D0944758

SHOCK AND VIBRATION HANDBOOK

McGRAW-HILL HANDBOOKS

ALJIAN · Purchasing Handbook
AMERICAN INSTITUTE OF PHYSICS · American Institute of Physics Handbook
AMERICAN SOCIETY OF MECHANICAL ENGINEERS · ASME Handbooks:
  Engineering Tables Metals Engineering—Processes
  Metals Engineering—Design Metals Properties
BAUMEISTER AND MARKS · Standard Handbook for Mechanical Engineers
BEEMAN · Industrial Power Systems Handbook
BRADY · Materials Handbook
BRATER AND KING · Handbook of Hydraulics
BURINGTON · Handbook of Mathematical Tables and Formulas
BURINGTON AND MAY · Handbook of Probability and Statistics with Tables
CONDON AND ODISHAW · Handbook of Physics
CONSIDINE · Process Instruments and Controls Handbook
CROCKER AND KING · Piping Handbook
CROFT, CARR, AND WATT · American Electricians' Handbook
DAVIS AND SORENSEN · Handbook of Applied Hydraulics
DEAN · Lange's Handbook of Chemistry
ETHERINGTON · Nuclear Engineering Handbook
FINK AND CARROLL · Standard Handbook for Electrical Engineers
HARRIS · Handbook of Noise Control
HICKS · Standard Handbook of Engineering Calculations
HUSKEY AND KORN · Computer Handbook
JURAN · Quality-control Handbook
KORN AND KORN · Mathematical Handbook for Scientists and Engineers
LA LONDE AND JANES · Concrete Engineering Handbook
LANDEE, DAVIS, AND ALBRECHT · Electronic Designers' Handbook
LE GRAND · The New American Machinist's Handbook
MANTELL · Engineering Materials Handbook
MAYNARD · Handbook of Business Administration
MAYNARD · Industrial Engineering Handbook
MERRITT · Building Construction Handbook
MORROW · Maintenance Engineering Handbook
PERRY · Chemical Engineers' Handbook
SOCIETY OF MANUFACTURING ENGINEERS · Die Design Handbook
SOCIETY OF MANUFACTURING ENGINEERS · Tool Engineers Handbook
STREETER · Handbook of Fluid Dynamics
STUBBS · Handbook of Heavy Construction
TRUXAL · Control Engineers' Handbook
URQUHART · Civil Engineering Handbook
WOODS · Highway Engineering Handbook

SHOCK AND VIBRATION HANDBOOK

SECOND EDITION

Edited by

CYRIL M. HARRIS

Professor of Electrical Engineering and Architecture
Columbia University

and

CHARLES E. CREDE

Late Professor of Mechanical Engineering and Applied Mechanics
California Institute of Technology

McGRAW-HILL BOOK COMPANY

New York St. Louis San Francisco Auckland Düsseldorf Johannesburg
Kuala Lumpur London Mexico Montreal New Delhi Panama
Paris Sao Paulo Singapore Sydney Tokyo Toronto

Library of Congress Cataloging in Publication Data

Harris, Cyril M ed.
Shock and vibration handbook.

Includes bibliographies.
1. Vibration—Handbooks, manuals, etc. 2. Shock (Mechanics)—Handbooks, manuals, etc. I. Crede, Charles E., joint ed. II. Title.
TA355.H35 1976 620.3 76-55
ISBN 0-07-026799-5

1234567890 KPKP 785432109876

The editors for this book were Harold B. Crawford, Ross J. Kepler, and Betty Gatewood, and the production supervisor was Teresa F. Leaden.
It was set in Modern by Bi-Comp.

It was printed and bound by Kingsport Press.

CONTENTS

PREFACE TO THE SECOND EDITION

There have been many important technical developments since the publication of the First Edition of the "Shock and Vibration Handbook" in 1961 and the untimely death of Charles E. Crede, friend and co-editor, in 1964. New shock and vibration measuring instruments are now available and others have become obsolete; many standards have been established in vibration measurement; remarkable advances in data processing techniques in shock and vibration analysis have accompanied the widespread use of large-scale computers. These and other developments in many specialized areas of shock and vibration have brought about the need for a Second Edition of the Handbook to reflect current engineering practice.

The First Edition of the "Shock and Vibration Handbook" recognized for the first time the full scope of the field of shock and vibration and brought togethei under one title classical vibration theory combined with modern applications of the theory to current engineering practice, including the topics of mechanical shock and the instrumentation of shock and vibration. Each chapter covered the particular subject matter in the most comprehensive manner possible. Many chapters on specialized topics contained more material than had been found collectively in all books previously published in the field; others contained considerable material that had not been summarized in the literature. As a result, the Handbook became a standard reference work throughout the world.

Comprehensive coverage has continued to be a primary objective in the Second Edition, but there have been major changes in this revision. All archival material has been deleted. Also deleted from this edition are those chapters whose importance has diminished because of subsequent technical developments: for example, Analog Methods of Analysis, and chapters that provided details in highly specialized areas such as shock and vibration in ships, aircraft, or missiles (although applicable theory, instrumentation, measurement techniques, data analysis, and methods of control that apply to these special areas are retained in other chapters). In contrast, the treatment of current engineering problems of general interest has been expanded. While greater coverage is now given practical application, the deletions have reduced the total number of pages of the original three-volume edition by one-third, making it possible to publish the new edition in the more convenient form of a single volume.

The Second Edition, a unified treatment of the subject of shock and vibration, is the equivalent in content of at least five textbooks of the usual size. The 44 chapters were written by 54 authorities from industry, government laboratories, and universities.

In the Second Edition, each chapter covers a particular topic. Chapters dealing with related topics are grouped together. The first group of chapters provides a theoretical basis for shock and vibration. The second group considers instrumentation and measurements. This is followed by a new chapter on vibration standards. The next group of chapters deals with analysis and testing, concepts in the treatment of data obtained from measurements, and procedures for analyzing and testing systems subjected to vibration and shock. To this group has been added a new chapter on the use of digital computers. The important subject of methods of controlling shock and vibration is discussed in a group of chapters dealing with isolation, damping, and balancing. This is followed by chapters devoted to equipment design, packaging, and the effects of shock and vibration on man. Duplication of material between chapters is avoided insofar as this is desirable, cross-references to other chapters being used frequently. There are extensive references to available technical literature.

Although this Handbook is not intended primarily as a textbook, many teachers will find the classical and rigorous treatment suitable for classroom use; in particular, the extensive discussion of practical examples will be of value as a supplement to the usual classroom theory. The control of shock and vibration is of practical importance in many aspects of engineering. The Handbook is particularly intended to be used as a working reference by engineers and scientists in the mechanical, civil, acoustical, aeronautical, electrical, air-conditioning, transportation and chemical fields. Engineers in manufacturing, including plant maintenance, measurement and control, environmental testing, and packing and shipping, will find much of value, as will those engaged in development and design work.

In the preparation of the Second Edition, many persons and organizations (for example, the Shock and Vibration Information Center at the Naval Research Laboratory) made useful suggestions and contributions; these are far too numerous to acknowledge individually, but I am grateful to them all. The contributors worked diligently with me toward the objective of making each chapter the definitive treatment in its field. Finally I wish to express my appreciation to the government agencies and industrial organizations with whom some of our contributors are associated for clearing the material presented in chapters written by these contributors.

Cyril M. Harris

SHOCK AND VIBRATION HANDBOOK

1

INTRODUCTION TO THE HANDBOOK

Cyril M. Harris
Columbia University

Charles E. Crede
California Institute of Technology

CONCEPTS OF SHOCK AND VIBRATION

Vibration is a term that describes oscillation in a mechanical system. It is defined by the frequency (or frequencies) and amplitude. Either the motion of a physical object or structure or, alternatively, an oscillating force applied to a mechanical system is vibration in a generic sense. Conceptually, the time-history of vibration may be considered to be sinusoidal or simple harmonic in form. The frequency is defined in terms of cycles per unit time, and the magnitude in terms of amplitude (the maximum value of a sinusoidal quantity). The vibration encountered in practice often does not have this regular pattern. It may be a combination of several sinusoidal quantities, each having a different frequency and amplitude. If each frequency component is an integral multiple of the lowest frequency, the vibration repeats itself after a determined interval of time and is called *periodic*. If there is no integral relation among the frequency components, there is no periodicity and the vibration is defined as *complex*.

Vibration may be described as *deterministic* or *random*. If it is deterministic, it follows an established pattern so that the value of the vibration at any designated future time is completely predictable from the past history. If the vibration is random, its future value is unpredictable except on the basis of probability. Random vibration is defined in statistical terms wherein the probability of occurrence of designated magnitudes and frequencies can be indicated. The analysis of random vibration involves certain physical concepts that are different from those applied to the analysis of deterministic vibration.

Vibration of a physical structure often is thought of in terms of a model consisting of a mass and a spring. The vibration of such a model, or system, may be "free" or "forced." In *free vibration*, there is no energy added to the system but rather the vibration is the continuing result of an initial disturbance. An *ideal system* may be considered undamped for mathematical purposes; in such a system the free vibration is assumed to continue indefinitely. In any *real system*, damping (i.e., energy dissipation) causes the amplitude of free vibration to decay continuously to a negligible value. Such free vibration sometimes is referred to as *transient vibration*. *Forced vibration*, in contrast to free vibration, continues under "steady-state" conditions because energy is supplied to the system continuously to compensate for that dissipated by damping in the system. In general, the frequency at which energy is supplied (i.e., the forcing frequency) appears in the vibration of the system. Forced vibration may be either deterministic or random. In either instance, the vibration of the system depends upon the relation of the excitation or forcing function to the properties of the system. This relationship is a prominent feature of the analytical aspects of vibration.

Shock is a somewhat loosely defined aspect of vibration wherein the excitation is nonperiodic, e.g., in the form of a pulse, a step, or transient vibration. The word "shock" implies a degree of suddenness and severity. These terms are relative rather than absolute measures of the characteristic; they are related to a popular notion of the characteristics of shock and are not necessary in a fundamental analysis of the applicable principles. From the analytical viewpoint, the important characteristic of shock is that the motion of the system upon which the shock acts includes both the frequency of the shock excitation and the natural frequency of the system. If the excitation is brief, the continuing motion of the system is free vibration at its own natural frequency.

The technology of shock and vibration embodies both theoretical and experimental facets prominently. Thus, methods of analysis and instruments for the measurement of shock and vibration are of primary significance. The results of analysis and measurement are used to evaluate shock and vibration environments, to devise testing procedures and testing machines, and to design and operate equipment and machinery. Shock and/or vibration may be either wanted or unwanted, depending upon circumstances. For example, vibration is involved in the primary mode of operation of such equipment as conveying and screening machines; the setting of rivets depends upon the application of impact or shock. More frequently, however, shock and vibration are unwanted. Then the objective is to eliminate or reduce their severity or, alternatively, to design equipment to withstand their influences. These procedures are embodied in the control of shock and vibration. Methods of control are emphasized throughout this Handbook.

CONTROL OF SHOCK AND VIBRATION

Methods of shock and vibration control may be grouped into three broad categories:

1. Reduction at the Source

a. Balancing of Moving Masses. Where the vibration originates in rotating or reciprocating members, the magnitude of a vibratory force frequently can be reduced or possibly eliminated by balancing or counterbalancing. For example, during the manufacture of fans and blowers, it is common practice to rotate each rotor and to add or subtract material as necessary to achieve balance.

b. Balancing of Magnetic Forces. Vibratory forces arising in magnetic effects of electrical machinery sometimes can be reduced by modification of the magnetic path. For example, the vibration originating in an electric motor can be reduced by skewing the slots in the armature laminations.

c. Control of Clearances. Vibration and shock frequently result from impacts involved in operation of machinery. In some instances, the impacts result from inferior design or manufacture, such as excessive clearances in bearings, and can be reduced by closer attention to dimensions. In other instances, such as the movable armature of a relay, the shock can be decreased by employing a rubber bumper to cushion motion of the plunger at the limit of travel.

2. Isolation

a. Isolation of Source. Where a machine creates significant shock or vibration during its normal operation, it may be supported upon isolators to protect other machinery and personnel from shock and vibration. For example, a forging hammer tends to create shock of a magnitude great enough to interfere with the operation of delicate apparatus in the vicinity of the hammer. This condition may be alleviated by mounting the forging hammer upon isolators.

b. Isolation of Sensitive Equipment. Equipment often is required to operate in an environment characterized by severe shock or vibration. The equipment may be protected from these environmental influences by mounting it upon isolators. For example, equipment mounted in ships of the navy is subjected to shock of great severity during naval warfare and may be protected from damage by mounting it upon isolators.

3. Reduction of the Response

a. Alteration of Natural Frequency. If the natural frequency of the structure of an equipment coincides with the frequency of the applied vibration, the vibration condition may be made much worse as a result of resonance. Under such circumstances, if the frequency of the excitation is substantially constant, it often is possible to alleviate the vibration by changing the natural frequency of such structure. For example, the vibration of a fan blade was reduced substantially by modifying a stiffener on the blade, thereby changing its natural frequency and avoiding resonance with the frequency of rotation of the blade. Similar results are attainable by modifying the mass rather than the stiffness.

b. Energy Dissipation. If the vibration frequency is not constant or if the vibration involves a large number of frequencies, the desired reduction of vibration may not be attainable by altering the natural frequency of the responding system. It may be possible to achieve equivalent results by the dissipation of energy to eliminate the severe effects of resonance. For example, the housing of a washing machine may be made less susceptible to vibration by applying a coating of damping material on the inner face of the housing.

c. Auxiliary Mass. Another method of reducing the vibration of the responding system is to attach an auxiliary mass to the system by a spring; with proper tuning the mass vibrates and reduces the vibration of the system to which it is attached. For example, the vibration of a textile-mill building subjected to the influence of several hundred looms was reduced by attaching large masses to a wall of the building by means of springs; then the masses vibrated with a relatively large motion and the vibration of the wall was reduced. The incorporation of damping in this auxiliary mass system may further increase its effectiveness.

CONTENT OF HANDBOOK

The chapters of this Handbook each deal with a discrete phase of the subject of shock and vibration. Frequent references are made from one chapter to another, to refer to basic theory in other chapters, to call attention to supplementary information, and to give illustrations and examples. Therefore, each chapter when read with other referenced chapters presents one complete facet of the subject of shock and vibration.

Chapters dealing with similar subject matter are grouped together. The first ten chapters following this introductory chapter deal with fundamental concepts of shock and vibration. Chapter 2 discusses the free and forced vibration of linear systems that can be defined by lumped parameters with similar types of coordinates. The properties of rigid bodies are discussed in Chap. 3, together with the vibration of resiliently supported rigid bodies wherein several modes of vibration are coupled. Nonlinear vibration is discussed in Chap. 4, and self-excited vibration in Chap. 5. Chapter 6 discusses two degree-of-freedom systems in detail—including both the basic theory and the application of such theory to dynamic absorbers and auxiliary mass dampers. The vibration of systems defined by distributed parameters, notably beams and plates, is discussed in Chap. 7. Chapters 8 and 9 relate to shock; Chap. 8 discusses the response of lumped parameter systems to step- and pulse-type excitation, and Chap. 9 discusses the effects of impact on structures. Chapter 10, entitled "Mechanical Impedance," discusses the concepts whereby the characteristics of a composite system may be determined from the characteristics of the component parts. Random vibration is discussed in Chap. 11 under the title of "Statistical Concepts in Vibration."

The second group of chapters deals with instrumentation for the measurement of shock and vibration. The general principles are discussed in Chap. 12. A wide range of instruments are described—from the compact and self-contained instruments discussed in Chap. 13 to elaborate instrumentation systems comprised of transducers and associated recording means. There are many types of transducers with a wide range of characteristics. The importance of this topic is indicated by the fact that four chapters (Chaps. 14 to 17) are devoted to discussions of the design, performance, and general characteristics of a large number of transducers. The calibration of instruments is discussed in Chap. 18. Chapter 19 describes standards in shock and vibration measurements. Field measurement techniques are discussed in Chap. 20.

Equipment intended for use under environmental conditions characterized by shock and vibration frequently is tested in the laboratory under simulated conditions. The importance of such testing is recognized by the group of chapters devoted to this subject. Chapter 21 introduces the concepts, and Chaps. 22 and 23 discuss the applicable concepts of data analysis. Chapter 22 is concerned with the analysis of data defining vibration conditions, and discusses the transformation of a time-history of a vibration measurement into more compact forms of data. Chapter 23 is an analogous presentation—treating the transformation of time-histories defining conditions of shock. Chapter 24 discusses environmental specifications and testing. Chapter 25 describes the construction and operation of vibration testing machines for laboratory use, and Chap. 26 is the analogous presentation dealing with shock testing machines.

The next group of chapters deals with computational methods. The use of computers is discussed in Chap. 27; the concepts which are presented are useful in both analytical and experimental work. Chapter 28 describes modern numerical methods in vibration analysis, dealing largely with the formulation of matrices for use with digital computers, and other numerical calculating methods. Chapter 29 discusses vibrations which are excited by ground motion and by the flow of air.

One of the most common techniques of shock and vibration control involves the concept of isolation. The theory of vibration isolation is discussed in detail in Chap. 30; an analogous presentation of shock isolation is included in Chap. 31. Chapter 30 includes treatment of vibration isolation when the vibration frequencies are so high that associated structures cannot be considered as rigid bodies, and when the vibration is random in nature. The more practical aspects of isolation are considered in Chap. 32, including the application and design of isolators. Chapter 33 discusses two particular classes of isolators—using pneumatic springs and involving servo control. Such principles often are applied to a single isolation system and are discussed in a single chapter. Chapters 34 and 35 describe the materials for and design of springs for use in isolators, metal springs being discussed in Chap. 34 and rubber springs being discussed in Chap. 35.

An important method of controlling shock and vibration involves the addition of damping or energy-dissipating means to structures that are susceptible to vibration. Chapter 36 discusses the general concepts of damping together with the application of such concepts to hysteresis and slip damping. The application of damping materials to structures is discussed in Chap. 37.

The latter chapters of the Handbook deal with the specific application of fundamentals of analysis, methods of measurement, and control techniques—where these are developed sufficiently to form a separate and discrete subject. Torsional vibration is discussed in Chap. 38, with particular application to internal-combustion engines. The balancing of rotating equipment is discussed in Chap. 39, and balancing machines are described. Chapter 40 describes the special vibration problems associated with the design and operation of machine tools. Among the most prominent occurrences of shock and vibration are those that arise during handling and shipping of merchandise. Packaging of equipment to protect against such shock and vibration is discussed in Chap. 41. Chapters 42 and 43 describe procedures for the design of equipment to withstand shock and vibration—the former considering primarily the theory of design and the latter considering practical aspects. A comprehensive discussion of the human aspect of shock and vibration is considered in Chap. 44 which describes the effect of shock and vibration on man.

SYMBOLS

This section includes a list of symbols with their usual English units as used generally in the Handbook; metric units are given as alternates in the text. In special circumstances, some of the following symbols have different meanings in certain chapters but are defined in those chapters. Other symbols of special or limited application are defined in the respective chapters as they are used.

Symbol	Meaning	Units
a	radius	in.
B	magnetic flux density	gauss
c	damping coefficient	lb-sec/in.
c	velocity of sound	in./sec
c_c	coefficient for critical damping	lb-sec/in.
C	capacitance	farads
D	diameter	in.
e	electrical voltage	volts
e	eccentricity	in.
E	energy	in.-lb
E	modulus of elasticity in tension and compression (Young's modulus)	lb/in.^{2}
f	frequency	cycles/sec (cps)
f_n	undamped natural frequency	cycles/sec (cps)
f_i	undamped natural frequencies in a multiple degree-of-freedom system, where $i = 1, 2, \ldots$	cycles/sec (cps)
f_d	damped natural frequency	cycles/sec (cps)
f_r	resonant frequency	cycles/sec (cps)
F	force	lb
f_f	Coulomb friction force	lb
g	acceleration of gravity	in./sec^2
G	modulus of elasticity in shear	lb/in.^{2}
h	height, depth	in.
H	magnetic field strength	oersteds
i	electric current	amperes
I_i	area or mass moment of inertia (subscript indicates axis)	in.^{4}
I_p	polar moment of inertia	in.^{4}
I_{ij}	area or mass product of inertia (subscripts indicate axes)	in.^{4}
$\mathcal{I}$	imaginary part of	
j	$\sqrt{-1}$	
J	inertia constant (weight moment of inertia)	lb-in.^{2}
J	impulse	lb-sec
k	linear stiffness	lb/in.
k_t	rotational (torsional) stiffness	lb-in./rad
l	length	in.
L	inductance	henrys
m	mass	lb-sec^2/in.

Symbol	Quantity	Unit
m_u	unbalanced mass	lb-sec²/in.
M	torque	lb-in.
M	mutual inductance	henrys
$\mathfrak{M}$	mobility	in./lb-sec
n	number of coils, supports, etc.	
p	alternating pressure	lb/in.²
p	probability density	
P	probability distribution	
P	static pressure	lb/in.²
q	electric charge	coulombs
Q	resonance factor (also ratio of reactance to resistance)	
r	resistance	ohms
R	radius	in.
$\Re$	real part of	
s	arc length	in.
S	area of diaphragm, tube, etc.	in.²
t	thickness	in.
t	time	sec
T	transmissibility	
T	kinetic energy	in.-lb
v	linear velocity	in./sec
V	potential energy	in.-lb
w	width	in.
W	weight	lb
W	power	in.-lb/sec
W_e	spectral density of the excitation	(rms)²/unit freq.
W_r	spectral density of the response	(rms)²/unit freq.
x	linear displacement in direction of X axis	in.
y	linear displacement in direction of Y axis	in.
z	linear displacement in direction of Z axis	in.
Z	impedance	lb-sec/in.
α	rotational displacement about X axis	radians
β	rotational displacement about Y axis	radians
γ	rotational displacement about Z axis	radians
γ	shear strain	
γ	weight density	lb/in.³
δ	deflection	in.
δ_{st}	static deflection	in.
Δ	logarithmic decrement	
ϵ	tension or compression strain	
ζ	fraction of critical damping	
η	stiffness ratio	
θ	phase angle	radians
λ	wavelength	in.
μ	coefficient of friction	
μ	mass density	lb-sec²/in.⁴
ν	Poisson's ratio	
ρ	mass density	lb-sec²/in.⁴
ρ_i	radius of gyration (subscript indicates axis)	in.
σ	Poisson's ratio	
σ	normal stress	lb/in.²
σ	root-mean-square (rms) value	
τ	period	sec
τ	shear stress	lb/in.²
ϕ	magnetic flux	maxwell
ψ	phase angle	radians
ω	forcing frequency—angular	rad/sec
ω_n	undamped natural frequency—angular	rad/sec

ω_i	undamped natural frequencies—angular—in a multiple degree-of-freedom system, where $i = 1, 2, \ldots$	rad/sec
ω_d	damped natural frequency—angular....................	rad/sec
ω_r	resonant frequency—angular..........................	rad/sec
Ω	rotational speed....................................	rad/sec
$\simeq$	approximately equal to	

CHARACTERISTICS OF HARMONIC MOTION

Harmonic functions are employed frequently in the analysis of shock and vibration. A body that experiences simple harmonic motion follows a displacement pattern defined by

$$x = x_0 \sin (2\pi f t) = x_0 \sin \omega t \tag{1.1}$$

where f is the *frequency* of the simple harmonic motion, $\omega = 2\pi f$ is the corresponding *angular frequency*, and x_0 is the *amplitude* of the displacement.

The velocity $\dot{x}$ and acceleration $\ddot{x}$ of the body are found by differentiating the displacement once and twice, respectively:

$$\dot{x} = x_0(2\pi f) \cos 2\pi f t = x_0\omega \cos \omega t \tag{1.2}$$

$$\ddot{x} = -x_0(2\pi f)^2 \sin 2\pi f t = -x_0\omega^2 \sin \omega t \tag{1.3}$$

The maximum absolute values of the displacement, velocity, and acceleration of a body undergoing harmonic motion occur when the trigonometric functions in Eqs. (1.1) to (1.3) are numerically equal to unity. These values are known, respectively, as displacement, velocity, and acceleration amplitudes; they are defined mathematically as follows:

$$x_0 = x_0 \qquad \dot{x}_0 = (2\pi f)x_0 \qquad \ddot{x}_0 = (2\pi f)^2 x_0 \tag{1.4}$$

It is common to express the displacement amplitude x_0 in inches when the English system of units is used and in centimeters or millimeters when the metric system is used. Accordingly, the velocity amplitude $\dot{x}_0$ is expressed in inches per second in the English system (centimeters per second or millimeters per second in the metric system). The acceleration amplitude $\ddot{x}_0$ usually is expressed as a dimensionless multiple of the gravitational acceleration g, where $g = 386$ in./sec^2 or 980 cm/sec^2. For example, an acceleration of 3,860 in./sec^2 is written $10g$.

Table 1.1. Conversion Factors for Translational Velocity and Acceleration

Multiply Value in → or → By ↘ To obtain value in ↓	g-sec, g	ft/sec ft/sec^2	in./sec in./sec^2	cm/sec cm/sec^2	m/sec m/sec^2
g-sec, g	1	0.0311	0.00259	0.00102	0.102
ft/sec ft/sec^2	32.16	1	0.0833	0.0328	3.28
in./sec in./sec^2	386	12.0	1	0.3937	39.37
cm/sec cm/sec^2	980	30.48	2.540	1	100
m/sec m/sec^2	9.80	0.3048	0.0254	0.010	1

Factors for converting values of rectilinear velocity and acceleration to different units are given in Table 1.1; similar factors for angular velocity and acceleration are given in Table 1.2. Displacement, velocity, and acceleration amplitudes (expressed in the English system of units) as a function of frequency are shown graphically in Fig. 1.1 and by the nomograph of Fig. 1.2.

For certain purposes in analysis, it is convenient to express the amplitude in terms of the average value of the harmonic function, the root-mean-square (rms) value, or two times the amplitude (i.e., peak-to-peak value). These terms are defined mathematically in Chap. 22; numerical conversion factors are set forth in Table 1.3 for ready reference.

Table 1.2. Conversion Factors for Rotational Velocity and Acceleration

Multiply Value in→ or→ By↘ To obtain value in ↓	rad/sec rad/sec²	degree/sec degree/sec²	rev/sec rev/sec²	rev/min rev/min/sec
rad/sec rad/sec²	1	0.01745	6.283	0.1047
degree/sec degree/sec²	57.30	1	360	6.00
rev/sec rev/sec²	0.1592	0.00278	1	0.0167
rev/min rev/min/sec	9.549	0.1667	60	1

Table 1.3. Conversion Factors for Simple Harmonic Motion

Multiply numerical value in terms of→ By↘ To obtain value in terms of ↓	Amplitude	Average value	Root-mean-square value (rms)	Peak-to-peak value
Amplitude	1	1.571	1.414	0.500
Average value	0.637	1	0.900	0.318
Root-mean-square value (rms)	0.707	1.111	1	0.354
Peak-to-peak value	2.000	3.142	2.828	1

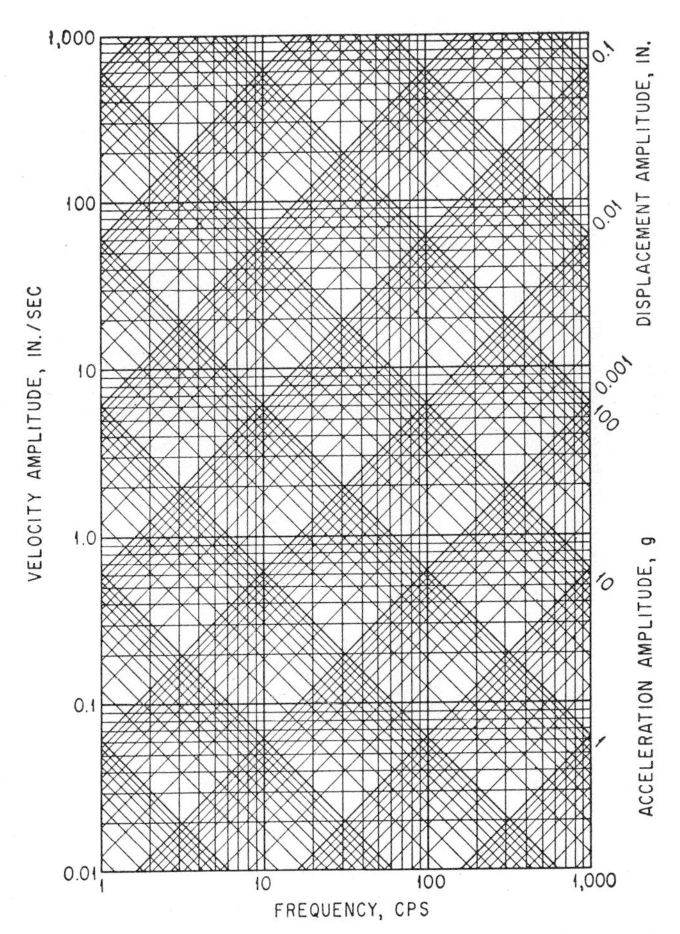

Fig. 1.1. Relation of frequency to the amplitudes of displacement, velocity, and acceleration in harmonic motion.

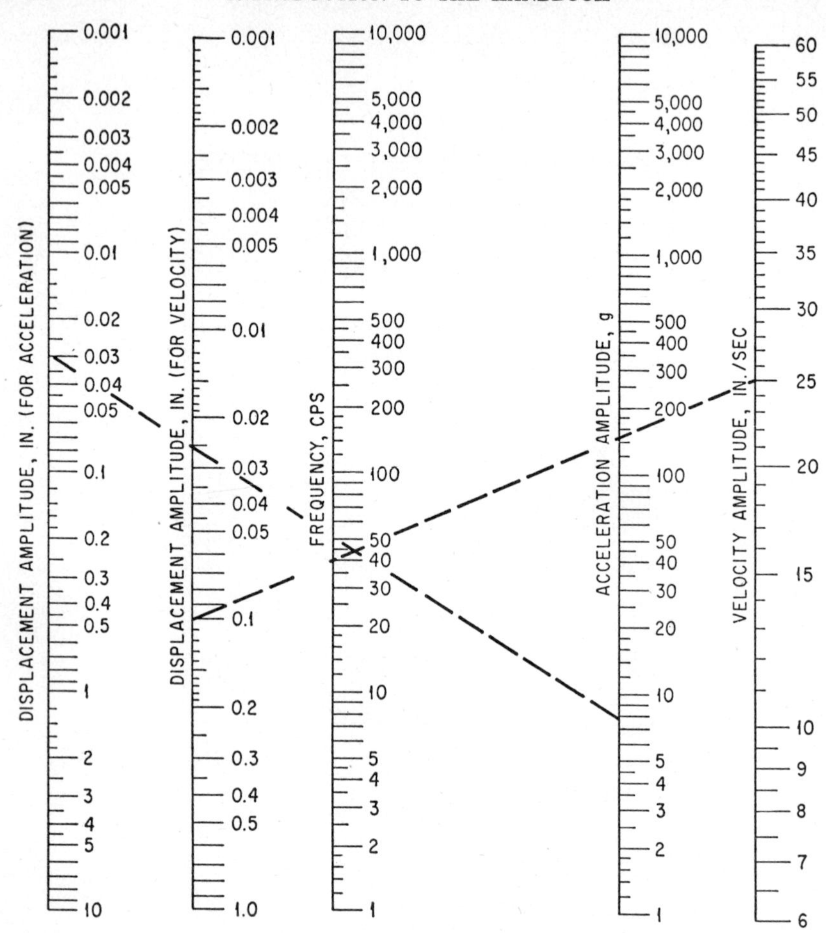

Fig. 1.2. Nomogram for harmonic motion showing relation of frequency and acceleration amplitude to displacement amplitude (extreme left scale) and relation of frequency and velocity amplitude to displacement amplitude (middle scale on left). Use of the nomogram is indicated by the broken lines. For example, harmonic motion with a displacement amplitude of 0.03 in. and a frequency of 50 cps has an acceleration amplitude of 7.7*g*; harmonic motion with a displacement amplitude of 0.10 in. and a frequency of 40 cps has a velocity amplitude of 25.1 in./sec.

APPENDIX 1.1

NATURAL FREQUENCIES OF COMMONLY USED SYSTEMS

The most important aspect of vibration analysis often is the calculation or measurement of the natural frequencies of mechanical systems. Natural frequencies are discussed prominently in many chapters of the Handbook. Appendix 1.1 includes in tabular form, convenient for ready reference, a compilation of frequently used expressions for the natural frequencies of common mechanical systems. The data for beams and plates are abstracted from Chap. 7.

MASS-SPRING SYSTEMS IN TRANSLATION
(RIGID MASS AND MASSLESS SPRING)

k = SPRING STIFFNESS, LB/IN.
m = MASS, LB-SEC²/IN.
ω_n = ANGULAR NATURAL FREQUENCY, RAD/SEC

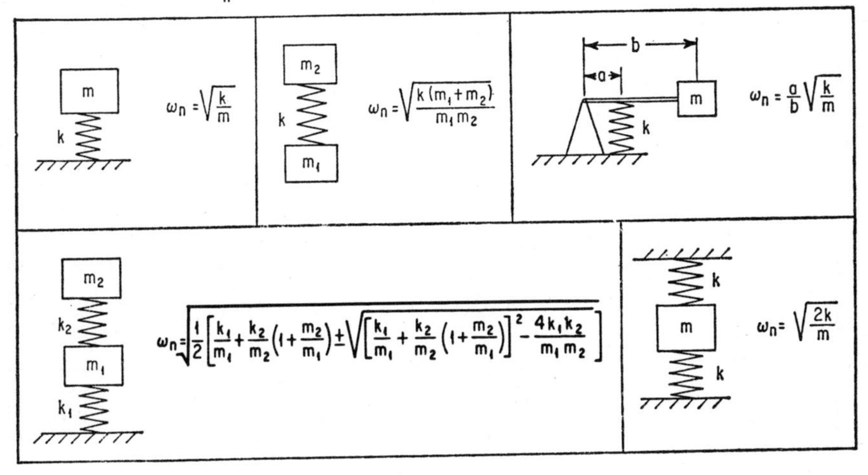

$\omega_n = \sqrt{\frac{k}{m}}$
$\omega_n = \sqrt{\frac{k(m_1+m_2)}{m_1 m_2}}$
$\omega_n = \frac{a}{b}\sqrt{\frac{k}{m}}$
$\omega_n = \sqrt{\frac{1}{2}\left[\frac{k_1}{m_1}+\frac{k_2}{m_2}\left(1+\frac{m_2}{m_1}\right)\pm\sqrt{\left[\frac{k_1}{m_1}+\frac{k_2}{m_2}\left(1+\frac{m_2}{m_1}\right)\right]^2-\frac{4k_1k_2}{m_1m_2}}\right]}$
$\omega_n = \sqrt{\frac{2k}{m}}$

SPRINGS IN COMBINATION

k_r = RESULTANT STIFFNESS OF COMBINATION

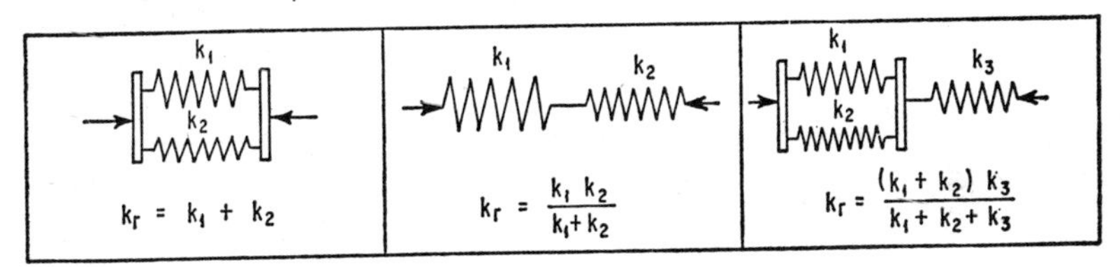

$k_r = k_1 + k_2$
$k_r = \frac{k_1 k_2}{k_1+k_2}$
$k_r = \frac{(k_1+k_2)k_3}{k_1+k_2+k_3}$

HELICAL SPRINGS

d = WIRE DIAMETER, IN
D = MEAN COIL DIAMETER, IN.
n = NUMBER OF ACTIVE COILS
E = YOUNG'S MODULUS, LB/IN.²
G = MODULUS OF ELASTICITY IN SHEAR, LB/IN.²

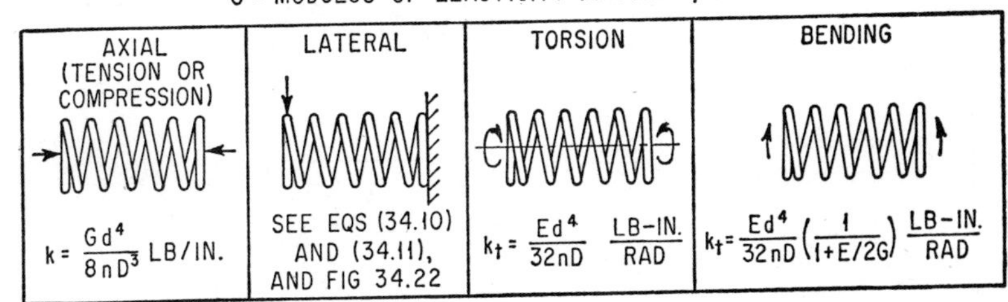

AXIAL (TENSION OR COMPRESSION)
$k = \frac{Gd^4}{8nD^3}$ LB/IN.
LATERAL
SEE EQS (34.10) AND (34.11), AND FIG 34.22
TORSION
$k_t = \frac{Ed^4}{32nD}$ $\frac{\text{LB-IN.}}{\text{RAD}}$
BENDING
$k_t = \frac{Ed^4}{32nD}\left(\frac{1}{1+E/2G}\right)$ $\frac{\text{LB-IN.}}{\text{RAD}}$

ROTOR-SHAFT SYSTEMS
(RIGID ROTOR AND MASSLESS SHAFT)

k_t = TORSIONAL STIFFNESS OF SHAFT, LB-IN./RAD
I = MASS MOMENT OF INERTIA OF ROTOR, LB-IN.-SEC^2
ω_n = ANGULAR NATURAL FREQUENCY, RAD/SEC

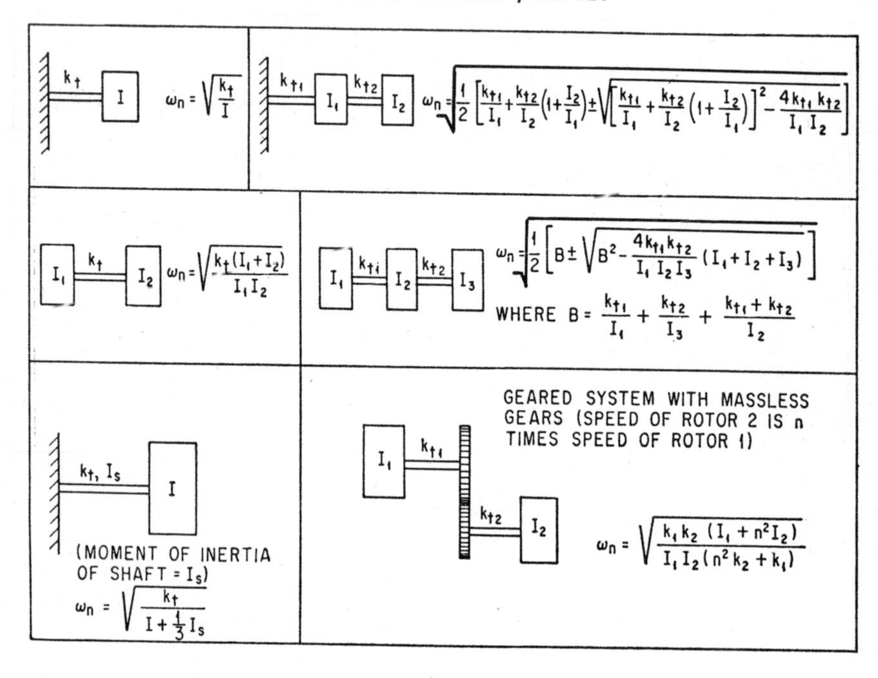

$\omega_n = \sqrt{\frac{k_t}{I}}$
$\omega_n = \sqrt{\frac{1}{2}\left[\frac{k_{t1}}{I_1}+\frac{k_{t2}}{I_2}\left(1+\frac{I_2}{I_1}\right) \pm \sqrt{\left[\frac{k_{t1}}{I_1}+\frac{k_{t2}}{I_2}\left(1+\frac{I_2}{I_1}\right)\right]^2 - \frac{4k_{t1}k_{t2}}{I_1 I_2}}\right]}$
$\omega_n = \sqrt{\frac{k_t(I_1+I_2)}{I_1 I_2}}$
$\omega_n = \sqrt{\frac{1}{2}\left[B \pm \sqrt{B^2 - \frac{4k_{t1}k_{t2}}{I_1 I_2 I_3}(I_1+I_2+I_3)}\right]}$
WHERE $B = \frac{k_{t1}}{I_1} + \frac{k_{t2}}{I_3} + \frac{k_{t1}+k_{t2}}{I_2}$
GEARED SYSTEM WITH MASSLESS GEARS (SPEED OF ROTOR 2 IS n TIMES SPEED OF ROTOR 1)
$\omega_n = \sqrt{\frac{k_1 k_2 (I_1 + n^2 I_2)}{I_1 I_2 (n^2 k_2 + k_1)}}$
(MOMENT OF INERTIA OF SHAFT = I_s)
$\omega_n = \sqrt{\frac{k_t}{I + \frac{1}{3} I_s}}$

STIFFNESS OF SHAFTS IN TORSION

G = MODULUS OF ELASTICITY IN SHEAR, LB/IN.^{2}
l = LENGTH OF SHAFT, IN.
I_p = POLAR MOMENT OF INERTIA OF SHAFT CROSS-SECTION, IN.^{4}.

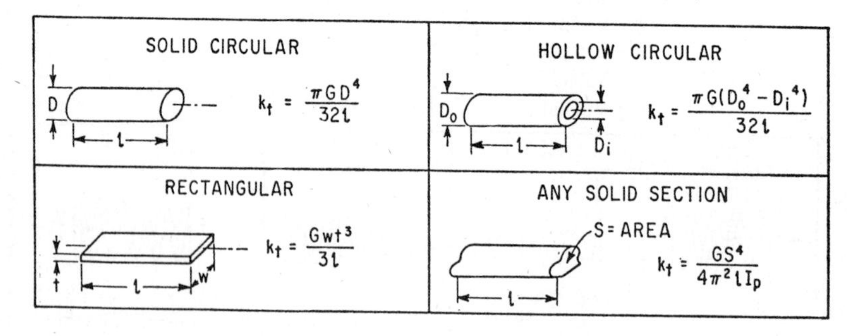

MASSLESS BEAMS WITH CONCENTRATED MASS LOADS

m = MASS OF LOAD, LB-SEC²/IN.
l = LENGTH OF BEAM, IN.
I = AREA MOMENT OF INERTIA OF BEAM CROSS SECTION, IN.⁴
E = YOUNG'S MODULUS, LB/IN.²
ω_n = ANGULAR NATURAL FREQUENCY, RAD/SEC

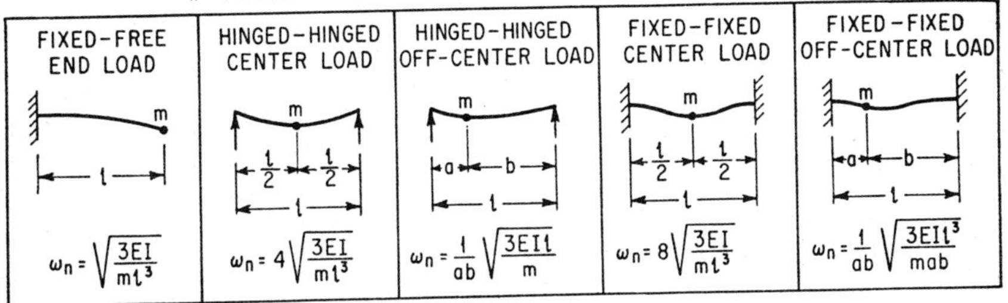

MASSIVE SPRINGS (BEAMS) WITH CONCENTRATED MASS LOADS

m = MASS OF LOAD, LB-SEC²/IN.
$m_s(m_b)$ = MASS OF SPRING (BEAM), LB-SEC²/IN.
k = STIFFNESS OF SPRING LB/IN.
l = LENGTH OF BEAM, IN.
I = AREA MOMENT OF INERTIA OF BEAM CROSS SECTION, IN.⁴
E = YOUNG'S MODULUS, LB/IN.²
ω_n = ANGULAR NATURAL FREQUENCY, RAD/SEC

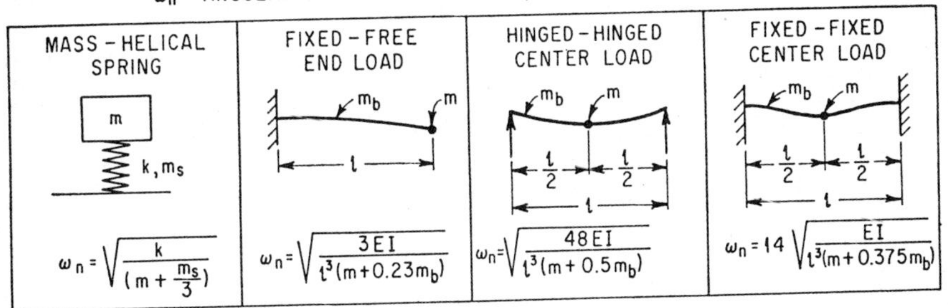

AREA MOMENT OF INERTIA OF BEAM SECTIONS
(WITH RESPECT TO AXIS a-a)

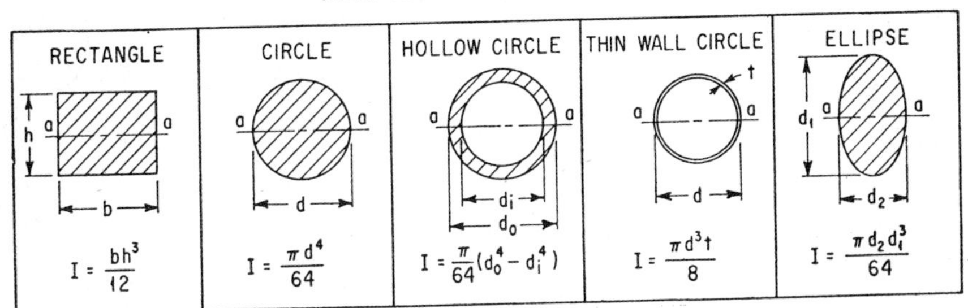

BEAMS OF UNIFORM SECTION AND UNIFORMLY DISTRIBUTED LOAD

ANGULAR NATURAL FREQUENCY $\omega_n = A\sqrt{\frac{EI}{\mu l^4}}$ RAD/SEC

WHERE E = YOUNG'S MODULUS, LB/IN.^{2}

I = AREA MOMENT OF INERTIA OF BEAM CROSS SECTION, IN.^{4}

l = LENGTH OF BEAM, IN.

μ = MASS PER UNIT LENGTH OF BEAM, LB-SEC^2/IN.^{2}

A = COEFFICIENT FROM TABLE BELOW

NODES ARE INDICATED IN TABLE BELOW AS A PROPORTION OF LENGTH l MEASURED FROM LEFT END

FIXED-FREE (CANTILEVER)	A = 3.52	0.774 A = 22.4	0.500 0.868 A = 61.7	0.356 0.644 0.906 A = 121.0	0.279 0.500 0.723 0.926 A = 200.0
HINGED-HINGED (SIMPLE)	A = 9.87	0.500 A = 39.5	0.333 0.667 A = 88.9	0.25 0.50 0.75 A = 158	0.20 0.40 0.60 0.80 A = 247
FIXED-FIXED (BUILT-IN)	A = 22.4	0.500 A = 61.7	0.359 0.641 A = 121	0.278 0.500 0.722 A = 200	0.227 0.409 0.591 0.773 A = 298
FREE-FREE	0.224 0.776 A = 22.4	0.132 0.500 0.868 A = 61.7	0.094 0.356 0.644 0.906 A = 121	0.073 0.277 0.500 0.723 0.927 A = 200	0.060 0.227 0.409 0.591 0.773 0.940 A = 298
FIXED-HINGED	A = 15.4	0.560 A = 50.0	0.384 0.692 A = 104	0.294 0.529 0.765 A = 178	0.238 0.429 0.619 0.810 A = 272
HINGED-FREE	0.736 A = 15.4	0.446 0.853 A = 50.0	0.308 0.616 0.898 A = 104	0.235 0.471 0.707 0.922 A = 178	0.190 0.381 0.581 0.763 0.937 A = 272

NATURAL FREQUENCIES OF THIN FLAT PLATES OF UNIFORM THICKNESS

$$\omega_n = B\sqrt{\frac{Et^2}{\rho a^4(1-\nu^2)}} \text{ RAD/SEC}$$

E = YOUNG'S MODULUS, LB/IN.^{2}
t = THICKNESS OF PLATE, IN.
ρ = MASS DENSITY, LB-SEC^2/IN.^{4}
a = DIAMETER OF CIRCULAR PLATE OR SIDE OF SQUARE PLATE, IN.
ν = POISSON'S RATIO

SHAPE OF PLATE	DIAGRAM	EDGE CONDITIONS	VALUE OF B FOR MODE:							
			1	2	3	4	5	6	7	8
CIRCULAR		CLAMPED AT EDGE	11.84	24.61	40.41	46.14	103.12			
CIRCULAR		FREE	6.09	10.53	14.19	23.80	40.88	44.68	61.38	69.44
CIRCULAR		CLAMPED AT CENTER	4.35	24.26	70.39	138.85				
CIRCULAR		SIMPLY SUPPORTED AT EDGE	5.90							
SQUARE		ONE EDGE CLAMPED– THREE EDGES FREE	1.01	2.47	6.20	7.94	9.01			
SQUARE		ALL EDGES CLAMPED	10.40	21.21	31.29	38.04	38.22	47.73		
SQUARE		TWO EDGES CLAMPED– TWO EDGES FREE	2.01	6.96	7.74	13.89	18.25			
SQUARE		ALL EDGES FREE	4.07	5.94	6.91	10.39	17.80	18.85		
SQUARE		ONE EDGE CLAMPED– THREE EDGES SIMPLY SUPPORTED	6.83	14.94	16.95	24.89	28.99	32.71		
SQUARE		TWO EDGES CLAMPED– TWO EDGES SIMPLY SUPPORTED	8.37	15.82	20.03	27.34	29.54	37.31		
SQUARE		ALL EDGES SIMPLY SUPPORTED	5.70	14.26	22.82	28.52	37.08	48.49		

MASSLESS CIRCULAR PLATE WITH CONCENTRATED CENTER MASS

CLAMPED EDGES

$$\omega_n = 4.09\sqrt{\frac{Eh^3}{ma^2(1-\nu^2)}}$$

SIMPLY SUPPORTED EDGES

$$\omega_n = 4.09\sqrt{\frac{Eh^3}{ma^2(1-\nu)(3+\nu)}}$$

NATURAL FREQUENCIES OF MISCELLANEOUS SYSTEMS

(ω_n = ANGULAR NATURAL FREQUENCY, RAD/SEC)

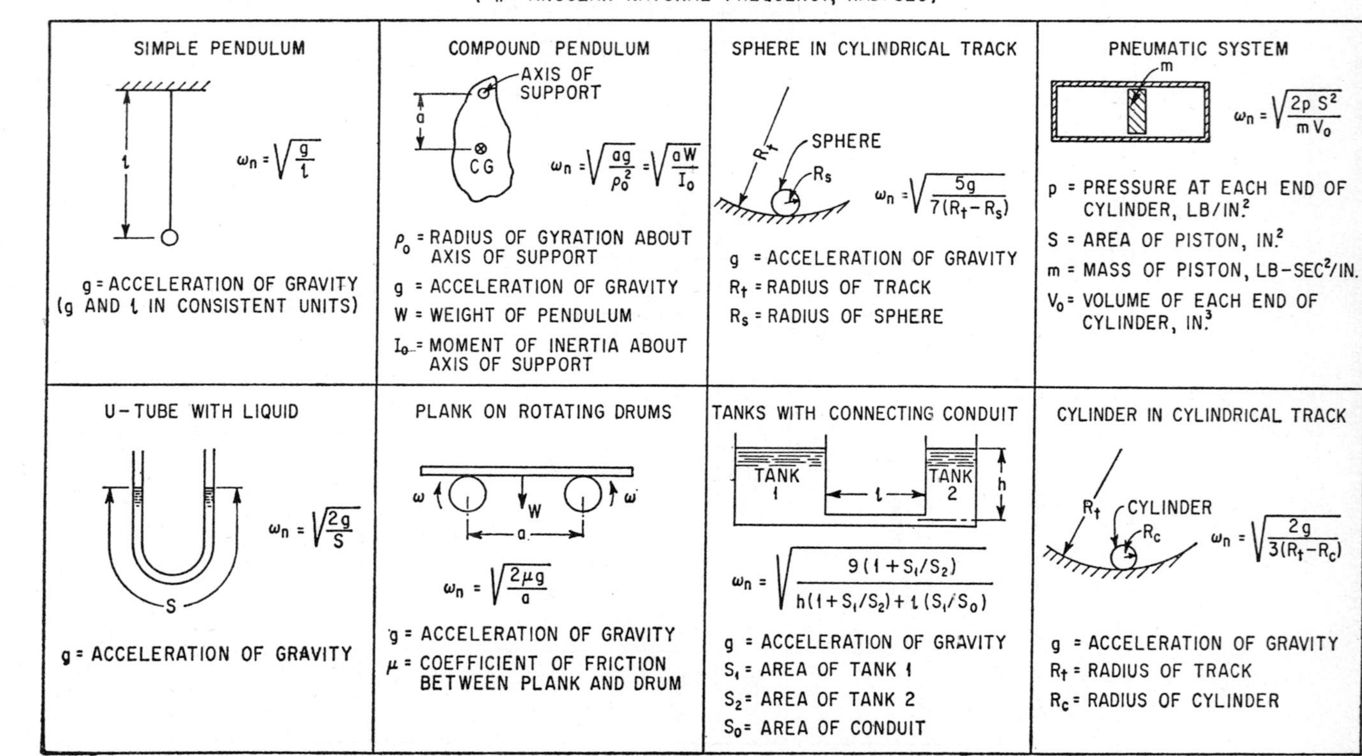

APPENDIX 1.2
TERMINOLOGY

For convenience, definitions of terms which are used frequently in the field of shock and vibration are assembled here. Many of the definitions in this appendix are identical with those contained in a document prepared by ISO, TC-108, Mechanical vibration and shock.† For terms which are not listed below, the reader is referred to the Index.

Acceleration. Acceleration is a vector quantity that specifies the time rate of change of velocity.

Amplitude. Amplitude is the maximum value of a sinusoidal quantity.

Analog. If a first quantity or structural element is analogous to a second quantity or structural element belonging in another field of knowledge, the second quantity is called the analog of the first, and vice versa.

Analogy. An analogy is a recognized relationship of consistent mutual similarity between the equations and structures appearing within two or more fields of knowledge, and an identification and association of the quantities and structural elements that play mutually similar roles in these equations and structures, for the purpose of facilitating transfer of knowledge of mathematical procedures of analysis and behavior of the structures between these fields.

Angular Frequency (Circular Frequency). The angular frequency of a periodic quantity, in radians per unit time, is the frequency multiplied by 2π.

Angular Mechanical Impedance (Rotational Mechanical Impedance). Angular mechanical impedance is the impedance involving the ratio of torque to angular velocity. (See *Impedance.*)

Antinode (Loop). An antinode is a point, line, or surface in a standing wave where some characteristic of the wave field has maximum amplitude.

Antiresonance. For a system in forced oscillation, antiresonance exists at a point when any change, however small, in the frequency of excitation causes an increase in the response at this point.

Audio Frequency. An audio frequency is any frequency corresponding to a normally audible sound wave.

Autocorrelation Coefficient. The autocorrelation coefficient of a signal is the ratio of the autocorrelation function to the mean-square value of the signal:

$$R(\tau) = \overline{x(t)x(t+\tau)}/\overline{[x(t)]^2}$$

Autocorrelation Function. The autocorrelation function of a signal is the average of the product of the value of the signal at time t with the value at time $t + \tau$:

$$R(\tau) = \overline{x(t)x(t+\tau)}$$

For a stationary random signal of infinite duration, the power spectral density (except for a constant factor) is the cosine Fourier transform of the autocorrelation function.

Auxiliary Mass Damper (Damped Vibration Absorber). An auxiliary mass damper is a system consisting of a mass, spring, and damper which tends to reduce vibration by the dissipation of energy in the damper as a result of relative motion between the mass and the structure to which the damper is attached.

Background Noise. Background noise is the total of all sources of interference in a system used for the production, detection, measurement, or recording of a signal, independent of the presence of the signal.

Balancing. Balancing is a procedure for adjusting the mass distribution of a rotor so that vibration of the journals, or the forces on the bearings at once-per-revolution, are reduced or controlled. (See *Chap. 39* for a complete list of definitions related to *balancing.*)

Band-pass Filter. A band-pass filter is a wave filter that has a single transmission band extending from a lower cutoff frequency greater than zero to a finite upper cutoff frequency.

Bandwidth, Effective. (See *Effective Bandwidth.*)

Beats. Beats are periodic variations that result from the superposition of two simple harmonic quantities of different frequencies f_1 and f_2. They involve the periodic increase and decrease of amplitude at the beat frequency $(f_1 - f_2)$.

† Copies of this document, as well as others in the field of mechanical shock and vibration, are available from the American National Standards Institute, Inc., 1430 Broadway, New York 10018.

Broad-band Random Vibration. Broad-band random vibration is random vibration having its frequency components distributed over a broad frequency band. (See *Random Vibration.*)

Circular Frequency. (See *Angular Frequency.*)

Complex Angular Frequency. As applied to a function $a = Ae^{\sigma t} \sin (\omega t - \phi)$, where σ, ω, and ϕ are constant, the quantity $\omega_c = \sigma + j\omega$ is the complex angular frequency where j is an operator with rules of addition, multiplication, and division as suggested by the symbol $\sqrt{-1}$. If the signal decreases with time, σ must be negative.

Complex Function. A complex function is a function having real and imaginary parts.

Complex Vibration. Complex vibration is vibration whose components are sinusoids not harmonically related to one another. (See *Harmonic.*)

Compliance. Compliance is the reciprocal of stiffness.

Continuous System (Distributed System). A continuous system is one that is considered to have an infinite number of possible independent displacements. Its configuration is specified by a function of a continuous spatial variable or variables in contrast to a discrete or lumped parameter system which requires only a finite number of coordinates to specify its configuration.

Correlation Coefficient. The correlation coefficient of two variables is the ratio of the correlation function to the product of the averages of the variables:

$$\overline{x_1(t) \cdot x_2(t)} / \overline{x_1(t)} \cdot \overline{x_2(t)}$$

Correlation Function. The correlation function of two variables is the average value of their product:

$$\overline{x_1(t) \cdot x_2(t)}$$

Coulomb Damping (Dry Friction Damping). Coulomb damping is the dissipation of energy that occurs when a particle in a vibrating system is resisted by a force whose magnitude is a constant independent of displacement and velocity, and whose direction is opposite to the direction of the velocity of the particle.

Coupled Modes. Coupled modes are modes of vibration that are not independent but which influence one another because of energy transfer from one mode to the other. (See *Mode of Vibration.*)

Coupling Factor, Electromechanical. The electromechanical coupling factor is a factor used to characterize the extent to which the electrical characteristics of a transducer are modified by a coupled mechanical system, and vice versa.

Critical Damping. Critical damping is the minimum viscous damping that will allow a displaced system to return to its initial position without oscillation.

Critical Speed. Critical speed is a speed of a rotating system that corresponds to a resonant frequency of the system.

Cycle. A cycle is the complete sequence of values of a periodic quantity that occur during a period.

Damped Natural Frequency. The damped natural frequency is the frequency of free vibration of a damped linear system. The free vibration of a damped system may be considered periodic in the limited sense that the time interval between zero crossings in the same direction is constant, even though successive amplitudes decrease progressively. The frequency of the vibration is the reciprocal of this time interval.

Damping. Damping is the dissipation of energy with time or distance.

Damping Ratio. (See *Fraction of Critical Damping.*)

Decibel (db). The decibel is a unit which denotes the magnitude of a quantity with respect to an arbitrarily established reference value of the quantity, in terms of the logarithm (to the base 10) of the ratio of the quantities. For example, in electrical transmission circuits a value of power may be expressed in terms of a power level in decibels; the power level is given by 10 times the logarithm (to the base 10) of the ratio of the actual power to a reference power (which corresponds to 0 db).

Degrees-of-freedom. The number of degrees-of-freedom of a mechanical system is equal to the minimum number of independent coordinates required to define completely the positions of all parts of the system at any instant of time. In general, it is equal to the number of independent displacements that are possible.

Deterministic Function. A deterministic function is one whose value at any time can be predicted from its value at any other time.

Displacement. Displacement is a vector quantity that specifies the change of position of a body or particle and is usually measured from the mean position or position of rest. In general, it can be represented as a rotation vector or a translation vector, or both.

Distortion. Distortion is an undesired change in waveform. Noise and certain desired changes in waveform, such as those resulting from modulation or detection, are not usually classed as distortion.

Distributed System. (See *Continuous System.*)

Driving Point Impedance. Driving point impedance is the impedance involving the ratio of force to velocity when both the force and velocity are measured at the same point and in the same direction. (See *Impedance.*)

Dry Friction Damping. (See *Coulomb Damping.*)

Duration of Shock Pulse. The duration of a shock pulse is the time required for the acceleration of the pulse to rise from some stated fraction of the maximum amplitude and to decay to this value. (See *Shock Pulse.*)

Dynamic Vibration Absorber (Tuned Damper). A dynamic vibration absorber is an auxiliary mass-spring system which tends to neutralize vibration of a structure to which it is attached. The basic principle of operation is vibration out-of-phase with the vibration of such structure, thereby applying a counteracting force.

Effective Bandwidth. The effective bandwidth of a specified transmission system is the bandwidth of an ideal system which (1) has uniform transmission in its pass band equal to the maximum transmission of the specified system and (2) transmits the same power as the specified system when the two systems are receiving equal input signals having a uniform distribution of energy at all frequencies.

Electromechanical Coupling Factor. (See *Coupling Factor, Electromechanical.*)

Electrostriction. Electrostriction is the phenomenon wherein some dielectric materials experience an elastic strain when subjected to an electric field, this strain being independent of the polarity of the field.

Ensemble. A collection of signals. (Also see *Process.*)

Environment. (See *Natural Environments* and *Induced Environment.*)

Equivalent System. An equivalent system is one that may be substituted for another system for the purpose of analysis. Many types of equivalence are common in vibration and shock technology: (1) equivalent stiffness; (2) equivalent damping; (3) torsional system equivalent to a translational system; (4) electrical or acoustical system equivalent to a mechanical system; etc.

Equivalent Viscous Damping. Equivalent viscous damping is a value of viscous damping assumed for the purpose of analysis of a vibratory motion, such that the dissipation of energy per cycle at resonance is the same for either the assumed or actual damping force.

Ergodic Process. An ergodic process is a random process that is stationary and of such a nature that all possible time averages performed on one signal are independent of the signal chosen and hence are representative of the time averages of each of the other signals of the entire random process.

Excitation (Stimulus). Excitation is an external force (or other input) applied to a system that causes the system to respond in some way.

Filter. A filter is a device for separating waves on the basis of their frequency. It introduces relatively small insertion loss to waves in one or more frequency bands and relatively large insertion loss to waves of other frequencies. (See *Insertion Loss.*)

Force Factor. The force factor of an electromechanical transducer is: (*a*) the complex quotient of the force required to block the mechanical system divided by the corresponding current in the electric system; (*b*) the complex quotient of the resulting open-circuit voltage in the electric system divided by the velocity in the mechanical system. Force factors (*a*) and (*b*) have the same magnitude when consistent units are used and the transducer satisfies the principle of reciprocity. It is sometimes convenient in an electrostatic or piezoelectric transducer to use the ratios between force and charge or electric displacement, or between voltage and mechanical displacement.

Forced Vibration (Forced Oscillation). The oscillation of a system is forced if the response is imposed by the excitation. If the excitation is periodic and continuing, the oscillation is steady-state.

Foundation (Support). A foundation is a structure that supports the gravity load of a mechanical system. It may be fixed in space, or it may undergo a motion that provides excitation for the supported system.

Fraction of Critical Damping. The fraction of critical damping (damping ratio) for a system with viscous damping is the ratio of actual damping coefficient c to the critical damping coefficient c_c.

Free Vibration. Free vibration of a system is vibration that occurs in the absence of forced vibration.

Frequency. The frequency of a function periodic in time is the reciprocal of the period. The unit is the cycle per unit time and must be specified; the unit *cycle per second* is called *Hertz* (Hz).

Frequency, Angular. (See *Angular Frequency.*)

Fundamental Frequency. (1) The fundamental frequency of a periodic quantity is the frequency of a sinusoidal quantity which has the same period as the periodic quantity. (2) The fundamental frequency of an oscillating system is the lowest natural frequency. The normal mode of vibration associated with this frequency is known as the fundamental mode.

Fundamental Mode of Vibration. The fundamental mode of vibration of a system is the mode having the lowest natural frequency.

***g*.** The quantity g is the acceleration produced by the force of gravity, which varies with the latitude and elevation of the point of observation. By international agreement, the value 980.665 cm/sec^2 = 386.087 in./sec^2 = 32.1739 ft/sec^2 has been chosen as the standard acceleration due to gravity.

Harmonic. A harmonic is a sinusoidal quantity having a frequency that is an integral multiple of the frequency of a periodic quantity to which it is related.

Harmonic Motion. (See *Simple Harmonic Motion.*)

Harmonic Response. Harmonic response is the periodic response of a vibrating system exhibiting the characteristics of resonance at a frequency that is a multiple of the excitation frequency.

High-pass Filter. A high-pass filter is a wave filter having a single transmission band extending from some critical or cutoff frequency, not zero, up to infinite frequency.

Image Impedances. The image impedances of a structure or device are the impedances that will simultaneously terminate all of its inputs and outputs in such a way that at each of its inputs and outputs the impedances in both directions are equal.

Impact. An impact is a single collision of one mass in motion with a second mass which may be either in motion or at rest.

Impedance. Mechanical impedance is the ratio of a force-like quantity to a velocity-like quantity when the arguments of the real (or imaginary) parts of the quantities increase linearly with time. Examples of force-like quantities are: force, sound pressure, voltage, temperature. Examples of velocity-like quantities are: velocity, volume velocity, current, heat flow. *Impedance* is the reciprocal of *mobility.* (Also see *Angular Mechanical Impedance, Linear Mechanical Impedance, Driving Point Impedance,* and *Transfer Impedance.*)

Impulse. Impulse is the product of a force and the time during which the force is applied; more specifically, the impulse is $\int_{t_1}^{t_2} F dt$ where the force F is time dependent and equal to zero before time t_1 and after time t_2.

Induced Environments. Induced environments are those conditions generated as a result of the operation of a structure or equipment.

Insertion Loss. The insertion loss, in decibels, resulting from insertion of an element in a transmission system is 10 times the logarithm to the base 10 of the ratio of the power delivered to that part of the system that will follow the element, before the insertion of the element, to the power delivered to that same part of the system after insertion of the element.

Isolation. Isolation is a reduction in the capacity of a system to respond to an excitation, attained by the use of a resilient support. In steady-state forced vibration, isolation is expressed quantitatively as the complement of transmissibility.

Jerk. Jerk is a vector that specifies the time rate of change of acceleration; jerk is the third derivative of displacement with respect to time.

Line Spectrum. A line spectrum is a spectrum whose components occur at a number of discrete frequencies.

Linear Mechanical Impedance. Linear mechanical impedance is the impedance involving the ratio of force to linear velocity. (See *Impedance.*)

Linear System. A system is linear if for every element in the system the response is proportional to the excitation. This definition implies that the dynamic properties of each element in the system can be represented by a set of linear differential equations with constant coefficients, and that for the system as a whole superposition holds.

Logarithmic Decrement. The logarithmic decrement is the natural logarithm of the ratio of any two successive amplitudes of like sign, in the decay of a single-frequency oscillation.

Longitudinal Wave. A longitudinal wave in a medium is a wave in which the direction of displacement at each point of the medium is normal to the wave front.

Low-pass Filter. A low-pass filter is a wave filter having a single transmission band extending from zero frequency up to some critical or cutoff frequency which is not infinite.

Magnetic Recorder. A magnetic recorder is equipment incorporating an electromagnetic transducer and means for moving a ferromagnetic recording medium relative to the transducer for recording electric signals as magnetic variations in the medium.

Magnetostriction. Magnetostriction is the phenomenon wherein ferromagnetic materials experience an elastic strain when subjected to an external magnetic field. Also, magnetostriction is the converse phenomenon in which mechanical stresses cause a change in the magnetic induction of a ferromagnetic material.

Mechanical Admittance. (See *Mobility.*)

Mechanical Impedance. (See *Impedance.*)

Mechanical Shock. Mechanical shock is a nonperiodic excitation (e.g., a motion of the foundation or an applied force) of a mechanical system that is characterized by suddenness and severity, and usually causes significant relative displacements in the system.

Mechanical System. A mechanical system is an aggregate of matter comprising a defined configuration of mass, stiffness, and damping.

Mobility (Mechanical Admittance). Mobility is the ratio of a velocity-like quantity to a force-like quantity when the arguments of the real (or imaginary) parts of the quantities increase linearly with time. *Mobility* is the reciprocal of *impedance.* The terms *Angular Mobility, Linear Mobility, Driving-point Mobility,* and *Transfer Mobility* are used in the same sense as corresponding impedances.

Modal Numbers. When the normal modes of a system are related by a set of ordered integers, these integers are called modal numbers.

Mode of Vibration. In a system undergoing vibration, a mode of vibration is a characteristic pattern assumed by the system in which the motion of every particle is simple harmonic with the same frequency. Two or more modes may exist concurrently in a multiple degree-of-freedom system.

Modulation. Modulation is the variation in the value of some parameter which characterizes a periodic oscillation. Thus, amplitude modulation of a sinusoidal oscillation is a variation in the amplitude of the sinusoidal oscillation.

Multiple Degree-of-freedom System. A multiple degree-of-freedom system is one for which two or more coordinates are required to define completely the position of the system at any instant.

Narrow-band Random Vibration (Random Sine Wave). Narrow-band random vibration is random vibration having frequency components only within a narrow band. It has the appearance of a sine wave whose amplitude varies in an unpredictable manner. (See *Random Vibration.*)

Natural Environments. Natural environments are those conditions generated by the forces of nature and whose effects are experienced when the equipment or structure is at rest as well as when it is in operation.

Natural Frequency. Natural frequency is the frequency of free vibration of a system. For a multiple degree-of-freedom system, the natural frequencies are the frequencies of the normal modes of vibration.

Noise.* Noise is any undesired signal. By extension, noise is any unwanted disturbance within a useful frequency band, such as undesired electric waves in a transmission channel or device.

Nominal Bandwidth. The nominal bandwidth of a filter is the difference between the nominal upper and lower cutoff frequencies. The difference may be expressed (1) in cycles per second; (2) as a percentage of the pass-band center frequency; or (3) in octaves.

Nominal Pass-band Center Frequency. The nominal passband center frequency is the geometric mean of the nominal cutoff frequencies.

Nominal Upper and Lower Cutoff Frequencies. The nominal upper and lower cutoff frequencies of a filter pass-band are those frequencies above and below the frequency of maximum response of a filter at which the response to a sinusoidal signal is 3 db below the maximum response.

Nonlinear Damping. Nonlinear damping is damping due to a damping force that is not proportional to velocity.

Normal Mode of Vibration. A normal mode of vibration is a mode of vibration that is uncoupled from (i.e., can exist independently of) other modes of vibration of a system. When vibration of the system is defined as an eigenvalue problem, the normal modes are the eigenvectors and the normal mode frequencies are the eigenvalues. The term "classical normal mode" is sometimes applied to the normal modes of a vibrating system characterized by vibration of each element of the system at the same frequency and phase. In general, classical normal modes exist only in systems having no damping or having particular types of damping.

Oscillation. Oscillation is the variation, usually with time, of the magnitude of a quantity with respect to a specified reference when the magnitude is alternately greater and smaller than the reference.

Partial Node. A partial node is the point, line, or surface in a standing-wave system where some characteristic of the wave field has a minimum amplitude differing from zero. The appropriate modifier should be used with the words "partial node" to signify the type that is intended; e.g., displacement partial node, velocity partial node, pressure partial node.

Peak-to-peak Value. The peak-to-peak value of a vibrating quantity is the algebraic difference between the extremes of the quantity.

Period. The period of a periodic quantity is the smallest increment of the independent variable for which the function repeats itself.

Periodic Quantity. A periodic quantity is an oscillating quantity whose values recur for certain increments of the independent variable.

Phase of a Periodic Quantity. The phase of a periodic quantity, for a particular value of the independent variable, is the fractional part of a period through which the independent variable has advanced, measured from an arbitrary reference.

Pickup. (See *Transducer.*)

Piezoelectric (Crystal) (Ceramic) Transducer. A piezoelectric transducer is a transducer that depends for its operation on the interaction between the electric charge and the deformation of certain asymmetric crystals having piezoelectric properties.

Piezoelectricity. Piezoelectricity is the property exhibited by some asymmetrical crystalline materials which when subjected to strain in suitable directions develop electric polarization proportional to the strain. Inverse piezoelectricity is the effect in which mechanical strain is produced in certain asymmetrical crystalline materials when subjected to an external electric field; the strain is proportional to the electric field.

Power Spectral Density. Power spectral density is the limiting mean-square value (e.g., of acceleration, velocity, displacement, stress or other random variable) per unit bandwidth, i.e., the limit of the mean-square value in a given rectangular bandwidth divided by the bandwidth, as the bandwidth approaches zero.

Power Spectral Density Level. The spectrum level of a specified signal at a particular frequency is the level in decibels of that part of the signal contained within a band 1 cycle per second wide, centered at the particular frequency. Ordinarily this has significance only for a signal having a continuous distribution of components within the frequency range under consideration.

Process. A process is a collection of signals. The word *process* rather than the word *ensemble* ordinarily is used when it is desired to emphasize the properties the signals have or do not have as a group. Thus, one speaks of a stationary process rather than a stationary ensemble.

Pulse Rise Time. The pulse rise time is the interval of time required for the leading edge of a pulse to rise from some specified small fraction to some specified larger fraction of the maximum value.

***Q* (Quality Factor).** The quantity Q is a measure of the sharpness of resonance or frequency selectivity of a resonant vibratory system having a single degree of freedom, either mechanical or electrical. In a mechanical system, this quantity is equal to one-half the reciprocal of the damping ratio. It is commonly used only with reference to a lightly damped system, and is then approximately equal to the following:

(1) Transmissibility at resonance,
(2) π/logarithmic decrement,
(3) $2\pi W/\Delta W$ where W is the stored energy and ΔW the energy dissipation per cycle, and
(4) $f_r/\Delta f$ where f_r is the resonant frequency and Δf is the bandwidth between the half-power points.

Quasi-ergodic Process. A quasi-ergodic process is a random process which is not necessarily stationary but of such a nature that some time averages performed on a signal are independent of the signal chosen.

Quasi-periodic Signal. A signal consisting only of quasi-sinusoids.

Quasi-sinusoid. A function of the form $a = A \sin (2\pi ft - \phi)$ where either A or f, or both, is not a constant but may be expressed readily as a function of time. Ordinarily ϕ is considered constant.

Random Sine Wave. (See *Narrow-band Random Vibration.*)

Random Vibration. Random vibration is vibration whose instantaneous magnitude is not specified for any given instant of time. The instantaneous magnitudes of a random vibration are specified only by probability distribution functions giving the probable fraction of the total time that the magnitude (or some sequence of magnitudes) lies within a specified range. Random vibration contains no periodic or quasi-periodic constituents. If random vibration has instantaneous magnitudes that occur according to the Gaussian distribution, it is called "Gaussian random vibration."

Ratio of Critical Damping. (See *Fraction of Critical Damping.*)

Rayleigh Wave. A Rayleigh wave is a surface wave associated with the free boundary of a solid, such that a surface particle describes an ellipse whose major axis is normal to the surface, and whose center is at the undisturbed surface. At maximum particle displacement away from the solid surface the motion of the particle is opposite to that of the wave.

Recording Channel. The term "recording channel" refers to one of a number of independent recorders in a recording system or to independent recording tracks on a recording medium.

Recording System. A recording system is a combination of transducing devices and associated equipment suitable for storing signals in a form capable of subsequent reproduction.

Relaxation Time. Relaxation time is the time taken by an exponentially decaying quantity to decrease in amplitude by a factor of $1/e = 0.3679$.

Re-recording. Re-recording is the process of making a recording by reproducing a recorded signal source and recording this reproduction.

Resonance. Resonance of a system in forced vibration exists when any change, however small, in the frequency of excitation causes a decrease in the response of the system.

Resonance Frequency (Resonant Frequency). A frequency at which resonance exists.

Response. The response of a device or system is the motion (or other output) resulting from an excitation (stimulus) under specified conditions.

Response Spectrum. (See *Shock Spectrum.*)

Rotational Mechanical Impedance. (See *Angular Mechanical Impedance.*)

Self-induced (Self-excited) Vibration.* The vibration of a mechanical system is self-induced if it results from conversion, within the system, of nonoscillatory excitation to oscillatory excitation.

Shake Table. (See *Vibration Machine.*)

Shear Wave (Rotational Wave). A shear wave is a wave in an elastic medium which causes an element of the medium to change its shape without a change of volume.

Shock. (See *Mechanical Shock.*)

Shock Absorber. A shock absorber is a device which dissipates energy to modify the response of a mechanical system to applied shock.

Shock Isolator (Shock Mount). A shock isolator is a resilient support that tends to isolate a system from a shock motion.

Shock Machine. A shock machine is a device for subjecting a system to controlled and reproducible mechanical shock.

Shock Motion. Shock motion is an excitation involving motion of a foundation. (See *Foundation* and *Mechanical Shock.*)

Shock Mount. (See *Shock Isolator.*)

Shock Pulse. A shock pulse is a substantial disturbance characterized by a rise of acceleration from a constant value and decay of acceleration to the constant value in a short period of time. Shock pulses are normally displayed graphically as curves of acceleration as functions of time.

Shock-pulse Duration. (See *Duration of Shock Pulse.*)

Shock Spectrum (Response Spectrum). A shock spectrum is a plot of the maximum response experienced by a single degree-of-freedom system, as a function of its own natural frequency, in response to an applied shock. The response may be expressed in terms of acceleration, velocity or displacement.

Signal. A signal is (1) a disturbance used to convey information; (2) the information to be conveyed over a communication system.

Simple Harmonic Motion. A simple harmonic motion is a motion such that the displacement is a sinusoidal function of time; sometimes it is designated merely by the term *harmonic motion.*

Single Degree-of-freedom System. A single degree-of-freedom system is one for which only one coordinate is required to define completely the configuration of the system at any instant.

Sinusoidal Motion. (See *Simple Harmonic Motion.*)

Snubber. A snubber is a device used to increase the stiffness of an elastic system (usually by a large factor) whenever the displacement becomes larger than a specified value.

Spectrum. A spectrum is a definition of the magnitude of the frequency components that constitute a quantity.

Spectrum Density. (See *Power Spectral Density.*)

Standard Deviation. Standard deviation is the square root of the variance; i.e., the square root of the mean of the squares of the deviations from the mean value of a vibrating quantity.

Standing Wave. A standing wave is a periodic wave having a fixed distribution in space which is the result of interference of progressive waves of the same frequency and kind. Such waves are characterized by the existence of nodes or partial nodes and antinodes that are fixed in space.

Stationary Process. A stationary process is an ensemble of signals such that an average of values over the ensemble at any given time is independent of time.

Stationary Signal. A stationary signal is a random signal of such nature that averages over samples of finite time intervals are independent of the time at which the sample occurs.

Steady-state Vibration. Steady-state vibration exists in a system if the velocity of each particle is a continuing periodic quantity.

Stiffness. Stiffness is the ratio of change of force (or torque) to the corresponding change in translational (or rotational) deflection of an elastic element.

Subharmonic. A subharmonic is a sinusoidal quantity having a frequency that is an integral submultiple of the fundamental frequency of a periodic quantity to which it is related.

Subharmonic Response. Subharmonic response is the periodic response of a mechanical system exhibiting the characteristic of resonance at a frequency that is a submultiple of the frequency of the periodic excitation.

Superharmonic Response. Superharmonic response is a term sometimes used to denote a particular type of harmonic response which dominates the total response of the system; it frequently occurs when the excitation frequency is a submultiple of the frequency of the fundamental resonance.

Transducer (Pickup). A transducer is a device which converts shock or vibratory motion into an optical, a mechanical, or most commonly to an electrical signal that is proportional to a parameter of the experienced motion.

Transfer Impedance. Transfer impedance between two points is the impedance involving the ratio of force to velocity when force is measured at one point and velocity at the other point. The term *transfer impedance* also is used to denote the ratio of force to velocity measured at the same point but in different directions. (See *Impedance.*)

Transient Vibration. Transient vibration is temporarily sustained vibration of a mechanical system. It may consist of forced or free vibration or both.

Transmissibility. Transmissibility is the nondimensional ratio of the response amplitude of a system in steady-state forced vibration to the excitation amplitude. The ratio may be one of forces, displacements, velocities, or accelerations.

Transmission Loss. Transmission loss is the reduction in the magnitude of some characteristic of a signal, between two stated points in a transmission system.

Transverse Wave. A transverse wave is a wave in which the direction of displacement at each point of the medium is parallel to the wave front.

Tuned Damper. (See *Dynamic Vibration Absorber.*)

Uncorrelated. Two signals or variables $a_1(t)$ and $a_2(t)$ are said to be uncorrelated if the average value of their product is zero: $\overline{a_1(t) \cdot a_2(t)} = 0$. If the correlation coefficient is equal to unity, the variables are said to be completely correlated. If the coefficient is less than unity but larger than zero, they are said to be partially correlated. (See *Correlation Coefficient.*)

Uncoupled Mode. An uncoupled mode of vibration is a mode that can exist in a system concurrently with and independently of other modes.

Undamped Natural Frequency. The undamped natural frequency of a mechanical system is the frequency of free vibration resulting from only elastic and inertial forces of the system.

Variance. Variance is the mean of the squares of the deviations from the mean value of a vibrating quantity.

Velocity. Velocity is a vector quantity that specifies the time rate of change of displacement with respect to a reference frame. If the reference frame is not inertial, the velocity is often designated "relative velocity."

Velocity Shock. Velocity shock is a particular type of shock motion characterized by a sudden velocity change of the foundation. (See *Foundation* and *Mechanical Shock.*)

Vibration. Vibration is an oscillation wherein the quantity is a parameter that defines the motion of a mechanical system. (See *Oscillation.*)

Vibration Isolator. A vibration isolator is a resilient support that tends to isolate a system from steady-state excitation.

Vibration Machine. A vibration machine is a device for subjecting a mechanical system to controlled and reproducible mechanical vibration.

Vibration Meter. A vibration meter is an apparatus for the measurement of displacement, velocity, or acceleration of a vibrating body.

Vibration Mount. (See *Vibration Isolator.*)

Vibration Pickup. (See *Transducer.*)

Viscous Damping. Viscous damping is the dissipation of energy that occurs when a particle in a vibrating system is resisted by a force that has a magnitude proportional to the magnitude of the velocity of the particle and direction opposite to the direction of the particle.

Viscous Damping, Equivalent. (See *Equivalent Viscous Damping.*)

Wave. A wave is a disturbance which is propagated in a medium in such a manner that at any point in the medium the quantity serving as measure of disturbance is a function of the time, while at any instant the displacement at a point is a function of the position of the point. Any physical quantity that has the same relationship to some independent variable (usually time) that a propagated disturbance has, at a particular instant, with respect to space, may be called a wave.

Wave Interference. Wave interference is the phenomenon which results when waves of the same or nearly the same frequency are superposed; it is characterized by a spatial or temporal distribution of amplitude of some specified characteristic differing from that of the individual superposed waves.

Wavelength. The wavelength of a periodic wave in an isotropic medium is the perpendicular distance between two wave fronts in which the displacements have a difference in phase of one complete period.

White Noise. White noise is a noise whose power spectral density is substantially independent of frequency over a specified range.

2

BASIC VIBRATION THEORY

Ralph E. Blake
Lockheed Aircraft Corporation

INTRODUCTION

This chapter presents the theory of free and forced steady-state vibration of single degree-of-freedom systems. Undamped systems and systems having viscous damping and structural damping are included. Multiple degree-of-freedom systems are discussed, including the normal-mode theory of linear elastic structures and Lagrange's equations.

ELEMENTARY PARTS OF VIBRATORY SYSTEMS

Vibratory systems are comprised of means for storing potential energy (spring), means for storing kinetic energy (mass or inertia), and means by which the energy is gradually lost (damper). The vibration of a system involves the alternating transfer of energy between its potential and kinetic forms. In a damped system, some energy is dissipated at each cycle of vibration and must be replaced from an external source if a steady vibration is to be maintained. Although a single physical structure may store both kinetic and potential energy, and may dissipate energy, this chapter considers only *lumped parameter systems* comprised of ideal springs, masses, and dampers wherein each element has only a single function. In translational motion, displacements are defined as linear distances; in rotational motion, displacements are defined as angular motions.

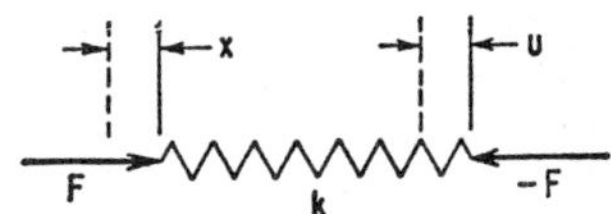

FIG. 2.1. Linear spring.

TRANSLATIONAL MOTION

SPRING. In the linear spring shown in Fig. 2.1, the change in the length of the spring is proportional to the force acting along its length:

$$F = k(x - u) \tag{2.1}$$

The ideal spring is considered to have no mass; thus, the force acting on one end is equal and opposite to the force acting on the other end. The constant of proportionality k is the *spring constant or stiffness.*

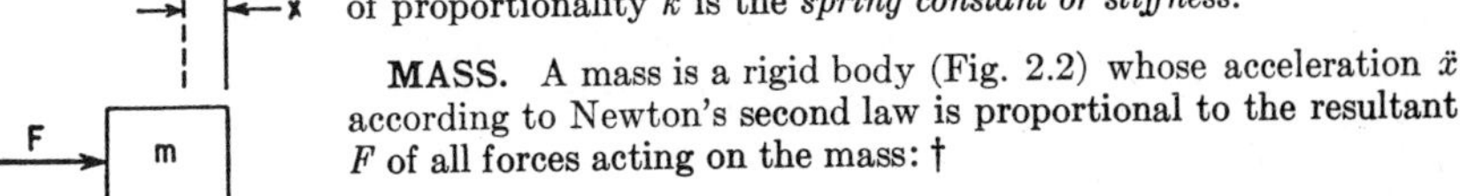

FIG. 2.2. Rigid mass.

MASS. A mass is a rigid body (Fig. 2.2) whose acceleration $\ddot{x}$ according to Newton's second law is proportional to the resultant F of all forces acting on the mass: †

$$F = m\ddot{x} \tag{2.2}$$

† It is common to use the word "mass" in a general sense to designate a rigid body. Mathematically, the mass of the rigid body is defined by m in Eq. (2.2).

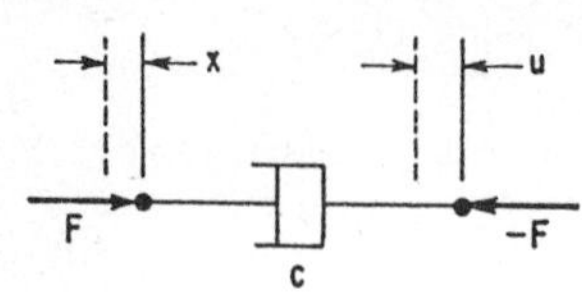

FIG. 2.3. Viscous damper.

DAMPER. In the viscous damper shown in Fig. 2.3, the applied force is proportional to the relative velocity of its connection points:

$$F = c(\dot{x} - \dot{u}) \tag{2.3}$$

The constant c is the *damping coefficient*, the characteristic parameter of the damper. The ideal damper is considered to have no mass; thus the force at one end is equal and opposite to the force at the other end. "Structural damping" is considered below and several other types of damping are considered in Chap. 30.

ROTATIONAL MOTION

The elements of a mechanical system which moves with pure rotation of the parts are wholly analogous to the elements of a system that moves with pure translation. The property of a rotational system which stores kinetic energy is inertia; stiffness and damping coefficients are defined with reference to angular displacement and angular velocity, respectively. The analogous quantities and equations are listed in Table 2.1.

Table 2.1. Analogous Quantities in Translational and Rotational Vibrating Systems

Translational quantity	*Rotational quantity*
Linear displacement x	Angular displacement α
Force F	Torque M
Spring constant k	Spring constant k_r
Damping constant c	Damping constant c_r
Mass m	Moment of inertia I
Spring law $F = k(x_1 - x_2)$	Spring law $M = k_r(\alpha_1 - \alpha_2)$
Damping law $F = c(\dot{x}_1 - \dot{x}_2)$	Damping law $M = c_r(\dot{\alpha}_1 - \dot{\alpha}_2)$
Inertia law $F = m\ddot{x}$	Inertia law $M = I\ddot{\alpha}$

Inasmuch as the mathematical equations for a rotational system can be written by analogy from the equations for a translational system, only the latter are discussed in detail. Whenever translational systems are discussed, it is understood that corresponding equations apply to the analogous rotational system, as indicated in Table 2.1.

SINGLE DEGREE-OF-FREEDOM SYSTEM

The simplest possible vibratory system is shown in Fig. 2.4; it consists of a mass m attached by means of a spring k to an immovable support. The mass is constrained to translational motion in the direction of the X axis so that its change of position from an initial reference is described fully by the value of a single quantity x. For this reason it is called a *single degree-of-freedom system.* If the mass m is displaced from its equilibrium position and then allowed to vibrate free from further external forces, it is said to have *free vibration.* The vibration also may be forced; i.e., a continuing force acts upon the mass or the foundation experiences a continuing motion. Free and forced vibration are discussed below.

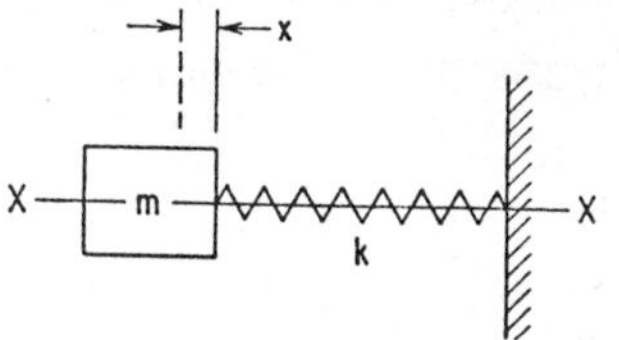

FIG. 2.4. Undamped single degree-of-freedom system.

FREE VIBRATION WITHOUT DAMPING [1, 2, 3]

Considering first the free vibration of the undamped system of Fig. 2.4, Newton's equation is written for the mass m. The force $m\ddot{x}$ exerted by the mass on the spring is equal and opposite to the force kx applied by the spring on the mass:

$$m\ddot{x} + kx = 0 \tag{2.4}$$

where $x = 0$ defines the equilibrium position of the mass.

The solution of Eq. (2.4) is

$$x = A \sin \sqrt{\frac{k}{m}}\, t + B \cos \sqrt{\frac{k}{m}}\, t \tag{2.5}$$

where the term $\sqrt{k/m}$ is the *angular natural frequency* defined by

$$\omega_n = \sqrt{\frac{k}{m}} \qquad \text{rad/sec} \tag{2.6}$$

The sinusoidal oscillation of the mass repeats continuously, and the time interval to complete one cycle is the *period:*

$$\tau = \frac{2\pi}{\omega_n} \tag{2.7}$$

The reciprocal of the period is the *natural frequency:*

$$f_n = \frac{1}{\tau} = \frac{\omega_n}{2\pi} = \frac{1}{2\pi}\sqrt{\frac{k}{m}} = \frac{1}{2\pi}\sqrt{\frac{kg}{W}} \tag{2.8}$$

where $W = mg$ is the weight of the rigid body forming the mass of the system shown in Fig. 2.4. The relations of Eq. (2.8) are shown by the solid lines in Fig. 2.5.

INITIAL CONDITIONS. In Eq. (2.5), B is the value of x at time $t = 0$, and the value of A is equal to $\dot{x}/\omega_n$ at time $t = 0$. Thus, the conditions of displacement and velocity which exist at zero time determine the subsequent oscillation completely.

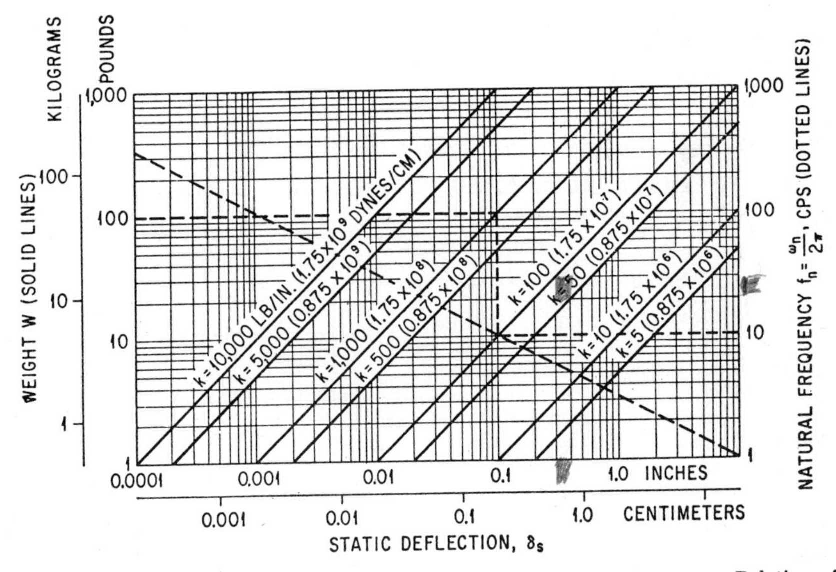

FIG. 2.5. Natural frequency relations for a single degree-of-freedom system. Relation of natural frequency to weight of supported body and stiffness of spring [Eq. (2.8)] is shown by solid lines. Relation of natural frequency to static deflection [Eq. (2.10)] is shown by diagonal-dashed line. Example: To find natural frequency of system with $W = 100$ lb and $k = 1{,}000$ lb/in., enter at $W = 100$ on left ordinate scale; follow the dashed line horizontally to solid line $k = 1{,}000$, then vertically down to diagonal-dashed line, and finally horizontally to read $f_n = 10$ cps from right ordinate scale.

PHASE ANGLE. Equation (2.5) for the displacement in oscillatory motion can be written, introducing the frequency relation of Eq. (2.6),

$$x = A \sin \omega_n t + B \cos \omega_n t = C \sin (\omega_n t + \theta) \tag{2.9}$$

where $C = (A^2 + B^2)^{1/2}$ and $\theta = \tan^{-1} (B/A)$. The angle θ is called the *phase angle.*

STATIC DEFLECTION. The static deflection of a simple mass-spring system is the deflection of spring k as a result of the gravity force of the mass, $\delta_{st} = mg/k$. (For example, the system of Fig. 2.4 would be oriented with the mass m vertically above the spring k.) Substituting this relation in Eq. (2.8),

$$f_n = \frac{1}{2\pi}\sqrt{\frac{g}{\delta_{st}}} \tag{2.10}$$

The relation of Eq. (2.10) is shown by the diagonal-dashed line in Fig. 2.5. This relation applies only when the system under consideration is both linear and elastic. For example, rubber springs tend to be nonlinear or exhibit a dynamic stiffness which differs from the static stiffness; hence, Eq. (2.10) is not applicable.

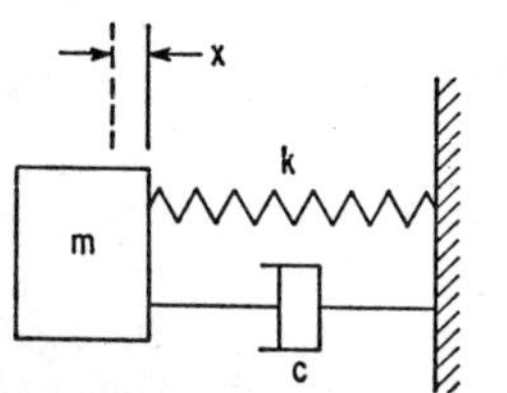

FIG. 2.6. Single degree-of-freedom system with viscous damper.

FREE VIBRATION WITH VISCOUS DAMPING [1, 2, 3]

Figure 2.6 shows a single degree-of-freedom system with a viscous damper. The differential equation of motion of mass m, corresponding to Eq. (2.4) for the undamped system, is

$$m\ddot{x} + c\dot{x} + kx = 0 \tag{2.11}$$

The form of the solution of this equation depends upon whether the damping coefficient is equal to, greater than, or less than the *critical damping coefficient* c_c:

$$c_c = 2\sqrt{km} = 2m\omega_n \tag{2.12}$$

The ratio $\zeta = c/c_c$ is defined as the *fraction of critical damping.*

LESS-THAN-CRITICAL-DAMPING. If the damping of the system is less than critical, $\zeta < 1$; then the solution of Eq. (2.11) is

$$x = e^{-ct/2m}(A \sin \omega_d t + B \cos \omega_d t)$$
$$= Ce^{-ct/2m} \sin (\omega_d t + \theta) \tag{2.13}$$

where C and θ are defined with reference to Eq. (2.9). The *damped natural frequency* is related to the undamped natural frequency of Eq. (2.6) by the equation

$$\omega_d = \omega_n(1 - \zeta^2)^{1/2} \quad \text{rad/sec} \tag{2.14}$$

Equation (2.14), relating the damped and undamped natural frequencies, is plotted in Fig. 2.7.

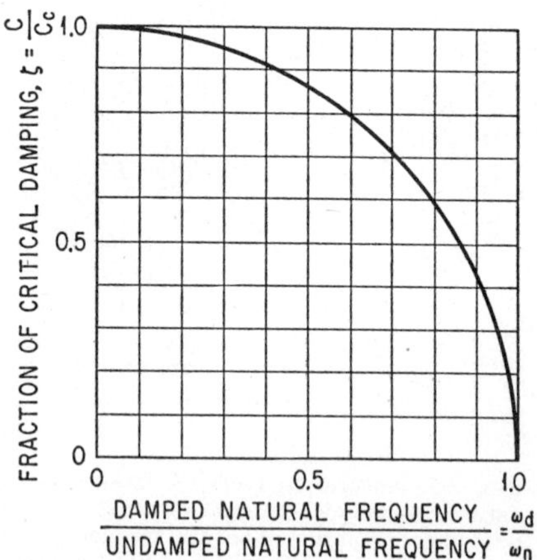

FIG. 2.7. Damped natural frequency as a function of undamped natural frequency and fraction of critical damping.

CRITICAL DAMPING. When $c = c_c$, there is no oscillation and the solution of Eq. (2.11) is

$$x = (A + Bt)e^{-ct/2m} \tag{2.15}$$

GREATER-THAN-CRITICAL-DAMPING. When $\zeta > 1$, the solution of Eq. (2.11) is

$$x = e^{-ct/2m}(Ae^{\omega_n\sqrt{\zeta^2-1}\,t} + Be^{-\omega_n\sqrt{\zeta^2-1}\,t}) \tag{2.16}$$

This is a nonoscillatory motion; if the system is displaced from its equilibrium position, it tends to return gradually.

LOGARITHMIC DECREMENT. The degree of damping in a system having $\zeta < 1$ may be defined in terms of successive peak values in a record of a free oscillation. Substituting the expression for critical damping from Eq. (2.12), the expression for free vibration of a damped system, Eq. (2.13), becomes

$$x = Ce^{-\zeta\omega_n t}\sin(\omega_d t + \theta) \tag{2.17}$$

Consider any two maxima (i.e., value of x when $dx/dt = 0$) separated by n cycles of oscillation, as shown in Fig. 2.8. Then the ratio of these maxima is

$$\frac{x_n}{x_0} = e^{-2\pi n\zeta/(1-\zeta^2)^{1/2}} \tag{2.18}$$

Values of x_n/x_0 are plotted in Fig. 2.9 for several values of n over the range of ζ from 0.001 to 0.10.

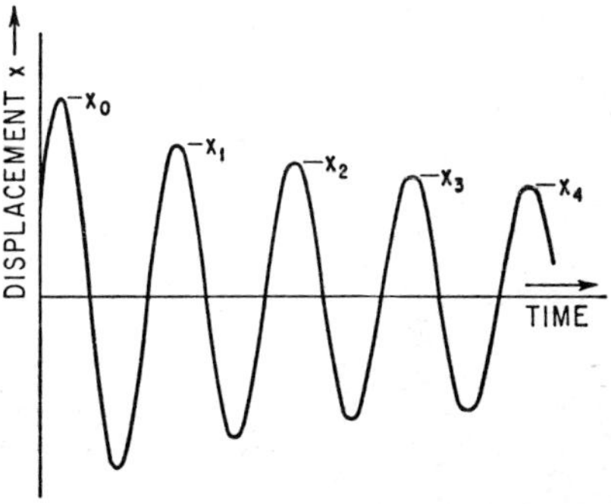

FIG. 2.8. Trace of damped free vibration showing amplitudes of displacement maxima.

The *logarithmic decrement* Δ is the natural logarithm of the ratio of the amplitudes of two successive cycles of the damped free vibration:

$$\Delta = \log\frac{x_1}{x_2} \qquad \text{or} \qquad \frac{x_2}{x_1} = e^{-\Delta} \tag{2.19}$$

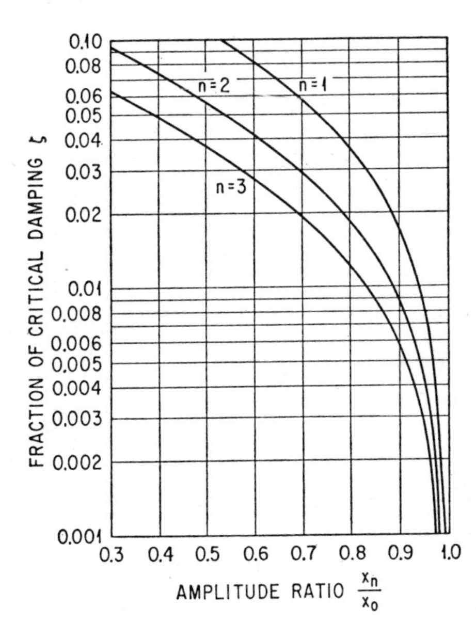

FIG. 2.9. Effect of damping upon the ratio of displacement maxima of a damped free vibration.

A comparison of this relation with Eq. (2.18) when $n = 1$ gives the following expression for Δ:

$$\Delta = \frac{2\pi\zeta}{(1-\zeta^2)^{1/2}} \tag{2.20}$$

The logarithmic decrement can be expressed in terms of the difference of successive amplitudes by writing Eq. (2.19) as follows:

$$\frac{x_1 - x_2}{x_1} = 1 - \frac{x_2}{x_1} = 1 - e^{-\Delta}$$

Writing $e^{-\Delta}$ in terms of its infinite series, the following expression is obtained which gives a good approximation for $\Delta < 0.2$:

$$\frac{x_1 - x_2}{x_1} = \Delta \tag{2.21}$$

For small values of ζ (less than about 0.10), an approximate relation between the fraction of critical damping and the logarithmic decrement, from Eq. (2.20), is

$$\Delta \simeq 2\pi\zeta \tag{2.22}$$

FORCED VIBRATION

Forced vibration in this chapter refers to the motion of the system which occurs in response to a continuing excitation whose magnitude varies sinusoidally with time. (See Chaps. 8 and 23 for a treatment of the response of a simple system to step, pulse, and transient vibration excitations.) The excitation may be, alternatively, force applied to the system (generally, the force is applied to the mass of a single degree-of-freedom system) or motion of the foundation that supports the system. The resulting response of the system can be expressed in different ways, depending upon the nature of the excitation and the use to be made of the result.

1. If the excitation is a force applied to the mass of the system shown in Fig. 2.4, the result may be expressed in terms of (*a*) the amplitude of the resulting motion of the mass or (*b*) the fraction of the applied force amplitude that is transmitted through the system to the support. The former is termed the *motion response* and the latter is termed the *force transmissibility*.

2. If the excitation is a motion of the foundation, the resulting response usually is expressed in terms of the amplitude of the motion of the mass relative to the amplitude of the motion of the foundation. This is termed the *motion transmissibility* for the system.

In general, the response and transmissibility relations are functions of the forcing frequency, and vary with different types and degrees of damping. Results are presented in this chapter for undamped systems, and for systems with either viscous or structural damping. Corresponding results are given in Chap. 30 for systems with Coulomb damping, and for systems with either viscous or Coulomb damping in series with a linear spring.

FORCED VIBRATION WITHOUT DAMPING

FORCE APPLIED TO MASS. When the sinusoidal force $F = F_0 \sin \omega t$ is applied to the mass of the undamped single degree-of-freedom system shown in Fig. 2.10, the differential equation of motion is

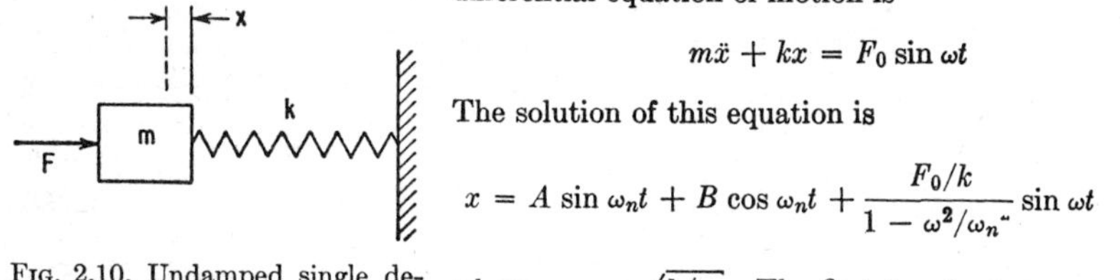

Fig. 2.10. Undamped single degree-of-freedom system excited in forced vibration by force acting on mass.

$$m\ddot{x} + kx = F_0 \sin \omega t \tag{2.23}$$

The solution of this equation is

$$x = A \sin \omega_n t + B \cos \omega_n t + \frac{F_0/k}{1 - \omega^2/\omega_n^2} \sin \omega t \tag{2.24}$$

where $\omega_n = \sqrt{k/m}$. The first two terms represent an oscillation at the undamped natural frequency ω_n. The coefficient B is the value of x at time $t = 0$, and the coefficient A may be found from the velocity at time $t = 0$. Differentiating Eq. (2.24) and setting $t = 0$,

$$\dot{x}(0) = A\omega_n + \frac{\omega F_0/k}{1 - \omega^2/\omega_n^2} \tag{2.25}$$

The value of A is found from Eq. (2.25).

The oscillation at the natural frequency ω_n gradually decays to zero in physical systems because of damping. The steady-state oscillation at forcing frequency ω is

$$x = \frac{F_0/k}{1 - \omega^2/\omega_n^2} \sin \omega t \tag{2.26}$$

This oscillation exists after a condition of equilibrium has been established by decay of the oscillation at the natural frequency ω_n and persists as long as the force F is applied.

The force transmitted to the foundation is directly proportional to the spring deflection: $F_t = kx$. Substituting x from Eq. (2.26) and defining transmissibility $T = F_t/F$,

$$T = \frac{1}{1 - \omega^2/\omega_n^2} \tag{2.27}$$

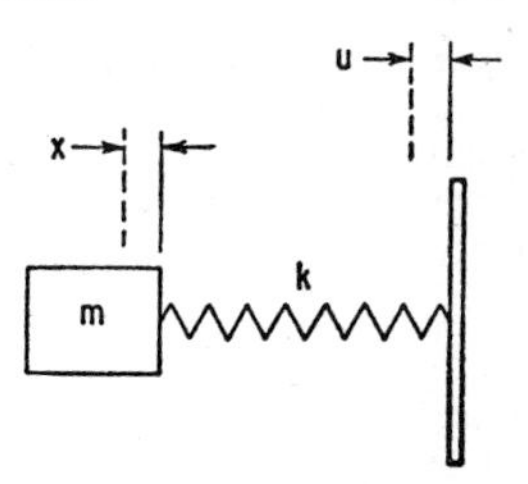

FIG. 2.11. Undamped single degree-of-freedom system excited in forced vibration by motion of foundation.

If the mass is initially at rest in the equilibrium position of the system (i.e., $x = 0$ and $\dot{x} = 0$) at time $t = 0$, the ensuing motion at time $t > 0$ is

$$x = \frac{F_0/k}{1 - \omega^2/\omega_n^2}\left(\sin \omega t - \frac{\omega}{\omega_n} \sin \omega_n t\right) \tag{2.28}$$

For large values of time, the second term disappears because of the damping inherent in any physical system, and Eq. (2.28) becomes identical to Eq. (2.26).

When the forcing frequency coincides with the natural frequency, $\omega = \omega_n$ and a condition of resonance exists. Then Eq. (2.28) is indeterminate and the expression for x may be written as

$$x = -\frac{F_0\omega}{2k} t \cos \omega t \tag{2.29}$$

According to Eq. (2.29), the amplitude x increases continuously with time, reaching an infinitely great value only after an infinitely great time.

MOTION OF FOUNDATION. The differential equation of motion for the system of Fig. 2.11 excited by a continuing motion $u = u_0 \sin \omega t$ of the foundation is

$$m\ddot{x} = -k(x - u_0 \sin \omega t)$$

The solution of this equation is

$$x = A_1 \sin \omega_n t + B_2 \cos \omega_n t + \frac{u_0}{1 - \omega^2/\omega_n^2} \sin \omega t$$

where $\omega_n = \sqrt{k/m}$ and the coefficients A_1, B_1 are determined by the velocity and displacement of the mass, respectively, at time $t = 0$. The terms representing oscillation at the natural frequency are damped out ultimately, and the ratio of amplitudes is defined in terms of transmissibility T:

$$\frac{x_0}{u_0} = T = \frac{1}{1 - \omega^2/\omega_n^2} \tag{2.30}$$

where $x = x_0 \sin \omega t$. Thus, in the forced vibration of an undamped single degree-of-freedom system, the motion response, the force transmissibility, and the motion transmissibility are numerically equal.

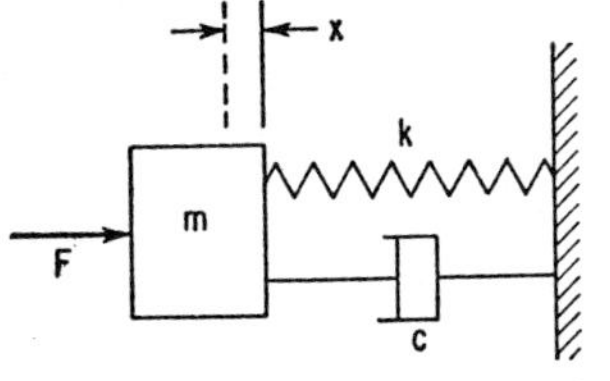

FIG. 2.12. Single degree-of-freedom system with viscous damping, excited in forced vibration by force acting on mass.

FORCED VIBRATION WITH VISCOUS DAMPING

FORCE APPLIED TO MASS. The differential equation of motion for the single degree-of-freedom system with viscous damping shown in Fig. 2.12, when the excitation is a force $F = F_0 \sin \omega t$ applied to the mass, is

$$m\ddot{x} + c\dot{x} + kx = F_0 \sin \omega t \tag{2.31}$$

Equation (2.31) corresponds to Eq. (2.23) for forced vibration of an undamped system; its solution would correspond to Eq. (2.24) in that it includes terms representing oscillation at the natural frequency. In a damped system, however, these terms are damped out rapidly and only the steady-state solution usually is considered. The resulting motion occurs at the forcing frequency ω; when the damping coefficient c is greater than zero, the phase between the force and resulting motion is different than zero. Thus, the

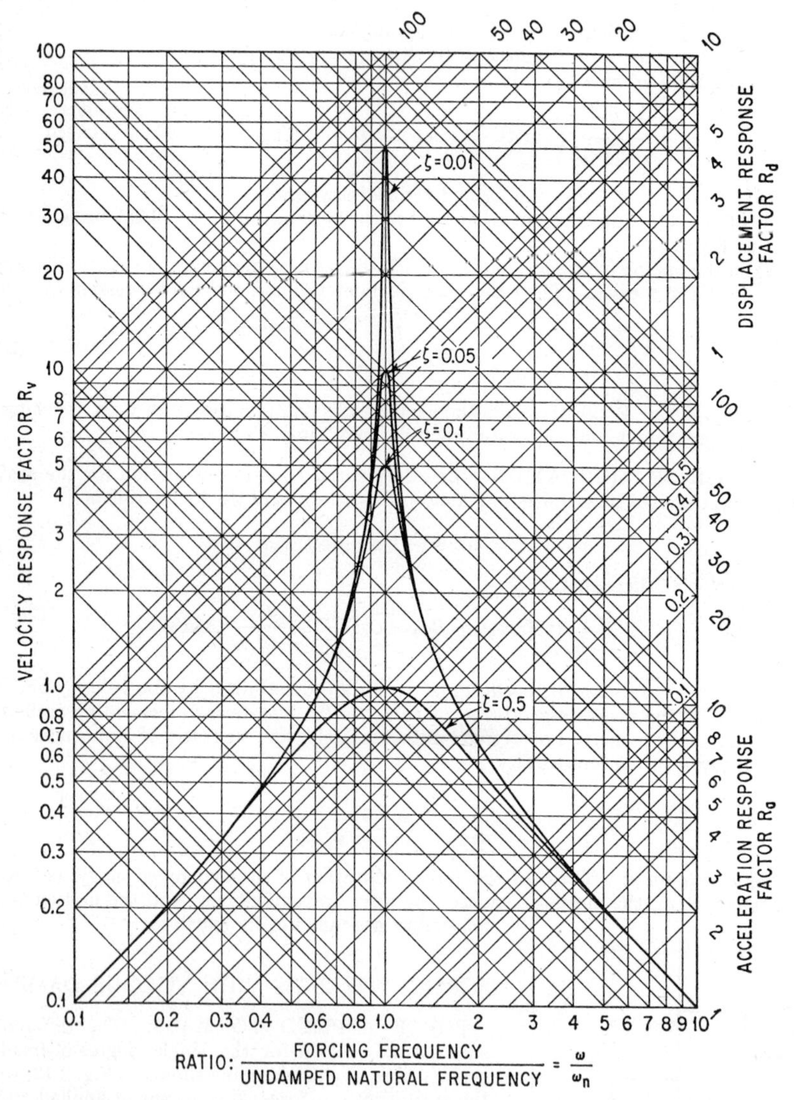

FIG. 2.13. Response factors for a viscous-damped single degree-of-freedom system excited in forced vibration by a force acting on the mass. The velocity response factor shown by horizontal lines is defined by Eq. (2.36); the displacement response factor shown by diagonal lines of positive slope is defined by Eq. (2.33); and the acceleration response factor shown by diagonal lines of negative slope is defined by Eq. (2.37).

response may be written

$$x = R \sin(\omega t - \theta) = A_1 \sin \omega t + B_1 \cos \omega t \tag{2.32}$$

Substituting this relation in Eq. (2.31), the following result is obtained:

$$\frac{x}{F_0/k} = \frac{\sin(\omega t - \theta)}{\sqrt{(1 - \omega^2/\omega_n^2)^2 + (2\zeta\omega/\omega_n)^2}} = R_d \sin(\omega t - \theta) \tag{2.33}$$

where
$$\theta = \tan^{-1}\left(\frac{2\zeta\omega/\omega_n}{1 - \omega^2/\omega_n^2}\right)$$

and R_d is a *dimensionless response factor* giving the ratio of the amplitude of the vibratory displacement to the spring displacement that would occur if the force F were applied statically. At very low frequencies R_d is approximately equal to 1; it rises to a peak near ω_n and approaches zero as ω becomes very large. The displacement response is defined at these frequency conditions as follows:

$$x \simeq \left(\frac{F_0}{k}\right) \sin \omega t \qquad [\omega \ll \omega_n]$$

$$x = \frac{F_0}{2k\zeta} \sin\left(\omega_n t + \frac{\pi}{2}\right) = -\frac{F_0 \cos \omega_n t}{c\omega_n} \qquad [\omega = \omega_n] \tag{2.34}$$

$$x \simeq \frac{\omega_n^2 F_0}{\omega^2 k} \sin(\omega t + \pi) = \frac{F_0}{m\omega^2} \sin \omega t \qquad [\omega \gg \omega_n]$$

For the above three frequency conditions, the vibrating system is sometimes described as *spring-controlled, damper-controlled,* and *mass-controlled,* respectively, depending on which element is primarily responsible for the system behavior.

Curves showing the dimensionless response factor R_d as a function of the frequency ratio ω/ω_n are plotted in Fig. 2.13 on the coordinate lines having a positive 45° slope. Curves of the phase angle θ are plotted in Fig. 2.14. A phase angle between 180 and 360°

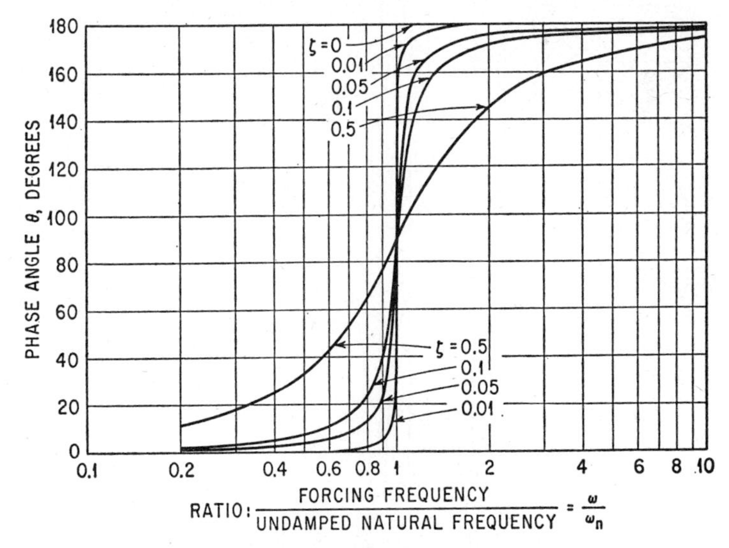

FIG. 2.14. Phase angle between the response displacement and the excitation force for a single degree-of-freedom system with viscous damping, excited by a force acting on the mass of the system.

cannot exist in this case since this would mean that the damper is furnishing energy to the system rather than dissipating it.

An alternative form of Eqs. (2.33) and (2.34) is

$$\frac{x}{F_0/k} = \frac{(1 - \omega^2/\omega_n^2) \sin \omega t - 2\zeta(\omega/\omega_n) \cos \omega t}{(1 - \omega^2/\omega_n^2)^2 + (2\zeta\omega/\omega_n)^2} \tag{2.35}$$

$$= (R_d)_x \sin \omega t + (R_d)_R \cos \omega t$$

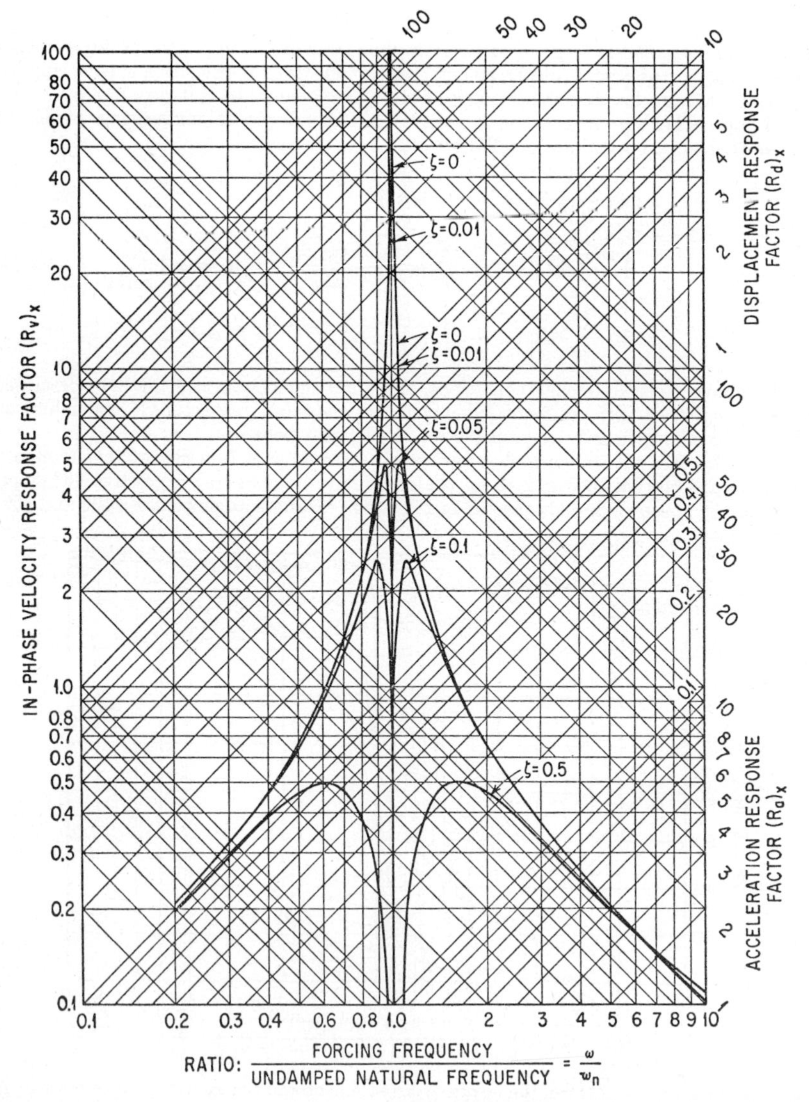

FIG. 2.15. In-phase component of response factor of a viscous-damped system in forced vibration. All values of the response factor for $\omega/\omega_n > 1$ are negative but are plotted without regard for sign. The fraction of critical damping is denoted by ζ.

This shows the components of the response which are in phase $[(R_d)_x \sin \omega t]$ and 90° out of phase $[(R_d)_R \cos \omega t]$ with the force. Curves of $(R_d)_x$ and $(R_d)_R$ are plotted as a function of the frequency ratio ω/ω_n in Figs. 2.15 and 2.16.

Velocity and Acceleration Response. The shape of the response curves changes distinctly if velocity $\dot{x}$ or acceleration $\ddot{x}$ is plotted instead of displacement x. Differentiating Eq. (2.33),

$$\frac{\dot{x}}{F_0/\sqrt{km}} = \frac{\omega}{\omega_n} R_d \cos(\omega t - \theta) = R_v \cos(\omega t - \theta) \tag{2.36}$$

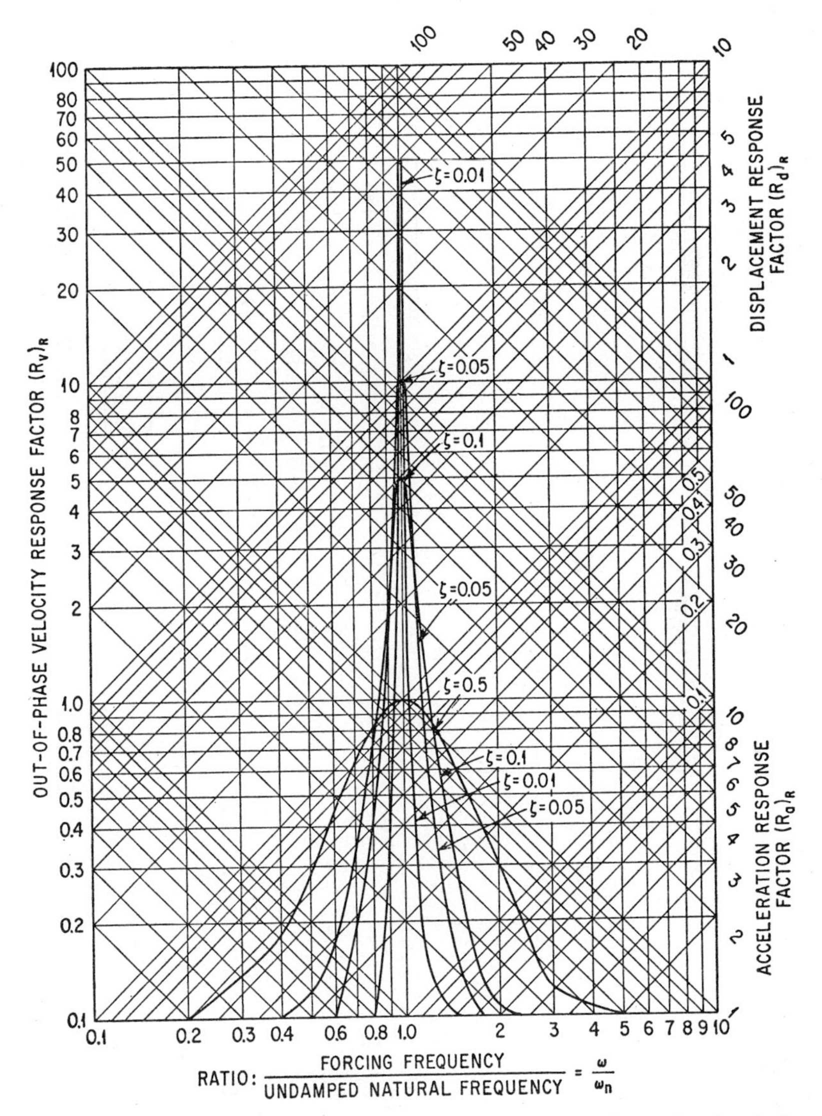

FIG. 2.16. Out-of-phase component of response factor of a viscous-damped system in forced vibration. The fraction of critical damping is denoted by ζ.

The acceleration response is obtained by differentiating Eq. (2.36):

$$\frac{\ddot{x}}{F_0/m} = -\frac{\omega^2}{\omega_n{}^2} R_d \sin(\omega t - \theta) = -R_a \sin(\omega t - \theta) \tag{2.37}$$

The velocity and acceleration response factors defined by Eqs. (2.36) and (2.37) are shown graphically in Fig. 2.13, the former to the horizontal coordinates and the latter to the coordinates having a negative 45° slope. Note that the velocity response factor approaches zero as $\omega \rightarrow 0$ and $\omega \rightarrow \infty$, whereas the acceleration response factor approaches 0 as $\omega \rightarrow 0$ and approaches unity as $\omega \rightarrow \infty$.

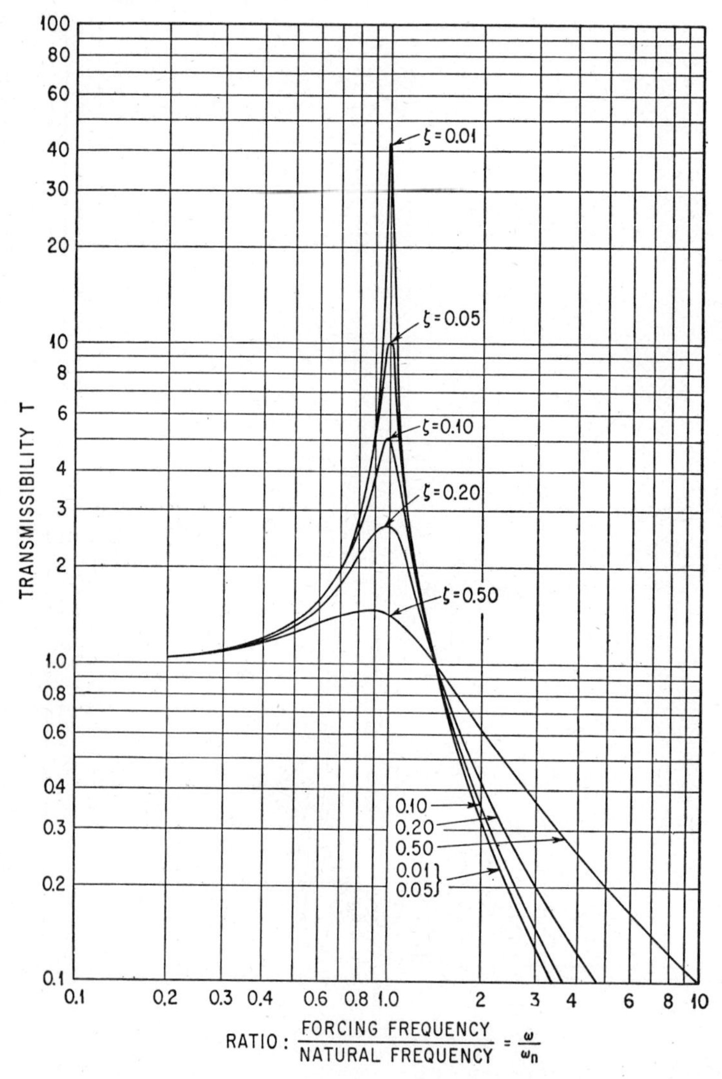

FIG. 2.17. Transmissibility of a viscous-damped system. Force transmissibility and motion transmissibility are identical numerically. The fraction of critical damping is denoted by ζ.

Force Transmission. The force transmitted to the foundation of the system is

$$F_T = c\dot{x} + kx \tag{2.38}$$

Since the forces $c\dot{x}$ and kx are 90° out of phase, the magnitude of the transmitted force is

$$|F_T| = \sqrt{c^2\dot{x}^2 + k^2x^2} \tag{2.39}$$

The ratio of the transmitted force F_T to the applied force F_0 can be expressed in terms of transmissibility T:

$$\frac{F_T}{F_0} = T \sin(\omega t - \psi) \tag{2.40}$$

where

$$T = \sqrt{\frac{1 + (2\zeta\omega/\omega_n)^2}{(1 - \omega^2/\omega_n^2)^2 + (2\zeta\omega/\omega_n)^2}} \tag{2.41}$$

and

$$\psi = \tan^{-1} \frac{2\zeta(\omega/\omega_n)^3}{1 - \omega^2/\omega_n^2 + 4\zeta^2\omega^2/\omega_n^2}$$

The transmissibility T and phase angle ψ are shown in Figs. 2.17 and 2.18, respectively, as a function of the frequency ratio ω/ω_n and for several values of the fraction of critical damping ζ.

Hysteresis. When the viscous damped, single degree-of-freedom system shown in Fig. 2.12 undergoes vibration defined by

$$x = x_0 \sin \omega t \tag{2.42}$$

the net force exerted on the mass by the spring and damper is

$$F = kx_0 \sin \omega t + c\omega x_0 \cos \omega t \tag{2.43}$$

Equations (2.42) and (2.43) define the relation between F and x; this relation is the ellipse

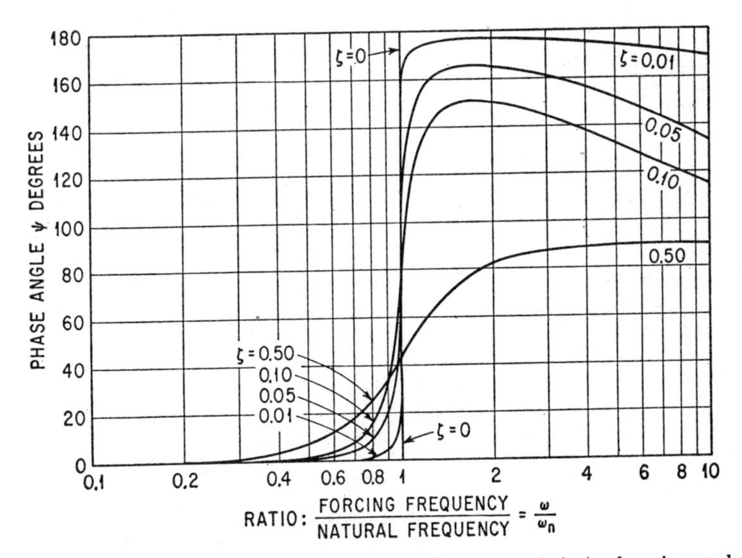

FIG. 2.18. Phase angle of force transmission (or motion transmission) of a viscous-damped system excited (1) by force acting on mass and (2) by motion of foundation. The fraction of critical damping is denoted by ζ.

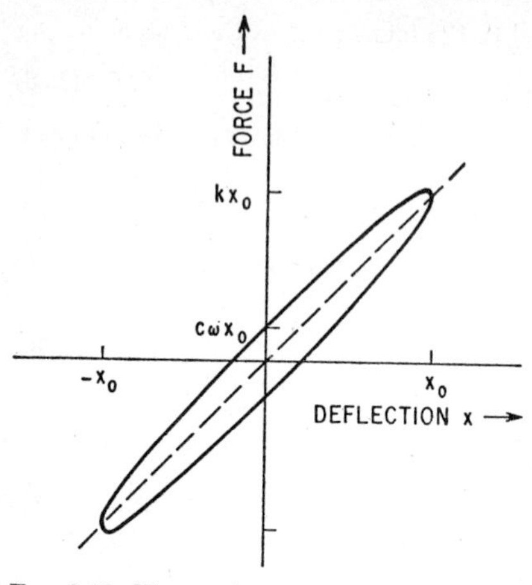

FIG. 2.19. Hysteresis curve for a spring and viscous damper in parallel.

shown in Fig. 2.19. The energy dissipated in one cycle of oscillation is

$$W = \int_T^{T+2\pi/\omega} F \frac{dx}{dt} dt = \pi c \omega x_0^2 \qquad (2.44)$$

MOTION OF FOUNDATION. The excitation for the elastic system shown in Fig. 2.20 may be a motion $u(t)$ of the foundation. The differential equation of motion for the system is

$$m\ddot{x} + c(\dot{x} - \dot{u}) + k(x - u) = 0 \qquad (2.45)$$

Consider the motion of the foundation to be a displacement that varies sinusoidally with time, $u = u_0 \sin \omega t$. A steady-state condition exists after the oscillations at the natural frequency ω_n are damped out, defined by the displacement x of mass m:

$$x = Tu_0 \sin (\omega t - \psi) \qquad (2.46)$$

where T and ψ are defined in connection with Eq. (2.40) and are shown graphically in Figs. 2.17 and 2.18, respectively. Thus, the motion transmissibility T in Eq. (2.46) is identical numerically to the force transmissibility T in Eq. (2.40). The motion of the foundation and of the mass m may be expressed in any consistent units, such as displacement, velocity, or acceleration, and the same expression for T applies in each case.

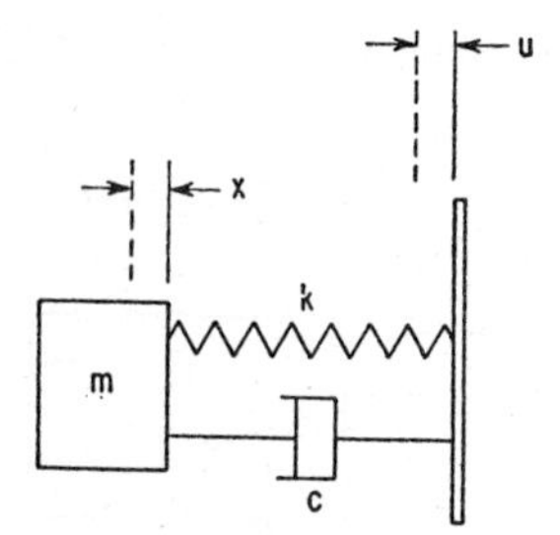

FIG. 2.20. Single degree-of-freedom system with viscous damper, excited in forced vibration by foundation motion.

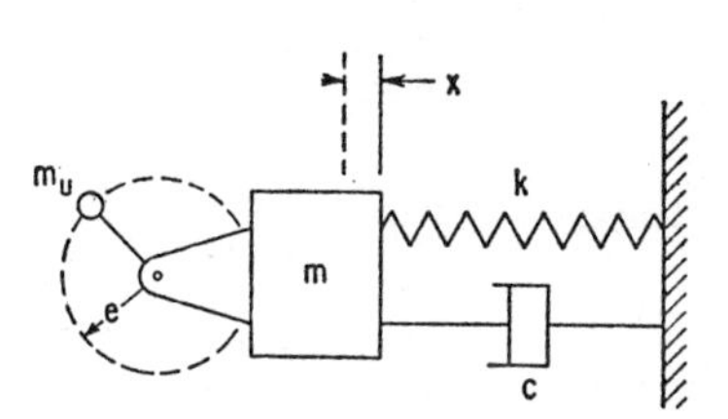

FIG. 2.21. Single degree-of-freedom system with viscous damper, excited in forced vibration by rotating eccentric weight.

VIBRATION DUE TO A ROTATING ECCENTRIC WEIGHT. In the mass-spring-damper system shown in Fig. 2.21, a mass m_u is mounted by a shaft and bearings to the mass m. The mass m_u follows a circular path of radius e with respect to the bearings. The component of displacement in the X direction of m_u relative to m is

$$x_3 - x_1 = e \sin \omega t \qquad (2.47)$$

where x_3 and x_1 are the absolute displacements of m_u and m, respectively, in the X direction; e is the length of the arm supporting the mass m_u; and ω is the angular velocity of the arm in radians per second. The differential equation of motion for the system is

$$m\ddot{x}_1 + m_u\ddot{x}_3 + c\dot{x}_1 + kx_1 = 0 \qquad (2.48)$$

Differentiating Eq. (2.47) with respect to time, solving for $\ddot{x}_3$, and substituting in Eq. (2.48):

$$(m + m_u)\ddot{x}_1 + c\dot{x}_1 + kx_1 = m_u e \omega^2 \sin \omega t \qquad (2.49)$$

Equation (2.49) is of the same form as Eq. (2.31); thus, the response relations of Eqs. (2.33), (2.36), and (2.37) apply by substituting $(m + m_u)$ for m and $m_u e\omega^2$ for F_0. The resulting displacement, velocity, and acceleration responses are

$$\frac{x_1}{m_u e\omega^2} = R_d \sin(\omega t - \theta)$$

$$\frac{\dot{x}_1\sqrt{km}}{m_u e\omega^2} = R_v \cos(\omega t - \theta) \qquad (2.50)$$

$$\frac{\ddot{x}_1 m}{m_u e\omega^2} = -R_a \sin(\omega t - \theta)$$

RESONANT FREQUENCIES. The peak values of the displacement, velocity, and acceleration response of a system undergoing forced, steady-state vibration occur at slightly different forcing frequencies. Since a *resonant frequency* is defined as the frequency for which the response is a maximum, a simple system has three resonant frequencies if defined only generally. The natural frequency is different from any of the resonant frequencies. The relations among the several resonant frequencies, the damped natural frequency, and the undamped natural frequency ω_n are:

Displacement resonant frequency: $\omega_n(1 - 2\zeta^2)^{1/2}$

Velocity resonant frequency: ω_n

Acceleration resonant frequency: $\omega_n/(1 - 2\zeta^2)^{1/2}$

Damped natural frequency: $\omega_n(1 - \zeta^2)^{1/2}$

For the degree of damping usually embodied in physical systems, the difference among the three resonant frequencies is negligible.

RESONANCE, BANDWIDTH, AND THE QUALITY FACTOR Q. Damping in a system can be determined by noting the maximum response, i.e., the response at the resonant frequency as indicated by the maximum value of R_v in Eq. (2.36). This is defined by the factor Q sometimes used in electrical engineering terminology and defined with respect to mechanical vibration as

$$Q = (R_v)_{\max} = 1/2\zeta$$

The maximum acceleration and displacement responses are slightly larger, being

$$(R_d)_{\max} = (R_a)_{\max} = \frac{(R_v)_{\max}}{(1 - \zeta^2)^{1/2}}$$

The damping in a system is also indicated by the sharpness or width of the response curve in the vicinity of a resonant frequency ω_n. Designating the width as a frequency increment $\Delta\omega$ measured at the "half-power point" (i.e., at a value of R equal to $R_{\max}/\sqrt{2}$), as illustrated in Fig. 2.22, the damping of the system is defined to a good approximation by

$$\frac{\Delta\omega}{\omega_n} = \frac{1}{Q} = 2\zeta \qquad (2.51)$$

for values of ζ less than 0.1.

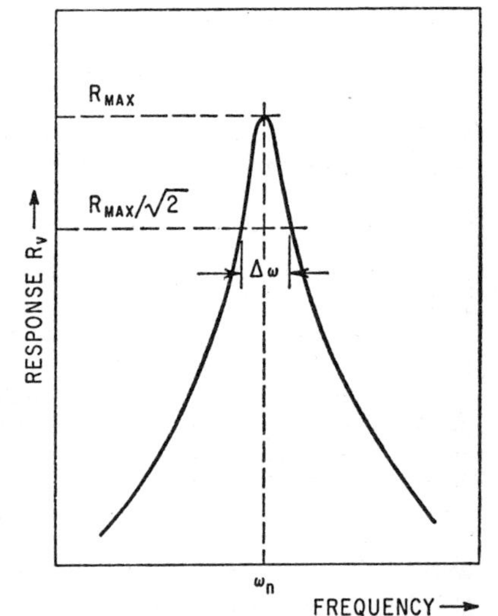

FIG. 2.22. Response curve showing bandwidth at "half-power point."

STRUCTURAL DAMPING. The energy dissipated by the damper is known as *hysteresis loss;* as indicated by Eq. (2.44), it is proportional to the forcing frequency ω. However, the hysteresis loss of many engineering structures has been found to be independent of frequency. To provide a better model for defining the *structural damping* experienced during vibration, an arbitrary damping term $k\mathfrak{g} = c\omega$ is introduced. In effect, this defines the damping force as being equal to the viscous damping force at some frequency, depending upon the value of $\mathfrak{g}$, but being invariant with frequency. The relation of the damping force F to the displacement x is defined by an ellipse similar to Fig. 2.19, and the displacement response of the system is described by an expression corresponding to Eq. (2.33) as follows:

$$\frac{x}{F_0/k} = R_g \sin(\omega t - \theta) = \frac{\sin(\omega t - \theta)}{\sqrt{(1 - \omega^2/\omega_n^2)^2 + \mathfrak{g}^2}} \tag{2.52}$$

where $\mathfrak{g} = 2\zeta\omega/\omega_n$. The resonant frequency is ω_n, and the value of R_g at resonance is $1/\mathfrak{g} = Q$.

The equations for the hysteresis ellipse for structural damping are

$$\begin{aligned} F &= kx_0 (\sin \omega t + \mathfrak{g} \cos \omega t) \\ x &= x_0 \sin \omega t \end{aligned} \tag{2.53}$$

UNDAMPED MULTIPLE DEGREE-OF-FREEDOM SYSTEMS

An elastic system sometimes cannot be described adequately by a model having only one mass but rather must be represented by a system of two or more masses considered to be *point masses* or *particles* having no rotational inertia. If a group of particles is bound together by essentially rigid connections, it behaves as a rigid body having both mass (significant for translational motion) and moment of inertia (significant for rotational motion). There is no limit to the number of masses that may be used to represent a system. For example, each mass in a model representing a beam may be an infinitely thin slice representing a cross section of the beam; a differential equation is required to treat this continuous distribution of mass.

DEGREES-OF-FREEDOM

The number of independent parameters required to define the distance of all the masses from their reference positions is called the number of *degrees-of-freedom N*. For example, if there are N masses in a system constrained to move only in translation in the X and Y directions, the system has $2N$ degrees-of-freedom. A continuous system such as a beam has an infinitely large number of degrees-of-freedom.

For each degree-of-freedom (each coordinate of motion of each mass) a differential equation can be written in one of the following alternative forms:

$$m_j\ddot{x}_j = F_{xj} \qquad\qquad I_k\ddot{\alpha}_k = M_{\alpha k} \tag{2.54}$$

where F_{xj} is the component in the X direction of all external, spring, and damper forces acting on the mass having the jth degree-of-freedom, and $M_{\alpha k}$ is the component about the α axis of all torques acting on the body having the kth degree-of-freedom. The moment of inertia of the mass about the α axis is designated by I_k. (This is assumed for the present analysis to be a principal axis of inertia, and product of inertia terms are neglected. See Chap. 3 for a more detailed discussion.) Equations (2.54) are identical in form and can be represented by

$$m_j\ddot{x}_j = F_j \tag{2.55}$$

where F_j is the resultant of all forces (or torques) acting on the system in the jth degree-of-freedom, $\ddot{x}_j$ is the acceleration (translational or rotational) of the system in the jth degree-of-freedom, and m_j is the mass (or moment of inertia) in the jth degree-of-freedom. Thus, the terms defining the motion of the system (displacement, velocity, and acceleration) and the deflections of structures may be either translational or rotational, depending upon the type of coordinate. Similarly, the "force" acting on a system may be either a force or a torque, depending upon the type of coordinate. For example, if a system has n bodies each free to move in three translational modes and three rotational modes, there would be $6n$ equations of the form of Eq. (2.55), one for each degree-of-freedom.

DEFINING A SYSTEM AND ITS EXCITATION

The first step in analyzing any physical structure is to represent it by a mathematical model which will have essentially the same dynamic behavior. A suitable number and distribution of masses, springs, and dampers must be chosen, and the input forces or foundation motions must be defined. The model should have sufficient degrees-of-freedom to determine the modes which will have significant response to the exciting force or motion.

The properties of a system that must be known are the natural frequencies ω_n, the normal mode shapes D_{jn}, the damping of the respective modes, and the mass distribution m_j. The detailed distributions of stiffness and damping of a system are not used directly but rather appear indirectly as the properties of the respective modes. The characteristic properties of the modes may be determined experimentally as well as analytically.

STIFFNESS COEFFICIENTS

The spring system of a structure of N degrees-of-freedom can be defined completely by a set of N^2 *stiffness coefficients*.[5] A stiffness coefficient K_{jk} is the change in spring force acting on the jth degree-of-freedom when only the kth degree-of-freedom is slowly displaced a unit amount in the negative direction. This definition is a generalization of the linear, elastic spring defined by Eq. (2.1). Stiffness coefficients have the characteristic of reciprocity, i.e., $K_{jk} = K_{kj}$. The number of independent stiffness coefficients is $(N^2 + N)/2$.

The total elastic force acting on the jth degree-of-freedom is the sum of the effects of the displacements in all of the degrees-of-freedom:

$$F_{el} = -\sum_{k=1}^{N} K_{jk}x_k \tag{2.56}$$

Inserting the spring force F_{el} from Eq. (2.56) in Eq. (2.55) together with the external forces F_j results in the n equations:

$$m_j\ddot{x}_j = F_j - \sum_k K_{jk}x_k \tag{2.56a}$$

FREE VIBRATION

When the external forces are zero, the preceding equations become

$$m_j\ddot{x}_j + \sum_k K_{jk}x_k = 0 \tag{2.57}$$

Solutions of Eq. (2.57) have the form

$$x_j = D_j \sin(\omega t + \theta) \tag{2.58}$$

Substituting Eq. (2.58) in Eq. (2.57),

$$m_j\omega^2 D_j = \sum_k K_{jk}D_k \tag{2.59}$$

This is a set of n linear algebraic equations with n unknown values of D. A solution of these equations for values of D other than zero can be obtained only if the determinant of the coefficients of the D's is zero:

$$\begin{vmatrix} (m_1\omega^2 - K_{11}) & -K_{12} & \cdot\ \cdot & -K_{in} \\ -K_{21} & (m_2\omega^2 - K_{22}) & \cdot\ \cdot & \cdot \\ \cdot & \cdot & \cdot\ \cdot & \cdot \\ \cdot & \cdot & \cdot\ \cdot & \cdot \\ -K_{ni} & \cdot & \cdot\ \cdot & (m_n\omega^2 - K_{nn}) \end{vmatrix} = 0 \tag{2.60}$$

Equation (2.60) is an algebraic equation of the nth degree in ω^2; it is called the *frequency equation* since it defines n values of ω which satisfy Eq. (2.57). The roots are all real; some may be equal, and others may be zero. These values of frequency determined from Eq. (2.60) are the frequencies at which the system can oscillate in the absence of external forces. These frequencies are the *natural frequencies* ω_n of the system. Depending upon the initial conditions under which vibration of the system is initiated, the oscillations may occur at any or all of the natural frequencies and at any amplitude.

Example 2.1. Consider the three degree-of-freedom system shown in Fig. 2.23; it consists of three equal masses m and a foundation connected in series by three equal springs k. The absolute displacements of the masses are x_1, x_2, and x_3. The stiffness coefficients (see section entitled *Stiffness Coefficients*) are thus $K_{11} = 2k$, $K_{22} = 2k$, $K_{33} = k$, $K_{12} = K_{21} = -k$, $K_{23} =$

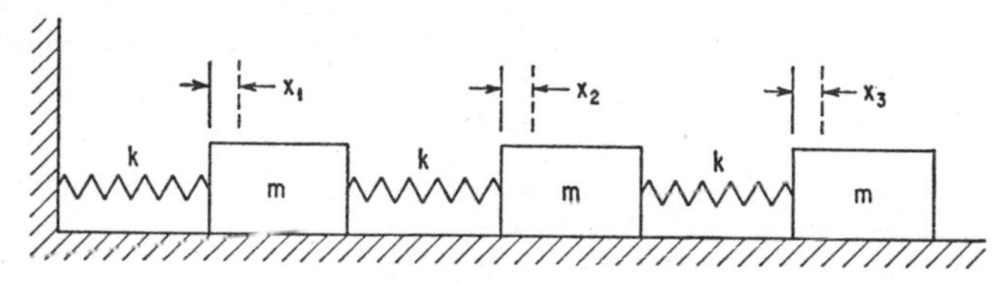

FIG. 2.23. Undamped three degree-of-freedom system on foundation.

$K_{32} = -k$, and $K_{13} = K_{31} = 0$. The frequency equation is given by the determinant, Eq. (2.60),

$$\begin{vmatrix} (m\omega^2 - 2k) & k & 0 \\ k & (m\omega^2 - 2k) & k \\ 0 & k & (m\omega^2 - k) \end{vmatrix} = 0$$

The determinant expands to the following polynomial:

$$\left(\frac{m\omega^2}{k}\right)^3 - 5\left(\frac{m\omega^2}{k}\right)^2 + 6\left(\frac{m\omega^2}{k}\right) - 1 = 0$$

Solving for ω,

$$\omega = 0.445\sqrt{\frac{k}{m}}, \quad 1.25\sqrt{\frac{k}{m}}, \quad 1.80\sqrt{\frac{k}{m}}$$

NORMAL MODES OF VIBRATION. A structure vibrating at only one of its natural frequencies ω_n does so with a characteristic pattern of amplitude distribution called a *normal mode of vibration.*[3,4,5] A normal mode is defined by a set of values of D_{jn} [see Eq. (2.58)] which satisfy Eq. (2.59) when $\omega = \omega_n$:

$$\omega_n^2 m_j D_{jn} = \sum_k K_{jn} D_{kn} \tag{2.61}$$

A set of values of D_{jn} which form a normal mode is independent of the absolute values of D_{jn} but depends only on their relative values. To define a mode shape by a unique set of numbers, any arbitrary *normalizing condition* which is desired can be used. A condition often used is to set $D_{1n} = 1$ but $\sum_j m_j D_{jn}^2 = 1$ and $\sum_j m_j D_{jn}^2 = \sum_j m_j$ also may be found convenient.

ORTHOGONALITY OF NORMAL MODES. The usefulness of normal modes in dealing with multiple degree-of-freedom systems is due largely to the orthogonality of the normal modes. It can be shown[3,4,5] that the set of inertia forces $\omega_n^2 m_j D_{jn}$ for one mode does no work on the set of deflections D_{jm} of another mode of the structure:

$$\sum_j m_j D_{jm} D_{jn} = 0 \qquad [m \neq n] \tag{2.62}$$

This is the *orthogonality condition.*

NORMAL MODES AND GENERALIZED COORDINATES. Any set of N deflections x_j can be expressed as the sum of normal mode amplitudes:

$$x_j = \sum_{n=1}^{N} q_n D_{jn} \tag{2.63}$$

The numerical values of the D_{jn}'s are fixed by some normalizing condition, and a set of values of the N variables q_n can be found to match any set of x_j's. The N values of q_n constitute a set of *generalized coordinates* which can be used to define the position coordinates x_j of all parts of the structure. The q's are also known as the amplitudes of the normal modes, and are functions of time. Equation (2.63) may be differentiated to obtain

$$\ddot{x}_j = \sum_{n=1}^{N} \ddot{q}_n D_{jn} \tag{2.64}$$

Any quantity which is distributed over the j coordinates can be represented by a linear transformation similar to Eq. (2.63). It is convenient now to introduce the parameter γ_n relating D_{jn} and F_j/m_j as follows:

$$\frac{F_j}{m_j} = \sum_n \gamma_n D_{jn} \tag{2.65}$$

where F_j may be zero for certain values of n.

FORCED MOTION

Substituting the expressions in generalized coordinates, Eqs. (2.63) to (2.65), in the basic equation of motion, Eq. (2.56*a*),

$$m_j \sum_n \ddot{q}_n D_{jn} + \sum_k k_{jk} \sum_n q_n D_{kn} - m_j \sum_n \gamma_n D_{jn} = 0 \tag{2.66}$$

The center term in Eq. (2.66) may be simplified by applying Eq. (2.61) and the equation rewritten as follows:

$$\sum_n (\ddot{q}_n + \omega_n^2 q_n - \gamma_n) m_j D_{jn} = 0 \tag{2.67}$$

Multiplying Eqs. (2.67) by D_{jm} and taking the sum over j (i.e., adding all the equations together),

$$\sum_n (\ddot{q}_n + \omega_n^2 q_n - \gamma_n) \sum_j m_j D_{jn} D_{jm} = 0$$

All terms of the sum over n are zero, except for the term for which $m = n$, according to the orthogonality condition of Eq. (2.62). Then since $\sum_j m_j D_{jn}^2$ is not zero, it follows that

$$\ddot{q}_n + \omega_n^2 q_n - \gamma_n = 0$$

for every value of n from 1 to N.

An expression for γ_n may be found by using the orthogonality condition again. Multiplying Eq. (2.65) by $m_j D_{jm}$ and taking the sum taken over j,

$$\sum_j F_j D_{jm} = \sum_n \gamma_n \sum_j m_j D_{jn} D_{jm} \tag{2.68}$$

All the terms of the sum over n are zero except when $n = m$, according to Eq. (2.62), and Eq. (2.68) reduces to

$$\gamma_n = \frac{\sum_j F_j D_{jn}}{\sum_j m_j D_{jn}^2} \tag{2.69}$$

Then the differential equation for the response of any generalized coordinate to the externally applied forces F_j is

$$\ddot{q}_n + \omega_n^2 q_n = \gamma_n = \frac{\sum_j F_j D_{jn}}{\sum_j m_j D_{jn}^2} \tag{2.70}$$

where $\Sigma F_j D_{jn}$ is the generalized force, i.e., the total work done by all external forces during a small displacement δq_n divided by δq_n, and $\Sigma m_j D_{jn}^2$ is the generalized mass.

Thus the amplitude q_n of each normal mode is governed by its own equation, independent of the other normal modes, and responds as a simple mass-spring system. Equation (2.70) is a generalized form of Eq. (2.23).

The forces F_j may be any functions of time. Any equation for the response of an undamped mass-spring system applies to each mode of a complex structure by substituting:

The *generalized coordinate* q_n for x

The *generalized force* $\sum_j F_j D_{jn}$ for F (2.71)

The *generalized mass* $\sum_j m_j D_{jn}$ for m

The *mode natural frequency* ω_n for ω_n

RESPONSE TO SINUSOIDAL FORCES. If a system is subjected to one or more sinusoidal forces $F_j = F_{0j} \sin \omega t$, the response is found from Eq. (2.26) by noting that $k = m\omega_n^2$ [Eq. (2.6)] and then substituting from Eq. (2.71):

$$q_n = \frac{\sum_j F_{0j} D_{jn}}{\omega_n^2 \sum_j m_j D_{jn}^2} \frac{\sin \omega t}{(1 - \omega^2/\omega_n^2)} \tag{2.72}$$

Then the displacement of the kth degree-of-freedom, from Eq. (2.63), is

$$x_k = \sum_{n=1}^{N} \frac{D_{kn} \sum_j F_{0j} D_{jn} \sin \omega t}{\omega_n^2 \sum_j m_j D_{jn}^2 (1 - \omega^2/\omega_n^2)} \tag{2.73}$$

This is the general equation for the response to sinusoidal forces of an undamped system of N degrees-of-freedom. The application of the equation to systems free in space or attached to immovable foundations is discussed below.

Example 2.2. Consider the system shown in Fig. 2.24; it consists of 3 equal masses m connected in series by 2 equal springs k. The system is free in space and a force $F \sin \omega t$ acts on the first mass. Absolute displacements of the masses are x_1, x_2, and x_3. Determine the acceleration $\ddot{x}_3$. The stiffness coefficients (see section entitled *Stiffness Coefficients*) are $K_{11} = K_{33} = k$, $K_{22} = 2k$, $K_{12} = K_{21} = -k$, $K_{13} = K_{31} = 0$, and $K_{23} = K_{32} = -k$. Substituting in Eq. (2.60), the frequency equation is

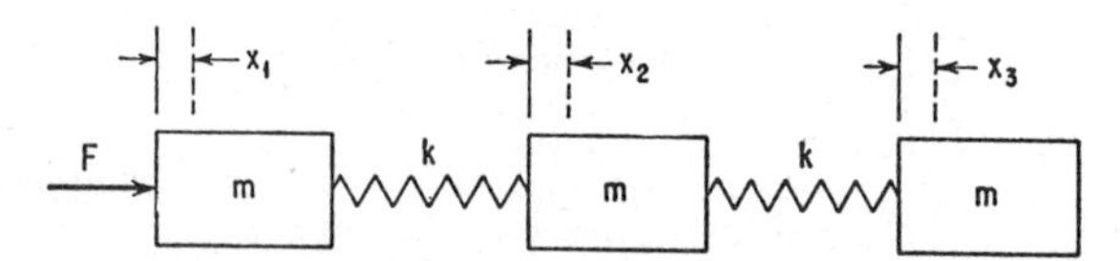

FIG. 2.24. Undamped three degree-of-freedom system acted on by sinusoidal force.

$$\begin{vmatrix} (m\omega^2 - k) & k & 0 \\ k & (m\omega^2 - 2k) & k \\ 0 & k & (m\omega^2 - k) \end{vmatrix} = 0$$

The roots are $\omega_1 = 0$, $\omega_2 = \sqrt{k/m}$, and $\omega_3 = \sqrt{3k/m}$. The zero value for one of the natural frequencies indicates that the entire system translates without deflection of the springs. The mode shapes are now determined by substituting from Eq. (2.58) in Eq. (2.57), noting that $\ddot{x} = -D\omega^2$, and writing Eq. (2.59) for each of the three masses in each of the oscillatory modes 2 and 3:

$$mD_{21}\left(\frac{k}{m}\right) = K_{11}D_{21} + K_{21}D_{22} + K_{31}D_{23}$$

$$mD_{22}\left(\frac{k}{m}\right) = K_{12}D_{21} + K_{22}D_{22} + K_{32}D_{23}$$

$$mD_{23}\left(\frac{k}{m}\right) = K_{13}D_{21} + K_{23}D_{22} + K_{33}D_{23}$$

$$mD_{31}\left(\frac{3k}{m}\right) = K_{11}D_{31} + K_{21}D_{32} + K_{31}D_{33}$$

$$mD_{32}\left(\frac{3k}{m}\right) = K_{12}D_{31} + K_{22}D_{32} + K_{32}D_{33}$$

$$mD_{33}\left(\frac{3k}{m}\right) = K_{13}D_{31} + K_{23}D_{32} + K_{33}D_{33}$$

where the first subscript on the D's indicates the mode number (according to ω_1 and ω_2 above) and the second subscript indicates the displacement amplitude of the particular mass. The values of the stiffness coefficients K are calculated above. The mode shapes are defined by the relative displacements of the masses. Thus, assigning values of unit displacement to the first mass (i.e., $D_{21} = D_{31} = 1$), the above equations may be solved simultaneously for the D's:

$$D_{21} = 1 \qquad D_{22} = 0 \qquad D_{23} = -1$$

$$D_{31} = 1 \qquad D_{32} = -2 \qquad D_{33} = 1$$

Substituting these values of D in Eq. (2.71), the generalized masses are determined: $M_2 = 2m$, $M_3 = 6m$.

Equation (2.73) then can be used to write the expression for acceleration $\ddot{x}_3$:

$$\ddot{x}_3 = \left[\frac{1}{3m} + \frac{(\omega^2/\omega_2^2)(-1)(+1)}{2m(1 - \omega^2/\omega_2^2)} + \frac{(\omega^2/\omega_3^2)(+1)(+1)}{6m(1 - \omega^2/\omega_3^2)}\right] F_1 \sin \omega t$$

FREE AND FIXED SYSTEMS. For a structure which is free in space, there are six "normal modes" corresponding to $\omega_n = 0$. These represent motion of the structure without relative motion of its parts; this is rigid body motion with six degrees-of-freedom.

The rigid body modes all may be described by equations of the form

$$D_{jm} = a_{jm}D_m \qquad [m = 1, 2, \ldots, 6]$$

where D_m is a motion of the rigid body in the m coordinate and a is the displacement of the jth degree-of-freedom when D_m is moved a unit amount. The geometry of the structure determines the nature of a_{jm}. For example, if D_m is a rotation about the Z axis, $a_{jm} = 0$ for all modes of motion in which j represents rotation about the X or Y axis and $a_{jm} = 0$ if j represents translation parallel to the Z axis. If D_{jm} is a translational mode of motion parallel to X or Y, it is necessary that a_{jm} be proportional to the distance r_j of m_j from the Z axis and to the sine of the angle between r_j and the jth direction. The above relations may be applied to an elastic body. Such a body moves as a rigid body in the gross sense in that all particles of the body move together generally but may experience relative vibratory motion. The orthogonality condition applied to the relation between any rigid body mode D_{jm} and any oscillatory mode D_{jn} yields

$$\sum_j m_j D_{jn} D_{jm} = \sum_j m_j a_{jm} D_{jn} = 0 \qquad \begin{bmatrix} m \leq 6 \\ n > 6 \end{bmatrix} \tag{2.74}$$

These relations are used in computations of oscillatory modes, and show that normal modes of vibration involve no net translation or rotation of a body.

A system attached to a fixed foundation may be considered as a system free in space in which one or more "foundation" masses or moments of inertia are infinite. Motion of the system as a rigid body is determined entirely by the motion of the foundation. The amplitude of an oscillatory mode representing motion of the foundation is zero; i.e., $M_jD_{jn}^2 = 0$ for the infinite mass. However, Eq. (2.73) applies equally well regardless of the size of the masses.

FOUNDATION MOTION. If a system is small relative to its foundation, it may be assumed to have no effect on the motion of the foundation. Consider a foundation of large but unknown mass m_0 having a motion $x_0 \sin \omega t$, the consequence of some unknown force

$$F_0 \sin \omega t = -m_0 x_0 \omega^2 \sin \omega t \tag{2.75}$$

acting on m_0 in the x_0 direction. Equation (2.73) is applicable to this case upon substituting

$$-m_0 x_0 \omega^2 D_{0n} = \sum_j F_{0j} D_{jn} \tag{2.76}$$

where D_{0n} is the amplitude of the foundation (the 0 degree-of-freedom) in the nth mode.

The oscillatory modes of the system are subject to Eqs. (2.74):

$$\sum_{j=0} m_j a_{jm} D_{jn} = 0$$

Separating the 0th degree-of-freedom from the other degrees-of-freedom:

$$\sum_{j=0} m_j a_{jm} D_{jn} = m_0 a_{0m} D_{0n} + \sum_{j=1} m_j a_{jm} D_{jn}$$

If m_0 approaches infinity as a limit, D_{0n} approaches zero and motion of the system as a rigid body is identical with the motion of the foundation. Thus, a_{0m} approaches unity for motion in which $m = 0$, and approaches zero for motion in which $m \neq 0$. In the limit:

$$\lim_{m_0 \to \infty} m_0 D_{0n} = -\sum_j m_j a_{j0} D_{jn} \tag{2.77}$$

Substituting this result in Eq. (2.76),

$$\lim_{m_0 \to \infty} \sum_j F_{0j} D_{jn} = x_0 \omega^2 \sum_j m_j a_{j0} D_{jn} \tag{2.78}$$

The generalized mass in Eq. (2.73) includes the term $m_0 D_{0n}^2$, but this becomes zero as m_0 becomes infinite.

The equation for response of a system to motion of its foundation is obtained by substituting Eq. (2.78) in Eq. (2.73):

$$x_k = \sum_{n=1}^{N} \frac{\omega^2}{\omega_n^2} D_{kn} \frac{\sum_j m_j a_{j0} D_{jn} x_0 \sin \omega t}{\sum_j m_j D_{jn}^2 (1 - \omega^2/\omega_n^2)} + x_0 \sin \omega t \tag{2.79}$$

DAMPED MULTIPLE DEGREE-OF-FREEDOM SYSTEMS

Consider a set of masses interconnected by a network of springs and acted upon by external forces, with a network of dampers acting in parallel with the springs. The viscous dampers produce forces on the masses which are determined in a manner analogous to that used to determine spring forces and summarized by Eq. (2.56). The damping force acting on the jth degree-of-freedom is

$$(F_d)_j = -\sum_k C_{jk} \dot{x}_k \tag{2.80}$$

where C_{jk} is the resultant force on the jth degree-of-freedom due to a unit velocity of the kth degree-of-freedom.

In general, the distribution of damper sizes in a system need not be related to the spring or mass sizes. Thus, the dampers may couple the normal modes together, allowing motion of one mode to affect that of another. Then the equations of response are not easily separable [6] into independent normal mode equations. However, there are two types of damping distribution which do not couple the normal modes.[6,7,8] These are known as *uniform viscous damping* and *uniform mass damping*.

UNIFORM VISCOUS DAMPING

Uniform damping is an appropriate model for systems in which the damping effect is an inherent property of the spring material. Each spring is considered to have a damper acting in parallel with it, and the ratio of damping coefficient to stiffness coefficient is the same for each spring of the system. Thus, for all values of j and k,

$$\frac{C_{jk}}{k_{jk}} = 2\mathcal{G} \tag{2.81}$$

where $\mathcal{G}$ is a constant.

Substituting from Eq. (2.81) in Eq. (2.80),

$$-(F_d)_j = \sum_k C_{jk}\dot{x}_k = 2\mathcal{G}\sum_k k_{jk}\dot{x}_k \tag{2.82}$$

Since the damping forces are "external" forces with respect to the mass-spring system, the forces $(F_d)_j$ can be added to the external forces in Eq. (2.70) to form the equation of motion:

$$\ddot{q}_n + \omega_n^2 q_n = \frac{\sum_j (F_d)_j D_{jn} + \sum_j F_j D_{jn}}{\sum_j m_j D_{jn}^2} \tag{2.83}$$

Combining Eqs. (2.61), (2.63), and (2.82), the summation involving $(F_d)_j$ in Eq. (2.83) may be written as follows:

$$\sum_j (F_d)_j D_{jn} = -2\mathcal{G}\omega_n^2 \dot{q}_n \sum_j m_j D_{jn}^2 \tag{2.84}$$

Substituting Eq. (2.84) in Eq. (2.83),

$$\ddot{q}_n + 2\mathcal{G}\omega_n^2\dot{q}_n + \omega_n^2 q_n = \gamma_n \tag{2.85}$$

Comparison of Eq. (2.85) with Eq. (2.31) shows that each mode of the system responds as a simple damped oscillator.

The damping term $2\mathcal{G}\omega_n^2$ in Eq. (2.85) corresponds to $2\zeta\omega_n$ in Eq. (2.31) for a simple system. Thus, $\mathcal{G}\omega_n$ may be considered the critical damping ratio of each mode. Note that the effective damping for a particular mode varies directly as the natural frequency of the mode.

FREE VIBRATION. If a system with uniform viscous damping is disturbed from its equilibrium position and released at time $t = 0$ to vibrate freely, the applicable equation of motion is obtained from Eq. (2.85) by substituting $2\zeta\omega$ for $2\mathcal{G}\omega_n^2$ and letting $\gamma_n = 0$:

$$\ddot{q}_n + 2\zeta\omega_n\dot{q}_n + \omega_n^2 q_n = 0 \tag{2.86}$$

The solution of Eq. (2.86) for less than critical damping is

$$x_j(t) = \sum_n D_{jn} e^{-\zeta\omega_n t}(A_n \sin \omega_d t + B_n \cos \omega_d t) \tag{2.87}$$

where $\omega_d = \omega_n(1 - \zeta^2)^{1/2}$.

The values of A and B are determined by the displacement $x_j(0)$ and velocity $\dot{x}_j(0)$ at time $t = 0$:

$$x_j(0) = \sum_n B_n D_{jn}$$

$$\dot{x}_j(0) = \sum_n (A_n \omega_{dn} - B_n \zeta \omega_n) D_{jn}$$

Applying the orthogonality relation of Eq. (2.62) in the manner used to derive Eq. (2.69),

$$B_n = \frac{\sum_j x_j(0) m_j D_{jn}}{\sum_j m_j D_{jn}^2}$$

$$A_n \omega_{dn} - B_n \zeta \omega_{dn} = \frac{\sum_j \dot{x}_j(0) m_j D_{jn}}{\sum_j m_j D_{jn}^2} \tag{2.88}$$

Thus each mode undergoes a decaying oscillation at the damped natural frequency for the particular mode, and the amplitude of each mode decays from its initial value, which is determined by the initial displacements and velocities.

UNIFORM STRUCTURAL DAMPING

To avoid the dependence of viscous damping upon frequency, as indicated by Eq. (2.85), the uniform viscous damping factor $\mathcal{G}$ is replaced by $\mathfrak{g}/\omega$ for uniform structural damping. This corresponds to the structural damping parameter $\mathfrak{g}$ in Eqs. (2.52) and (2.53) for sinusoidal vibration of a simple system. Thus, Eq. (2.85) for the response of a mode to a sinusoidal force of frequency ω is

$$\ddot{q}_n + \frac{2\mathfrak{g}}{\omega} \omega_n^2 \dot{q}_n + \omega_n^2 q_n = \gamma_n \tag{2.89}$$

The amplification factor at resonance ($Q = 1/\mathfrak{g}$) has the same value in all modes.

UNIFORM MASS DAMPING

If the damping force on each mass is proportional to the magnitude of the mass,

$$(F_d)_j = -B m_j \dot{x}_j \tag{2.90}$$

where B is a constant. For example, Eq. (2.90) would apply to a uniform beam immersed in a viscous fluid.

Substituting as $\dot{x}_j$ in Eq. (2.90) the derivative of Eq. (2.63),

$$\Sigma (F_d)_j D_{jn} = -B \sum_j m_j D_{jn} \sum_m \dot{q}_m D_{jm} \tag{2.91}$$

Because of the orthogonality condition, Eq. (2.62):

$$\Sigma (F_d)_j D_{jn} = -B \dot{q}_n \sum_j m_j D_{jn}^2$$

Substituting from Eq. (2.91) in Eq. (2.83), the differential equation for the system is

$$\ddot{q}_n + B \dot{q}_n + \omega_n^2 q_n = \gamma_n \tag{2.92}$$

where the damping term B corresponds to $2\zeta\omega$ for a simple oscillator, Eq. (2.31). Then $B/2\omega_n$ represents the fraction of critical damping for each mode, a quantity which diminishes with increasing frequency.

GENERAL EQUATION FOR FORCED VIBRATION

All the equations for response of a linear system to a sinusoidal excitation may be regarded as special cases of the following general equation:

$$x_k = \sum_{n=1}^{N} \frac{D_{kn}}{\omega_n^2} \frac{F_n}{m_n} R_n \sin(\omega t - \theta_n) \tag{2.93}$$

where x_k = displacement of structure in kth degree-of-freedom
N = number of degrees-of-freedom, including those of the foundation
D_{kn} = amplitude of kth degree-of-freedom in nth normal mode
F_n = generalized force for nth mode
m_n = generalized mass for nth mode
R_n = response factor, a function of the frequency ratio ω/ω_n (Fig. 2.13)
θ_n = phase angle (Fig. 2.14)

Equation (2.93) is of sufficient generality to cover a wide variety of cases, including excitation by external forces or foundation motion, viscous or structural damping, rotational and translational degrees-of-freedom, and from one to an infinite number of degrees-of-freedom.

LAGRANGIAN EQUATIONS

The differential equations of motion for a vibrating system sometimes are derived more conveniently in terms of kinetic and potential energies of the system than by the application of Newton's laws of motion in a form requiring the determination of the forces acting on each mass of the system. The formulation of the equations in terms of the energies, known as Lagrangian equations,[3,4,5] is expressed as follows:

$$\frac{d}{dt}\frac{\partial T}{\partial \dot{q}_n} - \frac{\partial T}{\partial q_n} + \frac{\partial V}{\partial q_n} = F_n \tag{2.94}$$

where T = total kinetic energy of system
V = total potential energy of system
q_n = generalized coordinate—a displacement
$\dot{q}_n$ = velocity at generalized coordinate q_n
F_n = generalized force, the portion of the total forces not related to the potential energy of the system (gravity and spring forces appear in the potential energy expressions and are not included here)

The method of applying Eq. (2.94) is to select a number of independent coordinates (generalized coordinates) equal to the number of degrees-of-freedom, and to write expressions for total kinetic energy T and total potential energy V. Differentiation of these expressions successively with respect to each of the chosen coordinates leads to a number of equations similar to Eq. (2.94), one for each coordinate (degree-of-freedom). These are the applicable differential equations and may be solved by any suitable method.

Example 2.3. Consider free vibration of the three degree-of-freedom system shown in Fig. 2.23; it consists of three equal masses m connected in tandem by equal springs k. Take as coordinates the three absolute displacements x_1, x_2, and x_3. The kinetic energy of the system is

$$T = \tfrac{1}{2}m(\dot{x}_1^2 + \dot{x}_2^2 + \dot{x}_3^2)$$

The potential energy of the system is

$$V = \frac{k}{2}[x_1^2 + (x_1 - x_2)^2 + (x_2 - x_3)^2] = \frac{k}{2}(2x_1^2 + 2x_2^2 + x_3^2 - 2x_1x_2 - 2x_2x_3)$$

Differentiating the expression for the kinetic energy successively with respect to the velocities,

$$\frac{\partial T}{\partial \dot{x}_1} = m\dot{x}_1 \qquad \frac{\partial T}{\partial \dot{x}_2} = m\dot{x}_2 \qquad \frac{\partial T}{\partial \dot{x}_3} = m\dot{x}_3$$

The kinetic energy is not a function of displacement; therefore, the second term in Eq. (2.94) is zero. The partial derivatives with respect to the displacement coordinates are

$$\frac{\partial V}{\partial x_1} = 2kx_1 - kx_2 \qquad \frac{\partial V}{\partial x_2} = 2kx_2 - kx_1 - kx_3 \qquad \frac{\partial V}{\partial x_3} = kx_3 - kx_2$$

In free vibration, the generalized force term in Eq. (2.93) is zero. Then, substituting the derivatives of the kinetic and potential energies from above into Eq. (2.94),

$$\begin{aligned} m\ddot{x}_1 + 2kx_1 - kx_2 &= 0 \\ m\ddot{x}_2 + 2kx_2 - kx_1 - kx_3 &= 0 \\ m\ddot{x}_3 + kx_3 - kx_2 &= 0 \end{aligned}$$

The natural frequencies of the system may be determined by placing the preceding set of simultaneous equations in determinant form, in accordance with Eq. (2.60):

$$\begin{vmatrix} (m\omega^2 - 2k) & k & 0 \\ k & (m\omega^2 - 2k) & k \\ 0 & k & (m\omega^2 - k) \end{vmatrix} = 0$$

The natural frequencies are equal to the values of ω that satisfy the preceding determinant equation.

Example 2.4. Consider the compound pendulum of mass m shown in Fig. 2.25, having its center-of-gravity located a distance l from the axis of rotation. The moment of inertia is I

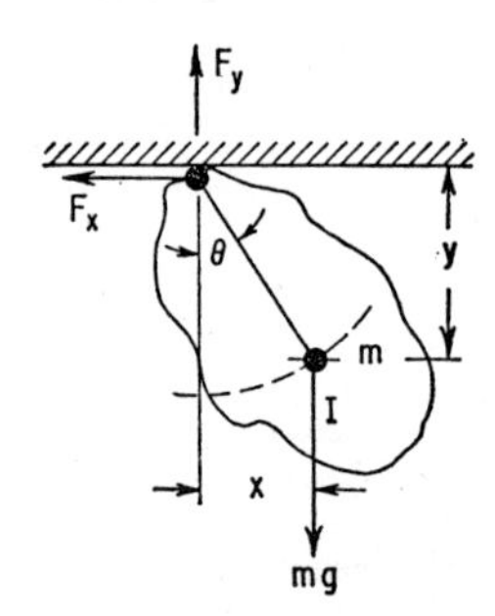

FIG. 2.25. Forces and motions of a compound pendulum.

about an axis through the center-of-gravity. The position of the mass is defined by three coordinates, x and y to define the location of the center-of-gravity, and θ to define the angle of rotation.

The *equations* of *constraint* are $y = l \cos \theta$; $x = l \sin \theta$. Each equation of constraint reduces the number of degrees-of-freedom by one; thus the pendulum is a one degree-of-freedom system whose position is defined uniquely by θ alone.

The kinetic energy of the pendulum is

$$T = \tfrac{1}{2}(I + ml^2)\dot{\theta}^2$$

The potential energy is

$$V = mgl(1 - \cos \theta)$$

Then

$$\frac{\partial T}{\partial \dot{\theta}} = (I + ml^2)\dot{\theta} \qquad \frac{d}{dt}\left(\frac{\partial T}{\partial \dot{\theta}}\right) = (I + ml^2)\ddot{\theta}$$

$$\frac{\partial T}{\partial \theta} = 0 \qquad \frac{\partial V}{\partial \theta} = mgl \sin \theta$$

Substituting these expressions in Eq. (2.94), the differential equation for the pendulum is

$$(I + ml^2)\ddot{\theta} + mgl \sin \theta = 0$$

Example 2.5. Consider oscillation of the water in the U tube shown in Fig. 2.26. If the displacements of the water levels in the arms of a uniform-diameter U tube are h_1 and h_2, then conservation of matter requires that $h_1 = -h_2$. The kinetic energy of the water flowing in the tube with velocity $\dot{h}_1$ is

$$T = \tfrac{1}{2}\rho S l \dot{h}_1^2$$

where ρ is the water density, S is the cross-section area of the tube, and l is the developed length of the water column. The potential energy (difference in potential energy between arms of tube) is

$$V = S\rho g h_1^2$$

Taking h_1 as the generalized coordinate, differentiating the expressions for energy, and substituting in Eq. (2.94),

$$S\rho l \ddot{h}_1 + 2\rho g S h_1 = 0$$

Dividing through by $\rho S l$,

$$\ddot{h}_1 + \frac{2g}{l} h_1 = 0$$

This is the differential equation for a simple oscillating system of natural frequency ω_n, where

$$\omega_n = \sqrt{\frac{2g}{l}}$$

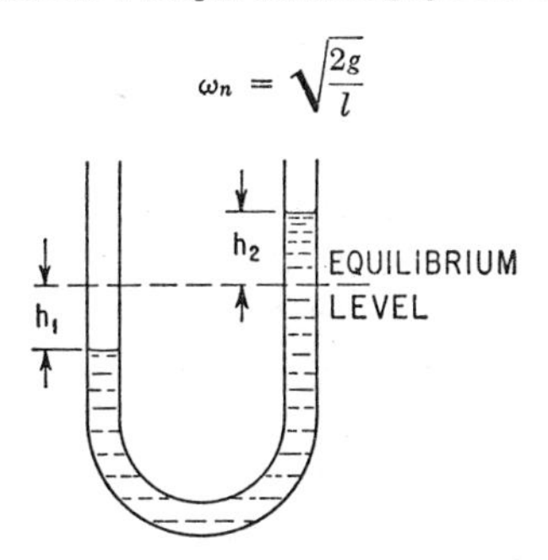

FIG. 2.26. Water column in a U tube.

REFERENCES

1. Timoshenko, S.: "Vibration Problems in Engineering," D. Van Nostrand Company, Inc., Princeton, N.J., 1937.
2. Hansen, H. M., and P. F. Chenea: "Mechanics of Vibration," John Wiley & Sons, Inc., New York, 1952.
3. Kármán, T. V., and M. A. Biot: "Mathematical Methods in Engineering," McGraw-Hill Book Company, Inc., New York, 1940.
4. Slater, J. C., and N. H. Frank: "Mechanics," McGraw-Hill Book Company, Inc., New York, 1947.
5. Pipes, L. A.: "Applied Mathematics for Engineers and Physicists," McGraw-Hill Book Company, Inc., New York, 1958.
6. Foss, K. A.: "Coordinates Which Uncouple the Equations of Motion of Damped Linear Dynamic Systems," *ASME Applied Mechanics Paper* 57-A-86, December, 1957.
7. Rayleigh: "The Theory of Sound," vol. I, Macmillan & Co., Ltd., London, 1894.
8. Crumb, S. F.: "A Study of the Effects of Damping on Normal Modes of Electrical and Mechanical Systems," California Institute of Technology, Pasadena, Calif., 1955.

3

VIBRATION OF A RESILIENTLY SUPPORTED RIGID BODY

Harry Himelblau, Jr.
Rockwell International Corporation

Sheldon Rubin
The Aerospace Corporation

INTRODUCTION

SCOPE

This chapter discusses the vibration of a rigid body on resilient supporting elements, including (1) methods of determining the inertial properties of a rigid body, (2) discussion of the dynamic properties of resilient elements, and (3) motion of a single rigid body on resilient supporting elements for various dynamic excitations and degrees of symmetry.

The general equations of motion for a rigid body on linear massless resilient supports are given; these equations are general in that they include any configuration of the rigid body and any configuration and location of the supports. They involve six simultaneous equations with numerous terms, for which a general solution is impracticable without the use of high-speed automatic computing equipment. Various degrees of simplification are introduced by assuming certain symmetry, and results useful for engineering purposes are presented. Several topics are considered: (1) determination of undamped natural frequencies and discussion of coupling of modes of vibration; (2) forced vibration where the excitation is a vibratory motion of the foundation; (3) forced vibration where the excitation is a vibratory force or moment generated within the body; and (4) free vibration caused by an instantaneous change in velocity of the system (velocity shock). Results are presented mathematically and, where feasible, graphically.

SYSTEM OF COORDINATES

The motion of the rigid body is referred to a fixed "inertial" frame of reference. The inertial frame is represented by a system of cartesian coordinates $\bar{X}$, $\bar{Y}$, $\bar{Z}$. A similar system of coordinates X, Y, Z fixed in the body has its origin at the center-of-mass. The two sets of coordinates are coincident when the body is in equilibrium under the action of gravity alone. The motions of the body are described by giving the displacement of the body axes relative to the inertial axes. The translational displacements of the center-of-mass of the body are x_c, y_c, z_c in the $\bar{X}$, $\bar{Y}$, $\bar{Z}$ directions, respectively. The rotational displacements of the body are characterized by the angles of rotation α, β, γ of

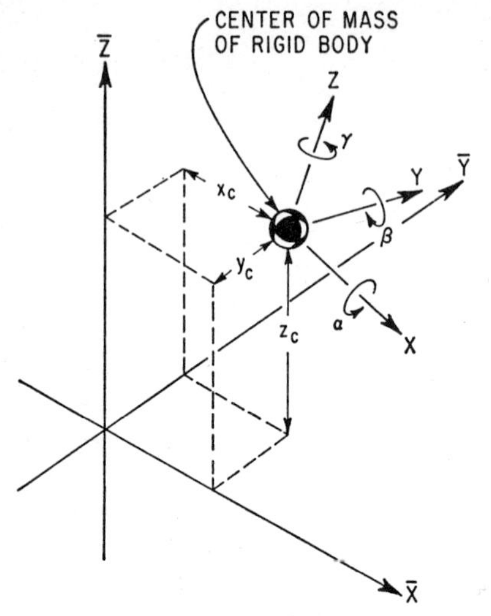

FIG. 3.1. System of coordinates for the motion of a rigid body consisting of a fixed inertial set of reference axes ($\bar{X}$, $\bar{Y}$, $\bar{Z}$), and a set of axes (X, Y, Z) fixed in the moving body with its origin at the center-of-mass. The axes $\bar{X}$, $\bar{Y}$, $\bar{Z}$ and X, Y, Z are coincident when the body is in equilibrium under the action of gravity alone. The displacement of the center-of-mass is given by the translational displacements x_c, y_c, z_c and the rotational displacements α, β, γ as shown. A positive rotation about an axis is one which advances a right-handed screw in the positive direction of the axis.

the body axes about the $\bar{X}$, $\bar{Y}$, $\bar{Z}$ axes, respectively. These displacements are shown graphically in Fig. 3.1.

Only small translations and rotations are considered. Hence, the rotations are commutative (i.e., the resulting position is independent of the order of the component rotations) and the angles of rotation about the body axes are equal to those about the inertial axes. Therefore, the displacements of a point b in the body (with the coordinates b_x, b_y, b_z in the X, Y, Z directions, respectively) are the sums of the components of the center-of-mass displacement in the directions of the $\bar{X}$, $\bar{Y}$, $\bar{Z}$ axes plus the tangential components of the rotational displacement of the body:

$$\begin{aligned} x_b &= x_c + b_z\beta - b_y\gamma \\ y_b &= y_c - b_z\alpha + b_x\gamma \\ z_b &= z_c - b_x\beta + b_y\alpha \end{aligned} \tag{3.1}$$

EQUATIONS OF SMALL MOTION OF A RIGID BODY

The equations of motion for the translation of a rigid body are

$$m\ddot{x}_c = \mathbf{F}_x \qquad m\ddot{y}_c = \mathbf{F}_y \qquad m\ddot{z}_c = \mathbf{F}_z \tag{3.2}$$

where m is the mass of the body, $\mathbf{F}_x$, $\mathbf{F}_y$, $\mathbf{F}_z$ are the summation of all forces acting on the body, and $\ddot{x}_c$, $\ddot{y}_c$, $\ddot{z}_c$ are the accelerations of the center-of-mass of the body in the $\bar{X}$, $\bar{Y}$, $\bar{Z}$ directions, respectively. The motion of the center-of-mass of a rigid body is the same as the motion of a particle having a mass equal to the total mass of the body and acted upon by the resultant external force.

The equations of motion for the rotation of a rigid body are

$$\begin{aligned} I_{xx}\ddot{\alpha} - I_{xy}\ddot{\beta} - I_{xz}\ddot{\gamma} &= \mathbf{M}_x \\ -I_{xy}\ddot{\alpha} + I_{yy}\ddot{\beta} - I_{yz}\ddot{\gamma} &= \mathbf{M}_y \\ -I_{xz}\ddot{\alpha} - I_{yz}\ddot{\beta} + I_{zz}\ddot{\gamma} &= \mathbf{M}_z \end{aligned} \tag{3.3}$$

where $\ddot{\alpha}$, $\ddot{\beta}$, $\ddot{\gamma}$ are the rotational accelerations about the X, Y, Z axes, as shown in Fig. 3.1; $\mathbf{M}_x$, $\mathbf{M}_y$, $\mathbf{M}_z$ are the summation of torques acting on the rigid body about the axes X, Y, Z, respectively; and I_{xx} . . ., I_{xy} . . . are the moments and products of inertia of the rigid body as defined below.

INERTIAL PROPERTIES OF A RIGID BODY

The properties of a rigid body that are significant in dynamics and vibration are the mass, the position of the center-of-mass (or center-of-gravity), the moments of inertia, the products of inertia, and the directions of the principal inertial axes. This section discusses the properties of a rigid body, together with computational and experimental methods for determining the properties.

MASS

COMPUTATION OF MASS. The mass of a body is computed by integrating the product of mass density $\rho(V)$ and elemental volume dV over the body:

$$m = \int_V \rho(V)\,dV \tag{3.4}$$

If the body is made up of a number of elements, each having constant or an average density, the mass is

$$m = \rho_1 V_1 + \rho_2 V_2 + \cdots + \rho_n V_n \tag{3.5}$$

where ρ_1 is the density of the element V_1, etc. Densities of various materials may be found in handbooks containing properties of materials.[1]

If a rigid body has a common geometrical shape, or if it is an assembly of subbodies having common geometrical shapes, the volume may be found from compilations of formulas. Typical formulas are included in Tables 3.1 and 3.2. Tables of areas of plane sections as well as volumes of solid bodies are useful.

If the volume of an element of the body is not given in such a table, the integration of Eq. (3.4) may be carried out analytically, graphically, or numerically. A graphical approach may be used if the shape is so complicated that the analytical expression for its boundaries is not available or is not readily integrable. This is accomplished by graphically dividing the body into smaller parts, each of whose boundaries may be altered slightly (without change to the area) in such a manner that the volume is readily calculable or measurable.

The weight W of a body of mass m is a function of the acceleration of gravity g at the particular location of the body in space:

$$W = mg \tag{3.6}$$

Unless otherwise stated, it is understood that the weight of a body is given for an average value of the acceleration of gravity on the surface of the earth. For engineering purposes, $g = 32.2$ ft/sec^2 or 386 in./sec^2 is usually used.

EXPERIMENTAL DETERMINATION OF MASS. Although Newton's second law of motion, $F = m\ddot{x}$, may be used to measure mass, this usually is not convenient. The mass of a body is most easily measured by performing a static measurement of the weight of the body and converting the result to mass. This is done by use of the value of the acceleration of gravity at the measurement location [Eq. (3.6)].

CENTER-OF-MASS

COMPUTATION OF CENTER-OF-MASS. The center-of-mass (or center-of-gravity) is that point located by the vector

$$\mathbf{r}_c = \frac{1}{m}\int_m \mathbf{r}(m)\,dm \tag{3.7}$$

where $\mathbf{r}(m)$ is the radius vector of the element of mass dm. The center-of-mass of a body in a cartesian coordinate system X, Y, Z is located at

$$\begin{aligned} X_c &= \frac{1}{m}\int_V X(V)\rho(V)\,dV \\ Y_c &= \frac{1}{m}\int_V Y(V)\rho(V)\,dV \\ Z_c &= \frac{1}{m}\int_V Z(V)\rho(V)\,dV \end{aligned} \tag{3.8}$$

where $X(V)$, $Y(V)$, $Z(V)$ are the X, Y, Z coordinates of the element of volume dV and m is the mass of the body.

If the body can be divided into elements whose centers-of-mass are known, the center-of-mass of the entire body having a mass m is located by equations of the following type:

$$X_c = \frac{1}{m}(X_{c1}m_1 + X_{c2}m_2 + \cdots + X_{cn}m_n), \text{ etc.} \tag{3.9}$$

where X_{c1} is the X coordinate of the center-of-mass of element m_1. Tables (see Tables 3.1 and 3.2) which specify the location of centers of area and volume (called centroids) for simple sections and solid bodies often are an aid in dividing the body into the submasses indicated in the above equation. The centroid and center-of-mass of an element are coincident when the density of the material is uniform throughout the element.

EXPERIMENTAL DETERMINATION OF CENTER-OF-MASS. The location of the center-of-mass is normally measured indirectly by locating the center-of-gravity of the body, and may be found in various ways. Theoretically, if the body is suspended by a flexible wire attached successively at different points on the body, all lines represented by the wire in its various positions when extended inwardly into the body intersect at the center-of-gravity. Two such lines determine the center-of-gravity, but more may be used as a check. There are practical limitations to this method in that the point of intersection often is difficult to designate.

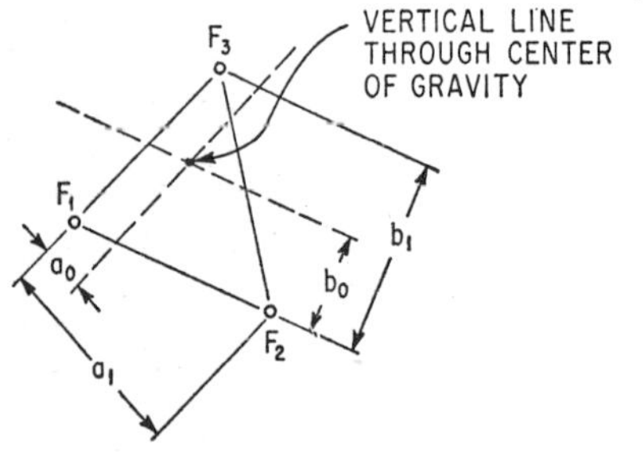

FIG. 3.2. Three-scale method of locating the center-of-gravity of a body. The vertical forces F_1, F_2, F_3 at the scales result from the weight of the body. The vertical line located by the distances a_o and b_o [see Eqs. (3.10)] passes through the center-of-gravity of the body.

Other techniques are based on the balancing of the body on point or line supports. A point support locates the center-of-gravity along a vertical line through the point; a line support locates it in a vertical plane through the line. The intersection of such lines or planes determined with the body in various positions locates the center-of-gravity. The greatest difficulty with this technique is the maintenance of the stability of the body while it is balanced, particularly where the height of the body is great relative to a horizontal dimension. If a perfect point or edge support is used, the equilibrium position is inherently unstable. It is only if the support has width that some degree of stability can be achieved, but then a resulting error in the location of the line or plane containing the center-of-gravity can be expected.

Another method of locating the center-of-gravity is to place the body in a stable position on three scales. From static moments the vector weight of the body is the resultant of the measured forces at the scales as shown in Fig. 3.2. The vertical line through the center-of-gravity is located by the distances a_0 and b_0:

$$\begin{aligned} a_0 &= \frac{F_2}{F_1 + F_2 + F_3} a_1 \\ b_0 &= \frac{F_3}{F_1 + F_2 + F_3} b_1 \end{aligned} \tag{3.10}$$

This method cannot be used with more than three scales.

MOMENT AND PRODUCT OF INERTIA

COMPUTATION OF MOMENT AND PRODUCT OF INERTIA.[2] The moments of inertia of a rigid body with respect to the orthogonal axes X, Y, Z fixed in the body are

$$I_{xx} = \int_m (Y^2 + Z^2)\, dm \qquad I_{yy} = \int_m (X^2 + Z^2)\, dm \qquad I_{zz} = \int_m (X^2 + Y^2)\, dm \tag{3.11}$$

where dm is the infinitesimal element of mass located at the coordinate distances X, Y, Z; and the integration is taken over the mass of the body. Similarly, the products of inertia are

$$I_{xy} = \int_m XY\, dm \qquad I_{xz} = \int_m XZ\, dm \qquad I_{yz} = \int_m YZ\, dm \tag{3.12}$$

It is conventional in rigid body mechanics to take the center of coordinates at the center-of-mass of the body. Unless otherwise specified, this location is assumed and the moments of inertia and products of inertia refer to axes through the center-of-mass of the body. For a unique set of axes, the products of inertia vanish. These axes are called the principal inertial axes of the body. The moments of inertia about these axes are called the principal moments of inertia.

The moments of inertia of a rigid body can be defined in terms of radii of gyration as follows:

$$I_{xx} = m\rho_x^{\,2} \qquad I_{yy} = m\rho_y^{\,2} \qquad I_{zz} = m\rho_z^{\,2} \tag{3.13}$$

where $I_{xx}, \ldots$ are the moments of inertia of the body as defined by Eqs. (3.11), m is the mass of the body, and $\rho_x, \ldots$ are the radii of gyration. The radius of gyration has the dimension of length, and often leads to convenient expressions in dynamics of rigid bodies when distances are normalized to an appropriate radius of gyration. Solid bodies of various shapes have characteristic radii of gyration which sometimes are useful intuitively in evaluating dynamic conditions.

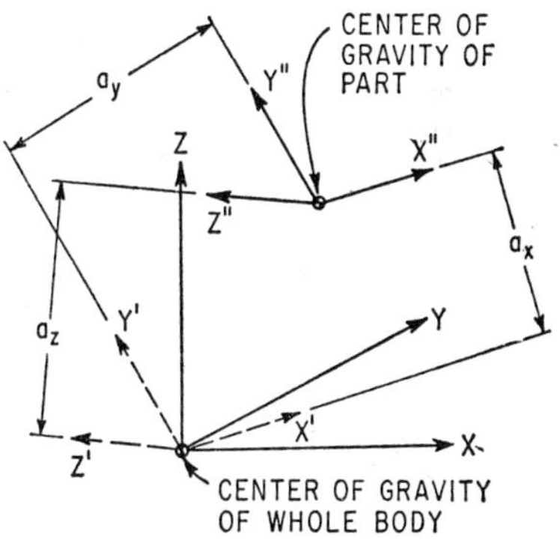

FIG. 3.3. Axes required for moment and product of inertia transformations. Moments and products of inertia with respect to the axes X'', Y'', Z'' are transferred to the mutually parallel axes X', Y', Z' by Eqs. (3.14) and (3.15), and then to the inclined axes X, Y, Z by Eqs. (3.16) and (3.17).

Unless the body has a very simple shape, it is laborious to evaluate the integrals of Eqs. (3.11) and (3.12). The problem is made easier by subdividing the body into parts for which simplified calculations are possible. The moments and products of inertia of the body are found by first determining the moments and products of inertia for the individual parts with respect to appropriate reference axes chosen in the parts, and then summing the contributions of the parts. This is done by selecting axes through the centers-of-mass of the parts, and then determining the moments and products of inertia of the parts relative to these axes. Then the moments and products of inertia are transferred to the axes chosen through the center-of-mass of the whole body, and the transferred quantities summed. In general, the transfer involves two sets of nonparallel coordinates whose centers are displaced. Two transformations are required as follows.

Transformation to Parallel Axes. Referring to Fig. 3.3, suppose that X, Y, Z is a convenient set of axes for the moment of inertia of the whole body with its origin at the center-of-mass. The moments and products of inertia for a part of the body are $I_{x''x''}$,

$I_{y''y''}$, $I_{z''z''}$, $I_{x''y''}$, $I_{x''z''}$, and $I_{y''z''}$, taken with respect to a set of axes X'', Y'', Z'' fixed in the part and having their center at the center-of-mass of the part. The axes X', Y', Z' are chosen parallel to X'', Y'', Z'' with their origin at the center-of-mass of the body. The perpendicular distance between the X'' and X' axes is a_x; between Y'' and Y' is a_y; between Z'' and Z' is a_z. The moments and products of inertia of the part of mass m_n with respect to the X', Y', Z' axes are

$$\begin{aligned} I_{x'x'} &= I_{x''x''} + m_n a_x^2 \\ I_{y'y'} &= I_{y''y''} + m_n a_y^2 \\ I_{z'z'} &= I_{z''z''} + m_n a_z^2 \end{aligned} \qquad (3.14)$$

The corresponding products of inertia are

$$\begin{aligned} I_{x'y'} &= I_{x''y''} + m_n a_x a_y \\ I_{x'z'} &= I_{x''z''} + m_n a_x a_z \\ I_{y'z'} &= I_{y''z''} + m_n a_y a_z \end{aligned} \qquad (3.15)$$

If X'', Y'', Z'' are the principal axes of the part, the product of inertia terms on the right-hand side of Eqs. (3.15) are zero.

Transformation to Inclined Axes. The desired moments and products of inertia witn respect to axes X, Y, Z are now obtained by a transformation theorem relating the properties of bodies with respect to inclined sets of axes whose centers coincide. This theorem makes use of the direction cosines λ for the respective sets of axes. For example, $\lambda_{xx'}$ is the cosine of the angle between the X and X' axes. The expressions for the moments of inertia are

$$\begin{aligned} I_{xx} &= \lambda_{xx'}^2 I_{x'x'} + \lambda_{xy'}^2 I_{y'y'} + \lambda_{xz'}^2 I_{z'z'} \\ &\quad - 2\lambda_{xx'}\lambda_{xy'} I_{x'y'} - 2\lambda_{xx'}\lambda_{xz'} I_{x'z'} - 2\lambda_{xy'}\lambda_{xz'} I_{y'z'} \\ I_{yy} &= \lambda_{yx'}^2 I_{x'x'} + \lambda_{yy'}^2 I_{y'y'} + \lambda_{yz'}^2 I_{z'z'} \\ &\quad - 2\lambda_{yx'}\lambda_{yy'} I_{x'y'} - 2\lambda_{yx'}\lambda_{yz'} I_{x'z'} - 2\lambda_{yy'}\lambda_{yz'} I_{y'z'} \\ I_{zz} &= \lambda_{zx'}^2 I_{x'x'} + \lambda_{zy'}^2 I_{y'y'} + \lambda_{zz'}^2 I_{z'z'} \\ &\quad - 2\lambda_{zx'}\lambda_{zy'} I_{x'y'} - 2\lambda_{zx'}\lambda_{zz'} I_{x'z'} - 2\lambda_{zy'}\lambda_{zz'} I_{y'z'} \end{aligned} \qquad (3.16)$$

The corresponding products of inertia are

$$\begin{aligned} -I_{xy} &= \lambda_{xx'}\lambda_{yx'} I_{x'x'} + \lambda_{xy'}\lambda_{yy'} I_{y'y'} + \lambda_{xz'}\lambda_{yz'} I_{z'z'} \\ &\quad - (\lambda_{xx'}\lambda_{yy'} + \lambda_{xy'}\lambda_{yx'}) I_{x'y'} - (\lambda_{xy'}\lambda_{yz'} + \lambda_{xz'}\lambda_{yy'}) I_{y'z'} \\ &\quad - (\lambda_{xz'}\lambda_{yx'} + \lambda_{xx'}\lambda_{yz'}) I_{x'z'} \\ -I_{xz} &= \lambda_{xx'}\lambda_{zx'} I_{x'x'} + \lambda_{xy'}\lambda_{zy'} I_{y'y'} + \lambda_{xz'}\lambda_{zz'} I_{z'z'} \\ &\quad - (\lambda_{xx'}\lambda_{zy'} + \lambda_{xy'}\lambda_{zx'}) I_{x'y'} - (\lambda_{xy'}\lambda_{zz'} + \lambda_{xz'}\lambda_{zy'}) I_{y'z'} \\ &\quad - (\lambda_{xx'}\lambda_{zz'} + \lambda_{xz'}\lambda_{zx'}) I_{x'z'} \\ -I_{yz} &= \lambda_{yx'}\lambda_{zx'} I_{x'x'} + \lambda_{yy'}\lambda_{zy'} I_{y'y'} + \lambda_{yz'}\lambda_{zz'} I_{z'z'} \\ &\quad - (\lambda_{yx'}\lambda_{zy'} + \lambda_{yy'}\lambda_{zx'}) I_{x'y'} - (\lambda_{yy'}\lambda_{zz'} + \lambda_{yz'}\lambda_{zy'}) I_{y'z'} \\ &\quad - (\lambda_{yz'}\lambda_{zx'} + \lambda_{yx'}\lambda_{zz'}) I_{x'z'} \end{aligned} \qquad (3.17)$$

EXPERIMENTAL DETERMINATION OF MOMENTS OF INERTIA. The moment of inertia of a body about a given axis may be found experimentally by suspending the body as a pendulum so that rotational oscillations about that axis can occur. The period of free oscillation is then measured, and is used with the geometry of the pendulum to calculate the moment of inertia.

Two types of pendulums are useful: the compound pendulum and the torsional pendulum. When using the compound pendulum, the body is supported from two overhead points by wires, illustrated in Fig. 3.4. The distance l is measured between the axis of support O–O and a parallel axis C–C through the center-of-gravity of the body. The moment of inertia about C–C is given by

$$I_{cc} = ml^2 \left[\left(\frac{\tau_0}{2\pi} \right)^2 \left(\frac{g}{l} \right) - 1 \right] \quad (3.18)$$

where τ_0 is the period of oscillation in seconds, l is the pendulum length in inches, g is the gravitational acceleration in in./sec^2, and m is the mass in lb-sec^2/in., yielding a moment of inertia in lb-in.-sec^2.

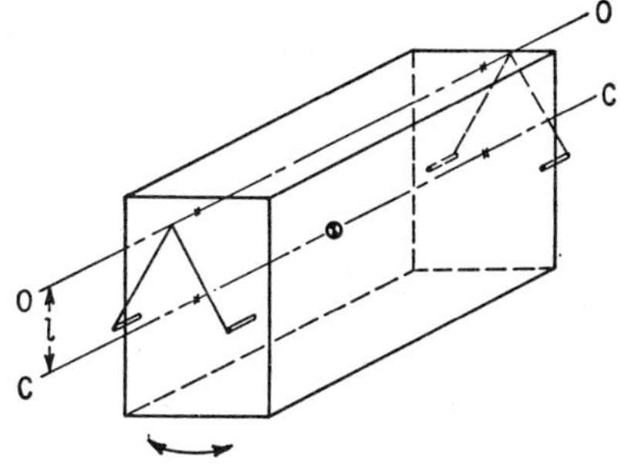

FIG. 3.4. Compound pendulum method of determining moment of inertia. The period of oscillation of the test body about the horizontal axis O–O and the perpendicular distance l between the axis O–O and the parallel axis C–C through the center-of-gravity of the test body give I_{cc} by Eq. (3.18).

The accuracy of the above method is dependent upon the accuracy with which the distance l is known. Since the center-of-gravity often is an inaccessible point, a direct measurement of l may not be practicable. However, a change in l can be measured quite readily. If the experiment is repeated with a different support axis O'–O', the length l becomes $l + \Delta l$ and the period of oscillation becomes τ_0'. Then, the distance l can be written in terms of Δl, and the two periods τ_0, τ_0':

$$l = \Delta l \left[\frac{(\tau_0'^2/4\pi^2)(g/\Delta l) - 1}{[(\tau_0^2 - \tau_0'^2)/4\pi^2][g/\Delta l] - 1} \right] \quad (3.19)$$

This value of l can be substituted into Eq. (3.18) to compute I_{cc}.

Note that accuracy is not achieved if l is much larger than the radius of gyration ρ_c of the body about the axis C–C ($I_{cc} = m\rho_c^2$). If l is large, then $(\tau_0/2\pi)^2 \simeq l/g$ and the expression in brackets in Eq. (3.18) is very small; thus, it is sensitive to small errors in the measurement of both τ_0 and l. Consequently, it is highly desirable that the distance l be chosen as small as convenient, preferably with the axis O–O passing through the body.

A torsional pendulum may be constructed with the test body suspended by a single torsional spring (in practice, a rod or wire) of known stiffness, or by three flexible wires. A solid body supported by a single torsional spring is shown in Fig. 3.5. From the known torsional stiffness k_t and the measured period of torsional oscillation τ, the moment of inertia of the body about the vertical torsional axis is

$$I_{cc} = \frac{k_t \tau^2}{4\pi^2} \quad (3.20)$$

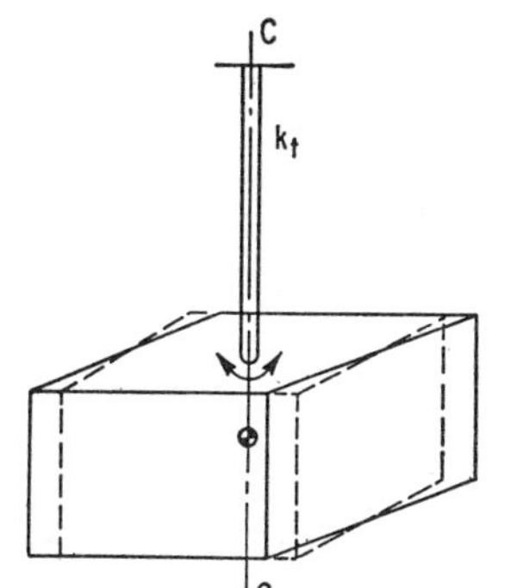

FIG. 3.5. Torsional pendulum method of determining moment of inertia. The period of torsional oscillation of the test body about the vertical axis C–C passing through the center-of-gravity and the torsional spring constant k_t give I_{cc} by Eq. (3.20).

A platform may be constructed below the torsional spring to carry the bodies to be measured, as shown in Fig. 3.6. By repeating the experiment with two different bodies placed on the platform, it becomes unnecessary to measure the torsional stiffness k_t. If a body with a *known* moment of inertia I_1 is placed on the platform and an oscillation period τ_1 results, the moment of inertia I_2 of a body which produces a period τ_2 is given by

$$I_2 = I_1 \left[\frac{(\tau_2/\tau_0)^2 - 1}{(\tau_1/\tau_0)^2 - 1} \right] \quad (3.21)$$

where τ_0 is the period of the pendulum comprised of platform alone.

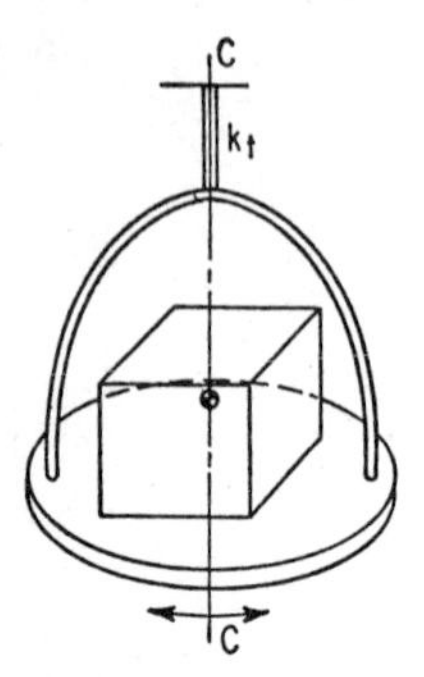

FIG. 3.6. A variation of the torsional pendulum method shown in Fig. 3.5 wherein a light platform is used to carry the test body. The moment of inertia I_{cc} is given by Eq. (3.20).

A body suspended by three flexible wires, called a trifilar pendulum, as shown in Fig. 3.7, offers some utilitarian advantages. Designating the perpendicular distances of the wires to the vertical axis C–C through the center-of-gravity of the body by R_1, R_2, R_3, the angles between wires by ϕ_1, ϕ_2, ϕ_3, and the length of each wire by l, the moment of inertia about axis C–C is

$$I_{cc} = \frac{mgR_1R_2R_3\tau^2}{4\pi^2 l} \frac{R_1 \sin \phi_1 + R_2 \sin \phi_2 + R_3 \sin \phi_3}{R_2R_3 \sin \phi_1 + R_1R_3 \sin \phi_2 + R_1R_2 \sin \phi_3} \tag{3.22}$$

Apparatus that is more convenient for repeated use embodies a light platform supported by three equally spaced wires. The body whose moment of inertia is to be measured is placed on the platform with its center-of-gravity equidistant from the wires. Thus $R_1 = R_2 = R_3 = R$ and $\phi_1 = \phi_2 = \phi_3 = 120°$. Substituting these relations in Eq. (3.22), the moment of inertia about the vertical axis C–C is

$$I_{cc} = \frac{mgR^2\tau^2}{4\pi^2 l} \tag{3.23}$$

where the mass m is the sum of the masses of the test body and the platform. The moment of inertia of the platform is subtracted from the test result to obtain the moment of inertia of the body being measured. It becomes unnecessary to know the distances R and l in Eq. (3.23) if the period of oscillation is measured with the platform empty, with the body being measured on the platform, and with a second body of known mass m_1 and known moment of inertia I_1 on the platform. Then the desired moment of inertia I_2 is

$$I_2 = I_1 \left[\frac{[1 + (m_2/m_0)][\tau_2/\tau_0]^2 - 1}{[1 + (m_1/m_0)][\tau_1/\tau_0]^2 - 1} \right] \tag{3.24}$$

where m_0 is the mass of the unloaded platform, m_2 is the mass of the body being measured, τ_0 is the period of oscillation with the platform unloaded, τ_1 is the period when loaded with known body of mass m_1, and τ_2 is the period when loaded with the unknown body of mass m_2.

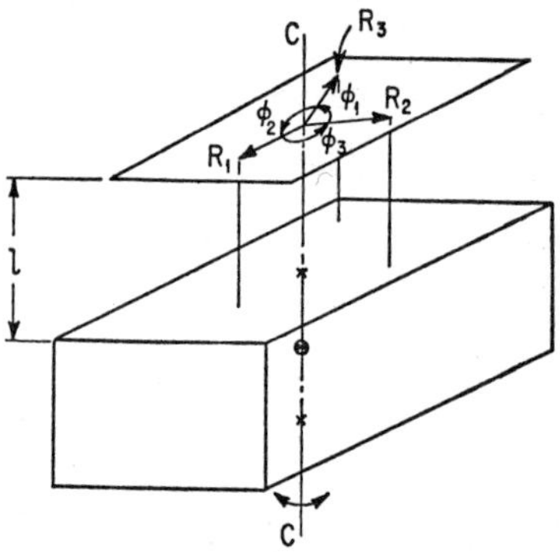

FIG. 3.7. Trifilar pendulum method of determining moment of inertia. The period of torsional oscillation of the test body about the vertical axis C–C passing through the center-of-gravity and the geometry of the pendulum give I_{cc} by Eq. (3.22); with a simpler geometry, I_{cc} is given by Eq. (3.23).

EXPERIMENTAL DETERMINATION OF PRODUCT OF INERTIA. The experimental determination of a product of inertia usually requires the measurement of moments of inertia. (An exception is the balancing machine technique described later.) If possible, symmetry of the body is used to locate directions of principal inertial axes, thereby simplifying the relationship between the moments of inertia as known and the products of inertia to be found. Several alternative procedures are described below, depending on the number of principal inertia axes whose directions are known. Knowledge of two principal axes implies a knowledge of all three since they are mutually perpendicular.

If the directions of all three principal axes (X', Y', Z') are known and it is desirable to use another set of axes (X, Y, Z), Eqs. (3.16) and (3.17) may be simplified because the products of inertia with respect to the principal directions are zero. First, the three principal moments of inertia ($I_{x'x'}$, $I_{y'y'}$, $I_{z'z'}$) are measured by one of the above techniques; then the moments of inertia with respect to the X, Y, Z axes are

$$\begin{aligned} I_{xx} &= \lambda_{xx'}^2 I_{x'x'} + \lambda_{xy'}^2 I_{y'y'} + \lambda_{xz'}^2 I_{z'z'} \\ I_{yy} &= \lambda_{yx'}^2 I_{x'x'} + \lambda_{yy'}^2 I_{y'y'} + \lambda_{yz'}^2 I_{z'z'} \\ I_{zz} &= \lambda_{zx'}^2 I_{x'x'} + \lambda_{zy'}^2 I_{y'y'} + \lambda_{zz'}^2 I_{z'z'} \end{aligned} \tag{3.25}$$

The products of inertia with respect to the X, Y, Z axes are

$$\begin{aligned} -I_{xy} &= \lambda_{xx'}\lambda_{yx'} I_{x'x'} + \lambda_{xy'}\lambda_{yy'} I_{y'y'} + \lambda_{xz'}\lambda_{yz'} I_{z'z'} \\ -I_{xz} &= \lambda_{xx'}\lambda_{zx'} I_{x'x'} + \lambda_{xy'}\lambda_{zy'} I_{y'y'} + \lambda_{xz'}\lambda_{zz'} I_{z'z'} \\ -I_{yz} &= \lambda_{yx'}\lambda_{zx'} I_{x'x'} + \lambda_{yy'}\lambda_{zy'} I_{y'y'} + \lambda_{yz'}\lambda_{zz'} I_{z'z'} \end{aligned} \tag{3.26}$$

The direction of one principal axis Z may be known from symmetry. The axis through the center-of-gravity perpendicular to the plane of symmetry is a principal axis. The product of inertia with respect to X and Y axes, located in the plane of symmetry, is determined by first establishing another axis X' at a counterclockwise angle θ from X, as shown in Fig. 3.8. If the three moments of inertia I_{xx}, $I_{x'x'}$, and I_{yy} are measured by any applicable means, the product of inertia I_{xy} is

$$I_{xy} = \frac{I_{xx}\cos^2\theta + I_{yy}\sin^2\theta - I_{x'x'}}{\sin 2\theta} \tag{3.27}$$

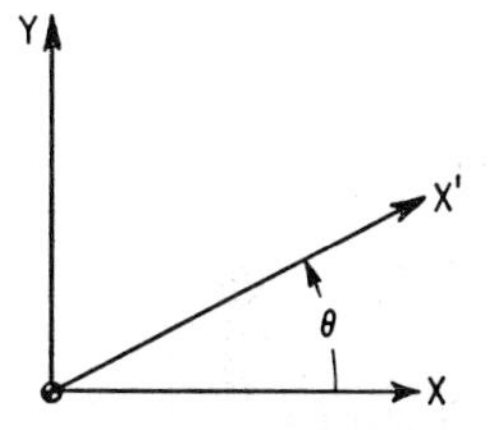

FIG. 3.8. Axes required for determining the product of inertia with respect to the axes X and Y when Z is a principal axis of inertia. The moments of inertia about the axes X, Y, and X', where X' is in the plane of X and Y at a counterclockwise angle θ from X, give I_{xy} by Eq. (3.27).

where $0 < \theta < \pi$. For optimum accuracy, θ should be approximately $\pi/4$ or $3\pi/4$. Since the third axis Z is a principal axis, I_{xz} and I_{yz} are zero.

Another method is illustrated in Fig. 3.9.[3,4] The plane of the X and Z axes is a plane of symmetry, or the Y axis is otherwise known to be a principal axis of inertia. For determining I_{xz}, the body is suspended by a cable so that the Y axis is horizontal and the Z axis is vertical. Torsional stiffness about the Z axis is provided by four springs acting in the Y direction at the points shown. The body is oscillated about the Z axis with various positions of the springs so that the angle θ can be varied. The spring stiffnesses and locations must be such that there is no net force in the Y direction due to a rotation about the Z axis. In general, there is coupling between rotations about the X and Z axes, with the result that oscillations about both axes occur as a result of an initial rotational displacement about the Z axis. At some particular value of $\theta = \theta_0$, the two rotations are uncoupled; i.e., oscillation about the Z axis does not cause oscillation about the X axis. Then

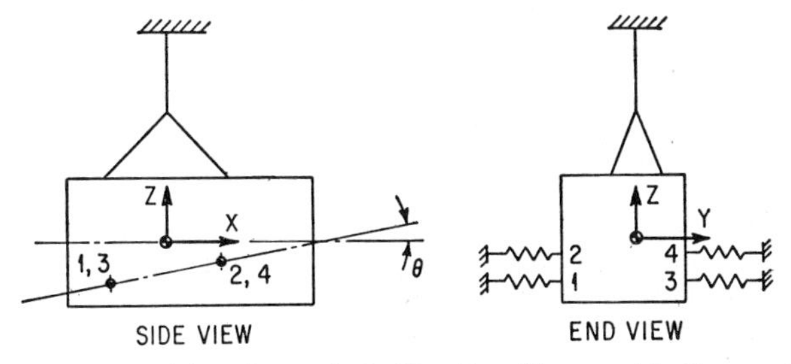

FIG. 3.9. Method of determining the product of inertia with respect to the axes X and Z when Y is a principal axis of inertia. The test body is oscillated about the vertical Z axis with torsional stiffness provided by the four springs acting in the Y direction at the points shown. There should be no net force on the test body in the Y direction due to a rotation about the Z axis. The angle θ is varied until, at some value of $\theta = \theta_0$, oscillations about X and Z are uncoupled. The angle θ_0 and the moment of inertia about the Z axis give I_{xz} by Eq. (3.28).

$$I_{xz} = I_{zz} \tan \theta_0 \tag{3.28}$$

The moment of inertia I_{zz} can be determined by one of the methods described under *Experimental Determination of Moments of Inertia.*

When the moments and product of inertia with respect to a pair of axes X and Z in a principal plane of inertia XZ are known. the orientation of a principal axis P is given by

$$\theta_p = \tfrac{1}{2} \tan^{-1} \left(\frac{2I_{xz}}{I_{zz} - I_{xx}} \right) \tag{3.29}$$

where θ_p is the counterclockwise angle from the X axis to the P axis. The second principal axis in this plane is at $\theta_p + 90°$.

Consider the determination of products of inertia when the directions of all principal axes of inertia are unknown. In one method, the moments of inertia about two independent sets of three mutually perpendicular axes are measured, and the direction cosines between these sets of axes are known from the positions of the axes. The values for the six moments of inertia and the nine direction cosines are then substituted into Eqs. (3.16) and (3.17). The result is six linear equations in the six unknown products of inertia, from which the values of the desired products of inertia may be found by simultaneous solution of the equations. This method leads to experimental errors of relatively large magnitude because each product of inertia is, in general, a function of all six moments of inertia, each of which contains an experimental error.

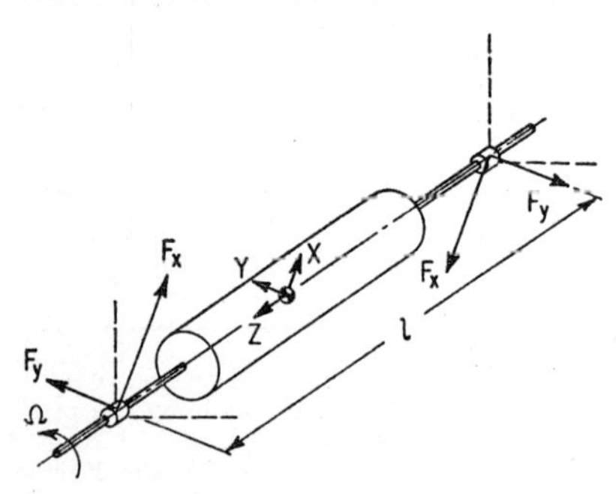

FIG. 3.10. Balancing machine technique for determining products of inertia. The test body is rotated about the Z axis with angular velocity Ω. The dynamic reactions F_x and F_y measured at the bearings, which are a distance l apart, give I_{xz} and I_{yz} by Eq. (3.30).

An alternative method is based upon the knowledge that one of the principal moments of inertia of a body is the largest and another is the smallest that can be obtained for any axis through the center-of-gravity. A trial-and-error procedure can be used to locate the orientation of the axis through the center-of-gravity having the maximum and/or minimum moment of inertia. After one or both are located, the moments and products of inertia for any set of axes are found by the techniques previously discussed.

The products of inertia of a body also may be determined by rotating the body at a constant angular velocity Ω about an axis passing through the center of gravity, as illustrated in Fig. 3.10. This method is similar to the balancing machine technique used to balance a body dynamically (see Chap. 39). If the bearings are a distance l apart and the dynamic reactions F_x and F_y are measured, the products of inertia are

$$I_{xz} = -\frac{F_x l}{\Omega^2} \qquad I_{yz} = -\frac{F_y l}{\Omega^2} \tag{3.30}$$

Limitations to this method are (1) the size of the body that can be accommodated by the balancing machine and (2) the angular velocity that the body can withstand without damage from centrifugal forces. If the angle between the Z axis and a principal axis of inertia is small, high rotational speeds may be necessary to measure the reaction forces accurately.

PROPERTIES OF RESILIENT SUPPORTS

A resilient support is considered to be a three-dimensional element having two terminals or end connections. When the end connections are moved one relative to the other in any direction, the element resists such motion. In this chapter, the element is considered to be massless; the force that resists relative motion across the element is considered to consist of a spring force that is directly proportional to the relative displacement (deflection across the element) and a damping force that is directly proportional to the relative velocity (velocity across the element). Such an element is defined as a *linear resilient support.* Nonlinear elements are discussed in Chap. 4; elements with mass are discussed in Chap. 30; and nonlinear damping is discussed in Chaps. 2 and 30.

In a single degree-of-freedom system or in a system having constraints on the paths of motion of elements of the system (Chap. 2), the resilient element is constrained to deflect in a given direction and the properties of the element are defined with respect to the force opposing motion in this direction. In the absence of such constraints, the application of a force to a resilient element generally causes a motion in a different direction. The *principal elastic axes* of a resilient element are those axes for which the element, when unconstrained, experiences a deflection colineal with the direction of the applied force.[5] Any axis of symmetry is a principal elastic axis.

In rigid body dynamics, the rigid body sometimes vibrates in modes that are coupled by the properties of the resilient elements as well as by their location. For example, if the body experiences a static displacement x in the direction of the X axis only, a resilient element opposes this motion by exerting a force $k_{xx}x$ on the body in the direction of the X axis where one subscript on the spring constant k indicates the direction of the force exerted by the element and the other subscript indicates the direction of the deflection. If the X direction is not a principal elastic direction of the element and the body experiences a static displacement x in the X direction, the body is acted upon by a force $k_{yx}x$ in the Y direction if no displacement y is permitted. The stiffnesses have reciprocal properties; i.e., $k_{xy} = k_{yx}$. In general, the stiffnesses in the directions of the coordinate axes can be expressed in terms of (1) principal stiffnesses and (2) the angles between the coordinate axes and the principal elastic axes of the element. (See Chap. 30 for a detailed discussion of a biaxial stiffness element.) Therefore, the stiffness of a resilient element can be represented pictorially by the combination of three mutually perpendicular, idealized springs oriented along the principal elastic directions of the resilient element. Each spring has a stiffness equal to the principal stiffness represented.

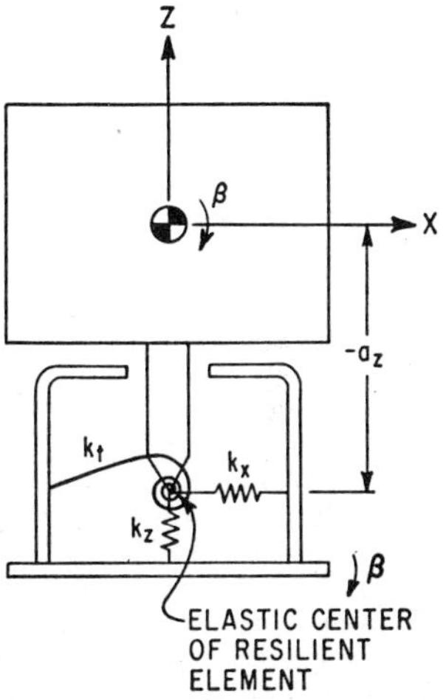

FIG. 3.11. Pictorial representation of the properties of an undamped resilient element in the XZ plane including a torsional spring k_t. An analysis of the motion of the supported body in the XZ plane shows that the torsional spring can be neglected if $k_t \ll a_z^2 k_x$.

A resilient element is assumed to have damping properties such that each spring representing a value of principal stiffness is paralleled by an idealized viscous damper, each damper representing a value of principal damping. Hence, coupling through damping exists in a manner similar to coupling through stiffness. Consequently, the viscous damping coefficient c is analogous to the spring coefficient k; i.e., the force exerted by the damping of the resilient element in response to a velocity $\dot{x}$ is $c_{xx}\dot{x}$ in the direction of the X axis and $c_{yx}\dot{x}$ in the direction of the Y axis if $\dot{y}$ is zero. Reciprocity exists; i.e., $c_{xy} = c_{yx}$.

The point of intersection of the principal elastic axes of a resilient element is designated as the *elastic center of the resilient element.* The elastic center is important since it defines the theoretical point location of the resilient element for use in the equations of motion of a resiliently supported rigid body. For example, the torque on the rigid body about the Y axis due to a force $k_{xx}x$ transmitted by a resilient element in the X direction is $k_{xx}a_zx$, where a_z is the Z coordinate of the elastic center of the resilient element.

In general, it is assumed that a resilient element is attached to the rigid body by means of "ball joints"; i.e., the resilient element is incapable of applying a couple to the body. If this assumption is not made, a resilient element would be represented not only by translational springs and dampers along the principal elastic axes but also by torsional springs and dampers resisting rotation about the principal elastic directions.

Figure 3.11 shows that the torsional elements usually can be neglected. The torque which acts on the rigid body due to a rotation β of the body and a rotation $\boldsymbol{\beta}$ of the support is $(k_t + a_z^2 k_x)(\beta - \boldsymbol{\beta})$, where k_t is the torsional spring constant in the β direction. The torsional stiffness k_t usually is much smaller than $a_z^2 k_x$ and can be neglected. Treatment of the general case indicates that if the torsional stiffnesses of the resilient element are small compared with the product of the translational stiffnesses times the square of distances from the elastic center of the resilient element to the center-of-gravity of the rigid body, the torsional stiffnesses have a negligible effect on the vibrational behavior of the body. The treatment of torsional dampers is completely analogous.

EQUATIONS OF MOTION FOR A RESILIENTLY SUPPORTED RIGID BODY

The differential equations of motion for the rigid body are given by Eqs. (3.2) and (3.3) where the **F**'s and **M**'s represent the forces and moments acting on the body, either directly or through the resilient supporting elements. Figure 3.12 shows a view of a rigid body at rest with an inertial set of axes $\bar{X}$, $\bar{Y}$, $\bar{Z}$ and a coincident set of axes X, Y, Z fixed in the rigid body, both sets of axes passing through the center-of-mass. A typical resilient element (2) is represented by parallel spring and viscous damper combinations arranged respectively parallel with the $\bar{X}$, $\bar{Y}$, $\bar{Z}$ axes. Another resilient element (1) is shown with its principal axes not parallel with $\bar{X}$, $\bar{Y}$, $\bar{Z}$.

The displacement of the center-of-gravity of the body in the $\bar{X}$, $\bar{Y}$, $\bar{Z}$ directions is in Fig. 3.1 indicated by x_c, y_c, z_c, respectively; and rotation of the rigid body about these axes is indicated by α, β, γ, respectively. In Fig. 3.12, each resilient element is represented by three mutually perpendicular spring-damper combinations. One end of each such combination is attached to the rigid body; the other end is considered to be attached to a foundation whose corresponding translational displacement is defined by u, v, w in the $\bar{X}$, $\bar{Y}$, $\bar{Z}$ directions, respectively, and whose rotational displacement about these axes is defined by $\boldsymbol{\alpha}$, $\boldsymbol{\beta}$, $\boldsymbol{\gamma}$, respectively. The point of attachment of each of the idealized resilient elements is located at the coordinate distances a_x, a_y, a_z of the elastic center of the resilient element.

Consider the rigid body to experience a translational displacement x_c of its center-of-gravity and no other displacement, and neglect the effects of the viscous dampers. The force developed by a resilient element has the effect of a force $-k_{xx}(x_c - u)$ in the X direction, a moment $k_{xx}(x_c - u)a_y$ in the γ coordinate (about the Z axis) and a moment $-k_{xx}(x_c - u)a_z$ in the β coordinate (about the Y axis). Furthermore, the coupling stiffness causes a force $-k_{xy}(x_c - u)$ in the Y direction and a force $-k_{xz}(x_c - u)$ in the Z direction. These forces have the moments: $k_{xy}(x_c - u)a_z$ in the α coordinate; $-k_{xy}(x_c - u)a_x$ in the γ coordinate; $k_{xz}(x_c - u)a_x$ in the β coordinate; and $-k_{xz}(x_c - u)a_y$ in the α coordinate. By considering in a similar manner the forces and moments developed by a resilient element for successive displacements of the rigid body in the three translational and three rotational coordinates, and summing over the number of resilient elements, the equations of motion are written as follows:[6, 7, 8]

$$\begin{aligned} m\ddot{x}_c &+ \Sigma k_{xx}(x_c - u) + \Sigma k_{xy}(y_c - v) + \Sigma k_{xz}(z_c - w) \\ &+ \Sigma(k_{xz}a_y - k_{xy}a_z)(\alpha - \boldsymbol{\alpha}) + \Sigma(k_{xx}a_z - k_{xz}a_x)(\beta - \boldsymbol{\beta}) \\ &+ \Sigma(k_{xy}a_x - k_{xx}a_y)(\gamma - \boldsymbol{\gamma}) = F_x \end{aligned} \tag{3.31a}$$

$$\begin{aligned} I_{xx}\ddot{\alpha} &- I_{xy}\ddot{\beta} - I_{xz}\ddot{\gamma} + \Sigma(k_{xz}a_y - k_{xy}a_z)(x_c - u) \\ &+ \Sigma(k_{yz}a_y - k_{yy}a_z)(y_c - v) + \Sigma(k_{zz}a_y - k_{yz}a_z)(z_c - w) \\ &+ \Sigma(k_{yy}a_z^2 + k_{zz}a_y^2 - 2k_{yz}a_ya_z)(\alpha - \boldsymbol{\alpha}) \\ &+ \Sigma(k_{xz}a_ya_z + k_{yz}a_xa_z - k_{zz}a_xa_y - k_{xy}a_z^2)(\beta - \boldsymbol{\beta}) \\ &+ \Sigma(k_{xy}a_ya_z + k_{yz}a_xa_y - k_{yy}a_xa_z - k_{xz}a_y^2)(\gamma - \boldsymbol{\gamma}) = M_x \end{aligned} \tag{3.31b}$$

$$
\begin{aligned}
m\ddot{y}_c &+ \Sigma k_{xy}(x_c - u) + \Sigma k_{yy}(y_c - v) + \Sigma k_{yz}(z_c - w) \\
&+ \Sigma(k_{yz}a_y - k_{yy}a_z)(\alpha - \boldsymbol{\alpha}) + \Sigma(k_{xy}a_z - k_{yz}a_x)(\beta - \boldsymbol{\beta}) \\
&+ \Sigma(k_{yy}a_x - k_{xy}a_y)(\gamma - \boldsymbol{\gamma}) = F_y
\end{aligned}
\tag{3.31c}
$$

$$
\begin{aligned}
I_{yy}\ddot{\beta} - I_{xy}\ddot{\alpha} - I_{yz}\ddot{\gamma} &+ \Sigma(k_{xx}a_z - k_{xz}a_x)(x_c - u) \\
&+ \Sigma(k_{xy}a_z - k_{yz}a_x)(y_c - v) + \Sigma(k_{xz}a_z - k_{zz}a_x)(z_c - w) \\
&+ \Sigma(k_{xz}a_ya_z + k_{yz}a_xa_z - k_{zz}a_xa_y - k_{xy}a_z^2)(\alpha - \boldsymbol{\alpha}) \\
&+ \Sigma(k_{xx}a_z^2 + k_{zz}a_x^2 - 2k_{xz}a_xa_z)(\beta - \boldsymbol{\beta}) \\
&+ \Sigma(k_{xy}a_xa_z + k_{xz}a_xa_y - k_{xx}a_ya_z - k_{yz}a_x^2)(\gamma - \boldsymbol{\gamma}) = M_y
\end{aligned}
\tag{3.31d}
$$

$$
\begin{aligned}
m\ddot{z}_c &+ \Sigma k_{xz}(x_c - u) + \Sigma k_{yz}(y_c - v) + \Sigma k_{zz}(z_c - w) \\
&+ \Sigma(k_{zz}a_y - k_{yz}a_z)(\alpha - \boldsymbol{\alpha}) + \Sigma(k_{xz}a_z - k_{zz}a_x)(\beta - \boldsymbol{\beta}) \\
&+ \Sigma(k_{yz}a_x - k_{xz}a_y)(\gamma - \boldsymbol{\gamma}) = F_z
\end{aligned}
\tag{3.31e}
$$

$$
\begin{aligned}
I_{zz}\ddot{\gamma} - I_{xz}\ddot{\alpha} - I_{yz}\ddot{\beta} &+ \Sigma(k_{xy}a_x - k_{xx}a_y)(x_c - u) \\
&+ \Sigma(k_{yy}a_x - k_{xy}a_y)(y_c - v) + \Sigma(k_{yz}a_x - k_{xz}a_y)(z_c - w) \\
&+ \Sigma(k_{xy}a_ya_z + k_{yz}a_xa_y - k_{yy}a_xa_z - k_{xz}a_y^2)(\alpha - \boldsymbol{\alpha}) \\
&+ \Sigma(k_{xy}a_xa_z + k_{xz}a_xa_y - k_{xx}a_ya_z - k_{yz}a_x^2)(\beta - \boldsymbol{\beta}) \\
&+ \Sigma(k_{xx}a_y^2 + k_{yy}a_x^2 - 2k_{xy}a_xa_y)(\gamma - \boldsymbol{\gamma}) = M_z
\end{aligned}
\tag{3.31f}
$$

where the moments and products of inertia are defined by Eqs. (3.11) and (3.12) and the stiffness coefficients are defined as follows:

$$
\begin{aligned}
k_{xx} &= k_p\lambda_{xp}^2 + k_q\lambda_{xq}^2 + k_r\lambda_{xr}^2 \\
k_{yy} &= k_p\lambda_{yp}^2 + k_q\lambda_{yq}^2 + k_r\lambda_{yr}^2 \\
k_{zz} &= k_p\lambda_{zp}^2 + k_q\lambda_{zq}^2 + k_r\lambda_{zr}^2 \\
k_{xy} &= k_p\lambda_{xp}\lambda_{yp} + k_q\lambda_{xq}\lambda_{yq} + k_r\lambda_{xr}\lambda_{yr} \\
k_{xz} &= k_p\lambda_{xp}\lambda_{zp} + k_q\lambda_{xq}\lambda_{zq} + k_r\lambda_{xr}\lambda_{zr} \\
k_{yz} &= k_p\lambda_{yp}\lambda_{zp} + k_q\lambda_{yq}\lambda_{zq} + k_r\lambda_{yr}\lambda_{zr}
\end{aligned}
\tag{3.32}
$$

where the λ's are the cosines of the angles between the principal elastic axes of the resilient supporting elements and the coordinate axes. For example, λ_{xp} is the cosine of the angle between the X axis and the P axis of principal stiffness.

The equations of motion, Eqs. (3.31), do not include forces applied to the rigid body by damping forces from the resilient elements. To include damping, appropriate damping terms analogous to the corresponding stiffness terms are added to each equation. For example, Eq. (3.31*a*) would become

$$
\begin{aligned}
m\ddot{x}_c &+ \Sigma c_{xx}(\dot{x}_c - \dot{u}) + \Sigma k_{xx}(x_c - u) + \cdots \\
&+ \Sigma(c_{xz}a_y - c_{xy}a_z)(\dot{\alpha} - \dot{\boldsymbol{\alpha}}) + \Sigma(k_{xz}a_y - k_{xy}a_z)(\alpha - \boldsymbol{\alpha}) + \cdots = F_x
\end{aligned}
\tag{3.31a'}
$$

where

$$
\begin{aligned}
c_{xx} &= c_p\lambda_{xp}^2 + c_q\lambda_{xq}^2 + c_r\lambda_{xr}^2 \\
c_{xy} &= c_p\lambda_{xp}\lambda_{yp} + c_q\lambda_{xq}\lambda_{yq} + c_r\lambda_{xr}\lambda_{yr}
\end{aligned}
$$

The number of degrees-of-freedom of a vibrational system is the minimum number of coordinates necessary to define completely the positions of the mass elements of the system in space. The system of Fig. 3.12 requires a minimum of six coordinates $(x_c,y_c,z_c,\alpha,\beta,\gamma)$ to define the position of the rigid body in space; thus, the system is said to vibrate in six degrees-of-freedom. Equations (3.31) may be solved simultaneously for the three components x_c, y_c, z_c of the center-of-gravity displacement and the three components α, β, γ of the rotational displacement of the rigid body. In most practical instances, the equations are simplified considerably by one or more of the following simplifying conditions:

1. The reference axes X, Y, Z are selected to coincide with the principal inertial axes of the body; then

$$I_{xy} = I_{xz} = I_{yz} = 0 \tag{3.33}$$

2. The resilient supporting elements are so arranged that one or more planes of symmetry exist; i.e., motion parallel to the plane of symmetry has no tendency to excite motion perpendicular to it, or rotation about an axis lying in the plane does not

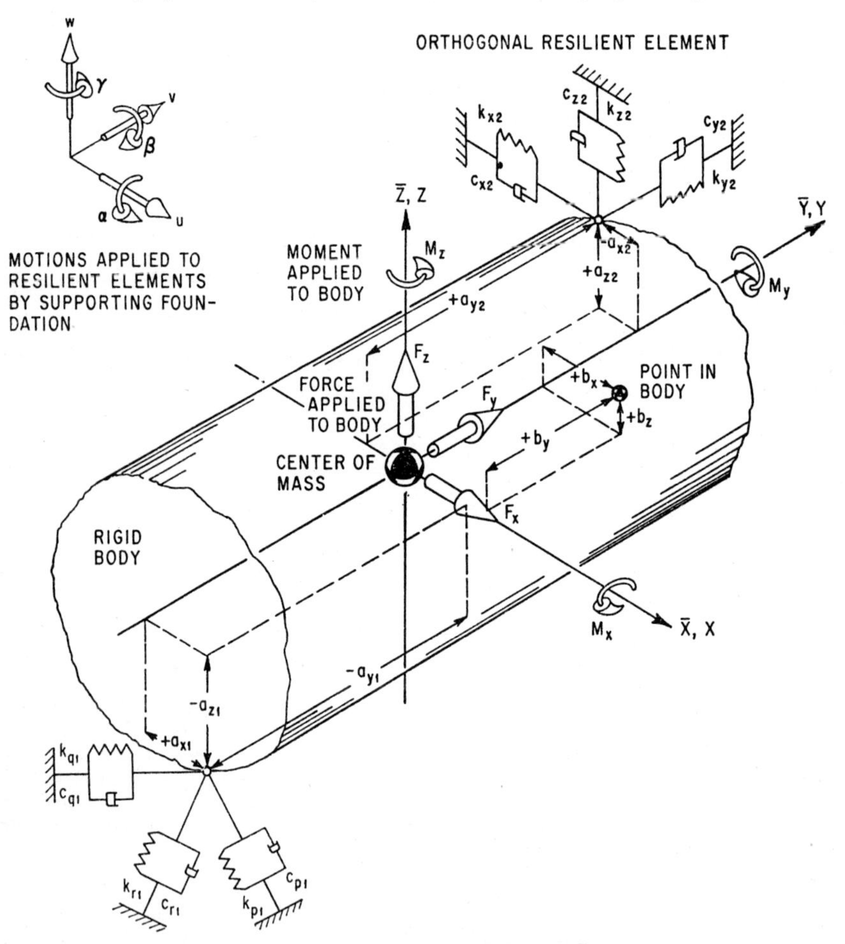

Fig. 3.12. Rigid body at rest supported by resilient elements, with inertial axes $\bar{X}$, $\bar{Y}$, $\bar{Z}$ and coincident reference axes X, Y, Z passing through the center-of-mass. The forces F_x, F_y, F_z and the moments M_x, M_y, M_z are applied directly to the body; the translations u, v, w and rotations $\boldsymbol{\alpha}$, $\boldsymbol{\beta}$, $\boldsymbol{\gamma}$ in and about the X, Y, Z axes, respectively, are applied to the resilient elements located at the coordinates a_x, a_y, a_z. The principal directions of resilient element (2) are parallel to the $\bar{X}$, $\bar{Y}$, $\bar{Z}$ axes (orthogonal) and those of resilient element (1) are not parallel to the $\bar{X}$, $\bar{Y}$, $\bar{Z}$ axes (inclined).

excite motion parallel to the plane. For example, in Eq. (3.31*a*), motion in the XY plane does not tend to excite motion in the XZ or YZ plane if Σk_{xz}, $\Sigma(k_{xz}a_y - k_{xy}a_z)$, and $\Sigma(k_{xx}a_z - k_{xz}a_x)$ are zero.

3. The principal elastic axes P, Q, R of all resilient supporting elements are orthogonal with the reference axes X, Y, Z of the body, respectively. Then, in Eqs. (3.32),

$$k_{xx} = k_p = k_x \qquad k_{yy} = k_q = k_y \qquad k_{zz} = k_r = k_z$$
$$k_{xy} = k_{xz} = k_{yz} = 0 \tag{3.34}$$

where k_x, k_y, k_z are defined for use when orthogonality exists. The supports are then called *orthogonal supports.*

4. The forces F_x, F_y, F_z and moments M_x, M_y, M_z are applied directly to the body and there are no motions ($u = v = w = \alpha = \beta = \gamma = 0$) of the foundation; or alternatively, the forces and moments are zero and excitation results from motion of the foundation.

In general, the effect of these simplifications is to reduce the numbers of terms in the equations and, in some instances, to reduce the number of equations that must be solved simultaneously. Simultaneous equations indicate coupled modes; i.e., motion cannot exist in one coupled mode independently of motion in other modes which are coupled to it.

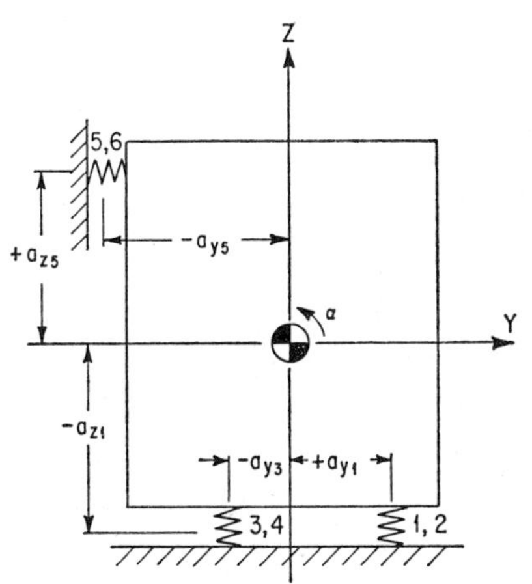

MODAL COUPLING AND NATURAL FREQUENCIES

Several conditions of symmetry resulting from zero values for the product of inertia terms in Eq. (3.33) are discussed in the following sections.

ONE PLANE OF SYMMETRY WITH ORTHOGONAL RESILIENT SUPPORTS

When the YZ plane of the rigid body system in Fig. 3.12 is a plane of symmetry, the following terms in the equations of motion are zero:

$$\Sigma k_{yy}a_x = \Sigma k_{zz}a_x = \Sigma k_{yy}a_xa_z = \Sigma k_{zz}a_xa_y = 0 \tag{3.35}$$

Introducing the further simplification that the principal elastic axes of the resilient elements are parallel with the reference axes, Eqs. (3.34) apply. Then the motions in the three coordinates y_c, z_c, α are coupled but are independent of motion in any of the other coordinates; furthermore, the other three coordinates x_c, β, γ also are coupled. For example, Fig. 3.13 illustrates a resiliently supported rigid body, wherein

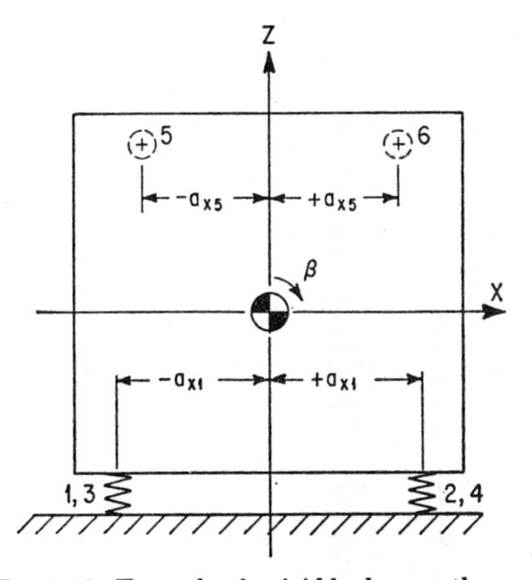

FIG. 3.13. Example of a rigid body on orthogonal resilient supporting elements with one plane of symmetry. The YZ plane is a plane of symmetry since each resilient element has properties identical to those of its mirror image in the YZ plane; i.e., $k_{x1} = k_{x2}$, $k_{x3} = k_{x4}$, $k_{x5} = k_{x6}$, etc. The conditions satisfied are Eqs. (3.33) to (3.35).

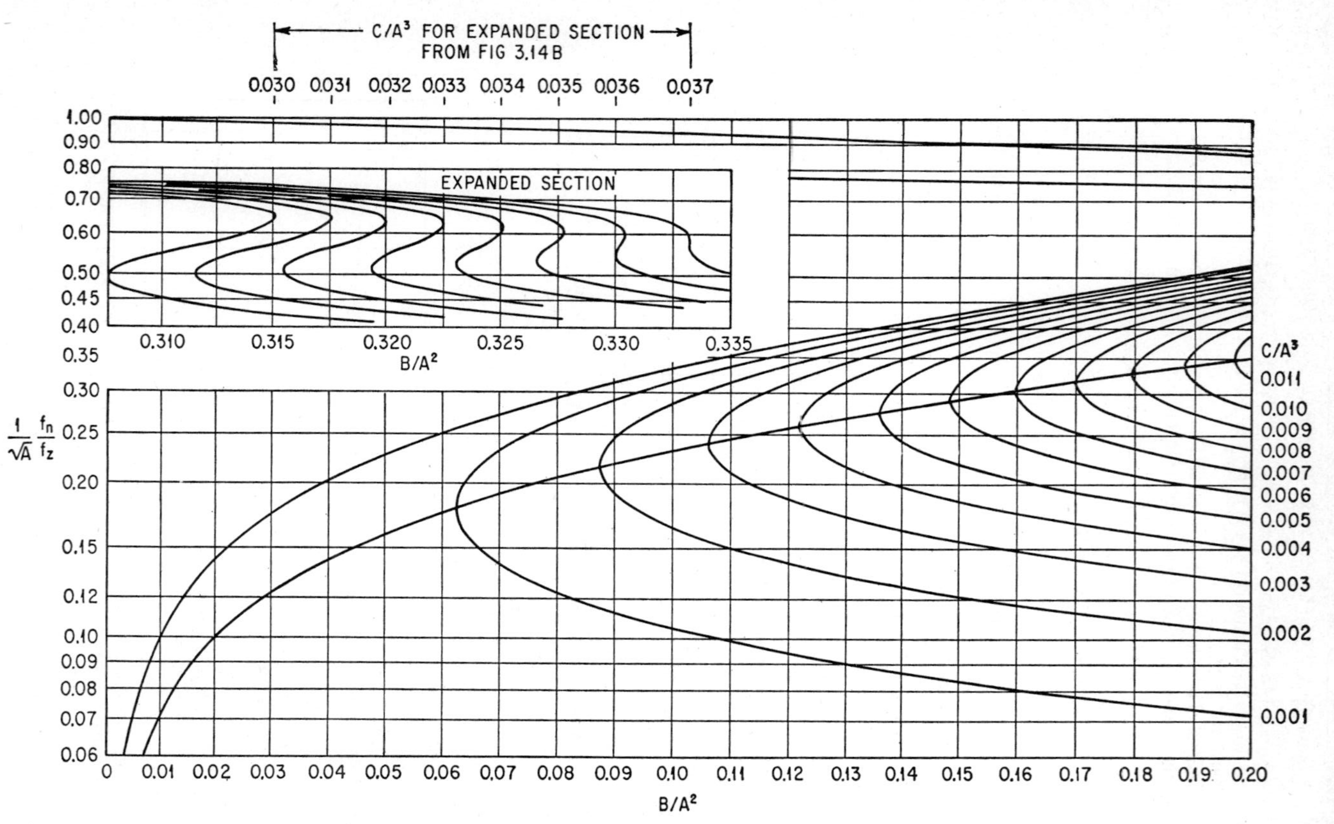

FIG. 3.14*A*. Graphical method of determining solutions of the cubic Eq. (3.36). Calculate A, B, C for the appropriate set of coupled coordinates, enter the abscissa at B/A^2 (values less than 0.2 on Fig. 3.14*A*, values greater than 0.2 on Fig. 3.14*B*), and read three values of $(f_n/f_z)/\sqrt{A}$ from the curve having the appropriate value of C/A^3.

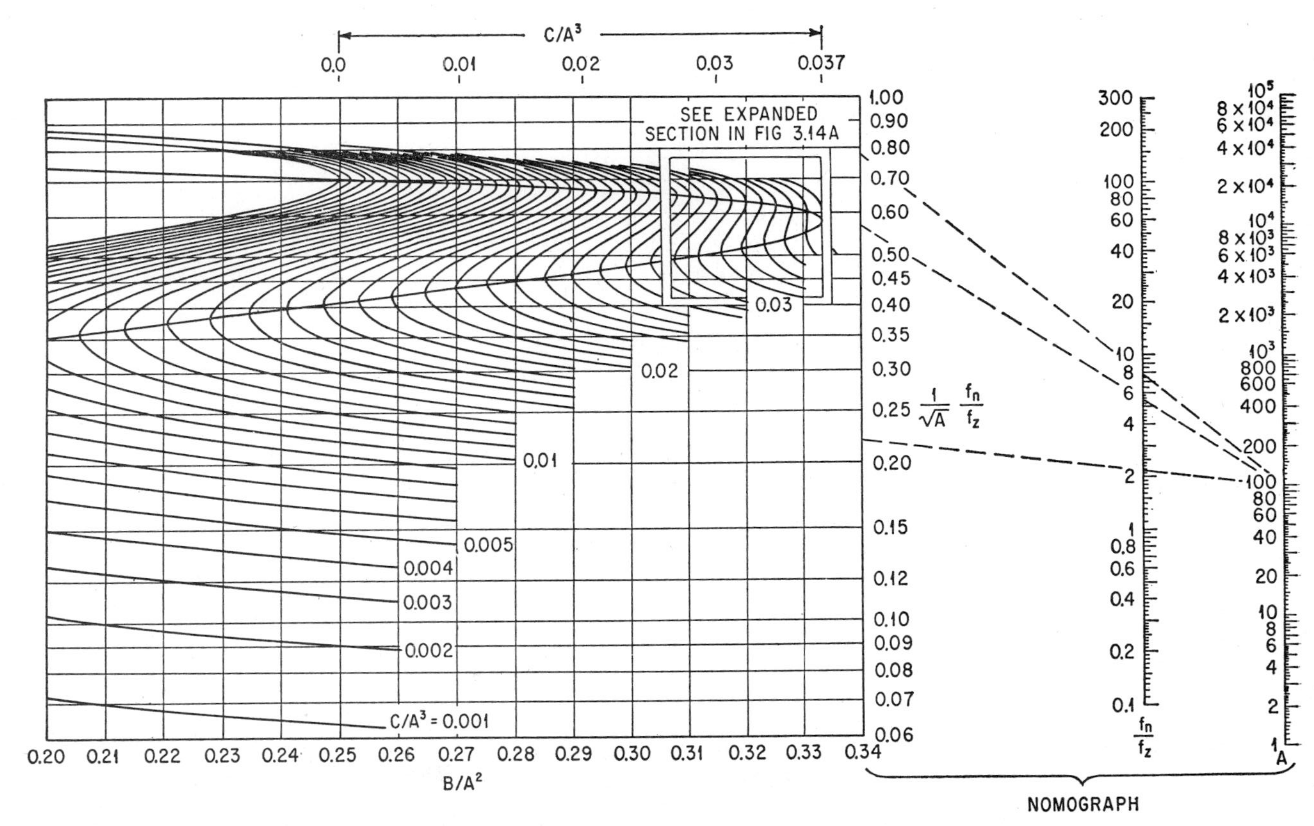

FIG. 3.14*B*. Using the above nomograph with values of $(f_n/f_z)/\sqrt{A}$ (see Fig. 3.14*A*), a diagonal line is drawn from each value of $(f_n/f_z)/\sqrt{A}$ on the left scale of the nomograph to the value of A on the right scale, as indicated by the dotted lines. The three roots f_n/f_z of Eq. (3.36) are given by the intercept of these dotted lines with the center scale of the nomograph. (*After F. F. Vane.*[9])

the YZ plane is a plane of symmetry that meets the requirements of Eq. (3.35). The three natural frequencies for the y_c, z_c, α coupled directions are found by solving Eqs. (3.31b), (3.31c), and (3.31e) [or Eqs. (3.31a), (3.31d), and (3.31f) for the x_c, β, γ coupled directions] simultaneously.[9, 10]

$$\left(\frac{f_n}{f_z}\right)^6 - A\left(\frac{f_n}{f_z}\right)^4 + B\left(\frac{f_n}{f_z}\right)^2 - C = 0 \tag{3.36}$$

where

$$f_z = \frac{1}{2\pi}\sqrt{\frac{\Sigma k_z}{m}} \tag{3.37}$$

is a quantity having mathematical rather than physical significance if translational motion in the direction of the Z axis is coupled to other modes of motion. (Such coupling exists for the system of Fig. 3.13.) The roots f_n represent the natural frequencies of the system in the coupled modes. The coefficients A, B, C for the coupled modes in the y_c, z_c, α coordinates are

$$A_{yz\alpha} = 1 + \frac{\Sigma k_y}{\Sigma k_z} + D_{zx}$$

$$B_{yz\alpha} = D_{zx} + \frac{\Sigma k_y}{\Sigma k_z}(1 + D_{zx}) - \frac{(\Sigma k_y a_z)^2 + (\Sigma k_z a_y)^2}{\rho_x^2(\Sigma k_z)^2}$$

$$C_{yz\alpha} = \frac{\Sigma k_y}{\Sigma k_z}\left(D_{zx} - \frac{(\Sigma k_z a_y)^2}{\rho_x^2(\Sigma k_z)^2}\right) - \frac{(\Sigma k_y a_z)^2}{\rho_x^2(\Sigma k_z)^2}$$

where

$$D_{zx} = \frac{\Sigma k_y a_z^2 + \Sigma k_z a_y^2}{\rho_x^2 \Sigma k_z}$$

and ρ_x is the radius of gyration of the rigid body with respect to the X axis.

The corresponding coefficients for the coupled modes in the x_c, β, γ coordinates are

$$A_{x\beta\gamma} = \frac{\Sigma k_x}{\Sigma k_z} + D_{zy} + D_{zz}$$

$$B_{x\beta\gamma} = \frac{\Sigma k_x}{\Sigma k_z}(D_{zy} + D_{zz}) + D_{zy}D_{zz} - \frac{(\Sigma k_x a_z)^2}{\rho_y^2(\Sigma k_z)^2} - \frac{(\Sigma k_x a_y)^2}{\rho_z^2(\Sigma k_z)^2} - \frac{(\Sigma k_x a_y a_z)^2}{\rho_y^2\rho_z^2(\Sigma k_z)^2}$$

$$C_{x\beta\gamma} = \frac{\Sigma k_x}{\Sigma k_z}\left[D_{zy}D_{zz} - \frac{(\Sigma k_x a_y a_z)^2}{\rho_y^2\rho_z^2(\Sigma k_z)^2}\right] - \frac{(\Sigma k_x a_y)^2}{\rho_z^2(\Sigma k_z)^2}D_{zy} - \frac{(\Sigma k_x a_z)^2}{\rho_y^2(\Sigma k_z)^2}D_{zz} + 2\frac{(\Sigma k_x a_y)(\Sigma k_x a_z)(\Sigma k_x a_y a_z)}{\rho_y^2\rho_z^2(\Sigma k_z)^3}$$

where

$$D_{zy} = \frac{\Sigma k_x a_z^2 + \Sigma k_z a_x^2}{\rho_y^2 \Sigma k_z} \qquad D_{zz} = \frac{\Sigma k_x a_y^2 + \Sigma k_y a_x^2}{\rho_z^2 \Sigma k_z}$$

and ρ_y, ρ_z are the radii of gyration of the rigid body with respect to the Y, Z axes.

The roots of the cubic equation Eq. (3.36) may be found graphically from Fig. 3.14.[9] The coefficients A, B, C are first calculated from the above relations for the appropriate set of coupled coordinates. Figure 3.14 is entered on the abscissa scale at the appropriate value for the quotient B/A^2. Small values of B/A^2 are in Fig. 3.14A, and large values in Fig. 3.14B. The quotient C/A^3 is the parameter for the family of curves. Upon selecting the appropriate curve, three values of $(f_n/f_z)/\sqrt{A}$ are read from the ordinate and transferred to the left scale of the nomograph in Fig. 3.14*B*. Diagonal lines are drawn for each root to the value of A on the right scale, as indicated by dotted lines, and the roots f_n/f_z of the equation are indicated by the intercept of these dotted lines with the center scale of the nomograph.

The coefficients A, B, C can be simplified if all resilient elements have equal stiffness in the same direction. The stiffness coefficients always appear to equal powers in numerator and denominator, and lead to dimensionless ratios of stiffness. For n resilient elements, typical terms reduce as follows:

$$\frac{\Sigma k_y}{\Sigma k_z} = \frac{k_y}{k_z} \qquad \frac{\Sigma k_z a_y^2}{\rho_x^2 \Sigma k_z} = \frac{\Sigma a_y^2}{n \rho_x^2}$$

$$\frac{(\Sigma k_x a_y a_z)^2}{\rho_y^2 \rho_z^2 (\Sigma k_z)^2} = \left(\frac{k_x}{nk_z} \frac{\Sigma a_y a_z}{\rho_y \rho_z}\right)^2, \text{ etc.}$$

TWO PLANES OF SYMMETRY WITH ORTHOGONAL RESILIENT SUPPORTS

Two planes of symmetry may be achieved if, in addition to the conditions of Eqs. (3.33) to (3.35), the following terms of Eqs. (3.31) are zero:

$$\Sigma k_{xx} a_y = \Sigma k_{zz} a_y = \Sigma k_{xx} a_y a_z = 0 \quad (3.38)$$

Under these conditions, Eqs. (3.31) separate into two independent equations, Eqs. (3.31*e*) and (3.31*f*), and two sets each consisting of two coupled equations [Eqs. (3.31*a*) and (3.31*d*); Eqs. (3.31*b*) and (3.31*c*)]. The planes of symmetry are the XZ and YZ planes. For example, a common system is illustrated in Fig. 3.15, where four identical resilient supporting elements are located symmetrically about the Z axis in a plane not containing the center-of-gravity.[6, 11, 12] Coupling exists between translation in the X direction and rotation about the Y axis (x_c,β), as well as between translation in the Y direction and rotation about the X axis (y_c,α). Translation in the Z direction (z_c) and rotation about the Z axis (γ) are each independent of all other modes.

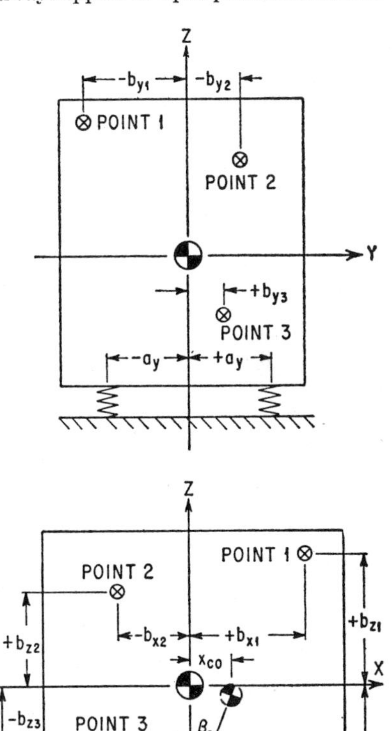

FIG. 3.15. Example of a rigid body on orthogonal resilient supporting elements with two planes of symmetry. The XZ and YZ planes are planes of symmetry since the four resilient supporting elements are identical and are located symmetrically about the Z axis. The conditions satisfied are Eqs. (3.33), (3.34), (3.35), and (3.38). At any single frequency, coupled vibration in the x_c, β direction due to X vibration of the foundation is equivalent to a pure rotation of the rigid body with respect to an axis of rotation as shown. Points 1, 2, and 3 refer to the example of Fig. 3.26.

The natural frequency in the Z direction is found by solving Eq. (3.31*e*) to obtain Eq. (3.37), where $\Sigma k_{zz} = 4k_z$. The rotational natural frequency f_γ about the Z axis is found by solving Eq. (3.31*f*); it can be expressed with respect to the natural frequency in the direction of the Z axis:

$$\frac{f_\gamma}{f_z} = \sqrt{\frac{k_x}{k_z}\left(\frac{a_y}{\rho_z}\right)^2 + \frac{k_y}{k_z}\left(\frac{a_x}{\rho_z}\right)^2} \quad (3.39)$$

where ρ_z is the radius of gyration with respect to the Z axis.

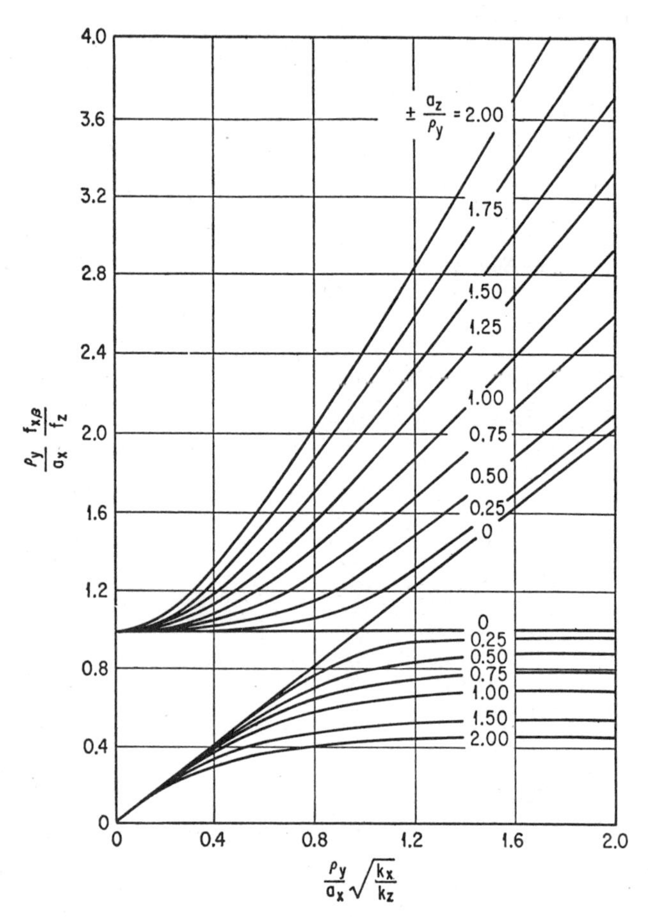

FIG. 3.16. Curves showing the ratio of each of the two coupled natural frequencies $f_{x\beta}$ to the decoupled natural frequency f_z, for motion in the XZ plane of symmetry for the system in Fig. 3.15 [see Eq. (3.40)]. Calculate the abscissa $(\rho_y/a_x)\sqrt{k_x/k_z}$ and the parameter a_z/ρ_y, where a_x, a_z are indicated in Fig. 3.15; k_x, k_z are the stiffnesses of the resilient supporting elements in the X, Z directions, respectively; and ρ_y is the radius of gyration of the body about the Y axis. The two values read from the ordinate when divided by ρ_y/a_x give the natural frequency ratios $f_{x\beta}/f_z$. (*After C. E. Crede.*[12])

The natural frequencies in the coupled x_c, β modes are found by solving Eqs. (3.31*a*) and (3.31*d*) simultaneously; the roots yield the following expression for natural frequency:

$$\frac{f_{x\beta}^2}{f_z^2} = \frac{1}{2}\left\{\frac{k_x}{k_z}\left(1 + \frac{a_z^2}{\rho_y^2}\right) + \frac{a_x^2}{\rho_y^2} \pm \sqrt{\left[\frac{k_x}{k_z}\left(1 + \frac{a_z^2}{\rho_y^2}\right) + \frac{a_x^2}{\rho_y^2}\right]^2 - 4\frac{k_x}{k_z}\frac{a_x^2}{\rho_y^2}}\right\} \quad (3.40)$$

Figure 3.16 provides a convenient graphical method for determining the two coupled natural frequencies $f_{x\beta}$. An expression similar to Eq. (3.40) is obtained for $f_{y\alpha}^2/f_z^2$ by

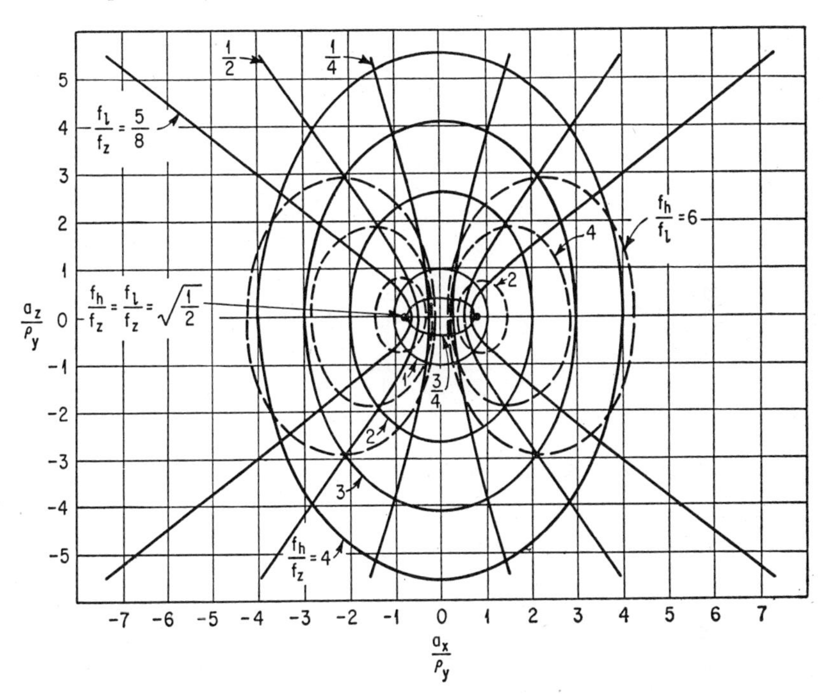

FIG. 3.17. Space-plot for the system in Fig. 3.15 when the stiffness ratio $k_x/k_z = 0.5$, obtained from Eqs. (3.40*a*) to (3.40*c*). With all dimensions divided by the radius of gyration ρ_y about the *Y* axis, superimpose the outline of the rigid body in the *XZ* plane on the plot; the center-of-gravity of the body is located at the coordinate center of the plot. The elastic centers of the resilient supporting elements give the natural frequency ratios f_l/f_z, f_h/f_z and f_h/f_l for x_c, β coupled motion, each ratio being read from one of the three families of curves as indicated on the plot. Replacing k_x, ρ_y, a_x with k_y, ρ_x, a_y, respectively, allows the plot to be applied to motions in the *YZ* plane.

solving Eqs. (3.31*b*) and (3.31*d*) simultaneously. By replacing ρ_y, a_x, k_x, $f_{x\beta}$ with ρ_x, a_y, k_y, $f_{y\alpha}$, respectively, Fig. 3.16 also can be used to determine the two values of $f_{y\alpha}$.

It may be desirable to select resilient element locations a_x, a_y, a_z which will produce coupled natural frequencies in specified frequency ranges, with resilient elements having specified stiffness ratios k_x/k_z, k_y/k_z. For this purpose it is convenient to plot solutions of Eq. (3.40) in the form shown in Figs. 3.17 to 3.19. These plots are termed *space-plots* and their use is illustrated in Example 3.1.[13, 14]

The space-plots are derived as follows: In general, the two roots of Eq. (3.40) are numerically different, one usually being greater than unity and the other less than unity. Designating the root associated with the positive sign before the radical (higher value)

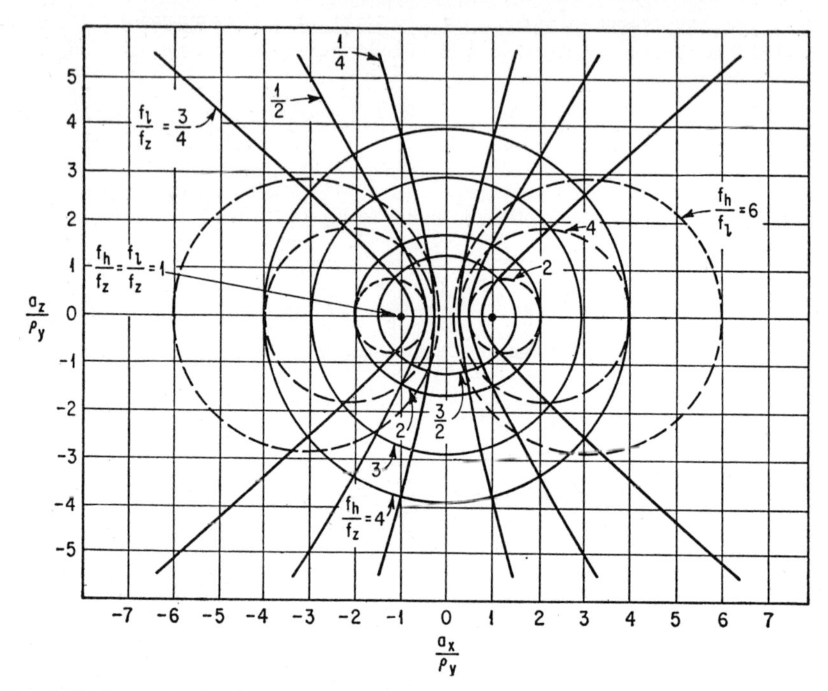

FIG. 3.18. Space-plot for the system in Fig. 3.15 when the stiffness ratio $k_x/k_z = 1$. See caption for Fig. 3.17.

as f_h/f_z, Eq. (3.40) may be written in the following form:

$$\frac{(a_x/\rho_y)^2}{(f_h/f_z)^2} + \frac{(a_z/\rho_y)^2}{(k_z/k_x)(f_h/f_z)^2 - 1} = 1 \tag{3.40a}$$

Equation (3.40a) is shown graphically by the large ellipses about the center of Figs. 3.17 to 3.19, for stiffness ratios k_x/k_z of ½, 1, and 2, respectively. A particular type of resilient element tends to have a constant stiffness ratio k_x/k_z; thus, Figs. 3.17 to 3.19 may be used by cut-and-try methods to find the coordinates a_x, a_z of such elements to attain a desired value of f_h.

Designating the root of Eq. (3.40) associated with the negative sign (lower value) by f_l, Eq. (3.40) may be written as follows:

$$\frac{(a_x/\rho_y)^2}{(f_l/f_z)^2} - \frac{(a_z/\rho_y)^2}{1 - (k_z/k_x)(f_l/f_z)^2} = 1 \tag{3.40b}$$

Equation (3.40b) is shown graphically by the family of hyperbolas on each side of the center in Figs. 3.17 to 3.19, for values of the stiffness ratio k_x/k_z of ½, 1, and 2.

The two roots f_h/f_z and f_l/f_z of Eq. (3.40) may be expressed as the ratio of one to the other. This relationship is given parametrically as follows:

$$\left[\frac{2\dfrac{a_x}{\rho_y} \pm \sqrt{\dfrac{k_x}{k_z}}\left(\dfrac{f_h}{f_l} + \dfrac{f_l}{f_h}\right)}{\sqrt{\dfrac{k_x}{k_z}}\left(\dfrac{f_h}{f_l} - \dfrac{f_l}{f_h}\right)}\right]^2 + \left[\frac{2\dfrac{a_z}{\rho_y}}{\dfrac{f_h}{f_l} - \dfrac{f_l}{f_h}}\right]^2 = 1 \tag{3.40c}$$

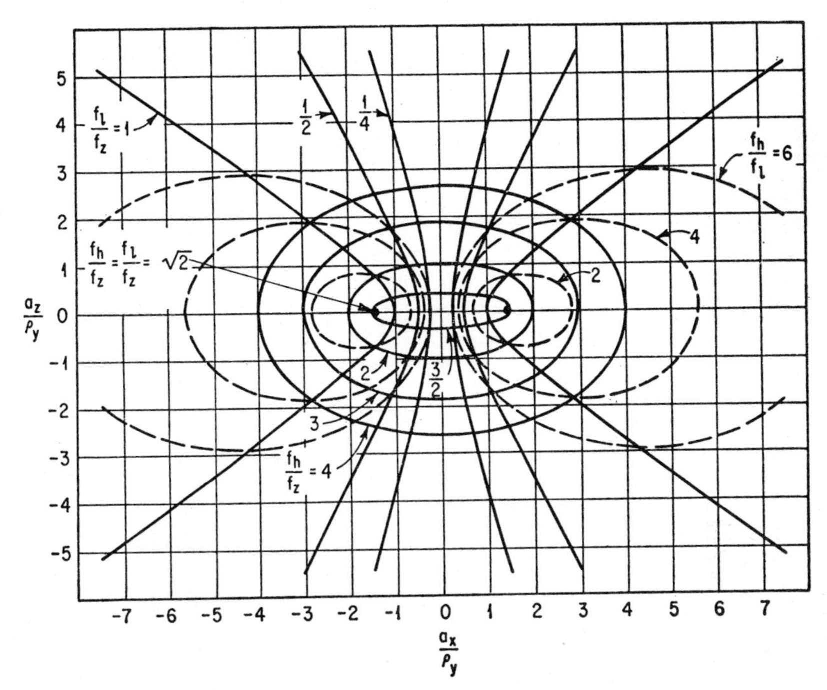

FIG. 3.19. Space-plot for the system in Fig. 3.15 when the stiffness ratio $k_x/k_z = 2$. See caption for Fig. 3.17.

Equation (3.40*c*) is shown graphically by the smaller ellipses (shown dotted) displaced from the vertical center line in Figs. 3.17 to 3.19.

Example 3.1. A rigid body is symmetrical with respect to the *XZ* plane; its width in the *X* direction is 13 in. and its height in the *Z* direction is 12 in. The center-of-gravity is 5½ in. from the lower side and 6¾ in. from the right side. The radius of gyration about the *Y* axis through the center-of-gravity is 5.10 in. Use a space-plot to evaluate the effects of the location for attachment of resilient supporting elements having the characteristic stiffness ratio $k_x/k_z = \frac{1}{2}$.

Superimpose the outline of the body on the space-plot of Fig. 3.20, with its center-of-gravity at the coordinate center of the plot. (Figure 3.20 is an enlargement of the central portion of Fig. 3.17.) All dimensions are divided by the radius of gyration ρ_y. Thus, the four corners of the body are located at coordinate distances as follows:

Upper right corner:
$$\frac{a_z}{\rho_y} = \frac{+6.50}{5.10} = +1.28 \qquad \frac{a_x}{\rho_y} = \frac{+6.75}{5.10} = +1.32$$

Upper left corner:
$$\frac{a_z}{\rho_y} = \frac{+6.50}{5.10} = +1.28 \qquad \frac{a_x}{\rho_y} = \frac{-6.25}{5.10} = -1.23$$

Lower right corner:
$$\frac{a_z}{\rho_y} = \frac{-5.50}{5.10} = -1.08 \qquad \frac{a_x}{\rho_y} = \frac{+6.75}{5.10} = +1.32$$

Lower left corner:
$$\frac{a_z}{\rho_y} = \frac{-5.50}{5.10} = -1.08 \qquad \frac{a_x}{\rho_y} = \frac{-6.25}{5.10} = -1.23$$

The resilient supports are shown in heavy outline at *A* in Fig. 3.20, with their elastic centers indicated by the solid dots. The horizontal coordinates of the resilient supports are $a_x/\rho_y =$

± 0.59, or $a_x = \pm 0.59 \times 5.10 = \pm 3$ in. from the vertical coordinate axis. The corresponding natural frequencies are $f_h/f_z = 1.25$ (from the ellipses) and $f_l/f_z = 0.33$ (from the hyperbolas). An alternative position is indicated by the hollow circles B. The natural frequencies for this position are $f_h/f_z = 1.43$ and $f_l/f_z = 0.50$. The natural frequency f_z in vertical translation is found from the mass of the equipment and the summation of stiffnesses in the Z direction, using Eq. (3.37). This example shows how space-plots make it possible to determine the locations of the resilient elements required to achieve given values of the coupled natural frequencies with respect to f_z.

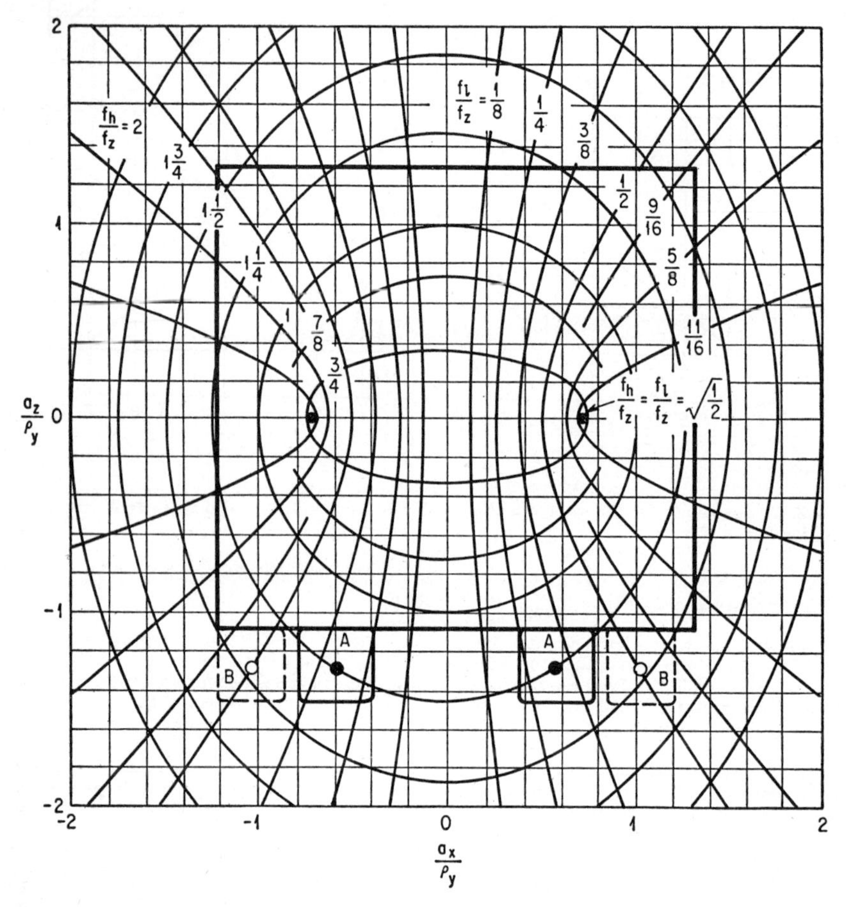

FIG. 3.20. Enlargement of the central portion of Fig. 3.17 with the outline of the rigid body discussed in Example 3.1.

THREE PLANES OF SYMMETRY WITH ORTHOGONAL RESILIENT SUPPORTS

A system with three planes of symmetry is defined by six independent equations of motion. A system having this property is sometimes called a *center-of-gravity system.* The equations are derived from Eqs. (3.31) by substituting, in addition to the conditions of Eqs. (3.33), (3.34), (3.35), and (3.38), the following condition:

$$\Sigma k_{xx}a_z = \Sigma k_{yy}a_z = 0 \tag{3.41}$$

The resulting six independent equations define six uncoupled modes of vibration, three in translation and three in rotation. The natural frequencies are:

Translation along X axis:

$$f_x = \frac{1}{2\pi}\sqrt{\frac{\Sigma k_x}{m}}$$

Translation along Y axis:

$$f_y = \frac{1}{2\pi}\sqrt{\frac{\Sigma k_y}{m}}$$

Translation along Z axis:

$$f_z = \frac{1}{2\pi}\sqrt{\frac{\Sigma k_z}{m}}$$

Rotation about X axis:

$$f_\alpha = \frac{1}{2\pi}\sqrt{\frac{\Sigma(k_y a_z^2 + k_z a_y^2)}{m\rho_x^2}} \tag{3.42}$$

Rotation about Y axis:

$$f_\beta = \frac{1}{2\pi}\sqrt{\frac{\Sigma(k_x a_z^2 + k_z a_x^2)}{m\rho_y^2}}$$

Rotation about Z axis:

$$f_\gamma = \frac{1}{2\pi}\sqrt{\frac{\Sigma(k_x a_y^2 + k_y a_x^2)}{m\rho_z^2}}$$

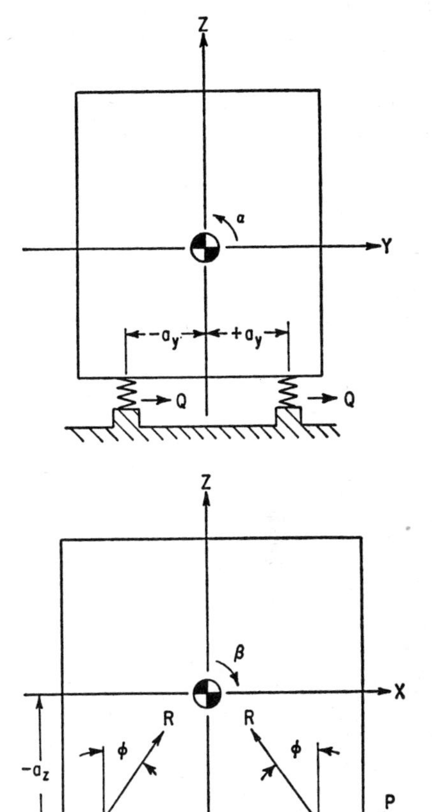

FIG. 3.21. Example of a rigid body on resilient supporting elements inclined toward the YZ plane. The resilient supporting elements are identical and are located symmetrically about the Z axis, making XZ and YZ planes of symmetry. The principal stiffnesses in the XZ plane are k_p and k_r. The conditions satisfied are Eqs. (3.33), (3.35), and (3.38).

TWO PLANES OF SYMMETRY WITH RESILIENT SUPPORTS INCLINED IN ONE PLANE ONLY

When the principal elastic axes of the resilient supporting elements are inclined with respect to the X, Y, Z axes, the stiffness coefficients k_{xy}, k_{xz}, k_{yz} are nonzero. This introduces elastic coupling which must be considered in evaluating the equations of motion. Two planes of symmetry may be achieved by meeting the conditions of Eqs. (3.33), (3.35), and (3.38). For example, consider the rigid body supported by four identical resilient supporting elements located symmetrically about the Z axis, as shown in Fig. 3.21. The XZ and the YZ planes are planes of symmetry and the resilient elements are inclined toward the YZ plane so that one of their principal elastic axes R is inclined at the angle ϕ with the Z direction as shown; hence $k_{yy} = k_q$, and $k_{xy} = k_{yz} = 0$.

Because of symmetry, translational motion z_c in the Z direction and rotation γ about the Z axis are each decoupled from the other modes. The pairs of translational and rotational modes in the x_c, β and y_c, α coordinates are coupled. The natural frequency in the Z direction is

$$\frac{f_z}{f_r} = \sqrt{\frac{k_p}{k_r}\sin^2\phi + \cos^2\phi} \tag{3.43}$$

where f_r is a fictitious natural frequency used for convenience only; it is related to Eq (3.37) wherein $4k_r$ is substituted for Σk_z:

$$f_r = \frac{1}{2\pi}\sqrt{\frac{4k_r}{m}}$$

Equation (3.43) is plotted in Fig. 3.22 where the angle ϕ is indicated by the upper of the abscissa scales.

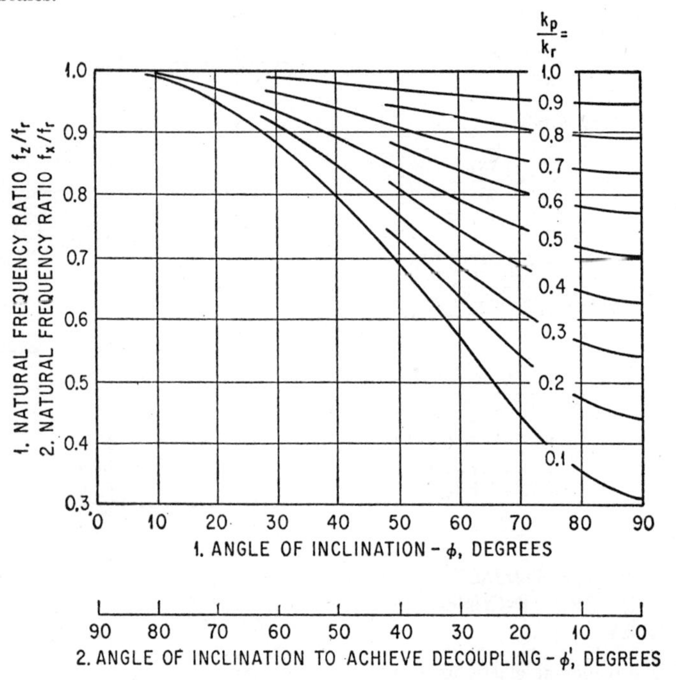

FIG. 3.22. Curves showing the ratio of the decoupled natural frequency f_z of translation z_c to the fictitious natural frequency f_r for the system shown in Fig. 3.21 [see Eq. (3.43)] when the resilient supporting elements are inclined at the angle ϕ. The curves also indicate the ratio of the decoupled natural frequency f_x of translation x_c to f_r when ϕ has a value ϕ' (use lower abscissa scale) which decouples x_c, β motions [see Eqs. (3.47) and (3.48)]. (*After C. E. Crede.*[15])

The rotational natural frequency about the Z axis is obtained from

$$\frac{f_\gamma}{f_r} = \sqrt{\left(\frac{k_p}{k_r}\cos^2\phi + \sin^2\phi\right)\left(\frac{a_y}{\rho_z}\right)^2 + \frac{k_q}{k_r}\left(\frac{a_x}{\rho_z}\right)^2} \tag{3.44}$$

For the x_c, β coupled mode, the two natural frequencies are

$$\frac{f_{x\beta}}{f_r} = \frac{1}{2}\left[A \pm \sqrt{A^2 - 4\frac{k_p}{k_r}\left(\frac{a_x}{a_y}\right)^2}\right] \tag{3.45}$$

where $A = \left(\frac{k_p}{k_r}\cos^2\phi + \sin^2\phi\right)\left[1 + \left(\frac{a_z}{\rho_y}\right)^2\right] + \left(\frac{k_p}{k_r}\sin^2\phi + \cos^2\phi\right)\left(\frac{a_x}{\rho_y}\right)^2$

$$+ 2\left(1 - \frac{k_p}{k_r}\right)\left|\frac{a_x}{\rho_y}\right|\sin\phi\cos\phi$$

For the y_c, α coupled mode, the natural frequencies are

$$\frac{f_{y\alpha}}{f_r} = \frac{1}{2}\left[B \pm \sqrt{B^2 - 4\frac{k_q}{k_r}\left(\frac{k_p}{k_r}\sin^2\phi + \cos^2\phi\right)\left(\frac{a_y}{\rho_x}\right)^2}\right] \tag{3.46}$$

where

$$B = \frac{k_q}{k_r}\left[1 + \left(\frac{a_z}{\rho_x}\right)^2\right] + \left(\frac{k_p}{k_r}\sin^2\phi + \cos^2\phi\right)\left(\frac{a_y}{\rho_x}\right)^2$$

DECOUPLING OF MODES IN A PLANE USING INCLINED RESILIENT SUPPORTS

The angle ϕ of inclination of principal elastic axes (see Fig. 3.21) can be varied to produce changes in the amount of coupling between the x_c and β coordinates. Decoupling

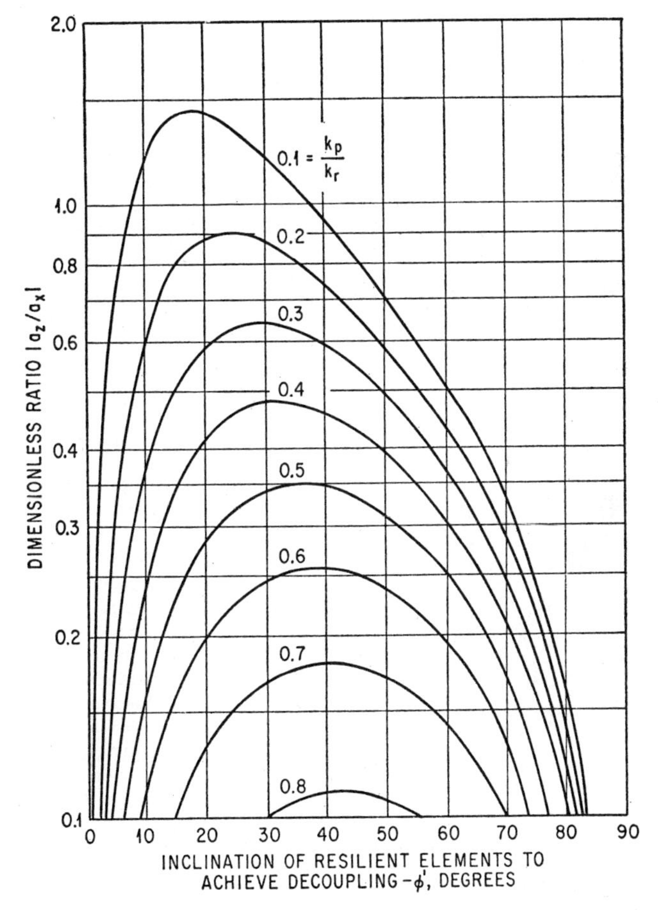

FIG. 3.23. Curves showing the angle of inclination ϕ' of the resilient elements which achieves decoupling of the x_c, β motions in Fig. 3.21 [see Eq. (3.47)]. Calculate the ordinate $|a_z/a_x|$ and with the stiffness ratio k_p/k_r determine two values of ϕ' for which decoupling is possible. Decoupling is not possible for a particular value of k_p/k_r if $|a_z/a_y|$ has a value greater than the maximum ordinate of the k_p/k_r curve.

of the x_c and β coordinates is effected if[15]

$$\left|\frac{a_z}{a_x}\right| = \frac{[1 - (k_p/k_r)] \cot \phi'}{1 + (k_p/k_r) \cot^2 \phi'} \tag{3.47}$$

where ϕ' is the value of the angle of inclination ϕ required to achieve decoupling. When Eq. (3.47) is satisfied, the configuration is sometimes called an "equivalent center-of-gravity system" in the YZ plane since all modes of motion in that plane are decoupled. Figure 3.23 is a graphical presentation of Eq. (3.47). There may be two values of ϕ' that decouple the x_c and β modes for any combination of stiffness and location for the resilient supporting elements.

The decoupled natural frequency for translation in the X direction is obtained from

$$\frac{f_x}{f_r} = \sqrt{\frac{k_p}{k_r} \cos^2 \phi' + \sin^2 \phi'} \tag{3.48}$$

The relation of Eq. (3.48) is shown graphically in Fig. 3.22 where the angle ϕ' is indicated by the lower of the abscissa scales. The natural frequency in the β mode is obtained from

$$\frac{f_\beta}{f_r} = \frac{a_x}{\rho_y} \sqrt{\frac{1}{(k_r/k_p) \sin^2 \phi' + \cos^2 \phi'}} \tag{3.49}$$

COMPLETE DECOUPLING OF MODES USING RADIALLY INCLINED RESILIENT SUPPORTS

In general, the analysis of rigid body motion with the resilient supporting elements inclined in more than one plane is quite involved. A particular case where sufficient symmetry exists to provide relatively simple yet useful results is the configuration illustrated in Fig. 3.24. From symmetry about the Z axis, $I_{xx} = I_{yy}$. Any number n of resilient supporting elements greater than 3 may be used. For clarity of illustration, the rigid body is shown as a right circular cylinder with $n = 3$.

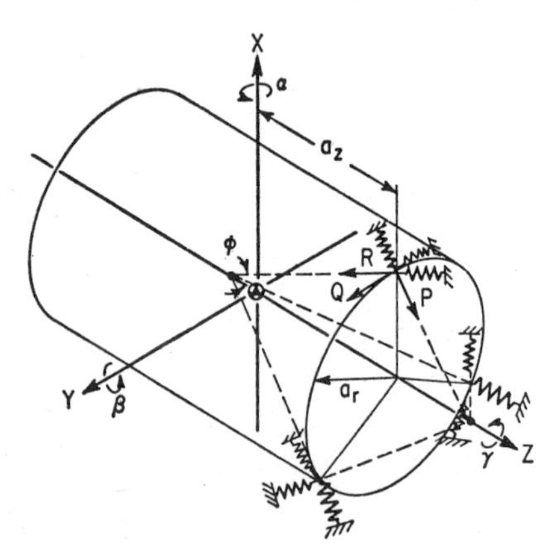

FIG. 3.24. Example of a rigid cylindrical body on radially inclined resilient supports. The resilient supports are attached symmetrically about the Z axis to one end face of the cylinder at a distance a_r from the Z axis and a distance a_z from the XY plane. The resilient elements are inclined so that their principal elastic axes R and P intersect the Z axis at common points. The angle between the R axes and the Z axis is ϕ; and the angle between the P axis and Z axis is $90° - \phi$. The Q principal elastic axes are each tangent to the circle of radius a_r.

The resilient supporting elements are arranged symmetrically about the Z axis; they are attached to one end face of the cylinder at a distance a_r from the Z axis and a distance a_z from the XY reference plane. The resilient elements are inclined so that their principal elastic axes R intersect at a common point on the Z axis; thus, the angle between the Z axis and the R axis for each element is ϕ. The principal elastic axes P also intersect at a common point on the Z axis, the angle between the Z axis and the P axis for each element being $90° - \phi$. Consequently, the Q principal elastic axes are each tangent to the circle of radius a_r which bounds the end face of the cylinder.

The use of such a configuration permits decoupling of all six modes of vibration of the rigid body. This complete decoupling is achieved if the angle of inclination ϕ has the value ϕ' which satisfies the following equation:

$$\left|\frac{a_z}{a_r}\right| = \frac{(\frac{1}{2})[1 - (k_p/k_r)] \sin 2\phi'}{(k_q/k_r) + (k_p/k_r) + [1 - (k_p/k_r)] \sin^2 \phi'} \tag{3.50}$$

Since complete decoupling is effected, the system may be termed an "equivalent center-of-gravity system."[16,17] The natural frequencies of the six decoupled modes are

$$\frac{f_x}{f_r} = \frac{f_y}{f_r} = \sqrt{\frac{1}{2}\left(\frac{k_p}{k_r}\cos^2 \phi' + \sin^2 \phi' + \frac{k_q}{k_r}\right)} \tag{3.51}$$

$$\frac{f_\alpha}{f_r} = \frac{f_\beta}{f_r} = \left\{\frac{a_r}{2\rho_x}\left[\frac{k_p}{k_r}\sin \phi'\left(\frac{a_r}{\rho_x}\sin \phi' + \frac{a_z}{\rho_x}\cos \phi'\right) + \cos \phi'\left(\frac{a_r}{\rho_x}\cos \phi' - \frac{a_z}{\rho_x}\sin \phi'\right)\right]\right\}^{1/2} \tag{3.52}$$

$$\frac{f_\gamma}{f_r} = \sqrt{\frac{k_q}{k_r}\frac{a_r}{\rho_z}} \tag{3.53}$$

The frequency ratio f_z/f_r is given by Eq. (3.43) or Fig. 3.22. The fictitious natural frequency f_r is given by

$$f_r = (1/2\pi)\sqrt{nk_r/m}$$

Similar solutions are also available for the configuration of four resilient supports located in a rectangular array and inclined to achieve complete decoupling.[18]

FORCED VIBRATION

Forced vibration results from a continuing excitation that varies sinusoidally with time. The excitation may be a vibratory displacement of the foundation for the resiliently supported rigid body (*foundation-induced vibration*), or a force or moment applied to or generated within the rigid body (*body-induced vibration*). These two forms of excitation are considered separately.

FOUNDATION-INDUCED SINUSOIDAL VIBRATION

This section includes an analysis of foundation-induced vibration for two different systems, each having two planes of symmetry. In one system, the principal elastic axes of the resilient elements are parallel to the X, Y, Z axes; in the other system, the principal elastic axes are inclined with respect to two of the axes but in a plane parallel to one of the reference planes. The excitation is translational movement of the foundation in its own plane, without rotation. No forces or moments are applied directly to the rigid body; i.e., in the equations of motion [Eqs. (3.31)], the following terms are equal to zero:

$$F_x = F_y = F_z = M_x = M_y = M_z = \alpha = \beta = \gamma = 0 \tag{3.54}$$

TWO PLANES OF SYMMETRY WITH ORTHOGONAL RESILIENT SUPPORTS

The system is shown in Fig. 3.15. The excitation is a motion of the foundation in the direction of the X axis defined by $u = u_0 \sin \omega t$. (Alternatively, the excitation may be the displacement $v = v_0 \sin \omega t$ in the direction of the Y axis, and analogous results are obtained.) The resulting motion of the resiliently supported rigid body involves translation x_c and rotation β simultaneously. The conditions of symmetry are defined by Eqs. (3.33), (3.34), (3.35), and (3.38); these conditions decouple Eqs. (3.31) so that only Eqs. (3.31*a*) and (3.31*d*), and Eqs. (3.31*b*) and (3.31*c*), remain coupled. Upon substituting $u = u_0 \sin \omega t$ as the excitation, the response in the coupled modes is of a form

$x_c = x_{c0} \sin \omega t$, $\beta = \beta_0 \sin \omega t$ where x_{c0} and β_0 are related to u_0 as follows:[19]

$$\frac{x_{c0}}{u_0} = \frac{\dfrac{k_x}{k_z}\left[\left(\dfrac{a_x}{\rho_y}\right)^2 - \left(\dfrac{f}{f_z}\right)^2\right]}{\left(\dfrac{f}{f_z}\right)^4 - \left[\dfrac{k_x}{k_z} + \dfrac{k_x}{k_z}\left(\dfrac{a_z}{\rho_y}\right)^2 + \left(\dfrac{a_x}{\rho_y}\right)^2\right]\left(\dfrac{f}{f_z}\right)^2 + \dfrac{k_x}{k_z}\left(\dfrac{a_x}{\rho_y}\right)^2} \tag{3.55}$$

$$\frac{\beta_0}{u_0/\rho_y} = \frac{-\dfrac{k_x}{k_z}\dfrac{a_z}{\rho_y}\left(\dfrac{f}{f_z}\right)^2}{\left(\dfrac{f}{f_z}\right)^4 - \left[\dfrac{k_x}{k_z} + \dfrac{k_x}{k_z}\left(\dfrac{a_z}{\rho_y}\right)^2 + \left(\dfrac{a_x}{\rho_y}\right)^2\right]\left(\dfrac{f}{f_z}\right)^2 + \dfrac{k_x}{k_z}\left(\dfrac{a_x}{\rho_y}\right)^2} \tag{3.56}$$

where $f_z = \frac{1}{2\pi}\sqrt{4k_z/m}$ in accordance with Eq. (3.37). A similar set of equations apply for vibration in the coupled y_c, α coordinates. There is no response of the system in the z_c or γ modes since there is no net excitation in these directions; that is, $\mathbf{F}_z$ and $\mathbf{M}_z$ are zero.

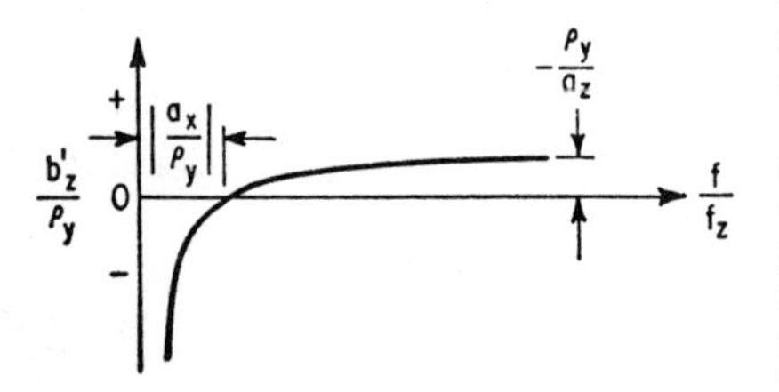

FIG. 3.25. Curve showing the position of the axis of pure rotation of the rigid body in Fig. 3.15 as a function of the frequency ratio f/f_z when the excitation is sinusoidal motion of the foundation in the X direction [see Eq. (3.57)]. The axis of rotation is parallel to the Y axis and in the XZ plane, and its coordinate along the Z axis is designated by b_z'.

As indicated by Eqs. (3.1), the displacement at any point in a rigid body is the sum of the displacement at the center-of-gravity and the displacements resulting from motion of the body in rotation about axes through the center-of-gravity. Equations (3.55) and (3.56) together with analogous equations for y_{c0}, α_0 provide the basis for calculating these displacements. Care should be taken with phase angles, particularly if two or more excitations u, v, w exist concurrently.

At any single frequency, coupled vibration in the x_c,β modes is equivalent to a pure rotation of the rigid body with respect to an axis parallel to the Y axis, in the YZ plane and displaced from the center-of-gravity of the body (see Fig. 3.15). As a result, the rigid body has zero displacement x in the horizontal plane containing this axis. Therefore, the Z coordinate of this axis b_z' satisfies $x_{c0} + b_z'\beta_0 = 0$, which is obtained from the first of Eqs. (3.1) by setting $x_b = 0$ (γ_0 motion is not considered). Substituting Eqs. (3.55) and (3.56) for x_{c0} and β_0, respectively, the axis of rotation is located at

$$\frac{b_z'}{\rho_y} = \frac{(a_x/\rho_y)^2 - (f/f_z)^2}{(a_z/\rho_y)(f/f_z)^2} \tag{3.57}$$

Figure 3.25 shows the relation of Eq. (3.57) graphically. At high values of frequency f/f_z, the axis does not change position significantly with frequency; b_z'/ρ_y approaches a positive value as f/f_z becomes large since a_z is negative (see Fig. 3.15).

When the resilient supporting elements have damping as well as elastic properties, the solution of the equations of motion [see Eq. (3.31a)] becomes too laborious for general use. Responses of systems with damping have been obtained for several typical cases using a digital computer.[20] Figures 3.26A, B, and C show the response at three points in the body of the system shown in Fig. 3.15, with the excitation $u = u_0 \sin \omega t$. The weight of the body is 45 lb; each of the four resilient supporting elements has a stiffness $k_z = 1{,}050$ lb/in. and stiffness ratios $k_x/k_z = k_y/k_z = \frac{1}{2}$. The critical damping coefficients in the X, Y, Z directions are taken as $c_{cx} = 2\sqrt{4k_x m}$, $c_{cy} = 2\sqrt{4k_y m}$, $c_{cz} = 2\sqrt{4k_z m}$, respectively, where the expression for c_{cz} follows from the single degree-of-freedom case defined by Eq. (2.12). The fractions of critical damping are $c_x/c_{cx} = c_y/c_{cy} = c_z/c_{cz} = c/c_c$, the parameter of the curves in Figs.

3.26*A*, *B*, and *C*. Coordinates locating the resilient elements are $a_x = \pm 5.25$ in., $a_y = \pm 3.50$ in., and $a_z = -6.50$ in. The radii of gyration with respect to the *X*, *Y*, *Z* axes are $\rho_x = 4.40$ in., $\rho_y = 5.10$ in., and $\rho_z = 4.60$ in.

Natural frequencies calculated from Eqs. (3.37) and (3.40) are $f_z = 30.0$ cps; $f_{x\beta} = 43.7$ cps, 15.0 cps; and $f_{y\alpha} = 43.2$ cps, 11.7 cps. The fraction of critical damping c/c_c varies between 0 and 0.25. Certain characteristic features of the response curves in Figs. 3.26*A*, *B*, and *C* are:

1. The relatively small response at the frequency of 24.2 cps in Fig. 3.26*C* occurs because point 3 lies near the axis of rotation of the rigid body at that frequency. Point 2 lies near the axis of rotation at higher frequencies and the response becomes correspondingly low, as shown in Fig. 3.26*B*. The position of the axis of rotation changes rapidly for small changes of fre-

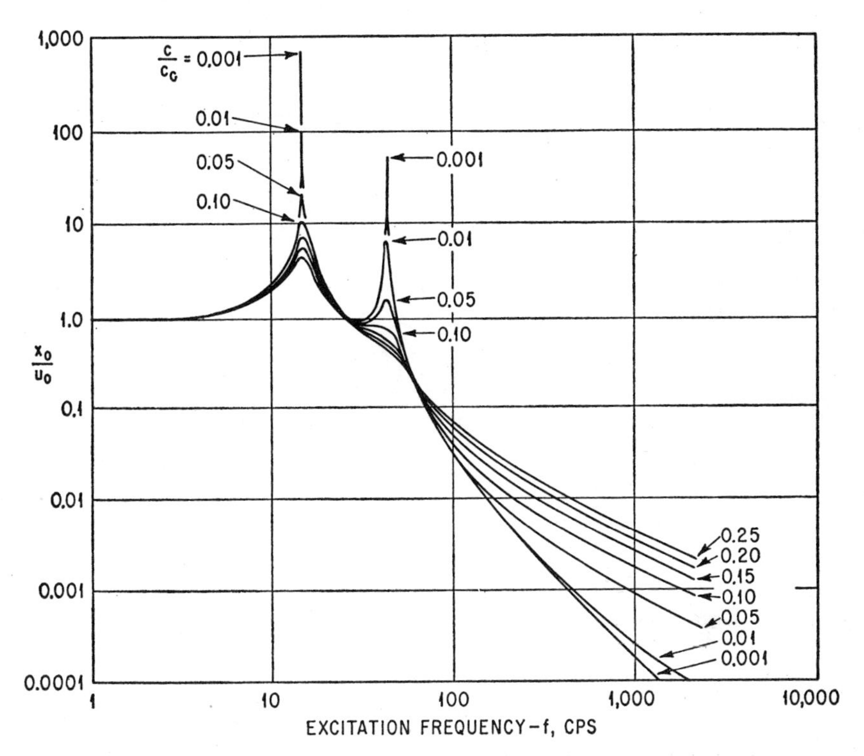

FIG. 3.26*A*. Response curves for point 1 with damping in the resilient supports in the system shown in Fig. 3.15. The response is the ratio of the amplitude at point 1 of the rigid body in the *X* direction to the amplitude of the foundation in the *X* direction (x_0/u_0). The fraction of critical damping c/c_c is the same in the *X*, *Y*, *Z* directions.

quency in the low- and intermediate-frequency range (indicated by the sharp dip in the curves for small damping in Fig. 3.26*C*) and varies asymptotically toward a final position as the forcing frequency increases (see Fig. 3.25).

2. The effect of damping on the magnitude of the response at the higher and lower natural frequencies in coupled modes is illustrated. When the fraction of critical damping is between 0.01 and 0.10, the response at the lower of the coupled natural frequencies is approximately 10 times as great as the response at the higher of the coupled natural frequencies. With greater damping ($c/c_c \geq 0.15$), the effect of resonance in the vicinity of the higher coupled natural frequency becomes so slight as to be hardly discernible.

TWO PLANES OF SYMMETRY WITH RESILIENT SUPPORTS INCLINED IN ONE PLANE ONLY. The system is shown in Fig. 3.21, and the excitation is $u = u_0 \sin \omega t$. The conditions of symmetry are defined by Eqs. (3.33), (3.35), and (3.38).

The response is entirely in the x_c,β coupled mode with the following amplitudes:

$$\frac{x_{c0}}{u_0} = \frac{\frac{k_p}{k_r}\left(\frac{a_x}{\rho_y}\right)^2 - \left(\frac{k_p}{k_r}\cos^2\phi + \sin^2\phi\right)\left(\frac{f}{f_r}\right)^2}{\left(\frac{f}{f_r}\right)^4 - A\left(\frac{f}{f_r}\right)^2 + \frac{k_p}{k_r}\left(\frac{a_x}{\rho_y}\right)^2}$$

$$\frac{\beta_0}{u_0/\rho_y} = \frac{-\left[\left(\frac{k_p}{k_r}\cos^2\phi + \sin^2\phi\right)\left(\frac{a_z}{\rho_y}\right) + \left(1 - \frac{k_p}{k_r}\right)\left|\frac{a_x}{\rho_y}\right|\sin\phi\cos\phi\right]\left(\frac{f}{f_r}\right)^2}{\left(\frac{f}{f_r}\right)^4 - A\left(\frac{f}{f_r}\right)^2 + \frac{k_p}{k_r}\left(\frac{a_x}{\rho_y}\right)^2} \tag{3.58}$$

where A is defined after Eq. (3.45). A similar set of expressions may be written for the response in the y_c,α coupled mode when the excitation is the motion $v = v_0 \sin \omega t$ of the foundation:

$$\frac{y_{c0}}{v_0} = \frac{\frac{k_q}{k_r}\left(\frac{k_p}{k_r}\sin^2\phi + \cos^2\phi\right)\left(\frac{a_y}{\rho_x}\right)^2 - \frac{k_q}{k_r}\left(\frac{f}{f_r}\right)^2}{\left(\frac{f}{f_r}\right)^4 - B\left(\frac{f}{f_r}\right)^2 + \frac{k_q}{k_r}\left(\frac{k_p}{k_r}\sin^2\phi + \cos^2\phi\right)\left(\frac{a_y}{\rho_x}\right)}$$

$$\frac{\alpha_0}{v_0/\rho_x} = \frac{\frac{k_q}{k_r}\frac{a_z}{\rho_x}\left(\frac{f}{f_r}\right)^2}{\left(\frac{f}{f_r}\right)^4 - B\left(\frac{f}{f_r}\right)^2 + \frac{k_q}{k_r}\left(\frac{k_p}{k_r}\sin^2\phi + \cos^2\phi\right)\left(\frac{a_y}{\rho_x}\right)} \tag{3.59}$$

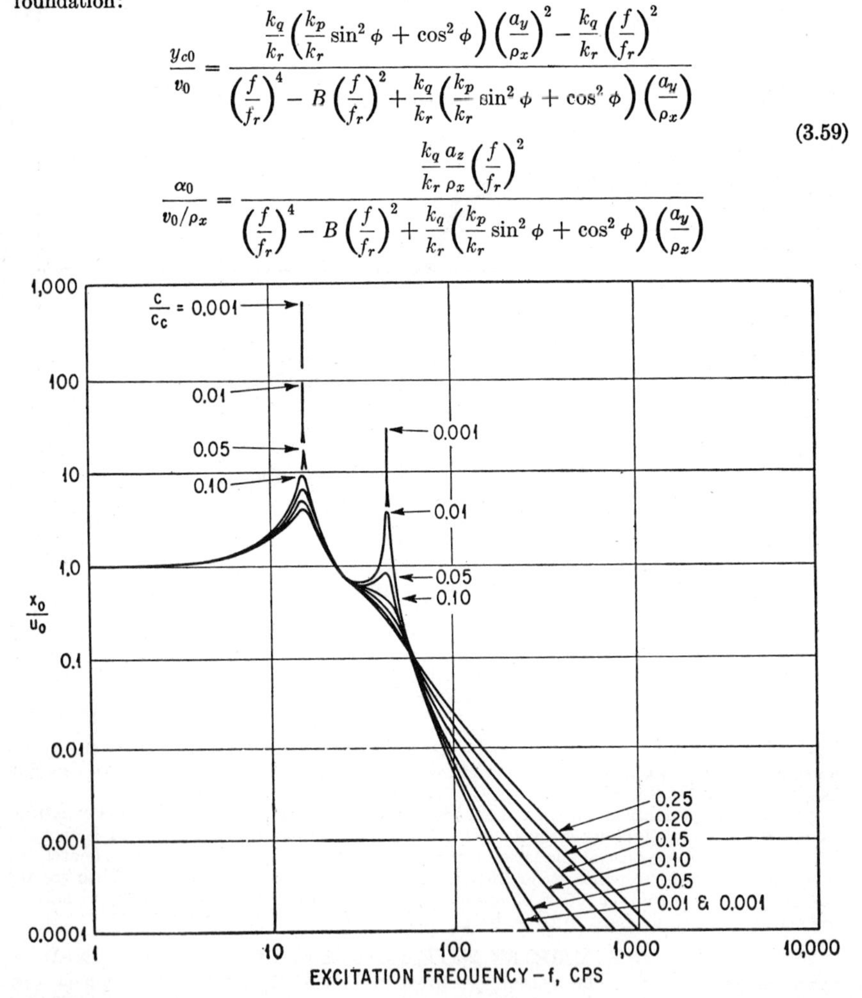

Fig. 3.26*B*. Response curves at point 2 in the system shown in Fig. 3.15. See caption for Fig. 3.26*A*.

where B is defined after Eq. (3.46). No motion occurs in the z_c or γ mode since the quantities F_z and M_z are zero in Eqs. (3.31*e*) and (3.31*f*).

Response curves for the system shown in Fig. 3.21 when damping is included are qualitatively similar to those shown in Figs. 3.26.[21] The significant advantage in the use of inclined resilient supports is the additional versatility gained from the ability to vary the angle of inclination ϕ which directly affects the degree of coupling in the x_c,β coupled mode. For example, a change in the angle ϕ produces a change in the position of the axis of pure rotation of the rigid body. In a manner similar to that used to derive Eq. (3.57), Eqs. (3.58) yield the following expression defining the location of the axis of rotation:

$$\frac{b_z'}{\rho_y} = \frac{\dfrac{k_p}{k_r}\left(\dfrac{a_x}{\rho_y}\right)^2 - \left(\dfrac{k_p}{k_r}\cos^2\phi + \sin^2\phi\right)\left(\dfrac{f}{f_r}\right)^2}{\left[\left(\dfrac{k_p}{k_r}\cos^2\phi + \sin^2\phi\right)\dfrac{a_z}{\rho_y} + \left(1 - \dfrac{k_p}{k_r}\right)\left|\dfrac{a_x}{\rho_y}\right|\sin\phi\cos\phi\right]\left(\dfrac{f}{f_r}\right)^2} \tag{3.60}$$

BODY-INDUCED SINUSOIDAL VIBRATION

This section includes the analysis of a resiliently supported rigid body wherein the excitation consists of forces and moments applied directly to the rigid body (or originating within the body). The system has two planes of symmetry with orthogonal resilient supports; the modal coupling and natural frequencies for such a system are considered above. Two types of excitation are considered: (1) a force rotating about an axis parallel to one of the principal inertial axes and (2) an oscillatory moment acting about one of the

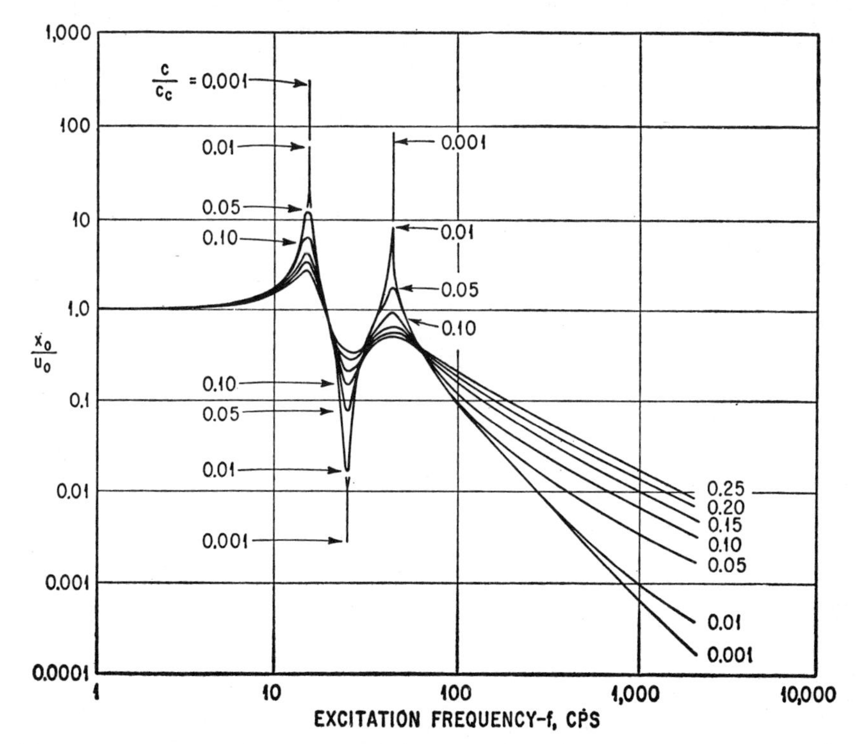

Fig. 3.26*C*. Response curves at point 3 in the system shown in Fig. 3.15. See caption for Fig. 3.26*A*.

principal inertial axes.[22,23] There is no motion of the foundation that supports the resilient elements; thus, the following terms in Eqs. (3.31) are equal to zero:

$$u = v = w = \alpha = \beta = \gamma = 0 \tag{3.61}$$

TWO PLANES OF SYMMETRY WITH ORTHOGONAL RESILIENT ELEMENTS EXCITED BY A ROTATING FORCE. The system excited by the rotating force is illustrated in Fig. 3.27. The force F_0 rotates at frequency ω about an axis parallel to the Y

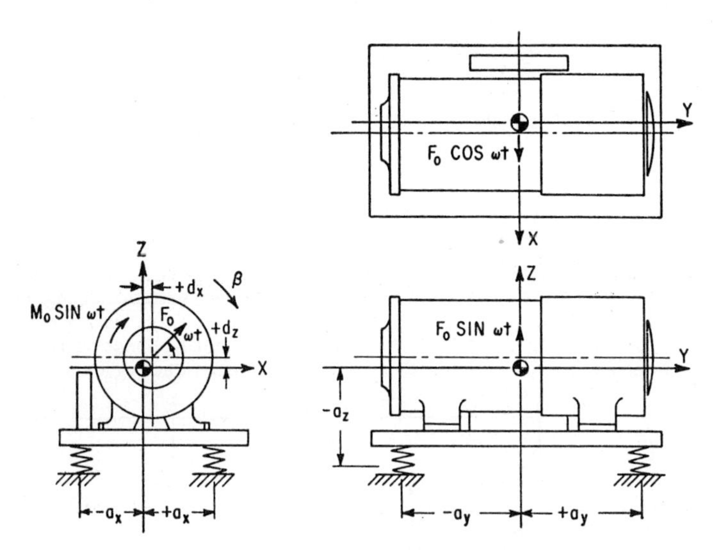

FIG. 3.27. Example of a rigid body on orthogonal resilient supports with two planes of symmetry, excited by body-induced sinusoidal excitation. Alternative excitations are (1) the force F_0 in the XZ plane rotating with angular velocity ωt about an axis parallel to the Y axis and (2) the oscillatory moment $M_0 \sin \omega t$ acting about the Y axis. There is no motion of the foundation that supports the resilient elements.

axis but spaced therefrom by the coordinate distances d_x, d_z; the force is in the XZ plane. The forces and moments applied to the body by the rotating force F_0 are

$$\begin{aligned} F_x &= F_0 \cos \omega t & M_x &= 0 \\ F_y &= 0 & M_y &= F_0(d_z \cos \omega t - d_x \sin \omega t) \\ F_z &= F_0 \sin \omega t & M_z &= 0 \end{aligned} \tag{3.62}$$

The conditions of symmetry are defined by Eqs. (3.33), (3.34), (3.35), and (3.38); and the excitation is defined by Eqs. (3.61) and (3.62). Substituting these conditions into the equations of motion, Eqs. (3.31) show that vibration response is not excited in the coupled y_c,α mode or in the γ mode. In the Z direction, the motion z_{c0} of the body and the force F_{tz} transmitted through the resilient elements can be found from Eq. (2.30) and Fig. 2.17 since single degree-of-freedom behavior is involved. The horizontal displacement amplitude x_{c0} of the center-of-gravity in the X direction and the rotational displacement amplitude β_0 about the Y axis are given by

$$\frac{x_{c0}}{F_0/4k_x} = \frac{k_x}{k_z} \frac{\sqrt{\left[\frac{k_x}{k_z}\frac{a_z}{\rho_y}\left(\frac{a_z}{\rho_y} - \frac{d_z}{\rho_y}\right) + \left(\frac{a_x}{\rho_y}\right)^2 - \left(\frac{f}{f_z}\right)^2\right]^2 + \left[\frac{k_x}{k_z}\frac{d_x}{\rho_y}\frac{a_z}{\rho_y}\right]^2}}{\left(\frac{f}{f_z}\right)^4 - \left[\frac{k_x}{k_z} + \frac{k_x}{k_z}\left(\frac{a_z}{\rho_y}\right)^2 + \left(\frac{a_x}{\rho_y}\right)^2\right]\left(\frac{f}{f_z}\right)^2 + \frac{k_x}{k_z}\left(\frac{a_x}{\rho_y}\right)^2} \tag{3.63}$$

$$\frac{\beta_0}{F_0/4k_x\rho_y} = \frac{k_x}{k_z} \frac{\sqrt{\left[\frac{k_x}{k_z}\left(\frac{a_z}{\rho_y} - \frac{d_z}{\rho_y}\right) + \frac{d_z}{\rho_y}\left(\frac{f}{f_z}\right)^2\right]^2 + \left[\frac{d_x}{\rho_y}\left(\frac{k_x}{k_z} - \frac{f^2}{f_z^2}\right)\right]^2}}{\left(\frac{f}{f_z}\right)^4 - \left[\frac{k_x}{k_z} + \frac{k_x}{k_z}\left(\frac{a_z}{\rho_y}\right)^2 + \left(\frac{a_x}{\rho_y}\right)^2\right]\left(\frac{f}{f_z}\right)^2 + \frac{k_x}{k_z}\left(\frac{a_x}{\rho_y}\right)^2}$$

where a_x, a_z are location coordinates of the resilient supports, and

$$f_z = \frac{1}{2\pi}\sqrt{\frac{4k_z}{m}} \tag{3.64}$$

The amplitude of the oscillating force F_{tx} in the X direction and the amplitude of the oscillating moment M_{ty} about the Y axis which are transmitted to the foundation by the combination of resilient elements are

$$\begin{aligned} F_{tx} &= 4k_x\sqrt{x_{c0}^2 - 2a_zx_{c0}\beta_0 \cos(\phi_x - \phi_\beta) + a_z^2\beta_0^2} \\ M_{ty} &= 4k_za_x^2\beta_0 \end{aligned} \tag{3.65}$$

where F_{tx} is the sum of the forces transmitted by the individual resilient elements and M_{ty} is a moment formed by forces in the Z direction of opposite sign at opposite resilient supports. The angles ϕ_x and ϕ_β are defined by

$$\tan\phi_x = \frac{\frac{k_x}{k_z}\frac{a_z}{\rho_y}\left(\frac{a_z}{\rho_y} - \frac{d_z}{\rho_y}\right) + \left(\frac{a_x}{\rho_y}\right)^2 - \left(\frac{f}{f_z}\right)^2}{\frac{k_x}{k_z}\frac{a_z}{\rho_y}\frac{d_x}{\rho_y}} \qquad [0° \leq \phi_x \leq 360°]$$

$$\tan\phi_\beta = \frac{\frac{k_x}{k_z}\left(\frac{a_z}{\rho_y} - \frac{d_z}{\rho_y}\right) + \frac{d_z}{\rho_y}\left(\frac{f}{f_z}\right)^2}{\frac{d_x}{\rho_y}\left[\frac{k_x}{k_z} - \left(\frac{f}{f_z}\right)^2\right]} \qquad [0° \leq \phi_\beta \leq 360°]$$

To obtain the correct value of $(\phi_x - \phi_\beta)$ in Eq. (3.65), the signs of the numerator and denominator in each tangent term must be inspected to determine the proper quadrant for ϕ_x and ϕ_β.

Example 3.2. Consider an electric motor which has an unbalanced rotor, creating a centrifugal force. The motor weighs 3,750 lb, and has a radius of gyration $\rho_y = 9.10$ in. The distances $d_x = d_y = d_z = 0$, that is, the axis of rotation is the Y principal axis and the center-of-gravity of the rotor is in the XZ plane. The resilient supports each have a stiffness ratio of $k_x/k_z = 1.16$, and are located at $a_z = -14.75$ in., $a_x = \pm 12.00$ in. The resulting displacement amplitudes of the center-of-gravity, expressed dimensionlessly, are shown in Fig. 3.28; the force and moment amplitudes transmitted to the foundation, expressed dimensionlessly, are shown in Fig. 3.29. The displacements of the center-of-gravity of the body are dimensionalized with respect to the displacements at zero frequency:

$$z_{c0}(0) = \frac{F_0}{4k_z}$$

$$x_{c0}(0) = \frac{F_0}{4k_x}\left[1 + \frac{k_x}{k_z}\left(\frac{a_z}{a_x}\right)^2\right] \tag{3.66}$$

$$\beta_0(0) = \frac{F_0}{4k_z a_z}\left(\frac{a_z}{a_x}\right)^2$$

At excitation frequencies greater than the higher natural frequency of the x_c,β coupled motion, the displacements, forces, and moment all continuously decrease as the frequency increases.

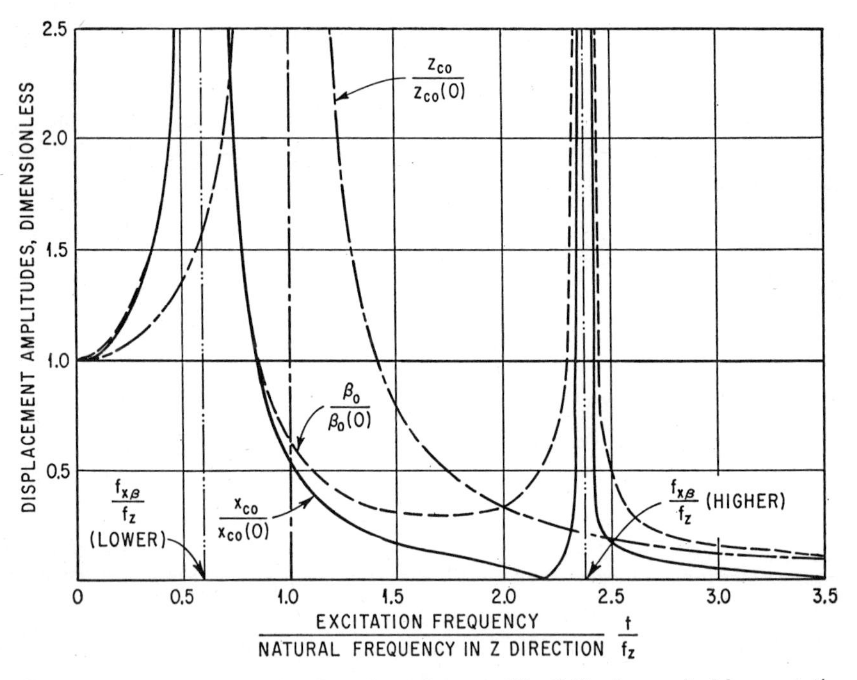

FIG. 3.28. Response curves for the system shown in Fig. 3.27 when excited by a rotating force F_0 acting about the Y axis. The parameters of the system are $k_x/k_z = 1.16$, $a_x/\rho_y = \pm 1.32$, $a_z/\rho_y = -1.62$. Only x_c, z_c, β displacements of the body are excited [see Eqs. (3.63)]. The displacements are expressed dimensionlessly by employing the displacements at zero frequency [see Eqs. (3.66)].

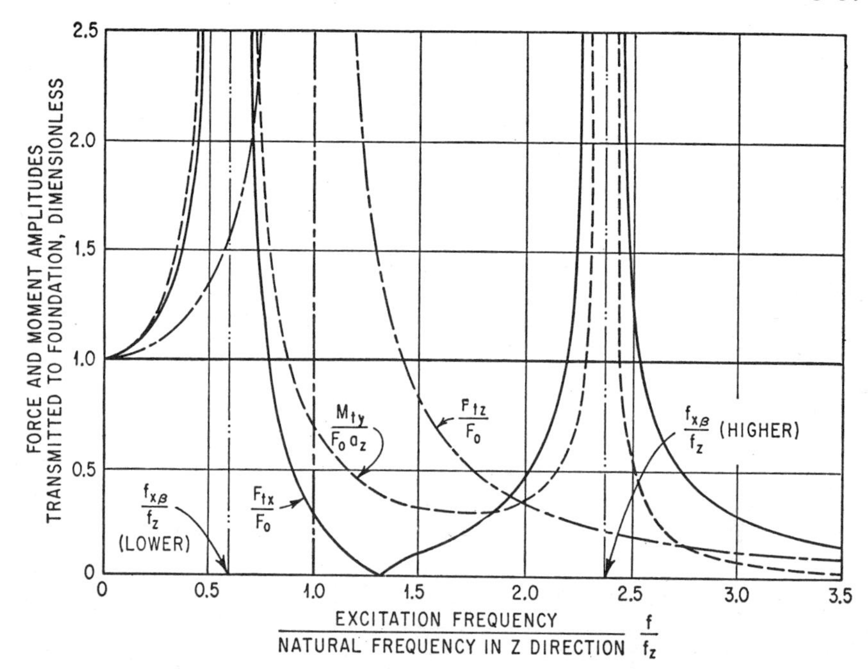

FIG. 3.29. Force and moment amplitudes transmitted to the foundation for the system shown in Fig. 3.27 when excited by a rotating force F_0 acting about the Y axis. The parameters of the system are $k_x/k_z = 1.16$, $a_x/\rho_y = \pm 1.32$, $a_z/\rho_y = -1.62$. The amplitudes of the oscillating forces in the X and Z directions transmitted to the foundation are F_{tx} and F_{tz}, respectively. The amplitude of the total oscillating moment about the Y axis transmitted to the foundation is M_{ty}.

TWO PLANES OF SYMMETRY WITH ORTHOGONAL RESILIENT ELEMENTS EXCITED BY AN OSCILLATING MOMENT. Consider the oscillatory moment M_0 acting about the Y axis with forcing frequency ω. The resulting applied forces and moments acting on the body are

$$\begin{aligned} M_y &= M_0 \sin \omega t \\ F_x = F_y = F_z &= M_x = M_z = 0 \end{aligned} \tag{3.67}$$

Substituting conditions of symmetry defined by Eqs. (3.33), (3.34), (3.35), and (3.38), and the excitation defined by Eqs. (3.61) and (3.67), the equations of motion [Eqs. (3.31)] show that oscillations are excited only in the x_c,β coupled mode. Solving for the resulting displacements,

$$\begin{aligned} \frac{x_{c0}}{M_0/4k_x\rho_y} &= \frac{\left(\frac{k_x}{k_z}\right)^2 \frac{a_z}{\rho_y}}{\left(\frac{f}{f_z}\right)^4 - \left[\frac{k_x}{k_z} + \frac{k_x}{k_z}\left(\frac{a_z}{\rho_y}\right)^2 + \left(\frac{a_x}{\rho_y}\right)^2\right]\left(\frac{f}{f_z}\right)^2 + \frac{k_x}{k_z}\left(\frac{a_x}{\rho_y}\right)^2} \\ \frac{\beta_0}{M_0/4k_x{\rho_y}^2} &= \frac{\frac{k_x}{k_z}\left[\frac{k_x}{k_z} - \left(\frac{f}{f_z}\right)^2\right]}{\left(\frac{f}{f_z}\right)^4 - \left[\frac{k_x}{k_z} + \frac{k_x}{k_z}\left(\frac{a_z}{\rho_y}\right)^2 + \left(\frac{a_x}{\rho_y}\right)^2\right]\left(\frac{f}{f_z}\right)^2 + \frac{k_x}{k_z}\left(\frac{a_x}{\rho_y}\right)^2} \end{aligned} \tag{3.68}$$

The amplitude of the oscillating force F_{tx} in the X direction and the amplitude of the oscillating moment M_{ty} about the Y axis transmitted to the foundation by the combination of resilient supports are

$$F_{tx} = 4k_x(x_{c0} - a_z\beta_0)$$
$$M_{ty} = 4k_z a_x^2 \beta_0 \qquad (3.69)$$

where F_{tx} and M_{ty} have the same meaning as in Eqs. (3.65). Low vibration transmission of force and moment to the foundation is decreased at the higher frequencies in a manner similar to that shown in Fig. 3.29.

FOUNDATION-INDUCED VELOCITY SHOCK

A velocity shock is an instantaneous change in the velocity of one portion of a system relative to another portion. In this section, the system is a rigid body supported by orthogonal resilient elements within a rigid container; the container experiences a velocity shock. The system has one plane of symmetry; modal coupling and natural frequencies for such a system are considered above. Two types of velocity shock are analyzed: (1) a sudden change in the translational velocity of the container and (2) a sudden change in the rotational velocity of the container. In both instances the change in velocity is from an initial velocity to zero. No forces or moments are applied directly to the resiliently supported body; i.e., only the forces transmitted by the resilient supports act. Thus, in the equations of motion, Eqs. (3.31):

$$F_x = F_y = F_z = M_x = M_y = M_z = 0 \qquad (3.70)$$

The modal coupling and natural frequencies for this system have been determined when the YZ plane is a plane of symmetry and the conditions of symmetry of Eqs. (3.33) to (3.35) apply. It is assumed that the velocity components of the body ($\dot{x}_c$, $\dot{y}_c$, $\dot{z}_c$, $\dot{\alpha}$, $\dot{\beta}$, $\dot{\gamma}$) and the velocity components of the supporting container ($\dot{u}$, $\dot{v}$, $\dot{w}$, $\dot{\mathbf{a}}$, $\dot{\boldsymbol{\beta}}$, $\dot{\boldsymbol{\gamma}}$) are respectively equal at time $t < 0$. At $t = 0$, all velocity components of the supporting container are brought to zero instantaneously. To determine the subsequent motion of the resiliently supported body, the natural frequencies f_n in the coupled modes of response are first calculated using Eq. (3.36). Then the response motion of the resiliently supported body to the two types of velocity shock can be found by the analyses which follow.[24]

ONE PLANE OF SYMMETRY WITH ORTHOGONAL RESILIENT SUPPORTS EXCITED BY A TRANSLATIONAL VELOCITY SHOCK. Figure 3.30 shows a rigid body supported within a rigid container by resilient supports in such a manner that the YZ plane is a plane of symmetry. The entire system moves with constant velocity $\dot{v}_0$ and without relative motion. At time $t = 0$, the container impacts inelastically against the rigid wall shown at the right. The following initial conditions of displacement and velocity apply at the instant of impact ($t = 0$):

$$\dot{y}_c(0) = \dot{v}_0$$
$$x_c(0) = y_c(0) = z_c(0) = \alpha(0) = \beta(0) = \gamma(0) = 0 \qquad (3.71)$$
$$\dot{x}_c(0) = \dot{z}_c(0) = \dot{\alpha}(0) = \dot{\beta}(0) = \dot{\gamma}(0) = 0$$

As a result of the impact, the velocity of the supported rigid body tends to continue and is responsible for excitation of the system in the coupled mode of the y_c, z_c, α motions. The maximum displacements of the center-of-gravity of the supported body are

$$\frac{y_{cm}}{2\pi\dot{v}_0/f_z} = \frac{1}{B}\sum_{n=1}^{3}\left(\frac{|A_n|}{f_n/f_z}\right)$$
$$\frac{z_{cm}}{2\pi\dot{v}_0/f_z} = \frac{1}{B}\sum_{n=1}^{3}\left(\frac{|M_nA_n|}{f_n/f_z}\right) \qquad (3.72)$$
$$\frac{\alpha_m}{2\pi\dot{v}_0/\rho_x f_z} = \frac{1}{B}\sum_{n=1}^{3}\left(\frac{|N_nA_n|}{f_n/f_z}\right)$$

The maximum accelerations of the center-of-gravity of the supported body are

$$\frac{\ddot{y}_{cm}}{2\pi f_z \dot{v}_0} = \frac{1}{B} \sum_{n=1}^{3} \left(|A_n| \frac{f_n}{f_z} \right)$$

$$\frac{\ddot{z}_{cm}}{2\pi f_z \dot{v}_0} = \frac{1}{B} \sum_{n=1}^{3} \left(|M_n A_n| \frac{f_n}{f_z} \right) \tag{3.73}$$

$$\frac{\ddot{\alpha}_m}{2\pi f_z \dot{v}_0/\rho_x} = \frac{1}{B} \sum_{n=1}^{3} \left(|N_n A_n| \frac{f_n}{f_z} \right)$$

where the subscript m denotes maximum value and

$$M_n = \frac{1}{1 - (f_n/f_z)^2} \left[\frac{\Sigma k_y}{\Sigma k_z} - \left(\frac{f_n}{f_z} \right)^2 \right] \frac{\Sigma k_z a_y}{\Sigma k_y a_z}$$

$$N_n = \left[\frac{\Sigma k_y}{\Sigma k_z} - \left(\frac{f_n}{f_z} \right)^2 \right] \frac{\rho_x \Sigma k_z}{\Sigma k_y a_z} \tag{3.74}$$

$$A_n = M_{n+1} N_{n+2} - M_{n+2} N_{n+1}$$

$$B = \left| \sum_{n=1}^{3} M_n (N_{n+1} - N_{n+2}) \right|$$

The fictitious natural frequency f_z is defined for mathematical purposes by Eq. (3.37). The numerical values of the subscript numbers n, $n + 1$, $n + 2$ denote the three natural frequencies in the coupled mode of the y_c, z_c, α motions determined from Eq. (3.36). These natural frequencies are arbitrarily assigned the values $n = 1, 2, 3$. When $n + 1$

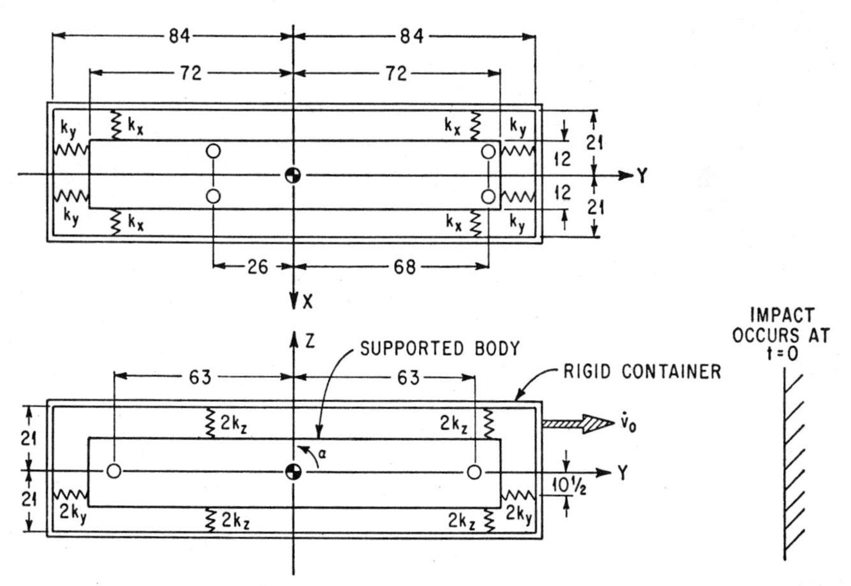

FIG. 3.30. Example of a rigid body supported within a rigid container by resilient elements with YZ a plane of symmetry. Excitation is by a translational velocity shock in the Y direction. Prior to impact the entire system moves with constant velocity $\dot{v}_0$ and without relative motion. The rigid container impacts inelastically against the wall shown at the right, and y_c, z_c, α motions of the internally supported body result, as described mathematically by Eqs. (3.72) and (3.73).

or $n + 2$ equals 4, use 1 instead; when $n + 2$ equals 5, use 2 instead. Maximum displacements and accelerations may be calculated for other points in the supported rigid body by using Eqs. (3.1) except that each of the terms must be made numerically additive. For example, the maximum value of the y displacement at the point b having the Z coordinate b_z is

$$y_{bm} = y_{cm} + |b_z|\alpha_m \tag{3.75}$$

since $\gamma = 0$.

Since the system is assumed undamped, the response of the suspended body in terms of displacement or acceleration consists of a superposition of three sinusoidal components at the three natural frequencies in the coupled y_c, z_c, α mode. The absolute values of terms appear in Eq. (3.75) because the maximum response is the sum of the amplitudes

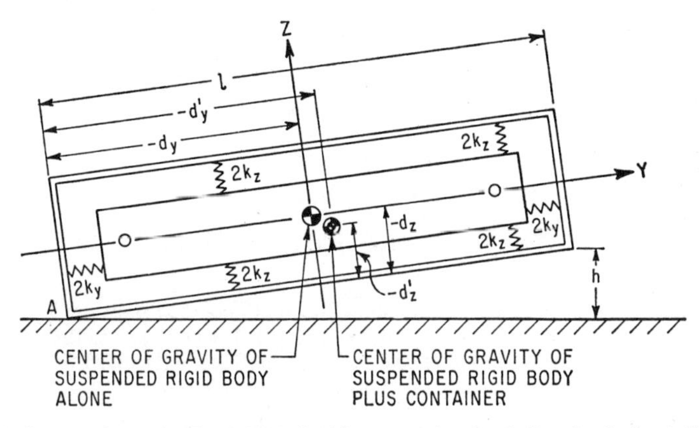

FIG. 3.31. System shown in Fig. 3.30 excited by a rotational velocity shock about the X axis. The shock is induced by lifting and dropping one end of the rigid container to make inelastic impact with the foundation. If the height of drop is h, the rotational velocity of the system about the corner A at the instant of impact is given by Eq. (3.79). The response of the resiliently supported body is described mathematically by Eqs. (3.77) and (3.78).

of the three component vibrations which comprise the over-all response. In general, the maximum response occurs when the three component vibrations reach their maximum positive or negative values at the same instant. Thus, the maximum values of response apply both in positive and negative directions.

ONE PLANE OF SYMMETRY WITH ORTHOGONAL RESILIENT SUPPORTS EXCITED BY A ROTATIONAL VELOCITY SHOCK. Alternative to the type of impact illustrated in Fig. 3.30, the system may be excited by imparting a rotational velocity shock (e.g., by lifting and dropping one end of the container), as illustrated in Fig. 3.31. It is assumed that the container impacts inelastically. The system has the same form of symmetry as that shown in Fig. 3.30, and only the y_c, z_c, α modes are excited. The initial conditions at the instant of impact ($t = 0$), based upon the angular velocity $\dot{\alpha}_0$ of the rigid container about point A in Fig. 3.31, are

$$\begin{gathered} \dot{y}_c(0) = -d_z\dot{\alpha}_0 \qquad \dot{z}_c(0) = d_y\dot{\alpha}_0 \qquad \dot{\alpha}(0) = \dot{\alpha}_0 \\ x_c(0) = y_c(0) = z_c(0) = \alpha(0) = \beta(0) = \gamma(0) = 0 \\ \dot{x}_c(0) = \dot{\beta}(0) = \dot{\gamma}(0) = 0 \end{gathered} \tag{3.76}$$

Note that d_y and d_z are negative quantities. The initial conditions in Eqs. (3.76) are based on the assumption that motion of the rigid body relative to the container during the fall is negligible compared to that which occurs after the impact. The maximum displacements of the center-of-gravity of the supported body are

$$\frac{y_{cm}}{2\pi\rho_x\dot{\alpha}_0/f_z} = \frac{1}{B}\sum_{n=1}^{3}\left[\left|\frac{d_z}{\rho_x}A_n + \frac{d_y}{\rho_x}(N_{n+1} - N_{n+2}) + (M_{n+2} - M_{n+1})\right|\frac{f_z}{f_n}\right]$$

$$\frac{z_{cm}}{2\pi\rho_x\dot{\alpha}_0/f_z} = \frac{1}{B}\sum_{n=1}^{3}\left[\left|M_n\left(\frac{d_z}{\rho_x}A_n + \frac{d_y}{\rho_x}(N_{n+1} - N_{n+2}) + (M_{n+2} - M_{n+1})\right)\right|\frac{f_z}{f_n}\right] \quad (3.77)$$

$$\frac{\alpha_m}{2\pi\dot{\alpha}_0/f_z} = \frac{1}{B}\sum_{n=1}^{3}\left[\left|N_n\left(\frac{d_z}{\rho_x}A_n + \frac{d_y}{\rho_x}(N_{n+1} - N_{n+2}) + (M_{n+2} - M_{n+1})\right)\right|\frac{f_z}{f_n}\right]$$

The maximum accelerations of the center-of-gravity of the supported body are

$$\frac{\ddot{y}_{cm}}{2\pi\rho_x f_z\dot{\alpha}_0} = \frac{1}{B}\sum_{n=1}^{3}\left[\left|\frac{d_z}{\rho_x}A_n + \frac{d_y}{\rho_x}(N_{n+1} - N_{n+2}) + (M_{n+2} - M_{n+1})\right|\frac{f_n}{f_z}\right]$$

$$\frac{\ddot{z}_{cm}}{2\pi\rho_x f_z\dot{\alpha}_0} = \frac{1}{B}\sum_{n=1}^{3}\left[\left|M_n\left(\frac{d_z}{\rho_x}A_n + \frac{d_y}{\rho_x}(N_{n+1} - N_{n+2}) + (M_{n+2} - M_{n+1})\right)\right|\frac{f_n}{f_z}\right] \quad (3.78)$$

$$\frac{\ddot{\alpha}_m}{2\pi f_z\dot{\alpha}_0} = \frac{1}{B}\sum_{n=1}^{3}\left[\left|N_n\left(\frac{d_z}{\rho_x}A_n + \frac{d_y}{\rho_x}(N_{n+1} - N_{n+2}) + (M_{n+2} - M_{n+1})\right)\right|\frac{f_n}{f_z}\right]$$

where d_z and d_y are the Z and Y coordinates, respectively, of the edges of the container, as shown in Fig. 3.31, and the other quantities are the same as those appearing in Eqs. (3.72) and (3.74). The maximum response at any point in the suspended body can be found in the manner of Eq. (3.75).

The rotational velocity $\dot{\alpha}_0$ of the container about the corner A in Fig. 3.31 may be induced by lifting the opposite end to a height h and dropping it. The resulting velocity $\dot{\alpha}_0$ is

$$\dot{\alpha}_0 = \left\{\frac{2g}{\rho_A{}^2}\left[\frac{h}{l}d_y' + d_z'\sqrt{1 - \left(\frac{h}{l}\right)^2} - d_z'\right]\right\}^{1/2} \quad (3.79)$$

where g is the acceleration of gravity, ρ_A is the radius of gyration of the rigid body plus container about the corner A, h is the initial elevation of the raised end of the container, l is the length of the container, and d_y' and d_z' are the Y and Z coordinates, respectively, of the edges of the container with respect to the center-of-gravity of the assembly of rigid body plus container (see Fig. 3.31).

Example 3.3. The rigid body shown in Fig. 3.31 weighs 1,500 lb and has a radius of gyration $\rho_x = 42$ in. with respect to the X axis. The resilient supporting elements apply forces parallel to their longitudinal axes *only*. Each element with its longitudinal axis in the X or Y direction has a stiffness of $k_x = k_y = 500$ lb/in. Each element whose longitudinal axis extends in the Z direction has a stiffness $k_z = 1{,}000$ lb/in. The resilient elements are positioned as shown in Fig. 3.30, and $l = 168$ in., $d_y = d_y' = -84$ in., $d_z = d_z' = -21$ in., $\rho_A = 308$ in. The rotational velocity shock results from a height of drop $h = 36$ in.

The fictitious natural frequency f_z is obtained from Eq. (3.37), yielding $f_z = 7.22$ cps. From Eq. (3.36) or Fig. 3.14, the natural frequencies in the y_c, z_c, α mode are $f_1 = 3.58$ cps, $f_2 = 6.02$ cps, and $f_3 = 9.75$ cps. From Eqs. (3.74), it is determined that $M_1 \simeq 0$, $M_2 = 11.7$, $M_3 = -15.3$, $N_1 = -0.1$, $N_2 = 7.1$, $N_3 = 25.1$, $A_1 = 402$, $A_2 = 2$, $A_3 = 1$, $B = 405$. Sample calculations for M_1 and A_1 are

$$M_1 = \frac{1}{1 - (3.58/7.22)^2}\left[\frac{4(500)}{8(1{,}000)} - \left(\frac{3.58}{7.22}\right)^2\right]\frac{4(1{,}000)(68 - 26)}{4(500)(-10.5)} = -0.04$$

$$A_1 = M_2N_3 - M_3N_2 = (11.7)(25.1) - (-15.3)(7.1) = 402$$

From Eq. (3.79), $\dot{\alpha}_0 = 0.38$ rad/sec. Then Eqs. (3.78) give the maximum acceleration of the center-of-gravity in the Y direction of the supported body as follows:

$$\ddot{y}_{cm} = \frac{2\pi\rho_x f_z \ddot{\alpha}_0}{B}\left[\begin{array}{l} \left|\frac{d_z}{\rho_x}A_1 + \frac{d_y}{\rho_x}(N_2 - N_3) + (M_3 - M_2)\right|\frac{f_1}{f_z} \\ + \left|\frac{d_z}{\rho_x}A_2 + \frac{d_y}{\rho_x}(N_3 - N_1) + (M_1 - M_3)\right|\frac{f_2}{f_z} \\ + \left|\frac{d_z}{\rho_x}A_3 + \frac{d_y}{\rho_x}(N_1 - N_2) + (M_2 - M_1)\right|\frac{f_3}{f_z} \end{array}\right]$$

$$= \frac{724 \text{ in./sec}^2}{405}\left[\begin{array}{l} \left|\frac{-21}{42}(402) + \frac{-84}{42}(7.1 - 25.1) + (-15.3 - 11.7)\right|\frac{3.58}{7.22} \\ + \left|\frac{-21}{42}(2) + \frac{-84}{42}(25.1 + 0.1) + (0 + 15.3)\right|\frac{6.02}{7.22} \\ + \left|\frac{-21}{42}(1) + \frac{-84}{42}(-0.1 - 7.1) + (11.7 - 0)\right|\frac{9.75}{7.22} \end{array}\right]$$

$$= 286 \text{ in./sec}^2 = 0.74g$$

In a similar manner:

$$z_{cm} = 1{,}580 \text{ in./sec}^2 = 4.09g$$
$$\ddot{\alpha}_m = 45.9 \text{ rad/sec}^2$$

REFERENCES

1. Marks, L. S., and T. Baumeister: "Mechanical Engineer's Handbook," 6th ed., p. **6**-6, McGraw-Hill Book Company, Inc., New York, 1958.
2. Housner, G. W., and D. E. Hudson: "Applied Mechanics-Dynamics," 2d ed., chap. 7, D. Van Nostrand Company, Inc., Princeton, N. J., 1959.
3. Boucher, R. W., D. A. Rich, H. L. Crane, and C. E. Matheny: *NACA Tech. Note* 3084, 1954.
4. Woodward, C. R.: "Handbook of Instructions for Experimentally Determining the Moments and Products of Inertia of Aircraft by the Spring Oscillation Method," *WADC Tech. Rept.* 55–415, June 1955.
5. Rubin, S.: *SAE Preprint* 197, 1957.
6. Macduff, J. N.: *Product Eng.*, **17**:106, 154 (1946).
7. Vane, F. F.: "A Guide for the Selection and Application of Resilient Mountings to Shipboard Equipment—Revised," *David W. Taylor Model Basin Rept.* 880, February, 1958, p. 98.
8. Smollen, L. E.: *J. Acoust. Soc. Amer.*, **40**:195 (1966).
9. Ref. 7, p. 50.
10. Crede, C. E.: "Vibration and Shock Isolation," p. 68, John Wiley & Sons, Inc., New York, 1951.
11. Ref. 7, pp. 37–49.
12. Ref. 10, pp. 53–58.
13. Lewis, R. C., and K. Unholtz: *Trans. ASME*, **69**:813 (1947).
14. Klein, E., R. S. Ayre, and I. Vigness: "Fundamentals of Guided Missile Packaging," *Dept. Defense (U.S.) Rept.* RD 219/3, July, 1955, appendix 8, pp. 49–52.
15. Ref. 10, p. 73.
16. Taylor, E. S., and K. A. Browne: *J. Aeronaut. Sci.*, **6**:43 (1938).
17. Browne, K. A.: *Trans. SAE*, **44**:185 (1939).
18. Derby, T. F.: "Decoupling the Three Translational Modes from the Three Rotational Modes of a Rigid Body Supported by Four Corner-Located Isolators," *Shock and Vibration Bull.* 43, pt. 4, June 1973, pp. 91–108.
19. Ref. 10, p. 50.
20. Himelblau, H.: "A Reliable Approach to Protecting Fragile Equipment from Aircraft Vibration," *North American Aviation, Inc., Rept.* NA-56-1030, 1957, pp. 16, 86.
21. Ref. 20, pp. 22, 95.
22. Ref. 10, pp. 43, 61.
23. Himelblau, H.: *Product Eng.*, **23**:151 (1952).
24. Ref. 14, chap. 11, November, 1955.
25. Ref. 2, appendix IV.

Table 3.1. Properties of Plane Sections

(After G. W. Housner and D. E. Hudson.[25])

The dimensions X_c, Y_c are the X, Y coordinates of the centroid, A is the area, $I_x \ldots$ is the area moment of inertia with respect to the $X \ldots$ axis, $\rho_x \ldots$ is the radius of gyration with respect to the $X \ldots$ axis; uniform solid cylindrical bodies of length l in the Z direction having the various plane sections as their cross sections have mass moment and product of inertia values about the Z axis equal to ρl times the values given in the table, where ρ is the mass density of the body; the radii of gyration are unchanged.

	Plane section	Area and centroid	Area moment of inertia	Square of radius of gyration	Area product of inertia
1	Right triangle	$A = \frac{1}{2}bh$ $X_c = \frac{2}{3}b$ $Y_c = \frac{1}{3}h$	$I_{x_c} = \frac{bh^3}{36}$ $I_{y_c} = \frac{b^3h}{36}$	$\rho_{x_c}^2 = \frac{1}{18}h^2$ $\rho_{y_c}^2 = \frac{1}{18}b^2$	$I_{x_cy_c} = \frac{A}{36}hb = \frac{h^2b^2}{72}$
2		$A = \frac{1}{2}bh$ $X_c = \frac{1}{3}b$ $Y_c = \frac{1}{3}h$	$I_{x_c} = \frac{bh^3}{36}$ $I_{y_c} = \frac{b^3h}{36}$	$\rho_{x_c}^2 = \frac{1}{18}h^2$ $\rho_{y_c}^2 = \frac{1}{18}b^2$	$I_{x_cy_c} = -\frac{A}{36}hb = -\frac{h^2b^2}{72}$
3	Triangle	$A = \frac{1}{2}bh$ $X_c = \frac{1}{3}(a+b)$ $Y_c = \frac{1}{3}h$	$I_{x_c} = \frac{bh^3}{36}$ $I_{y_c} = \frac{bh}{36}(b^2 - ab + a^2)$	$\rho_{x_c}^2 = \frac{1}{18}h^2$ $\rho_{y_c}^2 = \frac{1}{18}(b^2 - ab + a^2)$	$I_{x_cy_c} = \frac{Ah}{36}(2a - b) = \frac{bh^2}{72}(2a - b)$

Table 3.1 Properties of Plane Sections (Continued)

	Plane Section	Area and centroid	Area moment of inertia	Square of radius of gyration	Area product of inertia
4	Square (Y, X, C, a, a)	$A = a^2$ $X_c = \frac{1}{2}a$ $Y_c = \frac{1}{2}a$	$I_{x_c} = I_{y_c} = \frac{a^4}{12}$	${\rho_{x_c}}^2 = {\rho_{y_c}}^2 = \frac{1}{12}a^2$	$I_{x_c y_c} = 0$
5	Rectangle (Y, X, C, b, h)	$A = bh$ $X_c = \frac{1}{2}b$ $Y_c = \frac{1}{2}h$	$I_{x_c} = \frac{bh^3}{12}$ $I_{y_c} = \frac{b^3h}{12}$	${\rho_{x_c}}^2 = \frac{1}{12}h^2$ ${\rho_{y_c}}^2 = \frac{1}{12}h^2$	$I_{x_c y_c} = 0$
6	Parallelogram (Y, X, C, θ, a, b)	$A = ab \sin\theta$ $X_c = \frac{1}{2}(b + a\cos\theta)$ $Y_c = \frac{1}{2}(a\sin\theta)$	$I_{x_c} = \frac{a^3b}{12}\sin^3\theta$ $I_{y_c} = \frac{ab}{12}\sin\theta\,(b^2 + a^2\cos^2\theta)$	${\rho_{x_c}}^2 = \frac{1}{12}(a\sin\theta)^2$ ${\rho_{y_c}}^2 = \frac{1}{12}(b^2 + a^2\cos^2\theta)$	$I_{x_c y_c} = \frac{a^3b}{12}\sin^2\theta\cos\theta$
7	Trapezoid (Y, X, C, a, b, h)	$A = \frac{1}{2}h(a + b)$ $Y_c = \frac{1}{3}h\left(\frac{2a + b}{a + b}\right)$	$I_{x_c} = \frac{h^3(a^2 + 4ab + b^2)}{36(a + b)}$	${\rho_{x_c}}^2 = \frac{h^2(a^2 + 4ab + b^2)}{18(a + b)^2}$	

8 Y, C, a, X Circle	$A = \pi a^2$ $X_c = a$ $Y_c = a$	$I_{x_c} = I_{y_c} = \frac{1}{4}\pi a^4$	$\rho_{x_c}^2 = \rho_{y_c}^2 = \frac{1}{4}a^2$	$I_{x_c y_c} = 0$
9 Y, C, a, b, X Annulus	$A = \pi(a^2 - b^2)$ $X_c = a$ $Y_c = a$	$I_{x_c} = I_{y_c} = \frac{\pi}{4}(a^4 - b^4)$	$\rho_{x_c}^2 = \rho_{y_c}^2 = \frac{1}{4}(a^2 + b^2)$	$I_{x_c y_c} = 0$
10 Y, C, 2a, X Semicircle	$A = \frac{1}{2}\pi a^2$ $X_c = a$ $Y_c = \frac{4a}{3\pi}$	$I_{x_c} = \frac{a^4(9\pi^2 - 64)}{72\pi}$ $I_{y_c} = \frac{1}{8}\pi a^4$	$\rho_{x_c}^2 = \frac{a^2(9\pi^2 - 64)}{36\pi^2}$ $\rho_{y_c}^2 = \frac{1}{4}a^2$	$I_{x_c y_c} = 0$
11 Y, a, θ, θ, C, X Circular sector	$A = a^2\theta$ $X_c = \frac{2a}{3}\frac{\sin\theta}{\theta}$ $Y_c = 0$	$I_x = \frac{1}{4}a^4(\theta - \sin\theta\cos\theta)$ $I_y = \frac{1}{4}a^4(\theta + \sin\theta\cos\theta)$	$\rho_x^2 = \frac{1}{4}a^2\left(\frac{\theta - \sin\theta\cos\theta}{\theta}\right)$ $\rho_y^2 = \frac{1}{4}a^2\left(\frac{\theta + \sin\theta\cos\theta}{\theta}\right)$	$I_{x_c y_c} = 0$ $I_{xy} = 0$

Table 3.1. Properties of Plane Sections (*Continued*)

	Plane section	Area and centroid	Area moment of inertia	Square of radius of gyration	Area product of inertia
12	Circular segment	$A = a^2(\theta - \frac{1}{2}\sin 2\theta)$ $X_c = \frac{2a}{3}\left(\frac{\sin^3\theta}{\theta - \sin\theta\cos\theta}\right)$ $Y_c = 0$	$I_x = \frac{Aa^2}{4}\left[1 - \frac{2\sin^3\theta\cos\theta}{3(\theta - \sin\theta\cos\theta)}\right]$ $I_y = \frac{Aa^2}{4}\left[1 + \frac{2\sin^3\theta\cos\theta}{\theta - \sin\theta\cos\theta}\right]$	$\rho_x^2 = \frac{a^2}{4}\left[1 - \frac{2\sin^3\theta\cos\theta}{3(\theta - \sin\theta\cos\theta)}\right]$ $\rho_y^2 = \frac{a^2}{4}\left[1 + \frac{2\sin^3\theta\cos\theta}{\theta - \sin\theta\cos\theta}\right]$	$I_{x_c y_c} = 0$ $I_{xy} = 0$
13	Ellipse	$A = \pi ab$ $X_c = a$ $Y_c = b$	$I_{x_c} = \frac{\pi}{4}ab^3$ $I_{y_c} = \frac{\pi}{4}a^3b$	${\rho_{x_c}}^2 = \frac{1}{4}b^2$ ${\rho_{y_c}}^2 = \frac{1}{4}a^2$	$I_{x_c y_c} = 0$
14	Semiellipse	$A = \frac{1}{2}\pi ab$ $X_c = a$ $Y_c = \frac{4b}{3\pi}$	$I_{x_c} = \frac{ab^3}{72\pi}(9\pi^2 - 64)$ $I_{y_c} = \frac{\pi}{8}a^3b$	${\rho_{x_c}}^2 = \frac{b^2}{36\pi^2}(9\pi^2 - 64)$ ${\rho_{y_c}}^2 = \frac{1}{4}a^2$	$I_{x_c y_c} = 0$
15	Parabola	$A = \frac{4}{3}ab$ $X_c = \frac{3}{5}a$ $Y_c = 0$	$I_{x_c} = \frac{4}{15}ab^3$ $I_{y_c} = \frac{16}{175}a^3b$	${\rho_{x_c}}^2 = \frac{1}{5}b^2$ ${\rho_{y_c}}^2 = \frac{12}{175}a^2$	$I_{x_c y_c} = 0$
16	Semiparabola	$A = \frac{2}{3}ab$ $X_c = \frac{3}{5}a$ $Y_c = \frac{3}{8}b$	$I_x = \frac{2}{15}ab^3$ $I_y = \frac{2}{7}a^3b$	$\rho_x^2 = \frac{1}{5}b^2$ $\rho_y^2 = \frac{3}{7}a^2$	$I_{xy} = \frac{A}{4}ab = \frac{1}{6}a^2b^2$

17	$Y = \frac{h}{b^n} X^n$ *n*th-degree parabola	$A = \frac{bh}{n+1}$ $X_c = \frac{n+1}{n+2} b$ $Y_c = \frac{h}{2}\left(\frac{n+1}{2n+1}\right)$	$I_x = \frac{bh^3}{3(3n+1)}$ $I_y = \frac{hb^3}{n+3}$	$\rho_x^2 = \frac{h^2(n+1)}{3(3n+1)}$ $\rho_y^2 = \frac{n+1}{n+3} b^2$	
18	$Y = \frac{h}{b^{\frac{1}{n}}} X^{\frac{1}{n}}$ *n*th-degree parabola	$A = \frac{n}{n+1} bh$ $X_c = \frac{n+1}{2n+1} b$ $Y_c = \frac{n+1}{2(n+2)} h$	$I_x = \frac{n}{3(n+3)} bh^3$ $I_y = \frac{n}{3n+1} hb^3$	$\rho_x^2 = \frac{n+1}{3(n+3)} h^2$ $\rho_y^2 = \frac{n+1}{3n+1} b^2$	

Table 3.2. Properties of Homogeneous Solid Bodies

(*After G. W. Housner and D. E. Hudson.*[25])

The dimensions X_c, Y_c, Z_c are the X, Y, Z coordinates of the centroid, S is the cross-sectional area of the thin rod or hoop in cases 1 to 3, V is the volume, $I_x \ldots$ is the mass moment of inertia with respect to the $X \ldots$ axis, $\rho_x \ldots$ is the radius of gyration with respect to the $X \ldots$ axis, ρ is the mass density of the body.

	Solid body	Volume and centroid	Mass moment of inertia	Radius of gyration squared	Mass product of inertia
1	Thin rod	$V = Sl$ $X_c = \frac{1}{2}l$ $Y_c = 0$ $Z_c = 0$	$I_{x_c} = 0$ $I_{y_c} = I_{z_c} = \frac{\rho V}{12} l^2$	${\rho_{x_c}}^2 = 0$ ${\rho_{y_c}}^2 = {\rho_{z_c}}^2 = \frac{1}{12}l^2$	$I_{x_c y_c}$, etc. $= 0$
2	Thin circular rod	$V = 2SR\theta$ $X_c = \frac{R \sin \theta}{\theta}$ $Y_c = 0$ $Z_c = 0$	$I_x = I_{x_c}$ $= \frac{\rho V R^2(\theta - \sin \theta \cos \theta)}{2\theta}$ $I_y = \frac{\rho V R^2(\theta + \sin \theta \cos \theta)}{2\theta}$ $I_z = \rho V R^2$	${\rho_x}^2 = {\rho_{x_c}}^2 = \frac{R^2(\theta - \sin \theta \cos \theta)}{2\theta}$ ${r_y}^2 = \frac{R^2(\theta + \sin \theta \cos \theta)}{2\theta}$ ${\rho_z}^2 = R^2$	$I_{x_c y_c}$, etc. $= 0$ I_{xy}, etc. $= 0$
3		$V = 2\pi SR$ $X_c = R$ $Y_c = R$ $Z_c = 0$	$I_{x_c} = I_{y_c} = \frac{\rho V}{2} R^2$ $I_{z_c} = \rho V R^2$	${\rho_{x_c}}^2 = {\rho_{y_c}}^2 = \frac{1}{2}R^2$ ${\rho_{z_c}}^2 = R^2$	$I_{x_c y_c}$, etc. $= 0$

4	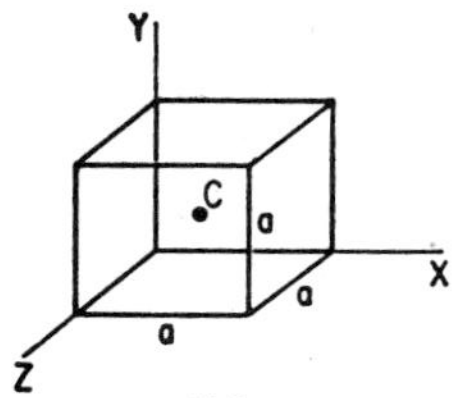 Cube	$V = a^3$ $X_c = \frac{1}{2}a$ $Y_c = \frac{1}{2}a$ $Z_c = \frac{1}{2}a$	$I_{x_c} = I_{y_c} = I_{z_c} = \frac{1}{6}\rho V a^2$	${\rho_{x_c}}^2 = {\rho_{y_c}}^2 = {\rho_{z_c}}^2 = \frac{1}{6}a^2$	$I_{x_c y_c}$, etc. $= 0$
5	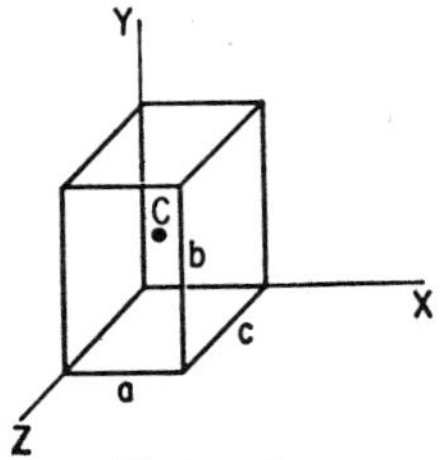 Rectangular prism	$V = abc$ $X_c = \frac{1}{2}a$ $Y_c = \frac{1}{2}b$ $Z_c = \frac{1}{2}c$	$I_{x_c} = \frac{1}{12}\rho V(b^2 + c^2)$	${\rho_{x_c}}^2 = \frac{1}{12}(b^2 + c^2)$	$I_{x_c y_c}$, etc. $= 0$
6	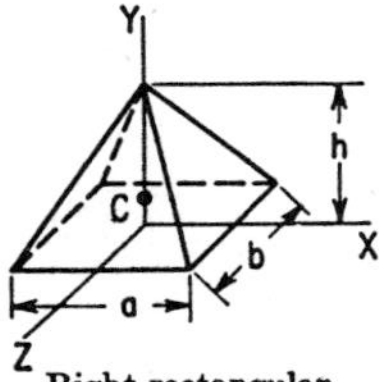 Right rectangular pyramid	$V = \frac{1}{3}abh$ $X_c = 0$ $Y_c = \frac{1}{4}h$ $Z_c = 0$	$I_{x_c} = \frac{1}{80}\rho V(4b^2 + 3h^2)$ $I_{y_c} = \frac{1}{20}\rho V(a^2 + b^2)$	${\rho_{x_c}}^2 = \frac{1}{80}(4b^2 + 3h^2)$ ${\rho_{y_c}}^2 = \frac{1}{20}(a^2 + b^2)$	$I_{x_c y_c}$, etc. $= 0$

Table 3.2. Properties of Homogeneous Solid Bodies (*Continued*)

	Solid body	Volume and centroid	Mass moment of inertia	Radius of gyration squared	Mass product of inertia
7	Right circular cone	$V = \frac{1}{3}\pi R^2 h$ $X_c = 0$ $Y_c = \frac{1}{4}h$ $Z_c = 0$	$I_{x_c} = I_{z_c} = \frac{3\rho V}{80}(4R^2 + h^2)$ $I_{y_c} = \frac{3}{10}\rho V R^2$	${\rho_{x_c}}^2 = {\rho_{z_c}}^2 = \frac{3}{80}(4R^2 + h^2)$ ${\rho_{y_c}}^2 = \frac{3}{10}R^2$	$I_{x_c y_c}$, etc. $= 0$
8	Right circular cylinder	$V = \pi R^2 h$ $X_c = 0$ $Y_c = \frac{1}{2}h$ $Z_c = 0$	$I_{x_c} = I_{z_c} = \frac{1}{12}\rho V(3R^2 + h^2)$ $I_{y_c} = \frac{1}{2}\rho V R^2$	${\rho_{x_c}}^2 = {\rho_{z_c}}^2 = \frac{1}{12}(3R^2 + h^2)$ ${\rho_{y_c}}^2 = \frac{1}{2}R^2$	$I_{x_c y_c}$, etc. $= 0$
9	Hollow right circular cylinder	$V = \pi h(R_1^2 - R_2^2)$ $X_c = 0$ $Y_c = \frac{1}{2}h$ $Z_c = 0$	$I_{x_c} = I_{z_c}$ $= \frac{1}{12}\rho V(3R_1^2 + 3R_2^2 + h^2)$ $I_{y_c} = \frac{1}{2}\rho V(R_1^2 + R_2^2)$	${\rho_{x_c}}^2 = {\rho_{z_c}}^2 = \frac{1}{12}(3R_1^2 + 3R_2^2 + h^2)$ ${\rho_{y_c}}^2 = \frac{1}{2}(R_1^2 + R_2^2)$	$I_{x_c y_c}$, etc. $= 0$

10 Sphere	$V = \frac{4}{3}\pi R^3$ $X_c = 0$ $Y_c = 0$ $Z_c = 0$	$I_{x_c} = \frac{2}{5}\rho V R^2$ $I_{y_c} = \frac{2}{5}\rho V R^2$ $I_{z_c} = \frac{2}{5}\rho V R^2$	${\rho_{x_c}}^2 = \frac{2}{5}R^2$ ${\rho_{y_c}}^2 = \frac{2}{5}R^2$ ${\rho_{z_c}}^2 = \frac{2}{5}R^2$	$I_{x_c y_c}$, etc. $= 0$
11 Hollow sphere	$V = \frac{4}{3}\pi({R_1}^3 - {R_2}^3)$ $X_c = 0$ $Y_c = 0$ $Z_c = 0$	$I_x = I_y = I_z$ $= \frac{2}{5}\rho V \frac{{R_1}^5 - {R_2}^5}{{R_1}^3 - {R_2}^3}$	${\rho_x}^2 = {\rho_y}^2 = {\rho_z}^2$ $= \frac{2}{5}\frac{{R_1}^5 - {R_2}^5}{{R_1}^3 - {R_2}^3}$	I_{xy}, etc. $= 0$
12 Hemisphere	$V = \frac{2}{3}\pi R^3$ $X_c = 0$ $Y_c = \frac{3}{8}R$ $Z_c = 0$	$I_x = I_y = I_z = \frac{2}{5}\rho V R^2$	${\rho_x}^2 = {\rho_y}^2 = {\rho_z}^2 = \frac{2}{5}R^2$	$I_{x_c y_c}$, etc. $= 0$ I_{xy}, etc. $= 0$

Table 3.2. Properties of Homogeneous Solid Bodies (*Continued*)

	Solid body	Volume and centroid	Mass moment of inertia	Radius of gyration squared	Mass product of inertia
13	Ellipsoid	$V = \frac{4}{3}\pi abc$ $X_c = 0$ $Y_c = 0$ $Z_c = 0$	$I_x = \frac{1}{5}\rho V(b^2 + c^2)$ $I_y = \frac{1}{5}\rho V(a^2 + c^2)$ $I_z = \frac{1}{5}\rho V(a^2 + b^2)$	$\rho_x^2 = \frac{1}{5}(b^2 + c^2)$ $\rho_y^2 = \frac{1}{5}(a^2 + c^2)$ $\rho_z^2 = \frac{1}{5}(a^2 + b^2)$	I_{xy}, etc. $= 0$
14	Paraboloid of revolution	$V = \frac{1}{2}\pi R^2 h$ $X_c = \frac{2}{3}h$ $Y_c = 0$ $Z_c = 0$	$I_{x_c} = \frac{1}{3}\rho V R^2$ $I_{y_c} = I_{z_c} = \frac{1}{18}\rho V(3R^2 + h^2)$	$\rho_{x_c}^2 = \frac{1}{3}R^2$ $\rho_{y_c}^2 = \rho_{z_c}^2 = \frac{1}{18}(3R^2 + h^2)$	$I_{x_c y_c}$, etc. $= 0$
15	Elliptic paraboloid	$V = \frac{1}{2}\pi abc$ $X_c = \frac{2}{3}a$ $Y_c = 0$ $Z_c = 0$	$I_{x_c} = \frac{1}{6}\rho V(b^2 + c^2)$ $I_{y_c} = \frac{1}{18}\rho V(3c^2 + a^2)$ $I_{z_c} = \frac{1}{18}\rho V(3b^2 + a^2)$	$\rho_{x_c}^2 = \frac{1}{6}(b^2 + c^2)$ $\rho_{y_c}^2 = \frac{1}{18}(3c^2 + a^2)$ $\rho_{z_c}^2 = \frac{1}{18}(3b^2 + a^2)$	$I_{x_c y_c}$, etc. $= 0$

4

NONLINEAR VIBRATION

H. Norman Abramson
Southwest Research Institute

INTRODUCTION

A vast body of scientific knowledge has been developed over a long period of time devoted to a description of natural phenomena. In the field of mechanics, rapid progress in the past two centuries has occurred, due in large measure to the ability of investigators to represent physical laws in terms of rather simple equations. In many cases the governing equations were not so simple; therefore, certain assumptions, more or less consistent with the physical situation, were employed to reduce the equations to types more easily soluble. Thus, the process of linearization has become an intrinsic part of the rational analysis of physical problems. An analysis based on linearized equations, then, may be thought of as an analysis of a corresponding but idealized problem.

In many instances the linear analysis is insufficient to describe the behavior of the physical system adequately. In fact, one of the most fascinating features of a study of nonlinear problems is the occurrence of new and totally unsuspected phenomena; i.e., new in the sense that the phenomena are not predicted, or even hinted at, by the linear theory. On the other hand, certain phenomena observed physically are unexplainable except by giving due consideration to nonlinearities present in the system.

The branch of mechanics that has been subjected to the most intensive attack from the nonlinear viewpoint is the theory of vibration of mechanical and electrical systems. Other branches of mechanics, such as incompressible and compressible fluid flow, elasticity, plasticity, wave propagation, etc., also have been studied as nonlinear problems, but the greatest progress has been made in treating vibration of nonlinear systems. The systems treated in this chapter are systems with a finite number of degrees-of-freedom which can be defined by a finite number of simultaneous ordinary differential equations; on the other hand, the mechanics of continua involves partial differential equations. Nonlinear ordinary differential equations are easier to handle than nonlinear partial differential equations. Interesting surveys of the entire realm of nonlinear mechanics are given in Refs. 1 and 2.

This chapter provides information concerning features of nonlinear vibration theory likely to be encountered in practice, and methods of nonlinear vibration analysis which find ready application.

EXAMPLES OF SYSTEMS POSSESSING NONLINEAR CHARACTERISTICS

SIMPLE PENDULUM

As a first example of a system possessing nonlinear characteristics, consider a simple pendulum of length l having a bob of mass m, as shown in Fig. 4.1. The well-known differential equation governing free vibration is

$$ml^2\ddot{\theta} + mgl\theta = 0 \tag{4.1}$$

This equation holds only for small oscillations about the position of equilibrium since the actual restoring moment is characterized by the quantity $\sin \theta$. Equation (4.1) thus employs the assumption $\sin \theta \simeq \theta$. The exact, but nonlinear, equation of motion is

$$ml^2\ddot{\theta} + mgl \sin \theta = 0 \tag{4.2}$$

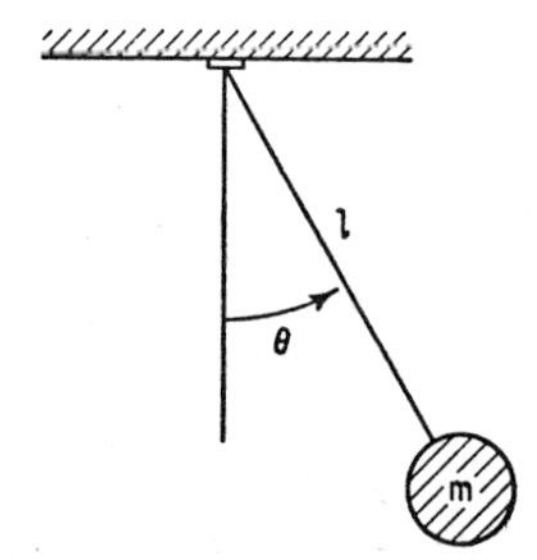

FIG. 4.1. Simple pendulum.

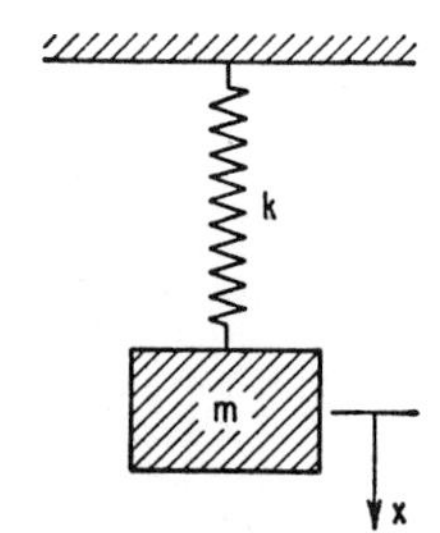

FIG. 4.2. Simple spring-mass system.

SIMPLE SPRING-MASS SYSTEM

A simple spring-mass system, as shown in Fig. 4.2, is characterized by the equation

$$m\ddot{x} + kx = 0$$

This equation is based on the assumption that the elastic spring obeys Hooke's law; i.e., the characteristic curve of restoring force versus displacement is a straight line. However, many materials do not exhibit such a linear characteristic. Further, in the case of a simple coil spring, a deviation from linearity occurs at large compression as the coils begin to close up, or conversely, when the extension becomes so great that the coils begin to lose their individual identity. In either case, the spring exhibits a characteristic such that the restoring force increases more rapidly than the displacement. Such a characteristic is called *hardening*. In a similar manner, certain systems (e.g., a simple pendulum) exhibit a *softening* characteristic. Both types of characteristic are shown in Fig. 4.3. A simple system with either softening or hardening restoring force may be described approximately by an

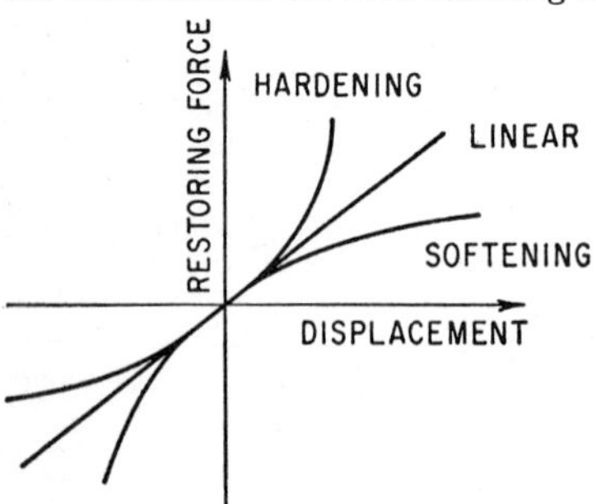

FIG. 4.3. Restoring force characteristic curves for linear, hardening, and softening vibration systems.

equation of the form

$$m\ddot{x} + k(x \pm \mu^2 x^3) = 0$$

where the upper sign refers to the hardening characteristic and the lower to the softening characteristic.

It is possible for a system with only linear components to exhibit nonlinear characteristics, by snubber action for example, as shown in Fig. 4.4. A system undergoing vibration of small amplitude also may exhibit nonlinear characteristics; for example, in the pendulum shown in Fig. 4.5, the length depends on the amplitude.

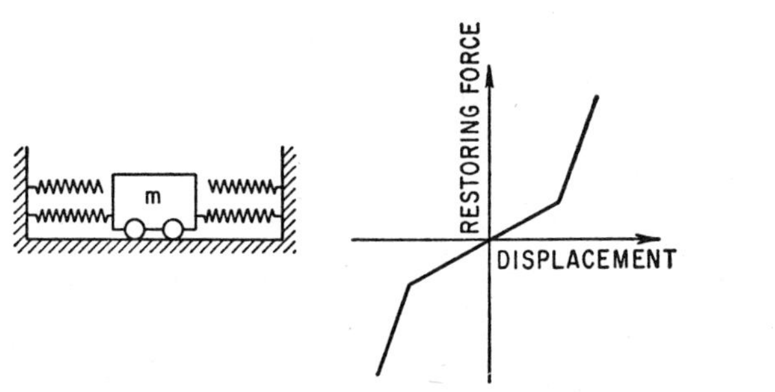

FIG. 4.4. Nonlinear mechanical system with snubber action showing piecewise linear restoring force characteristic curve.

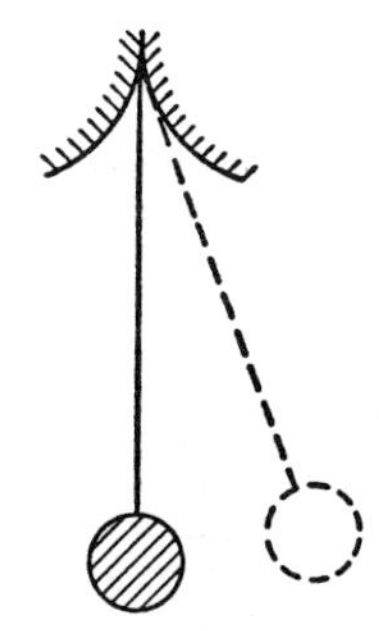

FIG. 4.5. Pendulum with nonlinear characteristic resulting from dependence of length on vibration amplitude.

STRETCHED STRING WITH CONCENTRATED MASS

The large amplitude vibration of a stretched string with a concentrated mass, as shown in Fig. 4.6, offers another example of a nonlinear system. The governing nonlinear differential equation is, approximately,

$$m\ddot{w} + F_0\left(\frac{l}{ab}\right)w + (SE - F_0)\left(\frac{a^3 + b^3}{2a^3b^3}\right)w^3 = 0$$

where F_0 is the initial tension, S is the cross-sectional area, and E is the elastic modulus of the string. Consider now the special case of $a = b$ and denote the unstretched length

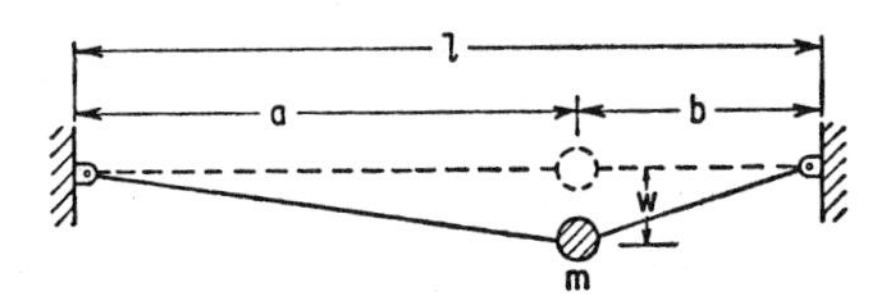

FIG. 4.6. Vibration of a weighted string as an example of a nonlinear system.

of the half string by l_0. Then the initial tension and the restoring force become

$$F_0 = SE\left(\frac{a - l_0}{l_0}\right)$$

$$F_r \simeq SE\left[2\left(\frac{a}{l_0} - 1\right)\left(\frac{w}{a}\right) + \left(2 - \frac{a}{l_0}\right)\left(\frac{w}{a}\right)^3\right]$$

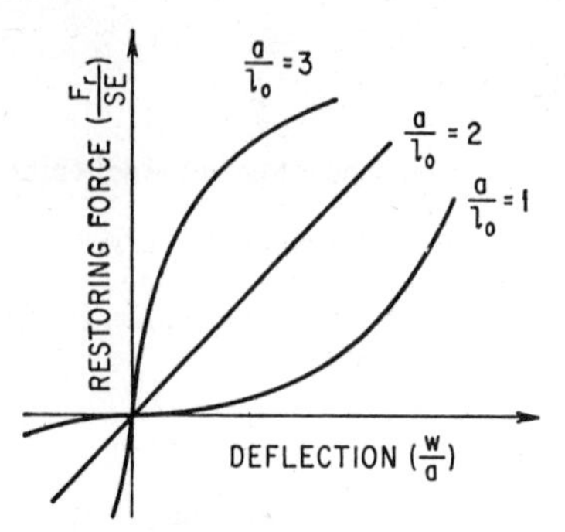

FIG. 4.7. Restoring force characteristics for the weighted string shown in Fig. 4.6.

An interesting feature of this system is that it exhibits a wide variety of either hardening or softening characteristics, depending upon the value of a/l_0, as shown in Fig. 4.7.

SYSTEM WITH VISCOUS DAMPING

The foregoing examples all involve nonlinearities in the elastic components, either as a result of appreciable amplitudes of vibration or as a result of peculiarities of the elastic element. Consider a simple spring-mass system which also includes a dashpot. The usual assumptions pertaining to this system are that the spring is linear and that the motion is sufficiently slow that the viscous resistance provided by the dashpot is proportional to the velocity; therefore, the governing equation of motion is linear. Frequently, the dashpot resistance is more correctly expressed by a term proportional to the square of the velocity. Further, the resistance is always such as to oppose the motion; therefore, the nonlinear equation of motion may be written

$$m\ddot{x} + c|\dot{x}|\dot{x} + kx = 0$$

BELT FRICTION SYSTEM

The system shown in Fig. 4.8 involves a nonlinearity depending upon the dry friction between the mass and the moving belt. The belt has a constant speed v_0, and the applicable equation of motion is

$$m\ddot{x} + F(\dot{x}) + kx = 0$$

where the friction force $F(\dot{x})$ is shown in Fig. 4.9. For large values of displacement, the damping term is positive, has positive slope, and removes energy from the system; for small values of displacement, the damping term is negative, has negative slope, and actually puts energy into the system. Even though there is no external stimulus, the system can have an oscillatory solution, and thus corresponds to a nonlinear *self-excited* system.

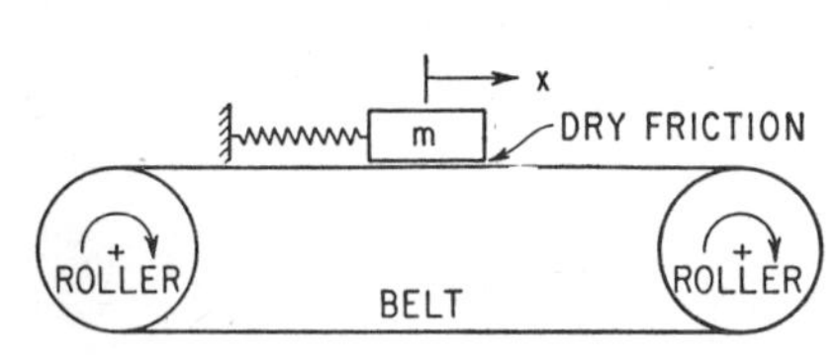

FIG. 4.8. Belt friction system which exhibits self-excited vibration.

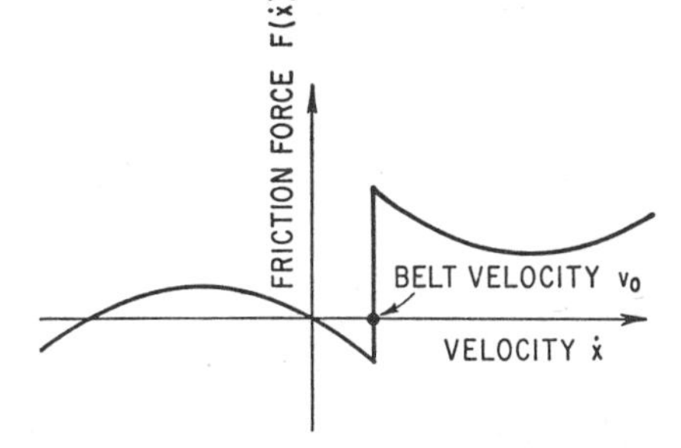

FIG. 4.9. Damping force characteristic curve for the belt friction system shown in Fig. 4.8.

Many other examples of nonlinear systems are given in the references of this chapter, particularly Refs. 3 to 5.

DESCRIPTION OF NONLINEAR PHENOMENA

This section describes briefly, largely in nonmathematical terms, certain of the more important features of nonlinear vibration. Further details and methods of analysis are given later.

FREE VIBRATION

In so far as the free vibration of a system is concerned, one distinguishing feature between linear and nonlinear behavior is the dependence of the period of the motion in nonlinear vibration on the amplitude. For example, the simple pendulum of Fig. 4.1 may be analyzed on the basis of the linearized equation of motion, Eq. (4.1), from which it is found that the period of the vibration is given by the constant value $\tau_0 = 2\pi/\omega_n$. An analysis on the basis of the nonlinear equation of motion, Eq. (4.2), leads to an expression for the period of the form

$$\frac{\tau}{\tau_0} = 1 + \tfrac{1}{4}(U)^2 + \tfrac{9}{64}(U)^4 + \tfrac{25}{256}(U)^6 + \cdots \tag{4.3}$$

where U is related to the amplitude of the vibration Θ by the relation $U = \sin(\Theta/2)$. The linear solution thus corresponds to the first term of Eq. (4.3). The dependence of

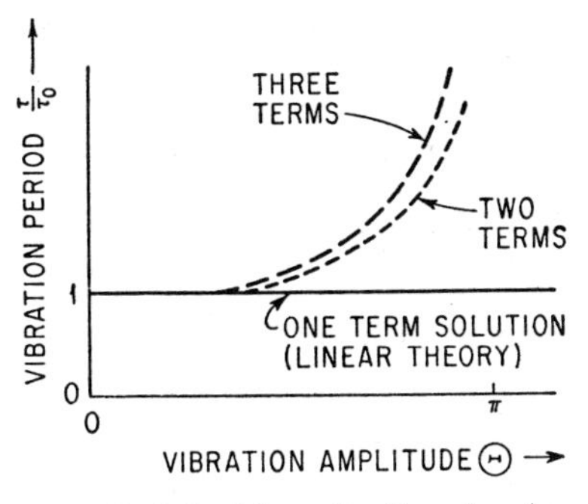

FIG. 4.10. Period of free vibration of a simple pendulum according to Eq. (4.3) and showing the effect of nonlinear terms.

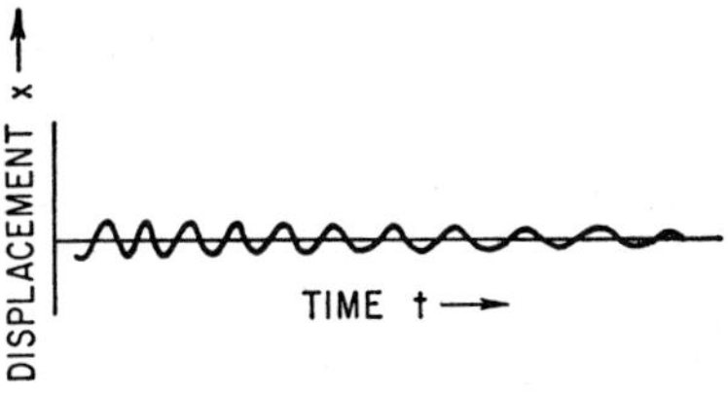

FIG. 4.11. Deflection time-history for free damped vibration of the nonlinear system described by Duffing's equation [Eq. (4.16)].

the period of vibration on amplitude is shown in Fig. 4.10. Systems in which the period of vibration is independent of the amplitude are called *isochronous*, while those in which the period τ is dependent on the amplitude are called *nonisochronous*.

The dependence of period on amplitude also may be seen from the vibration trace shown in Fig. 4.11, which corresponds to a solution of the equation

$$m\ddot{x} + c\dot{x} + k(x + \mu^2 x^3) = 0$$

CHARACTER OF THE RESPONSE CURVES FOR FORCED VIBRATION

Representations of vibration behavior in the form of curves of response amplitude versus exciting frequency are called *response curves*. The response curves for an undamped linear system acted on by a harmonic exciting force of amplitude p and frequency ω may be derived from the equation of motion

$$\ddot{x} + \omega_n^2 x = \frac{p}{m} \cos \omega t \tag{4.4}$$

The solution has the form shown in Fig. 4.12. The vertical line at $\omega = \omega_n$ corresponds not only to resonance, but also to free vibration ($p = 0$); the amplitude in this instance is determined by the initial conditions of the motion. In a nonlinear system the character of the motion is dependent upon the amplitude. This requires that the natural frequency likewise be amplitude-dependent; hence, it follows that the free vibration curve $p = 0$ for nonlinear systems cannot be a straight line. Figure 4.13 shows free vibration

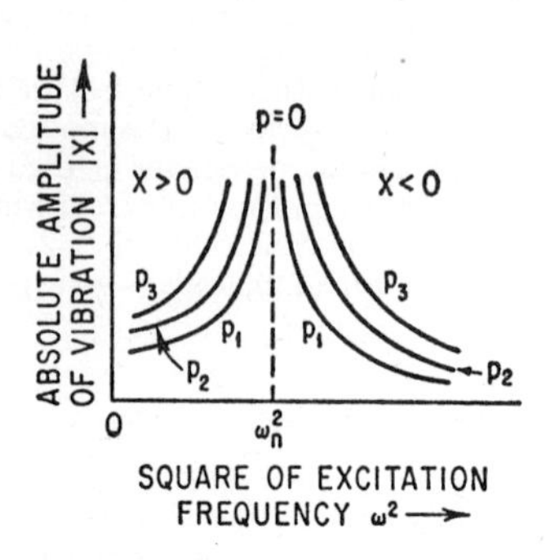

Fig. 4.12. Family of response curves for the undamped linear system defined by Eq. (4.4).

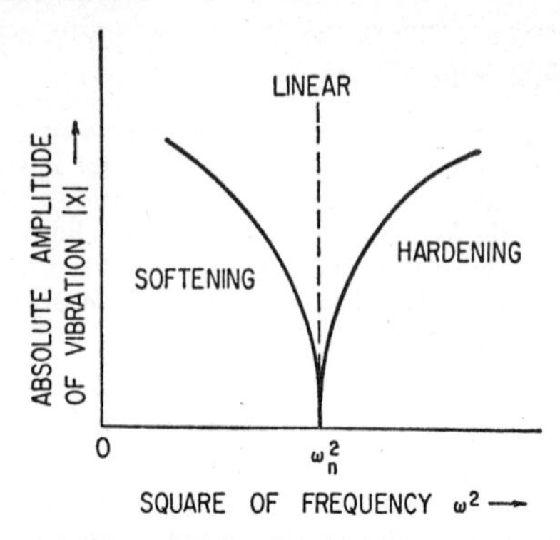

Fig. 4.13. Free vibration curves (natural frequency as a function of amplitude) in the response diagram for linear, hardening, and softening vibration systems [see Eq. (4.49)].

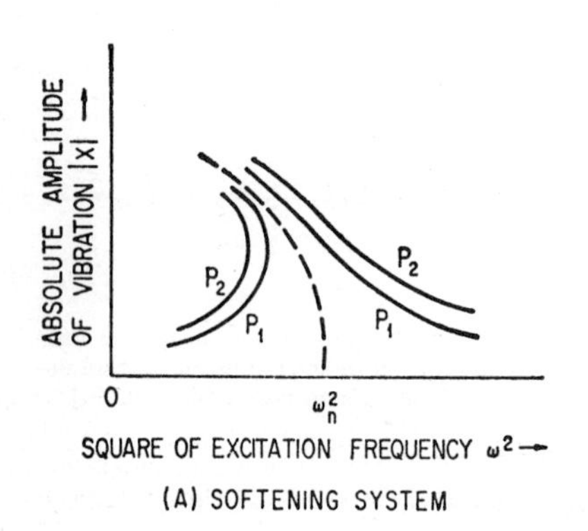

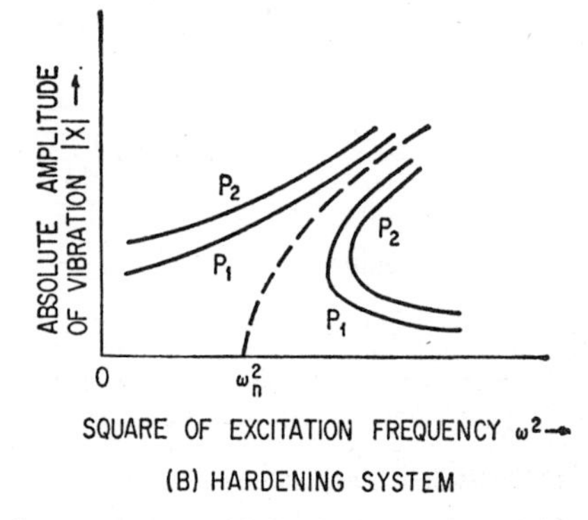

Fig. 4.14. Response curves for undamped nonlinear systems with hardening and softening restoring force characteristics [see Eq. (4.50)].

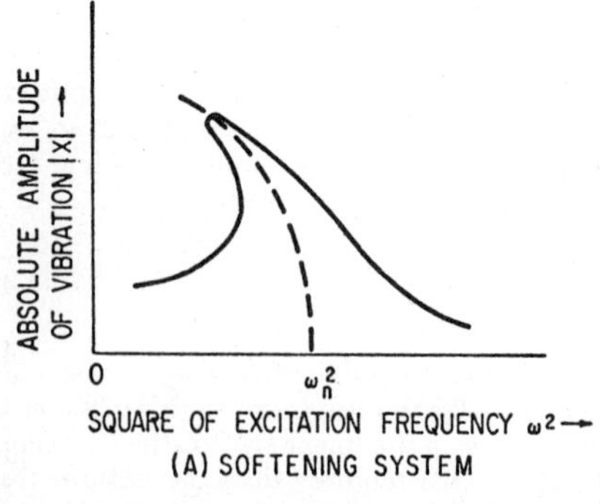

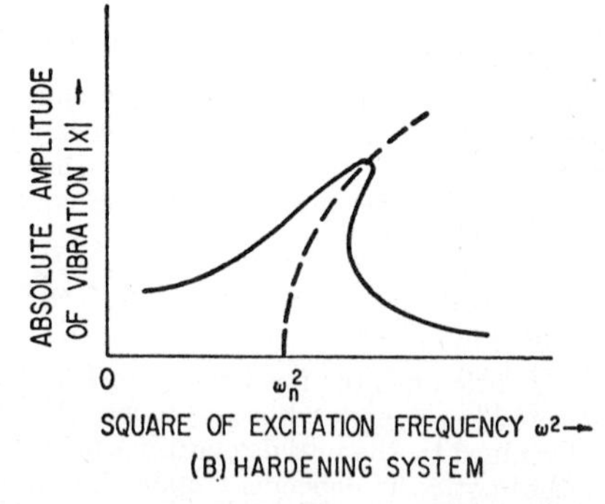

Fig. 4.15. Response curves for damped nonlinear systems with hardening and softening restoring force characteristics [see Eq. (4.52)].

curves (i.e., natural frequency as a function of amplitude) for hardening and softening systems.

Figures 4.12 and 4.13 suggest that the forced vibration response curves for systems with nonlinear restoring forces have the general form of those of a linear system, but are "swept over" to the right or left, depending on whether the system is hardening or softening. These are shown in Fig. 4.14. The principal effect of damping in forced vibration of a nonlinear system is to limit the amplitude at resonance, as shown in Fig. 4.15.

STABILITY

Consider a hardening system whose response curve is shown in Fig. 4.15*B*. Suppose that the exciting frequency starts at a low value, and increases continuously at a slow rate. The amplitude of the vibration also increases, but only up to a point. In particular, at

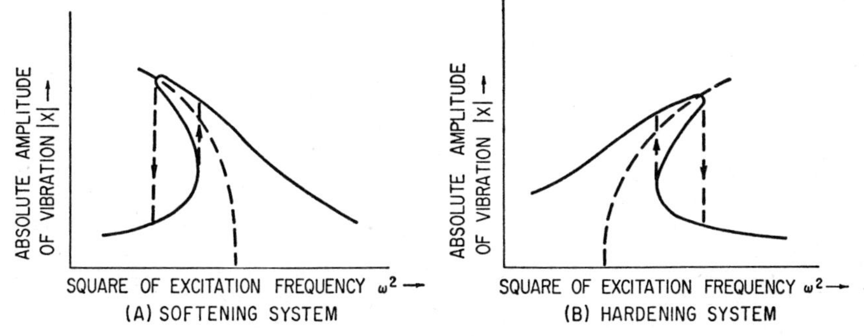

Fig. 4.16. Jump phenomenon in hardening and softening systems.

the point of vertical tangency of the response curve, a slight increase in frequency requires that the system perform in an unusual manner; i.e., that it "jump" down in amplitude to the lower branch of the response curve. This experiment may be repeated by starting with a large value of exciting frequency, but requiring that the forcing frequency be continuously reduced. A similar situation again is encountered; the system must jump up in amplitude in order to meet the conditions of the experiment. This *jump phenomenon* is shown in Fig. 4.16 for both the hardening and softening systems. The jump is not instantaneous in time, but requires a few cycles of vibration to establish a steady-state vibration at the new amplitude.

There is a portion of the response curve which is "unattainable"; it is not possible to obtain that particular amplitude by a suitable choice of forcing frequency. Thus, for certain values of ω there appear to be three possible amplitudes of vibration but only the upper and lower can actually exist. If by some means it were possible to initiate a steady-state vibration with just the proper amplitude and frequency to correspond to the middle branch, the condition would be unstable; at the slightest disturbance the motion would jump to either of the other two states of motion. The direction of the jump depends on the direction of the disturbance. Thus, of the three possible states of motion, one in phase and two out-of-phase with the exciting force, the one having the larger amplitude of the two out-of-phase motions is

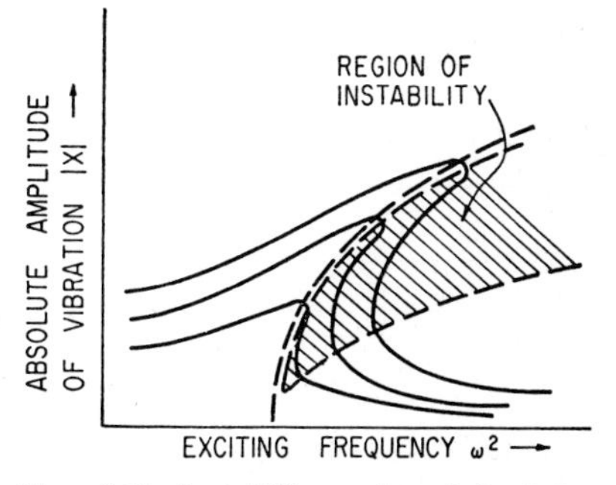

Fig. 4.17. Instability region defined by the loci of vertical tangents of the damped response curves for the hardening system.

unstable. This region of instability in the response diagram is defined by the loci of vertical tangents to the response curves, and is shown for a hardening system in Fig. 4.17.

SUBHARMONIC AND SUPERHARMONIC VIBRATIONS

In the preceding discussion, only the harmonic solutions of nonlinear differential equations were considered; i.e., the solutions described were those whose frequency is the same as that of the exciting force $F \cos \omega t$. However, permanent oscillations whose frequency is a fraction ($\frac{1}{2}$, $\frac{1}{3}$, ..., $1/n$) of that of the exciting force also can occur in nonlinear systems. The term *subharmonic vibration* usually is applied to this phenomenon, although the term *frequency demultiplication* is also used. In a similar manner, *superharmonic vibration* of frequency equal to some multiple (2, 3, ..., n) of the exciting frequency also can occur.

Although subharmonic and superharmonic oscillations occur in nonlinear systems, it is not a simple matter to give a plausible physical explanation for their occurrence. In a linear system, if the frequency of the free vibration is ω_n, then a periodic external force of frequency ω can excite free oscillation in addition to steady-state forced vibration of frequency ω. But since every physical system possesses some damping, the free vibration eventually disappears, leaving only the steady-state forced vibration. In a nonlinear system, the free vibration contains many higher harmonics; hence, it is possible that an exciting force with a frequency corresponding to one of these higher harmonics might be able to excite and sustain that particular harmonic component of the free vibration, in addition to the normal forced vibration.

Two essential differences between subharmonic and superharmonic oscillations may be noted: (1) subharmonics may occur under very special conditions while superharmonics always occur; (2) damping only diminishes the amplitudes of superharmonic vibrations, but may completely prohibit the existence of subharmonic vibrations if it is greater than a certain amount. Subharmonic and superharmonic vibrations are discussed analytically later in this chapter. Jump phenomena also can exist in both superharmonic and subharmonic vibrations.

OTHER PHENOMENA

SELF-EXCITED VIBRATION.* Consider the nonlinear equation of motion

$$m\ddot{x} + c(x^2 - 1)\dot{x} + kx = 0$$

This is known as Van der Pol's equation and may be written alternatively

$$\ddot{x} - \epsilon(1 - x^2)\dot{x} + \kappa^2 x = 0 \tag{4.5}$$

The principal feature of this self-excited system resides in the damping term; for small displacements the damping is negative, and for large displacements the damping is positive. Thus, even an infinitesimal disturbance causes the system to oscillate; however, when the displacement becomes sufficiently large, the damping becomes positive and limits further increase in amplitude. This is shown in Fig. 4.18. Such systems, which start in a spontaneous manner, often are called "soft" systems in contrast with "hard" systems which exhibit sustained oscillations only if a shock in excess of a certain level is applied. Note that stability questions arise here (which are different from those discussed earlier in connection with jump phenomena) concerning the existence of one or more limiting amplitudes, such as the one noted above in the Van der Pol oscillator.

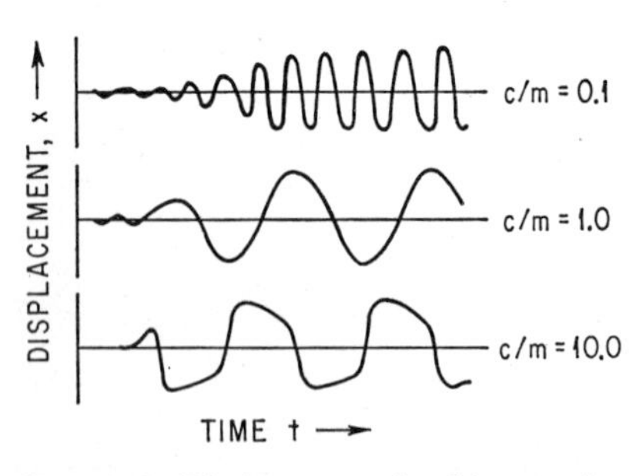

FIG. 4.18. Displacement time-histories for Van der Pol's equation [Eq. (4.5)] for various values of the damping.

* A general treatment of self-excited vibration is given in Chap. 5.

RELAXATION OSCILLATIONS. As shown in Fig. 4.18, the motion of the Van der Pol oscillator is very nearly harmonic for $c/m = 0.1$ while the motion is made up of relatively sudden transitions between deflections of opposite sign for $c/m = 10.0$. The period of the harmonic motion for $c/m = 0.1$ is determined essentially by the linear spring stiffness k and the mass m; the period of the motion corresponding to $c/m = 10.0$ is very much larger and depends also on c. Thus, it is possible to obtain an undamped periodic oscillation in a damped system as a result of the particular behavior of the damping term. Such oscillations are often called *relaxation oscillations.*

ASYNCHRONOUS EXCITATION AND QUENCHING. In linear systems, the principle of superposition is valid, and there is no interaction between different oscillations. Moreover, the mathematical existence of a periodic solution always indicates the existence of a periodic phenomenon. In nonlinear systems, there is an interaction between oscillations; the mathematical existence of a periodic solution is only a necessary condition for the existence of corresponding physical phenomena. When supplemented by the condition of stability, the conditions become both necessary and sufficient for the appearance of the physical oscillation. Therefore, it is conceivable that under these conditions the appearance of one oscillation may either create or destroy the stability condition for another oscillation. In the first case, the other oscillation appears (*asynchronous excitation*), and in the second case, disappears (*asynchronous quenching*). The term asynchronous is used to indicate that there is no relation between the frequencies of these two oscillations.

ENTRAINMENT OF FREQUENCY. According to linear theory, if two frequencies ω_1 and ω_2 are caused to beat in a system, the period of beating increases indefinitely as ω_2 approaches ω_1. In nonlinear systems, the beats disappear as ω_2 reaches certain values. Thus, the frequency ω_1 falls in synchronism with, or is entrained by, the frequency ω_2 within a certain range of values. This is called *entrainment of frequency,* and the band of frequencies in which entrainment occurs is called the zone of entrainment or the interval of synchronization. In this region, the frequencies ω_1 and ω_2 combine and only vibration at a single frequency ensues.

EXACT SOLUTIONS

It is possible to obtain exact solutions for only a relatively few second-order nonlinear differential equations. In this section, some of the more important of these exact solutions are listed. They are exact in the sense that the solution is given either in closed form or in an expression that can be evaluated numerically to any desired degree of accuracy. The examples given here are fairly general in nature; more specialized examples are given in Ref. 4.

FREE VIBRATION

Consider the free vibration of an undamped system with a general restoring force $f(x)$ as governed by the differential equation

$$\ddot{x} + \kappa^2 f(x) = 0$$

This can be rewritten as

$$\frac{d(\dot{x}^2)}{dx} + 2\kappa^2 f(x) = 0 \tag{4.6}$$

and integrated to yield

$$\dot{x}^2 = 2\kappa^2 \int_x^X f(\xi)\, d\xi$$

where ξ is an integration variable and X is the value of the displacement when $\dot{x} = 0$. Thus

$$|\dot{x}| = \kappa\sqrt{2}\sqrt{\int_x^X f(\xi)\, d\xi}$$

This may be integrated again to yield

$$t - t_0 = \frac{1}{\kappa\sqrt{2}} \int_0^x \frac{d\zeta}{\sqrt{\int_\zeta^X f(\xi)\, d\xi}} \tag{4.7}$$

where ζ is an integration variable and t_0 corresponds to the time when $x = 0$. The displacement-time relation may be obtained by inverting this result. Considering the restoring force term to be an odd function, i.e.,

$$f(-x) = -f(x)$$

and considering Eq. (4.7) to apply to the time from zero displacement to maximum displacement, the period τ of the vibration is

$$\tau = \frac{4}{\kappa\sqrt{2}} \int_0^X \frac{d\zeta}{\sqrt{\int_\zeta^X f(\xi)\, d\xi}} \tag{4.8}$$

Exact solutions can be obtained in all cases where the integrals in Eq. (4.8) can be expressed explicitly in terms of X.

CASE 1. PURE POWERS OF DISPLACEMENT. Consider the restoring force function

$$f(x) = x^n$$

Equation (4.8) then becomes

$$\tau = \frac{4}{\kappa}\sqrt{\frac{n+1}{2}} \int_0^X \frac{d\zeta}{\sqrt{X^{n+1} - \zeta^{n+1}}}$$

Setting $u = \zeta/X$,

$$\tau = \frac{4}{\kappa\sqrt{X^{n-1}}} \left(\sqrt{\frac{n+1}{2}} \int_0^1 \frac{du}{\sqrt{1 - u^{n+1}}} \right)$$

The expression within the parentheses depends only on the parameter n and is denoted by $\psi(n)$. Thus

$$\tau = \frac{4}{\kappa\sqrt{X^{n-1}}}\psi(n) \tag{4.9}$$

The factor $\psi(n)$ may be evaluated numerically to any desired degree of accuracy, and is tabulated in Table 4.1.

CASE 2. POLYNOMIALS OF DISPLACEMENT. Consider the binomial restoring force

$$f(x) = x^n + \mu x^m \qquad [m > n \geq 0]$$

Introducing this expression into Eq. (4.8) and performing the integrations:[5]

$$\tau = \frac{4}{\kappa\sqrt{X^{n+1}}} \left(\sqrt{\frac{n+1}{2}} \int_0^1 \frac{du}{\sqrt{(1 + \bar{\mu}) - (u^{n+1} + \bar{\mu} u^{m+1})}} \right) \tag{4.10}$$

where

$$\bar{\mu} = \mu X^{m-n} \left(\frac{n+1}{m+1} \right) \tag{4.11}$$

For particular values of n, m, and $\bar{\mu}$, the expression within the parentheses can be evaluated to any desired degree of accuracy by numerical methods. The extension of this method to higher-order polynomials can be made quite readily.

CASE 3. HARMONIC FUNCTION OF DISPLACEMENT. Consider now the problem of the simple pendulum which has a restoring force of the form

$$f(x) = \sin x$$

Introducing this relation into Eq. (4.7):

$$t - t_0 = \frac{1}{2\kappa}\int_0^x \frac{d\zeta}{\sqrt{\sin^2 \frac{X}{2} - \sin^2 \frac{\zeta}{2}}}$$

If $x = X$ and $t_0 = 0$, this integral can be reduced to the standard form of the complete elliptic integral of the first kind:

$$\hat{K}(\alpha) = \int_0^{\pi/2} \frac{dv}{\sqrt{1 - \sin^2 \alpha \sin^2 v}} \tag{4.12}$$

Thus, the period of vibration is

$$\tau = \frac{1}{\kappa}\hat{K}\left(\frac{X}{2}\right) \tag{4.13}$$

The displacement-time function can be obtained by inversion and leads to the inverse elliptic functions. Replacing $\sin \alpha$ by U in Eq. (4.12), expanding by the binomial theorem, and then integrating yields Eq. (4.3).

CASE 4. VELOCITY SQUARED DAMPING. As indicated by Eq. (4.6), the introduction of any other function of $\dot{x}^2$ does not complicate the problem. Thus, the differential equation *

$$\ddot{x} \pm \frac{\delta}{2}\dot{x}^2 + \kappa^2 f(x) = 0$$

can be reduced to

$$\frac{d(\dot{x}^2)}{dx} \pm \delta\dot{x}^2 = 2\kappa^2 f(x)$$

Integrating the above equation,

$$\dot{x}^2 = 2\kappa^2 e^{\mp\delta x}\int_x^X e^{\pm\delta\xi} f(\xi)\, d\xi$$

Integrating again,

$$t = \int_{x_0}^x \frac{d\eta}{\dot{x}(\eta)}$$

where η is an integration variable.

FORCED VIBRATION

Exact solutions for forced vibration of nonlinear systems are virtually nonexistent, except as the system can be represented in a stepwise linear manner. For example, consider a system with a stepwise linear symmetrical restoring force characteristic, as shown in Fig. 4.4. Denote the lower of the two stiffnesses by k_1, the upper by k_2, and the dis-

* The $\pm$ sign is employed here, and elsewhere in this chapter, to account for the proper direction of the resisting force. Consequently, reference frequently is made to upper or lower sign rather than to plus or minus.

placement at which the change in stiffness occurs by x_1. Thus, the problem reduces to the solution of two linear differential equations:

$$m\ddot{x}' + k_1 x' = \pm P \sin \omega t \qquad [x_1 \geq x' \geq 0] \tag{4.14a}$$

$$m\ddot{x}'' + (k_1 - k_2)x_1 + k_2 x'' = \pm P \sin \omega t \qquad [x'' \geq x_1] \tag{4.14b}$$

where the upper sign refers to in-phase exciting force and the lower sign to out-of-phase exciting force. The appropriate boundary conditions are

$$\begin{aligned} x'(t = 0) &= 0 \\ x'(t = t_1) &= x''(t = t_1) = x_1 \\ \dot{x}'(t = t_1) &= \dot{x}''(t = t_1) \\ \dot{x}''\left(t = \frac{\pi}{2\omega}\right) &= 0 \end{aligned} \tag{4.15}$$

The solutions of Eqs. (4.14) are

$$x' = \frac{\pm P/k_1}{1 - \omega^2/\omega_1^2} \sin \omega t + A_1 \cos \omega_1 t + B_1 \sin \omega_1 t$$

$$x'' = \frac{\pm P/k_2}{1 - \omega^2/\omega_2^2} \sin \omega t + A_2 \cos \omega_2 t + B_2 \sin \omega_2 t + \left(1 - \frac{k_1}{k_2}\right) x_1$$

where $\omega_1^2 = k_1/m$, $\omega_2^2 = k_2/m$, and the constants A_1, A_2, B_1, B_2 may be evaluated from the boundary conditions, Eq. (4.15).

This analysis also applies to the case of free vibration by setting $P = 0$. By assigning various values to k_1 and k_2, a wide variety of specific problems may be treated; a collection of such solutions is given in Refs. 7 and 8. It is not necessary to restrict the restoring forces to odd functions.

APPROXIMATE METHODS OF NONLINEAR VIBRATION ANALYSIS

Any second-order ordinary differential equation can be integrated numerically in a stepwise manner to yield a time-history of the motion. Known methods of stepwise integration [9] may be employed for this purpose and result in a solution of any desired degree of accuracy, depending on the size of the steps. Numerical methods apply only to specific equations, and are not useful for general studies of the behavior of nonlinear vibrating systems.

A large number of approximate analytical methods of nonlinear vibration analysis exist, each of which may or may not possess advantages for certain classes of problems. Some of these are restricted techniques which may work well with some types of equations, but not with others. The methods which are outlined below are among the better known and possess certain advantages as to ranges of applicability.[4–21]

DUFFING'S METHOD

Consider the nonlinear differential equation

$$\ddot{x} + \kappa^2(x \pm \mu^2 x^3) = p \cos \omega t \tag{4.16}$$

where the $\pm$ sign indicates either a hardening or softening system.* As a first approximation to a harmonic solution, assume that

$$x_1 = A \cos \omega t \tag{4.17}$$

* Equation (4.16) is known as Duffing's equation.

and rewrite Eq. (4.16) to obtain an equation for the second approximation:

$$\ddot{x}_2 = -(\kappa^2 A \pm \tfrac{3}{4}\kappa^2\mu^2 A^3 - p)\cos\omega t - \tfrac{1}{4}\kappa^2\mu^2 A^3 \cos\omega t$$

This equation may now be integrated to yield

$$x_2 = \frac{1}{\omega^2}(\kappa^2 A \pm \tfrac{3}{4}\kappa^2\mu^2 A^3 - p)\cos\omega t + \tfrac{1}{36}\kappa^2\mu^2 A^3 \cos 3\omega t \tag{4.18}$$

where the constants of integration have been taken as zero to insure periodicity of the solution.

This may be regarded as an iteration procedure by reinserting each successive approximation into Eq. (4.16) and obtaining a new approximation. For this iteration procedure to be convergent, the nonlinearity must be small; i.e., κ^2, μ^2, A, and p must be small quantities. This restricts the study to motions in the neighborhood of linear vibration (but not near $\omega = \kappa$, since A would then be large); thus, Eq. (4.17) must represent a reasonable first approximation. It follows that the coefficient of the cos ωt term in Eq. (4.18) must be a good second approximation, and should not be far different from the first approximation.[22] Since this procedure furnishes the exact result in the linear case, it might be expected to yield good results for the "slightly nonlinear" case. Thus, a relation between frequency and amplitude is found by equating the coefficients of the first and second approximations:

$$\omega^2 = \kappa^2(1 \pm \tfrac{3}{4}\mu^2 A^2) - \frac{p}{A} \tag{4.19}$$

This relation describes the response curves, as shown in Fig. 4.14.

The above method applies equally well when linear velocity damping is included. Further details concerning this method, and an analysis of its applicability when only μ^2 and p are considered small, may be found in Ref. 10.

RAUSCHER'S METHOD [23]

Duffing's method considered above is based on the idea of starting the iteration procedure from the linear vibration. More rapid convergence might be expected if the approximations were to begin with free nonlinear vibration; Rauscher's method [23] is based on this idea.

Consider a system with general restoring force described by the differential equation

$$\omega^2 x'' + \kappa^2 f(x) = p \cos\omega t \tag{4.20}$$

where primes denote differentiation with respect to ωt, and $f(x)$ is an odd function. Assume that the conditions at time $t = 0$ are $x(0) = A$, $x'(0) = 0$. Start with the free nonlinear vibration as a first approximation, i.e., with the solution of the equation

$$\omega_0^2 x'' + \kappa^2 f(x) = 0 \tag{4.21}$$

such that $x = x_0(\phi)$ (where $\omega t = \phi$) has the period 2π and $x_0(0) = A$, $x_0'(0) = 0$. Equation (4.21) may be solved exactly in the form of quadratures according to Eq. (4.7):

$$\phi = \phi_0(x) = \frac{\omega_0}{\kappa\sqrt{2}} \int_A^x \frac{d\zeta}{\sqrt{\int_\zeta^A f(\xi)\, d\xi}} \tag{4.22}$$

Since $f(x)$ is an odd function and noting that ωt varies from 0 to $\pi/2$ as x varies from 0 to A,

$$\frac{1}{\omega_0} = \frac{\sqrt{2}}{\kappa\pi} \int_0^A \frac{d\zeta}{\sqrt{\int_\zeta^A f(\xi)\, d\xi}} \tag{4.23}$$

With ω_0 and ϕ_0 determined by Eqs. (4.23) and (4.22), respectively, the next approximation may be found from the equation

$$\omega_1^2 x'' + \kappa^2 f(x) - p \cos \phi_0 = 0 \tag{4.24}$$

In the original differential equation, Eq. (4.20), ωt is replaced by its first approximation ϕ_0 and ω_0 (now known) is replaced by its second approximation ω_1, thus giving Eq. (4.24). This equation is again of a type which may be integrated explicitly; therefore, the next approximation ω_1 and ϕ_1 may be determined. In those cases where $f(x)$ is a complicated function, the integrals may be evaluated graphically.[23]

This method involves reducing nonautonomous systems to autonomous ones * by an iteration procedure in which the solution of the free vibration problem is used to replace the time function in the original equation which is then solved again for $t(x)$. The method is accurate and frequently two iterations will suffice.

THE PERTURBATION METHOD

In one of the most common methods of nonlinear vibration analysis, the desired quantities are developed in powers of some parameter which is considered small; then the coefficients of the resulting power series are determined in a stepwise manner. The method is straightforward, although it becomes cumbersome for actual computations if many terms in the perturbation series are required to achieve a desired degree of accuracy.

Consider Duffing's equation, Eq. (4.16), in the form

$$\omega^2 x'' + \kappa^2(x + \mu^2 x^3) - p \cos \phi = 0 \tag{4.25}$$

where $\phi = \omega t$ and primes denote differentiation with respect to ϕ. The conditions at time $t = 0$ are $x(0) = A$ and $x'(0) = 0$, corresponding to harmonic solutions of period $2\pi/\omega$. Assume that μ^2 and p are small quantities, and define $\kappa^2\mu^2 \equiv \epsilon$, $p \equiv \epsilon p_0$. The displacement $x(\phi)$ and the frequency ω may now be expanded in terms of the small quantity ϵ:

$$\begin{aligned} x(\phi) &= x_0(\phi) + \epsilon x_1(\phi) + \epsilon^2 x_2(\phi) + \cdots \\ \omega &= \omega_0 + \epsilon\omega_1 + \epsilon^2\omega_2 + \cdots \end{aligned} \tag{4.26}$$

The initial conditions are taken as $x_i(0) = x_i'(0) = 0$ $[i = 1, 2, \ldots]$.

Introducing Eq. (4.26) into Eq. (4.25) and collecting terms of zero order in ϵ gives the linear differential equation

$$\omega_0^2 x_0'' + \kappa^2 x_0 = 0$$

Introducing the initial conditions into the solution of this linear equation gives $x_0 = A \cos \omega t$ and $\omega_0 = \kappa$. Collecting terms of the first order in ϵ,

$$\omega_0^2 x_1'' + \kappa^2 x_1 - (2\omega_0\omega_1 A - \tfrac{3}{4}A^3 + p_0) \cos \phi + \tfrac{1}{4}A^3 \cos 3\phi = 0 \tag{4.27}$$

The solution of this differential equation has a nonharmonic term of the form $\phi \cos \phi$, but since only harmonic solutions are desired, the coefficient of this term is made to vanish so that

$$\omega_1 = \frac{1}{2\kappa}\left(\tfrac{3}{4}A^2 - \frac{p_0}{A}\right)$$

Using this result and the appropriate initial conditions, the solution of Eq. (4.27) is

$$x_1 = \frac{A^3}{32\kappa^2}(\cos 3\phi - \cos \phi)$$

* An autonomous system is one in which the time *does not* appear explicitly, while a nonautonomous system is one in which the time *does* appear explicitly.

To the first order in ϵ, the solution of Duffing's equation, Eq. (4.25), is

$$x = A \cos \omega t + \epsilon \frac{A^3}{32\kappa^2} (\cos 3\omega t - \cos \omega t)$$

$$\omega = \kappa + \frac{\epsilon}{2\kappa} \left(\tfrac{3}{4} A^2 - \frac{p_0}{A} \right)$$

This agrees with the results obtained previously [Eqs. (4.18) and (4.19)]. The analysis may be carried beyond this point, if desired, by application of the same general procedures.

As a further example of the perturbation method, consider the self-excited system described by Van der Pol's equation

$$\ddot{x} - \epsilon(1 - x^2)\dot{x} + \kappa^2 x = 0 \tag{4.5}$$

where the initial conditions are $x(0) = 0$, $\dot{x}(0) = A\kappa_0$. Assume that

$$x = x_0 + \epsilon x_1 + \epsilon^2 x_2 + \cdots$$

$$\kappa^2 = \kappa_0^2 + \epsilon \kappa_1^2 + \epsilon^2 \kappa_2^2 + \cdots$$

Inserting these series into Eq. (4.5) and equating coefficients of like terms, the result to the order ϵ^2 is

$$x = \left(2 - \frac{29\epsilon^2}{96\kappa_0^2}\right) \sin \kappa_0 t + \frac{\epsilon}{4\kappa_0} \cos \kappa_0 t + \frac{\epsilon}{4\kappa_0} \left(\frac{3\epsilon}{4\kappa_0} \sin 3\kappa_0 t - \cos 3\kappa_0 t \right) - \frac{5\epsilon^2}{124\kappa_0^2} \sin 5\kappa_0 t \tag{4.28}$$

An application of the perturbation method which employs operational calculus to solve the resulting linear differential equations is given in Ref. 21, while other applications of the method are given in Refs. 24 and 25. An application to the problem of subharmonic response is outlined in Ref. 1.

THE METHOD OF KRYLOFF AND BOGOLIUBOFF [11]

Consider the general autonomous differential equation

$$\ddot{x} + F(x,\dot{x}) = 0$$

which can be rewritten in the form

$$\ddot{x} + \kappa^2 x + \epsilon f(x,\dot{x}) = 0 \qquad [\epsilon \ll 1] \tag{4.29}$$

For the corresponding linear problem ($\epsilon \equiv 0$), the solution is

$$x = A \sin (\kappa t + \theta) \tag{4.30}$$

where A and θ are constants.

The procedure employed often is used in the theory of ordinary linear differential equations, and known variously as the method of variation of parameters or the method of Lagrange. In the application of this procedure to a nonlinear equation of the form of Eq. (4.29), assume the solution to be of the form of Eq. (4.30) but with A and θ as time-dependent functions rather than constants. This procedure, however, introduces an excessive variability into the solution; consequently, an additional restriction may be introduced. The assumed solution, of the form of Eq. (4.30), is differentiated once considering A and θ as time-dependent functions; this is made equal to the corresponding relation from the linear theory (A and θ constant) so that the additional restriction

$$\dot{A}(t) \sin [\kappa t + \theta(t)] + \dot{\theta}(t)A(t) \cos [\kappa t + \theta(t)] = 0 \tag{4.31}$$

is placed on the solution. The second derivative of the assumed solution is now formed and these relations are introduced into the differential equation, Eq. (4.29). Combining this result with Eq. (4.31),

$$\dot{A}(t) = -\left(\frac{\epsilon}{\kappa}\right) f[A(t)\sin\Phi,\ A(t)\kappa\cos\Phi]\cos\Phi$$

$$\dot{\theta}(t) = \frac{\epsilon}{\kappa A(t)} f[A(t)\sin\Phi,\ A(t)\kappa\cos\Phi]\sin\Phi$$

where

$$\Phi = \kappa t + \theta(t)$$

Thus, the second-order differential equation, Eq. (4.29), has been transformed into two first-order differential equations for $A(t)$ and $\theta(t)$.

The expressions for $\dot{A}(t)$ and $\dot{\theta}(t)$ may now be expanded in Fourier series:

$$\begin{aligned} \dot{A}(t) &= -\left(\frac{\epsilon}{\kappa}\right)\left\{K_0(A) + \sum_{n=1}^{r}[K_n(A)\cos n\Phi + L_n(A)\sin n\Phi]\right\} \\ \dot{\theta}(t) &= \frac{\epsilon}{\kappa A}\left\{P_0(A) + \sum_{n=1}^{r}[P_n(A)\cos n\Phi + Q_n(A)\sin n\Phi]\right\} \end{aligned} \tag{4.32}$$

where

$$K_0(A) = \frac{1}{2\pi}\int_0^{2\pi} f[A\sin\Phi,\ A\kappa\cos\Phi]\cos\Phi\, d\Phi$$

$$P_0(A) = \frac{1}{2\pi}\int_0^{2\pi} f[A\sin\Phi,\ A\kappa\cos\Phi]\sin\Phi\, d\Phi$$

It is apparent that A and θ are periodic functions of time of period $2\pi/\kappa$; therefore, during one cycle, the variation of $\dot{A}$ and $\dot{\theta}$ is small because of the presence of the small parameter ϵ in Eqs. (4.32). Hence, the average values of $\dot{A}$ and $\dot{\theta}$ are considered. Since the motion is over a single cycle, and since the terms under the summation signs are of the same period and consequently vanish, then approximately:

$$\dot{A} \simeq -\left(\frac{\epsilon}{\kappa}\right) K_0(A)$$

$$\dot{\theta} \simeq \frac{\epsilon}{\kappa A} P_0(A)$$

$$\dot{\Phi} \simeq \kappa + \frac{\epsilon}{\kappa A} P_0(A)$$

For example, consider Rayleigh's equation

$$\ddot{x} - (\alpha - \beta\dot{x}^2)\dot{x} + \kappa^2 x = 0 \tag{4.33}$$

By application of the above procedures:

$$\begin{aligned} \dot{A} &= -\left(\frac{1}{\kappa}\right) K_0(A) = -\frac{1}{\kappa}\left[\frac{1}{2\pi}\int_0^{2\pi}(-\alpha + \beta A^2\kappa^2\cos^2\Phi)A\kappa\cos^2\Phi\, d\Phi\right] \\ &= \frac{A}{2}(\alpha - \tfrac{3}{4}\beta A^2\kappa^2) \end{aligned} \tag{4.34}$$

Equation (4.34) may be integrated directly:

$$t = 2\int_{A_0}^{A}\frac{dA}{A(\alpha - \gamma A^2)} = \frac{1}{\alpha}\ln\frac{A^2}{\alpha - \gamma A^2}$$

Solving for A,

$$A = \frac{\alpha}{\gamma}\left[\frac{1}{1 + \left(\frac{\alpha}{\gamma A_0^2} - 1\right)e^{-\alpha t}}\right]^{1/2} \tag{4.35}$$

where

$$\gamma = \tfrac{3}{4}\beta^2\kappa^2$$

The application of the method to Van der Pol's equation, Eq. (4.5), is easily accomplished and leads to a solution in the first approximation of the form

$$x = 2 \cos o + \theta) \tag{4.36}$$

This may be compared with the perturbation solution given by Eq. (4.28).

The method may be applied equally well to nonautonomous systems.[5,19,20]

THE RITZ METHOD

In addition to methods of nonlinear vibration analysis stemming from the idea of small nonlinearities and from extensions of methods applicable to linear equations, other methods are based on such ideas as satisfying the equation at certain points of the motion [13] or satisfying the equation in the average. The Ritz method is an example of the latter method, and is quite powerful for general studies.

One method of determining such "average" solutions is to multiply the differential equation by some "weight function" $\psi_n(t)$ and then integrate the product over a period of the motion. If the differential equation is denoted by E, this procedure leads to

$$\int_0^{2\pi} E \cdot \psi_n(t)\, dt = 0 \tag{4.37}$$

A second method of obtaining such "average" solutions can be derived from the calculus of variations by seeking functions that minimize a certain integral:

$$I = \int_{t_0}^{t_1} F(\dot{x},x,t)\, dt = \text{minimum}$$

Consider a function of the form

$$\bar{x}(t) = a_1\psi_1(t) + a_2\psi_2(t) + \cdots + a_n\psi_n(t)$$

where the $\psi_k(t)$ are prescribed functions. If $\bar{x}$ is now introduced for x, then

$$I = I(a_1, a_2, \ldots, a_n)$$

and a necessary condition for I to be a minimum is

$$\frac{\partial I}{\partial a_1} = 0, \qquad \frac{\partial I}{\partial a_2} = 0, \ldots, \qquad \frac{\partial I}{\partial a_n} = 0 \tag{4.38}$$

This gives n equations of the form

$$\frac{\partial I}{\partial a_k} = \int_{t_0}^{t_1}\left(\frac{\partial F}{\partial \bar{x}}\psi_k + \frac{\partial F}{\partial \dot{\bar{x}}}\dot{\psi}_k\right) dt = 0 \tag{4.39}$$

for determining the n unknown coefficients. Integrating Eq. (4.39),

$$\frac{\partial I}{\partial a_k} = \left[\frac{\partial F}{\partial \dot{\bar{x}}}\psi_k\right]_{t_0}^{t_1} + \int_{t_0}^{t_1}\left[\frac{\partial F}{\partial \bar{x}} - \frac{d}{dt}\left(\frac{\partial F}{\partial \dot{\bar{x}}}\right)\right]\psi_k\, dt = 0$$

The first term is zero because ψ_k must satisfy the boundary conditions; the expression in brackets under the integral in the second term is Euler's equation. The conditions given in Eqs. (4.38) then reduce to

$$\int_{t_0}^{t} E(\bar{x})\psi_k \, dt = 0 \qquad [k = 1, 2, \ldots, n] \tag{4.40}$$

This is the same as Eq. (4.37); thus, it is not necessary to "know" the variational problem, but only the differential equation. The conditions given in Eqs. (4.40) then yield average solutions based on variational concepts.

EXAMPLES. As a first example of the application of the Ritz method, consider the equation

$$\ddot{x} + \kappa^2 x^n = 0$$

for which an exact solution was given earlier in this chapter [Eq. (4.9)]. Assume a single-term solution of the form

$$\bar{x} = A \cos \omega t$$

The Ritz procedure, defined by Eq. (4.40), gives

$$\int_0^{2\pi} (-\omega^2 A \cos^2 \omega t + \kappa^2 A^n \cos^{n+1} \omega t) \, d(\omega t) = 0$$

from which

$$\frac{\omega^2}{\kappa^2} = \frac{4}{\pi} A^{n-1} \int_0^{\pi/2} \cos^{n+1} \omega t \, d(\omega t) = A^{n-1} \varphi(n) \tag{4.41}$$

The comparable exact solution obtained previously by introducing in Eq. (4.9) the quantity $2\pi/\omega$ for the period τ is

$$\frac{\omega^2}{\kappa^2} = \left[\frac{\pi^2/4}{\psi^2(n)}\right] X^{n-1} = \Phi(n) X^{n-1} \tag{4.42}$$

Values of $\varphi(n)$ from the approximate analysis and $\Phi(n)$ from the exact analysis are compared directly in Table 4.1, affording an appraisal of the accuracy of the method.

Table 4.1. Values of the Functions $\psi(n)$, $\Phi(n)$, $\varphi(n)$ *

n	$\psi(n)$	$\Phi(n)$	$\varphi(n)$
0	1.4142	1.2337	1.2732
1	1.5708	1.0000	1.0000
2	1.7157	0.8373	0.8488
3	1.8541	0.7185	0.7500
4	1.9818	0.6282	0.6791
5	2.1035	0.5577	0.6250
6	2.2186	0.5013	0.5820
7	2.3282	0.4552	0.5469

* The mathematical expressions for $\psi(n)$, $\Phi(n)$, and $\varphi(n)$ and the equations to which they refer are:

$$\psi(n) = \sqrt{\frac{n+1}{2}} \int_0^1 \frac{du}{\sqrt{1 - u^{n+1}}} \qquad \text{[Eq. (4.9)]}$$

$$\Phi(n) = \frac{\pi^2/4}{\psi^2(n)} \qquad \text{[Eq. (4.42)]}$$

$$\varphi(n) = \frac{4}{\pi} \int_0^{\pi/2} \cos^{n+1} \sigma \, d\sigma \qquad \text{[Eq. (4.41)]}$$

Consider now the nonautonomous system described by Duffing's equation

$$E \equiv \ddot{x} + \kappa^2(x + \mu^2 x^3) - p \cos \omega t = 0$$

Assuming

$$\bar{x} = A \cos \phi, \qquad \phi = \omega t$$

the Ritz condition, Eq. (4.40), leads to

$$\int_0^{2\pi} \{[(1 - \eta^2)A - s] \cos \phi + \mu^2 A^3 \cos^3 \phi\} \cos \phi \, d\phi$$

from which the amplitude-frequency relation is

$$(1 - \eta^2)A + \tfrac{3}{4}\mu^2 A^3 = \pm s \tag{4.43}$$

where

$$s = \frac{p}{\kappa^2}, \qquad \eta^2 = \frac{\omega^2}{\kappa^2} \tag{4.44}$$

The upper sign indicates vibration in phase with the exciting force. Equation (4.43) describes the response curves shown in Fig. 4.14*A*, and corresponds to Eq. (4.19) obtained by Duffing's method.

Application of the Ritz method to Van der Pol's equation, Eq. (4.5), leads to the identical result given by Eq. (4.36).

GENERAL EQUATIONS FOR RESPONSE CURVES

The Ritz method has been applied extensively in studies of nonlinear differential equations.[6, 8] Some of the general equations for response curves thereby obtained are given here, both as a further example of the application of the method and as a collection of useful relations.

SYSTEM WITH LINEAR DAMPING AND GENERAL RESTORING FORCES

Consider a system with general elastic restoring force (an odd function) and described by the equation of motion

$$a\ddot{x} + b\dot{x} + cf(x) - P \cos \omega t = 0$$

A solution may be assumed in the form

$$\bar{x} = A \cos (\omega t - \theta) = B \cos \phi + C \sin \phi \tag{4.45}$$

where $\phi = \omega t$, $B = A \cos \theta$, $C = A \sin \theta$. Introducing Eq. (4.45) according to the Ritz conditions, and recalling that $f(x)$ is to be an odd function,

$$\begin{aligned} -a\omega^2 A \cos \theta + b\omega A \sin \theta + cAF(A) \cos \theta &= P \\ -a\omega^2 A \sin \theta - b\omega A \cos \theta + cAF(A) \sin \theta &= 0 \end{aligned} \tag{4.46}$$

where

$$F(A) = \frac{1}{\pi A} \int_0^{2\pi} f(A \cos \sigma) \cos \sigma \, d\sigma$$

and σ is simply an integration variable.

Some algebraic manipulations with Eqs. (4.46) give independent equations for the two unknowns A and θ:

$$[F(A) - \eta^2]^2 + 4D^2\eta^2 = \left(\frac{s}{A}\right)^2 \tag{4.47}$$

$$\tan \theta = \frac{2D\eta}{F(A) - \eta^2} \tag{4.48}$$

where η^2 and s are defined according to Eq. (4.44) and

$$\kappa^2 = \frac{c}{a} \qquad p = \frac{P}{a} \qquad D = \frac{b}{2\sqrt{ac}}$$

Equation (4.47) describes response curves of the form shown in Fig. 4.15, and Eq. (4.48) gives the corresponding phase angle relationships. These two equations also yield other special relations which describe various curves in the response diagram:

Undamped free vibration curve (Fig. 4.13),

$$\eta^2 = F(A) \tag{4.49}$$

Undamped response curves (Fig. 4.14),

$$\eta^2 = F(A) \mp \frac{s}{A} \tag{4.50}$$

Locus of vertical tangents of undamped response curves (Fig. 4.17),

$$\eta^2 = F(A) + A\frac{\partial F(A)}{\partial A} \tag{4.51}$$

Damped response curves (Fig. 4.15),

$$\eta^2 = [F(A) - 2D^2] \mp \sqrt{\left(\frac{s}{A}\right)^2 - 4D^2[F(A) - D^2]} \tag{4.52}$$

Locus of vertical tangents of damped response curves (Fig. 4.17),

$$[F(A) - \eta^2]\left[F(A) + A\frac{\partial F(A)}{\partial A} - \eta^2\right] = -4D^2\eta^2 \tag{4.53}$$

The maximum amplitude of vibration is of interest. The amplitude at the point at which a response curve crosses the free vibration curve is termed the *resonance amplitude*, and is determined in the nonlinear case by solving Eqs. (4.49) and (4.52) simultaneously. This leads to [27]

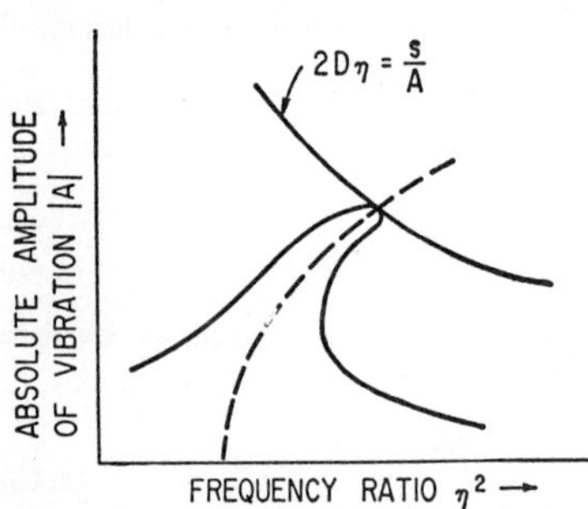

FIG. 4.19. Determination of the resonant amplitude in accordance with Eq. (4.54).

$$2D\eta = \frac{s}{A} \qquad \theta = \frac{\pi}{2} \tag{4.54}$$

The first of these two equations defines a hyperbola in the response diagram, describing the locus of crossing points, as shown in Fig. 4.19; hence, the intersection of this curve with the free vibration curve gives the resonance amplitude. The phase angle at resonance has the value $\pi/2$, as in the linear case. This result is of great help in computing response curves since the effect of damping (except for very large values) is negligible except in the neighborhood of resonance. Therefore, one may compute only the undamped curves (which is not difficult) and the hyperbola (which does not contain the nonlinearity); then, the effect of damping may be sketched in from knowledge of the crossing point.

SYSTEM WITH GENERAL DAMPING AND GENERAL RESTORING FORCES

The preceding analysis may be extended to include the more general differential equation

$$E \equiv \ddot{x} + 2D\kappa g(\dot{x}) + \kappa^2 f(x) - p\cos\omega t = 0$$

By procedures similar to those employed above:

$$[F(A) - \eta^2]^2 + 4D^2S^2(A) = \left(\frac{s}{A}\right)^2 \tag{4.55}$$

$$\tan\theta = \frac{2DS(A)}{F(A) - \eta^2} \tag{4.56}$$

where
$$S(A) = \frac{1}{\pi\kappa A}\int_0^{2\pi} g(\omega A \sin\sigma)\sin\sigma\, d\sigma$$

In the case of linear velocity damping, $S(A) = \eta$, and Eqs. (4.55) and (4.56) reduce to Eqs. (4.47) and (4.48). The results for various types of damping forces are:

Coulomb damping: $g(\dot{x}) = \pm\nu_0$ $\qquad S(A) = \dfrac{4}{\pi}\dfrac{\nu_0}{\kappa A}$

Linear velocity damping: $g(\dot{x}) = \nu_1\dot{x}$ $\qquad S(A) = \nu_1\eta$

Velocity squared damping: $g(\dot{x}) = \nu_2\dot{x}|\dot{x}|$ $\qquad S(A) = \dfrac{8}{3\pi}\nu_2\eta(A\omega)$

nth-power velocity damping: $g(\dot{x}) = \nu_n\dot{x}|\dot{x}|^{n-1}$ $\qquad S(A) = \nu_n\eta(A\omega)^{n-1}\varphi(n)$

where $\varphi(n)$ is defined in Eq. (4.41) and values are given in Table 4.1.

The locus of resonance amplitudes or crossing points is now given by [27]

$$2DS(A) = \frac{s}{A} \qquad \theta = \frac{\pi}{2}$$

GRAPHICAL METHODS OF INTEGRATION

INTRODUCTION

Graphical methods may be employed in the analysis of nonlinear vibration and often prove to be of great value both for general studies of the behavior of a particular system and for actual integration of the equation of motion.

A single degree-of-freedom system requires two parameters to describe completely the state of the motion. When these two parameters are used as coordinate axes, the graphical representation of the motion is called a *phase-plane* representation. In dealing with ordinary dynamical problems, these parameters frequently are taken as the displacement and velocity. First consider an undamped linear system having the equation of motion

$$\ddot{x} + \omega_n^2 x = 0 \tag{4.57}$$

and the solution

$$x = A\cos\omega_n t + B\sin\omega_n t$$
$$y = \frac{\dot{x}}{\omega_n} = -A\sin\omega_n t + B\cos\omega_n t \tag{4.58}$$

Eliminating time as a variable between Eqs. (4.58):

$$x^2 + y^2 = c^2$$

Thus, the phase-plane representation is a family of concentric circles with centers at the origin. Such curves are called *trajectories*. The necessary and sufficient conditions in the phase-plane for periodic motions are (1) closed trajectories and (2) paths described in finite time.

Now, suppose that the solution of Eq. (4.57) is not known. By introducing $y = \dot{x}/\omega_n$,

$$\frac{dx}{dt} = \omega_n y \qquad \frac{dy}{dt} = -\omega_n x$$

Therefore

$$\frac{dy}{dx} = -\frac{x}{y} \tag{4.59}$$

Thus, the path in the phase-plane is described by a simple first-order differential equa tion. This process of eliminating the time always can be done in principle, but frequently the problem is too difficult. Since the time is to be eliminated, *only autonomous systems can be treated by phase-plane methods.* When an equation of the type of Eq. (4.59) can

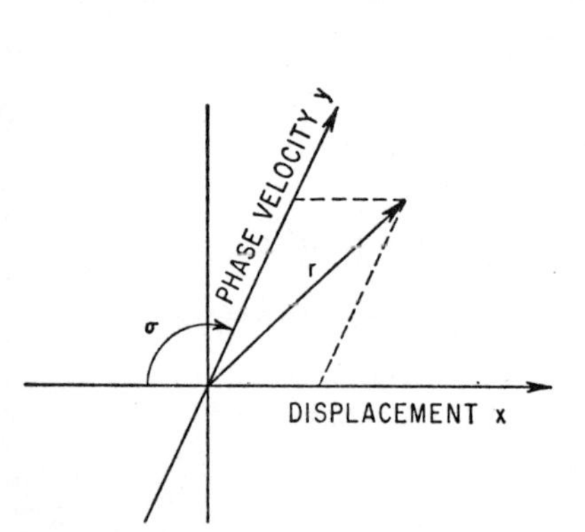

FIG. 4.20. Phase-plane using oblique coordinates which results in a logarithmic spiral trajectory for a linear system with viscous damping [Eq. (4.60)].

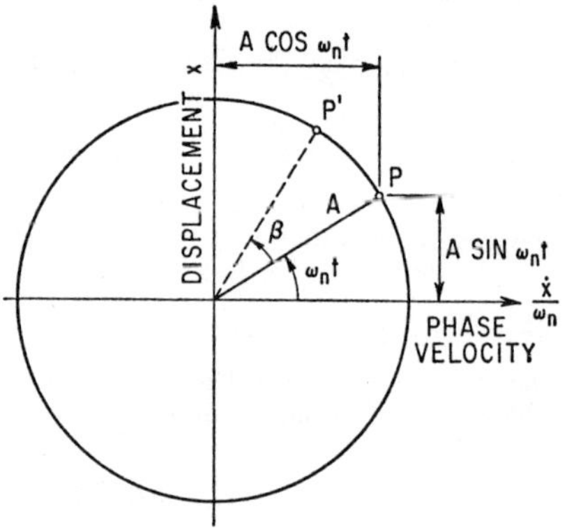

FIG. 4.21. Phase-plane solution for a linear undamped vibrating system.

be found, a direct solution of the problem follows since slopes of the trajectories can be sketched in the phase-plane and the trajectories determined by connecting the tangents; this is known as the method of *isoclines*. It sometimes happens that $dx = 0$, $dy = 0$ simultaneously so that there is no knowledge of the direction of the motion; such points in the phase-plane are called *singular* points. In the present example, the origin constitutes a singular point.

Consider now a damped linear system having the equation of motion

$$\ddot{x} + 2\zeta\omega_n\dot{x} + \omega_n^2 x = 0$$

and the solution

$$x = Ce^{-\delta t}\cos\Phi$$

$$y = \frac{\dot{x}}{\omega_n} = Ce^{-\delta t}\cos(\Phi + \sigma) \tag{4.60}$$

where

$$\delta = \zeta\omega_n = -\omega_n\cos\sigma \qquad \left[\sigma > \frac{\pi}{2}\right]$$

$$\Phi = \nu t + \theta$$

$$\nu = \omega_n\sqrt{1 - \zeta^2} = \omega_n\sin\sigma$$

Equations (4.60) indicate that the trajectories in the phase-plane are some form of spiral (one of the simplest known of which is the logarithmic spiral). By referring to the oblique coordinate system shown in Fig. 4.20, and recalling that $\sin\sigma$ is a constant and

$$r^2 = x^2 + y^2 - 2xy\cos\sigma$$

Eqs. (4.60) reduce to

$$r = Ce^{-\delta t} \sin \sigma$$

This is a form of a logarithmic spiral.

The trajectories also could be found in a rectangular coordinate system, by the method of isoclines, without knowledge of the solution [Eq. (4.60)]. The governing differential equation is

$$\frac{dy}{dx} = -\frac{2\zeta y + x}{y} \tag{4.61}$$

The resulting trajectories can be sketched in the phase-plane. On the other hand, Eq. (4.61) also can be integrated analytically by use of the substitution $z = y/x$ and separation of the variables:

$$y^2 + 2\zeta xy + x^2 = C \exp\left(\frac{2\zeta}{\sqrt{1-\zeta^2}} \tan^{-1} \frac{x\zeta + y}{x\sqrt{1-\zeta^2}}\right)$$

This is a spiral of the form of Eq. (4.60).

The method of isoclines is extremely useful in studying the behavior of solutions in the neighborhood of singular points and for the related questions of stability of solutions. In this sense, phase-plane methods may be thought of as topological methods.[5,10,12,19] However, it is desirable also to study the over-all solutions, rather than solutions in the neighborhood of special points, and preferably by some straightforward method of graphical integration. Such integration methods are given in the following sections of this chapter, while topological studies, with special reference to questions of stability, are treated in Chap. 5.

PHASE-PLANE INTEGRATION OF STEPWISE LINEAR SYSTEMS

Consider the undamped linear system described by Eq. (4.57). The known solution $x = A \sin \omega_n t$, $\dot{x} = A\omega_n \cos \omega_n t$ may be shown graphically in the phase-plane representation of Fig. 4.21. The point P moves with constant angular velocity ω_n, and the deflection increases to P' in the time β/ω_n.

If the system has a nonlinear restoring force composed of straight lines (as in Fig. 4.4), the motion within the region represented by any one linear segment can be described as above. For example, consider a system with the force-deflection characteristic shown at the top of Fig. 4.22. If the motion starts with initial velocity q_6 and zero initial deflection, the motion is described by a circular arc with center at 0 and angular velocity $\omega_{n_1} = \sqrt{\frac{1}{m}} \tan \alpha_1$, from q_6 to q_5. At the point q_5, it is seen that $\dot{x}_A/\omega_{n_1} = \overline{x_A q_5}$ and $\dot{x}_A/\omega_{n_2} = \overline{x_A q_4}$. Therefore

$$\frac{\overline{x_A q_4}}{\overline{x_A q_5}} = \sqrt{\frac{\tan \alpha_2}{\tan \alpha_1}}$$

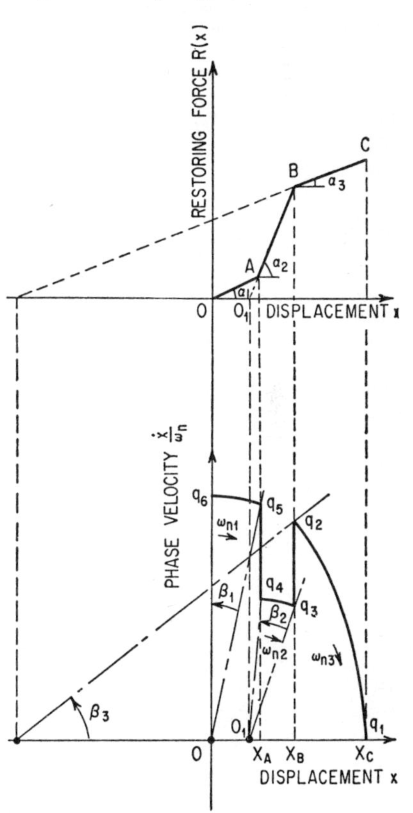

FIG. 4.22. Phase-plane solution for the stepwise linear restoring force characteristic curve shown at the top. The motion starts with zero displacement but finite velocity.

In this example, $\tan \alpha_1 < \tan \alpha_2$ so that $\overline{x_A q_4} < \overline{x_A q_5}$. The circular arc from q_4 to q_3 corresponds to the segment AB of the restoring force characteristic with center at the intersection 0_1 of the segment (extended) with the X axis. The radius of this circle is $\overline{0_1 q_4}$, where

$$q_4 = \frac{\dot{x}_B}{\omega_{n_2}} \qquad \text{and} \qquad \omega_{n_2} = \sqrt{\frac{1}{m} \tan \alpha_2}$$

The total time required to go from q_6 to q_1 is

$$t = \frac{\beta_1}{\omega_{n_1}} + \frac{\beta_2}{\omega_{n_2}} + \frac{\beta_3}{\omega_{n_3}}$$

For a symmetrical system this is one quarter of the period.

If the force-deflection characteristic of a nonlinear system is a smooth curve, it may be approximated by straight line segments and treated as above. It should be noted that the time required to complete one cycle is strongly influenced by the nature of the curve in regions where the velocity is low; therefore, linear approximations near the equilibrium position do not greatly affect the period.

The time-history of the motion (i.e., the x,t representation) may be obtained quite readily by projecting values from the X axis to an x,t plane.

Inasmuch as phase-plane methods are restricted to autonomous systems, only free vibration is discussed above. However, if a constant force were to act on the system, the nature of the vibration would be unaffected, except for a displacement of the equilibrium position in the direction of the force and equal to the static deflection produced by that force. Thus, the trajectory would remain a circular arc but with its center displaced from the origin. Therefore, *nonautonomous systems may be treated by phase-plane methods, if the time function is replaced by a series of stepwise constant values.* The degree of accuracy attained in such a procedure depends only on the number of steps assumed to represent the time function.

A system having a bilinear restoring force and acted upon by an external stepwise function of time, treated by the method described above, is shown in Fig. 4.23. Phase-plane methods therefore offer the possibility of treating transient as well as free vibrations.

Phase-plane methods have been widely used for the analysis of control mechanisms.[29] A comprehensive analysis of discontinuous-type systems possessing various types of nonlinearities is given in Ref. 30.

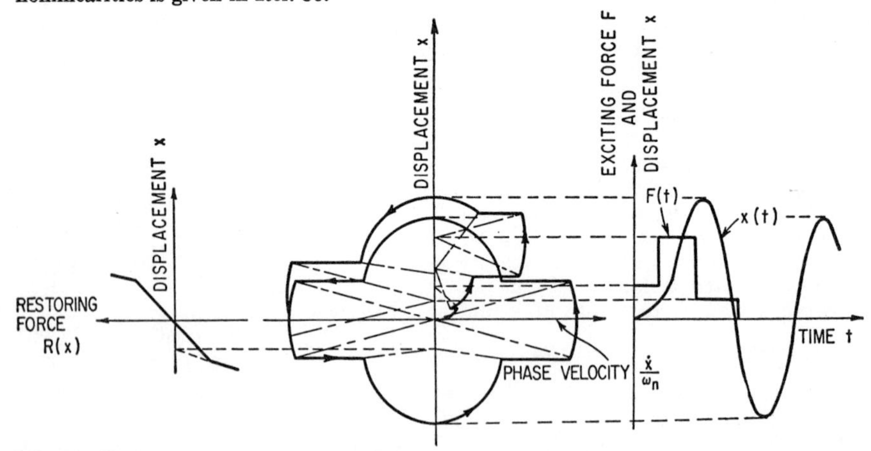

FIG. 4.23. Phase-plane solution for transient motion. The bilinear restoring force characteristic curve is shown at the left, and the exciting force $F(t)$ and the resulting motion of the system $X(t)$ are shown at the right. (*After Evaldson et al.*[28])

PHASE-PLANE INTEGRATION OF AUTONOMOUS SYSTEMS WITH NONLINEAR DAMPING

Consider the differential equation

$$\ddot{x} + g(\dot{x}) + \kappa^2 x = 0$$

Introducing $y = \dot{x}/\kappa$, the following isoclinic equation is obtained:

$$\frac{dy}{dx} = -\frac{g(y) + x}{y} \tag{4.62}$$

For points of zero slope in the phase-plane, the numerator of Eq. (4.62) must vanish; therefore, the condition for zero slope is

$$x_0 = -g(y)$$

Points of infinite slope correspond to the X axis. Singular points occur where the x_0 curve intersects the X axis.

To construct the trajectory, the slope at any point P_i must be determined first. This is done as illustrated in Fig. 4.24: A line is drawn parallel to the X axis through P_i. The intersection of this line with the x_0 curve determines a point S_i on the X axis. With S_i as the center, a circular arc of short length is drawn through P_i; the tangent to this arc is the required slope. The termination of this short arc may be taken as the point P_{i+1}, etc. The accuracy of the construction is dependent on the lengths of the arcs. This construction is known as Liénard's method.[5,10]

As an example of Liénard's method, consider Rayleigh's equation, Eq. (4.33), in the form

$$\ddot{x} + \epsilon\left(\frac{\dot{x}^3}{3} - \dot{x}\right) + x = 0$$

The corresponding isoclinic equation is

$$\frac{dy}{dx} = \frac{\epsilon(y - y^3/3) - x}{y}$$

The x_0 curve is given by

$$x_0 = \epsilon\left(y - \frac{y^3}{3}\right) \tag{4.63}$$

This is illustrated in Fig. 4.25.

A little experimentation shows that if a point P_1 is taken near the origin, the slope is such as to take the trajectory away from the origin (as compared with the undamped

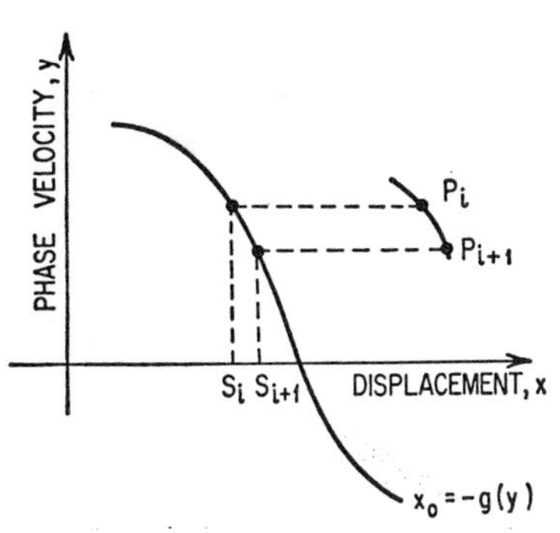

FIG. 4.24. Liénard's construction for phase-plane integration of autonomous systems with nonlinear damping.

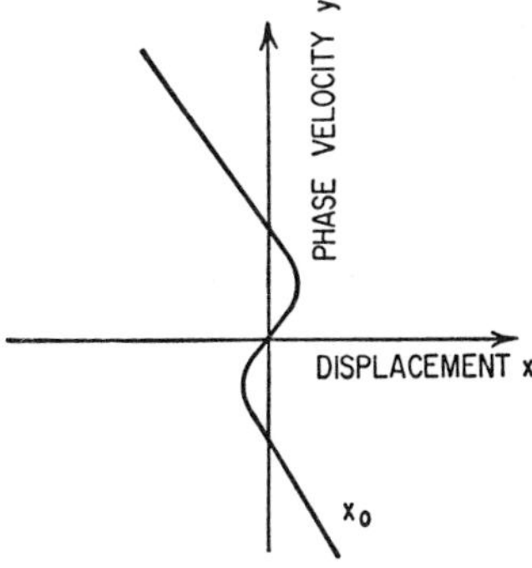

FIG. 4.25. Curve of x_0 for Rayleigh's equation [Eq. (4.33)] as given by Eq. (4.63).

vibration); by the same reasoning, a point P_2 far from the origin tends to take the trajectory toward the origin (again as compared with the undamped vibration). Therefore, there is some neutral curve, describing a periodic motion, toward which the trajectories tend; this neutral curve is called a *limit cycle* and is illustrated in Fig. 4.26. Such a limit cycle is obtained when x_0 has a different sign for different parts of the Y axis.

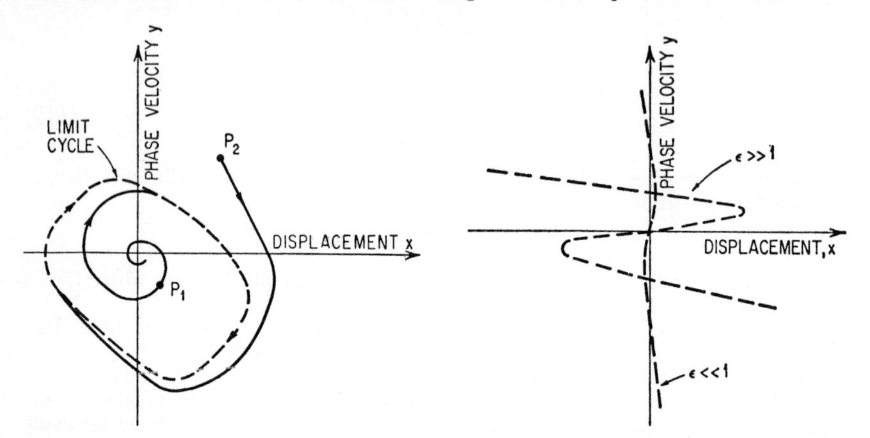

FIG. 4.26. Limit cycle for Rayleigh's equation [Eq. (4.33)].

FIG. 4.27. Curves of x_0 for extreme values of ϵ in Rayleigh's equation [Eq. (4.33)]. See Fig. 4.25 for a solution with a moderate value of ϵ.

For extreme values of ϵ, the x_0 curves would appear as shown in Fig. 4.27. For $\epsilon \gg 1$, introduce the notation $\xi = x/\epsilon$; then

$$\frac{dy}{d\xi} = \frac{\epsilon}{y}\left(y - \frac{y^3}{3} - \xi\right)$$

This leads to a trajectory as shown in Fig. 4.28. This type of motion is known as a *relaxation oscillation*. Note from Fig. 4.28 that for this case of large ϵ the slope changes quickly from horizontal to vertical. Hence, for a motion starting at some point P_i, a vertical

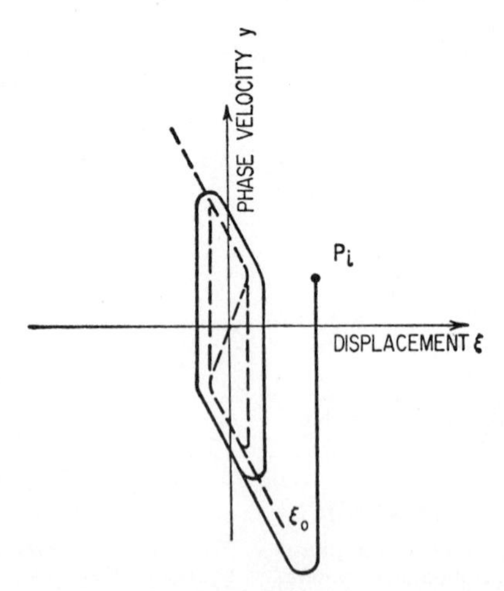

FIG. 4.28. Relaxation oscillations of Rayleigh's equation [Eq. (4.33)].

trajectory is followed until it intersects the ξ_0 curve; then, the trajectory turns and follows the ξ_0 curve until it enters the vertical field at the lower knee in the curve. The trajectory then moves straight up until it intersects ξ_0 again after which it swings right and down again. A few circuits bring the trajectory into the limit cycle.

There is a possibility that more than one limit cycle may exist. If the x_0 curve crosses the X axis more than three times, it can be shown that at least two limit cycles may exist. Further discussion of limit cycles is given in Chap. 5.

GENERALIZED PHASE-PLANE ANALYSIS

The following method of integrating second-order differential equations by phase-plane techniques has general application.[31, 32] Consider the general equation

$$\ddot{x} + F(x,\dot{x},t) = 0 \tag{4.64}$$

Equation (4.64) can be converted to the form

$$\ddot{x} + \kappa^2 x = g(x,\dot{x},t)$$

by adding $\kappa^2 x$ to both sides where

$$\kappa^2 x - F(x,\dot{x},t) = g(x,\dot{x},t)$$

Let

$$g(x_0,\dot{x}_0,t_0) = -\kappa^2 \Delta_0$$

where κ is chosen arbitrarily. At some point P_0 on the trajectory,

$$\ddot{x} + \kappa^2(x + \Delta_0) = 0$$

and

$$\frac{dy}{dx} = -\frac{x + \Delta_0}{y}$$

Referring to Fig. 4.29,

$$dt = \frac{1}{\kappa}\frac{dx}{y} = \frac{1}{\kappa}\,d\theta$$

Therefore, the time may be obtained by integration of the angular displacements. Thus, at a nearby point P_1 on the trajectory:

$$\begin{aligned} x_1 &= x_0 + dx \\ y_1 &= y_0 + dy \\ t_1 &= t_0 + dt \end{aligned}$$

Now, compute Δ_1 for the new center, and repeat the process.

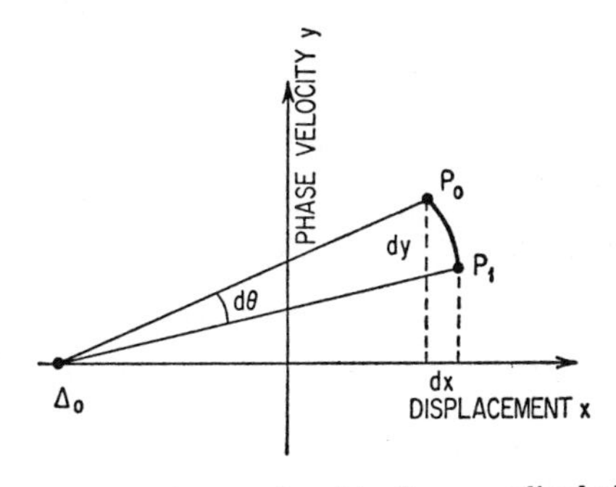

FIG. 4.29. Method of construction employed in the generalized phase-plane analysis.

This method has been applied to a very wide variety of linear and nonlinear equations.[31, 33] For example, Fig. 4.30 shows the solution of Bessel's equation

$$\ddot{x} + \frac{1}{t}\dot{x} + \left(p^2 - \frac{n^2}{t^2}\right)x = 0$$

of order zero. The angle (or time) projection of x yields $J_0(pt)$, while the $\dot{x}/p$ projection yields $J_1(pt)$; that is, the Bessel functions of the zeroth and first order of the first kind. Bessel functions of the second kind also can be obtained.

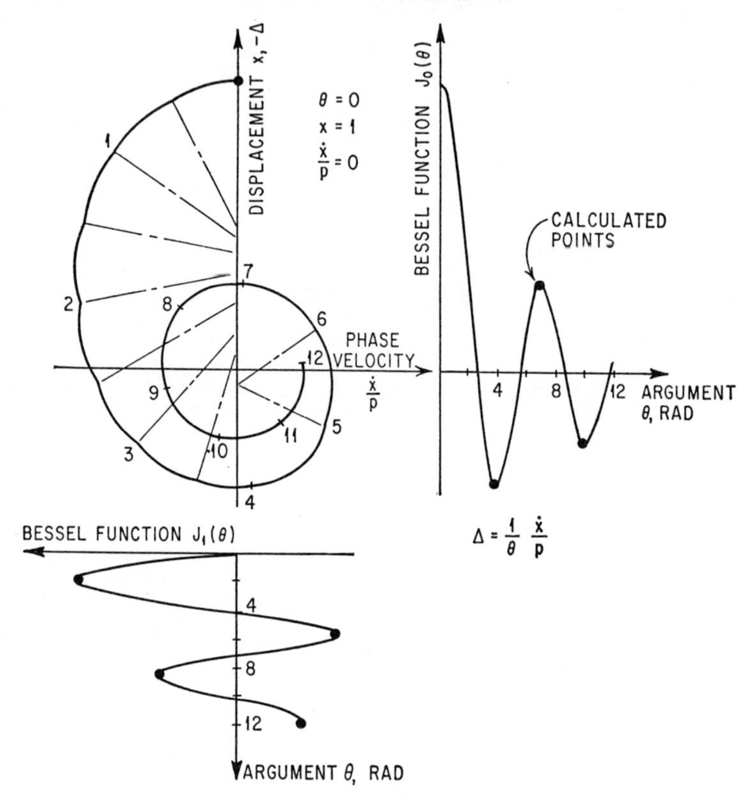

FIG. 4.30. Generalized phase-plane solution of Bessel's equation. (*Jacobsen.*[31])

STABILITY OF PERIODIC NONLINEAR VIBRATION

Certain systems having nonlinear restoring forces and undergoing forced vibration exhibit unstable characteristics for certain combinations of amplitude and exciting frequency. The existence of such an instability leads to the "jump phenomenon" shown in Fig. 4.16. To investigate the stability characteristics of the response curves, consider Duffing's equation

$$\ddot{x} + \kappa^2(x + \mu^2 x^3) = p \cos \omega t \tag{4.65}$$

Assume that two solutions of this equation exist and have slightly different initial conditions:

$$x_1 = x_0$$

$$x_2 = x_0 + \delta \qquad [\delta \ll x_0]$$

Introducing the second of these into Eq. (4.65) and employing the condition that x_0 is also a solution,

$$\ddot{\delta} + \kappa^2(1 + 3\mu^2 x_0^2)\delta = 0 \tag{4.66}$$

Now an expression for x_0 must be obtained; assuming a one-term approximation of the form $x_0 = A \cos \omega t$, Eq. (4.66) becomes

$$\frac{d^2\delta}{d\varphi^2} + (\lambda + \gamma \cos \varphi)\delta = 0 \tag{4.67}$$

where

$$\kappa^2(1 + \tfrac{3}{2}\mu^2 A^2) = 4\omega^2 \lambda$$

and

$$\tfrac{3}{2}\kappa^2 \mu^2 A^2 = 4\omega^2 \gamma \qquad 2\omega t = \varphi \tag{4.68}$$

Equation (4.67) is known as Mathieu's equation.

Mathieu's equation has appeared in this analysis as a variational equation characterizing small deviations from the given periodic motion whose stability is to be investigated; thus, the stability of the solutions of Mathieu's equation must be studied. A given periodic motion is stable if *all* solutions of the variational equation associated with it tend toward zero for all positive time and unstable if there is at least one solution which does not tend toward zero. The stability characteristics of Eq. (4.67) often are represented in a chart as shown in Fig. 4.31.[10]

From the response diagram of Duffing's equation, the out-of-phase motion having the larger amplitude appears to be unstable. This portion of the response diagram (Fig. 4.17) corresponds to unstable motion in the Mathieu stability chart (Fig. 4.31), and the locus of vertical tangents of the response curves (considering undamped vibration for simplicity) corresponds exactly to the boundaries between stable and unstable regions in the stability chart. Thus, the region of interest in the response diagram is described by the free vibration

$$\omega^2 = \kappa^2(1 + \tfrac{3}{4}\mu^2 A^2) \tag{4.69}$$

and the locus of vertical tangents

$$\tfrac{3}{2}\kappa^2 \mu^2 A^2 + \frac{p}{A} = 0 \tag{4.70}$$

The corresponding curves in the stability chart are taken as those for small positive values of γ and λ which have the approximate equations

$$\gamma = \tfrac{1}{2} - 2\lambda \tag{4.71}$$

$$\gamma = -\tfrac{1}{2} + 2\lambda \tag{4.72}$$

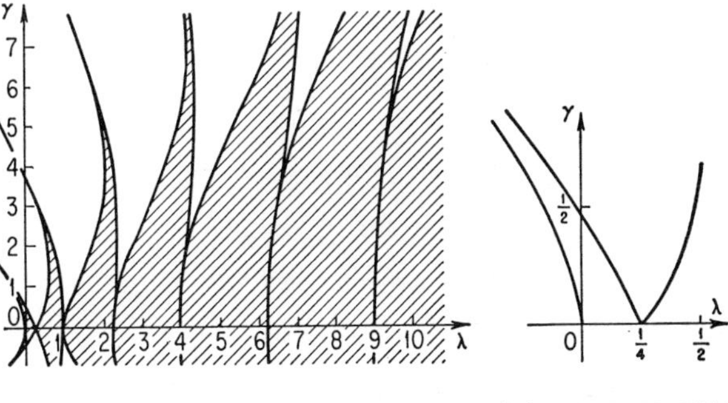

FIG. 4.31. Stability chart for Mathieu's equation [Eq. (4.67)].

Now, if Eq. (4.69) is introduced into Eqs. (4.68), the resulting equations expanded by the binomial theorem (assuming μ^2 small), and Eq. (4.72) introduced, the result is an identity. Therefore, the free vibration-response curve maps onto the curve of positive slope in the stability chart. The locus of vertical tangents to the response curves maps into the curve of negative slope in the stability chart; this may be seen from the identity obtained by introducing the equations obtained above by binomial expansion into Eq. (4.71) and then employing Eq. (4.70).

In any given case, it can be determined whether a motion is stable or unstable on the basis of the values of γ and λ, according to the location of the point in the stability chart.

The question of stability of response also can be resolved by means of a "stability criterion" developed from the Kryloff-Bogoliuboff procedures.[34] The differential equation of motion is considered in the form

$$\ddot{x} + \kappa^2 x + f(x,\dot{x}) = p \cos \omega t$$

Proceeding in the manner of the Kryloff-Bogoliuboff procedure described earlier,

$$\dot{A} = \frac{1}{\kappa} f(x,\dot{x}) \sin (\kappa t + \theta) - \frac{p}{\kappa} \cos \omega t \sin (\kappa t + \theta)$$

$$\dot{\theta} = \frac{1}{\kappa} f(x,\dot{x}) \cos (\kappa t + \theta) - \frac{p}{A\kappa} \cos \omega t \cos (\kappa t + \theta)$$

Expanding the last terms of these equations, the result contains motions of frequency κ, $\kappa + \omega$, and $\kappa - \omega$. The motion over a long interval of time is of interest, and the motions of frequencies $\kappa + \omega$ and $\kappa - \omega$ may be averaged out; this is accomplished by integrating over the period $2\pi/\omega$:

$$\dot{A} = S(A) - \frac{p}{2\kappa} \sin (\Phi - \omega t)$$

$$\dot{\theta} = \frac{C(A)}{A} - \frac{p}{2\kappa A} \cos (\Phi - \omega t)$$

where

$$S(A) = \frac{1}{2\pi\kappa} \int_0^{2\pi} f(A \cos \Phi, -A\kappa \sin \Phi) \sin \Phi \, d\Phi$$

$$C(A) = \frac{1}{2\pi\kappa} \int_0^{2\pi} f(A \cos \Phi, -A\kappa \sin \Phi) \cos \Phi \, d\Phi$$

The steady-state solution may be determined by employing the conditions $A = A_0$, $\psi = \Phi - \omega t = \psi_0$:

$$\frac{p^2}{4\kappa^2} = S^2(A_0) + [C(A_0) + A_0(\kappa - \omega)]^2$$

$$\tan \psi_0 = \frac{S(A_0)}{C(A_0) + A_0(\kappa - \omega)}$$

This steady-state solution will now be perturbed and the stability of the ensuing motion investigated. Let

$$A(t) = A_0 + \xi(t) \qquad [\xi \ll A_0]$$

$$\psi(t) = \psi_0 + \eta(t) \qquad [\eta \ll \psi_0]$$

By Taylor's series expansion:

$$\dot{\xi} = \xi S'(A_0) - \frac{p}{2\kappa} \eta \cos \psi_0$$

$$\dot{\eta} = \frac{\xi}{A_0} [(\kappa - \omega) + C'(A_0)] + \frac{p}{2\kappa A_0} \eta \sin \psi_0$$

where primes indicate differentiation with respect to A. These two differential equations are satisfied by the solutions

$$\xi = \mathcal{A}e^{zt} \qquad \eta = \mathcal{B}e^{zt}$$

where $\mathcal{A}$ and $\mathcal{B}$ are arbitrary constants and

$$z = \frac{1}{2A_0}\left\{[S(A_0) + A_0S'(A_0)] \pm \sqrt{[S(A_0) + A_0S'(A_0)]^2 - 4A_0\bar{p}\frac{d\bar{p}}{dA_0}}\right\}$$

and $\bar{p} = p/2\kappa$.

For stability, the real parts of z must be negative; hence, the following criteria can be established:[34]

$$[S(A_0) + A_0S'(A_0)] < 0, \frac{d\bar{p}}{dA_0} > 0, \text{ ensures stability}$$

$$[S(A_0) + A_0S'(A_0)] < 0, \frac{d\bar{p}}{dA_0} < 0, \text{ ensures instability}$$

$$[S(A_0) + A_0S'(A_0)] > 0, \frac{d\bar{p}}{dA_0} \gtrless 0, \text{ ensures instability}$$

$$[S(A_0) + A_0S'(A_0)] = 0, \frac{d\bar{p}}{dA_0} > 0, \text{ ensures stability}$$

These criteria can be interpreted in terms of response curves by reference to Fig. 4.14. For systems of this type, $[S(A_0) + A_0S'(A_0)] < 0$; when $\frac{d\bar{p}}{dA_0} > 0$, $\bar{p}$ increases as A_0 also increases. This does not hold for the middle branch of the response curves, thus confirming the earlier results. Other analyses of stability are found in Refs. 35 to 38.

SUPERHARMONIC AND SUBHARMONIC VIBRATIONS

INTRODUCTION

Steady-state response of a nonlinear system also may occur at some multiple or submultiple of the forcing frequency. The response depends on both the frequency and the amplitude of the excitation, and may vary with time as well as with different initial conditions. Such vibration can be analyzed by several of the methods discussed above: The Kryloff and Bogoliuboff method,[39] the Ritz method,[40] a method employing Fourier series with either a perturbation or an iteration procedure used to determine the coefficients,[10] etc. Other studies are given in Refs. 41 to 43, as well as in many of the other references of this chapter.

SUPERHARMONIC RESPONSE

Consider forced vibration of a system defined by the differential equation

$$\ddot{x} + 2\delta\kappa\dot{x} + \kappa^2(x + \mu^2x^3) = p\cos\omega t \tag{4.73}$$

The solution of Eq. (4.73) may be of the form

$$x = A_1\cos\omega t + A_3\cos 3\omega t \tag{4.74}$$

Typical values for the coefficients A_1 and A_3 are illustrated graphically in Fig. 4.32 [2] as a function of the frequency ratio ω/ω_n, where ω_n is the frequency of free vibration obtained from Eq. (4.73) by letting $\mu^2 = 0$.

If the term with frequency 3ω (i.e., the third harmonic) in Eq. (4.74) is neglected, the response can be described by plotting the coefficient A_1 against frequency, as shown by the broken lines *a* in Fig. 4.32. Such response is designated harmonic response; the resonance condition (at $\omega/\omega_n = 1$) is designated *harmonic resonance* and the peak is bent to the right as a consequence of the assumed hardening type of nonlinearity. When the third harmonic is included, the response is described by the two coefficients A_1 and A_3,

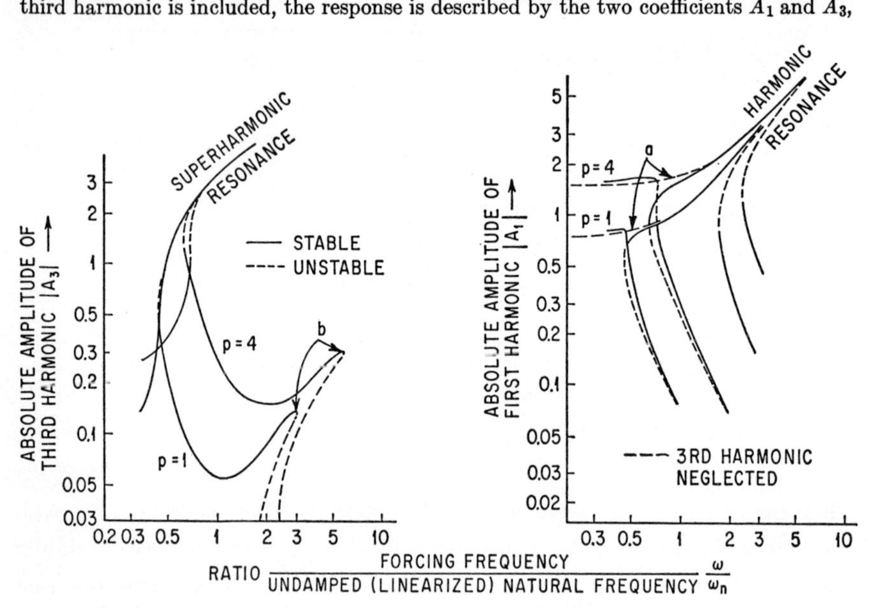

FIG. 4.32. Response curves for system defined by Eq. (4.73) with harmonic response. (*After Clauser.*[2])

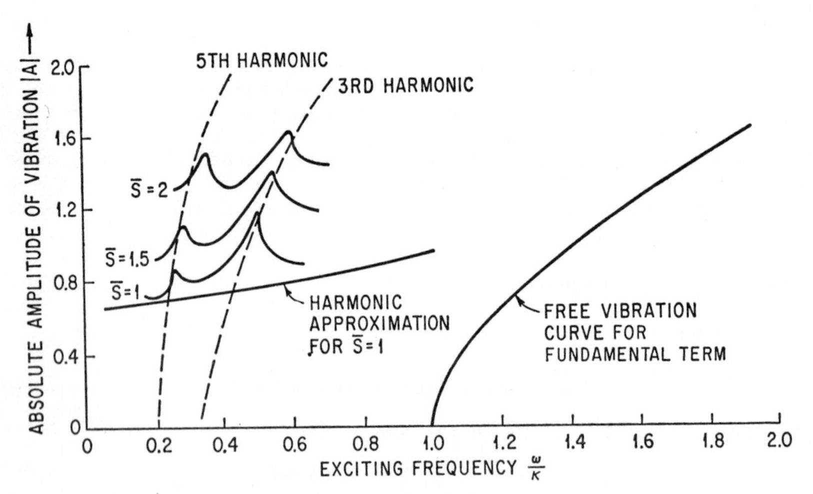

FIG. 4.33. Response curves for system with harmonic response in the third and fifth harmonics, where $\bar{S} = \frac{p}{\kappa^2}\sqrt{\frac{3}{4}\mu^2}$ and $\delta = 0.01$. (*After Atkinson.*[43])

plotted separately in Fig. 4.32. The latter is characterized by the following features:

1. A peak in the region $\omega/\omega_n = 1$; this represents a vibration at a frequency of approximately $3\omega_n$ superimposed upon the resonance of the harmonic response whose frequency is ω_n. This peak (b in Fig. 4.32) is designated the *third-order component of the harmonic resonance.*

2. A peak in the region $\omega/\omega_n = 1/3$; this represents vibration at a frequency of approximately ω_n when the forcing frequency is $\omega_n/3$. This region of the response is designated *superharmonic resonance.** It is accompanied by a decrease in A_1, as indicated in Fig. 4.32; thus, the value of A_3 at the frequency of the superharmonic resonance may exceed the value of A_1.

If terms of frequency 5ω, 7ω ... with corresponding coefficients A_5, A_7, ... had been included in Fig. 4.32, the same general pattern would have been repeated at the frequency ratios $\omega/\omega_n = 1/5, 1/7$ .... The vibration amplitude of the response may then be expressed as $A = A_1 + A_3 + \cdots$ and is illustrated graphically as a sequence of superharmonic resonances superposed on the approximate harmonic response, as shown in Fig. 4.33.[43] The jump phenomenon described with reference to Fig. 4.16 also may occur with superharmonic resonance. If the damping is very small, the superharmonic resonances may be of greater magnitude than the first-order component of the harmonic resonance because the nonlinearity provides a mechanism whereby energy from the excitation may be transferred to the higher harmonics.†

SUBHARMONIC RESPONSE

The subharmonic response of the system described by Eq. (4.73) may be studied by assuming a solution of the form

$$x = A_{1/3} \cos \tfrac{1}{3}\omega t + A_1 \cos \omega t \tag{4.75}$$

Typical values for the coefficients $A_{1/3}$ and A_1 are illustrated graphically in Fig. 4.34 as a function of the frequency ratio ω/ω_n, where ω_n is again the frequency of the free vibration obtained from Eq. (4.73) by letting $\mu^2 = 0$.

If the term with frequency $(1/3)\omega$ (i.e., the one-third harmonic) in Eq. (4.75) is neglected, the response can be described by the coefficient A_1 plotted against frequency, as shown by the lines c in Fig. 4.34.[2] Such response is designated harmonic response, and the resonance condition is designated harmonic resonance. When the one-third harmonic is included, the response is defined by the coefficients $A_{1/3}$ and A_1, plotted separately in Fig. 4.34. The response is characterized by a significant vibration $A_{1/3} \sin (1/3)\omega t$ in the region of $\omega/\omega_n = 3$; physically, this means that the system vibrates at the frequency $\omega = \omega_n$ when the forcing frequency is $3\omega_n$. This is the *subharmonic resonance.* The harmonic response of the system is modified by the subharmonic resonance and is characterized by the solid (stable condition) and dotted (unstable condition) lines D in Fig. 4.34. The subharmonic response offers an alternate mode of vibration rather than a modification of the fundamental response. Since the subharmonic response is unstable at the junction with the fundamental response, the subharmonic cannot simply arise but must be induced by a disturbance of sufficient magnitude. Again, jump phenomena and subharmonics of higher orders are possible.

Consider the undamped system defined by

$$\omega^2 x'' + \kappa^2(x + \mu^2 x^3) = p \cos \omega t \tag{4.76}$$

Introducing an assumed solution of the form of Eq. (4.75):

$$\begin{gathered}\left(\kappa^2 - \frac{\omega^2}{9}\right) A_{1/3} + \tfrac{3}{4}\kappa^2\mu^2(A_{1/3}^3 + A_{1/3}^2 A_1 + 2A_{1/3}A_1^2) = 0 \\ (\kappa^2 - \omega^2)A_1 + \tfrac{1}{4}\kappa^2\mu^2(A_{1/3}^3 + 6A_{1/3}^2 A_1 + 3A_1^3) = p\end{gathered} \tag{4.77}$$

* The term *ultraharmonic resonance* sometimes is used in place of the term superharmonic resonance.

† Even as well as odd harmonics may appear in the superharmonic response.[40, 43]

These equations can be studied by an iteration procedure starting from the linear forced vibration.[10] Since $A_{1/3}$ is presumed to exist, Eqs. (4.77) can be rewritten

$$\omega^2 = 9\kappa^2 + {}^{27}\!/\!_4\kappa^2\mu^2(A_{1/3}^2 + A_{1/3}A_1 + 2A_1^2)$$

$$-8\kappa^2 A_1 = p + (\omega^2 - 9\kappa^2)A_1 - {}^1\!/\!_4\kappa^2\mu^2(A_{1/3}^3 + 6A_{1/3}^2 A_1 + 3A_1^3)$$

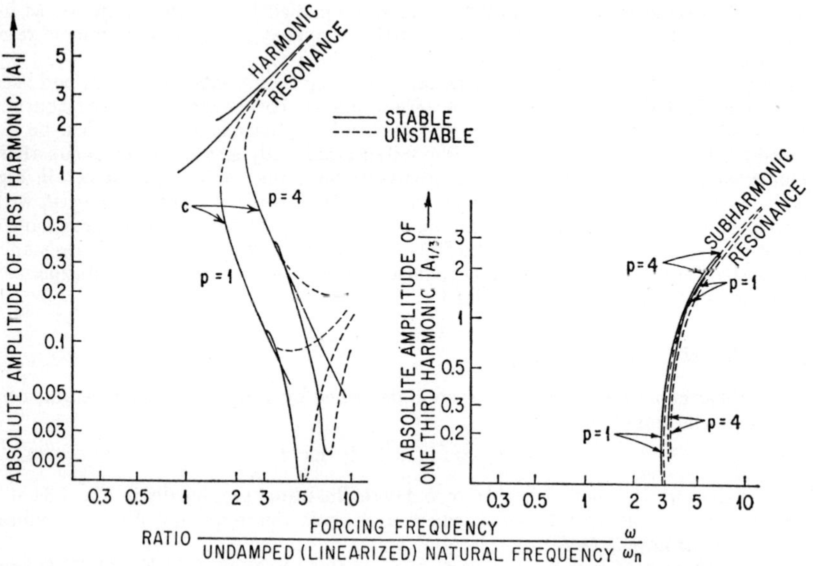

FIG. 4.34. Response curves for system defined by Eq. (4.74) with subharmonic response. (*After Clauser.*[2])

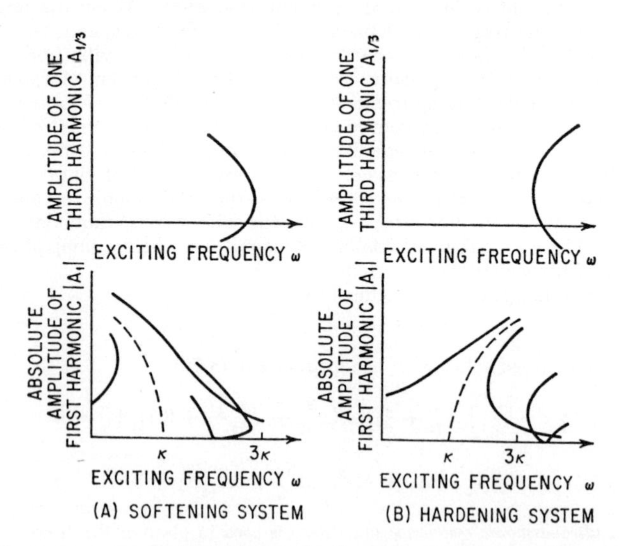

FIG. 4.35. Response curves for hardening and softening systems with subharmonic response, defined by Eq. (4.76). (*After Stoker.*[10])

Beginning the iteration with $\mu = 0$ (linear forced vibration) yields

$$\omega = 3\kappa \qquad A_1 = -\frac{p}{8\kappa}$$

The next step becomes

$$\omega^2 = 9\kappa + {}^{27}\!/_{4}\kappa^2\mu^2\left(A_{1/3}^2 - \frac{p}{8\kappa}A_{1/3} + \frac{p^2}{32\kappa^2}\right)$$

$$A_1 = -\frac{p}{8\kappa} + {}^{1}\!/_{32}\mu^2\left(A_{1/3}^3 + \frac{21}{8}\frac{p}{\kappa}A_{1/3}^2 - \frac{27}{64}\frac{p^2}{\kappa^2}A_{1/3} + \frac{51}{512}\frac{p^3}{\kappa^3}\right)$$

This solution applies to both the hardening system $(x + \mu^2x^3)$ and the softening system $(x - \mu^2x^3)$.

In the response diagram showing $A_{1/3}$ as a function of ω, ω^2 has a minimum for $+\mu^2$ and a maximum for $-\mu^2$ corresponding to

$$A_{1/3} = \frac{p}{16\kappa}$$

$$\omega^2 = 9[\kappa^2 + ({}^{21}\!/_{1024})\mu^2p^2]$$

Thus, the subharmonic vibration can exist only for

$$\omega < 9[\kappa^2 + ({}^{21}\!/_{1024})\mu^2p^2] \qquad \text{(hardening system)}$$

$$\omega > 9[\kappa^2 + ({}^{21}\!/_{1024})\mu^2p^2] \qquad \text{(softening system)}$$

Response curves for both systems are shown in Fig. 4.35.

This analysis can be extended to include linear damping [10]; then the subharmonic of order $1/3$ cannot occur unless the damping is very small.

SYSTEMS OF MORE THAN A SINGLE DEGREE-OF-FREEDOM

Interest in systems of more than one degree-of-freedom arises from the problem of the dynamic vibration absorber. The earliest studies of nonlinear two degree-of-freedom systems were those of vibration absorbers having nonlinear elements.[44–46]

The analysis of multiple degree-of-freedom systems can be carried out by various of the methods described earlier in this chapter; thus, a stepwise linear system is treated in Ref. 47, and more general systems are treated in Refs. 47 to 51. The extension of phase-plane methods to such systems is given in Refs. 29, 30, and 52.

All the essential features of nonlinear vibration of single degree-of-freedom systems described earlier occur in the multiple degree-of-freedom systems as well. An analysis which considers subharmonic vibration in such systems by an iteration procedure is given in Ref. 53 and is completely analogous to that given here for the single degree-of-freedom system, with analogous results.

REFERENCES

Additional citations to the literature are given in Refs. 4, 29, 32, 54, and 55, and in past and current issues of the journal *Applied Mechanics Reviews*.

1. von Kármán, T.: *Bull. Am. Math. Soc.*, **46**: 615 (1940).
2. Clauser, F. H.: *J. Aeronaut. Sci.*, **23**:411 (1956).
3. Davis, S. A.: *Product Eng.*, **25**:181 (1954).
4. McLachlan, N. W.: "Ordinary Nonlinear Differential Equations," Oxford University Press, New York, 1956.
5. Minorsky, N.: "Introduction to Nonlinear Mechanics," Edwards Bros., Ann Arbor, Mich., 1947.
6. Klotter, K.: *Proc. 1st U.S. Natl. Congr. Appl. Mechs.*, 1951.

7. Klotter, K.: *Stanford Univ., Rept.* 17, Contract N6 ONR-251-II, 1951.
8. Klotter, K.: *Proc. Symposium on Nonlinear Circuit Analysis*, 1953.
9. Levy, H., and E. A. Baggott: "Numerical Solutions of Differential Equations," Dover Publications, New York, 1950.
10. Stoker, J. J.: "Nonlinear Vibrations," Interscience Publishers, Inc., New York, 1950.
11. Kryloff, N., and N. Bogoliuboff: "Introduction to Nonlinear Mechanics," Princeton University Press, Princeton, N.J., 1943.
12. Andronow, A. A., and C. E. Chaikin: "Theory of Oscillations," Princeton University Press, Princeton, N.J., 1949.
13. Rudenberg, R.: *ZAMM*, **3**:454 (1923).
14. Brock, J. E.: *J. Appl. Mechanics*, **18** (1951).
15. Roberson, R. E.: *J. Appl. Mechanics*, **20**:237 (1953).
16. Wylie, C. R.: *J. Franklin Inst.*, **236**:273 (1943).
17. Young, D.: *Proc. 1st Midwest. Conf. Solid Mechanics*, 1953.
18. Fifer, S.: *J. Appl. Phys.*, **22**:1421 (1951).
19. Minorsky, N.: "Advances in Applied Mechanics," Vol. I, Academic Press, Inc., New York, 1948.
20. Bellin, A. I.: "Advances in Applied Mechanics," Vol. III, Academic Press, Inc., New York, 1953.
21. Pipes, L. A.: *J. Appl. Phys.*, **13**:117 (1942).
22. Duffing, G.: "Erzwungene Schwingungen bei veranderlicher Eigenfrequenz," F. Vieweg u Sohn, Brunswick, 1918.
23. Rauscher, M.: *J. Appl. Mechanics*, **5**:169 (1938).
24. Minorsky, N.: *J. Franklin Inst.*, **248**:205 (1949).
25. Carrier, G. F.: *Quart. Appl. Math.*, **3**:157 (1945).
26. Schwesinger, G.: *J. Appl. Mechanics*, **17**:202 (1950).
27. Abramson, H. N.: *Product Eng.*, **25**:179 (1954).
28. Evaldson, R. L., R. S. Ayre, and L. S. Jacobsen: *J. Franklin Inst.*, **248**:473 (1949).
29. Ku, Y. H.: "Analysis and Control of Nonlinear Systems," The Ronald Press Company, New York, 1958.
30. Flügge-Lotz, I.: "Discontinuous Automatic Control," Princeton University Press, Princeton, N.J., 1953.
31. Jacobsen, L. S.: *J. Appl. Mechanics*, **19**:543 (1952).
32. Jacobsen, L. S., and R. S. Ayre: "Engineering Vibrations," McGraw-Hill Book Company, Inc., New York, 1958.
33. Bishop, R. E. D.: *Proc. Inst. Mech. Engrs.*, **168**:299 (1954).
34. Klotter, K., and E. Pinney: *J. Appl. Mechanics*, **20**:9 (1953).
35. John, F.: "Studies in Nonlinear Vibration Theory," New York University, 1946.
36. Rosenberg, R. M.: *Proc. 2nd Natl. Congr. Appl. Mechanics*, 1954.
37. Young, D., and P. N. Hess: *Proc. 2nd Natl. Congr. Appl. Mechanics*, 1954.
38. Hayashi, C.: "Forced Oscillations in Nonlinear Systems," Nippon Pub. Co., Osaka, Japan, 1953.
39. Caughey, T. K.: *J. Appl. Mechanics*, **21**:327 (1954).
40. Burgess, J. C.: *Stanford Univ., Rept.* 27, Contract N6 ONR-251-II, 1954.
41. Wu, M. H. L.: *Proc. 1st Natl. Congr. Appl. Mechanics*, 1951.
42. Rosenberg, R. M.: *Proc. 2nd Midwest. Conf. Solid Mechanics*, 1955.
43. Atkinson, C. P.: *J. Appl. Mechanics*, **24**:520 (1957).
44. Roberson, R. E.: *J. Franklin Inst.*, **254**:205 (1952).
45. Pipes, L. A.: *J. Appl. Mechanics*, **20**:515 (1953).
46. Arnold, F. R.: *J. Appl. Mechanics*, **22**:487 (1955).
47. Soroka, W. W.: *J. Appl. Mechanics*, **17**:185 (1950).
48. Sethna, P. R.: *Proc. 2nd Natl. Congr. Appl. Mechanics*, 1954.
49. Huang, T. C.: *J. Appl. Mechanics*, **22**:107 (1955).
50. Klotter, K.: *Trans. IRE on Circuit Theory*, **CT-1** (4):13 (1954).
51. Stoker, J. J.: *Proc. 2nd Natl. Congr. Appl. Mechanics*, 1954.
52. Ku, Y. H.: *J. Franklin Inst.*, **259**:115 (1955).
53. Huang, T. C.: *Proc. 2nd Natl. Congr. Appl. Mechanics*, 1954.
54. Minorsky, N.: *Appl. Mechanics Revs.*, **4**:266 (1951).
55. Klotter, K.: *Appl. Mechanics Revs.*, **10**:495 (1957).

5

SELF-EXCITED VIBRATION

F. F. Ehrich
General Electric Co.

INTRODUCTION

GENERAL NATURE

Self-excited systems begin to vibrate of their own accord spontaneously, the amplitude increasing until some nonlinear effect limits any further increase. The energy supplying these vibrations is obtained from a uniform source of power associated with the system which, due to some mechanism inherent in the system, gives rise to oscillating forces. The nature of self-excited vibration compared to forced vibration is:[1]

In self-excited vibration the alternating force that sustains the motion is created or controlled by the motion itself; when the motion stops, the alternating force disappears.

In a forced vibration the sustaining alternating force exists independent of the motion and persists when the vibratory motion is stopped.

The occurrence of self-excited vibration in a physical system is intimately associated with the stability of equilibrium positions of the system. If the system is disturbed from a position of equilibrium, forces generally appear which cause the system to move either toward the equilibrium position or away from it. In the latter case the equilibrium position is said to be unstable; then the system may either oscillate with increasing amplitude or monotonically recede from the equilibrium position until nonlinear or limiting restraints appear. The equilibrium position is said to be stable if the disturbed system approaches the equilibrium position either in a damped oscillatory fashion or asymptotically.

The forces which appear as the system is displaced from its equilibrium position may depend on the displacement or the velocity, or both. If displacement-dependent forces appear and cause the system to move away from the equilibrium position, the system is said to be statically unstable. For example, an inverted pendulum is statically unstable. Velocity-dependent forces which cause the system to recede from a statically stable equilibrium position lead to dynamic instability.

Self-excited vibrations are characterized by the presence of a mechanism whereby a system will vibrate at its own natural or critical frequency, essentially *independent* of the *frequency* of any external stimulus. In mathematical terms, the motion is described by the unstable *homogeneous* solution to the homogeneous equations of motion. In contradistinction, in the case of "forced," or "resonant," vibrations, the *frequency* of the oscillation is *dependent* on (equal to, or a whole number ratio of) the frequency of a forcing function external to the vibrating system (e.g., shaft rotational speed in the case of rotating shafts). In mathematical terms, the forced vibration is the *particular* solution to the *nonhomogeneous* equations of motion.

Self-excited vibrations pervade all areas of design and operations of physical systems where motion or time-variant parameters are involved—aeromechanical systems (flutter, aircraft flight dynamics); aerodynamics (separation, stall, musical wind instruments, diffuser and inlet chugging); aerothermodynamics (flame instability, combustor screech); mechanical systems (machine-tool chatter); and feedback networks (pneumatic, hydraulic, and electromechanical servomechanisms).

ROTATING MACHINERY

One of the more important manifestations of self-excited vibrations, and the one that is the principal concern in this chapter, is that of rotating machinery, specifically, the self-excitation of lateral, or flexural, vibration of rotating shafts (as distinct from torsional, or longitudinal, vibration).

In addition to the description of a large number of such phenomena in standard vibrations textbooks (most typically and prominently, Ref. 1), the field has been subject to several generalized surveys.[2, 3, 4] The mechanisms of self-excitation which have been identified can be categorized as follows:

WHIRLING OR WHIPPING
- Hysteretic whirl
- Fluid trapped in the rotor
- Dry friction whip
- Fluid bearing whip
- Seal and blade-tip-clearance effect in turbomachinery
- Propeller and turbomachinery whirl

PARAMETRIC INSTABILITY
- Asymmetric shafting
- Pulsating torque
- Pulsating longitudinal loading

STICK-SLIP RUBS AND CHATTER

INSTABILITIES IN FORCED VIBRATIONS
- Bistable vibration
- Unstable imbalance

In each instance, the physical mechanism is described and aspects of its prevention or its diagnosis and correction are given. Some exposition of its mathematical analytic modeling is also included.

WHIRLING OR WHIPPING

ANALYTIC MODELING

In the most important subcategory of instabilities (generally termed whirling or whipping), the unifying generality is the generation of a tangential force, normal to an arbitrary radial deflection of a rotating shaft, whose magnitude is proportional to (or varies monotonically with) that deflection. At some "onset" rotational speed, such a force system will overcome the stabilizing external damping forces which are generally present and induce a whirling motion of ever-increasing amplitude, limited only by nonlinearities which ultimately limit deflections.

A close mathematical analogy to this class of phenomena is the concept of "negative damping" in linear systems with constant coefficients, subject to *plane* vibration.

A simple mathematical representation of a self-excited vibration may be found in the concept of negative damping. Consider the differential equation for a damped, free vibration:

$$m\ddot{x} + c\dot{x} + kx = 0 \tag{5.1}$$

This is generally solved by assuming a solution of the form

$$x = Ce^{st}$$

Substitution of this solution into Eq. (5.1) yields the characteristic (algebraic) equation

$$s^2 + \frac{c}{m}s + \frac{k}{m} = 0 \tag{5.2}$$

If $c < 2\sqrt{mk}$, the roots are complex:

$$s_{1,2} = -\frac{c}{2m} \pm iq$$

where

$$q = \sqrt{\frac{k}{m} - \left(\frac{c}{2m}\right)^2}$$

The solution takes the form

$$x = e^{-ct/2m}(A \cos qt + B \operatorname{sm} qt) \tag{5.3}$$

This represents a decaying oscillation because the exponential factor is negative, as illustrated in Fig. 5.1*A*. If $c < 0$, the exponential factor has a positive exponent and the vibration appears as shown in Fig. 5.1*B*. The system, initially at rest, begins to oscillate spontaneously with ever-increasing amplitude. Then, in any physical system, some nonlinear effect enters and Eq. (5.1) fails to represent the system realistically. Equation (5.4) defines a nonlinear system with negative damping at small amplitudes but with large positive damping at larger amplitudes, thereby limiting the amplitude to finite values:

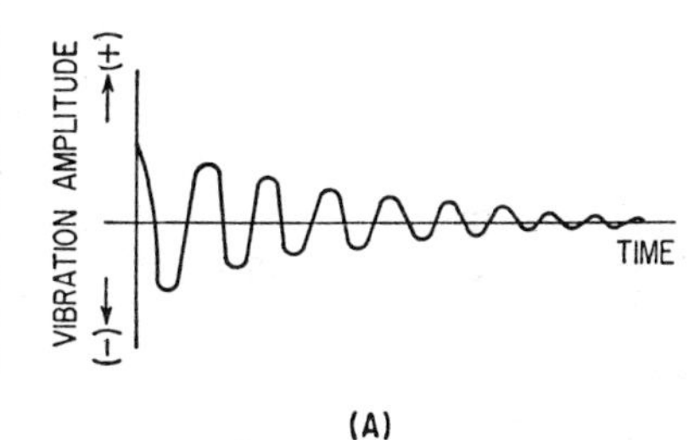

$$m\ddot{x} + (-c + ax^2)\dot{x} + kx = 0 \tag{5.4}$$

Thus, the fundamental criterion of stability in linear systems is that the roots of the characteristic equation have negative real parts, thereby producing decaying amplitudes.

In the case of a whirling or whipping shaft, the equations of motion (for an idealized shaft with a single lumped mass m) are more appropriately written in polar coordinates for the radial force balance,

$$-m\omega^2 r + m\ddot{r} + c\dot{r} + kr = 0 \tag{5.5}$$

and for the tangential force balance,

$$2m\omega\dot{r} + c\omega r - F_n = 0 \tag{5.6}$$

Fig. 5.1. (*A*) Illustration showing a decaying vibration (stable) corresponding to negative real parts of the complex roots. (*B*) Increasing vibration corresponding to positive real parts of the complex roots (unstable).

where we presume a constant rate of whirl ω.

In general, the whirling is predicated on the existence of some physical phenomenon which will induce a force F_n that is normal to the radial deflection r and is in the direction of the whirling motion—i.e., in opposition to the damping force, which tends to inhibit the whirling motion. Very often, this normal force can be characterized or approximated as being proportional to the radial deflection:

$$F_n = f_n r \tag{5.7}$$

The solution then takes the form

$$r = r_0 e^{at} \tag{5.8}$$

For the system to be stable, the coefficient of the exponent

$$a = \frac{f_n - c\omega}{2m\omega} \tag{5.9}$$

must be negative, giving the requirement for stable operation as

$$f_n < \omega c \tag{5.10}$$

As a rotating machine increases its rotational speed, the left-hand side of this inequality (which is generally also a function of shaft rotation speed) may exceed the right-hand side, indicative of the onset of instability. At this onset condition,

$$a = 0+ \tag{5.11}$$

so that whirl speed at onset is found to be

$$\omega = \left(\frac{k}{m}\right)^{1/2} \tag{5.12}$$

That is, the whirling speed at onset of instability is the shaft's natural or critical frequency, irrespective of the shaft's rotational speed (rpm). The direction of whirl may

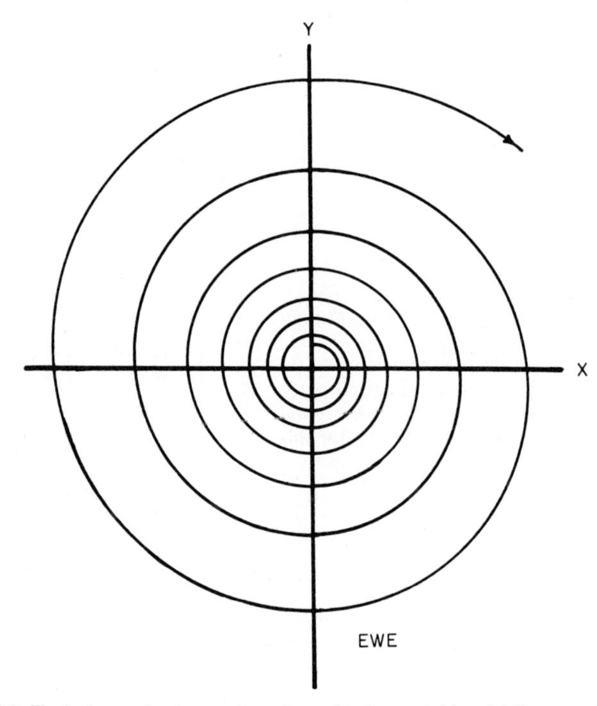

FIG. 5.2. Trajectory of rotor center of gravity in unstable whirling or whipping.

be in the same rotational direction as the shaft rotation (*forward* whirl) or opposite to the direction of shaft rotation (*backward* whirl) depending on the direction of the destabilizing force F_n.

When the system is unstable, the solution for the trajectory of the shaft's mass is, from Eq. (5.8), an exponential spiral as in Fig. 5.2. Any planar component of this two-dimensional trajectory takes the same form as the unstable planar vibration shown in Fig. 5.1*B*.

GENERAL DESCRIPTION

The most important examples of whirling and whipping instabilities are
- Hysteretic whirl
- Fluid trapped in the rotor
- Dry friction whip
- Fluid bearing whip
- Seal and blade-tip-clearance effect in turbomachinery
- Propeller and turbomachinery whirl

All these self-excitation systems involve friction or fluid energy mechanisms to generate the destabilizing force.

These phenomena are rarer than forced vibration due to unbalance or shaft misalignment, and they are difficult to anticipate before the fact or diagnose after the fact

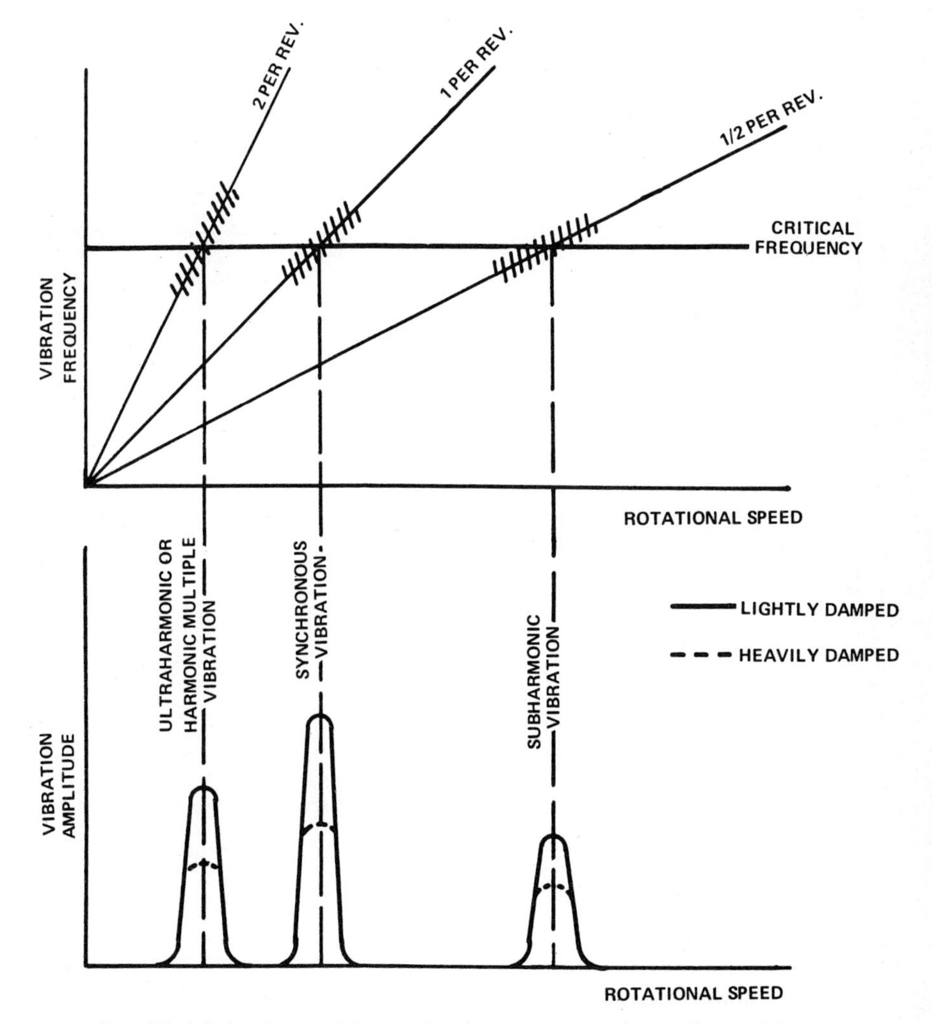

FIG. 5.3. (*A*) Attributes of forced vibration or resonance in rotating machinery.

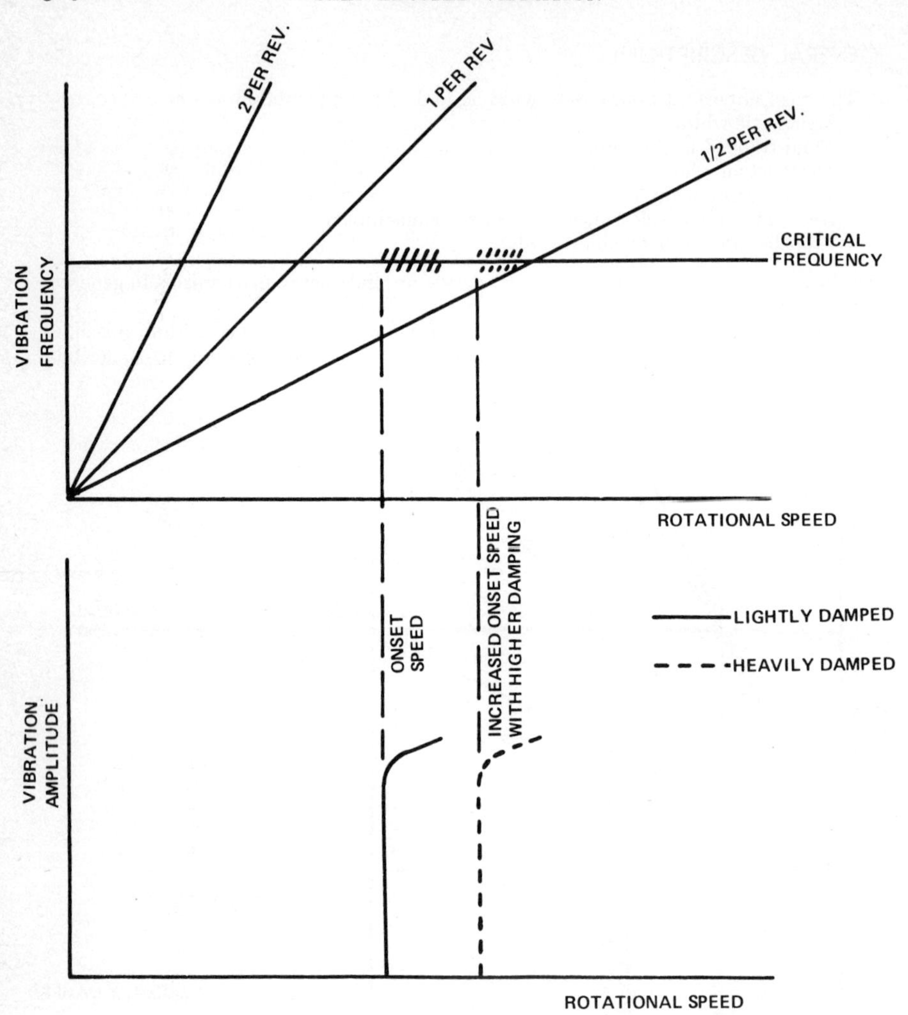

FIG. 5.3. (*B*) Attributes of whirling or whipping in rotating machinery.

because of their subtlety. Also, self-excited vibrations are potentially more destructive, since the asynchronous whirling of self-excited vibration induces alternating stresses in the rotor and can lead to fatigue failures of rotating components. Synchronous forced vibration typical of unbalance does not involve alternating stresses in the rotor and will rarely involve rotating element failure. The general attributes of these instabilities, insofar as they differ from forced excitations, are summarized in Table 5.1 and Fig. 5.3*A* and *B*.

HYSTERETIC WHIRL

The mechanism of hysteretic whirl, as observed experimentally,[5] defined analytically,[6] or described in standard texts,[7] may be understood from the schematic representation

Table 5.1. Characterization of Two Categories of Vibration of Rotating Shafts

	Forced or resonant vibration	Whirling or whipping
Vibration frequency–rpm relationship	Frequency is equal to (i.e., synchronous with) rpm or a whole number or rational fraction of rpm, as in Fig. 5.3*A*.	Frequency is nearly constant and relatively independent of rotor rotational speed or any external stimulus and is at or near one of the shaft critical or natural frequencies, as in Fig. 5.3*B*.
Vibration amplitude–rpm relationship	Amplitude will peak in a narrow band of rpm wherein the rotor's critical frequency is equal to the rpm or to a whole-number multiple or a rational fraction of the rpm or an external stimulus, as in Fig. 5.3*A*.	Amplitude will suddenly increase at an onset rpm and continue at high or increasing levels as rpm is increased, as in Fig. 5.3*B*.
Influence of damping	Addition of damping may reduce peak amplitude but not materially affect rpm at which peak amplitude occurs, as in Fig. 5.3*A*.	Addition of damping may defer onset to a higher rpm but not materially affect amplitude after onset, as in Fig. 5.3*B*.
System geometry	Excitation level and hence amplitude are dependent on some lack of axial symmetry in the rotor mass distribution or geometry, or external forces applied to the rotor. Amplitudes may be reduced by refining the system to make it more perfectly axisymmetric.	Amplitudes are independent of system axial symmetry. Given an infinitesimal deflection to an otherwise symmetric system, the amplitude will self-propagate.
Rotor fiber stress	For synchronous vibration, the rotor vibrates in frozen, deflected state, without oscillatory fiber stress.	Rotor fibers are subject to oscillatory stress at a frequency equal to the difference between rotor rpm and whirling speed.
Avoidance or elimination	1. Tune the system's critical frequencies to be out of the rpm operating range.	1. Restrict operating rpm to below instability onset rpm.
	2. Eliminate all deviations from axial symmetry in the system as built or as induced during operation (e.g., balancing).	2. Defeat or eliminate the instability mechanism.
	3. Introduce damping to limit peak amplitudes at critical speeds which must be traversed.	3. Introduce damping to raise the instability onset speed to above the operating speed range.

of Fig. 5.4. With some nominal radial deflection of the shaft, the flexure of the shaft would induce a neutral strain axis normal to the deflection direction. From first-order considerations of elastic-beam theory, the neutral axis of stress would be coincident with the neutral axis of strain. The net elastic restoring force would then be perpendicular to the neutral stress axis, i.e., parallel to and opposing the deflection. In actual fact, hysteresis, or internal friction, in the rotating shaft will cause a phase shift in the development of stress as the shaft fibers rotate around through peak strain to the neutral strain axis. The net effect is that the neutral stress axis is displaced in angle orientation from the neutral strain axis, and the resultant force is not parallel to the deflection. In particular, the resultant force has a tangential component *normal* to deflection which is the fundamental precondition for whirl. This tangential force component is in the direction of rotation and induces a *forward* whirling motion which in-

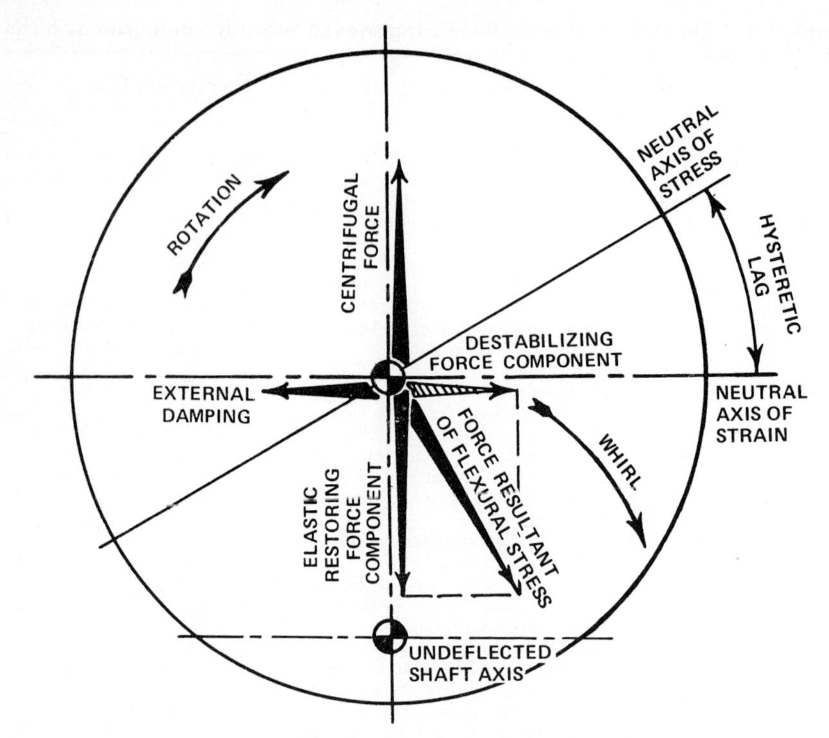

FIG. 5.4. Hysteretic whirl.

creases centrifugal force on the deflected rotor, thereby increasing its deflection. As a consequence, induced stresses are increased, thereby increasing the whirl-inducing force component.

Several surveys and contributions to the understanding of the phenomenon have been published in Refs. 8, 9, 10, and 11. It has generally been recognized that hysteretic whirl can occur only at rotational speeds above the first-shaft critical speed (the lower the hysteretic effect, the higher the attainable whirl-free operating rpm). It has been shown[12] that once whirl has started, the critical whirl speed that will be induced (from among the spectrum of criticals of any given shaft) will have a frequency approximately half the onset rpm.

A straightforward method for hysteretic whirl avoidance is that of limiting shafts to subcritical operation, but this is unnecessarily and undesirably restrictive. A more effective avoidance measure is to limit the hysteretic characteristic of the rotor. Most investigators (e.g., Ref. 5) have suggested that the essential hysteretic effect is caused by working at the interfaces of joints in a rotor rather than within the material of that rotor's components. Success has been achieved in avoiding hysteretic whirl by minimizing the number of separate elements, restricting the span of concentric rabbets and shrunk fitted parts, and providing secure lockup of assembled elements held together by tie bolts and other compression elements. Bearing-foundation characteristics also play a role in suppression of hysteretic whirl.[8]

WHIRL DUE TO FLUID TRAPPED IN ROTOR

There has always been a general awareness that high-speed centrifuges are subject to a special form of instability. It is now appreciated that the same self-excitation may be experienced more generally in high-speed rotating machinery where liquids (e.g., oil from bearing sumps, or steam condensate, etc.) may be inadvertently trapped in the internal cavity of hollow rotors. The mechanism of instability is shown schematically in Fig. 5.5. For some nominal deflection of the rotor, the fluid is flung out radially in

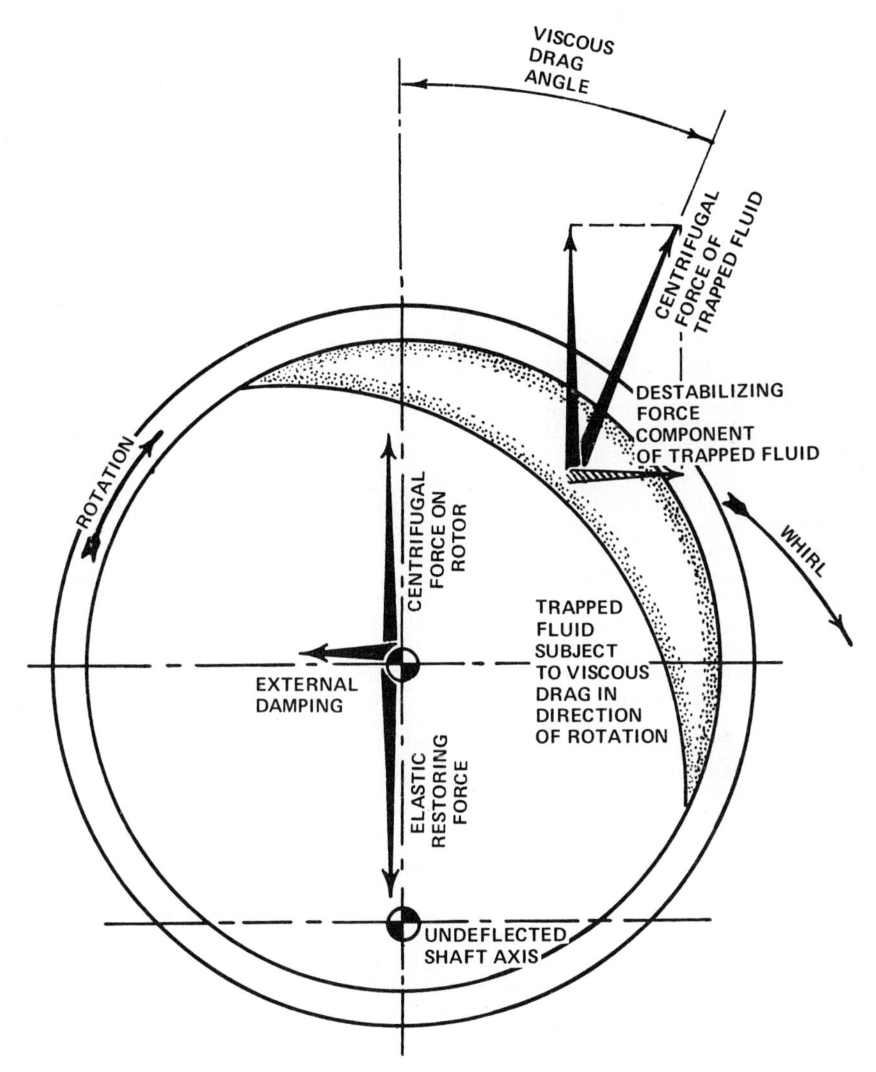

FIG. 5.5. Whirl due to fluid trapped in rotor.

the direction of deflection. But the fluid does not remain in simple radial orientation. The spinning surface of the cavity drags the fluid (which has some finite viscosity) in the direction of rotation. This angle of advance results in the centrifugal force on the fluid having a component in the tangential direction in the direction of rotation. This force then is the basis of instability, since it induces forward whirl which increases centrifugal force on the fluid and thereby increases the whirl-inducing force.

Contributions to the understanding of the phenomenon as well as a complete history of the phenomenon's study are available in Ref. 13. It has been shown[14] that onset speed for instability is always above critical rpm and below twice-critical rpm. Since the whirl is at the shaft's critical frequency, the ratio of whirl frequency to rpm will be in the range of 0.5 to 1.0.

Avoidance of this self-excitation can be accomplished by running shafting subcritically, although this is generally undesirable in centrifuge-type applications when further consideration is made of the role of trapped fluids as unbalance in forced vibration of rotating shafts (as described in Ref. 14). Where the trapped fluid is not fundamental to the machine's function, the appropriate avoidance measure, if the particular application permits, is to provide drain holes at the outermost radius of all hollow cavities where fluid might be trapped.

DRY FRICTION WHIP

As described in standard vibration texts (e.g., Ref. 15), dry friction whip is experienced when the surface of a rotating shaft comes in contact with an unlubricated stationary guide or shroud or stator system. This can occur in an unlubricated journal bearing or with loss of clearance in a hydrodynamic bearing or inadvertent closure and contact in the radial clearance of labyrinth seals or turbomachinery blading or power screws.[16]

The phenomenon may be understood with reference to Fig. 5.6. When radial contact is made between the surface of the rotating shaft and a static part, Coulomb friction will induce a tangential force on the rotor. Since the friction force is approximately proportional to the radial component of the contact force, we have the preconditions for instability. The tangential force induces a whirling motion which induces larger centrifugal force on the rotor, which in turn induces a large radial contact and hence larger whirl-inducing friction force.

It is interesting to note that this whirl system is *counter* to the shaft rotation direction (i.e., *backward* whirl). One may envision the whirling system as the rolling (accompanied by appreciable slipping) of the shaft in the stator system.

The same situation can be produced by a thrust bearing where angular deflection is combined with lateral deflection.[17] If contact occurs on the same side of the disc as the virtual pivot point of the deflected disc, then backward whirl will result. Conversely, if contact occurs on the side of the disc opposite to the side where the virtual pivot point of the disc is located, then forward whirl will result.

It has been suggested (but not concluded)[18] that the whirling frequency is generally less than the critical speed.

The vibration is subject to various types of control. If contact between rotor and stator can be avoided or the contact area can be kept well lubricated, no whipping will occur. Where contact must be accommodated, and lubrication is not feasible, whipping may be avoided by providing abradability of the rotor or stator element to allow disengagement before whirl. When dry friction is considered in the context of the dynamics of the stator system in combination with that of the rotor system,[19] it is found that whirl can be inhibited if the independent natural frequencies of the rotor and stator are kept dissimilar, that is, a very stiff rotor should be designed with a very soft mounted stator element that may be subject to rubs. No first-order interdependence of whirl speed with rotational speed has been established.

FLUID BEARING WHIP

As described in experimental and analytic literature,[20] and in standard texts (e.g., Ref. 21), fluid bearing whip can be understood by referring to Fig. 5.7. Consider some nominal radial deflection of a shaft rotating in a fluid (gas- or liquid-) filled clearance. The entrained, viscous fluid will circulate with an average velocity of about half the shaft's surface speed. The bearing pressures developed in the fluid will not be symmetric about the radial deflection line. Because of viscous losses of the bearing fluid

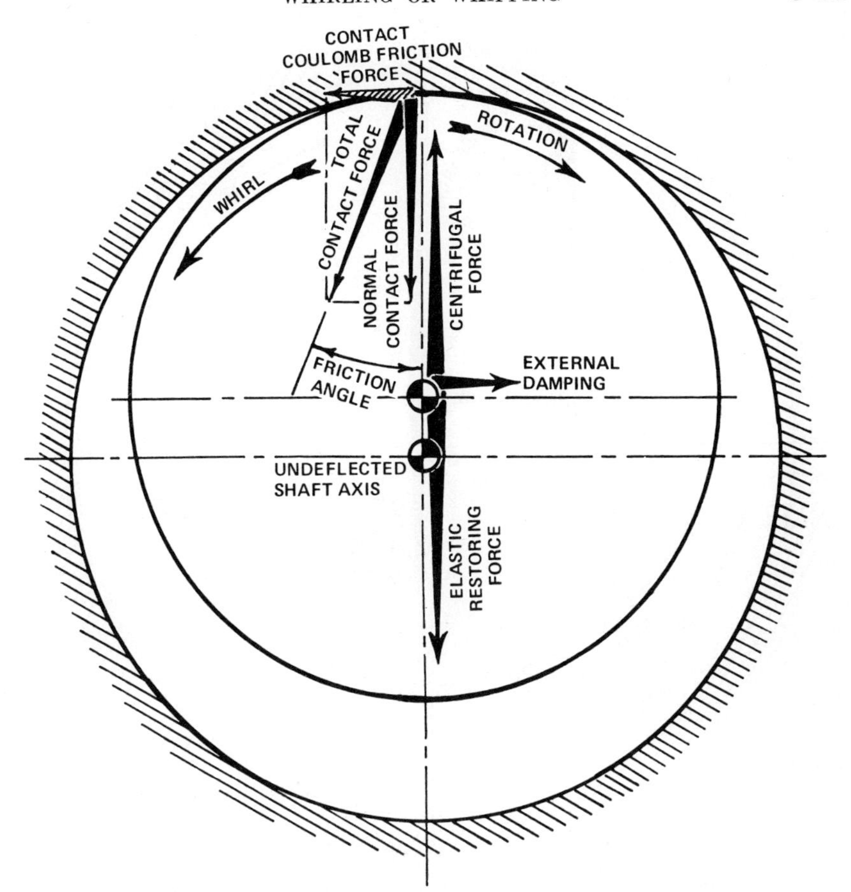

FIG. 5.6. Dry friction whip.

circulating through the close clearance, the pressure on the upstream side of the close clearance will be higher than that on the downstream side. Thus, the resultant bearing force will include a tangential force component in the direction of rotation which tends to induce *forward* whirl in the rotor. The tendency to instability is evident when this tangential force exceeds inherent stabilizing damping forces. When this happens, any induced whirl results in increased centrifugal forces; this, in turn, closes the clearance further and results in ever-increasing destabilizing tangential force. Detailed reviews of the phenomenon are available in Refs. 22 and 23.

These and other investigators have shown that to be unstable, shafting must rotate at an rpm equal to or greater than approximately twice the critical speed, so that one would expect the ratio of frequency to rpm to be equal to less than approximately 0.5.

The most obvious measure for avoiding fluid bearing whip is to restrict rotor maximum rpm to less than twice its lowest critical speed. Detailed geometric variations in the bearing runner design, such as grooving and tilt-pad configurations, have also been found effective in inhibiting instability. In extreme cases, use of rolling contact bearings instead of fluid film bearings may be advisable.

Various investigators (e.g., Ref. 24) have noted that fluid seals as well as fluid bearings are subject to this type of instability.

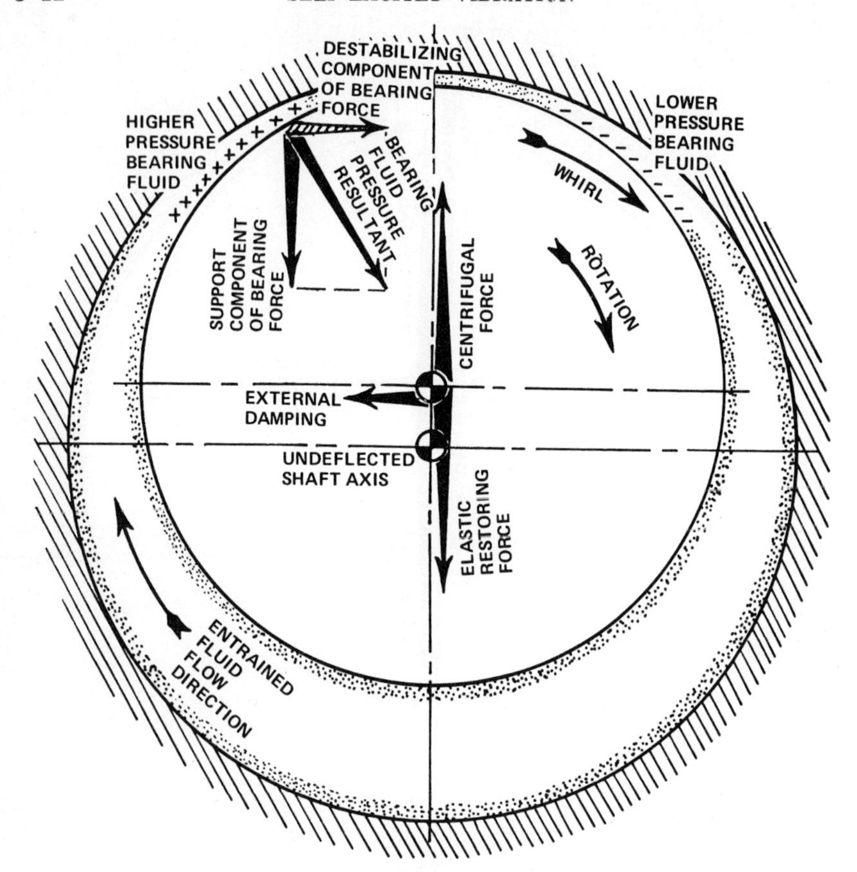

Fig. 5.7. Fluid bearing whip.

SEAL AND BLADE-TIP-CLEARANCE EFFECT IN TURBOMACHINERY

Axial-flow turbomachinery may be subject to an additional whirl-inducing effect by virtue of the influence of tip clearance on turbopump or compressor or turbine efficiency.[25] As shown schematically in Fig. 5.8, some nominal radial deflection will close the radial clearance on one side of the turbomachinery component and open the clearance 180° away on the opposite side. We would expect the closer clearance zone to operate more efficiently than the open clearance zone. For a compressor, the more efficient zone requires a lesser blade force to achieve the average pressure rise. For a turbine, a greater work extraction and blade force level is achieved in the more efficient region for a given average pressure drop. In either case, a resultant net tangential force may be generated to induce whirl in the direction of rotor rotation (i.e., forward whirl). In the case of radial-flow turbomachinery, it has been suggested[26] that destabilizing forces are exerted on an eccentric (i.e., dynamically deflected) impeller due to variations of loading of the diffuser vanes.

One text[27] describes several manifestations of this class of instability—in the thrust balance piston of a steam turbine; in the radial labyrinth seal of a radial-flow Ljungstrom counterrotating steam turbine; in the Kingsbury thrust bearing of a vertical-shaft hydraulic turbogenerator; and in the tip seals of a radial-inflow hydraulic Francis turbine.

A survey paper[3] includes a bibliography of several German papers on the subject from 1958 to 1969.

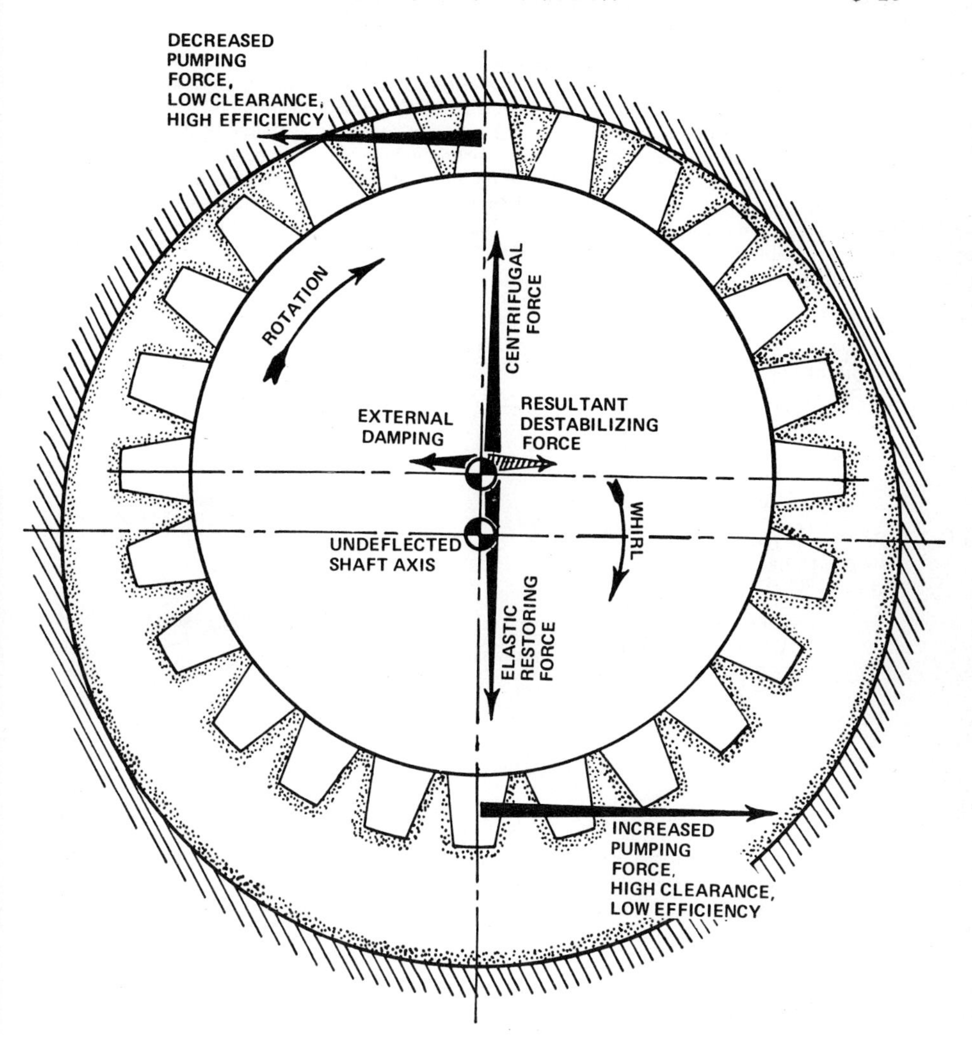

FIG. 5.8. Turbomachinery tip clearance effect's contribution to whirl.

An analysis is available[28] dealing with the possibility of stimulating flexural vibrations in the seals themselves, although it is not clear if the solutions pertain to gross deflections of the entire rotor.

It is reasonable to expect that such destabilizing forces may at least contribute to instabilities experienced on high-powered turbomachines. If this mechanism were indeed a key contributor to instability, one would conjecture that very small or very large initial tip clearances would minimize the influence of tip clearance on the unit's performance and, hence, minimize the contribution to destabilizing forces.

PROPELLER AND TURBOMACHINERY WHIRL

Propeller whirl has been identified both analytically[29] and experimentally.[30] In this instance of shaft whirling, a small deflection of the shaft will generally be accompanied by incremental velocities of the propeller blades. Where the plane of the propeller rotates as a result of the slope deflection of the shaft centerline, incremental axial velocities are induced, as shown in Fig. 5.9. The incremental velocities result in changes in relative velocities between the blades and the airstream which change the aerodynamic forces on the blades. At high airspeeds, the forces are destabilizing and can overcome damping forces to cause destructive whirling of the entire system. The whirling is generally found to be counter to the shaft-rotation direction. It has been suggested[31] that equivalent stimulation is possible in turbomachinery in the process of diagnosing and correcting whirl in an aircraft-engine compressor. An attempt has been made[32] to generalize the analysis for axial-flow turbomachinery. Although it has been shown that this analysis is in error, the general deduction seems appropriate that forward whirl may also be possible if the virtual pivot point of the deflected rotor is forward of the rotor (i.e., on the side of the approaching fluid).

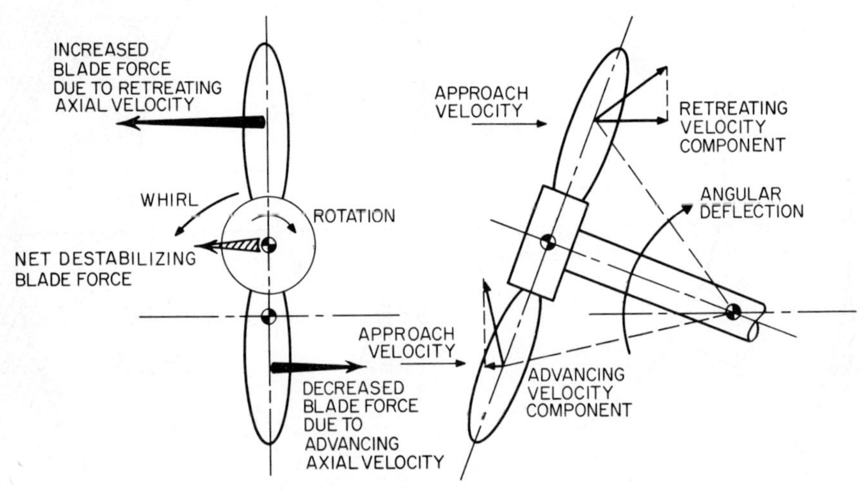

FIG. 5.9. Propeller whirl.

Instability is found to be load-sensitive in the sense of being a function of the velocity and density of the impinging flow. It is not thought to be sensitive to the torque level of the turbomachine since, for example, experimental work[30] was on an unloaded windmilling rotor. Corrective action is generally recognized to be stiffening the entire system and manipulating the effective pivot center of the whirling mode to inhibit angular motion of the propeller (or turbomachinery) disc as well as system damping.

PARAMETRIC INSTABILITY

ANALYTIC MODELING

There are systems in engineering and physics which are described by linear differential equations having periodic coefficients,

$$\frac{d^2y}{dz^2} + p(z)\frac{dy}{dz} + q(z)y = 0 \tag{5.13}$$

where $p(z)$ and $q(z)$ are periodic in z. These systems also may exhibit self-excited vibrations, but the stability of the system cannot be evaluated by finding the roots of a characteristic equation. A specialized form of this equation, which is representative of a variety of real physical problems in rotating machinery, is Mathieu's equation:

$$\frac{d^2f}{dz^2} + (a - 2q\cos 2z)f = 0 \tag{5.14}$$

A system that possesses periodically varying coefficients is a shaft whose bending rigidity differs about the two axes 1-1 and 2-2, as shown in Fig. 5.10. The shaft may develop vibration of large amplitude when it is run at submultiples of the usual critical speed. The differential equations describing the motion of the center-of-gravity of the disc carried by the shaft are

$$\ddot{y} + \left(\frac{\omega_1^2 + \omega_2^2}{2} - \frac{\omega_1^2 - \omega_2^2}{2} \cos 2\Omega t\right) y = -\frac{\omega_1^2 - \omega_2^2}{2} \sin 2\Omega t x - g$$

$$\ddot{x} + \left(\frac{\omega_1^2 + \omega_2^2}{2} + \frac{\omega_1^2 - \omega_2^2}{2} \cos 2\Omega t\right) x = -\frac{\omega_1^2 - \omega_2^2}{2} \sin 2\Omega t y$$

where ω_1 and ω_2 are the natural circular frequencies in the first bending mode about the axes 1-1 and 2-2 and Ω is the rotational speed. If the shaft is operated in the vicinity of the critical speed $\Omega = \sqrt{\frac{\omega_1^2 + \omega_2^2}{2}}$ or at one-half, one-fourth, one-sixth, etc., of the critical speed, violent vibration is likely to occur. If the motion is restricted to the x coordinate only, a Mathieu equation results.

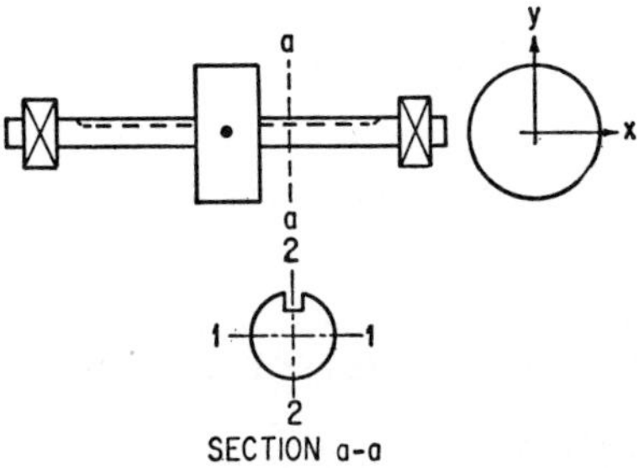

FIG. 5.10. Shaft system possessing unequal rigidities, leading to a pair of coupled inhomogeneous Mathieu equations.

An equation somewhat more general than Mathieu's equation is Hill's equation:

$$\frac{d^2y}{dz^2} + [a - 2q\psi(z)]y = 0 \qquad (5.15)$$

where $\psi(z)$ is any even function of z.

GENERAL THEORY

The following theory applies to any linear differential equation of the second order with single-valued periodic coefficients, such as Eq. (5.13).[33] If $y_1(z)$ and $y_2(z)$ are any two linearly independent fundamental solutions, the complete solution is

$$y(z) = Ay_1(z) + By_2(z) \qquad (5.16)$$

where A and B are arbitrary constants. If the period of the coefficients is τ, then $y_1(z + \tau)$ and $y_2(z + \tau)$ are also solutions and each can be expressed in terms of the fundamental solutions $y_1(z)$ and $y_2(z)$:

$$\begin{aligned} y_1(z + \tau) &= \alpha_1 y_1(z) + \alpha_2 y_2(z) \\ y_2(z + \tau) &= \beta_1 y_1(z) + \beta_2 y_2(z) \end{aligned} \qquad (5.17)$$

From Eq. (5.16),

$$y(z + \tau) = Ay_1(z + \tau) + By_2(z + \tau)$$

Substituting in this equation from Eq. (5.17),

$$y(z + \tau) = (A\alpha_1 + B\beta_1)y_1(z) + (A\alpha_2 + B\beta_2)y_2(z)$$

If $y(z + \tau) = \phi y(z)$, where ϕ is a constant, then

$$(A\alpha_1 + B\beta_1)y_1(z) + (A\alpha_2 + B\beta_2)y_2(z) = \phi Ay_1(z) + \phi By_2(z)$$

Equating the coefficients of $y_1(z)$ and $y_2(z)$ to zero,

$$A(\alpha_1 - \phi) + B\beta_1 = 0$$

$$A\alpha_2 + B(\beta_2 - \phi) = 0$$

Since $A \neq B \neq 0$,

$$\begin{vmatrix} (\alpha_1 - \phi), & \beta_1 \\ \alpha_2, & (\beta_2 - \phi) \end{vmatrix} = 0$$

Writing the determinant as a quadratic in ϕ,

$$\phi^2 - (\alpha_1 + \beta_2)\phi + \alpha_1\beta_2 - \alpha_2\beta_1 = 0 \tag{5.18}$$

Hence, ϕ can be determined if α_1, α_2, β_1, and β_2 are known. The parameters α_1, α_2, β_1, and β_2 are determined from Eqs. (5.17) and the following initial conditions:

$$y_1(0) = 1 \qquad y_1'(0) = 0 \qquad y_2(0) = 0 \qquad y_2'(0) = 1$$

It is assumed that $y_1(z)$ is an even function of z and $y_2(z)$ an odd function. Thus

$$\alpha_1 = y_1(\tau) \qquad \alpha_2 = y_1'(\tau)$$

$$\beta_1 = y_2(\tau) \qquad \beta_2 = y_2'(\tau)$$

Let $\phi \equiv e^{\mu\tau}$, where μ is a number depending on the parameters in Eq. (5.13), or on a and q in Eq. (5.14). Also, let $\Phi(z) \equiv e^{-\mu z}y(z)$. Then $\Phi(z + \tau) = e^{-\mu(z+\tau)}y(z + \tau)$. Since $y(z + \tau) = \phi y(z)$ and $\phi = e^{\mu\tau}$,

$$y(z + \tau) = e^{-\mu z}y(z) = \Phi(z)$$

Hence, $\Phi(z)$ is periodic in z with period τ and

$$y(z) = e^{\mu z}\Phi(z) \tag{5.19}$$

Consequently, a solution of Eq. (5.13) or (5.14), which itself may or may not be periodic, is of the form of the product of $e^{\mu z}$ and a periodic function of z, where $\mu = \alpha + i\beta$.

In the particular case of Mathieu's and Hill's equations where the coefficient is even in z,

$$y(-z) = e^{-\mu z}\Phi(-z) \tag{5.20}$$

is also a solution, independent of Eq. (5.19), provided either $\alpha \neq 0$ or, when $\alpha = 0$, β is nonintegral. Therefore, the complete solution of Mathieu's or Hill's equation is

$$y(z) = Ae^{\mu z}\Phi(z) + Be^{-\mu z}\Phi(-z)$$

where A and B are arbitrary constants.

Since $\Phi(z)$ is periodic, it may be expressed as

$$\Phi(z) = \sum_{r=-\infty}^{\infty} c_{2r}e^{2rzi} \qquad \Phi(-z) = \sum_{r=-\infty}^{\infty} c_{2r}e^{-2rzi}$$

Hence

$$y(z) = Ae^{\mu z}\sum_{r=-\infty}^{\infty} c_{2r}e^{2rzi} + Be^{-\mu z}\sum_{r=-\infty}^{\infty} c_{2r}e^{-2rzi} \tag{5.21}$$

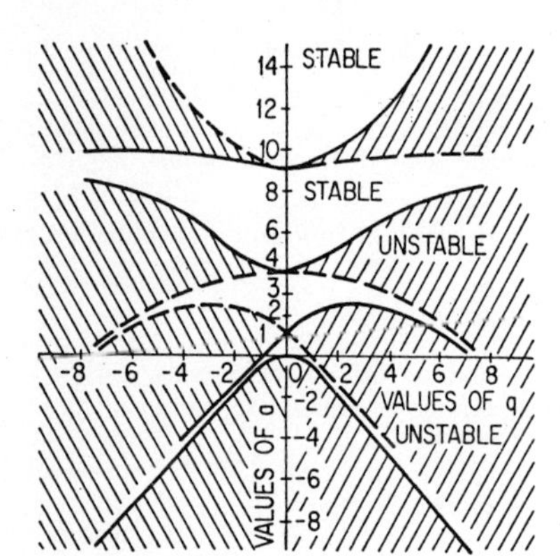

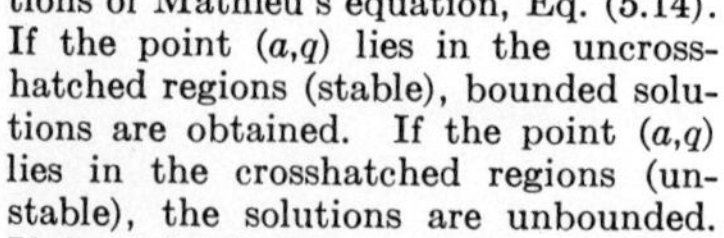

Fig. 5.11. Mathieu stability diagram showing the values of a and q which yield stable, unstable, and periodic solutions of Mathieu's equation, Eq. (5.14). If the point (a,q) lies in the uncrosshatched regions (stable), bounded solutions are obtained. If the point (a,q) lies in the crosshatched regions (unstable), the solutions are unbounded. If the point (a,q) lies on either the solid or dotted boundary curves, the solutions are periodic and are known as Mathieu functions of integral order.

MATHIEU'S STABILITY CRITERIA

The stability of the solutions to Mathieu's equation is dependent on the value of μ in Eq. (5.21), which in turn is a function of the parameters a and q in Eq. (5.14). If μ is a positive real number, the first term in Eq. (5.21) tends to infinity as $z \to \infty$; therefore, the complete solution is unstable. If μ is negative, the second term is unbounded and the complete solution also is unstable; if $\mu = \alpha + i\beta$ and $\alpha \neq 0$, the solution is unstable.

When $\mu = i\beta$ and β is nonintegral, Eq (5.21) gives the stable solution:

$$y(z) = A \sum_{r=-\infty}^{\infty} c_{2r}e^{(2r+\beta)zi} + B \sum_{r=-\infty}^{\infty} c_{2r}e^{-(2r+\beta)zi} \quad (5.22)$$

If β is a rational fraction p/s, where p and s are prime numbers, both terms in Eq. (5.22) are periodic with period $2\pi s$. When β is irrational, the solution is oscillatory, bounded but not periodic.

The regions of the a,q plane corresponding to stable and unstable solutions of Eq. (5.14) are mapped as shown in Fig. 5.11. If (a,q) lies in a stable area, Eq. (5.14) has bounded solutions. These solutions are referred to as Mathieu functions of fractional order. The limiting periodic solutions corresponding to the boundary curves of Fig. 5.11 are Mathieu functions of integral order. Figure 5.12 is an extended version of Fig. 5.11. Further details on Mathieu functions are given in Ref. 34.

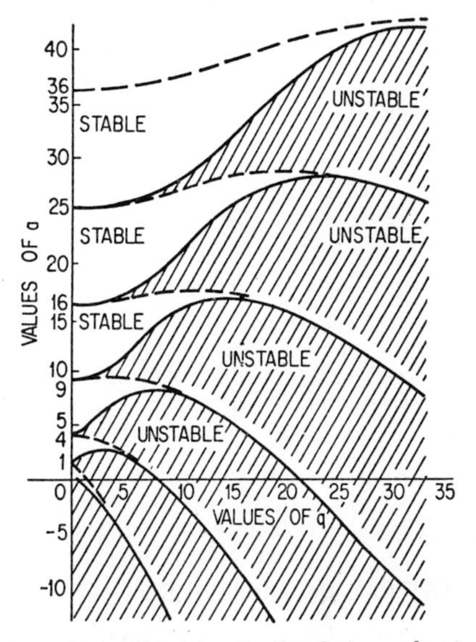

FIG. 5.12. Extension in the first quadrant of the Mathieu stability diagram of Fig. 5.11.

For example, consider a long pin-jointed column which sustains a steady axial force P_0, as shown in Fig. 5.13. Because of out of balance in associated rotating machinery, an alternating component of force $-2\gamma P_0 \cos 2\omega t$ is superimposed. The equation of lateral motion[34] is

$$EI \frac{\partial^4 y}{\partial x^4} - P_0(1 - 2\gamma \cos 2\omega t) \frac{\partial^2 y}{\partial x^2} + m_l \frac{\partial^2 y}{\partial t^2} = 0 \quad (5.23)$$

where y = lateral displacement at point x
m_l = mass per unit length
2γ = strength of alternating force component relative to sustained force

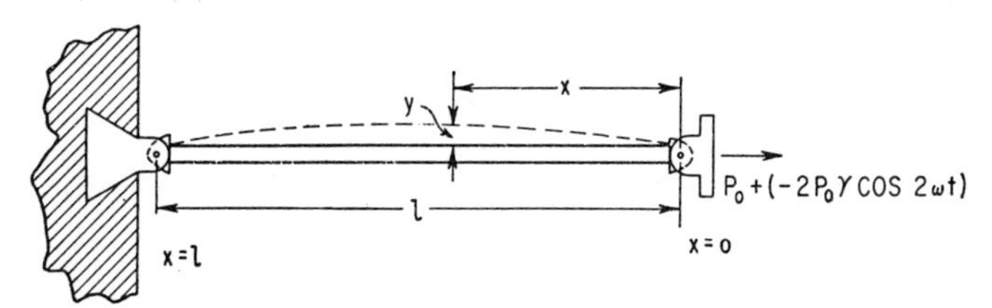

FIG. 5.13. Long column with pinned ends illustrating application of Figs. 5.11 and 5.12 for the determination of stability. A periodic force is superimposed upon a constant axial pull. (*After McLachlan.*[41])

The boundary conditions are $y = 0$, $\partial^2 y/\partial x^2 = 0$ at $x = 0$ and $x = l$. These conditions are satisfied by assuming

$$y = f(t) \sin \frac{n\pi x}{l}$$

Substituting the assumed solution in Eq. (5.23),

$$\frac{d^2 f}{dz^2} + (a - 2q \cos 2z)f = 0$$

where

$$z = \omega t$$

$$a = \frac{\left(\frac{n\pi}{l}\right)^2 \left[EI\left(\frac{n\pi}{l}\right)^2 + P_0\right]}{m_l\omega^2}$$

$$q = \frac{\gamma P_0 \left(\frac{n\pi}{l}\right)^2}{m_l\omega^2}$$

For a given value of n, the stability is determined by computing values of a and q, and observing their location in Fig. 5.11.

This general subcategory of self-excited vibrations is termed "parametric instability," since instability is induced by the effective periodic variation of the system's parameters (stiffness, inertia, natural frequency, etc.). Three particular instances of interest in the field of rotating machinery are

Lateral instability due to asymmetric shafting and/or bearing characteristics
Lateral instability due to pulsating torque
Lateral instabilities due to pulsating longitudinal compression

LATERAL INSTABILITY DUE TO ASYMMETRIC SHAFTING

As noted in the above example, if a rotor or its stator contains sufficient levels of asymmetry in the flexibility associated with its two principal axes of flexure, self-excited vibration may take place. This phenomenon is completely independent of any unbalance, and independent of the forced vibrations associated with twice-per-revolution excitation of such shafting mounted horizontally in a gravitational field.

As is described in standard vibration texts,[35] we find that presupposing a nominal whirl amplitude of the shaft at some whirl frequency, the rotation of the asymmetric shaft at an rpm different from the whirling speed will appear as periodic change in flexibility in the plane of the whirling shaft's radial deflection. This will result in an instability in certain specific ranges of rpm as a function of the degree of asymmetry. In general, instability is experienced when the rpm is approximately one-third and one-half the critical rpm and approximately equal to the critical rpm (where the critical rpm is defined with the average value of shaft stiffness), as in Fig. 5.14. The ratios of whirl frequency to rotational speed will then be approximately 3.0, 2.0, and 1.0. But with gross asymmetries, and with the additional complication of asymmetrical inertias with principal axes in arbitrary orientation to the shaft's principal axes' flexibility, no simple generalization is possible.

There is a considerable literature dealing with many aspects of the problem and substantial bibliographies.[36, 37, 38]

Stability is accomplished by minimizing shaft asymmetries and avoiding rpm ranges of instability.

LATERAL INSTABILITY DUE TO PULSATING TORQUE

Experimental confirmation[39] has been achieved that establishes the possibility of inducing first-order lateral instability in a rotor-disc system by the application of a proper combination of constant and pulsating torque. The application of torque to a shaft affects its natural frequency in lateral vibration so that the instability may also be characterized as "parametric." Analytic formulation and description of the phenomenon are available in Ref. 40 and in the bibliography of Ref. 3. The experimental work (Ref. 39) explored regions of shaft speed where the disc always whirled at the first critical speed of the rotor-disc system, regardless of the torsional forcing frequency or the rotor speed within the unstable region.

It therefore appears that combinations of ranges of steady and pulsating torque, which have been identified[40] as being sufficient to cause instability, should be avoided in the narrow-speed bands where instability is possible in the vicinity of twice the critical

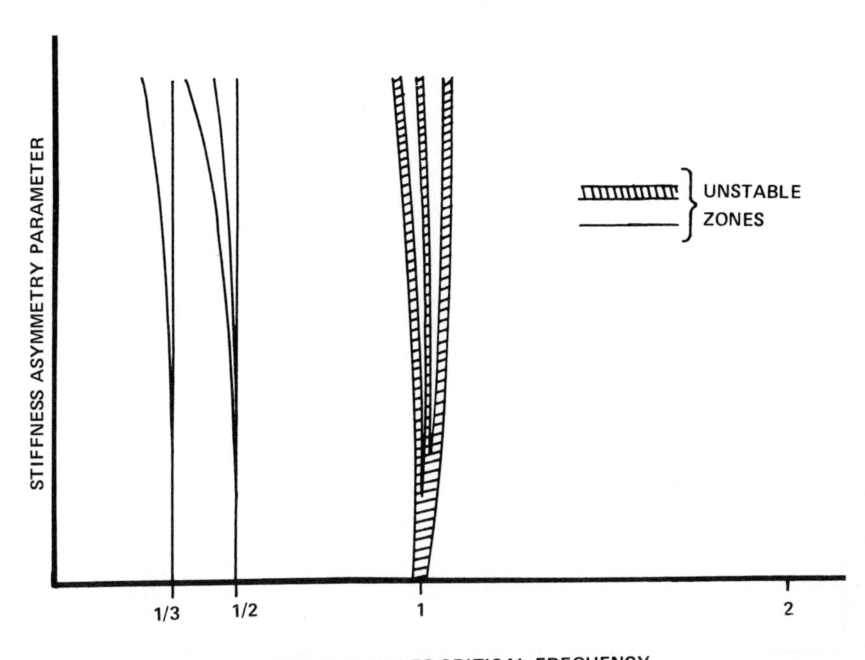

FIG. 5.14. Instability regimes of rotor system induced by asymmetric stiffness (Ref. 37).

speed and lesser instabilities at 2/2, 2/3, 2/4, 2/5, . . . times the critical frequency, as in Fig. 5.15, implying frequency/speed ratios of approximately 0.5, 1.0, 1.5, 2.0, 2.5, ...

LATERAL INSTABILITY DUE TO PULSATING LONGITUDINAL LOADS

Longitudinal loads on a shaft which are of an order of magnitude of the buckling will tend to reduce the natural frequency of that lateral, flexural vibration of the shaft. Indeed, when the compressive buckling load is reached, the natural frequency goes to zero. Therefore pulsating longitudinal loads effectively cause a periodic variation in stiffness, and they are capable of inducing "parametric instability" in rotating as well as stationary shafts,[41] as noted above in Fig. 5.13.

STICK-SLIP RUBS AND CHATTER

GENERAL DESCRIPTION

Mention is appropriate of another family of instability phenomena—stick-slip or chatter. Though the instability mechanism is associated with the dry friction contact force at the point of rubbing between a rotating shaft and a stationary element, it must not be confused with dry friction whip, previously discussed. In the case of stick-slip, as is described in standard texts (e.g., Ref. 42), the instability is caused by the irregular nature of the friction force developed at very low rubbing speeds.

At high velocities, the friction force is essentially independent of contact speed. But at very low contact speeds we encounter the phenomenon of "stiction," or breakaway friction, where higher levels of friction force are encountered, as in Fig. 5.16. Any periodic motion of the rotor's point of contact, superimposed on the basic relative contact velocity, will be self-excited. In effect, there is negative damping (as illustrated in Fig.

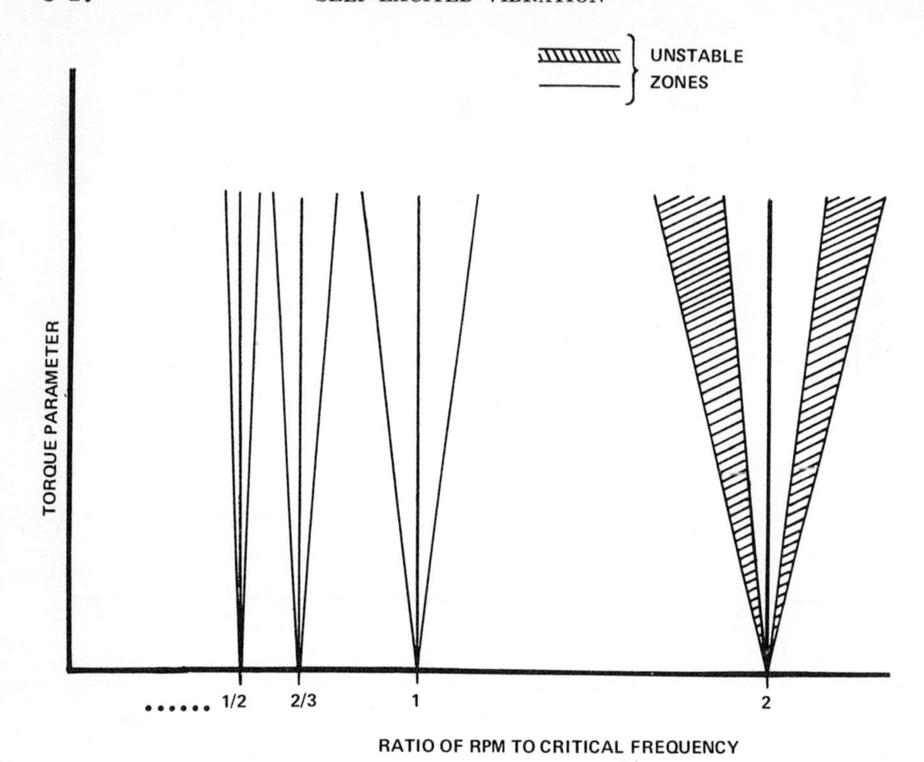

FIG. 5.15. Instability regimes of rotor system induced by pulsating torque (Ref. 40).

5.1*B*) since motion of the rotor's contact point in the direction of rotation will increase relative contact velocity and reduce stiction and the net force resisting motion. Rotor motion counter to the contact velocity will reduce relative velocity and increase friction force, again reinforcing the periodic motion. The ratio of vibration frequency to rotation speed will be much larger than unity.

While the vibration associated with stick-slip or chatter is often reported to be torsional, planar lateral vibrations can also occur. Surveys of the phenomenon are included in Refs. 43 and 44.

Measures for avoidance are similar to those prescribed for dry friction whip: avoid contact where feasible and lubricate the contact point where contact is essential to the function of the apparatus.

INSTABILITIES IN FORCED VIBRATIONS

GENERAL DESCRIPTION

In a middle ground between the generic categories of force vibrations and self-excited vibrations is the category of *instabilities in force vibrations*. These instabilities are characterized by forced vibration at a frequency equal to rotor rotation (generally induced by unbalance), but with the amplitude of that vibration being unsteady or unstable. Such unsteadiness or instability is induced by the interaction of the forced vibration on the mechanics of the system's response, or on the unbalance itself. Two manifestations of such instabilities and unsteadiness have been identified in the literature—bistable vibration and unstable imbalance.

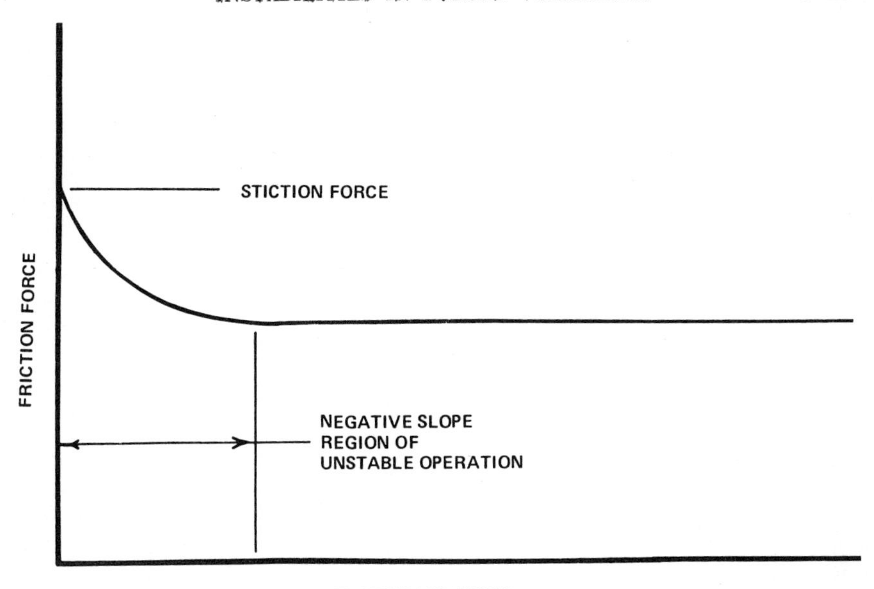

FIG. 5.16. Dry friction characteristic giving rise to stick-slip rubs or chatter.

BISTABLE VIBRATION

A classical model of one type of unstable motion is the "relaxation oscillator," or "*multivibrator*." A system subject to relaxation oscillation has *two* fairly stable states, separated by a zone where stable operation is impossible.[45] Furthermore, in each of the stable states, a mechanism exists which will induce the system to drift toward the unstable state. The system will develop a periodic motion of the general form shown in Fig. 5.17.

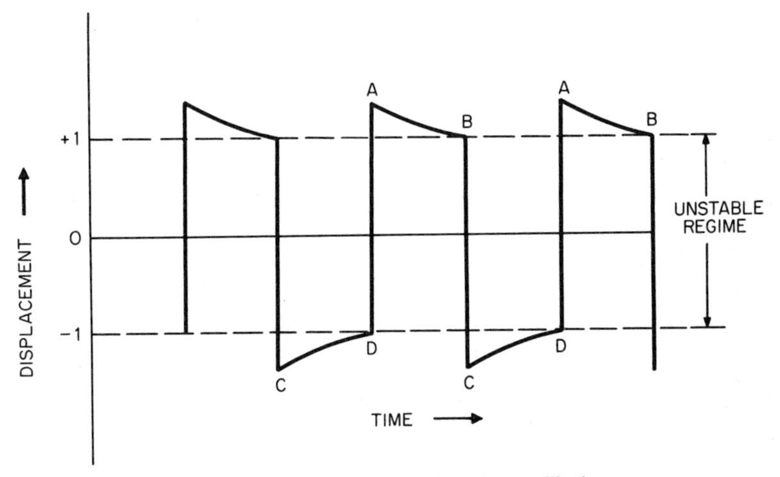

FIG. 5.17. General form of relaxation oscillations.

An idealized formulation of this class of vibration with nonlinear damping is[46]

$$m\ddot{x} + c(x^2 - 1)\dot{x} + kx = 0 \tag{5.24}$$

When the deflection amplitude x is greater than $+1$ or less than -1, as in *A-B* and *C-D*, the damping coefficient is positive, and the system is stable, although presence of a spring system k will always tend to drag the mass to a smaller absolute deflection amplitude. When the deflection amplitude lies between -1 and $+1$, as in *B-C* or *D-A*, the damping coefficient is negative and the system will move violently until it stabilizes in one of the damped stable zones.

While such systems are common in electronic circuitry, they are rather rare in the field of rotating machinery. One instance has been observed[47] in a rotor system supported by rolling element bearings with finite internal clearance. In this situation, the effective stiffness of the rotor is small for small deflections (within the clearance) but large for large deflections (when full contact is made between the rollers and the rotor and stator). Such a nonlinearity in stiffness causes a "leftward leaning" peak in the response curve when the rotor is operating in the vicinity of its critical speed and being stimulated by unbalance. In this region, two stable modes of operation are possible, as in Fig. 5.18. In region *A-B*, the rotor and stator are in solid contact through the rollers. In region *C-D*, the rotor is whirling within the clearance, out of contact. A jump in amplitude is experienced when operating from *B* to *C* or *D* to *A*. When operating at constant speed, either of the nominally stable states can drift toward instability by virtue of thermal effects on the rollers. When the rollers are unloaded, they will skid and heat up, thereby reducing the clearance. When the rollers are loaded, they will be cooled by lubrication and will tend to contract and increase clearance. In combination, these mechanisms are sufficient to cause a relaxation oscillation in the amplitude of the forced vibration.

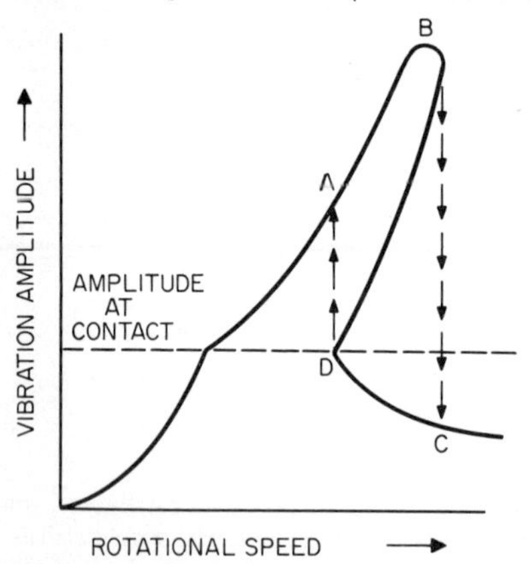

FIG. 5.18. Response of a rotor, in bearings with (constant) internal clearance, to unbalance excitation in the vicinity of its critical speed.

The remedy for this type of self-excited vibration is to eliminate the precondition of skidding rollers by reducing bearing geometric clearance, by preloading the bearing, or by increasing the temperature of any recirculating lubricant.

UNSTABLE IMBALANCE

A standard text[48] describes the occurrence of unstable vibration of steam turbines where the rotor "would vibrate with the frequency of its rotation, obviously caused by unbalance, but the intensity of the vibration would vary periodically and extremely slowly." The instability in the vibration amplitude is attributable to thermal bowing of the shaft, which is caused by the heat input associated with rubbing at the rotor's deflected "high spot," or by the mass of accumulated steam condensate in the inside of a hollow rotor at the rotor's deflected high spot. In either case, there is basis for continuous variation of amplitude, since unbalance gives rise to deflection and the deflection is, in turn, a function of that imbalance.

The phenomenon is sometimes referred to as the Newkirk effect in reference to its early recorded experimental observation.[49] A manifestation of the phenomenon in a steam turbine has been diagnosed and reported in Ref. 50 and a bibliography is available in Ref. 51. An analytic study[52] shows the possibility of both spiraling, oscillating, and constant modes of amplitude variability.

IDENTIFICATION OF SELF-EXCITED VIBRATION

Even with the best of design practice and application of the most effective methods of avoidance, the conditions and mechanisms of self-excited vibrations in rotating machinery are so subtle and pervasive that incidents continue to occur, and the major task for the vibrations engineer is diagnosis and correction.

Figure 5.3*B* suggests the forms for display of experimental data to perceive the patterns characteristic of whirling or whipping, so as to distinguish it from forced vibration, Fig. 5.3*A*. Table 5.2 summarizes particular quantitative measurements that can be made to distinguish between the various types of whirling and whipping, and other types of self-excited vibrations. The table includes the characteristic ratio of whirl speed to rotation speed at onset of vibration, and the direction of whirl with respect to the rotor rotation. The latter parameter can generally be sensed by noting the phase relation between two stationary vibration pickups mounted at 90° to one another at similar radial locations in a plane normal to the rotor's axis of rotation. Table 5.1 and specific prescriptions in the foregoing text and references suggest corrective action based on these diagnoses.

Table 5.2. Diagnostic Table of Rotating Machinery Self-excited Vibrations

	R, characteristic ratio: whirl frequency/rpm	Whirl direction
Whirling or whipping:		
Hysteretic whirl	$R \approx 0.5$	Forward
Fluid trapped in rotor	$0.5 < R < 1.0$	Forward
Dry friction whip	No functional relationship; whirl frequency a function of coupled rotor-stator system; onset rpm is a function of rpm at contact	Backward—axial contact on disc side nearest virtual pivot; Forward—axial contact on disc side opposite to virtual pivot; Backward—radial contact
Fluid bearing whip	$R < 0.5$	Forward
Seal and blade-tip-clearance effect in turbomachinery	Load-dependent	Forward—blade tip clearance; Unspecified—for seal clearance
Propeller and turbomachinery whirl	Load-dependent	Backward—virtual pivot aft of rotor; Forward—virtual pivot front of rotor (where front is source of impinging flow)
Parametric instability:		
Asymmetric shafting	$R \approx 1.0, 2.0, 3.0, \ldots$	Unspecified
Pulsating torque	$R \approx 0.5, 1.0, 1.5, 2.0, \ldots$	Unspecified
Pulsating longitudinal load	A function of pulsating load frequency rather than rpm	Unspecified
Stick-slip rubs and chatter	$R \ll 1$	Essentially planar rather than whirl motion
Instabilities in forced vibrations:		
Bistable vibration	$R = 1$ with periodic square wave fluctuations in amplitude of frequency much lower than rotation rate	Forward
Unstable imbalance	$R = 1$ with slow variation in amplitude	Forward

REFERENCES

1. Den Hartog, J. P.: "Mechanical Vibrations," 4th ed., p. 346, McGraw-Hill Book Company, Inc., New York, 1956.
2. Ehrich, F. F.: "Identification and Avoidance of Instabilities and Self-Excited Vibrations in Rotating Machinery," *ASME Paper* 72-DE-21, May 1972.
3. Kramer, E.: "Instabilities of Rotating Shafts," *Proc. Conf. Vib. Rotating Systems, Inst. Mech. Eng., London*, February 1972.
4. Vance, J. M.: "High Speed Rotor Dynamics—Assessment of Current Technology for Small Turboshaft Engines," USAAMRDL-TR-74-66, Ft. Eustis, Va., July 1974.
5. Newkirk, B. L.: *Gen. Elec. Rev.*, **27**:169 (1924).
6. Kimball, A. L.: *Gen. Elec. Rev.*, **17**:244 (1924).
7. Ref. 1, pp. 295–296.
8. Gunter, E. J.: "Dynamic Stability of Rotor-Bearing Systems," NASA SP-113, chap. 4, 1966.
9. Bolatin, V. V.: "Non-Conservative Problems of the Theory of Elastic Stability," Pergamon Press, New York, 1964.
10. Bentley, D. E.: "The Re-Excitation of Balance Resonance Regions by Internal Friction," *ASME Paper* 72-PET-49, September 1972.
11. Vance, J. M., and J. Lee: "Stability of High Speed Rotors with Internal Friction," *ASME Paper* 73-DET-127, September 1973.
12. Ehrich, F. F.: *J. Appl. Mech.*, (E) **31**(2):279 (1964).
13. Wolf, J. A.: "Whirl Dynamics of a Rotor Partially Filled with Liquids," *ASME Paper* 68-WA/APM-25, December 1968.
14. Ehrich, F. F.: *J. Eng. Ind.*, (B) **89**(4):806 (1967).
15. Ref. 1, pp. 292–293.
16. Sapetta, L. P., and R. J. Harker: "Whirl of Power Screws Excited by Boundary Lubrication at the Interface," *ASME Paper* 67-Vibr-37, March 1967.
17. Ref. 1, pp. 293–295.
18. Begg, I. C.: "Friction Induced Rotor Whirl—A Study in Stability," *ASME Paper* 73-DET-10, September 1973.
19. Ehrich, F. F.: "The Dynamic Stability of Rotor/Stator Radial Rubs in Rotating Machinery," *ASME Paper* 69-Vibr-56, April 1969.
20. Newkirk, B. L., and H. D. Taylor: *Gen. Elec. Rev.*, **28**:559–568 (1925).
21. Ref. 1, pp. 297–298.
22. Ref. 8, chap. 5.
23. Pinkus, O., and B. Sternlight: "Theory of Hydrodynamic Lubrication," chap. 8, McGraw-Hill Book Company, Inc., New York, 1961.
24. Black, H. F., and D. N. Jenssen: "Effects of High Pressure Seal Rings on Pump Rotor Vibrations," ASME Paper 71-WA/FE-38, December 1971.
25. Alford, J. S.: *J. Eng. Power*, **87**(4):333 October 1965.
26. Black, H. F.: "Calculation of Forced Whirling and Stability of Centrifugal Pump Rotor Systems," *ASME Paper* 73-DET-131, September 1973.
27. Ref. 1, pp. 317–321.
28. Ehrich, F. F.: *Trans. ASME*, (A) **90**(4):369 (1968).
29. Taylor, E. S., and K. A. Browne: *J. Aeronaut. Sci.*, **6**(2):43–49 (1938).
30. Houbolt, J. C., and W. H. Reed: "Propeller Nacelle Whirl Flutter," *I.A.S. Paper* 61–34, January 1961.
31. Trent, R., and W. R. Lull: "Design for Control of Dynamic Behavior of Rotating Machinery," *ASME Paper* 72-DE-39, May 1972.
32. Ehrich, F. F.: "An Aeroelastic Whirl Phenomenon in Turbomachinery Rotors," *ASME Paper* 73-DET-97, September 1973.
33. Floquet, G.: *Ann. l'école normale supérieure*, **12**:47 (1883).
34. McLachlan, N. W.: "Theory and Applications of Mathieu Functions," p. 40, Oxford University Press, New York, 1947.
35. Ref. 1, pp. 336–339.
36. Brosens, P. J., and Crandall, H. S.: *J. Appl. Mech.*, **83**(4):567 (1961).
37. Messal, E. E., and Bronthon, R. J.: "Subharmonic Rotor Instability Due to Elastic Asymmetry," *ASME Paper* 71-Vibr-57, September 1971.
38. Arnold, R. C., and E. E. Haft: "Stability of an Unsymmetrical Rotating Cantilever Shaft Carrying an Unsymmetrical Rotor," *ASME Paper* 71-Vibr-57, September 1971.
39. Eshleman, R. L., and R. A. Eubanks: "Effects of Axial Torque on Rotor Response: An Experimental Investigation," *ASME Paper* 70-WA/DE-14, December 1970.
40. Wehrli, V. C.: "Uber Kritische Drehzahlen unter Pulsierender Torsion," *Ing. Arch.*, **33**:73–84 (1963).

41. Ref. 34, p. 292.
42. Ref. 1, p. 290.
43. Conn, H.: *Tool Eng.*, **45**:61–65 (1960).
44. Sadowy, M.: *Tool Eng.*, **43**:99–103 (1959).
45. Ref. 1, pp. 365–368.
46. Van der Pol, B.: *Phil. Mag.*, **2**:978 (1926).
47. Ehrich, F. F.: "Bi-Stable Vibration of Rotors in Bearing Clearance," *ASME Paper* 65-WA/MD-1, November 1965.
48. Ref. 1, pp. 245–246.
49. Newkirk, B. L.: "Shaft Rubbing," *Mech. Eng.*, **48**:830 (1926).
50. Kroon, R. P., and W. A. Williams: "Spiral Vibration of Rotating Machinery," *5th Int. Congr. Appl. Mech.*, John Wiley & Sons, Inc., New York, 1939, p. 712.
51. Dimarogonas, A. D., and Sander, G. N.: *Wear*, **14**(3):153 (1969).
52. Dimarogonas, A. D.: "Newkirk Effect: Thermally Induced Dynamic Instability of High-Speed Rotors," *ASME Paper* 73-GT-26, April 1973.

6

DYNAMIC VIBRATION ABSORBERS AND AUXILIARY MASS DAMPERS

F. Everett Reed
Littleton Research and Engineering Corp.

INTRODUCTION

Auxiliary masses are frequently attached to vibrating systems by springs and damping devices to assist in controlling the amplitude of vibration of the system. Depending upon the application, these auxiliary mass systems fall into two distinct classes.

1. If the primary system is excited by a force or displacement that has a constant frequency, or in some cases by an exciting force that is a constant multiple of a rotational speed, then it is possible to modify the vibration pattern and to reduce its amplitude significantly by the use of an auxiliary mass on a spring tuned to the frequency of the excitation. When the auxiliary mass system has as little damping as possible, it is called a *dynamic absorber*.

2. If it is impossible to incorporate damping into a structure that vibrates excessively, it may be possible to provide the damping in an auxiliary system attached to the structure. When used in this manner, the auxiliary mass system is one form of a damper. (Other forms may be incorporated as an integral part of the system.) The names *damped absorber* or *auxiliary mass damper* are given to this type of system.

It is sometimes useful to analyze the auxiliary mass system in terms of its electrical analog.

FORMS OF DYNAMIC ABSORBERS AND AUXILIARY MASS DAMPERS

In its simplest form, as applied to a single degree-of-freedom system, the character of the auxiliary mass system is the same as that of the primary system. Thus a torsional system has a torsionally connected auxiliary mass; a linear system has a linear-spring connected mass; and a pendulum has an auxiliary pendulum. Examples of undamped auxiliary mass systems attached to single degree-of-freedom systems are shown in Figs. 6.1 and 6.2; examples of damped auxiliary mass systems are shown in Figs. 6.3 and 6.4. With multiple degree-of-freedom systems the attachment of the auxiliary masses is not as conventional as with the single degree-of-freedom system. For example, consider the two degree-of-freedom system shown in Fig. 6.5*A* consist-

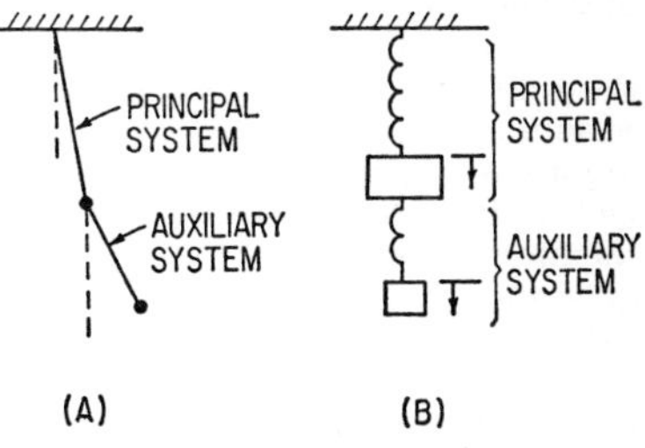

FIG. 6.1. Dynamic vibration absorbers in pendulum form (*A*) and linear form (*B*).

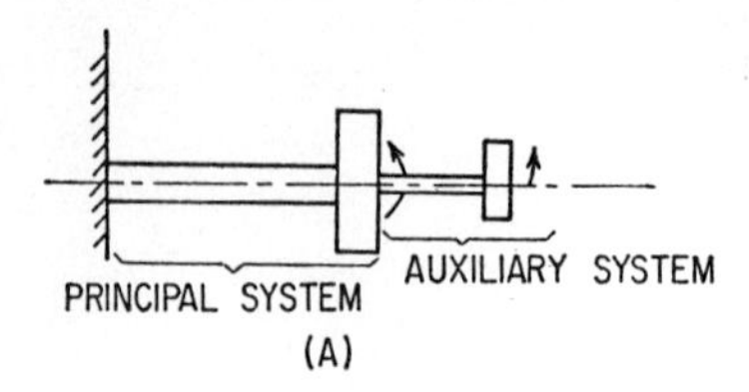

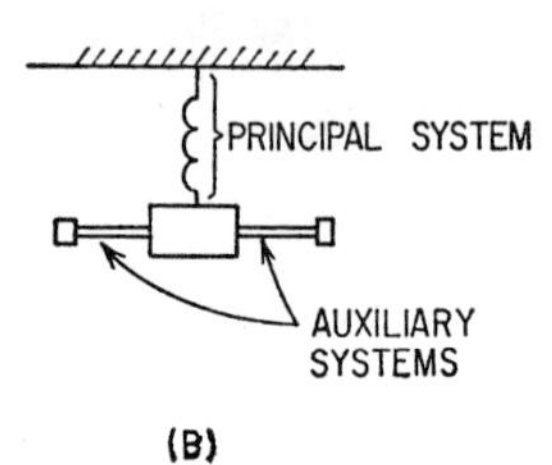

FIG. 6.2. Typical dynamic vibration absorbers. The principal and auxiliary systems vibrate in torsion in the arrangement at (*A*); the auxiliary system is in the form of masses and beams at (*B*).

ing of two masses m_1 and m_2 on a rigid, massless bar. A dynamic absorber of the type shown in Fig. 6.5*B* is effective for the vertical translational motion; however, if the auxiliary masses are on cantilever beams mounted on the rigid bar, as shown in Fig. 6.5*C*, the absorber can be made effective for both vertical translational motion and rotational motion about an axis normal to the page.

WAYS OF EXPRESSING THE EFFECTS OF AUXILIARY MASS SYSTEMS

Suppose a linear auxiliary mass system, consisting of one or more masses, springs, and dampers, is attached to a vibrating primary system. The reaction back on the primary system is proportional to the amplitude of motion at the point of attachment. It is a function of the frequency of excitation and of the masses, spring stiffnesses, and damping constants of the auxiliary mass system. If there is no damping in the auxiliary mass system, the reaction forces are either in phase or 180° out of phase with the displacement and the acceleration at the point of attachment. However, where there is damping in the auxiliary system, the reaction has a component that is 90° out of phase with the acceleration and the displacement.

Since the reaction is proportional to the amplitude of motion, it is possible to express the properties of the auxiliary mass system in terms of the motion at the point of attachment. This can be done in three ways: (1) the ratio of the reaction force to the displacement at the point of attachment, (2) the ratio of the reaction force to the velocity at the point of attachment, or (3) the ratio of the reaction force to the acceleration at the point of attachment. The first ratio can be considered equivalent to a spring whose stiffness changes with frequency. The second ratio can be considered equivalent to a damper; at any frequency it is equal in magnitude to the force-displacement ratio divided by the angular frequency. The phase angle between the force and the velocity is 90° from the phase angle between the force and the displacement. This force-velocity ratio is called the *mechanical impedance* Z of the auxiliary system. The third ratio corresponds to a mass and is designated *equivalent mass* m_{eq}. The equivalent mass of a system is $-1/\omega^2$ that of the equivalent spring k_{eq} of the system.

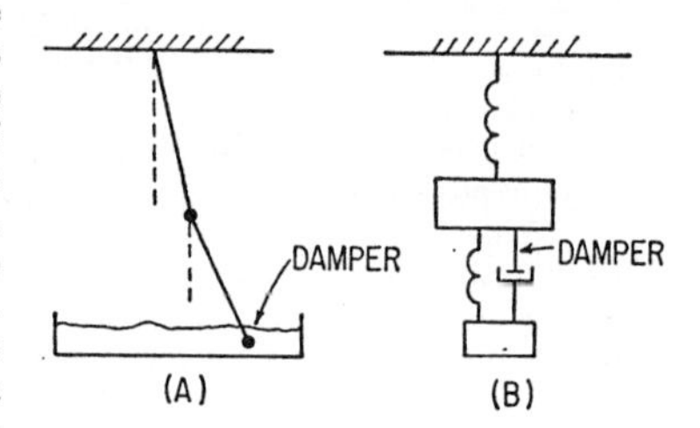

FIG. 6.3. Damped auxiliary mass systems corresponding to the undamped vibration absorbers shown in Fig. 6.1.

Because of the phase relations between the force and the displacement, velocity, and acceleration at the point of connection, it is customary to represent the ratios as complex quantities. Thus $Z = k_{eq}/j\omega = j\omega m_{eq}$. Most dynamic analyses of mechanical systems are made on purely reactive systems, i.e., systems having masses and stiffnesses only, and no damping. The effects of auxiliary mass systems are most easily understood if the effect of the auxiliary system is represented as a reactive subsystem. For this reason, and because the hypothetical addition of a mass to a system is often more easily comprehended than the addition of a spring, the effects of auxiliary mass systems are treated in terms of the equivalent masses in this chapter, i.e., in terms of the ratio of the force exerted by the auxiliary system upon the primary system to the acceleration at the point of attachment of the auxiliary system.

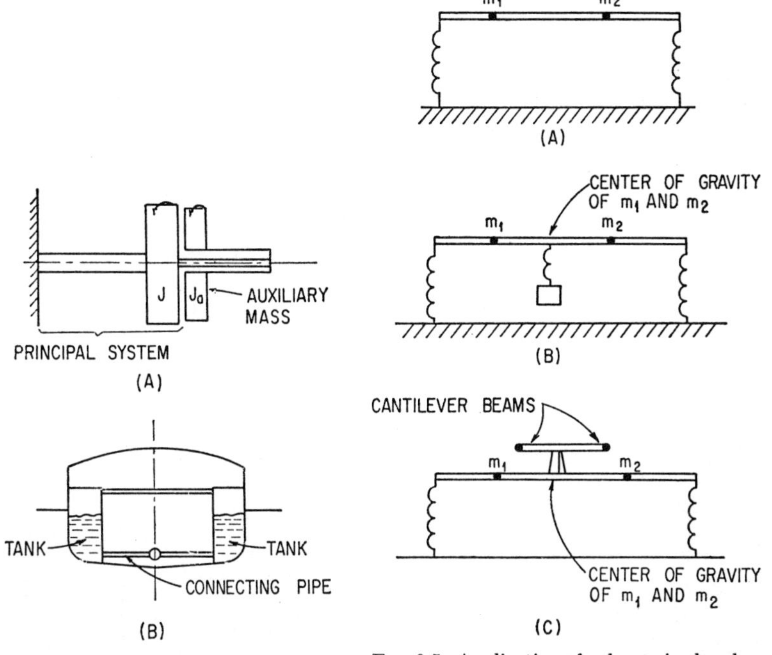

FIG. 6.4. Typical damped auxiliary mass systems. In the torsional system at (*A*), damping is provided by relative motion of the flywheels J, J_a. In the antiroll tanks for ships shown at (*B*), water flows from one tank to the other and damping is provided by a constriction in the connecting pipe.

FIG. 6.5. Application of a dynamic absorber to reduce the vibration of the spring-mounted bar at (*A*) in both vertical translational and rotational modes. The linear mass-spring system at (*B*) is effective for only translational motion, whereas the cantilever beams at (*C*) are effective for rotational as well as translational motion.

THE INFLUENCE OF A SIMPLE AUXILIARY MASS SYSTEM UPON A VIBRATING SYSTEM

The magnitude of the equivalent mass of a simple auxiliary mass system, consisting of a mass m_a, spring k_a, and viscous damper c_a, can be determined readily by evaluating the forces exerted by such a system upon a foundation vibrating at a frequency $f = \omega/2\pi$. The system with its assumed constants and displacements is shown in Fig. 6.6*A*. The spring and damping forces acting on m are shown in Fig. 6.6*B*, and the equation of motion is

$$(-k_a x_r - c_a j\omega x_r)e^{j\omega t} = -m_a(x_0 + x_r)\omega^2 e^{j\omega t}$$

Solving for x_r,

$$x_r = \frac{m_a\omega^2 x_0}{-m_a\omega^2 + jc_a\omega + k_a} \tag{6.1}$$

The force acting on the foundation is

$$Fe^{j\omega t} = (k_a + jc_a\omega)x_r e^{j\omega t}$$

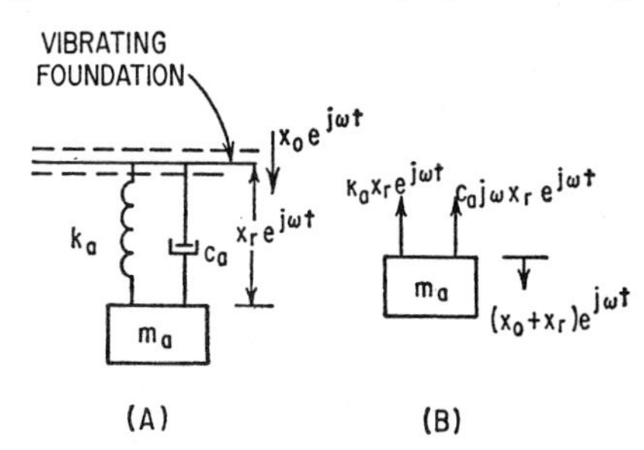

FIG. 6.6. Auxiliary mass damper. The arrangement of the damper is shown at (*A*), and the forces acting on the mass are indicated at (*B*).

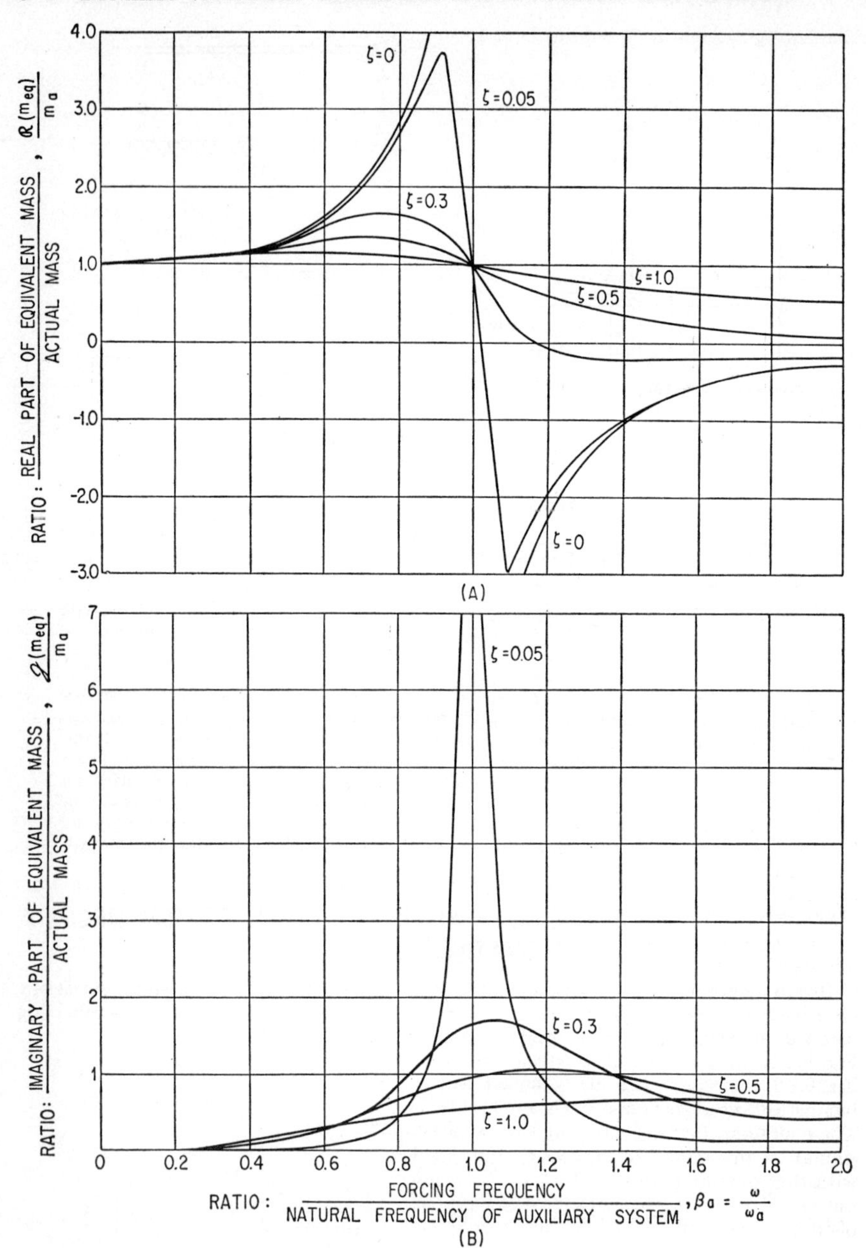

FIG. 6.7. Equivalent mass m_{eq} of the auxiliary-mass system shown in Fig. 6.6. The real part oft he equivalent mass is shown at (*A*) and the imaginary part at (*B*).

Eliminating x_r from the preceding equations,

$$F = \frac{(k_a + jc_a\omega)m_a\omega^2}{-m_a\omega^2 + jc_a\omega + k_a} x_0 \tag{6.2}$$

Since the force exerted by an equivalent mass m_{eq} rigidly attached to the moving foundation is $F = m_{eq}\omega^2 x_0$:

$$m_{eq} = \frac{(k_a + jc_a\omega)}{k_a + jc_a\omega - m_a\omega^2} m_a \tag{6.3}$$

Equation (6.3) can be written in terms of nondimensional quantities:

$$m_{eq} = \frac{1 + 2\zeta\beta_a j}{(1 - \beta_a^2) + 2\zeta\beta_a j} m_a \tag{6.4}$$

where $\beta_a = \dfrac{\omega}{\omega_a}$, a tuning parameter

$\omega_a^2 = \dfrac{k_a}{m_a}$, the natural frequency of the auxiliary system

$\zeta = \dfrac{c_a}{c_{ca}}$, a damping parameter

$c_{ca} = 2\sqrt{k_a m_a}$, critical damping of the auxiliary system

Equation (6.4) can be divided into the following real and imaginary components:

$$m_{eq} = \frac{(1 - \beta_a^2) + (2\zeta\beta_a)^2}{(1 - \beta_a^2)^2 + (2\zeta\beta_a)^2} m_a - \frac{2\zeta\beta_a^3}{(1 - \beta_a^2)^2 + (2\zeta\beta_a)^2} jm_a \tag{6.5}$$

The real and imaginary parts of m_{eq} are shown in Fig. 6.7*A* and Fig. 6.7*B*, respectively. If there is no damping, $\zeta = 0$ and

$$m_{eq} = \frac{1}{1 - \beta_a^2} m_a \tag{6.6}$$

If $\beta_a = 1$ in Eq. (6.6), m_{eq} becomes infinite and a finite force produces no displacement. Thus, the auxiliary mass enforces a point of no motion (i.e., a node) at its point of attachment.

This concept can be applied to reduce the amplitude of the forced vibration of a single degree-of-freedom system by attaching a damped absorber.[1,2] A sketch of the system with a damped auxiliary mass system is shown in Fig. 6.8*A*. In the equivalent system shown in Fig. 6.8*B*, there is no force acting on the mass m but instead the support is given a motion $ue^{j\omega t}$. The equations for the system of Fig. 6.8*B* are similar to those for the system of Fig. 6.8*A* with the value ku substituted for F. The amplitude of forced vibration of a single degree-of-freedom system, Eq. (2.24), is

$$x_0 = \frac{F/k}{1 - m\omega^2/k}$$

The effect of the auxiliary mass system is to increase the mass m of the primary system by the equivalent mass of the auxiliary system as

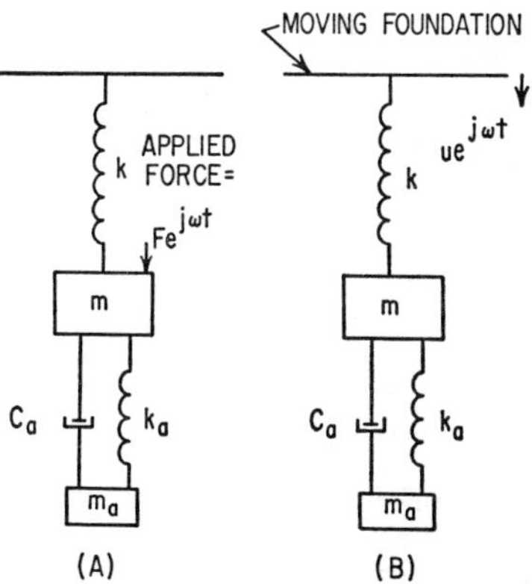

FIG. 6.8. Schematic diagram of auxiliary mass m_a coupled by a spring k_a and viscous damper c_a to a primary system k, m. The primary system is excited by the force F at (A), or alternatively by the foundation motion u at (B).

given by Eq. (6.4):

$$x_0 = \frac{F/k}{1 - \dfrac{\omega^2}{k}\left[m + m_a \dfrac{(1 + 2\zeta\beta_a j)}{(1 - \beta_a^2) + 2\zeta\beta_a j}\right]}$$

Substituting $\mu = m_a/m$, the mass ratio, $\delta_{st} = F/k$, the static deflection of the spring of the primary system, and $\beta = \sqrt{m\omega^2/k}$, the ratio of the forcing frequency to the natural frequency of the primary system, and writing in dimensionless form,

$$\frac{x_0}{\delta_{st}} = \frac{(1 - \beta_a^2) + 2\zeta\beta_a j}{(1 - \beta_a^2) + 2\zeta\beta_a j - \beta^2[(1 - \beta_a^2) + 2\zeta\beta_a j + \mu(1 + 2\zeta\beta_a j)]}$$

The amplitude of motion of the primary mass, without regard to phase, is

$$\frac{x_0}{\delta_{st}} = \left\{\frac{(1 - \beta_a^2)^2 + (2\zeta\beta_a)^2}{[(1 - \beta_a^2)(1 - \beta^2) - \beta^2\mu]^2 + (2\zeta\beta_a)^2[1 - \beta^2 - \beta^2\mu]^2}\right\}^{1/2} \tag{6.7}$$

If $\zeta = 0$ (no damping), then

$$\frac{x_0}{\delta_{st}} = \frac{1 - \beta_a^2}{(1 - \beta_a^2)(1 - \beta^2) - \beta^2\mu} \tag{6.8}$$

If $\beta_a = 1$, $x_0 = 0$; that is, the vibration of the primary system is eliminated entirely when the auxiliary system is undamped and is tuned to the forcing frequency.

THE DYNAMIC ABSORBER

If the auxiliary mass system has no damping and is tuned to the forcing frequency, it acts as a dynamic absorber and enforces a node at its point of attachment. The auxiliary mass must be sufficiently large so that it will not have an excessive amplitude.[3] For a dynamic absorber attached to the primary system at the point where the excitation is introduced, the required mass of the auxiliary body is easily determined. Since the primary mass is motionless, the force exerted by the absorber, when the amplitude of motion of the auxiliary mass is u_0, is equal and of opposite sign to the exciting force F. Hence

$$F = m_a\omega^2 u_0 \tag{6.9}$$

Since the frequency is known, the mass and amplitude of motion necessary to neutralize a given excitation force are determined by Eq. (6.9). The spring stiffness in the auxiliary system is determined by the requirement that the auxiliary system be tuned to the frequency of the exciting force:

$$k_a = m_a\omega^2 \tag{6.10}$$

Although the concept of tuning a dynamic absorber appears simple, practical considerations make it difficult to tune any system exactly. When the auxiliary mass is small relative to the mass of the primary system, its effectiveness depends upon accurate tuning. If the tuning is incorrect, the addition of the auxiliary mass may bring the composite system (primary and auxiliary systems) into resonance with the exciting force.

Consider the natural frequencies of the composite system. The natural frequency of the primary system is $\omega_0 = \sqrt{k/m}$. With this relation, Eq. (6.8) in which the damping is zero ($\zeta = 0$) becomes

$$\frac{x_0}{\delta_{st}} = \frac{1 - \omega^2/\omega_a^2}{(1 - \omega^2/\omega_a^2)(1 - \omega^2/\omega_0^2) - (\omega^2/\omega_0^2)\mu}$$

At resonance the denominator is zero and ω is designated ω_n:

$$(\omega_n^2 - \omega_a^2)(\omega_n^2 - \omega_0^2) - \omega_n^2\,\omega_a^2\mu = 0 \tag{6.11}$$

The natural frequencies are found from the roots ${\omega_n}^2$ of Eq. (6.11):

$$
{\omega_n}^2 = \frac{{\omega_a}^2(1+\mu) + {\omega_0}^2}{2} \pm \sqrt{\left[\frac{{\omega_a}^2(1+\mu) - {\omega_0}^2}{2}\right]^2 + {\omega_a}^2{\omega_0}^2\mu} \qquad (6.12)
$$

This last relation may be represented by Mohr's circle, Fig. 6.9.

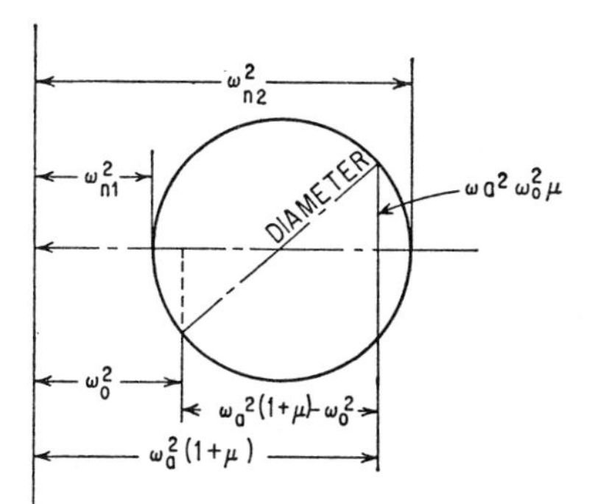

FIG. 6.9. Representation of the natural frequencies ω_n of the composite system by Mohr's circle. The circle is constructed on the diameter located by the natural frequencies ω_0, ω_a of the primary and auxiliary systems, respectively. The natural frequencies of the composite system are indicated by the intercept of the circle with the horizontal axis.

Since the absorber is nominally tuned to the frequency of the excitation, the root ω_{n2}^2 that is closer to the forcing frequency is of interest. The ratio ω_{n2}/ω_a is a measure of the sensitivity of the tuning required to avoid resonance. This is given as a function of μ for various ratios of ω_0/ω_a in Fig. 6.10. Dynamic absorbers are most generally used when the primary system without the absorber is nearly in resonance with the excitation. If the natural frequency of the primary system is less than the forcing frequency, it is preferable to tune the dynamic absorber to a frequency slightly lower than the forcing frequency to avoid the resonance that lies above the natural frequency of the primary system. Likewise if the natural frequency of the primary system is above the forcing frequency, it is well to tune the damper to a frequency slightly greater than the forcing frequency. Figure 6.10 shows that the tuning for a primary system with high natural frequency is more sensitive than that for a primary system with low natural frequency. Mohr's circle of Fig. 6.9 provides a useful graphical representation.

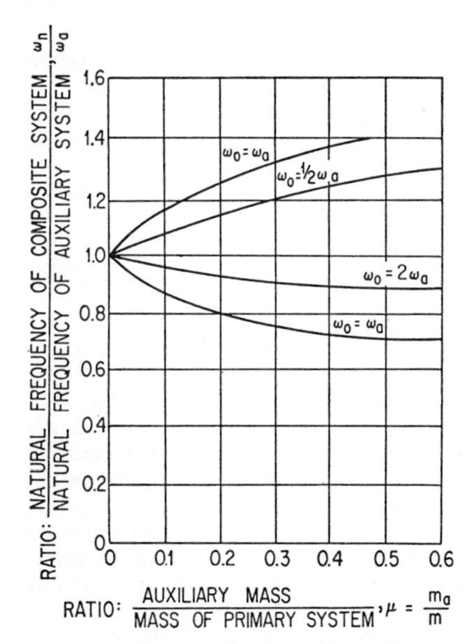

FIG. 6.10. Curves showing effect of mass ratio m_a/m on the natural frequencies ω_n of the composite system, for several ratios of the natural frequency ω_a of the auxiliary system to the natural frequency ω_0 of the primary system.

Where the natural frequency of the composite system is nearly equal to the tuned frequency of the absorber, the amplitude of motion of the primary mass at resonance is much smaller than that of the absorber. Consequently, the motion of the primary mass does not become large even at resonance; but the motion of the absorber, unless limited by damping, may become so large that failure occurs.

The use of the dynamic absorber is not restricted to single degree-of-freedom systems or to locations in simple systems where the exciting forces act. However, dynamic absorbers are most effective if located where the excitation force acts. For example, consider a dynamic absorber that is attached to the spring in the simple system shown in Fig. 6.11. When the absorber is tuned so that $\sqrt{k_a/m_a} = \omega$, the equivalent mass is infinite at its point of attachment and enforces a node at point A. If the stiffness of the spring between A and the mass m is k_1, then the force F' exerted by the absorber to

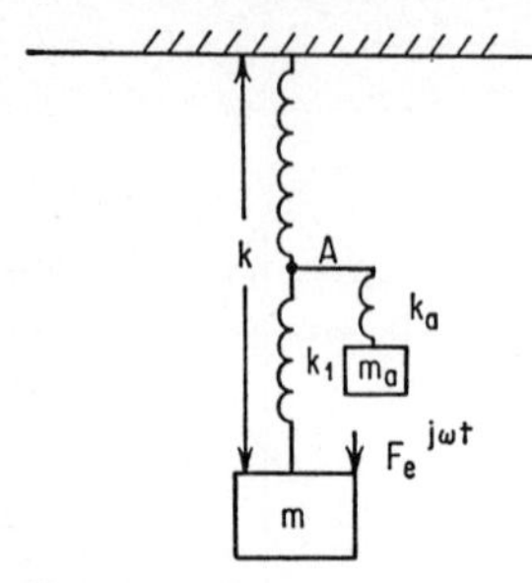

FIG. 6.11. Dynamic absorber attached to the spring of the primary system. The analysis shows that this is not as effective as if it were attached to the rigid body on which the force acts.

enforce the node is equal to that exerted by a system composed of the mass m and the spring k_1 attached to a fixed foundation at A and acted upon by the force $Fe^{j\omega t}$. The force F' is

$$F' = \frac{F}{1 - (m\omega^2/k_1)}$$

Thus the amplitude of motion of the auxiliary mass is

$$u_0 = \frac{F}{1 - (m\omega^2/k_1)} \times \frac{1}{m_a\omega^2} \tag{6.13}$$

The amplitude of motion of the primary mass is

$$x = \frac{F}{k_1}\left(1 - \frac{m\omega^2}{k_1}\right)^{-1} \tag{6.14}$$

Hence, an absorber attached to the spring is not as effective as one attached to the body where the force is acting. It is possible for the primary system to come into resonance about the new node at A.

AUXILIARY MASS DAMPERS

In general, the dynamic absorber is effective only for a system that is subjected to a constant frequency excitation. In the special case of a pendulum absorber (discussed later in this chapter), it is effective for an excitation that is a constant multiple of a rotating shaft speed. When excited at frequencies other than the frequency to which it is tuned, the absorber acts as an attached mass of positive value at frequencies below the tuned frequency and of negative value at frequencies above the tuned frequency. It introduces an additional degree-of-freedom and an additional natural frequency into the primary system.

In a multiple degree-of-freedom system, the introduction of an auxiliary mass system tends to lower those original natural frequencies of the primary system that are below the tuned frequency of the auxiliary system. This is because the auxiliary mass system adds a positive equivalent mass at frequencies below the tuned frequency. The original natural frequencies of the primary system that are higher than the tuned frequency of the auxiliary system are raised by adding the auxiliary mass system, because the equivalent mass of the auxiliary system is negative. A new natural mode of vibration corresponding to the vibration of the auxiliary mass system against the primary system is injected between the displaced initial natural frequencies of the primary system. Because the equivalent mass of the auxiliary mass system is large only at frequencies near the tuned frequency, those frequencies of the primary system that are closest to the tuned frequency are most strongly influenced by the auxiliary mass system. The addition of damping in the auxiliary mass system can be effective in reducing the amplitudes of motion of the primary system at the natural frequencies. For this reason auxiliary mass dampers are used quite commonly to reduce over-all vibration stresses and amplitudes.

Studies of the effects of a damped auxiliary mass system upon the amplitude of motion of an undamped, single degree-of-freedom system [1–5] have been applied to a multimass system.[6,7] In analyzing dampers utilizing auxiliary masses, it is desirable to consider a composite system in which the characteristics of both the primary and auxiliary systems are fixed. This composite system is excited by a harmonic force of varying frequency. It is desirable to express the tuned frequency of the auxiliary mass system in terms of the natural frequency of the primary system rather than the ratio β_a of the excitation frequency ω to the tuned frequency ω_a of the auxiliary system. Defining a new ratio α,

$$\alpha = \frac{\omega_a}{\omega_0} = \frac{\beta}{\beta_a}$$

Then Eq. (6.7) becomes

$$\frac{x_0}{\delta_{st}} = \left\{\frac{(\alpha^2 - \beta^2)^2 + (2\zeta\alpha\beta)^2}{[(\alpha^2 - \beta^2)(1 - \beta^2) - \alpha^2\beta^2\mu]^2 + (2\zeta\alpha\beta)^2(1 - \beta^2 - \beta^2\mu)^2}\right\}^{1/2} \tag{6.15}$$

This equation is plotted in Fig. 6.12. Note that all curves pass through two points A, B on the graph, independent of the damping parameter ζ. These points are known as *fixed points*. Their locations are independent of the value of ζ if the ratio of the coeffi-

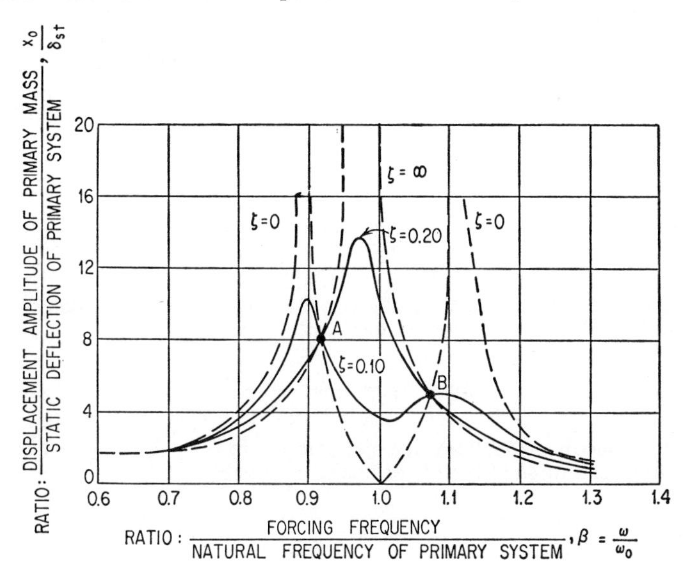

FIG. 6.12. Curves for auxiliary mass damper showing amplitude of vibration of mass of primary system, as given by Eq. (6.15), as a function of the ratio of forcing frequency ω to natural frequency of primary system $\omega = \sqrt{k/m}$. The mass ratio $m_a/m = 0.05$, and the natural frequency ω_a of the auxiliary mass system is equal to the natural frequency ω_0 of the primary system. Curves are included for several values of damping in the auxiliary system.

cient of ζ^2 to the term independent of ζ is the same in both numerator and denominator of Eq. (6.15):

$$\frac{(2\alpha\beta)^2}{(\alpha^2 - \beta^2)^2} = \frac{2\alpha\beta(1 - \beta^2 - \beta^2\mu)^2}{[(\alpha^2 - \beta^2)(1 - \beta^2) - \alpha^2\beta^2\mu]^2} \tag{6.16}$$

This equation is satisfied if

$$(2\alpha\beta)^2 = 0$$

$$\frac{1}{\alpha^2 - \beta^2} + \frac{(1 - \beta^2 - \beta^2\mu)}{(\alpha^2 - \beta^2)(1 - \beta^2) - \alpha^2\beta^2\mu} = 0$$

$$\frac{1}{\alpha^2 - \beta^2} - \frac{(1 - \beta^2 - \beta^2\mu)}{(\alpha^2 - \beta^2)(1 - \beta^2) - \alpha^2\beta^2\mu} = 0$$

The first two solutions are trivial. The third yields the equation

$$\beta^4\left(1 + \frac{\mu}{2}\right) - \beta^2(1 + \alpha^2 + \alpha^2\mu) + \alpha^2 = 0 \tag{6.17}$$

The solution of this equation gives two values of β, designated β_c, one corresponding to each fixed point.

The amplitude of motion at each fixed point may be found by substituting each value of β_c given by Eq. (6.17) into Eq. (6.15). Since the amplitude is independent of ζ, the value that gives the simplest calculation (namely, $\zeta = \infty$) can be used for the calculation:

$$\left.\frac{x_0}{\delta_{st}}\right|_c = \left[\frac{1}{(1 - \beta_c^2 - \beta_c^2\mu)^2}\right]^{1/2} \tag{6.18}$$

For the auxiliary mass damper to be most effective in limiting the value of x_0/δ_{st} over a full range of excitation frequencies, it is necessary to select the spring and damping constants of the system as given by the parameters α and ζ, respectively, so that the amplitude x_0 of the primary mass is a minimum. First consider the influence of the ratio α. As α is varied, the values of β_c computed from Eq. (6.17) are substituted in Eq. (6.18) to obtain values of x_0/δ_{st} for the fixed points A and B. The optimum value of α is that for which the amplitude x_0 at A is equal to that at B.

Let the two roots of Eq. (6.17) be β_1^2 and β_2^2, where β_1^2 is less than one and β_2^2 is greater than one. When x_0/δ_{st} has the same value for both β_1 and β_2 in Eq. (6.18),

$$\beta_1^2 + \beta_2^2 = \frac{2}{1 + \mu}$$

In an equation having unity for the coefficient of its highest power, the sum of the roots is equal to the coefficient of the second term with its sign changed:

$$\beta_1^2 + \beta_2^2 = \frac{1 + \alpha^2 + \alpha^2\mu}{1 + \mu/2}$$

From the two preceding equations, the optimum tuning (i.e., that required to give the same amplitude of motion at both fixed points) is obtained:

$$\alpha_{opt} = \frac{1}{1 + \mu} \tag{6.19}$$

where α is defined by the equation preceding Eq. (6.15).

If the effect of the damping is considered, it is possible to choose a value of the damping parameter ζ that will make the fixed points nearly the points of greatest amplitude of the motion. Consider Fig. 6.13, which represents the curves defining the motion of a single degree-of-freedom system to which an ideally tuned damped vibration absorber is attached (Fig. 6.8). The solid curves (1) represent the response of a system fitted with an undamped absorber. Curve 2 represents infinite damping of the auxiliary system. Curves 3 have horizontal tangents at the fixed points A and B, respectively. Since it is difficult to determine the required damping from maxima at the fixed points, the assumption is made that an optimum damping gives the same value of x_0/δ_{st} at a convenient point between A and B as at these fixed points. First find the values of β at A and B. This is done by solving Eq. (6.17) with the values of α as determined by Eq. (6.19) substituted:

$$\beta^4 - \frac{2\beta^2}{1 + \mu} + \frac{2}{(2 + \mu)(1 + \mu)^2} = 0$$

Solving for β to obtain the abscissas at the fixed points,

$$\beta^2 = \frac{1}{1 + \mu}\left(1 \pm \sqrt{\frac{\mu}{2 + \mu}}\right) \tag{6.20}$$

A convenient value for β lying between the two fixed points A and B is defined by

$$\beta_t^2 = \frac{1}{1 + \mu} \tag{6.21}$$

The frequency corresponding to this frequency ratio β_l is the natural frequency of the composite system when the damping is infinite; it is called the locked frequency.[7] The value of x_0/δ_{st} at the fixed points is found by substituting Eq. (6.20) into Eq. (6.18):

$$\frac{x_0}{\delta_{st}} \text{ at fixed point } = \sqrt{1 + \frac{2}{\mu}} \tag{6.22}$$

An approximate value for the maximum damping is obtained by solving for the value of ζ in Eq. (6.15) that gives a value of $x_0/\delta_{st} = \sqrt{1 + 2/\mu}$ when β_l^2 (the locked frequency)

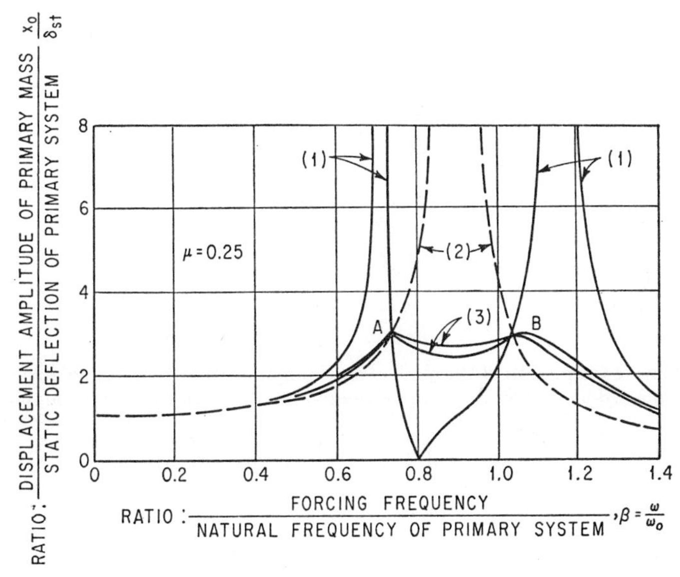

FIG. 6.13. Curves similar to Fig. 6.12 but with optimum tuning. Curves 1 apply to an undamped absorber, curve 2 represents infinite damping in the auxiliary system, and curves 3 have horizontal tangents at the fixed points A and B.

is given by Eq. (6.21) and α has the optimum value given by Eq. (6.19). This gives the following value for the optimum damping parameter:

$$\zeta_{opt} = \sqrt{\frac{\mu}{2(1+\mu)}} \tag{6.23}$$

It is possible to find the value of ζ that makes the fixed point A a maximum on the x_0/δ_{st} vs. β plot, Fig. 6.13, and also to find the value of ζ that makes the point B a maximum. The average of the two values so obtained indicates optimum damping:[4]

$$\zeta_{opt} = \sqrt{\frac{3\mu}{8(1+\mu)^3}} \tag{6.24}$$

OPTIMUM DAMPING FOR AN AUXILIARY MASS ABSORBER CONNECTED TO THE PRIMARY SYSTEM WITH DAMPING ONLY. In general, the most effective damping is obtained where the auxiliary mass damping system includes a spring in its connection to the primary system. However, such a design requires a calculation of the optimum stiffness of the spring. Sometimes it is more expedient to add an oversize mass, coupled only by damping to the primary system, than it is to compute the optimum system. However, if use is made of such a simplified damper by taking it from a list of standard dampers and applying it with a minimum of calculations, the stock dampers should be as efficient as the application will permit.

In computing the optimum damping characteristic for an auxiliary mass absorber, attached to a single degree-of-freedom system by damping only, from the relations that have been developed, note in Eq. (6.4) that $\zeta = \infty$ and $\beta_a = \infty$ when $k = 0$. Then $\alpha = \beta/\beta_a = 0$. However, the product $\zeta\alpha = \zeta\beta/\beta_a$ is finite; thus, substituting $\alpha = 0$ but retaining the product $\zeta\alpha$ in Eq. (6.15),

$$\frac{x_0}{\delta_{st}} = \sqrt{\frac{\beta^2 + 4(\zeta\alpha)^2}{\beta^2(1-\beta^2)^2 + 4(\zeta\alpha)^2[1-\beta^2(1+\mu)]^2}} \tag{6.25}$$

The value of x_0/δ_{st} is independent of $\zeta\alpha$ where the ratio of the coefficient of $\zeta\alpha$ to the term independent of $\zeta\alpha$ in the numerator is the same as the corresponding ratio in the denominator:

$$\frac{4}{\beta^2} = \frac{4[1-\beta^2(1+\mu)]^2}{\beta^2(1-\beta^2)^2}$$

The solution of this equation for β gives the fixed points

$$\beta^2 = 0 \qquad \text{and} \qquad \beta^2 = \frac{2}{2+\mu} \tag{6.26}$$

The amplitude of motion of the primary mass where $\beta^2 = 2/(2+\mu)$ is

$$\frac{x_0}{\delta_{st}} = \frac{2+\mu}{\mu} \tag{6.27}$$

Curves showing the motion of the mass of a primary system fitted with an auxiliary mass system connected by damping only are given in Fig. 6.14. The optimum damping is that

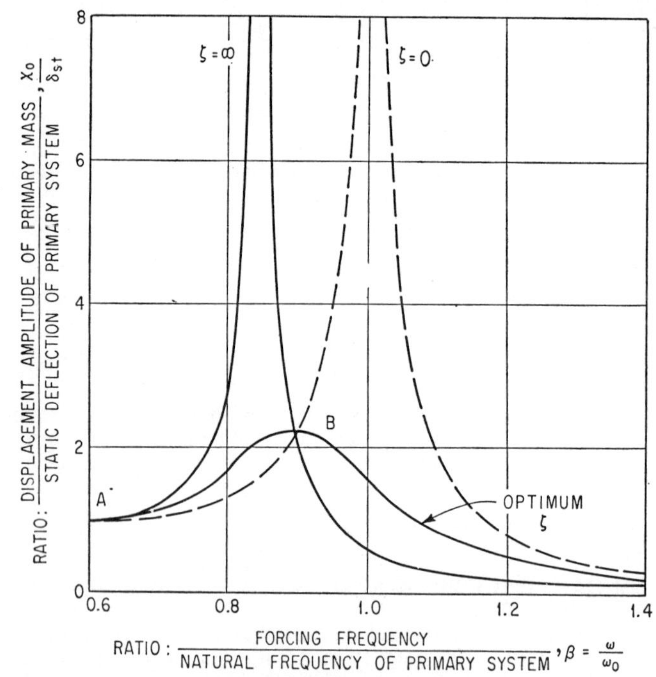

FIG. 6.14. Curves similar to Fig. 6.12 for system having auxiliary mass coupled by damping only. Several values of damping are included.

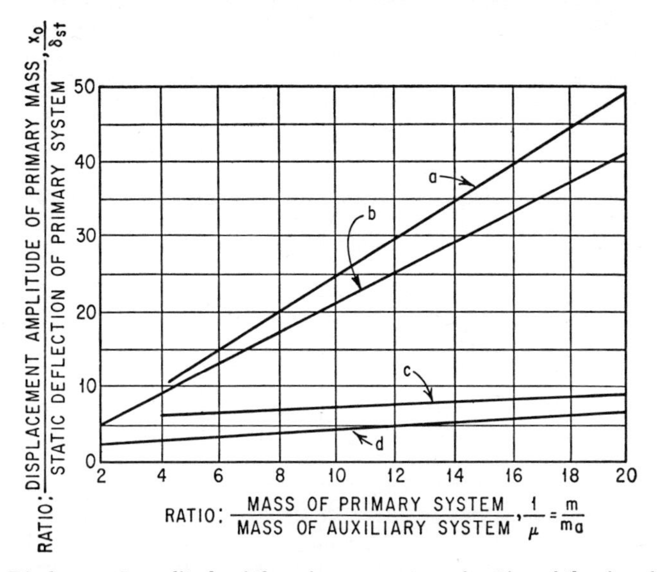

FIG. 6.15. Displacement amplitude of the primary mass as a function of the size of the auxiliary mass: (*a*) auxiliary system coupled only by Coulomb friction ($\alpha = 0$) with optimum damping; (*b*) auxiliary system coupled only by viscous damping ($\alpha = 0$) of optimum value; (*c*) auxiliary system coupled by spring and damper tuned to frequency of primary system ($\alpha = 1$) with optimum damping; (*d*) auxiliary system coupled by spring and damper with optimum tuning [$\alpha = 1/(1 + \mu)$] and optimum damping.

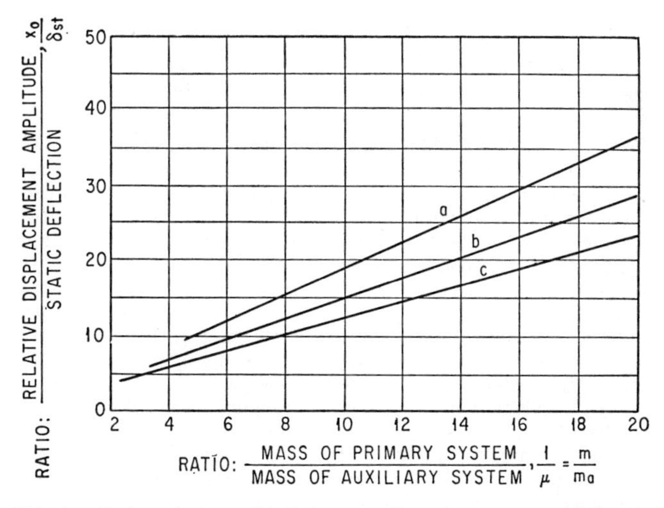

FIG. 6.16. Relative displacement amplitude between the primary mass and the auxiliary mass as a function of the size of the auxiliary mass: (*a*) auxiliary system coupled by spring and damper with optimum tuning [$\alpha = 1/(1 + \mu)$] and optimum damping; (*b*) auxiliary system coupled only by viscous damping ($\alpha = 0$) of optimum value; (*c*) auxiliary system coupled by spring and damper tuned to frequency of primary system ($\alpha = 1$) with optimum damping.

which makes the maximum amplitude occur at the fixed point B. By finding the value of $\zeta\alpha$ that makes the slope of x_0/δ_{st} versus β equal to zero at $\beta^2 = 2/(2 + \mu)$, the optimum damping is defined by

$$(\zeta\alpha)_{\text{opt}} = \sqrt{\frac{1}{2(2 + \mu)(1 + \mu)}} \tag{6.28}$$

The values for the amplitude of vibration of the primary mass, the relative amplitude between the primary and auxiliary masses, and the optimum damping constants are given in Figs. 6.15 to 6.17 as functions of the mass ratio $\mu = m_a/m$.

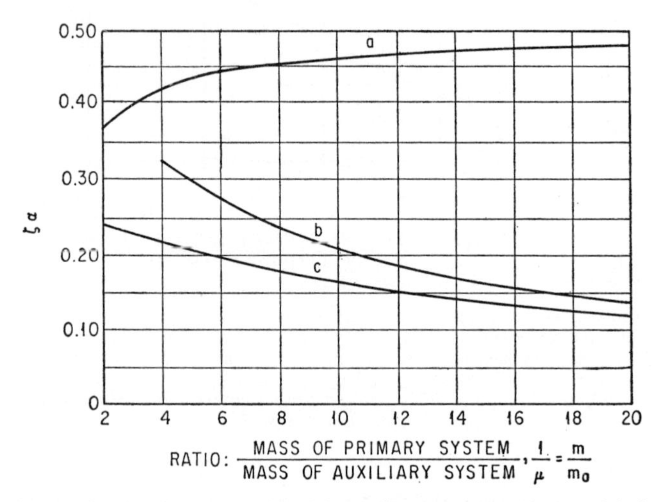

FIG. 6.17. Curves showing damping required in auxiliary mass systems to minimize vibration amplitude of primary system: (*a*) auxiliary mass coupled by viscous damping only ($\alpha = 0$); (*b*) auxiliary system coupled by spring and damper tuned to frequency of primary system ($\alpha = 1$); (*c*) auxiliary system coupled by spring and damper with optimum tuning [$\alpha = 1/(1 + \mu)$]. The ordinate of the curves is $\zeta\alpha$, where ζ is the fraction of critical damping in the auxiliary system [Eq. (6.4)] and α is the tuning parameter [Eq. (6.15)].

THE USE OF AUXILIARY MASS ABSORBERS FOR VIBRATION ENERGY DISSIPATION. When a complicated mass-spring system is analyzed for possible vibration troubles, it is customary to compute the natural frequencies of the several modes of vibration of the system. The vibration amplitudes and stresses are estimated by making an energy balance between the energy input from the various exciting forces and the energy dissipated in the system and external reactions. From this point of view, it is desirable to know how much energy is dissipated in auxiliary mass systems and what value the damping constant should have in an auxiliary mass system of limited size to give maximum energy absorption. This is not the best criterion for determining the optimum damping because it neglects the effects of damping upon the mode shapes and the frequencies of the system, but it is generally adequate when compared with the other uncertainties of the calculations. Methods of designing dampers for torsional systems are given in Chap. 38.

Optimum Viscous Damping to Give Large Energy Absorption in an Auxiliary Mass Absorber.[8] Suppose the amplitude of motion of the primary system is unaffected by the auxiliary mass system which is attached to it. All energy absorption occurs in the damping element of the auxiliary mass system and is obtained by integrating the differential work done in the damper over a vibration cycle. The force exerted by damping is $c\dot{x}_r$ (the subscripts a are dropped), where x_r is the relative motion and the increment of work is $c\dot{x}_r\,dx_r = c\dot{x}_r^2\,dt$. If $x_r = x_{r0}\cos\omega t$, the work done over a cycle is

$$V = \oint c\omega^2 x_{r0}^2 \sin^2 \omega t\,dt = \pi c x_{r0}^2 \omega \tag{6.29}$$

For a damper attached to a support moving in harmonic motion of amplitude x_0, the relative motion x_r is given by Eq. (6.1). The amplitude of relative motion is

$$x_{r0} = \frac{m\omega^2 x_0}{\sqrt{(k - m\omega^2)^2 + c^2\omega^2}} = \frac{\beta_a^2 x_0}{\sqrt{(1 - \beta_a^2)^2 + (2\zeta\beta_a)^2}}$$

Substituting the above value of x_{r0} in Eq. (6.29) and integrating,

$$V = \frac{\pi c \omega x_0^2 m^2 \omega^4}{(k - m\omega^2)^2 + c^2\omega^2} = \frac{\pi x_0^2 m\omega^2 (2\zeta\beta_a)\beta_a^2}{(1 - \beta_a^2)^2 + (2\zeta\beta_a)^2} \tag{6.30}$$

Equation (6.30) can be used to find the tuning and the damping that gives the maximum energy dissipation when the amplitude of the forcing motion remains constant. Placing $\partial V/\partial\beta_a = 0$, the optimum value of β_a for given values of ζ is found from

$$(\beta_a)^2_{\text{opt}} = (2\zeta^2 - 1) \pm 2\sqrt{1 - \zeta^2 + \zeta^4} \tag{6.31}$$

Placing $\partial V/\partial\zeta = 0$, the optimum value of ζ for a given value of β_a is

$$\zeta_{\text{opt}} = \frac{1 - \beta_a^2}{2\beta_a} \tag{6.32}$$

Where $k = 0$, the optimum damping is determined most conveniently by setting $\partial V/\partial c = 0$, using the dimensional form of Eq. (6.30), and determining c for maximum energy absorption:

$$c_{\text{opt}} = m\omega \tag{6.33}$$

AUXILIARY MASS DAMPER USING COULOMB FRICTION DAMPING.[9] Dampers relying on Coulomb friction (i.e., friction whose force is constant) have been widely used. A damper relying on dry friction and connected to its primary system with a spring is too complicated to be analyzed or to be adjusted by experiment. For this reason, a damper with Coulomb friction has been used with only friction damping connecting the seismic mass (usually in a torsional application) to the primary system.[1,2,9] Because the motion is irregular, it is necessary to use energy methods of analysis. The analysis given here applies to the case of linear vibration. By analogy, the application to torsional or other vibration can be made easily (see Table 2.1 for analogous parameters).

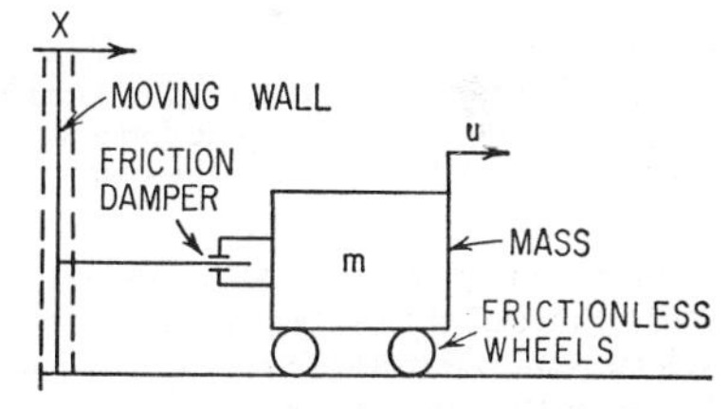

FIG. 6.18. Schematic diagram of auxiliary mass absorber with Coulomb friction damping.

Consider the system shown in Fig. 6.18. It consists of a mass resting on wheels that provide no resistance to motion and connected through a friction damper to a wall that is moving sinusoidally. The friction damper consists of two friction facings that are held on opposite sides of a plate by a spring that can be adjusted to give a desired clamping force. The maximum force that can be transmitted through each interface of the damper is the product of the normal force and the coefficient of friction; the maximum total force for the damper is the summation over the number of interfaces.

Consider the velocity diagrams shown in Fig. 6.19*A*, *B*, and *C*. In these diagrams the velocity of the moving wall, $\dot{x} = x_0\omega \sin\omega t$, is shown by curve 1; the velocity $\dot{u}$ of the mass is shown by curve 2. The force exerted by the damper when slipping occurs is F_s.

When $F_s \geq m\ddot{u}$, the mass moves sinusoidally with the wall. When $F_s < m\ddot{u}$, slipping occurs in the damper and the mass is accelerated at a constant rate. Since a constant acceleration produces a uniform change in velocity, the velocity of the mass when the damper is slipping is shown by straight lines. The relative velocity between the wall and the mass is shown by the vertical shading.

Figure 6.19*A* applies to a damper with a low friction force. The damper slips continuously. In Fig. 6.19*B* the velocities resulting from a larger friction force are shown. Slipping disappears for certain portions of the cycle. Where the wall and the mass have the same velocity, their accelerations also are equal. Slipping occurs when the force

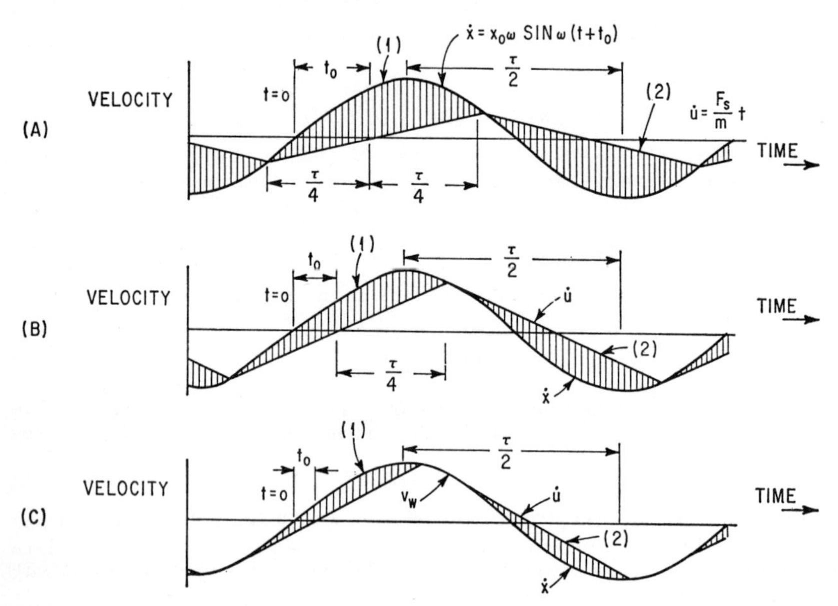

FIG. 6.19. Velocity-time diagrams for motion of wall (curve 1) and mass (curve 2) of Fig. 6.18. The conditions for a small damping force are shown at (*A*), for an intermediate damping force at (*B*), and for a large damping force at (*C*). The relative velocity between the wall and the mass is indicated by vertical shading.

transmitted by the damper is not large enough to keep the mass accelerating with the wall. Since at the breakaway point the accelerations of the wall and mass are equal, their velocity-time curves have the same slope; i.e., the curves are tangent at this point. In Fig. 6.19*C*, the damping force is so large that the mass follows the wall for a considerable portion of the cycle, and slips only where its acceleration becomes greater than the value of F_s/m. A slight increase in the clamping force or in the coefficient of friction locks the mass to the wall; then there is no relative motion and no damping.

Because of the nature of the damping force, the damping provided by the friction damper can be computed most practically in terms of energy. If the friction force exerted through the damper is F_s, the energy dissipated by the damper is the product of the friction force and the total relative motion between the mass and the moving wall. The time reference is taken at the moment when the auxiliary mass m has a zero velocity and is being accelerated to a positive velocity, Fig. 6.19*A*. Let the period of the vibratory motion of the wall be $\tau = 2\pi/\omega$, where ω is the angular frequency of the wall motion. By symmetry, the points of no slippage in the damper occur at times $-\tau/4$, $\tau/4$, and $3\tau/4$. Let the time when the velocity of the wall is zero be $-t_0$; then the velocity of the wall $\dot{x}$ is

$$\dot{x} = +x_0\omega \sin \omega(t + t_0)$$

The velocity $\dot{u}$ of the mass for $-\tau/4 < t < \tau/4$ is

$$\dot{u} = \ddot{u}t = \frac{F_s}{m}t$$

The velocities of the wall and the mass are equal at time $t = \tau/4$:

$$x_0\omega \sin \omega\left(\frac{\tau}{4} + t_0\right) = \frac{F_s}{m}\frac{\tau}{4}$$

Since $\omega\tau/4 = \pi/2$, $\sin \omega(\tau/4 + t_0) = \cos \omega t_0$. Therefore

$$\cos \omega t_0 = \frac{F_s}{m}\frac{\pi}{2x_0\omega^2}$$

The relative velocity between the moving wall and the mass is $\dot{x} - \dot{u}$, and the total relative motion is the integral of the relative velocity over a cycle. Note that the area between the two curves for the second half of the cycle is the same as for the first. Hence, the work V per cycle is

$$V = 2\int_{-\tau/4}^{\tau/4} F_s(\dot{x} - \dot{u})\, dt = 4F_s x_0\sqrt{1 - \left(\frac{F_s\pi}{2mx_0\omega^2}\right)^2} \tag{6.34}$$

Optimum damping occurs when the work per cycle is a maximum. It can be determined by setting the derivative of V with respect to F_s in Eq. (6.34) equal to zero and solving for F_s:

$$(F_s)_{\text{opt}} = \frac{\sqrt{2}}{\pi} m\omega^2 x_0 \tag{6.35}$$

Energy absorption per cycle with optimum damping is, from Eq. (6.34),

$$V_{\text{opt}} = \frac{4}{\pi} m\omega^2 x_0^2 \tag{6.36}$$

A comparison of the effectiveness of the Coulomb friction damper with other types is given in Fig. 6.15.

EFFECT OF NONLINEARITY IN THE SPRING OF AN AUXILIARY MASS DAMPER

It is possible to extend the range of frequency over which a dynamic absorber is effective by using a nonlinear spring.[10–12] When a nonlinear spring is used, the natural frequency of the absorber is a function of the amplitude of vibration; it increases or decreases, depending upon whether the spring stiffness increases or decreases with deflection. Figure 6.20*A* shows a typical response curve for a system with increasing spring stiffness; Fig. 6.20*B* illustrates types of systems having increasing spring stiffness and shows typical force-deflection curves. Figure 6.21*A* shows a typical response curve for a system of decreasing spring stiffness; Fig. 6.21*B* illustrates types of systems having decreasing stiffnesses and shows typical force-deflection curves.

To compute the equivalent mass at a given frequency when a nonlinear spring is used, it is necessary to use a trial-and-error procedure. By the methods given in Chap. 4, compute the natural frequency of the auxiliary mass system, assuming the point of attachment fixed, as a function of the amplitude of motion of the auxiliary mass. This will result in a curve similar to the dotted curves in Figs. 6.20*A* and 6.21*A*. At the given frequency, compute β_a in Eq. (6.4) in terms of the tuned frequency of the absorber at zero amplitude. (The tuned frequency will change with amplitude because the spring constant changes.) With this value of β_a compute the equivalent mass from Eq. (6.6). With this mass in the system, compute the amplitude of motion x_0 of the primary mass to which the auxiliary system is attached [Eq. (6.7)] and the amplitude of the relative motion $x_{r0} = \nu^2(1 - \nu^2)x_0$. Using this value of x_{r0}, ascertain the corresponding value of

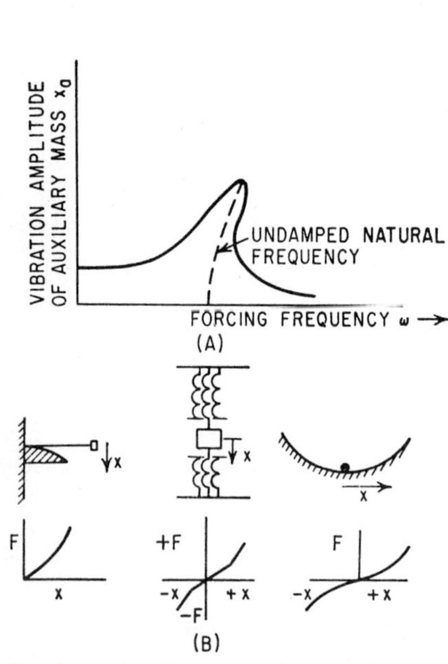

FIG. 6.20. Auxiliary mass damper with nonlinear spring having stiffness that increases as deflection increases. The response to forced vibration and the natural frequency are shown at (*A*). Several arrangements of nonlinear systems with the corresponding force-deflection curves are shown at (*B*).

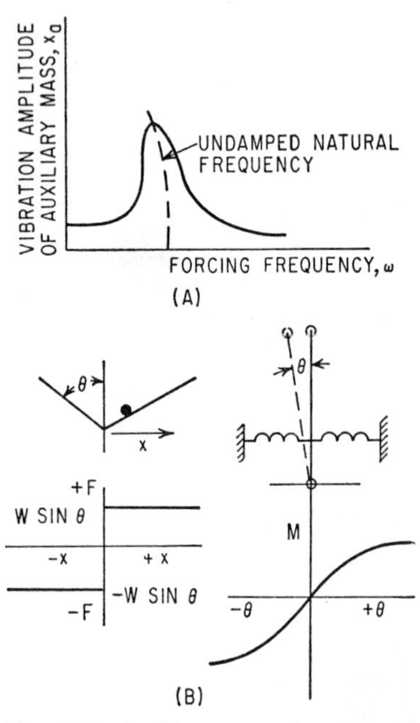

FIG. 6.21. Auxiliary mass damper with nonlinear spring having stiffness that decreases as deflection increases. The response to forced vibration and the natural frequency are shown at (*A*). Two arrangements of nonlinear systems with the corresponding force-deflection curves are shown at (*B*).

resonant frequency of the system from the computed curve, and compute the new value of β_a. Repeat the process until the value of β_a remains unchanged upon repeated calculation.

A dynamic absorber having a nonlinear characteristic can be used to introduce nonlinearity into a resonant system. This can be useful in the case where a machine passes through a resonance rapidly as the speed is increased but slowly as the speed is decreased. In bringing this machine up to speed, there is a natural frequency that comes into strong resonance, giving a critical speed. A strongly nonlinear dynamic absorber tuned at low amplitudes to the optimum frequency for the damped absorber can be used to reduce the effects of the critical speed. Two resonant peaks will be introduced, as shown on curve 1 of Fig. 6.13. By making the dynamic absorber nonlinear, so that the stiffness becomes greater as the amplitude of vibration is increased, the peaks are bent over to provide the response curve shown in Fig. 6.22. In starting, the machine is accelerated through the two critical speeds so fast that a resonance is unable to build up. In coasting to a stop, there would be ample time for significant amplitudes to build up if the nonlinearity did not exist. Because of the nonlinearity, the amplitude of vibration as a function of speed (since β is proportional to speed) follows the path A, B, C, D, E, F, G and never reaches the extreme amplitudes H_1 and H_2.

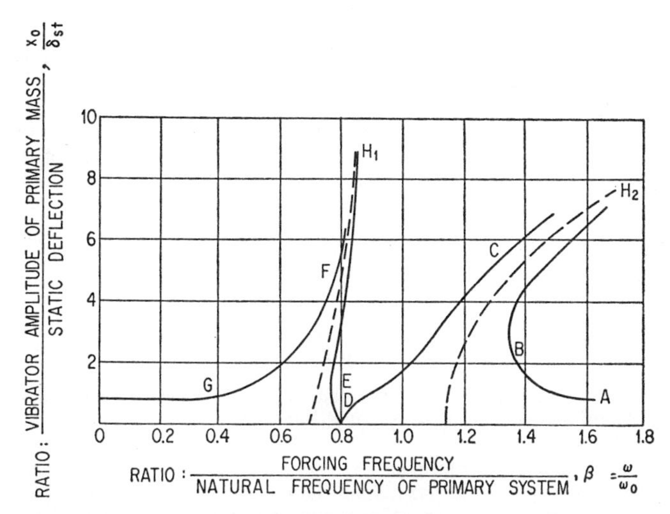

FIG. 6.22. Motion of the primary mass, as a function of forcing frequency, in a system having a nonlinear dynamic absorber whose natural frequency increases with amplitude. The mass of the absorber is 0.25 times the mass of the primary system ($\mu = 0.25$).

MULTIMASS ABSORBERS

In general, only one mass is used in a dynamic absorber. However, it is possible to provide a dynamic absorber that is effective for two or more frequencies by attaching an auxiliary mass system that resonates at the frequencies that are objectionable. The principle that would make such a dynamic absorber effective is utilized in the design of the elastic system of a ship's propulsion plant driven by independent high-pressure and low-pressure turbines. By making the frequencies of the two branches about the reduction gear identical, the gear becomes a node for one of the resonant modes. Then it is impossible to excite the mode of vibration where one turbine branch vibrates against the other as a result of excitation transmitted by the propeller shaft to that node.

DISTRIBUTED MASS ABSORBERS

It is possible to use distributed masses as vibration dampers. Consider an undamped rod of distributed mass and elasticity attached to a foundation that vibrates the rod axially, as shown in Fig. 6.23. The differential equation for the motion of this rod is derived in Chap. 7. The values of the constants are set by the boundary conditions:

$$\begin{aligned} \text{Stress} &= E\frac{\partial u}{\partial x} = 0 \qquad \text{where } x = l \\ u &= u_0 \cos t \qquad \text{where } x = 0 \end{aligned} \tag{6.37}$$

The solution of the equation of motion is

$$\begin{aligned} u &= u_0 \cos \omega t \left(\cos \sqrt{\frac{\gamma\omega^2}{Eg}}\, x + \tan \sqrt{\frac{\gamma\omega^2}{Eg}}\, l \sin \sqrt{\frac{\gamma\omega^2}{Eg}}\, x \right) \\ &= u_0 \cos \omega t \frac{\cos \sqrt{(\gamma\omega^2/Eg)}\,(1 - x)}{\cos \sqrt{(\gamma\omega^2/Eg)}\, l} \end{aligned} \tag{6.38}$$

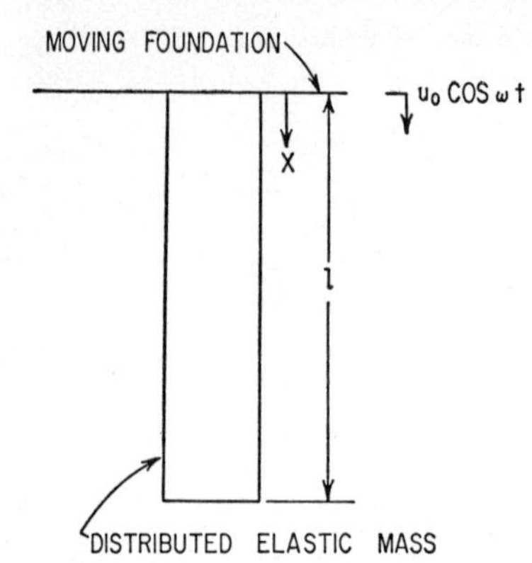

FIG. 6.23. Elastic body with distributed mass used as auxiliary mass damper.

where E is the modulus of elasticity and γ is the weight density of the material. When $x = 0$, the force F on the foundation is

$$F = SE \frac{\partial u}{\partial x}\bigg|_0 = SEu_0 \sqrt{\frac{\gamma\omega^2}{Eg}} \left(\tan \sqrt{\frac{\gamma\omega^2}{Eg}} l\right) \tag{6.39}$$

where S is the cross-sectional area of the bar. It is apparent that as the argument of the tangent has successive values of $\pi/2$, $3\pi/2$, $5\pi/2$, . . ., the force exerted on the foundation becomes infinite. The distributed mass acts as a dynamic absorber enforcing a node at its point of attachment. By tuning the mass so that

$$\sqrt{\frac{\gamma\omega^2}{Eg}} l = \frac{n\pi}{2} \qquad \text{or} \qquad l = \frac{2}{n\pi\omega} \sqrt{\frac{Eg}{\gamma}} \tag{6.40}$$

the distributed mass acts as a dynamic absorber for not only the fundamental frequency $\omega/2\pi$, but also for the third, fifth, seventh, . . . harmonics of the fundamental.

The above solution neglects damping. It is possible to consider the effect of damping by including a damping term in the differential equation. The stress in an element is assumed to be the sum of a deformation stress and a stress related to the velocity of strain:

$$\sigma = E\epsilon + \mu \frac{d\epsilon}{dt} \tag{6.41}$$

where $\epsilon = \partial u/\partial x$ is the strain. Then the differential equation becomes

$$E \frac{\partial^2 u}{\partial x^2} + \mu \frac{\partial^3 u}{\partial x^2\, \partial t} = \frac{\gamma}{g} \frac{\partial^2 u}{\partial t^2} \tag{6.42}$$

Since the absorber is excited by a foundation moving with a frequency $f = \omega/2\pi$, u may be expressed as $\Re u_1 e^{j\omega t}$ and the partial differential equation can be written as the ordinary linear differential equation

$$E \frac{d^2 u_1}{dx^2} + j\omega\mu \frac{d^2 u_1}{dx^2} + \frac{\gamma\omega^2 u_1}{g} = 0$$

This equation may be written

$$\left(1 + \frac{\mu\omega j}{E}\right) \frac{d^2 u_1}{dx^2} + \frac{\gamma\omega^2}{Eg} u_1 = 0 \tag{6.43}$$

Since Eq. (6.43) is a second-order linear differential equation, the solution may be written

$$u = A_1 e^{\beta_1 x} + A_2 e^{\beta_2 x} \tag{6.44}$$

where β_1 and β_2 are the two roots of the equation

$$\beta^2 = \frac{-(\gamma/Eg)\omega^2}{1 + (j\mu\omega/E)} \tag{6.45}$$

For small values of μ, by a binomial expansion of the denominator,

$$\pm \beta = \frac{1}{2}\sqrt{\frac{\gamma}{Eg}}\frac{\mu\omega^2}{E} + j\sqrt{\frac{\gamma}{Eg}}\,\omega \tag{6.46}$$

where μ is defined by Eq. (6.41).

The boundary conditions to be met by the damper are:

$$\text{At } x = 0: \qquad u = u_0 \qquad \text{therefore, } A_1 + A_2 = u_0$$
$$\text{At } x = l: \qquad \sigma = (E + j\omega\mu)\frac{\partial u}{\partial x} = 0 \qquad \text{therefore, } A_1 e^{\beta l} - A_2 e^{-\beta l} = 0 \tag{6.47}$$

Solving Eqs. (6.47) for A_1 and A_2 and substituting the result in (6.44),

$$u = u_0 \frac{\cosh \beta(l - x)}{\cosh \beta l} \tag{6.48}$$

Substituting from Eq. (6.48) in Eq. (6.39), the force exerted on the foundation by the damper is

$$F_{(x=0)} = S\sigma_{(x=0)} = -Su_0(E + j\omega\mu)\frac{\sinh \beta l}{\cosh \beta l} \tag{6.49}$$

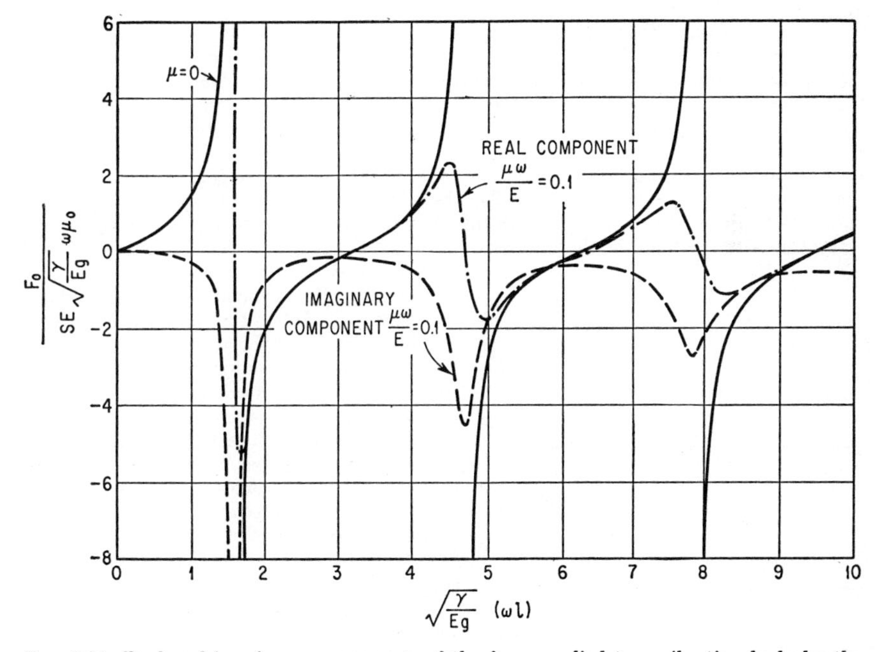

FIG. 6.24 Real and imaginary components of the force applied to a vibrating body by the distributed mass damper shown in Fig. 6.23. These relations are given mathematically by Eq. (6.50), and the terms are defined in connection with Eq. (6.38). The curves are for a value of the damping coefficient $\mu\omega/E = 0.1$, where μ is defined by Eq. (6.41).

where S is the cross-section area of the bar. When the complex value of β as given in Eq. (6.46) is substituted in Eq. (6.49), the following value for the dynamic force exerted on the foundation is obtained:

$$\frac{F_{(x=0)}}{SE\sqrt{\frac{\gamma}{Eg}}\,\omega u_0} = \frac{\left(1+\frac{\mu^2\omega^2}{2E^2}\right)\sin 2\sqrt{\frac{\gamma}{Eg}}\,\omega l + \frac{\mu\omega}{2E}\sinh\frac{\mu\omega}{E}\sqrt{\frac{\gamma}{Eg}}\,\omega l}{\cos\sqrt{\frac{\gamma}{Eg}}\,\omega l + \cosh\frac{\mu\omega}{E}\sqrt{\frac{\gamma}{Eg}}\,\omega l}$$
$$+\,j\,\frac{\frac{\mu\omega}{2E}\sin 2\sqrt{\frac{\gamma}{Eg}}\,\omega l - \left(1+\frac{\mu^2\omega^2}{2E^2}\right)\sinh\frac{\mu\omega}{E}\sqrt{\frac{\gamma}{Eg}}\,\omega l}{\cos\sqrt{\frac{\gamma}{Eg}}\,\omega l + \cosh\frac{\mu\omega}{E}\sqrt{\frac{\gamma}{Eg}}\,\omega l} \qquad (6.50)$$

A plot of the real and imaginary values of $F_{(x=0)}/SE\sqrt{\frac{\gamma}{Eg}}\,\omega u_0$ is given in Fig. 6.24 for zero damping and for a damping coefficient $\mu\omega/E = 0.1$ as a function of a tuning parameter $\sqrt{\gamma/Eg}\,(\omega l)$. Damping decreases the effectiveness of the distributed mass damper substantially, particularly for the higher modes.

PRACTICAL APPLICATIONS OF AUXILIARY MASS DAMPERS AND ABSORBERS TO SINGLE DEGREE-OF-FREEDOM SYSTEMS

THE DYNAMIC ABSORBER

The dynamic absorber, because of its tuning, can be used to eliminate vibration only where the frequency of the vibration is constant. Many pieces of equipment to which it is applied are operated by alternating current. So that it can be used for time keeping, the frequency of this a-c is held remarkably constant. For this reason, most applications of dynamic absorbers are made to mechanisms that operate in synchronism from an a-c power supply.

An application of a dynamic absorber to the pedestal of an a-c generator having considerable vibration is shown in Fig. 6.25, where the relative sizes of absorber and pedestal are shown approximately to scale. In this case, the application is made to a complicated structure and the mass of the absorber is much less than that of the primary system; however, since the frequency of the excitation is constant, the dynamic absorber reduces the vibration. When the mass ratio is small, it is important that the absorber be accurately tuned and that the damping be small. In this case, the excitation was the unbalance in the turbine rotor which was elastically connected to the pedestal through the flexibility of the shaft. If the absorber were ideally effective, there would be no forces at the frequency of the shaft speed; therefore, there would be no displacements from the pedestal where the force is neutralized through the remainder of the structure.

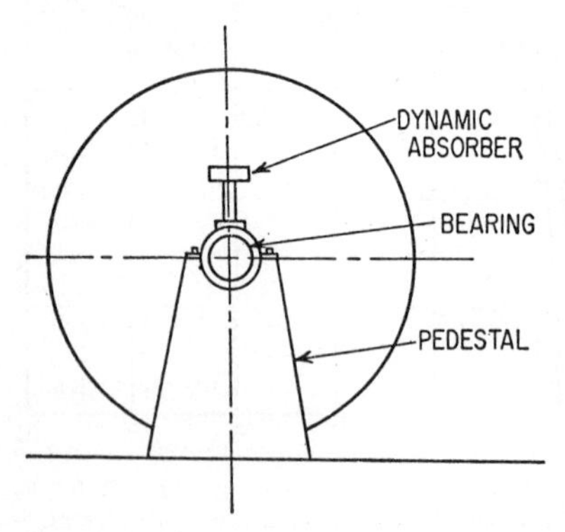

FIG. 6.25. Application of a dynamic absorber to the bearing pedestal of an a-c generator.

The dynamic absorber has been applied to the electric clipper shown in Fig. 6.26. The structure consisting of the cutter blade and its driving mechanism is actuated by the magnetic field at a frequency of 120 cps, as a result of the 60-cps a-c power supply. The forces and torques required to move the blade are balanced by reactions on the housing, causing it to vibrate. The dynamic absorber tuned to a frequency of 120 cps enforces a node at the location of its mass. Since this is approximately the center-of-gravity of the assembly of the cutter and its driving mechanism, the absorber effectively neutralizes the unbalanced force. The moment caused by the rotation of the moving parts is still unbalanced. A second very small dynamic absorber placed in the handle of the clipper could enforce a node at the handle and substantially eliminate all vibration. The design of these absorbers is simple after the unbalanced forces and torques generated by the cutter mechanism are computed. The sum of the inertia forces generated by the two absorbers, $m_1x_1\omega^2 + m_2x_2\omega^2$ (where m_1 and x_1 are the mass and amplitude of motion of the first absorber, m_2 and x_2 are the corresponding values for the second absorber, and $\omega = 240\pi$), must equal the unbalanced force generated by the clipper mechanism. The torque generated by the two absorbers must balance the torque of the mechanism. Since the value of ω^2 is known, the values of m_1x_1 and m_2x_2 can be determined. Weights that fit into the available space with adequate room to move are chosen, and a spring is designed of such stiffness that the natural frequency is 120 cps.

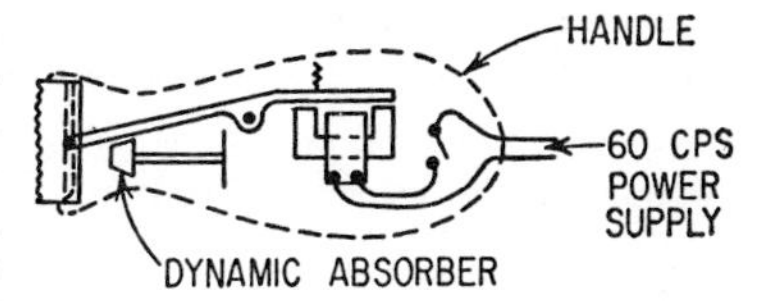

FIG. 6.26. Application of a dynamic absorber to a hair clipper.

Because of the desirable balancing properties of the simple dynamic absorber and the constancy of frequency of a-c power, it might be expected that devices operating at a frequency of 120 cps would be used more widely. However, their application is limited because the frequency of vibration is too high to allow large amplitudes of motion.

REDUCTION OF ROLL OF SHIPS BY AUXILIARY TANKS

An interesting application of auxiliary mass absorbers is found in the auxiliary tanks used to reduce the rolling of ships,[1, 13] as shown in Fig. 6.27. When a ship is heeled, the restoring moment $k_r\phi$ acting on it is proportional to the angle of heel (or roll). This restoring moment acts to return the ship (and the water that moves with it) to its equilibrium position. If I_s represents the polar moment of inertia of the ship and its entrained water, the differential equation for the rolling motion of the ship is

$$I_s\ddot{\phi} + k_r\phi = M_s \tag{6.51}$$

where M_s represents the rolling moments exerted on the ship, usually by waves.

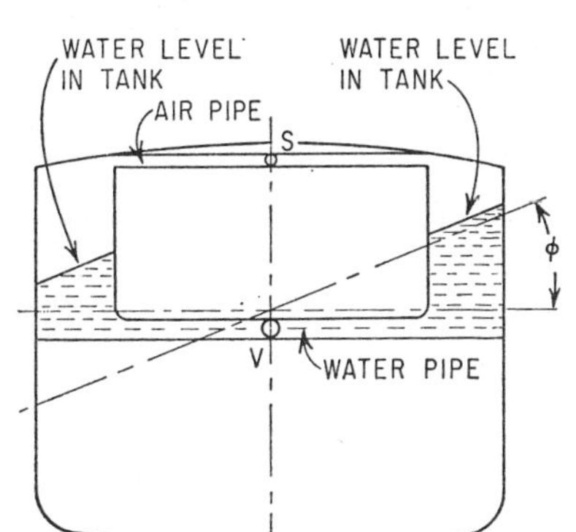

FIG. 6.27. Cross section of ship equipped with antiroll tanks. The flow of water from one tank to the other tends to counteract rolling of the ship.

To reduce rolling of the ship, auxiliary wing tanks connected by pipes are used. The water flowing from one tank to another has a natural frequency that is determined by the length and cross-sectional area of the tube connecting the tanks. The damping is controlled by restricting the flow of water, either with a valve S in the line that allows air to flow between the tanks (Fig. 6.27) or with a valve V in the water line. Since the tanks occupy valuable space, the mass ratio of the water in the tanks to the ship is small. Fortunately, the excitation from waves generally is not large relative to the restoring moments, and roll becomes objectionable only because the normal damping of a ship in rolling motion is not very large. The use of antirolling tanks in the German luxury liners *Bremen* and *Europa* reduced the maximum roll from 15 to 5°.

REDUCTION OF ROLL OF SHIPS BY GYROSCOPES

A large gyroscope may be used to reduce roll in ships, as shown in Fig. 6.28.[1,14] In response to the velocity of roll of a ship, the gyroscope precesses in the plane of symmetry of the ship. By braking this precession, energy can be dissipated and the roll reduced. The torque exerted by the gyroscope is proportional to the rate of change of the angular momentum about an axis perpendicular to the torque. Letting I represent the polar moment of inertia of the gyroscope about its spin axis and $\dot{\theta}$ the angular velocity of precession of the gyroscope, then the equation of motion of the ship is

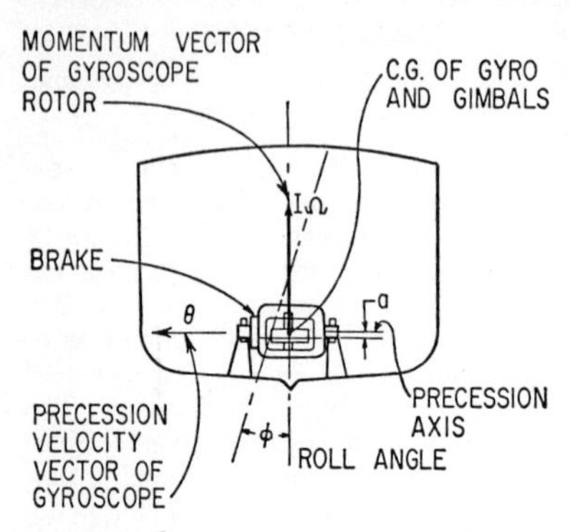

FIG. 6.28. Application of a gyroscope to a ship to reduce roll.

$$I_s\ddot{\phi} + k_r\phi + I\Omega\dot{\theta} = M_s \tag{6.52}$$

Assume that the gyroscope has (1) a moment of inertia about the precession axis of I_g, (2) a weight of W, and (3) that its center-of-gravity is below the gimbal axis (as it must be for the gyro to come to equilibrium in a working position) a distance a, as shown in Fig. 6.28. Then the equation of motion of the gyroscope is

$$I_g\ddot{\theta} + W_a\theta + c\dot{\theta} - I\Omega\dot{\phi} = 0 \tag{6.53}$$

where Ω is the spin velocity of the gyroscope. From Eq. (6.53), for a roll frequency of ω, the angle of precession of the gyroscope is

$$\theta = \frac{jI\Omega\omega\phi}{-I_g\omega^2 + Wa - jc\omega} \tag{6.54}$$

The torque exerted on the ship is

$$I\Omega\dot{\theta} = \frac{-(I\Omega)^2\omega^2\phi}{-I_g\omega^2 + Wa + jc\omega} \tag{6.55}$$

The equivalent moment of inertia of the gyroscope system in its reaction on the ship is

$$\frac{I\Omega^2}{-I_g\omega^2 + Wa + cj\omega} \tag{6.56}$$

By analogy with the steps of Eqs. (6.2) through (6.7), it follows that

$$\frac{\phi}{\phi_{st}} = \sqrt{\frac{(1-\beta_g^2)^2 + (2\zeta\beta_g)^2}{[(1-\beta_g^2)(1-\beta^2) - \beta^2\mu]^2 + (2\zeta\beta_g)^2(1-\beta^2)^2}} \tag{6.57}$$

where the parameters are defined in terms of ship and gyro constants as follows:

$$\beta_g = \frac{\omega}{\sqrt{Wa/I_g}} \qquad \beta = \frac{\omega}{\sqrt{k_r/I_s}} \qquad \zeta = \frac{c}{2\sqrt{WaI_g}} \qquad \mu = \frac{(I\Omega)^2}{WaI_s} \qquad \psi_{st} = \frac{M_s}{k_r}$$

Because $I\Omega$ can be made large by using a large gyro rotor and spinning it at a high speed, and Wa can be made small by choice of a design, the value of μ can be made quite large even though I_s is large. In one experimental ship, $\mu = 20$ was obtained. Even with this large value of μ, the precession angle of the gyroscope would become very large for optimum damping. Therefore it is necessary to use much more damping than optimum.

Gyro stabilizers were used on the Italian ship *Conte di Savoia;* they are sometimes installed on yachts.

Both antirolling tanks and gyro stabilizers are more effective if they are active rather than passive. Activated dampers are considered below.

AUXILIARY MASS DAMPERS APPLIED TO ROTATING MACHINERY

An important industrial use of auxiliary mass systems is to neutralize the unbalance of centrifugal machinery. A common application is the balance ring in the spin dryer of home washing machines. The operation of such a balancer is dependent upon the basket of the washer rotating at a speed greater than the natural frequency of its support. The balance ring is attached to the washing machine basket concentric with its axis of rotation, as shown in Fig. 6.29.

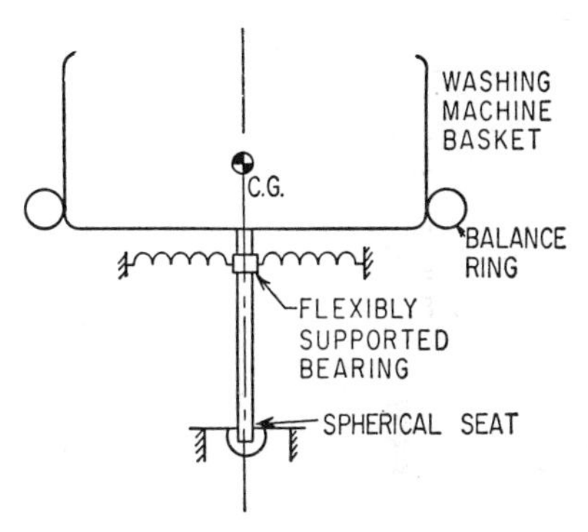

Fig. 6.29. Schematic diagram showing location of balance ring on basket of a spin dryer.

Consider the washing machine basket shown in Fig. 6.29. When its center-of-gravity does not coincide with its axis of rotation and it is rotating at a speed lower than its critical speed (corresponding to the natural frequency in rocking motion about the spherical seat), the centrifugal force tends to pull the rotational axis in the direction of the unbalance. This effect increases with an increase in rotational speed until the critical speed is reached. At this speed the amplitude would become infinite if it were not for the damping in the system. Above the critical speed, the phase position of the axis of rotation relative to the center-of-gravity shifts so that the basket tends to rotate about its center-of-gravity with the flexibly supported bearing moving in a circle about an axis through the center-of-gravity. The relative positions of the bearing center and the center-of-gravity are shown in Fig. 6.30*A* and *B*.

Since the balance ring is circular with a smooth inner surface, any weights or fluids contained in the ring can be acted upon only by forces directed radially. When the ring is rotated about a vertical axis, the weights or fluids will move within the ring in such a manner as to be concentrated on the side farthest from the axis of rotation. If this concentration occurs below the natural frequency (Fig. 6.30*A*), the weights tend to move further from the axis and the resultant center-of-gravity is displaced so as to give a greater eccentricity. The points A and G rotate about the axis O at the frequency ω. The initial eccentricity of the center-of-gravity of the washer basket and its load from the axis of rotation is represented by e, and ρ is the elastic displacement of this center of rotation due to the centrifugal force. Where the off-center rotating weight is W, the unbalanced force is $(W/g)(\rho + e)\omega^2$ [where $\rho = e/(1 - \beta^2)$ and $\beta^2 = \omega^2/\omega_n^2 < 1$] and acts in the direction from A to G.

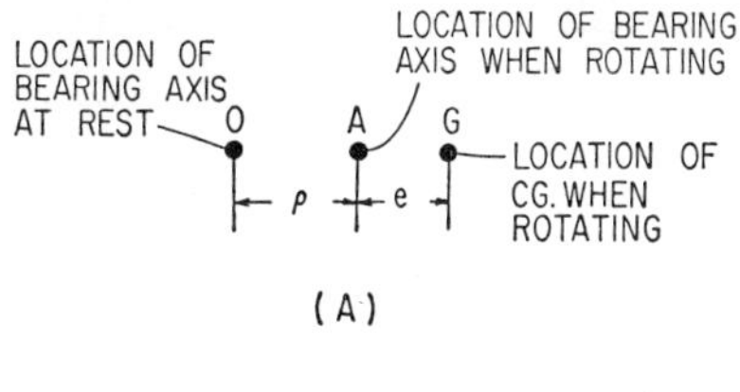

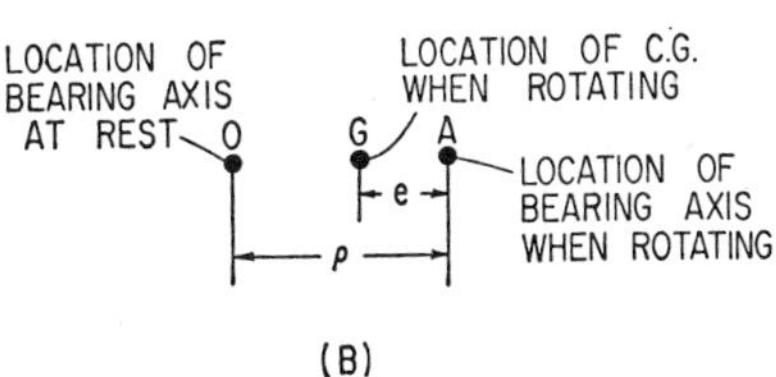

Fig. 6.30. Diagram in plane normal to axis of rotation of spin dryer in Fig. 6.29. Relative positions of axes when rotating speed is less than natural frequency are shown at (*A*); corresponding diagram for rotation speed greater than natural frequency is shown at (*B*).

If the displacement of the weights or fluids in the balance ring occurs above the natural frequency, the center-of-gravity tends to

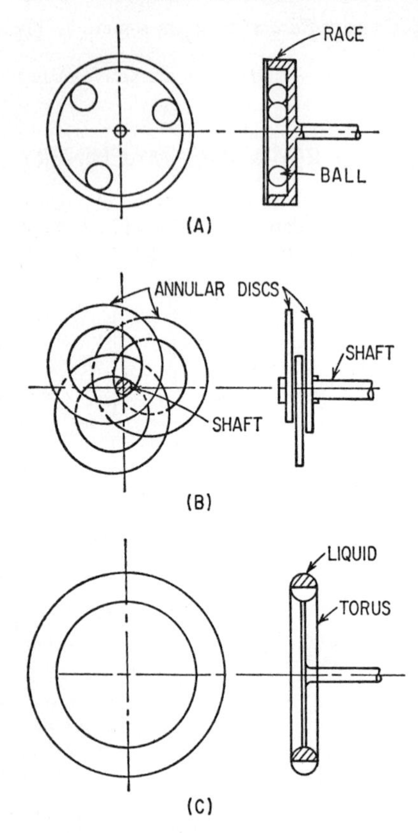

FIG. 6.31. Examples of balancing means for rotating machinery: (*A*) spheres (or cylinders) in a race; (*B*) annular discs rotating on shaft; (*C*) damping fluid in torus.

move closer to the dynamic location of the axis. The action in this case is shown in Fig. 6.30*B*. Then the points *A* and *G* rotate about *O* at the frequency ω. The unbalanced force is $(W/g)(\rho + e)\omega^2$ [where $\rho = e/(1 - \beta^2)$ and $\beta^2 = \omega^2/\omega_n{}^2 > 1$]. This gives a negative force that acts in a direction from *G* to *A*. Thus the eccentricity is brought toward zero and the rotor is automatically balanced. Because it is necessary to pass through the critical speed in bringing the rotor up to speed and in stopping it, it is desirable to heavily damp the balancing elements, either fluid or weights.

In practical applications, the balancing elements can take several forms. The earliest form consisted of two or more spheres or cylinders free to move in a race concentric with the axis of the rotor, as shown in Fig. 6.31*A*. A later modification consists of three annular discs that rotate about an enlarged shaft concentric with the axis, as indicated in Fig. 6.31*B*. These are contained in a sealed compartment with oil for lubrication and damping. A fluid type of damper is shown in Fig. 6.31*C*, the fluid usually being a high-density viscous material. With proper damping, mercury would be excellent, but it is too expensive. Therefore a more viscous, high-density halogenated fluid is used.

The balancers must be of sufficient weight and operate at such a radius that the product of their weight and the maximum eccentricity they can attain is equivalent to the unbalanced moment of the load. This requirement makes the use of the spheres or cylinders difficult because they cannot be made large; it makes the annular plates large because they are limited in the amount of eccentricity that can be obtained.

In a cylindrical volume 24 in. (61 cm) in diameter and 2 in. (5 cm) thick, seven spheres 2 in. (5 cm) in diameter can neutralize 98.6 lb-in. (114 kg-cm) of unbalance; three cylinders 4 in. (10 cm) in diameter by 2 in. (5 cm) thick can neutralize 255 lb-in. (295 kg-cm); three annular discs, each ⅝ in. (1.6 cm) thick with an outside diameter of 19.55 in. (50 cm) and an inside diameter of 10.45 in. (26.5 cm) [the optimum for a center post 6 in. (15.2 cm) in diameter], can neutralize 250 lb-in. (290 kg-cm); and half of a 2-in. (5-cm) diameter torus filled with fluid of density 0.2 lb/in³ (5.5 gm/cm³) can neutralize 609 lb-in. (700 kg-cm). Only the fluid-filled torus would be initially balanced.

AUXILIARY MASS DAMPERS APPLIED TO TORSIONAL VIBRATION

Dampers and absorbers are used widely for the control of torsional vibration of internal-combustion engines. The most common absorber is the viscous-damped, untuned auxiliary mass unit shown in Fig. 6.32. The device is comprised of a cylindrical housing carrying an inertia mass that is free to rotate. There is a preset clearance between the housing and the inertia mass that is filled with a silicone oil of proper viscosity. Silicone oil is used because of its high viscosity index; i.e., its viscosity changes relatively little with temperature. With the inertia mass and the damping medium contained, the housing is seal-welded to provide a leakproof and simple absorber. However, the silicone oil has poor boundary lubricating properties and if decomposed by a local hot spot (such as might be caused by a reduced clearance at some particular spot), the decomposed damping fluid is abrasive.

Because of the simplicity of this untuned damper, it is commonly used in preference to the more effective tuned absorber. However, it is possible to use the same construction methods for a tuned damper, as shown in Fig. 6.33. It is also possible to mount the standard damper with the housing for the unsprung inertia mass attached to the main

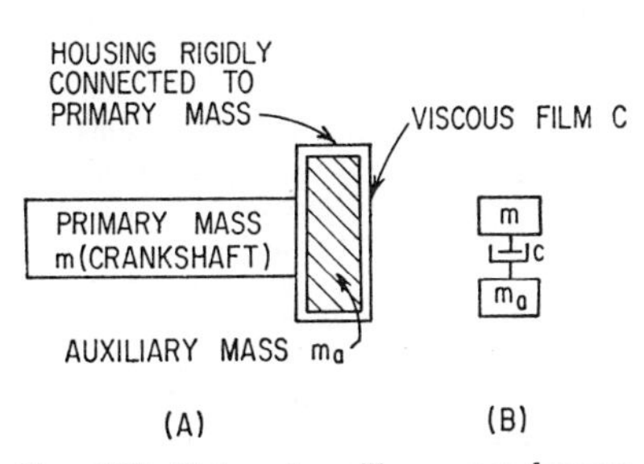

FIG. 6.32. Untuned auxiliary mass damper with viscous damping. The application to a torsional system is shown at (*A*), and the linear analog at (*B*).

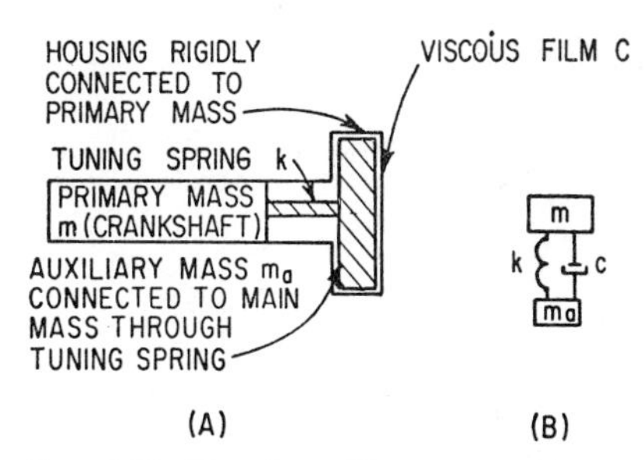

FIG. 6.33. Tuned auxiliary mass damper with viscous damping. The application to a torsional system is shown at (*A*), and the linear analog at (*B*).

mass by a spring, as shown in Fig. 6.34. If the viscosity of the oil and the dimensions of the masses and the clearance spaces are known, the damping effects of the dampers shown in Figs. 6.32 and 6.34 can be computed directly in terms of the equations previously developed. The damper in Fig. 6.34 can be analyzed by treating the spring and housing as additional elements in the main system and the untuned mass as a viscous damped auxiliary mass. If the inertia of the housing is negligible, the inertia mass is

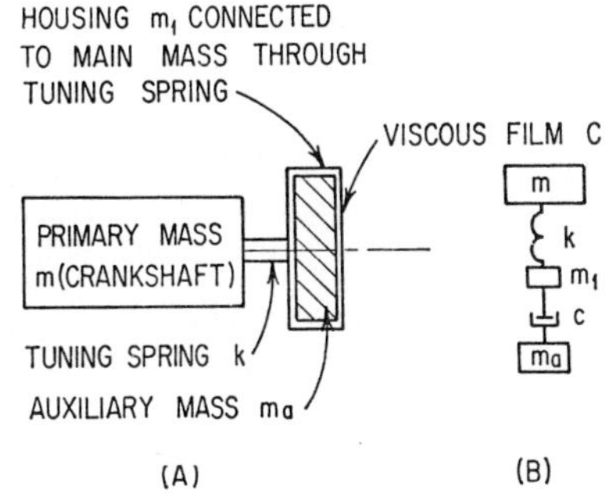

FIG. 6.34. Auxiliary mass damper with viscous damping and spring-mounted housing. The application to a torsional system is shown at (*A*), and the linear analog at (*B*).

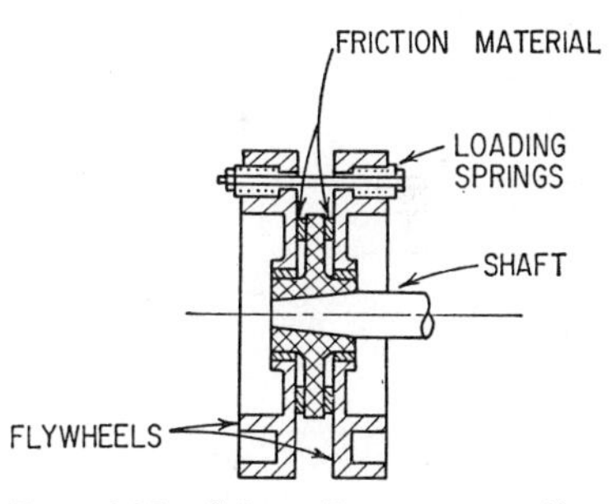

FIG. 6.35. Schematic cross section through Lanchester damper.

effectively connected to the main mass through a spring and a dashpot in series. The two elements in series can be represented by a complex spring constant equal to

$$\frac{1}{(1/jc\omega) + (1/k)} = \frac{kcj\omega}{k + cj\omega}$$

Where there is no damping in parallel with the spring, Eq. (6.3) becomes

$$m_{eq} = km/(k - m\omega^2)$$

Substituting the complex value of the spring constant, the effective mass is

$$m_{eq} = \frac{ckj\omega}{k + cj\omega}\left[\frac{m}{-m\omega^2 + cjk\omega/(k + cj\omega)}\right] \tag{6.58}$$

In terms of the nondimensional parameters defined in Eq. (6.4):

$$m_{eq} = \frac{(2\zeta\beta_a)^2(1 - \beta_a^2)}{\beta_a^4 - (2\zeta\beta_a)^2(1 - \beta_a^2)} m + \frac{-2\zeta\beta_a^3 m}{\beta_a^4 - (2\zeta\beta_a)^2(1 - \beta_a^2)} j \tag{6.59}$$

Before the advent of silicone oil with its chemical stability and relatively constant viscosity over service temperature conditions, the damper most commonly used for absorbing torsional vibration energy was the dry friction or Lanchester damper shown in Fig. 6.35. The damping is determined by the spring tension and the coefficient of friction at the sliding interfaces. Its optimum value is determined by the equation for a torsional system analogous to Eq. (6.35) for a linear system:

$$(T_s)_{opt} = \frac{\sqrt{2}}{\pi} I\omega^2\theta_0 \tag{6.60}$$

where T_s is the slipping torque, I is the moment of inertia of the flywheels, and θ_0 is the amplitude of angular motion of the primary system. The dry-friction-based Lanchester damper requires frequent adjustment, as the braking material wears, to maintain a constant braking force.

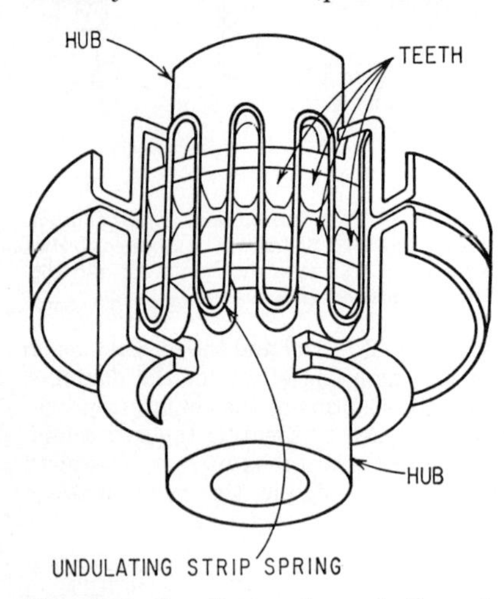

FIG. 6.36. Coupling used as elastic and damping element in auxiliary mass damper for torsional vibration. The torque is transmitted by an undulating strip of thin steel interposed between the teeth on opposite hubs. The stiffness of the strip is nonlinear, increasing as torque increases. Oil pumped between the strip and teeth dissipates energy.

It is possible to use torque-transmitting couplings that can absorb vibration energy, as the spring elements for tuned dampers. The Bibby coupling (Fig. 6.36) is used in this manner. Since the stiffness of this coupling is nonlinear, the optimum tuning of such an absorber is secured for only one amplitude of motion.

A discussion of dampers and of their application to engine systems is given in Chap. 38.

DYNAMIC ABSORBERS TUNED TO ORDERS OF VIBRATION RATHER THAN CONSTANT FREQUENCIES

In the torsional vibration of rotating machinery, it is generally found that exciting torques and forces occur at the same frequency as the rotational speed, or at multiples of this frequency. The ratio of the frequency of vibration to the rotational speed is called the *order of the vibration* q. Thus a power plant driving a four-bladed propeller may have a torsional vibration whose frequency is four times the rotational speed of the drive shaft; sometimes it may have a second torsional vibration whose frequency is eight times the rotational speed. These are called the fourth-order and eighth-order torsional vibrations.

If a dynamic absorber in the form of a pendulum acting in a centrifugal field is used, then its natural frequency increases linearly with speed. Therefore it can be used to neutralize an order of vibration.[15–19]

Consider a pendulum of length l and of mass m attached at a distance R from the center of a rotating shaft, as shown in Fig. 6.37. Since the pendulum is excited by torsional vibration in the shaft, let the radius R be rotating at a constant speed Ω with a superposed vibration $\theta = \theta_0 \cos q\Omega t$, where q represents the order of the vibration. Then the angle of R with respect to any desired reference is $\Omega t + \theta_0 \cos q\Omega t$. The angle of the pendulum with respect to the radius R is defined as $\psi = \psi_0 \cos q\Omega t$, as shown by Fig. 6.37.

The acceleration acting on the mass m at

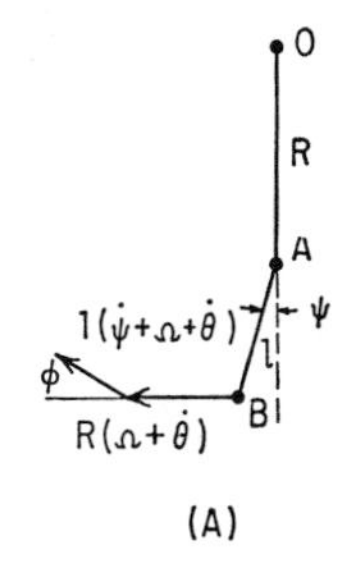

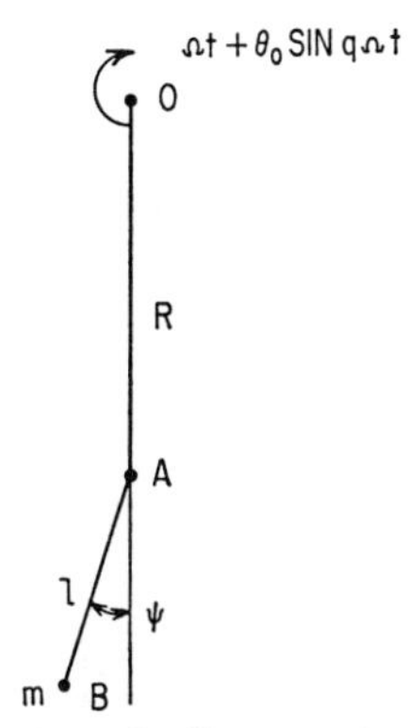

Fig. 6.37. Schematic diagram of pendulum absorber.

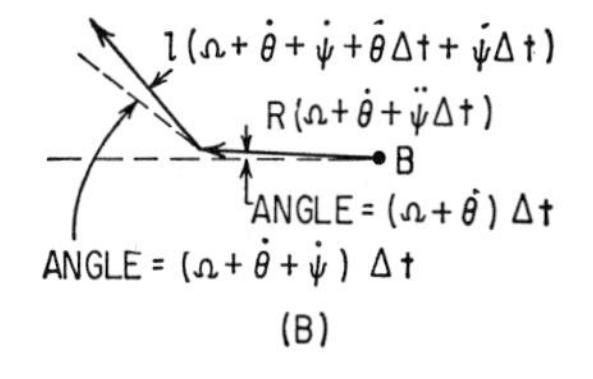

Fig. 6.38. Velocity vectors for the pendulum absorber: (A) velocities at time t; (B) velocities at time $t + \Delta t$; (C) change in velocities during time increment Δt.

position B is most easily ascertained by considering the change in velocity during a short increment of time Δt. The components of velocity of the mass m at time t are shown graphically in Fig. 6.38A; at time $t + \Delta t$, the corresponding velocities are shown in Fig. 6.38B. The change in velocity during the time interval Δt is shown in Fig. 6.38C. Since the acceleration is the change in velocity per unit of time, the accelerations along and perpendicular to l are:

Acceleration along l:

$$\frac{-l(\Omega + \dot\theta + \dot\psi^2)\,\Delta t - R(\Omega + \dot\theta)^2\,\Delta t \cos\psi + R\ddot\theta\,\Delta t \sin\psi}{\Delta t} \tag{6.61}$$

Acceleration perpendicular to l:

$$\frac{l(\ddot\theta + \ddot\psi)\,\Delta t + R(\Omega + \dot\theta)^2\,\Delta t \sin\psi + R\ddot\theta\,\Delta t \cos\psi}{\Delta t} \tag{6.62}$$

Only the force $-F$, directed along the pendulum, acts on the mass m. Therefore the equations of motion are

$$\begin{aligned} -F &= -ml(\Omega + \dot\theta + \dot\psi)^2 - mR(\Omega + \dot\theta)^2 \cos\psi + R\ddot\theta \sin\psi \\ 0 &= ml(\ddot\theta + \ddot\psi) + mR(\Omega + \dot\theta)^2 \sin\psi + mR\ddot\theta \cos\dot\psi \end{aligned} \tag{6.63}$$

Assuming that ψ and θ are small, Eqs. (6.63) simplify to

$$\begin{aligned} Ft &= m(R + l)\Omega^2 \\ l(\ddot\theta + \ddot\psi) &+ R\Omega^2\psi + R\ddot\theta = 0 \end{aligned} \tag{6.64}$$

The second of Eqs. (6.64) upon substitution of $\theta = \theta_0 \cos q\Omega t$ and $\psi = \psi_0 \cos q\Omega t$ yields

$$\frac{\psi_0}{\theta_0} = \frac{(q\Omega)^2(l + R)}{-(q\Omega)^2 l + \Omega^2 R} = \frac{q^2(l + R)}{R - q^2 l} \tag{6.65}$$

The torque M exerted at point 0 by the force F is

$$M = RF \sin \psi = RF\psi \qquad \text{when } \psi \text{ is small}$$

From Eqs. (6.64) and (6.65), when ψ is small,

$$M = \frac{mq^2R(R + l)^2\Omega^2}{R - q^2 l} \tag{6.66}$$

If a flywheel having a moment of inertia I is accelerated by a shaft having an amplitude of angular vibratory motion θ_0 and a frequency $q\Omega$, the torque amplitude exerted on the shaft is $I(q\Omega)^2\theta_0$. Therefore, the equivalent moment of inertia I_{eq} of the pendulum is

$$I_{eq} = \frac{mR(R + l)^2}{R - q^2 l} = \frac{m(R + l)^2}{1 - q^2 l/R} \tag{6.67}$$

When

$$\frac{R}{l} = q^2 \tag{6.68}$$

the equivalent inertia is infinite and the pendulum acts as a dynamic absorber by enforcing a node at its point of attachment.

Where the pendulum is damped, the equivalent moment of inertia is given by an equation analogous to Eqs. (6.4) and (6.5):

$$\begin{aligned} I_{eq} &= \frac{1 + 2\zeta\nu j}{(1 - \nu^2) + 2\zeta\nu j}\, m(R + l)^2 \\ &= m(R + l)^2 \left[\frac{1 - \nu^2 + (2\zeta\nu)^2}{(1 - \nu^2)^2 + (2\zeta\nu)^2} - \frac{2\zeta\nu^3 j}{(1 - \nu^2)^2 + (2\zeta\nu)^2}\right] \end{aligned} \tag{6.69}$$

where $\nu^2 = q^2 l/R$ and $\zeta = (c/2m\Omega)\sqrt{l/R}$.

When the pendulum is attached to a single degree-of-freedom system as is shown in Fig. 6.39, the amplitude of motion θ_a of the flywheel of inertia I is given, by analogy to Eq. (6.7), as

$$\frac{\theta_a}{\theta_{st}} = \sqrt{\frac{(1 - \nu^2)^2 + (2\zeta\nu)^2}{[(1 - \nu^2)(1 - \beta_p{}^2) - \beta_p{}^2\mu]^2 + (2\zeta\nu)^2[1 - \beta_p{}^2 - \beta_p{}^2\mu_p]^2}} \tag{6.70}$$

where

$$2\zeta\nu = \frac{cql}{mR}$$

$$\mu_p = \frac{m(R + l)^2}{I}$$

$$\beta_p = \frac{q}{k_r I}$$

$$\theta_{st} = \frac{m_0}{k_r}$$

The pendulum tends to detune when the amplitude of motion of the pendulum is large, thereby introducing harmonics of the torque that it neutralizes.[17] Suppose the shaft rotates at a constant speed Ω, i.e., $\dot{\theta}_0 = 0$, and consider the torque exerted on the shaft as m moves through a large amplitude ψ_0 about its equilibrium position. Equations (6.63) become

$$\begin{aligned} F &= ml(\Omega + \dot{\psi})^2 + mR\Omega^2 \cos \psi \\ l\ddot{\psi} &+ R\Omega^2 \sin \psi = 0 \end{aligned} \tag{6.71}$$

A solution for the second of Eqs. (6.71) is

$$\dot{\psi} = \sqrt{\frac{2\Omega^2 R}{l}} \sqrt{\cos \psi - \cos \psi_0} \tag{6.72}$$

The solution of Eq. (6.72) involves elliptic integrals and is given approximately by

$$\psi = \psi_0 \sin \omega t$$

where

$$\omega = \sqrt{\frac{R}{l}} \frac{\pi/2}{F(\psi_0/2, \pi/2)} \Omega$$

and $F(\psi_0/2, \pi/2)$ is an elliptic function of the first kind whose value may be obtained from tables.

Since $\omega/\Omega = q$ (the order of the disturbance), the tuning of the damper will be changed for large angles and becomes

$$q^2 = \frac{R}{l} \left(\frac{\pi/2}{F(\psi_0/2, \pi/2)} \right)^2 \tag{6.73}$$

The value of $q^2 l/R = \nu^2$ used in Eqs. (6.69) and (6.70) is given in Fig. 6.40 as a function of the amplitude of the pendulum.

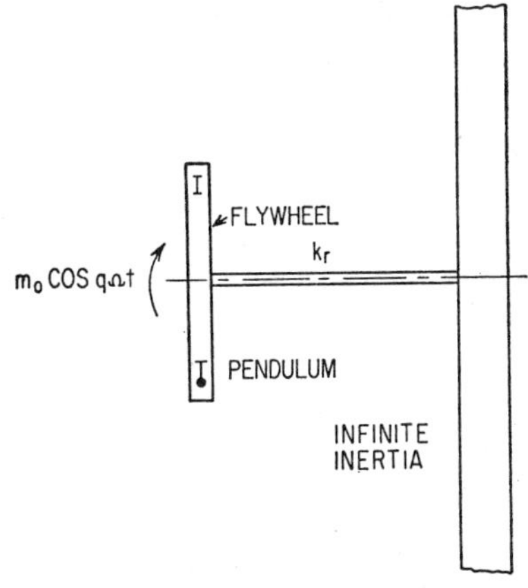

Fig. 6.39. Application of pendulum absorber to a rotational single degree-of-freedom system.

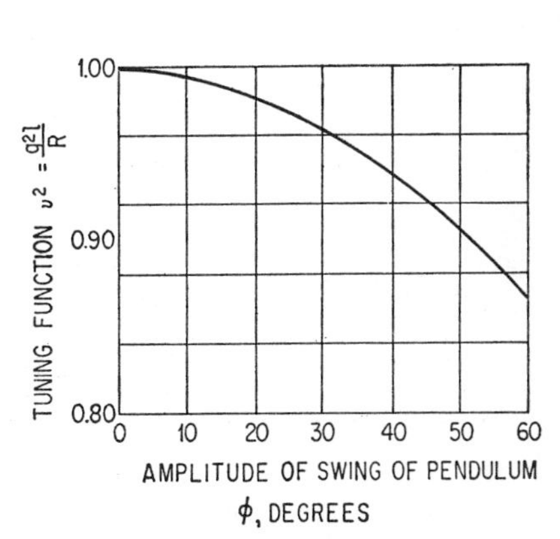

Fig. 6.40. Tuning function for a pendulum absorber used in Eqs. (6.69) and (6.70).

Since the force exerted by the mass m is directed along the rod connecting it to the pivot A (Fig. 6.37), the reactive torque on the shaft is

$$
\begin{aligned}
M &= FR \sin \psi \\
&= mR^2\Omega^2 \left[\frac{l}{R} \left(1 + \frac{\dot{\psi}}{\Omega}\right)^2 \sin \psi + \sin \psi \cos \psi \right] \\
&= mR^2\Omega^2 (A_1 \sin q\Omega t + A_2 \sin 2q\Omega t + A_3 \sin 3q\Omega t + \cdots) \qquad (6.74)
\end{aligned}
$$

The values of the fundamental torque corresponding to the tuned frequency and to the second and third harmonics of this tuned frequency are given in Fig. 6.41 as a function of the angle of swing of the pendulum, for a typical installation. In this case, the pendulum is tuned to the 4½ order of vibration. (The 4½ order of vibration is one whose frequency is 4½ times the rotational frequency and 9 times the fundamental frequency. The latter is called the half order and occurs at half of the rotational frequency. This is common in four-cycle engines.)

Two types of pendulum absorber are used. The one most commonly used is shown in Fig. 6.42. The counterweight, which also is used to balance rotating forces in the engine, is suspended from a hub carried by the crankshaft by pins that act through holes with clearance, Fig. 6.42*A*. By suspending the pendulum from two pins, the pendulum when oscillating does not rotate but rather moves as shown in Fig. 6.42*B*. Since it is not subjected to angular acceleration, it may be treated as a particle located at its center-

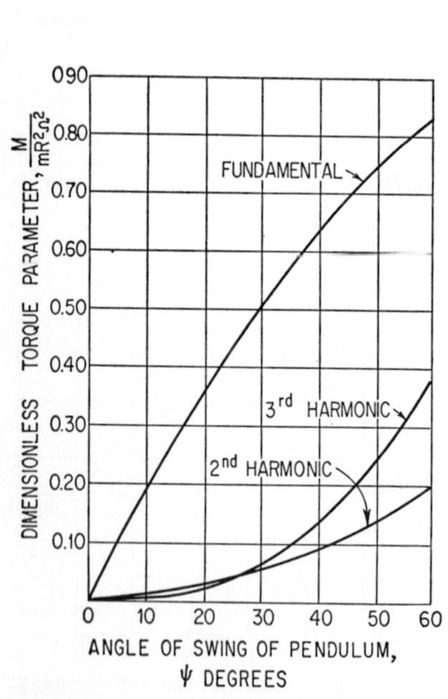

FIG. 6.41. Harmonic components of torque generated by a pendulum absorber as a function of its angle of swing. The torque is expressed by the parameters used in Eq. (6.74).

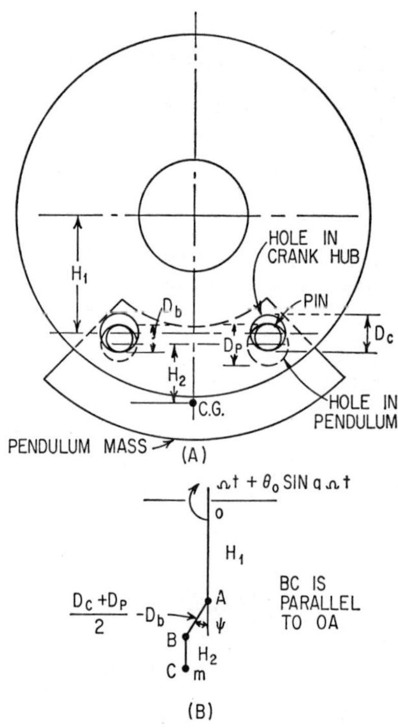

FIG. 6.42. Bifilar type of pendulum absorber. The mechanical arrangement is shown at (*A*), and a schematic diagram at (*B*).

of-gravity. Referring to Fig. 6.42*A* and *B*, the expression for acceleration [Eqs. (6.61) and (6.62)] and the equations of motion [Eqs. (6.63)] apply if

$$R = H_1 + H_2 \qquad l = \frac{D_c + D_p}{2} - D_b \tag{6.75}$$

where H_1 = distance from center of rotation to center of holes in crank hub

H_2 = distance from center of holes in pendulum to center-of-gravity of pendulum

D_c = diameter of hole in crank hub

D_p = diameter of hole in pendulum

D_b = diameter of pin

In practice, difficulty arises from the wear of the holes and the pin. Moreover, the motion on the pins generally is small and the loads due to centrifugal forces are large so that fretting is a problem. Because the radius of motion of the pendulum is short, only a small amount of wear can be tolerated. Hardened pins and bushings are used to reduce the wear.

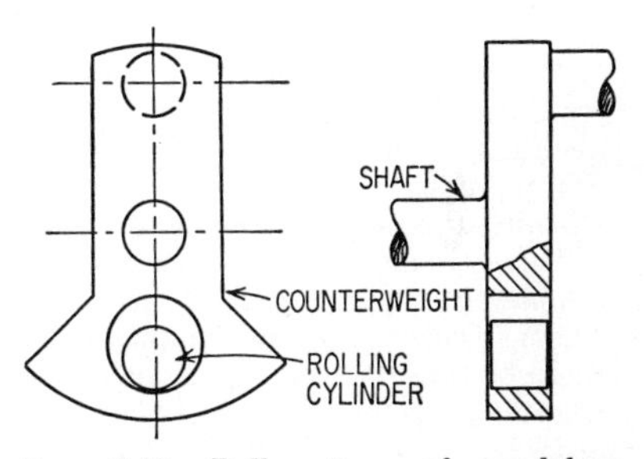

FIG. 6.43. Roller type of pendulum absorber.

The pendulum is most easily designed if it is recognized that the inertia torques generated by the pendulum must neutralize the forcing torques. Thus

$$m\omega^2 l\psi_0 R = M \tag{6.76}$$

The radii l and R are set by the design of the crank and the order of vibration to be neutralized. The original motion ψ_0 is generally limited to a small angle, approximately 20°. It is probable that the most stringent condition is at the lowest operating speed, although the absorber may be required only to avoid difficulty at some particular critical speed. Knowing the excitation M, it is possible to compute the required mass of the pendulum weight.

A second type of pendulum absorber is a cylinder that rolls in a hole in a counterweight, as shown in Fig. 6.43. In this type, the radius of the pendulum corresponds to the difference in the radii of the hole and of the cylinder. It is found, by observing tests and checking the tuning of actual systems using cylindrical pendulums, that the weight rotates with a uniform angular velocity. Therefore the tuning is independent of the moments of inertia of the cylinder. It is common to allow a larger amplitude of motion with the absorber of Fig. 6.43 than with the absorber of Fig. 6.40.

Applications of pendulum absorbers to torsional-vibration problems are given in Chap. 38.

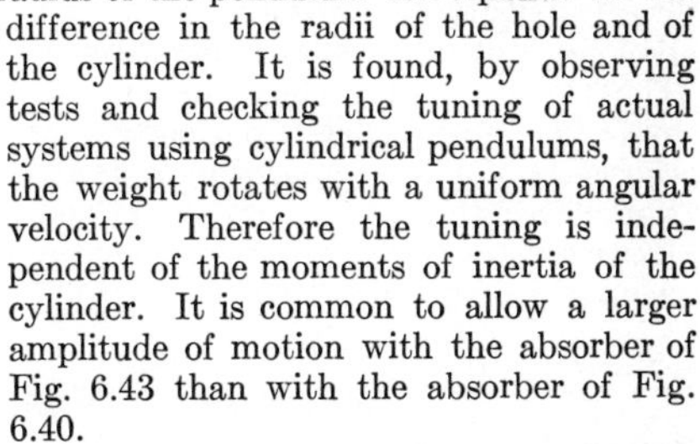

FIG. 6.44. Application of pendulum absorbers to counteract linear vibration.

PENDULUM ABSORBER FOR LINEAR VIBRATION

The principle of the pendulum absorber can be applied to linear vibration as well as to torsional vibration. To neutralize linear vibration, pendulums are rotated about an axis parallel to the direction of vibration, as shown in Fig. 6.44. This can be accomplished with an absorber mounted on the moving body. Two or more pendulums are used so that centrifugal forces are balanced. Free rotational movement of each pendulum in the plane of the axis allows the axial forces to be neutralized. The pendulum assembly must rotate about the axis at some submultiple of the frequency of vibration. The size of the absorber is determined by the condition that the components of the inertia forces of the weights in the axial direction $[\Sigma m\omega^2 r\theta]$ must balance the exciting forces. This device can be applied where the vibration is generated by the action of rotating members, but the magnitude of the vibratory forces is uncertain. A discussion of this absorber, including the influence of moments of inertia and damping of the pendulum, together with some applications to the elimination of vibration in special locations on a ship, is given in Ref. 20.

APPLICATIONS OF DAMPERS TO MULTIPLE DEGREE-OF-FREEDOM SYSTEMS

Auxiliary mass dampers as applied to systems of several degrees-of-freedom can be represented most effectively by equivalent masses or moments of inertia, as determined by Eq. (6.5) or Eq. (6.6). The choice of proper damping constants is more difficult. For the case of torsional vibration, the practical problems of designing dampers and selecting the proper damping are considered in Chap. 38.

There are many applications of dampers to vibrating structures that illustrate the use of different types of auxiliary mass damper. One such application has been to ships.[21] These absorbers had low damping and were designed to be filled with water so that they could be tuned to the objectionable frequencies. In one case, the absorber was located near the propeller (the source of excitation) and when properly tuned was found to be effective in reducing the resonant vibration of the ship. In another case, the absorber was located on an upper deck but was not as effective. It enforced a node at its point of attachment but, because of the flexibility between the upper deck and the bottom of the ship, there was appreciable motion in the vicinity of the propeller and vibratory energy was fed to the ship's structure. To operate properly, the absorbers must be closely tuned and the propeller speed closely maintained. Because the natural frequencies of the ship vary with the types of loading, it is not sufficient to install a fixed frequency absorber that is effective at only one natural frequency of the hull, corresponding to a particular loading condition.

An auxiliary mass absorber has been applied to the reduction of vibration in a heavy building that vibrated at a low frequency under the excitation of a number of looms.[22] The frequency of the looms was substantially constant. However, the magnitude of the excitation was variable as the looms came into and out of phase. The dynamic absorber, consisting of a heavy weight hung as a pendulum, was tuned to the frequency of excitation. Because the frequency was low and the forces large, the absorber was quite large. However, it was effective in reducing the amplitude of vibration in the building and was relatively simple to construct.

ACTIVATED VIBRATION ABSORBERS

The cost and space that can be allotted to ship antirolling devices are limited. Therefore it is desirable to activate the absorbers so that their full capacity is used for small amplitudes as well as large. Activated dampers can be made to deliver as large restoring forces for small amplitudes of motion of the primary body as they would be required to deliver if the motions were large. For example, the gyrostabilizer that is used in the ship is precessed by a motor through its full effective range, in the case of small angles as well as large. Thus, it introduces a restoring torque that is much larger than would be introduced by the normal damped precession.[14] In the same manner the water in antiroll tanks is always pumped to the tank where it will introduce the maximum torque to counteract the roll. By pumping, much larger quantities of water can be transferred and larger damping moments obtained than can be obtained by controlled gravity flows.

Devices for damping the roll are desirable for ships. It has been common practice to install bilge keels (long fins which extend into the water) in steel ships. Some ships are now fitted with activated, retractable hydrofoils located at the bilge at the maximum beam. Both these devices are effective only when the ship is in motion and add to the resistance of the ship.

Activated vibration absorbers are essentially servomechanisms designed to maintain some desired steady state. Steam and gas turbine speed governors, wicket gate controls for frequency regulation in water turbines, and temperature control equipment can be considered as special forms of activated vibration absorbers.[23]

THE USE OF AUXILIARY MASS DEVICES TO REDUCE TRANSIENT AND SELF-EXCITED VIBRATIONS

Where the vibration is self-excited or caused by repeated impact, it is necessary to have sufficient damping to prevent a serious build-up of vibration amplitude. This damping, which need not always be large, may be provided by a loosely coupled auxiliary mass. A simple application of this type is the ring fitted to the interior of a gear, as shown in Fig. 6.45. By fitting this ring with the proper small clearance so that relative motion occurs between it and the gear, it is possible to obtain enough energy dissipation to damp the high-frequency, low-energy vibration that causes the gear to ring. The rubbery coatings applied to large, thin-metal panels such as automobile doors to give them a solid rather than a "tinny" sound depend for their effectiveness on a proper balance of mass, elasticity, and damping (see Chap. 37).

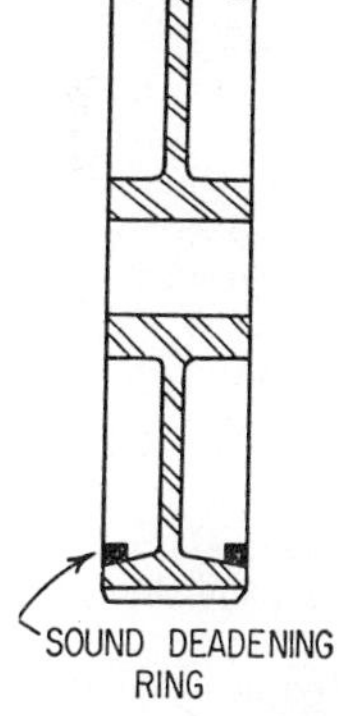

FIG. 6.45. Application of auxiliary mass damper to deaden noise in gear.

Another application where auxiliary mass dampers are useful is in the prevention of fatigue failures in turbines. At the high-pressure end of an impulse turbine, steam or hot gas is admitted through only a few nozzles. Consequently, as the blade passes the nozzle it is given an impulse by the steam and set into vibration at its natural frequency. It is a characteristic of alloy steels that they have very little internal damping at high operating temperature. For this reason the free vibration persists with only a slightly diminished amplitude until the blade again is subjected to the steam impulse. Some of these second impulses will be out of phase with the motion of the blade and will reduce its amplitude; however, successive impulses may increase the amplitude on subsequent passes until failure occurs. Damping can be increased by placing a number of loose wires in a cylindrical hole cut in the blade in a radial direction. The damping of a number of these wires has been computed in terms of the geometry of the application (number of wires, density of wires, size of the hole, radius of the blade, rotational speed, etc.) and the amplitude of vibration.[24] These computations show reasonable agreement with experimental results.

An auxiliary mass has been used to damp the cutting tool chatter set up in a boring bar.[25] Because of the characteristics of the metal-cutting process or of some coupling between motions of the tool parallel and perpendicular to the work face, it is sometimes found that a self-excited vibration is initiated at the natural frequency of the cutter system. Since the self-excitation energy is low, the vibration usually is initiated only if the damping is small. Chatter of the tool is most common in long, poorly supported tools, such as boring bars (see Chap. 40). To eliminate this chatter, a loose auxiliary mass is incor-

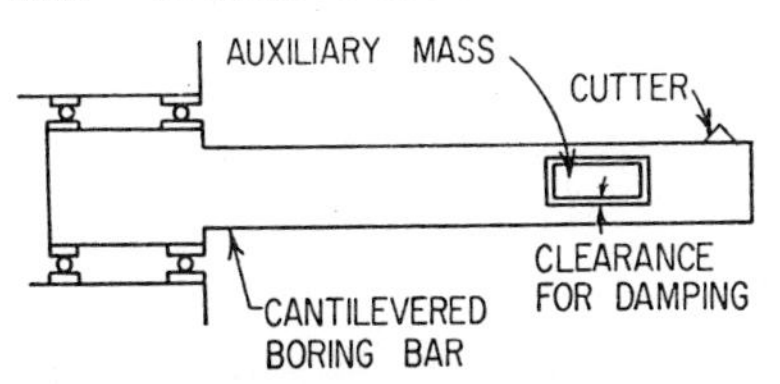

FIG. 6.46. Application of auxiliary mass damper to reduce chatter in boring bar.

porated in the boring bar, as shown in Fig. 6.46. This may be air-damped or fluid-damped. Since the excitation is at the natural frequency of the tool, the damping should be such that the tool vibrates with a minimum amplitude at this frequency. The damping requirement can be estimated by substituting $\beta = 1$ in Eq. (6.25),

$$\frac{x_0}{\delta_{st}} = \sqrt{\frac{1 + 4(\zeta\alpha)^2}{4(\zeta\alpha)^2\mu^2}} \tag{6.77}$$

The optimum value of the parameter $(\zeta\alpha)$ is infinity. Thus when the frequency of excitation is constant, a greater reduction in amplitude can be obtained by a shift in natural

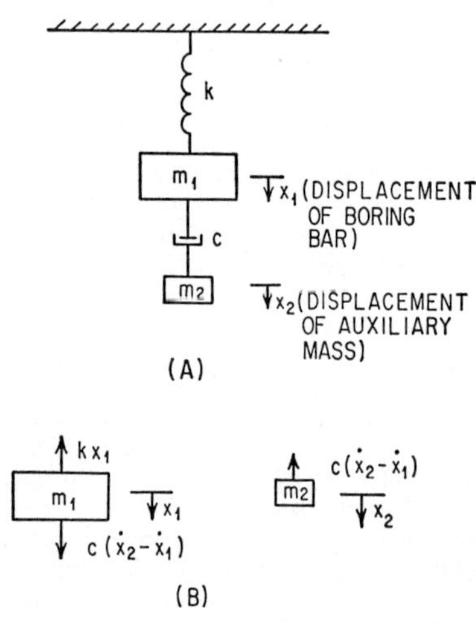

FIG. 6.47. Schematic diagram of damper shown in Fig. 6.46. The arrangement is shown at (*A*), and the forces acting on the boring bar and auxiliary mass are shown at (*B*).

frequency than by damping. However, such a shift cannot be attained because the frequency of the excitation always coincides with the natural frequency of the complete system. Instead, a better technique is to determine the damping that gives the maximum decrement of the free vibration.

Let the boring bar and damper be represented by a single degree-of-freedom system with a damper mass coupled to the main mass by viscous damping, as shown in Fig. 6.47*A*. The forces acting on the masses are shown in Fig. 6.47*B*. The equations of motion are

$$\begin{aligned} -kx_1 - c\dot{x}_1 + c\dot{x}_2 &= m_1\ddot{x}_1 \\ c\dot{x}_1 - c\dot{x}_2 &= m_2\ddot{x}_2 \end{aligned} \tag{6.78}$$

Substituting $x = Ae^{st}$, the resulting frequency equation is

$$s^3 + \frac{c(m_1 + m_2)}{m_1m_2}s^2 + \frac{k}{m_1}s + \frac{kc}{m_1m_2} = 0 \tag{6.79}$$

Where chatter occurs, this equation has three roots, one real and two complex. The complex roots correspond to decaying free vibrations. Let the roots be as follows:

$$\alpha_1, \quad \alpha_2 + j\beta, \quad \alpha_2 - j\beta$$

The value of β determines the frequency of the free vibration, and the value of α_2 determines the decrement (rate of decrease of amplitude) of the free vibration. The decrement α_2 is of primary interest; it is most easily found from the conditions that when the coefficient of s^3 is unity, (1) the sum of the roots is equal to the negative of the coefficient of s^2, (2) the sum of the products of the roots taken two at a time is the negative of the coefficient of s, and (3) the product of the roots is the negative of the constant term. The equations thus obtained are

$$\alpha_1 + 2\alpha_2 = -\frac{c(1+\mu)}{\mu m_1} \tag{6.80}$$

$$2\alpha_1\alpha_2 + \alpha_2^2 + \beta^2 = -\omega_n^2 \tag{6.81}$$

$$\alpha_1(\alpha_2^2 + \beta^2) = -\omega_n^2 \frac{c}{m_1\mu} \tag{6.82}$$

where $\omega_n^2 = k/m_1$ and $\mu = m_2/m_1$. It is not practical to find the optimum damping by solving these equations for α_2 and then setting the derivative of α_2 with respect to c equal to zero. However, it is possible to find the optimum damping by the following process. Eliminate $(\alpha_2^2 + \beta^2)$ between Eqs. (6.81) and (6.82) to obtain

$$2\alpha_1^2\alpha_2 = \omega_n^2\left(\frac{c}{\mu m_1} - \alpha_1\right) \tag{6.83}$$

Substituting the value of α_1 from Eq. (6.80) in Eq. (6.83),

$$2\alpha_2\left[2\alpha_2 + \frac{c(1+\mu)}{\mu m_1}\right]^2 = \frac{c\omega_n^2}{\mu m_1} + \omega_n^2\left[2\alpha_2 + \frac{c(1+\mu)}{\mu m_1}\right] \tag{6.84}$$

To find the damping that gives the maximum decrement, differentiate with respect to c and set $d\alpha_2/dc = 0$:

$$2\alpha_2\left[2\alpha_2 + \frac{c(1+\mu)}{\mu m_1}\right] = \tfrac{1}{2}\omega_n^2\frac{2+\mu}{1+\mu} \tag{6.85}$$

Solving Eqs. (6.84) and (6.85) simultaneously,

$$c_{\text{opt}} = \frac{\mu^2 m_1\omega_n}{2(1+\mu)^{3/2}} \tag{6.86}$$

$$(\alpha_2)_{\text{opt}} = -\frac{(2+\mu)\omega_n}{4(1+\mu)^{1/2}} \tag{6.87}$$

These values may be obtained by proper choice of clearance between the auxiliary mass and the hole in which it is located. Air damping is preferable to oil because it

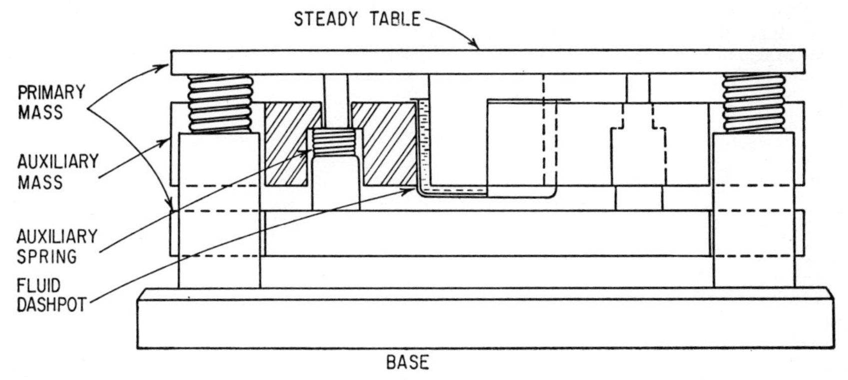

FIG. 6.48. Application of auxiliary mass to spring-mounted table to reduce vibration of table. (*Macinante.*[26])

requires less clearance. Therefore the plug is not immobilized by the centrifugal forces that, with the rotating boring bar, become larger as the clearance is increased.

In precision measurements, it is necessary to isolate the instruments from effects of shock and vibration in the earth and to damp any oscillations that might be generated in the measuring instruments. A heavy spring-mounted table fitted with a heavy auxiliary mass that is attached to the table by a spring and submerged in an oil bath (Fig. 6.48) has proved to be effective.[26] In this example the table has a top surface of 13½ in. (34 cm) by 13½ in. (34 cm) and a height of 6 in. (15 cm). Each auxiliary mass weighs about 70 lb (32 kg). The springs for both the primary table and the auxiliary system are designed to give a natural frequency between 2 and 4 cps in both the horizontal and vertical directions. By trying different fluids in the bath, suitable damping may be obtained experimentally.

REFERENCES

1. Timoshenko, S.: "Vibration Problems in Engineering," p. 240, D. Van Nostrand Company, Inc., Princeton, N.J., 1937.
2. Den Hartog, J. P.: "Mechanical Vibrations," chap. III, McGraw-Hill Book Company, Inc., New York, 1956.
3. Ormondroyd, J., and J. P. Den Hartog: *Trans. ASME*, **50**:A9 (1928).
4. Brock, J. E.: *J. Appl. Mechanics*, **13**(4):A-284 (1946).
5. Brock, J. E.: *J. Appl. Mechanics*, **16**(1):86 (1949).
6. Saver, F. M., and C. F. Garland: *J. Appl. Mechanics*, **16**(2):109 (1949).
7. Lewis, F. M.: *J. Appl. Mechanics*, **22**(3):377 (1955).
8. Georgian, J. C.: *Trans. ASME*, **16**:389 (1949).
9. Den Hartog, J. P., and J. Ormondroyd: *Trans. ASME*, **52**:133 (1930).
10. Roberson, R. E.: *J. Franklin Inst.*, **254**:205 (1952).
11. Pipes, L. A.: *J. Appl. Mechanics*, **20**:515 (1953).
12. Arnold, F. R.: *J. Appl. Mechanics*, **22**:487 (1955).
13. Hort, W.: "Technische Schwingungslehre," 2d ed., Springer-Verlag, Berlin, 1922.
14. Sperry, E. E.: *Trans. SNAME*, **30**:201 (1912).
15. Solomon, B.: *Proc. 4th Intern. Congr. Appl. Mechanics, Cambridge, England*, 1934.
16. Taylor, E. S.: *Trans. SAE*, **44**:81 (1936).
17. Den Hartog, J. P.: "Stephen Timoshenko 60th Anniversary Volume," The Macmillan Company, New York, 1939.
18. Porter, F. P.: "Evaluation of Effects of Torsional Vibration," SAE War Engineering Board, SAE, New York, 1945, p. 269.
19. Crossley, F. R. E.: *J. Appl. Mechanics*, **20**(1):41 (1953).
20. Reed, F. E.: *J. Appl. Mechanics*, **16**:190 (1949).
21. Constanti, M.: *Trans. Inst. of Naval Arch.*, **80**:181 (1938).
22. Crede, C. E.: *Trans. ASME*, **69**:937 (1947).
23. Brown, G. S., and D. P. Campbell: "Principles of Servomechanisms," John Wiley & Sons, Inc., New York, 1948.
24. DiTaranto, R. A.: *J. Appl. Mechanics*, **25**(1):21 (1958)
25. Hahn, R. S.: *Trans. ASME*, **73**:331 (1951).
26. Macinante, J. A.: *J. Sci. Instr.*, **35**:224 (1958)

7

VIBRATION OF SYSTEMS HAVING DISTRIBUTED MASS AND ELASTICITY

William F. Stokey
Carnegie-Mellon University

INTRODUCTION

Preceding chapters consider the vibration of lumped parameter systems; i.e., systems that are idealized as rigid masses joined by massless springs and dampers. Many engineering problems are solved by analyses based on ideal models of an actual system, giving answers that are useful though approximate. In general, more accurate results are obtained by increasing the number of masses, springs, and dampers; i.e., by increasing the number of degrees-of-freedom. As the number of degrees-of-freedom is increased without limit, the concept of the system with distributed mass and elasticity is formed. This chapter discusses the free and forced vibration of such systems. Types of systems include rods vibrating in torsional modes and in tension-compression modes, and beams and plates vibrating in flexural modes. Particular attention is given to the calculation of the natural frequencies of such systems for further use in other analyses. Numerous charts and tables are included to define in readily available form the natural frequencies of systems commonly encountered in engineering practice.

FREE VIBRATION

DEGREES-OF-FREEDOM. Systems for which the mass and elastic parts are lumped are characterized by a finite number of degrees-of-freedom. In physical systems, all elastic members have mass, and all masses have some elasticity; thus, all real systems have distributed parameters. In making an analysis, it is often assumed that real systems have their parameters lumped. For example, in the analysis of a system consisting of a mass and a spring, it is commonly assumed that the mass of the spring is negligible so that its only effect is to exert a force between the mass and the support to which the spring is attached, and that the mass is perfectly rigid so that it does not deform and exert any elastic force. The effect of the mass of the spring on the motion of the system may be considered in an approximate way, while still maintaining the assumption of one degree-of-freedom, by assuming that the spring moves so that the deflection of each of its elements can be described by a single parameter. A commonly used assumption is that the deflection of each section of the spring is proportional to its distance from the support, so that if the deflection of the mass is given, the deflection of any part of the spring is defined. For the exact solution of the problem, even though the mass is considered to be perfectly rigid, it is necessary to consider that the deformation of the spring can occur in any manner consistent with the requirements of physical continuity.

Systems with distributed parameters are characterized by having an infinite number of degrees-of-freedom. For example, if an initially straight beam deflects laterally, it may be necessary to give the deflection of each section along the beam in order to define completely the configuration. For vibrating systems, the coordinates usually are defined in such a way that the deflections of the various parts of the system from the equilibrium position are given.

NATURAL FREQUENCIES AND NORMAL MODES OF VIBRATION. The number of natural frequencies of vibration of any system is equal to the number of degrees-of-freedom; thus, any system having distributed parameters has an infinite number of natural frequencies. At a given time, such a system usually vibrates with appreciable amplitude at only a limited number of frequencies, often at only one. With each natural frequency is associated a shape, called the normal or natural mode, which is assumed by the system during free vibration at the frequency. For example, when a uniform beam with simply supported or hinged ends vibrates laterally at its lowest or fundamental natural frequency, it assumes the shape of a half sine wave; this is a normal mode of vibration. When vibrating in this manner, the beam behaves as a system with a single degree-of-freedom, since its configuration at any time can be defined by giving the deflection of the center of the beam. When any linear system, i.e., one in which the elastic restoring force is proportional to the deflection, executes free vibration in a single natural mode, each element of the system except those at the supports and nodes executes simple harmonic motion about its equilibrium position. All possible free vibration of any linear system is made up of superposed vibrations in the normal modes at the corresponding natural frequencies. The total motion at any point of the system is the sum of the motions resulting from the vibration in the respective modes.

There are always nodal points, lines, or surfaces, i.e., points which do not move, in each of the normal modes of vibration of any system. For the fundamental mode, which corresponds to the lowest natural frequency, the supported or fixed points of the system usually are the only nodal points; for other modes, there are additional nodes. In the modes of vibration corresponding to the higher natural frequencies of some systems, the nodes often assume complicated patterns. In certain problems involving forced vibrations, it may be necessary to know what the nodal patterns are, since a particular mode usually will not be excited by a force acting at a nodal point. Nodal lines are shown in some of the tables.

METHODS OF SOLUTION. The complete solution of the problem of free vibration of any system would require the determination of all the natural frequencies and of the mode shape associated with each. In practice, it often is necessary to know only a few of the natural frequencies, and sometimes only one. Usually the lowest frequencies are the most important. The exact mode shape is of secondary importance in many problems. This is fortunate, since some procedures for finding natural frequencies involve assuming a mode shape from which an approximation to the natural frequency can be found.

Classical Method. The fundamental method of solving any vibration problem is to set up one or more equations of motion by the application of Newton's second law of motion. For a system having a finite number of degrees-of-freedom, this procedure gives one or more ordinary differential equations. For systems having distributed parameters partial differential equations are obtained. Exact solutions of the equations are possible for only a relatively few configurations. For most problems other means of solution must be employed.

Rayleigh's and Ritz's Methods. For many elastic bodies, Rayleigh's method is useful in finding an approximation to the fundamental natural frequency. While it is possible to use the method to estimate some of the higher natural frequencies, the accuracy often is poor; thus, the method is most useful for finding the fundamental frequency. When any elastic system without damping vibrates in its fundamental normal mode, each part of the system executes simple harmonic motion about its equilibrium position. For example, in lateral vibration of a beam the motion can be expressed as $y = X(x) \sin \omega_n t$ where X is a function only of the distance along the length of the beam. For lateral vibration of a plate, the motion can be expressed as $w = W(x,y) \sin \omega_n t$ where x and y are the coordinates in the plane of the plate. The equations show that when the deflection from equilibrium is a maximum, all parts of the body are motionless. At that time all the energy associated with the vibration is in the form of elastic strain energy. When the body is passing through its equilibrium position, none of the vibrational energy is in the form of strain energy so that all of it is in the form of kinetic energy. For conservation of energy, the strain energy in the position of maximum deflection must equal the kinetic energy when passing through the equilibrium position. Rayleigh's method of finding the natural frequency is to compute these maximum energies, equate them, and solve for the frequency. When the kinetic-energy term is evaluated, the frequency always appears as a factor. Formulas for finding the strain and kinetic energies of rods, beams, and plates are given in Table 7.1.

If the deflection of the body during vibration is known exactly, Rayleigh's method gives the true natural frequency. Usually the exact deflection is not known, since its determination involves the solution of the vibration problem by the classical method. If the classical solution is available, the natural frequency is included in it, and nothing is gained by applying Rayleigh's method. In many problems for which the classical solution is not available, a good approximation to the deflection can be assumed on the basis of physical reasoning. If the strain and kinetic energies are computed using such an assumed shape, an approximate value for the natural frequency is found. The correctness of the approximate frequency depends on how well the assumed shape approximates the true shape.

In selecting a function to represent the shape of a beam or a plate, it is desirable to satisfy as many of the boundary conditions as possible. For a beam or plate supported at a boundary, the assumed function must be zero at that boundary; if the boundary is built in, the first derivative of the function must be zero. For a free boundary, if the conditions associated with bending moment and shear can be satisfied, better accuracy

Table 7.1. Strain and Kinetic Energies of Uniform Rods, Beams, and Plates

Member	Strain energy V	Kinetic energy T	
		General	Maximum *
Rod in tension or compression	$\frac{SE}{2}\int_0^l \left(\frac{\partial u}{\partial x}\right)^2 dx$	$\frac{S\gamma}{2g}\int_0^l \left(\frac{\partial u}{\partial t}\right)^2 dx$	$\frac{S\gamma\omega_n^2}{2g}\int_0^l V^2\,dx$
Rod in torsion	$\frac{GI_p}{2}\int_0^l \left(\frac{\partial \phi}{\partial x}\right)^2 dx$	$\frac{I_p\gamma}{2g}\int_0^l \left(\frac{\partial \phi}{\partial t}\right)^2 dx$	$\frac{I_p\gamma\omega_n^2}{2g}\int_0^l \Phi^2\,dx$
Beam in bending	$\frac{EI}{2}\int_0^l \left(\frac{\partial^2 y}{\partial x^2}\right)^2 dx$	$\frac{S\gamma}{2g}\int_0^l \left(\frac{\partial y}{\partial t}\right)^2 dx$	$\frac{S\gamma\omega_n^2}{2g}\int_0^l Y^2\,dx$
Rectangular plate in bending [1]	$\frac{D}{2}\int_S\int \left\{\left(\frac{d^2w}{dx^2}+\frac{d^2w}{dy^2}\right)^2 - 2(1-\mu)\left[\frac{\partial^2 w}{\partial x^2}\frac{\partial^2 w}{\partial y^2} - \left(\frac{\partial^2 w}{\partial x\,\partial y}\right)^2\right]\right\} dx\,dy$	$\frac{\gamma h}{2g}\int_S\int \left(\frac{\partial w}{\partial t}\right)^2 dx\,dy$	$\frac{\gamma h\omega_n^2}{2g}\int_S\int W^2\,dx\,dy$
Circular plate (deflection symmetrical about center) [1]	$\pi D\int_0^a \left\{\left(\frac{\partial^2 w}{\partial r^2}+\frac{1}{r}\frac{\partial w}{\partial r}\right)^2 - 2(1-\mu)\frac{\partial^2 w}{\partial r^2}\frac{1}{r}\frac{\partial w}{\partial r}\right\} r\,dr$	$\frac{\pi\gamma h}{g}\int_0^a \left(\frac{\partial w}{\partial t}\right)^2 r\,dr$	$\frac{\pi\gamma h\omega_n^2}{g}\int_0^a W^2 r\,dr$

u = longitudinal deflection of cross section of rod
ϕ = angle of twist of cross section of rod
y = lateral deflection of beam
w = lateral deflection of plate
Capitals denote values at extreme deflection for simple harmonic motion.
l = length of rod or beam
a = radius of circular plate
h = thickness of beam or plate

S = area of cross section
I_p = polar moment of inertia
I = moment of inertia of beam
γ = weight density
E = modulus of elasticity
G = modulus of rigidity
μ = Poisson's ratio
$D = \frac{Eh^3}{12(1-\mu^2)}$

* This is the maximum kinetic energy in simple harmonic motion.

usually results. It can be shown [2] that the frequency that is found by using any shape except the correct shape always is higher than the actual frequency. Therefore, if more than one calculation is made, using different assumed shapes, the lowest computed frequency is closest to the actual frequency of the system.

In many problems for which a classical solution would be possible, the work involved is excessive. Often a satisfactory answer to such a problem can be obtained by the application of Rayleigh's method. In this chapter several examples are worked using both the classical method and Rayleigh's method. In all, Rayleigh's method gives a good approximation to the correct result with relatively little work. Many other examples of solutions to problems by Rayleigh's method are in the literature.[3,4,5]

Ritz's method is a refinement of Rayleigh's method. A better approximation to the fundamental natural frequency can be obtained by its use, and approximations to higher natural frequencies can be found. In using Ritz's method, the deflections which are assumed in computing the energies are expressed as functions with one or more undetermined parameters; these parameters are adjusted to make the computed frequency a minimum. Ritz's method has been used extensively for the determination of the natural frequencies of plates of various shapes, and is discussed in the section on the lateral vibrations of plates.

Lumped Parameters. A procedure that is useful in many problems for finding approximations to both the natural frequencies and the mode shapes is to reduce the system with distributed parameters to one having a finite number of degrees-of-freedom. This is done by lumping the parameters for each small region into an equivalent mass and elastic element. Several formalized procedures for doing this and for analyzing the re-

Table 7.2. Approximate Formulas for Natural Frequencies of Systems Having Both Concentrated and Distributed Mass

TYPE OF SYSTEM	NATURAL FREQUENCY $\omega_n = 2\pi f_n$	STIFFNESS
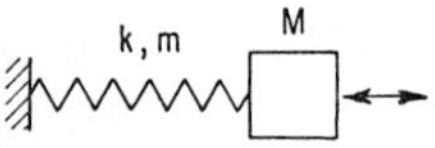 SPRING WITH MASS ATTACHED (k, m; M)	$\sqrt{\frac{k}{M + m/3}}$	$k = \frac{Gd^4}{8nD^3}$ D = COIL DIA d = WIRE DIA n = NUMBER OF TURNS
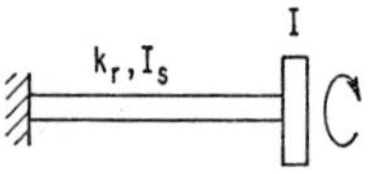 CIRCULAR ROD, WITH DISC ATTACHED, IN TORSION (k_r, I_s; I)	$\sqrt{\frac{k_r}{I + I_s/3}}$	$k_r = \frac{G\pi D^4}{32\,l}$ D = ROD DIAMETER l = ROD LENGTH
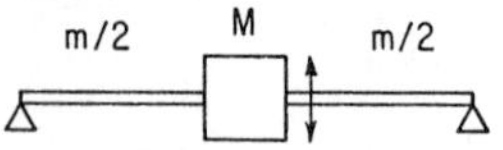 UNIFORM SIMPLY SUPPORTED BEAM WITH MASS IN CENTER ($m/2$, M, $m/2$)	$\sqrt{\frac{k}{M + m/2}}$	$k = \frac{48EI}{l^3}$ l = BEAM LENGTH I = MOMENT OF INERTIA
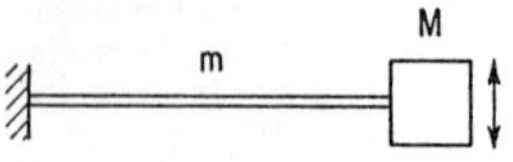 UNIFORM CANTILEVER BEAM WITH MASS ON END (m; M)	$\sqrt{\frac{k}{M + 0.23m}}$	$k = \frac{3EI}{l^3}$ l = BEAM LENGTH I = MOMENT OF INERTIA

sulting systems are described in Chap. 28. If a system consists of a rigid mass supported by a single flexible member whose mass is not negligible, the elastic part of the system sometimes can be treated as an equivalent spring; i.e., some of its mass is lumped with the rigid mass. Formulas for several systems of this kind are given in Table 7.2.

ORTHOGONALITY. It is shown in Chap. 2 that the normal modes of vibration of a system having a finite number of degrees-of-freedom are orthogonal to each other. For a system of masses and springs having n degrees-of-freedom, if the coordinate system is selected in such a way that X_1 represents the amplitude of motion of the first mass, X_2 that of the second mass, etc., the orthogonality relations are expressed by $(n - 1)$ equations as follows:

$$m_1 X_1^a X_1^b + m_2 X_2^a X_2^b + \cdots = \sum_{i=1}^{n} m_i X_i^a X_i^b = 0 \qquad [a \neq b]$$

where X_1^a represents the amplitude of the first mass when vibrating only in the ath mode, X_1^b the amplitude of the first mass when vibrating only in the bth mode, etc.

For a body such as a uniform beam whose parameters are distributed only lengthwise; i.e., in the X direction, the orthogonality between two normal modes is expressed by

$$\int_0^l \rho \phi_a(x) \phi_b(x)\, dx = 0 \qquad [a \neq b] \tag{7.1}$$

where $\phi_a(x)$ represents the deflection in the ath normal mode, $\phi_b(x)$ the deflection in the bth normal mode, and ρ the density.

For a system, such as a uniform plate, in which the parameters are distributed in two dimensions, the orthogonality condition is

$$\int_A \int \rho \phi_a(x,y) \phi_b(x,y)\, dx\, dy = 0 \qquad [a \neq b] \tag{7.2}$$

LONGITUDINAL AND TORSIONAL VIBRATIONS OF UNIFORM CIRCULAR RODS

Equations of Motion. A circular rod having a uniform cross section can execute longitudinal, torsional, or lateral vibrations, either individually or in any combination. The equations of motion for longitudinal and torsional vibrations are similar in form, and the solutions are discussed together. The lateral vibration of a beam having a uniform cross section is considered separately.

In analyzing the longitudinal vibration of a rod, only the motion of the rod in the longitudinal direction is considered. There is some lateral motion because longitudinal stresses induce lateral strains; however, if the rod is fairly long compared to its diameter, this motion has a minor effect.

Consider a uniform circular rod, Fig. 7.1*A*. The element of length dx, which is formed by passing two parallel planes *A–A* and *B–B* normal to the axis of the rod, is shown in Fig. 7.1*B*. When the rod executes only longitudinal vibration, the force acting on the

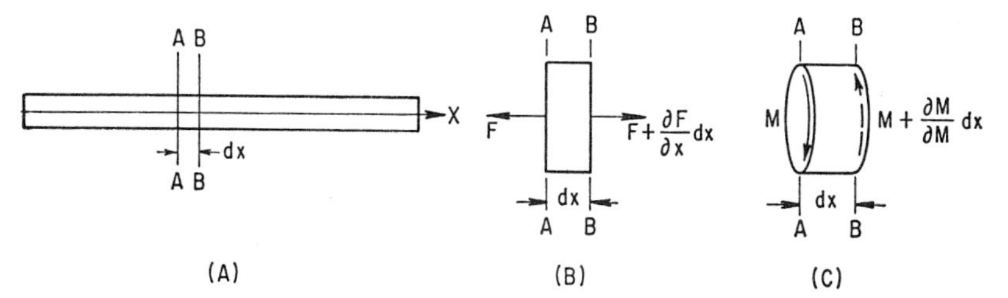

FIG. 7.1. (*A*) Rod executing longitudinal or torsional vibration. (*B*) Forces acting on element during longitudinal vibration. (*C*) Moments acting on element during torsional vibration.

face A–A is F, and that on face B–B is $F + (\partial F/\partial x)\,dx$. The net force acting to the right must equal the product of the mass of the element $(\gamma/g)S\,dx$ and its acceleration $\partial^2 u/\partial t^2$, where γ is the weight density, S the area of the cross section, and u the longitudinal displacement of the element during the vibration:

$$\left(F + \frac{\partial F}{\partial x}dx\right) - F = \frac{\partial F}{\partial x}dx = \left(\frac{\gamma}{g}\right)S\,dx\,\frac{\partial^2 u}{\partial t^2} \qquad \text{or} \qquad \frac{\partial F}{\partial x} = \frac{\gamma S}{g}\frac{\partial^2 u}{\partial t^2} \tag{7.3}$$

This equation is solved by expressing the force F in terms of the displacement. The elastic strain at any section is $\partial u/\partial x$, and the stress is $E\partial u/\partial x$. The force F is the product of the stress and the area, or $F = ES\,\partial u/\partial x$, and $\partial F/\partial x = ES\,\partial^2 u/\partial x^2$. Equation (7.3) becomes $Eu'' = \gamma/g\ddot{u}$, where $u'' = \partial^2 u/\partial x^2$ and $\ddot{u} = \partial^2 u/\partial t^2$. Substituting $a^2 = Eg/\gamma$,

$$a^2 u'' = \ddot{u} \tag{7.4}$$

The equation governing the torsional vibration of the circular rod is derived by equating the net torque acting on the element, Fig. 7.1C, to the product of the moment of inertia J and the angular acceleration $\ddot{\phi}$, ϕ being the angular displacement of the section. The torque on the section A–A is M and that on section B–B is $M + (\partial M/\partial x)\,dx$. By an analysis similar to that for the longitudinal vibration, letting $b^2 = Gg/\gamma$,

$$b^2\phi'' = \ddot{\phi} \tag{7.5}$$

Solution of Equations of Motion. Since Eqs. (7.4) and (7.5) are of the same form, the solutions are the same except for the meaning of a and b. The solution of Eq. (7.5) is of the form $\phi = X(x)T(t)$ in which X is a function of x only and T is a function of t only. Substituting this in Eq. (7.5) gives $b^2X''T = X\ddot{T}$. By separating the variables,[6]

$$T = A\cos(\omega_n t + \theta)$$
$$X = C\sin\frac{\omega_n x}{b} + D\cos\frac{\omega_n x}{b} \tag{7.6}$$

The natural frequency ω_n can have infinitely many values, so that the complete solution of Eq. (7.5) is, combining the constants,

$$\phi = \sum_{n=1}^{n=\infty}\left(C_n\sin\frac{\omega_n x}{b} + D_n\cos\frac{\omega_n x}{b}\right)\cos(\omega_n t + \theta_n) \tag{7.7}$$

The constants C_n and D_n are determined by the end conditions of the rod and by the initial conditions of the vibration. For a built-in or clamped end of a rod in torsion, $\phi = 0$ and $X = 0$ because the angular deflection must be zero. The torque at any section of the shaft is given by $M = (GI_p)\phi'$, where GI_p is the torsional rigidity of the shaft; thus, for a free end, $\phi' = 0$ and $X' = 0$. For the longitudinal vibration of a rod, the boundary conditions are essentially the same; i.e., for a built-in end the displacement is zero ($u = 0$) and for a free end the stress is zero ($u' = 0$).

Example 7.1. The natural frequencies of the torsional vibration of a circular steel rod of 2-in. diameter and 24-in. length, having the left end built in and the right end free, are to be determined.

Solution. The built-in end at the left gives the condition $X = 0$ at $x = 0$ so that $D = 0$ in Eq. (7.6). The free end at the right gives the condition $X' = 0$ at $x = l$. For each mode of vibration, Eq. (7.6) is $\cos\frac{\omega_n l}{b} = 0$ from which $\frac{\omega_n l}{b} = \frac{\pi}{2}, \frac{3\pi}{2}, \frac{5\pi}{2}, \ldots$. Since $b^2 = \frac{Gg}{\gamma}$, the natural frequencies for the torsional vibration are

$$\omega_n = \frac{\pi}{2l}\sqrt{\frac{Gg}{\gamma}},\ \frac{3\pi}{2l}\sqrt{\frac{Gg}{\gamma}},\ \frac{5\pi}{2l}\sqrt{\frac{Gg}{\gamma}},\ \ldots \qquad \text{rad/sec}$$

For steel, $G = 11.5 \times 10^6$ lb/in.^{2} and $\gamma = 0.28$ lb/in.^{3} The fundamental natural frequency is

$$\omega_n = \frac{\pi}{2(24)} \sqrt{\frac{(11.5 \times 10^6)(386)}{0.28}} = 8{,}240 \text{ rad/sec} = 1{,}311 \text{ cps}$$

The remaining frequencies are 3, 5, 7, etc., times ω_n.

Since Eq. (7.4), which governs longitudinal vibration of the bar, is of the same form as Eq. (7.5), which governs torsional vibration, the solution for longitudinal vibration is the same as Eq. (7.7) with u substituted for ϕ and $a = \sqrt{Eg/\gamma}$ substituted for b. The natural frequencies of a uniform rod having one end built in and one end free are obtained by substituting a for b in the frequency equations found above in Example 7.1:

$$\omega_n = \frac{\pi}{2l}\sqrt{\frac{Eg}{\gamma}},\ \frac{3\pi}{2l}\sqrt{\frac{Eg}{\gamma}},\ \frac{5\pi}{2l}\sqrt{\frac{Eg}{\gamma}},\ \ldots$$

The frequencies of the longitudinal vibration are independent of the lateral dimensions of the bar, so that these results apply to uniform noncircular bars. Equation (7.5) for torsional vibration is valid only for circular cross sections.

TORSIONAL VIBRATIONS OF CIRCULAR RODS WITH DISCS ATTACHED. An important type of system is that in which a rod which may twist has mounted on it one or more rigid discs or members that can be considered as the equivalents of discs. Many systems can be approximated by such configurations. If the moment of inertia of the rod is small compared to the moments of inertia of the discs, the mass of the rod may be neglected and the system considered to have a finite number of degrees-of-freedom. Then the methods described in Chaps. 2 and 38 are applicable. Even if the moment of inertia of the rod is not negligible, it usually may be lumped with the moment of inertia of the disc. For a shaft having a single disc attached, the formula in Table 7.2 gives a close approximation to the true frequency.

The exact solution of the problem requires that the effect of the distributed mass of the rod be considered. Usually it can be assumed that the discs are rigid enough that their elasticity can be neglected; only such systems are considered. Equation (7.5) and its solution, Eq. (7.7), apply to the shaft where the constants are determined by the end conditions. If there are more than two discs, the section of shaft between each pair of discs must be considered separately; there are two constants for each section. The constants are determined from the following conditions:

1. For a disc at an end of the shaft, the torque of the shaft at the disc is equal to the product of the moment of inertia of the disc and its angular acceleration.
2. Where a disc is between two sections of shaft, the angular deflection at the end of each section adjoining the disc is the same; the difference between the torques in the two sections is equal to the product of the moment of inertia of the disc and its angular acceleration.

Example 7.2. The fundamental frequency of vibration of the system shown in Fig. 7.2 is to be calculated and the result compared with the frequency obtained by considering that each half of the system is a simple shaft-disc system with the end of the shaft fixed. The system consists of a steel shaft 24 in. long and 4 in. in diameter having attached to it at each end a rigid steel disc 12 in. in diameter and 2 in. thick. For the approximation, add one-third of the moment of inertia of half the shaft to that of the disc (Table 7.2). (Because of symmetry, the center of the shaft is a nodal point; i.e., it does not move. Thus, each half of the system can be considered as a rod-disc system.)

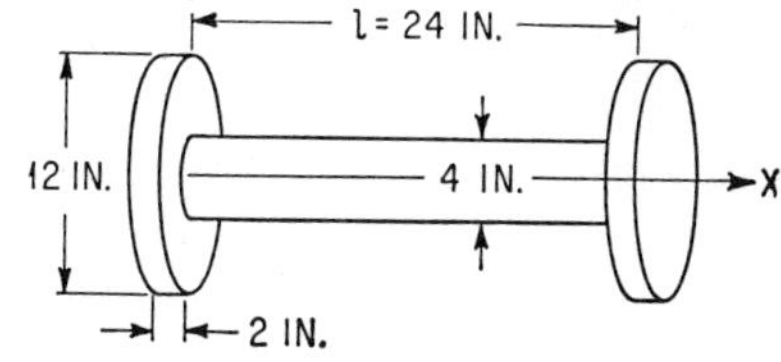

FIG. 7.2. Rod with disc attached at each end.

Exact Solution. The boundary conditions are: at $x = 0$, $M = GI_p\phi' = I_1\ddot{\phi}$; at $x = l$, $M = GI\phi' = -I_2\ddot{\phi}$, where I_1 and I_2 are the moments of inertia of the discs. The signs are opposite for the two boundary conditions because, if the shaft is twisted in a certain direction, it will tend to accelerate the disc at the left end in one direction and the disc at the right end in the other. In the present example, $I_1 = I_2$; however, the solution is carried out in general terms.

Using Eq. (7.7), the following is obtained for each value of n:

$$\phi' = \frac{\omega_n}{b}\left(C \cos\frac{\omega_n x}{b} - D \sin\frac{\omega_n x}{b}\right)\cos(\omega_n t + \theta)$$

$$\ddot{\phi} = \omega_n^2\left(C \sin\frac{\omega_n x}{b} + D \cos\frac{\omega_n x}{b}\right)[-\cos(\omega_n t + \theta)]$$

The boundary conditions give the following:

$$GI_p\frac{\omega_n}{b}C = -\omega_n^2 DI_1 \qquad \text{or} \qquad C = -\frac{b\omega_n I_1}{GI_p}D$$

$$\frac{\omega_n}{b}GI_p\left(C\cos\frac{\omega_n l}{b} - D\sin\frac{\omega_n l}{b}\right) = \omega_n^2 I_2\left(C\sin\frac{\omega_n l}{b} + D\cos\frac{\omega_n l}{b}\right)$$

These two equations can be combined to give

$$-\frac{\omega_n}{b}GI_p\left(\frac{b\omega_n I_1}{GI_p}\cos\frac{\omega_n l}{b} + \sin\frac{\omega_n l}{b}\right) = \omega_n^2 I_2\left(-\frac{b\omega_n I_1}{GI_p}\sin\frac{\omega_n l}{b} + \cos\frac{\omega_n l}{b}\right)$$

The preceding equation can be reduced to

$$\tan\alpha_n = \frac{(c + d)\alpha_n}{cd\alpha_n^2 - 1} \tag{7.8}$$

where $\alpha_n = (\omega_n l)/b$, $c = I_1/I_s$, $d = I_2/I_s$, and I_s is the polar moment of inertia of the shaft as a rigid body. There is a value for X in Eq. (7.6) corresponding to each root of Eq. (7.8) so that Eq. (7.7) becomes

$$\theta = \sum_{n=1}^{n=\infty} A_n\left(\cos\frac{\omega_n x}{b} - c\alpha_n \sin\frac{\omega_n x}{b}\right)\cos(\omega_n t + \theta_n)$$

For a circular disc or shaft, $I = \frac{1}{2}mr^2$ where m is the total mass; thus $c = d = \frac{D^4}{d^4}\frac{h}{l} = 6.75$. Equation (7.8) becomes $(45.56\alpha_n^2 - 1)\tan\alpha_n = 13.5\alpha_n$, the lowest root of which is $\alpha_n = 0.538$. The natural frequency is $\omega_n = 0.538\sqrt{Gg/\gamma l^2}$ rad/sec.

Approximate Solution. From Table 7.2, the approximate formula is

$$\omega_n = \left(\frac{k_r}{I + I_s/3}\right)^{1/2} \qquad \text{where } k_r = \frac{\pi d^4}{32}\frac{G}{l}$$

For the present problem where the center of the shaft is a node, the values of moment of inertia I_s and torsional spring constant for half the shaft must be used:

$$\tfrac{1}{2}I_s = \frac{\pi d^4}{32}\frac{\gamma}{g}\frac{l}{2} \qquad \text{and} \qquad k_r = 2\left[\frac{\pi d^4}{32}\frac{G}{l}\right]$$

From the previous solution:

$$I_1 = 6.75I_s \qquad I_1 + \frac{1}{2}\left(\frac{I_s}{3}\right) = \frac{\pi d^4}{32}\frac{\gamma}{g}\frac{l}{2}[2(6.75) + 0.333]$$

Substituting these values into the frequency equation and simplifying gives

$$\omega_n = 0.538\sqrt{\frac{Gg}{\gamma l^2}}$$

In this example, the approximate solution is correct to at least three significant figures. For larger values of I_s/I, poorer accuracy can be expected.

For steel, $G = 11.5 \times 10^6$ lb/in.^{2} and $\gamma = 0.28$ lb/in.^{3}; thus

$$\omega_n = 0.538\sqrt{\frac{(11.5 \times 10^6)(386)}{(0.28)(24)^2}} = 0.538 \times 5{,}245 = 2{,}822 \text{ rad/sec} = 449 \text{ cps}$$

LONGITUDINAL VIBRATION OF A ROD WITH MASS ATTACHED. The natural frequencies of the longitudinal vibration of a uniform rod having rigid masses attached to it can be solved in a manner similar to that used for a rod in torsion with discs attached. Equation (7.4) applies to this system; its solution is the same as Eq. (7.7) with a substituted for b. For each value of n,

$$u = \left(C_n \sin \frac{\omega_n x}{a} + D_n \cos \frac{\omega_n x}{a}\right) \cos (\omega_n t + \theta)$$

In Fig. 7.3, the rod of length l is fixed at $x = 0$ and has a mass m_2 attached at $x = l$. The boundary conditions are: at $x = 0$, $u = 0$ and at $x = l$, $SEu' = -m_2\ddot{u}$. The latter expresses the condition that the force in the bar equals the product of the mass and its acceleration at the end with the mass attached. The sign is negative because the force is tensile or positive when the acceleration of the mass is negative. From the first boundary condition, $D_n = 0$. The second boundary condition gives

$$\frac{\omega_n SE}{a} C_n \cos \frac{\omega_n l}{a} = m_2 \omega_n^2 C_n \sin \frac{\omega_n l}{a}$$

from which

$$\frac{SEl}{m_2 a^2} = \frac{\omega_n l}{a} \tan \frac{\omega_n l}{a}$$

Since $a^2 = Eg/\gamma$, this can be written

$$\frac{m_1}{m_2} = \frac{\omega_n l}{a} \tan \frac{\omega_n l}{a}$$

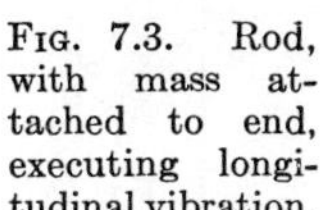

FIG. 7.3. Rod, with mass attached to end, executing longitudinal vibration.

where m_1 is the mass of the rod. This equation can be applied to a simple mass-spring system by using the relation that the constant k of a spring is equivalent to SE/l for the rod, so that $l/a = (m_1/k)^{1/2}$, where m_1 is the mass of the spring:

$$\frac{m_1}{m_2} = \omega_n \sqrt{\frac{m_1}{k}} \tan \omega_n \sqrt{\frac{m_1}{k}} \tag{7.9}$$

Rayleigh's Method. An accurate approximation to the fundamental natural frequency of this system can be found by using Rayleigh's method. The motion of the mass can be expressed as $u_m = u_0 \sin \omega t$. If it is assumed that the deflection u at each section of the rod is proportional to its distance from the fixed end, $u = u_0(x/l) \sin \omega_n t$. Using this relation in the appropriate equation from Table 7.1, the strain energy V of the rod at maximum deflection is

$$V = \frac{SE}{2} \int_0^l \left(\frac{\partial u}{\partial x}\right)^2 dx = \frac{SE}{2} \int_0^l \left(\frac{u_0}{l}\right)^2 dx = \frac{SEu_0^2}{2l}$$

The maximum kinetic energy T of the rod is

$$T = \frac{S\gamma}{2g} \int_0^l V^2_{\max}\, dx = \frac{S\gamma}{2g} \int_0^l \left(\omega_n u_0 \frac{x}{l}\right)^2 dx = \frac{S\gamma}{2g} \omega_n^2 u_0^2 \frac{l}{3}$$

The maximum kinetic energy of the mass is $T_m = m_2 \omega_n^2 u_0^2/2$. Equating the total maximum kinetic energy $T + T_m$ to the maximum strain energy V gives

$$\omega_n = \left(\frac{SE}{l(m_2 + m_1/3)}\right)^{1/2}$$

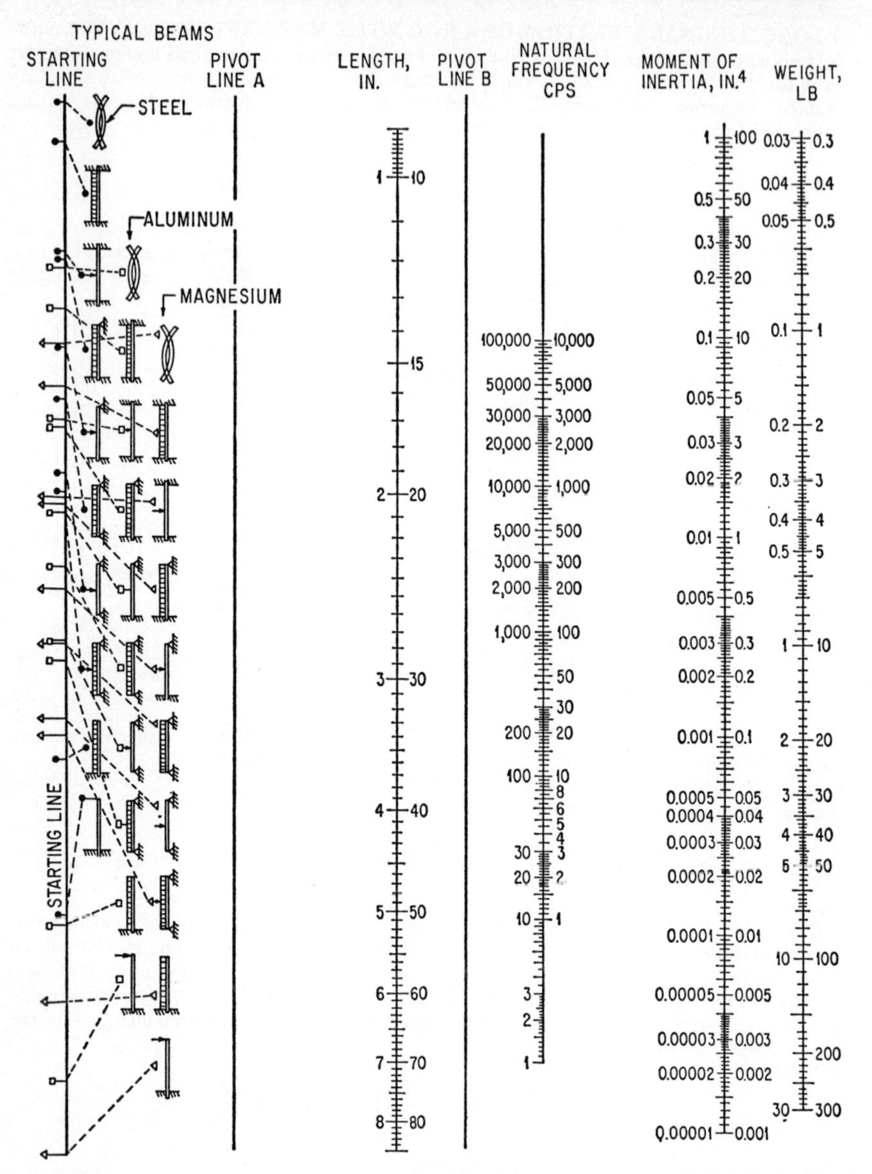

Fig. 7.4. Nomograph for determining fundamental natural frequencies of beams. From the point on the starting line which corresponds to the loading and support conditions for the beam, a straight line is drawn to the proper point on the length line. (If the length appears on the left side of this line, subsequent readings on all lines are made to the left; and if the length appears to the right, subsequent readings are made to the right.) From the intersection of this line with pivot line A, a straight line is drawn to the moment of inertia line; from the intersection of this line with pivot line B, a straight line is drawn to the weight line. (For concentrated loads, the weight is that of the load; for uniformly distributed loads, the weight is the total load on the beam, including the weight of the beam.) The natural frequency is read where the last line crosses the natural frequency line. (*J. J. Kerley.*[7])

where $m_1 = S\gamma l/g$ is the mass of the rod. Letting $SE/l = k$,

$$\omega_n = \sqrt{\frac{k}{M + m/3}} \tag{7.10}$$

This formula is included in Table 7.2. The other formulas in that table are also based on analyses by the Rayleigh method.

Example 7.3. The natural frequency of a simple mass-spring system for which the weight of the spring is equal to the weight of the mass is to be calculated and compared to the result obtained by using Eq. (7.10).

Solution. For $m_1/m_2 = l$, the lowest root of Eq. (7.9) is $\omega_n\sqrt{m/k} = 0.860$. When $m_2 = m_1$,

$$\omega_n = 0.860\sqrt{\frac{k}{m_2}}$$

Using the approximate equation,

$$\omega_n = \sqrt{\frac{k}{m_2(1 + \frac{1}{3})}} = 0.866\sqrt{\frac{k}{m_2}}$$

LATERAL VIBRATION OF STRAIGHT BEAMS

Natural Frequencies from Nomograph. For many practical purposes the natural frequencies of uniform beams of steel, aluminum, and magnesium can be determined with sufficient accuracy by the use of the nomograph, Fig. 7.4. This nomograph applies to many conditions of support and several types of load. Figure 7.4*A* indicates the procedure for using the nomograph.

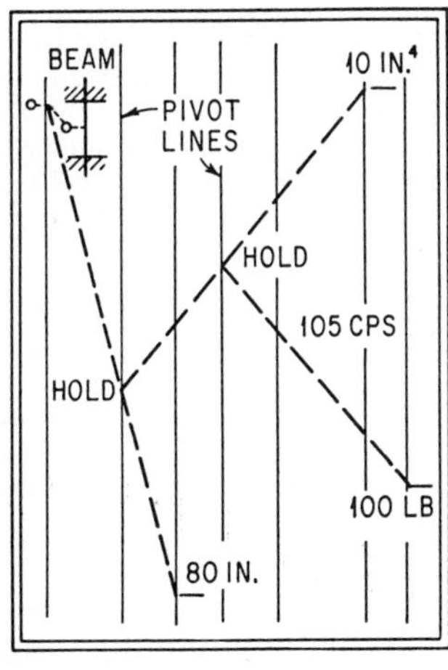

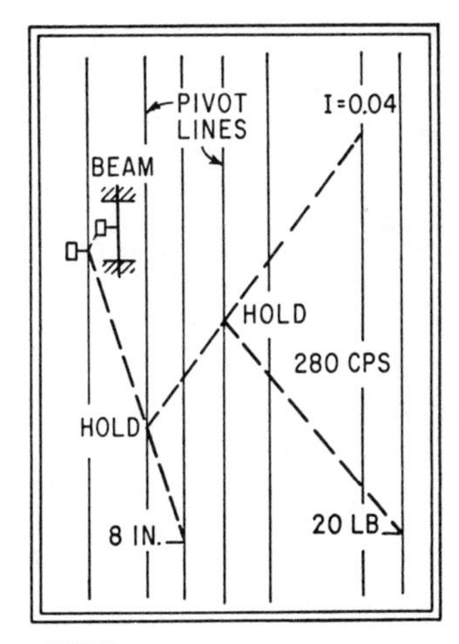

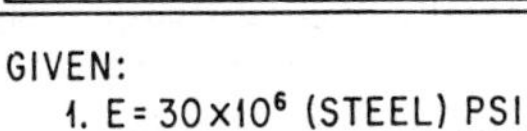

GIVEN:
1. E=10.5×10⁶ (ALUM) PSI
2. I=0.04 IN.⁴
3. l=8 IN.
4. W=20 LB
5. DOUBLE END FIXITY

Fig. 7.4*A*. Example of use of Fig. 7.4. The natural frequency of the steel beam is 105 cps and that of the aluminum beam is 280 cps. (*J. J. Kerley*.[7])

Classical Solution. In the derivation of the necessary equation, use is made of the relation

$$EI \frac{d^2y}{dx^2} = M \tag{7.11}$$

This equation relates the curvature of the beam to the bending moment at each section of the beam. This equation is based upon the assumptions that the material is homogeneous, isotropic and obeys Hooke's law, and that the beam is straight and of uniform cross section. The equation is valid for small deflections only, and for beams that are long compared to cross-sectional dimensions since the effects of shear deflection are neg-

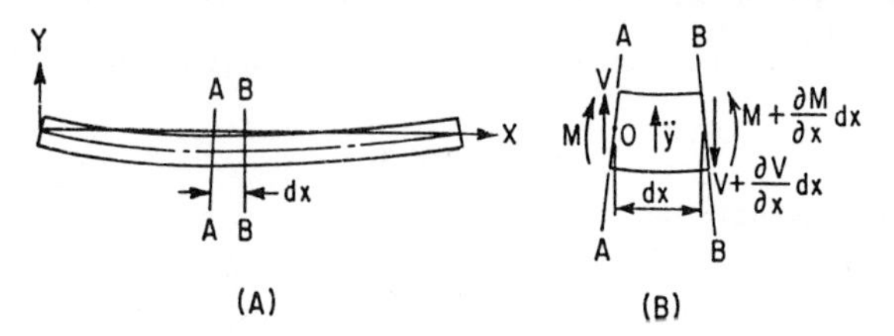

FIG. 7.5. (*A*) Beam executing lateral vibration. (*B*) Element of beam showing shear forces and bending moments.

lected. The effects of shear deflection and rotation of the cross sections are considered later.

The equation of motion for lateral vibration of the beam shown in Fig. 7.5*A* is found by considering the forces acting on the element, Fig. 7.5*B*, which is formed by passing two parallel planes *A–A* and *B–B* through the beam normal to the longitudinal axis. The vertical elastic shear force acting on section *A–A* is V, and that on section *B–B* is $V + (\partial V/\partial x)\,dx$. Shear forces acting as shown are considered to be positive. The total vertical elastic shear force at each section of the beam is composed of two parts: that caused by the static load including the weight of the beam and that caused by the vibration. The part of the shear force caused by the static load exactly balances the load, so that these forces need not be considered in deriving the equation for the vibration if all deflections are measured from the position of equilibrium of the beam under the static load. The sum of the remaining vertical forces acting on the element must equal the product of the mass of the element $S\gamma/g\,dx$ and the acceleration $\partial^2 y/\partial t^2$ in the lateral direction: $V + (\partial V/\partial x)\,dx - V = (\partial V/\partial x)\,dx = -(S\gamma/g)(\partial^2 y/\partial t^2)\,dx$, or

$$\frac{\partial V}{\partial x} = -\frac{\gamma S}{g}\frac{\partial^2 y}{\partial t^2} \tag{7.12}$$

If moments are taken about point 0 of the element in Fig. 7.5*B*, $V\,dx = (\partial M/\partial x)\,dx$, and $V = \partial M/\partial x$. Other terms contain differentials of higher order, and can be neglected. Substituting this in Eq. (7.12) gives $-\partial^2 M/\partial x^2 = (S\gamma/g)(\partial^2 y/\partial t^2)$. Substituting Eq. (7.11) gives

$$-\frac{\partial^2}{\partial x^2}\left(EI\frac{\partial^2 y}{\partial x^2}\right) = \frac{\gamma S}{g}\frac{\partial^2 y}{\partial t^2} \tag{7.13}$$

Equation (7.13) is the basic equation for the lateral vibration of beams. The solution of this equation, if EI is constant, is of the form $y = X(x)\,[\cos(\omega_n t + \theta)]$, in which X is a function of x only. Substituting

$$\kappa^4 = \frac{\omega_n^2 \gamma S}{EIg} \tag{7.14}$$

and dividing Eq. (7.13) by $\cos(\omega_n t + \theta)$:

$$\frac{d^4X}{dx^4} = \kappa^4 X \tag{7.15}$$

where X is any function whose fourth derivative is equal to a constant multiplied by the function itself. The following functions satisfy the required conditions and represent the solution of the equation:

$$X = A_1 \sin \kappa x + A_2 \cos \kappa x + A_3 \sinh \kappa x + A_4 \cosh \kappa x$$

The solution can also be expressed in terms of exponential functions, but the trigonometric and hyperbolic functions usually are more convenient to use.

For beams having various support conditions, the constants A_1, A_2, A_3, and A_4 are found from the end conditions. In finding the solutions, it is convenient to write the equation in the following form in which two of the constants are zero for each of the usual boundary conditions:

$$\begin{aligned} X = A(\cos \kappa x + \cosh \kappa x) + B(\cos \kappa x - \cosh \kappa x) \\ + C(\sin \kappa x + \sinh \kappa x) + D(\sin \kappa x - \sinh \kappa x) \end{aligned} \quad (7.16)$$

In applying the end conditions, the following relations are used where primes indicate successive derivatives with respect to x:

The deflection is proportional to X and is zero at any rigid support.
The slope is proportional to X' and is zero at any built-in end.
The moment is proportional to X'' and is zero at any free or hinged end.
The shear is proportional to X''' and is zero at any free end.

The required derivatives are:

$$\begin{aligned} X' &= \kappa[A(-\sin \kappa x + \sinh \kappa x) + B(-\sin \kappa x - \sinh \kappa x) \\ &\quad + C(\cos \kappa x + \cosh \kappa x) + D(\cos \kappa x - \cosh \kappa x)] \end{aligned}$$

$$\begin{aligned} X'' &= \kappa^2[A(-\cos \kappa x + \cosh \kappa x) + B(-\cos \kappa x - \cosh \kappa x) \\ &\quad + C(-\sin \kappa x + \sinh \kappa x) + D(-\sin \kappa x - \sinh \kappa x)] \end{aligned}$$

$$\begin{aligned} X''' &= \kappa^3[A(\sin \kappa x + \sinh \kappa x) + B(\sin \kappa x - \sinh \kappa x) \\ &\quad + C(-\cos \kappa x + \cosh \kappa x) + D(-\cos \kappa x - \cosh \kappa x)] \end{aligned}$$

For the usual end conditions, two of the constants are zero, and there remain two equations containing two constants. These can be combined to give an equation which contains only the frequency as an unknown. Using the frequency, one of the unknown constants can be found in terms of the other. There always is one undetermined constant, which can be evaluated only if the amplitude of the vibration is known.

Example 7.4. The natural frequencies and modes of vibration of the rectangular steel beam shown in Fig. 7.6 are to be determined and the fundamental frequency compared with that obtained from Fig. 7.4. The beam is 24 in. long, 2 in. wide, and ¼ in. thick, with the left end built in and the right end free.

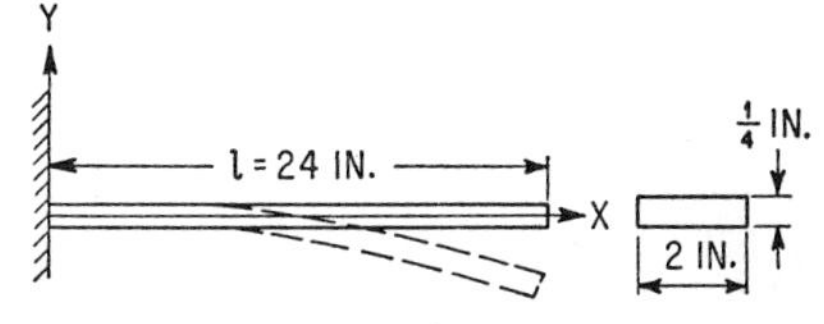

FIG. 7.6. First mode of vibration of beam with left end clamped and right end free.

Solution. The boundary conditions are: at $x = 0$, $X = 0$ and $X' = 0$; at $x = l$, $X'' = 0$ and $X''' = 0$. The first condition requires that $A = 0$ since the other constants are multiplied by zero at $x = 0$. The second condition requires that $C = 0$. From the third and fourth conditions, the following equations are obtained:

$$0 = B(-\cos \kappa l - \cosh \kappa l) + D(-\sin \kappa l - \sinh \kappa l)$$

$$0 = B(\sin \kappa l - \sinh \kappa l) + D(-\cos \kappa l - \cosh \kappa l)$$

Solving each of these for the ratio D/B and equating, or making use of the mathematical condition that for a solution the determinant of the two equations must vanish, the following equation results:

$$\frac{D}{B} = -\frac{\cos \kappa l + \cosh \kappa l}{\sin \kappa l + \sinh \kappa l} = \frac{\sin \kappa l - \sinh \kappa l}{\cos \kappa l + \cosh \kappa l} \quad (7.17)$$

Table 7.3. Natural Frequencies and Normal Modes of Uniform Beams

SUPPORTS	MODE n	(A) SHAPE AND NODES (NUMBERS GIVE LOCATION OF NODES IN FRACTION OF LENGTH FROM LEFT END)	(B) BOUNDARY CONDITIONS EQ (7.16)	(C) FREQUENCY EQUATION	(D) CONSTANTS EQ (7.16)	(E) kl EQ (7.14) $\omega_n = k^2\sqrt{\frac{EIg}{A\gamma}}$	(F) R RATIO OF NON-ZERO CONSTANTS COLUMN (D)
HINGED-HINGED	1	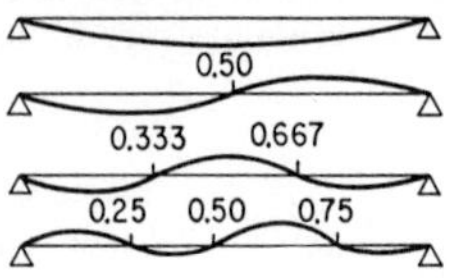	$x=0\begin{cases}X=0\\X''=0\end{cases}$	$\sin kl=0$	$A=0$	3.1416	1.0000
	2	0.50			$B=0$	6.283	1.0000
	3	0.333 0.667			$\frac{C}{D}=1$	9.425	1.0000
	4	0.25 0.50 0.75	$x=l\begin{cases}X=0\\X''=0\end{cases}$			12.566	1.0000
	n>4					$\approx n\pi$	1.0000
CLAMPED-CLAMPED	1	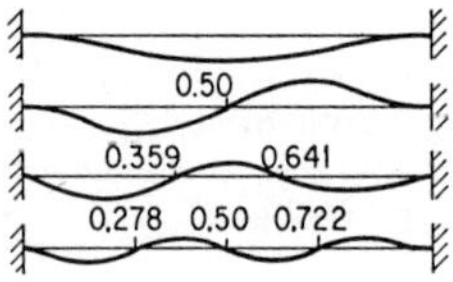	$x=0\begin{cases}X=0\\X'=0\end{cases}$	$(\cos kl)(\cosh kl)=1$	$A=0$	4.730	-0.9825
	2	0.50			$C=0$	7.853	-1.0008
	3	0.359 0.641			$\frac{D}{B}=R$	10.996	-1.0000-
	4	0.278 0.50 0.722	$x=l\begin{cases}X=0\\X'=0\end{cases}$			14.137	-1.0000+
	n>4					$\approx\frac{(2n+1)\pi}{2}$	-1.0000-
CLAMPED-HINGED	1	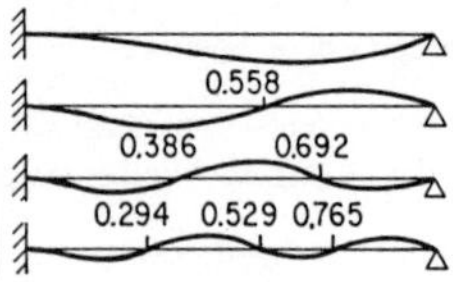	$x=0\begin{cases}X=0\\X'=0\end{cases}$	$\tan kl=\tanh kl$	$A=0$	3.927	-1.0008
	2	0.558			$C=0$	7.069	-1.0000+
	3	0.386 0.692			$\frac{D}{B}=R$	10.210	-1.0000
	4	0.294 0.529 0.765	$x=l\begin{cases}X=0\\X''=0\end{cases}$			13.352	-1.0000
	n>4					$\approx\frac{(4n+1)\pi}{4}$	-1.0000
CLAMPED-FREE	1	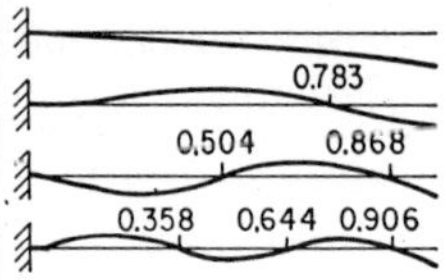	$x=0\begin{cases}X=0\\X'=0\end{cases}$	$(\cos kl)(\cosh kl)=-1$	$A=0$	1.875	-0.7341
	2	0.783			$C=0$	4.694	-1.0185
	3	0.504 0.868			$\frac{D}{B}=R$	7.855	-0.9992
	4	0.358 0.644 0.906	$x=l\begin{cases}X''=0\\X'''=0\end{cases}$			10.996	-1.0000+
	n>4					$\approx\frac{(2n-1)\pi}{2}$	-1.0000-
FREE-FREE	1	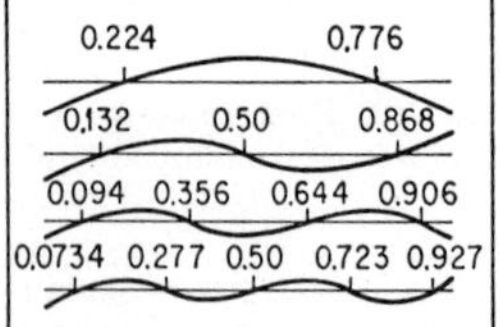	$x=0\begin{cases}X''=0\\X'''=0\end{cases}$	$(\cos kl)(\cosh kl)=1$	$B=0$	0 (REPRESENTS TRANSLATION)	
	2	0.224 0.776			$D=0$	4.730	-0.9825
	3	0.132 0.50 0.868			$\frac{C}{A}=R$	7.853	-1.0008
	4	0.094 0.356 0.644 0.906	$x=l\begin{cases}X''=0\\X'''=0\end{cases}$			10.996	-1.0000-
	5	0.0734 0.277 0.50 0.723 0.927				14.137	-1.0000+
	n>5					$\approx\frac{(2n-1)\pi}{2}$	-1.0000-

Equation (7.17) reduces to $\cos \kappa l \cosh \kappa l = -1$. The values of κl which satisfy this equation can be found by consulting tables of hyperbolic and trigonometric functions. The first five are: $\kappa_1 l = 1.875$, $\kappa_2 l = 4.694$, $\kappa_3 l = 7.855$, $\kappa_4 l = 10.996$, $\kappa_5 l = 14.137$. The corresponding frequencies of vibration are found by substituting the length of the beam to find each κ, and then solving Eq. (7.14) for ω_n:

$$\omega_n = \kappa_n^2 \sqrt{\frac{EIg}{\gamma S}}$$

For the rectangular section, $I = bh^3/12 = 1/384$ in.^{4} and $S = bh = 0.5$ in.^{2} For steel, $E = 30 \times 10^6$ lb/in.^{2} and $\gamma = 0.28$ lb/in.^{3} Using these values,

$$\omega_1 = \frac{(1.875)^2}{(24)^2}\sqrt{\frac{(30 \times 10^6)(386)}{(0.28)(384)(0.5)}} = 89.6 \text{ rad/sec} = 14.26 \text{ cps}$$

The remaining frequencies can be found by using the other values of κ. Using Fig. 7.4, the fundamental frequency is found to be about 12 cps.

To find the mode shapes, the ratio D/B is found by substituting the appropriate values of κl in Eq. (7.17). For the first mode:

$$\cosh 1.875 = 3.33710 \qquad \sinh 1.875 = 3.18373$$

$$\cos 1.875 = -0.29953 \qquad \sin 1.875 = 0.95409$$

Therefore, $D/B = -0.73410$. The equation for the first mode of vibration becomes

$$y = B_1[(\cos \kappa x - \cosh \kappa x) - 0.73410 (\sin \kappa x - \sinh \kappa x)] \cos (\omega_1 t + \theta_1)$$

in which B_1 is determined by the amplitude of vibration in the first mode. A similar equation can be obtained for each of the modes of vibration; all possible free vibration of the beam can be expressed by taking the sum of these equations.

Frequencies and Shapes of Beams. Table 7.3 gives the information necessary for finding the natural frequencies and normal modes of vibration of uniform beams having various boundary conditions. The various constants in the table were determined by computations similar to those used in Example 7.4. The table includes (1) diagrams showing the modal shapes including node locations; (2) the boundary conditions; (3) the frequency equation that results from using the boundary conditions in Eq. (7.16); (4) the constants that become zero in Eq. (7.16); (5) the values of κl from which the natural frequencies can be computed by using Eq. (7.14); and (6) the ratio of the nonzero constants in Eq. (7.16). By the use of the constants in this table, the equation of motion for any normal mode can be written. There always is a constant which is determined by the amplitude of vibration.

Values of characteristic functions representing the deflections of beams, at fifty equal intervals, for the first five modes of vibration have been tabulated.[8] Functions are given for beams having various boundary conditions, and the first three derivatives of the functions are also tabulated.

Rayleigh's Method. This method is useful for finding approximate values of the fundamental natural frequencies of beams. In applying Rayleigh's method, a suitable function is assumed for the deflection, and the maximum strain and kinetic energies are calculated, using the equations in Table 7.1. These energies are equated and solved for the frequency. The function used to represent the shape must satisfy the boundary conditions associated with deflection and slope at the supports. Best accuracy is obtained if other boundary conditions are also satisfied. The equation for the static deflection of the beam under a uniform load is a suitable function, although a simpler function often gives satisfactory results with less numerical work.

Example 7.5. The fundamental natural frequency of the cantilever beam in Example 7.4 is to be calculated using Rayleigh's method.

Solution. The assumed deflection $Y = (a/3l^4)[x^4 - 4x^3l + 6x^2l^2]$ is the static deflection of a cantilever beam under uniform load and having the deflection $Y = a$ at $x = l$. This deflection satisfies the conditions that the deflection Y and the slope Y' be zero at $x = 0$. Also, at $x = l$, Y'' which is proportional to the moment and Y''' which is proportional to the shear are zero. The second derivative of the function is $Y'' = (4a/l^4)[x^2 - 2xl + l^2]$. Using this in the expression from Table 7.1, the maximum strain energy is

$$V = \frac{EI}{2}\int_0^l \left(\frac{d^2Y}{dx^2}\right)^2 dx = \frac{8}{5}\frac{EIa^2}{l^3}$$

The maximum kinetic energy is

$$T = \frac{\omega_n^2 \gamma S}{2g}\int_0^l Y^2\, dx = \frac{52}{405}\frac{\omega_n^2 \gamma S l a^2}{g}$$

Equating the two energies and solving for the frequency,

$$\omega_n = \sqrt{\frac{162}{13} \times \frac{EIg}{\gamma S l^4}} = \frac{3.530}{l^2}\sqrt{\frac{EIg}{\gamma S}}$$

The exact frequency as found in Example 4 is $(3.516/l^2)\sqrt{EIg/\gamma S}$; thus, Rayleigh's method gives good accuracy in this example.

If the deflection is assumed to be $Y = a[1 - \cos(\pi x/2l)]$, the calculated frequency is $(3.66/l^2)\sqrt{EIg/\gamma S}$. This is less accurate, but the calculations are considerably shorter. With this function, the same boundary conditions at $x = 0$ are satisfied; however, at $x = l$, $Y'' = 0$ but Y''' does not equal zero. Thus, the condition of zero shear at the free end is not satisfied. The trigonometric function would not be expected to give as good accuracy as the static deflection relation used in the example, although for most practical purposes the result would be satisfactory.

EFFECTS OF ROTARY MOTION AND SHEARING FORCE. In the preceding analysis of the lateral vibration of beams it has been assumed that each element of the beam moves only in the lateral direction. If each plane section that is initially normal to the axis of the beam remains plane and normal to the axis, as assumed in simple beam theory, then each section rotates slightly in addition to its lateral motion when the beam deflects.[9] When a beam vibrates, there must be forces to cause this rotation, and for a complete analysis these forces must be considered. The effect of this rotation is small except when the curvature of the beam is large relative to its thickness; this is true either for a beam that is short relative to its thickness or for a long beam vibrating in a higher mode so that the nodal points are close together.

Another factor that affects the lateral vibration of a beam is the lateral shear force. In Eq. (7.11) only the deflection associated with the bending stress in the beam is included. In any beam except one subject only to pure bending, a deflection due to the shear stress in the beam occurs. The exact solution of the beam vibration problem requires that this deflection be considered. The analysis of beam vibration including both the effects of rotation of the cross section and the shear deflection is called the "Timoshenko beam theory." The following equation governs such vibration:[10]

$$a^2\frac{\partial^4 y}{\partial x^4} + \frac{\partial^2 y}{\partial t^2} - \rho^2\left(1 + \frac{E}{\varkappa G}\right)\frac{\partial^4 y}{\partial x^2 \partial t^2} + \rho^2\frac{\gamma}{g\varkappa G}\frac{\partial^4 y}{\partial t^4} = 0 \tag{7.18}$$

where $a^2 = EIg/S\gamma$, E = modulus of elasticity, G = modulus of rigidity, and $\rho = \sqrt{I/S}$, the radius of gyration; $\varkappa = F_s/GS\beta$, F_s being the total lateral shear force at any section and β the angle which a cross section makes with the axis of the beam because of shear deformation. Under the assumptions made in the usual elementary beam theory, $\varkappa$ is $2/3$ for a beam with a rectangular cross section and $3/4$ for a circular beam. More refined analysis shows[11] that, for the present purposes, $\varkappa = 5/6$ and $9/10$ are more accurate values for rectangular and circular cross sections, respectively. Using a solution of the form $y = C\sin(n\pi x/l)\cos\omega_n t$, which satisfies the necessary end conditions, the following frequency equation is obtained for beams with both ends simply supported:

$$a^2\frac{n^4\pi^4}{l^4} - \omega_n^2 - \omega_n^2\frac{n^2\pi^2\rho^2}{l^2} - \omega_n^2\frac{n^2\pi^2\rho^2}{l^2}\frac{E}{\varkappa G} + \frac{\rho^2\gamma}{g\varkappa G}\omega_n^4 = 0 \tag{7.18a}$$

If it is assumed that $nr/l \ll 1$, Eq. (7.18a) reduces to

$$\omega_n = \frac{a\pi^2}{(l/n)^2}\left[1 - \frac{\pi^2 n^2}{2}\left(\frac{\rho}{l}\right)^2\left(1 + \frac{E}{\varkappa G}\right)\right] \tag{7.18b}$$

When $nr/l < 0.08$, the approximate equation gives less than 5 per cent error in the frequency.[11]

Values of the ratio of ω_n to the natural frequency uncorrected for the effects of rotation and shear have been plotted,[11] using Eq. (7.18a) for three values of $E/\varkappa G$, and are shown in Fig. 7.7.

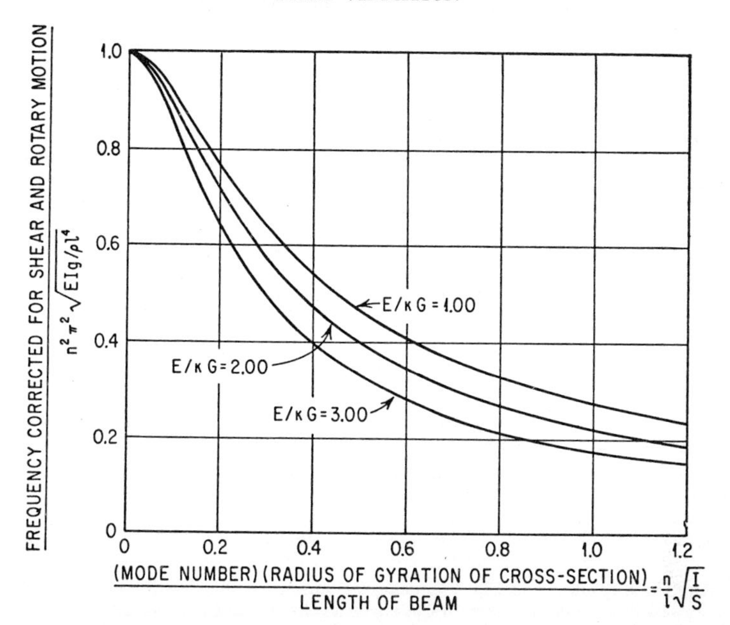

FIG. 7.7. Influence of shear force and rotary motion on natural frequencies of simply supported beams. The curves relate the corrected frequency to that given by Eq. (7.14). (*J. G. Sutherland and L. E. Goodman.*[11])

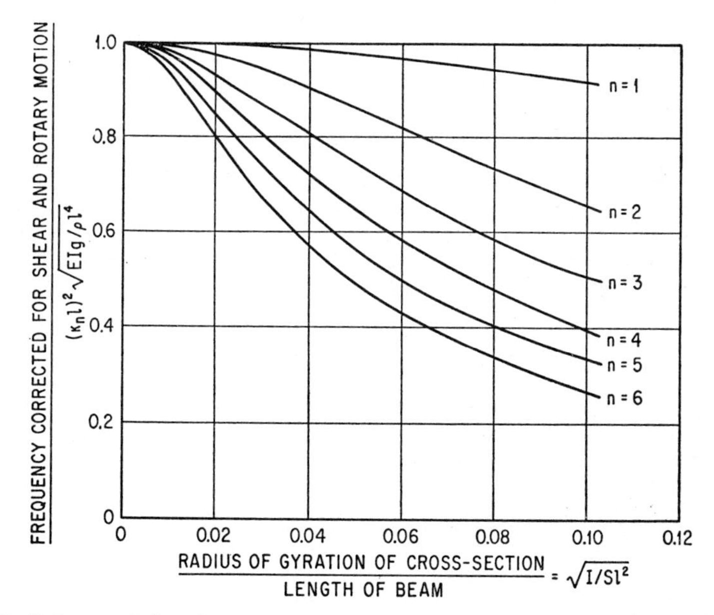

FIG. 7.8. Influence of shear force and rotary motion on natural frequencies of uniform cantilever beams ($E/\kappa G = 3.20$). The curves relate the corrected frequency to that given by Eq. (7.14). (*J. G. Sutherland and L. E. Goodman.*[11])

For a cantilever beam the frequency equation is quite complicated. For $E/\kappa G = 3.20$, corresponding approximately to the value for rectangular steel or aluminum beams, the curves in Fig. 7.8 show the effects of rotation and shear on the natural frequencies of the first six modes of vibration.

Example 7.6. The first two natural frequencies of a rectangular steel beam 40 in. long, 2 in. wide, and 6 in. thick, having simply supported ends, are to be computed with and without including the effects of rotation of the cross sections and shear deflection.

Solution. For steel $E = 30 \times 10^6$ lb/in.^{2}, $G = 11.5 \times 10^6$ lb/in.^{2}, and for a rectangular cross section $\kappa = 5/6$; thus $E/\kappa G = 3.13$. For a rectangular beam $\rho = h/\sqrt{12}$ where h is the thickness; thus $\rho/l = 6/(40\sqrt{12}) = 0.0433$. The approximate frequency equation, Eq. (7.18*b*), becomes

$$\omega_n = \frac{a\pi^2}{(l/n)^2}\left[1 - \frac{\pi^2}{2}(0.0433n)^2(1 + 3.13)\right]$$

$$= \frac{a\pi^2}{(l/n)^2}(1 - 0.038n^2)$$

Letting $\omega_0 = a\pi^2/(l/n)^2$ be the uncorrected frequency obtained by neglecting the effect of n in Eq. (7.18*b*):

For $n = 1$: $$\frac{\omega_n}{\omega_0} = 1 - 0.038 = 0.962$$

For $n = 2$: $$\frac{\omega_n}{\omega_0} = 1 - 0.152 = 0.848$$

Comparing these results with Fig. 7.7, using the curve for $E/\kappa G = 3.00$, the calculated frequency for the first mode agrees with the curve as closely as the curve can be read. For the second mode, the curve gives $\omega_n/\omega_0 = 0.91$; therefore the approximate equation for the second mode is not very accurate. The uncorrected frequencies are, since $I/S = \rho^2 = h^2/12$:

For $n = 1$: $$\omega_0 = \frac{\pi^2}{l^2}\sqrt{\frac{EIg}{S\gamma}} = \frac{\pi^2}{(40)^2}\sqrt{\frac{(30 \times 10^6)(36)386}{(12)(0.28)}} = 2{,}170 \text{ rad/sec} = 345 \text{ cps}$$

For $n = 2$: $$\omega_0 = 345 \times 4 = 1{,}380 \text{ cps}$$

The frequencies corrected for rotation and shear, using the value from Fig. 7.7 for correction of the second mode, are:

For $n = 1$: $$f_n = 345 \times 0.962 = 332 \text{ cps}$$

For $n = 2$: $$f_n = 1{,}380 \times 0.91 = 1{,}256 \text{ cps}$$

EFFECT OF AXIAL LOADS. When an axial tensile or compressive load acts on a beam, the natural frequencies are different from those for the same beam without such load. The natural frequencies for a beam with hinged ends, as determined by an energy analysis, assuming that the axial force F remains constant, are [12]

$$\omega_n = \frac{\pi^2 n^2}{l^2}\sqrt{\frac{EIg}{S\gamma}}\sqrt{1 \pm \frac{\alpha^2}{n^2}} = \omega_0\sqrt{1 \pm \frac{\alpha^2}{n^2}}$$

where $\alpha^2 = Fl^2/EI\pi^2$, n is the mode number, ω_0 is the natural frequency of the beam with no axial force applied, and the other symbols are defined in Table 7.1. The plus sign is for a tensile force and the minus sign for a compressive force.

For a cantilever beam with a constant axial force F applied at the free end, the natural frequency is found by an energy analysis [13] to be $[1 + \frac{5}{14}(Fl^2/EI)]^{1/2}$ times the natural frequency of the beam without the force applied. If a uniform axial force is applied along the beam, the effect is the same as if about seven-twentieths of the total force were applied at the free end of the beam.

If the amplitude of vibration is large, an axial force may be induced in the beam by the supports. For example, if both ends of a beam are hinged but the supports are rigid enough so that they cannot move axially, a tensile force is induced as the beam deflects. The force is not proportional to the deflection; therefore, the vibration is of the type characteristic of nonlinear systems in which the natural frequency depends on the amplitude of vibration. The natural frequency of a beam having immovable hinged ends is given in the following table where the axial force is zero at zero deflection of the beam [14] and where x_0 is the amplitude of vibration, I the moment of inertia, and S the area of the cross section; ω_0 is the natural frequency of the unrestrained bar.

$\frac{x_0}{\sqrt{I/S}}$	0	0.1	0.2	0.4	0.6	0.8
$\frac{\omega_n}{\omega_0}$	1	1.0008	1.0038	1.015	1.038	1.058
$\frac{x_0}{\sqrt{I/S}}$	1.0	1.5	2	3	4	5
$\frac{\omega_n}{\omega_0}$	1.089	1.190	1.316	1.626	1.976	2.35

BEAMS HAVING VARIABLE CROSS SECTIONS. The natural frequencies for beams of several shapes having cross sections that can be expressed as functions of the distance along the beam have been calculated.[15] The results are shown in Table 7.4. In the analysis, Eq. (7.13) was used, with EI considered to be variable.

Rayleigh's method or Ritz's method can be used to find approximate values for the frequencies of such beams. The frequency equation becomes, using the equations in Table 7.1, and letting $Y(x)$ be the assumed deflection,

$$\omega_n^2 = \frac{Eg}{\gamma} \frac{\int_0^l I \left(\frac{d^2Y}{dx^2}\right)^2 dx}{\int_0^l SY^2\, dx}$$

where $I = I(x)$ is the moment of inertia of the cross section and $S = S(x)$ is the area of the cross section. Examples of the calculations are in the literature.[18] If the values of $I(x)$ and $S(x)$ cannot be defined analytically, the beam may be divided into two or more sections, for each of which I and S can be approximated by an equation. The strain and kinetic energies of each section may be computed separately, using an appropriate function for the deflection, and the total energies for the beam found by adding the values for the individual sections.

CONTINUOUS BEAMS ON MULTIPLE SUPPORTS. In finding the natural frequencies of a beam on multiple supports, the section between each pair of supports is considered as a separate beam with its origin at the left support of the section. Equation (7.16) applies to each section. Since the deflection is zero at the origin of each section, $A = 0$ and the equation reduces to

$$X = B(\cos \kappa x - \cosh \kappa x) + C(\sin \kappa x + \sinh \kappa x) + D(\sin \kappa x - \sinh \kappa x)$$

There is one such equation for each section, and the necessary end conditions are as follows:

1. At each end of the beam the usual boundary conditions are applicable, depending on the type of support.
2. At each intermediate support the deflection is zero. Since the beam is continuous, the slope and the moment just to the left and to the right of the support are the same.

General equations can be developed for finding the frequency for any number of spans.[19,20] Table 7.5 gives constants for finding the natural frequencies of uniform continuous beams on uniformly spaced supports for several combinations of end supports.

Table 7.4. Natural Frequencies of Variable-section Steel Beams (*J. N. Macduff and R. P. Felgar.*[16,17])

Beam structure	b/b_0	h/h_0	$(f_n l^2/\rho)/10^4$, n = 1	n = 2	n = 3
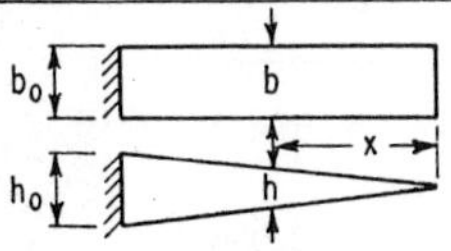 Cantilever: width b_0 constant, depth h_0 tapering to zero	1	$\frac{x}{l}$	17.09	48.89	96.57
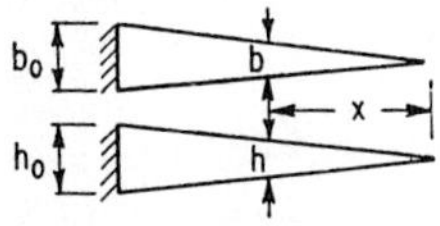 Cantilever: width b_0 and depth h_0 both tapering to zero	$\frac{x}{l}$	$\frac{x}{l}$	26.08	68.08	123.64
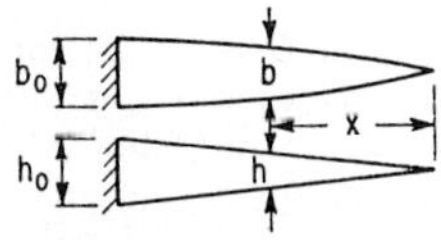 Cantilever: width b_0 parabolic taper, depth h_0 linear taper	$\left(\frac{x}{l}\right)^{\frac{1}{2}}$	$\frac{x}{l}$	22.30	58.18	109.90
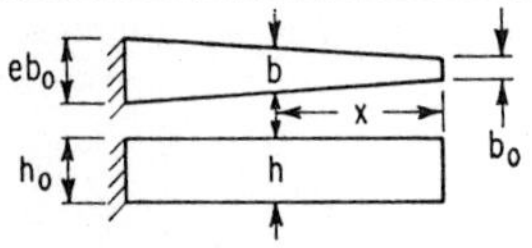 Cantilever: width from eb_0 to b_0, depth h_0 constant	$e^{x/l}$	1	15.23	77.78	206.07
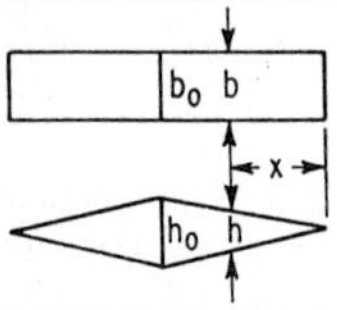 Free beam: width b_0 constant, depth h_0 double taper	1	$\frac{x}{l}$	21.21* 35.05†	56.97	
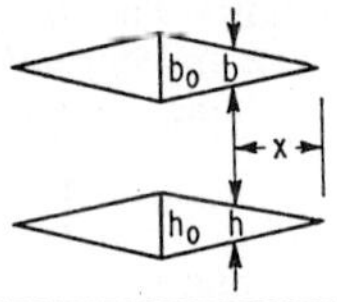 Free beam: width b_0 and depth h_0 double taper	$\frac{x}{l}$	$\frac{x}{l}$	32.73* 49.50†	76.57	
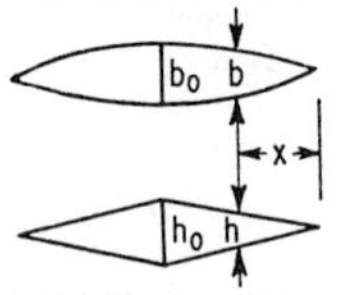 Free beam: width b_0 parabolic double taper, depth h_0 double taper	$\left(\frac{x}{l}\right)^{\frac{1}{2}}$	$\frac{x}{l}$	25.66* 42.02†	66.06	

*SYMMETRIC †ANTISYMMETRIC

f_n = natural frequency, cps
$\rho = \sqrt{I/S}$ = radius of gyration, in.
h = depth of beam, in.
l = beam length, in.
n = mode number
b = width of beam, in.

For materials other than steel: $f_n = f_{ns}\sqrt{\frac{E\gamma_s}{E_s\gamma}}$

E = modulus of elasticity, lb/in.^{2}
γ = density, lb/in.^{3}
Terms with subscripts refer to steel
Terms without subscripts refer to other material

Table 7.5. Natural Frequencies of Continuous Uniform Steel * Beams (*J. N. Macduff and R. P. Felgar.*[16,17])

Beam structure	$(f_n l^2/\rho)/10^4$					
	N	$n = 1$	$n = 2$	$n = 3$	$n = 4$	$n = 5$
	Extreme Ends Simply Supported					
	1	31.73	126.94	285.61	507.76	793.37
	2	31.73	49.59	126.94	160.66	285.61
	3	31.73	40.52	59.56	126.94	143.98
	4	31.73	37.02	49.59	63.99	126.94
	5	31.73	34.99	44.19	55.29	66.72
	6	31.73	34.32	40.52	49.59	59.56
	7	31.73	33.67	38.40	45.70	53.63
	8	31.73	33.02	37.02	42.70	49.59
	9	31.73	33.02	35.66	40.52	46.46
	10	31.73	33.02	34.99	39.10	44.19
	11	31.73	32.37	34.32	37.70	41.97
	12	31.73	32.37	34.32	37.02	40.52
	Extreme Ends Clamped					
	1	72.36	198.34	388.75	642.63	959.98
	2	49.59	72.36	160.66	198.34	335.20
	3	40.52	59.56	72.36	143.98	178.25
	4	37.02	49.59	63.99	72.36	137.30
	5	34.99	44.19	55.29	66.72	72.36
	6	34.32	40.52	49.59	59.56	67.65
	7	33.67	38.40	45.70	53.63	62.20
	8	33.02	37.02	42.70	49.59	56.98
	9	33.02	35.66	40.52	46.46	52.81
	10	33.02	34.99	39.10	44.19	49.59
	11	32.37	34.32	37.70	41.97	47.23
	12	32.37	34.32	37.02	40.52	44.94
	Extreme Ends Clamped-Supported					
	1	49.59	160.66	335.2	573.21	874.69
	2	37.02	63.99	137.30	185.85	301.05
	3	34.32	49.59	67.65	132.07	160.66
	4	33.02	42.70	56.98	69.51	129.49
	5	33.02	39.10	49.59	61.31	70.45
	6	32.37	37.02	44.94	54.46	63.99
	7	32.37	35.66	41.97	49.59	57.84
	8	32.37	34.99	39.81	45.70	53.63
	9	31.73	34.32	38.40	43.44	49.59
	10	31.73	33.67	37.02	41.24	46.46
	11	31.73	33.67	36.33	39.81	44.19
	12	31.73	33.02	35.66	39.10	42.70

* For materials other than steel, use equation at bottom of Table 7.4.

f_n = natural frequency, cps
$\rho = \sqrt{I/S}$ = radius of gyration, in.
l = span length, in.
n = mode number
N = number of spans

BEAMS WITH PARTLY CLAMPED ENDS. For a beam in which the slope at each end is proportional to the moment, the following empirical equation gives the natural frequency:[21]

$$f_n = f_0 \left[n + \frac{1}{2}\left(\frac{\beta_L}{5n + \beta_L}\right)\right]\left[n + \frac{1}{2}\left(\frac{\beta_R}{5n + \beta_R}\right)\right]$$

where f_0 is the frequency of the same beam with simply supported ends and n is the mode number. The parameters $\beta_L = k_L l/EI$ and $\beta_R = k_R l/EI$ are coefficients in which k_L and k_R are stiffnesses of the supports as given by $k_L = M_L/\theta_L$, where M_L is the moment and θ_L the angle at the left end, and $k_R = M_R/\theta_R$, where M_R is the moment and θ_R the angle at the right end. The error is less than 2 per cent except for bars having one end completely or nearly clamped ($\beta > 10$) and the other end completely or nearly hinged ($\beta < 0.9$).

LATERAL VIBRATION OF BEAMS WITH MASSES ATTACHED

The use of Fig. 7.4 is a convenient method of estimating the natural frequencies of beams with added loads.

Exact Solution. If the masses attached to the beam are considered to be rigid so that they exert no elastic forces, and if it is assumed that the attachment is such that the bending of the beam is not restrained, Eqs. (7.13) and (7.16) apply. The section of the beam between each two masses, and between each support and the adjacent mass, must be considered individually. The constants in Eq. (7.16) are different for each section. There are $4N$ constants, N being the number of sections into which the beam is divided. Each support supplies two boundary conditions. Additional conditions are provided by:

1. The deflection at the location of each mass is the same for both sections adjacent to the mass.
2. The slope at each mass is the same for each section adjacent thereto.
3. The change in the lateral elastic shear force in the beam, at the location of each mass, is equal to the product of the mass and its acceleration $\ddot{y}$.
4. The change of moment in the beam, at each mass, is equal to the product of the moment of inertia of the mass and its angular acceleration $(\partial^2/\partial t^2)(\partial y/\partial x)$.

Setting up the necessary equations is not difficult, but their solution is a lengthy process for all but the simplest configurations. Even the solution of the problem of a beam with hinged ends supporting a mass with negligible moment of inertia located anywhere except at the center of the beam is fairly long. If the mass is at the center of the beam, the solution is relatively simple because of symmetry, and is illustrated to show how the result compares with that obtained by Rayleigh's method.

Rayleigh's Method. Rayleigh's method offers a practical method of obtaining a fairly accurate solution of the problem, even when more than one mass is added. In carrying out the solution, the kinetic energy of the masses is added to that of the beam. The strain and kinetic energies of a uniform beam are given in Table 7.1. The kinetic energy of the ith mass is $(m_i/2)\omega_n{}^2Y^2(x_i)$, where $Y(x_i)$ is the value of the amplitude at the location of mass. Equating the maximum strain energy to the total maximum kinetic energy of the beam and masses, the frequency equation becomes

$$\omega_n{}^2 = \frac{EI\int_0^l (Y'')^2\,dx}{\frac{\gamma S}{g}\int_0^l Y^2\,dx + \sum_{i=1}^{n} m_i Y^2(x_i)} \tag{7.19}$$

where $Y(x)$ is the maximum deflection. If $Y(x)$ were known exactly, this equation would give the correct frequency; however, since Y is not known, a shape must be assumed. This may be either the mode shape of the unloaded beam or a polynomial that satisfies the necessary boundary conditions, such as the equation for the static deflection under a load.

Beam as Spring. A method for obtaining the natural frequency of a beam with a single mass mounted on it is to consider the beam to act as a spring, the stiffness of which is found by using simple beam theory. The equation $\omega_n = \sqrt{k/m}$ is used. Best accuracy is obtained by considering m to be made up of the attached mass plus some portion of the mass of the beam. The fraction of the beam mass to be used depends on the type of beam. The equations for simply supported and cantilevered beams with masses attached are given in Table 7.2.

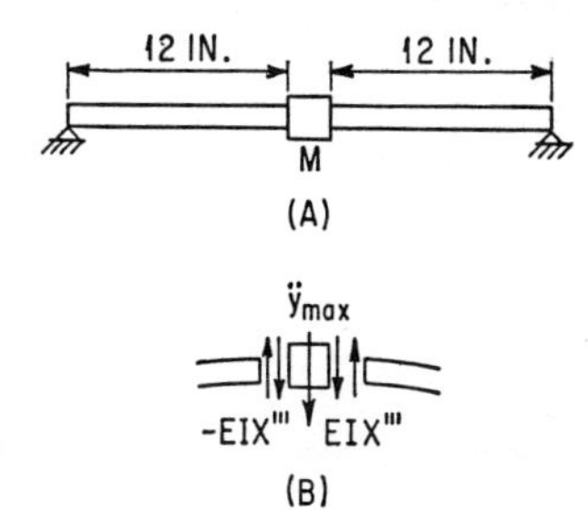

FIG. 7.9. (*A*) Beam having simply supported ends with mass attached at center. (*B*) Forces exerted on mass, at extreme deflection, by shear stresses in beam.

Example 7.7. The fundamental natural frequencies of a beam with hinged ends 24 in. long, 2 in. wide, and ¼ in. thick having a mass m attached at the center (Fig. 7.9) are to be calculated by each of the three methods, and the results compared for ratios of mass to beam mass of 1, 5, and 25. The result is to be compared with the frequency from Fig. 7.4.

Exact Solution. Because of symmetry, only the section of the beam to the left of the mass has to be considered in carrying out the exact solution. The boundary conditions for the left end are: at $x = 0$, $X = 0$, and $X'' = 0$. The shear force just to the left of the mass is negative at maximum deflection (Fig. 7.9*B*) and is $F_s = -EIX'''$; to the right of the mass, because of symmetry, the shear force has the same magnitude with opposite sign. The difference between the shear forces on the two sides of the mass must equal the product of the mass and its acceleration. For the condition of maximum deflection,

$$2EIX''' = m\ddot{y}_{\max} \tag{7.20}$$

where X''' and $\ddot{y}_{\max}$ must be evaluated at $x = l/2$. Because of symmetry the slope at the center is zero. Using the solution $y = X \cos \omega_n t$ and $\ddot{y}_{\max} = -\omega_n^2 X$, Eq. (7.20) becomes

$$2EIX''' = -m\omega_n^2 X \tag{7.21}$$

The first boundary condition makes $A = 0$ in Eq. (7.16) and the second condition makes $B = 0$. For simplicity, the part of the equation that remains is written

$$X = C \sin \kappa x + D \sinh \kappa x \tag{7.22}$$

Using this in Eq. (7.20) gives

$$2EI\left(-C\kappa^3 \cos \frac{\kappa l}{2} + D\kappa^3 \cosh \frac{\kappa l}{2}\right) = -m\omega_n^2 \left(C \sin \frac{\kappa l}{2} + D \sinh \frac{\kappa l}{2}\right) \tag{7.23}$$

The slope at the center is zero. Differentiating Eq. (7.22) and substituting $x = l/2$,

$$\kappa\left(C \cos \frac{\kappa l}{2} + D \cosh \frac{\kappa l}{2}\right) = 0 \tag{7.24}$$

Solving Eqs. (7.23) and (7.24) for the ratio C/D and equating, the following frequency equation is obtained:

$$2\frac{m_b}{m} = \frac{\kappa l}{2}\left(\tan \frac{\kappa l}{2} - \tanh \frac{\kappa l}{2}\right)$$

where $m_b = \gamma Sl/g$ is the total mass of the beam. The lowest roots for the specified ratios m/m_b are as follows:

m/m_b	1	5	25
$\kappa l/2$	1.1916	0.8599	0.5857

The corresponding natural frequencies are found from Eq. (7.14) and are tabulated, with the results obtained by the other methods, at the end of the example.

Solution by Rayleigh's Method. For the solution by Rayleigh's method it is assumed that $Y = B \sin (\pi x/l)$. This is the fundamental mode for the unloaded beam (Table 7.3). The terms in Eq. (7.19) become

$$\int_0^l (Y'')^2\,dx = B^2\left(\frac{\pi}{l}\right)^4 \int_0^l \sin^2\frac{\pi x}{l}\,dx = B^2\frac{l}{2}\left(\frac{\pi}{l}\right)^4$$

$$\int_0^l Y^2\,dx = B^2\int_0^l \sin^2\frac{\pi x}{l}\,dx = B^2\frac{l}{2}$$

$$Y^2(x_1) = B^2$$

Substituting these terms, Eq. (7.19) becomes

$$\omega_n = \sqrt{\frac{EIB^2\dfrac{l}{2}\left(\dfrac{\pi}{l}\right)^4}{\dfrac{S\gamma B^2 l}{2g} + mB^2}} = \frac{\pi^2}{\sqrt{1 + 2m/m_b}}\sqrt{\frac{EIg}{S\gamma l^4}}$$

The frequencies for the specified values of m/m_b are tabulated at the end of the example. Note that if $m = 0$, the frequency is exactly correct, as can be seen from Table 7.3. This is to be expected since, if no mass is added, the assumed shape is the true shape.

Lumped Parameter Solution. Using the appropriate equation from Table 7.2, the natural frequency is

$$\omega_n = \sqrt{\frac{48EI}{l^3(m + 0.5m_b)}}$$

Since $m_b = \gamma Sl/g$, this becomes

$$\omega_n = \sqrt{\frac{48}{(m/m_b) + 0.5}}\sqrt{\frac{EIg}{S\gamma l^4}}$$

Comparison of Results. The results for each method can be expressed as a coefficient α multiplied by $\sqrt{EIg/S\gamma l^4}$. The values of α for the specified values by m/m_b for the three methods of solution are:

m/m_b	1	5	25
Exact.....	5.680	2.957	1.372
Rayleigh ..	5.698	2.976	1.382
Spring	5.657	2.954	1.372

The results obtained by all the methods agree closely. For large values of m/m_b the third method gives very accurate results.

Numerical Calculations. For steel, $E = 30 \times 10^6$ lb/in.², $\gamma = 0.28$ lb/in.³; for a rectangular beam, $I = bh^3/12 = 1/384$ in.⁴ and $S = bh = \frac{1}{2}$ in.² The fundamental frequency using the value of α for the exact solution when $m/m_b = 1$ is

$$\omega_1 = \frac{\alpha}{l^2}\sqrt{\frac{EIg}{S\gamma}} = \frac{5.680}{576}\sqrt{\frac{(30 \times 10^6)(386)}{(0.5)(384)(0.28)}} = 145 \text{ rad/sec} = 23 \text{ cps}$$

Other frequencies can be found by using the other values of α. Nearly the same result is obtained by using Fig. 7.4, if half the mass of the beam is added to the additional mass.

LATERAL VIBRATION OF PLATES

General Theory of Bending of Rectangular Plates. For small deflections of an initially flat plate of uniform thickness (Fig. 7.10) made of homogeneous isotropic material and subjected to normal and shear forces in the plane of the plate, the following equation relates the lateral deflection w to the lateral loading:[22]

$$D\nabla^4 w = D\left(\frac{\partial^4 w}{\partial x^4} + 2\frac{\partial^4 w}{\partial x^2\,\partial y^2} + \frac{\partial^4 w}{\partial y^4}\right) = P + N_x\frac{\partial^2 w}{\partial x^2} + 2N_{xy}\frac{\partial^2 w}{\partial x\,\partial y} + N_y\frac{\partial^2 w}{\partial y^2} \quad (7.25)$$

where $D = Eh^3/12(1 - \mu^2)$ is the plate stiffness, h being the plate thickness and μ Poisson's ratio. The parameter P is the loading intensity, N_x the normal loading in the X direction per unit of length, N_y the normal loading in the Y direction, and N_{xy} the shear load parallel to the plate surface in the X and Y directions.

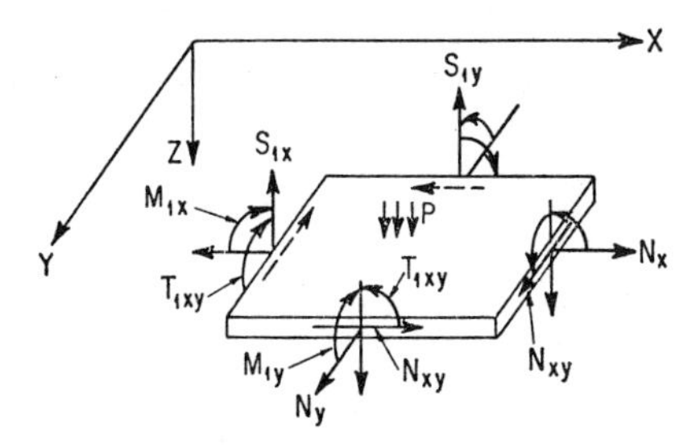

FIG. 7.10. Element of plate showing bending moments, normal forces, and shear forces.

The bending moments and shearing forces are related to the deflection w by the following equations:[23]

$$M_{1x} = -D\left(\frac{\partial^2 w}{\partial x^2} + \mu\frac{\partial^2 w}{\partial y^2}\right) \qquad M_{1y} = -D\left(\frac{\partial^2 w}{\partial y^2} + \mu\frac{\partial^2 w}{\partial x^2}\right)$$

$$T_{1xy} = D(1 - \mu)\frac{\partial^2 w}{\partial x\,\partial y} \tag{7.26}$$

$$S_{1x} = -D\left(\frac{\partial^3 w}{\partial x^3} + \frac{\partial^3 w}{\partial x\,\partial y^2}\right) \qquad S_{1y} = -D\left(\frac{\partial^3 w}{\partial y^3} + \frac{\partial^3 w}{\partial x^2\,\partial y}\right)$$

As shown in Fig. 7.10, M_{1x} and M_{1y} are the bending moments per unit of length on the faces normal to the X and Y directions, respectively, T_{1xy} is the twisting or warping moment on these faces, and S_{1x}, S_{1y} are the shearing forces per unit of length normal to the plate surface.

The boundary conditions that must be satisfied by an edge parallel to the X axis, for example, are as follows:

Built-in edge:

$$w = 0 \qquad \frac{\partial w}{\partial y} = 0$$

Simply supported edge:

$$w = 0 \qquad M_{1y} = -D\left(\frac{\partial^2 w}{\partial y^2} + \mu\frac{\partial^2 w}{\partial x^2}\right) = 0$$

Free edge:

$$M_{1y} = -D\left(\frac{\partial^2 w}{\partial y^2} + \mu\frac{\partial^2 w}{\partial x^2}\right) = 0 \qquad T_{1xy} = 0 \qquad S_{1y} = 0$$

which together give

$$\frac{\partial}{\partial y}\left[\frac{\partial^2 w}{\partial y^2} + (2 - \mu)\frac{\partial^2 w}{\partial x^2}\right] = 0$$

Similar equations can be written for other edges. The strains caused by the bending of the plate are

$$\epsilon_x = -z\frac{\partial^2 w}{\partial x^2} \qquad \epsilon_y = -z\frac{\partial^2 w}{\partial y^2} \qquad \gamma_{xy} = 2z\frac{\partial^2 w}{\partial x\,\partial y} \tag{7.27}$$

where z is the distance from the center plane of the plate.

Hooke's law may be expressed by the following equations:

$$\epsilon_x = \frac{1}{E}(\sigma_x - \mu\sigma_y) \qquad \sigma_x = \frac{E}{1-\mu^2}(\epsilon_x + \mu\epsilon_y)$$

$$\epsilon_y = \frac{1}{E}(\sigma_y - \mu\sigma_x) \qquad \sigma_y = \frac{E}{1-\mu^2}(\epsilon_y + \mu\epsilon_x) \tag{7.28}$$

$$\gamma_{xy} = \frac{\tau_{xy}}{G} \qquad \tau_{xy} = G\gamma_{xy}$$

Substituting the expressions giving the strains in terms of the deflections, the following equations are obtained for the bending stresses in terms of the lateral deflection:

$$\sigma_x = -\frac{Ez}{1-\mu^2}\left(\frac{\partial^2 w}{\partial x^2} + \mu\frac{\partial^2 w}{\partial y^2}\right) = \frac{12M_{1x}}{h^3}z$$

$$\sigma_y = -\frac{Ez}{1-\mu^2}\left(\frac{\partial^2 w}{\partial y^2} + \mu\frac{\partial^2 w}{\partial x^2}\right) = \frac{12M_{1y}}{h^3}z \tag{7.29}$$

$$\tau_{xy} = 2G\frac{\partial^2 w}{\partial x\,\partial y}z = \frac{12T_{1xy}}{h^3}z$$

Table 7.6 gives values of maximum deflection and bending moment at several points in plates which have various shapes and conditions of support, and which are subjected to uniform lateral pressure. The results are all based on the assumption that the deflections are small and that there are no loads in the plane of the plate. The bending stresses are found by the use of Eqs. (7.29). Bending moments and deflections for many other types of load are in the literature.[22]

The stresses caused by loads in the plane of the plate are found by assuming that the stress is uniform through the plate thickness. The total stress at any point in the plate is the sum of the stresses caused by bending and by the loading in the plane of the plate.

For plates in which the lateral deflection is large compared to the plate thickness but small compared to the other dimensions, Eq. (7.25) is valid. However, additional equations must be introduced because the forces N_x, N_y, and N_{xy} depend not only on the initial loading of the plate but also upon the stretching of the plate due to the bending. The equations of equilibrium for the X and Y directions in the plane of the plate are

$$\frac{\partial N_x}{\partial x} + \frac{\partial N_{xy}}{\partial y} = 0 \qquad \frac{\partial N_{xy}}{\partial x} + \frac{\partial N_y}{\partial y} = 0 \tag{7.30}$$

It can be shown [27] that the strain components are given by

$$\epsilon_x = \frac{\partial u}{\partial x} + \frac{1}{2}\left(\frac{\partial w}{\partial x}\right)^2 \qquad \epsilon_y = \frac{\partial v}{\partial y} + \frac{1}{2}\left(\frac{\partial w}{\partial y}\right)^2$$

$$\gamma_{xy} = \frac{\partial u}{\partial y} + \frac{\partial v}{\partial x} + \frac{\partial w}{\partial x}\frac{\partial w}{\partial y} \tag{7.31}$$

where u is the displacement in the X direction and v is the displacement in the Y direction. By differentiating and combining these expressions, the following relation is obtained:

$$\frac{\partial^2 \epsilon_x}{\partial y^2} + \frac{\partial^2 \epsilon_y}{\partial x^2} - \frac{\partial^2 \gamma_{xy}}{\partial x\,\partial y} = \left(\frac{\partial^2 w}{\partial x\,\partial y}\right)^2 - \frac{\partial^2 w}{\partial x^2}\frac{\partial^2 w}{\partial y^2} \tag{7.32}$$

If it is assumed that the stresses caused by the forces in the plane of the plate are uni-

Table 7.6. Maximum Deflection and Bending Moments in Uniformly Loaded Plates under Static Conditions

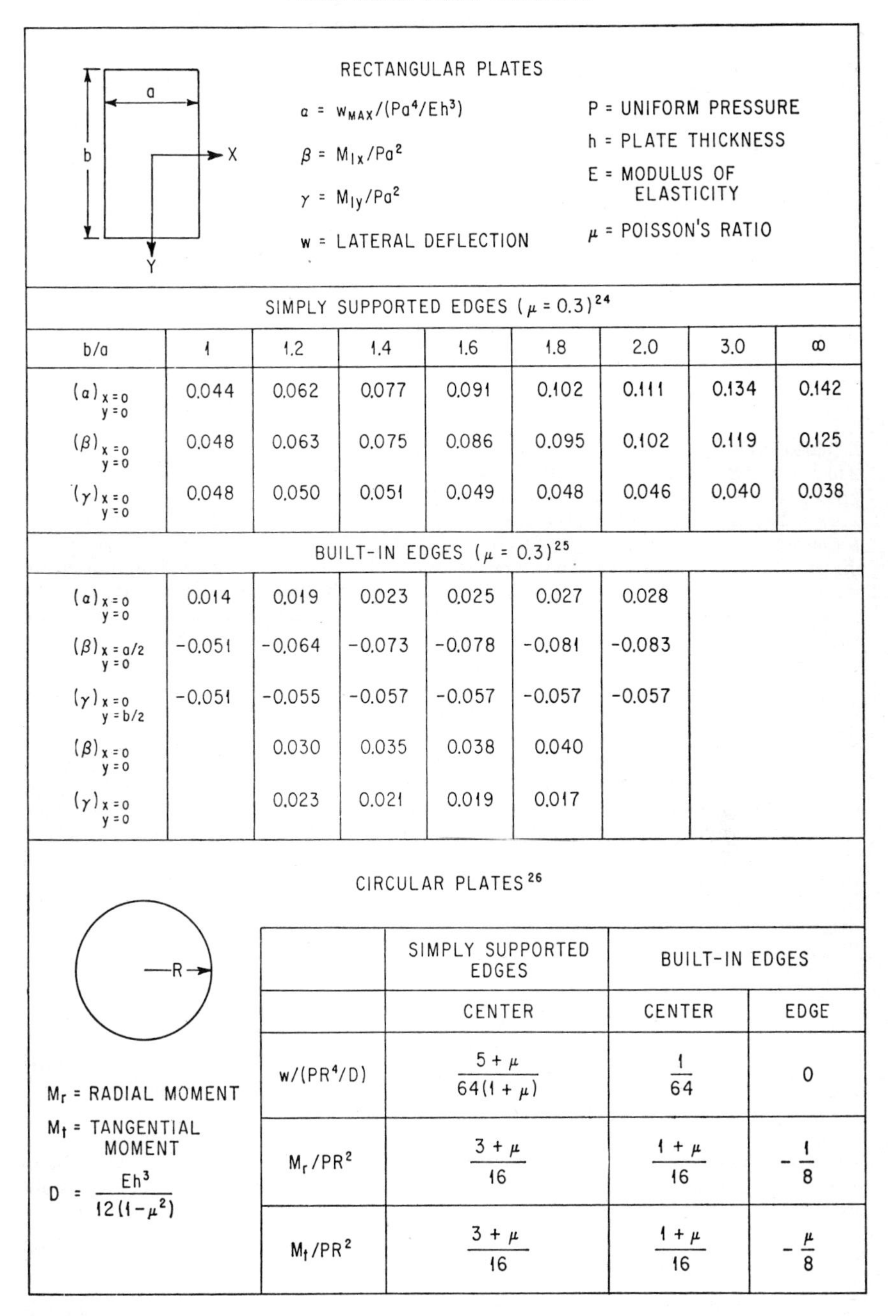

RECTANGULAR PLATES

$\alpha = w_{MAX}/(Pa^4/Eh^3)$

$\beta = M_{1x}/Pa^2$

$\gamma = M_{1y}/Pa^2$

w = LATERAL DEFLECTION

P = UNIFORM PRESSURE

h = PLATE THICKNESS

E = MODULUS OF ELASTICITY

μ = POISSON'S RATIO

SIMPLY SUPPORTED EDGES ($\mu = 0.3$)[24]

b/a	1	1.2	1.4	1.6	1.8	2.0	3.0	∞
$(\alpha)_{x=0,\ y=0}$	0.044	0.062	0.077	0.091	0.102	0.111	0.134	0.142
$(\beta)_{x=0,\ y=0}$	0.048	0.063	0.075	0.086	0.095	0.102	0.119	0.125
$(\gamma)_{x=0,\ y=0}$	0.048	0.050	0.051	0.049	0.048	0.046	0.040	0.038

BUILT-IN EDGES ($\mu = 0.3$)[25]

b/a	1	1.2	1.4	1.6	1.8	2.0
$(\alpha)_{x=0,\ y=0}$	0.014	0.019	0.023	0.025	0.027	0.028
$(\beta)_{x=a/2,\ y=0}$	-0.051	-0.064	-0.073	-0.078	-0.081	-0.083
$(\gamma)_{x=0,\ y=b/2}$	-0.051	-0.055	-0.057	-0.057	-0.057	-0.057
$(\beta)_{x=0,\ y=0}$		0.030	0.035	0.038	0.040	
$(\gamma)_{x=0,\ y=0}$		0.023	0.021	0.019	0.017	

CIRCULAR PLATES[26]

	SIMPLY SUPPORTED EDGES	BUILT-IN EDGES	
	CENTER	CENTER	EDGE
$w/(PR^4/D)$	$\dfrac{5+\mu}{64(1+\mu)}$	$\dfrac{1}{64}$	0
M_r/PR^2	$\dfrac{3+\mu}{16}$	$\dfrac{1+\mu}{16}$	$-\dfrac{1}{8}$
M_t/PR^2	$\dfrac{3+\mu}{16}$	$\dfrac{1+\mu}{16}$	$-\dfrac{\mu}{8}$

M_r = RADIAL MOMENT

M_t = TANGENTIAL MOMENT

$$D = \frac{Eh^3}{12(1-\mu^2)}$$

formly distributed through the thickness, Hooke's law, Eqs. (7.28), can be expressed:

$$\epsilon_x = \frac{1}{hE}(N_x - \mu N_y) \qquad \epsilon_y = \frac{1}{hE}(N_y - \mu N_x) \qquad \gamma_{xy} = \frac{1}{hG}N_{xy} \tag{7.33}$$

The equilibrium equations are satisfied by a stress function ϕ which is defined as follows:

$$N_x = h\frac{\partial^2\phi}{\partial y^2} \qquad N_y = h\frac{\partial^2\phi}{\partial x^2} \qquad N_{xy} = -h\frac{\partial^2\phi}{\partial x\,\partial y} \tag{7.34}$$

If these are substituted into Eqs. (7.33) and the resulting expressions substituted into Eq. (7.32), the following equation is obtained:

$$\frac{\partial^4\phi}{\partial x^4} + 2\frac{\partial^4\phi}{\partial x^2\,\partial y^2} + \frac{\partial^4\phi}{\partial y^4} = E\left[\left(\frac{\partial^2 w}{\partial x\,\partial y}\right)^2 - \frac{\partial^2 w}{\partial x^2}\frac{\partial^2 w}{\partial y^2}\right] \tag{7.35}$$

A second equation is obtained by substituting Eqs. (7.34) in Eq. (7.25):

$$D\nabla^4 w = P + h\left(\frac{\partial^2\phi}{\partial y^2}\frac{\partial^2 w}{\partial x^2} - 2\frac{\partial^2\phi}{\partial x\,\partial y}\frac{\partial^2 w}{\partial x\,\partial y} + \frac{\partial^2\phi}{\partial x^2}\frac{\partial^2 w}{\partial y^2}\right) \tag{7.36}$$

Equations (7.35) and (7.36), with the boundary conditions, determine ϕ and w, from which the stresses can be computed. General solutions to this set of equations are not known, but some approximate solutions can be found in the literature.[28]

Free Lateral Vibrations of Rectangular Plates. In Eq. (7.25), the terms on the left are equal to the sum of the rates of change of the forces per unit of length in the X and Y directions where such forces are exerted by shear stresses caused by bending normal to the plane of the plate. For a rectangular element with dimensions dx and dy, the net force exerted normal to the plane of the plate by these stresses is $D\nabla^4 w\,dx\,dy$. The last three terms on the right-hand side of Eq. (7.25) give the net force normal to the plane of the plate, per unit of length, which is caused by the forces acting in the plane of the plate. The net force caused by these forces on an element with dimensions dx and dy is $(N_x\,\partial^2 w/\partial x^2 + 2N_{xy}\,\partial^2 w/\partial x\,\partial y + N_y\,\partial^2 w/\partial y^2)\,dx\,dy$. As in the corresponding beam problem, the forces in a vibrating plate consist of two parts: (1) that which balances the static load P including the weight of the plate and (2) that which is induced by the vibration. The first part is always in equilibrium with the load and together with the load can be omitted from the equation of motion if the deflection is taken from the position of static equilibrium. The force exerted normal to the plane of the plate by the bending stresses must equal the sum of the force exerted normal to the plate by the loads acting in the plane of the plate; i.e., the product of the mass of the element $(\gamma h/g)\,dx\,dy$ and its acceleration $\ddot{w}$. The term involving the acceleration of the element is negative, because when the bending force is positive the acceleration is in the negative direction. The equation of motion is

$$D\nabla^4 w = -\frac{\gamma}{g}h\ddot{w} + \left(N_x\frac{\partial^2 w}{\partial x^2} + 2N_{xy}\frac{\partial^2 w}{\partial x\,\partial y} + N_y\frac{\partial^2 w}{\partial y^2}\right) \tag{7.37}$$

This equation is valid only if the magnitudes of the forces in the plane of the plate are constant during the vibration. For many problems these forces are negligible and the term in parentheses can be omitted.

When a system vibrates in a natural mode, all parts execute simple harmonic motion about the equilibrium position; therefore, the solution of Eq. (7.37) can be written as $w = AW(x,y)\cos(\omega_n t + \theta)$ in which W is a function of x and y only. Substituting this in Eq. (7.37) and dividing through by $A\cos(\omega_n t + \theta)$ gives

$$D\nabla^4 W = \frac{\gamma h\omega_n^2}{g}W + \left(N_x\frac{\partial^2 W}{\partial x^2} + 2N_{xy}\frac{\partial^2 W}{\partial x\,\partial y} + N_y\frac{\partial^2 W}{\partial y^2}\right) \tag{7.38}$$

The function W must satisfy Eq. (7.38) as well as the necessary boundary conditions.

The solution of the problem of the lateral vibration of a rectangular plate with all edges simply supported is relatively simple; in general, other combinations of edge conditions require the use of other methods of solution. These are discussed later.

Example 7.8. The natural frequencies and normal modes of small vibration of a rectangular plate of length a, width b, and thickness h are to be calculated. All edges are hinged and subjected to unchanging normal forces N_x and N_y.

Solution. The following equation, in which m and n may be any integers, satisfies the necessary boundary conditions:

$$W = A \sin \frac{m\pi x}{a} \sin \frac{n\pi y}{b} \tag{7.39}$$

Substituting the necessary derivatives into Eq. (7.38),

$$D\left[\left(\frac{m}{a}\right)^4 + 2\left(\frac{m}{a}\right)^2\left(\frac{n}{b}\right)^2 + \left(\frac{n}{b}\right)^4\right]\pi^4 \sin \frac{m\pi x}{a} \sin \frac{n\pi y}{b}$$

$$= \frac{\gamma h \omega_n^2}{g} \sin \frac{m\pi x}{a} \sin \frac{n\pi y}{b} - \pi^2\left[N_x\left(\frac{m}{a}\right)^2 + N_y\left(\frac{n}{b}\right)^2\right] \sin \frac{m\pi x}{a} \sin \frac{n\pi y}{b}$$

Solving for ω_n^2,

$$\omega_n^2 = \frac{g}{\gamma h}\left\{\pi^4 D\left[\left(\frac{m}{a}\right)^2 + \left(\frac{n}{b}\right)^2\right]^2 + \pi^2\left[N_x\left(\frac{m}{a}\right)^2 + N_y\left(\frac{n}{b}\right)^2\right]\right\} \tag{7.40}$$

By using integral values of m and n, the various frequencies are obtained from Eq. (7.40) and the corresponding normal modes from Eq. (7.39). For each mode, m and n represent the number of half sine waves in the X and Y directions, respectively. In each mode there are $m - 1$ evenly spaced nodal lines parallel to the Y axis, and $n - 1$ parallel to the X axis.

Rayleigh's and Ritz's Methods. The modes of vibration of a rectangular plate with all edges simply supported are such that the deflection of each section of the plate parallel to an edge is of the same form as the deflection of a beam with both ends simply supported. In general, this does not hold true for other combinations of edge conditions. For example, the vibration of a rectangular plate with all edges built in does not occur in such a way that each section parallel to an edge has the same shape as does a beam with both ends built in. A function that is made up using the mode shapes of beams with built-in ends obviously satisfies the conditions of zero deflection and slope at all edges, but it cannot be made to satisfy Eq. (7.38).

The mode shapes of beams give logical functions with which to formulate shapes for determining the natural frequencies, for plates having various edge conditions, by the Rayleigh or Ritz methods. By using a single mode function in Rayleigh's method an approximate frequency can be determined. This can be improved by using more than one of the modal shapes and using Ritz's method as discussed below.

The strain energy of bending and the kinetic energy for plates are given in Table 7.1. Finding the maximum values of the energies, equating them, and solving for ω_n^2 gives the following frequency equation:

$$\omega_n^2 = \frac{V_{\max}}{\dfrac{\gamma h}{2g}\displaystyle\int_A\int W^2\,dx\,dy} \tag{7.41}$$

where V is the strain energy.

In applying the Rayleigh method, a function W is assumed that satisfies the necessary boundary conditions of the plate. An example of the calculations is given in the section on circular plates. If the shape assumed is exactly the correct one, Eq. (7.41) gives the exact frequency. In general, the correct shape is not known and a frequency greater than the natural frequency is obtained. The Ritz method involves assuming W to be of the form $W = a_1W_1(x,y) + a_2W_2(x,y) + \ldots$ in which $W_1, W_2, \ldots$ all satisfy the boundary conditions, and $a_1, a_2, \ldots$ are adjusted to give a minimum frequency. Reference 28a is an extensive compilation, with references to sources, of calculated and experimental results for plates of many shapes. Some examples are cited in the following sections.

Table 7.7. Natural Frequencies and Nodal Lines of Square Plates with Various Edge Conditions (*After D. Young.*[29])

	1ST MODE	2ND MODE	3RD MODE	4TH MODE	5TH MODE	6TH MODE
$\omega_n / \sqrt{Dg/\gamma h a^4}$	3.494	8.547	21.44	27.46	31.17	
NODAL LINES	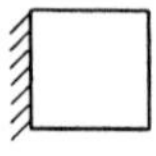	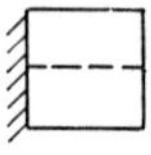	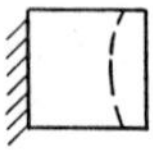	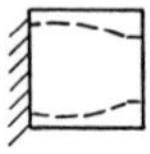	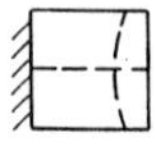	
$\omega_n / \sqrt{Dg/\gamma h a^4}$	35.99	73.41	108.27	131.64	132.25	165.15
NODAL LINES	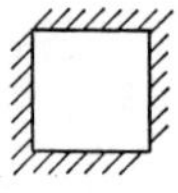	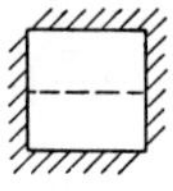	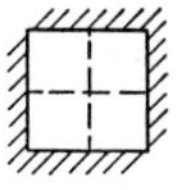	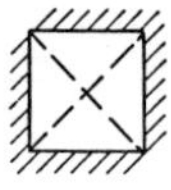	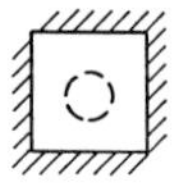	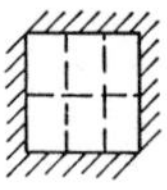
$\omega_n / \sqrt{Dg/\gamma h a^4}$	6.958	24.08	26.80	48.05	63.14	
NODAL LINES	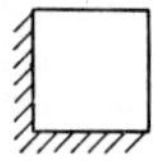	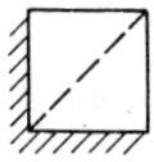	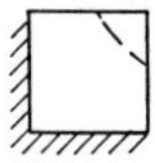	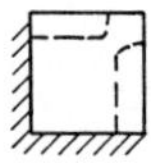	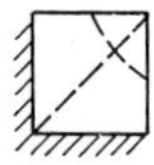	

$\omega_n = 2\pi f_n$ h = PLATE THICKNESS

$D = Eh^3/12(1-\mu^2)$ a = PLATE LENGTH

γ = WEIGHT DENSITY

SQUARE, RECTANGULAR, AND SKEW RECTANGULAR PLATES. Tables of the functions necessary for the determination of the natural frequencies of rectangular plates by the use of the Ritz method are available,[29] these having been derived by using the modal shapes of beams having end conditions corresponding to the edge conditions of the plates. Information is included from which the complete shapes of the vibrational modes can be determined. Frequencies and nodal patterns for several modes of vibration of square plates having three sets of boundary conditions are shown in Table 7.7. By the use of functions which represent the natural modes of beams, the frequencies and nodal patterns for rectangular and skew cantilever plates have been determined [30] and are shown in Table 7.8. Comparison of calculated frequencies with experimentally determined values shows good agreement. Natural frequencies of rectangular plates having other boundary conditions are given in Table 7.9.

TRIANGULAR AND TRAPEZOIDAL PLATES. Nodal patterns and natural frequencies for triangular plates have been determined [32] by the use of functions derived from the mode shapes of beams, and are shown in Table 7.10. Certain of these have been compared with experimental values and the agreement is excellent. Natural frequencies and nodal patterns have been determined experimentally for six modes of vibration of a number of cantilevered triangular plates [33] and for the first six modes of cantilevered trapezoidal plates derived by trimming the tips of triangular plates parallel to the clamped edge.[34] These triangular and trapezoidal shapes approximate the shapes of various delta wings for aircraft and of fins for missiles.

CIRCULAR PLATES. The solution of the problem of small lateral vibration of circular plates is obtained by transforming Eq. (7.38) to polar coordinates and finding the solution that satisfies the necessary boundary conditions of the resulting equation.

Table 7.8. Natural Frequencies and Nodal Lines of Cantilevered Rectangular and Skew Rectangular Plates ($\mu = 0.3$)* (*M. V. Barton.*[30])

MODE \ a/b	1/2	1	2	5
FIRST 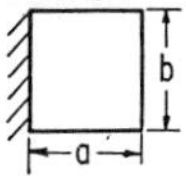	3.508	3.494	3.472	3.450
SECOND 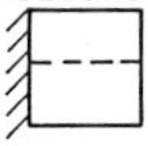	5.372	8.547	14.93	34.73
THIRD 	21.96	21.44	21.61	21.52
FOURTH 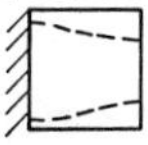	10.26	27.46	94.49	563.9
FIFTH 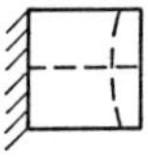	24.85	31.17	48.71	105.9

MODE	FIRST	SECOND	FIRST	SECOND	FIRST	SECOND
$\omega_n/\sqrt{Dg/\gamma h a^4}$	3.601	8.872	3.961	10.190	4.824	13.75
NODAL LINES	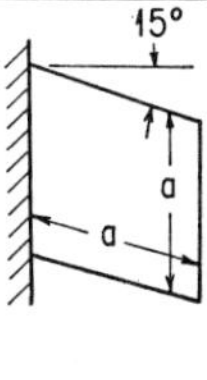 15°	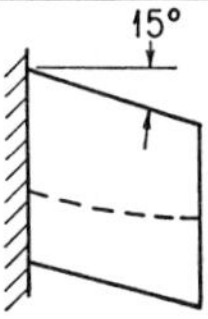 15°	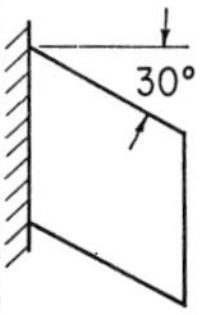 30°	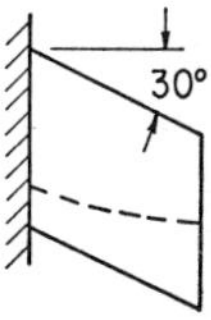 30°	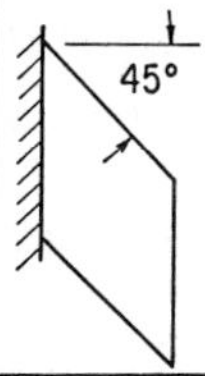 45°	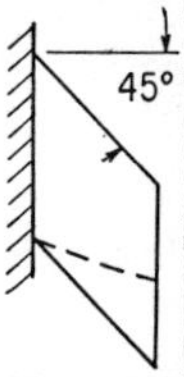 45°

* For terminology, see Table 7.7.

Table 7.9. Natural Frequencies of Rectangular Plates (*R. F. S. Hearmon.*[31])

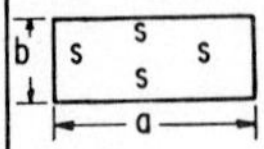 S S S S (b × a)	b/a	1.0	1.5	2.0	2.5	3.0	∞
	$\omega_n / \sqrt{Dg/\gamma h a^4}$	19.74	14.26	12.34	11.45	10.97	9.87
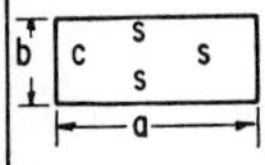 C S S S (b × a)	b/a	1.0	1.5	2.0	2.5	3.0	∞
	$\omega_n / \sqrt{Dg/\gamma h a^4}$	23.65	18.90	17.33	16.63	16.26	15.43
	a/b	1.0	1.5	2.0	2.5	3.0	∞
	$\omega_n / \sqrt{Dg/\gamma h b^4}$	23.65	15.57	12.92	11.75	11.14	9.87
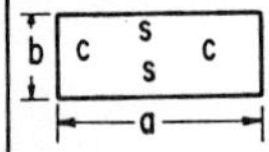 C S C S (b × a)	b/a	1.0	1.5	2.0	2.5	3.0	∞
	$\omega_n / \sqrt{Dg/\gamma h a^4}$	28.95	25.05	23.82	23.27	22.99	22.37
	a/b	1.0	1.5	2.0	2.5	3.0	∞
	$\omega_n / \sqrt{Dg/\gamma h b^4}$	28.95	17.37	13.69	12.13	11.36	9.87
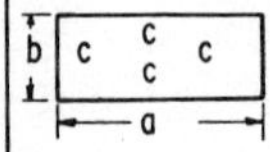 C C C C (b × a)	b/a	1.0	1.5	2.0	2.5	3.0	∞
	$\omega_n / \sqrt{Dg/\gamma h a^4}$	35.98	27.00	24.57	23.77	23.19	22.37

S DENOTES SIMPLY SUPPORTED EDGE

C DENOTES BUILT-IN OR CLAMPED EDGE

a = LENGTH OF PLATE

b = WIDTH OF PLATE

FOR OTHER TERMINOLOGY SEE TABLE 7.7

Omitting the terms involving forces in the plane of the plate,[35]

$$\left(\frac{\partial^2}{\partial r^2} + \frac{1}{r}\frac{\partial}{\partial r} + \frac{1}{r}\frac{\partial^2}{\partial \theta^2}\right)\left(\frac{\partial^2 W}{\partial r^2} + \frac{1}{r}\frac{\partial W}{\partial r} + \frac{1}{r}\frac{\partial^2 W}{\partial \theta^2}\right) = \varkappa^4 W \tag{7.42}$$

where

$$\varkappa^4 = \frac{\gamma h {\omega_n}^2}{gD}$$

The solution of Eq. (7.42) is[35]

$$W = A \cos (n\theta - \beta)[J_n(\varkappa r) + \lambda J_n(i\varkappa r)] \tag{7.43}$$

where J_n is a Bessel function of the first kind. When $\cos (n\theta - \beta) = 0$, a mode having a nodal system of n diameters, symmetrically distributed, is obtained. The term in

Table 7.10. Natural Frequencies and Nodal Lines of Triangular Plates (*B. W. Anderson*.[32])

MODE \ k	2	4	8	14
FIRST (a/k, a)	7.194	7.122	7.080	7.068
SECOND	30.803	30.718	30.654	30.638
THIRD	61.131	90.105	157.70	265.98
FOURTH	148.8	259.4	493.4	853.6
k	2	4	7	
FIRST	5.887	6.617	6.897	
SECOND (a/k, a)	25.40	28.80	30.28	

a = LENGTH OF TRIANGLE
k = RATIO OF LENGTH TO WIDTH OF TRIANGLE
FOR OTHER TERMINOLOGY SEE TABLE 7.7

brackets represents modes having concentric nodal circles. The values of $\varkappa$ and λ are determined by the boundary conditions, which are, for radially symmetrical vibration:

Simply supported edge:

$$W = 0 \qquad M_{1r} = D\left(\frac{d^2W}{dr^2} + \frac{\mu}{a}\frac{dW}{dr}\right) = 0$$

Fixed edge:

$$W = 0 \qquad \frac{dW}{dr} = 0$$

Free edge:

$$M_{1r} = D\left(\frac{d^2W}{dr^2} + \frac{\mu}{a}\frac{dW}{dr}\right) = 0 \qquad \frac{d}{dr}\left(\frac{d^2W}{dr^2} + \frac{1}{r}\frac{dW}{dr}\right) = 0$$

Example 7.9. The steel diaphragm of a radio earphone has an unsupported diameter of 2.0 in. and is 0.008 in. thick. Assuming that the edge is fixed, the lowest three frequencies for the free vibration in which only nodal circles occur are to be calculated, using the exact method and the Rayleigh and Ritz methods.

Exact Solution. In this example $n = 0$, which makes $\cos(n\theta - \beta) = 1$; thus, Eq. (7.43) becomes

$$W = A[J_0(\varkappa r) + \lambda I_0(\varkappa r)]$$

where $J_0(i\varkappa r) = I_0(\varkappa r)$ and I_0 is a modified Bessel function of the first kind.

At the boundary where $r = a$,

$$\frac{\partial W}{\partial r} = A\varkappa[-J_1(\varkappa a) + \lambda I_1(\varkappa a)] = 0 \qquad -J_1(\varkappa a) + \lambda I_1(\varkappa a) = 0$$

The deflection at $r = a$ is also zero:

$$J_0(\varkappa a) + \lambda I_0(\varkappa a) = 0$$

The frequency equation becomes

$$\lambda = \frac{J_1(\varkappa a)}{I_1(\varkappa a)} = -\frac{J_0(\varkappa a)}{I_0(\varkappa a)}$$

The first three roots of the frequency equation are: $\varkappa a = 3.196, 6.306, 9.44$. The corresponding natural frequencies are, from Eq. (7.42),

$$\omega_n = \frac{10.21}{a^2}\sqrt{\frac{Dg}{\gamma h}}, \quad \frac{39.77}{a^2}\sqrt{\frac{Dg}{\gamma h}}, \quad \frac{88.9}{a^2}\sqrt{\frac{Dg}{\gamma h}}$$

For steel, $E = 30 \times 10^6$ lb/in.², $\gamma = 0.28$ lb/in.³, and $\mu = 0.28$. Hence

$$D = \frac{Eh^3}{12(1-\mu^2)} = \frac{30 \times 10^6(0.008)^3}{12(1-0.078)} = 1.38 \text{ lb-in.}$$

Thus, the lowest natural frequency is

$$\omega_1 = 10.21\sqrt{\frac{(1.38)(386)}{(0.28)(0.008)}} = 4{,}960 \text{ rad/sec} = 790 \text{ cps}$$

The second frequency is 3,070 cps, and the third is 6,880 cps.

Solution by Rayleigh's Method. The equations for strain and kinetic energies are given in Table 7.1. The strain energy for a plate with clamped edges becomes

$$V = \pi D \int_0^a \left(\frac{\partial^2 W}{\partial r^2} + \frac{1}{r}\frac{\partial W}{\partial r}\right)^2 r\,dr$$

The maximum kinetic energy is

$$T = \frac{\omega_n^2 \pi \gamma h}{g}\int_0^a W^2 r\,dr$$

An expression of the form $W = a_1[1 - (r/a)^2]^2$, which satisfies the conditions of zero deflection and slope at the boundary, is used. The first two derivatives are $\partial W/\partial r = a_1(-4r/a^2 + 4r^3/a^4)$ and $\partial^2 W/\partial r^2 = a_1(-4/a^2 + 12r^2/a^4)$. Using these values in the equations for strain and kinetic energy, $V = 32\pi D a_1^2/3a^2$ and $T = \omega_n^2 \pi \gamma h a^2 a_1^2/10g$. Equating these values and solving for the frequency,

$$\omega_n = \sqrt{\frac{320\,Dg}{3a^4\,\gamma h}} = \frac{10.33}{a^2}\sqrt{\frac{Dg}{\gamma h}}$$

This is somewhat higher than the exact frequency.

Solution by Ritz's Method. Using an expression for the deflection of the form

$$W = a_1[1 - (r/a)^2]^2 + a_2[1 - (r/a)^2]^3$$

and applying the Ritz method, the following values are obtained for the first two frequencies:

$$\omega_1 = \frac{10.21}{a^2}\sqrt{\frac{Dg}{\gamma h}} \qquad \omega_2 = \frac{43.04}{a^2}\sqrt{\frac{Dg}{\gamma h}}$$

The details of the calculations giving this result are in the literature.[36] The first frequency agrees with the exact answer to four significant figures, while the second frequency is somewhat high. A closer approximation to the second frequency and approximations of the higher frequencies could be obtained by using additional terms in the deflection equation.

The frequencies of modes having n nodal diameters are:[36]

$n = 1$: $$\omega_1 = \frac{21.22}{a^2}\sqrt{\frac{Dg}{\gamma h}}$$

$n = 2$: $$\omega_2 = \frac{34.84}{a^2}\sqrt{\frac{Dg}{\gamma h}}$$

For a plate with its center fixed and edge free, and having m nodal circles, the frequencies are:[37]

m	0	1	2	3
$\omega_n a^2 \Big/ \sqrt{\frac{Dg}{\gamma h}}$	3.75	20.91	60.68	119.7

STRETCHING OF MIDDLE PLANE. In the usual analysis of plates, it is assumed that the deflection of the plate is so small that there is no stretching of the middle plane. If such stretching occurs, it affects the natural frequency. Whether it occurs depends

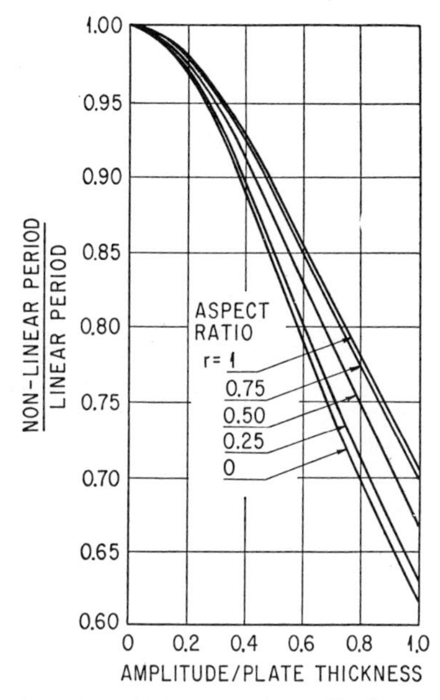

FIG. 7.11. Influence of amplitude on period of vibration of uniform rectangular plates with immovable hinged edges. The aspect ratio r is the ratio of width to length of the plate. (*H. Chu and G. Herrmann.*[38])

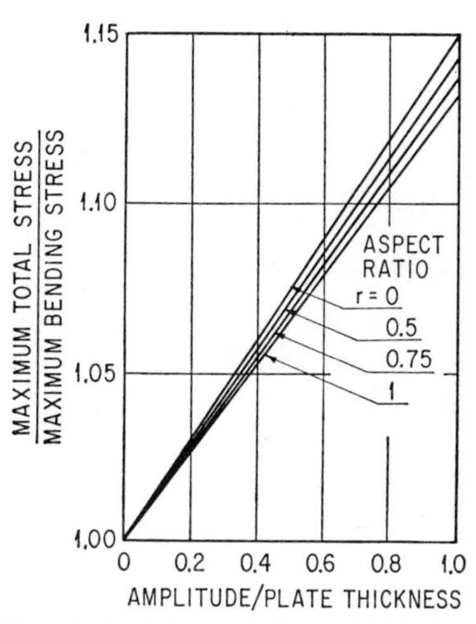

FIG. 7.12. Influence of amplitude on maximum total stress in rectangular plates with immovable hinged edges. The aspect ratio r is the ratio of width to length of the plate. (*H. Chu and G. Herrmann.*[38])

on the conditions of support of the plate, the amplitude of vibration, and possibly other conditions. In a plate with its edges built in, a relatively small deflection causes a significant stretching. The effect of stretching is not proportional to the deflection; thus, the elastic restoring force is not a linear function of deflection. The natural frequency is not independent of amplitude, but becomes higher with increasing amplitudes. If a plate is subjected to a pressure on one side, so that the vibration occurs about a deflected position, the effect of stretching may be appreciable. The effect of stretching in rectangular plates with immovable hinged supports has been discussed.[38] The effect of the amplitude on the natural frequency is shown in Fig. 7.11; the effect on the total stress in the plate is shown in Fig. 7.12. The natural frequency increases rapidly as the amplitude of vibration increases.

ROTATIONAL MOTION AND SHEARING FORCES. In the foregoing analysis, only the motion of each element of the plate in the direction normal to the plane of the plate is considered. There is also rotation of each element, and there is a deflection associated with the lateral shearing forces in the plate. The effects of these factors becomes

Table 7.11. Natural Frequencies of Complete Circular Rings Whose Thickness in Radial Direction Is Small Compared to Radius

TYPE OF VIBRATION	SHAPE OF LOWEST MODE	RECTANGULAR CROSS SECTION	CIRCULAR CROSS SECTION
		ω_n	ω_n
FLEXURAL IN PLANE OF RING WITH $\underline{n}$ COMPLETE WAVE-LENGTH IN CIRCUMFERENCE	R; n = 2	$\sqrt{\frac{Eg}{\gamma}\frac{I}{AR^4}\frac{n^2(n^2-1)^2}{n^2+1}}$ $\underline{n}$ ANY INTEGER > 1	$\sqrt{\frac{E\pi r^4}{4mR^4}\frac{n^2(n^2-1)^2}{n^2+1}}$ $\underline{n}$ ANY INTEGER > 1
FLEXURAL NORMAL TO PLANE OF RING	R; n = 2		$\sqrt{\frac{E\pi r^4}{4mR^4}\frac{n^2(n^2-1)^2}{n^2+1+\mu}}$ $\underline{n}$ ANY INTEGER > 1
TORSIONAL	R	FIRST MODE $\sqrt{\frac{Eg}{\gamma R^2}\frac{I_x}{I_p}}$	$\sqrt{\frac{G\pi r^2}{mR^2}(n^2+1+\mu)}$ n = 0, OR ANY INTEGER
EXTENSIONAL	R	$\sqrt{\frac{Eg}{\gamma R^2}}$	$\sqrt{\frac{E\pi r^2}{mR^2}(1+n^2)}$ $\underline{n}$ = 0, OR ANY INTEGER

E = MODULUS OF ELASTICITY
G = MODULUS OF RIGIDITY
γ = WEIGHT DENSITY
n : DEFINED FOR EACH TYPE OF VIBRATION
R = RADIUS OF RING
μ = POISSON'S RATIO

PROPERTIES OF CROSS SECTIONS
I = MOMENT OF INERTIA WITH RESPECT TO AXIS OF SECTION
I_x = MOMENT OF INERTIA WITH RESPECT TO RADIAL LINE
I_p = POLAR MOMENT OF INERTIA
A = AREA
r = RADIUS
m = MASS PER UNIT OF LENGTH

significant if the curvature of the plate is large relative to its thickness, i.e., for a plate in which the thickness is large compared to the lateral dimensions or when the plate is vibrating in a mode for which the nodal lines are close together. These effects have been analyzed for rectangular plates [39] and for circular plates.[40]

COMPLETE CIRCULAR RINGS. Equations have been derived [41, 42] for the natural frequencies of complete circular rings for which the radius is large compared to the thickness of the ring in the radial direction. Such rings can execute several types of free vibration, which are shown in Table 7.11 with the formulas for the natural frequencies.

TRANSFER MATRIX METHOD

In some assemblies which consist of various types of elements, e.g., beam segments, the solution for each element may be known. The transfer matrix method[43, 44] is a procedure by means of which the solution for such elements can be combined to yield a frequency equation for the assembly. The associated mode shapes can then be determined. The method is an extension to distributed systems of the Holzer method, described in Chap. 38, in which torsional problems are solved by dividing an assembly into lumped masses and elastic elements, and of the Myklestad method,[45] in which a similar procedure is applied to beam problems. The method has been used[46] to find the natural frequencies and mode shapes of the internals of a nuclear reactor by modeling the various elements of the system as beam segments.

The method will be illustrated by setting up the frequency equation for a cantilever beam, Fig. 7.13, composed of three segments, each of which has uniform section properties. Only the effects of bending will be considered, but the method can be extended to include other effects, such as shear deformation and rotary motion of the cross section.[44] Application to other geometries is described in Ref. 44.

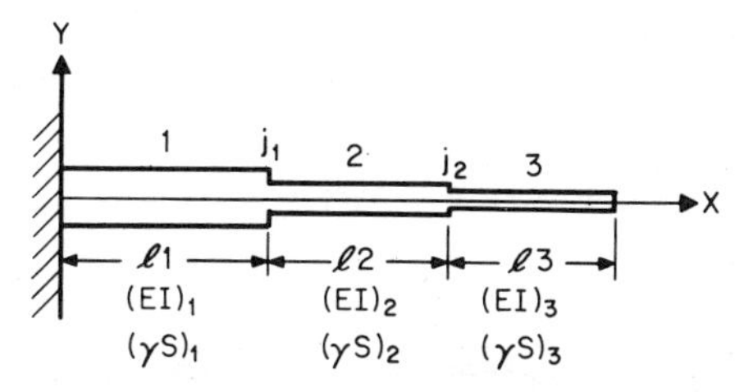

FIG. 7.13. Cantilever beam made up of three segments having different section properties.

Depending on the type of element being considered, the values of appropriate parameters must be expressed at certain sections of the piece in terms of their values at other sections. In the beam problem, the deflection and its first three derivatives must be used.

Transfer Matrices. Two types of transfer matrix are used. One, which for the beam problem is called the **R** matrix (after Lord Rayleigh[43]), yields the values of the parameters at the right end of a uniform segment of the beam in terms of their values at the left end of the segment. The other type of transfer matrix is the point matrix, which yields the values of the parameters just to the right of a joint between segments in terms of their values just to the left of the joint.

As can be seen by looking at the successive derivatives, the coefficients in Eq. (7.16) are equal to the following, where the subscript 0 indicates the value of the indicated parameter at the left end of the beam:

$$A = \frac{X_0}{2} \qquad C = \frac{X_0'}{2\kappa} \qquad B = \frac{-X_0''}{2\kappa^2} \qquad D = \frac{-X_0'''}{2\kappa^3}$$

Using the following notation, X and its derivatives at the right end of a beam segment can be expressed, by the matrix equation, in terms of the values at the left end of the segment. The subscript n refers to the number of the segment being considered, the subscript l to the left end of the segment and the subscript r to the right end.

$$C_{0n} = \frac{\cos \kappa_n l_n + \cosh \kappa_n l_n}{2}$$

$$S_{1n} = \frac{\sin \kappa_n l_n + \sinh \kappa_n l_n}{2\kappa_n}$$

$$C_{2n} = \frac{-(\cos \kappa_n l_n - \cosh \kappa_n l_n)}{2\kappa_n^2}$$

$$S_{3n} = \frac{-(\sin \kappa_n l_n - \sinh \kappa_n l_n)}{2\kappa_n^3}$$

where κ_n takes the value shown in Eq. (7.14) with the appropriate values of the parameters for the segment and l_n is the length of the segment.

$$\begin{bmatrix} X \\ X' \\ X'' \\ X''' \end{bmatrix}_{rn} = \begin{bmatrix} C_{0n} & S_{1n} & C_{2n} & S_{3n} \\ \kappa_n^4 S_{3n} & C_{0n} & S_{1n} & C_{2n} \\ \kappa_n^4 C_{2n} & \kappa_n^4 S_{3n} & C_{0n} & S_{1n} \\ \kappa_n^4 S_{1n} & \kappa_n^4 C_{2n} & \kappa_n^4 S_{3n} & C_{0n} \end{bmatrix} \begin{bmatrix} X \\ X' \\ X'' \\ X''' \end{bmatrix}_{ln}$$

or $\mathbf{x}_{rn} = \mathbf{R}_n\mathbf{x}_{ln}$, where the boldface capital letter denotes a square matrix and the boldface lowercase letters denote column matrices. Matrix operations are discussed in Chap. 28.

At a section where two segments of a beam are joined, the deflection, the slope, the bending moment, and the shear must be the same on the two sides of the joint. Since $M = EI \cdot X''$ and $V = EI \cdot X'''$, the point transfer matrix for such a joint is as follows, where the subscript jn refers to the joint to the right of the nth segment of the beam.

$$\begin{bmatrix} X \\ X' \\ X'' \\ X''' \end{bmatrix}_{rjn} = \begin{bmatrix} 1 & 0 & 0 & 0 \\ 0 & 1 & 0 & 0 \\ 0 & 0 & (EI)_l/(EI)_r & 0 \\ 0 & 0 & 0 & (EI)_l/(EI)_r \end{bmatrix} \begin{bmatrix} X \\ X' \\ X'' \\ X''' \end{bmatrix}_{ljn}$$

or $\mathbf{x}_{rjn} = \mathbf{J}_n\mathbf{x}_{ljn}$.

The Frequency Equation. For the cantilever beam shown in Fig. 7.13, the coefficients relating the values of X and its derivatives at the right end of the beam to their values at the left end are found by successively multiplying the appropriate **R** and **J** matrices, as follows:

$$\mathbf{x}_{r3} = \mathbf{R}_3\mathbf{J}_2\mathbf{R}_2\mathbf{J}_1\mathbf{R}_1\mathbf{x}_{l1}$$

Carrying out the multiplication of the square **R** and **J** matrices, and calling the resulting matrix **P** yields

$$\begin{bmatrix} X \\ X' \\ X'' \\ X''' \end{bmatrix}_{r3} = \begin{bmatrix} P_{11} & P_{12} & P_{13} & P_{14} \\ P_{21} & P_{22} & P_{23} & P_{24} \\ P_{31} & P_{32} & P_{33} & P_{34} \\ P_{41} & P_{42} & P_{43} & P_{44} \end{bmatrix} \begin{bmatrix} X \\ X' \\ X'' \\ X''' \end{bmatrix}_{l1}$$

The boundary conditions at the fixed left end of the cantilever beam are $X = X' = 0$. Using these, and performing the multiplication of **P** by $\mathbf{x}_{l1}$ yields the following:

$$\begin{aligned} X_{r3} &= P_{13}X_{l1}'' + P_{14}X_{l1}''' \\ X_{r3}' &= P_{23}X_{l1}'' + P_{24}X_{l1}''' \\ X_{r3}'' &= P_{33}X_{l1}'' + P_{34}X_{l1}''' \\ X_{r3}''' &= P_{43}X_{l1}'' + P_{44}X_{l1}''' \end{aligned} \tag{7.44}$$

The boundary conditions for the free right end of the beam are $X'' = X''' = 0$. Using these in the last two equations results in two simultaneous homogeneous equations, so that the following determinant, which is the frequency equation, results.

$$\begin{vmatrix} P_{33} & P_{34} \\ P_{43} & P_{44} \end{vmatrix} = 0$$

It can be seen that for a beam consisting of only one segment, this determinant yields a result which is equivalent to Eq. (7.17).

While in theory it would be possible to multiply the successive **R** and **J** matrices and obtain the **P** matrix in literal form, so that the transcendental frequency equation could be written, the process, in all but the simplest problems, would be long and time-consuming. A more practicable procedure is to perform the necessary multiplications with numbers, using a digital computer, and finding the roots by trial and error.

Mode Shapes. Either of the last two equations of Eq. (7.44) may be used to find the ratio X_{l1}''/X_{l1}'''. These are used in Eq. (7.16), with $\kappa = \kappa_1$ to find the shape of the first segment. By the use of the **R** and **J** matrices the values of the coefficients in Eq. (7.16) are found for each of the other segments.

With intermediate rigid supports or pinned connections, numerical difficulties occur in the solution of the frequency equation. These difficulties are eliminated by the use of delta matrices, the elements of which are combinations of the elements of the **R** matrix. These delta matrices, for various cases, are tabulated in Refs. 43 and 44.

In Ref. 46 transfer matrices are developed and used for structures which consist, in part, of beams that are parallel to each other.

FORCED VIBRATION

CLASSICAL SOLUTION. The classical method of analyzing the forced vibration that results when an elastic system is subjected to a fluctuating load is to set up the equation of motion by the application of Newton's second law. During the vibration, each element of the system is subjected to elastic forces corresponding to those experienced during free vibration; in addition, some of the elements are subjected to the disturbing force. The equation which governs the forced vibration of a system can be obtained by adding the disturbing force to the equation for free vibration. For example, in Eq. (7.13) for the free vibration of a uniform beam, the term on the left is due to the elastic forces in the beam. If a force $F(x,t)$ is applied to the beam, the equation of motion is obtained by adding this force to Eq. (7.13), which becomes, after rearranging terms,

$$EI\frac{\partial^4 y}{\partial x^4} + \frac{\gamma S}{g}\frac{\partial^2 y}{\partial t^2} = F(x,t)$$

where EI is a constant. The solution of this equation gives the motion that results from the force F. For example, consider the motion of a beam with hinged ends subjected to a sinusoidally varying force acting at its center. The solution is obtained by representing the concentrated force at the center by its Fourier series:

$$EIy'''' + \frac{\gamma S}{g}\ddot{y} = \frac{2F}{l}\left[\sin\frac{\pi x}{l} - \sin\frac{3\pi x}{l} + \sin\frac{5\pi x}{l}\cdots\right]\sin\omega t$$

$$= \frac{2F}{l}\sum_{n=1}^{n=\infty}\left(\sin\frac{n\pi}{2}\sin\frac{n\pi x}{l}\right)\sin\omega t \tag{7.45}$$

where $\sin(n\pi/2)$, which appears in each term of the series, makes the nth term positive, negative, or zero. The solution of Eq. (7.45) is

$$y = \sum_{n=1}^{n=\infty}\left[A_n\sin\frac{n\pi x}{l}\sin\omega_n t + B_n\sin\frac{n\pi x}{l}\cos\omega_n t + \sin\frac{n\pi}{2}\frac{2Fg/S\gamma l}{(n\pi/l)^4(EIg/S\gamma) - \omega^2}\sin\frac{n\pi x}{l}\sin\omega t\right] \tag{7.46}$$

The first two terms of Eq. (7.46) are the values of y which make the left side of Eq. (7.45) equal to zero. They are obtained in exactly the same way as in the solution of the free-vibration problem and represent the free vibration of the beam. The constants are determined by the initial conditions; in any real beam, damping causes the free vibration to die out. The third term of Eq. (7.46) is the value of y which makes the left-hand side of Eq. (7.45) equal the right-hand side; this can be verified by substitution. The third term represents the forced vibration. From Table 7.3, $\kappa_n l = n\pi$ for a beam with hinged ends; then from Eq. (7.14), $\omega_n^2 = n^4\pi^4 EIg/S\gamma l^4$. The term representing the forced vibration in Eq. (7.46) can be written, after rearranging terms,

$$y = \frac{2Fg}{S\gamma l}\sum_{n=1}^{n=\infty}\frac{\sin(n\pi/2)}{\omega_n^2[1-(\omega/\omega_n)^2]}\sin\frac{n\pi x}{l}\sin\omega t \tag{7.47}$$

From Table 7.3 and Eq. (7.16), it is evident that this deflection curve has the same shape as the nth normal mode of vibration of the beam since, for free vibration of a beam with hinged ends, $X_n = 2C\sin\kappa x = \sin(n\pi x/l)$.

The equation for the deflection of a beam under a distributed static load $F(x)$ can be obtained by replacing $-(\gamma S/g)\ddot{y}$ with F in Eq. (7.12); then Eq. (7.13) becomes

$$y_s'''' = \frac{F(x)}{EI} \tag{7.48}$$

where EI is a constant. For a static loading $F(x) = 2F/l \sin \frac{n\pi}{2} \sin \frac{n\pi x}{l}$ corresponding to the nth term of the Fourier series in Eq. (7.45), Eq. (7.48) becomes $y_{sn}'''' = 2F/EIl \sin \frac{n\pi}{2} \sin \frac{n\pi x}{l}$. The solution of this equation is

$$y_{sn} = \frac{2F}{EIl}\left(\frac{l}{n\pi}\right)^4 \sin \frac{n\pi}{2} \sin \frac{n\pi x}{l}$$

Using the relation $\omega_n^2 = n^4\pi^4EIg/S\gamma l^4$, this can be written

$$y_{sn} = \frac{2Fg}{\omega_n^2 S\gamma l} \sin \frac{n\pi x}{l} \sin \frac{n\pi}{2}$$

Thus, the nth term of Eq. (7.47) can be written

$$y_n = y_{sn} \frac{1}{1 - (\omega/\omega_n)^2} \sin \omega t$$

Thus, the amplitude of the forced vibration is equal to the static deflection under the Fourier component of the load multiplied by the "amplification factor" $1/[1 - (\omega/\omega_n)^2]$. This is the same as the relation that exists, for a system having a single degree-of-freedom, between the static deflection under a load F and the amplitude under a fluctuating load $F \sin \omega t$. Therefore, in so far as each mode alone is concerned, the beam behaves as a system having a single degree-of-freedom. If the beam is subjected to a force fluctuating at a single frequency, the amplification factor is small except when the frequency of the forcing force is near the natural frequency of a mode. For all even values of n, $\sin (n\pi/2) = 0$; thus, the even-numbered modes are not excited by a force acting at the center, which is a node for those modes. The distribution of the static load that causes the same pattern of deflection as the beam assumes during each mode of vibration has the same form as the deflection of the beam. This result applies to other beams since a comparison of Eqs. (7.15) and (7.48) shows that if a static load $F = (\omega_n^2\gamma S/g)y$ is applied to any beam it will cause the same deflection as occurs during the free vibration in the nth mode.

The results for the simply supported beam are typical of those which are obtained for all systems having distributed mass and elasticity. Vibration of such a system at resonance is excited by a force which fluctuates at the natural frequency of a mode, since nearly any such force has a component of the shape necessary to excite the vibration. Even if the force acts at a nodal point of the mode, vibration may be excited because of coupling between the modes.

METHOD OF WORK. Another method of analyzing forced vibration is by the use of the theorem of virtual work and D'Alembert's principle. The theorem of virtual work states that when any elastic body is in equilibrium, the total work done by all external forces during any virtual displacement equals the increase in the elastic energy stored in the body. A virtual displacement is an arbitrary small displacement that is compatible with the geometry of the body, and which satisfies the boundary conditions.

In applying the principle of work to forced vibration of elastic bodies, the problem is made into one of equilibrium by the application of D'Alembert's principle. This permits a problem in dynamics to be considered as one of statics by adding to the equation of static equilibrium an "inertia force" which, for each part of the body, is equal to the product of the mass and the acceleration. Using this principle, the theorem of virtual work can be expressed in the following equation:

$$\Delta V = \Delta(F_I + F_E) \tag{7.49}$$

in which V is the elastic strain energy in the body, F_I is the inertia force, F_E is the external disturbing force, and Δ indicates the change of the quantity when the body undergoes a virtual displacement. The various quantities can be found separately.

For example, consider the motion of a uniform beam having hinged ends with a sinusoidally varying force acting at the center, and compare the result with the solution obtained by the classical method. All possible motions of any beam can be represented by a series of the form

$$y = q_1X_1 + q_2X_2 + q_3X_3 + \cdots = \sum_{n=1}^{n=\infty} q_nX_n \tag{7.50}$$

in which the X's are functions representing displacements in the normal modes of vibration and the q's are coefficients which are functions of time. The determination of the values of q_n is the problem to be solved. For a beam having hinged ends, Eq. (7.50) becomes

$$y = \sum_{n=1}^{n=\infty} q_n \sin \frac{n\pi x}{l} \tag{7.51}$$

This is evident by using the values of $\kappa_n l$ from Table 7.3 in Eq. (7.16). A virtual displacement, being any arbitrary small displacement, can be assumed to be

$$\Delta y = \Delta q_m X_m = \Delta q_m \sin \frac{m\pi x}{l}$$

The elastic strain energy of bending of the beam is

$$V = \frac{EI}{2} \int_0^l \left(\frac{\partial^2 y}{\partial x^2}\right)^2 dx = \frac{EI}{2} \sum_{n=1}^{n=\infty} q_n{}^2 \int_0^l \left[\frac{\partial^2}{\partial x^2}\left(\sin \frac{n\pi x}{l}\right)\right]^2 dx$$

$$= \frac{EI}{2} \sum_{n=1}^{n=\infty} q_n{}^2 \left(\frac{n\pi}{l}\right)^4 \int_0^l \left(\sin \frac{n\pi x}{l}\right)^2 dx = \frac{EI}{2} \sum_{n=1}^{n=\infty} q_n{}^2 \left(\frac{n\pi}{l}\right)^4 \frac{l}{2}$$

For the virtual displacement, the change of elastic energy is

$$\Delta V = \frac{\partial V}{\partial q_m} \Delta q_m = \frac{EI}{2l^3} (n\pi)^4 q_m \, \Delta q_m = \frac{EI}{2l^3} (\kappa_n l)^4 q_m \, \Delta q_m$$

The value of the "inertia force" at each section is

$$F_I = -\frac{\gamma S}{g} \ddot{y} = -\frac{\gamma S}{g} \sum_{n=1}^{n=\infty} \frac{d^2 q_n}{dt^2} \sin \frac{n\pi x}{l}$$

The work done by this force during the virtual displacement Δy is

$$\Delta F_I = F_I \, \Delta y = -\frac{\gamma S}{g} \sum_{n=1}^{n=\infty} \frac{d^2 q_n}{dt^2} \Delta q_m \int_0^l \sin \frac{n\pi x}{l} \sin \frac{m\pi x}{l} \, dx$$

$$= -\frac{\gamma S l}{2g} \frac{d^2 q_m}{dt^2} \Delta q_m$$

The orthogonality relation of Eq. (7.1) is used here, making the integral vanish when $n = m$. For a disturbing force F_E, the work done during the virtual displacement is

$$\Delta F_E = F_E \, \Delta y = F(X_m)_{x=c} \, \Delta q_m$$

in which $(X_m)_{x=c}$ is the value of X_m at the point of application of the load. Substituting the terms into Eq. (7.49),

$$\frac{\gamma S l}{2g} \ddot{q}_m + \frac{EI}{2l^3} (\kappa_m l)^4 q_m = F(X_m)_{x=c}$$

Rearranging terms and letting $EI/S\gamma = a^2$,

$$\ddot{q}_m + \kappa_m{}^4 a^2 q_m = \frac{2g}{\gamma S l} F(X_m)_{x=c} \tag{7.52}$$

If F_E is a force which varies sinusoidally with time at any point $x = c$,

$$F(X_m)_{x=c} = \bar{F} \sin \frac{m\pi c}{l} \sin \omega t$$

and Eq. (7.52) becomes

$$\ddot{q}_m + \kappa_m{}^4 a^2 q_m = \frac{2g\bar{F}}{\gamma Sl} \sin \frac{m\pi c}{l} \sin \omega t$$

The solution of this equation is

$$q_m = A_m \sin \kappa_m{}^2 at + B_m \cos \kappa_m{}^2 at + \frac{2\bar{F}g}{\gamma Sl} \frac{\sin (m\pi c/l)}{\kappa_m{}^4 a^2 - \omega^2} \sin \omega t$$

Since $\kappa_m{}^2 a = \omega_m$,

$$q_m = A_m \sin \omega_m t + B_m \cos \omega_m t + \frac{2\bar{F}g}{\gamma Sl} \frac{\sin (m\pi c/l)}{\omega_m{}^2 - \omega^2} \sin \omega t$$

when the force acts at the center $c/l = \frac{1}{2}$. Substituting the corresponding values of q in Eq. (7.51), the solution is identical to Eq. (7.46), which was obtained by the classical method.

VIBRATION RESULTING FROM MOTION OF SUPPORT. When the supports of an elastic body are vibrated by some external force, forced vibration may be induced in the body.[47] For example, consider the motion that results in a uniform beam, Fig. 7.14,

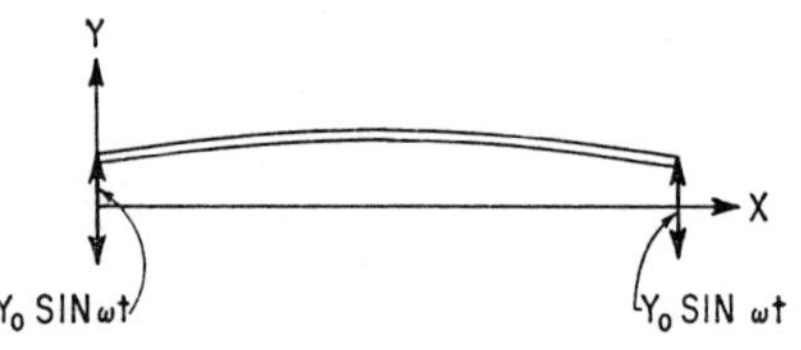

FIG. 7.14. Simply supported beam undergoing sinusoidal motion induced by sinusoidal motion of the supports.

when the supports are moved through a sinusoidally varying displacement $(y)_{x=0,l} = Y_0 \sin \omega t$. Although Eq. (7.13) was developed for the free vibration of beams, it is applicable to the present problem because there is no force acting on any section of the beam except the elastic force associated with the bending of the beam. If a solution of the form $y = X(x) \sin \omega t$ is assumed and substituted into Eq. (7.13):

$$X'''' = \frac{\omega^2 \gamma S}{EIg} X \tag{7.53}$$

This equation is the same as Eq. (7.15) except that the natural frequency $\omega_n{}^2$ is replaced by the forcing frequency ω^2. The solution of Eq. (7.53) is the same except that κ is replaced by $\kappa' = (\omega^2 \gamma S/EIg)^{1/4}$:

$$X = A_1 \sin \kappa' x + A_2 \cos \kappa' x + A_3 \sinh \kappa' x + A_4 \cosh \kappa' x \tag{7.54}$$

The solution of the problem is completed by finding the constants, which are determined by the boundary conditions. Certain boundary conditions are associated with the supports of the beam and are the same as occur in the solution of the problem of free vibration. Additional conditions are supplied by the displacement through which the supports are forced. For example, if the supports of a beam having hinged ends are moved sinusoidally, the boundary conditions are: at $x = 0$ and $x = l$, $X'' = 0$, since the moment exerted by a hinged end is zero, and $X = Y_0$, since the amplitude of vibration is prescribed at each end. By the use of these boundary conditions, Eq. (7.54) becomes

$$X = \frac{Y_0}{2} \left[\tan \frac{\kappa' l}{2} \sin \kappa' x + \cos \kappa' x - \tanh \frac{\kappa' l}{2} \sinh \kappa' x + \cosh \kappa' x \right] \tag{7.55}$$

The motion is defined by $y = X \sin \omega t$. For all values of κ', each of the coefficients except the first in Eq. (7.55) is finite. The tangent term becomes infinite if $\kappa' l = n\pi$, for odd values of n. The condition for the amplitude to become infinite is $\omega = \omega_n$ because $\kappa'/\kappa = \omega^2/\omega_n^2$ and, for natural vibration of a beam with hinged ends, $\kappa_n l = n\pi$. Thus, if the supports of an elastic body are vibrated at a frequency close to a natural frequency of the system, vibration at resonance occurs.

DAMPING. The effect of damping on forced vibration can be discussed only qualitatively. Damping usually decreases the amplitude of vibration, as it does in systems having a single degree-of-freedom. In some systems, it may cause coupling between modes, so that motion in a mode of vibration that normally would not be excited by a certain disturbing force may be induced.

REFERENCES

1. Timoshenko, S.: "Vibration Problems in Engineering," 3d ed., pp. 442, 448, D. Van Nostrand Company, Inc., Princeton, N.J., 1955.
2. Den Hartog, J. P.: "Mechanical Vibrations," 4th ed., p. 161, McGraw-Hill Book Company, Inc., New York, 1956.
3. Ref. 2, p. 152.
4. Hansen, H. M., and P. F. Chenea: "Mechanics of Vibration," p. 274, John Wiley & Sons, Inc., New York, 1952.
5. Jacobsen, L. S., and R. S. Ayre: "Engineering Vibrations," p. 73, McGraw-Hill Book Company, Inc., New York, 1958.
6. Ref. 4, p. 256.
7. Kerley, J. J.: *Prod. Eng., Design Digest Issue*, Mid-October, 1957, p. F34.
8. Young, D., and R. P. Felgar: "Tables of Characteristic Functions Representing Normal Modes of Vibration of a Beam," *Univ. Texas Bur. Eng. Research Bull.* 44, July 1, 1949.
9. Rayleigh, Lord: "The Theory of Sound," 2d rev. ed., vol. 1, p. 293; reprinted by Dover Publications, New York, 1945.
10. Timoshenko, S.: *Phil. Mag.* (ser. 6), **41**:744 (1921); **43**:125 (1922).
11. Sutherland, J. G., and L. E. Goodman: "Vibrations of Prismatic Bars Including Rotatory Inertia and Shear Corrections," Department of Civil Engineering, University of Illinois, Urbana, Ill., April 15, 1951.
12. Ref. 1, p. 374.
13. Ref. 1, 2d ed., p. 366.
14. Woinowsky-Krieger, S.: *J. Appl. Mechanics*, **17**:35 (1950).
15. Cranch, E. T., and A. A. Adler: *J. Appl. Mechanics*, **23**:103 (1956).
16. Macduff, J. N., and R. P. Felgar: *Trans. ASME*, **79**:1459 (1957).
17. Macduff, J. N., and R. P. Felgar: *Machine Design*, **29**(3):109 (1957).
18. Ref. 1, p. 386.
19. Darnley, E. R.: *Phil. Mag.*, **41**:81 (1921).
20. Smith, D. M.: *Engineering*, **120**:808 (1925).
21. Newmark, N. M., and A. S. Veletsos: *J. Appl. Mechanics*, **19**:563 (1952).
22. Timoshenko, S.: "Theory of Plates and Shells," p. 301, McGraw-Hill Book Company, Inc., New York, 1940.
23. Ref. 22, p. 88.
24. Ref. 22, p. 133.
25. Evans, T. H.: *J. Appl. Mechanics*, **6**:A-7 (1939).
26. Ref. 22, p. 58.
27. Ref. 22, p. 304.
28. Ref. 22, p. 344.

28a. Leissa, A. W.: "Vibration of Plates," NASA SP-160, 1969.

29. Young, D.: *J. Appl. Mechanics*, **17**:448 (1950).
30. Barton, M. V.: *J. Appl. Mechanics*, **18**:129 (1951).
31. Hearmon, R. F. S.: *J. Appl. Mechanics*, **19**:402 (1952).
32. Anderson, B. W.: *J. Appl. Mechanics*, **21**:365 (1954).
33. Gustafson, P. N., W. F. Stokey, and C. F. Zorowski: *J. Aeronaut. Sci.*, **20**:331 (1953).
34. Gustafson, P. N., W. F. Stokey, and C. F. Zorowski: *J. Aeronaut. Sci.*, **21**:621 (1954).
35. Ref. 9, p. 359.
36. Ref. 1, p. 449.
37. Southwell, R. V.: *Proc. Roy. Soc. (London)*, **A101**:133 (1922).
38. Chu, Hu-Nan, and G. Herrmann: *J. Appl. Mechanics*, **23**:532 (1956).

39. Mindlin, R. D., A. Schacknow, and H. Deresiewicz: *J. Appl. Mechanics*, **23**:430 (1956).
40. Deresiewicz, H., and R. D. Mindlin: *J. Appl. Mechanics*, **22**:86 (1955).
41. Love, A. E. H.: "A Treatise on the Mathematical Theory of Elasticity," 4th ed., p. 451, reprinted by Dover Publications, New York, 1944.
42. Ref. 1, p. 425.
43. Marguerre, K.: *J. Math. Phys.*, **35**:28 (1956).
44. Pestel, E. C., and F. A. Leckie: "Matrix Methods in Elastomechanics," McGraw-Hill Book Company, Inc., New York, 1963.
45. Myklestad, N. O.: *J. Aeronaut. Sci.*, **11**:153 (1944).
46. Bohm, G. J.: *Nucl. Sci. Eng.*, **22**:143 (1965).
47. Mindlin, R. D., and L. E. Goodman: *J. Appl. Mech.*, **17**:377 (1950).

8

TRANSIENT RESPONSE TO STEP AND PULSE FUNCTIONS *

Robert S. Ayre
University of Colorado

INTRODUCTION

In analyses involving shock and transient vibration, it is essential in most instances to begin with the time-history of a quantity that describes a motion, usually displacement, velocity, or acceleration. The method of reducing the time-history depends upon the purpose for which the reduced data will be used. When the purpose is to compare shock motions, to design equipment to withstand shock, or to formulate a laboratory test as means to simulate an environmental condition, the *response spectrum* is found to be a useful concept. This concept in data reduction is discussed in Chap. 23, and its application to the simulation of environmental conditions is discussed in Chap. 24.

This chapter deals briefly with methods of analysis for obtaining the response spectrum from the time-history, and includes in graphical form certain significant spectra for various regular step- and pulse-type excitations. The usual concept of the response spectrum is based upon the single degree-of-freedom system, usually considered linear and undamped, although useful information sometimes can be obtained by introducing nonlinearity or damping. The single degree-of-freedom system is considered to be subjected to the shock or transient vibration, and its response determined.

The *response spectrum* is a graphical presentation of a selected quantity in the response taken with reference to a quantity in the excitation. It is plotted as a function of a dimensionless parameter that includes the natural period of the responding system and a significant period of the excitation. The excitation may be defined in terms of various physical quantities, and the response spectrum likewise may depict various characteristics of the response.

LINEAR, UNDAMPED, SINGLE DEGREE-OF-FREEDOM SYSTEMS

DIFFERENTIAL EQUATION OF MOTION

It is assumed that the system is linear and undamped. The excitation, which is a known function of time alone, may be a force function $F(t)$ acting directly on the mass of the system (Fig. 8.1*A*) or it may be a ground motion, i.e., foundation or base motion, acting on the spring anchorage. The ground motion may be expressed as a ground dis-

* Chapter 8 is based on Chaps. 3 and 4 of "Engineering Vibrations," by L. S. Jacobsen and R. S. Ayre, McGraw-Hill Book Company, Inc., 1958, and on the report, "A Comparative Study of Pulse and Step-type Loads on a Simple Vibratory System," by L. S. Jacobsen and R. S. Ayre, U.S. Navy Contract N6-ori-154, T. O. 1, Stanford University, 1952.

placement function $u(t)$ (Fig. 8.1*B*). In many cases, however, it is more useful to express it as a ground acceleration function $\ddot{u}(t)$ (Fig. 8.1*C*).

The differential equation of motion, written in terms of each of the types of excitation, is given in Eqs. (8.1*a*), (8.1*b*), and (8.1*c*).

$$m\ddot{x} = -kx + F(t) \quad \text{or} \quad \frac{m\ddot{x}}{k} + x = \frac{F(t)}{k} \tag{8.1a}$$

$$m\ddot{x} = -k[x - u(t)] \quad \text{or} \quad \frac{m\ddot{x}}{k} + x = u(t) \tag{8.1b}$$

$$m[\ddot{\delta}_x + \ddot{u}(t)] = -k\delta_x \quad \text{or} \quad \frac{m\ddot{\delta}_x}{k} + \delta_x = -\frac{m\ddot{u}(t)}{k} \tag{8.1c}$$

where x is the displacement (absolute displacement) of the mass relative to a *fixed reference* and δ_x is the displacement relative to a *moving anchorage* or ground. These displacements are related to the ground displacement by $x = u + \delta_x$. Similarly, the accelerations are related by $\ddot{x} = \ddot{u} + \ddot{\delta}_x$.

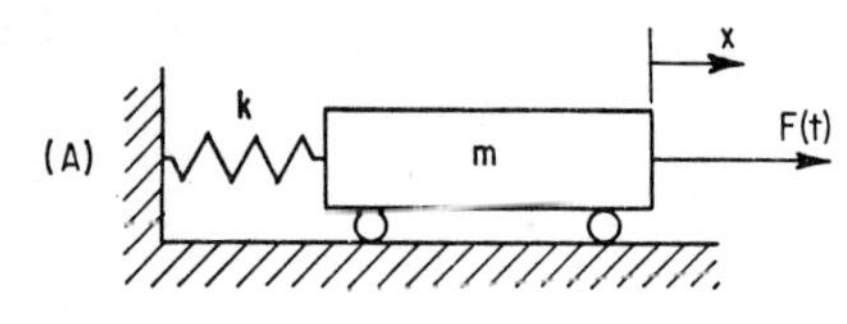

Furthermore, if Eq. (8.1*b*) is differentiated twice with respect to time, a differential equation is obtained in which ground acceleration $\ddot{u}(t)$ is the excitation and the absolute acceleration $\ddot{x}$ of the mass m is the variable. The equation is

$$\frac{m}{k}\frac{d^2\ddot{x}}{dt^2} + \ddot{x} = \ddot{u}(t) \tag{8.1d}$$

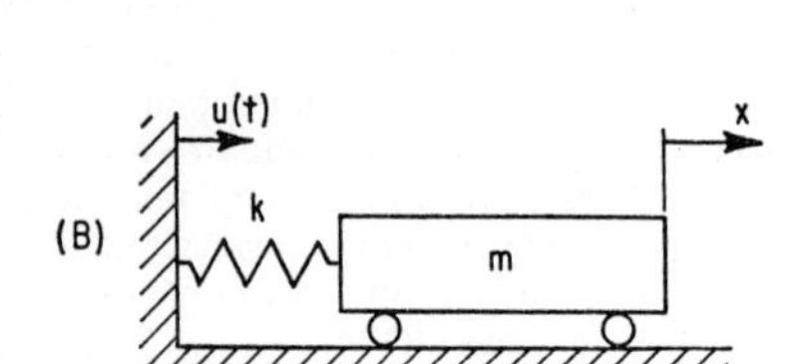

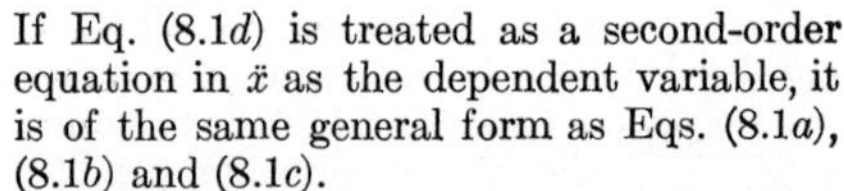

If Eq. (8.1*d*) is treated as a second-order equation in $\ddot{x}$ as the dependent variable, it is of the same general form as Eqs. (8.1*a*), (8.1*b*) and (8.1*c*).

Occasionally, the excitation is known in terms of ground velocity $\dot{u}(t)$. Differentiating Eq. (8.1*b*) once with respect to time, the following second-order equation in $\dot{x}$ is obtained:

$$\frac{m}{k}\frac{d^2\dot{x}}{dt^2} + \dot{x} = \dot{u}(t) \tag{8.1e}$$

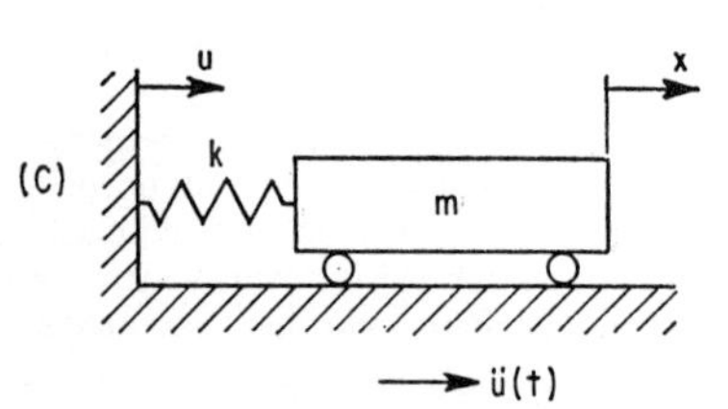

FIG. 8.1. Simple oscillator acted upon by known excitation functions of time: (*A*) force $F(t)$, (*B*) ground displacement $u(t)$, (*C*) ground acceleration $\ddot{u}(t)$.

The analogy represented by Eqs. (8.1*b*), (8.1*d*), and (8.1*e*) may be extended further since it is generally possible to differentiate Eq. (8.1*b*) any number of times n:

$$\frac{m}{k}\frac{d^2}{dt^2}\left(\frac{d^n x}{dt^n}\right) + \left(\frac{d^n x}{dt^n}\right) = \left(\frac{d^n u}{dt^n}\right)(t) \tag{8.1f}$$

This is of the same general form as the preceding equations if it is considered to be a second-order equation in $(d^n x/dt^n)$ as the response variable, with $(d^n u/dt^n)$ (t), a known function of time, as the excitation.

ALTERNATE FORMS OF THE EXCITATION AND OF THE RESPONSE

The foregoing equations are alike, mathematically, and a solution in terms of one of them may be applied to any of the others by making simple substitutions. Therefore, the equations may be expressed in the single general form:

$$\frac{m}{k}\ddot{\nu} + \nu = \xi(t) \tag{8.2}$$

where ν and ξ are the *response* and the *excitation*, respectively, at time t.

A general notation (ν and ξ) is desirable in the presentation of response functions and response spectra for general use. However, in the discussion of examples of solution, it sometimes is preferable to use more specific notations. Both types of notation are used in this chapter. For ready reference, the alternate forms of the excitation and the response are given in Table 8.1 where $\omega_n^2 = k/m$.

Table 8.1. Alternate Forms of Excitation and Response in Eq. (8.2)

Excitation $\xi(t)$		Response ν	
Force	$\frac{F(t)}{k}$	Absolute displacement	x
Ground displacement	$u(t)$	Absolute displacement	x
Ground acceleration	$\frac{-\ddot{u}(t)}{\omega_n^2}$	Relative displacement	δ_x
Ground acceleration	$\ddot{u}(t)$	Absolute acceleration	$\ddot{x}$
Ground velocity	$\dot{u}(t)$	Absolute velocity	$\dot{x}$
nth derivative of ground displacement	$\frac{d^n u}{dt^n}(t)$	nth derivative of absolute displacement	$\frac{d^n x}{dt^n}$

METHODS OF SOLUTION OF THE DIFFERENTIAL EQUATION

A brief review of four methods of solution is given in the following sections.

CLASSICAL SOLUTION. The complete solution of the linear differential equation of motion consists of the sum of the *particular integral* x_1 and the *complementary function* x_2, that is, $x = x_1 + x_2$. Since the differential equation is of second order, two constants of integration are involved. They appear in the complementary function and are evaluated from a knowledge of the initial conditions.

Example 8.1. Versed-sine Force Pulse. In this case the differential equation of motion, applicable for the duration of the pulse, is

$$\frac{m\ddot{x}}{k} + x = \frac{F_p}{k}\frac{1}{2}\left(1 - \cos\frac{2\pi t}{\tau}\right) \qquad [0 \leq t \leq \tau] \tag{8.3a}$$

where, in terms of the general notation, the excitation function $\xi(t)$ is

$$\xi(t) \equiv \frac{F(t)}{k} = \frac{F_p}{k}\frac{1}{2}\left(1 - \cos\frac{2\pi t}{\tau}\right)$$

and the response ν is displacement x. The maximum value of the pulse excitation force is F_p.

The particular integral (particular solution) for Eq. (8.3*a*) is of the form

$$x_1 = M + N\cos\frac{2\pi t}{\tau} \tag{8.3b}$$

By substitution of the particular solution into the differential equation, the required values of the coefficients M and N are found.

The complementary function is

$$x_2 = A\cos\omega_n t + B\sin\omega_n t \tag{8.3c}$$

where A and B are the constants of integration. Combining x_2 and the explicit form of x_1 gives the complete solution:

$$x = x_1 + x_2 = \frac{F_p/2k}{1 - \tau^2/T^2}\left(1 - \frac{\tau^2}{T^2} + \frac{\tau^2}{T^2}\cos\frac{2\pi t}{\tau}\right) + A\cos\omega_n t + B\sin\omega_n t \quad (8.3d)$$

If it is assumed that the system is initially at rest, $x = 0$ and $\dot{x} = 0$ at $t = 0$, and the constants of integration are

$$A = -\frac{F_p/2k}{1 - \tau^2/T^2} \quad \text{and} \quad B = 0 \quad (8.3e)$$

The complete solution takes the following form:

$$\nu \equiv x = \frac{F_p/2k}{1 - \tau^2/T^2}\left(1 - \frac{\tau^2}{T^2} + \frac{\tau^2}{T^2}\cos\frac{2\pi t}{\tau} - \cos\omega_n t\right) \quad (8.3f)$$

If other starting conditions had been assumed, A and B would have been different from the values given by Eqs. (8.3e). It may be shown that if the starting conditions are general, namely, $x = x_0$ and $\dot{x} = \dot{x}_0$ at $t = 0$, it is necessary to superimpose on the complete solution already found, Eq. (8.3f), only the following additional terms:

$$x_0\cos\omega_n t + \frac{\dot{x}_0}{\omega_n}\sin\omega_n t \quad (8.3g)$$

For values of time equal to or greater than τ, the differential equation is

$$m\ddot{x} + kx = 0 \qquad [\tau \le t] \quad (8.4a)$$

and the complete solution is given by the complementary function alone. However, the constants of integration must be redetermined from the known conditions of the system at time $t = \tau$. The solution is

$$\nu \equiv x = \frac{F_p}{k}\frac{\sin(\pi\tau/T)}{1 - \tau^2/T^2}\sin\omega_n\left(t - \frac{\tau}{2}\right) \qquad [\tau \le t] \quad (8.4b)$$

The additional terms given by expressions (8.3g) may be superimposed on this solution if the conditions at time $t = 0$ are general.

DUHAMEL'S INTEGRAL. The use of Duhamel's integral (convolution integral or superposition integral) is a well-known approach to the solution of transient vibration problems in linear systems. Its development [7, 33] is based on the *superposition* of the responses of the system to a sequence of impulses.

A general excitation function is shown in Fig. 8.2, where $F(t)$ is a known force function of time, the variable of integration is t_v between the limits of integration 0 and t and the elemental impulse is $F(t_v)\,dt_v$. It may be shown that the complete solution of the differential equation is

$$x = \left(x_0 - \frac{1}{m\omega_n}\int_0^t F(t_v)\sin\omega_n t_v\,dt_v\right)\cos\omega_n t + \left(\frac{\dot{x}_0}{\omega_n} + \frac{1}{m\omega_n}\int_0^t F(t_v)\cos\omega_n t_v\,dt_v\right)\sin\omega_n t \quad (8.5)$$

where x_0 and $\dot{x}_0$ are the initial conditions of the system at zero time.

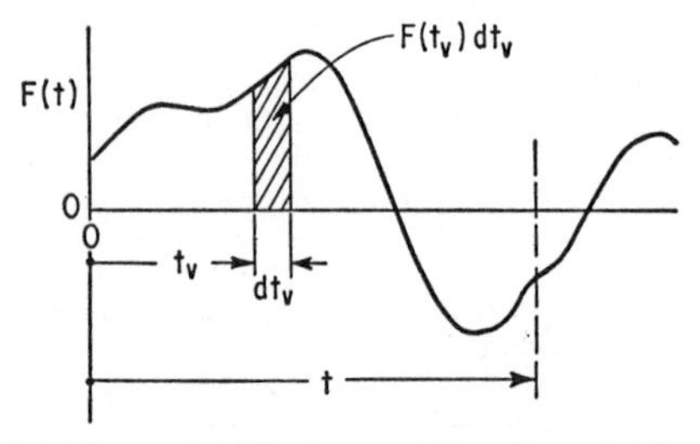

FIG. 8.2. General excitation and the elemental impulse.

Example 8.2. Half-cycle Sine, Ground Displacement Pulse. Consider the following excitation:

$$\xi(t) \equiv u(t) = \begin{cases} u_p \sin \dfrac{\pi t}{\tau} & [0 \leq t \leq \tau] \\ 0 & [\tau \leq t] \end{cases}$$

The maximum value of the excitation displacement is u_p. Assume that the system is initially at rest, so that $x_0 = \dot{x}_0 = 0$. Expressing the excitation function in terms of the variable of integration t_v, Eq. (8.5) may be rewritten for this particular case in the following form:

$$x = \frac{ku_p}{m\omega_n}\left(-\cos\omega_n t \int_0^t \sin\frac{\pi t_v}{\tau}\sin\omega_n t_v\,dt_v + \sin\omega_n t\int_0^t \sin\frac{\pi t_v}{\tau}\cos\omega_n t_v\,dt_v\right) \tag{8.6a}$$

Equation (8.6a) may be reduced, by evaluation of the integrals, to

$$\nu \equiv x = \frac{u_p}{1 - T^2/4\tau^2}\left(\sin\frac{\pi t}{\tau} - \frac{T}{2\tau}\sin\omega_n t\right) \qquad [0 \leq t \leq \tau] \tag{8.6b}$$

where $T = 2\pi/\omega_n$ is the natural period of the responding system.

For the second era of time, where $\tau \leq t$, it is convenient to choose a new time variable $t' = t - \tau$. Noting that $u(t) = 0$ for $\tau \leq t$, and that for continuity in the system response the initial conditions for the second era must equal the closing conditions for the first era, it is found from Eq. (8.5) that the response for the second era is

$$x = x_\tau \cos\omega_n t' + \frac{\dot{x}_\tau}{\omega_n}\sin\omega_n t' \tag{8.7a}$$

where x_τ and $\dot{x}_\tau$ are the displacement and velocity of the system at time $t = \tau$ and hence at $t' = 0$. Equation (8.7a) may be rewritten in the following form:

$$\nu \equiv x = u_p \frac{(T/\tau)\cos(\pi\tau/T)}{(T^2/4\tau^2) - 1}\sin\omega_n\left(t - \frac{\tau}{2}\right) \qquad [\tau \leq t] \tag{8.7b}$$

PHASE-PLANE GRAPHICAL METHOD. Several numerical and graphical methods,[18, 23, 27, 37] all related in general but differing considerably in the details of procedure, are available for the solution of linear transient vibration problems. Of these methods, the phase-plane graphical method[25] is one of the most useful. The procedure is basically

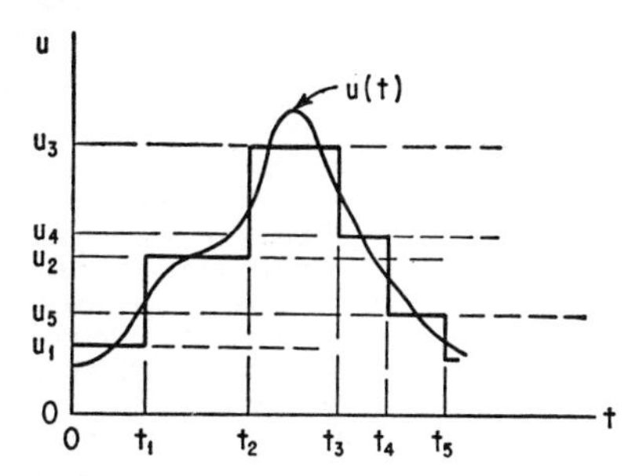

FIG. 8.3. General excitation approximated by a sequence of finite rectangular steps.

very simple, it gives a clear physical picture of the response of the system, and it may be applied readily to some classes of *nonlinear* systems.[3, 5, 6, 8, 13, 15, 21, 22]

In Fig. 8.3 a general excitation in terms of ground displacement is represented, approximately, by a sequence of finite steps. The ith step has the total height u_i, where u_i is constant for the duration of the step. The differential equation of motion and its complete solution, applying for the duration of the step, are

$$\frac{m\ddot{x}}{k} + x = u_i \qquad [t_{i-1} \leq t \leq t_i] \tag{8.8a}$$

$$x - u_i = (x_{i-1} - u_i)\cos\omega_n(t - t_{i-1}) + \frac{\dot{x}_{i-1}}{\omega_n}\sin\omega_n(t - t_{i-1}) \tag{8.8b}$$

where x_{i-1} and $\dot{x}_{i-1}$ are the displacement and velocity of the system at time t_{i-1}; consequently, they are the initial conditions for the ith step. The system velocity (divided by ω_n) during the ith step is

$$\frac{\dot{x}}{\omega_n} = -(x_{i-1} - u_i)\sin\omega_n(t - t_{i-1}) + \frac{\dot{x}_{i-1}}{\omega_n}\cos\omega_n(t - t_{i-1}) \tag{8.8c}$$

Squaring Eqs. (8.8*b*) and (8.8*c*) and adding them,

$$\left(\frac{\dot{x}}{\omega_n}\right)^2 + (x - u_i)^2 = \left(\frac{\dot{x}_{i-1}}{\omega_n}\right)^2 + (x_{i-1} - u_i)^2 \tag{8.8d}$$

This is the equation of a circle in a rectangular system of coordinates $\dot{x}/\omega_n$, x. The center is at 0, u_i; and the radius is

$$R_i = \left[\left(\frac{\dot{x}_{i-1}}{\omega_n}\right)^2 + (x_{i-1} - u_i)^2\right]^{1/2} \tag{8.8e}$$

The solution for Eq. (8.8*a*) for the ith step may be shown, as in Fig. 8.4, to be the arc of the circle of radius R_i and center 0, u_i, subtended by the angle $\omega_n(t_i - t_{i-1})$ and starting at the point $\dot{x}_{i-1}/\omega_n$, x_{i-1}. Time is positive in the counterclockwise direction.

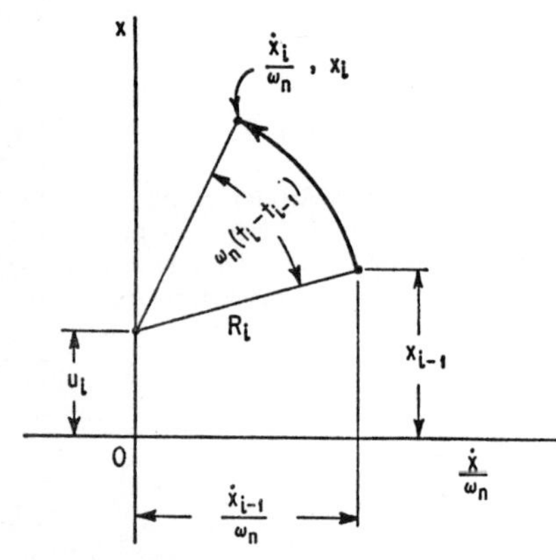

Fig. 8.4. Graphical representation in the phase-plane of the solution for the ith step.

Example 8.3. Application to a General Pulse Excitation. Figure 8.5 shows an application of the method for the general excitation $u(t)$ represented by seven steps in the time-displacement plane. Upon choice of the step heights u_i and durations $(t_i - t_{i-1})$, the arc-center locations can be projected onto the X axis in the phase-plane and the arc angles $\omega_n(t_i - t_{i-1})$ can be computed. The graphical construction of the sequence of circular arcs, the *phase trajectory*, is then carried out, using the system conditions at zero time (in this example, 0,0) as the starting point.

Projection of the system displacements from the phase-plane into the time-displacement plane at once determines the time-displacement response curve. The time-velocity response can also be determined by projection as shown. The velocities and displacements at particular instants of time can be found directly from the phase trajectory coordinates without the necessity for drawing the time-response curves. Furthermore, the times of occurrence and the magnitudes of all the maxima also can be obtained directly from the phase trajectory.

Good accuracy is obtainable by using reasonable care in the graphical construction and in the choice of the steps representing the excitation. Usually, the time intervals should not be longer than about one-fourth the natural period of the system.[22]

THE LAPLACE TRANSFORMATION. The Laplace transformation provides a powerful tool for the solution of linear differential equations. The following discussion of the technique of its application is limited to the differential equation of the type applying to the undamped linear oscillator. Application to the linear oscillator with viscous damping is illustrated in a later part of this chapter.

Definitions. The Laplace transform $F(s)$ of a known function $f(t)$, where $t > 0$, is defined by

$$F(s) = \int_0^\infty e^{-st} f(t)\,dt \tag{8.9a}$$

where s is a complex variable. The transformation is abbreviated as

$$F(s) = \mathcal{L}[f(t)] \tag{8.9b}$$

The limitations on the function $f(t)$ are not discussed here. For the conditions for existence of $\mathcal{L}[f(t)]$, for complete accounts of the technique of application, and for extensive tables of function-transform pairs, the references should be consulted.[16, 17, 34, 36]

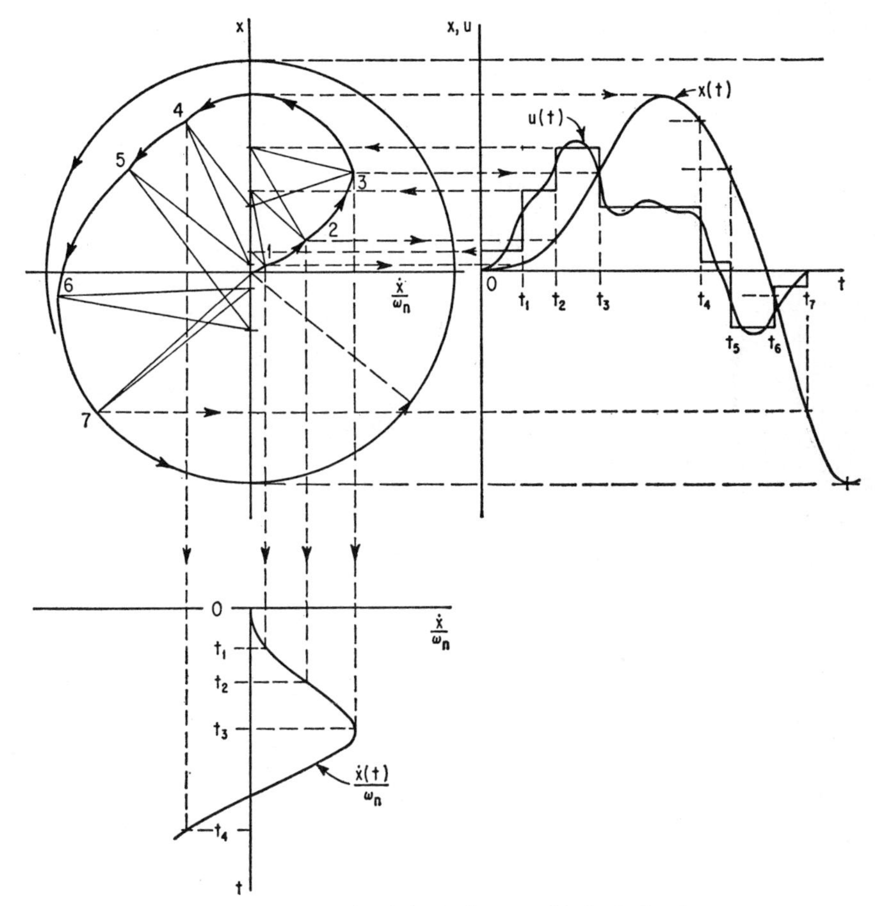

FIG. 8.5. Example of phase-plane graphical solution.[2]

General Steps in Solution of the Differential Equation. In the solution of a differential equation by Laplace transformation, the first step is to transform the differential equation, in the variable t, into an algebraic equation in the complex variable s. Then, the algebraic equation is solved, and the solution of the differential equation is determined by an *inverse* transformation of the solution of the algebraic equation. The process of inverse Laplace transformation is symbolized by

$$\mathcal{L}^{-1}[F(s)] = f(t) \tag{8.10}$$

Tables of Function-transform Pairs. The processes symbolized by Eqs. (8.9b) and (8.10) are facilitated by the use of tables of function-transform pairs. Table 8.2 is a brief

Table 8.2. Pairs of Functions $f(t)$ and Laplace Transforms $F(s)$

	$f(t)$	$F(s)$
	Operation Transforms	
1	Definition, $f(t)$	$F(s) = \int_0^\infty e^{-st} f(t)\, dt$
2	First derivative, $f'(t)$	$sF(s) - f(0)$
3	nth derivative, $f^{(n)}(t)$	$s^n F(s) - s^{n-1} f(0) - s^{n-2} f'(0) - \cdots - s f^{(n-2)}(0) - f^{(n-1)}(0)$ †
4	Superposition, $C_1 f_1(t) + C_2 f_2(t) + \cdots + C_n f_n(t)$	$C_1 F_1(s) + C_2 F_2(s) + \cdots + C_n F_n(s)$
5	Shifting in s plane, $e^{at} f(t)$	$\int_0^\infty e^{-st} e^{at} f(t)\, dt = \int_0^\infty e^{-(s-a)t} f(t)\, dt = F(s-a)$
6	Shifting in t plane $\begin{cases} f(t-b) \text{ when } t > b, \\ 0 \text{ when } t < b \end{cases}$	$e^{-bs} F(s)$
	Function Transforms	
7	1	$\frac{1}{s}$
8	$\frac{t^{n-1}}{(n-1)!}$	$\frac{1}{s^n}$, for $n = 1, 2, \ldots$
9	e^{-at}	$\frac{1}{s+a}$
10	$\frac{1}{a}(1 - e^{-at})$	$\frac{1}{s(s+a)}$
11	te^{-at}	$\frac{1}{(s+a)^2}$
12	$\frac{1}{a} \sin at$	$\frac{1}{s^2 + a^2}$
13	$\frac{1}{a^2}(1 - \cos at)$	$\frac{1}{s(s^2 + a^2)}$
14	$\frac{1}{a^3}(at - \sin at)$	$\frac{1}{s^2(s^2 + a^2)}$
15	$\frac{1}{(b-a)}(e^{-at} - e^{-bt})$	$\frac{1}{(s+a)(s+b)}$
16	$\frac{1}{ab} + \frac{be^{-at} - ae^{-bt}}{ab(a-b)}$	$\frac{1}{s(s+a)(s+b)}$
17	$\frac{a \sin bt - b \sin at}{ab(a^2 - b^2)}$	$\frac{1}{(s^2 + a^2)(s^2 + b^2)}$
18	$e^{-at}(1 - at)$	$\frac{s}{(s+a)^2}$
19	$\cos at$	$\frac{s}{s^2 + a^2}$
20	Rectangular pulse (height 1 from 0 to τ)	$\frac{1 - e^{-s\tau}}{s}$
21	Sine pulse (height 1 from 0 to τ)	$\frac{\pi/\tau}{s^2 + \pi^2/\tau^2}(1 + e^{-s\tau})$

† $f(t)$ and its derivatives through $f^{(n-1)}(t)$ must be continuous.

example. Transforms for general operations, such as differentiation, are included as well as transforms of explicit functions.

In general, the transforms of the explicit functions can be obtained by carrying out the integration indicated by the definition of the Laplace transformation. For example:

For $f(t) = 1$:

$$F(s) = \int_0^\infty e^{-st}\,dt = -\frac{1}{s}e^{-st}\Big]_0^\infty = \frac{1}{s}$$

Transformation of the Differential Equation. The differential equation for the undamped linear oscillator is given in general form by

$$\frac{1}{\omega_n^2}\ddot{\nu} + \nu = \xi(t) \tag{8.11}$$

Applying the operational transforms (items 1 and 3, Table 8.2), Eq. (8.11) is transformed to

$$\frac{1}{\omega_n^2}s^2F_r(s) - \frac{1}{\omega_n^2}sf(0) - \frac{1}{\omega_n^2}f'(0) + F_r(s) = F_e(s) \tag{8.12a}$$

where $F_r(s)$ = the transform of the unknown response $\nu(t)$, sometimes called the *response transform*

$s^2F_r(s) - sf(0) - f'(0)$ = the transform of the second derivative of $\nu(t)$

$f(0)$ and $f'(0)$ = the known *initial values* of ν and $\dot{\nu}$, i.e., ν_0 and $\dot{\nu}_0$

$F_e(s)$ = the transform of the known excitation function $\xi(t)$, written $F_e(s) = \mathcal{L}[\xi(t)]$, sometimes called the *driving transform*

It should be noted that the initial conditions of the system are explicit in Eq. (8.12a).

The Subsidiary Equation. Solving Eq. (8.12a) for $F_r(s)$,

$$F_r(s) = \frac{sf(0) + f'(0) + \omega_n^2F_e(s)}{s^2 + \omega_n^2} \tag{8.12b}$$

This is known as the *subsidiary equation* of the differential equation. The first two terms of the transform derive from the initial conditions of the system, and the third term derives from the excitation.

Inverse Transformation. In order to determine the response function $\nu(t)$, which is the solution of the differential equation, an inverse transformation is performed on the subsidiary equation. The entire operation, applied explicitly to the solution of Eq. (8.11), may be abbreviated as follows:

$$\nu(t) = \mathcal{L}^{-1}[F_r(s)] = \mathcal{L}^{-1}\left[\frac{s\nu_0 + \dot{\nu}_0 + \omega_n^2\mathcal{L}[\xi(t)]}{s^2 + \omega_n^2}\right] \tag{8.13}$$

Example 8.4. Rectangular Step Excitation. In this case $\xi(t) = \xi_c$ for $0 \leq t$ (Fig. 8.6A). The Laplace transform $F_e(s)$ of the excitation is, from item 7 of Table 8.2,

$$\mathcal{L}[\xi_c] = \xi_c\mathcal{L}[1] = \xi_c\frac{1}{s}$$

Assume that the starting conditions are general, that is, $\nu = \nu_0$ and $\dot{\nu} = \dot{\nu}_0$ at $t = 0$. Substituting the transform and the starting conditions into Eq. (8.13), the following is obtained:

$$\nu(t) = \mathcal{L}^{-1}\left[\frac{s\nu_0 + \dot{\nu}_0 + \omega_n^2\xi_c(1/s)}{s^2 + \omega_n^2}\right] \tag{8.14a}$$

The foregoing may be rewritten as three separate inverse transforms:

$$\nu(t) = \nu_0\mathcal{L}^{-1}\left[\frac{s}{s^2 + \omega_n^2}\right] + \dot{\nu}_0\mathcal{L}^{-1}\left[\frac{1}{s^2 + \omega_n^2}\right] + \xi_c\omega_n^2\mathcal{L}^{-1}\left[\frac{1}{s(s^2 + \omega_n^2)}\right] \tag{8.14b}$$

The inverse transforms in Eq. (8.14*b*) are evaluated by use of items 19, 12 and 13, respectively, in Table 8.2. Thus, the time-response function is given explicitly by

$$\nu(t) = \nu_0 \cos \omega_n t + \frac{\dot{\nu}_0}{\omega_n} \sin \omega_n t + \xi_c(1 - \cos \omega_n t) \tag{8.14c}$$

The first two terms are the same as the starting condition response terms given by expressions (8.20*a*). The third term agrees with the response function shown by Eq. (8.22), derived for the case of a start from rest.

Example 8.5. Rectangular Pulse Excitation. The excitation function, Fig. 8.6*B*, is given by

$$\xi(t) = \begin{cases} \xi_p & \text{for } 0 \le t \le \tau \\ 0 & \text{for } \tau \le t \end{cases}$$

For simplicity, assume a start from rest, i.e., $\nu_0 = 0$ and $\dot{\nu}_0 = 0$ when $t = 0$.

During the first time interval, $0 \le t \le \tau$, the response function is of the same form as Eq. (8.14*c*) except that, with the assumed start from rest, the first two terms are zero.

During the second interval, $\tau \le t$, the transform of the excitation is obtained by applying the delayed-function transform (item 6, Table 8.2) and the transform for the rectangular step function (item 7) with the following result:

$$F_e(s) = \mathfrak{L}[\xi(t)] = \xi_p \left(\frac{1}{s} - \frac{e^{-s\tau}}{s}\right)$$

This is the transform of an excitation consisting of a rectangular step of height $-\xi_p$ starting at time $t = \tau$, superimposed on the rectangular step of height $+\xi_p$ starting at time $t = 0$.

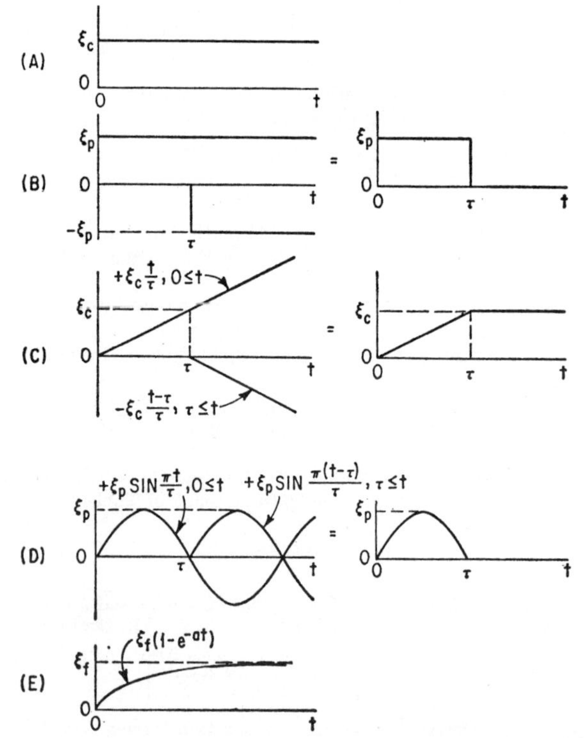

FIG. 8.6. Excitation functions in examples of use of the Laplace transform: (*A*) rectangular step, (*B*) rectangular pulse, (*C*) step with constant-slope front, (*D*) sine pulse, and (*E*) step with exponential asymptotic rise.

Substituting for $\mathcal{L}[\xi(t)]$ in Eq. (8.13),

$$\nu(t) = \xi_p \omega_n^2 \left\{ \mathcal{L}^{-1} \left[\frac{1}{s(s^2 + \omega_n^2)} \right] - \mathcal{L}^{-1} \left[\frac{e^{-s\tau}}{s(s^2 + \omega_n^2)} \right] \right\} \tag{8.15a}$$

The first inverse transform in Eq. (8.15*a*) is the same as the third one in Eq. (8.14*b*) and is evaluated by use of item 13 in Table 8.2. However, the second inverse transform requires the use of items 6 and 13. The function-transform pair given by item 6 indicates that when $t < b$ the inverse transform in question is zero, and when $t > b$ the inverse transform is evaluated by replacing t by $t - b$ (in this particular case, by $t - \tau$). The result is as follows:

$$\begin{aligned} \nu(t) &= \xi_p \omega_n^2 \left\{ \frac{1}{\omega_n^2} (1 - \cos \omega_n t) - \frac{1}{\omega_n^2} [1 - \cos \omega_n (t - \tau)] \right\} \\ &= 2\xi_p \sin \frac{\pi\tau}{T} \sin \omega_n \left(t - \frac{\tau}{2} \right) \qquad [\tau \leq t] \end{aligned} \tag{8.15b}$$

Theorem on the Transform of Functions Shifted in the Original (t) Plane. In Example 8.5, use is made of the theorem on the transform of functions shifted in the original plane. The theorem (item 6 in Table 8.2) is known variously as the second shifting theorem, the theorem on the transform of delayed functions, and the time-displacement theorem. In determining the transform of the excitation, the theorem provides for shifting, i.e., displacing the excitation or a component of the excitation in the positive direction along the time axis. This suggests the term *delayed function.* Examples of the shifting of component parts of the excitation appear in Fig. 8.6*B*, 8.6*C*, and 8.6*D*. Use of the theorem also is necessary in determining, by means of inverse transformation, the response following the delay in the excitation. Further illustration of the use of the theorem is shown by the next two examples.

Example 8.6. Step Function with Constant-slope Front. The excitation function (Fig. 8.6*C*) is expressed as follows:

$$\xi(t) = \begin{cases} \xi_c \dfrac{t}{\tau} & [0 \leq t \leq \tau] \\ \xi_c & [\tau \leq t] \end{cases}$$

Assume that $\nu_0 = 0$ and $\dot{\nu}_0 = 0$.

The driving transforms for the first and second time intervals are

$$\mathcal{L}[\xi(t)] = \begin{cases} \xi_c \dfrac{1}{\tau} \dfrac{1}{s^2} & [0 \leq t \leq \tau] \\ \xi_c \dfrac{1}{\tau} \left(\dfrac{1}{s^2} - \dfrac{e^{-s\tau}}{s^2} \right) & [\tau \leq t] \end{cases}$$

The transform for the second interval is the transform of a *negative* constant slope excitation, $-\xi_c(t - \tau)/\tau$, starting at $t = \tau$, superimposed on the transform for the positive constant slope excitation, $+\xi_c t/\tau$, starting at $t = 0$.

Substituting the transforms and starting conditions into Eq. (8.13), the responses for the two time eras, in terms of the transformations, are

$$\nu(t) = \begin{cases} \xi_c \dfrac{\omega_n^2}{\tau} \mathcal{L}^{-1} \left[\dfrac{1}{s^2(s^2 + \omega_n^2)} \right] & [0 \leq t \leq \tau] \\ \xi_c \dfrac{\omega_n^2}{\tau} \left\{ \mathcal{L}^{-1} \left[\dfrac{1}{s^2(s^2 + \omega_n^2)} \right] - \mathcal{L}^{-1} \left[\dfrac{e^{-s\tau}}{s^2(s^2 + \omega_n^2)} \right] \right\} & [\tau \leq t] \end{cases} \tag{8.16a}$$

Evaluation of the inverse transforms by reference to Table 8.2 [item 14 for the first of Eqs. (8.16*a*), items 6 and 14 for the second] leads to the following:

$$\nu(t) = \begin{cases} \xi_c \dfrac{\omega_n^2}{\tau} \dfrac{1}{\omega_n^3} (\omega_n t - \sin \omega_n t) & [0 \leq t \leq \tau] \\ \xi_c \dfrac{\omega_n^2}{\tau} \left\{ \dfrac{1}{\omega_n^3} (\omega_n t - \sin \omega_n t) - \dfrac{1}{\omega_n^3} [\omega_n (t - \tau) - \sin \omega_n (t - \tau)] \right\} & [\tau \leq t] \end{cases}$$

Simplifying,

$$\nu(t) = \begin{cases} \xi_c \dfrac{1}{\omega_n \tau} (\omega_n t - \sin \omega_n t) & [0 \le t \le \tau] \\ \xi_c \left[1 + \dfrac{2}{\omega_n \tau} \sin \dfrac{\omega_n \tau}{2} \cos \omega_n \left(t - \dfrac{\tau}{2} \right) \right] & [\tau \le t] \end{cases} \tag{8.16b}$$

Example 8.7. Half-cycle Sine Pulse. The excitation function (Fig. 8.6*D*) is

$$\xi(t) = \begin{cases} \xi_p \sin \dfrac{\pi t}{\tau} & [0 \le t \le \tau] \\ 0 & [\tau \le t] \end{cases}$$

Let the system start from rest. The driving transforms are

$$\mathcal{L}[\xi(t)] = \begin{cases} \xi_p \dfrac{\pi}{\tau} \dfrac{1}{s^2 + \pi^2/\tau^2} & [0 \le t \le \tau] \\ \xi_p \dfrac{\pi}{\tau} \left(\dfrac{1}{s^2 + \pi^2/\tau^2} + \dfrac{e^{-s\tau}}{s^2 + \pi^2/\tau^2} \right) & [\tau \le t] \end{cases}$$

The driving transform for the second interval is the transform of a sine wave of *positive* amplitude ξ_p and frequency π/τ starting at time $t = \tau$, superimposed on the transform of a sine wave of the same amplitude and frequency starting at time $t = 0$.

By substitution of the driving transforms and the starting conditions into Eq. (8.13), the following are found:

$$\nu(t) = \begin{cases} \xi_p \dfrac{\pi}{\tau} \omega_n^2 \mathcal{L}^{-1} \left[\dfrac{1}{s^2 + \pi^2/\tau^2} \cdot \dfrac{1}{s^2 + \omega_n^2} \right] & [0 \le t \le \tau] \\ \xi_p \dfrac{\pi}{\tau} \omega_n^2 \left\{ \mathcal{L}^{-1} \left[\dfrac{1}{s^2 + \pi^2/\tau^2} \cdot \dfrac{1}{s^2 + \omega_n^2} \right] + \mathcal{L}^{-1} \left[\dfrac{e^{-s\tau}}{s^2 + \pi^2/\tau^2} \cdot \dfrac{1}{s^2 + \omega_n^2} \right] \right\} & [\tau \le t] \end{cases} \tag{8.17a}$$

Determining the inverse transforms from Table 8.2 [item 17 for the first of Eqs. (8.17*a*), items 6 and 17 for the second]:

$$\nu(t) = \begin{cases} \xi_p \dfrac{\pi}{\tau} \omega_n^2 \dfrac{\omega_n \sin (\pi t/\tau) - (\pi/\tau) \sin \omega_n t}{(\pi \omega_n/\tau)(\omega_n^2 - \pi^2/\tau^2)} & [0 \le t \le \tau] \\ \xi_p \dfrac{\pi}{\tau} \omega_n^2 \left[\dfrac{\omega_n \sin (\pi t/\tau) - (\pi/\tau) \sin \omega_n t}{(\pi \omega_n/\tau)(\omega_n^2 - \pi^2/\tau^2)} \right. \\ \qquad \left. + \dfrac{\omega_n \sin [\pi(t - \tau)/\tau] - (\pi/\tau) \sin \omega_n (t - \tau)}{(\pi \omega_n/\tau)(\omega_n^2 - \pi^2/\tau^2)} \right] & [\tau \le t] \end{cases}$$

Simplifying,

$$\nu(t) = \begin{cases} \xi_p \dfrac{1}{1 - T^2/4\tau^2} \left(\sin \dfrac{\pi t}{\tau} - \dfrac{T}{2\tau} \sin \omega_n t \right) & [0 \le t \le \tau] \\ \xi_p \dfrac{(T/\tau) \cos (\pi \tau/T)}{(T^2/4\tau^2) - 1} \sin \omega_n \left(t - \dfrac{\tau}{2} \right) & [\tau \le t] \end{cases} \tag{8.17b}$$

where $T = 2\pi/\omega_n$ is the natural period of the responding system. Equations (8.17*b*) are equivalent to Eqs. (8.6*b*) and (8.7*b*) derived previously by the use of Duhamel's integral.

Example 8.8. Exponential Asymptotic Step. The excitation function (Fig. 8.6*E*) is

$$\xi(t) = \xi_f (1 - e^{-at}) \qquad [0 \le t]$$

Assume that the system starts from rest. The driving transform is

$$\mathcal{L}[\xi(t)] = \xi_f \left(\frac{1}{s} - \frac{1}{s + a} \right) = \xi_f a \frac{1}{s(s + a)}$$

It is found by Eq. (8.13) that

$$\nu(t) = \xi_f a\omega_n^2 \mathfrak{L}^{-1}\left[\frac{1}{s(s+a)(s^2+\omega_n^2)}\right] \qquad [0 \le t] \tag{8.18a}$$

It frequently happens that the inverse transform is not readily found in an available table of transforms. Using the above case as an example, the function of s in Eq. (8.18a) is first *expanded in partial fractions*; then the inverse transforms are sought, thus:

$$\frac{1}{s(s+a)(s^2+\omega_n^2)} = \frac{\kappa_1}{s} + \frac{\kappa_2}{s+a} + \frac{\kappa_3}{s+j\omega_n} + \frac{\kappa_4}{s-j\omega_n} \tag{8.18b}$$

where $j = \sqrt{-1}$

$$\kappa_1 = \left[\frac{1}{(s+a)(s+j\omega_n)(s-j\omega_n)}\right]_{s=0} = \frac{1}{a\omega_n^2}$$

$$\kappa_2 = \left[\frac{1}{s(s+j\omega_n)(s-j\omega_n)}\right]_{s=-a} = \frac{1}{-a(a^2+\omega_n^2)}$$

$$\kappa_3 = \left[\frac{1}{s(s+a)(s-j\omega_n)}\right]_{s=-j\omega_n} = \frac{1}{-2\omega_n^2(a-j\omega_n)}$$

$$\kappa_4 = \left[\frac{1}{s(s+a)(s+j\omega_n)}\right]_{s=+j\omega_n} = \frac{1}{-2\omega_n^2(a+j\omega_n)}$$

Consequently, Eq. (8.18a) may be rewritten in the following expanded form:

$$\nu(t) = \xi_f \left\{\mathfrak{L}^{-1}\left[\frac{1}{s}\right] - \frac{\omega_n^2}{a^2+\omega_n^2}\mathfrak{L}^{-1}\left[\frac{1}{s+a}\right] - \frac{a}{2(a-j\omega_n)}\cdot \mathfrak{L}^{-1}\left[\frac{1}{s+j\omega_n}\right] - \frac{a}{2(a+j\omega_n)}\mathfrak{L}^{-1}\left[\frac{1}{s-j\omega_n}\right]\right\} \tag{8.18c}$$

The inverse transforms may now be found readily (items 7 and 9, Table 8.2):

$$\nu(t) = \xi_f\left[1 - \frac{\omega_n^2}{a^2+\omega_n^2}e^{-at} - \frac{a}{2(a-j\omega_n)}e^{-j\omega_n t} - \frac{a}{2(a+j\omega_n)}e^{j\omega_n t}\right]$$

Rewriting,

$$\nu(t) = \xi_f\left[1 - \frac{\omega_n^2 e^{-at} + a^2 \tfrac{1}{2}(e^{j\omega_n t} + e^{-j\omega_n t}) - aj\omega_n \tfrac{1}{2}(e^{j\omega_n t} - e^{-j\omega_n t})}{a^2+\omega_n^2}\right]$$

Making use of the relations, $\cos z = (\tfrac{1}{2})(e^{jz} + e^{-jz})$ and $\sin z = -j(\tfrac{1}{2})(e^{jz} - e^{-jz})$, the equation for $\nu(t)$ may be expressed as follows:

$$\nu(t) = \xi_f\left[1 - \frac{(a/\omega_n)[\sin\omega_n t + (a/\omega_n)\cos\omega_n t] + e^{-at}}{1 + a^2/\omega_n^2}\right] \tag{8.18d}$$

Partial Fraction Expansion of $F(s)$. The partial fraction expansion of $F_r(s)$, illustrated for a particular case in Eq. (8.18b), is a necessary part of the technique of solution. In general $F_r(s)$, expressed by the subsidiary equation (8.12b) and involved in the inverse transformation, Eqs. (8.10) and (8.13), is a quotient of two polynomials in s, thus

$$F_r(s) = \frac{A(s)}{B(s)} \tag{8.19}$$

The purpose of the expansion of $F_r(s)$ is to divide it into simple parts, the inverse transforms of which may be determined readily. The general procedure of the expansion is to factor $B(s)$ and then to rewrite $F_r(s)$ in partial fractions.[16, 17, 34, 36]

INITIAL CONDITIONS OF THE SYSTEM

In all the solutions for response presented in this chapter, unless otherwise stated, it is assumed that the initial conditions (ν_0 and $\dot{\nu}_0$) of the system are both zero. Other starting conditions may be accounted for merely by superimposing on the time-response functions given the additional terms

$$\nu_0 \cos \omega_n t + \frac{\dot{\nu}_0}{\omega_n} \sin \omega_n t \tag{8.20a}$$

These terms are the complete solution of the homogeneous differential equation, $m\ddot{\nu}/k + \nu = 0$. They represent the free vibration resulting from the initial conditions.

The two terms in Eq. (8.20a) may be expressed by either one of the following combined forms:

$$\sqrt{\nu_0^2 + \left(\frac{\dot{\nu}_0}{\omega_n}\right)^2} \sin (\omega_n t + \theta_1) \qquad \text{where } \tan \theta_1 = \frac{\nu_0 \omega_n}{\dot{\nu}_0} \tag{8.20b}$$

$$\sqrt{\nu_0^2 + \left(\frac{\dot{\nu}_0}{\omega_n}\right)^2} \cos (\omega_n t - \theta_2) \qquad \text{where } \tan \theta_2 = \frac{\dot{\nu}_0}{\nu_0 \omega_n} \tag{8.20c}$$

where $\sqrt{\nu_0^2 + \left(\frac{\dot{\nu}_0}{\omega_n}\right)^2}$ is the resultant amplitude and θ_1 or θ_2 is the phase angle of the *initial-condition free vibration.*

PRINCIPLE OF SUPERPOSITION

When the system is linear, the *principle of superposition* may be employed. Any number of component excitation functions may be superimposed to obtain a prescribed total excitation function, and the corresponding component response functions may be superimposed to arrive at the total response function. However, the superposition must be carried out on a time basis and with complete regard for algebraic sign. The superposition of maximum component responses, disregarding time, may lead to completely erroneous results. For example, the response functions given by Eqs. (8.31) to (8.34) are defined completely with regard to time and algebraic sign, and may be superimposed for any combination of the excitation functions from which they have been derived.

COMPILATION OF RESPONSE FUNCTIONS AND RESPONSE SPECTRA; SINGLE DEGREE-OF-FREEDOM, LINEAR UNDAMPED SYSTEMS

STEP-TYPE EXCITATION FUNCTIONS

CONSTANT-FORCE EXCITATION (SIMPLE STEP IN FORCE). The excitation is a constant force applied to the mass at zero time, $\xi(t) \equiv F(t)/k = F_c/k$. Substituting this excitation for $F(t)/k$ in Eq. (8.1a) and solving for the absolute displacement x,

$$x = \frac{F_c}{k}(1 - \cos \omega_n t) \tag{8.21a}$$

CONSTANT-DISPLACEMENT EXCITATION (SIMPLE STEP IN DISPLACEMENT). The excitation is a constant displacement of the ground which occurs at zero time, $\xi(t) \equiv u(t) = u_c$. Substituting for $u(t)$ in Eq. (8.1*b*) and solving for the absolute displacement x,

$$x = u_c(1 - \cos \omega_n t) \tag{8.21b}$$

CONSTANT-ACCELERATION EXCITATION (SIMPLE STEP IN ACCELERATION). The excitation is an instantaneous change in the ground acceleration at zero time, from zero to a constant value $\ddot{u}(t) = \ddot{u}_c$. The excitation is thus

$$\xi(t) \equiv -m\ddot{u}_c/k = -\ddot{u}_c/\omega_n^2.$$

Substituting in Eq. (8.1*c*) and solving for the *relative* displacement δ_x,

$$\delta_x = \frac{-\ddot{u}_c}{\omega_n}(1 - \cos \omega_n t) \tag{8.21c}$$

When the excitation is defined by a function of acceleration $\ddot{u}(t)$, it is often convenient to express the response in terms of the absolute acceleration $\ddot{x}$ of the system. The force acting on the mass in Fig. 8.1*C* is $-k\ \delta_x$; the acceleration $\ddot{x}$ is thus $-k\ \delta_x/m$ or $-\delta_x\omega_n^2$. Substituting $\delta_x = -\ddot{x}/\omega_n^2$ in Eq. (8.21*c*),

$$\ddot{x} = \ddot{u}_c(1 - \cos \omega_n t) \tag{8.21d}$$

The same result is obtained by letting $\xi(t) \equiv \ddot{u}(t) = \ddot{u}_c$ in Eq. (8.1*d*) and solving for $\ddot{x}$. Equation (8.21*d*) is similar to Eq. (8.21*b*) with acceleration instead of displacement on both sides of the equation. This analogy generally applies in step- and pulse-type excitations.

The absolute displacement of the mass can be obtained by integrating Eq. (8.21*d*) twice with respect to time, taking as initial conditions $x = \dot{x} = 0$ when $t = 0$,

$$x = \frac{\ddot{u}_c}{\omega_n^2}\left[\frac{\omega_n^2 t^2}{2} - (1 - \cos \omega_n t)\right] \tag{8.21e}$$

Equation (8.21*e*) also may be obtained from the relation $x = u + \delta_x$, noting that in this case $u(t) = \ddot{u}_c t^2/2$.

CONSTANT-VELOCITY EXCITATION (SIMPLE STEP IN VELOCITY). This excitation, when expressed in terms of ground or spring anchorage motion, is equivalent to prescribing, at zero time, an instantaneous change in the ground velocity from zero to a constant value $\dot{u}_c$. The excitation is $\xi(t) \equiv u(t) = \dot{u}_c t$, and the solution for the differential equation of Eq. (8.1*b*) is

$$x = \frac{\dot{u}_c}{\omega_n}(\omega_n t - \sin \omega_n t) \tag{8.21f}$$

For the velocity of the mass,

$$\dot{x} = \dot{u}_c(1 - \cos \omega_n t) \tag{8.21g}$$

The result of Eq. (8.21*g*) could have been obtained directly by letting $\xi(t) \equiv \dot{u}(t) = \dot{u}_c$ in Eq. (8.1*e*) and solving for the *velocity* response $\dot{x}$.

GENERAL STEP EXCITATION. A comparison of Eqs. (8.21*a*), (8.21*b*), (8.21*c*), (8.21*d*), and (8.21*g*) with Table 8.1 reveals that the response ν and the excitation ξ are related in a common manner. This may be expressed as follows:

$$\nu = \xi_c(1 - \cos \omega_n t) \tag{8.22}$$

where ξ_c indicates a constant value of the excitation. The excitation and response of the system are shown in Fig. 8.7.

ABSOLUTE DISPLACEMENT RESPONSE TO VELOCITY-STEP AND ACCELERATION-STEP EXCITATIONS. The absolute displacement responses to the velocity-step and the acceleration-step excitations are given by Eqs. (8.21*f*) and (8.21*e*) and are shown in Figs. 8.8 and 8.9, respectively. The comparative effects of displacement-step, velocity-step, and acceleration-step excitations, in terms of *absolute displacement* response, may be seen by comparing Figs. 8.7 to 8.9.

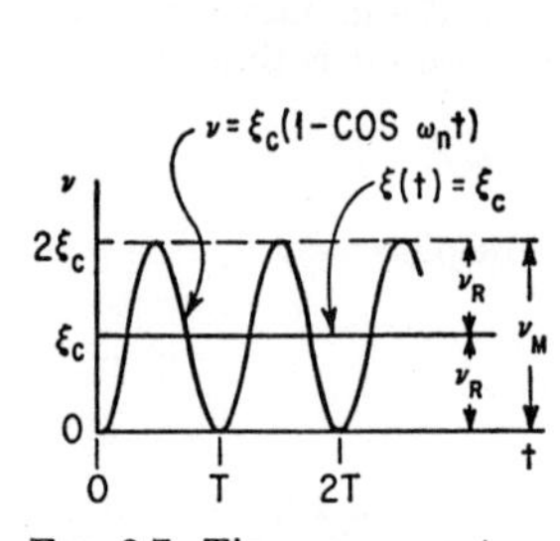

Fig. 8.7. Time response to a simple step excitation (general notation).

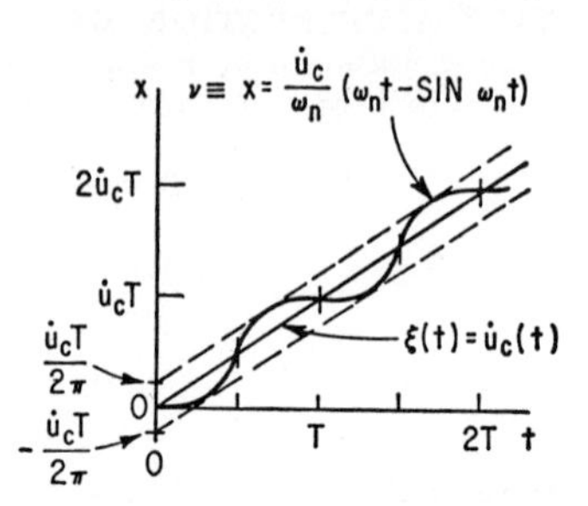

Fig. 8.8. Time-displacement response to a constant-velocity excitation (simple step in velocity).

In the case of the velocity-step excitation, the *velocity* of the system is always positive, except at $t = 0, T, 2T, \ldots$, when it is zero. Similarly, an acceleration-step excitation results in system *acceleration* that is always positive, except at $t = 0, T, 2T, \ldots$, when it is zero. The natural period of the responding system is $T = 2\pi/\omega_n$.

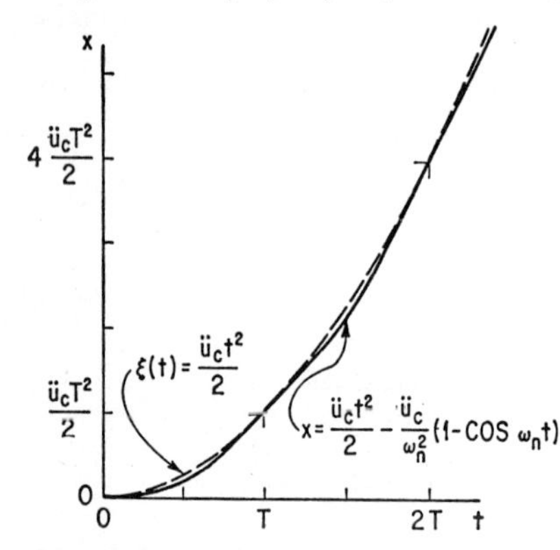

Fig. 8.9. Time-displacement response to a constant-acceleration excitation (simple step in acceleration).

RESPONSE MAXIMA. In the response of a system to step or pulse excitation, the maximum value of the response often is of considerable physical significance. Several kinds of maxima are important. One of these is the *residual response amplitude,* which is the amplitude of the free vibration about the final position of the excitation as a base. This is designated ν_R, and for the response given by Eq. (8.22):

$$\nu_R = \pm\xi_c \tag{8.22a}$$

Another maximum is the *maximax response,* which is the greatest of the maxima of ν attained at *any time* during the response. In general, it is of the same sign as the excitation. For the response given by Eq. (8.22), the maximax response ν_M is

$$\nu_M = 2\xi_c \tag{8.22b}$$

ASYMPTOTIC STEP. In the exponential function $\xi(t) = \xi_f(1 - e^{-at})$, the maximum value ξ_f of the excitation is approached asymptotically. This excitation may be defined alternatively by $\xi(t) = (F_f/k)(1 - e^{-at})$; $u_f(1 - e^{-at})$; $(-\ddot{u}_f/\omega_n^2)(1 - e^{-at})$; etc. (see Table 8.1). Substituting the excitation $\xi(t) = \xi_f(1 - e^{-at})$ in Eq. (8.2), the response ν is

$$\nu = \xi_f \left[1 - \frac{(a/\omega_n)[\sin \omega_n t + (a/\omega_n) \cos \omega_n t] + e^{-at}}{1 + a^2/\omega_n^2}\right] \tag{8.23a}$$

The excitation and the response of the system are shown in Fig. 8.10. For large values of the exponent at, the motion is nearly simple harmonic. The residual amplitude, relative to the final position of equilibrium, approaches the following value asymptotically.

$$\nu_R \rightarrow \xi_f \frac{1}{\sqrt{1 + \omega_n^2/a^2}} \tag{8.23b}$$

The maximax response $\nu_M = \nu_R + \xi_f$ is plotted against ω_n/a to give the response spectrum in Fig. 8.11.

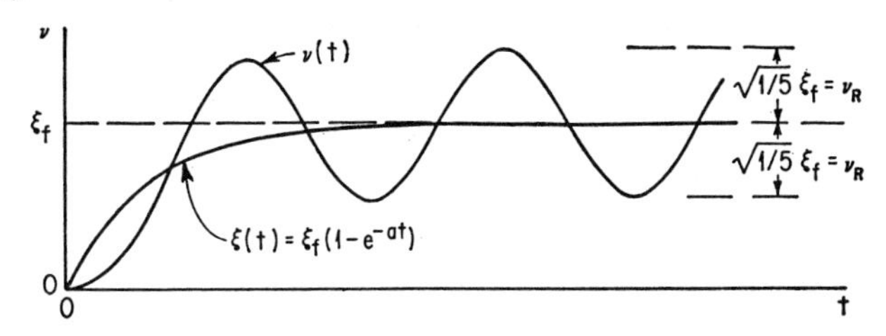

FIG. 8.10. Time response to an exponentially asymptotic step for the particular case $\omega_n/a = 2$.

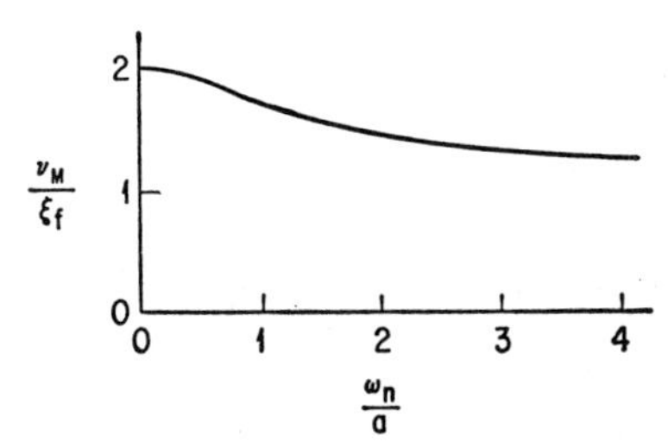

FIG. 8.11. Spectrum for maximax response resulting from exponentially asymptotic step excitation.

STEP-TYPE FUNCTIONS HAVING FINITE RISE TIME. Many step-type excitation functions rise to the constant maximum value ξ_c of the excitation in a finite length of time τ, called the *rise time*. Three such functions and their first three time derivatives are shown in Fig. 8.12. The step having a *cycloidal* front is the only one of the three that does not include an infinite third derivative; i.e., if the step is a ground displacement, it does not have an infinite rate of change of ground acceleration (infinite "jerk").

The excitation functions and the expressions for maximax response are given by the following equations:

Constant-slope front:

$$\xi(t) = \begin{cases} \xi_c \dfrac{t}{\tau} & [0 \leq t \leq \tau] \\ \xi_c & [\tau \leq t] \end{cases} \tag{8.24a}$$

$$\frac{\nu_M}{\xi_c} = 1 + \left| \frac{T}{\pi\tau} \sin \frac{\pi\tau}{T} \right| \tag{8.24b}$$

Versed-sine front:

$$\xi(t) = \begin{cases} \dfrac{\xi_c}{2}\left(1 - \cos \dfrac{\pi t}{\tau}\right) & [0 \leq t \leq \tau] \\ \xi_c & [\tau \leq t] \end{cases} \tag{8.25a}$$

$$\frac{\nu_M}{\xi_c} = 1 + \left| \frac{1}{(4\tau^2/T^2) - 1} \cos \frac{\pi\tau}{T} \right| \tag{8.25b}$$

Cycloidal front:

$$\xi(t) = \begin{cases} \dfrac{\xi_c}{2\pi}\left(\dfrac{2\pi t}{\tau} - \sin\dfrac{2\pi t}{\tau}\right) & [0 \le t \le \tau] \\ \xi_c & [\tau \le t] \end{cases} \tag{8.26a}$$

$$\frac{\nu_M}{\xi_c} = 1 + \left|\frac{T}{\pi\tau(1-\tau^2/T^2)}\sin\frac{\pi\tau}{T}\right| \tag{8.26b}$$

where $T = 2\pi/\omega_n$ is the natural period of the responding system.

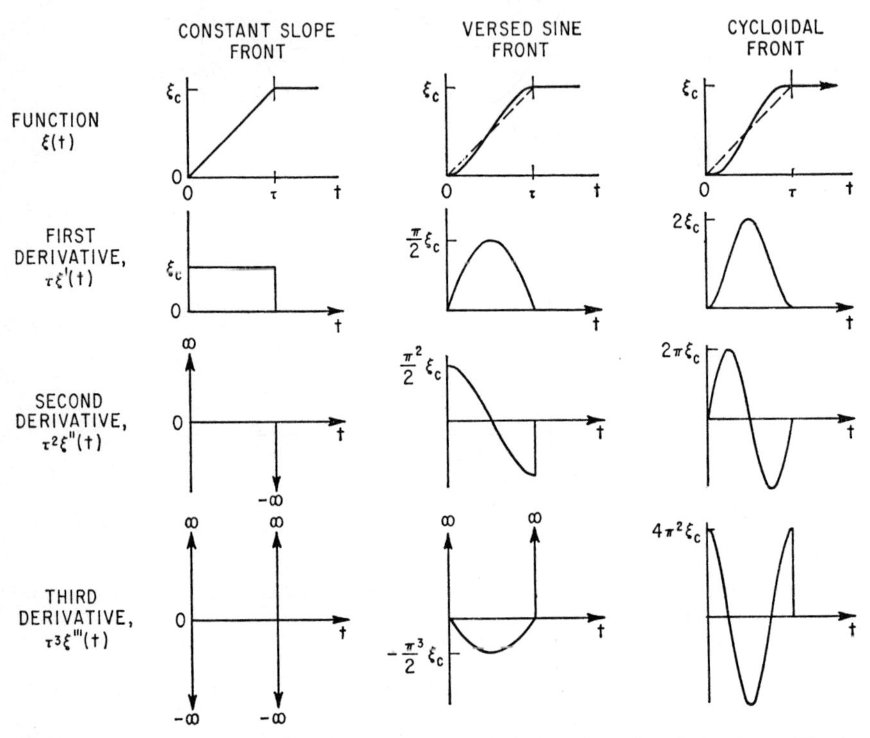

FIG. 8.12. Three step-type excitation functions and their first three time-derivatives. (*Jacobsen and Ayre.*[22])

In the case of step-type excitations, the maximax response occurs after the excitation has reached its constant maximum value ξ_c and is related to the residual response amplitude by

$$\nu_M = \nu_R + \xi_c \tag{8.27}$$

Figure 8.13 shows the spectra of maximax response versus step rise time τ expressed relative to the natural period T of the responding system. In Fig. 8.13*A* the comparison is based on *equal rise times*, and in Fig. 8.13*B* it relates to *equal maximum slopes of the step fronts.* The residual response amplitude has values of zero ($\nu_M/\xi_c = 1$) in all three cases; for example, the step excitation having a constant-slope front results in zero residual amplitude at $\tau/T = 1, 2, 3, \ldots$.

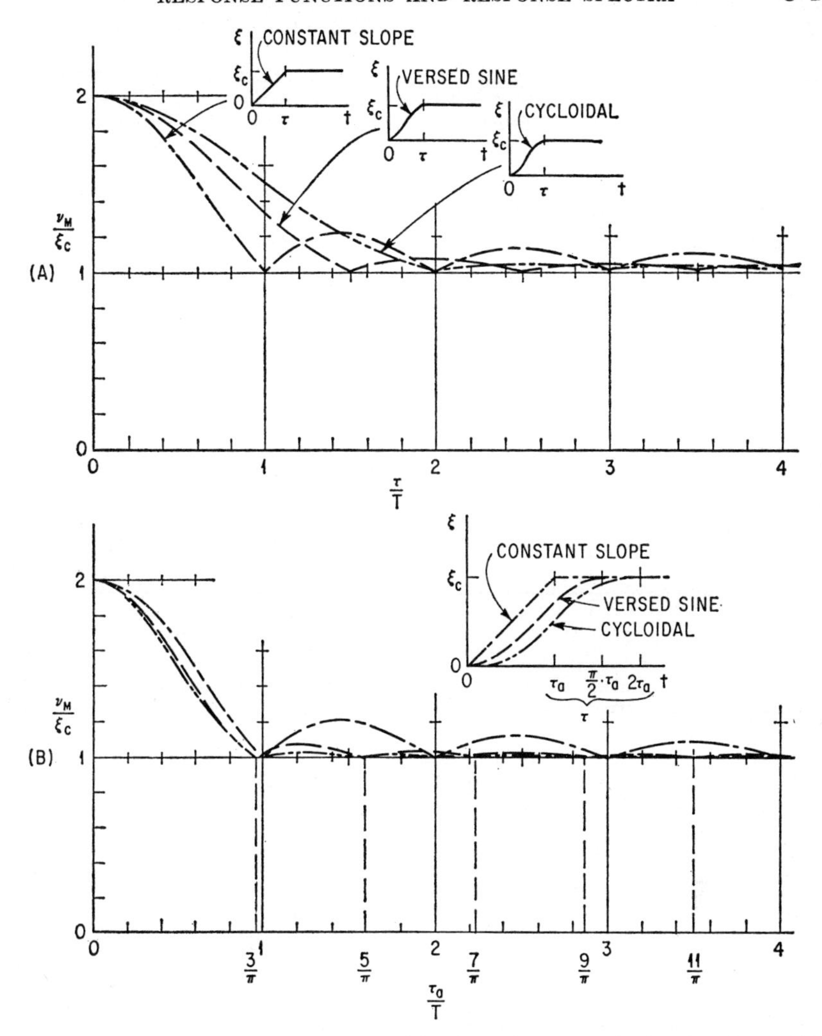

FIG. 8.13. Spectra of maximax response resulting from the step excitation functions of Fig. 8.12. (*A*) For step functions having equal rise time τ. (*B*) For step functions having equal maximum slope ξ_c/τ_a. (*Jacobsen and Ayre.*[22])

A FAMILY OF EXPONENTIAL STEP FUNCTIONS HAVING FINITE RISE TIME. The inset diagram in Fig. 8.14 shows and Eqs. (8.28*a*) define a family of step functions having fronts which rise exponentially to the constant maximum ξ_c in the rise time τ. Two limiting cases of vertically fronted steps are included in the family: When $a \to -\infty$, the vertical front occurs at $t = 0$; when $a \to +\infty$, the vertical front occurs at $t = \tau$. An intermediate case has a constant-slope front ($a = 0$). The maximax responses are given

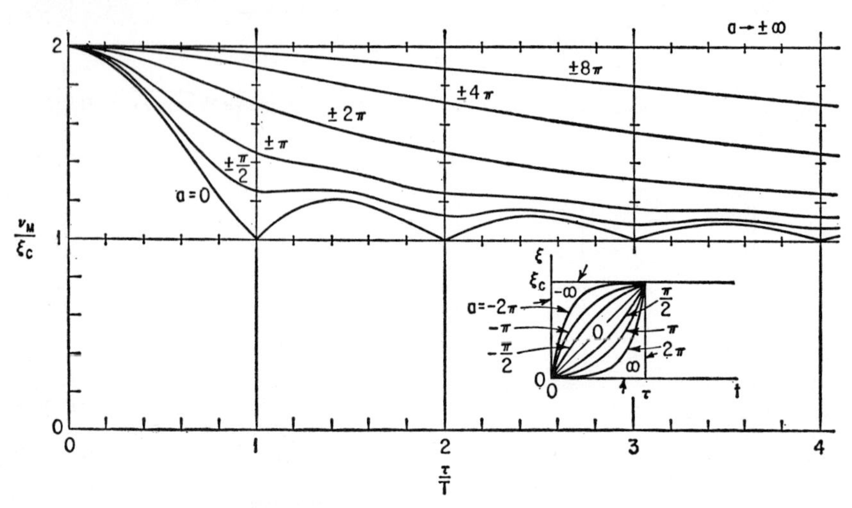

Fig. 8.14. Spectra of maximax response for a family of step functions having exponential fronts, including the vertical fronts $a \to \pm\infty$, and the constant-slope front $a = 0$, as special cases. (*Jacobsen and Ayre.*[22])

by Eq. (8.28*b*) and by the response spectra in Fig. 8.14. The values of the maximax response are independent of the sign of the parameter a.

$$\xi(t) = \begin{cases} \xi_c \dfrac{1 - e^{at/\tau}}{1 - e^a} & [0 \le t \le \tau] \\ \xi_c & [\tau \le t] \end{cases} \tag{8.28a}$$

$$\frac{\nu_M}{\xi_c} = 1 + \left| \frac{a}{1 - e^a} \left[\frac{1 - 2e^a \cos(2\pi\tau/T) + e^{2a}}{a^2 + 4\pi^2\tau^2/T^2} \right]^{1/2} \right| \tag{8.28b}$$

where T is the natural period of the responding system.

There are zeroes of residual response amplitude ($\nu_M/\xi_c = 1$) at finite values of τ/T only for the constant-slope front ($a = 0$). Each of the step functions represented in Fig. 8.13 results in zeroes of residual response amplitude, and each function has antisymmetry with respect to the half-rise time $\tau/2$. This is of interest in the selection of cam and control-function shapes, where one of the criteria of choice may be *minimum residual amplitude of vibration* of the driven system.

PULSE-TYPE EXCITATION FUNCTIONS

THE SIMPLE IMPULSE. If the duration τ of the pulse is short relative to the natural period T of the system, the response of the system may be determined by equating the impulse J, i.e., the force-time integral, to the momentum $m\dot{x}_J$:

$$J = \int_0^\tau F(t)\,dt = m\dot{x}_J \tag{8.29a}$$

Thus, it is found that the impulsive velocity $\dot{x}_J$ is equal to J/m. Consequently, the velocity-time response is given by $\dot{x} = \dot{x}_J \cos \omega_n t = (J/m) \cos \omega_n t$. The displacement-time response is obtained by integration, assuming a start from rest,

$$x = x_J \sin \omega_n t$$

where

$$x_J = \frac{J}{m\omega_n} = \omega_n \int_0^\tau \frac{F(t)\,dt}{k} \tag{8.29b}$$

The impulse concept, used for determining the response to a short-duration force pulse, may be generalized in terms of ν and ξ by referring to Table 8.1. The *generalized impulsive response* is

$$\nu = \nu_J \sin \omega_n t \tag{8.30a}$$

where the amplitude is

$$\nu_J = \omega_n \int_0^\tau \xi(t)\,dt \tag{8.30b}$$

The impulsive response amplitude ν_J and the generalized impulse $k \int_0^\tau \xi(t)\,dt$ are used in comparing the effects of various pulse shapes when the pulse durations are short.

SYMMETRICAL PULSES. In the following discussion a comparison is made of the responses caused by single symmetrical pulses of rectangular, half-cycle sine, versed-sine, and triangular shapes. The excitation functions and the time-response equations are given by Eqs. (8.31) to (8.34). Note that the residual response amplitude factors are set in brackets and are identified by the time interval $\tau \leq t$.

Rectangular:

$$\left.\begin{aligned} \xi(t) &= \xi_p \\ \nu &= \xi_p(1 - \cos \omega_n t) \end{aligned}\right\} \qquad [0 \leq t \leq \tau] \tag{8.31a}$$

$$\left.\begin{aligned} \xi(t) &= 0 \\ \nu &= \xi_p \left[2 \sin \frac{\pi\tau}{T}\right] \sin \omega_n \left(t - \frac{\tau}{2}\right) \end{aligned}\right\} \qquad [\tau \leq t] \tag{8.31b}$$

Half-cycle sine:

$$\left.\begin{aligned} \xi(t) &= \xi_p \sin \frac{\pi t}{\tau} \\ \nu &= \frac{\xi_p}{1 - T^2/4\tau^2}\left(\sin \frac{\pi t}{\tau} - \frac{T}{2\tau} \sin \omega_n t\right) \end{aligned}\right\} \qquad [0 \leq t \leq \tau] \tag{8.32a}$$

$$\left.\begin{aligned} \xi(t) &= 0 \\ \nu &= \xi_p \left[\frac{(T/\tau) \cos (\pi\tau/T)}{(T^2/4\tau^2) - 1}\right] \sin \omega_n \left(t - \frac{\tau}{2}\right) \end{aligned}\right\} \qquad [\tau \leq t] \tag{8.32b}$$

Versed-sine:

$$\left.\begin{aligned} \xi(t) &= \frac{\xi_p}{2}\left(1 - \cos \frac{2\pi t}{\tau}\right) \\ \nu &= \frac{\xi_p/2}{1 - \tau^2/T^2}\left(1 - \frac{\tau^2}{T^2} + \frac{\tau^2}{T^2} \cos \frac{2\pi t}{\tau} - \cos \omega_n t\right) \end{aligned}\right\} \qquad [0 \leq t \leq \tau] \tag{8.33a}$$

$$\left.\begin{aligned} \xi(t) &= 0 \\ \nu &= \xi_p \left[\frac{\sin \pi\tau/T}{1 - \tau^2/T^2}\right] \sin \omega_n \left(t - \frac{\tau}{2}\right) \end{aligned}\right\} \qquad [\tau \leq t] \tag{8.33b}$$

Triangular:

$$\left.\begin{aligned} \xi(t) &= 2\xi_p \frac{t}{\tau} \\ \nu &= 2\xi_p \left(\frac{t}{\tau} - \frac{T}{\tau}\frac{\sin \omega_n t}{2\pi}\right) \end{aligned}\right\} \qquad \left[0 \le t \le \frac{\tau}{2}\right] \qquad (8.34a)$$

$$\left.\begin{aligned} \xi(t) &= 2\xi_p \left(1 - \frac{t}{\tau}\right) \\ \nu &= 2\xi_p \left(1 - \frac{t}{\tau} - \frac{T}{\tau}\frac{\sin \omega_n t}{2\pi} + \frac{T}{\tau}\frac{\sin \omega_n (t - \tau/2)}{\pi}\right) \end{aligned}\right\} \qquad \left[\frac{\tau}{2} \le t \le \tau\right] \qquad (8.34b)$$

$$\left.\begin{aligned} \xi(t) &= 0 \\ \nu &= \xi_p \left[2 \frac{\sin^2 (\pi\tau/2T)}{\pi\tau/2T}\right] \sin \omega_n (t - \tau/2) \end{aligned}\right\} \qquad [\tau \le t] \qquad (8.34c)$$

where T is the natural period of the responding system.

Equal Maximum Height of Pulse as Basis of Comparison. Examples of time response, for six different values of τ/T, are shown separately for the rectangular, half-cycle sine, and versed-sine pulses in Fig. 8.15, and for the triangular pulse in Fig. 8.22B. The basis of comparison is equal maximum height of excitation pulse ξ_p.

Residual Response Amplitude and Maximax Response. The spectra of maximax response ν_M and residual response amplitude ν_R are given in Fig. 8.16 by (A) for the rectangular pulse, by (B) for the sine pulse, and by (C) for the versed-sine pulse. The maximax response may occur either within the duration of the pulse or after the pulse function has dropped to zero. In the latter case the maximax response is equal to the residual response amplitude. In general, the maximax response is given by the residual response amplitude only in the case of short-duration pulses; for example, see the case $\tau/T = 1/4$ in Fig. 8.15 where T is the natural period of the responding system. The response spectra for the triangular pulse appear in Fig. 8.24.

Maximax Relative Displacement When the Excitation Is Ground Displacement. When the excitation $\xi(t)$ is given as *ground displacement* $u(t)$, the response ν is the absolute displacement x of the mass (Table 8.1). It is of practical importance in the investigation of the maximax *distortion* or *stress* in the elastic element to know the maximax value of the relative displacement. In this case the relative displacement is a *derived quantity* obtained by taking the difference between the response and the excitation, that is, $x - u$ or, in terms of the general notation, $\nu - \xi$.

If the excitation is given as ground acceleration, the response is determined directly as relative displacement and is designated δ_x (Table 8.1). To avoid confusion, relative displacement determined as a *derived quantity*, as described in the first case above, is designated by $x - u$; relative displacement determined directly as the *response variable* (second case above) is designated by δ_x. The distinction is made readily in the general notation by use of the symbols $\nu - \xi$ and ν, respectively, for relative response and for response. The maximax values are designated $(\nu - \xi)_M$ and ν_M, respectively.

The maximax relative response may occur *either* within the duration of the pulse or during the residual vibration era ($\tau \le t$). In the latter case the maximax relative response is equal to the residual response amplitude. This explains the discontinuities which occur in the spectra of maximax relative response shown in Fig. 8.16 and elsewhere.

The meaning of the relative response $\nu - \xi$ may be clarified further by a study of the time-response and time-excitation curves shown in Fig. 8.15.

Equal Area of Pulse as Basis of Comparison. In the preceding section on the comparison of responses resulting from pulse excitation, the pulses are assumed of equal maximum height. Under some conditions, particularly if the pulse duration is short relative to the natural period of the system, it may be more useful to make the comparison on the basis of equal pulse area; i.e., equal impulse (equal time integral).

The *areas* for the pulses of maximum height ξ_p and duration τ are as follows: rectangle, $\xi_p\tau$; half-cycle sine, $(2/\pi)\xi_p\tau$; versed-sine $(1/2)\xi_p\tau$; triangle, $(1/2)\xi_p\tau$. Using the area of the

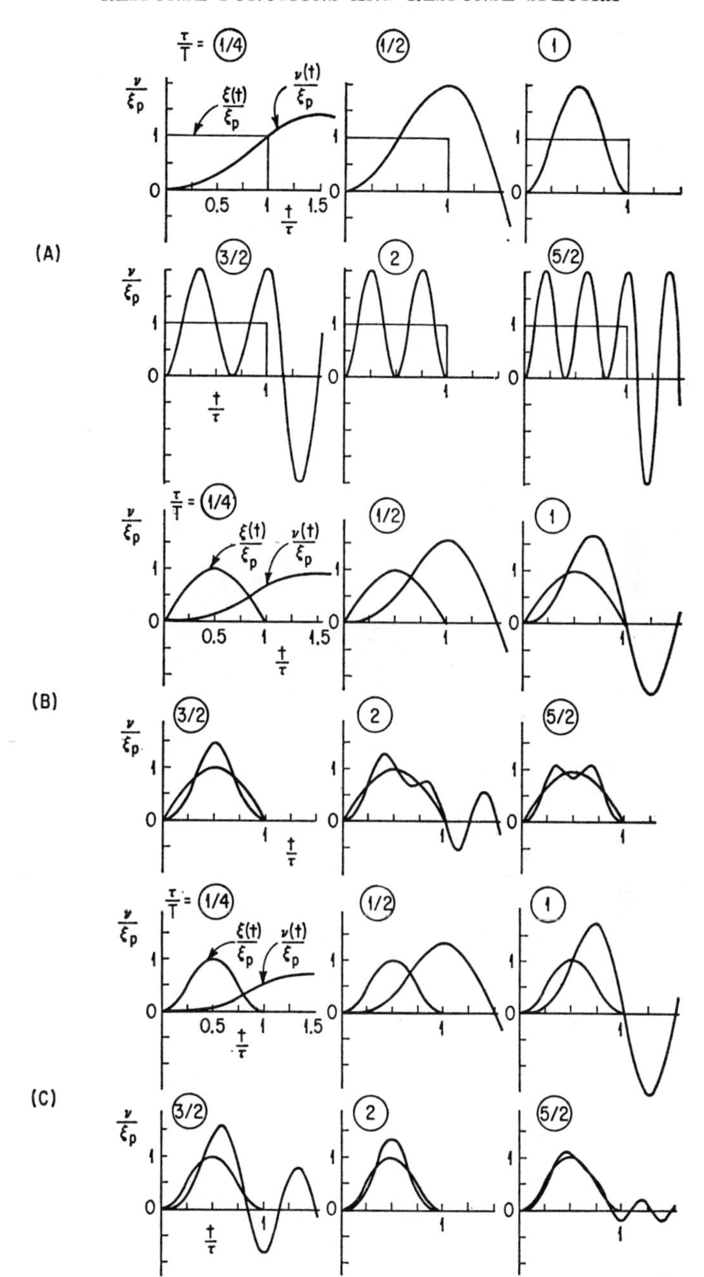

FIG. 8.15. Time response curves resulting from single pulses of (*A*) rectangular, (*B*) half-cycle sine, and (*C*) versed-sine shapes.[19]

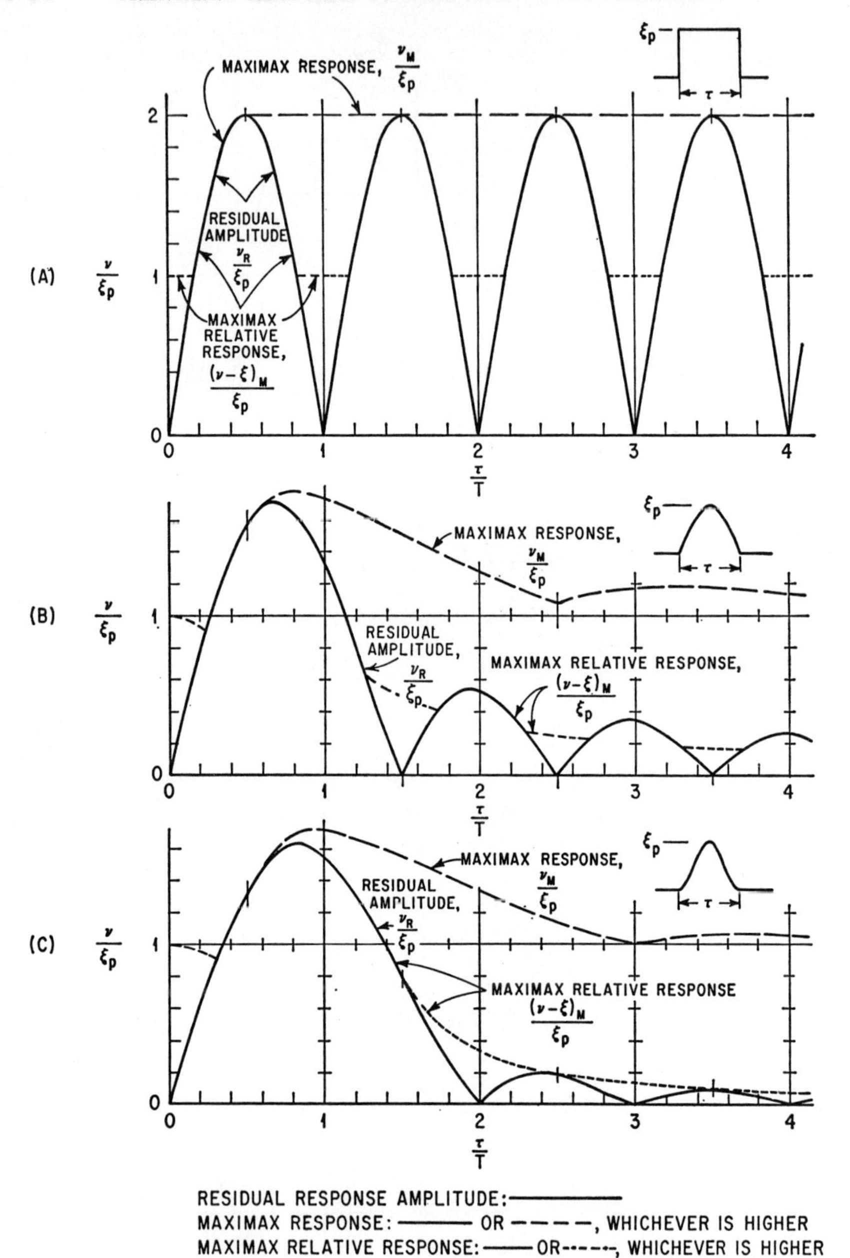

Fig. 8.16. Spectra of maximax response, residual response amplitude, and maximax relative response resulting from single pulses of (*A*) rectangular, (*B*) half-cycle sine, and (*C*) versed-sine shapes.[19] The spectra are shown on another basis in Fig. 8.18.

triangular pulse as the basis of comparison, and requiring that the areas of the other pulses be equal to it, it is found that the pulse *heights*, in terms of the height ξ_{po} of the *reference triangular pulse*, must be as follows: rectangle, $(\frac{1}{2})\xi_{po}$; half-cycle sine, $(\pi/4)\xi_{po}$; versed-sine, ξ_{po}.

Figure 8.17 shows the time responses, for four values of τ/T, redrawn on the basis of *equal pulse area* as the criterion for comparison. Note that the response reference is the constant ξ_{po}, which is the height of the triangular pulse. To show a direct comparison, the response curves for the various pulses are superimposed on each other. For the shortest duration shown, $\tau/T = \frac{1}{4}$, the response curves are nearly alike. Note that the responses to two different rectangular pulses are shown, one of duration τ and height $\xi_{po}/2$, the other of duration $\tau/2$ and height ξ_{po}, both of area $\xi_{po}\tau/2$.

The response spectra, plotted on the basis of equal pulse area, appear in Fig. 8.18. The residual response spectra are shown altogether in (*A*), the maximax response spectra in (*B*), and the spectra of maximax relative response in (*C*).

Since the pulse area is $\xi_{po}\tau/2$, the generalized impulse is $k\xi_{po}\tau/2$, and the amplitude of vibration of the system computed on the basis of the generalized impulse theory, Eq. (8.30*b*), is given by

$$\nu_J = \omega_n \xi_{po} \frac{\tau}{2} = \pi \frac{\tau}{T} \xi_{po} \tag{8.35}$$

A comparison of this straight-line function with the response spectra in Fig. 8.18*B* shows that *for values of* τ/T *less than one-fourth the shape of the symmetrical pulse is of little concern.*

Family of Exponential, Symmetrical Pulses. A continuous variation in shape of pulse may be investigated by means of the family of pulses represented by Eqs. (8.36*a*) and shown in the inset diagram in Fig. 8.19*A*:

$$\xi(t) = \begin{cases} \xi_p \dfrac{1 - e^{2at/\tau}}{1 - e^a} & \left[0 \le t \le \dfrac{\tau}{2}\right] \\ \xi_p \dfrac{1 - e^{2a(1-t/\tau)}}{1 - e^a} & \left[\dfrac{\tau}{2} \le t \le \tau\right] \\ 0 & [\tau \le t] \end{cases} \tag{8.36a}$$

The family includes the following special cases:

$a \to -\infty$: rectangle of height ξ_p and duration τ;
$a = 0$: triangle of height ξ_p and duration τ;
$a \to +\infty$: spike of height ξ_p and having zero area.

The residual response amplitude of vibration of the system is

$$\frac{\nu_R}{\xi_p} = \frac{2aT}{\pi\tau} \left(\frac{e^a - \cos(\pi\tau/T) - (aT/\pi\tau)\sin(\pi\tau/T)}{(1 - e^a)(1 + a^2T^2/\pi^2\tau^2)} \right) \tag{8.36b}$$

where T is the natural period of the responding system. Figure 8.19*A* shows the spectra for residual response amplitude for seven values of the parameter a, compared on the basis of *equal pulse height*. The zero-area spike ($a \to +\infty$) results in zero response.

The area of the general pulse of height ξ_p is

$$A_p = \xi_p \frac{\tau}{a} \left(\frac{1 - e^a + a}{1 - e^a} \right) \tag{8.36c}$$

If a comparison is to be drawn on the basis of *equal pulse area* using the area $\xi_{po}\tau/2$ of the

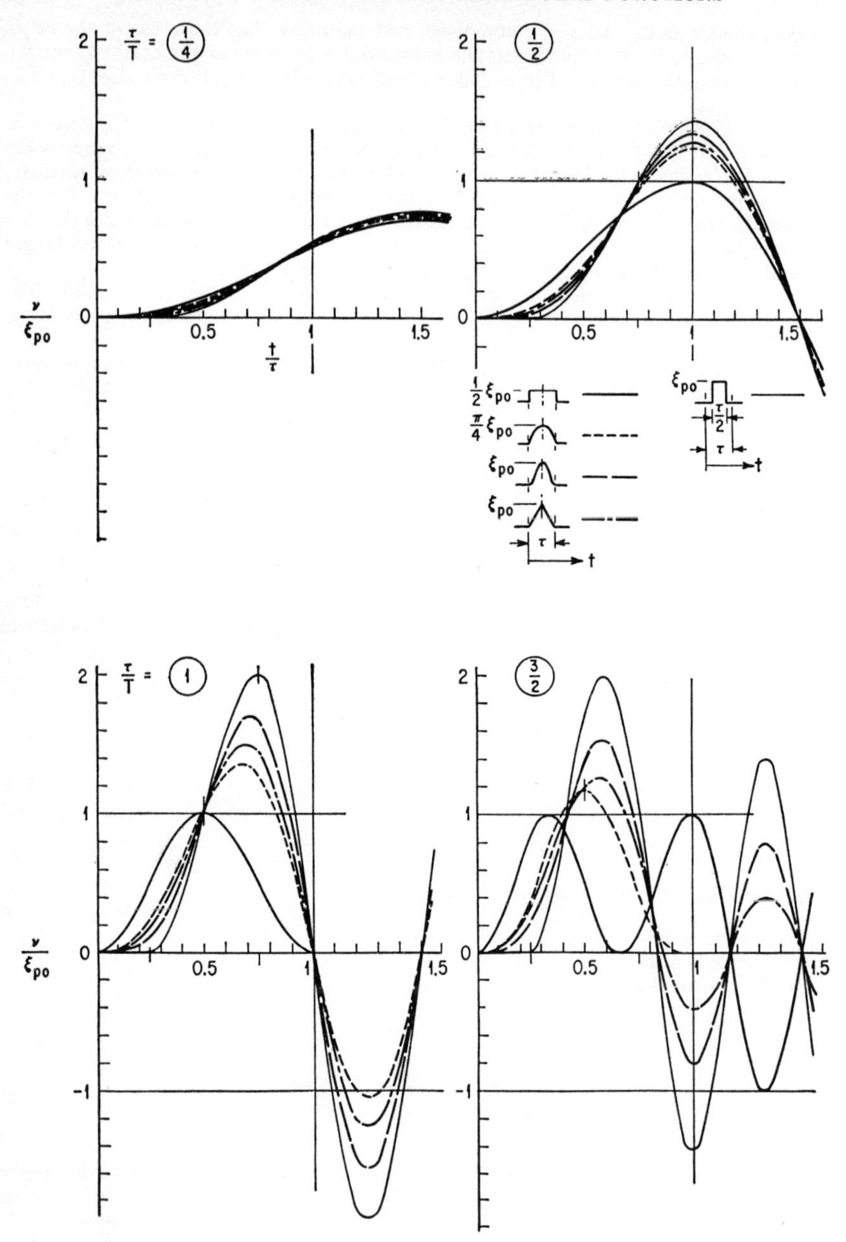

FIG. 8.17. Time response to various symmetrical pulses having equal pulse area, for four different values of τ/T. (*Jacobsen and Ayre.*[22])

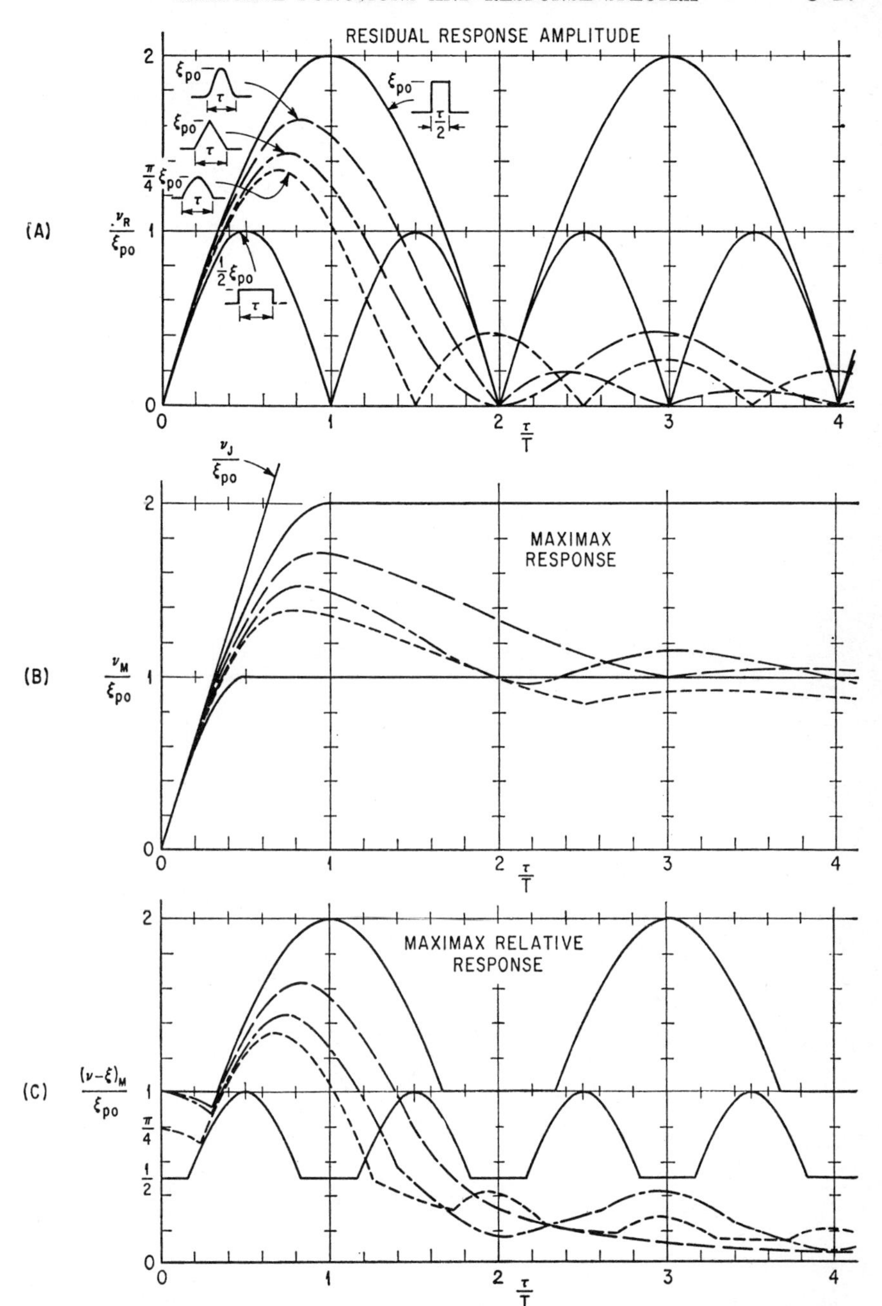

FIG. 8.18. Response spectra for various symmetrical pulses having equal pulse area: (*A*) residual response amplitude, (*B*) maximax response, and (*C*) maximax relative response. (*Jacobsen and Ayre.*[22])

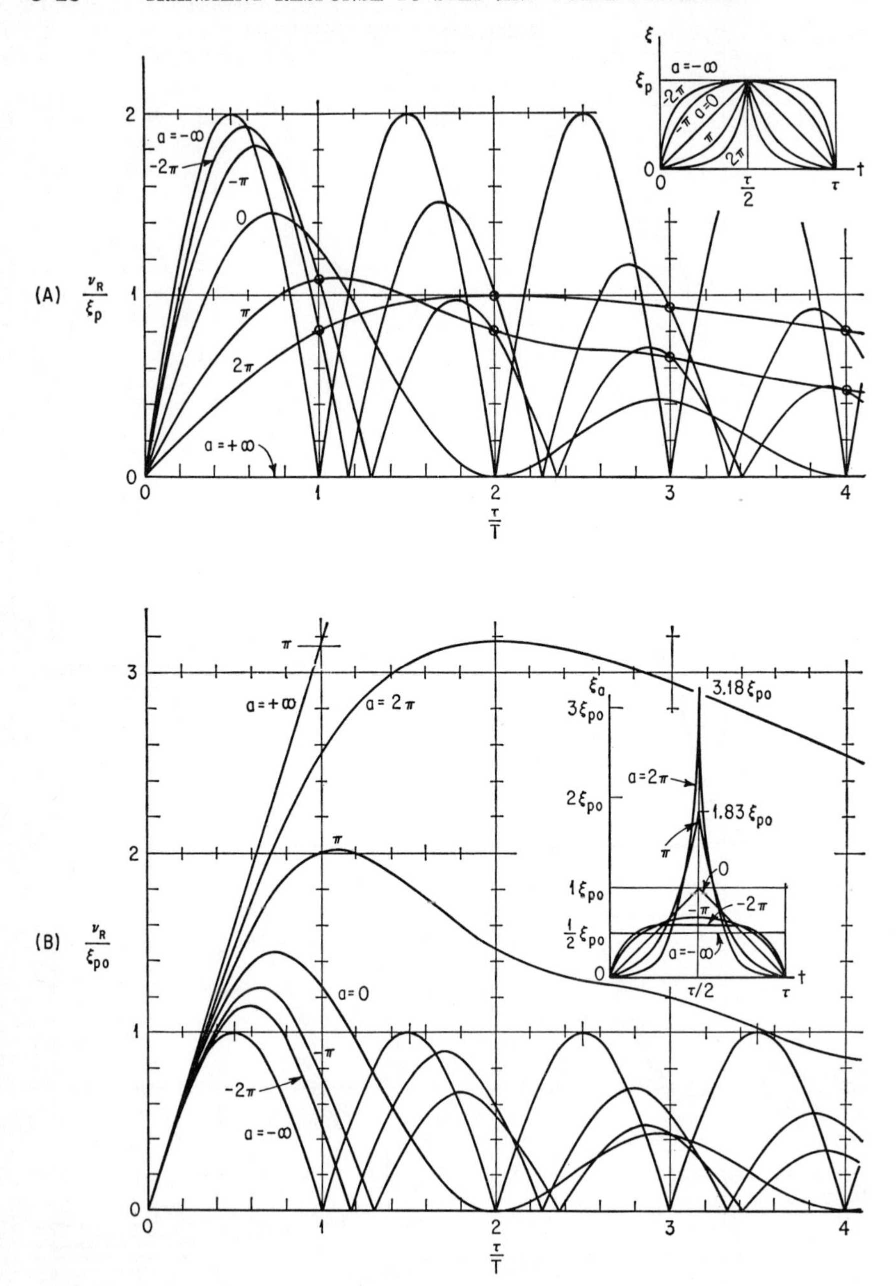

FIG. 8.19. Spectra for residual response amplitude for a family of exponential, symmetrical pulses: (*A*) pulses having equal height; (*B*) pulses having equal area. (*Jacobsen and Ayre.*[22])

triangular pulse as the reference, the height ξ_{pa} of the general pulse is

$$\xi_{pa} = \xi_{po} \frac{a}{2} \left(\frac{1 - e^a}{1 - e^a + a} \right) \tag{8.36d}$$

The residual response amplitude spectra, based on the equal-pulse-area criterion, are shown in Fig. 8.19*B*. The case $a \rightarrow +\infty$ is equivalent to a generalized impulse of value $k\xi_{po}\tau/2$ and results in the straight-line spectrum given by Eq. (8.35).

Symmetrical Pulses Having a Rest Period of Constant Height. In the inset diagrams of Fig. 8.20 each pulse consists of a rise, a central rest period or "dwell" having constant height, and a decay. The expressions for the pulse *rise* functions may be obtained from Eqs. (8.24*a*), (8.25*a*), and (8.26*a*) by substituting $\tau/2$ for τ. The pulse *decay* functions are available from symmetry.

If the *rest* period is long enough for the maximax displacement of the system to be reached during the duration τ_r of the pulse rest, the maximax may be obtained from Eqs. (8.24*b*), (8.25*b*), and (8.26*b*) and, consequently, from Fig. 8.13. The substitution of $\tau/2$ for τ is necessary.

Equations (8.37) to (8.39) give the residual response amplitudes. The spectra computed from these equations are shown in Fig. 8.20.

Constant-slope rise and decay:

$$\frac{\nu_R}{\xi_p} = \frac{2T}{\pi\tau} \left[1 - \cos\frac{\pi\tau}{T} + \frac{1}{2}\cos\frac{2\pi\tau_r}{T} - \cos\frac{\pi(\tau + 2\tau_r)}{T} + \frac{1}{2}\cos\frac{2\pi(\tau + \tau_r)}{T} \right]^{1/2} \tag{8.37}$$

Versed-sine rise and decay:

$$\frac{\nu_R}{\xi_p} = \frac{1}{1 - \tau^2/T^2} \left[1 + \cos\frac{\pi\tau}{T} - \frac{1}{2}\cos\frac{2\pi\tau_r}{T} - \cos\frac{\pi(\tau + 2\tau_r)}{T} - \frac{1}{2}\cos\frac{2\pi(\tau + \tau_r)}{T} \right]^{1/2} \tag{8.38}$$

Cycloidal rise and decay:

$$\frac{\nu_R}{\xi_p} = \frac{2T/\pi\tau}{1 - \tau^2/4T^2} \left[\cos\frac{\pi\tau_r}{T} - \cos\frac{\pi(\tau + \tau_r)}{T} \right] \tag{8.39}$$

Note that τ in the abscissa is the sum of the rise time and the decay time and is *not* the total duration of the pulse. Attached to each spectrum is a set of values of τ_r/T where T is the natural period of the responding system.

When $\tau_r/T = 1, 2, 3, \ldots$, the residual response amplitude is equal to that for the case $\tau_r = 0$, and the spectrum starts at the origin. If $\tau_r/T = \frac{1}{2}, \frac{3}{2}, \frac{5}{2}, \ldots$, the spectrum has the maximum value 2.00 at $\tau/T = 0$. The *envelopes* of the spectra are of the same forms as the *residual-response-amplitude* spectra for the related *step functions*; see the spectra for $[(\nu_M/\xi_c) - 1]$ in Fig. 8.13*A*. In certain cases, for example, at $\tau/T = 2, 4, 6, \ldots$, in Fig. 8.20*A*, $\nu_R/\xi_p = 0$ for all values of τ_r/T.

UNSYMMETRICAL PULSES. Pulses having only slight asymmetry may often be represented adequately by symmetrical forms. However, if there is considerable asymmetry, resulting in appreciable steepening of either the rise or the decay, it is necessary to introduce a parameter which defines the *skewing* of the pulse.

The ratio of the rise time to the pulse period is called the *skewing constant*, $\sigma = t_1/\tau$. There are three special cases:

$\sigma = 0$: the pulse has an *instantaneous* (vertical) *rise*, followed by a decay having the duration τ. This case may be used as an elementary representation of a *blast pulse*.

$\sigma = \frac{1}{2}$: the pulse may be *symmetrical*.

$\sigma = 1$: the pulse has an *instantaneous decay*, preceded by a rise having the duration τ.

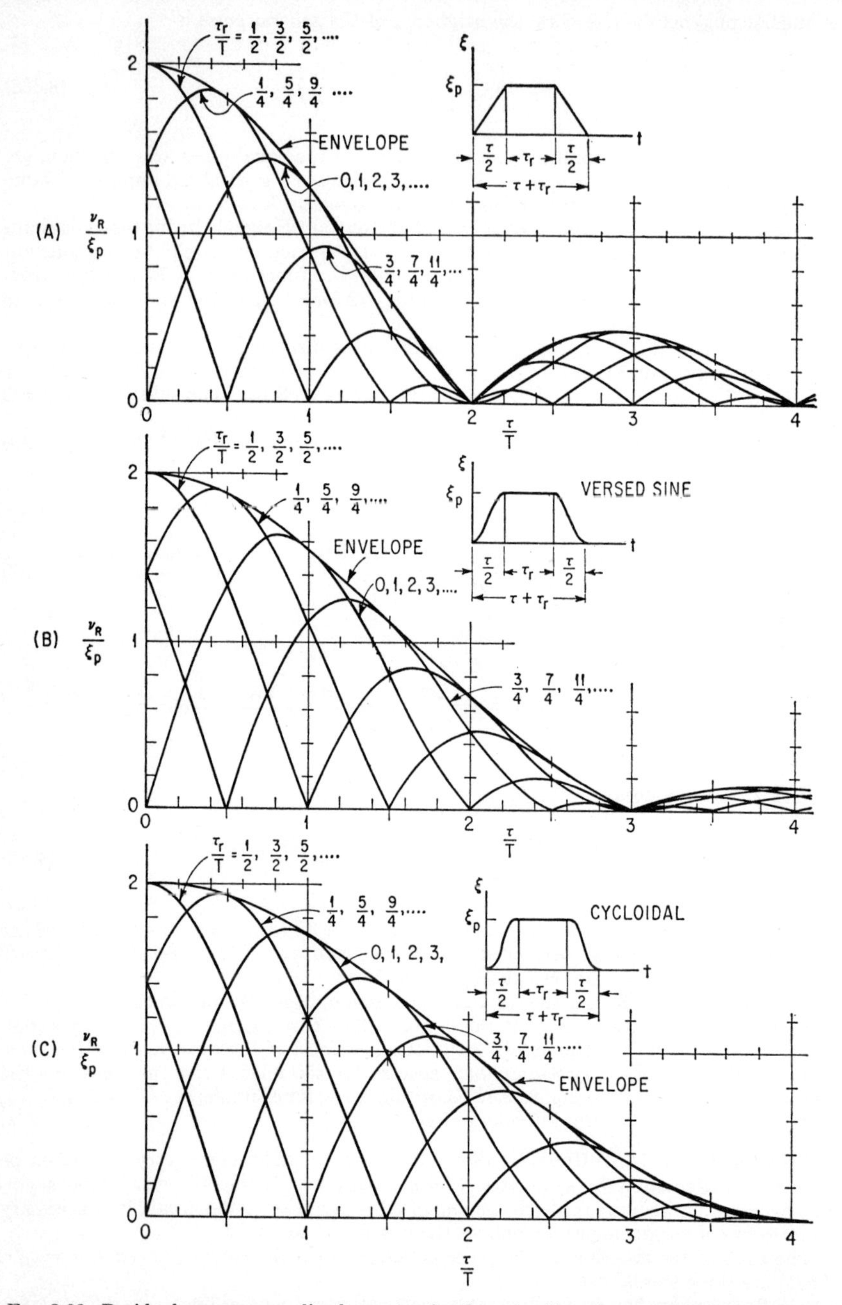

FIG. 8.20. Residual response amplitude spectra for three families of symmetrical pulses having a central rest period of constant height and of duration τ_r. Note that the abscissa is τ/T, where τ is the sum of the rise time and the decay time. (*A*) Constant-slope rise and decay. (*B*) Versed-sine rise and decay. (*C*) Cycloidal rise and decay. (*Jacobsen and Ayre.*[22])

Triangular Pulse Family. The effect of asymmetry in pulse shape is shown readily by means of the family of triangular pulses (Fig. 8.21). Equations (8.40) give the excitation and the time response.

Rise era: $0 \leq t \leq t_1$

$$\xi(t) = \xi_p \frac{t}{t_1}$$

$$\nu = \xi_p \left(\frac{t}{t_1} - \frac{T}{2\pi t_1} \sin \omega_n t \right) \tag{8.40a}$$

Decay era: $0 \leq t' \leq t_2$, where $t' = t - t_1$

$$\xi(t) = \xi_p \left(1 - \frac{t'}{t_2} \right)$$

$$\nu = \xi_p \left[1 - \frac{t'}{t_2} + \frac{T}{2\pi t_2} \left(1 + 4 \frac{t_2}{t_1} \frac{\tau}{t_1} \sin^2 \frac{\pi t_1}{T} \right)^{1/2} \sin (\omega_n t' + \theta') \right] \tag{8.40b}$$

where

$$\tan \theta' = \frac{\sin (2\pi t_1/T)}{\cos (2\pi t_1/T) - \tau/t_2}$$

Residual-vibration era: $0 \leq t''$, where $t'' = t - \tau = t - t_1 - t_2$

$$\xi(t) = 0$$

$$\nu = \xi_p \frac{1}{\pi} \left[\frac{T}{t_1} \frac{T}{t_2} \left(\frac{\tau}{t_1} \sin^2 \frac{\pi t_1}{T} + \frac{\tau}{t_2} \sin^2 \frac{\pi t_2}{T} - \sin^2 \frac{\pi \tau}{T} \right) \right]^{1/2} \sin (\omega_n t'' + \theta_R) \tag{8.40c}$$

where

$$\tan \theta_R = \frac{(\tau/t_2) \sin (2\pi t_2/T) - \sin (2\pi \tau/T)}{(\tau/t_2) \cos (2\pi t_2/T) - \cos (2\pi \tau/T) - t_1/t_2}$$

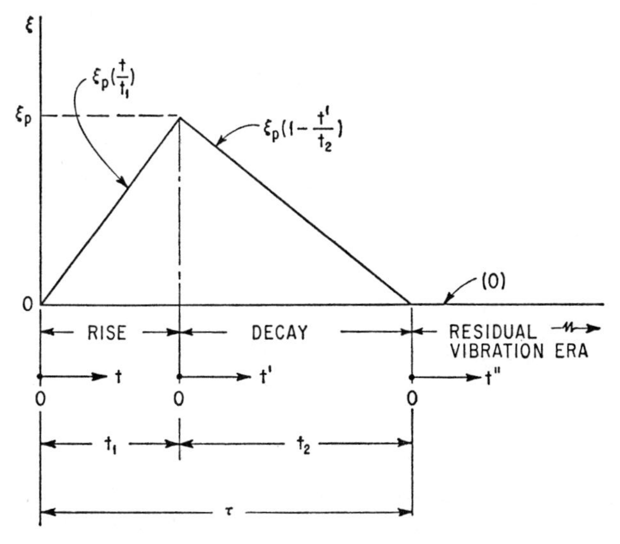

FIG. 8.21. General triangular pulse.

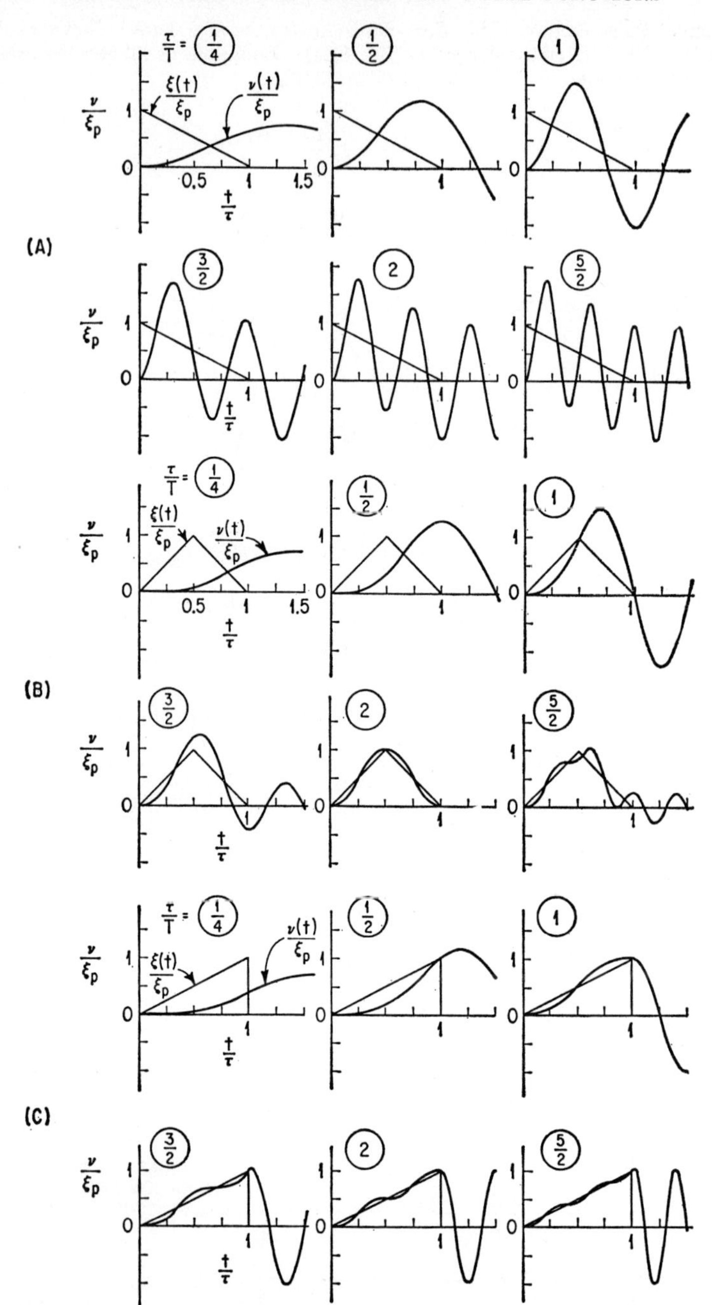

Fig. 8.22. Time response curves resulting from single pulses of three different triangular shapes: (*A*) vertical rise (elementary blast pulse), (*B*) symmetrical, and (*C*) vertical decay.[19]

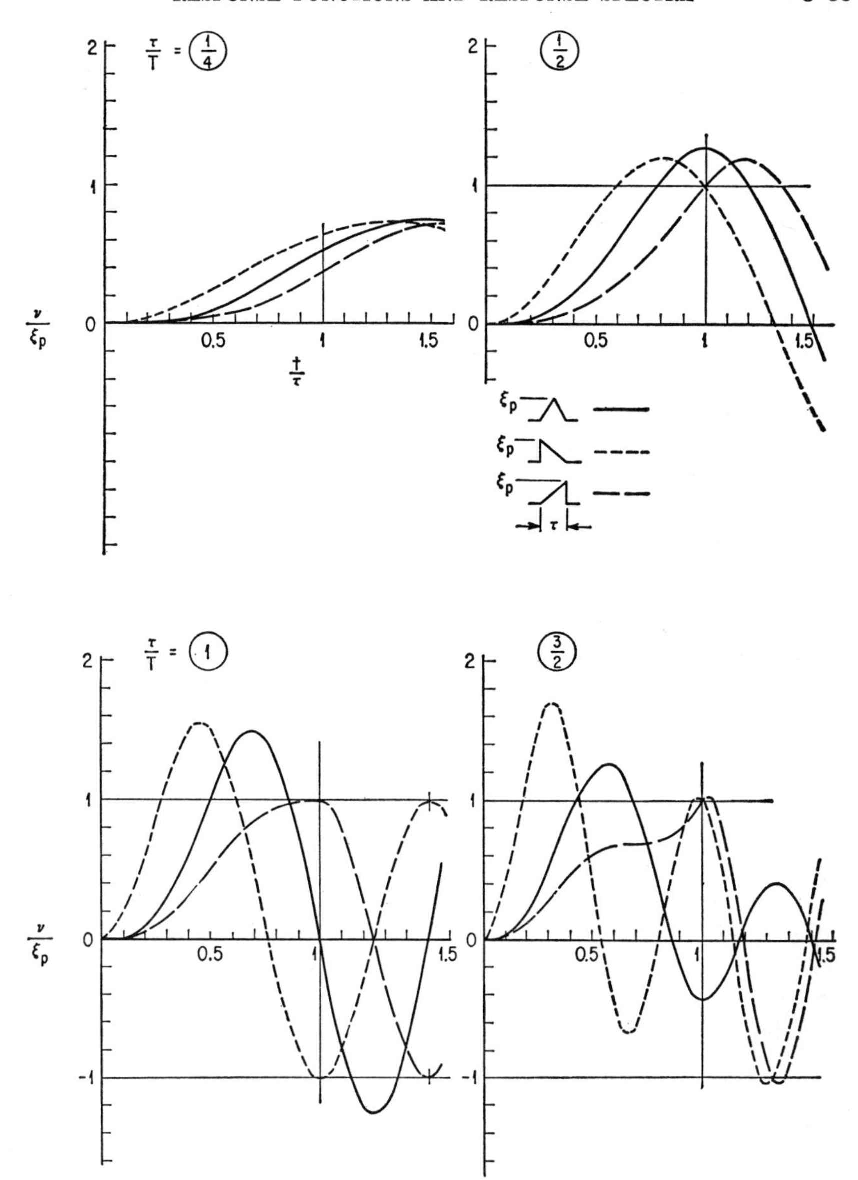

FIG. 8.23. Time response curves of Fig. 8.22 superposed, for four values of τ/T. (*Jacobsen and Ayre.*[22])

For the special cases $\sigma = 0$, ½ and 1, the time responses for six values of τ/T are shown in Fig. 8.22, where T is the natural period of the responding system. Some of the curves are superposed in Fig. 8.23 for easier comparison. The response spectra appear in Fig. 8.24. The straight-line spectrum ν_J/ξ_p for the amplitude of response based on the impulse theory also is shown in Fig. 8.24*A*. In the two cases of extreme skewing, $\sigma = 0$ and

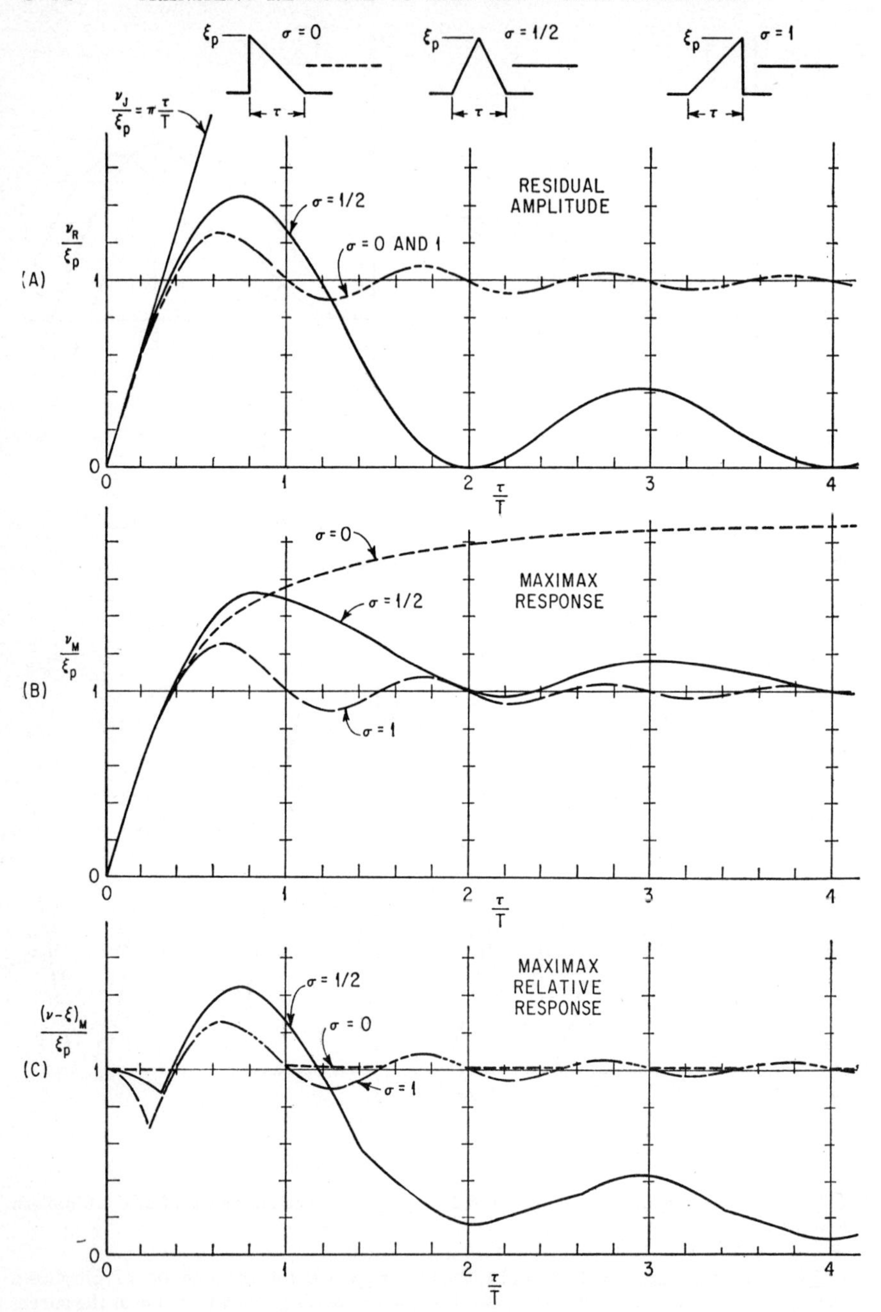

FIG. 8.24. Response spectra for three types of triangular pulse: (*A*) Residual response amplitude. (*B*) Maximax response. (*C*) Maximax relative response. (*Jacobsen and Ayre.*[22])

$\sigma = 1$, the residual amplitudes are *equal* and are given by Eq. (8.41*a*). For the symmetrical case, $\sigma = \frac{1}{2}$, ν_R is given by Eq. (8.41*b*).

$$\sigma = 0 \text{ and } 1: \quad \frac{\nu_R}{\xi_p} = \left[1 - \frac{T}{\pi\tau}\sin\frac{2\pi\tau}{T} + \left(\frac{T}{\pi\tau}\right)^2 \sin^2\frac{\pi\tau}{T}\right]^{\frac{1}{2}} \tag{8.41a}$$

$$\sigma = \tfrac{1}{2}: \quad \frac{\nu_R}{\xi_p} = 2\,\frac{\sin^2(\pi\tau/2T)}{\pi\tau/2T} \tag{8.41b}$$

The residual response amplitudes for other cases of skewness may be determined from the amplitude term in Eqs. (8.40*c*); they are shown by the response spectra in Fig. 8.25. The residual response amplitudes resulting from single pulses that are mirror images of

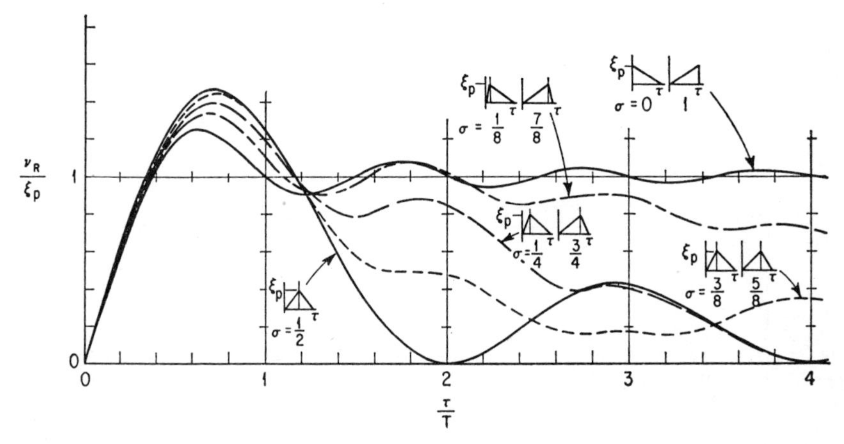

FIG. 8.25. Spectra for residual response amplitude for a family of triangular pulses of varying skewness. (*Jacobsen and Ayre.*[22])

each other in time are equal. In general, the phase angles for the residual vibrations are unequal.

Note that in the cases $\sigma = 0$ and $\sigma = 1$ for *vertical rise* and *vertical decay*, respectively, there are *no zeroes of residual amplitude*, except for the trivial case, $\tau/T = 0$.

The family of triangular pulses is particularly advantageous for investigating the effect of varying the skewness, because both criteria of comparison, equal pulse height and equal pulse area, are satisfied simultaneously.

Various Pulses Having Vertical Rise or Vertical Decay. Figure 8.26 shows the spectra of residual response amplitude plotted on the basis of *equal pulse area*. The rectangular pulse is included for comparison. The expressions for residual response amplitude for the rectangular and the triangular pulses are given by Eqs. (8.31*b*) and (8.41*a*), and for the quarter-cycle sine and the half-cycle versed-sine pulses by Eqs. (8.42) and (8.43).

Quarter-cycle "sine":

$$\xi(t) = \xi_p \begin{cases} \sin\dfrac{\pi t}{2\tau} & \text{for vertical decay} \\ \text{or} & \\ \cos\dfrac{\pi t}{2\tau} & \text{for vertical rise} \end{cases} \qquad [0 \le t \le \tau]$$

$$\xi(t) = 0 \qquad [\tau \le t]$$

$$\frac{\nu_R}{\xi_p} = \frac{4\tau/T}{(16\tau^2/T^2) - 1}\left(1 + \frac{16\tau^2}{T^2} - \frac{8\tau}{T}\sin\frac{2\pi\tau}{T}\right)^{\frac{1}{2}} \tag{8.42}$$

Half-cycle "versed-sine":

$$\xi(t) = \xi_p \begin{cases} \dfrac{1}{2}\left(1 - \cos\dfrac{\pi t}{\tau}\right) & \text{for vertical decay} \\ \text{or} & \\ \dfrac{1}{2}\left(1 + \cos\dfrac{\pi t}{\tau}\right) & \text{for vertical rise} \end{cases} \qquad [0 \le t \le \tau]$$

$$\xi(t) = 0 \qquad [\tau \le t]$$

$$\frac{\nu_R}{\xi_p} = \frac{1/2}{(4\tau^2/T^2) - 1}\left[1 + \left(1 - \frac{8\tau^2}{T^2}\right)^2 - 2\left(1 - \frac{8\tau^2}{T^2}\right)\cdot\cos\frac{2\pi\tau}{T}\right]^{1/2} \tag{8.43}$$

where T is the natural period of the responding system.

Note again that the residual response amplitudes, caused by single pulses that are mirror images in time, are equal. Furthermore, it is seen that the unsymmetrical pulses,

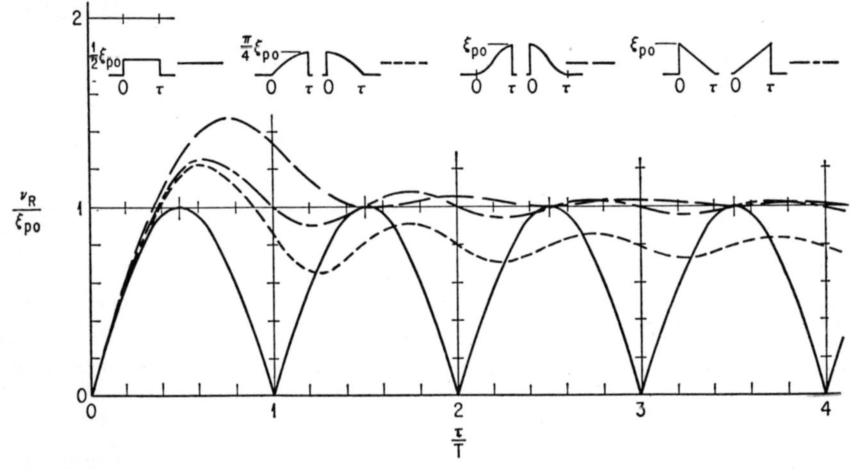

FIG. 8.26. Spectra for residual response amplitude for various unsymmetrical pulses having either vertical rise or vertical decay. Comparison on the basis of equal pulse area.[19]

having either vertical rise or vertical decay, result in no zeroes of residual response amplitude, except in the trivial case $\tau/T = 0$.

Exponential Pulses of Finite Duration, Having Vertical Rise or Vertical Decay. Families of exponential pulses having either a vertical rise or a vertical decay, as shown in the inset diagrams in Fig. 8.27, can be formed by Eqs. (8.44*a*) and (8.44*b*).

Vertical rise with exponential decay:

$$\xi(t) = \begin{cases} \xi_p\left(\dfrac{1 - e^{a(1-t/\tau)}}{1 - e^a}\right) & [0 \le t \le \tau] \\ 0 & [\tau \le t] \end{cases} \tag{8.44a}$$

Exponential rise with vertical decay:

$$\xi(t) = \begin{cases} \xi_p\left(\dfrac{1 - e^{at/\tau}}{1 - e^a}\right) & [0 \le t \le \tau] \\ 0 & [\tau \le t] \end{cases} \tag{8.44b}$$

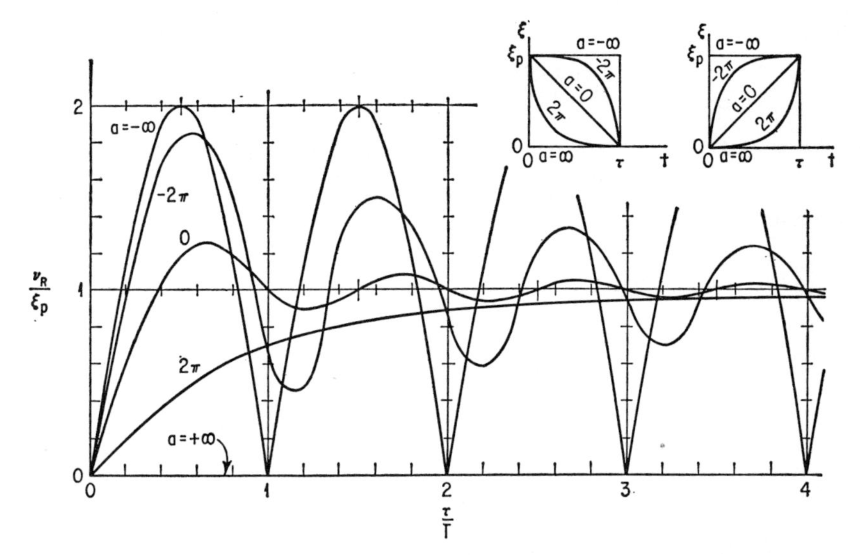

FIG. 8.27. Spectra for residual response amplitude for unsymmetrical exponential pulses having either vertical rise or vertical decay. Comparison on the basis of equal pulse height.[19]

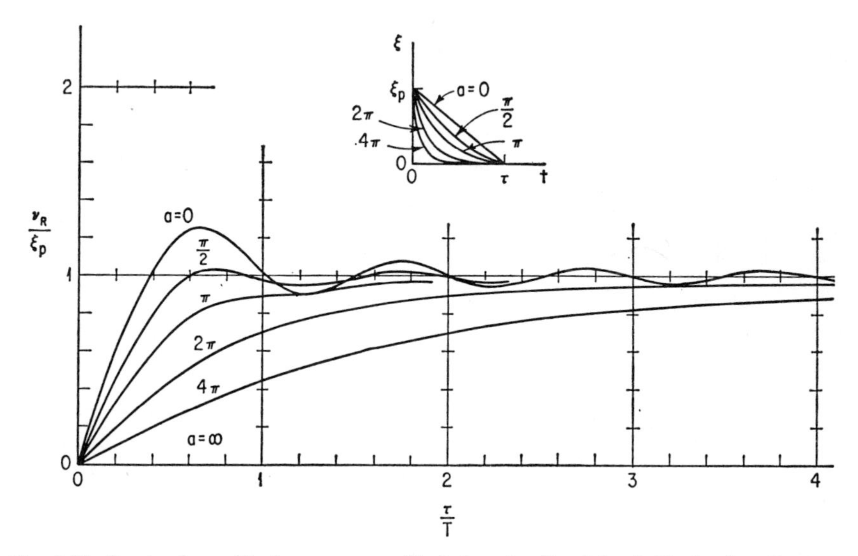

FIG. 8.28. Spectra for residual response amplitude for a family of simple blast pulses, the same family shown in Fig. 8.27 but limited to positive values of the exponential decay parameter *a*. Comparison on the basis of equal pulse height. (These spectra also apply to mirror-image pulses having vertical decay.) (*Jacobsen and Ayre.*[22])

Residual response amplitude for *either form of pulse:*

$$\frac{\nu_R}{\xi_p} = \frac{a}{1 - e^a} \left\{ \frac{[(2\pi\tau/T)(1 - e^a)/a + \sin(2\pi\tau/T)]^2 + [1 - \cos(2\pi\tau/T)]^2}{a^2 + 4\pi^2\tau^2/T^2} \right\}^{1/2} \tag{8.44c}$$

When $a = 0$, the pulses are triangular with vertical rise or vertical decay. If $a \to +\infty$ or $-\infty$, the pulses approach the shape of a zero-area spike or of a rectangle, respectively. The spectra for residual response amplitude, plotted on the basis of *equal pulse height,* are shown in Fig. 8.27.

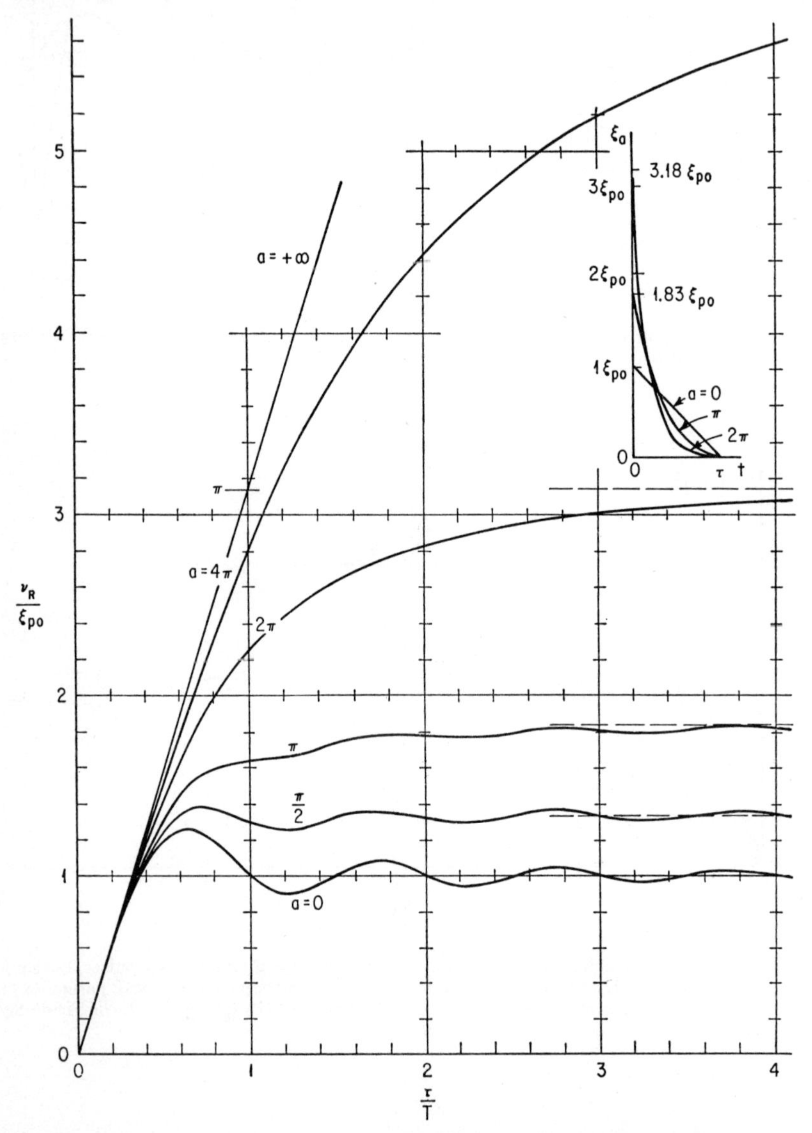

FIG. 8.29. Spectra for residual response amplitude for the family of simple blast pulses shown in Fig. 8.28, compared on the basis of equal pulse area. (*Jacobsen and Ayre.*[22])

Figure 8.28 shows the spectra of residual response amplitude in greater detail for the range in which the parameter a is limited to positive values. This group of pulses is of interest in studying the effects of a simple form of blast pulse, in which the peak height and the duration are constant but the rate of decay is varied.

The areas of the pulses of equal height ξ_p, and the heights of the pulses of equal area $\xi_{po}\tau/2$ are the same as for the symmetrical exponential pulses [see Eqs. (8.36*c*) and (8.36*d*)]. If the spectra in Fig. 8.28 are redrawn, using *equal pulse area* as the criterion for comparison, they appear as in Fig. 8.29. The limiting pulse case $a \to +\infty$ represents a generalized impulse of value $k\xi_{po}\tau/2$. The asymptotic values of the spectra are equal to the peak heights of the equal area pulses and are given by

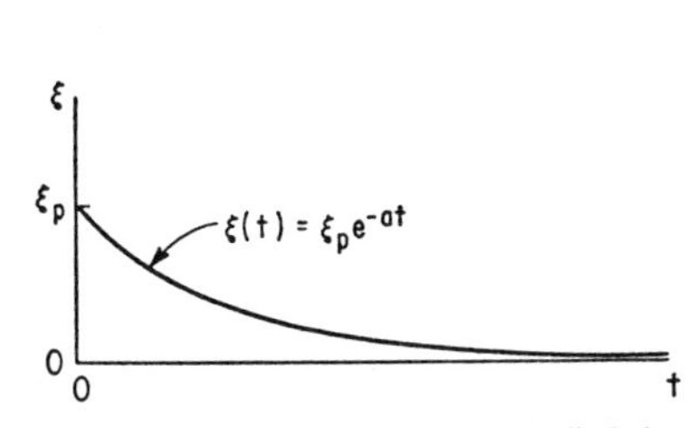

FIG. 8.30. Pulse consisting of vertical rise followed by exponential decay of infinite duration.

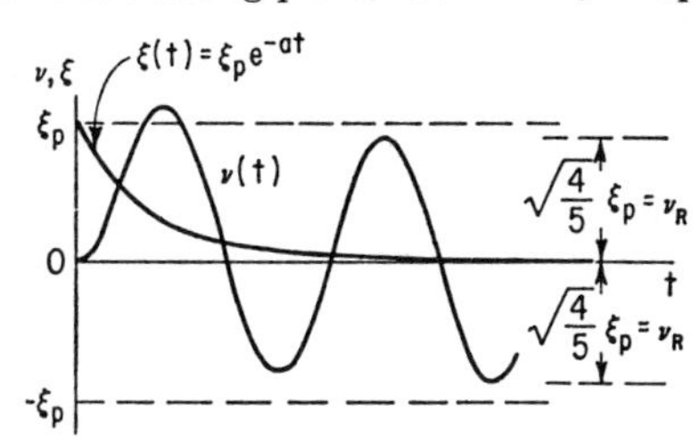

FIG. 8.31. Time response to the pulse having a vertical rise and an exponential decay of infinite duration (Fig. 8.30), for the particular case $\omega_n/a = 2$.

$$\frac{\nu_R}{\xi_{po}} \to \frac{a(1 - e^a)}{2(1 - e^a + a)} \qquad \text{as } \frac{\tau}{T} \to \infty \tag{8.44d}$$

Exponential Pulses of Infinite Duration. Five different cases are included as follows:

1. The excitation function, consisting of a vertical rise followed by an exponential decay, is

$$\xi(t) = \xi_p e^{-at} \qquad [0 \le t] \tag{8.45a}$$

It is shown in Fig. 8.30. The response time equation for the system is

$$\nu = \xi_p \frac{(a/\omega_n)\sin \omega_n t - \cos \omega_n t + e^{-at}}{1 + a^2/\omega_n^2} \tag{8.45b}$$

and the asymptotic value of the residual amplitude is given by

$$\frac{\nu_R}{\xi_p} \to \frac{1}{\sqrt{1 + a^2/\omega_n^2}} \tag{8.45c}$$

The maximax response is the first maximum of ν. The time response, for the particular case $\omega_n/a = 2$, and the response spectra are shown in Figs. 8.31 and 8.32, respectively.

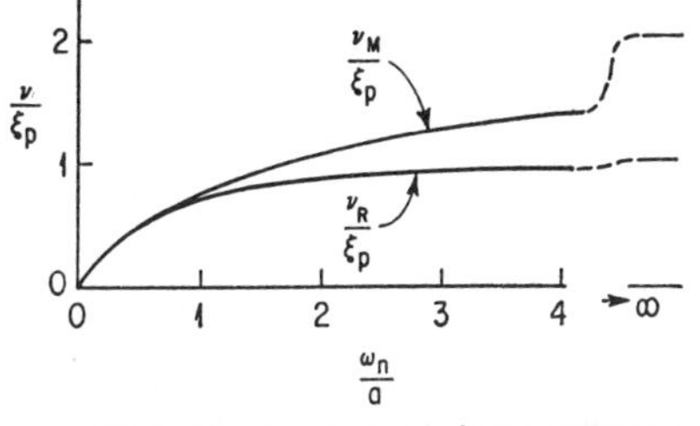

FIG. 8.32. Spectra for maximax response and for asymptotic residual response amplitude, for the pulse shown in Fig. 8.30.

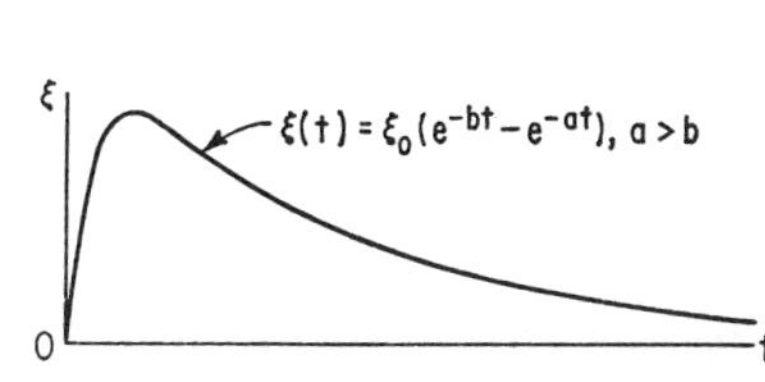

FIG. 8.33. Pulse formed by taking the difference of two exponentially decaying functions.

2. The *difference of two exponential functions*, of the type of Eq. (8.45*a*), results in the pulse given by Eq. (8.46*a*):

$$\xi(t) = \xi_0(e^{-bt} - e^{-at}) \qquad (8.46a)$$

$$a > b \qquad [0 \leq t]$$

The shape of the pulse is shown in Fig. 8.33. Note that ξ_0 is the ordinate of each of the exponential functions at $t = 0$; it is *not* the pulse maximum. The asymptotic residual response amplitude is

$$\nu_R \to \xi_0 \frac{(b/\omega_n) - (a/\omega_n)}{[(1 + a^2/\omega_n^2)(1 + b^2/\omega_n^2)]^{1/2}} \qquad (8.46b)$$

3. The *product of the exponential function* e^{-at} *by time* results in the excitation given by Eq. (8.47*a*) and shown in Fig. 8.34.

$$\xi(t) = C_0 te^{-at} \qquad (8.47a)$$

where C_0 is a constant. The peak height of the pulse ξ_p is equal to C_0/ae, and occurs at the time $t_1 = 1/a$. Equations (8.47*b*) and (8.47*c*) give the time response and the asymptotic residual response amplitude:

$$\nu = \xi_p \frac{ae/\omega_n}{(1 + a^2/\omega_n^2)^2} \left\{ \left[\frac{2a}{\omega_n} + \left(1 + \frac{a^2}{\omega_n^2}\right) \omega_n t \right] e^{-at} - \frac{2a}{\omega_n} \cos \omega_n t - \left(1 - \frac{a^2}{\omega_n^2}\right) \sin \omega_n t \right\} \qquad (8.47b)$$

$$\frac{\nu_R}{\xi_p} \to \frac{e}{(a/\omega_n) + (\omega_n/a)} \qquad (8.47c)$$

The *maximum* value of ν_R occurs in the case $a/\omega_n = 1$, and is given by

$$(\nu_R)_{\max} = \xi_p e/2 = 1.36\xi_p$$

Both of the excitation functions described by Eqs. (8.46*a*) and (8.47*a*) include finite times of rise to the pulse peak. These rise times are dependent on the exponential decay constants.

4. The rise time may be made independent of the decay by inserting a separate rise function before the decay function, as in Fig. 8.35, where a *straight-line rise precedes the exponential decay.* The response-time equations are as follows:

Pulse rise era:

$$\nu = \xi_p \frac{\omega_n t - \sin \omega_n t}{\omega_n t_1} \qquad [0 \leq t \leq t_1] \qquad (8.48a)$$

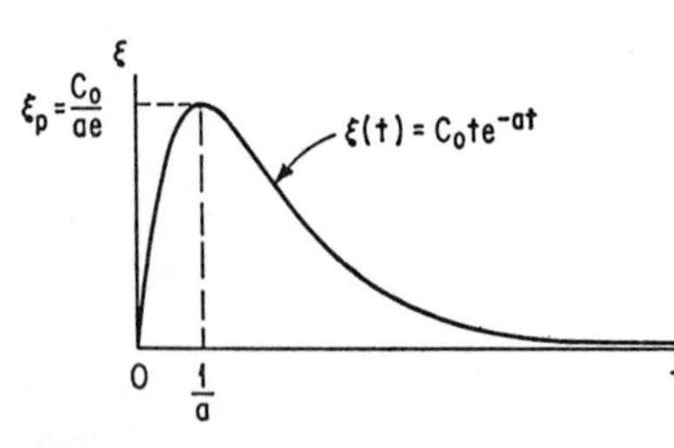

Fig. 8.34. Pulse formed by taking the product of an exponentially decaying function by time.

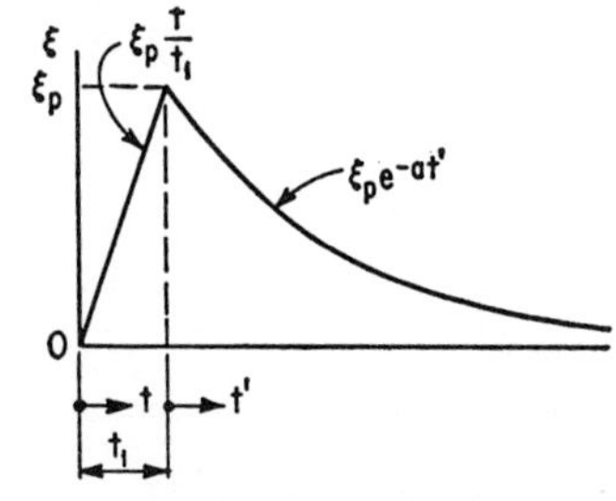

Fig. 8.35. Pulse formed by a straight-line rise followed by an exponential decay asymptotic to the time axis.

Pulse decay era:

$$\nu = \xi_p \left[\frac{e^{-at'}}{1 + a^2/\omega_n^2} + \left(\frac{a^2/\omega_n^2}{1 + a^2/\omega_n^2} - \frac{\sin \omega_n t_1}{\omega_n t_1} \right) \cos \omega_n t' + \left(\frac{a/\omega_n}{1 + a^2/\omega_n^2} + \frac{1 - \cos \omega_n t_1}{\omega_n t_1} \right) \sin \omega_n t' \right] \quad (8.48b)$$

where $t' = t - t_1$ and $0 \leq t'$.

5. Another form of pulse, which is a more complete representation of a blast pulse since it includes the possibility of a negative phase of pressure,[14] is shown in Fig. 8.36. It consists of a straight-line rise, followed by an exponential decay through the positive phase, into the negative phase, finally becoming asymptotic to the time axis. The rise time is t_1 and the duration of the positive phase is $t_1 + t_2$.

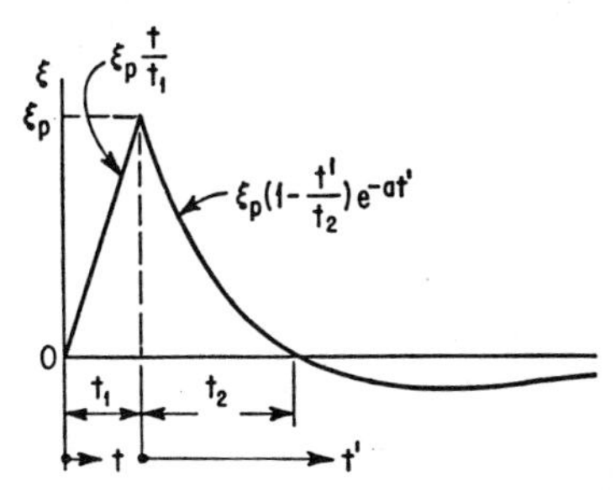

Fig. 8.36. Pulse formed by a straight-line rise followed by a continuous exponential decay through positive and negative phases. (*Frankland*.[14])

Unsymmetrical Exponential Pulses with Central Peak. An interesting family of unsymmetrical pulses may be formed by using Eqs. (8.36*a*) and changing the sign of the exponent of e in both the numerator and the denominator of the second of the equations. The resulting family consists of pulses whose maxima occur at the mid-period time and which satisfy simultaneously both criteria for comparison (equal pulse height and equal pulse area).

Figure 8.37 shows the spectra of residual response amplitude and, in the inset diagrams, the pulse shapes. The limiting cases are the symmetrical triangle of duration τ and height ξ_p, and the rectangles of duration $\tau/2$ and height ξ_p. All pulses in the family have the area $\xi_p \tau/2$. Zeroes of residual response amplitude occur for all values of a, at

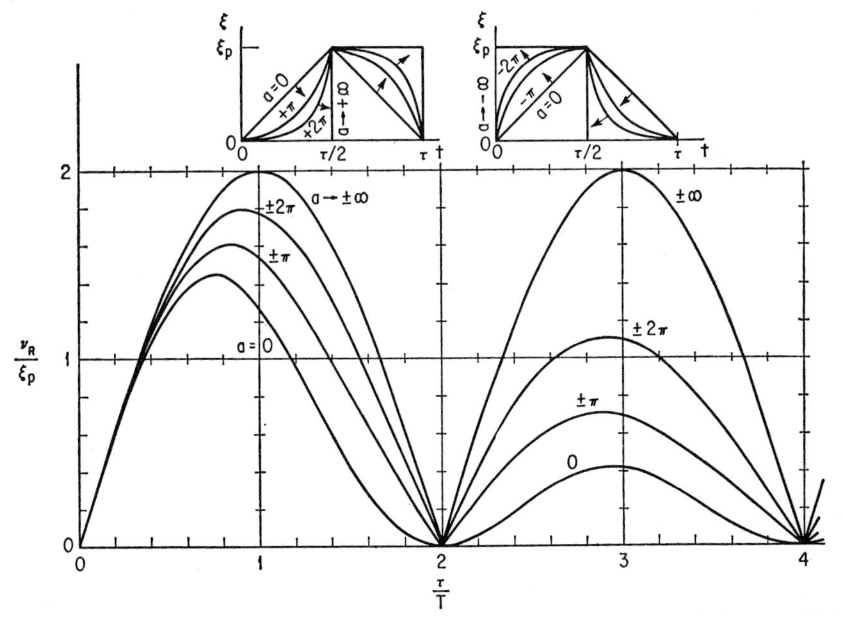

Fig. 8.37. Spectra of residual response amplitude for a family of unsymmetrical exponential pulses of equal area and equal maximum height, having the pulse peak at the mid-period time. (*Jacobsen and Ayre*.[22])

even integer values of τ/T. The residual response amplitude is

$$\frac{\nu_R}{\xi_p} = \frac{aT/\pi\tau}{1 - \cosh a} \cdot \left[\frac{\cosh 2a - \cosh a - (1 - \cosh a)\cos(2\pi\tau/T) + (1 - \cosh 2a)\cos(\pi\tau/T)}{1 + a^2T^2/\pi^2\tau^2}\right]^{1/2} \tag{8.49}$$

Pulses which are mirror images of each other in time result in equal residual amplitudes.

Skewed Versed-sine Pulse. By taking the product of a decaying exponential and the versed-sine function, a family of pulses with varying skewness is obtained.[13,22] The

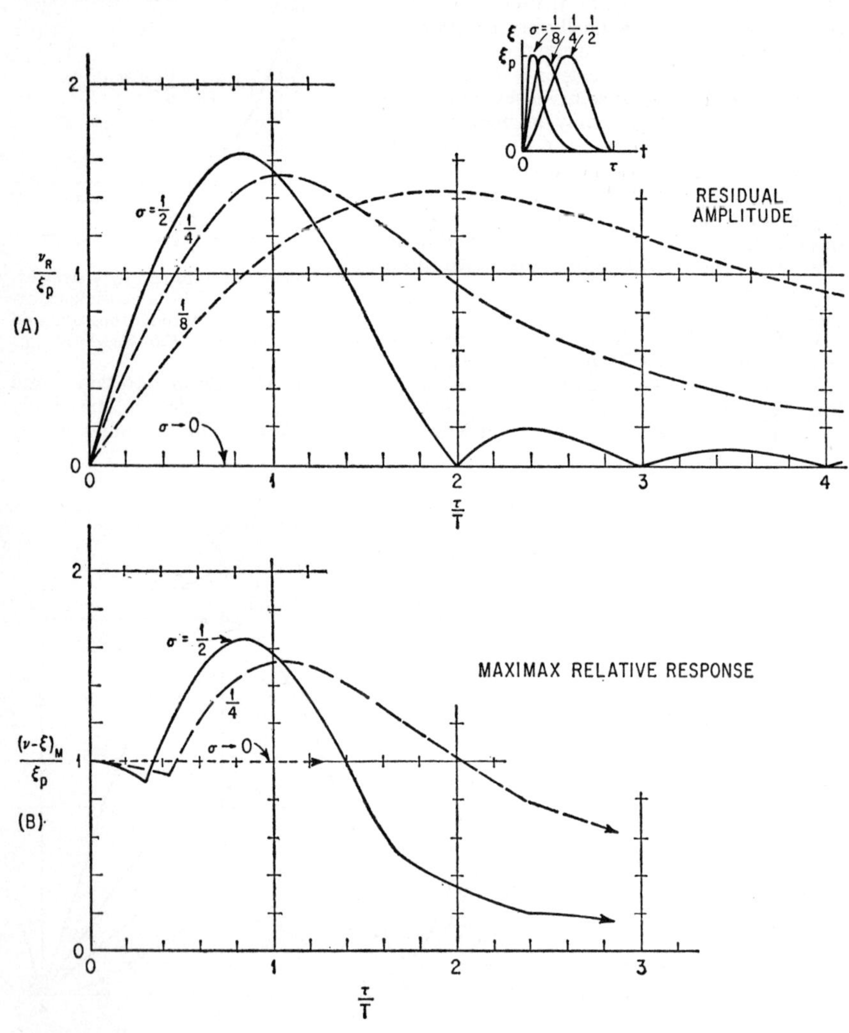

FIG. 8.38. Response spectra for the skewed versed-sine pulse, compared on the basis of equal pulse height: (*A*) Residual response amplitude. (*B*) Maximax relative response. (*Jacobsen and Ayre.*[22])

family is described by the following equation:

$$\xi(t) = \begin{cases} \xi_p \dfrac{e^{2\pi(\sigma - t/\tau)\cot \pi\sigma}}{1 - \cos 2\pi\sigma}(1 - \cos 2\pi t/\tau) & [0 \leq t \leq \tau] \\ 0 & [\tau \leq t] \end{cases} \tag{8.50}$$

These pulses are of particular interest when the excitation is a ground displacement function because they have continuity in both velocity and displacement; thus, they do not involve theoretically infinite accelerations of the ground. When the skewing constant σ equals one-half, the pulse is the symmetrical versed sine. When $\sigma \to 0$, the front of the pulse approaches a straight line with infinite slope, and the pulse area approaches zero.

The spectra of residual response amplitude and of maximax relative response, plotted on the basis of equal pulse height, are shown in Fig. 8.38 for several values of σ. The residual response amplitude spectra are reasonably good approximations to the spectra of maximax relative response except at the lower values of τ/T.

Figure 8.39 compares the residual response amplitude spectra on the basis of equal pulse area. The required pulse heights, for a constant pulse area of $\xi_{po}\tau/2$, are shown in the inset diagram. On this basis, the pulse for $\sigma \to 0$ represents a generalized impulse of value $k\xi_{po}\tau/2$.

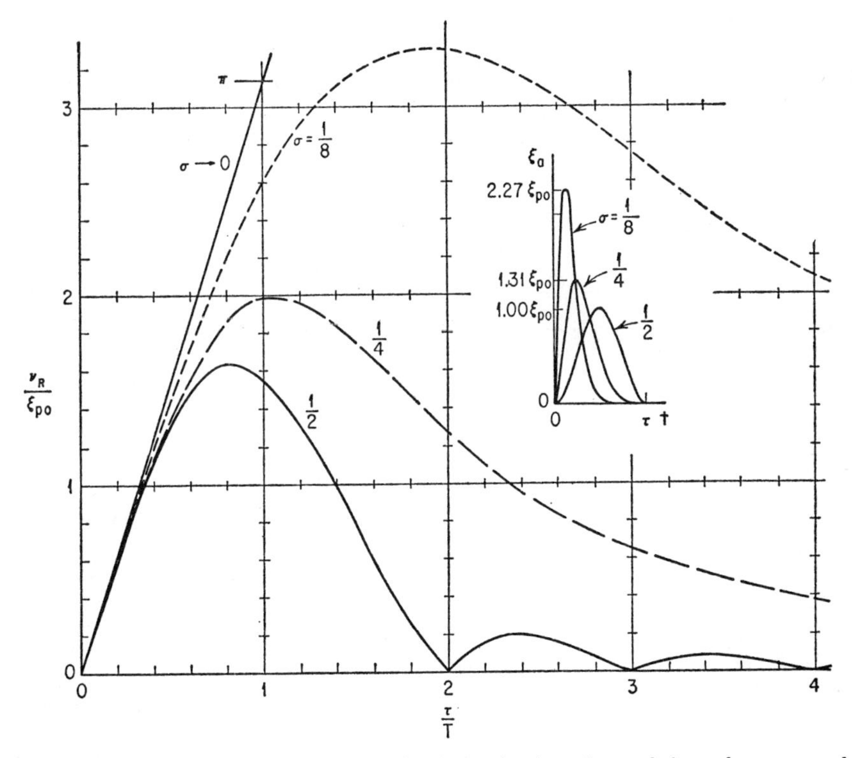

Fig. 8.39. Spectra of residual response amplitude for the skewed versed-sine pulse, compared on the basis of equal pulse area.[19]

FULL-CYCLE PULSES (FORCE-TIME INTEGRAL = 0). The residual response amplitude spectra for three groups of *full-cycle pulses* are shown as follows: in Fig. 8.40 for the rectangular, the sinusoidal, and the symmetrical triangular pulses; in Fig. 8.41 for three types of pulse involving sine and cosine functions; and in Fig. 8.42 for three forms of triangular pulse. The pulse shapes are shown in the inset diagrams. Expressions for the residual response amplitudes are given in Eqs. (8.51) to (8.53).

Full-cycle rectangular pulse:

$$\frac{\nu_R}{\xi_p} = 2 \sin \frac{\pi\tau}{T}\left[2 \sin \frac{\pi\tau}{T}\right] \tag{8.51}$$

Full-cycle "sinusoidal" pulses:
Symmetrical half cycles

$$\frac{\nu_R}{\xi_p} = 2 \sin \frac{\pi\tau}{T}\left[\frac{T/\tau}{(T^2/4\tau^2) - 1} \cos \frac{\pi\tau}{T}\right] \tag{8.52a}$$

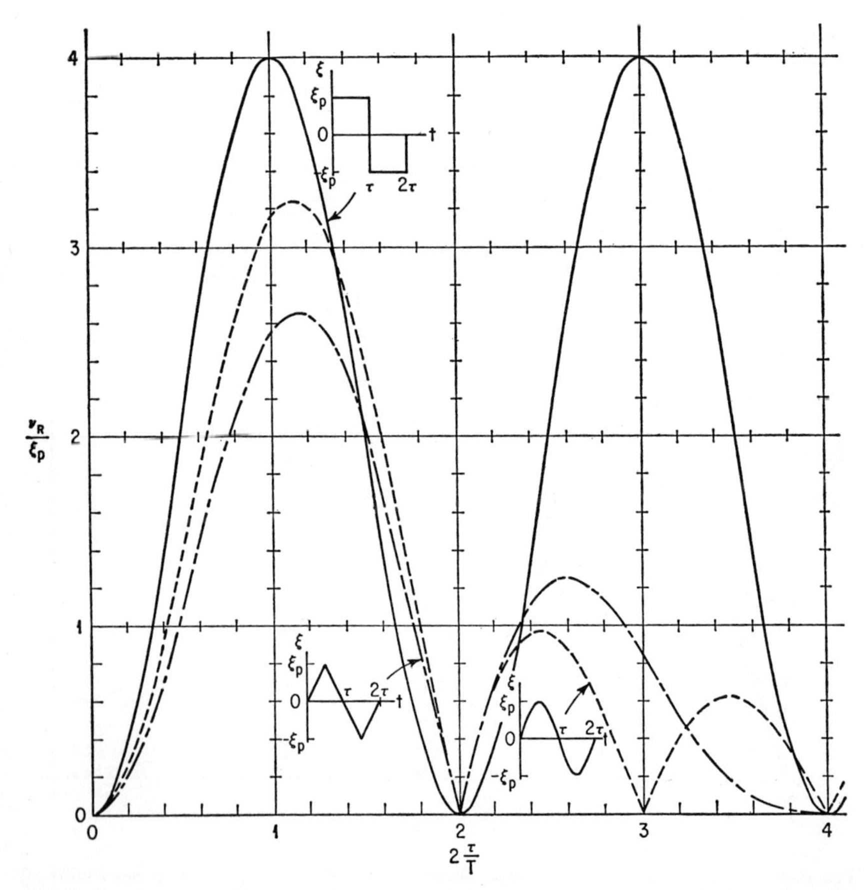

Fig. 8.40. Spectra of residual response amplitude for three types of full-cycle pulses. Each half cycle is symmetrical.[19]

Vertical front and vertical ending

$$\frac{\nu_R}{\xi_p} = \frac{2}{1 - T^2/16\tau^2} \cos \frac{2\pi\tau}{T} \tag{8.52b}$$

Vertical jump at mid-cycle

$$\frac{\nu_R}{\xi_p} = \frac{2}{1 - T^2/16\tau^2} \left(1 - \frac{T}{4\tau} \sin \frac{2\pi\tau}{T}\right) \tag{8.52c}$$

Full-cycle triangular pulses:

Symmetrical half cycles

$$\frac{\nu_R}{\xi_p} = 2 \sin \frac{\pi\tau}{T} \left[\frac{4T}{\pi\tau} \sin^2 \frac{\pi\tau}{2T}\right] \tag{8.53a}$$

Vertical front and vertical ending

$$\frac{\nu_R}{\xi_p} = 2 \left(\frac{T}{2\pi\tau} \sin \frac{2\pi\tau}{T} - \cos \frac{2\pi\tau}{T}\right) \tag{8.53b}$$

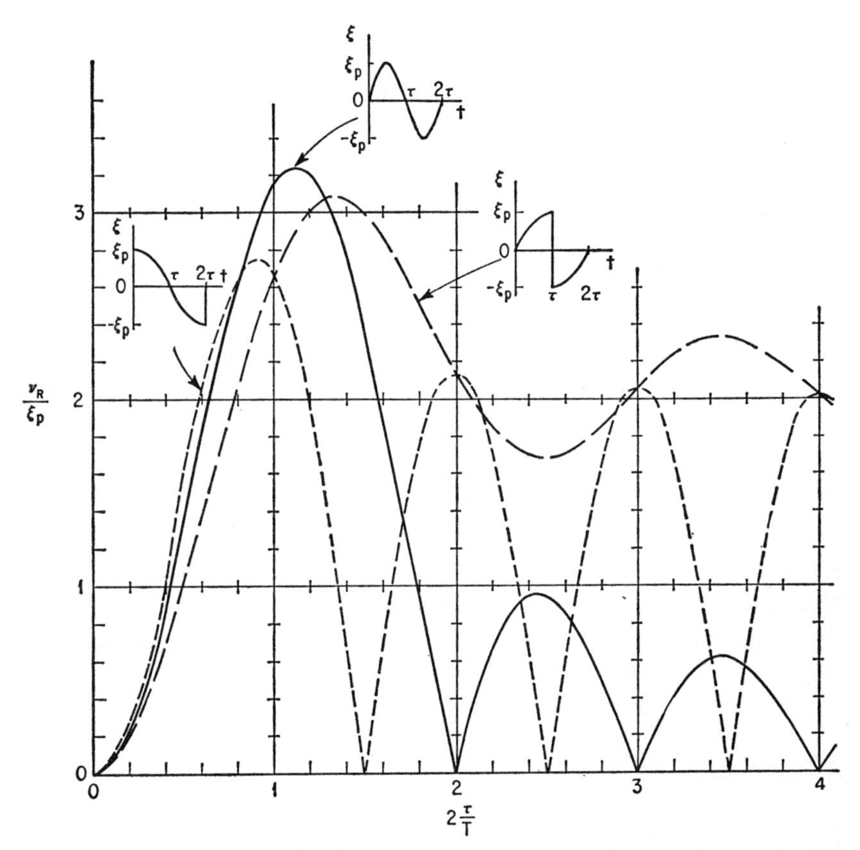

Fig. 8.41. Spectra of residual response amplitude for three types of full-cycle "sinusoidal" pulses.[19]

Vertical jump at mid-cycle

$$\frac{\nu_R}{\xi_p} = 2\left(1 - \frac{T}{2\pi\tau}\sin\frac{2\pi\tau}{T}\right) \tag{8.53c}$$

In the case of full-cycle pulses having symmetrical half cycles, note that the residual response amplitude equals the residual response amplitude of the symmetrical one-half-cycle pulse of the same shape, multiplied by the dimensionless residual response amplitude function $2\sin(\pi\tau/T)$ for the single rectangular pulse. Compare the bracketed functions in Eqs. (8.51), (8.52*a*), and (8.53*a*) with the bracketed functions in Eqs. (8.31*b*), (8.32*b*), and (8.34*c*), respectively.

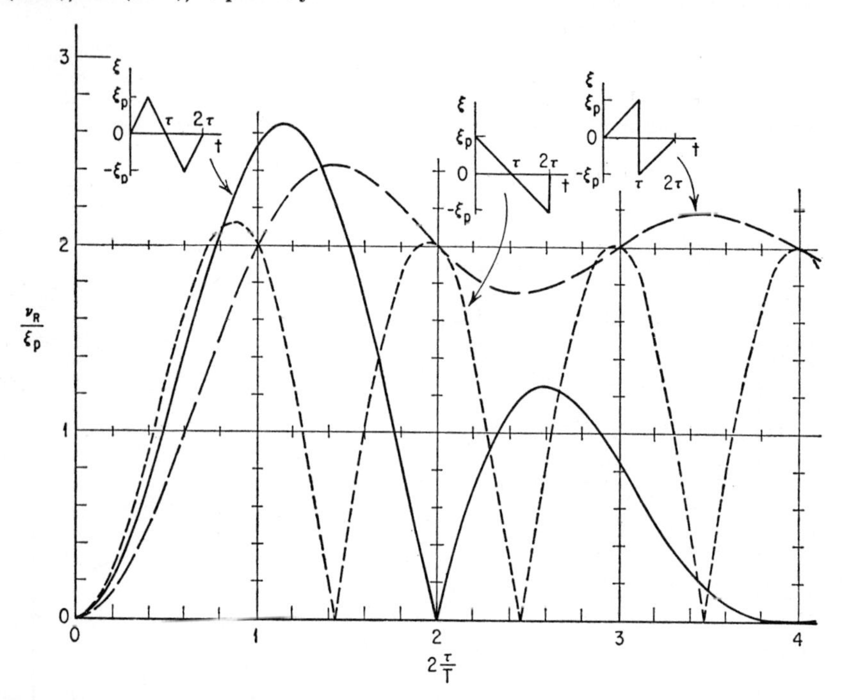

FIG. 8.42. Spectra of residual response amplitude for three types of full-cycle triangular pulses.[19]

SUMMARY OF TRANSIENT RESPONSE SPECTRA FOR THE SINGLE DEGREE-OF-FREEDOM, LINEAR, UNDAMPED SYSTEM

INITIAL CONDITIONS. The following conclusions are based on the assumption that the system is initially at rest.

STEP-TYPE EXCITATIONS:

1. The maximax response ν_M occurs *after* the step has risen (monotonically) to full value ($\tau \leq t$, where τ is the step rise time). It is equal to the residual response amplitude plus the constant step height ($\nu_M = \nu_R + \xi_c$).

2. The extreme values of the ratio of maximax response to step height ν_M/ξ_c are 1 and 2. When the ratio of step rise time to system natural period τ/T approaches zero, the step approaches the simple rectangular step in shape and ν_M/ξ_c approaches the upper extreme of 2. If τ/T approaches infinity, the step loses the character of a dynamic ex-

citation; consequently, the inertia forces of the system approach zero and ν_M/ξ_c approaches the lower extreme of 1.

3. For some particular shapes of step rise, ν_M/ξ_c is equal to 1 at certain finite values of τ/T. For example, for the step having a constant-slope rise, $\nu_M/\xi_c = 1$ when $\tau/T = 1, 2, 3, \ldots$. The lowest values of $\tau/T = (\tau/T)_{\min}$, for which $\nu_M/\xi_c = 1$, are, for three shapes of step rise: constant-slope, 1.0; versed-sine, 1.5; cycloidal, 2.0. The lowest possible value of $(\tau/T)_{\min}$ is 1.

4. In the case of step-type excitations, when $\nu_M/\xi_c = 1$ the residual response amplitude ν_R is zero. Sometimes it is of practical importance in the design of cams and dynamic control functions to achieve the smallest possible residual response.

SINGLE-PULSE EXCITATIONS:

1. When the ratio τ/T of pulse duration to system natural period is less than ½, the time shapes of certain types of equal area pulses are of secondary significance in determining the maxima of system response [maximax response ν_M, maximax relative response $(\nu - \xi)_M$, and residual response amplitude ν_R]. If τ/T is less than ¼, the pulse shape is of little consequence in almost all cases and the system response can be determined to a fair approximation by use of the simple impulse theory. If τ/T is larger than ½, the pulse shape may be of great significance.

2. The maximum value of maximax response for a given shape of pulse, $(\nu_M)_{\max}$, usually occurs at a value of the period ratio τ/T between ½ and 1. The maximum value of the ratio of maximax response to the reference excitation, $(\nu_M)_{\max}/\xi_p$, is usually between 1.5 and 1.8.

3. If the pulse has a *vertical rise*, ν_M is the first maximum occurring, and $(\nu_M)_{\max}$ is an asymptotic value approaching $2\xi_p$ as τ/T approaches infinity. In the special case of the rectangular pulse, $(\nu_M)_{\max}$ is equal to $2\xi_p$ and occurs at values of τ/T equal to or greater than ½.

4. If the pulse has a *vertical decay*, $(\nu_M)_{\max}$ is equal to the maximum value $(\nu_R)_{\max}$ of the residual response amplitude.

5. The maximum value $(\nu_R)_{\max}$ of the residual response amplitude, for a given shape of pulse, often is a reasonably good approximation to $(\nu_M)_{\max}$, except if the pulse has a steep rise followed by a decay. A few examples are shown in Table 8.3. Furthermore, if $(\nu_M)_{\max}$ and $(\nu_R)_{\max}$ for a given pulse shape are approximately equal in magnitude, they occur at values of τ/T not greatly different from each other.

6. Pulse shapes that are mirror images of each other in time result in equal values of residual response amplitude.

7. The residual response amplitude ν_R generally has zero values for certain finite values of τ/T. However, if the pulse has either a vertical rise or a vertical decay, but not both, there are no zero values except the trivial one at $\tau/T = 0$. In the case of the rectangular

Table 8.3. Comparison of Greatest Values of Maximax Response and Residual Response Amplitude

Pulse shape	$(\nu_M)_{\max}/(\nu_R)_{\max}$
Symmetrical:	
Rectangular	1.00
Sine	1.04
Versed sine	1.05
Triangular	1.06
Vertical-decay pulses	1.00
Vertical-rise pulses:	
Rectangular	1.00
Triangular	1.60
Asymptotic exponential decay	2.00

pulse, $\nu_R = 0$ when $\tau/T = 1, 2, 3, \ldots$. For several shapes of pulse the values of $(\tau/T)_{\min}$ (lowest values of τ/T for which $\nu_R = 0$) are as follows: rectangular, 1.0; sine, 1.5; versed-sine, 2.0; symmetrical triangle, 2.0. The lowest possible value of $(\tau/T)_{\min}$ is 1.

8. In the formulation of pulse as well as of step-type excitations, it may be of practical consequence for the residual response to be as small as possible; hence, attention is devoted to the case, $\nu_R = 0$.

SINGLE DEGREE-OF-FREEDOM LINEAR SYSTEM WITH DAMPING

The calculation of the effects of damping on transient response may be laborious. If the investigation is an extensive one, use should be made of an analog computer.

DAMPING FORCES PROPORTIONAL TO VELOCITY (VISCOUS DAMPING)

In the case of steady forced vibration, even very small values of the viscous damping coefficient have great effect in limiting the system response at or near resonance. If the excitation is of the single step- or pulse-type, however, the effect of damping on the maximax response may be of relatively less importance, unless the system is highly damped.

For example, in a system under steady sinusoidal excitation at resonance, a tenfold increase in the fraction of critical damping c/c_c from 0.01 to 0.1 results in a theoretical tenfold decrease in the magnification factor from 50 to 5. In the case of the same system, initially at rest and acted upon by a half-cycle sine pulse of "resonant duration" $\tau = T/2$, the same increase in the damping coefficient results in a decrease in the maximax response of only about 9 per cent.

HALF-CYCLE SINE PULSE EXCITATION. Figure 8.43 shows the spectra of maximax response for a viscously damped system excited by a half-cycle sine pulse.[12,29] The

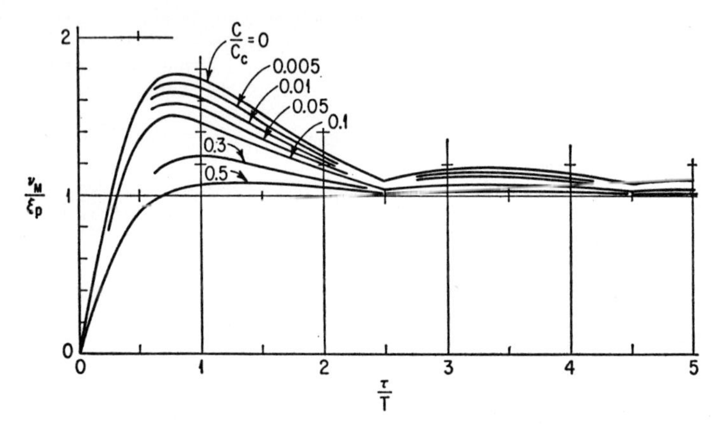

FIG. 8.43. Spectra of maximax response for a viscously damped single degree-of-freedom system acted upon by a half-cycle sine pulse. (*R. D. Mindlin, F. W. Stubner, and H. L. Cooper.*[29])

system is initially at rest. The results apply to the cases indicated by the following differential equations of motion:

$$\frac{m\ddot{x}}{k} + \frac{c\dot{x}}{k} + x = \frac{F_p}{k} \sin \frac{\pi t}{\tau} \tag{8.54a}$$

$$\frac{m\ddot{x}}{k} + \frac{c\dot{x}}{k} + x = u_p \sin \frac{\pi t}{\tau} \tag{8.54b}$$

$$\frac{m\ddot{\delta}_x}{k} + \frac{c\dot{\delta}_x}{k} + \delta_x = \frac{-m\ddot{u}_p}{k} \sin \frac{\pi t}{\tau} \tag{8.54c}$$

and in general

$$\frac{m\ddot{\nu}}{k} + \frac{c\dot{\nu}}{k} + \nu = \xi_p \sin \frac{\pi t}{\tau} \tag{8.54d}$$

where $0 \leq t \leq \tau$.

For values of t greater than τ, the excitation is zero. The distinctions among these cases may be determined by referring to Table 8.1. The fraction of critical damping c/c_c in Fig. 8.43 is the ratio of the damping coefficient c to the critical damping coefficient $c_c = 2\sqrt{mk}$. The damping coefficient must be defined in terms of the velocity $(\dot{x}, \dot{\delta}_x, \dot{\nu})$ appropriate to each case. For $c/c_c = 0$, the response spectrum is the same as the spectrum for maximax response shown for the undamped system in Fig. 8.16*B*.

OTHER FORMS OF EXCITATION; METHODS. Qualitative estimates of the effects of viscous damping in the case of other forms of step or pulse excitation may be made by the use of Fig. 8.43 and of the appropriate spectrum for the undamped response to the excitation in question.

Quantitative calculations may be effected by extending the methods described for the undamped system. If the excitation is of general form, given either numerically or graphically, the *phase-plane-delta* [21,22] method described in a later section of this chapter may be used to advantage. Of the analytical methods, the *Laplace transformation* is probably the most useful. A brief discussion of its application to the viscously damped system follows.

LAPLACE TRANSFORMATION. The differential equation to be solved is

$$\frac{m\ddot{\nu}}{k} + \frac{c\dot{\nu}}{k} + \nu = \xi(t) \tag{8.55a}$$

Rewriting Eq. (8.55*a*),

$$\frac{\ddot{\nu}}{\omega_n^2} + \frac{2\zeta\dot{\nu}}{\omega_n} + \nu = \xi(t) \tag{8.55b}$$

where $\zeta = c/c_c$ and $\omega_n^2 = k/m$.

Applying the operation transforms of Table 8.2 to Eq. (8.55*b*), the following algebraic equation is obtained:

$$\frac{1}{\omega_n^2}[s^2F_r(s) - sf(0) - f'(0)] + \frac{2\zeta}{\omega_n}[sF_r(s) - f(0)] + F_r(s) = F_e(s) \tag{8.56a}$$

The *subsidiary* equation is

$$F_r(s) = \frac{(s + 2\zeta\omega_n)f(0) + f'(0) + \omega_n^2F_e(s)}{s^2 + 2\zeta\omega_n s + \omega_n^2} \tag{8.56b}$$

where the initial conditions $f(0)$ and $f'(0)$ are to be expressed as ν_0 and $\dot{\nu}_0$, respectively.

By performing an *inverse* transformation of Eq. (8.56*b*), the response is determined in the following operational form:

$$\begin{aligned} \nu(t) &= \mathcal{L}^{-1}[F_r(s)] \\ &= \mathcal{L}^{-1}\left[\frac{(s + 2\zeta\omega_n)\nu_0 + \dot{\nu}_0 + \omega_n^2F_e(s)}{s^2 + 2\zeta\omega_n s + \omega_n^2}\right] \end{aligned} \tag{8.57}$$

Example 8.9. Rectangular Step Excitation. Assume that the damping is less than critical ($\zeta < 1$), that the system starts from rest ($\nu_0 = \dot{\nu}_0 = 0$), and that the system is acted upon by the rectangular step excitation: $\xi(t) = \xi_c$ for $0 \leq t$. The transform of the excitation is given by

$$F_e(s) = \mathcal{L}[\xi(t)] = \mathcal{L}[\xi_c] = \xi_c \frac{1}{s}$$

Substituting for ν_0, $\dot{\nu}_0$ and $F_e(s)$ in Eq. (8.57), the following equation is obtained:

$$\nu(t) = \mathcal{L}^{-1}[F_r(s)] = \xi_c\omega_n{}^2\mathcal{L}^{-1}\left[\frac{1}{s(s^2 + 2\zeta\omega_n s + \omega_n{}^2)}\right] \tag{8.58a}$$

Rewriting,

$$\nu(t) = \xi_c\omega_n{}^2\mathcal{L}^{-1}\left[\frac{1}{s[s + \omega_n(\zeta - j\sqrt{1 - \zeta^2})][s + \omega_n(\zeta + j\sqrt{1 - \zeta^2})]}\right] \tag{8.58b}$$

where $j = \sqrt{-1}$.

To determine the inverse transform $\mathcal{L}^{-1}[F_r(s)]$, it may be necessary to expand $F_r(s)$ in partial fractions as explained previously. However, in this particular example the transform pair is available in Table 8.2 (see item 16). Thus, it is found readily that $\nu(t)$ is given by the following:

$$\nu(t) = \xi_c\omega_n{}^2\left[\frac{1}{ab} + \frac{be^{-at} - ae^{-bt}}{ab(a - b)}\right] \tag{8.59a}$$

where $a = \omega_n(\zeta - j\sqrt{1 - \zeta^2})$ and $b = \omega_n(\zeta + j\sqrt{1 - \zeta^2})$. By using the relations, $\cos z = (\tfrac{1}{2})(e^{jz} + e^{-jz})$ and $\sin z = -(\tfrac{1}{2})j(e^{jz} - e^{-jz})$, Eq. (8.59a) may be expressed in terms of *cosine* and *sine* functions:

$$\nu(t) = \xi_c\left[1 - e^{-\zeta\omega_n t}\left(\cos\omega_d t + \frac{\zeta}{\sqrt{1 - \zeta^2}}\sin\omega_d t\right)\right] \qquad [\zeta < 1] \tag{8.59b}$$

where the *damped* natural frequency $\omega_d = \omega_n\sqrt{1 - \zeta^2}$.

If the damping is negligible, $\zeta \rightarrow 0$ and Eq. (8.59b) reduces to the form of Eq. (8.22) previously derived for the case of zero damping:

$$\nu(t) = \xi_c(1 - \cos\omega_n t) \qquad [\zeta = 0] \tag{8.22}$$

CONSTANT (COULOMB) DAMPING FORCES; PHASE-PLANE METHOD

The phase-plane method is particularly well suited to the solving of transient response problems involving Coulomb damping forces.[21,22] The problem is truly a stepwise linear one, provided the usual assumptions regarding Coulomb friction are valid. For example, the differential equation of motion for the case of ground displacement excitation is

$$m\ddot{x} \pm F_f + kx = ku(t) \tag{8.60a}$$

where F_f is the Coulomb friction force. In Eq. (8.60b) the friction force has been moved to the right side of the equation and the equation has been divided by the spring constant k:

$$\frac{m\ddot{x}}{k} + x = u(t) \mp \frac{F_f}{k} \tag{8.60b}$$

The effect of friction can be taken into account readily in the construction of the phase trajectory by modifying the ordinates of the stepwise excitation by amounts equal to $\mp F_f/k$. The quantity F_f/k is the Coulomb friction "displacement," and is equal to one-fourth the decay in amplitude in each cycle of a *free* vibration under the influence of Coulomb friction. The algebraic sign of the friction term changes when the velocity changes sign. When the friction term is placed on the right-hand side of the differential equation, it must have a *negative* sign when the velocity is positive.

Example 8.10. Free Vibration. Figure 8.44 shows an example of free vibration with the initial conditions $x = x_0$ and $\dot{x} = 0$. The locations of the arc centers of the phase trajectory alternate each half cycle from $+F_f/k$ to $-F_f/k$.

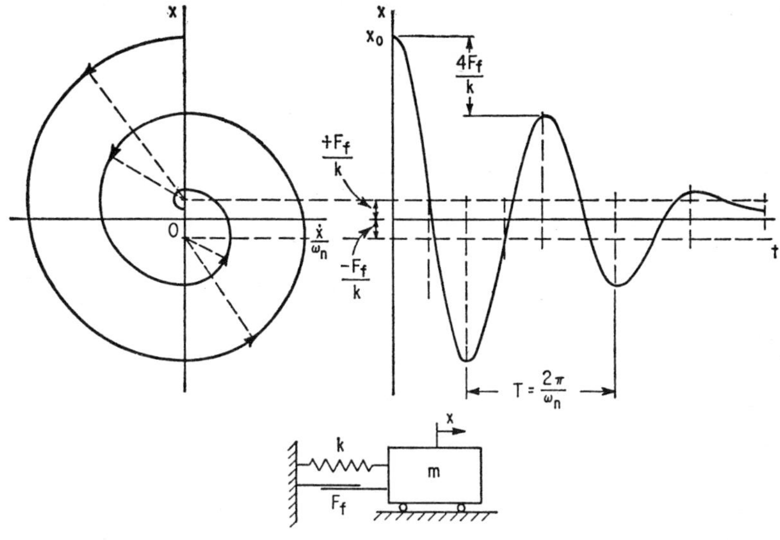

FIG. 8.44. Example of phase-plane solution of free vibration with Coulomb friction;[2] the natural frequency is $\omega_n = \sqrt{k/m}$.

Example 8.11. General Transient Excitation. A general stepwise excitation $u(t)$ and the response x of a system under the influence of a friction force F_f are shown in Fig. 8.45. The case of zero friction is also shown. The initial conditions are $x = 0$, $\dot{x} = 0$. The arc centers are located at ordinates of $u(t) \mp F_f/k$. During the third step in the excitation, the velocity of the system changes sign from positive to negative (at $t = t_2'$); consequently, the friction displacement must also change sign, but from negative to positive.

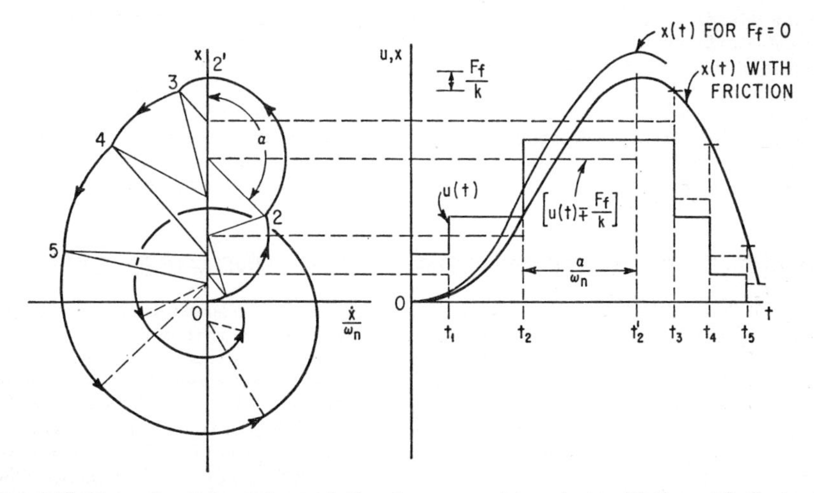

FIG. 8.45. Example of phase-plane solution for a general transient excitation with Coulomb friction in the system.[2]

SINGLE DEGREE-OF-FREEDOM NONLINEAR SYSTEMS

PHASE-PLANE-DELTA METHOD

The transient response of damped linear systems and of nonlinear systems of considerable complexity can be determined by the *phase-plane-delta* method.[21,22] Assume that the differential equation of motion of the system is

$$m\ddot{x} = G(x,\dot{x},t) \tag{8.61a}$$

where $G(x,\dot{x},t)$ is a general function of x, $\dot{x}$, and t to any powers. The coefficient of $\ddot{x}$ is constant, either inherently or by a suitable division.

In Eq. (8.61*a*) the general function may be replaced by another general function minus a linear, constant-coefficient, restoring force term:

$$G(x,\dot{x},t) = g(x,\dot{x},t) - kx$$

By moving the linear term kx to the left side of the differential equation, dividing through by m, and letting $k/m = \omega_n{}^2$, the following equation is obtained:

$$\ddot{x} + \omega_n{}^2 x = \omega_n{}^2 \delta \tag{8.61b}$$

where the *operative displacement* δ is given by

$$\delta = \frac{1}{k} g(x,\dot{x},t) \tag{8.61c}$$

The separation of the kx term from the G function does not require that the kx term exist physically. Such a term can be separated by first adding to the G function the *fictitious* terms, $+kx - kx$.

With the differential equation of motion in the δ form, Eq. (8.61*b*), the response problem can now be solved readily by stepwise linearization. The left side of the equation represents a simple, undamped, linear oscillator. Implicit in the δ function on the right side of the equation are the nonlinear restoration terms, the linear or nonlinear dissipation terms, and the excitation function.

If the δ function is held constant at a value δ for an interval of time Δt, the response of the linear oscillator in the phase-plane is an arc of a circle, with its center on the X axis at δ and subtended by an angle equal to $\omega_n \Delta t$. The graphical construction may be similar in general appearance to the examples already shown for linear systems in Figs. 8.5, 8.44, and 8.45. Since in the general case the δ function involves the dependent variables, it is necessary to estimate, before constructing each step, appropriate average values of the system displacement and/or velocity to be expected during the step. In some cases, more than one trial may be required before suitable accuracy is obtained.

Many examples of solution for various types of systems are available in the literature.[3,5,6,8,13,15,20,21,22,25]

MULTIPLE DEGREE-OF-FREEDOM, LINEAR, UNDAMPED SYSTEMS

Some of the transient response analyses, presented for the single degree-of-freedom system, are in complete enough form that they can be employed in determining the responses of linear, undamped, multiple degree-of-freedom systems. This can be done by the use of *normal (principal) coordinates*. A system of normal coordinates is a system of generalized coordinates chosen in such a way that vibration in each normal mode involves only one coordinate, a normal coordinate. The differential equations of motion, when written in normal coordinates, are all independent of each other. Each differential equation is related to a particular normal mode and involves only one coordinate. The differential equations are of the same general form as the differential equation of motion for the single degree-of-freedom system. The response of the system in terms of the physical coordinates, for example, displacement or stress at various locations in the system, is determined by superposition of the normal coordinate responses. Normal coordinates are discussed in Chaps. 2 and 7, and in Refs. 37 and 38.

Example 8.12. Sine Force Pulse Acting on a Simple Beam. Consider the flexural vibration of a prismatic bar with simply supported ends, Fig. 8.46. A sine-pulse concentrated force $F_p \sin(\pi t/\tau)$ is applied to the beam at a distance c from the left end (origin of coordinates). Assume that the beam is initially at rest. The displacement response of the beam, during the time of action of the pulse, is given by the following series:

$$y = \frac{2F_p l^3}{\pi^4 EI} \sum_{i=1}^{i=\infty} \frac{1}{i^4} \sin\frac{i\pi c}{l} \sin\frac{i\pi x}{l} \left[\frac{1}{1 - T_i^2/4\tau^2}\left(\sin\frac{\pi t}{\tau} - \frac{T_i}{2\tau}\sin\omega_i t\right)\right] \qquad [0 \le t \le \tau] \qquad (8.62a)$$

where

$$i = 1, 2, 3, \ldots; \; T_i = \frac{2\pi}{\omega_i} = \frac{2l^2}{i^2\pi}\sqrt{\frac{A\gamma}{EIg}} = \frac{T_1}{i^2}, \text{ sec}$$

A comparison of Eqs. (8.62*a*) and (8.32*a*) shows that the time function $[\sin(\pi t/\tau) - (T_i/2\tau) \sin\omega_i t]$ for the ith term in the beam-response series is of exactly the same form as the time function $[\sin(\pi t/\tau) - (T/2\tau)\sin\omega_n t]$ in the response of the single degree-of-freedom system. Furthermore, the magnification factors $1/(1 - T_i^2/4\tau^2)$ and $1/(1 - T^2/4\tau^2)$ in the two equations have identical forms.

Following the end of the pulse, beginning at $t = \tau$, the vibration of the beam is expressed by

$$y = \frac{2F_p l^3}{\pi^4 EI} \sum_{i=1}^{i=\infty} \frac{1}{i^4} \sin\frac{i\pi c}{l} \sin\frac{i\pi x}{l} \left[\frac{(T_i/\tau)\cos(\pi\tau/T_i)}{(T_i^2/4\tau^2) - 1}\sin\omega_i\left(t - \frac{\tau}{2}\right)\right] \qquad [\tau \le t] \qquad (8.62b)$$

A comparison of Eqs. (8.62*b*) and (8.32*b*) leads to the same conclusion as found above for the time era $0 \le t \le \tau$.

Excitation and Displacement at Mid-span. As a specific case, consider the displacement at mid-span when the excitation is applied at mid-span ($c = x = l/2$). The even-numbered terms of the series now are all zero and the series take the following forms:

$$y_{l/2} = \frac{2F_p l^3}{\pi^4 EI} \sum_{i=1,3,5,\ldots}^{\infty} \frac{1}{i^4}\left[\frac{1}{1 - T_i^2/4\tau^2}\left(\sin\frac{\pi t}{\tau} - \frac{T_i}{2\tau}\sin\omega_i t\right)\right] \qquad [0 \le \;\; \le \tau] \qquad (8.63a)$$

$$y_{l/2} = \frac{2F_p l^3}{\pi^4 EI} \sum_{i=1,3,5,\ldots}^{\infty} \frac{1}{i^4}\left[\frac{(T_i/\tau)\cos(\pi\tau/T_i)}{(T_i^2/4\tau^2) - 1}\sin\omega_i\left(t - \frac{\tau}{2}\right)\right] \qquad [\tau \le t] \qquad (8.63b)$$

Assume, for example, that the pulse period τ equals two-tenths of the fundamental natural period of the beam ($\tau/T_1 = 0.2$). It is found from Fig. 8.16*B*, by using an abscissa value of 0.2, that the maximax response in the *fundamental* mode ($i = 1$) occurs in the residual vibration era ($\tau \le t$). The value of the corresponding ordinate is 0.75. Consequently, the maximax response for $i = 1$ is 0.75 $(2F_p l^3/\pi^4 EI)$.

In order to determine the maximax for the *third* mode ($i = 3$), an abscissa value of $\tau/T_i = i^2\tau/T_1 = 3^2 \times 0.2 = 1.8$, is used. It is found that the maximax is greater than the residual amplitude and consequently that it occurs during the time era $0 \le t \le \tau$. The value of the corresponding ordinate is 1.36; however, this must be multiplied by $1/3^4$, as indicated by the series. The maximax for $i = 3$ is thus 0.017 $(2F_p l^3/\pi^4 EI)$.

The maximax for $i = 5$ also occurs in the time era $0 \le t \le \tau$ and the ordinate may be estimated to be about 1.1. Multiplying by $1/5^4$, it is found that the maximax for $i = 5$ is approximately 0.002 $(2F_p l^3/\pi^4 EI)$, a negligible quantity when compared with the maximax value for $i = 1$.

To find the maximax total response to a reasonable approximation, it is necessary to sum on a time basis several terms of the series. In the particular example above, the maximax total response occurs in the residual vibration era and a reasonably accurate value can be obtained by considering only the first term ($i = 1$) in the series, Eq. (8.63*b*).

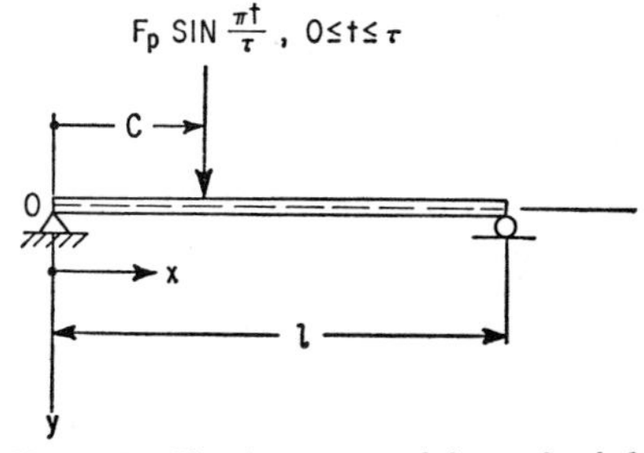

FIG. 8.46. Simply supported beam loaded by a concentrated force sine pulse of half-cycle duration.

GENERAL INVESTIGATION OF TRANSIENTS

An extensive (and efficient) investigation of transient response in multiple degree-of-freedom systems requires the use of an automatic computer. In some of the simpler cases, however, it is feasible to employ numerical or graphical methods. For example, the phase-plane method may be applied to multiple degree-of-freedom linear systems [1,2] through the use of normal coordinates. This involves independent phase-planes having the coordinates q_i and $\dot{q}_i/\omega_i$, where q_i is the ith normal coordinate.

REFERENCES

1. Ayre, R. S.: *J. Franklin Inst.*, **253**:153 (1952).
2. Ayre, R. S.: *Proc. World Conf. Earthquake Eng.*, 1956, p. 13–1.
3. Ayre, R. S., and J. I. Abrams: *Proc. ASCE*, EM 2, Paper 1580, 1958.
4. Biot, M. A.: *Trans. ASCE*, **108**:365 (1943).
5. Bishop, R. E. D.: *Proc. Inst. Mech. Engrs.* (*London*), **168**:299 (1954).
6. Braun, E.: *Ing.-Arch.*, **8**:198 (1937).
7. Bronwell, A.: "Advanced Mathematics in Physics and Engineering," McGraw-Hill Book Company, Inc., New York, 1953.
8. Bruce, V. G.: *Bull. Seismol. Soc. Amer.*, **41**:101 (1951).
9. Cherry, C.: "Pulses and Transients in Communication Circuits," Dover Publications, New York, 1950.
10. Crede, C. E.: "Vibration and Shock Isolation," John Wiley & Sons, Inc., New York, 1951.
11. Crede, C. E.: *Trans. ASME*, **77**:957 (1955).
12. Criner, H. E., G. D. McCann, and C. E. Warren: *J. Appl. Mechanics*, **12**:135 (1945).
13. Evaldson, R. L., R. S. Ayre, and L. S. Jacobsen: *J. Franklin Inst.*, **248**:473 (1949).
14. Frankland, J. M.: *Proc. Soc. Exptl. Stress Anal.*, **6**:2, 7 (1948).
15. Fuchs, H. O.: *Product Eng.*, August, 1936, p. 294.
16. Gardner, M. F., and J. L. Barnes: "Transients in Linear Systems," vol. I, John Wiley & Sons, Inc., New York, 1942.
17. Hartman, J. B.: "Dynamics of Machinery," McGraw-Hill Book Company, Inc., New York, 1956.
18. Hudson, G. E.: *Proc. Soc. Exptl. Stress Anal.*, **6**:2, 28 (1948).
19. Jacobsen, L. S., and R. S. Ayre: "A Comparative Study of Pulse and Step-type Loads on a Simple Vibratory System," *Tech. Rept.* N16, under contract N6-ori-154, T. O. 1, U.S. Navy, Stanford University, 1952.
20. Jacobsen, L. S.: *Proc. Symposium on Earthquake and Blast Effects on Structures*, 1952, p. 94.
21. Jacobsen, L. S.: *J. Appl. Mechanics*, **19**:543 (1952).
22. Jacobsen, L. S., and R. S. Ayre: "Engineering Vibrations," McGraw-Hill Book Company, Inc., New York, 1958.
23. Kelvin, Lord: *Phil. Mag.*, **34**:443 (1892).
24. Kornhauser, M.: *J. Appl. Mechanics*, **21**:371 (1954).
25. Lamoen, J.: *Rev. universelle mines*, ser. 8, **11**:7, 3 (1935).
26. McCann, G. D., and J. M. Kopper: *J. Appl. Mechanics*, **14**:A127 (1947).
27. Meissner, E.: *Schweiz. Bauzt.*, **99**:27, 41 (1932).
28. Mindlin, R. D.: *Bell System Tech. J.*, **24**:353 (1945).
29. Mindlin, R. D., F. W. Stubner, and H. L. Cooper: *Proc. Soc. Exptl. Stress Anal.*, **5**:2, 69 (1948).
30. Morrow, C. T.: *J. Acoust. Soc. Amer.*, **29**:596 (1957).
31. Muller, J. T.: *Bell System Tech. J.*, **27**:657 (1948).
32. Rothbart, H. A.: "Cams—Design, Dynamics and Accuracy," John Wiley & Sons, Inc., New York, 1956.
33. Salvadori, M. G., and R. J. Schwarz: "Differential Equations in Engineering Problems," Prentice-Hall, Inc., Englewood Cliffs, N.J., 1954.
34. Scott, E. J.: "Transform Calculus," Harper & Brothers, New York, 1955.
35. Shapiro, H., and D. E. Hudson: *J. Appl. Mechanics*, **20**:422 (1953).
36. Thomson, W. T.: "Laplace Transformation," Prentice-Hall, Inc., Englewood Cliffs, N.J., 1950.
37. Timoshenko, S. P., and D. H. Young: "Advanced Dynamics," McGraw-Hill Book Company, Inc., New York, 1948.
38. Timoshenko, S. P., and D. H. Young: "Vibration Problems in Engineering," 3d ed., D. Van Nostrand Company, Inc., Princeton, N.J., 1955.
39. Walsh, J. P., and R. E. Blake: *Proc. Soc. Exptl. Stress Anal.*, **6**:2, 151 (1948).
40. Williams, H. A.: *Trans. ASCE*, **102**:838 (1937).

9

EFFECTS OF IMPACT ON STRUCTURES

W. H. Hoppmann II
University of South Carolina

INTRODUCTION

This chapter discusses a particular phenomenon in the general field of shock and vibration usually referred to as impact.[1] An impact occurs when two or more bodies collide. An important characteristic of an impact is the generation of relatively large forces at points of contact for relatively short periods of time. Such forces sometimes are referred to as *impulse-type* forces.

Three general classes of impact are considered in this chapter: (1) impact between spheres or other rigid bodies, where a body is considered to be rigid if its dimensions are large relative to the wavelengths of the elastic stress waves in the body; (2) impact of a rigid body against a beam or plate that remains substantially elastic during the impact; and (3) impact involving yielding of structures.

DIRECT CENTRAL IMPACT OF TWO SPHERES

The elementary analysis of the central impact of two bodies is based upon an experimental observation of Newton.[2] According to that observation, the relative velocity of two bodies after impact is in constant ratio to their relative velocity before impact and is in the opposite direction. This constant ratio is the *coefficient of restitution;* usually it is designated by e.[3]

Let $\dot{u}$ and $\dot{x}$ be the components of velocity along a common line of motion of the two bodies before impact, and $\dot{u}'$ and $\dot{x}'$ the component velocities of the bodies in the same direction after impact. Then, by the observation of Newton,

$$\dot{u}' - \dot{x}' = -e(\dot{u} - \dot{x}) \tag{9.1}$$

Now suppose that a smooth sphere of mass m_u and velocity $\dot{u}$ collides with another smooth sphere having the mass m_x and velocity $\dot{x}$ moving in the same direction. Let the coefficient of restitution be e, and let $\dot{u}'$ and $\dot{x}'$ be the velocities of the two spheres, respectively, after impact. Figure 9.1 shows the condition of the two spheres just before collision. The only force acting on the spheres during impact is the force at the point of contact, acting along the line through the centers of the spheres.

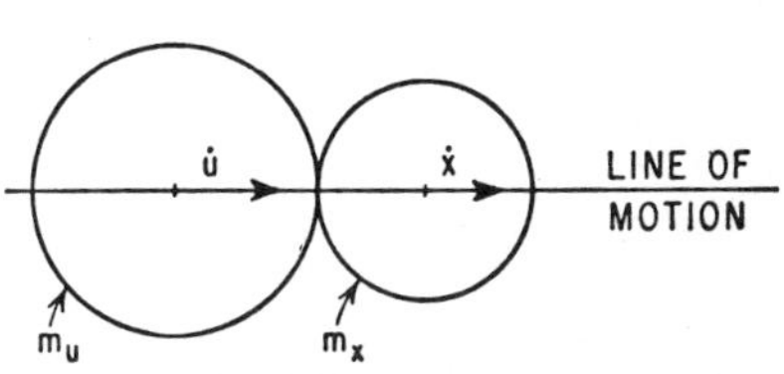

FIG. 9.1. Positions of two solid spheres at instant of central impact.

According to the law of conservation of linear momentum:

$$m_u\dot{u}' + m_x\dot{x}' = m_u\dot{u} + m_x\dot{x} \tag{9.2}$$

Solving Eqs. (9.1) and (9.2) for the two unknowns, the velocities $\dot{u}'$ and $\dot{x}'$ after impact,

$$\dot{u}' = \frac{(m_u\dot{u} + m_x\dot{x}) - em_x(\dot{u} - \dot{x})}{m_u + m_x}$$
$$\dot{x}' = \frac{(m_u\dot{u} + m_x\dot{x}) + em_u(\dot{u} - \dot{x})}{m_u + m_x} \qquad (9.3)$$

This analysis yields the resultant velocities for the two spheres on the basis of an experimental law and the principle of the conservation of momentum, without any specific reference to the force of contact F. A similar result is obtained for a ballistic pendulum used to measure the muzzle velocity of a bullet. A bullet of mass m_u and velocity $\dot{u}$ is fired into a block of wood of mass m_x which is at rest initially and finally assumes a velocity $\dot{x}'$ after the impact. Using only the principle of the conservation of momentum,

$$\dot{u} = \frac{(m_u + m_x)\dot{x}'}{m_u} \qquad (9.4)$$

No knowledge of the complicated pattern of force acting on the bullet and the pendulum during the embedding process is required.

These simple facts are introductory to the more complicated problem involving the vibration of at least one of the colliding bodies, as discussed in a later section.

HERTZ THEORY OF IMPACT OF TWO SOLID SPHERES

The theory of two solid elastic spheres which collide with one another is based upon the results of an investigation of two elastic bodies pressed against one another under purely statical conditions.[4] For these static conditions, the relations between the sum of the displacements at the point of contact in the direction of the common line of motion and the resultant total pressure have been derived. The sum of these displacements is equal to the relative approach of the centers of the spheres, assuming that the spheres act as rigid bodies except for elastic compression at the point of contact. The relative approach varies as the two-thirds power of the total pressure; a formula is given for the time of duration of the contact.[4] The theory is valid only if the duration of contact is long in comparison with the period of the fundamental mode of vibration of either sphere.

The range of validity of the Hertz theory is related to the possibility of exciting vibration in the spheres.[5] The dimensionless ratio of the maximum kinetic energy of vibration to the sum of the kinetic energies of the two spheres just before collision is approximately

$$R = \frac{1}{50}\frac{\dot{u} - \dot{x}}{\sqrt{E/\rho}} \qquad (9.5)$$

where $\dot{u} - \dot{x}$ = relative velocity of approach, in./sec
E = Young's modulus of elasticity, assumed to be the same for each sphere, lb/in.^{2}
ρ = density of each sphere, lb-sec^2/in.^{4}
$\sqrt{E/\rho}$ = approximate velocity of propagation of dilatational waves, in./sec

The ratio R usually is a very small quantity; thus, the theory of impact set forth by Eq. (9.5) has wide application because vibration is not generated in the spheres to an appreciable degree under ordinary conditions. The energy of the colliding spheres remains translational, and the velocities after impact are deducible from the principles of energy and of momentum. The important point of plastic deformation at the point of contact is discussed in a later section.

Formulas for force between the spheres, the radius of the circular area of contact, and the relative approach of the centers of the spheres, all as functions of time, can be determined for any two given spheres.[6]

IMPACT OF A SOLID SPHERE ON AN ELASTIC PLATE

An extension of the Hertz theory of impact to include the effect of vibration of one of the colliding bodies involves a study of the transverse impact of a solid sphere upon an infinitely extended plate.[7] The plate has the role of the vibrating body. The coefficient of restitution is an important element in any analysis of the motion ensuing after the collision of two bodies.

The analysis is based on the assumption that the principal elastic waves of importance are flexural waves of half-period equal to the duration of impact. Let $2h$ and $2D$ be the thickness of plate and diameter of sphere, respectively, ρ_1, ρ_2 their densities, E_1, E_2 their Young's moduli, ν_1, ν_2 their values of Poisson's ratio, and τ_H the duration of impact. The velocity c of long flexural waves of wavelength λ in the plate is given by

$$c^2 = \frac{4\pi^2}{3}\frac{h^2}{\lambda^2}\frac{E_1}{\rho_1(1 - \nu_1^2)} \tag{9.6}$$

The radius a of the circle on the plate over which the disturbance has spread at the termination of impact is given by

$$a = c\tau_H = \frac{\lambda}{2} \tag{9.7}$$

Combining Eqs. (9.6) and (9.7),

$$a^2 = \pi\tau_H h \sqrt{\frac{E_1}{3\rho_1(1 - \nu_1^2)}} \tag{9.8}$$

The next step is to find the kinetic and potential energies of the wave motion of the plate. The kinetic energy may be determined from the transverse velocity of the plate at each point over the circle of radius a covered by the wave. Figure 9.2 shows an approximate distribution of velocity over the circle of radius a at the end of impact.[8] The direction of the impact also is shown. The kinetic energy in the wave at the end of impact is

$$T = \int_0^a \tfrac{1}{2}\cdot 2h\cdot \rho_1 \cdot 2\pi R \cdot \dot{w}^2\, dR \tag{9.9}$$

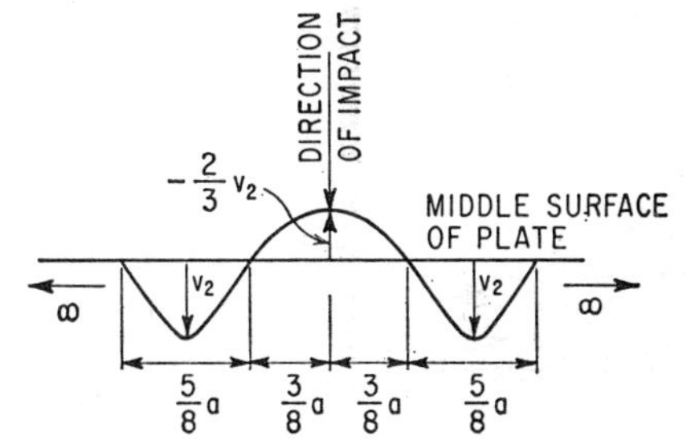

FIG. 9.2. Distribution of transverse velocities in plate as a result of impact by a moving body. (*After Lamb.*)

where $\dot{w}$ is the transverse velocity at distance R from the origin. As an approximation it is assumed that the sum of the potential energy and the kinetic energy in the wave is $2T$. With considerable effort these energies can be calculated in terms of the motion of the plate, although the calculation may be laborious.

The impulse in the plate produced by the colliding body is

$$J = \int_0^a \tfrac{1}{2}\cdot 2h \cdot \rho_1 \cdot 2\pi R \cdot \dot{w}\, dr \tag{9.10}$$

The integration should be carried out with due regard to the sign of velocity. If m_u is the mass of colliding body, $\dot{u}$ its velocity before impact, and e the coefficient of restitution, the following relations are obtained on the assumption that the energy is conserved:

$$\tfrac{1}{2}m_u\dot{u}^2(1 - e^2) = 2T \tag{9.11}$$

$$m_u\dot{u}(1 + e) = J \tag{9.12}$$

Equation (9.11) represents the energy lost to the moving sphere as a result of impact and Eq. (9.12) represents the change in momentum of the sphere.

The coefficient of restitution e is determined by evaluating the integrals for T and J and substituting their values in Eq. (9.12). The necessary integrations can be performed by taking the function for transverse velocity in Fig. 9.2 as arcs of sine curves. The resultant expression for e is

$$e = \frac{h\rho_1 a^2 - 0.56 m_u}{h\rho_1 a^2 + 0.56 m_u} \tag{9.13}$$

where a, the radius of the deformed region, is given by Eq. (9.8) and τ_H, the time of contact between sphere and plate, is given by Hertz's theory of impact to a first approximation.[4] The mass of the sphere is m_u; the mass of the plate is assumed to be infinite. Large discrepancies between theory and experiment occur when the diameter of the sphere is large compared with the thickness of the plate. The duration of impact τ_H is

$$\tau_H = 2.94 \frac{\alpha}{\dot{u}} \tag{9.14}$$

where $$\alpha = \left[\frac{15}{16} \nu_1^2 \left(\frac{1 - \nu_1^2}{E_1} + \frac{1 - \nu_2^2}{E_2}\right) m_u\right]^{2/5} R_s^{-1/5}$$

The radius of the striking sphere is R_s and its velocity before impact is $\dot{u}$. Subscripts 1 and 2 represent the properties of the sphere and plate, respectively. The value of τ_H may be substituted in Eq. (9.8) above.

Experimental results verify the theory when the limitations of the theory are not violated. The velocity of impact must be sufficiently small to avoid plastic deformation. When the collision involves steel on steel, the velocity usually must be less than 1 ft/sec. However, useful engineering results can be obtained with this approach even though plastic deformation does occur locally.[9, 10]

TRANSVERSE IMPACT OF A MASS ON A BEAM

If $F(t)$ is the force acting between the sphere and the beam during contact, the distance traveled by the sphere in time t after collision is [11]

$$\dot{u}t - \frac{1}{m_u} \int_0^t F(t_v)(t - t_v)\, dt_v \tag{9.15}$$

where $\dot{u}$ = velocity of sphere before collision (beam assumed to be at rest initially)
m_u = mass of solid sphere

The beam is assumed to be at rest initially.

For example, the deflection of a simply supported beam under force $F(t_v)$ at its center is

$$\sum_{1,3,5\ldots}^{\infty} \frac{1}{m_b} \int_0^t F(t_v) \frac{\sin \omega_n (t - t_v)}{\omega_n}\, dt_v \tag{9.16}$$

where m_b = one-half of mass of beam
ω_n = angular frequency of the nth mode of vibration

Equation (9.16) represents the transverse vibration of a beam. While the present case is only for direct central impact, the cases for noncentral impact depend only on the corresponding solution for transverse vibration. Oblique impact also is treated readily.

The expression for the relative approach of the sphere and beam; i.e., penetration of beam by sphere, is [11]

$$\alpha = \kappa_1 F(t)^{2/3} \tag{9.17}$$

where κ_1 is a constant depending on the elastic and geometrical properties of the sphere and the beam at the point of contact, and α is given by Eq. (9.14). Consequently, the equation that defines the problem is

$$\alpha = K_1 F^{2/3} = \dot{u}t - \frac{1}{m_u} \int_0^t F(t_v)(t - t_v)\, dt_v - \sum_{1,3,5}^{\infty} \frac{1}{m_b} \int_0^t F(t_v) \frac{\sin \omega_n (t - t_v)}{\omega_n}\, dt_v \tag{9.18}$$

Equation (9.18) has been solved numerically for two specific problems by subdividing the time interval 0 to t into small elements and calculating, step by step, the displacements of the sphere.[11] The results are not general but rather apply only to the cases of beam and sphere.

For the impact of a mass on a beam, the sum of the kinetic and the potential energies may be expressed in terms of the unknown contact force.[12] Also, the impulse integral J in terms of the contact force may be expressed as

$$J = \int_0^t F(t)\, dt = m_u \dot{u}(1 + e) \tag{9.19}$$

A satisfactory approximation to $F(t)$ is defined in terms of a normalized force $\bar{F}$:

$$F(t) = m_u \dot{u}(1 + e)\bar{F}(t) \tag{9.20}$$

Thus, from Eqs. (9.19) and (9.20),

$$\int_0^t \bar{F}\, dt = 1 \tag{9.21}$$

The value of this integral is independent of the shape of $F(t)$. The normalized force is defined such that its maximum value equals the maximum value of the corresponding normalized Hertz force.[12] To perform the necessary integrations, a suitable function for defining $F(t)$ is chosen as follows:

$$\begin{aligned} \bar{F}(t) &= \frac{\pi}{2\tau_L} \sin \frac{\pi}{\tau_L} t \qquad [0 < t < \tau_L] \\ \bar{F}(t) &= 0 \qquad [|t| > \tau_L] \end{aligned} \tag{9.22}$$

Results for particular problems solved in this manner agree well with those obtained for the same problems by the numerical solution of the exact integral equation.[12]

To apply these results to a specific beam impact problem, it is necessary to express the deflection equation for the beam in terms of known quantities. One of these quantities is the coefficient of restitution; a formula must be provided for its determination in terms of known functions. This is given by Eq. (9.31).

IMPACT OF A RIGID BODY ON A DAMPED ELASTICALLY SUPPORTED BEAM

For the more general case of impact of a rigid body on a damped, elastically supported beam, it is assumed that there is external damping, damping determined by the Stokes' law of stress-strain, and an elastic support attached to the beam along its length in such a manner that resistance is proportional to deflection.[13] The differential equation for the deflection of the beam is

$$EI \frac{\partial^4 w}{\partial x^4} + c_1 I \frac{\partial^5 w}{\partial x^4\, \partial t} + c_2 \frac{\partial w}{\partial t} + kw + \rho S \frac{\partial^2 w}{\partial t^2} = F(x,t) \tag{9.23}$$

where w = deflection, in.
E = Young's modulus, lb/in.^{2}
I = moment of inertia for cross section (constant), in.^{4}
c_1 = internal damping coefficient, lb/in.^{2}-sec (Stokes' law)
c_2 = external damping coefficient, lb/in.^{2}-sec
k = foundation modulus, lb/in.^{2}
ρ = density, lb-sec^2/in.^{4}
S = area of cross section (constant), in.^{2}
$\frac{\partial^2 w}{\partial t^2}$ = acceleration, in./sec^2
t = time, sec
$F(x,t)$ = driving force per unit length of beam, lb/in.

For example, to illustrate the application of specific boundary conditions, consider a simply supported beam of length l. The moments and deflections must vanish at the ends. The beam is assumed undeflected and at rest just before impact, and central impact is assumed although with some additional computation this restriction may be dropped. The solution may be written as follows:

$$w(x,t) = \sum^{\infty} \sin\frac{n\pi x}{l} \sin\frac{n\pi}{2}\frac{1}{m}\frac{1}{\sqrt{\omega_n^2 - \delta_n^2}}$$

$$\times \int_0^t e^{-\delta_n(t-\tau)} \sin\left[\sqrt{\omega_n^2 - \delta_n^2}\cdot(t-\tau)\right] F_1(\tau)\, d\tau \quad (9.24)$$

where e = base of natural logarithms

δ_n = damping numbers $= \frac{1}{2}\left(r_i \frac{n^4\pi^4}{l^4} + r_e\right)$

$r_i = \frac{c_1 I}{\rho S}$

$r_e = \frac{c_2}{\rho S}$

ω_n = angular frequencies

$m = \frac{1}{2}\rho A l$

A satisfactory analytical expression for the contact force $F_1(t)$, a particularization of $F(x,t)$ in Eq. (9.23), must be developed. Although $F_1(t)$ is assumed to act at the center of the beam, the methods apply with only minor alterations if the impact occurs at any other point of the beam.

One of the conditions which the contact force must satisfy is that its time integral for the duration of impact equal the change in momentum of the striking body. The change of momentum is

$$m\dot{z} - m\dot{z}' = m\dot{z}\left(1 - \frac{\dot{z}'}{\dot{z}}\right) \quad (9.25)$$

where m = mass of rigid body, lb-sec^2/in.

$\dot{z}$ = velocity of rigid body just before collision, in./sec

$\dot{z}'$ = velocity of rigid body just after collision, in./sec

When the velocity of the beam is zero, Eq. (9.1) may be written

$$e = -\frac{\dot{z}'}{\dot{z}} \quad (9.26)$$

Equation (9.26) may be written

$$m\dot{z}\left(1 - \frac{\dot{z}'}{\dot{z}}\right) = m\dot{z}(1 + e) \quad (9.27)$$

From the equivalence of impulse and momentum:

$$\int_0^{\tau_0} F_1(t)\, dt = m\dot{z}(1 + e) \quad (9.28)$$

where τ_0 is the time of contact.

It can then be shown [13] that the impact force may be written

$$F_1(t) = m\dot{z}(1 + e)\frac{\pi}{2\tau_L}\sin\frac{n\pi t}{\tau_L} \qquad [0 < t < \tau_L]$$

$$F_1 = 0 \qquad [t > \tau_L] \quad (9.29)$$

It can be shown further [13] that

$$\tau_L = 3.28 \left[\frac{m^2}{\dot{z}R} \cdot \frac{(1-\nu^2)}{E^2} \right]^{1/5} \tag{9.30}$$

where R = radius of sphere, in.
ν = Poisson's ratio

The time interval τ_L is a special value of the time of contact T_0. It agrees well with experimental results.

The coefficient of restitution e is [13]

$$e = \frac{1 - \frac{m}{m_b}\sum_1^\infty \Phi_n - \frac{m}{m_b}\sum_1^\infty \Psi_n}{1 + \frac{m}{m_b}\sum_1^\infty \Phi_n + \frac{m}{m_b}\sum_1^\infty \Psi_n} \tag{9.31}$$

where m = mass of sphere
m_b = half mass of beam

The functions Φ_n and Ψ_n are given in the form of curves in Figs. 9.3 and 9.4; the symbol $\beta_n = \delta_n/\omega_n$ represents fractional damping and $Q_n = \omega_n \tau_L / 2\pi$ is a dimensionless frequency where ω_n = angular frequency of nth mode of vibration of undamped vibration

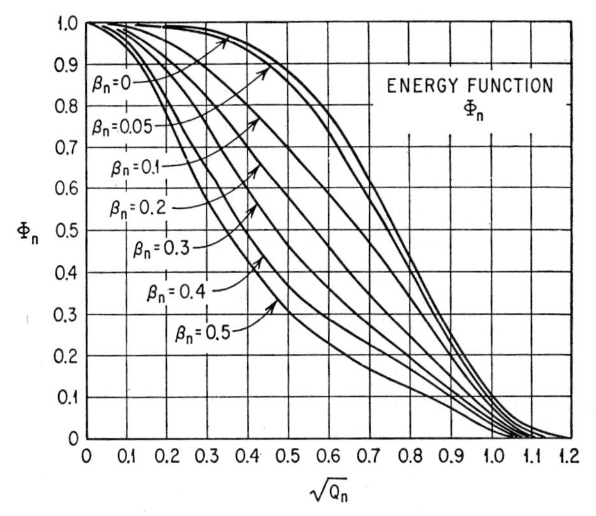

FIG. 9.3. Energy functions ϕ_n used with Eq. (9.31) to determine the coefficient of restitution from the impact of a rigid body on a damped elastically supported beam.

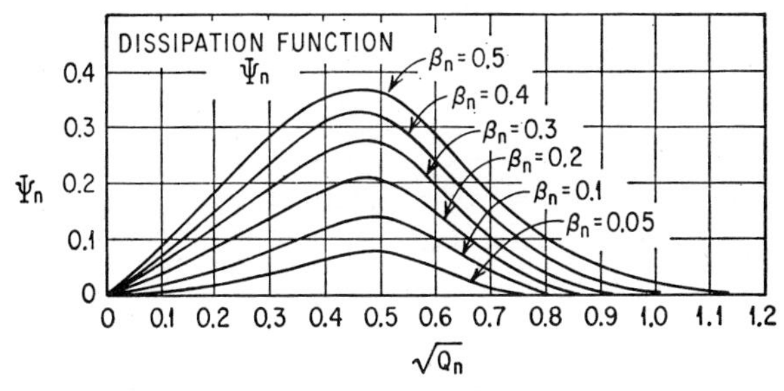

FIG. 9.4. Dissipative (damping) functions ψ_n used with Eq. (9.31) to determine the coefficient of restitution from the impact of a rigid body on a damped elastically supported beam.

of beam, rad/sec, and τ_L = length of time the sinusoidal pulse is assumed to act on beam [see Eq. (9.30)]. If damping is neglected, the functions Ψ_n vanish from Eq. (9.31).

The above theory may be generalized to apply to the response of plates to impact. The deflection equation of a plate subjected to a force applied at a point is required. The various energy distributions at the end of impact are arrived at in a manner analogous to that for the beam.

The theory has been applied to columns and continuous beams [14,15] and also could be applied to transverse impact on a ring. Measurement of the force of impact illustrates the large number of modes of vibration that can be excited by an impact.[16,17,22]

Principal qualitative results of the foregoing analysis are:

1. Impacts by bodies of relatively small mass moving with low velocities develop significant bending strains in beams.
2. External damping of the type assumed above has a rapidly decreasing effect on reducing deflection and strain as the number of the mode increases.
3. Internal damping of the viscous type here assumed reduces deflection and strain appreciably in the higher modes. For a sufficiently high mode number, the vibration becomes aperiodic.
4. Increasing the modulus for an elastic foundation reduces the energy absorbed by the structure from the colliding body.
5. Impacts from collision produce sharp initial rises in strain which are little influenced by damping.
6. Because of result 5, the fatigue problem for machines and structures, in which the impact conditions are repeated many times, can be serious. Ordinary damping affords little protection.
7. The structure seldom can be treated as a single degree-of-freedom system with any degree of reliability in predicting strain.[13,19]

LONGITUDINAL AND TORSIONAL IMPACT ON BARS

If a mass strikes the end of a long bar, the response may be investigated by means of the Hertz contact theory.[11] The normal modes of vibration must be known so the displacement at each part of the bar can be calculated in terms of a contact force. In a similar manner, the torsional vibration of a long bar can be studied, using the normal modes of torsional vibration.

PLASTIC DEFORMATION RESULTING FROM IMPACT

Many problems of interest involve plastic deformation rather than elastic deformation as considered in the preceding analyses. Using the concept of the plastic hinge, the large plastic deformation of beams under transverse impact [23] and the plastic deformation of free rings under concentrated dynamic loads [24] have been studied. In such analyses, the elastic portion of the vibration usually is neglected. To make further progress in analyses of large deformations as a result of impact, a realistic theory of material behavior in the plastic phase is required.

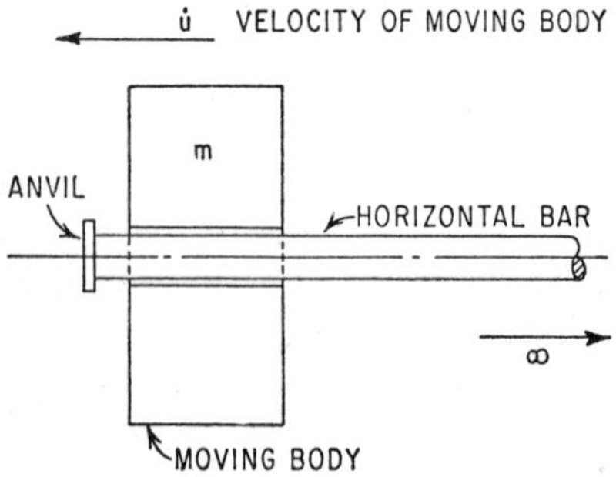

An attempt to solve the problem for the longitudinal impact on bars has been made using the static engineering-type stress-strain curve as a part of the analysis.[25] An extension of the work to transverse impact also was attempted.[26]

Figure 9.5 illustrates the impact of a large body m colliding axially with a long rod. The body m has an initial velocity $\dot{u}$ and is sufficiently large that the end of the rod may be assumed to move with constant

FIG. 9.5. Longitudinal impact of moving body on end of rod.

velocity $\dot{u}$. At any time t a stress wave will have moved into the bar a definite distance; by the condition of continuity (no break in the material), the struck end of the bar will have moved a distance equal to the total elongation of the end portion of the bar:

$$\dot{u}t = \epsilon \cdot l \tag{9.32}$$

The velocity c of a stress wave is $c = l/t$, and Eq. (9.32) becomes

$$\epsilon = \frac{\dot{u}}{c} \tag{9.33}$$

The stress and strain in an elastic material are related by Young's modulus. Substituting for strain from Eq. (9.33),

$$\sigma = \epsilon \cdot E = E\frac{\dot{u}}{c} \tag{9.34}$$

where $\dot{u}$ = velocity of end of rod, in./sec
l = distance stress wave travels in time t, in.
t = time, sec
σ = stress, lb/in.^{2}
ϵ = strain (uniform), in./in.
E = Young's modulus, lb/in.^{2}
c = velocity of stress wave (dilatational), in./sec

When the yield point of the material is exceeded, Eq. (9.34) is inapplicable. Extensions of the analysis, however, lead to some results in the case of plastic deformation.[25] The differential equation for the elastic case is

$$E\frac{\partial^2 u}{\partial x^2} = \rho\frac{\partial^2 u}{\partial t^2} \tag{9.35}$$

where u = displacement, in.
x = coordinate along rod, in.
t = time, sec
E = Young's modulus, lb/in.^{2}
ρ = mass density, lb-sec^2/in.^{4}

The velocity of the elastic dilatational wave obtained from Eq. (9.35) is

$$c = \sqrt{\frac{E}{\rho}}$$

The modulus E is the slope of the stress-strain curve in the initial linear elastic region. Replacing E by $\partial\sigma/\partial\epsilon$ for the case in which plastic deformation occurs, the slope of the static stress-stress curve can be determined at any value of the strain ϵ.[25] Equation (9.35) then becomes

$$\frac{\partial\sigma}{\partial\epsilon}\frac{\partial^2 u}{\partial x^2} = \rho\frac{\partial^2 u}{\partial t^2} \tag{9.36}$$

Equation (9.36) is nonlinear; its general solution never has been obtained. For the simple type of loading discussed above and an infinitely long bar, the theory predicts a so-called critical velocity of impact because the velocities of the plastic waves are much smaller than those for the elastic waves and approach zero as the strain is indefinitely increased.[25] Since the impact velocity $\dot{u}$ is an independent quantity, it can be made larger and larger while the wave velocities are less than the velocity for elastic waves. Hence a point must be reached at which the continuity of the material is violated. Experimental data illustrate this point.[27]

ENERGY METHOD

Many problems in the design of machines and structures require knowledge of the deformation of material in the plastic condition. In statical problems the method of limit design [28] may be used. In dynamics, the most useful corresponding concept is less theoretical and may be termed the energy method; it is based upon the impact test used for the investigation of brittleness in metals. Originally, the only purpose of this test was to break a standard specimen as an index of brittleness or ductility. The general method, using a tension specimen, may be used in studying the dynamic resistance of materials.[27] An axial force is applied along the length of the specimen and causes the material to rupture ultimately. The energy of absorption is the total amount of energy taken out of the loading system and transferred to the specimen to cause the plastic deformation. The elastic energy and the specific mode of build-up of stress to the final plastic state are ignored. Such an approach has value only to the extent that the material has ductility. For example, in a long tension-type specimen of medium steel, the energy absorbed before neck-down and rupture is of the order of 500 ft-lb per cubic inch of material. Thus, if the moving body in Fig. 9.5 weighs 200 lb and has an initial velocity of 80 ft/sec, it represents 20,000 ft-lb of kinetic energy. If the tension bar subjected to the impact is 10 in. long and 0.5 in. in diameter, it will absorb approximately 1,000 ft-lb of energy. Under these circumstances it will rupture. On the other hand, if the moving body m weighs only 50 lb and has an initial velocity of 30 ft/sec, its kinetic energy is approximately 700 ft-lb and the bar will not rupture.

If the tension specimen were severely notched at some point along its length, it would no longer absorb 500 ft-lb per cubic inch to rupture. The material in the immediate neighborhood of the notch would deform plastically; a break would occur at the notch with the bulk of the material in the specimen stressed below the yield stress for the material. A practical structural situation related to this problem occurs when a butt weld is located at some point along an unnotched specimen. If the weld is of good quality, the full energy absorption of the entire bar develops before rupture; with a poor weld, the rupture occurs at the weld and practically no energy is absorbed by the remainder of the material. This is an important consideration in applying the energy method to design problems.

REFERENCES

1. Love, A. E. H.: "The Mathematical Theory of Elasticity," p. 25, Cambridge University Press, New York, 1934.
2. Timoshenko, S., and D. H. Young: "Engineering Mechanics," p. 334, McGraw-Hill Book Company, Inc., New York, 1940.
3. Loney, S. L.: "A Treatise on Elementary Dynamics," Cambridge University Press, p. 199, New York, 1900.
4. Hertz, H.: *J. Math.* (*Crelle*), pp. 92, 155, 1881.
5. Rayleigh, Lord: *Phil. Mag.* (ser. 6), **11**:283 (1906).
6. Timoshenko, S.: "Theory of Elasticity," p. 350, McGraw-Hill Book Company, Inc., New York, 1934.
7. Raman, C. V.: *Phys. Rev.*, **15,** 277 (1920).
8. Lamb, H.: *Proc. London Math. Soc.*, **35,** 141 (1902).
9. Hoppmann, II, W. H.: *Proc. SESA*, **9**:2, 21 (1952).
10. Hoppmann, II, W. H.: *Proc. SESA*, **10**:1, 157 (1952).
11. Timoshenko, S.: "Vibration Problems in Engineering," 3d ed., p. 413, D. Van Nostrand Company, Inc., Princeton, N.J., 1955.
12. Zener, C., and H. Feshbach: *Trans. ASME*, **61**:a-67 (1939).
13. Hoppmann, II, W. H.: *J. Appl. Mechanics*, **15**:125 (1948).
14. Hoppmann, II, W. H.: *J. Appl. Mechanics*, **16**:370 (1949).
15. Hoppmann, II, W. H.: *J. Appl. Mechanics*, **17**:409 (1950).
16. Goldsmith, W., and D. M. Cunningham: *Proc. SESA*, **14**:1, 179 (1956).
17. Barnhart, Jr., K. E., and Werner Goldsmith: *J. Appl. Mechanics*, **24**:440 (1957).
18. Emschermann, H. H., and K. Ruhl: *VDI-Forschungsheft* 443, Ausgabe B, Band 20, 1954.
19. Hoppmann, II, W. H.: *J. Appl. Mechanics*, **19** (1952).
20. Wenk, E., Jr.: Dissertation, The Johns Hopkins University, 1950, and *David W. Taylor Model Basin Rept.* 704, July, 1950.

21. Compendium, "Underwater Explosion," O. N. R., Department of the Navy, 1950.
22. Prager, W.: James Clayton Lecture, *The Institution of Mechanical Engineers, London*, 1955.
23. Lee, E. H., and P. S. Symonds: *J. Appl. Mechanics*, **19**:308 (1952).
24. Owens, R. H., and P. S. Symonds: *J. Appl. Mechanics*, **22** (1955).
25. Von Kármán, T.: *NDRC Rept.* A-29, 1943.
26. Duwez, P. E., D. S. Clark, and H. F. Bohnenblust: *J. Appl. Mechanics*, **17**, 27 (1950).
27. Hoppmann, II, W. H.: *Proc. ASTM*, **47**:533 (1947).
28. Symposium on the Plastic Theory of Structures, Cambridge University, September, 1956, *British Welding J.*, **3**(8) (1956); **4**(1) (1957).

10

MECHANICAL IMPEDANCE

Elmer L. Hixson
The University of Texas at Austin

INTRODUCTION

The differential equations of motion of a linear mechanical system can be expressed in terms of the driving force and the acceleration, velocity, or displacement of the elements of the system. With a sinusoidal force driving a linear system, the steady-state response exhibits sinusoidal acceleration, velocity, and displacement at the driving frequency with a fixed phase relation to the driving force. The relationship between driving force and the motion resulting at various points of the system can be expressed by algebraic equations involving complex numbers. The force and motion characteristics of the system are defined as described in the following paragraphs.

Force. A sinusoidal force can be represented as a rotating phasor (to differentiate from a vector having magnitude and direction) on the complex plane, as shown in Fig. 10.1. The phasor has a length or magnitude F_0; it rotates counterclockwise with an angular velocity ω, and at some time t subtends an angle ωt with the (+) real axis. The instantaneous force is the projection of the phasor on the real axis, $F_0 \cos \omega t$. The complex representation of the force in terms of real and imaginary components is

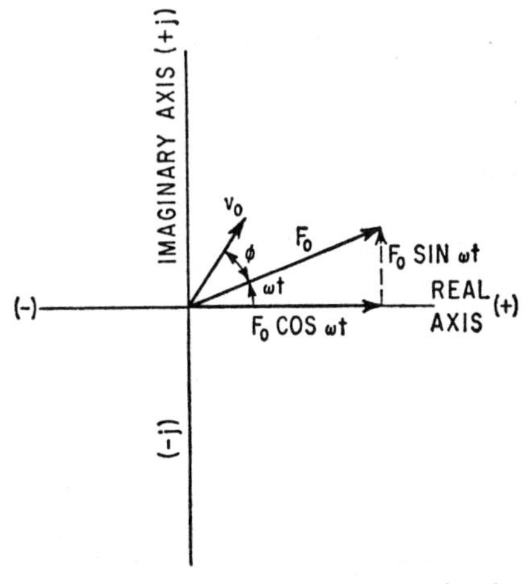

FIG. 10.1. Complex-plane representation for sinusoidal force and velocity.

$$F = F_0 \cos \omega t + jF_0 \sin \omega t \tag{10.1}$$

The force also can be expressed in terms of magnitude and phase angle:

$$F = F_0 e^{j\omega t} \tag{10.2}$$

Velocity. A sinusoidal velocity also can be represented as a phasor, as shown in Fig. 10.1. When v_0 is rotating at the same angular velocity ω as F_0, it can be represented as

$$v = v_0[\cos (\omega t + \phi) + j \sin (\omega t + \phi)] = v_0 e^{j(\omega t + \phi)} \tag{10.3}$$

where ϕ is the phase angle between F and v. The phase reference in Eq. (10.3) is the (+)

real axis. When F is taken as the phase reference, v may be written

$$v = v_0 e^{j\phi} \tag{10.4}$$

Displacement. Displacement in sinusoidal motion is readily determined from velocity by expressing velocity in the form of Eq. (10.2), $v = v_0 e^{j\omega t}$, and noting that $x = \int v\,dt$:

$$x = \frac{v}{j\omega} \tag{10.5}$$

Acceleration. Acceleration is derived from velocity by using the relation $\ddot{x} = dv/dt$:

$$\ddot{x} = j\omega v \tag{10.6}$$

Mechanical Driving-point Impedance. The ratio of the driving force acting on a system to the resulting velocity of the system is *mechanical impedance.** If the velocity is measured at the point of application of the driving force, the ratio of force to velocity is *mechanical driving-point impedance* Z defined as follows: [1,2,3]

$$Z = \frac{F}{v} \tag{10.7}$$

where F is the applied force and v is the resultant velocity in the direction of the force. The quantities F and v are represented by complex numbers, as indicated by Eqs. (10.1) to (10.3). When F is taken as the phase reference, $\phi = \theta$ where θ is defined as the *impedance angle.*

Mechanical Transfer Impedance. The ratio of the driving force to the velocity at another point in the system (or in a direction different from the direction of the applied force) may be expressed as an impedance:

$$Z_{12} = \frac{F_1}{v_2} \tag{10.8}$$

where F_1 is the sinusoidal force applied at one point in a system and v_2 is the resulting sinusoidal velocity at another point in the system, F_1 and v_2 being complex quantities.

Mechanical Driving-point Mobility. Mobility, the inverse of impedance, is determined as follows: [4,5,6]

$$\mathfrak{M} = \frac{v}{F} \tag{10.9}$$

where F and v are as defined by Eqs. (10.1) and (10.3). The concept of mobility is useful in representing electromechanical systems. Mobility sometimes is called *mechanical admittance.*

Dynamic Modulus. Dynamic modulus is defined as follows: [7,8]

$$D = \frac{F}{x} \tag{10.10}$$

where F is the applied sinusoidal force and x is the resulting displacement in the direction of the force, both expressed as complex quantities.

Receptance. Receptance is the reciprocal of dynamic modulus; it is defined as follows: [9,10]

$$R = \frac{x}{F} \tag{10.11}$$

where F and x are defined as above.

* The ratio of driving force to resulting displacement sometimes is termed "mechanical impedance." This is not recommended usage.

Table 10.1. Units of Parameters Used in Analysis by Mechanical Impedance and Mobility Methods

Quantity	Units		
	cgs	mks	English gravitational
Velocity, v	cm/sec	meters/sec	in./sec
Force, F	dynes	newtons	lb
Mass, m	grams	kilograms	lb-sec²/in.
Spring stiffness, k	dynes/cm	newtons/meter	lb/in.
Damping coefficient, c	dynes/cm/sec	newtons/meter/sec	lb/in./sec
Impedance, Z	dyne-sec/cm grams/sec	newton-sec/meter kilograms/sec	lb-sec/in.
Mobility, $\mathfrak{M}$	cm/dyne/sec sec/gram	meters/newton/sec sec/kilogram	in./lb-sec
Power, P	dyne-cm/sec ergs/sec	newton-meters/sec joules/sec	in.-lb/sec

In the ensuing discussion, only impedance and mobility are used; the discussion is limited to linear systems in which the motion is translational in the direction of the driving force. Other systems that are analogous to translational systems may be analyzed by the same methods using analogous parameters. The several quantities used in impedance methods may be expressed in any consistent system of units. Three such systems are given in Table 10.1.

Impedance methods allow a complete analysis of the motions and forces acting on each part of a physical system; however, the differential equations of Newtonian mechanics become algebraic equations. In addition, a "black box" concept is introduced in which the forces and motions at one or two points of major interest in a system are determined without a complete analysis of the entire system. These methods are adaptable to the use of results of physical measurements on systems that are so complicated that a complete analysis becomes impractical. Typical curves illustrating the quantitative aspect of impedance as measured on a wide range of structure types are presented.

IMPEDANCE OF MECHANICAL ELEMENTS

Three idealized mechanical system elements with lumped constants are conventionally assembled to form linear physical systems. These elements are resistance (or damping), spring, and mass.

MECHANICAL RESISTANCE (DAMPER). A mechanical resistance is a device in which the relative velocity between its end points is proportional to the force applied to the end points. Such a device can be represented by the dashpot of Fig. 10.2 in which the force resisting the extension (or compression) of the dashpot is the result of viscous friction. An ideal resistance is assumed to be made of massless, infinitely rigid elements.

In Fig. 10.2 the velocity of point A with respect to point B is

$$v = (v_1 - v_2) = \frac{F_a}{c} \tag{10.12}$$

where c is the constant of proportionality called the *mechanical resistance* or *damping constant.* For there to be a relative velocity v as a result of force at A, there must be an equal reaction force at B. Thus, it may be considered that a transmitted force F_b is

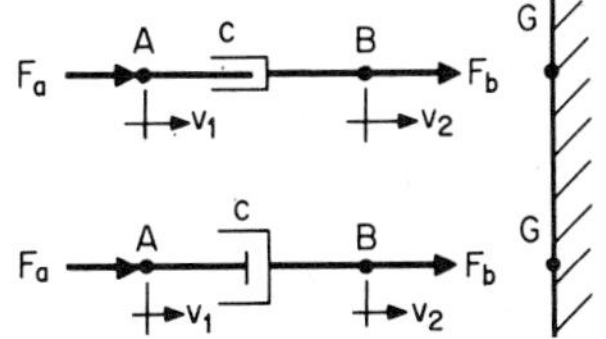

FIG. 10.2. Two alternate schematic representations of an ideal mechanical resistance element. The upper one is that recommended by the American National Standards Institute, ANSI Y-32.18 (1972).

equal to F_a. The velocities v_1 and v_2 are measured with respect to the stationary reference G; their difference is the relative velocity between the end points of the resistance.

With the sinusoidal force of Eq. (10.2) applied to point A of Fig. 10.2 and point B attached to a fixed (immovable) point, the velocity v_1 is obtained from Eq. (10.12):

$$v_1 = \frac{F_0 e^{j\omega t}}{c} = v_0 e^{j\omega t} \tag{10.13}$$

The force and velocity phasors thus are rotating at the same angular frequency and are said to be "in phase," as shown in Fig. 10.3.

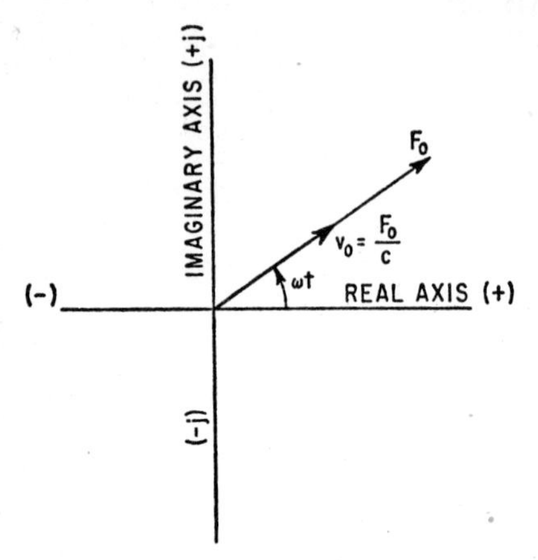

Fig. 10.3. Force and velocity relationships at the connections of an ideal resistance element.

The mechanical impedance of the resistance is obtained by substituting from Eqs. (10.2) and (10.13) in Eq. (10.7):

$$Z_c = \frac{F}{v} = \frac{F_0 e^{j\omega t}}{\dfrac{F_0}{c} e^{j\omega t}} = c \tag{10.14}$$

The mechanical impedance of a resistance is the value of its damping constant c; i.e., when a resistance is placed in a mechanical system, the ratio of the transmitted force to the relative velocity of its end points is its impedance c.

SPRING. A linear spring is a device for which the relative displacement between its end points is proportional to the force applied. It is illustrated in Fig. 10.4 and can be represented mathematically as follows:

$$\delta x = x_1 - x_2 = \frac{F_a}{k} \tag{10.15}$$

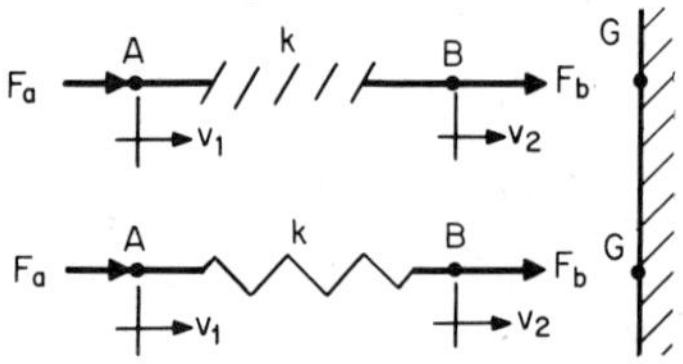

Fig. 10.4. Two alternate schematic representations of an ideal spring. The upper one is that recommended by the American National Standards Institute, ANSI Y-32.18 (1972).

where k is the *spring stiffness* and x_1, x_2 are displacements relative to the reference point G. The stiffness k can be expressed alternately in terms of a *compliance* $C = 1/k$. The spring transmits the applied force so that $F_b = F_a$.

With the force of Eq. (10.2) applied to point A and with point B fixed, the displacement of point A is given by Eq. (10.15):

$$x_1 = \frac{F_0 e^{j\omega t}}{k} = x_0 e^{j\omega t} \tag{10.16}$$

The displacement is thus sinusoidal and in phase with the force. The relative velocity of the end connections is required for impedance calculations and is given by the differentiation of x with respect to time:

$$\dot{x} = v = \frac{j\omega F_0 e^{j\omega t}}{k} = j v_0 e^{j\omega t} \tag{10.17}$$

From Eqs. (10.2) and (10.17) the impedance of the spring is

$$Z_k = \frac{F_0 e^{j\omega t}}{j\omega F_0 e^{j\omega t}/k} = \frac{k}{j\omega} = -\frac{jk}{\omega} \tag{10.18}$$

The ratio of force to velocity for a spring thus depends on the stiffness and frequency, and is an imaginary number.

To find the phase relation between v and F for a spring, Eq. (10.17) may be written in the form of Eq. (10.1):

$$v = \frac{F_0}{k}(-\omega \sin \omega t + j\omega \cos \omega t) = \frac{\omega}{k} F_0 e^{j(\omega t + 90°)}$$

The velocity is always 90° ahead of, or leading, the applied force, as shown in Fig. 10.5. This is indicated by the factor j in Eq. (10.17).

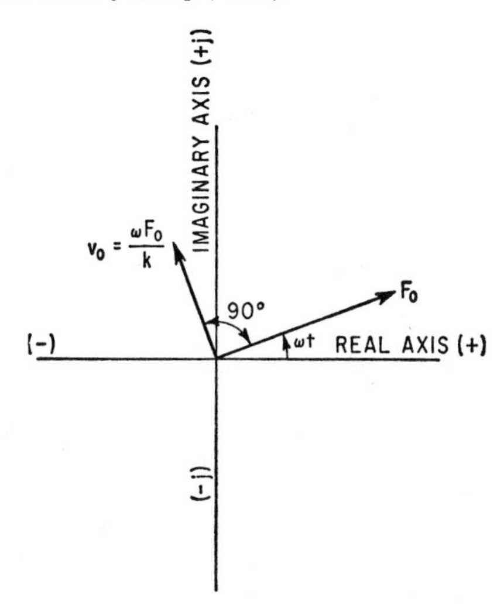

Fig. 10.5. Force and velocity relationships at the connections of an ideal spring.

MASS. In the ideal mass illustrated in Fig. 10.6*A*, the acceleration $\ddot{x}$ of the rigid body is proportional to the applied force F:

$$\ddot{x}_1 = \frac{F_a}{m} \tag{10.19}$$

where m is the mass of the body and $\ddot{x}_1 = \ddot{x}_2$ since the element is rigid. By Eq. (10.19), the force F_a is required to give the mass the acceleration $\ddot{x}_1$ and the transmitted force F_b is zero. The representations of Fig. 10.6*B* and *C* are used commonly to indicate that one

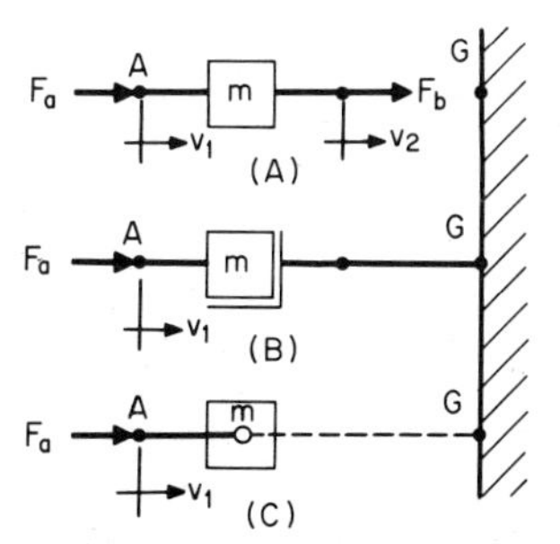

Fig. 10.6. Schematic representations of an ideal mass. The one shown in *C* is that recommended by the American National Standards Institute, ANSI Y-32.18 (1972).

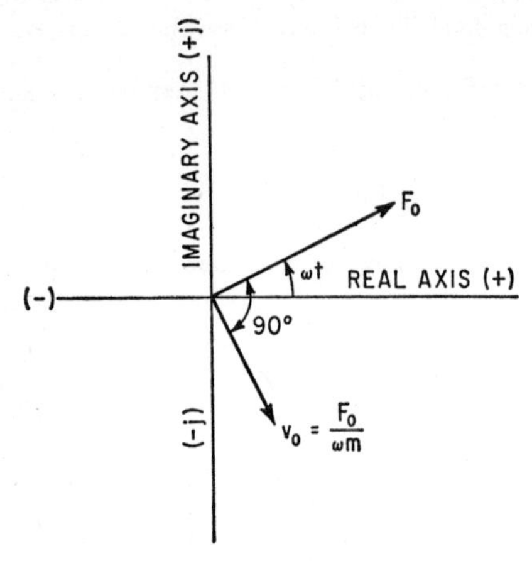

FIG. 10.7. Force and velocity relationships for an ideal mass.

end of a mass is connected rigidly to an inertial reference point, leaving it with only one end which is free to move. When a sinusoidal force is applied, Eq. (10.19) becomes

$$\ddot{x}_1 = \frac{F_0 e^{j\omega t}}{m} \tag{10.20}$$

The acceleration is sinusoidal and in phase with the applied force.

Integrating Eq. (10.20) to find velocity

$$\dot{x} = v = \frac{F_0 e^{j\omega t}}{j\omega m} = -jv_0 e^{j\omega t} \tag{10.21}$$

The mechanical impedance of the mass is the ratio of F to v:

$$Z_m = \frac{F_0 e^{j\omega t}}{F_0 e^{j\omega t}/j\omega m} = j\omega m \tag{10.22}$$

Thus, the impedance of a mass is an imaginary quantity that depends on the magnitude of the mass and on the frequency.

To find the phase relation between F and v, Eq. (10.21) may be written:

$$v = \frac{F_0}{\omega m}(\sin \omega t - j \cos \omega t) = \frac{F_0}{\omega m} e^{j(\omega t - 90°)}$$

Thus the velocity of the mass lags behind the applied force by 90° as shown in Fig. 10.7, and indicated by the $-j$ in Eq. (10.21).

The impedances of a resistance, a spring, and a mass at any particular frequency may be represented on the complex plane as shown in Fig. 10.8. Their variation with frequency is indicated in Fig. 10.9. The impedances of elements and several combinations of elements are given in Table 10.2.

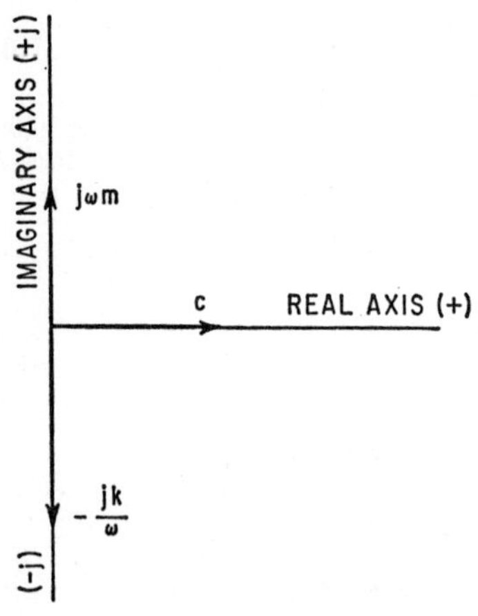

FIG. 10.8. A complex-plane representation for the impedance of ideal resistance, spring, and mass at a fixed frequency.

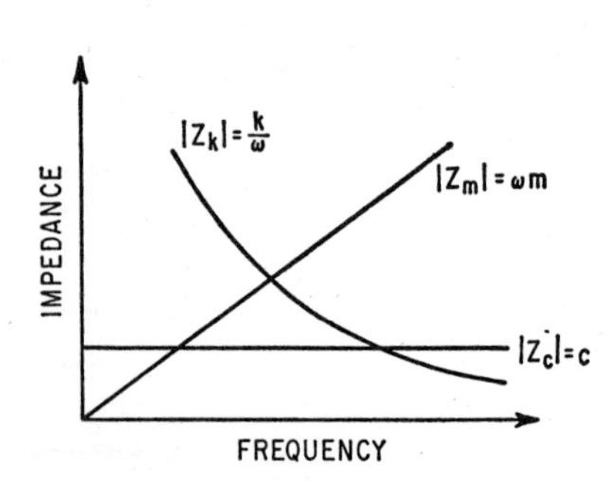

FIG. 10.9. The variation with frequency of the magnitude of the impedance of ideal resistance, spring, and mass.

Table 10.2. Driving-point Impedance and Mobility of Ideal Mechanical Elements and Lumped Parameter Systems

When the system of elements shown includes a mass, the impedance or mobility given is the relationship between force and velocity at the one available connection, the other connection being attached to the inertial reference plane. When no mass is included, the impedance or mobility describes the relation between the force applied to the two system connections and the resulting relative velocity between these connections. Graphs of magnitude of impedance vs. ω and magnitude of mobility vs. ω are plotted on a log-log scale.

DIAGRAM OF SYSTEM	MATHEMATIC FORMULAS: IMPEDANCE Z - EQ. (10.7) MOBILITY $\mathcal{M}$ - EQ. (10.9)	IMPEDANCE IN THE COMPLEX PLANE	MOBILITY IN THE COMPLEX PLANE	MAGNITUDE OF IMPEDANCE	MAGNITUDE OF MOBILITY	IMPEDANCE ANGLE θ FIG. 10.34
1. 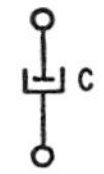 c	$Z = c$ $\mathcal{M} = 1/c$	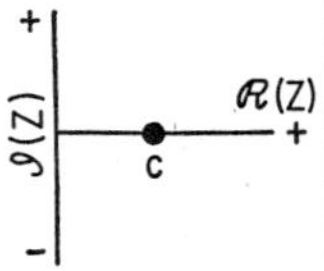 $\mathcal{I}(Z)$, $\mathcal{R}(Z)$, +, −, c	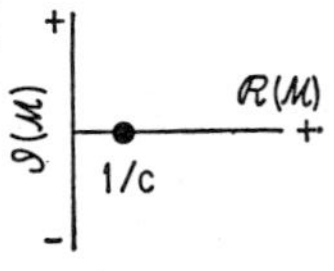 $\mathcal{I}(\mathcal{M})$, $\mathcal{R}(\mathcal{M})$, +, −, 1/c	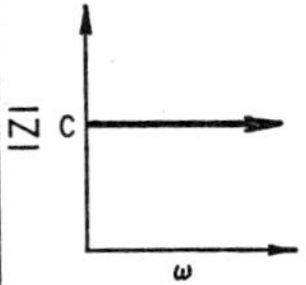 $\lvert Z \rvert$, c, ω	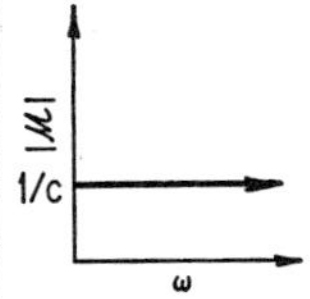 $\lvert \mathcal{M} \rvert$, 1/c, ω	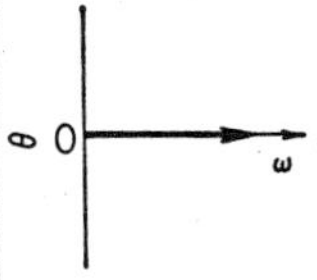 θ, 0, ω
2.  k	$Z = \frac{k}{j\omega}$ $\mathcal{M} = \frac{j\omega}{k}$	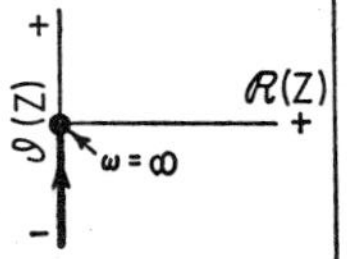 $\mathcal{I}(Z)$, $\mathcal{R}(Z)$, +, −, $\omega = \infty$	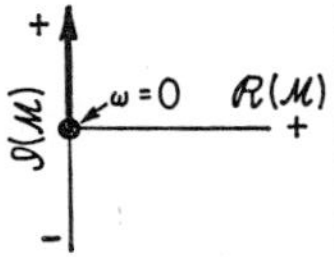 $\mathcal{I}(\mathcal{M})$, $\mathcal{R}(\mathcal{M})$, +, −, $\omega = 0$	 $\lvert Z \rvert$, SLOPE=-1, ω	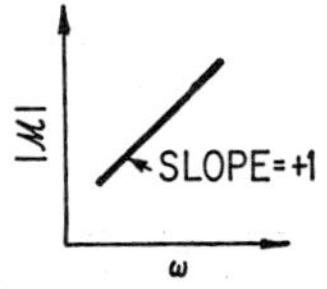 $\lvert \mathcal{M} \rvert$, SLOPE=+1, ω	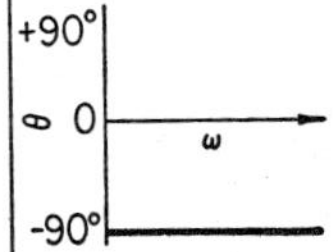 θ, +90°, 0, -90°, ω
3. 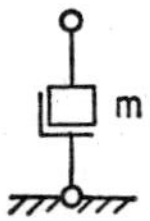 m	$Z = j\omega m$ $\mathcal{M} = \frac{1}{j\omega m}$	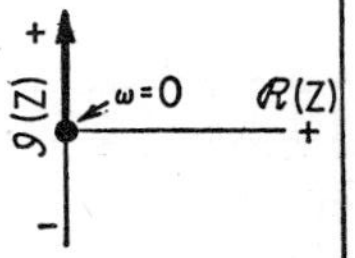 $\mathcal{I}(Z)$, $\mathcal{R}(Z)$, +, −, $\omega = 0$	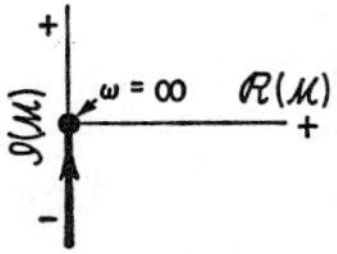 $\mathcal{I}(\mathcal{M})$, $\mathcal{R}(\mathcal{M})$, +, −, $\omega = \infty$	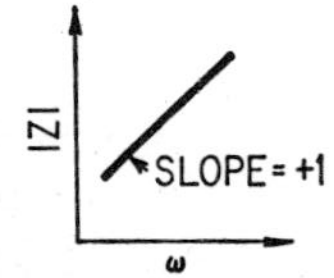 $\lvert Z \rvert$, SLOPE=+1, ω	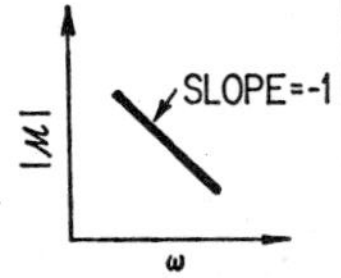 $\lvert \mathcal{M} \rvert$, SLOPE=-1, ω	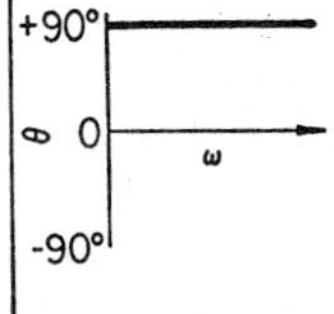 θ, +90°, 0, -90°, ω

Table 10.2. (*Continued*)

DIAGRAM OF SYSTEM	MATHEMATIC FORMULAS: IMPEDANCE Z - EQ. (10.7) MOBILITY $\mathcal{M}$ - EQ. (10.9)	IMPEDANCE IN THE COMPLEX PLANE	MOBILITY IN THE COMPLEX PLANE	MAGNITUDE OF IMPEDANCE	MAGNITUDE OF MOBILITY	IMPEDANCE ANGLE θ FIG. 10.34
4. 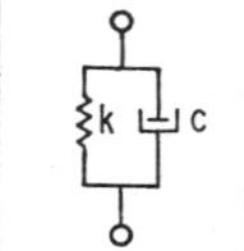 k, c	$Z = c + \frac{k}{j\omega}$ $\mathcal{M} = \frac{c - \frac{k}{j\omega}}{c^2 + (k/\omega)^2}$ $\omega_1 = \frac{k}{c}$	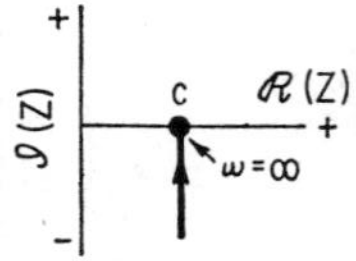 $\mathcal{I}(Z)$, $\mathcal{R}(Z)$, +, −, c, $\omega=\infty$	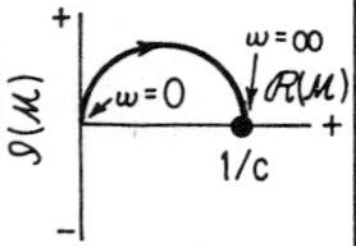 $\mathcal{I}(\mathcal{M})$, $\mathcal{R}(\mathcal{M})$, +, −, $\omega=0$, $\omega=\infty$, 1/c	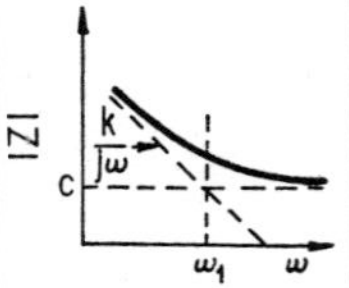 $\lvert Z\rvert$, c, $\frac{k}{j\omega}$, ω_1, ω	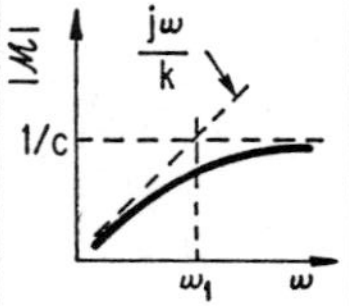 $\lvert\mathcal{M}\rvert$, 1/c, $\frac{j\omega}{k}$, ω_1, ω	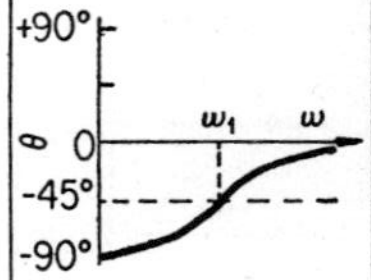 +90°, θ 0, -45°, -90°, ω_1, ω
5.  c, k	$Z = \frac{1/c - j\omega/k}{(1/c)^2 + (\omega/k)^2}$ $\mathcal{M} = \frac{1}{c} + \frac{j\omega}{k}$ $\omega_1 = \frac{k}{c}$	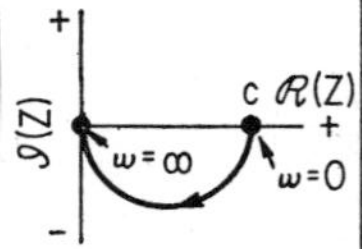 $\mathcal{I}(Z)$, $\mathcal{R}(Z)$, +, −, c, $\omega=\infty$, $\omega=0$	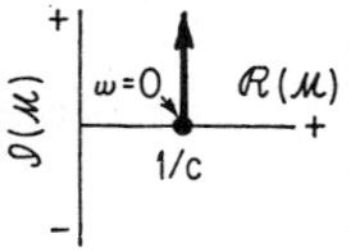 $\mathcal{I}(\mathcal{M})$, $\mathcal{R}(\mathcal{M})$, +, −, $\omega=0$, 1/c	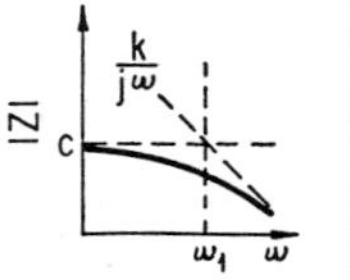 $\lvert Z\rvert$, c, $\frac{k}{j\omega}$, ω_1, ω	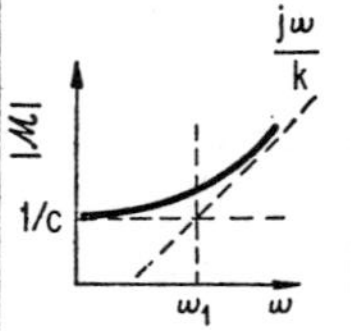 $\lvert\mathcal{M}\rvert$, 1/c, $\frac{j\omega}{k}$, ω_1, ω	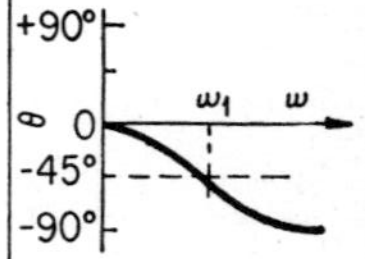 +90°, θ 0, -45°, -90°, ω_1, ω
6. 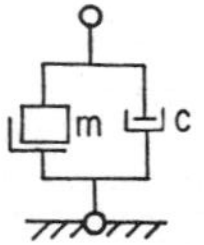 m, c	$Z = c + j\omega m$ $\mathcal{M} = \frac{c - j\omega m}{c^2 + \omega^2 m^2}$ $\omega_1 = \frac{c}{m}$	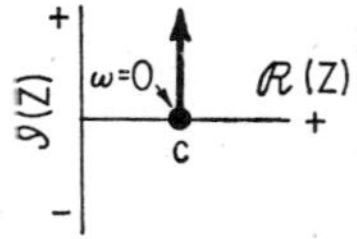 $\mathcal{I}(Z)$, $\mathcal{R}(Z)$, +, −, c, $\omega=0$	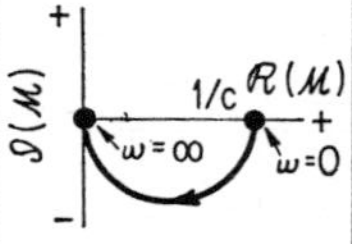 $\mathcal{I}(\mathcal{M})$, $\mathcal{R}(\mathcal{M})$, +, −, $\omega=\infty$, $\omega=0$, 1/c	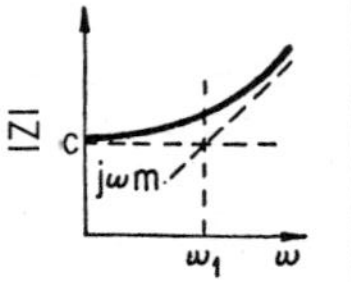 $\lvert Z\rvert$, c, $j\omega m$, ω_1, ω	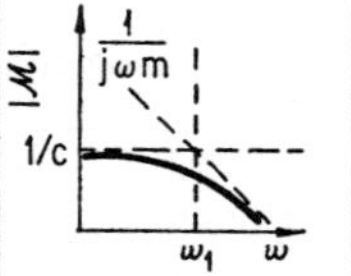 $\lvert\mathcal{M}\rvert$, 1/c, $\frac{1}{j\omega m}$, ω_1, ω	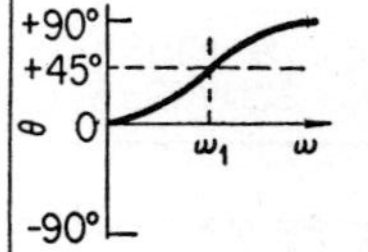 +90°, +45°, θ 0, -90°, ω_1, ω

DIAGRAM OF SYSTEM	MATHEMATIC FORMULAS: IMPEDANCE Z - EQ.(10.7) MOBILITY $\mathcal{M}$ - EQ.(10.9)	IMPEDANCE IN THE COMPLEX PLANE	MOBILITY IN THE COMPLEX PLANE	MAGNITUDE OF IMPEDANCE	MAGNITUDE OF MOBILITY	IMPEDANCE ANGLE θ FIG.10.34
7. 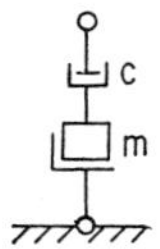	$Z = \frac{1/c + j/\omega m}{(1/c)^2 + (1/\omega m)^2}$ $\mathcal{M} = 1/c + 1/j\omega m$ $\omega_1 = \frac{c}{m}$	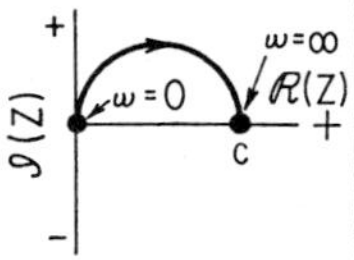	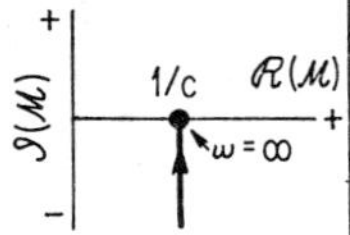	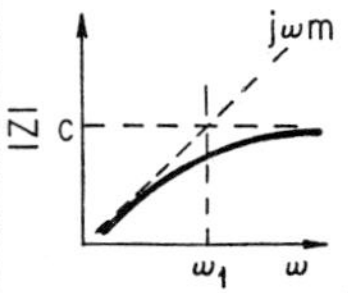	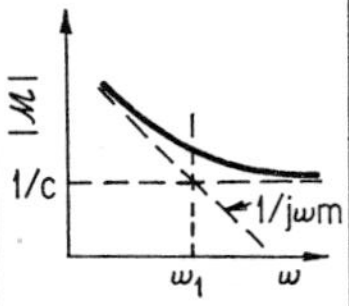	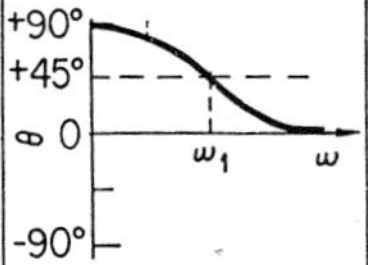
8. 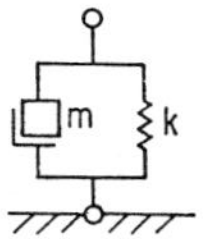	$Z = j\omega m + \frac{k}{j\omega}$ $\mathcal{M} = \frac{-j}{\omega m - k/\omega}$ $\omega_0 = \sqrt{\frac{k}{m}}$	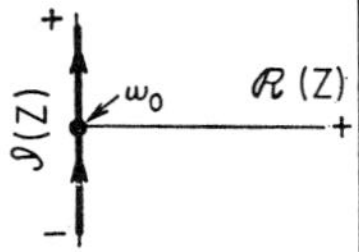	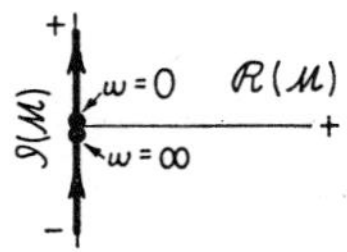	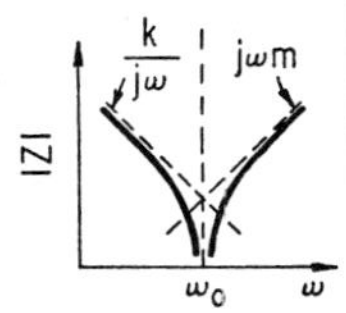	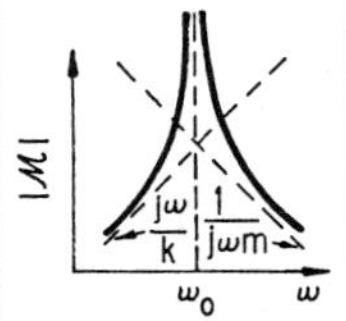	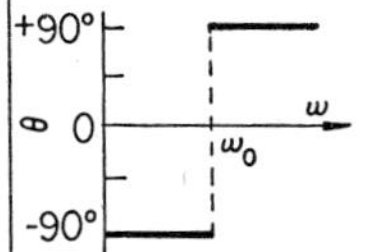
9. 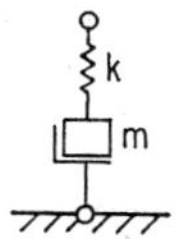	$Z = \frac{-j}{\omega/k - 1/\omega m}$ $\mathcal{M} = j\omega/k + 1/j\omega m$ $\omega_0 = \sqrt{\frac{k}{m}}$	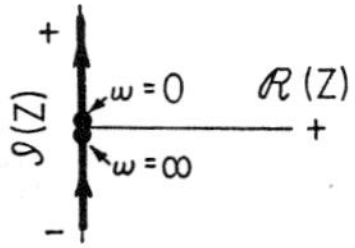	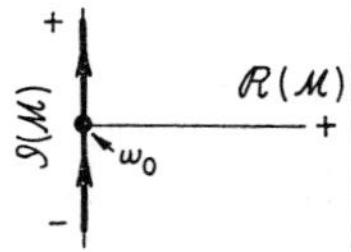	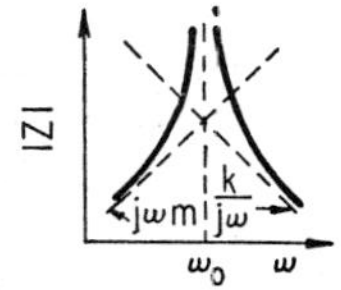	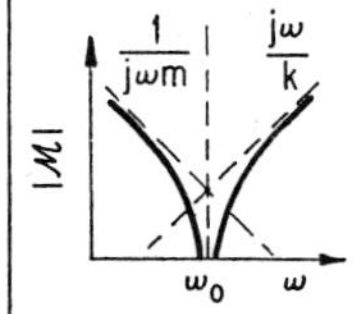	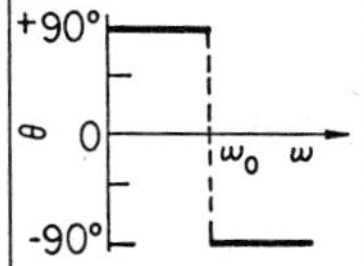

Table 10.2. (Continued)

DIAGRAM OF SYSTEM	MATHEMATIC FORMULAS: IMPEDANCE Z - EQ.(10.7) MOBILITY $\mathcal{M}$ - EQ.(10.9)	IMPEDANCE IN THE COMPLEX PLANE	MOBILITY IN THE COMPLEX PLANE	MAGNITUDE OF IMPEDANCE	MAGNITUDE OF MOBILITY	IMPEDANCE ANGLE θ FIG. 10.34
10. 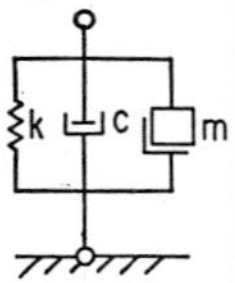 k c m	$Z = c + j(\omega m - k/\omega)$ $\mathcal{M} = \dfrac{c - j(\omega m - k/m)}{c^2 + (\omega m - k/m)^2}$ $\omega_0 = \sqrt{\dfrac{k}{m}}$	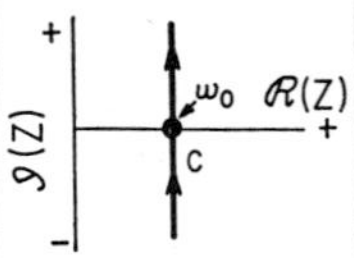 + $\mathcal{I}(Z)$ − ω_0 $\mathcal{R}(Z)$ + c	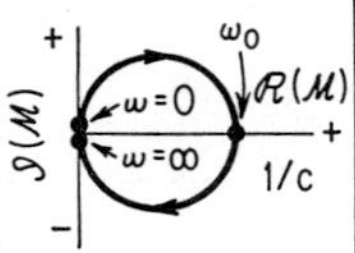 + $\mathcal{I}(\mathcal{M})$ − ω_0 $\mathcal{R}(\mathcal{M})$ + $\omega=0$ $\omega=\infty$ 1/c	$\lvert Z \rvert$ $\dfrac{k}{j\omega}$ $j\omega m$ c ω_0 ω	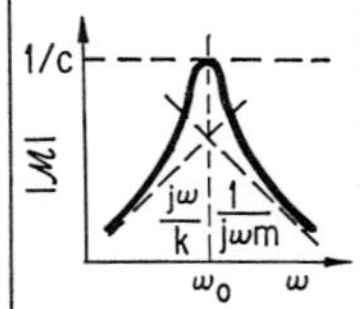 1/c $\lvert \mathcal{M} \rvert$ $\dfrac{j\omega}{k}$ $\dfrac{1}{j\omega m}$ ω_0 ω	+90° θ 0 −90° ω_0 ω
11. 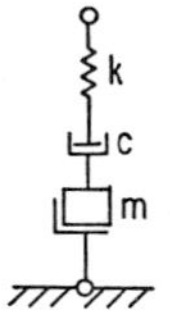 k c m	$Z = \dfrac{1/c - j(\omega/k - 1/\omega m)}{(1/c)^2 + (\omega/k - 1/\omega m)^2}$ $\mathcal{M} = 1/c + j(\omega/k - 1/\omega m)$ $\omega_0 = \sqrt{\dfrac{k}{m}}$	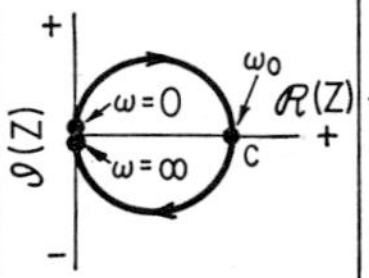 + $\mathcal{I}(Z)$ − ω_0 $\mathcal{R}(Z)$ + $\omega=0$ $\omega=\infty$ c	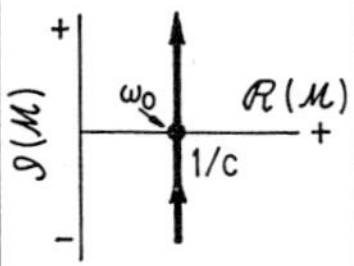 + $\mathcal{I}(\mathcal{M})$ − ω_0 $\mathcal{R}(\mathcal{M})$ + 1/c	$\lvert Z \rvert$ c $k/j\omega$ $j\omega m$ ω_0 ω	$\lvert \mathcal{M} \rvert$ $\dfrac{1}{j\omega m}$ $\dfrac{j\omega}{k}$ 1/c ω_0 ω	+90° θ 0 −90° ω_0 ω
12. c_1 c_2 m	$Z = \dfrac{\dfrac{c_1 + c_2}{c_1^2} + \dfrac{c_2}{\omega^2 m^2} + \dfrac{j}{\omega m}}{(1/c_1)^2 + (1/\omega m)^2}$ $\mathcal{M} = \dfrac{\dfrac{c_1 + c_2}{c_1^2} + \dfrac{c_2}{\omega^2 m^2} - \dfrac{j}{\omega m}}{\left(\dfrac{c_1 + c_2}{c_1}\right)^2 + \left(\dfrac{c_2}{\omega m}\right)^2}$	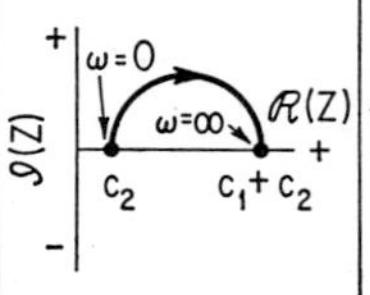 + $\mathcal{I}(Z)$ − $\omega=0$ $\omega=\infty$ $\mathcal{R}(Z)$ + c_2 c_1+c_2	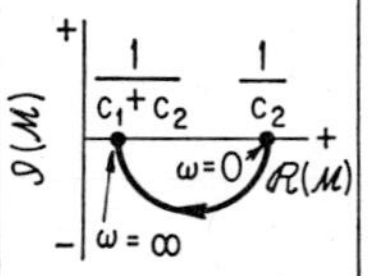 + $\mathcal{I}(\mathcal{M})$ − $\dfrac{1}{c_1+c_2}$ $\dfrac{1}{c_2}$ + $\omega=0$ $\mathcal{R}(\mathcal{M})$ $\omega=\infty$	$\lvert Z \rvert$ c_1+c_2 c_2 ω	$\lvert \mathcal{M} \rvert$ $\dfrac{1}{c_2}$ $\dfrac{1}{c_1+c_2}$ ω	+90° θ 0 −90° ω

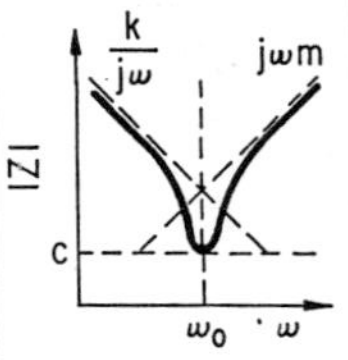

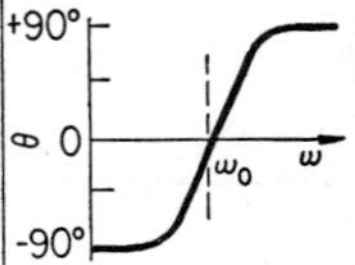

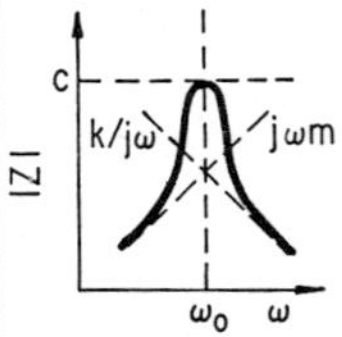

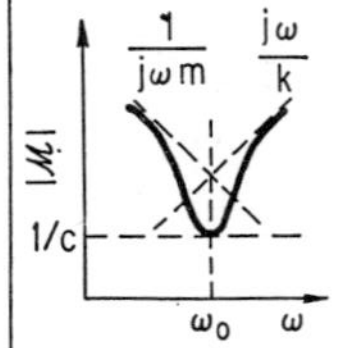

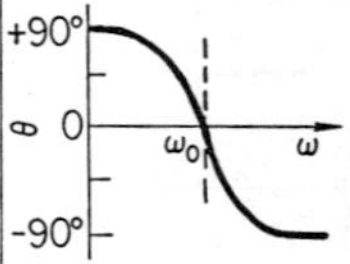

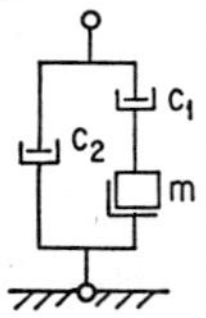

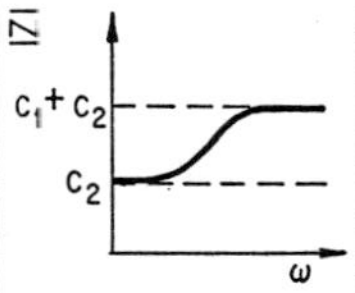

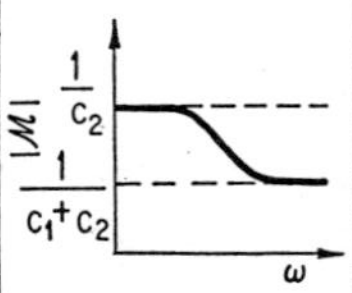

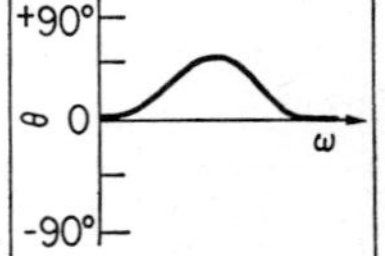

DIAGRAM OF SYSTEM	MATHEMATIC FORMULAS: IMPEDANCE Z - EQ. (10.7) MOBILITY $\mathcal{M}$ - EQ. (10.9)	IMPEDANCE IN THE COMPLEX PLANE	MOBILITY IN THE COMPLEX PLANE	MAGNITUDE OF IMPEDANCE	MAGNITUDE OF MOBILITY	IMPEDANCE ANGLE θ FIG. 10.34
13. 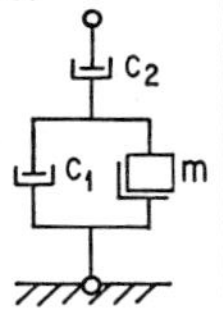 c_2, c_1, m	$Z = \dfrac{c_1\left(\dfrac{c_1+c_2}{c_2}\right) + \dfrac{\omega^2 m^2}{c_2} + j\omega m}{\left(\dfrac{c_1+c_2}{c_2}\right)^2 + \left(\dfrac{\omega m}{c_2}\right)^2}$ $\mathcal{M} = \dfrac{c_1\left(\dfrac{c_1+c_2}{c_2}\right) + \dfrac{\omega^2 m^2}{c_2} - j\omega m}{c_1^2 + \omega^2 m^2}$	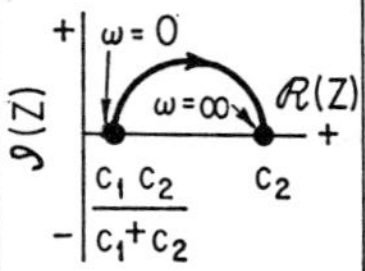 $\mathcal{I}(Z)$, +, −, $\omega = 0$, $\omega = \infty$, $\dfrac{c_1 c_2}{c_1 + c_2}$, c_2, $\mathcal{R}(Z)$	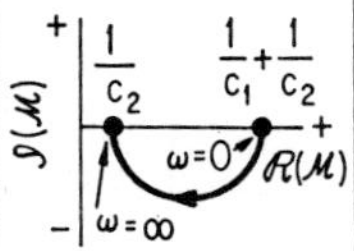 $\mathcal{I}(\mathcal{M})$, +, −, $\dfrac{1}{c_2}$, $\dfrac{1}{c_1} + \dfrac{1}{c_2}$, $\omega = 0$, $\omega = \infty$, $\mathcal{R}(\mathcal{M})$	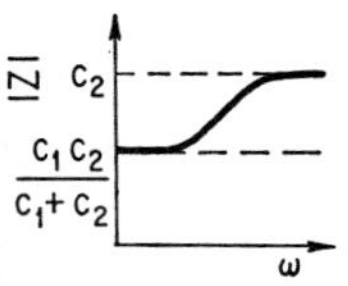 $\lvert Z \rvert$, c_2, $\dfrac{c_1 c_2}{c_1 + c_2}$, ω	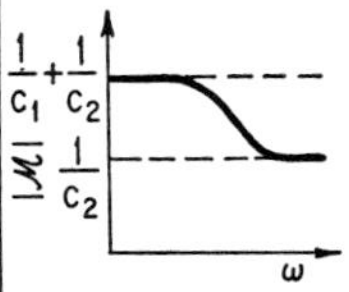 $\lvert \mathcal{M} \rvert$, $\dfrac{1}{c_1} + \dfrac{1}{c_2}$, $\dfrac{1}{c_2}$, ω	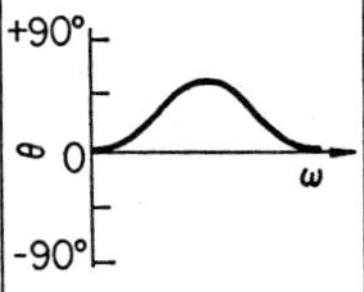 +90°, θ 0, −90°, ω
14. 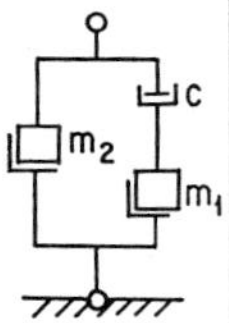 c, m_2, m_1	$Z = \dfrac{\dfrac{1}{c} + j\left[\dfrac{\omega m_2}{c^2} + \dfrac{1}{\omega m_1}\left(\dfrac{m_1+m_2}{m_1}\right)\right]}{(1/c)^2 + (1/\omega m_1)^2}$ $\mathcal{M} = \dfrac{\dfrac{1}{c} - j\left[\dfrac{\omega m_2}{c^2} + \dfrac{1}{\omega m_1}\left(\dfrac{m_1+m_2}{m_1}\right)\right]}{\left(\dfrac{m_1+m_2}{m_1}\right)^2 + \left(\dfrac{\omega m_2}{c}\right)^2}$ $\omega_1 = \dfrac{c}{m_2}$ $\qquad \omega_2 = \dfrac{c}{m_1}\left(\dfrac{m_1+m_2}{m_1}\right)$	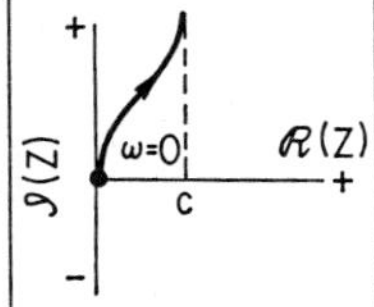 $\mathcal{I}(Z)$, +, −, $\omega = 0$, c, $\mathcal{R}(Z)$	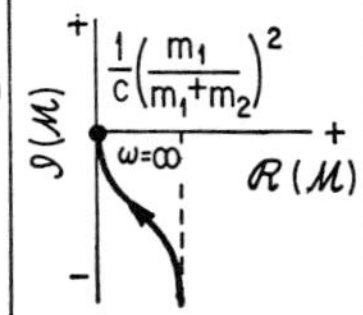 $\mathcal{I}(\mathcal{M})$, +, −, $\dfrac{1}{c}\left(\dfrac{m_1}{m_1+m_2}\right)^2$, $\omega = \infty$, $\mathcal{R}(\mathcal{M})$	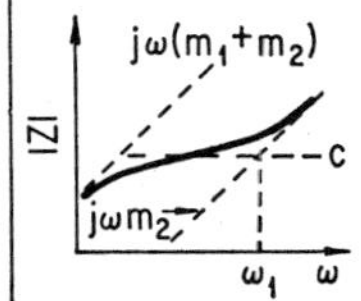 $\lvert Z \rvert$, $j\omega(m_1+m_2)$, c, $j\omega m_2$, ω_1, ω	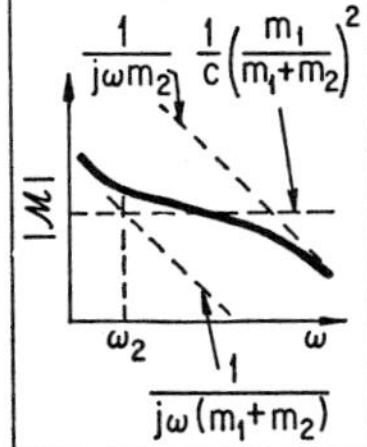 $\lvert \mathcal{M} \rvert$, $\dfrac{1}{j\omega m_2}$, $\dfrac{1}{c}\left(\dfrac{m_1}{m_1+m_2}\right)^2$, ω_2, ω, $\dfrac{1}{j\omega(m_1+m_2)}$	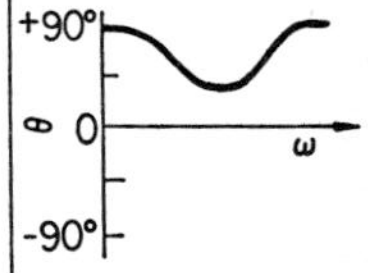 +90°, θ 0, −90°, ω
15. 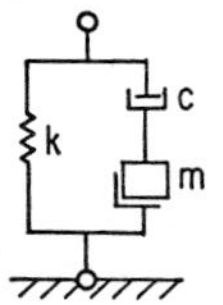 c, k, m	$Z = \dfrac{\dfrac{1}{c} + \dfrac{j}{\omega}\left[\dfrac{1}{m} - k\left(\dfrac{1}{c^2} + \dfrac{1}{\omega^2 m^2}\right)\right]}{(1/c)^2 + (1/\omega m)^2}$ $\mathcal{M} = \dfrac{\dfrac{\omega^2 m^2}{c} + j\left(\dfrac{\omega m^2 k}{c} + \dfrac{k}{\omega} - \omega m\right)}{(mk/c)^2 + (\omega m - k/\omega)^2}$ $\omega_0 = \sqrt{\dfrac{k}{m - \dfrac{km^2}{c^2}}}$	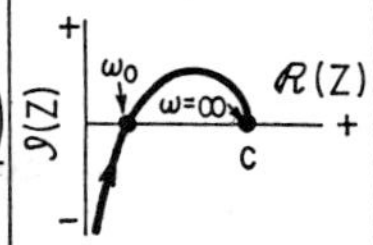 $\mathcal{I}(Z)$, +, −, ω_0, $\omega = \infty$, c, $\mathcal{R}(Z)$	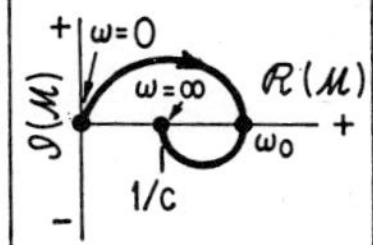 $\mathcal{I}(\mathcal{M})$, +, −, $\omega = 0$, $\omega = \infty$, 1/c, ω_0, $\mathcal{R}(\mathcal{M})$	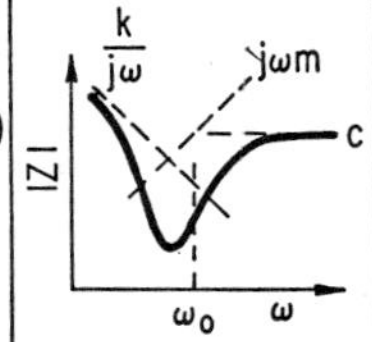 $\lvert Z \rvert$, $\dfrac{k}{j\omega}$, $j\omega m$, c, ω_0, ω	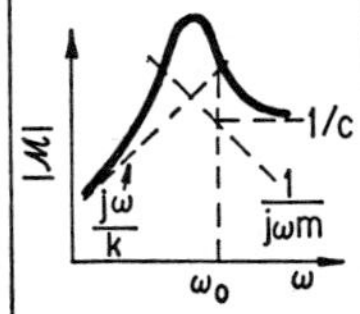 $\lvert \mathcal{M} \rvert$, 1/c, $\dfrac{j\omega}{k}$, $\dfrac{1}{j\omega m}$, ω_0, ω	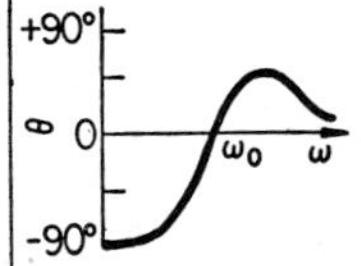 +90°, θ 0, −90°, ω_0, ω

Table 10.2. *(Continued)*

DIAGRAM OF SYSTEM	MATHEMATIC FORMULAS: IMPEDANCE Z - EQ. (10.7) MOBILITY $\mathcal{M}$ - EQ. (10.9)	IMPEDANCE IN THE COMPLEX PLANE	MOBILITY IN THE COMPLEX PLANE	MAGNITUDE OF IMPEDANCE	MAGNITUDE OF MOBILITY	IMPEDANCE ANGLE θ FIG. 10.34
16. k, c, m	$Z=\dfrac{\dfrac{ck^2}{\omega^2}-jkm\left[(\omega m-k/\omega)+\dfrac{c^2k}{\omega m}\right]}{c^2+(\omega m-k/\omega)^2}$ $\mathcal{M}=\dfrac{c+j\omega\left(\dfrac{c^2+\omega^2m^2}{k}-m\right)}{c^2+\omega^2m^2}$ $\omega_0=\sqrt{\dfrac{k}{m}-\dfrac{c^2}{m^2}}$	$\mathcal{I}(Z)$, $\mathcal{R}(Z)$, $\omega=\infty$, $\omega=0$, c, ω_0	$\mathcal{I}(\mathcal{M})$, $\mathcal{R}(\mathcal{M})$, ω_0, $\omega=0$, 1/c	$\lvert Z\rvert$, c, $j\omega m$, $\dfrac{k}{j\omega}$, ω_0, ω	$\lvert\mathcal{M}\rvert$, 1/c, $\dfrac{1}{j\omega m}$, $\dfrac{j\omega}{k}$, ω_0, ω	θ: +90°, 0, -90°; ω_0, ω
17. c_1, k, c_2	$Z=\dfrac{\dfrac{c_1+c_2}{c_1^2}+\dfrac{c_2\omega^2}{k^2}-j\dfrac{\omega}{k}}{(1/c_1)^2+(\omega/k)^2}$ $\mathcal{M}=\dfrac{\dfrac{c_1+c_2}{c_1^2}+\dfrac{c_2\omega^2}{k^2}+\dfrac{j\omega}{k}}{\left(\dfrac{c_1+c_2}{c_1}\right)^2+\left(\dfrac{c_2\omega}{k}\right)^2}$	$\mathcal{I}(Z)$, $\mathcal{R}(Z)$, c_2, c_1+c_2, $\omega=0$, $\omega=\infty$	$\mathcal{I}(\mathcal{M})$, $\mathcal{R}(\mathcal{M})$, $\omega=0$, $\omega=\infty$, $\dfrac{1}{c_1+c_2}$, $\dfrac{1}{c_2}$	$\lvert Z\rvert$, c_1+c_2, c_2, ω	$\lvert\mathcal{M}\rvert$, $\dfrac{1}{c_2}$, $\dfrac{1}{c_1+c_2}$, ω	θ: +90°, 0, -90°; ω
18. c_2, k, c_1	$Z=\dfrac{c_1\left(\dfrac{c_1+c_2}{c_2}\right)+\dfrac{k^2}{c_2\omega^2}-j\dfrac{k}{\omega}}{\left(\dfrac{c_1+c_2}{c_2}\right)^2+\left(\dfrac{k}{\omega c_2}\right)^2}$ $\mathcal{M}=\dfrac{c_1\left(\dfrac{c_1+c_2}{c_2}\right)+\dfrac{k^2}{c_2\omega^2}+\dfrac{jk}{\omega}}{c_1^2+(k/\omega)^2}$	$\mathcal{I}(Z)$, $\mathcal{R}(Z)$, $\dfrac{c_1c_2}{c_1+c_2}$, c_2, $\omega=0$, $\omega=\infty$	$\mathcal{I}(\mathcal{M})$, $\mathcal{R}(\mathcal{M})$, $\omega=0$, $\omega=\infty$, $\dfrac{1}{c_2}$, $\dfrac{1}{c_1}+\dfrac{1}{c_2}$	$\lvert Z\rvert$, c_2, $\dfrac{c_1c_2}{c_1+c_2}$, ω	$\lvert\mathcal{M}\rvert$, $\dfrac{1}{c_1}+\dfrac{1}{c_2}$, $\dfrac{1}{c_2}$, ω	θ: +90°, 0, -90°; ω

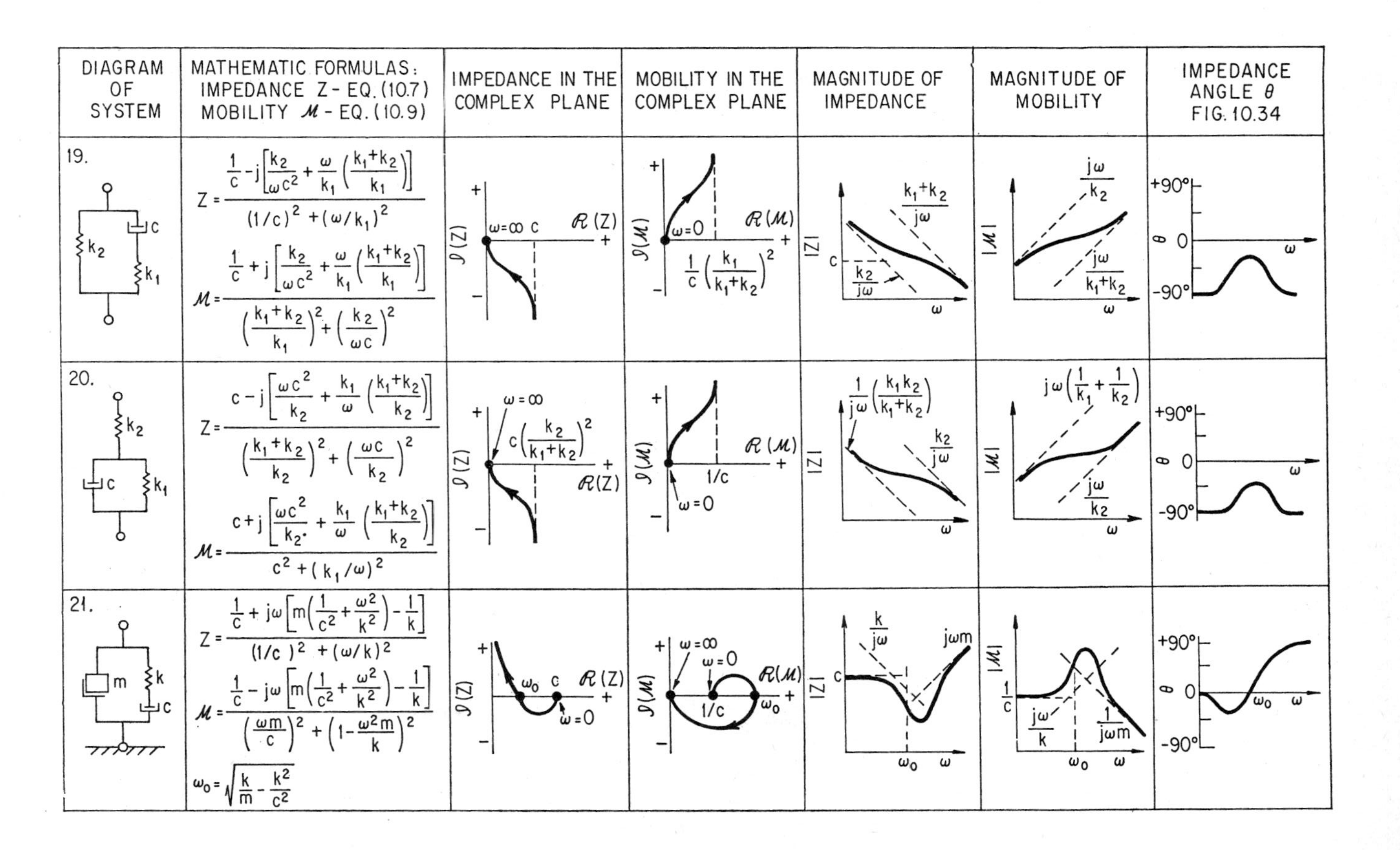
DIAGRAM OF SYSTEM
MATHEMATIC FORMULAS: IMPEDANCE Z - EQ. (10.7) MOBILITY $\mathcal{M}$ - EQ. (10.9)
IMPEDANCE IN THE COMPLEX PLANE
MOBILITY IN THE COMPLEX PLANE
MAGNITUDE OF IMPEDANCE
MAGNITUDE OF MOBILITY
IMPEDANCE ANGLE θ FIG. 10.34
19.
$Z = \dfrac{\frac{1}{c} - j\left[\frac{k_2}{\omega c^2} + \frac{\omega}{k_1}\left(\frac{k_1+k_2}{k_1}\right)\right]}{(1/c)^2 + (\omega/k_1)^2}$
$\mathcal{M} = \dfrac{\frac{1}{c} + j\left[\frac{k_2}{\omega c^2} + \frac{\omega}{k_1}\left(\frac{k_1+k_2}{k_1}\right)\right]}{\left(\frac{k_1+k_2}{k_1}\right)^2 + \left(\frac{k_2}{\omega c}\right)^2}$
20.
$Z = \dfrac{c - j\left[\frac{\omega c^2}{k_2} + \frac{k_1}{\omega}\left(\frac{k_1+k_2}{k_2}\right)\right]}{\left(\frac{k_1+k_2}{k_2}\right)^2 + \left(\frac{\omega c}{k_2}\right)^2}$
$\mathcal{M} = \dfrac{c + j\left[\frac{\omega c^2}{k_2} + \frac{k_1}{\omega}\left(\frac{k_1+k_2}{k_2}\right)\right]}{c^2 + (k_1/\omega)^2}$
21.
$Z = \dfrac{\frac{1}{c} + j\omega\left[m\left(\frac{1}{c^2} + \frac{\omega^2}{k^2}\right) - \frac{1}{k}\right]}{(1/c)^2 + (\omega/k)^2}$
$\mathcal{M} = \dfrac{\frac{1}{c} - j\omega\left[m\left(\frac{1}{c^2} + \frac{\omega^2}{k^2}\right) - \frac{1}{k}\right]}{\left(\frac{\omega m}{c}\right)^2 + \left(1 - \frac{\omega^2 m}{k}\right)^2}$
$\omega_0 = \sqrt{\frac{k}{m} - \frac{k^2}{c^2}}$

Table 10.2. (Continued)

DIAGRAM OF SYSTEM	MATHEMATIC FORMULAS: IMPEDANCE Z - EQ. (10.7) MOBILITY $\mathcal{M}$ - EQ. (10.9)	IMPEDANCE IN THE COMPLEX PLANE	MOBILITY IN THE COMPLEX PLANE	MAGNITUDE OF IMPEDANCE	MAGNITUDE OF MOBILITY	IMPEDANCE ANGLE θ FIG. 10.34
22. c, k, m	$Z = c + \frac{j\omega mk}{k-\omega^2 m}$ $\mathcal{M} = \frac{c - \frac{j\omega mk}{k-\omega^2 m}}{c^2 + \left(\frac{\omega mk}{k-\omega^2 m}\right)^2}$ $\omega_0 = \sqrt{\frac{k}{m}}$	$\mathscr{I}(Z)$, +, −, $\omega=0$, $\omega=\infty$, c, $\mathscr{R}(Z)$, +	$\mathscr{I}(\mathcal{M})$, +, −, $\omega=\infty$, ω_0, $\mathscr{R}(\mathcal{M})$, +, $\omega=0$, 1/c	$\lvert Z\rvert$, $\frac{k}{j\omega}$, $j\omega m$, c, ω_0, ω	$\lvert\mathcal{M}\rvert$, $\frac{1}{j\omega m}$, $\frac{j\omega}{k}$, 1/c, ω_0, ω	+90°, θ 0, −90°, ω_0, ω
23. c, m, k	$Z = \frac{\frac{1}{c} - \frac{j\omega}{k-\omega^2 m}}{\left(\frac{1}{c}\right)^2 + \left(\frac{\omega}{k-\omega^2 m}\right)^2}$ $\mathcal{M} = \frac{1}{c} + \frac{j\omega}{k-\omega^2 m}$ $\omega_0 = \sqrt{\frac{k}{m}}$	$\mathscr{I}(Z)$, +, −, $\omega=\infty$, ω_0, $\mathscr{R}(Z)$, +, $\omega=0$, c	$\mathscr{I}(\mathcal{M})$, +, −, $\omega=0$, $\omega=\infty$, 1/c, $\mathscr{R}(\mathcal{M})$, +	$\lvert Z\rvert$, $\frac{k}{j\omega}$, $j\omega m$, c, ω_0, ω	$\lvert\mathcal{M}\rvert$, $\frac{1}{j\omega m}$, $\frac{j\omega}{k}$, 1/c, ω_0, ω	+90°, θ 0, −90°, ω_0, ω
24. k_1, k_2, m	$Z = \frac{-j\left(\frac{k_1+k_2}{k_1} - \frac{k_2}{\omega^2 m}\right)}{\omega/k_1 - 1/\omega m}$ $\mathcal{M} = \frac{+j(\omega/k_1 - 1/\omega m)}{\left(\frac{k_1+k_2}{k_1}\right) - \frac{k_2}{\omega^2 m}}$ $\omega_1 = \sqrt{\frac{1}{m}\left(\frac{k_1 k_2}{k_1+k_2}\right)}$ $\quad \omega_2 = \sqrt{\frac{k_1}{m}}$	$\mathscr{I}(Z)$, +, −, $\omega_1, \omega=\infty$, $\mathscr{R}(Z)$, +	$\mathscr{I}(\mathcal{M})$, +, −, $\omega=0, \omega_2$, $\mathscr{R}(\mathcal{M})$, +	$\lvert Z\rvert$, $\frac{1}{j\omega}\left(\frac{k_1k_2}{k_1+k_2}\right)$, $j\omega m$, $\frac{k_1}{j\omega}$, ω_1, ω_2, ω	$\lvert\mathcal{M}\rvert$, $j\omega\left(\frac{k_1+k_2}{k_1k_2}\right)$, $\frac{j\omega}{k_1}$, $\frac{1}{j\omega m}$, ω_1, ω_2, ω	+90°, θ 0, −90°, ω_1, ω_2, ω

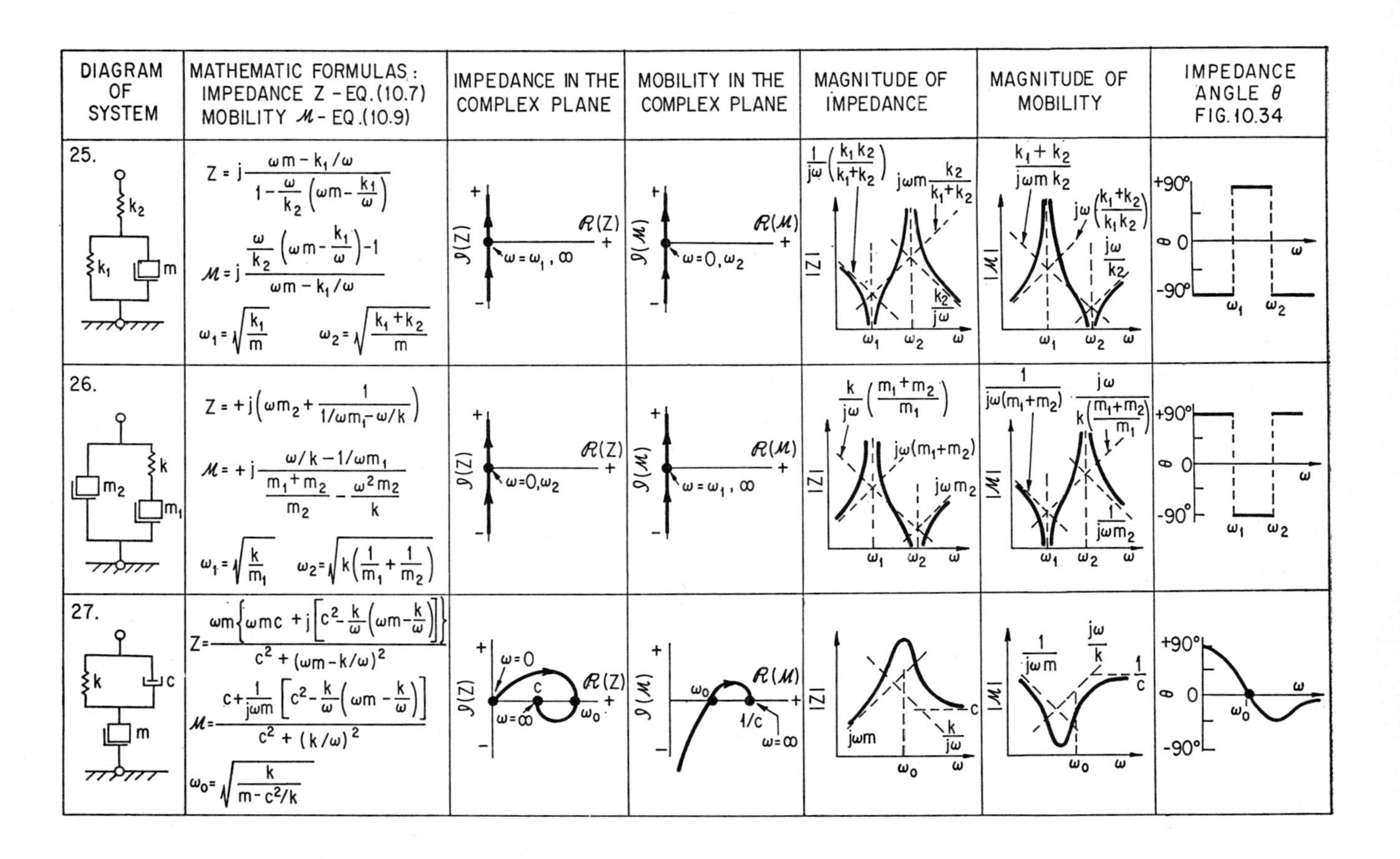
DIAGRAM OF SYSTEM
MATHEMATIC FORMULAS: IMPEDANCE Z - EQ. (10.7) MOBILITY $\mathcal{M}$ - EQ. (10.9)
IMPEDANCE IN THE COMPLEX PLANE
MOBILITY IN THE COMPLEX PLANE
MAGNITUDE OF IMPEDANCE
MAGNITUDE OF MOBILITY
IMPEDANCE ANGLE θ FIG. 10.34
25.
k_2
k_1
m
$Z = j\dfrac{\omega m - k_1/\omega}{1 - \dfrac{\omega}{k_2}\left(\omega m - \dfrac{k_1}{\omega}\right)}$
$\mathcal{M} = j\dfrac{\dfrac{\omega}{k_2}\left(\omega m - \dfrac{k_1}{\omega}\right) - 1}{\omega m - k_1/\omega}$
$\omega_1 = \sqrt{\dfrac{k_1}{m}}$
$\omega_2 = \sqrt{\dfrac{k_1 + k_2}{m}}$
$\mathcal{I}(Z)$
$\mathcal{R}(Z)$
$\omega = \omega_1, \infty$
$\mathcal{I}(\mathcal{M})$
$\mathcal{R}(\mathcal{M})$
$\omega = 0, \omega_2$
$\dfrac{1}{j\omega}\left(\dfrac{k_1 k_2}{k_1 + k_2}\right)$
$j\omega m \dfrac{k_2}{k_1 + k_2}$
$\dfrac{k_2}{j\omega}$
$|Z|$
$\dfrac{k_1 + k_2}{j\omega m k_2}$
$j\omega\left(\dfrac{k_1 + k_2}{k_1 k_2}\right)$
$\dfrac{j\omega}{k_2}$
$|\mathcal{M}|$
$+90°$
θ 0
$-90°$
ω_1
ω_2
ω
26.
k
m_2
m_1
$Z = +j\left(\omega m_2 + \dfrac{1}{1/\omega m_1 - \omega/k}\right)$
$\mathcal{M} = +j\dfrac{\omega/k - 1/\omega m_1}{\dfrac{m_1 + m_2}{m_2} - \dfrac{\omega^2 m_2}{k}}$
$\omega_1 = \sqrt{\dfrac{k}{m_1}}$
$\omega_2 = \sqrt{k\left(\dfrac{1}{m_1} + \dfrac{1}{m_2}\right)}$
$\omega = 0, \omega_2$
$\omega = \omega_1, \infty$
$\dfrac{k}{j\omega}\left(\dfrac{m_1 + m_2}{m_1}\right)$
$j\omega(m_1 + m_2)$
$j\omega m_2$
$\dfrac{1}{j\omega(m_1 + m_2)}$
$\dfrac{j\omega}{k\left(\dfrac{m_1 + m_2}{m_1}\right)}$
$\dfrac{1}{j\omega m_2}$
27.
c
$Z = \dfrac{\omega m\left\{\omega m c + j\left[c^2 - \dfrac{k}{\omega}\left(\omega m - \dfrac{k}{\omega}\right)\right]\right\}}{c^2 + (\omega m - k/\omega)^2}$
$\mathcal{M} = \dfrac{c + \dfrac{1}{j\omega m}\left[c^2 - \dfrac{k}{\omega}\left(\omega m - \dfrac{k}{\omega}\right)\right]}{c^2 + (k/\omega)^2}$
$\omega_0 = \sqrt{\dfrac{k}{m - c^2/k}}$
$\omega = 0$
$\omega = \infty$
ω_0
$1/c$
$j\omega m$
$\dfrac{k}{j\omega}$
$\dfrac{1}{j\omega m}$
$\dfrac{j\omega}{k}$
$\dfrac{1}{c}$

MOBILITY OF MECHANICAL ELEMENTS

Mobility as defined by Eq. (10.9) is the reciprocal of impedance. Expressions for the mobility of the three basic elements can be written directly from the respective expressions for impedance:

Resistance:
$$\mathfrak{M}_c = \frac{1}{c} \tag{10.23}$$

Spring:
$$\mathfrak{M}_k = \frac{j\omega}{k} \tag{10.24}$$

Mass:
$$\mathfrak{M}_m = \frac{1}{j\omega m} = \frac{-j}{\omega m} \tag{10.25}$$

These mobilities are plotted on the complex plane in Fig. 10.10*A*, and represented as functions of frequency in Fig. 10.10*B*. Mobility of various elements and combinations of elements are summarized in Table 10.2.

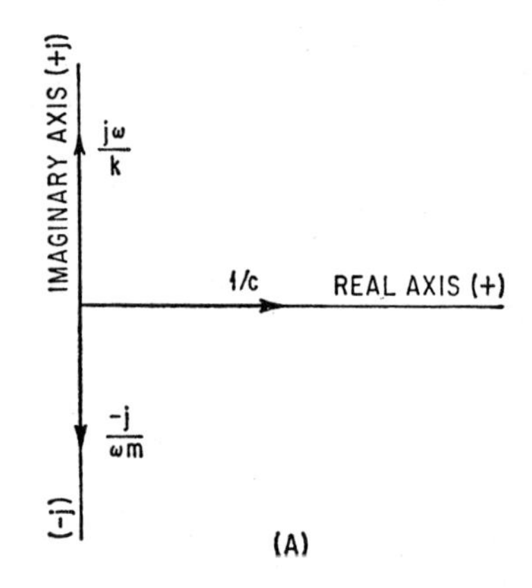

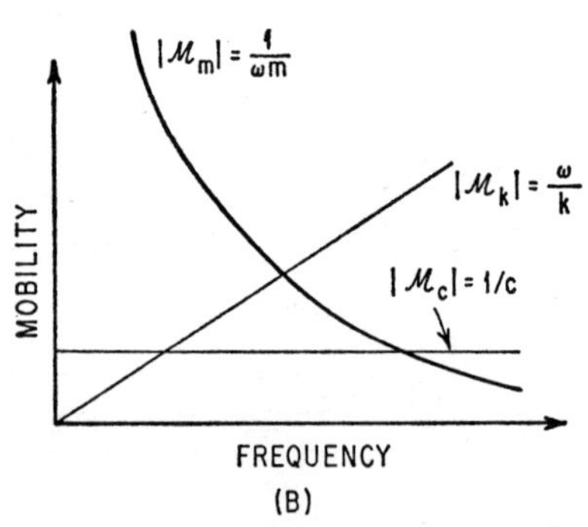

FIG. 10.10. The complex-plane representation of the mobility of an ideal resistance, spring, and mass at a fixed frequency is shown in (*A*); the variation of the mobility magnitude with frequency is shown in (*B*).

MECHANICAL SYSTEMS

The properties of a mechanical system that are used in analyses involving mechanical impedance and mobility are points of force application, paths for transmitting forces, and points of common velocities. The velocity at each connection point and the forces exerted on each element may be calculated by the proper application of Eq. (10.7) or Eq. (10.9) if the impedance of each element in the system and the forces or velocities imposed on points in the system are known. In many applications, only the motion at a few points is of primary interest. Then it is advantageous to combine groups of elements into single impedances. Methods for calculating the impedance of such combined elements are given in this section.

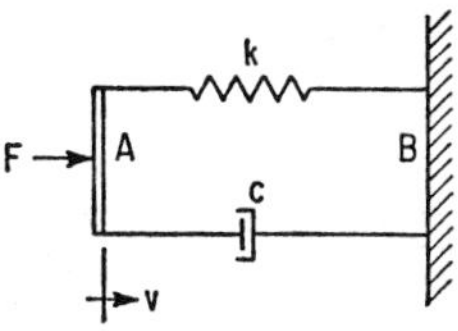

Fig. 10.11. Schematic representation of a parallel spring-resistance combination.

PARALLEL ELEMENTS

To illustrate the concept of parallel elements only ideal elements, multiply connected across two points, will be considered. Two ideal elements connected in parallel are shown in Fig. 10.11. Both are constrained to have the same relative velocities between their connections. The force required to give the resistance the velocity v is found from Eqs. (10.7) and (10.14):

$$F_c = vZ_c = vc$$

The force required to give the spring this same velocity is, from Eq. (10.18),

$$F_k = vZ_k = \frac{vk}{j\omega}$$

The total force F is

$$F = F_c + F_k = v\left(c + \frac{k}{j\omega}\right)$$

Since $Z = F/v$,

$$Z = c + \frac{k}{j\omega}$$

Thus, the over-all impedance is the sum of the impedances of the two elements.

By extending this concept to any number of parallel elements, the driving force F equals the sum of the resisting forces:

$$F = \sum_{i=1}^{n} vZ_i = v\sum_{i=1}^{n} Z_i \qquad \text{and} \qquad Z_p = \sum_{i=1}^{n} Z_i \tag{10.26}$$

where Z_p is the total mechanical impedance of the parallel combination of the individual elements Z_i.

When the properties of the parallel elements are expressed as mobilities, the total mobility of the combination follows from Eqs. (10.9) and (10.26):

$$\frac{1}{\mathfrak{M}_p} = \sum_{i=1}^{n} \frac{1}{\mathfrak{M}_i} \tag{10.27}$$

SERIES ELEMENTS

In determining the impedance presented by the end of a number of series-connected elements, consider the arrangement shown in Fig. 10.12. Elements Z_1 and Z_2 must have no mass since a mass always has one end connected to a stationary inertial reference. However, the impedance Z_3 may be a mass. The relative velocities between the end connections of each element are indicated by v_a, v_b, and v_c; the velocities of the connections with respect to the stationary reference point G are indicated by v_1, v_2, and v_3:

$$v_3 = v_c \qquad v_2 = v_3 + (v_2 - v_3) = v_c + v_b$$

$$v_1 = v_2 + (v_1 - v_2) = v_a + v_b + v_c$$

The impedance at point 1 is F/v_1, and the force F is transmitted to all three elements. The relative velocities are

$$v_a = \frac{F}{Z_1} \qquad v_b = \frac{F}{Z_2} \qquad v_c = \frac{F}{Z_3}$$

Thus, the total impedance is defined by

$$\frac{1}{Z} = \frac{F/Z_1 + F/Z_2 + F/Z_3}{F} = \frac{1}{Z_1} + \frac{1}{Z_2} + \frac{1}{Z_3}$$

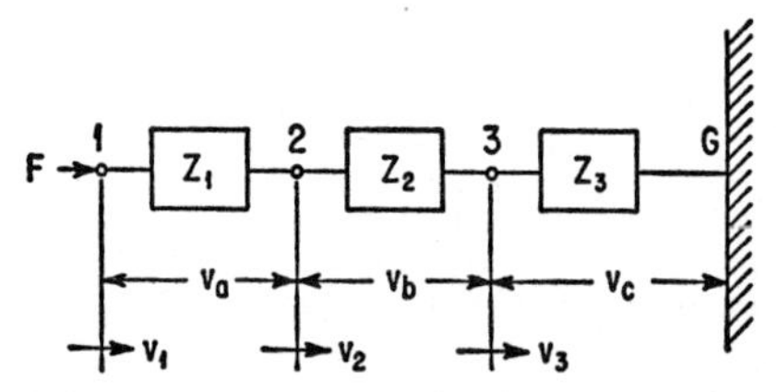

FIG. 10.12. Generalized three-element system of series-connected mechanical impedances.

Extending this principle to any number of massless series elements,

$$\frac{1}{Z_s} = \sum_{i=1}^{n} \frac{1}{Z_i} \tag{10.28}$$

where Z_s is the total mechanical impedance of the elements Z_i connected in series.

Using Eq. (10.9), the total mobility of series connected elements (expressed as mobilities) is

$$\mathfrak{M}_s = \sum_{i=1}^{n} \mathfrak{M}_i \tag{10.29}$$

SYSTEM SIMPLIFICATION

When one point of attachment of a multiple element system is of interest, the above methods for calculating the impedance of combined elements may be used to give one expression for the impedance at that point. When a force is applied to the point of attachment, the resulting velocity can be calculated from Eq. (10.7). In general, such an expression involves real and imaginary terms. When an *impedance* expression is obtained, the two terms can be considered to represent a purely resistive element connected in *parallel* with a purely imaginary or reactive element. A *mobility* expression may be considered to represent a resistive element in *series* with a reactive element. Either of these two representations has the characteristics of the original multiple element system and is called an *equivalent system.*

The calculation of the impedance of a multielement system using Eqs. (10.26) through (10.29) requires considerable manipulation of complex quantities.

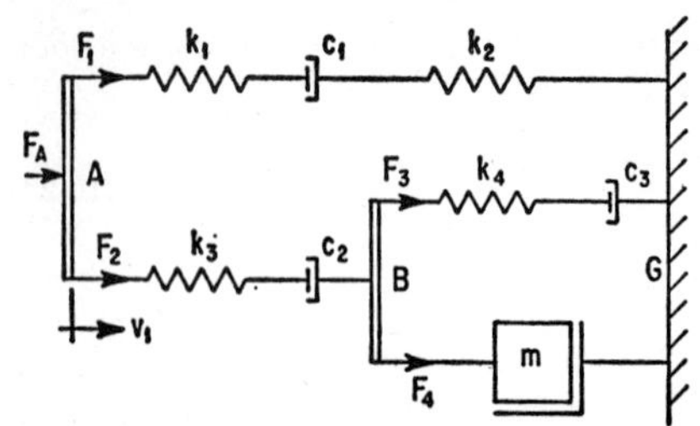

FIG. 10.13. System of several ideal elements analyzed in Example 10.1.

Example 10.1. For the system of Fig. 10.13, calculate the impedance at point A so that the velocity v_1 as a result of the application of force F_A may be determined. The procedure to be followed is basically as follows:

1. When series elements such as k_4 and c_3 are paralleled by other elements (such as m), add the mobilities of the series elements and convert the resulting expression to an impedance which may be added directly to the impedance of the parallel elements.

2. When parallel connected elements such as k_4, c_3, and m are connected in series with other elements (such as k_3 and c_2), convert the impedance of the parallel elements to a mobility and add to the mobility of the series elements.

In accordance with (1), add the mobilities of k_4 and c_3 to obtain the mobility of the branch at B:

$$\mathfrak{M}_{B1} = \frac{1}{c_3} + \frac{j\omega}{k_4}$$

The impedance of this branch is then

$$Z_{B1} = \frac{1}{\mathfrak{M}_{B1}} = \frac{1/c_3 - j\omega/k_4}{(1/c_3)^2 + (\omega/k_4)^2}$$

$$= \frac{1/c_3}{(1/c_3)^2 + (\omega/k_4)^2} - \frac{j\omega/k_4}{(1/c_3)^2 + (\omega/k_4)^2}$$

The system at B is then the parallel combination shown in Fig. 10.14A where $\mathcal{R}(Z_{B1})$ represents the real part of Z_{B1}, and $\mathcal{I}(Z_{B1})$ is the imaginary part of Z_{B1}. The parallel impedance at B including the impedance of mass m is

$$Z_B = \frac{1/c_3}{(1/c_3)^2 + (\omega/k_4)^2} + j\left[\omega m - \frac{\omega/k_4}{(1/c_3)^2 + (\omega/k_4)^2}\right]$$

The corresponding mobility is

$$\mathfrak{M}_B = \frac{\mathcal{R}(Z_B)}{|Z_B|^2} - \frac{j\mathcal{I}(Z_B)}{|Z_B|^2}$$

where $|Z_B|$ is the magnitude of Z_B which is represented by Fig. 10.14B. Adding the mobilities of c_2 and k_3, the mobility presented at A by this branch is

$$\mathfrak{M}_{A2} = \mathcal{R}(\mathfrak{M}_B) + \frac{1}{c_2} + j\left(\mathcal{I}(\mathfrak{M}_B) + \frac{\omega}{k_3}\right)$$

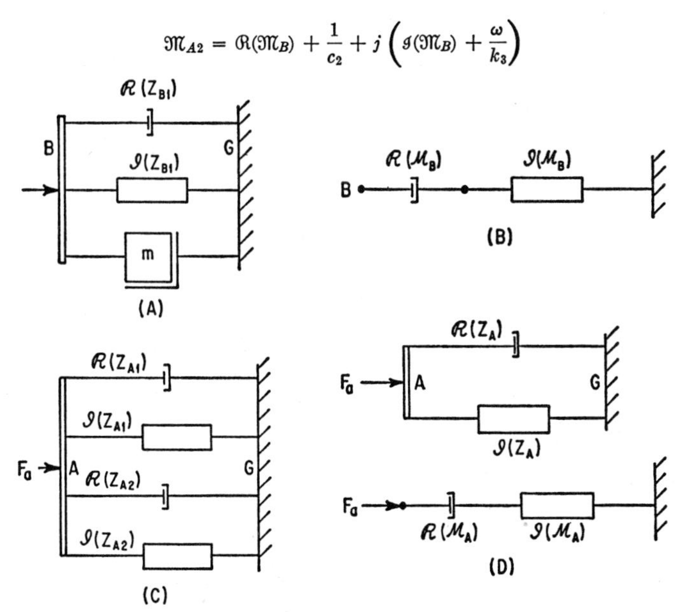

FIG. 10.14. Equivalent representations of parts of the system of Fig. 10.13 used in reducing that system to a representation equivalent to a single impedance or mobility.

To add the effect of this branch to that of the parallel branch consisting of k_1, k_2, and c_1, the impedances of both branches must be found:

$$Z_{A2} = \frac{1}{\mathfrak{M}_{A2}} = \frac{\mathcal{R}(\mathfrak{M}_{A2})}{|\mathfrak{M}_{A2}|^2} - j\frac{\mathcal{I}(\mathfrak{M}_{A2})}{|\mathfrak{M}_{A2}|^2}$$

$$\mathfrak{M}_{A1} = \frac{1}{c_1} + \frac{k_1 + k_2}{j\omega}$$

$$Z_{A1} = \frac{1/c_1}{(1/c_1)^2 + [(k_1 + k_2)/\omega]^2} + j\frac{(k_1 + k_2)/\omega}{(1/c_1)^2 + [(k_1 + k_2)/\omega]^2}$$

The impedances Z_{A1} and Z_{A2} then can be represented by the four parallel elements of Fig. 10.14*C*. The total impedance at A is

$$Z_A = Z_{A1} + Z_{A2} = \mathcal{R}(Z_{A1}) + \mathcal{R}(Z_{A2}) + j[\mathcal{I}(Z_{A1}) + \mathcal{I}(Z_{A2})]$$

and can be represented by the two parallel elements of Fig. 10.14*D*. The mobility expression at A is $1/Z_A$; it can be represented by the series elements of Fig. 10.14*D*.

As an aid to system simplification problems, the impedances and mobilities of several combinations of elements are given in Table 10.2.

EQUIVALENT SYSTEMS

The ratio of an applied force to a resulting velocity at the point A of the eight-element system in Fig. 10.13 is finally expressed analytically by an impedance having real and imaginary terms. The impedance characteristics at A are represented by the two-element systems of Fig. 10.14*D*. The two elements are not necessarily the ideal resistance, spring, or mass. The resistance term may vary with frequency and the reactance term may appear springlike, masslike, or be zero—depending on frequency. Even the impedance Z_{B1} of the two series elements c_3 and k_4 in Example 10.1 exhibits some of these characteristics.

The advantage of the two-element equivalent representation is that the response of complex systems to single frequency excitation may be calculated easily and the system may be represented by an ideal mass or spring and a resistance. In many cases there are ranges of frequencies over which complex systems can be represented by one or two ideal elements. Thus in Example 10.1, when $1/c_3 \gg \omega/k_4$ (which occurs when $\omega \ll k_4/c_3$), $Z_{B1} = c_3$, and the two elements can be represented by the resistance c_3. At higher frequencies, $\omega \gg k_4/c_3$, $Z_{B1} = k_4/j\omega$, and the impedance is springlike.

A further advantage of the equivalent-system representation lies in considering the effect of other elements to be attached to the point of interest. Qualitative considerations become apparent immediately. For example, when a mass is attached to a structure that appears springlike at the excitation frequency, the possibility of system resonance exists. This may or may not be desirable, but in either case the size of the mass is very important. The quantitive determination of the response can be determined by combining the added element with the equivalent system.

PHYSICAL ELEMENTS

In general, the characteristics of real masses, springs, and mechanical resistance elements differ from those of ideal elements in three respects:

1. A spring may have a nonlinear force-deflection characteristic; a mass may suffer plastic deformation with motion; and the force presented by a resistance may not be exactly proportional to velocity.

2. All materials have some mass; thus, a perfect spring or resistance cannot be made. Some compliance or spring effect is inherent in all elements. Energy can be dissipated in a system in several ways: friction, acoustic radiation, hysteresis, etc. Such a loss can be represented as a resistive component of the element impedance.

3. Elements can differ from the ideal elements at high frequencies or when long connecting elements are used because the element length becomes comparable to a wavelength of stress waves in the material. Then a lumped system analysis does not apply, and analysis on the basis of wave motion in a transmitting medium must be made.

LUMPED ELEMENTS

A piece of high-density metal of spherical or cubical shape provides a nearly ideal mass. Wave motion effects do not occur until very high frequencies are reached, and a rigid attachment is feasible. However, supporting means for the mass may involve friction, and the attaching elements may have significant compliance. A parallel representation of the impedance of a physical mass at low frequencies is

$$Z_m = c + j\omega m + \frac{k}{j\omega}$$

If the mass is freely suspended by fairly rigid supports, the $j\omega m$ term predominates over a wide frequency range and Eqs. (10.22) and (10.25) may be used.

A helical wound spring can be made to have a nearly linear force-displacement characteristic, thereby attaining a constant value of k. But, under dynamic conditions the mass of the spring must be considered. When one end of a spring is rigidly fixed and the other end attached to a vibrating point, one-third of the mass of the spring may be considered to be lumped at the vibrating end of the spring. When both ends are vibrating, from one-third to one-half the mass of the spring may be considered lumped at each end. At low frequencies the mass reactance may be small compared to the compliant effect so that either approximation for the mass of the spring may be adequate.

Hysteresis effects in the spring metal can cause energy loss resulting in a resistive component in the spring impedance. The impedance of a spring can be represented by a parallel connection of the three ideal elements, but in most cases the resistive term is negligible. At low frequencies, the impedance of a spring when one end is fixed is approximately

$$Z_k = \frac{k}{j\omega} + j\omega\frac{m_k}{3} \tag{10.30}$$

In many cases a mechanical resistance represents damping which is inherent in physical systems. In the case of sliding friction the reaction force may not be proportional to velocity but may be a constant, independent of velocity. Such a characteristic can be represented by a constant-force generator, 180° out-of-phase with the relative velocity across the sliding elements.

Resistive effects also may arise from the losses in rubber, felt, cork, or other vibration isolation materials. With such materials the damping coefficient may vary with the amplitude of motion.

One type of resistive element can be constructed by an arrangement of parts so arranged that the applied force pushes a body through a liquid or subjects it to other viscous forces. Because of the mass and compliance of the parts, such a device will not constitute an ideal resistance. Elements can be made sufficiently rigid so that the spring term is negligible; however, the effects of the mass of the elements may not be negligible.

Although ideal elements do not actually exist, it is common in the use of impedance methods to represent physical members by one or more ideal elements. Figure 10.15*A* shows a system of five physical elements as represented, for analysis at low frequency, by ideal elements. Many of the masses are essentially in parallel and can be added together as follows to produce the system of Fig. 10.15*B*:

$$m' = \frac{m_{k1}}{2} + \frac{m_{c1}}{2} \qquad m'' = m + \frac{m_{c2}}{2} + \frac{m_{k2}}{3} + \frac{m_{k1}}{2} + \frac{m_{c1}}{2}$$

Since one end of k_2 is rigidly fixed, one-third of its mass is lumped on the moving end. The masses on the fixed ends of c_2 and k_2 are not included in Fig. 10.15*B* because they do not enter into the dynamic problem.

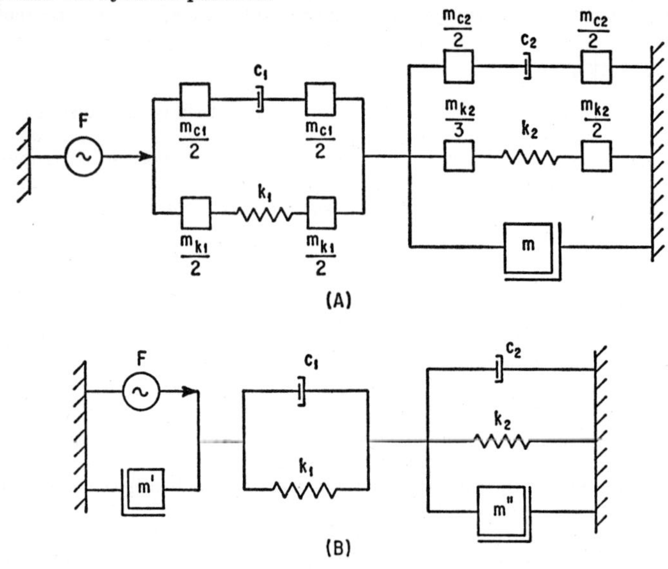

FIG. 10.15. The schematic representation applicable at low frequencies of a system of physical elements is shown in (*A*); the redrawn system suitable for impedance method analysis is shown in (*B*).

ELEMENTS WITH DISTRIBUTED MASS

All mechanical elements have distributed mass since mass cannot be concentrated at a point. If the vibration frequency is low enough that the time of travel of stress waves in an element of length l is very small compared to a period (i.e., if $l/\lambda \ll 1/2\pi$), the mass can be considered as a lumped element.[11]

The velocity of compressional stress waves in an elastic bar is $c_s = \sqrt{E/\rho}$, where E is Young's modulus and ρ is mass density. In long thin rods the velocity is c_s if the diameter is small compared to a wavelength, but when the diameter is as great as 2λ, the velocity drops to about $0.6c_s$.[12]

The force transmitted by and the velocities at the ends of a distributed mass element may be determined by transmission line methods.[13, 14, 15] For example, consider a uniform bar, as shown in Fig. 10.16*A*. (Other element shapes may be approximated by sections of uniform bars.) The bar has a length l, a cross-sectional area S, a density ρ, and a stress-wave velocity c_s. The force and velocity at the driven end (vibrational source attachment) are represented by F_i and v_i; F_r and v_r are the force and velocity at the load end. The material

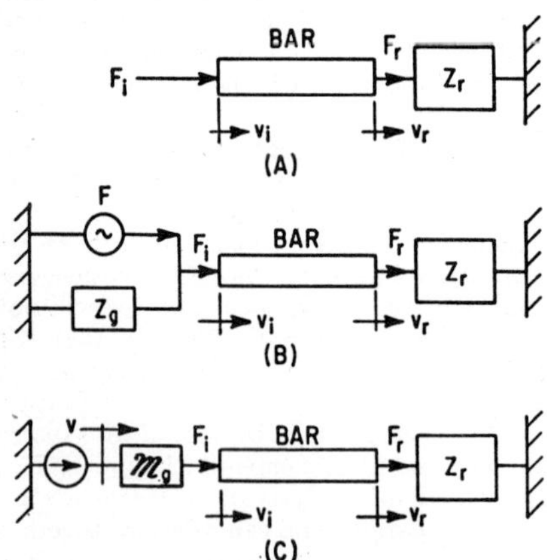

FIG. 10.16. The schematic representation of an element having distributed mass and its representation when it is connected to a source of vibration energy.

is assumed to be lossless, and the bar is characterized by a *propagation constant* $\gamma = j\beta = j\omega/c_s$ (where $\beta = \omega/c_s$) and a *characteristic impedance* (the impedance seen at the driven end for an infinitely long bar) $Z_0 = \rho c_s S$. Analysis by transmission line methods with the load Z_r attached gives the following relationships:

Load velocity in terms of the driven end velocity:

$$v_r = \frac{v_i}{\cos \beta\ell + j(Z_r/Z_0) \sin \beta\ell} \tag{10.31}$$

Load force in terms of the driving end force:

$$F_r = \frac{F_i}{\cos \beta\ell + j(Z_0/Z_r) \sin \beta\ell} \tag{10.32}$$

Driving end impedance:

$$Z_i = Z_0 \frac{Z_r \cos \beta\ell + jZ_0 \sin \beta\ell}{Z_0 \cos \beta\ell + jZ_r \sin \beta\ell} \tag{10.33}$$

When the bar is driven by a force generator with a paralleled impedance Z_g the schematic representation is that of Fig. 10.16*B*. The load element force and velocity, and the driven end force and velocity, may be determined in terms of the known force F by the following expressions:

$$F_i = \frac{F}{1 + Z_i/Z_g} \qquad V_i = \frac{F}{Z_i(1 + Z_g/Z_i)}$$

$$F_r = \frac{F}{(1 + Z_g/Z_i)[\cos \beta\ell + j(Z_0/Z_r) \sin \beta\ell]} \tag{10.34}$$

$$v_r = \frac{F}{(Z_g + Z_i)[\cos \beta\ell + j(Z_r/Z_0) \sin \beta\ell]}$$

where Z_g is the impedance of the generator.

When the bar is driven from a velocity generator with a series mobility $\mathfrak{M}_g$, as in Fig. 10.16*C*, the quantities of Eq. (10.34) are given in terms of the known velocity v and the mobility:

$$\mathfrak{M}_0 = \frac{1}{Z_0} \qquad \mathfrak{M}_r = \frac{1}{Z_r} \qquad \mathfrak{M}_g = \frac{1}{Z_g} \qquad \text{and} \qquad \mathfrak{M}_i = \frac{1}{Z_i}$$

$$v_i = \frac{v}{1 + \mathfrak{M}_g/\mathfrak{M}_i} \qquad F_i = \frac{v}{\mathfrak{M}_g + \mathfrak{M}_i}$$

$$v_r = \frac{v}{(1 + \mathfrak{M}_g/\mathfrak{M}_i)[\cos \beta\ell + j(\mathfrak{M}_0/\mathfrak{M}_r) \sin \beta\ell]} \tag{10.35}$$

$$F_r = \frac{v}{(\mathfrak{M}_g + \mathfrak{M}_i)[\cos \beta\ell + j(\mathfrak{M}_r/\mathfrak{M}_0) \sin \beta\ell]}$$

Equation (10.33) can be applied to a mass at frequencies for which the lumped approximation is no longer valid. In this case $Z_r = 0$ and

$$Z_m = jZ_0 \tan \beta\ell \tag{10.36}$$

where $Z_0 = \rho c_s S$. This expression may be compared with Eq. (10.22), the impedance of an ideal mass, $Z_m = j\omega m$.

THEOREMS

The following theorems are statements of basic principles (or combination of them) which apply to elements of a mechanical system; they are in a form convenient to analysis by the impedance method. In all but Kirchhoff's laws, these theorems apply only to systems made up of linear, bilateral elements. Linearity implies that the system elements can be represented by combinations of ideal elements in which c, k, and m are constants regardless of motion amplitude. A bilateral element is one in which forces are transmitted equally well in either direction across its connections.

Kirchhoff's Laws. 1. *The sum of all the forces acting at a point (common connection of several elements) is zero:*

$$\sum_i^n F_i = 0 \qquad \text{(at a point)} \tag{10.37}$$

This follows directly from the consideration leading to Eq. (10.26).

2. *The sum of the relative velocities across the connections of series mechanical elements taken around a closed loop is zero:*

$$\sum_i^n v_i = 0 \qquad \text{(around a closed loop)} \tag{10.38}$$

This follows from the considerations leading to Eq. (10.28).

Kirchhoff's laws apply to any system even when the elements are not linear or bilateral.

Superposition Theorem. *If a mechanical system of linear bilateral elements includes more than one vibration source, the force or velocity response at a point in the system can be determined by adding the response to each source, taken one at a time (the other sources supplying no energy but replaced by their internal impedances).*

The internal impedance of a vibrational generator is that impedance presented at its connection point when the generator is supplying no energy. This theorem finds useful application in systems having several sources. A very important application arises when the applied force is nonsinusoidal but can be represented by a Fourier series. Each term in the series can be considered a separate sinusoidal generator. The response at any point in the system can be calculated for each generator by using the impedance values at that frequency. Each response term becomes a term in the Fourier series representation of the total response function. The over-all response as a function of time then can be synthesized from the series.

Reciprocity Theorem. *If a force generator operating at a particular frequency at some point* (1) *in a system of linear bilateral elements produces a velocity at another point* (2), *the generator can be removed from* (1) *and placed at* (2); *then the former velocity at* (2) *will exist at* (1), *provided the impedances at all points in the system are unchanged.* This theorem also can be stated in terms of a vibration generator that produces a certain velocity at its point of attachment (1), regardless of force required, and the force resulting on some element at (2).

Reciprocity is an important characteristic of linear bilateral elements. It indicates that a system of such elements can transmit energy equally well in both directions. It further simplifies calculation on two-way energy transmission systems since the characteristics need be calculated for only one direction.

Thévenin's Equivalent System. *If a mechanical system of linear bilateral elements contains vibration sources and produces an output to a load at some point at any particular frequency, the whole system can be represented at that frequency by a single constant-force generator F_c in parallel with a single impedance Z_i connected to the load.* Thévenin's equivalent-system representation for a physical system may be determined by the following experimental procedure: Denote by F_c the force which is transmitted by the attachment point of the system to an infinitely rigid fixed point; this is called the *clamped force.* When the load connection is disconnected and perfectly free to move, a free velocity v_f is measured. Then the parallel impedance Z_i is F_c/v_f. The impedance Z_i also can be determined by measuring the internal impedance of the system when no source is supplying motional energy.

If the values of all the system elements in terms of ideal elements are known, F_c and Z_i may be determined analytically. A great advantage is derived from this representation in that attention is focused on the characteristics of a system at its output point and not on the details of the elements of the system. This allows an easy prediction of the response when different loads are attached to the output connection. After a final load condition has been determined, the system may be analyzed in detail for strength considerations.

Norton's Equivalent System. *A mechanical system of linear bilateral elements having vibration sources and an output connection may be represented at any particular frequency by a single constant-velocity generator v_f in series with an internal impedance Z_i.*

This is the series system counterpart of Thévenin's equivalent system where v_f is the free velocity and Z_i is the impedance as defined above. The same advantages in analysis exist as with Thévenin's parallel representation. The most advantageous one to use depends upon the type of structure to be analyzed. In the experimental determination of an equivalent system, it is usually easier to measure the free velocity than the clamped force on large heavy structures, while the converse is true for light structures. In any case, one representation is easily derived from the other. When v_f and Z_i are determined, $F_c = v_f Z_i$.

T and π Equivalent System. *Any mechanical system having an input and output connection and which is composed of linear bilateral elements may be represented by a three-element system of the T or π configuration.* The required equivalent system impedances

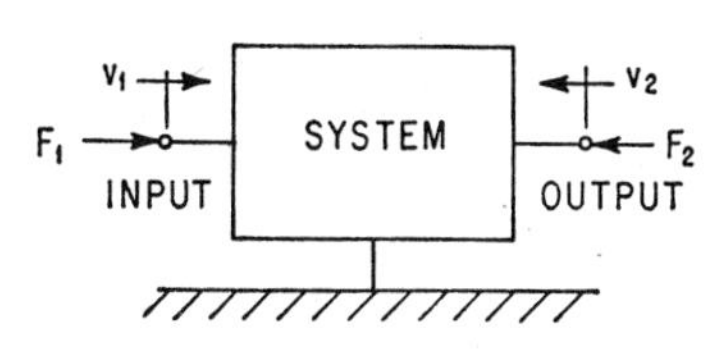

FIG. 10.17. Generalized two-connection mechanical system.

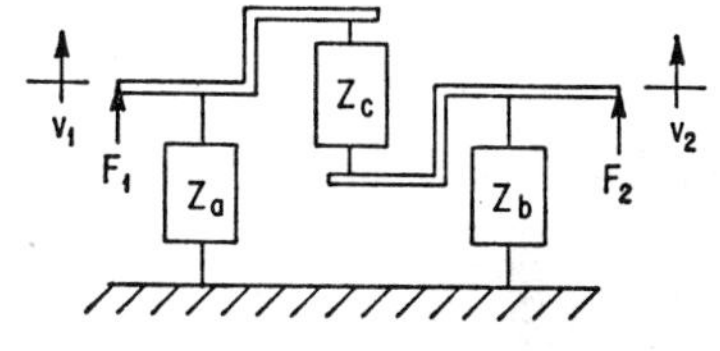

FIG. 10.18. A π equivalent system representation for a passive two-connection system.

determined below may not be obtainable physically, but the mathematical model represents the original system accurately.

π System. A generalized system is shown in Fig. 10.17. If a force F_1 is applied to the input connection, a velocity v_1 results. The ratio of F_1 to v_1 is the *input impedance.* If a force F_2 is applied to the output connection, a velocity v_2 results; the ratio of F_2 to v_2 is the *output impedance.* When the force F_1 is applied to the input, a velocity v_2 at the output results; the ratio of F_1 to v_2 is the *reverse transfer impedance.* If the force F_2 is applied to the output and a velocity v_1 results, the ratio of F_2 to v_1 is the *forward transfer impedance.*

These definitions can be used to find the impedances of an equivalent π system, as shown in Fig. 10.18, that are required to represent any general system, such as Fig. 10.17. First, find the input, output, and transfer impedances of the π system under the conditions specified below:

$$Z_{11} = \left. \frac{F_1}{v_1} \right|_{v_2=0} \tag{10.39}$$

where Z_{11} is the input impedance with the output connected to a rigid point so that

$$v_2 = 0$$

i.e., output clamped.

$$Z_{22} = \left. \frac{F_2}{v_2} \right|_{v_1=0} \tag{10.40}$$

where Z_{22} is the output impedance with the input clamped.

$$Z_{12} = \left. \frac{F_1}{v_2} \right|_{v_1=0} \tag{10.41}$$

where Z_{12} is the reverse transfer impedance with the input clamped and F_1 is the force required to maintain the input velocity $v_1 = 0$.

$$Z_{21} = \frac{F_2}{v_1}\bigg|_{v_2=0} \tag{10.42}$$

where Z_{21} is the forward transfer impedance with the output clamped and F_2 is the force required to clamp the output.

Applying Eq. (10.39) to the system of Fig. 10.18, the input impedance when the output is clamped consists of Z_a and Z_c in parallel:

$$Z_{11} = Z_a + Z_c \tag{10.43}$$

Applying Eq. (10.40),

$$Z_{22} = Z_b + Z_c \tag{10.44}$$

When the input is clamped and a force F_2 is applied to the output, the velocity of the output v_2 results. The force F_1 then is transmitted by Z_c to the clamping point, and v_2 is the relative velocity of the connections of Z_c. By Eq. (10.7),

$$\frac{F_1}{v_2} = Z_c = Z_{12} \tag{10.45}$$

By the same procedure, $Z_{21} = Z_c$. Solving Eqs. (10.43) to (10.45) for Z_a, Z_b, and Z_c,

$$Z_a = Z_{11} - Z_{12} \qquad Z_b = Z_{22} - Z_{21} \qquad Z_c = Z_{12} = Z_{21} \tag{10.46}$$

If the impedances defined by Eqs. (10.39) to (10.42) can be determined for a system by either analytical or experimental means, the three impedances of an equivalent π system can be determined by Eqs. (10.46).

T System. An equivalent T system is shown in Fig. 10.19. The system values are most conveniently found in terms of mobilities. The necessary mobilities for a generalized system such as Fig. 10.17 are as follows:

The input mobility with the output "free" (no restraining force):

$$\mathfrak{M}_{11} = \frac{v_1}{F_1}\bigg|_{F_2=0} \tag{10.47}$$

The output mobility with the input "free":

$$\mathfrak{M}_{22} = \frac{v_2}{F_2}\bigg|_{F_1=0} \tag{10.48}$$

The reverse transfer mobility with the input "free":

$$\mathfrak{M}_{12} = \frac{v_1}{F_2}\bigg|_{F_1=0} \tag{10.49}$$

where v_1 is the velocity of the "free" input point when the force F_2 is applied to the output.

The forward transfer mobility with the output "free":

$$\mathfrak{M}_{21} = \frac{v_2}{F_1}\bigg|_{F_2=0} \tag{10.50}$$

Finding these quantities for the T system of Fig. 10.19,

$$\mathfrak{M}_{11} = \mathfrak{M}_x + \mathfrak{M}_z \qquad \mathfrak{M}_{22} = \mathfrak{M}_y + \mathfrak{M}_z \qquad \mathfrak{M}_{12} = \mathfrak{M}_{21} = \mathfrak{M}_z \tag{10.51}$$

Solving Eqs. (10.51) for $\mathfrak{M}_x$, $\mathfrak{M}_y$, and $\mathfrak{M}_z$,

$$\mathfrak{M}_x = \mathfrak{M}_{11} - \mathfrak{M}_{12} \qquad \mathfrak{M}_y = \mathfrak{M}_{22} - \mathfrak{M}_{21} \qquad \mathfrak{M}_z = \mathfrak{M}_{12} = \mathfrak{M}_{21} \tag{10.52}$$

If the mobilities required by Eqs. (10.47) to (10.50) can be determined analytically or experimentally for a system, the three mobilities of an equivalent T system can be obtained from Eqs. (10.52).

The forward and reverse transfer impedances and mobilities are equal for both π and T systems. This is a characteristic of even very complicated systems when the system elements are linear and bilateral; it is not true if the system contains sources of vibration energy. However, systems with vibration sources can be represented by π and T equivalent systems of a slightly more complex form.

A system of mechanical elements with one connection point can be represented by a single impedance. Thévenin's or Norton's equivalent systems represent systems having vibration sources and an output connection by a source with a constant output and a single impedance element. The π or T equivalent systems can represent connecting elements between a vibration source and some load system. Combining the three representations gives a system that is simple to analyze; it focuses attention on the force and motion at the load connection and at the output connection of the source. The effects of changes in the source, load, or connecting elements then can be readily calculated.

Another advantage of the π and T representations is that the equivalent system can be obtained from direct measurements on complicated systems that are almost impossible to represent by a collection of idealized mechanical elements. An analytical expression for the system thus is obtained that allows an analysis by the methods discussed in the next section. After the values of the equivalent elements are determined at the operating frequencies, it may become evident that the original system may be replaced by a much simpler system of physical elements, based on the equivalent representation.

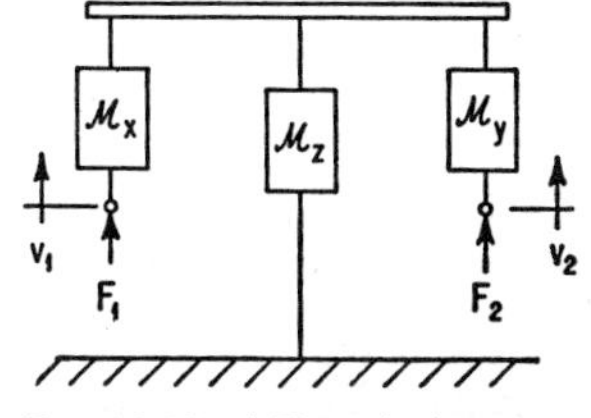

FIG. 10.19. A T equivalent system representation for a passive two-connection system.

Despite the many advantages of equivalent system representations, the complete schematic representation of a system often is desired. This representation is necessary for a complete analysis to determine the stresses on all of the parts of the system. To draw such a representation, a detailed study is required that can lead to a better understanding of the operation of the system.

In all but Kirchhoff's laws, the above theorems require linear, bilateral elements. This does not prevent their use in nonlinear, nonbilateral systems. Equivalent system representations with linear bilateral elements can be determined in many cases that represent accurately the nonlinear system over a limited range of operation.

ANALYSIS METHODS

The analysis of a system by impedance methods leads to the determination of forces acting on elements and the velocities of their connection points. From these quantities stresses may be determined, and accelerations and displacements may be calculated from Eqs. (10.5) and (10.6). When the motion is nonsinusoidal, the forces and velocities can be determined with the impedance approach, using the methods of analysis presented in this section. These methods are not the only ones available, but they are basic and adequate for many problems. Other methods may be found in the literature on electric circuit theory.

COMPLETE ANALYSIS

Kirchhoff's laws provide a sufficient number of simultaneous equations to determine all forces and velocities in a mechanical system. Equation (10.37) allows a force equation to be written for each independent connection point (force node). These equations usually are written in terms of impedances and velocities since a connection point ensures that one end of each element connected there has the same velocity when referred to a stationary reference. Equation (10.38) allows a velocity equation to be written for each independent loop, usually in terms of mobilities and forces. Velocity equations, best suited to series systems, do not find much usage in physical systems. Because the masses associated with physical elements have one connection effectively attached to a stationary reference (inertial reference), such a system seldom includes ideal series connected elements.

Example 10.2. Find the velocity of all the connection points and the forces acting on the elements of the system shown in Fig. 10.20. The system contains two velocity generators v_1 and v_6. Their magnitudes are known, their frequencies are the same, and they are 180° out-of-phase.

A. Using Eq. (10.37), write a force equation for each connection point except *a* and *e*. At point *b*: $F_1 - F_2 - F_3 = 0$. In terms of velocities and impedances:

$$(v_1 - v_2)Z_1 - (v_2 - v_3)Z_2 - (v_2 - v_4)Z_4 = 0 \tag{a}$$

At *c*, the two series elements have the same force acting: $F_2 - F_2 = 0$. In terms of velocities and impedances:

$$(v_2 - v_3)Z_2 - (v_3 - v_4)Z_3 = 0 \tag{b}$$

At *d*: $F_2 + F_3 - F_4 - F_5 = 0$. In terms of velocities and impedances:

$$(v_3 - v_4)Z_3 + (v_2 - v_4)Z_4 - (v_4 + v_6)Z_5 - (v_4 - v_5)Z_6 = 0 \tag{c}$$

Note that v_6 is (+) because of the 180° phase relation to v_1.
At *f*: $F_5 - F_5 = 0$. In terms of velocities and impedances:

$$(v_4 - v_5)Z_6 - v_5 Z_7 = 0 \tag{d}$$

Since v_1 and v_6 are known, the four unknown velocities v_2, v_3, v_4, and v_5 may be determined by solving the four simultaneous equations above. After the velocities are obtained, the forces may be determined from the following:

$$F_1 = (v_1 - v_2)Z_1 \qquad F_2 = (v_2 - v_3)Z_2 = (v_3 - v_4)Z_3$$

$$F_3 = (v_2 - v_4)Z_4 \qquad F_4 = (v_4 + v_6)Z_5$$

$$F_5 = (v_4 - v_5)Z_6 = v_5 Z_7$$

B. The method of *node forces.* Equations (*a*) through (*d*) above can be rewritten as follows:

$$v_1 Z_1 = (Z_1 + Z_2 + Z_3)v_2 - Z_2 v_3 - Z_4 v_4 \tag{a'}$$

$$0 = -Z_2 v_2 + (Z_2 + Z_3)v_3 - Z_3 v_4 \tag{b'}$$

$$0 = -Z_4 v_2 - Z_3 v_3 + (Z_3 + Z_4 + Z_5 + Z_6)v_4 - Z_6 v_5 \tag{c'}$$

$$-v_6 Z_5 = -Z_6 v_4 + (Z_6 + Z_7)v_5 \tag{d'}$$

These equations can be written by inspection of the schematic diagram by the following rule: *At each point with a common velocity (force node), equate the force generators to the sum of the impedances attached to the node multiplied by the velocity of the node, minus the impedances multiplied by the velocities of their other connection points.*

When the equations are written so that the unknown velocities form columns, the equations are in the proper form for a determinant solution for any of the unknowns. Note that the

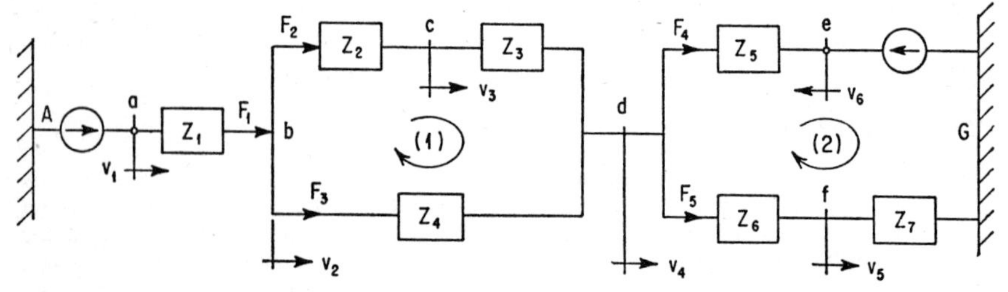

Fig. 10.20. System of mechanical elements and vibration sources analyzed in Example 10.2 to find the velocity of each connection and the force acting on each element.

determinant of the Z's is symmetrical about the main diagonal. This condition always exists and provides a check for the correctness of the equations.

C. Using Eq. (10.38), write a velocity equation in terms of force and mobility around enough closed loops to include each element at least once. In Fig. 10.20, note that

$$F_3 = F_1 - F_2 \quad \text{and} \quad F_5 = F_1 - F_4$$

Around loop (1):

$$F_2(\mathfrak{M}_2 + \mathfrak{M}_3) - (F_1 - F_2)\mathfrak{M}_4 = 0 \tag{e}$$

The minus sign preceding the second term results from going across the element 4 in a direction opposite to the assumed force acting on it.

Around loop (2):

$$F_4\mathfrak{M}_5 - v_6 - (F_1 - F_4)(\mathfrak{M}_6 + \mathfrak{M}_7) = 0 \tag{f}$$

A summation of velocities from A to G along the upper path forms the following closed loop:

$$v_1 + F_1\mathfrak{M}_1 + F_2(\mathfrak{M}_2 + \mathfrak{M}_3) + F_4\mathfrak{M}_5 - v_6 = 0 \tag{g}$$

Equations (*e*), (*f*), and (*g*) then may be solved for the unknown forces F_1, F_2, and F_4. The other forces are $F_3 = F_1 - F_2$ and $F_5 = F_1 - F_4$. The velocities are:

$$v_2 = v_1 - F_1\mathfrak{M}_1 \qquad v_3 = v_2 - F_2\mathfrak{M}_2 \qquad v_4 = v_2 - F_3\mathfrak{M}_4 \qquad v_5 = F_5\mathfrak{M}_7$$

When a system includes more than one source of vibration energy, a Kirchhoff's law analysis with impedance methods can be made only if all the sources are operating at the same frequency. This is the case because sinusoidal forces and velocities can add as phasors only when their frequencies are identical. However, they may differ in magnitude and phase. Kirchhoff's laws still hold for instantaneous values and can be used to write the differential equations of motion for any system.

SUPERPOSITION

The superposition method of system analysis is advantageous when there are several vibration sources in a system. This method is illustrated by an example involving two force generators.

Example 10.3. In the system of Fig. 10.21*A*, the force generators F_1 and F_2 operate at the same frequency and their magnitudes are known. Determine the velocity of the common connection point *c* by the superposition theorem. First, determine the velocity v_c by Kirchhoff's laws for comparison with the results obtained by superposition. Writing the force-node equations at *a*, *b*, and *c*,

At *a*: $F_1 = Z_1v_a - Z_1v_c$

At *c*: $0 = -Z_1v_a + (Z_1 + Z_2 + Z_3)v_c - Z_2v_b$

At *b*: $F_2 = -Z_2v_c + Z_3v_b$

The impedance Z_1 is c_1 and k_1 in parallel, Z_2 is c_2 and k_2 in parallel, and Z_3 is c_3, k_3, and m_3 in parallel. Solving the equations for forces at *a*, *b*, and *c* to obtain v_c,

$$v_c = \frac{F_1 + F_2}{Z_3}$$

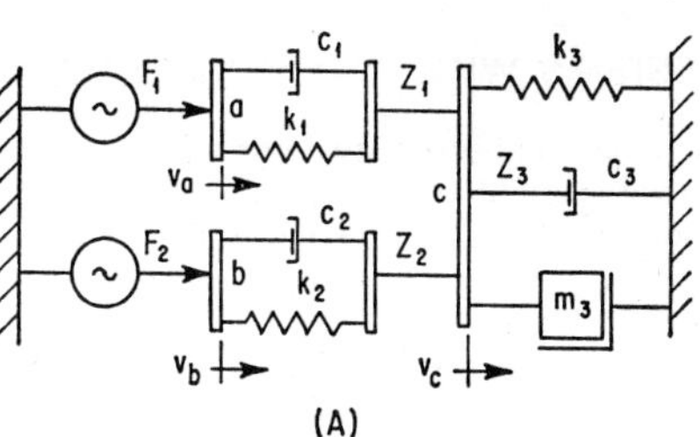

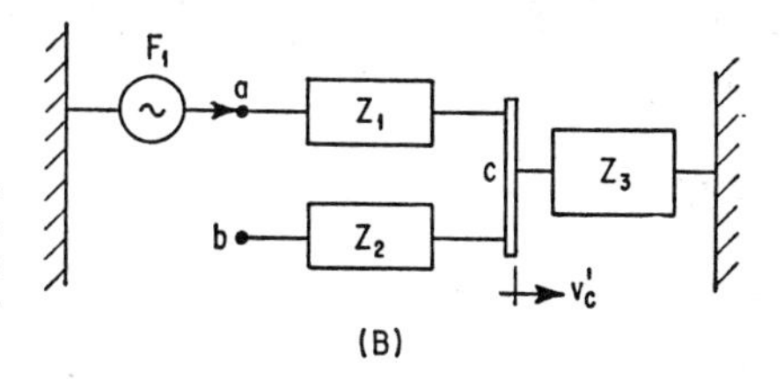

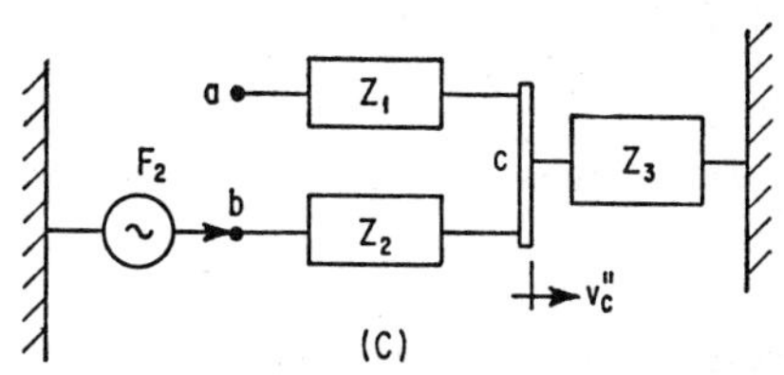

FIG. 10.21. System of mechanical elements including two force generators used to illustrate the principle of superposition in Example 10.3.

Using the superposition theorem, let the force F_2 be zero, and redraw the system as in Fig. 10.21*B*. Then let the velocity at point *c* be v_c' when only F_1 is applied. Since the force F_1 is transmitted by Z_1 to Z_3, v_c' becomes

$$v_c' = \frac{F_1}{Z_3}$$

Now letting $F_1 = 0$, applying F_2 as in Fig. 10.21*C*, and denoting the velocity of point *c* by v_c'',

$$v_c'' = \frac{F_2}{Z_3}$$

By the superposition theorem,

$$v_c = v_c' + v_c'' = \frac{F_1 + F_2}{Z_3}$$

If the force generators in Example 10.3 operate at different frequencies, the velocities are $v_c' = F_1/Z_3'$ and $v_c'' = F_2/Z_3''$, where Z_3' is the value of Z_3 at the frequency of force generator 1 and Z_3'' is the value of Z_3 at the frequency of generator 2. The instantaneous value of v_c, when phase angles are neglected, is

$$v_c = |v_c'| \sin \omega_1 t + |v_c''| \sin \omega_2 t$$

The mean square value is

$$v_c^2 = |v_c'|^2 + |v_c''|^2$$

Since the solution for a velocity or force by the superposition method involves only one source at a time, Kirchhoff's laws and impedance methods can be used in finding the responses to each source.

SYSTEMS WITH ONE CONNECTION

In one-connection systems considered here, one connection is attached to a rigid immovable point or inertial reference and another connection is available for external connection. These systems differ from multiple-connection systems in that they do not transmit a force or velocity from one system to another. One-connection systems can be divided into two types:

1. Passive systems, in which no sources of vibration energy are included. These systems can be represented by a single impedance or mobility function as shown in the section on *system simplification.* Such systems often are connected to, and considered as loads on, active systems as defined below.
2. Active systems, in which one or more sources of vibration energy is included. A complete schematic representation of such a system and its external connections may be drawn and analyzed by Kirchhoff's laws and superposition. The simplified representations of Thévenin's and Norton's equivalent systems are considered here.

THÉVENIN'S EQUIVALENT SYSTEM. Example 10.4. The system of Fig. 10.22*A* has one force generator, a number of elements, and a connection point *b*. Find Thévenin's equivalent system for this system. First combine the parallel elements c_1, k_1, and m_1 into one impedance Z_1, combine c_2 and k_2 into Z_2, and express m_3 in terms of Z_3. The schematic representation is shown in Fig. 10.22*B*. Determine the value of the force transmitted to a rigid immovable structure when point *b* is clamped to it. Point *b* is shown clamped in Fig. 10.22*C*, and Z_3 is omitted because it has no motion. Impedances Z_1 and Z_2 then are in parallel, and the force *F* divides between them to give the force transmitted to Z_2:

$$F_c = \frac{FZ_2}{Z_1 + Z_2} \qquad (a)$$

The impedance required for the Thévenin equivalent system can be determined from the system as rearranged in Fig. 10.22*D*. The force generator is removed and the impedance at *b* is calculated:

$$Z_i = Z_3 + \frac{1}{1/Z_1 + 1/Z_2} = \frac{Z_1Z_2 + Z_1Z_3 + Z_2Z_3}{Z_1 + Z_2} \qquad (b)$$

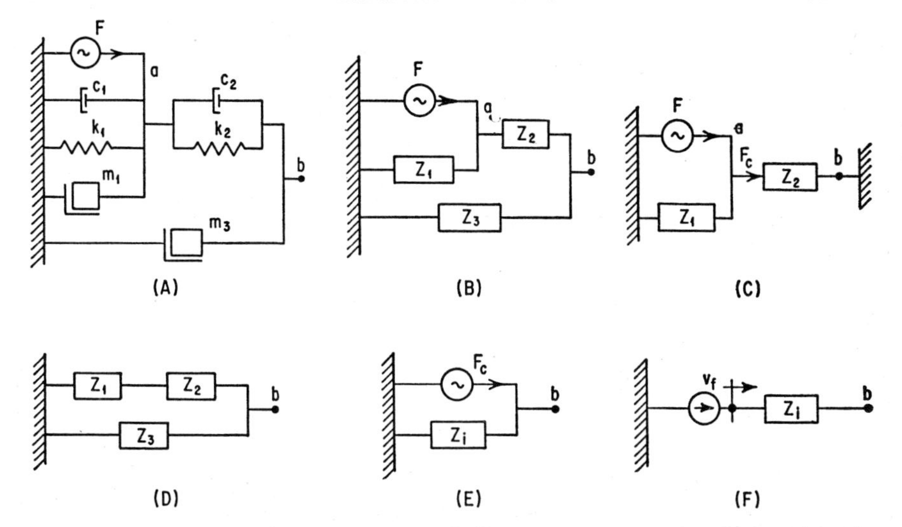

FIG. 10.22. System of mechanical elements including a force generator for which a Thévenin equivalent system is determined in Example 10.4 and a Norton equivalent system is determined in Example 10.5.

The clamped force is given by Eq. (a) above, and the internal impedance is given by Eq. (b). The Thévenin equivalent system is shown in Fig. 10.22E.

When a load Z_4 is connected to point b, the equivalent representation is shown in Fig. 10.23. Under these conditions the velocity v_b of point b and the force applied to Z_4 can be calculated. Writing the force-node equation at b,

$$F_c = (Z_i + Z_4)v_b$$

Solving for v_b,

$$v_b = \frac{F_c}{Z_i + Z_4} \tag{c}$$

The force at b is then $F_b = v_b Z_4$, or

$$F_b = \frac{F_c Z_4}{Z_i + Z_4} \tag{d}$$

For any load impedance Z_4, the force and velocity at b can be calculated from Eqs. (c) and (d).

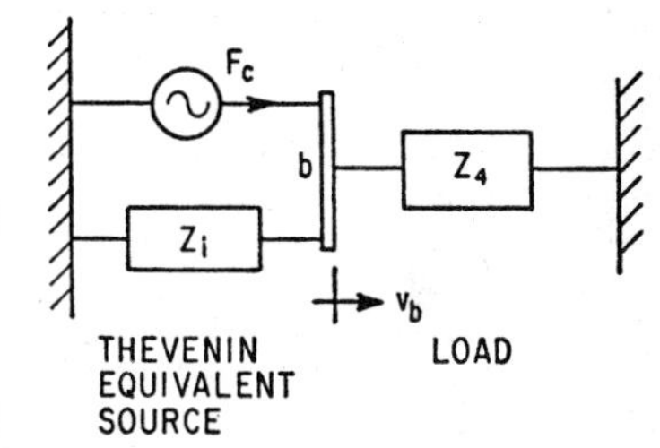

FIG. 10.23. A Thévenin equivalent system connected to a load impedance.

NORTON'S EQUIVALENT SYSTEM. Example 10.5. Find Norton's equivalent representation for the system of Fig. 10.22A, and find the force and velocity of point b when a load is attached to that point.

For this representation the free velocity at point b is required. This velocity v_f can be calculated from the system as drawn in Fig. 10.22B. Write the force-node equations at a and b, and let v_b be called v_f:

$$F = (Z_1 + Z_2)v_a - Z_2 v_f$$
$$0 = -Z_2 v_a + (Z_2 + Z_3)v_f$$

Solving for v_f,

$$v_f = \frac{FZ_2}{Z_1Z_2 + Z_1Z_3 + Z_2Z_3} \tag{a}$$

The internal impedance is given by Eq. (b) of Example 10.4, the free velocity is given by Eq. (a) above, and the Norton equivalent representation is shown in Fig. 10.22F.

When the load Z_4 is connected to b, the equivalent system is shown in Fig. 10.24.

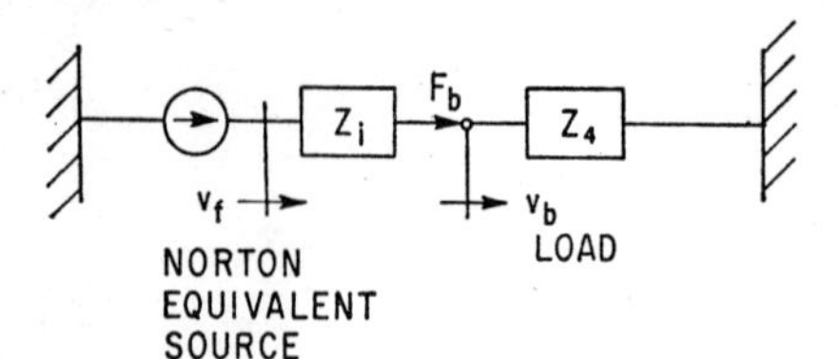

FIG. 10.24. A Norton equivalent system connected to a load mobility.

Since the elements are in series, they all transmit the same force F_b:

$$F_b = \frac{v_f}{\mathfrak{M}_i + \mathfrak{M}_4} = \frac{v_f}{1/Z_i + 1/Z_4} = \frac{v_f Z_i Z_4}{Z_i + Z_4} \qquad (b)$$

The velocity at b is

$$v_b = \frac{F_b}{Z_4} = \frac{v_f Z_i}{Z_i + Z_4} \qquad (c)$$

The force and velocity of b can be determined for any load Z_4 in terms of the constant velocity source and internal impedance of the equivalent system representation.

SYSTEMS WITH TWO CONNECTIONS

Only two-connection systems used to transmit vibration energy are considered here. Active and passive systems of this type are possible. Systems that include vibration sources that are unrelated to the force or velocity to be transmitted are not considered active in the usual sense. They are best analyzed using superposition. This involves determining the responses at the two connections due to the internal sources by Kirchhoff's laws, finding the transmitted motion and force by the methods that follow, and adding the results.

An active two-connection system by the usual definition is one in which an input function applied to one connection controls an energy source so as to produce an output function of the same shape as the input, but with an increased amplitude. Such a device is called an *amplifier*. The hydraulically powered vibration generator that has its control valve driven by a small electrodynamic vibration exciter is an example of this type of system. Many amplifying systems produce mechanical output functions, but they involve other forms of energy and dynamic media. In such cases, analysis is possible by impedance methods when compatible dynamical analogies and suitable coupling factors between dynamic media are used.

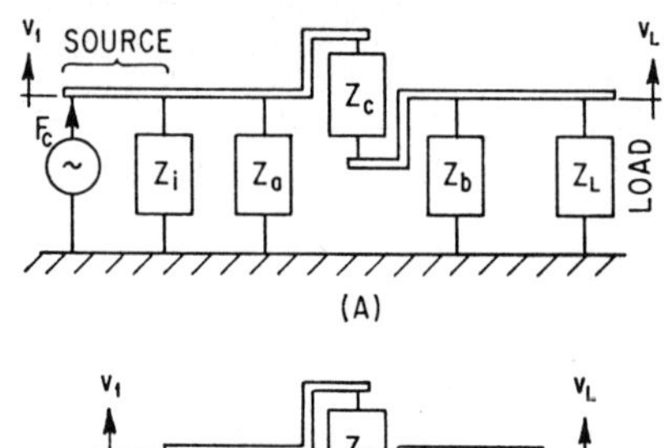

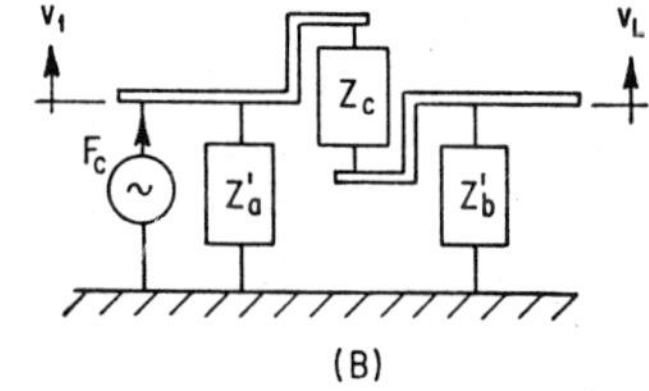

FIG. 10.25. A π equivalent system representation for a two-connection element used to transmit vibration energy, as analyzed in Example 10.6.

EQUIVALENT π SYSTEMS. Example 10.6. When a passive two-connection system is represented by an equivalent π system as shown in Fig. 10.18, its operation as a device to transmit vibration energy may be analyzed by adding the vibration source and the load impedance to the schematic representation. With the equivalent π system, it is convenient to use a Thévenin equivalent system for the source and a single impedance Z_c to represent the load. The complete system then is represented as shown in Fig. 10.25*A*. Since Z_i and Z_a as well as Z_b and Z_L are in parallel, they may be combined as in Fig. 10.25*B* where

$$Z_a' = Z_i + Z_a \qquad \text{and} \qquad Z_b' = Z_L + Z_b$$

Writing force node equations in terms of v_1 and v_L,

$$F_c = (Z_a' + Z_c)v_1 - Z_c v_L$$

$$0 = -Z_c v_1 + (Z_b' + Z_c)v_L$$

Solving for v_1 and v_L,

$$v_1 = \frac{F_c(Z_b' + Z_c)}{Z_a'Z_b' + Z_a'Z_c + Z_b'Z_c}$$

$$v_L = \frac{F_cZ_c}{Z_a'Z_b' + Z_a'Z_c + Z_b'Z_c}$$

The force applied to the load is $F_L = v_LZ_L$.

EQUIVALENT T SYSTEM. Example 10.7. When the T equivalent system representation is used, it is convenient to use the Norton equivalent representation for the vibration source and a mobility representation for the load. The complete representation is shown in Fig. 10.26*A*. Mobilities $\mathfrak{M}_i$ and $\mathfrak{M}_x$ as well as $\mathfrak{M}_2$ and $\mathfrak{M}_y$ are in series and can be combined as in Fig. 10.26*B* in which $\mathfrak{M}_x' = \mathfrak{M}_x + \mathfrak{M}_i$ and $\mathfrak{M}_y' = \mathfrak{M}_y + \mathfrak{M}_L$. A determination of the velocity v of the common connection allows a simple determination of F_1 and F_L. Since $\mathfrak{M}_y$ and $\mathfrak{M}_L$ are in series, the force F_L is the force applied to the load. The force F_1 is supplied by the vibration source. Proceeding to find v, write a force equation at the common connection by Eq. (10.37):

$$\frac{v_f - v}{\mathfrak{M}_x'} = \frac{v}{\mathfrak{M}_y'} + \frac{v}{\mathfrak{M}_z}$$

$$v = \frac{v_f}{1 + \mathfrak{M}_x'/\mathfrak{M}_y' + \mathfrak{M}'_x/\mathfrak{M}_z}$$

The force acting on the load is then $F_L = v/\mathfrak{M}_y'$, and the load velocity is

$$v_L = F_L\mathfrak{M}_L = v_L\mathfrak{M}_L/\mathfrak{M}_y'$$

The force produced by the source which acts at the input connection is

$$F_1 = (v_f - v)/\mathfrak{M}_x'$$

and the velocity of the input connection v_1 is

$$v_1 = v + F_1\mathfrak{M}_x = \frac{v\mathfrak{M}_i + v_f\mathfrak{M}_x}{\mathfrak{M}_x'}$$

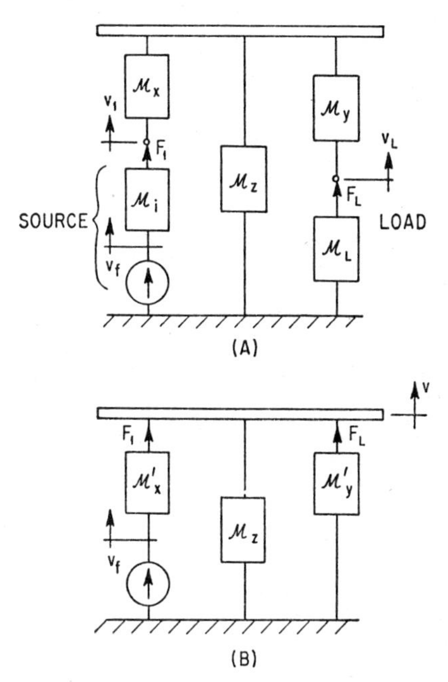

FIG. 10.26. A T equivalent system representation for the two-connection system analyzed in Example 10.7.

IMPEDANCE AND MOBILITY PARAMETERS. In Examples 10.6 and 10.7 the actual system is represented by an equivalent system made up of fictitious impedance or mobility elements. Another approach is to represent a two-connection passive system by a "black box," attached to an inertial reference, that has input and output connections. Such a "black box" is illustrated in Fig. 10.27. When the elements in the system are linear and bilateral, and force generators are attached to the connections, the relationship between the velocities and forces may be expressed:

$$\begin{aligned} F_1 &= Z_{11}v_1 + Z_{12}v_2 \\ F_2 &= Z_{21}v_1 + Z_{22}v_2 \end{aligned} \tag{10.53}$$

where the Z's are defined by Eqs. (10.39) to (10.42) and are called the *impedance parameters*. When the impedance parameters and two forces are known, the two velocities may be obtained by solving Eqs. (10.53).

The forces and velocities of a system considered as a "black box" also may be expressed as

$$
\begin{aligned}
v_1 &= \mathfrak{M}_{11}F_1 + \mathfrak{M}_{12}F_2 \\
v_2 &= \mathfrak{M}_{21}F_1 + \mathfrak{M}_{22}F_2
\end{aligned}
\tag{10.54}
$$

Here the $\mathfrak{M}$'s are defined by Eqs. (10.47) to (10.50) and are called the *mobility parameters*. This representation is useful when the velocities at connections (1) and (2) are known.

When a "black box" system is used to transmit vibration energy in only one direction, there is a source attached to one connection and a load connected to the other. Such a case is shown in Fig. 10.27.

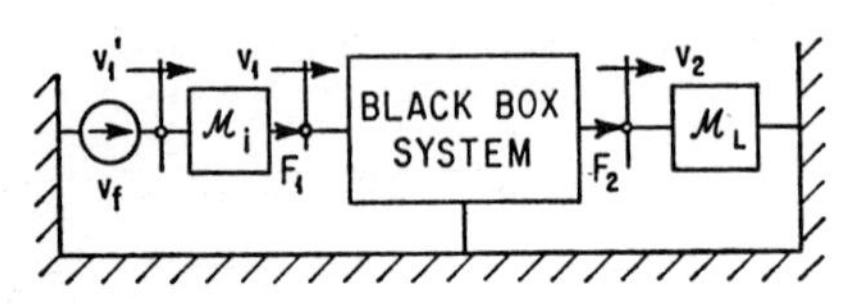

FIG. 10.27. A "black box" representation of a two-connection system analyzed by the mobility parameter method in Example 10.8.

Example 10.8. Find the force and velocity at connections 1 and 2 for the system of Fig. 10.27. A Norton equivalent system is used to represent the vibration source and the load is represented by a mobility. The mobility parameters are thus the most advantageous to use. The load and internal mobility of the source may be included with the "black box" system by measuring or calculating the mobility parameters with $\mathfrak{M}_L$ and $\mathfrak{M}_i$ in place. Equations (10.54) then become

$$
\begin{aligned}
v_f &= \mathfrak{M}_{11}'F_1 + \mathfrak{M}_{12}F_2 \\
0 &= \mathfrak{M}_{21}F_1 + \mathfrak{M}_{22}'F_2
\end{aligned}
$$

Forces and velocities are considered at points 1′ and 2. The velocity v_1 becomes v_f, and v_2 is zero since no external force is applied. The mobility $\mathfrak{M}_{11}'$ is $\mathfrak{M}_{11}$ of Eq. (10.47) determined at point 1′, and $\mathfrak{M}_{22}'$ is $\mathfrak{M}_{22}$ of Eq. (10.48) with $\mathfrak{M}_L$ in place. Mobilities $\mathfrak{M}_{12}$ and $\mathfrak{M}_{21}$ are not dependent on the external connections. Solving for F_1 and F_2:

$$
F_1 = \frac{v_f\mathfrak{M}_{22}'}{\mathfrak{M}_{11}'\mathfrak{M}_{22}' - \mathfrak{M}_{12}\mathfrak{M}_{21}}
$$

$$
F_2 = \frac{-v_f\mathfrak{M}_{21}}{\mathfrak{M}_{11}'\mathfrak{M}_{22}' - \mathfrak{M}_{12}\mathfrak{M}_{21}}
$$

The force applied to the load is thus $F_L = F_2$, and the load velocity is $v_L = F_2\mathfrak{M}_L$. The force at connection 1 is F_1, since $\mathfrak{M}_i$ transmits this force and the input connection velocity is

$$
v_1 = v_f - F_1\mathfrak{M}_i
$$

FOUR-POLE PARAMETERS. The relationship between input and output forces and velocities can be written in still another form:

$$
\begin{aligned}
F_1 &= \alpha_{11}F_2 + \alpha_{12}v_2 \\
v_1 &= \alpha_{21}F_2 + \alpha_{22}v_2
\end{aligned}
\tag{10.55}
$$

The α's are called *four-pole parameters* and are defined as follows:

$$
\alpha_{11} = \left.\frac{F_1}{F_2}\right|_{v_2=0} \qquad \alpha_{12} = \left.\frac{F_1}{v_2}\right|_{F_2=0}
$$

$$
\alpha_{21} = \left.\frac{v_1}{F_2}\right|_{v_2=0} \qquad \alpha_{22} = \left.\frac{v_1}{v_2}\right|_{F_2=0}
$$

The notation $v_2 = 0$ indicates that the output connection (2) is *clamped* and $F_2 = 0$ indicates the output is *free*. The quantities α_{11} and α_{22} are the force and velocity transfer functions, while α_{12} is an impedance and α_{21} is a mobility. "Black box" systems may be analyzed by the use of these parameters and Eqs. (10.55). In addition, the system may be analyzed by the method that uses the four-pole parameters of the ideal elements and rules for combining their parameters when they are series or parallel connected.[16]

To establish the four-pole parameters for the ideal elements, the relationships between applied and transmitted forces and the velocities of the connections are written in the form of Eqs. (10.55); then the α's are noted as the coefficients of the F_2 and v_2 terms.

Example 10.9. Determine the four-pole parameters for a mass, spring, and resistance.

Mass. Since a mass may be considered as a rigid body, the velocities of its connections are equal. The force required to give the mass a velocity v_1 is $j\omega m v_1$ (or $j\omega m v_2$). If a force F_2 is transmitted, this must be added to $j\omega m v_2$ to determine F_1. Writing these relationships in the form of Eqs. (10.55),

$$F_1 = F_2 + j\omega m v_2$$

$$v_1 = v_2$$

The α's are $\alpha_{11} = 1$, $\alpha_{12} = j\omega m = Z_m$, $\alpha_{21} = 0$, and $\alpha_{22} = 1$.

Spring. The ideal spring transmits an applied force so that $F_1 = F_2$. The relative velocity is $(v_1 - v_2) = j\omega F_2/k$. Expressing these relationships in the form of Eqs. (10.55),

$$F_1 = F_2$$

$$v_1 = \frac{j\omega}{k} F_2 + v_2$$

The α's are $\alpha_{11} = 1$, $\alpha_{12} = 0$, $\alpha_{21} = j\omega/k = \mathfrak{M}_k$, and $\alpha_{22} = 1$.

Resistance. For the resistance, $F_1 = F_2$ and the relative velocity is $(v_1 - v_2) = F_2/c$. In the form of Eqs. (10.55):

$$F_1 = F_2$$

$$v_1 = \frac{F_2}{c} + v_2$$

The α's are $\alpha_{11} = 1$, $\alpha_{12} = 0$, $\alpha_{21} = 1/c = \mathfrak{M}_c$, and $\alpha_{22} = 1$.

Note that the determinant of the α's in each case in Example 10.9 is equal to unity. This is a characteristic of all systems made up of linear bilateral elements and is another indication of reciprocity. The unity value of the α determinant greatly simplifies the solution of Eqs. (10.55) by the determinant or matrix method.

Matrix methods are convenient in system analysis by the four-pole parameter methods. The parameters for a series-connected system, expressed in terms of the individual element parameters, are determined as follows:

SERIES-CONNECTED SYSTEMS. Two such elements are shown in Fig. 10.28*A*. The four-pole parameter equations for both elements in matrix form are

$$\begin{bmatrix} F_1 \\ v_1 \end{bmatrix} = \begin{bmatrix} \alpha_{11} & \alpha_{12} \\ \alpha_{21} & \alpha_{22} \end{bmatrix} \begin{bmatrix} F_2 \\ v_2 \end{bmatrix}$$

$$\begin{bmatrix} F_2 \\ v_2 \end{bmatrix} = \begin{bmatrix} \alpha_{11}' & \alpha_{12}' \\ \alpha_{21}' & \alpha_{22}' \end{bmatrix} \begin{bmatrix} F_3 \\ v_3 \end{bmatrix}$$

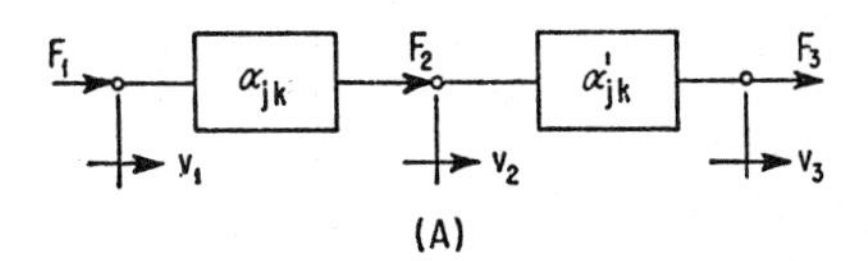

Combining these to find the output at (3) in terms of the input at (1):

$$\begin{bmatrix} F_1 \\ v_1 \end{bmatrix} = \begin{bmatrix} \alpha_{11} & \alpha_{12} \\ \alpha_{21} & \alpha_{22} \end{bmatrix} \begin{bmatrix} \alpha_{11}' & \alpha_{12}' \\ \alpha_{21}' & \alpha_{22}' \end{bmatrix} \begin{bmatrix} F_3 \\ v_3 \end{bmatrix}$$

$$\begin{bmatrix} F_1 \\ v_1 \end{bmatrix} = \begin{bmatrix} \alpha_{11}'' & \alpha_{12}'' \\ \alpha_{21}'' & \alpha_{22}'' \end{bmatrix} \begin{bmatrix} F_3 \\ v_3 \end{bmatrix}$$

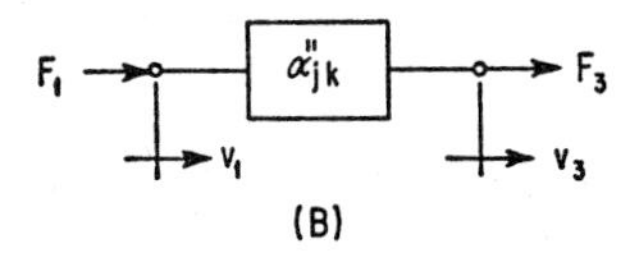

FIG. 10.28. Series-connected mechanical elements considered in the four-pole parameter method.

The single system of Fig. 10.28*B* is represented by the α'''s and replaces the two series elements. The final four-pole parameters in terms of the original element parameters are

$$\alpha_{11}'' = \alpha_{11}\alpha_{11}' + \alpha_{12}\alpha_{21}' \qquad \alpha_{12}'' = \alpha_{11}\alpha_{12}' + \alpha_{12}\alpha_{22}'$$
$$\alpha_{21}'' = \alpha_{21}\alpha_{11}' + \alpha_{22}\alpha_{21}' \qquad \alpha_{22}'' = \alpha_{22}\alpha_{22}' + \alpha_{21}\alpha_{12}' \tag{10.56}$$

This process can be continued to combine any number of elements in a series string. Note that this method allows for a series mass and any number of elements to be considered series-connected.

PARALLEL-CONNECTED ELEMENTS. The canonical equations for parallel connected elements, as shown in Fig. 10.29*A*, can be written in the form of Eqs. (10.55)

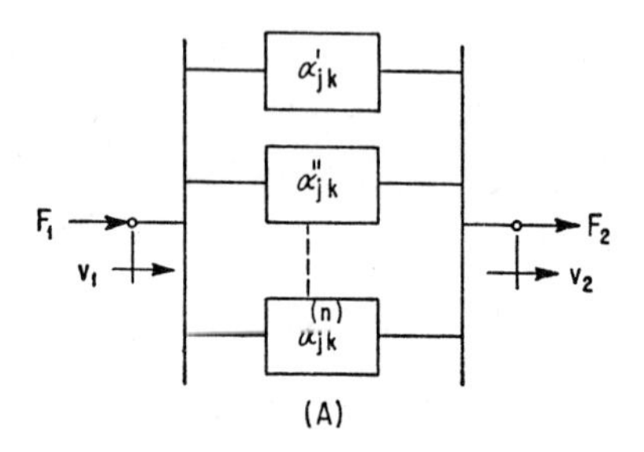

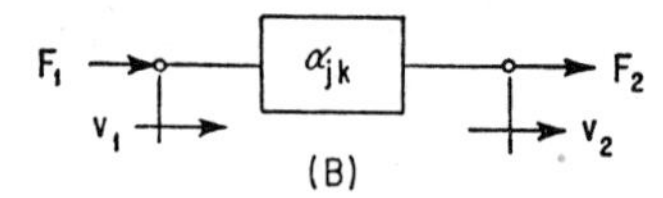

FIG. 10.29. Parallel-connected mechanical elements considered in the four-pole parameter method.

where the α's for the single resultant element are defined in terms of the individual elements in the system:

$$\alpha_{11} = \frac{A}{B} \qquad \alpha_{12} = \frac{AC}{B} - B \qquad \alpha_{21} = \frac{1}{B} \qquad \alpha_{22} = \frac{C}{B} \tag{10.57}$$

where $$A = \sum_{i=1}^{n} \frac{\alpha_{11}^{(i)}}{\alpha_{21}^{(i)}} \qquad B = \sum_{i=1}^{n} \frac{1}{\alpha_{21}^{(i)}} \qquad C = \sum_{i=1}^{n} \frac{\alpha_{22}^{(i)}}{\alpha_{21}^{(i)}}$$

The factors A, B, and C depend on the parameters of the individual elements as indicated by $i = 1, 2, \ldots, n$. The parallel combination then is represented by a single element as in Fig. 10.29*B*.

VIBRATION SOURCES. Thévenin's and Norton's equivalent system representations are convenient in this method. Consider a Thévenin representation such as Fig. 10.23. The force F_1 applied to a load and the velocity of the load input v_1 are related to the clamped force F_c and internal impedance Z_i as follows:

$$F_1 = F_c - Z_i v_1 \tag{10.58}$$

For the Norton representation in Fig. 10.24, the force F_1 is related to the v_1 as follows:

$$F_1 = Z_i v_c - Z_i v_1 \tag{10.59}$$

Example 11.10. By the four-pole parameter method, determine the velocity of a mass driven by a vibration source. The mass is shown driven by a Thévenin equivalent source in

Fig. 10.30. When there is no force transmitted by the mass, the four-pole parameter equations for the mass are

$$\begin{bmatrix} F_1 \\ v_1 \end{bmatrix} = \begin{bmatrix} 1 & j\omega m \\ 0 & 1 \end{bmatrix} \begin{bmatrix} 0 \\ v_2 \end{bmatrix}$$

In conventional form, $F_1 = j\omega m v$ and $v_1 = v_2$. The source equation $F_1 = F_c - Z_i v_1$ is combined with the mass equation to give

$$F_c - Z_i v_1 = j\omega m v_2 = j\omega m v_1$$

Solving for v_1,

$$v_1 = \frac{F_c}{Z_i + j\omega m}$$

When the mass is driven by a Norton equivalent system with free velocity v_f, the combination of equations yields

$$Z_i v_f - Z_i v_1 = j\omega m v_1$$

Solving for v_1,

$$v_1 = \frac{Z_i v_f}{Z_i + j\omega m}$$

The "black box" concept is evident in each of the above methods of analysis for two-connection systems. This approach focuses attention on two points in a system, usually the more important points. In the case of vibration isolators, the performance is measured in terms of the velocities or forces at the two connections when the devices are used to connect two systems. The above methods allow a calculation of the response of such a device in any application when the impedances of the systems to be connected are known, and one of the sets of the above parameters is known. All the parameters lend themselves to direct measurement. Vibration response characteristics of very complex systems under many different conditions may be determined with a minimum of analysis time when one of the sets of parameters is known. All the sets of parameters are defined in terms of sinusoidal forces and velocities, and they must be measured with this type of driving function.

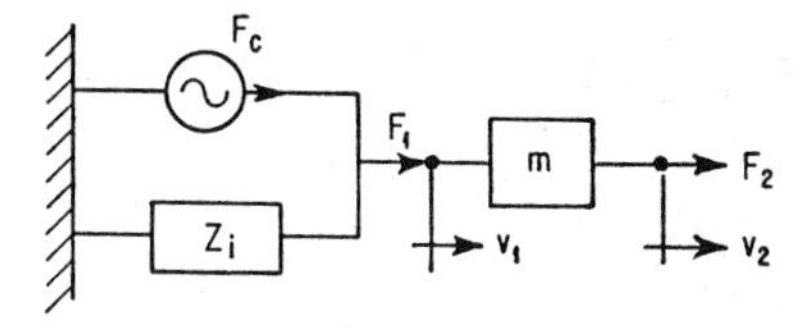

Fig. 10.30. A mass driven by a Thévenin equivalent source, as analyzed by the four-pole parameter method in Example 10.10.

NONSINUSOIDAL FUNCTIONS

The forces generated by the sources of vibration energy encountered in the normal operation of many mechanical systems are not sinusoidal in form. The responses of systems to nonsinusoidal driving functions can be determined by the impedance method. The Fourier analysis methods and the principle of superposition make this possible.

Excitations of three types frequently encountered are:

1. Continuous, periodic but nonsinusoidal force or velocity functions. Such functions can be represented by a Fourier series and the response to each term found by the impedance method.

2. Nonrepeated transient excitation, usually called shock excitation. The Fourier integral transform of the function yields a continuous frequency spectrum. When impedance expressions are considered as frequency functions, the frequency spectrum of the response may be determined. By the inverse Fourier integral transform the response as an instantaneous time function may be obtained.

3. Continuous, nonperiodic random excitation. The Fourier integral transform applies, and the methods are similar to those for (2) above. Random vibration is discussed in Chap. 11.

PRESENTATION AND EVALUATION OF IMPEDANCE DATA

At a particular frequency the mechanical impedance (or mobility) of a system is a complex number which can be expressed as the sum of a real and an imaginary component or as a magnitude and impedance or mobility angle.

$$Z = \mathcal{R}(Z) + j\mathcal{J}(Z) = |Z|e^{j\theta}$$

The magnitude $|Z|$ and angle θ are

$$|Z| = \sqrt{[\mathcal{R}(Z)]^2 + [\mathcal{J}(Z)]^2} \qquad \theta = \tan^{-1} \frac{\mathcal{J}(Z)}{\mathcal{R}(Z)}$$

These relations also may be expressed

$$\mathcal{R}(Z) = |Z| \cos \theta \qquad \mathcal{J}(Z) = |Z| \sin \theta$$

Impedance at a particular frequency may be represented as a directed line on the complex plane, as shown in Fig. 10.31. Thus, two quantities are required to specify completely an impedance or mobility.

The impedances of the three ideal elements (mass, spring, and resistance) are functions of frequency as indicated by Eqs. (10.14), (10.18), and (10.22). It is convenient to plot

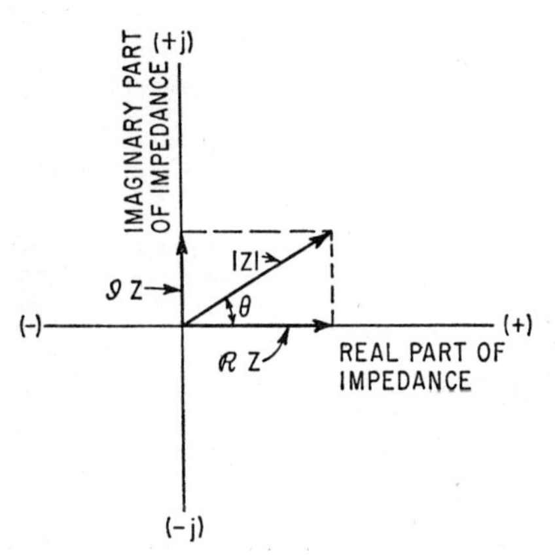

Fig. 10.31. The complex-plane representation of magnitude and angle of impedance, by real and imaginary parts.

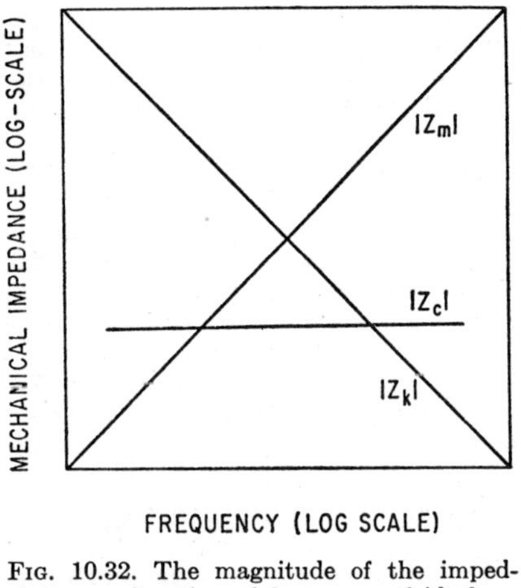

Fig. 10.32. The magnitude of the impedance as a function of frequency of ideal resistance, spring, and mass plotted to a log-log scale.

impedance data on graph paper with logarithmic impedance and frequency scales (usually referred to as log-log paper). The impedance curves of Fig. 10.9 are replotted on log-log paper in Fig. 10.32. Curves for impedance of masses have a slope of $(+1)$; they cross the line $f = 1$ cps at $Z_m = 2\pi m$. Thus $m = Z_m/2\pi$. The value of m for an unknown mass thus can be obtained from the impedance curve. The curves for impedance of ideal springs have slopes of (-1) and they intercept the line $f = 1$ cps at $Z_k = k/2\pi$; thus, $k = 2\pi Z_k$. Resistance is a constant with frequency; thus, the impedance Z_c is represented by a horizontal line.

The impedances of the ideal elements can be represented by single curves because they are entirely real or imaginary; however, for combinations of elements it often is necessary to keep the real and imaginary parts separate. For example, consider the parallel

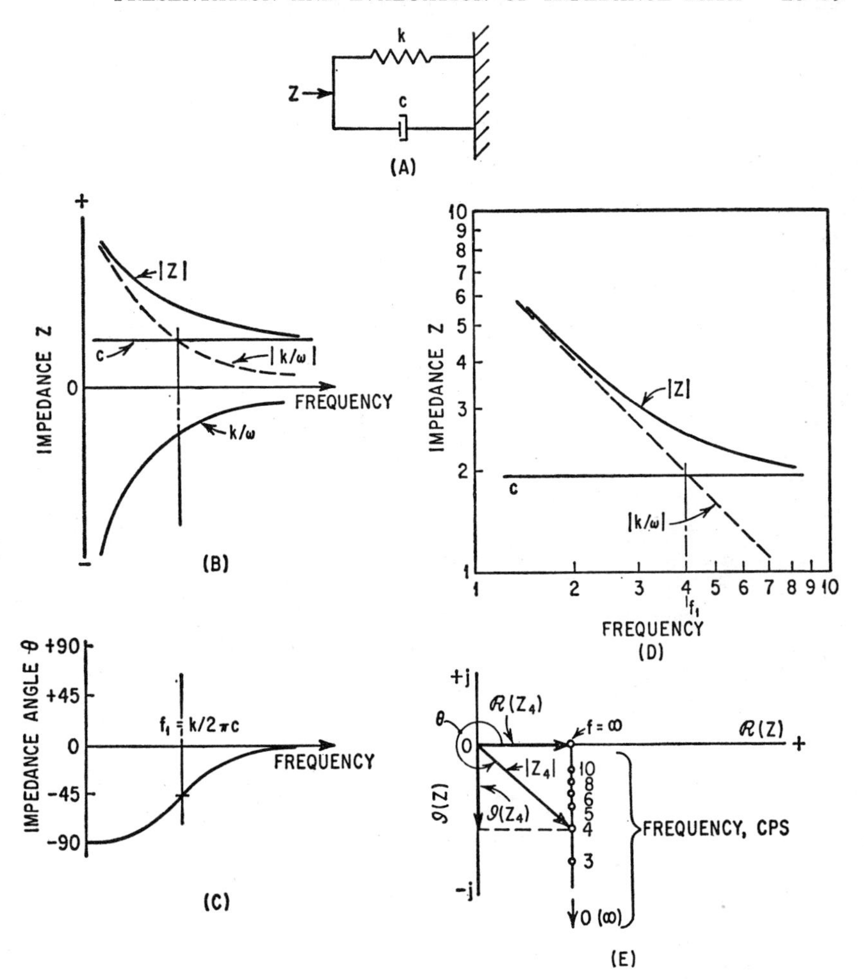

FIG. 10.33. Impedance characteristics of a parallel resistance-spring combination.

spring-dashpot arrangement of Fig. 10.33*A*. The mechanical impedance of the system is

$$Z = c - \frac{jk}{2\pi f}$$

Expressed as magnitude and angle:

$$|Z| = \sqrt{c^2 + \left(\frac{k}{2\pi f}\right)^2} \qquad \theta = \tan^{-1}\left(-\frac{k}{2\pi fc}\right)$$

where the real part $\mathcal{R}(Z) = c$ and the imaginary part $\mathcal{I}(Z) = -k/2\pi f$ are plotted in Fig. 10.33*B* to linear coordinates. The magnitude $|Z|$ is plotted on the same set of coordinates. The imaginary part $\mathcal{I}(Z)$ is shown dotted in the positive impedance region to indicate that $|Z|$ approaches the imaginary term at low frequencies and the real term at high frequencies. The angle of the impedance is plotted in Fig. 10.33*C*. The curves

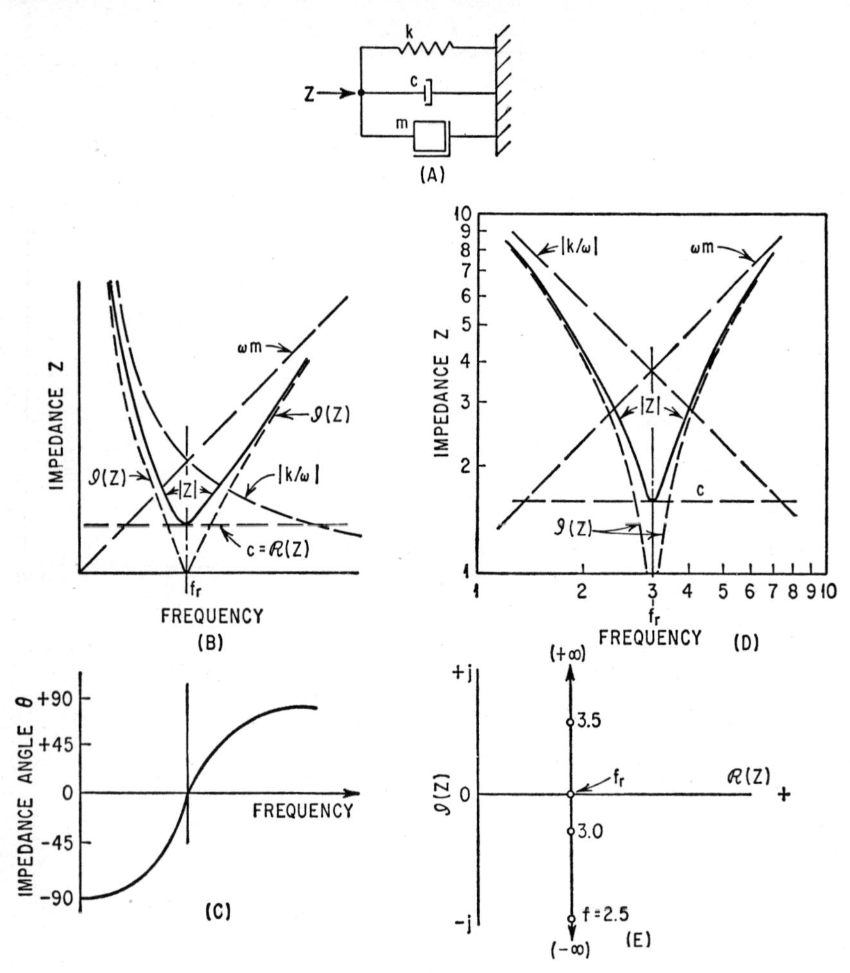

Fig. 10.34. Impedance characteristics of a parallel resistance-spring-mass system.

shown in Fig. 10.33*B* to linear coordinates are shown in Fig. 10.33*D* to logarithmic coordinates.

An alternate method for presenting impedance data is shown in Fig. 10.33*E*. All four factors required to specify impedance are presented on one curve. At each frequency the real and imaginary parts of impedance are plotted as a point on the complex plane, corresponding to Fig. 10.31. These points are connected to form the impedance curve. As shown in Fig. 10.33*E* for a frequency of 4 cps, the coordinates of the point are the real and imaginary parts of Z. The length of the radial line from the origin to the point is then $|Z|$, and the angle that this line makes with the (+) real axis is the angle of impedance. Frequency appears as a parameter in this presentation.

As shown in Fig. 10.33*E*, the impedance becomes real at high frequencies since $|Z|$ approaches c and θ approaches 360° (0°). At low frequencies $|Z|$ becomes large, θ approaches 270° (−90°), and the negative imaginary impedance indicates springlike action. The asymptotes of $|Z|$ indicate the characteristics of the system at frequency extremes.

The addition of a mass to the parallel system of Fig. 10.33*A* gives the system of Fig.

10.34A. The impedance at the driving point is

$$Z = c + j\left(2\pi fm - \frac{k}{2\pi f}\right)$$

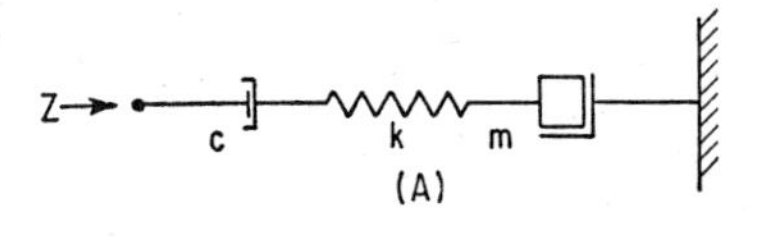

The magnitude and angle are

$$|Z| = \sqrt{c^2 + \left(2\pi fm - \frac{k}{2\pi f}\right)^2}$$

$$\theta = \tan^{-1}\frac{2\pi fm - k/2\pi f}{c}$$

The impedance of each element is plotted without regard for sign on the linear coordinates of Fig. 10.34B and the logarithmic coordinates of Fig. 10.34D. At the frequency of

$$f_r = \frac{\sqrt{k/m}}{2\pi}$$

the imaginary term is zero. This frequency is the *resonant frequency;* the impedance Z is real and equal to c. At very low frequencies, $|Z|$ approaches the spring impedance and the system is said to be "stiffness controlled." At high frequencies, $|Z|$ approaches the mass impedance and the system is "mass controlled." The impedance angle is plotted on a linear plot in Fig. 10.34C. The asymptotes of this curve are $\theta = -90°$ at low frequencies, and $\theta = +90°$ at high frequencies. At $f = f_r$, $\theta = 0$.

A plot in the complex plane is given in Fig. 10.34E. The curve, parallel to the imaginary axis, indicates a constant real term. The imaginary term approaches infinity at both extremes and becomes zero at f_r.

In Figs. 10.33 and 10.34, the real part of impedance is constant with frequency. This is not the case for the series elements of Fig. 10.35A. The mobility of this system is

$$\mathfrak{M} = \frac{1}{c} + j\left(\frac{2\pi f}{k} - \frac{1}{2\pi fm}\right)$$

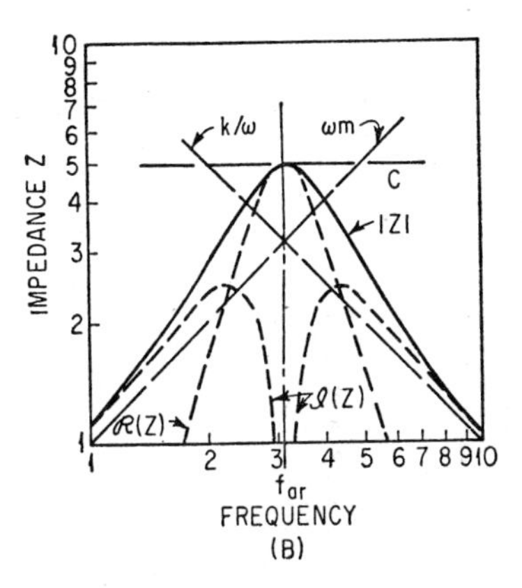

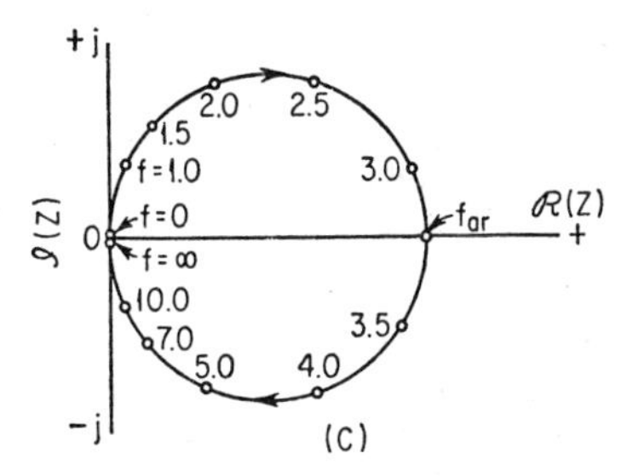

FIG. 10.35. Impedance characteristics of a series resistance-spring-mass system.

The impedance is the reciprocal of the mobility:

$$Z = \frac{1}{\mathfrak{M}} = \frac{1/c}{1/c^2 + (2\pi f/k - 1/2\pi fm)^2} - j\frac{(2\pi f/k - 1/2\pi fm)}{1/c^2 + (2\pi f/k - 1/2\pi fm)^2}$$

The impedances of the individual elements are plotted to logarithmic coordinates in Fig. 10.35B, together with the magnitude of the impedance $|Z|$ and its real and imaginary parts. The impedance $|Z|$ reaches a maximum at the *antiresonant frequency*

$$f_{ar} = \frac{\sqrt{k/m}}{2\pi}$$

When the system is driven by a sinusoidal excitation of constant-force amplitude, the velocity amplitude at the driving point is at a minimum at the antiresonant frequency.

The complex plane representation is shown in Fig. 10.35*C*. At low frequencies the impedance is small, the imaginary term is positive, and the angle of the impedance is nearly +90°. Masslike action is indicated. At high frequencies, the impedance is small but the imaginary term is negative and the angle is −90°. The system is then "stiffness controlled." At intermediate frequencies the locus of the impedance graph is a circle having a diameter *c* with its center on the + real axis. At f_{ar} the impedance is real and equal to *c*.

Note that the mobility of the series-connected elements of Fig. 10.35*A* is given by an expression of the same form as that for the impedance of the parallel-connected elements of Fig. 10.34*A*. Thus, when mobility is plotted in the same way as impedance, the shapes of the resulting curves for the series elements (Fig. 10.36) are the same as for the impedance of the parallel elements shown in Fig. 10.34.

The mobility of the parallel elements of Fig. 10.34*A* takes the same form as the impedance of the series elements of Fig. 10.35*A*. Conversely, the mobility curves for the

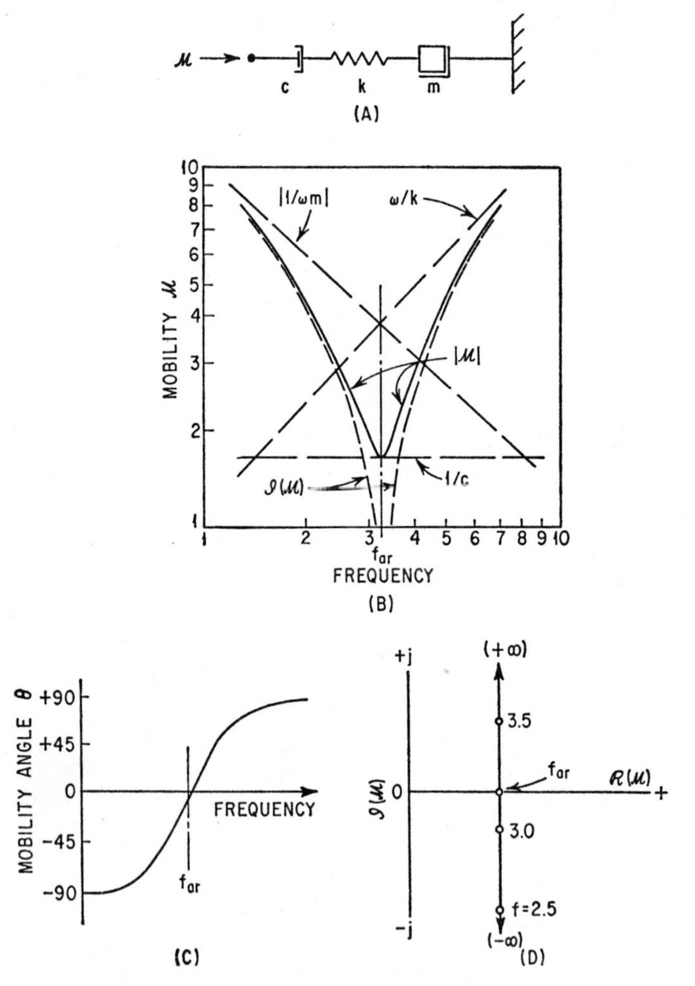

FIG. 10.36. Mobility characteristics of a series resistance-spring-mass system.

parallel-connected elements have the same form as the impedance curves for the series-connected elements of Fig. 10.35.

Complicated mechanical structures may have many resonances at different frequencies. If the impedance is measured and plotted on log-log paper, results similar to those presented in Figs. 10.34 and 10.36 are useful in analyzing the system qualitatively. Positive peaks on the impedance curve indicate the resonance of series elements, and negative peaks indicate the resonance of parallel elements. Frequency regions where the impedance curve has a constant positive slope of +1 indicate masslike action with a positive impedance angle. Similarly, regions of the curve with a slope of −1 indicate springlike action with a negative impedance angle.

On the complex plane, each crossing of the real axis indicates a resonance since the reactance $\mathscr{I}(Z)$ is zero at that frequency. In general, circles such as Fig. 10.35*C* on the impedance curve indicate series element resonances even if they do not cross the real axis. Fig. 10.37*A* shows a system in which there are parallel resonant elements but one parallel branch has a series resonance at a frequency at which the imaginary part of Z is masslike. The arrows in Fig. 10.37*B* indicate the direction the locus takes as frequency is increased. The parallel resonance is indicated by the crossing of the real axis, and the circle indicates the resonance of the series branch.

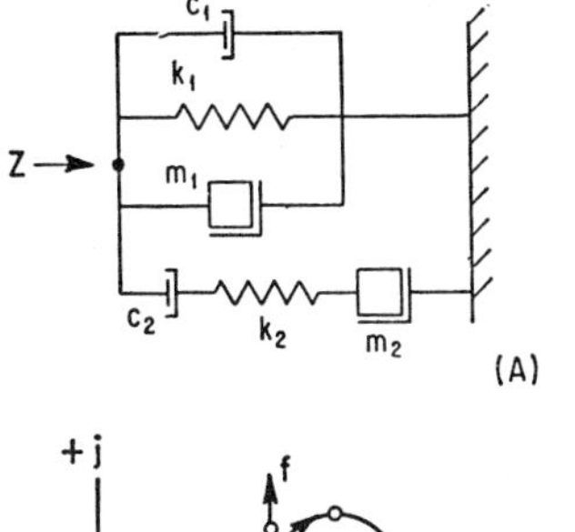

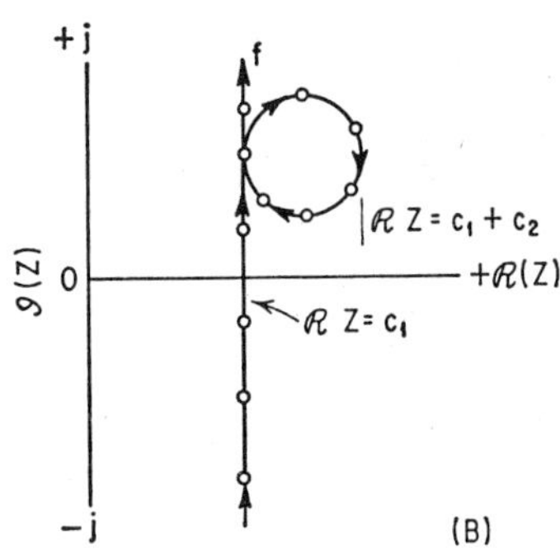

Fig. 10.37. Impedance characteristics of a system having series and parallel resonances.

The characteristics of the mobility plotted on the complex plane take on a form that is essentially opposite to the form of the impedance plot. Since mobility and impedance are related by the transformation $Z = 1/\mathfrak{M}$, straight lines on the Z plot transform to circles on the $\mathfrak{M}$ plot, and vice versa. Thus, real axis crossings on the $\mathfrak{M}$ plot indicate series resonances and circles indicate parallel resonances.

The graphical representation of impedance and mobility is useful in predicting the response of physical systems. If a vibration generator with a constant velocity with respect to frequency is applied to a system at the point at which the impedance is known, the force required to produce this velocity is directly proportional to the impedance. Thus, the impedance plot is a force plot when multiplied by velocity. The mobility plot serves a similar purpose in becoming a velocity plot when a constant force is applied to the system.

The graphical representation of measured impedance data is useful in the study of complicated systems. Dips in the curves indicate parallel resonances. At these frequencies large motions can be obtained with small forces. Peaks in the impedance curves indicate antiresonances. At these frequencies large forces are required to produce significant motion. In some cases the curve shape as determined by measurement may be approximated by the impedance curve of a simple two- or three-element equivalent system. Such an equivalent system of ideal elements can represent a complicated system quite accurately if the frequency range is small. Once the equivalent system is determined, an analytical expression for impedance can be obtained from known theory and the response predicted when the original system is combined with another system of known impedance.

MEASUREMENT OF MECHANICAL IMPEDANCE

In the measurement of mechanical impedance a sinusoidal force is applied to the point at which the impedance is desired. Then a measurement is made of the applied force and resulting motion at the point of application. From this information, the value of the impedance is calculated.[17] Every point of attachment in a system has six possible directions of motion with respect to any set of reference coordinates: three translational and three rotational. A measured impedance in each of these directions is necessary for a complete description of a system. The process is complex and time-consuming and the technology for rotational impedance measurements is undeveloped, so that such a complete description is seldom attempted. The description here is therefore limited to measurement in one dimension. In many cases this is adequate and provides a powerful tool for dynamic analysis and design. Experience is a great aid in making accurate mechanical impedance measurements. Good experience may be obtained by making measurements on simple structures having known characteristics.

FORCE OR MOTION GENERATION

Since impedance is defined in terms of sinusoidal variables, the force or motion applied should have, as nearly as possible, a perfectly sinusoidal waveshape. Its frequency should be easily controllable over the range of interest and its amplitude conveniently variable. The amplitude of force and motion should be kept within the linear range. The most widely used means of excitation is an electromagnetic shaker driven by an electronic oscillator and power amplifier. In general, large forces are not required;[18] a peak force of 10 lb is usually sufficient. The mechanical excitor should be mounted in such a way that the application of the exciting force or motion is restricted to the point at which the impedance is desired.

MEASUREMENT OF FORCE AND MOTION

Force transducers and accelerometers are available commercially for use in impedance measurements.* Usually they are employed in an arrangement such as that shown in

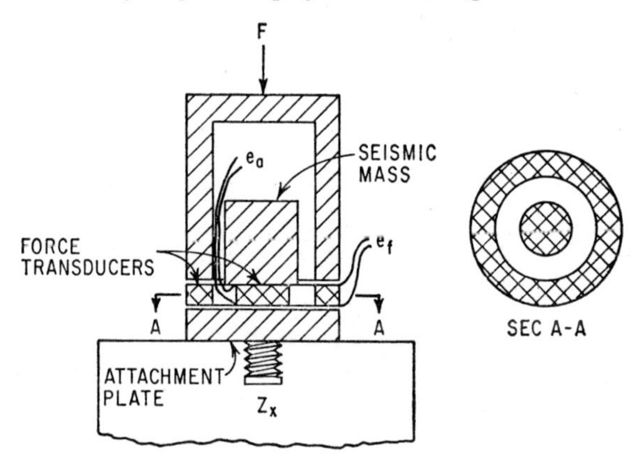

FIG. 10.38. Device for the measurement of mechanical impedance in which force and acceleration are measured. (*Plunkett.*[19])

Fig. 10.38. Here the force is applied through a force transducer and the resulting motion of the point to which it is attached is measured by an accelerometer consisting of a force transducer and seismic mass. The electrical outputs of the force transducer and the accelerometer may be filtered electrically by very narrow bandpass filters, centered at the excitation frequency, to ensure that only the signals at the driving frequency are measured and to reduce interfering noise. One disadvantage of the arrangement of Fig. 10.38 is that the force required to move the acclerometer and the attachment plate is also measured by the force transducer. Thus, the measured force may be in error.

* For example, from B & K Instruments, Inc., Cleveland, Ohio 44142; Wilcoxon Research, Bethesda, Md. 20014.

However, such an error can be compensated to some extent, as indicated in the next section.

IMPEDANCE CALCULATIONS

Mechanical impedance is defined as the ratio of force to velocity [see Eq. (10.7)]. Since the impedance transducer of Fig. 10.38 measures force and acceleration, the velocity must be determined from the acceleration measurement. For sinusoidal motion, the velocity v is related to acceleration a as follows:

$$v = \frac{a}{j\omega}$$

so that Eq. (10.7) becomes

$$Z = \frac{j\omega F}{a} \tag{10.60}$$

For the device of Fig. 10.38, the voltage produced by the force transducer is

$$e_f = K_f F$$

where K_f is the force-transducer calibration constant in volts per unit force.* The voltage produced by the acceleration transducer is

$$e_a = K_a a$$

where K_a is the accelerometer calibration constant in volts per unit acceleration. The impedance then is obtained from Eq. (10.60) by substituting the values of force and acceleration obtained from these voltage measurements, taking phase angle into account; the transducer constants must be known.

If the accelerometer and attachment plate in Fig. 10.38 had no mass, the unknown impedance Z_x would be given by

$$Z_x = \frac{\omega |e_f|/K_f}{|e_a|/K_a} \exp j\left(\theta + \frac{\pi}{2}\right) \tag{10.61}$$

Here θ is the phase angle between the voltage produced by the force and acceleration transducers; the angle $\pi/2$ accounts for the 90° phase shift between acceleration and velocity.

Suppose the mass of the accelerometer and attachments in Fig. 10.38 is represented by m_0. This additional mass may then be taken into account in Eq. (10.61) by subtracting the impedance of the additional mass. Then the unknown impedance becomes

$$Z_x = \omega \frac{|e_f|/K_f}{|e_a|/K_a} \exp j\left(\theta + \frac{\pi}{2}\right) - j\omega m_0 \tag{10.62}$$

The correction term, $j\omega m_0$, in Eq. (10.62) can be accounted for electronically by subtracting a voltage equal to $m_0(K_f/K_a)e_a$ from the voltage produced by the force transducer; this subtraction is known as *mass cancellation.*† Then a new force voltage is obtained:

$$e_f' = e_f - m_0 \frac{K_f}{K_a} e_a \tag{10.63}$$

* When K_f is expressed as volts/newton, and K_a is volts/meter/sec^2 the calculated impedance is in units of newton-sec/meter. For K_f in volts/lb and K_a in volts/in./sec^2, the units of Z are lb-sec/in.

† Mass cancellation is performed electrically by subtracting a variable fraction of the voltage produced by the acceleration transducer from the voltage produced by the force transducer *without* the unknown impedance attached; this fraction is adjusted until e_f' is zero. An exact null is generally not possible over a large range of frequencies because of differential phase shifts in the two electrical circuits and because the accelerometer plus its attachments do not behave as an ideal mass.

When this corrected value of force voltage is substituted in Eq. (10.62), the corrected value of Z_x (taking the effects of m_0 into account) is

$$Z_x = \frac{\omega|e_f'|/K_f}{|e_a|/K_a} \exp j\left(\theta + \frac{\pi}{2}\right) \tag{10.64}$$

CALIBRATION

In order to determine the unknown impedance by means of Eq. (10.64), the transducer calibration constants K_f and K_a must be known. These constants can be determined from measurements made on a known impedance, such as on a spherical or cubical block of stiff, dense material, (steel or tungsten is nearly ideal). The mass m_s of such a block is given by W/g, where W is the weight of the block of material and g is the gravitational constant. Well below resonance of the mass, its impedance is given by $j\omega m_s$. Then Eq. (10.64) becomes

$$j\omega m_s = j\omega \frac{e_{f0}K_a}{e_{a0}K_f}$$

so that

$$\frac{K_a}{K_f} = m_s \frac{e_{a0}}{e_{f0}}$$

where e_{f0} and e_{a0} are the measured low-frequency values of e_f and e_a with the calibrated mass attached.

MEASUREMENTS SYSTEMS

Complete systems are commercially available which provide the automatic computation and plotting of mechanical impedance. (Such systems generally provide a sinusoidal excitation voltage which can be swept over the frequency range of interest.) Firms which manufacture such systems include: Bafco, Inc., Warminster, Pa. 18971; B & K Instruments, Inc., Cleveland, Ohio 44142; Hewlett-Packard, Inc., Palo Alto, Calif. 94306; Nicolet Scientific Corp., Northvale, N.J. 07647; Spectral Dynamics Corp., San Diego, Calif. 92112.

REFERENCES

1. Thomson, W. T.: "Mechanical Vibrations," p. 70, Prentice-Hall, Inc., Englewood Cliffs, N.J., 1948.
2. Rocard, Y.: "Dynamique générale des vibrations," p. 62, Masson et Cie, Paris, 1949.
3. McLachlan, N. W.: "Theory of Vibrations," p. 27, Dover Publications, New York, 1941.
4. Firestone, F. A.: *J. Appl. Phys.*, **9**:373 (1938).
5. Freeberg, C. R., and E. N. Kemler: "Elements of Mechanical Vibrations," p. 179, John Wiley & Sons, Inc., New York, 1949.
6. Firestone, F. A.: *J. Acoust. Soc. Amer.*, **28**:1117 (1956).
7. Biot, M. A.: *J. Aeronaut. Sci.*, **7**:376 (1940).
8. Von Kármán, T., and M. A. Biot: "Mathematical Methods in Engineering," McGraw-Hill Book Company, Inc., New York, 1940.
9. Duncan, W. J.: *British Aeronautical Research Committee R & M* 2000, 1947.
10. Bishop, R. E. D.: *J. Roy. Aeronaut. Soc.*, **58**:703 (1954).
11. Harrison, M., A. O. Sykes, and P. G. Marcotti: *J. Acoust. Soc. Amer.*, **24**:384 (1952).
12. Bancroft, D.: *Phys. Revs.*, **59**:588 (1941).
13. Sykes, A. O.: "The Effects of Machine and Foundation Resilience and of Wave Propagation on the Isolation Provided by Vibration Mounts," SAE National Aeronautic Meeting, Los Angeles, Calif., Oct. 2, 1957.
14. Wright, D. V.: *Colloq. Mechanical Impedance Methods Mechanical Vibrations, ASME*, Dec. 2, 1958, pp. 19–42.
15. Chenea, P. F.: *J. Appl. Mechanics*, **20**:233 (1953).
16. Molloy, C. T.: *J. Acoust. Soc. Amer.*, **29**:842 (1957).
17. "Nomenclature and Symbols for Specifying the Mechanical Impedance of Structures," American National Standards Institute, S2.6–1963 (R 1971):
18. Muster, D. F.: *Colloq. Mechanical Impedance Methods Mechanical Vibrations, ASME*, Dec. 2, 1958, p. 87.
19. Plunkett, R.: ASME Paper 53-A-45, *J. Appl. Mechanics*, **21**:250 (1954).

11

STATISTICAL CONCEPTS IN VIBRATION

John W. Miles
University of California, San Diego

W. T. Thomson
University of California, Santa Barbara

INTRODUCTION

A random vibration is one whose instantaneous value is not predictable. Such vibration is generated, for example, by the rocket engines of guided missiles and by aerodynamic turbulence and gusts. Figure 11.1*A* is an acceleration-time record of a vibration measured in the vicinity of a rocket engine. Clearly there are no well-defined periodicities, and the value of the acceleration at any instant of time is unrelated to that at any other instant and cannot be predicted.

Although the time variation of a random vibration cannot be predicted, the probability of a given value being within a certain range is predictable on a statistical basis. For example, consider a short sample of the above record. The acceleration $\ddot{x}$, at any instant of time t, varies in a random manner about a mean value $\bar{\ddot{x}}$. Suppose the vertical scale is divided into small divisions $\Delta\ddot{x}$. Now determine the statistical probability that the value will be within $\Delta\ddot{x}$. This can be done by measuring the time Δt_n during which the signal has amplitude values between $\ddot{x}$ and $\ddot{x} + \Delta\ddot{x}$ relative to the total length of time T over which the phenomenon is studied. Thus *amplitude probability* P in this example is given by

$$P(\ddot{x}, \ddot{x} + \Delta\ddot{x}) = \frac{\sum_i \Delta t_n}{T}$$

and the *amplitude density* p is given by

$$p(\ddot{x}) = \lim_{\Delta\ddot{x}\to 0} \frac{P(\ddot{x}, \ddot{x} + \Delta\ddot{x})}{\Delta\ddot{x}}$$

The probability of the instantaneous value being infinite is zero; the probability of its peak value being zero or infinite is zero; and the probability of any other value may be distributed as indicated in Fig. 11.1*B*.

This chapter deals with the problem of a linear, vibrating system subjected to a random excitation, including (1) a statistical description of the random input (*excitation*), (2) a description of the linear system in terms that permit the calculation of its response over the frequency spectrum of the input, and (3) a statistical description of the random output (*response*). Mathematical concepts are discussed first, followed by illustrative applications to mechanical systems characterized by one or more resonances and applica-

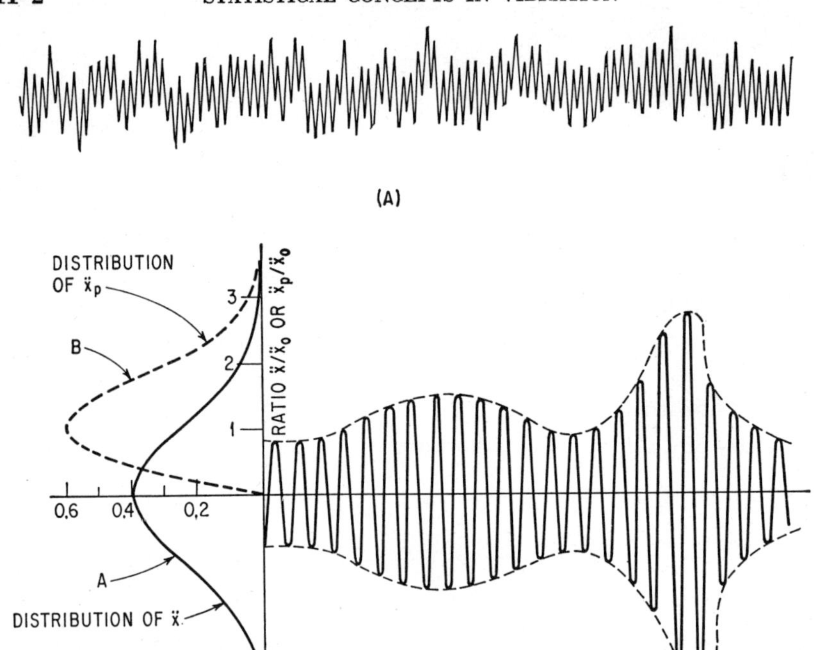

FIG. 11.1. Examples of random vibration. (*A*) Acceleration vs. time record of vibration measured in the vicinity of a rocket engine. (*B*) Example of narrow-band random vibration; the curves along the vertical axis give the probability distribution of the instantaneous (dashed curve) and peak (solid curve) values.

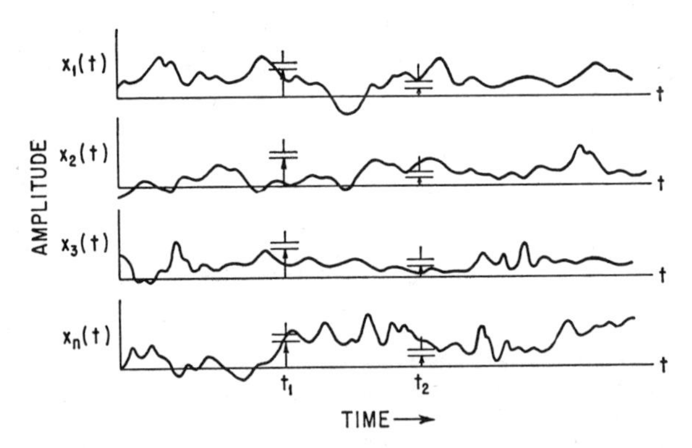

FIG. 11.2. Assembly of random time functions representing a random phenomenon $x(t)$.

tions to the problem of fatigue. A study of the statistical properties of the spectra of a vibrating system often is useful in leading to an understanding of the physical mechanism producing the vibration. Furthermore, it provides information which is useful in the simulation of vibration and useful for design purposes.

STATISTICAL CONCEPTS

FIRST PROBABILITY DISTRIBUTION

The statistical properties of a random time function, such as the acceleration-time record shown in Fig. 11.1*A*, are related to its so-called "probability distribution." For the determination of the statistical properties of a random function, it is generally necessary to consider a large assembly of random time records, as shown in Fig. 11.2. In this illustration, x is a function that may represent displacement, velocity, or acceleration. Let n represent the number of the members of this family of functions (called an *assembly* or *ensemble*). This chapter is restricted to a consideration of "stationary random processes."

A random process (an ensemble of time functions) is said to be *stationary* if any translation of the time origin leaves its statistical properties unaffected. The *ergodic hypothesis* states that assembly averaging in stationary random processes can be replaced by time averaging over a single record of very long duration. Time averaging of random processes greatly simplifies the task of data analysis.

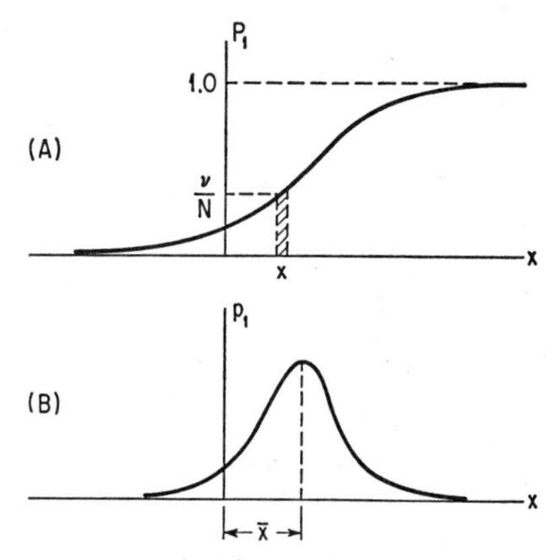

FIG. 11.3. (*A*) First probability distribution of the assembly of time functions shown in Fig. 11.2 at time *t*. (*B*) The probability density function corresponding to (*A*).

At any time t_1, the value of $x(t_1)$ will differ in the various records; however, it is possible to count the fraction ν/n of the total number of records for which $x(t_1)$ will be less than some specified value x and plot it as shown in Fig. 11.3*A*. Such a curve is known as the *first probability distribution*, $P_1(x,t_1)$, at time t_1, and it will tend to a stationary value as $n \to \infty$. The first probability distribution $P_1(x,t_1)$ is a nondecreasing function of x with the following bounds (see Fig. 11.3):

The fraction having a value less than ∞ at time t_1 is 1, i.e., the probability distribution is given by

$$P_1(\infty,t_1) = 1$$

The fraction having a value less than $-\infty$ at time t_1 is 0:

$$P_1(-\infty,t_1) = 0$$

The fraction having a value less than some intermediate value x at time t_1 is given by

$$0 \leq P_1(x,t_1) \leq 1$$

If the derivative of $P_1(x,t_1)$ exists, the density of the *first probability distribution* $p_1(x,t_1)$ is given by

$$p_1(x,t_1) = \frac{d}{dx} P_1(x,t_1) \tag{11.1}$$

as shown in Fig. 11.3*B*.

The probability of $x(t_1)$ having a value between x_1 and $x_1 + dx$, as shown in Fig. 11.3, is

$$dP_1 = P_1(x_1 + dx,\, t_1) - P_1(x_1,t_1) = p_1(x_1,t_1)\, dx \tag{11.2}$$

It follows that

$$P_1(x,t_1) = \int_{-\infty}^{x} p_1(x,t_1)\,dx \tag{11.3}$$

$$P_1(\infty,t_1) = \int_{-\infty}^{\infty} p_1(\xi,t_1)\,d\xi = 1 \tag{11.4}$$

SECOND PROBABILITY DISTRIBUTION

The *second probability distribution* P_2 determines the likelihood of pairs of values occurring a specified time interval apart. Suppose this time interval to be τ sec, so that $t_2 = t_1 + \tau$. Then a count of the number of records in which $x(t)$ lies between x_1 and $(x_1 + dx_1)$ at t_1 and between x_2 and $(x_2 + dx_2)$ at t_2, as shown in Fig. 11.2, defines the probability of a pair of values in any record being in the specified ranges at the specified times as

$$dP_2 = p_2(x_1,t_1,\, x_2,t_2)\,dx_1\,dx_2 \tag{11.5}$$

In a similar manner, higher-order probability functions can be determined.

AVERAGE VALUE

Given the first probability density $p_1(x)$, the average value $\bar{x}$ of the random function $x(t)$ can be determined from

$$\bar{x} = \int_{-\infty}^{\infty} x p_1(x,t_1)\,dx \tag{11.6}$$

The nth moment $\overline{x^n}$ is given by

$$\overline{x^n} = \int_{-\infty}^{\infty} x^n p_1(x,t_1)\,dx \tag{11.7}$$

For stationary random processes, Eqs. (11.6) and (11.7) reduce to the following time averages by letting $n = 2$ in Eq. (11.7):

Average value:

$$\bar{x} = \lim_{T\to\infty} \frac{1}{2T} \int_{-T}^{T} x(t)\,dt \tag{11.8}$$

Mean-square value:

$$\overline{x^2} = \lim_{T\to\infty} \frac{1}{2T} \int_{-T}^{T} x^2(t)\,dt \tag{11.9}$$

GAUSSIAN (NORMAL) DISTRIBUTION

The probability distribution that is most frequently encountered (at least as an acceptable approximation) in random vibration problems and the only such distribution that is amenable to extensive analysis is the *Gaussian or normal distribution.* Suppose the amplitude-density curve is normalized:

$$\int_{-\infty}^{\infty} p(x)\,dx = 1$$

where $p(x)$ is the amplitude density. Then for a Gaussian distribution the amplitude density at time t_1 is given by

$$p_1(x,t_1) = \frac{1}{\sigma\sqrt{2\pi}}\, e^{-\frac{(x-\bar{x})^2}{2\sigma^2}} \tag{11.10}$$

where $\bar{x}$ is the mean value of x and σ, the *standard deviation* (i.e., the rms value), is defined by

$$\sigma^2 = \int_{-\infty}^{\infty} (x - \bar{x})^2 p_1(x,t_1)\, dx = \overline{x^2} - (\bar{x})^2 \tag{11.11}$$

The quantity σ^2 is a measure of the dispersion of the random function and is called the *variance.*

The Gaussian distribution defined by Eq. (11.10) is plotted in Fig. 11.4. It is the only distribution (for instantaneous values) considered in this chapter. The assumption that a distribution is Gaussian automatically excludes such properties as skewness (asymmetry of the distribution). Conversely, the existence of some degree of skewness in the plot of probability density vs. $(x - \bar{x})$ is a priori evidence of departure from a Gaussian distribution (see Figs. 11.5 and 22.9).

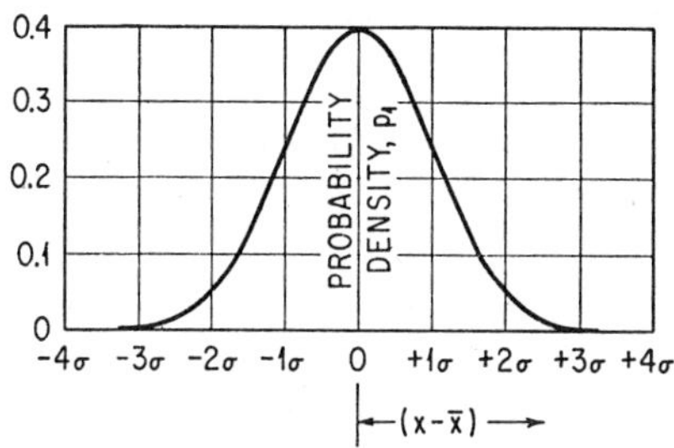

FIG. 11.4. Gaussian (also called normal) distribution. This represents the probability that the magnitude of the instantaneous value of x exceeds its rms value. For example, the probability of x exceeding 2σ is 4.6 per cent.

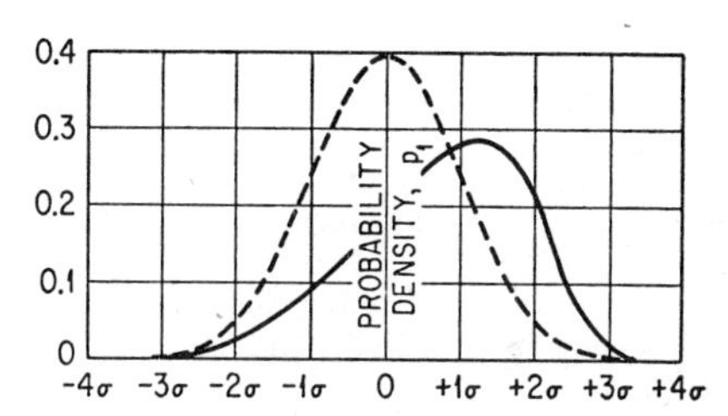

FIG. 11.5. Example of skewed distribution compared with normal distribution.

POWER SPECTRAL DENSITY

Power, which is the rate of doing work, is proportional to the square of the amplitude of a harmonic vibration. If two frequencies are present in a vibration, the power is proportional to the sum of the squares of the individual amplitudes associated with the two frequencies.

A random vibration can be considered to be a sum of a large number (tending to infinity) of harmonic vibrations of appropriate amplitudes and phase. The total power is again the sum of the power of the component harmonic vibrations, and it is of interest to know how this power is distributed as a function of frequency. Therefore *power spectral density* is defined as the power per unit frequency interval. A plot of this quantity indicates the frequency distribution of power [see Eq. (22.17)].

DETERMINATION OF POWER SPECTRAL DENSITY

The analysis of random vibration always involves a record of finite length. Assume that such a record has been recorded on magnetic tape and is to be analyzed by an electronic analyzer as a stationary process; the tape can be formed into a loop of arbitrary length and made to repeat indefinitely on the analyzer. By such a procedure, an arbitrary fundamental period corresponding to the length of the loop is established, and the contents of the record may be defined in terms of the Fourier coefficients of integral multiple harmonics of the loop frequency ω_0.

The function $x(t)$ then can be represented by the real part of the Fourier series,

$$x(t) = \mathcal{R} \sum_{n=-\infty}^{\infty} C_n e^{jn\omega_0 t} = C_0 + 2 \sum_{n=1}^{\infty} |C_n| \cos (n\omega_0 t - \alpha_n) \tag{11.12}$$

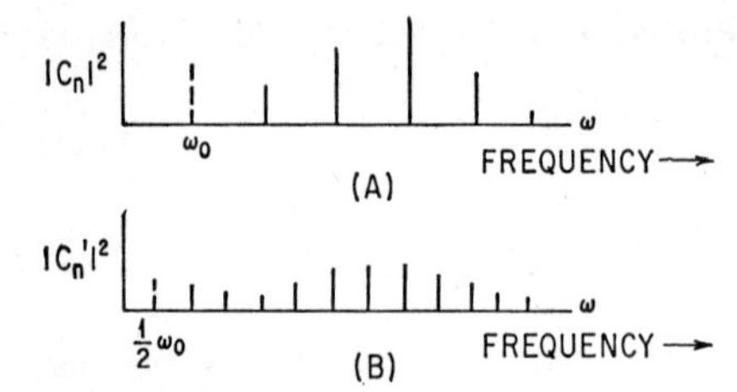

FIG. 11.6. Spectrum analysis of random vibration, as determined from a magnetic-tape loop. In (*A*) the period is T. In (*B*) the period is $2T$. Note that doubling loop period doubles number of spectral lines and reduces $|C_n|^2$ to one-half the former value.

where the complex amplitude C_n is defined by the equation

$$C_n = \frac{1}{2T}\int_{-T}^{T} x(\xi)e^{-jn\omega_0\xi}\,d\xi \quad (11.13)$$

and $2T$ is the loop period.

The mean-square value $\overline{x^2}$ is a stationary property of $x(t)$ that is determined by Eq. (11.9) as

$$\overline{x^2} = \frac{1}{2T}\int_{-T}^{T} x^2(t)\,dt = C_0{}^2 + 2\sum_{n=1}^{\infty}|C_n|^2 \quad (11.14)$$

This equation indicates that the contribution to the mean-square value in any frequency interval is the summation of all the components within that frequency band.

If now the length of the loop and the loop period are doubled, the number of spectral lines will be doubled. Since the mean-square value of the entire spectrum must remain essentially constant, the values of the new coefficients $|C_n'|^2$ must decrease to approximately one-half their former values, as illustrated in Fig. 11.6. The sum of the lengths of the $|C_n|^2$ lines, up to any frequency, is the same in each diagram of Fig. 11.6, and the increment in $\overline{x^2}$ divided by the increment in frequency is essentially the same in each case,

$$\frac{\Delta\overline{x^2}}{\Delta\omega} = \frac{2|C_n|^2}{\omega_0} = \frac{2|C_n'|^2}{\frac{1}{2}\omega_0} = W(\omega_n) \quad (11.15)$$

The quantity $W(\omega_n)$ is the *power spectral density* at $\omega = \omega_n$. The dimensions are the square of the parameter represented by $\Delta\overline{x^2}$ in Eq. (11.15) per unit of frequency.

As the length of the loop is increased, more and more spectral lines are introduced; the magnitude of $|C_n|^2$ decreases correspondingly, but $W(\omega_n)$ remains finite. In the limiting case, as $T\to\infty$ or $\Delta\omega\to 0$, the spectrum becomes continuous, and the discrete values of $W(\omega_n)$ approach a smooth function $W(\omega)$:

$$W(\omega) = \lim_{\Delta\omega\to 0}\frac{\Delta\overline{x^2}}{\Delta\omega} = \frac{d\overline{x^2}}{d\omega} \quad (11.16)$$

The mean-square value can be expressed by the integral

$$\overline{x^2} = \int_0^{\infty} W(\omega)\,d\omega \quad (11.17)$$

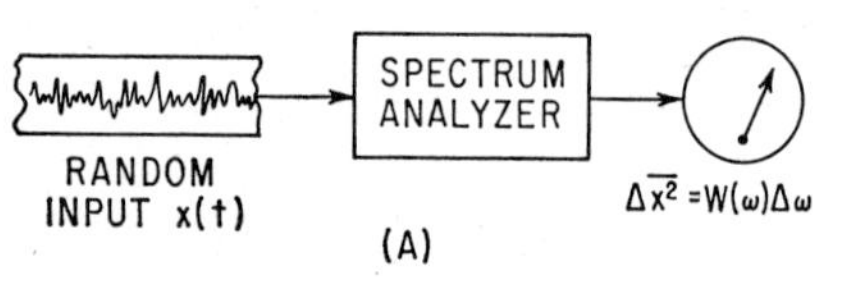

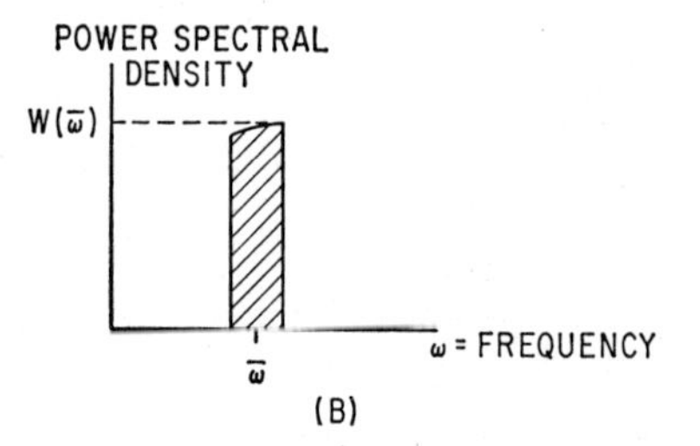

FIG. 11.7. The experimental determination of the power spectral density $W(\omega)$ of a random function $x(t)$. The function $x(t)$ is recorded on a magnetic tape and fed into a spectrum analyzer that transmits only those frequencies within the passband $\bar{\omega} \pm \frac{1}{2}\Delta\omega$. The output in the passband, namely, $\Delta\overline{x^2}$, is recorded on a mean-square meter, and the mean power spectral density over the passband is determined by dividing the meter reading by $\Delta\omega$. Thus, as shown in (*B*), the meter reading, $\Delta\overline{x^2}$, appears in the power spectral density vs. frequency plane as an increment of area of width $\Delta\omega$ and mean ordinate $f(\bar{\omega})$.

The power spectral density of a random time function may be determined by (see Fig. 11.7) feeding the function $x(t)$ into a spectrum analyzer that transmits only those frequency components within the passband of the analyzer, $\bar{\omega} \pm \frac{1}{2}\Delta\omega$. The output in the passband, namely, $\Delta\overline{x^2}$ is then indicated by a "mean-square" meter. The mean power spectral density $W(\bar{\omega})$ over the passband, at $\bar{\omega}$, is determined by dividing the meter reading by $\Delta\omega$. This division may be incorporated directly in the calibration of the meter.

The complete distribution of $W(\omega)$ then is determined by changing the frequency $\bar{\omega}$ in increments of $\Delta\omega$.

TYPICAL POWER SPECTRAL DENSITY

Typical power spectral densities may be characterized *broadly* by two parameters: the mean-square value, given by Eq. (11.17), and some characteristic frequency—say ω_1. For example, the *white noise* * distribution with a cutoff frequency ω_1 is described by the equation

$$W(\omega) = \begin{cases} \overline{x^2}/\omega_1 & [\omega < \omega_1] \\ 0 & [\omega > \omega_1] \end{cases} \tag{11.18}$$

If $W(\omega)$ is (at least approximately) a monotonically decreasing function of ω, it is convenient to choose $\omega_1 = \omega_{1/2}$, where $\omega_{1/2}$ defines (at least roughly) the *half-power point* according to

$$\int_0^{\omega_{1/2}} W(\omega)\, d\omega \doteq \int_{\omega_{1/2}}^{\infty} W(\omega)\, d\omega \tag{11.19}$$

If $W(\omega)$ exhibits a definite peak at some frequency appreciably greater than zero, say ω_m, it is convenient to choose $\omega_1 = \omega_m$.

Examples of these two possible types of power spectral density are given by

Monotonic:

$$W_A(\omega) = \left(\frac{2\overline{x^2}}{\pi^{1/2}\omega_1}\right) e^{-(\omega/\omega_1)^2} \tag{11.20}$$

Peaked:

$$W_B(\omega) = \left(\frac{4\overline{x^2}}{\pi^{1/2}\omega_1}\right) \left(\frac{\omega}{\omega_1}\right)^2 e^{-(\omega/\omega_1)^2} \tag{11.21}$$

and are plotted in Fig. 11.8. The actual half-power point for $W_A(\omega)$ occurs at $\omega = 0.48\omega_1$, while the peak of $W_B(\omega)$ occurs at $\omega = \omega_1$. It should be emphasized that these examples are hypothetical; although they are representative of the more important types, most spectra are so complicated that they can be defined only graphically.

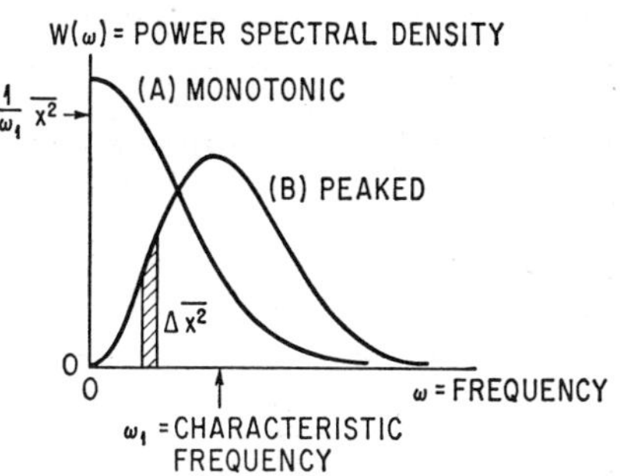

FIG. 11.8. Typical power spectral densities: (*A*) as given by Eq. (11.20) and (*B*) by Eq. (11.21). Note that $\overline{x^2} = \int_0^\infty W(\omega)\, d\omega$.

Whether or not either of the (closely related) methods discussed in this section is the most expedient for the actual determination of the power spectral density from a given random function depends on the equipment available, the accuracy, and the ultimate application of the data. Another procedure for obtaining $W(\omega)$ is based on the autocorrelation of $x(t)$, making use of Eq. (11.60).

INPUT-OUTPUT RELATIONSHIP

The relationship between the input and output of a linear system for any time function $x(t)$, whether random or not, may be expressed in terms of the integral

$$y(t) = \int_{-\infty}^{t} x(\tau)h(t-\tau)\, d\tau = \int_0^\infty x(t-\xi)h(\xi)\, d\xi \tag{11.22}$$

where $h(t)$ is the response of the system to a unit impulse $\delta(t)$ and $x(t)$ is the input, which may have existed from $t = -\infty$. In the usual form of the above integral, the input is

* The power spectral density of true white noise is a constant for all ω.

assumed to be zero for $t < 0$, in which case the range of integration becomes 0 to t in both integrals.

The spectral aspects of the system are defined by a response function which is directly significant only for harmonic excitation. Letting $x(t) = e^{j\omega t}$ represent such a harmonic excitation, and noting that there is no provision for starting or stopping such a function, its substitution into Eq. (11.22) results in the following:

$$y(t) = \int_0^\infty e^{j\omega(t-\tau)}h(\tau)\,d\tau = e^{j\omega t}\int_0^\infty e^{-j\omega\tau}h(\tau)\,d\tau$$

$$= A(j\omega)e^{j\omega t} \tag{11.23}$$

where

$$A(j\omega) = \int_0^\infty e^{-j\omega\tau}h(\tau)\,d\tau \tag{11.24}$$

is defined as the response function of the system. Thus the response to a harmonic excitation is the excitation multiplied by the response function $A(j\omega)$ which then serves to define the spectral characteristics of the system. However, a sinusoidal input is given by either the real or the imaginary part of $e^{j\omega t}$, and either the real or the imaginary part of the response $y(t)$ represents the solution to such an excitation. Since $h(\tau)$ is zero for $\tau < 0$, the lower limit of Eq. (11.24) can be extended to $-\infty$ without altering the integral. Thus $A(j\omega)$ is the Fourier transform of the impulsive response function $h(\tau)$.

Suppose that a random time function (over a finite time) is expressible in terms of a Fourier series. Determine the spectral form of the output after the function has passed through a linear system having the response function $A(j\omega)$. Let Eq. (11.12) represent the input, and substitute into the integral of Eq. (11.22) for the response:

$$y(t) = \int_0^\infty h(\tau) \sum_{n=-\infty}^{\infty} C_n e^{jn\omega_0(t-\tau)}\,d\tau$$

$$= \sum_{n=-\infty}^{\infty} C_n e^{jn\omega_0 t}\int_0^\infty h(\tau)e^{-jn\omega_0\tau}\,d\tau$$

$$= \sum_{n=-\infty}^{\infty} C_n e^{jn\omega_0 t}A(jn\omega_0) \tag{11.25}$$

Note that Eq. (11.25) is merely a superposition of harmonic components, one of which is given by Eq. (11.23). For a real input, the output must be real; then Eq. (11.25) becomes

$$y(t) = C_0A(0) + 2\sum_{n=1}^{\infty}|C_n|\,|A(jn\omega_0)|\cos(n\omega_0 t + \beta_n) \tag{11.26}$$

From Eq. (11.9), the mean-square value of the response is

$$\overline{y^2} = \frac{1}{2T}\int_{-T}^{T} y^2(t)\,dt = C_0^2A^2(0) + 2\sum_{n=1}^{\infty}|C_n|^2|A(jn\omega_0)|^2 \tag{11.27}$$

The increment of the mean-square value of the response at $\omega = n\omega_0$ is

$$\Delta\overline{y^2} = 2|C_n|^2|A(jn\omega_0)|^2 \tag{11.28}$$

and the power spectral density of the output for the discrete spectrum, with the frequency increment of $\omega_0 = \pi/T$, is

$$W_r(\omega_n) = \frac{2|C_n|^2|A(jn\omega_0)|^2}{\omega_0} \tag{11.29}$$

Since the power spectral density of the input $W_e(\omega_n)$ is given by Eq. (11.15), the input and output spectra are related by the equation

$$W_r(\omega_n) = W_e(\omega_n)\,|A(jn\omega_0)|^2 \tag{11.30}$$

If the period $2T$ is increased without limit, the discrete spectrum becomes continuous; then $n\omega_0 = \omega_n = \omega$, so that Eq. (11.30) may be written as

$$W_r(\omega) = W_e(\omega)\,|A(j\omega)|^2 \tag{11.31}$$

The mean-square output is given by the integral

$$\overline{y^2} = \int_0^\infty W_r(\omega)\,d\omega = \int_0^\infty W_e(\omega)\,|A(j\omega)|^2\,d\omega \tag{11.32}$$

Thus the mean-square output (which is a statistical property of the output) is determined by the integral, over frequency, of the power spectral density of the input as modified by the response function of the system. When the time variation of the input is random and unknown, the time variation of the output is also random and unknown, and the relationship between the output and input is the statistical one given by Eqs. (11.31) and (11.32).

RESPONSE OF A SINGLE DEGREE-OF-FREEDOM SYSTEM

According to Eq. (2.27), the differential equation for a mechanical system of one degree-of-freedom (referred to as a "simple oscillator") is

$$m\ddot{y} + c\dot{y} + ky = F_0 \sin \omega t \tag{11.33}$$

Let

$\omega_n = \sqrt{\dfrac{k}{m}}$ = natural angular frequency of the oscillator

$c_c = 2\sqrt{km}$ = critical damping

$\zeta = \dfrac{c}{c_c}$ = fraction of critical damping

If the excitation is due to the motion $u(t)$ of the support instead of the force $F_0 \sin \omega t$, according to Eq. (2.47), the equation of motion of the mass has the same general form:

$$m\ddot{z} + c\dot{z} + kz = -m\ddot{u} \tag{11.34}$$

where $z = (y - u)$ is the relative displacement.

The response function of the simple oscillator can be determined from its Fourier transform, or more simply from the relation

$$z(t) = \ddot{u}_0 e^{j\omega t} A_1(j\omega) \tag{11.35}$$

which expresses its steady-state response to the harmonic input $\ddot{u}_0 e^{j\omega t}$. The square of the absolute value of the response function for Eqs. (11.33) and (11.34) is

$$|A_1(j\omega)|^2 = \frac{1}{\omega_n{}^4 \left\{ \left[1 - \left(\dfrac{\omega}{\omega_n} \right)^2 \right]^2 + \left(2\zeta \dfrac{\omega}{\omega_n} \right)^2 \right\}} \tag{11.36}$$

The power spectral density of the response then may be obtained from Eq. (11.32). In general this integration can be performed graphically, but it sometimes may be possible to effect analytical approximations, as in the following paragraph.

RESPONSE OF A NARROW-BAND OSCILLATOR

An important case, illustrated in Fig. 11.9, arises when the bandwidth of the oscillator, $\zeta\omega_n$, is small compared with the frequency interval over which the power spectral density is substantial and when $W_e(\omega_n)$, the power spectral density of the input at the resonant frequency of the oscillator, is of the same order of magnitude as $W_e(\omega_1)$. The response

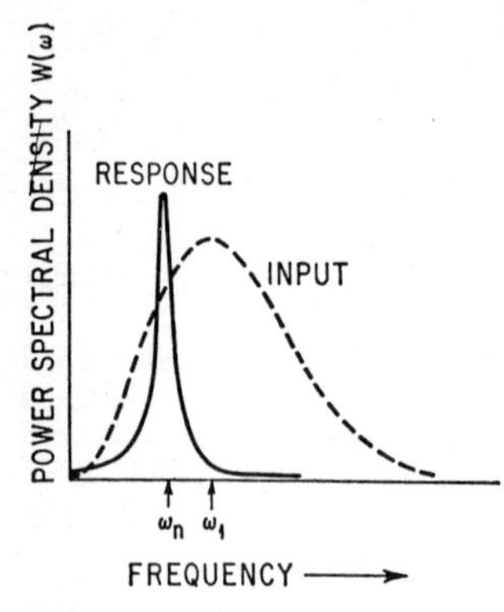

FIG. 11.9. Power spectral density of the output response of a simple oscillator for an input having the power spectral density shown by the dashed curve [see Eq. (11.37)].

of the oscillator then is essentially resonant, and it receives energy only in the neighborhood of its resonant frequency. Then its power spectral density of relative displacement may be approximated by

$$W_r(\omega) = W_e(\omega_n)\,|A_1(j\omega)|^2 \tag{11.37}$$

where $W_r(\omega)$ is the mean-square displacement density of relative motion z; $W_e(\omega)$ is the mean-square acceleration density of the input u; and $A_1(j\omega)$ is the response function defined by Eq. (11.36). The corresponding mean-square response then is given by Eq. (11.32) as

$$\overline{z^2} = W_e(\omega_n)\int_0^\infty |A_1(j\omega)|^2\,d\omega = \frac{\pi}{4\zeta}\frac{W_e(\omega_n)}{\omega_n{}^3} \tag{11.38}$$

where the integral has been evaluated by substituting $|A(j\omega)|^2$ from Eq. (11.36) and assuming $\zeta \ll 1$ *after* carrying out the integration (which is given in integral tables). It may be assumed, without loss of generality, that the mean response $\bar{z}$ is zero (since, if $\bar{z} \neq 0$, it is necessary only to measure z from $\bar{z}$, i.e., replace z in the following analysis by $z - \bar{z}$); it then follows that $\sigma^2 = \overline{z^2}$.

The resonant response of Eq. (11.38) is simple in that, whatever the probability distribution of the input $x(t)$, the probability distribution of the response $z(t)$ tends to the Gaussian distribution of Eq. (11.10) as $\zeta\omega_n/\omega_1$ tends to zero, i.e., as the bandwidth of the oscillator becomes small compared with that of the input.* (Note that this last qualification excludes a periodic input.) The statistical properties of this essentially resonant response then are *completely* characterized by its resonant frequency ω_n and its mean-square value (compare the *broad* characterization of the input by its characteristic frequency ω_1 and mean-square value $\overline{x^2}$).

Note that if a random input is applied to a lightly damped oscillator, it acts as a narrow bandpass filter—passing only those frequencies in the neighborhood $\omega_n \pm \zeta\omega_n$ [cf. Eq. (11.37) and Fig. 11.9]. Because of the phenomenon of *beats* (wherein the addition of two harmonic waves of approximately the same frequency gives rise to a wave oscillating with the mean of the two frequencies and having an amplitude envelope that fluctuates at a rate equal to the difference frequency), the addition of a large number (tending to infinity) of harmonic waves having frequencies in the approximate band $\omega_n \pm \zeta\omega_n$ and randomly distributed phases will give rise to a wave oscillating with frequency ω_n and having an amplitude envelope that exhibits a random fluctuation, the rapidity of which must be of the order of $2\zeta\omega_n$. Such a response (illustrated in Fig. 11.1*B*) is called a *random sine wave*.

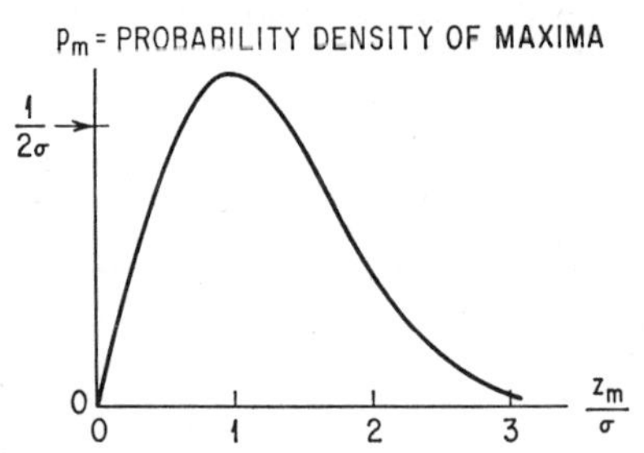

FIG. 11.10. Probability density distribution of an idealized random amplitude vibration of narrow bandwidth (nearly single frequency). This curve, known as a "Rayleigh distribution," gives the probability density of the maxima for essentially resonant loading as defined by Eq. (11.39).

The probability distribution of this envelope (as opposed to the distribution of the instantaneous response) may be determined by synthesizing it from the randomly oriented vectors representing the individual harmonic waves.†

* It is generally argued that this result follows directly from the "central-limit theorem."

† This problem, known as the two-dimensional *random walk* problem, was first solved by Lord Rayleigh.

RAYLEIGH DISTRIBUTION

The probability that the envelope $z_m(t)$ lies between z and $z + dz$ is

$$P\{z < z_m(t) < z + dz\} = p_m\,dz = \sigma^{-2}e^{-\frac{1}{2}(z/\sigma)^2} z\,dz \tag{11.39}$$

where σ, the standard deviation or rms value, is defined by Eq. (11.11). The quantity p_m, known as a *Rayleigh distribution* density, is plotted in Fig. 11.10. Also see Fig. 22.12.

The Rayleigh distribution of Eq. (11.39) describes the variation in amplitudes of the envelope of the random sine wave; the probability distribution of the instantaneous value is given by the Gaussian distribution of Eq. (11.10). Thus, the probability that $|z|$ will exceed a specified value z_1 is determined by

$$P\{|z| > z_1\} = \frac{2}{\sqrt{2\pi}}\int_{z_1}^{\infty} e^{-\frac{1}{2}(z/\sigma)^2} d\left(\frac{z}{\sigma}\right) = erfc\,\frac{z_1}{\sigma\sqrt{2}} \tag{11.40a}$$

$$P\{|z| > z_1\} \simeq \sqrt{\frac{2}{\pi}}\left(\frac{\sigma}{z_1}\right) e^{-\frac{1}{2}(z_1/\sigma)^2}\left[1 - \left(\frac{\sigma}{z_1}\right)^2 + \cdots\right] \tag{11.40b}$$

where *erfc* denotes the complementary error function and Eq. (11.40*b*) gives its asymptotic approximation for $z_1 \gg \sigma$. Equations (11.40*a*) and (11.40*b*) may be used to determine the probability of exceeding the ultimate stress, whereas the Rayleigh distribution P_m is employed in studies of fatigue failure.

RESPONSE OF MULTIPLE DEGREE-OF-FREEDOM AND CONTINUOUS SYSTEMS

Every elastic system has associated with it certain normal-mode properties. Because of the orthogonal character of normal modes, such a system can be treated in terms of simple oscillators.[2]

The displacement response of a linear structure can be expressed in terms of its normal modes $\varphi_n(x,y,z)$: *

$$u(x,y,z,t) = \sum_n q_n(t)\varphi_n(x,y,z) \tag{11.41}$$

Assuming structural damping (damping in phase with the velocity and proportional to the generalized displacement of each mode), the Lagrange equations of the system have the form

$$\ddot{q}_n + (1 + j\gamma_n)\omega_n^2 q_n = \frac{Q_n(t)}{M_n} \tag{11.42}$$

where γ_n = coefficient of structural damping

ω_n = nth normal angular frequency

$$M_n = \int \varphi_n^2(x,y,z)\,dm = \text{generalized mass} \tag{11.43}$$

$$Q_n(t) = \int f(x,y,z,t)\varphi_n(x,y,z)\,dv = \text{generalized force} \tag{11.44}$$

$dv = dx\,dy\,dz$

The generalized mass and generalized force are obtained by integration over the structure.

* The concept of normal modes is discussed in detail under *Normal Modes of Vibration* in Chaps. 2 and 7. In particular, see Eq. (2.52) for a discussion of structural damping and Eq. (2.71) for a discussion of generalized mass and generalized force.

To establish the response function, a harmonic force can be applied. In this case the generalized force becomes

$$Q_n(t) = e^{j\omega t}\int F(x,y,z)\varphi_n(x,y,z)\,dv = e^{j\omega t}w_0 w_n \tag{11.45}$$

where
$$w_0 = \int F(x,y,z)\,dv = \text{amplitude of the total applied force} \tag{11.46}$$

and
$$w_n = \frac{1}{w_0}\int F(x,y,z)\varphi_n(x,y,z)\,dv = \text{mode participation factor of load} \tag{11.47}$$

The steady-state solution for q_n is

$$q_n = \frac{w_0 w_n e^{j(\omega t+\theta_n)}}{M_n\omega_n{}^2\left\{\left[1-\left(\frac{\omega}{\omega_n}\right)^2\right]^2+\gamma_n{}^2\right\}^{1/2}} \tag{11.48}$$

which results in the response function

$$A(j\omega) = \sum_n \frac{w_n\varphi_n(x,y,z)e^{j\theta_n}}{M_n\omega_n{}^2\left\{\left[1-\left(\frac{\omega}{\omega_n}\right)^2\right]^2+\gamma_n{}^2\right\}} \tag{11.49}$$

The square of the response function, required for the evaluation of the mean-square response, is then

$$|A(j\omega)|^2 = A(j\omega)A(-j\omega)$$
$$= \sum_n \frac{w_n{}^2\varphi_n{}^2(x,y,z)}{M_n{}^2\omega_n{}^4\left\{\left[1-\left(\frac{\omega}{\omega_n}\right)^2\right]^2+\gamma_n{}^2\right\}} \tag{11.50}$$

Utilizing again the approximations allowed for small damping, the mean-square response for the structure becomes

$$\overline{u^2} = \frac{\pi}{2}\sum_n \frac{w_n{}^2\varphi_n{}^2(x,y,z)W(\omega_n)}{\gamma_n M_n{}^2\omega_n{}^3} \tag{11.51}$$

Equation (11.51) is expressed in terms of the normal modes of the structure and the power spectral density of the excitation, which allows a simple calculation for any linear structure.

CORRELATION FUNCTIONS

In statistical studies, the correlation function indicates the correlation between two sets of numbers. The autocorrelation, $R(t_1,\tau)$, of a random function $x(t)$ is a measure of the correlation between two values of $x(t)$ spaced τ apart. Crosscorrelation indicates the correlation between sets of numbers in two quantities $x(t)$ and $y(t)$.

AUTOCORRELATION FUNCTIONS*

Autocorrelation is found by averaging the product of the two values $x(t_1)$ and $x(t_1+\tau)$ of each record over the assembly of records. The autocorrelation function $R(t,\tau)$ is related to the second probability density as follows:

$$R(t_1,\tau) = \overline{x_i(t_1)x_i(t_1+\tau)}$$
$$= \int_{-\infty}^{\infty}\int_{-\infty}^{\infty} x_1x_2p_2(x_1,t_1,x_2,t_2)\,dx_1\,dx_2 \tag{11.52}$$

* Also see Chap. 22.

For a stationary function, the autocorrelation function $R(\tau)$ is determined by time averaging.

$$R(\tau) = \lim_{T\to\infty} \frac{1}{2T} \int_{-T}^{T} x(t)x(t + \tau)\, dt \tag{11.53}$$

From Eq. (11.12) the two terms to be multiplied and averaged are

$$x(t) = C_0 + 2 \sum_{n=1}^{\infty} |C_n| \cos(n\omega_0 t - \alpha_n) \tag{11.54}$$

$$x(t + \tau) = C_0 + 2 \sum_{n=1}^{\infty} |C_n| \{\cos(n\omega_0 t - \alpha_n) \cos n\omega_0\tau - \sin(n\omega_0 t - \alpha_n) \sin n\omega_0\tau\} \tag{11.55}$$

and the correlation function $R(\tau_n)$ for a finite period τ_n becomes

$$R(\tau_n) = \frac{1}{2T} \int_{-T}^{T} x(t)x(t + \tau)\, dt = C_0^2 + 2 \sum_{n=1}^{\infty} |C_n|^2 \cos n\omega_0\tau \tag{11.56}$$

The increment in $R(\tau_n)$ divided by the increment in frequency is then

$$\frac{\Delta R(\tau_n)}{\Delta\omega} = \frac{2|C_n|^2}{\omega_0} \cos n\omega_0\tau = W(\omega_n) \cos n\omega_0\tau \tag{11.57}$$

As $T \to \infty$, $n\omega_0 = \omega$, $W(\omega_n) \to W(\omega)$, $R(\tau_n) \to R(\tau)$, and

$$\frac{dR(\tau)}{d\omega} = W(\omega) \cos \omega\tau \tag{11.58}$$

Thus, by integration, the relationship between the power spectral density $W(\omega)$ and the autocorrelation function $R(\tau)$ is

$$R(\tau) = \int_0^{\infty} W(\omega) \cos \omega\tau\, d\omega \tag{11.59}$$

It is evident from this expression that $R(\tau)$ is an even function of τ. In any practical mechanical system $W(\omega)$ approaches zero as $\omega \to \infty$, and $R(\infty) \to 0$. Also if $\tau \to 0$, the value of the function at each time increment is multiplied by itself; then Eq. (11.59) and Eq. (11.17) are identical, indicating that $R(0)$ is equal to the mean-square value $\overline{x^2}$. For $0 < \tau < \infty$, $R(\tau)$ may experience a change in sign and the curve for $R(\tau)$ may appear as in Fig. 11.11.

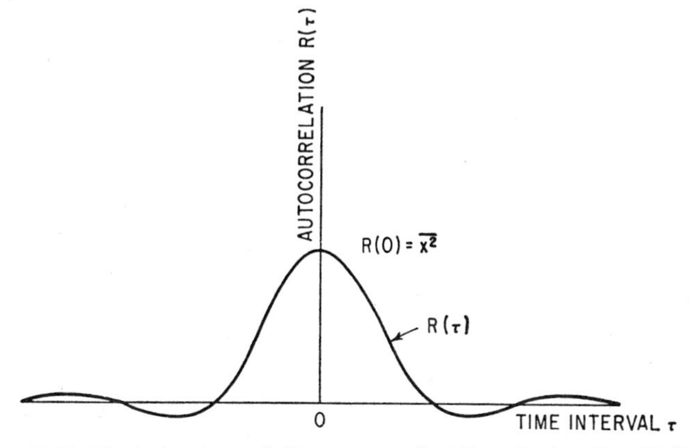

FIG. 11.11. Typical autocorrelation, an even function of τ [see Eq. (11.59)].

Equation (11.59) may be identified as a Fourier cosine transform, whence the power spectral density may be determined from the autocorrelation function by taking the inverse transform to obtain

$$W(\omega) = \frac{2}{\pi} \int_0^\infty R(\tau) \cos \omega\tau \, d\tau \tag{11.60}$$

It is often desirable to obtain $W(\omega)$ for a given record by first obtaining $R(\tau_n)$ according to Eq. (11.56), with T finite, and then evaluating $W(\omega)$ from Eq. (11.60). These operations are mathematically equivalent to those discussed under *power spectral density,* and the ultimate choice depends on the equipment available, the required accuracy, and the intended application of the results as indicated in Chap. 22.

REFERENCES

1. Rice, S. O.: *Bell System Tech. J.*, **23**:282 (1944) and **24**:44 (1945); reprinted in "Noise and Stochastic Processes," N. Wax (ed.), Dover Publications, New York, 1954.
2. Thomson, W. T., and M. V. Barton: *J. Appl. Mechanics*, **24**:248 (1957).

12

INTRODUCTION TO SHOCK AND VIBRATION MEASUREMENTS

Eldon E. Eller
Endevco Corporation

Robert W. Conrad
National Aeronautics and Space Administration

INTRODUCTION

This chapter is the first in a group of nine chapters on the measurement of shock and vibration. The following chapters describe in detail various types of instruments and measuring systems, and set forth primary considerations in the calibration and field use of such instruments and systems. This chapter defines the terms and describes the general principles of shock- and vibration-measuring instruments; it also sets forth the mathematical basis for the measurement of shock and vibration, and includes a brief description of important instruments and instrumentation systems.

DEFINITION OF TERMS

A *transducer* (or "*pickup*") is a device which converts shock or vibratory motion into an optical, a mechanical, or, most commonly, an electrical signal that is proportional to a parameter of the experienced motion.

A *transducing element* is the part of the transducer that accomplishes the conversion of motion into the signal.

A *measuring instrument or system* converts shock and vibratory motion into an *observable* form that is directly proportional to a parameter of the experienced motion. It may consist of a transducer with transducing element, signal-conditioning equipment, and device for displaying the signal. An *instrument* contains all of these elements in one package (see Chap. 13), while a *system* utilizes separate packages.

An *accelerometer* is a transducer whose output is proportional to the acceleration input.

A *velocity pickup* is a transducer whose output is proportional to the velocity input.

A *displacement pickup* is a transducer whose output is proportional to the displacement input.

CLASSIFICATION OF INSTRUMENTS

In principle, shock and vibration are measured with reference to a point fixed in space.* This may be accomplished by either of two fundamentally different types of instruments:

* For special purposes, the motion of one point may be measured relative to another point. Then one terminal of the instrument is attached to each of the points.

1. In a *fixed reference instrument,* one terminal of the instrument is attached to a point that is fixed in space and the other terminal is attached (e.g., mechanically, electrically, optically, etc.) to the point whose motion is to be measured.
2. In a *mass-spring instrument* (*seismic instrument*), the only terminal is the base of a mass-spring system; this base is attached at the point where the shock or vibration is to be measured. The motion at the point is inferred from the motion of the mass relative to the base.

FIXED REFERENCE INSTRUMENTS

Figure 12.1 illustrates schematically the principle of the fixed reference instrument for measuring the motion of the "vibrating part" relative to the "fixed reference." By attaching a scale to the fixed reference and a pointer to the vibrating part, as shown in

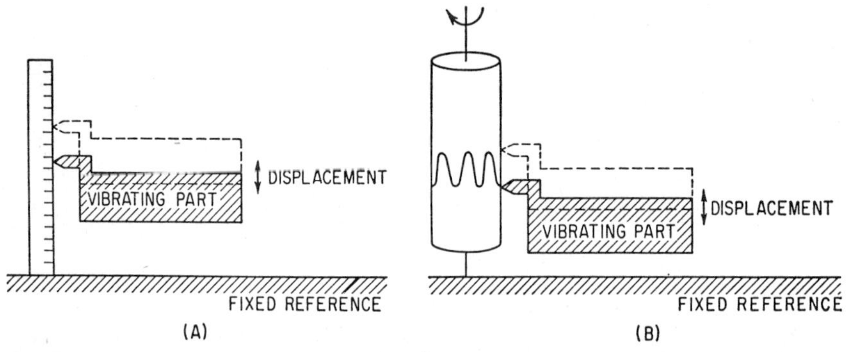

FIG. 12.1. Measurement of vibration with a fixed reference system: (*A*) Displacement of the vibrating part is indicated for direct observation, with or without optical enlargement techniques. (*B*) Displacements are recorded on a rotating drum, describing the waveform as a function of time.

Fig. 12.1*A*, the limits of motion of the vibrating part can be determined visually. If the pointer inscribes a mark upon the scale, the peak-to-peak displacement can be determined by noting the length of the inscribed line. Additional information can be obtained by substituting a rotating drum for the scale, as illustrated in Fig. 12.1*B*. Then the inscribed trace represents the time-history of the displacement of the vibrating part relative to the fixed reference.

Various means are used to perform the functions indicated schematically in Fig. 12.1. Usually, the displacement associated with vibratory motion is relatively small; thus, magnifying means must be used to obtain useful sensitivity and accuracy of measurement. This may be done mechanically by incorporating a lever with the pointer or scriber to amplify the motion to a more readily observed size (see *Vibrographs* in Chap. 13). Because of mechanical limitations, a lever is useful only for the measurement of shock and vibration having a relatively large displacement and a relatively low frequency. Electrical means may be employed to indicate the motion of the vibrating part relative to the fixed reference; such means are described in other chapters. Optical or visual methods also are used wherein the observer occupies the position of the scale or drum in Fig. 12.1 and is provided with means to determine the displacement of the vibrating part.

VISUAL DISPLACEMENT INDICATORS. The accuracy of an optical system at a high frequency of vibration is limited by the resolution and rigidity of the fixed reference used in the system. With good technique, it is possible to achieve accurate measurements at frequencies as high as 2,000 cps. Under highly specialized conditions, good results may be obtained at frequencies as high as 20,000 cps. Inasmuch as displacements at high frequency usually are measured in microinches or microns, it is essential that the optical system be secured firmly to the fixed reference.

Fixed reference instruments of the visual observation type are described in Chap. 13. Such an instrument employs an appropriate "target" which is attached to the surface of the vibrating part in position for visual observation. A number of different types of targets are utilized, several of which are shown in Fig. 12.2. (Also see Fig. 18.8, which illustrates a method of vibration displacement measurement that uses a vibrating wedge.)

Figure 12.2*A* shows a "target" made of a material containing reflecting spots *b* as it appears through a microscope equipped with a reticle scale *c*. Typical reflective materials are Scotchlite, fine garnet paper, or a blackened mirror with a pinhole through the blacking. Each of these materials reflects a fine pinpoint of light when illuminated by a light source. When such a reflector is vibrated, persistence of vision causes each reflecting point to appear as a straight line *d* in Fig. 12.2*A*. The length of the line may be determined by the use of the microscope reticle, and indicates the peak-to-peak displacement of the vibrating part.

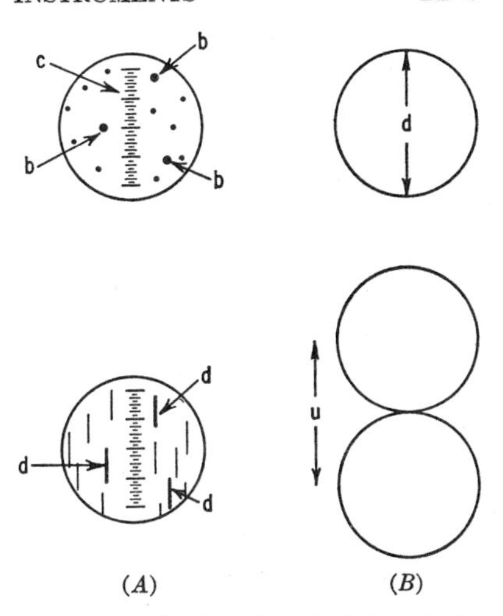

FIG. 12.2. Visual and optical targets for measurement of vibratory displacement. The appearance of the targets when stationary is shown in the upper views; the corresponding appearance when vibrating is shown in the lower views.

(*A*) Illuminated target containing small reflecting spots observed through microscope equipped with reticle. Under vibration the spots become lines. The observer selects the most prominent line and measures its length by the scale in the reticle. This is a measure of the peak-to-peak displacement *u* multiplied by the fixed power of the optical system.

(*B*) A circular target of diameter *d* equal to a desired peak-to-peak displacement *u* will appear as two touching circles when *u* equals *d*.

When it is desired to set a vibratory displacement to a known and fixed value, a target of the type shown in Fig. 12.2*B* is useful. This target is a circle with its diameter *d* equal to the desired peak-to-peak displacement. During vibration two circles appear because the velocity of the moving target is zero at times of maximum displacement. These circles move further apart as the displacement is increased. The desired value of displacement is attained when the circles appear to just touch one another, as shown in the lower view of Fig. 12.2*B*. Because of persistence of vision, the target appears motionless at these points. The required size of target circles may be too small to be drawn accurately to size; then the circle may be drawn on a larger scale and reduced photographically.

STROBOSCOPE. A stroboscope is a light source that can be adjusted to flash at a desired rate. It may be used to illuminate the vibrating surface in a fixed reference system (see Chap. 13). If the flashing frequency is the same as the vibration frequency or some submultiple thereof, the vibrating surface is in the same position each time it is illuminated. Therefore, the vibrating surface appears motionless due to persistence of vision. If the flashing frequency is slightly different from the vibration frequency, the vibratory displacement appears in slow motion. The stroboscope may be calibrated and used to measure the frequency of vibration.

MASS-SPRING TRANSDUCERS (SEISMIC TRANSDUCERS)

In many applications, such as moving vehicles or missiles, it is impossible to establish a fixed reference for shock and vibration measurements. Therefore, many transducers use the response of a mass-spring system to measure shock and vibration. A mass-spring transducer is shown schematically in Fig. 12.3; it consists of a mass *m* suspended from the transducer case *a* by a spring of stiffness *k*. The motion of the mass within the case may be damped by a viscous fluid or electric current symbolized by a dashpot with damp-

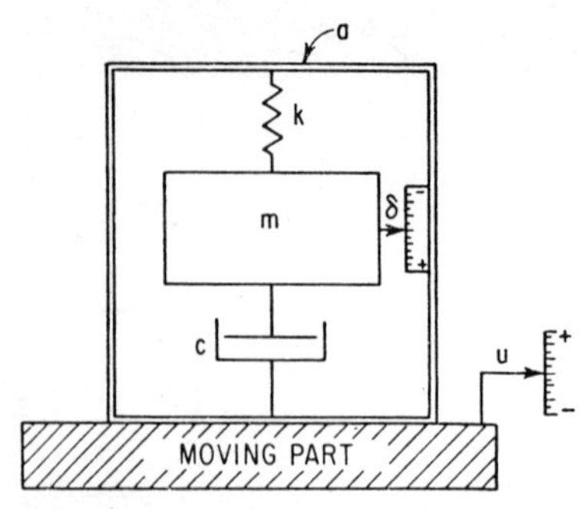

FIG. 12.3. Mass-spring type of vibration-measuring instrument consisting of a mass m supported by spring k and viscous damper c. The case a of the instrument is attached to the moving part whose vibratory motion u is to be measured. The motion u is inferred from the relative motion δ between the mass m and the case a.

ing coefficient c. It is desired to measure the motion of the moving part whose displacement with respect to fixed space is indicated by u. When the transducer case is attached to the moving part, the transducer may be used to measure displacement, velocity, or acceleration, depending on the portion of the frequency range which is utilized and whether the relative displacement δ or relative velocity $d\delta/dt$ is sensed by the transducing element. The typical response of the mass-spring system is analyzed in the following paragraphs and applied to the interpretation of transducer output.

Consider an instrument whose case experiences a motion u, and let the relative displacement between the mass and the case be δ. Then the motion of the mass with respect to a reference fixed in space is $\delta + u$, and the force causing its acceleration is $m[d^2(\delta + u)/dt^2]$. Thus, the force applied by the mass to the spring and dashpot assembly is $-m[d^2(\delta + u)/dt^2]$. The force applied by the spring is $-k\delta$, and the force applied by the damper is $-c(d\delta/dt)$. Adding all force terms and equating the sum to zero,

$$-m\frac{d^2(\delta + u)}{dt^2} - c\frac{d\delta}{dt} - k\delta = 0 \tag{12.1}$$

Equation (12.1) may be rearranged:

$$m\frac{d^2\delta}{dt^2} + c\frac{d\delta}{dt} + k\delta = -m\frac{d^2u}{dt^2} \tag{12.2}$$

Assume that the motion u is sinusoidal, $u = u_0 \cos \omega t$, where $\omega = 2\pi f$ is the angular frequency in radians per second and f is expressed in cycles per second. Neglecting transient terms, the response of the instrument is defined by $\delta = \delta_0 \cos(\omega t - \theta)$; then the solution of Eq. (12.2) is

$$\frac{\delta_o}{u_o} = \frac{\omega^2}{\sqrt{\left(\frac{k}{m} - \omega^2\right)^2 + \left(\omega\frac{c}{m}\right)^2}} \tag{12.3}$$

$$\theta = \tan^{-1}\frac{\omega\frac{c}{m}}{\frac{k}{m} - \omega^2} \tag{12.4}$$

The undamped natural frequency f_n of the instrument is the frequency at which

$$\frac{\delta_o}{u_o} = \infty$$

when the damping is zero ($c = 0$), or the frequency at which $\theta = 90°$. From Eqs. (12.3) and (12.4), this occurs when the denominators are zero:

$$\omega_n = 2\pi f_n = \sqrt{\frac{k}{m}} \quad \text{rad/sec} \tag{12.5}$$

Thus, a stiff spring and/or light mass produces an instrument with high natural frequency. A heavy mass and/or compliant spring produces an instrument with a low natural frequency.

The damping in a transducer is specified as a *fraction of critical damping.* Critical damping c_c is the minimum level of damping that prevents a mass-spring transducer from oscillating when excited by a step function or other transient. It is defined by

$$c_c = 2\sqrt{km} \tag{12.6}$$

Thus, the fraction of critical damping ζ is

$$\zeta = \frac{c}{c_c} = \frac{c}{2\sqrt{km}} \tag{12.7}$$

It is convenient to define the excitation frequency ω for a transducer in terms of the undamped natural frequency ω_n by using the dimensionless frequency ratio ω/ω_n. Substituting this ratio and the relation defined by Eq. (12.7), Eqs. (12.3) and (12.4) may be written:

$$\frac{\delta_o}{u_o} = \frac{\left(\dfrac{\omega}{\omega_n}\right)^2}{\sqrt{\left[1 - \left(\dfrac{\omega}{\omega_n}\right)^2\right]^2 + \left(2\zeta\dfrac{\omega}{\omega_n}\right)^2}} \tag{12.8}$$

$$\theta = \tan^{-1}\frac{2\zeta\dfrac{\omega}{\omega_n}}{1 - \left(\dfrac{\omega}{\omega_n}\right)^2} \tag{12.9}$$

The response of the mass-spring transducer given by Eq. (12.8) may be expressed in terms of the acceleration $\ddot{u}$ of the moving part by substituting $\ddot{u}_0 = -u_0\omega^2$. Then the ratio of the relative displacement amplitude δ_0 between the mass m and transducer case a to the impressed acceleration amplitude $\ddot{u}_0$ is

$$\frac{\delta_o}{\ddot{u}_o} = -\frac{1}{\omega_n^2}\left[\frac{1}{\sqrt{\left[1 - \left(\dfrac{\omega}{\omega_n}\right)^2\right]^2 + \left[2\zeta\dfrac{\omega}{\omega_n}\right]^2}}\right] \tag{12.10}$$

The relation between δ_0/u_0 and the frequency ratio ω/ω_n is shown graphically in Fig. 12.4 for several values of the fraction of critical damping ζ. Corresponding curves for $\delta_o/\ddot{u}_o$ are shown in Fig. 12.5. The phase angle θ defined by Eq. (12.9) is shown graphically in Fig. 12.6, using the scale at the left side of the figure. Corresponding phase angles between the relative displacement δ, and the velocity $\dot{u}$ and acceleration $\ddot{u}$, are indicated by the scales at the right side of the figure.

APPLICATION OF MASS-SPRING TRANSDUCERS

The mass-spring transducer may be adapted for the measurement of displacement, velocity, or acceleration, depending upon the natural frequency of the transducer and the type of transducing element used. Considerable versatility frequently is available in instrumentation systems for transforming one parameter to another. For example, velocity may be measured with an accelerometer by integrating the signal from the transducer. Often, this is done electrically and the recorded result indicates velocity directly.

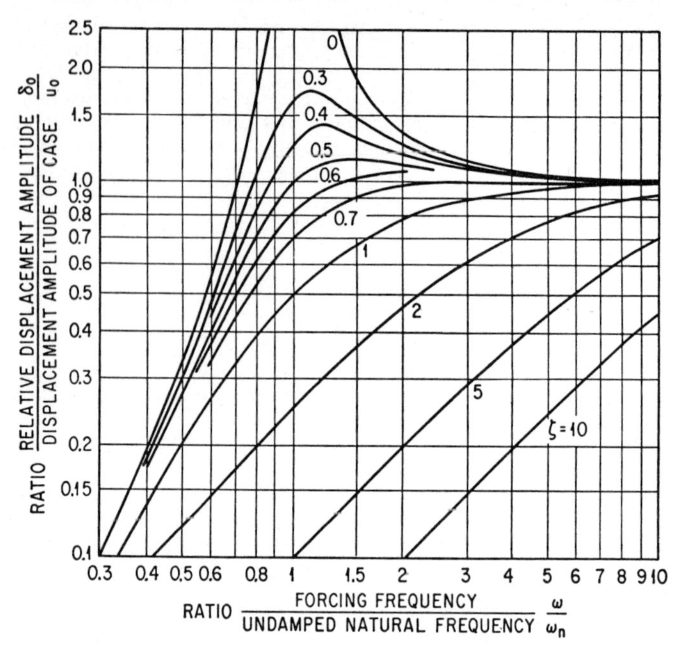

FIG. 12.4. Displacement response δ_o/u_o of a mass-spring system subjected to a sinusoidal displacement $u = u_o \sin \omega t$. The fraction of critical damping ζ is indicated for each curve.

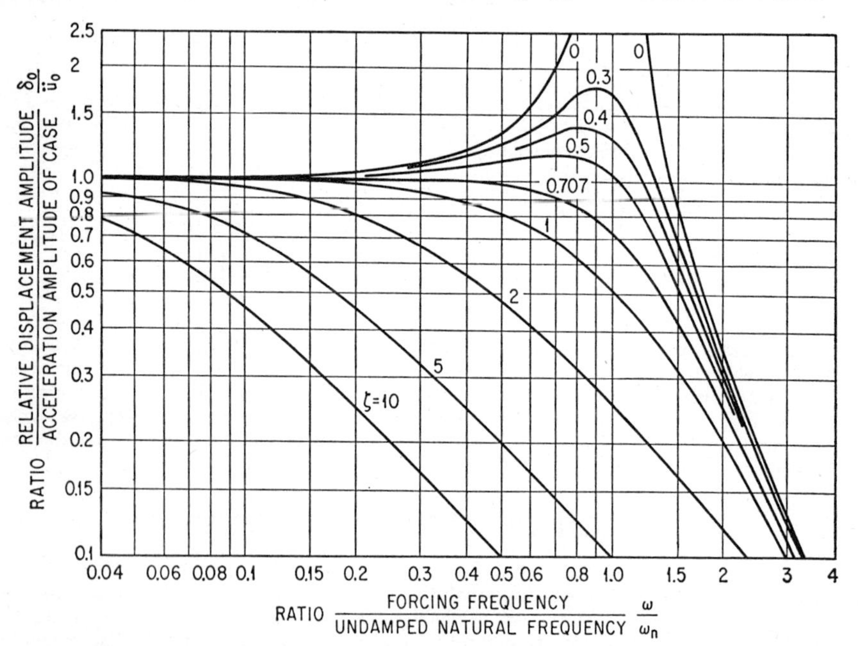

FIG. 12.5. Relationship between the relative displacement amplitude δ_o of a mass-spring system and the acceleration amplitude $\ddot{u}_o$ of the case. The fraction of critical damping ζ is indicated for each response curve.

DISPLACEMENT-MEASURING TRANSDUCERS

When ω/ω_n is substantially greater than 1.0, the ratio δ_o/u_o approaches the constant value of 1.0, approximately independent of frequency. If the signal obtained from the transducing element is directly proportional to δ_o, the transducer measures the displacement amplitude u_o to the approximation and over the frequency range indicated by Fig. 12.4. Therefore, the instrument is a displacement-measuring instrument. The addition of substantial damping makes the transducer useful for indicating displacement amplitude over a wider range of frequencies, but introduces limitations in the form of wave distortion (see *Phase Shift* in this chapter).

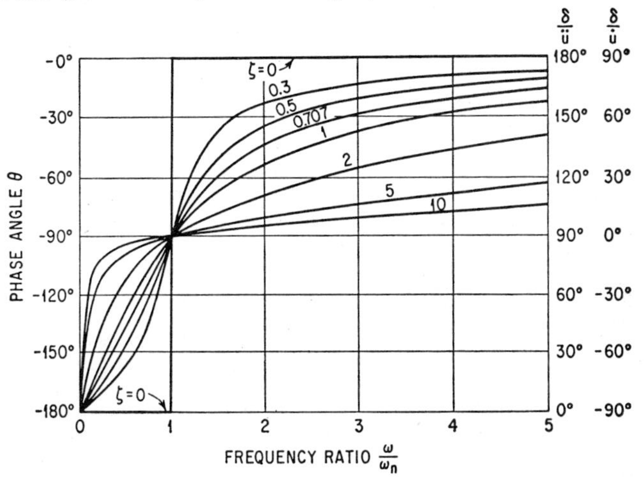

FIG. 12.6. Phase angle of a mass-spring transducer when used to measure sinusoidal vibration. The phase angle θ on the left-hand scale relates the relative displacement δ to the impressed displacement, as defined by Eq. (12.9). The right-hand scales relate the relative displacement δ to the impressed velocity and acceleration.

DISPLACEMENT LIMITS. Mass-spring transducers operating above their natural frequencies have several general characteristics in common, independent of the type of transducing element employed. The size of the transducer is determined by the requirement that the mass remain nearly stationary in space while the transducer case is moved about it. Thus, the internal clearance in both plus and minus directions of the sensitive axis must be larger than the largest displacement amplitude to be measured. Most transducers of this type contain "stops" to limit the relative displacement between case and mass, thereby defining the displacement limits of the transducer.

HIGH-FREQUENCY RESPONSE. The curves of Fig. 12.4 indicate that the response δ_o/u_o of the transducer is constant to infinitely high frequencies. This assumes that the mass and transducer case are ideal rigid structures and that the spring and dashpot are massless. In practical transducer designs, however, an upper limit of frequency is imposed by secondary resonances in the spring. A lower limit of displacement is imposed by noise or lack of resolution in the measuring system; this limit becomes significant at high frequency because displacements tend to become small as the frequency increases.

VELOCITY-MEASURING TRANSDUCERS *

A mass-spring transducer for measuring velocity usually is a transducer with a low natural frequency, utilizing the frequency range above its natural frequency. The rel-

* Several velocity-measuring transducers are described in Chap. 15.

ative velocity $d\delta/dt$ between the mass and case is measured with a velocity-sensing transducing element, such as a coil moving in a magnetic field. Since the mass-spring system of a velocity transducer operates in the same frequency range as that of a displacement transducer, it has the same limitation of size, input displacement, frequency range, and phase shift as a displacement-measuring instrument.

ACCELERATION-MEASURING TRANSDUCERS

As indicated in Fig. 12.5, the relative displacement amplitude δ_o is directly proportional to the acceleration amplitude $\ddot{u}_o = -u_o\omega^2$ of the sinusoidal vibration being measured, at small values of the frequency ratio ω/ω_n. Thus, when the natural frequency ω_n of the transducer is high, the transducer is an accelerometer. If the transducer is undamped, the response curve of Fig. 12.5 is substantially flat when $\omega/\omega_n < 0.2$, approximately. Consequently, an undamped accelerometer can be used for the measurement of acceleration when the vibration frequency does not exceed approximately 20 per cent of the natural frequency of the accelerometer. The range of measurable frequency increases as the damping of the accelerometer is increased, up to an optimum value of damping. When the fraction of critical damping is approximately 0.65, an accelerometer gives reasonably accurate results in the measurement of vibration at frequencies as great as approximately 60 per cent of the natural frequency of the accelerometer.

As indicated in Fig. 12.5, the useful frequency range of an accelerometer increases as its natural frequency ω_n increases. However, the deflection of the spring in an accelerometer is inversely proportional to the square of the natural frequency; i.e., for a given value of $\ddot{u}_o$, the relative displacement is directly proportional to $1/\omega_n^2$ [see Eq. (12.10)]. As a consequence, the signal from the transducing element may be very small, thereby requiring a large amplification to increase the signal to a level at which recording is feasible. For this reason, a compromise usually is made between high sensitivity and highest attainable natural frequency, depending upon the desired application.

The usefulness of a particular instrument in the measurement of vibration is determined by decomposing the vibration into a series of sinusoidal or harmonic components. For steady-state periodic vibration, these components are defined by the Fourier coefficients which give the amplitudes at the discrete frequencies of the components (see Chap. 22). Then it becomes possible to determine which of the components can be measured effectively by a particular instrument. A nonperiodic motion can be defined by its Fourier spectrum, which, in general, is a continuous function of frequency; i.e., all frequencies within a given range are present (see Chap. 23). A comparison of the Fourier spectrum with the frequency-response characteristics of the instrument, such as Figs. 12.4 and 12.5, indicates the extent to which the instrument can measure the nonperiodic motion effectively. If the spectrum includes frequency components beyond the range of capability of the instrument, these are deleted or modified as indicated by the instrument-response characteristics. Thus, Figs. 12.4 and 12.5 indicate in general the capability of an instrument in measuring both periodic and nonperiodic vibration.

RESPONSE TO SHOCK. The capability of an accelerometer in the measurement of shock may be evaluated by noting the response of the accelerometer to acceleration pulses. Ideally, the response of the accelerometer (i.e., the output of the transducing element) should correspond identically with the pulse. In general, this result may be approached but not attained exactly. Three typical pulses and the corresponding responses [1] of accelerometers are shown in Figs. 12.7 to 12.9. The pulses are shown in dotted lines, a sinusoidal pulse in Fig. 12.7, a triangular pulse in Fig. 12.8, and a rectangular pulse in Fig. 12.9. Curves of the response of the accelerometer are shown in solid lines. For each of the three pulse shapes, the response is given for ratios τ_n/τ of 1.014, 0.338, and 0.203, where τ is the pulse duration and $\tau_n = 1/f_n$ is the natural period of the accelerometer. These response curves, computed for the fraction of critical damping $\zeta = 0$, 0.4, 0.7, and 1.0, indicate the following general relationships:

1. The response of the accelerometer follows the pulse most faithfully when the natural period of the accelerometer is smallest relative to the period of the pulse. For example, the responses at A in Figs. 12.7 to 12.9 show considerable deviaton between the pulse

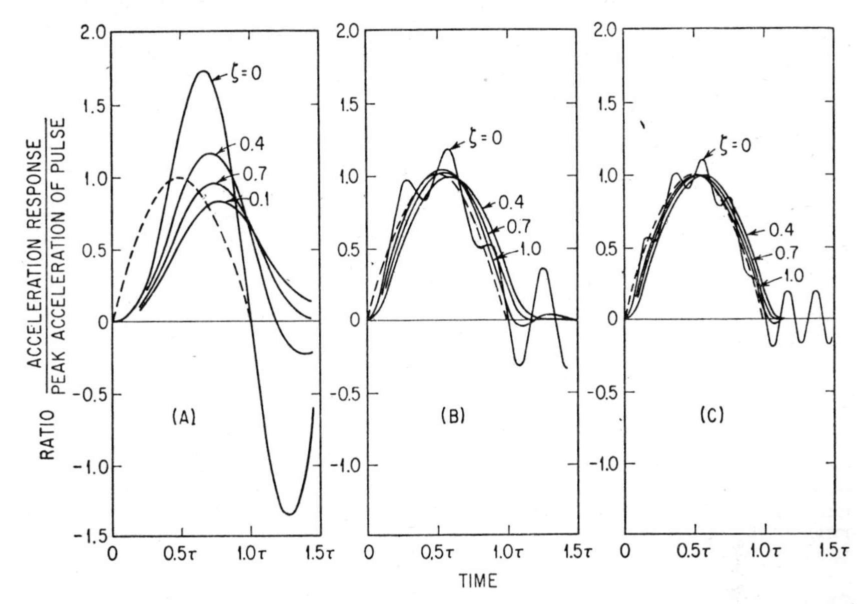

FIG. 12.7. Acceleration response to a half-sine pulse of acceleration of duration τ (dashed curve) of a mass-spring transducer whose natural period τ_n is equal to: (*A*) 1.014 times the duration of the pulse, (*B*) 0.338 times the duration of the pulse, and (*C*) 0.203 times the duration of the pulse. The fraction of critical damping ζ is indicated for each response curve. (*Levy and Kroll.*[1])

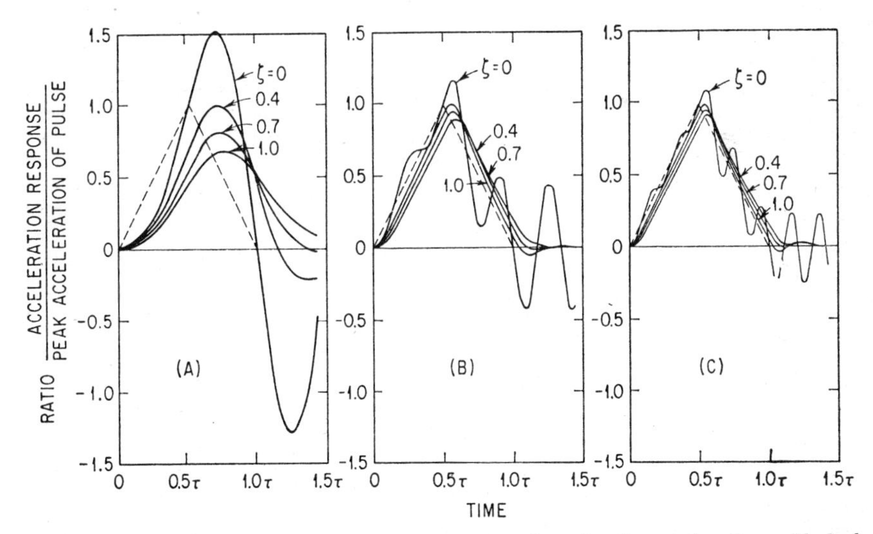

FIG. 12.8. Acceleration response to a triangular pulse of acceleration of duration τ (dashed curve) of a mass-spring transducer whose natural period τ_n is equal to: (*A*) 1.014 times the duration of the pulse, (*B*) 0.338 times the duration of the pulse, and (*C*) 0.203 times the duration of the pulse. The fraction of critical damping ζ is indicated for each response curve. (*Levy and Kroll.*[1])

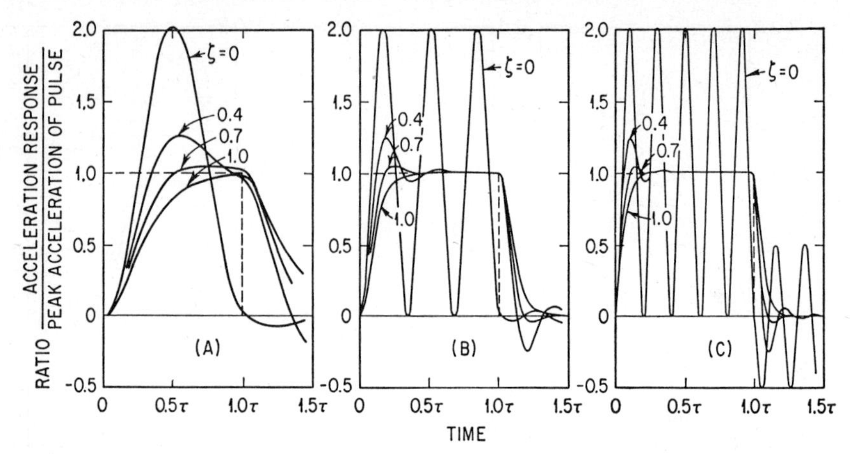

FIG. 12.9. Acceleration response to a rectangular pulse of acceleration of duration τ (dashed curve) of a mass-spring transducer whose natural period τ_n is equal to: (*A*) 1.014 times the duration of the pulse, (*B*) 0.338 times the duration of the pulse, and (*C*) 0.203 times the duration of the pulse. The fraction of critical damping ζ is indicated for each response curve. (*Levy and Kroll.*[1])

and the response; this occurs when τ_n is approximately equal to τ. However, when τ_n is small relative to τ (Figs. 12.7*C* to 12.9*C*), the deviation between the pulse and the response is much smaller.

2. Damping in the transducer reduces the response of the transducer at its own natural frequency; i.e., it reduces the transient vibration superimposed upon the pulse. Damping also reduces the maximum value of the response to a value lower than the actual pulse in the case of large damping. For example, in some cases a fraction of critical damping $\zeta = 0.7$ provides an instrument response that does not reach the peak value of the acceleration pulse.

LOW-FREQUENCY RESPONSE. The measurement of shock requires that the transducer and its associated equipment have good response at low frequencies because pulses and other types of shock motions characteristically include low-frequency components. The frequency content of nonperiodic functions is indicated by the Fourier spectrum. For example, see the Fourier spectra of typical pulses in Fig. 23.3. Such pulses can be measured accurately only with an instrumentation system whose response is flat down to the lowest frequency of the spectrum; in general, this lowest frequency is zero for pulses.

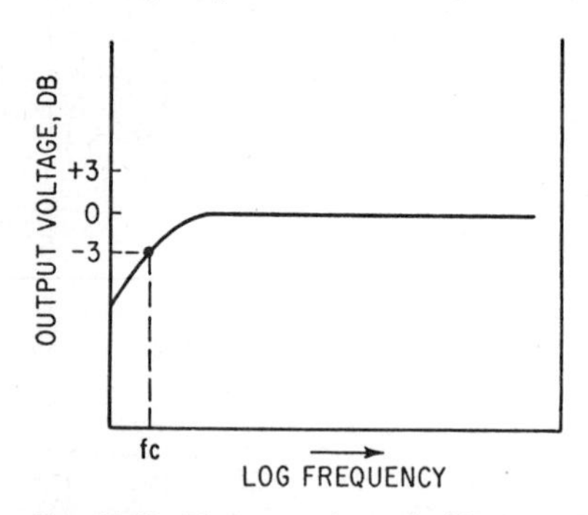

FIG. 12.10. Response of a typical instrument for measuring shock, illustrating drop-off at low frequencies. The low-frequency cutoff point f_c is the frequency at which the response is down 3 db or 30 per cent below the flat portion of the response curve.

The response of an instrumentation system is defined by a plot of output voltage vs. excitation frequency; the response of a typical system is shown in Fig. 12.10. For purposes of shock measurement, the decrease in response at low frequencies is significant. The decrease is defined quantitatively by the frequency f_c at which the response is down 3 db or approximately 30 per cent below the flat response which exists at the higher frequencies. The distortion which occurs in the measurement of a pulse is related to the frequency f_c in a manner illustrated with respect to the example[2] of Fig. 12.11.

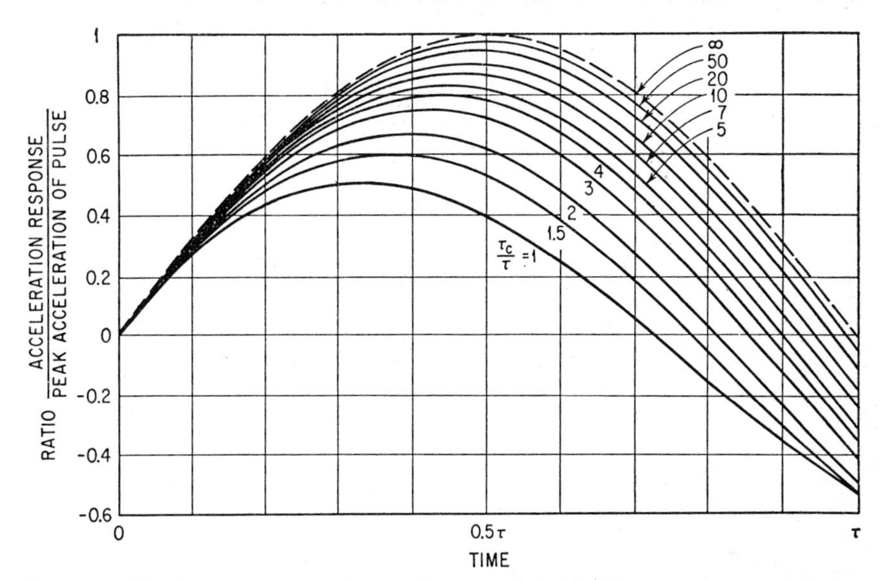

FIG. 12.11. Signal from a mass-spring accelerometer (solid line) in response to a half-sine pulse of acceleration (dotted line). The parameter of the family of curves is τ_c/τ, where τ is the duration of the acceleration pulse and $\tau_c = 1/f_c$ where f_c is the frequency defined in Fig. 12.10. The signal is given relative to the signal obtained when $f_c = 0$; i.e., when τ_c/τ is infinite. (*Lawrence.*[2])

Figure 12.11 shows by the dotted line a half-sinusoidal pulse of acceleration having a duration τ, and in solid lines the signal from a transducer having various values of the frequency f_c defined in Fig. 12.10. The frequency f_c is expressed in terms of its period $\tau_c = 1/f_c$; the parameter for the family of signal curves is τ_c/τ, where τ is the duration of the pulse shown by the dotted line. When $f_c = 0$, τ_c is infinitely great and the signal from the instrumentation system corresponds to the pulse. For values of f_c significantly greater than zero, the signal deviates from the pulse in the following respects: (1) The maximum value of the signal is lower than the maximum value of the pulse and (2) the signal exhibits a negative value near the termination of the pulse. As indicated in Fig. 12.11, this distortion becomes severe for a half-sinusoidal pulse when the period τ_c is of the same order of magnitude as the period τ of the pulse. Values of τ_c in excess of fifty times τ are desirable.

IMPORTANT CHARACTERISTICS OF SHOCK- AND VIBRATION-MEASURING INSTRUMENTS

SENSITIVITY

The *sensitivity* of a shock- and vibration-measuring instrument is the ratio of its electrical output to its mechanical input. The output usually is expressed in terms of voltage per unit of displacement, velocity, or acceleration. This specification of sensitivity is sufficient for instruments which generate their own voltage independent of an external power source. However, the sensitivity of an instrument requiring an external voltage usually is specified in terms of output voltage per unit of voltage supplied to the instrument per unit of displacement, velocity, or acceleration, e.g., millivolts per volt per g of acceleration. It is important to note the terms in which the respective parameters are

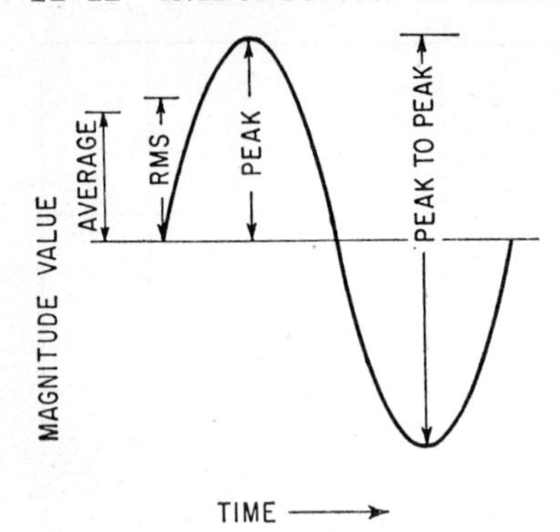

FIG. 12.12. Relationships between average, rms, peak, and peak-to-peak values for a simple sine wave. These values are used in specifying sensitivities of shock and vibration transducers (e.g., peak millivolts per peak g, or rms millivolts per peak-to-peak displacement). These relationships do *not* hold true for other than simple sine waves.

expressed; e.g., average, rms, or peak. The relation between these terms is shown in Fig. 12.12. Also see Table 1.3.

RESOLUTION

The *resolution* of a measuring instrument is the smallest change in mechanical input, i.e., displacement, velocity, or acceleration, for which a change in the electrical output is discernible. Neither mass-spring nor fixed reference instruments have a theoretical limit of resolution. The resolution usually is a function of the transducing element.

Recording equipment, indicating equipment, and other auxiliary equipment used with the vibration-measuring instruments often establish the resolution of the over-all measurement system. If the electrical output of an instrument is indicated by a meter, the resolution may be established by the smallest increment that can be read from the meter. Resolution can be limited by noise levels in the instrument or in the system. In general, any signal change smaller than the noise level will be obscured by the noise, thus determining the resolution of the system.

TRANSVERSE SENSITIVITY*

If a transducer is subjected to vibration of unit amplitude along its axis of maximum sensitivity, the amplitude of the voltage output e_{max} is the sensitivity. The sensitivity e_θ along the X axis, inclined at an angle θ to the axis of e_{max}, is $e_\theta = e_{max} \cos \theta$, as illus-

* Also called *lateral* or *cross-axis sensitivity*.

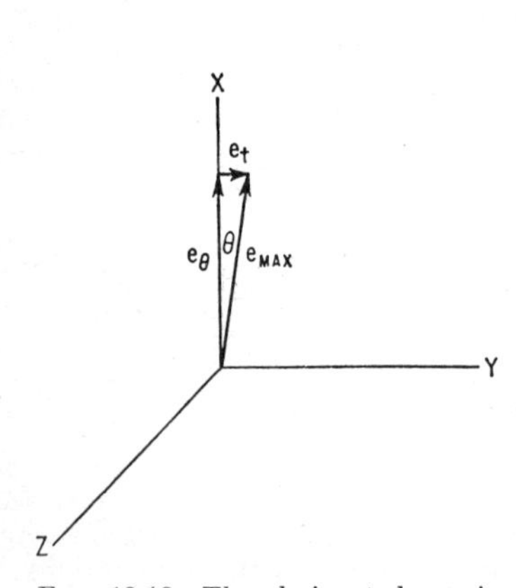

FIG. 12.13. The designated sensitivity e_θ and cross-axis sensitivity e_t that result when the axis of maximum sensitivity e_{max} is not aligned with the axis of e_θ.

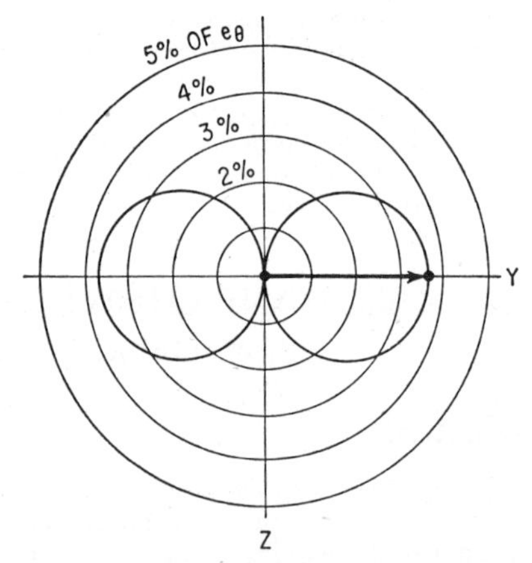

FIG. 12.14. Plot of transducer sensitivity in all axes normal to the designated axis e_θ, plotted according to axes shown in Fig. 12.23. Cross-axis sensitivity reaches a maximum e_t along the Y axis and a minimum value along the Z axis.

trated in Fig. 12.13. Similarly, the sensitivity along the Y axis is $e_t = e_{max} \sin \theta$. In general, the sensitive axis of a transducer is designated. Ideally, the X axis would be designated the sensitive axis, and the angle θ would be zero. Practically, θ can be made only to approach zero because of manufacturing tolerances and/or unpredictable variations in the characteristics of the transducing element. Then the transverse sensitivity is expressed as the tangent of the angle θ; i.e., the ratio of e_t to e_θ:

$$\frac{e_t}{e_\theta} = \tan \theta$$

In practice, $\tan \theta$ is between 0.01 and 0.10, and is expressed as a percentage. For example, if $\tan \theta = 0.05$, the transducer is said to have a transverse sensitivity of 5 per cent. Figure 12.14 is a typical polar plot of transverse sensitivity.

AMPLITUDE LINEARITY AND LIMITS

When the ratio of the electrical output of an instrument to the mechanical input (i.e., the sensitivity) remains constant within specified limits, the instrument is said to be "linear" within those limits, as illustrated in Fig. 12.15. A vibration-measuring instrument is linear only over a certain range of amplitude values. The lower end of this range is determined by either (1) the resolution of the instrument or (2) the nonlinear behavior of the instrument. For example, an instrument may include frictional forces which determine the operating characteristics of the instrument at a very low input level. It may be necessary that the mechanical input exceed some threshold value before static friction is overcome and the instrument begins to operate linearly.

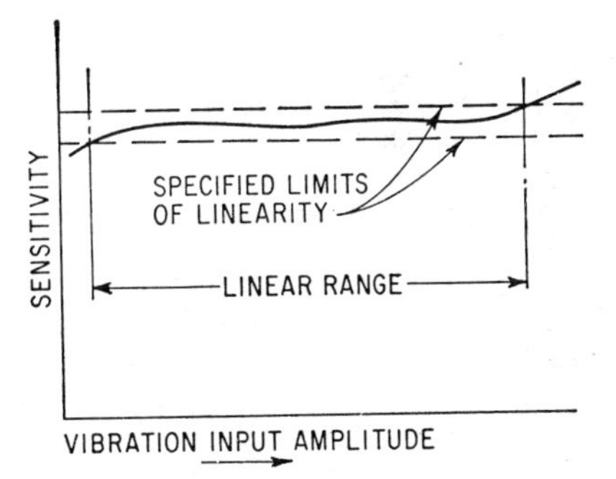

FIG. 12.15. Typical plot of sensitivity as a function of amplitude for a shock and vibration transducer. The *linear range* is established by the intersection of the sensitivity curve and the specified limits (dashed lines).

The upper limit of linearity may be imposed by the electrical characteristics of the transducing element, and by the size or the fragility of the instrument. For example, if the spring-supported mass is displaced so far that it strikes the pickup case or mechanical "stops," the instrument becomes nonlinear. Similarly, for very large acceleration values, the large forces produced by the spring of the mass-spring system may exceed the yield strength of a part of the instrument, causing nonlinear behavior or complete failure.

FREQUENCY RANGE

The operating frequency range is the range over which the sensitivity of the transducer does not vary more than a stated percentage from the rated sensitivity. This range may be limited by the electrical or mechanical characteristics of the pickup or by its associated auxiliary equipment. These limits can be added to amplitude linearity limits to define completely the operating ranges of the instrument, as illustrated in Fig. 12.16.

LOW-FREQUENCY LIMIT. The *mechanical* response of a mass-spring transducer does not impose a low-frequency limit for an acceleration transducer because the transducer responds to vibration with frequencies less than the natural frequency of the transducer; however, there is a low-frequency limit for a displacement (or velocity) transducer because it responds significantly only to vibration whose frequency is greater than the natural frequency of the transducer.

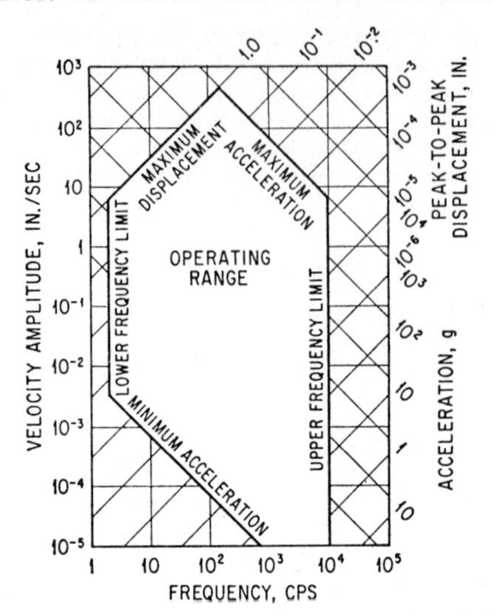

FIG. 12.16. Linear operating range of a transducer. Amplitude linearity limits are shown as a combination of displacement and acceleration values. The lower amplitude limits usually are expressed in acceleration values as shown.

In evaluating the low-frequency limit, it is necessary to consider the electrical characteristics of both the transducer and the associated equipment. In general, a transducing element that utilizes external power or a carrier voltage does not have a lower frequency limit, whereas a self-generating transducing element is not operative at zero frequency. The frequency response of amplifiers and other circuit components may limit the lowest usable frequency of an instrumentation system; frequently, an instrumentation system may be used to very low frequencies with a-c amplifiers through the use of a carrier voltage together with a discriminator placed after the amplifiers in the circuit.

HIGH-FREQUENCY LIMIT. In principle, the mechanical response of a mass-spring transducer does not impose a high-frequency limit for a displacement (or velocity) transducer because the transducer responds to all vibration with frequencies greater than the natural frequency of the transducer. In practice, however, the transducer is usable only at frequencies below the natural frequencies of mechanical components of the transducer.

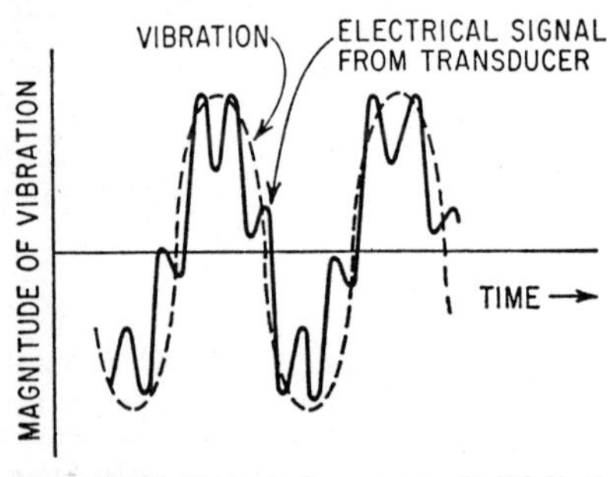

FIG. 12.17. Distorted response (solid line) of a lightly damped ($\zeta < 0.1$) mass-spring accelerometer to vibration (dashed line) containing a small harmonic content of the same frequency as the natural frequency of the accelerometer.

An acceleration transducer (accelerometer) has an upper usable frequency limit because it responds to vibration whose frequency is less than the natural frequency of the transducer. The limit is a function of (1) the natural frequency and (2) the damping of the transducer, as discussed with reference to Fig. 12.5. An attempt to use such a transducer beyond this frequency limit may result in distortion of the signal, as illustrated in Fig. 12.17.

The upper frequency limit for slightly damped vibration-measuring instruments is important because these instruments exaggerate the small amounts of harmonic content that may be contained in the motion, even when the operating frequency is well within the operating range of the instrument. The result of exciting an undamped instrument at its natural frequency may either damage the instrument or obscure the desired measurement. Figure 12.17 shows how a small amount of harmonic distortion in the vibratory motion may be exaggerated by an undamped transducer.

The upper frequency limit of transducers may be established by the electrical characteristics of the transducing element. For example, in some transducers a fixed frequency carrier voltage is required to excite the transducing element. The signal from the transducing element can be separated from the carrier signal only if the frequency of the vibration is a small percentage of the carrier frequency. This establishes an upper limit for the operating frequency range.

PHASE SHIFT

Phase shift is the time delay between the mechanical input and the electrical output signal of the instrumentation system. Unless the phase-shift characteristics of an instrumentation system meet certain requirements, a distortion may be introduced that consists of the superposition of vibration at several different frequencies. Consider first an accelerometer, for which the phase angle θ_1 is given by Fig. 12.6. If the accelerometer is undamped, $\theta_1 = 0$ for values of ω/ω_n less than 1.0; thus, the phase of the relative displacement δ is equal to that of the acceleration being measured, for all values of frequency within the useful range of the accelerometer. Therefore, an undamped accelerometer measures acceleration without distortion of phase. If the fraction of critical damping ζ for the accelerometer is 0.65, the phase angle θ_1 increases approximately linearly with the frequency ratio ω/ω_n, within the useful frequency range of the accelerometer. Then the expression for the relative displacement may be written

$$\delta = \delta_0 \cos (\omega t - \theta) = \delta_0 \cos (\omega t - a\omega) = \delta_0 \cos \omega(t - a)$$

Thus, the relative motion δ of the instrument is displaced in phase relative to the acceleration $\ddot{u}$ being measured; however, the increment along the time axis is a constant independent of frequency. Consequently, the waveform of the accelerometer output is undistorted but has a phase difference with respect to the waveform of the vibration being measured. As indicated by Fig. 12.6, any value of damping in an accelerometer other than $\zeta = 0$ or $\zeta = 0.65$ (approximately) results in nonlinear shift of phase with frequency and consequent distortion of the waveform.

When a transducer is used in the frequency range where ω/ω_n is substantially greater than 1.0, it measures displacement or velocity. Then the phase relations are less favorable. If $\zeta = 0$, the phase angle θ is $-180°$, independent of frequency; this represents only a change of sign and there is no distortion of the waveform. However, there is no value of damping greater than zero for which the phase angle varies linearly with frequency throughout the range of frequency for which the transducer is useful unless the damping is very large. Thus, a displacement or velocity-measuring instrument with a moderate degree of damping always introduces distortion of complex or superimposed waveforms, depending on the degree of damping in the instrument.

CALIBRATION REQUIREMENTS

The calibration of shock- and vibration-measuring instruments consists in determining the relationship of the output (e.g., electrical or mechanical) of the transducer to the input (displacement, velocity, or acceleration), i.e., determining its sensitivity. Calibration methods are described in detail in Chap. 18. The type and amount of calibration information required depend both on the type of instrument and on its intended application; in general, the following information is required:

1. *The sensitivity over the frequency range of interest.* It is necessary to know the ratio of the output to the input at all frequencies where measurements are to be made. Usually this ratio is measured at a frequency near the center of the range; then a plot is made of the deviation of this value as the frequency is varied.

2. *The sensitivity over an environmental range of interest.* If accurate results are to be obtained over a wide range of environmental conditions, it is necessary to determine the

effects upon the characteristics of the transducer of temperature, supply voltage variation, radiation, acoustic noise, electromagnetic field, altitude, and humidity.

3. *The sensitivity over an amplitude range of interest.* If the vibration (or shock) level to be measured is expected to have a large range of magnitude, the characteristics of the transducer should be measured at both high and low vibration (or shock) levels to determine the effects of noise, resolution, and nonlinearity.

4. *Stability of calibration with time.* The sensitivity of most vibration- and shock-measuring instruments is relatively stable with time. However, instruments should be recalibrated on a regular basis to ensure continued accuracy and reliability.

ENVIRONMENTAL EFFECTS

TEMPERATURE. The sensitivity, natural frequency, and damping of a transducer may be affected by temperature. The specific effects produced depend on the type of transducer and the details of its design. For example, Fig. 12.18 shows the variation of damping with temperature for several different damping media. Either of two methods

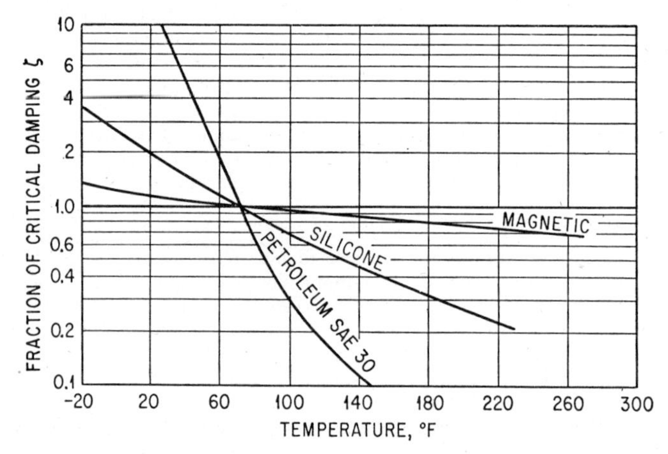

FIG. 12.18. Variation of damping with temperature for different damping means. The ordinate indicates the fraction of critical damping ζ at various temperatures assuming $\zeta = 1$ at 70°F.

may be employed to compensate for temperature effects: (1) the temperature of the pickup may be held constant by local heating or cooling or (2) the pickup characteristics may be measured as a function of temperature; if necessary, the appropriate corrections can then be applied to the measured data.

HUMIDITY. Humidity may affect the characteristics of certain types of vibration instruments. In general, a transducer which operates at a high electrical impedance is affected by humidity more than a transducer which operates at a low electrical impedance. It usually is impractical to correct the measured data for humidity effects. However, instruments that might otherwise be adversely affected by humidity often are sealed hermetically to protect them from the effects of moisture.

ACOUSTIC NOISE. Acoustic energy of high intensity often accompanies vibration. If structures of the transducer or auxiliary equipment can be excited acoustically to vibrate, there may be large error signals. Sometimes the vibration to be measured is the source of the acoustic energy; in other cases, the acoustic energy induces the vibration to be measured. Only in the latter case is the ratio of the acoustic energy to the vibration energy high enough to produce serious errors. In general, an accelerometer subjected to acoustic noise does not produce an electrical output equivalent to the voltage produced by an acceleration of $1g$ until the sound pressure level of the acoustic noise exceeds 150 db. With such a high sound pressure level it is likely that the vibration level is large and the error introduced by the acoustic noise is not apt to be important. Figure 16.20 shows the output of several types of piezoelectric accelerometers as a function of the sound pressure level of the sound field to which they are exposed. Where it is necessary to measure vibration of small magnitude in a high-intensity acoustic field, the response of the instrument to the acoustic field alone should be measured to determine the effect of the noise on the measurement of vibration. In performing such a test, the transducer is mounted so that its mounting will not be set into vibration by the acoustic field.

PHYSICAL PROPERTIES

Size and weight of the transducer are very important considerations in many vibration and shock measurements. A large instrument may require a mounting structure that will change the local vibration characteristics of the structure whose vibration is being measured. Similarly, the added mass of the transducer may produce substantial changes in the vibratory response of such structure. Generally, the natural frequency of a structure is lowered by the addition of mass; specifically, for a simple spring-mass structure:

$$\Delta f_n = f_n \left(1 - \sqrt{\frac{1}{1 + (\Delta m/m)}}\right)$$

where f_n is the natural frequency of the structure, Δf_n is the change in natural frequency, m is the mass of the structure, and Δm is the increase in mass resulting from the addition of the transducer.

In general, for a given type of transducing element, the sensitivity increases approximately in proportion to the mass of the transducer. In most applications, it is more important that the transducer be small in size than that it have high sensitivity because amplification of the signal increases the output to a usable level.

Mass-spring-type transducers for the measurement of displacement usually are larger and heavier than similar transducers for the measurement of acceleration. In the former, the mass must be free to remain substantially stationary in space while the instrument case moves about it; this requirement does not exist with the latter.

For the measurement of shock and vibration in aircraft or missiles, the size and weight of not only the transducer but also the auxiliary equipment are important. In these applications, self-generating instruments that require no external power may have a significant advantage.

SIGNAL TRANSMISSION CABLES

The electrical cable connecting a transducer to the electronic components with which it is associated is a very important element of the over-all measurement system. Cable types and applications range from well-shielded coaxial cables employed over the long distances encountered in some types of field tests to short lengths of unshielded twisted-pair wires which are widely used in strain-gage measurements in the laboratory. Choice of a suitable cable depends on the particular application, transducer, cable length, and environmental conditions.

Cables normally are considered passive elements. However, under certain conditions of use and environment, spurious signals may be induced in, or generated by, the connecting cables. The signals at the receiving end of the line then contain components that were not initially present in the transducer output. The most common cause of interference is electrical "pickup" from strong electromagnetic fields associated with nearby power lines, electrical machinery, switching transients, or radar and commercial radio transmitters. "Ground loops," i.e., circulating ground currents, also can produce spurious signals very similar in appearance to electrical pickup, but by an entirely different mechanism. In addition to these external sources of interference, untreated cables operated at high-impedance levels are prone to exhibit a self-generated type of "cable noise" when mechanically agitated or distorted.

These departures from the normal passive nature of signal transmission cables play an important part in selecting a suitable cable for a particular application, and affect the quality and reliability of the recorded data. The following sections discuss the circumstances under which these cable effects arise, and describe practical measures and techniques which negate their consequences.

TRANSMISSION-LINE CHARACTERISTICS

CABLE PARAMETERS. Characteristics of cables and transmission lines are determined by their fundamental physical properties.* For the idealized transmission line of Fig. 19.1 having distributed parameters, the series impedance Z and shunt admittance Y per unit length of line are given by

$$Z = R + j\omega L \qquad \text{ohms/unit length} \tag{12.11a}$$

$$Y = G + j\omega C \qquad \text{mhos/unit length} \tag{12.11b}$$

where R = series resistance, ohms/unit length
L = series inductance, henrys/unit length
G = shunt conductance, mhos/unit length
C = shunt capacitance, farads/unit length

As a circuit element, the electrical properties of a cable are more conveniently expressed in terms of its *characteristic* or *surge* impedance Z_0 and its *propagation constant* γ. These, derived from Eq. (12.11*a*), are defined as

$$Z_0 = \left(\frac{Z}{Y}\right)^{1/2} \qquad \text{ohms}$$

$$\gamma = (ZY)^{1/2} = \alpha + j\beta$$

where α = attenuation constant, nepers
β = phase constant, radians

In the general case, both Z_0 and γ are complex quantities. However, on high-quality low-loss cables or at radio frequencies where the reactive components dominate the expression of Eqs. (12.11), the characteristic impedance becomes resistive and is independent of frequency, i.e.,

$$Z_0 = (L/C)^{1/2} \qquad \text{ohms}$$

This is the impedance seen looking into the input terminals of an infinitely long cable. For finite lengths the input impedance depends on the value of terminating impedance relative to the characteristic impedance. When a cable is terminated by a resistance equal to its characteristic impedance, its input impedance also is equal to Z_0, and all the energy in the incident wave will be absorbed in the terminating resistance. For all other values of terminating impedance, the receiving end represents a discontinuity which is responsible for propagating a reflected wave back to the sending end. Standing-wave patterns then appear on the line, the amplitude of which depend on the amount of mismatch and line losses.

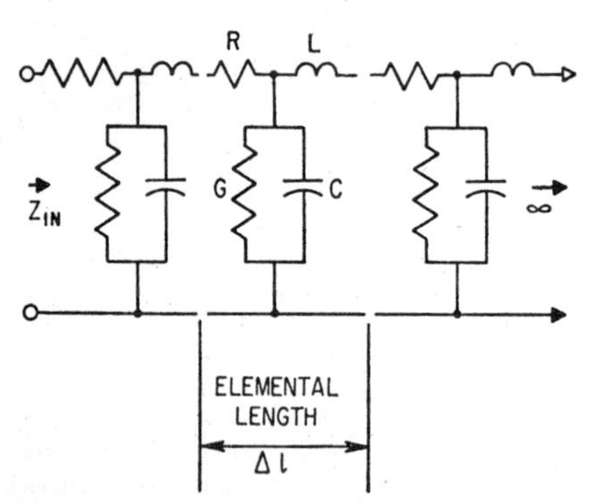

FIG. 12.19. Equivalent circuit of an infinitely long transmission line with distributed constants.

* The material presented here is restricted to elementary fundamentals of transmission-line theory. It is developed only to the extent required to cover most shock and vibration applications. The reader desiring more comprehensive treatment of transmission-line theory is directed to one of the numerous texts on the subject.[3–5]

The propagation constant γ contains a real part α, the *attenuation constant*, and an imaginary part β, the *phase constant*. These constants represent the rate of loss in amplitude and the rate of phase variation, respectively, as the wave is propagated down the line.

Nominal values of characteristic impedance range from about 35 to 150 ohms for coaxial cables, and considerably higher for open or twisted pair. The attenuation and phase constants are frequency-dependent, and they become quite small for frequencies below 1 megacycle.

Cable parameters and characteristics, outlined above, provide the basis of transmission-line theory, which is usually concerned with circumstances where the cable is many wavelengths long. Voice frequencies over long-distance telephone lines, and television or other UHF frequencies over short cables are typical applications. In shock and vibration work, cable lengths seldom exceed a small fraction of a wavelength for the highest frequencies of interest, even with the long cable lengths required with some field tests. Except for a few special cases, general transmission-line theory is not applicable. The following sections deal with the physical properties of cables as they affect the performance of transducers and circuits normally encountered in shock and vibration applications.

LOW-FREQUENCY CHARACTERISTICS OF SHORT CABLES

CABLE CONSTANTS. When operated in the audio-frequency range the series inductance L and the shunt leakage G of good-quality, short cables are negligibly small in comparison with the other parameters and may be neglected. Figure 12.20*A* shows

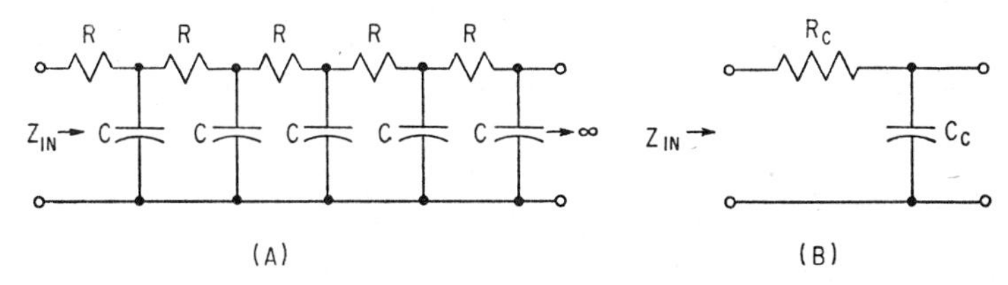

FIG. 12.20. Successive approximations in the representation of a short high-quality transmission line at audio frequencies. (*A*) Distributed constant configuration, neglecting series inductance and shunt leakage. (*B*) Lumped-constant configuration.

the equivalent low-frequency representation of a cable with distributed constants. For most purposes in shock and vibration applications, the simpler lumped-constant arrangement of Fig. 12.20*B* is sufficiently accurate. The quantities R_c and C_c are now the total ohmic resistance of the conductors and total capacity between them, respectively. Values for a typical coaxial cable, such as RG-58A/U,* are $R_c = 0.01$ ohm/ft, $C_c =$ 29 mmfd/ft. The nominal characteristic impedance of 50 ohms for this cable has no significance in these short low-frequency applications. The open-circuit input impedance of the cable is almost exclusively capacitive. When terminated, it takes on the impedance of the load, modified by the series and shunt parameters.

When used as a circuit element to connect a transducer to its associated electronic equipment, or to couple two pieces of equipment together, the capacity of the cable has, by far, the most detrimental effects on system performance. The extent of this shunt loading depends on the impedance levels involved in relation to the magnitude of the cable capacity, and the maximum frequencies of interest. The presence of the cable series resistance can be neglected except when it becomes comparable to either the source or the terminating impedance.

TRANSDUCER CABLES. A generalized circuit consisting of a cable connecting a signal source to its load is given by Fig. 12.21*A*. The generator may be a transducer or

* Coaxial cables of this type are manufactured by numerous companies, among which are Amphenol Cable and Wire Div., Tensolite Insulated Wire Co., Inc.

some preceding electronic stage having an open-circuit output voltage e_1. Its internal impedance Z_1 may be resistive or reactive. The load R_L is usually resistive, i.e., the input of an electronic circuit or a galvanometer. Typical applications are given below.

Piezoelectric Accelerometer Cables. As indicated in Chap. 16, piezoelectric accelerometers are charge generators with a very high (capacitive) internal impedance. Typical values of capacitance for barium-titanate accelerometers range around 500 mmfd. Such pickups generally operate into a cathode follower whose input impedance is so high that the loading effect of the input impedance may be neglected. The shunt

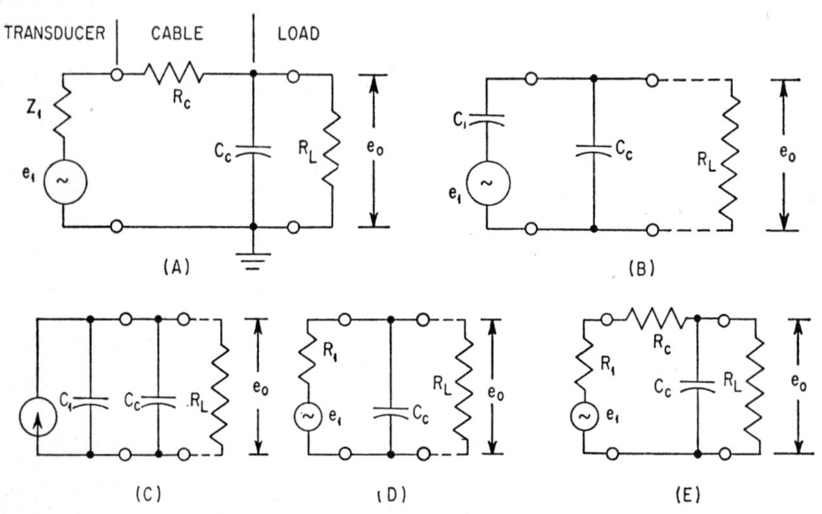

FIG. 12.21. Generalized circuits of a cable connecting a signal source to its load. (*A*) Generalized circuit. (*B*) Piezoelectric accelerometer as a voltage generator. (*C*) Piezoelectric accelerometer as a charge generator. (*D*) Electrodynamic velocity transducer operating into a high-impedance load. (*E*) Electrodynamic velocity transducer operating into a low-impedance load.

cable capacity C_c then forms a capacitive voltage divider with the internal accelerometer impedance C_1 (Fig. 12.21*B*) to yield an output voltage

$$e_0 = \frac{C_1}{C_1 + C_c} e_1$$

Since the open-circuit voltage depends on the accelerometer's charge sensitivity according to

$$\frac{e_1}{g} = \frac{Q/g}{C_1}$$

a substitution can be made yielding

$$\frac{e_0}{g} = \frac{Q/g}{C_1 + C_c} \tag{12.12}$$

for which the revised circuit diagram of Fig. 19.3*C* applies.

Equation (12.12) shows that the *voltage sensitivity* of a piezoelectric accelerometer is inversely proportional to the total shunt capacitance across its terminals. For this reason the accelerometer's sensitivity must be recalculated each time its cable length is changed. This method of adding shunt capacitors is sometimes used at the amplifier input to trim the voltage sensitivity to a particular value.

Cable loading effects on piezoelectric accelerometers are quite large. For example, the voltage sensitivity of a typical small barium-titanate accelerometer having an internal impedance of 500 mmfd, including a short length of cable, is reduced to two-thirds of its original value by the addition of only 8 ft of RG-58A/U cable. Because of these extreme variations, voltage-sensitivity figures for piezoelectric accelerometers are only approximate unless exact loading conditions are specified.

Velocity Transducer Cables. Small velocity pickups usually are of the electrodynamic type and have a resistive internal impedance R_1 of about 600 to 1,000 ohms. Usually they are connected to a cathode-ray oscillograph or other recorder whose input impedance R_L is several orders of magnitude greater than their own internal impedance. Output loading, if any, then will be due principally to the cable shunt capacitance C_c.

Figure 12.21*D* shows a circuit representing an electrodynamic velocity pickup operating into a high-impedance load. Neglecting R_L and the series cable resistance R_c, the ratio of output voltage to open-circuit voltage is

$$\frac{e_0}{e_1} = \frac{Z_c}{Z_1 + Z_c} = \frac{1}{1 + \omega R_1 C_c}$$

which indicates that the error increases with capacitive loading and frequency. For example, if typical conditions are assumed:

$$R_1 = 1{,}000 \text{ ohms}$$

$$f_{\max} = 2{,}000 \text{ cps}$$

$$\frac{e_0}{e_1} = 0.95 \text{ (5 per cent error)}$$

the capacitance C_c can be as great as 4,200 mmfd, equivalent to approximately 137 ft of RG-58A/U cable. The relatively long cable length permitted by these transducers, due to their low internal impedance, is the principal reason for their popularity in field measurements. No corrections need be made for normal laboratory lengths of connecting cable.

Occasionally a very low-impedance velocity pickup is operated directly into a magnetic oscillograph galvanometer or is terminated by a low-valued resistor to provide proper damping. In these instances, Fig. 12.21*E*, the cable series resistance R_c and connector contact resistances become more important than the shunt capacitance C_c. For example, if $R_1 = 0.5$ ohm and $R_L = 25$ ohms, the output voltage

$$\frac{e_0}{e_1} = \frac{R_L}{R_1 + R_c + R_L}$$

is 5 per cent down when the cable resistance R_c equals 0.8 ohm, amounting to approximately 80 ft of RG-58A/U cable.

Cables for Other Transducers. Cable effects for strain gages, potentiometers, and most other types of transducers, having relatively low-resistive internal impedances, are generally intermediate to the two cases cited above for velocity transducers. Transducer impedance, load impedance, and cable characteristics will determine which parameters are important and which can be neglected. In multichannel arrangements, phase shift and time delays caused by the reactive components of the input network and cable should be considered in establishing a common time base for different transducers.

INTERFERENCE AND NOISE IN TRANSMISSION CABLES

In an electrical sense, any unintentional or undesirable component of an electrical signal is described as "noise." Noise may be produced by many causes, but the types considered here are restricted to those which arise out of the use of connecting cables.

ELECTRICAL INTERFERENCE. The most common and troublesome sources of noise in instrumentation systems result from electrical interference or "pickup." These are noise components superimposed on the desired signal due to the proximity of the connecting cable to the electromagnetic field of an electrical disturbance. Hum from power lines, pickup from radio-frequency transmitters, and "static-type" disturbances from switching transients are the worst offenders. In general, precautionary procedures against electrical interference take the form of short cables, shielding, grounding, and, where permissible, the lowering of the line impedance.

Shielding. Shielding is accomplished by completely surrounding a susceptible circuit by a conductive surface which keeps the enclosed space free of external fields. Requirements for magnetic and electrostatic shielding are somewhat different. *Magnetic shields* are effective partially due to short circuiting of magnetic flux lines by low-reluctance paths and partially from self-annihilation due to opposing fields set up by eddy currents. Accordingly, they are made from high-permeability materials such as Permalloy, and are as thick as possible and contain a minimum of joints, holes, etc. A good magnetic shield is also a good electrostatic shield, but the converse is not true. *Electrostatic shields* provide a conducting surface for the termination of electrostatic flux lines, but need not be magnetic material. Stranded braid, mesh, and screens of good electrical conductors such as aluminum or copper are good electrostatic shields. Most shielded cables use copper braid as the outer conductor and shield.

Grounds. A circuit is *grounded* when one terminal is connected to the "earth" by a low-impedance path. Grounding removes the potential difference between that side of the circuit and earth, and eliminates the variable stray capacitances which tend to induce voltages in "floating" (i.e., ungrounded) systems. A good ground has the ability to absorb electrons without change in potential. In the laboratory, water and steam pipes make good ground connections because of their intimate contact with the earth. Metal building frames and power cable armor may be less satisfactory because of poorer bonding at junctions. A metal pipe driven several feet into moist soil makes a good ground in the field. At sea, the ship's metal hull and piping make good grounds.

Line Impedance. The magnitude of interfering components is proportional to cable impedance. High-impedance lines are more susceptible to undesirable pickup than low-impedance lines. Cable impedances therefore should be kept as low as time constants and loading effects will permit. If the interfering signals are considerably higher in frequency than the desired components, as is usually the case with radio-frequency transmitter interference, a shunt capacitor across the line will serve to short out the high-frequency pickup. The capacitor should be chosen small enough not to attenuate the highest frequencies desired, but large enough to be an effective short at the interfering frequency. Shunt capacitors are also useful for removing sharp spikes associated with interference from switching transients.

INTERFERENCE-SUPPRESSION TECHNIQUES. Double Shielding. Single-shielded cable is sufficient under ordinary operating conditions to keep electrical pickup levels to an acceptable minimum. For those installations where the cable lengths are long, where the impedance levels are high, or the interference is overly objectionable, a double-shielded cable is available. In this type cable, a second shielding braid is woven over the cable jacket, electrically insulated from the inner shield. The inner braid furnishes additional shielding against electrostatic fields which penetrate the first shield. For best results the shields should be connected together and grounded at only one point, preferably at the input to the associated electronic equipment.

Magnetic Fields. Magnetic fields associated with current-carrying power lines, electronic equipment power transformers, and line regulators are the most troublesome sources of magnetic interference in instrumentation setups. Interference is chiefly audio "hum" at the power-line frequency and its harmonics. Fields from power transformers produce a readily recognizable waveform containing a large third-harmonic component. At power-line frequencies, cable shielding provides rather ineffective protection. Since these fields attenuate rapidly away from the offending source, the most practical solution for this type of interference is to keep signal cables as far removed as possible. Occasionally, movable equipment can be reoriented to produce a null in induced voltage in the affected cable.

Ground Loops. Ground loops are formed when a common connection in a system is grounded at more than one point, Fig. 12.22*A*. If circulating ground currents cause a potential difference e_{gnd} to develop between these grounding points, the normal input signal will be modulated by hum and pickup from such a potential difference. The voltage seen by the input amplifier is then the sum of the normal input signal e_1 in series with the hum voltage e_{gnd} as given by the equivalent circuit, Fig. 12.22*B*. It is not uncommon for ground potential differences of several volts to exist between grounded electrical items only a few feet apart.

The principal source of ground loops in instrumentation setups results from the established practice of connecting the "low" side of transducers and electronic signal circuits to the instrument case. When several such items in a system are remotely located, Fig. 12.23*A*, and each housing is connected to the common signal lead and to a local earth ground, circulating ground currents can flow in the signal circuit if ground potential differences exist. The situation is further complicated if a number of inde-

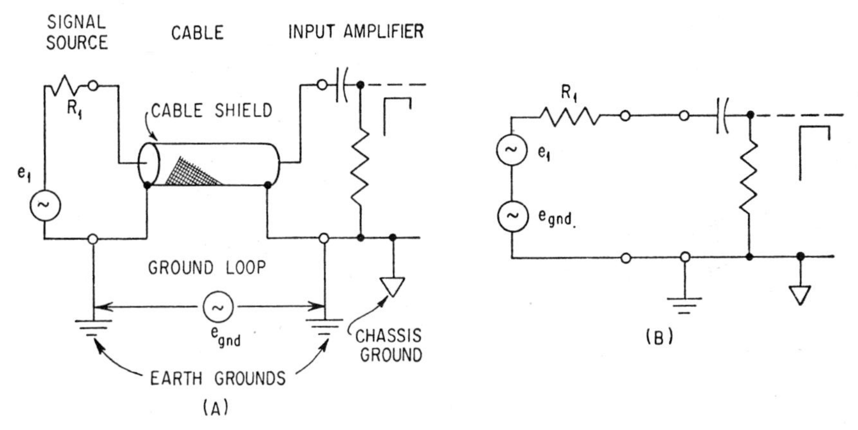

Fig. 12.22. Ground loops. (*A*) Multiple grounds in a system form ground loops which can cause interference if a ground potential difference exists. (*B*) Equivalent circuit of a system with a ground loop.

pendent data channels culminate in a single, multichannel recorder. Common lead grounding generally does not cause trouble if the separate equipments are physically close together or otherwise situated so that ground potential differences do not occur.

Ground-loop interference can be eliminated only by avoiding multiple grounds on the signal circuit; *the circuit should be grounded at only one point.* When it is possible to disconnect the low-signal side from the various housings, the arrangement of Fig. 12.23*B* is satisfactory. Although the case of each unit is connected to its own local ground, the signal circuit is grounded only at the recorder input and the previous ground loops are open-circuited. When single-conductor shielded cable is used with this arrangement, care must be taken to ensure that cable and connector shields do not inadvertently come into contact with grounded objects.

Some equipment, for example small piezoelectric accelerometers, Fig. 12.24*A*, are constructed so that the case is irrevocably part of the signal circuit. Ground loops can be broken with this type equipment by electrically insulating the housing from ground, as in Fig. 12.24*B*. Since potential differences are seldom over several volts, electrical breakdown is not a problem, and virtually any nonconductor will suffice. For transducers, the insulating material should be quite stiff so as not to filter out the higher-frequency components of the motion being measured. Dental cement is quite useful for this purpose; not only is it a rigid insulator but it sets up quickly and eliminates drilling holes in the test structure. Preamplifiers can be wrapped in electrical tape or merely placed on a pad of paper or cardboard, if they need not be tied down. Sponge-rubber pads also reduce microphonics from structure-borne vibration.

Two-conductor shielded cables should preferably be used when the signal circuit can be completely isolated from the transducer case, Fig. 12.24*C*. This relieves the

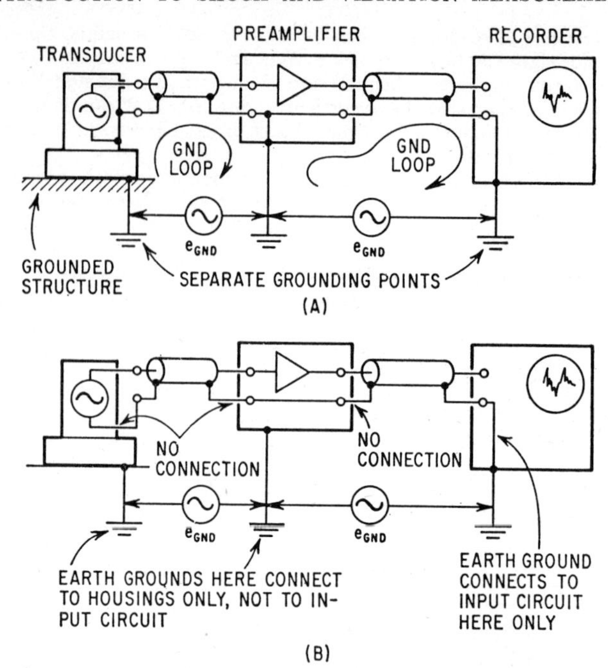

FIG. 12.23. Ground loops and preventive arrangements in a system of several series elements. (*A*) Transducers and preamplifiers with one side of their signal circuit connected to their cases or chassis are likely to suffer from ground-loop interference. (*B*) Proper arrangement for isolating the low side from individual chassis to open-circuit ground-loop potentials.

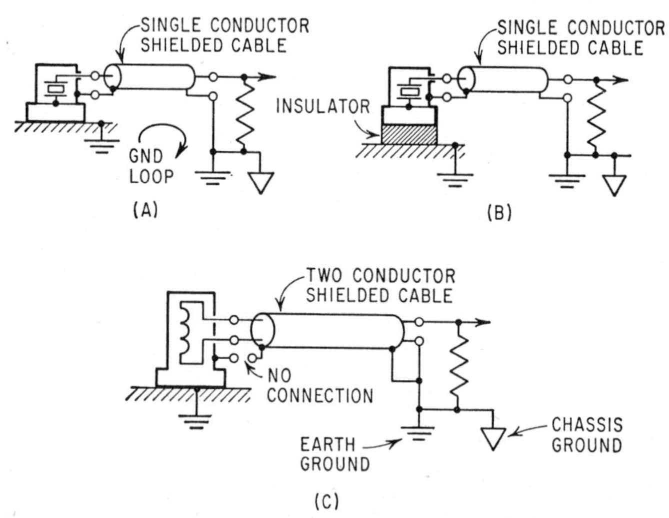

FIG. 12.24. Transducer cable connections for eliminating ground loops. (*A*) Many small piezoelectric accelerometers use the transducer case as the ground return and form ground loops when mounted on grounded structures. (*B*) Ground loops can be broken on grounded transducers by insulating the case of the transducer from ground. (*C*) Proper connections for transducers with floating terminals. Shield does not carry signal currents and is grounded at equipment input only.

shield of the responsibility for one side of the signal circuit and permits both sides of the line to be shielded. Normally the shield is connected to ground at one end only. However, little interference would result if it were accidentally grounded at a second place since it is not part of the signal circuit.

In all these circuits, care should be taken to maintain complete electrostatic shielding of the cable wires against radiated interference. With multiconductor cables this can be done, while preventing ground-loop currents through the shield, by connecting the shield to the connector shell at one end only. At the other end the shield is carried into the connector, but electrically insulated from it.

The most satisfactory place to make the system ground is at the recorder input. This is the only location in multichannel circuits that will not cause ground loops between channels. Grounding the recorder, which is frequently a-c operated, shorts out the large stray capacity to the power-line ground and greatly reduces system noise and hum. It also reduces the danger of shock to operating personnel who must make adjustments

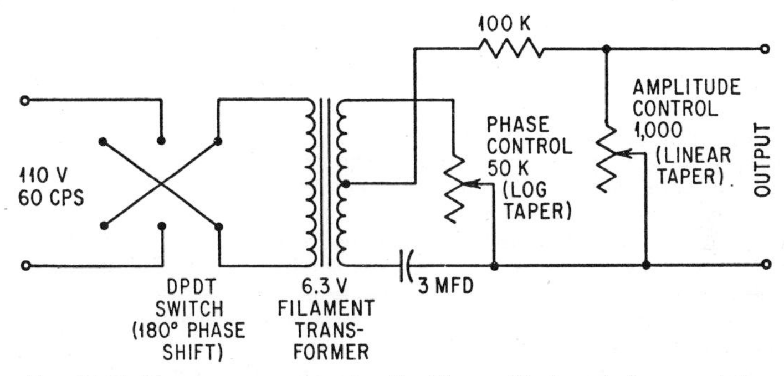

FIG. 12.25. Hum-compensating circuit with amplitude and phase controls.

to the equipment. Preamplifiers ahead of the recorder are often battery-operated and remote, and may be left "floating" without introducing appreciable capacitive hum pickup or becoming a shock hazard.

In some special cases, the system ground must be made at a point other than at the recorder input—for example, when measuring the line current across a shunt with an unbalanced input circuit. Operator safety dictates that the shunt should be in the line with the lowest potential to ground. The shunt then becomes the ground point of the entire system, and may make the recorder "hot" (i.e., at a high potential) with respect to other grounded objects. Consequently, it should be ungrounded and insulated. If it is part of a multichannel system it is a good precaution to fuse the line between shunt and recorder low side in the event an accidental earth ground develops on any other part of the system. With the capacity of the generator available, short-circuit current can be very damaging to electronic circuit parts and wiring.

Hum Compensators. Ungrounded transducers, or those not readily adapted to shielding, pickup, and leakage within a-c operated equipment, are common sources of hum interference which are difficult or inconvenient to eliminate. When the residual noise in a system cannot be reduced to an acceptable level by careful application of interference-suppression techniques, and consists mostly of hum at the power-line frequency, considerable improvement in signal-to-noise ratio can be obtained with *hum compensators* or *hum buckers*. These networks introduce into the circuit an additional signal derived from the power line, whose amplitude and phase can be adjusted to exactly cancel the residual hum interference. A typical circuit is shown in Fig. 12.25. The circuit permits independent adjustment of amplitude and phase. A filament transformer supplies low-voltage power to the phase-shift network and isolates it from the power-line ground. The hum compensator can be connected either in series or parallel with the signal, depending on the components selected, local d-c levels, and impedance at the point of injection.

TRIBOELECTRIC CABLE NOISE. One of the early difficulties encountered with accelerometer systems in shock work was a spurious "cable noise" generated by the cable whenever it was suddenly squeezed, bent, struck, or mechanically distorted. Peak noise voltages from this source were frequently as large as the actual acceleration signals being recorded. Little was then known of the mechanisms which produced cable noise, except that it was found experimentally that cables of certain type and manufacture were better than others and that good cables deteriorated rather rapidly in shock test use. One homemade design, consisting of a center conductor encased in ¼-in. copper tubing and then filled with beeswax, proved far superior to any of the then available commercial microphone or coaxial cables. Its principal disadvantage was its rigidity. The same objection pertained to a thick underwater cable used somewhat later, but which had acceptably low-noise characteristics due to a conducting layer of graphite between the dielectric and shield.

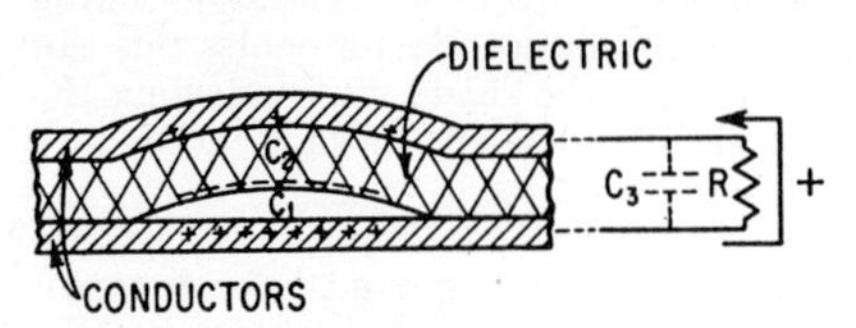

FIG. 12.26. A section of cable during distortion showing how the separation of triboelectric charges produces "cable noise" across the terminating resistor. (*After T. A. Perls.*[6])

As a result of extensive investigations in this field,[6] it was shown that cable noise was the result of the separation of triboelectric charges when the dielectric momentarily lost intimate contact with either the center conductor or the shield because of mechanical distortion. A simplified representation of this condition is shown in Fig. 12.26. In this case it is assumed that negative charges have been left on the dielectric, where they are trapped by the low conductivity of the dielectric surface. The excess charge on the conductor, being free to move, is neutralized by a transfer of electrons between conductors through the terminating impedance at a rate determined by the time constant. The transfer produces a voltage pulse on the input of the following electronic equipment. On reestablishment of dielectric-conductor contact, the excess negative charge on the conductor-dielectric interface is again neutralized by charge redistribution, resulting in a pulse of the opposite polarity. Peak amplitudes of the voltages developed depend on the affected area, charge separation, total cable capacity, and the ratio of the two local capacities formed by separation.

From the above analysis, it is apparent that flexible low-noise cables can be produced if the dielectric surfaces are covered by a conducting coating. The conductivity of the coating need not be high, just as long as it provides a leakage path on the dielectric surface for charges which would otherwise be immobilized there during mechanical separation. The charges then are redistributed through a local path rather than through the terminating impedance where they are indistinguishable from accelerometer signals.

Miniaturized, flexible, low-noise cables using this coating technique are available from several manufacturers. These treated cables exhibit very low cable-noise characteristics and are generally capable of withstanding considerable abuse before the coatings develop enough holes to become noisy. In this respect it is important in assembling these cables, as in splicing or fitting them with connectors, that none of the conducting material be allowed to form a leakage path between conductors. Carbon tetrachloride and xylene are satisfactory solvents and cleaning agents.

REFERENCES

1. Levy, S., and W. D. Kroll: *Research Paper* 2138, *J. Research Natl. Bur. Standards*, **45**:4 (1950).
2. Lawrence, A. F.: "Crystal Accelerometer Response to Mechanical Shock Impulses," *Rept.* 168–56–16, Sandia Corp., Albuquerque, N.M.
3. Landee, R. E., D. C. Davis, and A. P. Albrecht: "Electronic Designers' Handbook," McGraw-Hill Book Company, Inc., New York, 1957.
4. Terman, F. E.: "Electronic and Radio Engineering," 4th ed., McGraw-Hill Book Company, Inc., New York, 1955.
5. Skilling, H. H.: "Electric Transmission Lines," McGraw-Hill Book Company, Inc., New York, 1951.
6. Perls, T. A.: *J. Appl. Phys.*, **23**(6):674 (1952).

13

MECHANICAL INSTRUMENTS FOR MEASURING SHOCK AND VIBRATION

George Stathopoulos
Naval Surface Weapons Center

Charles E. Fridinger
Naval Surface Weapons Center

INTRODUCTION

This chapter describes mechanical instruments used in the measurement of shock and vibration. Such instruments produce limited information and may thus lack the precision and accuracy obtained with electromechanical transducers and electronic systems, but they have features which make their use attractive under certain conditions: they are self-contained, rugged, simple to use, highly reliable, and low in cost. Normally a technically oriented person can become proficient in their use and operation with only a minimum of training. Many of these instruments can be used with equal facility in the laboratory and in the field. With one exception, the instruments described are available commercially.

CLASSIFICATION OF INSTRUMENTS

The instruments described in this chapter are classified as follows:

VIBRATION-MEASURING INSTRUMENTS
- Vibrographs
- Vibrometers
- Reed-type frequency-indicating vibrometers
  - Multiple-reed frequency gages
  - Variable-length reed vibrometers

SHOCK-MEASURING INSTRUMENTS
- Shock recorders
- Shock-overload indicators
- Peak-reading instruments

VIBRATION-MEASURING INSTRUMENTS

PRINCIPLES OF OPERATION

An ideal mass-spring type pickup (sometimes called a "seismic instrument"), described in Chap. 12, is shown in Fig. 12.3. It consists of a mass supported on, or suspended from, the case by a spring; the motion of the mass within the case may be damped by a viscous fluid or electric current. The pickup case is attached to the vibrating surface by suitable means and its motion is essentially the same as that of the surface. The relative motion between the case and the spring-supported mass is the response of the instrument. The natural frequency of the mass-spring system determines the instrument's response characteristics.

DISPLACEMENT MEASUREMENT

For an applied sinusoidal motion, if the frequency of applied motion is much higher than the natural frequency of the mass-spring system, then the response displacement amplitude is approximately equal to the displacement amplitude of applied motion. This is the characteristic of a displacement pickup of this type. Instruments using this principle for measuring displacement are no longer available commercially. Those which are commercially available are designed to be hand-held rather than fastened to the vibrating surface. In a device of this type, the instrument has a prod which is held against the surface whose vibration is to be measured (see Fig. 13.1). The prod is forced against the surface by a spring fastened to the case of the instrument. The spring, which pushes against the case also, acts as the spring element of the mass-spring system—the mass of which is provided by the remainder of the instrument. In principle, the hands of the operator support the case in such a way that the case acts as the mass. In practice, the operator's hands introduce an indeterminate and variable amount of mass, stiffness, and damping. Consequently, the hand vibration pickup is not a true mass-spring type of instrument.

ACCELERATION MEASUREMENT

The acceleration pickup has essentially the same features of design, construction, and behavior as the mass-spring type of displacement pickup described above. In contrast to the latter, however, the acceleration pickup has a relatively stiff spring so that its natural frequency is comparatively high. Physically, the behavior of the accelerometer pickup may be described as follows. Consider a mass-spring system which is less than critically damped. Suppose that the system is subjected to a sinusoidal vibration and has a natural frequency several times greater than the frequency of applied vibration. Then, the spring experiences an alternating force of amplitude equal to the mass multiplied by the acceleration amplitude; consequently, the spring deflection (i.e., the response of the system) is proportional to the applied acceleration. Several instruments for measuring acceleration at low frequencies, associated with packaged-item response, are included in this chapter under the heading *Shock-Measuring Instruments*.

VIBROGRAPHS

A vibrograph is a mass-spring type of displacement pickup with self-contained means for recording the relative motion between the prod and the case and consisting of a recording medium contained on a supply spool and a transport mechanism for the recording medium.* The principle of operation of the vibrograph is shown in the schematic diagram of Fig. 13.1. The relative motion between the prod and the case is inscribed on a recording medium by the stylus, being magnified by the lever ratio $(c - g)/(e - c)$. The ratio between the displacement of the stylus and the relative motion between the prod and case is called the "magnification ratio."

In use, the operator holds the case and pushes the prod against the surface whose vibration is being measured, applying enough force to center the stylus on the paper. This introduces a significant limitation in the use of the instrument; any unsteadiness of the operator in holding the instrument is recorded on the paper. It is sometimes difficult to distinguish this unsteadiness from low-frequency vibration, particularly if the measure-

* A commercially available vibrograph of this type, the Davey Hand Vibrograph, is manufactured by the Vibroscope Co., Inc., Kingston, N.Y.

ments are made in a moving vehicle or under other conditions where steadiness is difficult to maintain. A steady, vibration-free support for the instrument is advantageous but often unavailable.

The principal advantages of the vibrograph are: (1) a permanent record of vibratory displacement as a function of time is available immediately, thus giving an indication of the waveshape and frequency content of both steady-state and transient vibration; and (2) the instrument is portable, convenient, and ready for immediate use without assembling auxiliary equipment.

There are pronounced disadvantages. (1) If the vibratory accelerations are high, because the motion is of large amplitude or high frequency, the prod may not be able to follow the motion; instead it may chatter. (2) The form of the record is not well adapted to detailed data reduction except by laborious graphical methods, although estimates of frequency and displacement may be obtained by inspection. (3) The instrument is limited to the measurement of vibration whose frequency is relatively low and whose displacement is relatively large, although the range can be extended to a lower value by the use of a microscope to analyze the record. The improved resolution obtained by use of a microscope is generally limited somewhat by the "fuzziness" of the trace at moderate magnification.

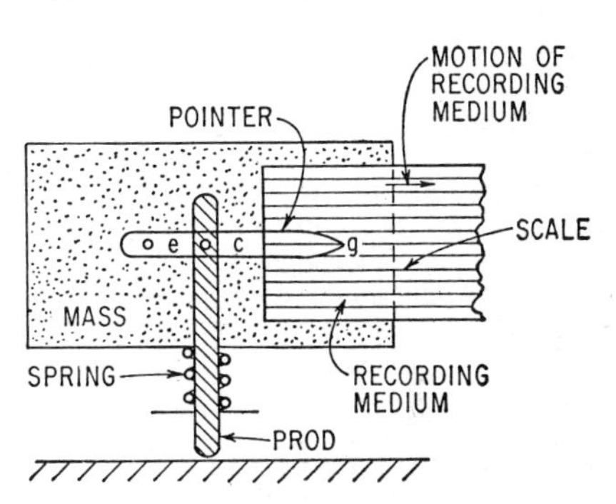

FIG. 13.1. Schematic diagram of the hand vibrograph. The pointer arm is supported by the housing (mass) and is coupled to a prod. The end of the prod is held against the vibrating surface. Motion of the prod is magnified by the ratio $(e - g)/(e - c)$ and inscribed on a moving recording medium. The recording medium and the associated drive mechanism are a part of the mass.

Another significant disadvantage of the vibrograph is the relatively large loading imposed on the vibrating surface which usually makes it impractical for use in recording vibration of relatively light structures. Most vibrographs weigh several pounds, although some are many times heavier. The loading imposed on the structure being measured is that due to the "unsprung" weight of the vibrograph, plus that due to the reaction forces in the springs supporting the mass of the mass-spring system. For hand vibrographs, only the spring-reaction loading is imposed via the prod; however, even this can modify the vibration of a light structure, thereby introducing error into the measurement.

The upper limit of the useful frequency range frequently is determined by the dynamics of the mechanical linkage and the recorder stylus.

VIBROMETERS

A vibrometer is essentially the same as a vibrograph, except that the response of the instrument is displayed as a nonpermanent reading rather than as a chart tracing. One manufacturer* provides vibrometers in which the prod drives a small mirror system through a sensitive leverage system for amplification. A bulb lights the mirror, and the light is then reflected onto a calibrated ground-glass scale. The vibration causes the light beam to form a ribbon of light on the scale that represents the peak-to-peak displacement of the vibrating body. Models are available which are designed to be used by supporting them on a firm base near the vibrating object and for hand-held use.

REED-TYPE FREQUENCY-INDICATING VIBROMETERS

In measurement problems involving relatively steady vibration, one of the quantities of major interest is the dominant frequency. Determination of this frequency often

* The Vibroscope Co., Inc., Kingston, N.Y.

provides the information needed to identify an unknown source of vibration. For example, the source of vibration in a building can be traced to an offending compressor. Measurement of vibration frequency also can be used conveniently to determine the rotational speed of an engine or motor.

Reed-type frequency-indicating vibrometers are useful devices for such measurements. Such instruments depend on the resonant vibration of a cantilever reed with low damping to indicate the frequency of the exciting vibration; low damping is achieved by careful design and assembly to minimize losses at the clamping end. These devices are of two types: (1) multiple-reed gages which employ a number of reeds of different natural frequencies that are closely spaced and (2) single-reed, adjustable-length gages.

The fundamental natural frequency f_n of a cantilever reed of the type used in these instruments is given by

$$f_n = \frac{3.52}{2\pi} \sqrt{\frac{EI}{\mu l^4}} \quad \text{cps} \tag{13.1}$$

where E is modulus of elasticity, I the section moment of inertia, μ the mass per unit length of the reed, and l the length of the reed. In the multiple-reed type of frequency indicator, each reed generally has a small mass on the end. For a cantilever with a mass on the end, the fundamental natural frequency is given by the expression

$$f_n = \frac{1}{2\pi} \sqrt{3EI/l^3(m_2 + 0.23m_1)} \quad \text{cps} \tag{13.2}$$

where m_1 is the mass of the reed and m_2 is the value of the mass added to the end.

In order to determine the frequency of vibration, the reed vibrometer is held or placed with its base in contact with the vibrating surface. A multiple-reed instrument contains a number of reeds of different natural frequency, usually progressing in uniform increments of frequency between adjacent reeds from the lowest to the highest. If the vibration being observed is steady and has a major frequency component within the range covered by the reeds, several of the reeds closest in natural frequency to this component will vibrate with relatively large excursions, the one which is closest in frequency having the largest excursion. The instrument has a frequency scale indicating the natural frequency of each reed, so that the vibration frequency may be determined visually.

In vibrometers having a single, adjustable reed, the clamped end of the reed is constrained between two spring-loaded rollers; its length (and therefore its natural frequency) is adjusted by extending the reed from the body of the instrument. In use, with the instrument held against the vibrating surface, the length of the reed is varied slowly until the reed vibrates steadily with its greatest excursion. The nonvibrating portion of the reed has an index mark on its very end, which moves along a scale calibrated in frequency as the free length of the reed is changed. The natural frequency of the reed is read on this scale.

Although this type of instrument is convenient to use because it is extremely light and compact and can be used anywhere the observer can reach, it possesses a serious limitation. Adjustment of the reed length, in the process of tuning for resonance, as well as inadvertent motions of the operator's hand, may excite the reed into vibration at its natural frequency; since the reed is very lightly damped, such spurious oscillations continue for some time. This often makes tuning to the frequency of applied vibration time-consuming—particularly when the vibration is varying in amplitude. However, for relatively pure sinusoidal vibration, there is a marked indication at resonance, and the frequency can be determined readily.

MULTIPLE-REED FREQUENCY GAGES. A typical multiple-reed gage is shown schematically in Fig. 13.2. The reeds all have different natural frequencies, designed to give continuous frequency coverage throughout a given range of frequencies. This indication is given by resonant vibration of the reed at, or closest to, the frequency of the vibrating body. The reeds have white marks on their ends to aid in determining visually which reed is in resonance. If a multiple-reed gage of this type is subject to vibration of discrete frequency, the reed corresponding most nearly in natural frequency to the exciting frequency will show the greatest displacement amplitude; if the exciting frequency is midway between the natural frequencies of two adjacent reeds, they will show equal displacement amplitudes.

These instruments have the advantage of being small, portable, reasonably accurate and reliable; they do not require a source of power. They are particularly useful for indicating speeds of rotating machinery. A major limitation is that they are effective only for rather steady vibration. In addition, a relatively limited frequency range is covered by individual units.

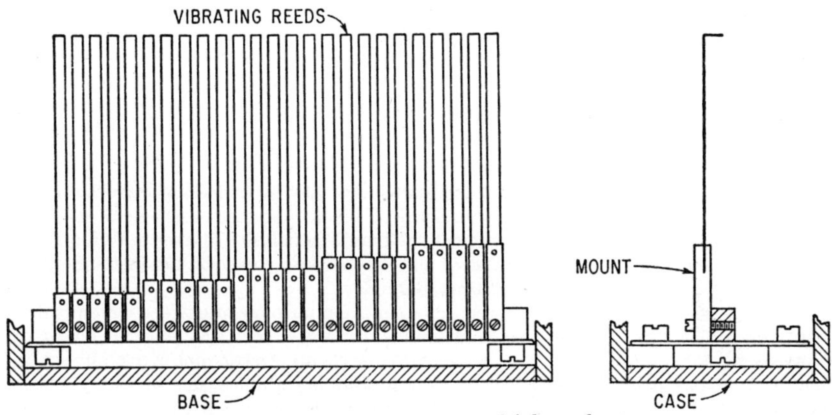

Fig. 13.2. Schematic diagram, multiple-reed gage.

The minimum displacement amplitude which gives an observable response is about 0.1 mil.

One manufacturer* has a multiple-reed gage which covers the frequency range from 10 to 1,500 cps. Another manufacturer† has an instrument of this type which covers the range from 10 to 1,000 cps.

VARIABLE-LENGTH REED VIBROMETER

This vibrometer is a single-reed, adjustable-length frequency indicator. A typical device of this type, shown schematically in Fig. 13.3, consists of a thin steel cantilever reed, clamped between rollers. Rotation of one of the rollers, by means of a thumb nut,

* James G. Biddle Co., Plymouth Meeting, Pa.
† Herman H. Sticht Co., Inc., New York, N.Y.

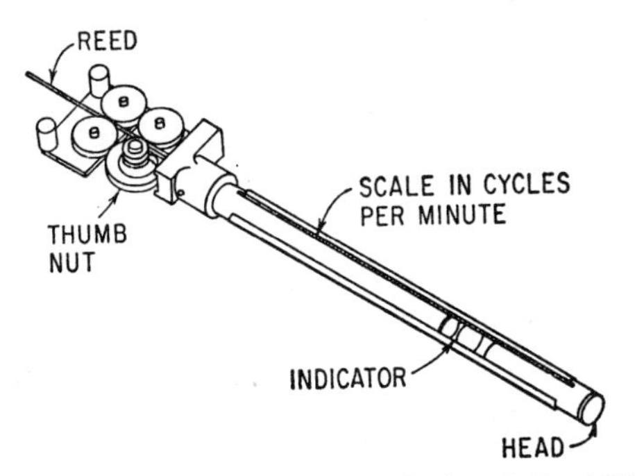

Fig. 13.3. Schematic diagram of a variable-length Reed Vibrometer.

changes the length of the free reed, resulting in a corresponding change in natural frequency. The frequency is indicated by the position of a pointer on the fixed end of the reed along a calibrated frequency scale on the frame. To measure vibration frequency, the head of the instrument is held against the vibrating surface and the reed length is slowly extended until the reed vibrates at maximum amplitude. One manufacturer* provides vibrometers covering the range from 200 to 21,000 cpm. A second manufacturer† provides vibrometers covering a range from 250 to 20,000 cpm.

SHOCK-MEASURING INSTRUMENTS

Instruments for measuring mechanical shock may be classified as
Shock recorders
Shock-overload indicators
Peak-reading instruments

SHOCK RECORDERS

A "shock recorder" is a self-contained instrument having a mass-spring system which is intended to provide a time-history record of shock motions. Such devices are used principally in the transportation and package-handling fields. The natural frequency of the mass-spring system is a function of the acceleration range of the instrument—the higher the range, the higher the natural frequency.

A stylus (attached to the mass) records its motion on a paper surface which is driven continuously or intermittently. "Continuous-drive systems" are powered either by a slow-speed clock motor (usually ½ to 3 in./hr) or by an electric motor (usually up to 12 in./sec). Clock drives are used in long-term shipping applications in which peak acceleration and time of occurrence are to be determined; motor drives are used in short-term shipping and laboratory applications in which magnitude-pulse shape information is to be determined. "Intermittent-drive systems" are powered by a motor which advances the writing surface when a preset value of acceleration is exceeded, thereby actuating a ratchet. This allows a large number of shock recordings on a relatively small roll of paper but does not provide information about time of occurrence. The useful frequency range which is determined by the natural frequency of the mass-spring system is limited.

Some instruments of this type are labeled "accelerometers" by their manufacturers even though they are not accelerometers in the strict sense of the term. Nevertheless, they can produce amplitude information that is indicative of the maximum value of acceleration. Other instruments provide shock magnitude information in terms of "zones"; each zone is associated with a certain severity (usually defined in terms of g) of shock or impact. The response data can be interpreted as a velocity change for those cases in which the pulse duration is considerably shorter (at least five times) than the natural period of the mass-spring system. The governing relationship is

$$\Delta V = \frac{A_{pk}}{2\pi f_n} \tag{13.3}$$

where A_{pk} is the peak recorded acceleration in feet per second squared, f_n is the natural frequency of the mass-spring system, and ΔV is the velocity change in feet per second.

Most shock recorders are friction-damped, having an equivalent damping ratio of approximately 0.01. Special models are damped pneumatically, having a damping ratio of 0.65. For most impact situations, peak response errors of less than 5 per cent can be expected for lightly damped instruments having a natural frequency at least five times that of the highest significant frequency of the input pulse.[1] For instruments having a damping ratio of 0.65, the natural frequency need be only 1.7 times the highest frequency of the input pulse for comparable error. The two damping ratios meet the phase-shift requirements, zero or linear phase shift, for accurately recording pulse-shape information.

* Martin Engineering Co., Neponset, Ill.
† Vibroscope Co., Inc., Kingston, N.Y.

One type* of shock recorder is used to record the severity of freight-car end impacts (longitudinal) and the times at which the impacts occur. Different models of this recorder with greater sensitivity are used in the same manner in freight cars equipped with longitudinal cushioning devices, in rail passenger cars, and in laboratory package testing. The sensing unit consists of a weight mounted on two rods riding on rollers, the movement of which is restrained by springs on either side. The natural frequency for this system is about 6.5 cps.

The direct-recording stylus is mounted in the center of the moving weight. Longitudinal impacts are recorded from both sides of the centerline in five zones of impact; the width of each zone after the first is 0.47 in. The instrument has a static calibration of $2g$ per zone.

A second shock recorder† employs three-axis sensing elements, each of which is an angular vibration system with one degree-of-freedom in the direction in which it monitors. The sensing elements are mounted on a "stylus bracket" with an amplifying linkage. Recording of stylus motion is on pressure-sensitive, preprinted, plastic- or wax-coated chart paper. The natural frequencies of the sensing elements are sufficiently low that most of the shocks being monitored appear impulsively applied. Thus frequency response, in the conventional sense, is not a consideration. Maximum sensitivity ratings of these instruments are 2, 4, 6, 10, 15, 20, 30, 50, 70, 100, 200, or $300g$.

A third type‡ of shock recorder has available sensing elements in a wide range of frequencies and amplitudes. The recorders are used to measure accelerations in automotive vehicles, railroad cars, ships, aircraft, missiles, and laboratory impact testing. Two models are available; one has one-axis sensitivity and the other has three-axis sensitivity. The one-axis unit is electrically driven with drive speeds up to 2 in./sec and has stylus deflections to 4 in. The three-axis unit has clock drives and electrical drives; drive speeds up to 2 in./sec; and stylus deflections up to 2 in. At the lower chart speeds the recordings are of acceleration peaks only; however, at the higher speeds low-frequency shock and vibration acceleration-time oscillograms can be obtained.

A fourth type§ of shock recorder is used to record acceleration in three axes for a variety of transportation and impact-testing applications. Each stylus records its respective impact on a 1-in. space on the chart. Two types of damping are available: friction and pneumatic. The friction-damping ratio is 0.01 and the pneumatic-damping ratio is 0.65. The recorder ratings are 1, 2, 4, 5, 6, 10, 15, 20, 25, 30, 50, 70, and $100g$. Their natural frequencies are a function of the g rating and range from 10 cps for the $1g$ instrument to 88 cps for the $100g$ instrument. The operating or flat response ranges of the pneumatic-damped and friction-damped instruments are 50 and 25 per cent, respectively, of the natural frequency. When used with high-speed electric drives, shock and vibration acceleration-time oscillograms are produced.

The smallest in a line of ruggedized shock recorders‖ contains sensors of the mass-spring type with a compound spring system consisting of two primary coiled springs and two secondary cantilever springs. The springs are secured in such a manner that the g range, frequency response, stylus pressure, and zero setting can be changed. An eccentric system provides a 16:1 amplification between the sensing mass and the sapphire recording stylus. The standard sensitivity settings range from 2.5 to $100g$ with corresponding natural frequencies of approximately 14 to 86 cps. The system incorporates friction-damping which permits a distortionless recording response from zero frequency to a frequency equal to 25 per cent of the natural frequency. Recording is on plastic- or wax-coated charts.

SHOCK-OVERLOAD INDICATORS

These instruments are used widely in the transportation and package-handling fields to determine whether the peak of an applied shock exceeds a value of acceleration which has been preset on the instrument. They are low in cost, reusable, and available in a

* Savage Impact Recorder, Impact Register Co., Champaign, Ill.
† Impact-O-Graph, Impact-O-Graph Corp., Bedford, Ohio.
‡ Model 1A and 3A Accelerometers, Impact Register Co., Champaign, Ill.
§ Model R-M Three Way Accelerometer, Impact Register Co., Champaign, Ill.
‖ GEMM (G-Environmental Mini-Monitor) Recorder, Impact Register Co., Champaign, Ill.

wide range of actuation levels, but their performance varies both with pulse duration and orientation.[2, 3] To obtain the most accurate data with these devices, they should be calibrated under conditions anticipated in actual service.

Many different schemes are used in shock-overload indicators for sensing the applied shock and indicating that it has exceeded the preset level. Basically, the sensing elements can be considered as masses and springs in an undamped seismic system. The motion of the sensing system can be approximated by Eq. (12.2) with one important restraint, viz., no motion is generated until the input acceleration exceeds the preset acceleration value of the indicator; then an indicator is actuated—provided the duration of the pulse is greater than a threshold value (threshold values of 10 to 30 millisec are typical).

Many types of shock-overload indicators are available commercially. The manufacturer's instructions should be followed in attaching the indicators to the specimen being monitored. Inadequate or excessive attachment forces can lead to erroneous results. Indicators having definite axes of sensitivity should be mounted so they are properly aligned with the anticipated direction of applied acceleration.

One type* of shock-overload indicator contains a spring-loaded mass with an integral trigger which engages a spring-loaded sleeve. If the shock pulses are of sufficient magnitude, properly oriented, and at least 30 millisec in duration, the inertia force of the mass compresses the spring. This action disengages the trigger from the sleeve. Once free, the trigger is forced by its spring to move along a rod, exposing a red warning band which can be viewed through a transparent plastic dome. This device is sensitive only in a hemispherical direction, but two units may be mounted back to back so that omnidirectional sensitivity can be obtained. They are rated from 5 to 50g in 5g increments.

A second type† of shock-overload indicator employs two identical sensing elements oriented at right angles to each other. Each sensing element consists of a spring with a steel ball at each end. The spring-ball assemblies are compressed and nested within a transparent impact-resistant plastic housing. The springs are calibrated for the actuation range of the indicator. This device is sensitive to acceleration pulses in any direction if the pulse has an amplitude in excess of the rating and the pulse is at least 8.4 millisec in duration. Actuation is indicated by the dislodging of at least one spring-ball assembly. These devices are rated from 1 to 150g.

A third type‡ of shock-overload indicator is a uniaxial, bidirectional unit. A mass-spring system is used for sensing acceleration in each direction. The mass is a steel weight free to slide within the housing guide and designed to bear against the spring. The spring is a leaf-type, snap-action spring supported at its ends. The mass-spring parameters are chosen to provide a snap action at the g rating of the unit. The triggered spring causes an arrow to move into view. Three units mounted in mutually perpendicular orientations will provide three-axis coverage, although higher levels are required to trigger it in directions other than the principal axes. These devices are rated at 5, 10, 15, 25, or 50g. The minimum pulse duration for accurate operation varies somewhat with the setting of the instrument—the higher the setting, the lower the duration of the pulse. The 25g indicator requires a shock duration of at least 11 millisec.

PEAK-READING INSTRUMENTS

Peak-reading instruments are used in the transportation and package-handling fields and in high-shock applications such as missile ground- and water-impact shocks. They are also used in shock application for which it is difficult or expensive to use electronic systems requiring trailing wires or telemetry. They are low in cost, reusable, and available in a wide range of maximum ranges. The principal disadvantage is that time information is not provided, making it difficult to interpret the readings. These instruments are of the mass-spring type. They provide an indication of the peak acceleration of the applied shock; for an ideal instrument it is the maximum response of Eq. (12.2).

* Shockmaster Shock-Overload Indicator, Arizona Gear and Manufacturing Co., Tucson, Ariz.

† Omni-G All Direction Shock Indicator, Impact-O-Graph Corp., Bedford, Ohio.

‡ Model 130 Impact Recorder, Vexilar, Inc., Minneapolis, Minn.

Clusters of these instruments, each having a different natural frequency, are used to obtain shock spectra. Care must be exercised to use these instruments in their linear range in order to minimize errors in interpretation.

One type* of peak-reading instrument has as its principal elements an ink-coated silicone sphere, a frosted Mylar cube liner, and a molded 2-in. clear Lexan cube. The liner covers the inside faces of the cube and has a graduated scale, on each of its six faces, which can be viewed through the cube. The sphere is placed within the liner, making point contact against the center of each of its six faces. The sphere acts as the mass-spring system; the liner acts as the recording surface; and the cube acts as the housing. Under shock, the sphere deforms against one or more faces of the liner, leaving a permanent circular imprint. This action is similar to that of pressing a wet ball against a wall: the harder the ball is pressed, the larger the impression left on the wall. The imprint diameters are related to peak acceleration by calibration data furnished by the manufacturer. The unit has omnidirectional sensitivity, and the direction of shock can be inferred from the surfaces which are inked. Its range is 3 to 150*g*. It operates within its calibrated range for shock pulses as short as 11 millisec.

Another type† of peak-recording instrument is a uniaxial, bidirectional unit. The sensing elements are reeds equipped with diamond-tipped styli. Under shock the reeds deflect and the motions of their styli are inscribed on a gold-plated recording hub. The inscribed deflections are measured with the aid of a microscope and related to accelerations by calibration data provided by the manufacturer. The recording hub is replaced for each shock. Two models are available. One has reeds ranging in natural frequency from 310 to 870 cps in 80-cps increments; and the other has reeds ranging in natural frequency from 950 to 2,000 cps in 150-cps increments. The *g* range of each reed is a function of its natural frequency. The overall *g* range for the two models is 5.6 to 10,200*g*. The two models may be used simultaneously by gang-mounting them. The peak acceleration of each reed can be plotted as a function of its natural frequency to obtain a shock spectrum of the pulse.

COPPER-BALL ACCELEROMETER. Copper-ball accelerometers[4] are a type of peak-reading instrument. The components of these accelerometers are a cylindrical mass (inertia weight), an annealed copper ball, a ball holder, and a housing as shown in Fig. 13.4. Under shock the mass permanently deforms the ball, and the deflection is

* Recorda-g Impact Recorder, H. L. Boardman, Inc., Clayton, Mo.

† Peak Shock Recorder, Engdahl Enterprises, Costa Mesa, Calif.

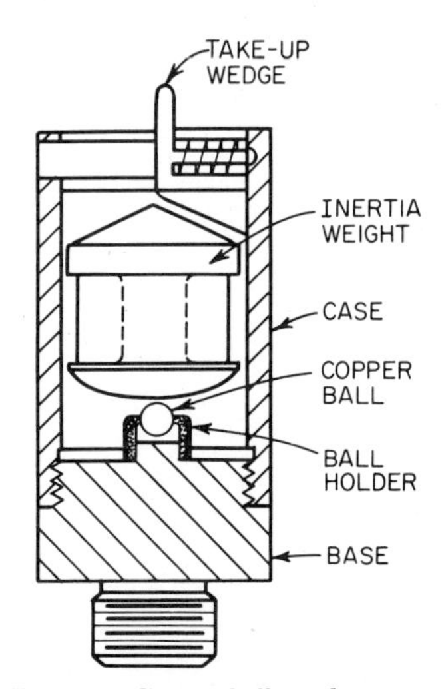

FIG. 13.4. Copper-ball accelerometer.

interpreted as the peak acceleration. The deflections are read with either micrometers or microscopes. The near-linear, force-deflection characteristics of the copper balls permit their use as the sensing element of an accelerometer.

REFERENCES

1. Stathopoulos, G., W. W. Mebane, and W. W. Wassmann: "Shock and Vibration Instrumentation," *Instr. Soc. Amer. Paper* ASIT 74202 (15–64), May, 1974.
2. Isaacson, A., and I. Zuckerman: "Report of Investigation of Alternate Accelerometers for Monitoring Shock Levels During Transportation of LM Flight Components," *NASA Rept.* LRP400-1, October, 1970.
3. Brown, R. V.: "Analysis of Shock Indicator Devices for Packaging Applications," *Wright-Patterson AFB Rept.* 73–52, July, 1973.
4. DeVost, V. F.: "Shock Spectra Measurements Using Multiple Mechanical Gages," *U.S. Naval Ordnance Lab. Tech. Rept.* 67–151, September, 1967.

14

SELF-CONTAINED ELECTRONIC VIBRATION METERS AND SPECIAL-PURPOSE TRANSDUCERS

John D. Ramboz
National Bureau of Standards

INTRODUCTION

This chapter describes: (1) commercially available electronic vibration meters which are "self-contained" (i.e., instruments whose components are confined to a single portable case, usually with batteries, and connected to a vibration pickup by a cable), and (2) special-purpose and miscellaneous vibration transducers and measurement systems of the following types:

Capacitance-type transducer
Potentiometer-type transducer
Servo accelerometer
Optical-electronic
Reflected wave (microwave or ultrasonic)
Optical interferometer

SELF-CONTAINED VIBRATION METERS

The term "vibration meter" is applied to a self-contained measurement system comprised of a vibration pickup, adjustable attenuator, self-contained amplifier, direct-reading meter or oscilloscope, and an internal power source, usually rechargeable batteries. Many are equipped with an electrical output for magnetic-tape recording, headphone listening, and oscilloscopic display, and for connection to a frequency analyzer or some other type of data-processing equipment. Some vibration analyzers include electrical filters.

Vibration meters provide the following advantages over the mechanical vibrographs described in Chap. 13. (1) Their frequency ranges are larger. (2) Their amplitude ranges are greater. (3) Their use of a separate vibration pickup (usually small and light, e.g., 2 to 4 cm or less in maximum dimension, and 20 to 200 gm in mass) results in a smaller load on the object or structure being measured and thereby a smaller influence on its motion. (4) Their output signals can be transmitted conveniently to a remote recorder. Vibration meters can therefore be used for measurements at relatively inaccessible, inconvenient, or hazardous locations, or they can be used to measure vibration at many points simultaneously with a multiplicity of pickups. (5) Their outputs can be filtered electrically, thereby unwanted signal components are rejected, or the outputs can be integrated or differentiated so as to provide displacement, velocity, acceleration, or jerk data. (6) By the use of damping of the output meter, rapid variations in amplitude can be averaged, providing a visual indication of a short-time average of the meter reading. (7) By proper electrical scaling, the vibration parameters can be converted or indicated in peak-to-peak, peak, average, or root-mean-square values. (8) Peak-sensing holding circuits are available on some vibration meters which provide an indication of the true maximum values of vibration.

COMMERCIALLY-AVAILABLE VIBRATION METERS*

Most vibration meters tabulated below are operated by replaceable or rechargeable batteries. Models employing solid-state circuitry are lighter than the earlier vacuum-tube models and they are not subject to microphonically induced noise when measurements are made in high noise levels. Models with amplitude ranges indicated in the English system of units are usually available from the manufacturers in the metric system of units. Most, but not all, of the units are provided with an electrical output. The manufacturers should be consulted for further details.

Bently Nevada Corporation, Model TK-7 portable vibration indicator
Overall frequency range: 1.7 to 5,000 cps
Amplitude ranges
Displacement: 0.15 to 15 mils peak-to-peak, full scale
Velocity: 0.15 to 15 in./sec peak, full scale

B & K Instruments, Model 2510 portable vibration meter
Overall frequency range: 10 to 1,000 cps
Amplitude range
Velocity: 0.3 to 30 mm/sec, full scale

B & K Instruments, Model 2511 portable vibration meter
Overall frequency range: 0.3 to 15,000 cps
Amplitude ranges
Displacement: 0.003 to 10 mm, full scale (rms or peak-to-peak)
Velocity: 0.02 to 1,000 mm/sec, full scale (rms or peak-to-peak)
Acceleration: 0.002 to 100 mm/sec^2, full scale (rms or peak-to-peak)

CEC/Bell and Howell, Model 1-157-0002 vibration meter
Overall frequency range: 5 to 2,000 cps (displacement); 5 to 20,000 cps (velocity)
Amplitude ranges
Displacement: 0.0005 to 0.500 in. peak-peak, full scale
Velocity: 0.05 to 50 in./sec average, full scale

Columbia Research Labs, Model VM-103 vibration meter
Overall frequency range: 15 to 1,000 cps
Amplitude ranges
Displacement: 0.001 to 10 in., full scale
Velocity: 0.1 to 1,000 in./sec, full scale
Acceleration: 0.01 to 100g, full scale

Dytronics Company, Model 4000 vibration analyzer
Overall frequency range: 0.8 to 8,300 cps
Amplitude ranges
Displacement: 0.1 to 100 mils peak-to-peak, full scale
Acceleration: 0.01 to 10g, peak, full scale

Dytronics Company, Model 4400B vibration surveyor
Overall frequency range: 0.8 to 8,300 cps
Amplitude ranges
Displacement: 0.1 to 100 mils peak-to-peak, full scale
Acceleration: 0.01 to 10g, peak, full scale

General Radio Company, Model 1553-A vibration meter
Overall frequency range: 2 to 2,000 cps
Amplitude ranges
Displacement: 0.003 to 3,000 mils average, full scale
Velocity: 0.003 to 3,000 in./sec average, full scale
Acceleration: 0.03 to 30,000 in./sec^2 average, full scale
Jerk: 3 to 30,000 in./sec^3 average, full scale

IRD Mechanalysis, Model 320 vibration selector
Overall frequency range: 5 to 5,000 cps

* The units tabulated in this chapter are representative of commercially available devices The listings are not necessarily complete, nor are the units given necessarily recommended by the author.

Amplitude ranges
Displacement: 0.1 to 100 mils peak-to-peak, full scale
Velocity: 0.1 to 100 in./sec peak, full scale

IRD Mechanalysis, Models 330 and 350 vibration analyzer and dynamic balancer
Overall frequency range: 0.8 to 8,300 cps
Amplitude ranges
Displacement: 0.03 to 100 mils peak-to-peak, full scale
Velocity: 0.03 to 100 in./sec peak, full scale

Metrix Instrument Company, Model 5115B vibration analyzer and dynamic balancer
Overall frequency range: 0.8 to 1,700 cps
Amplitude ranges
Displacement: 0.1 to 10 mils per division (oscilloscope display)
Velocity: 0.01 to 1 in./sec per division
Acceleration: 0.1 to 10g per division

Metrix Instrument Company, Model 5160 miniature vibration meter
Overall frequency range: 10 to 1,000 cps
Amplitude ranges
Displacement: 1, 10, 100 mils, peak-to-peak, full scale
Velocity: 0.1, 1, 10 in./sec, peak, full scale
Acceleration: 0.1, 1, 10g, peak, full scale

CAPACITANCE-TYPE TRANSDUCER

The capacitance-type transducer is basically a displacement-sensitive device. Its output is proportional to the change in capacitance between two plates caused by the change of relative displacement between them as a result of the motion to be measured.

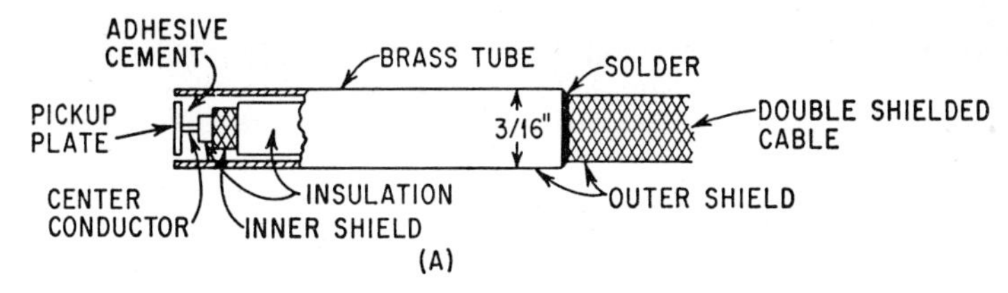

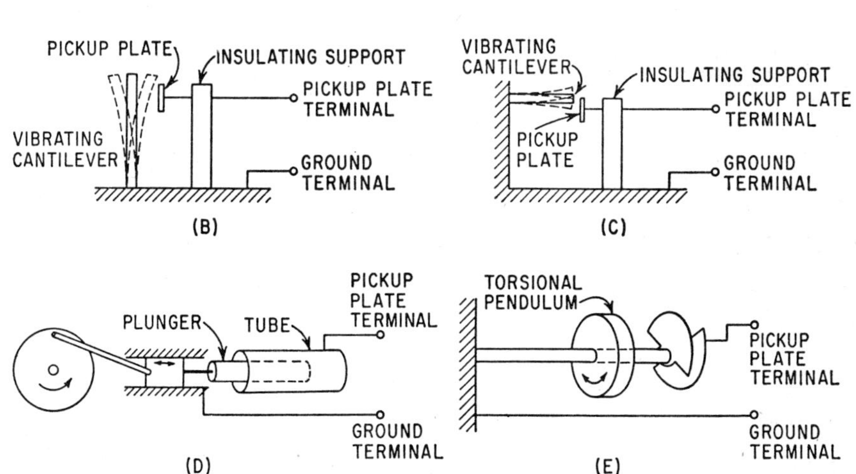

FIG. 14.1. Capacitance-type transducers and their application: (*A*) construction of typical assembly, (*B*) gap length or spacing sensitive pickup for transverse vibration, (*C*) area sensitive pickup for transverse vibration, (*D*) area sensitive pickup for axial vibration, and (*E*) area sensitive pickup for torsional vibration.

Appropriate electronic equipment is used to generate a voltage corresponding to the change in capacitance.

Transducers based upon the variation of capacitance are primarily special-purpose devices. This is principally due to the fact that auxiliary electronic equipment required for their operation is inherently more complicated than that required for the more popular types of transducers. The capacitance-type transducer's main advantages are (1) its simplicity in installation, (2) its negligible effect on the operation of the vibrating system since it is a proximity-type pickup which adds no mass or restraints, (3) its extreme sensitivity, (4) its wide displacement range, due to its low background noise, and (5) its wide frequency range, which is limited only by the electric circuit used.

The capacitance-type transducer often is applied to a conducting surface of a vibrating system by using this surface as the ground plate of the capacitor. In this arrangement, the insulated plate of the capacitor should be supported on a rigid structure close to the vibrating system. Figure 14.1*A* shows the construction of a typical capacitance pickup; Figs. 14.1*B*, *C*, *D*, and *E* show a number of possible methods of applying this type of transducer. In each of these, the metallic vibrating system is the ground plate of the capacitor. Where the vibrating system at the point of instrumentation is an electrical insulator, the surface can be made slightly conducting and grounded by using a metallic paint or by rubbing the surface with graphite.

The maximum operating temperature of the transducer is limited by the insulation breakdown of the plate supports and leads. Bushings made of alumina No. 38900 are commercially available and provide adequate insulation at temperatures as high as 2000°F (1093°C).

THE CAPACITOR

DISPLACEMENT-TO-CAPACITANCE SENSITIVITY. The capacitance C between two parallel conducting plates that are insulated from each other is directly proportional to the effective area S of the plates and is inversely proportional to the distance d between the plates, i.e.,

$$C = 0.225 \frac{S(\text{in.}^2)}{d(\text{in.})} = 0.0885 \frac{S(\text{cm}^2)}{d(\text{cm})} \quad \text{mmfd} \tag{14.1}$$

Table 14.1. Formulas for Calculating Capacitance and Change in Capacitance for Various Geometries of Air-dielectric-type Capacitors

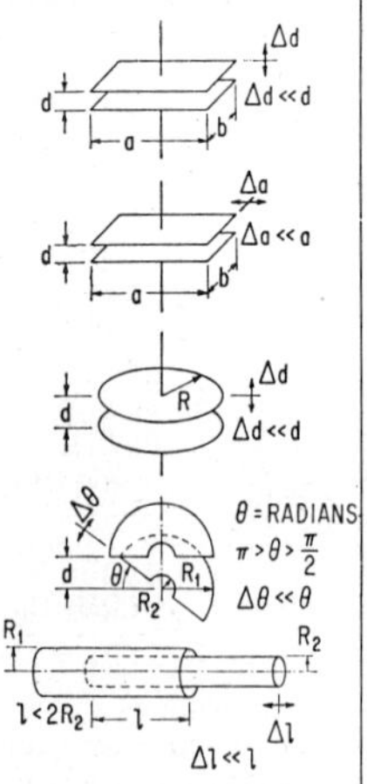

	Capacitance in air, mmfd		Change in capacitance due to change in position ΔC, mmfd
	Dimensions, in.	Dimensions, cm	
d, a, b, Δd, Δd ≪ d	$C = 0.225 \frac{ab}{d}$	$C = 0.0885 \frac{ab}{d}$	$\Delta C \simeq C \frac{\Delta d}{d}$
d, a, b, Δa, Δa ≪ a	$C = 0.225 \frac{ab}{d}$	$C = 0.0885 \frac{ab}{d}$	$\Delta C \simeq C \frac{\Delta a}{a}$
d, R, Δd, Δd ≪ d	$C = 0.225 \frac{\pi R^2}{d}$	$C = 0.0885 \frac{\pi R^2}{d}$	$\Delta C \simeq C \frac{\Delta d}{d}$
Δθ, d, θ, R_1, R_2, θ = RADIANS, $\pi > \theta > \frac{\pi}{2}$, Δθ ≪ θ	$C = 0.353 \frac{R_1^2 - R_2^2}{d} \left(\frac{\Theta}{\pi}\right)$	$C = 0.139 \frac{R_1^2 - R_2^2}{d} \left(\frac{\Theta}{\pi}\right)$	$\Delta C \simeq C \frac{\Delta\Theta}{\Theta}$
R_1, R_2, $l < 2R_2$, l, Δl, Δl ≪ l	$C = \frac{0.613l}{\log \frac{R_1}{R_2}}$	$C = \frac{0.242l}{\log \frac{R_1}{R_2}}$	$\Delta C \simeq C \frac{\Delta l}{l}$

for an air-dielectric capacitor at standard conditions. Equation (14.1) shows that the capacitance variation can be obtained either by changing the effective area S or by changing the spacing d. Table 14.1 shows this equation applied to various geometric shapes of capacitors; it also gives the relationship for the change in capacitance ΔC due to a small change in plate position. (Fringing effects are neglected.)

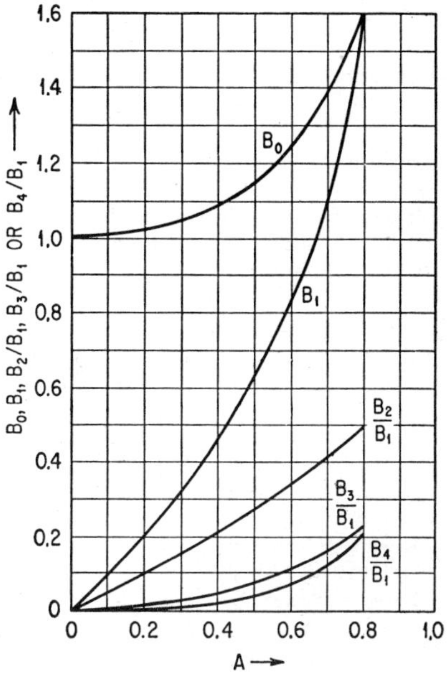

CAPACITANCE VERSUS DISPLACEMENT NONLINEARITY. Variable-capacitance transducers have three principal sources of nonlinearity in the relationship of output voltage to mechanical displacement which should be considered in any application of the transducer: (1) the effect of fringing in the electric field between the capacitor plates, (2) the fact that the capacitance is inversely, and not directly, proportional to the spacing, and (3) the electric circuit associated with the transducer. The effect of fringing depends to such an extent upon the transducer plate design and its shielding that it is difficult to determine analytically; fringing effects should be checked experimentally in each installation.

FIG. 14.2. Harmonic coefficients in Eq. (14.5) as a function of the variation in spacing of the condenser plates.

The capacitance variation, which is obtained when the spacing between the plates is changed, is not a linear function of the spacing. This can lead to serious distortion unless the percentage change in the spacing is limited. For example, assume a sinusoidal variation in spacing d:

$$d = d_0 + d_s \sin (2\pi ft) \tag{14.2}$$

where d_0 = average plate spacing
d_s = amplitude of spacing variation
f = vibration frequency, cps
t = time, sec

Equation (14.2) can be rewritten as

$$d = d_0(1 + A \sin \omega t) \tag{14.3}$$

where $A = d_s/d_0$ and $\omega = 2\pi f$. Substituting Eq. (14.3) into Eq. (14.1), the capacitance can be expressed as

$$C = \frac{0.225S}{d_0(1 + A \sin \omega t)} \quad \text{mmfd} \tag{14.4}$$

Expanding Eq. (14.4) in a Fourier series,

$$C = 0.225 \frac{S}{d_0} \left[B_0 + B_1 \left(\sin \omega t + \frac{B_2}{B_1} \cos 2\omega t + \frac{B_3}{B_1} \sin 3\omega t + \frac{B_4}{B_1} \cos 4\omega t + \cdots \right) \right] \quad \text{mmfd} \tag{14.5}$$

where the values of B_0, B_1, B_2, B_3, B_4 ... are functions of the variation of the spacing ratio A. The coefficients for the first four harmonics as a function of A are given in Fig. 14.2. The deviation of B_0 from the value 1.0 indicates the amount of zero shift obtained as the value of A increases. The conditions for the capacitance change to be proportional to the vibration amplitude are $B_1 = A$, and $B_2 = B_3 = B_4 = 0$. The harmonic distortion is indicated by the curves for B_2/B_1, B_3/B_1, and B_4/B_1 for the second, third, and fourth harmonics, respectively.

ELECTRICAL CIRCUITS

There are many electric circuits in which a small change in the capacitance will result in a corresponding change in output voltage. Descriptions of a number of these circuits are given in Refs. 1, 2, and 3.

COMMERCIALLY AVAILABLE CAPACITANCE-TYPE TRANSDUCERS

Commercially available vibration-measuring equipment that uses the capacitance-variation principle includes:

B & K Instruments, capacitance-type transducer Model MM 0004
Sensitivity: 25 V/mm

Dressen-Barnes Corporation (Omega Instruments), Type 40
Sensitivity: 100 mV/μ in.

Dressen-Barnes Corporation (Omega Instruments), Types 70 and 80

Acceleration range, peak g	10	50	100	500	1000	2000
Sensitivity, mV/g	500	100	50	10	5	1

Dressen-Barnes Corporation (Omega Instruments) Type 75

Acceleration range, peak g	1	2	5	10
Sensitivity, mV/g	5,000	2,500	1,000	500

Dynasciences Corporation (Photocon), proximity transducers, Models PT-3, 4, 5, 6, 7, and 14
Displacement range: 1μ in. (2.54×10^{-6} cm) to 0.005 in. (0.013 cm)

Sprengnether Instrument Co., Inc. Model VCT-200
Sensitivity: 300 mV/μ in.

Wayne Kerr Corporation, Model B-731A
Sensitivity: 10,000 to 10 V/in.

POTENTIOMETER-TYPE TRANSDUCER

The potentiometer-type transducer is a variable-voltage divider in which the output voltage is a function (usually linear) of a displacement. If the displacement varies with time, then the output voltage will vary with time in a similar manner.

APPLICATION TO MOTION MEASUREMENT

This type of transducer is used in the study of the relative motion of one part of a structure with respect to another. It is also used as the sensing element in force gages, pressure gages, etc., and in remote position-indicating systems.

The potentiometer-type transducer may incorporate a seismic system consisting of a spring-mounted mass whose motion with respect to the base is measured. In this application, the transducer output voltage is proportional to the vibration acceleration of the base, provided that the vibration frequency is substantially below the natural frequency of the seismic system (see Chap. 12). If the vibration frequency is substantially above the natural frequency, the transducer output voltage is proportional to the vibration displacement.

GENERAL PRINCIPLES

The potentiometer-type transducer circuit is shown schematically in Fig. 14.3*A* and its equivalent circuit in Fig. 14.3*B*. The potentiometer is energized with the d-c voltage e. Its total circuit resistance is r_p, which includes the resistance of the d-c source, the potentiometer, and the line between the source and potentiometer. The potentiometer senses a displacement by moving a variable contact through a distance which is proportional to the displacement; this movement causes a corresponding change in resistance r_2, or in the resistance ratio $\alpha = r_2/r_p$ in Fig. 14.3*A*.

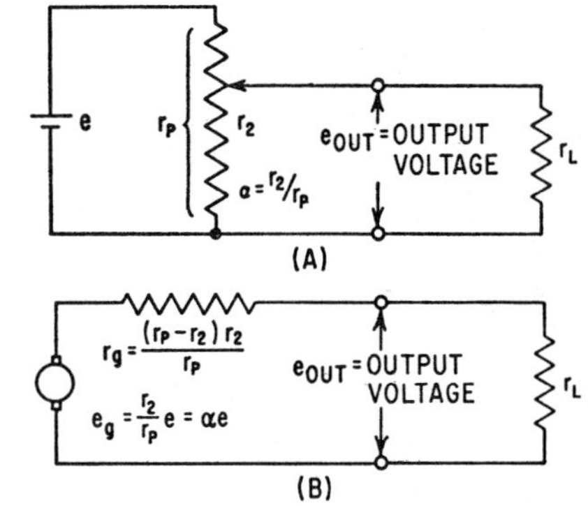

When there is no electrical circuit load on the output terminal of the potentiometer, the output voltage $e_{out} = e\alpha$ (the equivalent generator voltage of Fig. 14.3*B*). If an indicating instrument or load having a resistance r_L is connected across the potentiometer output terminals, the loading effect[4-6] introduces an error or nonlinearity in the transducer action as shown by Eq. (14.6). Then the output voltage e_{out} is not a linear function of α, but is given by

FIG. 14.3. Potentiometer-type transducer circuits: (*A*) typical circuit and (*B*) equivalent circuit.

$$e_{out} = \frac{\dfrac{r_L}{r_p}\dfrac{r_2}{r_p}e}{\dfrac{r_L}{r} + \dfrac{r_2}{r_p} - \dfrac{(r_2)^2}{(r_p)^2}} \tag{14.6}$$

FACTORS AFFECTING USE

SENSITIVITY. The full-scale travel of the movable contact in the potentiometer depends upon the particular transducer. In some accelerometers, a full-scale travel of less than 0.1 in. is used. If the transducer is used in the study of the motion of controls, control surfaces, actuators, etc., a travel up to 10 in. may be required.

The sensitivity in terms of output voltage per unit travel of the movable contact is seldom listed by the manufacturer. The general practice is to give the operating limitations, and to let the user calculate the sensitivity according to his specific operating conditions. Manufacturers recommend either the maximum energizing voltage, the maximum energizing current, or the maximum power input to the potentiometer. Knowing these and the full-scale travel, it is possible to determine the sensitivity.

RESOLUTION. The resolution of potentiometers used in transducers is usually between 0.1 and 0.5 per cent of the full range. It represents the step change in voltage obtained when the movable contact moves from one turn to the next in the potentiometer resistance winding.

LINEARITY. The potentiometer can be linear to within 0.1 per cent full scale in the more accurate transducers, but the average value is about 1.0 per cent full scale.

OUTPUT-CIRCUIT LOADING. Typical potentiometers in such transducers have a resistance r_p between 100 and 10,000 ohms. The actual value of r_p which is selected is a compromise between current drain on the battery and the resulting heat dissipation within the potentiometer, and the nonlinearity due to the loading effect of r_L in the indicating instrument.

HYSTERESIS OR MECHANICAL DRAG. Friction is a serious drawback in potentiometers when they are used with seismic instruments. The sticking friction determines the threshold sensitivity, while both sticking and rubbing friction result in a hysteresis loop.

COMMERCIALLY AVAILABLE POTENTIOMETER-TYPE TRANSDUCERS

Most accelerometers of this type have natural frequencies below 100 cps. A typical unit is fluid, air, or magnetically damped to 0.7 of critical damping and is relatively heavy, approximately ½ to 2 lb (about 200 to 1,000 gm). The principal advantage of these units is their high sensitivity (up to 20 volts per g) and their flat response from half of their natural frequency down to zero frequency. Because they respond to d-c, the calibration of most units of this type can be checked by inverting their sensitive axis. This reversal is equivalent to a $2g$ change in acceleration. Manufacturers include Bourns Labs., Cedar Engineering, Conrac Corp., Edcliff Instruments, Honeywell, Inc. (Aerospace Division), Humphrey, Inc.

SERVO ACCELEROMETER

The servo accelerometer (sometimes called a "force-balance accelerometer") contains a seismically suspended mass which has a displacement sensor (e.g., a capacitance-type transducer) attached to it. If the accelerometer is subject to motion, the mass is displaced with respect to the accelerometer case; this displacement generates a servo-loop error signal, establishing a restoring force that is applied to the mass in such a way as to reduce the initial error signal to zero. The acceleration, which is proportional to the restoring force, may be determined from the voltage drop across a resistance in series with a coil which generates the restoring force.

Servo accelerometers can be made very sensitive, some having threshold sensitivities on the order of a few micro-g.[7] Excellent amplitude linearity is attainable, usually on the order of a few hundredths of one per cent. Typical frequency ranges are from zero cps to 50 cps, although some units operate at frequencies up to several hundred cycles per second. Eddy current, gas, fluid, and electrical damping may be employed to provide a wide range of damping characteristics. Some types of servo accelerometers are designed to measure angular acceleration.

Manufacturers of servo accelerometers include Bell Aerospace, Columbia Research Labs, Gulton Industries, Larson Aero, Schaevitz Engineering, Singer (Kearfott Division), Sunstrand (Endevco), Sunstrand (Kistler), Sunstrand (Palomar), Sunstrand (United Controls), and Systron-Donner.

OPTICAL-ELECTRONIC TRANSDUCER

This device (also called an "optical tracker") is a commercially available noncontacting transducer for the measurement of displacement, which incorporates an optical system in an electronic circuit. The exact waveshape of the motion showing all transients and other characteristics of the displacement are displayed on an oscilloscope. This device can be used to measure amplitude and frequency of shake tables and accelerometers, runout of rotating equipment, or the response of cam followers. Its main advantage is that it does not require contact with the work. For example, it can measure the vibration of individual blades of an impeller while running; as each blade comes past, the device can be "locked in" on the motion of the blade, following it through a segment of its revolution.

In the optical-electronic transducer, an optical "light-dark" discontinuity is detected by a photosensitive device. This device generates an electrical output proportional to the displacement of the optical discontinuity. A wide range of displacement amplitudes can be measured by the selection of an appropriate lens; typical amplitudes may be as little as 0.02 mm or as great as 15 meters. Some models are designed to measure biaxial or angular displacements.

Manufacturers of optical-electronic transducers include Universal Technology (Optron Division), and PhysiTech Inc.

REFLECTED-WAVE TRANSDUCER SYSTEMS

The motion of some vibrating systems can be detected by reflections from the vibrating surface of electromagnetic or ultrasonic waves which impinge on the surface. In each of the systems described below, there is no physical contact between the vibrating surface and the source of wave motion. Therefore, an instrument using the reflected-wave principle offers several advantages: (1) the instrument will not load or disturb the vibrating surface under observation, (2) a large number of regions can be examined in rapid succession, and (3) an extremely wide bandwidth can be achieved, since there are no mechanical resonances.

ELECTROMAGNETIC (MICROWAVE) SYSTEMS

WAVE-GUIDE SYSTEM. In this system, microwave radio and wave-guide components are employed in the measurement of the motions of a surface.[8, 12–14] The open end of the wave guide is directed toward the vibrating surface. By tuning the wave guide and adjusting the microwave frequency, it is possible to establish a standing-wave pattern in the wave guide that is very sensitive to the position of the vibrating surface. As the reflecting surface vibrates, a corresponding modulation can be detected in the standing-wave pattern within the wave guide which, when demodulated, provides an output voltage proportional to the motion of the surface.

One type of microdisplacement meter employs a wave-guide transmission line having a length of one-quarter wavelength.[12] In this transmission line, the vibrating surface forms a termination at one end of the line. An oscillator circuit is used such that only the quarter-wave mode is excited. Since the frequency of the quarter-wave mode is a function of the line length, the motion of the vibrating surface causes a corresponding modulation in the oscillator frequency. This measurement system is capable of detecting and measuring the amplitude of vibration of small surfaces (such as the vibrating faces of piezoelectric crystal assemblies) when the displacements are less than 1 microinch; acoustical loading is negligible. The usable frequency range is 100 to 100,000 cps.

Another wave-guide technique has been used to study surface irregularities of commutators while spinning in a noncontact "tracing" system,[13, 14] an example of which is shown in Fig. 14.4. In setting up this system, it is necessary to position the open end of a microwave "magic tee"[8] carefully with respect to the commutator. It is also necessary to adjust the shorting plug and the klystron oscillator frequency properly in order to obtain the proper sensitivity in the standing-wave pattern at the detector section of the tee. For the system described, the maximum displacement that can be measured is in the order of one-eighth wavelength of the microwave carrier used, or 0.050 in. (0.13 cm), while displacements of 100 microinches (25×10^{-6} cm) give readable deflections (0.1 in.) in a high-gain cathode-ray oscilloscope. The frequency range over which measurement can be made is determined primarily by the crystal detector and its associated amplifiers. The technique may be used from 0 cps to approximately one-hundredth of the carrier frequency.

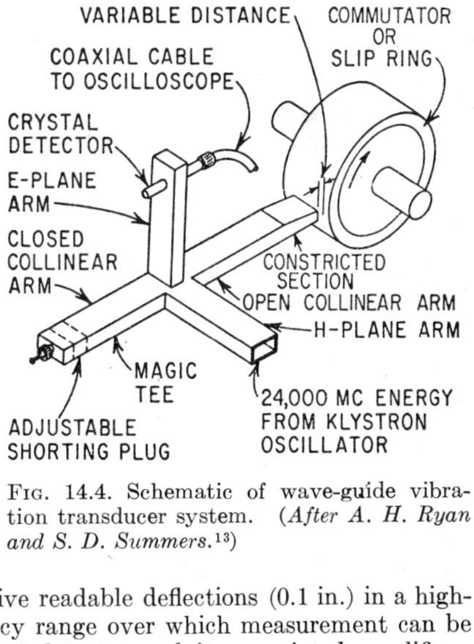

FIG. 14.4. Schematic of wave-guide vibration transducer system. (*After A. H. Ryan and S. D. Summers.*[13])

Another measurement system,[9] similar in principle to that described above, is shown in Fig. 14.5*A*. A reflex-type klystron oscillator supplies approximately 30 mw of power at a wavelength of 3.2 cm to an electromagnetic horn-type wave guide. The open end of the horn contains a dielectric lens that focuses the microwave energy on the vibrating object. The vibrating surface reflects some of the energy back into

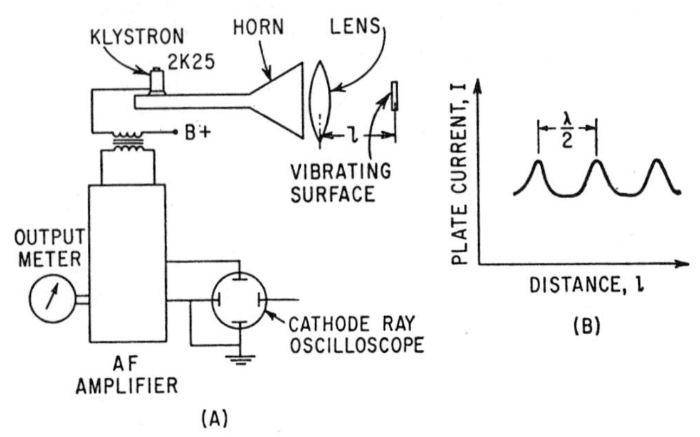

FIG. 14.5. Horn-type wave-guide vibration transducer: (*A*) circuit diagram; (*B*) change in plate current with vibration displacement. (*After S. Satomura et al.*[9])

the wave guide, modulating the standing-wave amplitude in the wave guide and thus the output of the klystron oscillator. This change in the efficiency of the oscillator causes a corresponding change in the klystron plate current, as shown in Fig. 14.5*B*. The output of the transformer in the plate circuit is then a voltage which is proportional to the vibration of the object. With a spacing (l in Fig. 14.5*A*) of 1.26 in. (3.2 cm), it is possible to measure vibration displacement amplitudes as large as 0.15 in. (0.38 cm) to approximately 40×10^{-6} in. (1×10^{-4} cm) at vibration frequencies from 5 to over 100,000 cps, depending primarily upon the type of transformer used.

MICROWAVE-BEAM REFLECTING SYSTEM. If a narrow beam of microwave electromagnetic energy is transmitted toward a vibrating surface, the phase of the re-

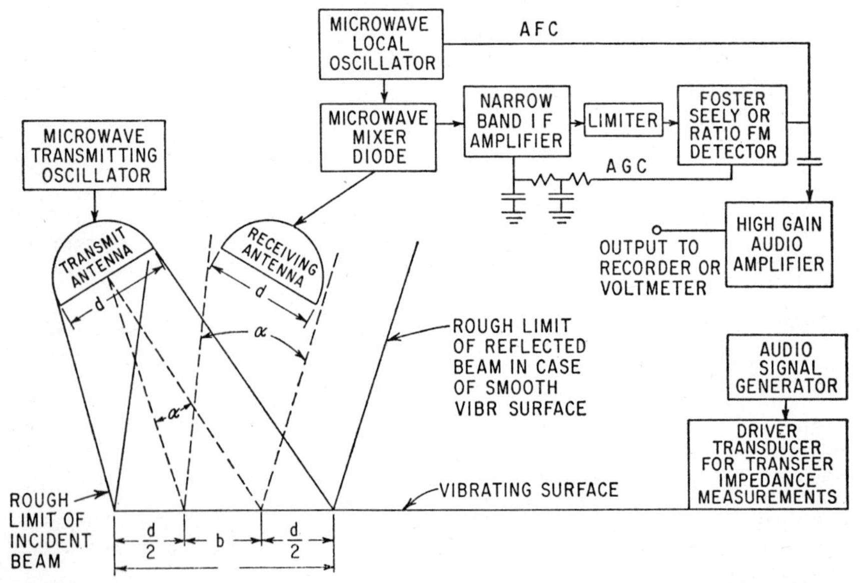

FIG. 14.6. Block diagram of microwave reflected-beam system. (*After C. Stewart.*[15])

flected wave will be modulated in proportion to the vibration amplitude and at the vibration frequency. By detecting this phase modulation, an electrical voltage is obtained which is proportional to the vibration velocity of the vibrating surface. This measurement is free from effects of mechanical loading and surface reflectivity variation, without requiring close proximity to the surface.

A design, based on this principle, is shown in the block diagram of Fig. 14.6.[15] The 20-milliwatt klystron oscillator and transmitting antenna direct a beam of microwave energy toward the vibrating surface. This beam strikes the surface at an angle so that only the reflected wave is picked up by the receiving antenna. The microwave signal is compared to a signal from a fixed local oscillator. As the vibrating surface varies the path length of the received signal, a phase-shift modulation between the received signal and the local oscillator signal is observed. By the use of an FM detector, an electrical signal proportional to the phase modulation and consequently to the surface vibration is obtained. For this system the observation spot for 10 in. clearance (25 cm) will have a diameter approximately twice the 2-in. (5-cm) diameter of the antenna. The maximum displacement that can be measured is of the order of a wavelength of the microwave beam or 0.34 in. (0.85 cm). For a smooth surface the minimum displacement that can be measured with 1 per cent accuracy is 0.1×10^{-6} in. (0.25×10^{-6} cm), determined primarily by the signal-to-noise ratio.

ULTRASONIC SYSTEMS

REFLECTED ENERGY SYSTEM. The total power radiated by an ultrasonic transducer is affected by both the amount and phase of the ultrasonic energy reflected back

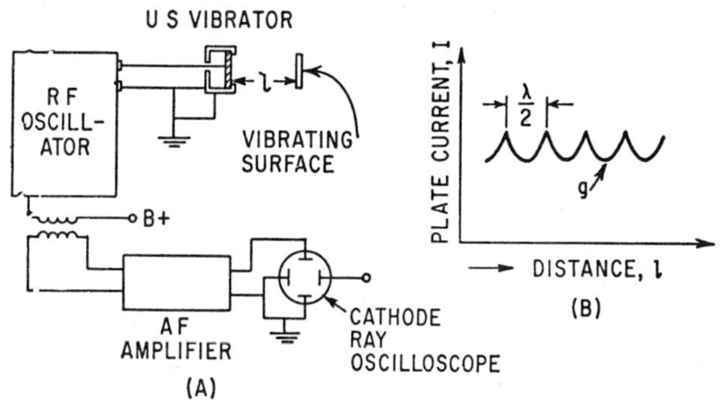

FIG. 14.7. Ultrasonic vibration transducer: (*A*) circuit diagram; (B) change in plate current of RF oscillator with vibration displacement. (*After S. Satomura et al.*[9])

upon the transducer. When the reflecting surface is vibrating, there is a corresponding modulation in the radiated energy. A sensitive vibration-measuring system using this principle is shown in the block diagram of Fig. 14.7*A*.[8] The plate current of the RF oscillator is modulated by the reflected energy seen by the transducer. The change in plate current with relative position of the reflecting surface is illustrated by the curve in Fig. 14.7*B*. The sensitivity and selectivity of this technique can be increased if the transducer is concave so as to concentrate or focus the ultrasonic energy on a small area; it is possible to measure vibration displacements as small as 10×10^{-6} in. (0.25×10^{-4} cm) and as large as approximately 1/16 wavelength at frequencies from 5 to over 100,000 cps, depending primarily upon the type of transformer used.

PULSE SYSTEM. The length of time required for a pulse of ultrasonic energy to travel from the source to a reflecting surface and return can be used as a measure of the position of that surface. The actual travel time (which is an indication of the distance from the source to the reflecting surface) may be indicated by the position of a spike on the horizontal axis (i.e., time axis) of a cathode-ray oscilloscope. Vibration of the reflecting surface causes a corresponding variation in the position of this spike. This technique can be used in vibration measurements having peak displacements as small as ±0.01 in. (0.025 cm).

ULTRASONIC DOPPLER-SHIFT SYSTEM. In this system, the velocity of a vibrating surface is measured by determining the Doppler-frequency shift in an ultrasonic carrier-frequency beam reflected from a vibrating surface.[10, 11] It is a frequency-modulation system, so that its calibration does not depend upon the distance between the ultrasonic pickup transducer and the surface being measured; its output is proportional to the velocity of motion of the surface.

For example, in the design of one system, the ultrasonic source and receiving transducers are barium titanate crystals operating at 110 kc.[10] The ultrasonic beam width is approximately 15°. The unit can be operated at distances from about 2 in. to 6 ft (i.e., from about 5 to 180 cm). The frequency response of the system is flat between 0 and 700 cps. The minimum velocity that can be measured is approximately 0.01 in./sec (0.025 cm/sec).

The reflected-energy system described in Ref. 8 also can be used as a Doppler-shift system since there is the same shift in the frequency of the reflected wave proportional to the normal velocity of the reflecting surface.

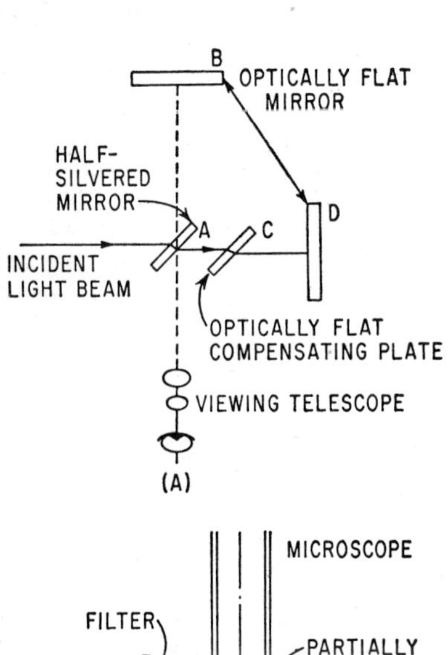

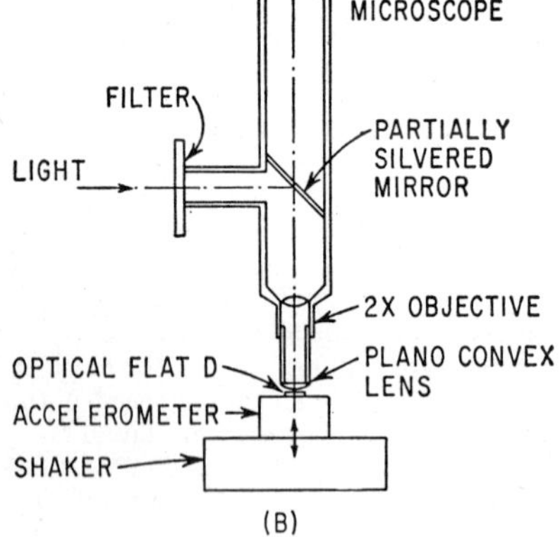

FIG. 14.8. Schematic diagram of optical paths of interferometers used for vibration-amplitude measurement. (*A*) Michelson type. (*After E. I. Feder and A. M. Gillen.*[19]) (*B*) Fizeau type. (*After S. Edelman, E. Jones, and E. R. Smith.*[20])

OPTICAL INTERFEROMETER

An interferometer divides a beam of light into two parts which travel different paths and recombine to form interference fringes. The form of these fringes is determined by the difference in the optical path traveled by the two beams. This optical method is adaptable as a means of absolute calibration above vibration frequencies of 1,000 cps where the vibration displacements encountered are low (Chap. 18).

The optical interferometer method of vibration measurement[16–18] is best suited for measuring sinusoidal displacements at amplitudes below ±30 microinches (about 80×10^{-6} cm) in the frequency range from 30 to over 10,000 cps. When used with monochromatic light of 6,328 angstroms (from a HeNe laser), the smallest displacement that can be detected is about 5 microinches (12×10^{-6} cm).

GENERAL PRINCIPLE OF OPERATION

In the interferometers shown schematically in Fig. 14.8*A* and *B*, a beam of light from a monochromatic source is reflected in part from one surface and the balance from another; the resulting beam of light is recombined and viewed. If the effective length of the light path from the two mirrors is equal, then the resulting light beam has the same intensity as the original or incident beam. If the light is split into equal parts and totally reflected from the two mirrors, and if the effective light path length for one mirror is a half wave-length longer than that for the other, then when the two light beams are again recombined, cancellation will cause a dark field. If one of the two mirrors is not perfectly perpendicular to its incident beam of light, then the combined beam of light will have dark and light bands, as shown in Fig. 14.9*A*.

The interferometer when used for vibration measurements is arranged so that mirror *D* (Fig. 14.8*A* and *B*) is mounted on the vibrating surface whose motion is essentially rectilinear. When the vibration amplitude of this mirror is slowly increased, the fringes will disappear very sharply at set amplitudes. As the amplitude is further increased, the

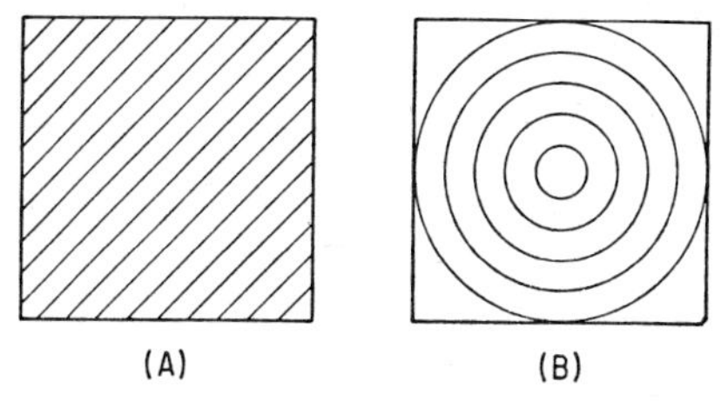

FIG. 14.9. Optical interferometer interference patterns: (*A*) Michelson type; (*B*) Fizeau type.

interference bands will again appear, and as they disappear the second time, a second known displacement amplitude is determined. The order of the disappearance and the resultant displacement amplitude are given in Table 18.1 for monochromatic light of 6,328 angstroms (632.8 nm).

ANALYSIS OF OPERATING PRINCIPLES

The intensity of the fringe pattern I produced by an interferometer, if the mirrors are stationary, is given by

$$I = K\left(1 + \cos 2\pi \frac{x}{h}\right) \tag{14.7}$$

where K = constant
x = distance measured perpendicularly across the fringes
h = fringe width

If the mirror D in Fig. 14.8 is vibrated sinusoidally in a direction perpendicular to its plane, the fringe pattern is set into simple harmonic motion. If the frequency is greater than 30 cps, or faster than the eye can follow, then according to Eq. (18.15) the time average of the light intensity is[16]

$$I = K\left[1 + J_0\left(\frac{2\pi d}{\lambda}\right) \cos \frac{2\pi x}{h}\right] \tag{14.8}$$

where $J_0(2\pi d/\lambda)$ is a Bessel function of order zero and
where λ = wavelength of light used
d = peak-to-peak amplitude of the vibrating surface

The value of $J_0(2\pi d/\lambda)$ is dependent only upon the amplitude of vibration and the wavelength λ of the light source. As the value of d is increased, the value of $J_0(2\pi d/\lambda)$ will go through zero successively at the values given in Table 18.1. The value of d is tabulated for each of the successive zeros (for λ = 6,328Å); it represents those values of d for which the fringe pattern disappears and the illumination is uniform. Since the Bessel function $J_0(2\pi d/\lambda)$ has a zero value only at discrete points, this fringe disappearance is very sharp.

MICHELSON-TYPE INTERFEROMETER

In the Michelson interferometer,[16–18, 21] Fig. 14.15*A*, a beam of light from a monochromatic source is directed toward the half-silvered mirror *A* which is set at an angle of 45°. The half-silvered mirror A has a coating which is so thin that about half the light is transmitted through it and the rest is reflected. The beam thus divides into two parts, one of which is reflected to the completely silvered mirror B and is reflected back through mirror A toward the viewing telescope. The other part of the beam passes through A, through the compensating plate C, and is reflected by mirror D back to mirror A, then toward the viewing telescope.

If the incident and reflected beams from both B and D are parallel and involve exactly the same time delay, then the observer will see a uniformly illuminated field. If one mirror, either B or D, is moved in the direction of the light beam a distance of a quarter wavelength, then the field of view is dark. If either of the two mirrors is set at a slight angle to the incident light, then a type of fringe pattern is viewed which contains a series of parallel lines, Fig. 14.9A.

FIZEAU-TYPE INTERFEROMETER

The Fizeau-type interferometer,[22–24] shown in Fig. 14.8B, can be made by modifying a microscope. This is done by mounting the microscope vertically with a 2X objective in position and with a plano-convex lens (focal length about 1,000 mm) at the focus of the objective, curved side down. The microscope should be mounted so that an optical flat mounted rigidly to the vibrating surface and normal to the direction of motion is approximately 0.1 mm (approximately 0.004 in.) from the plano-convex lens. Monochromatic light enters the microscope through the vertical illuminator and is deflected down the axis of the microscope by the partially silvered mirror. It is then partially reflected from the curved surface of the plano-convex lens and from the optical flat D mounted on the vibrating surface. As a result of interference between these two reflected beams (similar to that described for the Michelson interferometer), a set of Newton's rings or fringes, Fig. 14.9B, is formed which can be viewed through the eyepiece of the microscope.

PHOTOELECTRIC FRINGE SENSING

Using the interferometers described above, or others,[25] it is possible to detect and measure fringe characteristics with photoelectric sensors. At lower vibration frequencies, fringe-counting techniques are useful. A slit or pinhole is placed in front of the photodetector so that only the light from a portion of one fringe illuminates the detector. As vibration occurs, the fringes will pass the slit, thus generating an electrical signal each time a fringe is passed. By counting the number of fringes that pass the slit per vibration cycle, the total peak-to-peak displacement can be computed from the relationship shown by Eq. (18.12). Because the vibration displacement amplitude generally decreases with increasing frequency, electronic fringe counting is only useful to about several hundred cycles per second.

If the output of the photodetector is processed through a narrow bandpass filter, tuned to the vibration frequency, fringe disappearances may be counted; these fringe disappearances are related to the vibration amplitude by a Bessel function. With proper mirror-separation adjustments, it can be shown[26] that fringe disappearances occur whenever $J_1(4\pi d/\lambda) = 0$, where J_1 is the Bessel function of the first kind, first order, d is the amplitude of vibration, and λ is the wavelength of light being used.

If the upper mirror (plano-convex lens) of the Fizeau-type interferometer shown in Fig. 14.8B is modulated at a frequency at least ten times less than the vibration, and the photodetector output is processed through a bandpass filter tuned to the modulation frequency, the fringe disappearances can be counted and will occur whenever $J_0(4\pi d/\lambda) = 0$,[27] where J_0 is the zero-order Bessel function of the first kind and d and λ are as described above. This is the same relationship as that for visual fringe disappearance as described in Chap. 18. The fringe-disappearance methods are useful at the higher frequencies, beginning at about 1,000 cps, and can be used successfully to as high as 50,000 cps.[27, 28]

HOLOGRAPHIC INTERFEROMETRY

Laser-illuminated vibrating surfaces can be examined by the use of time-average holography or stroboscopic holography. A three-dimensional view can be made to detect mode shapes and estimates of vibration amplitude. Holographic setups are generally considered as laboratory tools and are not practical to implement in typical industrial working areas.[29, 30]

REFERENCES

1. Roberts, H. C.: "Mechanical Measurements by Electrical Methods," The Instruments Publishing Co., Inc., Pittsburgh, Pa., 1951.
2. Attree, V. H.: *Electronic Engr.*, **27**:308 (1955).
3. Shafer, S. N., and R. Plunkett: *Proc. Soc. Exptl. Stress Anal.*, **13**(1):123 (1955).
4. Nettleton, L. A., and F. E. Dole: *Rev. Sci. Instr.*, **11**(5):332 (1947).
5. Kaufman, A. B.: *Radio-Electronic Engr.*, **44**(3):12 (1952).
6. Post, R. E., and J. Zdzieborski: *Electrical Design News*, **4**(6):52 (1959).
7. Pope, K. E.: *Control Eng.*, **5**(11):97 (1958).
8. Pound, R. V., and E. Durand: "Microwave Mixers," MIT Radiation Laboratory Series, vol. 16, p. 259, McGraw-Hill Book Company, Inc., New York, 1948. Also see: Gik, L. D., L. Ya. Mizuk, and K. B. Karandeyev, *Byulleten izobreteniy* (*USSR*), **11**:76 (1958).
9. Satomura, S., S. Matsubara, and M. Yoshioka: *Mem. Inst. Sci. Ind. Research, Osaka Univ.*, **13**:125, 1956.
10. Hardy, H. C., H. H. Hall, D. B. Callaway, and D. S. Schorer: *J. Acoust. Soc. Amer.*, **27**:1004 (1955).
11. Hardy, H. C.: U.S. Patent 2,733,597, February, 1956, assigned to Armour Research Foundation.
12. Post, R. F., and R. L. Howard: *J. Acoust. Soc. Amer.*, **19**(1):283 (1947).
13. Ryan, A. H., and S. D. Summers: *Elect. Engr.*, **73**:251 (1945).
14. Cohen, G. I., and B. Ebstein: *Proc. Natl. Electronics Conf.*, **12**:982 (1956).
15. Stewart, C.: *J. Acoust. Soc. Amer.*, **30**:644 (1958).
16. Ostenberg, H.: *J. Opt. Soc. Amer.*, **22**:19 (1932).
17. Kennedy, W. J.: *J. Opt. Soc. Amer.*, **31**:99 (1941).
18. Ziegler, C. A.: *J. Acoust. Soc. Amer.*, **25**:135 (1953).
19. Feder, E. I., and A. M. Gillen: *IRE Trans. on Instrumentation*, **I-6**(2):98 (1957).
20. Edelman, S., E. Jones, and E. R. Smith: *J. Acoust. Soc. Amer.*, **27**(4):732 (1955).
21. Thomas, H. A., and G. W. Warren: *Phil. Mag.*, **5**:1125 (1928).
22. Smith, D. H.: *Proc. Phil. Soc.* (*London*), **57**:534 (1945).
23. Hunton, R. D., A. Weis, and W. Smith: *J. Opt. Soc. Amer.*, **44**:264 (1954).
24. Orlacchio, A. W.: *Elect. Mfg.*, **59**(1):78 (1957).
25. Smith, E. R., S. Edelman, V. A. Schmidt, and E. Jones: *J. Acoust. Soc. Amer.*, **30**:867 (1948).
26. Schmidt, V. A., S. Edelman, E. R. Smith, and E. Jones: *J. Acoust. Soc. Amer.*, **33**(6):748 (1961).
27. Schmidt, V. A., S. Edelman, E. R. Smith, and E. T. Pierce: *J. Acoust. Soc. Amer.*, **34**(4): 455 (1962).
28. Pierce, E. T., S. Edelman, L. D. Ballard, and W. S. Epstein: "Low Amplitude Vibration Measurement System," *Natl. Bur. Std.* (*U.S.*) *Rept.* 10, Rept. 10836, 1972.
29. Ross, M.: "Laser Applications," vol. 1, p. 62, Academic Press, Inc., New York, 1971.
30. Collier, R. J., C. B. Burckhardt, L. H. Lin: "Optical Holography," chap. 15, p. 418, Academic Press, Inc., New York, 1971.

15

INDUCTIVE-TYPE PICKUPS

R. R. Bouche
Bouche Laboratories

INTRODUCTION

An inductive-type shock and vibration pickup is an electromechanical device in which the magnetic characteristics of its electric circuit change in response to the motion of an object. Inductive-type pickups have several distinct advantages. Inherently, they are extremely stable and consequently can be employed in very accurate measurements. They have low electrical impedance so that they usually can be used with very long cables without significantly affecting their output voltages.

The *self-generating type* of inductive pickup generates an electromotive force in the pickup in response to the motion of an object. There are four different self-generating inductive types: electrodynamic, electromagnetic, eddy-current, and magnetostrictive. These pickups operate on the principle that a voltage is generated when a conductor cuts through a magnetic field or when the magnetic field is varied about a stationary conductor. The electrodynamic velocity types have the advantage of producing very large output voltages, so that they can be used without amplifiers. It is usually not practical, however, to use them to measure vibration above several thousand cycles per second. The output voltages of electrodynamic-type pickups are proportional to velocity; their outputs, corresponding to given accelerations, therefore decrease with increasing frequency. For accelerations several times the acceleration of gravity, the outputs usually are very small at frequencies above several thousand cycles per second.

In the *passive type* of inductive pickup, a change in the inductance of the coils in the pickup is caused by the motion of an object. There are three basically different passive inductive types: mutual inductance, differential transformer, and variable reluctance. Their principle of operation is that the inductance of the coils is changed by varying the configuration of the magnetic flux path.

TYPES OF PICKUPS USING INDUCTIVE PRINCIPLES

An inductive-type pickup may be referred to as a displacement, velocity, acceleration, or jerk pickup, depending upon whether its electrical output throughout a specified range of frequencies is proportional to displacement, velocity, acceleration, or jerk. Inductive-type pickups also may be classified as proximity pickups, movable-core pickups, or seismic pickups, as indicated in Table 15.1.

PROXIMITY PICKUPS

No part of a proximity pickup touches the object whose motion it is measuring. The pickup is usually attached to a fixture so that there is a small gap between the pickup

Table 15.1. Types of Shock and Vibration Pickups Using Inductive Principles

	Proximity	Movable-core	Seismic	
			Used above natural frequency	Used below natural frequency
Displacement	Mutual inductance	Differential transformer Variable reluctance		
Velocity	Electromagnetic	Electrodynamic	Electrodynamic	
Acceleration			Eddy-current	Differential transformer Variable reluctance
Jerk				Magnetostrictive

and the object. The proximity pickup has the advantage that it does not mechanically load the vibrating object. For this reason, it can be used to measure the vibration of very thin diaphragms, for example. The mutual inductance proximity pickup measures the displacement of the vibrating object. The electromagnetic proximity pickup measures the velocity of the vibrating object. These pickups are described in the sections entitled *Mutual-inductance Pickups* and *Electromagnetic Pickups.*

MOVABLE-CORE PICKUPS

In a movable-core pickup, only the core is mechanically attached to the vibrating object. The movable-core pickup can be used without modification to measure very-low-frequency vibratory motions; it also can be constructed into a seismic pickup (see section on *Seismic Pickups,* below). Movable-core pickups are described in sections on the differential transformer, variable reluctance, and electrodynamic type of pickups, according to the inductive principles which they employ.

SEISMIC PICKUPS

Ideally a seismic pickup consists of a mass element connected to the housing (or case) of the pickup by a massless spring and a massless damping device. Equation (12.8) expresses the displacement amplitude of the mass element relative to the housing of the pickup. As indicated in Fig. 12.4 at high frequencies, the mass element remains fixed in space as the housing of the pickup vibrates. If the output of the pickup is proportional to the displacement amplitude of the mass δ_0, the displacement sensitivity of the pickup is constant over a considerable range of frequencies above the natural frequency of the pickup; the pickup is then a displacement pickup. If the output is proportional to $\delta_0\omega$, it is a velocity pickup because the sensitivity of the pickup is the output divided by the applied velocity $u_0\omega$. Table 15.1 indicates that the electrodynamic principle is employed in the seismic velocity pickup normally used above its natural frequency. Pickups whose outputs are proportional to $\delta_0\omega^2$ and $\delta_0\omega^3$ are acceleration and jerk pickups, respectively; such pickups operate at frequencies *above* their natural frequencies. For example, the eddy-current pickup listed in Table 15.1 is an acceleration pickup of this type.

Many inductive-type pickups are designed to be used at frequencies well *below* their natural frequencies. Thus, below its natural frequency a pickup whose transducing element output is proportional to δ_0 is called an "acceleration pickup" because its acceleration sensitivity (its output divided by the applied acceleration $u_0\omega^2$) is constant over a considerable range of frequencies. Differential-transformer and variable-reluctance principles are used in this type of acceleration pickup.

The so-called "jerk pickup" has an output which is proportional to $\delta_0\omega$; its sensitivity (output divided by $u_0\omega^3$) is constant in the frequency range below its natural frequency. The magnetostrictive pickup, Table 15.1, is a jerk pickup. It is interesting to note that if the electrodynamic velocity seismic pickup (normally used *above* its natural frequency) is used *below* its natural frequency, it acts as a jerk pickup. This is because its sensitivity, output divided by $u_0\omega^3$, is constant at frequencies below about one-third of the natural frequency.

SELF-GENERATING-TYPE PICKUPS

Self-generating inductive pickups make use of either a steady magnetic field or a variable magnetic field. That is, a voltage can be induced in a conductor (1) by changing the position of the conductor in a steady magnetic field or (2) by using a fixed conductor about an electromagnet of variable field strength. The electrodynamic and eddy-current pickups are based on the steady magnetic field principle; the voltage induced in the conductor is proportional to the velocity of the conductor relative to the magnet. Electromagnetic and magnetostrictive pickups generate a voltage proportional to the time rate of change of the magnetic field about the conductor.

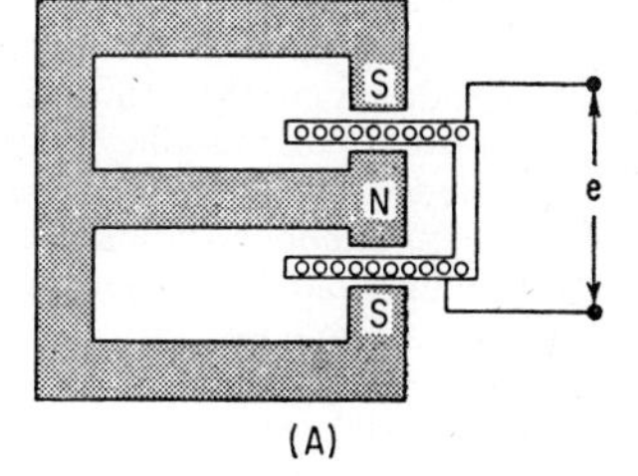

(A)

ELECTRODYNAMIC PICKUPS

The output voltage of the electrodynamic pickup is proportional to the relative velocity between the coil and the magnetic flux lines being cut by the coil. For this reason it is commonly called a velocity pickup. The principle of operation of the device is illustrated in Fig. 15.1*A*. A magnet has an annular gap in which a coil wound on a hollow cylinder of nonmagnetic material moves. Usually a permanent magnet is used, although an electromagnet may be used. The pickup also can be designed with the coil stationary and the magnet movable. The open-circuit voltage e generated in the coil is [1,2]

$$e = -Blv(10^{-8}) \quad \text{volts} \tag{15.1}$$

where B is the flux density in gausses; l is the total length in centimeters of the conductor in the magnetic field; and v is the relative velocity in centimeters per second between the coil and magnetic field. The magnetic field decreases sharply outside the space between the pole pieces; therefore, the length of coil wire outside the gap generates only a very small portion of the total voltage.

(B)

FIG. 15.1. Electrodynamic principle: (*A*) rectilinear; (*B*) rotational. The voltage e generated in the coil is proportional to the velocity of the coil relative to the magnet.

The basis of the rotational electrodynamic principle is illustrated in Fig. 15.1*B*. A soft-iron core is placed within the coil so that the magnetic field in the gaps remains sufficiently intense. The output voltage e is

$$e = -Blr \frac{d\phi}{dt} (10^{-8}) \qquad \text{volts} \tag{15.2}$$

where r is half the distance in centimeters between the lengths of wire in the two gaps, and ϕ is the angle of the rotation in radians of the rectangular-shaped coil about the geometric axis of the iron core.

MOVABLE-CORE ELECTRODYNAMIC PICKUPS. The electrodynamic pickup is the only velocity-type movable-core transducer in use. The pickup consists of a permanent bar magnet that moves in the core of a long coil wound on a form of nonmagnetic material. The self-generated voltage is proportional to the relative velocity between the magnet core and coil. For a constant velocity, a d-c voltage is generated. For vibratory motions, a-c voltage is generated which is an accurate analog of the mechanical motion. The performance [3] characteristics of the pickup are indicated in Table 15.3. It has good stability, and it has a large output which is a linear function of velocity if the magnet is kept well within the coil. Considerable off-centering of the longitudinal axis of the magnet relative to the longitudinal axis of the coil does not significantly affect the sensitivity or linearity nor does the presence of metallic objects outside the coil have any appreciable effect.[3] In fact, it is sometimes desirable to enclose the coil in a soft-iron shield to reduce the stray magnetic field from the magnet that may affect other instruments. The output resistance and number of turns are kept reasonably small so that the output is not excessively reduced by the electrical loading of the input impedance of the voltage-measuring equipment used with the pickup. Pickups which are designed for shorter displacement amplitude ranges than those listed in Table 15.3 have smaller output impedances. If the coil is wound on metallic forms, the sensitivity is reduced somewhat at high frequencies due to eddy-current flow. Some improvement in this respect can be achieved if the coil form contains longitudinal slots. The pickup normally is used below the elastic body longitudinal natural frequency of the bar or coil-form assembly, depending upon which is vibrated. The sensitivity of the pickup would be affected if the bar or coil-form assembly were vibrated at their longitudinal resonances. This is usually not a problem because the first axial resonance may be as high as 1,000 cps or 10,000 cps, depending upon the size of the pickup. Normally transverse resonance will not affect the performance if the transverse motion is small compared to the gap between the magnet and coil form.

A large percentage of all vibration pickup calibrators incorporate the movable-core electrodynamic principle. Instead of using a movable magnet, the calibrator usually is constructed with a coil that vibrates within the annular gap of a permanent magnet, seismically mounted relative to the frame of the calibrator. At frequencies above the natural frequency of the seismic magnet, the coil measures the absolute velocity of the motion. Absolute calibrations [4] of these coils can be made up to 5,000 cps.

SEISMIC RECTILINEAR ELECTRODYNAMIC PICKUPS. The electrodynamic principle is used in the velocity-type seismic pickup. There are many variations in the design of the spring-mass system used in the pickup. Figure 15.2 illustrates three of the designs in common use. The mass element of the pickup in Fig. 15.2*A* consists of an extremely light coil attached to a support arm which is pivoted on a shaft with bearings. A spiral spring attached to the arm and the housing of the pickup completes the seismic system. The Alnico permanent magnet rigidly attached to the housing of the pickup provides the steady magnetic flux field about the coil. When the housing of the pickup is vibrated above its natural frequency, the support arm rotates about a point near its center-of-gravity. Depending upon the center of rotation of the support arm, the displacement of the coil may be more or less than the displacement applied to the housing. In either case, the output voltage of the coil is proportional to the rectilinear velocity of the case. The permanent magnet of the pickup shown in Fig. 15.2*B* also is attached to the housing. The coil is suspended by springs from the housing. Above its natural frequency, the coil stands still in space when the housing is vibrated. In another common design, the coil is attached rigidly to the housing, as shown in Fig. 15.2*C*, and the permanent magnet is mounted seismically through a spring to the housing of the pickup.

Some of the performance characteristics of such pickups are listed in Table 15.4. Usually the pickup is used only at frequencies above its natural frequency. It is not very useful at frequencies above several thousand cycles per second. The sensitivity of most pickups of this type is quite large, particularly at low frequencies where their output voltage is larger than that of many other types of pickups. The coil impedance is quite small even at relatively high frequencies so that the output voltage can be measured directly with a high-impedance voltmeter.

This type of pickup is designed to measure quite large displacement amplitudes. Its use in some applications is limited by the effect of its relatively large weight on the structural response. For a pickup in which the coil is seismically mounted, the mechanical loading on the structure corresponds closely to the total weight of the pickup. When the permanent magnet is the seismic mass element, the effective mass attached to the vibrating structure is less than the total weight of the pickup. The actual effect on the structure is dependent on the mechanical impedance of the pickup. The pickup must be attached quite rigidly to the structure, particularly for measurements at very high frequencies. The contact resonance, i.e., the resonance between the pickup and the structure, should be considerably higher than the highest frequency of motion to be measured. The contact resonance of a typical pickup, for example, is about 3,000 cps when attached to the structure by a screw of approximately 0.4 in. diameter.[5]

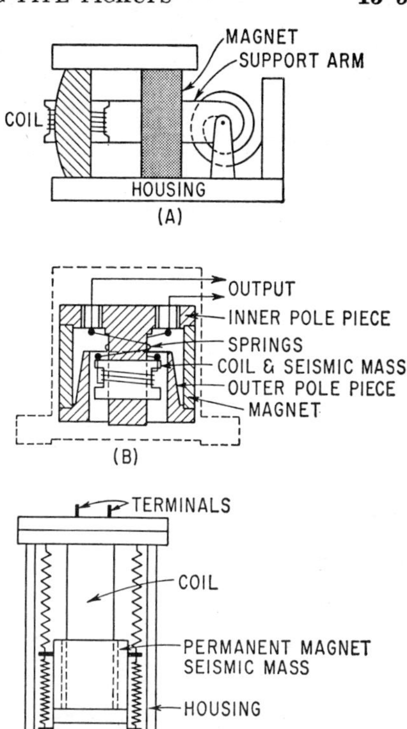

FIG. 15.2. Rectilinear velocity pickups based on the electrodynamic principle. The pickup is used at frequencies above its natural frequency. [(*A*) *Courtesy of MB Electronics.* (*B*, *C*) *Courtesy of Consolidated Electrodynamic Corp.*]

Most of the pickups can be used to measure vibration in either the vertical or horizontal direction. In some cases it is necessary to adjust the springs in the pickup to compensate for the deflection of the mass element due to its own weight when the position of the pickup is changed. Regardless of position, the transverse sensitivity of the pickup is usually near zero. However, for a pickup in which frictional contact can exist on the moving parts inside the pickup, the output may not indicate very accurately the motion along the sensitive axis when the pickup is vibrated in another direction.

The damping in a pickup is a factor in determining the shape of its frequency-response curve. Figure 15.3 shows typical response curves for an undamped pickup and for other pickups having various types of damping. The sensitivity of the undamped pickup, Fig. 15.3*A*, is constant for frequencies above several times its natural frequency. However, the sensitivity increases to a very large value near its natural frequency. Figure 15.3*B* illustrates eddy-current damping. For this pickup, the damping is achieved by eddy-current flow in the coil form (see section on *Eddy-current Pickups* below); the amount of damping increases at the higher frequencies, and the sensitivity drops off significantly. Eddy-current damping also can be achieved by placing a relatively small resistive shunt on the coil so that appreciable current flows in the coil. The effects of shunts of different resistance are shown in Fig. 15.3*C*. If the resistance of the shunt is reduced, the damping is increased; also, the sensitivity drops off at high frequencies

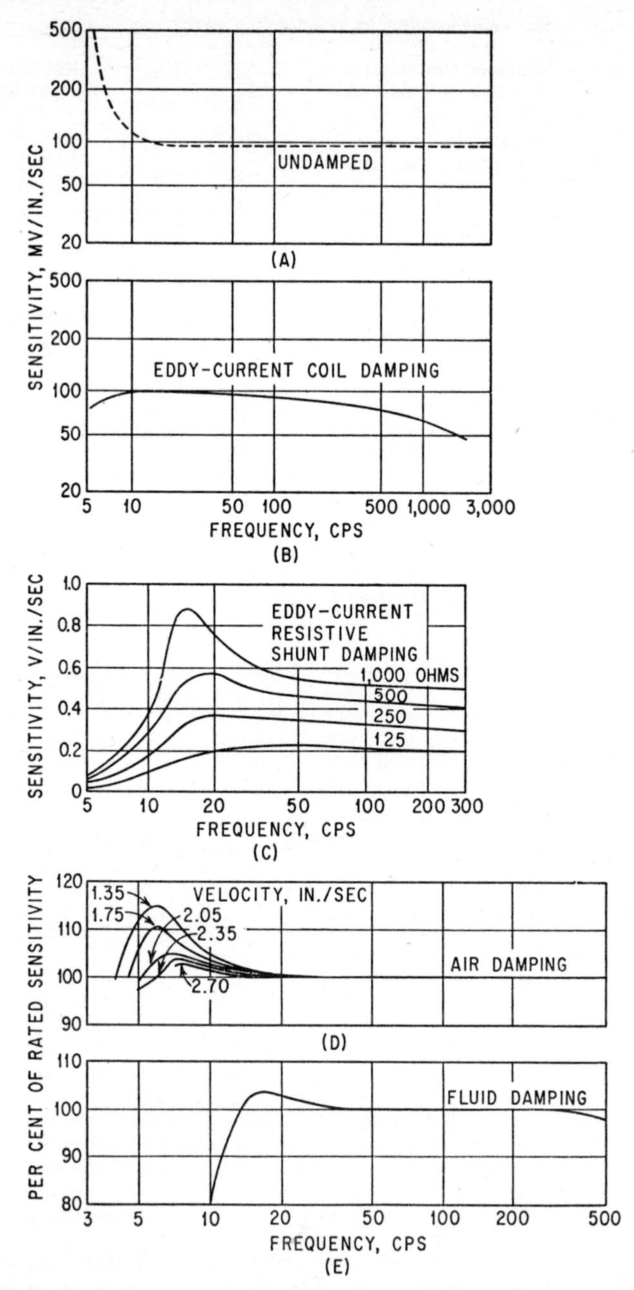

Fig. 15.3. Typical response curves of rectilinear velocity pickups with various types of damping. (*A*) MB Model 125; (*B*) MB Model 124. (*Courtesy of MB Electronics.*) (*C*) Electro-Tech. Model EVS-2. (*Courtesy of Electro-Technical Labs.*) (*D*) Sperry Model 1777291. (*Courtesy of Sperry Gyroscope Co.*) (*E*) I. R. D. Model 545. (*Courtesy of International Research and Development Corp.*) The sensitivity of the undamped pickup is constant at frequencies above the natural frequency. The undamped pickup is used in applications where the motion does not excite the natural frequency of the pickup. The damped pickups are used in rugged laboratory tests.

since the impedance of the coil is increased and the effect of the resistive shunt is greater.[6] The response of a pickup with air damping is shown in Fig. 15.3*D*. The damping is not proportional to the velocity. Over an appreciable range of velocities in the operating range of the pickup, the sensitivity of the pickup is nearly constant. A typical response characteristic of a fluid-damped pickup is shown in Fig. 15.3*E*. The sensitivity does not change greatly at high frequencies. However, fluid damping has the disadvantage

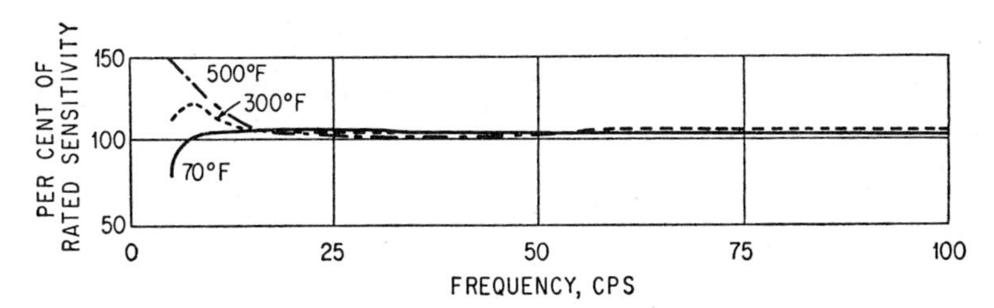

FIG. 15.4. Effect of temperature on response of a rectilinear velocity pickup. (*Courtesy of MB Electronics.*)

that it changes considerably with temperature.[7] The effect of temperature on a pickup having eddy-current damping is shown in Fig. 15.4. There is little change in sensitivity within the operating frequency range for temperatures up to 500°F (260°C). Another eddy-current damped pickup[5] experiences a 1 per cent change in the sensitivity for a temperature change of approximately 18°F (10°C).

The variation of phase lag of the output voltage relative to the applied velocity is shown in Fig. 15.5 for the two pickups whose sensitivities are shown in Fig. 15.3*A* and Fig. 15.3*B*. The phase lag is constant for the undamped pickup and varies linearly with frequency for the damped pickup at frequencies well above the natural frequency of the pickup.

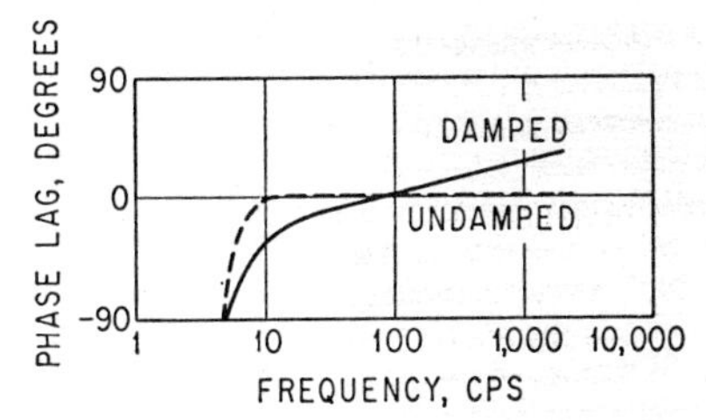

FIG. 15.5. Typical phase-lag curves of rectilinear velocity pickups. Damped, MB Model 124; undamped, MB Model 125. (*Courtesy of MB Electronics.*) The phase lag is 90° at the undamped natural frequency of the pickup. For measuring transient motions, it is necessary for the phase lag to be zero or directly proportional to frequency. The phase-lag response of these pickups is linear at frequencies significantly above the natural frequency.

SEISMIC ANGULAR ELECTRODYNAMIC PICKUPS. The construction of an electrodynamic angular vibration pickup is basically that shown in Fig. 15.1*B*. A permanent magnet is seismically mounted to the housing and the coil is rigidly attached to the housing. The pickup is used in torsional-vibration applications, e.g., on the arbor of a milling machine. At frequencies above its natural frequency, the voltage output of the pickup is proportional to the angular velocity of torsional oscillations superimposed on the rotational speed of the shaft to which the pickup is attached. The weight of the pickup, 7 lb (3.2 kg), limits its use to vibration testing of large machinery.

ELECTROMAGNETIC PICKUPS

The operation of electromagnetic pickups is based on the principle that a voltage is induced in a conductor when the magnetic flux about it is varied. This is illustrated in Fig. 15.6. When a metallic object is moved relative to the permanent magnet, the magnetic flux density changes and a voltage is induced in the coil. The voltage induced e is proportional to the time rate of change of the flux, and is proportional to the

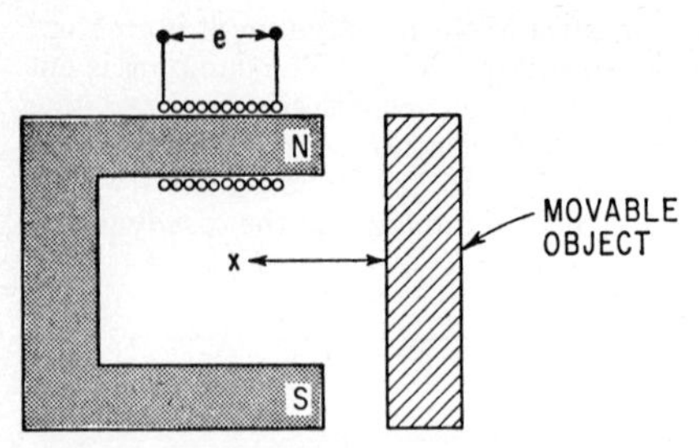

FIG. 15.6. Electromagnetic principle. The voltage generated in the coil wound on the magnet is proportional to the time rate of change of the flux. The flux varies as the gap changes between the magnet and metallic objects.

velocity $\dot{x}$ of the object, in centimeters per second, i.e.,

$$e = -n \frac{d\phi}{dx} \dot{x}(10^{-8}) \qquad \text{volts} \qquad (15.3)$$

where n is the number of turns in the coil and ϕ is the flux in maxwells. Over a limited range, the rate of change of the flux as the gap changes is nearly a constant for small changes in position, x, of the movable object. However, since the device is not perfectly linear, its application is limited.

The primary advantage of a pickup based on this principle is that it can be used to measure the motion of an object without touching it. By placing the electromagnet in the proximity of the object, a voltage is induced when the gap is changed. If the object is fabricated of magnetic material, the reluctance of the flux path is increased when the gap is decreased; the flux increases and thus generates a voltage in the coil. If the object is of nonmagnetic material, currents are set up in the object which oppose the magnetic field, thereby reducing the flux and inducing a voltage in the coil. Actually both effects may be present; if so, they tend to cancel each other. For best results, the material should either have a high permeability or be very nearly nonmagnetic.

The last two pickups listed in Table 15.2 are typical proximity pickups of the electromagnetic type. They are used for a variety of applications; for example, in speed measurements they are placed about 0.005 in. from a rotating object such as a rotating toothed gear. The output voltage increases as the speed of the gear increases. In vibration measurements, the voltage generated by the pickup is nearly proportional to the velocity of a metallic object in its proximity. Figure 15.7 shows the variation in sensitivity of a pickup for different spacings of the gap between the pickup and the vibrating object. The sensitivity changes as a function of the displacement of the vibrating object. This nonlinear behavior introduces some distortion in the output voltage. However, the distortion is small for small displacements if the average spacing of the gap is several times the displacement amplitude.

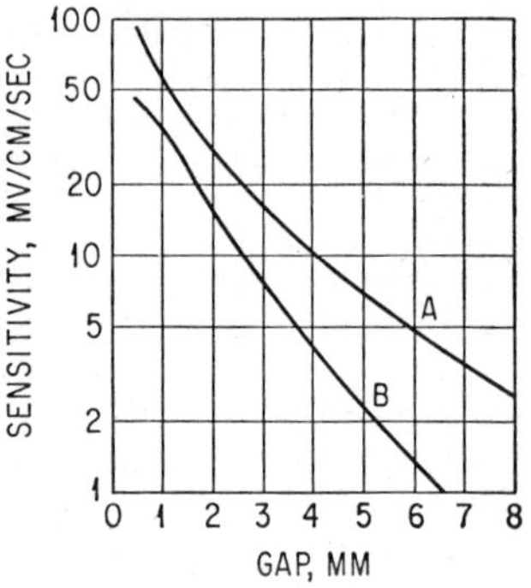

FIG. 15.7. Variation in sensitivity for Philips electromagnetic pickup Model PR 9262. The curve A is plotted for large ferromagnetic objects; curve B for nonmagnetic objects to which silicon-iron discs have been attached. (*After Anon.*[5])

EDDY-CURRENT PICKUPS

Consider a nonferrous plate of low resistivity which is moved in a direction perpendicular to the flux lines of a magnet, as indicated in Fig. 15.8. A current is generated in the plate which is proportional to the velocity of the plate; these eddy currents set up a magnetic field in a direction opposing the magnetic field that creates them. Thus, the resulting magnetic field varies as the generated current changes. The output voltage e of a coil of wire wound on the magnet is proportional to the time rate of change of the eddy currents, and therefore to the acceleration of the motion of the plate relative to the flux field. The above principle forms the basis of the eddy-current pickup. Such a device has an advantage over an electromagnetic pickup in that the voltage is linearly related to the motion since the gap between the plate and electromagnet remains constant.

Table 15.2. Typical Characteristics * of Proximity Pickups

Manufacturer	Model	Principle	Frequency range, cps	Amplitude limit, in.	Temperature range, °F	Output coil impedance, ohms
Tel-Instrument Electronics Corp.	501	Mutual inductance	10–20,000	0.020		
Electro Products Laboratories	3055	Electromagnetic	....	.....	−400 to 500	90
Philips (Holland)	PR 9262	Electromagnetic	0–2,000	0.3	....	1,260

* Taken from manufacturers' literature which gives detailed specifications and information on additional models available.

A significant feature of a pickup based on the eddy-current principle is that the force produced by the flux field from the eddy currents opposes the motion of the plate and is proportional to its velocity. Thus mechanical damping is achieved. Such a pickup has inherent damping in the same plate that is used to cause a voltage to be generated in the coil.

In the pickup which is based on the eddy-current principle, a permanent magnet and coil are rigidly attached to the housing of the pickup. A rectangular plate is seismically mounted on the housing. The eddy currents generated in the plate induce a voltage in the coil and at the same time provide damping in the pickup. Above the natural frequency of the pickup, the generated voltage is proportional to the acceleration applied to the pickup. The frequency response of the pickup is given in Fig. 15.9. This type

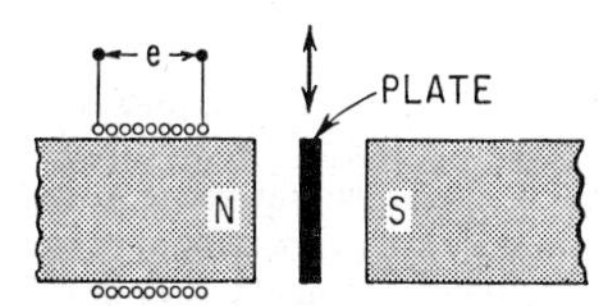

FIG. 15.8. Eddy-current principle. Current generated in the plate is proportional to the velocity of the plate as it cuts the flux field in the gap of the magnet. The current reduces the flux field, and the voltage e generated in the coil is proportional to the acceleration of the plate.

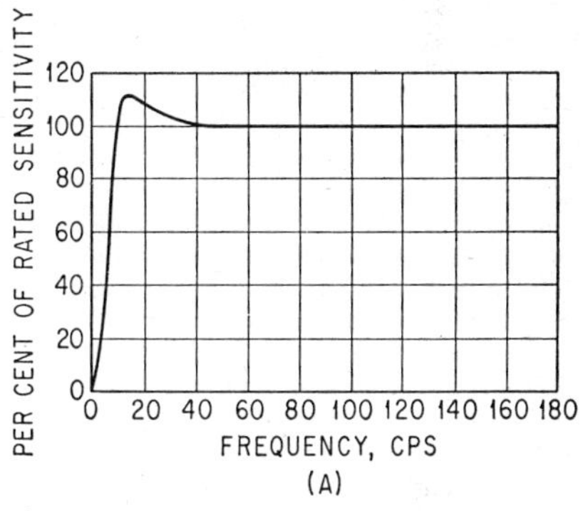

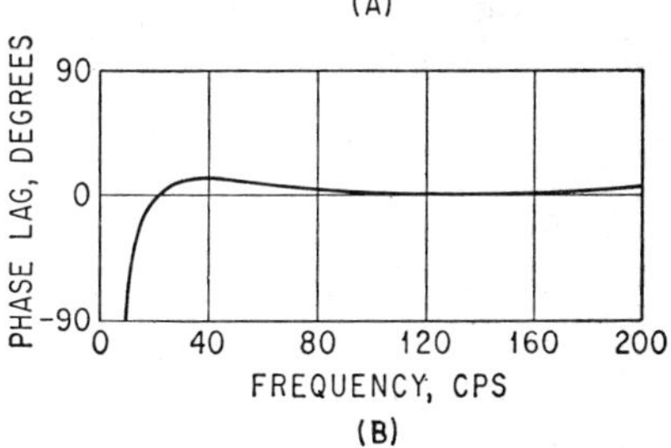

FIG. 15.9. Response curves of a prototype model of an acceleration pickup based on the eddy-current principle; (*A*) sensitivity and (*B*) phase angle. The pickup is used above its natural frequency. The undamped natural frequency of the pickup is 11 cps. (*Courtesy of The Calidyne Company.*)

of response is nearly ideal for measuring transient motion. The phase lag between the output voltage and acceleration applied to the pickup is close to 0° at frequencies above its natural frequency, and the percentage increase in the sensitivity is very small near the natural frequency of the pickup. These characteristics reduce the distortion in the output voltage of the pickup.

MAGNETOSTRICTIVE PICKUPS

The magnetic flux density in a ferromagnetic material changes when its length is altered due to the application of stress. This characteristic is used in magnetostrictive pickups. The flux density in some materials increases with the application of tensile

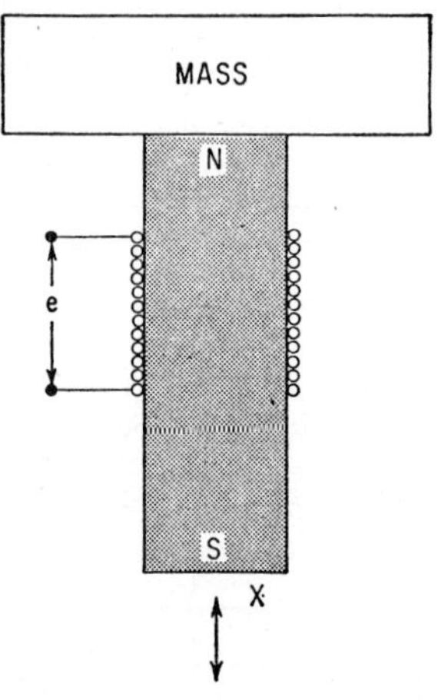

FIG. 15.10. Magnetostrictive principle. The force resulting from the mass motion changes the stress and flux density in the magnet, and a voltage e is generated in the coil.

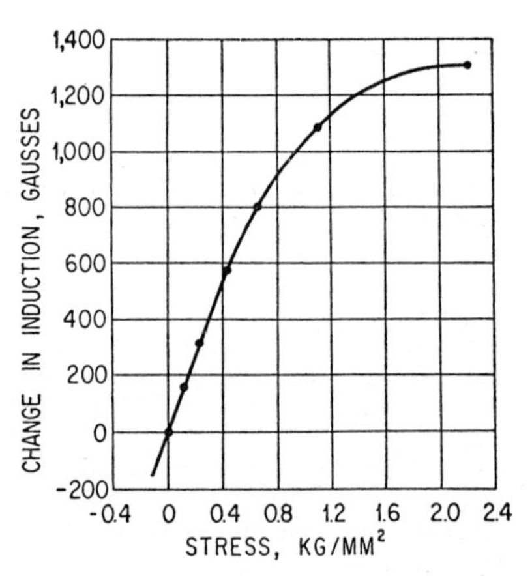

FIG. 15.11. Magnetostrictive effect in 45 Permalloy. The change in inductive flux density is linear only over a limited stress range. (*After R. M. Bozarth.*[8])

stress and it decreases with compressive stress; the opposite is true in other materials. For example, the flux density in Permalloy [8] is increased when tensile stresses are applied, whereas in nickel it is decreased. The application of the magnetostrictive principle to the measurement of mechanical motion is illustrated in Fig. 15.10. If mechanical motion is applied to the bar magnet, a force which is proportional to the acceleration of the attached mass is exerted on the magnet, the stress in the magnet varies with its acceleration, and the changes in the flux density are proportional to the stress. Therefore the voltage e generated in a coil which surrounds the magnet is proportional to the time rate of change of the flux density and is a function of the time rate of change of the acceleration:

$$e = CG(\sigma)\frac{d^3x}{dt^3} \qquad \text{volts} \tag{15.4}$$

where C is a constant and $G(\sigma)$ is the slope of the flux density vs. applied stress. The slope $G(\sigma)$ is constant only over a limited range of stress, as indicated in Fig. 15.11. In the range of constant slope, the generated voltage is proportional to the time rate of change of the acceleration. Therefore magnetostrictive pickups are linear over only a small range of stresses in the magnet.

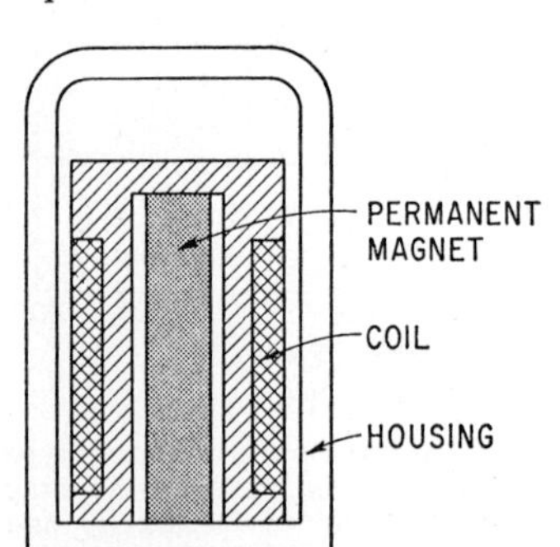

FIG. 15.12. Jerk pickup based on magnetostrictive principle. The voltage output of the coil is proportional to the time rate of change of the acceleration applied to the housing of the pickup. (*Courtesy of Sperry Gyroscope Company.*)

A cross-sectional view of a magnetostrictive jerk pickup is shown in Fig. 15.12. These pickups are used primarily for measuring the vibration of internal-combustion-engine cylinder heads in evaluating engine knock, valve timing, ignition, and mechanical malfunction, etc. The sensitivity of one of these pickups is approximately 51×10^{-9} mv/in./sec^3 (see Table 15.5). A half-sine-wave shock pulse of 100g with a time duration of 1 millisecond produces an output voltage of about 6 millivolts. Another magnetostrictive jerk pickup [9] is designed for the measurement of acceleration pulses of 1 millisecond duration up to more than 1,000g. However, because of its low sensitivity, frequency components at low accelerations and frequencies are difficult to measure.

PASSIVE INDUCTIVE-TYPE PICKUPS

There are three types of passive inductive pickups: mutual-inductance, differential-transformer, and variable-reluctance. In these devices the inductance changes as a component of the pickup changes its position. This results in a corresponding change in output voltage. Pickups of this type can be constructed which have a usable response down to zero frequency.

MUTUAL-INDUCTANCE PICKUPS

The operation of a pickup based on the mutual-inductance principle is illustrated in Fig. 15.13. Two coils are wound on a form which is fabricated of nonmetallic material. The primary coil is energized by an alternating current i_p. The magnetic field produced by the primary coil sets up a field which is opposed by that produced by eddy-current flow in the metal surface. The resulting magnetic field induces a voltage in the secondary coil. As the metal surface moves closer to the secondary coil, the opposing eddy-current field increases, and the output of the secondary coil is reduced. The voltage e in the secondary coil is

$$e = M\omega i_p \tag{15.5}$$

where M is the mutual inductance in henries, ω the angular frequency in radians per second, and i_p the alternating current in the primary coil in amperes. Figure 15.14 shows the variation in mutual inductance for a nonmagnetic metal surface. The mutual inductance can be made nearly linear over a surface displacement range up to approximately 5 per cent of the coil diameter. For a metal surface of ferromagnetic material, the intercept of the curve is greater and its slope is less because the high flux in the surface creates a field that opposes the field due to the eddy currents. Therefore, a nonmagnetic vibrating surface induces a larger output in the secondary coil per unit of displacement of the surface than does a surface of magnetic material.

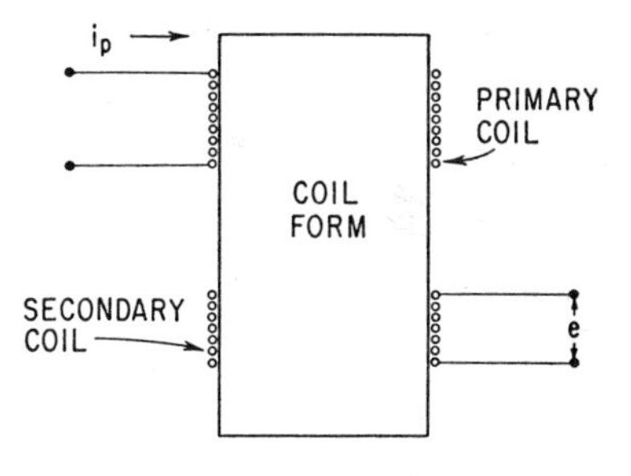

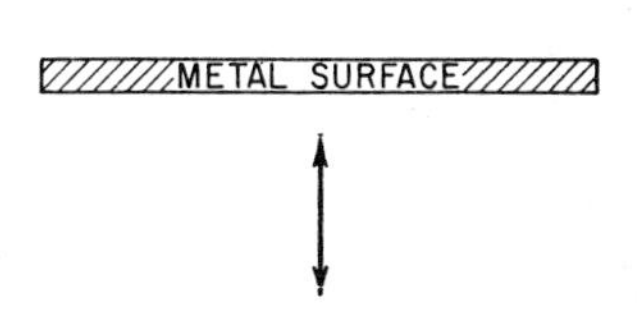

FIG. 15.13. Mutual-inductance principle. The mutual inductance of the coils changes as the gap between the coil form and metal surface is varied. The output voltage e of the secondary coil is proportional to the mutual inductance when the input current i_p is maintained constant. (*After M. L. Greenough.*[10])

Under static conditions, the output of the secondary coil is a voltage having the frequency of the primary current. If the surface vibrates, this voltage is frequency-modulated at a frequency corresponding to the motion of the metal surface, and the output voltage varies in proportion to the surface-to-coil spacing. The distinct advantage of a pickup of this type is that it can measure motions from d-c to very high frequencies without touching the vibrating object.

The properties of a typical mutual-inductance

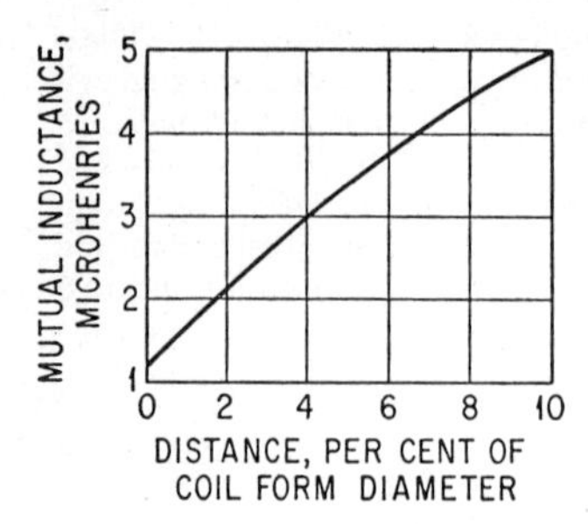

FIG. 15.14. Typical variation of mutual inductance for the displacement-measuring device. The variation is nearly linear for changes in the distance between the coil form and metal surface up to 5 per cent of the diameter of the coil form.

pickup are listed in Table 15.2. This pickup is used to measure the vibration of nonmagnetic metallic objects by bringing the two-coil probe to within 0.03 in. of the object; for other materials it is recommended that a disc at least 1 in. in diameter and 0.01 in. thick be attached to the vibrating object.[11, 12] The output of the pickup is sufficiently linear so that it is practicable to measure vibration up to 0.020 in. amplitude. This limits its practical usefulness at very low frequencies where large displacements may occur. Electronic instrumentation is available for the pickup which includes a regulated 2.5-megacycle oscillator for the primary coil and a voltmeter to read the output. The variation in mutual inductance can also be used to vary the frequency of an oscillator.[13]

This type of pickup can be used to measure the extraneous motion of linear vibration exciters.[14] It also can be used for many other vibration-measuring applications, particularly when it is undesirable to load mechanically the vibrating object with a seismic-type vibration pickup. Since it is a zero-frequency displacement-type pickup, it may be calibrated statically with a dial micrometer or other accurate static displacement device, although it is desirable to calibrate it throughout the frequency range of intended use.

DIFFERENTIAL-TRANSFORMER PICKUPS

The operation of a pickup of the differential-transformer type is similar to one of the mutual-inductance type in that its output depends upon the mutual inductance between a primary and secondary coil. The basic components are shown in Fig. 15.15. The pickup consists of a core of magnetic material, a primary coil, and two secondary coils. The voltage induced in each secondary coil is equivalent to that given by Eq. (15.5). When the core is exactly in the center, the same length of core is within both secondary coils. Therefore the mutual inductances of both secondary coils are identical, and the voltages across both secondary coils are equal in magnitude. However, they are connected in series opposition so that the output voltage is zero. As the core is moved up or down, the inductance and induced voltage of one secondary coil are increased while those of the other are decreased. The output voltage is the difference between these two induced voltages. In this type of transducer, the output voltage is proportional to the displacement of the core over an appreciable range. A typical response characteristic is shown in Fig. 15.16. In practice, the output voltage at the carrier frequency of the primary current is not exactly zero when the core is centered, and the output near the center position is not exactly linear. When the core is vibrated, the output voltage is a carrier wave, modulated at a frequency and amplitude corresponding to the motion of the core relative to the coils.

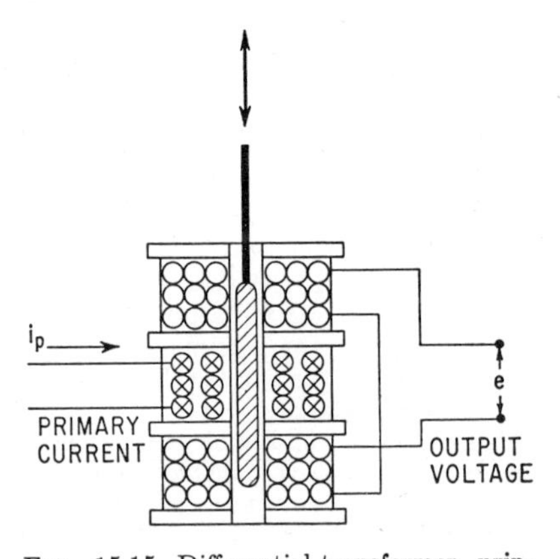

FIG. 15.15. Differential-transformer principle. The inductance of the coils changes as the core is moved. For constant input current i_p to the primary coil, the output voltage e is the difference of the voltages in the two secondary coils which are wound in series opposition. (*Courtesy of Automatic Timing and Controls, Inc.*)

There are a considerable number of ways [15, 16, 17] that these cores and coils can be arranged. For example, the core may be attached to one object and the coils to another, i.e., as a movable-core pickup. It also is common to use the core as the mass element of a seismic acceleration pickup operated below its natural frequency.

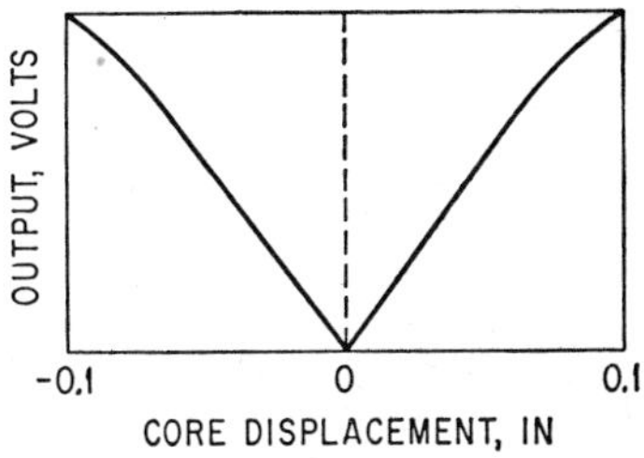

Fig. 15.16. Variation in output voltage with core displacement for a differential-transformer pickup. (*After H. Schaevitz.*[17])

MOVABLE-CORE DIFFERENTIAL-TRANSFORMER PICKUPS. Table 15.3 lists a typical movable-core differential-transformer pickup and some of its important characteristics. These pickups are used for very-low-frequency measurements. The sensitivity listed in Table 15.3 gives an approximate indication of the output voltages obtainable. Actually the sensitivity varies with the carrier frequency of the current in the primary coil. A typical plot of the sensitivity vs. carrier frequency is given in Fig. 15.17. The carrier frequency should be at least ten times the highest frequency of the mechanical motion to be measured in order that the output voltage, when demodulated, be an accurate representation of the mechanical motion. Therefore the carrier frequency usually is above 500 cps. The last column in Table 15.3 gives the d-c resistance of the secondary coils. At high carrier frequencies, the reactive component of the coil impedances becomes large, and the output impedance of the coil is significantly greater than its d-c resistance. The load impedance of the voltage-measuring instrument should be very large compared to the output impedance of the differential transformer in order to avoid excessive reduction in the output voltage.

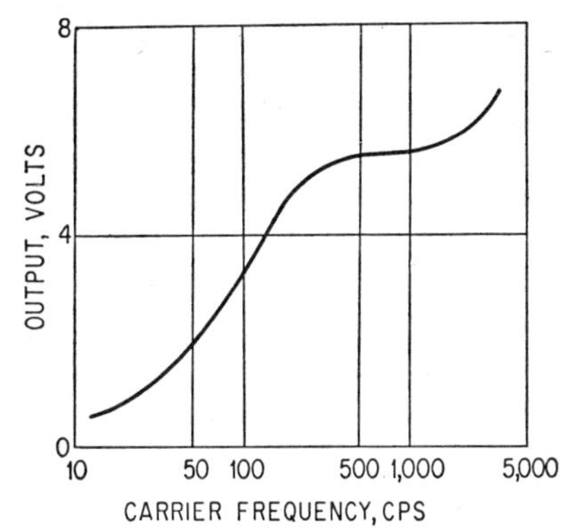

Fig. 15.17. Variation in output voltage with carrier frequency of the input current for a differential-transformer pickup. Usually a carrier frequency of several hundred cycles per second is used. (*After H. Schaevitz.*[17])

Note that if the primary coil is energized with direct current, the differential-transformer type of pickup becomes an electrodynamic pickup. However, few pickups are used this way since the sensitivity of the resulting electrodynamic pickup usually would be less than that of a permanent-magnet electrodynamic pickup.

RECTILINEAR SEISMIC DIFFERENTIAL-TRANSFORMER PICKUPS. A typical design of a differential-transformer seismic pickup is shown in Fig. 15.18. The flat diaphragm spring permits the mass-element iron core to move axially in the differential transformer. A pickup of this type is used in the frequency range from 0 cps up to about two-thirds of its natural frequency. Table 15.5 lists typical acceleration ranges. Pickups with the higher acceleration ranges have larger usable frequency ranges; however, their sensitivities are correspondingly smaller.

Eddy-current damping is used frequently in pickups with small frequency ranges. The effect of temperature on these pickups is small. Fluid damping is used frequently in pickups with large frequency ranges, but is significantly affected by temperature. If the pickup were used only for sinusoidal motion measurements at frequencies up to one-third of its undamped natural frequency, it would be preferable to use no damping. However, in general-purpose testing and in measuring transient motions when significant components of the motion are present at frequencies near the natural frequency, damping is necessary in order to avoid excessive voltage outputs and to avoid contact between the mass element and housing of the pickup as a result of excessive displacements. For ac-

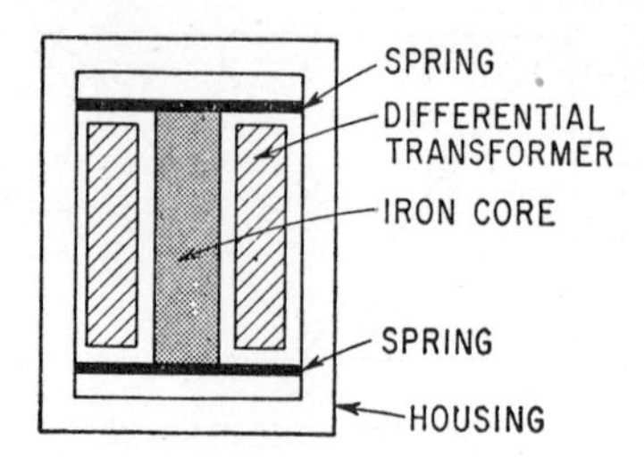

Fig. 15.18. Typical differential-transformer acceleration pickup used below its natural frequency. The output voltage of the differential transformer is proportional to the acceleration of the housing in the operating frequency range of the pickup. (*Courtesy of Gulton Industries, Inc.*)

curate measurements of transient motion, it is also necessary for the phase angle of the output voltage and input motion to be zero or to be directly proportional to frequency. According to the curves of Fig. 12.26 the phase response of such a device should be nearly proportional to frequency within its operating range.

ANGULAR SEISMIC DIFFERENTIAL-TRANSFORMER PICKUPS. A differential-transformer seismic angular-acceleration pickup (used below its natural frequency) is listed in Table 15.3. Other models with natural frequencies as high as 100 cps are available. The performance characteristics are similar to those of the rectilinear pickup. A differential-transformer seismic angular-displacement pickup (used above its natural frequency) has been built [18] which has a natural frequency of 2.5 cps and constant sensitivity to 100 cps.

VARIABLE-RELUCTANCE PICKUPS

The operation of a variable-reluctance type of pickup is based on the variation of inductance that is produced in a coil when the reluctance of the magnetic flux path about the coil is changed. The inductance of a coil may be changed either by changing the length of a magnetic core which is inserted into the coil or by changing the size of the air gap of an electromagnet.

MOVABLE-CORE VARIABLE-RELUCTANCE PICKUPS. Consider the inductance of a simple coil of small diameter which is wound on a cylinder. The inductance L of such a coil is, approximately,

$$L = 4\pi \mathfrak{n}^2 \mu S l (10^{-9}) \qquad \text{henrys} \tag{15.6}$$

where $\mathfrak{n}$ is the number of turns of wire per unit length, μ is the permeability of the cylinder, l is the length of the cylinder, and S is the area of the cylinder. If the total length of the coil is nearly filled with a magnetic core, the inductance of the length of coil over the core contributes the major part of the total inductance of the coil. If the remainder of the coil is filled with air or a core of low-permeability material, it contributes little to

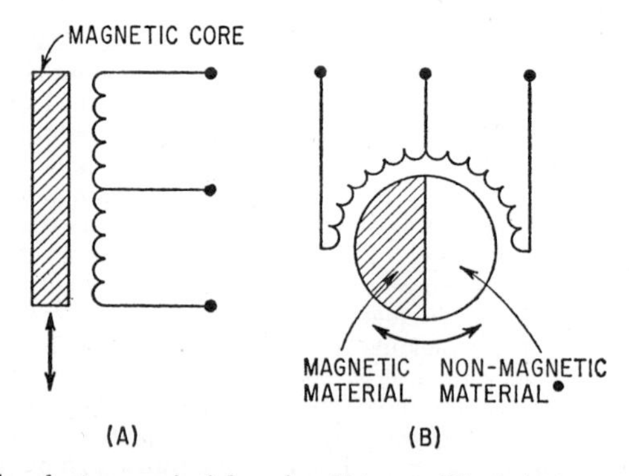

Fig. 15.19. Variable-reluctance principle, solenoid type. The inductance of the coils changes as the reluctance of the flux path is changed due to motion of the core. The core is made of magnetic material. The coils are usually arranged as part of a Wheatstone bridge circuit. Design (*A*) is for measuring rectilinear motion and (*B*) for rotational motion. In the latter, only half of the cylinder is made of magnetic material.

the total inductance of the coil. Therefore, for small displacements of the core, the inductance of the coil is nearly proportional to the length of core within the coil. The coil usually is center-tapped as shown in Fig. 15.19*A*. When the core is moved up or down, the inductance of half the coil is increased while that of the other half is decreased. The two halves of the coil are arranged as two arms of a Wheatstone bridge and the output is linear [19] over an appreciable range. Because the use of this type of pickup is limited to the very low-frequency range, input carrier frequencies as low as 60 cps are used, although the recommended carrier frequency for most models is between 600 and 50,000 cps. The effects of the carrier frequency on the performance of the pickup are similar to those given for differential transformers. This type of pickup has a very low output impedance.

Table 15.3 lists a commercial pickup of this type. Other models are available for the measurement of both rectilinear and angular motion. In such commercial units, the core which moves within the coil usually is made of high-permeability stainless-steel rod. A nonmagnetic extension is attached to the rod so that the core can be mechanically attached to an object. The output from the bridge is proportional to the relative displacement between the core and coil. One of the designs [19] of this type of pickup (in which solid ceramic potting is used) may be operated at temperatures as high as 1300°F (704°C).

Angular displacement can be measured by using a cylindrical core with only half of its cross section made of magnetic material as indicated by the shaded area in Fig. 15.19*B*. The core is normal to the axis of the coil. Its principle of operation is similar to the rectilinear coil in that the inductances of the two halves of the coil depend upon the amount of magnetic material in their flux paths.

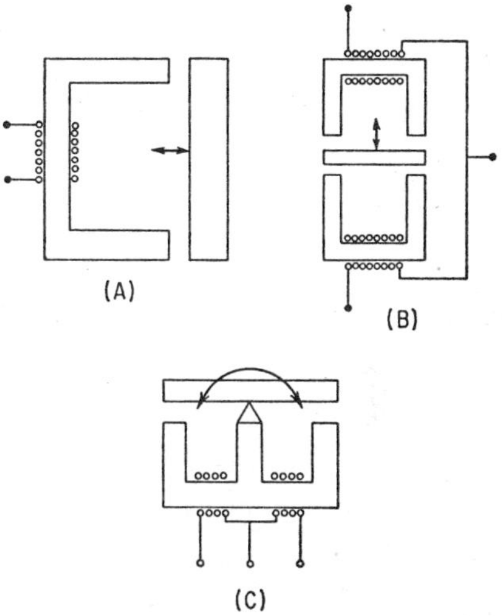

(A)

(B)

(C)

FIG. 15.20. Variable-reluctance principle, air-gap type. The inductance of the coils is changed by varying the air gap between the movable and stationary parts of the magnets. Designs including (*A*) a single magnet, (*B*) a double magnet, and (*C*) an E-core magnet are used. The coils are usually arranged as part of a Wheatstone bridge circuit.

SEISMIC VARIABLE-RELUCTANCE PICKUPS. The variable-reluctance principle also can be applied in pickups by varying the thickness of a small air gap in the magnetic flux path of an electromagnet. In the pickups illustrated in Fig. 15.20*A*, *B*, and *C*, the air gap is very small compared to the total length of magnetic core material through which the intense flux field passes. For such an arrangement the inductance of the coil L is approximately

$$L = \frac{4\pi n^2 S(10^{-9})}{l} \quad \text{henrys} \tag{15.7}$$

where S is the area of the air gap perpendicular to the direction of flux flow, l is its thickness, and n is the total number of coil turns. For small air gaps, the inductance of the coil is proportional to the change in air-gap thickness over an appreciable range. Usually one movable armature is used between two electromagnets, each of which has coils wound on it as shown in Fig. 15.20*B*, or an armature is pivoted on an E-core magnet as shown in Fig. 15.20*C*. These two coils form two arms of a Wheatstone bridge. Figure 15.21 illustrates the linear range of such an arrangement of coils.

The armature of the air-gap variable-reluctance pickup is usually made the mass element of a seismic system. Variable-reluctance pickups of the rectilinear type are in most general use, although angular-acceleration pickups have been built.[21] The characteristics of an acceleration pickup of the rectilinear type are listed in Table 15.3. A diagram of a typical design is shown in Fig. 15.22. The mass element and the magnetic armature lever rotate and change the gap dimensions when the pickup is subjected to vibration in the direction of the arrows. Single-coil variable-reluctance pickups with the magnetic armature suspended by diaphragm springs, and other designs are used.[22] This type of variable-reluctance pickup is used at frequencies up to about two-thirds

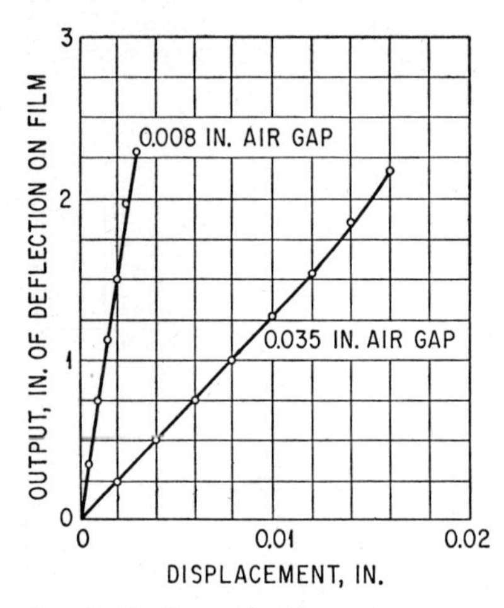

FIG. 15.21. Curve showing output measured from oscillographic recording of an air-gap type of variable-reluctance displacement-measuring device. The output is nearly linear for applied displacements up to about one-fourth of the air gap. (*After B. F. Langer.*[20])

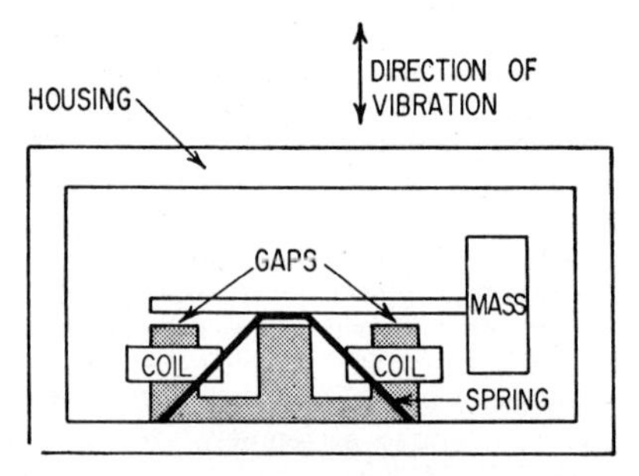

FIG. 15.22. Typical variable-reluctance acceleration pickup used below its natural frequency. The inductance of the coils changes as the gaps vary when the pickup is vibrated. (*Courtesy of Wiancko Engineering Company.*)

of its natural frequency. In addition to the one listed in Table 15.3, pickups with acceleration ranges as small as $\pm 0.5g$ and as large as $\pm 1{,}000g$ are available. A carrier frequency of several thousand cycles per second is fed into the input when the coils are

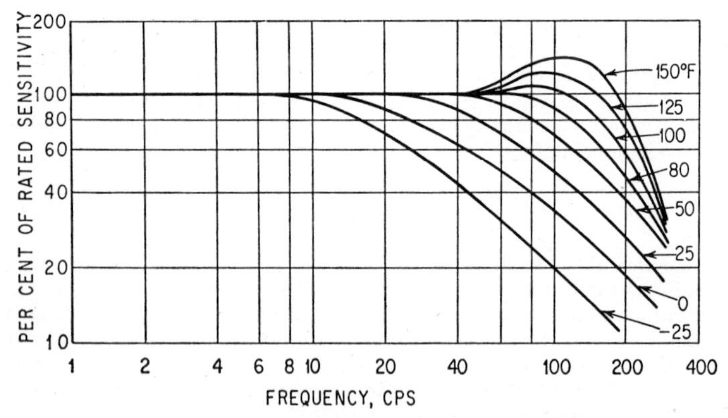

FIG. 15.23. Effect of temperature on response of CEC Model 4-260 variable-reluctance acceleration pickup. The operating frequency range within which the sensitivity is constant is reduced at low temperatures by an increase in the damping. (*Courtesy of Consolidated Electrodynamics Corp.*)

arranged in a Wheatstone bridge. The output voltage is amplitude-modulated at the frequency of the vibratory motion.

Both eddy-current and fluid damping are used. The effect of temperature on a pickup with fluid damping is indicated in Fig. 15.23. Near room temperature, the damping is about 0.6 of critical damping and the frequency range of flat response is large compared to that which would exist were no damping present. At very high and low temperatures, the damping is changed so that the usable frequency range is much less than that obtained at room temperature. The considerations regarding damping and phase angle discussed in the section *Rectilinear Seismic Differential-transformer Pickups* also apply to this type of pickup.

REFERENCES

1. Mason, W. P.: "Electromechanical Transducers and Wave Filters," p. 187, D. Van Nostrand Company, Inc., Princeton, N.J., 1948.
2. Nelson, H. M.: "Moving Coil Vibration Generators and Pickups," *Instruments Note* 54, Aeronautical Research Laboratories, Australia, 1954.
3. Perls, T. A., and E. Buchmann: *Rev. Sci. Instr.*, **22**:475 (1951).
4. Levy, S., and R. R. Bouche: *J. Research Natl. Bur. Standards*, **57**:227 (1956).
5. Anon.: "Philips Vibration Measuring and Exciting Instruments," Philips, Eindhoven, Holland.
6. Leslie, C. B., J. M. Kendall, and J. L. Jones: *J. Acoust. Soc. Amer.*, **28**:711 (1956).
7. White, G. E.: *ASME Symposium on Shock and Vibration Instr.*, vol. 10, 1952.
8. Bozarth, R. M., and H. J. Williams: *Revs. Modern Phys.*, **17**(1):72 (1945).
9. Wilde, H., and E. Eisele: *Z. angew. Phys.*, **1**(8):359 (1949).
10. Greenough, M. L.: *Trans. Am. Inst. Elect. Engineers*, **67**(I):589 (1948).
11. Yates, W., and M. Davidson: *Electronics*, **26**(9):183 (1953).
12. Schacher, D. L.: *Instr. and Automation*, **30**:470 (1957).
13. Clark, H. F.: *Trans. AIEE*, **74**(I): 186 (1955).
14. Elliott, W. R.: *Proc. Instr. Soc. Amer.*, **10**(55-21-1) (1955).
15. Boggis, A. G.: *Proc. Soc. Exptl. Stress Analysis*, **9**(2):171 (1952).
16. MacGeorge, W. D.: *Instruments*, **23**(6):610 (1950).
17. Schaevitz, H.: *Proc. Soc. Exptl. Stress Analysis*, **4**(2):79 (1947).
18. DeMichele, D. J.: *Proc. Instr. Soc. Amer.*, **10**(55-12-1) (1955).
19. Sawyer, E. V.: *Proc. Instr. Soc. Amer.*, **10**(55-7-2) (1955).
20. Langer, B. F.: *Rev. Sci. Instr.*, **2**:336 (1931).
21. Wiancko, T. H.: U.S. Patent 2,759,157, 1956.
22. Dranetz, A. I.: *Machine Design*, **30**(1):120 (1958).

Table 15.3. Typical Characteristics of Inductive Pickups

Type	Principle	Sensitivity*	Natural frequency, cps	Frequency range, cps	Amplitude limit, in.†	Temperature range, °F	Damping fraction of critical	Type	Weight, oz.	Output coil impedance, ohms
Movable core	Differential transformer	300	....	0–60	1	....	....	....	8	600
Velocity	Electrodynamic	105	30	50–500	0.1	−65–300	....	Magnetic	2.2	800
Velocity	Electrodynamic	1,080	16	10–1,000	30*g*	−40–500	0.7	Magnetic	21	2,000
Acceleration	Reluctance	190	14	0–10	1*g*	−65–200	0.65	Gas		200

* Units are millivolts per volt per inch for movable-core type; millivolts per inch per second for seismic velocity type used above natural frequency; and millivolts per volt per gram for seismic acceleration type used below natural frequency.

† Units are inches unless otherwise specified.

16

PIEZOELECTRIC AND PIEZORESISTIVE PICKUPS

Abraham I. Dranetz
Dranetz Engineering Laboratories, Inc.

Anthony W. Orlacchio
Gulton Industries, Inc.

INTRODUCTION

Certain solid-state materials are electrically responsive to mechanical force; they often are used as the mechanical-to-electrical transduction devices in shock and vibration pickups. Generally exhibiting high elastic stiffness, these materials can be divided into two categories: *the self-generating type,* in which electric charge is generated as a direct result of applied force, and the *passive-circuit type,* in which applied force causes a change in the electrical characteristics of the material. *Piezoelectric* materials are of the self-generating type. In the linear elastic range, a piezoelectric material produces an electric charge proportional to stress. Passive-circuit types of materials include *magnetostrictive* and *piezoresistive* materials. A magnetostrictive material has a magnetic permeability which depends upon applied force; a piezoresistive material has an electrical resistance which depends upon applied force.

A solid-state material whose electrical resistance depends upon the applied magnetic field is said to be *magnetoresistive.* Their use is restricted to vibration pickups having highly compliant spring-mass combinations, and hence low resonant frequency.

PIEZOELECTRIC ACCELEROMETERS

PRINCIPLE OF OPERATION OF SIMPLE SEISMIC SYSTEM

An accelerometer of the type shown in Fig. 16.1*A* is a linear seismic transducer (i.e., pickup) utilizing a piezoelectric element in such a way that an electric charge is produced which is proportional to the applied acceleration. This "ideal" seismic piezoelectric pickup can be represented (over most of its frequency range) by the elements shown in Fig. 16.1*B*. A mass is supported on a linear spring which is fastened to the frame of the instrument. The piezoelectric crystal which produces the charge acts as the spring. Viscous damping between the mass and the frame is represented by dashpot c. In Fig. 16.1*C* the frame is given an acceleration upward to a displacement of u, thereby producing a compression in the spring equal to δ. The displacement of the mass relative to the frame is dependent upon the applied acceleration of the frame, the spring stiffness, the mass and the viscous damping between the mass and the frame, as indicated in Eq. (12.10) and illustrated in Fig. 12.5.

For frequencies far below the resonant frequency of the mass and spring, this displacement is directly proportional to the acceleration of the frame and is independent of frequency. At low frequencies, the phase angle of the relative displacement x, with respect to the applied acceleration, is proportional to frequency. As indicated in Fig. 12.6, for low fractions of critical damping which is characteristic of many piezoelectric pickups, the phase angle is proportional to frequency at frequencies below 30 per cent of the resonant frequency.

In Fig. 16.1, inertial force of the mass causes a mechanical strain in the piezoelectric element which produces an electric charge proportional to the stress* and, hence, proportional to strain and acceleration.[1] If the dielectric constant of the piezoelectric material does not change with electric charge, the voltage generated also is proportional to

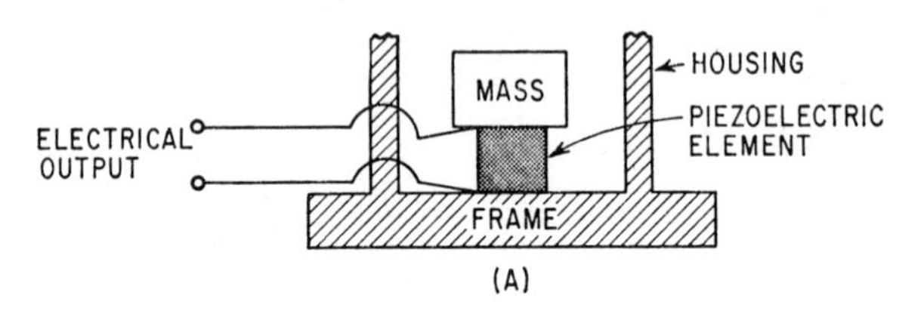

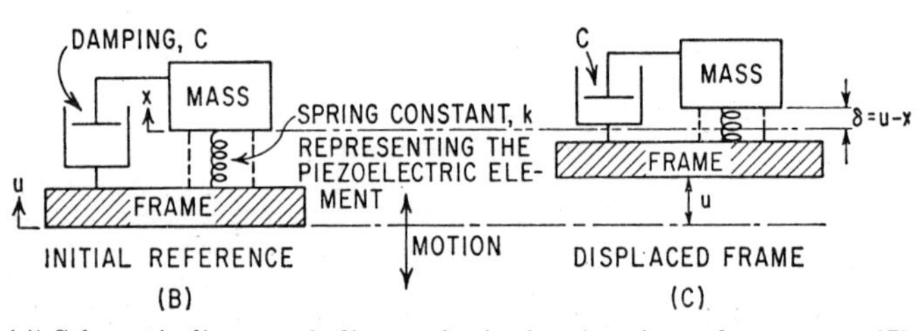

FIG. 16.1. (*A*) Schematic diagram of a linear seismic piezoelectric accelerometer. (*B*) A simplified representation of the accelerometer shown in (*A*) which applies over most of the useful frequency range. A mass m rests on the piezoelectric element which acts as a spring having a spring constant k. The damping in the system, represented by the dashpot, has a damping coefficient c. (*C*) The frame is accelerated upward, producing a displacement u of the frame, moving the mass from its initial position by an amount x, and compressing the spring by an amount δ.

acceleration. Metallic electrodes are applied to the piezoelectric element and electrical leads are connected to the electrodes for measurement of the electrical output of the piezoelectric element.

In the ideal seismic system shown in Fig. 16.1, the mass has infinite stiffness, the spring has zero mass, viscous damping exists only between mass and frame, and the frame has infinite stiffness. In practical piezoelectric pickups, these assumptions cannot be fulfilled. For example, the mass may have as much compliance as the piezoelectric element. In some seismic elements (particularly those which produce voltage from bending of the piezoelectric element) the mass and spring are inherently a single structure. Furthermore, in many practical designs where the frame is used to hold the mass and piezoelectric element, distortion of the frame may produce mechanical forces upon the seismic element.[2] All these factors may change the performance of the seismic system from those calculated using equations based on an ideal system. In particular, the resonant frequency of the piezoelectric combination may be substantially lower than that indicated by theory. Nevertheless, the equations for an ideal system are useful both in design and application of piezoelectric accelerometers.

* The piezoelectric element contained in an accelerometer normally is used within its linear elastic range; hence, one may consider that the charge is a result of the stress applied or of the strain which results from the stress.

ELECTRICAL ANALOG OF PIEZOELECTRIC ACCELEROMETER. The electrical analog of a mechanical system often offers a useful representation, particularly to those familiar with electric circuit analysis. An *electrical analog* is particularly useful in describing an electromechanical device, such as a piezoelectric pickup, since it provides a description for the entire device of a consistent set of electrical parameters.

Table 16.1 Analogous Parameters for an Electric Circuit Which Is the Direct Analogy of a Mechanical System

Electrical parameters		Mechanical parameters	
Symbol	Description	Symbol	Description
q	Charge	x	Displacement
$\dot{q}, i$	Current	$\dot{x}$	Velocity
$\ddot{q}, \frac{di}{dt}$	Rate of change of current	$\ddot{x}$	Acceleration
e	Voltage	F	Force
L	Inductance	m	Mass
R	Resistance	c	Damping coefficient
C	Capacitance	$1/k$	Compliance

Employing Eq. (12.1), a simple substitution of mechanical parameters by analogous electrical parameters in the equation of motion for the simple seismic system shown in Fig. 16.1 yields

$$L(\ddot{q}_2 - \ddot{q}_1) + R\dot{q}_2 + \frac{1}{C}q_2 = 0 \tag{16.1}$$

where inductance L is analogous to mass m, resistance R is analogous to damping coefficient c, capacitance C is analogous to the reciprocal of spring constant k, charge q_1 is analogous to input frame displacement u, and charge q_2 is analogous to spring compression δ. The analogous quantities are summarized in Table 16.1. This type of analog is called a "direct analog" or "force-voltage" analog. It can be shown that Eq. (16.1) is satisfied by the electric circuit of Fig. 16.2A, in which the frame acceleration is represented by $\ddot{q}_1$ and the spring distortion is represented by q_2.

In a simple seismic accelerometer, the deflection of the spring is a measure of the acceleration. However, in the analysis of a seismic accelerometer sometimes it is easier to relate the piezoelectric constants to the applied force rather than to the deflection. Assuming linear elastic characteristics of the piezoelectric element, the force F and strain δ are related by

$$\delta = \frac{F}{k} \quad \text{meters} \tag{16.2}$$

where k = stiffness of the piezoelectric element in newtons/meter and F = force in newtons.

As a result of mechanical stress in the piezoelectric element, an electric charge q is generated and built up across the internal electrical capacitance C_E of the piezoelectric element. The electric charge is given by

$$q = A\delta = \frac{AF}{k} \quad \text{coulombs} \tag{16.3}$$

where A is a constant (expressed in coulombs per meter) determined by the size, shape, and material properties of the piezoelectric element and by the position of the electrodes. For certain simple configurations, the constant A is given by

$$A = d_{ij}k \qquad \text{coulombs/meter} \tag{16.4}$$

where d_{ij} is the piezoelectric constant of the material relating applied stress to generated electric charge. Subscripts i and j refer to orientation of electrodes and direction of applied stress [see Eq. (16.8)].

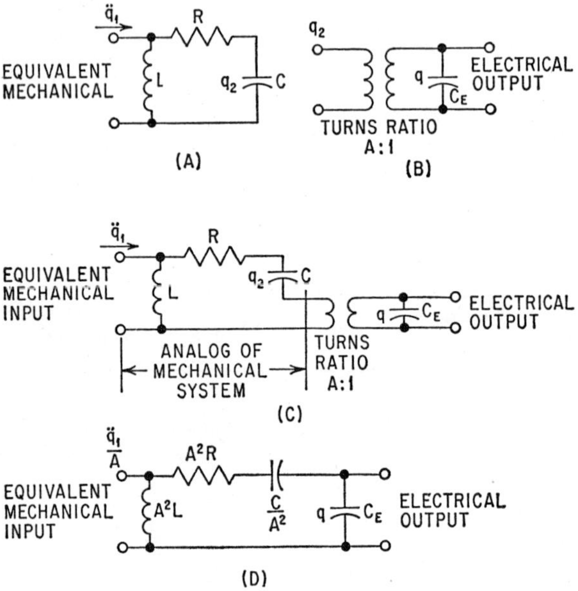

FIG. 16.2. Direct analog and equivalent circuit of a piezoelectric accelerometer employing the seismic system of Fig. 16.1*A*. (*A*) Direct analog of simple seismic system of Fig. 16.1*A*. (*B*) Electrical equivalent of a piezoelectric element. (*C*) Electrical equivalent circuit of a piezoelectric accelerometer. (*D*) Electrical equivalent circuit of a piezoelectric accelerometer which is an alternate form of the circuit shown in (*C*).

The equivalent circuit representing the generation of electric charge across the piezoelectric element is shown in Fig. 16.2*B*. The transformer is a perfect transformer having no loss between zero frequency and the highest frequency for which the piezoelectric element is operative; it has a turns ratio of $A:1$.

The electric equivalent circuit of a piezoelectric accelerometer (assuming a simple seismic system of the type indicated in Fig. 16.1*B*) is shown in Fig. 16.2*C* which represents the combination of Figs. 16.2*A* and 16.2*B*. Another form of this electric equivalent circuit is shown in Fig. 16.2*D*. Often the viscous damping coefficient is negligible in piezoelectric accelerometers so that the circuit element, R in Fig. 16.2*C* and *D*, is small.

At frequencies well below the resonant frequency of the pickup, the electric charge generated is directly proportional to acceleration, as given, for example, by Eq. (16.10). Therefore, at these low frequencies, the equivalent circuit of the pickup, as seen at the electrical output terminals, is that given in Fig. 16.3. A *charge equivalent circuit* is shown in Fig. 16.3*A*, where the electric charge generated by acceleration is represented by q and the electrical capacitance of the piezoelectric element is represented by C_E. The open-circuit output voltage generated by the pickup is given by

$$e = \frac{q}{C_E} \qquad \text{volts} \tag{16.5}$$

The *charge equivalent circuit* of Fig. 16.3*A* is equivalent to the *voltage equivalent circuit* of Fig. 16.3*B*. These equivalent circuits are particularly useful in determining the effect of coupling the output of a pickup to other electronic equipment.

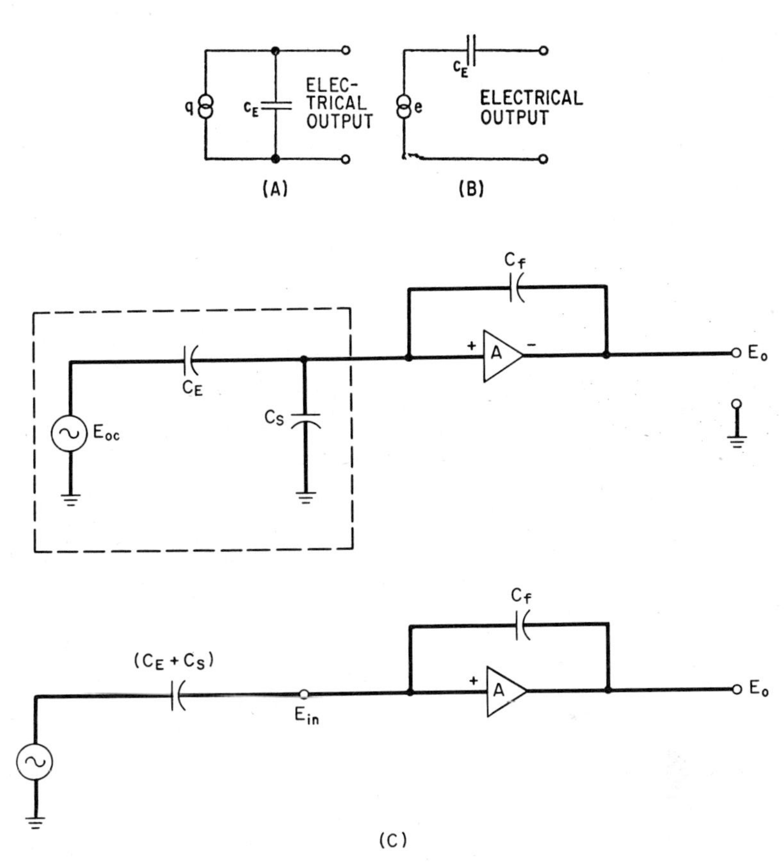

Fig. 16.3. Simplified equivalent circuit of a piezoelectric seismic accelerometer when the frequency of applied acceleration is far below the resonant frequency. (*A*) Charge generating equivalent circuit. (*B*) Voltage generating equivalent circuit in which $e = q/C_E$. (*C*) A charge amplifier circuit which permits lengths of cable to be used without affecting the charge sensitivity of the transducer system.

A charge accelerometer, symbolized by a voltage source, E_{oc}, and two capacitors, C_E and C_S (C_E represents the accelerometer capacitance, and C_S the cable capacitance), is connected to an operational amplifier that has capacitive feedback (Fig. 16.3*C*). E_o represents the output voltage. The input voltage E_{in} is related to the generator voltage E_{oc} by the equation

$$E_{\text{in}} = E_{oc} \frac{C_E}{C_E + C_S} \tag{16.5A}$$

and the gain of the amplifier is given by

$$\frac{E_o}{E_{\text{in}}} = \frac{C_E + C_S}{C_f} \tag{16.5B}$$

The output voltage E_o is obtained by substituting Eq. (16.5*A*) in Eq. (16.5*B*):

$$E_o = E_{oc}\frac{C_E}{C_f}$$

The generator voltage, E_{oc}, may be equated to the charge on capacitor C_E:

$$E_{oc} = \frac{Q}{C_E}$$

so that $E_o = Q/C_f$. Cable capacitance has been eliminated as a factor. The output of the operational amplifier may thus be expressed in volts per picocoulomb. For example, if the feedback capacitor has a value of 1,000 picofarad, the gain of the amplifier is 1 millivolt/picocoulomb.

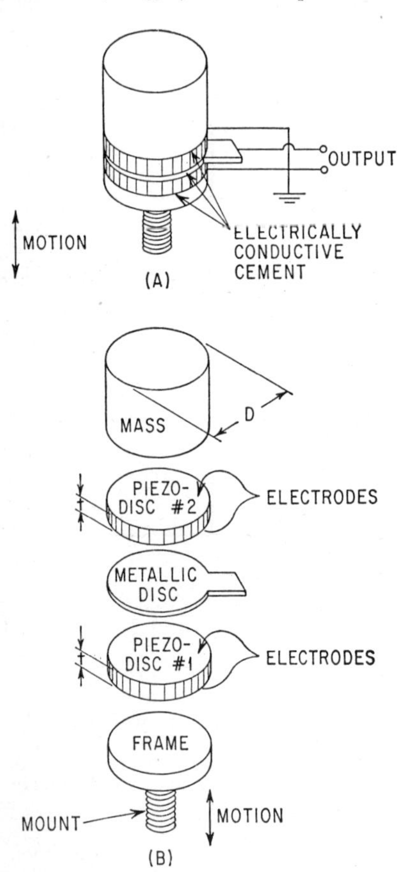

Fig. 16.4. Compression-type piezoelectric accelerometer without housing: (*A*) assembled and (*B*) "exploded" view.

CALCULATION OF PERFORMANCE CHARACTERISTICS OF AN ACCELEROMETER. The approximate performance characteristics of a practical piezoelectric accelerometer can be calculated, if it is assumed that the behavior is that of an ideal simple seismic system. For example, consider the pickup shown in Fig. 16.4. This piezoelectric accelerometer consists of a frame to which is fastened piezoelectric disc 1. This disc is provided with metallic electrodes on its upper and lower faces. A metallic disc (for electrical contact) is fastened to the upper face of piezoelectric disc 1; also bonded to the metallic disc is piezoelectric disc 2 (electroded on upper and lower faces); a mass is bonded to piezoelectric disc 2.* Acceleration in the direction indicated causes compressive or tensile forces on the piezoelectric elements. For this reason, such a pickup is termed a *compression-type accelerometer*. The piezoelectric elements are electrically connected so as to add the electric charges which are created by each element when compressed. Assuming negligible damping, by methods outlined below, one can calculate (1) the resonant frequency f_n, (2) *charge* and *voltage sensitivities*† at frequencies far below the resonant frequency (e.g., below $0.3f_n$), and (3) the electromechanical conversion efficiency. From such calculations one may determine reasonable design compromises among sensitivity, frequency response, and weight.

* Often only one piezoelectric disc is used. The use of two piezoelectric discs, however, makes it possible to connect the mass to the housing electrically, without the use of electrical insulators necessary in a single disc design.

† *Charge sensitivity* is defined as the electric charge generated per unit of acceleration (given by $S_q = q/\ddot{x}$). *Voltage sensitivity* is defined as the open-circuit voltage generated per unit of acceleration (given by $S_v = e/\ddot{x}$).

Approximate Formulas for Transducer Characteristics. The resonant frequency of the pickup shown in Fig. 16.4 is given by

$$f_n = \frac{1}{2\pi}\left(\frac{k}{m}\right)^{1/2} \quad \text{cps} \tag{16.6}$$

where k, the combined stiffness of the two piezoelectric discs (one-half that of each disc), is given by

$$k = \frac{E\pi D^2}{8t} \quad \text{newtons/meter} \tag{16.7}$$

where E is the modulus of elasticity of the piezoelectric material in newtons per square meter, D is the diameter of the discs in meters, and t is the thickness of each disc in meters.

The total charge q generated by both discs due to vibration at frequencies far below the resonant frequency f_n is

$$q = 2d_{33}F \quad \text{coulombs} \tag{16.8}$$

where d_{33} is the piezoelectric constant* of the piezoelectric material and F is the force produced on the piezoelectric discs as a result of acceleration.

The force F is given by

$$F = m\ddot{x} \quad \text{newtons} \tag{16.9}$$

where m is the mass in kilograms and $\ddot{x}$ is the applied acceleration in meters per second per second. The electric charge generated thus is

$$q = 2d_{33}m\ddot{x} \quad \text{coulombs} \tag{16.10}$$

The electrical capacitance C_E of the pickup (due to the two discs in parallel) is given by

$$C_E = \frac{\epsilon\pi D^2}{2t} \quad \text{farads} \tag{16.11}$$

where ϵ is the dielectric constant of the piezoelectric material in farads per meter.

The open-circuit voltage e generated at frequencies well below f_n is

$$e = \frac{q}{C_E} = \frac{2d_{33}m\ddot{x}}{C_E} \quad \text{volts} \tag{16.12}$$

Substitution from Eq. (16.11) for C_E yields

$$e = \frac{4d_{33}mt\ddot{x}}{\epsilon\pi D^2} = \frac{4g_{33}mt\ddot{x}}{\pi D^2} \quad \text{volts} \tag{16.13}$$

where $g_{33} = \frac{d_{33}}{\epsilon}$ is a piezoelectric constant relating open-circuit voltage to applied stress.

The electrical energy generated by the pickup is given by

$$E = \frac{q^2}{2C_E} = \frac{(2d_{33}m\ddot{x})^2 t}{\epsilon\pi D^2} \quad \text{joules} \tag{16.14}$$

* The piezoelectric constant d_{ij} relates electric charge to the applied stress. Subscripts 33 refer to the simple case of a piezoelectric ceramic disc undergoing a compressive (or tensile) force in the direction of the created electrical field (i.e., perpendicular to the metallic electrodes). Other piezoelectric configurations also are used. For example, if the stress is applied parallel to (rather than perpendicular to) the electrodes, the d_{31} constant is used. In some cases, a mechanical shear stress may be applied, in which case the d_{15} constant is applicable.

Example of Calculations for an Accelerometer. Consider a seismic accelerometer, similar in design to that shown in Fig. 16.4, which utilizes a polycrystalline ceramic composed of 96 per cent barium titanate, and 4 per cent lead titanate. However, instead of mechanical bonding, a rugged housing is utilized to clamp the entire seismic assembly under a heavy preloading force (up to 400 lb/in.^{2}). Assume the following constants:

$$
\begin{aligned}
D &= 0.175 \text{ in.} = 4.45 \times 10^{-3} \text{ meter} \\
t &= 0.020 \text{ in.} = 5.09 \times 10^{-4} \text{ meter} \\
E &= 1.13 \times 10^{12} \text{ dynes/cm}^2 = 1.13 \times 10^{11} \text{ newtons/meter}^2 \\
m &= 1.1 \text{ gm} = 1.1 \times 10^{-3} \text{ kg} \\
\epsilon &= 900\, \epsilon_0 = 7.95 \times 10^{-9} \text{ farad/meter} \\
d_{33} &= 280 \times 10^{-8} \text{ statcoulomb/dyne} = 9.32 \times 10^{-11} \text{ coulomb/newton} \\
g &= 9.8 \text{ meter/sec}^2
\end{aligned}
$$

From Eqs. (16.6), (16.10), (16.11), and (16.13) the following characteristics of the pickup may be calculated: natural frequency, 198,000 cps; capacitance, 490 mmfd; charge sensitivity, 2.0×10^{-12} coulomb/g; and voltage sensitivity, 3.1×10^{-3} volt/g. In practical designs, the calculated values may differ considerably from the measured values because the calculations neglect effects such as clamping by the housing and because of resonances introduced by the distributed parameter system.

TYPICAL ACCELEROMETER CONSTRUCTIONS

Piezoelectric accelerometers utilize a variety of seismic element configurations. Most pickups are constructed of polycrystalline ceramic piezoelectric materials because of their ease of manufacture, high piezoelectric sensitivity, and excellent time and temperature stability. These seismic devices may be classified in three modes of operation: compression-, bender- or shear-type accelerometers.

COMPRESSION-TYPE ACCELEROMETER. The compression-type seismic accelerometer, in its simplest form, consists of a piezoelectric disc and a mass placed on a frame as shown in Fig. 16.4. Motion in the direction indicated causes compressive (or tensile) forces to act on the piezoelectric element, producing an electrical output proportional to acceleration. In this example, the mass is cemented with a conductive material to the piezoelectric element which, in turn, is cemented to the frame. The components must be cemented firmly so as to avoid being separated from each other by the applied acceleration.

In the unit shown in Fig. 16.5 the mass is held in place by means of a stud extending from the frame through the ceramic. This accelerometer (NRL Type C-4), employing

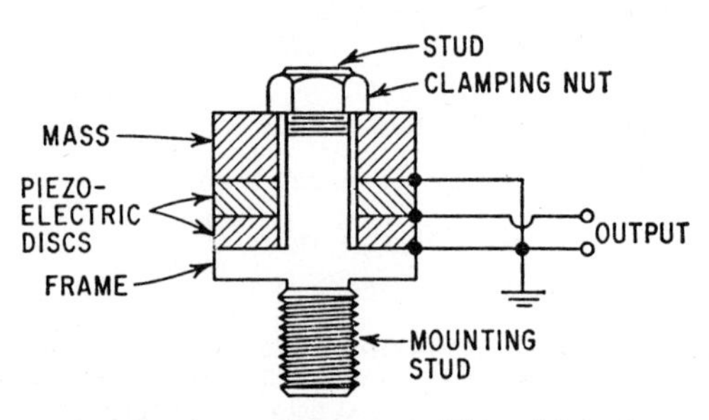

FIG. 16.5. NRL-type piezoelectric accelerometer. This pickup has a sensitivity of 100 millivolts/g, a resonant frequency of 14,000 cps, and a useful acceleration up to 5,000g. (*Courtesy of the U.S. Naval Research Laboratory.*)

a ceramic element, has the following characteristics:[3]

Diameter = 0.80 in. = 2.03 cm
Thickness = 0.30 in. = 0.76 cm
$d_{33} = 2.5 \times 10^{-6}$ coulombs/newton
$\epsilon/\epsilon_0 = 1{,}000$ (where ϵ_0 is dielectric constant of free space)

Sensitivity = 100 millivolts/g (open circuit)
Acceleration range = 1 to 5,000g
Resonant frequency = 14,000 cps
Useful frequency range = 10 to 4,000 cps
Temperature range (depends on piezoelectric material)

A typical commercial accelerometer is shown in Fig. 16.6. This unit consists of a housing, one piezoelectric ceramic element, a seismic mass, and a spring and cap for clamping the mass against the ceramic element. The compression cap applies the necessary preloading through the case isolation spring. The resonant frequency of the pickup is always lower than that of the mass-ceramic combination alone. This type of accelerometer must be attached to the structure with care in order to minimize distortion of

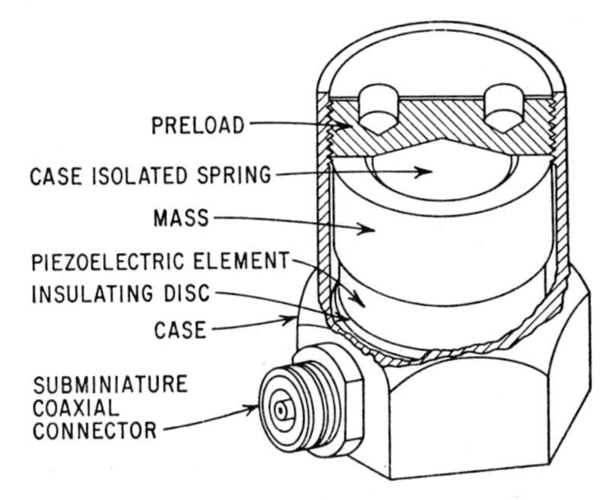

FIG. 16.6. Commercial compression-type accelerometer. The spring provides the necessary preload for the mass and piezoelectric element, so as to maintain intimate contact throughout the rated range of acceleration. (*Courtesy of Endevco Corp.*)

the housing which may affect the sensitivity of the accelerometer. For this reason a recommended value of mounting torque in attaching the unit and mounting surface often is specified.[4]

In the accelerometer shown in Fig. 16.6, an insulating material is placed between the housing base and the lower electrode of the piezoelectric element in order to provide an insulated output lead. The upper electrode of the ceramic is connected electrically to the housing through the mass-spring combination. The preloading of the spring is adjusted so as to prevent "chattering" of the mass for values of acceleration up to the rated acceleration range of the instrument. If a clamp or cementing technique is not used for maintaining the mass in contact with the ceramic element in the above type of construction, the output will be proportional to acceleration only if the forces due to the applied acceleration do not exceed the bias force. An unclamped seismic system is sometimes used to calibrate vibrating systems, by observing the occurrence of chatter,[4] which takes place whenever the downward acceleration exceeds 1g (see Chap. 18).

The temperature characteristics of compression-type accelerometers have been improved greatly in recent years; it is now possible to measure acceleration over a temperature range of −425°F (−254°C) to +1400°F (+760°C). This wider range has been primarily a result of the use of two new ceramic materials: sodium bismuth titanate and lithium niobate.

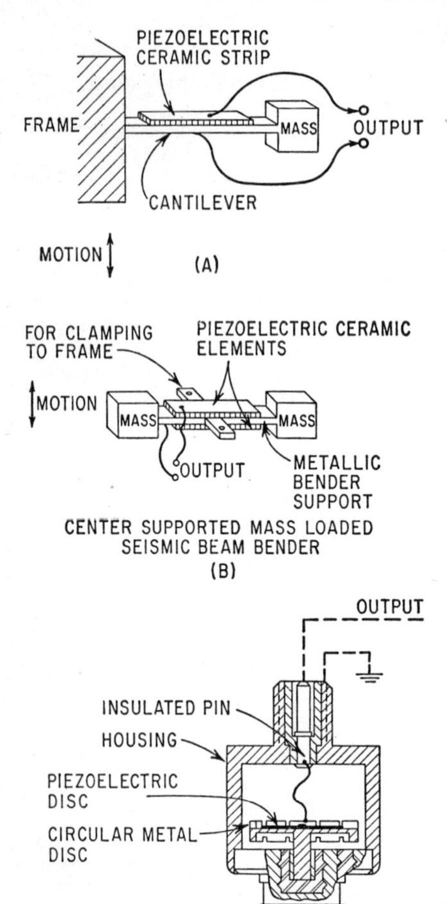

FIG. 16.7. (*A*) A typical piezoelectric seismic element employing the bending principle, showing the operation of a seismic system utilizing a mass-loaded cantilever. Below the natural frequency f_n of the cantilever the amplitude of bending strain is proportional to the applied acceleration. The strain in the piezoelectric strip generates an electrical output voltage proportional to the applied acceleration. (*B*) Two ceramic strips are connected in parallel so that they create a charge of the same polarity when strained. (*C*) A commercial bender-type seismic accelerometer. Acceleration in the axial direction produces a strain due to bending of the metal disc, resulting in an output proportional to acceleration. (*Gulton Industries, Inc.*)

BENDER-TYPE ACCELEROMETER. A typical piezoelectric seismic element operating on a bending principle is shown in Fig. 16.7*A*. The cantilever construction shown consists of a frame to which a flat mass-loaded cantilever strip is attached. Acceleration causes this cantilever to bend—the magnitude of the deflection depending on its stiffness and the mass loading; at frequencies well below the resonant frequency f_n the bending strain is proportional to the applied acceleration of the frame. A strip of piezoelectric ceramic having electrodes on both top and bottom is bonded to the cantilever. Upward curvature of the cantilever causes compression in length of the piezoelectric ceramic, and downward curvature causes tension. The resulting strain in the piezoelectric strip generates an electrical output voltage which is proportional to the applied acceleration.

Another typical bender-type element used in accelerometers is shown in Fig. 16.7*B*; this type is supported at its center and has a mass at each end with a piezoelectric strip bonded to each side. A means is provided for mounting the element to the frame. In this configuration the metallic support for the element is connected to the output ground; the two outer piezoelectric surfaces are silvered and are connected together to the "high" side of the output. When acceleration is applied to the frame, tensile stresses on one ceramic element and compressive stresses on the other are produced simultaneously. The two piezoelectric ceramic strips are oriented so as to generate electric charge of the same polarity when one is compressed and the other is extended so that the generated charges are additive. Typical characteristics of such a device are: length, 1 in.; mass loading, 3 gm; sensitivity, 150 mv/g; capacitance, 1,000 mmfd; and resonant frequency, 3,600 cps. It also is possible to connect the output of the ceramic strips in series rather than in parallel; this doubles the sensitivity and reduces the capacitance by a factor of 4.

In Fig. 16.7*C* a circular bender-type accelerometer is shown. This unit has an element which consists of a circular metal disc which is supported at its center by a mounting stud; a circular ceramic disc is cemented to the flat circular portion of the metal. The sensitivity and resonant frequency of this seismic element may be changed by mass loading in the form of a peripheral ring. Acceleration along the axial direction, as shown, results in a bending of the disc which strains the ceramic in a radial direction. The ceramic produces an output voltage proportional to stress. The base of the unit is comprised of two molded ceramic insulating rings, an outer metal and inner metal ring, and the mounting stud. The piezoelectric element is clamped in the inner metal ring and electrical connection is made from the inner metal ring to the protective housing, which acts as the negative electrical lead. The positive electrical lead is attached to the top of the ceramic and is connected to the insulated pin of the coaxial connector. The purpose of electrically insulating the seismic element from the mounting stud is to isolate *ground-loop* voltages from the electrical output terminals of the accelerometer (see *Ground Loops*, Chap. 11).

SHEAR-TYPE ACCELEROMETERS

One shear-type accelerometer employs a ceramic cylinder fitted around a middle mounting post; a loading ring is cemented to the outer dimension of the piezoelectric element (see Fig. 16.8). The ceramic cylinder is polarized along its length; the output voltage of the accelerometer is taken from the inner and outer walls.

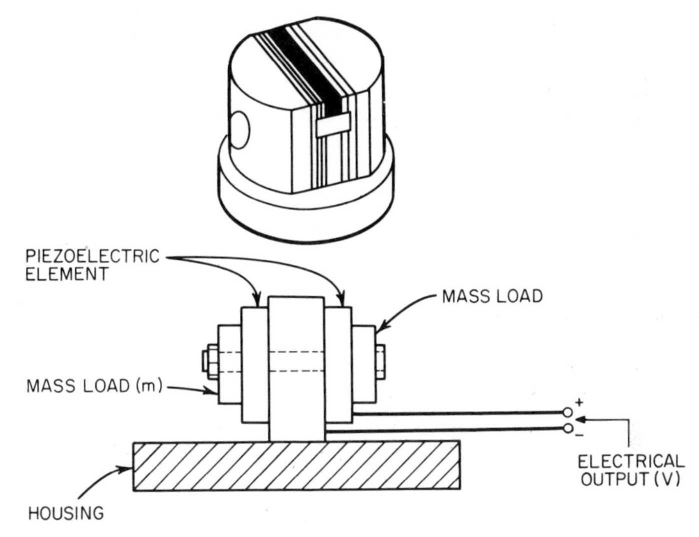

FIG. 16.8. A shear-type piezoelectric accelerometer showing the seismic elements. When subject to acceleration, a strain is generated by the mass m, resulting in a shear stress on the piezoelectric ceramic. (*Courtesy of Gulton Industries, Inc.*)

Another shear-type accelerometer provides better characteristics than either the compression or bender seismic systems, having a lower cross-axis response, excellent temperature characteristics, and negligible output from strain sensitivity or base bending. The temperature range of the bolted shear design is from −425°F (−254°C) to +1400°F (+760°C). The specifications are: sensitivity, 100 picocoulombs/g; acceleration range, 1 to 500g; resonant frequency, 25,000 cps; useful frequency range, 3 to 5,000 cps; temperature range, −425°F to +1400°F (−254°C to +760°C); transverse response, 2 per cent.

UNIAXIAL AND TRIAXIAL ACCELEROMETERS

The majority of piezoelectric accelerometers manufactured are of the uniaxial type, i.e., for measuring acceleration along a single axis. This axis is usually perpendicular to the plane of the mounting surface.

A triaxial accelerometer contains three uniaxial seismic elements which are orthogonal to each other and from which three independent electrical outputs are obtained. By precision manufacture in a single housing, a triaxial accelerometer is more compact and ensures more precise orthogonality in comparison with three individual uniaxial accelerometers. In the simultaneous measurement of acceleration components along three orthogonal directions, unless the transverse sensitivity* of each piezoelectric seismic system is relatively low compared with the maximum sensitivity,[1] erroneous indications will be obtained regardless of whether three uniaxial pickups or a single triaxial pickup is used.

Consider a triaxial accelerometer, each element of which is identical. Suppose the transverse response of these elements is 10 per cent of the value in the direction of maximum response. Then if there is an acceleration of $10g$ in a transverse direction, the orthogonal elements will produce an output equivalent to $1g$; an error of 10 per cent may result.

An *omnidirectional accelerometer* is defined as one which produces a voltage output that is dependent on the magnitude of the applied acceleration, but which is independent of the direction of the applied acceleration. Usually, such accelerometers are fabricated only for special applications.

PHYSICAL CHARACTERISTICS OF PIEZOELECTRIC PICKUPS. Shape, Size, and Weight. Commercially available piezoelectric accelerometers usually are cylindrical in shape. They are available with both attached and detachable mounting studs at the bottom of the cylinder. A coaxial cable connector is provided at either the top or side of the pickup.

Most commercially available piezoelectric pickups† are relatively light in weight, ranging from approximately ½ gm to 60 gm. Usually, the larger the accelerometer, the higher is its sensitivity and the lower is its resonant frequency. The smallest-size units have a diameter of about ¼ in. and a height of ¼ in., exclusive of the mounting stud and connector; the larger units have a diameter of about 2 in. and a height of about 2 in.

Resonant Frequency. The highest fundamental resonant frequency of an accelerometer, due to the combination of the mass and piezoelectric element, may be above 100,000 cps. The higher the resonant frequency, the lower will be the capacitance or sensitivity, and the more difficult it will be to provide mechanical damping. Therefore, some compromise must be made in design of a piezoelectric accelerometer. In compression-type units, an additional resonant frequency (which is lower in frequency and of low Q) may be introduced when the accelerometer is mounted because of the seismic mass and the compliance of the housing. An additional resonant frequency is not introduced in a bender-type pickup if the suspension is of the type shown in Fig. 16.7*C*.

Damping. The *amplification ratio* of an accelerometer is defined as the voltage sensitivity at its resonant frequency to the voltage sensitivity in the frequency band in which sensitivity is independent of frequency. This ratio depends on the amount of damping in the seismic system; it decreases with increasing damping. Most piezoelectric accelerometers are essentially undamped, having magnification ratios between 5 and 50.

The frequency response of a typical piezoelectric accelerometer using a bender-type center-supported cantilever is shown in Fig. 16.9. The dashed curve shows the frequency-response characteristics for the accelerometer without damping. The solid curve shows the characteristics when damping is added. The piezoelectric element has an undamped resonant frequency of 10,000 cps; its magnification ratio of 15 is reduced to 5 by the addition of 100,000-centistoke damping oil, and its resonant frequency is reduced from about 10,000 cps to 9,000 cps.

* Also called *lateral sensitivity* or *cross-axis sensitivity*.

† Manufacturers of commercially available piezoelectric accelerometers include: Clevite Brush Development Co., Cleveland, Ohio; Columbia Research Laboratories, Woodlyn, Pa.; Endevco Corp., Pasadena, Calif.; Gulton Industries, Inc., Metuchen, N.J.; Kistler Instrument Co., Clarence, N.Y.; Bruel and Kjaer, Naerum, Denmark; Philips, N. V., Eindhoven, Netherlands.

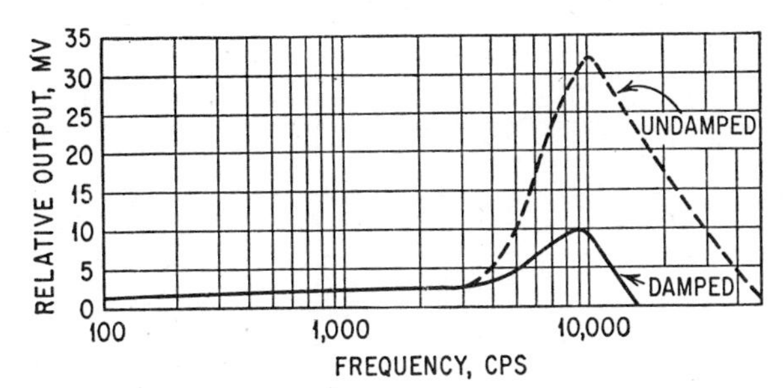

FIG. 16.9. The frequency response of a typical accelerometer is shown with damping (solid curve) and without damping (dashed curve). Damping with a fluid having a kinematic viscosity of 100,000 centistokes greatly reduces the amplification ratio and shifts the undamped resonant frequency to a slightly lower value.

Damping affects the phase-shift characteristics which may be important in complex vibration or shock measurements.

ELECTRICAL CHARACTERISTICS OF PIEZOELECTRIC ACCELEROMETERS

DEPENDENCE OF SENSITIVITY ON SHUNT CAPACITANCE. The *sensitivity* of a pickup is defined as the electrical output per unit of applied acceleration. The sensitivity of a piezoelectric accelerometer can be expressed either as a *charge sensitivity* or *voltage sensitivity*. Charge sensitivity usually is expressed in units of coulombs generated per g of applied acceleration; voltage sensitivity usually is expressed in volts per g (where g is the acceleration of gravity). Voltage sensitivity often is expressed as *open-circuit voltage* sensitivity, i.e., in terms of the voltage produced across the electrical terminals per unit acceleration, when the electrical load impedance is infinitely high. Open-circuit voltage sensitivity may be given either with or without the connecting cable.

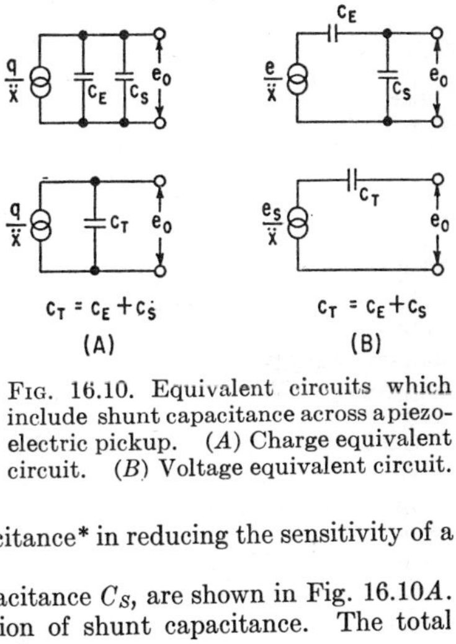

$C_T = C_E + C_S$ (A) $\qquad$ $C_T = C_E + C_S$ (B)

FIG. 16.10. Equivalent circuits which include shunt capacitance across a piezoelectric pickup. (*A*) Charge equivalent circuit. (*B*) Voltage equivalent circuit.

An electrical capacitance often is placed across the output terminals of a piezoelectric pickup. This added capacitance (called *shunt capacitance*) may result from the connection of an electrical cable between the pickup and other electrical equipment (all electrical cables exhibit interlead capacitance). The effect of shunt capacitance* in reducing the sensitivity of a pickup is shown in Figs. 16.10 and 16.11.

The charge equivalent circuits, with shunt capacitance C_S, are shown in Fig. 16.10*A*. The charge sensitivity is not changed by addition of shunt capacitance. The total capacitance of the pickup including shunt is given by

$$C_T = C_E + C_S \quad \text{farads} \tag{16.15}$$

where C_E is the capacitance of the pickup without shunt capacitance.

The voltage equivalent circuits are shown in Fig. 16.10*B*. With the shunt capacitance C_S, the total capacitance is given by Eq. (16.15) and the open-circuit voltage sensitivity is given by

$$\frac{e_s}{\ddot{x}} = \left(\frac{q}{\ddot{x}}\right)\left(\frac{1}{C_E + C_S}\right) \quad \frac{\text{volt-sec}^2}{\text{meter}} \tag{16.16}$$

* In order to eliminate the effect of cable shunt capacitance, accelerometers are available with the coupling electronic circuits built into the same housing as the seismic element.

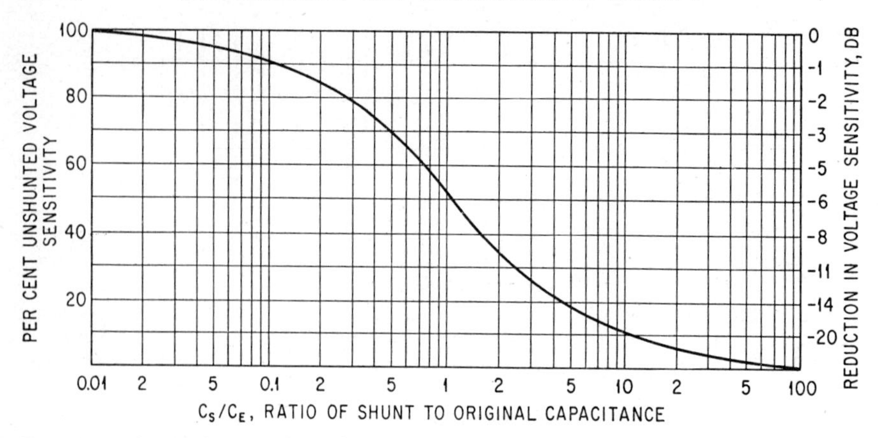

FIG. 16.11. Curve showing the reduction of open-circuit voltage sensitivity due to shunt capacitance across a pickup; the charge output from the transducer remains constant, regardless of the capacitive loading.

where $q/\ddot{x}$ is the charge sensitivity. The voltage sensitivity without shunt capacitance is given by

$$\frac{e}{\ddot{x}} = \left(\frac{q}{\ddot{x}}\right)\left(\frac{1}{C_E}\right) \qquad \frac{\text{volt-sec}^2}{\text{meter}} \tag{16.17}$$

Therefore the effect of the shunt capacitance is to reduce the voltage sensitivity by a factor:

$$\frac{e_s/\ddot{x}}{e/\ddot{x}} = \frac{C_E}{C_E + C_S} \tag{16.18}$$

Figure 16.11 shows a plot of the factor $C_E/(C_E + C_S)$ vs. C_S/C_E.

Because the addition of a shunt capacitance reduces voltage sensitivity of a piezoelectric pickup, a shunt capacitance sometimes is added to adjust the sensitivity of a pickup to a value which is set by the dynamic range of the auxiliary equipment associated with the pickup.

OUTPUT RESPONSE TO VIBRATION. The sensitivity of a piezoelectric accelerometer is dependent upon the frequency of the applied vibration as shown in Fig. 12.5. Over a given frequency region, the sensitivity is independent of the frequency (this is called the *mid-band range*); however, at frequencies below and above this band the sensitivity is dependent upon the frequency of the applied vibration.

Low-frequency Response. At low frequencies (i.e., below the mid-band range), the frequency response is dependent upon the electrical capacitance C_T of the accelerometer (including its cable capacitance and other shunt capacitance) and the input resistance R of the coupling amplifier (or cathode follower) to which the pickup output is connected. The magnitude of the sensitivity at low frequencies is given by

$$\frac{e_f}{\ddot{x}} = \frac{1}{\left[1 + \left(\frac{1}{2\pi f R C_T}\right)^2\right]^{1/2}} \qquad \frac{\text{volt-sec}^2}{\text{meter}} \tag{16.19}$$

where f is the frequency and $e_s/\ddot{x}$ is the open-circuit voltage sensitivity.

Equation (16.19) shows that the frequency dependence of sensitivity is a function of the product RC. This product is expressed in units of seconds and is called the *RC time constant.* Thus the combination R = 10 megohms and C = 10,000 mmfd provides the same low frequency response as the combination R = 100 megohms and C = 1,000 mmfd. Plots of e_f/e_s (i.e., open-circuit voltage sensitivity) vs. frequency are given in Fig. 16.12 for three values of RC. Because an increase in the RC time constant improves low-frequency response, a common technique used to obtain better low-frequency response is to shunt the accelerometer with a large capacitor. This has the disadvantage of reducing voltage sensitivity throughout the mid-band frequency range.

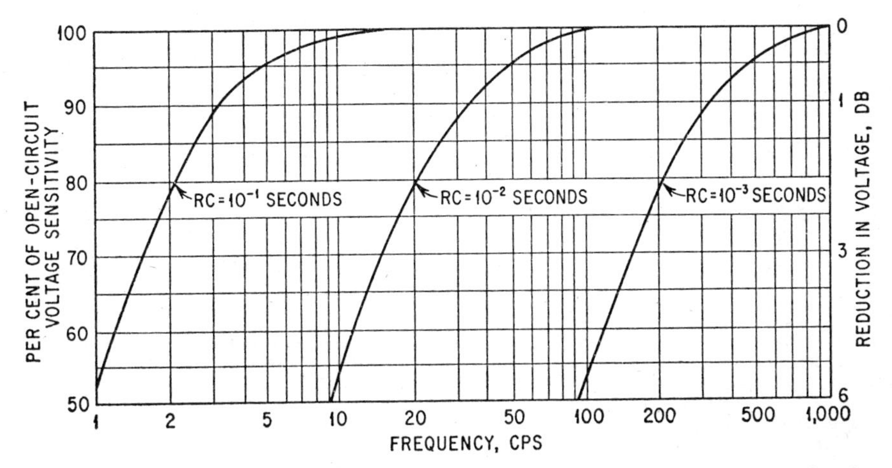

FIG. 16.12. Ratio of voltage output to open-circuit voltage vs. frequency for a piezoelectric pickup having RC time constants of 0.1, 0.01, and 0.001 sec.

The low-frequency response may be limited by the following factors in addition to the load resistance R of the coupling amplifier:

1. The internal shunt resistance of the piezoelectric or insulating materials within the accelerometer. Normally this shunt resistance is extremely high (100,000 megohms or more), and its effect can be neglected; however, at high temperatures it may be less than 10 megohms.

2. The low-frequency response of the coupling amplifier. The a-c coupled stages of an amplifier may cause a greater drop in low-frequency response than that which depends on the RC time constant described above. Some piezoelectric pickups may generate a slowly varying voltage as a result of temperature change* due to the pyroelectric effect. Therefore it is desirable in such cases to use a coupling amplifier that cuts off at very low frequencies in order to attenuate voltages from this source.[8]

High-frequency Response. At frequencies above the mid-band range, where sensitivity is independent of frequency, the response depends upon the mechanical characteristics of the accelerometer. This response depends primarily upon the resonant frequency f_n and viscous damping c. Most commercially available piezoelectric accelerometers have a fraction of critical damping $\zeta = c/c_c$ of 0.05 or less (where c_c represents the value of critical damping); they may be used at frequencies up to as high as 0.2 to $0.3f_n$. At $f = 0.3f_n$ the sensitivity is approximately 12 per cent higher than the mid-band sensitivity. Accelerometers having a greater amount of damping may be used above $0.3f_n$. For example, an accelerometer which has a fraction of critical damping ζ of 0.3 may be used up to approximately $0.4f_n$; and an accelerometer which has a fraction of critical damping ζ of 0.7 may be used to $0.7f_n$.

* The generation of electric charge by temperature change in a piezoelectric material is called the *pyroelectric effect;* it is a property of certain piezoelectric materials. (Natural crystals that are nonpolar do not exhibit this effect.) In a piezoelectric accelerometer the temperature changes within the piezoelectric material often occur at considerably slower rate than 0.01°F/sec, which allows much of the output to be attenuated in an a-c amplifier.

A typical frequency response curve for an accelerometer, Fig. 16.13, is flat (+5 per cent) from 5 cps to 5,000 cps. The accelerometer in this example has a resonant frequency of 35,000 cps. At resonance, the unit has an amplification factor of about 20; thus its output at resonance is 20 times larger than its output at frequencies over the flat portion of the curve.

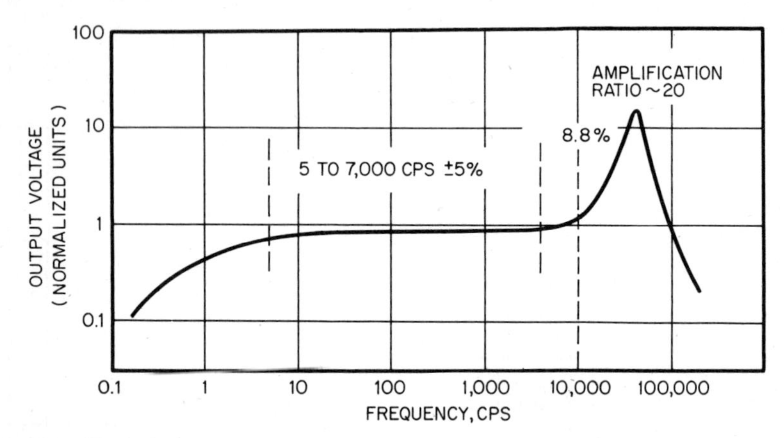

FIG. 16.13. Typical output response of a piezoelectric accelerometer is very poor near direct current, at input frequencies in the bandpass it is flat, and at resonance it increases sharply.

The following formula may be used to estimate the percentage rise in the output in the upper frequency range:

$$\text{Per cent rise} = \frac{1}{1 - (f/f_n)^2} \tag{16.20}$$

where f_n is the resonant frequency of the accelerometer, and f is the frequency at which the rise is measured. For example, an accelerometer having a resonant frequency of 35,000 cps has a rise in output voltage of 8.8 per cent at 10,000 cps.

OUTPUT RESPONSE TO SHOCK. Faithful reproduction of a shock pulse may be attained with a piezoelectric accelerometer if the duration of the shock pulse is sufficiently short and the rise time of the shock pulse is sufficiently long.

Dependence of Response on Pulse Duration. When an amplifier is connected to a piezoelectric accelerometer, an electrical resistance path is placed across the pickup through which the electric charge may leak off. This leaking off of charge results in a continuous decay of the output voltage with time. For example, if an acceleration is applied suddenly to the pickup and is maintained at this fixed level, the output voltage e of the pickup is given as a function of time by

$$e = e_0 e^{-t/RC_T} \tag{16.21}$$

where e_0 is the voltage generated immediately after application of the acceleration, C_T is the combined electrical capacitance of the accelerometer and connecting cable, R is the input resistance of the coupling amplifier, and t is the time after application of the acceleration. The voltage decay is dependent upon the ratio of time t to the RC time constant, as indicated by Eq. (16.21). Over extended durations of time, the decay can be a large percentage of the voltage e_0. For example, when t exceeds $0.1RC$, the voltage e is less than $0.9e_0$.

While the allowable decay and overshoot depend upon the specific requirement of measurement accuracy, it is recommended that for most applications the RC time constant be at least ten times the pulse length in seconds.

As examples of the dependence of the response on pulse duration, consider the responses of an accelerometer to square-wave and half-sine-wave pulses. A square-wave acceleration pulse is one which exhibits the following characteristic: Prior to time $t = 0$, the acceleration is zero; at time $t = 0$, the acceleration suddenly is increased to a fixed level and is maintained at this level for a time $t = \tau$, at which time the acceleration instantaneously returns to zero and is maintained at zero thereafter. The response of a piezoelectric pickup to a square-wave pulse is shown in Fig. 16.14 for four values of RC. During the acceleration pulse, the output exhibits an exponential decay (except for the curve which corresponds to $RC = \infty$); when the acceleration returns to zero, the output decreases to a negative value (by an amount equal to the decay at $t = \tau$). This is followed by a second exponential decay toward zero. The negative pulse is called *negative overshoot;* the smaller the ratio t/RC, the less will be the negative overshoot. The curves

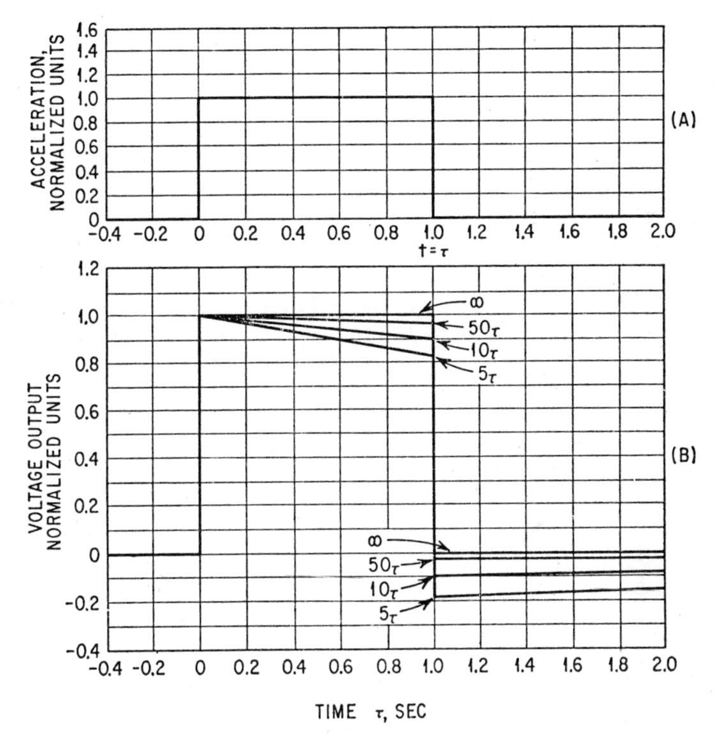

FIG. 16.14. Response of a piezoelectric accelerometer to a square acceleration pulse for RC time constants of 5τ, 10τ, 50τ, and ∞. (*A*) Input acceleration vs. time and (*B*) output voltage vs. time.

of Fig. 16.14 can be used to determine the required input resistance to maintain the voltage decay and "negative overshoot" within specified limits. For example, assume a square-wave pulse having a duration of 10 milliseconds, and assume the decay and negative overshoot are to be maintained to less than 10 percent for an accelerometer which has a capacitance of 1,000 mmfd. The curves of Fig. 16.14 indicate that an RC product of $10\ \tau$ or higher is required. Thus R must be greater than $10\ \tau/C$, or greater than 100 megohms. If the decay and overshoot are to be within 2 per cent of the steady-state voltages, the input impedance must be 500 megohms or higher. Shunting the accelerometer with a capacitance of 10,000 mmfd reduces the input resistance requirement of the amplifier by a factor of approximately 10.

The response to a half-sine pulse is indicated in Fig. 16.15 for several values of the RC time constant.[9] These curves indicate that the response is a faithful reproduction of a half-sine pulse if $RC = \infty$, but for finite values of RC, the output decays with time and may become negative before the acceleration pulse returns to zero.

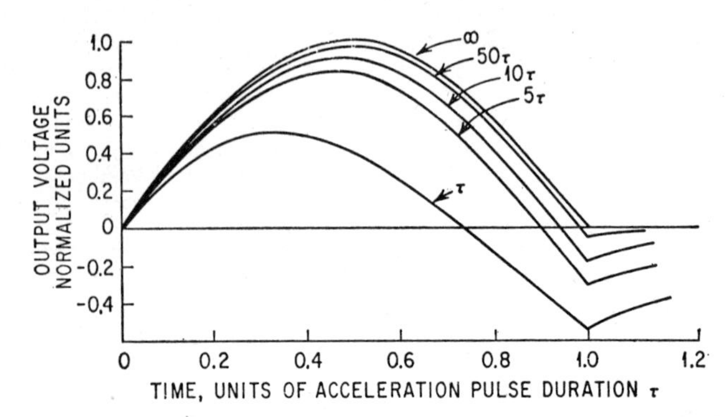

FIG. 16.15. Response of a piezoelectric accelerometer to a half-sine acceleration pulse for RC time constants equal to τ, 5τ, 10τ, 50τ, and ∞, where τ is equal to the duration of the half-sine pulse.[10]

Dependence of Response on Rise and Decay Times of a Shock Pulse. The faithful response of an accelerometer to shock is dependent not only on the pulse duration but also on the ability of the seismic system to follow rapid changes of acceleration. The response of the seismic system is dependent on the rise and decay times of the shock pulse and on the resonant frequency and viscous damping of the seismic element. In general, faithful seismic response to pulses is attained when (1) the natural period τ_n of the accelerometer is short compared to the rise and decay time of the pulse (the natural period τ_n of a pickup is the reciprocal of its natural frequency f_n, and is expressed in seconds per cycle); and (2) the damping coefficient c of the pickup is approximately $0.7c_c$, where c_c is the coefficient of critical damping.

Curves of the response of an ideal seismic system to half-sine, triangular, and square-wave pulses of acceleration are shown in Figs. 12.7, 12.8, and 12.9. For each of the three pulse shapes, the response is given for ratios τ_n/τ of 1.014, 0.338, and 0.203 where τ is the pulse duration.

ACCELERATION-AMPLITUDE CHARACTERISTICS. Piezoelectric accelerometers are generally useful for the measurement of acceleration of magnitudes of from $10^{-4}g$ to more than 10^4g. The lowest value of acceleration which can be measured is approximately that which will produce an output voltage equivalent to the electrical input noise of the coupling amplifier connected to the accelerometer when the pickup is at rest. Over its useful operating range, the output of a piezoelectric pickup is directly and continuously proportional to the input acceleration. A single accelerometer often can be used to provide measurements over a dynamic amplitude range of 10,000 to 1 (80 db), which is substantially greater than the dynamic range of most of the associated transmission, recording, and analysis equipment.

Commercial accelerometers generally exhibit excellent linearity of output voltage vs. input acceleration and virtually no mechanical hysteresis* under normal usage. A linearity of better than 2 per cent of actual reading up to $4{,}000g$ and better than 5 per cent of actual reading up to $20{,}000g$ is possible.[11]

* Mechanical hysteresis in this context is the change in sensitivity of a pickup at one level of acceleration after being subjected to another level of acceleration within the acceleration-measuring range of the instrument.

At very high values of acceleration (depending upon the design characteristics of the particular pickup), nonlinearity, hysteresis, or damage may occur. For example, large dynamic forces may produce voltage outputs sufficient to reduce permanently the sensitivity of the piezoelectric material. Further, if the dynamic forces exceed the biasing or clamping forces, the seismic element may "chatter" or fracture, although such a fracture might not be observed in subsequent low-level acceleration calibrations. High dynamic accelerations also may cause a slight physical shift in position of the piezoelectric element in the accelerometer—sometimes sufficient to cause a change in sensitivity. The upper limit of acceleration measurements depends upon the specific design and construction details of the pickup, and may vary considerably from one accelerometer to another, even though the design is the same. In certain designs the upper limit is dependent upon the machining accuracy of the parts. It is not always possible to calculate the upper acceleration limit of a pickup. Therefore one cannot assume linearity at acceleration levels for which calibration data cannot be obtained.

DIRECTIONAL SENSITIVITY—TRANSVERSE SENSITIVITY. The output of an ideal piezoelectric pickup depends upon the direction of applied acceleration and is given by

$$e_\theta = e_{\max} \cos \theta \tag{16.22}$$

where $e_{\max}$ is the output in the direction of maximum sensitivity and θ is the angle between the direction of maximum sensitivity and the direction of acceleration. The directional dependence of sensitivity of an ideal pickup is shown in Fig. 16.16. If the maximum sensitivity is in the Y direction, the sensitivity in the XY or YZ planes is described by a *figure-eight* locus. The sensitivity at $\theta = 90°$ is called the *transverse sensitivity.** In an ideal pickup the transverse sensitivity is zero.

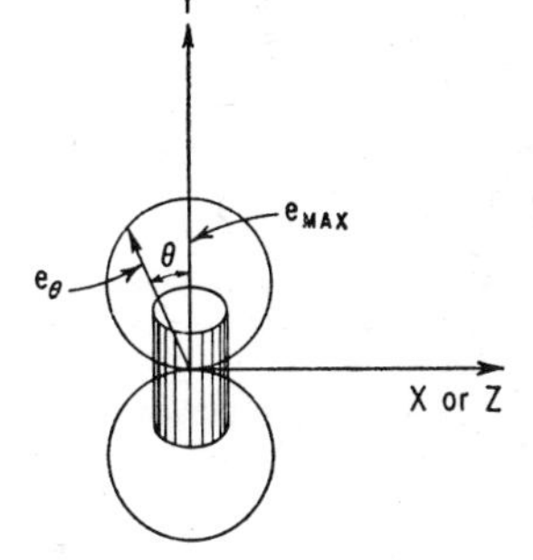

FIG. 16.16. Directional sensitivity of an ideal seismic piezoelectric pickup.

It is not practical to construct an accelerometer having a transverse sensitivity of zero. Because of practical fabrication tolerances, the piezoelectric element in a pickup may not seat tightly against the frame, the mass may have an axial tilt, or the mounting stud or flat may vary slightly from the ideal direction. These factors create an effective angular tilt of the axis of maximum sensitivity.

The effect of tilting the axis of maximum sensitivity by an angle θ_1 in the XY plane is shown in Fig. 16.17. As indicated in Fig. 16.17*A*, in the XY plane this results in a voltage output of

$$e_\theta = e_{\max} \cos (\theta - \theta_1) \tag{16.23}$$

Along the Y axis, the following voltage is produced:

$$e_y = e_{\max} \cos \theta_1 \tag{16.24}$$

Along the X axis, the following voltage results:

$$e_x = e_{\max} \sin \theta_1 \tag{16.25}$$

In the YZ plane a typical figure-eight locus is obtained (see Fig. 16.17*B*) which has a maximum sensitivity (represented by the diameter of each locus circle) given by Eq. (16.24). The output in the transverse plane (XZ) is shown in Fig. 16.17*C*. In the transverse plane, the sensitivity at an angle ϕ is given by

$$e_\phi = e_{\max} \sin \theta_1 \cos \phi \tag{16.26}$$

* The terms cross-axis sensitivity and lateral sensitivity are used as synonyms for transverse sensitivity.

In an ideal pickup, $e_\phi = 0$. Output also can be produced by transverse acceleration which creates small shear forces in the piezoelectric material; these forces are independent of the direction of the transverse acceleration in the XZ plane, i.e., independent of ϕ.

As the result of tilt and related effects, the transverse sensitivity usually is given by the equation:

$$e_\phi = e_{\max} (A + B \cos \phi) \tag{16.27}$$

where A is a constant depending upon effects such as the shear effect described above, B is the effect of the angle of tilt, and ϕ is the angle in the transverse plane. Sometimes A is greater than B; in this case the transverse sensitivity may not pass through zero even though the angle ϕ is rotated through 360°.

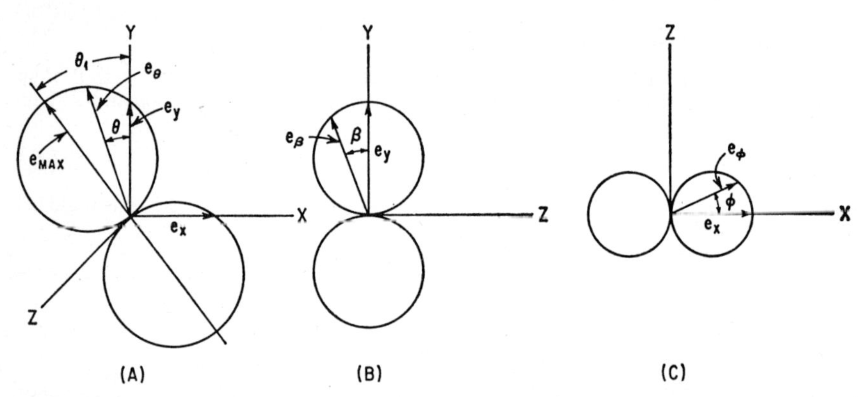

Fig. 16.17. Directional sensitivity of a pickup tilted from the Y axis by an angle θ_1. (A) Output in the XY plane, (B) output in the YZ plane, and (C) output in the (transverse) XZ plane.

In commercial accelerometers, the maximum transverse sensitivity (which occurs when $\phi = 0°$) normally is specified in per cent of the maximum sensitivity and is usually between 0.5 and 15 per cent. The maximum transverse sensitivity varies between units of a single design as well as between different designs. The directional sensitivity of a pickup must be determined with extreme care. If a vibration machine is used for this purpose, the presence of lateral motion in the vibration table may produce significant errors (see Chap. 18).

EFFECTS OF MOUNTING ON PICKUP CHARACTERISTICS. The torque used to mount an accelerometer in place should be controlled to prevent damage to the pickup. Excessive torque can strip or break mounting threads and studs. The application of insufficient torque will result in a mounting that is not secure, thereby causing large errors in high-frequency vibration measurements. Mounting torques of the order of 6 to 18 in.-lb usually are recommended by manufacturers.

Because of design limitations, some compression-type accelerometers have sensitivities that are a function of mounting torque.[5] This characteristic is called *torque sensitivity*. Since *torque sensitivity* generally is attributed to bending stresses induced in the housing of the accelerometer, the sensitivity may also depend upon the location of intimate contact between the pickup and the mounting surface. If an accelerometer exhibits *torque sensitivity*, one should use the same type of mounting and the same mounting torque in each application in order to obtain reproducible and accurate data.

The condition of the mounting surface to which the accelerometer is attached is important (see Fig. 20.2). If the surface is rough, a tight joint cannot be made between the accelerometer and the structure. As a result, the accelerometer will be decoupled from the vibratory motion of the structure. Any mounting plate, jig, or fixture between the accelerometer and the structure should be examined to ensure that a tight joint is obtained. The mounting jig itself is effectively a spring which becomes softer with increasing thickness and stiffer with increasing diameter. Many mounting fixtures have their first resonance below 2,000 cps and therefore can severely affect the accuracy of the measurement being made with the accelerometer. Consideration also should be given to the characteristic of any mounting fixture subject to transverse acceleration. Transverse motion may excite resonances which usually occur at lower frequencies than resonances in the axial direction.

Mounting. Misalignment in the mounting angle of an accelerometer will produce an effect equivalent to a component of transverse sensitivity, as indicated by Eq. (16.23) and Fig. 16.17. For example, a misalignment of 4° from the true perpendicular will produce an effect equivalent to a transverse sensitivity of 7 per cent along one axis.

EFFECTS OF TEMPERATURE ON A PICKUP. Piezoelectric pickups are available which may be used in the temperature range from −425°F (−254°C) to above +1400°F (+760°C) without the aid of external cooling. The voltage sensitivity, charge sensitivity, capacitance, and frequency response of a pickup depend upon the ambient

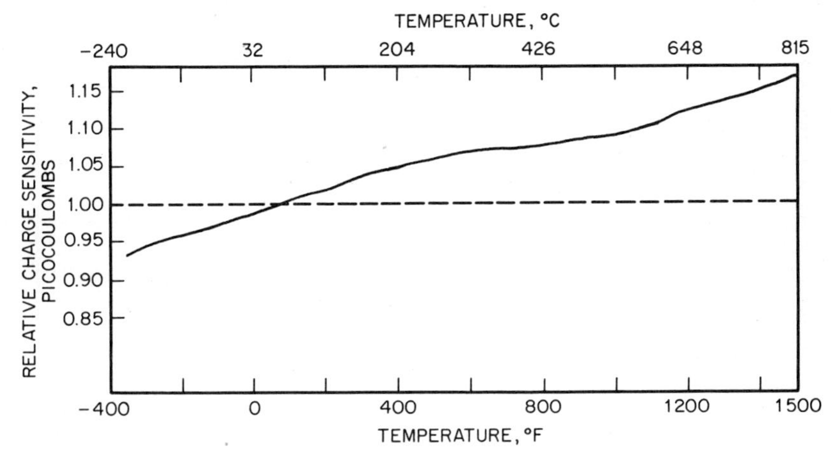

Fig. 16.18. Temperature dependence on charge sensitivity for lithium niobate measured at room temperature, 70°F. *(Gulton Industries, Inc.)*

temperature of the pickup. This temperature dependence is due primarily to variations in the characteristics of the piezoelectric material, of which several types are described in the second half of this chapter, but it also may be due to variations in the insulation resistance of cables and connectors—especially at high temperatures.[12] Consider the following example of the temperature dependence of a pickup based upon the use of a typical piezoelectric material—lithium niobate.

Effects of Temperature on Charge Sensitivity. The charge sensitivity of a piezoelectric accelerometer is directly proportional to the d_{ij} piezoelectric constant of the material used in the piezoelectric element. The d_{ij} constants of most piezoelectric materials vary with temperature. The change in charge sensitivity with temperature for an accelerometer using lithium niobate as the piezoelectric ceramic is illustrated in Fig. 16.18.[15]

Effects of Temperature on Voltage Sensitivity. The open-circuit voltage sensitivity of an accelerometer is the ratio of its charge sensitivity to its total capacitance ($C_S + C_E$). Hence, the temperature variation in voltage sensitivity depends on the temperature dependence of both charge sensitivity and capacitance. With zero shunt capacitance ($C_S = 0$), the open-circuit voltage sensitivity, for the piezoelectric material illustrated in Fig. 16.19, varies by less than ±10 per cent from the sensitivity at 70°F (21°C), over a temperature range from −425°F (−254°C) to +1400°F (+760°C). This is owing to the fact that both the charge sensitivity and capacitance have similar temperature dependence over this temperature range. This characteristic is typical of many piezoelectric materials within a limited temperature range. At temperatures above 1000°F (+538°C), the voltage sensitivity becomes highly dependent, at low frequencies, on the resistivity of the material. Accelerometer manufacturers should be consulted for data regarding individual models.

Effects of Temperature on Low-frequency Response. As indicated in the section on *Low-frequency Response*, see Fig. 16.12, the voltage output at low frequencies depends on the shunt resistance across the pickup, both external and internal. Since the insulation resistance of all insulating materials and piezoelectric materials decreases exponentially with increasing temperature, the low-frequency response may be affected at very high temperatures.

Effects of Temperature on High-frequency Response. To some extent, the temperature of a pickup affects its damping coefficient and (to a much lesser extent) its resonant frequency. The resonant frequency is affected because the elastic coefficients of the materials of which it is constructed vary with temperature. Damping is typically dependent upon temperature. Because most piezoelectric pickups are used at frequencies far below their lowest natural frequency (e.g., below $0.3f_n$), changes in resonant frequency and damping coefficient due to temperature variations have little effect upon their usable high-frequency response.

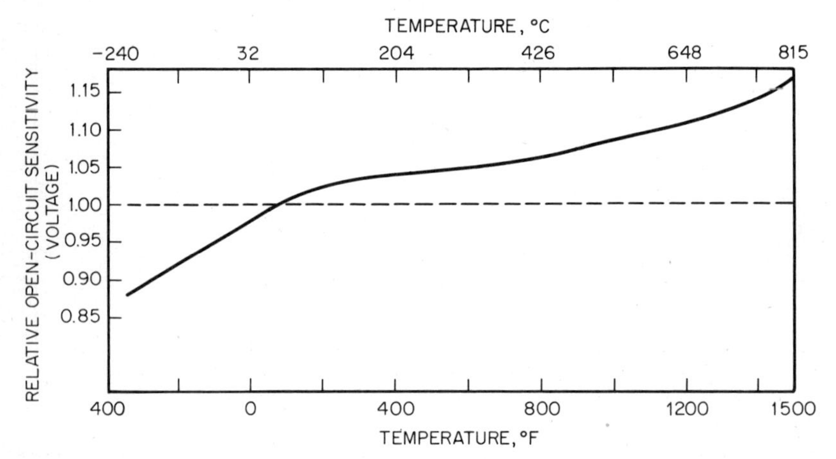

FIG. 16.19. Effect of temperature on the open-circuit voltage sensitivity of a piezoelectric pickup.

Effects of Temperature on Shock Pulse Response. The response of a pickup to a shock pulse may be temperature dependent because of variation in capacitance and insulation resistance with temperature. These variations affect the electrical *RC* time constant, and hence may affect the ability of the pickup to reproduce long shock pulses faithfully. The response also may be temperature-dependent because of the variation in damping coefficient with temperature, which may affect the short rise-time response. Results of these variations are described in previous sections of this chapter.

Effects of Transient Temperature. The voltage sensitivity of a piezoelectric pickup is not affected by the *rate of change* of the temperature of the instrument. However, electrical voltages (as large as 10 volts) due to a temperature fluctuation may be produced at the output terminals of the pickup in the following ways: (1) Differential expansion or compression of the mechanical and piezoelectric elements of the pickup may produce slowly varying forces on the piezoelectric element, and hence produce an output. (2) Many piezoelectric materials exhibit a *pyroelectric effect,* which is the generation of electric charge when the temperature of the piezoelectric material is changed. In general, the charge produced is proportional to the temperature change.

The voltage produced by temperature fluctuations occurs at a very slow rate and follows the temperature changes. Thermal insulation of the seismic element often is used to increase the time lag. Such pyroelectrically generated voltages may be filtered out electrically by means of a high-pass filter. Such a filter (having a cutoff frequency of the order of 5 cps) may be incorporated in the coupling amplifier. However, if the input voltage which results from this source is sufficiently high, the operation of the amplifier may be affected even though such a filter is used.

EFFECTS OF HIGH-INTENSITY SOUND. High-intensity sound can be an important source of vibratory excitation. Thus when vibration measurements are made on a structure that is exposed to such sound, the structure may vibrate with a considerable acceleration amplitude as a result of this source of excitation. The acoustic excitation also may affect the operation of an accelerometer used in the measurements by mechanical excitation of the piezoelectric element used in the accelerometer.

An illustration of the effect of high-intensity sound in producing an electrical output is shown in Fig. 16.20. In this example, a group of pickups are exposed to a random noise sound source having a uniform pressure spectral density in the frequency range from 150 to 9,600 cps. The output of the accelerometer resulting from the acoustic excitation is expressed in terms of the acceleration required to produce an equivalent voltage. These data are given for various models of a piezoelectric pickup.[13] For example, if one of these pickups is exposed to a sound pressure level of 170 db, the acoustical excitation produces an electrical signal equivalent to that produced by an acceleration of $10g$; at

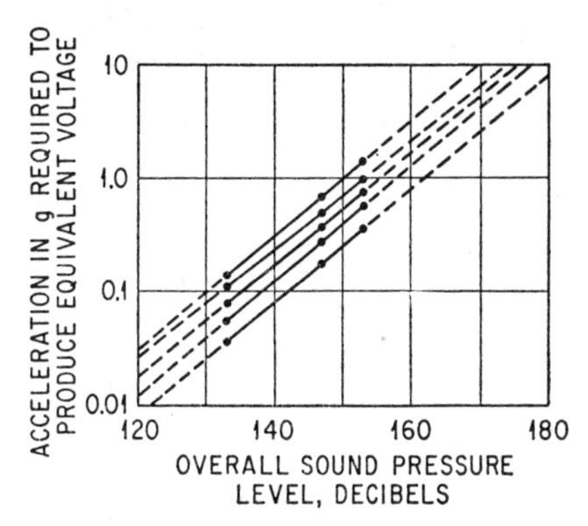

FIG. 16.20. The electrical outputs of several representative types of piezoelectric accelerometers as a result of acoustic excitation. The output voltage is expressed in terms of the acceleration in g required to produce an equivalent voltage. The sound source is random noise having uniform spectral density in the frequency range from 150 to 9,600 cps. The over-all sound-pressure level of this noise is expressed in decibels referred to 0.0002 dyne/cm². (*After W. Bradley, Jr.*[13])

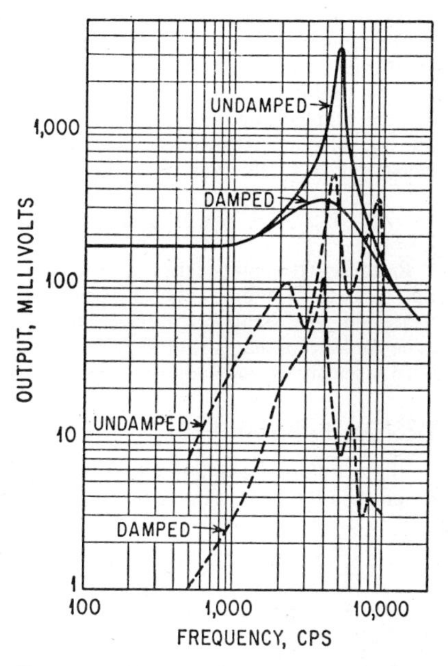

FIG. 16.21. Frequency-response curves for a piezoelectric accelerometer, illustrating the effects of damping on acoustic sensitivity. The solid lines show the usual sensitivity vs. frequency-response curves for $1g$ acceleration for the undamped and damped accelerometer; damping reduces the amplification ratio. The dashed curves show the voltage output when the accelerometers are exposed to a sound field of 150 db (referred to 0.0002 dyne/cm²), for sine-wave excitation.

150 db, the acoustic excitation produces an electrical signal equivalent to an acceleration of $1g$; and at 130 db, the corresponding value is only $0.1g$.

When a piezoelectric accelerometer is exposed to noise, the electrical output which is produced will be greatest at the resonant frequency of the seismic system. This is illustrated by the dashed curves in Fig. 16.21 for an undamped seismic system which is exposed to (essentially) a 150-db sine-wave acoustic excitation. For acoustic excitation at 3,000 cps at a sound pressure level of 150 db, a voltage is generated which is greater than that which would be produced by a vibratory acceleration of approximately $1g$.

Figure 16.21 shows that damping in the accelerometer reduces the effect of acoustic excitation, which effect can also be obtained by substantially increasing the resonant frequency of the accelerometer.

EFFECTS OF NUCLEAR RADIATION

An important characteristic of a piezoelectric accelerometer used in a nuclear power plant is its resistance to deterioration under nuclear bombardment. Such accelerometers may be used in the reactor, where they are exposed both to high temperature and to high neutron and gamma radiation levels. Special accelerometers have been developed which are capable of withstanding neutron bombardment of 10^{17} n/cm^2 and total gamma dosages of 4×10^{10} rads.[17]

PIEZORESISTIVE ACCELEROMETERS

Piezoresistive solid-state materials can be employed as strain elements in accelerometers so as to provide a much higher sensitivity (e.g., by a factor of 50 to 100) than conventional wire strain-gage accelerometers.* A major advantage of the piezoresistive-type accelerometers is that they have good frequency response down to d-c (0 cps) along with a relatively good high-frequency response.

DESIGN PARAMETERS

Many different configurations are possible for an accelerometer of this type (for example, see Ref. 16). For purposes of illustration, the design parameters are considered

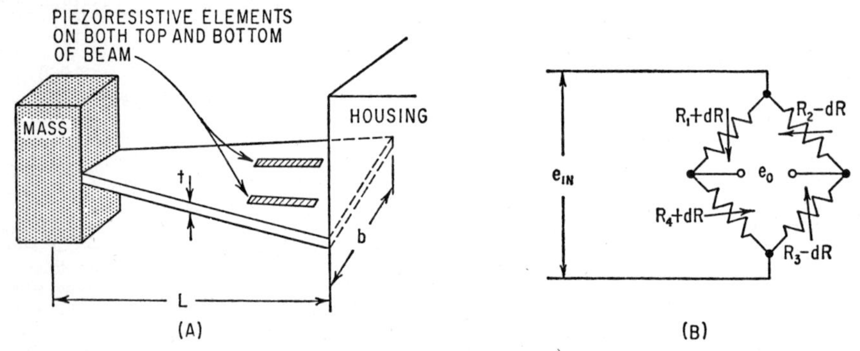

Fig. 16.22. (*A*) Schematic drawing of a piezoresistive accelerometer of the cantilever beam type. (*B*) Four piezoresistive elements are used—two cemented to each side of the uniformly stressed beam. These elements are connected in the above bridge circuit.

for a piezoresistive accelerometer which has a cantilever arrangement as shown in Fig. 16.22*A*. This uniformly stressed cantilever beam is loaded at its end with mass m. In this arrangement, four identical piezoresistive elements are used—two cemented to each

* Manufacturers include: Gulton Industries, Inc., Metuchen, N.J.; Statham Instruments, Inc., Los Angeles, Calif.; Consolidated Electrodynamics Corp., Pasadena, Calif.; Endevco Corp., Pasadena, Calif.

side of the beam whose length is L in. These elements, whose resistance is R, form the active arms of the balanced bridge shown in Fig. 16.22B. A change of length ΔL of the beam produces a change in resistance ΔR in each element. The gage factor K for each of the elements [defined by Eq. (17.1)] is

$$K = \frac{\Delta R/R}{\Delta L/L} = \frac{\Delta R/R}{\epsilon}$$

where ϵ is the strain induced in the beam, expressed in inches/inch, at the surface where the elements are cemented. If the resistances in the four arms of the bridge are equal, then the ratio of the output voltage of the bridge circuit to the input is[16]

$$\frac{e_o}{e_{\text{in}}} = \frac{dr}{R} \simeq \frac{\Delta R}{R} = K\epsilon$$

For the accelerometer shown in Fig. 16.22, suppose that
b = width of beam at the support, in.
t = thickness of beam, in.
E = Young's modulus of beam, lb/in.^{2}
m = mass of load at end of beam, lb-sec^2/ft

Then the following characteristics of the *uniformly stressed* beam, which is approximated by the configuration shown in Fig. 16.22A, may be derived which affect the response of this piezoresistive accelerometer:

Natural frequency:

$$f_n \simeq \frac{1}{2\pi}\sqrt{\frac{bEt^3}{6L^3m}} \qquad \text{cps}$$

Deflection x of end of beam for mass m:

$$x \simeq \frac{mgL^3}{2EI} = \frac{6mg}{Eb}\left(\frac{L^3}{t^3}\right) = \epsilon\left(\frac{L^2}{t}\right) \qquad \text{in.}$$

Strain at the surface to which the elements are bonded:

$$\epsilon \simeq \frac{6mgL}{Ebt^2}$$

The specifications for one example of a piezoresistive accelerometer of this type are given in Table 16.2.

Table 16.2. Specifications for One Example of a Piezoresistive Accelerometer

Frequency response	0 to 2,000 cps
Dynamic acceleration range	$2g$ to $1{,}000g$
Full-scale output	30 millivolts/(volt input)
Input voltage	10 volts (maximum)
Sensitivity	0.03 millivolts/volt/g
Temperature range	−65°F to 212°F (−54°C to 100°C)
Temperature sensitivity	0.02 per cent full scale per °F
Zero drift with temperature	0.0015 per cent full scale per °F
Dimensions	1 in. diameter by 2 in. length
Weight	2 oz
Output impedance	250 to 350 ohms
Transverse response	Less than 2 per cent of the maximum sensitivity
Linearity (including hysteresis)	±1 per cent

REFERENCES

1. Guttwein, G. K., and A. I. Dranetz: *Electronics*, **24**:120 (1951).
2. Orlacchio, A. W., and G. Hieber: *Electronic Inds.*, **11**:75 (1957).
3. Peters, R. J.: "NRL Type C-4 Barium Titanate Accelerometers," Proceedings of Symposium on Barium Titanate Accelerometers, *NBS Rept.* 2654, 1953, p. 57.
4. Perls, T. A.: "Determination of Sinusoidal Acceleration at Peak Levels Near That of Gravity by the Chatter Method," *NBS Rept.* 3399, September, 1954.
5. Perls, T. A., and C. W. Kissinger: "A Barium Titanate Accelerometer with Wide Frequency and Acceleration Ranges," *NBS Rept.* 2390, April, 1953.
6. Kissinger, C. W.: "Transverse Response," Proceedings of Symposium on Barium Titanate Accelerometers, *NBS Rept.* 2654, 1953, p. 119.
7. Dranetz, A. I.: "Natural Frequency and Frequency Response," Proceedings of Symposium on Barium Titanate Accelerometers, *NBS Rept.* 2654, 1953, p. 95.
8. Fleming, S. T.: "Cathode Follower Design," Proceedings of Symposium on Barium Titanate Accelerometers, *NBS Rept.* 2654, 1953, p. 175.
9. Lawrence, A. F.: "Crystal Accelerometer Response to Mechanical Shock Impulses," *Shock and Vibration Bull.* 24, Office of Secretary of Defense, February, 1957, p. 298.
10. Levy, S., W. D. Knoll, and D. Wilhelmia: *J. Research Natl. Bur. Standards*, **45**:303/RP2138 (1950).
11. Perls, T. A.: "Tests of Accelerometer Linearity at High Accelerations," Proceedings of Symposium on Barium Titanate Accelerometers, *NBS Rept.* 2654, 1953, p. 133.
12. Orlacchio, A. W.: *Elec. Mfg.*, **59**:78 (1957).
13. Bradley, W., Jr.: "Effects of High Intensity Acoustic Fields on Crystal Vibration Pickups," *Proc. 26th Shock and Vibration Symposium*, Office of Secretary of Defense, 1958.
14. Padgett, E. D., and W. V. Wright: "Silicon Piezoresistive Devices," *Proc. Instr. Soc. Amer.*, vol. 15, *Paper* 42-NY60, 1960. Also, Xavier, M. A., and C. O. Vogt: "Characteristics and Applications of a Semiconductor Strain Gage," *Proc. Instr. Soc. Amer.*, vol. 15. *Paper* 16-NY60, 1960.
15. Mason, W. P.: "American Institute of Physics Handbook," sec. 3g, McGraw-Hill Book Company, Inc., New York, 1957.
16. Mason, W. P., J. J. Forst, and L. M. Tornillo: "Recent Developments in Semiconductor Strain Transducers," *Instr. Soc. Amer., Preprint* 15 NY 60, 1960.
17. Aerojet General, "Evaluation of Irradiation Test Results for Candadate NERVA Piezoelectric Accelerometers Ground Test," Reactor Irradiation Test 22, December, 1971.

17

STRAIN-GAGE INSTRUMENTATION

Earl J. Wilson
National Aeronautics and Space Administration

INTRODUCTION

The resistance strain gage may be employed in shock or vibration instrumentation in either of two ways. The strain gage may be the active element in a commercial or special-purpose transducer or pickup, or it may be bonded directly to a critical area on a vibrating member. Both of these applications are considered in this chapter, together with a discussion of strain-gage types and characteristics, cements and bonding techniques, circuitry for signal enhancement and temperature compensation, and related aspects of strain-gage technology.

The electrical resistance strain gage discussed in this chapter is basically a piece of very thin foil or fine wire which exhibits a change in resistance proportional to the mechanical strain imposed on it. In order to handle such a delicate filament, it is either mounted on or bonded to some type of carrier material and is known as the *bonded strain gage*.

The strain gage is used universally by stress analysts in the experimental determination of stresses. Since strain always accompanies vibration, the strain gage or the principle by which it works is broadly applicable in the field of shock and vibration measurement. Here it serves to determine not only the magnitude of the strains produced by the shock or vibration, but also the entire time-history of the event, no matter how great the frequency of the phenomenon.

BASIC STRAIN-GAGE THEORY AND PROPERTIES

The relationship between resistance change and strain in the foil or wire used in strain-gage construction can be expressed as

$$\frac{\Delta L}{L} = \frac{1}{K}\frac{\Delta R}{R}$$

or

$$K = \frac{\Delta R/R}{\Delta L/L} \tag{17.1}$$

where K is defined as the *gage factor* of the foil or wire, ΔR is the resistance change due to strain, R is the initial resistance, ΔL is the change in length, L is the original length of the wire or foil, and $\Delta L/L$ is the unit strain to which the wire or foil is subjected.

All materials do not exhibit this strain-sensitivity effect and different materials have different gage factors. Filament materials in common use in strain gages are Constantan (Ni 0.45, Cu 0.55), which has a gage factor of approximately +2.0, Iso-elastic (Ni 0.36, Cu 0.08, Fe 0.52, and Mo 0.005), which has a gage factor of about +3.5, and modified Karma (Ni 0.75, Cr 0.20, plus additions), which has a gage factor of +2.1.

STRAIN-GAGE CONSTRUCTION

Since the foil used in a strain gage must be very fine or thin to have a relatively high electrical resistance, it is difficult to handle. For example, the foil used in gages is about 0.1 mil in thickness. In the past, many strain gages have been constructed with fine-wire filaments, but this type of gage is now rarely used except in special or high-temperature applications.

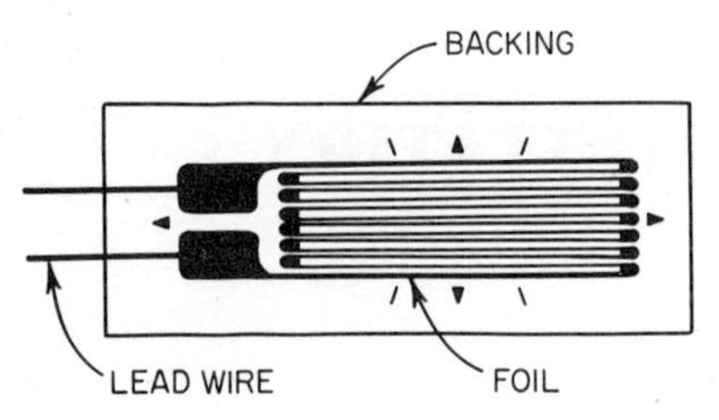

Fig. 17.1. Typical construction of a foil strain gage.

In order to handle this foil, it must be provided with a "carrier medium," usually a piece of paper or plastic to which the filament is cemented. Lead wires are often provided on foil gages. The cement and paper or plastic sandwich perform another very important function in addition to providing ease of handling and simplicity of application. The cement provides so much lateral resistance to the foil that it can be shortened significantly without buckling; then compressive as well as tensile strains can be measured. Foil gages are formed in a grid as shown in Fig. 17.1.

TRANSVERSE SENSITIVITY

Because of its construction, a portion of the foil in each gage lies in the transverse direction and will respond to transverse strain. Therefore the gage factor K of a gage* is always slightly smaller than the gage factor of the material of which it is fabricated. One of the desirable features of foil-type gages is their low transverse sensitivity. In this case, the gage consists of a flat foil grid; a sufficiently large amount of the foil is

* In determining the *gage factor* of the gage it is assumed that the gage is mounted on a material having a Poisson's ratio of 0.285 and subjected to uniaxial stress in the direction of the gage axis.

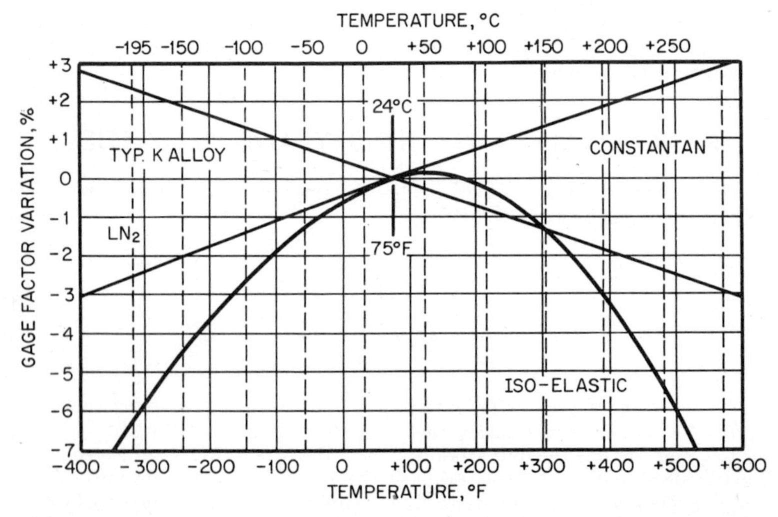

Fig. 17.2. Typical variation in the gage factor of strain-gage alloys as function of temperature.

left at the ends of each strand to reduce the transverse sensitivity of the gage to one-half the value for wire gages for some types and to essentially zero for others.

TEMPERATURE EFFECTS

The effects of temperature on the gage factor of several alloys are illustrated in Fig. 17.2. When a bonded strain gage is used in measurements, any change in resistance in the strain-gage measurement system is interpreted as resulting from a strain. If thermal expansion is not induced, then this change will result from a mechanical strain. However, if thermal expansion is induced, then there will be a change in resistance resulting from the mechanical strain; in addition, there will be a change in resistance resulting from the response of the strain gage to changes in temperature. The strain indication which results from such a temperature effect is known as an "apparent strain." Figure 17.3 shows typical apparent strain for three commonly used alloys. This effect is

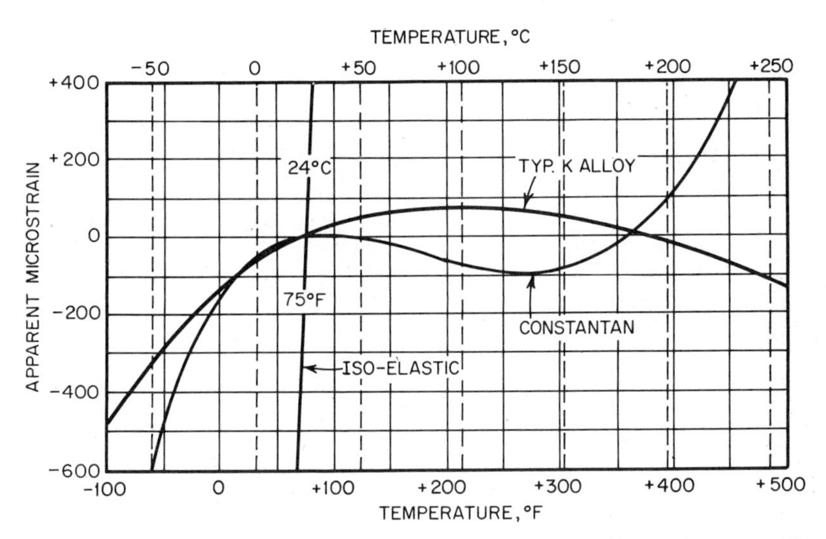

FIG. 17.3. Typical apparent strain for three alloys commonly used in strain gages. These data are based on an instrument gage factor of 2.00.

usually negligible in the measurement of dynamic strains, since the readout instrument associated with the strain gage usually does not respond to static or slow changes in its resistance. However, in the measurement of static strains, the effects of temperature represent the largest potential source of error and require some form of temperature compensation.[2]

STRAIN-GAGE CLASSIFICATIONS

Strain gages are classified in several ways. One classification cites the purpose for which the gage is to be used, that is, for static or dynamic strain measurement. Static gages are made up with Constantan foil, which has a minimum change of resistance with temperature. Dynamic strain gages are made up with Iso-elastic foil, which provides a greater gage factor than Constantan. The dynamic gages, while having a much greater resistance change for a given strain than the static gages, also are much more sensitive to changes in temperature. They are used only where the phenomenon to be measured is so short in time duration that no temperature change of any consequence can occur during the time of measurement. Gages also are available for the measurement of very large strains (up to 20 per cent) occurring in the plastic region of the material as distinguished from the more common gages which are used to measure elastic strains (up to 1 per cent).

STRAIN-GAGE SELECTION CONSIDERATIONS

In the case of shock measurements, a transient may be applied to the structure that is under investigation only once or it may be repetitive. Shock is of very short time duration, and the problem of temperature compensation is nonexistent because in most cases the temperature does not have time to change during the impact. For this reason a dynamic-type gage usually can be employed for the measurement of shock. This type of gage has the advantage of a higher gage factor than the static gage so that it will provide the greatest possible electrical signal for a given strain.

For vibration measurement the type of gage selected is dependent on the kind of information desired. If only the frequency of vibration and the magnitude of the cyclic stresses are desired, dynamic-type gages can be used since temperature changes will not affect the results obtained unless the temperature fluctuates at the same rate as the stress. If, however, a measurement of the static or slowly varying component of the stress is also to be determined (i.e., if the absolute values of the stresses are desired), a static-type gage must be employed. Since changes in temperature will affect the gage reading, temperature compensation must be incorporated to obtain true values of stress.

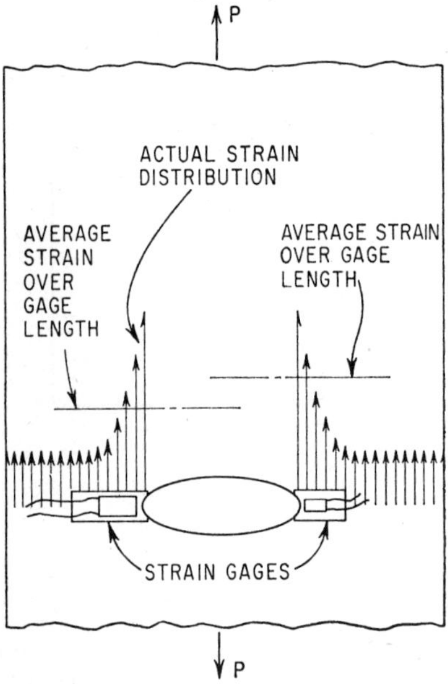

Gage selection is dependent on the space limitation and steepness of strain gradient in any region. The strain gage indicates the average strain over the length of the gage; in a region of steep strain gradient, this indicated value may be much less than the maximum strain. The shorter the gage used in such a region, the closer is the gage indication to the maximum strain (Fig. 17.4). However, two possible objectives must be considered quite carefully in selecting a gage for a particular installation: (1) the determination of the frequency of vibration, or comparison of relative amplitudes and frequencies with different conditions of excitation, and (2) the determination of the maximum stress pattern resulting from the vibration set up. In the first case there is considerable freedom with regard to the location of the gage on the structure, and therefore with the selection of the gage itself. In the second case severe restrictions exist in regard to the region of application of the gage and its possible dimensions. In general, very short gage-length gages are more difficult to apply properly. Therefore it is desirable to employ gage lengths of ¼ in. whenever possible. When the actual magnitude of the maximum stress resulting from shock or vibration is to be determined, a much more complicated system of gages must be employed. A single gage can be used in only the very limited case where a stress exists in one direction only, and that direction must be known. If stresses exist in several directions, or if the direction of a singly existing stress is unknown, a strain-gage rosette consisting of three or more gages must be employed.[3]

FIG. 17.4. Effect of gage length on indicated strain in the presence of a severe strain gradient. The shorter gage on the right indicates a higher strain. An infinitesimal gage length would be necessary to indicate the peak strain.

PHYSICAL ENVIRONMENT

The physical environment of the applied gage is an important factor which must be considered in gage selection and protective treatment. Temperature, pressure, humidity, oil, corrosive acid, abrasive action, and possible electromagnetic, neutron, and radiation fields are conditions which affect the choice of gage and its required protection. If high temperatures (up to 500°F or 260°C) are to be encountered, a Bakelite or other high-temperature-type gage must be selected. If even higher temperatures must be withstood, a ceramic-type gage should be employed. Gages of this sort are used at temperatures as high as 2000°F (1100°C). If the temperature never exceeds 200°F (95°C), however, any type of gage can be used. Most gages operate satisfactorily at very low temperatures.

ACCURACY CONSIDERATIONS

Gages must be selected with regard to the desired precision of the results. If only the frequency of the vibration or the duration of a shock wave is required, almost any gage, properly chosen for the temperature and humidity conditions to be encountered, gives quite satisfactory results. However, if the magnitude of the stresses produced is to be determined in addition, then considerable care must be exercised to select the proper gage to obtain the desired results. Not only must the gage be the proper one to portray the encountered strain faithfully, but precautions must be taken to install the gage correctly.

The testing "environment" can affect strain-gage accuracy in many ways. Magnetostrictive effects,[4] hydrostatic pressure,[5] nuclear radiation,[6] and high humidity are examples of conditions that may cause large strain-gage errors. Creep, drift, and fatigue life in the gages themselves may be important. In most normal environments these errors are either small or undetectable. Whenever unusual or harsh environments are encountered, it is wise to consult the strain-gage manufacturer to obtain recommendations for gage systems and estimates of expected accuracies.

BONDING TECHNIQUES

The proper functioning of a strain gage is completely dependent on the bond which holds it to the structure undergoing test. If the bond does not faithfully transmit the strain from the test piece to the wire or foil of the gage, the results obtained cannot be accurate. Failure to bond over even a minute area of the gage will result in incorrect strain indications. The greatest weakness in the entire technique of strain measurement by means of wire or foil gages is in the bonding of the gage to the test piece. Usually, the manufacturer of the strain gages will recommend cements which are compatible with their use and will provide instructions for their proper installation.

STRAIN-GAGE MEASUREMENTS

The resistance strain gage, because of its inherent linearity, very small mass, wide frequency response (from zero to more than 50,000 cps),[7] general versatility, and ease of installation in a variety of applications, is an ideal sensitive component for electrical transducers for use in shock and vibration instrumentation.[8] The Wheatstone bridge circuit, described in a subsequent section, can be used to extend the versatility of the strain gage to still broader applications by performing mathematical operations on the strain-gage output signals. The combination of these two devices can be used effectively for the measurement of acceleration, displacement, force, torque, pressure, and similar mechanical variables. Other useful attributes include the capacity for separation of forces and moments, vector resolution of forces and accelerations, and cancellation of undesired vector components.

The usual technique for employing a strain gage as a transducing element is to attach the gage to some form of mechanical member which is loaded or deformed in such a manner as to produce a signal in the strain gage proportional to the variable being measured. The mechanical member can be utilized in tension, compression, bending, torsion, or any combination of these. All strain-gage-actuated transducers can be considered as either force- or torque-measuring instruments. Any mechanical variable which can be predictably manifested as a force or a couple can be instrumented with strain gages.

There are a number of precautions which should be observed in the design and construction of custom-made strain-gage transducers.[9] First, the elastic member on which the strain gage is to be mounted should be characterized by very low mechanical hysteresis and should have a high ratio of proportional limit to modulus of elasticity (i.e., as large an elastic strain as possible). Although aluminum, bronze, and other metals are often employed for this purpose, steel is the most common material. An alloy steel such as SAE 4340, heat-treated to a hardness of RC 40, will ordinarily function very satisfactorily.

The physical form of the elastic member, and the location of the strain gages thereon, are not subject to specific recommendation, but vary with the special requirements of each individual instrumentation task. When no such requirements exist, a standard commercial transducer ordinarily should be used. In general, the shape of the member should be such as to (1) allow adequate space for mounting strain gages (preferably in regions of zero or near-zero strain gradient), (2) provide the desired natural frequency, (3) produce a strain in the gages which is great enough at low values of the measured variable to result in an output signal readily subject to accurate indication or recording, and not so great as to cause nonlinearities or abbreviated gage life at peak load values, (4) provide temperature compensation and/or signal augmentation (as described in a subsequent section) whenever feasible, and (5) allow for simplicity of machining, ease of gage attachment and wiring, and, if necessary, protection of the gages.

The strain gages should be cemented to the elastic member with the usual care and cleanliness necessary to all strain-gage applications, special attention being given to minimizing the bulk of the installation if the added mass is significant to the frequency response of the instrument. Other considerations vital to successful strain-gage-application technique are described elsewhere in this chapter.

DISPLACEMENT MEASUREMENT

Measurement of displacement with strain gages can be accomplished by exploiting the fact that the deflection of a beam or other loaded mechanical member is ordinarily proportional to the strain at every point in the member as long as all strains are within the elastic limit.

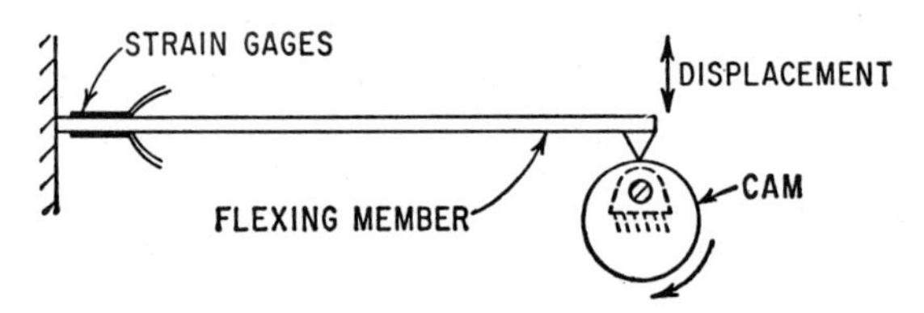

FIG. 17.5. Strain gages mounted on a cantilever beam for displacement measurement produce electrical signal proportional to cam motion.

For small displacements at low frequencies, a cantilever beam arranged as shown in Fig. 17.5 can be employed. The beam should be mounted with sufficient preload on the moving surface that continuous contact at the maximum operating frequency is assured. In the case of higher frequency applications the beam can be held in contact with the moving surface magnetically or by a fork or yoke arrangement, as illustrated in Fig. 17.6. It is necessary to make certain that the measuring beam will not affect the displacement to be instrumented, and that no natural mode of vibration of the beam itself will be excited.[10]

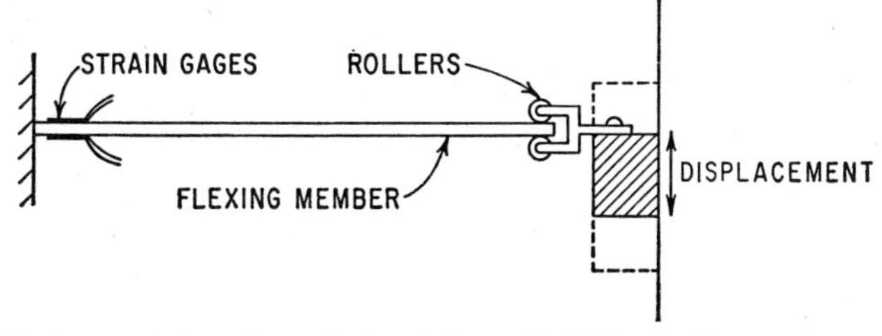

FIG. 17.6. Displacement transducer designed for continuous, positive contact with moving object.

The measurable displacement magnitude can be increased above that for the cantilever beam by employing other schemes such as the "clip gage" shown in Fig. 17.7. This gage is constructed by bonding strain gages to the upper and lower sides of a piece of channel-shaped spring steel, as shown in Fig. 17.7. The assembly is then clipped or otherwise mounted on the test specimen so that the legs deflect as the specimen is strained, thus straining the backbone of the clip gage to a greater or lesser extent. Any desired reduction in strain magnitude can be obtained in this manner by merely altering the proportions of the clip gage. Unfortunately, the maximum allowable frequency generally decreases as the displacement amplitude increases, since stiffness and natural frequency tend to change together. Displacement also can be measured through the use of the relative motion of a seismically mounted mass of much lower natural frequency than the applied frequency.

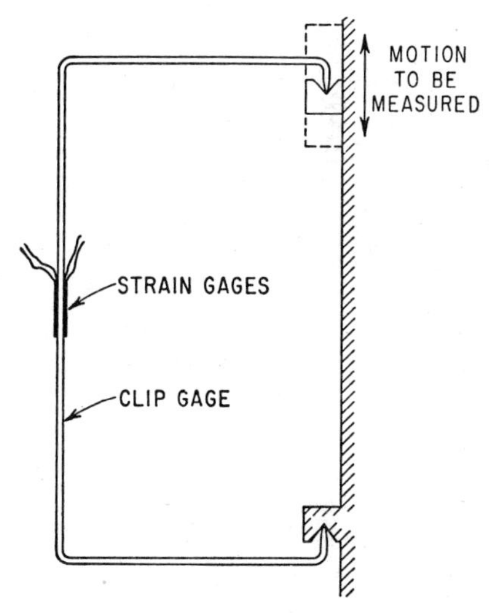

FIG. 17.7. Clip gage for instrumenting large displacements. Proportions of clip gage are designed to keep strain well within the proportional limit of the material.

VELOCITY

Velocities can be measured directly with strain-gage transducers only by producing a force such as viscous damping or hydro- or aerodynamic drag force which is uniquely related to velocity. Velocity indication also can be obtained with strain gages by differentiation of a displacement function or integration of an acceleration function. In either case, the transducer-design considerations correspond to those for force measurement described in the following section.

FORCE MEASUREMENT

The principle of force measurement with strain-gage-actuated transducers is very similar to that for displacement.[9] The procedure consists of placing a strain-gage-instrumented elastic member in series with the force to be measured. The strain in the transducer, and thus the output signal, is proportional to the force if all stresses are kept within the elastic limit. The proportionality constant between strain and force must be obtained by calibration if precise results are desired. Otherwise, tolerances on the gage factor of the strain gage, and uncertainty as to the elastic properties of the instrumented member, can produce errors of 5 per cent or greater—even for transducer configurations with readily calculable strain distributions.

Figure 17.8 illustrates a common form of force transducer, the cantilever beam. Strain gages are mounted on the top and bottom of the beam, producing double sensitivity (output) and virtually complete temperature compensation. While this type of transducer is probably best suited to static or quasi-static measurements such as reaction forces, it also can be used very successfully for many shock and vibration problems as long as the natural frequency of the beam is higher than the frequency of the force being measured. The ring gage (Fig. 17.9) can be categorized with the cantilever beam, and is equally applicable to static or dynamic force measurement within the limitations imposed by its comparatively low natural frequency.

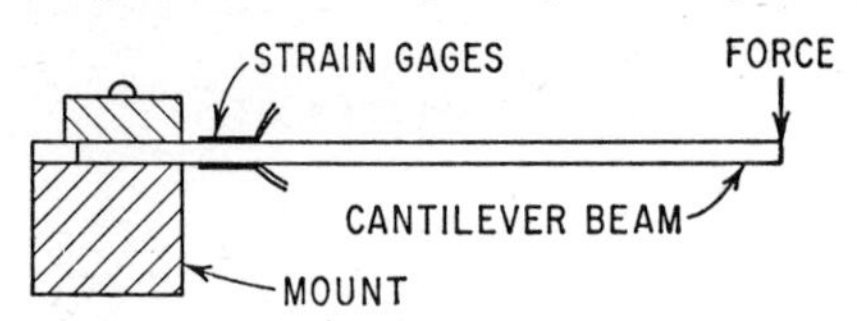

FIG. 17.8. Cantilever force-measuring transducer, consisting of beam with load applied at free end. Gage strain is a linear function of the force if the proportional limit is not exceeded.

For most dynamic force-instrumentation problems a small compression or tension member (Fig. 17.10) is ordinarily employed. If the load is characterized by alternation between compression and tension, the transducer must be designed for a rigid, integral connection, with no backlash or clearance. This can be accomplished by employing threaded ends with lock nuts for joining the transducer to the remainder of the assembly. In many problems involving machine parts or other mechanical components it is possible to measure loads by applying strain gages to the machine member itself, necessitating calibration of the member to determine the relationship between force and strain.

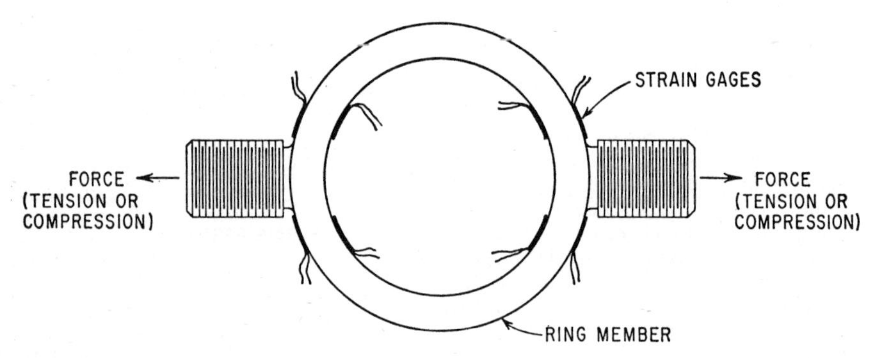

FIG. 17.9. Ring gage for force measurement. This type of gage provides sensitive axial load measurement without undue loss of rigidity or ruggedness.

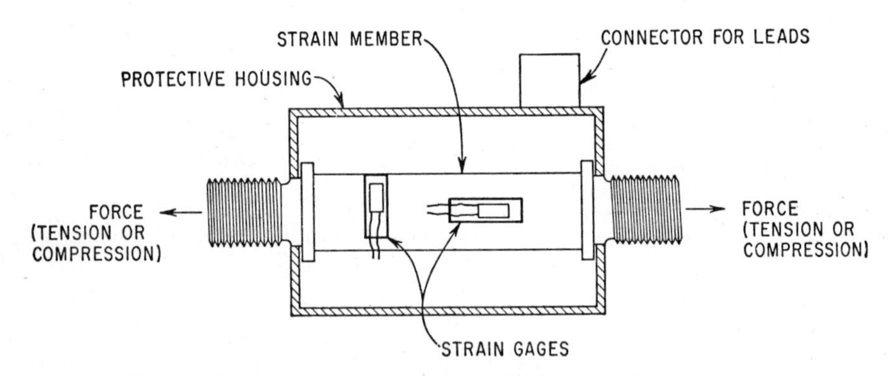

FIG. 17.10. Widely used commercial form of axial force transducer for large loads.

PRESSURE

In hydraulic and aerodynamic devices, pressure fluctuations are often associated with vibration phenomena—either as cause or effect. Strain-gage transducers are widely used in such situations.[11]

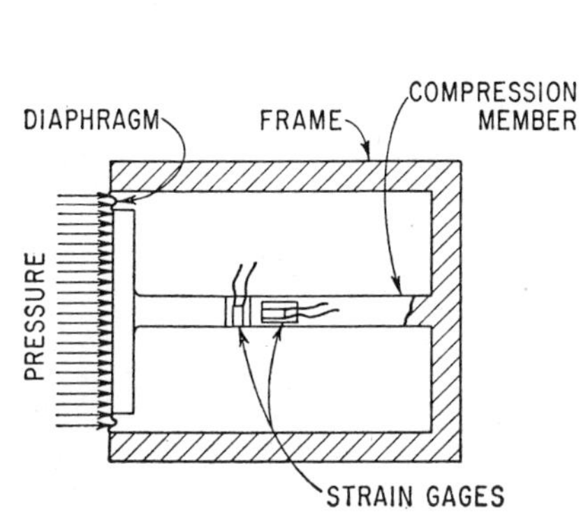

FIG. 17.11. Piston-type pressure transducer with diaphragm seal for piston. Pressure load on piston head is sensed by strain gages on supporting column.

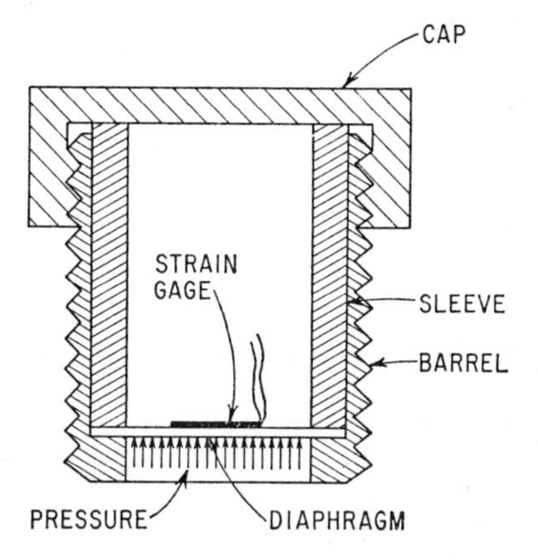

FIG. 17.12. Pressure pickup whose output is a function of diaphragm strain. As diaphragm deforms under pressure, strain is transmitted to gage to produce electrical signal.

Pressure pickups based on strain gages are commonly one of three principal types: piston, diaphragm, or tube. In the piston type the pressure acts against a freely movable flat surface (which may be either a piston or diaphragm), the motion of which is inhibited by an elastic member instrumented with strain gages to measure the force (Fig. 17.11).

Diaphragm-type pressure transducers, shown in Fig. 17.12, have the strain gages applied directly to the back surface of the diaphragm so that diaphragm strain is a measure of pressure.[12] The simplest form of pressure transducer to construct is the tube type, shown in Fig. 17.13. In this type, strain gages are applied to the outer surface of a tube which has the fluid pressure acting on its inner surface. It is sometimes necessary to thin the wall of the tube or to use a longitudinally crimped tube in order to increase the strain magnitude to a measurable level. As a convenient alternative, the bourdon tube in a conventional mechanical pressure gage can serve as the transducing element if strain gages are attached to it. The compressibility of the fluid contained in the tube must be considered for its effect on the frequency response of this type of unit. Pressure pickups should be calibrated statically, and preferably dynamically, prior to use.

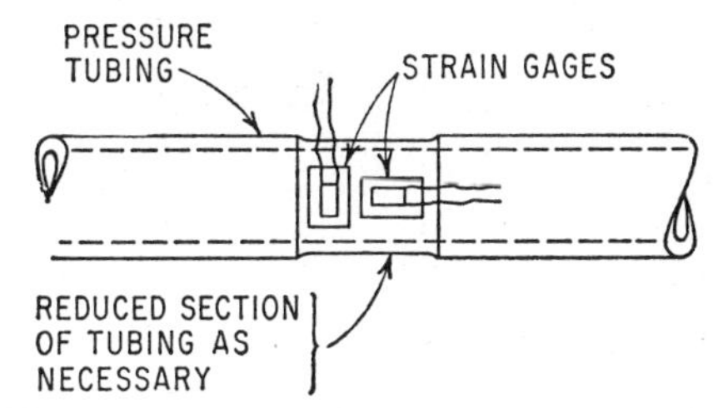

FIG. 17.13. Readily made pressure transducer consisting of length of tubing with strain gages attached. Dilation of the tube with pressure creates strain in gages.

ACCELERATION

Strain-gage accelerometers are similar in design to force-sensing units. Ordinarily they consist of a mass mounted so as to deform an elastic member in an amount proportional to the inertia force.[13] Strain-gage accelerometers, in contrast to the more commonly used piezoelectric types, are suitable for use at very low frequencies. However, the output signal from any strain-gage device is considerably smaller than that from a piezoelectric transducer.

Two types of strain-gage accelerometers are shown in Fig. 17.14 in greatly simplified form. In addition to the requirement for a high natural frequency, the other criteria for successful operation include selection of fatigue-resistant strain gages, placement of

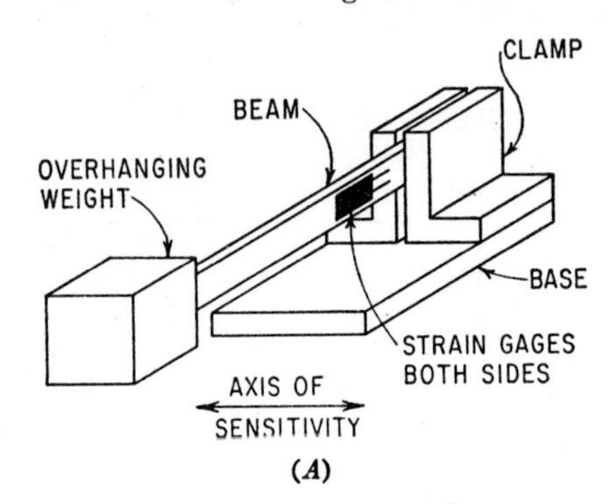

(A)

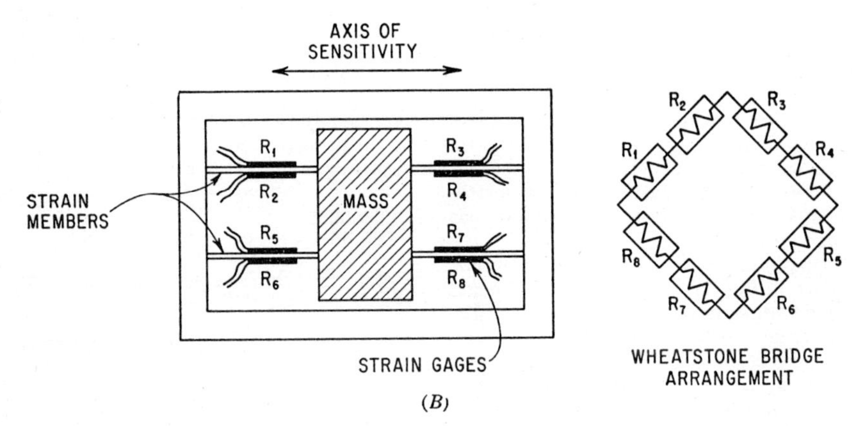

(B)

Fig. 17.14. Common forms of strain-gage accelerometer. (A) Weight on free end of cantilever beam. Inertia forces from acceleration cause beam to bend and strain the gages proportionally. (B) High-g accelerometer with seismic mass producing axial strain in instrumented members.

gages whenever possible for augmented output and temperature compensation, limiting maximum strain so as to enhance linearity and minimize hysteresis, and calibration of the completed transducer. Damping of the accelerometer can be accomplished by enclosing the unit in a case filled with a dielectric damping fluid. Except for details associated with the strain gages themselves, the over-all design considerations for strain-gage accelerometers are the same as those for other seismic accelerometers (see Chap. 12).

STRAIN-GAGE CIRCUITRY AND INSTRUMENTATION

The output of a resistance strain gage usually is from 10 to 1,000 microvolts for practicable strain ranges. In order to study the detailed cyclic nature of vibration problems or the transient phenomena commonly associated with mechanical shock, it is necessary to obtain some form of oscillographic record of the events. Oscillographic recording equipment commonly requires an input signal several orders of magnitude greater than the strain-gage output. Thus, some form of electrical amplification is necessary between these two components. The strain gage also needs a stable source of electric current, or excitation, to produce an output voltage proportional to resistance change. These two factors are of primary importance in determining the nature of the electrical instrumentation system which can be used satisfactorily with the resistance strain gage.

THE POTENTIOMETER CIRCUIT

The simplest circuit arrangement for supplying a strain gage with excitation current, and obtaining a signal corresponding to deformation of the gage, is known as the *potentiometer circuit* (Fig. 17.15). It is very well suited to the instrumentation of dynamic or fluctuating strains, but is totally unsuited for the measurement of static strains or the static component of a combined static and dynamic strain.

In Fig. 17.15, R_B represents a ballast resistor, the principal function of which is to maintain the current flow in the circuit relatively constant and independent of small resistance changes in the strain gage R_G.

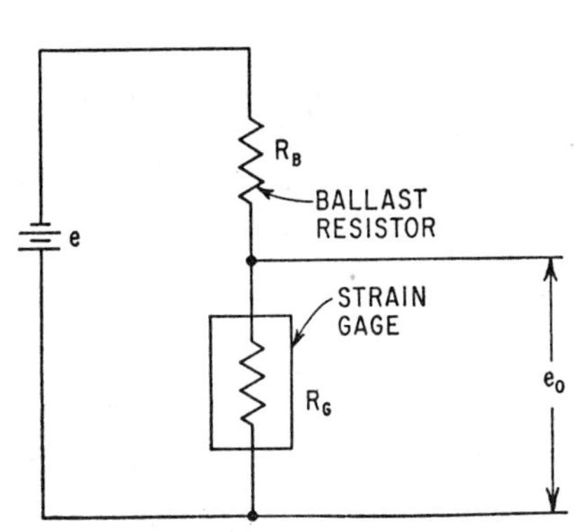

FIG. 17.15. Potentiometer circuit for dynamic strain signals. Nearly constant current through the circuit, combined with varying gage resistance, produces output signal.

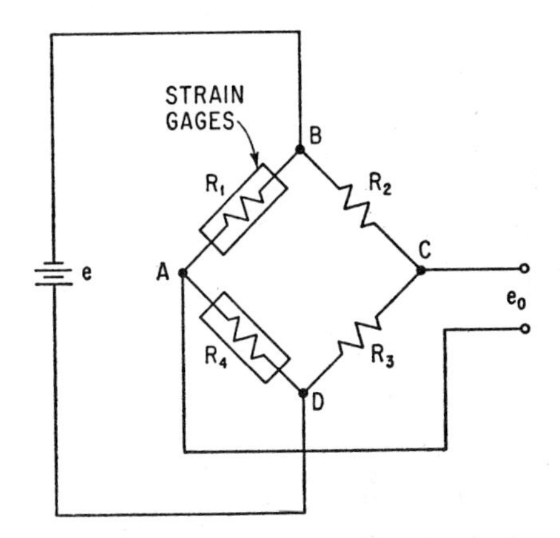

FIG. 17.16. Wheatstone-bridge circuit for static and dynamic strain measurement.

THE WHEATSTONE BRIDGE

In the potentiometer circuit it is necessary to block the d-c component of the output voltage with a capacitor before feeding the signal to the input of an amplifier. The same effect can be achieved by suppressing the d-c component of the signal by connecting two potentiometer circuits in parallel and taking the output signal from corresponding points in the two branches of the resulting network as shown in Fig. 17.16. This circuit arrangement is generally referred to as a *Wheatstone bridge*, and represents one of the most precise methods known for measuring (or comparing) resistances. Advantages of the Wheatstone bridge over the potentiometer circuit are (1) much greater flexibility in circuit arrangements for signal augmentation, temperature compensation, and cancellation or separation of variables, (2) capacity for accurately indicating combined static and dynamic strains, and (3) virtually complete freedom from error due to resistive changes in the conductors connecting the supply voltage to the network. As an example of the significance of the last point, consider the effect of the contact resistance variations which might occur in a set of slip rings being used in conjunction with a test of torsional vibration in a rotating shaft.

SELECTION OF INSTRUMENTS FOR STRAIN MEASUREMENT[14]

The output voltage from a strain-gage potentiometer circuit or Wheatstone bridge is, for elastic strain magnitudes in metals, very small. Electrical amplification is required to bring the signal to a level where it can be used conveniently for indication or recording. To assure satisfactory performance and precision, the entire instrument system, from power supply to recording instrument, should be considered as a unit. Figure 17.17 illustrates in block form the basic elements of a strain-gage instrumentation

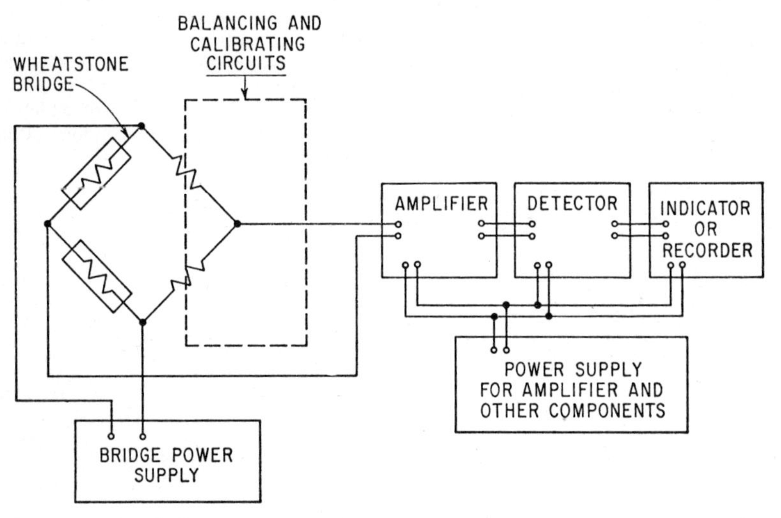

FIG. 17.17. Block diagram of basic elements of strain-gage instrumentation system.

system. The criteria for selecting the individual components of such a system are fixed by the nature of the strain being studied, the type of information required from the system, and the mutual compatibility of the various system components. Consideration should be given to the required frequency response, the input and output impedances of the units in the system, the signal amplitudes being dealt with, and the accuracy of measurement desired. In general, it is safe to assume that the strain gage will respond to frequencies considerably higher than any mechanical device to which it may be attached. In the case of small members vibrating at high frequencies the limitation is more apt to arise from the change in mass due to the presence of the gage and its lead wires.

Commercial instruments for use with strain gages usually combine into a single unit several if not all of the components of Fig. 17.17. The limitations of such devices should be investigated prior to purchase. For example, the instrument may include an alternating-frequency source of power for the Wheatstone bridge. This can lead to difficulties in the measurement of high-frequency strains. The frequency of the strain being measured (which will modulate the power supply in the bridge circuit) is limited to approximately 10 to 20 per cent of the carrier frequency. If the carrier frequency is high enough to overcome this objection, the capacitive unbalance and pickup in the strain-gage leads are apt to be excessive.

Alternating-current bridge supply sources usually range from 500 to 25,000 cps, and impose corresponding limitations on the strain frequency which can be studied with precision. Commercial instruments usually include convenient, built-in circuits for balancing and calibration. These features should be investigated and tested for the distortion and gage desensitization effects which they may introduce into the system. The carrier systems have several advantages: a high degree of stability, capacity for handling low-frequency and static components of strain as well as higher-frequency signals, and (in conjunction with a phase-sensitive demodulator) the ability to preserve the sign (tensile or compressive) of the strain. For studying mechanical vibration at frequencies above 1,000 cps it is generally advisable to employ a battery or regulated d-c source to supply the bridge circuit, a direct-coupled amplifier, and a cathode-ray oscilloscope.

Table 17.1. Manufacturers of Bonded Strain Gages*

Manufacturer	Type of gage
BLH Electronics, Inc., Waltham, Mass.	Wire, foil; static, dynamic; room temperature, high temperature, very high temperature
Dentronics, Inc., Hackensack, N.J.	Die cut foil; static, dynamic; room temperature, high temperature
Hottinger Baldwin Messtechnik GmbH, Germany	Wire, foil; static, dynamic; room temperature, high temperature, very high temperature
Hitec Corporation, Monson, Mass.	Wire; static, dynamic; high temperature, very high temperature
Kyowa Electronic Instrument Co., Ltd., Japan	Wire, foil; static, dynamic; room temperature, high temperature
Magnaflux Corporation, Chicago, Ill.	Foil; static; dynamic; room temperature, high temperature
Micro-Measurements Division, Romulus, Mich.	Foil; static, dynamic; room temperature, high temperature, very high temperature
Micro-Strain, Inc., Spring City, Pa.	Foil, static, dynamic; room temperature, high temperature
Philips Company, Netherlands	Wire, foil; static, dynamic; room temperature, high temperature, very high temperature
Shinkoh Company, Japan	Wire, foil; static, dynamic; room temperature, elevated temperature
Showa Sokki, Ltd., Japan	Wire, foil; static, dynamic; room temperature, high temperature
Tinsley Telecon, Ltd., England	Wire, foil; static, dynamic; room temperature, high temperature, very high temperature
Toko Sokki Kenkyujo Co., Ltd., Japan	Wire, foil; static, dynamic; room temperature, high temperature
Toyo Baldwin Company, Ltd., Japan	Wire, foil; static, dynamic; room temperature, high temperature, very high temperature

* This list is not necessarily complete, but it is representative of firms manufacturing such equipment.

COMMERCIAL STRAIN-MEASURING GAGES AND INSTRUMENTS*

Table 17.1 lists commercially available bonded strain gages. Such units can be used to construct custom-made transducers to perform specific jobs, but it is usually more satisfactory to employ commercially available transducers of this type because of their

* Of the many instruments available for strain measurement, only a few representative types are discussed here. The instruments described were selected for purposes of illustration, and are not necessarily those recommended by the author.

better linearity and greater accuracy. A list of commercially available strain-gage transducers is given in Table 17.2. Manufacturers of strain-measuring equipment are listed in Table 17.3.

Table 17.2. Manufacturers of Strain-gage Transducers*
(For the measurement of acceleration, force, torque, pressure, etc.)

BLH Electronics, Inc., Waltham, Mass.
Bell & Howell Corp., Monrovia, Calif.
Dynisco, Westwood, Mass.
Electro Development Corp., Lynnwood, Wash.
GSE Associates, Farmington Hills, Mich.
Gentran, Inc., Sunnyvale, Calif.
Interface, Inc., Scottsdale, Ariz.
Instron Corp., Canton, Mass.
Kistler Instrument Corp., Redmond, Wash.
Lebow Associates, Inc., Troy, Mich.
Revere Electronics Divn., Neptune Meter Co., Wallingford, Conn.
Sensotec, Inc., Columbus, Ohio
Standard Controls, Inc., Seattle, Wash.
Strainsert Co., Bryn Mawr, Pa.
Structural Instrumentation, Inc., Tukila, Wash.
Taber Instruments, North Tonawanda, N.Y.
Toledo Scale Corp., Toledo, Ohio
Toroid Corp., Huntsville, Ala.
Viatran Corp., Grand Island, N.Y.
Weigh-Tronix Divn., Arts-Way Manufacturing Co., Armstrong, Iowa

* This list is not necessarily complete, but it is representative of firms manufacturing such equipment.

Table 17.3. Manufacturers of Strain-gage Instruments*
(Amplifiers, Wheatstone Bridges, Strain Indicators, etc.)

Acurex Corp., Mountain View, Calif.
Automation-Peekel, N.V., Rotterdam, Netherlands
B & F Instruments, Cornwell Heights, Pa.
BLH Electronics, Inc., Waltham, Mass.
Beckman Instruments, Inc., Offner Division, Shiller Park, Ill.
Bell & Howell Company, Pasadena, Calif.
Bruel & Kjaer, Naerum, Denmark
Dynamics Electronic Products, Inc., Santa Fe Springs, Calif.
Endevco, San Juan Capistrano, Calif.
Gould, Inc., Instrument Systems Division; Cleveland, Ohio
S. Himmelstein & Co., Elk Grove Village, Ill.
Honeywell, Inc., Test Instruments Division, Denver, Colo.
Hottinger-Baldwin Mess-Technik GmbH, Darmstadt, Germany
Ithaco, Inc., Ithaca, N.Y.
Kelvin & Hughes, Ltd., London, England
Kyowa Electronic Instruments Co., Ltd., Tokyo, Japan
Philips, N.V., Eindhoven, Netherlands
Sedeme, Paris, France
Shinkoh Communication Industry, Kanagawa-ken, Japan
Solartron Electronics Group, Ltd., Farnborough, Hants, England
Strainsert Co., Bryn Mawr, Pa.
H. Tinsley & Company, Ltd., London, England
Tokyo Sokki Kenkyujo Co., Ltd., Tokyo, Japan
Vishay Instruments Division, Malvern, Pa.

* This list is not necessarily complete, but it is representative of firms manufacturing such equipment.

The functional block diagram for a typical static strain indicator is shown in Fig. 17.18. This unit consists of a Wheatstone-bridge circuit, an oscillator to power the bridge with alternating current, an a-c amplifier, followed by a phase-sensitive demodulator, and a meter for indicating bridge balance.

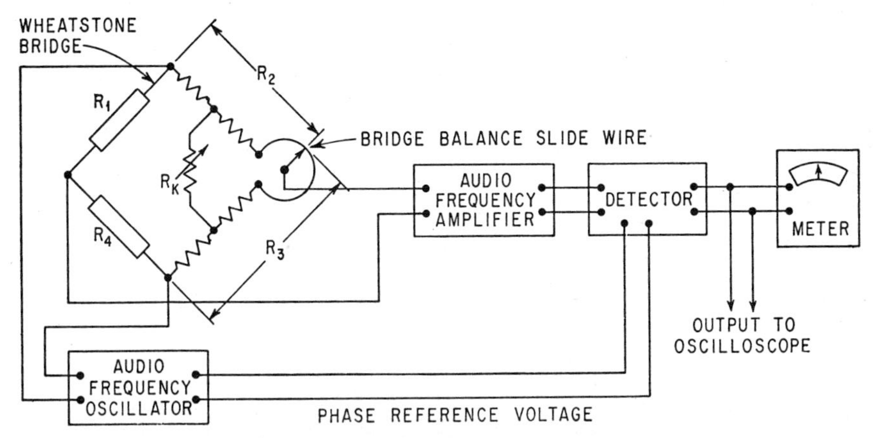

FIG. 17.18. Block diagram of commercial static strain indicator. It employs a-c bridge powered by oscillator at approximately 1,000 cps. Bridge output is amplified and then rectified in a phase-sensitive detector to give a d-c signal proportional to strain, and of a corresponding sign. This indicator can be used with an oscilloscope for dynamic strains at frequencies up to approximately 100 cps.

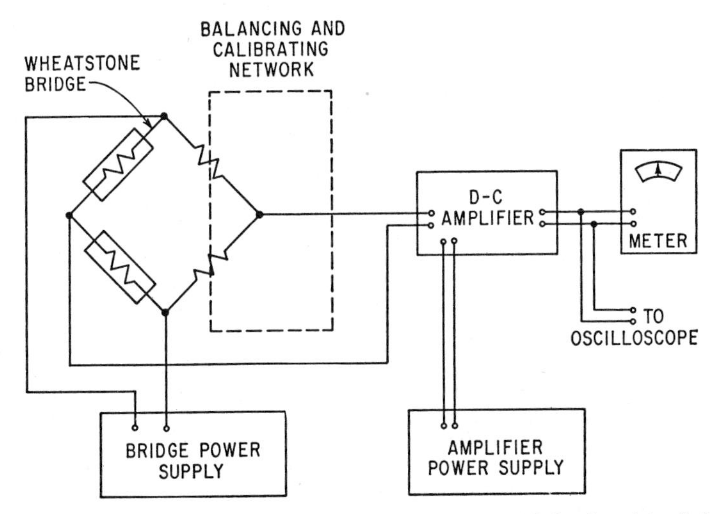

FIG. 17.19. Block diagram of strain amplifier with direct-connected circuitry, intended primarily for use with an oscilloscope.

The instrument is battery-powered and transistorized for portability. It has provisions for employing one, two, or four active strain gages in the bridge circuit, and is equipped with a finely calibrated balancing slide wire for accurate null-balance operation. Although this unit is basically a null-balance static strain indicator, it includes an output jack for supplying a signal to an oscilloscope, and thus can be used for dynamic strains as well. Limited by a carrier frequency of approximately 1,000 cps, the instrument is satisfactory for cyclic strains at frequencies below 100 cps.

Another form of strain amplifier, primarily for use with an oscilloscope, is shown in functional block form in Fig. 17.19. This transistorized instrument employs a stabilized direct-coupled amplifier to obtain a frequency response to 20,000 cps or above. The battery-supplied d-c Wheatstone bridge is arranged to accept one, two, or four active strain gages.

The circuit of Figure 17.19 can also be used with a direct-writing recorder. For frequencies up to perhaps 100 cps, pen-and-ink or hot stylus (thermal) recorders may be used. The gain and/or power output of the d-c amplifier may have to be improved.

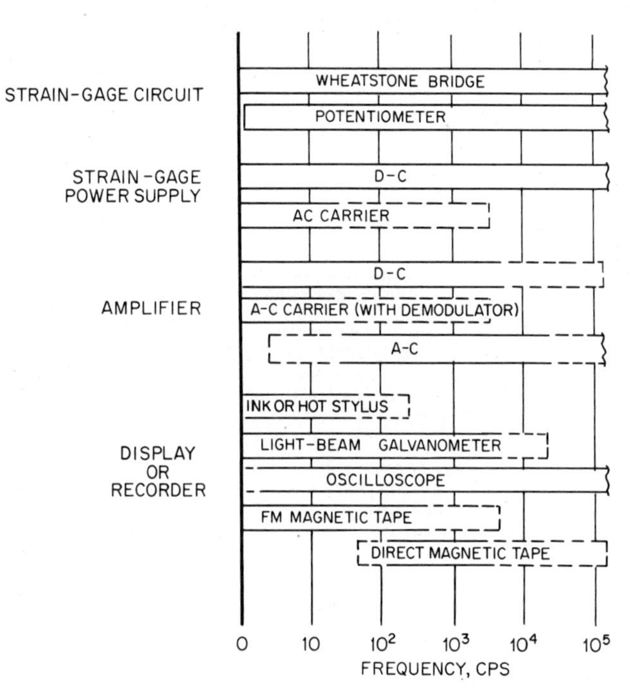

FIG. 17.20. Chart showing the frequency range of components in a strain-gage measurement system. The solid portion of the blocks indicates typical frequency ranges for such components; the dashed portions of blocks indicate extended ranges, possible with careful choice of equipment.

Higher frequency strain recording instruments (up to 5,000 cps) employ mirror galvanometers which record on photosensitive paper with an ultraviolet beam. Generally these galvanometers require a d-c amplifier with quite high current output capacity.

In selecting instrumentation for dynamic strain measurements, careful consideration should be given to the various components in the measurement system so as to obtain a system capable of operating over the desired frequency range. The chart shown in Fig. 17.20 indicates typical frequency ranges of many types of components.

REFERENCES

1. Wu, C. T.: "Transverse Sensitivity of Bonded Strain Gages," *Experimental Mechanics, J. Soc. for Exptl. Stress Anal.*, November, 1962.
2. Hines, F. H., and L. J. Weymouth: Practical Aspects of Temperature Effects on Resistance Strain Gages, in M. Dean, III and R. D. Douglas (eds.), "Semiconductor and Conventional Strain Gages," Academic Press, Inc., New York, 1962.

3. Vigness, I.: "Magnetostrictive Effects in Wire Strain Gages," *Proc. Soc. Exptl. Stress Anal.*, **14**(2):139 (1957).
4. Milligan, R. V.: The Gross Hydrostatic-Pressure Effects as Related to Foil and Wire Strain Gages, "Experimental Mechanics," pp. 67ff, Society for Experimental Stress Analysis, Westport, Conn., February, 1967.
5. Vulliet, P. (ed.): *Proc. Western Regional Strain Gage Comm.*—1968 Spring Meeting, Marina Del Rey, Calif., Society for Experimental Stress Analysis, Westport, Conn.
6. Perry, C. C., and H. R. Lissner: "The Strain Gage Primer," pp. 117ff., McGraw-Hill Book Company, Inc., New York, 1955.
7. Bickle, L. W.: The Use of Strain Gages for the Measurement of Propagating Strain Waves, Proc. Tech. Comm. Strain Gages, Oct. 23, 1970, Society for Experimental Stress Analysis, Westport, Conn.
8. Norton, H. N.: "Handbook of Transducers for Electronic Measuring Systems," pp. 42ff., Prentice-Hall, Inc., Englewood Cliffs, N.J., 1969.
9. Motsinger, R. N.: Flexural Devices in Measurement Systems, in P. K. Stein (ed.), "Measurement Engineering," vol. 1 chap. 11, Stein Engineering Services, Phoenix, Ariz., 1964.
10. Cleveland, A. W.: *J. Soc. Auto. Eng.*, Vol. **59**:34ff., (1951).
11. Jasper, N. H.: *Proc. Soc. Exptl. Stress Anal.*, **8**(2):83 (1951).
12. Perry, C. C.: "Design Considerations for Diaphragm Pressure Transducers," *Tech. Note* TN-129, Micro-Measurements Division, Romulus, Mich., 1974.
13. Dove, R. C., and P. H. Adams: "Experimental Stress Analysis and Motion Measurement," chap. 9, Charles E. Merrill Publishing Co., Columbus, Ohio, 1964.
14. "Advances in Test Measurement," *Proc. 5th Ann. ISA Test Measurement Symp.*, vol. 5, October, 1968.

18

CALIBRATION OF PICKUPS

John D. Ramboz
National Bureau of Standards

INTRODUCTION

This chapter describes various methods of calibrating shock and vibration pickups. Each method is subject to inherent mechanical and electrical limitations such as the frequency and amplitude of vibration, electrical noise, and the mass and size of the pickup. For these reasons, it is often necessary to apply several calibration methods in order to determine the sensitivity of a single vibration pickup. A calibration method should be selected which provides the required information in the range of frequencies and amplitude levels in which the pickup will operate.

The chapter is divided into four major parts: calibration methods suitable for low frequency and low acceleration; calibration of pickups by direct measurement; high-acceleration calibration methods; and calibration by comparison methods.

Only the fundamental basis for the given calibration methods is generally discussed, with the frequency, amplitude, and uncertainty ranges indicated. Supplemental calibrations of transverse sensitivity ratio, temperature response, and other spurious sensitivities are not discussed.[1] The reader is guided to the references, some of which discuss these tests in detail.

An appreciation of the difficulty of accurate calibration of vibration pickups is obtained when one considers that vibration at an acceleration amplitude of $5g$ at 1,000 cps corresponds to a peak-to-peak displacement of only 25 micrometers. If this calibration is to be done by measuring the peak-to-peak displacement and if an accuracy of greater than 1 per cent is to be attained, the accuracy of measurement of the peak-to-peak displacement must be to somewhat better than 25 nanometers. Correspondingly, when considering a shock pulse with a peak value of acceleration of $5{,}000g$ and a duration of 100 microseconds, extreme accuracy of time measurement is required, which may present a problem.

SENSITIVITY, CALIBRATION FACTOR

The "sensitivity factor" of a linear vibration pickup is defined[2] as the ratio of electrical output to the mechanical input applied to a specified axis. The sensitivity factor is specified at a discrete frequency. In general, the sensitivity factor includes both amplitude and phase information and is a complex quantity. When the phase information is given, it is common practice to express the sensitivity factor in polar notation, i.e., amplitude and the phase lag in degrees.

The sensitivity factor for all pickups varies with frequency. A typical response for a piezoelectric accelerometer is shown in Fig. 18.1, and the phase lag is shown in Fig. 18.2. If the sensitivity factor is substantially independent of frequency over a specified range of frequencies, it is referred to the calibration factor[2] for that range.

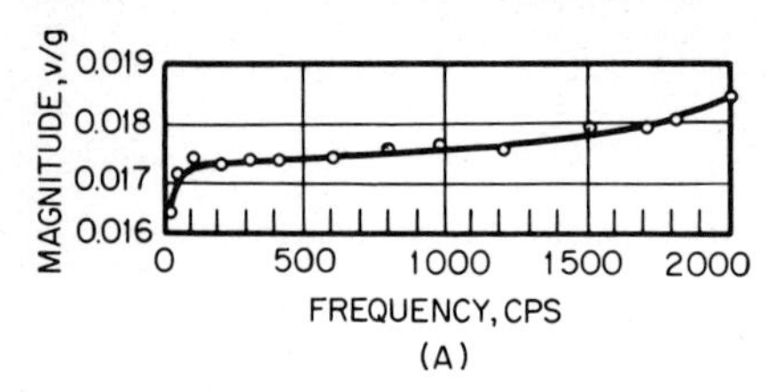

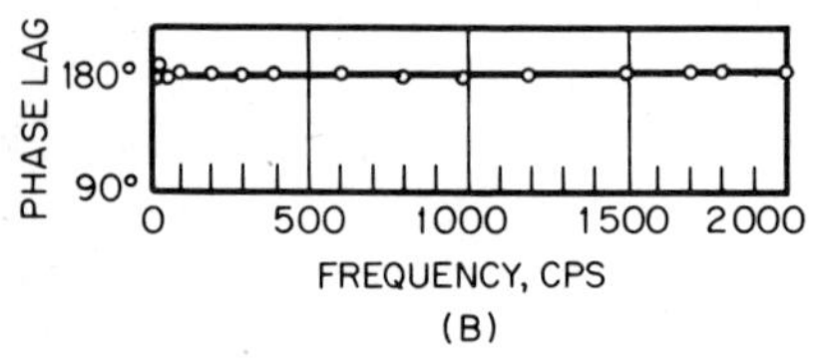

FIG. 18.1. Sensitivity factor of a vibration pickup (piezoelectric accelerometer) calibrated using a reference which had been established by the reciprocity calibration method as described in this chapter. (*A*) The magnitude is plotted as a function of frequency. (*B*) Phase lag versus frequency. (*After S. Levy and R. R. Bouche.*[22])

There are many reasons why the sensitivity varies with frequency. Most important among these is the effect of resonance of the seismic element, resulting in a rising response as the vibration period approaches the natural period of the pickup. Other factors which cause the sensitivity to vary with frequency are resonances in the pickup case and external electrical loading resulting from the limitations of auxiliary circuits. In some instances, damping fluid may develop springlike properties at high frequencies or the fluid may experience a change in temperature, thereby altering the damping rate. Joints which behave nearly elastically at low frequencies can develop appreciable damping at higher frequencies. Generally these effects upon frequency response occur simultaneously, adding to the variation of the sensitivity factor.

The mechanical input to the vibration pickup may be displacement, velocity, or acceleration depending on the type of pickup. (*Jerk*, the time deviative of acceleration, is seldom used.) Displacement is usually expressed as single- or double-amplitude (peak or peak-to-peak) values. Velocity is usually expressed as peak, root-mean-square (rms), or average values. Acceleration is usually expressed as peak or rms values.

The electrical output of the vibration pickup may be expressed as peak, root-mean-square, or average value.

The sensitivity factor (or calibration factor) is commonly given in similarly expressed units, i.e., the numerator and denominator are both peak, or both rms, etc., units. For example, a typical sensitivity factor for an accelerometer may be expressed in units of millivolts per unit acceleration, or simply mV/g. The factor of $\sqrt{2}$ may have to be used to express the sensitivity factor as desired.

For some special applications it may be desirable to express the sensitivity factor in mixed values, such as rms millivolts per peak unit acceleration, or mV rms/g peak. In such instances, care should be exercised in stating the "type" of units used in expressing the sensitivity factor.

For primary vibration standards, quite often a "system" calibration is performed. The system may consist of an accelerometer, cable assembly, and associated amplifier. The output of the amplifier is considered as the system sensitivity factor. This is done to achieve a lower measurement uncertainty for the standard. Thereafter, care must be taken always to use the same components together as a system.

AMPLITUDE RANGE

In the recommended frequency range of operation of a pickup, the sensitivity factor is usually nearly constant for a wide range of amplitudes. At amplitudes above this range, the factor may vary because of nonlinear performance of the mechanical or electrical elements, amplitude limitations set by mechanical stops, or other reasons. At amplitudes below this range, the sensitivity factor may vary because of the limited resolving power of the sensing element, electrical noise, or sticking where rubbing friction is present.

PHASE LAG

The phase lag for a vibration pickup is a measure of the time lag between the mechanical input and the electrical output. A distortion-free phase condition exists if the phase shift is 0 or 180°, or if it is proportional to the frequency. As indicated in Chap. 12, accelerometers with damping about 0.65 of critical have a phase shift which is almost proportional to frequency over their usable range. Therefore pickups that are intended for use in measuring either transient disturbances or vibrations containing several frequency components simultaneously should be calibrated to determine that the phase-angle distortion is acceptably small.

When a mechanical vibration containing several frequency components is applied to a vibration-measuring instrument and it is found that the waveform of the electrical output is undistorted from the vibration input waveform, phase distortion does not exist. Phase-angle distortion of the pickup may be compensated by associated electric circuits.

CALIBRATION METHODS SUITABLE FOR LOW FREQUENCY AND LOW ACCELERATION

This section describes several methods of calibrating vibration pickups in a frequency range from 0 to 2,000 cps, in the acceleration amplitude range from a fraction of $1g$ up to approximately $100g$. The methods make use of the earth's gravitational field, centrifuges, mechanical exciters, and electrodynamic exciters.

EARTH'S GRAVITATIONAL FIELD

The earth's gravitational field provides a convenient means of applying small constant acceleration levels to a pickup. It is particularly useful in calibrating accelerometers whose frequency range extends to 0 cps. A $2g$ change in acceleration may be obtained by first orienting the accelerometer with the positive direction of its sensing axis up, and then rotating the accelerometer through 180° so that the positive direction is down.

Increments of acceleration within the $1g$ range can be applied by means of a tilting-support calibrator with an accuracy better than $0.004g$ near the horizontal and $0.0003g$ near the vertical.[2] The pickup to be calibrated is fastened as shown in Fig. 18.2 to a platform at the end of an arm. The arm may be set at any angle ϕ between 0 and 180° relative to the vertical. It is furnished with a pointer to indicate the angle ϕ. Care should be taken to level the base when the arm is set at $\phi = 0°$. Positioning of the arm angle ϕ to $\pm 0.2°$ or less is possible with an accurately divided circle. The component of acceleration along the arm is given by

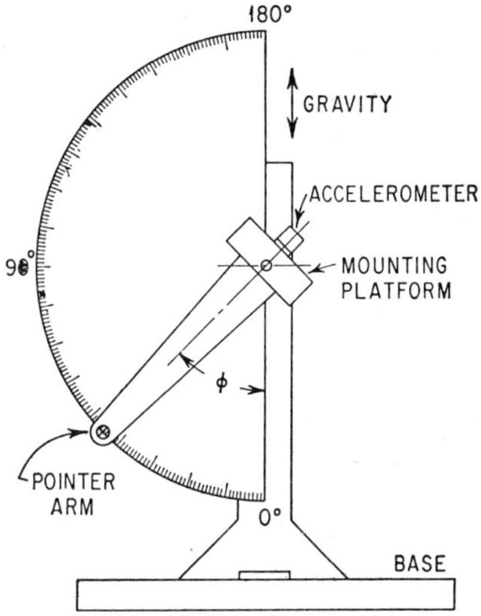

$$a = g \cos \phi \tag{18.1}$$

FIG. 18.2. Tilting-support calibrator used to supply incremental static accelerations of $1g$ or less to accelerometers which have zero frequency response such as strain-gage accelerometers. (*After ANSI Standard S 2.2-1959.*[2])

The pickup is subjected to a component of acceleration at right angles to its sensitive direction equal to

$$a_t = g \sin \phi \qquad \text{(transverse)} \tag{18.2}$$

Because of the transverse component, this method is not recommended for calibration of pickups for which the transverse sensitivity is significant.

The calibration factor is obtained by plotting the response of the accelerometer as a function of the acceleration a, given by Eq. (18.1), for successive values of ϕ and determining the slope of the straight line fitted through the points.

PHYSICAL PENDULUM CALIBRATOR

The physical pendulum calibrator is a simple device for imparting to a pickup a transient pulse of acceleration of as great as $10g$ with a duration of the order of a second.[2,3] It imparts a transverse and an angular acceleration as well as the acceleration inward from its center of rotation. Therefore it should be used with caution in calibrating pickups sensitive to these extraneous motions.

The physical pendulum calibrator in Fig. 18.3 consists of a beam and platform with its center-of-gravity pivoted so as to swing about a horizontal axis. The pickup to be calibrated is attached to the platform, and is carefully aligned so that the pickup axis of sensitivity is along a radius of the beam. Electrical connections can be brought from the pickup along the beam bridging to the support frame. A well-constructed pendulum has negligible damping. If the pendulum is released to swing after being raised to an angle ϕ above the vertical, the increment a in acceleration along the radius vector, from the time of release to the bottom of the swing, is

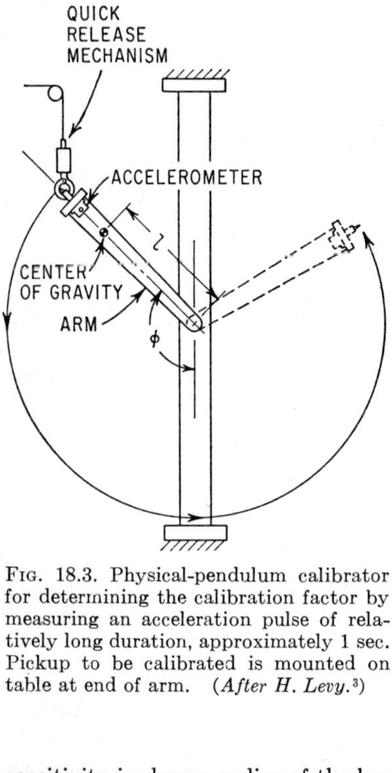

FIG. 18.3. Physical-pendulum calibrator for determining the calibration factor by measuring an acceleration pulse of relatively long duration, approximately 1 sec. Pickup to be calibrated is mounted on table at end of arm. (*After H. Levy.*[3])

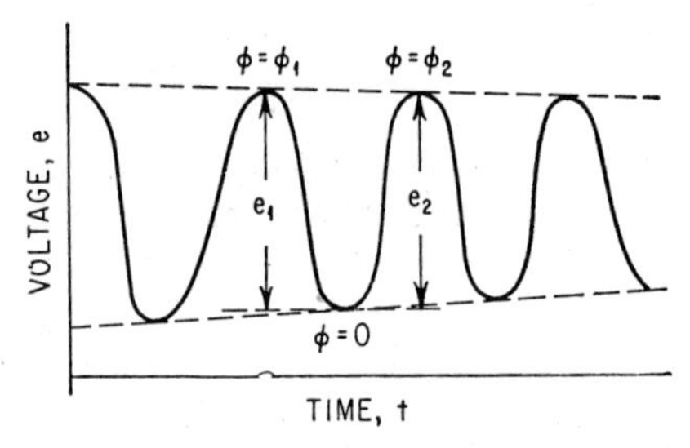

FIG. 18.4. Typical oscillogram of an accelerometer response to the physical-pendulum calibrator in Fig. 18.3. The release angle is ϕ_1, and the angle to which the pendulum rises after passing the bottom of its swing is ϕ_2. The average angle $(\phi_1 + \phi_2)/2$ corresponds to the average response $(e_1 + e_2)/2$ (*After H. Levy.*[3])

$$a = \left(g + \frac{8\pi^2 l}{\tau^2}\right)(1 - \cos\phi) \tag{18.3}$$

where g is the acceleration of gravity, l is the distance from the center-of-gravity of the mass element to the pivot, τ is the period of the pendulum for small amplitudes, and ϕ is equal to the average of the release angle and the angle to which the pendulum rises after passing the bottom of its swing.

The calibration factor normally is determined by plotting the peak response

$$R = \tfrac{1}{2}(e_1 + e_2)$$

of the pickup as a function of the acceleration a as given by Eq. (18.3) for successive values of ϕ, and determining the slope of the straight line fitted through the data. The error is ordinarily below 1 per cent, where care is exercised. A typical oscillogram record of an accelerometer response as a function of time, obtained from a calibration using the physical pendulum, is shown in Fig. 18.4; a block diagram of an instrumentation recording system used in this method is shown in Fig. 18.5.

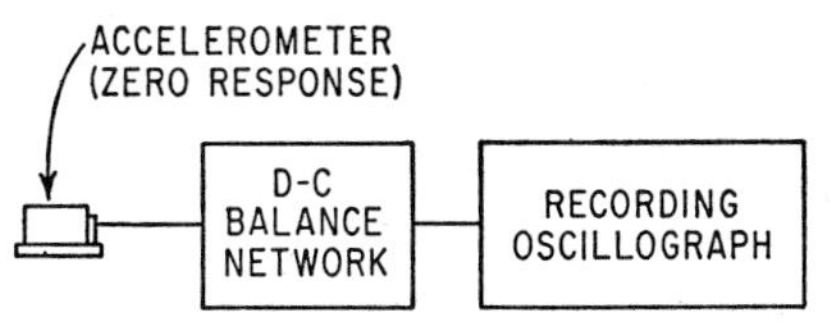

FIG. 18.5. Block diagram of the instrumentation recording system used to obtain the oscillogram for Fig. 18.4. (*After H. Levy.*[3])

CENTRIFUGE

A centrifuge provides a convenient means of applying constant acceleration to a pickup. Centrifuges can be obtained readily for acceleration levels up to $100g$ and have been made for use at much higher values. They are particularly useful in calibrating rectilinear accelerometers whose frequency range extends down to 0 cps and whose sensitivity to rotation is negligible. Centrifuges are mounted so as to rotate about a vertical axis. Cable leads from the pickup, as well as power leads, usually are brought to the table of the centrifuge through specially selected "low-noise" slip rings and brushes.

To perform a calibration, the accelerometer is mounted on the centrifuge with its axis of sensitivity carefully aligned along a radius of the circle of rotation. If the centrifuge rotates with an angular velocity of ω radians/sec, the acceleration acting on the pickup is

$$a = \omega^2 r \tag{18.4}$$

where r is the distance from the center-of-gravity of the mass element of the pickup to the axis of rotation. Where the exact location of the center-of-gravity of the mass in the pickup is not known, mount the pickup with its positive sensing axis first outward and then inward; then compare the average response with the average acceleration acting on the pickup computed from Eq. (18.4), where r is taken as the mean of the radii to a given point on the pickup base. Care should be taken to mount the pickup at a distance from the axis of rotation such that the deflection of the mass element is negligible in the determination of r.

The calibration factor usually is determined by plotting the output e of the pickup as a function of the acceleration a given by Eq. (18.4) for successive values of ω and determining the slope of the straight line fitted through the data. Error results primarily in measuring the angular velocity accurately and in holding the angular velocity constant during the time required to take a reading.

ROTATING-TABLE METHOD (CENTRIFUGAL FIELD METHOD)

In the rotating-table method of calibration, the accelerometer is rotated at a uniform angular rate about a horizontal axis.[4-6] The experimental arrangement is shown in Fig. 18.6. The pickup is mounted so as to rotate about a horizontal axis; the sensitive axis of the accelerometer rotates in a vertical plane. Thus it is possible to apply very low-frequency sinusoidal motion in the region where sinusoidal linear translation of large amplitude would otherwise be required to give useful accelerations. As a result, a sinusoidal acceleration, having a peak amplitude of $1g$ at the rotation frequency, is superposed on the centrifugal acceleration. By this method it is possible to obtain the response of the accelerometer under both static and dynamic conditions in the same test

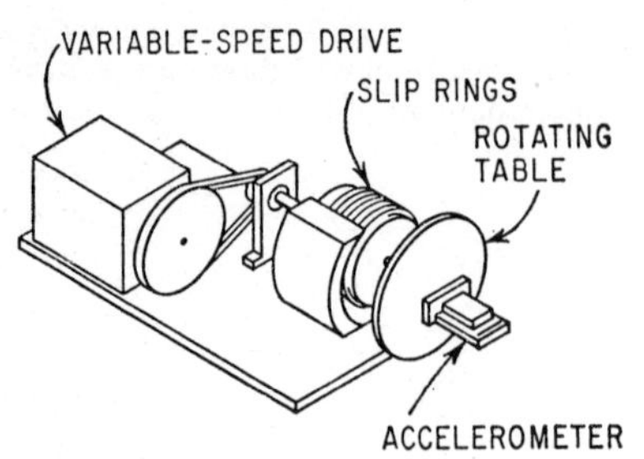

FIG. 18.6. Rotating-table calibrator which is used for both static and dynamic calibration with a peak acceleration amplitude of $1g$. Table rotates in a vertical plane. Pickup mounted on table is alternately subjected to plus and minus $1g$ as table rotates. (*After W. A. Wildback and R. O. Smith.*[5])

setup. The rotating-table method not only provides excitation at very low frequencies but also provides an accurate constant level of acceleration. The basic consideration in applying this method is that the vibration pickup be sufficiently sensitive to have adequate output in the range of acceleration up to $1g$. The threshold sensitivity and the repeatability of the pickup under test must be such that a periodic excitation of $\pm 1g$ or less produces a response that is representative of the instrument's behavior. Excitation at less than $\pm 1g$ can be provided by inclining the axis of rotation. The useful frequency range over which this calibration technique applies is from about 0.5 to 45 cps.

Accurate calibration using electrodynamic exciters is generally difficult below frequencies of about 10 cps; this rotating table method can provide good results in this range within the limitations of $\pm 1g$.

Bearing noise and dynamic unbalance limits the upper frequency range. Improved performance from that obtained from the calibrator shown in Fig. 18.6 can be obtained by incorporating air bearings and a magnetic clutch-drive arrangement;[6] the position of the pickup is adjusted by a fine lead screw arrangement and a series of adjustable counterbalance masses are arranged about the table's circumference.

The chief limitations of the rotating-table method of dynamic calibration are (1) the rotation speed of a table is limited; (2) the acceleration level is limited to $1g$; (3) the response to the centrifugal acceleration tends to become infinite at the natural frequency of the pickup. Measuring the output of the pickup must be done with care. At frequencies below about 5 cps, errors introduced by voltmeters become a limiting factor in the overall measurement uncertainty. Overall pickup calibration uncertainties of ± 1 per cent in the frequency range from 2 to 8 cps are attainable, and ± 0.5 per cent in the frequency range from 8 to 45 cps can be maintained.

MECHANICAL EXCITERS

Rectilinear motion can be produced by mechanical exciter systems of the type described under *Direct-drive Mechanical Vibration Machine* in Chap. 25. At low frequencies, a mechanical drive consisting of an electric motor which turns a cam or drives a linkage can be used to produce nearly sinusoidal motion. Rotating eccentric weights also can be used to produce sinusoidal motion of a spring-supported table.[7] This equipment is described under *Reaction-type Mechanical Vibration Machine* in Chap. 25. In both of these machines, the amplitude of the table is nearly independent of frequency. The usable frequency range is from a few cycles per second to about 100 cps. Although the displacement is essentially sinusoidal, appreciable distortion of acceleration from a pure sinusoid usually is present as a result of bearing tolerances and the geometry of linkage systems.

ELECTRODYNAMIC CALIBRATOR

The rectilinear electrodynamic exciter (see Chap. 25) is widely used in calibration equipment. It provides a means for applying sinusoidal motion over a range of displacement amplitudes from about 2.5 cm at low frequencies to about the amplitude which is equivalent to an acceleration amplitude of approximately $50g$ at high frequencies. Over a frequency range from about 5 to 10,000 cps electrodynamic calibrators are available which provide relatively undistorted sinusoidal waveform.[2, 8–11] Ordinarily, several exciters are needed to cover this broad frequency range since it is necessary to avoid resonant frequencies of the exciter when using it for calibration of pickups. Freedom from harmonics in the motion is largely dependent on the pure sinusoidal waveform of the driving current. In some cases ripple in the direct current used to energize the field coil of the exciter introduces small amplitude spurious motions by transformer coupling with the armature. (The latter is evident only when the armature circuit is closed.)

The construction of a typical electrodynamic vibration pickup calibrator is illustrated in Fig. 18.7. The pickup to be calibrated is attached to the calibrator table. The armature and the table are as rigid as practical to minimize table bending and to increase the first axial resonance to as high a frequency as possible. Flexures or air

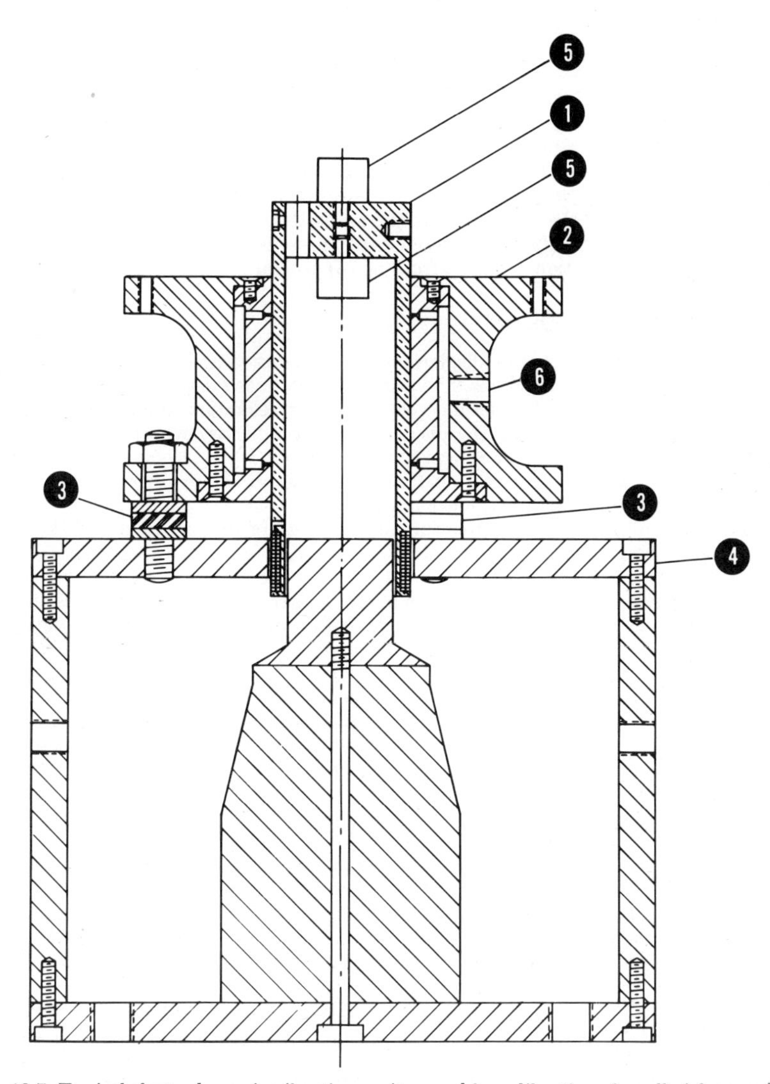

FIG. 18.7. Typical electrodynamic vibration exciter used for calibration of small pickups. The major components are (1) ceramic moving element, (2) air bearing, (3) vibration absorbers, (4) permanent magnet, (5) vibration pickups, and (6) air inlet. (*After T. Dimoff.*[8])

bearings usually are used to constrain undesirable lateral motions. Sinusoidal motion is obtained when a sinusoidal current is passed through the drive coil.

Frequently, calibrators of this type are mounted on trunnions to allow use in either vertical or horizontal directions. Magnetic shielding or bucking magnetic coils can be used to reduce the stray magnetic field at the mounting table to nearly zero. This is important when calibrating pickups sensitive to magnetic fields, e.g., electrodynamic velocity-type pickups. To achieve isolation from building vibration, a calibrator may be mounted on a heavy rigid mass which in turn is supported by soft springs.

Calibration exciters often have a built-in accelerometer or velocity coil intended as a reference standard. Such internal standards are calibrated by methods described in this chapter.

CALIBRATION OF PICKUPS BY DIRECT MEASUREMENT

The pickup to be calibrated is attached rigidly to the calibrator table with its center-of-gravity centered on the table. Pickup lead wires are secured tightly to avoid whipping. Particular care must be taken in securing cables from a pickup that is sensitive to distortion of its case. The pickup output voltage then is fed into suitable auxiliary instrumentation. The oscillator frequency is adjusted to desired values throughout the frequency range of the pickup, while the amplifier gain is set to give desired amplitudes. Waveform normally is monitored with an oscilloscope or measured by a wave analyzer to avoid distortion resulting from nonsinusoidal motion of the armature.

DISPLACEMENT MEASUREMENT

The measurement of dynamic displacement is a common and valuable technique for the calibration of vibration pickups.[2] A variety of methods exist, and refined techniques have proven reliable. Both dynamic velocity and acceleration amplitudes can be accurately calculated by knowing the displacement time-history waveform. Velocity is the first time derivative of displacement and acceleration is the second time derivative of displacement. These fundamental relationships are expressed in Chap. 1 [see Eqs. (1.1) through (1.4)] for sinusoidal rectilinear motion.

The calibration procedure requires the measurement of displacement and the output of the pickup being calibrated. The pickup sensitivity factor can then be obtained by calculating the ratio of the pickup output divided by the motion input. Using a factor of ω or ω^2 to obtain velocity or acceleration parameters from a displacement measurement is only valid for sinusoidal motion. If harmonic distortion is present, errors will result. The magnitude of these errors will depend upon the order of the harmonic and the phase relationships. Experience has shown that if the total harmonic distortion is less than a few per cent in the acceleration waveform, calibration errors will be small and generally of no concern.

Some of the displacement measuring methods discussed below yield double-amplitude displacements (peak-to-peak), while others give peak. Exercise care in determining the units of sensitivity factor, as discussed earlier in this chapter.

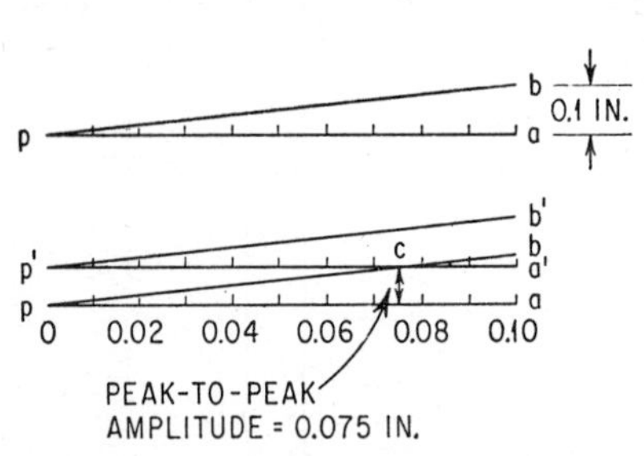

FIG. 18.8. Method of displacement measurement of relatively large amplitudes by the vibrating-wedge technique. A card is attached to the body whose motion is to be measured. The card is oriented so that at rest the line *p-a* is perpendicular to the expected motion as illustrated in the upper figure. When the body vibrates, two angles are seen. Apparent intercept *c* permits direct reading of peak-to-peak amplitude *pp'*. (*ANSI Standard S 2.2-1959.*[2])

VIBRATING WEDGE. One method of measuring relatively large displacement amplitudes is shown in Fig. 18.8.[2] A card is attached to the body whose vibratory motion is to be measured. This card contains two lines that intersect to form an angle as shown in the upper portion of the illustration. With the body at rest, the card is oriented so that a straight line *p–a* is perpendicular to the expected motion. Line *p–b* intersects *p–a* and forms a small angle with it. A scale is marked along *p–a* at each tenth of its length and the length of *a–b* is set at 0.1 in. When vibrating, two angles *apb* and *a′p′b′* are seen, corresponding to the extremes of motion, and an apparent intercept *c* appears. In the example the intercept *c* falls 0.75 of the distance from *p* to *a* so that the peak-to-peak displacement amplitude *p–p′* is read as 0.075 in. (The use of this method with the triangular area *apb* filled in to form a wedge results in a penumbra error, depending on the observer, and would require calibration.)

When calibrating displacement pickups, the peak-to-peak amplitude $2x_0$ is measured directly. Half this value is generally used in calculating the pickup sensitivity factor.

Velocity can be calculated from the equation

$$\dot{x} = 3.1416(2x_0)f \tag{18.5}$$

where the term $2x_0$ in parentheses is the peak-to-peak displacement. The velocity assumes the units of the displacement parameter. For example, if the displacement x_0 is in meters and the frequency f is in cycles per second, then the velocity $\dot{x}$ is in units of peak meters per second.

Acceleration can be calculated from the equation

$$\alpha = 19.739(2x_0)f^2 \tag{18.6}$$

where the term $2x_0$ in parentheses is the peak-to-peak displacement measured. The acceleration α assumes the peak units of the displacement parameter per unit time squared. For example, if the displacement is measured in meters and the frequency in cycles per second, the acceleration is in units of peak meters per second squared. By dividing Eq. (18.6) by the standard acceleration of gravity (9.80665 m/sec² or 386.089 in./sec²), the peak acceleration can be expressed as

$$\alpha = 2.0128(2x_0)f^2 \qquad g(x_0 \text{ in meters}) \tag{18.7a}$$

$$\alpha = 0.051125(2x_0)f^2 \qquad g(x_0 \text{ in inches}) \tag{18.7b}$$

MICROSCOPE-STREAK METHOD. The displacement amplitude of the input vibration can be measured optically by observing a spot of high-intensity reflected light. A good source is an individual bright reflection from a field of emery cloth cemented to the edge of the calibrator table. Peak-to-peak displacement amplitudes of 10 mm can be measured with an accuracy of ±0.25 mm using conventional microscopes. With a high-quality microscope of high magnification and equipped with an optical micrometer, a peak-to-peak displacement of the order of 0.001 mm can be measured if care is taken to isolate the microscope from the vibration exciter. Stroboscopic illumination can be used to slow or stop the image. Use of Eqs. (18.5) through (18.7) will yield velocity and acceleration.

RONCHI RULING METHOD. Electronic sensing of vibration displacement can be achieved by the use of two matched Ronchi rulings.[12, 13] One ruling is rigidly mounted on the vibration exciter table and the second ruling is attached onto a seismically stable and adjustable mount. It is important that vibration from the vibration exciter does not cause vibration on the fixed ruling. Figure 18.9 shows a block diagram of a typical setup.

A laser may be used to illuminate the rulings, although collimated light may be used. When properly aligned, as the rulings are moved with respect to one another due to applied vibration, a first-order Moire fringe causes light pulses to fall upon the photodetector. These pulses may be counted with an electronic counter. An output pulse will be generated for every displacement equal to the width of one line pair. (A line pair is one transparent line and one opaque line on a ruling.) The number of output pulses per vibration cycle is directly proportional to the vibration-displacement amplitude. If we assume sinusoidal motion, the peak acceleration α can be calculated from

$$\alpha = \frac{\pi^2 f^2 \eta}{ng} \qquad g \tag{18.8}$$

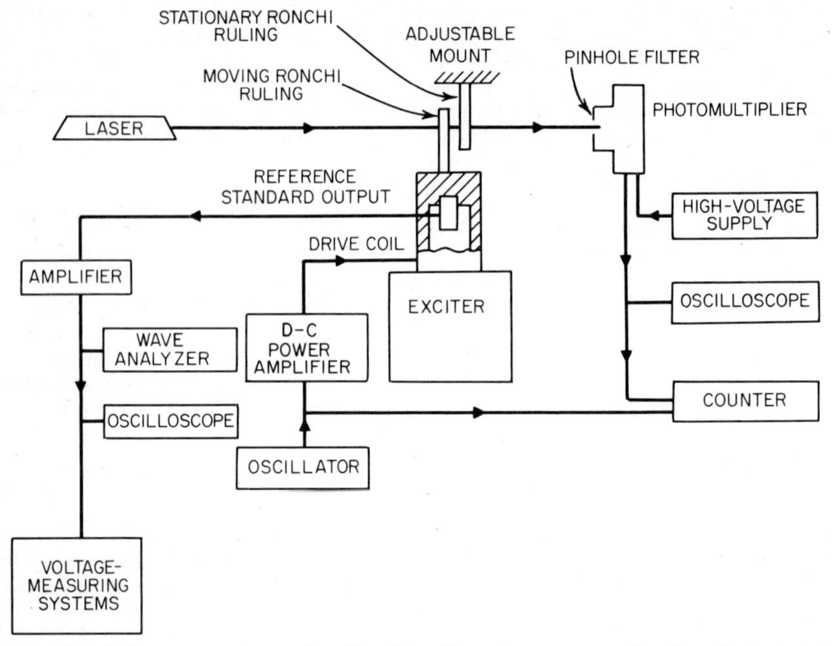

FIG. 18.9. Laboratory setup using Ronchi rulings to measure vibration displacement. Displacement is displayed on the gated counter. (*After R. S. Koyanagi.*[13])

where f is the vibration frequency in cycles per second, η is the number of light pulses per vibration cycle, n is the Ronchi ruling constant in line pairs per unit length, and g is the acceleration of gravity.

It is crucial that the rulings be precisely aligned with each other. The line pairs on the stationary ruling must be parallel to the line pairs on the moving ruling. Additionally, the ruled surfaces must be parallel. Micrometer adjustments for rotation and translation of the stationary ruling are useful.

Ronchi rulings are commercially available with 200 line pairs per millimeter. By using such rulings, accelerometer calibration can be done to frequencies as high as several hundred cycles per second, depending on the ratings of the vibration exciter. The measurement errors are chiefly due to displacement resolution, voltage measurement, and nonsinusoidal motion. Overall uncertainties can generally be maintained less than ± 1 per cent.

OPTICAL INTERFEROMETRIC FRINGE COUNTING METHOD. An optical interferometer can be used to measure vibration displacement. The Michelson and Fizeau interferometers are the popular configuration most often used. These are discussed in Chap. 14. A modified Michelson interferometer is shown in Fig. 18.10.[13, 14] A corner cube retroreflector is mounted on the vibration-exciter table. A helium-neon laser is used as an illumination source. The photodiode and its amplifier must have sufficient bandwidth (as high as 50 mc) to accommodate the Doppler frequency shift due to the high vibration peak velocities. An electrical pulse is generated by the photodiode for each optical fringe passing. The vibration-displacement amplitude is directly proportional to the number of fringes counted. The vibration peak acceleration can be calculated by

$$\alpha = \frac{\lambda \eta \pi^2 f^2}{2g} \qquad g \tag{18.9}$$

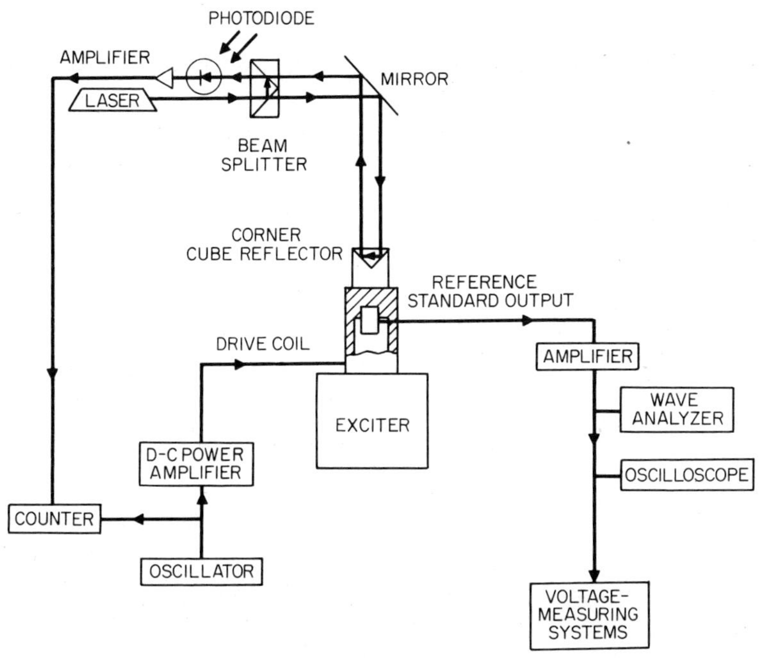

FIG. 18.10. Typical laboratory setup for interferometric fringe counting of vibration displacement. Displacement is displayed by the gated counter. (*After R. S. Koyanagi.*[13])

where λ = wavelength of the illumination light source
η = number of fringes (output pulses of the photodiode) per vibration cycle
f = vibration frequency, cps
g = acceleration of gravity

When a helium-neon laser is used (λ = 632.8 nm), Eq. (18.9) becomes

$$\alpha = 3.1843\eta f^2 \times 10^{-7} g \tag{18.10}$$

Interferometric fringe counting is useful for vibration-displacement measurement in the lower frequency ranges, perhaps to several hundred cycles per second, depending on the rating of the vibration exciter being used.[15] Very good accuracies can be achieved because a large number of fringes can be counted accurately and the wavelength of light is well known.

OPTICAL INTERFEROMETRIC FRINGE DISAPPEARANCE METHOD. The phenomenon of interference band disappearance in an optical interferometer can be used as a precision means of determining amplitude of motion. Figure 18.11 shows the principle of operation of the optical interferometer employed in this technique.[16] One of the mirrors D, Fig. 18.11A, is attached to the mounting plate of the calibrator. It is possible to obtain interference fringe patterns which are composed of straight lines, or bands, as shown in Fig. 18.11B. The intensity I of light for 50 per cent transmission through the silvered surface when K is the illuminating intensity is given by

$$I = K\left(1 + \cos\frac{2\pi x}{h}\right) \tag{18.11}$$

where h is the separation of bands and x is the lateral displacement from a point midway between two bands.

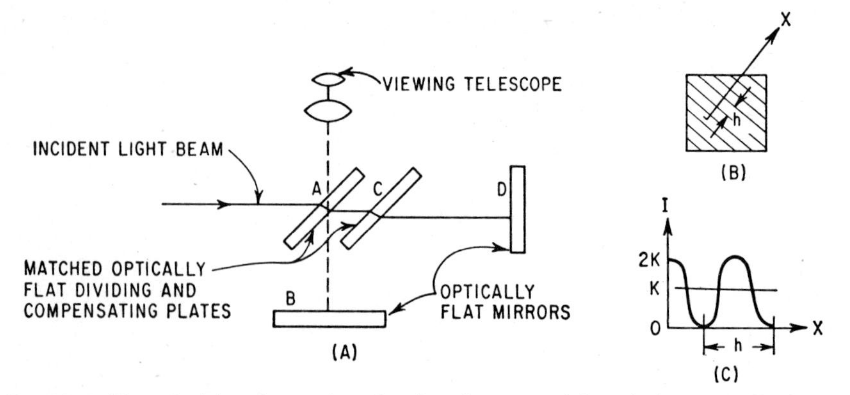

FIG. 18.11. The principles of operation of an interferometer; (*A*) optical system, (*B*) the observed interference pattern, and (*C*) the light intensity along the *X* axis. The system is used in the fringe disappearance technique for measuring displacement.

A movement of one of the mirrors *D* in the direction of the light path by an amount r causes a change in the length of the light path of the amount $2r$ and a shift in the position of the pattern by the amount x'. When $2r$ is equal to the wavelength λ of the light used, the shift x' equals h. When $2r$ has other values, the shift is given by

$$x' = \frac{2rh}{\lambda} \tag{18.12}$$

When the mirror *D* vibrates sinusoidally with a frequency f and a peak-to-peak displacement amplitude of $2x_0$, its position is given by

$$r = x_0 \sin \omega t \tag{18.13}$$

and therefore

$$x' = \frac{2x_0 h}{\lambda} \sin \omega t \tag{18.14}$$

where $\omega = 2\pi f$.

Table 18.1. Optical Interferometric Displacements, J_0 Nulls

Values of peak-to-peak displacement amplitude d for which fringe pattern vanishes for a HeNe laser source (λ = 632.8 nm).

Root number	Displacement, d	
1	9.536×10^{-6} in.	242.2 nm
2	21.887	555.93
3	34.313	871.57

The time average of the light intensity at position x when t is zero, as a result of the varying band shift x', is given by

$$I = \frac{K\omega}{2\pi} \int_0^{2\pi/\omega} \left[1 + \cos \frac{2\pi(x - x')}{h} \right] dt$$

Substituting for x' its value as given by Eq. (18.14) and evaluating the integral,

$$I = K\left[1 + J_0\left(\frac{2\pi d}{\lambda}\right)\cos\left(\frac{2\pi x}{h}\right)\right] \tag{18.15}$$

where J_0 is a Bessel function of zero order.

For certain values of the argument, the Bessel function of zero order is zero; then the fringe pattern disappears and only the illuminating intensity K is present. Table 18.1 shows the calculated values of displacement for fringe pattern disappearance for a helium-neon laser source ($\lambda = 632.8$ nm). Additional disappearances occur at half-wavelength increments in d, that is, 316.4 nm (12.46×10^{-6} in.).

The fringe disappearance may be sensed by an optoelectronic method.[14, 17] A portion of the fringe pattern is focused onto a photodetector whose output is fed into a narrow bandpass filter. (A wave analyzer is useful as a tuned voltmeter.) When the filter is tuned to the vibration frequency, the J_1 Bessel function nulls can be sensed. The vibration amplitude is slowly increased until the Bessel function zero is observed as a null on the analyzer output. The photodetector output shown in Eq. (18.16) applies, and values of displacement are those shown in Table 18.2.

$$I \propto J_1 \frac{4\pi d}{\lambda} \tag{18.16}$$

Table 18.2. Optical Interferometric Displacements, J_1 Nulls

Values of peak-to-peak displacement amplitude d for which fringe pattern vanishes for a HeNe laser source ($\lambda = 632.8$ nm) for J_1 Bessel function nulls.

Root number	Displacement, d	
1	385.9 nm	15.194×10^{-6} in.
2	706.6	27.818
3	1024.6	40.336
4	1341.9	52.830
5	1658.8	65.308
6	1975.5	77.777
7	2292.2	90.244
8	2608.8	102.709
9	2925.4	115.173
10	3241.8	127.630

When using interferometric setups, background displacement noise must be minimized. Fringe counting arrangements are not as sensitive to these noises as are the fringe disappearance setups. For the latter, it is usually necessary to provide good seismic isolation from typical floors and to work on rigid massive surfaces. Even acoustic noises can sometimes be a problem.

The use of piezoelectric exciters is common for high-frequency calibration of accelerometers.[18] They provide good motion, but more important to the interferometric setup, they are structurally stiff at the lower frequencies where displacement noise is bothersome. When electrodynamic exciters are used with fringe disappearance methods, it is generally necessary to stiffen the armature suspensions to reduce the background displacement noise.[14] This can usually be done with rubber pieces which are selected to be dynamically stiff at low frequencies and soft at the vibration frequencies where measurements are being made. These pieces can usually be wedged around the existing flexure or armature to block it loosely.

If the interferometer fixed mirror (mirror D in Fig. 18.11A, or the plane-convex mirror of Fig. 14.16B) is modulated at a frequency 10 or more times removed from the vibration frequency, the fringe disappearances can also be sensed by a photodetector.[19] A narrow bandpass filter is tuned to the mirror modulation frequency and nulls occur at the J_0 Bessel function zeros as shown by Eq. (18.15). Displacement amplitude will have values shown in Table 18.1 for helium-neon laser illumination ($\lambda = 632.8$ nm).

OPTOELECTRONIC TRACKERS. Optoelectronic trackers, described in Chap. 14, can be used to measure vibration displacements. Care must be taken to determine the displacement sensitivity of the tracker. This is a function of the lens used, the tracker-to-vibrating target distance, target geometry, and electronic gains within the tracker circuitry. Optoelectronic trackers provide a convenient noncontacting, massless sensor which is useful at the lower range of vibration frequencies where displacements are not extremely small.

"CHATTER" METHOD OF DETERMINING 1*g*

The chatter method utilizes the earth's gravitational field as a means of determining 1*g*. If a small mass is placed upon a vibrating surface which is perpendicular to the gravity vector and the mass is permitted to rest loosely on it, separation will occur whenever the downward acceleration exceeds 1*g*. Separation will be maintained until the deceleration portion of the cycle. Separation may be sensed electrically and contact may be detected electrically or audibly.[20]

This method is useful only for vertical vibration at frequencies less than about 100 cps and can indicate the 1*g* acceleration level to within about ± 1 to ± 5 per cent, depending on the mass, the vibrating surface, the frequency, and the means used to detect separation or contact. Special-purpose accelerometers[20] and small exciters[21] have been built which employ this chatter method.

RECIPROCITY CALIBRATION

The reciprocity calibration method is an absolute means for calibrating vibration exciters having a velocity coil or reference accelerometer.[2, 22–24] This method relates the pickup sensitivity to measurements of voltage ratio, resistance, frequency, and mass. For this method to be applicable, it is necessary that the exciter system be linear, i.e., that displacement, velocity, acceleration, and driver coil current each increase linearly with force and driver coil voltage. It is used chiefly with electrodynamic exciters. Neither the internal exciter structure, the magnet structure, the armature, nor mounting need be perfectly rigid. However, all joints must be tight.

Reciprocity calibration procedures can generally be applied only under controlled laboratory conditions. Many precautions must be taken and the process is usually time-consuming. It is used at national standards laboratories such as the National Bureau of Standards and some other laboratories engaged in high-level standardization measurements. Several methods have evolved over the past years, the most widely used and accepted being the Bouche-Levy reciprocity method,[2, 22–24] discussed in this chapter. For other methods, see Refs. 25 and 26.

The Bouche-Levy reciprocity calibration method consists of two laboratory experiments: (1) the measurement of the transfer admittance between the exciter's driver coil and the attached velocity coil or accelerometer, and (2) the measurement of the voltage ratio of the open-circuit velocity coil or accelerometer and the driving coil while the exciter is driven by a second external exciter. In recent years a high-quality piezoelectric accelerometer has been incorporated into reciprocity calibration techniques rather than the velocity-coil-type transducers because of the frequency ranges of use, stability, and convenience. The use of an accelerometer is assumed here. A simple block diagram in Fig. 18.12 shows the circuit used for the transfer admittance measurements.[24] The relationship for the transfer admittance is

$$\bar{Y} = \frac{\bar{I}}{\bar{E}_{12}} \tag{18.17}$$

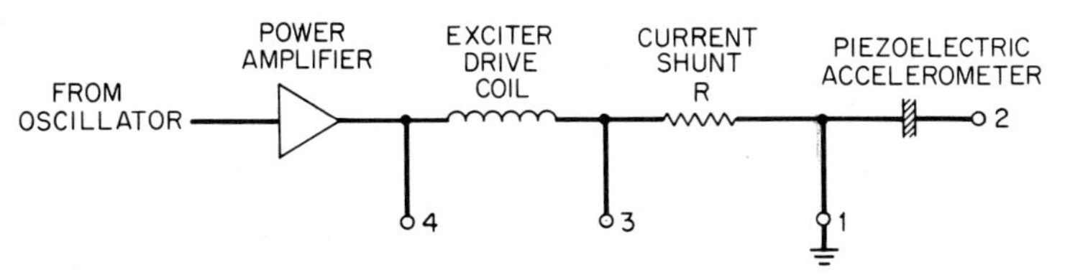

FIG. 18.12. Simplified diagram of driver coil and accelerometer circuits used in the Bouche-Levy reciprocity calibration method. The voltage drop across the shunt resistor R is used to measure drive-coil current. (*After B. F. Payne.*[24])

where $\bar{Y}$ = transfer admittance
$\bar{I}$ = driver coil current
$\bar{E}_{12}$ = output of the accelerometer

(The voltage $\bar{E}_{12}$ can be an amplified accelerometer output.) The bar over each of the parameters indicates complex quantities. The transfer admittances measurements are made with a series of masses attached one at a time to the table of the exciter. Also, the zero-load transfer admittance is generally measured before and after each admittance measurement is made with the various masses. (The zero-load transfer admittance is denoted as $\bar{Y}_0$.)

Voltage measuring ratiometers provide a convenient and accurate means of measurement. By using a ratiometer, the magnitude of the transfer admittance, $|Y|$, can be measured by obtaining the ratio of the voltages E_{13} and E_{12}, as shown in Fig. 18.12, and dividing by the value of R. The phase of the transfer admittance, ϕ, can be determined by the use of a phase meter, which leads to the relationship shown in Eq. (18.17).

$$|Y| = \frac{|E_{13}/E_{12}|}{R} \tag{18.18}$$

Using the measured values of the transfer admittance, both the real and imaginary parts of the ratio of $W/\bar{Y} - \bar{Y}_0$ are plotted against W, where W is the value of the mass used in determining the admittance. When plotted against the mass W the ordinate intercepts J and the slopes Q (real and imaginary of each) of the function are needed subsequently in determining the calibration factor.

The second experiment for the Bouche-Levy reciprocity method requires that a second exciter be used to provide a sinusoidal motion to the exciter being calibrated. (Some electrodynamic exciters are made with two electrically independent driver coils so that a second exciter is not essential.) The purpose of these measurements is to determine the ratio $\bar{K}$ of the open-circuited voltages from the accelerometer and driver coil, such that

$$\bar{K} = \frac{\bar{E}_{12}}{\bar{E}_{14}} \tag{18.19}$$

Figure 18.12 shows the points within the circuit where the ratio and phase angles are measured. The shunt resistor can be removed for this measurement and replaced with a short, left in place; because zero current is flowing in the driver coil, no voltage drop will occur across the shunt resistor. A ratiometer can be used to measure the ratio and a phase meter can be used to measure the phase. Using the relationship

$$\bar{Y}' = \frac{\bar{J}(\bar{Y} - \bar{Y}_0)}{1 - \bar{Q}(\bar{Y} - \bar{Y}_0)} \tag{18.20}$$

the sensitivity of the accelerometer is

$$\bar{S} = 2634.95 \left(\frac{\bar{K}\bar{J}}{jf}\right)^{1/2} \left(1 + \frac{\bar{Y}'\bar{Q}}{\bar{J}}\right) \quad \text{millivolts}/g \tag{18.21}$$

Hence the accelerometer sensitivity $\bar{S}$ is determined from $\bar{K}$, $\bar{J}$, $\bar{Q}$, $\bar{Y}$, and f for the masses attached to the table. Because the quantities are complex, the use of a computer is helpful, especially if the calibration is to be done at a number of frequencies.

Temperature, armature position, and a number of other conditions must be carefully controlled. With experience and the proper hardware, vibration exciters can be calibrated by the Bouche-Levy reciprocity method with uncertainties less than ± 1 per cent. At very low frequencies, less than perhaps 10 to 20 cps, problems may be encountered because of low acceleration levels or harmonic distortion. At frequencies greater than about 3 to 10 kc, it is difficult to obtain satisfactory results because of problems related to the dynamic characteristics of the masses. See Refs. 2, 14, and 22 to 35 for a full discussion of the reciprocity method.

HIGH-ACCELERATION CALIBRATION METHODS

Some applications in shock or vibration measurement require that high amplitudes be determined accurately. To ensure that the pickups used in such applications meet certain performance criteria, it is necessary to be able to calibrate at these high amplitudes. Several methods are available, depending on application. For pickups which respond to zero frequency (typically piezoresistive or other strain-gage-type accelerometers), centrifuge calibration as discussed earlier can provide a check on the amplitude linearity of the pickup. Accelerations in excess of several hundred g are possible. The calibration methods discussed in this portion are dynamic methods and are divided into sinusoidal methods and impact (shock) methods.

SINUSOIDAL METHODS FOR HIGH-ACCELERATION CALIBRATION

The use of a resonant bar, Figs. 18.13 and 18.14, to apply sinusoidal accelerations for calibration purposes has several advantages:[36–38] (1) The frequency is inherently

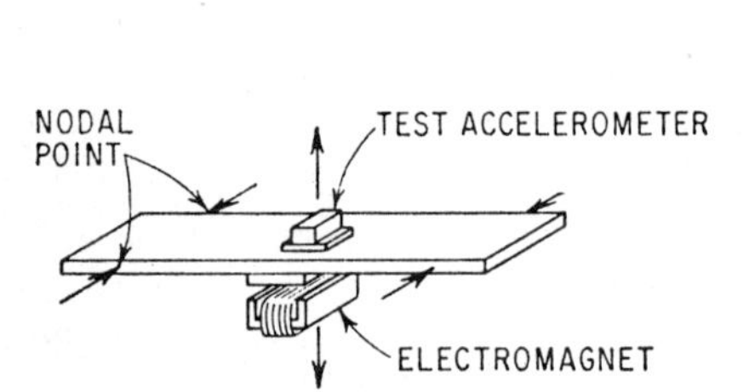

FIG. 18.13. Resonant-bar calibrator supported at the nodal points for free-free bending vibration. The bar at its natural frequency is driven by an electromagnet. The pickup is mounted at the mid-length of the bar. (*After ANSI Standard S 2.2-1959.*[2])

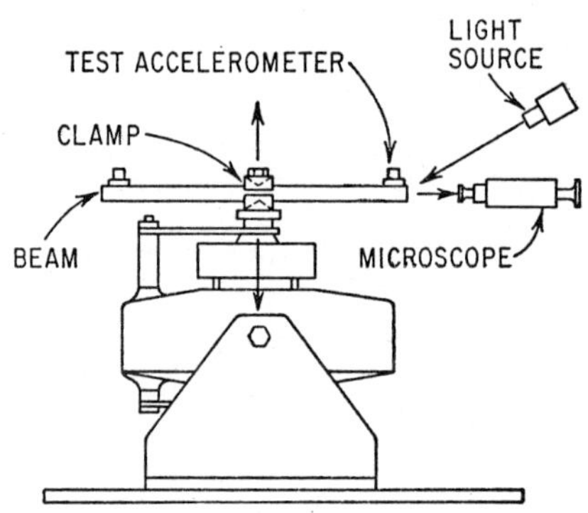

FIG. 18.14. Resonant-bar calibrator driven by an electrodynamic shaker. The pickup is mounted at one end and a counterbalancing weight is at the other. (*After E. I. Feder and A. M. Gillen.*[38])

constant and (2) very large amplitudes of acceleration (as high as 4,000g) with very little distortion in the waveform can be attained with moderate driving force. A disadvantage of this type of calibrator is that only the resonant frequencies are available for use.

In a typical mounting, Fig. 18.13, the bar is supported at its nodal points and the pickup to be calibrated is mounted at the mid-length of the bar.[36, 37] The bar is energized by an electromagnet. The magnet current is obtained by superposing on a polarizing direct current the output of a power amplifier energized by an oscillator. Self-excitation is possible if the oscillator is replaced by the output of a second pickup mounted on the beam, in conjunction with a phase shifter to form a closed servo loop. Beams of about 2-in. thickness vary in natural frequency from about 400 cps for a 31-in.-length beam to about 2,000 cps for a 14-in.-length beam. Acceleration amplitudes of several thousand g are obtained. By employing smaller beams, higher frequencies and acceleration levels can be excited. For example, calibrations can be performed at acceleration levels of $12{,}000g$ employing bars in axial resonance.

The displacement at the point of attachment of the pickup can be measured optically since displacements encountered are adequately large. The response R and the peak-to-peak displacement $2x_0$ are read for a series of values of driving force. The values of R are then plotted against $4\pi^2 f^2 x_0$ and the slope of the line through the points is taken as the pickup sensitivity factor for the frequency f of resonance of the bar. To calibrate a pickup at other frequencies, the bar must ordinarily be changed and the procedure repeated.

The resonant-bar calibrator shown in Fig. 18.14 is primarily limited in amplitude by the fatigue resistance of the bar.[38] Using aluminum bars, levels up to $500g$ have been attained without special design criteria. For levels up to $4{,}000g$ a bar material such as tempered vanadium steel machined to have a mounting boss at its mid-length is more suitable. The resonant bar is mounted at its mid-length on a conventional electrodynamic shaker. The accelerometer being calibrated is mounted at one end of the bar and an equivalent balance weight is mounted at the opposite end in the same relative position. The calibration procedure uses an optical means for measuring displacement amplitude from which the applied acceleration amplitude is computed as $4\pi^2 f^2 x_0$, where f is the frequency and $2x_0$ is the peak-to-peak displacement. (It should be noted that a pickup mounted at the end of a resonant bar is subjected to a rocking motion in addition to the desired translation. High-frequency pickups generally are unaffected by this extraneous rocking motion.)

Axial resonances of long rods and tubes have been used to generate motion for accurate calibration of vibration pickups over a frequency range from about 1 to 20 kc and at acceleration levels up to $12{,}000g$.[39–41]

The resonant rods can be driven by electrodynamic exciters at low frequencies and with piezoelectric exciters at high frequencies. The use of axially driven rods has an advantage over the beams discussed above in that no bending or lateral motion is present.[41] This minimizes errors from the pickup response to such unwanted modes and also from direct measurement of the displacement having nonrectilinear motion.

The vibration amplitude can be measured by an optical method, discussed previously. As with many measurement methods, laboratory skills and techniques must be developed in order to achieve satisfactory results. When performing sinusoidal calibrations at very high levels, damage to the pickup or cable assembly may occur. Experience has shown that fatigue damage can be sustained by accelerometers at sinusoidal accelerations considerably less than the shock performance rating.[41] Furthermore, it is common that under some circumstances severe cable heating may occur—enough sometimes to melt the cable insulation.

IMPACT METHODS FOR HIGH-ACCELERATION CALIBRATION

There are several methods by which sudden velocity change may be applied to pickups designed for high-frequency acceleration measurement. Any method which generates a reproducible velocity change and time duration can be used to obtain the calibration factor.[2] Impactive techniques can be employed to obtain calibrations over an amplitude range from a few to over $30{,}000g$. An accurate determination of shock performance of an accelerometer depends not only upon frequency response, and mechanical and electrical characteristics designed into the transducer, but also upon the characteristics of the instrumentation and recording equipment. It is often best to perform system calibrations to determine the linearity of the vibration pickup as well as the linearity of the recording instrumentation in the approximate range under observation. Several of the following methods make use of the fact that the velocity change during a transient pulse is equal to the time integral of acceleration:

$$v = \int_{t_1}^{t_2} a\,dt \tag{18.22}$$

where the initial or final velocity is taken as reference zero and the integration is performed to or from the time at which the velocity is constant.[42]

In this section, several methods are presented for applying known velocity changes v to a pickup. The voltage output e of a pickup of calibration factor $\mathcal{S}_1$ for an acceleration a is

$$e = \frac{\mathcal{S}_1 a}{g} \tag{18.23}$$

where $\mathcal{S}_1$ is the calibration factor in volts per g (g being the acceleration of gravity). Combining the preceding two equations,

$$\mathcal{S}_1 = \frac{\int_{t_1}^{t_2} eg\,dt}{v} \tag{18.24}$$

Methods are presented in this section for measuring the integral in this equation. Having the value of both the integral and v, the calibration factor $\mathcal{S}_1$ can be computed. The linear range of a pickup is determined by noting at what magnitude of the velocity change v the value of the calibration factor $\mathcal{S}_1$ deviates appreciably from its previous values. The minimum pulse duration is similarly found by shortening the pulse duration and noting when $\mathcal{S}_1$ changes appreciably from previous values. Typical auxiliary circuit considerations are described in this section and typical calibrations are given.

BALLISTIC PENDULUM CALIBRATOR. The ballistic pendulum calibrator provides a means for applying a sudden velocity change to a vibration pickup. The calibrator consists of two masses which are suspended by wires or metal ribbons. These ribbons restrict the motion of the masses to a common vertical plane.[42–44] This arrangement, shown in Fig. 18.15, maintains horizontal alignment of the principal axes of the masses in the direction parallel to the direction of motion at impact. The velocity attained by the anvil mass as the result of the sudden impact is determined.

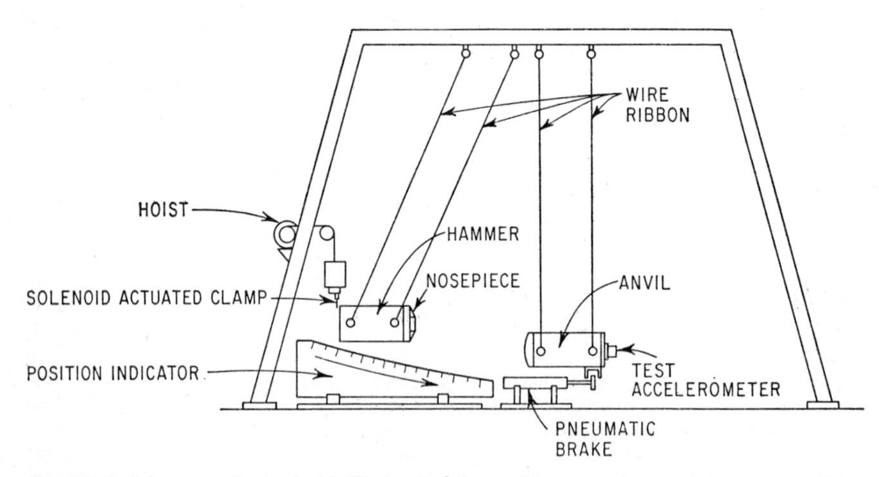

FIG. 18.15. Schematic of a typical ballistic pendulum calibrator system used to apply a sudden velocity change to a vibration pickup. (*After E. I. Feder and A. M. Gillen.*[38])

The accelerometer to be calibrated is mounted to an adapter which attaches to the forward face of the anvil. The hammer is raised to a predetermined height and held in the release position by a solenoid-actuated clamp. Since the anvil is at rest prior to impact, recording and measurement of the change in velocity of the anvil and the transient waveform on a calibrated time base are required. One method of measurement of the velocity change is performed by focusing a light beam through a grating attached to the anvil, as shown in Fig. 18.16. The slots modulate the light beam intensity, thus varying the

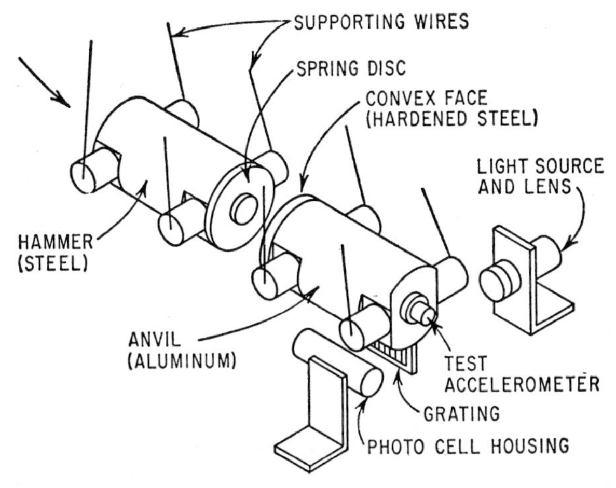

Fig. 18.16. Schematic arrangement of the ballistic pendulum with phototube and light grating to determine the anvil-velocity change during impact. (*After R. W. Conrad and I. Vigness.*[42])

phototube output which is recorded with the pickup output. Since the distance between grating lines is known, the velocity of the anvil is calculated directly, assuming that the velocity is essentially constant over the distance between successive grating lines. Other schemes may be used to measure velocity change, such as a sliding contactor or any suitable linear displacement device which can be timed as it crosses a sensitive region whose dimensions are known. The velocity of the anvil in each case is determined directly; the time relation between initiation of the velocity and the pulse at the output of the pickup is obtained by recording both signals on the same time base. The most frequently used method infers the anvil velocity from its vertical rise by measuring the maximum horizontal displacement and making use of the geometry of the pendulum system. This method has been proved quite satisfactory and provides results that correlate with the more precise techniques described above to within a few per cent.

The duration of the pulse, which is the time during which the hammer and anvil are in contact, can be varied within close limits.[42] In Fig. 18.16 the hammer nosepiece is a disc with a raised spherical surface. It develops a contact time of 0.55 millisecond. For larger periods, ranging up to 1 millisecond, the stiffness of the nosepiece is decreased by bolting a hollow ring between it and the hammer. Pulses longer than 1 millisecond may be obtained by placing various compliant materials, such as lead, between the contacting surfaces.

DROP-TEST CALIBRATOR. Another frequently used impulsive device is the drop tester shown in Fig. 18.17. The pickup is attached to the hammer using a suitable adapter plate. An impact is produced as the guided hammer falls under the influence of gravity and strikes the fixed anvil. To determine the velocity change, measurement is made of the time required for a contactor to pass over a known region just prior to and after impact. The pickup output and the contactor indication are recorded simultaneously in conjunction with a calibrated time base. The velocity change also

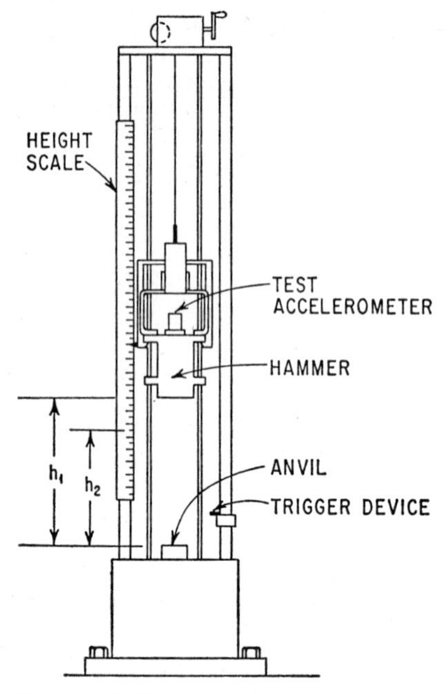

FIG. 18.17. Schematic of a conventional drop tester used to apply a sudden velocity change to a vibration pickup. (*After R. W. Conrad and I. Vigness.*[42])

may be determined by measuring the height h_1 of hammer drop before rebound and the height h_2 of hammer rise after rebound. The total velocity is calculated from the following relationship:

$$v = (2gh_1)^{1/2} + (2gh_2)^{1/2} \tag{18.25}$$

A total velocity change of 40 ft/sec (1,219 cm/sec) is a typical value which has been achieved by this type of machine.

DROP-BALL SHOCK CALIBRATOR. Figure 18.18 shows a drop-ball shock calibrator.[45–47] The accelerometer to be calibrated is mounted on an anvil which is held in position by a magnet assembly. A large steel ball is dropped from the top of the calibrator impacting the anvil. The anvil (and mounted accelerometer) are accelerated in a short free-flight path. A cushion catches the anvil and accelerometer. Shortly after impact, the anvil passes through an optical timing gate of a known distance. From this, the velocity after impact can be calculated.

Acceleration amplitudes and pulse durations can be varied by selecting the mass of the anvil, mass of the impacting ball, and resilient pads on top of the anvil where the ball strikes. Common accelerations and durations are 100g at 3 milliseconds, 500g at 1 millisecond, 1,000g at 1 millisecond, 5,000g at 0.2 millisecond, and 10,000g at 0.1 millisecond.[47] With experience and care, shock calibrations can be performed with an uncertainty of about ±5 to ±10 per cent.

HIGH-IMPACT CALIBRATIONS

Methods for calibrating a pickup by applying a sudden velocity change to it at a higher acceleration level than obtainable by methods previously mentioned have been developed using specially modified ballistic pendulums, air guns, inclined troughs, and other devices.[43] Successful calibrations have been performed on these at accelerations up to 40,000g for pulse durations of 23 to 70 microseconds with an approximate scatter of 16 per cent.

INTEGRATION OF ACCELERATION FROM RECORD. An example[42] of a typical acceleration vs. time record describing the characteristic response of a vibration pickup to a pulse input is shown in Fig. 18.19. The following linear relationship holds between acceleration a and voltage output e:

$$\mathcal{S}_1 = \frac{e}{a/g} \tag{18.26}$$

where $\mathcal{S}_1$ is the accelerometer calibration factor expressed in output volts per g acceleration (g is the acceleration of gravity). The calibration factor k_1 for the recording system ordinate scale is obtained by applying a known calibration voltage at the input terminals and noting the resulting deflection y:

$$k_1 = \frac{e}{y} \tag{18.27}$$

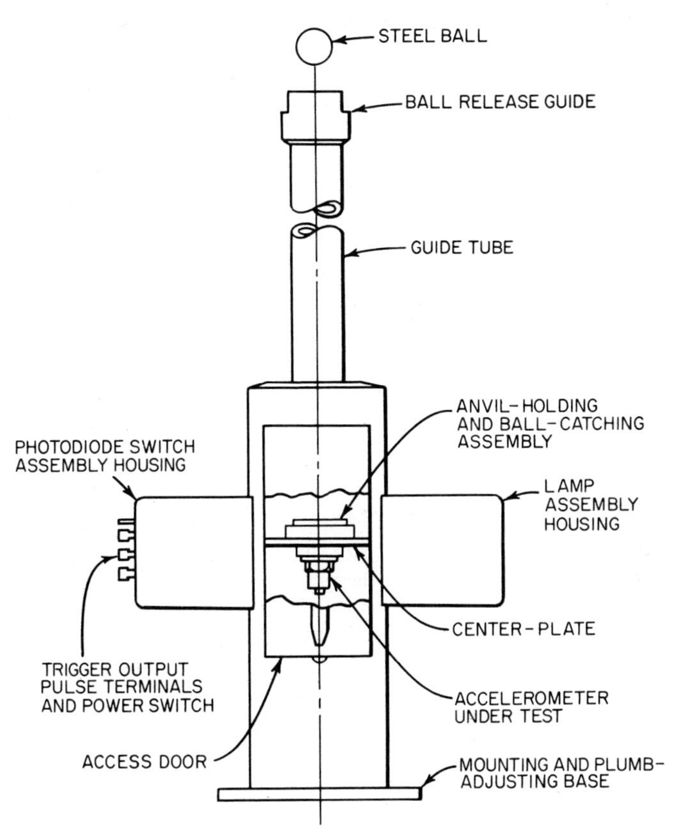

FIG. 18.18. Diagram of a drop-ball shock calibrator. The accelerometer being calibrated is mounted on an anvil which is held in place by small magnets. A steel ball is dropped and impacts the anvil, thereby generating a shock acceleration. (*After R. R. Bouche.*[45, 47])

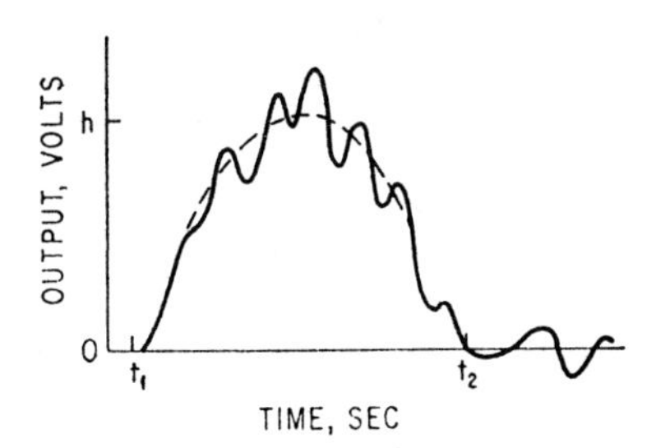

FIG. 18.19. Typical acceleration-time record of a shock pulse resulting from a conventional impact calibrator. The dashed curve closely resembles a half-sine pulse. (*After R. W. Conrad and I. Vigness.*[42])

The calibration factor k_2 for the abscissa scale is obtained from a precision timing trace introduced at the terminals and the corresponding deflection x:

$$k_2 = \frac{t}{x} \tag{18.28}$$

By substitution Eq. (18.24) becomes

$$\mathcal{S}_1 = \left(\frac{k_1 k_2 g}{v}\right) \int_{x_1}^{x_2} y\,dx \tag{18.29}$$

and since

$$\int_{x_1}^{x_2} y\,dx = S \tag{18.30}$$

where S is the area under the acceleration vs. time curve, the calibration factor for the test accelerometer expressed in terms of volts per unit gravity is

$$\mathcal{S}_1 = \frac{k_1 k_2 S g}{v} \tag{18.31}$$

The area S can be determined by use of a planimeter or by numerical integration. If the output closely resembles a half-sine pulse, the area is equal to approximately $2hx/\pi$, where h is the height of the pulse and x its width.

ELECTRICAL INTEGRATION OF PICKUP RESPONSE. The integration of the area response S may be performed by electrical integration. It is necessary to select constants for the integrating network that will not appreciably load the voltage e. If an RC network consisting of a series resistance R followed by a parallel capacitance C, as in Fig. 18.20, is used to integrate the accelerometer output voltage, the maximum integrated output voltage is

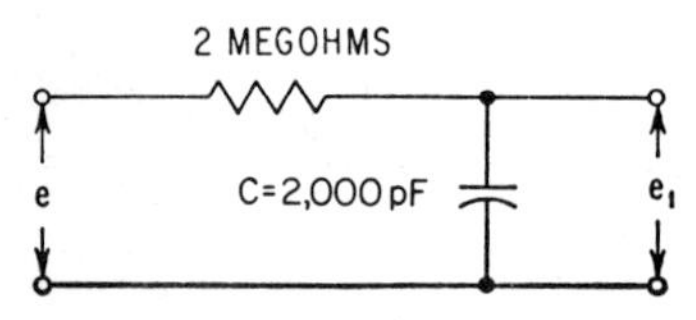

Fig. 18.20. Single-stage integration network. The resistance $R = 2$ megohms; the parallel capacitance $C = 2{,}000$ pF. (*After T. A. Perls and C. W. Kissinger.*[43])

$$e_1 = \frac{1}{RC} \int_{t_1}^{t_2} e\,dt \tag{18.32}$$

It follows from Eq. (18.24) that

$$\mathcal{S}_1 = \frac{gRCe_1}{v} \tag{18.33}$$

Several restrictions are imposed by this method of calibration and generally are related to the low-frequency response of the electrical integrator and the natural period of the accelerometer. In the case of piezoelectric pickups, the distortion in response at low frequency is limited by the time constant derived from the total capacitance of the accelerometer and cables and the input impedance of the conversion system. The response at high frequency is limited by the natural period of the accelerometer and the nonlinearity of the integration network.

The theoretical limit of calibration accuracy[42] for accelerometers with little or no damping is 2 per cent for ratios of pulse duration to natural period greater than 6. In practice, calibration errors may vary from as little as 5 per cent to as much as 20 per cent, depending upon the region of frequency response that is explored. Experimental transient studies[43] conducted for two accelerometers over a range of pulse duration greater and less than the natural period are shown in Fig. 18.21. The variations in relative response become greater for those values of the ratio of pulse duration to natural period greater than 6.

AUXILIARY CIRCUIT EFFECTS ON VELOCITY-CHANGE CALIBRATION METHOD. Cathode followers and amplifying and recording equipment ordinarily are used with high-frequency, high-acceleration pickups. Distortion in these auxiliary circuits can be more important in high-level pickups than in low-level pickups. Therefore, the accurate determination of the performance of a pickup by the velocity-change method depends on the lack of frequency distortion and nonlinearity in the auxiliary circuits for the range of frequencies and amplitudes excited by the shock motion. A low-pass filter sometimes can be used, as shown in Fig. 18.22, to eliminate unwanted high-frequency high-amplitude portions of the response. Where this is done, it is important that the linear range not be exceeded in portions of the circuit preceding the filter and that the upper cutoff frequency of the filter be well above the frequency range over which the calibration is desired.

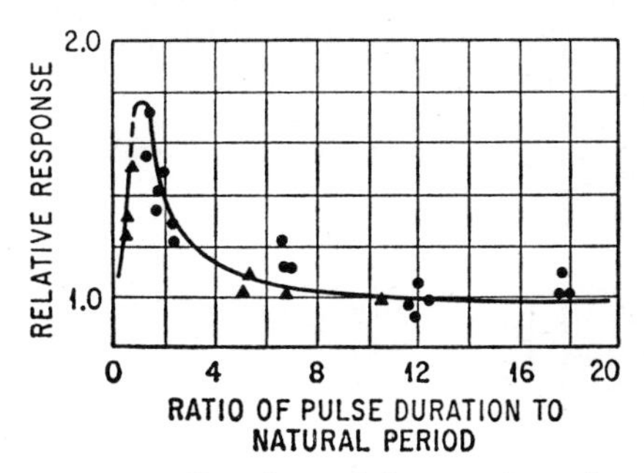

FIG. 18.21. Experimental data on the variation of relative response with relative pulse duration for two commercial piezoelectric accelerometers. (*After T. A. Perls and C. W. Kissinger.*[43])

Wherever possible it is desirable to determine that the response of circuit elements is unaffected by frequency distortion in the auxiliary circuits, and that the expected voltages will not exceed the linear range of any element. A more detailed and general discussion of calibration of auxiliary circuits is given later in this chapter.

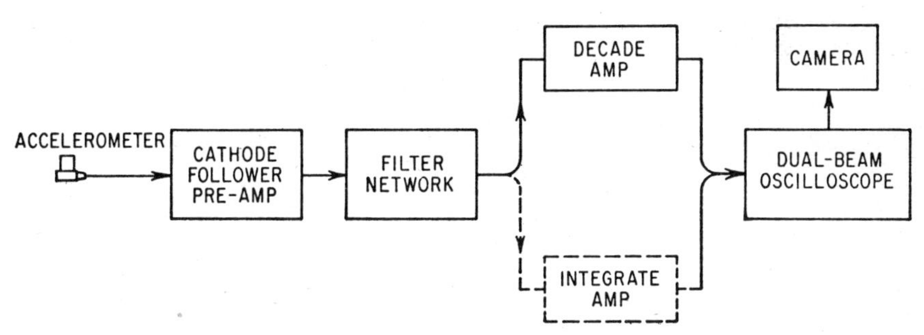

FIG. 18.22. Schematic of a typical circuit arrangement used to calibrate piezoelectric accelerometers by the velocity-change method. Dual-beam oscilloscope is used to record velocity before and after impact on one channel and pickup output on the other.

IMPACT-FORCE SHOCK CALIBRATOR

The impact-force shock calibrator has a free-falling carriage and a quartz load cell.[46, 48] The accelerometer to be calibrated is mounted onto the top of the carriage, as shown in Fig. 18.23. The carriage is suspended about one-half to one meter above the load cell and allowed to fall freely onto the cell.[49] The carriage's path is guided by a plastic tube. Cushion pads are attached at the top of the load cell to lengthen the impulse duration and to shape the pulse. Approximate haversines are generated by this calibrator.

The outputs of the accelerometer and load cell are fed to two nominally identical charge amplifiers. The outputs of the two charge amplifiers connect to some means of measuring the transient outputs; peak-holding meters can be used, or a storage-type oscilloscope. Figure 18.24 shows a typical setup.

During impact, the voltage produced at the output of the acceleration-charge amplifier $e_a(t)$ is

$$e_a(t) = a(t)S_aH_a \tag{18.34}$$

where $a(t)$ = acceleration
S_a = calibration factor for the accelerometer
H_a = charge amplifier gain

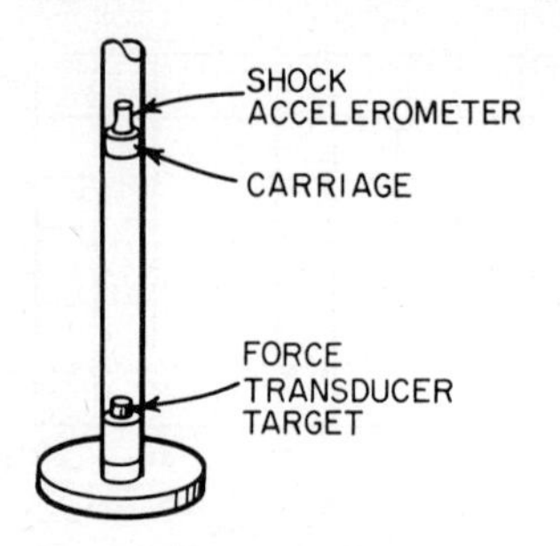

FIG. 18.23. Impact-force calibrator for shock accelerometers. The shock accelerometer to be calibrated is mounted on a carriage. It is raised and then dropped to impact a force transducer (load cell). (*After W. P. Kistler.*[48])

The output of the load cell-charge amplifier $e_f(t)$ is

$$e_f(t) = F(t)S_fH_f \tag{18.35}$$

where $F(t)$ = force
S_f = calibration factor for the load cell
H_f = charge amplifier gain

By using the relationship $F(t) = ma(t)$, where m is the falling mass, and combining Eqs. (18.34) and (18.35),

$$\frac{e_a(t)}{e_f(t)} = \frac{a(t)S_aH_a}{ma(t)S_fH_f} \tag{18.36}$$

so that

$$S_a = \left(\frac{e_a(t)}{e_f(t)}\right)\left(\frac{H_f}{H_a}\right)\frac{m}{g}S_f \qquad \text{picocoulomb}/g \tag{18.37}$$

In Eq. (18.37) the amplifier gains H_f and H_a are in units of volts per unit charge, mass m is in kilograms, g is the acceleration of gravity (9.80665 m/sec²), and the load cell calibration factor is in units of charge per unit force (picocoulomb/newton).

When calculating the mass, it is necessary to know the mass of the carriage, accelerometer, mounting stud, cable connector, and a short portion of the accelerometer cable. Experience has shown that for small coaxial cables, a length of about 2 to 4 cm is correct. When setting the value of charge gain H_f for the load cell charge amplifier, the terms mS_f/g of Eq. (18.37) can be computed and made a part of H_f to normalize the load cell

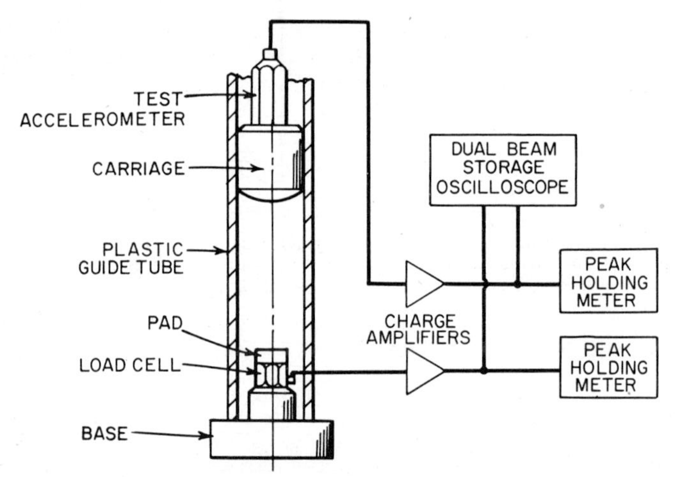

FIG. 18.24. Impact-force calibrator with the charge amplifiers and peak-holding meters. A dual-beam storage oscilloscope is also helpful. (*After W. P. Kistler.*[48])

channel output. By doing so, the accelerometer calibration factor can be read directly as $(1/H_a)$ on the accelerometer charge amplifier when adjusted such that $e_a(t) = e_f(t)$. This is done by successive trial drops and by adjusting H_a until the readings of the two peak meters are the same. Calibrations by this method can be accomplished with uncertainties generally between 2 to 5 per cent.

FOURIER TRANSFORM SHOCK CALIBRATION

The previously discussed shock calibration methods yield the "calibration factor" for the accelerometer being tested, that is, a single value which ideally characterizes the

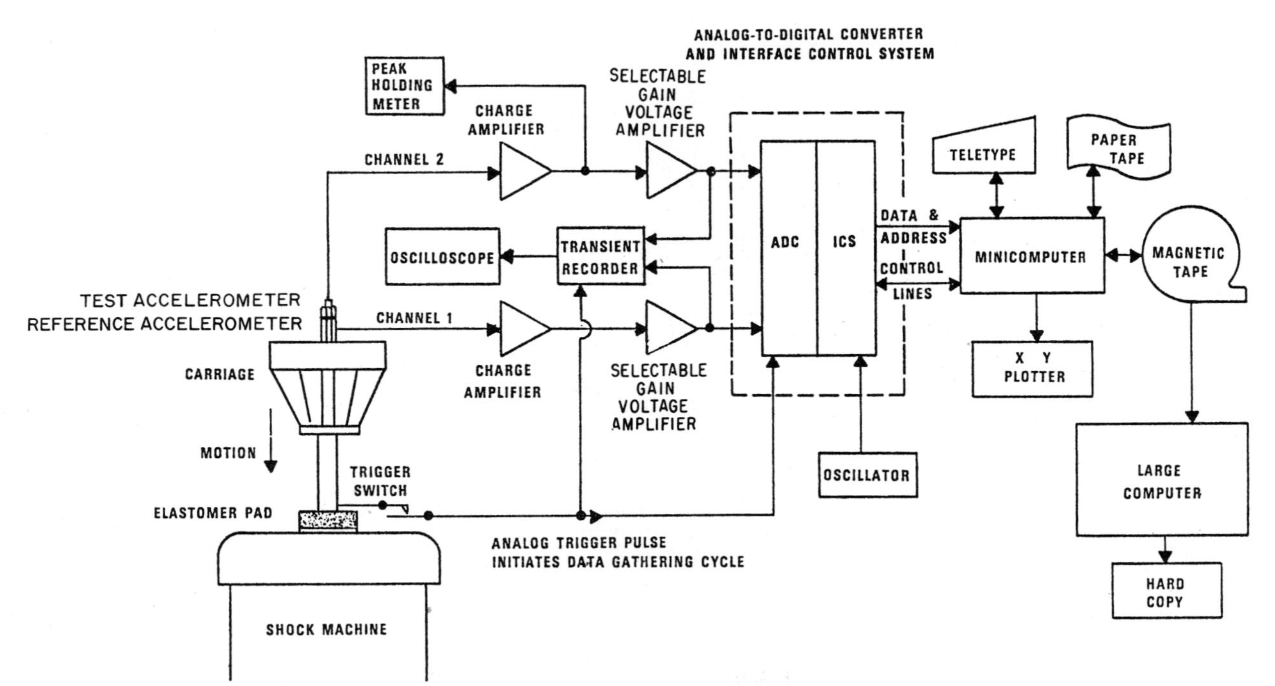

FIG. 18.25. Block diagram of a system using Fourier techniques for shock accelerometer calibration. Electrical analog shock signals are digitized and stored in a computer. Fourier analysis transforms the time-domain data to frequency-domain data. Sensitivity factor and phase are calculated versus frequency. (*After J. D. Ramboz and C. Federman.*[52])

"sensitivity factor." The latter is specified as a function of frequency, while the former is a single constant value which applies over a range of frequencies. (See the discussion under *Sensitivity, Calibration Factor* at the beginning of this chapter.) For many applications, a knowledge of the calibration factor may be sufficient. However, for shock standards or other critical applications, it is desirable to know the sensitivity factor (both amplitude and phase).

The shock calibration method described in this section employs Fourier techniques to *transform* time-domain data into frequency-domain data. By doing so, the frequency-dependent sensitivity factor can be determined both in amplitude and phase.[50–53]

The laboratory equipment consists of a mechanical shock-generating machine, two accelerometers, a data-transfer system, and a small computer for data storage and processing. Figure 18.25 shows[52, 53] a block diagram of a typical setup. The calibration of the test accelerometer is in terms of a reference standard. It is convenient to employ a piggyback standard accelerometer which permits the test accelerometer to be mounted directly on its top in a back-to-back configuration.[54–56] The signals from each accelerometer pass through charge amplifiers into selectable-gain voltage amplifiers and into an analog-to-digital converter (ADC). Digital data is then stored in a minicomputer core or other suitable device. Data can then be transferred to a large computer where Fourier transforms can be done quickly and economically.

A peak-holding meter, transient recorder, and oscilloscope are used to measure approximate acceleration peak values and to examine the time-domain waveshape. These instruments are not used in the Fourier analysis. An oscillator serves as clock for the conversion rate of the ADC; this frequency is a necessary parameter in the Fourier analysis. A digital 10-bit word is sufficient for most work, and conversion rates of at least 100 kc are desirable.

The computer performs a Fourier analysis by use of a fast Fourier transform (FFT).[53] Ratios of the test accelerometer's Fourier amplitudes with respect to the standard accelerometer's amplitude, taken frequency by frequency, lead to the test accelerometer sensitivity in terms of the standard. Phase calibration data is obtained by taking the phase difference between the test and standard Fourier outputs.

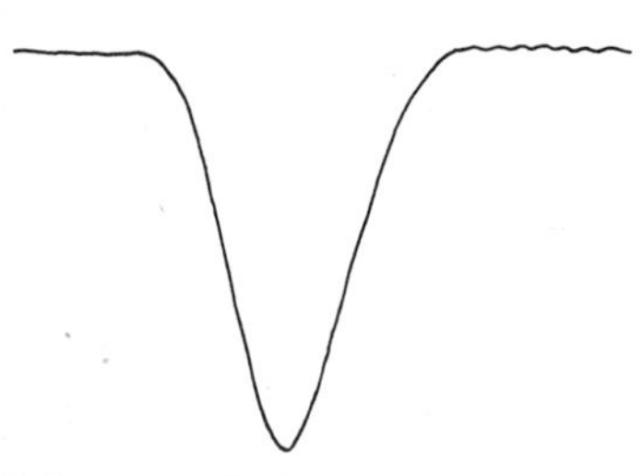

FIG. 18.26. A typical half-sine shock pulse generated by a pneumatic shock machine. Deceleration amplitude of 900*g* and a pulse duration of 1 microsecond. The Fourier transform is shown in Fig. 18.27*C*. (*After J. D. Ramboz and C. Federman.*[52])

Figure 18.26 shows a typical half-sine shock pulse. The Fourier transform is shown in Fig. 18.27. The range of frequency is limited by the pulse shape and duration, sampling rate, and Fourier analysis. Generally, this type of calibration can yield complete calibration results over a frequency range from 0 to 10 kc or more. Uncertainties are usually less than ±5 per cent. Figure 18.28 shows a typical calibration result for a piezoelectric accelerometer specifically made for shock measurements. Note that the sensitivity factor changes as function of frequency by several per cent. Also shown is a sinusoidal calibration response. At the left, peak-comparison calibration factors are shown which were derived from time-domain data as a function of the applied shock pulse. The limitations of time-domain peak-comparison calibrations are illustrated by the 50*g* point being lower than the remaining data.

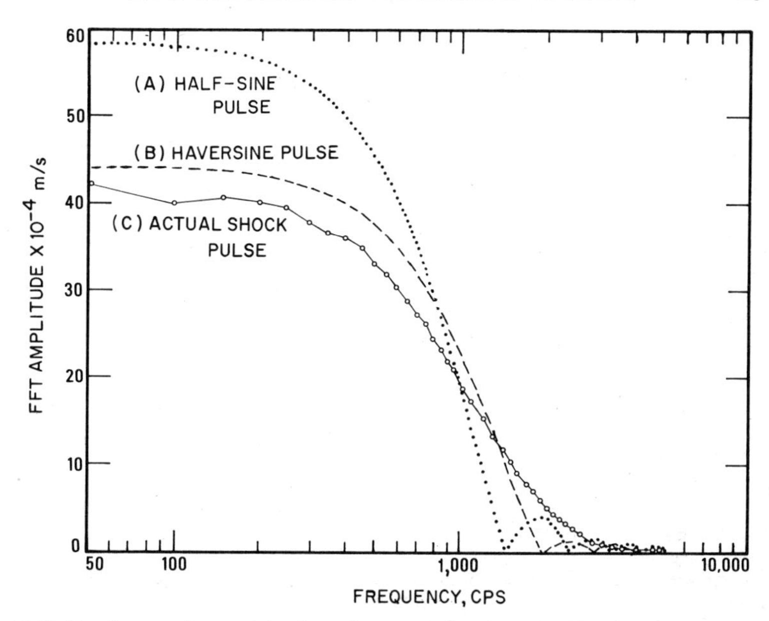

FIG. 18.27. Fourier transforms of shock accelerometer signals. (*A*) Ideal half-sine. (*B*) Ideal haversine. (*C*) An actual shock pulse of 900*g* peak and a duration of 1 millisecond. The ideal pulses have the same peak acceleration and time durations as the actual pulse as that shown by Fig. 18.26. (*Courtesy National Bureau of Standards.*)

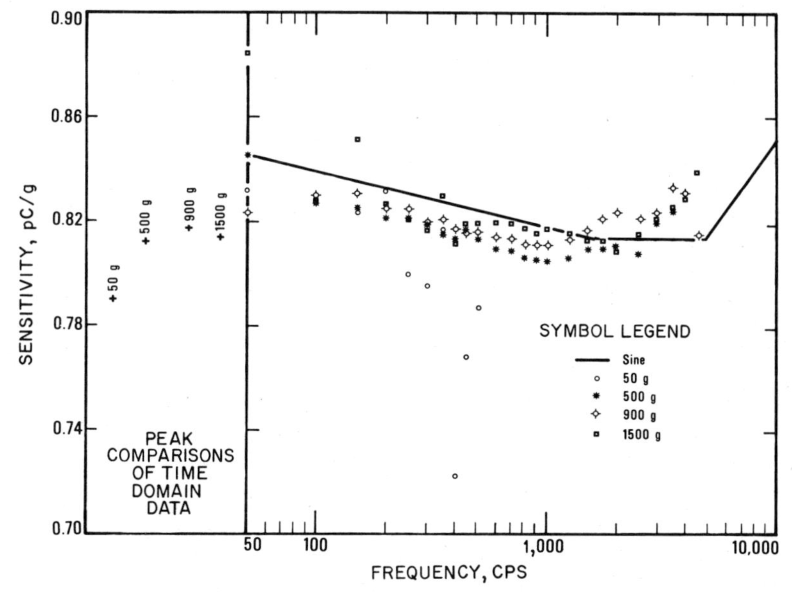

FIG. 18.28. Results of shock calibration of an accelerometer using Fourier techniques. Four acceleration amplitudes are showing and are compared with the estimated true sensitivity (sine). Time-domain calibration factors are shown at the left side for comparison purposes. (*After W. H. Walston and C. Federman.*[53])

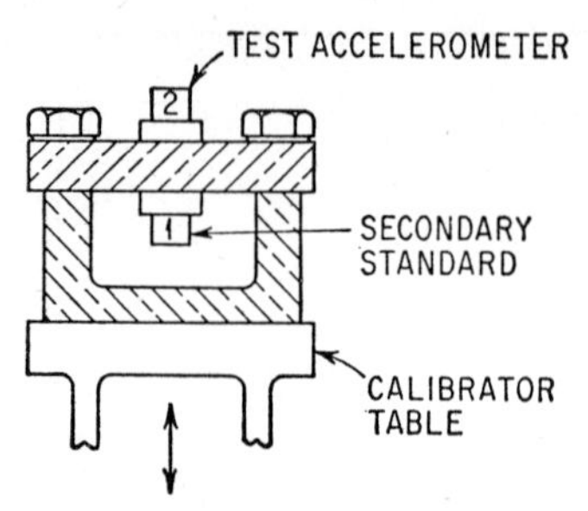

FIG. 18.29. Comparison method of calibration. Pickup 2 is calibrated against pickup 1 as the secondary standard. The two pickups are excited by any of the means described in this chapter. (*After ANSI Standard S 2.2-1959.*[2])

CALIBRATION BY THE COMPARISON METHOD

A rapid and convenient method of obtaining the sensitivity factor of a vibration pickup is by direct comparison of the pickup voltage to that of a second pickup used as a secondary standard and calibrated for frequency and phase response by one of the methods described in this chapter.[2]

Pickups are mounted back-to-back as shown in Fig. 18.29. The most important consideration is to ensure that each pickup experiences the same motion. If both pickups are rectilinear and are placed on the calibrator table, the angular rotation of the table should be small to avoid any difference in excitation between the two pickup locations. The error due to rotation may be reduced by carefully locating the pickups firmly on opposite faces with the center-of-gravity of the pickups located at the center of the table. Relative differences in pickup excitation may be observed by reversing the pickup locations and observing if the voltage ratio is the same in both positions.

Calibration by the comparison method is limited to the range of frequencies and amplitudes for which the secondary pickup has been calibrated. If both pickups are linear, the amplitude sensitivity $\mathcal{S}_2$ of a vibration pickup can be determined by

$$\mathcal{S}_2 = \frac{e_2}{e_1} \mathcal{S}_1 \tag{18.38}$$

where $\mathcal{S}_1$ is the sensitivity of the calibrated pickup and e_2 and e_1 are the corresponding measured voltages. The phase relationship is obtained by phase measurement between output voltages and phase calibration of the secondary-standard pickup.

SINUSOIDAL FREQUENCY-DWELL METHOD

A simple and convenient way to perform comparison calibrations is to fix the frequency at a desired value, adjust the vibration amplitude, and then make the necessary ratio measurements. The circuit shown in Fig. 18.30 is a simple and direct method of comparison calibration. The pickups are assumed to experience identical motion. Their outputs are processed through amplifiers (either voltage or charge amplifiers). The amplifier in the test channel may have a variable gain which has been calibrated. It is common practice to calibrate a gain adjustment inversely proportional to gain, i.e., a dial reading $\propto 1/H_t$. The sensitivity factor for the reference accelerometer channel is commonly a "system" sensitivity factor such that the sensitivity is specified at the output of the amplifier, thus accounting for gain H_r. An attenuator is sometimes used in the reference channel (assuming this voltage output is greater than that of the test channel). A single voltmeter is employed to indicate the output of both channels, as shown in Fig. 18.30.

In the first instance, assume the amplifier in the test channel has a fixed gain, H_t, and the attenuater is used in the reference channel. The calibration procedure is to switch

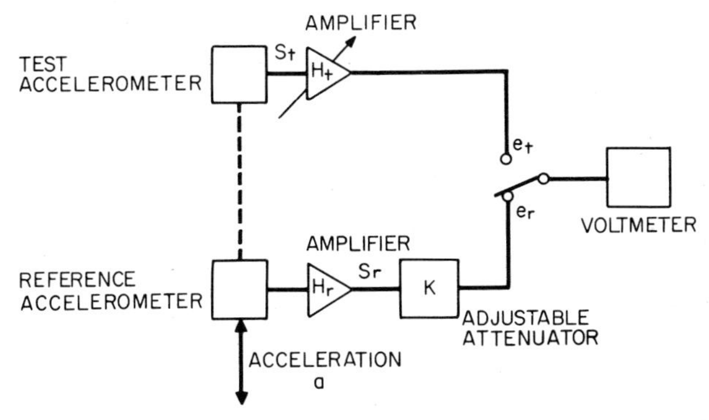

FIG. 18.30. Simplified diagram of a simple ratio-comparison calibration setup. The circuit is alternately switched so that the voltages e_t and e_r are measured. (*After K. Unholtz.*[58])

the voltmeter alternately between the channels and adjust the attenuator until the voltmeter gives the same indication. The two voltages, e_t and e_r, can be expressed as

$$e_t = aS_tH_t \tag{18.39}$$

$$e_r = aS_rK \tag{18.40}$$

When the two voltages are equal, as indicated by the voltmeter, then [by equating Eqs. (18.39) and 18.40)]

$$S_t = \frac{K}{H_t} S_r \tag{18.41}$$

Thus, the test accelerometer can be calibrated in terms of a measurable attenuation ratio, amplifier gain, and the sensitivity factor of the reference pickup.

Another common method involves less computation when the product KS_r is adjusted to be a power of 10, such as

$$KS_r = 10^n \tag{18.42}$$

Substituting Eq. (18.42) into Eq. (18.41), the test accelerometer sensitivity factor becomes (where n is an integer)

$$S_t = \frac{1}{Ht} 10^n \tag{18.43}$$

If the amplifier gain in the test channel is calibrated such that the dial reading is $1/H_t$, then the test accelerometer's sensitivity factor can be read directly as the dial reading times 10^n. This process is easy and speeds the manual calibration process considerably.

A significant advantage of using a circuit such as is shown in Fig. 18.30 is that the voltages e_t and e_r need not be measured accurately. If the voltmeter gives the same indication for both channels, errors in the voltmeter tend to divide to zero [see Eq. (18.38)]. This is not the case if two separate voltmeters are used, or if a single voltmeter is used and the voltages e_t and e_r are widely different in magnitude. Additionally, when $e_t = e_r$ and some harmonic distortion is present, the single voltmeter tends to make the same voltage-measurement error for both readings; hence this error tends to divide to zero.

Minicomputer control and the use of electronic voltage ratiometers can also be utilized for comparison calibrations.[57] Correction factors for frequency response and statistical measurements can be performed to provide high accuracy.

SINUSOIDAL SWEPT-FREQUENCY METHOD

A graphical plot of the test accelerometer sensitivity factor (or deviation from a reference) versus frequency can provide a quick and convenient method of calibration. This gives a continuous sensitivity factor so that minor discontinuities or deviations can be detected.[58, 59]

A circuit such as is shown in Fig. 18.31 provides good resolution and a stable comparison system. The system configuration is much the same as for the previous method,

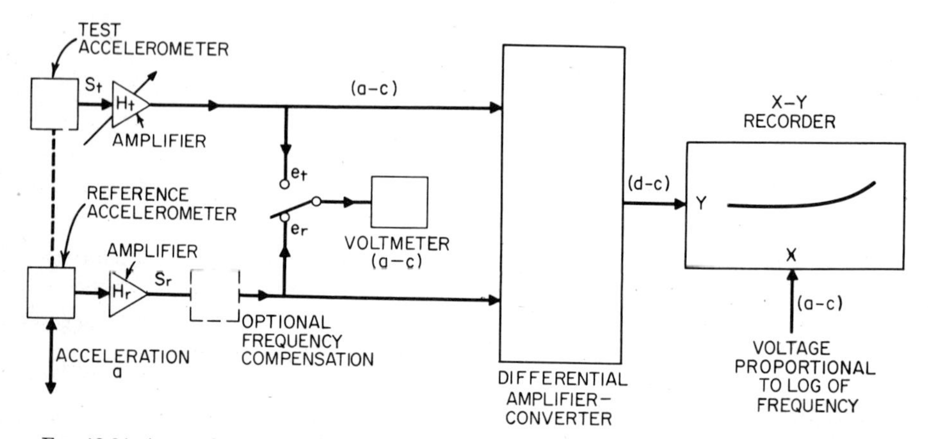

FIG. 18.31. An accelerometer calibration plotting system using a differential amplifier, a-c–to–d-c converter and an X-Y recorder. The two a-c voltages are made equal at some reference condition by adjusting amplifier gains. Recorder gain (ordinate) is adjusted so that a given difference equals a percentage of the test accelerometer calibration factor. (*After K. Unholtz.*[58])

however, a differential amplifier is used to measure the voltage differences between the two channels. The a-c voltages are converted to d-c voltages and fed to an X-Y plotter. The ordinate gain is adjusted such that, for example, a 1 per cent change in difference causes a 2-cm deflection on the recorder. The other recorder axis is fed a sweep voltage proportional to frequency (or log frequency). The plotting systems can save calibration time if large numbers of similar accelerometers are being calibrated.

RANDOM-EXCITATION–TRANSFER-FUNCTION METHOD

The use of random vibration excitation and transfer-function analysis techniques can provide quick and accurate comparison calibrations.[60] Two pickups are mounted back-to-back on a suitable vibration calibration exciter. Figure 18.32 shows a typical setup. The Fourier analyzer is a key piece of equipment for this method. The two accelerometer channels are fed into the analyzer through a pair of low-pass (antialiasing) filters. The random signal to drive the exciter is generated under control of the analyzer.

This method provides a nearly continuous calibration over the desired frequency spectra, the resulting sensitivity factors having both amplitude and phase information. Purely sinusoidal motion is not a requirement as in other calibration methods. This lessens the requirements for the power amplifier and exciter to maintain low values of harmonic distortion; in fact, harmonic distortion tends to have little consequence. A very useful measure of process quality is obtained by computing a statistical parameter called the "coherence function." This statistically relates the input and output such that the quality of the transfer function can be indicated.

Four fundamental parameters must be determined for this method: the power spectrum, the cross power spectrum, the transfer function, and the coherence function.

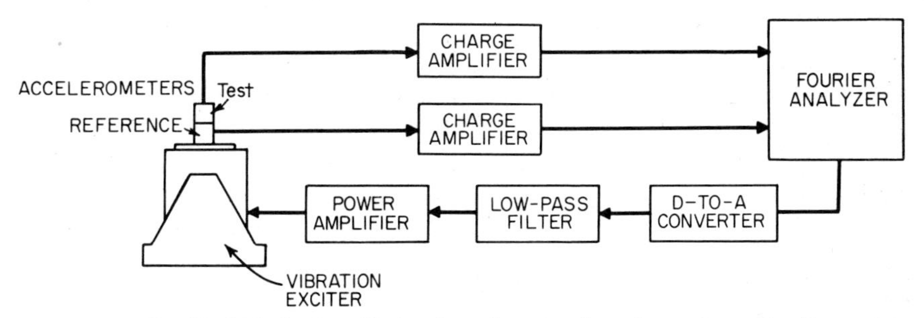

FIG. 18.32. Simplified block of a wide-band random signal accelerometer calibration system using Fourier analysis. The Fourier analyzer and low-pass filter control the frequency content of the power spectrum as experienced by the two accelerometers. (*After K. Ramsey.*[60])

The power spectrum of the system input can be defined and computed as

$$G_{xx} = A_x A_x^* \tag{18.44}$$

where G_{xx} is the power spectrum input of $a(t)$, and A_x^* is the complex conjugate of A_x. The output power spectrum G_{yy} can be defined and computed by

$$G_{yy} = A_y A_y^* \tag{18.45}$$

where A_y^* is the complex conjugate of A_y. The cross power spectrum between the input and output is denoted by G_{yx}, where

$$G_{yx} = A_y A_x^* \tag{18.46}$$

Using these relationships, the transfer function (i.e., sensitivity factor) can be described by

$$H(f) = \frac{\bar{G}_{yx}}{\bar{G}_{xx}} \tag{18.47}$$

where $\bar{G}_{yx}$ denotes the ensemble average of the cross power spectrum and $\bar{G}_{xx}$ represents the ensemble average of the input power spectrum.

The question of whether the system output is totally caused by the system input is important. Noise or nonlinear effects can create large errors at various frequencies. The coherence function gives an indication of such errors and is defined as

$$\gamma^2 = \frac{G_{yx}^2}{G_{xx}G_{yy}} \tag{18.48}$$

where $0 \leq \gamma^2 \leq 1$. When $\gamma = 1$ at any frequency, the system transfer function has perfect "causality" at that frequency; the output was caused totally by the input. (Other sources which are perfectly coherent with the input will also contribute to $\gamma = 1$; however, the likelihood of such occurring in this application is very small.) A low value of coherence function at given frequencies indicates that the transfer function has inaccuracies at those frequencies. When $\gamma = 0$, the output is caused totally by sources other than the input. A thorough treatment of the mathematics and theory of spectrum analysis, transfer functions, and coherence functions is given in Ref. 61.

CALIBRATION OF AUXILIARY CIRCUITS

The calibration of auxiliary circuits for vibration pickups is as important as the calibration of the pickups themselves. Calculation of the performance of various circuit elements individually and in combination in a system is difficult and in most cases impossible. Very often nonlinearities detected in the calibration of pickups can be attributed to limitations imposed by the transformation and recording system. Variations in low-frequency response may be directly related to an incorrectly selected matching amplifier or to the loss of necessary high insulation resistance as the result of excessive moisture or poor wiring techniques. Loss in sensitivity frequently occurs because of cable loading. Variations in high-frequency response may be caused by limitations of integrator response or galvanometer response or, in the case of magnetic tape, by head gap, tape speed, or the allowable deviation of center frequency. In order to avoid inadvertent errors in circuit selection and construction, system checks may be performed by calibrating through the system to be used during the test or by electrically simulating physical quantities based upon the predetermined calibration factor and the anticipated range of operation. In those instances where the pickup output is a voltage or charge, the auxiliary circuits can be calibrated by applying appropriate voltages or charges at selected frequencies and recording the response, thus establishing the scale factor for the instrumentation system.

A simple illustration of a system calibration consists in shunting a precision resistor across one arm of a resistance-bridge-type vibration pickup.[62] The pickup output is essentially in the form of a resistance change. The effect simulates the combined resistance changes of the active bridge arms due to acceleration. Electrical calibration in this manner does not remove the necessity for a precise calibration of the system. A calculated value of resistance is introduced during the physical calibration of such magnitude that the resulting indication is equal to that resulting from direct excitation. Since the resistance change appears as a step function, it is important that the auxiliary circuits have a flat response down to 0 cps. The circuit schematic of a typical system is shown in Fig. 18.33.

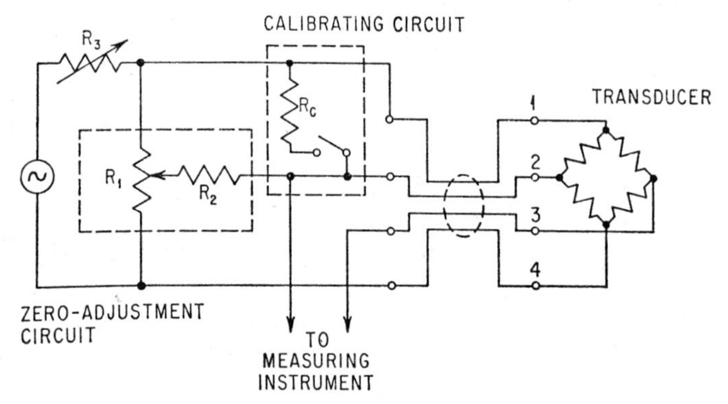

FIG. 18.33. Schematic diagram of circuit details in the resistance-change calibration of a Wheatstone-bridge unbonded-strain-gage accelerometer. (*After B. B. Helfand and J. Burns.*[62])

The calculation of the calibrating resistor value corresponding to an equivalent acceleration is given as

$$R_c = \frac{10^6 r}{2N\mathcal{S}} \tag{18.49}$$

where R_c is the calibrating resistor value in ohms, r is the resistance of each of the four bridge arms in ohms, $\mathcal{S}$ is the calibration factor in microvolts open-circuit output per volt input applied across the bridge per unit acceleration determined by physical calibration, and N is the mechanical acceleration input simulated by the calibrating resistor. The accuracy of the calibrating resistor method depends upon the stability of the calibrating resistor R_c, the bridge-arm resistance r, and the calibration factor $\mathcal{S}$.

A complete system checkout is important in order to provide assurance that the vibration-measuring system will function properly under test conditions. It is desirable to supplement the laboratory calibration of a vibration pickup and measuring circuit with at least a one-point calibration of the field installation. Small portable exciters may be used to perform this function. By such means an over-all transducer-to-read-out calibration can be made of conventional or telemetered channels assigned to vibration measurement.

REFERENCES

1. American National Standard for the Selection of Calibrations and Tests for Electrical Transducers used for Measuring Shock and Vibration, S2.11-1969, American National Standards Institute.
2. Methods for the Calibration of Shock and Vibration Pickups, S2.2-1959, American National Standards Institute.
3. Levy, H.: "Description of Pendulum Accelerometer Calibrator," *Naval Air Mat. Cen. (Philadelphia) Rept.* AML. NAM 2425, part V, Aug. 25, 1945.
4. White, G. E., and S. Kerstner: "Basic Methods for Accelerometer Calibration," *Statham Lab. Instr. Notes* 17 and 18, October, 1950.
5. Wildback, W. A., and R. O. Smith: *Proc. Instr. Soc. Amer.*, vol. 9, part 5, Paper 54-40-3, 1954.
6. Hilten, J. S.: "Accelerometer Calibration With the Earth's Field Dynamic Calibrator," Natl. Bur. Std. (U.S.) *Tech. Note* 517, March, 1970.
7. Levy, S., A. E. McPherson, and E. V. Hobbs: *J. Res. Natl. Bur. Std.*, **41**(5):359, 1948.
8. Dimoff, T.: *J. Acoust. Soc. Amer.*, **40**(3):671, 1966.
9. Model 2901 Primary Vibration Standard Shaker, Endevco Corp. Data Sheet, 1973.
10. Series 300 Vibration Calibration Systems, *Unholtz-Dickie Bull.* C300-1-65-2M, 1965.
11. Model CS75 Calibration System, *Gilmore Industries, Inc. (MB Electronics Division) Bull. 235A.*
12. Model DO-5000 Digital Optical Vibration Calibration Sensor, *Unholtz-Dickie Bull. DO5000-7-66*, 1966.
13. Koyanagi, R. S.: "Development of a Low-Frequency Vibration Calibration System," *Natl. Bur. Std. (U.S.) Rept.* 10-529, 1971.
14. Payne, B. F.: "Accelerometer Calibration at the National Bureau of Standards," *Instr. Soc. Amer. Preprint* ASI-75255, 1975.
15. Logue, S. H.: "A Laser Interferometer and its Applications to Length, Displacement and Angle Measurement," *Proc. Inst. Environ. Sci.*, p. 465, 1968.
16. Orlacchio, A. W.: *Elec. Mfg.*, January, 1957, p. 78.
17. Schmidt, V. A., S. Edelman, E. R. Smith, and E. Jones: *J. Acoust. Soc. Amer.*, **33**(6):748, 1961.
18. Jones, E., W. B. Yelon, and S. Edelman: *J. Acoust. Soc. Amer.*, **45**(6):1556, 1969.
19. Schmidt, V. A., S. Edelman, E. R. Smith, and E. T. Pierce: *J. Acoust. Soc. Amer.*, **34**(4): 455, 1962.
20. Kissinger, C. W.: *Proc. Instr. Soc. Amer.*, vol. 10, Paper 54-40-1, 1955.
21. Bruel and Kjaer: Instruction and Application No. 1606, January, 1958.
22. Levy, S., and R. R. Bouche: *J. Res. Natl. Bur. Std.*, **57**:227, 1956.
23. Bouche, R. R., and L. C. Ensor: "Use of Reciprocity Calibrated Accelerometer Standards for Performing Routine Laboratory Comparison Calibrations," *Shock and Vibration Bull.* 34, pt. 4, 1965.
24. Payne, B. F.: "Absolute Calibration of Vibration Generators with Time-Sharing Computer as Integral Part of System," *Shock and Vibration Bull.*, 36, pt. 6, 1967.
25. Harrison, M., A. O. Sykes, and P. G. Marcotte: "The Reciprocity Calibration of Piezoelectric Accelerometers," *David W. Taylor Model Basin Rept.* 811, March, 1952.
26. Harrison, M., A. O. Sykes, and P. G. Marcotte: *J. Acoust. Soc. Amer.*, **24**(4), 1952.
27. Ramboz, J. D.: "Absolute Calibration of Vibration Standards by the Three-Mass Reciprocity Method," *Natl. Bur. Std. (U.S.) Rept.* NBSIR 74-481, 1974.
28. Trent, H. M.: *J. Appl. Mech.*, **15**(1):49, 1948.
29. London, A.: "The Absolute Calibration of Vibration Pickups," *Natl. Bur. Std. (U.S.) Tech. News Bull.* 32, January, 1948.
30. Thompson, S. P.: "Reciprocity Calibration of Primary Vibration Standards," *NRL Rept.* F-3337, 1948.
31. Thompson, S. P.: *J. Acoust. Soc. Amer.*, **20**(5), 1948.
32. Thompson, S. P.: "Theoretical Aspects of the Reciprocity Calibration of Electromechanical Transducers," *NRL Rept.* F-3371, 1948.
33. Camm, J. C.: "The Reciprocity Calibration of a Vibration Pickup Calibrator," *Natl. Bur. Std. (U.S.) Rept.* 2651, 1953.
34. Sheeks, O. P.: "Methods for the Practical Reciprocity Calibration of Piezoelectric Accelerometers," *Instr. Soc. Amer. Reprint* 68-581, 1968.

35. Kühl, R.: "A Portable Calibrator for Accelerometers," *Bruel and Kjaer Tech. Rev.*, **1**:26, 1971.
36. Perls, T. A., C. W. Kissinger, and D. R. Paquette: *Bull. Amer. Phys. Soc.*, **30**(3):34 (1955).
37. Tyzzer, F. G., and H. C. Hardy: *J. Acoust. Soc. Amer.*, **22**:454 (1950).
38. Feder, E. I., and A. M. Gillen: *IRE Trans. Instr.*, **1–6**(2) (June, 1957).
39. Brennan, J. N.: *J. Acoust. Soc. Amer.*, **25**(4):610, 1953.
40. Nisbet, J. S., J. N. Brennan, H. I. Tarpley: *J. Acoust. Soc. Amer.*, **32**(1):71, 1960.
41. Jones, E., S. Edelman, K. S. Sizemore: *J. Acoust. Soc. Amer.*, **33**(11):1462, 1961.
42. Conrad, R. W., and I. Vigness: *Proc. Instr. Soc. Amer.*, vol. 8, Paper 53-11-3, 1953.
43. Perls, T. A., and C. W. Kissinger: *Proc. Instr. Soc. Amer.*, vol. 9, Paper 54-40-2, 1954.
44. Perls, T. A., C. W. Kissinger, and D. R. Paquette: *Bull. Amer. Phys. Soc.*, **30**(3):34 (1955).
45. Bouche, R. R.: "The Absolute Calibration of Pickups on a Drop-Ball Shock Machine of the Ballistic Type," *Endevco Corp. Tech. Paper* TP 206, April, 1961.
46. Kelly, R. W.: "Calibration of Shock Accelerometers," *Instr. Soc. Amer. Preprint* M18-2-MISTIND-67, 1967.
47. Model 2965C Comparison Shock Calibrator, Endevco Corp. Data Sheet, 1969.
48. Kistler, W. P.: "New Precision Calibration Techniques for Vibration Transducers," *Shock and Vibration Bull.* 35, pt. 4, 1966.
49. Model 894K, Impact-Force Calibrator, Kistler Instr. Corp. Data Sheet, 1967.
50. Favour, J. D.: "Accelerometer Calibration by Impulse Excitation Techniques," *Instr. Soc. Amer. Preprint* P13-1-PHYMMID-67, 1967.
51. Favour, J. D.: "Calibration of Accelerometers by Impulse Excitation and Fourier Integral Transform Techniques," *Shock and Vibration Bull.* 37, pt. 2, p. 17, 1968.
52. Ramboz, J. D., C. Federman: "Evaluation and Calibration of Mechanical Shock Accelerometers by Comparison Methods," *Natl. Bur. Std. (U.S.) Rept.* NBSIR 74-480, March, 1974.
53. Walston, W. H., C. Federman: "Fourier Transform Techniques for the Calibration of Shock Accelerometers," *Instr. Soc. Amer. Preprint* ASI-75257, 1975.
54. Model 2270, Primary Standard Accelerometer, Endevco Corp. Data Sheet, 1968.
55. Data Sheet Models 808K, 809K, 819K Vibration Calibration Standards, Kistler Instr. Corp.
56. Model 8305, Vibration Transducer, Bruel and Kjaer Master Catalog, p. 177, 1974.
57. Payne, B. F.: "An Automated Precision Calibration System for Accelerometers," paper presented at 17th National Aerospace Instrumentation Symposium, *Instr. Soc. Amer.*, May, 1971.
58. Unholtz, K.: "Vibration Transducer Calibration Using the Comparison Method," *Instr. Soc. Amer. Preprint* M18-3-MESTIND-67, 1967.
59. Chernoff, R.: "Comparison Calibration Techniques for Vibration Transducers," *Instr. Soc. Amer. Preprint* M18-3-MESTIND-67, 1967.
60. Ramsey, K.: "Accelerometer Calibration Using Random Noise and Transfer Function Measurements," *Instr. Soc. Amer. Preprint* ASI 75256, 1975.
61. Fourier Analyzer Training Manual, 02-5952-0651, Hewlett-Packard Co., 1970.
62. Helfand, B. B., and J. Burns: "Calibration of Resistance-bridge Transducer Circuits under Temperature Extremes," *Statham Lab. Instr. Notes* 14.

19

VIBRATION STANDARDS

Ronald L. Eshleman
The Vibration Institute

INTRODUCTION

A good standard must represent a consensus of opinion among users, be simple to understand, be easy to use, and contain no loopholes or ambiguities. Any standard must contain vital information that leads to common measurement and evaluation of data that are compared with agreed-upon criteria. In general, a vibration standard should establish classifications for equipment which is being rated and indicate how measurements are to be made and how the data, so obtained, are to be analyzed; it may indicate how the equipment is to be operated during the test. This chapter is mainly concerned with such classifications. Details on measurement techniques and instrumentation are covered in other chapters of this handbook (especially in Chaps. 12 through 18). Environmental specifications are considered in Chap. 24.

Standards are used: (1) to assure public safety and to eliminate misunderstandings between manufacturer and purchaser, (2) to aid in the selection of the proper product for a particular need, and (3) to assist in comparison of structural safety, performance, and maintenance qualities of similar pieces of equipment. The nature of a standard developed for use in a factory to control the quality of a product is usually very different from standards used for national and international trade. Most vibration standards are relatively new. One reason is that the problems associated with the development of high-speed machinery are complex; the establishment of associated standards is difficult. Among the earliest standards were those established by the insurance industry and by the U.S. government.

The objectives of vibration standards (e.g., to establish and control quality, performance, and safety; to establish and implement maintenance programs; and to perform fault diagnosis of equipment) are interrelated even though procedures and criteria for evaluation sometimes differ considerably. Consider the following examples: Vibration standards may be adopted to control quality of performance, but they are also useful in considerations of safety. In addition, safety and production maintenance are interrelated: preventive maintenance through vibration monitoring protects property and life, but it can also ensure the maintenance of production schedules. Alarms and shutdown devices set according to vibration standards are often used on vibration monitoring instruments to protect equipment and personnel. The quality of new or rebuilt equipment, from the point of view of vibrations, can be judged by comparing measured vibration levels to a standard. Vibration standards are included in specifications for new and rebuilt equipment and in more general codes and standards to guarantee performance and to permit the correction of malfunctions in equipment component performance before a catastrophic failure occurs.

STANDARDS ORGANIZATIONS AND GROUPS

In the field of vibration, the two recognized international organizations are the International Standards Organization (ISO), which is technology-oriented, and the International Electrical Commission (IEC), which is product-oriented. The ISO works in cooperation with national organizations, such as the American National Standards Institute (ANSI)—a nongovernmental institution that coordinates the development of voluntary national standards in the United States; outside the United States, national standards organizations usually are government institutions. Vibration standards activity in the ISO is guided by Technical Committee 108, Mechanical Vibration and Shock; vibration standards in ANSI are guided by Technical Committee S2, Mechanical Vibration and Shock. The secretariat for both organizations is held by the Acoustical Society of America. Vibration Standards activity in the IEC is guided by Technical Committee 50; and in the United States, the Institute of Environmental Sciences is the associated technical society. IEC works with trade associations, such as the National Electrical Manufacturers Association; other trade organizations that have adopted formal vibration standards include, for example, the American Petroleum Institute, the Hydraulic Institute, and the Compressed Air and Gas Association. Various technical societies have been instrumental in the development of codes and standards concerned with vibration; such documents have been established by consensus of consumers and manufacturers and their use is voluntary. In addition, standards and specifications have been developed by the Department of Defense to ensure the quality of equipment which it procures.

CLASSIFICATION OF SEVERITY OF MACHINERY VIBRATION

Should measurements of vibration displacement, velocity, or acceleration be used as a basis for the establishment of standards for machinery vibration? This depends on the type of standard, the frequency range, and other factors. In classifying machinery vibration in the 10 to 1,000 cps range, vibration velocity usually is used because vibration velocity is independent of frequency in this frequency range, yielding a simple measure of the severity of vibration of a new or operating machine.

There is no general agreement on whether machinery vibration may best be classified in terms of peak or root-mean-square velocity measurements. For simple harmonic motion, either peak or rms values may be used; however, for machines whose motion is complex, the use of these two indices provide distinctly different results—mainly because the higher frequency harmonics are given different weights. In the United States and Canada, the peak value of the velocity has been used to characterize machinery vibration. In contrast, in Europe the rms value has been used. The International Standards Organization has adopted a special quantity—*vibration severity*[1]—for this purpose. The rms value of vibration velocity of a machine is measured (at prescribed measuring points and in prescribed directions); the largest such measured value is said to characterize the vibration severity.

USE OF RMS VALUES—VIBRATION SEVERITY

An ISO standard[2] for classifying the severity of vibration of large rotating machines in situ, operating at speeds from 600 to 12,000 rpm, is shown in Table 19.1. Root-mean-square velocity measurements were made on the bearing housings in three orthogonal directions. The rating of a machine depends on the classification of the supports for the machine, i.e., whether they are "hard" or "soft." The supports are said to be soft if the fundamental frequency of the machine on its supports is lower than its main excitation frequency. The supports are said to be hard if the fundamental frequency of the machine on its supports is higher than its main excitation frequency.

Table 19.2 shows a classification scheme[1] which applies to rotating machines in the operating speed range of 600 to 12,000 rpm. It may be used to compare similar machines

Table 19.1. Quality Judgment of Vibration Severity

(*After ISO Std IS 3945.*[2])

Vibration severity		Support classification	
in./sec	mm/sec	Hard supports	Soft supports
0.017	0.45		
0.028	0.71	Good	Good
0.044	1.12		
0.071	1.8		
0.11	2.8	Satisfactory	
0.18	4.5		Satisfactory
0.28	7.1	Unsatisfactory	
0.44	11.2		Unsatisfactory
0.71	18.0		
1.10	28.0	Impermissible	
2.80	71.0		Impermissible

Table 19.2. Vibration Severity Ranges and Examples of Their Application

(*After ISO Std IS 2372.*[1])

Range of vibration severity	Examples of quality judgment for separate classes of machines			
Limits of range, mm/sec	Small machines, class I	Medium machines, class II	Large machines, class III	Turbo-machines, class IV
0.28	A			
0.45		A	A	
0.71				A
1.12	B			
1.8		B		
2.8	C		B	
4.5		C		B
7.1	D		C	
11.2				C
18				
28		D	D	
45				D

The letters A, B C and D represent machine vibration quality grades, ranging from good (A) to unacceptable (D).

or to compare the vibration of "normal" machines with respect to their reliability, safety, and effects on the environment.

European national standards are available on allowable limits for using unbalanced solid machine elements,[3] judging the influence of mechanical vibration on machine quality,[4] evaluating vibration in machinery,[5] and measuring machine vibration.[6] A standard[7] that specifies requirements for instruments used to measure vibration severity in rotating and reciprocating machinery has been adopted to reduce measurement inaccuracies.

USE OF PEAK VALUES

A number of rating systems have been developed for characterizing the state of machinery vibration which make use of measurements of peak values of the velocity of vibration. Such rating systems can act as a guide to classifying machinery vibration; for example, see Ref. 8. Another guide, developed as part of a plant maintenance program for bolted-down steady-rotating machinery, makes use of the maximum value of peak velocity, which is measured in three directions on the bearing housing.[9] Another guide,[10] for the classification of vibration of rotating and reciprocating machines, is based on the peak values of velocity of vibration measured on the bearing housings in three orthogonal directions, at speeds from 190 to 12,000 rpm.

COMPRESSORS

The vibration of compressors is influenced by many factors, such as speed, flow rate, pressure, and temperature. Therefore compressor vibration must be measured under carefully controlled conditions. An ISO document has proposed the use of vibration severity for the evaluation of reciprocating compressors. Guides which are based on measurements of peak velocity are given in Ref. 10 for reciprocating and rotating compressors, and in Ref. 8 for centrifugal compressors.

CRITICAL SPEEDS.* According to an industry standard, the first lateral critical speed of rigid-shaft compressors[11] must be at least 20 per cent higher than the maximum continuous speed. Flexible-shaft compressors must operate with the first critical speed at least 15 per cent below any operating speed; the second lateral critical speed must be 20 per cent above the maximum continuous speed. No actual torsional natural frequency may be within 10 per cent of the first or second harmonic of the operating speed range nor within 10 per cent of gear-tooth contact frequencies encountered.

SHAFT VIBRATION. Data provided by the measurement of shaft vibration (near the bearings) with a proximity-type transducer form the basis for standards[11,12] for the judgment of quality of compressors. The vibration is measured on the shaft relative to each radial bearing. According to these standards, the assembled compressor, operating at maximum continuous speed (or at any other speed within the specified operating speed range) may not have a double-amplitude vibration in any plane that exceeds the value of Eq. (19.1) or 2.0 mils, whichever is less:

$$\text{Peak-to-peak}\dagger\ \text{displacement} = \sqrt{\frac{12{,}000}{\text{rpm}}} \tag{19.1}$$

Guidelines[13] for vibration-measuring equipment, measurements, and vibration limits for maintenance decisions based on shaft vibration measurement are available for centrifugal compressors.

ELECTRIC MOTORS

Standards that establish classifications systems for permissible vibration in electric motors have been adopted by various trade associations, industries, and countries, and by the ISO. These classification systems are not identical; some standards are based on peak-to-peak displacement measurements, whereas others are based on rms or peak velocity measurements. Usually each standard specifies the method of test mounting, provides a guide on instrumentation to be used in the test, establishes measurement procedures, and provides maintenance information.

* Critical speeds are rotor speeds that induce a condition of resonance in rotor-bearing systems. Resonance occurs when a system natural frequency coincides with a forcing phenomenon frequency which may be rotor speed or some multiple thereof.

† Includes the runout in mils. Shaft runout is the total indicator reading in a radial direction when the shaft is rotated in its bearings.

TRADE ASSOCIATION STANDARDS. A standard[14] for integral horsepower alternating-current and direct-current motors is summarized in Table 19.3, which lists the

Table 19.3. Maximum Permissible Vibration for Integral Horsepower Electric Motors

(*After NEMA MG1-12.05.*[14])

Speed, rpm*	Peak-to-peak displacement amplitude, in.
3,000–4,000	0.001
1,500–2,999	0.0015
1,000–1,499	0.002
999 and below	0.0025

* For alternating-current motors, use the highest synchronous speed. For direct-current motors, use the highest rated speed. For series and universal motors, use the operating speed.

maximum permissible values of displacement amplitude. A standard procedure for mounting the motor, for measuring its vibration, and for operating it is specified.[15]

A standard[16] for large induction motors is summarized in Table 19.4, which lists the maximum permissible values of displacement amplitude. A standard procedure[17] for mounting the motor, for measuring its vibration, and for operating it is specified.

Table 19.4. Maximum Permissible Vibration for Large Induction Motors

(*After NEMA MG1-20.52.*[16])

Speed, rpm	Peak-to-peak displacement amplitude, in.
3,000 and above	0.001
1,500–2,999	0.002
1,000–1,499	0.0025
999 and below	0.003

A standard which specifies the maximum permissible vibration for form-wound squirrel-cage induction motors (200 hp and larger) is summarized in Table 19.5. The measurement procedures outlined in Refs. 15 and 17 are used for elastic and rigidly mounted motors, respectively.

INDUSTRY STANDARDS. A guide for classifying the severity of vibration in uninstalled electric motors that are new, repaired and balanced, or repaired without balancing is given in Ref. 19, this guide makes use of peak-to-peak displacement amplitude measurements taken at the bearing cap.

NATIONAL STANDARDS. Measurement and evaluation of the quality of electrical rotating machines with respect to mechanical vibration are described in several European national standards.[20–23]

INTERNATIONAL STANDARDS. For three-phase alternating current motors and direct-current machines with shaft heights between 80 and 400 mm, an ISO standard[24] has been established for evaluating motor quality. This standard specifies the machine mounting, where the measurements are made (bearing housing), and the operating condi-

Table 19.5. Maximum Permissible Vibration for Form-wound Squirrel-cage Induction Motors

(*After API STD 541.*[18])

Synchronous speed, rpm	Peak-to-peak displacement amplitude, in.	
	Motor on elastic mount	Motor on rigid mount
720 to 1,499	0.002	0.0025
1,500 to 3,000	0.0015	0.002
3,000 and above	0.001	0.001

Table 19.6. Quality Judgment of Electric Motors Based on Vibration Severity

(*After ISO IS 2373.*[24])

Quality grade	Rotational speed	Maximum rms values of velocity amplitude for the shaft height H in mm*					
		$80 < H < 132$		$132 < H < 225$		$225 < H < 400$	
	rpm	mm/sec	in./sec	mm/sec	in./sec	mm/sec	in./sec
N (normal)	600 to 3,600	1.8	0.044	2.8	0.071	4.5	0.11
R (reduced)	600 to 1,800	0.71	0.018	1.12	0.028	1.8	0.044
	<1,800 to 3,600	1.12	0.028	1.8	0.044	2.8	0.071
S (special)	600 to 1,800	0.45	0.011	0.71	0.018	1.12	0.028
	<1,800 to 3,600	0.71	0.018	1.12	0.028	1.8	0.044

* A single set of values, such as those applicable to the 132- to 225-mm shaft height, may be used if shown by experience to be required.

tions during tests; the true rms value of the velocity must be used. The recommended limits of vibration are given in Table 19.6 for various degrees of quality, speed, and shaft height.

FANS

Unbalanced fans may cause structural damage, undesirable noise, and vibration. A guide for rating the unbalance of cooling-tower fans based on economic and safety conditions is given in Ref. 25. One European country has a national standard for the measurement of vibration and noise from ventilators.[26]

GEAR UNITS

A standard[27] specifying permissible limits of vibration in gear units is summarized in Table 19.7. Usually measurements are made with a proximity-type transducer which does not touch the gear unit. According to this standard, lateral critical speeds

Table 19.7. Vibration Limits for Gear Units

(*After API STD 613.*[27])

Maximum continuous speed, rpm	Double amplitude including runout, mils	
	Shop test unloaded	Shop test loaded
Up to 8,000	2.0	1.5
8,000 to 12,000	1.5	1.0
Over 12,000	Less than 1.5	Less than 1.0

of the gear unit should be at least 20 per cent below, or 20 per cent above, the operating speed range. Another standard, developed by the American Gear Manufacturers Association (AGMA), specifies the vibration limits shown in Fig. 19.1; this standard, based on peak-to-peak displacement measurements, uses damage avoidance as a criterion.[28]

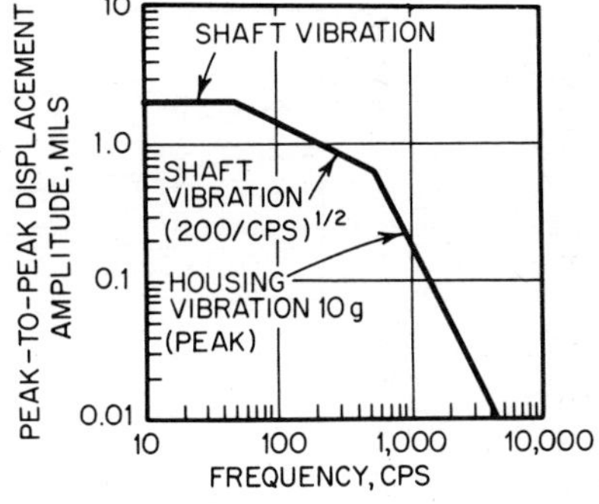

Fig. 19.1. Vibration limits for gear units (*AGMA.*[28]).

MACHINE TOOLS

The vibration of machine tools affects the quality of the work produced, the machine structure, and the surrounding environment. Reference 29 includes a guide which indicates the tolerable range for such vibration.

PUMPS

A standard of the Hydraulic Institute specifies the maximum permissible amplitude of displacement in any plane for clean-liquid-handling pumps[30] (see Fig. 19.2). A similar curve for nonclog centrifugal pumps is shown in Fig. 19.3. Vibration-displacement measurements made at the bearing housings are compared with the appropriate curve when the pump is operating at rated speed and within 10 per cent above or below rated capacity. Vibration measurements are made at the top motor bearing and bearing housing, respectively, for vertical and horizontal pumps. It is the Institute's recommendation that if the vibration exceeds the limits shown in Figs. 19.2 and 19.3, a pump should be examined for defects.

TURBOMACHINERY

Standards which establish permissible vibration in turbomachinery are useful for evaluating quality and establishing maintenance action. Most turbomachines operate at a high speed, above their first or second critical speed. Hence measurements of shaft vibration provide a better indication of conditions of vibration of the machinery than do other vibration measurements—for example vibration measurements at the bearing caps.

CRITICAL SPEEDS.[31–33] According to standards for single- and multistage steam turbines or special-purpose steam turbines, to avoid excessive vibration the critical speed (see Chap. 39) of a rigid-shaft turbine rotor should be at least 10 per cent higher than

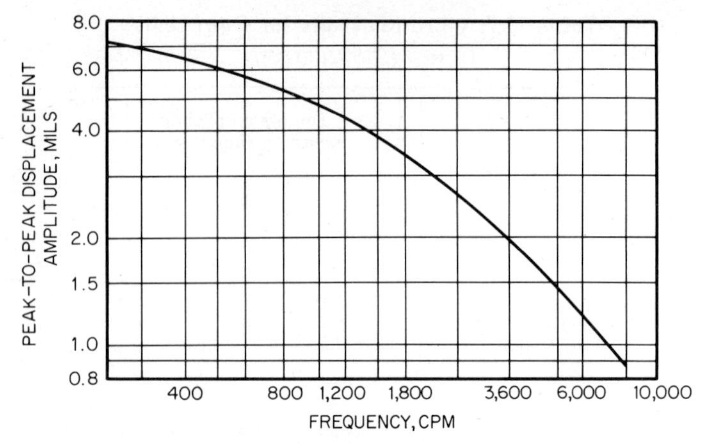

FIG. 19.2. Vibration classification for clean-liquid-handling centrifugal pumps. (*After Hydraulic Institute.*[30])

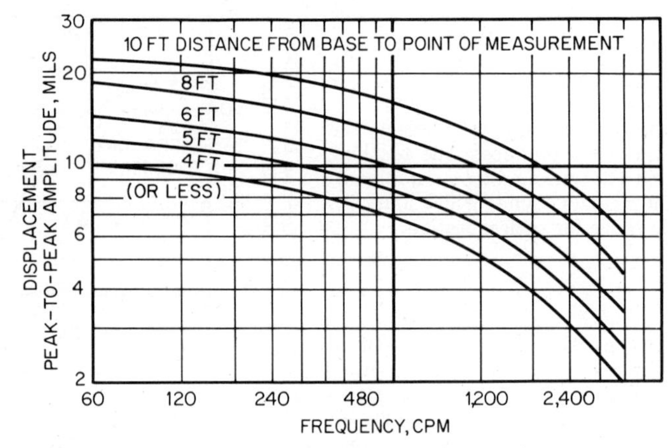

FIG. 19.3. Vibration classification for centrifugal pumps, vertical or horizontal nonclog. (*After Hydraulic Institute.*[30])

its maximum speed; the value of the first critical speed of a flexible-shaft turbine rotor should not exceed 60 per cent of the maximum continuous speed, nor should it be within 10 per cent of any operating speed.

ACCEPTABLE LEVELS OF VIBRATION. Standards for acceptable levels of vibration exist only for restricted classes of turbomachinery. Most of the standards are based on measured values of vibration displacement (relative to the shaft) of turbomachinery operating in situ. Acceptable limits of vibration for various types of turbomachines are shown in Fig. 19.4. For example, these curves indicate that for a single-stage steam turbine,[31] the vibration displacement should not exceed 2 mils double amplitude for rated speeds of 4,000 rpm or less; for operating speeds above 4,000 rpm, the double amplitude measured on the shaft should not exceed 1.5 mils. According to this standard, if it is not possible to measure the vibration on the shaft, a measurement of vibration displacement at the bearing housing should be made; this value should not exceed 50 per cent of the value of the displacement of the shaft.

The acceptable vibration level for multistage steam turbines[32] for mechanical drive

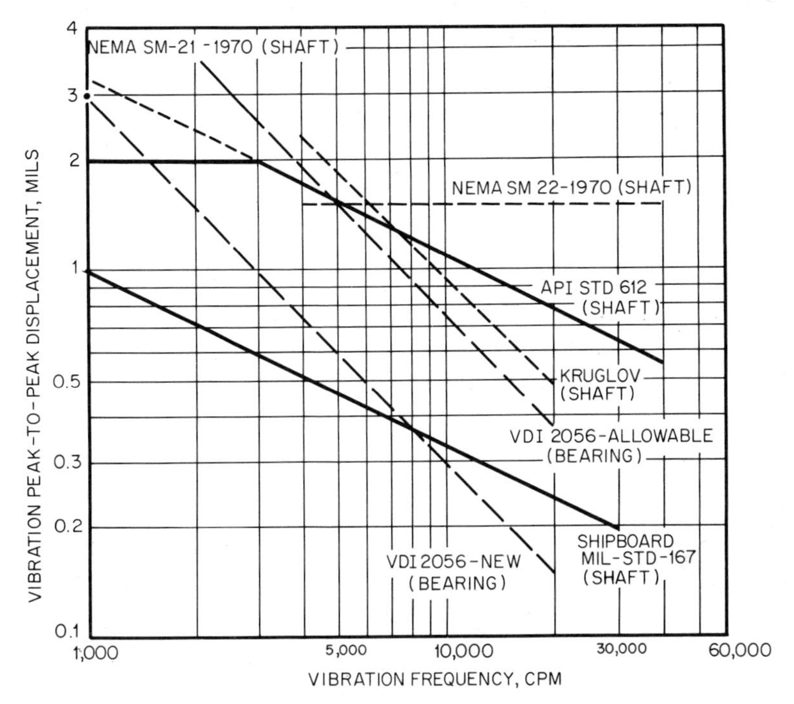

FIG. 19.4. Turbomachinery shaft vibration limits.

service (Fig. 19.4) can be expressed by Eq. (19.1), which is given for centrifugal compressors. According to Fig. 19.4 for special-purpose steam turbines[33] for refinery service, below 3,000 rpm an increase in vibration displacement amplitude greater than 2 mils is not permitted.

A guide to the classification of vibration limits based on quality of performance for turboalternators operating at 3,000 rpm is contained in Ref. 34.

Acceptable values[35] for flexible-rotor machines mounted aboard a ship are an order of magnitude lower than those accepted by trade associations. The values shown in Fig. 19.5 are for new turbomachines mounted aboard ship and for equipment being evaluated for quality after overhaul; they are not an indication of maintenance action.

A USSR standard[36] for high-speed turbomachinery using shaft measurements is given in Fig. 19.6. Some agreement may be noted with acceptable limits given in Fig. 19.4 established by trade associations in the United States.

A German standard[37], based on bearing vibration measurements, is shown in Fig. 19.7. A vibration velocity of 2.8 mm/sec is allowed for new machinery; a vibration velocity in excess of 7 mm/sec is allowed for short running periods only if the source of the vibration is known. German standard VDI 2059[38] involves the measurement of shaft vibration on machines.

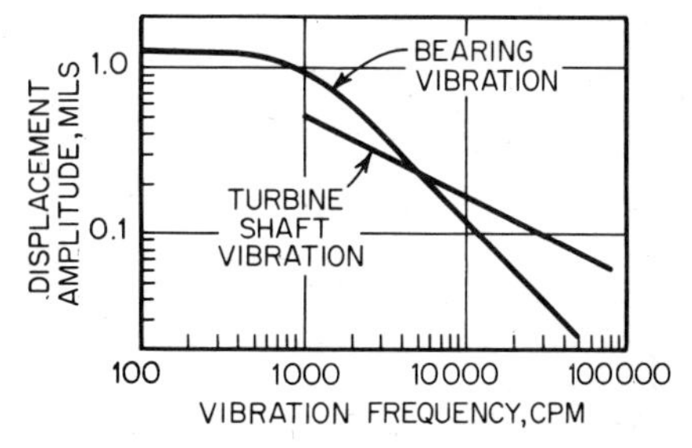

FIG. 19.5. Maximum allowable vibration for shipboard-mounted machinery. (*After MIL-STD*-167-I *SHIPS*.[35])

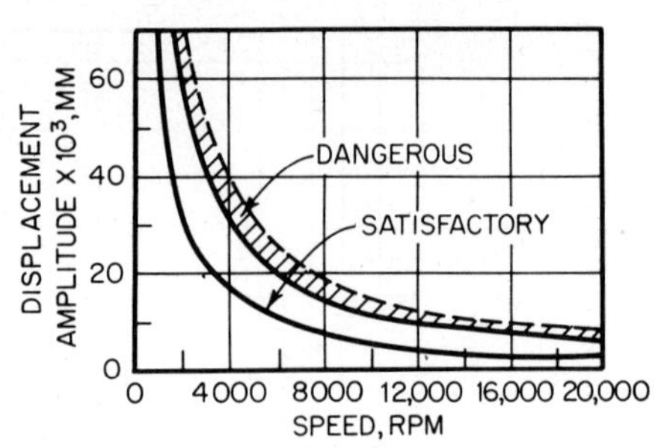

Fig. 19.6. Vibration standards of high-speed machines. (*After Kruglov.*[36])

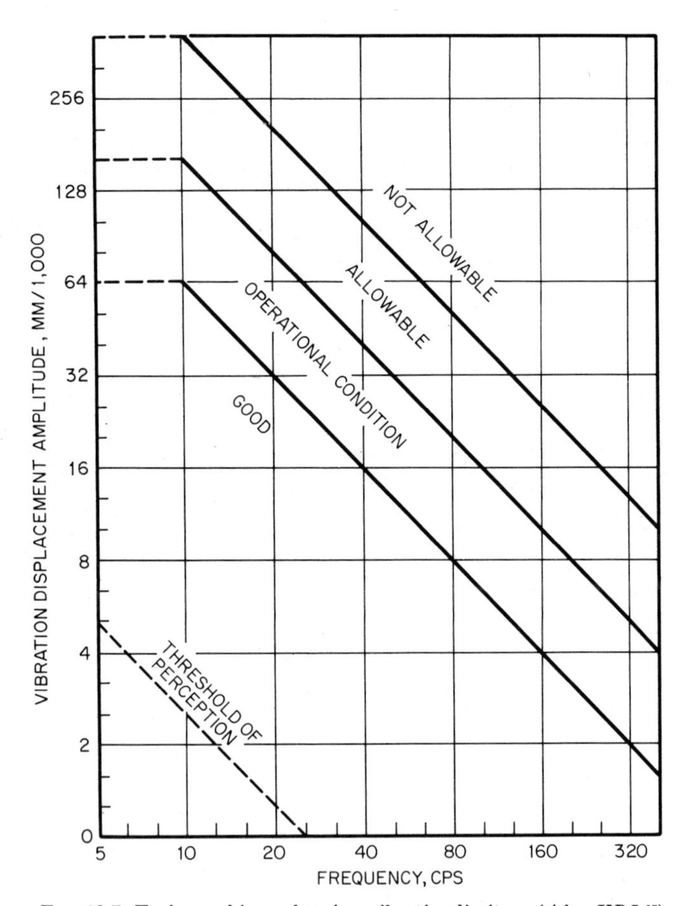

Fig. 19.7. Turbomachinery bearing vibration limits. (*After VDI.*[37])

In the United States it is industrial practice[39] to limit vibration on high-speed machines to values of 2.2 and 4.2 mils peak-to-peak for warning and shutdown, respectively. Rotor-thrust displacement monitors are set with 15-mils-over-normal load deflection for automatic shutdown.

MACHINERY COMPONENTS

ANTIFRICTION BEARINGS. An assessment of the state of bearings by vibration measurement techniques is essential to the evaluation of bearing quality and maintenance requirements. A guide to the classification of vibration of installed bearings is given in Ref. 8.

MACHINE FOUNDATIONS. A guide indicating the safe limits of vibration for machine foundations is given in Ref. 40.

VIBRATION STANDARDS FOR SHIPS

Vibration standards for ships are used by the military to ensure structural integrity and to minimize undesirable radiation of noise. According to one such specification for naval equipment, the residual vibration of shipboard-mounted equipment must have a magnitude less than that shown in Fig. 19.5; these limits apply to systems having an elastic mounting with a natural frequency less than 0.25 of the minimum rotational frequency of the unit. Vibration-displacement amplitudes are measured on the bearing housing in the direction of maximum amplitudes (except for turbines, for which measurements are made on the rotating shaft adjacent to the bearings). On constant-speed machines, measurements are made at operating speed; on variable-speed machines, measurements are made at maximum speed and at all critical speeds within the operating range. In the case of complex machinery such as reduction gearing and impellers, the peak vibration velocity must not exceed 0.15 in./sec; the displacement amplitude must not exceed the values given in Fig. 19.5 and not exceed 1.25 mils in any case.

Hull vibration is excited by the propeller shaft and the propeller blade. Such vibration is greatest in the afterpart of the ship (except when one of the hull resonant frequencies coincides with a driving frequency). The displacement amplitudes often exceed 0.1 in. At higher speeds of the propeller shaft, the amplitudes of the high modes of vibration of the hull tend to be smaller. Vibration data indicate that the maximum displacement amplitude of the main structural members of the hull (measured at the antinodes and at the end of the ship) is smaller at the higher excited frequencies.[41] When shipboard machinery and equipment are fastened to the hull, deck, and bulkhead stiffeners, the vibration amplitudes at the equipment supports are substantially the same as the main structural members of the hull.

The amplitudes of vibration of the bow and stern are used as a measure of hull vibration. In some types of ships, the maximum displacement amplitude of hull vibration is less than 0.01 in. at propeller-blade frequencies. A displacement amplitude of less than 0.02 in. is considered satisfactory; if this value is exceeded, corrective measures are usually taken.

VIBRATION STANDARDS FOR AIRCRAFT, SPACECRAFT, AND MISSILES

Aircraft vibration may be excited by many sources, including engines, propellers, airflow over the external surfaces of the aircraft, and auxiliary mechanical equipment. Military standard MIL-STD-810D is used to determine the acceptability of equipment with respect to vibrationally induced stresses and to ensure that performance degradations or malfunctions will not be produced by the service vibration environment whether it be aircraft or other vehicles such as missiles, spacecraft, or helicopters.

The effects of gun-blast pressure impinging on aircraft structure from high-speed repetitive firing of installed guns and general aircraft motions caused by such factors as runway roughness, landing, and gusts have been incorporated into criteria of MIL-STD-810D.

An international standard has been established[43] with test procedures similar to those described above. The criteria in this IEC document are based upon the ability of equipment to function and endure specified magnitudes of vibration. Sinusoidal and random vibration testing, as well as shock testing, are included in this document, which is applicable to wide classes of equipment. The severity of vibration that acceptable equipment is required to withstand is specified by a combination of parameters including frequency range, vibration amplitude, and endurance duration. For sinusoidal testing below a specified frequency called the crossover frequency (57 to 62 cps), all amplitudes are specified as constant displacement, whereas above this frequency, amplitudes are specified as constant acceleration. The test procedures include an initial resonance search followed by endurance conditioning by sweeping, endurance conditioning at resonance frequencies, and/or endurance conditioning at predetermined frequencies for

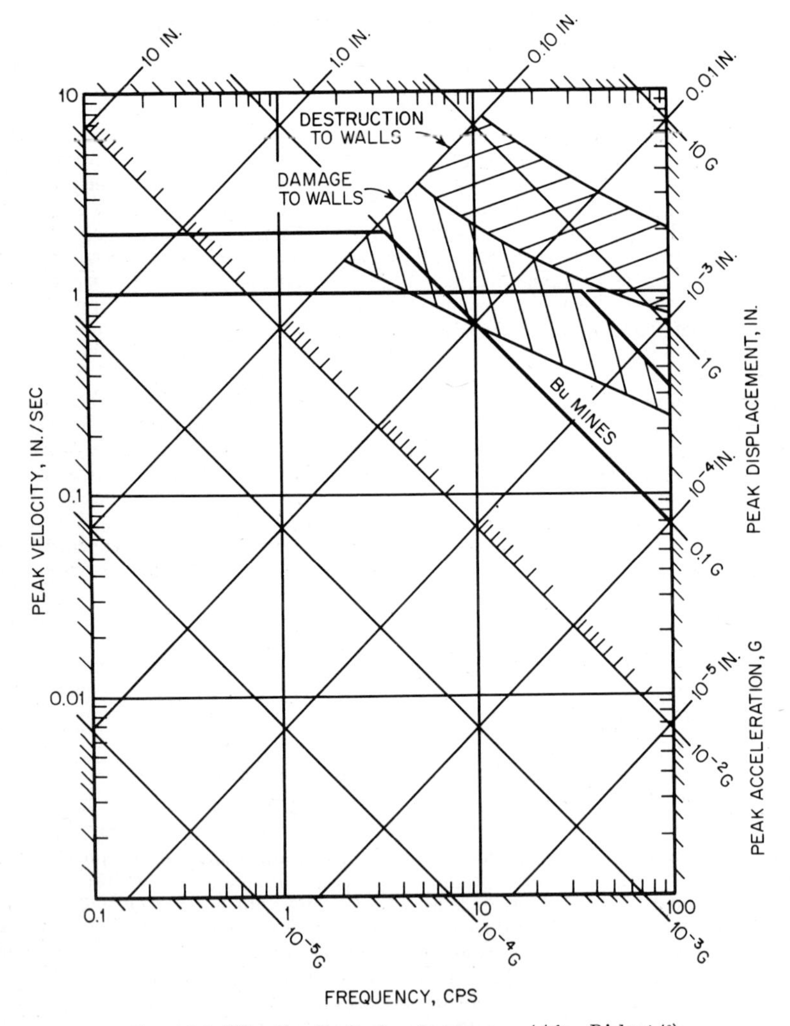

FIG. 19.8. Vibration limits for structures. (*After Richart.*[40])

specified levels and time. To qualify, equipment must not malfunction or sustain significant damage as a result of the test.

STRUCTURES

Structures such as public buildings, offices, factories, bridges, and power plants are subject to vibration as a result of forces generated by ground motion, wind, traffic, and machinery. Vibration limits for structures subject to blasting are given in Fig. 19.8. Although the lower limit for the safe zone (denoted "caution to structures") represents a peak velocity of 3 in./sec, it is general practice[44] to limit the peak velocity to 2 in./sec. For example, the U.S. Bureau of Mines criterion[40] for structural safety against damage from blasting, shown in Fig. 19.8, limits the peak velocity to 2 in./sec below 3 cps; above this frequency the peak acceleration is limited to $0.1g$. Another chart showing the safe limits of vibration in structures, for the frequency range from 0 to 50 cps, is given in Fig. 19.9.[45]

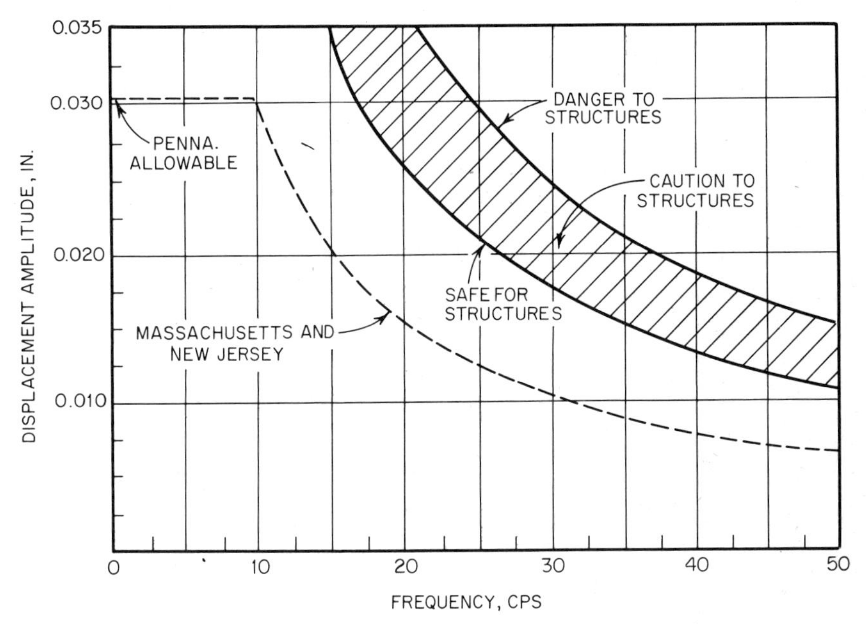

FIG. 19.9. Safe limits for structures. (*After L. D. Leet.*[45])

To facilitate the evaluation and comparison of data and to give provisional guidance on acceptable levels of vibration, an ISO document[46] suggests limits for vibration caused by quarry blasting or the like, and for floor vibration from disturbance or machinery (see Fig. 19.10). In this document, vibration is expressed in terms of maximum rms velocity amplitude as calculated or measured from the orthogonal components of vibration in the structure; the greatest value, determined at the prescribed measuring points, characterizes the vibration severity of the building.

An IEEE standard[47] serves as a guide for the qualification of class I electrical equipment* which is installed in nuclear power plants that are subject to earthquakes. This standard establishes qualification procedures on an experimental, analytical, or com-

* Class I electrical equipment is electrical equipment that is essential to the safe shutdown and isolation of the reactor or whose failure or damage could result in significant release of radioactive material.

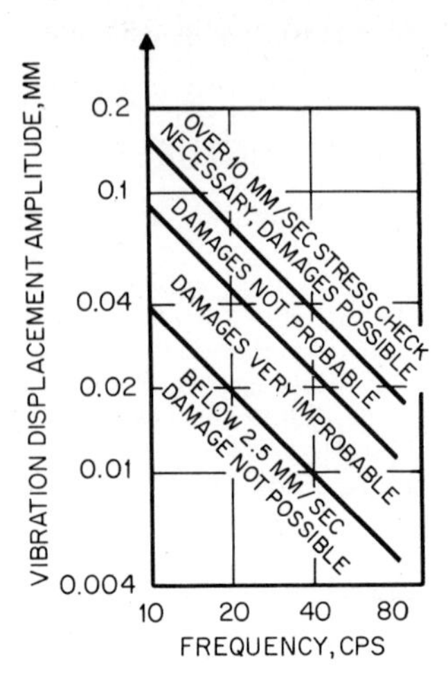

FIG. 19.10. Rough evaluation of stationary-floor vibrations by measurement of vibration displacement amplitude and frequency. (*ISO.*[46])

bined analytical-experimental basis. Qualification of class I electric equipment requires that it meet performance criteria during and after one safe shutdown earthquake.*

REFERENCES

1. International Standards Organization Standard: "Mechanical Vibration of Machines with Operating Speeds from 10 to 200 rps—Basis for Specifying Evaluation Standards," ISO/IS 2732, 1972.
2. International Standards Organization Standard: "The Measurement and Evaluation of Vibration Severity of Large Rotating Machine in Situ, Operating at Speeds from 10 to 200 rps," ISO/IS 3945, 1975.
3. Czechoslovakian National Standard: "Permitted Limits for Unbalanced Solid Machine Elements," CSN 011410, 1971.
4. German National Standard: "Beurterlung der Einwirkung Mechanischen Schwingungen auf den Menschen," VDI 2057.
5. British National Standard: "A Basis for Comparative Evaluation of Vibration in Machinery," BS4675.
6. Austrian National Standard: "Schwingungsgedämmte Maschinenaufstellung," ÖAL-Richtlinie Nr. 7.
7. International Standards Organization Standard: "Mechanical Vibration of Rotating and Reciprocating Machinery—Requirements for Instruments for Measuring Vibration Severity," ISO/IS 2954, 1973.
8. Blake, M. P.: "Standards and Tables: Vibration and Acoustic Measurement Handbook," Spartan Books, New York, 1972.
9. Maten, S.: "Velocity Criteria for Machine Vibration: Vibration and Acoustic Measurement Handbook," p 267, Spartan Books, New York, 1972.
10. Lortie, L. J.: "Philosophy of Diagnosis," Technology Interchange—Research 1, The Vibration Institute, Chicago, April, 1974.
11. American Petroleum Institute Standard, "Centrifugal Compressor for General Refinery Services," API Standard 617 Third Edition, October, 1973.
12. Compressed Air and Gas Institute Standard: "Standards for Centrifugal Air Compressors," 1975.
13. Clark Turbo Products Engineering Department: "General Guide Lines for Vibration on Clark Centrifugal Compressors," Clark Turbo: Compressor Division, Dresser Industrial, Olean, N.Y., 1968.

* Safe shutdown earthquake is an earthquake producing the maximum vibratory ground motion that the nuclear-power-generating station is designed to withstand without functional impairment of those features necessary to shut down the reactor, maintain the station in a safe condition, and prevent undue risk to the health and safety of the public.

14. National Electrical Manufacturers Association Standard, "MG1-12.05, Dynamic Balance of Motor," Dec. 16, 1971.
15. National Electrical Manufacturers Association Standard: "MG1-12.06, Method of Measuring the Motor Vibration," December, 1971.
16. National Electrical Manufacturers Association Standards: "MG1-20.52, Balance of Machines," July 16, 1969.
17. National Electrical Manufacturers Association Standard: "MG1-20.53, Method of Measuring the Motor Vibration," July, 1969.
18. American Petroleum Institute Standard, "Recommended Practice for Form-Wound Squirrel-Cage Induction Motors," API Standard 541.
19. Blake, M. P.: *Hydrocarbon Process.*, **45**(3):126 (1966).
20. Bulgarian National Standard: "Measurement of Vibration on Electrical Rotating Machines," BDS 5626-65.
21. German National Standard: "Schwingstärke von Rotierenden Elektrischen Maschinen den Baugrössen 80bis 315 Messverfahren und Grenzwerte," DIN 45665.
22. Polish National Standard: "Measurement of Vibration on Electrical Rotating Machines," PN-65/E-04255, 1965.
23. Russian National Standard: "Measurement of Vibration on Electrical Rotating Machines," GOST 12379-66.
24. International Standards Organization: "Mechanical Vibration of Certain Rotating Electrical Machinery with Shaft Heights between 80 and 400 mm—Measurement and Evaluation of the Vibration Severity," ISO/IS 2373, 1971.
25. Blake, M. P.: "Balance Cooling Tower Fans in Place," *Hydrocarbon Process.*, **46**:6(150) (1967).
26. Czechoslovakian National Standard: "Measurement of Noise and Vibration from Ventilators," ČSN 123062, 1961.
27. American Petroleum Institute Standard: "High-Speed, Special Purpose Gear Units for Refinery Services," API STD 613, August, 1968.
28. AGMA Standard: "Specification for Measurement of Lateral Vibration on High Speed Helical and Herringbone Gear Units," Standard 426.01.
29. Baxter, R. L., and Bernhard, D. L.: "Vibration Tolerances For Industry," *ASME* Paper 67-PEM-14, 1967.
30. Hydraulic Institute Standard: "Vibration Limits of Centrifugal Pumps," 11th ed., Centrifugal Pump Section, VI, Application Standards B-74-1, May, 1967.
31. National Electrical Manufacturers Association Standard: "Single Stage Steam Turbines for Mechanical Drive Service," SM 22-1970.
32. National Electrical Manufacturers Association Standard: "Multistage Steam Turbines For Mechanical Drive Service," SM21-1970.
33. American Petroleum Institute Standard: "Special-Purpose Steam Turbines For Refinery Services," API Standard 612, 1st ed., November, 1969.
34. Parvis, E., and Appendino, M.: "Large Size Turbogenerator Foundations—Dynamic Problems and Consideration on Designing," Ente Nationale per l'Energia Electrica, Department of Milan, Steam Power Plant Design, Engineering, and Construction Division, Italy, 1966.
35. Military Standard: "Mechanical Vibrations of Shipboard Equipment (Type I—Environmental and Type II—Internally Excited)," MIL-STD-167-I (Ships), May, 1974.
36. Kruglov, N. V., "Turbomachine Vibration Standards," *Teploenerg*, **8**:85 (1959).
37. Verlin Deutscher Ingenieure: "Beurteilung der Einwirkung mechanischer Schwingungen auf den Menschen," VDI Standard 2056, October, 1964.
38. German National Standard: "Wellenschwingungsmessungen zur Überwachung von Turbomachimen," VDI 2059.
39. Jackson, C.: "Optimize Your Vibration Analysis Procedures, Turbomachinery Handbook, Hydrocarbon Processing," p. 111, Gulf Publishing Company, Houston, 1974.
40. Richart, Jr., F. E., et al.: "Vibrations of Soils and Foundations," Prentice Hall Inc., Englewood Cliffs, N.J., 1970.
41. Vigness, I., and Hardy, V. S.: "Vibration on Ships," *Shock and Vibration Confidential, Bull.* 23, Office of Secretary of Defense, 1956.
42. Military Standard: "Environmental Test Methods," MIL-STD-810D, 1975.
43. International Electrical Commission Standard: "Basic Environmental Testing Procedures," IEC 68-2.
44. Wiss, J. F.: *Civil Eng.*, **38**(7) pp 46–48 (July, 1968).
45. Leet, L. D.: "Vibrations from Blasting Rock," p. 96, Harvard University Press, Cambridge, Mass., 1960.
46. International Standards Organization Standard: "Evaluation and Measurement of Vibration in Buildings," ISO/TC108/SC2/Wg3-9, in preparation.
47. IEEE Standard: "Guide for Seismic Qualification of Class I Electrical Equipment for Nuclear Power Generating Stations," IEEE 344:1973.

20

MEASUREMENT TECHNIQUES

Richard D. Baxter
Convair, Division of General Dynamics

John J. Beckman
Convair, Division of General Dynamics

Harold A. Brown
Convair, Division of General Dynamics

INTRODUCTION

This chapter outlines many of the techniques employed in shock and vibration measurement. It includes a discussion of planning the test objectives, selecting types of measurements to implement them, selecting the measurement system best suited to these requirements, installing the components of the system, and testing and calibrating the system prior to field measurements. In addition, other factors must be considered, such as the method of data analysis to be employed. Many of the measurement techniques referred to here are treated in detail in preceding chapters.

DEFINING THE PROBLEM

The first step toward measurement is to define the nature of the test and what is to be measured. Careful pretest planning may save much time in making the measurements and obtaining the most useful information from the test data. Planning should start with a clear, written definition of the test objectives. The next step is to establish the various measurement requirements. Examples of such requirements are listed in Table 20.1. In the more simple vibration measurement problems, only a few of these factors need be considered. On the basis of this information, one can select specific instrumentation for the test.

SELECTION OF EQUIPMENT—PRETEST PLANNING

The selection of a measurement system for a particular test depends on many different requirements, for example, on the number and characteristics of variables to be measured. A close examination of these requirements may indicate that the desired data may be obtained from any of several different sets of measurements of variables. For example, a determination of the amount of damping in a single degree-of-freedom mass-spring sys-

Table 20.1. List of Measurement Requirements and Considerations

Measurement location
Measurement direction
Frequency range
Amplitude range
Required accuracy
Test condition for recording data (e.g., from 10 sec before to 10 sec after wheel touchdown on landing of an aircraft)
Total length of recording time required
Time correlation between channels and with test conditions
Method of structural excitation to obtain vibration
Method of record recovery
Space available for installation
Accessibility for service or calibration
Electrical power available
Total number of channels of recording
Number of vibration measurement channels of similar range
Redundancy requirement to ensure against loss of data
Specific automatic or manual data reduction operations
Type of playback equipment already available or needed
Calibration method and special fixtures needed (pretest, test, and posttest)
Percentage of the units under test that are required as spare units (or procedure to be followed in case of component failure)
Environment of operation
- Temperature
- Humidity
- Corrosiveness
- Magnetic and radio-frequency fields
- Acoustic fields
- Nuclear radiation
- Pressure (altitude)

Expected calendar duration of test program and total expected operating hours of equipment
Schedule considerations
Financial considerations

tem which is less than critically damped can be made from measurements of any of the following sets of variables:

1. Mass; spring constant; instantaneous values of driving force, displacement, velocity and acceleration of the mass.
2. Peak amplitudes of successive cycles (logarithmic decrement) and the frequency of oscillation when the mass is released at rest from a displaced position.
3. Mass, spring constant, steady-state values of driving force, displacement, and frequency.

Each possible measurement method should be examined to determine whether it satisfies the requirements which have been established and an appropriate choice should be made. Thus, in the above example, choice 3 probably is easiest to instrument and to employ.

In a proposed measurement system, the number of variables to be recorded is a most important consideration because the number may exceed the available number of recording channels. If the maximum available frequency response of the recorder is greater than that required for the data, multiplexing techniques may be used which increase the effective number of recording channels by restricting the frequency response of each channel. This may be accomplished by the frequency-sharing or time-division techniques.

In any multichannel recording system, it is desirable that the total number of recording channels include some spare channels to cover additional requirements for data not envisioned when the test run was planned and to anticipate breakdowns in the recording system.

In certain types of measurements, one must decide whether on-board recording or remote transmission is to be used. The flight testing of a long-range missile over a considerable distance, for example, requires a choice between radio telemetry or a recoverable on-board recorder. Short-range missiles can use either of these methods or a "trailing-wire" transmission system. The latter technique is generally useful during the period shortly after launch. On the other hand, radio telemetry usually is more satisfactory than wire transmission from any moving vehicle requiring ranges greater than a few hundred feet. This is because the volume of wire required is large for long distances and the handling of the wire becomes excessively difficult.

One advantage that the telemetering or trailing-wire transmission systems have in the testing of controlled vehicles, compared to the on-board recorder, is that instantaneous monitoring can be performed. For example, this allows the engineers to detect unsafe values or to redirect the test plans as the test progresses.

The choice of recorders usually is dependent on such factors as the magnitude of the data analysis task contemplated, the required frequency range, and the weight of the recorder. Magnetic-tape recordings are reproducible in a form suitable for high-speed automatic analyzers. In contrast, photographic records usually are more difficult to analyze, but they are relatively simple.

SELECTION OF A TRANSDUCER

The various characteristics of transducers are outlined in Chap. 12, and various types of transducers are described in Chaps. 13 to 17. The engineer must select the most appropriate transducer for the specific application from the many that are available. The selection may be determined by size, weight, electrical characteristics, environment in which the transducer is to operate, or limitations imposed by required auxiliary equipment. In addition, the selection is determined by the vibratory characteristics of the member to be studied. Some a priori knowledge of the motion is important in deciding whether to select a displacement, velocity, acceleration, or strain-measuring transducer. The following considerations often are of importance in determining the type of measurement to be made:

Displacement measurements may be useful:

1. Where the amplitude of displacement is particularly important (for example, in assemblies where vibrating parts must not touch).
2. Where the magnitude of the measured displacement may be an indication of stresses to be analyzed.
3. In studying low-frequency vibration where corresponding velocity and acceleration measurements may yield outputs which are too small for practical use.

Velocity measurements may be useful:

1. At intermediate frequencies when displacement amplitudes are too small to measure conveniently.
2. In correlating acoustic and vibration measurements, because a vibrating member may produce sound pressure in air which is proportional to velocity.

Acceleration measurements may be useful:

1. At high frequencies, where the highest signal output usually can be obtained from such measurements.
2. Where forces, loads, and stresses must be analyzed, since force is proportional to acceleration.
3. Where suitable displacement or velocity pickups would be too large because of clearance requirements.

Strain measurements may be useful:

1. Where a portion of the specimen that is being vibration tested has an appreciable variation in strain caused by vibration.

AUXILIARY EQUIPMENT

When the full-scale output voltage of the transducer is insufficient to drive the recording equipment with satisfactory accuracy, a signal amplifier is required. Many commercially available units are miniaturized, and have excellent characteristics with respect to noise and sensitivity.

The amplified or direct transducer output may feed into a visual or a photographic recorder in simple systems, or into a commutator or a subcarrier oscillator for tape recording or radio telemetry.

In mobile installations serious limitations frequently are imposed on the size and weight of instruments and by the availability of power. All these factors are more critical on smaller vehicles. In experimental vehicles, it usually is important to select instruments that will operate satisfactorily when the electrical voltage or frequency is considerably outside the design tolerance. Battery power is advisable in many cases, but it carries the penalty of additional volume and weight. Sometimes vehicle power "backed up" by emergency batteries is the best choice. The final selection depends, to a large extent, on the balance between weight and reliability requirements. Volume limitations may be largely overcome by using modules specially shaped to fit into space which may otherwise be wasted. Such installations are more costly as well as more difficult to manufacture and maintain. Equipment maintenance should not be overlooked during system design or installation.

TIMING AND DATA CORRELATION METHODS

In tests involving observation of many variables, it is important to be able to compare the data on all recorded channels at a given instant of time and to correlate these data with external events. This usually is accomplished by providing a direct timing channel on the recording system and a visual indication of the timing channel for the system operator.

The choice of time coding depends on length of the test and ultimate use of the data. For very short tests, a continuous periodic pulse of known frequency will suffice. For longer tests, a coded pulse whose code indicates the elapsed time from a starting reference often proves more satisfactory for reasons given below. For tests which run continuously for days, or where correlation with time of day is important, a pulse-coded time of day in hours, minutes, and seconds may be useful.* The coded pulse has the following advantages over a continuous periodic pulse:

1. A specific value of time t on a long-duration record may be located much more rapidly, since it is unnecessary to count the elapsed time from zero to t.

2. Loss of a number of coded time signals does not affect the accuracy of time measurement of those remaining, while loss of a few of the continuous periodic pulses results in the wrong value of time from point of loss.

Code time signals are particularly useful when a correlation must be provided between different recorders and when the tests are so long that the recorders must be stopped for servicing or to save recording medium during the test. Coded time signals also may be used to provide a knowledge of the "off" time. "Off" time sometimes is related to instrumentation drifts; hence this technique can be used as an indirect measure of confidence one should place in data anomalies separated by periods of "no record."

It is desirable that all timing channels of the recording system be driven from a common timing mechanism, so that synchronization between channels and recorders is maintained even if absolute time is lost because of a malfunction of the timing mechanism. The ability to synchronize data from a number of channels at a given instant of time is limited by the accuracy with which time delays in the various channels of the recording or playback system are known.

ANALYSIS CONSIDERATIONS. The method of data analysis to be employed should be considered in selecting measurement instrumentation.

CATALOGING METHODS. Instrumentation selection in a large test organization may be aided considerably by a carefully indexed catalog (such as punched cards) of each test and test installation that has been made over a period of years. When a prob-

* Such a coded pulse may be synchronized with signals from radio station WWV when time power is first applied and periodically thereafter; this radio station, in Washington, transmits frequency standard signals and Greenwich Mean Time signals at frequent intervals.

lem is presented, a quick search can be made to see whether a similar measurement has been performed previously. Any such information may be very helpful in the new test, occasionally supplying the complete answer.

INSTRUMENTATION INSTALLATIONS

In studying the motion of a specimen or vehicle one must consider the effect of the added mass and the change in stiffness of the specimen which is introduced by the measuring instruments. It is important in testing to choose measurement equipment which will not affect the characteristics of the system under test. Thus, such considerations may dictate that test equipment weigh less than a specified value, or that special mountings be provided so that the test equipment does not increase the stiffness of the system under test.

TRANSDUCER MOUNTING TECHNIQUES

The basic problem in designing a mounting is to couple the transducer to the system under test so that the transducer accurately follows the motion of the surface to which it is attached. This requires that the effective stiffness of the transducer mounting be large in the frequency range of interest; otherwise the mounting will deflect under the inertia load of the transducer mass.

Many mounting fixtures or brackets have resonant frequencies which are below 2,000 cps and have little damping. The use of such a bracket may result in significant measurement errors as a result of resonant amplification or attenuation of vibration in the bracket. This is illustrated in Fig. 20.1, which shows the frequency response of a stand-

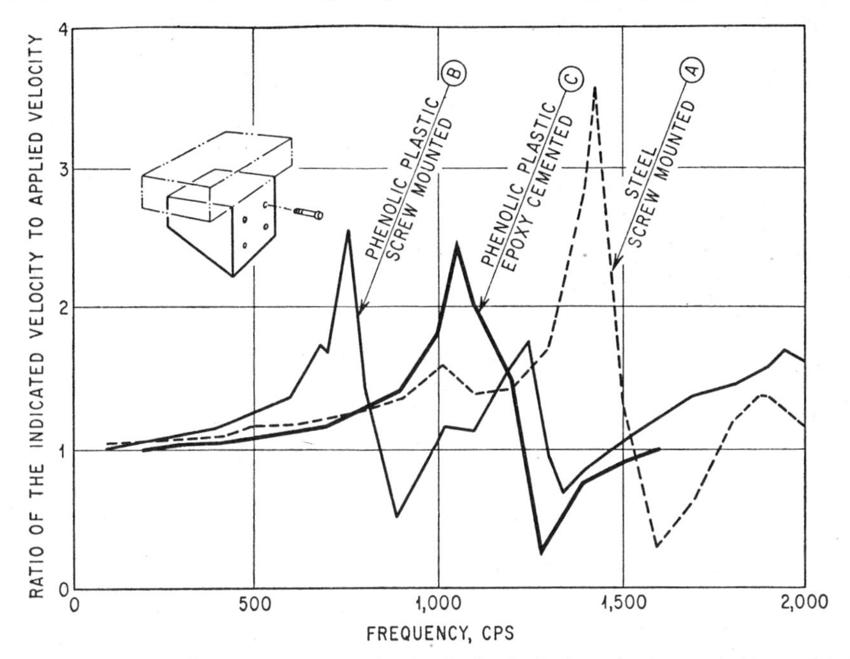

FIG. 20.1. Relative frequency response of a standard velocity transducer mounted on various brackets having identical geometry but fabricated of different materials. Curve *A* shows the response using a steel bracket attached to the specimen with four screws. Curve *B* shows the response using a bracket fabricated of a cloth-reinforced phenolic plastic and attached to the specimen with four screws. Curve *C* shows the response using a bracket fabricated of cloth-reinforced phenolic plastic and attached to the specimen with an epoxy resin adhesive.

ard velocity pickup mounted on brackets which are identical in geometry but which are fabricated from different materials. Note that a change in bracket material from steel to a cloth-reinforced plastic halves the resonant frequency of the mounting. A change in bracket attachment, from four screws to an epoxy resin adhesive bond, increases the frequency of the mounting resonance 60 or 70 per cent. Although these results are not of a general nature, they illustrate that minor variations in the transducer mounting bracket may produce significant changes in the output characteristics. Examples of the effects of various types of mountings on the frequency response of a transducer are shown in Fig. 20.2.

In order to design a transducer mounting properly, one must know the nature of its use, the frequency range and the maximum acceleration of the measurements, and the mechanical specifications of the test object. Specially designed mountings may be required for each test setup. Not only must mountings be designed carefully, but they should be tested under conditions closely approximating the service environment.

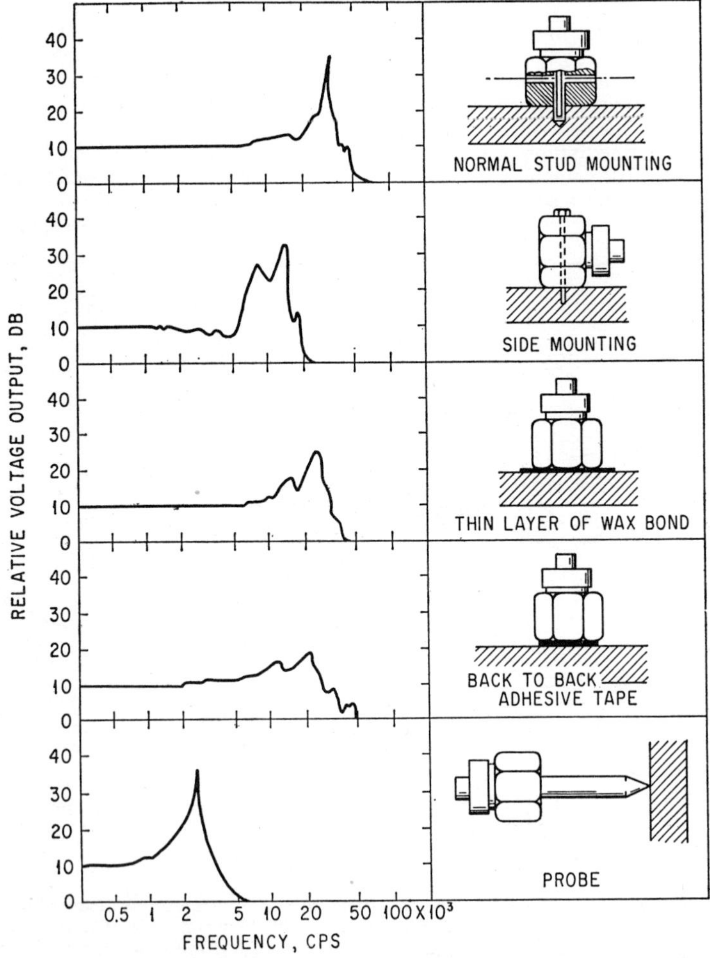

FIG. 20.2. Typical methods of coupling a crystal accelerometer to a test item and corresponding frequency-response curves. (*Courtesy of Brüel and Kjaer.*[1])

The effect of the transducer-mounting system can be estimated if it is assumed that it behaves as a simple spring-mass system driven at the end of the spring. Then, the acceleration of the transducer is:

$$\ddot{x} = \ddot{u}\frac{k}{k + m(2\pi f)^2}$$

where $\ddot{u}$ is the specimen acceleration, m is the transducer mass, k is the spring constant of the mounting, and f is the frequency of vibration. For $\ddot{x}$ to be within 1 per cent of $\ddot{u}$, $k > 100m(2\pi f)^2$. Since $f_n = (1/2\pi)(k/m)^{1/2}$ is the undamped natural frequency of the transducer-mounting system, then $f_n > 10f$. For example, suppose an accelerometer weighs 0.1 lb. In order that the data obtained with this transducer be accurate to 1 per cent at a frequency of 100 cps, the stiffness of the mounting must be such that $k > 10{,}000$ lb/in., i.e., the mounting must have a resonant frequency greater than 1,000 cps. Because the mass of the mounting is not negligible, the values of k calculated in this manner represent a lower limit of the required stiffness.

GENERAL RULES FOR TRANSDUCER MOUNTING DESIGN

Several types of mounting brackets are illustrated in Fig. 20.3. Some general rules to observe in the mechanical design of transducer mounts follow: *

1. The mounting must be rigid, but should be as light as possible.
2. The use of long, thin structural members and long bolts should be avoided. (Such members have relatively low spring constants and contribute to a low resonant frequency of the mounting.)
3. The major resonant frequency of the mounting must be well above the test frequency range.
4. In order to obtain maximum damping in the transducer mounting, cast mountings are preferred; next in order of preference are welded constructions and machined assemblies.
5. Methods of mechanical attachment of the transducer, in order of preference, are:
   a. Transducer bolted directly to the structure (subject to surface mounting conditions).
   b. Transducer bolted to a mounting which is itself attached to the structure.
   c. Transducer attached to a multiple mounting bracket (where *a* and *b* are not possible).
6. Employ flat machined mating surfaces between the transducer and the specimen or bracket. (This helps to avoid mechanical distortion of the transducer, when it is attached to the test article, with consequent effect on the response of the transducer.)
7. Avoid locating transducers on thin skins or delicate structural members in the test specimen. (This reduces the possibility of introducing a spurious low-frequency resonance due to the combination of the transducer mass and the stiffness of the skin.)

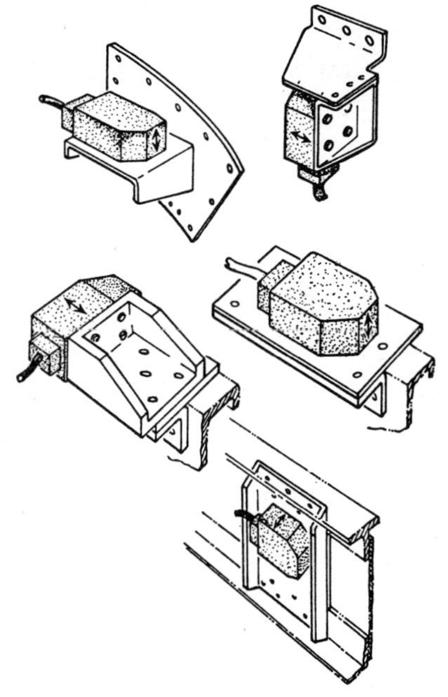

FIG. 20.3. Typical mounting brackets for velocity-type transducers. Arrows on the transducers indicate the direction of sensed motion.

ADHESIVE TRANSDUCER MOUNTING. Adhesive bonding of transducers to test

* Also see *Effects of Mounting on Pickup Characteristics*, Chap. 16.

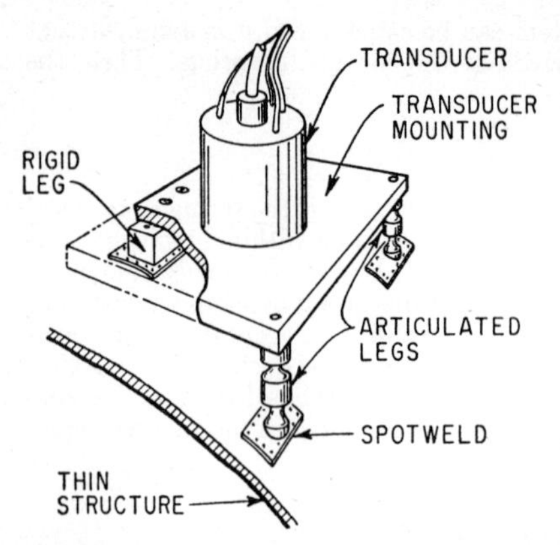

FIG. 20.4. A transducer mounting bracket for use on thin structure. This tripod arrangement is comprised of one rigid leg and two articulated legs, allowing the structure to flex without distorting the transducer.

specimens usually requires little preparation of the surface to which the transducer is to be attached. Often, a bracket is not required. However, the success of such adhesive bonds largely depends on the care with which the bond is made. Adequate cleaning of the mating surfaces with solvents to remove grease and wax is absolutely essential; furthermore, a thin layer of glue must be used to prevent decoupling resulting from elasticity of the cement. Mechanically hard, catalytic, or thermosetting cements are suitable for transducer mountings. Solvent-drying cements are undesirable because metal surfaces prevent reasonably rapid escape of the solvent which leaves the cement in a plastic condition. Nonhardening cements are not suitable because they act like soft springs, thereby decoupling the transducer from the specimen.

A very fast method of transducer bonding (suitable for use with lightweight pickups which are used at low acceleration levels) employs double-backed pressure-sensitive tape as the bonding agent. Such tape consists of a thin plastic ribbon, coated with pressure-sensitive adhesive on both sides. It is available from several manufacturers. The highest frequency usable with tape-bonded transducers often is limited by the elasticity of the tape and adhesive.

SPECIAL MOUNTING METHODS. Thin Structures. When a transducer is to be mounted on a thin structure, the mounting design must avoid producing stress concen-

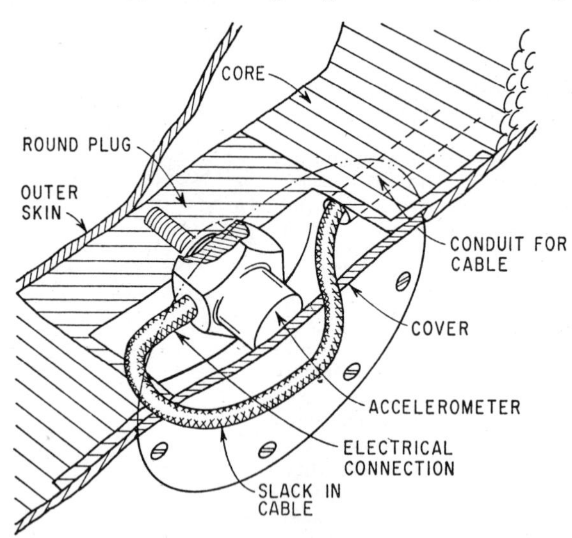

FIG. 20.5. Preferred installation of a transducer in a honeycomb material requires the incorporation of a suitable transducer mounting base in the core material prior to assembly of the honeycomb skin.

trations in the structure and modifying the characteristics of the structure. The mass of the transducer bracket chosen for such an application should be as small as practical, to reduce loading on the specimen. A typical bracket design, shown in Fig. 20.4, employs a tripod arrangement with one rigid leg and two articulated legs connecting the pickup to the thin structure. The articulated legs allow the structure to flex without distorting the transducer.

Honeycomb Structures. The mounting of a transducer on a honeycomb structure can be difficult because honeycomb structures have very little local strength. The best solution is to install a mounting base for the transducer during manufacture of the honeycomb, as shown in Fig. 20.5. When circumstances prohibit a solution of this type, some other method of distributing the mounting loads over a considerable area of the honeycomb skin must be employed. For example, an adapter plate tapering in thickness to a feather edge can be bonded to the skin as shown in Fig. 20.6. To avoid possible delamination of the honeycomb skin, the thickened part of such an adapter should be reduced to the smallest practical size.

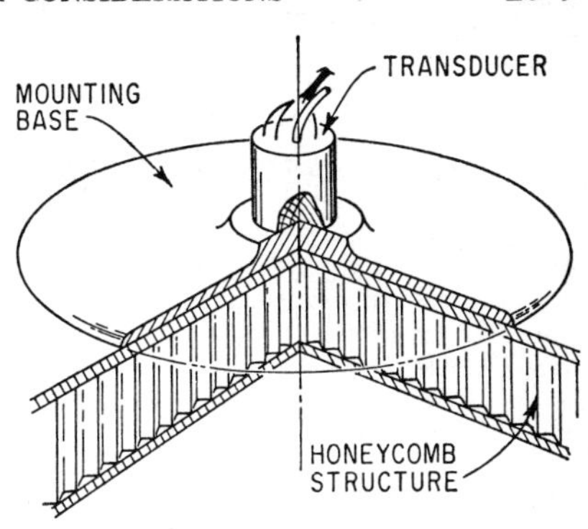

FIG. 20.6. A typical transducer base bonded to honeycomb skin. Such a base is required to distribute the mounting loads of the transducer to avoid producing stress concentrations and local failure. To avoid delamination of the honeycomb skin from the core, the thickened part of the transducer mounting base should be as small as practical.

SPECIAL TRANSDUCER CONSIDERATIONS

Various classes of transducers have characteristic features which must be considered in mounting design. Some of these are given in the following sections.

SOFT-SPRING TRANSDUCERS

In mounting a transducer that employs a mass-spring system having a low natural frequency (as found in some velocity and displacement transducers), the attitude of the transducer relative to gravity or other static acceleration fields should be considered. The effect of gravity will deflect the mass toward one of the internal stops in the transducer, thus limiting the maximum vibration amplitude which can be applied. Some commercially available transducers incorporate spring fixture adjustments which may be used to restore the mass of the mass-spring system to its normal position in the "installed" pickup attitude. In using other types of transducers, judicious orientation of the transducer and test article can reduce the effect of gravity.

In transducers employing a pivoted arm as the support, orientation of the pivoted arm parallel to the static acceleration vector, or in the opposite direction, respectively increases or decreases the effective restoring force applied to the mass of the mass-spring system. This effect is particularly important in velocity pickups having soft springs. For example, in one common pickup, the resonant frequency changes ± 12 per cent, depending on its orientation in the earth's gravitational field. This results in a sensitivity change of ± 7 per cent at 7.5 cps, decreasing to ± 2 per cent at 20 cps. Since this effect cannot be eliminated, it must be accounted for in transducer calibration.

INDUCTIVE-TYPE PICKUPS

When using a pickup which incorporates a magnet as part of the transducing element or for magnetic damping, avoid using magnetic materials in the transducer supports or mounting bolts.

The use of such materials often results in shunting the magnetic gaps in the transducer, thereby changing its characteristics.

PIEZOELECTRIC ACCELEROMETERS

The sensitivity of some piezoelectric accelerometers varies if the case is subject to distortion. Therefore, such pickups must be provided with distortion-free mountings. Piezoelectric transducers frequently are provided with a single mounting screw or stud threaded into the bottom of the instrument case. Then it is convenient to provide a flat mounting bed by spot facing on the test specimen a circle just larger than the instrument case. Alternatively, a special stud having a small diameter shoulder can be used to raise the transducer case a few hundredths of an inch above a specimen whose surface is rough,

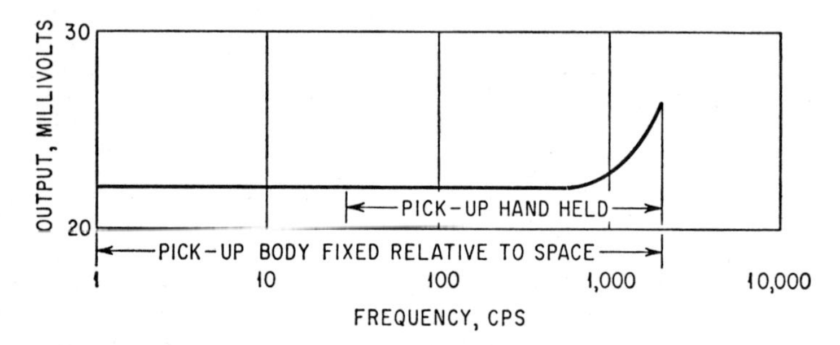

Fig. 20.7. Typical frequency response of a velocity pickup employing a probe. A pickup of this type will follow the typical curve within 10 per cent when hand-held or within 5 per cent when the case is rigidly mounted. The output voltage shown is for a sinusoidal excitation at 0.22 in./sec rms velocity.

irregular, or under flexure. Shoulder studs should not be used for very high levels of acceleration since they limit the area of contact between the specimen and the pickup.

The length of the screw or stud used in installing a piezoelectric pickup should be selected with care, since it is possible to bottom a long stud in the pickup, thereby causing the case to distort and possibly causing permanent damage. The torque applied to a mounting stud should follow the manufacturer's recommendations.

HAND-HELD VELOCITY PICKUPS

The application of seismic velocity pickups sometimes is limited by the amount of mass that can be added to the test specimen when the pickup is installed. Velocity pickups of the probe type are commercially available and largely overcome this limitation. The body of the transducer is hand-held and the motion of the test surface is sensed by a probe protruding from the transducer. In this case, the only mass added to the specimen is that of the probe itself. Such transducers find wide application in vibration survey work due to their low effective mass and the ease with which they can be moved about. Figure 20.7 shows a frequency response curve of a typical hand-held velocity pickup. Note that the response at low frequencies is limited if the transducer is hand-held. As indicated in Chap. 13, an extraneous signal is introduced by any motion of the hand.

INSTRUMENTATION WIRING CONSIDERATIONS

The use of electromechanical transducers requires the installation of a certain amount of wiring in the equipment under test. Where a number of pickups are required, the amount of wire needed may become quite large. The installation of this wiring should take account of the following facts.

EFFECTS OF THE MECHANICAL STIFFNESS AND MASS

The mechanical stiffness and mass of a bundle of wire may be considerable. For example, a well-laced bundle of 50 No. 22 copper conductors, weighing about ¼ lb/ft, has elastic properties similar to a 0.25-in.-diameter aluminum tube with 0.040-in. wall thickness. This wire has structural properties which, if neglected, can appreciably affect the natural frequencies of a test object. Mass loading effects of the wire on mode shapes can be minimized by distributing the wire over the specimen surface. This also reduces the cross-sectional moment of inertia of the wires, thereby reducing their bending stiffness. Routing of wires along the neutral axes of beams will further reduce specimen stiffening effects.

ELECTRICAL NOISE SUSCEPTIBILITY

Wiring installations should be designed to minimize self-generated and induced noise. The induced noise results when electrical energy from local electrical equipment is coupled into the measurement circuits by one or more of the following mechanisms:

1. Ground loops
2. Varying magnetic fields
3. Varying electric fields
4. Resistive leakage paths

Noise introduced by these effects can be limited significantly by judicious application of the following techniques:

1. Use separate signal and power current return leads to reduce ground-loop coupling between signal and power circuits.
2. Use common-current return leads of sufficiently low resistance where an inadequate number of current return leads prevent separate returns for each signal current.
3. Use twisted pair or coaxial cable to reduce inductive pickup in signal leads.
4. Reorient or relocate circuits to reduce inductive or capacitive coupling between circuits.
5. Use twisted-pair power-distribution leads to reduce the magnetic fields responsible for inductive pickup.
6. Use coaxial cable or shielding to reduce capacitive coupling between circuits.
7. Use single-point ground bonding to aid in location of unintentional ground loops and/or leakage paths.

Electrical noise may be generated by motion of some part of the wiring because of variations in contact resistance in connectors, because of changes in geometry of the wire (particularly coaxial cable) or as a result of voltages induced by motion of the conductors through magnetic fields which may be present. In general, such electrical noise will be reduced if cable harnesses securely fasten the wire cable to a structure at frequent intervals, and if connectors are provided with mechanical locks and strain-relief loops in their cables. Cables are available commercially which are especially constructed to minimize the noise resulting from flexure. Such cable should be employed when the wiring will be subject to high accelerations or to high shock loads. Noise generated by wire motion in a magnetic field can best be eliminated by removing or reorienting the field.

WIRING INSTALLATION AND ELECTRICAL CHECKOUT PROVISIONS

The mechanical problems associated with wiring installation, especially when making a measurement in a complex vehicle such as an airplane, often result in the cutting of the cable into sections for ease of handling. It should be noted that the wires are much more prone to failure at connectors or terminals than at the intermediate points. When cables must be spliced, soldered connections provided with adequate stress reliefs are preferable to mechanical plugs or connectors. This is because the electrical resistance is lower and the possibility of the connectors being opened (thereby stressing the mechanical contacts) for trouble shooting is removed.

Electrical checkout considerations dictate that test points be available at various points in the circuits. To maintain high circuit reliability, such test points should not introduce any additional joints into the wiring system. Ordinarily, test point locations and installation splice locations can be chosen so that a single break in the cable will serve both purposes. Electrical test points should be of a parallel type, allowing connection of the test equipment to circuits without disturbing circuit continuity. Such connections allow observation of the circuit in operation if desired, and do not require the making and breaking of mechanical circuit connections.

FIELD CHECKOUT

Ordinarily a complete field calibration of a vibration measurement system is impractical. Calibration in the field usually is restricted to functional checkout of the system and of the auxiliary equipment such as the recorders and amplifier. Methods of calibration and channel identification are described in the next section. This section considers cross talk and electrical noise in the over-all measurement system, and the problem of spurious transducer-mounting resonances.

CROSS TALK

Cross talk, i.e., signal interference between channels in a measurement system, usually is due to coupling between measurement channels occurring in some common element, for example, a power supply having high internal impedance circuits. The source of coupling must be eliminated. Decoupling in common power supplies sometimes can be achieved by adding isolation networks such as those shown in Fig. 20.8. If other elements couple several measurement channels, often the simplest solution is to replace the common element by separate components in each channel.

ELECTRICAL NOISE

Significant electrical noise may be evident from the output of the measurement system even though no input signal is supplied. Such noise results: (1) from coupling between circuits in the measurement system and those of the power circuits, (2) from mechanical strain or vibration sensitive elements other than transducers, or (3) from improper equipment design. Often it is most convenient to locate the source of noise by

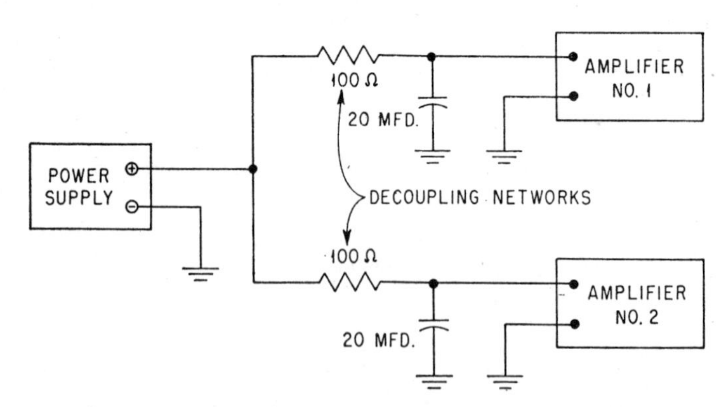

FIG. 20.8. Block diagram showing decoupling networks used to reduce cross talk between two amplifiers connected to a common high-voltage power supply having a high source impedance.

using an oscilloscope which is first connected to the transducer output with no vibration applied. Then the connection to the oscilloscope is moved component by component through the measurement system toward the recorder until the noise is observed. It is necessary that the system be electrically intact during this operation or the source of noise may be disconnected while the search for it is conducted. Another approach is to short-circuit the signal path at various points in the system, one at a time until the system noise disappears. Usually, this pin-points the source as the component next nearest the transducer from the last short circuit.

After locating the position of the noise source in the measurement system, the mechanism of coupling must be determined before corrective action can be taken. Often, elimination of the coupling mechanism eliminates the problem of cross talk.

Mechanical or acoustical sources must be eliminated or controlled if they result in noise in the measurement system. The component which is excited and which produces the noise must be decoupled from the driving source by isolating techniques or by physical separation. The latter is usually the more successful method.

SPURIOUS TRANSDUCER-MOUNTING RESONANCES

On initial installation, and whenever transducers are removed and remounted, it is advisable to check the transducer installation for spurious mounting resonances. These resonances may result from the selection of a transducer bracket which is not stiff enough. They also may result from faulty fasteners (such as loose rivets or bolts) or from improper seating of the transducer. Hence a careful visual check of the transducer mounting is helpful. It often is very useful to excite the transducer-mounting system by a blow while observing the transducer's output to detect resonances other than the resonant frequency of the transducer. The other resonant frequencies which appear may be due to (1) resonances in the test specimen or (2) resonances in the transducer-mounting system. Loose mountings usually produce "noisy" records and may produce audible buzzing sounds or clattering noises. Often it is difficult to determine the difference between resonances in the mounting and the resonances in the actual test specimen. If serious doubt exists, a different type of transducer mounting can be substituted and the test repeated. If the resonant frequencies are identical for the new mounting, the resonances are probably due to the test specimen, and original mount probably was satisfactory.

FIELD CALIBRATION TECHNIQUES

EARTH'S GRAVITATIONAL FIELD METHOD

This calibration technique is useful for calibrating accelerometers having useful sensitivity down to 0 cps. To employ this technique, the accelerometer is dismounted, but left electrically connected. Its output is observed for a $2g$ change in acceleration. Such a change is obtained by first orienting the accelerometer with the positive direction up (along the vertical) and then rotating the accelerometer through 180°. It is also possible to employ increments of acceleration within the $1g$ range [see Eq. (18.1)].

This technique is not applicable to crystal accelerometers or velocity pickups, or any instrument sensitive only to vibratory motion. It is not recommended for the calibration of accelerometers having a significant transverse sensitivity.

CHATTER METHOD

This calibration technique for accelerometers provides a $1g$ calibration point at any desired frequency. Such calibrations can be performed by mounting the accelerometer in a vertical position and placing a small mass on top of the transducer. The accelerometer is then placed on a small "shake table" which is vibrated. When the downward acceleration exceeds $1g$, the mass will become separated from the accelerometer. At this level, the test mass will "chatter" or bounce on the accelerometer case—providing

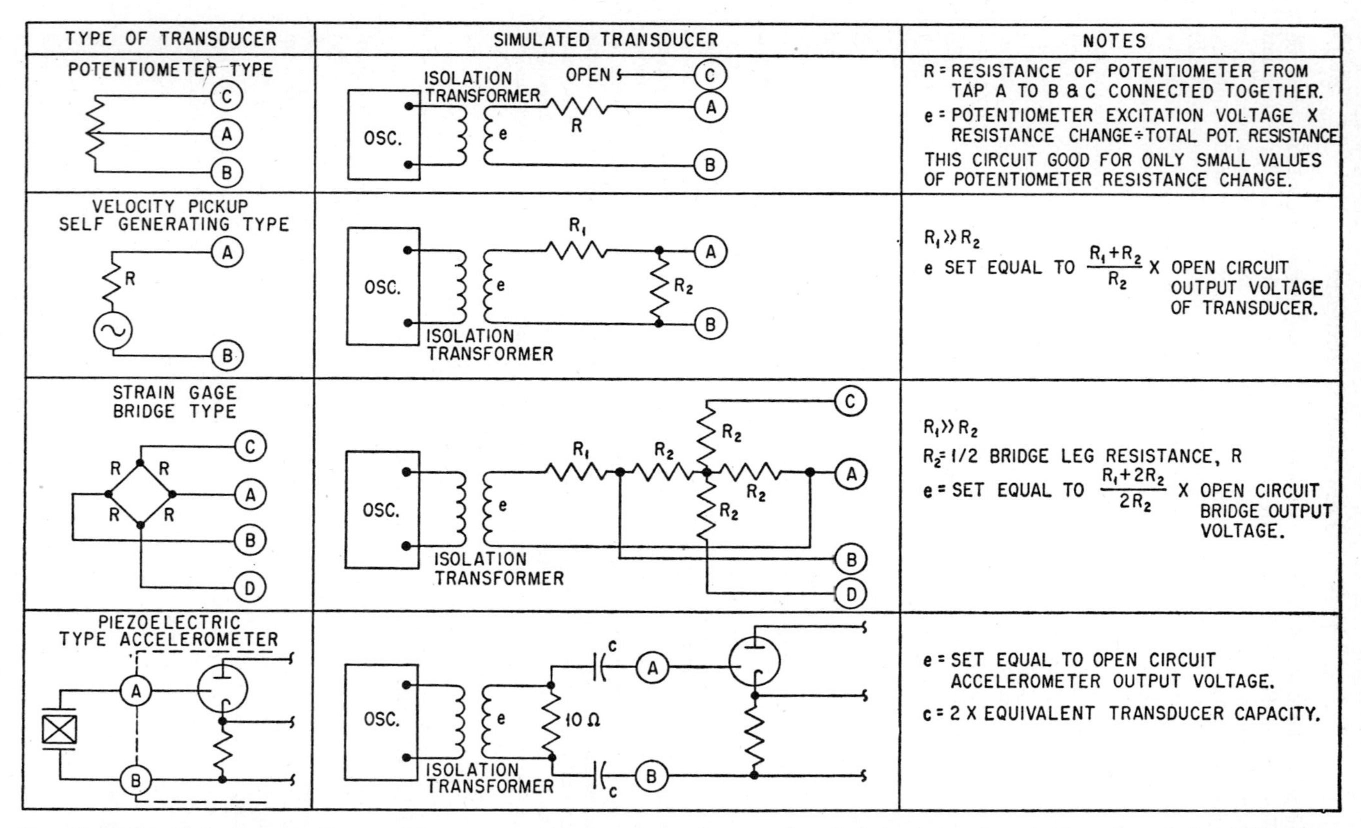

TYPE OF TRANSDUCER
SIMULATED TRANSDUCER
NOTES
POTENTIOMETER TYPE
ISOLATION TRANSFORMER
OPEN
OSC.
R = RESISTANCE OF POTENTIOMETER FROM TAP A TO B & C CONNECTED TOGETHER.
e = POTENTIOMETER EXCITATION VOLTAGE X RESISTANCE CHANGE ÷ TOTAL POT. RESISTANCE
THIS CIRCUIT GOOD FOR ONLY SMALL VALUES OF POTENTIOMETER RESISTANCE CHANGE.
VELOCITY PICKUP SELF GENERATING TYPE
ISOLATION TRANSFORMER
$R_1 \gg R_2$
e SET EQUAL TO $\frac{R_1+R_2}{R_2}$ X OPEN CIRCUIT OUTPUT VOLTAGE OF TRANSDUCER.
STRAIN GAGE BRIDGE TYPE
ISOLATION TRANSFORMER
$R_1 \gg R_2$
R_2 = 1/2 BRIDGE LEG RESISTANCE, R
e = SET EQUAL TO $\frac{R_1+2R_2}{2R_2}$ X OPEN CIRCUIT BRIDGE OUTPUT VOLTAGE.
PIEZOELECTRIC TYPE ACCELEROMETER
10 Ω
ISOLATION TRANSFORMER
e = SET EQUAL TO OPEN CIRCUIT ACCELEROMETER OUTPUT VOLTAGE.
c = 2 X EQUIVALENT TRANSDUCER CAPACITY.

FIG. 20.9. Electrical schematic diagrams of some common types of transducers and typical circuits used to simulate them during field calibration. Terminals labeled *A* and *B* are the signal lead connections to which either the transducer or simulated transducer is connected.

an audible indication and producing a spike on the accelerometer output which may be observed on an oscilloscope (see Figs. 18.12 and 18.13).

SYSTEM CALIBRATION BY COMBINING THE CHARACTERISTICS OF THE SYSTEM COMPONENTS

An over-all system calibration can be determined by combining the measured electrical characteristics of all components in the system from transducer to recorder. Obtaining a system calibration in this way circumvents the difficulties of precise field calibration, but it requires that each element in the system be calibrated in the laboratory with extreme care and that the effects of the source and load impedances be completely accounted for. Thus, a system calibration is subject to the sum of the experimental errors introduced by the calibration of each element in addition to any errors resulting from improper simulation of, or accounting for, loading effects. In general, the calibration of each element is performed before the system is assembled, and so this method is subject to (1) error because of the possibility of undetected damage to components between calibration and use and (2) errors resulting from improper connections, misidentification, or confusion in polarity. While this type of calibration saves time and conserves trained personnel, an independent check (such as that provided by the method given below) usually is advisable.

VOLTAGE SUBSTITUTION METHOD OF SYSTEM CALIBRATION

A suitable "simulated transducer" for use in field checkout must duplicate the electrical outputs of the actual transducer for the various vibration conditions to be simulated. To do this, it must either (1) reproduce the electrical voltage or current-generating characteristics of the actual transducer and must have the same output impedance or (2) duplicate the electrical quantity generated by the actual transducer when connected to its load. Failure to meet these conditions will result in different electrical loading of the actual and simulated transducers, and will probably cause calibration errors. It also is important that the simulated transducer have the same electrical grounding configuration as the actual transducer; otherwise electric-circuit noise and cross-talk effects will not be represented accurately when the simulated transducer is in use. Typical examples of circuits which simulate transducers are shown in Fig. 20.9. The "simulated transducer" introduces an electrical signal into the measurement system, thereby simulating the response of the actual transducer. For checks of phase shift and polarity, it is necessary to observe the output of both the simulated transducer and the over-all system output of a signal measurement channel, as functions of time. This can often be done by temporarily connecting the simulated transducer reference output to some other recorder channel with known phase-shift characteristics.

If a simulated transducer output is connected to only one measurement channel and the output of the other measurement channels is observed, the amount of interference or cross talk between channels can be observed. Such cross talk may appear as a reproduction of the simulated transducer output or as some "noise" of unrecognized form. It can be distinguished from other electrical disturbances because it is a function of the output of the simulated transducer and disappears when the simulated transducer output is removed.

Small, self-contained, self-excited vibration calibrators that can generate a known vibration level are available for field use. Such devices can be used to compare transducers or to calibrate working transducers against a laboratory standard.

REFERENCES

1. Brüel and Kjaer: "*Instructions and Applications, Accelerometer Sets 4308 and 4309,*" Naerum, Denmark, May, 1957.

21

INTRODUCTION TO TESTING AND DATA REDUCTION

Charles T. Morrow
Advanced Technology Center, Inc.

INTRODUCTION

This chapter sets in practical perspective various concepts and procedures related to testing and data reduction which are defined more precisely or discussed in greater detail in other parts of this handbook. Some engineering cautions are suggested that would be difficult to emphasize in more fundamental treatments. Whereas most of the other related chapters start primarily from fundamentals and appropriately are concerned primarily with fundamental information to be applied at the readers' discretion according to the problems they encounter, this chapter works backward from the most prominent applications and returns for further examination at the end. The treatment of the mathematics is primarily descriptive because derivations are available in other chapters.

SPECIFICATIONS

Shock and vibration engineering has a legal and managerial as well as technical foundation,[1] the legal aspects of which have important influences on all actions taken. In its most distinct and recognizable scope, shock and vibration engineering primarily is associated with government development and production contracts. In contrast to engineering and marketing for commercial sales, this situation provides only one customer, except to the extent that more than one service or more than one branch of the government supports development or buys the product. It involves only one source except to the extent that the government provides for alternate contractors. The incentives of competition are not effective beyond award of contract. Therefore, the contract defines at great length and often necessarily in cumbersome detail what the customers expect to get for their money. The contract in particular defines certain tests and other actions necessary to qualify a design or a product for acceptance.

If environmental influences may have adverse effects on performance or reliability, the contract calls out certain government design and test specifications or standards as binding requirements on the contractor or, in some instances, directs the contractor to devise specifications for approval, which then remain as binding requirements unless renegotiated. The documents have a different status from that of technical society standards, which are applied more or less at the discretion of the individual engineer. They will henceforth be referred to simply as "specifications" or "controlling specifications." They commonly become effective before the acquisition of any environmental data for the specific application. They are the counterparts of the government laws and regulations that provide a legal foundation for noise control. In turn, they lead to test specifications that are, in contrast, detailed technical procedures.

It is important that the controlling environmental specifications be recognized as legal and managerial and not technical documents. As much technical insight as possible should indeed be applied in their formulation. But any other interpretation leads to logical inconsistencies and to less-than-optimum use of both specification and environmental data. An environmental specification is an instruction to take certain actions—to consider certain stated environments in design and incorporate them quantitatively into the design process to the extent that is feasible, or to test to certain stated environmental conditions. Unless a waiver is requested by the contractor and granted by the customer, the specification remains mandatory in all respects until revision. It determines the acceptance or rejection of the design or product unless a variation (on the grounds that the discrepancy does not affect function) or a deviation (on the grounds that the effect is not significant) is requested by the contractor and granted by a customer representative.

In particular, an environmental specification is not a technical statement of the environment in which a deliverable item will be transported or used. Otherwise, there would be a conflict of status between its details and those of any accumulating data bank.

An environmental specification may be influenced by practical considerations other than the character of anticipated actual environments. A specification for design may be influenced by the effectiveness with which a design team can utilize particular modes of presentation of an environment. A specification for test (which indirectly affects initial design by imposing a risk of having to redesign after test) may be influenced by what is feasible and economical in test apparatus and procedure. A test condition such as a shock may even be prescribed, not because of a certainty that a device will be subject to shock after delivery but because of a feeling that a shock test facility that might prove necessary would not otherwise be purchased in time. There may be an intent to prescribe a test condition somewhat more severe than any anticipated in order to ensure with a limited number of tests that there is a reasonable margin of safety in the design. For such reasons there may be apparent inconsistencies between the environmental test requirements for airborne equipment and the design requirements for the structure to which the equipment will be attached.

Caution should be taken in stating any statistical relationship between a prescribed environment and an environment that it may be intended to simulate, in the legally mandatory portion of a specification, especially if there are inadequate data to support such a statement.

However, it can be beneficial for a specification to include some clarification of the intent and rationale of its contents. Those clarifications that are technical should be put in a nonmandatory section for information only.

Specifications are usually prepared by committee in accordance with negotiations of various people who will be affected in one way or another by the contents. Consequently, it is not always possible to include in the appropriate section of a specification a reasonable explanation of technical rationale. In such cases, it would be helpful for the person with the primary responsibility (who may or may not be the chairman of a committee) to prepare, as a memorandum for his or her file for future reference, a succinct record of the important considerations and compromises that affected the result.

DEVELOPMENT AND PRODUCTION

For major structures such as bridges and airframes, it is possible to anticipate and correct for most of the potential difficulties during the initial design process. For electromechanical, electrohydraulic, and sophisticated electronic equipment, the initial design process is commonly inadequate to ensure proper function in the laboratory bench environment or survival and function in extreme environments. Therefore initial design is followed by a period of intensive development—by construction, inspection, test, evaluation, and redesign, and by some feasible demonstration intended to show that the objectives have been achieved. The environmental specifications apply most directly to the last step but serve in an even more important way as a guide to developmental testing.

For hardware to be made in any quantity, the development phase is followed by a production phase with suitable changes in tooling and processing and with more detailed instructions so that manufacture will be less dependent on personal skill. The drawings are redone. Environmental specifications continue to be essential in this phase, to help ensure that the design does not deteriorate as a consequence of change, that any improvements in quality hoped for as a result of production methods of fabrication are realized, and that a suitable standard of workmanship is maintained.

PERFORMANCE AND RELIABILITY

The environments have two extreme effects—deterioration of performance not necessarily associated with permanent damage, and unreliability commonly but not always associated with permanent damage. A badly drifting inertial gyroscope, as a direct result of damage to its rotor bearing or as a consequence of subsequent wear, would be an example of the former effect. So would an electronic amplifier exhibiting excessive electrical noise (but not overload) induced by vibration or shock. A gyroscope inoperative because of fatigue and breakage of an electrical power wire would be an example of the latter effect. So also would probably be a relay making or breaking the wrong connections during exposure to vibration or shock. Performance deterioration is more or less proportionate to the severity of the environment and is relatively reproducible. Therefore, any environmental tests for it should ideally be at the actual environmental severity the device will experience. Unreliability implies that the environment has exceeded some critical threshold of the test item, which may, however, vary in manufacture. Therefore, if reliability is to be ensured by a small number of environmental tests, the test environment should ideally be somewhat more severe than the actual, to ensure that there is some margin of safety in the design. The practices of basing specified environments on worst conditions and enveloped spectra tend to be consistent with the latter objective. The relation of specifications to tests is discussed further in Chap. 24.

GLOSSARY

Much of the character of environmental data and its applications can be clarified by defining a number of terms.

Data measurement—a measurement of the environment, not necessarily as part of an orderly or extensive plan, that provides either raw data or results of simple data reduction.

Raw data—essentially complete data in the time domain for a particular measurement, without data reduction to deemphasize unimportant or redundant detail or to provide a more useful mode of presentation.

Data acquisition—the acquisition of environmental data by literature search, measurement, and/or data reduction according to some orderly plan and usually for a specific purpose.

Data collection—the collection of measured and/or reduced environmental data for some intended purpose.

Data reduction—the processing of data so as to present it in more useful form and focus on the more important aspects for an intended purpose. For vibration and shock, the processing usually involves a conversion from the time domain to the frequency domain—i.e., the computation of one or more spectra. Particular aspects in the time domain, such as velocity change or peak acceleration, are sometimes noted and recorded.

Data analysis—the analysis of environmental data, usually in reduced form, for some intended purpose, such as prescription of design criteria or test environments or estimation of the degree of environmental hazard in a new situation.

Actual environment—the environment (complete or reduced) that a device and/or its container may actually experience in transportation or use, not necessarily identical with design or test environments, for which, however, the environment commonly supplies supporting data. The design and test environments are legal requirements for engineering action and are influenced by the constraints of their intended purposes.

The actual environment may be natural (as a high ambient temperature) or induced (as a shock induced by a pyrotechnic shock) or a combination (as a combination of the two examples, or a device overheating because of natural and generated heat).

Design environment—an environment prescribed for design purposes, or certain simpler derived criteria (e.g., equivalent static acceleration) for more direct involvement in the design process.

Test environment—an environment prescribed for or actually used in an environmental test.

Environmental test—a test of a device to a specified environment (or sometimes the environment generated by a specified machine as in certain Navy shock-test machines) to determine whether there is any deterioration of performance or any damage that could affect reliability.

Test specification—a more comprehensive and detailed procedure and set of requirements than in the environmental specification from which it may be derived. It covers preparation of the test item, identification of all test equipment and instrumentation, description of any necessary test fixture, instructions for mounting any accelerometers and other sensors, and step-by-step procedures for operating the equipment and test item (if operation of the latter is required) and for recording data.

Analytic test—a test in which an environment is applied to a device to determine properties or parameters (e.g., resonance frequencies) of that device.

Diagnostic test—an analytic or environmental test carried out as an aid in diagnosis of some observed or suspected design or workmanship deficiency.

Qualification test—an environmental test applied as an aid in determining whether sufficient development or production engineering has been carried out to ensure a satisfactory design for the intended purpose.

Developmental test—a test similar in requirements to those of a qualification test, but applied in an exploratory manner earlier in the program as an aid to development.

Production or quality-control test—an environmental test, usually but not necessarily less severe than a qualification test, applied as an aid in ensuring that a satisfactory quality or workmanship is being maintained.

This chapter is concerned primarily with some aspects of data acquisition or measurement, with data reduction, with developmental or qualification (environmental) tests, and with some aspects of reliability theory.

ENVIRONMENTS

The environments that must be considered for rather generally exposed hardware include those of Table 21.1. In principle, all, or most, can occur simultaneously and

Table 21.1. Environments

Atmospheric pressure	Salt spray
Temperature	Fungus
Wind	Vibration
Rain, hail, and snow	Shock
Humidity	Acceleration (sustained)
Sand and dust	Acoustic noise

have their interactions. In practice, for economy of cost and effort, design and test are usually based on individual environments, with perhaps some supplementary design or environmental tests for interactions among effects of sustained acceleration, vibration, shock, temperature, and possibly atmospheric pressure.

In this handbook, we are primarily concerned with the last four environments. Their effects are not independent. For example, when a structure is immersed in a fluid (gas), there is always an interchange between mechanical energy and energy of waves in the fluid. The boosting of a missile after launch generates rocket noise, which shakes the whole missile, for subsonic airspeeds. A pyrotechnic blast may induce transient vibration (shock) in a structure and any equipment supported by it.

ENVIRONMENTAL DISTINCTIONS

Practical distinctions between vibration, shock, and sustained acceleration are not sharp and cannot be made absolute—independent of the dynamical system subject to the environment. For a tall building, an earthquake has much the character of a shock, even though by most standards, the excitation is low in frequency and long in duration. Indeed, the most-common spectral measures of shock severity originated in studies of excitation by earthquakes.

Each more or less linear dynamical system subject to the environments is characterized by its own resonance frequencies and by a set of corresponding transient (decaying) responses after any momentary excitation. An excitation that is of duration comparable to or less than the response times (or decay times) of the transients of a system is properly called a "shock," for that particular mechanical system, to identify the appropriate set of rules for any calculation of response. A dynamic excitation that is of duration long by comparison with the response times is properly called a "vibration," for that particular mechanical system, as a partial identification of the set of rules for any calculation of response. An acceleration that is unchanged for a time interval long by comparison with the response times is properly called a "sustained acceleration," for that particular mechanical system, and leads only to a problem in statics. Thus these environments are distinguished according to the appropriate set of rules necessary to calculate responses of particular mechanical systems. We may not intend to make the calculations, but a different set of rules implies a possible gross difference in magnitude or character of effect and points to different criteria for severity of the environment.

Typically an accurate knowledge of the resonances and response times in any specific case may not be available. The practical distinctions among the environments must be somewhat arbitrary, with a touch of conservatism so that design deficiencies do not result from the decisions. For example, if a brief acceleration is *likely* to excite some resonances, the environment is better considered to be a shock rather than a steady acceleration. Any simulation will then include a shock test rather than simply a centrifuge test.

Environments intermediate between those identified and defined can be visualized. But the practical techniques of simulation in test and response calculation for complicated hardware are not so refined as to require at this time more meticulous classification, except to point to two special examples of vibration. If a *periodic* vibration does not change its character significantly within the response times of the hardware, it can be considered to be *steady-state* or *quasi-steady-state*. If a *random* vibration does not change significantly (on a statistical basis) within the response times, it can be considered to be *stationary* or *quasi-stationary*.

One other special environment, the swept sine wave, will be introduced later.

MULTIPLE ENVIRONMENTS

On occasion in practice, shock, vibration, and steady acceleration may be superposed in the actual environment and exert their influences more or less simultaneously. This may be compensated for by a higher intensity in one or more of the test environments or, when the complication is warranted, may lead to an attempt to simulate all three environments simultaneously as a final check on design adequacy.

DESCRIPTIONS OF THE ENVIRONMENTS

Environmental specifications focus on levels of severity. Such levels are essential to most environmental tests as normally conceived and also on occasion when it is necessary to compute responses. Therefore quantitative descriptions of the environments are necessary, preferably correlating to a high degree with the damage potential for hardware, even when detailed data on dynamics of the hardware are unavailable and too costly and time-consuming to obtain. The following brief presentation defines the basic concepts, in a preliminary way, sets them in perspective, and extends some of the distinctions already discussed.

PERIODIC VIBRATION

Steady-state periodic vibration repeats its waveshape throughout all time or consists of a sum of motions that so repeat. It may be expressed as a trigonometric function of time or as a sum of such trigonometric functions. For vibration expressed by acceleration,

$$a = \Sigma A_k \cos (\omega_k t - \phi_k) = \mathrm{Re}\, \Sigma A_k e^{j(\omega_k t - \phi_k)}$$

$$= \Sigma A_k \sin (\omega_k t - \theta_k) = \mathrm{Im}\, \Sigma A_k e^{j(\omega_k t - \theta_k)} \tag{21.1}$$

where

$$\omega_k = 2\pi f_k \tag{21.2}$$

is the kth angular frequency in radians per second, f_k is the corresponding frequency in cps, and Euler's relationship is

$$e^{j(\omega_k t - \phi_k)} = \cos (\omega_k t - \phi_k) + j \sin (\omega_k t - \phi_k) \tag{21.3}$$

or

$$e^{j(\omega_k t - \phi_k)} = \cos (\omega_k t - \theta_k) + j \sin (\omega_k t - \theta_k) \tag{21.4}$$

If all the f_ks are in a harmonic relationship (i.e., multiples of some fundamental frequency), the composite waveshape repeats itself over all time.

Equations (21.3) and (21.4) express as a complex exponential function of time a counterclockwise (in accordance with the traditions of electrical and electronic engineering as opposed to quantum physics) rotating unit vector in the complex plane, whose projection on the x (real) axis is a cosine and whose projection on the y (imaginary) axis is a sine. The complex exponential is used for theoretical analyses in preference to the trigonometric functions, partly because its time derivative is simpler and partly because the introduction of complex numbers provides a simple method of expressing changes in phase as well as magnitude of a vibration in the course of its propagation.

The standard *fundamental* method of predicting response, when the resonances of the hardware are known, starts with computation of a ratio of response to excitation, versus frequency, or a measurement, if the hardware is available and a point where the response is of interest is accessible. On the assumption of linearity, each sinusoid of interest may then be multiplied by the appropriate factor, according to its frequency, to yield a response, and the instantaneous responses at the various frequencies can be summed. The response/excitation ratio is often expressed as a complex number so as to contain phase-shift information. If the response and excitation are in the same units (e.g., acceleration), the ratio is known as a "transmissibility." If the response is a motion (or, more strictly, a velocity) and the excitation is a force, the ratio is known as a "mobility" or, inversely, a "mechanical impedance." If the response point of interest and the excitation point differ, the more precise terms are "transfer mobility" and "transfer impedance." Acceleration is obtained from velocity by multiplication by $2\pi f$, in accordance with formulas for differentiation, or by $j2\pi f$ if complex notation is used. The unit imaginary j expresses a 90° difference in phase.

Steady-state vibration is a mathematical concept which does not exist in the real world where everything must have a beginning and an end. If A_k, ϕ_k, and θ_k are slow functions of time, Eq. (21.1) expresses quasi-periodic vibration as a summation of quasi sinusoids. The methods of computing response or describing severity at any particular time remain unaltered, provided that the vibration does not change significantly within the response times of the hardware excited.

As descriptions of severity of periodic vibration, only the magnitudes A_k or the corresponding root-mean-square values are considered. The phase angles ϕ_k or θ_k are ignored. However, instantaneous accelerations add arithmetically and magnitudes of sinusoids at the same frequency add vectorially. Phase *shift* is important theoretically in the analysis of mechanical systems and practically as a symptom of resonance when resonance frequencies are being identified experimentally. Near a center frequency, the phase of a response changes more rapidly than the magnitude with frequency. This is clearly evident in a Lissajous figure obtained from excitation and response signals.

SWEPT SINE WAVE

The swept sine wave is a vibration excitation occurring primarily in test. It has been used extensively because it can be controlled by relatively simple instrumentation, by comparison with the general case of Eq. (21.1) or with random vibration, to be discussed later. Equation (21.1) reduces to

$$a = A(t) \cos [\omega(t)t] \tag{21.5}$$

or

$$a = A(t) \sin [\omega(t)t] \tag{21.6}$$

where $f = \omega/2\pi$ is swept upward and downward slowly at a prescribed rate over some prescribed interval, frequently from 2 to 2,000 cps. The swept sinusoid tends in some respects to be a decelerated test. Any one resonator in a test item responds significantly only a small portion of the test time. If it is lightly damped, it may not attain full steady-state response unless the sweep rate is very slow.

A multifrequency periodic excitation in accordance with the general case of Eq. (21.1) is almost never used in test; but sometimes the swept sine wave is supplemented by a dwell for a prescribed time interval at each resonance frequency of the test item. As applied in practice, this tends frequently to result in overtest.

Both the swept sine wave and the dwell suggest a somewhat arbitrary but simple and potentially useful means of ensuring quality of a product rather than an attempt to simulate the actual environment. The swept sine wave is seldom a prominent feature of the actual environment. The dwell is, by common practice, at frequencies determined by the test item rather than by the actual environment.

MECHANICAL SHOCK*

In contrast to vibration, mechanical shock is of relatively short duration and either is terminated abruptly (as in some environmental tests) or dies away rapidly. Its most *fundamental* representation, without recourse to complex frequency, may be interpreted as a limiting case of Eq. (21.1), with the summation replaced by a continuous integration over frequency.

$$a(t) = 2 \operatorname{Re} \int_0^\infty F(f) e^{j2\pi ft}\, df = \int_{-\infty}^{\infty} F(f) e^{j2\pi ft}\, df \tag{21.7}$$

where the *Fourier spectrum* or *Fourier transform*

$$F(f) = \int_{-\infty}^{\infty} a(t) e^{-j2\pi ft}\, dt \tag{21.8}$$

is a complex function of frequency incorporating a phase angle $\phi(f)$. The complex exponential in the integrand of Eq. (21.8) may be regarded as a mathematical probe inserted by the analyst to indicate what frequency is of interest for the moment, so that integration will yield the value of the transform for that frequency.

If $Y(f)$ is the complex transmissibility or transfer mobility of a mechanical system from an excitation point to a point of interest, the response is given by

$$a_r(t) = 2 \operatorname{Re} \int_0^\infty F_r(f) e^{j2\pi ft}\, df \tag{21.9}$$

where

$$F_r(f) = Y(f)F(f) \tag{21.10}$$

* Also see Chap. 23.

The *fundamental* procedure for computing response to shock, closely similar to that for periodic vibration, is thus to convert from the time domain to the frequency domain by taking the transform, to multiply by the complex transmissibility, and to return to the time domain by taking the inverse transform. Therefore the magnitude $|F(f)|$ of the Fourier transform may be interpreted as a measure of shock severity, in much the same way that amplitudes of sinusoids are taken as measures of severity of periodic vibration.

The procedure also involves a generalization to complex numbers at the beginning for mathematical convenience and, if one is more comfortable with real answers or wishes to know instantaneous numerical values, a restriction to real numbers at the end.

In addition to the complex exponential transforms, there are cosine and sine Fourier transforms. The cosine transform

$$F_c = \int_{-\infty}^{\infty} a(t) \cos 2\pi ft \, dt \tag{21.11}$$

is used primarily, not in shock theory, but as an intermediate step in digital computation of the power spectrum of a random vibration, to be discussed later.

More powerful approaches to shock theory become possible if the integration in Eq. (21.8) is restricted to positive time, and $j\omega$ is replaced by s, which is permitted to have a real part, expressing a decay rate, to yield the *Laplace transform* [see Eq. (8.9*a*)]:

$$L(s) = \int_0^{\infty} a(t)e^{-st} \, dt \tag{21.12}$$

Restricting the integration range is equivalent to assuming that the shock acceleration is zero for negative time, for the proof of general theorems. The real part of s, depending on its sign, describes an exponential rate of decrease or increase of a sinusoid. Any linear passive damped mechanical system exhibits certain transients, with particular frequencies and decay rates, after an excitation. The Laplace transform of the response increases beyond limit as s approaches the complex frequency (including decay rate and angular frequency) of any of these transients, thereby providing an opportunity to characterize the system simply by means of its transient response as an alternate to its steady-state response.

Control systems, such as active isolation systems, may, on first trial, exhibit an exponentially *increasing* output (up to the limits of linearity) at one or more complex frequencies. Consequently, the Laplace transform has become the foundation for control-system stability theory. However, in shock and vibration engineering, active isolation systems are used only when necessary, because they consume power, are expensive, and must be tailored to the specific application by relatively sophisticated engineering methods in order to perform satisfactorily at all. Accordingly, shock theory is limited primarily to the effects of excitation of passive systems.

The spectral descriptions most commonly used for shock appear, at first glance, to be simpler in concept or at least less abstract than the Fourier and Laplace transforms. The study of excitation of tall buildings by earthquakes led to the use of the peak response of a simple mechanical resonator as a function of its resonance frequency, now known to shock and vibration engineers as the "shock spectrum," as a measure of shock severity (see Chap. 23).

The various forms of the shock spectrum have been interpreted alternately as representative responses of practical hardware or as indirect descriptions of the excitation. In the course of years, the term shock spectrum has been distinguished as undamped or damped according to the amount of damping assumed in the resonator, and as positive, negative, or composite according to the direction in which the peak response is observed. The actual shock environment, as exemplified by pyrotechnic shock in aerospace vehicles, tends to consist of transients characterizing the structure conveying the excitation. Laboratory test shocks more usually resemble simple acceleration pulse shapes such as square (rectangular) wave, half-sine wave, and sawtooth, partly because of relative apparent ease of description. Especially for test shocks, the shock spectrum has been distinguished as initial (the term primary, also used, tends to convey a false impression of importance), residual, and maximax, according to whether the time interval in which the peak response is observed is during the pulse, after the pulse, or unrestricted. The undamped residual shock spectrum (which is identical for positive and negative directions) is simply related to the magnitude of the Fourier transform at the same frequency by

$$A_p = 2\pi f\,|F(f)| \tag{21.13}$$

so that it can also be regarded as a fundamental measure of excitation of even complicated mechanical systems.

The shock spectrum is usually expressed in terms of peak acceleration, but the undamped residual spectrum is converted to displacement simply by division by $\omega^2 = 4\pi^2 f^2$. At low frequencies, where displacement is more important, this can be accomplished by slanted grid lines against the same spectral plot.

The terminal peak sawtooth, noted for its relative smoothness of spectra, originated as a means of attaining a prescribed minimum shock spectrum.[5] Since then there has been a strong tendency to specify a nominal spectrum and tolerances without mention of pulse shape or, following previous practice, to specify the nominal pulse shape and tolerances without mention of spectra. Of the two, the latter has the virtue that it better defines the excitation for any *rigid* hardware that transmits shock without modification. The former has the virtue that it maintains better control of the response of any mechanical *resonator*. However, the spectral magnitude versus frequency does not uniquely define the severity of a shock excitation for all possible test items.

In contrast to the Fourier transform, the undamped residual shock spectrum, as presently defined, contains no phase information. There are indications that for laboratory test shocks the phase-versus-frequency characteristic influences the response of coupled mechanical resonators. Therefore there may be benefit from placing tolerances on phase, preferably referred to some standard frequency more or less in the middle of the range of interest, and on phase rate, as an alternative to specifying general pulse shape in addition to spectral magnitude and tolerances.

RANDOM VIBRATION*

Like shock, random vibration can be regarded as a limiting case of Eq. (21.1), with an infinite number of infinitesimal sinusoids in random relationships in the frequency domain. Alternately, it can be regarded as the sum of an infinite number of infinitesimal shocks occurring randomly in the time domain. It is predictable and concisely describable only in a statistical sense. Yet either point of view leads to the same fundamental description and measure of severity. The first point of view leads to a less abstract development of spectral concepts, whereas the second is perhaps more vividly allied with the practical orgins of random vibration or random time functions in general.

Any finite-duration sample of random vibration can be approximated, within its duration, by Eq. (21.1), with a finite number of sinusoids. In fact, if the sample is recorded on magnetic tape, a signal of the form of Eq. (21.1) can be generated by joining the two ends of the sample, so as to make a tape loop, and playing the sample back repeatedly. The sinusoids will be in a harmonic relationship with a fundamental frequency that is the inverse of the duration but will have random amplitude and phase relationships. If successive samples of the same length are compared, sinusoids at the same frequency will have a random relationship to each other. If the duration is increased, the sinusoids will increase in number and, on the average, decrease in magnitude. For any one sample, from considerations of power dissipation (in a standard load) statistics and simple algebra, the mean-square values for the various sinusoids are directly additive, so that the root-mean-square value (or in statistical terms, the standard deviation from any mean or steady acceleration) must be proportional on the average to the square root of the number of sinusoids, or to the square root of any bandwidth through which the playback signal is examined. Dividing the mean-square value for the effective bandwidth by the effective bandwidth leads to the *power spectral density* or

* Also see Chap. 11.

power spectrum as a spectral measure that is, to a first approximation, independent of the bandwidth. The overall rms value can be found by integrating overall frequencies and taking the square root.

The term "white random vibration" refers to a power spectrum constant over a significant frequency range and implies a power spectrum of acceleration. But it could be a flat power spectrum of velocity, acceleration, force, or any other pertinent random variable. To convert from acceleration to displacement power spectral density, divide by $\omega^4 = 16\pi^4 f^4$.

More precisely, we can define an *estimate* of power spectral density at the frequency f, for a sample of duration T, as observed through a bandwidth B, centering on f, to be

$$w_e(f,B,T) = \frac{1}{2B} \sum_B A_k^2 \tag{21.14}$$

Then the power spectral density is defined to be, for a random vibration sustained through all time,

$$w(f) = \lim_{\substack{T \to \infty \\ B \to 0}} w_e(f,B,T) \tag{21.15}$$

subject to the constraint that T increases more rapidly than B decreases so that the number of spectral lines within B increases beyond limit, leaving $w(f)$ as a continuous function of frequency. This implies that we examine successive samples, while gradually increasing T and decreasing B, subject to the constraint.

For a stationary random vibration, $w(f)$ is found to be a mean about which the estimates must vary. For any fixed T and B, the standard deviation of $w_e(f,B,T)$, divided by $w(f)$, is of the order of

$$\sigma = \frac{1}{\sqrt{BT}} \tag{21.16}$$

depending slightly on the details of the selectivity curve and method of time-averaging.[6] Consequently any estimate of $w(f)$ from a finite sample, especially at low frequencies, is a compromise between resolution (small B) and statistical significance.

If the complex transmissibility or transfer mobility of a linear mechanical system is $Y(f)$, the response power spectrum corresponding to an excitation $w(f)$ is given by

$$w_r(f) = |Y|^2 \, w(f) \tag{21.17}$$

and the overall rms acceleration may be obtained by integrating this and taking the square root. Therefore any calculation of response of a linear system to random excitation involves the same data about the system as for periodic vibration or shock, but the data on the magnitude of transmissibility are used in a slightly different way.

The alternate point of view of considering a sample of random vibration to be a summation of shocks suggests the use of a Fourier transform and leads to an alternate definition for the estimate of a power spectrum:

$$w_e(f,B,T) = \frac{2}{BT} \int_B |F(f)|^2 \, df \tag{21.18}$$

where $F(f)$ is the Fourier transform of a (t), and this involves[7] averaging the square of the magnitude over the bandwidth B. Equation (21.18) can be shown to be consistent with Eq. (21.14) and to lead to the same compromises between resolution (small B) and statistical significance for any sample of finite duration.

Phase-versus-frequency relationships are generally unimportant for random vibration.

ENVIRONMENTAL TEST

Many of the considerations pertinent to environmental test have been set in perspective in previous sections of this chapter. It remains to discuss briefly the test apparatus for the environments of greatest interest and the concept of equivalence.

TEST APPARATUS

Steady acceleration is usually simulated by a centrifuge.

In the early days of vibration testing, the swept sine wave and the dwell were applied by variable-speed mechanical shakers (for low frequencies only) and by electromagnetic shakers powered by motor generator sets with variable-speed motors and several frequency ranges available by switching connections on the generators. The magnitude of excitation was controlled, in the case of the electromagnetic shaker, by a simple servomechanism which altered the field current in the generator, so as to hold an accelerometer signal to predetermined values. For the dwell test, the sweep mechanism was inactivated and the vibration tuned by hand successively to the resonance frequencies of the test item.

With the advent of electronically powered shakers for random vibration testing, the swept sine wave was not abandoned entirely. Advantage was taken of the flexibility and precision of electronic oscillators and electronic automatic gain controls. It became possible to sweep up to 2,000 cps without switching frequency ranges.

Random vibration testing required an electronic signal supply for an electromechanical or (sometimes) electrohydraulic shaker. The complete electronic system consisted of a multikilowatt power amplifier, a limiter to prevent bottoming of the shaker armature, a quick shutdown device to protect shaker and test item in the event of a major malfunction of the power amplifier, an equalizer to compensate for resonances in the shaker and test item fixture and shape the power spectrum as required, and a power-spectrum measuring instrument so that the equalizer could be properly adjusted. The first equalizers were sets of manually adjustable analog peak-notch filters. These gave way to sets of narrow-band filters with manually adjusted attenuators whose output signals were summed, and in turn to automatically adjusted sets. Finally, digital circuitry was applied.

Most shock testing has been done in drop towers, with the test item fastened to a fixture and a rigid carriage, and with a provision for decelerating the carriage at the bottom of its fall so as to apply a prescribed pulse shape or spectrum. In recent years there has been an emphasis on electronic synthesis of shock signals for electrodynamic shakers. With such an approach, if both vibration and shock tests are required, they can be accomplished in a single setup and the shock test can be somewhat better controlled. The U.S. Navy has usually specified, instead of shock environment, a shock testing machine that to some degree simulates the structure of a ship.

One awkward question is what is the upper frequency limit which must be prescribed in a specification of a spectrum. In principle, the acceleration the test item can survive should increase with the frequency of a bending mode, because of the decreased effective mass in the neighborhood of an antinode (point of maximum motion), so that at sufficiently high frequencies the excitation must become unimportant. Furthermore, at sufficiently high frequencies, the excitation must become negligible by any standard. But neither consideration yields a quantitative upper limit. Because of resonances in shaker armatures, shock carriages, and fixtures, it is difficult to control excitation at frequencies much higher than 2,000 cps except for very small test items. It is common to taper off a spectrum as 2,000 cps is approached to avoid high-frequency overtest and then to truncate the spectrum altogether.

Much effort is expended in making fixtures rigid. The equalization problem would be alleviated if fixtures were more highly damped, even at the expense of somewhat increased flexibility.

A somewhat easier question is what is the lower frequency limit which must be prescribed in a specification of a spectrum. Unless the actual environment is entirely negligible at low frequencies, the test spectrum should extend below the lowest resonance frequency the designer is likely to incorporate into the test item.

Testing is further discussed in Chaps. 24, 25, and 26.

EQUIVALENCE

The term equivalence was first introduced in a negative way, to show that one could not establish an equivalence between broad-band random vibration and a dwell test at resonance, with respect to damage potential, unless one knew the damping of the resonance.[8]

However, random-vibration test equipment did not become generally available for the greater part of a decade. Even since then it has often been necessary to try to compare a swept sine qualification test with a new severe random environment as a basis for engineering decisions to save the cost of requalification. Engineers have been forced to make the best possible use of the equivalence concept on the basis of estimated ranges of damping and other characteristics of resonance. For such reasons, increasingly sophisticated formulas have been devised for equivalence between random vibration, the resonant dwell, the swept sine wave, and shock.[9] The simple resonator, which had long been the favored idealized model for the theoretical investigation of shock effects, became the favored idealized model for theoretical investigation of equivalence.

DATA ACQUISITION AND REDUCTION

Many of the considerations pertinent to data acquisition and reduction have been set in perspective in earlier sections of this chapter. The reduction process is usually a matter of expressing in terms of numbers the quantitative spectral descriptions already defined, except when a simpler criterion, such as height of fall, is sufficient.[10, 11] It remains to discuss some of the instrumentation and planning for data acquisition, and instrumentation for data reduction.

DATA ACQUISITION AND MEASUREMENT

In the acquisition of data on the actual vibration and shock environments, it is customary now to record on magnetic tape for later data reduction, although peak reading devices are sometimes used. In missiles, the data are relayed by telemetry to the ground for recording. For airplanes, and land vehicles and ships, which have more room and are recoverable, the recorders are usually on board.

The sensing device is customarily a piezoelectric accelerometer (containing a mass-loaded piezoelectric ceramic with foil electrodes) mounted close to equipment whose environment is of interest. A force gage could be used and might be preferable under some circumstances, but it would have to be inserted between the structure and the equipment and might alter the environment. A strain gage on the equipment near a mounting point could serve much the same purpose as the force gage, but it would be difficult to use and interpret unless attached to a rather flexible member of known properties.

The low-frequency cutoff of a piezoelectric accelerometer is determined by its electrical relationship to its preamplifier, but other cutoffs in the preamplifier and other electronics may be the controlling factors. Limited low-frequency response eliminates the mean from the data. Some steady or low-frequency accelerations may have to be measured by inertial guidance equipment, if on board, or estimated from advance knowledge of a flight profile.

What is the upper frequency limit one should select in measuring an environment? Piezoelectric accelerometers are commonly advertised as being flat to 50,000 cps or more, but it is difficult to mount them without altering the apparent environment above 2,000 cps or so unless they are very light or the structure is very rigid. It is difficult to utilize high-frequency environmental data quantitatively in design or in test above 2,000 cps. In missiles, extended frequency response is obtained only at the expense of fewer environmental telemetry channels. Extended frequency response places greater demands on recording and data-reduction equipment. For such practical reasons, measurements of vibration and shock are usually limited to 2,000 cps—and sometimes 5,000 cps.

Data acquisition implies some plan, with respect to both measurement points and directions and the selection of samples in the time domain for data reduction, if not also for recording before data reduction. For full benefit in supporting and supplementing specification requirements, the details of the plan should remain available and accessible so that the significance of the reduced data is not misunderstood.

Because of the shortage of telemetry channels in missiles and of recording channels in airplanes, acquisition will be scanty for both measurement points and directions unless there is a major emergency with respect to some particular item of hardware.

Even so, if data reduction were to be carried out profusely and indiscriminately, there would be more reduced data than could be utilized. The samples selected for data reduction are usually taken at times when something of interest is happening, preferably when the environment can be classified distinctly as vibration or shock and when conditions suggest that vibration may be steady-state or stationary.

Because of its variability and potentially long duration, the transportation environment presents the most extreme example of need for planning. Sampling must be done not only for data reduction but earlier for recording. Random sampling is not cost-effective. In addition, the measurements must be at points where cargo is likely to be stored and representative of the environment prevailing when the cargo is stored.

DATA REDUCTION

For engineering decisions, acceleration-time histories as recorded on tape must be reduced to concise criteria of severity. This usually involves a transition from the time to the frequency domain and evaluation of appropriate quantitative criteria, discussed earlier in this chapter.

Periodic vibration is usually reduced simply by a wave analyzer that contains amplifiers, a tunable narrow bandpass filter, and a meter.

At first, power spectra for random vibration were obtained by a wave analyzer with a resistance-capacitance averaging circuit inserted before the meter. Because of the conflict between resolution and required averaging time, plotting a wide frequency range spectrum was tedious and time-consuming. The wave analyzer gave way to narrow-band filter sets, with automatic scanning for readout, and automatic plotting.

More recently, digital circuitry has been applied to the problem.[12] The continuous acceleration function is converted by a digital-analog converter to a series of data points. From these an estimate

$$R_e(\tau) = \frac{1}{T} \int_0^T a(t)\, a(t + \tau)\, dt \tag{21.19}$$

of the *autocorrelation function*, where τ is a relative time delay inserted before multiplication and integration is computed [see Eqs. (11.52) and (22.21)]. The cosine Fourier transform of this estimate is computed and averaged over sufficient bandwidth to provide statistical significance. Many computations are performed more or less simultaneously to yield the entire spectrum very quickly. The fast Fourier transform (FFT) algorithm, derived from matrix theory, greatly simplifies the numerical computation and permits a smaller, simpler computer to be adequate.[12, 13] Digital computers are discussed in Chap. 27.

In the early exploratory investigations of shock, measurement and data reduction were combined in a *reed gage*. The lengths of marks made on paper by the various reeds were taken as measures of the shock spectrum at different frequencies. Later analog electronic circuitry was used to simulate sets of simple mechanical resonators, and peak responses in appropriate time intervals were read out automatically. It should be noted that the response acceleration is completely determined by frequency and damping, so that sets of masses and corresponding spring constants for the same frequency need not be considered. Again, in time, digital circuitry was applied.

It thus appears that there are now available not only essentially suitable mathematical measures of environmental severity as a function of frequency, except for some limitations in case of shock, but essentially satisfactory data-reduction instrumentation for computing them. To ensure that reduced data on the actual environment serve as well as possible to support and supplement the specification requirements, which tend to focus on level of excitation, with rather smooth spectra, attention should be called not only to levels but to frequencies at which there are particular concentrations of energy.

Data reduction is discussed further in Chaps. 22, 23, and 24.

STATISTICS

The theory of statistics and probability seems to have originated largely in games of chance. If a card dealer deals enough rounds from an unstacked deck of cards without cheating, the chance of any one player having a king (for example) in his or her hand is found to be proportional to the number of kings in the pack and inversely proportional to the total number of cards. The chance of the player having two kings is found to be proportional to the chance of having one king times the chance of receiving a king from the remainder of the pack. Any one hand is an example of a statistical *sample,* and the set of hands dealt a player in the course of a game or several games is an example of a statistical *ensemble.* From such considerations, numerous useful formulas have been obtained, on the basis of the proportion of a group in a larger population and rules for combination and permutation.

At one extreme, the formulas may be regarded as empirical formulas interconnected by logic. At another extreme, if probability theory is to be rigorous and dependent as little as possible on the real world, the simpler formulas may be taken as axiomatic or as definitions of what is meant by probability in mathematics. The engineer, however, must maintain an awareness of relationships of probability to the real world.

Earlier in this chapter, the concept of stationarity was introduced rather loosely in terms of successive samples from a single acceleration-time record. Theoreticians prefer to postulate an infinite ensemble of records of infinite duration, as from an infinite number of extended missile flights, and to average over the complete ensemble for successive instants in testing stationarity. If they presume to find stationarity, they call the ensemble a *stationary process.* If they presume to find that time averages are identical with ensemble averages, they call the process *ergodic.* The theoreticians' approach makes the concept of stationarity a little sharper by avoiding the statistics of small samples. However, it is not a panacea for all lack of clarity. The qualities prescribed for the ensemble or process do not prohibit individual records from exhibiting short-term behavior that conflicts with our intuitive feeling for stationarity. In practice, it is small ensembles (frequently an ensemble of one), *piecewise stationarity* and *piecewise ergodicity,* that concern us. For example, if two ICBM flights are for different ranges, the reentry environments will be compared directly, with no concern for the fact that the total flight durations were different.

Nevertheless statistics is an important branch of mathematics for the engineer. In mechanical design, it has been traditional to focus on deterministic concepts and absolute survivability up to predetermined environmental levels. The actual behavior of hardware under static and dynamic stress is more statistical in nature.

Statistical concepts have been implied by the discussions of stationarity, the use of the rms value as an overall measure of random vibration, the definitions of the power spectrum and an estimate of the power spectrum, the compromise between statistical significance and resolution, and the use of the autocorrelation function as a step in digital computation of the power spectrum. It remains now to introduce several further concepts applicable to random time series and to discuss statistical reliability.

Although the statistics of material fatigue will be omitted in this chapter, it is necessary to point out that, even with specially prepared specimens tested in the laboratory, the spread of lifetimes for the same excitation level is an order of magnitude or more. This makes it difficult to rank-order the accuracy of various theories for the mean. Furthermore, as an alternative to fatigue-life prediction, it may be preferable in the presentation of data to emphasize the environmental level at which the material has a prescribed probability of exceeding a prescribed lifetime. This would be more nearly what a designer needs and would focus on the variable with the least scatter.

DISTRIBUTIONS

The root-mean-square value (from the mean) or standard deviation has already been introduced as a measure of statistical spread, but nothing has been said about more detailed descriptions.

The probability of finding an acceleration within da of a prescribed acceleration a is the number of such accelerations found in a large number of trials divided by the number of trials. It follows that the integral over all accelerations for such a probability, or probability distribution, must be unity.

The most-common distribution assumed for instantaneous accelerations of a random vibration is the *Gaussian*, or *normal*, distribution[7, 10–11]

$$p(a)\,da = \frac{1}{\sigma\sqrt{2\pi}}\,e^{-a^2/2\sigma^2}\,da \tag{21.20}$$

with the mean or steady acceleration omitted for simplicity and for correspondence with the typical accelerometer signal and with σ as the standard deviation [see Eqs. (11.10) and (22.10)]. Deviations from the Gaussian distribution are seldom significant unless measured at an actual failure point, which is not usually accessible. As random vibration propagates through a linear mechanical system, it tends to become more precisely Gaussian. Nonlinearity in the failure mechanism may produce deviations, but these are not predictable from those of the external excitation. Pseudo-random vibration may not become more Gaussian as it propagates, but distributions at failure points are not simply predictable.

Another matter of interest is the distribution of the positive peaks of a Gaussian vibration. As the bandwidth through which the vibration is examined decreases below an octave so that harmonics are eliminated, the positive peaks become confined to the positive side of the (zero) mean. They approach the *Rayleigh* distribution

$$P(A_p)\,dp = \frac{A_p}{\sigma^2}\exp\frac{-A_p{}^2}{2\sigma^2}\,dp \tag{21.21}$$

where σ is the same standard distribution [see Eq. (11.39) and Chap. 22].

Failure mechanisms tend to absorb energy from the higher peaks so as to modify the tails of these distributions.

CORRELATION

The autocorrelation function of a random vibration is a measure of how predictable, on the average, the acceleration at any particular moment is from another acceleration a time τ earlier or later. It is obtained, in theory, by delaying the random time function by τ, multiplying by the original function, and integrating over all time.

The correlation function of two random time variables is a measure of how predictable, on the average, one is at any instant from a measurement of the other at the same instant. It is obtained, in theory, by multiplying the two variables together and integrating over all time. The cross-correlation function is more general and permits one signal to be delayed by an amount τ [see Eq. (22.27)].

We are concerned with random vibration of resonant hardware and ultimately with the response of resonant structure to random pressure fields such as rocket noise and external turbulence. Consequently, the correlations of greatest interest in themselves are expressed on a narrow-band basis. The cospectrum can be obtained by multiplying two signals, within the same narrow band, together and integrating over all time. For identical signals, it reduces to the power spectrum. The quadspectrum can be obtained by shifting one signal first by 90°. The cross-power spectrum, a complex sum of cospectrum and quadspectrum, expresses any dominant phase angle.[10, 11] Such a phase angle is related in concept to the delay τ at which the cross-correlation function may have a maximum.

The cross-power spectrum of fluctuating pressures at pairs of points on a structure affects the magnitude and character of the mechanical response. Rocket noise tends to remain correlated for significant separations of the points. For boundary-layer turbulence, in contrast, the correlation distance across the flow is of the order of the thickness of the layer and may be only a small fraction of a wavelength. This is impossible for radiated sound.

RELIABILITY OF SHORT LIFETIME DEVICES

We will first discuss reliability theory in connection with relatively high production rate, short use-time devices (such as guided missiles). No maintenance during flight and no re-use are possible. Finally, in the next section, we will look briefly at the longer lifetime problem.

The reliability of a weapon system for a designated time, such as from missile launch to closest approach to the target, is the number of systems exhibiting no failure divided by the total number tried, for a large total number. Subsystem reliability can be defined in the same way.

Lusser's product formula relating the reliability of a system to the reliabilities of a complete set of non-redundant subsystems is

$$R = R_1R_2R_3, \ldots, R_n \tag{21.22}$$

According to this, if the subsystem reliabilities are equal, which for subsystems of comparable complexity would represent an optimum design,

$$R_n = \sqrt[n]{R} \text{ or } R^{1/n} \tag{21.23}$$

and, more generally, the subsystem reliabilities must be of this order, as tabulated in Table 21.2.

Table 21.2. Subsystem Reliability R_n

R	n								
	2	3	4	5	6	7	8	9	10
0.5	0.71	0.79	0.84	0.87	0.89	0.91	0.92	0.93	0.93
0.6	0.77	0.84	0.88	0.90	0.92	0.93	0.94	0.94	0.95
0.7	0.84	0.89	0.91	0.93	0.94	0.95	0.96	0.96	0.96
0.8	0.89	0.93	0.95	0.96	0.96	0.97	0.97	0.98	0.98
0.9	0.95	0.97	0.97	0.98	0.98	0.99	0.99	0.99	0.99

The product formula serves as a beneficial reminder that reliability tends to be more difficult to obtain as the complexity of a system increases.

At the beginning of a weapon-system development program, it is almost standard practice to use the product formula and a principle of equal engineering challenge to apportion reliability objectives among subsystems and again among smaller assemblies. It is common to prescribe numbers of test items and tests to verify these objectives according to a statistical trade-off between chances of accepting an unsatisfactory subsystem design versus chances of rejecting a satisfactory subsystem design. This is known as statistical design of tests. The reliability objectives usually survive contractual negotiations. The statistical test program usually does *not*, except possibly for some critical major subsystems.

Why is this so? Let us explore more deeply what may at first appear to be a very simple logical application of statistics.

Examination of Table 21.2 quickly reveals a dilemma for technical management in both the customer and contractor organizations. As a representative example, assume a system reliability objective $R = 0.9$ and a number of subsystems $n = 10$. The typical subsystem reliability must be of the order of $R_n = 0.99$; or, in other words, there must be approximately 100 test items and 100 tests for an even chance of observing one failure. Moreover, the single failure would not provide a statistically significant estimate of the subsystem reliability. More nearly 1,000 test items and 1,000 tests would be required. Without carrying the analysis further, we can conclude that the statistical test program, if carried through for any but major subsystems, would require many more test items and many more tests than would normally be called for in a development program. Consequently the program could be expected to involve overwhelming penalties in cost and schedule.

Consider a program for the development of a guided missile. The program usually involves an extended experimental flight test schedule to identify system difficulties that might not be disclosed by laboratory tests, to verify precision of guidance and other system-performance design objectives, and to monitor improvement in reliability. The flights are system tests in the *actual* environments of use. A running sample of a few missiles is adequate to maintain current and statistically significant estimates of system reliability up to the typical system objective. If the system reliability comes close to the objective, it is satisfactory for system production, and the subsystem designs are presumably ready for production except for possible difficulties in high-rate manufacture. If the system reliability does not come close, more development is needed, and effort must be directed toward areas that provide the greatest potential benefit. The subsystem design short-comings are usually all too obvious from the various assembly operations, inspections, and tests. The primary contractor challenge is to maintain customer confidence while introducing effective design changes in an orderly manner, after extensive testing and evaluation, so that the reliability will indeed increase as planned. Why, then, should customer or contractor management want an expensive program to *verify subsystem reliability* objectives at any one point in development?

We can think of three extreme examples that warrant some sort of extensive program of this general nature, and there may be others. Two relate to critical items that are intended for more than one system, perhaps eventually if not immediately. One is a new item and of such critical importance that an extensive evaluation can be funded. Another is a critical device of such simplicity that it can be manufactured and tested rapidly in large quantity at low cost. The third is a device critical to safety.

The statistically designed test program in its purest sense is an attempt at absolute elimination of personal bias. Its intent is to shuffle the cards so that the dealer cannot cheat or to eliminate the dealer altogether whenever possible. Its limitation is that, at potentially high cost, it ignores existing knowledge that may be useful if management can be trusted to make satisfactory judgments of engineering bias.

But other subtle difficulties appear if we look more closely at the product formula and the nature of the failures that may occur. A fair deal may be of little help to the competent player if the deck is already stacked so its makeup will alter the statistics of the game.

The product formula is often taken to be axiomatic—almost a definition of subsystem reliability. However, it can be derived from the basic definitions only on the assumption of independent failure rates for different modes of failure. What modes are suitable for consideration in connection with the product formula?

Clearly, except in special situations such as control-system instability induced by change of gain, performance parameters out of tolerance are *not suitable.* Such defects seldom abort a mission. In the absence of blunders, performance parameters are well-behaved statistically in manufacture, generally following Gaussian distributions, and are verifiable by economical statistical sampling. Their change with environment is highly reproducible, so that only a few environmental tests are needed for their investigation, and their probabilities of being driven out of tolerance are not independent.

Failure modes that are sensitive to environmental levels do not have independent failure rates unless the environmental transmission paths or the vulnerabilities at the failure points are very variable. Otherwise, in their connection, the subsystem reliability objectives are pessimistic and virtual, but not real objectives that are useful as reference points for making judgments but not accurate as literal definitions of the maximum permissible failure rates.

Subsystem reliabilities are sometimes defined in terms of failure rates in environmental test as opposed to failure rates in use—but in what environment? A single overall level of vibration or shock, which has been useful during development in disclosing design deficiencies with minimal testing, is too severe for use in a statistically designed test program. For environmentally sensitive failure modes, especially in view of the limited funding that can be devoted to quantity and quality of data on the actual environment, the limitations of test apparatus, and the problems of equivalence, it is seldom clear how the test environment should be decreased.

In principle, the manufacturing blunder and the defect, such as the loose screw or the intermittent electrical connection, which can be detected in a mild vibration or shock environment and which are not sensitive to severity level (but which can incapacitate an essential portion of a system), satisfy the assumptions for applicability of the product formula. But, where possible, it is better to spend money on elimination of defects than on measurement of their rate of occurrence. Low-level vibration tests are often used during production as an aid in 100 per cent inspection to eliminate such defects. The sticking hydraulic valve is another example of the compliance with the product formula. Especially allied to the product formula are defects of potentially catastrophic consequence that develop in time from random degradation. Fatigue life in the presence of *low-level* alternating stress is a special example, but this is more allied to airplanes and ships than to missiles, and it is more economically investigated by accelerated tests.

There is a branch of statistics known as "Bayesian statistics" that makes a compromise on the absence of bias. It incorporates an estimate based on existing information and uses future tests to correct the estimate. It is not widely popular with statisticians; as mathematicians, they prefer analyses to be as independent of preconceived notions as possible. In a practical sense, it is not clear what advantages the compromise has over that of going from one extreme or the other.

Reliability theory is not simply academic to shock and vibration engineers. They are often participants in negotiations that involve reliability concepts or they must often adjust their approach according to the outcomes. The difficulties discussed here are indicative of the problems customer and contractor managements have had, under pressure for high reliability, in defining the responsibility for reliability monitoring without often implying an obligation for collecting and perfecting information they do not really want.

Reliability engineering and shock and vibration engineering have developed independently, with different preoccupations. Shock and vibration engineering, in part because of pressures from designers and managers,[1] has often focused on environmental simulation even when this is not feasible or will have no beneficial effect on design. Reliability engineering often has focused on statistical verification of reliability even when this is costly and time consuming and may even hamper development rather than serve as a motivation for adequate design.[1] It has generally assumed that environmental tests are precise simulations, while often placing little importance on details of test conditions. But the two fields are closely interrelated. With better definition of objectives, a more unified point of view may evolve for both, with more benefits per dollar and more satisfaction to the people involved in each field.

RELIABILITY OF LONG LIFETIME DEVICES

Consider now a long lifetime system, consisting of an airplane or ship, and including all equipment necessary for its function, but excluding missiles and other short lifetime expendable armament.

The reliability concept becomes blurred by the indefiniteness of mission duration, the practice of at least limited maintenance or of equipment replacement during a mission, greater equipment redundancy and variable mission objectives. Mean time between failure (MTBF) tends to become a more important criterion.

However, many of the problems of reliability verification persist or become intensified. Whereas reliability objectives for missile subsystems imply many test items and many tests, those for airplane or ship subsystems often are construed as requiring tests of duration comparable to the useful lifetime of the system—or even longer. Shock and vibration test realism is even more questionable than for missile subsystems. Accelerated tests are economical for guiding design and development, and for rapid qualification procedures, but not for a direct, accurate measurement of reliability. A complete and uncompromising test for more realistic environments may not be completed before the system or subsystem is on the verge of obsolesence.

REFERENCES

1. Morrow, C. T.: *J. Acous. Soc. Amer.*, **55**:695 (1974).
2. Morrow, C. T.: "Shock and Vibration Engineering," vol. 1, John Wiley & Sons, Inc., New York, 1963.
3. Brown, G. S., and D. P. Campbell: "Principles of Servomechanisms," John Wiley & Sons, Inc., New York, 1948.
4. Biot, M. A.: *Trans. ASCE*, No. 108, p. 365, 1943.
5. Morrow, C. T., and H. I. Sergeant: *J. Acous. Soc. Amer.*, **28**:959 (1956).
6. Morrow, C. T.: *J. Acous. Soc. Amer.*, **30**:456 (1958).
7. Rice, S. O.: *Bell Sys. Tech. J.*, **23**:282 (1944); **24**:44 (1945). Also reprinted in N. Wax, "Selected Papers on Noise and Stochastic Processes," Dover Publications, Inc., New York, 1954.
8. Morrow, C. T., and R. B. Muchmore: *J. Appl. Mech.*, **22**:367 (1955).
9. Fackler, W. C.: "Equivalence Techniques for Vibration Testing," *Shock and Vibration Mon.* 9, Shock and Vibration Information Center, 1972.
10. Bendat, J. S., and A. G. Piersol: "Measurement and Analysis of Random Data," John Wiley & Sons, Inc., New York, 1966.
11. Bendat, J. S.: "Principles and Applications of Random Noise Theory," John Wiley & Sons, Inc., New York, 1958.
12. Enochson, L. D., and R. K. Otnes: "Programming and Analysis for Digital Time Series Data," *Shock and Vibration Mon.* 3, Shock and Vibration Information Center, 1968.
13. Cooley, J. W., and J. W. Tukey: *Math. Comput.*, **19**:297 (1965).
14. Lusser, R. K., *NAMTC Tech. Rept.* 75, Naval Air Missile Test Center, Point Mugu, Calif., July 1950.
15. Lusser, R. K.: *NAMTC Tech. Rept.* 84, Naval Air Missile Test Center, Point Mugu, California, July 1951.

22

CONCEPTS IN VIBRATION DATA ANALYSIS

Allen J. Curtis
Hughes Aircraft Company

INTRODUCTION

This chapter discusses the mathematical concepts involved in the analysis of vibration data which are essentially steady-state or slowly time-varying processes. It provides the mathematical basis for data reduction procedures by which the original time-history of vibration is transformed to other forms required for particular applications. Classes of data discussed in this chapter include sinusoidal, periodic, complex, and random (shock and transient vibration are discussed in Chap. 23).

The basic objectives of vibration data analysis, discussed in Chap. 21, may be achieved by various means. No one method of analysis can be considered *the* correct one, since both the type of data and the use to be made of the results influence the appropriateness of a particular type of analysis. This chapter is restricted to a description of the basic principles and techniques of data analysis which will enable the reader to make a logical choice of the type of analysis and data presentation most suited to the particular problem at hand.

The techniques of data analysis are illustrated in this chapter by functional block diagrams and are necessarily described in terms associated with analog processing of the signals. With the advent of minicomputers and implementation of the fast Fourier transform algorithm, it became practical to perform these analyses digitally. Chapter 27 describes the use of digital computers for performing the types of vibration data analyses described in principle in this chapter.

The complete definition of a vibration condition requires the description of magnitude,* and its variation with both frequency and time. Various types of vibration are described mathematically, and the respective magnitude parameters defined explicitly or statistically. The determination of the statistical properties of random vibration is discussed, including the probability density and probability distribution functions. The determination of the frequency or spectral characteristics of the vibration magnitude at a given time is discussed, followed by a discussion of means of determining the time variation of the magnitude of the vibration. Finally, autocorrelation and crosscorrelation are discussed as means to determine both the statistical and spectral characteristics of vibration.

* The word "magnitude" is used in this chapter to indicate the severity of vibration without relating it specifically to one of the various mathematical expressions used to describe vibration quantitatively.

QUALITATIVE DESCRIPTION OF VIBRATION

This section considers various means which are commonly used in vibration analysis to describe time-varying functions such as displacement, velocity, acceleration, or force, and describes the principal characteristics of these functions.

Figure 22.1 illustrates three typical vibration waveforms or time-histories. The solid curve of Fig. 22.1*A* is the sum of the two sinusoidal vibrations shown by the dotted lines. Figure 22.1*B* represents broad-band random vibration, while Fig. 22.1*C* illustrates the waveform of a narrow-band random vibration. The latter has the appearance of a sine wave whose amplitude, represented by the dotted lines, varies in an unpredictable manner. If the envelope of the waveform varies sinusoidally, the waveform is the sum of two sinusoids whose frequency difference is small compared to the sum, resulting in beats.

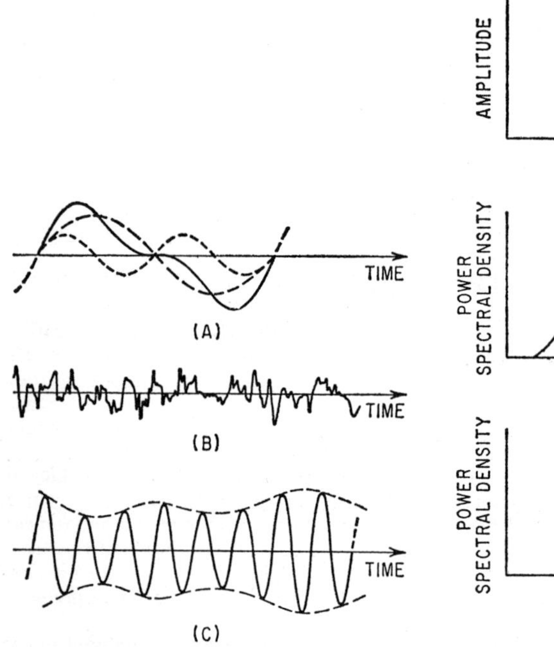

FIG. 22.1. Typical vibration time-histories: (*A*) Sum of two sinusoids; (*B*) broad-band random vibration; (*C*) narrow-band random vibration.

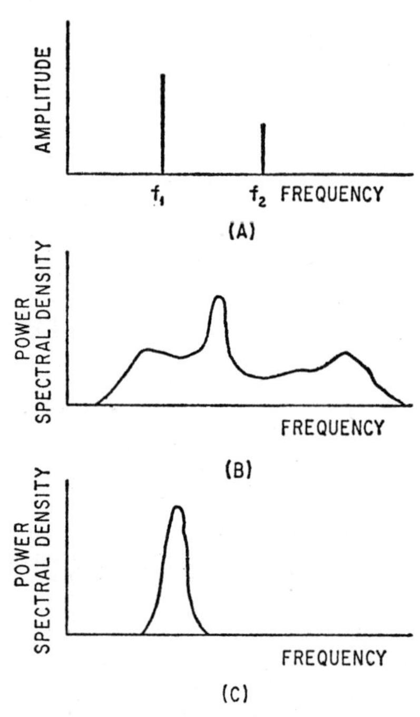

FIG. 22.2. Results of data analysis of time-histories of Fig. 22.1: (*A*) Sum of two sinusoids; (*B*) broad-band random vibration; (*C*) narrow-band random vibration.

Figure 22.2 illustrates typical forms for describing, after suitable data analysis, the time-histories shown in Fig. 22.1. In Fig. 22.2*A*, the two sine waves are represented by vertical lines at their respective frequencies; the height of each line is proportional to the amplitude of the respective sine wave. This plot is known as a *line spectrum* or *discrete frequency spectrum*. The information also can be presented as a tabulation of frequencies and associated amplitudes. For complete definition of the waveform of Fig. 22.1*A*, it is necessary to specify the phase angle between the two sine waves at a convenient time origin. However, phase information seldom is required in vibration data analysis and the plot of Fig. 22.2*A* is the extent of the information usually required.

Figure 22.2*B* is a plot of the power spectral density* vs. frequency of broad-band random vibration showing the frequency content to be spread over a wide frequency range. Figure 22.2*C* illustrates the power spectral density of narrow-band random vibration showing the concentration of the frequency content in a narrow band of frequencies.

In general, three coordinates are necessary to describe a vibration time-history: (1) the magnitude of the vibration; e.g., the amplitude of sine waves or the power spectral density; (2) the variation of magnitude with frequency, i.e., spectral characteristics; and (3) the variation of the magnitude of the vibration at a particular frequency as a function of time.

The magnitude of a vibration may be represented in terms of its spectral characteristics and its variation with time by the surface shown in Fig. 22.3. The height of a point

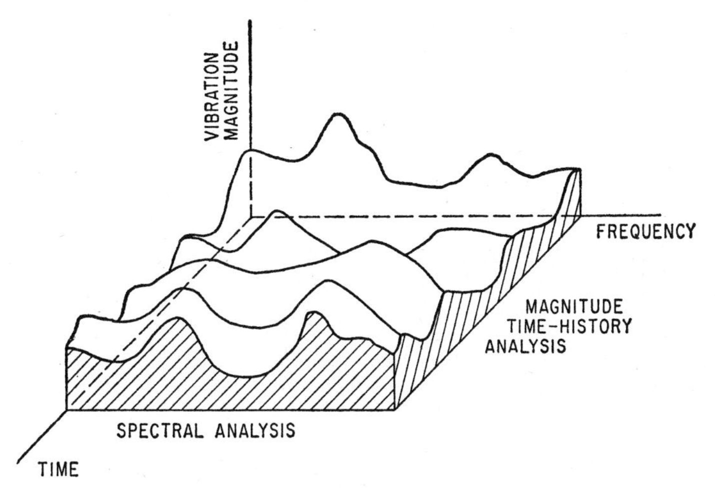

Fig. 22.3. Vibration magnitude surface: the height of the surface at any time and frequency represents the vibration magnitude at that time and frequency. Units of vibration magnitude depend on type of vibration, e.g., random, complex, or periodic.

on the surface raised over the time-frequency plane defines the magnitude of the vibration at a particular time and frequency. For example, the surface could represent the vibration encountered during the flight of an aircraft from take-off to landing. As the flight conditions change, both the vibration magnitude and the frequency characteristics of the vibration may change.

At a specific instant of time, say t_0, the vibration is described by its variation of magnitude with frequency in the plane normal to the time axis passing through t_0, i.e., in the plane labeled "Spectral Analysis" in Fig. 22.3 and shown in the two-dimensional plot of Fig. 22.4.

The variation of the magnitude of vibration at a particular frequency, say f_0, as a function of time is shown by passing a plane normal to the frequency axis through f_0, i.e., in the plane labeled "Magnitude–Time-history Analysis" in Fig. 22.3 and shown in the two-dimensional plot of Fig. 22.5.

In the case of truly steady-state vibration, i.e., vibration independent of time as during some laboratory vibration tests, the surface of Fig. 22.3 can be represented by the magnitude-frequency plot of Fig. 22.4. Similarly, a vibration consisting of a single sine

* Power spectral density is defined under *Spectral Analysis* [see Eq. (22.17) in this chapter; also see Eq. (11.16)]. It is a generic term used regardless of the physical quantity represented by the time-history. However, it is preferable to indicate the physical quantity involved. For example, the terms "mean-square acceleration density" or "acceleration spectral density" are used when the time-history of acceleration is to be described.

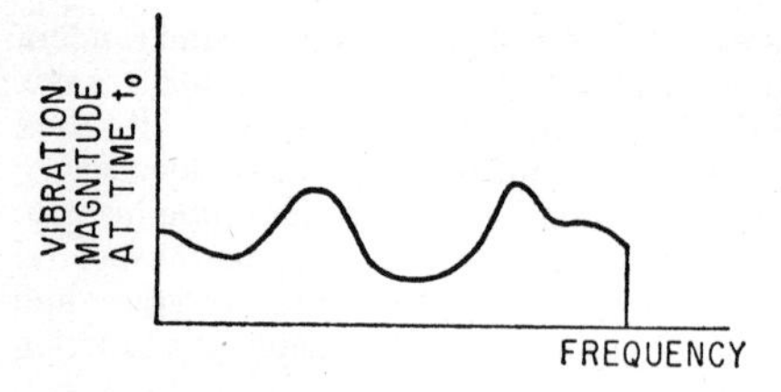

FIG. 22.4. Typical plot of spectral analysis.

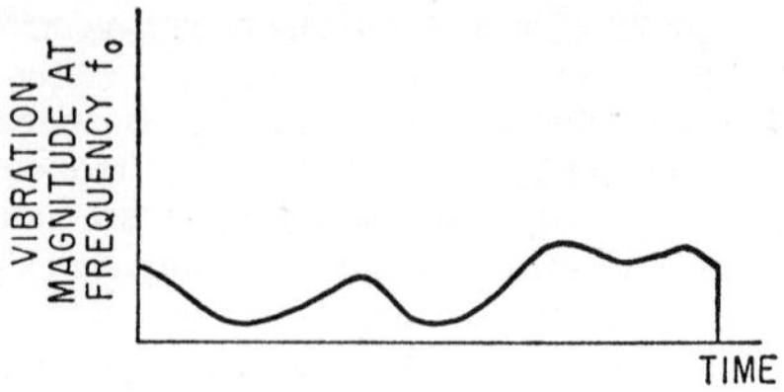

FIG. 22.5. Typical plot of magnitude time-history analysis.

wave at a constant frequency whose amplitude varies slowly with time may be represented by a single amplitude-time plot such as Fig. 22.5.

BASIC TYPES OF TIME-VARYING FUNCTIONS

This section considers certain relations used to describe specific time-varying functions in vibration analysis and describes their principal characteristics. Types of functions discussed are sinusoidal, periodic,[1–3] phase-coherent,[4,5] complex, and random.[6–8,29,32] In the following sections, $\xi(t)$ is the function which represents the time variation of any physical parameter such as displacement, force, or acceleration.

SINUSOIDAL FUNCTIONS. The simplest time-varying quantity of concern in vibration analysis is a sinusoidal function having constant amplitude ξ_0 and frequency f:

$$\xi(t) = \xi_0 \sin (\omega t + \theta) \tag{22.1}$$

where $\omega = 2\pi f$, and θ is the phase angle with respect to the time origin. The objective of data analysis on this waveform is the determination of ξ_0, ω, and, less frequently, θ.

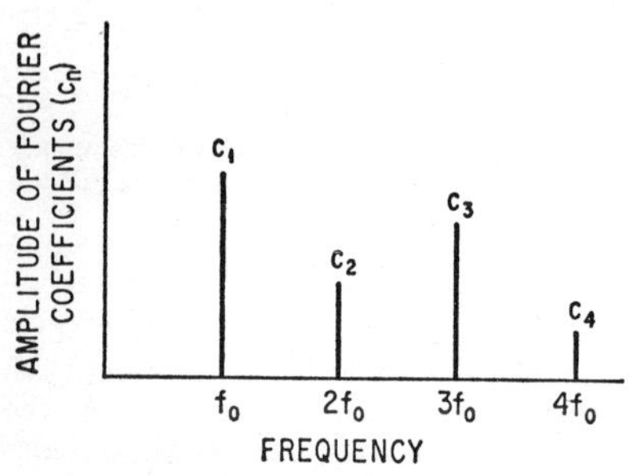

FIG. 22.6. Line spectrum to represent Fourier series of periodic function. Fundamental frequency $f_0 = 1/\tau$, where τ is period of function [see Eq. (22.8)].

PERIODIC FUNCTIONS. If a function has a waveform which repeats itself exactly in consecutive time periods τ:

$$\xi(t) = \xi(t \pm n\tau) \tag{22.2}$$

where n is any integer; then it is said to be a *periodic function*[1–3] having a fundamental frequency $f = 1/\tau$. For example, a wing panel adjacent to a propeller may vibrate periodically at a fundamental frequency equal to the cyclic rate at which propeller blades pass the panel and also at higher frequencies equal to multiples of the blade passage frequency. A periodic function may be represented by the line spectra shown in Fig. 22.6.

PHASE-COHERENT FUNCTIONS. A phase-coherent function[4,5] is defined as a sine wave whose amplitude and frequency vary smoothly with time. It is assumed that the changes in amplitude and frequency are significant in magnitude only over a time duration equal to a number of cycles of the waveform. Mathematically, it is expressed by

$$\xi(t) = \xi_0(t) \sin [\omega(t)t + \theta] \tag{22.3}$$

where ξ_0 and ω are functions of time and θ is the phase angle of the waveform at the time origin. A function of this type is useful in a description of vibration measured in a system whose natural frequencies vary with time, such as a rocket engine during the burning of the propellant. Equation (22.3) defines the input excitation to the test object during a sweep frequency vibration test.

COMPLEX FUNCTIONS. A time-history which is composed of a sum (or in the limit, integral) of sinusoidal functions which is not necessarily periodic is defined as a complex function. Such functions may be represented graphically by line spectra as shown in Fig. 22.6 except that the lines are not equally spaced along the frequency axis; when the complex function consists of an integral or continuous distribution of sinusoidal functions, it may be represented by a Fourier spectrum as discussed in Chap. 23.

TRANSIENT FUNCTION—SHOCK. A transient function is a time-varying function which is nonzero only over a given restricted time interval. Transients are discussed in Chap. 23.

RANDOM FUNCTIONS. A random time-function[6–9] (see Chap. 11) consists of a continuous distribution of sine waves at all frequencies, the amplitudes and phase angles of which vary in an unpredictable manner as a function of time. This unpredictability distinguishes random functions from "deterministic" or coherent functions (e.g., sinusoidal, periodic, and complex functions) in the following respect. Knowledge of the instantaneous value of a random function at a given time does not provide any information on the value which it will have at some specific later time. It is known only that there is a certain probability that the value will lie within a certain range of values, i.e., it can be described only in statistical terms. In contrast, if the value of a deterministic function is known at any time, the value at any other time can be determined.

DEFINITION OF MEAN, AVERAGE, AND MEAN-SQUARE VALUES.
Mean Value. The mean value of a function $\xi(t)$ is defined by

$$\bar{\xi} = \frac{1}{T} \int_0^T \xi(t)\, dt \tag{22.4}$$

where T is the "averaging" time.
Average Value. The term "average value" of a function $\xi(t)$ is employed to denote the time average of the magnitude of a function with zero mean value:

$$\overline{|\xi|} = \frac{1}{T} \int_0^T [\xi(t)]\, dt \tag{22.5}$$

Mean-square Value. The mean-square value of a function $\xi(t)$ is the time average of the square of the function:

$$\overline{[\xi]^2} = \frac{1}{T} \int_0^T [\xi(t)]^2\, dt \tag{22.6}$$

The rms value is the square root of the mean-square value.
For example, the mean, average, and mean-square values of a sine wave [Eq. (22.1)] averaged over an integer number of periods, are

$$\begin{aligned} \bar{\xi} &= 0 \\ \overline{|\xi|} &= \frac{2}{\pi}\, \xi_0 \\ \overline{[\xi]^2} &= \tfrac{1}{2}\xi_0^2 \end{aligned} \tag{22.7}$$

ANALYSIS OF PERIODIC FUNCTIONS

FOURIER SERIES

Periodic functions encountered in vibration analysis may be represented by a Fourier series[1–3] which consists of a sum of sine waves whose frequencies are all multiples of the fundamental frequency. The amplitudes of the sine waves are known as the Fourier coefficients c_n. An additional constant term equal to the mean value of the waveform during the period τ must also be included. The function is expressed mathematically as

$$\xi(t) = \sum_{n=1}^{\infty} a_n \sin n\omega t + \sum_{n=0}^{\infty} b_n \cos n\omega t = \sum_{n=0}^{\infty} c_n \sin (n\omega t + \theta_n) \tag{22.8}$$

where

$$c_n = +[a_n{}^2 + b_n{}^2]^{1/2} \qquad \theta_n = \sin^{-1} \frac{b_n}{c_n} = \cos^{-1} \frac{a_n}{c_n}$$

$$a_n = \frac{2}{\tau} \int_{-\tau/2}^{\tau/2} \xi(t) \sin n\omega t \, dt$$

$$b_n = \frac{2}{\tau} \int_{-\tau/2}^{\tau/2} \xi(t) \cos n\omega t \, dy$$

$$b_0 = \frac{1}{\tau} \int_{-\tau/2}^{\tau/2} \xi(t) \, dt = \bar{\xi}$$

If the function $\xi(t)$ is an *even function*, i.e., $\xi(t)$ equals $\xi(-t)$, the series consists of cosine terms only; for an *odd function*, $\xi(t)$ equals $-\xi(-t)$ and the series consists of sine terms only.

OBJECTIVE AND RESULTS OF ANALYSIS

The objective of the analysis of periodic functions is to determine the Fourier coefficients c_n and the phase angles θ_n of Eq. (22.8). By describing a function as a Fourier series, it is implied that the waveform during the period τ is repeated exactly in all successive periods between negative and positive infinite times and that all frequencies associated with the waveform are determined by τ, as shown by Eq. (22.8). No measured vibration data completely fulfill these requirements for description by a Fourier series. When conducting a Fourier series analysis, it is common practice to select what appears to be a typical section of the time-history and to treat it as if it were the time-history during one period of a periodic function. The consequences of this selection are (1) the time period chosen fixes the frequencies of the resulting line spectra, whether these are the actual frequencies present in the function or not, and (2) the signal is assumed, a priori, to consist of a sum of pure sinusoids.

The results of data analysis involving periodic vibration may be summarized by plotting the discrete frequencies and the corresponding amplitudes. For example, Fig. 22.2*A* applies to two superimposed sinusoidal functions of arbitrary frequencies f_1, f_2; Fig. 22.6 is an analogous plot of the Fourier coefficients of a periodic quantity. This form of data presentation may be abbreviated by plotting only points representing the upper ends of the spectrum lines in Figs. 22.2*A* and 22.2*B*. Representation of the data by such points is particularly advantageous for summarizing the results of analysis of the vibration of a number of data samples. For example, the vibration at a number of different flight conditions or on a number of different aircraft may be shown on a single plot, so that each point represents one measurement of a sinusoidal function at the given amplitude and frequency. If the number of points becomes excessive, different-sized symbols are used to represent given numbers, say 1, 10, 50, or 100 readings at the particular amplitude and frequency. The density of points in various frequency regions or at various amplitudes indicates, at least qualitatively, the frequency of occurrence of those values.

DETERMINATION OF FOURIER COEFFICIENTS

The Fourier coefficients of a periodic function are defined by Eq. (22.8) and may be determined by several alternate means. If an oscillographic recording of the function $\xi(t)$ during one complete period is obtained and the instantaneous value of $\xi(t)$ at N equally spaced time intervals is measured, the Fourier coefficients c_n and phase angles θ_n from $n = 0$ to $n = (N - 1)/2$ may be calculated by tabular methods.[1, 3, 10, 11] A detailed description of these tabular methods is given with respect to Fig. 38.12. Similar computations also may be carried out by digital computers.

A machine known as a Coradi* analyzer may be used to determine the first five Fourier coefficients from an oscillographic record, provided the time scale is expanded to yield 40 cm of record for one period of the function. The machine solves Eq. (22.8) by the use of planimeter-type mechanical integrators while the operator traces the time-history with a stylus. A similar mechanical analyzer known as the Harvey Harmonic Analyzer[11] is capable of determining the first fourteen coefficients. However, because of their complexity, such machines are rarely used. Instead, electrical analysis techniques are employed.

When the time-history is available as an analog voltage, filters may be used to determine the Fourier components as described later in this chapter under *Spectral Analysis*.

STATISTICAL ANALYSIS

Statistical analysis[13–20, 29–32] of a vibration time-history is carried out to determine the characteristics, e.g., sinusoidal and/or random, of the time-history so that the appropriate mathematical function for description of the vibration magnitude may be chosen. This section describes the statistical properties of random and sinusoidal functions, and discusses concepts of data analysis for determining the significant statistical parameters of the time-histories.

STATISTICAL PARAMETERS OF RANDOM FUNCTIONS

Statistical parameters used frequently in data analysis in addition to the mean and mean-square values defined by Eqs. (22.4) and (22.6) are the probability density function and probability distribution function.

PROBABILITY DENSITY FUNCTION. The probability density function $p(\xi/\sigma)$, illustrated in Fig. 22.7, defines the probability (or fraction of time, on the average) that

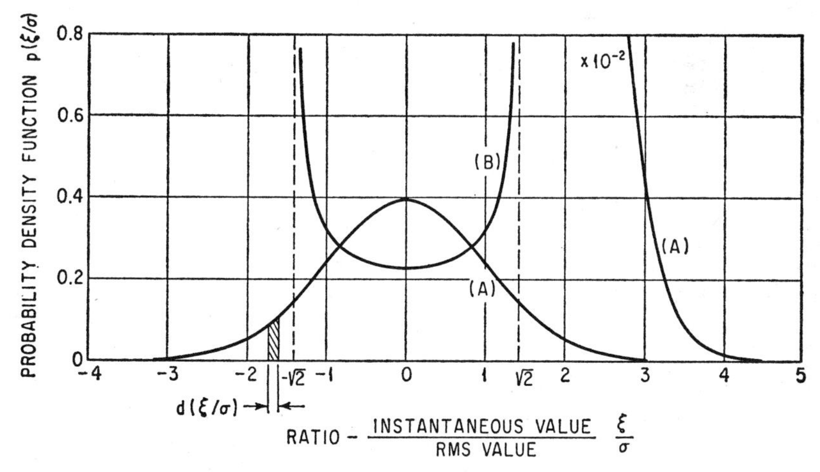

Fig. 22.7. Normalized probability density functions: (*A*) Gaussian or normal distribution [Eq. (22.10)]; (*B*) distribution of instantaneous values of a sine wave [Eq. (22.11)]. The curve (*A*) marked $\times 10^{-2}$ indicates hundredfold expansion of ordinate scale.

* The Henrici analyzer is a very similar device.

the magnitude of the quantity $\xi(t)$ will lie between two values. It is customary to normalize the curve by plotting the magnitude divided by the rms value σ as the abscissa. Then the probability that the magnitude lies between ξ/σ and $(\xi + d\xi)/\sigma$ is equal to $p(\xi/\sigma)d(\xi/\sigma)$, i.e., the shaded area shown in Fig. 22.7. Since it is certain, with probability 1.0, that the function $\xi(t)$ lies between plus and minus infinity, the area under the entire curve is unity.

PROBABILITY DISTRIBUTION FUNCTION. The probability distribution function (or *cumulative distribution function*) $P(\xi/\sigma\geq)$* defines the probability that the magnitude of ξ/σ will exceed a certain value. However, from the definition of the probability density function, $P(\xi/\sigma\geq)$ is equal to the area under the plot of $p(\xi/\sigma)$ as a function of ξ/σ between ξ/σ and infinity:

$$P(\xi/\sigma\geq) = \int_{\xi/\sigma}^{\infty} p\left(\frac{\xi'}{\sigma}\right) d\left(\frac{\xi'}{\sigma}\right) \tag{22.9}$$

where, for this equation and Eq. (22.12) only, ξ' is used as a dummy variable of integration for ξ.

The value of the distribution function $P(\xi/\sigma\geq)$ at the minimum possible value of ξ is $P(\xi_{\min}/\sigma) = 1.0$ since ξ is never less than $\xi_{\min}$; for the maximum possible value of ξ, $P(\xi_{\max}/\sigma)$ is zero since ξ never exceeds $\xi_{\max}$. Further, $P(\xi/\sigma\geq)$ must decrease monotonically between $\xi_{\min}$ and $\xi_{\max}$, as illustrated in Fig. 22.11. Since the probability of exceeding a value is one minus the probability of not exceeding that value, the ordinate of Fig. 22.11 may be changed to $P(\xi/\sigma\leq)$ if the scale is changed to vary from 1 to 0 instead of 0 to 1.

COMPARISON OF PARAMETERS FOR SINUSOIDAL AND RANDOM FUNCTIONS

Although a sinusoidal function is deterministic, the probability density and probability distribution of a sine wave may be determined for comparison with those of a random function. The comparison may be made on the basis of (1) the instantaneous values $\xi(t)$ of the function and (2) the peak values or maxima $\xi_p(t)$ of the function.

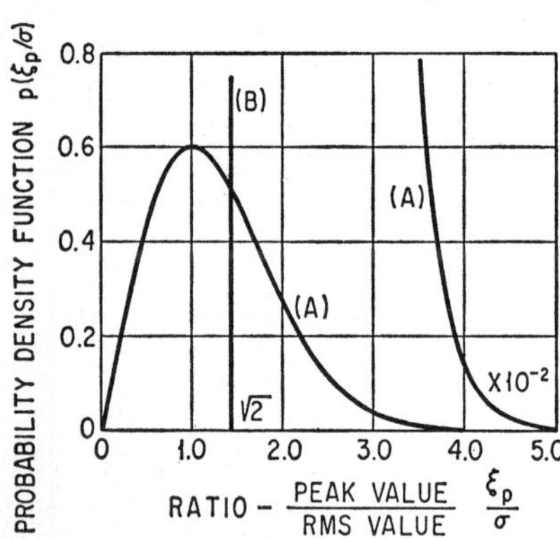

FIG. 22.8. Normalized probability density functions: (*A*) Rayleigh distribution for peaks of narrow-band Gaussian vibration [Eq. (22.14)]; (*B*) distribution for peaks of sine-wave delta function at $\xi_p/\sigma = \sqrt{2}$. The curve (*A*) marked $\times 10^{-2}$ indicates hundredfold expansion of ordinate scale.

DISTRIBUTION OF INSTANTANEOUS VALUES OF $\xi(t)$. A comparison of the identifying characteristics of the probability densities and distributions for the instantaneous values of a sinusoidal function and the particular case of Gaussian random vibration is shown in Figs. 22.7, 22.8, and 22.9. The term "Gaussian" is used to describe a random function whose *instantaneous* value is defined by the Gaussian or normal probability density function [see Eq. (11.10)] given by

$$p(\xi/\sigma) = \frac{1}{\sqrt{2\pi}} e^{-(\xi^2/2\sigma^2)} \tag{22.10}$$

where σ is the rms value. Equation (22.10) is shown by curve *A* of Fig. 22.7. The prob-

* The probability that ξ/σ is "greater than" is written $P(\xi/\sigma\geq)$; conversely, the probability that ξ/σ is "less than" is written $P(\xi/\sigma\leq)$. $P(\xi/\sigma\geq) = 1 - P(\xi/\sigma\leq)$.

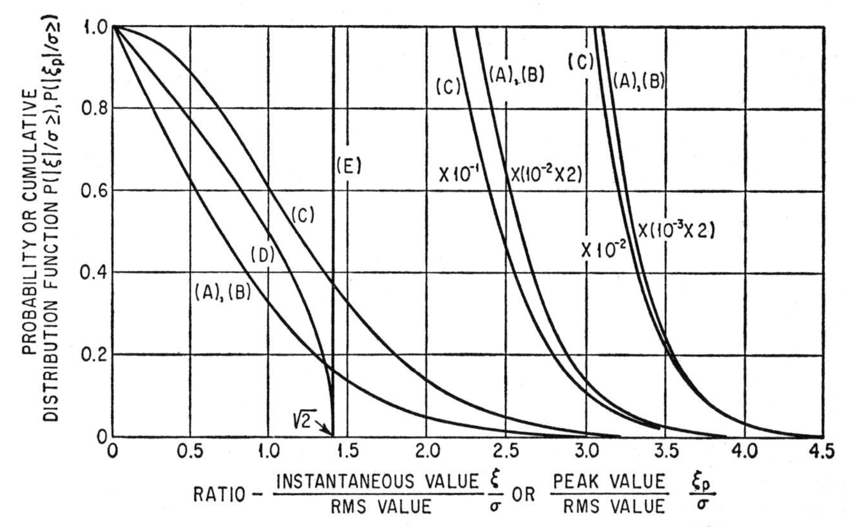

Fig. 22.9. Probability distribution functions: (*A*) Instantaneous values of broad-band and narrow-band random vibration—Gaussian distribution [Eq. (22.12)]; (*B*) peaks of broad-band random vibration [Eq. (22.12)]; (*C*) peaks of narrow-band random vibration [Eq. (22.15)]; (*D*) instantaneous values of a sine wave [Eq. (22.13)]; (*E*) peak values of a sine wave. Multiply ordinate scale by factors marked adjacent to curves for large values of ξ/σ.

ability density function of the *instantaneous* value of a sinusoid is shown by curve *B* of Fig. 22.7 and is defined by

$$p(\xi/\sigma) = \frac{1}{\pi\sqrt{2 - (\xi/\sigma)^2}} \tag{22.11}$$

The probability density functions of Eqs. (22.10) and (22.11) are symmetrical about a mean assumed to be zero; then the probability that ξ exceeds a given absolute value (or magnitude) $|\xi|$ is twice the probability that it exceeds the same absolute value in either the positive or negative sense [see Eq. (22.9) or Fig. 22.7]. Therefore, it is convenient to plot the probability distribution function in terms of the absolute value of ξ, i.e., $P(|\xi|/\sigma \geq)$, as shown in Fig. 22.9.

The probability distribution functions for the Gaussian and sinusoidal functions are obtained by integration of Eqs. (22.10) and (22.11):

Gaussian:

$$P(|\xi|/\sigma \geq) = \frac{2}{\sqrt{2\pi}} \int_{\xi/\sigma}^{\infty} e^{-(\xi'^2/2\sigma^2)} d\left(\frac{\xi'}{\sigma}\right) \tag{22.12}$$

Sinusoid:

$$P(|\xi|/\sigma \geq) = \frac{2}{\pi} \cos^{-1} \frac{\xi}{\sigma\sqrt{2}} \tag{22.13}$$

The relations of Eqs. (22.12) and (22.13) are plotted as curves *A* and *B* in Fig. 22.9.

DISTRIBUTION OF PEAK VALUES (MAXIMA) OF $\xi(t)$. When the peak values or maxima of a function are considered, the two statistical functions differ from those found for the instantaneous values. For a sine wave, all maxima are of equal magnitude and the probability density function $p(\xi_p/\sigma)$ becomes a Dirac delta function[2] (see Chap. 23) as shown by curve *B* of Fig. 22.8. For broad-band Gaussian vibration, i.e., vibration with nonzero spectral density over a frequency bandwidth which is not small compared to the average or center frequency of the bandwidth, the distribution of peak values is normal, as shown by curves *A* of Fig. 22.7 and *B* of Fig. 22.9. However, for narrow-band Gaussian noise, i.e., noise with negligible spectral density except in a frequency bandwidth which is small compared to the center frequency, the distribution of peak values becomes the *Rayleigh distribution.* The probability density and distribution functions for the Rayleigh distribution are defined by:

Probability density:

$$p(|\xi_p|/\sigma) = \frac{\xi_p}{\sigma} e^{-(\xi_p^2/2\sigma^2)} \tag{22.14}$$

Probability distribution:

$$P(|\xi_p|/\sigma) = e^{-\xi_p^2/2\sigma^2} \tag{22.15}$$

The relations given by Eqs. (22.14) and (22.15) are shown graphically by curve *A* of Fig. 22.8 and curve *C* of Fig. 22.9, respectively.

Figures 22.7 to 22.9 are plotted on the basis that the mean value of the function $\xi(t)$ is zero. In the more general case of nonzero mean value, the abscissae of these figures are $(\xi - \bar{\xi})/\sigma$, where $\bar{\xi}$ is the mean value of the function $\xi(t)$. However, in vibration data analysis, it is common practice to constrain the mean value to zero by use of a high-pass or bandpass filter prior to statistical analysis. Thus these figures show the form usually encountered.

Figures 22.7 to 22.9 show that the probability density and probability distribution curves for sinusoidal and random functions differ considerably, thus providing identifying characteristics for each type of function. When the time-history $\xi(t)$ is a combination of sinusoidal and random functions, the shape of the probability density and distribution curves depends on the relative magnitudes of each type of function.

RELATIONSHIP BETWEEN AVERAGE AND RMS VALUES. The average value may be determined from the probability density function since $\overline{|\xi|}$ is represented by the distance from the vertical axis to the center-of-gravity of the area under the probability density function, for either positive or negative values (assuming a symmetrical density function):

$$\overline{|\xi|} = 2\sigma \int_0^\infty p\left(\frac{\xi}{\sigma}\right) \frac{\xi}{\sigma}\, d\left(\frac{\xi}{\sigma}\right) \tag{22.16}$$

where $\overline{|\xi|}/\sigma$ is indicated in Fig. 22.10 for the probability density functions for the positive values of (*A*) a Gaussian distribution and (*B*) a sinusoid. For a sine wave, the average

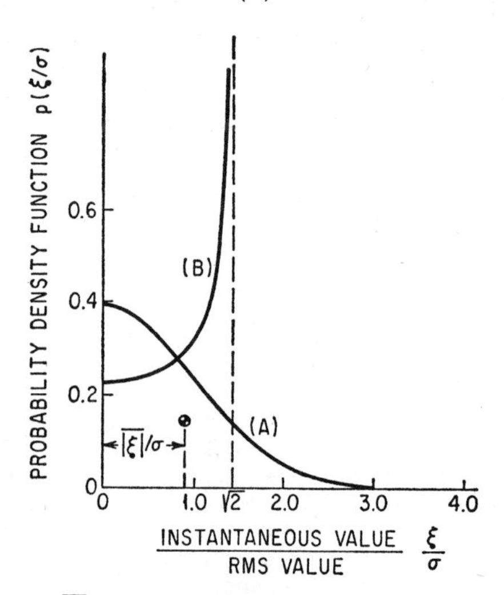

Fig. 22.10. Average value $\overline{|\xi|}$ of function with zero mean value: (*A*) Gaussian distribution $\overline{|\xi|}/\sigma = 0.798$; (*B*) sine wave $\overline{|\xi|}/\sigma = 0.900$.

value $\overline{|\xi|} = 0.900\sigma$; this is 0.636 times the amplitude. The average value of a Gaussian function $\overline{|\xi|} = 0.798\sigma$; it cannot be defined in terms of a peak value since the peak values vary from zero to infinity. However, physical processes usually are only approximately Gaussian in that the peak values are limited to some value between three and ten times the rms value.

PROBABILITY GRAPH PAPER

The characteristics of the probability density and distribution functions shown in Figs. 22.7 to 22.9 are emphasized by plotting the functions on probability paper.

GAUSSIAN DISTRIBUTION. Gaussian probability paper* is graph paper having a set of coordinates such that when the Gaussian distribution function [Eq. (22.12)] is plotted on it, a straight line is obtained, as shown in Fig. 22.11. For example, if the

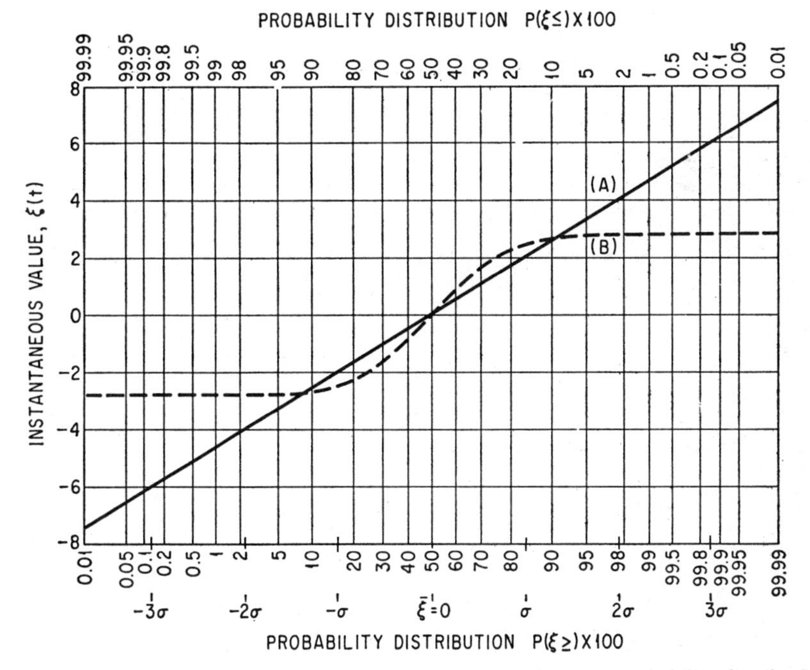

FIG. 22.11. Probability paper for Gaussian distribution: (*A*) Gaussian probability distribution function gives a straight line where mean value $\bar{\xi}$ is at $P\ (\xi \geq) = 0.50$ and rms value σ is at 0.159 or 0.841; (*B*) probability distribution of a sine wave gives a nonlinear plot.

solid straight line in this figure represents the measured distribution function of the instantaneous value of $\xi(t)$, then $\xi(t)$ is random having a Gaussian distribution. In contrast, a sinusoidal function, for example, has a distribution function of the form shown by the dotted line of Fig. 22.11. Therefore, this type of paper is helpful in vibration analysis problems to determine the extent to which a time function has a normal distribution by observing the linearity of the plotted distribution function. To be useful in determining the randomness of $\xi(t)$, the probability distribution must be plotted when $\xi(t)$ is the *output of a relatively narrow-band filter* and the distribution function of the *instantaneous* values has been measured. A close approximation to the Gaussian dis-

* Available commercially, for example: Probability Scale Graph Paper No. 359-23, Keuffel & Esser Co., New York.

tribution can be synthesized by as few as seven appropriately chosen sinusoids,[21] and any complex function with broad frequency content will approximate a Gaussian distribution in accordance with the central-limit theorem.[9] Thus the probability distribution function of a broad-band signal is an insensitive measure of randomness.

The symmetry of the probability density function about the mean value also may be determined readily by use of this paper. If the shape of the curve is unchanged when it is rotated 180° about its intersection with the ordinate $P(\xi \geq) = 50$ per cent, the density function is symmetrical about the mean and the *mean value* $\bar{\xi}$ is the value of $\xi(t)$ at $P(\xi \geq) = 50$ per cent. The mean value of zero is plotted in Fig. 22.11. If the distribution function is linear as well as symmetrical when plotted on Gaussian probability paper, the *rms value* σ may be read from the curve at the 15.9 or 84.1 per cent points. The 2σ points at 2.3, or 97.7 per cent, and the 3σ points at 0.13, or 99.87 per cent, also may be used in Fig. 22.11. In Fig. 22.11, the rms value plotted is 2.0.

RAYLEIGH DISTRIBUTION. Rayleigh probability paper* is graph paper having a set of coordinates such that when the Rayleigh distribution function [Eq. (22.15)] is plotted on it, a straight line is obtained. This is shown in Fig. 22.12. The solid curve shows the distribution function of the *peak values* ξ_p of a narrow-band random function which has an rms value of 2.0. The analogous distribution function of the peak values of a sine wave is a horizontal line at the amplitude of the sine wave, as shown by the dotted curve of Fig. 22.12 for rms value of 2.0. Thus, the paper shown in Fig. 22.12 is useful in determining the randomness or coherence of $\xi(t)$ when $\xi(t)$ is the output of a relatively narrow-band filter and the distribution function of the *peak* values has been measured.

When data are presented in the form shown in Fig. 22.12, one curve is required for each narrow band of the signal which is analyzed. Since this may lead to a large number of plots, a technique of data presentation which includes all frequency bands and, at least qualitatively, a picture of the distribution of maxima is illustrated in Fig. 22.13. The abscissa is the center frequency of each band, and the ordinate is the magnitude of the maxima. Lines are then drawn between the values at each center frequency which represent the magnitude at or below which the indicated percentages of the peaks occur. For sinusoidal functions, all lines are coincident since all peaks are of equal magnitude.

DETERMINATION OF DISTRIBUTION FUNCTIONS

In practice, the characteristics of a signal $\xi(t)$ at a specific time and at a specific frequency cannot be determined, since only the instantaneous value of $\xi(t)$ is known at a specific time. Instead, the characteristics of the signal in a given restricted frequency bandwidth during a short period of time are determined. Thus the first step in a statistical analysis is to select a time interval short enough so that the nature and magnitude of $\xi(t)$ may be assumed constant, but long enough to give a statistically significant result.[6] Often, it is convenient to store the data sample in the form of a recording on a continuous loop of magnetic tape so that it may be played back repeatedly. Then a time sample of the signal may be analyzed to determine the probability distribution of either the in-

* Rayleigh probability paper can be constructed as follows:
The probability distribution function for the Rayleigh distribution, Eq. (22.15), can be written:

$$P(|\xi_p|/\sigma \leq) = 1 - e^{-\xi_p{}^2/2\sigma^2} \qquad \xi_p/\sigma = \left[2 \log_e \left(\frac{1}{1-P}\right)\right]^{1/2}$$

Let $x(P)$ be the distance from the origin of a point on the abscissa corresponding to the ordinate ξ_p/σ. For the above equation to be a straight line

$$x(P) = a\frac{\xi_p}{\sigma} + b = a\left[2 \log_e \left(\frac{1}{1-P}\right)\right]^{1/2} + b$$

The values of a and b are determined by the choice of origin and scale. If $x(0)$ is zero, then b also is zero and a is determined by the distance between $x(0)$ and $x(0.9999)$.

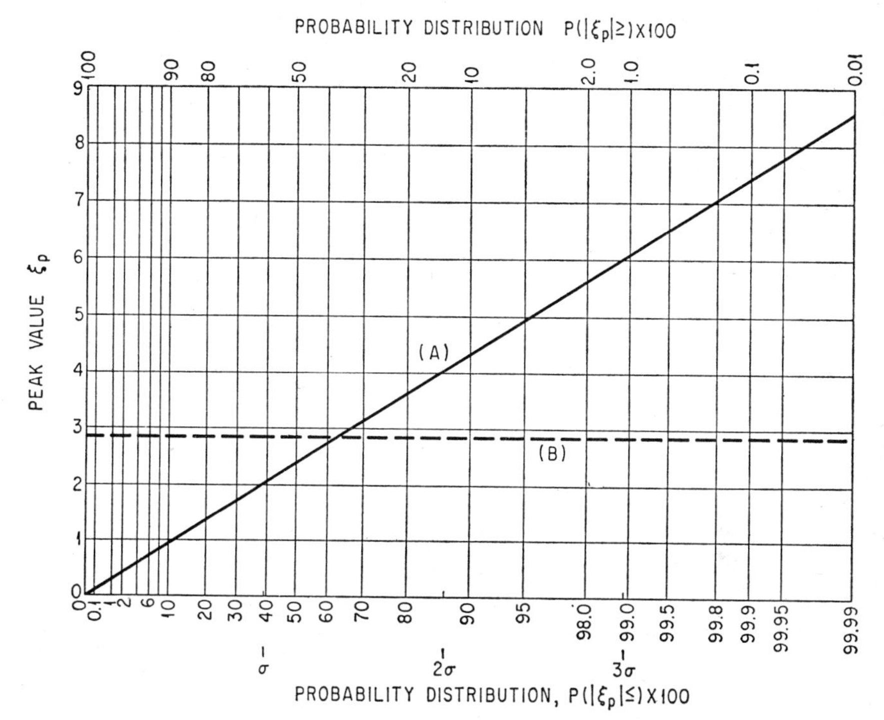

FIG. 22.12. Probability paper on which Rayleigh distribution function gives a straight line: (*A*) Rayleigh distribution for peaks of narrow-band random vibration with $\sigma = 2.0$; (*B*) distribution function of peaks of a sinusoid $- \sigma = 2.0$.

stantaneous or the peak values of the signal. The concepts involved in this determination are considered separately in the next two sections dealing with instantaneous and peak values, respectively. The instantaneous values of the *filtered* signal are indicated by $\xi'(t)$ while the peak values of the *filtered* signal are indicated by $\xi_p'(t)$.

INSTANTANEOUS VALUES. The probability distribution of instantaneous values is obtained for a set of discrete values of $\xi'(t)$ rather than as the continuous function that is illustrated in Fig. 22.9. The probability distribution function at a value ξ_1' is the probability $P(\xi' \geq)$ that $\xi'(t)$ exceeds ξ_1'.* Referring to Fig. 22.14, this is equivalent to summing all the Δt's during which $\xi'(t)$ exceeds ξ_1' or level L_1 and dividing by the total elapsed time. If this is done for a number of selected levels, $L_1, L_2, \ldots, L_n$, the probability distribution function at these levels is obtained. Figure 22.15*A* illustrates the block diagram of a system to make these measurements. The filtered signal $\xi'(t)$ is fed to an array of "discriminators" which determine whether a signal is greater or less than a preselected value. The levels of the discriminators are set at L_0, L_1, L_2, etc., usually in equal increments of voltage. The detail or resolution obtained is governed by the number of discriminator levels selected for measurement. The detail and accuracy yielded by ten levels for each polarity are sufficient for many engineering purposes. Since the resolution obtained improves with the number of counters registering, a variable gain

* The filtered signal $\xi'(t)$ may be passed first through a mechanism for changing the polarity (or sign) of the signal so that the positive and negative values of $\xi(t)$ can be analyzed separately, thus halving the number of discriminators and counters required for a given resolution. The time-history must be played back for twice as long but this usually will cause a rather small increase in the total analysis time.

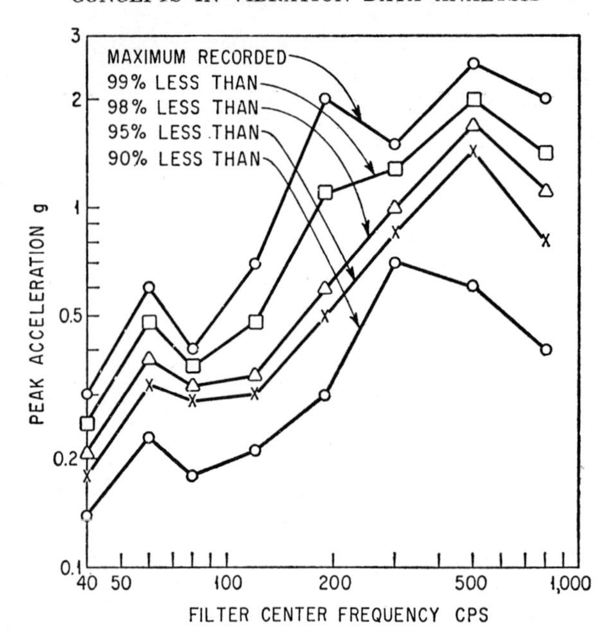

Fig. 22.13. Distribution of peak accelerations as a function of frequency. Each curve represents the value below which the indicated percentage of peak values occurs in the response of the filter to $\xi(t)$, for the indicated center frequency of the filter.

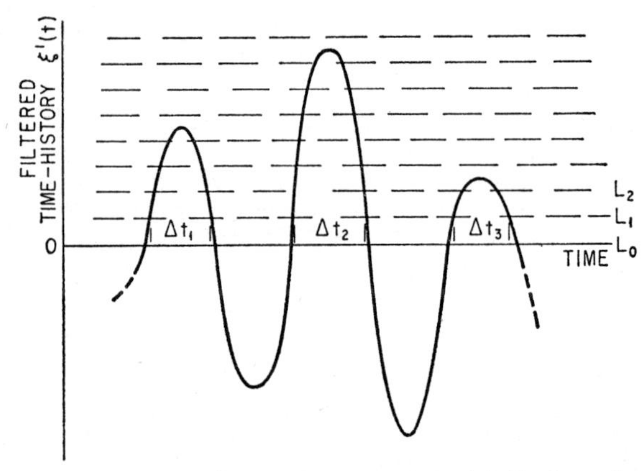

Fig. 22.14. Measurements required for statistical analysis: (A) To determine distribution of instantaneous values, sum of time increments Δt_i during which $\xi'(t)$ exceeds L_0, L_1, L_2 . . . is measured; (B) to determine distribution of peak values, number of times that $\xi'(t)$ exceeds L_0, L_1, L_2 . . . is measured.

control is needed to adjust the signal level, by a known amount, until an adequate number of counters are registering.

When the signal $\xi'(t)$ exceeds the level of a particular discriminator, a clock is started. It is stopped when $\xi'(t)$ next falls below that level and the counter counts the elapsed time between these events. At the end of the data sample, the readings of the counters are recorded by the read-out mechanism, a digital printer, for example. If each reading is divided by the total elapsed time of the data sample, the values obtained are the values

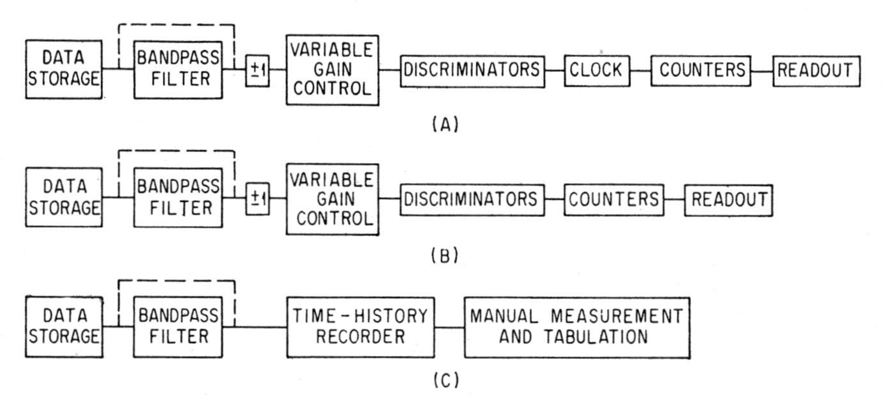

Fig. 22.15. Block diagrams showing steps in statistical analysis: (*A*) Instantaneous value analysis—automatic; (*B*) peak value analysis—automatic; (*C*) peak value analysis—manual. For each analysis, the center frequency of the filter remains constant.

of the probability distribution function at the values of the levels L_0, L_1, L_2, etc. If the values at adjacent levels are subtracted, the average probability density function between these two levels is obtained. The polarity of the signal is now changed and the process is repeated. However, using positive polarity, the distribution function $P(\xi' \geq)$ is obtained while the negative polarity yields $P(\xi' \geq)$. Since $P(\xi' \leq) = 1 - P(\xi' \geq)$, the two sets of values may be combined to give the complete distribution function. The character of the signal is then evaluated by the use of the probability paper illustrated in Fig. 22.11.

In addition to determining the characteristics of the signal, this analysis can be used to determine the mean square value of $\xi'(t)$, either by use of the probability paper if the signal is Gaussian or by computing the moment of inertia of the probability density function about the mean value [see Eq. (11.11)].

PEAK VALUES. The probability distribution of peak values or maxima $\xi'_p(t)$ of the filtered signal $\xi'(t)$ is obtained for a set of discrete values of $\xi'_p(t)$ rather than as the continuous function illustrated in Fig. 22.9. The probability distribution function at a value ξ'_{p1} is the probability $P(\xi'_p \geq)$ that a peak or maximum of $\xi'(t)$ exceeds ξ'_{p1}. Referring to Fig. 22.14, this is equivalent to counting the *number* of times that $\xi'(t)$ exceeds ξ'_{p1}(i.e., level L_1) and dividing by the total number of times that $\xi'(t)$ exceeds L_0. If this count is made for a number of levels, $L_1, L_2, \ldots$, the probability distribution function at these values is obtained.

Figure 22.15*B* illustrates the block diagram of a system for making such measurements. The filtered signal $\xi'(t)$ passes through a bank of level discriminators and causes the associated counters to register one count each time the level of the particular discriminator is exceeded. (Sometimes an array of discriminator-counter combinations is called a "pulse-height analyzer.") At the end of the data sample, the readings of the counters are recorded. Division of each count by the zero level count yields the probability distribution function $P(\xi'_p \geq)$ at the values equivalent to the levels $L_1, L_2, \ldots$. If the values at adjacent levels are subtracted, the average probability density function between these two levels is obtained. The character of the signal then is evaluated by the use of the probability paper illustrated in Fig. 22.12.

For some end uses of the data, such as fatigue analysis, the probability distribution of the total or unfiltered signal $\xi(t)$ rather than the probability distribution of the filtered signal $\xi'(t)$ is desired. Then the distribution of the peaks of $\xi(t)$ is obtained by shunting out the filter in the block diagram of Fig. 22.15*B* as indicated by the dotted lines.

The systems illustrated by the block diagrams of Fig. 22.15*A* and *B* are relatively complex, and unless a large amount of analysis of this type is to be carried out, the cost of such systems may not be justified. A much simpler method is illustrated in Fig. 22.15*C*.

In this case a direct-writing recorder is employed and the number of peaks which exceed various levels are counted manually (see Fig. 22.14). In principle, the distribution of instantaneous values may be measured in this way, but practical difficulties make this method difficult to employ.

SPECTRAL ANALYSIS

The objective of spectral analysis[6, 14, 20, 22, 23, 24, 29–32] is to determine the variation of vibration magnitude with frequency. The narrowness of the bandwidth of the filter employed in the analysis determines the frequency resolution of the analysis and, therefore, the ability to detect the "fine-grain" variation of magnitude with frequency. The magnitude obtained at a particular frequency will be the average magnitude over the short time interval of data analyzed, during which the nature and magnitude of the vibration may be considered constant (see Figs. 22.3 and 22.4).

In preparing for spectral analysis, it is helpful if the characteristics of the time-history $\xi(t)$ are known from statistical analysis (or are known from previous experience) so that the most appropriate units to be employed for describing the vibration magnitude can be selected. In practice, the spectral analysis often is carried out first and the necessity for statistical analysis is judged from the appearance of the resulting spectrum.

DEFINITION OF POWER SPECTRAL DENSITY*

Power spectral density[6–9, 29] is defined as the limiting value of the mean-square response $\overline{[\xi']^2}$ of an ideal bandpass filter† to $\xi(t)$, divided by the bandwidth B of the filter, as the bandwidth of the filter approaches zero.

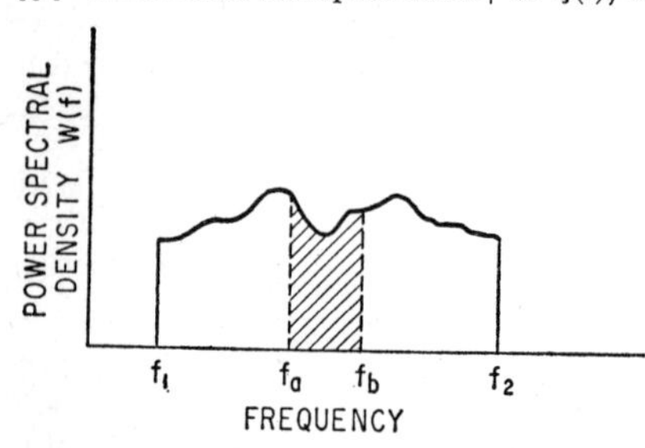

FIG. 22.16. Typical power spectral density plot of broad-band random function $\xi(t)$. Mean-square value of frequency content of $\xi(t)$ between f_a and f_b is equal to shaded area [Eq. (22.18)]. For "white noise" or flat spectrum, $W(f)$ is a constant.

An alternative definition[7] is as follows. If the function $\xi(t)$ is passed through an ideal low-pass filter‡ with cutoff frequency f_c, the mean-square response of the filter $\overline{[\xi']^2}$ will increase or decrease as f_c is increased or decreased, i.e., more or less of the function will be passed by the filter (assuming f_c is varied in a frequency range where the power spectral density is nonzero). The power spectral density $W(f)$ is the rate of change of $\overline{[\xi']^2}$ with respect to f_c, i.e.,

$$W(f_c) = \frac{d}{df_c}\{\overline{[\xi']^2}\} \tag{22.17}$$

Figure 22.16 illustrates a plot of power spectral density as a function of frequency obtained, for example, from spectral analysis of a random function. The mean-square value, or variance, of the frequency content of $\xi(t)$ between the frequencies f_a and f_b is equal to the shaded area of Fig. 22.16:

$$\sigma^2(f_a \leq f \leq f_b) = \int_{f_a}^{f_b} W(f)\,df \tag{22.18}$$

* Power spectral density is defined by Eq. (22.17) and also by Eq. (11.16). It is a generic term used regardless of the physical quantity represented by the time-history. However, it is preferable to indicate the physical quantity involved. For example, the term "mean-square acceleration density" or "acceleration spectral density" is used when the time-history of acceleration is to be described.

† An ideal bandpass filter has a transmission characteristic which is rectangular in shape so that all frequency components within the filter bandwidth are passed with unity gain and zero phase distortion, while frequency components outside the bandwidth are completely removed. If the transmission characteristic is H instead of unity in the bandwidth B, the spectral density is obtained by dividing the mean-square response by $B \times H^2$ instead of B.

‡ An ideal low-pass filter is an ideal bandpass filter having a lower cutoff frequency of zero.

The rms value is σ. The mean-square value of the complete function $\overline{[\xi]^2}$ is given by Eq. (22.18) when f_a and f_b are zero and infinity, respectively; it is equal to the area under the entire spectrum. In the case of *white noise*, for which the spectral density is independent of frequency, i.e., $W(f) = W$, Eq. (22.18) simplifies to

$$\sigma^2 = W(f_2 - f_1) \tag{22.19}$$

where f_1 and f_2 are the limiting frequencies of the noise.

RELATIONSHIP OF POWER SPECTRAL DENSITY TO LINE SPECTRUM. The mathematical relationship between power spectral density $W(f)$ and the Fourier coefficients c_n is

$$\begin{aligned} W(f) &= 2\pi W(\omega) = 2\pi \lim_{B\to 0} \frac{\overline{[\xi']^2}}{B} = 2\pi \frac{d}{d\omega} \overline{[(\xi')^2]} \\ &= \frac{1}{2} - \sum_{n=1}^{\infty} c_n{}^2\, \delta(f - f_n) \qquad [b_0 = 0] \end{aligned} \tag{22.20}$$

where $\overline{[\xi']^2}$ is the mean-square response of the ideal filter with bandwidth B and center frequency $f = \omega/2\pi$; $\delta(f - f_n)$ is the Dirac delta function[2] (see Chap. 23).

The power spectral density of a sinusoidal function is a line spectrum since such a function has effectively zero bandwidth. This may be seen from Eq. (22.16) since $\overline{[\xi']^2}$ will change instantaneously as the cutoff frequency f_c increases through the frequency of the sine wave giving, theoretically, an infinite value for $W(f)$.

DETERMINATION OF POWER SPECTRAL DENSITY

One technique of obtaining a spectral analysis of a signal is to record it on a continuous loop of magnetic tape. As described in Chap. 11, the loop is played back repeatedly. Each time, an analysis is made at a slightly different frequency using, for example, a tunable filter. To obtain the complete spectrum, the process is repeated at enough frequencies to give a plot of magnitude vs. frequency. These steps are illustrated in the first two blocks of Figs. 22.17 and 22.18, which are block diagrams depicting examples of

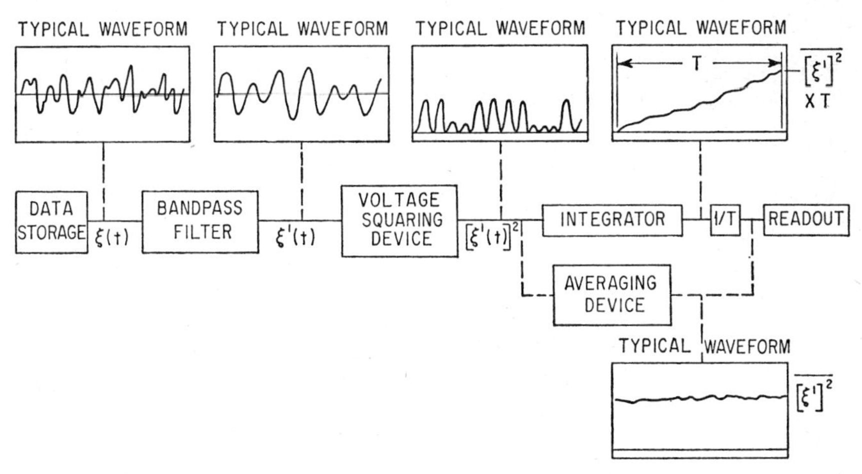

FIG. 22.17. Block diagram showing spectral analysis by the mean-square method. Dotted lines indicate alternative use of averaging device in place of integrator. Insets show typical waveforms at successive stages of analysis. Signal $\xi(t)$ is filtered to give $\xi'(t)$; squared to give $[\xi'(t)]^2$; integrated and divided by time of integration T to give $\overline{[\xi']^2}$, the mean-square value. Division by effective bandwidth of filter yields power spectral density of random function.

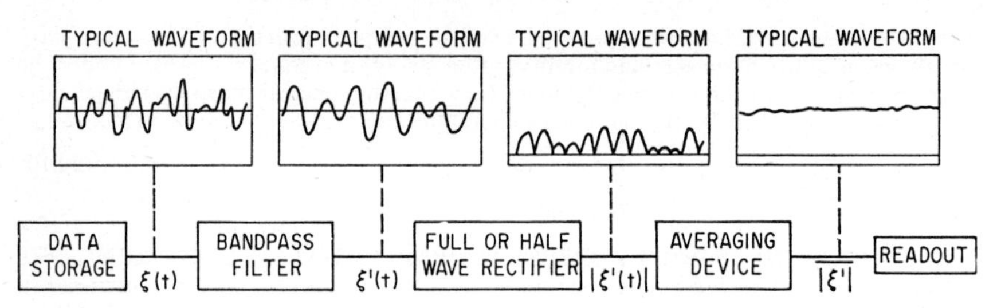

FIG. 22.18. Block diagram showing spectral analysis by averaging method. Insets show typical waveforms at successive stages of analysis. Signal $\xi(t)$ is filtered to give $\xi'(t)$; rectified (half or full wave) to give $|\xi'(t)|$; and averaged to yield $\overline{|\xi'|}$, the average value of $\xi'(t)$.

techniques employed in spectral analysis. Typical waveforms for one value of the center frequency of the filter are shown in the insets above the blocks. Techniques also are available for spectral analysis in real time.[12]

When a sample of a time-history of finite duration is employed to compute the power spectral density of a random function, it is assumed that (1) the function is ergodic, i.e., that averaging one time-history with respect to time yields the same result as averaging over an ensemble of time-histories at a given instant of time, and (2) that the function is *stationary*, i.e., that the power spectral density is independent of the sample of the time-history chosen. Further, the averaging time or sample duration must be long enough to yield a statistically significant value. Thus, the mean-square value obtained should not vary appreciably with a change in averaging time. The time over which a vibration record may be considered a stationary process and the need for a sufficiently long averaging time often are conflicting requirements.

MEASUREMENT OF POWER SPECTRAL DENSITY BY THE USE OF FILTERS

If a random function is applied to the input of a filter, the spectral density of the output is obtained by multiplying the spectral density of the input by the square of the transmission characteristic curve* at every frequency. First, consider the response of a filter to *white noise* (constant spectral density W). The output of an ideal bandpass filter is constant spectral density within the bandwidth B, and zero elsewhere. From Eq. (22.18), the mean-square output will be $W \times B$. The output of a practical filter (i.e., a filter having practically attainable characteristics in contrast to an ideal filter) has a spectral density which is directly proportional to the square of the transmission characteristic; the mean-square output is W multiplied by the area under the curve of the squared transmission characteristic. This area is defined as the effective bandwidth of the filter. However, due to the nonlinearities inherent in many filters, the effective bandwidth of a filter should be determined by measurement of the mean-square response of the filter to known white noise.

Now consider a random signal whose spectral density varies with frequency, as shown in Fig. 22.19*B*. It is desired to measure the spectral density by the use of filters, i.e., to perform a spectral analysis. In the case of an ideal filter, the mean-square output divided by B yields the mean (or average) value of the spectral density within the filter bandwidth [see Eq. (22.18) and Fig. 22.16]. When the signal is applied to a practical filter and the mean-square output is divided by the effective bandwidth, an approximation to the average spectral density within the filter bandwidth is obtained.

EFFECTS OF FILTER CHARACTERISTICS. The suitability of practical filters for the measurement of power spectral density depends on the rate of variation of the

* The ratio of the response amplitude to the input amplitude when the input is a sinusoidal function is the *transmission characteristic* or *frequency-response characteristic* of the filter. If this ratio is independent of the input amplitude, the filter is linear. The ratio usually is normalized so that the value at the center frequency is unity.

spectral density with frequency of the signal being analyzed, the shape of the transmission characteristic of the filter, and the time constant of the filter.

Figure 22.19*F* shows the spectral densities which would be measured using filters with the transmission characteristics shown in Fig. 22.19*C* and 22.19*D*, if the input power spectral density is actually as shown in Fig. 22.19*B*. The smoothing effect on the measured spectral density of the filter with the wider bandwidth is evident, particularly where the magnitude of the power spectral density is changing rapidly with frequency. Sharp peaks or valleys in the power spectral density are masked when the effective filter bandwidth approaches the bandwidth of the peak or valley. The power spectral density measured by the use of filters is exact only when an ideal filter is used, in the limit as the bandwidth approaches zero. Furthermore the time constant of the filter may affect the measurements by time-averaging the output signal.

MEASUREMENT OF LINE SPECTRA BY THE USE OF FILTERS

In the measurement of sinusoidal functions by the use of filters,[25, 26] the output amplitude is measured and plotted in place of the power spectral density used with random functions. The accuracy with which such measurements can be made depends upon the bandwidth of the filter. For example, Fig. 22.19*E* shows in solid lines the measured spec-

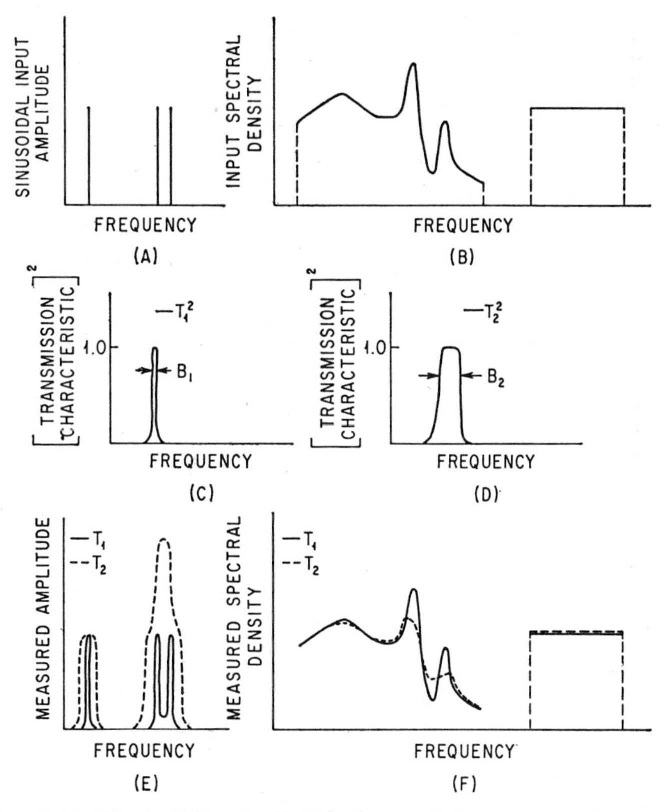

FIG. 22.19. Effects of filter bandwidth characteristics on measured spectra.

trum when the line spectrum of Fig. 22.19*A* is the input to the filters having the characteristics shown in Fig. 22.19*C*; the dotted lines in Fig. 22.19*E* show the spectrum obtained with the filter of Fig. 22.19*D*. The error introduced by the filter with the wider bandwidth is particularly evident when the sine waves are closely spaced in frequency. Assume now that the input $\xi(t)$ is a complex function comprised of several sinusoidal functions at unrelated frequencies. The output of the filter is the mean-square value of one or the sum of many sinusoids, depending on the spacing of the lines of the input spectrum, the breadth of the filter bandwidth, and the shape of the filter transmission characteristic. However, the output is often considered to be the mean-square value of a single sinusoidal function at the center frequency of the filter.

LINE SPECTRUM SUPERIMPOSED ON RANDOM VIBRATION SPECTRUM

The input signal $\xi(t)$ may be a "mixture" of a random function plus a single sinusoidal function within the bandwidth of the filter, as shown in Fig. 22.20. The mean-square output of the filter $\overline{[\xi']^2}$ is the sum of the mean-square values of the sinusoid and the frequency content of the random function within the filter bandwidth. Depending on the end use of the data and the relative magnitudes of the sinusoidal and random functions, it *may* be satisfactory to take the spectral density computed from this mean-square value $\overline{[\xi']^2}$ as a description of the vibration magnitude. Conversely, it may be preferable to consider $\overline{[\xi']^2}$ to be the mean-square value of a sinusoid at the filter center frequency. However, power spectral density should be used only to describe a random function. Periodic and complex functions should be represented by Fourier spectra, either line or continuous.

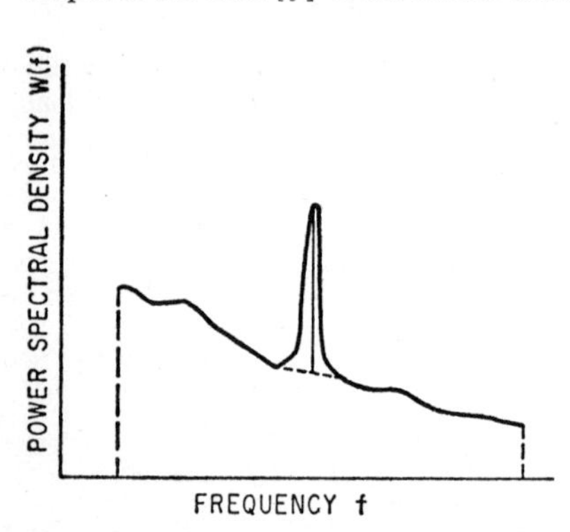

FIG. 22.20. Example of spectral analysis when signal contains narrow-band peak of random vibration or line spectrum superimposed on random vibration spectrum as shown by dotted line. Because of limitations of analysis equipment, line spectrum may give same appearance as narrow-band random vibration.

It is difficult to separate the random and coherent parts of a vibration time-history when such a combination occurs. For example, if the power spectral density of an acceleration time-history has been computed using a system such as illustrated in Fig. 22.17, and the spectrum has a peak in a narrow frequency band as shown in Fig. 22.20, it is necessary to know whether this peak represents a truly random function of increased intensity in this frequency band or a sinusoidal function superimposed on the smoother spectral density plot indicated by the dotted line in Fig. 22.20. Fundamentally, a sinusoidal function is represented by a line spectrum but may give the appearance of narrow-band random vibration because of the limitations of the spectral analysis techniques.[25] The probability density or probability distribution at the frequency of the peak in the spectrum may be used to determine the characteristics of the signal in this frequency region.

MEASUREMENT OF FILTER OUTPUT

The magnitude of the filtered signal is obtained in terms of (1) the mean-square value or (2) the average value.*

MEAN-SQUARE VALUE. Assume that statistical analysis has shown $\xi'(t)$ to be a random function. Then the mean-square value $\overline{[\xi']^2}$ must be obtained. Figure 22.17 shows one method of accomplishing this. The function $\xi'(t)$ is squared, integrated over the period of the data sample, and divided by the time of integration T to yield $\overline{[\xi']^2}$.

* See definition of average value in Eq. (22.5).

Typical waveforms during this process are shown. Alternatively, the integrator may be replaced by an averaging device, as shown by the dotted lines. The terminal value of the integrator output, divided by T, or the output of the averaging device is read out; division by the effective bandwidth of the filter yields the average power spectral density within the filter bandwidth.

AVERAGE VALUE.* Figure 22.18 is a block diagram showing an example of a system in which the filter output $\xi'(t)$ is half-wave (or full-wave) rectified and averaged. Thus the quantity measured by the read-out is $\overline{|\xi'|}$, the average value of $\xi'(t)$. If $\xi'(t)$ is known to be a random function, the relationships discussed under *Statistical Analysis* or empirical measurement of the response to a known spectral density can be used to relate the average value to the mean-square value, and thus to obtain the power spectral density.

EFFECT OF SIGNAL CHARACTERISTICS. When the magnitude of the filter output is obtained in terms of the mean-square value, as shown in Fig. 22.17, the read-out may be directly converted to the appropriate magnitude quantity, for example, power spectral density, the amplitude of a sine wave, etc. If the average value is obtained as shown in Fig. 22.18, the conversion of the read-out to an appropriate magnitude quantity is not direct, and may not even be possible. For example, if the filter output is the sum of two sinusoidal functions, the average value is a function of the phase angle between the sine waves, while the mean-square value is independent of phase angle. When the filter output is a mixture of a random function and a sinusoidal function, a similar indeterminacy exists in the average value. Thus, detailed knowledge of the characteristics of the function is required to interpret correctly the magnitude of $\xi(t)$ from its average value.†

MAGNITUDE TIME-HISTORY ANALYSIS

In magnitude time-history analysis, the variation of the magnitude‡ of the vibration as a function of time is examined. The result may be plotted as shown in Fig. 22.5 where the magnitude is computed as the average over a few seconds while the vibration data record has a duration of perhaps many minutes. In some cases, it may be desired to determine the time variation of the magnitude of the complete signal, e.g., the variation of the mean-square value $\overline{[\xi]^2}$ or, alternatively, the time-history of the magnitude of the vibration in a restricted frequency band. For example, suppose that spectral analysis of a number of time samples of the record shows that the vibration magnitude is concentrated in several narrow frequency bands, and that these bands are the only important components of the vibration. Then a magnitude time-history analysis would be carried out for these frequencies. However, if spectral analysis of a number of time samples revealed that the shape of the spectrum was approximately constant even though the magnitude varied, then a time-history of the magnitude of the complete signal would be appropriate. The amount of analysis carried out is determined by the completeness with which the vibration magnitude surface of Fig. 22.3 must be defined.

DETERMINATION OF MAGNITUDE TIME-HISTORY

Figure 22.21 shows block diagrams for carrying out a magnitude time-history analysis. The data storage contains the complete vibration record, rather than the short-duration samples used in spectral and statistical analysis. The signal $\xi(t)$ from the record is

* See definition of average value in Eq. (22.5).

† Similar effects must be taken into account in the use of voltmeters which, for example, respond to the average value of the input signal and have a scale which is graduated in terms of rms value.

‡ The word "magnitude" is used in this chapter to indicate the severity of vibration without specifically relating it to one of the various mathematical expressions used to describe vibration quantitatively.

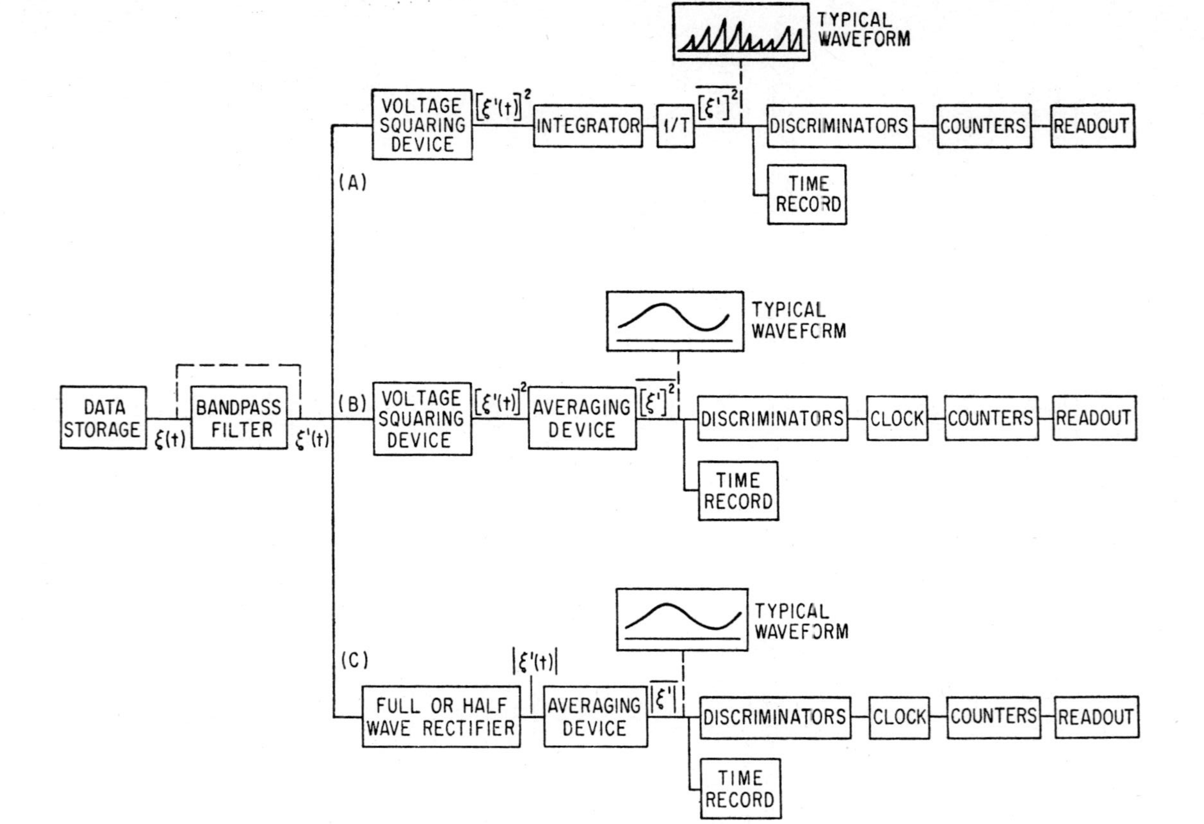

FIG. 22.21. Block diagrams showing magnitude time-history analysis. To determine time-history of magnitude, "time record" blocks are used; magnitude distribution function is determined from discriminators and counters. Typical waveforms of magnitude time-history are shown by insets. For analysis of variation of magnitude of complete signal, filter is shunted as shown by dotted line. Filter center frequency is maintained constant during playback of entire vibration record. (Refer to *Statistical Analysis* and *Spectral Analysis* regarding counting and averaging techniques.) Diagrams show (*A*) mean-square value by successive integration; (*B*) mean-square value by averaging; (*C*) average value.

passed through a filter whose center frequency is held constant for the entire record. The mean-square value $\overline{[\xi']^2}$ or the average value $\overline{|\xi'|}$ of the output of the filter $\xi'(t)$ is obtained by one of the three alternative systems outlined in Fig. 22.21: (*A*) the mean-square value may be obtained by integration of $[\xi'(t)]^2$ during successive time intervals T; (*B*) the mean-square value may be obtained by averaging of $[\xi'(t)]^2$; or (*C*) the average value $\overline{|\xi'|}$ may be obtained. The results are recorded at the time record stations in Fig. 22.21. The filter may be shunted as shown by dotted lines to analyze the time variation of $\xi(t)$ rather than that of the filtered signal $\xi'(t)$.

When a running mean-square or running average value is obtained, as indicated in Fig. 22.21*B* and *C*, the value obtained at any instant is affected by the values at times previous to the time of observation, depending on the time constant of the averaging device. This tends to smooth out the time variation of the magnitude and is avoided when integration over consecutive periods is used because each integrand then is independent. Typical time records of the value of $\overline{[\xi']^2}$ or $\overline{|\xi'|}$ are shown in the insets adjoining the systems *A*, *B*, and *C* in Fig. 22.21. Such a time record permits correlation of the magnitude of the vibration with physical parameters. If the time-histories of the vibration magnitude are obtained for a number of filter bands and laid side by side, the vibration magnitude surface of Fig. 22.3 is synthesized.

MAGNITUDE DISTRIBUTION FUNCTION. In addition to the magnitude time-history, the total time that the magnitude is equal to or greater than a particular magnitude or set of magnitudes, i.e., the magnitude distribution function, may be determined, as indicated in Fig. 22.21. If $\overline{[\xi']^2}$ is obtained by integration [system (*A*)], the mean-square value is fed into a bank of discriminator-counter combinations which sense the number of times that the integrand exceeds various levels. If $\overline{[\xi']^2}$ is obtained by averaging [system (*B*)] or if $\overline{|\xi'|}$ is obtained [system (*C*)], the time that these quantities exceed various levels is measured by the bank of discriminator-clock-counter combinations. The function of discriminators and related equipment is discussed with reference to Figs. 22.18 and 22.19.

Figure 22.22 illustrates a manner of plotting the readings from the counters. The abscissa is normalized to indicate the percentage of the total time that the level is exceeded, while the ordinate scale corresponds to the magnitude of the level which is exceeded; e.g., magnitude L_1 corresponding to discriminator 1 is exceeded t_1 per cent of the time. The vibration magnitude distribution function illustrated in Fig. 22.22 is analogous to the probability distribution function shown in Fig. 22.9. The filter frequency and bandwidth must be specified unless the analysis represents the total (i.e., the unfiltered) signal. The time of occurrence of the various vibration magnitudes is lost and cannot be correlated with physical conditions. If the magnitude distribution functions are obtained for a number of frequencies and placed side by side, the vibration magnitude distribution surface is formed. This describes the total vibration signal in terms of magnitude, frequency, and duration in the most concise form possible.

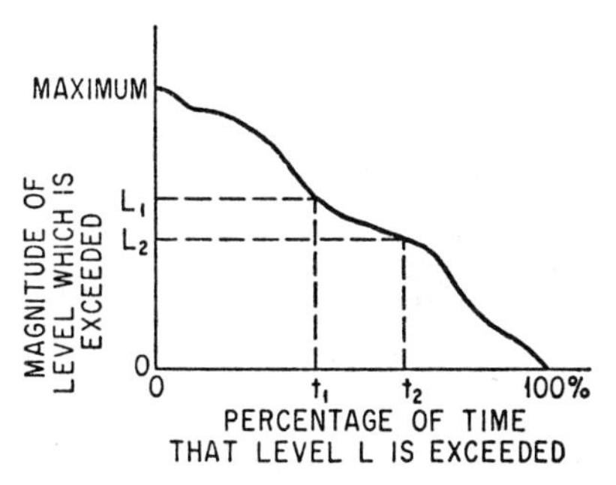

Fig. 22.22. Distribution function of vibration magnitude for a filter having a center frequency f_0.

CORRELATION ANALYSIS

CORRELATION FUNCTIONS AND THEIR USE

A correlation function[6, 7, 8, 9, 13, 29–32] defines the correlation between two parameters as a function of the times at which the parameters are observed. If the two parameters are the same except for the time of observation, i.e., $\xi(t)$ and $\xi(t + \tau)$, the function is known as an *autocorrelation function*. For example, the autocorrelation function of a single acceleration measurement may be used to determine both the nature of $\xi(t)$ and its power spectral density. If the two parameters $\xi_1(t)$ and $\xi_2(t + \tau)$ are physically distinct, the function is known as a *crosscorrelation function*. For example, the crosscorrelation between acceleration measurements at two different points of a structure may be determined for the purpose of studying the propagation of vibration through the structure. Crosscorrelation functions are not restricted to correlation of parameters with the same physical units; for example, one might determine the crosscorrelation between the applied force and the acceleration response to that force.

Correlation analysis of vibration data has many uses. For example, crosscorrelation data may be used to identify the normal modes of a vibrating structure; the autocorrelation function may be used to determine the power spectral density, and thus may be considered to be an alternative approach to spectral analysis. In the latter case, this technique has the advantage of yielding both the spectral density and a means of determining the characteristics of the signal, e.g., sinusoidal vs. random, etc., from the form of the autocorrelation function. Correlation techniques also are useful as a powerful method of detecting and analyzing a weak signal in the presence of noise.

Correlation analysis is concerned with stationary processes; in vibration analysis, it usually is carried out on a short interval of record during which physical parameters are assumed unchanged, as in spectral analysis.

AUTOCORRELATION FUNCTION

From Eq. (11.53), the autocorrelation function $R_\xi(\tau)$ of a time-varying parameter $\xi(t)$ is given by

$$R_\xi(\tau) = R_\xi(-\tau) = \lim_{T\to\infty} \frac{1}{2T} \int_{-T}^{T} \xi(t)\xi(t - \tau)\, dt = \overline{\xi(t)\xi(t - \tau)} \tag{22.21}$$

Physically, $R_\xi(\tau)$ may be considered as the time average of the product of the instantaneous values of the function measured at two instants separated by a time interval τ. Setting $\tau = 0$ in Eq. (22.21) yields the mean-square value

$$R_\xi(0) = \overline{[\xi]^2} \tag{22.22}$$

Since a parameter cannot be more closely related to another parameter than it is to itself, i.e., be better correlated, it follows that the maximum value of $R_\xi(\tau)$ occurs when $\tau = 0$.

The relationship between the autocorrelation function $R_\xi(\tau)$ and the power spectral density $W(f)$ is

$$W(f) = 4 \int_0^\infty R_\xi(\tau) \cos \omega\tau \, d\tau \tag{22.23}$$

From Eq. (11.59), the inverse transformation is given by

$$R_\xi(\tau) = \frac{1}{2\pi} \int_0^\infty W(f) \cos \omega\tau \, d\omega \tag{22.24}$$

Equation (22.23) is employed in the determination of the power spectral density of a random function from the autocorrelation function.

AUTOCORRELATION FUNCTION OF A SINE WAVE. Assume that $\xi(t)$ is a sine wave[13] as defined in Eq. (22.1). Then $R_\xi(\tau)$ is given by

$$R_\xi(\tau) = \frac{\xi_0^2}{2} \cos \omega\tau \tag{22.25}$$

The relation given by Eq. (22.25) is shown by curve A in Fig. 22.23. This curve shows that the autocorrelation function of a sine wave is periodic with a frequency equal to the frequency of the sine wave. The physical interpretation is that a sine wave is unchanged by a translation of the time origin by an integer number of periods $2\pi/\omega$. The maxima of the autocorrelation function occur at $\tau = 2\pi n/\omega$ and are all equal to $R_\xi(0)$, i.e., the correlation at time delays τ equal to an integer number of periods $2\pi/\omega$ is the same as it is for zero time delay.

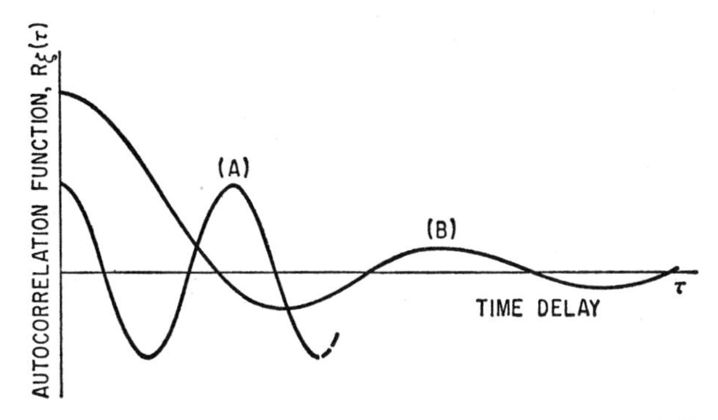

FIG. 22.23. Typical autocorrelation functions: (A) Sine wave [Eq. (22.25)]; (B) band-limited white noise [Eq. (22.26)].

AUTOCORRELATION FUNCTION OF WHITE NOISE. Suppose $\xi(t)$ represents band-limited white noise which has constant spectral density W up to a frequency f_c and zero for higher frequencies. According to Eq. (22.24), the autocorrelation function[13] is

$$R_\xi(\tau) = Wf_c \frac{\sin 2\pi f_c \tau}{2\pi f_c \tau}$$
$$R_\xi(0) = Wf_c = \overline{[\xi]^2} \qquad (22.26)$$

Equations (22.26) define the decaying sinusoidal function shown by curve B in Fig. 22.23. As the cutoff frequency f_c approaches infinity, $R_\xi(\tau)$ approaches a delta function at the origin.

The autocorrelation function of the random process decreases rapidly as τ increases, and as the cutoff frequency f_c increases. Thus there is little correlation between the instantaneous values of $\xi(t)$ at different times. In the limit, as f_c approaches infinity, there is no correlation at all; i.e., the value of $\xi(t)$ is unpredictable from knowledge of its previous value $\xi(t - \tau)$. The autocorrelation function for a random function which has variable spectral density will have the same properties as curve B of Fig. 22.23, i.e., decreasing correlation for increasing τ and increasing bandwidth. Thus, the form of the autocorrelation function can be used to determine the character of $\xi(t)$, i.e., whether it is a random, complex, or periodic time function.

DETERMINATION OF AUTOCORRELATION FUNCTION. Figure 22.24 illustrates one method of obtaining a correlation analysis by analog means.[14, 20, 27, 28] (The arrangement is similar for either autocorrelation or crosscorrelation analyses.) The stored signal $\xi(t)$ is fed into a multiplier directly and through a variable time-delay τ. The time average of the multiplicand $\overline{\xi(t)\xi(t - \tau)}$ for various time-delays is obtained either by integration or averaging as shown by the alternative blocks in Fig. 22.24. The plot of the output vs. the time-delay is the autocorrelation function.

Another method of obtaining a correlation function from an analog record is to convert the analog record to digital form. Then a standard high-speed digital computer can be programmed to calculate the autocorrelation function and the power spectral density.

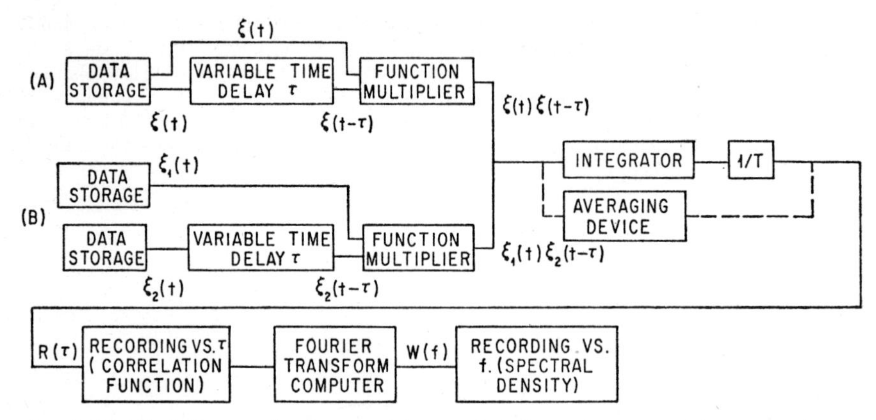

FIG. 22.24. Block diagram showing correlation analysis by analog means: (*A*) autocorrelation analysis; (*B*) crosscorrelation analysis. Dotted line shows alternative method of averaging multiplicand (see Fig. 22.17).

CROSSCORRELATION FUNCTION

The crosscorrelation function of two distinct time-varying parameters $\xi_1(t)$ and $\xi_2(t)$ is given by

$$R_{\xi_1\xi_2}(\tau) = \lim_{T\to\infty} \frac{1}{2T} \int_{-T}^{T} \xi_1(t)\xi_2(t-\tau)\, dt = \overline{\xi_1(t)\xi_2(t-\tau)} \tag{22.27}$$

The magnitude of $R_{\xi_1\xi_2}(\tau)$ is a measure of the correlation of ξ_1 and ξ_2 when observed at times separated by a time τ.

Example 22.1. Assume ξ_1 is directly proportional to ξ_2 except for a time lag τ_1:

$$\xi_1(t) = A\xi_2(t-\tau_1) \tag{22.28}$$

Then, by Eq. (22.27),

$$R_{\xi_1\xi_2}(\tau_1) = A\overline{[\xi_2]^2} = AR_{\xi_2}(0) \tag{22.29}$$

Example 22.2. Assume ξ_1 and ξ_2 are sinusoidal functions at the same frequency:

$$\xi_1(t) = A\xi_0 \sin\omega(t-\tau_1) \tag{22.30}$$

$$\xi_2(t) = \xi_0 \sin\omega t$$

Then the crosscorrelation function is

$$R_{\xi_1\xi_2}(\tau) = \frac{A[\xi_0]^2}{2} \cos\omega(\tau-\tau_1) \tag{22.31}$$

This is the same form as the autocorrelation function for a sine wave given in Eq. (22.25) except for the time shift τ_1 (see Fig. 22.23). Thus, for example, the increase or decrease in the vibration displacement amplitude between points 1 and 2 may be determined from the crosscorrelation function given by Eq. (22.31).

DETERMINATION OF CROSSCORRELATION FUNCTION. One method of obtaining a crosscorrelation function by analog means is illustrated at *B* in Fig. 22.24. One function $\xi_1(t)$ is fed directly to the multiplier and the second function $\xi_2(t)$ is fed through the variable time-delay to the multiplier. The time average of the multiplicand $\overline{\xi_1(t)\xi_2(t-\tau)}$ plotted vs. τ is the crosscorrelation function.

Crosscorrelation analysis also may be carried out by the use of standard high-speed digital computers which have been programmed for this purpose. Crosscorrelation analysis by digital means is almost identical to autocorrelation analysis since the only difference is that two distinct functions, $\xi_1(t)$ and $\xi_2(t)$, are employed instead of the one function $\xi(t)$ used in autocorrelation analysis.

REFERENCES

1. von Kármán, T., and M. A. Biot: "Mathematical Methods in Engineering," McGraw-Hill Book Company, Inc., New York, 1940.
2. Pipes, L. A.: "Applied Mathematics for Engineers and Physicists," 2d ed., McGraw-Hill Book Company, Inc., New York, 1958.
3. Campbell, G. A., and R. M. Foster: "Fourier Series for Practical Applications," D. Van Nostrand Company, Inc., Princeton, N.J., 1940. Previously published as No. B-584 of the Bell System Series of Monographs, 1931.
4. Bradford, R. S.: "Phase-coherent Vibration in Missiles," *External Publ.* 349, Jet Propulsion Laboratory, Pasadena, Calif., October, 1956.
5. Bradford, R. S.: *Shock and Vibration Bull.* 23, p. 315. Office of Secretary of Defense, Research and Development, Washington, D.C., February, 1957.
6. Crandall, S. H.: "Random Vibration," Technology Press of MIT, Cambridge, Mass., 1958.
7. Bendat, J. S.: "Principles and Applications of Random Noise Theory," John Wiley & Sons, Inc., New York, 1958.
8. Davenport, W. B., Jr., and W. L. Root: "An Introduction to the Theory of Random Signals and Noise," McGraw-Hill Book Company, Inc., New York, 1958.
9. Miller, K. S.: "Engineering Mathematics," Holt, Winston & Rinehart, Inc., New York, 1957.
10. Scarborough, J. B.: "Numerical Mathematical Analysis," 2d ed., The Johns Hopkins Press, Baltimore, 1950.
11. Manley, R. G.: "Waveform Analysis," John Wiley & Sons, Inc., New York, 1945.
12. Bickel, H. J.: "Spectrum Analysis with Delay-line Filters," IRE Wescon Convention Record (Part 8), 1959.
13. Knudtzon, N. H.: "Experimental Study of Statistical Characteristics of Filtered Random Noise," *MIT Research Lab. Electronics, Tech. Rept.* 115, July, 1949.
14. Symposium on Applications of Autocorrelation Analysis to Physical Problems, Woods Hole, Mass., June 13–14, 1949, published by U.S. Office of Naval Research, Washington, D.C., ATI No. 86556.
15. Whiteley, T. B.: "A Method for Amplitude Analysis of Vibration Data," *Tech. Memo.* 65–100, U.S. Naval Ordnance Lab., Corona, Calif., January, 1956.
16. Fine, A., T. B. Whiteley, D. Bell, and M. Buus: *Shock and Vibration Bull.* 23, p. 184. Office of Secretary of Defense, Research and Development, Washington, D.C., June, 1956.
17. White, H. E.: "An Analog Probability Density Analyzer," *MIT Research Lab. Electronics, Tech. Rept.* 326, April, 1957.
18. Meyer, D. D.: *Shock and Vibration Bull.* 23, p. 229. Office of Secretary of Defense, Research and Development, Washington, D.C., June, 1956.
19. Baldwin, F. L.: *Shock and Vibration Bull.* 25, part I, p. 151. Office of Secretary of Defense, Research and Development, Washington, D.C., December, 1957.
20. Carlson, E. R., C. C. Conger, J. C. Laurence, E. H. Meyn, and R. A. Yocke: *Proc. IRE,* **44**:956 (1959).
21. Baruch, J. J.: *Shock and Vibration Bull.* 25, part II, p. 25. Office of Secretary of Defense, Research and Development, Washington, D.C., December, 1957.
22. Stallard, R. L.: *Shock and Vibration Bull.* 24, p. 70. Office of Secretary of Defense, Research and Development, Washington, D.C., February, 1957.
23. Stallard, R. L.: "The Validity of the Single Degree of Freedom System Analog for Vibration Analysis," *Johns Hopkins Univ. Appl. Phys. Lab. Rept.* APL/JHU CF-2518, May, 1956.
24. Gruen Applied Science Laboratories, Inc.: "Methods of Analyzing Shock and Vibration," *Rept.* 10002-F, July, 1957.
25. Granick, N.: "Status Report on Random Vibration Simulation," *WADC Tech. Note* 58–274, *ASTIA Document* No. AD 203125, March, 1959.
26. Haase, K. H., and F. Vilrig: "Errors in Spectrum Analysis by a Set of Narrowband Selecting Filters," *Air Force Cambridge Research Center Tech. Rept.* AFCRC—TR—56–121, *ASTIA Document* No. AD 110197, November, 1956.
27. Zabusky, N.: "The Mechanical Correlation Computer," *MIT Servomechanisms Lab., Eng. Rept.* 32, October, 1951. *ASTIA Document* No. 45.
28. Howes, W. L., E. E. Callahan, W. D. Coles, and H. R. Mull: "Near Noise Field of a Jet-engine Exhaust," *NACA Rept.* 1338, 1957. (Supersedes *NACA TN*'s 3763 and 3764.)

29. Bendat, Julius S., and Allan G. Piersol: "Random Data: Analysis and Measurement Procedures," Wiley-Interscience, New York, 1971.
30. Enochson, Loren D., and Robert K. Otnes: "Programming and Analysis for Digital Time Series Data," *Shock and Vibration Mon.* 3, Shock and Vibration Information Center, Washington, D. C., 1968.
31. Otnes, Robert K., and Loren Enochson: "Digital Time Series Analysis," Wiley-Interscience, New York, 1972.
32. Methods for Analysis and Presentation of Shock and Vibration Data, American National Standard, ANSI S2.10–1971.

23

CONCEPTS IN SHOCK DATA ANALYSIS

Sheldon Rubin

The Aerospace Corporation

INTRODUCTION

SCOPE

This chapter discusses the interpretation of shock measurements and the reduction of data to a form adapted to further engineering use. Methods of data reduction also are discussed. A shock measurement is a trace giving the time-history of a shock parameter over the duration of the shock. The shock parameter may define motion (such as acceleration, velocity, or displacement) or loading (such as force, pressure, stress, or torque). It is assumed that any corrections that should be applied to eliminate distortions resulting from the instrumentation have been made. The trace may be a pulse or transient vibration. The interpretation of periodic and random vibration measurements is discussed in Chap. 22.

Examples of sources of shock to which this discussion applies are aircraft landing, braking and gust loading; missile launching and staging; transportation of fragile equipment; accidental collision of vehicles; gunfire; explosions; and high-speed fluid entry.

ENGINEERING USES OF SHOCK MEASUREMENTS

Often, a shock measurement in the form of a time-history of a motion or loading parameter is not useful directly for engineering purposes. Reduction to a different form is then necessary, the type of data reduction employed depending upon the ultimate use of the data.

COMPARISON OF MEASURED RESULTS WITH THEORETICAL PREDICTION. The correlation of experimentally determined and theoretically predicted results by comparison of records of time-histories is difficult. Generally, it is impractical in theoretical analyses to give consideration to all the effects which may influence the experimentally obtained results. For example, the measured shock often includes the vibrational response of the structure to which the shock-measuring device is attached. Such vibration obscures the determination of the shock input for which an applicable theory is being tested; thus, data reduction is useful in minimizing or eliminating the irrelevancies of the measured data to permit ready comparison of theory with corresponding aspects of the experiment. It often is impossible to make such comparisons on the basis of original time-histories.

CALCULATION OF STRUCTURAL RESPONSE. In the design of equipment to withstand shock, the required strength of the equipment is indicated by its response to the shock. The response may be measured in terms of the deflection of a member of the equipment relative to another member or by the magnitude of the dynamic loads imposed upon the equipment. The structural response can be calculated from the time-history by known means; however, certain techniques of data reduction result in descriptions of the shock that are related directly to structural response.

As a design procedure it is convenient to represent the equipment by an appropriate model that is better adapted to analysis.* A typical model is shown in Fig. 23.1; it consists of a secondary structure supported by a primary structure. Depending upon the ultimate objective of the design work, certain characteristics of the response of the model must be known:

1. If design of the secondary structure is to be effected, it is necessary to know the time-history of the motion of the primary structure. Such motion constitutes the excitation for the secondary structure.
2. In the design of the primary structure, it is necessary to know the deflection of such structure as a result of the shock, either the time-history or the maximum value.

By selection of suitable data reduction methods, response information useful in the design of the equipment is obtained from the original time-history.

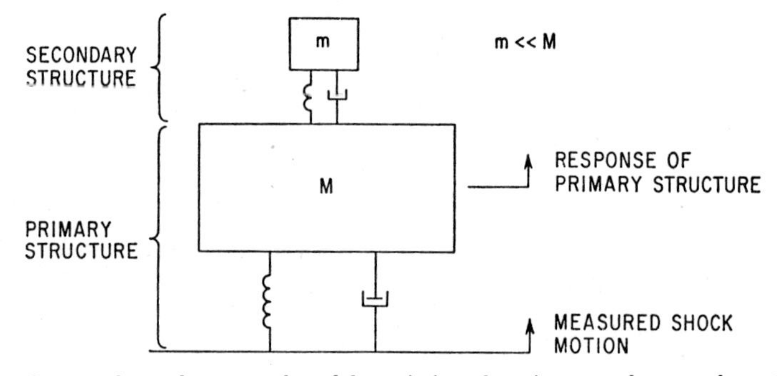

Fig. 23.1. Commonly used structural model consisting of a primary and a secondary structure. Each structure is represented as a lumped-parameter single degree-of-freedom system with the secondary mass m much smaller than the primary mass M so that the response of the primary mass is unaffected by the response of the secondary mass. The response of the primary mass to an input shock motion is the input shock motion to the secondary structure.

LABORATORY SIMULATION OF MEASURED SHOCK. Because of the difficulty of using analytical methods in the design of equipment to withstand shock, it is common practice to prove the design of equipments by laboratory tests that simulate the anticipated actual shock conditions. Unless the shock can be defined by one of a few simple functions, it is not feasible to reproduce in the laboratory the complete time-history of the actual shock experienced in service. Instead, the objective is to synthesize a shock having the characteristics and severity considered significant in causing damage to equipment. Then, the data reduction method is selected so that it extracts from the original time-history the parameters that are useful in specifying an appropriate laboratory shock test. The considerations involved in simulation are discussed in Chap. 24. Shock testing machines are discussed in Chap. 26.

EXAMPLES OF SHOCK MOTIONS

Five examples of shock motions are illustrated in Fig. 23.2 to show typical characteristics and to aid in the comparison of the various techniques of data reduction. The acceleration impulse and the acceleration step are the classical limiting cases of shock motions. The half-sine pulse of acceleration, the decaying sinusoidal acceleration, and the complex oscillatory type motion typify shock motions encountered frequently in practice.

* See Chap. 42 for a more complete discussion of models.

In selecting data reduction methods to be used in a particular circumstance, the applicable physical conditions must be considered. The original record, usually a time-history, may indicate any of several physical parameters; e.g., acceleration, force, velocity, or pressure. Data reduction methods discussed in subsequent sections of this chapter are applicable to a time-history of any parameter. For purposes of illustration in the following examples, the primary time-history is that of acceleration; time-histories of velocity and displacement are derived therefrom by integration. These examples are included to show characteristic features of typical shock motions and to demonstrate data reduction methods.

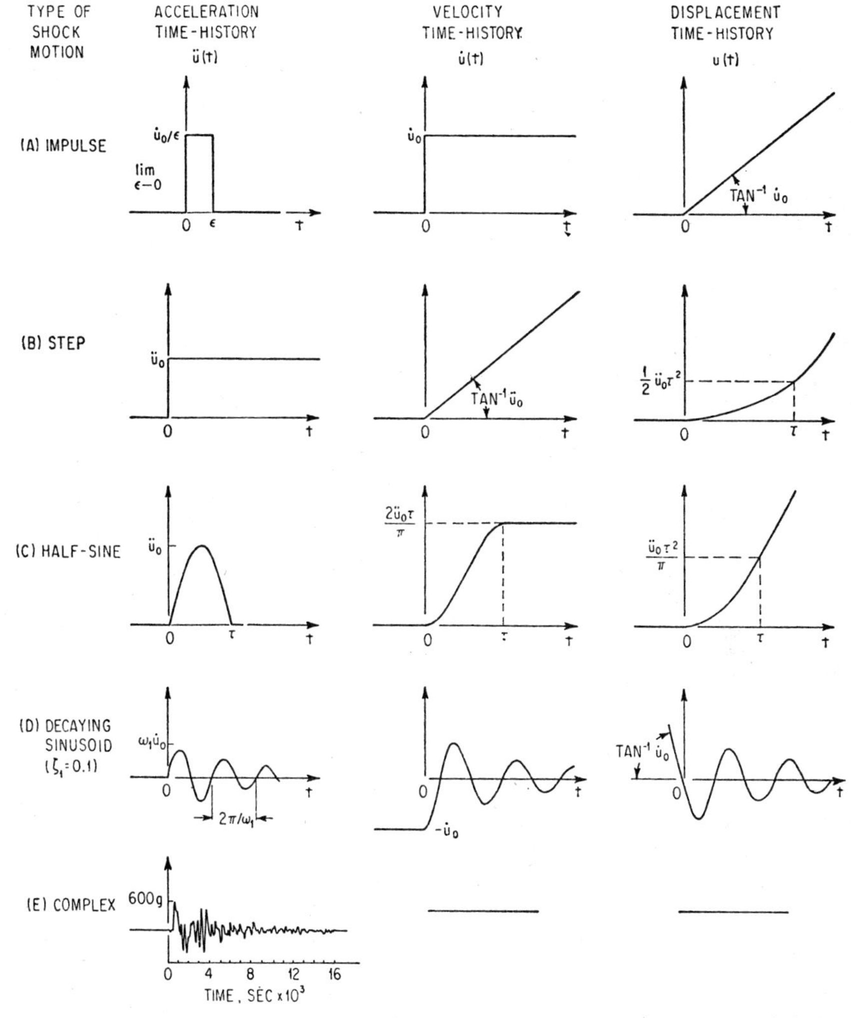

Fig. 23.2. Five examples of shock motions. These examples are used to show typical characteristics of shock motions and to aid in the comparison of various techniques of data reduction. The acceleration time-history is considered to be the primary description; and corresponding velocity and displacement time-histories are shown. [Mathematical descriptions appear in Eqs. (23.1) to (23.14).]

ACCELERATION IMPULSE OR STEP VELOCITY

The *delta function* $\delta(t)$ is defined mathematically as a function consisting of an infinite ordinate (acceleration) occurring in a vanishingly small interval of abscissa (time) at time $t = 0$ such that the area under the curve is unity. An acceleration time-history of this form is shown diagrammatically in Fig. 23.2*A*. If the velocity and displacement are zero at time $t = 0$, the corresponding velocity time-history is the velocity step and the corresponding displacement time-history is a line of constant slope, as shown in the figure. The mathematical expressions describing these time histories are

$$\ddot{u}(t) = \dot{u}_0\delta(t) \tag{23.1}$$

where $\delta(t) = 0$ when $t \neq 0$, $\delta(t) = \infty$ when $t = 0$, and $\int_{-\infty}^{\infty} \delta(t)\,dt = 1$. The acceleration can be expressed alternatively as

$$\ddot{u}(t) = \lim_{\epsilon\to 0} \dot{u}_0/\epsilon \qquad [0 < t < \epsilon] \tag{23.2}$$

where $\ddot{u}(t) = 0$ when $t < 0$ and $t > \epsilon$. The corresponding expressions for velocity and displacement for the initial conditions $u = \dot{u} = 0$ when $t < 0$ are

$$\dot{u}(t) = \dot{u}_0 \qquad [t > 0] \tag{23.3}$$

$$u(t) = \dot{u}_0 t \qquad [t > 0] \tag{23.4}$$

ACCELERATION STEP

The *unit step function* $\mathbf{1}(t)$ is defined mathematically as a function which has a value of zero at time less than zero ($t < 0$) and a value of unity at time greater than zero ($t > 0$). The mathematical expressions describing the acceleration step are

$$\ddot{u}(t) = \ddot{u}_0\mathbf{1}(t) \tag{23.5}$$

where $\mathbf{1}(t) = 1$ for $t > 0$ and $\mathbf{1}(t) = 0$ for $t < 0$. An acceleration time-history of the unit step function is shown in Fig. 23.2*B*; the corresponding velocity and displacement time-histories are also shown for the initial conditions $u = \dot{u} = 0$ when $t = 0$.

$$\dot{u}(t) = \ddot{u}_0 t \qquad [t > 0] \tag{23.6}$$

$$u(t) = \tfrac{1}{2}\ddot{u}_0 t^2 \qquad [t > 0] \tag{23.7}$$

The unit step function is the time integral of the delta function:

$$\mathbf{1}(t) = \int_{-\infty}^{t} \delta(t)\,dt \qquad [t > 0] \tag{23.8}$$

HALF-SINE ACCELERATION

A half-sine pulse of acceleration of duration τ is shown in Fig. 23.2*C*; the corresponding velocity and displacement time-histories also are shown, for the initial conditions $u = \dot{u} = 0$ when $t = 0$. The applicable mathematical expressions are

$$\begin{aligned} \ddot{u}(t) &= \ddot{u}_0 \sin\left(\frac{\pi t}{\tau}\right) && [0 < t < \tau] \\ \ddot{u}(t) &= 0 && \text{when } t < 0 \quad \text{and } t > \tau \end{aligned} \tag{23.9}$$

$$\begin{aligned} \dot{u}(t) &= \frac{\ddot{u}_0\tau}{\pi}\left(1 - \cos\frac{\pi t}{\tau}\right) && [0 < t < \tau] \\ \dot{u}(t) &= \frac{2\ddot{u}_0\tau}{\pi} && [t > \tau] \end{aligned} \tag{23.10}$$

$$u(t) = \frac{\ddot{u}_0\tau^2}{\pi^2}\left(\frac{\pi t}{\tau} - \sin\frac{\pi t}{\tau}\right) \qquad [0 < t < \tau]$$
$$u(t) = \frac{\ddot{u}_0\tau^2}{\pi}\left(\frac{2t}{\tau} - 1\right) \qquad [t > \tau] \qquad (23.11)$$

This example is typical of a class of shock motions in the form of acceleration pulses not having infinite slopes.

DECAYING SINUSOIDAL ACCELERATION

A decaying sinusoidal trace of acceleration is shown in Fig. 23.2*D*; the corresponding time-histories of velocity and displacement also are shown for the initial conditions $\dot{u} = -\dot{u}_0$ and $u = 0$ when $t = 0$. The applicable mathematical expressions are

$$\ddot{u}(t) = \frac{\dot{u}_0\omega_1}{\sqrt{1-\zeta_1^2}} e^{-\zeta_1\omega_1 t} \sin\left(\sqrt{1-\zeta_1^2}\,\omega_1 t + \sin^{-1}\left(2\zeta_1\sqrt{1-\zeta_1^2}\right)\right) \qquad [t > 0] \qquad (23.12)$$

where ω_1 is the frequency of the vibration and ζ_1 is the fraction of critical damping corresponding to the decrement of the decay. Corresponding expressions for velocity and displacement are

$$\dot{u}(t) = -\frac{\dot{u}_0}{\sqrt{1-\zeta_1^2}} e^{-\zeta_1\omega_1 t} \cos\left(\sqrt{1-\zeta_1^2}\,\omega_1 t + \sin^{-1}\zeta_1\right) \qquad [t > 0] \qquad (23.13)$$

where $\dot{u}(t) = -\dot{u}_0$ when $t < 0$.

$$u(t) = -\frac{\dot{u}_0}{\omega_1\sqrt{1-\zeta_1^2}} e^{-\zeta_1\omega_1 t} \sin\left(\sqrt{1-\zeta_1^2}\,\omega_1 t\right) \qquad [t > 0] \qquad (23.14)$$

where $u(t) = -\dot{u}_0 t$ when $t < 0$.

COMPLEX SHOCK MOTION

The trace shown in Fig. 23.2*E* is an acceleration time-history representing typical field data. It cannot be defined by an analytic function; consequently, the corresponding velocity and displacement time-histories can be obtained only by numerical, graphical, or analog integration of the acceleration time-history.

CONCEPTS OF DATA REDUCTION

Consideration of the engineering uses of shock measurements indicates two basically different methods for describing a shock: (1) a description of the shock in terms of its inherent properties, in the time domain or in the frequency domain; and (2) a description of the shock in terms of the effect on structures when the shock acts as the excitation. The latter is designated reduction to the response domain. The following sections discuss concepts of data reduction to the frequency and response domains.

Whenever practical, the original time-history should be retained even though the information included therein is reduced to another form. The purpose of data reduction is to make the data more useful for some particular application. The reduced data usually have a more limited range of applicability than the original time-history. These limitations must be borne in mind if the data are to be applied intelligently.

DATA REDUCTION TO THE FREQUENCY DOMAIN

Any nonperiodic function can be represented as the superposition of sinusoidal components, each with its characteristic amplitude and phase.[1] This superposition is the Fourier spectrum, a plot of the amplitude and phase of the sinusoidal components into which the function can be decomposed. It is analogous to the Fourier components of a periodic function (Chap. 22). The Fourier components of a periodic function occur at discrete frequencies and the composite function is obtained by superposition of components. By contrast, the Fourier spectrum for a nonperiodic function is a continuous function of frequency and the composite function is achieved by integration. Applicable mathematical properties are given in the Appendix; the following sections discuss the application of the Fourier spectrum to describe the shock motions illustrated in Fig. 23.2.

ACCELERATION IMPULSE. Using the definition of the acceleration pulse given by Eq. (23.2) and substituting this for $f(t)$ in Eq. (23.57) of the Appendix,

$$\mathbf{F}(\omega) = \lim_{\epsilon \to 0} \int_0^{\epsilon} \frac{\dot{u}_0}{\epsilon} e^{-j\omega t}\, dt \tag{23.15}$$

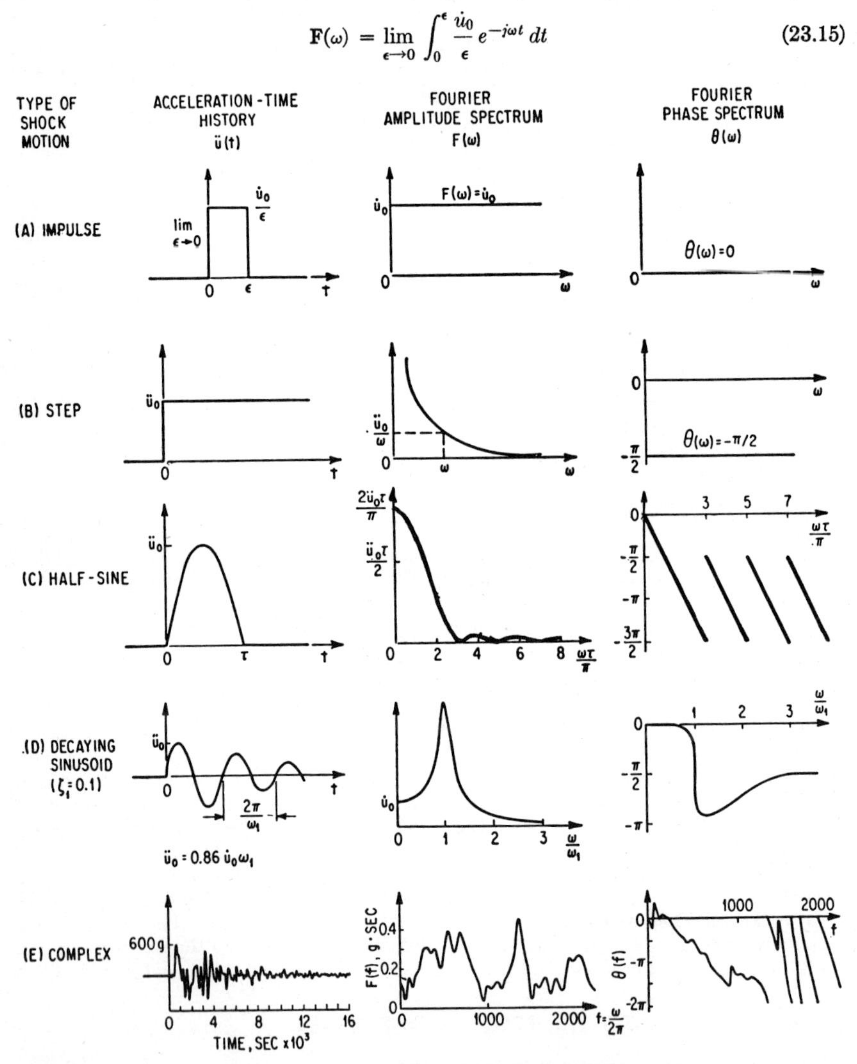

FIG. 23.3. Fourier amplitude and phase spectra for the shock motions in Fig. 23.2. These spectra represent the amplitude and phase of the continuous distribution of frequency components into which the shock motions can be decomposed. [Mathematical descriptions appear in Eqs. (23.15) to (23.27).]

Carrying out the integration,

$$\mathbf{F}(\omega) = \lim_{\epsilon \to 0} \frac{\dot{u}_0(1 - e^{-j\omega\epsilon})}{j\omega\epsilon} = \dot{u}_0 \qquad (23.16)$$

The corresponding amplitude and phase spectra, from Eqs. (23.63) and (23.64) of the Appendix, are

$$F(\omega) = \dot{u}_0; \qquad \theta(\omega) = 0 \qquad (23.17)$$

These spectra are shown in Fig. 23.3*A*. The magnitude of the Fourier amplitude spectrum is a constant, independent of frequency, equal to the area under the acceleration-time curve.

The physical significance of the spectra in Fig. 23.3*A* is shown in Fig. 23.4 where the rectangular acceleration pulse of magnitude $\dot{u}_0/\epsilon$ and duration $t = \epsilon$ is shown as approximated by superposed sinusoidal components for several different upper limits of frequency for the components. With the frequency limit $\omega_l = 4/\epsilon$, the pulse has a noticeably rounded contour formed by the superposition of all components whose frequencies are less than ω_l. These components tend to add in the time interval $0 < t < \epsilon$ and, though existing for all time from $-\infty$ to $+\infty$, cancel each other outside this interval so that $\ddot{u}$ approaches zero. When $\omega_l = 16/\epsilon$, the pulse is more nearly rectangular and $\ddot{u}$ approaches zero more rapidly for time $t < 0$ and $t > \epsilon$. When $\omega_l = \infty$, the superposition of sinusoidal components gives $\ddot{u} = \dot{u}_0/\epsilon$ for the time interval of the pulse, and $\ddot{u} = \dot{u}_0/2\epsilon$ at $t = 0$ and $t = \epsilon$. The components cancel completely for all other times. As $\epsilon \to 0$ and $\omega_l \to \infty$, the infinitely large number of superimposed frequency components gives $\ddot{u} = \infty$ at $t = 0$. The same general result is obtained when the Fourier components of other forms of $\ddot{u}(t)$ are superimposed.

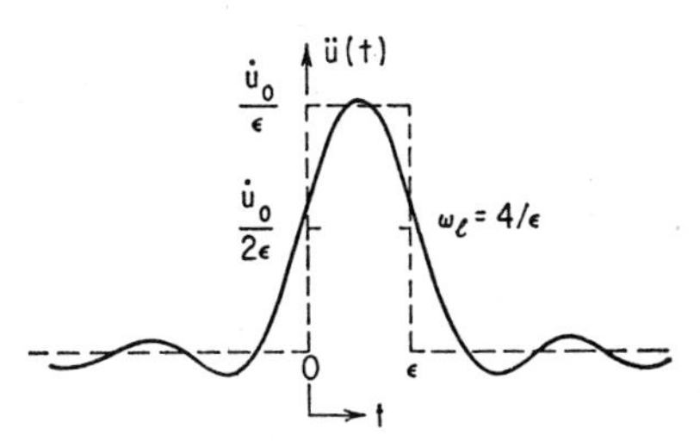

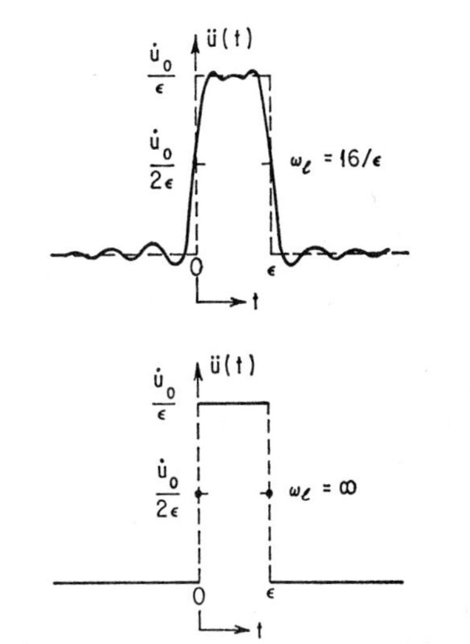

FIG. 23.4. Time-histories which result from the superposition of the Fourier components of a rectangular pulse for several different upper limits of frequency ω_l of the components. The upper time-history ($\omega_l = 4/\epsilon$) has a noticeably rounded contour; the middle time-history ($\omega_l = 16/\epsilon$) is more nearly rectangular; and the lower time-history ($\omega_l = \infty$) is the complete Fourier representation of a rectangular pulse.

ACCELERATION STEP. The Fourier spectrum of the acceleration step does not exist in the strict sense (see Appendix) since the integrand of Eq. (23.56) does not tend to zero as $\omega \to \infty$. Using a convergence factor, the Fourier transform is found by substituting $\ddot{u}(t)$ for $f(t)$ in Eq. (23.73) of the Appendix:

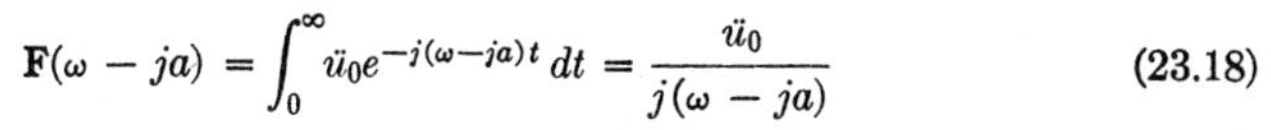

$$\mathbf{F}(\omega - ja) = \int_0^{\infty} \ddot{u}_0 e^{-j(\omega - ja)t}\, dt = \frac{\ddot{u}_0}{j(\omega - ja)} \qquad (23.18)$$

Taking the limit as $a \to 0$,

$$\mathbf{F}(\omega) = \frac{\ddot{u}_0}{j\omega} \qquad (23.19)$$

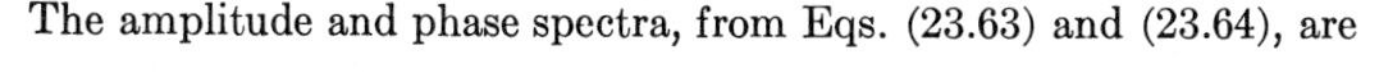

The amplitude and phase spectra, from Eqs. (23.63) and (23.64), are

$$F(\omega) = \frac{\ddot{u}_0}{\omega}; \qquad \theta(\omega) = -\frac{\pi}{2} \qquad (23.20)$$

These spectra are shown in Fig. 23.3B; the amplitude spectrum decreases as frequency increases, whereas the phase is a constant independent of frequency. Note that the spectrum of Eq. (23.19) is $1/j\omega$ times the spectrum for the impulse, Eq. (23.16), in accordance with Eq. (23.65) of the Appendix.

HALF-SINE ACCELERATION. Substitution of the half-sine acceleration time-history, Eq. (23.9), into Eq. (23.57) gives

$$\mathbf{F}(\omega) = \int_0^\tau \ddot{u}_0 \sin \frac{\pi t}{\tau} e^{-j\omega t}\, dt \tag{23.21}$$

Performing the indicated integration gives

$$\begin{aligned} \mathbf{F}(\omega) &= \frac{\ddot{u}_0\tau/\pi}{1 - (\omega\tau/\pi)^2}(1 + e^{-j\omega\tau}) && [\omega \neq \pi/\tau] \\ \mathbf{F}(\omega) &= -\frac{j\ddot{u}_0\tau}{2} && [\omega = \pi/\tau] \end{aligned} \tag{23.22}$$

Applying Eqs. (23.63) and (23.64) to find expressions for the spectra of amplitude and phase,

$$\begin{aligned} F(\omega) &= \frac{2\ddot{u}_0\tau}{\pi}\left|\frac{\cos(\omega\tau/2)}{1 - (\omega\tau/\pi)^2}\right| && [\omega \neq \pi/\tau] \\ F(\omega) &= \frac{\ddot{u}_0\tau}{2} && [\omega = \pi/\tau] \end{aligned} \tag{23.23}$$

$$\theta(\omega) = -\frac{\omega\tau}{2} + n\pi \tag{23.24}$$

where n is the smallest integer that prevents $|\theta(\omega)|$ from exceeding $3\pi/2$. The Fourier spectra of the half-sine pulse of acceleration are plotted in Fig. 23.3C.

DECAYING SINUSOIDAL ACCELERATION. The application of Eq. (23.57) to the decaying sinusoidal acceleration defined by Eq. (23.12) gives the following expression for the Fourier spectrum:

$$\mathbf{F}(\omega) = \dot{u}_0 \frac{1 + j2\zeta_1\omega/\omega_1}{(1 - \omega^2/\omega_1^2) + j2\zeta_1\omega/\omega_1} \tag{23.25}$$

This can be converted to a spectrum of absolute values by applying Eq. (23.63):

$$F(\omega) = \dot{u}_0 \sqrt{\frac{1 + (2\zeta_1\omega/\omega_1)^2}{(1 - \omega^2/\omega_1^2)^2 + (2\zeta_1\omega/\omega_1)^2}} \tag{23.26}$$

A spectrum of phase angle is obtained from Eq. (23.64):

$$\theta(\omega) = -\tan^{-1}\frac{2\zeta_1(\omega/\omega_1)^3}{(1 - \omega^2/\omega_1^2) + (2\zeta_1\omega/\omega_1)^2} \tag{23.27}$$

These spectra are shown in Fig. 23.3D for a value of $\zeta = 0.1$. The peak in the amplitude spectrum near the frequency ω_1 indicates a strong concentration of Fourier components near the frequency of occurrence of the oscillations in the shock motion.

COMPLEX SHOCK. The complex shock motion shown in Fig. 23.3E is the result of actual measurements; hence, its functional form is unknown. Its Fourier spectrum must be computed numerically. The Fourier spectrum shown in Fig. 23.3E was evaluated digitally using 100 time increments of 0.00015 sec duration. The peaks in the amplitude spectrum indicate concentrations of sinusoidal components near the frequencies of various oscillations in the shock motion. The portion of the phase spectrum at the high frequencies creates an appearance of discontinuity. If the phase angle were not returned to zero each time it passes through −360°, as a convenience in plotting, the curve would be continuous.

APPLICATION OF THE FOURIER SPECTRUM. The Fourier spectrum description of a shock is useful in linear analysis when the properties of a structure on which the shock acts are defined as a function of frequency. Such properties are designated by the general term *transfer function;* in the shock and vibration technology, commonly used transfer functions are mechanical impedance, mobility, and transmissibility.

When a shock acts on a structure, the structure responds in a manner that is essentially oscillatory. The frequencies that appear predominantly in the response are (1) the preponderant frequencies of the shock and (2) the natural frequencies of the structure. The Fourier spectrum of the response $\mathbf{R}(\omega)$ is the product of the Fourier spectrum of the shock $\mathbf{F}(\omega)$ and an appropriate transfer function for the structure. This is indicated by Eq. (23.74) of the Appendix. For example, if $\mathbf{F}(\omega)$ and $\mathbf{R}(\omega)$ are Fourier spectra of acceleration, the transfer function is the transmissibility of the structure, i.e., the ratio of acceleration at the responding station to the acceleration at the driving station, as a function of frequency. However, if $\mathbf{R}(\omega)$ is a Fourier spectrum of velocity and $\mathbf{F}(\omega)$ is a Fourier spectrum of force, the transfer function is mobility as a function of frequency.

The Fourier spectrum also finds application in evaluating the effect of a load upon a shock source. A source of shock generally consists of a means of shock excitation and a resilient structure through which the excitation is transmitted to a load. Consequently, the character of the shock delivered by the resilient structure of the shock source is influenced by the nature of the load being driven. The characteristics of the source and load may be defined in terms of mechanical impedance or mobility (see Chap. 10). If the shock motion at the source output is measured with no load and expressed in terms of its Fourier spectrum, the effect of the load upon this shock motion can be determined by Eq. (23.79) of the Appendix. The resultant motion with the load attached is described by its Fourier spectrum.

The transfer function of a structure may be determined by applying a transient force to the structure and noting the response. This is analogous to the more commonly used method of applying a sinusoidally varying force whose frequency can be varied over a wide range and noting the sinusoidally varying motion at the frequency of the force application. In some circumstances, it may be more convenient to apply a transient. From the measured time-histories of the force and the response, the corresponding Fourier spectra can be calculated. The transfer function is the quotient of the Fourier spectrum of the force divided by the Fourier spectrum of the response, as indicated by Eq. (23.82) of the Appendix.

DATA REDUCTION TO THE RESPONSE DOMAIN

A structure or physical system has a characteristic response to a particular shock applied as an excitation to the structure. The magnitudes of the response peaks can be used to define certain effects of the shock by considering systematically the properties of the system and relating the peak responses to such properties. This is in contrast to the Fourier spectrum description of a shock in the following respects:

1. Whereas the Fourier spectrum defines the shock in terms of the amplitudes and phase relations of its frequency components, the response spectrum describes only the effect of the shock upon a structure in terms of peak responses. This effect is of considerable significance in the design of equipments (Chap. 42), and in the specification of laboratory tests (Chap. 24).

2. The time-history of a shock cannot be determined from the knowledge of the peak responses of a system excited by the shock; i.e., the calculation of peak responses is an irreversible operation. This contrasts with the Fourier spectrum where the Fourier spectrum can be determined from the time-history, and vice versa.

By limiting consideration to the response of a linear, viscously damped single degree-of-freedom structure with lumped parameters (hereafter referred to as a simple structure and illustrated in Fig. 23.5), there are only two structural parameters upon which the response depends: (1) the undamped natural frequency and (2) the fraction of critical

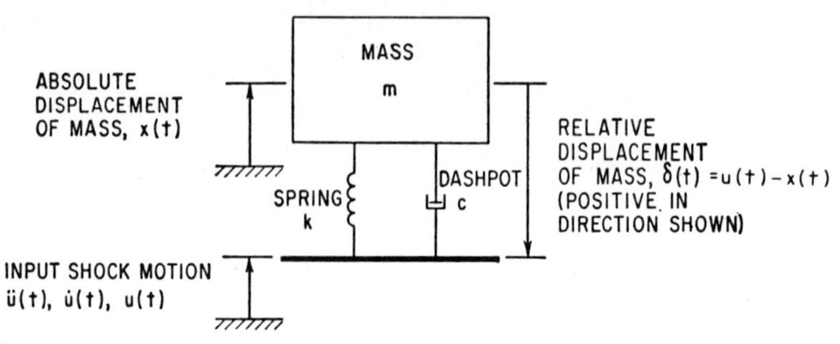

FIG. 23.5. Representation of a simple structure used to accomplish the data reduction of a shock motion to the response domain. The differential equation of motion is given by Eq. (23.31).

damping. With only two parameters involved, it is feasible to obtain from the shock measurement a systematic presentation of the peak responses of many simple structures. This process is termed *data reduction to the response domain.* This type of reduced data applies directly to a system that responds in a single degree-of-freedom; it is useful to some extent by normal-mode superposition to evaluate the response of a linear system that responds in more than one degree-of-freedom. The conditions of a particular application determine the magnitude of errors resulting from superposition.[2–4]

SHOCK SPECTRUM.* The response of a system to a shock can be expressed as the time-history of a parameter that describes the motion of the system. For a simple system, the magnitudes of the response peaks can be summarized as a function of the natural frequency or natural period of the responding system, at various values of the fraction of critical damping. This type of presentation is termed a shock spectrum. In the shock spectrum, or more specifically the two-dimensional shock spectrum, only the maximum value of the response found in a single time-history is plotted. The three-dimensional shock spectrum takes the form of a surface and shows the distribution of response peaks throughout the time-history. The two-dimensional spectrum is more common and is discussed in considerable detail in the immediately following section; for convenience, it is referred to hereafter simply as the *shock spectrum.* The three-dimensional spectrum is discussed in less detail in a later section.

Parameters for the Shock Spectrum. The peak response of the simple structure may be defined, as a function of natural frequency, in terms of any one of several parameters that describe its motion. The parameters often are related to each other by the characteristics of the structure. However, inasmuch as one of the advantages of the shock spectrum method of data reduction and presentation is convenience of application to physical situations, it is advantageous to give careful consideration in advance to the particular parameter that is best adapted to the attainment of particular objectives. Referring to the simple structure shown in Fig. 23.5, the following significant parameters may be determined directly from measurements on the structure:

1. Absolute displacement $x(t)$ of mass m. This indicates the displacement of the responding structure with reference to an inertial reference plane, i.e., coordinate axes fixed in space.

2. Relative displacement $\delta(t)$ of mass m. This indicates the displacement of the responding structure relative to its support, a quantity useful for evaluating the distortions and strains within the responding structure.

3. Absolute velocity $\dot{x}(t)$ of mass m. This quantity is useful for determining the kinetic energy of the structure.

4. Relative velocity $\dot{\delta}(t)$ of mass m. This quantity is useful for determining the stresses generated within the responding structure due to viscous damping and the maximum energy dissipated by the responding structure.

* This spectrum sometimes is designated the *response spectrum.*

5. Absolute acceleration $\ddot{x}(t)$ of mass m. This quantity is useful for determining the stresses generated within the responding structure due to the combined elastic and damping reactions of the structure.

The *equivalent static acceleration* is that steadily applied acceleration, expressed as a multiple of the acceleration of gravity, which distorts the structure to the maximum distortion resulting from the action of the shock.[5] For the simple structure of Fig. 23.5, the relative displacement response δ indicates the distortion under the shock condition. The corresponding distortion under static conditions, in a $1g$ gravitational field, is

$$\delta_{st} = \frac{mg}{k} = \frac{g}{\omega_n^2} \tag{23.28}$$

By analogy, the maximum distortion under the shock condition is

$$\delta_{\text{max}} = \frac{A_{\text{eq}}g}{\omega_n^2} \tag{23.29}$$

where A_{eq} is the equivalent static acceleration in units of gravitational acceleration. From Eq. (23.29),

$$A_{\text{eq}} = \frac{\delta_{\text{max}}\omega_n^2}{g} \tag{23.30}$$

The maximum relative displacement δ_{max} and the equivalent static acceleration A_{eq} are directly proportional.

If the shock is a loading parameter, such as force, pressure, or torque, as a function of time, the corresponding equivalent static parameter is an equivalent static force, pressure, or torque, respectively. Since the supporting structure is assumed to be motionless when a shock loading acts, the relative response motions and absolute response motions become identical.

The differential equation of motion for the system shown in Fig. 23.5 is

$$-\ddot{x}(t) + 2\zeta\omega_n\dot{\delta}(t) + \omega_n^2\delta(t) = 0 \tag{23.31}$$

where ω_n is the undamped natural frequency and ζ is the fraction of critical damping. When $\zeta = 0$, $\ddot{x}_{\text{max}} = A_{\text{eq}}g$; this follows directly from the relation of Eq. (23.29). When $\zeta \neq 0$, the acceleration $\ddot{x}$ experienced by the mass m results from forces transmitted by the spring k and the damper c. Thus, in a damped system, the maximum acceleration of mass m is not exactly equal to the equivalent static acceleration. However, in most mechanical structures, the damping is relatively small; therefore, the equivalent static acceleration and the maximum absolute acceleration often are interchangeable with negligible error.

Referring to the model in Fig. 23.1, suppose the equivalent static acceleration A_{eq} and the maximum absolute acceleration $\ddot{x}_{\text{max}}$ are known for the primary structure. Then A_{eq} is useful directly for calculating the maximum relative displacement response of the primary structure. When the natural frequency of the secondary structure is much higher than the natural frequency of the primary structure, the maximum acceleration $\ddot{x}_{\text{max}}$ of M is useful for calculating the maximum relative displacement of m with respect to M. The secondary structure then responds in a "static manner" to the acceleration of the mass M; i.e., the maximum acceleration of m is approximately equal to that of M. Consequently, both A_{eq} and $\ddot{x}_{\text{max}}$ can be used for design purposes to calculate equivalent static loads on structures of equipment (see Chap. 42).

If the damping in the responding structure is large ($\zeta > 0.2$), the values of A_{eq} and $\ddot{x}_{\text{max}}$ are significantly different. Because the maximum distortion of primary structures often is the type of information required and the equivalent static acceleration is an expression of this response in terms of an equivalent static loading, the following discussion is limited to shock spectra in terms of A_{eq}.

The response of a simple structure with small damping to oscillatory-type shock excitation often is substantially sinusoidal at the natural frequency of the structure, i.e., the envelope of the oscillatory response varies in a relatively slow manner as depicted in Fig. 23.6. The maximum relative displacement δ_{max}, the maximum relative velocity $\dot{\delta}_{max}$, and the maximum absolute acceleration $\ddot{x}_{max}$ are related approximately as follows:

$$\dot{\delta}_{max} = \omega_n \delta_{max}; \qquad \ddot{x}_{max} = \omega_n \dot{\delta}_{max}; \qquad \ddot{x}_{max} = \omega_n^2 \delta_{max} \tag{23.32}$$

where the sign may be neglected since the positive and negative maxima are approximately equal. When applicable, these relations may be used to convert from a spectrum expressed in one parameter to a spectrum expressed in another parameter.

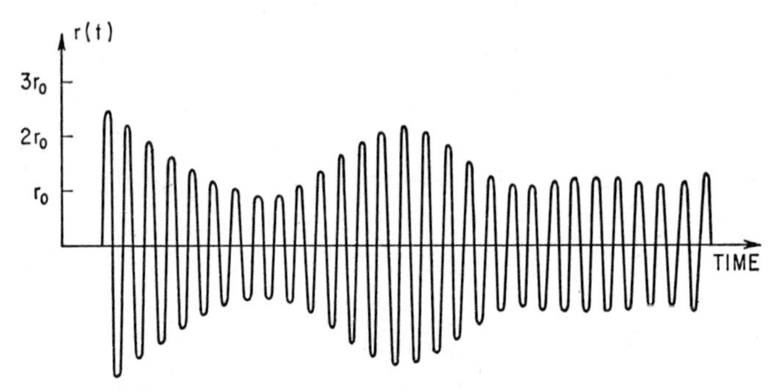

FIG. 23.6. Example of an oscillatory response time-history $r(t)$ for which the envelope of the response varies in a relatively slow manner. This type of response often results when an oscillatory-type shock motion acts on a simple structure with small damping. When such a response occurs, the response parameters of the simple structure are related in accordance with Eq. (23.32).

For idealized shock motions which often are approximated in practice, it is desirable to use a dimensionless ratio for the ordinate of the shock spectrum. Some of the more common dimensionless ratios are

$$\frac{gA_{eq}}{\ddot{u}_{max}} = \frac{\omega_n^2 \delta_{max}}{\ddot{u}_{max}}; \qquad \frac{\ddot{x}_{max}}{\ddot{u}_{max}}; \qquad \frac{\omega_n \delta_{max}}{\Delta \dot{u}}; \qquad \frac{\dot{\delta}_{max}}{\Delta \dot{u}}; \qquad \frac{\delta_{max}}{u_{max}}$$

where $\ddot{u}_{max}$ and u_{max} are the maximum acceleration and displacement, respectively, of the shock motion and $\Delta \dot{u}$ is the velocity change of the shock motion (equal to the area under the acceleration time-history). Sometimes these ratios are referred to as *shock amplification factors.*

Calculation of Shock Spectrum. The relative displacement response of a simple structure (Fig. 23.5) resulting from a shock defined by the acceleration $\ddot{u}(t)$ of the support is given by the Duhamel integral[6]

$$\delta(t) = \frac{1}{\omega_d} \int_0^t \ddot{u}(t_v) e^{-\zeta\omega_n(t-t_v)} \sin \omega_d(t - t_v)\, dt_v \tag{23.33}$$

where $\omega_n = (k/m)^{1/2}$ is the undamped natural frequency, $\zeta = c/2m\omega_n$ is the fraction of critical damping and $\omega_d = \omega_n(1 - \zeta^2)^{1/2}$ is the damped natural frequency. The excitation $\ddot{u}(t_v)$ is defined as a function of the variable of integration t_v and the response $\delta(t)$ is a function of time t. The relative displacement δ and relative velocity $\dot{\delta}$ are considered to be zero when $t = 0$. The equivalent static acceleration, defined by Eq. (23.30), as a function of ω_n and ζ is

$$A_{eq}(\omega_n,\zeta) = \frac{\omega_n^2}{g} \delta_{max}(\omega_n,\zeta) \tag{23.34}$$

If a shock loading such as the input force $F(t)$ rather than an input motion acts on the simple structure, the response is

$$\delta(t) = \frac{1}{m\omega_d}\int_0^t F(t_v)e^{-\zeta\omega_n(t-t_v)}\sin\omega_d(t-t_v)\,dt_v \tag{23.35}$$

and an equivalent static force is given by

$$F_{eq}(\omega_n,\zeta) = k\delta_{max}(\omega_n,\zeta) = m\omega_n{}^2\delta_{max}(\omega_n,\zeta) \tag{23.36}$$

The equivalent static force is related to equivalent static acceleration by

$$F_{eq}(\omega_n,\zeta) = mA_{eq}(\omega_n,\zeta) \tag{23.37}$$

It is often of interest to determine the maximum relative displacement of the simple structure in Fig. 23.5 in both a positive and a negative direction. If $\ddot{u}(t)$ is positive as shown, positive values of $\ddot{x}(t)$ represent upward acceleration of the mass m. Initially, the spring is compressed and the positive direction of $\delta(t)$ is taken to be positive as shown. Conversely, negative values of $\delta(t)$ represent extension of spring k from its original position. It is possible that the ultimate use of the reduced data would require that both extension and compression of spring k be determined. Correspondingly, a positive and a negative sign may be associated with an equivalent static acceleration A_{eq} of the support, so that $A_{eq}{}^+$ is an upward acceleration producing a positive deflection δ and $A_{eq}{}^-$ is a downward acceleration producing a negative deflection δ.

For some purposes it is desirable to distinguish between the maximum response which occurs during the time in which the measured shock acts and the maximum response which occurs during the free vibration existing after the shock has terminated. The shock spectrum based on the former is called a *primary shock spectrum* and that based on the latter is called a *residual shock spectrum*. For instance, the response $\delta(t)$ to the half-sine pulse in Fig. 23.2*C* occurring during the period $(t < \tau)$ is the primary response and the response $\delta(t)$ occurring during the period $(t > \tau)$ is the residual response. Reference is made to primary and residual shock spectra in the next section on examples of the shock spectrum and in the section on the relationship between the shock spectrum and the Fourier spectrum.

Examples of Shock Spectra.* In this section the shock spectra are presented for the five acceleration time-histories in Fig. 23.2. These spectra, shown in Fig. 23.7, are expressed in terms of equivalent static acceleration for the undamped responding structure, for $\zeta = 0.1$, 0.5, and other selected fractions of critical damping. Both the maximum positive and the maximum negative responses are indicated. In addition, a number of relative displacement response time-histories $\delta(t)$ are plotted to show the nature of the responses.

Acceleration Impulse. The application of Eq. (23.33) to the acceleration impulse shown in Fig. 23.2*A* and defined by Eq. (23.1) yields

$$\delta(t) = \frac{\dot{u}_0}{\omega_d}e^{-\zeta\omega_n t}\sin\omega_d t \qquad [\zeta < 1] \tag{23.38}$$

This response is plotted in Fig. 23.7*A* for $\zeta = 0$, 0.1, and 0.5. The response peaks are reached at the times $t = (\cos^{-1}\zeta)/\omega_d$, $\cos^{-1}\zeta$ increasing by π for each succeeding peak. The values of response at the peaks are

$$\delta_{max}^{(i)}(\omega_n,\zeta) = \frac{\dot{u}_0}{\omega_n}\exp\left(-\frac{\zeta}{\sqrt{1-\zeta^2}}[\cos^{-1}\zeta + (i-1)\pi]\right) \quad [0 < \cos^{-1}\zeta \le \pi/2] \tag{23.39}$$

where i is the number of the peak ($i = 1$ for the first positive peak, $i = 2$ for the first negative peak, etc.).

* A large number of shock spectra, based on various response parameters, are given in Chap. 8.

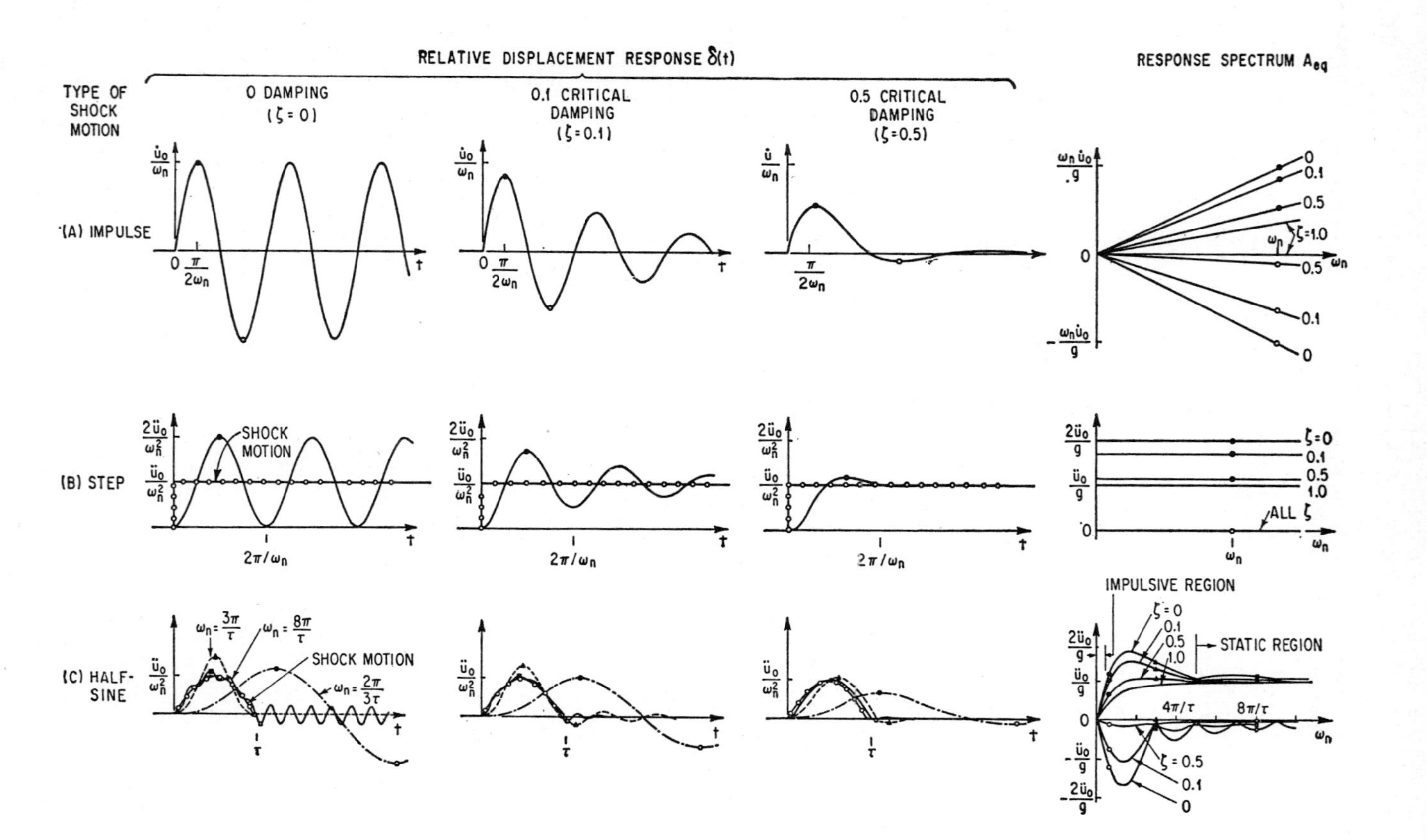

RELATIVE DISPLACEMENT RESPONSE $\delta(t)$
RESPONSE SPECTRUM A_{eq}
TYPE OF SHOCK MOTION
O DAMPING ($\zeta = 0$)
0.1 CRITICAL DAMPING ($\zeta = 0.1$)
0.5 CRITICAL DAMPING ($\zeta = 0.5$)
(A) IMPULSE
(B) STEP
(C) HALF-SINE
SHOCK MOTION
$\omega_n = \frac{3\pi}{\tau}$
$\omega_n = \frac{8\pi}{\tau}$
$\omega_n = \frac{2\pi}{3\tau}$
$2\pi/\omega_n$
$\frac{\pi}{2\omega_n}$
ALL ζ
IMPULSIVE REGION
STATIC REGION
$4\pi/\tau$
$8\pi/\tau$

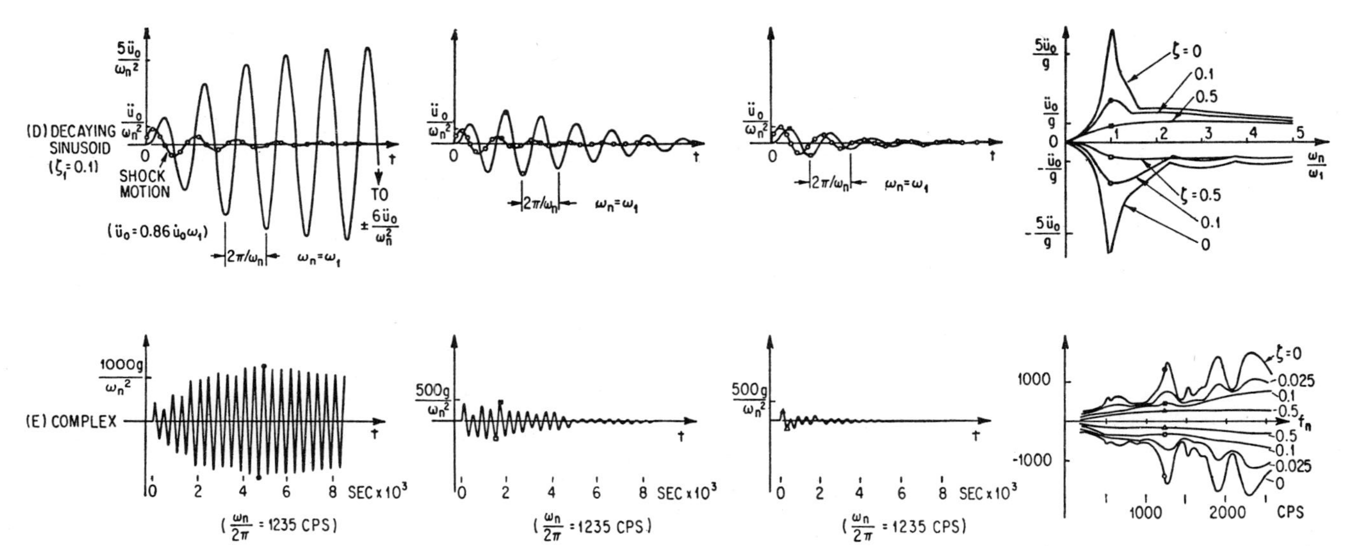

FIG. 23.7. Time-histories of response to shock motions defined in Fig. 23.2 and corresponding shock spectra. Time-histories for several fractions of critical damping in the responding system are included, and the spectra are for corresponding fractions of critical damping. The spectra represent maximum values of response, as indicated by similar symbols. When the shock motions shown in Fig. 23.2 are superimposed on the time-histories of response, the shock motions are indicated by lines with hollow circles. Natural frequencies corresponding to the shock spectra are indicated where they have a significant relation to the frequency of the shock motion.

The largest positive response occurs at the first peak, i.e., when $i = 1$, and is shown by the solid dots in Fig. 23.7*A*. Hence, the equivalent static acceleration in the positive direction is obtained by substitution of Eq. (23.39) into Eq. (23.34) with $i = 1$:

$$A_{eq}^{+}(\omega_n,\zeta) = \frac{\omega_n \dot{u}_0}{g} \exp\left(-\frac{\zeta}{\sqrt{1-\zeta^2}}\cos^{-1}\zeta\right) \tag{23.40}$$

The equivalent static acceleration in the negative direction is calculated from the maximum relative deflection at second peak, i.e., when $i = 2$, and is shown by the hollow dots in Fig. 23.7*A*:

$$A_{eq}^{-}(\omega_n,\zeta) = \frac{\omega_n \dot{u}_0}{g} \exp\left(-\frac{\zeta}{\sqrt{1-\zeta^2}}(\cos^{-1}\zeta + \pi)\right) \tag{23.41}$$

The resulting shock spectrum is shown in Fig. 23.7*A* with curves for $\zeta = 0, 0.1, 0.5$, and 1.0. At any value of damping, a shock spectrum is a straight line passing through the origin. The peak distortion of the structure δ_{max} is inversely proportional to frequency. Thus, the relative displacement of the mass increases as the natural frequency decreases, whereas the equivalent static acceleration has an opposite trend.

Acceleration Step. The response of a simple structure to the acceleration step in Fig. 23.2*B* is found by substituting from Eq. (23.5) in Eq. (23.33) and integrating:

$$\delta(t) = \frac{\ddot{u}_0}{\omega_n^2}\left[1 - \frac{e^{-\zeta\omega_n t}}{\sqrt{1-\zeta^2}}\cos(\omega_d t - \sin^{-1}\zeta)\right] \qquad [\zeta < 1] \tag{23.42}$$

The responses $\delta(t)$ are shown in Fig. 23.7*B* for $\zeta = 0, 0.1$, and 0.5. The response overshoots the value $\ddot{u}_0/\omega_n^2$ and then oscillates about this value as a mean with diminishing amplitude as energy is dissipated by damping. An overshoot to $2\ddot{u}_0/\omega_n^2$ occurs for zero damping. A response $\delta = \ddot{u}_0/\omega_n^2$ would result from a steady application of the acceleration $\ddot{u}_0$.

The response maxima and minima occur at the times $t = i\pi/\omega_d$, $i = 0$ providing the first minimum and $i = 1$ the first maximum. The maximum values of the relative displacement response are

$$\delta_{max}(\omega_n,\zeta) = \frac{\ddot{u}_0}{\omega_n^2}\left[1 + \exp\left(-\frac{\zeta i\pi}{\sqrt{1-\zeta^2}}\right)\right] \qquad [i \text{ odd}] \tag{23.43}$$

The largest positive response occurs at the first maximum; i.e., where $i = 1$, and is shown by the solid symbols in Fig. 23.7*B*. The equivalent static acceleration in the positive direction is obtained by substitution of Eq. (23.43) into Eq. (23.34) with $i = 1$:

$$A_{qe}^{+}(\omega_n,\zeta) = \frac{\ddot{u}_0}{g}\left[1 + \exp\left(-\frac{\zeta\pi}{\sqrt{1-\zeta^2}}\right)\right] \tag{23.44a}$$

The greatest negative response is zero; it occurs at $t = 0$, independent of the value of damping, as shown by open symbols in Fig. 23.7*B*. Thus, the equivalent static acceleration in the negative direction is

$$A_{eq}^{-}(\omega_n,\zeta) = 0 \tag{23.44b}$$

Since the equivalent static acceleration is independent of natural frequency, the shock spectrum curves shown in Fig. 23.7*B* are horizontal lines. The symbols shown on the shock spectra correspond to the responses shown.

The equivalent static acceleration for an undamped simple structure is twice the value of the acceleration step $\ddot{u}_0/g$. As the damping increases, the overshoot in response decreases; there is no overshoot when the structure is critically damped.

Half-sine Acceleration. The expressions for the response of the damped simple structure to the half-sine acceleration of Eq. (23.9) are too involved to have general usefulness.

For an undamped system, the response $\delta(t)$ is

$$\delta(t) = \frac{\ddot{u}_0}{\omega_n^2}\left(\frac{(\omega_n\tau/\pi)}{1-(\omega_n\tau/\pi)^2}\right)[\sin\omega_n t - (\omega_n\tau/\pi)\sin(\pi t/\tau)] \qquad [0 < t \leq \tau]$$
$$\delta(t) = \frac{\ddot{u}_0}{\omega_n^2}\left(\frac{(\omega_n\tau/\pi)}{1-(\omega_n\tau/\pi)^2}\right)2\cos\left(\frac{\omega_n\tau}{2}\right)\sin\left[\omega_n\left(t-\frac{\tau}{2}\right)\right] \qquad [t > \tau] \tag{23.45}$$

For zero damping the residual response is sinusoidal with constant amplitude. The first maximum in the response of a simple structure with natural frequency less than π/τ occurs during the residual response; i.e., after $t = \tau$. As a result, the magnitude of each succeeding response peak is the same as that of the first maximum. Thus the positive and negative shock spectrum curves are equal for $\omega_n \leq \pi/\tau$. The dot-dash curve in Fig. 23.7*C* is an example of the response at a natural frequency of $2\pi/3\tau$. The peak positive response is indicated by a solid circle; the peak negative response by an open circle. The positive and negative shock spectrum values derived from this response are shown on the undamped ($\zeta = 0$) shock spectrum curves at the right-hand side of Fig. 23.7*C*, using the same symbols.

At natural frequencies below $\pi/2\tau$, the shock spectra for an undamped system are very nearly linear with a slope $\pm 2\ddot{u}_0\tau/\pi g$. In this low-frequency region the response is essentially impulsive; i.e., the maximum response is approximately the same as that due to an ideal acceleration impulse (Fig. 23.7*A*) having a velocity change $\dot{u}_0$ equal to the area under the half-sine acceleration time-history.

The response at the natural frequency $3\pi/\tau$ is the dotted curve in Fig. 23.7*C*. The displacement and velocity response are both zero at the end of the pulse and hence no residual response occurs. The solid and open triangles indicate the peak positive and negative response, the latter being zero. The corresponding points appear on the undamped shock spectrum curves. As shown by the negative undamped shock spectrum curve, the residual spectrum goes to zero for all odd multiples of π/τ above $3\pi/\tau$.

As the natural frequency increases above $3\pi/\tau$, the response attains the character of relatively low amplitude oscillations occurring with the half-sine pulse shape as a mean. An example of this type of response is shown by the solid curve for $\omega_n = 8\pi/\tau$. The largest positive response is slightly higher than $\ddot{u}_0/\omega_n^2$ and the residual response occurs at a relatively low level. The solid and open square symbols indicate the largest positive and negative response.

As the natural frequency becomes extremely high, the response follows the half-sine shape very closely. In the limit, the natural frequency becomes infinite and the response approaches the half-sine wave shown in Fig. 23.7*C*. For natural frequencies greater than $5\pi/\tau$, the response tends to follow the input and the largest response is within 20 per cent of the response due to a static application of the peak input acceleration. This portion of the shock spectrum is sometimes referred to as the "static region" (see *Limiting Values of Shock Spectrum* below).

The equivalent static acceleration without damping for the positive direction is

$$A^+_{\text{eq}}(\omega_n,0) = \frac{\ddot{u}_0}{g}\left(\frac{2(\omega_n\tau/\pi)}{1-(\omega_n\tau/\pi)^2}\right)\cos\left(\frac{\omega_n\tau}{2}\right) \qquad \left[\omega_n \leq \frac{\pi}{\tau}\right]$$
$$A^+_{\text{eq}}(\omega_n,0) = \frac{\ddot{u}_0}{g}\left(\frac{(\omega_n\tau/\pi)}{(\omega_n\tau/\pi)-1}\right)\sin\left(\frac{2i\pi}{(\omega_n\tau/\pi)+1}\right) \qquad \left[\omega_n > \frac{\pi}{\tau}\right] \tag{23.46}$$

where i is the positive integer which maximizes the value of the sine term while the argument remains less than π. In the negative direction the peak response always occurs during the residual response; thus, it is given by the absolute value of the first of the expressions in Eq. (23.46):

$$A^-_{\text{eq}}(\omega_n,0) = \frac{\ddot{u}_0}{g}\left(\frac{2(\omega_n\tau/\pi)}{1-(\omega_n\tau/\pi)^2}\right)\cos\left(\frac{\omega_n\tau}{2}\right) \tag{23.47}$$

Shock spectra for damped systems can be found by use of an electrical analog or digital computer. Spectra for $\zeta = 0.1$ and 0.5 are shown in Fig. 23.7*C*.

The response of a damped structure whose natural frequency is less than $\pi/2\tau$ is essentially impulsive; i.e., the shock spectra in this frequency region are substantially identical to the spectra for the acceleration impulse in Fig. 23.7*A*. Except near the zeros in the negative spectrum for an undamped system, damping reduces the peak response. For the positive spectra, the effect is small in the static region since the response tends to follow the input for all values of damping. The greatest effect of damping is seen in the negative spectra because it affects the decay of response oscillations at the natural frequency of the structure.

Decaying Sinusoidal Acceleration. Although analytical expressions for the response of a simple structure to the decaying sinusoidal acceleration shown in Fig. 23.2*D* are available, calculation of spectra is impractical without use of a computer. Figure 23.7*D* shows spectra for several values of damping in the decaying sinusoidal acceleration. In the low-frequency region ($\omega_n < 0.2\omega_1$), the response is essentially impulsive. The area under the acceleration time-history of the decaying sinusoid is $\dot{u}_0$; hence, the response of a very low-frequency structure is similar to the response to an acceleration impulse of magnitude $\dot{u}_0$.

When the natural frequency of the responding system approximates the frequency ω_1 of the oscillations in the decaying sinusoid, a resonant type of build-up tends to occur in the response oscillations. The region in the neighborhood of $\omega_1 = \omega_n$ may be termed a quasi-resonant region of the shock spectrum. Responses for $\zeta = 0$, 0.1, and 0.5 and $\omega_n = \omega_1$ are shown in Fig. 23.7*D*. In the absence of damping in the responding system, the rate of build-up diminishes with time and the amplitude of the response oscillations levels off as the input acceleration decays to very small values. Small damping in the responding system; e.g., $\zeta = 0.1$, reduces the initial rate of build-up and causes the response to decay to zero after a maximum is reached. When damping is as large as $\zeta = 0.5$, no build-up occurs.

Complex Shock. The shock spectra for the complex shock of Fig. 23.2*E* are shown in Fig. 23.7*E*, as obtained with a direct-analog-type shock spectrum analyzer.[7] Time-histories of the response of a system with a natural frequency of 1,250 cps also are shown. The ordinate of the spectrum plot is equivalent static acceleration and the abscissa is the natural frequency in cycles per second. Three pronounced peaks appear in the spectra for zero damping, at approximately 1,250 cps, 1,900 cps, and 2,350 cps. Such peaks indicate a concentration of frequency content in the shock, similar to the spectra for the decaying sinusoid in Fig. 23.7*D*. Other peaks in the shock spectra for an undamped system indicate less significant oscillatory behavior in the shock. The two lower frequencies at which the pronounced peaks occur correlate with the peaks in the Fourier spectrum of the same shock, as shown in Fig. 23.3*E*. The highest frequency at which a pronounced peak occurs is above the range for which the Fourier spectrum was calculated.

Because of frequency-response limitations of the spectrum analyzer, the shock spectra do not extend below 200 cps. Since the duration of the complex shock of Fig. 23.2*E* is about 0.016 sec, an impulsive-type response occurs only for natural frequencies well below 200 cps. As a result, no impulsive region appears in the shock spectra. There is no static region of the spectra shown because calculations were not extended to a sufficiently high frequency.

In general, the equivalent static acceleration A_{eq} is reduced by additional damping in the responding structure system except in the region of valleys in the shock spectra where damping may increase the magnitude of the spectrum. Positive and negative spectra tend to be approximately equal in magnitude at any value of damping; thus, the spectra for a complex oscillatory type of shock may be based on peak response independent of sign to a good approximation.

Limiting Values of Shock Spectrum. The response data provided by the shock spectrum sometimes can be abstracted to simplified parameters that are useful for certain purposes. In general, this cannot be done without definite information on the ultimate use of the reduced data, particularly the natural frequencies of the structures upon which the shock acts. Two important cases are discussed in the following sections.

Impulse or Velocity Change. The duration of a shock sometimes is much smaller than the natural period of a structure upon which it acts. Then the entire response of the structure is essentially a function of the area under the time-history of the shock, described in terms of acceleration or a loading parameter such as force, pressure, or torque. Consequently, the shock has an effect which is equivalent to that produced by an impulse of infinitesimally short duration; i.e., an ideal impulse.

The shock spectrum of an ideal impulse is shown in Fig. 23.7*A*. All equivalent static acceleration curves are straight lines passing through the origin. The portion of the spectrum exhibiting such straight-line characteristics is termed the *impulsive region.*

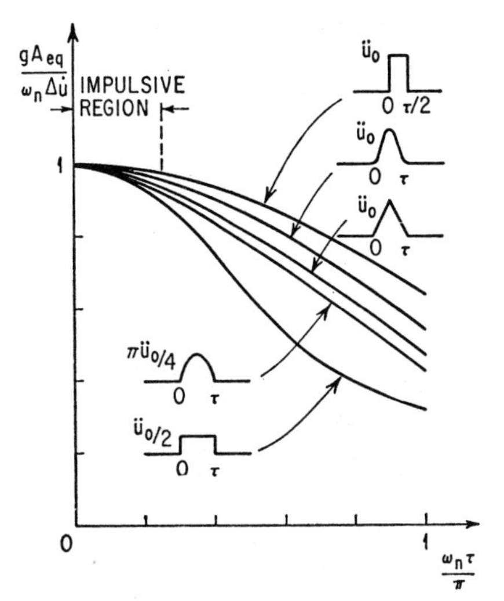

FIG. 23.8. Portions adjacent to the origin of the positive spectra of an undamped system for several single pulses of acceleration as shown. The ordinate is equivalent static acceleration A_{eq} normalized with respect to the peak impulsive response in g's, $\omega_n \Delta\dot{u}/g$, where $\Delta\dot{u}$ is the area under the corresponding acceleration time-history. An impulsive region is shown for which the deviation from impulsive response, $gA_{eq}/\omega_n \Delta\dot{u} = 1$, is less than 10 per cent.

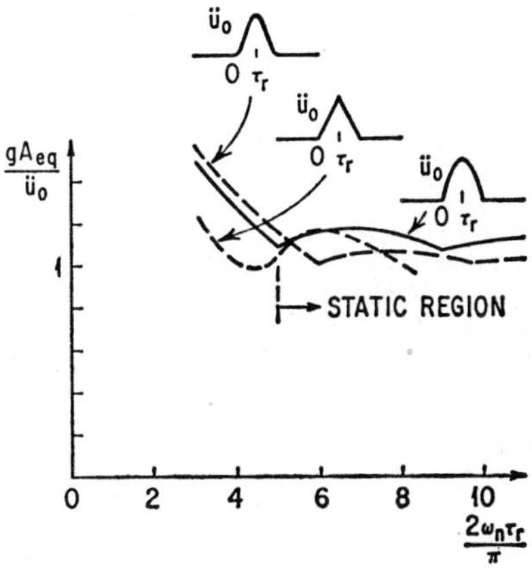

FIG. 23.9. Portions of the positive shock spectra of an undamped system with high natural frequencies for several single pulses of acceleration. The ordinate is equivalent static acceleration A_{eq} normalized with respect to the peak acceleration of the pulse, $\ddot{u}_0/g$. A static region is shown for which the deviation from static response, $gA_{eq}/\ddot{u}_0 = 1$, is less than 20 per cent.

The shock spectrum of the half-sine acceleration pulse has an impulsive region when ω_n is less than approximately $\pi/2\tau$, as shown in Fig. 23.7*C*. If the area under a time-history of acceleration or shock loading is not zero or infinite, an impulsive region exists in the shock spectrum. The extent of the region on the natural frequency axis depends on the shape and duration of the shock.

The portions adjacent to the origin of the positive shock spectra of an undamped system for several single pulses of acceleration are shown in Fig. 23.8. To illustrate the impulsive nature, each spectrum is normalized with respect to the peak impulsive response $\omega_n \Delta\dot{u}/g$, where $\Delta\dot{u}$ is the area under the corresponding acceleration time-history. Hence, the spectra indicate an impulsive response where the ordinate is approximately 1. The response to a single pulse of acceleration is impulsive within a tolerance of 10 per cent if $\omega_n < 0.25\pi/\tau$; i.e., $f_n < 0.4\tau^{-1}$, where f_n is the natural frequency of the responding structure in cps and τ is the pulse duration in seconds. This result also applies when the responding system is damped. Thus, it is possible to reduce the description of a shock pulse to a designated velocity change when the natural frequency of the responding structure is less than a specified value. The magnitude of the velocity change is the area under the acceleration pulse:

$$\Delta \dot{u} = \int_0^{\tau} \ddot{u}(t)\, dt \tag{23.48}$$

Peak Acceleration or Loading. The natural frequency of a structure responding to a shock sometimes is sufficiently high that the response oscillations of the structure at its natural frequency have a relatively small amplitude. Examples of such responses are shown in Fig. 23.7*C* for $\omega_n = 8\pi/\tau$ and $\zeta = 0, 0.1, 0.5$. As a result, the maximum response of the structure is approximately equal to the maximum acceleration of the shock and is termed *equivalent static response.* The magnitude of the spectra in such a static region is determined principally by the peak value of the shock acceleration or loading. Portions of the positive spectra of an undamped system in the region of high natural frequencies are shown in Fig. 23.9 for a number of acceleration pulses. Each spectrum is normalized with respect to the maximum acceleration of the pulse. If the ordinate is approximately 1, the shock spectrum curves behave approximately in a static manner.

The limit of the static region in terms of the natural frequency of the structure is more a function of the slope of the acceleration time-history than the duration of the pulse. Hence, the horizontal axis of the shock spectra in Fig. 23.9 is given in terms of the ratio of the rise time τ_r to the maximum value of the pulse. As shown in Fig. 23.9, the peak response to a single pulse of acceleration is approximately equal to the maximum acceleration of the pulse, within a tolerance of 20 per cent, if $\omega_n > 2.5\pi/\tau_r$; i.e., $f_n > 1.25\tau_r^{-1}$, where f_n is the natural frequency of the responding structure in cps and τ_r is the rise time to the peak value in seconds. The tolerance of 20 per cent applies to an undamped system; for a damped system, the tolerance is lower, as indicated in Fig. 23.7*C*.

The concept of the static region also can be applied to complex shocks. Suppose the shock is oscillatory, as shown in Fig. 23.2*E*. If the response to such a shock is to be nearly static, the response to each of the succession of pulses that make up the shock must be nearly static. This is most significant for the pulses of large magnitude because they determine the ordinate of the spectrum in the static region. Therefore, the shock spectrum for a complex shock in the static region is based upon the pulses of greatest magnitude and shortest rise time.

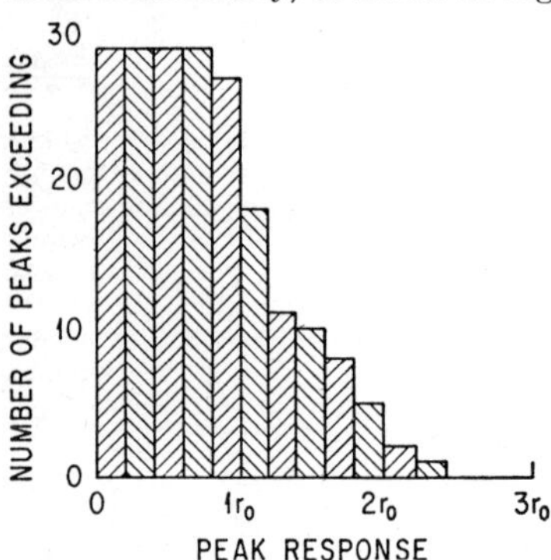

FIG. 23.10. Bar chart for the response of a system to a shock excitation. The chart represents a plane normal to the natural frequency axis in Fig. 23.11 and is obtained by counting the number of response maxima above various discrete increments of maximum response. In concept, the width of the increments approaches zero and a line faired through the ends of the bars is a smooth curve.

THREE-DIMENSIONAL SHOCK SPECTRUM.[8] In general, the response of a structure to a shock is oscillatory and continues for an appreciable number of oscillations. At each oscillation, the response has an interim maximum value that differs, in general, from the preceding or following maximum value. For example, a typical time-history of response of a simple system of given natural frequency is shown in Fig. 23.6; the characteristics of the response may be summarized by the block diagram of Fig. 23.10. The abscissa of Fig. 23.10 is the peak response at the respective cycles of the oscillation, and the ordinate is the number of cycles at which the peak response exceeds the indicated value. Thus, the time-history of Fig. 23.6 has 29 cycles of oscillation at which the peak response of the oscillation exceeds $0.6r_0$, but only 2 cycles at which the peak response exceeds $2.0r_0$.

In accordance with the concept of the shock spectrum, the natural frequency of the responding system is modified by discrete increments and the response determined at each increment. This leads to a number of time-histories of response corresponding to

Fig. 23.6, one for each natural frequency, and a similar number of block diagrams corresponding to Fig. 23.10. This group of block diagrams can be assembled to form a surface that shows pictorially the characteristics of the shock in terms of the response of a simple system. The axes of the surface are peak response, natural frequency of the responding system, and number of response cycles exceeding a given peak value. The block diagram of Fig. 23.10 is arranged on this set of axes at *A*, as shown in Fig. 23.11, at the appropriate position along the natural frequency axis. Other corresponding block diagrams are shown at *B*. The three-dimensional shock spectrum is the surface faired through the ends of the bars; the intercept of this surface with the planes of the block diagrams is indicated at *C* and with the maximum response–natural frequency plane at *D*. Surfaces are obtainable for both positive and negative values of the response, and a separate surface is obtained for each fraction of critical damping in the responding system.

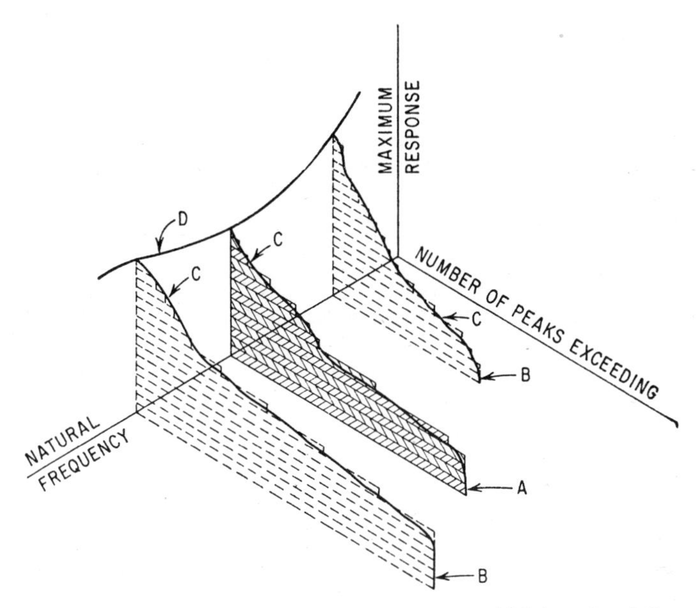

Fig. 23.11. Example of a three-dimensional shock spectrum, which is a plot of the maximum value of individual response oscillations as a function of the number of occurrences of these maxima and the natural frequency of the responding system. Bar charts are shown properly disposed with respect to the coordinate planes.

The two-dimensional shock spectrum is a special case of the three-dimensional surface. The former is a plot of the maximum response as a function of the natural frequency of the responding system; hence, it is a projection on the plane of the response and natural frequency axes of the maximum height of the surface. However, the height of the surface never exceeds that at one response cycle. Thus, the two-dimensional shock spectrum is the intercept of the surface with a plane normal to the "number of peaks exceeding" axis at the origin.

The response surface is a useful concept and illustrates a physical condition; however, it is not well adapted to quantitative analysis because the distances from the surface to the coordinate planes cannot be determined readily. A group of bar charts, each corresponding to Fig. 23.10, is more useful for quantitative purposes. The differences in lengths of the bars are discrete increments; this corresponds to the data reduction method in which the axis of response magnitudes is divided into discrete increments for purposes of counting the number of peaks exceeding each magnitude. In concept, the width of the increment may be considered to approach zero and the line faired through the ends of the bars represents the smooth intercept with the surface.

RELATIONSHIP BETWEEN SHOCK SPECTRUM AND FOURIER SPECTRUM. Although the shock spectrum and the Fourier spectrum are fundamentally different, there is a partial correlation between them. A direct relationship exists between a running Fourier spectrum, to be defined subsequently, and the response of an undamped simple structure. A consequence is a simple relationship between the Fourier spectrum of absolute values and the peak residual response of an undamped simple structure.

For the case of zero damping, Eq. (23.33) provides the relative displacement response

$$\delta(\omega_n,t) = \frac{1}{\omega_n} \int_0^t \ddot{u}(t_v) \sin \omega_n(t - t_v)\, dt_v \tag{23.49}$$

A form better suited to our needs here is

$$\delta(\omega_n,t) = \frac{1}{\omega_n} \mathcal{I} \left[e^{j\omega_n t} \int_0^t \ddot{u}(t_v)\, e^{-j\omega_n t_v}\, dt_v \right] \tag{23.50}$$

The integral above is seen to be the Fourier spectrum of the portion of $\ddot{u}(t)$ which lies in the time interval from zero to t, evaluated at the natural frequency ω_n. Such a time-dependent spectrum can be termed a "running Fourier spectrum" and denoted by $\mathbf{F}(\omega,t)$:

$$\mathbf{F}(\omega,t) = \int_0^t \ddot{u}(t_v) e^{-j\omega t_v}\, dt_v \tag{23.51}$$

It is assumed that the excitation vanishes for $t < 0$. The integral in Eq. (23.50) can be replaced by $\mathbf{F}(\omega_n,t)$; and after taking the imaginary part

$$\delta(\omega_n,t) = \frac{1}{\omega_n} F(\omega_n,t) \sin [\omega_n t + \theta(\omega_n,t)] \tag{23.52}$$

where $F(\omega_n,t)$ and $\theta(\omega_n,t)$ are the magnitude and phase of the running Fourier spectrum, corresponding to the definitions in Eqs. (23.63) and (23.64). Equation (23.52) provides the previously mentioned direct relationship between undamped structural response and the running Fourier spectrum.

When the running time t exceeds τ, the duration of $\ddot{u}(t)$, the running Fourier spectrum becomes the usual spectrum as given by Eq. (23.57), with τ used in place of the infinite upper limit of the integral. Consequently, Eq. (23.52) yields the sinusoidal residual relative displacement for $t > \tau$:

$$\delta_r(\omega_n,t) = \frac{1}{\omega_n} F(\omega_n) \sin [\omega_n t + \theta(\omega_n)] \tag{23.53}$$

The amplitude of this residual deflection and the corresponding equivalent static acceleration are

$$\begin{aligned} (\delta_r)_{\max} &= \frac{1}{\omega_n} F(\omega_n) \\ (A_{\text{eq}})_r &= \frac{\omega_n{}^2(\delta_r)_{\max}}{g} = \frac{\omega_n}{g} F(\omega_n) \end{aligned} \tag{23.54}$$

This result is clearly evident for the Fourier spectrum and undamped shock spectrum of the acceleration impulse. The Fourier spectrum is the horizontal line (independent of frequency) shown in Fig. 23.3*A* and the shock spectrum is the inclined straight line (increasing linearly with frequency) shown in Fig. 23.7*A*. Since the impulse exists only at $t = 0$, the entire response is residual. The undamped shock spectra in the impulsive region of the half-sine pulse and the decaying sinusoidal acceleration, Fig. 23.7*C* and *D*, respectively, also are related to the Fourier spectra of these shocks, Fig. 23.3*C* and *D*, in a similar manner. This results from the fact that the maximum response occurs in the residual motion for systems with small natural frequencies. Another example is the entire negative shock spectrum with no damping for the half-sine pulse in Fig. 23.7*C* whose values are ω_n/g times the values of the Fourier spectrum in Fig. 23.3*C*.

METHODS OF DATA REDUCTION

Even though preceding sections of this chapter include several analytic functions as examples of typical shocks, data reduction in general is applied to measurements of shock that are not definable by analytic functions. The following sections outline data reduction methods that are adapted for use with any general type of function. Standard forms for presenting the analysis results are given in Ref. 9.

FOURIER SPECTRUM

The Fourier spectrum can be evaluated by either analog or digital techniques.[4] Block diagrams of two basic approaches to analog computation are shown in Fig. 23.12. Such

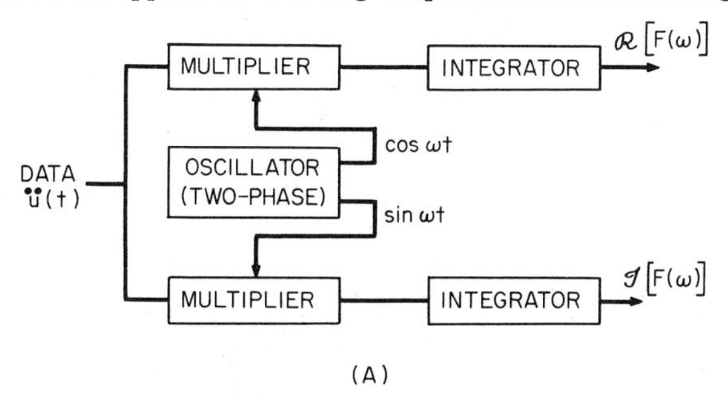

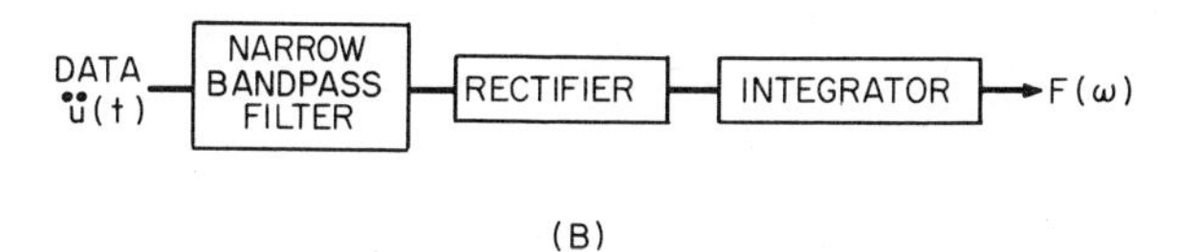

FIG. 23.12. Block diagrams for analog computation of Fourier spectra. (*A*) Direct implementation of Eqs. (23.60) and (23.61) and (*B*) bandpass filtering to determine spectrum magnitude.[4]

analyses can be performed in either a *parallel* or a *swept* fashion. In a parallel analysis, an array of oscillators or filters is implemented to determine all the spectral values of interest simultaneously. In a swept analysis, the computation is performed for one frequency at a time; the frequency is stepped or continuously varied slowly to determine the spectral values of interest in a serial fashion. For purposes of the data analysis the shock time-history can be made artificially periodic by repeating the shock at intervals of time τ. The Fourier spectrum then becomes a Fourier series involving discrete components at integer multiples of the fundamental frequency $1/\tau$.* The magnitude of the true continuous Fourier spectrum at the frequency of such a discrete component is

$$F\left(\frac{2\pi n}{\tau}\right) = \left(\frac{\tau}{2}\right) c_n \tag{23.55}$$

where c_n is the amplitude of the nth discrete component at the frequency $\omega = 2\pi n/\tau$, as defined by Eq. (22.8).

* See Chap. 22 for the defining relationships of a Fourier series and a discussion of data-reduction techniques.

Digital evaluation of the Fourier spectrum is most efficiently accomplished by means of the fast Fourier transform (FFT). Discussion of this approach appears in Chap. 27.

SHOCK SPECTRUM

Both analog and digital computation techniques can be employed for evaluation of the shock spectrum.[4] Block diagrams for analog computation are shown in Fig. 23.13. As with the Fourier spectrum, analyses can be performed in either a *parallel* or a *stepped* manner. In a parallel analysis, the shock-spectrum values are obtained from simultaneous determinations of the responses of simple structures corresponding to all natural frequencies of interest. A stepped analysis involves the determination of shock-spectrum results for one natural frequency and then a repeat of the determination for each successive value of the natural frequency.

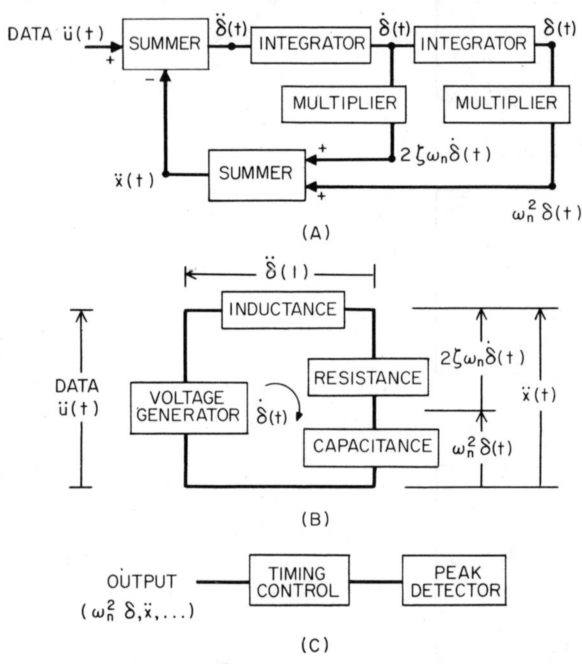

FIG. 23.13. Block diagrams for analog computation of shock spectra using the simple system shown in Fig. 23.5. (*A*) Active electronic analog, (*B*) passive electrical analog, (*C*) peak detection (positive, negative, or absolute) with timing control as required for primary, residual, or overall analysis.

Digital evaluation of the shock spectrum involves the use of a digital computer to determine the response of simple structures, followed by the detection of peak response. The response time-histories can be evaluated by the following techniques: (1) direct numerical or recursive integration of the Duhamel integral, Eq. (23.33), or (2) convolution or recursive filtering procedures. These techniques are discussed in detail in Ref. 4.

In principle, both analog and digital techniques can be modified to determine the three-dimensional shock spectrum. The modification involves the counting of the number of response maxima above various discrete increments of maximum response to obtain the results depicted in Fig. 23.11.

Reed Gage. The shock spectrum may be measured directly by a mechanical instrument that responds to shock in a manner analogous to the data reduction techniques used to obtain shock spectra from time-histories. The instrument includes a number of flexible mechanical systems that are considered to respond as single degree-of-freedom systems; each system has a different natural frequency and means are provided to indicate the maximum deflection of each system as a result of the shock. The instrument often is referred to as a *reed gage* because the flexible mechanical systems are small cantilever beams carrying end masses; these have the appearance of reeds.[10] The reed gage is described in Chap. 13.

The response parameter indicated by the reed gage is maximum deflection of the reeds relative to the base of the instrument; generally, this deflection is converted to equivalent static acceleration by applying the relation of Eq. (23.30). The reed gage offers a convenience in the indication of a useful quantity immediately and in the elimination of auxiliary electronic equipment. Also, it has important limitations: (1) the information is limited to the determination of a shock spectrum; (2) the deflection of a reed is inversely proportional to its natural frequency squared, thereby requiring high equivalent static accelerations to achieve readable records at high natural frequencies; (3) the means to indicate maximum deflection of the reeds (styli inscribing on a target surface) tend to introduce an undefined degree of damping; and (4) size and weight limitations on the reed gage for a particular application often limit the number of reeds which can be used and the lowest natural frequency for a reed. In spite of these limitations, the instrument sees continued use and has provided significant shock spectra where more elaborate instruments have failed.

APPENDIX 23.1

FOURIER INTEGRAL

Let the time-history of any motion or loading parameter (time-history of a shock) be designated $f(t)$. The corresponding Fourier spectrum or frequency spectrum $\mathbf{F}(\omega)$ is*

$$\mathbf{F}(\omega) = \int_{-\infty}^{\infty} f(t)e^{-j\omega t}\,dt \tag{23.56}$$

The Fourier spectrum $\mathbf{F}(\omega)$ is a complex number used to describe the amplitude and phase of a sinusoid at the frequency ω. The complex amplitude of the Fourier component at the frequency ω is $\mathbf{F}(\omega)\,d\omega$, the area under the Fourier spectrum curve in the interval $d\omega$. The quantity $\mathbf{F}(\omega)e^{j\omega t}\,d\omega$ is the sinusoidal variation at the frequency ω; the integral sums these sinusoids. If it is assumed that the shock starts at $t = 0$, Eq. (23.56) becomes

$$\mathbf{F}(\omega) = \int_{\infty}^{0} f(t)e^{-j\omega t}\,dt \tag{23.57}$$

If the duration of the time-history $f(t)$ has a finite limit, this limit is used in place of ∞ in Eq. (23.57). An alternate notation for $\mathbf{F}(\omega)$ is $\mathbf{F}[f(t)]$, which denotes the "Fourier transform of $f(t)$."

When $\omega = 0$, Eq. (23.56) becomes

$$\mathbf{F}(0) = \int_{-\infty}^{\infty} f(t)\,dt \tag{23.58}$$

Thus, the zero frequency component of the Fourier spectrum of a shock is equal to the area under the time-history of the shock.

Since $\mathbf{F}(\omega)$ represents a distribution of amplitudes expressed in complex form, it may be written in terms of the real and imaginary parts:

$$\mathbf{F}(\omega) = \mathcal{R}[\mathbf{F}(\omega)] + j\mathcal{I}[\mathbf{F}(\omega)] \tag{23.59}$$

where

$$\mathcal{R}[\mathbf{F}(\omega)] = \int_{0}^{\infty} f(t)\cos\omega t\,dt \tag{23.60}$$

$$\mathcal{I}[\mathbf{F}(\omega)] = -\int_{0}^{\infty} f(t)\sin\omega t\,dt \tag{23.61}$$

* Mathematical limitations on the function which permit this type of analysis do exist but these do not hinder engineering applications. See Refs. 11 to 13 for discussions of these limitations.

Alternatively, the Fourier spectrum may be defined in terms of its absolute value and phase angle:

$$\mathbf{F}(\omega) = F(\omega)e^{j\theta(\omega)} \tag{23.62}$$

where $F(\omega)$ is the absolute value of $\mathbf{F}(\omega)$ and $\theta(\omega)$ is the phase angle. The absolute value and phase angle are related to the real and imaginary parts by

$$F(\omega) = \sqrt{\mathcal{R}^2[\mathbf{F}(\omega)] + \mathcal{I}^2[\mathbf{F}(\omega)]} \tag{23.63}$$

$$\theta(\omega) = \tan^{-1}\left\{\frac{\mathcal{I}[\mathbf{F}(\omega)]}{\mathcal{R}[\mathbf{F}(\omega)]}\right\} \tag{23.64}$$

Figure 23.14 shows the vector $\mathbf{F}(\omega)$, its magnitude $F(\omega)$ and phase angle $\theta(\omega)$, and its real and imaginary parts.

Equation (23.56) may be used with any shock parameter, such as acceleration, velocity, displacement, force, or pressure. The area under the curve of Fourier spectrum magnitude versus frequency has the same units as the time-history of the shock parameter.

The Fourier spectrum of the time-derivative of $f(t)$ is the product of $j\omega$ and the Fourier spectrum of $f(t)$:

$$\mathbf{F}\left[\frac{df(t)}{dt}\right] = j\omega\mathbf{F}[f(t)] \tag{23.65}$$

Thus if $\ddot{u}(t)$, $\dot{u}(t)$, $u(t)$ are the acceleration, velocity, and displacement time-histories, respectively, of a given shock motion, their Fourier spectra are related as follows:

$$\mathbf{F}[\ddot{u}(t)] = j\omega\mathbf{F}[\dot{u}(t)] = -\omega^2\mathbf{F}[u(t)] \tag{23.66}$$

$$\mathbf{F}[\dot{u}(t)] = j\omega\mathbf{F}[u(t)] \tag{23.67}$$

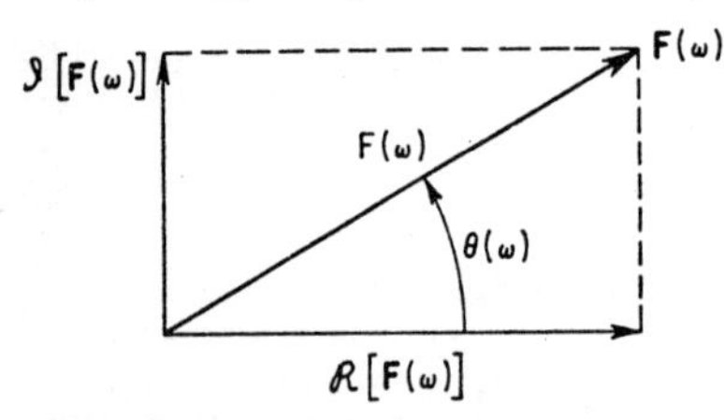

FIG. 23.14. Representation of the complex Fourier component at the frequency ω, its magnitude $F(\omega)$ and phase angle $\theta(\omega)$, and its real and imaginary parts, $\mathcal{R}[\mathbf{F}(\omega)]$ and $\mathcal{I}[\mathbf{F}(\omega)]$.

The time-history of a function can be found by integrating its Fourier spectrum components:

$$f(t) = \frac{1}{2\pi}\int_{-\infty}^{\infty} \mathbf{F}(\omega)e^{j\omega t}\,d\omega \tag{23.68}$$

Equations (23.68) and (23.56) are known as the inverse and direct Fourier transforms, respectively. The use of the factor $1/2\pi$ in Eq. (23.68) represents one formulation of the pair of Eqs. (23.56) and (23.68). In other formulations, the $1/2\pi$ appears in Eq. (23.56) and not in Eq. (23.68), or a $1/\sqrt{2\pi}$ appears in each. Caution should be exercised, when consulting treatises and tables of the Fourier transform, in determining which formulation of the transform is used. A useful compilation of direct and inverse Fourier transform pairs is given in Ref. 14.

The Fourier spectrum of a constant is a delta function (infinitely sharp spike) at zero frequency; hence, the addition of a constant to a function only alters the Fourier spectrum of that function at $\omega = 0$. Consequently, the relations for the spectra of acceleration, velocity, and displacement, Eqs. (23.66) and (23.67), are valid for all $\omega \neq 0$. This limitation is of little significance for engineering applications because the addition of a constant to a displacement time-history only changes the origin of the coordinate system. Similarly, the addition of a constant to a velocity time-history introduces another frame of reference which does not accelerate.

Equation (23.68) requires integration over negative frequencies. For engineering purposes, negative frequencies lack physical significance and a more desirable form of Eq. (23.68) is

$$\begin{aligned} f(t) &= \frac{2}{\pi}\int_0^{\infty} \mathcal{R}[\mathbf{F}(\omega)] \cos \omega t \, d\omega \qquad [t > 0] \\ f(t) &= 0 \qquad [t < 0] \end{aligned} \tag{23.69}$$

Equation (23.69) is derived from (23.68) by including the restriction that $f(t) = 0$ for $t < 0$. For this to be true, $\int_0^\infty \mathcal{R}[\mathbf{F}(\omega)] \cos \omega t \, d\omega$ must equal $-\int_0^\infty \mathcal{I}[\mathbf{F}(\omega)] \sin \omega t \, d\omega$.

COMPARISON OF FOURIER AND LAPLACE TRANSFORMS*

The Laplace transform of $f(t)$ is defined, as a function of the complex variable p, by the integral

$$\mathbf{L}[f(t)] = \int_0^\infty f(t)e^{-pt} \, dt \tag{23.70}$$

where $p = a + j\omega$ for $a > 0$ and $\mathbf{L}[f(t)]$ denotes the "Laplace transform of $f(t)$." Equation (23.70) sometimes is written $\mathbf{L}[f(t)] = p \int_0^\infty f(t)e^{-pt} \, dt$. Caution should be exercised in determining the formulation of the transform. A comparison of Eqs. (23.70) and (23.57) reveals that the Laplace transform of a function approaches the Fourier transform of that function as $a \to 0$ or $p \to j\omega$, if the function is zero for all negative time.[12] The time-history is retrieved by the use of the inverse Laplace transform, involving integration in the complex plane:

$$f(t) = \frac{1}{2\pi j} \int_{a-j\infty}^{a+j\infty} \mathbf{L}[f(t)]e^{pt} \, dp \tag{23.71}$$

where a is chosen so that $f(t) = 0$ for $t < 0$. If $\mathbf{L}[f(t)]$ approaches $\mathbf{F}(\omega)$ as $p \to j\omega$ for *all* values of the frequency ω, then Eq. (23.71) provides a result identical to that of Eq. (23.68).

For problems associated with physically achievable shocks, the Fourier and Laplace transforms give identical results. The Fourier transform has certain mathematical disadvantages which the Laplace transform overcomes, and more extensive tables of the Laplace transform are available.[14] A detailed discussion of both transform methods and their limitations is found in Chap. 1 of Ref. 13.

A disadvantage of the Fourier transform method is that the integral defining the transform sometimes does not converge unless an extra convergence factor is provided. For example, the acceleration step in Fig. 23.2*B* has no Fourier transform in the strict sense since the integrand of Eq. (23.56) is $e^{-j\omega t}$ which does not have a well-defined limit as $t \to \infty$. The Fourier transform can be made useful in such instances by introducing a convergence factor e^{-at} and modifying the function $f(t)$ as follows:

$$f(t,a) = e^{-at}f(t) \tag{23.72}$$

The Fourier spectrum is

$$\mathbf{F}(\omega - ja) = \int_{-\infty}^\infty f(t)e^{-j(\omega-ja)t} \, dt \tag{23.73}$$

If the limit exists as $a \to 0$ of $\mathbf{F}(\omega - ja)$ $\mathbf{F}(\omega)$ can be called the Fourier transform of $f(t)$. The Laplace transform incorporates an appropriate convergence factor in its definition, thus guaranteeing the existence of the transform for practically all functions of possible interest in engineering applications.

Another disadvantage of the Fourier transform is the difficulty in obtaining the inverse transform; i.e., the application of Eq. (23.68). The Laplace transform achieves the inversion by integration in the complex p plane so that all the background of complex-function theory is available.

On the basis of these factors it may appear that the Laplace transform is the broader and more basic concept, and that the Fourier transform represents a special case of the Laplace transform. However, for purposes of physical interpretation of actual measured data, the Fourier spectrum has a greater intuitive appeal as a logical extension of the Fourier series concepts for periodic functions.

* The Laplace transform is discussed in detail in Chap. 8.

APPLICATIONS OF FOURIER INTEGRAL

DETERMINATION OF STRUCTURAL RESPONSE

If a shock loading parameter whose Fourier spectrum $\mathbf{F}(\omega)$ is defined by Eq. (23.56) acts upon a structure, the frequency content of the structural response is found from

$$\mathbf{R}(\omega) = \mathbf{H}(\omega)\mathbf{F}(\omega) \tag{23.74}$$

where $\mathbf{R}(\omega)$ is the Fourier spectrum of the response and $\mathbf{H}(\omega)$ is the applicable transfer function. The transfer function is the steady-state response per unit sinusoidal input as a function of the frequency ω. If $\mathbf{R}(\omega)$ represents a velocity response and $\mathbf{F}(\omega)$ represents a force input, $\mathbf{H}(\omega)$ is given the special designation of mobility $\mathfrak{M}(\omega)$; the reciprocal of $\mathbf{H}(\omega)$ is the mechanical impedance $\mathbf{Z}(\omega)$. [*Mobility* is sometimes called *mechanical admittance.*] Mechanical impedance and mobility are discussed in detail in Chap. 10. Equation (23.74) is applicable when the velocity and displacement of the structure are zero at $t = 0$. If the initial conditions are not zero, additional terms are required in Eq. (23.74). These terms can be obtained from the governing differential equation. For example, if the differential equation has the form

$$a\ddot{r} + b\dot{r} + cr = f(t) \tag{23.75}$$

the corresponding equation in the frequency domain is found by applying the Fourier transform to both the response $r(t)$ and the excitation $f(t)$[11, 13]

$$a[-\omega^2\mathbf{R}(\omega) - j\omega r(0+) - \dot{r}(0+)] + b[j\omega\mathbf{R}(\omega) - r(0+)] + c[\mathbf{R}(\omega)] = \mathbf{F}(\omega)$$

Solving for $\mathbf{R}(\omega)$,

$$\mathbf{R}(\omega) = \frac{\mathbf{F}(\omega) + a\dot{r}(0+) + [b + j\omega a]r(0+)}{-\omega^2 a + j\omega b + c}$$

The quantities $r(0+)$ and $\dot{r}(0+)$ are the values of the response and its derivative at $t = 0+$, the plus sign allowing for any discontinuity at $t = 0$. If $r(0+)$ and $\dot{r}(0+)$ are zero, the transfer function $\mathbf{H}(\omega)$ is found from Eq. (23.74):

$$\mathbf{H}(\omega) = \frac{1}{-\omega^2 a + j\omega b + c} \tag{23.76}$$

The response in the time domain may be found by the use of Eq. (23.69):

$$r(t) = \frac{2}{\pi}\int_0^\infty \mathfrak{R}[\mathbf{R}(\omega)] \cos \omega t \, d\omega \qquad [t > 0]$$

Substituting for $\mathbf{R}(\omega)$ from Eq. (23.74),

$$r(t) = \frac{2}{\pi}\int_0^\infty \mathfrak{R}[\mathbf{H}(\omega)\mathbf{F}(\omega)] \cos \omega t \, d\omega \qquad [t > 0] \tag{23.77}$$

Example 23.1. Consider the response of the single degree-of-freedom structure in Fig. 23.5 to an input acceleration shock $\ddot{u}(t)$. The transfer function for the relative displacement response per unit input sinusoidal acceleration is

$$\mathbf{H}(\omega) = \frac{\bar{\delta}e^{j\omega t}}{e\ddot{\mathbf{U}}^{j\omega t}} = \frac{1}{(\omega_n{}^2 - \omega^2) + j(2\zeta\omega\omega_n)}$$

where $\bar{\delta}$ and $\ddot{\mathbf{U}}$ are, respectively, complex amplitudes of relative displacement response and input sinusoidal acceleration at the frequency ω. From Eq. (23.77), the time-history of the response is

$$\delta(t) = \frac{2}{\pi}\int_0^\infty \mathfrak{R}\, \frac{\mathbf{F}(\omega)}{(\omega_n{}^2 - \omega^2) + j(2\zeta\omega\omega_n)} \cos \omega t \, d\omega \tag{23.78}$$

where $\mathbf{F}(\omega)$ is the Fourier spectrum of $\ddot{u}(t)$.

CALCULATION OF THE EFFECT OF STRUCTURAL LOAD CHANGE ON A SHOCK SOURCE

An important application of the Fourier spectrum representation of a shock concerns the transmission of a shock from one structure to another structure. The combination of the structure from which the shock emanates and the process by which the shock is generated may be termed a mechanical shock source.[15] The other structure can be considered a load on this source; any change in the character of the load modifies the shock output of the source.

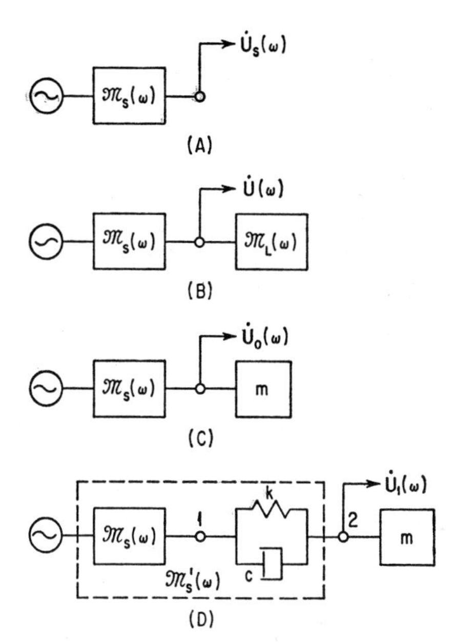

FIG. 23.15. Schematic representation of a mechanical source composed of a generator and an internal mobility.[16] The source is shown driving (*A*) no load, (*B*) a load with input mobility $\mathfrak{M}_L(\omega)$, (*C*) a mass load, and (*D*) a resiliently mounted mass.

A schematic representation of a simple mechanical source that is free at its output side is shown in Fig. 23.15*A*. The mobility of the source determined at its output junction is designated $\mathfrak{M}_s(\omega)$. The Fourier spectrum of the shock velocity delivered by the unloaded or free source is $\dot{\mathbf{U}}_s(\omega)$.

A source driving a load having an input mobility $\mathfrak{M}_L(\omega)$ is shown in Fig. 23.15*B*. The output of the source in this condition is

$$\dot{\mathbf{U}}(\omega) = \frac{1}{1 + [\mathfrak{M}_s(\omega)/\mathfrak{M}_L(\omega)]}\dot{\mathbf{U}}_s(\omega) \tag{23.79}$$

as $\mathfrak{M}_s/\mathfrak{M}_L$ approaches zero; i.e., when the impedance is large compared to the load impedance, the actual output shock approaches the free output of the source.

Example 23.2. Consider the effect of interposing a resilient mounting between a rigid mass and a source that drives the mass.[16] Figure 23.15*C* represents the direct mounting of a mass m to the source; the mobility of the mass m is $1/j\omega m$ (Chap. 10). Thus, the shock input to the mass is found by substituting in Eq. (23.79):

$$\dot{\mathbf{U}}_0(\omega) = \frac{1}{1 + j\omega m \mathfrak{M}_s(\omega)} \dot{\mathbf{U}}_s(\omega) \qquad (23.80)$$

Figure 23.15*D* shows the resiliently mounted mass, the resilient mount being represented by a stiffness k and a viscous damping coefficient c. Since the shock input to the mass is of interest, the output junction of the source can be considered to be moved from point 1 to point 2; i.e., the resilient mount becomes part of the source. The modified source mobility is

$$\mathfrak{M}_s'(\omega) = \mathfrak{M}_s(\omega) + \frac{1}{c + (k/j\omega)}$$

The output of the free source $\dot{\mathbf{U}}_s(\omega)$ remains unchanged, and the load mobility remains $1/j\omega m$. The shock input to the mass is

$$\dot{\mathbf{U}}_1(\omega) = \frac{1}{1 + j\omega m \mathfrak{M}_s'(\omega)} \dot{\mathbf{U}}_s(\omega) \qquad (23.81)$$

MEASUREMENT OF STRUCTURAL TRANSFER FUNCTION

The most straightforward technique for the measurement of a structural transfer function involves the application of a sinusoidal input motion and the measurement of the resulting sinusoidal output motion. The ratio of output to input as a function of frequency is, by definition, the transfer function. It may be desirable to determine the transfer function by applying an input shock and recording the output shock.

Given a structure which is initially at rest, a known shock input $f(t)$ is applied, producing a shock response $r(t)$. The direct application of Eq. (23.74) gives

$$\mathbf{H}(\omega) = \frac{\mathbf{F}(\omega)}{\mathbf{R}(\omega)} \qquad (23.82)$$

where $\mathbf{F}(\omega)$ and $\mathbf{R}(\omega)$ are the Fourier spectra of $f(t)$ and $r(t)$, respectively. Accurate values of $\mathbf{H}(\omega)$ are obtained only in those frequency ranges where $f(t)$ has a significant frequency content. If the absolute value of $\mathbf{F}(\omega)$ is zero at certain frequencies, the value of $\mathbf{H}(\omega)$ is indeterminate at those frequencies.

REFERENCES

1. von Kármán, T., and M. A. Biot: "Mathematical Methods in Engineering," p. 388, McGraw-Hill Book Company, Inc., New York, 1940.
2. Rubin, S.: *J. Appl. Mechanics*, **25**:501 (1958).
3. Fung, Y. C., and M. V. Barton: *J. Appl. Mechanics*, **25**:365 (1958).
4. Kelly, R. D., and G. Richman: "Principles and Techniques of Shock Data Analysis," Shock and Vibration Information Center, Washington, D.C., 1969.
5. Walsh, J. P., and R. E. Blake: *Proc. Soc. Exptl. Stress Anal.*, **6**(2):150 (1948).
6. Timoshenko, S., and D. H. Young: "Advanced Dynamics," p. 49, McGraw-Hill Book Company, Inc., New York, 1948.
7. Caughey, T. K., and D. E. Hudson: *Proc. Soc. Exptl. Stress Anal.*, **13**(1):199 (1956).
8. Lunney, E. J., and C. E. Crede, *WADC Tech. Rept.* 57-75, 1958
9. American National Standards Institute: Methods for Analysis and Presentation of Shock and Vibration Data, ANSI S2.10-1971.
10. Rubin, S.: *Proc. Soc. Exptl. Stress Anal.*, **16**(2):97 (1956).
11. Pipes, L. A.: "Applied Mathematics for Engineers and Physicists," 2d ed., McGraw-Hill Book Company, Inc., New York, 1958.
12. Campbell, G. A., and R. M. Foster: "Fourier Series for Practical Applications," D. Van Nostrand Company, Inc., Princeton, N.J., 1940. Previously published as No. B-584 of the Bell System Series of Monographs, 1931.
13. Weber, E.: "Linear Transient Analysis," vol. II, chap. 1, John Wiley & Sons, Inc., New York, 1956.
14. Erdelyi, A.: "Tables of Integral Transforms," vol. I, McGraw-Hill Book Company, Inc., New York, 1954.
15. Molloy, C. T.: *J. Acoust. Soc. Amer.*, **29**:842 (1957).
16. Rubin, S.: *SAE Trans.*, **68**:318 (1960).

24

ENVIRONMENTAL SPECIFICATIONS AND TESTING

Charles T. Morrow
Advanced Technology Center, Inc.

Environmental specifications are intended to provide incentives for products to be developed in such a way that they will survive transportation storage and use, and operate satisfactorily. The specifications lead first to developmental tests that can guide the designers in revealing marginal features that might require redesign. They also lead to qualification tests and to more formal requirements, that provide evidence that the developmental model or product is satisfactory or that it should not be accepted without further development. As indicated in Chap. 21, specifications are commonly negotiated and established before there is any extensive knowledge of the actual environments.

All environmental specifications exhibit an element of arbitrariness because of a need to simplify legal and managerial considerations and test procedures.[1] For example, consider a relatively simple deviation from the laboratory environment in terms of temperature. Storage at any one ambient temperature can be simulated exactly. The operation of equipment at any one ambient temperature for a sufficient length of time to establish equilibrium can likewise be simulated exactly. There may be temperature gradients within the product because of its operation, but they will remain in equilibrium and will not change unless the mode of operation is changed. But it is not economical to simulate all ambient temperatures. Therefore steady-temperature testing is usually limited to room temperature and upper and lower extremes. Likewise temperature cycling tests must be restricted to a small number of the possibilities even though realism of effects may be impaired.

Similarly a salt-spray test of practical duration may be an overtest for the housing of a product and an undertest for its interior parts.

Specifications generally serve their purposes without requiring perfect simulations of the actual environments. From this point on, this chapter will be concerned only with mechanical vibration and shock and the fluid-pressure variations that may induce them. These are the most complicated of the environments, because of the importance of their spectral characteristics, but they must still be dealt with in simple terms. Testing machines are discussed at length in Chaps. 25 and 26.

OBJECTIVES OF A TEST

After a nonoperational environmental test, the test item will be inspected for damage and its operation will be checked. During an operational environmental test, the test item should be monitored for any symptoms of malfunction or damage. It is not to be presumed that the design or test engineers have comprehensive knowledge of interior dynamics before the test. However, once a malfunction or failure has been identified, limited dynamical investigation may be an important precursor to redesign. Diagnostic testing may become important.

In the specification of a test environment and execution of the test, there are two possible objectives:

1. Designer motivation
2. Simulation of the actual environment

The first objective is the more important, but more conscious and explicit attention is commonly given to the second, even when that objective may be unattainable. It is sometimes possible to achieve the first objective while paying little attention to the second.

Consider a vibration environment that is confined to extremely low frequencies, as are many shipping environments and land and ship mobile environments. Assume that the product under test will not be mechanically isolated. One strategy that is very

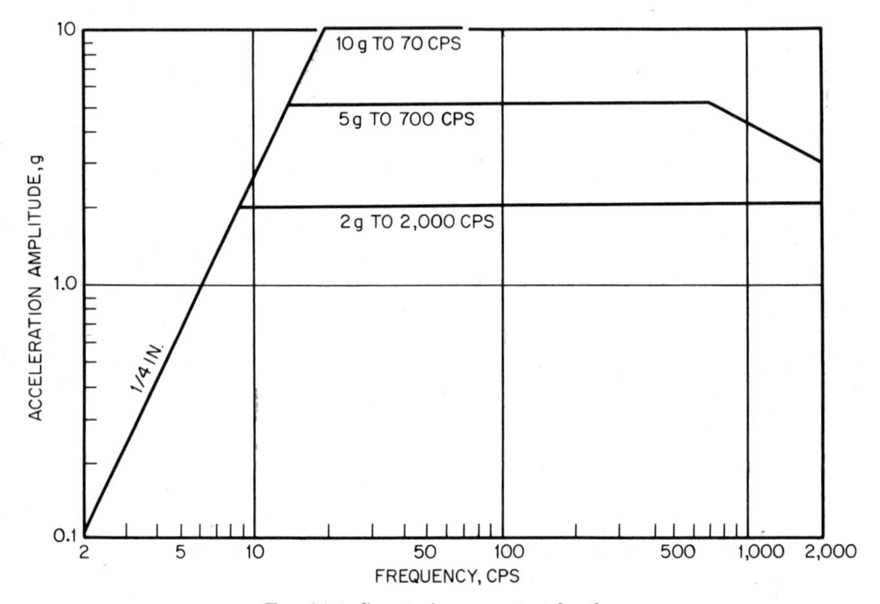

FIG. 24.1. Swept sine-wave test levels.

effective in ensuring its survival and operation is to raise the frequencies of all its mechanical resonances above the highest frequency of real importance in the actual environment. If this is feasible but is not explicitly incorporated into the design approach, ensurance that it will be may be provided by specifying a low-frequency vibration test, as in the upper curve of Fig. 24.1, with a suitable definite high-frequency limit. The test level at these low frequencies need not have any definite relation to the actual level. It must merely be sufficiently severe to result in failure at any resonance in the forbidden range, or at least to cause clattering, without requiring, in the absence of resonances, a design to a static stress so high as to be impractical.

ENVIRONMENT VERSUS MACHINE

In specifications, an attempt is usually made to prescribe a suitable environment while placing little restriction on the apparatus to be used in applying the environment to the product. For shipboard shock, an almost opposite approach is taken. A machine is specified which contains a plate for mounting the test item and a pivoted hammer that strikes the plate. Within the limitations imposed by its size, the mounting plate is intended to be representative of shipboard structure. Reduction of the environmental level by dynamic loading of the plate by the test item tends to be more realistic than are tests for which the environment is specified. Overtest may be less likely to occur.

An extension of this approach is the Navy barge test for simulating the "near-miss environment" of a submarine or surface ship. Explosives in the water are set off progressively closer to the barge, which is a low-cost substitute for the ship or submarine. Specification of the test environment would, in principle, permit more direct control for purposes of realism and would induce less demand for realism. But for very large test items, it is difficult to control the environment precisely according to any simple prescription; there appears to be little practical alternative to specifying a machine rather than an environment.

IN-USE SPECIFICATIONS

Many vibration and shock specifications relate to conditions at the locations of use in that they provide design and development incentives for survival and operation of the product in such conditions. Usually these specifications attempt to require the environmental tests to simulate such conditions. The specifications may be divided into at least three types: (1) general purpose, (2) formula, and (3) application. The first consists of a general set of requirements for a particular category of equipment (e.g., electronic communication) without regard to where the equipment may be used in any particular case; little refinement of simulation is possible. The second type of specification is provided by the customer in the form of formulas or curves. The contractor uses these and supplementary information to determine test conditions. For example, consider missiles stored under the wing of an airplane. The test condition is influenced by the type and location of engine, location of the store on the wing, etc., according to customer instructions. The third type of specification, often delegated to the contractor subject to customer approval (particularly applicable to airborne equipment for space vehicles and guided missiles), is a complete prescription of test conditions based on location in the airframe. Such a specification is intended to be refined as data about actual environment accumulate.

LOW-FREQUENCY VIBRATION IN TEST

Typical vibration-test conditions, expressed as acceleration amplitudes for swept sine waves, are shown in Fig. 24.1. (A complete test requirement includes a prescription of sweep rate and total test time, but these matters are not important for the immediate discussion.) The low-frequency slanted portion of the curves represents, not necessarily an anticipated actual environmental condition, but a round number for displacement amplitude not greater than the permissible displacement of available shakers. For a random vibration test, the power spectral density at low frequencies is similarly contoured so that the root-mean-square displacement is less than three times the permissible displacement. To protect the shaker armature and the test item from bottoming in response to occasional large-signal amplitudes, an electronic limiter is employed in the test system.

If the test item is not to be mounted on an isolator for vibration or shock, the low-frequency portion of the test-vibration curve is a satisfactory design motivation without further investigation or care in simulation. Its practical effect is to motivate the designer to keep resonances out of that frequency region. With an excitation amplitude of $\frac{1}{4}$ in., the relative motion at any resonance in this frequency range within the test item would be at least $\frac{3}{4}$ in. Neither the resultant clattering nor the extra space to prevent it can be tolerated.

On the other hand, if the test is to have an influence on isolator design or on government packaging for shipment, engineering decision making often becomes very difficult. The isolator or container resonance frequency eventually chosen is commonly 10 to 15 cps. The specified amplitude serves neither as an indication of the actual low-frequency environment nor as a direct measure of the low-frequency fragility. Decision making can be facilitated by estimating the actual low-frequency environment—especially the frequencies at which vibration may be most severe and the nature of the actual shocks.

It is common practice to accept any test conditions already used as indications of equipment fragility. But at low frequencies, in the absence of any resonances in this range, an equipment can stand an acceleration not only greater than that corresponding to the prescribed low-frequency displacement amplitude but many times greater than the acceleration prescribed for higher frequencies. A well-designed isolator providing ample protection without excessive sway space may require a waiver of some test requirements. Government representatives who must rule on any request for such a waiver should be given sufficient authority and have adequate background to exercise good judgment.

PACKAGING SPECIFICATIONS

At present there are three approaches to packaging requirements for equipment to be delivered to the government. One is to permit commercial packaging. This minimizes the negotiations between customer and contractor and permits quick responses to material shortages or excessive material cost increases. Another is to prescribe the container in detail, even to the point of supplying drawings. This makes the utmost use of existing designs if they are suitable for the problem at hand. The third approach, especially favored for equipment considered to be fragile, is to place only general requirements on the construction of the container but to specify drop and vibration tests and require that the transmitted environments be compared with the fragility of the contents.[2–4]

The environmental qualification test following from the last type of specification is usually a test for the container, not for the contents. Usually the shipping environment is not considered in the design of the contents. Dummy contents, instrumented, are commonly permitted in the container tests. Dummy contents present possible advantages in any event. In principle, they can permit container design to proceed concurrently with contents design and container design to be accomplished independently of the manufacturing schedule for the contents.

The prescribed drop-test height is usually dependent on the weight of the contents, in inverse relationship. The container must survive without damage and transmit no acceleration in excess of the estimated contents fragility. However, no limit is placed on the potential damage to container or contents in the event of a higher drop as a consequence of mishandling. In actual shipping and mishandling (with actual contents in the container) extent of damage tends to increase gradually with drop height. The difficulties of estimating fragility are similar to those encountered in estimating equipment to be used on isolators.

Packaging is discussed in greater detail in Chap. 41.

ENVELOPING

Usually, the spectra of an actual environment are expressed in terms of acceleration amplitude or displacement amplitude—in other words, in terms of motion rather than force. Enveloping the maxima to arrive at test conditions carries a tacit assumption that the peaks and valleys of the actual spectrum are entirely characteristic of the structure to which the equipment is mounted, that these may vary, and that the equipment should not be so designed as to be critical to the frequencies of these resonances. Sometimes it leads to impossible design criteria for shipboard equipment such as motor-generator sets. If the resonating members of the equipment are very massive, they decrease the deck motion, so that their resonance frequencies tend to coincide with the minima of the actual spectrum of deck vibration or shock. This suggests that an envelope of the minima would provide a more realistic test condition. However, this may not be adequate to ensure reliability if there are also some resonating members that are not so massive and therefore do not decrease the deck motion by dynamic loading. An alternative is to measure an actual spectrum of force at the mounting and envelope the maxima of this force spectrum.

The enveloping of spectra is one of the most common operations on reduced data to determine what test environments should be or to evaluate their realism retroactively as data accumulate. Enveloping usually leads to a test somewhat more severe than that of the actual environment, which is helpful in revealing design deficiencies without excessive test time, but the amount of severity increase is typically difficult to estimate.

The effect of spectral enveloping on test severity is influenced by numerous factors, including spectral resolution of data-reduction equipment and possible wild points in the data. For shock spectra, if obtained from responses of simulated undamped mechanical resonators, resolution is primarily a matter of the number of frequencies for which data points are computed. However, for computation of random vibration power spectra from samples of finite duration, a nonzero bandwidth must be used, because there must be a compromise between resolution and statistical significance, as explained in Chap. 21. Limited resolution results in the raising of apparent valleys and lowering of apparent peaks in the reduced data on actual environment.

Poor statistical significance tends to lower the envelope of the minima and raise the envelope of the maxima.

When there are many peaks in the actual spectra in the frequency range of interest, enveloping seems quite natural and the severity increase minimal. The operation is more questionable when there are only a few isolated peaks.

Enveloping is particularly sensitive to wild points and tends to give them undue weight. Data involving inappropriate accelerometer locations or faulty instrumentation are more likely to be too high or too low than to conform to the general trend. Before enveloping, the nature and quality of the data must be examined critically.

THE FIXTURE PROBLEM

The vibration and shock requirements for an equipment to be tested are specified in terms of their characteristics at the equipment mounting points. Neither a shaker armature nor a shock drop-tower carriage is likely to fit the equipment mounting points without a fixture or adapter. When the excitation is to be applied parallel to the plane of the mounting points, the fixture may become a major item of structure with a tendency to be flexible. It is intended to be as rigid as possible over the frequency range important to the test.

Much effort is expended in making fixtures rigid and little effort is taken in damping them. An undamped fixture resonance, which introduces a sharp peak and notch in the applied spectrum, is difficult to equalize electronically. Furthermore, it may be impossible to supply the electronic power that a specification requires at the frequencies in a deep notch in the response without overloading amplifiers. It would often be better to damp fixture resonances even at the expense of lowering them somewhat in frequency. One of the most effective methods of damping is to make a cut through a nodal region or region of relatively high stress concentration and bond the new surfaces to a sheet of dissipative plastic. If bolts are preferred to bonding, damping material must be incorporated so as to avoid metal-to-metal contact at the bolt heads. Damping can also be inserted at the interface between the fixture and shaker armature or drop-tower carriage.

EQUALIZATION

Every random vibration-test system includes an electronic equalizer to compensate for resonances in the fixture and shaker armature and to provide a spectrum in compliance with whatever is specified.[5] Early equalizers were manually adjusted peak and notch filters. These compensated phase as well as magnitude but were very difficult to adjust. Then sets of narrow-band filter sets were used with manually adjustable attenuators in the path to the output mixing circuit. Later automatically adjustable attenuators were employed with the bandpass filters. Equivalent digital circuitry is commonly used at present. The bandpass filters compensate approximately for magnitude but not for phase (which is not important for random vibration testing but has an influence on severity if the same system is to be used to apply a shock).

For large test items, there has been a tendency to use several shakers, one at each mounting point, rather than one large shaker. This makes the fixture design problem easier but makes an equalization policy problem more obvious. It raises a question of how the excitation should be apportioned among the several mounting points or if there should be as many equalizers as mounting points. With a single shaker, one does not have any practical direct control over the apportionment. With multiple shakers, as with a single shaker, there is little justification for making anything much more elaborate than an average of the accelerations or corresponding spectra conform to specification, especially as there will seldom be much data on the actual apportionment when the test item is in use.[5]

Further discussion of equalization and random vibration testing will be found in Chaps. 25 and 27.

PROBABILITY DISTRIBUTIONS

In random vibration data reduction and testing, details of probability distributions other than the Gaussian (or Rayleigh for narrow-band peaks) are usually not important.

The vibration at failure points is usually more Gaussian than at equipment mounting points, where the prescribed environment is applied.[6] There is, however, one special case where the reverse may be true. In some approaches to digital synthesis of random vibration, the first step is the generation of non-Gaussian narrow-band signals. It is claimed that mixing these signals results in wide-band Gaussian random vibration. The claim is true enough but not entirely relevant. Failure points are usually associated with responses of resonant parts. Their vibration is narrow-band, with distributions more like those of the initial signals generated. This is not to say that tests using such methods of synthesis are never useful. However, hasty surmises about the distributions at the failure points may be in serious error.

RESPONSE TO SWEPT SINE WAVE

Failure and malfunction of a test item are usually associated with its mechanical resonances. To gain insight into the effects of vibration and shock on typical test items, it is customary to examine a simple viscous damped resonator, and sometimes two coupled resonators, as test cases.

Consider a simple resonator of resonance frequency f_0 and

$$Q = \frac{2\pi f_0}{c} = \tfrac{1}{2}\eta \approx \frac{f_0}{B} \tag{24.1}$$

where c is the damping coefficient, η is its ratio to critical damping, and B is the bandwidth in cycles per second between the half-power points. Within B, the steady-state response is at least 0.707 of the response at f_0. Except when deliberate attempts are made to damp a resonator, as in isolator design, Q for an equipment resonance will be at least 10 and B will be at most $f_0/10$.

If the sweep rate of a swept sine wave is df/dt in the vicinity of f_0, the length of time spent within B for each sweep is

$$\Delta t_B = \frac{B}{df/dt} \tag{24.2}$$

and the number of complete cycles during this time interval is approximately

$$\Delta N_B = \frac{Bf_0}{df/dt} \tag{24.3}$$

If excitation is suddenly terminated, the response acceleration decays according to the formula

$$a = A_0 e^{-t/T} \cos 2\pi f_0 t \tag{24.4}$$

where

$$T = \frac{2m}{c} = \frac{Q}{\pi f_0} \tag{24.5}$$

For steady-state response during the sweep, it is necessary that the time within the bandwidth be sufficiently long

$$\Delta t_B = \beta T = \frac{\beta Q}{\pi f_0} \tag{24.6}$$

so that

$$\frac{df}{dt} = \frac{B}{\beta T} = \frac{\pi f_0^2}{\beta Q^2} \tag{24.7}$$

with

$$\beta \gg 1 \tag{24.8}$$

If the minimum Q is taken to be Q_0, independent of frequency, integration of Eq. (24.7) with f_0 generalized to f yields an inverse relationship between frequency and time. Assume that f is increasing from an initial value f_1.

$$\frac{1}{f_1} - \frac{1}{f} = \frac{2\pi t}{\beta Q_0^2} \tag{24.9}$$

The time for a sweep from f_1 to f_2 is

$$t_s = \frac{\beta Q_0^2(f_2 - f_1)}{2\pi f_1 f_2} \tag{24.10}$$

The fraction of time spent within the bandwidth centering on f_0 is

$$\frac{\Delta t_B}{t_s} = \frac{2 f_1 f_2}{Q_0 f_0 (f_2 - f_1)} \tag{24.11}$$

Alternately, this ratio can be held independent of frequency by letting β in Eq. (24.7) be proportional to frequency

$$\beta = \gamma f_0 \tag{24.12}$$

to yield a logarithmic sweep

$$\ln \frac{f}{f_1} = \frac{\pi t}{\gamma Q_0^2} \tag{24.13}$$

The time for a sweep from f_1 to f_2 is

$$t_s = \frac{\gamma Q_0^2}{\pi} \ln \frac{f_2}{f_1} \tag{24.14}$$

Instead of the equalities of Eq. (24.6),

$$\Delta t_B = \gamma f_0 T = \frac{\gamma Q_0}{\pi} \tag{24.15}$$

and instead of Eq. (24.11),

$$\frac{\Delta t_b}{t_s} = \frac{1}{Q_0 \ln (f_2/f_1)} \tag{24.16}$$

which makes the test deceleration approximately constant for all frequencies.

For sweep rates so slow that steady-state response is never violated, the response is simply the resonance curve times the excitation, attaining very closely Q times the excitation when $f = f_0$. The effect of fast sweep rates is illustrated[7] in Figs. 24.2 and 24.3.

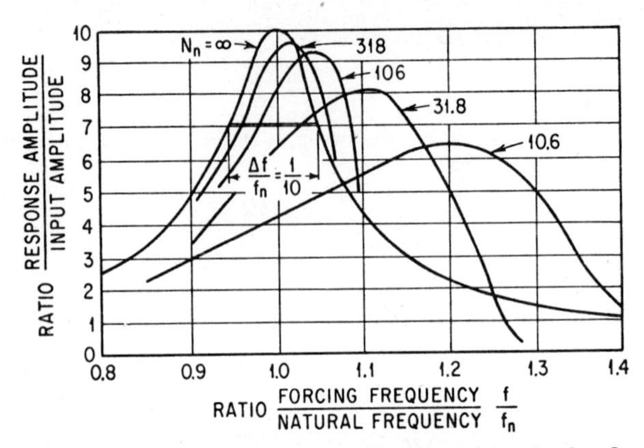

FIG. 24.2. Maximum response of a single degree-of-freedom system having $Q = 10$ when the forcing frequency is increased continuously. The parameter N_n is inversely proportional to the time rate of change of the forcing frequency. (*After F. M. Lewis.*[7])

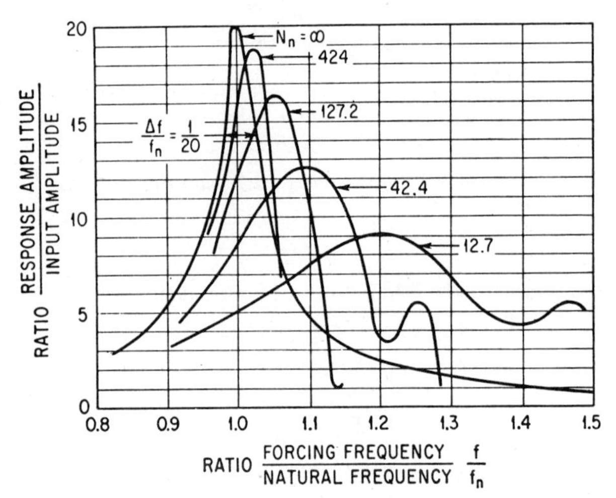

FIG. 24.3. Maximum response of a single degree-of-freedom system having $Q = 20$ when the forcing frequency is increased continuously. The parameter N_n is inversely proportional to the time rate of change of the forcing frequency. (*After F. M. Lewis.*[7])

RESPONSE TO RANDOM VIBRATION

For a power spectral density $w = w_0 = w(f_0)$ of input acceleration, constant in the neighborhood of the resonance frequency f_0, the root-mean-square acceleration of the mass is

$$(\overline{a^2})^{1/2} = \left(\frac{\pi f_0 Q w_0}{2}\right)^{1/2} \tag{24.17}$$

RESPONSE TO SHOCK

The peak acceleration of the mass is given by the shock spectrum. For simplicity, zero damping or infinite Q [as defined by Eq. (24.1)] is commonly assumed. Conservative practice uses the maximax spectrum, defined in Chap. 21, in describing the actual environment but prescribes the test environment in terms of the residual spectrum. However, most shock-spectrum computers offer several choices of Q and permit examination of entire response time-histories when desired.

FATIGUE

One of the most important effects of vibration and shock is cumulative damage by cyclic stress. This is a highly variable process involving the generation and propagation of a crack to the point that function is impaired or structural integrity is destroyed.

For experimental investigation of fatigue as a property of materials, carefully prepared polished specimens are used. The data are summarized in *S-N* curves as in Fig. 24.4, showing peak stress vs. number of cycles to failure. For most materials, the frequency at which the cyclic stress is applied is a secondary consideration. Some steels exhibit an endurance limit—a stress below which fatigue may be negligible but above which the rate of fatigue rapidly increases. Not surprisingly, the effective damping tends to increase with stress above the endurance limit.

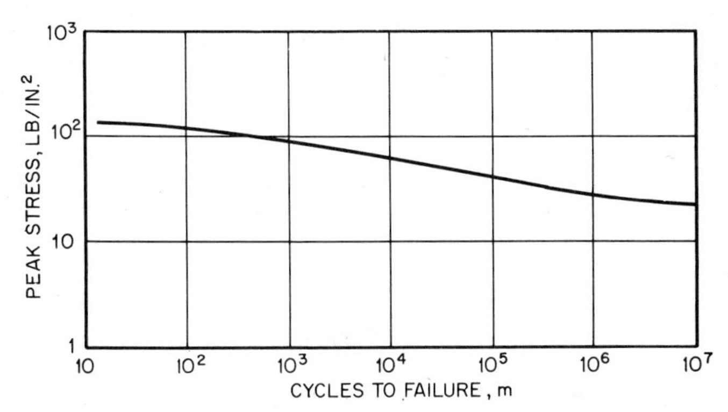

FIG. 24.4. Illustrative *S-N* curve.

When a material is subject to cyclic stresses of varying peak value, as in response of a resonator to random excitation, fatigue occurs, according to Miner's hypothesis,[8] when

$$\sum \frac{n_k}{N_k} = C \tag{24.18}$$

where n_k is the number of cycles experienced at the kth peak stress level and N_k would be the number of cycles to failure, for that constant stress level, according to an *S-N* curve. For any constant stress level, $C = 1$. Ideally it should also be unity for varying peak stresses, but experimentally it is found to vary at least from 0.3 to 3.0. For rough comparison of fatigue from random vibration to damage from other vibrations and from shocks, a value of 0.5 to 1.0 is reasonable.

More elaborate formulas than Eq. (24.18) have been proposed, but they are seldom used for investigation of equivalence of environments. Likewise, in recent years there has been an intensive theoretical and experimental investigation of material properties related to fracture, crack propagation, and impact, which has led to many criteria for optimum selection of materials. However, these criteria are seldom useful in precise lifetime prediction or in assessment of realism of environmental tests. Lifetimes are erratic at constant peak stress and very sensitive to stress level. Dynamic stress is seldom known precisely in advance.

SINE-RANDOM EQUIVALENCE

The response of a simple resonator to random excitation resembles a sine wave except for its randomly fluctuating amplitude. If the instantaneous values are Gaussian, the peaks or amplitudes are in a Rayleigh distribution. Fatigue is produced more rapidly by the larger peaks than the smaller. Analysis of Eqs. (24.17) and (24.18) suggests that a sinusoidal dwell at the resonance frequency (discussed in Chap. 21) may produce the same fatigue damage as random excitation if its rms response is about 1.5 times that given by Eq. (24.17). This is referred to the input by dividing by Q:

$$a_{\text{rms}} = 1.5 \left(\frac{\pi f_0 w_0}{2Q} \right)^{1/2} \tag{24.19}$$

which corresponds to a peak acceleration of about

$$A = 2.7 \left(\frac{f_0 w_0}{Q} \right)^{1/2} \tag{24.20}$$

The sine sweep is more generally used than the dwell in environmental testing. Comparison of the energies dissipated by the slow sine sweep and random excitations suggests an equivalent sinusoidal peak acceleration

$$A = \left(\frac{2 t_t S w_0}{N} \right)^{1/2} \tag{24.21}$$

where $S = df/dt$ is the sweep rate, t_t is the duration of the random test, and N is the number of sweeps past the resonance ferquency during the sweep test. On the other hand, a derivation based on energy absorption suggests for the resonant dwell, discussed in Chap. 21,

$$a_{\text{rms}} = \left(\frac{\pi f_0 w_0}{2Q} \right)^{1/2} \tag{24.22}$$

which is lower than Eq. (24.19). Therefore, it is not unreasonable to increase Eq. (24.21) by a factor of 1.5:

$$A = 1.5 \left(\frac{2 t_t S w_0}{N} \right)^{1/2} \tag{24.23}$$

But environmental tests are also intended to reveal malfunctions and performance degradations not necessarily involving damage, as, for example, an excessive electrical noise because of microphonics. For such situations, two vibrations may be equivalent if they produce the same root-mean-square responses, a condition not necessarily compatible with Eq. (24.20), (24.21), or (24.23).

Equivalences are important primarily when the actual environment is known but the test is or has been different in kind. A sine sweep may be used for a qualification test because random vibration test equipment is unavailable. A device may have been qualified by a sine sweep test and now face a new application with a severe random vibration environment. However it is done, requalification is costly and should be avoided if possible.

There is no standardization[9] on equivalence formulas for such decisions. Quite arbitrary rules are often used by practicing engineers.

Equation (24.20) requires the assumption of a typical value for Q. Equations (24.21) and (24.23) do not. For performance problems not involving damage, it is desirable that the environment be adjusted to a more suitable equivalence or that the undesirable effect be corrected for lack of equivalence. It is not reasonable to insist on detailed data on test-item dynamics in advance of test. However, once several failures, malfunctions, and performance degradations have been identified, economical diagnostic testing may reveal either feasible design changes or better measures of the realism of the test. On the basis of the latter, the customer may approve deviations from some required test results or waivers of some test conditions.

ACCELERATED TESTS

Airborne equipment of guided missiles, and space vehicle portions used only during a single boost period, can easily be tested for durations equal to or greater than their useful operational life. For equipment for ships or airplanes, an accelerated vibration test is more compatible with schedule and funding requirements. In other words, it is desirable to exaggerate the level of excitation to compensate for a shortening of time.

In an *S-N* plot as illustrated by Fig. 24.4, the horizontal scale may be converted to test duration by multiplication of n by the resonance frequency f_0. An exaggeration factor for constant amplitude sinusoidal excitation can be estimated from a typical *S-N* curve by sliding the test time to the left. For random excitation, if it is assumed that the response peaks between two and three times the rms acceleration produce most of the fatigue, an exaggeration factor can be estimated in a similar way.

However, advance estimation of exaggeration factors for a test item with a set of failures as yet undetermined in a variety of materials (as a consequence of transmission paths of undetermined characteristics) cannot be an exact engineering procedure. There is no universally accepted approach.

Inasmuch as specifications are usually written in advance of data on the actual environment, the usual question is not how to select an exaggeration factor as a basis for the specification but what exaggeration factor is indicated after the fact by accumulating data. Again, there is no universally accepted approach, although the problem is easier.

In any event, accelerated tests do not provide realistic environments for malfunctions and performance degradations not involving damage.

THE SUBASSEMBLY PROBLEM

In the development of equipment containing several important subassemblies, it is not advisable to delay developmental environmental testing until the subassemblies can be incorporated into the whole. If a subassembly is supplied by a vendor, shock and vibration are usually prescribed as a qualification test for acceptance. The subassembly environment of the complete equipment is seldom known, in either actual or in equipment environmental test.

The most common practice is to prescribe the same environment for the subassembly as for the whole equipment. This is likely to be a sort of spectral average of the environment applied by the equipment—certainly not an upper envelope unless additional precautions are taken during design and development. The subassembly test by itself provides only partial assurance that the subassembly will be reliable in use.

Figure 24.5 shows two extreme relationships between the resonances of the subassembly and the resonances of the equipment where it is to be mounted. For simplicity for illustrative purposes, dynamic loading interactions are neglected. For both relationships, the spectral peaks applied to the subassembly by the equipment may exceed the prescribed environment for the subassembly.

However, for relationship (*a*), the peaks occur at frequencies where the subassembly can stand little more than its prescribed environment; failure of the subassembly in equipment test is likely. For condition (*b*), the equipment spectral peaks occur at frequencies where the subassembly can stand significantly more than its prescribed environment; failure of the subassembly in equipment test would be unlikely even if the equipment environment were made somewhat more severe.

The prescribed subassembly environment could be increased until it corresponds more to a spectral envelope of the environment applied by the equipment. A better compromise in respect to both schedule and cost is to supplement a moderate subassembly environmental test by at least crude estimates of frequency relationships as development progresses. The subassembly environmental test is best regarded as a partial

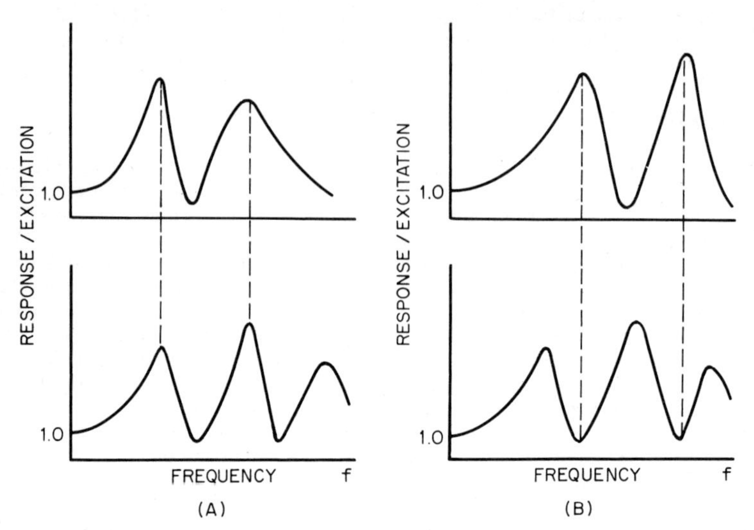

Fig. 24.5. Relationships of subassembly resonances to resonances of equipment without subassembly.

motivation for and guide to acceptable design, not as an attempt at precise environmental simulation, nor by itself a guarantee of adequate design.

LEVEL VS. FREQUENCY

For simplicity, environmental specifications focus more critically on environmental level than on frequencies where the actual environment may be concentrated. The subassembly problem is merely one example of the need to supplement formal compliance with specifications by informal attention to frequency relationships.

Attention to frequency seldom occurs when the equipment and subassembly are developed by different organizations, even within the same company, unless the eventual customer encourages it. Customer representatives should be provided with handbooks for background on such matters and to assist their judgment on requested waivers and deviations with respect to specified tests.

EQUIVALENCE, MORE GENERALLY

When simulation is the primary objective of environmental test, degree of equivalence is an important aspect, partly to assess test realism and partly to determine whether both vibration and shock or more than one kind of vibration should be simulated.[8] Among the equivalences to be considered are

1. Specified (enveloped) vs. actual, along single axis, without alteration of general type of environment
2. Excitation sequentially along orthogonal axes vs. actual excitation simultaneously
3. Dwell vs. random vibration
4. Swept sine vs. random vibration
5. Dwell vs. shock
6. Swept sine vs. shock
7. Random vibration vs. shock
8. Accelerated vs. unaccelerated

Numerous equivalence formulas, some of them rather elaborate, have been derived on the basis of various assumptions.[9] These are useful and should be among the resources of every specification writer and every environmental test engineer. However, they are only guides to judgment; and for some of the equivalence problems encountered in testing, there is little alternative but to make a choice and take a risk.

When environmental responses desired as clues to equivalence are difficult to obtain in closed mathematical form, numerical analysis by computer is an alternative. However, one must guard against letting the computer program or any unnecessarily elaborate results distract attention from assumptions, primary objectives, and significant conclusions.

The shock-spectrum computer can be a useful tool for the study of equivalences of excitations (along a single axis at a time) if it is sufficiently versatile or permits access so auxiliary electrical connections can be made. The basic computer amounts to a collection of simulated mechanical resonators at various frequencies, for generating responses corresponding to applied electrical inputs, with decision devices for measuring peak responses in selected time intervals. In other words, it is designed to provide measures of possible responses of test items. With the addition of squaring and integrating devices, it can be made to yield measures of energy absorbed by single degree-of-freedom systems. If the response of one resonator can be applied to a second computer or to another resonator in the same computer, insights can be gained into the behavior of two degree-of-freedom systems. If the difference of responses can be squared and integrated, the energy absorbed by the second resonator can be studied.

HIGH-INTENSITY NOISE TEST

Tests flights are part of the development program for an airplane. Extensive test firings are part of the development program for a guided missile. For a space vehicle, often made in quantities of only two or three, the first flight must be a highly reliable operational flight. Because of the limitations of vibration testing with mechanical excitations, and because of the problems of verifying system adequacy solely by tests of parts, it is common to subject major sections or airframe to high-intensity noise resembling that encountered in flight.[10, 11] The test vehicle is usually in a reverberation chamber. The airborne equipment is operated and monitored for malfunction and failure.

REFERENCES

1. Morrow, C. T.: *J. Acoust. Soc. Amer.*, **55**(4):695 (1974).
2. Mustin, G. S.: "Theory and Practice of Cushion Design," *Shock and Vibration Monograph* 2, Shock and Vibration Information Center, 1968.
3. Sevin, E., and W. D. Pilkey: "Optimum Shock and Vibration Isolation," *Shock and Vibration Monograph* 6, Shock and Vibration Information Center, 1971.
4. Ruzicka, J. E., and T. F. Derby: "Influence of Damping in Vibration Isolation," *Shock and Vibration Monograph* 7, Shock and Vibration Information Center, 1971.
5. Curtis, A. J., N. G. Tinling, and H. T. Abstein, Jr.: "Selection and Performance of Vibration Tests," *Shock and Vibration Monograph* 8, Shock and Vibration Information Center, 1971.
6. Morrow, C. T.: *Shock and Vibration Bull.* 28, pt. IV, p. 171, Shock and Vibration Information Center, 1960.
7. Lewis, F. M.: *Trans. ASME*, **54**:253 (1932).
8. Miner, M. A.: *J. Appl. Mech.*, **12**(3):A-159, 1945.
9. Fackler, W. C.: "Equivalence Techniques for Vibration Testing," *Shock and Vibration Monograph* 9, Shock and Vibration Information Center, 1972.
10. Lyon, R. H.: "Random Noise and Vibration in Space Vehicles," *Shock and Vibration Monograph* 1, Shock and Vibration Information Center, 1967.
11. Morrow, C. T.: *J. Acoust. Soc. Amer.*, **48**(1):162 (1970).

25

VIBRATION TESTING MACHINES

Karl Unholtz
Unholtz-Dickie Corporation

INTRODUCTION

This chapter describes some of the more common types of vibration testing machines* which are used for developmental, simulation, production, or exploratory vibration tests for the purpose of studying the effects of vibration or of evaluating physical properties of materials or structures. (*Fixed-frequency* vibrators, driven directly by the power line and used for sorting, sifting, conveying, compacting, hammering, processing, etc., are *not* included in this chapter.) A summary of the prominent features of each machine is given. These features should be kept in mind when selecting a vibration testing machine for a specific test.

A vibration testing machine (sometimes called a *shake table* or *shaker* and referred to here as a *vibration machine*) is distinguished from a vibration exciter in that it is complete with a mounting table which includes provisions for bolting the test article directly to it. A *vibration exciter*, also called a *vibration generator*, may be part of a vibration machine or it may be a device suitable for transmitting a vibratory force to a structure. A *constant-displacement* vibration machine attempts to maintain constant-displacement amplitude while the frequency is varied. Similarly, a *constant-acceleration* vibration machine attempts to maintain a constant-acceleration amplitude as the frequency is changed.

The *load* of a vibration machine includes the item under test and the supporting structures that are not normally a part of the vibration machine. In the case of equipment mounted on a vibration table, the load is the material supported by the table. In the case of objects separately supported, the load includes the test item and all fixtures partaking of the vibration. The load is frequently expressed as the weight of the material. The *test load* refers specifically to the item under test exclusive of supporting fixtures. A *dead-weight load* is a rigid load with rigid attachments. For nonrigid loads the reaction of the load on the vibration machine is a function of frequency. The vector force exerted by the load, per unit of acceleration amplitude expressed in units of gravity of the driven point at any given frequency, is the *effective load* for that frequency. The term *load capacity*, which is descriptive of the performance of reaction and direct-drive types of mechanical vibration machines, is the maximum dead-weight load that can be vibrated at the maximum acceleration rating of the vibration machine. The *load couple* for a dead-weight load is equal to the product of the force exerted on the load and the distance of the center-of-mass from the line-of-action of the force, or from some arbitrarily selected location (such as a table surface). The static and dynamic load couples are generally different for nonrigid loads.

* Patents are owned by various manufacturers of vibration machines which cover many aspects of design discussed here. Such discussion does not imply freedom from patent restrictions.

The term *force capacity,* which is descriptive of the performance of electrodynamic shakers, is defined as the maximum rated force generated by the machine. This force is usually specified, for continuous rating, as the maximum vector amplitude of a sinusoid that can be generated throughout a usable frequency range. A corresponding maximum rated acceleration, in units of gravity, can be calculated as the quotient of the force capacity divided by the total weight of the coil table assembly and the attached dead-weight loads. The *effective force* exerted by the load is equal to the effective load multiplied by the (dimensionless) ratio $\mathfrak{g}$, which represents the number of units of gravity acceleration of the driven point.

DIRECT-DRIVE MECHANICAL VIBRATION MACHINES

The direct-drive vibration machine consists of a rotating eccentric or cam driving a positive linkage connection which forces a displacement between the base and table of the machine. Except for the bearing clearances and strain in the load-carrying members,

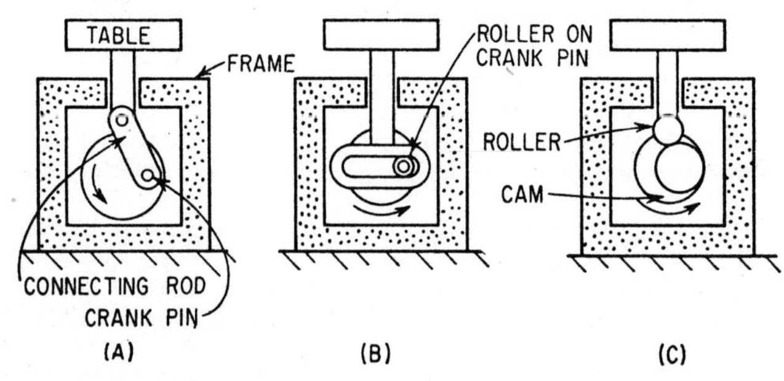

FIG. 25.1. Elementary direct-drive mechanical vibration machines: (*A*) Eccentric and connecting link. (*B*) Scotch yoke. (*C*) Cam and follower.

the machine tends to develop a displacement between the base and the table which is independent of the forces exerted by the load against the table. If the base is held in a fixed position, the table tends to generate a vibratory displacement of constant amplitude, independent of the operating rpm. Figure 25.1 shows the direct-drive mechanical machine in its simplest forms. This type of machine is sometimes referred to as a *brute force machine* since it will develop any force necessary to produce the table motion corresponding to the crank or cam offset, short of breaking the load-carrying members or stalling the driving shaft.

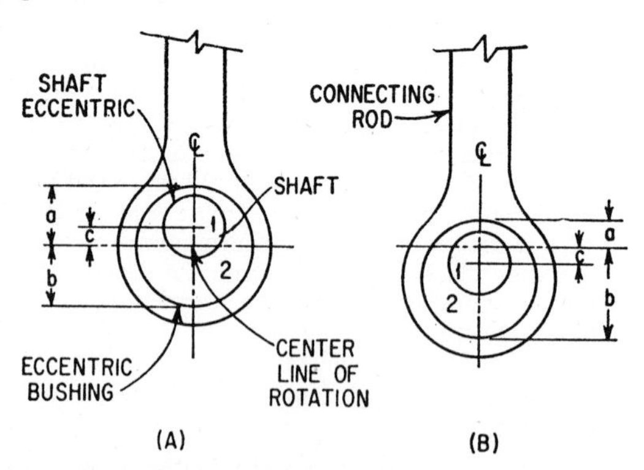

FIG. 25.2. Displacement adjusting method for a direct-drive vibration machine:
(*A*) $a = b$. Displacement $= (b - a) = 0$.
(*B*) $a \neq b$. Displacement $= 2c$ (peak-peak).

The direct-drive machines usually are rated in terms of pounds of test item that can be vibrated through the maximum displacement and frequency operating limits of the machine. (On some of the larger machines, the rating is based on a maximum value of the product of load weight plus table weight in pounds and the maximum acceleration, for a specified frequency range.) The operating limits are determined by allowable bearing loads based on the assumption that the test item behaves as a rigid mass and the bearing loads can be calculated as the product of mass and acceleration.

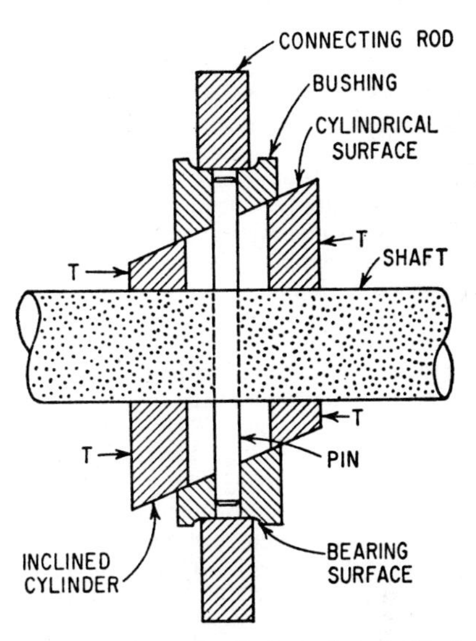

Fig. 25.3. A sectional view of an inclined cylinder-type adjustable amplitude eccentric. (*Courtesy L.A.B. Corp.*)

Figure 25.2 shows a method of providing stepless adjustability of crank offset between zero and full displacement. The boundary surfaces between shaft eccentric 1 and eccentric bushing 2 and between 2 and the connecting rod are circular cylinders. For a fixed-displacement setting, the eccentric 1 and the eccentric bushing 2 remain fixed relative to each other but rotate about the rotation center of the driving shaft. The displacement adjustment is made by rotating eccentric bushing 2 relative to shaft eccentric 1. In some machines it is necessary to stop the machine, unlock the cylindrical elements, reposition them according to a calibrated displacement scale of the desired displacement, and relock the position before resuming operation. In other machines, this adjustment can be made during machine operation.

Another basic mechanism for producing direct-drive vibration is an eccentric consisting of a cylinder inclined at a fixed angle with the main drive shaft on which it is mounted, as shown in Fig. 25.3. A bushing, which rotates with the shaft, and a connecting-rod bearing on the rotating bushing are assembled around the inclined cylinder. The bushing and the inclined cylinder are each pinned to the shaft, thereby being forced to turn with the shaft. The cylinder has a slot in the axial direction allowing axial motion of the cylinder while the axial position of the bushing and connecting rod is fixed. Thrust bearings (at T) position the inclined cylinder axially according to the setting of the amplitude control, thereby providing a stepless adjustment which can be made while the machine is operating.

The simplest direct-drive mechanical vibration machine is driven by a constant-speed motor in conjunction with a belt-driven speed changer and a frequency-indicating tachometer. Table displacement is set during shutoff and is assumed to hold during operation. An auxilary motor driving a cam may be included to provide frequency cycling between adjustable limits. More elaborate systems employ a direct-coupled variable-speed motor with electronic speed control, as well as amplitude adjustment from a control station. Machines have been developed which provide rectilinear, circular, and three-dimensional table movements—the latter giving complete, independent adjustment of magnitude and phase in the three directions.

PROMINENT FEATURES

1. Low operating frequencies and large displacements can be provided conveniently.

2. Theoretically, the machine maintains constant displacement regardless of the mechanical impedance of the table-mounted test item within force and frequency limits of the machine. However, in practice, the departure from this theoretical ideal is considerable, due to the elastic deformation of the load-carrying members with change in

output force. The output force changes in proportion to the square of the operating frequency and in proportion to the increased displacement resulting therefrom. Because the load-carrying members cannot be made infinitely stiff, the machines do not hold constant displacement with increasing frequency with a bare table. This characteristic is further emphasized with heavy table mass loads. Accordingly, some of the larger capacity machines which operate up to 60 cps include automatic adjustment of the crank offset as a function of operating frequency in order to hold displacement more nearly constant throughout the full operating range of frequency.

3. The machine must be designed to provide a stiff connection between the ground or floor support and the table. If accelerations greater than 1g are contemplated, the vibratory forces generated between the table and ground will be greater than the weight of the test item. Hence, all mass loads within the rating of the machine can be directly attached to the table without recourse to external supports.

4. The allowable range of operating frequencies is small in order to remain within bearing load ratings. Therefore, the direct-drive mechanical vibration machine can be designed to have all mechanical resonances removed from the operating frequency range. In addition, relatively heavy tables can be used in comparison to the weight of the test item. Consequently, misplacing the center-of-gravity of the test item relative to the table center for vibration normal to the table surface and the generation of moments by the test item (due to internal resonances) usually has less influence on the table motions for this type of machine than would other types which are designed for wide operational frequency bands.

5. Simultaneous rectilinear motion normal to the table surface and parallel to the table surface in two principal directions is practicable to achieve. It may be obtained with complete independent control of magnitude and phase in each of the three directions.

6. Displacement of the table is generated directly by a positive drive rather than by a generated force acting on the mechanical impedance of the table and load. Consequently, impact loads in the bearings, due to the necessary presence of some bearing clearance, result in the generation of relatively high impact forces which are rich in harmonics. Accordingly, although the waveform of displacement might be tolerated as such, the waveform of acceleration is normally sufficiently distorted to preclude recognition of the fundamental driven frequency, when displayed on a time base.

REACTION-TYPE MECHANICAL VIBRATION MACHINE

A vibration machine using a rotating shaft carrying a mass whose center-of-mass is displaced from the center-of-rotation of the shaft for the generation of vibration, is called a *reaction-type vibration machine*. The product of the mass and the distance of its center from the axis of rotation is referred to as the *mass unbalance*, the *rotating unbalance*, or simply the *unbalance*. The force resulting from the rotation of this unbalance is referred to as the *unbalance force*.

The *reaction-type vibration machine* consists of at least one[1] rotating-mass unbalance directly attached to the vibrating table. The table and rotating unbalance are suspended from a base or frame by soft springs which isolate most of the vibration forces from the supporting base and floor. The rotating unbalance generates an oscillating force which drives the table. The unbalance consists of a weight on an arm which is relatively long by comparison to the desired table displacement. The unbalance force is transmitted through bearings directly to the table mass, causing a vibratory motion without reaction of the force against the base. A vibration machine employing this principle is referred to as a reaction machine since the reaction to the unbalance force is supplied by the table itself rather than through a connection to the floor or ground.

CIRCULAR-MOTION MACHINE

The reaction-type machine, in its simplest form, uses a single rotating-mass unbalance which produces a force directed along the line connecting the center-of-rotation and the center-of-mass of the displaced mass. Referred to stationary coordinates, this force appears normal to the axis of rotation of the driven shaft, rotating about this axis at the rotational speed of the shaft. The transmission of this force to the vibration-machine table causes the table to execute a circular motion in a plane normal to the axis of the rotating shaft.

Figure 25.4 shows, schematically, a machine employing a single unbalance producing circular motion in the plane of the vibration-table surface. The unbalance is driven at various rotational speeds, causing the table and test item to execute circular motion at various frequencies. The counterbalance weight is adjusted to equal the test item mass moment calculated from *d*, the plane of the unbalance force, thereby keeping the combined center-of-gravity coincident with the generated force. Keeping the generated force acting through the combined center-of-gravity of the spring-mounted assembly eliminates vibratory moments which, in turn, would generate unwanted rotary motions in addition to the motion parallel to the test mounting surface. The vibration isolator supports the vibrating parts with minimum transmission of the vibration to the supporting floor.

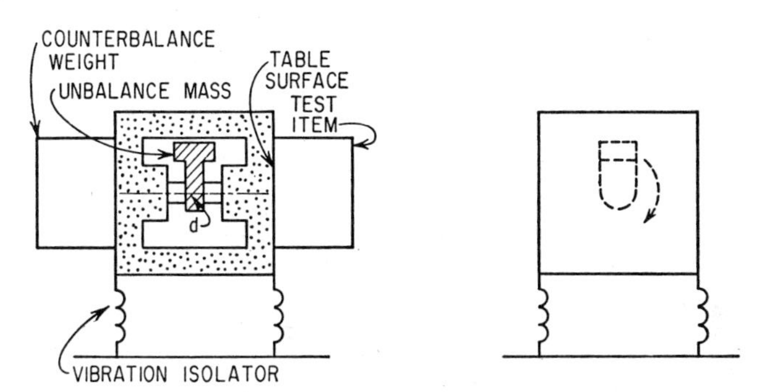

Fig. 25.4. Circular-motion reaction-type mechanical vibration machine.

Reference to the section on dynamic considerations for the reaction machine which appears later in this chapter shows that for a fixed amount of unbalance and for the case of the table and test item acting as a rigid mass, the displacement of motion tends to remain constant if there are no resonances in or near the operating frequency range. If constant table acceleration is desired, the unbalance force must remain constant, requiring the amount of unbalance to change with shaft speed.

RECTILINEAR-MOTION MACHINE

Rectilinear motion rather than circular motion can be generated by means of a reciprocating mass, as shown in Fig. 25.5. The shaft has an eccentric pin which engages the reciprocating mass which bears on guides in the frame. Rotation of the shaft causes reciprocating motion of the mass. The inertia forces caused by the reciprocating mass are transmitted to the frame in the direction shown, through the bearings supporting the shaft. This mechanism is referred to as a *Scotch yoke*. The frame may be attached to a vibration table or directly to a structure in order to excite the resonances in the structure. The crankpin can be counterbalanced, thereby eliminating most of the unwanted excitations at 90° to the motion of the reciprocating mass.

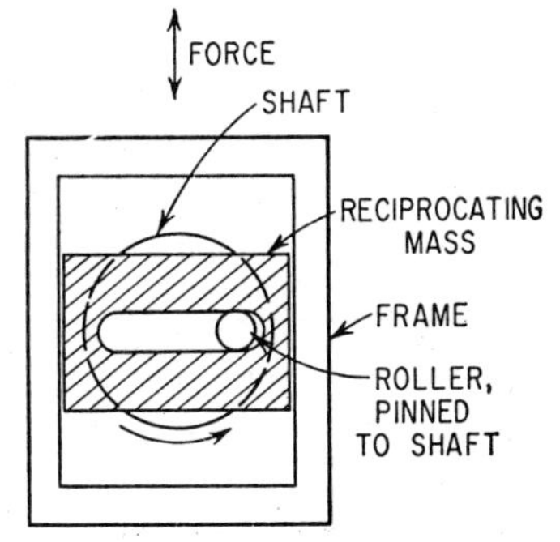

Fig. 25.5. Rectilinear-motion reaction-type mechanical vibration exciter using a reciprocating mass.

Rectilinear motions also can be produced with a single rotating unbalance by constraining the table to move in one direction. Two examples are shown in Fig. 25.6. Some fatigue testing machines use these configurations with springs *a* and *b* in Fig. 25.6*A* and spring *a* in Fig. 25.6*B* selected to have the proper stiffness in the direction of vibration so that the table and associated mass are resonant at a fixed operating frequency. Under these conditions, the force transmitted by the table to the item under test is approximately constant and independent of any changes in the mechanical mobility of the

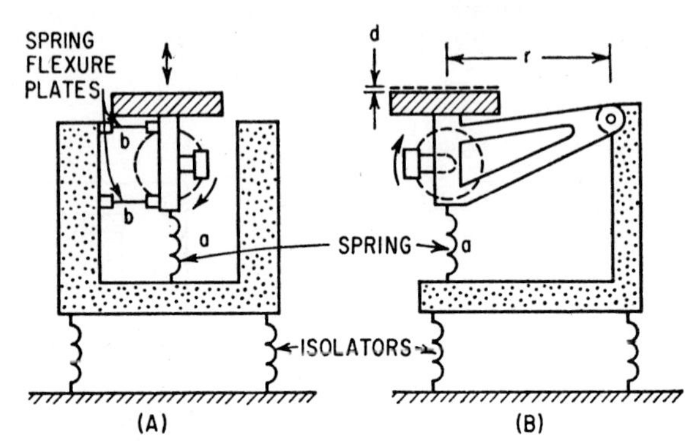

Fig. 25.6. Rectilinear-motion reaction-type vibration machine: (*A*) Flexure constraint. (*B*) Radius arm with $d \ll r$.

load—provided the length of the eccentric arm is large compared to the table displacement. If the resonant frequency is below the operating frequencies and the rotating unbalance remains constant, then the displacement of the table tends to remain constant through the operating frequency range for rigid dead-mass table loads.

TWO ROTATING UNBALANCES. The most common rectilinear reaction-type vibration machine consists of two rotating unbalances, turning in opposite directions and phased so that the unbalance forces add in the desired direction and cancel in other

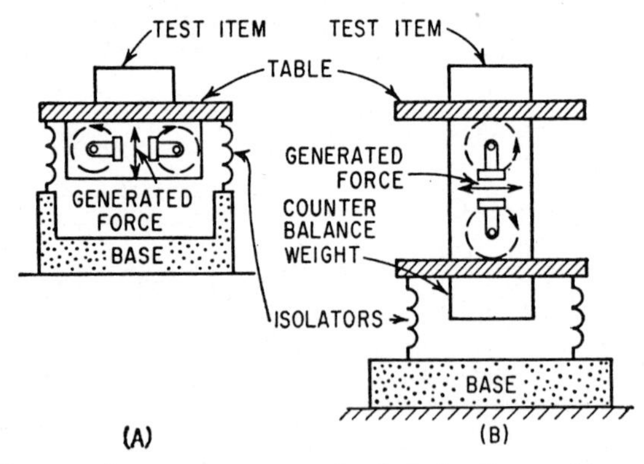

Fig. 25.7. Rectilinear-motion reaction-type mechanical vibration machine using two rotating unbalances: (*A*) Vibration perpendicular to table surface. (*B*) Vibration parallel to table surface.

directions. Figure 25.7 shows schematically how rectilinear motion perpendicular and parallel to the vibration table is generated. The effective generated force from the two rotating unbalances is midway between the two axes of rotation and is normal to a line connecting the two. In the case of motion perpendicular to the surface of the table, simply locating the center-of-gravity of the test item over the center of the table gives a proper load orientation. Tables are designed so that the resultant force always passes through this point. This results in collinearity of generated forces and inertia forces, thereby avoiding the generation of moments which would otherwise rock the table. In the case of motion parallel to the table surface, no simple orientation of the test item will achieve collinearity of the generated force and inertia force of the table and test item. Various methods are used to make the generated force pass through the combined center-of-gravity of the table and test item.

THREE ROTATING UNBALANCES. If a machine is desired which can be adjusted to give vibratory motion either normal to the plane of the table or parallel to the plane of the table, a minimum of three rotating unbalances is required. Inspection of Fig. 25.8 shows how rotating the two smaller mass unbalances relative to the single larger

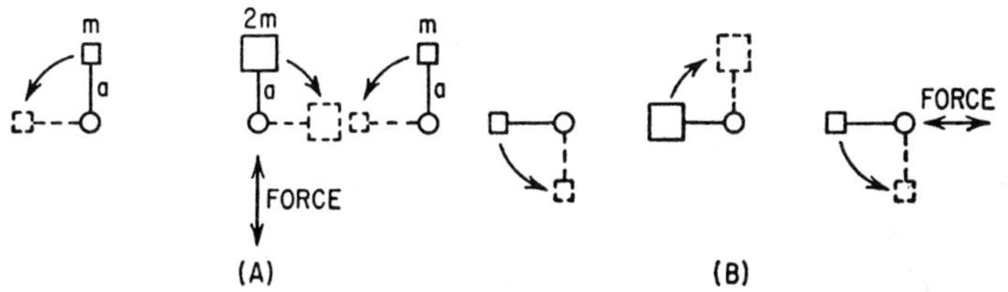

Fig. 25.8. Adjustment of direction of generated force in a reaction-type mechanical vibration exciter: (*A*) Vertical force. (*B*) Horizontal force.

unbalance results in the addition of forces in any desired direction, with cancellation of forces and force couples at 90° to this direction. Although parallel shafts are usually used as illustrated, occasionally the three unbalances may be mounted on collinear shafts, the two smaller unbalances being placed on either side of the single larger unbalance to conserve space and to eliminate the bending moments and shear forces imposed on the structure connecting the individual shafts.

PROMINENT FEATURES

1. The forces generated by the rotating unbalances are transmitted directly to the table without dependence upon a reactionary force against a heavy base or rigid ground connection.

2. Because the length of the arm which supports the unbalance mass can be large, relative to reasonable bearing clearances and the generation of a force which does not reverse its direction relative to the rotating unbalance arm, the generated waveform of motion imparted to the vibration machine table is superior to that attainable in the direct-drive type of vibration machine.

3. The generated vibratory force can be made to pass through the combined center-of-gravity of the table and test item in both the normal and parallel directions relative to the table surface, thereby minimizing vibratory moments giving rise to table rocking modes.

4. The attainable rpm and load ratings on bearings currently limit performance to a frequency of approximately 60 cps and a generated force of 300,000 lb, respectively, although in special cases frequencies up to 120 cps and higher can be obtained for smaller machines.

ELECTRODYNAMIC VIBRATION MACHINE

GENERAL DESCRIPTION

A complete electrodynamic vibration test system is comprised of an electrodynamic vibration machine, electrical power equipment which drives the vibration machine, and electrical controls and vibration monitoring equipment.

The electrodynamic vibration machine derives its name from the method of force generation. The force which causes motion of the table is produced electrodynamically by the interaction between a current flow in the armature coil and the intense magnetic d-c field which passes through the coil, as illustrated in Fig. 25.9. The table is structurally attached to a force-generating coil which is concentrically located (with radial clearances) in the annular air gap of the d-c magnet circuit. The assembly of the armature coil and the table is usually referred to as the "driver coil-table" or "armature." The magnetic circuit is made from soft iron which also forms the *body* of the vibration machine. The body is magnetically energized, usually by two field coils as shown in Fig. 25.9*C*, generating a radially directed field in the air gap, which is perpendicular to the direction of current flow in the armature coil. Alternatively, in small shakers, the magnetic field is generated by permanent magnets. The generated force in the armature coil is in the direction of the axis of the coil, perpendicular to the table surface. The direction of the force is also perpendicular to the armature-current direction and to the air-gap field direction.

The table and armature coil assembly is supported by elastic means from the machine body, permitting rectilinear motion of the table perpendicular to its surface, corresponding in direction to the axis of the armature coil. Motion of the table in all other directions is resisted by stiff restraints. Table motion results when an a-c current passes through the armature coil. The body of the machine is usually supported by a base with a trunnion shaft centerline passing horizontally through the center of gravity of the body assembly, permitting the body to be rotated about its center, thereby giving a vertical or horizontal orientation to the machine table. The base usually includes an elastic support of the body, providing vibration isolation between the body and the supporting floor.

Where a very small magnetic field is required at the vibration machine table due to the effect of the magnetic field on the item under test, *degaussing* may be provided. Magnetic fields of 5 to 30 gauss several inches above the table are normal for modern machines with double-ended, center air-gap magnet designs, Fig. 25.9*C*, without degaussing accessories; in contrast, with degaussing accessories, magnetic fields of 2 to 5 gauss can be achieved.

Because of copper and iron losses in the electrodynamic unit, provision must be made to carry off the dissipated heat. Cooling by convection air currents, compressed air, or a motor-driven blower is used and, in some cases, a recirculating fluid is used in conjunction with a heat exchanger. Fluid cooling is particularly useful under extremes of hot or cold environments or altitude conditions where little air pressure is available.

MAGNET CIRCUIT CONFIGURATIONS

Three magnet circuit configurations which are used in the electrodynamic machines are shown schematically in Fig. 25.9. In Fig. 25.9*A*, the table and driver coil are located at opposite ends of the magnet circuit. The advantage of this configuration is that the location of the annular air gap, the region of high magnetic leakage flux, is spaced from the table and the body itself acts as a magnetic shield, resulting in lower magnetic flux density at the table. The disadvantage lies in the loss of rigidity in the connecting structure between the driver coil and the table because of its length. This configuration is usually cooled by convection air currents or by forced air from a motor-driven blower.

In Fig. 25.9*B*, the table is connected directly to the driver coil. This eliminates the length of structure passing through the magnet structure, thereby increasing the rigidity of the driver coil-table assembly and allowing higher operating frequencies. The leakage magnetic field in the vicinity of the table is high in this configuration. It is therefore difficult, if not impossible, to reduce the leakage to acceptable levels without adding extra length to the driver coil assembly, elevating the table above the air gap. The configuration in Fig. 25.9*C* has a complete magnet circuit above and below the annular air gap, thereby reducing external leakage magnetic field to a minimum. This configuration also increases the total magnetic flux in the air gap by a factor of almost 2 for the same diameter driver coil, giving greater force generation and a more symmetrical magnetic flux density along the axis of the coil. Hence more uniform force generation results when the driver coil is moved axially throughout its total stroke. All high-efficiency and high-performance electrodynamic vibration machines use the configuration shown in Fig. 25.9*C*.

Configurations *B* and *C* may use air cooling throughout, or an air-cooled driver coil and liquid-cooled field coil(s) or total liquid cooling.

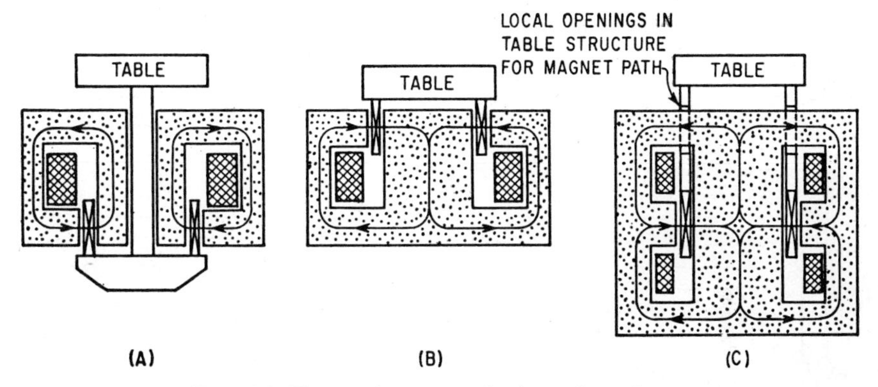

FIG. 25.9. Three main magnet circuit configurations.

The main magnetic circuit uses d-c field coils for generating the high-intensity magnetic flux in the annular gap in all of the larger sizes and most of the smaller units. Permanent magnet excitation is used in small portable units and in some general-purpose units up to about 500 lb generated force.

FREQUENCY-RESPONSE CONSIDERATIONS

Testing procedures which call for sinusoidal motion of a vibration-machine table can be performed even though the frequency-response curve of the electrodynamic vibration machine is far from flat. For a test at a fixed frequency, the driving voltage e_d is adjusted until the table motion is equal in amplitude to that required by the test specifications. If the procedure calls for cycling the frequency between two frequency limits while keeping a constant displacement or acceleration, an automatic regulating device or servo control adjusts the driver-coil voltage as required to maintain the desired vibration machine table motion independent of the frequency of operation. This regulating device provides a correction at any frequency of operation within the testing frequency limits but it can correct for only one operating frequency at any instant of time. The closer the frequency response is to the desired variation in acceleration with frequency, the smaller the corrections in driver-coil voltage will be from the automatic regulator—thereby improving the attainable accuracy of the control.

Testing procedures which call for a random vibration source, covering a broad frequency band, require that the frequency-response characteristic of the system be adjusted. The frequency response can be adjusted to be constant throughout the band of frequencies of the test with attenuation above and below this band. In this case, a "white noise" generator at the input of the system produces an acceleration spectral density which is constant throughout the frequency range of the test. Alternatively, the frequency-response characteristic can be adjusted so as to approximate the variation in acceleration spectral density found under actual service conditions. In this case, a white noise generator at the input of the system produces a variation in acceleration spectral density at the table as a function of frequency duplicating the service conditions.

CONTROL SYSTEMS

SINUSOIDAL VIBRATION TESTING

A typical analog control system for sinusoidal vibration testing is shown in Fig. 25.10 (also see Fig. 27.2). Voltage is supplied by a low-distortion sine-wave generator to an amplitude servo and programmer which adjusts the magnitude of this voltage; the adjustment is controlled by a voltage which is generated by the accelerometers. The sine-wave signal, so modified, is fed to a low-distortion power amplifier which drives the vibration machine.

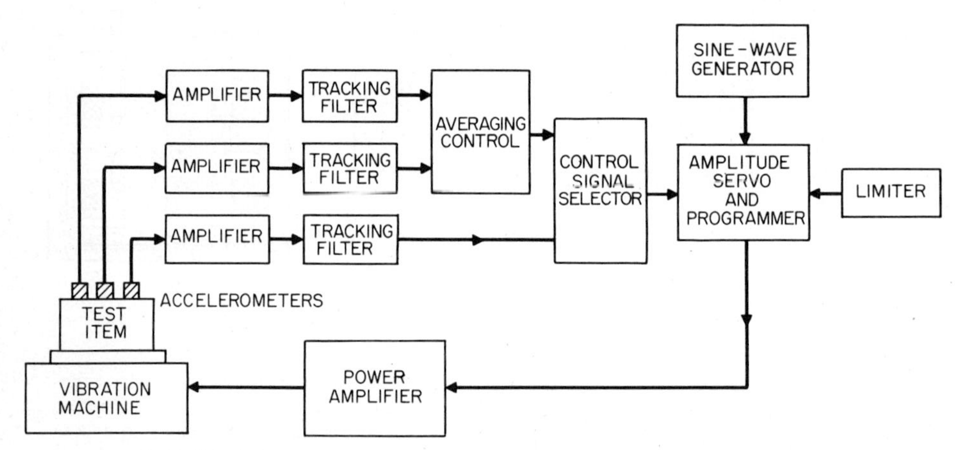

Fig. 25.10. Block diagram of a typical control system for sinusoidal vibration testing.

Sinusoidal motion of the vibration machine may induce harmonics, rattles, and other nonlinear response in the test item. The tracking filters (narrow bandpass filters whose center frequency always corresponds exactly to the frequency of the sine-wave generator) connected to the outputs of the accelerometers reject any signals other than those corresponding to the driving frequency of the oscillator. The function of the averaging control is to provide servo control on the arithmetic average of the individual accelerometer magnitudes. The control signal selector limits the test to preset values. The averager and/or the selector provide protection against excessive amplitude overtesting or undertesting.

RANDOM VIBRATION TESTING

A typical control system for random vibration testing is shown in Fig. 25.11. The voltage, supplied by a random noise generator, is fed to an equalizer which shapes the spectrum of the random noise. The equalizer is usually comprised of a number of narrow bandpass filters, each with an adjustable gain that divides the total frequency range of interest into contiguous narrowband segments.[10] Digital techniques may be employed to achieve similar results (see Fig. 27.3). The output of the equalizer provides the desired signal input to the power amplifier[2, 3] that drives the vibration machine.

Accelerometers are mounted on the item under test. The output of these accelerometers is amplified and fed into an averaging control whose input switches between the signals from the various accelerometers. Precautions must be observed in the use of such averaging controls.[5] The mean-square acceleration density, provided by the spectrum analyzer, is displayed by any of several convenient techniques—for example, by means of an oscilloscopic display.

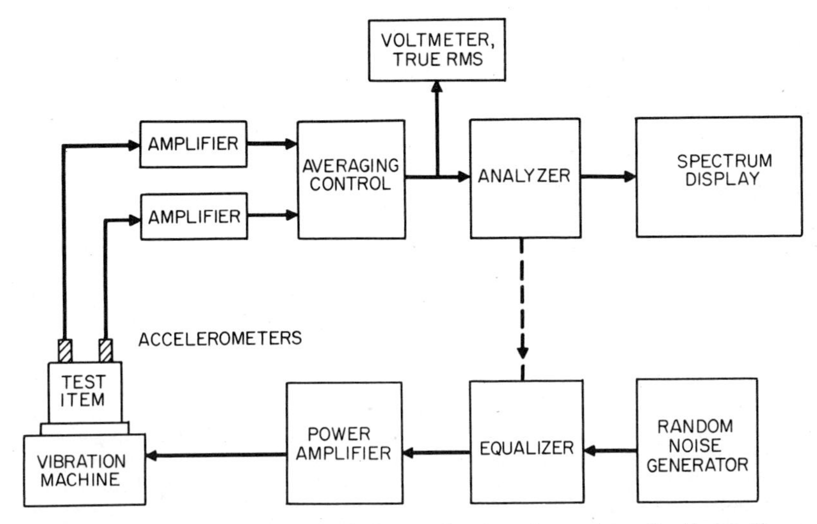

FIG. 25.11. Block diagram of a typical control system for random vibration testing.

SYSTEM RATINGS

The electrodynamic vibration machine system is rated: (1) in terms of the *peak value* of the sinusoidal generated force for *sinusoidal* vibration testing, and (2) in terms of the *rms* and *instantaneous* values of the maximum force generated under *random* vibration testing. In order to determine the acceleration rating of the system with a test load on the vibration table, the weight of the test load, assumed to be effective at all frequencies, must be known and used in the following expressions:

$$\mathfrak{g} = \frac{F}{W_L + W_T}$$

$$\mathfrak{g}_{\text{rms}} = \frac{F_{\text{rms}}}{W_L + W_T}$$

where $\mathfrak{g}$ = a/g, a number expressing the ratio of the peak sinusoidal acceleration to the acceleration due to gravity (i.e., the peak sinuosidal acceleration in g's)

$\mathfrak{g}_{\text{rms}}$ = a_{rms}/g, a number expressing the ratio of the rms value of random acceleration to the acceleration due to gravity

W_L = weight of load, lb

W_T = equivalent weight of table driver-coil assembly and associated moving parts, lb

F = rated peak value of sinusoidal generated force, lb

F_{rms} = rated rms value of random generated force, lb

The *force rating* of an electrodynamic vibration machine is the value of force which can be used to calculate attainable accelerations for any rigid-mass table load equal to (or greater than) the driver coil weight. It is not necessarily the force generated by the driver coil. These two forces are identical only if the operating frequencies are sufficiently below the axial resonant frequency of the armature assembly, where it acts as a rigid body. As the axial resonant frequency is approached, a mechanical magnification of the force generated electrically by the driver coil results. The design of the driving power supply takes into account the possible reduction in driver-coil current at frequencies approaching the armature axial resonant frequency, since full current in this range cannot be used without exceeding the rated value of transmitted force at the table —possibly causing structural damage.

In those cases where the test load dissipates energy mechanically, the system performance should be analyzed for each specific load since normal ratings are based on a dead-mass, nondissipative type of load. This consideration is particularly significant in resonant-type fatigue tests at high stress levels.[4]

PROMINENT FEATURES

1. A wide range of operating frequencies is possible, with a properly selected electric power source, from 0 to above 30,000 cps. Small, special-purpose machines have been made with the first axial resonant mode above 26,000 cps, giving inherently a resonance-free, flat response to 10,000 cps.

2. Frequency and amplitude are easily controlled by adjusting the power-supply frequency and voltage.

3. Pure sinusoidal table motion can be generated at all frequencies and amplitudes. Inherently, the table acceleration is the result of a generated force proportional to the driving current. If the electric power supply generates pure sinusoidal voltages and currents, the waveform of the acceleration of the table will be sinusoidal, and background noise will not be present. Operation with table acceleration waveform distortion of less than 10 per cent through a displacement range of 10,000-to-1 is common, even in the largest machines. Velocity and displacement waveforms obtained by the single and double integration of acceleration, respectively, will have even less distortion.

4. Random vibration, as well as sinusoidal vibration, or a combination of both, can be generated by supplying an appropriate input voltage.

5. A unit occupying a small volume, and powered from a remote source, can be used to generate small vibratory forces. A properly designed unit adds little mass at the point of attachment and can have high mobility without mechanical damping.

6. Leakage magnetic flux is present around the main magnet circuit. This leakage flux can be minimized by proper design and the use of degaussing coil techniques.

SPECIFICATIONS

DESIGN FACTORS

Force Output. The maximum vector-force output for sinusoidal excitation shall be given for continuous duty and may additionally be given for intermittent duty. When nonsinusoidal motions are involved, the force may additionally be given in terms of an rms value together with a maximum instantaneous value. The latter value is especially significant when random type of excitation is required.

In some cases of wide-frequency-band operation of the electrodynamic vibration machine, the upper frequencies are sufficiently near to the axial mechanical resonance of the coil-table assembly to provide some amplification of the generated force. Most system designs account for this magnification, when present, by reducing the capacity of the electrical driving power accordingly.

The peak values of the input electrical signal, for random excitation, may extend to indefinitely large values. In order that the armature coil voltage and generated force may be limited to reasonable values the peak values of excitation are clipped so that no maxima shall exceed a given multiple of the rms value. The magnitude of the maximum clipped output shall be specified preferably as a multiple of the rms value. If adjustments are possible, the range of magnitudes shall be given.

Weight of Vibrating Assembly. The weight of the vibration coil-table assembly shall be given. It shall include all parts which move with the table and an appropriate percentage of the weight of those parts connecting the moving and stationary parts giving an effective over-all weight.

Vibration Direction. The directions of vibration shall be specified with respect to the surface of the vibration table and with respect to the horizontal or vertical direction. Provisions for changing the direction of vibration shall be stated.

Unsupported Load. The maximum allowable weight of a load not requiring external supports shall be given for horizontal and vertical orientations of the vibration table. This load in no way relates to dyanmic performance, but is a design limitation, the basis of which may be stated by the manufacturer.

Static Moments and Torques. Static moments and torques may be applied to the coil-table assembly of a vibration machine by the tightening of bolts and by the overhang of the center-of-gravity of an unsupported load during horizontal vibration. The maximum permissible values of these moments and torques shall be specified. These loads in no way relate to the dynamic performance but are design limitations the basis for which may be stated by the manufacturer.

Total Excursion Limit. The maximum table motion between mechanical stops shall be given together with the maximum vibrational excursion permissible with no load and with maximum load supportable by the table.

Acceleration Limit. The maximum allowable table acceleration shall be given. (These large maxima may be involved in the drive of resonant systems.)

Stiffness of Coil-Table Assembly Suspension System

Axial Stiffness. The stiffness of the suspension system for axial deflections of the coil-table assembly shall be given in terms of pounds per inch of deflection. The natural frequency of the unloaded vibrating assembly may also be given. Provisions, if any, to adjust the table position to compensate for position changes caused by different loads shall be described.

Suspension Resonances. Resonances of the suspension system should be described together with means for their adjustment where applicable.

Axial Coil-Table Resonance. The resonant frequency of the lowest axial mode of vibration of the coil-table assembly shall be given for no load and for an added dead-weight load equal to one and to three times the coil-table assembly weight. If this resonant frequency is not obvious from measurements of the table amplitude vs. frequency, it may be taken to be approximately equal to the lowest frequency, above the rigid-body resonance of the table-coil assembly on its suspension system, at which the phase difference between the armature coil current and the acceleration of the center of the table is 90°.

Impedance Characteristics. When an exciter or vibration machine is considered independent of its power supply, information concerning the electrical impedance characteristics of the machine shall be given in sufficient detail to permit matching of the power-supply output to the vibration-machine input. It is suggested that consideration be given to providing schematic circuit diagrams (electrical and mechanical, or equivalent electrical) together with corresponding equations which contain the principal features of the machine.

Environmental Extremes. When it is anticipated that the vibration machine will be used under conditions of abnormal pressure and temperature, the following information shall be supplied as may be applicable: maximum simulated altitude (or minimum pressure) under which full performance ratings can be applied; maximum simulated altitude under which reduced performance ratings can be applied; maximum ambient temperature for rated output; low-temperature limitations; humidity limitations.

PERFORMANCE. The performance relates in part to the combined operation of the vibration generator and its power supply.

Amplitude-Frequency Relations. Data on sinusoidal operation shall be given as a series of curves for several table loads, including zero load, and for a load at least three times the weight of the coil-table assembly. Maximum loads corresponding to $20g$ and $10g$ table acceleration under full rated force output would be preferred. These curves should give amplitudes of table displacement, velocity, or acceleration, whichever is limiting, throughout the complete range of operating frequencies corresponding to maximum continuous ratings of the system. Additionally, the maximum rated force should be given. If this force is frequency-dependent, it should be presented as a curve with the ordinate representing the force and the abscissa the frequency.

If the system is for broad-band use, necessarily employing an electronic power amplifier, the exciting voltage signal applied to the input of the system shall be held constant and the output acceleration shall be plotted as a function of frequency with and without peak-notch filters or other compensating devices for the loads and accelerations indicated above. If the vibrator is used only for sinusoidal vibrations, and employs servo amplitude control, the curves should be obtained under automatic frequency sweeping conditions with the control system included. If cycling the exciter through its complete frequency range involves switching operations, such as may be required for matching impedances or for changing power sources, these discontinuities should be noted on the amplitude-frequency curves.

Waveform. Total rms distortion of the acceleration waveform at the center of the vibration table, or at the center on top of the added test weight, shall be furnished to show at least the frequencies of worst waveform under the test conditions specified under the above paragraph. The pickup type, and frequency range, shall be given together with the frequency range of associated equipment. It is desirable to have the over-all frequency range at least ten times the frequency of the fundamental being recorded. Tabular data on harmonic analysis may alternatively or additionally be given.

Magnetic Fields. The maximum values of constant and alternating magnetic fields, due to the vibration exciter, in the region over the surface of the vibration table should be indicated. If degaussing coils are furnished, these values should be given with and without the use of the degaussing coils.

Frequency Range. The over-all frequency range, and a division of frequency ranges for different alternators that supply the exciting current, if applicable, shall be given. A group of frequency ranges shall also be given for electronic power supplies if they require changes of their output impedance for the different ranges.

Automatic Frequency Control. The following factors shall be stated or provided: The range, or ranges, of the frequency bands that can be automatically cycled at full rated output or at reduced output if applicable; the minimum and maximum time for each frequency cycle; a curve of frequency vs. time for a maximum and a minimum cycle period and for the principal frequency ranges; the dwell time at the points of reversal of the direction of frequency change.

Frequency Drift. The probable drift of a set frequency shall be stated, together with factors that contribute to the drift. This shall apply for nonresonant loads.

Signal Generator. A vibration pickup, if built into the vibration machine, shall have calibrations furnished over a specified frequency and amplitude range.

INSTALLATION REQUIREMENTS. Recommendations shall be given as to suitable methods for installing the vibration machine and auxiliary equipment. Electrical and other miscellaneous requirements shall be stated.

HYDRAULIC VIBRATION MACHINE

The *hydraulic vibration machine* is a device which transforms power in the form of a high-pressure flow of fluid from a pump to a reciprocating motion of the table of the vibration machine. A schematic diagram of a typical machine is shown in Fig. 25.12. In this example, a two-stage electrohydraulic valve is used to deliver high-pressure fluid, first to one side of the piston in the actuator and then to the other side, forcing the actuator to move with a reciprocating motion. This valve consists of a pilot stage and power stage, the former being driven with a reciprocating motion by the electrodynamic driver. At the time the actuator moves under the force of high-pressure fluid on one side of the piston, the fluid on the other side of the piston is forced back through the valve at reduced pressure, and returned to the pump.

The electrohydraulic valve is usually mounted directly on the side of the actuator cylinder, forming a close-coupled assembly of massive steel parts. The proximity of the valve and cylinder is desirable in order to reduce the volume and length of the connecting fluid paths between the several spools and the actuator, thereby minimizing the effects of the compliance of the fluid and the friction to its flow. (Many types of electrohydraulic valves exist, all of which fail to meet the requirement of sufficient flow at high frequencies to give vibration machine performance equivalent to existing electrodynamic machine performance at 2,000 cps.)

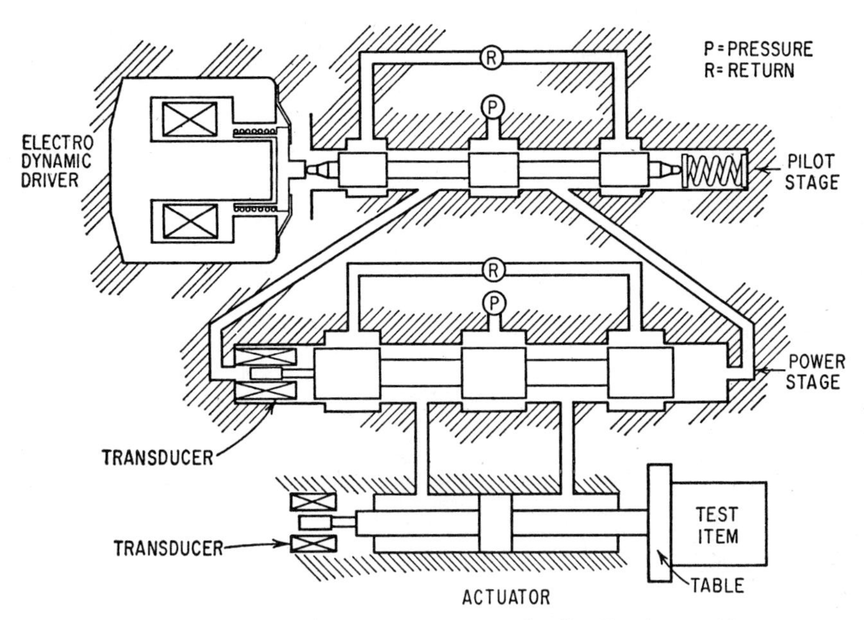

FIG. 25.12. Schematic diagram of a typical hydraulic vibration machine.

OPERATING PRINCIPLE

In Fig. 25.12, the *pilot* and *power spools* of a hydraulic vibration machine are shown in the "middle" or "balanced" position, blocking both the pump high-pressure flow P and the return low-pressure flow R. Correspondingly, the piston of the actuator must be stationary since there can be no fluid flow either to or from the actuator cylinder. If the pilot spool is displaced to the right of center by a force from the electrodynamic driver, then high-pressure fluid P will flow through the passage from the pilot spool to the left end of the power spool, causing it to move to the right also. This movement forces the trapped fluid from the right-hand end of the power spool through the connecting passage, back to the pilot stage, and then through the opening caused by the displacement of the pilot spool to the right, to the chamber R connected to the return to the pump. Correspondingly, if the pilot spool moves to the left, the flow to and from the power spool is reversed, causing it to move to the left. For a given displacement of the pilot spool, a flow results which causes a corresponding velocity of the power spool. A displacement of the power spool to the right allows the flow of high-pressure fluid P from the pump to the left side of the piston in the actuator, causing it to move to the right and forcing the trapped fluid on the right of the piston to be expelled through the connecting passage to the power spool and out past the right-hand restrictions to the return fluid chamber R. The transducers shown on the power spool and the actuator shaft are of the differential transformer type* and are used in the feedback circuit to improve system operation and provide electrical control of the average (i.e., stationary) position of the actuator shaft relative to the actuator cylinder.

A block diagram of the complete hydraulic vibration machine system is shown in Fig. 25.13. The pump, in conjunction with accumulators in the pressure and return lines at the hydraulic valve, should be capable of variable flow while maintaining a fixed pressure. Most systems to date have required an operating pump pressure of 3,000 lb/in.^{2} The upper limit of efficiency of the hydraulic valve is approximately 60 per cent, the losses being dissipated in the form of heat. Mechanical loads are seldom capable of dis-

* See Chap. 15 for a discussion of such transducers.

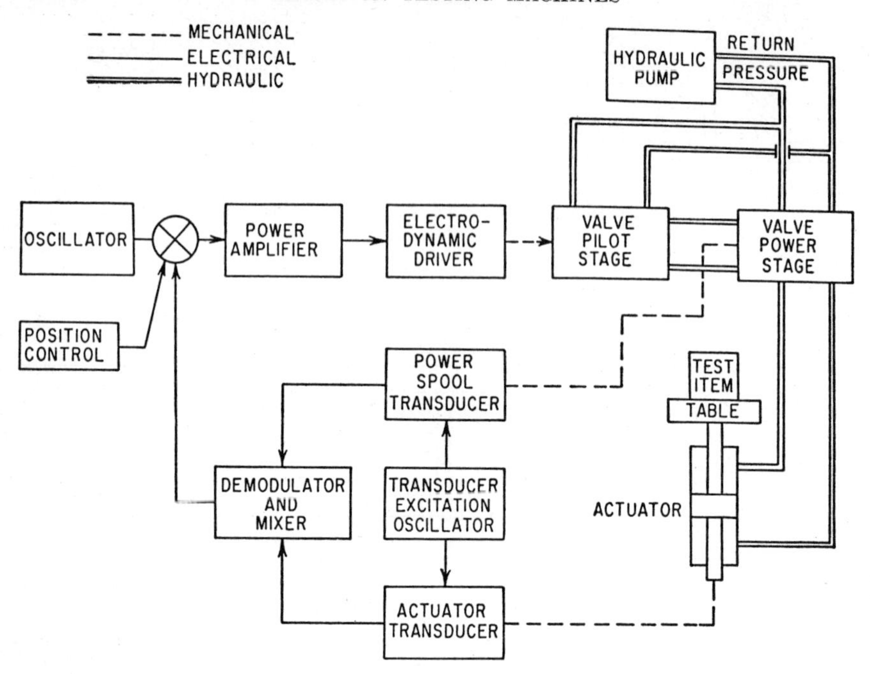

Fig. 25.13. Block diagram—hydraulic vibration machine system.

sipating appreciable power; most of the power in the pump discharge is converted to a temperature rise in the fluid. Therefore a heat exchanger limiting the fluid temperature must be included as part of the system.

PROMINENT FEATURES

1. Large generated forces or large strokes can be provided relatively easily. Large forces and large velocities of motion, made possible with a large stroke, determine the power capacity of the system. For example, one hydraulic vibration machine has a peak output power of 450,000 lb-in./sec (519,000 kg-cm/sec) or approximately 34 hp with a single electrohydraulic valve. This power can be increased by the installation of several valves on a single actuator. Appreciable increases in valve flow can be realized by sacrificing high-frequency performance. Hence, the hydraulic vibration machine excels at low frequencies where large force, stroke, and power capacity are required.

2. The hydraulic machine is small in weight, relative to the forces attainable; therefore, a rigid connection to firm ground or a large massive base is necessary to anchor the machine in place and to attenuate the vibration transmitted to the surrounding area.

3. The main power source is hydraulic, which is essentially d-c in character from available pumps. The electrical driving power for controlling the valve is small. Therefore, the operating frequency range can be extended down to zero cps.

4. The magnetic leakage flux in the region of the table is insignificant by comparison with the electrodynamic-type vibration machine.

5. The machine, with little modification, is suitable for use in high- and low-temperature, humidity, and altitude environments.

6. The machine is inherently nonlinear with amplitude in terms of electrical input and output flow or velocity.

7. The waveform of table motion in general is not as good as that of the electrodynamic type, because of distortion introduced by the valve itself and friction in the actuator which is most noticeable at low drive levels.

8. The hydraulic machine actuator consists of a cylinder, piston, and shaft which are simple in construction and not easily damaged. Forces are generated hydraulically—no primary loads are carried through bearings.

9. Sufficient force capacity generally is available to support all rated test loads hydraulically in the vertical position without resorting to external supports.

10. Hydraulic power requires larger hose sizes than an electrical cable of equal capacity. Hydraulic fluid cleanliness, seepage, and leakage are problems not encountered in electrical power systems.

APPLICATION AND SELECTION OF VIBRATION MACHINES AND EXCITERS

Some of the most common types of vibration tests are listed in this section and the desirable characteristics of the test equipment are indicated. The type of vibration exciter or machine best fitted for a given test is suggested.

THE GENERATION OF PURE SINUSOIDAL RECTILINEAR MOTION FOR THE CALIBRATION OF VIBRATION PICKUPS

The generation of vibratory motion of suitable quality for calibration work, and the determination of the amplitude, as well as the frequency, of the generated vibration, is difficult. The deviations from ideal in the form of harmonics, unwanted lateral and rotational motions superposed on the desired rectilinear motion, lack of infinite stiffness in the vibration table, etc., which can be tolerated in the general vibration test, may be intolerable in calibration work. Therefore, a vibration machine designed specifically for calibration work may have limited usefulness for general vibration testing because it necessarily must be designed to accommodate small test loads to minimize design compromises. (See Chap. 18 for calibration techniques.)

DESIRED CALIBRATOR CHARACTERISTICS. A vibration machine for calibration work should provide the following:

1. Distortion-free sinusoidal motion.
2. True rectilinear motion normal to the vibration-table surface without the presence of any other motion.
3. A rigid table for all design loads at all operating frequencies.
4. Stepless variation of frequency and amplitude of motion within specified limits, easily adjustable during operation.
5. A table that remains at ambient temperature and does not provide either a source or sink for heat regardless of the ambient temperature.
6. A mounting area at the table free from electromagnetic disturbances.
7. Horizontal as well as vertical motion, if required.
8. Features allowing the use of absolute calibration methods when calibrating the calibrator.
9. Data which can be used to define the vibration imposed on the pickup.

If a range of calibrating frequencies from 2 to 10,000 cps is desired, the electrodynamic-type machine more nearly satisfies all the requirements of the ideal calibrator. For high-frequency (1,000 to 20,000 cps) and high-acceleration calibrations using steady sinusoidal vibration, the piezoelectric type has advantages over other types. The piezoelectric type is limited to very small displacements, making low-frequency calibrations difficult, and displacement must be measured by interferometric methods or external displacement-measuring pickups, such as the capacitance or variable-reluctance types. Mechanical direct-drive calibrators are no longer used because they have an upper-frequency limit of less than 100 cps, and because of the presence of background mechanical bearing noise and the difficulty of adjusting speed and amplitude.

DETERMINING THE NATURAL MODES OF A STRUCTURE

Vibration exciters are used to determine experimentally the natural modes of a structure. The frequency, the deflection form including nodal lines, and the damping factor associated with the natural mode are usually determined for all modes of vibration of interest. Usually, the dynamic characteristics of a structure are determined by operating at vibration levels which are small compared to damaging levels. If, in the exceptional case, nonlinear characteristics are suspected and are to be evaluated, then testing levels approaching operating levels may be desirable.

Structures are vibrated by introducing a vibratory force by a vibration exciter at one or more points on the structure while the structure is supported in an appropriate manner resembling the service configuration. An excellent example of natural mode testing is found in the "ground vibration testing" of winged aircraft where the results of the tests are used to check the calculated natural modes of the aircraft wings and tail as well as the complete aircraft.

In some cases, several natural modes may be sufficiently close in frequency so that they will be excited simultaneously from a single input-force location. Since the motions from the several modes add as vector quantities throughout the structure to give the actual structural response at any point, recognition of the individual natural modes becomes difficult. A method using multiple exciters appropriately distributed throughout the structure has been developed[6] and can be used to excite individual natural modes in a complex structure. An iteration process is used to arrive at the correct magnitude and phase of output force at each exciter to overcome locally the dissipation forces associated with the mode in question. In this manner, the excitation of modes closely adjacent in frequency is minimized.

DESIRED VIBRATION EXCITER CHARACTERISTICS. The desirable characteristics of a vibration exciter for determining the natural modes of vibration of a structure are:

1. Insignificant dynamic changes in the vibration characteristics of the structure should result from the addition of the exciter to the structure. For this condition to be met, the mass, stiffness, and damping introduced by the exciter must be small compared to the effective values of the structure at the point of attachment.
2. Force and frequency should be easily controlled by the operator at a position removed from the exciter. If the structure is large, a portable control station for frequency and amplitude is sometimes helpful. With a set position of the force control, it is desirable that the force remain constant, independent of frequency and mechanical impedance of the structure.
3. Stepless variation in force and frequency should be available to the operator. The operating frequency should correspond to the set frequency, independent of the mechanical structure characteristics.
4. It should be possible to interrupt the generation of force abruptly. Furthermore, the exciter should introduce negligible damping into the vibrating structure.
5. For the testing of complex structures, a number of exciters may be required—each with the force individually controllable in magnitude and phase characteristics.
6. The exciter should be portable and easily supported adjacent to the structure under test. Heavy and stiff connections to drive the exciter should be avoided where possible.
7. The exciter should generate good waveform in order to avoid the excitation of higher-frequency modes occurring at harmonics of the driven frequency.
8. Sufficient displacement capacity should be provided in the exciter to allow for acceptable vibratory displacements at the point of attachment, as well as inadvertent positional shifts between the structure and the supports of the stationary part of the exciter.

FATIGUE TESTING MACHINES

Many types of commercial equipment are available that are designed specifically for fatigue testing. These machines usually use (1) mechanical direct-drive, rotating unbalances, or (2) a hydraulic means of displacement or force generation. However, general-purpose vibration testing machines described earlier in this chapter may be employed. In some test setups, force amplification is obtained by mechanical resonance.

The operating frequency of the testing machine should be as high as possible in order to shorten the time of the test, but the frequency is limited by the temperature rise of the test item if the endurance limits are dependent upon temperature.

A typical fatigue testing machine includes some or all of the following general features:

1. It should be capable of maintaining stress or strain in the test item within stated limits for the duration of the test.
2. Steady loads, applied simultaneously with vibratory loads, should be available when required by the test procedure.
3. Provisions should be available for maintaining elevated or reduced temperatures in the test item.
4. The exciting unit should operate without inadvertent transients or other malfunctions which could be transmitted to the test item, thereby complicating the analysis of the test results.

APPLICATION OF VIBRATION MACHINES AS FATIGUE TESTING MACHINES. In selecting a vibration machine to perform a fatigue test, it is best to select a type that inherently produces the output desired. For example, if a fixed displacement is desired, it is advantageous to select a direct-drive-type machine. If a constant force is desired, it is advantageous to select a machine that incorporates a rotating unbalance at a fixed frequency since this type of machine generates a constant force against a point of high mechanical impedance. If the mechanical impedance is not high, then a machine of the type shown in Fig. 25.6 can provide a constant force.

THE USE OF MECHANICAL RESONANCE. In fatigue testing, vibration machines are used which inherently give a constant output; the machines are operated at frequencies removed from mechanical resonances; and they require only occasional manual adjustments for maintaining a constant output. By making use of mechanical resonance for force amplification, with a given capacity of vibration exciter, a large increase in performance can be realized in some tests. This increase is possible if the test fixture is carefully designed, and if a continuous control on the output of the exciter is provided. In addition, the damping of the test item must be small to allow amplification of force by resonance. Additional advantages from a resonant-system test include:

1. The generation of the desired motion, to the exclusion of all other motions and with good waveform.
2. The generation of bending stresses, direct tension-compression stresses, or torsion stresses with a general-purpose vibration exciter.
3. The reduction of damaging impact loads transmitted to the exciter which originate from free play or chatter in the test item, by allowing the use of a relatively large mass between the exciter and the test item.
4. The concentration of high stress in local areas of choice in order to reproduce service failures under laboratory testing procedures. This is possible in many instances by selecting the proper fixture and natural mode of vibration.

For example, in Fig. 25.14, a resonant system for producing bending moments is shown. The massive legs provide a large mass moment of inertia at each end of the test item. The bending stiffness of the test item and the mass properties of the legs determine the resonant frequency. Some axial force, in addition to the moments, is placed on the test item. The axial force can be eliminated by extending the length of the legs equally on both sides of the test item. A spring between the extremities of the legs can be used for a preload moment. The exciter and test system are flexibly supported from the floor to eliminate the transmission of unwanted disturbances to the surrounding area and to eliminate unpredictable power losses through the supports. A single vibration exciter is connected to one leg—although two may be used, one connected to each leg, if more power is required than is available from a single unit. If two exciters are used, they are normally driven from a common power source to ensure synchronism. A connection is used between the exciter and leg which gives axial stiffness and yet is soft to bending and

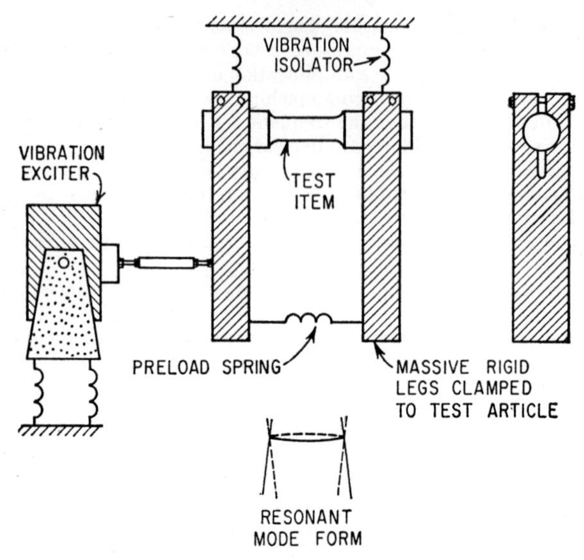

FIG. 25.14. A resonant fatigue test: bending stress.

shearing deflections, which is necessary to allow the leg to assume its pendulous motion while the exciter moves with rectilinear motion. Necked-down connectors on either end of a stiff member are satisfactory for this purpose. Moving the exciter toward (or away from) the test item along the leg increases (or decreases) the mechanical impedance, respectively, at the point of attachment, allowing maximum values of both force and velocity to be generated in the exciter for maximum power transfer to the test item. With good design, the desired mode of operation can be made high in frequency—consistent with dissipation and power considerations; the resonant frequency can be selected so that it is not near the natural frequencies of all other modes of the system and only a single mode of vibration is excited.

Figure 25.15 is similar to Fig. 25.14 except tension-compression forces are generated in the test item. In this case, the plate at the top of the legs serves as an ideal hinge.

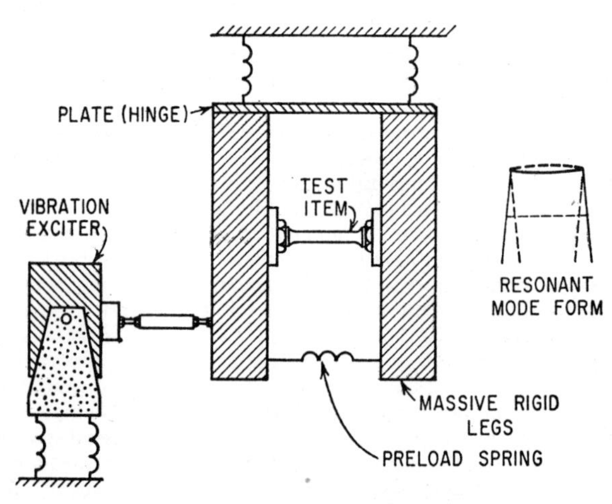

FIG. 25.15. A resonant fatigue test: tension-compression stress.

It should be stiff in its own plane, but soft to bending. Figure 25.16 shows a suggested configuration for torsion motion. The mechanical impedance seen by the exciter in this case can be adjusted by changing the radial location of the point of exciter attachment and the ratio of the top and bottom mass moments of inertia. Reducing the top inertia, with respect to the bottom inertia, reduces the amplitude of rotational displacement at the bottom for a given rotational strain in the test item—thereby increasing the mechanical impedance at that end.

Figure 25.17 illustrates how bending moments in the plane of the paper can be maximized at several locations along the length of the crankshaft. In Fig. 25.17*A*, with the exciter connected in the center of the crankshaft, the natural mode shown gives a zero bending moment at the ends and a maximum moment at the center. In Fig. 25.17*B*, using large mass moments of inertia at both ends, and driving horizontally at one end,

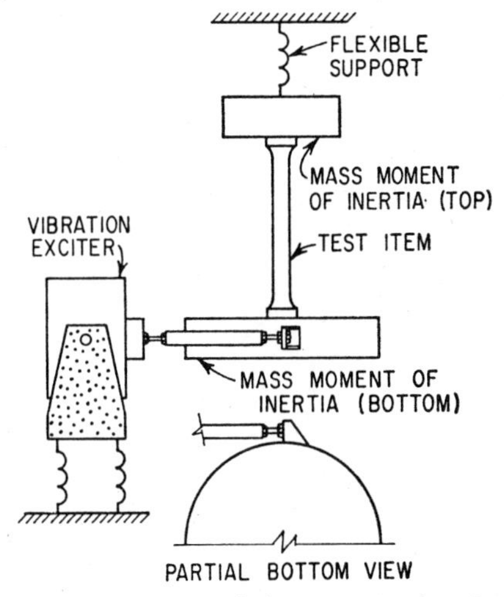

FIG. 25.16. A resonant fatigue test: torsion stress.

gives a nearly constant moment throughout the length of the shaft if the end masses are large compared to the distributed mass of the shaft. In Fig. 25.17*C*, driving at one of the crankthrow bearings and operating at the mode shown gives a zero moment at each end and at the center, with the maximum moments occurring at the two positions between the three zero moment locations. By adding large end-mass moments of inertia and exciting the mode in Fig. 25.17*D*, the maximum bending moment occurs at the two ends, with a zero moment in the center. Tests of this kind permit failures to be duplicated in the laboratory which have been experienced in the field. They also allow one to determine the effect of oil holes, fillets, section changes, materials, etc., on the endurance limit of the complete shaft.

In fatigue tests of the type described, it is usually necessary to measure stress or strain at an appropriate point on the test piece, and to use the voltage generated by the measuring device to control the operation of the system, so as to maintain, automatically, the set operating level. Control can be performed manually by an operator. In this case the operator must maintain a constant meter reading of stress or strain in the test article—this can be tedious.

In the examples shown, a large amplification of force can be achieved. For example, in Fig. 25.15, amplifications of the vibration exciter force of 50 to 250 times in the test item are achieved readily. However, if the item is tested to failure, within a relatively few cycles of stress near or beyond the yield point, a large part of the energy expended in deforming the test item is not returned to the system in the form of kinetic energy upon removal of the load. In this case, large power may be dissipated, and the amplification of mechanical force attainable at resonance is reduced accordingly.

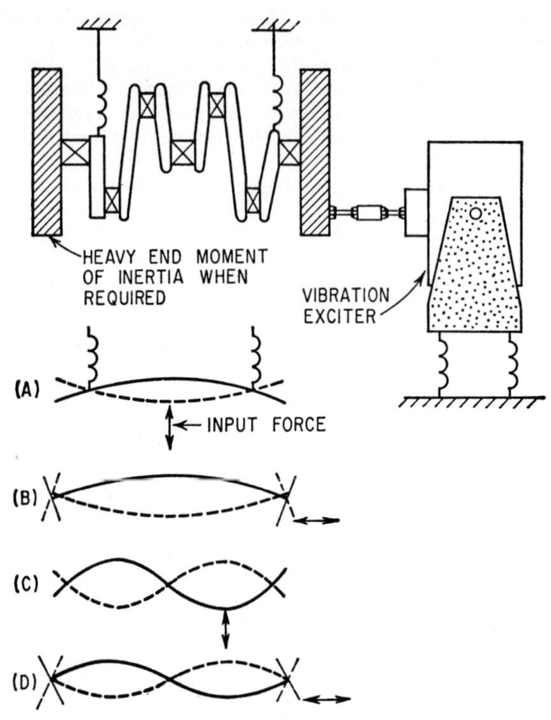

FIG. 25.17. A resonant fatigue test: selective location of maximum bending moment.

Types of Exciters Employed in Resonant Fatigue Testing Machines. Vibration exciters which are applied to resonant fatigue testing include the following types: rotating unbalance, electrodynamic, hydraulic, electromagnetic, and pneumatic. In general, for this application, force-generating machines are employed rather than displacement-generating machines. The direct-drive mechanical-type exciter can be included if a spring is added between the output of the exciter and the input to the structure. By comparison with other types, the application of this exciter is awkward and difficult. The piezoelectric type is excluded because of its small displacements and small power capacity. The electromagnetic and pneumatic types offer the advantage of not requiring physical contact with the resonant member. The rotating-unbalance and hydraulic types can be applied to resonant fatigue tests although their characteristics of constant output force (with fixed frequency) and large output force potential, respectively, particularly suit them for nonresonant-type fatigue tests. The electrodynamic vibration exciter offers some advantages for resonant fatigue tests in the form of a wide frequency range of operation and ease of control.

RESONANT-LOAD EFFECTS ON THE VIBRATION EXCITER. In applying vibration exciters to resonant fatigue testing machines the following considerations are important:

1. With an approximately constant input force applied to a resonant structure, an increase in amplitude occurs at resonance, as a result of which the power absorbed by the structure is increased.

2. If the power originates from a variable-speed motor (e.g., from a rotating-unbalance or motor-alternator type of supply for the electrodynamic exciter), it may be impossible to adjust the operating frequency exactly to the resonant frequency. This is because of

the rapidly increasing torque placed on the motor with increasing frequency below resonance, and the rapidly decreasing torque with increasing frequency above resonance. Unless sufficient speed regulation exists, the motor may tend to avoid the speed corresponding to a structure resonance (1) by slowing down below resonance or (2) (with a slight upward adjustment in frequency setting) by increasing its speed and passing through the resonant frequency—stabilizing at a speed in excess of resonance. The use of an electronic amplifier and frequency source will avoid this problem since the mechanical load impedance is not reflected back to the oscillator which controls the frequency of operation.

3. If a measurement of the amplitude of motion at the structure is employed to control the amplitude of output of the exciter, note that a system (described in item 2 above) which is stable with respect to frequency and operable at resonance with fixed or manual amplitude control, may become unstable with respect to frequency if automatic amplitude control is added.

4. If the operating frequency is adjusted for maximum amplitude at resonance, assuming a force amplification of at least several times, any variations in the properties of the test item or vibration exciter frequency will cause the amplitude to diminish. It is desirable to maintain constant amplitude conditions for a fatigue test, but if inadvertent variations occur, decreases in amplitude can be corrected far more easily than increases in amplitude.

5. The electrodynamic vibration exciter will tend to adjust its driving force as required to maintain a constant amplitude of motion at a given operating frequency with a fixed position of the amplitude control, (*a*) if an electrodynamic vibration exciter is used below electrical resonance, (*b*) if the part of the impedance measured at the electrical terminals of the driver coil, due to the motion of the driver coil, is large compared to the impedance of the driver-coil electrical inductance and resistance, and (*c*) if the internal impedance of the power supply is small. By making use of this characteristic and by adjusting for the peak amplitude at the resonant frequency, as described in item 4, many different types of specimens can be fatigue-tested in a resonant system without automatic controls and with only occasional minor adjustments to the operating controls.

6. Many vibration exciters are rated in terms of rigid-mass loads. In resonant tests, the exciter may easily attain accelerations in excess of those corresponding to full rated force, even with no mechanical load. These accelerations should be maintained below the manufacturer's stated maximum. In the case of Fig. 25.15, for example, for a given strain across the test article the acceleration at the exciter can be varied by moving the exciter along the length of the leg. Operation at resonance can absorb more power from the exciter than that anticipated by the manufacturer, who usually designs the system only (*a*) for mass load conditions, (*b*) for overloading of the driving motor, in the case of the rotating unbalance, or (*c*) for overloading of the motor-alternator, in the case of the electrodynamic exciter.[4] In the case of an electronic amplifier, sufficient voltage may not be provided to generate maximum velocity and force simultaneously in the electrodynamic exciter. For example, consider one commercially available electrodynamic vibration machine which is driven by an electronic amplifier power source. The system is rated at 1,200 lb peak sinusoidal generated force for any mass load on the table, the amplifier capacity is 2,900 watts, and the maximum table displacement is 1 in. peak-to-peak. Suppose that this vibration machine were attached to a resonant load and its acceleration were limited to 100*g* for structural strength reasons. If the mechanical impedance of the resonant load has a value such that a 1,200 lb peak sinuosidal force produces a 1-in. peak-to-peak displacement at approximately 44 cps (the frequency of maximum velocity at 100*g* acceleration), the mechanical power absorbed by the mechanical impedance would be 9,400 watts, or more than three times the power normally provided by the power source. This excludes the power loss in the electrodynamic exciter itself, which is significant.

7. If the driver-coil electrical impedance of an electrodynamic exciter is determined mainly by its mechanical motion, the driver-coil circuit must be protected against large currents and forces generated by the exciter should the mechanical connection between it and the structure fail, or should changes in the structure destroy the resonant operating condition as a result of incipient failure.

FIXED-FREQUENCY VIBRATION TEST

Some test specifications for equipment require that the test item be vibrated for extended periods at a frequency corresponding to the frequency of the most serious resonance in the equipment or, if no resonances are found, at some arbitrary nonresonant frequency. The amplitude of motion of the vibration table is specified as a function of frequency for the test. When searching for resonant frequencies, the frequency is slowly changed by adjusting the frequency manually. This search may be conducted at reduced amplitudes compared to those specified for the endurance test.

It may be possible to use a machine which has insufficient capacity when used to vibrate the loads directly. First search for resonances at reduced vibration levels which this machine will provide; then obtain the required output vibration force at the fixed-frequency endurance run by force amplification from a mechanical resonant system. A free-free uniform beam employed as a resonant system is simple to construct and has some advantage compared to many other types of resonant systems. As an example of such a system, see Fig. 25.18. Some adjustment in frequency can be obtained by adding weights at the extremities of the beam, as shown schematically in Fig. 25.19.

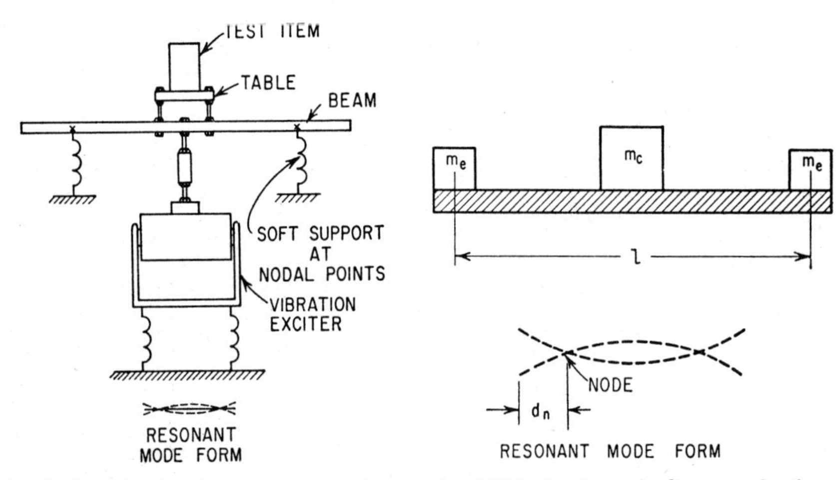

FIG. 25.18. A free-free beam resonant system.

FIG. 25.19. A schematic diagram of a free-free beam with concentrated weights at the center and at the ends.

The natural frequency of the lowest mode of a free-free beam system of Fig. 25.19 is, approximately,

$$f_n \approx \frac{\pi}{2}\left[\frac{EI(m_c + 2m_e + \mu l)}{l^3[2m_e(2m_c + \mu l) + m_c\mu l(3 - 8/\pi) + (\mu l)^2(1 - 8/\pi^2)]}\right]^{1/2} \tag{25.1}$$

where f_n = natural frequency, cps
E = Young's modulus of beam material, lb/in.^{2}
I = area moment of inertia of beam cross section, in.^{4}
m_c = mass at center of beam, including that of test article and vibration exciter element attached to beam, lb sec^2/in.
m_e = mass at each end of beam, lb sec^2/in.
l = total separation of end masses m_e if m_e is large compared to μl over the same spanwise distance; otherwise, the length of the beam, in.
μ = beam mass per inch of length, lb sec^2/in.

The nodal lines pass through the beam parallel to the end and midway between the upper and lower surfaces at a distance from each end approximately equal to

$$d_n \approx \frac{l}{\pi} \sin^{-1} \left(\frac{[2\mu l/\pi] + m_c}{\mu l + 2m_e + m_c} \right) \tag{25.2}$$

where the angle is expressed in radians. Equations (25.1) and (25.2) are based on an assumed sinusoidal deflection form of the beam, and only translational motions are considered.

If the concentrated masses m_c and m_e are very small compared to the mass of the beam, the beam can be treated as a uniform beam without added loads. In this case, if the material of the beam is steel and it has a length l and a thickness t, both expressed in inches, the natural frequency is approximately 209,000 t/l^2 cps; the nodal line is located approximately $0.23l$ in. from each end. The frequency is not dependent upon the width of the beam since both the mass and stiffness are directly proportional to the width. The test item is assumed to be a rigid mass connected to the center of the beam. The beam should be designed for vibratory stresses well below the endurance limit of its material at the maximum desired deflection.

Electromagnetic, hydraulic, or rotating-unbalanced exciters may be employed in this system. However, the electrodynamic vibration exciter is particularly well suited for this type of system because of its wide range of operating frequencies and ease of control. The system is awkward in that mechanical changes must be made to change the resonant frequency, but it has the advantage of imposing almost pure sinusoidal, rectilinear motion on the test item at its mounting points. This motion may be imparted to the test item even though its center-of-gravity is displaced horizontally from the center of its mounting, since the beam can be made quite massive and of large dimension by comparison to a typical vibration machine. A resonant-beam system[7] has been constructed which is adjustable in frequency from 5 to 500 cps, with mass loads ranging up to 200 lb, and which will impart $20g$ acceleration to the mass load with the maximum input force not exceeding 200 lb. This corresponds to a force amplification of 20 times throughout the frequency range. If the test item adds sufficient damping to the resonant beam mode or if it acts as a tuned vibration absorber at the resonant mode frequency, little or no amplification of vibration exciter force may be provided by a resonant beam system.

ENVIRONMENTAL VIBRATION TESTING

Specifications usually require that rectilinear motion be applied in each of three mutually perpendicular directions relative to the test item. With the exception of some of the mechanical types of vibration machines, all others generate motion normal to the vibration table surface, making it necessary for the test item to be reoriented relative to the table in order to produce vibratory motion in each of the three directions. This requires a versatile attachment between the table and test item or possibly three different attaching fixtures. If the direction of gravity is important to the test item and the vibration machine generates motion in one direction only, the complete machine must be rotated from vertical to horizontal, or vice versa, to meet the test conditions.

In order to obtain horizontal motion in the plane of a horizontal table, stiff tables, supported by elastic restraints which permit motion only in the horizontal direction, are used, driven usually by an electrodynamic machine rotated to give horizontal motion. For example, see Fig. 25.20A. Vibration in the plane of the table is more difficult to achieve than vibration normal to the plane of the table because the test article necessarily has dimensions of height separating its center-of-gravity vertically from the table surface. Accordingly, reactionary moments as well as forces are transmitted to the table. Tables for test items in excess of 50 lb weight are difficult to construct for use above 500 cps with acceptable performance; the large weight added to the vibration machine reduces the attainable vibration levels, reduces the frequency of the inevitable flexing and in-plane resonant modes of the horizontal table, and emphasizes the requirement of attachment to the vibration machine with sufficient rigidity. (Laminated Sitka-spruce-wood tables, having a depth approaching the horizontal dimension, and

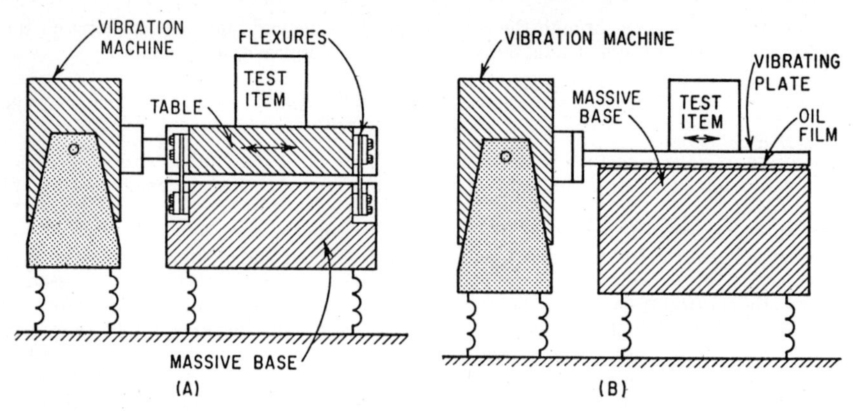

Fig. 25.20. An auxiliary horizontal vibration table with vibration in the plane of the table: (*A*) Elastic support. (*B*) Fluid film support.

having metal inserts at attaching points, are used by one environmental testing laboratory.) Good performance is claimed by several users[8, 9] of a horizontal plate driven by an electrodynamic vibration machine. The plate is separated from a flat opposing surface of a massive block by an oil film, as shown in Fig. 25.20*B*. The thin oil film provides little shearing restraint but great stiffness normal to the surface, the stiffness being distributed uniformly over the complete horizontal area. Accordingly, a relatively light moving plate can be vibrated which has the properties of the massive rigid block in the direction normal to its plane.

In general, in the low-frequency category the mechanical direct-drive and hydraulic vibration machines are suitable for the lowest frequencies and largest displacements. If displacements are less than 1 in. and frequencies above 5 cps, the mechanical reaction-type machine may be considered. All three of these machines can be fitted with large massive tables, simplifying the attachment of the test article to the vibration table. The electrodynamic vibration machine can be used for displacements below 1 in. and frequencies above 2 to 5 cps. However, large, heavy test items in this case may require external flexible supports to avoid forcing the table against its stops. In the intermediate-frequency range, the electrodynamic and hydraulic vibration machines may be used. In the wide-band range, the electrodynamic machine is most widely used.

REFERENCES

1. Bernhard, R. K.: *Proc. ASTM*, **49**:1016 (1949).
2. Usher, T.: *AIEE Paper* No 58:1237, 1958.
3. Oleson, M. W., and R. E. Blake: *Publication Board Rept.* 132741, Washington, D.C. 1957.
4. Reen, G. K.: *MB Vibration Notebook*, **2**(1):3 (1956). MB Manufacturing Co., New Haven, Connecticut (1956).
5. Curtis, A. J.: *Shock and Vibration Bulletin No.* 42, *part III*, p. 1, U.S. Naval Research Laboratory, Washington, D.C., January, 1972.
6. Lewis, R. C., and D. L. Wrisley: *J. Aeronaut. Sci.*, **17**:705 (1950).
7. Granick, N., D. C. Kennard, and K. Unholtz: *Shock and Vibration Bulletin 17*, p. 84, U.S. Naval Research Laboratory, Washington, D.C., March, 1951.
8. Hansen, W. O.: 25th *Shock and Vibration Bulletin, part II*, p. 93, U.S. Naval Research Laboratory, Washington, D.C., December, 1957.
9. Klein, F. P.: *Rept.* GM-TN-0165-00141, Space Technology Laboratories, Inc., Los Angeles, May, 1958.
10. Klapper, J., and C. M. Harris: *J. Audio Eng. Soc.*, **8**:177 (1960).

26

SHOCK TESTING MACHINES

Jerry R. Sullivan
U.S. Naval Ship Engineering Center

Irwin Vigness
U.S. Naval Research Laboratory

INTRODUCTION

Equipment must be built sufficiently rugged to operate satisfactorily in the shock and vibration environments to which it will be exposed and to survive transportation to the site of ultimate use. To ensure that the equipment is sufficiently rugged and to determine what its mechanical faults are, it is subjected to controlled mechanical shocks on shock testing machines. The severity and nature of the applied shocks are usually intended to simulate environments expected in later use, or to be similar to important components of those environments. However, a principal characteristic of shocks encountered in the field is their variety. These field shocks cannot be defined exactly. Therefore shock simulation can never exactly duplicate shock conditions that occur in the field.

There is no general requirement that a shock testing machine reproduce field conditions. All that is required is that the shock testing machine provide a shock test such that equipment which survives be acceptable under service conditions. Assurance that this condition exists requires a comparison of shock test results and field experience extending over long periods of time. This comparison is not possible for newly developed items. It is generally accepted that shocks that occur in field environments be measured; and that shock machines simulate the important characteristics of shocks that occur in field environments, or that the shock machines have a damage potential which by analysis is shown to be similar to that of a composite field shock environment against which protection is required.

A *shock testing machine* (frequently called a *shock machine*) is a mechanical device that applies a mechanical shock to an equipment under test. The nature of the shock is determined from an analysis of the field environment.* Tests by means of shock machines usually are preferable to tests under actual field conditions for four principal reasons: (1) The nature of the shock is under good control, and the shock can be repeated with reasonable exactness. This permits a comparative evaluation of equipment under test, and allows exact performance specifications to be written. (2) The intensity and nature of shock motions can be produced which represents an average condition for which protection is practical, whereas a field test may involve only a specific condition that is contained in this average. (3) The shock machine can be housed at a convenient location with suitable facilities available for monitoring the test. (4) The shock machine is relatively inexpensive to operate so that it is practical to perform a great number of developmental tests on components and subassemblies in a manner not otherwise practical.

* A mechanical shock occurs when the position of an item is significantly changed in a relatively short time in a nonperiodic manner. It is characterized by suddenness and develops significant inertial forces in the item.

SHOCK-MACHINE CHARACTERISTICS

DAMAGE POTENTIAL AND SHOCK SPECTRA

The damage potential of a shock motion is dependent upon the nature of an equipment subjected to shock, as well as upon the nature and intensity of the shock motion. To describe the damage potential, a description of what the shock does to an equipment must be given—a description of the shock motion is not sufficient. To obtain a comparative measure of the damage potential of a shock motion, it is customary to determine the effect of the motion on simple mechanical systems. This is done by determining the maximum responses of a series of single degree-of-freedom systems to the shock motion, and considering the magnitude of the response of each of these systems as indicative of the damage potential of the shock motion to these "standard" systems. The responses are plotted as a function of these natural frequencies. A curve representing these responses is called a *shock spectrum*, or *response spectrum* (see Chap. 23). Its magnitude at any given frequency is a quantitative measure of the damage potential of a particular shock motion to a single degree-of-freedom system of that natural frequency. This concept of the shock spectrum originally was applied only to undamped single degree-of-freedom systems, but the concept has been extended to include systems in which any specified amount of damping exists.

The response of a simple system can be expressed in terms of the relative displacement, velocity, or acceleration of the system. It is customary to define velocity and acceleration responses as $2\pi f$ and $(2\pi f)^2$ times the maximum displacement response, where f is frequency in cycles per second. The corresponding response curves are called *displacement, velocity*, or *acceleration shock spectra*. A more detailed discussion of shock spectra is given in Chap. 23.

MODIFICATION OF CHARACTERISTICS BY REACTIONS OF TEST ITEM

The shock motion produced by a shock machine may depend upon the mass and frequency characteristics of the item under test. However, if the effective weight of the item is small compared with the weight of the moving parts of the shock machine, its influence is relatively unimportant. Generally, however, the reaction of the test item on the shock machine is appreciable and it is not possible to specify the test in terms of the shock motions unless large tolerances are permissible. The test item acts like a dynamic vibration absorber (see Chap. 6) for fixed-base * natural frequencies of the test item (antiresonant frequencies of the shock-machine test-item system). If the item is relatively heavy, this causes the shock spectra of the exciting shock to have minima at these frequencies; it also causes its mounting foundation to have these minima during shock excitation at field installations. Shock tests and design factors are sometimes established on the basis of an envelope of the maximum values of shock spectra. However, maximum stresses in the test item will most probably occur at the antiresonant frequencies where the shock spectrum exhibits minimum values. To require that the item withstand the upper limit of spectra at these frequencies may result in overtesting and overdesign. Considerable judgment is therefore required both in the specification of shock tests and in the establishment of theoretical design factors on the basis of field measurements. See Chap. 42 for a more complete discussion of this subject.

DOMINANT FREQUENCIES OF SHOCK MACHINES

The shock motion produced by a shock machine may exhibit frequencies that are characteristic of the machine. The frequencies may be affected by the equipment under test. The probability that these particular frequencies will occur in the field is no greater than the probability of other frequencies in the general range of interest. A shock test,

* Fixed-base natural frequencies of a flexibly mounted item are those which would be calculated if the base on which the item is mounted were rigid and of infinite mass. They correspond to antiresonant frequencies of the system.

therefore, discriminates against equipment having elements whose natural frequencies coincide with frequencies introduced by the shock machine. This may cause failures to occur in relatively good equipment whereas other equipment, having different natural frequencies, may pass the test even though of poorer quality. Because of these factors, there is an increasing tendency to design shock machines to be as rigid as possible, so that their natural frequencies are above the range of frequencies that might be strongly excited in the equipment under test. The shock motion is then designed to be the simplest shape pulse that will give a desired shock motion or spectrum.

CALIBRATION

A *shock-machine calibration* is a determination of the shock motions, or spectra, generated by the machine under standard specified conditions of load, mounting arrangements, methods of measurement, and machine operation. The purpose of the calibration is not to present a complete study of the characteristics of the machine, but rather to present a sufficient measure of its performance to ensure the user that the machine is in a satisfactory condition. Measurements should therefore be made under a limited number of significant conditions that can be accurately specified and easily duplicated. Calibrations are usually performed with dead-weight loads rigidly attached to the shock machine.

The statement of calibration results must include information relative to all factors that may affect the nature of the motion. These include the magnitude, dimensions, and type of load; the location and method of mounting of the load; factors related to the operation of the shock machine; the locations and mounting arrangements of pickups; and the frequency range over which the measurements extend.

SPECIFYING A SHOCK TEST

Two methods of specification are employed in defining a shock test: (1) a specification of the shock motions (or spectra) to which the item under test is subjected, and (2) a specification of the shock machine, the method of mounting the test item, and the procedure for operating the machine.[1]

The first method of specification can be used only when the shock motion can be defined in a reasonably simple manner, and when the application of forces is not so sudden as to excite structural vibration of significant amplitude in the shock machine. If equipment under test is relatively heavy, and if its normal modes of vibration are excited with significant amplitude, the shock motions are affected by the load; then the specified shock motions should be regarded as nominal. If comparable results are to be obtained for tests on different machines of the same type, the method of mounting and operational procedures must be the same.

The second method of specification for a shock test assumes that it is impractical to specify a shock motion because of its complexity; instead, the specification states that the shock test shall be performed in a given manner on a particular machine. The second method permits a machine to be developed and specified as a standard shock testing machine. Those who are responsible for the specification then should ensure that the shock machine generates appropriate shock motions. This method avoids a difficulty that arises in the first method when measurements show that the shock motions differ from those specified. These differences are to be expected if load reactions are appreciable and complex.

A shock testing machine must be capable of reproducing shock motions with good precision for purposes of comparative evaluation of equipment and for the determination as to whether a manufacturer has met contractual obligations. Moreover, different machines of the same type must be able to provide shocks of equivalent damage potential to the same types of equipment under test. Precision in machine performance, therefore, is required on the basis of contractual obligations and for the comparative evaluation of equipments even though it is not justified on the basis of knowledge of field conditions.

Sometimes equipment under test may fail consistently to meet specification requirements on one shock machine, but may be acceptable when tested on a different shock machine of the same type. The reason for this is that small changes of natural frequencies and of internal damping, of either the equipment or the shock machine, may cause large changes in the likelihood of failure of the item. Results of this kind do not necessarily mean that a test has been performed on a faulty machine; normal variations of natural frequencies and internal damping from machine to machine make such changes possible. However, standard calibrations of shock machines should be made from time to time to ensure that significant changes in the machines have not occurred.

SHOCK TESTING MACHINES[2]

CHARACTERISTIC TYPES OF SHOCKS

The shock machines described below are grouped according to types of shocks they produce. When a machine can be classified under several headings, it is placed in the one for which it is primarily intended. The types of shock, shown in Fig. 26.1, are classified as: (*A*), velocity shock; (*B*) through (*D*), simple shock pulses; (*E*), single complex shock; and (*F*), multiple shock. Shocks not included in this classification are grouped under a miscellaneous category.

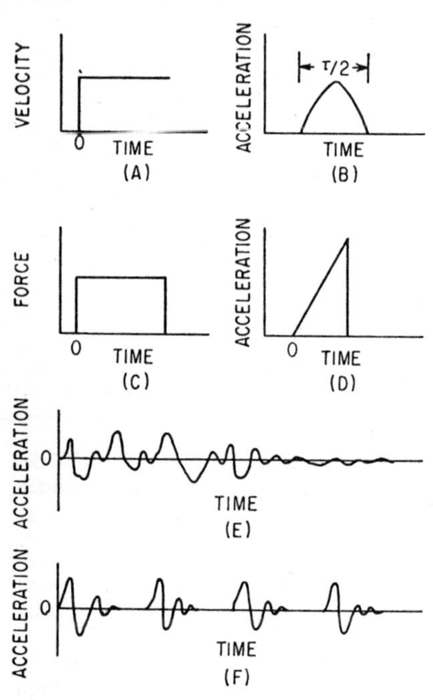

FIG. 26.1. Characteristic types of shocks. (*A*) Velocity shock, or step velocity change. (*B*) Simple half-sine acceleration shock pulse. (*C*) Rectangular force pulse. (*D*) Sawtooth acceleration pulse. (*E*) Single complex shock. (*F*) Multiple shock.

These types of shock are defined as follows: A *velocity shock* consists of a sudden change of velocity. If the change of velocity takes place in a time that is short compared with principal natural periods of vibration of the item subjected to the shock, it is considered to be a sudden change of velocity. For a velocity shock, the magnitude of the acceleration involved in the velocity change is unimportant as the item is isolated from the maximum acceleration by the flexibility of its parts. The magnitude of the velocity change defines the intensity of the shock. A *simple shock pulse* is said to exist if the shape of the shock motion-time, or force-time, curve is known and can be expressed in practical mathematical form. Usual shapes include the half sine-wave acceleration pulse, the rectangular force pulse, and other simple shaped motions or force functions. A *single complex shock* is any transient motion, or force function, caused by a single primary excitation, and that cannot be expressed practically in mathematical terms. A *multiple shock* consists of a number of individual shocks repeated over a relatively long time interval. Shocks which do not fall into the above categories are labeled "miscellaneous."

VELOCITY SHOCK

DROP TESTS. A *drop tester* may be either of two types. It may consist of a hammer assembly that drops upon a test item in order that the resilience of the item (usually a cushioning material) can be determined; or it may be a device by means of which an item under test (a package or a packaged item) can be dropped in a prescribed manner onto a suitable surface.

DROP TESTS FOR SHIPPING CONTAINERS. Drop testers, in contrast to most shock machines, can be made in a variety of ways without appreciably altering the results of tests performed on them. The American Society for Testing Materials has stated, in part, that the shipping container drop tester shall be any suitable apparatus that conforms to the following requirements:[3]

1. Permits accurate pre-positioning of the container to ensure a free fall and impact at the exact places and in the direction desired.
2. Permits accurate and convenient control of the height of drop.
3. Utilizes lifting devices that will not damage the containers.

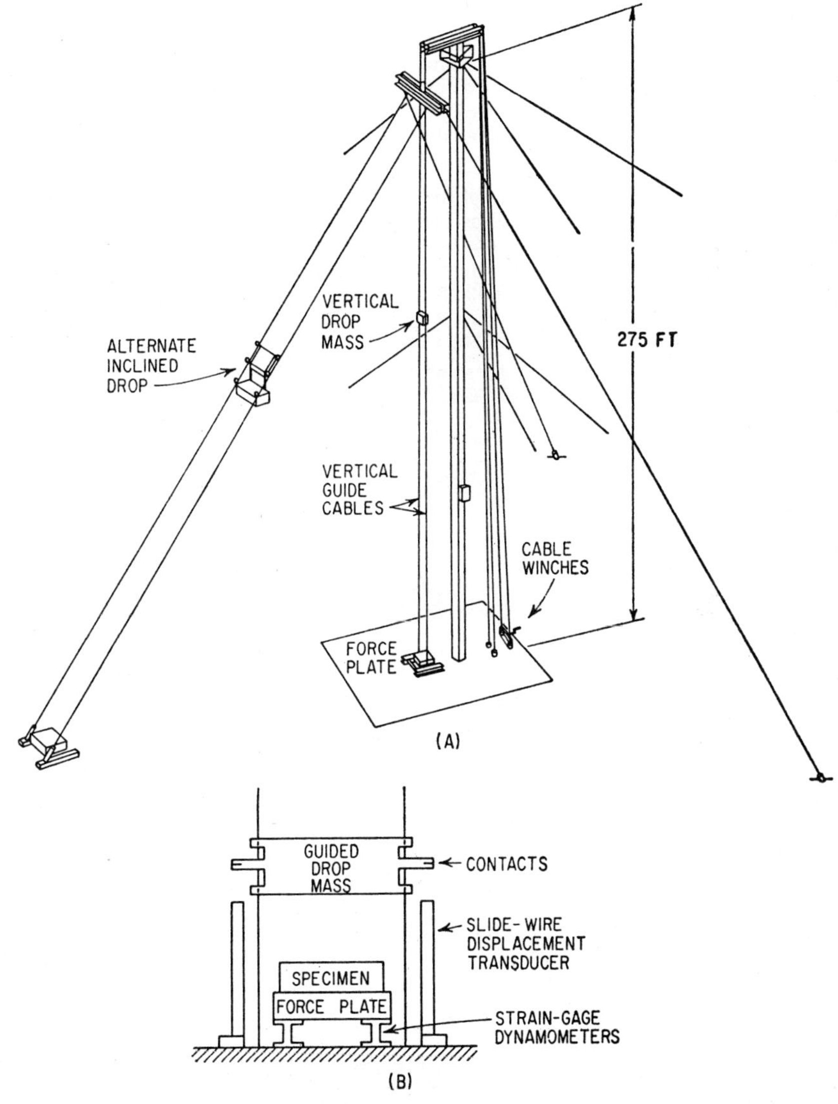

FIG. 26.2. (*A*) 275-ft drop-test facility. (*B*) Force plate for force-displacement measurements. (*After Matlock, Ripperger, Turnbow, and Thompson.*[4])

4. Permits a free unobstructed fall.
5. Provides for variations in height of drops within limits of anticipated requirements.
6. Provides a solid surface of concrete, stone, or steel of sufficient mass to absorb all shock without appreciable deflection.

DROP TESTS FOR AIR-DROP DELIVERY. The military services sometimes require that supplies be dropped from airplanes. Parachutes are used only if required for equipment survival, and if used are such as to permit maximum tolerable impact velocities. A simulation of these conditions requires impact velocities up to several hundred feet per second. In order to obtain such velocities within a relatively small distance, the item must be accelerated by a force in addition to that caused by gravity (this type of tester is considered in a later section), or very large drop heights must be employed.

A drop tester is shown in Fig. 26.2*A* where a drop height up to 275 ft can be obtained.[4] The installation illustrated is used primarily to test package cushioning materials. Weights, guided by cables, are dropped onto the cushioning material and measurements of the force-deflection curve are made by means of the arrangement shown in Fig. 26.2*B*. This installation also permits an inclined drop so that a horizontal component of velocity is obtainable.

INCLINED PLANE TEST. Some of the shocks received by packages because of careless handling, or encountered in railroad trains during switching, are simulated by the inclined plane test. One such test has been standardized as an *Incline Impact Test for Shipping Containers.*[5] This is also referred to as a *Conbur Test.* The testing machine employed for this test, shown in Fig. 26.3, consists of a rolling carriage and an inclined

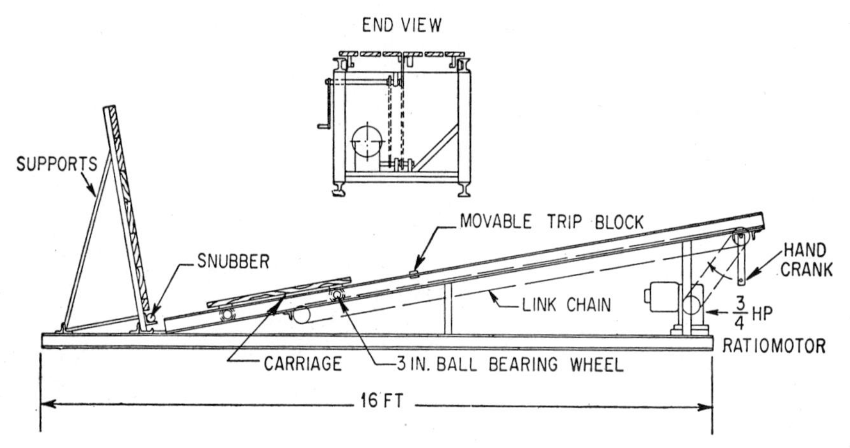

FIG. 26.3. Conbur tester; apparatus for incline-plane impact test. The carriage, which is 42 in. long and 36 in. wide, has a maximum run of 116 in. and a maximum vertical drop of 20 in. The test item is placed on the carriage so that it strikes a 55- by 58½-in. barrier, consisting of 1⅝-in. planking, when the carriage is rolled down the incline. (*L.A.B. Corp.*)

track which is terminated by a barrier at its lower end. The item under test is placed on the carriage so that it will strike the barrier when the carriage is rolled down the incline. This testing machine has been standardized so that different laboratories can obtain comparable results, and so that the test may be cited in specifications. The standard provides detailed instructions relating to the construction, calibration, and use of this testing machine.

MACHINES FOR PRODUCING A SIMPLE SHOCK PULSE

Although shocks encountered in the field are usually complex in nature (for example, see Fig. 26.1*E*), it is frequently advantageous to simulate a field shock by a shock of mathematically simple form. This permits designers to calculate equipment response more easily and allows tests to be performed that can check these calculations. This technique is additionally justifiable if the pulses are shaped so as to provide shock spectra similar to those obtained for a suitable average of a given type of field condition. Machines are therefore built to provide these simple shock motions. However, note that the motions provided by actual machines are only ideally simple. The ideal outputs may be given as nominal values; the actual outputs can only be determined by measurement.

SPRING DROP TABLE. If a rigid body is dropped onto a weightless linear spring, the body will be subjected to an acceleration pulse that has the shape of a sine wave of half-period duration. The maximum acceleration of this pulse is

$$A_0 = \sqrt{\frac{2hk}{W}}\, g$$

and its duration is

$$t = \frac{\tau}{2} = \pi \sqrt{\frac{W}{kg}} \quad \text{sec}$$

where A_0 is the magnitude of the acceleration pulse in units of gravity, W is the weight of the body (lb or gm), h is the height of drop (in. or cm), k is the spring constant (lb/in. or gm/cm); g is the acceleration of gravity (386 in./sec^2 or 980 cm/sec^2), τ is the period of the sine-wave function, and t is the duration of the pulse.

A machine of this type is illustrated in Fig. 26.4*A*. The carriage is dropped from some height so that a flat, simply supported leaf spring on the bottom of the carriage impacts on its center against a rigid anvil structure. The calculated and measured accelerations experienced by the carriage are shown in Fig. 26.4*B*. Because of distributed weight in the spring and lack of rigidity of the carriage, the half-sine acceleration will have high-frequency components superimposed upon it. The filtered value of accelerometer output, which excludes the high-frequency component shown in the figure, and the pulse time agree quite exactly with the values calculated from the above equations. Machines involving this principle are simple to design, construct, and build. Their shapes and sizes, spring types and orientations, can be adjusted to give the acceleration magnitude and duration of the pulse required by a particular application.

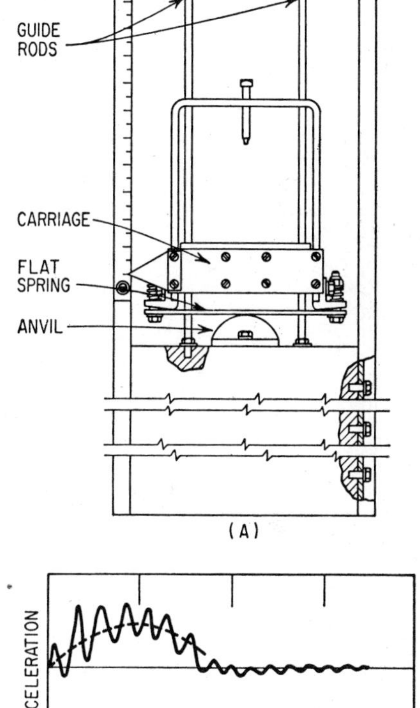

FIG. 26.4. (*A*) Typical spring drop table. (*B*) Calculated (dotted) and actual acceleration of carriage during impact.

SAND DROP TABLE. Certain military general shock specifications include requirements that an equipment be subjected to accelerations of about 15 to 30g lasting for about 11 milliseconds; other values of acceleration are additionally included. A

Sanddrop Shock Testing Machine[6] (also known as the *Medium-impact Shock Machine*) is normally specified for these tests. This machine, shown in Fig. 26.5, consists of a drop table whose fall is arrested by dropping into a sandbox which forms the base of the machine. An adjustable number of blocks, attached to the underside of the table, penetrate the sand. The magnitude and duration of the stopping acceleration are determined by the height of drop and number of blocks. A refined 30–40 grit* sand is used, which must be carefully maintained as to depth, packing, and surface condition if consistent results are to be obtained. The machines are made of several different sizes so that different load ranges can be accommodated. Figure 26.6 illustrates nominal values of maximum acceleration and pulse duration of shocks delivered by a machine of this type having a 150-lb load rating. The different block arrangements used are shown in the upper part of this figure. The values illustrated represent the results of measurements for a specified location on the drop table, for a specified load, and for a specified frequency-response range of the instrumentation. This machine should be operated according to procedures prescribed in the applicable American Standard.[6]

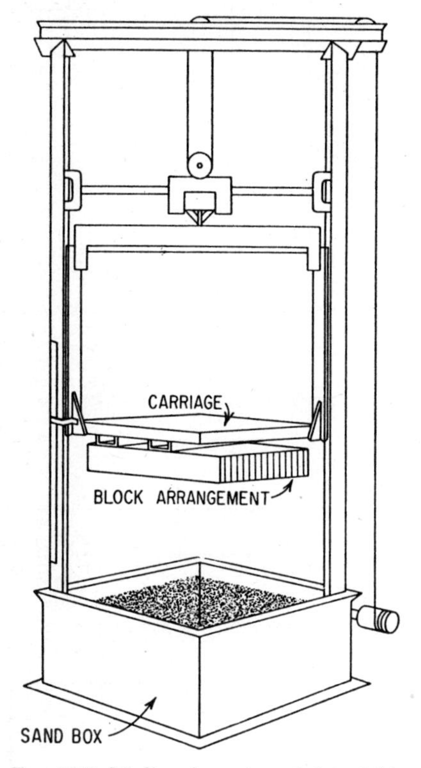

Fig. 26.5. Medium-impact sand-drop table.

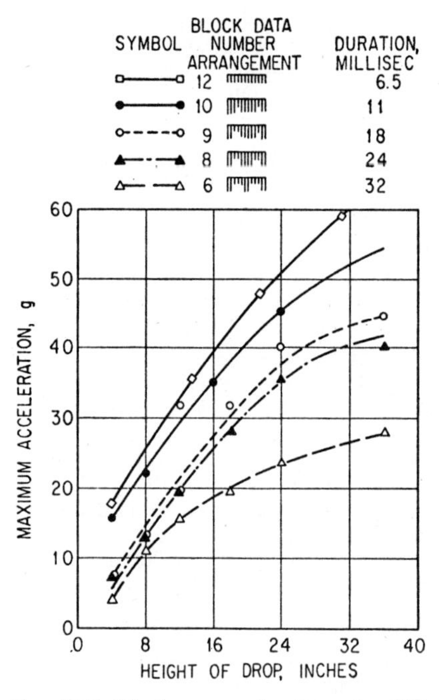

Fig. 26.6. Maximum acceleration of a 150-lb medium-impact shock machine using a 100-cps low-pass filter. The upper section of the figure shows the block arrangement and nominal duration of the pulse.

PLASTIC PELLET DROP TABLES. A great variety of drop testers[2] are used to obtain acceleration pulses having magnitudes ranging from 80,000*g* down to a few *g*.

* 100 per cent of a 30 grit sand should pass through a No. 18 sieve of the U.S. Standard Sieve Series and not more than 3 per cent should pass through a No. 40 sieve.

The machines each include a carriage (or table) on which the item under test is mounted; the carriage can be hoisted up to some required height and dropped onto an anvil. Guides are provided to keep the carriage properly oriented.

A plastic material, placed upon the point of impact of the anvil, can be used to control the acceleration vs. time characteristic and the shape of the acceleration pulse. These plastic pellets are usually made of lead, but a variety of materials and shapes have been used generated by the impact of a drop on the pellet. When rubber-like materials are used, the acceleration pulse can be made to approach the shape of a half-sine curve.

When large velocity changes are required, the carriage may be accelerated downward by means other than gravity. Frequently parts of the carriage, associated with its lifting and guiding mechanism, are flexibly mounted to the rigid part of the carriage structure that receives the impact. This is to isolate the main carriage structure from its flexible appendages so as to retain the simple pulse structure of the stopping acceleration.

A *sawtooth acceleration pulse*,[7] Fig. 26.1*D*, has an acceleration shock spectrum that is relatively smooth and with no extreme values which would make it discriminate for or against an item under test because of the excessive test severity at certain frequencies. Theoretically and experimentally obtained shock spectra for such a pulse are shown in Fig. 26.7*A* and *B*. Figure 26.7*C* illustrates the time-history of acceleration of a point on a beam when subjected to the sawtooth pulse shown on the same time scale. This type of pulse can be obtained relatively easily by dropping a rigid carriage onto a cylindrical lead pellet with a conical top. This form of excitation provides a satisfactory test for many types of shock environments where it is required that the shock spectrum rise to a maximum value within the first 100 cps and remain constant thereafter. The amplitude and duration of the pulse may be modified by changing the dimensions of the pellet, so as to provide an appropriate response level and rise time for the spectral curves.

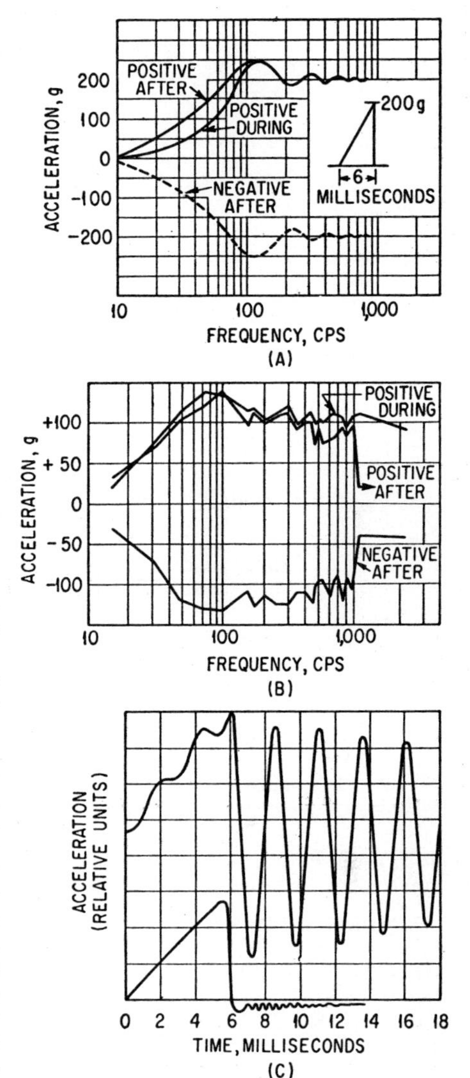

FIG. 26.7. (*A*) Theoretical acceleration shock spectra for the sawtooth acceleration pulse shown in inset. The response during and after the pulse interval is shown separately. (*After Morrow and Sargeant.*[7]) (*B*) Experimental shock-response spectrum for a 100*g*, 6-millisecond, sawtooth shock pulse. (*C*) Response of a beam of 400-cps natural frequency to the inset sawtooth shock pulse. (*After Jensen.*[8])

A typical machine of this type is shown in Fig. 26.8. The carriage consists of a rigid platform on which the specimen is mounted and which embodies the impacting surface

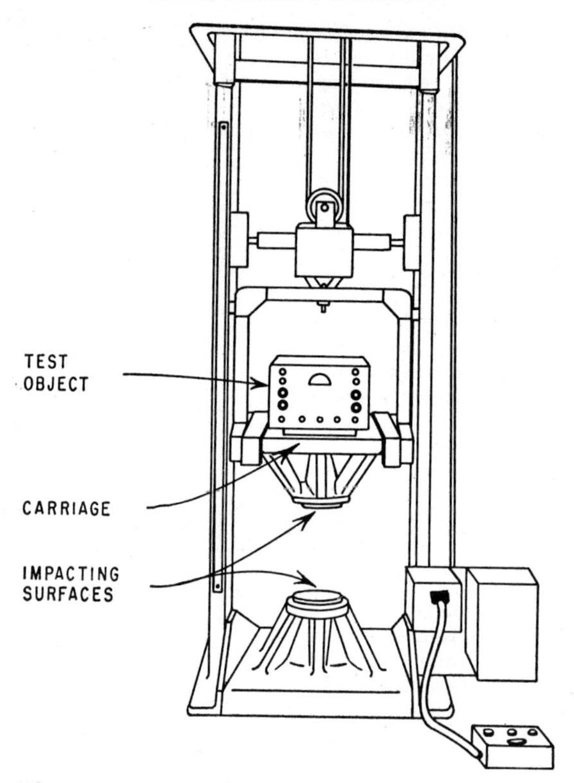

FIG. 26.8. Drop-table arrangement capable of being used with plastic pellets on the impacting surfaces so as to produce a pulse of desired shape. (*Barry Wright Corporation.*)

on its lower side. The platform is usually cast of aluminum and designed to have no natural frequencies below about 1,000 cps. The impacting surfaces are of hardened steel.

MISCELLANEOUS DROP TABLES. A great variety of drop-table machines are in existence.[2] One such machine, a *100-foot Portable Drop Tester*, is shown in Fig. 26.9*A*. While the impact velocity of the carriage corresponds to that attained by a free fall of 100 ft, the height of the machine is only about 11 ft. The carriage is accelerated downward by rubber shock cords, and is arrested by a sand-filled rubber bag. Figure 26.9*B* indicates the complete acceleration-time curve. Bags of different shapes and stiffnesses permit the shape of the stopping pulse to be varied. The extremely large accelerations shown on the table are obtained for direct impact (the sand-filled rubber bag is omitted) of the impacting surfaces.

AIR GUNS. Air guns frequently are used to impart large accelerations to pistons on which items under test can be attached. The piston is mechanically retained in position near the breech end of the gun while air pressure is built up within the breech. A quick-release mechanism suddenly releases the piston and the air pressure projects the piston down the gun barrel. The muzzle end of the gun is closed so that the piston is stopped by compressing the air in the muzzle end. Air bleeder holes may be placed in the gun barrel to absorb energy and to prevent an excessive number of oscillations of the piston between its two ends.

A variety of such guns, used by the Naval Surface Weapon Center, can provide acceleration pulses as shown in Figs. 26.10*A* and *B*. As shown in Table 26.1, the peak acceleration may extend from a maximum of about 1,000*g* for the large-diameter (21

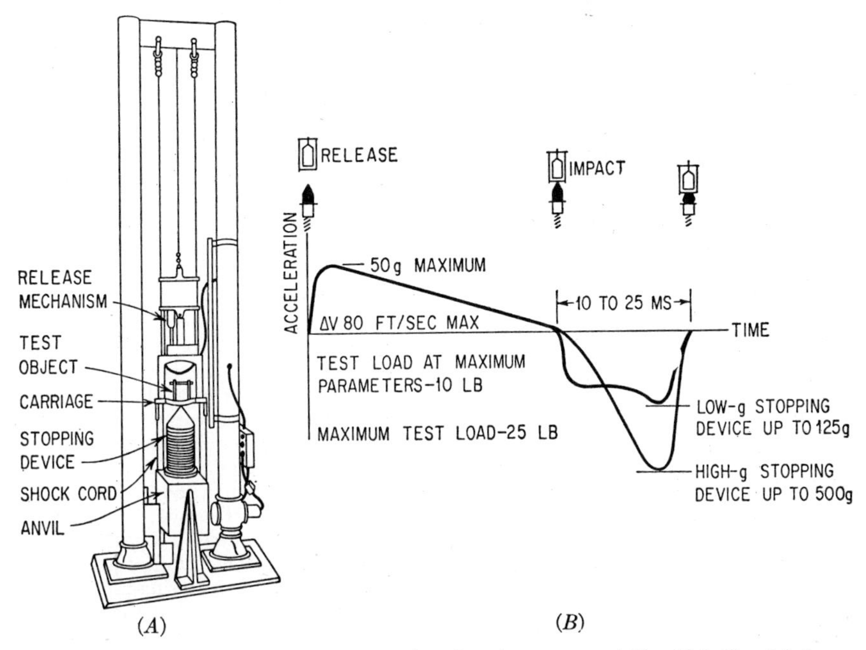

FIG. 26.9. (*A*) 100-ft drop tester. (*B*) Acceleration-time curve. (*After U.S. Naval Ordnance Laboratory Report.*)

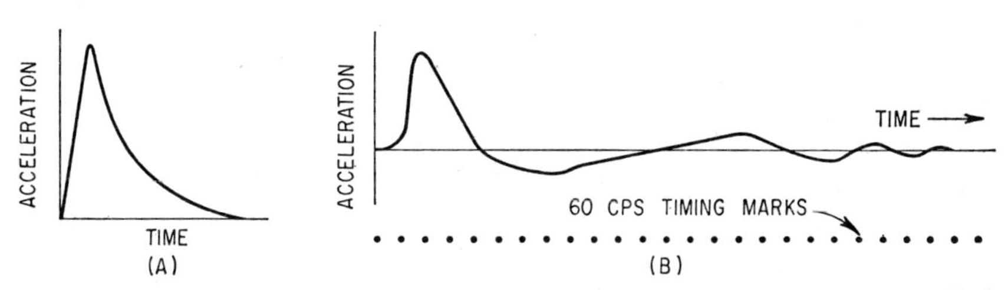

FIG. 26.10. Typical acceleration-time curves for (*A*) 5-in. air gun; (*B*) 21-in. air gun. Peak accelerations extend to 5,000*g*, with pulse lengths between 3 and 6 milliseconds for the 5-in. gun. The accelerations for the 21-in. gun extend to about 1,000*g*. (*After U.S. Naval Ordnance Laboratory*.[10])

in.) guns up to 200,000*g* for small-diameter (2 in.) guns. The pulse length varies correspondingly from about 50 to 3 milliseconds. The maximum piston velocity varies from about 400 to 750 ft/sec. The maximum velocities are not dependent upon piston diameter.

High-acceleration gas guns have been developed for testing electronic tubes. The items under test are attached to the piston. The gun consists of a barrel (cylinder) that is closed at the muzzle end, but which has large openings to the atmosphere a short distance from the muzzle end. The piston is held in place while a relatively low pressure gas (usually air or nitrogen) is applied at the breech end of the gun. The piston is then released, whereby it is accelerated over a relatively long distance until it reaches the position along the length of the cylinder that is open to the atmosphere. This initial acceleration is of relatively small magnitude. After the piston has passed these openings it is stopped by the compression of gas in the short closed end of the cylinder. This results in a reverse acceleration of relatively large magnitude. (Sometimes an inert gas, such as nitrogen, is used in the closed end to prevent explosions which might be caused by oil particles igniting under the high temperatures incident to the compression.) Thus, in contrast to the previously described devices, the major acceleration pulse is delivered during stopping rather than starting. An advantage of this latter technique is that the difficult problem of constructing a quick-release mechanism for the piston, which will work satisfactorily under the large forces exerted by the piston, is greatly simplified.

HYGE SHOCK TESTER.[9] The *Hyge Tester* is a device that can generate a large force through a given displacement and so can provide an acceleration to some object against which it acts. A typical device of this kind is shown on Fig. 26.11. The pressure in the lower chamber is greater than that in the upper chamber, but prior to operation the upward force exerted on the thrust column is less than the downward force because of the smaller area exposed to the lower chamber. As the pressure difference is increased, the upward force becomes greater than the downward force and the ring seal, which prevents the pressure of the lower chamber from acting over the entire bottom area of the piston, breaks. There is then a sudden increase in the upward thrust. The shape of the acceleration pulse experienced by the thrust column and its load is determined principally by orifice sizes and metering-pin shapes. Reactive loads will affect the shape of the pulse. A liquid usually is used, as shown, in the region surrounding the metering pins; the liquid is forced through the orifices by the gas pressure of the accumulator and involved in the pulse-shaping mechanism. Acceleration pulses of many shapes can be made by proper design of the metering pins, consideration being taken of the mass of the load. However, since load reactions affect the pulse shape, simple pulse shapes only can be obtained when these reactions are simple or unimportant. Some models have a maximum thrust output of 250,000 lbs and, depending on the load,[2] can provide sustained accelerations in excess of 800g.

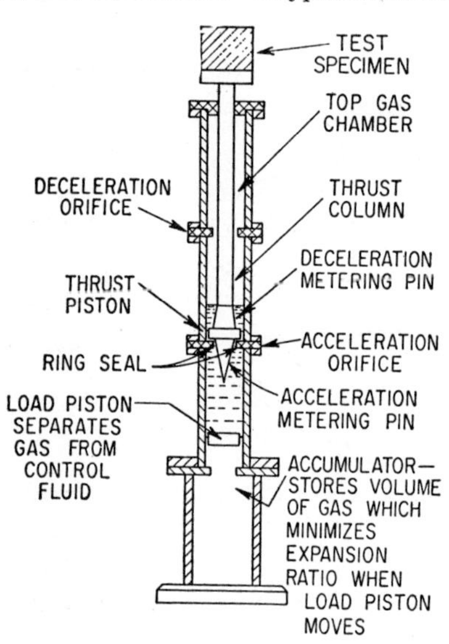

FIG. 26.11. Schematic drawing of a typical Hyge Shock Exciter.

CONSIDERATIONS RELATIVE TO SIMPLE SHOCK-PULSE MACHINES

In the above description of the output of shock machines designed to deliver simple shock pulses of adjustable shapes, it is assumed that the load imposed on the machine by the item under test has little effect upon the shock motions. This is true only when the effective weight of the load is negligibly small compared with that of the shock-machine mounting platform. If the effective weight of the load is independent of frequency, i.e., if it behaves as a rigid body, it is simple to compensate for the effect of the load by adjusting machine parameters. However, when the load is flexible and the reactions of excited vibrations are appreciable, the motions of the shock-machine platform are complex. Specifications involving the use of these types of machines should require that the mounting platform have no significant natural frequencies below a specified frequency. The weight of this platform together with that of all rigidly attached elements, exclusive of the test load, also should be specified. Pulse shapes then may be specified for motions of this platform or for the platform together with given dead-weight loads. These may be specified as nominal values for test loads, but it is neither practical nor desirable to require that the pulse shape be maintained in simple form for complex loads of considerable mass.

SHOCK MACHINES FOR SINGLE COMPLEX SHOCK

Because of the infinite variety of shock motions possible under field conditions it is not practical or desirable to construct a shock machine to reproduce a particular shock that may be encountered in the field. However, it is sometimes desirable to simulate some average of a given type of shock motion. To accomplish this may require that the shock machine deliver a complex motion. A shock of this type cannot be specified easily in terms of the shock motions, as the motions are very complex and dependent upon the nature and mounting of the load. It is customary, therefore, to specify a test in terms of a shock machine, the conditions for its operation, and a method of mounting the item under test.

NAVY HIGH-IMPACT SHOCK MACHINE FOR LIGHTWEIGHT EQUIPMENT.[1, 10, 12, 14] The *Navy High-impact Shock Machine for Lightweight Equipment* is designed to subject equipment to shocks of the nature and intensity that might occur on board ship that is exposed to severe, but sublethal, noncontact underwater explosions. Such severe shocks produce motions that extend throughout the ship. The machine is for equipment secured or mounted to bulkheads or decks. A machine to produce shocks of this type is shown in Fig. 26.12. The anvil A is struck on the back side by the pendu-

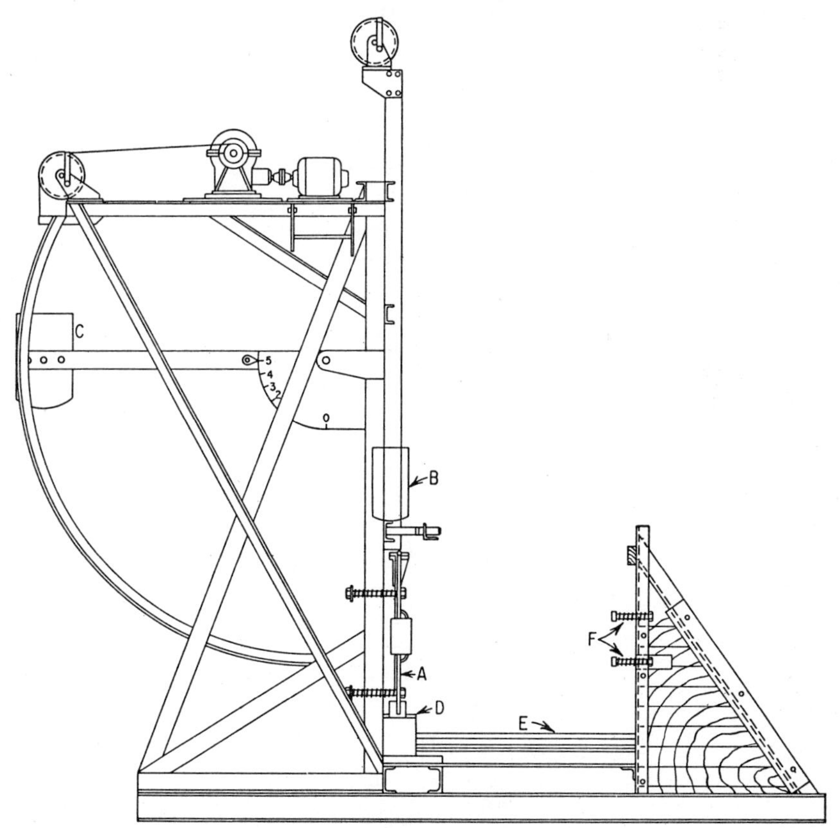

FIG. 26.12. Navy High-impact Shock Machine for Lightweight Equipment. *A*: Anvil plate. The anvil plate can also be oriented in the plane of the paper for an "end" hammer blow. *B*: Hammer for vertical blow. *C*: Hammer for horizontal blow. *D*: Restoring springs for vertical blow. *E*: Rail support for end blow. An upper support angle, not shown, is also required. *F*: Positioning springs for end blow.

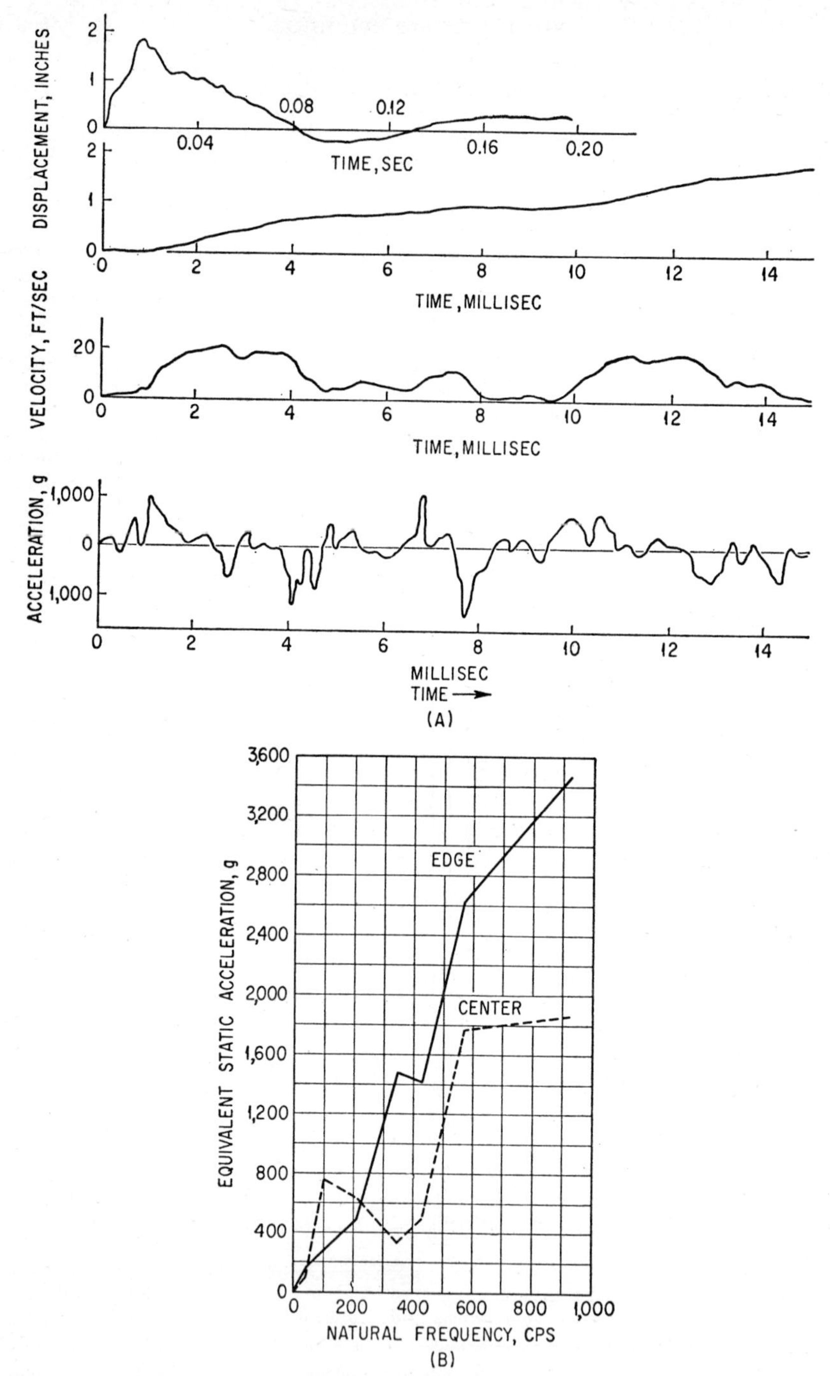

FIG. 26.13. (*A*) Shock motions of the center of an unloaded 4A mounting plate of the Shock Machine for Lightweight Equipment for a 5-ft back blow. All curves represent the same motion. (*After Oliver.*[11]) (*B*) Shock spectra for lightweight high-impact shock machine. Dotted line, center of 4A plate. Solid line, edge of 4A plate over channel. Five-foot back blow. (*After Walsh and Blake.*[14])

lum hammer *C*, or the anvil is rotated 90° on a vertical axis and struck on the end by the pendulum hammer. The drop hammer *B* can be made to strike the top of the anvil, thus providing principal shock motions in the third orthogonal direction. Acceleration, velocity, and displacement measured at the center of an unloaded mounting plate (4*A* plate) are plotted in Fig. 26.13*A*. The weights of equipment tested on this machine are limited to 250 lb.

It can be seen that acceleration magnitudes extend to about 1,000*g*, velocities to 20 ft/sec, and displacements to 1.7 in. Shock spectra of shock motions generated by this machine at two different locations on the mounting plate are shown on Fig. 26.13*B*. The spectrum for motion at the edge of the plate is typical of that for a relatively rigid structure—a straight line having a slope proportional to the velocity change resulting from the impact. The spectrum for the motion at the center of the plate illustrates the amplification of spectrum level at a natural frequency of the plate (about 100 cps) and some attenuation at higher frequencies.

THE HIGH-IMPACT SHOCK MACHINE FOR MEDIUM-WEIGHT EQUIPMENT.[1, 12, 13, 14] This machine is used to test equipment that, together with its supporting structures, weighs up to 7,400 lb. The machine, shown in Fig. 26.14, consists princi-

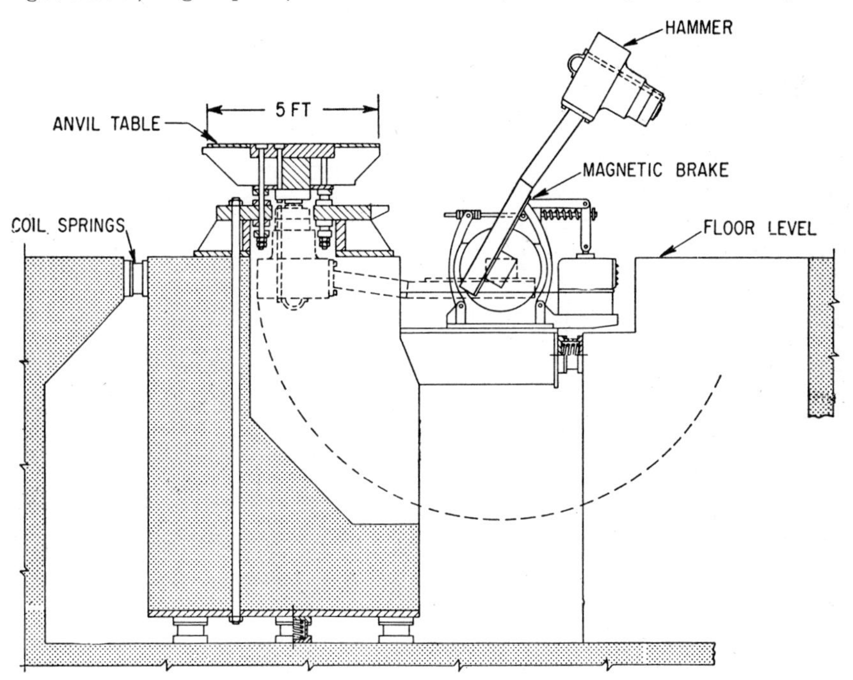

FIG. 26.14. High-impact Shock Machine for Medium Weight Equipment.

pally of a 3,000-lb hammer and 4,000-lb anvil. Loads are not attached directly to the rigid anvil structure but rather to a group of steel channel beams which are supported at their ends by steel members, which in turn are attached to the anvil table. The number of channels employed is dependent upon the weight of the load and is such as to cause the natural frequency of the load on these channels to be about 60 cps. The hammer can be dropped from a maximum effective height of 5.5 ft. It swings around on its axle so as to strike the anvil on the bottom, giving it an upward velocity. The anvil is permitted to travel a distance up to 3 in. before being stopped by a ring of retaining bolts. The machine is secured to a large block of concrete which is mounted on springs to isolate the surrounding area from shock motions.

The general nature of the shock is similar to that of the lightweight machine. Curves illustrating some shock motions generated by the machine are shown in Fig. 26.15. Figure 26.15*A* indicates the complex nature of the shock motions generated by this machine. Figure 26.15*B* shows that much of the high-amplitude, high-frequency components of the shock motions is not transmitted to the load. Figures 26.15*C* and *D*

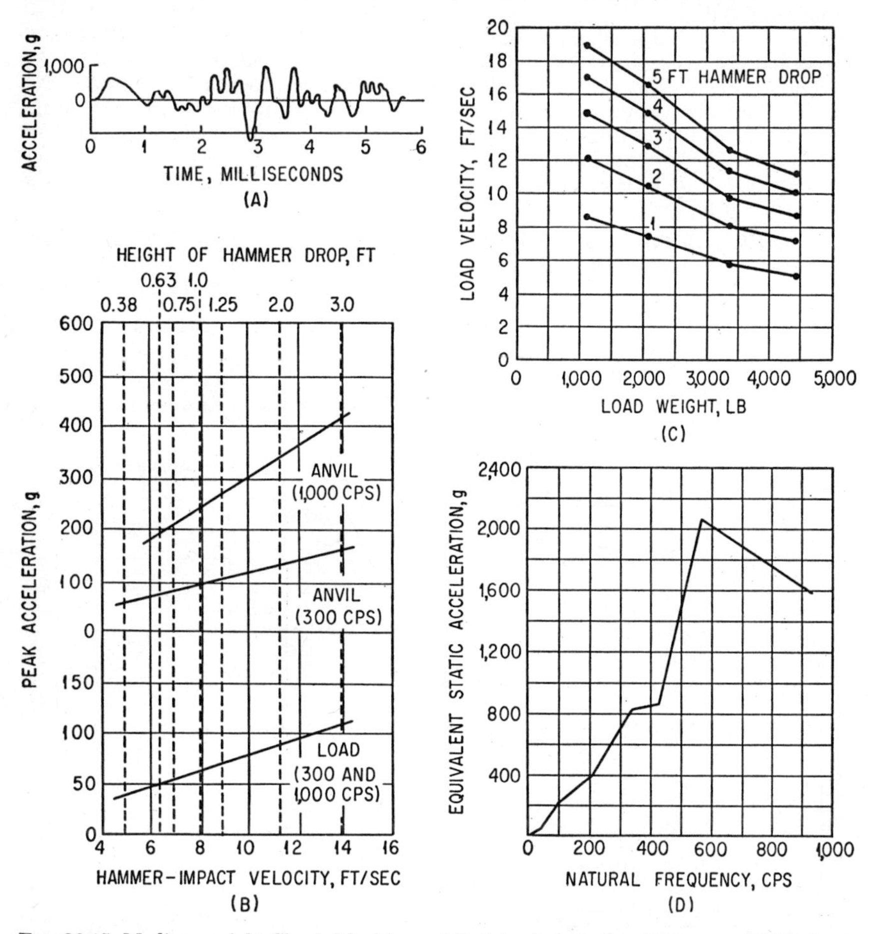

FIG. 26.15. Medium-weight Shock Machine. (*A*) A typical acceleration curve (400-lb load, 3-ft hammer drop, 5,000-cps low-pass filter). (*B*) Maximum accelerations of the anvil and of a dead-weight load of 1,115 lb mounted on the anvil; according to specifications, 1,000- and 300-cps low-pass filters were used. (*C*) Maximum velocities attained by different dead-weight loads. (*D*) A typical shock spectrum delivered by the shock machine for a 1-ft hammer drop and 500-lb load.

illustrate, respectively, the velocity attained by the load and a shock spectrum of the motions near the base of a load. A complete study of the performance of this machine is given in Ref. 14.

MULTIPLE IMPACT SHOCK

Many environments, particularly those involving transportation, subject equipment to a relatively large number of shocks. These are of lesser severity than the shocks of major intensity that have been considered above, but their cumulative effect can be just as damaging. It has been observed that the components of an equipment that are damaged as a result of a large number of shocks of relatively low intensity are usually different than those that are damaged as a result of a few shocks of relatively high intensity. The damage effects of a large number of shocks of low intensity cannot generally be produced by a small number of shocks of high intensity. Separate tests are therefore required in order that the multiple number of low-intensity shocks be properly simulated.

PACKAGE TESTER.[2] Packages that are not securely fastened may bounce while being transported by a vehicle. To simulate this mistreatment, a table is caused to vibrate in a vertical plane at a low frequency and at an acceleration amplitude slightly greater than $1g$. The package then loses contact with the table at each cycle, subjecting the package to repeated impacts as it strikes the table. One of the commercially available package testers has been designed for various load ranges and for operating at $1.25g$ acceleration amplitude (0.5 in. displacement amplitude) at about 5 cps. The table is of steel-frame construction with a wood top which is bordered by a fence in order to prevent the packages from falling off.

REVOLVING-DRUM TUMBLING TEST.[2, 15] A steel drum, having a cross-sectional shape of a regular hexagon and having circular end flanges so that the drum can be rotated about its axis, is shown in Fig. 26.16. Baffles are located on the inside faces of the drum so that a container under test is carried around with the drum as it rotates, until a position is attained such that the container rolls over the baffle onto the area below. The construction and operation of these drums have been standardized. Drums having diameters of 7 or 14 ft are used; the former is rotated at 1⅚ rpm and the latter at 1 rpm. Tests generally are continued until a failure develops in the container. The relative qualities of containers are compared by their relative survival times.

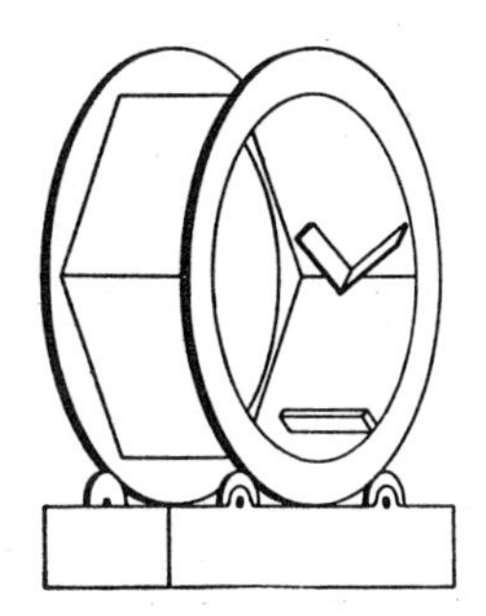

FIG. 26.16. Revolving-steel drum for testing of shipping containers.

BUMP TEST MACHINE. A machine has been designed in England which subjects equipment to repeated bumps at the rate of several bumps per second. Although the test conditions achieved by these machines are not necessarily intended to simulate the field conditions, experience has shown that items surviving the test conditions are satisfactory for transportation environments.

Machines are designed for loads up to 51 lb and up to 250 lb. Each machine includes a table on which equipment is mounted. The table is raised by a cam and allowed to fall freely onto a rubber anvil, the drop height of the table and the hardness of the anvil being selected to produce a specified maximum acceleration and waveform. The drop height is adjustable between about ¼ in. and about 1¼ in.; it is normally adjusted so that the maximum acceleration is about $40g$ (determined by acceleration measurements). The waveform of the acceleration pulse is approximately that of a half-period sine wave having a pulse duration in the order of 5 to 6 milliseconds.

A similar set of machines is described by the International Electrotechnical Commission.[16] The principal difference is that the impacting members for these latter machines are made of wood rather than rubber. The machines are calibrated by means of a Brinell type test. The acceleration causes an iron cylinder, weighing 1 kg, to press a steel ball (diameter 3.97 mm) onto an aluminum plate of 3 mm thickness. The diameter of the indentation is determined. The ball is then statically pressed onto another part of the plate so as to give the same diameter impression. The nominal acceleration, in g, is equal to the static load in kilograms that gives the same diameter impression as the bump. The specified accelerations are given in terms of these nominal values. It is, perhaps, unfortunate that the measure of shock intensity is given in terms of acceleration; it would seem that the specification could as well be in terms of impression diameter and thus prevent confusion with accelerations which might be measured by more accurate means.

ELECTRODYNAMIC BUMP TESTER. An electrodynamic vibration testing machine of the type described in Chap. 25 can provide multiple low-level shocks. Such a machine can be excited periodically by a condenser discharge or by electric current pulses of desired shapes. Maximum accelerations up to about $100g$ are obtainable for test items that are light compared with the armature of the machine.

MISCELLANEOUS

PENDULUM TAPPER. A *pendulum tapper* has been developed to provide an impulsive force type of excitation to electron tubes. The tapper, shown in Fig. 26.17*A*,

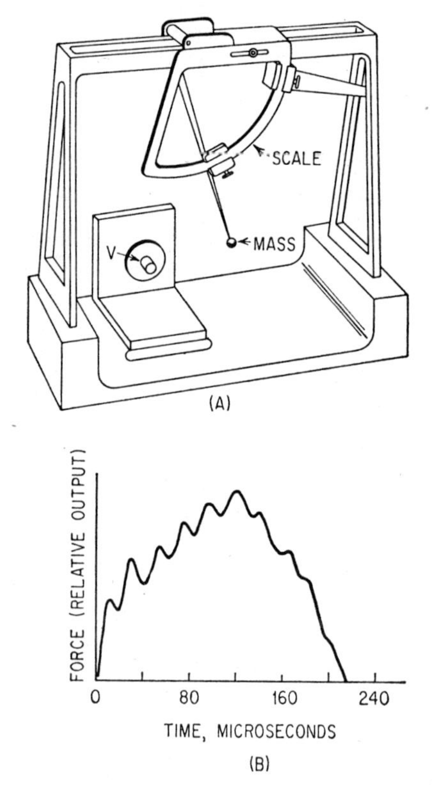

FIG. 26.17. (*A*) Schematic of NML Pendulum Tube Tapper. A small mass suspended by two light rods swings through an angle indicated by the scale and strikes an electron tube *V*. (*B*) The force pulse delivered to a metal octal tube. The curve represents the output of an accelerometer located on the mass.

consists of a light hammer which strikes the envelope of the electron tube. The impulse delivered, Fig. 26.17*B*, can excite all frequencies from those of very low value to about 100,000 cps. The test is therefore useful for comparing the relative level of microphonics, or of audio- and radio-frequency noise, generated by this test for different tubes.

QUICK-STARTING CENTRIFUGE OR ROTARY ACCELERATOR.[2] A *quick-starting centrifuge* is used to attain quickly and to maintain an acceleration for a long period of time. The accelerator consists of a rotating arm which is suddenly set into rotation by an air-operated piston assembly. The test object is mounted on a table attached to the outer end of the arm. The table swings on a pivot so that the resultant direction of acceleration is always along a fixed table axis. Initially the resultant acceleration is caused largely by angular acceleration of the arm, so that this axis is in a circumferential direction. As the centrifuge attains its full speed, the acceleration is caused primarily by centrifugal forces so that this table axis assumes a radial direction. These machines are built in several sizes. They require between 5 and 60 milliseconds to reach the maximum value of acceleration. For small test items (8 lb), a maximum acceleration of $450g$ is attainable; for heavy test items (100 lb), the maximum value is $40g$.

REFERENCES

1. Kennard, D. D., and I. Vigness: "Shock and Vibration Instrumentation," an ASME Publication, June, 1956, p. 127.
2. Shock and Vibration Information Center, U.S. Naval Research Laboratory: Index of Environmental Test Equipment in Government Establishments, November, 1967.
3. ASTM Standards, part 7, p. 1087, 1955.
4. Matlock, H., and E. A. Ripperger, J. Turnbow, and J. N. Thompson: *Shock and Vibration Bull.* 25, part 2, p. 144, December, 1957, Office of Dept. of Defense R & D. (Unclassified.)
5. ASTM Standards, part 7, p. 1101, 1955.
6. American Standard Specification for Design, Construction and Operation of Variable Duration, Medium-impact Shock-testing Machine for Lightweight Equipment, Publication *ASA Publ.* S2.1–1961.
7. Morrow, C. T., and I. Sargeant: *J. Acoust. Soc. Amer.*, **28**:959 (1956).
8. Jensen, G. A.: *Shock and Vibration Bull.* 25, part 2, pp. 137–143, December, 1957. Office of Secretary of Defense Research and Engineering, Washington 25, D.C.
9. The Hyge Shock Tester, *Bull.* 4-70, Consolidated Electrodynamics Corp., Rochester, N.Y., February, 1957.
10. The American National Standards Institute Specification for the Design, Construction and Operation of Class HI (High Impact) Shock-testing Machines for Lightweight Equipment. *ASA Publ.* S2.15-1972.
11. Oliver, R.: *Shock and Vibration Bull.* 3, *U.S. Naval Research Lab. Rept.* S-3106, May, 1947, p. 10. (Unclassified.)
12. MIL-S-901C (NAVY): "Military Specification. Shock Tests, HI (High-Impact); Shipboard Machinery, Equipment, and Systems, Requirements for," January 15, 1963.
13. MacDuff, J. N., and J. R. Curreri: "Vibration Control," chap. 7, McGraw-Hill Book Company, Inc., New York, 1958.
14. Clements, E. W.: "Shipboard Shock and Navy Devices for its Simulation," *U.S. Naval Res. Lab. Rept.* 7396, July, 1972.
15. Standard Method of Test for Shipping Containers in Revolving Hexagonal Drum, *ASTM Standards*, part 7, p. 1097, 1955.
16. Basic Climatic and Mechanical Robustness Testing Procedure for Components. Published by the Central Office of the International Electrotechnical Commission, Geneva, Switzerland, 1954.

27

APPLICATIONS OF DIGITAL COMPUTERS

Allen J. Curtis
Hughes Aircraft Company

INTRODUCTION

This chapter identifies and describes in general terms several basic categories of applications of digital computers to shock and vibration problems. Such categories are readily identifiable as (1) the use of computers to carry out numerical analyses of dynamic systems which are too complex and/or time-consuming to perform by hand calculation, and (2) the use of computers to perform other than analytical tasks which were performed previously by analog equipment or perhaps were not practical to perform. These categories tend to distinguish between analytical and experimental tasks. The major emphasis of this chapter is directed toward experimental applications.

FOR WHAT PROBLEMS SHOULD DIGITAL COMPUTATION BE USED?

The decision to employ a digital computer in the solution of a shock or vibration problem should be made with considerable care. Before a program of digital computation is decided on, the following questions should be answered.

1. Is existing software available to perform the required task?
2. If not, to what extent must the task or the software be modified in order to perform the task?
3. If no applicable software exists, what is the extent of the software development needed?
4. What are the detailed assumptions inherent in the software program (e.g., linearity, proportional damping, etc.)?
5. Is the software able to compute and output the information required (e.g., absolute vs. relative motion, phase relationships, etc.)?
6. What are the detailed input and output limitations of the program (e.g., types of input excitations, graphic outputs, etc.)?
7. How much computer time will be used to perform the task?

After satisfactory answers are obtained to such questions, the user must realize that the results obtained from the computer output can be no better than the input to the computer. For example, the quality of the natural frequencies and mode shapes obtained from a structural analysis computer program does not depend on the computer program but on the degree to which the mathematical model employed is representative of the mass, stiffness, and damping properties of the physical structuıe. Likewise, a spectral analysis of a signal with poor signal-to-noise ratio will provide an accurate spectrum of the signal *plus noise*, not of the signal.

ANALYTICAL APPLICATIONS

The development of large-scale computers having a very short cycle time (i.e., the time required to perform a single operation such as adding two numbers) and a very large memory now permits detailed analyses of structural responses to shock and vibration excitation which are many orders of magnitude more complex than was previously practical. A number of software programs have been developed to perform these analyses. In this chapter, they are categorized as *general-purpose programs* and *special-purpose programs*. References 1 and 2 contain extensive anthologies of both general-purpose and special-purpose analytical programs.

GENERAL-PURPOSE PROGRAMS

Programs may be classed as "general purpose" if they are applicable to a wide range of structures and permit the user to select a number of options, such as damping (viscous or structural), various types of excitation (sinusoidal or random vibration, transients), etc. Almost all general-purpose programs are limited to the analysis of linear systems.

FINITE-ELEMENT PROGRAMS. The most numerous programs are classed as "finite-element" or "lumped-parameter" programs.

In lumped-parameter programs, the structure to be analyzed is represented in a model as a number of point masses (or inertias) connected by massless, springlike elements. The points at which these elements are connected, and at which a mass may or may not be located, are the "nodes" of the system. Each node may have up to six degrees-of-freedom at the option of the analyst. The size of the model is determined by the sum of the degrees-of-freedom at the nodes and by the *dynamic* degrees-of-freedom (which is the number of degrees-of-freedom for which the mass or inertia is nonzero). The number of natural frequencies and modes which may be computed is equal to the number of dynamic degrees-of-freedom. However, the number of frequencies and modes which reliably represent the physical structure is generally only a fraction of the number which can be computed. Each program is limited in capacity to some combination of dynamic and zero mass degrees-of-freedom. The springlike elements are chosen to represent the stiffness of the physical structure between the selected nodes and generally may be represented by springs, beams, or plates of specified shapes. The material properties, geometric properties, and boundary conditions for each element are selected by the analyst.

In the more general finite-element programs, the springlike elements are not necessarily massless but may have distributed mass properties. In addition, lumped masses may be used at any of the nodes of the system.

The equations of motion of the finite-element model can be expressed in matrix form and solved by the methods described in Chap. 28. Regardless of the computational algorithms employed, the program computes the set of natural frequencies and orthogonal mode shapes of the finite-element system. These modes and frequencies are stored for future use in computing the response of the system to a specified excitation. For the latter computations, a damping factor must be specified. Depending on the program, this damping factor may have to be equal for all modes, or it may have a selected value for each mode.

The most sophisticated finite-element program, known as NASTRAN, was developed by the National Aeronautics and Space Administration (NASA).[3,4,5,6] NASTRAN can be employed on most large-scale computer facilities. Many similar programs with fewer options and limited to fewer degrees-of-freedom are also available on commercial computing facilities. In addition, many large firms have developed their own programs.

Component Mode Synthesis.[7,8,9] The method of modeling described above leads to the prescribing of models with a very large number of degrees-of-freedom compared with the number of modes and frequencies actually of interest. Not only is this expensive, but it rapidly exceeds the capacity of many programs. To overcome these problems, component mode synthesis techniques have been developed. Instead of the modeling of an entire physical system, several models are developed, each representing a distinct identifiable region of the total structure and within the capacity of the computer program. The modes and frequencies of interest in each of these models are computed independently. A model of the entire structure is then obtained by joining these several models, using the component mode synthesis technique. This model retains the essential features of each substructure model and thus of the entire structure while using a greatly reduced number of degrees-of-freedom to do so.

Reduction of Model Complexity. Companion techniques, developed to reduce the cost of analysis and to permit the joining of several substructure models, are known as "reduction" methods, of which the Guyan reduction method[10, 11] is an example. Again, the objective is to reduce the mass and stiffness matrices to the minimum size consistent with retaining the modes and frequencies of interest together with other dynamic characteristics such as base impedance. It should be noted, however, that the Guyan reduction method yields a mass matrix which is nondiagonal (see Chap. 28) and which may be unacceptable for some computer programs. It is also of interest that the rigid-body mass properties, i.e., total masses and inertias of the structure, are not identifiable in the reduced mass matrix.

DISTRIBUTED (CONTINUOUS) SYSTEM PROGRAMS.[1, 2, 4] While still classified as general-purpose programs, a number of more specialized programs treating the analysis of distributed or continuous structural systems have been developed. These programs deal with beams, plates, shells, rings, etc., that is, standard structural members. They are general purpose in the sense that each program can be applied for a broad, selectable range of physical properties and dimensions of the particular structural shape. Not all programs will employ the same theory of elasticity; the user must thus examine the theoretical basis on which the program was developed. For example, the user must determine if the program includes such effects as rotary inertia or shear deformation.

SPECIAL-PURPOSE PROGRAMS[1, 2, 4]

The need for a special-purpose program may arise in several ways. First, for an engineering activity engaged in the design, on a repetitive basis, of what amounts analytically to the same structure, it may be economical to develop an analysis program which efficiently analyzes that particular structure. The analysis of vibration isolator systems or automobile suspension systems may be an example. Similarly, parametric studies of a particular structure, either to gain understanding or to optimize the design, may require a sufficient number of computer runs to justify the software development. A second type of special-purpose program includes programs which in some way perform an unusual type of analysis—for example, that of nonlinear systems.

Access to existing special-purpose programs is generally more restricted than is access to general-purpose programs, in part because of their frequent proprietary nature and in part because of the investment in software required to develop them.

EXPERIMENTAL APPLICATIONS

The classification "experimental applications" covers uses of computers which involve, in some way, the processing of shock and vibration information originally obtained during test or field operation of physical equipment. When this handbook was first published in 1961, experimental applications of digital computers were essentially nonexistent. Two developments were crucial to the enabling of many of the applications described in later sections: (1) the recognition of the computational efficiency of the fast Fourier transform [FFT] algorithm,[12] and (2) the development of hardware FFT processors coupled with the development of minicomputers.

These developments now permit use of digital computers for such tasks as vibration data analysis, shock data analysis, and shock, vibration, and modal testing—all described in later sections. The information resulting from such applications is in digital form, which permits more sophisticated engineering evaluation of the information through further efficient digital processing, e.g., regression analysis, averaging, etc.

FUNDAMENTAL MATHEMATICAL RELATIONSHIPS[13, 14, 15]

A number of fundamental mathematical relationships between functions in the time and/or frequency domains are common to most experimental applications of digital computers to shock and vibration problems. These are summarized below. For clarity, the relationships are generally presented in terms of *continuous* functions though the reader must bear in mind that digital processing dictates that the functions can be computed only at a *discrete number* of frequencies or instants of time.

FOURIER TRANSFORM. The basic relationships used to transform information from the time domain to the frequency domain, or vice versa, are the Fourier transform pair (see Chap. 23).

$$S_x(f) = \int_{-\infty}^{\infty} x(t)e^{-i2\pi ft}\,dt \tag{27.1}$$

$$x(t) = \int_{-\infty}^{\infty} S_x(f)e^{i2\pi ft}\,df \tag{27.2}$$

where $x(t)$ is the time-history of a function x, and $S_x(f)$ is the Fourier transform of x, a complex quantity.

DISCRETE FOURIER TRANSFORM. In practical digital applications, $x(t)$ is known only for a finite time interval or record length T, at a total of N equally spaced time intervals Δt apart. The discrete Fourier transform (DFT) pair, equivalent to Eqs. (27.1) and (27.2), become

$$S_x(m\,\Delta f) = \Delta t \sum_{i=0}^{N-1} x(n\,\Delta t)e^{-i2\pi m\,\Delta f\,n\,\Delta t} \qquad m = 0, \ldots, \left(\frac{N}{2}\right) \tag{27.3}$$

$$x(n\,\Delta t) = \Delta f \sum_{m=0}^{N/2} S_x(m\,\Delta f)e^{i2\pi m\,\Delta f\,n\,\Delta t} \qquad n = 0, \ldots, (N-1) \tag{27.4}$$

The DFT Eq. (27.3) is correct only if $x(t)$ is periodic with period T. The set of S_x are the coefficients of the Fourier series expansion of $x(t)$ [see Eq. (22.8)]. If $x(t)$ is not periodic, the DFT treats $x(t)$ as if it were so and is thus only an estimate of the true Fourier transform of $x(t)$.

As a consequence of Shannon's theorem,[16] the DFT of $x(t)$ is defined only at $N/2$ frequencies at equally spaced intervals Δf apart, up to a maximum frequency $F_{\max}$. The following fundamental relationships apply:

Block size:

$$N = \frac{T}{\Delta t} \tag{27.5}$$

Frequency range:

$$F_{\max} = 1/2\Delta t = \Delta f\left(\frac{N}{2}\right) \tag{27.6}$$

Sampling rate:

$$\frac{1}{\Delta t} = 2\,F_{\max} \tag{27.7}$$

Frequency resolution:

$$\Delta f = \frac{1}{T} \tag{27.8}$$

Record length:

$$T = \frac{1}{\Delta f} = N\,\Delta t \tag{27.9}$$

In performing a DFT analysis, assuming no restrictions on block size N, the parameters which may be selected are limited. The frequency range or the sampling rate, but not both, may be selected at will. Similarly, the frequency resolution or the record length may be chosen. If the block size is fixed, then only one of these four parameters may be selected.

The DFT, and thus any other quantities derived therefrom, is susceptible to certain errors, the most important of which is known as "aliasing."[13, 14, 15] Aliasing error arises when the signal $x(t)$ contains frequency components or "energy" above F_{max} of the DFT. Owing to the sampling or digitizing process, this energy will appear to be within the DFT frequency range. Therefore, to avoid aliasing error, either F_{max} must be large enough to include all significant frequency components of $x(t)$ or, for a given F_{max}, the components above F_{max} must be removed by analog filtering before digitization.

SPECTRAL DENSITY—FOURIER TRANSFORM. The concept of *power spectral density,* also called *auto spectral density,* is defined in analog terms in Chaps. 11 and 22. This quantity describes the frequency or spectral properties of a single time-history. A companion quantity, the *cross spectral density* describes the joint spectral properties of two time-histories. Both quantities are related to the Fourier transforms of the time histories through the following relationships:

Auto spectral density:

$$G_{xx}(f) = S_x(f)S_x^*(f) = |S_x(f)|^2 \tag{27.10}$$

where $G_{xx}(f)$ is the one-sided auto spectral density and $S_x^*(f)$ is the complex conjugate of $S_x(f)$.

Cross spectral density:

$$G_{xy}(f) = S_x(f)S_y^*(f) \tag{27.11}$$

$$G_{yx}(f) = S_y(f)S_x^*(f) \tag{27.12}$$

where $G_{xy}(f)$ and $G_{yx}(f)$ are one-sided cross spectral densities and are complex quantities, and $S_x(f)$ and $S_y(f)$ are the Fourier transforms of $x(t)$ and $y(t)$, respectively.

When the Fourier transforms of $x(t)$ and $y(t)$ are obtained from digital data, i.e., DFT's, the spectral density functions are defined only at the frequencies for which the DFT is calculated. Further properties of the spectral density functions so obtained are described under *Vibration Data Analysis.*

SPECTRAL DENSITY—CORRELATION FUNCTIONS. The spectral density functions, which describe signal characteristics in the frequency domain, and the correlation functions, which describe signal characteristics in the time domain (see Chap. 22), are related through the Fourier transform as follows:

Auto:

$$G_{xx}(f) = 2\int_{-\infty}^{\infty} R_{xx}(\tau)e^{-i2\pi f\tau}\,d\tau \qquad 0 \le f < \infty \tag{27.13}$$

$$R_{xx}(\tau) = \tfrac{1}{2}\int_{-\infty}^{\infty} G_{xx}(f)e^{i2\pi f\tau}\,df \qquad -\infty < \tau < \infty \tag{27.14}$$

Cross:

$$G_{xy}(f) = 2\int_{-\infty}^{\infty} R_{xy}(\tau)e^{-i2\pi f\tau}\,d\tau \qquad 0 \le f < \infty \tag{27.15}$$

$$R_{xy}(\tau) = 2\int_{-\infty}^{\infty} G_{xy}(f)e^{i2\pi f\tau}\,df \qquad -\infty < \tau < \infty \tag{27.16}$$

where $R_{xx}(\tau)$ is the auto correlation function at $x(t)$, and $R_{xy}(\tau)$ is the cross correlation function at $x(t)$ and $y(t)$.

FREQUENCY-RESPONSE FUNCTIONS. Frequency-response functions, synonymously known as *transfer functions*, describe in the frequency domain the relationship between the input and output or response of a linear system. This is shown schematically in Fig. 27.1 for a single input–single output system. (See Ref. 13 for multiple input-output systems.) The transfer function $H(f)$ describes the magnitude and phase of the response per unit sinusoidal input as a function of the input frequency. It is thus a complex quantity. Transfer functions can also be determined from the Fourier

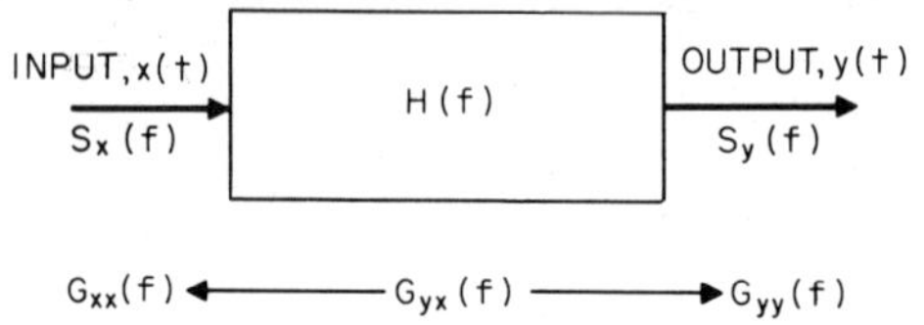

FIG. 27.1. Schematic representation of transfer function $H(f)$.

transforms of the input and response time-histories and from the spectral densities of the input and response signals when the input is a random process. The governing relationships are

Fourier transforms:

$$S_y(f) = S_x(f)H(f) \qquad \text{or} \qquad H(f) = \frac{S_y(f)}{S_x(f)} \tag{27.17}$$

Auto spectral densities:

$$G_{yy}(f) = G_{xx}(f)|H(f)|^2 \qquad \text{or} \qquad |H(f)|^2 = \frac{G_{yy}(f)}{G_{xx}(f)} \tag{27.18}$$

Cross spectral densities:

$$G_{yx}(f) = G_{xx}(f)H(f) \qquad \text{or} \qquad H(f) = \frac{G_{yx}(f)}{G_{xx}(f)} \tag{27.19}$$

The relationships above assume that no extraneous signal or noise is present in either $x(t)$ or $y(t)$ (see *Coherence Function and Modal Testing*).

It is seen that only the transfer function magnitude can be determined from auto spectral densities; that is, phase information is unobtainable. This follows because the auto spectral densities themselves are void of phase information. In using cross spectral density, in which phase information is retained, both magnitude and phase of the transfer function is obtained.

COHERENCE FUNCTION. The coherence function, γ_{xy}^2, is a measure of the quality of the input, response and cross spectral densities, and the causality of input to response for a system as shown in Fig. 27.1.

$$\gamma_{xy}^2(f) = \frac{|G_{xy}(f)|^2}{G_{xx}(f)G_{yy}(f)} \tag{27.20}$$

From Eqs. (27.18) and (27.19),

$$\gamma_{xy}^2(f) = \frac{|G_{xx}(f)H(f)|^2}{G_{xx}(f)|H(f)|^2G_{xx}(f)} = 1 \tag{27.21}$$

assuming $x(t)$ and $y(t)$ are noise-free.

It can be shown[13] that, in practical applications,

$$0 \leq \gamma_{xy}^2 \leq 1.0$$

Values of γ_{xy}^2 less than unity indicate that the response is not attributable to the input, due, for example, to extraneous noise or nonlinearity of the system. In the frequency domain, the coherence function is analogous to the correlation coefficient in the time domain.

DIGITAL VIBRATION DATA ANALYSIS

The basic principles of vibration data analysis are described in Chap. 22. These principles are illustrated in terms of analog methods of processing data. The principles of analysis are not changed by a switch to digital methods and are not repeated. Only the basic digital methods are described.

A number of hybrid, i.e., combined analog and digital, vibration analyzers have been developed, primarily for spectral density analysis of random vibration. For example, in one system[17] the computer is essentially used as a 52-channel detector, i.e., voltmeter, to detect the output of an analog comb filter. These types of computer applications have enhanced the capabilities of traditional analog analyzers but do not constitute a fundamental change of method.

Types of data analysis which may be performed entirely by digital computation parallel those performed by analog methods. Three basic types are spectral analysis, correlation analysis, and statistical or probability analysis. Common to all digital vibration data analysis is the implicit assumption that the data are sampled from a stationary or steady-state process; or that at least the data are varying slowly enough so the data sample may be considered stationary during the sample duration.

ANALOG-TO-DIGITAL CONVERSION AND DATA PREPARATION. The first step in any digital method is the analog-to-digital conversion (ADC) of the time-history or time-histories. This operation is generally built into self-contained digital analysis systems. However, when the digital processing is performed on a general-purpose scientific computer, this operation must usually be carried out at another facility.

A prime advantage of digital analysis methods is the requirement to play back the analog time-history of the signal no more than once for ADC, or alternatively, to obviate the necessity for recording the time-history if on-line analysis is desired. However, analog recording in parallel may be prudent for backup data purposes. Thus, for example, the problems associated with preparing a satisfactory tape-loop for repeated playback of the data sample to be analyzed by tunable analog filters are avoided.

For each of the digital data samples which constitute the digital record to be processed, the ADC process consists of two steps: (1) the analog signal is sampled by a sample-and-hold device, and (2) the analog voltage is converted to a binary digital number. Each sample is subject to two errors, which appear as digital noise on the record. First, the sample-and-hold device samples over a finite though small time interval. If the analog signal can vary significantly during this time interval, the "held" voltage will be inaccurate. Second, further error may be generated if the digital word length, i.e., number of bits, is insufficient. Since there is always an uncertainty equal to the least significant bit, this may represent a significant error, particularly for samples taken near a zero crossing.

When possible, it is desirable to preprocess the digital record prior to analysis to remove any wild points and any trends.[15] "Wild points" are defined as any individual digital samples (or possibly several successive samples) which are either unreasonably large or essentially zero for some generally unknown reason, such as analog tape spikes or dropouts or ADC momentary failure. "Trends" are defined as slowly varying changes in the data, such as a DC bias or a slow change in the running mean value of the signal. Trends can often be eliminated by the use of high-pass filtering of the signal prior to ADC. Alternatively, the digital record can be preprocessed to remove the calculated trend from each digital sample and a new record created.

The data must be preprocessed and digitized in a manner which avoids frequency aliasing. Aliasing errors may be avoided by using a sampling rate which is at least twice that of the highest significant frequency component in the signal. However, since this may be difficult before analysis, which has the objective of determining the frequency components up to some desired maximum frequency of interest, it is generally preferable to avoid aliasing errors by low-pass filtering prior to ADC.

SPECTRAL ANALYSIS. Spectral analysis (see Chap. 22) is performed to determine the frequency characteristics of a time-history. For deterministic signals, a line spectrum is desired; for random processes, a spectral density function is desired.

Deterministic Signal. Determination of the line spectrum of a deterministic signal $[x(t)]$ by digital computation requires only computation of the Fourier transform of the signal $S(x)$. The computed transform may be presented either in terms of magnitude $|S(x)|$ and phase or in terms of the real and imaginary parts for each computation frequency, or both.

Equations (27.5) through (27.9) define the relationships for selection of the analysis parameters. It should be noted that the Fourier transform treats the data as if it were periodic with period T. Thus, for a deterministic signal, the frequencies at which $S(x)$ is computed may not and probably will not coincide exactly with the actual frequency components of the signal. It is important, therefore, to select a frequency resolution, Δf, which is sufficiently small to minimize errors in both frequency and amplitude owing to this misalignment of actual and computed frequencies.

Random Signal. Determination of the spectral density function, either auto or cross, of a random signal may be performed in either of two alternative methods. The older and traditional method, known as the Blackman-Tukey method,[18] arrives at the spectral density by first computing the correlation function [Eq. (22.21) or (22.27)], followed by Fourier transformation of the correlation function to obtain the spectral density [Eq. (27.13) or (27.15)]. The newer method (using the FFT algorithm), known as the Cooley-Tukey method,[12] obtains the spectral density directly from the Fourier transform of the time-history [Eqs. (27.10), (27.11), and (27.12)]. For applications requiring real-time, or more exactly, almost real-time, analysis the latter method is mandatory.

Equations (27.5) through (27.9) define the relationships for selection of the analysis parameters, using the FFT method. Indirectly, these equations also define the analysis parameters for correlation analysis when using the Blackman-Tukey method. Assuming the correlation functions are computed at incremental delay times, $\Delta\tau$ equal to the sampling interval, Δt, up to a maximum delay, m $\Delta\tau$, where m is the maximum lag number, then

$$\Delta f = \frac{1}{m\,\Delta\tau} \tag{27.22}$$

It should be noted that for spectral density analysis, these methods determine the average spectral density within each bandwidth, Δf wide, centered at the $N/2$ computation frequencies.

The maximum lag number m should be small compared with the total number of samples N. The effective record length for spectral analysis is thus $m\,\Delta t$ rather than the $N\,\Delta t$ needed to compute the correlation function.

Statistical Error. Spectral density functions computed by either of the methods described above are only estimates of the true spectral densities of the original analog time-histories. These estimates are subject to statistical error very similar to the statistical errors in analog processing. The normalized standard error ϵ of the spectral density estimate computed from a single FFT is 1.0 and is governed by the chi-squared distribution for two degrees-of-freedom. To obtain spectral density estimates with a more reasonable statistical error, further averaging is necessary. This may be accomplished by averaging over frequency and/or averaging over segments or records. In the first case, the spectral estimates in l adjacent frequency intervals are averaged to obtain an estimate of the average spectral density in a frequency bandwidth, $l\,\Delta f$. The standard error of the resultant spectral density then becomes $\epsilon = 1/\sqrt{l}$. Thus, statistical accuracy is gained at the expense of frequency resolution. In the second case, the spectral estimates obtained from a series of records, each $N\,\Delta t = T$ in duration, are averaged. This process is known as "ensemble averaging." The standard error for an ensemble average of q records is $1/\sqrt{q}$. Therefore, statistical accuracy is gained without loss of frequency resolution, for a constant block size N, by repeated spectral analysis. If both frequency and ensemble averaging are employed, the standard error becomes $1/\sqrt{ql}$. The error is chi-squared distributed with $2l$, $2q$, and $2ql$ degrees-of-freedom, respectively, for the three types of averaging.

Leakage. Spectral analysis by digital methods is subject to error known as "leakage."[13, 14, 15] Leakage arises from the truncation which occurs when either the FFT is computed over a finite frequency range, 0 to F_{max}, or the correlation function is computed over a finite maximum time delay, $m\,\Delta\tau$. This truncation causes undesirable side lobes in the effective filter shape of the transmission characteristics of a filter. These side lobes can be minimized by the use of one of several types of tapering functions instead of truncating in the digital processing. Commonly used functions are known as Parzen[19, 13, 15] and Hamming[18, 13, 15] windows.

STATISTICAL ANALYSIS. Chapter 22 described, in analog terms, the basic methods of determining the probability distribution functions of both instantaneous and peak values of a signal. Performance of these types of analyses using digital computer techniques is no different in principle but is much more straightforward to implement. References 13, 14, and 15 contain detailed descriptions of the required digital operations.

An advantage gained by use of digital techniques is the ability to perform goodness-of-fit tests on the resultant probability density functions. These tests determine (to some selected confidence level, and based on the number of digital samples) whether the distribution should or should not be considered Gaussian.

CORRELATION ANALYSIS. Auto- and cross-correlation functions may be determined either from Fourier-transforming the auto- or cross-spectral density functions [Eqs. (27.14) and (27.16)] or directly from the time-histories. The latter is generally the case during computation of PSD by the Blackman-Tukey method. The principles of the latter method for digital processing are the same as described in Chap. 22 for analog processing. Detailed digital procedures are described in Refs. 13, 14, and 15, and are clearly very simple, consisting merely of summing the products of pairs of samples. The choice of record length, digitizing rate, and number of lags determines the statistical error of the correlation function estimate. In addition, these choices define the frequency resolution, frequency range, and statistical error of density spectra, later computed from the correlation functions.

DIGITAL CONTROL SYSTEMS FOR VIBRATION TESTING

The vibratory motions prescribed for the majority of vibration tests consist of either sinusoidal, with fixed or variable frequency, or random motions. The system employed to control the vibration level during the test utilizes the output signal from an accelerometer mounted at an appropriate location on the vibration exciter as the feedback signal to a servo system. The servo system adjusts the driving signal to the power amplifier to maintain the desired test level. Figure 27.2 shows this closed-loop system schematically. Vibration testing machines are described in Chap. 25, and Ref. 20 describes the

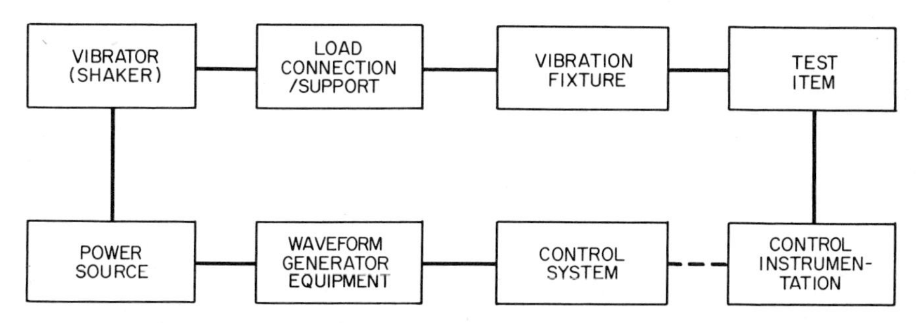

FIG. 27.2. Functional block diagram—closed-loop vibration system.

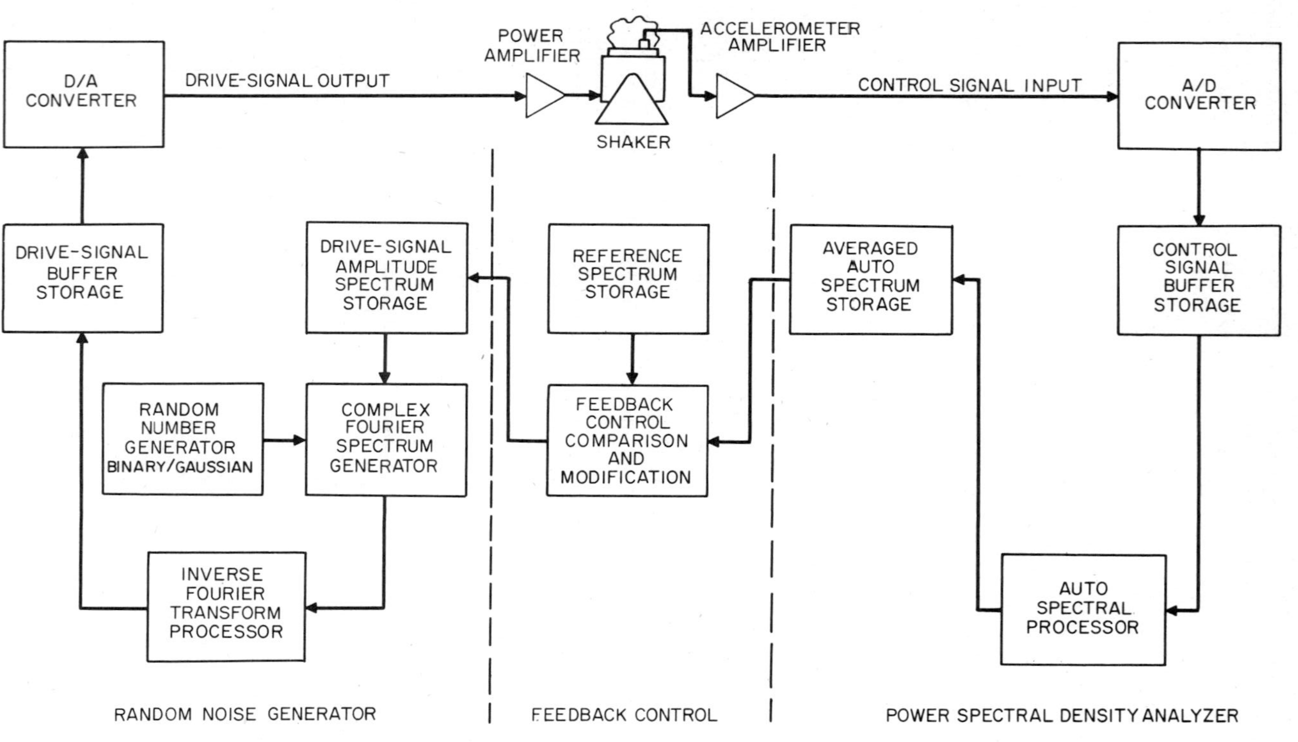

Fig. 27.3. Functional block diagram of random vibration digital control system.

performance of a number of types of vibration tests in detail when using analog control systems.

With the advent of FFT processors and minicomputers, it became possible to perform spectral analysis of random processes rapidly enough to permit the use of digital control systems for random vibration testing. An extension of software also enables the system to control sinusoidal testing.

RANDOM VIBRATION TESTS.[21, 22, 23, 24, 25, 26, 27] Figure 27.3 shows a functional block diagram of a digitally controlled random vibration system. Starting with the random number generator, which is the digital equivalent of an analog random noise generator, a complex Fourier spectrum is generated. The inverse Fourier transform converts this spectrum to a digital time-history which is used, after digital-to-analog conversion, to drive the power amplifier–electrodynamic shaker. The control accelerometer feedback signal is digitized and the auto spectral density computed by an FFT processor. After sufficient ensemble averaging, the auto-spectrum is compared with the specified test spectrum and a correction or modification of the driving spectrum is generated, thus closing the loop.

Two types of drive signals are employed for random vibration testing. (1) A purely random signal uses a completely new random time-history for each succeeding time-frame of data, obtained from the random number generator. In this case, the response time of the servo system is dependent on the time to ensemble-average the feedback spectrum sufficiently to obtain the required statistical accuracy. Typically, this time is 1 to 2 seconds, depending primarily on the chosen frequency resolution. (2) A pseudo-random signal (zero-variance signal) may be employed in order to achieve shorter servo response times, particularly during initial startup of the test. This type of signal consists of repeated use of the same time-frame of data, sometimes alternately played backward and forward, and is thus periodic. However, the time-frame itself displays the desired random characteristics. Since the time-frames are repetitive, the need to ensemble-average is obviated and the driving signal can be corrected every other time-frame.

SINUSOIDAL VIBRATION TESTS.[28, 29, 30] To perform sinusoidal vibration tests, the random noise generator in Fig. 27.3 is replaced by a digital sine-wave synthesizer which can be programmed. In addition, the feedback signal is analyzed to determine the amplitude of the sinusoidal signal rather than the spectral density of the signal for comparison to the reference (i.e., desired) amplitude.

DIGITAL CONTROL SYSTEMS FOR SHOCK TESTING

(The analysis of shock data is described in Chap. 23.) The shock-test machines described in Chap. 26 may be looked upon as devices for storing kinetic energy which can be suddenly released in order to create the desired shock or transient. For a number of reasons, it is desirable to perform shock or transient testing using electrodynamic or electrohydraulic vibration systems to induce the desired transient motion at the mounting points of the test item. The ability to employ this method is dependent on such parameters as the stroke or maximum allowable motion of the vibration exciter, the transient waveform, the mass of the test item, etc. Given that the required test is within the performance capability of an available vibration system, the ability to obtain and control the desired motion has been greatly expanded by the use of digital control equipment.

Shock-test requirements are specified in one of two alternative methods. The first, and most direct, method specifies a certain acceleration waveform, such as a half sine wave of specified duration and maximum acceleration. Most requirements of this type suffer from the following deficiency or indefiniteness. Most specified acceleration waveforms do not have a zero mean value and thus indirectly specify a certain velocity change. Since any laboratory test must start and eventually stop at rest, i.e., zero velocity change, additional accelerations, positive or negative, are applied before and/or after the specified waveform in order to give zero velocity change. These additional accelerations are minimized in amplitude by maximizing their duration as limited by the shock-test machine. The second method of specifying shock-test requirements employs shock spectra (see Chap. 23). The requirements must specify the frequency range, damping factor, type of spectrum, and either minimum or nominal values with an allowable tolerance on spectrum values.[31]

WAVEFORM CONTROL TESTS.[32, 33, 34, 35] Shock tests, defined in terms of waveform, are performed as follows:

1. With the test item installed on the vibration exciter, the complex transfer function between the drive signal input and the control accelerometer is measured. A short burst of random vibration or a suitable transient excitation may be used.
2. The complex Fourier spectrum of the desired waveform is calculated.
3. The Fourier spectrum of the desired waveform is divided by the complex transfer function of the system to obtain the complex Fourier spectrum of the required drive signal.
4. The inverse Fourier transform of this spectrum is computed to obtain the required drive signal time-history.

Cited references describe in detail the application of digital control systems employing FFT processors and minicomputers to perform the steps above. If the specified waveform does not have a zero mean value, i.e., nonzero velocity change, the Fourier spectrum in step 2 must be calculated for a desired waveform which includes any additional accelerations required to give a zero velocity change or the vibration exciter may well become a launching pad.

SHOCK-SPECTRUM CONTROL TESTS.[36, 37, 38] When the shock test is specified in terms of a shock spectrum, no explicit requirements are placed on the acceleration waveform. Thus, a number of different waveforms may all meet the requirements. Waveforms such as a burst of random noise, a chirp (i.e., a very rapidly swept sinusoidal sweep), and combinations of exponentially growing and decaying sine waves have been used. To control shock tests in terms of a required shock spectrum the following strategy is employed.

1. Select the type of acceleration time-history to be employed, i.e., random burst, etc.
2. Calculate the amplitude and frequency characteristics of this waveform which will give the desired shock spectrum.
3. Perform an exploratory test, using the four steps to calculate the drive signal described above for waveform control test.
4. Calculate the shock spectrum of the resulting acceleration waveform measured by the control accelerometer.
5. Repeat steps 2, 3, and 4 as necessary until the desired shock spectrum is obtained.

Cited references describe in detail the application of digital systems for the shock-spectrum computations and test control of these types of tests.

MODAL TESTING[39, 40, 41, 42, 43, 44]

Modal tests, also known as ground vibration tests in the aircraft industry, are tests conducted to determine experimentally the natural frequencies, mode shapes, and associated damping factors of a structure. For many years, these tests have been conducted by exciting the system with sinusoidal excitation at a number of points of the structure. The driving frequency, the relative amplitudes, and phases of the excitations are adjusted to obtain the "purest" possible excitation of a single mode. The responses at locations throughout the structure then define the mode shape for that frequency. Damping factors can be measured by measurement of the bandwidth of the mode.

Mode shapes may also be measured by the use of broadband vibration or transient excitations applied at a single point of the structure for any one measurement. The responses to these excitations are measured at all points of interest. From these data, a set of transfer functions between the excitation and each response measurement location may be calculated for each excitation employed. The frequencies, mode shapes, and damping factors are then obtained from these measured transfer functions.

Either of the approaches given above involves the acquisition and analysis of a vast amount of data. Digital processing of these data on line, i.e., during the test, has greatly improved the efficiency and accuracy of this type of test, as well as the post-test data processing of the data to determine the desired information.

Digital computers have been applied in two distinct ways. First, for sinusoidal excitation,[39, 40, 44] computers have been employed as an aid in obtaining the desired purity of the modal excitation as well as in acquiring and processing data. Second, and more essentially, for broadband excitation[41, 42, 43, 44] (either random vibration or transient) FFT processing of the data is used to determine transfer functions essentially on line. In addition, the quality of the transfer function can be assessed immediately by computation and display of the coherence function [Eq. (27.20)]. If an unsatisfactory coherence function is obtained, the excitation can be repeated immediately until satisfactory data quality is achieved. However, care must be exercised in computing the coherence function. The coherence function obtained from raw spectral density estimates using the FFT method, i.e., from a single frame of data, will always be unity.[13] Reliable frequency-response-function and coherence-function estimates may be obtained either from smoothed spectral estimates obtained by FFT processes (Cooley-Tukey) or from spectral estimates obtained from correlation functions (Blackman-Tukey).

REFERENCES

1. Pilkey, W., K. Saczalski, and H. Schaeffer (eds.): "Structural Mechanics Computer Programs, Surveys, Assessments and Availability," University Press of Virginia, Charlottesville, Virginia, 1974.
2. Pilkey, W., and B. Pilkey: "Shock Vibration Computer Programs—Reviews and Summaries," SVM-10, Shock and Vibration Information Center, Washington, D.C., 1975.
3. NASA SP-222 (01): "The NASTRAN Users Manual," June, 1972 (and later updated editions).
4. NASA SP-221 (01): "The NASTRAN Theoretical Manual," June, 1972 (and later updated editions).
5. NASA SP-223: "The NASTRAN Programmers Manual," June, 1972 (and later updated editions).
6. MacNeal, R. H., and C. W. McCormick: "The NASTRAN Computer Program for Structural Analysis," *SAE Paper* 690612, October, 1969.
7. MacNeal, R. H.: *Computers Struct.*, **1**:581–601 (1971).
8. Hurty, W. C., J. D. Collins, and G. C. Hart: *Computers Struct.*, **1**:535–563 (1971).
9. Rubin, S.: "An Improved Component-Mode Representation," *AIAA Paper* 74-386, April, 1974.
10. Guyan, R. J.: *AIAA*, **3**(2):380 (February, 1965).
11. Kuhar, E. J., and C. V. Stahle: "A Dynamic Transformation Method for Modal Synthesis," *AIAA Paper* 73-396, March, 1973.
12. Cooley, J. W., and J. W. Tukey: *Math. Computing*, **19**:297 (April, 1965).
13. Bendat, Julius S., and Allan G. Piersol: "Random Data: Analysis and Measurement Procedures," John Wiley & Sons, Inc., New York, 1971.
14. Enochson, Loren D., and Robert K. Otnes: "Programming and Analysis for Digital Time Series Data," SVM-3, Shock and Vibration Information Center, Washington, D.C., 1968.
15. Otnes, Robert K., and Loren Enochson: "Digital Time Series Analysis," John Wiley & Sons, Inc., New York, 1972.
16. Shannon, C. E.: *Proc. IRE*, **37**:10 (1949).
17. Curtis, A. J., and J. G. Herrera: "Random-Vibration Test Level Control Using Input and Test Item Response Spectra," *Shock and Vibration Bull.* 37, pt. 3, p. 47, Shock and Vibration Information Center, Washington, D.C., January, 1968.
18. Blackman, R. B., and Tukey, J. W.: "The Measurement of Power Spectra," Dover Publications, Inc., New York, 1958.
19. Parzen, E.: "Stochastic Processes," Holden-Day, Inc., Publisher, San Francisco, 1962.
20. Curtis, A. J., N. G. Tinling, and H. T. Abstein, Jr.: "Selection and Performance of Vibration Tests," SVM-8, Shock and Vibration Information Center, Washington, D.C., 1971.
21. Chapman, C. P., J. Shipley, and C. L. Heizman: "A Digitally Controlled Vibration or Acoustics Testing System," *IES 1969 Proc.*, pts. I, II, and III, pp. 387–409.
22. Ratz, A. G., "Random Vibration Test Systems Using Digital Equalizers," *IES 1970 Proc.*, p. 75.
23. Nelson, D. B.: "Some Considerations in Design, Specification and Evaluation of Digital Control System for Random Vibration Testing," *ASME Paper* 71-Vibr-30, September, 1971.

24. Chapman, C. P.: "Computer-Controlled Vibration Test Systems: Criteria for Selection, Installation and Maintenance," *SAE Paper* 720819, October, 1972.
25. Heizman, C. L., and E. A. Sloane: "Evolution of Digital Vibration System," *SAE Paper* 720820, October, 1972.
26. Tebbs, J. D., and N. F. Hunter, Jr.: "Digital Control of Random Vibration Tests Using a Sigma V Computer," *IES 1974 Proc.*, pp. 36–43.
27. Tebbs, J. D., and D. O. Smallwood: "Extension of Control Techniques for Digital Control of Random Vibration Tests," *Shock and Vibration Bull.* 45, 1975.
28. Reiner, R. L., "A Digitally-Controlled Sweep Oscillator," *IES 1970 Proc.*, pp. 52–57.
29. Bosso, F. C.: "Sine-Test Electrodynamic Vibration Systems Using Digital Control," *IES 1970 Proc.*, pp. 61–69.
30. Norin, R. S.: "A Multi-Channel, Multi-Strategy Sinusoidal Vibration Control System," *IES 1974 Proc.*, pp. 222–228.
31. American National Standards Institute: "Methods for Analysis and Presentation of Shock and Vibration Data," ANSI S2.10-1971.
32. Favour, J. D., and J. M. LeBrun: "Final Report, Feasibility and Conceptual Design Study—Vibration Generator Transient Waveform Control System," NAS5-15171, NASA, 1969.
33. Favour, J. D., J. M. LeBrun, and J. P. Young: "Transient Waveform Control of Electromagnetic Vibration Test Equipment," *Shock and Vibration Bull.* 40, pt. 2, pp. 157–171, December, 1969.
34. Barthmaier, J. P.: "Shock Testing Under Minicomputer Control," *IES 1974 Proc.*, pp. 207–215.
35. Hunter, N. F., Jr.: "Transient Waveform Reproduction on Hydraulic Actuators Using a Nonlinear Gain Estimation Technique," *IES 1974 Proc.*, pp. 202–206.
36. Kao, G. C., K. Y. Chang, and W. W. Holbrook: "Digital Control Technique for Seismic Simulation," *Shock and Vibration Bull.* 43, pt. I, pp. 109–117, June, 1973.
37. Smallwood, D. O., and A. F. Wilte: "The Use of Shaker Optimized Periodic Transients in Matching Field Shock Spectra," *Shock and Vibration Bull.* 43, pt. 1, pp. 139–150, June, 1973.
38. Yang, R. C., and H. R. Saffell: "Development of a Waveform Synthesis Technique—A Supplement to Response Spectrum as a Definition of Shock Environment," *Shock and Vibration Bull.* 42, pt. 2, pp. 45–53, January, 1972.
39. Salyer, Robert A.: "Hybrid Techniques for Modal Survey Control and Data Appraisal," *Shock and Vibration Bull.* 41, pt. 3, pp. 25–42, December, 1970.
40. Budd, R. W.: "MODAPS—A Real Time Data Processing System for Modal Vibration Testing," *IES 1971 Proc.*, pp. 169–177.
41. Favour, J. D., M. C. Mitchell, and N. L. Olson: "Transient Test Techniques for Mechanical Impedance and Modal Survey Testing," *Shock and Vibration Bull.* 42, pt. 1, pp. 71–82, January 1972.
42. Richardson, M., and R. Potter, "Identification of the Modal Properties of an Elastic Structure from Measured Transfer Function Data," *20th Intern. Instr. Symp.*, Albuquerque, New Mexico, May, 1974.
43. Potter, R., and M. Richardson, "Mass, Stiffness, and Damping Matrices from Measured Modal Parameters," *Proc. 1974 ISA Intern. Instr. Automation Conf.*, October, 1974.
44. Smith, S., R. C. Stroud, G. A. Hamma, W. L. Hallaver, and R. C. Yee: "MODALAB A, A New System for Structural Dynamic Testing," *Shock and Vibration Bull.* 45, 1975.

28

MATRIX METHODS OF ANALYSIS

Stephen H. Crandall
Massachusetts Institute of Technology

Robert B. McCalley, Jr.
General Electric Company

The mathematical language which is most convenient for analyzing multiple degree-of-freedom vibratory systems is that of *matrices*. Matrix notation simplifies the preliminary analytical study and in situations where particular numerical answers are required, matrices provide a standardized format for organizing the data and the computations. Computations with matrices can be carried out by hand or by digital computers. The availability of large high-speed digital computers makes the solution of many complex problems in vibration analysis a matter of routine.

This chapter describes how matrices are used in vibration analysis. It begins with definitions and rules for operating with matrices. The formulation of vibration problems in matrix notation then is treated. This is followed by general matrix solutions of several important types of vibration problems including free and forced vibrations of both undamped and damped linear multiple degree-of-freedom systems. The computational problems involved in handling matrices then are discussed.

MATRICES

Matrices[1] are mathematical entities which facilitate the handling of simultaneous equations. They are applied to the differential equations of a vibratory system as follows:

A single degree-of-freedom system of the type in Fig. 28.1 has the differential equation

$$m\ddot{x} + c\dot{x} + kx = F$$

where m is the mass, c is the damping coefficient, k is the stiffness, F is the applied force,

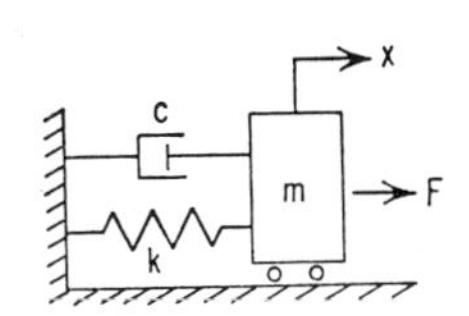

FIG. 28.1. Single degree-of-freedom system.

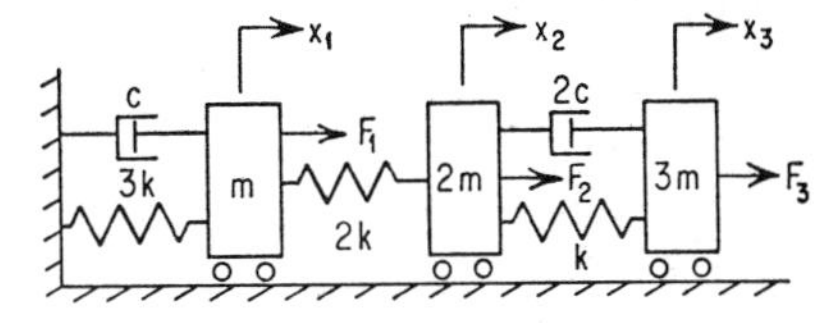

FIG. 28.2. Three degree-of-freedom system.

x is the displacement coordinate, and dots denote time derivatives. In Fig. 28.2 a similar three degree-of-freedom system is shown. The equations of motion may be obtained by applying Newton's second law to each mass in turn:

$$\begin{aligned} m\ddot{x}_1 \quad &+ c\dot{x}_1 &&+ 5kx_1 - 2kx_2 &&= F_1 \\ 2m\ddot{x}_2 \quad &+ 2c\dot{x}_2 - 2c\dot{x}_3 &&- 2kx_1 + 3kx_2 - kx_3 &&= F_2 \\ 3m\ddot{x}_3 \quad &- 2c\dot{x}_2 + 2c\dot{x}_3 &&- kx_2 + kx_3 &&= F_3 \end{aligned} \tag{28.1}$$

The accelerations, velocities, displacements, and forces may be organized into columns, denoted by single boldface symbols:

$$\ddot{\mathbf{x}} = \begin{bmatrix} \ddot{x}_1 \\ \ddot{x}_2 \\ \ddot{x}_3 \end{bmatrix} \qquad \dot{\mathbf{x}} = \begin{bmatrix} \dot{x}_1 \\ \dot{x}_2 \\ \dot{x}_3 \end{bmatrix} \qquad \mathbf{x} = \begin{bmatrix} x_1 \\ x_2 \\ x_3 \end{bmatrix} \qquad \mathbf{f} = \begin{bmatrix} F_1 \\ F_2 \\ F_3 \end{bmatrix} \tag{28.2}$$

The inertia, damping, and stiffness coefficients may be organized into square arrays:

$$\mathbf{M} = \begin{bmatrix} m & 0 & 0 \\ 0 & 2m & 0 \\ 0 & 0 & 3m \end{bmatrix} \qquad \mathbf{C} = \begin{bmatrix} c & 0 & 0 \\ 0 & 2c & -2c \\ 0 & -2c & 2c \end{bmatrix} \qquad \mathbf{K} = \begin{bmatrix} 5k & -2k & 0 \\ -2k & 3k & -k \\ 0 & -k & k \end{bmatrix} \tag{28.3}$$

By using these symbols, it is shown below that it is possible to represent the three equations of Eq. (28.1) by the following single equation:

$$\mathbf{M}\ddot{\mathbf{x}} + \mathbf{C}\dot{\mathbf{x}} + \mathbf{K}\mathbf{x} = \mathbf{f} \tag{28.4}$$

Note that this has the same form as the differential equation for the single degree-of-freedom system of Fig. 28.1. The notation of Eq. (28.4) has the advantage that in systems of many degrees of freedom it clearly states the physical principle that at every coordinate the external force is the sum of the inertia, damping, and stiffness forces. Equation (28.4) is an abbreviation for Eq. (28.1). It is necessary to develop the rules of operation with symbols such as those in Eqs. (28.2) and (28.3) to ensure that no ambiguity is involved. The algebra of *matrices* is devised to facilitate manipulations of simultaneous equations such as Eq. (28.1). Matrix algebra does not in any way simplify individual operations such as multiplication or addition of numbers, but it is an organizational tool which permits one to keep track of a complicated sequence of operations in an optimum manner.

DEFINITIONS

A *matrix* is an array of elements arranged systematically in rows and columns. For example, a rectangular matrix **A**, of elements a_{jk}, which has m rows and n columns is

$$\mathbf{A} = [a_{jk}] = \begin{bmatrix} a_{11} & a_{12} & \cdots & a_{1n} \\ a_{21} & a_{22} & \cdots & a_{2n} \\ \cdots & \cdots & \cdots & \cdots \\ a_{m1} & a_{m2} & \cdots & a_{mn} \end{bmatrix}$$

The elements a_{jk} are usually numbers or functions but, in principle, they may be any well-defined quantities. The first subscript j on the element refers to the row number while the second subscript k refers to the column number. The array is denoted by the single symbol **A** which can be used as such during operational manipulations in which it is not necessary to specify continually all the elements a_{jk}. When a numerical calculation is finally required, it is necessary to refer back to the explicit specifications of the elements a_{jk}.

A rectangular matrix with m rows and n columns is said to be of order (m,n). A matrix of order (n,n) is a *square matrix* and is said to be simply a square matrix of order n. A matrix of order $(n,1)$ is a *column matrix* and is said to be simply a column matrix of order n. A column matrix is sometimes referred to as a *column vector*. Similarly, a matrix of order $(1,n)$ is a *row matrix* or a *row vector*. Boldface *capital* letters are used here to represent square matrices and *lower-case* boldface letters to represent column matrices or vectors. For example, the matrices in Eq. (28.2) are column matrices of order three and the matrices in Eq. (28.3) are square matrices of order three.

Some special types of matrices are:

1. A *diagonal matrix* is a square matrix $\mathbf{A}$ whose elements a_{jk} are zero when $j \neq k$. The only nonzero elements are those on the *main diagonal* where $j = k$. In order to emphasize that a matrix is diagonal, it is often written with small ticks in the direction of the main diagonal:

$$\mathbf{A} = \left[\diagdown a_{jj} \diagdown\right]$$

2. A *unit matrix* or *identity matrix* is a diagonal matrix whose main diagonal elements are each equal to unity. The symbol $\mathbf{I}$ is used to denote a unit matrix. Examples are

$$\begin{bmatrix} 1 & 0 \\ 0 & 1 \end{bmatrix} \qquad \begin{bmatrix} 1 & 0 & 0 \\ 0 & 1 & 0 \\ 0 & 0 & 1 \end{bmatrix}$$

3. A *null matrix* or *zero matrix* has all its elements equal to zero and is simply written as zero.

4. The *transpose* $\mathbf{A}^T$ of a matrix $\mathbf{A}$ is a matrix having the same elements but with rows and columns interchanged. Thus, if the original matrix is

$$\mathbf{A} = [a_{jk}]$$

the transpose matrix is

$$\mathbf{A}^T = [a_{jk}]^T = [a_{kj}]$$

For example:

$$\mathbf{A} = \begin{bmatrix} 3 & 2 \\ -1 & 4 \end{bmatrix} \qquad \mathbf{A}^T = \begin{bmatrix} 3 & -1 \\ 2 & 4 \end{bmatrix}$$

The transpose of a square matrix may be visualized as the matrix obtained by rotating the given matrix about its main diagonal as an axis.

The transpose of a column matrix is a row matrix. For example,

$$\mathbf{x} = \begin{bmatrix} 3 \\ 4 \\ -2 \end{bmatrix} \qquad \mathbf{x}^T = [3 \quad 4 \quad -2]$$

Throughout this chapter a row matrix is referred to as the transpose of the corresponding column matrix.

5. A *symmetric matrix* is a square matrix whose off-diagonal elements are symmetric with respect to the main diagonal. A square matrix $\mathbf{A}$ is symmetric if, for all j and k,

$$a_{jk} = a_{kj}$$

A symmetric matrix is equal to its transpose. For example, all three of the matrices in Eq. (28.3) are symmetric. In addition the matrix $\mathbf{M}$ is a diagonal matrix.

MATRIX OPERATIONS

EQUALITY OF MATRICES. Two matrices of the same order are equal if their corresponding elements are equal. Thus two matrices $\mathbf{A}$ and $\mathbf{B}$ are equal if, for every j and k,

$$a_{jk} = b_{jk}$$

MATRIX ADDITION AND SUBTRACTION. Addition or subtraction of matrices of the same order is performed by adding or subtracting corresponding elements. Thus, $\mathbf{A} + \mathbf{B} = \mathbf{C}$, if for every j and k,

$$a_{jk} + b_{jk} = c_{jk}$$

For example, if

$$\mathbf{A} = \begin{bmatrix} 3 & 2 \\ -1 & 4 \end{bmatrix} \qquad \mathbf{B} = \begin{bmatrix} -1 & 2 \\ 5 & 6 \end{bmatrix}$$

then

$$\mathbf{A} + \mathbf{B} = \begin{bmatrix} 2 & 4 \\ 4 & 10 \end{bmatrix} \qquad \mathbf{A} - \mathbf{B} = \begin{bmatrix} 4 & 0 \\ -6 & -2 \end{bmatrix}$$

MULTIPLICATION OF A MATRIX BY A SCALAR. Multiplication of a matrix by a scalar c multiplies each element of the matrix by c. Thus

$$c\mathbf{A} = c[a_{jk}] = [ca_{jk}]$$

In particular, the negative of a matrix has the sign of every element changed.

MATRIX MULTIPLICATION. If $\mathbf{A}$ is a matrix of order (m,n) and $\mathbf{B}$ is a matrix of order (n,p), then their *matrix product* $\mathbf{AB} = \mathbf{C}$ is defined to be a matrix $\mathbf{C}$ of order (m,p) where, for every j and k,

$$c_{jk} = \sum_{r=1}^{n} a_{jr}b_{rk} \tag{28.5}$$

The product of two matrices can be obtained only if they are *conformable*, i.e., if the number of columns in A is equal to the number of rows in B. The symbolic equation

$$(m,n) \times (n,p) = (m,p)$$

indicates the orders of the matrices involved in a matrix product. Matrix products are not commutative, i.e., in general,

$$\mathbf{AB} \neq \mathbf{BA}$$

The matrix products which appear in this chapter are of the following types:
Square matrix × square matrix = square matrix
Square matrix × column vector = column vector
Row vector × square matrix = row vector
Row vector × column vector = scalar
Column vector × row vector = square matrix
In all cases, the matrices must be conformable. Numerical examples are given below.

$$\mathbf{AB} = \begin{bmatrix} 3 & 2 \\ -1 & 4 \end{bmatrix} \begin{bmatrix} -1 & 2 \\ 5 & 6 \end{bmatrix} = \begin{bmatrix} -(3 \times 1) + (2 \times 5) & (3 \times 2) + (2 \times 6) \\ (1 \times 1) + (4 \times 5) & -(1 \times 2) + (4 \times 6) \end{bmatrix}$$

$$= \begin{bmatrix} 7 & 18 \\ 21 & 22 \end{bmatrix}$$

$$\mathbf{Ax} = \begin{bmatrix} 3 & 2 \\ -1 & 4 \end{bmatrix} \begin{bmatrix} 5 \\ 3 \end{bmatrix} = \begin{bmatrix} (3 \times 5) + (2 \times 3) \\ -(1 \times 5) + (4 \times 3) \end{bmatrix} = \begin{bmatrix} 21 \\ 7 \end{bmatrix}$$

$$\mathbf{y}^T\mathbf{A} = [-2 \quad 1] \begin{bmatrix} 3 & 2 \\ -1 & 4 \end{bmatrix} = [-(2 \times 3) - (1 \times 1) \quad - (2 \times 2) + (1 \times 4)] = [-7 \quad 0]$$

$$\mathbf{y}^T\mathbf{x} = [-2 \quad 1] \begin{bmatrix} 5 \\ 3 \end{bmatrix} = (-10 + 3) = -7$$

$$\mathbf{xy}^T = \begin{bmatrix} 5 \\ 3 \end{bmatrix} [-2 \quad 1] = \begin{bmatrix} -(5 \times 2) & (5 \times 1) \\ -(3 \times 2) & (3 \times 1) \end{bmatrix} = \begin{bmatrix} -10 & 5 \\ -6 & 3 \end{bmatrix}$$

The last product always results in a matrix with proportional rows and columns.

The operation of matrix multiplication is particularly suited for representing systems of simultaneous linear equations in a compact form in which the coefficients are gathered into square matrices and the unknowns are placed in column matrices. For example, it is the operation of matrix multiplication which gives unambiguous meaning to the matrix abbreviation in Eq. (28.4) for the three simultaneous differential equations of Eq. (28.1). The two sides of Eq. (28.4) are column matrices of order three whose corresponding elements must be equal. On the right, these elements are simply the external forces at the three masses. On the left, Eq. (28.4) states that the resulting column is the sum of three column matrices, each of which results from the matrix multiplication of a square matrix of coefficients defined in Eq. (28.3) into a column matrix defined in Eq. (28.2). The rules of matrix operation just given ensure that Eq. (28.4) is exactly equivalent to Eq. (28.1).

Premultiplication or postmultiplication of a square matrix by the identity matrix leaves the original matrix unchanged; i.e.,

$$\mathbf{IA} = \mathbf{AI} = \mathbf{A}$$

Two symmetrical matrices multiplied together are generally not symmetric. The product of a matrix and its transpose is symmetric.

Continued matrix products such as **ABC** are defined, provided the number of columns in each matrix is the same as the number of rows in the matrix immediately preceding it. From the definition of matrix products, it follows that the *associative law* holds for continued products:

$$(\mathbf{AB})\mathbf{C} = \mathbf{A}(\mathbf{BC})$$

A square matrix **A** multiplied by itself yields a square matrix which is called the *square of the matrix* **A** and is denoted by $\mathbf{A}^2$. If $\mathbf{A}^2$ is in turn multiplied by **A**, the resulting matrix is $\mathbf{A}^3 = \mathbf{A}(\mathbf{A}^2) = \mathbf{A}^2(\mathbf{A})$. Extension of this process gives meaning to $\mathbf{A}^m$ for any positive integer *power m*. The extension to fractional and negative powers is described in connection with Eq. (28.44). Powers of symmetric matrices are themselves symmetric.

The rule for *transposition* of matrix products is

$$(\mathbf{AB})^T = \mathbf{B}^T\mathbf{A}^T$$

INVERSE OR RECIPROCAL MATRIX. If, for a given square matrix **A**, a square matrix $\mathbf{A}^{-1}$ can be found such that

$$\mathbf{A}^{-1}\mathbf{A} = \mathbf{AA}^{-1} = \mathbf{I} \tag{28.6}$$

then $\mathbf{A}^{-1}$ is called the *inverse* or *reciprocal* of **A**. Not every square matrix **A** possesses an inverse. If the determinant constructed from the elements of a square matrix is zero, the matrix is said to be *singular* and there is no inverse. Every nonsingular matrix possesses a unique inverse. The inverse of a symmetric matrix is symmetric. The rule for the *inverse of a matrix product* is

$$(\mathbf{AB})^{-1} = (\mathbf{B}^{-1})(\mathbf{A}^{-1})$$

The solution to the set of simultaneous equations

$$\mathbf{Ax} = \mathbf{c}$$

where **x** is the unknown vector and **c** is a known input vector can be indicated with the aid of the inverse of **A**. The formal solution for **x** proceeds as follows:

$$\mathbf{A}^{-1}\mathbf{Ax} = \mathbf{A}^{-1}\mathbf{c}$$

$$\mathbf{Ix} = \mathbf{x} = \mathbf{A}^{-1}\mathbf{c}$$

When the inverse $\mathbf{A}^{-1}$ is known, the solution vector **x** is obtained by a simple matrix multiplication of $\mathbf{A}^{-1}$ into the input vector **c**.

The problem of calculating the inverse of given matrix **A** is one of the central computational problems associated with matrices. Routine procedures for both hand and machine computation exist.

QUADRATIC FORMS

A general quadratic form Q of order n may be written as

$$Q = \sum_{j=1}^{n} \sum_{k=1}^{n} a_{jk} x_j x_k$$

where the a_{jk} are constants and the x_j are the n variables. The form is quadratic since it is of the second degree in the variables. The laws of matrix multiplication permit Q to be written as

$$Q = [x_1 x_2 \dots x_n] \begin{bmatrix} a_{11} & a_{12} & \dots & a_{1n} \\ a_{21} & a_{22} & \dots & a_{2n} \\ \dots & \dots & \dots & \dots \\ a_{n1} & a_{n2} & \dots & a_{nn} \end{bmatrix} \begin{bmatrix} x_1 \\ x_2 \\ \dots \\ x_n \end{bmatrix}$$

which is

$$Q = \mathbf{x}^T \mathbf{A} \mathbf{x}$$

Any quadratic form can be expressed in terms of a symmetric matrix. If the given matrix $\mathbf{A}$ is not symmetric, it can be replaced by the symmetric matrix

$$\mathbf{B} = \tfrac{1}{2}(\mathbf{A} + \mathbf{A}^T)$$

without changing the value of the form.

As an example of a quadratic form, the *potential energy* V for the system of Fig. 28.2 is given by

$$\begin{aligned} 2V &= 3kx_1^2 + 2k(x_2 - x_1)^2 + k(x_3 - x_2)^2 \\ &= 5kx_1x_1 - 2kx_1x_2 \\ &\quad - 2kx_2x_1 + 3kx_2x_2 - kx_2x_3 \\ &\quad - kx_3x_2 + kx_3x_3 \end{aligned}$$

Using the displacement vector $\mathbf{x}$ defined in Eq. (28.2) and the stiffness matrix $\mathbf{K}$ in Eq. (28.3), the potential energy may be written as

$$V = \tfrac{1}{2}\mathbf{x}^T \mathbf{K} \mathbf{x}$$

Similarly, the *kinetic energy* T is given by

$$2T = m\dot{x}_1^2 + 2m\dot{x}_2^2 + 3m\dot{x}_3^2$$

In terms of the inertia matrix $\mathbf{M}$ and the velocity vector $\dot{\mathbf{x}}$ defined in Eqs. (28.3) and (28.2) the kinetic energy may be written as

$$T = \tfrac{1}{2}\dot{\mathbf{x}}^T \mathbf{M} \dot{\mathbf{x}}$$

The *dissipation function* D for the system is given by

$$\begin{aligned} 2D &= c\dot{x}_1^2 + 2c(\dot{x}_3 - \dot{x}_2)^2 \\ &= c\dot{x}_1\dot{x}_1 \\ &\quad + 2c\dot{x}_2\dot{x}_2 - 2c\dot{x}_2\dot{x}_3 \\ &\quad - 2c\dot{x}_3\dot{x}_2 + 2c\dot{x}_3\dot{x}_3 \end{aligned}$$

In terms of the velocity vector $\dot{\mathbf{x}}$ and the damping matrix $\mathbf{C}$ defined in Eqs. (28.2) and (28.3) the dissipation function may be written as

$$D = \tfrac{1}{2}\dot{\mathbf{x}}^T \mathbf{C} \dot{\mathbf{x}}$$

The dissipation function gives half the rate at which energy is being dissipated in the system.

While quadratic forms assume positive and negative values in general, the three physical forms just defined are intrinsically *positive* for a vibrating system with linear springs, constant masses, and viscous damping; i.e., they can never be negative for a real motion of the system. Kinetic energy is zero only when the system is at rest. The same thing is not necessarily true for potential energy or the dissipation function.

Depending upon the arrangement of springs and dashpots in the system, there may exist motions which do not involve any potential energy or dissipation. For example, in vibratory systems where rigid body motions are possible (crankshaft torsional systems, free-free beams, etc.) no elastic energy is involved in the rigid body motions. Also, in Fig. 28.2, if x_1 is zero while x_2 and x_3 have the same motion, there is no energy dissipated and the dissipation function is zero. To distinguish between these two possibilities, a quadratic form is called *positive definite* if it is never negative and if the only time it vanishes is when all the variables are zero. Kinetic energy is always positive definite, while potential energy and the dissipation function are positive but not necessarily positive definite. It depends upon the particular configuration of a given system whether or not the potential energy and the dissipation function are positive definite or only positive. The terms positive and positive definite are applied also to the matrices from which the quadratic forms are derived. For example, of the three matrices defined in Eq. (28.3), the matrices **M** and **K** are positive definite but **C** is only positive. It can be shown that a matrix which is positive but not positive definite is *singular*.

DIFFERENTIATION OF QUADRATIC FORMS. In forming Lagrange's equations of motion for a vibrating system,* it is necessary to take derivatives of the potential energy V, the kinetic energy T, and the dissipation function D. When these quadratic forms are represented in matrix notation, it is convenient to have matrix formulas for differentiation. In this paragraph rules are given for differentiating the slightly more general *bilinear form*

$$F = \mathbf{x}^T\mathbf{A}\mathbf{y}$$

where $\mathbf{x}^T$ is a row vector of n variables x_j, $\mathbf{A}$ is a square matrix of constant coefficients, and $\mathbf{y}$ is a column matrix of n variables y_j. In a quadratic form the x_j are identical with the y_j.

For generality it is assumed that the x_j and the y_j are functions of n other variables u_j. In the formulas below the notation $\mathbf{X}_u$ is used to represent the following square matrix:

$$\mathbf{X}_u = \begin{bmatrix} \dfrac{\partial x_1}{\partial u_1} & \dfrac{\partial x_2}{\partial u_1} & \cdots & \dfrac{\partial x_n}{\partial u_1} \\ \dfrac{\partial x_1}{\partial u_2} & \dfrac{\partial x_2}{\partial u_2} & \cdots & \dfrac{\partial x_n}{\partial u_2} \\ \cdots & \cdots & \cdots & \cdots \\ \dfrac{\partial x_1}{\partial u_n} & \dfrac{\partial x_2}{\partial u_n} & \cdots & \dfrac{\partial x_n}{\partial u_n} \end{bmatrix}$$

Now letting $\partial/\partial\mathbf{u}$ stand for the column vector whose elements are the partial differential operators with respect to the u_j, the general differentiation formula is

$$\frac{\partial F}{\partial \mathbf{u}} = \begin{bmatrix} \dfrac{\partial F}{\partial u_1} \\ \dfrac{\partial F}{\partial u_2} \\ \cdots \\ \dfrac{\partial F}{\partial u_n} \end{bmatrix} = \mathbf{X}_u\mathbf{A}\mathbf{y} + \mathbf{Y}_u\mathbf{A}^T\mathbf{x}$$

* See Chap. 2 for a detailed discussion of Lagrange's equations.

For a quadratic form $Q = \mathbf{x}^T\mathbf{A}\mathbf{x}$ the above formula reduces to

$$\frac{\partial Q}{\partial \mathbf{u}} = \mathbf{X}_u(\mathbf{A} + \mathbf{A}^T)\mathbf{x}$$

Thus whether **A** is symmetric or not this kind of differentiation produces a *symmetrical* matrix of coefficients $(\mathbf{A} + \mathbf{A}^T)$. It is this fact which ensures that vibration equations in the form obtained from Lagrange's equations always have symmetrical matrices of coefficients. If **A** is symmetrical to begin with, the previous formula becomes

$$\frac{\partial Q}{\partial \mathbf{u}} = 2\mathbf{X}_u\mathbf{A}\mathbf{x}$$

Finally, in the important special case where the x_j are identical with the u_j the matrix $\mathbf{X}_x$ reduces to the identity matrix yielding

$$\frac{\partial Q}{\partial \mathbf{x}} = 2\mathbf{A}\mathbf{x} \tag{28.7}$$

which is employed in the following section in developing Lagrange's equations.

FORMULATION OF VIBRATION PROBLEMS IN MATRIX FORM

This section begins with a generalized derivation of vibration equations in matrix form making use of Lagrange's equations.[2] This is followed by several particular examples illustrating the theory and indicating important areas of application. For simplicity of exposition, the examples have only 2 to 6 degrees-of-freedom. In practice, systems with 30 degrees-of-freedom are common and systems with over 100 degrees-of-freedom have been successfully treated.

Consider a holonomic linear mechanical system with n degrees-of-freedom which vibrates about a stable equilibrium configuration. Let the motion of the system be described by n *generalized displacements* $x_j(t)$ which vanish in the equilibrium position. The potential energy V can then be expressed in terms of these displacements as a quadratic form. The kinetic energy T and the dissipation function D can be expressed as quadratic forms in the generalized velocities $\dot{x}_j(t)$.

The equations of motion are obtained by applying Lagrange's equations

$$\frac{d}{dt}\left(\frac{\partial T}{\partial \dot{x}_j}\right) + \frac{\partial D}{\partial \dot{x}_j} + \frac{\partial V}{\partial x_j} = f_j(t) \qquad [j = 1, 2, \ldots, n]$$

The *generalized external force* $f_j(t)$ for each coordinate may be an active force in the usual sense or a force generated by prescribed motion of the coordinates.

If each term in the foregoing equation is taken as the jth element of a column matrix, all n equations can be considered simultaneously and written in matrix form as follows:

$$\frac{d}{dt}\left(\frac{\partial T}{\partial \dot{\mathbf{x}}}\right) + \frac{\partial D}{\partial \dot{\mathbf{x}}} + \frac{\partial V}{\partial \mathbf{x}} = \mathbf{f}$$

The quadratic forms can be expressed in matrix notation as

$$T = \tfrac{1}{2}(\dot{\mathbf{x}}^T\mathbf{M}\dot{\mathbf{x}})$$

$$D = \tfrac{1}{2}(\dot{\mathbf{x}}^T\mathbf{C}\dot{\mathbf{x}})$$

$$V = \tfrac{1}{2}(\mathbf{x}^T\mathbf{K}\mathbf{x})$$

where the *inertia matrix* **M**, the *damping matrix* **C**, and the *stiffness matrix* **K** may be taken as symmetric square matrices of order n. Then the differentiation rule (28.7) yields

$$\frac{d}{dt}(\mathbf{M}\dot{\mathbf{x}}) + \mathbf{C}\dot{\mathbf{x}} + \mathbf{K}\mathbf{x} = \mathbf{f}$$

or simply

$$\mathbf{M}\ddot{\mathbf{x}} + \mathbf{C}\dot{\mathbf{x}} + \mathbf{K}\mathbf{x} = \mathbf{f} \tag{28.8}$$

as the equations of motion in matrix form for a general linear vibratory system with n degrees-of-freedom. This is a generalization of Eq. (28.4) for the three degree-of-freedom system of Fig. 28.2. Equation (28.8) applies to all linear constant parameter vibratory systems. The specifications of any particular system are contained in the *coefficient matrices* **M**, **C**, and **K**. The type of excitation is described by the column matrix **f**. The individual terms in the coefficient matrices have the following significance:

m_{jk} is the momentum component at j due to a unit velocity at k,
c_{jk} is the damping force at j due to a unit velocity at k,
k_{jk} is the elastic force at j due to a unit displacement at k.

In certain applications it is more convenient to deal with $\mathbf{K}^{-1}$, the inverse of the stiffness matrix, than with **K**. The elements α_{jk} of $\mathbf{K}^{-1}$ are called the *influence coefficients* and $\mathbf{K}^{-1}$ itself is sometimes called the *flexibility matrix*. The influence coefficients have the following significance:

α_{jk} is the elastic displacement at j due to unit force statically applied at k.

When **K** is singular its inverse does not exist. This occurs in systems where rigid body motions are possible. A unit force cannot be statically applied to such a system. The result of applying a force is acceleration.

The general solution to Eq. (28.8) contains $2n$ constants of integration which are usually fixed by the n displacements $x_j(t_0)$ and the n velocities $\dot{x}_j(t_0)$ at some initial time t_0. When the excitation matrix **f** is zero, Eq. (28.8) is said to describe the *free vibration* of the system. When **f** is nonzero, Eq. (28.8) describes a *forced vibration*. When the time behavior of **f** is periodic and steady, it is sometimes convenient to divide the solution into a *steady-state response* plus a *transient response* which decays with time. The steady-state response is independent of the initial conditions.

COUPLING OF THE EQUATIONS

The off-diagonal terms in the coefficient matrices are known as *coupling terms*. In general, the equations have inertia, damping, and stiffness coupling; however, it is often possible to obtain equations that have no coupling terms in one or more of the three

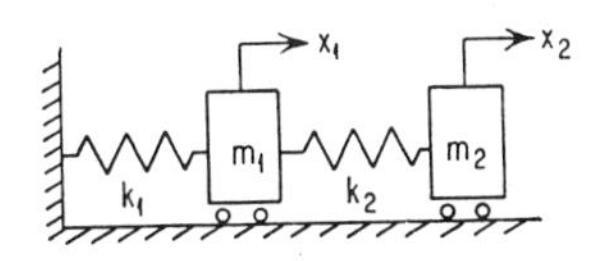

FIG. 28.3. Coordinates (x_1,x_2) with uncoupled inertia matrix.

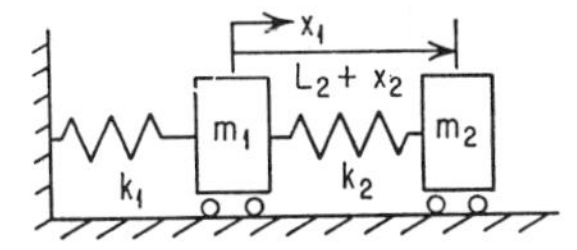

FIG. 28.4. Coordinates (x_1,x_2) with uncoupled stiffness matrix. The equilibrium length of the spring k_2 is L_2.

matrices. If the coupling terms vanish in all three matrices; i.e., if all three square matrices are diagonal matrices, the system of Eq. (28.8) becomes a set of independent uncoupled differential equations for the n generalized displacements $x_j(t)$. Each displacement motion is a single degree-of-freedom vibration independent of the motion of the other displacements.

The coupling in a system depends on the choice of coordinates used to describe the motion. For example, Figs. 28.3 and 28.4 show the same physical system with two different choices for the displacement coordinates.

The coefficient matrices corresponding to the coordinates shown in Fig. 28.3 are

$$\mathbf{M} = \begin{bmatrix} m_1 & 0 \\ 0 & m_2 \end{bmatrix} \qquad \mathbf{K} = \begin{bmatrix} k_1 + k_2 & -k_2 \\ -k_2 & k_2 \end{bmatrix}$$

Here the inertia matrix is uncoupled because the coordinates chosen are the absolute displacements of the masses. The elastic force in the spring k_2 is generated by the relative displacement of the two coordinates which accounts for the coupling terms in the stiffness matrix.

The coefficient matrices corresponding to the alternative coordinates shown in Fig. 28.4 are

$$\mathbf{M} = \begin{bmatrix} m_1 + m_2 & m_2 \\ m_2 & m_2 \end{bmatrix} \qquad \mathbf{K} = \begin{bmatrix} k_1 & 0 \\ 0 & k_2 \end{bmatrix}$$

Here the coordinates chosen relate directly to the extensions of the springs so that the stiffness matrix is uncoupled. The absolute displacement of m_2 is, however, the sum of the coordinates which accounts for the coupling terms in the inertia matrix.

A fundamental procedure for solving vibration problems in undamped systems may be viewed as the search for a set of coordinates which simultaneously uncouples both the stiffness and inertia matrices. This is always possible. In systems with damping (i.e., with all three coefficient matrices) there exist coordinates which uncouple two of these, but in general it is not possible to uncouple all three matrices simultaneously.

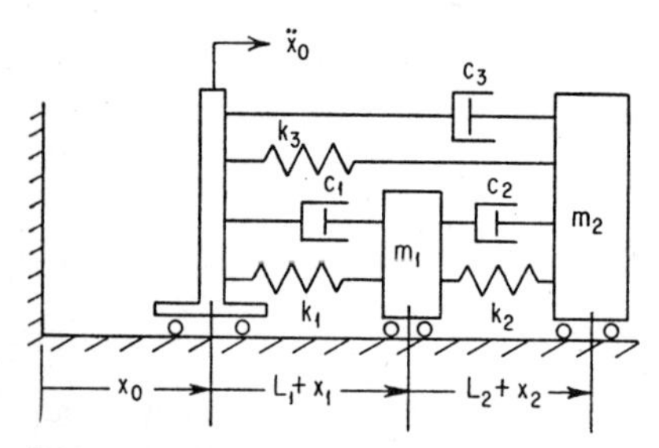

FIG. 28.5. Two degree-of-freedom vibratory system. The equilibrium length of the spring k_1 is L_1 and the equilibrium length of the spring k_2 is L_2.

The system of Fig. 28.2 provides an example of a three degree-of-freedom system with damping. The coefficient matrices are given in Eq. (28.3). The inertia matrix is uncoupled, but the damping and stiffness matrices are coupled.

Another example of a system with damping is furnished by the two degree-of-freedom system shown in Fig. 28.5. The excitation here is furnished by acceleration $\ddot{x}_0(t)$ of the base. This system is used as the basis for most of the numerical examples which follow. With the coordinates chosen as indicated in the figure, all three coefficient matrices have coupling terms. The equations of motion can be placed in the standard form of Eq. (28.8) where the coefficient matrices and the excitation column are as follows:

$$\mathbf{M} = \begin{bmatrix} m_1 + m_2 & m_2 \\ m_2 & m_2 \end{bmatrix} \qquad \mathbf{C} = \begin{bmatrix} c_1 + c_3 & c_3 \\ c_3 & c_2 + c_3 \end{bmatrix}$$

$$\mathbf{K} = \begin{bmatrix} k_1 + k_3 & k_3 \\ k_3 & k_2 + k_3 \end{bmatrix} \qquad \mathbf{f} = -\ddot{x}_0 \begin{bmatrix} m_1 + m_2 \\ m_2 \end{bmatrix} \tag{28.9}$$

VIBRATION ISOLATION SYSTEMS

In Fig. 28.6 the motions in the xy plane of the symmetrical rigid body supported by four symmetrically located equal isolator springs are considered. Each of the four springs has spring constants k_x and k_y in the x and y directions, respectively. The mass of the body is m and the moment of inertia about the z axis is I_z. The free vibration equations of motion are derived elsewhere [3] in algebraic form. In matrix format they are

$$\begin{bmatrix} m & 0 & 0 \\ 0 & m & 0 \\ 0 & 0 & I_z \end{bmatrix} \begin{bmatrix} \ddot{x} \\ \ddot{y} \\ \ddot{\theta} \end{bmatrix} + \begin{bmatrix} 4k_x & 0 & 4k_x a \\ 0 & 4k_y & 0 \\ 4k_x a & 0 & 4(k_x a^2 + k_y b^2) \end{bmatrix} \begin{bmatrix} x \\ y \\ \theta \end{bmatrix} = 0$$

Note that the y equation is uncoupled from the other two but that the horizontal and rocking motions are coupled. This coupling could be removed by redesigning the spring attachments so that $a = 0$.

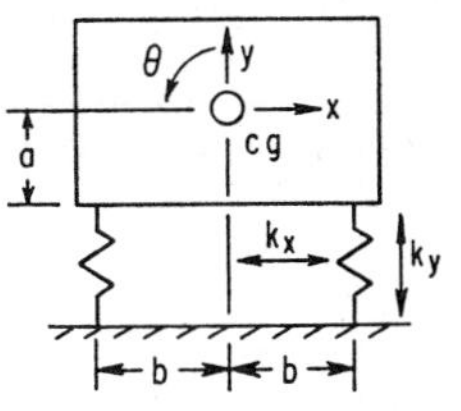

FIG. 28.6. Rigid body on isolator springs.

As a further extension of the suspension problem, consider a completely general rigid body, elastically supported at n points P_i by springs which have three mutually perpendicular stiffnesses k_{xi}, k_{yi}, k_{zi} for $i = 1, \ldots, n$, as shown in Fig. 28.7. The xyz axes are fixed in the body parallel to these directions through an arbitrary origin O. The inertia properties of the body are given by the coordinates $(\bar{x},\bar{y},\bar{z})$ of the center-of-mass G, the mass m, and the moments I_x, I_y, I_z and products I_{xy}, I_{yz}, I_{xz} of inertia with respect to the xyz axes. The equations of motion for small free vibration can be placed in the form

$$\mathbf{M}\ddot{\mathbf{x}} + \mathbf{K}\mathbf{x} = 0$$

by using the following sixth-order displacement vector

$$\mathbf{x} = \begin{bmatrix} x_0 \\ y_0 \\ z_0 \\ \theta_x \\ \theta_y \\ \theta_z \end{bmatrix}$$

where the first three elements describe the displacement of O and the second three elements describe the angular displacement of the $Oxyz$ frame. The matrices **K** and **M** are sixth-order symmetric square matrices.

$$\mathbf{K} = \begin{bmatrix}
\sum_{i=1}^{n} k_{xi} & 0 & 0 & 0 & \sum_{i=1}^{n} k_{xi}z_i & -\sum_{i=1}^{n} k_{xi}y_i \\
0 & \sum_{i=1}^{n} k_{yi} & 0 & -\sum_{i=1}^{n} k_{yi}z_i & 0 & \sum_{i=1}^{n} k_{yi}x_i \\
0 & 0 & \sum_{i=1}^{n} k_{zi} & \sum_{i=1}^{n} k_{zi}y_i & -\sum_{i=1}^{n} k_{zi}x_i & 0 \\
0 & -\sum_{i=1}^{n} k_{yi}z_i & \sum_{i=1}^{n} k_{zi}y_i & \sum_{i=1}^{n} (k_{yi}z_i^2 + k_{zi}y_i^2) & -\sum_{i=1}^{n} k_{zi}x_iy_i & -\sum_{i=1}^{n} k_{yi}x_iz_i \\
\sum_{i=1}^{n} k_{xi}z_i & 0 & -\sum_{i=1}^{n} k_{zi}x_i & -\sum_{i=1}^{n} k_{zi}x_iy_i & \sum_{i=1}^{n} (k_{zi}x_i^2 + k_{xi}z_i^2) & -\sum_{i=1}^{n} k_{xi}y_iz_i \\
-\sum_{i=1}^{n} k_{xi}y_i & \sum_{i=1}^{n} k_{yi}x_i & 0 & -\sum_{i=1}^{n} k_{yi}z_ix_i & -\sum_{i=1}^{n} k_{xi}y_iz_i & \sum_{i=1}^{n} (k_{xi}y_i^2 + k_{yi}x_i^2)
\end{bmatrix}$$

$$\mathbf{M} = \begin{bmatrix}
m & 0 & 0 & 0 & m\bar{z} & -m\bar{y} \\
0 & m & 0 & -m\bar{z} & 0 & m\bar{x} \\
0 & 0 & m & m\bar{y} & -m\bar{x} & 0 \\
0 & -m\bar{z} & m\bar{y} & I_x & -I_{xy} & -I_{xz} \\
m\bar{z} & 0 & -m\bar{x} & -I_{xy} & I_y & -I_{yz} \\
-m\bar{y} & m\bar{x} & 0 & -I_{xz} & -I_{yz} & I_z
\end{bmatrix}$$

Much of the elastic coupling is eliminated when the springs are located symmetrically. The inertia couplings are reduced if the origin 0 is taken through the center-of-mass G. If in addition the axes are principal axes of inertia, then there is no inertia coupling.

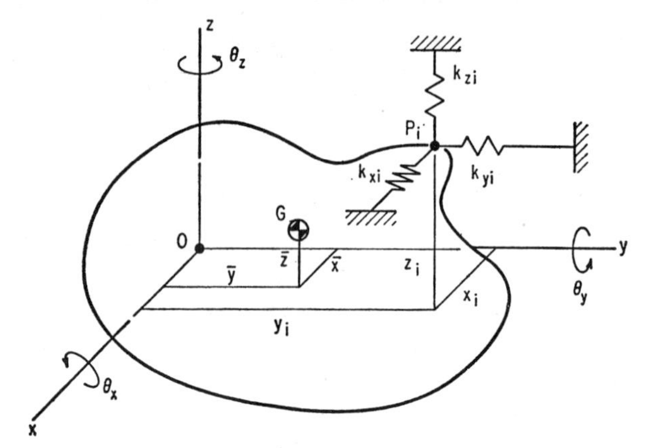

FIG. 28.7. General rigid body showing typical elastic suspension at point P_i.

TORSIONAL VIBRATION

Figure 28.8 shows a crankshaft of an engine; a mathematically idealized model is portrayed in Fig. 28.9. The torsional spring constants k_j represent the equivalent torsional stiffness of the complex crankshaft structure when supported in its bearings. The mass moments of inertia J_j represent the equivalent inertia of all the rotating and re-

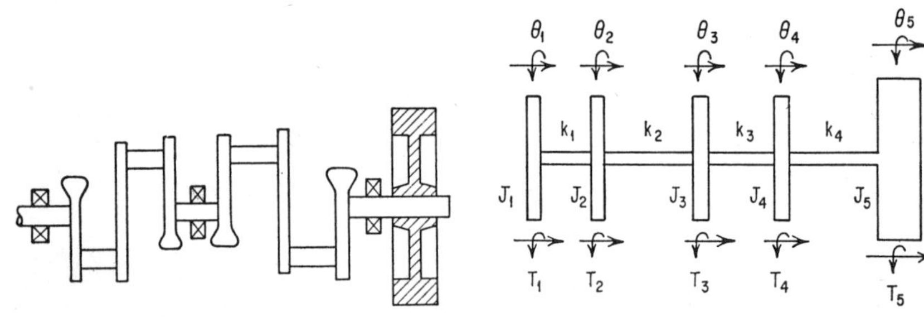

FIG. 28.8. Crankshaft with flywheel.

FIG. 28.9. Idealized equivalent torsional system for crankshaft.

ciprocating parts associated with the jth cylinder. The angles of rotation are θ_j and the applied torques are T_j. The equations of motion may be written in matrix notation as follows:

$$\begin{bmatrix} J_1 & 0 & 0 & 0 & 0 \\ 0 & J_2 & 0 & 0 & 0 \\ 0 & 0 & J_3 & 0 & 0 \\ 0 & 0 & 0 & J_4 & 0 \\ 0 & 0 & 0 & 0 & J_5 \end{bmatrix} \begin{bmatrix} \ddot{\theta}_1 \\ \ddot{\theta}_2 \\ \ddot{\theta}_3 \\ \ddot{\theta}_4 \\ \ddot{\theta}_5 \end{bmatrix} + \begin{bmatrix} k_1 & -k_1 & 0 & 0 & 0 \\ -k_1 & k_1 + k_2 & -k_2 & 0 & 0 \\ 0 & -k_2 & k_2 + k_3 & -k_3 & 0 \\ 0 & 0 & -k_3 & k_3 + k_4 & -k_4 \\ 0 & 0 & 0 & -k_4 & k_4 \end{bmatrix} \begin{bmatrix} \theta_1 \\ \theta_2 \\ \theta_3 \\ \theta_4 \\ \theta_5 \end{bmatrix} = \begin{bmatrix} T_1 \\ T_2 \\ T_3 \\ T_4 \\ T_5 \end{bmatrix}$$

As the whole system can rotate as a rigid body without elastic deformation of the crankshaft, the potential energy is positive rather than positive definite and the stiffness matrix is singular.

THE MATRIX EIGENVALUE PROBLEM

In the following sections the solutions to both free and forced vibration problems are given in terms of solutions to a specialized algebraic problem known as the matrix eigenvalue problem. In the present section a general theoretical discussion of the matrix eigenvalue problem is given. Later, in the section on numerical methods, practical procedures for solving the eigenvalue problem are described.

The free-vibration equation for an undamped system,

$$\mathbf{M\ddot{x}} + \mathbf{Kx} = 0 \tag{28.10}$$

follows from Eq. (28.8) when the excitation $\mathbf{f}$ and the damping $\mathbf{C}$ vanish. If a solution for $\mathbf{x}$ is assumed in the form

$$\mathbf{x} = \mathcal{R}\{\mathbf{v}e^{j\omega t}\}$$

where $\mathbf{v}$ is a column vector of unknown amplitudes, ω is an unknown frequency, j is the square root of -1, and $\mathcal{R}\{\quad\}$ signifies "the real part of," it is found on substituting in Eq. (28.10) that it is necessary for $\mathbf{v}$ and ω to satisfy the following algebraic equation:

$$\mathbf{Kv} = \omega^2\mathbf{Mv} \tag{28.11}$$

This algebraic problem is called the *matrix eigenvalue problem.* Where necessary it is called the *real* eigenvalue problem to distinguish it from the *complex* eigenvalue problem described in the section on vibration of systems with damping. Equation (28.11) defines the central problem of this chapter. The coefficient matrices $\mathbf{K}$ and $\mathbf{M}$ are known; the problem is to determine the scalar ω^2 and vector $\mathbf{v}$. This is not a simple problem, nor does it have a simple easily understood solution; however, the fact that the matrices $\mathbf{K}$ and $\mathbf{M}$ are always symmetric in vibration problems and that $\mathbf{M}$ is always positive definite does guarantee that the nature of the solution is a good deal simpler than it otherwise might be.

To indicate the formal solution to Eq. (28.11), it is rewritten as

$$(\mathbf{K} - \omega^2\mathbf{M})\mathbf{v} = 0 \tag{28.12}$$

which can be interpreted as a set of n homogeneous algebraic equations for the n elements v_j. This set always has the trivial solution

$$\mathbf{v} = 0$$

It also has nontrivial solutions if the determinant of the matrix multiplying the vector v is zero, i.e., if

$$\det(\mathbf{K} - \omega^2\mathbf{M}) = 0 \tag{28.13}$$

When the determinant is expanded, a polynomial of order n in ω^2 is obtained. Equation (28.13) is known as the *characteristic equation* or *frequency equation.* The restrictions that $\mathbf{M}$ and $\mathbf{K}$ be symmetric and that $\mathbf{M}$ be positive definite are sufficient to ensure[4] that there are n real roots for ω^2. If $\mathbf{K}$ is singular, at least one root is zero. If $\mathbf{K}$ is positive definite, all roots are positive. The n roots determine the n *natural frequencies* $\omega_r (r = 1, \ldots, n)$ of free vibration. These roots of the characteristic equation are also known as *normal values, characteristic values, proper values, latent roots,* or *eigenvalues.* When a natural frequency ω_r is known, it is possible to return to Eq. (28.12) and solve for the corresponding vector $\mathbf{v}_r$ to within a multiplicative constant. The eigenvalue problem does not fix the absolute amplitude of the vectors $\mathbf{v}$, only the relative amplitudes of the n coordinates. In the sections which follow, the absolute amplitudes are determined in both free and forced vibration problems. There are n independent vectors $\mathbf{v}_r$ corresponding to the n natural frequencies which are known as *natural modes.* These vectors are also known as *normal modes, characteristic vectors, proper vectors, latent vectors,* or *eigenvectors.*

MODAL AND SPECTRAL MATRICES

The complete solution to the eigenvalue problem of Eq. (28.11) consists of n eigenvalues and n corresponding eigenvectors. These can be assembled compactly into matrices. Let the eigenvector $\mathbf{v}_r$ corresponding to the eigenvalue ω_r^2 have elements v_{jr} (the first subscript indicates which row, the second subscript indicates which eigenvector). The n eigenvectors then can be displayed in a single square matrix $\mathbf{V}$, each column of which is an eigenvector

$$\mathbf{V} = [v_{jk}] = \begin{bmatrix} v_{11} & v_{12} & \cdots & v_{1n} \\ v_{21} & v_{22} & \cdots & v_{2n} \\ \cdots & \cdots & \cdots & \cdots \\ v_{n1} & v_{n2} & \cdots & v_{nn} \end{bmatrix}$$

The matrix $\mathbf{V}$ is called the *modal matrix* for the eigenvalue problem, Eq. (28.11).

The n eigenvalues ω_r^2 can be assembled into a diagonal matrix $\mathbf{\Omega}^2$ which is known as the *spectral matrix* of the eigenvalue problem, Eq. (28.11)

$$\mathbf{\Omega}^2 = \left[\diagdown \omega_r^2 \diagdown\right] = \begin{bmatrix} \omega_1^2 & 0 & \cdots & 0 \\ 0 & \omega_2^2 & \cdots & 0 \\ \cdots & \cdots & \cdots & \cdots \\ 0 & 0 & \cdots & \omega_n^2 \end{bmatrix}$$

Each eigenvector and corresponding eigenvalue satisfy a relation of the following form:

$$\mathbf{K}\mathbf{v}_r = \mathbf{M}\mathbf{v}_r\omega_r^2$$

By using the modal and spectral matrices it is possible to assemble all of these relations into a single matrix equation

$$\mathbf{K}\mathbf{V} = \mathbf{M}\mathbf{V}\mathbf{\Omega}^2 \tag{28.14}$$

Equation (28.14) provides a compact display of the complete solution to the eigenvalue problem Eq. (28.11).

PROPERTIES OF THE SOLUTION

The eigenvectors corresponding to different eigenvalues can be shown[5] to satisfy the following *orthogonality relations*. When $\omega_r^2 \neq \omega_s^2$,

$$\mathbf{v}_r^T\mathbf{K}\mathbf{v}_s = 0 \qquad \mathbf{v}_r^T\mathbf{M}\mathbf{v}_s = 0 \tag{28.15}$$

In case the characteristic equation has a p-fold multiple root for ω^2, then there is a p-fold infinity of corresponding eigenvectors. In this case, however, it is always possible[4] to choose p of these vectors which mutually satisfy Eq. (28.15) and to express any other eigenvector corresponding to the multiple root as a linear combination of the p vectors selected. If these p vectors are included with the eigenvectors corresponding to the other eigenvalues, a set of n vectors is obtained which satisfies the orthogonality relations of Eq. (28.15) for any $r \neq s$.

The orthogonality of the eigenvectors with respect to $\mathbf{K}$ and $\mathbf{M}$ implies that the following square matrices are *diagonal*.

$$\begin{aligned} \mathbf{V}^T\mathbf{K}\mathbf{V} &= \left[\diagdown \mathbf{v}_r^T\mathbf{K}\mathbf{v}_r \diagdown\right] \\ \mathbf{V}^T\mathbf{M}\mathbf{V} &= \left[\diagdown \mathbf{v}_r^T\mathbf{M}\mathbf{v}_r \diagdown\right] \end{aligned} \tag{28.16}$$

The elements $\mathbf{v}_r^T\mathbf{K}\mathbf{v}_r$ along the main diagonal of $\mathbf{V}^T\mathbf{K}\mathbf{V}$ are called the *modal stiffnesses* k_r, and the elements $\mathbf{v}_r^T\mathbf{M}\mathbf{v}_r$ along the main diagonal of $\mathbf{V}^T\mathbf{M}\mathbf{V}$ are called the *modal masses* m_r. Since $\mathbf{M}$ is positive definite, all modal masses are guaranteed to be positive. When

$\mathbf{K}$ is singular, at least one of the modal stiffnesses will be zero. Each eigenvalue ω_r^2 is the quotient of the corresponding modal stiffness divided by the corresponding modal mass; i.e.,

$$\omega_r^2 = \frac{k_r}{m_r}$$

In numerical work it is sometimes convenient to normalize each eigenvector so that its largest element is *unity*. In other applications it is common to normalize the eigenvectors so that the modal masses m_r all have the *same* value m where m is some convenient value such as the total mass of the system. In this case,

$$\mathbf{V}^T\mathbf{M}\mathbf{V} = m\mathbf{I} \tag{28.17}$$

and it is possible to express the inverse of the modal matrix $\mathbf{V}$ simply as

$$\mathbf{V}^{-1} = \frac{1}{m}\mathbf{V}^T\mathbf{M}$$

An interpretation of the modal matrix $\mathbf{V}$ can be given by showing that it defines a set of generalized coordinates for which both the inertia and stiffness matrices are uncoupled. Let $\mathbf{y}(t)$ be a column of displacements related to the original displacements $\mathbf{x}(t)$ by the following simultaneous equations:

$$\mathbf{y} = \mathbf{V}^{-1}\mathbf{x} \qquad \text{or} \qquad \mathbf{x} = \mathbf{V}\mathbf{y}$$

The potential and kinetic energies then take the forms

$$V = \tfrac{1}{2}\mathbf{x}^T\mathbf{K}\mathbf{x} = \tfrac{1}{2}\mathbf{y}^T(\mathbf{V}^T\mathbf{K}\mathbf{V})\mathbf{y}$$

$$T = \tfrac{1}{2}\dot{\mathbf{x}}^T\mathbf{M}\dot{\mathbf{x}} = \tfrac{1}{2}\dot{\mathbf{y}}^T(\mathbf{V}^T\mathbf{M}\mathbf{V})\dot{\mathbf{y}}$$

where, according to Eq. (28.16), the square matrices in parentheses on the right are *diagonal*; i.e., in the y_j coordinate system there is neither stiffness nor inertia coupling.

An alternative method for obtaining the same interpretation is to start from the eigenvalue problem of Eq. (28.11). Consider the structure of the related eigenvalue problem for $\mathbf{w}$ where again $\mathbf{w}$ is obtained from $\mathbf{v}$ by the transformation involving the modal matrix $\mathbf{V}$.

$$\mathbf{w} = \mathbf{V}^{-1}\mathbf{v} \qquad \text{or} \qquad \mathbf{v} = \mathbf{V}\mathbf{w}$$

Substituting in Eq. (28.11), premultiplying by $\mathbf{V}^T$ and using Eq. (28.14),

$$\mathbf{K}\mathbf{v} = \omega^2\mathbf{M}\mathbf{v}$$

$$\mathbf{K}\mathbf{V}\mathbf{w} = \omega^2\mathbf{M}\mathbf{V}\mathbf{w}$$

$$\mathbf{V}^T\mathbf{K}\mathbf{V}\mathbf{w} = \omega^2\mathbf{V}^T\mathbf{M}\mathbf{V}\mathbf{w}$$

$$(\mathbf{V}^T\mathbf{M}\mathbf{V})\Omega^2\mathbf{w} = \omega^2(\mathbf{V}^T\mathbf{M}\mathbf{V})\mathbf{w}$$

Now, since $\mathbf{V}^T\mathbf{M}\mathbf{V}$ is a diagonal matrix of positive elements it is permissible to cancel it from both sides, which leaves a simple diagonalized eigenvalue problem for $\mathbf{w}$:

$$\Omega^2\mathbf{w} = \omega^2\mathbf{w}$$

A modal matrix for $\mathbf{w}$ is the identity matrix $\mathbf{I}$ and the eigenvalues for $\mathbf{w}$ are the same as those for $\mathbf{v}$.

EIGENVECTOR EXPANSIONS

Any set of n independent vectors can be used as a basis for representing any other vector of order n. In the following sections, the eigenvectors of the eigenvalue problem of Eq. (28.11) are used as such a basis. An eigenvector expansion of an arbitrary vector $\mathbf{y}$

has the form

$$\mathbf{y} = \sum_{r=1}^{n} \mathbf{v}_r a_r \tag{28.18}$$

where the a_r are scalar *mode multipliers*. When $\mathbf{y}$ and the $\mathbf{v}_r$ are known, it is possible to evaluate the a_r by premultiplying both sides by $\mathbf{v}_s^T\mathbf{M}$. Because of the orthogonality relations of Eq. (28.15), all the terms on the right vanish except the one for which $r = s$. Inserting the value of the mode multiplier so obtained, the expansion can be rewritten as

$$\mathbf{y} = \sum_{r=1}^{n} \mathbf{v}_r \frac{\mathbf{v}_r^T\mathbf{M}\mathbf{y}}{\mathbf{v}_r^T\mathbf{M}\mathbf{v}_r} \tag{28.19}$$

or alternatively as

$$\mathbf{y} = \sum_{r=1}^{n} \frac{\mathbf{v}_r\mathbf{v}_r^T\mathbf{M}}{\mathbf{v}_r^T\mathbf{M}\mathbf{v}_r}\mathbf{y} \tag{28.20}$$

The form of Eq. (28.19) emphasizes the decomposition into eigenvectors since the fraction on the right is just a scalar. The form of Eq. (28.20) is convenient when a large number of vectors $\mathbf{y}$ are to be decomposed, since the fractions on the right, which are now square matrices, must be computed only once. The form of Eq. (28.20) becomes more economical of computation time when more than n vectors $\mathbf{y}$ have to be expanded. A useful check on the calculation of the matrices on the right of Eq. (28.20) is provided by the identity

$$\sum_{r=1}^{n} \frac{\mathbf{v}_r\mathbf{v}_r^T\mathbf{M}}{\mathbf{v}_r^T\mathbf{M}\mathbf{v}_r} = \mathbf{I} \tag{28.21}$$

which follows from Eq. (28.20) because $\mathbf{y}$ is completely arbitrary.

An alternative expansion which is useful for expanding the excitation vector $\mathbf{f}$ is

$$\mathbf{f} = \sum_{r=1}^{n} \omega_r^2\mathbf{M}\mathbf{v}_r a_r = \sum_{r=1}^{n} \mathbf{M}\mathbf{v}_r \frac{\mathbf{v}_r^T\mathbf{f}}{\mathbf{v}_r^T\mathbf{M}\mathbf{v}_r} \tag{28.22}$$

This may be viewed as an expansion of the excitation in terms of the *inertia force* amplitudes of the natural modes. The mode multiplier a_r has been evaluated by premultiplying by $\mathbf{v}_r^T$. A form analogous to Eq. (28.20) and an identity corresponding to Eq. (28.21) can easily be written.

The above properties of the matrix eigenvalue problem of Eq. (28.11) are employed in the following section to obtain free and forced vibration solutions for systems without damping. Further properties of the eigenvalue problem including perturbation formulas and Rayleigh's quotient are described in the section on *Numerical Methods*.

VIBRATIONS OF SYSTEMS WITHOUT DAMPING

In this section the damping matrix $\mathbf{C}$ is neglected in Eq. (28.8), leaving the general formulation in the form

$$\mathbf{M}\ddot{\mathbf{x}} + \mathbf{K}\mathbf{x} = \mathbf{f} \tag{28.23}$$

Solutions are outlined for the following three cases: free vibration ($\mathbf{f} = 0$), steady-state forced sinusoidal vibration ($\mathbf{f} = \Re\{\mathbf{d}e^{j\omega t}\}$, where $\mathbf{d}$ is a column vector of driving-force amplitudes), and the response to general excitation ($\mathbf{f}$ an arbitrary function of time). The first two cases are contained in the third, but for the sake of clarity each is described separately.

FREE VIBRATION WITH SPECIFIED INITIAL CONDITIONS

It is desired to find the solution $\mathbf{x}(t)$ of Eq. (28.23) when $\mathbf{f} = 0$ which satisfies the initial conditions

$$\mathbf{x} = \mathbf{x}(0) \qquad \dot{\mathbf{x}} = \dot{\mathbf{x}}(0) \tag{28.24}$$

at $t = 0$ where $\mathbf{x}(0)$ and $\dot{\mathbf{x}}(0)$ are columns of prescribed initial displacements and velocities. The differential equation to be solved is identical with Eq. (28.10) which led to the matrix eigenvalue problem in the preceding section. Assuming that the solution of the eigenvalue problem is available, the general solution of the differential equation is given by an arbitrary superposition of the natural modes

$$\mathbf{x} = \sum_{r=1}^{n} \mathbf{v}_r(a_r \cos \omega_r t + b_r \sin \omega_r t)$$

where the $\mathbf{v}_r$ are the eigenvectors or natural modes, the ω_r are the natural frequencies, and the a_r and b_r are $2n$ constants of integration. The corresponding velocity is

$$\dot{\mathbf{x}} = \sum_{r=1}^{n} \mathbf{v}_r\omega_r(-a_r \sin \omega_r t + b_r \cos \omega_r t)$$

Setting $t = 0$ in these expressions and substituting in the initial conditions of Eq. (28.24) provides $2n$ simultaneous equations for determination of the constants of integration.

$$\sum_{r=1}^{n} \mathbf{v}_r a_r = \mathbf{x}(0) \qquad \sum_{r=1}^{n} \mathbf{v}_r \omega_r b_r = \dot{\mathbf{x}}(0)$$

These equations may be interpreted as eigenvector expansions of the initial displacement and velocity. The constants of integration can be evaluated by the same technique used to obtain the mode multipliers in Eq. (28.19). Using the form of Eq. (28.20), the solution of the free vibration problem then becomes

$$\mathbf{x}(t) = \sum_{r=1}^{n} \frac{\mathbf{v}_r\mathbf{v}_r^T\mathbf{M}}{\mathbf{v}_r^T\mathbf{M}\mathbf{v}_r} \left\{\mathbf{x}(0) \cos \omega_r t + \frac{1}{\omega_r}\dot{\mathbf{x}}(0) \sin \omega_r t\right\} \tag{28.25}$$

EXAMPLE 28.1. Consider the system of Fig. 28.5 with the following particular values:

$$m_1 = 1 \text{ lb-sec}^2/\text{in.} \qquad m_2 = 2 \text{ lb-sec}^2/\text{in.} \qquad c_1 = c_2 = c_3 = 0$$

$$k_1 = 3 \text{ lb/in.} \qquad k_2 = 0.5 \text{ lb/in.} \qquad k_3 = 1 \text{ lb/in.}$$

Substitution of these values into Eq. (28.9) yields the following coefficient matrices

$$\mathbf{M} = \begin{bmatrix} 3 & 2 \\ 2 & 2 \end{bmatrix} \qquad \mathbf{C} = 0 \qquad \mathbf{K} = \begin{bmatrix} 4 & 1 \\ 1 & 1.5 \end{bmatrix}$$

Both **M** and **K** are positive definite.

By use of numerical methods given later in this chapter, the solution of the eigenvalue problem can be found to be

$$\mathbf{V} = \begin{bmatrix} 0.2179 & -0.9179 \\ 1.0000 & 1.0000 \end{bmatrix} \qquad \Omega^2 = \begin{bmatrix} 0.7053 & 0 \\ 0 & 3.5447 \end{bmatrix}$$
$$\omega_1 = 0.8398 \text{ rad/sec} \qquad \omega_2 = 1.8827 \text{ rad/sec} \tag{28.26}$$

One of the terms needed for substitution in Eq. (28.25) is calculated below:

$$\frac{\mathbf{v}_1\mathbf{v}_1^T\mathbf{M}}{\mathbf{v}_1^T\mathbf{M}\mathbf{v}_1} = \frac{\begin{bmatrix} 0.2179 \\ 1.0000 \end{bmatrix} [0.2179 \quad 1.0000] \begin{bmatrix} 3 & 2 \\ 2 & 2 \end{bmatrix}}{[0.2179 \quad 1.0000] \begin{bmatrix} 3 & 2 \\ 2 & 2 \end{bmatrix} \begin{bmatrix} 0.2179 \\ 1.0000 \end{bmatrix}} = \frac{\begin{bmatrix} 0.5782 & 0.5307 \\ 2.6537 & 2.4358 \end{bmatrix}}{3.0140}$$

$$= \begin{bmatrix} 0.1918 & 0.1761 \\ 0.8805 & 0.8082 \end{bmatrix}$$

Three similar matrices must be computed according to Eq. (28.25) to provide the complete solution which follows:

$$\begin{bmatrix} x_1(t) \\ x_2(t) \end{bmatrix} = \begin{bmatrix} 0.1918 & 0.1761 \\ 0.8805 & 0.8082 \end{bmatrix} \begin{bmatrix} x_1(0) \\ x_2(0) \end{bmatrix} \cos 0.8398t$$

$$+ \begin{bmatrix} 0.2284 & 0.2097 \\ 1.0484 & 0.9623 \end{bmatrix} \begin{bmatrix} \dot{x}_1(0) \\ \dot{x}_2(0) \end{bmatrix} \sin 0.8398t$$

$$+ \begin{bmatrix} 0.8082 & -0.1761 \\ -0.8805 & 0.1918 \end{bmatrix} \begin{bmatrix} x_1(0) \\ x_2(0) \end{bmatrix} \cos 1.8827t$$

$$+ \begin{bmatrix} 0.4292 & -0.0935 \\ -0.4676 & 0.1019 \end{bmatrix} \begin{bmatrix} \dot{x}_1(0) \\ \dot{x}_2(0) \end{bmatrix} \sin 1.8827t$$

The identity of Eq. (28.21) permits a partial check on the calculations. According to Eq. (28.21) the sum of the two square matrices multiplying the initial displacement vector $\mathbf{x}(0)$ should be the unit matrix $\mathbf{I}$. Note that this is actually the case.

STEADY-STATE FORCED SINUSOIDAL VIBRATION

It is desired to find the steady-state solution to Eq. (28.23) for single-frequency sinusoidal excitation $\mathbf{f}$ of the form

$$\mathbf{f} = \Re\,\{\mathbf{d}e^{j\omega t}\}$$

where $\mathbf{d}$ is a column vector of driving force amplitudes (these may be complex to permit differences in phase for the various components). The solution obtained is a useful approximation for lightly damped systems providing the forcing frequency ω is not too close to a natural frequency ω_r. For resonance and near-resonance conditions it is necessary to include the damping as indicated in the section which follows the present discussion.

The steady-state solution desired is assumed to have the form

$$\mathbf{x} = \Re\,\{\mathbf{a}e^{j\omega t}\}$$

where $\mathbf{a}$ is an unknown column vector of response amplitudes. When $\mathbf{f}$ and $\mathbf{x}$ are inserted in Eq. (28.23), the following set of simultaneous equations for the elements of $\mathbf{a}$ is obtained

$$(\mathbf{K} - \omega^2\mathbf{M})\mathbf{a} = \mathbf{d} \tag{28.27}$$

If ω is not a natural frequency, the square matrix $\mathbf{K} - \omega^2\mathbf{M}$ is nonsingular and may be inverted to yield

$$\mathbf{a} = (\mathbf{K} - \omega^2\mathbf{M})^{-1}\mathbf{d}$$

as a complete solution for the response amplitudes in terms of the driving-force amplitudes. This solution is useful if several force amplitude distributions are to be studied while the excitation frequency ω is held constant. The process requires repeated inversions if a range of frequencies is to be studied.

An alternate procedure which permits a more thorough study of the effect of frequency variation is available if the natural modes and frequencies are known. The driving-force vector $\mathbf{d}$ is represented by the eigenvector expansion of Eq. (28.22) and the response vector $\mathbf{a}$ is represented by the eigenvector expansion of Eq. (28.18):

$$\mathbf{d} = \sum_{r=1}^{n} \frac{\mathbf{M}\mathbf{v}_r\mathbf{v}_r^T}{\mathbf{v}_r^T\mathbf{M}\mathbf{v}_r}\,\mathbf{d} \qquad \mathbf{a} = \sum_{r=1}^{n} \mathbf{v}_r c_r$$

where the c_r are unknown coefficients. Substituting these into Eq. (28.27), and making use of the fundamental eigenvalue relation of Eq. (28.11) leads to

$$\sum_{r=1}^{n} (\omega_r^2 - \omega^2)\mathbf{M}\mathbf{v}_r c_r = \sum_{r=1}^{n} \frac{\mathbf{M}\mathbf{v}_r\mathbf{v}_r^T}{\mathbf{v}_r^T\mathbf{M}\mathbf{v}_r}\,d$$

This equation can be uncoupled by premultiplying both sides by $\mathbf{v}_r^T$ and using the orthogonality condition of Eq. (28.15) to obtain

$$(\omega_r^2 - \omega^2)\mathbf{v}_r^T\mathbf{M}\mathbf{v}_r c_r = \mathbf{v}_r^T\mathbf{d}$$

$$c_r = \frac{1}{\omega_r^2 - \omega^2}\frac{\mathbf{v}_r^T\mathbf{d}}{\mathbf{v}_r^T\mathbf{M}\mathbf{v}_r}$$

The final solution is then assembled by inserting the c_r back into **a** and **a** back into **x**.

$$\mathbf{x} = \Re\left\{\sum_{r=1}^{n}\frac{e^{j\omega t}}{\omega_r^2 - \omega^2}\frac{\mathbf{v}_r\mathbf{v}_r^T}{\mathbf{v}_r^T\mathbf{M}\mathbf{v}_r}\mathbf{d}\right\} \tag{28.28}$$

This form clearly indicates the effect of frequency on the response.

EXAMPLE 28.2. Equation (28.28) is illustrated below for the system of Fig. 28.5. The eigenvectors and eigenvalues are given in Eq. (28.26). The square matrix for $r = 1$ is evaluated below.

$$\frac{\mathbf{v}_1\mathbf{v}_1^T}{\mathbf{v}_1^T\mathbf{M}\mathbf{v}_1} = \frac{\begin{bmatrix}0.2179\\1.0000\end{bmatrix}[0.2179 \quad 1.0000]}{[0.2179 \quad 1.0000]\begin{bmatrix}3 & 2\\2 & 2\end{bmatrix}\begin{bmatrix}0.2179\\1.0000\end{bmatrix}} = \frac{\begin{bmatrix}0.0475 & 0.2179\\0.2179 & 1.0000\end{bmatrix}}{3.0140}$$

$$= \begin{bmatrix}0.0158 & 0.0723\\0.0723 & 0.3318\end{bmatrix}$$

Inserting this, and a similar matrix for $r = 2$, in Eq. (28.28) leads to the following representation for the steady state forced vibration:

$$\begin{bmatrix}x_1\\x_2\end{bmatrix} = \Re\left\{\frac{e^{j\omega t}}{0.7053 - \omega^2}\begin{bmatrix}0.0158 & 0.0723\\0.0723 & 0.3318\end{bmatrix}\begin{bmatrix}d_1\\d_2\end{bmatrix} + \frac{e^{j\omega t}}{3.5447 - \omega^2}\begin{bmatrix}0.9842 & -1.0723\\-1.0723 & 1.1682\end{bmatrix}\begin{bmatrix}d_1\\d_2\end{bmatrix}\right\}$$

where d_1 and d_2 are the excitation amplitudes.

RESPONSE TO GENERAL EXCITATION

It is now desired to obtain the solution to Eq. (28.23) for the general case in which the excitation $\mathbf{f}(t)$ is an arbitrary vector function of time and for which initial displacements $\mathbf{x}(0)$ and velocities $\dot{\mathbf{x}}(0)$ are prescribed. This kind of problem can be integrated directly as it stands by a step-by-step procedure[6] using a digital computer, or the system can be modelled on a large differential analyzer and the required excitations supplied by function generators. If the natural modes and frequencies of the system are available, it is again possible to split the problem up into n single degree-of-freedom response problems and to indicate a formal solution.

Following a procedure similar to that just used for steady-state forced sinusoidal vibrations, an eigenvector expansion of the solution is assumed

$$\mathbf{x}(t) = \sum_{r-1}^{n}\mathbf{v}_r c_r(t)$$

where the c_r are unknown functions of time and the known excitation $\mathbf{f}(t)$ is expanded according to Eq. (28.22). Inserting these into Eq. (28.23) yields

$$\sum_{r=1}^{n}(\mathbf{M}\mathbf{v}_r\ddot{c}_r + \mathbf{K}\mathbf{v}_r c_r) = \sum_{r=1}^{n}\frac{\mathbf{M}\mathbf{v}_r\mathbf{v}_r^T}{\mathbf{v}_r^T\mathbf{M}\mathbf{v}_r}\mathbf{f}(t)$$

Using Eq. (28.11) to eliminate $\mathbf{K}$ and premultiplying by $\mathbf{v}_r^T$ to uncouple the equation,

$$\ddot{c}_r + \omega_r^2 c_r^2 = \frac{\mathbf{v}_r^T \mathbf{f}(t)}{\mathbf{v}_r^T \mathbf{M} \mathbf{v}_r} \tag{28.29}$$

is obtained as a single second-order differential equation for the time behavior of the rth mode multiplier. The initial conditions for c_r can be obtained by making eigenvector expansions of $\mathbf{x}(0)$ and $\dot{\mathbf{x}}(0)$ as was done previously for the free vibration case. Formal solutions to Eq. (28.29) can be obtained by a number of methods including Laplace transforms and variation of parameters. When these mode multipliers are substituted back to obtain $\mathbf{x}$ the general solution has the following appearance:

$$\mathbf{x}(t) = \sum_{r=1}^{n} \frac{\mathbf{v}_r \mathbf{v}_r^T \mathbf{M}}{\mathbf{v}_r^T \mathbf{M} \mathbf{v}_r} \left\{ \mathbf{x}(0) \cos \omega_r t + \frac{1}{\omega_r} \dot{\mathbf{x}}(0) \sin \omega_r t \right\}$$

$$+ \sum_{r=1}^{n} \frac{\mathbf{v}_r \mathbf{v}_r^T}{\omega_r \mathbf{v}_r^T \mathbf{M} \mathbf{v}_r} \int_0^t \mathbf{f}(t') \sin \{\omega_r (t - t')\}\, dt' \tag{28.30}$$

This solution, while very general, is useful only if the integrals involving the excitation can be evaluated. If the elements $f_j(t)$ of $\mathbf{f}(t)$ are simple (e.g., step functions, ramps, single sine pulses, etc.), closed form solutions are available.[7] When the $f_j(t)$ are more complicated, it may be possible to evaluate the integrals numerically using a generalized Simpson's rule for trigonometric integrals.[8] It also may be possible to solve Eq. (28.29) on a small differential analyzer using a single function generator.

EXAMPLE 28.3. Consider the system of Fig. 28.5. Determine the response for a general excitation and also particularize this for the case of a velocity shock input. Since the treatment of initial conditions is given above, the initial displacements $\mathbf{x}(0)$ and velocities $\dot{\mathbf{x}}(0)$ are taken to be zero.

The natural modes and natural frequencies are given in Eq. (28.26). The square matrix for $r = 1$ required in Eq. (28.30) is calculated below.

$$\frac{\mathbf{v}_1 \mathbf{v}_1^T}{\omega_1 \mathbf{v}_1^T \mathbf{M} \mathbf{v}_1} = \frac{\begin{bmatrix} 0.2179 \\ 1.0000 \end{bmatrix} [0.2179 \quad 1.000]}{0.8398 [0.2179 \quad 1.0000] \begin{bmatrix} 3 & 2 \\ 2 & 2 \end{bmatrix} \begin{bmatrix} 0.2179 \\ 1.0000 \end{bmatrix}}$$

$$= \begin{bmatrix} 0.0188 & 0.0861 \\ 0.0861 & 0.3951 \end{bmatrix}$$

A similar calculation is required for $r = 2$. Substituting in Eq. (28.30) yields the following formal solution for the response, starting from rest at $t = 0$, due to arbitrary excitation components $f_1(t)$ and $f_2(t)$:

$$\begin{bmatrix} x_1(t) \\ x_2(t) \end{bmatrix} = \begin{bmatrix} 0.0188 & 0.0861 \\ 0.0861 & 0.3951 \end{bmatrix} \int_0^t \begin{bmatrix} f_1(t') \\ f_2(t') \end{bmatrix} \sin \{0.8398(t - t')\}\, dt'$$

$$+ \begin{bmatrix} 0.5228 & -0.5695 \\ -0.5695 & 0.6205 \end{bmatrix} \int_0^t \begin{bmatrix} f_1(t') \\ f_2(t') \end{bmatrix} \sin \{1.8827(t - t')\}\, dt'$$

In Fig. 28.5, consider that the excitation is due to a prescribed base acceleration $\ddot{x}_0(t)$; then the excitation vector $\mathbf{f}$ is, according to Eq. (28.9),

$$\mathbf{f} = -\ddot{x}_0(t) \begin{bmatrix} m_1 + m_2 \\ m_2 \end{bmatrix} = -\ddot{x}_0(t) \begin{bmatrix} 3 \\ 2 \end{bmatrix}$$

With this substituted in the solution above and the matrix multiplication completed, the response is

$$\begin{bmatrix} x_1(t) \\ x_2(t) \end{bmatrix} = -\begin{bmatrix} 0.2284 \\ 1.0484 \end{bmatrix} \int_0^t \ddot{x}_0(t) \sin \{0.8398(t - t')\} \, dt'$$

$$- \begin{bmatrix} 0.4292 \\ -0.4676 \end{bmatrix} \int_0^t \ddot{x}_0(t) \sin \{1.8827(t - t')\} \, dt'$$

If the base sustains a *velocity shock* of magnitude V_0, so that $\ddot{x}_0$ vanishes at all t except at $t = 0$ where it is infinite in such a way that its integral is V_0, the solution can be further particularized as follows:

$$\begin{bmatrix} x_1(t) \\ x_2(t) \end{bmatrix} = -\begin{bmatrix} 0.2284 \\ 1.0484 \end{bmatrix} V_0 \sin (0.8398t) - \begin{bmatrix} 0.4292 \\ -0.4676 \end{bmatrix} V_0 \sin (1.8827t)$$

VIBRATION OF SYSTEMS WITH DAMPING

In this section solutions to the complete governing equation, Eq. (28.8), are discussed. The results of the preceding section for systems without damping are adequate for many purposes. There are, however, important problems in which it is necessary to include the effect of damping, e.g., problems concerned with resonance, random vibration,[9] etc. Information concerning damping in vibratory systems is surveyed in Ref. 10.

COMPLEX EIGENVALUE PROBLEM

When there is no excitation, Eq. (28.8) becomes

$$\mathbf{M}\ddot{\mathbf{x}} + \mathbf{C}\dot{\mathbf{x}} + \mathbf{K}\mathbf{x} = 0$$

which describes the free vibration of the system. As in the undamped case there are $2n$ independent solutions which can be superposed to meet $2n$ initial conditions. Assuming a solution in the form

$$\mathbf{x} = \mathbf{u}e^{pt}$$

leads to the following algebraic problem:

$$(p^2\mathbf{M} + p\mathbf{C} + \mathbf{K})\mathbf{u} = 0 \tag{28.31}$$

for the determination of the vector $\mathbf{u}$ and the scalar p. This is a *complex eigenvalue problem* because the *eigenvalue* p and the elements of the *eigenvector* $\mathbf{u}$ are, in general, complex numbers. Let a superscript C denote the complex conjugate of a quantity. If $\mathbf{u}$ and p satisfy Eq. (28.31), then so also do $\mathbf{u}^C$ and p^C. There are $2n$ roots in the negative half-plane which occur in pairs of complex conjugates or as real negative numbers. When the damping is absent, all roots lie on the imaginary axis; for small damping the roots lie near the imaginary axis. The corresponding $2n$ eigenvectors satisfy[11] the following *orthogonality* relations:

$$\begin{aligned} (p_r + p_s)\mathbf{u}_r{}^T\mathbf{M}\mathbf{u}_s + \mathbf{u}_r{}^T\mathbf{C}\mathbf{u}_s &= 0 \\ \mathbf{u}_r{}^T\mathbf{K}\mathbf{u}_s - p_r p_s \mathbf{u}_r{}^T\mathbf{M}\mathbf{u}_s &= 0 \end{aligned} \tag{28.32}$$

whenever $p_r \neq p_s$; they can be made to hold for repeated roots by suitable choice of the eigenvectors associated with a multiple root. Some indication of the relative computational effort required for systems with damping as compared to systems without damping can be had by comparing these orthogonality relations with Eq. (28.15), keeping in mind that the individual operations in Eq. (28.32) involve addition and multiplication of complex numbers. It is sometimes convenient to display a complex eigenvalue p_r in the form

$$p_r = \omega_r(-\zeta_r + j\sqrt{1 - \zeta_r{}^2}) \tag{28.33}$$

which is commonly employed in treating single degree-of-freedom systems. The parameter ω_r, called the *undamped natural frequency,* and the parameter ζ_r, called the *ratio of critical damping* (related to *quality factor* Q_r by the equation $Q_r = 1/2\zeta_r$), can be determined from the following quotients involving the eigenvector $\mathbf{u}_r$ and its conjugate $\mathbf{u}_r^C$:

$$\omega_r^2 = \frac{\mathbf{u}_r^T\mathbf{K}\mathbf{u}_r^C}{\mathbf{u}_r^T\mathbf{M}\mathbf{u}_r^C} \qquad 2\zeta_r\omega_r = \frac{\mathbf{u}_r^T\mathbf{C}\mathbf{u}_r^C}{\mathbf{u}_r^T\mathbf{M}\mathbf{u}_r^C} \tag{28.34}$$

FORMAL SOLUTIONS

If the solution to the eigenvalue problem of Eq. (28.31) is available, it is possible to exhibit a general solution to the governing equation

$$\mathbf{M}\ddot{\mathbf{x}} + \mathbf{C}\dot{\mathbf{x}} + \mathbf{K}\mathbf{x} = \mathbf{f} \tag{28.8}$$

for arbitrary excitation $\mathbf{f}(t)$ which meets prescribed initial conditions for $\mathbf{x}(0)$ and $\dot{\mathbf{x}}(0)$ at $t = 0$. The solutions given below apply to the case where the $2n$ eigenvalues occur as n pairs of complex conjugates (which is usually the case when the damping is light). This does, however, restrict the treatment to systems with nonsingular stiffness matrices $\mathbf{K}$ because if $\omega_r^2 = 0$ is an undamped eigenvalue the corresponding eigenvalues in the presence of damping are real. All quantities in the solutions below are *real.* These forms have been obtained by breaking down complex solutions[12] into real and imaginary parts and recombining. Introducing the notation

$$p_r = -\alpha_r + j\beta_r \qquad \mathbf{u}_r = \mathbf{v}_r + j\mathbf{w}_r$$

for the real and imaginary parts of eigenvalues and eigenvectors, note that

$$\alpha_r = \zeta_r\omega_r \qquad \beta_r = \omega_r\sqrt{1 - \zeta_r^2}$$

The general solution to Eq. (28.8) is then

$$\begin{aligned}
\mathbf{x}(t) = &\sum_{r=1}^{n} \frac{2}{a_r^2 + b_r^2} \{\mathbf{G}_r\mathbf{M}\dot{\mathbf{x}}(0) + (-\alpha_r\mathbf{G}_r\mathbf{M} + \beta_r\mathbf{H}_r\mathbf{M} + \mathbf{G}_r\mathbf{C})\mathbf{x}(0)\} e^{-\alpha_r t} \cos \beta_r t \\
&+ \sum_{r=1}^{n} \frac{2}{a_r^2 + b_r^2} \{\mathbf{H}_r\mathbf{M}\dot{\mathbf{x}}(0) + (-\beta_r\mathbf{G}_r\mathbf{M} - \alpha_r\mathbf{H}_r\mathbf{M} + \mathbf{H}_r\mathbf{C})\mathbf{x}(0)\} e^{-\alpha_r t} \sin \beta_r t \\
&+ \sum_{r=1}^{n} \frac{2}{a_r^2 + b_r^2} \mathbf{G}_r \int_0^t \mathbf{f}(t') e^{-\alpha_r(t-t')} \cos \beta_r(t - t')\, dt' \\
&+ \sum_{r=1}^{n} \frac{2}{a_r^2 + b_r^2} \mathbf{H}_r \int_0^t \mathbf{f}(t') e^{-\alpha_r(t-t')} \sin \beta_r(t - t')\, dt'
\end{aligned} \tag{28.35}$$

where

$$\begin{aligned}
a_r &= -2\alpha_r(\mathbf{v}_r^T\mathbf{M}\mathbf{v}_r - \mathbf{w}_r^T\mathbf{M}\mathbf{w}_r) - 4\beta_r\mathbf{v}_r^T\mathbf{M}\mathbf{w}_r + \mathbf{v}_r^T\mathbf{C}\mathbf{v}_r - \mathbf{w}_r^T\mathbf{C}\mathbf{w}_r \\
b_r &= 2\beta_r(\mathbf{v}_r^T\mathbf{M}\mathbf{v}_r - \mathbf{w}_r^T\mathbf{M}\mathbf{w}_r) - 4\alpha_r\mathbf{v}_r^T\mathbf{M}\mathbf{w}_r + 2\mathbf{v}_r^T\mathbf{C}\mathbf{w}_r \\
\mathbf{A}_r &= \mathbf{v}_r\mathbf{v}_r^T - \mathbf{w}_r\mathbf{w}_r^T \qquad \mathbf{B}_r = \mathbf{v}_r\mathbf{w}_r^T + \mathbf{w}_r\mathbf{v}_r^T \\
\mathbf{G}_r &= a_r\mathbf{A}_r + b_r\mathbf{B}_r \qquad \mathbf{H}_r = b_r\mathbf{A}_r - a_r\mathbf{B}_r
\end{aligned}$$

The solution of Eq. (28.35) should be compared with the corresponding solution of Eq. (28.30) for systems without damping. When the damping matrix $\mathbf{C} = 0$, Eq. (28.35) reduces to Eq. (28.30).

For the important special case of steady-state forced sinusoidal excitation of the form

$$\mathbf{f} = \mathcal{R}\{\mathbf{d}e^{j\omega t}\}$$

where $\mathbf{d}$ is a column of driving-force amplitudes, the steady-state portion of the response can be written as follows, using the above notation:

$$\mathbf{x}(t) = \mathcal{R}\left\{\sum_{r=1}^{n} \frac{2e^{j\omega t}}{a_r^2 + b_r^2} \frac{\alpha_r \mathbf{G}_r + \beta_r \mathbf{H}_r + j\omega \mathbf{G}_r}{\omega_r^2 - \omega^2 + j2\zeta_r\omega_r\omega} \mathbf{d}\right\} \tag{28.36}$$

This result reduces to Eq. (28.28) when the damping matrix $\mathbf{C}$ is set equal to zero.

EXAMPLE 28.4. Consider the system of Fig. 28.5 with the same mass and stiffness values as used heretofore but now including damping coefficients

$$c_1 = 0.10 \text{ lb-sec/in.} \qquad c_2 = 0.02 \text{ lb-sec/in.} \qquad c_3 = 0.04 \text{ lb-sec/in.}$$

The coefficient matrices of Eq. (28.9) thus have the following numerical values:

$$\mathbf{M} = \begin{bmatrix} 3 & 2 \\ 2 & 2 \end{bmatrix} \qquad \mathbf{C} = \begin{bmatrix} 0.14 & 0.04 \\ 0.04 & 0.06 \end{bmatrix} \qquad \mathbf{K} = \begin{bmatrix} 4 & 1 \\ 1 & 1.5 \end{bmatrix}$$

The complex eigenvalues p_r and eigenvectors $\mathbf{u}_r$ are given below, correct to four decimal places.

$$p_r = -\alpha_r + j\beta_r \qquad \mathbf{u}_r = \mathbf{v}_r + j\mathbf{w}_r$$

$$2\alpha_1 = 0.0279 \qquad \alpha_1 = \zeta_1\omega_1 = 0.0139 \qquad \zeta_1 = 0.0166$$

$$\beta_1 = 0.8397 \qquad \omega_1 = 0.8398 \qquad \omega_1^2 = 0.7053$$

$$2\alpha_2 = 0.1221 \qquad \alpha_2 = \zeta_2\omega_2 = 0.0611 \qquad \zeta_2 = 0.0324$$

$$\beta_2 = 1.8818 \qquad \omega_2 = 1.8828 \qquad \omega_2^2 = 3.5449$$

$$\mathbf{V} = \begin{bmatrix} 0.2179 & -0.9179 \\ 1.0000 & 1.0000 \end{bmatrix} \qquad \mathbf{W} = \begin{bmatrix} 0.0016 & 0.0010 \\ 0 & 0 \end{bmatrix}$$

Note that this is a lightly damped system. The damping ratios in the two modes are 1.66 per cent and 3.24 per cent respectively. Alternatively, the quality factors are $Q_1 = 30.1$ and $Q_2 = 15.4$. Using the above values to calculate a_r, b_r, $\mathbf{A}_r$, $\mathbf{B}_r$, $\mathbf{G}_r$, and $\mathbf{H}_r$ for $r = 1$ and $r = 2$, on substitution into Eq. (28.35) the following general solution is found for vibration due to excitation $\mathbf{f}(t)$ which satisfies prescribed initial conditions for $\mathbf{x}(0)$ and $\dot{\mathbf{x}}(0)$ at $t = 0$:

$$\begin{aligned}
\begin{bmatrix} x_1 \\ x_2 \end{bmatrix} = &\left\{\begin{bmatrix} 0.0014 & 0.0012 \\ -0.0012 & -0.0014 \end{bmatrix}\begin{bmatrix} \dot{x}_1(0) \\ \dot{x}_2(0) \end{bmatrix}\right. \\
&+ \left.\begin{bmatrix} 0.1919 & 0.1761 \\ 0.8805 & 0.8081 \end{bmatrix}\begin{bmatrix} x_1(0) \\ x_2(0) \end{bmatrix}\right\} e^{-0.0140t}\cos 0.8397t \\
&+ \left\{\begin{bmatrix} -0.0014 & -0.0012 \\ 0.0011 & 0.0014 \end{bmatrix}\begin{bmatrix} \dot{x}_1(0) \\ \dot{x}_2(0) \end{bmatrix}\right. \\
&+ \left.\begin{bmatrix} 0.8081 & -0.1761 \\ -0.8805 & 0.1919 \end{bmatrix}\begin{bmatrix} x_1(0) \\ x_2(0) \end{bmatrix}\right\} e^{-0.0611t}\cos 1.8818t \\
&+ \left\{\begin{bmatrix} 0.2285 & 0.2097 \\ 1.0485 & 0.9624 \end{bmatrix}\begin{bmatrix} \dot{x}_1(0) \\ \dot{x}_2(0) \end{bmatrix}\right. \\
&+ \left.\begin{bmatrix} 0.0017 & 0.0020 \\ 0.0141 & 0.0149 \end{bmatrix}\begin{bmatrix} x_1(0) \\ x_2(0) \end{bmatrix}\right\} e^{-0.0140t}\sin 0.8397t \\
&+ \left\{\begin{bmatrix} 0.4294 & -0.0936 \\ -0.4679 & 0.1020 \end{bmatrix}\begin{bmatrix} \dot{x}_1(0) \\ \dot{x}_2(0) \end{bmatrix}\right. \\
&+ \left.\begin{bmatrix} 0.0269 & -0.0053 \\ -0.0283 & 0.0056 \end{bmatrix}\begin{bmatrix} x_1(0) \\ x_2(0) \end{bmatrix}\right\} e^{-0.0611t}\sin 1.8818t
\end{aligned}$$

$$+ \begin{bmatrix} 0.0002 & 0.0004 \\ 0.0004 & -0.0011 \end{bmatrix} \int_0^t [\mathbf{f}(t')e^{-0.0140(t-t')} \cos 0.8397(t - t')]\, dt'$$

$$+ \begin{bmatrix} -0.0002 & -0.0004 \\ -0.0004 & 0.0011 \end{bmatrix} \int_0^t [\mathbf{f}(t')e^{-0.0611(t-t')} \cos 1.8818(t - t')]\, dt'$$

$$+ \begin{bmatrix} 0.0188 & 0.0861 \\ 0.0861 & 0.3953 \end{bmatrix} \int_0^t [\mathbf{f}(t')e^{-0.0140(t-t')} \sin 0.8397(t - t')]\, dt'$$

$$+ \begin{bmatrix} 0.5230 & -0.5698 \\ -0.5698 & 0.6208 \end{bmatrix} \int_0^t [\mathbf{f}(t')e^{-0.0611(t-t')} \sin 1.8818(t - t')]\, dt'$$

Inserting these same values into Eq. (28.36), the following expression is obtained for the steady-state forced sinusoidal vibration of the same system:

$$\begin{bmatrix} x_1 \\ x_2 \end{bmatrix} = \mathcal{R} \left\{ \frac{e^{j\omega t}\left\{\begin{bmatrix} 0.0158 & 0.0723 \\ 0.0723 & 0.3318 \end{bmatrix} + j\omega \begin{bmatrix} 0.0002 & 0.0004 \\ 0.0004 & -0.0011 \end{bmatrix}\right\}}{0.7053 - \omega^2 + 0.0279j\omega} \begin{bmatrix} d_1 \\ d_2 \end{bmatrix} \right.$$

$$\left. + \frac{e^{j\omega t}\left\{\begin{bmatrix} 0.9842 & -1.0724 \\ -1.0724 & 1.1683 \end{bmatrix} + j\omega \begin{bmatrix} -0.0002 & -0.0004 \\ -0.0004 & 0.0011 \end{bmatrix}\right\}}{3.5449 - \omega^2 + 0.1221j\omega} \begin{bmatrix} d_1 \\ d_2 \end{bmatrix} \right\}$$

APPROXIMATE SOLUTIONS

For a lightly damped system the exact solutions of Eq. (28.35) and Eq. (28.36) can be abbreviated considerably by making approximations based on the smallness of the damping. A systematic method of doing this is to consider the system without damping as a base upon which an infinitesimal amount of damping is superposed as a perturbation. Using the calculus, it is possible to obtain the infinitesimal perturbation of the base solution which results. While only strictly valid for infinitesimal perturbations, this procedure provides useful approximations for small amounts of damping. In the section on numerical methods which follows, this process is carried out in detail to obtain complex eigenvectors and eigenvalues. It is found there that the effect of a small amount of damping on a natural vibration is (1) to introduce small exponential decay without altering the frequency and (2) to split the original real eigenvector into a pair of complex vectors each having the same real part as the undamped mode but having small conjugate imaginary parts. Thus in the perturbation solution the imaginary part of the eigenvalue p_r is simply $j\omega_r$, where ω_r is the corresponding undamped natural frequency and the real part of the eigenvector $\mathbf{u}_r$ is simply the corresponding undamped natural mode $\mathbf{v}_r$. The real part of the eigenvalue, i.e., $\alpha_r = \zeta_r\omega_r$, and the imaginary part of the eigenvector, i.e., $j\mathbf{w}_r$, are small quantities (strictly speaking, they are first-order infinitesimals).

This perturbation approximation can be continued into Eqs. (28.35) and (28.36) by simply neglecting all squares and products of the small quantities α_r, ζ_r, $\mathbf{w}_r$, and $\mathbf{C}$. When this is done it is found that the formulas of Eqs. (28.35) and (28.36) may still be used if the parameters therein are obtained from the simplified expressions below.

$$\left.\begin{aligned}
&\alpha_r = \zeta_r\omega_r \qquad \beta_r = \omega_r \\
&a_r = -4\omega_r\mathbf{v}_r^T\mathbf{M}\mathbf{w}_r \qquad b_r = 2\omega_r\mathbf{v}_r^T\mathbf{M}\mathbf{v}_r \\
&a_r^2 + b_r^2 = 4\omega_r^2(\mathbf{v}_r^T\mathbf{M}\mathbf{v}_r)^2 \\
&\mathbf{A}_r = \mathbf{v}_r\mathbf{v}_r^T \qquad \mathbf{B}_r = \mathbf{v}_r\mathbf{w}_r^T + \mathbf{w}_r\mathbf{v}_r^T \\
&\mathbf{G}_r = -4\omega_r(\mathbf{v}_r^T\mathbf{M}\mathbf{w}_r)\mathbf{v}_r\mathbf{v}_r^T + 2\omega_r(\mathbf{v}_r^T\mathbf{M}\mathbf{v}_r)(\mathbf{v}_r\mathbf{w}_r^T + \mathbf{w}_r\mathbf{v}_r^T) \\
&\mathbf{H}_r = 2\omega_r(\mathbf{v}_r^T\mathbf{M}\mathbf{v}_r)\mathbf{v}_r\mathbf{v}_r^T
\end{aligned}\right\} \qquad (28.37)$$

For example, the steady-state forced sinusoidal solution of Eq. (28.36) takes the following explicit form in the perturbation approximation:

$$\mathbf{x}(t) = \mathcal{R}\left\{\sum_{r=1}^{n} \frac{e^{j\omega t}}{\mathbf{v}_r^T\mathbf{M}\mathbf{v}_r} \frac{\mathbf{v}_r\mathbf{v}_r^T + \dfrac{j\omega}{\omega_r}\left[\mathbf{v}_r\mathbf{w}_r^T - \left(2\dfrac{\mathbf{v}_r^T\mathbf{M}\mathbf{w}_r}{\mathbf{v}_r^T\mathbf{M}\mathbf{v}_r}\right)\mathbf{v}_r\mathbf{v}_r^T + \mathbf{w}_r\mathbf{v}_r^T\right]}{\omega_r^2 - \omega^2 + j2\zeta_r\omega_r\omega}\,\mathbf{d}\right\} \tag{28.38}$$

When this approximation is applied to the numerical example previously given it is found that the results are almost identical. There are a few entries which differ in the fourth decimal place. The above formula simplifies somewhat if $\mathbf{w}_r$ is chosen to be orthogonal to $\mathbf{v}_r$ with respect to $\mathbf{M}$ since this causes the coefficient of $\mathbf{v}_r\mathbf{v}_r^T$ to vanish. In the perturbation procedure described in the section on numerical methods, the eigenvectors possess this orthogonality property before they are normalized.

A cruder approximation, which is often used, is based on accepting the complex eigenvalue $p_r = -\alpha_r + j\omega_r$ but completely neglecting the imaginary part $j\mathbf{w}_r$ of the eigenvector $\mathbf{u}_r = \mathbf{v}_r + j\mathbf{w}_r$. It is thus assumed that the undamped mode $\mathbf{v}_r$ still applies for the system with damping. The approximate parameter values of Eq. (28.37) are further simplified by this assumption; e.g., $a_r = 0$, $\mathbf{B}_r = \mathbf{G}_r = 0$. The steady forced sinusoidal response of Eq. (28.38) reduces to

$$\mathbf{x}(t) = \mathcal{R}\left\{\sum_{r=1}^{n} \frac{e^{j\omega t}}{\omega_r^2 - \omega^2 + j2\zeta_r\omega_r\omega} \frac{\mathbf{v}_r\mathbf{v}_r^T}{\mathbf{v}_r^T\mathbf{M}\mathbf{v}_r}\,\mathbf{d}\right\} \tag{28.39}$$

This approximation should be compared with the undamped solution of Eq. (28.28), as well as with the exact solution of Eq. (28.36) and the perturbation approximation of Eq. (28.38). An idea of the error committed in the cruder approximation is provided by comparing the numerical solutions to the examples used to illustrate Eq. (28.28) and Eq. (28.36). Although large errors are possible at high frequencies, the discrepancies when $\omega = \omega_2$ (the highest natural frequency) are only of the order of 1 per cent.

NUMERICAL METHODS

THE REAL EIGENVALUE PROBLEM

Given the symmetric matrices $\mathbf{K}$ and $\mathbf{M}$ with $\mathbf{M}$ positive definite, the problem is to determine the eigenvectors $\mathbf{v}_r$ and the eigenvalues ω_r^2 which satisfy

$$\mathbf{K}\mathbf{v} = \omega^2\mathbf{M}\mathbf{v} \tag{28.11}$$

Earlier in this chapter the basic properties of the solution are presented. Here a description is given of some additional properties of computational interest: Rayleigh's quotient, perturbation formulas, and Jacobi's method.

RAYLEIGH'S QUOTIENT. If Eq. (28.11) is premultiplied by $\mathbf{v}^T$ the following scalar equation is obtained:

$$\mathbf{v}^T\mathbf{K}\mathbf{v} = \omega^2\mathbf{v}^T\mathbf{M}\mathbf{v}$$

The positive definiteness of $\mathbf{M}$ guarantees that $\mathbf{v}^T\mathbf{M}\mathbf{v}$ is nonzero so that it is permissible to solve for ω^2.

$$\omega^2 = \frac{\mathbf{v}^T\mathbf{K}\mathbf{v}}{\mathbf{v}^T\mathbf{M}\mathbf{v}} \tag{28.40}$$

This quotient is called "Rayleigh's quotient." It also may be derived by equating time averages of potential and kinetic energy under the assumption that the vibratory system is executing simple harmonic motion at frequency ω with amplitude ratios given by $\mathbf{v}$ or by equating the maximum value of kinetic energy to the maximum value of potential energy under the same assumption. Rayleigh's quotient has the following interesting properties.

1. When $\mathbf{v}$ is an eigenvector $\mathbf{v}_r$ of Eq. (28.11), then Rayleigh's quotient is equal to the corresponding eigenvalue ω_r^2.

2. If $\mathbf{v}$ is an approximation to $\mathbf{v}_r$ with an error which is a *first-order* infinitesimal, then Rayleigh's quotient is an approximation to ω_r^2 with an error which is a *second-order* infinitesimal; i.e., Rayleigh's quotient is *stationary* in the neighborhoods of the true eigenvectors.

3. As $\mathbf{v}$ varies through all of n-dimensional vector space, Rayleigh's quotient remains bounded between the smallest and largest eigenvalues.

A common engineering application of Rayleigh's quotient involves simply evaluating Eq. (28.40) for a trial vector $\mathbf{v}$ which is selected on the basis of physical insight. When eigenvectors are obtained by approximate methods, Rayleigh's quotient provides a means of improving the accuracy in the corresponding eigenvalue. If the elements of an approximate eigenvector whose largest element is unity are correct to k decimal places, then Rayleigh's quotient can be expected to be correct to about $2k$ significant decimal places. Several interesting numerical procedures[13,14] for solving eigenvalue problems make use of the stationary property of Rayleigh's quotient.

PERTURBATION FORMULAS. The perturbation formulas which follow provide the basis for estimating the changes in the eigenvalues and the eigenvectors which result from *small* changes in the stiffness and inertia parameters of a system. The formulas are strictly accurate only for infinitesimal changes but are useful approximations for *small* changes. They may be used by the designer to estimate the effects of a proposed change in a vibratory system and may also be used to analyze the effects of minor errors in the measurement of the system properties. Iterative procedures for the solution of eigenvalue problems[15–17] can be based on these formulas. They are employed here to obtain approximations to the complex eigenvalues and eigenvectors of a lightly damped vibratory system in terms of the corresponding solutions for the same system without damping.

Suppose that the modal matrix $\mathbf{V}$ and the spectral matrix Ω^2 for the eigenvalue problem

$$\mathbf{KV} = \mathbf{MV}\Omega^2 \tag{28.14}$$

are known. Consider the perturbed eigenvalue problem

$$\mathbf{K}_*\mathbf{V}_* = \mathbf{M}_*\mathbf{V}_*\Omega_*{}^2$$

where

$$\mathbf{K}_* = \mathbf{K} + dK \qquad \mathbf{M}_* = \mathbf{M} + d\mathbf{M}$$

$$\mathbf{V}_* = \mathbf{V} + d\mathbf{V} \qquad \Omega_*{}^2 = \Omega^2 + d\Omega^2$$

The perturbation formula for the elements $d\omega_r^2$ of the diagonal matrix $d\Omega^2$ is

$$d\omega_r^2 = \frac{\mathbf{v}_r{}^T\, d\mathbf{K}\, \mathbf{v}_r - \omega_r{}^2\mathbf{v}_r{}^T\, d\mathbf{M}\, \mathbf{v}_r}{\mathbf{v}_r{}^T\mathbf{M}\mathbf{v}_r} \tag{28.41}$$

Thus in order to determine the change in a single eigenvalue due to changes in $\mathbf{M}$ and $\mathbf{K}$, it is only necessary to know the corresponding unperturbed eigenvalue and eigenvector. To determine the change in a single eigenvector, however, it is necessary to know *all* the unperturbed eigenvalues and eigenvectors. The following algorithm may be used to evaluate the perturbations of both the modal matrix and the spectral matrix. Calculate

$$\mathbf{F} = \mathbf{V}^T d\mathbf{K}\, \mathbf{V} - \mathbf{V}^T\, d\mathbf{M}\, \mathbf{V}\Omega^2$$

and

$$\mathbf{L} = \mathbf{V}^T\mathbf{M}\mathbf{V}$$

The matrix $\mathbf{L}$ is a diagonal matrix of positive elements and hence is easily inverted. Continue calculating

$$\mathbf{G} = \mathbf{L}^{-1}\mathbf{F} = [g_{jk}] \qquad \text{and} \qquad \mathbf{H} = [h_{jk}]$$

where

$$h_{jk} = \begin{cases} 0 & \text{if } \omega_j{}^2 = \omega_k{}^2 \\ \dfrac{g_{jk}}{\omega_k{}^2 - \omega_j{}^2} & \text{if } \omega_j{}^2 \neq \omega_k{}^2 \end{cases}$$

Then, finally, the perturbations of the modal matrix and the spectral matrix are given by

$$d\mathbf{V} = \mathbf{V}\mathbf{H} \qquad d\Omega^2 = \left[\diagdown g_{jj} \diagdown \right] \tag{28.42}$$

These formulas are derived by taking the total differential of Eq. (28.14), premultiplying each term by $\mathbf{V}^T$, and using a relation derived by taking the transpose of Eq. (28.14). Since eigenvectors are undetermined to the extent of a scale factor, the vector changes in the eigenvectors have a degree of indeterminateness. The particular selection in the above algorithm makes the change in each eigenvector orthogonal with respect to **M** to the corresponding unperturbed eigenvector, i.e.,

$$\mathbf{v}_j{}^T\mathbf{M}\, d\mathbf{v}_j = 0$$

JACOBI METHOD. The most satisfactory general-purpose method for obtaining the complete solution to the real eigenvalue problem of Eq. (28.14) appears to be based on *diagonalization by successive rotation*.[18–20] This procedure involves a simple repetitive algorithm which is relatively easy to program. No special provisions for equal or nearly equal eigenvalues are required. The procedure has proved to be remarkably immune to serious round-off errors. The computational time is only proportional to n^3 although the proportionality constant is so large (roughly about 60) that the method is quite unsuited to hand computation even for small n.

The basic computational operation in the method is the resolution of a single symmetric matrix **A** into its modal matrix **U** and its spectral matrix Λ according to the relation

$$\mathbf{A} = \mathbf{U}\Lambda\mathbf{U}^T \tag{28.43}$$

where Λ is a diagonal matrix of the eigenvalues λ_r, and the columns of **U** are the eigenvectors $\mathbf{u}_r$ for the eigenvalue problem

$$\mathbf{A}\mathbf{u} = \lambda\mathbf{u} \qquad \text{or} \qquad \mathbf{A}\mathbf{U} = \mathbf{U}\Lambda$$

The modal matrix **U** in Eq. (28.43) has the additional property that its columns are normalized so that $\mathbf{u}_r{}^T\mathbf{u}_r = 1$, i.e., $\mathbf{U}^T = \mathbf{U}^{-1}$.

Before describing how the resolution of Eq. (28.43) is arrived at by successive rotations, we indicate how the solution to the real eigenvalue problem of Eq. (28.14) with *two* symmetrical matrices **K** and **M** can be obtained by two successive applications of Eq. (28.43).

It is first observed that real integer powers of **A** can be resolved in the form

$$\mathbf{A}^m = \mathbf{U}\Lambda^m\mathbf{U}^T \tag{28.44}$$

where Λ^m is a diagonal matrix of elements $\lambda_r{}^m$; i.e., the eigenvectors of $\mathbf{A}^m$ are identical with the eigenvectors of **A** and the eigenvalues of $\mathbf{A}^m$ are simply the mth powers of the corresponding eigenvalues of **A**. The resolution of Eq. (28.44) then can be used to *define* fractional powers of **A**. Negative powers can likewise be defined, provided none of the λ_r equal zero, i.e., provided **A** is nonsingular.

Returning to the eigenvalue problem of Eq. (28.14) it is supposed that the resolution of Eq. (28.43) can be applied to the inertia matrix **M** and in particular that the matrix $\mathbf{M}^{-1/2}$ can be constructed according to Eq. (28.44). The problem of Eq. (28.14) now is reduced to an eigenvalue problem for a single symmetric matrix $\mathbf{A} = \mathbf{M}^{-1/2}\mathbf{K}\mathbf{M}^{1/2}$ with modal matrix $\mathbf{W} = \mathbf{M}^{1/2}\mathbf{V}$ as follows:

$$\mathbf{K}\mathbf{V} = \mathbf{M}\mathbf{V}\Omega^2$$

$$\mathbf{K}\mathbf{M}^{-1/2}\mathbf{M}^{1/2}\mathbf{V} = \mathbf{M}^{1/2}\mathbf{M}^{1/2}\mathbf{V}\Omega^2$$

$$(\mathbf{M}^{-1/2}\mathbf{K}\mathbf{M}^{-1/2})(\mathbf{M}^{1/2}\mathbf{V}) = (\mathbf{M}^{1/2}\mathbf{V})\Omega^2$$

$$\mathbf{A}\mathbf{W} = \mathbf{W}\Omega^2$$

The solution of this last eigenvalue problem provides the spectral matrix Ω^2 of the original problem. To obtain the modal matrix **V** of the original problem, it is necessary to perform the matrix multiplication

$$\mathbf{V} = \mathbf{M}^{-\frac{1}{2}}\mathbf{W}$$

which follows from inverting the definition of $\mathbf{W}$.

It remains to indicate how the resolution of Eq. (28.43) is obtained by successive rotations. The procedure is an iterative process in which the matrix $\mathbf{A}$ is gradually transformed into its spectral matrix Λ by a sequence of two-dimensional rotation-of-axes transformations $\mathbf{T}_i$. Starting with $\mathbf{A}_0 = \mathbf{A}$, the successive iterates are constructed as follows:

$$\mathbf{A}_1 = \mathbf{T}_1{}^T\mathbf{A}_0\mathbf{T}_1$$

$$\cdot \quad \cdot \quad \cdot \quad \cdot \quad \cdot \quad \cdot \quad \cdot$$

$$\mathbf{A}_i = \mathbf{T}_i{}^T\mathbf{A}_{i-1}\mathbf{T}_i$$

Each transformation matrix $\mathbf{T}_i$ has the following form:

$$\mathbf{T}_i = \begin{array}{c} \begin{array}{ccccccccc} & & & p & & & q & & \end{array} \\ \left[\begin{array}{ccccccccc} 1 & 0 & \cdots & 0 & 0 & 0 & 0 & \cdots & 0 \\ 0 & 1 & \cdots & 0 & 0 & 0 & 0 & \cdots & 0 \\ \cdot & \cdot & \cdot & \cdot & \cdot & \cdot & \cdot & \cdot & \cdot \\ 0 & 0 & \cdots & c & 0 & 0 & -s & \cdots & 0 \\ 0 & 0 & \cdots & 0 & 1 & 0 & 0 & \cdots & 0 \\ 0 & 0 & \cdots & 0 & 0 & 1 & 0 & \cdots & 0 \\ 0 & 0 & \cdots & s & 0 & 0 & c & \cdots & 0 \\ \cdot & \cdot & \cdot & \cdot & \cdot & \cdot & \cdot & \cdot & \cdot \\ 0 & 0 & \cdots & 0 & 0 & 0 & 0 & \cdots & 1 \end{array}\right] \end{array} \begin{array}{c} \\ \\ \\ \\ p \\ \\ \\ q \\ \\ \\ \end{array}$$

where $\qquad c = \cos\theta \qquad s = \sin\theta$

The matrix $\mathbf{T}_i$ represents a rotation through the angle θ in the (p,q) plane. By suitably selecting θ, the off-diagonal elements in the (p,q) and (q,p) positions of $\mathbf{A}_i$ can be made to vanish. Thus in each step a single pair of off-diagonal elements is reduced to zero at the expense of altering all the elements in the 2 rows and 2 columns involved. The $\mathbf{T}_i$ may be chosen so that at each step the largest off-diagonal pair is annihilated, although satisfactory results have been obtained by systematically treating the off-diagonal elements in turn without regard to their magnitude. The process converges[21] in the limit with $\mathbf{A}_i$ approaching the spectral matrix Λ and the continued product of the $\mathbf{T}_i$ approaching the modal matrix $\mathbf{U}$:

$$\lim_{i\to\infty} \mathbf{A}_i = \Lambda \qquad\qquad \lim_{i\to\infty} (\mathbf{T}_i\mathbf{T}_2 \ldots \mathbf{T}_i) = \mathbf{U}$$

For most matrices $\mathbf{A}$, convergence to about eight decimal places is obtained after[18, 20, 22] six to ten complete sweeps through the off-diagonal elements. The individual $\mathbf{T}_i$ possess the property that $\mathbf{T}_i{}^{-1} = \mathbf{T}_i{}^T$ and so does their continued product. Thus the transpose of $\mathbf{U}$ is also its inverse. The computation can be programmed so that the major storage requirements are $(n^2 + n)/2$ locations for the elements of $\mathbf{A}_i$ and n^2 locations for the elements of the continued product of the T_i.

Several noniterative procedures for solving eigenvalue problems have been programmed for digital computers. One of the most attractive[18, 23] involves using the above rotation transformation matrices $\mathbf{T}_i$, not to diagonalize $\mathbf{A}$, but to put it in a *tri-diagonal* form in which the main diagonal together with the two adjacent diagonals contain the only nonzero elements. This reduction can be performed exactly in a fixed number of steps without any iteration. The tridiagonal form is especially convenient for obtaining the roots of the characteristic equation by inserting successive trial eigenvalues and applying the method of false position to the value of the determinant of the tridiagonal matrix. The torsional-vibration system of Fig. 28.9 possesses a tridiagonal matrix $\mathbf{A} = \mathbf{M}^{-1}\mathbf{K}$. The most popular hand computation method for such a system is exactly this process of selecting trial frequencies and evaluating the remainder in a Holzer table.[24]

THE COMPLEX EIGENVALUE PROBLEM

Given the symmetric inertia, damping, and stiffness matrices **M**, **C**, and **K**, the problem is to determine the complex eigenvalues $p_r = -\alpha_r + j\beta_r$ and the complex eigenvectors $\mathbf{u}_r = \mathbf{v}_r + j\mathbf{w}_r$ which satisfy Eq. (28.31).

$$(p^2\mathbf{M} + p\mathbf{C} + \mathbf{K})\mathbf{u} = 0 \tag{28.31}$$

This problem is considerably more difficult than the real eigenvalue problem of Eq. (28.11), and much less attention has been given to efficient computational procedures. See, however, Refs. 25 and 26. Here only a *perturbation method* is given which furnishes a useful approximation solution for a lightly damped system when the solution to the corresponding undamped system is available.

When $\mathbf{C} = 0$ in Eq. (28.31), the complex eigenvalue problem reduces to the real eigenvalue problem of Eq. (28.11) with $p^2 = -\omega^2$. Suppose that the real eigenvalues ω_r^2 and real eigenvectors $\mathbf{v}_r$ are known. The perturbation formulas of Eqs. (28.41) and (28.42) can be used to estimate the perturbation of the rth mode due to the addition of small damping **C** by considering the damping to be a perturbation of the stiffness matrix of the form

$$d\mathbf{K} = j\omega_r\mathbf{C}$$

In this way it is found that the perturbed solution corresponding to the rth mode consists of a pair of complex conjugate eigenvalues

$$p_r = -\alpha_r + j\omega_r \qquad p_r^C = -\alpha_r - j\omega_r$$

and a pair of complex conjugate eigenvectors

$$\mathbf{u}_r = \mathbf{v}_r + j\mathbf{w}_r \qquad \mathbf{u}_r^C = \mathbf{v}_r - j\mathbf{w}_r$$

where ω_r and $\mathbf{v}_r$ are taken directly from the undamped system, and α_r and $\mathbf{w}_r$ are small perturbations which are given below. The real part of the eigenvalue which describes the rate of decay of the corresponding free motion is given by the following quotient:

$$2\alpha_r = 2\zeta_r\omega_r = \frac{\mathbf{v}_r^T\mathbf{C}\mathbf{v}_r}{\mathbf{v}_r^T\mathbf{M}\mathbf{v}_r} \tag{28.45}$$

The decay rate α_r and damping ratio ζ_r for a particular r depend only on the rth mode undamped solution. The imaginary part of the eigenvector $j\mathbf{w}_r$ which describes the perturbations in phase is more difficult to obtain. All the undamped eigenvalues and eigenvectors must be known. Let **W** be a square matrix whose columns are the $\mathbf{w}_r$. The following algorithm may be used to evaluate **W** when the undamped modal matrix **V** is known. Calculate

$$\mathbf{F} = \mathbf{V}^T\mathbf{C}\mathbf{V}$$

and

$$\mathbf{L} = \mathbf{V}^T\mathbf{M}\mathbf{V}$$

The matrix **L** is a diagonal matrix of positive elements and hence is easily inverted. Continue calculating

$$\mathbf{G} = \mathbf{L}^{-1}\mathbf{F} = [g_{jk}] \qquad \text{and} \qquad \mathbf{H} = [h_{jk}]$$

where

$$h_{jk} = \begin{cases} 0 & \text{if } \omega_j^2 = \omega_k^2 \\ \dfrac{g_{jk}\omega_k}{\omega_k^2 - \omega_j^2} & \text{if } \omega_j^2 \neq \omega_k^2 \end{cases}$$

Then, finally, the eigenvector perturbations are given by

$$\mathbf{W} = \mathbf{V}\mathbf{H} \tag{28.46}$$

The main diagonal elements g_{rr} of the matrix **G** are just the quotients of Eq. (28.45) and hence equal $2\alpha_r$.

EXAMPLE 28.6. To illustrate Eqs. (28.45) and (28.46), consider the system of Fig. 28.5 where

$$\mathbf{M} = \begin{bmatrix} 3 & 2 \\ 2 & 2 \end{bmatrix} \qquad \mathbf{C} = \begin{bmatrix} 0.14 & 0.04 \\ 0.04 & 0.06 \end{bmatrix} \qquad \mathbf{K} = \begin{bmatrix} 4 & 1.0 \\ 1 & 1.5 \end{bmatrix}$$

and the undamped modal matrix $\mathbf{V}$ and spectral matrix Ω^2 are known.

$$\mathbf{V} = \begin{bmatrix} 0.2179 & -0.9179 \\ 1.0000 & 1.0000 \end{bmatrix} \qquad \Omega^2 = \begin{bmatrix} 0.7053 & 0 \\ 0 & 3.5447 \end{bmatrix}$$

The decay rate α_1 can be obtained according to Eq. (28.45) as follows:

$$2\alpha_1 = \frac{\mathbf{v}_1{}^T\mathbf{C}\mathbf{v}_1}{\mathbf{v}_1{}^T\mathbf{M}\mathbf{v}_1} = \frac{[0.2179 \quad 1.0000]\begin{bmatrix} 0.14 & 0.04 \\ 0.04 & 0.06 \end{bmatrix}\begin{bmatrix} 0.2179 \\ 1.0000 \end{bmatrix}}{[0.2179 \quad 1.0000]\begin{bmatrix} 3 & 2 \\ 2 & 2 \end{bmatrix}\begin{bmatrix} 0.2179 \\ 1.0000 \end{bmatrix}} = 0.0279$$

In a similar fashion, $2\alpha_2 = 0.1221$. These agree (to four decimal places) with the exact solution given in the section on *Vibration of Systems with Damping*. There is, however, a small error in the imaginary parts of the perturbed eigenvalues. These do not remain precisely at their undamped values although the error is of second order in comparison with the decay rates.

Turning next to the algorithm leading to Eq. (28.46), the following are obtained in sequence:

$$\mathbf{F} = \begin{bmatrix} 0.2179 & 1.0000 \\ -0.9179 & 1.0000 \end{bmatrix}\begin{bmatrix} 0.14 & 0.04 \\ 0.04 & 0.06 \end{bmatrix}\begin{bmatrix} 0.2179 & -0.9179 \\ 1.0000 & 1.0000 \end{bmatrix} = \begin{bmatrix} 0.0841 & 0.0040 \\ 0.0040 & 0.1045 \end{bmatrix}$$

$$\mathbf{L} = \begin{bmatrix} 0.2179 & 1.0000 \\ -0.9179 & 1.0000 \end{bmatrix}\begin{bmatrix} 3 & 2 \\ 2 & 2 \end{bmatrix}\begin{bmatrix} 0.2179 & -0.9179 \\ 1.0000 & 1.0000 \end{bmatrix} = \begin{bmatrix} 3.0140 & 0 \\ 0 & 0.8560 \end{bmatrix}$$

$$\mathbf{L}^{-1} = \begin{bmatrix} 0.3318 & 0 \\ 0 & 1.1682 \end{bmatrix}$$

$$\mathbf{G} = \begin{bmatrix} 0.3318 & 0 \\ 0 & 1.1682 \end{bmatrix}\begin{bmatrix} 0.0841 & 0.0040 \\ 0.0040 & 0.1045 \end{bmatrix} = \begin{bmatrix} 0.0279 & 0.0013 \\ 0.0047 & 0.1221 \end{bmatrix}$$

The main diagonal elements of $\mathbf{G}$ are $2\alpha_1$ and $2\alpha_2$ as previously determined. Continuing the algorithm with $\omega_1 = (0.7053)^{1/2} = 0.8398$ and $\omega_2 = (3.5447)^{1/2} = 1.8827$, the following result is obtained:

$$\mathbf{H} = \begin{bmatrix} 0 & \dfrac{(0.0013)(1.8827)}{3.5447 - 0.7053} \\ \dfrac{(0.0047)(0.8398)}{0.7053 - 3.5447} & 0 \end{bmatrix} = \begin{bmatrix} 0 & 0.0009 \\ -0.0014 & 0 \end{bmatrix}$$

so that finally

$$\mathbf{W} = \begin{bmatrix} 0.2179 & -0.9179 \\ 1.0000 & 1.0000 \end{bmatrix}\begin{bmatrix} 0 & 0.0009 \\ -0.0014 & 0 \end{bmatrix} = \begin{bmatrix} 0.0013 & 0.0002 \\ -0.0014 & 0.0009 \end{bmatrix}$$

The complete modal matrix $\mathbf{U}$ is given by $\mathbf{V} + j\mathbf{W}$ in unnormalized form. After amplitude normalization,

$$\mathbf{U} = \begin{bmatrix} 0.2179 + j0.0016 & -0.9179 + j0.0010 \\ 1.0000 & 1.0000 \end{bmatrix}$$

is obtained as the approximate complex modal matrix which agrees (to four decimal places) with the exact solution given in the section on *Vibration of Systems with Damping*.

REFERENCES

1. Hildebrand, F. B.: "Methods of Applied Mathematics," Chap. 1, Prentice-Hall, Inc., Englewood Cliffs, N.J., 1965.
2. Timoshenko, S., and D. H. Young: "Advanced Dynamics," Chap. 3, McGraw-Hill Book Company, New York, 1948.
3. Crede, C. E.: "Vibration and Shock Isolation," p. 45, John Wiley & Sons, Inc., New York, 1951.
4. Ref. 1, p. 76.
5. Ref. 1, p. 75.
6. Crandall, S. H.: "Engineering Analysis, A Survey of Numerical Procedures," p. 191, McGraw-Hill Book Company, New York, 1956.
7. See Chap. 8.
8. Tranter, C. J.: "Integral Transforms in Mathematical Physics," 2d ed., p. 67, Methuen & Co., Ltd., London, and John Wiley & Sons, Inc., New York, 1956.
9. Crandall, S. H.: "Random Vibration," Technology Press, M.I.T., Cambridge, Mass., 1958.
10. Ruzicka, J. E. (ed.): "Structural Damping," ASME, New York, 1959.
11. McCalley, R. B.: "Discussion of Ref. 12," *J. Appl. Mechanics*, **26**:306 (1959).
12. Foss, K. A.: *J. Appl. Mechanics*, **25**:361 (1958).
13. Crandall, S. H.: *Proc. Roy. Soc. (London)*: **A207**:416 (1951).
14. Crandall, S. H., and W. G. Strang: *J. Appl. Mechanics*, **24**:228 (1957).
15. Jahn, H. A.: *Quart. J. Mech. and Appl. Math.*, **1**:131 (1948).
16. Collar, A. R.: *Quart. J. Mech. and Appl. Math.*, **1**:145 (1948).
17. Mahalingham, S.: *J. Appl. Mechanics*, **25**:618 (1958).
18. White, P. A.: *J. Soc. Ind. Appl. Math.*, **6**:393 (1958).
19. Ref. 6, p. 118.
20. Gregory, R. T.: "Computing Eigenvalues and Eigenvectors of a Symmetric Matrix on the Illiac," *Math. Tables and Other Aids to Comp.*, **7**:215 (1953).
21. Henrici, P.: *J. Soc. Ind. Appl. Math.*, **6**:144 (1958).
22. Pope, D. A., and C. B. Tompkins: *J. Assoc. Comput. Mach.*, **4**:459 (1957).
23. Paige, L. J., and O. Taussky: "Simultaneous Linear Equations and the Determination of Eigenvalues," National Bureau of Standards, Applied Mathematics Series, No. 29, 1953.
24. See Chap. 38.
25. Wayland, H.: *Quart. Appl. Math.*, **2**:277 (1945).
26. Tarnove, I.: *J. Soc. Ind. Appl. Math.*, **6**:163 (1958).

ADDITIONAL REFERENCES

Bathe, Klaus-Jurgen, and E. L. Wilson: "Eigensolution of Large Structural Systems with Small Bandwidth," *Proc. ASCE*, **99**(EM 3):467–479 (June, 1973).

Bodewig, E.: "Matrix Calculus," North-Holland Publishing Company, Amsterdam, and Interscience Publishers, Inc., New York, 1956.

Buckingham, R. A.: "Numerical Methods," Sir Isaac Pitman & Sons, Ltd., London, 1957.

Courant, R., and D. Hilbert: "Methods of Mathematical Physics," vol. I, first English edition, Interscience Publishers, Inc., New York and London, 1953.

Den Hartog, J. P.: "Mechanical Vibrations," Fourth Edition, McGraw-Hill Book Company, New York, 1956.

Faddeev, D. K., and V. N. Faddeeva: "Computational Methods of Linear Algebra," W. H. Freeman and Company, San Francisco, 1963.

Ferrar, W. L.: "Algebra," Oxford University Press, London, 1941.

Frazer, R. A., W. J. Duncan, and A. R. Collar: "Elementary Matrices," reprinted, Cambridge University Press, London, 1950.

Gupta, K. K.: "Eigenproblem Solution of Damped Structural Systems," *Intern. J. Numerical Methods Eng.*, **8**:877–911, 1974.

Hartree, D. R.: "Numerical Analysis," 2d ed., Oxford University Press, London, 1958.

Hildebrand, F. B.: "Introduction to Numerical Analysis," McGraw-Hill Book Company, Inc., New York, 1956.

Hildebrand, F. B.: "Methods of Applied Mathematics," Prentice-Hall, Inc., Englewood Cliffs, N.J., 1965.

Householder, A. S.: "Principles of Numerical Analysis," McGraw-Hill Book Company, Inc., New York, 1953.

Jacobsen, L. S., and R. S. Ayre: "Engineering Vibrations, with Applications to Structures and Machinery," McGraw-Hill Book Company, Inc., New York, 1958.

Lanczos, C.: "Applied Analysis," Prentice-Hall, Inc., Englewood Cliffs, N.J., 1956.
Milne, W. E.: "Numerical Calculus," Princeton University Press, Princeton, N.J., 1949.
Milne, W. E.: "Numerical Solution of Differential Equations," John Wiley & Sons, Inc., New York, 1953.
Paige, L. J., and Olga Taussky (eds.): "Simultaneous Linear Equations and the Determination of Eigenvalues," National Bureau of Standards, Applied Mathematics Series, No. 29. U.S. Government Printing Office, Washington, D.C., 1953.
Pipes, L. A.: "Applied Mathematics for Engineers and Physicists," 2d ed., McGraw-Hill Book Company, Inc., New York, 1958.
Plunkett, R. (ed.): "Mechanical Impedance Methods for Mechanical Vibrations," ASME, New York, 1958.
Przemieniecki, J. S.: "Theory of Matrix Structural Analysis," McGraw-Hill Book Company, Inc., New York, 1968.
Scarborough, J. B.: "Numerical Mathematical Analysis," 4th ed., Johns Hopkins Press, Baltimore, 1958.
Strutt, J. W. (Lord Rayleigh): "The Theory of Sound," vols. I and II, reprinted, Dover Publications, New York, 1945.
Taussky, O. (ed.): "Contributions to the Solution of Systems of Linear Equations and the Determination of Eigenvalues," National Bureau of Standards, Applied Mathematics Series, No. 39, U.S. Government Printing Office, Washington, D.C., 1954.
Timoshenko, S., and D. H. Young: "Vibration Problems in Engineering," 3d ed., D. Van Nostrand Company, Inc., Princeton, N.J., 1955.
von Kármán, T., and M. A. Biot: "Mathematical Methods in Engineering," McGraw-Hill Book Company, Inc., New York, 1940.
Webster, A. G.: "The Dynamics of Particles and of Rigid, Elastic, and Fluid Bodies," reprinted, Stechert-Hafner, Inc., New York, 1942.
Whittaker, E. T.: "Analytical Dynamics," 4th ed., reprinted, Dover Publications, New York, 1944.
Wilkinson, J. H.: "The Algebraic Eigenvalue Problem," Oxford University Press, New York, 1965.
Zurmühl, R.: "Matrizen und Ihre Technischen Anwendungen," 4th ed., Springer-Verlag OHG, Berlin, 1964.

29

PART I: VIBRATION OF STRUCTURES INDUCED BY GROUND MOTION

N. M. Newmark and W. J. Hall
University of Illinois, Urbana

INTRODUCTION

RESPONSE OF SIMPLE STRUCTURES TO GROUND MOTIONS

Four structures of varying size and complexity are shown in Fig. 29.1: (*a*) a simple, relatively compact machine anchored to a foundation, (*b*) a 15-story building, (*c*) a 40-story building, and (*d*) an elevated water ank.

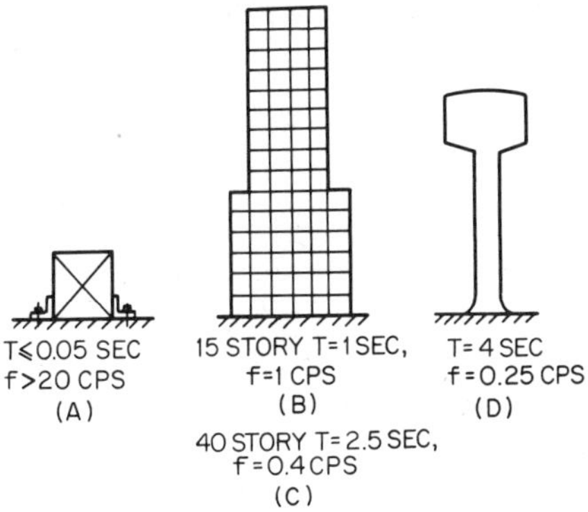

The dynamic response of each of the structures shown in Fig. 29.1 can be approximated by representing each as a simple mechanical oscillator consisting of a single mass supported by a spring and a dashpot (Fig. 29.2). The relation between the circular frequency of vibration $\omega = 2\pi f$, the natural period f, and the period T is given by the following equation in terms of the spring constant k and the mass m:

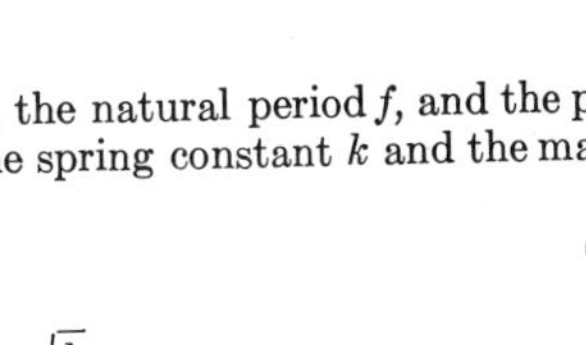

T≤0.05 SEC
f>20 CPS
(A)
15 STORY T=1 SEC,
f=1 CPS
(B)
40 STORY T=2.5 SEC,
f=0.4 CPS
(C)
T=4 SEC
f=0.25 CPS
(D)

FIG. 29.1. Structures subjected to earthquake ground motions.

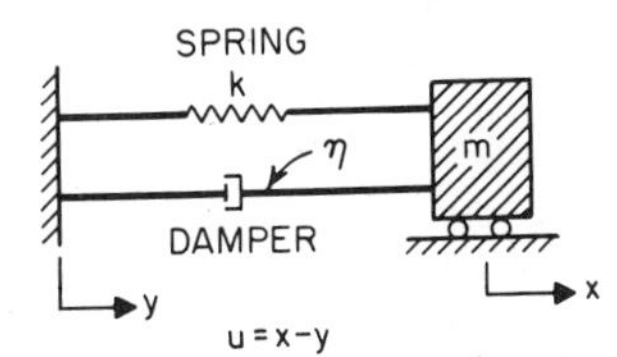

FIG. 29.2. System considered.

$$\omega^2 = \frac{k}{m} \tag{29.1}$$

$$f = \frac{1}{T} = \frac{\omega}{2\pi} = \frac{1}{2\pi}\sqrt{\frac{k}{m}} \tag{29.2}$$

In general, the effect of the dashpot is to produce damping of free vibrations or to reduce the amplitude of forced vibrations. The damping force is assumed to be equal to a damping coefficient c times the velocity $\dot{u}$ of the mass relative to the ground. The value of c at which the motion loses its vibratory character in free vibration is called the critical damping coefficient, for example, $c_c = 2m\omega$. The amount of damping is most conveniently considered in terms of the proportion of ζ of critical damping,

$$\zeta = \frac{c}{c_c} = \frac{c}{2m\omega} \tag{29.3}$$

For most practical structures ζ is relatively small, in the range of 0.5 to 10 or 20 per cent, and does not appreciably affect the natural period or frequency of vibration.

GROUND MOTIONS

Strong-motion earthquake acceleration records with respect to time have been obtained for a number of earthquakes. Ground motions from other sources of disturbance, such as quarry blasting and nuclear blasting, are also available and show many of the same characteristics. As an example of the application of such time-history records, the recorded accelerogram for the El Centro, California, earthquake of May 18, 1940, in the north-south component of horizontal motion, is shown in Fig. 29.3. On the same

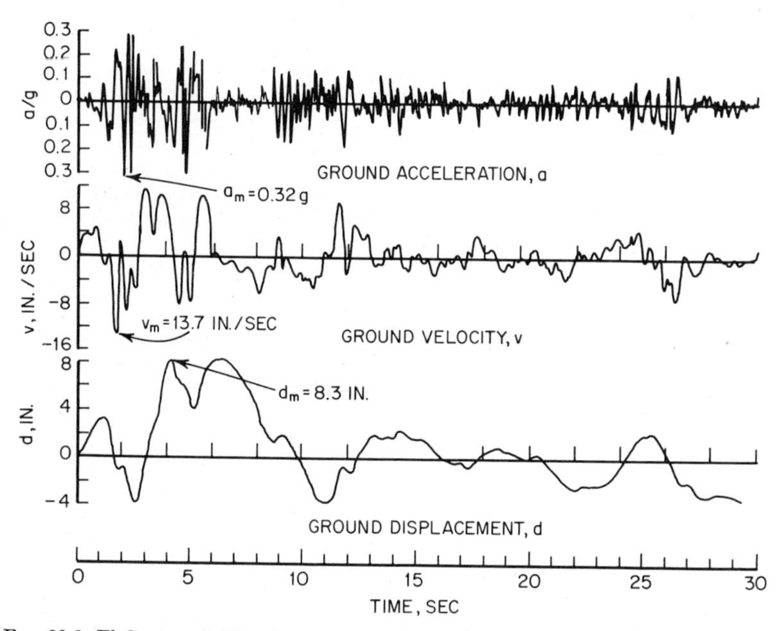

FIG. 29.3. El Centro, California, earthquake of May 18, 1940, north-south component.

figure are shown integration of the ground acceleration a to give the variation of ground velocity v with time, and the integration of velocity to give the variation of ground displacement d with time. These integrations normally require base-line corrections of various sorts, and the magnitude of the maximum displacement may vary depending on how the corrections are made. The maximum velocity is relatively insensitive to the corrections, however. For this earthquake, with the integrations shown in Fig. 29.3, the maximum ground acceleration is $0.32g$, the maximum ground velocity 13.7 in./sec, and the maximum ground displacement 8.3 in. These three maximum values are of particular interest because they help to define the response motions of the various structures considered in Fig. 29.1 most accurately if all three maxima are taken into account.

RESPONSE SPECTRA

ELASTIC SYSTEMS

The response of the simple oscillator shown in Fig. 29.2 to any type of ground motion can be readily computed as a function of time. The maximum values of the response are of particular interest. These maxima can be stated in terms of the maximum strain in the spring $u_m = D$, the maximum spring force, the maximum acceleration of the mass (which is related to the maximum spring force directly when there is no damping), or a quantity, having the dimensions of velocity, which gives a measure of the maximum energy absorbed in the spring. This quantity, designated the pseudovelocity V, is defined in such a way that the energy absorption in the spring is $\frac{1}{2}mV^2$. The relations among the maximum relative displacement of the spring D, the pseudovelocity V, and the pseudoacceleration A, which is a measure of the force in the spring, are as follows:

$$V = \omega D \tag{29.4}$$

$$A = \omega V = \omega^2 D \tag{29.5}$$

The pseudovelocity V is nearly equal to the maximum relative velocity for systems with moderate or high frequencies but may differ considerably from the maximum relative velocity for very low frequency systems. The pseudoacceleration A is exactly equal to the maximum acceleration for systems with no damping and is not greatly different from the maximum acceleration for systems with moderate amounts of damping, over the whole range of frequencies from very low to very high values.

Typical plots of the response of the system as a function of period or frequency are called response spectra. Plots for acceleration and for relative displacement, for a system with a moderate amount of damping, subjected to an input similar to that of Fig. 29.3 can be made. This arithmetic plot of maximum response is simple and convenient to use.

A somewhat more useful plot, which indicates at one and the same time the values

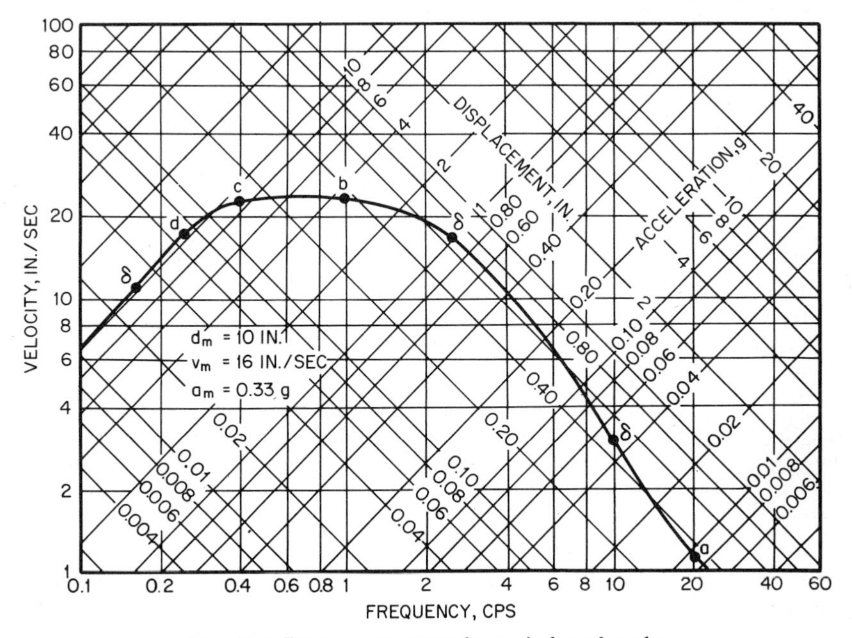

FIG. 29.4. Response spectrum for typical earthquake.

for D, V, and A, is shown in Fig. 29.4. This has the virtue that it also indicates more clearly the extreme or limiting values of the various parameters defining the response. The frequency is plotted on a logarithmic scale. Since the frequency is the reciprocal of the period, the logarithmic scale for period would have exactly the same spacing of the points, or in effect the scale for period would be turned end for end. The pseudovelocity is plotted on a vertical scale, also logarithmically. Then on diagonal scales along an axis that extends upward from right to left are plotted values of the displacement, and along an axis that extends upward from left to right the pseudoacceleration, in such a way that any one point defines for a given frequency the displacement D, the pseudovelocity V, and the pseudoacceleration A. Points are indicated in Fig. 29.4 for the several structures of Fig. 29.1, plotted at their fundamental frequencies.

A wide variety of motions have been considered in Refs. 9 and 10, ranging from simple pulses of displacement, velocity, or acceleration of the ground, through more complex motions such as those arising from nuclear-blast detonations, and for a variety of earthquakes as taken from available strong-motion records. Response spectra for the El Centro earthquake are shown in Fig. 29.5. The spectrum for small amounts of

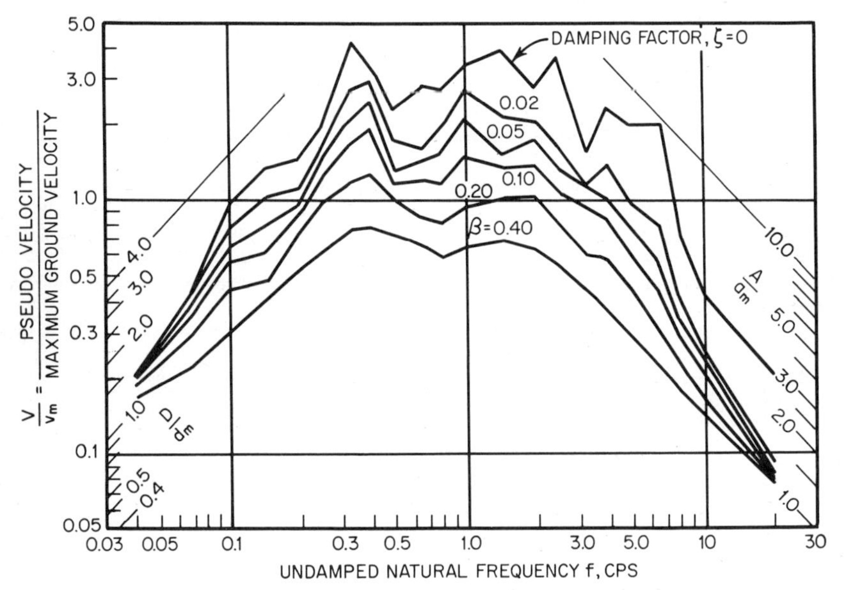

FIG. 29.5. Deformation spectra for elastic systems subjected to the El Centro earthquake.

damping is much more jagged than indicated by Fig. 29.4, but for the higher amounts of damping the response curves are relatively smooth. The scales are chosen in this instance to represent the amplifications of the response relative to the ground-motion values of displacement, velocity, or acceleration.

The spectra shown in Fig. 29.5 are typical of response spectra for nearly all types of ground motion. It is noted that on the extreme left, corresponding to very low frequency systems, the response for all degrees of damping approaches an asymptote corresponding to the value of the maximum ground displacement. A low-frequency system corresponds to one having a very heavy mass and a very light spring. When the ground moves relatively rapidly, the mass does not have time to move, and therefore the maximum strain in the spring is precisely equal to the maximum displacement of the ground. On the other hand, for a very high frequency system, the spring is relatively stiff and the mass very light. Therefore, when the ground moves, the stiff spring forces the mass to move in the same way the ground moves, and the mass therefore must have the same acceleration as the ground at every instant. Hence, the force in the spring is that required to move the mass with the same acceleration as the ground, and the maximum acceleration of the mass is precisely equal to the maximum acceleration of the ground. This is shown by the fact that all the lines on the extreme right-hand side of the figure approach as an asymptote the maximum ground-acceleration line.

For intermediate-frequency systems, there is an amplification of motion. In general, the amplification factor for displacement is less than that for velocity, which in turn is less than that for acceleration. Amplification factors for the undamped system ($\zeta = 0$) in Fig. 29.5 are of the order of about 3.5 for displacement, 4.2 for velocity, and 9.5 for acceleration.

The results of similar calculations for other ground motions are quite consistent with those in Fig. 29.5, even for simple motions. The general nature of the response spectrum (Fig. 29.6) consists of a central region of amplified response and two limiting

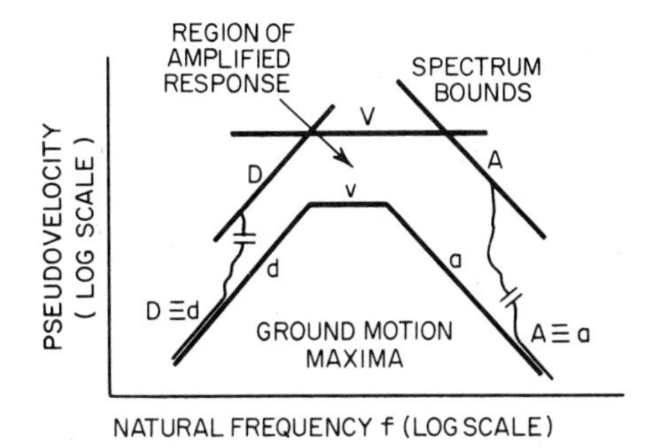

FIG. 29.6. Typical tripartite logarithmic plot of response spectrum bounds compared with maximum ground motions.

regions of response in which, for low-frequency systems, the response displacement is equal to the maximum ground displacement, and for high-frequency systems, the response acceleration is equal to the maximum ground acceleration. For damping of the order of about 5 to 10 per cent critical, the amplification factors for displacement, velocity, and acceleration are only slightly over 1, 1.5, and 2.0, respectively, for a wide variety of earthquake and ground shock motions. These amplification factors increase quite rapidly, however, as the damping decreases, and are approximately doubled for about 2 per cent damping. On the other hand, they decrease relatively slowly as the damping factor increases above the values of 5 to 10 per cent.

DESIGN RESPONSE SPECTRA

A response spectrum developed to give *design coefficients* is called a "design spectrum." As an example of its use in seismic design, for any given site, estimates are made of the maximum ground acceleration, maximum ground velocity, and maximum gound displacement. The lines representing these values can be drawn on the tripartite logarithmic chart of which Fig. 29.7 is an example. The lines showing the ground motion maxima in Fig. 29.7 are drawn for a maximum ground acceleration a of $1.0g$ velocity v of 48 in./sec, and displacement d of 36 in. These data represent motions more intense than those generally considered for any postulated design earthquake hazard. They are, however, approximately in correct proportion for a number of areas of the world, where earthquakes occur either on firm ground, soft rock, or competent sediments of various kinds. For relatively soft sediments, the velocities and displacements might require increases above the values corresponding to the given acceleration as scaled from Fig. 29.7. It is not likely that maximum ground velocities in excess of 4 to 5 ft/sec are obtainable under any circumstances.

As part of studies made by the authors, values have been determined for the hori-

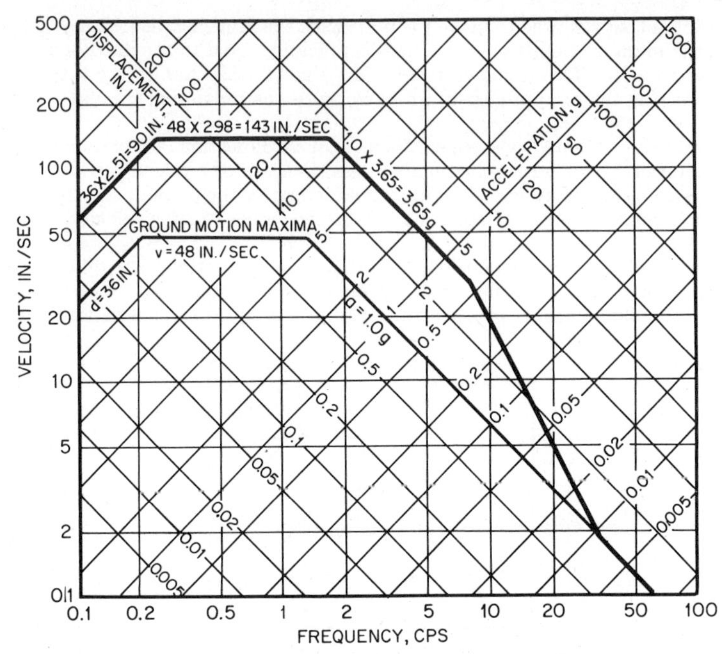

FIG. 29.7. Basic design spectrum normalized to 1.0g for 2 per cent damping, 84.1 percentile level.

zontal and vertical directions of excitation for various degrees of damping. Representative amplification levels for 50 and 84.1 percentile levels of horizontal response are presented in Table 29.1. A value of 84.1 percentile means that one could expect 84.1

Table 29.1. Values of Spectrum Amplification Factors[2]

Percentile	Damping, per cent	Amplification		
		D	*V*	*A*
50	0.5	1.97	2.58	3.67
	2.0	1.68	2.06	2.76
	5.0	1.40	1.66	2.11
	10.0	1.15	1.34	1.65
84.1	0.5	2.99	3.81	5.12
	2.0	2.51	2.98	3.65
	5.0	2.04	2.32	2.67
	10.0	1.62	1.81	2.01

per cent of the values to fall at or below that particular amplification. With these amplification values, and noting points *B* and *A* to fall at about 8 and 33 cps, the spectra may be constructed as shown in Fig. 29.4. Further information on construction of elastic response spectra may be found in Refs. 1 through 8.

RESPONSE SPECTRA FOR INELASTIC SYSTEMS

It is convenient to consider an elastoplastic resistance-displacement relation because one can draw response spectra for such a relation in generally the same way as the spectra were drawn for elastic conditions. A simple resistance-displacement relationship for a spring is shown by the light line in Fig. 29.8, where the yield point is indicated,

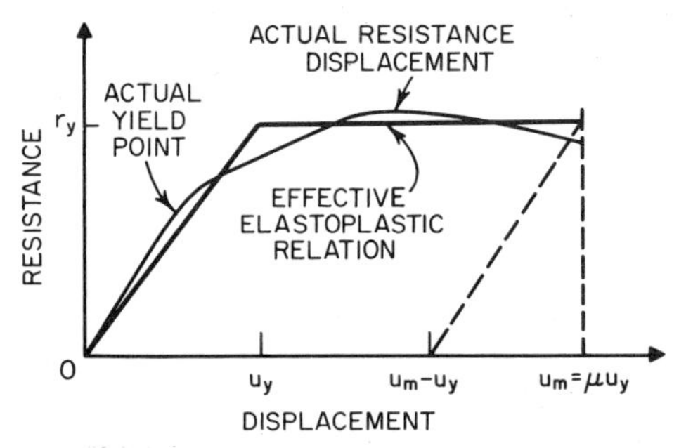

FIG. 29.8. Resistance-displacement relationship.

with a curved relationship showing a rise to a maximum resistance and then a decay to a point of maximum useful limit or failure at a displacement u_m. An equivalent elastoplastic resistance curve is shown by the heavy line in the figure, rising on a straight line to a point where the yield displacement is u_y and the resistance r_y, and then extending without appreciable increase in resistance to the maximum displacement u_m. The effective resistance curve is drawn so as to have the same area between the origin and u_y as the actual curve, and again the same area to the maximum displacement point. The ductility factor μ is defined as the ratio between the maximum permissible or useful displacement to the yield displacement for the effective curve.

The ductility factors for various types of construction are difficult to characterize briefly. They depend on the use of the building, the hazard involved in its failure, the material used, the framing or layout of the structure, and above all on the method of construction and the details of fabrication of joints and connections. A discussion of these topics is given in Ref. 1. In Fig. 29.9, there are shown acceleration spectra for elastoplastic systems having 2 per cent of critical damping for the El Centro 1940 earthquake. Here the symbol D_y represents the elastic component of the response displacement, but it is not the total displacement. Hence, the curves also give the elastic component of maximum displacement as well as the maximum acceleration A, but they do not give proper value of maximum pseudovelocity. This is designated by the use of the symbol V' for the pseudovelocity drawn in the figure. The figure is drawn for ductility factors ranging from 1 to 10.

One can also draw a response spectrum for total displacement. This is drawn for the same conditions as Fig. 29.9 and is obtained from Fig. 29.9 by multiplying each curve's ordinates by the value of ductility factor μ shown on that curve. It will be seen from such a figure that the maximum total displacement is virtually the same for all ductility factors, actually perhaps decreasing even slightly for the larger ductility factors in the low-frequency region, for frequencies below about 2 cps. Moreover, it appears from Fig. 29.9 that the maximum acceleration is very nearly the same for frequencies greater than about 20 or 30 cps for all ductility factors. In between, there is a transition. These remarks are applicable to the spectra for other earthquakes also. One can generalize about them in the following way for general nonlinear relations between resistance and displacement for single degree-of-freedom structures.

For low and intermediate frequencies, corresponding to something on the order of about 2 cps as an upper limit, total relative displacements are preserved and are very nearly the same for all ductility factors. As a matter of fact, inelastic systems have

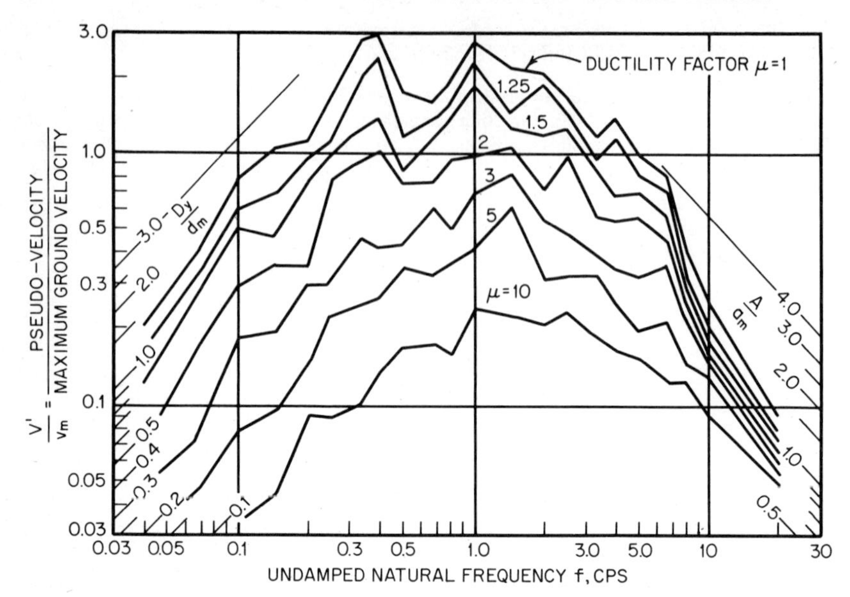

FIG. 29.9. Deformation spectra for elastoplastic systems with 2 per cent critical damping subjected to the El Centro earthquake.

perhaps even a smaller displacement than elastic systems for frequencies below about 0.3 cps. For frequencies between about 2 up to about 8 cps, the best relationship appears to be to equate the energy in the various curves, or to say that energy is preserved, with a corresponding relationship between deflections and accelerations or forces. There is a transition region between 8 and 30 to 33 cps, depending on the damping ratio. Above 33 cps, the force or acceleration is nearly the same for all ductility ratios.

To use the design spectrum to approximate inelastic behavior, the following suggestions are made. In the amplified displacement region of the spectra, the left-hand side, and in the amplified velocity region, at the top, the spectrum remains unchanged for total displacement and is divided by the ductility factor to obtain yield displacement or acceleration. The upper right-hand portion sloping down at 45°, or the amplified acceleration region of the spectrum, is relocated for an elastoplastic resistance curve, or for any other resistance curve for actual structural materials, by choosing it at a level which corresponds to the same energy absorption for the elastoplastic curve as for an elastic curve for the same period of vibration. The extreme right-hand portion of the spectrum, where the response is governed by the maximum ground acceleration, remains at the same acceleration level as for the elastic case, and therefore at a corresponding increased total displacement level. The frequencies at the corners are kept at the same values as in the elastic spectrum. The acceleration transition region of the response spectrum is now drawn also as a straight-line transition from the newly located amplified acceleration line and the ground acceleration line, using the same frequency points of intersection as in the elastic response spectrum.

In all cases the "inelastic maximum acceleration" spectrum and the "inelastic maximum displacement" spectrum differ by the factor μ at the same frequencies. The design spectrum so obtained is shown in Fig. 29.10.

The solid line $DVAA_0$ shows the elastic response spectrum. The heavy circles at the intersections of the various branches show the frequencies which remain constant in the construction of the inelastic design spectrum.

The dashed line $D'V'A'A_0$ shows the inelastic acceleration, and the line $DVA''A''_0$ shows the inelastic displacement. These two differ by a constant factor $\mu = 5$ for the

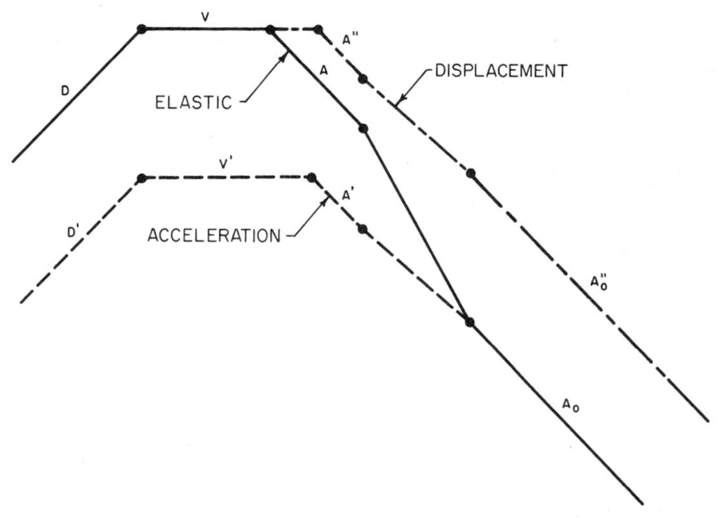

FIG. 29.10. Design spectra.

construction shown, but A and A' differ by the factor $\sqrt{2\mu - 1} = 3$, since this is the factor that corresponds to constant energy for an elastoplastic resistance.

Of course, the elastoplastic or other inelastic response spectra can be used only as an approximation for multi-degree-of-freedom systems. Additional information on development of inelastic design response spectra may be found for example in Refs. 1, 3, 4, 7, and 9–17.

MULTI-DEGREE-OF-FREEDOM SYSTEMS

USE OF RESPONSE SPECTRA

A multi-degree-of-freedom system has a number of different modes of vibration. For example, for the shear beam of Fig. 29.11*a* there are shown the fundamental mode of lateral oscillation (*b*), the second mode (*c*), and the third mode (*d*). The number of modes equals the number of degrees of freedom, five in this case. In a system that has independent (uncoupled) modes, which condition is usually satisfied for buildings, each mode responds to the base motion as an independent single-degree-of-freedom system. Thus, the modal responses are nearly independent functions of time. However, the maxima do not necessarily occur at the same time.

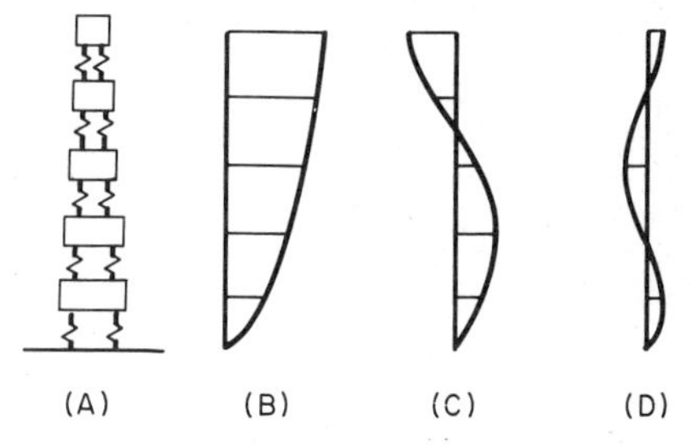

FIG. 29.11. Modes of vibration of shear beam.

For multi-degree-of-freedom systems, the concept of the response spectrum can also be used in most cases, although the use of the inelastic response spectrum is only approximately valid as a design procedure.[1, 16–18] For a system with a number of masses at nodes in a flexible framework, the equation of motion can be written in matrix form as follows:

$$M\ddot{u} + C\dot{u} + Ku = -M(\ddot{y})\{1\} \qquad (29.6)$$

in which the last symbol on the right represents a unit column vector. The mass matrix M is usually diagonal, but in all cases both M and the stiffness matrix K are symmetrical. When the damping matrix C satisfies certain conditions, the simplest of which is when it is a linear combination of M and K, then the system has normal modes of vibration, with modal displacement vectors u_n.

FUNDAMENTAL MODE*

Procedures are available for the computation of the periods of vibration of undamped† multi-degree-of-freedom systems (see Chap. 2) and applications to actual buildings are described in several papers.[19, 20]

Consider a system with a number of masses "lumped" at particular points, letting m_n represent the nth mass of the system, and assume that the system is vibrating in the jth mode. If the system is vibrating in a steady-state condition, without damping, the displacement u_{nj} of the nth mass can be written in the form

$$u_{nj} \sin \omega_j t \tag{29.7}$$

The acceleration experienced by the mass during its oscillatory motion is given by the second derivative with respect to time:

$$-\omega_j^2 u_{nj} \sin \omega_j t \tag{29.8}$$

The negative value of this acceleration, multiplied by the mass m_n, is considered to be a reversed effective force or inertial force applied at the point n. The inertial forces $\bar{Q}_{nj} \sin \omega_j t$ are considered to be applied to the structure at each mass point, where the coefficient of the sine term in the inertial-force expression has the form

$$\bar{Q}_{nj} = m_n \omega_j^2 u_{nj} \tag{29.9}$$

Since the inertial forces take account of the mass effects, the displacements of the structure due to the forces $\bar{Q}_{nj}$ must be precisely equal to the quantities u_{nj}. Consequently, in order to find the square of the circular frequency for the jth mode ω_j^2, it is necessary merely to find a set of displacements u_{nj} of such magnitudes that forces corresponding to each displacement multiplied by the local mass m_n, and by the square of the circular frequency for the jth mode ω_j^2, give rise to the displacements u_{nj}. Any procedure that will establish this condition will give both the modal frequencies and the modal deflection shapes. Multiplying the magnitudes of the modal deflections by a constant does not change the situation, since all the forces, and consequently all the deflections consistent with those forces, will be multiplied by the same constant.

However, it is not possible without other knowledge of the situation to write down directly a correct set of displacements for the jth mode. Therefore, the calculations must make it possible to arrive at these deflections as a result of a systematic method of computation. The most useful procedures, at least for the determination of the fundamental mode, are Rayleigh's method, or modifications thereof, or methods based on a procedure of successive approximations developed originally by Stodola. A description of the successive-approximations procedure follows:

1. Assume a set of deflections at each mass point of magnitude u_{na}. Compute for these deflections an inertial force $\bar{Q}_{na}$ given by

$$\bar{Q}_{na} = m_n u_{na} \omega^2 \tag{29.10}$$

where the quantity ω^2 is an unknown circular frequency which may be carried in the calculations as an unknown.

2. Apply these forces to the system and compute the corresponding deflections, designated by the symbol u_{nb},

$$u_{nb} = \bar{u}_{nb} \omega^2 \tag{29.11}$$

* The following article is based on material from Ref. 18, by permission of the Portland Cement Association.

† Damping of less than 20 per cent critical affects the computed periods by less than 2 per cent. Hence, nominal damping does not affect the period appreciably.

3. The problem is to make u_{nb} and u_{na} as nearly equal as possible. To do this, ω may be varied. The value of ω that gives the best fit is a good approximation to the circular frequency for the mode that corresponds to the deflection u_{nb}, which in general will be an approximation to the fundamental mode. In general, u_{nb} will be a better approximation to the fundamental mode shape than was u_{na}.

4. A repetition of the calculations using u_{nb} as the starting point will lead to a new derived deflection that will be an even better approximation.

In most cases, even with a very poor first assumption for the fundamental mode deflection, the process will converge with negligible errors to ω^2 in at most two or three cycles, and one can obtain a good approximation in only one cycle. However, the mode shape will not be so accurately determined unless the calculation is repeated several times.

If the quantity shown in Eq. (29.12) is made a minimum (in effect minimizing the square of the error between the derived deflection and the assumed deflection) the "best" value of ω^2 consistent with the assumed deflection curve can be determined:

$$\sum_n m_n(u_{nb} - u_{na})^2 = \text{minimum} \tag{29.12}$$

Substituting Eq. (29.11) into Eq. (29.12) and equating to zero the derivative with respect to ω^2 gives

$$\omega^2 = \frac{\Sigma m_u u_{na}\bar{u}_{nb}}{\Sigma m_n \bar{u}_{nb}{}^2} \tag{29.13}$$

The value of ω^2 given by Eq. (29.13) exceeds, generally only slightly, the true value for the fundamental mode.

Rayleigh's method is probably the most widely used engineering procedure for computing the period of the fundamental mode. Without modification, however, it does not generally give accurate values of the mode shape. Rayleigh's method for calculating the fundamental frequency of a building frame can be related to the procedure described above by equating u_{na} to unity throughout the structure. In the case of a vertical or horizontal beamlike structure, the derived displacements $\bar{u}_{nb}$ will be proportional to the deflections of the structure due to forces equal to the weight of the structure. Since Rayleigh's procedure using $u_{na} = 1$ gives a quite accurate determination of the fundamental frequency, it is obvious that Eq. (29.13) will yield a highly accurate value if any more reasonable deflection shape is assumed for the first mode.

The effect of foundation rotation, column shortening, or other contributions to deflection can be readily taken into account in both the successive-approximations procedure and the Rayleigh method.

HIGHER MODES

Several methods are available for computing the frequencies of modes higher than the fundamental mode for a multi-degree-of-freedom system. Two such procedures are described in Ref. 18. However, the high-speed digital computer can determine all the modes and frequencies of even highly complex systems in only a few minutes by use of standard programs, and detailed hand-calculation methods are no longer of interest except for approximating the fundamental frequency or period.

COMBINATIONS OF MODAL RESPONSES

When the modes and frequencies of the system are obtained, the modal responses are determined for each mode considering the "participation" factors, defined as follows for earthquake and general ground motion:

$$c_n = \frac{u_n{}^T M\{1\}}{u_n{}^T M u_n} \tag{29.14}$$

If the particular quantity desired—say the stress at a particular point, the relative displacement between two reference points, or any other effect—is designated by ω, then the modal values ω_n are determined for each mode and combined by use of the relations:

$$\alpha_{\max} \leq \sum_n |c_n \alpha_n D_n| \tag{29.15}$$

$$\alpha_{\text{prob}} \approx \sqrt{\sum_n (c_n \alpha_n D_n)^2} \tag{29.16}$$

$$V_n = \omega_n D_n \qquad D_n = \frac{V_n}{\omega_n} \tag{29.17}$$

$$A_n = \omega_n^2 D_n = \omega_n V_n \qquad D_n = \frac{A_n}{\omega_n^2} \tag{29.18}$$

For inelastic response, the quantities to be used are D_n', V_n', or A_n' from calculations such as those leading to Fig. 29.10. Equation (29.15) gives an upper bound to the value of α; and Eq. (29.16), the most probable value.

DESIGN

GENERAL CONSIDERATIONS

In the design of a building to resist earthquake motions, the designer works within certain constraints, such as the architectural configuration of the building, the foundation conditions, the nature and extent of the hazard should failure or collapse occur, the possibility of an earthquake, the possible intensity of earthquakes in the region, the cost or available capital for construction, and similar factors. He must have some basis for the selection of the strength and the proportions of the building and of the various members in it. The required strength depends on factors such as the intensity of earthquake motions to be expected, the flexibility of the structure, and its ductility or reserve strength before damage occurs. Because of the interrelations among flexibility and strength of a structure, and the forces generated in it by earthquake motions, the dynamic design procedure must take these various factors into account. The ideal to be achieved is one involving flexibility and energy-absorbing capacity which will permit the earthquake displacements to take place without generating unduly large forces. To achieve this end, control of the construction procedures and appropriate inspection practices are necessary. The attainment of the ductility required to resist earthquake motions must be emphasized.

EFFECTS OF DESIGN ON BEHAVIOR*

A structure designed for very much larger horizontal forces than are ordinarily prescribed will have a shorter period of vibration because of its greater stiffness. The shorter period results in higher spectral accelerations, so that the stiffer structure may attract more horizontal force. Thus, a structure designed for too large a force will not necessarily be safer than a similar structure based on smaller forces. On the other hand, a design based on too small a force makes the structure more flexible and will increase the relative deflections of the floors.

In general, yielding occurs first in the story that is weakest compared with the magnitudes of the shearing forces to be transmitted. In many cases this will be near the base of the structure. If the system is essentially elastoplastic, the forces transmitted through the yielded story cannot exceed the yield shear for that story. Thus, the shears, accelerations, and relative deflections of the portion of the structure above the yielded floor are reduced compared with those for an elastic structure subjected to the same base motion. Consequently, if a structure is designed for a base shear which is less than the maximum value computed for an elastic system, the lowest story will

* This section is based, by permission, on material from Ref. 18.

yield and the shears in the upper stories will be reduced. This means that, with proper provision for energy absorption in the lower stories, a structure will, in general, have adequate strength provided the design shearing forces for the upper stories are consistent with the design base shear. The Uniform Building Code recommendations are intended to provide such a consistent set of shears.

A significant inelastic deformation in a structure inhibits the higher modes of oscillation. Therefore, the major deformation is in the mode in which the inelastic deformation predominates, which is usually the fundamental mode. The period of vibration is effectively increased, and in many respects the structure responds almost as a single-degree-of-freedom system corresponding to its entire mass supported by the story which becomes inelastic. Therefore the base shear can be computed for the modified structure, with its fundamental period defining the modified spectrum on which the design should be based. The fundamental period of the modified structure *generally* will not be materially different from that of the original elastic structure in the case of framed structures. In the case of shear-wall structures it will be longer.

It is partly because of these facts that it is usually appropriate in design recommendations to use the frequency of the fundamental mode, without taking direct account of the higher modes. However, it is desirable to consider a shearing-force distribution which accounts for higher-mode excitations of the portion above the plastic region. This is implied in the UBC and SEAOC recommendations by the provision for lateral-force coefficients which vary with height. The distribution over the height corresponding to an acceleration varying uniformly from zero at the base to a maximum at the top takes into account the fact that local accelerations at higher levels in the structure are greater than those at lower levels, because of the larger motions at the higher elevations, and accounts quite well for the moments and shears in the structure.

DESIGN LATERAL FORCES

Although the complete response of multi-degree-of-freedom systems subjected to earthquake motions can be calculated, it should not be inferred that it is generally necessary to make such calculations as a routine matter in the design of multistory buildings. There are a great many uncertainties about the input motions and about the structural characteristics that can affect the computations. Moreover, it is not generally necessary or desirable to design tall structures to remain completely elastic under severe earthquake motions, and considerations of inelastic behavior lead to further discrepancies between the results of routine methods of calculation and the actual response of structures.

The Uniform Building Code recommendations for earthquake lateral forces are, in general, consistent with the forces and displacements determined by more elaborate procedures. A structure designed according to these recommendations will remain elastic, or nearly so, under moderate earthquakes of frequent occurrence, but must be able to yield locally without serious consequences if it is to resist an El Centro–type earthquake. Thus, design for the required ductility is an important consideration.

The ductility of the material itself is not a direct indication of the ductility of the structure. Laboratory and field tests and data from operational use of nuclear weapons indicate that structures of practical configurations having frames of ductile materials, or a combination of ductile materials, exhibit ductility factors μ ranging from a minimum of 3 to a maximum of 8. A minimum ductility factor of about 4 to 6 is a reasonable criterion for ordinary structures designed to UBC earthquake requirements.

As a result of the numerous earthquakes that have occurred throughout the world in the past two decades and of the resulting loss of life and property, seismic design codes are undergoing major revisions. These revisions are based in part on the results of research and design study, and in part on observational experience.

For example, the 1973 Uniform Building Code[21] recommends that the minimum total lateral seismic force V_s, assumed to act nonconcurrently in the direction of each of the main axes of the building, be determined by

$$V_s = ZKCW \tag{29.19}$$

whereas the 1974 SEAOC code[22] and proposed change to the UBC call for the base shear to be determined by

$$V_s = ZKCSIW \tag{29.20}$$

where Z = numerical coefficient related to expected severity of earthquakes in various regions of United States
K = numerical coefficient depending on type or arrangement of resisting elements
C = numerical coefficient dependent on period of structure
W = total dead load plus applicable portions of other loads as specified
S = numerical coefficient reflecting site-structure resonance conditions (intended to reflect foundation conditions)
I = occupancy importance factor

In general the seismic coefficients have been increased in comparison to earlier values and it should be obvious that the approaches being adopted attempt to take more factors into consideration in arriving at the design base shear. The newer values are sometimes 1.5 to 2 times the values used previously.

Other analytical approaches are under study for possible use in codes, including the response spectrum method and the time-history method. These methods have been used extensively in the design of special structures such as nuclear reactors, pipelines, and military facilities in recent years and it is expected they will be adopted for use in building systems in the near future, at least in special cases. It should be noted that the appropriate seismic coefficient can be determined from a response spectrum for use in design rather than from Eqs. (29.19) and (29.20) with the code design procedures followed thereafter.

Other references or parameters and methods of importance in seismic design are to be found in Refs. 23 to 29.

SEISMIC FORCES FOR OVERTURNING MOMENT AND SHEAR DISTRIBUTION

In general when modal analysis techniques are not used, in a complex structure or in one having several degrees of freedom, it is necessary to have a method of defining the seismic design forces at each mass point of the structure in order to be able to compute the shears and moments to be used for design throughout the structure. The method described in the SEAOC code[22] is preferable for this purpose. It is essentially the following:

1. Compute the total base shear corresponding to the seismic coefficient for the structure multiplied by the total weight.
2. Assign a proportion of the total base shear not exceeding 15 per cent to the top of the structure (see SEAOC code).
3. Assume a linear variation of acceleration in the structure from zero at the base to a maximum at the top.
4. Multiply the acceleration assumed in 1 by the mass at each elevation to find an inertial force acting at each level.
5. Adjust the assumed value of acceleration at the top of the structure in 3 so that the total distributed lateral forces add up to the total base shear computed in step 1.
6. Use the resulting seismic forces, assigned to the various masses at each elevation, to compute shears and moments throughout the structure.
7. The overturning moment at each elevation and at the base, so computed, may give rise to tensions and compressions in the columns and walls of the structure. Provision must be made for accommodating these forces.
8. The overturning moment at the base should be considered as causing a tilting of the base consistent with the foundation compliance and also may cause a partial uplift at one edge of the base. The increased foundation compression due to such tilting should be considered in the foundation design.

DAMPING

The damping in structural elements and components and in supports and foundations of the structure is a function of the intensity of motion and of the stress or strain levels introduced within the structural component or structure as well as being highly dependent on the makeup of the structure and the energy absorption mechanisms within it. For further detail see Refs. 1, 6, 16, 24, and 25.

GRAVITY LOADS

The effect of gravity loads, when the structures deform laterally by a considerable amount, can be of importance. In accordance with the general recommendations of most extant codes, the effects of gravity loads are to be added directly to the primary and earthquake effects. In general in computing the effect of gravity loads, one must take into account the actual deflection of the structure, not that corresponding to reduced seismic coefficients.

VERTICAL AND HORIZONTAL EXCITATION

Usually the stresses or strains at a particular point are affected primarily by the earthquake motions in only one direction; the second direction produces little if any influence. However, this is not always the case; and certainly not so for a simple square building supported on four columns where the stress in a corner column is in general affected equally by the earthquakes in the two horizontal directions and may be affected also by the vertical earthquake forces. Since the ground moves in all three directions in an earthquake, and even tilts and rotates, consideration of the combined effects of all these motions must be included in the design. When the response in the various directions may be considered to be uncoupled, then consideration can be given separately to the various components of base motion, and individual response spectra can be determined for each component of direction or of transient base displacement. Calculations have been made for the elastic response spectra in all directions for a number of earthquakes. Recent studies made by the authors (not yet reported) indicate that the vertical response spectrum is not more than about two-thirds the horizontal response spectrum for all frequencies, and it is recommended that a ratio of 2:3 for vertical response compared with horizontal response be used in design.

For parts of structures or components that are affected by motions in various directions in general, the response may be computed by either one of two methods. The first method involves computing the response for each of the directions independently and then taking the square root of the sums of the squares of the resulting stresses in the particular direction at a particular point as a combined response. Alternatively, one can use the procedure of taking the seismic forces corresponding to 100 per cent of the motion in one direction combined with 40 per cent of the motions in the other two orthogonal directions, then adding the absolute values of the effects of these to obtain the maximum resultant forces in a member or at a point in a particular direction, and computing the stresses corresponding to the combined effects. In general, this alternative method is slightly conservative.

A related matter that merits attention in design is the provision for relative motion of parts or elements having supports at different locations.

UNSYMMETRICAL STRUCTURES IN TORSION

In design, consideration should be given to the effects of torsion on unsymmetrical structures and even on symmetrical structures where torsions may arise accidentally because of various reasons, including lack of homogeneity of the structures or presence of the wave motions developed in earthquakes. Most codes provide values of accidental eccentricity to use in design, but in the event that analyses indicate values greater than those recommended by the code, the analytical values should be used in design.

SIMULATION TESTING

Simulation testing to create various vibration environments is not new; it has been employed for years in connection with the development of equipment that must withstand vibration. Over the years such testing of small components has been accomplished on shake tables and involves many different types of input functions.[30] As a result of improved development of electromechanical rams, large shake tables have been developed in the past decade, which can simulate the excitation that may be experienced in a building, structural component, or items of equipment, from various types of ground motions, including earthquake motions, nuclear ground motions, nuclear blast motions induced in the ground or in a structure, and traffic vibrations. Some of these devices are able to provide simultaneous motion in three orthogonal directions.

The matter of simulation testing became of great importance with regard to earthquake excitation because of the development of nuclear power plants and the necessity for components in these plants to remain operational for purposes of safe shutdown and containment and also because of the observed loss of lifeline items in recent earthquakes as, for example, communication and control equipment, utilities, and fire-fighting systems.

In testing a piece of equipment on a shake table, it is common for the natural frequencies of the equipment, as well as the system damping, to be determined through continuous-sweep frequency searches using a relatively small amplitude input. High levels of excitation are normally not used, in order to avoid damage to the equipment and components.

Afterward various types of excitation can be used to investigate in more detail the response of items of equipment and subassemblies. One particularly effective approach that has been developed in recent years is that of sine-beat seismic testing.[30–32]

For seismic testing of equipment to resist earthquake motions, for example, the vibration test input should reproduce the worst features of the base floor motion on which the equipment is mounted. Such floor motions are commonly obtained through normal-mode analysis procedures made with computers, and from these computations, so called floor response spectra are obtained. These curves do not represent the motion or acceleration of the floor, but rather the peak acceleration of a special class of objects attached to the floor in a particular manner. An example of horizontal floor response spectra for 5 per cent damping is shown in Fig. 29.12. The common procedure there-

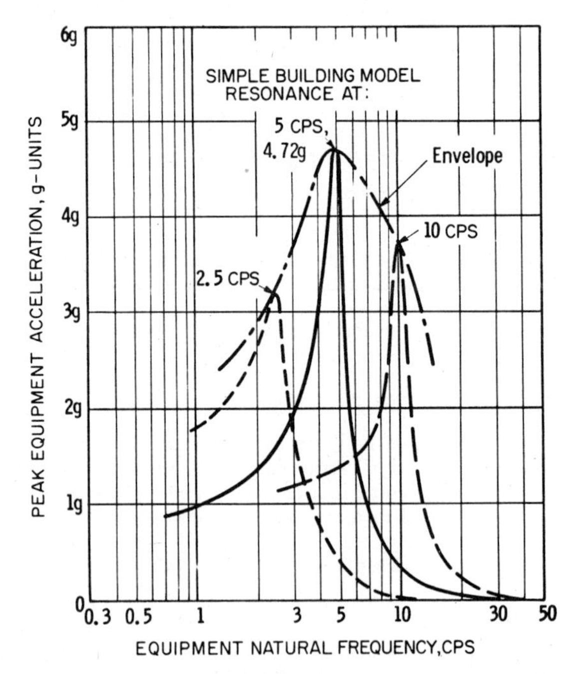

FIG. 29.12. Horizontal floor response spectra, $\xi = 5$ per cent, for simple building models excited by random earthquake motion (El Centro, 1940).

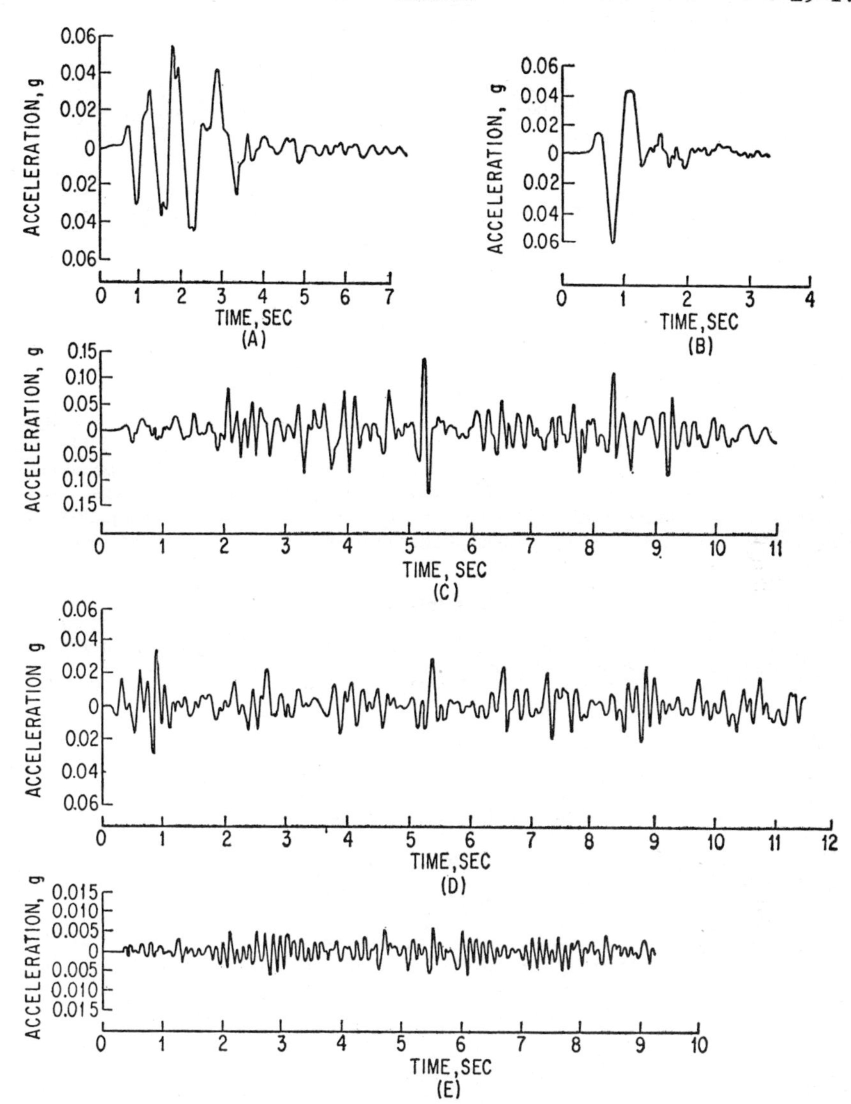

FIG. 29.13. Ground-acceleration–time curves for typical machine and vehicle excitations: (*A*) vertical acceleration measured on a concrete floor on sandy loam soil, at a point 6 ft from the base of a drop hammer, and (*B*) horizontal acceleration 50 ft from drop hammer. The weight of the drop hammer head was approximately 15,000 lb, and the hammer was mounted on three layers of 12- by 12-in. oak timbers on a large concrete base. (*C*) Vertical acceleration 6 ft from a railroad track on the well-maintained right-of-way of a major railroad, during passing of luxury-type passenger cars at a speed of approximately 20 mph. The accelerometer was bolted to a 2- by 2- by 2½-in. steel block which was firmly anchored to the ground. (*D*) Horizontal acceleration of the ground at 46 ft from the above railroad track, with a triple diesel-electric power unit passing at a speed of approximately 20 mph. (*E*) Horizontal acceleration of the ground 6 ft from the edge of a relatively smooth highway, with a large tractor and trailer unit passing on outside lane at approximately 35 mph with a full load of gravel.

after is to develop a time-history of motion which leads to such response and this is quite often in the form of a modified sine-beat or other type of excitation which can be applied through the test table. Equipment is often tested and evaluated on the basis of the response obtained while the equipment is in an operating mode on the test table.

Because of the importance of this subject, various standards have been developed for qualification testing.[32]

GROUND MOTIONS CAUSED BY MACHINERY AND VEHICLES*

The location of a precision machine shop near a railroad track, or the installation of delicate laboratory apparatus in a plant area containing heavy machinery, are examples of situations in which ground transmitted vibrations may pose serious problems. The variables involved in problems of this kind are exceedingly numerous, and few specific measurements are available to serve as a guide in estimating the ground motions that might be involved in particular cases. Figure 29.13 shows a number of acceleration-time curves for typical ground motions caused by the operation of machines and vehicles.[33] These curves were recorded with a variable-reluctance seismic-type accelerometer and recorder system having a flat frequency response from 0 to 100 cps. Thus the data can be compared directly with the similar measurements of earthquakes and generally handled in the same manner. Other examples of such motions are described in Ref. 35.

REFERENCES

1. Newmark, N. M., and E. Rosenblueth: "Fundamentals of Earthquake Engineering," Prentice-Hall, Inc., Englewood Cliffs, N.J., 1971.
2. Newmark, N. M., J. A. Blume, and K. K. Kapur: *J. Power Div. Am. Soc. Civil Engrs.*, **99**(PO2): 287 (November, 1973).
3. Newmark, N. M., and W. J. Hall: "Building Practices for Disaster Mitigation," *Nat. Bur. Std. Bldg. Sci. Ser.* 46, **1**:209–236 (February, 1973).
4. Newmark, N. M., and W. J. Hall: *Proc. 5th World Conf. Earthquake Eng., Intern. Assoc. Earthquake Eng., Rome,* **2**:2266–2275 (1974).
5. R. L. Wiegel (ed.): "Earthquake Engineering," Prentice-Hall, Inc., Englewood Cliffs, N.J., 1970.
6. Newmark, N. M., and W. J. Hall: *Proc. 4th World Conf. Earthquake Eng.*, Santiago, Chile, **II**, B4-37 to B4-50 (1969).
7. Newmark, N. M.: *Nucl. Eng. Design,* **20**(2):303–322 (July, 1972).
8. Newmark, N. M., and A. S. Veletsos: "Design Procedures for Shock Isolation Systems of Underground Protective Structures," Report for Air Force Weapons Laboratory, vol. III, General American Transportation Corporation, RTD TDR 63-3096, June, 1964.
9. Veletsos, A. S., and N. M. Newmark: *Proc. 2d World Conf. Earthquake Eng., Tokyo,* **II**:895–912 (1960).
10. Veletsos, A. S., N. M. Newmark, and C. V. Chelapati: *Proc. 3d World Congr. Earthquake Eng., New Zealand,* **2**:II-663–II-682 (1965).
11. *Proc. World Conf. Earthquake Eng. Earthquake Eng. Research Inst.*, San Francisco, (1956).
12. *Proc. 2d World Conf. Earthquake Eng.*, **I-III**, Science Council of Japan, 1960.
13. *Proc. 3d World Conf. Earthquake Eng.*, New Zealand Institution of Engineers, 1965.
14. *Proc. 4th World Conf. Earthquake Eng.*, Chilean Association on Seismology and Earthquake Engineering, Santiago, 1969.
15. *Proc. 5th World Conf. Earthquake Eng.*, EDIGRAF—Editrice Libraria, Rome, Italy, 1973.
16. Newmark, N. M.: Current Trends in the Seismic Analysis and Design of High Rise Structures, 403, in R. L. Wiegel (ed.), "Earthquake Engineering," chap. 16, Prentice-Hall, Inc., Englewood Cliffs, N.J., 1970.
17. Newmark, N. M.: *Proc. Symp. Earthquake Eng., University of British Columbia,* VI-1–VI-55 (September, 1965).
18. Blume, J. A., N. M. Newmark, and L. Corning: "Design of Multistory Reinforced Concrete Buildings for Earthquake Motions," Portland Cement Association, Chicago, 1961.

* This section is taken largely from Ref. 34.

19. J. A. Blume: *Proc. World Conf. Earthquake Eng., Berkeley, Calif.*, 11-1–11-27 (1956).
20. Zeevaert, L., and N. M. Newmark: *Proc. World Conf. Earthquake Eng., Berkeley, Calif.*, 35-1–35-11 (1956).
21. "Uniform Building Code—1973 Edition," International Conference of Building Officials, Whittier, Calif., 1973.
22. "Recommended Lateral Force Requirements and Commentary," Seismology Committee, Structural Engineers Association of California, 1974.
23. Newmark, N. M., and W. J. Hall: "Dynamic Behavior of Reinforced and Prestressed Concrete Buildings under Horizontal Forces and the Design of Joints (Including Wind, Earthquake, Blast Effects)," Preliminary Publication, 8 Congress, pp. 585–613, International Association Bridge and Structural Engineering, New York, 1968.
24. Newmark, N. M.: *Proc. IAEA Panel Aseismic Design and Testing of Nuclear Facilities*, pp. 90–113, Japan Earthquake Promotion Society, Tokyo, 1969.
25. Newmark, N. M., and W. J. Hall: *Proc. IAEA Panel Aseismic Design and Testing of Nuclear Facilities*, pp. 114–119, Japan Earthquake Engineering Promotion Society, Tokyo, 1969.
26. Blume, J. A.: *Trans. ASCE*, **125**:1088–1139 (1960).
27. Derecho, A. T., D. M. Schultz, and M. Fintel: "Analysis and Design of Small Reinforced Concrete Buildings for Earthquake Forces," Portland Cement Association, Engineering Bulletin, 1974.
28. Housner, G. W.: *Proc. 2d World Conf. Earthquake Eng.*, **1**:133 (1960).
29. Sozen, M. A., N. M. Newmark, and G. W. Housner: "Implications of Seismic Structural Design of the Evaluation of Damage to the Sheraton-Macato," *Proc. 4th World Conf. Earthquake Eng.*, **III**:J-2-137–J-2-150 (1969).
30. Gertel, Maurice: Specification of Laboratory Tests, in C. M. Harris and C. E. Crede (eds.), "Shock and Vibration Handbook," chap. 24, McGraw-Hill Book Co., New York, 1961.
31. Fischer, E. G., and F. H. Wolff: *Exp. Mech.*, **13**(12):531–538 (December, 1973).
32. Skreiner, K. M., E. G. Fischer, S. N. Hou, and G. Shipway: "New Seismic Requirements for Class I Electrical Equipment," *Trans. IEEE Paper* T 74 048-5, 1974.
33. Flygare, R. W.: "An Investigation of Ground Accelerations Produced by Machines," Mechanical's Thesis, California Institute of Technology, 1955.
34. Hudson, D. E.: Vibration of Structures Induced by Seismic Waves, in C. M. Harris and C. E. Crede (eds.), "Shock and Vibration Handbook," chap. 50, Vol. III, McGraw-Hill Book Company, New York, 1961.
35. Liu, T. K., E. B. Kinner, and M. K. Yegian: *Sound and Vibration*, 26–32 (October, 1974).

29

PART II: VIBRATION OF STRUCTURES INDUCED BY WIND

A. G. Davenport and M. Novak
University of Western Ontario

INTRODUCTION

Vibration of significant magnitude may be induced by wind in a wide variety of structures including buildings, television and cooling towers, chimneys, bridges, transmission lines, and radio telescopes. No structure exposed to wind seems entirely immune from such excitation. The material presented here describes several mechanisms causing these oscillations and suggests a few simpler approaches that may be taken in design to reduce vibration of structures induced by wind. There is an extensive literature[1–5] giving a more detailed treatment of the subject matter.

FORMS OF AERODYNAMIC EXCITATION

The types of structure referred to above are generally unstreamlined in shape. Such shapes are termed "bluff bodies" in contrast to streamlined "aeronautical" shapes. The distinguishing feature is that when the air flows around such a bluff body, a significant wake forms downstream, as illustrated in Fig. 29.14. The wake is separated from the outside flow region by a shear layer. With a sharp-edged body (such as a building or structural number) as in Fig. 29.14, this shear layer emanates from the corner. With oval bodies such as the cylinder in Fig. 29.14, the shear layer commences at a so-called

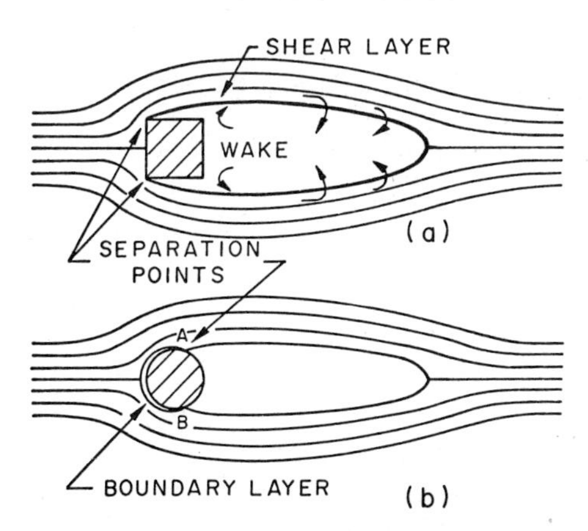

FIG. 29.14. Wake formation past bluff bodies: (*a*) sharp-edged body, (*b*) circular cylinder.

boundary layer on the upstream surface at points A and B (the separation points) and becomes a free shear layer. The exact position of these separation points depends on a wide variety of factors, such as the roughness of the cylinder, the turbulence in the flow, and the Reynolds number $R = VD/\nu$, where V = flow velocity, D = diameter of the body, and ν = kinematic viscosity.

The flow illustrated in Fig. 29.14 represents the time-average picture which would be obtained by averaging the movements of the fluid particles over a time interval that is long compared with the "transit time" D/V. The instantaneous picture of the flow may be quite different, as indicated in Fig. 29.15, for two reasons.

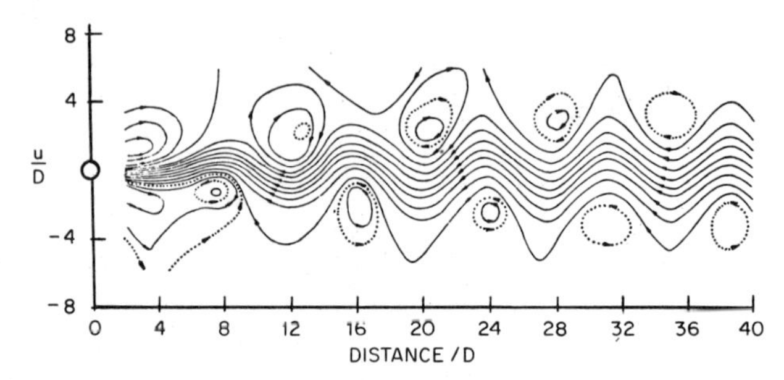

FIG. 29.15. Vortex street past circular cylinder (R = 56). (*After Kovasznay, Proc. Roy. Soc. London, 198, 1949.*)

First, if the flow is the wind, it is under almost all practical circumstances strongly turbulent; the oncoming flow will be varying continuously in direction and speed in an irregular manner. These fluctuating motions will range over a wide range of frequencies and scales (i.e., eddy sizes).

Second, the wake also will take on a fluctuating character. Here, however, the size of the dominant eddies (vortices) will be of a similar size to the body. The vortices tend to start off their career by curling up at the separation point and then are carried off downstream. Sometimes these eddies are fairly regular in character and are shed alternately from either side; if made visible by smoke or other means, they can be seen to form a more or less regular stepping-stone pattern until they are broken up by the turbulence or dissipate themselves. In a strongly turbulent flow, the regularity is disrupted.

The flow characteristics of the oncoming flow and the wake are the direct causes of the forces on the bodies responsible for their oscillation. The forms of the resulting oscillation are as follows.

1. Turbulence-induced oscillations. Certain types of oscillation of structures can be attributed almost exclusively to turbulence in the oncoming flow. In the wind these may be described as "gust-induced oscillations" (or turbulence-induced, oscillations). The gusts may cause longitudinal, transverse, or torsional oscillations of the structure, which increase with wind velocity (Fig. 29.16).

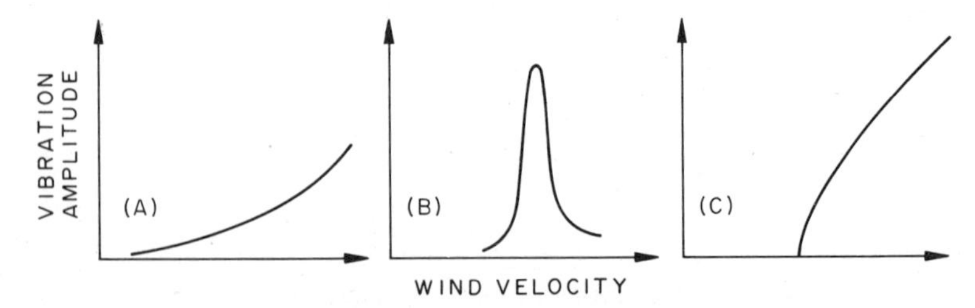

FIG. 29.16. Main types of wind-induced oscillations: (*a*) vibration due to turbulence, (*b*) vibration due to vortex shedding, and (*c*) aerodynamic instability.

2. Wake-induced oscillations. In other instances, the fluctuations in the wake may be the predominant agency. Since these fluctuations are generally characterized by alternating flow, first around one side of the body then around the other, the most significant pressure fluctuations act on the sides of the body in the wake behind the separation point (the so called after body); they act mainly laterally or torsionally and to a much lesser extent longitudinally. The resultant motion is known as "vortex-induced oscillation." Oscillation in the direction perpendicular to that of the wind is the most important type. It often features a pronounced resonance peak (Fig. 29.16*b*).

While these distinctions between gust-induced and wake-induced forces are helpful, they often strongly interact; the presence of free-stream turbulence, for example, may significantly modify the wake.

3. Buffeting by the wake of an upstream structure. A further type of excitation is that induced by the wake of an upstream structure (Fig. 29.17). Such an arrangement

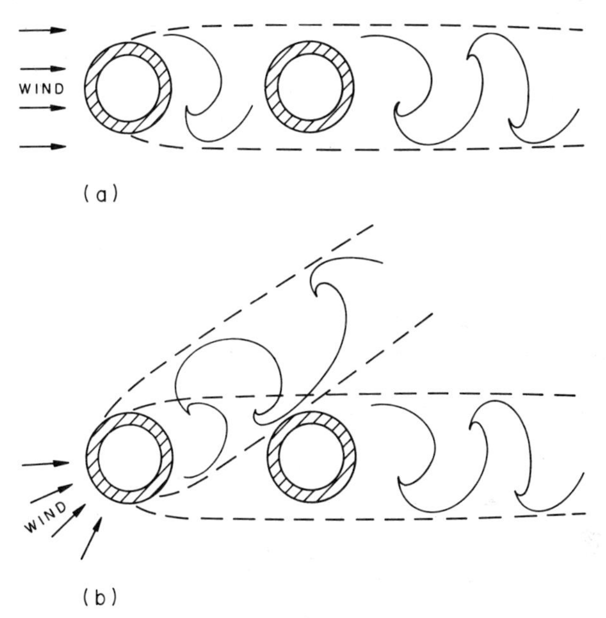

FIG. 29.17. Buffeting by the wake of an upstream structure.

of structures produces several effects. The turbulent wake containing strong vortices shed from the upstream structure can buffet the downstream structure. In addition, if the oncoming wind is very turbulent, it can cause the wake of the upstream structure to veer, subjecting the downstream structure successively to the free flow and the wake flow. This frequently occurs with chimneys in line, as well as with tall buildings.

4. Galloping and flutter mechanisms. The final mechanism for excitation is associated with the movements of the structure itself. As the structure moves relative to the flow in response to the forces acting, it changes the flow regime surrounding it. In so doing, the pressures change, and these changes are coupled with the motion. A pressure change coupled to the velocity (either linearly or nonlinearly) may be termed an "aerodynamic damping" term. It may be either positive or negative. If positive, it adds to the mechanical damping and leads to higher effective damping and a reduced tendency to vibrate; if negative, it can lead to instability and large amplitudes of movement. This type of excitation occurs with a wide variety of rectangular building shapes as well as bridge cross sections and common structural shapes such as angles and I sections.

In other instances, the coupling may be with either the displacement or acceleration, in which case they are described as either aerodynamic stiffness or mass terms, the effect of which is to modify the mass or stiffness terms in the equations of motion. Such modification can lead to changes in the apparent frequency of the structure. If the aerodynamic stiffness is negative, it can lead to a reduction in the effective stiffness of the structure and eventually to a form of instability known as "divergence." All types of instability feature a sudden start at a critical wind velocity and a rapid increase of violent displacements with wind velocity (Fig. 29.16*c*).

These various forms of excitation are briefly discussed in this chapter. Because all types of oscillations are influenced strongly by the properties of the wind, some basic wind characteristics are described first.

BASIC WIND CHARACTERISTICS

Wind is caused by differences in atmospheric pressure. At great altitudes, the air motion is independent of the roughness of the ground surface and is called the "geostrophic," or "gradient" wind. Its velocity is reached at a height called "gradient height," which lies between about 1,000 and 2,000 ft.

Below the gradient height, the flow is affected by surface friction, by the action of which the flow is retarded and turbulence is generated. In this region, known as the "planetary boundary layer," the three components of wind velocity resemble the traces shown in Fig. 29.18.

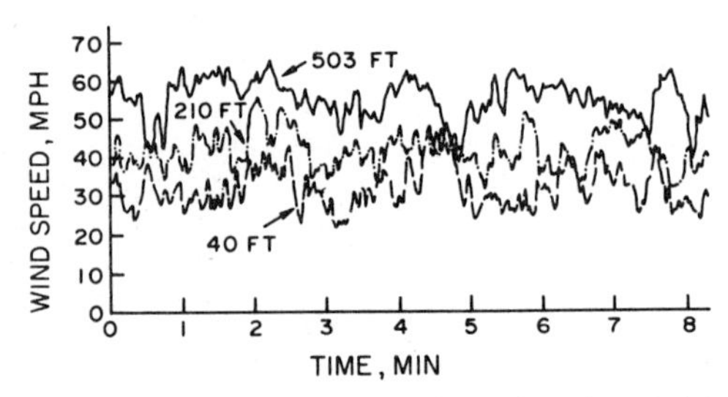

FIG. 29.18. Record of horizontal component of wind speed at three heights on 500 ft mast in open terrain. (*Courtesy of E. L. Deacon.*)

The longitudinal component consists of a mean plus an irregular turbulent fluctuation; the lateral and vertical components consist of similar fluctuations. These turbulent motions can be characterized in a number of different ways.

The longitudinal motion at height z can be expressed as

$$V_z(t) = \bar{V}_z + v(t) \tag{29.21}$$

where $\bar{V}_z$ = mean wind velocity (the bar denotes time average) and $v(t)$ = fluctuating component.

MEAN WIND VELOCITY. The mean wind velocity $\bar{V}_z$ varies with height z as represented by the mean wind velocity profile (Fig. 29.19). The profiles observed in the field can be matched by a logarithmic law, for which there are theoretical grounds, or by an empirical power law

$$\frac{\bar{V}_z}{\bar{V}_G} = \left[\frac{z}{z_G}\right]^\alpha \tag{29.22}$$

where $\bar{V}_G$ = gradient wind velocity, z_G = gradient height, and α = an exponent <1. Gradient height z_G and exponent α depend on the surface roughness, which can be characterized by the surface drag coefficient κ (here referenced to the wind speed at 10 meters).

A few typical values of these parameters are given in Fig. 29.19. The mean wind

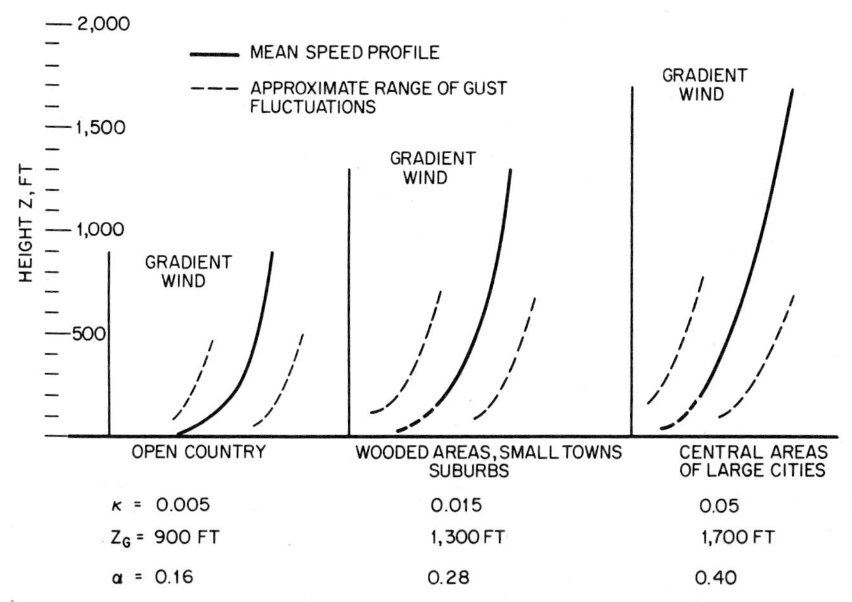

FIG. 29.19. Vertical profiles of mean wind velocity for three typical terrains.

profiles shown are characteristic of level terrain. They can significantly change, particularly in the lower region, when the air flow meets an abrupt change in surface roughness or terrain contour. A sudden increase in roughness reduces the wind speed near the ground while a hill accelerates the flow over its crest.

The mean wind profiles are useful when predicting the wind speed at a particular site. The gradient wind speed is estimated using data registered by the nearest meteorological stations at their standard height, which is usually 33 ft (10 meters).

The mean wind velocity generally depends on the period over which the wind speed is averaged. Periods from 10 to 60 min appear adequate for engineering considerations and usually yield reasonably steady mean values. The same duration is suitable to define the fluctuating wind component.

FLUCTUATING COMPONENTS OF THE WIND. The fluctuating components of the wind change with height less than the mean wind and are random both in time and space. The random nature of the wind requires the application of statistical concepts.*

The basic statistical characteristics of the velocity fluctuations are the intensity of turbulence, the power spectral density (power spectrum), the correlation between velocities at different points, and the probability distribution.

The intensity of turbulence is defined as $\sigma_v/\bar{V}_z$, where $\sigma_v = \sqrt{\overline{v^2(t)}}$ is the root-mean-square (rms) fluctuation in the longitudinal direction. The intensity of the lateral and vertical fluctuations can be described similarly. For wind, the intensity of turbulence is between 5 and 25 per cent. The magnitude σ_v also defines the probability distribution of the fluctuations which may be assumed to be Gaussian (normal).

The energy of turbulent fluctuations (gustiness) is distributed over a range of frequencies. This distribution of energy with frequency f can be described by the spectrum of turbulence (power spectral density) $S_v(f)$. The relationship between the spectrum and the variance is

$$\int_0^\infty S_v(f)\,df = \sigma_v^2$$

* Statistical approaches to random processes are described in Chap. 11 and Chap. 21.

which leads to another form of the spectrum known as the "logarithmic spectrum" $fS_v(f)/\sigma_v^2$. This form of the spectrum is dimensionless and preserves the relative contributions to the variance at different frequencies represented on a logarithmic scale; and its integral is

$$\int_0^\infty \frac{fS_v(f)}{\sigma_v^2}\, d\ln f = 1$$

The two forms of spectra are sketched in Fig. 29.20. A generalization of wind spectra for different wind velocities is possible if the frequency scale is so modified that

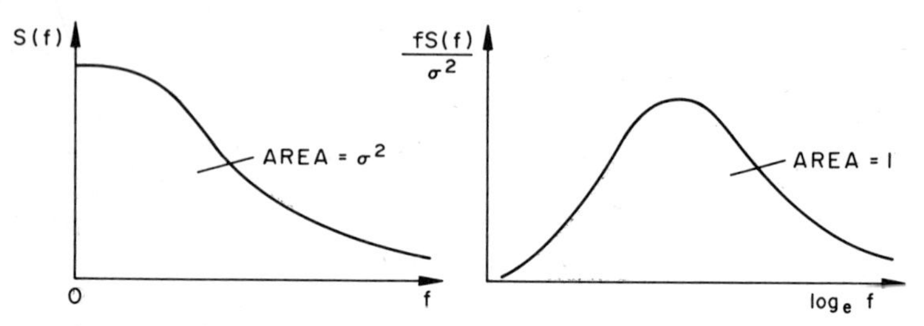

FIG. 29.20. Two different ways of presenting power spectral densities.

it too is dimensionless. The ratio $f/\bar{V}$ is the so-called "inverse wavelength" related to the "size" of atmospheric eddies. This may be expressed as a ratio to a representative length scale L, such as the wavelength of the eddies at the peak of the spectrum. The dimensionless frequency or inverse wavelength may now be written

$$\bar{f} = fL/\bar{V}$$

Under certain circumstances this relationship is also known as the Strouhal number or the reduced frequency.

It is generally found that while the length scale L in the oncoming flow corresponds to that of the turbulence itself (this in the natural wind is of the order of thousands of

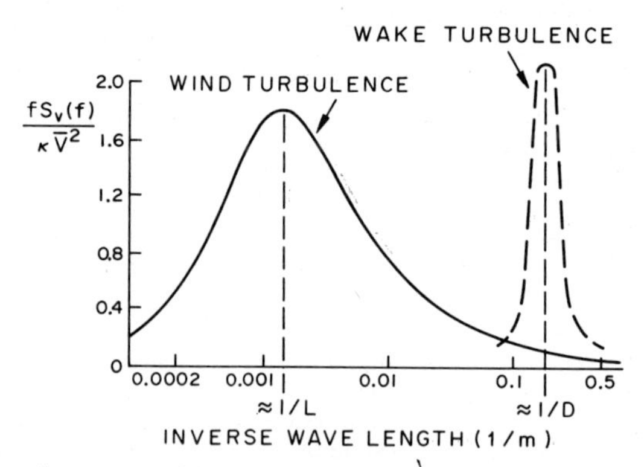

FIG. 29.21. Universal spectrum of horizontal gustiness in strong winds and example of spectrum of fluctuations in wake.

feet), in the wake the governing length scale is of the same order as the diameter of the body D. This is illustrated in Fig. 29.21.

The spectrum of horizontal gustiness in strong winds is largely independent of height above the ground, is proportional to both the surface drag coefficient κ and the square of the mean velocity at the standard height of 10 meters, $\bar{V}_{10}$, and can be represented, with some approximations, as[6, 7]

$$S_v(f) = 4\kappa \bar{V}_{10}^2 \frac{L/\bar{V}_{10}}{(2 + \bar{f}^2)^{5/6}} \tag{29.23}$$

in which f = frequency, cps, $\bar{f} = fL/\bar{V}_{10}$ where L = scale length $\approx$4,000 ft, and κ is given in Fig. 29.19. This spectrum is shown in Fig. 29.21.

The variance of the velocity fluctuations is

$$\sigma_v{}^2 = \int_0^\infty S_v(f)\, df = 6.68\kappa \bar{V}_{10}^2 \tag{29.24}$$

It can be seen from Eqs. (29.23) and (29.24) that large velocity fluctuations can be expected in rough terrain where coefficient κ is large.

The spatial correlation of wind speeds at two different stations is described by the coherence function (coherence),

$$\gamma_{12}{}^2(f) = \frac{|S_{12}(f)|^2}{S_1(f)S_2(f)} \leq 1 \tag{29.25}$$

where $S_{12}(f)$ = cross spectrum (generally complex) between stations 1 and 2; $S_1(f)$ and $S_2(f)$ are power spectra of the two stations. The coherence function depends primarily on the parameter $\Delta z f/\bar{V}$, where Δz = separation and $\bar{V} = 1/2(\bar{V}_1 + \bar{V}_2)$ is the average wind speed. A suitable approximate function is

$$\sqrt{\text{Coherence}} = e^{-c(\Delta z f/\bar{V})}$$

where c is a constant having a value of approximately 7 for vertical separation and approximately 15 for horizontal separation. Coherence decreases with both separation and frequency. A more detailed discussion of wind characteristics is given in Refs. 1 and 7.

EXCITATION DUE TO TURBULENCE

When a structure is exposed to the effects of wind, the fluctuating wind velocity translates into fluctuating pressures, which in turn produce time-variable response (deflection) of the structure. This response is random and represents the basic type of wind-induced oscillations. The theoretical prediction of this oscillation is rather complex but can be reduced to a simple procedure suitable for design purposes. The discussion of the oscillation is therefore presented in two parts. In the first part, the basic theoretical steps are outlined. In the second part, the design procedure known as the gust-factor approach is given in more detail.

FUNDAMENTALS OF RESPONSE PREDICTION

If the area A of the structure exposed to wind is small relative to the significant turbulent eddies, the so-called quasi-steady theory for turbulence can be used to estimate aerodynamic forces. In the drag direction, the drag force

$$\begin{aligned} D(t) &= \frac{1}{2}\rho C_D A V^2(t) \\ &= \frac{1}{2}\rho C_D A \bar{V}^2\left[1 + 2\frac{v(t)}{\bar{V}} + \frac{v^2(t)}{\bar{V}^2}\right] \end{aligned}$$

where ρ = air density (normally equal to 0.0024 slugs/cu ft), and C_D = drag coefficient. If $v(t) \ll \bar{V}$, the squared term is ignored. The spectra of the fluctuating drag and velocity are then related as

$$\frac{S_D(f)}{\bar{D}^2} = 4\frac{S_v(f)}{\bar{V}^2} \tag{29.26}$$

where the mean drag (static component of the drag) is

$$\bar{D} = \frac{1}{2}\rho C_D A \bar{V}^2 \tag{29.27}$$

and $S_v(f)$ is given by Eq. (29.23).

With large bodies the wavelength is comparable to the size of the body itself (that is, $f\sqrt{A}/\bar{V} \approx 1$) and it is necessary to modify the drag spectrum by the so-called "aerodynamic admittance function" $|X_{\text{aero}}(f)|^2$. This function[6] describes the modifying influence of any changes in effective drag coefficient, as well as the decrease in correlation of the eddies as the wavelength of the eddies approaches the diameter of the body. Thus, the modified drag spectrum is

$$\frac{S_D(f)}{\bar{D}^2} = 4|X_{\text{aero}}(f)|^2 \frac{S_v(f)}{\bar{V}^2}$$

If these forces act on an elastic spring-mass-damper system, the response of this system u will have a spectrum

$$\frac{S_u(f)}{\bar{u}^2} = |X_{\text{aero}}|^2 |X_{\text{mech}}|^2 \frac{4S_v(f)}{\bar{V}^2}$$

where static deflection $\bar{u} = \bar{D}/k$, k = stiffness constant, and the mechanical admittance function

$$|X_{\text{mech}}|^2 = \frac{1}{[1 - (f/f_0)^2]^2 + 4\zeta^2(f^2/f_0^2)}$$

where ζ = critical damping ratio, and f_0 = natural frequency of the system.

The transition from the spectrum of the wind-velocity fluctuations to the spectrum

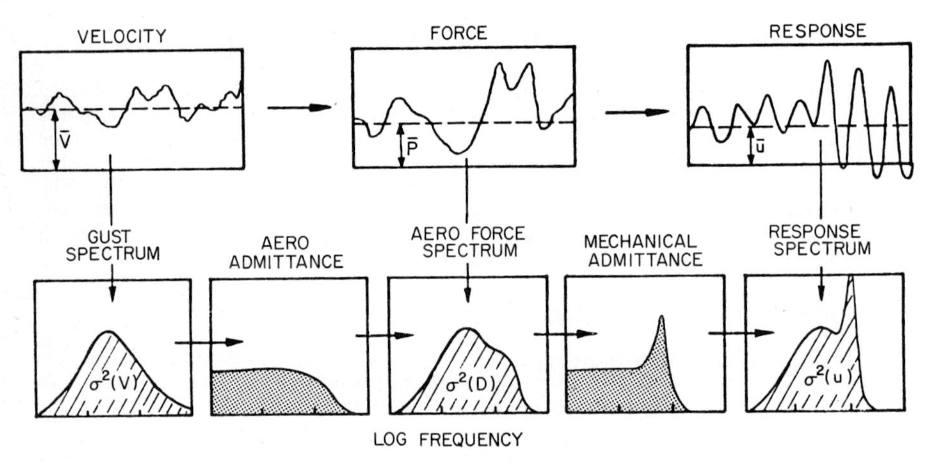

FIG. 29.22. Transition from gust spectrum to response spectrum.

of the response is shown diagrammatically in Fig. 29.22. The variance of the response σ_u^2 is obtained from the spectrum of the response,

$$\sigma_u^2 = \int_0^\infty S_u(f)\, df \tag{29.28}$$

The relationships above describe the mean and the variance of the response. For engineering purposes, it is also useful to define extreme values. It is often satisfactory to assume that the process in question is Gaussian with probability density function given by

$$p(u) = \frac{1}{\sqrt{2\pi}\,\sigma_u} e^{-(u-\bar{u})^2/2\sigma_u{}^2}$$

This distribution is fully described by the mean and the variance. Maximum values of the response during time T can be written as

$$u_{\max} = \bar{u} + g\sigma_u \tag{29.29}$$

where g = peak factor. The average largest value of the peak factor in a period T can be estimated from[6]

$$g = \sqrt{2 \ln \nu T} + \frac{0.5772}{\sqrt{2 \ln \nu T}} \tag{29.30}$$

where ν is an effective cycling rate of the process, generally close to the natural frequency. The relationship of the distribution of the largest peak value to the distribution of all values is shown in Fig. 29.23. As can be seen, when the period T or the

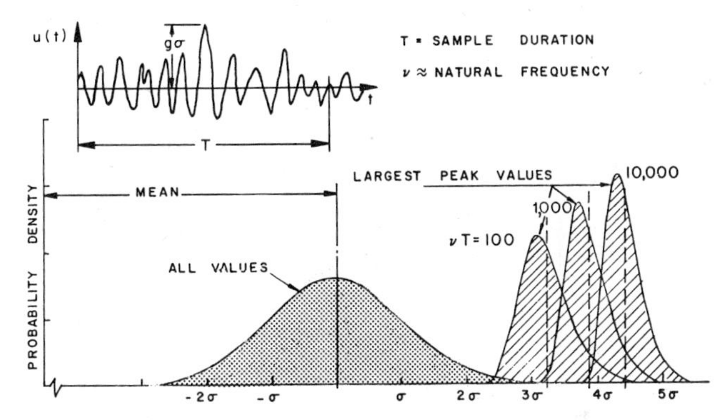

FIG. 29.23. Relationship of distribution of largest peak value to distribution of all values (for a stationary random process).

natural frequency increases, the expected peak displacement also increases. The factor g usually ranges between 3 and 5.

Further extension of the concept includes the cross correlation of the wind loads at different stations (e.g., heights), the shape of the vibration mode, and the nonuniformity of the mean flow. These factors can be included into the solution formulated in terms of modal analysis.*

With a prismatic structure, the displacement may be expressed in the form

$$u(z,t) = \sum_{j=1}^{\infty} q_j(t)\phi_j(z) \tag{29.31}$$

where $q_j(t)$ = the generalized coordinate of the jth mode, and $\phi_j(z)$ = the jth mode of natural vibrations to an arbitrary scale.

With damping small and natural frequencies well separated, the cross correlation of the generalized coordinate can be neglected and the mean-square displacement (the variance) is

$$\overline{u^2(z,t)} = \sum_{j=1}^{\infty} \overline{q_j{}^2}\,\phi_j{}^2(z) \tag{29.32}$$

* For detailed discussion of modal analysis, see Chaps. 2 and 7.

The variance of the generalized coordinate $\overline{q_j^2}$ is determined by the power spectrum of the generalized force Q_j. When the lateral dimension of the structure is small, only cross correlation in direction z need be considered. Then the power spectrum of the generalized force is

$$S_{Q_j}(f) = \int_0^H \int_0^H S_{12}(z_1,z_2,f)\phi_j(z_1)\phi_j(z_2)\,dz_1\,dz_2 \tag{29.33}$$

where $S_{12}(z_1,z_2,f)$ = cross spectrum of the wind loads at heights z_1 and z_2, and H = height of the structure. With respect to Eq. (29.26), the cross spectrum of the wind loads can be expressed in terms of the power spectrum of the wind speed [Eq. (29.23)] and the coherence function, Eq. (29.25).

The variance of q_j is

$$\begin{aligned}\overline{q_j^2} &= \int_0^\infty \frac{1}{(2\pi f_j)^4 M_j^2}\,\frac{1}{[1-(f/f_j)^2]^2 + 4\zeta^2(f/f_j)^2}\,S_{Q_j}(f)\\ &\approx \frac{1}{64\pi^3\zeta f_j^3 M_j^2}\,S_{Q_j}(f_j) + \frac{1}{(2\pi f_j)^4 M_j^2}\int_0^{f_j} S_{Q_j}(f)\,df\end{aligned} \tag{29.34}$$

where f_j = jth natural frequency and generalized mass

$$M_j = \int_0^H m(z)\phi_j^2(z)\,dz \tag{29.35}$$

where $m(z)$ = mass of the structure per unit length. The approximate integration* of Eq. (29.34) yields the response composed of two parts, the resonance effect (the first term) and the background turbulence effect (the second term) (Fig. 29.24). The vari-

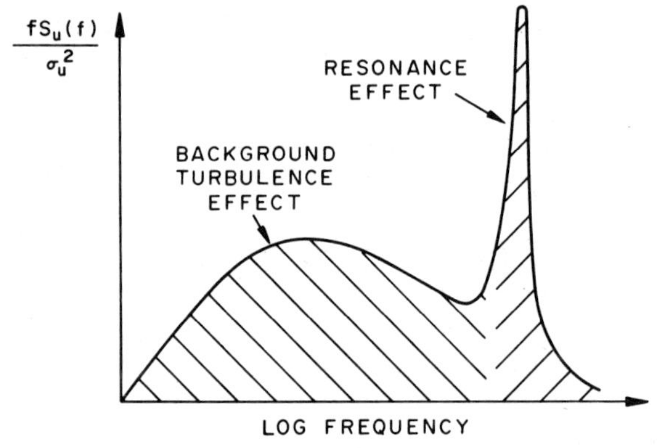

Fig. 29.24. Spectrum of structural response with indication of resonance effect and background turbulence effect.

ance of the displacement follows from Eq. (29.32), and its standard deviation (rms dynamic displacement) is $\sigma_u(z) = \sqrt{\overline{u^2(z,t)}}$. The peak response is established from Eq. (29.29) by means of the peak factor g [Eq. (29.30)] as in one degree-of-freedom. The mean deflection $\bar{u}(z)$ is the static deflection due to the mean wind $\bar{V}_z$.

Other analyses of slender structures are also available.[9, 10, 11] In applications to buildings and free-standing towers, the analysis can usually be limited to the first modal component in Eq. (29.32).

Application to buildings and structures with significant lateral dimension requires the incorporation of the horizontal cross correlation as well. A complete solution established by means of simplifying assumptions and numerical integrations is given below.

* The integral is similar to that appearing in Eq. (29.28) and can be evaluated accurately using the theory of residua. An example is given in Ref. 8.

GUST FACTOR APPROACH

The gust-factor approach is a design procedure derived on the basis of the theory above by means of a few simplifying assumptions. The approach given here is a modified version of the method described in Ref. 12 and adopted in Ref. 13. It considers only the response in the first vibration mode which is assumed to be linear. These assumptions are particularly suitable for buildings. The method yields all the data needed in design: the maximum response, the equivalent static wind load that would produce the same maximum response, and the maximum acceleration needed for the evaluation of the physiological effects of strong winds (human comfort).

The gust factor G is defined as the ratio of the expected peak displacement (load) in a period T to the mean displacement (load) $\bar{u}$. Hence, maximum expected response

$$u_{\max} = G\bar{u} = \left(1 + g\frac{\sigma_u}{\bar{u}}\right)\bar{u} \tag{29.36}$$

The gust factor is given as

$$G = 1 + g\sqrt{\frac{K}{C_e}\left(B + \frac{sF}{\zeta}\right)} \tag{29.37}$$

where ζ = damping ratio and K = factor related to the surface roughness; this factor is equal to

0.08 for open terrain (zone A)
0.10 for suburban, urban, or wooded terrain (zone B)
0.14 for concentrations of tall buildings (zone C)

All the other parameters appearing in Eq. (29.37) can be obtained from Fig. 29.25. C_e = exposure factor based on the mean wind speed profile (coefficient α) and thus on surface roughness. For the three zones, the exposure factor is obtained from Fig. 29.25*a* for the height of the building H. C_e relates to wind pressure rather than speed. Hence, the mean wind speed at the top of the building is given by

$$\bar{V}_H = \bar{V}_{10}\sqrt{C_e}$$

where $\bar{V}_{10}$ = reference wind speed at the standard height of 10 meters. $\bar{V}_{10}$ can be obtained from meteorological stations. Velocity $\bar{V}_H$ is needed for determination of parameters s and F. Factors B, s, F, and g are given in Fig. 29.25*c* to *f* as a function of parameters indicated; D = width of the frontal area, and f_0 = the first natural frequency of the structure in cycles per second. The average fluctuation rate ν, on which the peak factor g depends, is evaluated from formula

$$\nu = f_0\sqrt{\frac{sF/\zeta}{B + sF/\zeta}} \tag{29.38}$$

The peak factor g is plotted in Fig. 29.25*f*, assuming a period of observation T = 3,600 sec; it can also be calculated from Eq. (29.30).

The parameters given also yield the design wind pressure p, which produces displacement $u_{\max}$ if applied as a static load. This design pressure

$$p = qC_eGC_p \tag{29.39}$$

where $q = \frac{1}{2}\rho\bar{V}_{10}^2$ is the reference mean-velocity pressure, and C_e = exposure factor. In this case, C_e varies continuously with the elevation according to Fig. 29.25*a* for pressures acting on the windward face of the structure; for the leeward face, C_e is constant and evaluated at one-half the height of the building. The quantity C_p = average pressure coefficient, which depends on the shape of the structure and the flow pattern around it. For a typical building with a flat roof and a height greater than twice the width, the coefficients are given for the windward and leeward faces in Fig. 29.25*b* together with the pressure distribution.

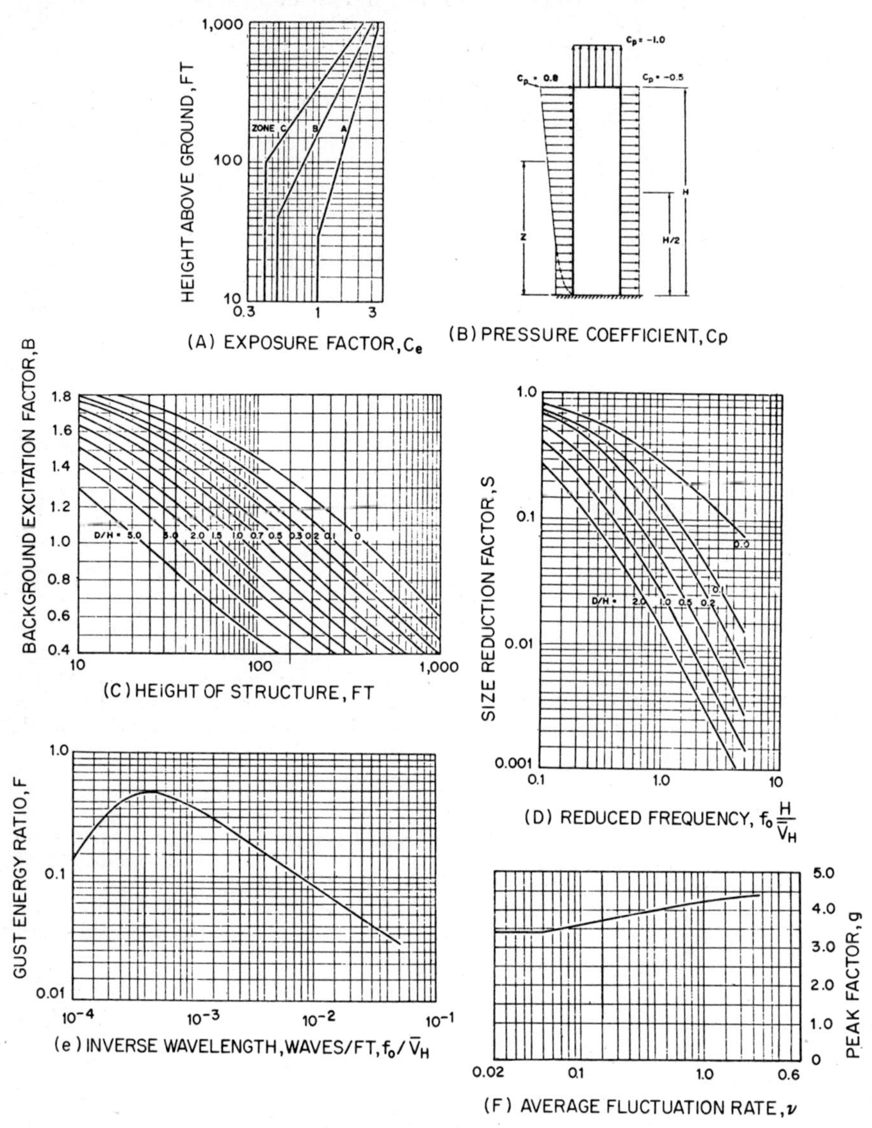

FIG. 29.25. Components of gust factor.

The peak acceleration A of a structure due to gusting wind is given by

$$A = u_{\max} \frac{4\pi^2 f_0^2 g}{G} \sqrt{\frac{KsF}{C_e \zeta}}$$

where $u_{\max}$ = maximum deflection under the design pressure p. The other parameters are equal to those used in Eq. (29.37). When the acceleration exceeds about 1 per cent of gravity, the motion is usually perceptible. However, there are large differences in the perceptibility of motions having very low frequencies.[14, 15]

Similar approaches are given in Refs. 16 to 18.

EFFECT OF GUSTS ON CLADDING AND WINDOWS

Wind gusts produce local pressure on cladding and window panels of a building. Because the natural frequency of such a panel is very high compared with the frequency components of the wind-speed fluctuations, the panel displacement is essentially static. Its design may be based on the static displacement resulting from maximum expected pressure, which is the algebraic sum of the height and time-dependent exterior pressure (or suction) and the constant interior pressure (or suction). If the fluctuating component of the pressure $p(t)$ is considered to be a stationary random process, the exterior expected maximum pressure is

$$p_{\max} = \bar{p}\left(1 + g\frac{\sigma_p}{\bar{p}}\right) = \bar{p}G \tag{29.40}$$

where $\bar{p} = \frac{1}{2}\rho C_p \bar{V}^2$ = mean pressure
C_p = local pressure coefficient
σ_p = standard deviation of the fluctuating pressure component
g = peak factor given by Eq. (29.30)
G = gust factor

To account for the sensitivity of glass to both static and dynamic fatigue, it has been suggested[19, 20] that g or G in Eq. (29.40) be multiplied by a wind-on-glass effect factor.

Factors g, $\sigma_p/\bar{p}$, and C_p are most reliably determined from wind-tunnel experiments. They strongly depend on location of the panel, wind direction, turbulence intensity and the local flow pattern determined by the shape of the building and its immediate environment. In full-scale experiments, values of g in excess of 10 have been observed in highly intermittent flow. Largest local pressure coefficients C_p (actually suctions) appear with skew wind at the leading edge of the building where a typical value is $C_p = -1.5$. In that part of the building exposed to free flow, a gust factor $G \approx 2.5$ is a reasonable estimate.[13, 21]

The interior pressure is not very high, but its magnitude and sign depend on openings and leakage.

Damage to windows may result from local wind pressure, but it also depends on material properties of glass and its fatigue. The fatigue limit of glass is only about 20 per cent of the instantaneous strength.[20]

VIBRATION DUE TO VORTEX SHEDDING

Vortex shedding represents the second most important mechanism for wind-induced oscillations. Unlike the gusts, vortex shedding produces forces which originate in the wake behind the structure, act mainly in the across-wind direction, and are, in general, rather regular. The resultant oscillation is resonant in character (Fig. 29.16*b*), is often almost periodic, and usually appears in the direction perpendicular to that of the wind. Lightly damped structures such as chimneys and towers are particularly susceptible to vortex shedding. Many failures attributed to vortex shedding have been reported.

When a bluff body is exposed to wind, vortices shed from the sides of the body creating a pattern in its wake often called the "Karman vortex street" (Fig. 29.15). The frequency of the shedding, nearly constant in many cases, depends on the shape and size of the body, the velocity of the flow, and to a lesser degree on the surface roughness and the turbulence of the flow. If the cross section of the body is noncircular, it also depends on the wind direction. The dominant frequency of vortex shedding, f_s, is given by

$$f_s = S\frac{\bar{V}}{D} \quad \text{cps} \tag{29.41}$$

where S = dimensionless constant called the Strouhal number, $\bar{V}$ = mean wind velocity, and D = width of the frontal area. The second dimensionless parameter is the Reynolds number $R = \bar{V}D/\nu$, where ν = kinematic viscosity. For air under normal conditions, $\nu = 1.6 \times 10^{-4}$ ft²/sec.

For a body having a rectangular or square cross section, the Strouhal number is almost independent of the Reynolds number.

For a body having a circular cross section, the Strouhal number varies with the regime of the flow as characterized by the Reynolds number. There are three major regions: the subcritical region for $R \lesssim 3 \times 10^5$, the supercritical region for $3 \times 10^5 \lesssim R \lesssim 3 \times 10^6$, and the transcritical region for $R \gtrsim 3 \times 10^6$. Approximate values of the Strouhal number for typical cross sections are given in Table 29.2. The numbers

Table 29.2. Aerodynamic Data for Prediction of Vortex-induced Oscillations in Turbulent Flow

Cross section	Strouhal number S	Rms lift coefficient σ_L	Bandwidth B	Correlation length L (diameters)
Circular: region				
Subcritical	0.2	0.5	0.1	2.5
Supercritical	Not marked	0.14	Not marked	1.0
Transcritical	0.25	0.25	0.3	1.5
Square:				
Wind normal to face	0.11	0.6	0.2	3

given in this table are based on Refs. 1, 22, 23, and 24 and other measurements and may be used for turbulent shear flow.

PREDICTION OF VORTEX-INDUCED OSCILLATION

Although the mechanism of vortex shedding and the character of the lift forces have been the subject of a great number of studies,[25] the available information does not permit an accurate prediction of these oscillations.

The motion is most often viewed as forced oscillation due to the lift force, which, per unit length, may be written as

$$F_L = \frac{1}{2}\rho D\bar{V}^2 C_L(t) \tag{29.42}$$

where $C_L(t)$ is a lift coefficient fluctuating in a harmonic or random way. (Some authors[26, 27] consider vortex shedding to be self-excitation, which does not seem necessary, however, for relatively small motions.) Hence, the solution of the response depends on the time-history assumed for $C_L(t)$.

HARMONIC EXCITATION OF PRISMATIC CYLINDERS BY VORTICES

Harmonic excitation represents a traditional model for vortex excitation, but it is really justified only for very low Reynolds numbers ($\lesssim$ 300) or possibly for large vibration where the motion starts controlling both the wake and the lift forces in the form of the "locking-in" phenomenon. Strongest oscillations arise at that wind velocity for which the frequency of vortex shedding f_s is equal to one of the natural frequencies of the structure f_j. This resonant wind velocity is, from Eq. (29.41),

$$V_c = \frac{1}{S} f_j D \tag{29.43}$$

With free-standing towers and stacks, resonance in the first two modes is met most often; resonance with higher modes has been observed as well with guyed towers (Fig. 29.26).

At the resonant wind velocity, the lift force is given by Eq. (29.42) in which $C_L(t) =$

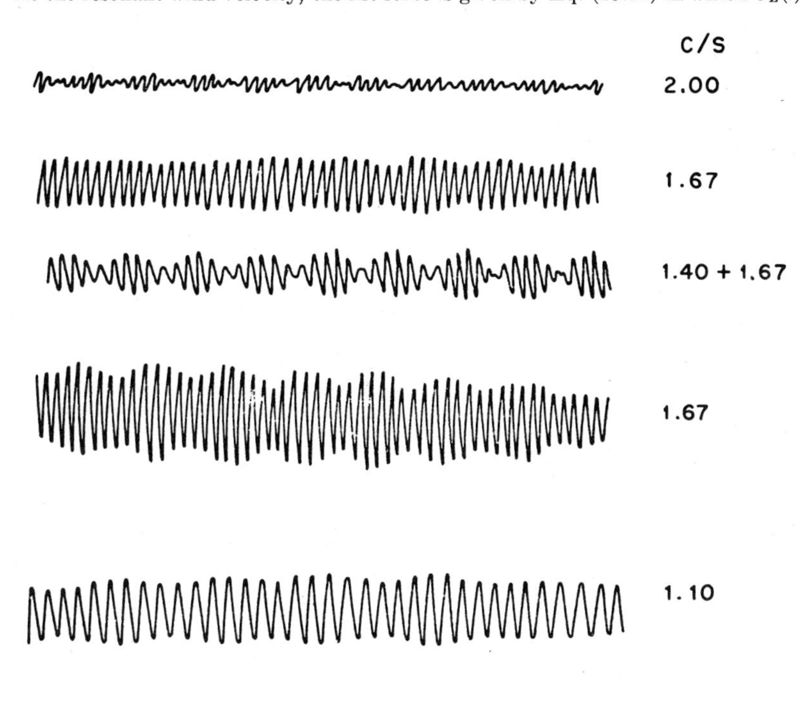

FIG. 29.26. Vortex-induced oscillations in different modes measured on 1,000 ft guyed tower.[28]

$C_L \sin 2\pi f_j t$, and C_L = amplitude of lift coefficient. Assuming a uniform wind profile and a constant diameter D, the resonant amplitude of mode j at the critical wind velocity V_c is, from Eq. (29.31),

$$u_j(z) = \frac{\rho C_L}{16\pi^2 S^2} \frac{D^3}{\zeta M_j} \phi_j(z) \int_0^H \phi_j(z)\, dz \tag{29.44}$$

where M_j is given by Eq. (29.35) and ζ = structural damping ratio. The formula can be further simplified if it is assumed that the lift force is distributed along the structure in proportion to the mode $\phi_j(z)$. (This assumption reflects the loss of spanwise correlation of the forces.) Then, with constant mass per unit length $m(z) = m$, the resonant amplitude at the height where the modal displacement is maximum:

$$u_j = \frac{\rho C_L}{16\pi^2 S^2} \frac{D^3}{\zeta m} \tag{29.45}$$

For the first mode of a free-standing structure, this occurs at the tip. In higher modes, this amplitude appears at the height where local resonance takes place. For circular cylinders, a design value of the lift coefficient C_L is about $\sqrt{2}\sigma_L$. This simple formula can be used for the first estimate of the amplitudes that are likely to represent the upper bound. It is also indicative of the role of the diameter, mass, and damping of the structure. Approximate values of σ_L are given in Table 29.1.

RANDOM EXCITATION OF PRISMATIC CYLINDERS BY VORTICES

Even when vortex shedding appears very regular, the lift force and thus $C_L(t)$ are not purely harmonic but random. The power spectrum of the lift force per unit length is from Eq. (29.42).

$$S_L(f) = \left(\frac{1}{2}\rho D\bar{V}^2\sigma_L\right)^2 S_L'(f) \tag{29.46}$$

where $\sigma_L = \sqrt{\overline{C_L{}^2(t)}}$ is the standard deviation of the lift coefficient and $S_L'(f)$ = normalized power spectrum of $C_L(t)$ for which

$$\int_0^\infty S_L'(f)\,df = 1 \tag{29.47}$$

With circular cylinders, the lift force is narrow-band random in the subcritical and transcritical[22, 23] ranges where the energy is distributed about the dominant frequency

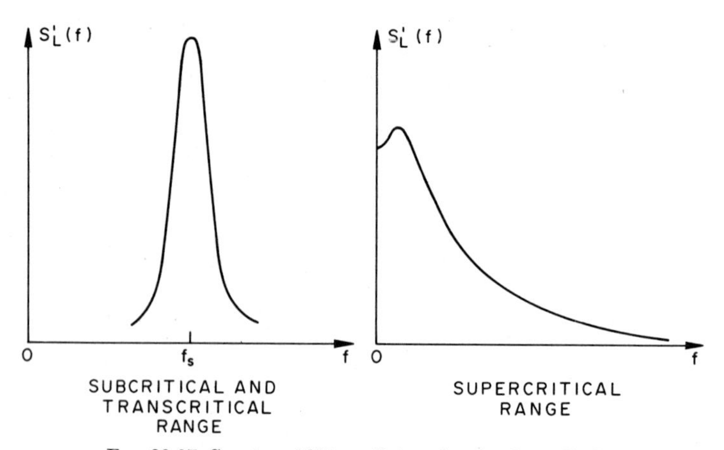

FIG. 29 27. Spectra of lift coefficient for circular cylinder.

f_s, given by Eq. (29.41) (Fig. 29.27*a*). Such spectra can be described by a Gaussian-type curve,

$$S_L'(f) = \frac{1}{\sqrt{\pi}\,Bf_s}\exp\left[-\left(\frac{1-f/f_s}{B}\right)^2\right] \tag{29.48}$$

A few design values of bandwidth B are given in Table 29.2. In the supercritical range, the power spectrum is broad (Fig. 29.27*b*) and can be expressed as[29]

$$S_L'(f) = 4.8\,\frac{1 + 682.2(fD/\bar{V})^2}{[1 + 227.4(fD/\bar{V})^2]^2}\,\frac{D}{\bar{V}} \tag{29.49}$$

Because the vortices are three-dimensional, a realistic treatment also requires the inclusion of the spanwise cross correlation of the lift forces. This can be done in terms of the "correlation length" L given in number of diameters.

Approximate values of L are given in Table 29.3. The correlation length decreases with turbulence[30] and shear and increases with aspect ratio $2H/D$ and the amplitude of the motion (Fig. 29.28).

Using the correlation length, the spectral density of the lift force, Eqs. (29.48) and (29.49), and a few further approximations, the vibration can be evaluated from Eqs.

(29.32) to (29.34). The root-mean-square (rms) displacement at height z in mode j is approximately[1]

$$\sqrt{\overline{u_j^2(z,t)}} = \frac{\pi^{1/4}\sigma_L \rho D^4 \phi_j(z/H)}{\sqrt{B\zeta}\,(4\pi S)^2 M_j} C$$

where

$$C^2 = \frac{(H/D)^2}{1 + (H/2LD)} \int_0^1 \left(\frac{z}{H}\right)^{3\alpha} \phi_j^2 \left(\frac{z}{H}\right) d\left(\frac{z}{H}\right)$$

where α = wind profile exponent (Fig. 29.19), and parameters S, σ_L, B, and L are given in Table 29.1. The mode $\phi_j(z/H)$ is dimensionless, and consequently M_j is in slugs in this

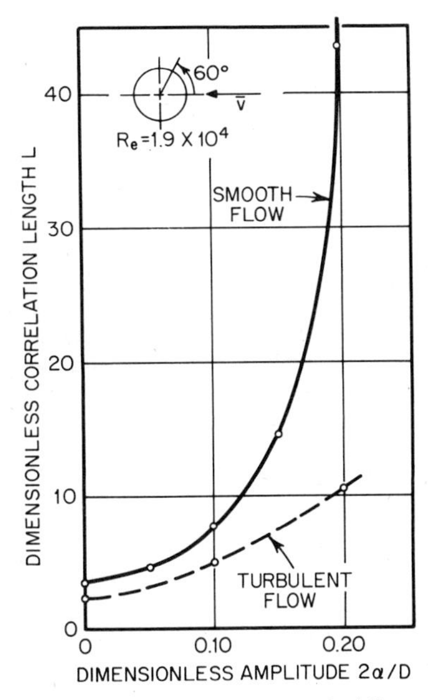

FIG. 29.28. Variation of correlation length of vortex shedding with amplitude of motion and turbulence ($2a$ = double amplitude, turbulence intensity 10 per cent).

case. The peak response is $g\sqrt{\overline{u_j^2(z,t)}}$, where the peak factor g is given by Eq. (29.30). If it is larger than about 2 per cent of diameter D, locking-in may develop and the analysis should be repeated assuming harmonic excitation or at least random excitation with a significantly increased correlation length, as Fig. 29.28 indicates.

RANDOM EXCITATION OF TAPERED CYLINDERS BY VORTICES

Tapered cylinders, such as stacks, also vibrate due to vortex shedding; but less is known about the mechanism of excitation. It appears that the lift forces are narrow-band random with a rather small correlation length L and with the dominant frequency f_s given by Eq. (29.41). As the diameter is variable, local resonance between f_s and the natural frequency f_j takes place at different heights z_r. As the wind speed increases, the resonance first appears at the tip and shifts downward. The critical wind speed for each height follows from Eq. (29.43) with $D = D(z_r)$. The rms displacements at height H due to local resonance at height z_r can be obtained from an approximate formula,[32]

$$\sqrt{\overline{u_j^2(H,t)}} = \sqrt{\frac{L}{2\pi^3\zeta\Psi}\,\frac{\sigma_L\rho D^4(z_r)\phi_j(z_r)}{8S^2M_j}}\;\phi_j(H)$$

where

$$\Psi = \frac{dD(z_r)}{dz} + \frac{\alpha D(z_r)}{z_r}$$

or with a constant taper

$$\Psi = \frac{t}{H} + \frac{\alpha D(z_r)}{z_r}$$

where $t = D(0) - D(H)$ and α = the wind-profile exponent.

The other parameters can be taken from Table 29.2. The values listed for the trans-critical region may be adequate, inasmuch as most tapered stacks are large. The peak displacement is again obtained by means of the peak factor given by Eq. (29.30).

Maximum response of chimneys in the first mode usually results from local resonance at about $\frac{3}{4}$ H. The height of maximum excitation follows from condition $d[D^4(z)\phi_j(z)]/dz = 0$.

SUPPRESSION OF VORTEX-INDUCED VIBRATIONS

Vortex shedding may induce severe vibration of a cylindrical structure such as a chimney, free-standing tower, guyed mast, bridge columns, etc. Very strong oscillations have been observed[28, 31] in all-welded structures where the damping ratio is extremely low, sometimes less than 0.005.[8, 28] Welded structures are particularly prone to fatigue failure, as the endurance limit may be only a fraction of the strength if heavy notches, flaws, attachments, or other adverse details are present. In other cases, the motion is intolerable because of its physiological effects or swaying of antennas. For these reasons, suppression of vibration is often desirable.

In some cases, vibration can be reduced by increasing the structural damping. This can be accomplished by additional dampers attached to an independent support[28] or to a special mass suspended from the structure and suitably tuned or by hanging chains.[33] Columns of a few bridges were filled with gravel, sand, or plastic balls partly filled with oil. The increase in mass may be favorable but can reduce the original structural damping.

Another successful method of vibration control is to break down the wake pattern by providing the surface by helical "strakes" or "spoilers."[28, 31, 34] A suitable height of the spoilers is about $0.1D$ or more with a pitch of about $5D$. A significant drawback of the spoilers is that they considerably increase the drag, sometimes as much as 100 per cent or even more.[31, 35]

WAKE BUFFETING

If one structure is located in the wake of another, vortices shed from the upstream structure may cause oscillation of the downstream structure[36, 37] (Fig. 29.17). If the two structures differ greatly in size or shape, this excitation is usually not significant. Strong vibration of the downstream structure may arise when two or more structures are identical and less than about 10 diameters apart. Then the structure in the wake is efficiently excited by well-tuned wake buffeting and its own vortex shedding. Such excitation has been observed with stacks and bridges and to a certain degree with hyperbolic cooling towers.[36]

GALLOPING OSCILLATIONS

Vibrations due to turbulence and vortices discussed above are induced by aerodynamic forces which are, to a high degree, independent of the motion and act even on stationary bodies. Quite a different kind of oscillation is induced by the aerodynamic forces generated by the motion itself. Such forces may result from oscillatory changes in pressure distribution brought about by the continuous change in the angle under which the wind strikes the structure ("angle of attack"). This kind of oscillation often has a tendency to diverge; it is called, summarily, "aerodynamic instability," "flutter," or "self-excited oscillation." Sudden start and violent amplitudes are typical of such phenomena (Fig. 29.16*c*).

The mechanism of this oscillation is, in general, complex. The aerodynamic forces may be a function of the displacements (translation and rotation), vibration velocity, or both, and they may interact with turbulence and vortex shedding. The basic type of the self-excited oscillations is the lateral (across-wind) oscillation induced by aerodynamic forces which are related to vibration velocity alone. Such oscillation is referred to as "galloping." Typical features of galloping oscillation are motion in the direction perpendicular to that of the wind, sudden onset, large steady amplitudes increasing with wind velocity, and a frequency equal to the natural frequency. Galloping oscillation occurs in transmission lines and in a variety of structures having square, rectangular, or other sharp-edged cross sections.

The origin of galloping oscillation depends on the relation between lift and drag. If a body moves with a velocity $\dot{u}$ in a flow having velocity $\bar{V}$ perpendicular to its direction

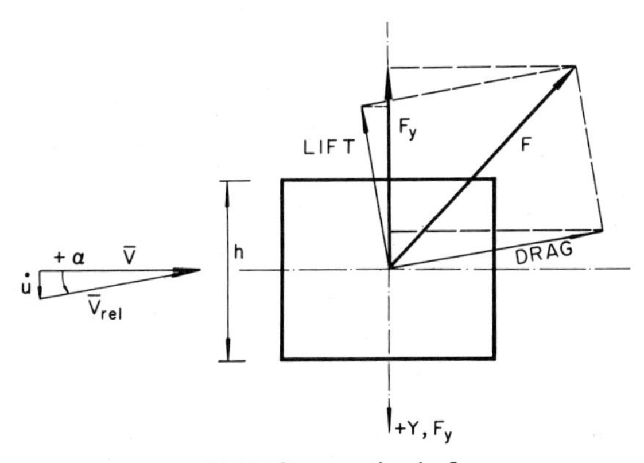

FIG. 29.29. Cross section in flow.

(Fig. 29.29), the aerodynamic force acting on the body is produced by relative wind velocity $\bar{V}_{\text{rel}}$. The angle of attack of relative wind is

$$\alpha = \arctan \frac{\dot{u}}{\bar{V}} \tag{29.50}$$

The drag and lift components D and L of the aerodynamic force F are

$$D = C_D \frac{1}{2} \rho h l \bar{V}_{\text{rel}}^2$$

$$L = C_L \frac{1}{2} \rho h l \bar{V}_{\text{rel}}^2$$

where C_D and C_L are drag and lift coefficients at angle α (Fig. 29.30), h = depth of the cross section, and l = length of the body.

The component of force F into the direction of axis Y, therefore, is

$$F_y = -(C_D \sin \alpha + C_L \cos \alpha) \frac{1}{2} \rho h l \bar{V}^2 \sec^2 \alpha = C_{F_y} \frac{1}{2} \rho h l \bar{V}^2 \tag{29.51}$$

where

$$C_{F_y} = -(C_L + C_D \tan \alpha) \sec \alpha \tag{29.52}$$

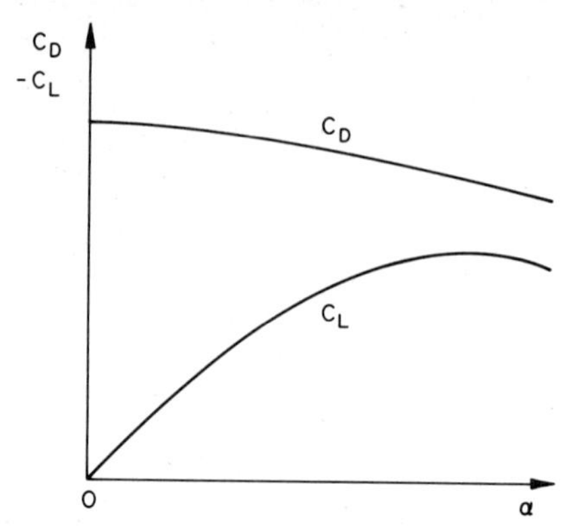

FIG. 29.30. Lift and drag as function of angle of attack.

The lateral force excites the vibration if the first derivative of C_{F_y} at $\alpha = 0$ is >0, hence

$$A_1 = \left.\frac{dC_{F_y}}{d\alpha}\right|_{\alpha=0} = -\left(\frac{dC_L}{d\alpha} + C_D\right) > 0 \tag{29.53}$$

This condition for aerodynamic instability is known as "Den Hartog's criterion."[38] Substitution of Eq. (29.50) into Eq. (29.52) indicates that the aerodynamic forces depend on vibration velocity and thus actually represent the aerodynamic damping. This damping is negative if $A_1 > 0$. Because the system also has structural damping ζ, which is positive, the vibration will start only if the total available damping becomes less than 0. This condition yields the onset (minimum) wind velocity for galloping from the equilibrium (or zero displacement) position,

$$\bar{V}_0 = \zeta \frac{2\pi f_j h}{n A_1} \tag{29.54}$$

where f_j = natural frequency, $n = \rho h^2/(4m)$ = mass parameter, and m = mass of the body per unit length. Some values of coefficient A_1 are given in Table 29.3.

Table 29.3. Coefficients A_1 for Determination of Galloping Onset Wind Velocity (Infinite Prisms)

	Cross section (Side ratio)					
	Unstable in smooth flow			Stable in smooth flow		
	Square	Rect.	Rect.	Rect.	Rect.	D-section*
$V \rightarrow$ Flow	1 × 1	2 × 3	1 × 2	3 × 2	2 × 1	2 × 1
Smooth	2.7	1.91	2.8	0	−0.03	−0.1
Turbulent ≈10 per cent intensity	2.6	1.83	−2.0	0.74	0.17	0

* Varies with Reynolds number.

Galloping oscillations starting from zero initial displacement can occur only when the cross section has $A_1 > 0$. Cross sections having $A_1 \leq 0$ are generally considered stable even though galloping may sometimes arise if triggered by a large initial amplitude.[41]

The response and the onset velocity are often very sensitive to turbulence. Some cross sections such as a flat rectangle or a D section are stable in smooth flow but can become unstable in turbulent flow.[41, 42] With other cross sections, turbulence may stabilize a shape that is unstable in smooth flow (see Table 29.2).

From Eqs. (29.51) and (29.52) the nonlinear, negative aerodynamic damping can be calculated[43] for inclusion in the treatment of the across-wind response due to atmospheric turbulence.

The prediction of oscillations for wind velocities greater than V_0 depends on the shape of the C_{F_y} coefficient and requires the application of nonlinear theory.[39-42] A few typical cases are shown in Fig. 29.31. The cases are typical of a square cross sec-

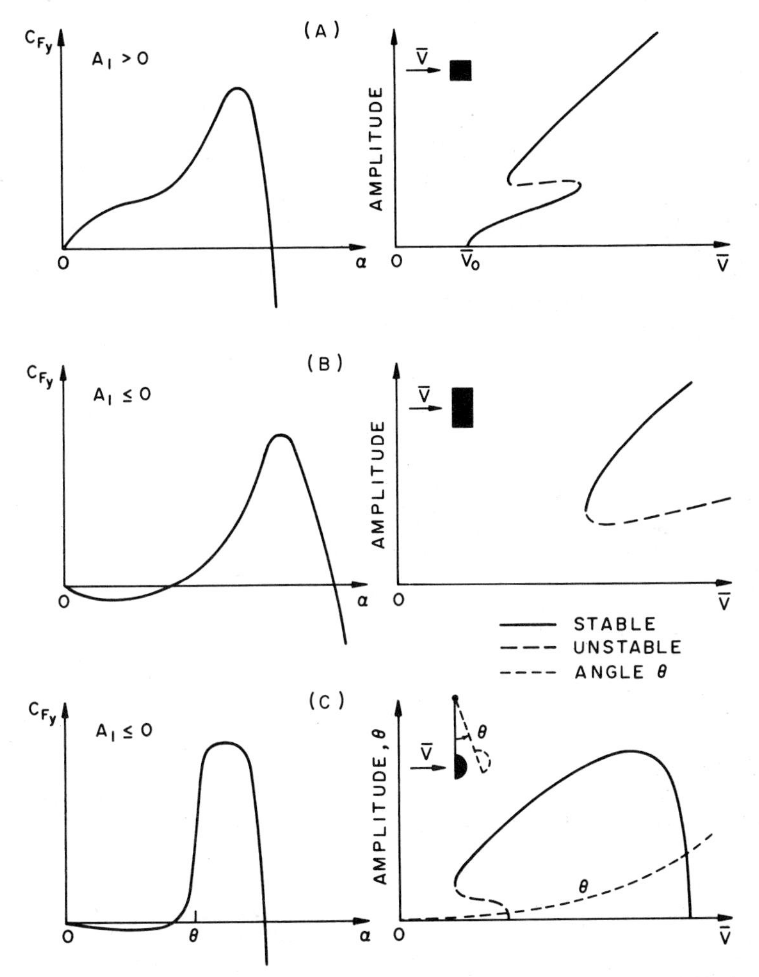

FIG. 29.31. Typical lateral force coefficients C_{F_y} and corresponding galloping oscillations: (*a*) vibration from equilibrium position, (*b*) vibration triggered by initial amplitudes, and (*c*) vibration with variable angle of attack.

tion, a flat rectangular section, and a D section whose angle of attack is allowed to change due to drag. Similar response can be expected with other cross sections.

Torsion can also participate in galloping oscillations and play an important part in the vibration. This is the case with angle cross sections[44] and bundled conductors.[45] The quasi-steady theory of pure torsional galloping can be found in Ref. 46. A solution of coupled galloping is presented in Ref. 47.

Galloping often appears in overhead conductors which also vibrate due to vortex shedding. Vortex shedding produces resonant vibration in a high-vibration mode. Galloping usually involves the fundamental mode and is known to occur when the conductor is ice-coated or free of ice. The vibration often leads to fatigue failures, and various techniques are therefore used to reduce the amplitude. This can be achieved by means of resonant dampers[48] consisting of auxiliary masses suspended on short lengths of cable which dissipate energy through the bending or aerodynamic dampers[49] consisting of perforated shrouds. Vibrations of bundled conductors can be eliminated by twisting the bundle[45] and thereby changing the aerodynamic characteristics in the spanwise direction.

VIBRATION OF SPECIAL STRUCTURES

The basic types of vibration discussed above are common in many structures. However, there are some special structures which would require individual treatment. A few examples are cited below.

Guyed towers experience complicated vibration patterns because of the nonlinearity of the guys, the three-dimensional character of the response, the interaction between the guys and the tower, and other factors.[28, 50–52]

Hyperbolic cooling towers can suffer from some of the effects of wake buffeting[36] and are susceptible to turbulence.[53] Information on vibration of a number of special structures can be found in Refs. 2 to 5.

REFERENCES

1. Davenport, A. G., et al.: "New Approaches to Design against Wind Action," to be published by the Faculty of Engineering Science, University of Western Ontario.
2. *Proc. IUTAM-IAHR Symp. Karlsruhe*, 1972.
3. *Proc. Conf. National Physical Laboratory, Teddinton, Middlesex*, 1963.
4. *Proc. Intern. Res. Seminar*, Ottawa, 1967.
5. *Proc. 3d Intern. Conf.*, Tokyo, 1971.
6. Davenport, A. G.: *Inst. Civil Eng. Paper* No. 6480, 449–472 (August, 1961).
7. Harris, R. I.: Seminar of Construction Industry Research and Information Association, paper 3, Institution of Civil Engineers, 1970.
8. Novak, M.: *Acta Tech. Czechoslovak Acad. Sci.*, **4**:375–404 (1967).
9. Vickery, B. J.: *J. Struct. Div. Am. Soc. Civil Engrs.*, **98**, 21–36 (January, 1972).
10. Davenport, A. G.: *Proc. Inst. Civil Engs.*, **23**:449–472 (1962).
11. Etkin, B.: Meeting on Ground Wind Load Problems in Relation to Launch Vehicles, pp. 21.1–15, Langley Research Center, NASA, June, 1966.
12. Davenport, A. G.: *J. Struct. Div., Am. Soc. Civil Engrs.*, **93**, 11–34 (June, 1967).
13. "Canadian Structural Design Manual 1970," Suppl. 4, National Research Council of Canada, 1970.
14. Chen, P. W., and L. E. Robertson: *J. Struct. Div. Am. Soc. Civil Engrs.*, **98**, 1681–1695 (August, 1972).
15. Hansen, R. T., J. W. Reed, and E. H. Vanmarcke: *J. Struct. Div. Am. Soc. Civil Engrs.*, **99**, 1589–1605 (July, 1973).
16. Vellozzi, Y., and E. Cohen: *J. Struct. Div. Am. Soc. Civil Engrs.*, **94**, 1295–1313 (June, 1968).
17. Vickery, B.: U.S. Dept. of Commerce, *Nat. Bur. Std. Bldg. Sci., Ser.* **30**:93–104.
18. Simiu, E.: *J. Struct. Div. Am. Soc. Civil Engrs.*, **100**, 1897–1910 (September, 1974).
19. Allen, D. E., and W. A. Dalgliesh: Preliminary Publication of IABSE Symposium on Resistance and Ultimate Deformability of Structures, pp. 279–285, Lisbon, 1973.
20. Dalgliesh, W. A.: *Proc. U.S.-Japan Res. Seminar Wind Effects Structures*, Kyoto, Japan, September, 1974.
21. Dalgliesh, W. A.: *J. Struct. Div. Am. Soc. Civil Engrs.*, **97**, 2173–2187 (September, 1971).

22. Roshko, A.: *J. Fluid Mech.*, **10**:345–356 (1961).
23. Cincotta, J. J., G. W. Jones, and R. W. Walker: Meeting on Ground Wind Load Problems in Relation to Launch Vehicles, pp. 20.1–35, Langley Research Center, NASA, 1966.
24. Novak, M.: Ref. 5, pp. 799–809.
25. Morkovin, M. V.: *Proc. Symp. Fully Separated Flows*, pp. 102–118, ASME, 1964.
26. Nakamura, Y.: *Rept. Res. Inst. Appl. Mech., Kyushu University*, **17**(59):217–234 (1969).
27. Hartlen, R. T., and I. G. Currie: *J. Eng. Mech. Div. Am. Soc. Civil Engrs.*, **70,** 577–591 (October, 1970).
28. Novak, M.: *Proc. IASS Symp. Tower-Shaped Steel r.c. Structures*, Bratislava, 1966.
29. Fung, Y. C.: *J. Aerospace Sci.*, **27**(11):801–814 (November, 1960).
30. Surry, D.: *J. Fluid Mech.*, **52**(3):543–563 (1972).
31. Wotton, L. R., and C. Scruton: Construction Industry Research and Information Association Seminar, Paper 5, June, 1970.
32. Vickery, B. J., and A. W. Clark: *J. Struct. Div. Am. Soc. Civil Engrs.*, **98,** 1–20 (January, 1972).
33. Reed, W. H.: Ref. 4, Paper 36, pp. 283–321.
34. Scruton, C.: National Physical Laboratory Note 1012, April, 1963.
35. Novak, M.: Ref. 4, pp. 429–457.
36. Scruton, C.: Ref. 4, pp. 115–161.
37. Cooper, K. R., and Wardlaw, R. L.: Ref. 5, pp. 647–655.
38. Den Hartog: *Trans., AIEE*, **51**:1074 (1932).
39. Parkinson, G. V., and J. D. Smith: *Quart. J. Mech. Appl. Math.*, **17**(2):225–239 (1964).
40. Novak, M.: *J. Eng. Mech. Div. Am. Soc. Civil Engrs.*, **95,** 115–142 (February, 1969).
41. Novak, M.: *J. Eng. Mech. Div. Am. Soc. Civil. Engrs.*, **98,** 27–46 (February, 1972).
42. Novak, M., and Tanaka, H.: *J. Eng. Mech. Div. Am. Soc. Civil Engrs.*, **100,** 27–47 (February, 1974).
43. Novak, M., and Davenport, A. G.: *J. Eng. Mech. Div. Am. Soc. Civil Engrs.*, **96,** 17–39 (February, 1970).
44. Wardlaw, R. L.: Ref. 4, pp. 739–772.
45. Wardlaw, R. L., K. R. Cooper, R. G. Ko, and J. A. Watts: *Trans. IEEE*, 1975.
46. Modi, V. J., and J. E. Slater: Ref. 2, pp. 355–372.
47. Blevins, R. D., and W. D. Iwan: *J. Appl. Mech.*, **41,** no. 4, 1974.
48. "Overhead Conductor Vibration," Alcoa Aluminum Overhead Conductor Engineering Data, no. 4, 1974.
49. Hunt, J. C. R., and D. J. W. Richards: *Proc. Inst. Elec. Engrs.*, **116**(11):1869–1874 (November, 1969).
50. Davenport, A. G., and G. N. Steels: *J. Struct. Div. Am. Soc. Civil Engrs.*, **91,** 43–70 (April, 1965).
51. Davenport, A. G.: Engineering Institute of Canada, **3**:119–141 (1959).
52. McCaffrey, R. J., and Hartmann, A. J.: *J. Struct. Div. Am. Soc. Civil Engrs.*, **98,** 1309–1323 (June, 1972).
53. Hashish, M. G., and Abu-Sitta, S. H., *J. Struct. Div. Am. Soc. Civil Engrs.*, **100,** 1037–1051 (May, 1974).

30

THEORY OF VIBRATION ISOLATION

Charles E. Crede
California Institute of Technology

Jerome E. Ruzicka
Bose Corporation

INTRODUCTION

Vibration isolation concerns means to bring about a reduction in a vibratory effect. A vibration isolator in its most elementary form may be considered as a resilient member connecting the equipment and foundation. The function of an isolator is to reduce the magnitude of motion transmitted from a vibrating foundation to the equipment, or to reduce the magnitude of force transmitted from the equipment to its foundation.

CONCEPT OF VIBRATION ISOLATION

The concept of vibration isolation is illustrated by consideration of the single degree-of-freedom system illustrated in Fig. 30.1. This system consists of a rigid body representing an equipment connected to a foundation by an isolator having resilience and energy-

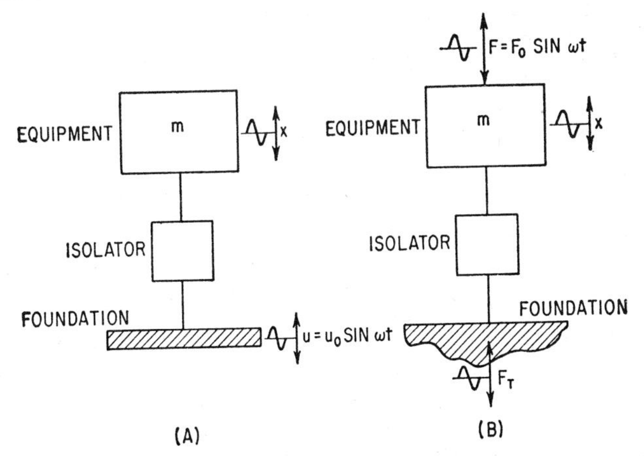

FIG. 30.1. Schematic diagrams of vibration isolation systems: (*A*) vibration isolation where motion u is imposed at the foundation and motion x is transmitted to the equipment; (*B*) vibration isolation where force F is applied by the equipment and force F_T is transmitted to the foundation.

dissipating means; it is unidirectional in that the body is constrained to move only in vertical translation. The performance of the isolator may be evaluated by the following characteristics of the response of the equipment-isolator system of Fig. 30.1 to steady-state sinusoidal vibration:

Absolute Transmissibility. Transmissibility is a measure of the reduction of transmitted force or motion afforded by an isolator. If the source of vibration is an oscillating motion of the foundation (motion excitation), transmissibility is the ratio of the vibration amplitude of the equipment to the vibration amplitude of the foundation. If the source of vibration is an oscillating force originating within the equipment (force excitation), transmissibility is the ratio of the force amplitude transmitted to the foundation to the amplitude of the exciting force.

Relative Transmissibility. Relative transmissibility is the ratio of the relative deflection amplitude of the isolator to the displacement amplitude imposed at the foundation. A vibration isolator effects a reduction in vibration by permitting deflection of the isolator. The relative deflection is a measure of the clearance required in the isolator. This characteristic is significant only in an isolator used to reduce the vibration transmitted from a vibrating foundation.

Motion Response. Motion response is the ratio of the displacement amplitude of the equipment to the quotient obtained by dividing the excitation force amplitude by the static stiffness of the isolator. If the equipment is acted on by an exciting force, the resultant motion of the equipment determines the space requirements for the isolator, i.e., the isolator must have a clearance at least as great as the equipment motion.

FORM OF ISOLATOR

The essential features of an isolator are resilient load-supporting means and energy-dissipating means. In certain types of isolators, the functions of the load-supporting means and the energy-dissipating means may be performed by a single element, e.g., natural or synthetic rubber. In other types of isolators, the resilient load-carrying means may lack sufficient energy-dissipating characteristics, e.g., metal springs; then separate and distinct energy-dissipating means (dampers) are provided. For purposes of analysis, it is assumed that the springs and dampers are separate elements. In general, the springs are assumed to be linear and massless. The effects of nonlinearity and mass of the load-supporting means upon vibration isolation are considered in later sections of this chapter.

Various types of dampers are shown in combination with ideal springs in the following idealized models of isolators illustrated in Table 30.1. Practical aspects of isolator design are considered in Chap. 32.

Rigidly Connected Viscous Damper. A viscous damper c is connected rigidly between the equipment and its foundation as shown in Table 30.1*A*. The damper has the characteristic property of transmitting a force F_c that is directly proportional to the relative velocity $\dot{\delta}$ across the damper, $F_c = c\dot{\delta}$. This damper sometimes is referred to as a *linear damper*.

Rigidly Connected Coulomb Damper. An isolation system with a rigidly connected Coulomb damper is indicated schematically in Table 30.1*B*. The force F_f exerted by the damper on the mass of the system is constant, independent of position or velocity, but always in a direction that opposes the relative velocity across the damper. In a physical sense, Coulomb damping is approximately attainable from the relative motion of two members arranged to slide one upon the other with a constant force holding them together.

Elastically Connected Viscous Damper. The elastically connected viscous damper is shown in Table 30.1*C*. The viscous damper c is in series with a spring of stiffness k_1; the load-carrying spring k is related to the damper spring k_1 by the parameter $N = k_1/k$. This type of damper system sometimes is referred to as a *viscous relaxation system*.

Elastically Connected Coulomb Damper. The elastically connected Coulomb damper is shown in Table 30.1*D*. The friction element can transmit only that force which is developed in the damper spring k_1. When the damper slips, the friction force F_f is independent of the velocity across the damper, but always is in a direction that opposes it.

Table 30.1. Types of Idealized Vibration Isolators

(A) RIGIDLY CONNECTED VISCOUS DAMPER	(B) RIGIDLY CONNECTED COULOMB DAMPER	(C) ELASTICALLY CONNECTED VISCOUS DAMPER	(D) ELASTICALLY CONNECTED COULOMB DAMPER
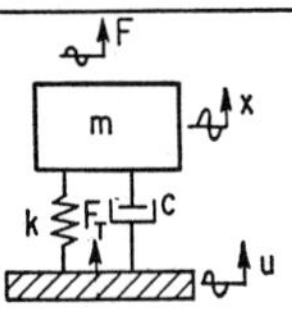 F, m, x, k, F_T, c, u	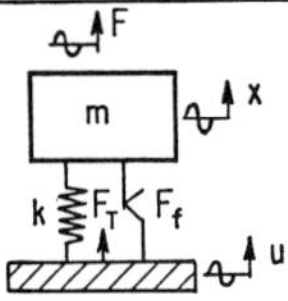 F, m, x, k, F_T, F_f, u	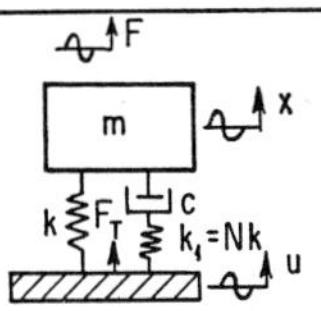 F, m, x, k, F_T, c, $k_1 = Nk$, u	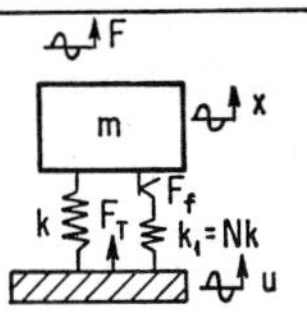 F, m, x, k, F_T, F_f, $k_1 = Nk$, u
EXCITATION			
$u = u_0 \text{SIN}\,\omega t$ $F = F_0 \text{SIN}\,\omega t$	$u = u_0 \text{SIN}\,\omega t$ * OR $\ddot{u} = \ddot{u}_0 \text{SIN}\,\omega t$ * $F = F_0 \text{SIN}\,\omega t$	$u = u_0 \text{SIN}\,\omega t$ $F = F_0 \text{SIN}\,\omega t$	$u = u_0 \text{SIN}\,\omega t$ * OR $\ddot{u} = \ddot{u}_0 \text{SIN}\,\omega t$ * $F = F_0 \text{SIN}\,\omega t$
RESPONSE			
$x = x_0 \text{SIN}(\omega t + \theta^{+})$ OR $\delta = \delta_0 \text{SIN}(\omega t + \theta^{+})$ WHERE $\delta = x - u$ $F_T = (F_T)_0 \text{SIN}(\omega t + \theta^{+})$	$x = x_0 \text{SIN}(\omega t + \theta^{+})$ OR $\delta = \delta_0 \text{SIN}(\omega t + \theta^{+})$ WHERE $\delta = x - u$ $F_T = (F_T)_0 \text{SIN}(\omega t + \theta^{+})$	$x = x_0 \text{SIN}(\omega t + \theta^{+})$ OR $\delta = \delta_0 \text{SIN}(\omega t + \theta^{+})$ WHERE $\delta = x - u$ $F_T = (F_T)_0 \text{SIN}(\omega t + \theta^{+})$	$x = x_0 \text{SIN}(\omega t + \theta^{+})$ OR $\delta = \delta_0 \text{SIN}(\omega t + \theta^{+})$ WHERE $\delta = x - u$ $F_T = (F_T)_0 \text{SIN}(\omega t + \theta^{+})$
FREQUENCY PARAMETERS			
$\omega_0 = \sqrt{k/m}\ (c = 0)$	$\omega_0 = \sqrt{k/m}\ (F_f = 0)$	$\omega_0 = \sqrt{k/m}\ (c = 0)$ $\omega_\infty = \sqrt{(N+1)\frac{k}{m}}\ (c = \infty)$	$\omega_0 = \sqrt{k/m}\ (F_f = 0)$ $\omega_\infty = \sqrt{(N+1)\frac{k}{m}}\ (F_f = \infty)$
DAMPING PARAMETERS			
$c_c = 2\sqrt{km}$ $\zeta = c/c_c$	$\eta = \frac{F_f}{ku_0}$ $\xi = \frac{F_f}{m\ddot{u}_0}$ $\xi_F = \frac{F_f}{F_0}$	$c_c = 2\sqrt{km}$ $\zeta = c/c_c$	$\eta = \frac{F_f}{ku_0}$ $\xi = \frac{F_f}{m\ddot{u}_0}$ $\xi_F = \frac{F_f}{F_0}$

* PHYSICALLY, THESE FORMS OF EXCITATION ARE IDENTICAL. THEY ARE EXPRESSED IN TWO DIFFERENT MATHEMATICAL FORMS FOR CONVENIENCE IN DEFINING THE DAMPING PARAMETER FOR COULOMB DAMPING.

† IN VIBRATION ISOLATION, ONLY THE MAGNITUDE OF THE RESPONSE IS OF INTEREST; THUS, THE PHASE ANGLE USUALLY IS NEGLECTED.

Table 30.2. Transmissibility and Motion Response

Where the equation is shown graphically, the applicable figure is

Type of damper	Absolute transmissibility
Rigidly connected viscous damper	(a) $T_A = \frac{x_0}{u_0} = \frac{F_T}{F_0} = \sqrt{\frac{1 + \left(2\zeta \frac{\omega}{\omega_0}\right)^2}{\left(1 - \frac{\omega^2}{\omega_0^2}\right)^2 + \left(2\zeta \frac{\omega}{\omega_0}\right)^2}}$ Fig. 30.2
Rigidly connected Coulomb damper (see Note 1)	(d) $(T_A)_D = \frac{x_0}{u_0} = \sqrt{\frac{1 + \left(\frac{4}{\pi}\eta\right)^2 (1 - 2\omega_0^2/\omega^2)}{\left(1 - \frac{\omega^2}{\omega_0^2}\right)^2}}$ Fig. 30.5 (see Note 2)
Elastically connected viscous damper	(g) $T_A = \frac{x_0}{u_0} = \frac{F_T}{F_0} = \sqrt{\frac{1 + 4\left(\frac{N+1}{N}\right)^2 \zeta^2 \frac{\omega^2}{\omega_0^2}}{\left(1 - \frac{\omega^2}{\omega_0^2}\right)^2 + \frac{4}{N^2}\zeta^2 \frac{\omega^2}{\omega_0^2}\left(N + 1 - \frac{\omega^2}{\omega_0^2}\right)^2}}$ Fig. 30.10 (see Note 3)
Elastically connected Coulomb damper (see Note 1)	(j) $(T_A)_D = \frac{x_0}{u_0} = \sqrt{\frac{1 + \left(\frac{4}{\pi}\eta\right)^2 \left[\left(\frac{N+2}{N}\right) - 2\left(\frac{N+1}{N}\right)\left(\frac{\omega_0}{\omega}\right)^2\right]}{\left(1 - \frac{\omega^2}{\omega_0^2}\right)^2}}$ Fig. 30.13 (see Notes 4 and 5)

NOTE 1: These equations apply only when there is relative motion across the damper.
NOTE 2: This equation applies only when excitation is defined in terms of displacement amplitude.
NOTE 3: These curves apply only for optimum damping [see Eq. (30.15)]; curves for other values of damping are given in Ref. 4.

INFLUENCE OF DAMPING IN VIBRATION ISOLATION

The nature and degree of vibration isolation afforded by an isolator is influenced markedly by the characteristics of the damper. This aspect of vibration isolation is evaluated in this section in terms of the single degree-of-freedom concept; i.e., the equipment and the foundation are assumed rigid and the isolator is assumed massless. The performance is defined in terms of absolute transmissibility, relative transmissibility, and motion response for isolators with each of the four types of dampers illustrated in Table 30.1. A system with a rigidly connected viscous damper is discussed in detail in Chap. 2, and important results are reproduced here for completeness; isolators with other types of dampers are discussed in detail in this chapter.

The characteristics of the dampers and the performance of the isolators are defined in terms of the parameters shown on the schematic diagrams in Table 30.1. Absolute transmissibility, relative transmissibility, and motion response are defined analytically in Table 30.2 and graphically in the figures referenced in Table 30.2. For the rigidly connected viscous and Coulomb-damped isolators, the graphs generally are explicit and complete. For isolators with elastically connected dampers, typical results are included and references are given to more complete compilations of dynamic characteristics.

for Isolation Systems Defined in Table 30.1

indicated below the equation. See Table 30.1 for definition of terms.

Relative transmissibility	Motion response
(*b*) $T_R = \frac{\delta_0}{u_0} = \sqrt{\frac{\left(\frac{\omega}{\omega_0}\right)^4}{\left(1-\frac{\omega^2}{\omega_0^2}\right)^2+\left(2\zeta\frac{\omega}{\omega_0}\right)^2}}$ Fig. 30.3	(*c*) $\frac{x_0}{F_0/k} = \sqrt{\frac{1}{\left(1-\frac{\omega^2}{\omega_0^2}\right)^2+\left(2\zeta\frac{\omega}{\omega_0}\right)^2}}$ Fig. 30.14
(*e*) $(T_R)_D = \frac{\delta_0}{u_0} = \sqrt{\frac{\left[\left(\frac{\omega}{\omega_0}\right)^4-\left(\frac{4}{\pi}\eta\right)^2\right]}{\left(1-\frac{\omega^2}{\omega_0^2}\right)^2}}$ Fig. 30.6 (see Note 2)	(*f*) $\frac{x_0}{F_0/k} = \sqrt{\frac{1-\left(\frac{4}{\pi}\xi\right)^2}{\left(1-\frac{\omega^2}{\omega_0^2}\right)^2}}$ (See Note 2)
(*h*) $T_R = \frac{\delta_0}{u_0} = \sqrt{\frac{\frac{\omega^2}{\omega_0^2}+\frac{4}{N^2}\zeta^2\frac{\omega^6}{\omega_0^6}}{\left(1-\frac{\omega^2}{\omega_0^2}\right)^2+\frac{4}{N^2}\zeta^2\frac{\omega^2}{\omega_0^2}\left(N+1-\frac{\omega^2}{\omega_0^2}\right)^2}}$ Fig. 30.11 (see Note 3)	(*i*) $\frac{x_0}{F_0/k} = \sqrt{\frac{1+\frac{4}{N^2}\zeta^2\frac{\omega^2}{\omega_0^2}}{\left(1-\frac{\omega^2}{\omega_0^2}\right)^2+\frac{4}{N^2}\zeta^2\frac{\omega^2}{\omega_0^2}\left(N+1-\frac{\omega^2}{\omega_0^2}\right)^2}}$ Fig. 30.9 (see Note 4)
(*k*) $(T_R)_D = \frac{\delta_0}{u_0} = \sqrt{\frac{\left(\frac{\omega}{\omega_0}\right)^4+\left(\frac{4}{\pi}\eta\right)^2\left[\frac{2}{N}\frac{\omega^2}{\omega_0^2}-\left(\frac{N+2}{N}\right)\right]}{\left(1-\frac{\omega^2}{\omega_0^2}\right)^2}}$ Fig. 30.14 (see Notes 4 and 5)	

NOTE 4: These curves apply only for $N = 3$.
NOTE 5: This equation applies only when excitation is defined in terms of displacement amplitude; for excitation defined in terms of force or acceleration, see Eq. (30.18).

RIGIDLY CONNECTED VISCOUS DAMPER

Absolute and relative transmissibility curves are shown graphically in Figs. 30.2 and 30.3, respectively.* As the damping increases, the transmissibility at resonance decreases and the absolute transmissibility at the higher values of the forcing frequency ω increases; i.e., reduction of vibration is not as great. For an undamped isolator, the absolute transmissibility at higher values of the forcing frequency varies inversely as the square of the forcing frequency. When the isolator embodies significant viscous damping, the absolute transmissibility curve becomes asymptotic at high values of forcing frequency to a line whose slope is inversely proportional to the first power of the forcing frequency.

The maximum value of absolute transmissibility associated with the resonant condition is a function solely of the damping in the system, taken with reference to critical damping. For a lightly damped system, i.e., for $\zeta < 0.1$, the maximum absolute transmissibility [see Eq. (2.51)] of the system is[1]

$$T_{\max} = \frac{1}{2\zeta} \tag{30.1}$$

where $\zeta = c/c_c$ is the fraction of critical damping defined in Table 30.1.

* For linear systems, the absolute transmissibility $T_A = x_0/u_0$ in the motion-excited system equals F_T/F_0 in the force-excited system. The relative transmissibility $T_R = \delta_0/u_0$ applies only to the motion-excited system.

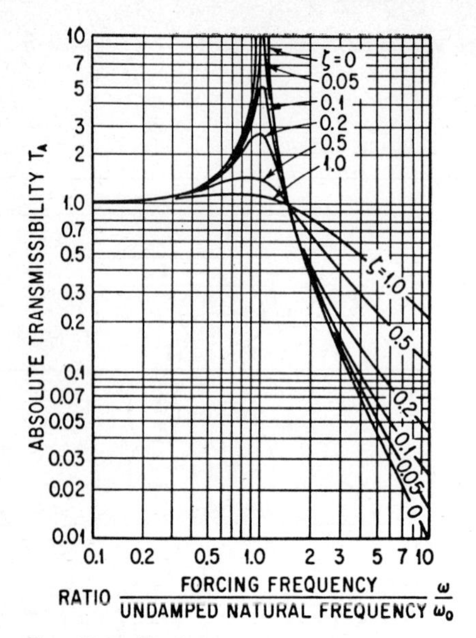

FIG. 30.2. Absolute transmissibility for the rigidly connected, viscous-damped isolation system shown at A in Table 30.1 as a function of the frequency ratio ω/ω_o and the fraction of critical damping ζ. The absolute transmissibility is the ratio (x_o/u_o) for foundation motion excitation (Fig. 30.1A) and the ratio (F_T/F_o) for equipment force excitation (Fig. 30.1B).

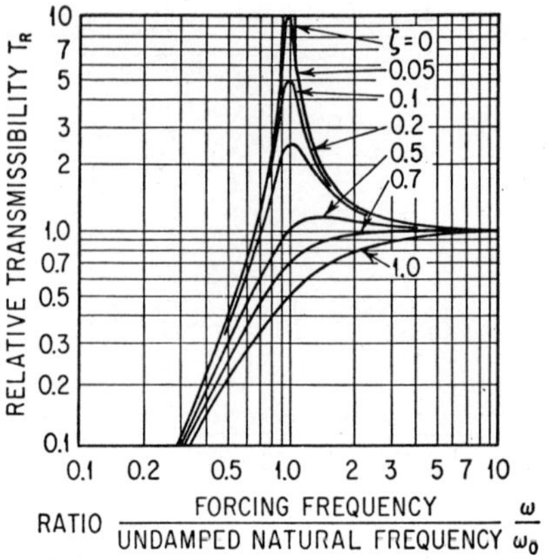

FIG. 30.3. Relative transmissibility for the rigidly connected, viscous-damped isolation system shown at A in Table 30.1 as a function of the frequency ratio ω/ω_o and the fraction of critical damping ζ. The relative transmissibility describes the motion between the equipment and the foundation (i.e., the deflection of the isolator).

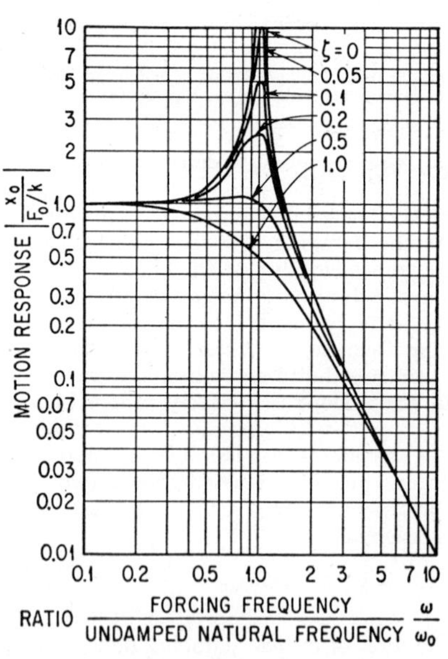

FIG. 30.4. Motion response for the rigidly connected viscous-damped isolation system shown at A in Table 30.1 as a function of the frequency ratio ω/ω_o and the fraction of critical damping ζ. The curves give the resulting motion of the equipment x in terms of the excitation force F and the static stiffness of the isolator k.

The motion response is shown graphically in Fig. 30.4. A high degree of damping limits the vibration amplitude of the equipment at all frequencies, compared to an undamped system. The single degree-of-freedom system with viscous damping is discussed more fully in Chap. 2.

RIGIDLY CONNECTED COULOMB DAMPER

The differential equation of motion for the system with Coulomb damping shown in Table 30.1*B* is

$$m\ddot{x} + k(x - u) \pm F_f = F_0 \sin \omega t \tag{30.2}$$

The discontinuity in the damping force that occurs as the sign of the velocity changes at each half cycle requires a step-by-step solution of Eq. (30.2).[2] An approximate solution based on the equivalence of energy dissipation involves equating the energy dissipation per cycle for viscous-damped and Coulomb-damped systems: [3]

$$\pi c \omega \delta_0^2 = 4F_f \delta_0 \tag{30.3}$$

where the left side refers to the viscous-damped system and the right side to the Coulomb-damped system; δ_0 is the amplitude of relative displacement across the damper. Solving Eq. (30.3) for c,

$$c_{eq} = \frac{4F_f}{\pi \omega \delta_0} = j\left(\frac{4F_f}{\pi \dot{\delta}_0}\right) \tag{30.4}$$

where c_{eq} is the *equivalent viscous damping coefficient* for a Coulomb-damped system having equivalent energy dissipation. Since $\dot{\delta}_0 = j\omega\delta_0$ is the relative velocity, the equivalent linearized dry friction damping force can be considered sinusoidal with an amplitude $j(4F_f/\pi)$. Since $c_c = 2k/\omega_0$ [see Eq. (2.12)],

$$\zeta_{eq} = \frac{c_{eq}}{c_c} = \frac{2\omega_0 F_f}{\pi \omega k \delta_0} \tag{30.5}$$

where ζ_{eq} may be defined as the *equivalent fraction of critical damping.* Substituting δ_0 from the relative transmissibility expression [(*b*) in Table 30.2] in Eq. (30.5) and solving for ζ_{eq}^2,

$$\zeta_{eq}^2 = \frac{\left(\frac{2}{\pi}\eta\right)^2 \left(1 - \frac{\omega^2}{\omega_0^2}\right)^2}{\frac{\omega^2}{\omega_0^2}\left[\frac{\omega^4}{\omega_0^4} - \left(\frac{4}{\pi}\eta\right)^2\right]} \tag{30.6}$$

where η is the Coulomb damping parameter for displacement excitation defined in Table 30.1.

The equivalent fraction of critical damping given by Eq. (30.6) is a function of the displacement amplitude u_0 of the excitation since the Coulomb damping parameter η depends on u_0. When the excitation is defined in terms of the acceleration amplitude $\ddot{u}_0$, the fraction of critical damping must be defined in corresponding terms. Thus, it is convenient to employ separate analyses for displacement transmissibility and acceleration transmissibility for an isolator with Coulomb damping.

DISPLACEMENT TRANSMISSIBILITY. The absolute displacement transmissibility of an isolation system having a rigidly connected Coulomb damper is obtained by substituting ζ_{eq} from Eq. (30.6) for ζ in the absolute transmissibility expression for viscous damping, (*a*) in Table 30.2. The absolute displacement transmissibility is shown graphically in Fig. 30.5, and the relative displacement transmissibility is shown in Fig. 30.6. The absolute displacement transmissibility has a value of unity when the forcing frequency is low and/or the Coulomb friction force is high. For these conditions, the friction damper is locked in, i.e., it functions as a rigid connection, and there is no relative motion across the isolator. The frequency at which the damper breaks loose, i.e., permits relative motion across the isolator, can be obtained from the relative displacement transmissibility expression, (*e*) in Table 30.2. The relative displacement is imaginary when $\omega^2/\omega_0^2 \leq (4/\pi)\eta$. Thus, the "break-loose" frequency ratio is *

$$\left(\frac{\omega}{\omega_0}\right)_L = \sqrt{\frac{4}{\pi}\eta} \tag{30.7}$$

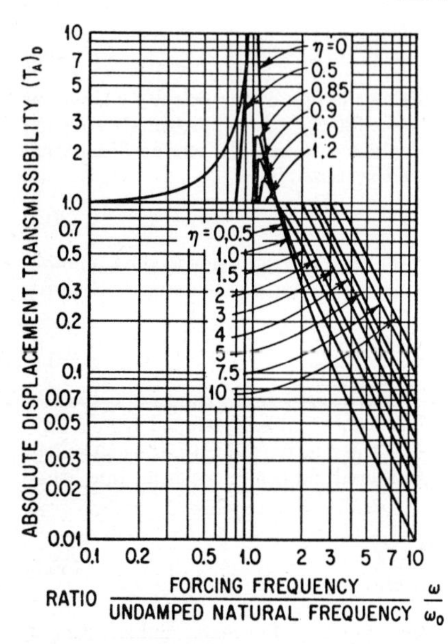

FIG. 30.5. Absolute displacement transmissibility for the rigidly connected, Coulomb-damped isolation system shown at *B* in Table 30.1 as a function of the frequency ratio ω/ω_0 and the displacement Coulomb-damping parameter η.

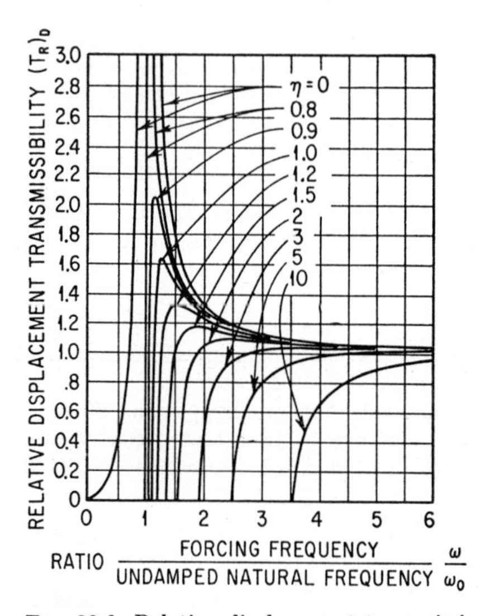

FIG. 30.6. Relative displacement transmissibility for the rigidly connected, Coulomb-damped isolation system shown at *B* in Table 30.1 as a function of the frequency ratio ω/ω_0 and the displacement Coulomb-damping parameter η.

The displacement transmissibility can become infinite at resonance, even though the system is damped, if the Coulomb damping force is less than a critical minimum value. The denominator of the absolute and relative transmissibility expressions becomes zero for a frequency ratio ω/ω_0 of unity. If the "break-loose" frequency is lower than the undamped natural frequency, the amplification of vibration becomes infinite at resonance. This occurs because the energy dissipated by the friction damping force increases linearly with the displacement amplitude, and the energy introduced into the system by the excitation source also increases linearly with the displacement amplitude. Thus, the energy dissipated at resonance is either greater or less than the input energy for *all* amplitudes of vibration. The minimum dry-friction force which prevents vibration of infinite magnitude at resonance is

$$(F_f)_{\min} = \frac{\pi k u_0}{4} = 0.79\, k u_0 \tag{30.8}$$

where k and u_0 are defined in Table 30.1.

* This equation is based upon energy considerations and is approximate. Actually, the friction damper breaks loose when the inertia force of the mass equals the friction force, $mu_0\omega^2 = F_f$. This gives the exact solution $(\omega/\omega_0)_L = \sqrt{\eta}$. A numerical factor of $4/\pi$ relates the Coulomb damping parameters in the exact and approximate solutions for the system.

As shown in Fig. 30.5, an increase in η decreases the absolute displacement transmissibility at resonance and increases the resonant frequency. All curves intersect at the point $(T_A)_D = 1$, $\omega/\omega_0 = \sqrt{2}$. With optimum damping force, there is no motion across the damper for $\omega/\omega_0 \leq \sqrt{2}$; for higher frequencies the displacement transmissibility is less than unity. The friction force that produces this "resonance-free" condition is

$$(F_f)_{\text{op}} = \frac{\pi k u_0}{2} = 1.57\, k u_0 \tag{30.9}$$

For high forcing frequencies, the absolute displacement transmissibility varies inversely as the square of the forcing frequency, even though the friction damper dissipates energy. For relatively high damping ($\eta > 2$), the absolute displacement transmissibility, for frequencies greater than the "break-loose" frequency, is approximately $4\eta\omega_0^2/\pi\omega^2$.

ACCELERATION TRANSMISSIBILITY. The absolute displacement transmissibility $(T_A)_D$ shown in Fig. 30.5 is the ratio of response of the isolator to the excitation, where each is expressed as a displacement amplitude in simple harmonic motion. The damping parameter η is defined with reference to the displacement amplitude u_0 of the excitation. Inasmuch as all motion is simple harmonic, the transmissibility $(T_A)_D$ also applies to acceleration transmissibility when the damping parameter is defined properly. When the excitation is defined in terms of the acceleration amplitude $\ddot{u}_0$ of the excitation,

$$\eta_{\ddot{u}_0} = \frac{F_f \omega^2}{k \ddot{u}_0} \tag{30.10}$$

where ω = forcing frequency, rad/sec
$\ddot{u}_0$ = acceleration amplitude of excitation, in./sec^2
k = isolator stiffness, lb/in.
F_f = Coulomb friction force, lb

For relatively high forcing frequencies, the acceleration transmissibility approaches a constant value $(4/\pi)\xi$, where ξ is the Coulomb damping parameter for acceleration excitation defined in Table 30.1. The acceleration transmissibility of a rigidly connected Coulomb damper system becomes asymptotic to a constant value because the Coulomb damper transmits the same friction force regardless of the amplitude of the vibration.

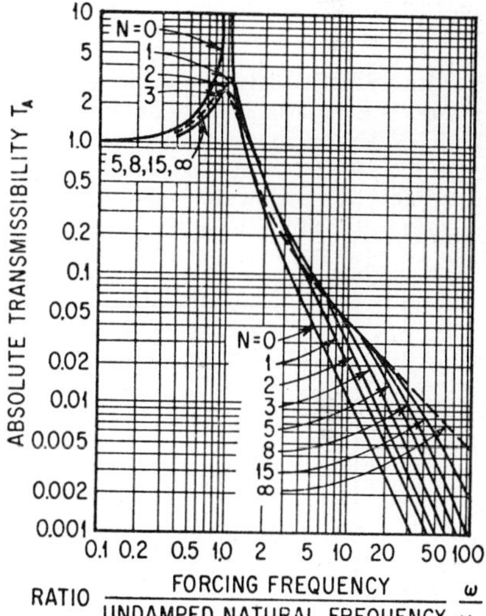

FIG. 30.7. Comparison of absolute transmissibility for rigidly and elastically connected, viscous damped isolation systems shown at A and C, respectively, in Table 30.1, as a function of the frequency ratio ω/ω_0. The solid curves refer to the elastically connected damper, and the parameter N is the ratio of the damper spring stiffness to the stiffness of the principal support spring. The fraction of critical damping $\zeta = c/c_c$ is 0.2 in both systems. The transmissibility at high frequencies decreases at a rate of 6 db per octave for the rigidly connected damper and 12 db per octave for the elastically connected damper.

ELASTICALLY CONNECTED VISCOUS DAMPER

The general characteristics of the elastically connected viscous damper shown at C in Table 30.1 may best be understood by successively assigning values to the viscous damper coefficient c while keeping the stiffness ratio N constant. For zero damping, the mass is supported by the isolator of stiffness k. The transmissibility curve has the characteristics typical of a transmissibility curve for an undamped system having the natural frequency

$$\omega_0 = \sqrt{\frac{k}{m}} \tag{30.11}$$

When c is infinitely great, the transmissibility curve is that of an undamped system having the natural frequency

$$\omega_\infty = \sqrt{\frac{k + k_1}{m}} = \sqrt{N + 1}\,\omega_0 \tag{30.12}$$

where $k_1 = Nk$. For intermediate values of damping, the transmissibility falls within the limits established for zero and infinitely great damping. The value of damping which produces the minimum transmissibility at resonance is called *optimum damping.*

All curves approach the transmissibility curve for infinite damping as the forcing frequency increases. Thus, the absolute transmissibility at high forcing frequencies is inversely proportional to the square of the forcing frequency. General expressions for absolute and relative transmissibility are given in Table 30.2.

A comparison of absolute transmissibility curves for the elastically connected viscous damper and the rigidly connected viscous damper is shown in Fig. 30.7. A constant viscous damping coefficient of $0.2c_c$ is maintained while the value of the stiffness ratio N is varied from zero to infinity. The transmissibilities at resonance are comparable, even for relatively small values of N, but a substantial gain is achieved in the isolation characteristics at high forcing frequencies by elastically connecting the damper.

TRANSMISSIBILITY AT RESONANCE. The maximum transmissibility (at resonance) is a function of the damping ratio ζ and the stiffness ratio N, as shown in Fig. 30.8.[4] The maximum transmissibility is nearly independent of N for small values of ζ. However, for $\zeta > 0.1$, the coefficient N is significant in determining the maximum transmissibility. The lowest value of the maximum absolute transmissibility curves corresponds to the conditions of optimum damping.

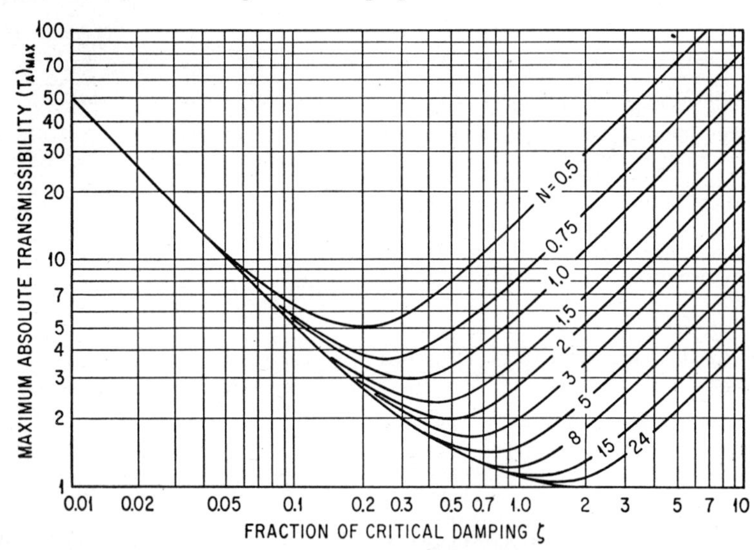

FIG. 30.8. Maximum absolute transmissibility for the elastically connected, viscous-damped isolation system shown at C in Table 30.1 as a function of the fraction of critical damping ζ and the stiffness of the connecting spring. The parameter N is the ratio of the damper spring stiffness to the stiffness of the principal support spring.

MOTION RESPONSE. A typical motion response curve is shown in Fig. 30.9 for the stiffness ratio $N = 3$. For small damping, the response is similar to the response of an isolation system with rigidly connected viscous damper. For intermediate values of damping, the curves tend to be flat over a wide frequency range before rapidly decreasing in value at the higher frequencies. For large damping, the resonance occurs near the natural frequency of the system with infinitely great damping. All response curves approach a high-frequency asymptote for which the attenuation varies inversely as the square of the excitation frequency.

OPTIMUM TRANSMISSIBILITY. For a system with optimum damping, maximum transmissibility coincides with the intersections of the transmissibility curves for zero and infinite damping. The frequency ratios $(\omega/\omega_0)_{op}$ at which this occurs are different for absolute and relative transmissibility:

Absolute transmissibility:

$$\left(\frac{\omega}{\omega_0}\right)_{op}^{(A)} = \sqrt{\frac{2(N+1)}{N+2}}$$

(30.13)

Relative transmissibility:

$$\left(\frac{\omega}{\omega_0}\right)_{op}^{(R)} = \sqrt{\frac{N+2}{2}}$$

The optimum transmissibility at resonance, for both absolute and relative motion, is

$$T_{op} = 1 + \frac{2}{N} \qquad (30.14)$$

The optimum transmissibility as determined from Eq. (30.14) corresponds to the minimum points of the curves of Fig. 30.8.

The damping which produces the optimum transmissibility is obtained by differentiating the general expressions for transmissibility [(g) and (h) in Table 30.2] with respect to the frequency ratio, setting the result equal to zero, and combining it with Eq. (30.13),

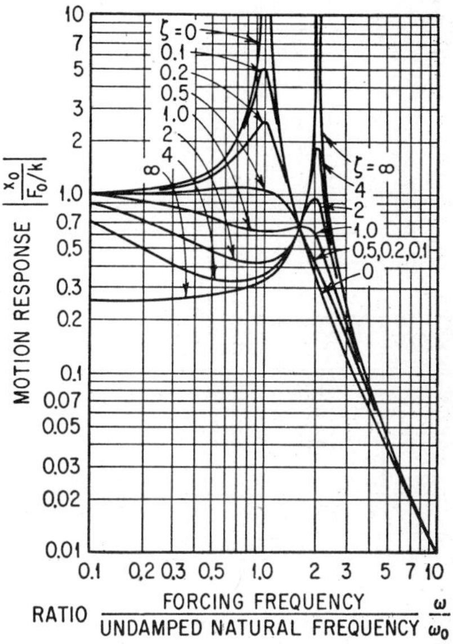

FIG. 30.9. Motion response for the elastically connected, viscous-damped isolation system shown at C in Table 30.1 as a function of the frequency ratio ω/ω_0 and the fraction of critical damping ζ. For this example, the stiffness of the damper connecting spring is three times as great as the stiffness of the principal support spring ($N = 3$). The curves give the resulting motion of the equipment in terms of the excitation force F and the static stiffness of the isolator k.

$$\text{Absolute transmissibility:} \quad (\zeta_{op})_A = \frac{N}{4(N+1)}\sqrt{2(N+2)}$$

$$\text{Relative transmissibility:} \quad (\zeta_{op})_R = \frac{N}{\sqrt{2(N+1)(N+2)}} \qquad (30.15)$$

Values of optimum damping determined from the first of these relations correspond to the minimum points of the curves of Fig. 30.8. By substituting the optimum damping ratios from Eqs. (30.15) into the general expressions for transmissibility given in Table 30.2, the optimum absolute and relative transmissibility equations are obtained, as shown graphically by Figs. 30.10 and 30.11, respectively.[5] For low values of the stiffness ratio N, the transmissibility at resonance is large but excellent isolation is obtained at high frequencies. Conversely, for high values of N, the transmissibility at resonance is lowered, but the isolation efficiency also is decreased.

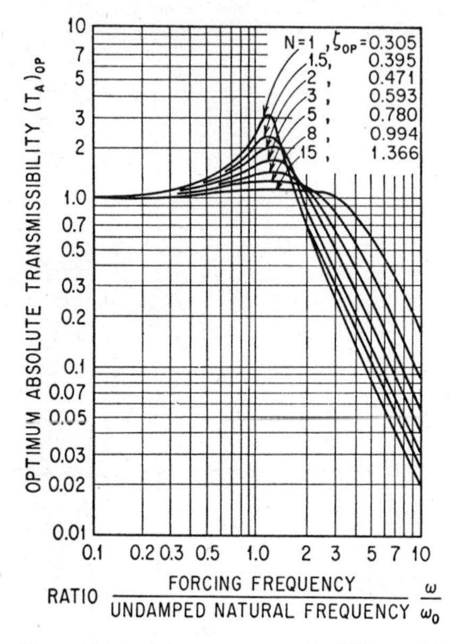

FIG. 30.10. Absolute transmissibility with optimum damping in elastically connected, viscous-damped isolation system shown at C in Table 30.1 as a function of the frequency ratio ω/ω_o and the fraction of critical damping ζ. These curves apply to elastically connected, viscous-damped systems having optimum damping for absolute motion. The transmissibility $(T_A)_{op}$ is $(x_o/u_o)_{op}$ for the motion-excited system and $(F_T/F_o)_{op}$ for the force-excited system.

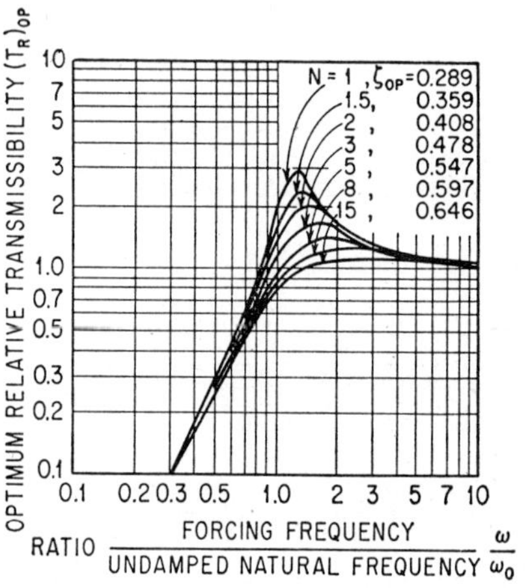

FIG. 30.11. Relative transmissibility with optimum damping in the elastically connected, viscous-damped isolation system shown at C in Table 30.1 as a function of the frequency ratio ω/ω_o and the fraction of critical damping ζ. These curves apply to elastically connected, viscous-damped systems having optimum damping for relative motion. The relative transmissibility $(T_R)_{op}$ is $(\delta_o/u_o)_{op}$ for the motion-excited system.

ELASTICALLY CONNECTED COULOMB DAMPER

Force-deflection curves for the isolators incorporating elastically connected Coulomb dampers, as shown at D in Table 30.1, are illustrated in Fig. 30.12. Upon application of the load, the isolator deflects; but since insufficient force has been developed in the spring k_1, the damper does not slide and the motion of the mass is opposed by a spring of stiffness $(N + 1)k$. The load is now increased until a force is developed in spring k_1 which equals the constant friction force F_f; then the damper begins to slide. When the load is increased further, the damper slides and reduces the effective spring stiffness to k. If the applied load is reduced after reaching its maximum value, the damper no longer displaces because the force developed in the spring k_1 is diminished. Upon completion of the load cycle, the damper will have been in motion for part of the cycle and at rest for the remaining part to form the hysteresis loops shown in Fig. 30.12.

Because of the complexity of the applicable equations, the equivalent energy method is used to obtain the transmissibility and motion response functions. Applying frequency, damping, and transmissibility expressions for the elastically connected viscous damped system to the elastically connected Coulomb damped system, the transmissibility expressions tabulated in Table 30.2 for the latter are obtained.[6]

If the coefficient of the damping term in each of the transmissibility expressions vanishes, the transmissibility is independent of damping. By solving for the frequency ratio ω/ω_0 in the coefficients that are thus set equal to zero, the frequency ratios obtained define the frequencies of optimum transmissibility. These frequency ratios are given by Eqs. (30.13) for the elastically connected viscous damped system and apply equally well to the elastically connected Coulomb damped system because the method of equivalent viscous damping is employed in the analysis. Similarly, Eq. (30.14) applies for optimum transmissibility at resonance.

The general characteristics of the system with an elastically connected Coulomb damper may be demonstrated by successively assigning values to the damping force

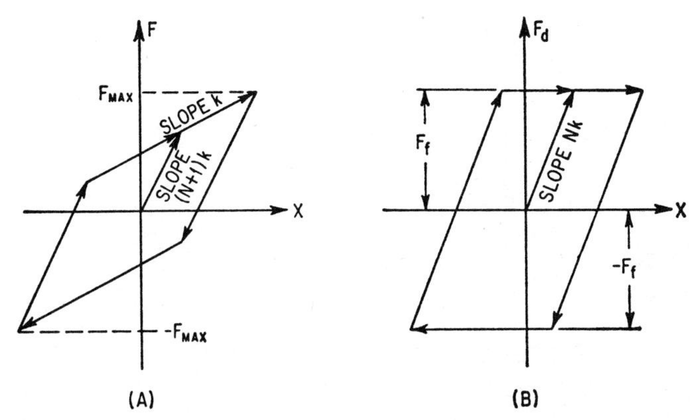

FIG. 30.12. Force-deflection characteristics of the elastically connected, Coulomb-damped isolation system shown at *D* in Table 30.1. The force-deflection diagram for a cyclic deflection of the complete isolator is shown at *A* and the corresponding diagram for the assembly of Coulomb damper and spring $k_1 = Nk$ is shown at *B*.

while keeping the stiffness ratio N constant. For zero and infinite damping, the transmissibility curves are those for undamped systems and bound all solutions. Every transmissibility curve for $0 < F_f < \infty$ passes through the intersection of the two bounding transmissibility curves. For low damping (less than optimum), the damper "breaks loose" at a relatively low frequency, thereby allowing the transmissibility to increase to a maximum value and then pass through the intersection point of the bounding transmissibility curves. For optimum damping, the maximum absolute transmissibility has a value given by Eq. (30.14); it occurs at the frequency ratio $\left(\frac{\omega}{\omega_0}\right)_{op}^{(A)}$ defined by Eq. (30.13). For high damping, the damper remains "locked-in" over a wide frequency range because insufficient force is developed in the spring k_1 to induce slip in the damper. For frequencies greater than the "break-loose" frequency, there is sufficient force in spring k_1 to cause relative motion of the damper. For a further increase in frequency, the damper remains broken loose and the transmissibility is limited to a finite value. When there is insufficient force in spring k_1 to maintain motion across the damper, the damper locks-in and the transmissibility is that of a system with the infinite damping.

The "break-loose" and "lock-in" frequencies are determined by requiring the motion across the Coulomb damper to be zero. Then the "break-loose" and "lock-in" frequency ratios are

$$\left(\frac{\omega}{\omega_0}\right)_L = \sqrt{\frac{\left(\frac{4}{\pi}\eta\right)(N+1)}{\left(\frac{4}{\pi}\eta\right) \pm N}} \tag{30.16}$$

where η is the damping parameter defined in Table 30.1 with reference to the displacement amplitude u_0. The plus sign corresponds to the break-loose frequency while the minus sign corresponds to the lock-in frequency. Damping parameters for which the denominator of Eq. (30.16) becomes negative correspond to those conditions for which the damper never becomes locked-in again after it has broken loose. Thus, the damper eventually becomes locked-in only if $\eta > (\pi/4)N$.

DISPLACEMENT TRANSMISSIBILITY. The absolute displacement transmissibility curve for the stiffness ratio $N = 3$ is shown in Fig. 30.13 where $(T_A)_D = x_0/u_0$.

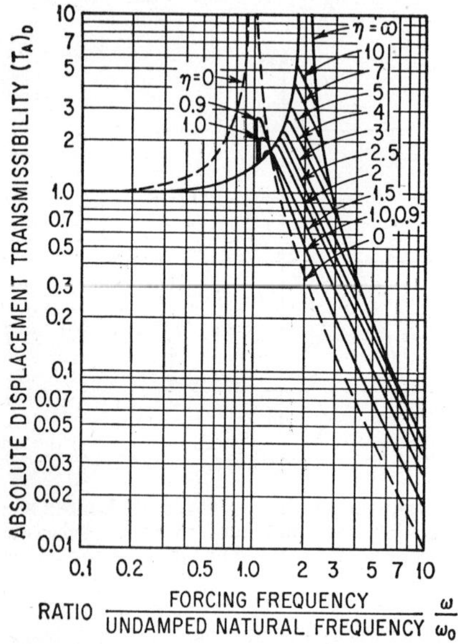

FIG. 30.13. Absolute displacement transmissibility for the elastically connected, Coulomb-damped isolation system illustrated at D in Table 30.1, for the damper spring stiffness defined by $N = 3$. The curves give the ratio of the absolute displacement amplitude of the equipment to the displacement amplitude imposed at the foundation, as a function of the frequency ratio ω/ω_0 and the displacement Coulomb-damping parameter η.

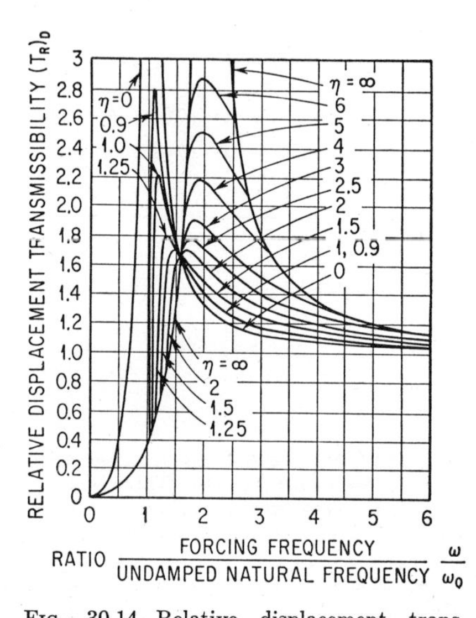

FIG. 30.14. Relative displacement transmissibility for the elastically connected, Coulomb-damped isolation system illustrated at D in Table 30.1 for the damper spring stiffness defined by $N = 3$. The curves give the ratio of the relative displacement amplitude (maximum isolator deflection) to the displacement amplitude imposed at the foundation, as a function of the frequency ratio ω/ω_0 and the displacement Coulomb-damping parameter η.

A small decrease in damping force F_f below the optimum value causes a large increase in the transmitted vibration near resonance. However, a small increase in damping force F_f above optimum causes only small changes in the maximum transmissibility. Thus, it is good design practice to have the damping parameter η equal to or greater than the optimum damping parameter η_{op}.

The relative transmissibility for $N = 3$ is shown in Fig. 30.14 where $(T_R)_D = \delta_0/u_0$. All curves pass through the intersection of the curves for zero and infinite damping. For optimum damping, the maximum relative transmissibility has a value given by Eq. (30.14); it occurs at the frequency ratio $\left(\frac{\omega}{\omega_0}\right)_{op}^{(R)}$ defined by Eq. (30.13).

ACCELERATION TRANSMISSIBILITY. The acceleration transmissibility can be obtained from the expression for displacement transmissibility by substitution of the effective displacement damping parameter in the expression for transmissibility of a system whose excitation is constant acceleration amplitude. If $\ddot{u}_0$ represents the acceleration amplitude of the excitation, the corresponding displacement amplitude is $u_0 = -\ddot{u}_0/\omega^2$. Using the definition of the acceleration Coulomb damping parameter ξ given in Table 30.1, the equivalent displacement Coulomb damping parameter is

$$\eta_{eq} = -\left(\frac{\omega}{\omega_0}\right)^2 \xi \tag{30.17}$$

Substituting this relation in the absolute transmissibility expression given at j in Table 30.2, the following equation is obtained for the acceleration transmissibility:

$$(T_A)_A = \frac{\ddot{x}_0}{\ddot{u}_0} = \sqrt{\frac{1 + \left(\frac{4}{\pi}\xi\right)^2 \left(\frac{\omega^2}{\omega_0^2}\right)\left[\left(\frac{N+2}{N}\right)\left(\frac{\omega^2}{\omega_0^2}\right) - 2\left(\frac{N+1}{N}\right)\right]}{\left(1 - \frac{\omega^2}{\omega_0^2}\right)^2}} \tag{30.18}$$

Equation (30.18) is valid only for the frequency range in which there is relative motion across the Coulomb damper. This range is defined by the "break-loose" and "lock-in" frequencies which are obtained by substituting Eq. (30.17) into Eq. (30.16):

$$\left(\frac{\omega}{\omega_0}\right)_L = \sqrt{\frac{\left(\frac{4}{\pi}\xi\right)(N+1) \pm N}{\frac{4}{\pi}\xi}} \tag{30.19}$$

where Eqs. (30.16) and (30.19) give similar results, damping being defined in terms of displacement and acceleration excitation, respectively. For frequencies not included in the range between "break-loose" and "lock-in" frequencies, the acceleration transmissibility is that for an undamped system. Equation (30.18) indicates that infinite acceleration occurs at resonance unless the damper remains locked-in beyond a frequency ratio of unity. The coefficient of the damping term in Eq. (30.18) is identical to the corresponding coefficient in the expression for $(T_A)_D$ at j in Table 30.2. Thus, the frequency ratio at the optimum transmissibility is the same as that for displacement excitation.

An acceleration transmissibility curve for $N = 3$ is shown by Fig. 30.15. Relative motion at the damper occurs in a limited frequency range; thus, for relatively high frequencies, the acceleration transmissibility is similar to that for infinite damping.

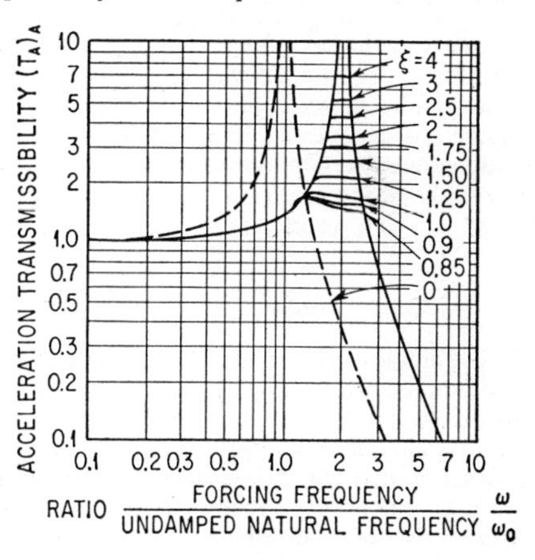

FIG. 30.15. Acceleration transmissibility for the elastically connected, Coulomb-damped isolation system illustrated at D in Table 30.1, for the damper spring stiffness defined by $N = 3$. The curves give the ratio of the acceleration amplitude of the equipment to the acceleration amplitude imposed at the foundation, as a function of the frequency ratio ω/ω_0 and the acceleration Coulomb-damping parameter ξ.

OPTIMUM DAMPING PARAMETERS. The optimum Coulomb damping parameters are obtained by equating the optimum viscous damping ratio given by Eq. (30.15) to the equivalent viscous damping ratio for the elastically supported damper system, and replacing the frequency ratio by the frequency ratio given by Eq. (30.13). The optimum value of the damping parameter η in Table 30.1 is

$$\eta_{op} = \frac{\pi}{2}\sqrt{\frac{N+1}{N+2}} \tag{30.20}$$

To obtain the optimum value of the damping parameter ξ in Table 30.1, Eq. (30.17) is substituted in Eq. (30.20):

$$\xi_{op} = \frac{\pi}{4}\sqrt{\frac{N+2}{N+1}} \tag{30.21}$$

FORCE TRANSMISSIBILITY. The force transmissibility $(T_A)_F = F_T/F_0$ is identical to $(T_A)_A$ given by Eq. (30.18) if $\xi = \xi_F$, where ξ_F is defined as

$$\xi_F = \frac{F_f}{F_0} \tag{30.22}$$

Thus, the transmissibility curve shown in Fig. 30.15 also gives the force transmissibility for $N = 3$. By substituting Eq. (30.22) into Eq. (30.21), the transmitted force is optimized when the friction force F_f has the following value:

$$(F_f)_{op} = \frac{\pi F_0}{4}\sqrt{\frac{N+2}{N+1}} \tag{30.23}$$

To avoid infinite transmitted force at resonance, it is necessary that $F_f > (\pi/4)F_0$.

COMPARISON OF RIGIDLY CONNECTED AND ELASTICALLY CONNECTED COULOMB-DAMPED SYSTEMS. A principal limitation of the rigidly connected Coulomb-damped isolator is the nature of the transmissibility at high forcing frequencies. Because the isolator deflection is small, the force transmitted by the spring is negligible; then the force transmitted by the damper controls the motion experienced by the equipment. The acceleration transmissibility approaches the constant value $(4/\pi)\xi$, independent of frequency. The corresponding transmissibility for an isolator with an elastically connected Coulomb damper is $(N + 1)/(\omega/\omega_0)^2$. Thus, the transmissibility varies inversely as the square of the excitation frequency and reaches a relatively low value at large values of excitation frequency.

MULTIPLE DEGREE-OF-FREEDOM SYSTEMS

The single degree-of-freedom systems discussed previously are adequate for illustrating the fundamental principles of vibration isolation, but are an oversimplification in so far as many practical applications are concerned. The condition of unidirectional motion of an elastically mounted mass is not consistent with the requirements in many applications. In general, it is necessary to consider freedom of movement in all directions, as dictated by existing forces and motions and by the elastic constraints. Thus, in the general isolation problem, the equipment is considered as a rigid body supported by resilient supporting elements or isolators. This system is arranged so that the isolators effect the desired reduction in vibration. Various types of symmetry are encountered, depending upon the equipment and arrangement of isolators.

NATURAL FREQUENCIES—ONE PLANE OF SYMMETRY

A rigid body supported by resilient supports with one vertical plane of symmetry has three coupled natural modes of vibration and a natural frequency in each of these modes. A typical system of this type is illustrated in Fig. 30.16; it is assumed to be symmetrical with respect to a plane parallel with the plane of the paper and extending through the center-of-gravity of the supported body. Motion of the supported body in horizontal and vertical translational modes and in the rotational mode, all in the plane of the paper, are coupled. The equations of motion of a rigid body on resilient supports with six degrees-of-freedom are given by Eq. (3.31). By introducing certain types of symmetry and setting the excitation equal to zero, a cubic equation defining the free vibration of the system shown in Fig. 30.16 is derived, as given by Eqs. (3.36). This equation may be solved graphically for the natural frequencies of the system by use of Fig. 3.14.

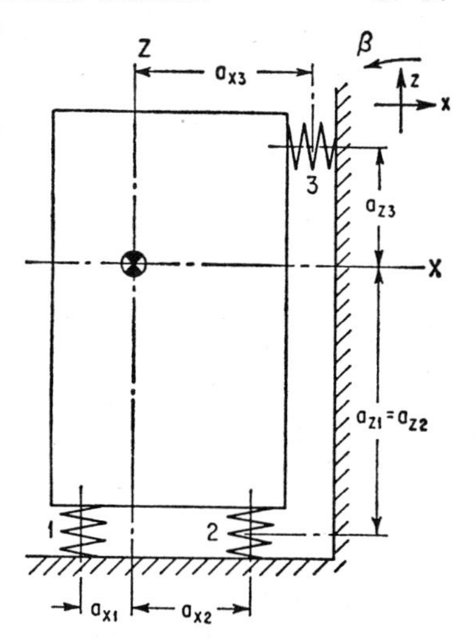

FIG. 30.16. Schematic diagram of a rigid equipment supported by an arbitrary arrangement of vibration isolators, symmetrical with respect to a plane through the center-of-gravity parallel with the paper.

SYSTEM WITH TWO PLANES OF SYMMETRY

A common arrangement of isolators is illustrated in Fig. 30.17; it consists of an equipment supported by four isolators located adjacent to the four lower corners. It is symmetrical with respect to two coordinate vertical planes through the center-of-gravity of the equipment, one of the planes being parallel with the plane of the paper. Because of this symmetry, vibration in the vertical translational mode is decoupled from vibration in the horizontal and rotational modes. The natural frequency in the vertical translational mode is $\omega_z = \sqrt{\Sigma k_z/m}$, where Σk_z is the sum of the vertical stiffnesses of the isolators.

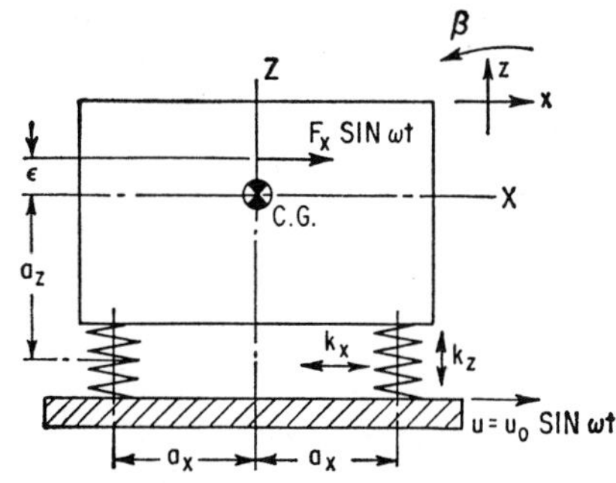

FIG. 30.17. Schematic diagram in elevation of a rigid equipment supported upon four vibration isolators. The plane of the paper extends vertically through the center-of-gravity; the system is symmetrical with respect to this plane and with respect to a vertical plane through the center-of-gravity perpendicular to the paper. The moment of inertia of the equipment with respect to an axis through the center-of-gravity and normal to the paper is I_y. Excitation of the system is alternatively a vibratory force $F_x \sin \omega t$ applied to the equipment or a vibratory displacement $u = u_0 \sin \omega t$ of the foundation.

Consider excitation by a periodic force $F = F_x \sin \omega t$ applied in the direction of the X axis at a distance ϵ above the center-of-gravity and in one of the planes of symmetry. The differential equations of motion for the equipment in the coupled horizontal translational and rotational modes are obtained by substituting in Eq. (3.31) the conditions of symmetry defined by Eqs. (3.33), (3.34), (3.35), and (3.38). The resulting equations of motion are

$$\begin{aligned} m\ddot{x} &= -4k_x x + 4k_x a\beta + F_x \sin \omega t \\ I_y\ddot{\beta} &= 4k_x a x - 4k_x a^2\beta - 4k_y b^2\beta \\ &\quad - F_x \epsilon \sin \omega t \end{aligned} \tag{30.24}$$

Making the common assumption that transients may be neglected in systems undergoing forced vibration, the translational and rotational displacements of the supported body are assumed to be harmonic at the excitation frequency. The differential equations of motion then are solved simultaneously to give the following expressions for the displacement amplitudes x_0 in horizontal translation and β_0 in rotation:

$$x_0 = \frac{F_x}{4k_z}\left(\frac{A_1}{D}\right) \qquad \beta_0 = \frac{F_x}{4\rho_y k_z}\left(\frac{A_2}{D}\right) \tag{30.25}$$

where
$$A_1 = \left(\frac{1}{\rho_y^2}\right)(\eta a_z^2 + a_x^2 - \eta\epsilon a_z) - \left(\frac{\omega}{\omega_z}\right)^2$$

$$A_2 = \frac{\epsilon}{\rho_y}\left(\frac{\omega}{\omega_z}\right)^2 + \frac{\eta}{\rho_y}(a_z - \epsilon) \tag{30.26}$$

$$D = \left(\frac{\omega}{\omega_z}\right)^4 - \left(\eta + \eta\frac{a_z^2}{\rho_y^2} + \frac{a_x^2}{\rho_y^2}\right)\left(\frac{\omega}{\omega_z}\right)^2 + \eta\left(\frac{a_x}{\rho_y}\right)^2$$

In the above equations, $\eta = k_x/k_z$ is the dimensionless ratio of horizontal stiffness to vertical stiffness of the isolators, $\rho_y = \sqrt{I_y/m}$ is the radius of gyration of the supported body about an axis through its center-of-gravity and perpendicular to the paper, $\omega_z = \sqrt{\Sigma k_z/m}$ is the undamped natural frequency in vertical translation, ω is the forcing frequency, a_z is the vertical distance from the effective height of spring (mid-height if symmetrical top to bottom) * to center-of-gravity of body m, and the other parameters are as indicated in Fig. 30.17.

Forced vibration of the system shown in Fig. 30.17 also may be excited by periodic motion of the support in the horizontal direction, as defined by $u = u_0 \sin \omega t$. The differential equations of motion for the supported body are

$$\begin{aligned} m\ddot{x} &= 4k_x(u - x - a_z\beta) \\ I_y\ddot{\beta} &= -4a_zk_x(u - x - a_z\beta) - 4k_za_x^2\beta \end{aligned} \tag{30.27}$$

Neglecting transients, the motion of the mounted body in horizontal translation and in rotation is assumed to be harmonic at the forcing frequency. Equations (30.27) may be solved simultaneously to obtain the following expressions for the displacement amplitudes x_0 in horizontal translation and β_0 in rotation:

$$x_0 = \frac{u_0 B_1}{D} \qquad \beta_0 = \frac{u_0 B_2}{\rho_y D} \tag{30.28}$$

where the parameters B_1 and B_2 are

$$B_1 = \eta\left(\frac{a_x^2}{\rho_y^2} - \frac{\omega^2}{\omega_z^2}\right) \qquad B_2 = \frac{\eta a_z}{\rho_y}\left(\frac{\omega}{\omega_z}\right)^2 \tag{30.29}$$

and D is given by Eq. (30.26).

NATURAL FREQUENCIES—TWO PLANES OF SYMMETRY. In forced vibration, the amplitude becomes a maximum when the forcing frequency is approximately equal to a natural frequency. In an undamped system, the amplitude becomes infinite at resonance. Thus, the natural frequency or frequencies of an undamped system may be determined by writing the expression for the displacement amplitude of the system in forced vibration and finding the excitation frequency at which this amplitude becomes infinite. The denominators of Eqs. (30.25) and (30.28) include the parameter D defined by Eq. (30.26). The natural frequencies of the system in coupled rotational and horizon-

* The distance a_z is taken to the mid-height of the spring to include in the equations of motion the moment applied to the body m by the fixed-end spring. If the spring is hinged to body m, the appropriate value for a_z is the distance from the X axis to the hinge axis.

tal translational modes may be determined by equating D to zero and solving for the forcing frequencies:[7]

$$\frac{\omega_{x\beta}}{\omega_z} \times \frac{\rho_y}{a_x} = \frac{1}{\sqrt{2}} \sqrt{\eta \left(\frac{\rho_y}{a_x}\right)^2 \left(1 + \frac{a_z^2}{\rho_y^2}\right) + 1 \pm \sqrt{\left[\eta \left(\frac{\rho_y}{a_x}\right)^2 \left(1 + \frac{a_z^2}{\rho_y^2}\right) + 1\right]^2 - 4\eta \left(\frac{\rho_y}{a_x}\right)^2}} \tag{30.30}$$

where $\omega_{x\beta}$ designates a natural frequency in a coupled rotational (β) and horizontal translational (x) mode, and ω_z designates the natural frequency in the decoupled vertical translational mode. The other parameters are defined in connection with Eq. (30.26). Two numerically different values of the dimensionless frequency ratio $\omega_{x\beta}/\omega_z$ are obtained from Eq. (30.30), corresponding to the two discrete coupled modes of vibration. Curves computed from Eq. (30.30) are given in Fig. 30.18.

The ratio of a natural frequency in a coupled mode to the natural frequency in the vertical translational mode is a function of three dimensionless ratios, two of the ratios relating the radius of gyration ρ_y to the dimensions a_z and a_x while the third is the ratio η of horizontal to vertical stiffnesses of the isolators. In applying the curves of Fig. 30.18, the applicable value of the abscissa ratio is first determined directly from the constants of the system. Two appropriate numerical values then are taken from the ordinate scale, as determined by the two curves for applicable values of a_z/ρ_y; the ratios of natural frequencies in coupled and vertical translational modes are determined by dividing these values by the dimensionless ratio ρ_y/a_x. The natural frequencies in coupled modes then are determined by multiplying the resulting ratios by the natural frequency in the decoupled vertical translational mode.

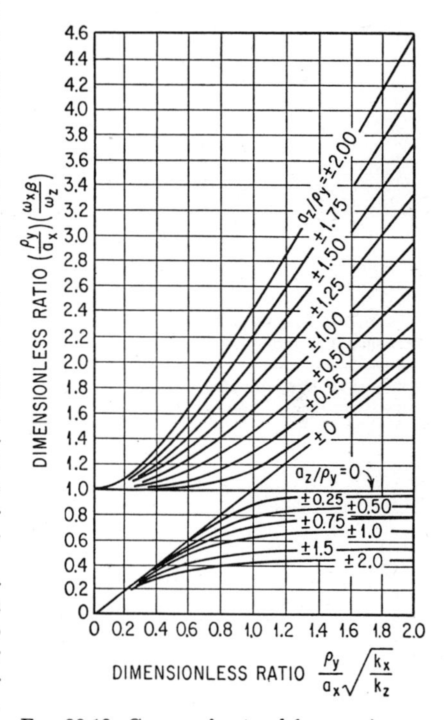

FIG. 30.18. Curves of natural frequencies $\omega_{x\beta}$ in coupled modes with reference to the natural frequency in the decoupled vertical translational mode ω_z, for the system shown schematically in Fig. 30.17. The isolator stiffnesses in the X and Z directions are indicated by k_x and k_z, respectively, and the radius of gyration with respect to the Y axis through the center-of-gravity is indicated by ρ_y.

The two straight lines in Fig. 30.18 for $a_z/\rho_y = 0$ represent natural frequencies in decoupled modes of vibration. When $a_z = 0$, the elastic supports lie in a plane passing through the center-of-gravity of the equipment. The horizontal line at a value of unity on the ordinate scale represents the natural frequency in a rotational mode. The inclined straight line for the value $a_z/\rho_y = 0$ represents the natural frequency of the system in horizontal translation.

Calculation of the coupled natural frequencies of a rigid body on resilient supports from Eq. (30.30) is sufficiently laborious to encourage the use of graphical means. For general purposes, both coupled natural frequencies can be obtained from Fig. 30.18. For a given type of isolators, $\eta = k_x/k_z$ is a constant and Eq. (30.30) may be evaluated in a manner that makes it possible to select isolator positions to attain optimum natural frequencies.[8–10] This is discussed under *Space-Plots* in Chap. 3. The convenience of the approach is partially offset by the need for a separate plot for each value of the stiffness ratio k_x/k_z. Applicable curves are plotted for several values of k_x/k_z in Figs. 3.17 to 3.19.

The preceding analysis of the dynamics of a rigid body on resilient supports includes the assumption that the principal axes of inertia of the rigid body are respectively parallel with the principal elastic axes of the resilient supports. This makes it possible to neglect the products of inertia of the rigid body. The coupling introduced by the product of inertia is not strong unless the angle between the above-mentioned inertia and elastic axes is substantial. It is convenient to take the coordinate axes through the center-of-

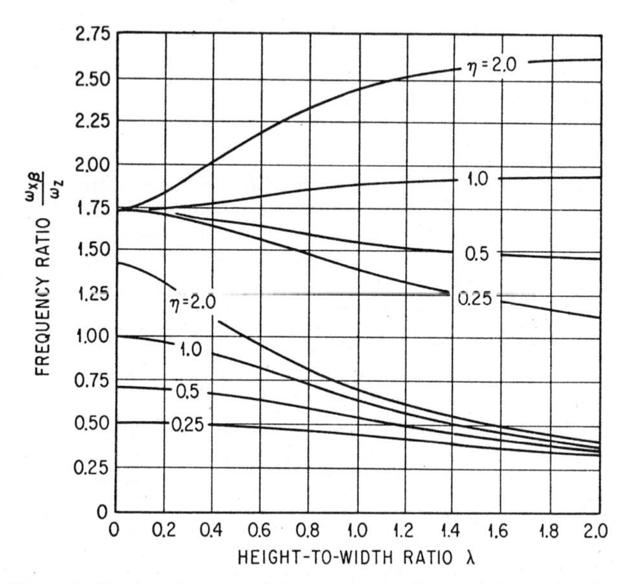

FIG. 30.19. Curves indicating the natural frequencies $\omega_{x\beta}$ in coupled rotational and horizontal translational modes with reference to the natural frequency ω_z in the decoupled vertical translational mode, for the system shown in Fig. 30.17. The ratio of horizontal to vertical stiffness of the isolators is η, and the height-to-width ratio for the equipment is λ. These curves are based upon the assumption that the mass of the equipment is uniformly distributed and that the isolators are attached precisely at the extreme lower corners thereof.

gravity of the supported body, parallel with the principal elastic axes of the isolators. If the moments of inertia with respect to these coordinate axes are used in Eqs. (30.24) to (30.30), the calculated natural frequencies usually are correct within a few per cent without including the effect of product of inertia. When it is desired to calculate the natural frequencies accurately or when the product of inertia coupling is strong, a calculation procedure is available that may be used for certain conventional arrangements using four isolators.[11]

The procedure for determining the natural frequencies in coupled modes summarized by the curves of Fig. 30.18 represents a rigorous analysis where the assumed symmetry exists. The procedure is somewhat indirect because the dimensionless ratio ρ_y/a_x appears in both ordinate and abscissa parameters, and because it is necessary to determine the radius of gyration of the equipment. The relations may be approximated in a more readily usable form if (1) the mounted equipment can be considered a cuboid having uniform mass distribution, (2) the four isolators are attached precisely at the four lower corners of the cuboid, and (3) the height of the isolators may be considered negligible. The ratio of the natural frequencies in the coupled rotational and horizontal translational modes to the natural frequency in the vertical translational mode then becomes a function of only the dimensions of the cuboid and the stiffnesses of the isolators in the several coordinate directions. Making these assumptions and substituting in Eq. (30.30),

$$\frac{\omega_{x\beta}}{\omega_z} = \frac{1}{\sqrt{2}}\sqrt{\frac{4\eta\lambda^2 + \eta + 3}{\lambda^2 + 1} \pm \sqrt{\left(\frac{4\eta\lambda^2 + \eta + 3}{\lambda^2 + 1}\right)^2 - \frac{12\eta}{\lambda^2 + 1}}} \tag{30.31}$$

where $\eta = k_x/k_z$ designates the ratio of horizontal to vertical stiffness of the isolators and $\lambda = 2a_z/2a_x$ indicates the ratio of height to width of mounted equipment. This relation is shown graphically in Fig. 30.19. The curves included in this figure are useful for calculating approximate values of natural frequencies and for indicating trends in natural frequencies resulting from changes in various parameters as follows:

1. Both of the coupled natural frequencies tend to become a minimum, for any ratio of height to width of the mounted equipment, when the ratio of horizontal to vertical stiffness k_x/k_z of the isolators is low. Conversely, when the ratio of horizontal to vertical stiffness is high, both coupled natural frequencies also tend to be high. Thus, when the isolators are located underneath the mounted body, a condition of low natural frequencies is obtained using isolators whose stiffness in a horizontal direction is less than the stiffness in a vertical direction. However, low horizontal stiffness may be undesirable in applications requiring maximum stability. A compromise between natural frequency and stability then may lead to optimum conditions.

2. As the ratio of height to width of the mounted equipment increases, the lower of the coupled natural frequencies decreases. The trend of the higher of the coupled natural frequencies depends on the stiffness ratio of the isolators. One of the coupled natural frequencies tends to become very high when the horizontal stiffness of the isolators is greater than the vertical stiffness, and when the height of the mounted equipment is approximately equal to or greater than the width. When the ratio of height to width of mounted equipment is greater than 0.5, the spread between the coupled natural frequencies increases as the ratio k_x/k_z of horizontal to vertical stiffness of the isolators increases.

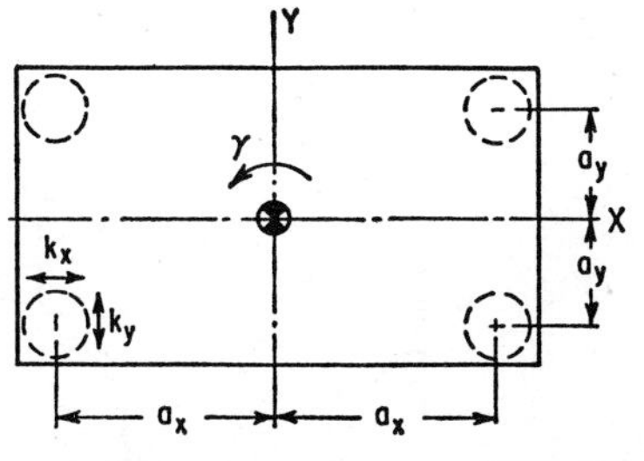

FIG. 30.20. Plan view of the equipment shown schematically in Fig. 30.17, indicating the uncoupled rotational mode specified by the rotation angle γ.

NATURAL FREQUENCY—UNCOUPLED ROTATIONAL MODE. Figure 30.20 is a plan view of the body shown in elevation in Fig. 30.17. The distances from the isolators to the principal planes of inertia are designated by a_x and a_y. The horizontal stiffnesses of the isolators in the directions of the coordinate axes X and Y are indicated by k_x and k_y, respectively. When the excitation is the applied couple $M = M_0 \sin \omega t$, the differential equation of motion is

$$I_z\ddot{\gamma} = -4\gamma a_x^2 k_y - 4\gamma a_y^2 k_x + M_0 \sin \omega t \tag{30.32}$$

where I_z is the moment of inertia of the body with respect to the Z axis. Neglecting transient terms, the solution of Eq. (30.32) gives the displacement amplitude γ_0 in rotation:

$$\gamma_0 = \frac{M_0}{4(a_x^2k_y + a_y^2k_x) - I_z\omega^2} \tag{30.33}$$

where the natural frequency ω_γ in rotation about the Z axis is the value of ω that makes the denominator of Eq. (30.33) equal to zero:

$$\omega_\gamma = 2\sqrt{\frac{a_x^2k_y + a_y^2k_x}{I_z}} \tag{30.34}$$

VIBRATION ISOLATION IN COUPLED MODES

When the equipment and isolator system has several degrees-of-freedom and the isolators are located in such a manner that several natural modes of vibration are coupled, it becomes necessary in evaluating the isolators to consider the contribution of the several modes in determining the motion transmitted from the support to the mounted equipment, or the force transmitted from the equipment to the foundation. Methods for determining the transmissibility under these conditions are best illustrated by examples.

For example, consider the system shown schematically in Fig. 30.21 wherein a machine is supported by relatively long beams which are in turn supported at their opposite ends by vibration isolators. The isolators are assumed to be undamped, and the excitation is

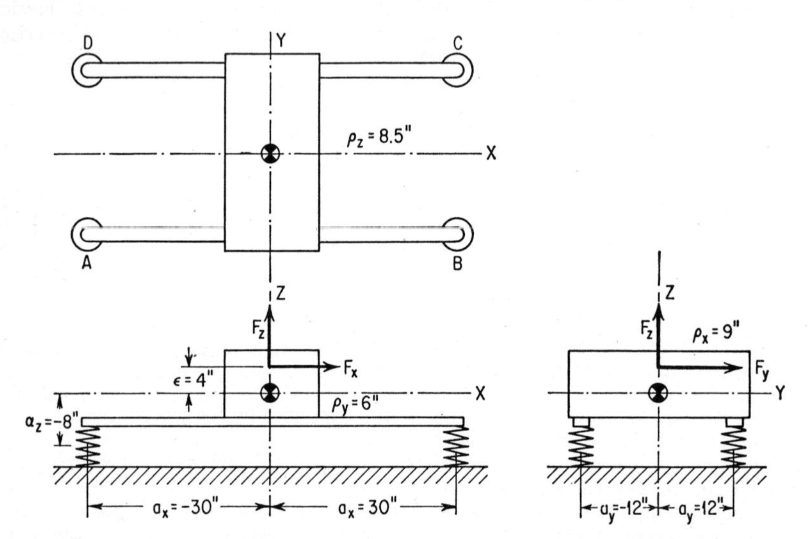

FIG. 30.21. Schematic diagram of an equipment mounted upon relatively long beams which are in turn attached at their opposite ends to vibration isolators. Excitation for the system is alternatively (1) the vibratory force $F_x = F_0 \cos \omega t$, $F_Z = F_0 \sin \omega t$ in the XZ plane or (2) the vibratory force $F_y = F_0 \cos \omega t$, $F_Z = F_0 \sin \omega t$ in the YZ plane.

considered to be a force applied at a distance $\epsilon = 4$ in. above the center-of-gravity of the machine-and-beam assembly. Alternatively, the force is (1) $F_x = F_0 \cos \omega t$, $F_z = F_0 \sin \omega t$ in a plane normal to the Y axis or (2) $F_y = F_0 \cos \omega t$, $F_z = F_0 \sin \omega t$ in a plane normal to the X axis. This may represent an unbalanced weight rotating in a vertical plane. A force transmissibility at each of the four isolators is determined by calculating the deflection of each isolator, multiplying the deflection by the appropriate isolator stiffness to obtain transmitted force, and dividing it by $F_0/4$.

When the system is viewed in a vertical plane perpendicular to the Y axis, the transmissibility curves are as illustrated in Fig. 30.22. The solid line defines the transmissibility at each of isolators B and C in Fig. 30.21, and the dotted line defines the transmissibility at each of isolators A and D. Similar transmissibility curves for a plane perpendicular to the X axis are shown in Fig. 30.23 wherein the solid line indicates the transmissibility at each of isolators C and D, and the dotted line indicates the transmissibility at each of isolators A and B.

Note the comparison of the transmissibility curves of Figs. 30.22 and 30.23 with the diagram of the system in Fig. 30.21. Figure 30.23 shows the three resonance conditions which are characteristic of a coupled system of the type illustrated. The transmissibility remains equal to or greater than unity for all excitation frequencies lower than the highest resonant frequency in a coupled mode. At greater excitation frequencies, vibration isolation is attained, as indicated by values of force transmissibility smaller than unity.

The transmissibility curves in Fig. 30.22 show somewhat similar results. The long horizontal beams tend to spread the resonant frequencies by a substantial frequency in-

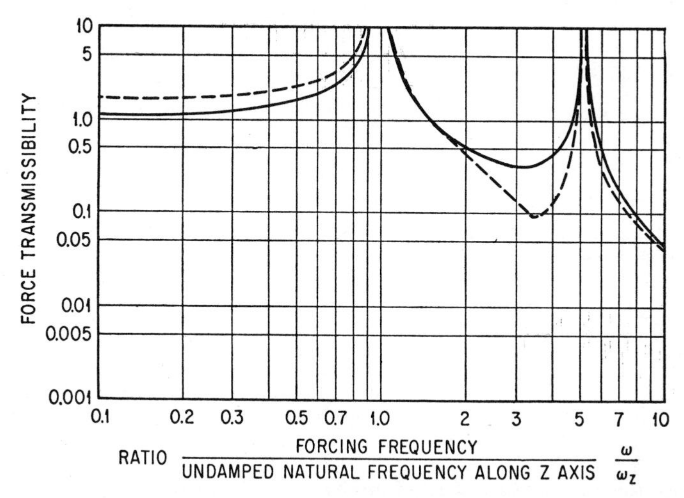

FIG. 30.22. Transmissibility curves for the system shown in Fig. 30.21 when the excitation is in a plane perpendicular to the Y axis. The solid line indicates the transmissibility at each of isolators B and C, whereas the dotted line indicates the transmissibility at each of isolators A and D.

crement, and merge the resonant frequency in the vertical translational mode with the resonant frequency in one of the coupled modes. A low transmissibility is again attained at excitation frequencies greater than the highest resonant frequency. Note that the transmissibility drops to a value slightly less than unity over a small frequency interval

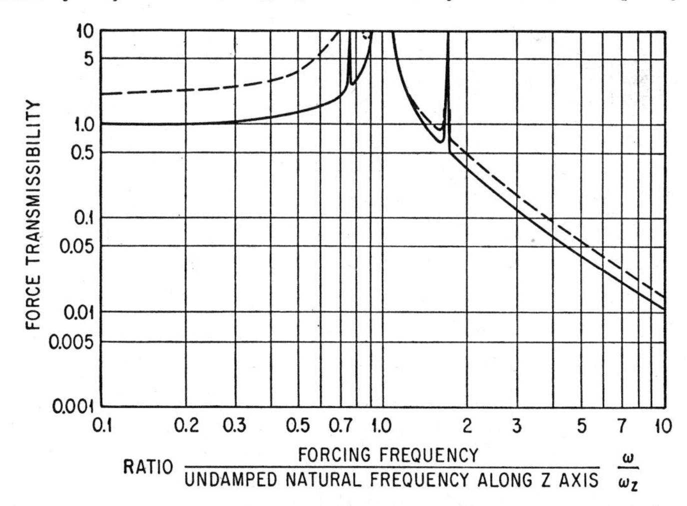

FIG. 30.23. Transmissibility curves for the system illustrated in Fig. 30.21 when the excitation is in a plane perpendicular to the X axis. The solid line indicates the transmissibility at each of isolators C and D, whereas the dotted line indicates the transmissibility at each of isolators A and B.

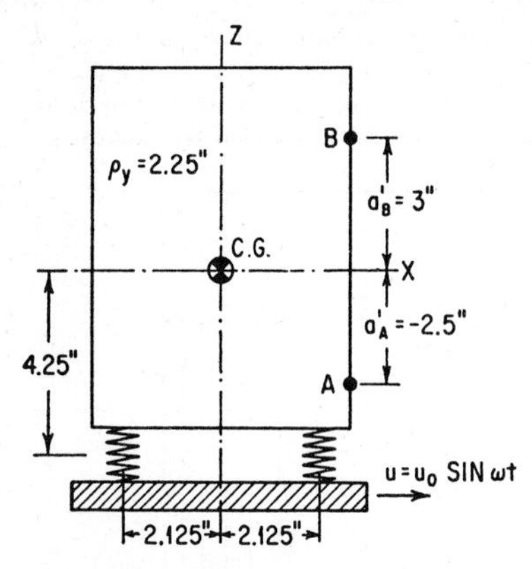

FIG. 30.24. Schematic diagram of an equipment supported by vibration isolators. Excitation is a vibratory displacement $u = u_0 \sin \omega t$ of the foundation.

between the predominant resonant frequencies. This is a force reduction resulting from the relatively long beams, and constitutes an acceptable condition if the magnitude of the excitation force in this direction is relatively small. Thus, the natural frequencies of the isolators could be somewhat higher with a consequent gain in stability; it is necessary, however, that the excitation frequency be substantially constant.

Consider the equipment illustrated in Fig. 30.24 when the excitation is horizontal vibration of the support. The effectiveness of the isolators in reducing the excitation vibration is evaluated by plotting the displacement amplitude of the horizontal vibration at points A and B with reference to the displacement amplitude of the support. Transmissibility curves for the system of Fig. 30.24 are shown in Fig. 30.25. The solid line in Fig. 30.25 refers to point A and the dotted line to point B. Note that there is no significant reduction of amplitude except when the forcing frequency exceeds the maximum resonant frequency of the system.

A general rule for the calculation of necessary isolator characteristics to achieve the results illustrated in Figs. 30.22, 30.23, and 30.25 is that the forcing frequency should be not less than 1.5 to 2 times the maximum natural frequency in any of six natural modes of vibration. In exceptional cases, such as illustrated in Fig. 30.22, the forcing frequency may be interposed between resonant frequencies if the forcing frequency is a constant.

Example 30.1. Consider the machine illustrated in Fig. 30.21. The force that is to be isolated is harmonic at the constant frequency of 8 cps; it is assumed to result from the rotation of an unbalanced member whose plane of rotation is alternatively (1) a plane perpendicular to the Y axis and (2) a plane perpendicular to the X axis. The distance between isolators is 60

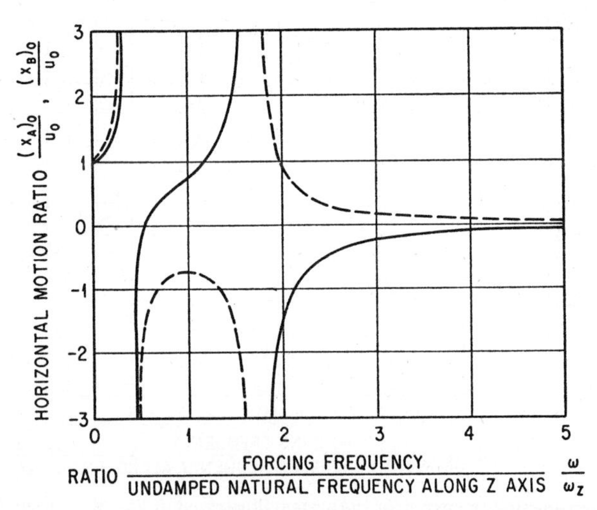

FIG. 30.25. Displacement transmissibility curves for the system of Fig. 30.24. Transmissibility between the foundation and point A is shown by the solid line; transmissibility between the foundation and point B is shown by the dotted line.

in. in the direction of the X axis and 24 in. in the direction of the Y axis. The center of coordinates is taken at the center-of-gravity of the supported body, i.e., at the center-of-gravity of the machine-and-beams assembly. The total weight of the machine and supporting beam assembly is 100 lb, and its radii of gyration with respect to the three coordinate axes through the center-of-gravity are $\rho_x = 9$ in., $\rho_z = 8.5$ in., and $\rho_y = 6$ in. The isolators are of equal stiffnesses in the directions of the three coordinate axes:

$$\eta = \frac{k_x}{k_z} = \frac{k_y}{k_z} = 1$$

The following dimensionless ratios are established as the initial step in the solution:

$$a_z/\rho_y = -1.333 \qquad a_z/\rho_x = -0.889$$

$$a_x/\rho_y = \pm 5.0 \qquad a_y/\rho_x = \pm 1.333$$

$$(a_z/\rho_y)^2 = 1.78 \qquad (a_z/\rho_x)^2 = 0.790$$

$$(a_x/\rho_y)^2 = 25.0 \qquad (a_y/\rho_x)^2 = 1.78$$

$$\eta(\rho_y/a_x)^2 = 0.04 \qquad \eta(\rho_x/a_y)^2 = 0.561$$

The various natural frequencies are determined in terms of the vertical natural frequency ω_z. Referring to Fig. 30.18, the coupled natural frequencies for vibration in a plane perpendicular to the Y axis are determined as follows:

First calculate the parameter

$$\frac{\rho_y}{a_x}\sqrt{\frac{k_x}{k_z}} = 0.2$$

For $a_z/\rho_y = -1.333$, $(\omega_{x\beta}/\omega_z)(\rho_y/a_x) = 0.19$; 1.03. Note the signs of the dimensionless ratios a_z/ρ_y and a_x/ρ_y. According to Eq. (30.30), the natural frequencies are independent of the sign of a_z/ρ_y. With regard to the ratio a_x/ρ_y, the sign chosen should be the same as the sign of the radical on the right side of Eq. (30.30). The frequency ratio $(\omega_{x\beta}/\omega_z)$ then becomes positive. Dividing the above values for $(\omega_{x\beta}/\omega_z)(\rho_y/a_x)$ by $\rho_y/a_x = 0.2$, $\omega_{x\beta}/\omega_z = 0.96$; 5.15.

Vibration in a plane perpendicular to the X axis is treated in a similar manner. It is assumed that exciting forces are not applied concurrently in planes perpendicular to the X and Y axes; thus, vibration in these two planes is independent. Consequently, the example entails two independent but similar problems and similar equations apply for a plane perpendicular to the X axis:

$$\frac{\rho_x}{a_y}\sqrt{\frac{k_z}{k_y}} = 0.75$$

For $a_z/\rho_x = 0.889$, $(\omega_{y\alpha}/\omega_z)(\rho_x/a_y) = 0.57$; 1.29. Dividing by $\rho_x/a_y = 0.75$, $\omega_{y\alpha}/\omega_z = 0.76$; 1.72.

The natural frequency in rotation with respect to the Z axis is calculated from Eq. (30.34) as follows, taking into consideration that there are two pairs of springs and that $k_x = k_y = k_z$:

$$\omega_\gamma = \sqrt{\left(\frac{a_x{}^2 + a_y{}^2}{\rho_z{}^2}\right)\left(\frac{4k_z g}{W}\right)} = 3.8\omega_z$$

The six natural frequencies are as follows:

1. Translational along Z axis: ω_z
2. Coupled in plane perpendicular to Y axis: $0.96\omega_z$
3. Coupled in plane perpendicular to Y axis: $5.15\omega_z$
4. Coupled in plane perpendicular to X axis: $0.76\omega_z$
5. Coupled in plane perpendicular to X axis: $1.72\omega_z$
6. Rotational with respect to Y axis: $3.8\omega_z$

Considering vibration in a plane perpendicular to the Y axis, the two highest natural frequencies are the natural frequency ω_y in the translational mode along the Z axis and the natural frequency $5.15\omega_z$ in a coupled mode. In a similar manner, the two highest natural frequencies in a plane perpendicular to the X axis are the natural frequency ω_z in translation along the Z axis and the natural frequency $1.72\omega_z$ in a coupled mode. The natural frequency in rotation about the Z axis is $3.80\omega_z$. The widest frequency increment which is void of natural frequencies is between $1.72\omega_z$ and $3.80\omega_z$. This increment is used for the forcing frequency which is taken as $2.5\omega_z$. Inasmuch as the forcing frequency is established at 8 cps, the vertical natural frequency is 8 divided by 2.5, or 3.2 cps. The required vertical stiffnesses of the isolators are calculated from Eq. (30.11) to be 105 lb/in. for the entire machine, or 26.2 lb/in. for each of the four isolators.

INCLINED ISOLATORS

Advantages in vibration isolation sometimes result from inclining the principal elastic axes of the isolators with respect to the principal inertia axes of the equipment, as illustrated in Fig. 30.26. The coordinate axes X and Z are respectively parallel with the principal inertia axes of the mounted body, but the center of coordinates is taken at the elastic axis. The location of the elastic axis is determined by the elastic properties of the system. If a force is applied to the body along a line extending through the elastic axis, the body is displaced in translation without rotation; if a couple is applied to the body, the body is displaced in rotation without translation.

The principal elastic axes r, p of the isolators are parallel with the paper and inclined with respect to the coordinate axes, as indicated in Fig. 30.26. The stiffness of each isolator in the direction of the respective principal axis is indicated by k_r, k_p. The principal elastic axis of an isolator is the axis along which a force must be applied to cause a deflection colinear with the applied force (see the section *Properties of a Biaxial Stiffness Isolator*).

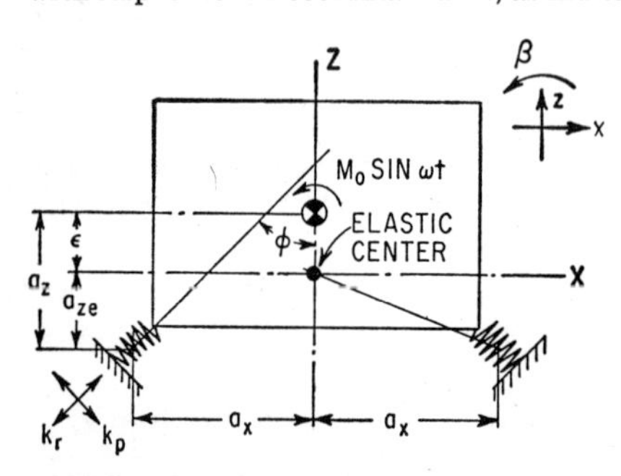

FIG. 30.26. Schematic diagram of an equipment supported by isolators whose principal elastic axes are inclined to the principal inertia axes of the equipment.

Assume the excitation for the system shown in Fig. 30.26 to be a couple $M_0 \sin \omega t$ acting about an axis normal to the paper. The equations of motion for the body in the horizontal translational and rotational modes may be written by noting that the displacement of the center-of-gravity in the direction of the X axis is $x - \epsilon\beta$; thus, the corresponding acceleration is $\ddot{x} - \epsilon\ddot{\beta}$. A translational displacement x produces only an external force $-k_x x$, whereas a rotational displacement β produces only an external couple $-k_\beta \beta$. The equations of motion are

$$
\begin{aligned}
m(\ddot{x} - \epsilon\ddot{\beta}) &= -k_x x \\
m\rho_e^2\ddot{\beta} - m\epsilon\ddot{x} &= -k_\beta\beta + M_0 \sin \omega t
\end{aligned}
\tag{30.35}
$$

where ρ_e is the radius of gyration of the mounted body with respect to the elastic axis. The radius of gyration ρ_e is related to the radius of gyration ρ_y with respect to a line through the center-of-gravity by $\rho_e = \sqrt{\rho_y^2 + \epsilon^2}$, where ϵ is the distance between the elastic axis and a parallel line passing through the center-of-gravity. In the equations of motion, k_x and k_β represent the translational and rotational stiffness of the isolators in the x and β coordinate directions, respectively.

By assuming steady-state harmonic motion for the horizontal translation x and rotation β, the following displacement amplitudes are obtained by solving Eqs. (30.35):

$$
\begin{aligned}
x_0 &= \frac{-M_0\epsilon\omega^2}{m[\rho_e^2(\omega^2 - \omega_\beta^2)(\omega^2 - \omega_x^2) - \epsilon^2\omega^4]} \\
\beta_0 &= \frac{-M_0}{m\left[\rho_e^2(\omega^2 - \omega_\beta^2) - \dfrac{\epsilon^2\omega^4}{\omega^2 - \omega_x^2}\right]}
\end{aligned}
\tag{30.36}
$$

where $\omega_x = \sqrt{k_x/m}$ and $\omega_\beta = \sqrt{k_\beta/m\rho_e^2}$ are hypothetical natural frequencies defined for convenience. The natural frequencies $\omega_{x\beta}$ in the coupled x,β modes are determined by equating the denominator of Eqs. (30.36) to zero and solving for ω (now identical to $\omega_{x\beta}$):

$$\frac{\omega_{x\beta}}{\omega_x} = \sqrt{\frac{1 + \lambda_1^2 \pm \sqrt{(1 + \lambda_1^2)^2 - 4\lambda_1^2[1 - (\epsilon/\rho_e)^2]}}{2[1 - (\epsilon/\rho_e)^2]}} \tag{30.37}$$

where λ_1 is a dimensionless ratio given by

$$\lambda_1 = \frac{(a_x/\rho_e)\sqrt{k_r/k_p}}{\cos^2 \phi + (k_r/k_p) \sin^2 \phi} \tag{30.38}$$

The hypothetical natural frequency ω_x is

$$\omega_x = \sqrt{\frac{4k_p}{m}\left[\cos^2 \phi + \frac{k_r}{k_p} \sin^2 \phi\right]} \tag{30.39}$$

The relation given by Eq. (30.37) is shown graphically by Fig. 30.27. The parameters needed to evaluate the natural frequencies by using this graph are calculated from the physical properties of the system and the relations of Eqs. (30.38) and (30.39). In addition, the distance ϵ between a parallel line passing through the center-of-gravity and the elastic axis must be known. The distance ϵ is determined by effecting a small horizontal displacement of the equipment in the X direction, and equating the resulting summation of elastic couples to zero:

$$\epsilon = a_z - \frac{a_x (1 - k_p/k_r) \cot \phi}{(k_p/k_r) \cot^2 \phi + 1} \tag{30.40}$$

where a_z is the distance between the parallel planes passing through the center-of-gravity of the body and the mid-height of the isolators, as shown in Fig. 30.26.

DECOUPLING OF MODES

The natural modes of vibration of a body supported by isolators may be decoupled one from another by proper orientation of the isolators. Each mode of vibration then exists independently of the others, and vibration in one mode does not excite vibration in other modes. The necessary conditions for decoupling may be stated as follows: The resultant of the forces applied to the mounted body by the isolators when the mounted body is displaced in translation must be a force directed through the center-of-gravity; or, the resultant of the couples applied to the mounted body by the isolators when the mounted body is displaced in rotation must be a couple about an axis through the center-of-gravity.

In general, the natural frequencies of a multiple degree-of-freedom system can be made equal only by decoupling the natural modes of vibration, i.e., by making $a_z = 0$ in Fig. 30.17. The natural frequencies in decoupled modes are indicated by the two straight lines in Fig. 30.18 marked $a_z/\rho_y = 0$. The natural frequencies in translation along the X axis and in rotation about the Y axis become

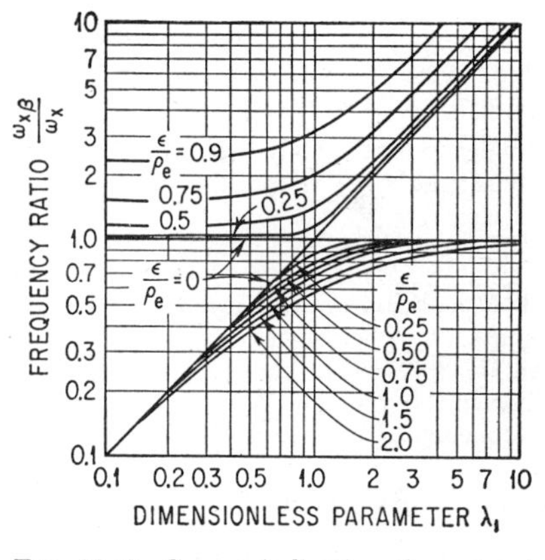

FIG. 30.27. Curves indicating the natural frequencies $\omega_{x\beta}$ in coupled modes with reference to the natural frequency in the decoupled (fictitious) horizontal translational mode ω_x for the system shown schematically in Fig. 30.26. The radius of gyration with respect to the elastic axis is indicated by ρ_e, and the distance between the center-of-gravity and the elastic center is ϵ. The dimensionless parameter λ_1 is defined by Eq. (30.38) and ω_x is defined by Eq. (30.39).

equal at the intersection of these lines; i.e., when $a_z/\rho_y = 0$, $k_x/k_z = 1$ and $\rho_y/a_x = 1$. The physical significance of these mathematical conditions is that the isolators be located in a plane passing through the center-of-gravity of the equipment, that the distance between isolators be twice the radius of gyration of the equipment, and that the stiffness of each isolator in the directions of the X and Z axes be equal.

When the isolators cannot be located in a plane which passes through the center-of-gravity of the equipment, decoupling can be achieved by inclining the isolators, as illustrated in Fig. 30.26. If the elastic axis of the system is made to pass through the center-of-gravity, the translational and rotational modes are decoupled because the inertia force of the mounted body is applied through the elastic center and introduces no

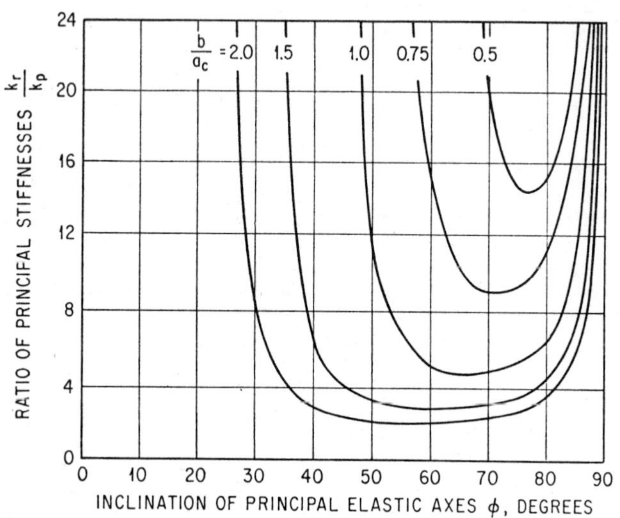

FIG. 30.28. Ratio of stiffnesses k_r/k_p along principal elastic axes required for decoupling the natural modes of vibration of the system illustrated in Fig. 30.26.

tendency for the body to rotate. The requirements for a decoupled system are established by setting $\epsilon = 0$ in Eq. (30.40) and solving for k_r/k_p:

$$\frac{k_r}{k_p} = \frac{(a_x/a_z) + \cot \phi}{(a_x/a_z) - \tan \phi} \tag{30.41}$$

The conditions for decoupling defined by Eq. (30.41) are shown graphically in Figs. 30.28 and 3.23. The decoupled natural frequencies are indicated by the straight lines $\epsilon/\rho_e = 0$ in Fig. 30.27. The horizontal line refers to the decoupled natural frequency ω_x in translation in the direction of the X axis, while the inclined line refers to the decoupled natural frequency ω_β in rotation about the Y axis.

PROPERTIES OF A BIAXIAL STIFFNESS ISOLATOR

A biaxial stiffness isolator is represented as an elastic element having a single plane of symmetry; all forces act in this plane and the resultant deflections are limited by symmetry or constraints to this plane. The characteristic elastic properties of the isolator may be defined alternatively by sets of influence coefficients as follows:

1. If the two coordinate axes in the plane of symmetry are selected arbitrarily, three stiffness parameters are required to define the properties of the isolator. These are the axial influence coefficients * along the two coordinate axes, and a characteristic coupling influence coefficient * between the coordinate axes.

* The influence coefficient κ is a function only of the isolator properties and not of the constraints imposed by the system in which the isolator is used. Both positive and negative values of the influence coefficient κ are permissible.

2. If the two coordinate axes in the plane of symmetry are selected to coincide with the principal elastic axes of the isolator, two influence coefficients are required to define the properties of the isolator. These are the principal influence coefficients. If the isolator is used in a system, a third parameter is required to define the orientation of the principal axes of the isolator with the coordinate axes of the system.

PROPERTIES OF ISOLATOR WITH RESPECT TO ARBITRARILY SELECTED AXES

A schematic representation of a linear biaxial stiffness element is shown in Fig. 30.29 where the X and Y axes are arbitrarily chosen to define a plane to which all forces and motions are restricted. In general, the deflection of an isolator resulting from an applied load is not in the same direction as the load, and a coupling influence coefficient is required to define the properties of the isolator in addition to the influence coefficients along the X and Y axes. The three characteristic stiffness coefficients that uniquely describe the load-deflection properties of a biaxial stiffness element are:

1. The influence coefficient of the element in the X coordinate direction is κ_x. It is the ratio of the component of the applied force in the X direction to the resulting deflection when the isolator is constrained to deflect in the X direction.

2. The influence coefficient of the element in the Y coordinate direction is κ_y. It is the ratio of the component of the applied force in the Y direction to the resulting deflection when the isolator is constrained to deflect in the Y direction.

3. The coupling influence coefficient is κ_{xy}. It represents the force required in the X direction to produce a unit displacement in the Y direction when the isolator is constrained to deflect only in the Y direction. (By Maxwell's reciprocity principle, the same force is required in the Y direction to produce a unit displacement in the X direction; i.e., $\kappa_{xy} = \kappa_{yx}$.)

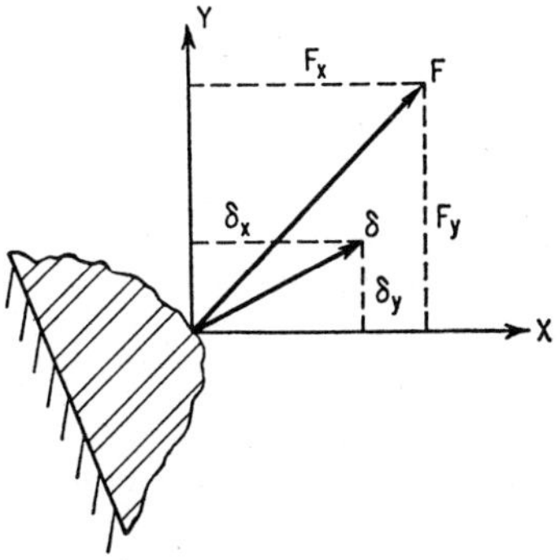

Y
F_x
F
δ_x
δ
F_y
δ_y
X

FIG. 30.29. Schematic diagram of a linear biaxial stiffness element.

Consider the isolator shown in Fig. 30.29 where the applied force F has components F_x and F_y; the resulting displacement has components δ_x and δ_y. From the above definitions of influence coefficients, the forces in the X and Y coordinate directions required to effect a displacement δ_x are

$$F_{xx} = \kappa_x\delta_x \qquad F_{yx} = \kappa_{yx}\delta_x \tag{30.42}$$

The forces required to effect a displacement δ_y in the Y direction are

$$F_{xy} = \kappa_{xy}\delta_y \qquad F_{yy} = \kappa_y\delta_y \tag{30.43}$$

The force components F_x and F_y required to produce the deflection having components δ_x, δ_y are the sums from Eqs. (30.42) and (30.43):

$$\begin{aligned} F_x &= \kappa_x\delta_x + \kappa_{xy}\delta_y \\ F_y &= \kappa_{yx}\delta_x + \kappa_y\delta_y \end{aligned} \tag{30.44}$$

If the three influence stiffness coefficients κ_x, κ_y and $\kappa_{xy} = \kappa_{yx}$ are known for a given stiffness element, the load-deflection properties are given by Eq. (30.44).

The deflections of the isolator in response to forces F_x, F_y are determined by solving Eqs. (30.44) simultaneously:

$$\begin{aligned} \delta_x &= \frac{F_x\kappa_y - F_y\kappa_{xy}}{\kappa_x\kappa_y - \kappa_{xy}^2} \\ \delta_y &= \frac{F_y\kappa_x - F_x\kappa_{xy}}{\kappa_x\kappa_y - \kappa_{xy}^2} \end{aligned} \tag{30.45}$$

These expressions give the orthogonal components of the displacement δ for any load having the components F_x and F_y applied to a biaxial stiffness isolator. By substituting the relations of Eqs. (30.45) into Eq. (30.44), the following alternate forms of the force-deflection equations are obtained:

$$F_x = \left(\kappa_x - \frac{\kappa_{xy}^2}{\kappa_y}\right)\delta_x + \frac{\kappa_{xy}}{\kappa_y}F_y$$
$$F_y = \left(\kappa_y - \frac{\kappa_{xy}^2}{\kappa_x}\right)\delta_y + \frac{\kappa_{xy}}{\kappa_x}F_x \qquad (30.46)$$

The specific force-deflection equations for a given situation are obtained from these general load-deflection expressions by applying the proper constraint conditions.

UNCONSTRAINED MOTION. The general force-deflection equations can be used to obtain the effective stiffness coefficients when the forces F_x and F_y shown in Fig. 30.29 are applied independently. The resulting deflection of the isolator is unconstrained motion, i.e., the isolator is free to deflect out of the line of force application. The force divided by that component of deflection along the line of action of the force is the effective stiffness k. When $F_y = 0$, the effective stiffness k_x resulting from the applied force F_x is obtained from Eq. (30.46):

$$k_x = \frac{F_x}{\delta_x} = \left(\kappa_x - \frac{\kappa_{xy}^2}{\kappa_y}\right) \qquad (30.47)$$

When $F_x = 0$, the effective stiffness k_y in response to the applied force F_y is

$$k_y = \frac{F_y}{\delta_y} = \left(\kappa_y - \frac{\kappa_{xy}^2}{\kappa_x}\right) \qquad (30.48)$$

For unconstrained motion, $k_x/k_y = \kappa_x/\kappa_y$; i.e., the ratio of the effective stiffnesses in two mutually perpendicular directions is equal to the ratio of the corresponding influence coefficients for the same directions.

CONSTRAINED MOTION. When the isolator is constrained either by the symmetry of a system or by structural constraints to deflect only along the line of the applied force, the effective stiffness is obtained directly by letting appropriate deflections be zero in Eq. (30.44):

$$k_x = \frac{F_x}{\delta_x} = \kappa_x \qquad k_y = \frac{F_y}{\delta_y} = \kappa_y \qquad (30.49)$$

The force required to maintain constrained motion is found by letting appropriate deflections be zero in Eqs. (30.46). For example, the force that must be applied in the X direction to insure that the isolator deflects in the Y direction in response to a force F_y is

$$F_x = \frac{\kappa_{xy}}{\kappa_y}F_y \qquad (30.50)$$

INFLUENCE COEFFICIENT TRANSFORMATION

Assume the influence coefficients κ_x, κ_y, and κ_{xy} are known in the X, Y coordinate system. It may be convenient to work with isolator influence coefficients in the X', Y' coordinate system as shown in Fig. 30.30. The X', Y' coordinate system is obtained by rotating the coordinate axes counterclockwise through an angle θ from the X, Y system. The influence coefficients with respect to the X', Y' axes are related to the influence coefficients with respect to the X, Y axes as follows:

$$\begin{aligned} \kappa_{x'} &= \frac{\kappa_x + \kappa_y}{2} + \frac{\kappa_x - \kappa_y}{2} \cos 2\theta + \kappa_{xy} \sin 2\theta \\ \kappa_{x'y'} &= \frac{\kappa_y - \kappa_x}{2} \sin 2\theta + \kappa_{xy} \cos 2\theta \\ \kappa_{y'} &= \frac{\kappa_x + \kappa_y}{2} - \frac{\kappa_x - \kappa_y}{2} \cos 2\theta - \kappa_{xy} \sin 2\theta \end{aligned} \tag{30.51}$$

The influence coefficient transformation of a biaxial stiffness isolator from one set of arbitrarily chosen coordinate axes to another arbitrarily chosen set of coordinate axes is described by the two-dimensional Mohr circle.[12] Since the influence coefficient is a

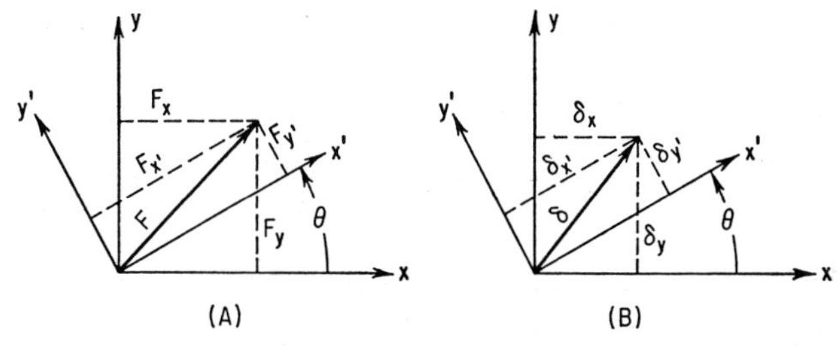

FIG. 30.30. Force and displacement transformation diagrams for a linear biaxial stiffness element.

tensor quantity, the following invariants of the influence coefficient tensor give additional relations between the influence coefficients in the X, Y and the X', Y' set of axes:

$$\begin{aligned} \kappa_x + \kappa_y &= \kappa_{x'} + \kappa_{y'} \\ \kappa_x \kappa_y - \kappa_{xy}^2 &= \kappa_{x'} \kappa_{y'} - \kappa_{x'y'}^2 \end{aligned} \tag{30.52}$$

PRINCIPAL INFLUENCE COEFFICIENTS

The set of axes for which there exists no coupling influence coefficient are the principal axes of stiffness (*principal elastic axes*). These axes can be found by requiring $\kappa_{x'y'}$ to be zero in Eq. (30.51) and solving for the rotation angle corresponding to this condition. Letting θ' represent the angle of rotation for which $\kappa_{x'y'} = 0$:

$$\tan 2\theta' = \frac{2\kappa_{xy}}{\kappa_x - \kappa_y} \tag{30.53}$$

By substituting this value of the angle of rotation into the general influence coefficient expressions, Eqs. (30.51), the following relation is obtained for the principal influence coefficients:

$$\kappa_p, \kappa_q = \frac{\kappa_x + \kappa_y}{2} \pm \sqrt{\left(\frac{\kappa_x - \kappa_y}{2}\right)^2 + \kappa_{xy}^2} \tag{30.54}$$

where p and q represent the principal axes of stiffness. The principal influence coefficients are the maximum and minimum influence coefficients that exist for a linear biaxial stiffness isolator. In Eq. (30.54), the plus sign gives the maximum influence coefficient whereas the minus sign gives the minimum influence coefficient. Either κ_p or κ_q can be the maximum influence coefficient, depending on the degree of axis rotation and the relative values of κ_x, κ_y, and κ_{xy}.

INFLUENCE COEFFICIENT TRANSFORMATION FROM THE PRINCIPAL AXES

The influence coefficient transformation from the principal axes p, q is of practical interest. The influence coefficients in the XY frame of reference are determined from Eq. (30.51) by setting $\kappa_{x'y'} = \kappa_{pq} = 0$, $\kappa_{x'} = \kappa_p$, $\kappa_{y'} = \kappa_q$, and $\theta = \theta'$. The influence coefficients in the XY frame-of-reference may be expressed in terms of the principal

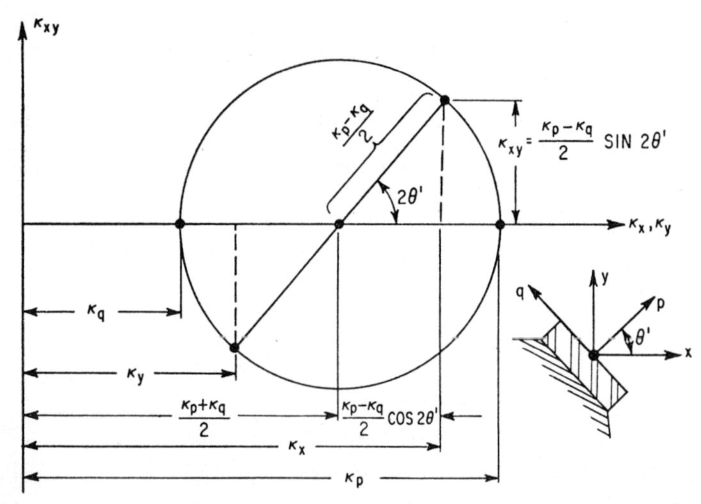

FIG. 30.31. Mohr-circle representation of the stiffness transformation from the principal axes of stiffness of a biaxial stiffness element. The p, q axes represent the principal stiffness axes and the X, Y axes are any arbitrary set of axes separated from the p, q axes by a rotation angle θ'.

influence coefficients as follows:

$$\begin{aligned}
\kappa_x &= \kappa_p \cos^2\theta' + \kappa_q \sin^2\theta' = \frac{\kappa_p + \kappa_q}{2} + \frac{\kappa_p - \kappa_q}{2}\cos 2\theta' \\
\kappa_{xy} &= (\kappa_p - \kappa_q)\sin\theta'\cos\theta' = \frac{\kappa_p - \kappa_q}{2}\sin 2\theta' \qquad (30.55) \\
\kappa_y &= \kappa_p \sin^2\theta' + \kappa_q \cos^2\theta' = \frac{\kappa_p + \kappa_q}{2} - \frac{\kappa_p - \kappa_q}{2}\cos 2\theta'
\end{aligned}$$

The transformation from the principal axes in the form of a two-dimensional Mohr's circle is shown by Fig. 30.31. This circle provides quick graphical determination of the three influence coefficients κ_x, κ_y, and κ_{xy} for any angle θ' between the P and X axes, where θ' is positive in the sense shown in the inset to Fig. 30.31.

Example 30.2. Consider the system shown schematically by Fig. 30.26. The transformation theory for the influence coefficient of a biaxial stiffness element may be applied to develop the effective stiffness coefficients for this system. The center of coordinates for the XZ axes is at the elastic center of the system. The principal elastic axes of the isolators p, r are oriented at an angle ϕ with the coordinate axes X, Z, respectively.* The position of the elastic center is determined by effecting a small horizontal displacement δ_x of the body, letting δ_z be zero and equating the summation of couples resulting from the isolator forces. The

* The properties of a biaxial stiffness element may be defined with respect to any pair of coordinate axes. In Fig. 30.26, the principal elastic axis q is parallel with the coordinate axis Y; then the analysis considers the principal elastic axes p, r which lie in the plane defined by the XZ coordinate axes.

forces F_x and F_z are determined from Eqs. (30.44):

$$F_x = \kappa_x \delta_x = k_x \delta_x \qquad F_z = \kappa_{zx} \delta_x = k_{zx} \delta_x$$

Each of the forces F_x acts at a distance $-a_{ze}$ from the elastic center; the force F_z at the right-hand isolator is positive and acts at a distance a_x from the elastic center whereas the force F_z at the left-hand isolator is negative and acts at a distance $-a_x$ from the elastic center. Taking a summation of the moments:

$$-2a_{ze}F_x + 2a_xF_z = 0$$

Substituting the above relations between the forces F_x, F_z and the influence coefficients κ_x, κ_{zx} into Eqs. (30.55), and noting that $\theta' = -\phi$ (compare Figs. 30.30 and 30.26), the following result is obtained in terms of principal stiffnesses:

$$\frac{a_{ze}}{a_x} = \frac{F_z}{F_x} = \frac{\kappa_{zx}}{\kappa_x} = \frac{(k_r - k_p)\sin\phi\cos\phi}{k_r\sin^2\phi + k_p\cos^2\phi}$$

Substituting $\epsilon = a_z - a_{ze}$ in the preceding equation, the relation for ϵ given by Eq. (30.40) is obtained.

Since the equations of motion are written in a coordinate system passing through the elastic center, all displacements in this frame-of-reference are constrained. Therefore, the effective stiffness coefficients for a single isolator may be obtained from Eq. (30.55) as follows [see Eq. (30.49)]:

$$k_x = \kappa_x = k_r\sin^2\phi + k_p\cos^2\phi$$

$$k_z = \kappa_z = k_r\cos^2\phi + k_p\sin^2\phi$$

These effective stiffness coefficients define the hypothetical natural frequency ω_x given by Eq. (30.39) as well as the uncoupled vertical natural frequency ω_z. Since four isolators are used in the problem represented by Fig. 30.26, the translational stiffnesses given by the above expressions for k_x and k_z must be multiplied by 4 to obtain the total translational stiffness.

The effective rotational stiffness of a single isolator k_β can be obtained by determining the sum of the restoring moments for a constrained rotation β. When the body is rotated through an angle β, the displacements at the right isolator are $\delta_x = -a_{ze}\beta$ and $\delta_z = a_x\beta$, where a_{ze} is a negative distance since it is measured in the negative Z direction. The sum of the restoring moments is $(F_za_x - F_xa_{ze})$, where F_x and F_z are the forces acting on the right isolator in Fig. 30.26. The forces F_x and F_z may be written in terms of the influence coefficients and the displacements δ_x and δ_z by use of Eq. (30.44) to produce the following moment equation:

$$M_\beta = k_\beta\beta = \beta[k_xa_{ze}^2 - 2k_{xz}a_{ze}a_x + k_za_x^2]$$

where the effective rotational stiffness k_β of a single isolator is

$$k_\beta = k_xa^2 - 2k_{xz}a_{ze}a_x + k_za_x^2$$

The distance a_{ze} can be eliminated from the expression for rotational stiffness by substituting $a_{ze} = a_xF_z/F_x$ obtained from the summation of couples about the elastic center:

$$k_\beta = a_x^2\left(\frac{k_xk_z - k_{xz}^2}{k_x}\right)$$

The numerator of this expression can be replaced by k_rk_p [see Eq. (30.52)] where the r, p axes are the principal elastic axes of the isolator and $k_{rp} = 0$. Also, k_x can be replaced by its equivalent form given by Eq. (30.55). Making these substitutions, the effective rotational stiffness for one isolator in terms of the principal stiffness coefficients of the isolator becomes

$$k_\beta = \frac{a_x^2k_p}{\sin^2\phi + (k_p/k_r)\cos^2\phi}$$

Since four isolators are used in the problem represented by Fig. 30.26, the rotational stiffness given by the above expression for k_β must be multiplied by 4 to obtain the total rotational stiffness of the system.

NONLINEAR VIBRATION ISOLATORS

In vibration isolation, the vibration amplitudes generally are small and linear vibration theory usually is applicable with sufficient accuracy.* However, the static effects of nonlinearity should be considered. Even though a nonlinear isolator may have approximately constant stiffness for small incremental deflections, the nonlinearity becomes important when large deflections of the isolator occur due to the effects of equipment weight and sustained acceleration. A vibration isolator often exhibits a stiffness that increases with applied force or deflection. Such a nonlinear stiffness is characteristic, for example, of rubber in compression or a conical spring.

In Eq. (30.11) for natural frequency, the stiffness k for a linear stiffness element is a constant. However, for a nonlinear isolator, the stiffness k is the slope of the force-deflection curve and Eq. (30.11) may be written

$$\omega_n = 2\pi f_n = \sqrt{\frac{g(dF/d\delta)}{W}} \tag{30.56}$$

where W is the total weight supported by the isolator, g is the acceleration of gravity, and $dF/d\delta$ is the slope of the line tangent to the force-deflection curve at the static equilibrium position. Vibration is considered to be small variations in the position of the supported equipment above and below the static equilibrium position, as indicated in Fig. 30.32. Thus, the natural frequency is determined solely by the stiffness characteristics in the region of the isolator deflection.

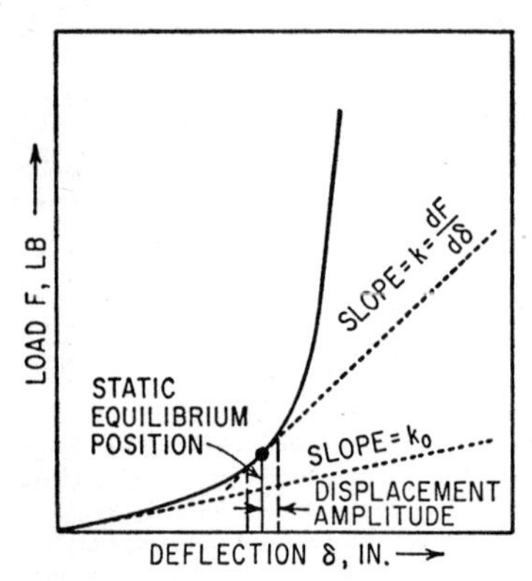

FIG. 30.32. Typical force-deflection characteristic of a tangent hardening isolator.

NATURAL FREQUENCY

In determining the natural frequency of a nonlinear isolator, it is important to note whether or not all the load results from the dead weight of a massive body. The force F on the isolator may be greater than the weight W because of a belt pull or sustained acceleration of a missile. Then the load on the isolator is

$$F = n_g W \tag{30.57}$$

where n_g is some multiple of the acceleration of gravity. For example, n_g may indicate the absolute value of the sustained acceleration of a missile measured in "number of g's."

CHARACTERISTIC OF TANGENT ISOLATOR. It is convenient to define the force-deflection characteristics of a nonlinear isolator having increasing stiffness (hardening characteristic) by a tangent function:[13]

$$F = \frac{2k_0 h_c}{\pi} \tan\left(\frac{\pi\delta}{2h_c}\right) \tag{30.58}$$

where F is the total force applied to the isolator, k_0 is the stiffness of the isolator at zero deflection, δ is the deflection of the isolator, and h_c is the characteristic height of the isolator. The force-deflection characteristic defined by Eq. (30.58) is shown graphically in Fig. 30.33A. The characteristic height h_c represents a height or thickness characteristic of the isolator which may be adjusted empirically to obtain optimum agreement, over the deflection range of interest, between Eq. (30.58) and the actual force-deflection curve for the isolator.

The stiffness of the tangent isolator is obtained by differentiation of Eq. (30.58) with

* If the vibration amplitude is large, nonlinear vibration theory as discussed in Chap. 4 is applicable.

respect to δ:

$$k = \frac{dF}{d\delta} = k_0 \sec^2\left(\frac{\pi\delta}{2h_c}\right) = k_0\left[1 + \left(\frac{F\pi}{2k_0h_c}\right)^2\right] \tag{30.59}$$

The stiffness-deflection relation defined by Eq. (30.59) is shown graphically in Fig. 30.33*B*.

Replacing the load F by n_gW in Eq. (30.59) and substituting the resulting stiffness relation into Eq. (30.56):

$$f_n\sqrt{h_c} = 3.13\sqrt{2.46n_g^2\left(\frac{W}{k_0h_c}\right) + \left(\frac{k_0h_c}{W}\right)} \tag{30.60}$$

The relation defined by Eq. (30.60) is shown graphically in Fig. 30.34. The ordinate is

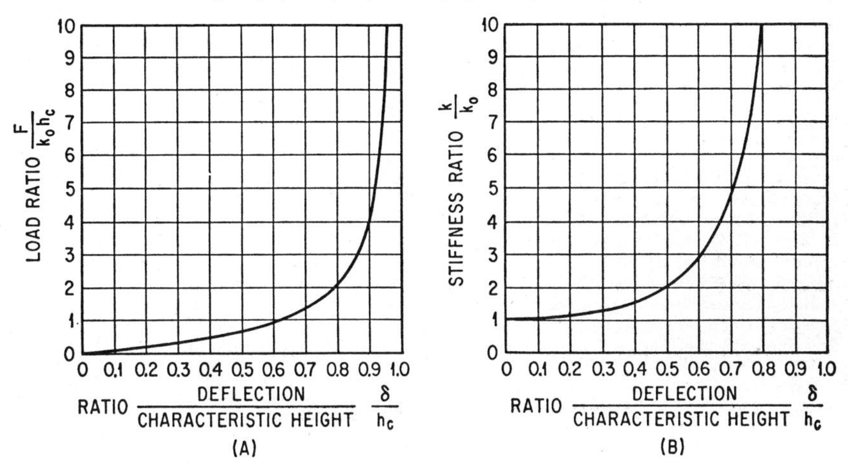

FIG. 30.33. Elastic properties of a tangent isolator in terms of its characteristic height h_c and stiffness k_o at zero deflection: (*A*) dimensionless force-deflection curve; (*B*) dimensionless stiffness-deflection curve.

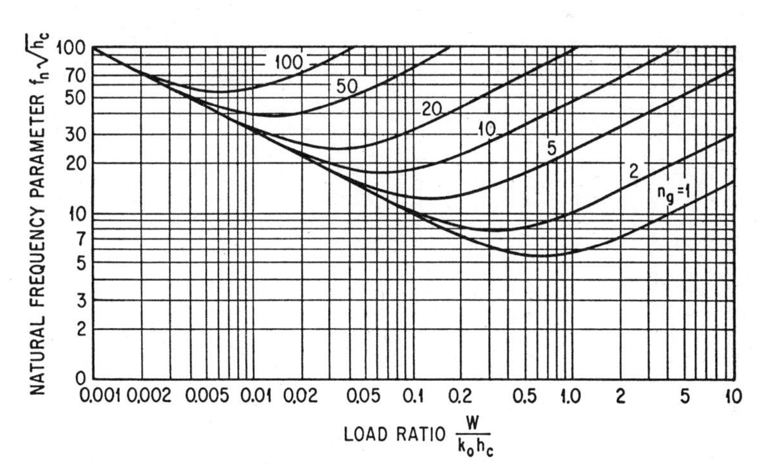

FIG. 30.34. Natural frequency f_n of a tangent isolator system when a portion of the total load applied to the isolator is nonmassive. The weight carried by the isolator is W and the sustained acceleration parameter is n_g, a multiple of the gravitational acceleration. The characteristic height is h_c and the stiffness at zero deflection is k_o.

the natural frequency f_n (cps) times the square root of the characteristic height of the isolator (in.). The theoretical and experimental force-deflection curves for the isolator are matched to establish the numerical value of the characteristic height. For a given value of the acceleration parameter n_g, the natural frequency of the isolation system is determined by h_c and W/k_0h_c.

The deflection of the isolator under a sustained acceleration loading is obtained by substituting Eq. (30.57) into the general force-deflection expression, Eq. (30.58), and solving for the dimensionless ratio δ/h_c:

$$\frac{\delta}{h_c} = \frac{2}{\pi}\tan^{-1}\left(\frac{\pi n_g}{2}\cdot\frac{W}{k_0h_c}\right) = \frac{2}{\pi}\tan^{-1}\left[15.37\left(\frac{n_g}{h_cf_{n0}^2}\right)\right] \tag{30.61}$$

A reference natural frequency f_{n0} is the natural frequency that occurs when the isolator is not deflected by the dead-weight load; i.e., $n_g = 0$. The nomograph of Fig. 30.35 gives the deflection ratio δ/h_c and the frequency ratio f_n/f_{n0}.[14] The value of the parameter $15.37(n_g/h_cf_{n0}^2)$ is transferred by a horizontal projection to the coordinate system for the curves. Values for the natural frequency ratio f_n/f_{n0} are read from the lower abscissa scale and values for the deflection ratio δ/h_c are read from the upper abscissa scale.

Example 30.3. A rubber isolator having a characteristic height $h_c = 0.5$ in. (determined experimentally for the particular isolator design) has a natural frequency $f_n = 10$ cps for small deflections and a fraction of critical damping $\zeta = 0.2$. The equipment supported by the isolator is subjected to a sustained acceleration of $11g$. It is desired to determine the absolute transmissibility of the isolation system when the forcing frequency is 100 cps, and to determine the deflection of the isolator under the sustained acceleration.

Referring to the nomograph of Fig. 30.35, a straight line is drawn from a value of 10 on the f_{n_0} scale to 0.5 on the h_c scale. A second straight line is drawn from the intersection of the first line with the R scale through the value $n_g = 11$. The second line intersects the left side

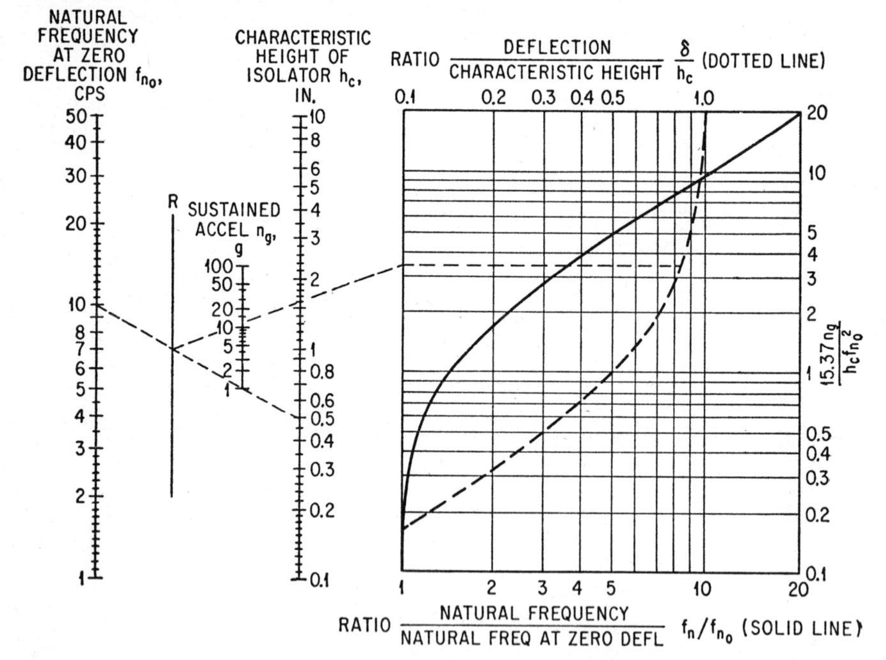

FIG. 30.35. Nomograph and curve for determining the natural frequency and deflection of an isolation system incorporating a tangent isolator when a portion of the total load applied to the isolator is nonmassive.

of the coordinate system and is extended horizontally so that it intersects the solid and dotted curves. The intersection points indicate that the natural frequency ratio $f_n/f_{n_0} = 3.5$ and the deflection ratio $\delta/h_c = 0.81$. The deflection of the isolator at equilibrium as a result of the sustained acceleration is $0.81h_c = 0.405$ in. The undamped natural frequency for the sustained acceleration of $11g$ is $f_n = 3.5 \times 10 = 35$ cps. The natural frequency also can be obtained from Fig. 30.34 by noting that $\frac{W}{k_0 h_c} = \frac{g/h_c}{(2\pi f_{n0})^2} = 0.196$ [see Eq. (30.60) when $n_g = 0$]. Then for $n_g = 11$, $f_n = 24.5/\sqrt{0.5} = 35$ cps.

From Fig. 30.2 the transmissibility for $\zeta = 0.2$, $f/f_n = 100/35 = 2.88$ is 0.22. In the absence of the sustained acceleration, the corresponding transmissibility would be 0.042 as obtained from Fig. 30.2 at $f/f_n = 100/10 = 10$. Thus, the transmissibility at 100 cps under a sustained acceleration of $11g$ is five times as great as that which would exist for a dead-weight loading of the isolator.

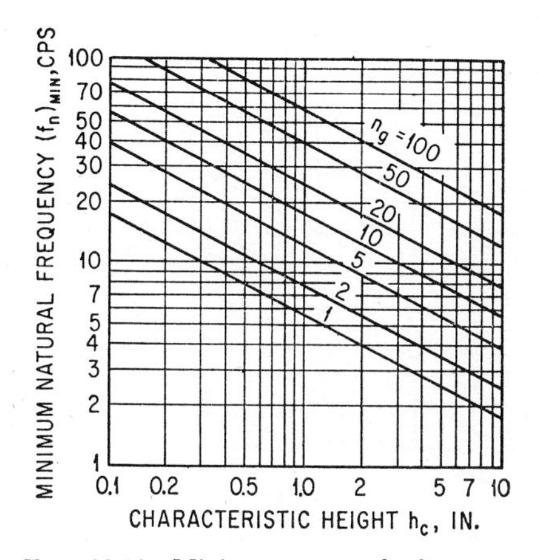

FIG. 30.36. Minimum natural frequency $f_{n(\min)}$ of a tangent isolator system as a function of (1) the characteristic height h_c of the isolator and (2) the sustained acceleration n_g expressed as a multiple of the gravitational acceleration.

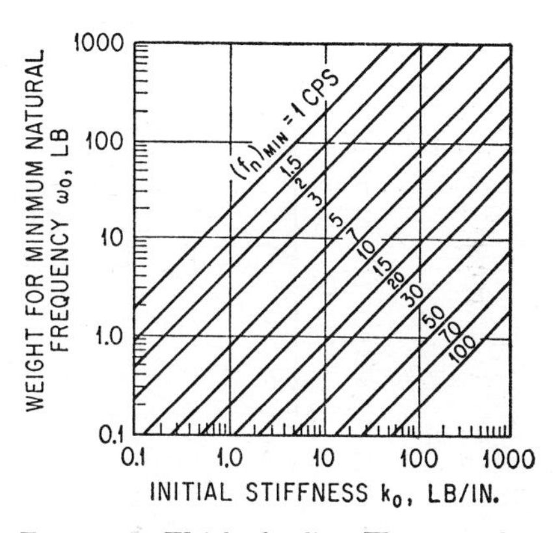

FIG. 30.37. Weight loading W_o required to cause a tangent isolator to have a minimum natural frequency $f_{n(\min)}$, as a function of the stiffness k_o at zero deflection.

Minimum Natural Frequency. The weight W_0 for which a given tangent isolator has a minimum natural frequency is [15]

$$W_0 = \frac{2k_0h_c}{\pi n_g} = \frac{k_0 g}{2\pi^2 (f_n)_{\min}^2} \qquad [f_n = \text{minimum}] \tag{30.62}$$

where the minimum natural frequency $(f_n)_{\min}$ is defined by

$$(f_n)_{\min} = \frac{1}{2}\sqrt{\frac{n_g g}{\pi h}} \tag{30.63}$$

The minimum natural frequency is shown graphically in Fig. 30.36 as a function of the characteristic height h_c and the sustained acceleration parameter n_g. The weight W_0 required to produce the minimum natural frequency $(f_n)_{\min}$ is shown graphically in Fig. 30.37 as a function of the initial stiffness k_0 and the minimum natural frequency $(f_n)_{\min}$. When the isolator is loaded to produce the minimum natural frequency, the isolator deflection is one-half the characteristic height ($\delta = h_c/2$) and the stiffness under load is twice the initial stiffness ($k = 2k_0$).

ISOLATION OF RANDOM VIBRATION

In random vibration, all frequencies exist concurrently, and the amplitude and phase relations are distributed in a random manner. A trace of random vibration is illustrated in Fig. 11.1*A*. The equipment-isolator assembly responds to the random vibration with the substantially single-frequency pattern shown in Fig. 11.1*B*. This response is similar to a sinusoidal motion with a continuously and irregularly varying envelope; it is described as narrow-band random vibration or a random sine wave.

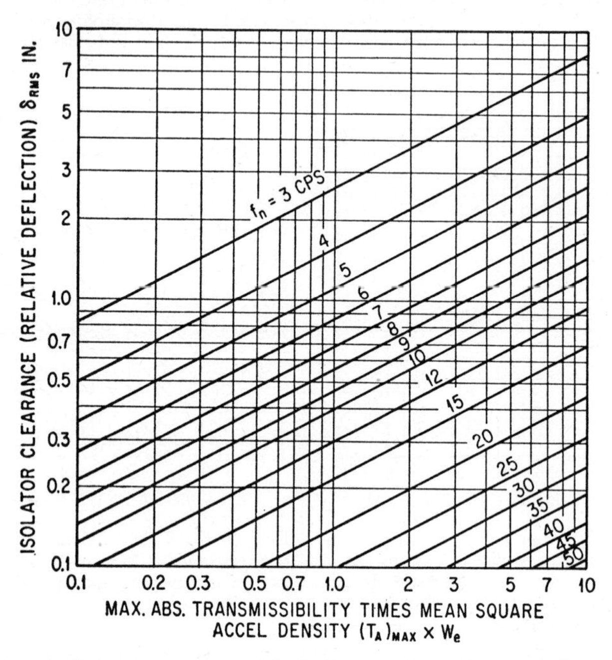

FIG. 30.38. Required clearance expressed in inches rms for a damped isolator with viscous damping when subjected to random vibration defined by a flat spectrum of mean-square acceleration density $W_e(f)$. The natural frequency of the isolator system in cycles per second is f_n.

The characteristics of random vibration are defined by a frequency spectrum of power spectral density (see Chaps. 11 and 22). This is a generic term used to designate the mean-square value of some magnitude parameter passed by a filter, divided by the bandwidth of the filter, and plotted as a spectrum of frequency. The magnitude is commonly measured as acceleration in units of g; then the particular expression to use in place of power spectral density is mean-square acceleration density, commonly expressed in units of g^2/cps. When the spectrum of mean-square acceleration density is substantially flat in the frequency region extending on either side of the natural frequency of the isolator, the response of the isolator may be determined in terms of (1) the mean-square acceleration density of the isolated equipment and (2) the deflection of the isolator at successive cycles of vibration.

The mean-square acceleration densities of the foundation and the isolated equipment are related by the absolute transmissibility that applies to sinusoidal vibration:

$$W_r(f) = W_e(f) T_A^2 \qquad (30.64)$$

where $W_r(f)$ and $W_e(f)$ are the mean-square acceleration densities of the equipment and the foundation, respectively, in units of g^2/cps and T_A is the absolute transmissibility for the vibration-isolation system [see Eq. (11.31)].

The severity of the vibration experienced by the isolated equipment may be expressed in terms of the rms value of acceleration at the foundation by integrating the mean-square acceleration density given by Eq. (30.64):

$$\ddot{x}_{\text{rms}} = \sqrt{\int W_e(f) T_A{}^2 \, df} \tag{30.65}$$

where $\ddot{x}_{\text{rms}}$ is the rms acceleration of the equipment and the integration is carried out over the frequency interval for which $W_e(f)$ is defined.

The clearance in the isolator is obtained from Eq. (11.38) by substituting (1) $W_e(f)$ in units of g^2/cps for $2\pi W_e(\omega_n)/g^2$ where $W_e(\omega_n)$ is in units of (in./sec^2)^2/(rad/sec), (2) $\omega_n = 2\pi f_n$, and (3) $(T_A)_{\text{max}} = 1/2\zeta$. The rms relative deflection δ_{rms} for a rigidly connected viscous damped isolator is

$$\delta_{\text{rms}} = \sqrt{\frac{g^2}{32\pi^3} \frac{(T_A)_{\text{max}} W_e(f)}{f_n{}^3}} = 12.25 \sqrt{\frac{(T_A)_{\text{max}} W_e(f)}{f_n{}^3}} \quad \text{in.} \tag{30.66}$$

where $W_e(f)$ is the mean-square acceleration density of the foundation measured in g^2/cps. Equation (30.66) applies only if the fraction of critical damping $\zeta = c/c_c$ for the isolation system is relatively small; i.e., $(T_A)_{\text{max}}$ is relatively large. The rms isolator clearance (relative motion between the equipment and the foundation) given by Eq. (30.66) is shown graphically by Fig. 30.38.

Example 30.4. Suppose the vibration of the foundation is defined by a flat spectrum of mean-square acceleration density of $0.2g^2$/cps over a frequency band from 10 to 500 cps (wide relative to the width of the absolute transmissibility curve of the isolator in the region of resonance). The isolator has a natural frequency of 25 cps and damping defined by $(T_A)_{\text{max}} = 5$ ($\zeta = 0.1$). Entering Fig. 30.38 at $(T_A)_{\text{max}} W_e(f) = 5 \times 0.2 = 1.0$ on the abscissa scale, the rms isolator clearance as read from the ordinate is $\delta_{\text{rms}} = 0.093$ in.

The deflection of the isolator varies from cycle to cycle. If the vibration has truly normal (Gaussian) characteristics, as discussed in Chaps. 11 and 22, very large values of amplitude occur occasionally. Then bottoming of the isolator cannot be prevented while maintaining the clearance reasonably small; rather, it can be made less frequent by increasing the isolator clearance. For example,* if the clearance of the isolator (in plus and minus directions along axis of vibration) is made $3\delta_{\text{rms}}$, it is probable that the isolator will bottom once in each 100 cycles of vibration; if the clearance is $4.3\delta_{\text{rms}}$, bottoming is probable once in 10^4 cycles; and if the clearance is $5.3\delta_{\text{rms}}$, bottoming is probable once in 10^6 cycles. It is common in certain testing procedures to limit the maximum amplitude to three times the rms value; then it is unlikely that bottoming will occur during tests if the clearance is made somewhat greater than $3\delta_{\text{rms}}$.

When the spectrum of mean-square acceleration density defining vibration of the support is not flat in the region of isolator natural frequency, the integration leading to the simple result of Eq. (30.66) cannot be carried out analytically. An equivalent result can be obtained by graphical integration of Eq. (11.32) where $A(j\omega)$ is a transfer function relating the acceleration of the support to the deflection of the isolator. When the mean-square acceleration density of the foundation $W_e(f)$ is not flat in the region of resonance, the rms isolator clearance is given approximately by

$$\delta_{\text{rms}} = 9.75 \sqrt{\frac{\int_{f_1}^{f_2} W_e(f) T_A{}^2 \frac{df_n}{f_n}}{f_n{}^3}} \quad \text{in.} \tag{30.67}$$

where T_A† is the absolute transmissibility of a rigidly connected viscous-damped isolation

* These probabilities are calculated from Eq. (11.39). Also, see Fig. 11.10.

† Equation (30.67) is approximate since the absolute transmissibility function T_A is used in place of a transfer response function (determined from a rigorous analysis) that has a frequency dependency similar to T_A in the region of resonance. It is desirable to use the absolute transmissibility T_A in the relation defined by Eq. (30.67) since this property of an isolation system is most frequently available by experimental means.

system; f_n is the natural frequency of the isolation system, cps; and $W_e(f)$ is the mean-square acceleration density of the foundation, in units of g^2/cps between the frequency limits f_1 and f_2.

A procedure for evaluating Eq. (30.67) graphically is shown in Fig. 30.39. The square of the absolute transmissibility curve is plotted on the left side of the figure to a dimensionless frequency scale that is divided, in the example shown, into increments $\Delta f_n/f_n = 0.25$. Smaller increments may be used for more accurate results. In the example shown, the area of each block is $1 \times 0.25 = 0.25$, where the differential df_n/f_n is replaced by the finite incremental value $\Delta f_n/f_n$. The spectrum of mean-square acceleration density is drawn at the right side

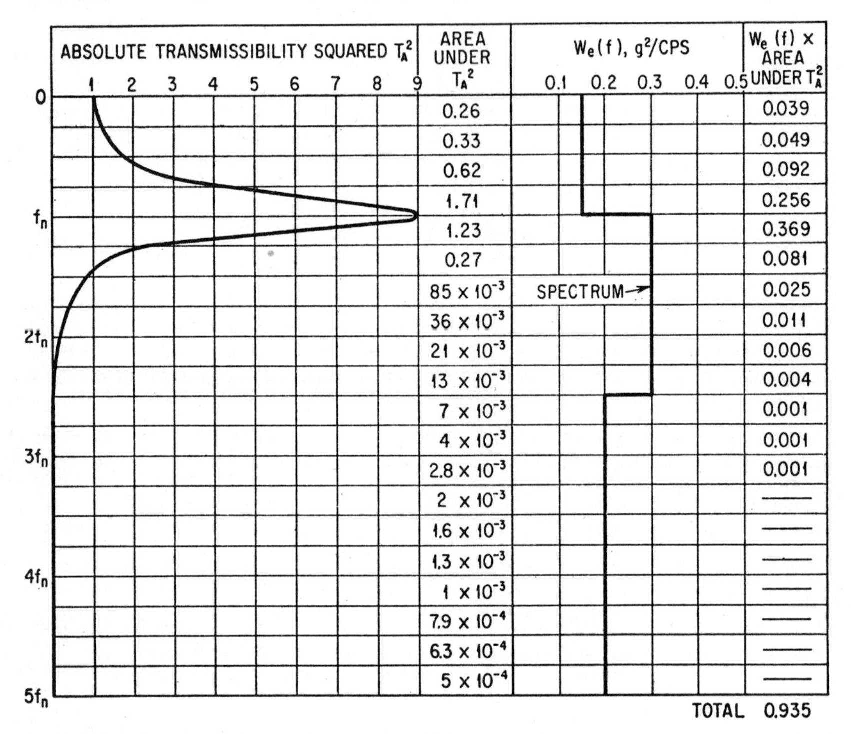

ABSOLUTE TRANSMISSIBILITY SQUARED T_A^2
1 2 3 4 5 6 7 8 9
0
f_n
$2f_n$
$3f_n$
$4f_n$
$5f_n$
AREA UNDER T_A^2
0.26
0.33
0.62
1.71
1.23
0.27
85 x 10^-3
36 x 10^-3
21 x 10^-3
13 x 10^-3
7 x 10^-3
4 x 10^-3
2.8 x 10^-3
2 x 10^-3
1.6 x 10^-3
1.3 x 10^-3
1 x 10^-3
7.9 x 10^-4
6.3 x 10^-4
5 x 10^-4
$W_e(f)$, g^2/CPS
0.1 0.2 0.3 0.4 0.5
SPECTRUM
$W_e(f)$ x AREA UNDER T_A^2
0.039
0.049
0.092
0.256
0.369
0.081
0.025
0.011
0.006
0.004
0.001
0.001
0.001
TOTAL 0.935

FIG. 30.39. Procedure for determining graphically the required isolator clearance when the excitation is random vibration. The example is for an isolation system with absolute transmissibility $T_A = 3$ ($\zeta = 0.17$) and natural frequency $f_n = 10$ cps; the mean-square acceleration density $W_e(f)$ varies with frequency according to the spectrum shown in the figure.

of the figure. For each frequency increment $\Delta f_n/f_n$, the product of $W_e(f)$ and the area under the curve of transmissibility squared is entered in the right-hand column. The sum of this column is $\int_0^\infty W_e(f) T_A^2 \, df_n/f_n$; δ_{rms} may be calculated from Eq. (30.67) using the appropriate value of the isolator natural frequency f_n. The necessary clearance and probability of bottoming can be determined as discussed above with reference to a flat spectrum of mean-square acceleration density.

VIBRATION ISOLATION OF NONRIGID BODY SYSTEMS

This section discusses the effect of nonrigidity of structures to which isolators are attached, in contrast to the rigid body theory discussed in the preceding sections of this chapter. Two aspects of vibration isolation are discussed in detail:

1. The equipment is attached by an isolator to a vibrating foundation which contains a power source. The vibration transferred to the equipment is determined.

2. A force-generating piece of equipment is connected by an isolator to a foundation which is at rest initially. The response of the equipment, the force transmitted to the foundation, and the resulting vibration of the foundation are determined.

The excitation in both cases is steady-state sinusoidal vibration. All forces and velocities are considered colinear, and all elements are considered linear. The equipment is represented by a model having one input terminal and one output terminal; the foundation is represented by a similar model.

MECHANICAL POWER SOURCES

A *mechanical power source* is a means to generate sinusoidally varying forces and motions, and an associated elastic system through which the forces and motions are transmitted to the output terminal. Such a source is shown schematically in Fig. 30.40. Practically, measurements can be made only at the power source output. The force applied by the source to any system attached to the source output is represented by F_s and v_s represents the resulting velocity at the source output; F_s and v_s are related by [16]

$$F_s = F_s^b - Z_s v_s = Z_s(v_s^f - v_s) \tag{30.68}$$

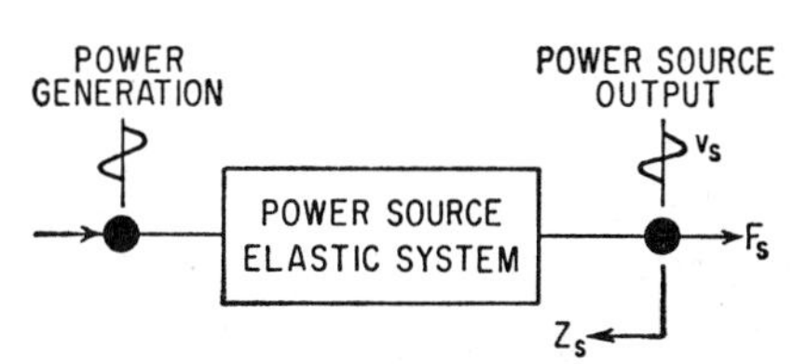

FIG. 30.40. Schematic diagram of a power source characterized by the point impedance Z_s at the power source output terminal.

where F_s^b is the *blocked output force* * of the power source, v_s^f is the *free output velocity* * of the power source, and Z_s is the point mechanical impedance of the power source output. (See Chap. 10 for a detailed discussion of mechanical impedance.) The blocked output force of the source is the force transmitted to a body of infinitely great mechanical impedance attached at the source output. The free output velocity of the source is the velocity at the output with no load attached. The point mechanical impedance of the power source output Z_s is the ratio of a force applied at the source output to the resulting velocity at the output, with the source inactive. The force, velocity, and impedance are related by

$$Z_s = \frac{F_s^b}{v_s^f} \tag{30.69}$$

IMPEDANCE PROPERTIES OF MECHANICAL ELEMENTS

A mechanical element may be represented by a generalized linear mechanical system having an input terminal i and an output terminal j, as shown by Fig. 30.41. The dynamic properties of the element may be described conveniently by the mechanical impedances (see Chap. 10) associated with the element:

Z_i^{jf} = point impedance at terminal i with terminal j free ($F_j = 0$)

Z_i^{jb} = point impedance at terminal i with terminal j blocked ($v_j = 0$)

$Z_{ij}^f = Z_{ji}^f$ = free transfer impedance between terminal i and terminal j; i.e., ratio of force applied at either one of the terminals to resulting velocity at the other terminal when the latter is free

$Z_{ij}^b = Z_{ji}^b$ = blocked transfer impedance between terminal i and terminal j; i.e., ratio of force developed at either one of the terminals when blocked to the velocity applied at the other terminal

Other impedance terms can be defined by substitution of the proper superscripts and subscripts in the above definitions.

* These quantities are the force and velocity *amplitudes* (complex) of sinusoidal quantities but are referred to as "force" and "velocity" for convenience.

Definitions for velocity and force transmissibility are:

$(T_v)_{ij}$ = velocity transmissibility from terminal i to terminal j; i.e., ratio of resulting velocity at terminal j to imposed velocity at terminal i when no load is attached at terminal j

$(T_F)_{ij}$ = force transmissibility from terminal i to terminal j; i.e., ratio of resulting force at terminal j to imposed force at terminal i when a body of infinitely great mechanical impedance is attached at terminal j

Other velocity and force transmissibilities can be defined by substitution of the proper subscripts in the above definitions. Note that the order of the subscripts in the transmissibility terms describes the direction in which the vibration is being transmitted. This

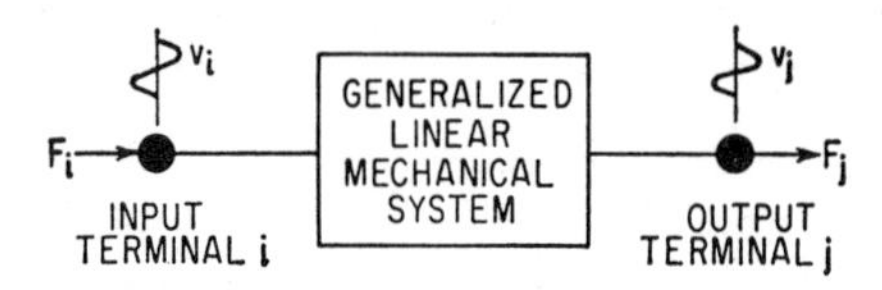

FIG. 30.41. Schematic diagram of a generalized linear mechanical system having an input at terminal i and an output at terminal j. The velocities v_i and v_j are of the same sign when in the same direction.

order is not significant when writing transfer impedance terms since the transfer impedance between any two terminals in opposite directions is equal.

Mechanical impedance identities used in the following analyses are:

$$Z_i^{jf} Z_j^{ib} = Z_i^{jb} Z_j^{if} = Z_{ij}^{f} Z_{ij}^{b}$$

$$\frac{Z_{ij}^{b}}{Z_{ij}^{f}} + 1 = \frac{Z_j^{ib}}{Z_j^{if}} = \frac{Z_i^{jb}}{Z_i^{jf}} \tag{30.70}$$

VELOCITY ISOLATION

In velocity isolation, a nonrigid equipment is mounted on a vibrating foundation by a generalized vibration isolator as shown in Fig. 30.42. The foundation is considered to have the properties of the power source shown in Fig. 30.40, and its point impedance measured at point 1 is designated Z_F. Without the equipment attached, the output of the foundation is vibration with a velocity v_F^f, the free source velocity of the foundation (Fig. 30.42*A*). When the equipment-isolator combination is attached to the foundation (Fig. 30.42*B*), the velocity at the foundation output changes to v_1, and velocity v_2 is transmitted to the input terminal of the equipment. In general, the equipment and the foundation are nonrigid; the isolator is characterized by mass, stiffness, and damping. The velocity at any other point on the equipment is designated by v_3. Transmissibility is referred to as velocity transmissibility; this is equal to displacement or acceleration transmissibility because the system is linear and the vibration sinusoidal.

GENERAL MODIFIED VELOCITY TRANSMISSIBILITY. The force F_1 applied to the isolation system at terminal 1 in Fig. 30.42*B* is first determined by considering the foundation as the power source and using Eq. (30.68):

$$F_1 = Z_F(v_F^f - v_1) \tag{30.71}$$

The *modified velocity transmissibility* $(T_v)_{F3}$ is the ratio of the equipment velocity v_3 to the foundation free velocity v_F^f. The velocity ratio $(T_V)_{F3} = v_3/v_F^f$ is designated a "modified" transmissibility since it is not a ratio of two velocities which exist simultaneously but rather the ratio of a resultant velocity at the equipment to the velocity of the foundation without equipment attached. When the foundation in Fig. 30.42 is rigid, $v_F^f = v_1$; then the modified transmissibility becomes the ratio of two concurrent velocity amplitudes and is a true transmissibility expression. The expression for modified velocity transmissibility is [17]

$$(T_v)_{F3} = \frac{v_3}{v_F^f} = \frac{1}{\frac{Z_{23}^f}{Z_{12}^b}\left[1 + \frac{Z_1^{2b}}{Z_F}\right] + \frac{Z_{23}^f Z_{12}^f}{Z_2^{3f}}\left[\frac{1}{Z_1^{2f}} + \frac{1}{Z_F}\right]} \tag{30.72}$$

where Z_{23}^f = free transfer impedance of equipment between terminals 2 and 3. (This may be determined by applying a known vibratory force at 2 and noting the resulting velocity at 3.)

Z_{12}^b = blocked transfer impedance of the isolator

Z_{12}^f = free transfer impedance of the isolator

Z_1^{2b} = point impedance of isolator at terminal 1 with a body of infinitely great impedance attached at terminal 2

Z_1^{2f} = point impedance of isolator at terminal 1 with no load at terminal 2

Z_F = point impedance of foundation measured at terminal 1. (In general, this determines the impedance of the foundation uniquely because any other terminal is inaccessible.)

Z_2^{3f} = point impedance of equipment at terminal 2 with no load at terminal 3

Only three of the four isolator impedances appearing in Eq. (30.72) must be measured, the fourth being determined by Eq. (30.70). If the actual velocity v_3 at terminal 3 of the equipment is desired, the free velocity v_F^f (see Fig. 30.42*A*) of the foundation must be measured. Thus, seven measurements are required to obtain the velocity v_3 at a point on the nonrigid equipment. In the following sections, the generalities of Eq. (30.72) are removed and corresponding expressions are derived for more restricted conditions.

MODIFIED VELOCITY TRANSMISSIBILITY FOR A MASSLESS ISOLATOR. If the isolator is massless, the point and transfer impedances Z_1^{2f}, Z_{12}^f of the isolator become zero; the point and transfer impedances Z_1^{2b}, Z_{12}^b become equal and are denoted by

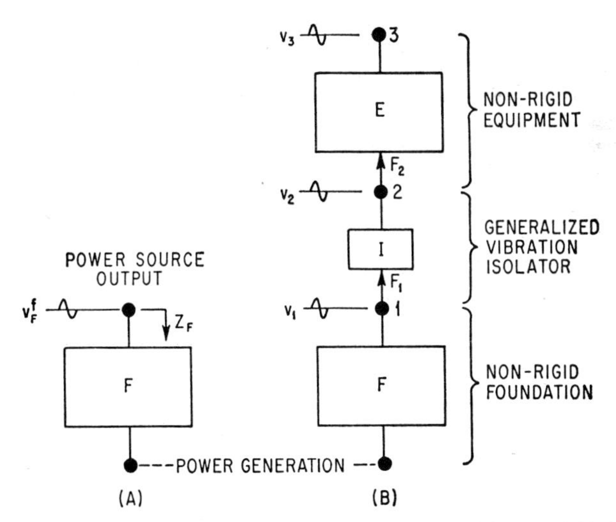

FIG. 30.42. Schematic diagram illustrating velocity isolation. The nonrigid foundation is considered a power source and is characterized by the impedance Z_F; initially it is free as illustrated at *A* and vibrates with a velocity v_F^f. A generalized system comprised of a nonrigid equipment and a generalized isolator is then attached to the foundation; this changes the foundation velocity to v_1 and transmits velocities v_2 and v_3 to the equipment as illustrated at *B*. The velocities are of the same sign when in the same direction.

Z_I. Then Eq. (30.72) becomes

$$(T_v)_{F3} = \frac{v_3}{v_F^f} = \frac{1}{Z_{23}^f(1/Z_I + 1/Z_F + 1/Z_2^{3f})} \tag{30.73}$$

Equation (30.73) can be written more conveniently in terms of mechanical mobility (see Chap. 10):

$$(T_v)_{F3} = \frac{v_3}{v_F^f} = \frac{\mathfrak{M}_{23}^f}{\mathfrak{M}_I + \mathfrak{M}_F + \mathfrak{M}_2^{3f}} \tag{30.74}$$

where $\mathfrak{M}_{23}^f = 1/Z_{23}^f$ = transfer mobility of equipment between terminals 2 and 3 with no load at terminal 3

$\mathfrak{M}_I = 1/Z_I^{2b} = 1/Z_{12}^b$ = mobility of isolator when measured with a load of infinite impedance attached at either terminal

$\mathfrak{M}_F = 1/Z_F$ = mobility of foundation

$\mathfrak{M}_2^{3f} = 1/Z_2^{3f}$ = mobility of equipment at terminal 2 with no load attached at terminal 3

The impedances are defined explicitly with reference to Eq. (30.72). Four measurements of impedance or mobility are required to determine the modified velocity transmissibility when the isolation system is massless.

MODIFIED VELOCITY TRANSMISSIBILITY TO INPUT TERMINAL OF NON-RIGID EQUIPMENT. The modified velocity transmissibility between the foundation and terminal 2 of the equipment may be determined from Eq. (30.72) by requiring terminals 2 and 3 to coincide and using the impedance identities stated by Eq. (30.70):

$$(T_v)_{F2} = \frac{v_2}{v_F^f} = \frac{Z_{12}^b}{(Z_1^{2b}/Z_F)(Z_E + Z_2^{1f}) + Z_E + Z_2^{1b}} \tag{30.75}$$

where $Z_E = Z_2^{3f} = Z_{23}^f$ represents the input point impedance of the equipment at terminal 2 for this case. This equation differs from Eq. (30.77) only in that the isolator may have mass. By use of the impedance identities stated by Eq. (30.70), the four isolator impedances indicated in Eq. (30.75) may be obtained from three measurements. In addition, the point impedances of the equipment and the foundation must be known to determine the velocity ratio v_2/v_F^f. However, if the velocity v_2 is required, v_F^f also must be measured, making a total of six measurements required. If the isolator has mass but the foundation is infinitely rigid, Eq. (30.75) becomes

$$(T_v)_{F2} = \frac{v_2}{v_F^f} = \frac{Z_{12}^b}{Z_E + Z_{12}^b} \tag{30.76}$$

If the isolator is massless, the modified velocity transmissibility between the foundation and terminal 2 of the equipment is

$$(T_v)_{F2} = \frac{v_2}{v_F^f} = \frac{Z_I}{Z_I + Z_E(1 + Z_I/Z_F)} = \frac{\mathfrak{M}_E}{\mathfrak{M}_I + \mathfrak{M}_F + \mathfrak{M}_E} \tag{30.77}$$

where the point impedance and mobility of the equipment at terminal 2 are Z_E and $\mathfrak{M}_E$, respectively, and the other parameters are defined with reference to Eqs. (30.72) to (30.74).

FORCE ISOLATION

Force isolation occurs where a power-generating equipment having the properties of a power source is attached to an initially motionless foundation by means of an isolator, as indicated schematically in Fig. 30.43. The equipment vibrates with a velocity v_E^f when unattached. After the equipment is attached to the isolator, the velocity of the equipment changes and vibration is transmitted to the foundation. The force transmitted to the foundation, the resulting velocity of the foundation, and the resulting velocity of the equipment are determined in the following sections.

The equipment containing the power source is characterized by its blocked output force $F_E{}^b$ (i.e., by the force applied at the output terminal when this terminal is attached to a body with an infinitely great impedance) and its point impedance Z_E as indicated in Fig. 30.43*A*.

MODIFIED FORCE TRANSMISSIBILITY. The *modified force transmissibility* $(T_F)_{E1}$ is defined as the ratio of the force F_1 transmitted to the foundation to the output

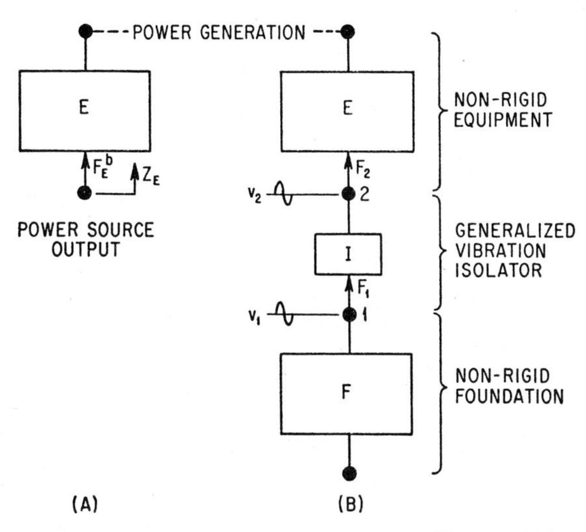

Fig. 30.43. Schematic diagram illustrating force isolation. The nonrigid equipment is considered a power source and is characterized by the impedance Z_E; initially it is attached to a rigid foundation and transmits a force $F_E{}^b$, as illustrated at *A*. The equipment is then attached by means of a generalized isolator to a nonrigid foundation; this changes the output force from the equipment to F_2 and transmits a force F_1 to the foundation. The velocities v_1, v_2 are of the same sign when in the same direction.

force $F_E{}^b$ of the equipment with point 2 blocked. The force ratio $(T_F)_{E1} = F_1/F_E{}^b$ is considered a "modified" transmissibility since it is not a true transfer force ratio of two forces that exist simultaneously; rather, it is the ratio of the force experienced by the foundation to the force generated by the equipment if rigidly constrained. Since the systems considered in this section are linear, the modified force transmissibility $(T_F)_{E1}$ is equal to the modified velocity transmissibility $(T_v)_{F2}$ given by Eq. (30.75). This relation applies for a power-generating nonrigid equipment, a generalized isolator, and a nonrigid foundation.

Massless Isolator. If the isolator is massless, all impedances (point and transfer impedances) of the isolation system with the output blocked are replaced by Z_I; the isolator impedances with the output free become zero. Then the modified force transmissibility $(T_F)_{E1}$ is equal to the modified velocity transmissibility $(T_v)_{F2}$ given by Eq. (30.77).

Rigid Foundation. If the isolator has mass, stiffness, and damping but the foundation is rigid, $Z_F = \infty$ and the modified force transmissibility $(T_F)_{E1}$ is equal to $(T_v)_{F2}$ given by Eq. (30.76).

VELOCITY RESPONSE OF FOUNDATION. In general, the nonrigid foundation experiences a velocity in response to force applied by the power-generating equipment,

The foundation velocity v_1 is

$$v_1 = \frac{F_1}{Z_F} \tag{30.78}$$

where Z_F is the impedance of the foundation and F_1 is the force applied by the isolator. Since $(T_F)_{E1} = F_1/F_E^b$, the foundation velocity v_1 is

$$v_1 = \frac{F_E^b}{Z_F}(T_F)_{E1} \tag{30.79}$$

By substitution of the proper modified force transmissibility function for the conditions of a problem, this relation gives the foundation velocity resulting from the vibration generated by the equipment. According to Eq. (30.69), the blocked output force F_E^b may be expressed as

$$F_E^b = Z_E v_E^f \tag{30.80}$$

where Z_E is the point impedance of the equipment measured at terminal 2. Substituting Eq. (30.80) into Eq. (30.79), the foundation velocity response v_1 can be written in dimensionless form as follows:

$$\frac{v_1}{v_E^f} = \frac{Z_E}{Z_F}(T_F)_{E1} \tag{30.81}$$

where v_E^f is the velocity of the equipment at terminal 2 when no load is attached. For example, velocity response of the foundation when the isolator is massless is obtained by substituting $(T_v)_{F2}$ from Eq. (30.77) for $(T_F)_{E1}$ in Eq. (30.79):

$$v_1 = \frac{F_E^b}{Z_E + Z_F(1 + Z_E/Z_I)} \tag{30.82}$$

Equation (30.82) can be written nondimensionally by replacing F_E^b by its equivalent form defined by Eq. (30.80):

$$\frac{v_1}{v_E^f} = \frac{Z_I}{Z_I + Z_F(1 + Z_I/Z_E)} \tag{30.83}$$

VELOCITY RESPONSE OF EQUIPMENT. The output velocity v_E^f of the equipment which exists before the attachment of the isolator is changed to the velocity v_2 when the attachment is made because the equipment is nonrigid:

$$\frac{v_2}{v_E^f} = 1 - (T_F)_{E1}\left(\frac{Z_2^{1b}}{Z_{12}^b} + \frac{Z_{12}^f}{Z_F}\right) \tag{30.84}$$

where Z_{12}^b = transfer impedance across isolator from terminal 1 to terminal 2, with terminal 2 blocked
Z_{12}^f = transfer impedance across isolator from terminal 1 to terminal 2, with terminal 2 free
Z_2^{1b} = point impedance of isolator at terminal 2 with terminal 1 blocked
Z_F = point impedance of the foundation at terminal 1

If the isolator is massless, $Z_{12}^b = Z_2^{1b} = Z_I$ and $Z_{12}^f = 0$; then Eq. (30.84) becomes

$$\frac{v_2}{v_E^f} = 1 - (T_F)_{E1} = \frac{Z_I + Z_F}{Z_I + Z_F(1 + Z_I/Z_E)} \tag{30.85}$$

where $(T_F)_{E1}$ for a massless isolator is given by Eq. (30.79).

ISOLATOR EFFECTIVENESS

The effectiveness of a vibration isolator is a measure of the reduction of vibration which it effects. In concept, effectiveness may be indicated in terms of vibratory velocity or vibratory force. Effectiveness in terms of velocity is the ratio of the velocity $v_2^{(U)}$ to the velocity $v_2^{(I)}$ (see Fig. 30.44) where $v_2^{(U)}$ is the (unisolated) velocity transmitted to an equipment attached directly to the foundation and $v_2^{(I)}$ is the corresponding (isolated) velocity when an isolator is interposed between equipment and foundation; effectiveness in terms of force is the ratio of the force $F_1^{(U)}$ to the force $F_1^{(I)}$, as indicated in Fig. 30.45.[18,19]

The velocity v_2 at the input terminal of the equipment (terminal 2) is given by Eq. (30.75) in terms of the free velocity of the foundation v_F^f and several characteristic im-

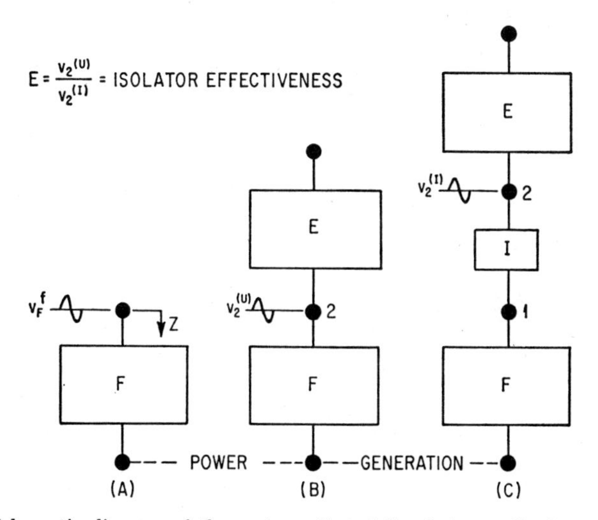

FIG. 30.44. Schematic diagram of the systems that define isolator effectiveness in velocity isolation as illustrated by Fig. 30.42. The isolator effectiveness is the ratio of the velocity $v_2^{(U)}$ transmitted to the unisolated equipment, as illustrated at B, to the velocity $V_2^{(I)}$ transmitted to the equipment through the isolator, as illustrated at C. The foundation without the equipment attached is shown at A.

pedances. This velocity is designated as $v_2^{(I)}$ when the isolator is effective; when the isolator is considered as a rigid, massless connection, $Z_{12}^b = Z_2^{1b} = Z_1^{2b} = \infty$, $Z_2^{1f} = 0$ and the corresponding velocity at the equipment is designated $v_2^{(U)}$. Then the expression for effectiveness is

$$E = \frac{v_2^{(U)}}{v_2^{(I)}} = \frac{Z_E + Z_2^{1b} + (Z_1^{2b}/Z_F)(Z_E + Z_2^{1f})}{Z_{12}^b(1 + Z_E/Z_F)} \tag{30.86}$$

where Z_E is the point impedance of the equipment at terminal 2, Z_1^{2b} is the point impedance of the isolator at terminal 1 with terminal 2 blocked, Z_2^{1f} is the point impedance of the isolator at terminal 2 with terminal 1 free, and the other parameters are defined in connection with Eq. (30.84). Inasmuch as the expression for force transmissibility is identical to Eq. (30.75) for velocity transmissibility, Eq. (30.86) for effectiveness applies to effectiveness of force transmissibility as follows:

$$E = \frac{F_1^{(U)}}{F_1^{(I)}} \tag{30.87}$$

where $F_1^{(U)}$ is the force transmitted to the foundation with the equipment attached directly thereto (Fig. 30.45*B*) and $F_1^{(I)}$ is the corresponding force with the isolator interposed therebetween (Fig. 30.45*C*).

EFFECTIVENESS OF MASSLESS ISOLATOR. If the isolator is massless, $Z_{12}^b = Z_2^{1b} = Z_1^{2b} = Z_I$ and $Z_2^{1f} = 0$; then the isolator effectiveness defined by Eq. (30.86) becomes

$$E = \frac{Z_E + Z_F(1 + Z_E/Z_I)}{Z_E + Z_F} = \frac{\mathfrak{M}_E + \mathfrak{M}_I + \mathfrak{M}_F}{\mathfrak{M}_E + \mathfrak{M}_F} \tag{30.88}$$

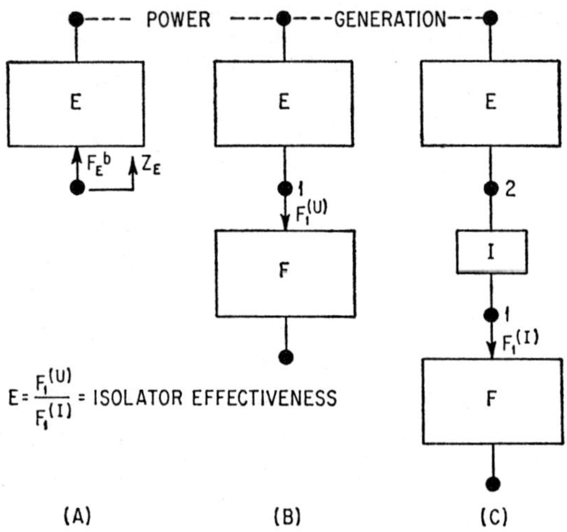

FIG. 30.45. Schematic diagram of the systems that define isolator effectiveness in force isolation as illustrated by Fig. 30.43. The isolator effectiveness is the ratio of the force $F_1^{(U)}$ transmitted to the unisolated foundation, as illustrated at *B*, to the force $F_1^{(I)}$ transmitted to the foundation through the isolator, as illustrated at *C*. The equipment is shown at *A* before being attached to the foundation.

NONRIGIDITY OF STRUCTURES

When the isolator is attached to a nonrigid structure of the equipment or foundation, the structure may vibrate with a relatively large amplitude if the excitation frequency coincides with a natural frequency of the structure. Thus, the isolator appears to be relatively ineffective at such frequencies and may afford little or no isolation. This ef-

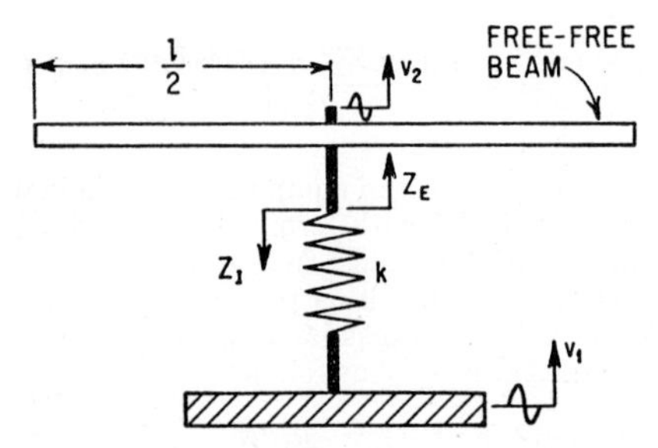

FIG. 30.46. Schematic diagram of a free-free beam of length l supported by an isolator in the form of a massless linear spring. The beam is an example of a nonrigid equipment.

fect can be evaluated quantitatively if the mechanical impedance or mobility of the equipment and foundation are known at point of attachment of the isolator.

For example, consider a free-free beam of mass m_B attached to a vibrating support by a linear, massless spring of stiffness k, as shown in Fig. 30.46. There is no damping in the system. The expression for transmissibility v_2/v_1 given by Eq. (30.77) becomes applicable by setting the foundation impedance Z_F equal to infinity:

$$(T_v)_{12} = \frac{v_2}{v_1} = \frac{Z_I}{Z_E + Z_I} = \frac{1}{1 + Z_E/Z_I} \tag{30.89}$$

where Z_I is the mechanical impedance of the massless isolator with one end blocked and Z_E is the mechanical impedance at the center point of the free-free beam. Substituting the impedance of a spring $Z_I = -jk/\omega$ (see Table 10.2) and the impedance Z_E of a free-free beam at its center point, Eq. (30.89) becomes [20]

$$(T_v)_{12} = \frac{1}{1 - \left(\frac{f}{f_0}\right)^2 \left[1 - \sum_{n=1}^{\infty} \frac{[\phi_n(l/2)]^2 (f/f_0)^2}{(f_n/f_0)^2 - (f/f_0)^2}\right]^{-1}} \tag{30.90}$$

where $\phi_n(l/2)$ = value of nth mode function at center point of free-free beam (see Chap. 7), dimensionless

f_n = natural frequency of nth mode of beam, cps

$f_0 = \frac{1}{2\pi}\sqrt{k/m_B}$ = natural frequency of beam (considered as a rigid body) on isolators, cps

$f = \omega/2\pi$ = forcing frequency, cps

The transmissibility as calculated from Eq. (30.90) and as determined experimentally is illustrated in Fig. 30.47 for three systems whose fundamental natural frequencies f_1 are two, five, and ten times the rigid body natural frequency f_0. The transmissibility in rigid body theory, Eq. (*a*) of Table 30.2 with $\zeta = 0$, is plotted in dash-dot lines for comparison with nonrigid body conditions.

As shown in Fig. 30.47, the rigid body theory is useful only for frequencies lower than the natural frequency of the beam in its fundamental mode. For example, when the natural frequency of the beam (considered as a rigid body) on the isolator is 0.2 times the fundamental natural frequency of the beam ($f_1/f_0 = 5$), as illustrated in Fig. 30.47*B*, the rigid body theory gives an accurate indication of transmissibility only for frequency ratios f/f_0 less than 2.5. This corresponds to a forcing frequency equal to 50 per cent of the lowest natural frequency of the beam. At higher frequencies, resonances of the beam become important and the rigid body theory is inapplicable. Generally similar results are obtained for other values of f_1/f_0, as indicated by Fig. 30.47*A* and *C*.

A result similar to that shown in Fig. 30.47 is obtained for force transmissibility when the foundation is nonrigid. The foundation responds with a large amplitude at its resonant frequencies; thus, the isolator appears to transmit large forces at such frequencies.

VIBRATION ISOLATION OF DAMPED STRUCTURES. The relatively large values of transmissibility in Fig. 30.47 result from vibration of the beam at its resonant frequencies; corresponding transmissibility curves are shown in Fig. 30.48 for a beam with significantly greater damping and with the same resonant frequencies.[21,22] The transmissibility curves for the undamped beams and the rigid body are repeated from Fig. 30.47 for comparison. The damped beams consist of two laminates separated by a thin layer of viscoelastic damping material.

Even though the damping in the beam reduces the amplification at structural resonances, the over-all isolation at the higher forcing frequencies is not as efficient as the rigid body theory predicts. The mean value of transmissibility for all beams, regardless of the amount of damping, has an attenuation rate that is less than that predicted by rigid body theory. Therefore, structural resonances, whether highly damped or not, reduce the over-all efficiency of a vibration isolation system at high forcing frequencies.

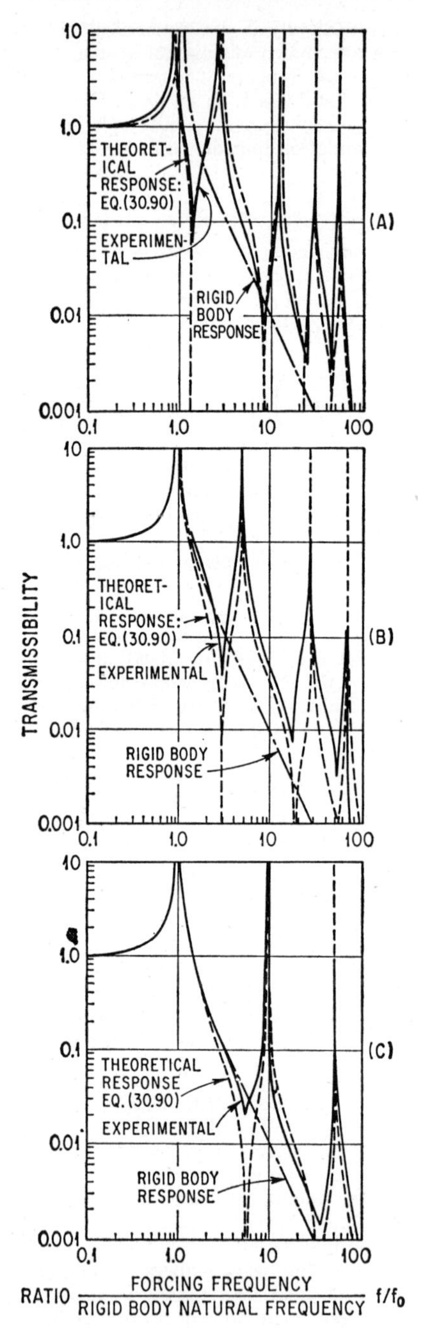

FIG. 30.47. Acceleration transmissibility for the system shown in Fig. 30.46 for three degrees of equipment nonrigidity. The fundamental natural frequency f_1 of the beam is twice the natural frequency f_0 of the beam (considered as a rigid body supported by the isolator) at *A*. At *B* and *C*, the ratio f_1/f_0 is 5 and 10, respectively.

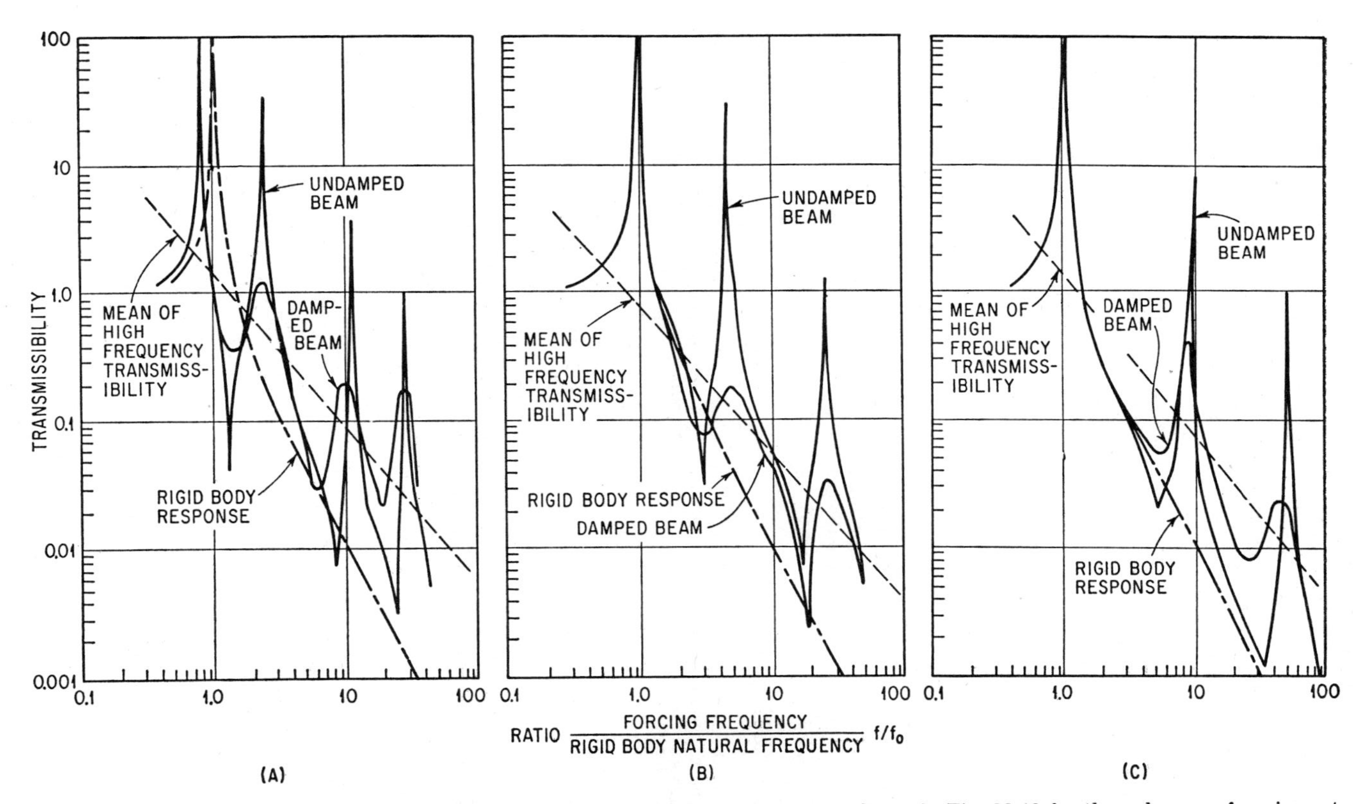

FIG. 30.48. Acceleration transmissibility as determined experimentally for the system shown in Fig. 30.46 for three degrees of equipment nonrigidity corresponding, respectively, to A, B, and C of Fig. 30.47. The damped beam is comprised of two steel members separated by a viscoelastic damping layer; the undamped beam is solid steel. The fundamental natural frequency f_1 of the beam is twice the natural frequency f_o of the beam considered as a rigid body supported by the isolator, in the curves at A; corresponding ratios of f_1/f_o are 5 and 10 at B and C, respectively.

WAVE EFFECTS IN ISOLATORS

When the forcing frequency becomes relatively high, standing waves tend to occur in an isolator and the classical theory of vibration isolation based upon a massless resilient element may not give acceptable results. The transmissibility may become relatively great at the standing-wave frequencies. It is difficult to determine by analytical means standing-wave frequencies of isolators which incorporate irregularly shaped metal or rubber springs. However, the principle can be demonstrated by a simplified model.

Consider a rigid equipment of mass m_E supported by a linear unidirectional isolator having a mass m_I as shown in Fig. 30.49. The transmissibility can be written in terms of displacement by using Eq. (30.76) and noting the equivalence of velocity and displacement transmissibility:

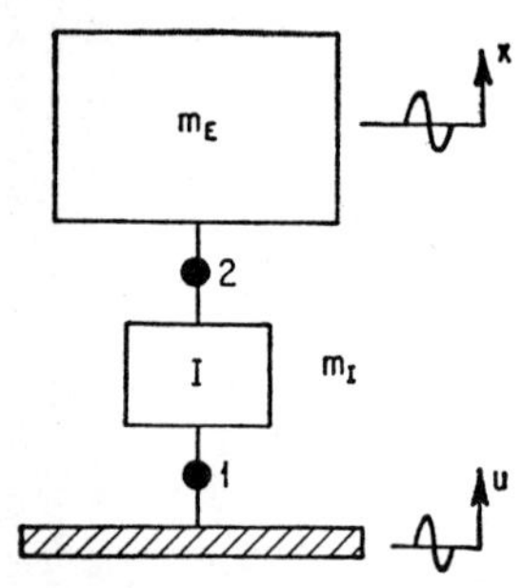

FIG. 30.49. Schematic diagram of a system used to demonstrate the effects on transmissibility of standing waves in the isolator. The rigid equipment having a mass m_E is supported by a unidirectional isolator having mass m_I.

$$(T_v)_{12} = \frac{x_0}{u_0} = \frac{Z_{12}^b}{Z_E + Z_2^{1b}} \tag{30.91}$$

where Z_{12}^b is the transfer impedance of the isolator with one end blocked, Z_2^{1b} is the point impedance of the isolator at station 2 with station 1 blocked, and Z_E is the point impedance of the equipment at terminal 2. Assuming the isolator to be a one-dimensional mechanical transmission line, the required impedances are [16,23]

$$Z_{12}^b = \frac{Z_c}{\sinh \gamma l}$$
$$Z_2^{1b} = Z_c\left(\frac{\cosh \gamma l}{\sinh \gamma l}\right) = Z_c \coth \gamma l \tag{30.92}$$

where the characteristic impedance Z_c is

$$Z_c = \frac{S\gamma}{\omega}\,[\omega\mu - j(c_0{}^2\rho)] \tag{30.93}$$

and ρ is the mass density of the isolator material, S is the isolator cross section area undergoing vibration, μ is the viscosity of the resilient material of the isolator * and c_0 is the undamped phase velocity or classical velocity of sound for the resilient material of the isolator when the wavelength of vibration is much larger than the lateral dimension of the isolator. The complex propagation function γ is

$$\gamma = \alpha + j\beta \tag{30.94}$$

The attenuation per unit length α (measure of damping in the material) and the phase function β may be determined by solving the general equations of motion that apply for the one-dimensional isolator.[24,25] For low damping, α and β are defined by [26]

$$\alpha = \frac{\omega^2\mu}{2\rho c_0{}^3} \qquad \beta = \frac{\omega}{c_0} \tag{30.95}$$

If the impedance relations for the isolator given by Eq. (30.92) are substituted into the transmissibility expression given by Eq. (30.91), the following result is obtained:

$$(T_v)_{12} = \frac{1}{\cosh \gamma l + \dfrac{Z_E}{Z_c}\sinh \gamma l} \tag{30.96}$$

* The coefficient of viscosity μ defines an effective viscous damping coefficient having a value $\mu S/l$ where S is the cross-section area and l is the length of the resilient material of the isolator.

Table 30.3. Characteristic Parameters Relating to Wave Effects in Undamped Isolators

DIAGRAMS OF ISOLATION SYSTEMS		m_E; ρ, S_C, E	I_E; ρ, D, I_p, G	m_E; ρ, S_S, G	m_E; D; ρ, N, G; d	m_E; $\rho, S_F \kappa$	m_E; ρ, S_A, V_0, n
PROPERTY	GENERAL	COMPRESSION	TORSION	SHEAR	HELICAL SPRING	FLUID	AIR
CROSS-SECTION AREA	S	S_C	$\frac{\pi D^2}{4}$	S_S	$\frac{\pi^2 N D d^2}{4 l}$	S_F	S_A
STATIC STIFFNESS	k_{st}	$\frac{S_C E}{l}$	$\frac{G I_p}{l}$	$\frac{S_S G}{l}$	$\frac{G d^4}{8 N D^3}$	$\frac{S_F \kappa}{l}$	$\frac{P_0 S_A^2 n}{V_0}$
MASS OF RESILIENT MATERIAL	m_I	$\rho S_C l$	$\frac{\pi \rho D^2 l}{4}$	$\rho S_S l$	$\frac{\pi^2 \rho N D d^2}{4}$	$\rho S_F l$	$\rho_0 V_0$
ELASTIC WAVE PROPAGATION VELOCITY c_0	$l \sqrt{\frac{k_{st}}{m_I}}$	$\sqrt{\frac{E}{\rho}}$	$\sqrt{\frac{G}{\rho}}$	$\sqrt{\frac{G}{\rho}}$	$\frac{l d}{\sqrt{2} \pi N D^2} \sqrt{\frac{G}{\rho}}$	$\sqrt{\frac{\kappa}{\rho}}$	$\sqrt{\frac{P_0 n}{\rho_0}}$
CHARACTERISTIC IMPEDANCE Z_C	$\sqrt{k_{st} m_I}$	$S_C \sqrt{E \rho}$	$I_p \sqrt{G \rho}$	$S_S \sqrt{G \rho}$	$\frac{\sqrt{2} \pi d^3}{8 D} \sqrt{G \rho}$	$S_F \sqrt{\kappa \rho}$	$S_A \sqrt{P_0 \rho_0 n}$

DIMENSIONS OF THE RESILIENT ELEMENTS ARE INDICATED IN THE SKETCHES; RELEVANT PROPERTIES OF THE RESILIENT MATERIAL ARE:

E = YOUNG'S MODULUS, LB/IN.^{2}

G = MODULUS OF ELASTICITY IN SHEAR, LB/IN.^{2}

κ = BULK MODULUS, LB/IN.^{2}

(SUBSCRIPT $_0$ INDICATES INITIAL CONDITIONS)

ρ = MASS DENSITY, LB-SEC^2/IN.^{4}

V = VOLUME OF RESILIENT MATERIAL, IN.^{3}

n = RATIO OF SPECIFIC HEAT AT CONSTANT PRESSURE TO SPECIFIC HEAT AT CONSTANT VOLUME

P = PRESSURE, LB/IN.^{2}

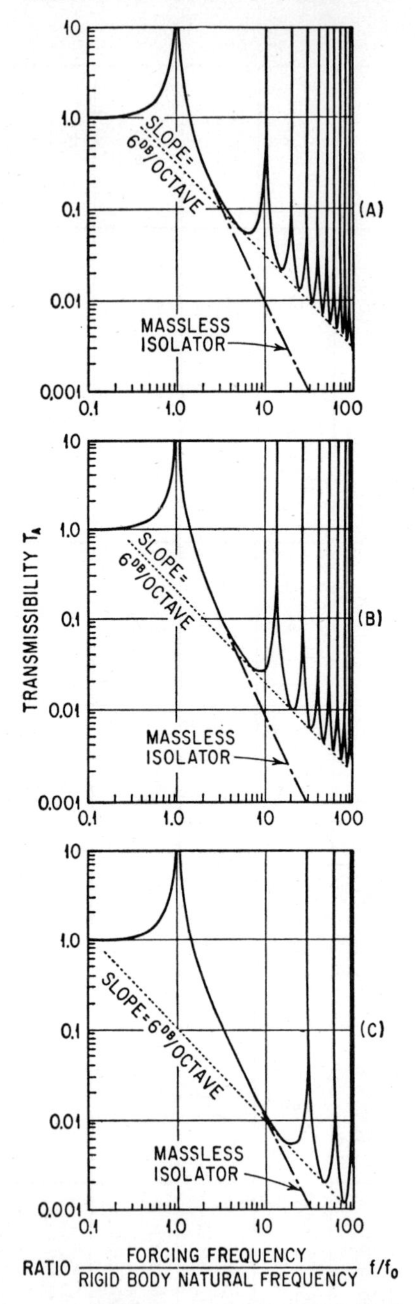

Fig. 30.50. Theoretical transmissibility curves for the system illustrated in Fig. 30.49. The ratio of the mass m_E of the equipment to the mass m_I of the isolator is 10 at A, 20 at B, and 100 at C. The isolator is undamped.

When the equipment is considered a rigid mass m_E, $Z_E = j\omega m_E$ and Eq. (30.96) becomes

$$(T_v)_{12} = \frac{1}{\cosh \gamma l + j \dfrac{\omega m_E}{Z_c} \sinh \gamma l} \tag{30.97}$$

The transmissibility expressions given by Eqs. (30.96) and (30.97) apply to unidirectional isolators for which the force-displacement equation is

$$F(x) = S(c_0^2 \rho + j\omega\mu) \frac{\partial \xi(x)}{\partial x} \tag{30.98}$$

where $F(x)$ is the force in the X coordinate direction, and $\xi(x)$ is the displacement of any cross section of the isolator from the equilibrium position in the X coordinate direction. Therefore, Eqs. (30.96) and (30.97) may be applied to isolators that can be classified as (1) resilient material strained in compression, torsion, or shear; (2) helical spring; and (3) fluid or air isolators. Expressions for c_0 and Z_c are given in Table 30.3 for the several classes of undamped isolators.[27]

For an undamped isolator, $\alpha = 0$ and $Z_c = \rho S c_0 = \sqrt{k_{st} m_I}$. Then Eq. (30.97) becomes

$$(T_v)_{12} = \frac{1}{\cos \dfrac{f}{f_0}\sqrt{\dfrac{m_I}{m_E}} - \dfrac{f}{f_0}\sqrt{\dfrac{m_E}{m_I}} \sin \dfrac{f}{f_0}\sqrt{\dfrac{m_I}{m_E}}} \tag{30.99}$$

The transmissibility defined by Eq. (30.99) is shown graphically in Fig. 30.50 as a function of the ratio of the forcing frequency f to the undamped natural frequency $f_0 = \dfrac{1}{2\pi}\sqrt{\dfrac{k_{st}}{m_E}}$ for three values of the ratio m_E/m_I. The mass m_I is the mass of that portion of the isolator which contributes the resilience. Since no damping exists in the isolator, the transmissibility approaches infinity when the forcing frequency f approaches one of the standing-wave frequencies f_w. The valleys in the transmissibility curve between the standing-wave frequencies have minimum values that indicate a decrease in transmissibility at the rate of 6 db/octave. The lowest frequency at which standing-wave resonances occur increases as the mass of the equipment increases relative to the mass of the isolator.

The transmissibility curves shown in Fig. 30.51 are experimental results obtained from tests on isolators made from cylindrical samples of various rubber materials for a mass ratio $m_E/m_I = 20$.[28] A comparison of the experimental curves of Fig. 30.51 with the theoretical curves of Fig. 30.50*B* indicates the effects that damping has on the standing-wave resonances. Damping in the isolator limits the transmissibility at the standing-wave resonances to finite values. If sufficient damping exists in the isolator, the standing-wave

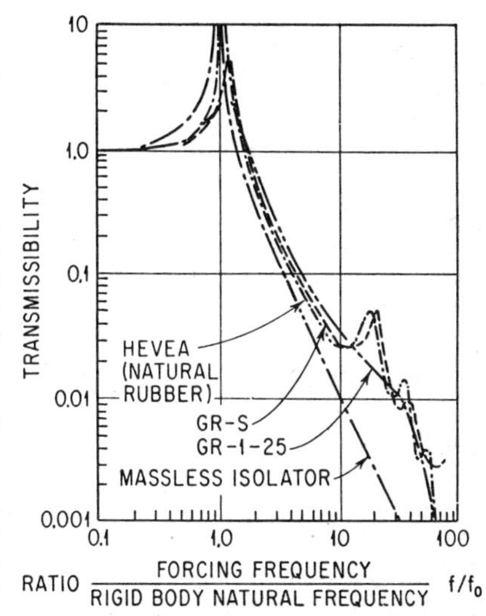

FIG. 30.51. Transmissibility curves for the system illustrated in Fig. 30.49 using various rubber compounds, as obtained experimentally to show the effect of isolator damping. The mass m_E of the equipment is 20 times as great as the mass m_I of the isolator. (*After A. O. Sykes.*[28])

effects of the isolator cause the isolation efficiency at high frequency to decrease without exhibiting any obvious resonant peaks.

The standing-wave frequencies do not occur exactly at the frequencies predicted by the analysis because the dynamic stiffness of a rubber isolator is greater than the static stiffness used to calculate the standing-wave frequencies. Thus, the experimental transmissibility curves shift to the right on the dimensionless frequency plot of Fig. 30.51. The different rubber compounds exhibit different dynamic-to-static stiffness ratios; thus, the curves shift by different increments along the horizontal axis.

The standing-wave frequency f_w is given approximately by the relation

$$\frac{f_w}{f_0} = n\pi \sqrt{\frac{m_E}{m_I}} \qquad [n = 1, 2, 3, \ldots] \tag{30.100}$$

where $f_0 = \frac{1}{2\pi}\sqrt{\frac{k_{st}}{m_E}}$ is the undamped natural frequency, n represents the mode of the standing wave, and m_E, m_I are the mass of the equipment and isolator, respectively. The relation given by Eq. (30.100) is shown graphically by Fig. 30.52 where the ratio of the standing-wave frequency f_w to the undamped natural frequency f_0 is given as a function of the mass ratio m_E/m_I for the first ten standing-wave modes. Standing-wave frequencies determined from this graph agree with the resonances in the transmissibility curves shown in Fig. 30.50.

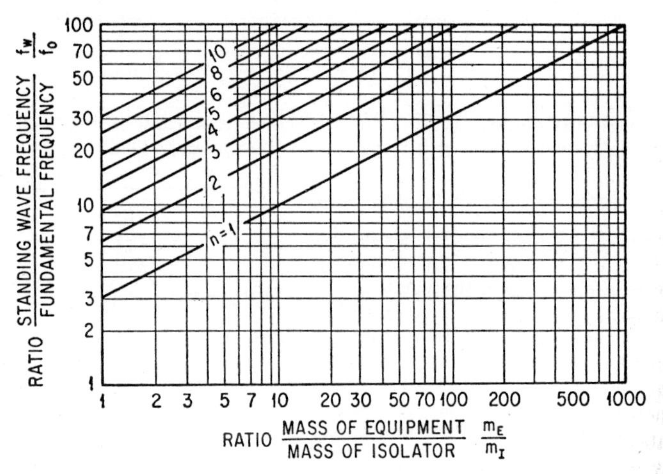

FIG. 30.52. Standing-wave frequencies for the system illustrated in Fig. 30.49. The standing-wave frequency f_w, which is referenced to the fundamental natural frequency f_o of the isolator system, is given as a function of the mass ratio m_E/m_I and the mode number n of the standing-wave vibration.

REFERENCES

1. Crede, C. E.: "Vibration and Shock Isolation," John Wiley & Sons, Inc., New York, 1951.
2. Den Hartog, J. P.: *Trans. ASME*, APM-53-9, 1932.
3. Jacobsen, L. S.: *Trans. ASME*, APM-52-15, 1931.
4. Ruzicka, J. E., and R. D. Cavanaugh: *Machine Design*, Oct. 16, 1958, p. 114.
5. Ruzicka, J. E.: Unpublished work.
6. Ruzicka, J. E.: "Forced Vibrations in Systems with Elastically Supported Dampers," Master's Thesis, Massachusetts Institute of Technology, Cambridge, Mass., 1957.
7. Crede, C. E., and J. P. Walsh: *J. Appl. Mechanics*, **14**:1A-7 (1947).
8. Lewis, R. C., and K. Unholtz: *Trans. ASME*, **69**:8 (1947).
9. Macduff, J. N.: *Prod. Eng.*, July, August, 1946.

10. de Gruben, K.: *VDI Zeitschrift*, **6**:41–42 (1942).
11. Crede, C. E.: *J. Appl. Mechanics*, **25**:541 (1958).
12. Timoshenko, S., and G. H. MacCullough: "Elements of Strength of Materials," 3d ed., p. 64, D. Van Nostrand Company, Inc., Princeton, N.J., 1949.
13. Mindlin, R. D.: *Bell System Tech. J.*, **24**(3–4):353 (1945).
14. Crede, C. E.: *Trans. ASME*, **76**(1):117 (1954).
15. Ruzicka, J. E.: *J. Eng. Industry* (*Trans. ASME*), **83B**(1):53 (1961).
16. Molloy, C. T.: *J. Acoust. Soc. Amer.*, **29**:842 (1957).
17. Ruzicka, J. E.: Unpublished work.
18. Leedy, H. A.: *J. Acoust. Soc. Amer.*, **11**:341 (1940).
19. Sykes, A. O.: "Shock and Vibration Instrumentation," *ASME*, 1956, p. 1.
20. Ruzicka, J. E., and R. D. Cavanaugh: "Mechanical Impedance Methods for Mechanical Vibrations," *ASME*, 1958, p. 109.
21. Ruzicka, J. E.: Paper No. 100Y, SAE National Aeronautic Meeting, October, 1959.
22. Ruzicka, J. E.: *J. Eng. Industry* (*Trans. ASME*), **83B**(4):403 (1961).
23. Sykes, A. O.: *Trans. SAE*, **66**:533 (1958).
24. Rayleigh, Lord: "The Theory of Sound," 2d ed., part II, p. 315, Dover Publications, New York, 1945.
25. Nolle, A. W.: *J. Acoust. Soc. Amer.*, **19**:194 (1947).
26. Harrison, M., A. O. Sykes, and M. Martin: "Wave Effects in Isolation Mounts," *David W. Taylor Model Basin Rept.* 766, 1952.
27. Ruzicka, J. E.: Unpublished work.
28. Sykes, A. O.: "A Study of Compression Noise Isolation Mounts Constructed from Cylindrical Samples of Various Natural and Synthetic Rubber Materials," *David W. Taylor Model Basin Rept.* 845, 1953.

31

THEORY OF SHOCK ISOLATION

R. E. Newton
Naval Postgraduate School

INTRODUCTION

This chapter presents an analytical treatment of the isolation of shock. Two classes of shock are considered: (1) shock characterized by motion of a support or foundation where a shock isolator reduces the severity of the shock experienced by equipment mounted on the support and (2) shock characterized by forces applied to or originating within a machine where a shock isolator reduces the severity of shock experienced by the support. In the simplified concept of shock isolation, the equipment and support are considered rigid bodies, and the effectiveness of the isolator is measured by the forces transmitted through the isolator (resulting in acceleration of equipment if assumed rigid) and by the deflection of the isolator. Linear isolators, both damped and undamped, together with isolators having special types of nonlinear elasticity are considered. When the equipment or floor is not rigid, the deflection of nonrigid members is significant in evaluating the effectiveness of isolators. Analyses of shock isolation are included which consider the response of nonrigid components of equipment and floor.

IDEALIZATION OF THE SYSTEM

In the application of shock isolators to actual equipments, the locations of the isolators are determined largely by practical mechanical considerations. In general, this results in types of nonsymmetry and coupled modes not well adapted to analysis by simple means. It is convenient in the design of shock isolators to idealize the system to a hypothetical one having symmetry and uncoupled modes of motion.

UNCOUPLED MOTIONS. The first step in idealizing the physical system is to separate the various translational and rotational modes, i.e., to *uncouple* the system. Consider the system of Fig. 31.1 consisting of a *homogeneous* block attached at the corners, by eight identical springs, to a movable rigid frame. The block and frame are constrained to move in the plane of the paper. With the system at rest, the frame is given a sudden vertical translation. Because of the symmetry of both mass and stiffness relative to a vertical plane perpendicular to the paper, the response motion of the block is pure vertical translation. Similarly, a sudden horizontal translation of the frame excites

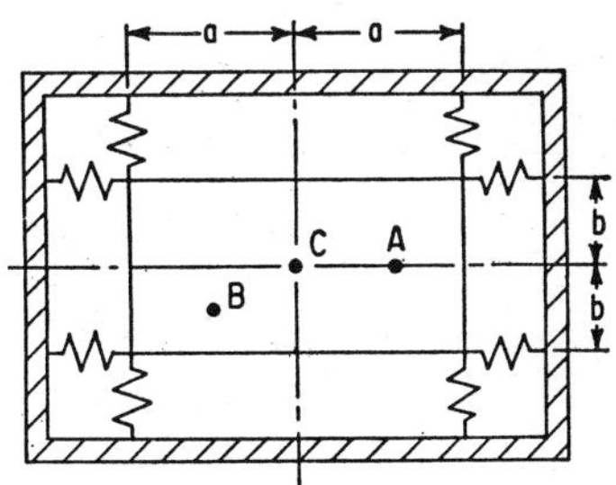

FIG. 31.1. Schematic diagram of three degree-of-freedom mounting. Block and frame are constrained to move in plane of paper.

pure horizontal translation of the block. A sudden rotation about an axis through the geometric center of the block produces pure rotation of the block about this axis. This set of response behaviors is characteristic of an *uncoupled* system.

If the block of Fig. 31.1 is not homogeneous, the mass center (or center-of-gravity) may be at A or B instead of C. Consider the response to a sudden vertical translation of the frame if the mass center is at A. If the response were pure vertical translation of the block, the dynamic forces induced in the vertical springs would have a resultant acting vertically through C. However, the "inertia force" of the block must act through the mass center at A. Thus, the response cannot be pure vertical translation, but must also include rotation. Then the motions of vertical translation and rotation are said to be *coupled*. A sudden horizontal translation of the frame would still excite only a horizontal translation of the block because A is symmetrical with respect to the horizontal springs; thus this horizontal motion remains *uncoupled*. If the mass center were at B, i.e., in neither the vertical nor the horizontal plane of symmetry, then a sudden vertical translation of the frame would excite both vertical and horizontal translations of the block, together with rotation. In this case, all three motions are said to be *coupled*.

It is not essential that a system have any kind of geometric symmetry in order that its motions be uncoupled but rather that the resultant of the spring forces be either a force directed through the center-of-gravity of the block or a couple. If the motions are completely uncoupled, there are three mutually orthogonal directions such that translational motion of the base in any one of these directions excites only a translation of the body in the same direction. Similarly there are three orthogonal axes, concurrent at the mass center, having the property that a pure rotation of the base about any one of these axes will excite a pure rotation of the body about the same axis. The idealized systems considered in this chapter are assumed to have uncoupled rigid body motions.

ANALOGY BETWEEN TRANSLATION AND ROTATION. If the motions in translational and rotational modes are uncoupled, motion in the rotational mode may be inferred by analogy from motion in the translational mode, and vice versa. Consider the system of Fig. 31.1 and assume that the mass center is at C. For horizontal motion the differential equation of motion is *

$$m\ddot{\delta} + 4k\delta = -m\ddot{u} \tag{31.1}$$

where δ = horizontal displacement of mass center of block relative to center-of-frame, in.
m = mass of block, lb-sec^2/in.
k = spring stiffness for each spring, lb/in.
u = absolute horizontal displacement of center-of-frame, in.†

Equation (31.1) may be written

$$\ddot{\delta} + \omega_n^2\delta = -\ddot{u} \tag{31.2}$$

where $\omega_n = \sqrt{4k/m}$, rad/sec, is the angular natural frequency in horizontal vibration.

For rotation of the block the corresponding equation of motion is

$$I\ddot{\gamma}_r + 4k(a^2 + b^2)\gamma_r = -I\ddot{\Gamma} \tag{31.3}$$

where I = mass moment of inertia of block about axis through C, perpendicular to plane of paper, lb-in.-sec^2
a, b = distances of spring center lines from mass center (see Fig. 31.1), in.
γ_r = rotation of block relative to frame in plane of paper, rad
Γ = absolute rotation of frame in plane of figure, rad

Equation (31.3) may be written

$$\ddot{\gamma}_r + \omega_{n1}^2\gamma_r = -\ddot{\Gamma} \tag{31.4}$$

where $\omega_{n1} = \sqrt{4k(a^2 + b^2)/I}$ is the angular natural frequency in rotation.

* It is assumed that forces in the four vertical springs have a negligible horizontal component at all times.

† In the equilibrium position the point C lies at the frame center.

Equations (31.2) and (31.4) are analogous; γ_r corresponds to δ, Γ corresponds to u, and ω_{n1} corresponds to ω_n. Because of this analogy, only the horizontal motion described by Eq. (31.2) is considered in subsequent sections; corresponding results for rotational motion may be determined by analogy.

CLASSIFICATION OF SHOCK ISOLATION PROBLEMS

It is convenient to divide shock isolation problems into two major classifications according to the physical conditions:

Class I. Mitigation of effects of foundation motion

Class II. Mitigation of effects of force generated by equipment

Isolators in the first class include such items as the draft gear on a railroad car, the shock strut of an aircraft landing gear, the mounts on air-borne electronic equipment, and the corrugated paper used to package light bulbs. The second class includes the recoil cylinders on gun mounts and the isolators on drop hammers, looms, and reciprocating presses.

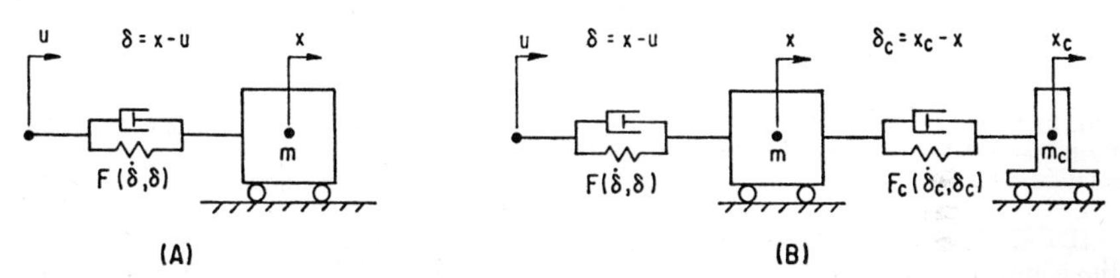

FIG. 31.2. Idealized systems showing use of isolator with transmitted force $F(\dot{\delta},\delta)$ to protect equipment of mass m from effects of support motion u. In (*A*) the equipment is rigid and in (*B*) there is a flexible component having stiffness-damping characteristics $F_c(\dot{\delta}_c, \delta_c)$ and mass m_c.

The objectives in the two classes of problems are allied, but distinct. In Class I the objective is to limit the shock-induced stresses in critical components of the protected equipment. In Class II the purpose is to limit the forces transmitted to the support for the equipment in which the shock originates.

IDEALIZED SYSTEMS—CLASS I. The simplest approach to problems of Class I is through a study of single degree-of-freedom systems.[1] Consider the system of Fig. 31.2*A*. The basic elements are a mass and a spring-dashpot unit attached to the mass at one end. The block may be taken to represent the equipment,* and the spring-dashpot unit to represent the shock isolator. The displacement of the support is u. The equation of motion is

$$m\ddot{\delta} + F(\dot{\delta},\delta) = -m\ddot{u} \tag{31.5}$$

where m = mass of block, lb.-sec²/in.

δ = deflection of spring ($\delta = x - u$; see Fig. 31.2), in.

$F(\dot{\delta},\delta)$ = force exerted on mass by spring-dashpot unit (positive when tensile), lb

u = absolute displacement of left-hand end of spring-dashpot unit, in.

In the typical shock isolation problem, the system of Fig. 31.2*A* is initially at rest ($\dot{u} = \dot{\delta} = 0$) in an equilibrium position ($u = \delta = 0$). An external shock causes the support to move. The corresponding movement of the left end of the shock isolator is described in terms of the support acceleration $\ddot{u}$. Then Eq. (31.5) may be solved for the resulting extreme values of δ and $F(\dot{\delta},\delta)$, and these values may be compared with the permissible deflection and force transmission limits of the shock isolator. It also is necessary to determine whether the internal stresses developed in the equipment are excessive. If the equipment is sufficiently rigid that all parts have substantially equal accelerations, then the internal stresses are proportional to $\ddot{x}$ where $-m\ddot{x} = F(\dot{\delta},\delta)$.

* In this simplified and idealized analysis, the equipment is considered a rigid body. The effect of flexibility of elements comprising the equipment is considered in a later section.

A critical component of the equipment may be sufficiently flexible to have a substantially different acceleration than that determined by assuming the equipment rigid. If the total mass of such components is small in comparison with the equipment mass, the above analysis may be extended to cover this case. Equation (31.5) is first solved to determine not merely the extreme value of $F(\dot{\delta},\delta)$ but its time-history. Then the acceleration $\ddot{x}$ may be determined from the relation $\ddot{x} = -F(\dot{\delta},\delta)/m$. Now consider the system shown in Fig. 31.2*B* having a component of mass m_c and stiffness-damping characteristics $F_c(\dot{\delta}_c,\delta_c)$. The force $F_c(\dot{\delta}_c,\delta_c)$ transmitted to the mass m_c and the resulting acceleration $\ddot{x}_c = -F_c(\dot{\delta}_c,\delta_c)/m_c$ may be found by solving an equation that is analogous to Eq. (31.5) where $\ddot{x}$ is substituted for $\ddot{u}$, $\ddot{\delta}_c$ for $\ddot{\delta}$, and $F_c(\dot{\delta}_c,\delta_c)$ for $F(\dot{\delta},\delta)$.

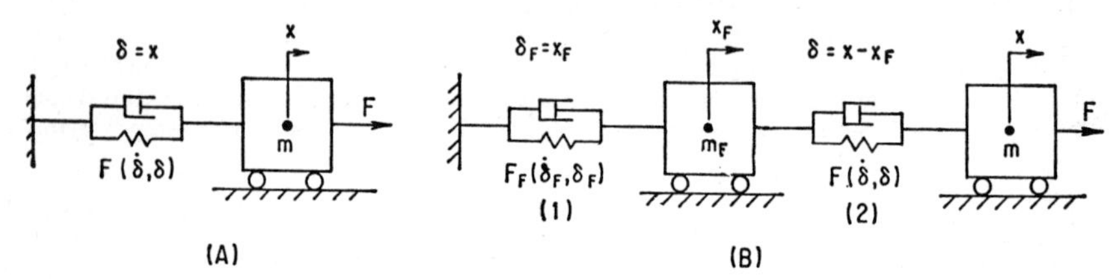

FIG. 31.3. Idealized systems showing use of isolator with transmitted force $F(\dot{\delta},\delta)$ to reduce force transmitted to foundation when force F is applied to equipment of mass m. In (*A*) the foundation is rigid and in (*B*) it has mass m_F and stiffness-damping characteristics $F_F(\dot{\delta}_F,\delta_F)$.

IDEALIZED SYSTEMS—CLASS II. Consider the system of Fig. 31.3*A* to represent the equipment (mass m) attached to its support by the shock isolator (spring-dashpot unit). The left end of the spring-dashpot unit is fixed to the supporting structure and there is a force F applied externally to the mass. The force F may be a real external force or it may be an "inertia force" generated by moving parts of the equipment. The equation of motion may be written

$$m\ddot{\delta} + F(\dot{\delta},\delta) = F \tag{31.6}$$

where F is the external force applied to the mass in pounds and the relative displacement δ of the ends of the spring-dashpot unit is equal to the absolute displacement x of the mass. Assuming the system to be initially in equilibrium ($\dot{\delta} = 0$, $\delta = 0$), Eq. (31.6) is solved for extreme values of δ and $F(\dot{\delta},\delta)$ since F is a known function of time. These are to be com-

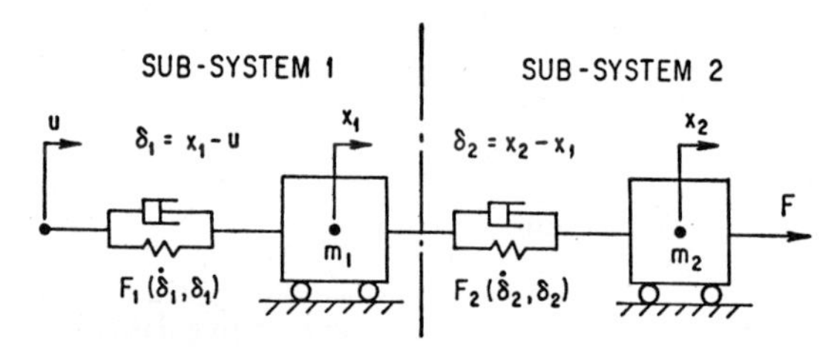

FIG. 31.4. General two degree-of-freedom system. Figures 31.2*B* and 31.3*B* represent particular cases of the general system.

pared with the displacement and force limitations of the shock isolator. Often the supporting structure is sufficiently rigid that the maximum force in the isolator may be considered as a force applied statically to the support. Then the foregoing analysis is adequate for determining the stress in the support.

The load on the floor may be treated as dynamic instead of static by a simple analysis if the displacement and velocity of the support are negligible in comparison with those of the equipment. Consider the system of Fig. 31.3*B* where the supporting structure is represented as a mass m_F and a spring-dashpot unit in place of the rigid support shown in

Fig. 31.3*A*. The force acting on the supporting structure is a known function of time $F(\dot{\delta},\delta)$ as found from the previous solution of Eq. (31.6). To find the maximum force *within* the support structure requires a solution of an equation analogous to Eq. (31.6) where $\ddot{\delta}_F$ is substituted for $\ddot{\delta}$, m_F for m, $F_F(\dot{\delta}_F,\delta_F)$ for $F(\dot{\delta},\delta)$ and $F(\dot{\delta},\delta)$ for F. For engineering purposes it suffices to find the extreme values of δ_F and $F_F(\dot{\delta}_F,\delta_F)$. The first is needed to verify the assumption that support motion is negligible compared with equipment motion, and can be used to determine the maximum stress in the support. The second is the maximum force applied by the support structure to its base.

MATHEMATICAL EQUIVALENCE OF CLASS I AND CLASS II PROBLEMS. The similarity of shock isolation principles in Class I and Class II is indicated by the similar form of Eqs. (31.5) and (31.6). The right-hand side ($-m\ddot{u}$ or F) is given as a function of time, and the extreme values of δ and $F(\dot{\delta},\delta)$ are desired. When the actual system is represented by two separate single degree-of-freedom systems as shown in Figs. 31.2*B* and 31.3*B*, the time-history of $F(\dot{\delta},\delta)$ is also required. Figure 31.4 may be considered a generalized form of the applicable system. In Class I, $F = 0$, $F_1(\dot{\delta}_1,\delta_1)$ represents the properties of the isolator and $m_2,F_2(\dot{\delta}_2,\delta_2)$ represents the component to be protected. In Class II, $u = 0$, $F_2(\dot{\delta}_2,\delta_2)$ represents the properties of the isolator, and $m_1,F_1(\dot{\delta}_1,\delta_1)$ represents the supporting structure.

The system of Fig. 31.4, with the spring-dashpot units nonlinear, requires the use of a digital or analog computer to investigate performance characteristics. Analytical methods are feasible if the system is linearized by assuming that each spring-dashpot unit has a force characteristic in the form

$$F(\dot{\delta},\delta) = c\dot{\delta} + k\delta \tag{31.7}$$

where c = damping coefficient, lb-sec/in., and k = spring stiffness, lb/in. Even with this simplification, the number of parameters (m_1,c_1,k_1,m_2,c_2,k_2) is so great that it is necessary to confine the analysis to a particular system. If the damping may be neglected [let $c = 0$ in Eq. (31.7)], then it is feasible to obtain equations in a form suitable for routine use.[2] Use of this idealization is described in the section on *Response of Equipment with a Flexible Component.*

A different form of idealization is indicated when the "equipment" is flexible; e.g., a large, relatively flexible aircraft subjected to landing shock. Then it is important to represent the aircraft as a system with several degrees-of-freedom. To find resulting stresses, it is necessary to superimpose the responses in the various modes of motion that are excited.

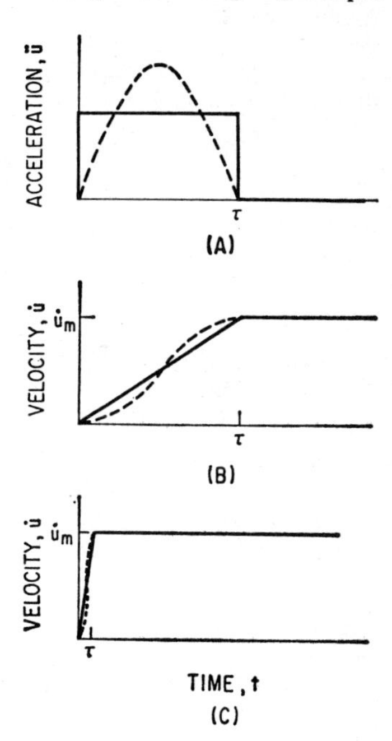

FIG. 31.5. Acceleration-time curves (*A*) and velocity-time curves (*B*) and (*C*) for rectangular acceleration pulse (solid curves) and half-sine acceleration pulse (dashed curves). In (*C*) the time scale is compressed to one-tenth that used in (*A*) and (*B*).

RESPONSE OF A RIGID BODY SYSTEM TO A VELOCITY STEP

PHYSICAL BASIS FOR VELOCITY STEP

The idealization of a shock motion as a simple change in velocity (velocity step) may form an adequate basis for designing a shock isolator and for evaluating its effectiveness. Consider the two types of acceleration $\ddot{u}$ vs. time t curves illustrated in Fig. 31.5*A*. The solid line represents a rectangular pulse of acceleration and the dashed line represents a half-sine pulse of acceleration. Each pulse has a duration τ. In Fig. 31.5*B*, the corresponding velocity-time curves are shown. Each of these curves is defined completely by specifying the type of acceleration pulse (rectangular or half-sine), the duration τ, and the velocity change $\dot{u}_m$. The curves of Fig. 31.5*B* are repeated in Fig. 31.5*C* with the time scale shrunk to one-tenth. If τ is *sufficiently short*, the only significant remaining characteristic of the velocity step is the velocity change $\dot{u}_m$. The idealized velocity step, then, is taken to be a discontinuous change of $\dot{u}$ from zero to $\dot{u}_m$. A shock isolator characteristically has a low natural frequency (long period), and this idealization leads to good results even when the pulse duration τ is significantly long.

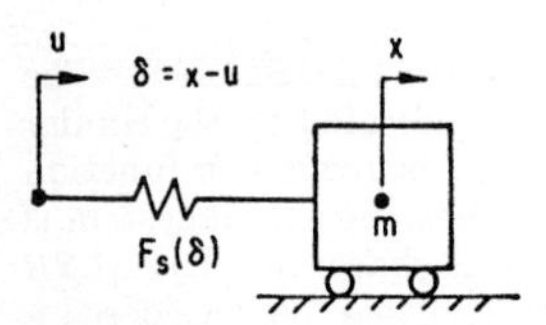

FIG. 31.6. Idealized system showing use of undamped isolator to protect equipment from effects of support motion u. The force transmitted by the isolator is $F_s(\delta)$.

GENERAL FORM OF ISOLATOR CHARACTERISTICS

The differential equation of motion for the undamped, single degree-of-freedom system shown in Fig. 31.6 is

$$m\ddot{\delta} + F_s(\delta) = -m\ddot{u} \tag{31.8}$$

where m represents the mass of the equipment considered as a rigid body, u represents the motion of the support which characterizes the condition of shock, and $F_s(\delta)$ is the force developed by the isolator at an extension δ (positive when tensile). Equation (31.8) differs from Eq. (31.5) in that $F_s(\delta)$, which does not depend upon $\dot{\delta}$, replaces $F(\dot{\delta},\delta)$ because the isolator is undamped. The effect of a velocity step of magnitude $\dot{u}_m$ at $t = 0$ is considered by choosing the initial conditions: * At $t = 0$, $\delta = 0$ and $\dot{\delta} = \dot{u}_m$. A first integration of Eq. (31.8) yields

$$\dot{\delta}^2 = \dot{u}_m{}^2 - \frac{2}{m}\int_0^\delta F_s(\delta)\,d\delta \tag{31.9}$$

At the extreme value of isolator deflection, $\delta = \delta_m$ and the velocity $\dot{\delta}$ of deflection is zero. Then from Eq. (31.9),

$$\int_0^{\delta_m} F_s(\delta)\,d\delta = \tfrac{1}{2}m\dot{u}_m{}^2 \tag{31.10}$$

The right side of Eq. (31.10) represents the initial kinetic energy of the equipment relative to the support, and the integral on the left side represents the work done on the isolator. The latter quantity is equal to the elastic potential energy stored in the isolator, since there is no damping.

For the special case of a rigid body mounted on an undamped isolator, Eq. (31.10) suffices to determine all important results. In particular, the quantities of engineering significance are:

1. The maximum deflection of the isolator δ_m
2. The maximum isolator force, $F_m = F_s(\delta_m) = m\ddot{x}_m$ †
3. The corresponding velocity change $\dot{u}_m$

The interrelations of these three quantities are shown graphically in Fig. 31.7. The curve *OAB* represents the spring force $F_s(\delta)$ as a function of deflection δ. If point *A* corresponds to the extreme excursion, then its abscissa represents the maximum deflection δ_m.

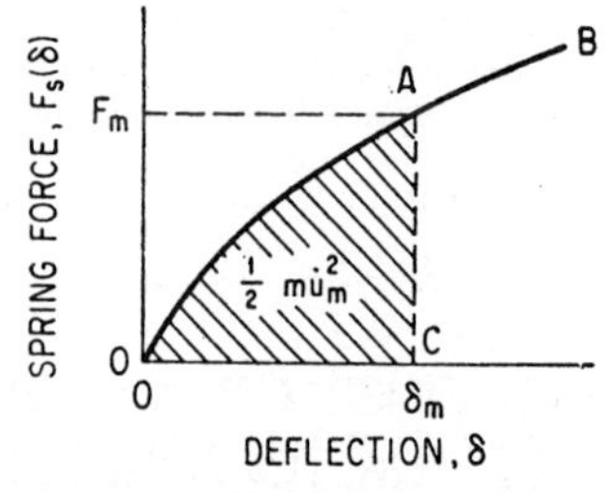

FIG. 31.7. Typical force-deflection curve for undamped isolator. Shaded area represents kinetic energy of equipment following a velocity step $\dot{u}_m$ [Eq. (31.10)].

* These conditions correspond to a negative velocity step. This choice is made to avoid dealing with negative values of δ and $\dot{\delta}$. If $F_s(\delta)$ is not an odd function of δ, a positive velocity step requires a separate analysis.

† The maximum absolute acceleration $\ddot{x}_m$ of the equipment is related to the maximum force F_m by $F_m = m\ddot{x}_m$. It sometimes is convenient to express results in terms of $\ddot{x}_m$ instead of F_m.

The shaded area OAC is proportional to the potential energy stored by the isolator; according to Eq. (31.10), this is equal to the initial kinetic energy $m\dot{u}_m^2/2$. The maximum ordinate (at A) represents the maximum spring force F_m. [It is possible to have a spring force $F_s(\delta)$ which attains a maximum value at $\delta = \delta_f < \delta_m$. Then $F_m = F_s(\delta_f)$.]

The design requirements for the isolator usually include as a specification one or more of the following quantities:

1. Maximum allowable deflection δ_a
2. Maximum allowable transmitted force F_a
3. Maximum expected velocity step $\dot{u}_a$

It is important to observe that the limits 1 and 2 establish an upper limit $F_a\delta_a$ on the work done on the mass. It follows that $\dot{u}_a$ must satisfy the relation

$$F_a\delta_a \geq m\dot{u}_a^2/2$$

or the specifications are impossible to meet. The specifications may be expressed mathematically as follows:

$$\delta_m \leq \delta_a \qquad F_m \leq F_a \qquad \dot{u}_m \geq \dot{u}_a \tag{31.11}$$

In many instances it is advantageous to eliminate explicit reference to the mass m. Then the allowable absolute acceleration $\ddot{x}_a$ of the mass is specified instead of the allowable force F_a where $F_a = m\ddot{x}_a$. With this substitution the second of Eqs. (31.11) is replaced by

$$\ddot{x}_m \leq \ddot{x}_a \tag{31.12}$$

The acceleration $\ddot{x}$ is determined as a function of time by using $\dot{\delta}$ from Eq. (31.9) and finding the time t corresponding to a given value of δ:

$$t = \int_0^\delta \frac{d\delta}{\dot{\delta}} \tag{31.13}$$

From Eq. (31.13) and the relation $\ddot{x} = F_s(\delta)/m$, the acceleration time-history is found.

The integrations required by Eqs. (31.9) and (31.13) sometimes are difficult to perform, and it is necessary to use numerical or graphical methods. Then a difficulty arises with the integral in Eq. (31.13): As δ approaches the extreme value δ_m, the velocity $\dot{\delta}$ in the denominator of the integrand approaches zero. The difficulty is circumvented by first using Eq. (31.13) to integrate up to some intermediate displacement δ_b less than δ_m; then the alternate form, Eq. (31.14), may be used in the region of $\delta = \delta_m$:

$$t = t_b + \int_{\dot{\delta}_b}^{\dot{\delta}} \frac{d\dot{\delta}}{\ddot{\delta}} \tag{31.14}$$

where t_b is the time at which $\delta = \delta_b$, as determined from Eq. (31.13).

In the next three sections three different kinds of spring force-deflection characteristics $F_s(\delta)$ are considered. Equation (31.10) is applied to find the relation between $\dot{u}_m$ and δ_m. Curves relating $\dot{u}_m$, δ_m, and $\ddot{x}_m$ in a form useful for design or analysis are presented.

EXAMPLES OF PARTICULAR ISOLATOR CHARACTERISTICS

LINEAR SPRING. The force-deflection characteristic of a linear spring is

$$F_s(\delta) = k\delta \tag{31.15}$$

where k = spring stiffness, lb/in. Using the notation

$$\omega_n = \sqrt{\frac{k}{m}} \qquad \text{rad/sec} \tag{31.16}$$

the maximum acceleration is

$$\ddot{x}_m = \omega_n^2\delta_m \tag{31.17}$$

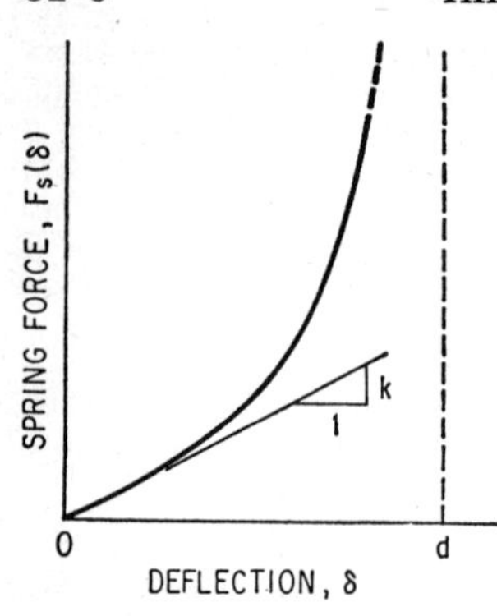

FIG. 31.8. Typical force-deflection curve for hardening spring (tangent elasticity) as given by Eq. (31.20). There is a vertical asymptote at the limiting deflection $\delta = d$.

From Eqs. (31.10) and (31.16), the relation between velocity change $\dot{u}_m$ and maximum deflection δ_m is

$$\dot{u}_m = \omega_n \delta_m \tag{31.18}$$

Combining Eqs. (31.18) and (31.17),

$$\ddot{x}_m = \omega_n \dot{u}_m \tag{31.19}$$

HARDENING SPRING (TANGENT ELASTICITY). The isolator spring may be nonlinear with a "hardening" characteristic; i.e., the slope of the curve representing spring force vs. deflection increases with increasing deflection. Rubber in compression has this behavior. A representative curve[1] having this characteristic is defined by

$$F_s(\delta) = \frac{2kd}{\pi} \tan \frac{\pi\delta}{2d} \tag{31.20}$$

where the constant k is the *initial* slope of the curve (lb/in.) and a vertical asymptote is defined by $\delta = d$ (in.). Such a curve is shown graphically in Fig. 31.8. Using the notation of Eq. (31.16) and the relation $m\ddot{x}_m = F_s(\delta_m)$, Eq. (31.20) gives the following relation between maximum acceleration and maximum deflection:

$$\frac{\ddot{x}_m}{\omega_n^2 d} = \frac{2}{\pi} \tan \frac{\pi\delta_m}{2d} \tag{31.21}$$

Note that ω_n, the angular natural frequency for a linear system, has the same meaning for small amplitude (small δ_m) motions of the nonlinear system. For large amplitudes the natural frequency depends on δ_m. Using Eq. (31.16), substituting for $F_s(\delta)$ from Eq. (31.20) in Eq. (31.10), and performing the indicated integration, the relation between velocity change and maximum displacement is

$$\frac{\dot{u}_m^2}{\omega_n^2 d^2} = \frac{8}{\pi^2} \log_e \left(\sec \frac{\pi\delta_m}{2d} \right) \tag{31.22}$$

A graphical presentation relating the important variables $\dot{u}_m$, $\ddot{x}_m$, and δ_m is convenient for design and analysis. Such data are presented compactly as relations among the dimensionless parameters δ_m/d, $\dot{u}_m/\omega_n d$ and $\ddot{x}_m\delta_m/\dot{u}_m^2$. The physical significance of the ratio $\ddot{x}_m\delta_m/\dot{u}_m^2$ is interpreted by multiplying both numerator and denominator by m. Then the numerator represents the product of the maximum spring force $F_m(= m\ddot{x}_m)$ and the maximum spring deflection δ_m. This product is the maximum energy that *could* be stored in the spring. The denominator $m\dot{u}_m^2$ is *twice* the energy that *is* stored in the spring. The minimum *possible* value of the ratio $\ddot{x}_m\delta_m/\dot{u}_m^2$ is 1/2. Actual values of the ratio, always greater than 1/2, may be considered to be a measure of the departure from optimum capability.

In Fig. 31.9 the solid curve represents $\dot{u}_m/\omega_n d$ as a function of δ_m/d and the dashed curve shows the corresponding result for a linear spring [see Eq. (31.18)]. In Fig. 31.10 the solid curve shows $\ddot{x}_m\delta_m/\dot{u}_m^2$ vs. δ_m/d for an isolator with tangent elasticity. The dashed curve in Fig. 31.10 shows $\ddot{x}_m\delta_m/\dot{u}_m^2$ for a linear spring [see Eqs. (31.17) and (31.18)]; the ratio is constant at a value of unity because a linear spring is 50 per cent efficient in storage of energy, independent of the deflection.

SOFTENING SPRING (HYPERBOLIC TANGENT ELASTICITY). A nonlinear isolator also may have a "softening" characteristic; i.e., the slope of the curve represent-

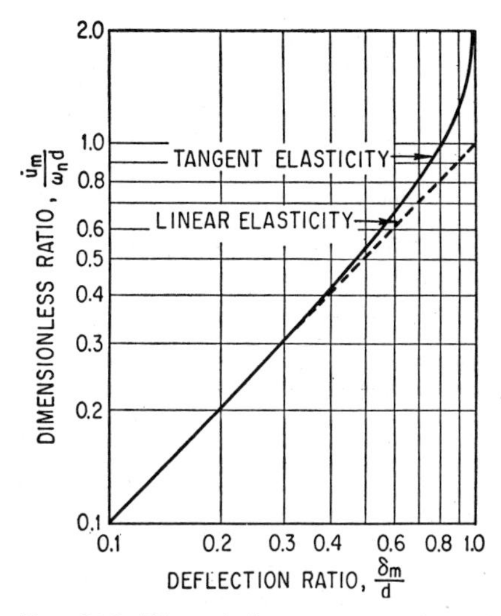

FIG. 31.9. Dimensionless representation of relation between velocity step $\dot{u}_m$ and maximum isolator deflection δ_m for undamped isolators having tangent elasticity (solid curve) and linear elasticity (dashed curve). The natural angular frequency for small oscillations is ω_n and the limiting deflection is d (Fig. 31.8).

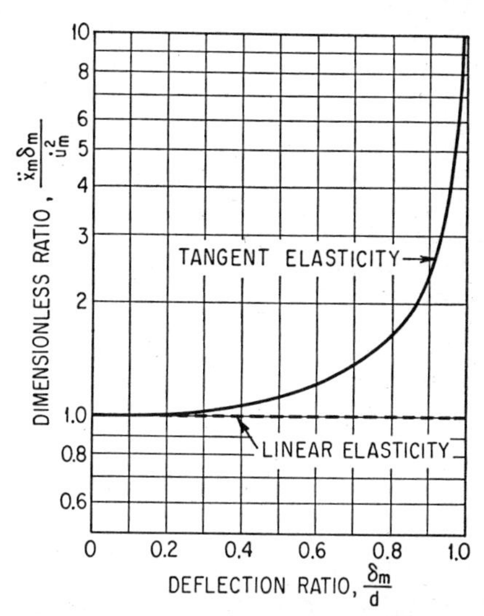

FIG. 31.10. Dimensionless representation of relation among velocity step $\dot{u}_m$, maximum transmitted acceleration $\ddot{x}_m$, and maximum isolator deflection δ_m for undamped isolators having tangent elasticity (solid curve) and linear elasticity (dashed curve). Ordinate is an inverse measure of energy-storage efficiency. The natural angular frequency for small oscillations is ω_n and the limiting deflection is d (Fig. 31.8).

ing force vs. deflection decreases with increasing deflection. The force-deflection characteristic for a typical "softening" isolator is [1]

$$F_s(\delta) = kd_1 \tanh \frac{\delta}{d_1} \tag{31.23}$$

where k is the initial slope of the curve. Figure 31.11 shows the form of this curve where the meaning of d_1 is evident from the figure. If $F_s(\delta)$ is replaced by $m\ddot{x}_m$, δ by δ_m, and k by $m\omega_n^2$, Eq. (31.23) becomes

$$\frac{\ddot{x}_m}{\omega_n^2 d_1} = \tanh \frac{\delta_m}{d_1} \tag{31.24}$$

where δ_m and $\ddot{x}_m$ are maximum values of deflection and acceleration, respectively, and ω_n may be interpreted as the angular natural frequency for small values of δ_m. To relate $\dot{u}_m$ to δ_m, substitute $F_s(\delta)$ from Eq. (31.23) in Eq. (31.10), let $\omega_n^2 = k/m$, and integrate:

$$\frac{\dot{u}_m^2}{\omega_n^2 d_1^2} = \log_e \left(\cosh^2 \frac{\delta_m}{d_1}\right) \tag{31.25}$$

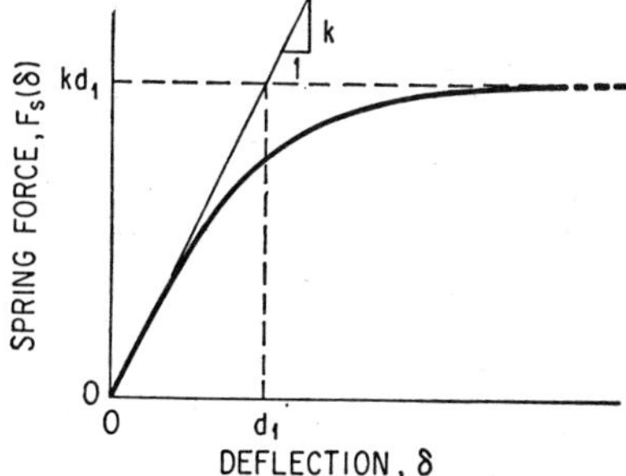

FIG. 31.11. Typical force-deflection curve for softening spring (hyperbolic tangent elasticity) as given by Eq. (31.23). There is a horizontal asymptote at the limiting force kd_1.

A graphical presentation of the relation between $\dot{u}_m/\omega_n d_1$ and δ_m/d_1 is given by the solid curve of Fig. 31.12. The dashed curve shows the corresponding relation for a linear spring. In Fig. 31.13 the solid curve represents $\ddot{x}_m \delta_m / \dot{u}_m^2$ as a function of δ_m/d_1. Note that, for large values of δ_m/d_1, the ordinate approaches the minimum value 1/2 attainable with an isolator of optimum energy storage efficiency. The dashed curve shows the same relation for a linear spring.

LINEAR SPRING AND VISCOUS DAMPING. The addition of viscous damping can almost double the energy absorption capability of a linear shock isolator. Consider the system of Fig. 31.2*A*, with both spring and dashpot linear as defined by Eq. (31.7). Substituting $F(\dot{\delta},\delta)$ from Eq. (31.7) in Eq. (31.5) gives the equation of motion. The initial conditions are $\dot{\delta} = \dot{u}_m$, $\delta = 0$, when $t = 0$; for $t > 0$, $\ddot{u} = 0$. Letting $c_c = 2m\omega_n$ and $\zeta = c/c_c$ [see Eq. (2.12)], the equation of motion becomes

$$\ddot{\delta} + 2\zeta\omega_n\dot{\delta} + \omega_n^2\delta = 0 \qquad (31.26)$$

Solutions of Eq. (31.26) for maximum deflection δ_m and maximum acceleration $\ddot{x}_m$ as

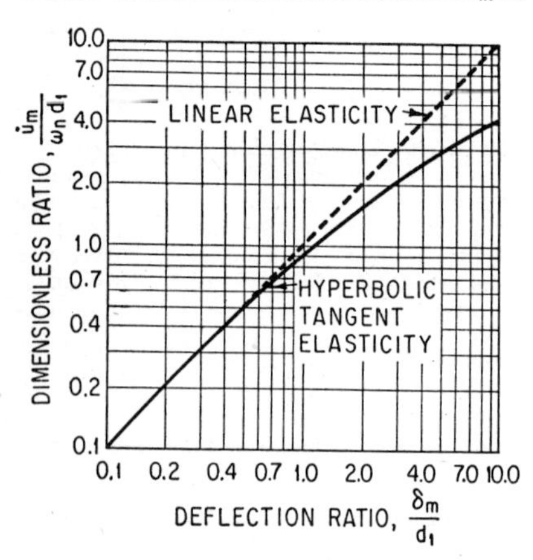

FIG. 31.12. Dimensionless representation of relation between velocity step $\dot{u}_m$ and maximum isolator deflection δ_m for undamped isolators having hyperbolic tangent elasticity (solid curve) and linear elasticity (dashed curve). The natural angular frequency for small oscillations is ω_n and the characteristic deflection d_1 is defined in Fig. 31.11.

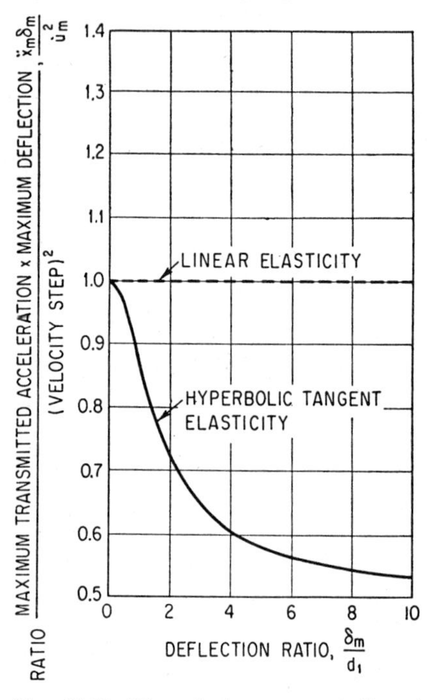

FIG. 31.13. Dimensionless representation of energy-storage capabilities of undamped isolators having hyperbolic tangent elasticity (solid curve) and linear elasticity (dashed curve). The ordinate is an inverse measure of energy-storage efficiency. The characteristic deflection d_1 is defined in Fig. 31.11.

functions of ζ are available [1] in analytical form; the solutions are shown graphically in Figs. 31.14 and 31.15. In Fig. 31.14, the dimensionless ratio $\ddot{x}_m/\dot{u}_m\omega_n$ is plotted as a function of the fraction of critical damping ζ. Note that the presence of small damping reduces the maximum acceleration. As ζ is increased beyond 0.25, the maximum acceleration increases again. For $\zeta > 0.50$, the maximum acceleration occurs at $t = 0$ and exceeds that for no damping; it is accounted for solely by the damping force $c\dot{\delta} = c\dot{u}_m$.

In Fig. 31.15 the parameter $\ddot{x}_m\delta_m/\dot{u}_m^2$ is plotted as a function of ζ. (As pointed out with reference to Fig. 31.10, $\ddot{x}_m\delta_m/\dot{u}_m^2$ is an inverse measure of shock isolator effectiveness.) Figure 31.15 shows that the presence of damping improves the energy storage effectiveness of the isolator even beyond $\zeta = 0.50$. In the neighborhood $\zeta = 0.40$, the parameter $\ddot{x}_m\delta_m/\dot{u}_m^2$ attains a minimum value of 0.52—only slightly above the theoretical minimum of 0.50. This parameter has the value 1.00 for an undamped linear

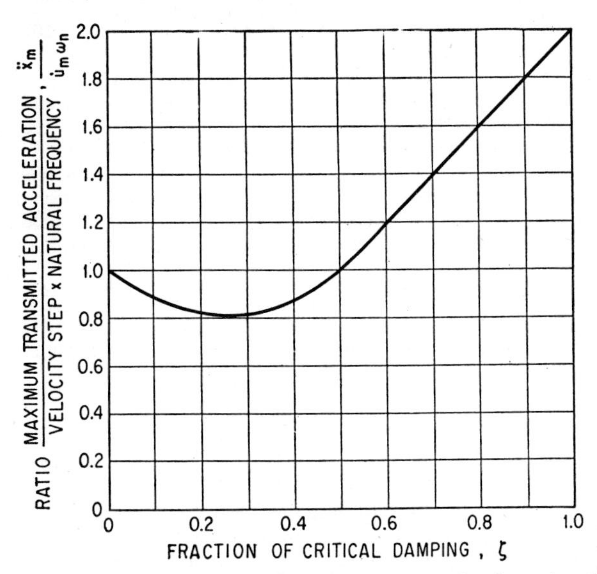

FIG. 31.14. Dimensionless representation of maximum transmitted acceleration $\ddot{x}_m$ for an isolator having a linear spring and viscous damping.

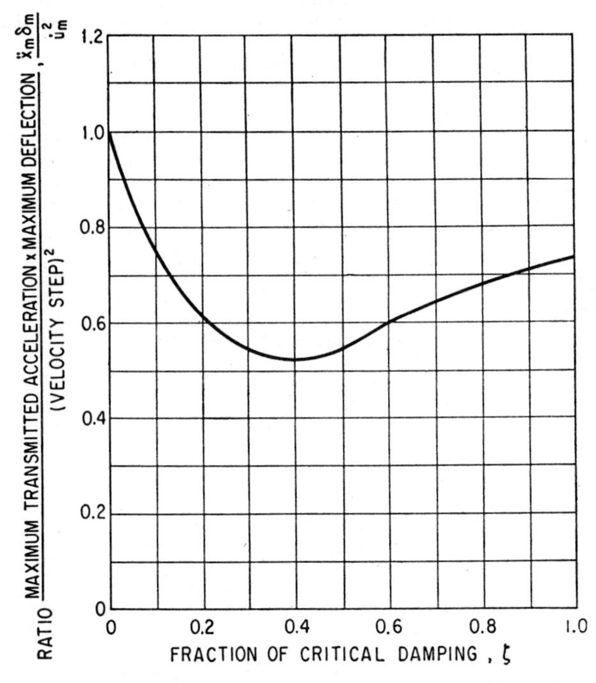

FIG. 31.15. Dimensionless representation of energy absorption capability of an isolator having a linear spring and viscous damping. Ordinate is an inverse measure of energy absorption capability.

system, and even higher values for a hardening spring (see Fig. 31.10). On the other hand, $\ddot{x}_m \delta_m / \dot{u}_m^2$ may approach 0.50 when a softening spring is used.

True viscous damping of the type considered above is difficult to attain except in electrical or magnetic form. Fluid dampers which depend upon orifices or other constricted passages to throttle the flow are likely to produce damping forces that vary more nearly as the square of the velocity. Dry friction tends to provide damping forces which are virtually independent of velocity. The analysis of response to a velocity step in the presence of Coulomb friction is similar to that described in the section entitled *General Formulas—No Damping.* A general analytic method for finding the behavior of systems with nonlinear springs in the presence of velocity-squared damping is also available.[3]

Example 31.1. Equipment weighing 40 lb and sufficiently stiff to be considered rigid is to be protected from a shock consisting of a velocity step $\dot{u}_a = 70$ in./sec. The maximum allowable acceleration is $\ddot{x}_a = 21\,g$ (g is the acceleration of gravity) and available clearance limits the deflection to $\delta_a = 0.70$ in. Find isolator characteristics for: linear spring, hardening spring, softening spring, and linear spring with viscous damping.

Linear Spring. Taking the maximum velocity $\dot{u}_m$ equal to the expected velocity $\dot{u}_a$ and using Eqs. (31.18) and (31.11),

$$\delta_m = \frac{\dot{u}_m}{\omega_n} \leq \delta_a; \qquad \text{or} \qquad \omega_n \geq \frac{70 \text{ in./sec}}{0.70 \text{ in.}} = 100 \text{ rad/sec}$$

From Eqs. (31.19) and (31.12), $\ddot{x}_m = \omega_n \dot{u}_m \leq \ddot{x}_a$. Then

$$\omega_n \leq \frac{\ddot{x}_a}{\dot{u}_m} = \frac{21 \times 386 \text{ in./sec}^2}{70 \text{ in./sec}} = 116 \text{ rad/sec}$$

Selecting a value in the middle of the permissible range gives $\omega_n = 108$ rad/sec [17.2 cps]. The corresponding maximum isolator deflection is $\delta_m = 0.65$ in. and the maximum acceleration of the equipment is $\ddot{x}_m = 7{,}580$ in./sec^2 = 19.6g. The isolator stiffness given by Eq. (31.16) is

$$k = m\omega_n^2 = \frac{40 \text{ lb}}{386 \text{ in./sec}^2} \times (108 \text{ rad/sec})^2 = 1{,}210 \text{ lb/in.}$$

If, as is usually the case, the isolation is provided by several individual isolators in parallel, then the above value of k represents the sum of the stiffnesses of the individual isolators.

Hardening Spring. The tangent elasticity represented by Eq. (31.20) is assumed. Since the linear spring meets the specifications with only a small margin of safety, it is inferred that the poorer energy storage capacity of the hardening spring shown by Fig. 31.10 will severely limit the permissible nonlinearity. Using the specified values as maxima,

$$\frac{\ddot{x}_m \delta_m}{\dot{u}_m^2} = \frac{\ddot{x}_a \delta_a}{\dot{u}_a^2} = \frac{(21 \times 386) \times 0.70}{(70)^2} = 1.16$$

From Fig. 31.10:

$$\frac{\delta_m}{d} = 0.54; \text{ thus } d = \frac{0.70}{0.54} = 1.30 \text{ in.}$$

From Fig. 31.9:

$$\frac{\dot{u}_m}{\omega_n d} = 0.58; \text{ thus } \omega_n = \frac{70}{1.30 \times 0.54} = 93 \text{ rad/sec [14.8 cps]}$$

The initial spring stiffness k from Eq. (31.16) is

$$k = \frac{40}{386}(93)^2 = 896 \text{ lb/in.}$$

Because the selected linear spring provides a small margin of safety and the hardening spring provides none, superficial comparison suggests that the former is superior. Various other considerations, such as compactness and stiffness along other axes, may offset the apparent advantage of the linear spring. Moreover, a shock more severe than that specified could cause the linear spring to bottom abruptly and cause much greater acceleration of the equipment.

Softening Spring. The hyperbolic tangent elasticity represented by Eq. (31.23) is assumed. The softening spring has high energy-storage capacity as shown by Fig. 31.13. By working to sufficiently high values of δ_m/d_1, it is possible to utilize this storage capacity to afford considerable overload capability. Choose $\ddot{x}_m = 20g$ and $\delta_m/d_1 = 3$. From Fig. 31.13, $\ddot{x}_m \delta_m / \dot{u}_m^2 = 0.645$ at $\delta_m/d_1 = 3$. Then

$$\delta_m = 0.645 \frac{(70)^2}{20 \times 386} = 0.41 \text{ in.} \qquad d_1 = \frac{\delta_m}{3} = 0.137 \text{ in.}$$

From Fig. 31.12, $\dot{u}_m/\omega_n d_1 = 2.15$ at $\delta_m/d_1 = 3$.

Then
$$\omega_n = \frac{70}{2.15 \times 0.137} = 238 \text{ rad/sec [37.9 cps]}$$

The initial spring stiffness k from Eq. (31.16) is

$$k = \frac{40}{386}(238)^2 = 5{,}870 \text{ lb/in.}$$

This initial stiffness is much greater than those found for the linear spring and hardening spring. Accordingly, for small shocks (small $\dot{u}_m$) the isolator with the softening spring will induce much higher acceleration of the equipment than will those with linear or hardening springs. This poorer performance for small shocks is unavoidable if the isolator with the softening spring is designed to take advantage of the large energy-storage capability under extreme shocks.

Linear Spring and Viscous Damping. The introduction of viscous damping in combination with a linear spring [Eq. (31.7)] affords the possibility of large energy dissipation capacity without deterioration of performance under small shocks. From Fig. 31.15, the best performance is obtained at the fraction of critical damping $\zeta = 0.40$ where $\ddot{x}_m\delta_m/\dot{u}_m^2 = 0.52$. If the maximum isolator deflection is chosen as $\delta_m = 0.47$ in. (67 per cent of δ_a), then

$$\ddot{x}_m = 0.52 \frac{\dot{u}_m^2}{\delta_m} = 5{,}450 \text{ in./sec}^2 = 14.1g$$

(This acceleration is 67 per cent of $\ddot{x}_a$.) From Fig. 31.14:

$$\frac{\ddot{x}_m}{\dot{u}_m\omega_n} = 0.86 \text{ at } \zeta = 0.40$$

Then
$$\omega_n = \frac{5{,}450}{0.86 \times 70} = 90 \text{ rad/sec [14.3 cps]}$$

The spring stiffness k from Eq. (31.16) is

$$k = \frac{40}{386}(90)^2 = 840 \text{ lb/in.}$$

The dashpot constant c is

$$c = 2\zeta m\omega_n = 2 \times 0.40 \times \frac{40}{386} \times 90 = 7.46 \text{ lb-sec/in.}$$

Example 31.2. The procedure for the analysis of a given isolator subjected to a specified velocity step differs somewhat from that used in design. For the isolators designed in Example 31.2, it is specified that each is subjected to a velocity step $\dot{u}_m = 84$ in./sec (20 per cent greater than in Example 31.1). Determine the maximum isolator deflection δ_m and maximum equipment acceleration $\ddot{x}_m$.

Linear Spring

From Eq (31.18):

$$\delta_m = \frac{\dot{u}_m}{\omega_n} = \frac{84}{108} = 0.78 \text{ in.}$$

From Eq. (31.17):

$$\ddot{x}_m = \omega_n^2\delta_m = (108)^2 \times 0.78 = 9{,}110 \text{ in./sec}^2 = 23.6g$$

The influence of the 20 per cent increase of velocity for this (linear) system is to increase both δ_m and $\ddot{x}_m$ by 20 per cent over the values from Example 31.1.

Hardening Spring. From Example 31.1, $d = 1.30$ in. and $\omega_n = 93$ rad/sec. Then

$$\frac{\dot{u}_m}{\omega_n d} = \frac{84}{93 \times 1.30} = 0.69$$

From Fig. 31.9, the corresponding value of δ_m/d is 0.63. Thus, $\delta_m = 0.63 \times 1.30 = 0.82$ in. From Fig. 31.10, at $\delta_m/d = 0.63$, $\ddot{x}_m\delta_m/\dot{u}_m^2 = 1.26$. Therefore

$$\ddot{x}_m = \frac{1.26 \times (84)^2}{0.82} = 10{,}850 \text{ in./sec}^2 = 28.1g$$

Compared with the results of Example 31.1, δ_m and $\ddot{x}_m$ are 17 per cent and 34 per cent greater, respectively.

Softening Spring. From the given data,

$$\frac{\dot{u}_m}{\omega_n d_1} = \frac{84}{238 \times 0.137} = 2.58$$

From Fig. 31.12 the corresponding value of δ_m/d_1 is 4.0; thus, $\delta_m = 4.0 \times 0.137 = 0.55$ in. From Fig. 31.13, at $\delta_m/d_1 = 4.0$, $\ddot{x}_m\delta_m/\dot{u}_m^2 = 0.603$; thus

$$\ddot{x}_m = \frac{0.603 \times (84)^2}{0.55} = 7{,}730 \text{ in./sec}^2 = 20g$$

Comparing results with those of Example 31.1, δ_m is increased by 34 per cent and $\ddot{x}_m$ is unchanged.

Linear Spring—Viscous Damping. For $\zeta = 0.4$, Fig. 31.14 gives $\ddot{x}_m/\dot{u}_m\omega_n = 0.86$; thus, $\ddot{x}_m = 0.86 \times 84 \times 90 = 6{,}500$ in./sec² $= 16.9g$. From Fig. 31.15, at $\zeta = 0.4$:

$$\frac{\ddot{x}_m\delta_m}{\dot{u}_m^2} = 0.522; \text{ thus, } \delta_m = \frac{0.522 \times (84)^2}{6{,}500} = 0.57 \text{ in.}$$

Again comparing with the results of Example 31.1, both δ_m and $\ddot{x}_m$ are increased by 20 per cent. This is the result of the linearity of the system.

Of the four isolator designs determined in Example 31.1, only the softening spring and the linear spring with viscous damping can meet the original deflection and acceleration limits (0.70 in. and 21g) when the velocity step is increased in magnitude by 20 per cent. This limitation is illustrated by using the specified acceleration (21g) and deflection (0.70 in.) limits, together with the increased velocity (84 in./sec), and calculating

$$\frac{\ddot{x}_m\delta_m}{\dot{u}_m^2} = \frac{(21 \times 386) \times 0.70}{(84)^2} = 0.804$$

According to Fig. 31.10, neither a linear spring nor a hardening spring can be found to meet a required value of $\ddot{x}_m\delta_m/\dot{u}_m^2$ less than unity. On the other hand, Fig. 31.13 shows that hyperbolic tangent springs operating beyond $\delta_m/d_1 = 1.4$ may be found to meet or exceed the requirements. Similarly, Fig. 31.15 shows that viscous damped linear systems with ζ greater than 0.08 can be designed to meet or exceed the requirements.

RESPONSE OF RIGID BODY SYSTEM TO ACCELERATION PULSE

The response of a spring-mounted rigid body to various acceleration pulses provides useful information. For example, it establishes limitations upon the use of the velocity step in place of an acceleration pulse and is significant in determining the response of an equipment component when the equipment support is subjected to a velocity step. Additional useful information is afforded by comparing the responses to acceleration pulses of different shapes.

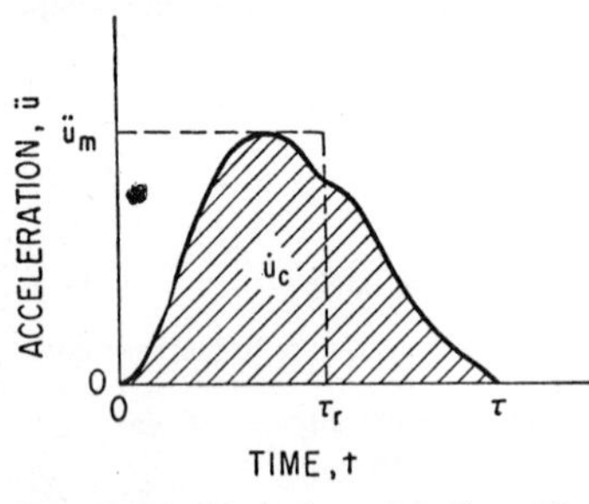

FIG. 31.16. Typical acceleration pulse with maximum acceleration $\ddot{u}_m$ and duration τ. The shaded area represents the resulting velocity change $\dot{u}_c$. The dashed rectangle represents an *equivalent rectangular pulse* of duration τ_r having the same velocity change $\dot{u}_c$.

For positive pulses ($\ddot{u} > 0$) having a single maximum value and finite duration, three basic characteristics of the pulse are of importance: maximum acceleration $\ddot{u}_m$, duration τ, and velocity change $\dot{u}_c$. A typical pulse is shown in Fig. 31.16. The relation among acceleration, duration and velocity change is

$$\dot{u}_c = \int_0^\tau \ddot{u}\, dt \tag{31.27}$$

where the value of the integral corresponds to the shaded area of the figure. The *equivalent rectangular pulse* is characterized by (*a*) the same maximum acceleration $\ddot{u}_m$ and (*b*) the same velocity change $\dot{u}_c$. In Fig. 31.16, the horizontal and vertical dashed lines outline the equivalent rectangular pulse corresponding to the shaded pulse. From condition (*b*) above and Eq. (31.27), the *effective duration* τ_r of the equivalent rectangular pulse is

$$\tau_r = \frac{1}{\ddot{u}_m} \int_0^\tau \ddot{u}\, dt \qquad (31.28)$$

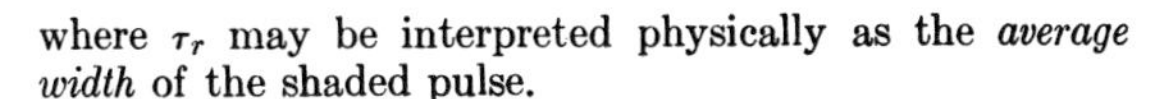

where τ_r may be interpreted physically as the *average width* of the shaded pulse.

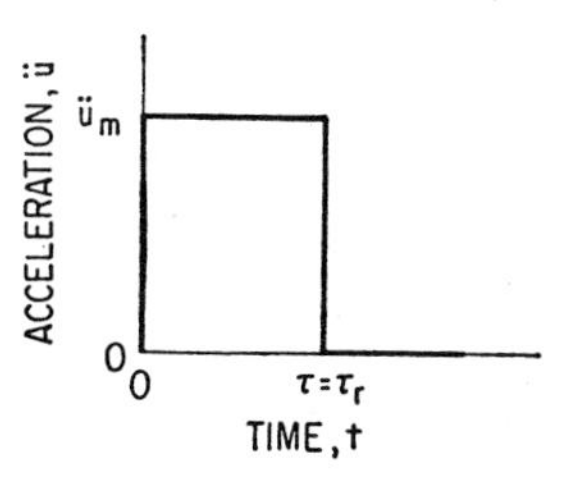

FIG. 31.17. Rectangular acceleration pulse.

RESPONSE TO A RECTANGULAR PULSE

The rectangular pulse shown in Fig. 31.17 has a maximum acceleration $\ddot{u}_m$ and duration τ; the velocity change is $\dot{u}_c = \ddot{u}_m\tau$. The response of an undamped, linear, single degree-of-freedom system (see Fig. 31.6) to this pulse is found from the differential equation obtained by substituting in Eq. (31.8) $F_s(\delta) = k\delta$ from Eq. (31.15) and $\omega_n^2 = k/m$ from Eq. (31.16):

$$\ddot{\delta} + \omega_n^2\delta = -\ddot{u}_m \qquad [0 \le t \le \tau] \qquad (31.29)$$

$$\ddot{\delta} + \omega_n^2\delta = 0 \qquad [t > \tau] \qquad (31.30)$$

Using the initial conditions $\dot{\delta} = 0$, $\delta = 0$ when $t = 0$, the solution of Eq. (31.29) is

$$\delta = \frac{\ddot{u}_m}{\omega_n^2}(\cos \omega_n t - 1) \qquad [0 \le t \le \tau] \qquad (31.31)$$

For the solution of Eq. (31.30), it is necessary to find as initial conditions the values of $\dot{\delta}$ and δ given by Eq. (31.31) for $t = \tau$. Using these values the solution of Eq. (31.30) is

$$\delta = \frac{\ddot{u}_m}{\omega_n^2}[(\cos \omega_n\tau - 1)\cos \omega_n(t - \tau) - \sin \omega_n\tau \sin \omega_n(t - \tau)] \qquad [t > \tau] \qquad (31.32)$$

The motion defined by Eqs. (31.31) and (31.32) is shown graphically in Fig. 31.18 for $\tau = \pi/2\omega_n$, π/ω_n and $3\pi/2\omega_n$.

In the isolation of shock, the extreme absolute acceleration $\ddot{x}_m$ of the mass is important. Since $\ddot{x}_m = \omega_n^2\delta_m$ [Eq. (31.17)], $\ddot{x}_m$ is found directly from the extreme value of δ. As indicated by Fig. 31.18, for values of τ greater than π/ω_n, the extreme (absolute) value of

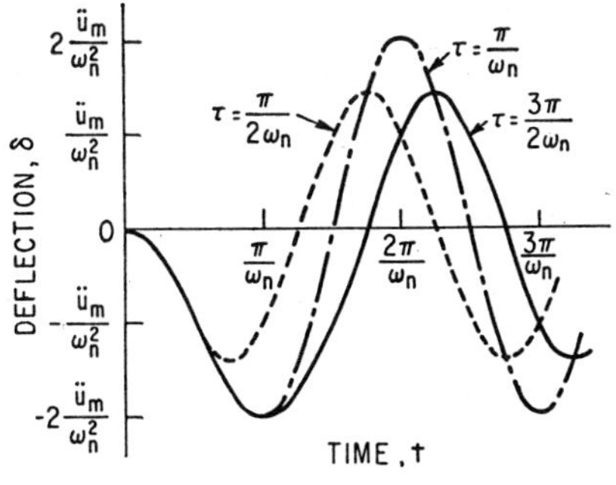

FIG. 31.18. Response curves for an undamped linear system subjected to rectangular acceleration pulses of height $\ddot{u}_m$ and various durations τ. The angular natural frequency of the system is ω_n.

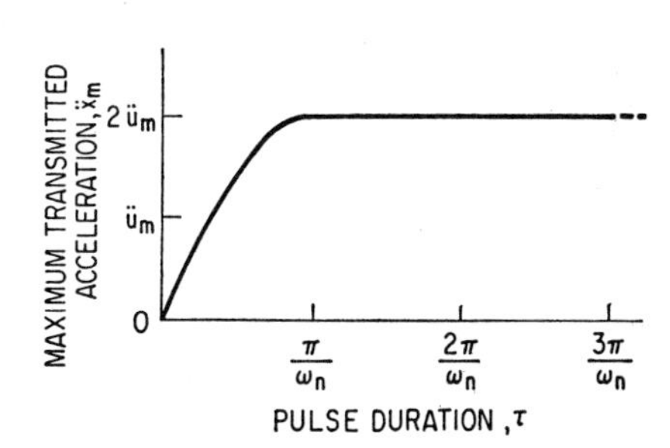

FIG. 31.19. Maximum acceleration spectrum for a linear system of angular natural frequency ω_n. Support motion is a rectangular acceleration pulse of height $\ddot{u}_m$.

δ encountered at $t = \pi/\omega_n$ is never exceeded. For values of τ less than π/ω_n, the extreme value occurs after the pulse has ended ($t > \tau$) and is the amplitude of the motion represented by Eq. (31.32). This amplitude may be written

$$\delta_m = 2\frac{\ddot{u}_m}{\omega_n^2}\sin\frac{\omega_n\tau}{2} \tag{31.33}$$

The extreme absolute values of the acceleration $\ddot{x}_m$ are plotted as a function of τ in Fig. 31.19. Note that the extreme value of acceleration is twice that of the acceleration of the rectangular pulse.

HALF-SINE PULSE

Consider the "half-sine" acceleration pulse (Fig. 31.20*A*) of amplitude $\ddot{u}_m$ and duration τ:

$$\begin{aligned} \ddot{u} &= \ddot{u}_m \sin\frac{\pi t}{\tau} \qquad [0 \le t \le \tau] \\ \ddot{u} &= 0 \qquad [t > \tau] \end{aligned} \tag{31.34}$$

From Eq. (31.28), the effective duration is

$$\tau_r = \frac{2}{\pi}\tau \tag{31.35}$$

The response of a single degree-of-freedom system to the half-sine pulse of acceleration, corresponding to Eqs. (31.31) and (31.32) for the rectangular pulse, is defined by Eq. (8.32).

VERSED SINE PULSE

The versed sine pulse (Fig. 31.20*B*) is described by

$$\begin{aligned} \ddot{u} &= \frac{\ddot{u}_m}{2}\left(1 - \cos\frac{2\pi t}{\tau}\right) = \ddot{u}_m \sin^2\frac{\pi t}{\tau} \qquad [0 \le t \le \tau] \\ \ddot{u} &= 0 \qquad [t > \tau] \end{aligned} \tag{31.36}$$

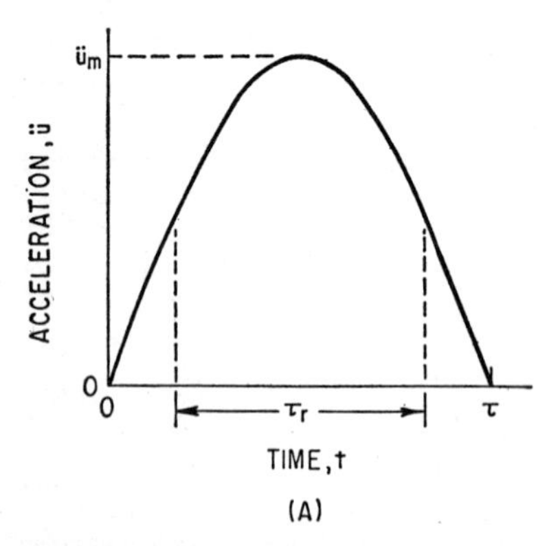

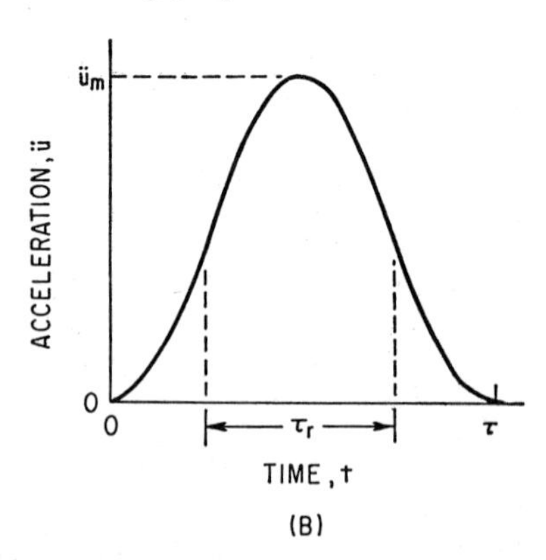

FIG. 31.20. Half-sine acceleration pulse (*A*) and versed sine acceleration pulse (*B*). The *effective duration* is τ_r [Eq. (31.28)].

The effective duration τ_r given by Eq. (31.28) is

$$\tau_r = (\tfrac{1}{2})\tau \tag{31.37}$$

The response of a single degree-of-freedom system to a versed sine pulse is defined by Eq. (8.33). The responses to a number of other types of pulse and step excitation also are defined in Chap. 8.

COMPARISON OF MAXIMUM ACCELERATIONS

VELOCITY STEP APPROXIMATION. A comparison of values of $\ddot{x}_m$ resulting from various acceleration pulses with that resulting from a velocity step is shown in Fig. 31.21.

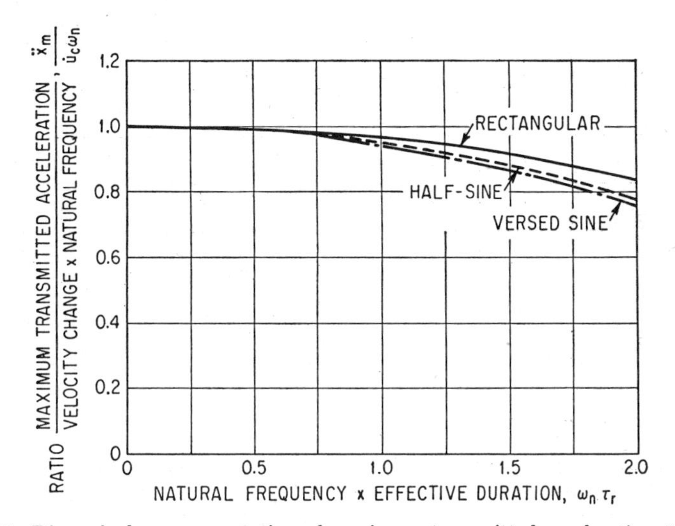

FIG. 31.21. Dimensionless representation of maximum transmitted acceleration $\ddot{x}_m$ for the undamped linear system of Fig. 31.6. Support motion is a rectangular, half-sine, or versed sine acceleration pulse.

The maximum acceleration induced by a velocity step is $\omega_n \dot{u}_m$ [see Eq. (31.19)]. The abscissa $\omega_n\tau_r$ is a dimensionless measure of pulse duration. The effect of pulse shape is imperceptible for values of $\omega_n\tau_r < 0.6$. For pulses of duration $\omega_n\tau_r < 1.0$, the effect of pulse shape is small and the maximum possible error resulting from use of the velocity step approximation is of the order of 5 per cent.

EFFECTS OF PULSE SHAPE. The effects of pulse shape upon the maximum response acceleration $\ddot{x}_m$ for values of $\omega_n\tau_r > 1.0$ are shown in Fig. 31.22. The ordinate $\ddot{x}_m/\ddot{u}_m$ is the ratio of maximum acceleration induced in the responding system to maximum acceleration of the pulse. All three pulses produce the highest value of response acceleration when $\omega_n\tau_r \simeq \pi$. Physically, this corresponds to an effective duration τ_r of one-half of the natural period of the spring-mass system. For longer pulse durations the curves for half-sine and versed sine pulses are similar. For pulse durations beyond the range of Fig. 31.22 ($\omega_n\tau_r > 16$), the half-sine and versed sine curves approach the limiting ordinate $\ddot{x}_m/\ddot{u}_m = 1$. This corresponds physically to approximating a static loading of the spring-mass system. A limiting acceleration ratio $\ddot{x}_m/\ddot{u}_m = 2$ is encountered for all rectangular pulses of duration greater than the half-period of the spring-mass system. A more extensive study of responses to a variety of pulse shapes is included in Chap. 8.

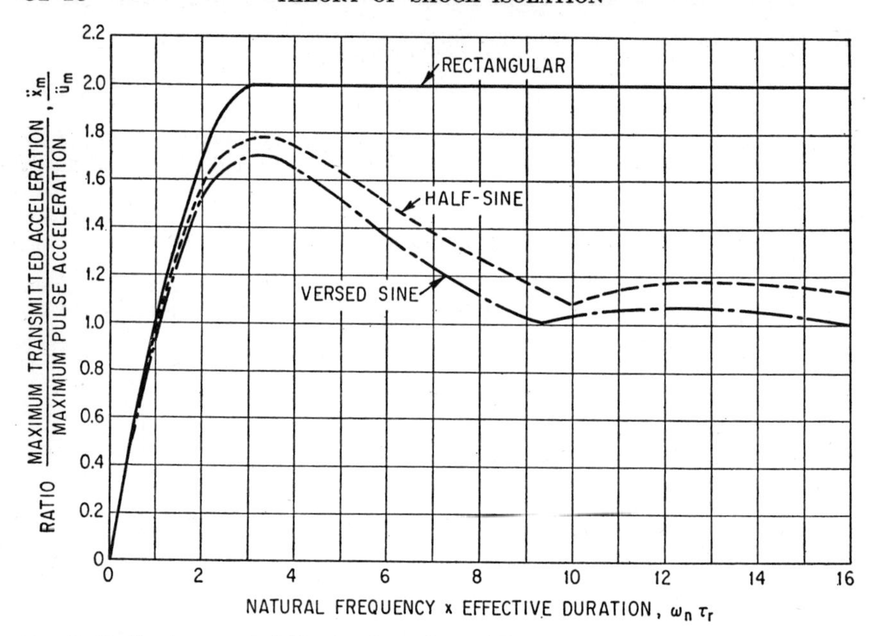

FIG. 31.22. Shock transmissibility for the undamped linear system of Fig. 31.6 as a function of angular natural frequency ω_n and effective pulse duration τ_r. Support motion is a rectangular, half-sine, or versed sine acceleration pulse.

SHOCK SPECTRUM

The abscissa $\omega_n\tau_r$ in Fig. 31.22 may be treated as a measure of pulse duration (proportional to τ_r) for a given spring-mass system with ω_n fixed. Alternatively, the pulse duration may be considered fixed; then the curves show the effect of varying the natural frequency ω_n of the spring-mass system. Each of the curves of Fig. 31.22 shows the maximum acceleration induced by a given acceleration pulse upon spring-mass systems of various natural frequencies ω_n; thus, Fig. 31.22 may be used to determine the required natural frequency of the isolator if $\ddot{x}_m$ and $\ddot{u}_m$ are known, and the pulse shape is defined.

Each curve shown in Fig. 31.22 may be interpreted as a description of a pulse, in terms of the response induced in a system subjected to the pulse. The curve of maximum response as a function of the natural frequency of the responding system is called a shock spectrum or response spectrum. This concept is discussed more fully in Chap. 23. A pulse is a particular form of a shock motion; thus, each shock motion has a characteristic shock spectrum. A shock motion has a characteristic effective value of time duration τ_r which need not be defined specifically; instead, the spectra are made to apply explicitly to a given shock motion by using the natural frequency ω_n as a dimensional parameter on the abscissa. By taking the isolator-and-equipment assembly to be the responding system, the natural frequency of the isolator may be chosen to meet any specified maximum acceleration $\ddot{x}_m$ of the equipment supported by isolators. Spectra of maximum isolator deflection δ_m also may be drawn, and are useful in predicting the maximum isolator deflection when the natural frequency of the isolator is known.

When damping is added to the isolator, the analysis of the response becomes much more complex. In general, it is possible to determine the maximum value of the response acceleration $\ddot{x}_m$ only by calculating the time-history of response acceleration over the entire time interval suspected of including the maximum response. A digital computer has been used to find shock spectra for "half-sine" acceleration pulses with various

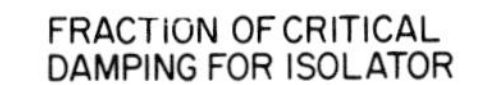

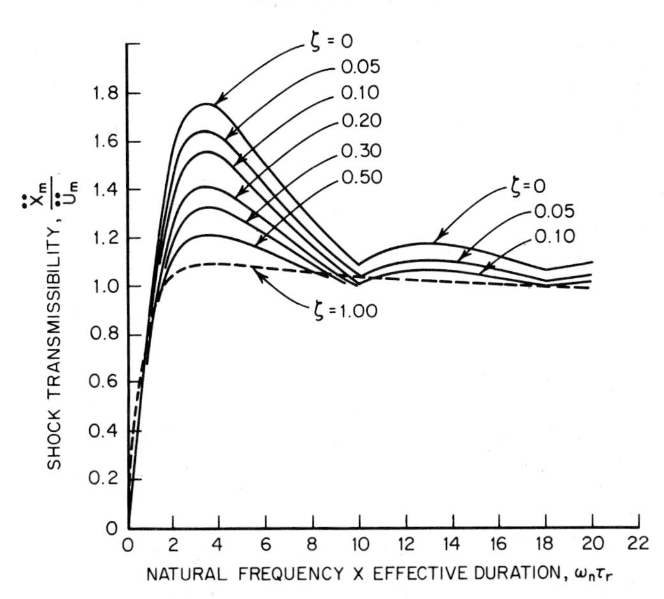

FIG. 31.23. Shock transmissibility for the system of Fig. 31.2*A* with linear spring and viscous damping. Support motion is a half-sine acceleration pulse of height $\ddot{u}_m$ and effective duration τ_r. Curves are for discrete values of the fraction of critical damping ζ in the isolator as indicated.

fractions of critical damping in the responding system, as shown in Fig. 31.23. Similar spectra could be obtained to indicate maximum values of isolator deflection. In selecting a shock isolator for a specified application, it may be necessary to use both maximum acceleration and maximum deflection spectra. This is illustrated in the following example.

Example 31.3. A piece of equipment weighing 230 lb is to be isolated from the effects of a vertical shock motion defined by the spectra of acceleration and deflection shown in Fig. 31.24. It is required that the maximum induced acceleration not exceed $7g$ (2,700 in./sec²). Clearances available limit the isolator deflection to 2.25 in. The curves in Fig. 31.24*A* represent maximum response acceleration $\ddot{x}_m$ as a function of the angular natural frequency ω_n of the equipment supported on the shock isolators. The isolator springs are assumed linear and viscously damped, and separate curves are shown for values of the damping ratio $\zeta = 0$, 0.1, 0.2, and 0.3. The curves in Fig. 31.24*B* represent the maximum isolator deflection δ_m as a function of ω_n for the same values of ζ.

Consider first the requirement that $\ddot{x}_m < 2{,}700$ in./sec². In Fig. 31.24*A*, the horizontal dashed line indicates this limiting acceleration. If the damping ratio $\zeta = 0.3$, then the angular natural frequency ω_n may not exceed 38.5 rad/sec on the criterion of maximum acceleration. The dashed horizontal line of Fig. 31.24*B* represents the deflection limit $\delta_m = 2.25$ in. For $\zeta = 0.3$, the minimum natural frequency is 30 rad/sec on the criterion of deflection. Considering both acceleration and deflection criteria, the angular natural frequency ω_n must lie between 30 rad/sec and 38.5 rad/sec. The spectra indicate that both criteria may be just met with $\zeta = 0.2$ if ω_n is 35 rad/sec. Smaller values of damping do not permit the satisfaction of both requirements.

Conservatively, a suitable choice of parameters is $\zeta = 0.3$, $\omega_n = 35$ rad/sec. This limits

$\ddot{x}_m$ to 2,500 in./sec² and δ_m to 2.0 in. The spring stiffness k is

$$k = \omega_n^2 m = (35)^2 \times \frac{230}{386} = 730 \text{ lb/in.}$$

If the equipment is to be supported by four like isolators, then the required stiffness of each isolator is $k/4 = 182.5$ lb/in.

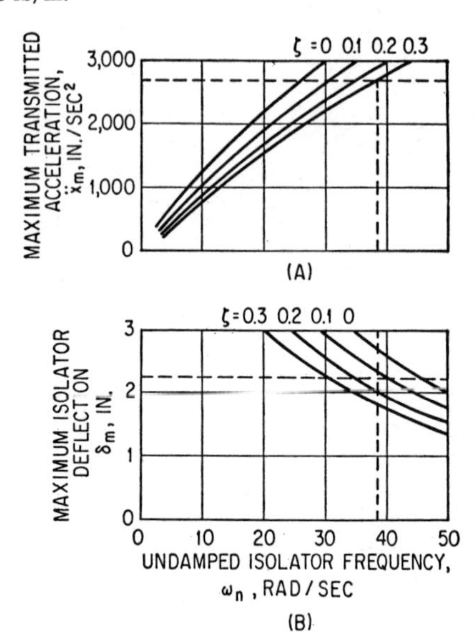

FIG. 31.24. Shock spectra: (*A*) maximum acceleration and (*B*) maximum isolator deflection for Example 31.3.

RESPONSE OF EQUIPMENT WITH A FLEXIBLE COMPONENT

IMPACT WITH REBOUND

Consider the system of Fig. 31.4. The block of mass m_1 represents the equipment and m_2 with its associated spring-dashpot unit represents a critical component of the equipment. The left spring-dashpot unit represents the shock isolator. It is assumed here that $m_1 \gg m_2$ so that the motion of m_1 is not sensibly affected by m_2; larger values of m_2 are considered in a later section. Consider the entire system to be moving to the left at uniform velocity when the left-hand end of the isolator strikes a fixed support (not shown). The isolator will be compressed until the equipment is brought to rest. Following this the compressive force in the isolator will continue to accelerate the equipment toward the right until the isolator loses contact with the support and the rebound is complete. This type of shock is called *impact with rebound*. Practical examples include the shock experienced by a single railroad car striking a bumper at the end of a siding and that experienced by packaged equipment, shock-mounted inside a container of small mass, when the container is dropped upon a hard surface and then rebounds.

The procedure for finding the maximum acceleration $\ddot{x}_{2m}$ of the component, assuming the component stiffness to be linear and neglecting component damping, is:

1. Using the known striking velocity determine, from velocity step results (Figs. 31.9, 31.10, 31.12 to 31.15), the maximum deflection δ_{1m} of the isolator and the maximum acceleration $\ddot{x}_{1m}$ of the equipment.

2. From Eq. (31.28), find the effective duration τ_r for the acceleration time-history $\ddot{x}_1(t)$ of the equipment.

3. From the shock spectra corresponding to the acceleration pulse $\ddot{x}_1(t)$, find the maximum acceleration $\ddot{x}_{2m}$ of the component.

Details of the procedure using the isolators of Example 31.1 are considered in Example 31.4.

Example 31.4. Let the equipment of Example 31.1 weighing 40 lb have a flexible component weighing 0.2 lb. By vibration testing, this component is found to have an angular natural frequency $\omega_n = 260$ rad/sec and to possess negligible damping. For the isolators of Example 31.1, it is desired to determine the maximum acceleration $\ddot{x}_{2m}$ experienced by the mass m_2 of the component if the equipment, traveling at a velocity of 70 in./sec, is arrested by the free end of the isolator striking a fixed support. The four cases are considered separately. It is assumed that the component has a negligible effect on the motion of the equipment because $m_2 \ll m_1$.

Linear Spring. From the results of Example 31.1, it is known that $\omega_n = 108$ rad/sec and that the maximum acceleration of the equipment as found from Eq. (31.19) is

$$\ddot{x}_{1m} = 7{,}580 \text{ in./sec}^2 = 19.6g$$

This acceleration occurs at the instant when the isolator deflection has the extreme value $\delta_{1m} = 0.65$ in.* Subsequently the isolator spring continues to accelerate the equipment until the isolator force is zero and the rebound is complete. Since there is no damping, the rebound velocity equals the striking velocity (with opposite sign). The velocity change $\dot{x}_{1c}$ is twice the striking velocity and the effective duration τ_r [Eq. (31.28)] is

$$\tau_r = \frac{\dot{x}_{1c}}{\ddot{x}_{1m}} = \frac{2 \times 70}{7{,}580} = 0.0185 \text{ sec}$$

The acceleration time-history of the equipment is a half-sine pulse as represented in Fig. 31.20 (the ordinate is $\ddot{x}_1$ instead of $\ddot{u}$).

Since the equipment is the "support" for the component, the response of the latter may be found from results developed for the response of a rigid body whose support experiences a half-sine pulse of acceleration. The half-sine curve of Fig. 31.22 gives the desired information if the following interpretations are made: For $\ddot{x}_m/\ddot{u}_m$ read $\ddot{x}_{2m}/\ddot{x}_{1m}$; for $\omega_n\tau_r$ read $\omega_{n2}\tau_r$. Now $\omega_{n2}\tau_r = 260 \times 0.0185 = 4.80$. From Fig. 31.22, $\ddot{x}_{2m}/\ddot{x}_{1m} = 1.66$, and $\ddot{x}_{2m} = 1.66 \times 7{,}580 = 12{,}600$ in./sec^2 $= 32.6g$.

Hardening Spring. From Example 31.1, the maximum equipment acceleration is $\ddot{x}_{1m} = 21g = 8{,}100$ in./sec^2. Since the velocity change $\dot{x}_{1c}$ is twice the striking velocity, the effective duration τ_r [Eq. (31.28)] is

$$\tau_r = \frac{\dot{x}_{1c}}{\ddot{x}_{1m}} = \frac{2 \times 70}{8{,}100} = 0.0173 \text{ sec}$$

With a hardening isolator spring, the shape of the acceleration pulse $\ddot{x}_1(t)$ experienced by the equipment varies considerably as the maximum deflection δ_{1m} approaches the upper limit d. Up to $\delta_{1m}/d = 0.5$, the shape is closely approximated by a half-sine pulse. For $\delta_{1m}/d = 0.8$, a symmetric triangular pulse is a good approximation. For higher values of δ_{1m}/d, the pulse is very sharply peaked. The maximum response curve for a half-sine pulse is given in Fig. 31.22. The corresponding curve for a symmetric triangular pulse (Fig. 8.18*b*) is similar to that for the versed sine pulse, though lying generally below the latter. Inasmuch as the curve for the versed sine pulse is below that for the half-sine pulse, it is conservative to use the half-sine pulse for all values of δ_{1m}/d. Accordingly, $\omega_{n2}\tau_r = 260 \times 0.0173 = 4.50$. From the half-sine curve of Fig. 31.22, $\ddot{x}_{2m}/\ddot{x}_{1m} = 1.69$, and $\ddot{x}_{2m} = 1.69 \times 8{,}1000 = 13{,}700$ in./sec^2 $= 36.4g$.

Softening Spring. From Example 31.1, the maximum equipment acceleration $\ddot{x}_{1m}$ is

$$\ddot{x}_{1m} = 20g = 7{,}720 \text{ in./sec}^2$$

The effective duration τ_r [Eq. (31.28)] is

$$\tau_r = \frac{\dot{x}_{1c}}{\ddot{x}_{1m}} = \frac{2 \times 70}{7{,}720} = 0.0181 \text{ sec}$$

The shape of the acceleration pulse $\ddot{x}_1(t)$ for the equipment varies markedly as the departure from linearity increases (increasing values of δ_{1m}/d_1). The pulse shape is found by first performing the integration of Eq. (31.9) with $F_s(\delta)$ as given by Eq. (31.23). The result supplies

* If the equipment (Fig. 31.4) is moving toward the left when the isolator contacts the support, the extreme value of δ_{1m} is negative. It suffices to deal here with absolute values.

the integrand required for Eq. (31.13) The integration of the latter equation is performed numerically. Results [1] show that the pulse shape undergoes a rapid transition from the half-sine pulse at very small values of δ_{1m}/d_1 to shapes that are closely approximated by the trapezoidal pulse of Fig. 31.25. Note that the pulse of Fig. 31.25 requires three parameters to fix it completely: the maximum acceleration $\ddot{x}_{1m}$; the effective duration τ_r; and the ratio τ_r/τ where τ is the actual duration and $\tau_r = \tau - \tau_1$. From results of the numerical integra-

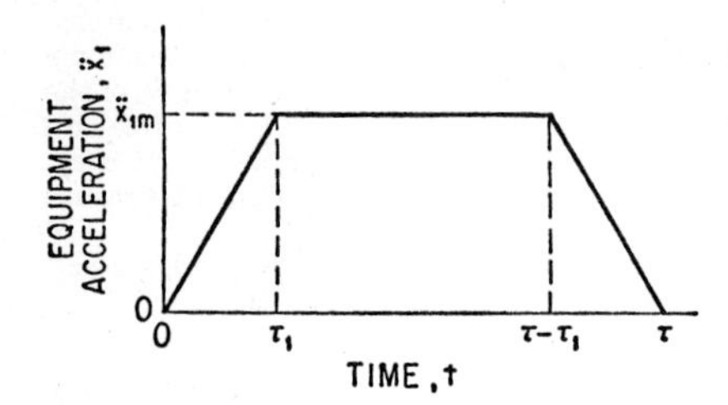

FIG. 31.25. Symmetric trapezoidal acceleration pulse.

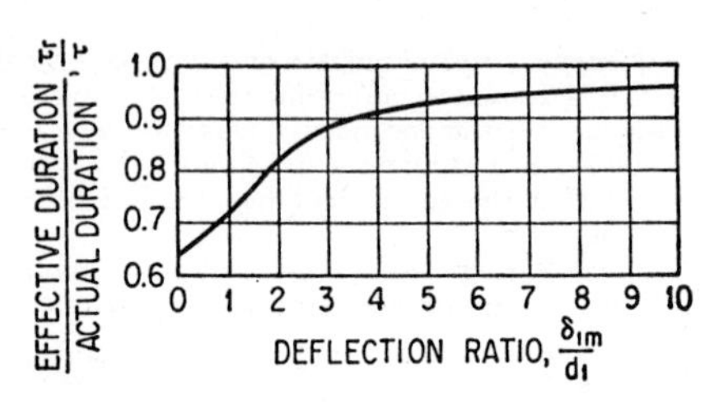

FIG. 31.26. Dimensionless representation of effective duration τ_r of acceleration pulse experienced by equipment during impact with rebound. Isolator has undamped hyperbolic tangent elasticity. Maximum isolator deflection is δ_{1m} and characteristic deflection d_1 is defined in Fig. 31.11.

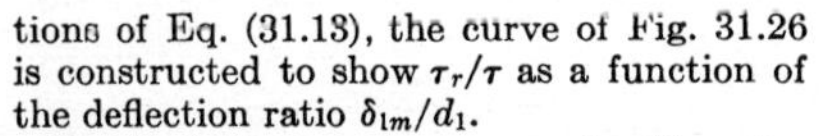

tions of Eq. (31.13), the curve of Fig. 31.26 is constructed to show τ_r/τ as a function of the deflection ratio δ_{1m}/d_1.

To find the maximum acceleration $\ddot{x}_{2m}$ of the component, the maximum response curves (shock spectra) of Fig. 31.27 are used. These curves are constructed for symmetric trapezoidal pulses (Fig. 31.25). The top curve ($\tau_r/\tau = 1.0$) corresponds to the limiting (rectangular) form. The dashed curve ($\tau_r/\tau = 0.64$) represents response to a half-sine pulse.

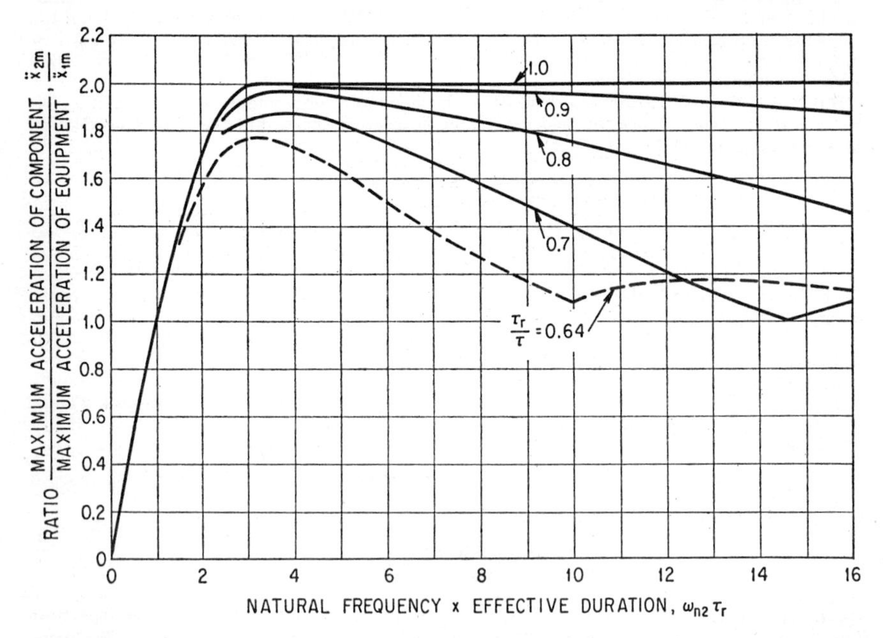

FIG. 31.27. Shock spectra for component having undamped linear elasticity with angular natural frequency ω_{n2}. Equipment motion is symmetric trapezoidal acceleration pulse (Fig. 31.25). Curves are for five discrete values of the ratio of effective pulse duration to actual duration τ_r/τ.

The value of δ_{1m}/d_1 corresponding to the maximum acceleration $\ddot{x}_{1m}$ of the equipment is (from Example 31.1) $\delta_{1m}/d_1 = 3$. From Fig. 31.26: $\tau_r/\tau = 0.88$. Now $\omega_n\tau_r = 260 \times 0.0181 = 4.7$. Using Fig. 31.27, linear interpolation between the curves for $\tau_r/\tau = 0.8$ and $\tau_r/\tau = 0.9$ gives $\ddot{x}_{2m}/\ddot{x}_{1m} = 1.98$ and $\ddot{x}_{2m} = 1.98 \times 7{,}720 = 15{,}300$ in./sec^2 = $39.6g$.

Linear Spring and Viscous Damping. The presence of damping in the isolator adds several complications: (1) the rebound velocity is no longer equal to the striking velocity; (2) the acceleration pulse of the equipment is not symmetrical and returns to zero before the isolator deformation δ_{1m} returns to zero; and (3) the pulse shape varies greatly with damping ratio ζ_1. Shock spectra for acceleration pulse shapes corresponding to damping ratios of particular interest ($0.10 < \zeta_1 < 0.40$) are not available. However, for single acceleration pulses which do not change sign, it is conservative to assume that the maximum acceleration $\ddot{x}_{2m}$ of the component is twice the maximum acceleration $\ddot{x}_{1m}$ of the equipment. Using the results of Example 31.1, the maximum acceleration of the component is $\ddot{x}_{2m} = 2\ddot{x}_{1m} = 2 \times 5{,}450 = 10{,}900$ in./sec^2 = $28.2g$.

IMPACT WITHOUT REBOUND

When impact of the isolator occurs without rebound, it must be recognized that the equipment-and-isolator system continues to oscillate until the initial kinetic energy is dissipated. Consider the system of Fig. 31.4; it consists of equipment m_1, shock isolator (left spring-dashpot unit), and flexible component (subsystem 2). The system is initially at rest. The left end of the shock isolator is attached to a support (not shown) which is given a velocity step of magnitude $\dot{u}_m$ at $t = 0$. The subsequent motion of the support is $u = \dot{u}_m t$. Determine the maximum force F_{1m} transmitted by the isolator, the maximum isolator deflection δ_{1m}, and the maximum acceleration $\ddot{x}_{2m}$ of the component.

Solutions are available only for linear systems, i.e., linear springs and viscous damping. Two such simplified analyses of this problem are included in the following sections: (1) The influence of damping is considered, but the component mass m_2 is assumed of negligible size relative to m_1 and (2) damping is neglected but the effect of the mass m_2 of the component upon the motion of the system is considered.

COMPONENT MASS NEGLIGIBLE. Assume that $m_1 \gg m_2$ so that the motion x_1 of the equipment may be determined by neglecting the effect of the component. Then the extreme value of the force F_{1m} transmitted by the isolator and the extreme deflection δ_{1m} of the isolator occur during the first quarter-cycle of the equipment motion; they may be found from Figs. 31.14 and 31.15 in the section on *Response of a Rigid Body System to a Velocity Step*. The subsequent motion of the equipment is an exponentially decaying sinusoidal oscillation or, if there is no damping in the isolator, a constant-amplitude oscillation. If the component also is undamped, an analytic determination of the component response is not difficult. The motion consists of harmonic oscillation at the frequency ω_{n1} of the equipment oscillation and a superposed oscillation at the frequency ω_{n2} of the component system. Since the oscillations are assumed to persist indefinitely in the absence of damping, the extreme acceleration of the component is the sum of the absolute values of the maximum accelerations associated with the oscillations at frequencies ω_{n1} and ω_{n2}. In the particular case of resonance ($\omega_{n1} = \omega_{n2}$), the vibration amplitude of the component increases indefinitely with time. Because actual systems always possess damping (usually a considerable amount in the isolator), solutions of this type tend to be unduly conservative for engineering applications.

The equation of motion for the viscous damped component is a special case of Eq. (31.5) with $F(\dot{\delta},\delta)$ as given by Eq. (31.7). If appropriate subscripts are supplied and customary substitutions are made, the equation is

$$\ddot{\delta}_2 + 2\zeta_2\omega_{n2}\dot{\delta}_2 + \omega_{n2}^2\delta_2 = -\ddot{x}_1 \tag{31.38}$$

Analytic solutions of Eq. (31.38) to find the acceleration $\ddot{x}_2 = \ddot{x}_1 + \ddot{\delta}_2$ of the component are too laborious to be practical. However, results have been obtained by analog computation [1,4] and are shown in Fig. 31.28. The ordinate is the ratio of the maximum acceleration $\ddot{x}_{2m}$ of the component to the maximum acceleration $\dot{u}_m\omega_{n2}$ [see Eq. (31.19)] that the component would experience if the shock isolator were rigid. The abscissa is the ratio of the undamped natural frequency ω_{n2} of the component to the undamped natural frequency ω_{n1} of the equipment on the isolator spring. Curves are given for several different values of the fraction of critical damping ζ_1 for the isolator. For all curves the fraction of critical damping for the component is $\zeta_2 = 0.01$. The effect of isolator damping in reducing the maximum acceleration $\ddot{x}_{1m}$ of the component is great in the neighborhood of $\omega_{n2}/\omega_{n1} = 1$. Above $\omega_{n2}/\omega_{n1} = 2$, small damping ($\zeta_1 \leq 0.1$) in the isolator has little effect and large damping may significantly increase the maximum acceleration of the component.

The ordinate in Fig. 31.28 represents the ratio of the maximum acceleration of the component to that which would be experienced with the isolator rigid (absent); thus, it may properly be called *shock transmissibility*. If shock transmissibility is less than unity, the isolator is beneficial (for the component considered). An isolator must have a natural frequency significantly less than that of the critical component in order to reduce the transmitted acceleration. If there are several critical components having different natural frequencies ω_{n2}, each must be considered separately and the natural frequency of the isolator must be significantly lower than the lowest natural frequency of a component.

TWO DEGREES-OF-FREEDOM—NO DAMPING. This section includes an analysis of the transient response of the two degree-of-freedom system shown in Fig. 31.4, neglecting the effects of damping but assuming the equipment mass m_1 and the component mass m_2 to be of the same order of magnitude. The equations of motion are

$$\begin{aligned} m_1\ddot{\delta}_1 + k_1\delta_1 &= k_2\delta_2 - m_1\ddot{u} \\ m_2\ddot{\delta}_2 + k_2\delta_2 &= -m_2\ddot{\delta}_1 - m_2\ddot{u} \end{aligned} \tag{31.39}$$

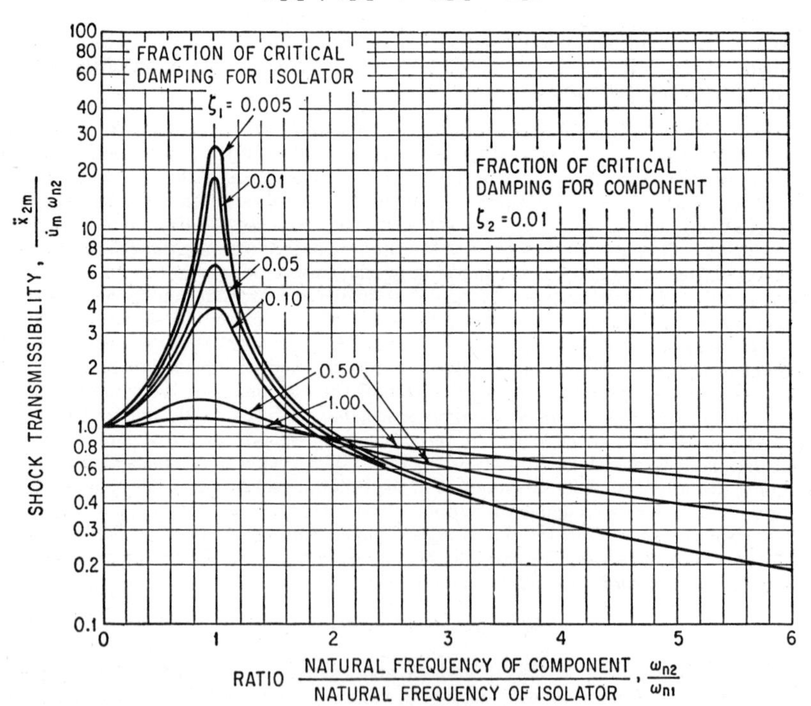

FIG. 31.28. Shock transmissibility for a component of a viscously damped system with linear elasticity. The maximum acceleration of the component is $\ddot{x}_{2m}$ and the velocity step (support motion) is $\dot{u}_m$. The effect of the component on the equipment motion is neglected. (*After R. D. Mindlin.*[1])

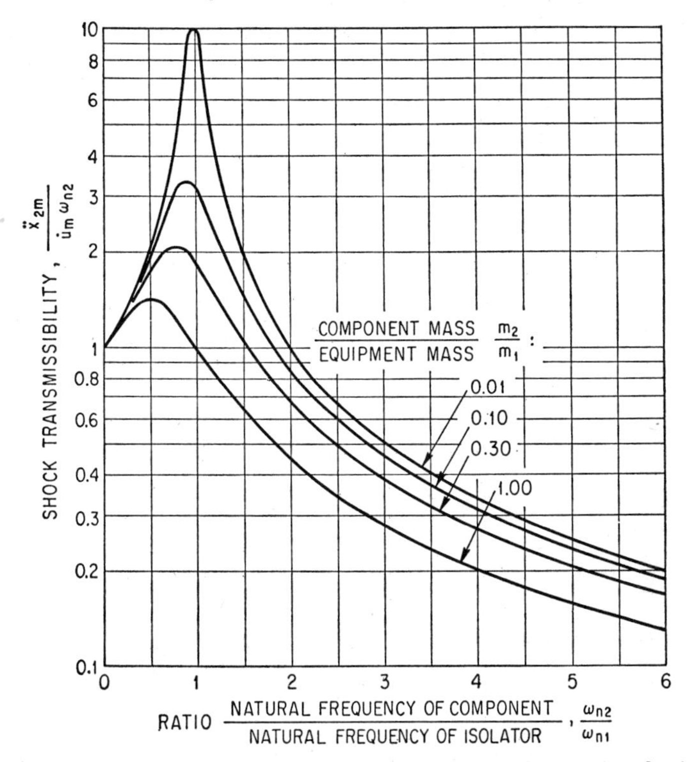

FIG. 31.29. Shock transmissibility for component of system of Fig. 31.4 under impact at velocity $\dot{u}_m$ without rebound. Component and isolator have undamped linear elasticity. Ordinate is ratio of maximum acceleration of component $\ddot{x}_{2m}$ to maximum acceleration which would be experienced by component with a rigid isolator. Curves are for four discrete values of the mass ratio m_2/m_1 as indicated.

where k_1 = stiffness of isolator spring, lb/in., and k_2 = stiffness of component, lb/in. The system is initially in equilibrium; at time $t = 0$, the left end of the isolator spring is given a velocity step of magnitude $\dot{u}_m$. Initial conditions are: $\dot{\delta}_1 = \dot{u}_m$, $\dot{\delta}_2 = 0$, $\delta_1 = \delta_2 = 0$. Equations (31.39) may be solved simultaneously [2] for maximum values of the acceleration $\ddot{x}_{2m}$ of the component and maximum deflection δ_{1m} of the isolator:

$$\ddot{x}_{2m} = \frac{\dot{u}_m \omega_{n2}}{\left[\left(\frac{\omega_{n2}}{\omega_{n1}} - 1\right)^2 + \frac{m_2}{m_1}\left(\frac{\omega_{n2}}{\omega_{n1}}\right)^2\right]^{1/2}} \tag{31.40}$$

$$\delta_{1m} = \frac{\dot{u}_m}{\omega_{n1}} \frac{1 + \frac{\omega_{n2}}{\omega_{n1}}\left(1 + \frac{m_2}{m_1}\right)}{\left[\left(\frac{\omega_{n2}}{\omega_{n1}} + 1\right)^2 + \frac{m_2}{m_1}\left(\frac{\omega_{n2}}{\omega_{n1}}\right)^2\right]^{1/2}} \tag{31.41}$$

where $\ddot{x}_{2m}$ = maximum absolute acceleration of component mass, in./sec^2; δ_{1m} = maximum deflection of isolator spring, in.; ω_{n1} * = angular natural frequency of isolator $(k_1/m_1)^{1/2}$, rad/sec; and ω_{n2} * = angular natural frequency of component $(k_2/m_2)^{1/2}$, rad/sec. Equation (31.40) is shown graphically in Fig. 31.29. The dimensionless ordi-

* The natural frequencies ω_{n1} and ω_{n2} are hypothetical in the sense that they do not consider the coupling between the subsystems.

nate is the ratio of maximum acceleration $\ddot{x}_{2m}$ of the component to the maximum acceleration $\dot{u}_m\omega_{n2}$ which the component would experience with no isolator present. The abscissa is the ratio of component natural frequency ω_{n2} to isolator natural frequency ω_{n1}. Separate curves are given for mass ratios $m_2/m_1 = 0.01$, 0.1, 0.3, and 1.0. Equation (31.41) is shown graphically in Fig. 31.30. The ordinate is the ratio of the maximum isolator deflection δ_{1m} to the deflection $\dot{u}_m(1 + m_2/m_1)^{1/2}/\omega_{n1}$ which would occur if component stiffness k_2 were infinite. The abscissa is the ratio of natural frequencies ω_{n2}/ω_{n1}, and curves are given for values of $m_2/m_1 = 0.1$ and 1.0.

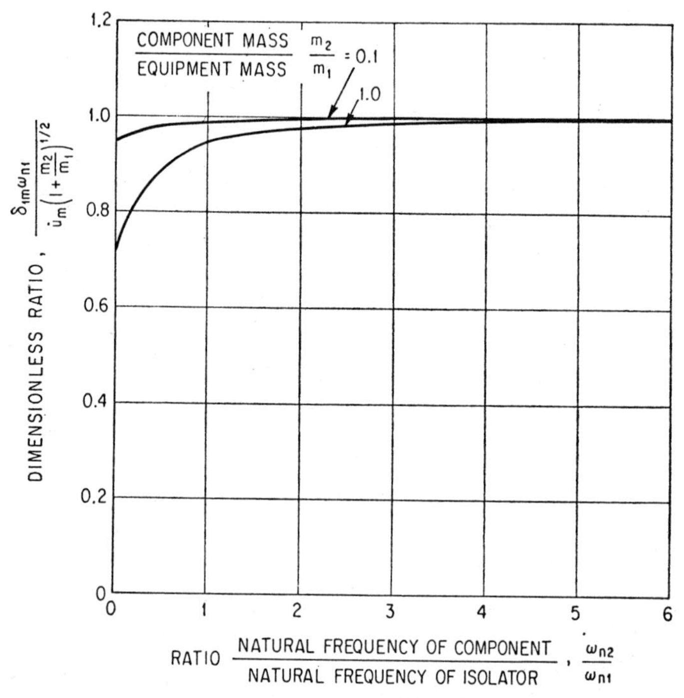

FIG. 31.30. Dimensionless representation of maximum isolator deflection in system of Fig. 31.4 under impact at velocity $\dot{u}_m$ without rebound. Component and isolator have undamped linear elasticity. Ordinate is ratio of maximum isolator deflection δ_{1m} to the isolator deflection which would result if the component stiffness were infinite. Curves are for two discrete values of the mass ratio m_2/m_1 as indicated.

Figure 31.29 shows that the effect of the mass ratio m_2/m_1 upon the maximum component acceleration $\ddot{x}_{2m}$ is very great near resonance ($\omega_{n2}/\omega_{n1} \simeq 1$). As ω_{n2}/ω_{n1} increases above resonance, the effect of finite component mass steadily decreases. Figure 31.30 shows that except for small values of ω_{n2}/ω_{n1} the effect of finite component mass on the maximum isolator deflection δ_{1m} is slight. As ω_{n2}/ω_{n1} increases, the curves for all mass ratios asymptotically approach the ordinate 1.0.

The factor $(1 + m_2/m_1)^{1/2}$ in the ordinate parameter of Fig. 31.30 is introduced because the total equipment mass is $m_1 + m_2$. For the limiting case of rigid equipment (k_2 infinite), the natural frequency ω_n is given by

$$\omega_n^2 = \frac{k_1}{m_1 + m_2} \qquad \omega_n = \frac{\omega_{n1}}{(1 + m_2/m_1)^{1/2}}$$

Substituting this relation in Eq. (31.18) and solving for δ_{1m}:

$$\delta_{1m} = \dot{u}_m(1 + m_2/m_1)^{1/2}/\omega_{n1}$$

This is in agreement with the result given by Eq. (31.41) as ω_{n2}/ω_{n1} approaches infinity.

Example 31.5. Equipment weighing 152 lb has a flexible component weighing 3 lb. The angular natural frequency of the component is $\omega_{n2} = 130$ rad/sec. The equipment is mounted on a shock isolator with a linear spring $k_1 = 2{,}400$ lb/in. and having a fraction of critical damping $\zeta_1 = 0.10$. Find the maximum isolator deflection δ_{1m} and the maximum component acceleration $\ddot{x}_{2m}$ which result when the base experiences a velocity step $\dot{u}_m = 55$ in./sec.

Consider first a solution assuming that m_2 has a negligible effect on the equipment motion:

$$m_1 = \frac{152 \text{ lb}}{386 \text{ in./sec}^2} = 0.393 \text{ lb-sec}^2\text{/in.}$$

$$\omega_{n1} = \sqrt{\frac{k_1}{m_1}} = \sqrt{\frac{2{,}400}{0.393}} = 78.1 \text{ rad/sec [12.4 cps]}$$

Figure 31.14 gives $\ddot{x}_{1m}/\dot{u}_m\omega_{n1} = 0.88$ and Fig. 31.15 gives $\ddot{x}_{1m}\delta_{1m}/\dot{u}_m^2 = 0.76$ for $\zeta_1 = 0.1$. Then

$$\delta_{1m} = \frac{0.76}{0.88} \times \frac{\dot{u}_m}{\omega_{n1}} = \frac{0.76 \times 55}{0.88 \times 78.1} = 0.61 \text{ in.}$$

In finding $\ddot{x}_{2m}$ it is assumed that damping of the component has the typical value $\zeta_2 = 0.01$. Using $\omega_{n1}/\omega_{n2} = 130/78.1 = 1.67$, Fig. 31.28 gives $\ddot{x}_{2m}/\dot{u}_m\omega_{n2} = 1.15$; then $\ddot{x}_{2m} = 1.15 \times 55 \times 130 = 8{,}230$ in./sec^2 = 21.3g.

A second solution, taking into consideration the mass m_2 of the component, may be obtained if the damping is neglected. From Eq. (31.41),

$$\delta_{1m} = \frac{\dot{u}_m}{\omega_{n1}} \frac{1 + \frac{\omega_{n2}}{\omega_{n1}}\left(1 + \frac{m_2}{m_1}\right)}{\left[\left(\frac{\omega_{n2}}{\omega_{n1}} + 1\right)^2 + \frac{m_2}{m_1}\left(\frac{\omega_{n2}}{\omega_{n1}}\right)^2\right]^{1/2}}$$

$$= \frac{55}{78.1} \times \frac{1 + 1.67(1 + 3/152)}{[(2.67)^2 + 3/152\,(1.67)^2]^{1/2}} = 0.71 \text{ in.}$$

From Eq. (31.40):

$$\ddot{x}_{2m} = \dot{u}_m\omega_{n2}\left[\left(\frac{\omega_{n2}}{\omega_{n1}} - 1\right)^2 + \frac{m_2}{m_1}\left(\frac{\omega_{n2}}{\omega_{n1}}\right)^2\right]^{-1/2}$$

$$= 55 \times 130[(0.67)^2 + 3/152(1.67)^2]^{-1/2}$$

$$= 10{,}070 \text{ in./sec}^2 = 26.1g$$

This example is too complex for a practicable solution when damping and the mass effects are considered together. However, the two above solutions may be taken conservatively as limiting conditions; it is unlikely that the actual acceleration and deflection would exceed the maxima of the limiting conditions.

SUPPORT PROTECTION

This section considers conditions in which the shock originates within the equipment (e.g., guns and drop hammers). Attention is first given to determining the response of the support for such equipment in the absence of a shock isolator. The effect of a shock isolator introduced to protect the support from excessive loads is considered later.

EQUIPMENT RIGIDLY ATTACHED TO SUPPORT

If the equipment is rigidly attached to the support, the support and equipment may be idealized as a single degree-of-freedom system for purposes of a simplified analysis. Consider the system of Fig. 31.3*B* with the spring-dashpot unit 2 assumed to be rigid. The mass m represents the equipment, and the mass m_F represents, with spring and dashpot assembly (1), the support. The force F, applied externally to the equipment, is taken to be a known function of time. The equation of motion is

$$(m_F + m)\ddot{\delta} + F(\dot{\delta},\delta) = F$$

Considering only force-time relations $F(t)$ in the form of a single pulse, the analogous mathematical relations of Eqs. (31.5) and (31.6) are used by defining the impulse J applied by the force F as

$$J = \int_0^\tau F\,dt \tag{31.42}$$

where τ is the duration of the pulse.

SHORT-DURATION IMPULSES. If τ is short compared with the half-period of free oscillation of the system, then the results derived in the section on *Response of a Rigid Body System to a Velocity Step* may be applied directly. An impulse J of negligible duration acting on the mass m produces a velocity change $\dot{u}_m$ given by

$$\dot{u}_m = \frac{J}{m} \tag{31.43}$$

The subsequent relative motion of the system is identical with that resulting from a velocity step of magnitude $\dot{u}_m$.

If the damping capacity of the support is small, then velocity step results derived for linear springs, hardening springs, and softening springs are applicable. If the damping of the support may be represented as viscous and the stiffness as linear, then the linear-spring viscous damping results apply. In most installations it is sufficiently accurate to consider the support an undamped linear system.

A structure used to support an equipment generally has distributed mass and elasticity; thus the application of an impulse tends to excite the structure to vibrate not only in its fundamental mode but also in higher modes of vibration. The mass-spring-dashpot system shown in Fig. 31.3*B* to represent the structure would have equivalent mass and stiffness suitable to simulate only the fundamental mode of vibration. In many applications, such simulation is adequate because the displacements and strains are greater in the fundamental mode than in higher modes. The vibration of members having distributed mass is discussed in Chap. 7, and the formulation of models suitable for use in the analysis of systems subjected to shock is discussed in Chap. 42.

LONG-DURATION IMPULSES. If the duration τ of the applied impulse exceeds about one-third of the natural period of the equipment-support system, application of velocity step results may be unduly conservative. Then the results developed in the section on *Response of Rigid Body System to Acceleration Pulse* are applicable. The mathematical equivalence of Eqs. (31.5) and (31.6) is based on identifying $-m\ddot{u}$ in the former with F in the latter. Accordingly, if the shape of the force F vs. time curve is similar to the shape of the curve of acceleration $\ddot{u}$ vs. time, then the response of a system to an acceleration pulse may be used by analogy to find the response to a force pulse by making the following substitutions:

$$\ddot{u}_m = \frac{F_m}{m}^{*} \qquad \tau_r = \frac{J}{F_m}$$

where F_m is the maximum value of F, $\ddot{u}_m$ is the maximum value of $\ddot{u}$ and τ_r is the effective duration.

EQUIPMENT SHOCK ISOLATED

IDEALIZED SYSTEM. When a shock isolator is used to reduce the magnitude of the force transmitted to the support, the idealized system is as shown in Fig. 31.4. Sub-

* If the mathematical equivalence is literally applied, F_m/m is analogous to $-\ddot{u}_m$, not $\ddot{u}_m$. Since acceleration pulse results are given in terms of extreme *absolute* values, the sign is not important.

system 2 represents the equipment (mass m_2) mounted on the shock isolator (right-hand spring-dashpot unit). Subsystem 1 is an idealized representation of the support with effective mass m_1 and with stiffness and damping capacity represented by the left spring-dashpot unit. The free end of the latter unit is taken to be fixed ($u = 0$).

It is assumed that the system is initially in equilibrium ($\dot{\delta}_1 = \dot{\delta}_2 = 0$; $\delta_1 = \delta_2 = 0$) and that force F (positive in the $+X$ direction) applies an impulse J to m_2. Analysis is simplified by treating the duration τ of impulse J as negligible. This assumption, always conservative, usually is warranted if the natural frequency of the shock isolator is small relative to the natural frequency of the support.

SYSTEM SEPARABLE. In many applications the support motion $x_1(= \delta_1)$ is sufficiently small compared with the equipment motion x_2 that the equipment acceleration $\ddot{x}_2$ is closely approximated by $\ddot{\delta}_2$ where $\ddot{x}_2 = \ddot{\delta}_2 + \ddot{x}_1$. Using this approximation, the analysis is resolved into two separate parts, each dealing with a single degree-of-freedom system.

If the system consists only of linear elements as defined by Eq. (31.7), the equation of motion of the equipment mounted on the shock isolator (subsystem 2 of Fig. 31.4) is

$$\ddot{\delta}_2 + 2\zeta_2\omega_{n2}\dot{\delta}_2 + \omega_{n2}^2\delta_2 = 0 \qquad (31.44)$$

where $\omega_{n2}^2 = k_2/m_2$ and $\zeta_2 = c_2/2m_2\omega_{n2}$. The initial conditions are: $\delta_2 = 0$, $\dot{\delta}_2 = \dot{u}_m = J/m_2$ when $t = 0$. Because of the similarity of Eqs. (31.26) and (31.44), and the respective initial conditions, the maximum equipment acceleration $\ddot{x}_{2m}$ and the maximum isolator deflection δ_{2m} may be found from Figs. 31.14 and 31.15.

The differential equation for the motion of the support in Fig. 31.4 is

$$\ddot{\delta}_1 + 2\zeta_1\omega_{n1}\dot{\delta}_1 + \omega_{n1}^2\delta_1 = -\frac{m_2}{m_1}\ddot{x}_2 \qquad (31.45)$$

where $\omega_{n1}^2 = k_1/m_1$ and $\zeta_1 = c_1/2m_1\omega_{n1}$. The initial conditions are $\dot{\delta}_1 = 0$, $\delta_1 = 0$.

The solution of Eq. (31.45) is formally identical with that of Eq. (31.38) because the equations differ only by the interchange of the numerical subscripts and the presence of the factor m_2/m_1 on the right-hand side of Eq. (31.45). The solutions of Eq. (31.45) as obtained by analog computer [1] are shown in Fig. 31.31. The ordinate is the ratio of the maximum force F_{1m} in the support to the quantity $J\omega_{n1}$. The latter quantity is the maximum force which would be developed in an undamped, linear, single degree-of-freedom support of mass m_1 and stiffness k_1 if the impulse J were applied directly to m_1. The abscissa in Fig. 31.31 is the ratio of the undamped support natural frequency ω_{n1} to the undamped isolator natural frequency ω_{n2}. Curves are drawn for various values of the fraction of critical damping ζ_2 for the isolator, assuming that the fraction of critical damping ζ_1 for the support is constant at $\zeta_1 = 0.01$.

Figure 31.31 appears to show that the presence of an isolator increases the maximum force F_{1m} transmitted by the support if the natural frequencies of isolator and support are nearly equal. This conclusion is misleading because the analysis assumes that the support deflection δ_1 is small compared with the isolator deflection δ_2—a condition which is not met in the neighborhood of unity frequency ratio. A more realistic analysis involves the two degree-of-freedom system discussed in the next section.

TWO DEGREE-OF-FREEDOM ANALYSIS. This section includes an analysis of the system of Fig. 31.4 considered as a coupled two degree-of-freedom system where both the support and isolator are linear and undamped [$F_1(\dot{\delta}_1,\delta_1) = k_1\delta_1$, $F_2(\dot{\delta}_2,\delta_2) = k_2\delta_2$]. This analysis makes it possible to consider the effect of deflection of the support on the motion of the equipment. Fixing the support base ($u = 0$), the equations of motion may be written

$$\begin{aligned} \ddot{\delta}_1 + \omega_{n1}^2\delta_1 &= \frac{m_2}{m_1}\omega_{n2}^2\delta_2 \\ \ddot{\delta}_2 + \omega_{n2}^2\delta_2 &= -\ddot{\delta}_1 \end{aligned} \qquad (31.46)$$

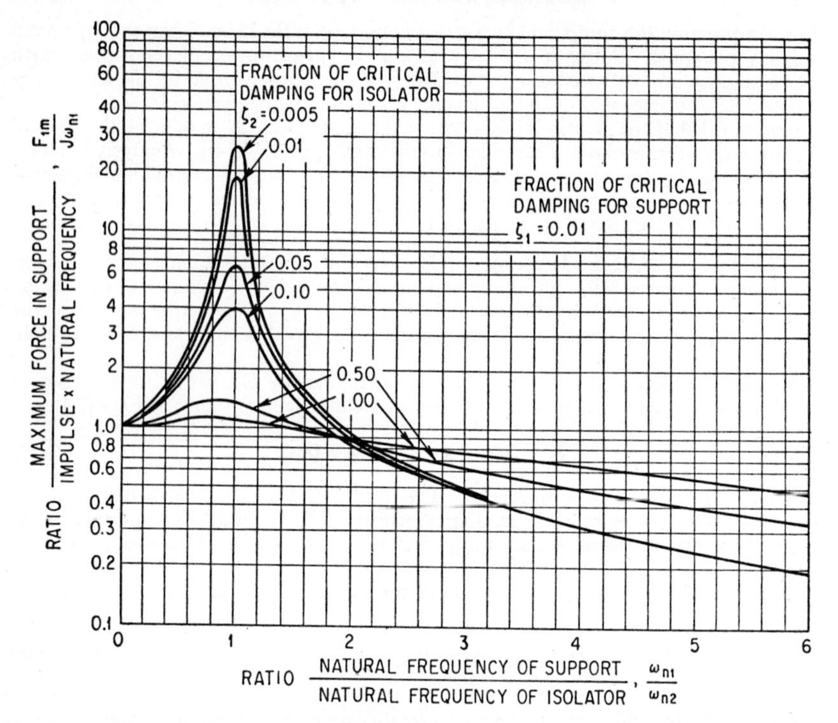

FIG. 31.31. Dimensionless representation of maximum force in support F_{1m} resulting from action of impulse J on equipment. Isolator and support have linear elasticity and fraction of critical damping ζ_2 and ζ_1, respectively. Ordinate is the ratio of maximum force in support to that which would result from direct application of the impulse to the support with equipment absent. (*After R. D. Mindlin.*[1])

Assuming that the impulse J has negligible duration, the initial conditions are: $\dot{\delta}_1 = 0$, $\dot{\delta}_2 = J/m_2$, $\delta_1 = \delta_2 = 0$.

The solution of Eqs. (31.46) parallels that of Eqs. (31.39); the resulting expressions for the maximum isolator deflection δ_{2m} and force F_{1m} applied to the support are

$$\delta_{2m} = \frac{J}{m_2\omega_{n2}}\left[1 + \frac{m_2/m_1}{(1 + \omega_{n1}/\omega_{n2})^2}\right]^{-1/2} \tag{31.47}$$

$$F_{1m} = J\omega_{n1}\left[\left(1 - \frac{\omega_{n1}}{\omega_{n2}}\right)^2 + \frac{m_2}{m_1}\right]^{-1/2} \tag{31.48}$$

The maximum deflection of the isolator given in Eq. (31.47) is shown graphically in Fig. 31.32. For small values of the ratio of support natural frequency to isolator natural frequency, the flexibility of the support may significantly reduce the maximum isolator deflection, especially if the mass of the support is small relative to the mass of the equipment. For large values of the frequency ratio, the effect of the mass ratio is small.

Maximum values of force in the support, given by Eq. (31.48), are shown in Fig. 31.33. The maximum deflection of the floor is the maximum force F_{1m} divided by the stiffness of the floor. The effect of mass ratio is profound for small values of the frequency ratio. The curves of Figs. 31.31 and 31.33 show corresponding results, the former including damping and the latter including the coupling effect between the two systems. The analysis which ignores the coupling effect may grossly overestimate the maximum force applied to the support at low values of the frequency ratio. At high values of the fre-

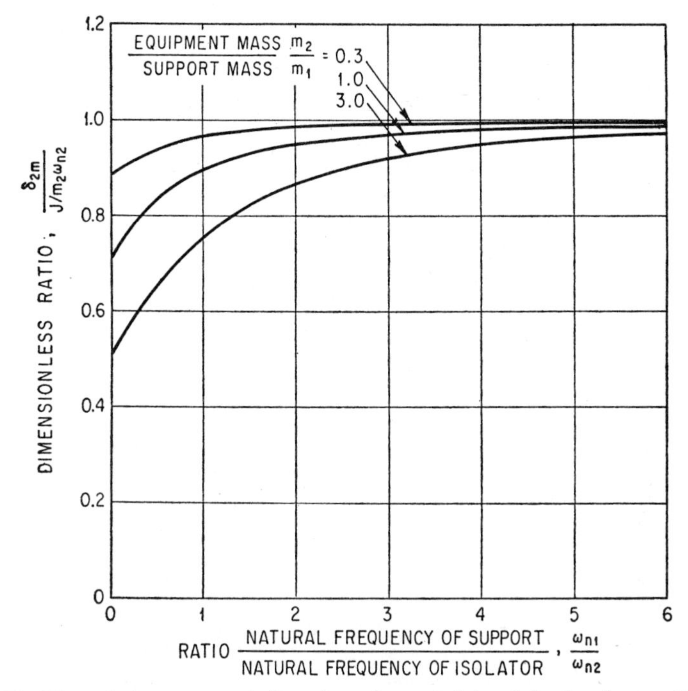

FIG. 31.32. Dimensionless representation of maximum isolator deflection δ_{2m} resulting from action of impulse J on equipment. Isolator and support have undamped linear elasticity. Ordinate is the ratio of maximum isolator deflection to that which would occur with a rigid support. Curves are for three discrete values of the mass ratio m_2/m_1 as indicated.

quency ratio, the two analyses yield like results if the fraction of critical damping in the isolator is less than about $\zeta_2 = 0.10$. The two methods are compared in Example 31.6.

Example 31.6. A forging machine weighs 7,000 lb exclusive of the 600-lb hammer. It is mounted at the center of a span formed by two 12-in., 50 lb/ft I beams having hinged ends and a span $l = 18$ ft. The hammer falls freely from a height of 60 in. before striking the work. Determine:

a. Maximum force F_{1m} in the beams and maximum deflection δ_{1m} of the beams if the machine is rigidly bolted to the beams.

b. The maximum force F_{1m} in the beams and the maximum deflection δ_{2m} of an isolator interposed between machine and beams.

Solution.

a. When the machine is bolted rigidly to the beams, the system may be considered to have only a single degree-of-freedom. The mass is that of the machine, plus the hammer, plus the effective mass of the beams. For the machine: $m_2 = (7{,}000 + 600)/386 = 19.2$ lb-sec²/in. The effective mass of the beams is taken as one-half of the actual mass:

$$m_1 = \frac{2(0.5)(18)(50)}{386} = 2.33 \text{ lb-sec}^2/\text{in.}$$

$$m = m_1 + m_2 = 21.5 \text{ lb-sec}^2/\text{in.}$$

The stiffness of the beams is

$$k = 2\frac{48EI}{l^3} = 2\frac{48 \times (30 \times 10^6) \times 302}{(18 \times 12)^3} = 123{,}000 \text{ lb/in.}$$

The natural frequency of the machine-and-beams system is

$$\omega_n = \sqrt{\frac{k}{m}} = \sqrt{\frac{123{,}000}{21.5}} = 75.6 \text{ rad/sec [12.0 cps]}$$

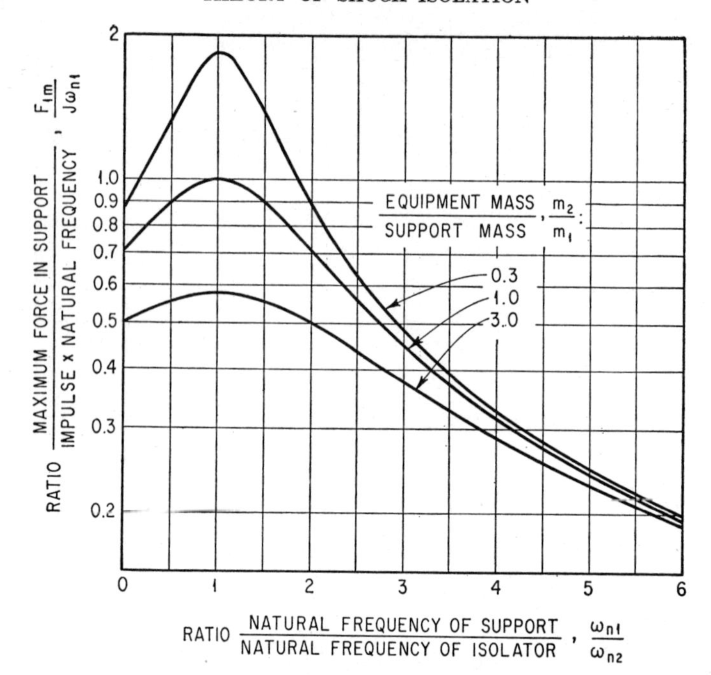

FIG. 31.33. Dimensionless representation of maximum force in support F_{1m} resulting from action of impulse J on equipment. Isolator and support have undamped linear elasticity. Ordinate is the ratio of maximum force in support to that which would result from direct application of the impulse to the support with equipment absent. Curves are for three discrete values of the mass ratio m_2/m_1.

If the impact between the hammer and work is inelastic and its duration is negligible, the resulting velocity $\dot{u}_m$ of the machine may be found from conservation of momentum. The impulse J is the product of weight of hammer and time of fall:

$$J = (600)\left(\frac{2 \times 60}{386}\right)^{1/2} = 335 \text{ lb-sec}$$

Then $\dot{u}_m = J/m = 335/21.5 = 15.6$ in./sec. If the damping of the beams is neglected, the maximum beam deflection is found from Eq. (31.18):

$$\delta_{1m} = \frac{\dot{u}_m}{\omega_n} = \frac{15.6}{75.6} = 0.21 \text{ in.}$$

The maximum force in the beams is the product of beam stiffness and maximum deflection:

$$F_{1m} = k\delta_{1m} = 25{,}300 \text{ lb}$$

b. An isolator having a stiffness $k_2 = 36{,}000$ lb/in. and a fraction of critical damping $\zeta_2 = 0.10$ is interposed between the machine and beams. The "uncoupled natural frequencies" defined in connection with Eqs. (31.40) and (31.41) are

$$\omega_{n2} = \sqrt{\frac{k_2}{m_2}} = \sqrt{\frac{36{,}000}{19.2}} = 43.3 \text{ rad/sec [6.9 cps]}$$

$$\omega_{n1} = \sqrt{\frac{k_1}{m_1}} = \sqrt{\frac{123{,}000}{2.33}} = 230 \text{ rad/sec [36.6 cps]}$$

Consider first that the system is separable. Figures 31.14 and 31.15 give, respectively: $\ddot{x}_{2m}/\dot{u}_m\omega_{n2} = 0.88$; $\ddot{x}_{2m}\delta_{2m}/\dot{u}_m{}^2 = 0.76$. Substituting $\dot{u}_m = J/m_2 = 17.4$ in./sec and solving for δ_{2m},

$$\delta_{2m} = \frac{0.76 \times 17.4}{0.88 \times 43.3} = 0.35 \text{ in.}$$

Entering Fig. 31.31 at $\omega_{n1}/\omega_{n2} = 5.3$, $F_{1m}/J\omega_{n1} = 0.23$. Then

$$F_{1m} = 17{,}700 \text{ lb}$$

Thus, the effect of the isolator is to reduce the maximum load in the beams from 25,300 lb to 17,700 lb. An isolator with less stiffness would permit a further reduction of this force at the expense of greater machine motion.

Consider now that the floor and machine-isolator systems are coupled, and use the two degree-of-freedom analysis which neglects damping. From Eq. (31.47):

$$\delta_{2m} = \frac{J}{m_2\omega_{n2}} \left[1 + \frac{m_2/m_1}{\left(1 + \frac{\omega_{n1}}{\omega_{n2}}\right)^2}\right]^{-1/2}$$

$$= \frac{335}{19.2 \times 43.3} \left[1 + \frac{19.2/2.33}{(1 + 5.3)^2}\right]^{-1/2} = 0.37 \text{ in.}$$

From Eq. (31.48):

$$F_{1m} = J\omega_{n1} \left[\left(1 - \frac{\omega_{n1}}{\omega_{n2}}\right)^2 + \frac{m_2}{m_1}\right]^{-1/2}$$

$$= 335 \times 230 \left[(1 - 5.3)^2 + \frac{19.2}{2.33}\right]^{-1/2} = 14{,}900 \text{ lb}$$

Thus, the two results for the isolator deflection δ_{2m} differ only slightly, but the two degree-of-freedom analysis gives a maximum load in the beams about 16 per cent smaller than that obtained by assuming the systems to be separable.

REFERENCES

1. Mindlin, R. D.: *Bell System Tech. J.*, **24**:353 (1945).
2. McCalley, R. B., Jr.: "Velocity Shock Transmission in Two Degree Series Mechanical Systems," Atomic Energy Commission Contract W-31-109 Eng-52, Knolls Atomic Power Laboratory, Schenectady, March, 1956.
3. Klotter, K.: *J. Appl. Mechanics*, **22**:493 (1955).
4. Crede, C. E.: "Vibration and Shock Isolation," John Wiley & Sons, Inc., New York, 1951.

32

APPLICATION AND DESIGN OF ISOLATORS

Charles E. Crede
California Institute of Technology

INTRODUCTION

This chapter discusses the practical aspects of isolation and isolators, supplementing the theoretical treatment of vibration isolation and shock isolation in Chaps. 30 and 31, respectively. Data are appended on the characteristics of typical commercially available isolators. These data include both dimensional and performance characteristics adequate for most application purposes. If commercially available isolators are not adequate, an isolator may be designed. General information is included on the application of springs as vibration and shock isolators, drawing upon the detailed material on metal and rubber springs in Chaps. 34 and 35, respectively. Dampers of various types are discussed in detail to provide design and application data where the damping attainable from internal hysteresis of the spring is not great enough.

This chapter also discusses the application of isolators. Application is considered in two categories: applications to particular machines and applications in particular environments. In the first category, special considerations introduced into isolator application and design by the peculiar characteristics of various types of machines are discussed. Application to particular environments refers primarily to the conditions in vehicles and to the consequent requirements imposed on isolators; the inferences are particularly useful for vehicles for military use.

APPLICATION OF ISOLATORS TO PARTICULAR MACHINES

This section discusses various types of machines and equipment to which isolators sometimes are applied, and describes the application of isolators. In general, the vibration or shock is created by operation of the machine; the objective in using isolators is to reduce transmission to adjacent structures or buildings.

ELECTRIC MOTORS

The torque delivered by an electric motor is the result of the attraction of the magnetic field of the stator upon current-carrying conductors of the armature. The nominally steady torque has superimposed fluctuations at a frequency that is a function of the motor speed and number of poles in the stator. Vibration results from two sources:

1. Fluctuation in the torque, characterized as an angular reaction upon the stator and thus upon the supports for the motor
2. Unbalance of the rotating parts, characterized as a rotating, radially directed force vector

Isolation of the vibration resulting from fluctuations in the torque is achieved by mounting the stator upon isolators having relatively low stiffness in rotation about the shaft center, the natural frequency of the stator upon the isolators being substantially lower than the frequency of the torque fluctuation. Several methods are employed to achieve the required low torsional stiffness of the isolator. One of these is illustrated in Fig. 32.1.[1] The isolator consists of a rubber ring which encircles the hub of the stator and is attached at its periphery to the motor support. The modulus of elasticity of rubber in shear is relatively low; the location of the rubber ring close to the axis of rotation thus makes possible the attainment of the low natural frequency in a rotational mode. An equivalent result is obtained from the arrangement of flat metal springs illustrated in Fig. 32.2.[2] The two legs of the spring form an included angle of 90° which has its apex at the shaft axis. The stator is free to rotate about the shaft axis with but little restraint since the springs exhibit a characteristic low stiffness when deflected in flexure.

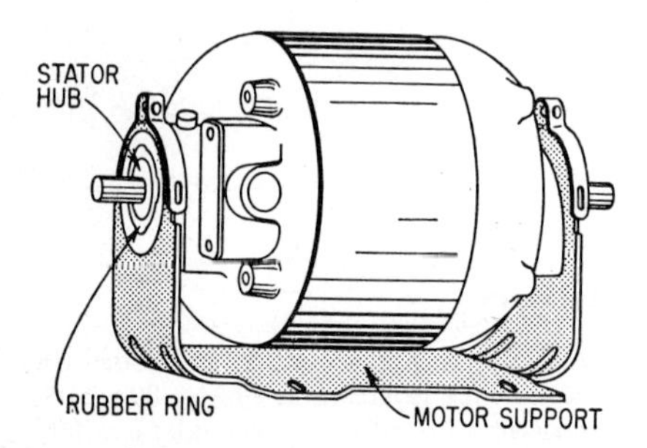

FIG. 32.1. Electric motor equipped with vibration isolators in the form of rubber rings surrounding the hubs of the motor end bells. (*General Electric Co.*)

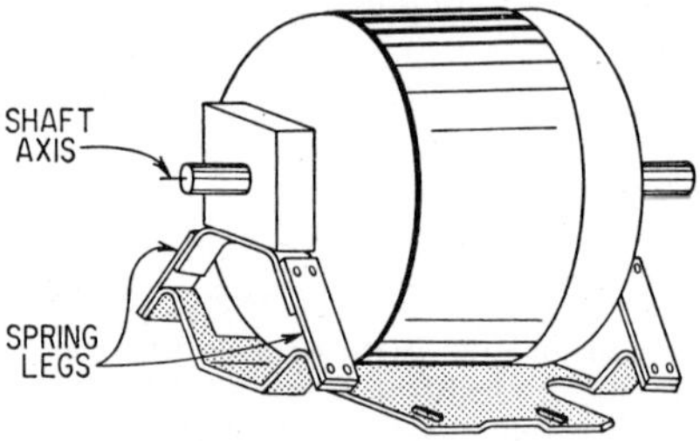

FIG. 32.2. Electric motor equipped with vibration isolators consisting of diagonally disposed leaf springs. (*Holtzer-Cabot.*)

The isolators shown in Figs. 32.1 and 32.2 are relatively stiff with respect to forces applied perpendicular to the shaft. In the former, this results from the relatively high modulus of rubber in compression and, in the latter, from the great stiffness of the flat springs in tension and compression. Any radial force on the shaft can be resolved into component forces along the two spring legs. Thus, the isolators are adapted to exhibit rigidity with respect to forces applied by belts and gears without sacrificing the low torsional stiffness necessary to attain isolation of the torque fluctuations.

Vibration resulting from the unbalance of rotating parts occurs at the rotational frequency of the motor. Except for certain high-speed motors, the effective isolation of such vibration would require isolators having relatively low stiffness in the radial direction. This low stiffness would interfere with the effective operation of belt and gear drives. The rotating parts of electric motors generally are well balanced; thus, it usually is unnecessary to isolate the resulting vibration.

RECIPROCATING MACHINES

Among the more troublesome machines from a vibration viewpoint are those involving reciprocating motion, such as internal-combustion engines, air compressors, vacuum pumps, etc. The vibration in a reciprocating machine results from two unrelated sources, namely, torque impulses and mass unbalance:

1. In the operation of a reciprocating machine, a series of impulses inherently are applied to the piston, whether the machine creates a rotational motion or is driven by rotation of the crankshaft. An equal and opposite series of impulses are applied to the piston heads, and the chassis of the machine receives a series of torque pulses equal to those acting upon the piston. The frequency of these torque pulses is a function of the rotational speed and the type of cycle, ranging from one cycle per two revolutions to several cycles per revolution. Vibration ascribable to torque pulses often is less important than

that resulting from inertia forces. In most applications isolators applied to isolate the vibration resulting from inertia forces also isolate the vibration resulting from torque pulses.

2. Vibration resulting from mass unbalance of a reciprocating machine is unrelated to the torque pulsation. Whereas the latter results from gas pressure on the piston, the former is due to the inertia of moving members of the machine. Each piston moves in a plane normal to the crankshaft; forces acting in such plane result from inertia forces associated with the motion of the piston, connecting rod, and crank. Vibration resulting from unbalance always occurs at the rotational frequency of the machine and harmonics thereof, the harmonics being caused by the finite length of the connecting rod. The level of vibration created by a particular machine is a function of the extent to which these forces and the couples resulting therefrom are unbalanced.

A machine having only one or two cylinders may exhibit extreme unbalance because it is impossible to balance the forces and couples. Often the total mass of the machine is not sufficiently large relative to the vibratory forces, and the machine may have excessive vibratory displacement on the isolators. In a machine having four or more cylinders, certain forces and couples can be balanced and the displacement usually is smaller. An obstacle to the effective isolation of vibration created by reciprocating machines may be the frequency of the machine. Reciprocating machines (particularly those of large size) often operate at relatively slow speeds, thereby requiring that the isolators have a large static deflection. This may lead to an unstable condition unless precautions are taken to ensure stability.

The vibratory force in a direction parallel to the crankshaft generally is not great * and isolation may not be necessary, provided the isolator system avoids having a natural frequency that coincides with the fundamental frequency of crankshaft rotation or a harmonic thereof. The crankshafts of some one- and two-cylinder machines extend in the short direction of the base, and isolators having a low stiffness in this direction tend to introduce a degree of instability. This instability can be largely overcome by extending the supporting beams as illustrated in Fig. 30.21, thereby introducing a relatively high natural frequency about an axis perpendicular to the crankshaft axis. Such an arrangement often is found to be acceptable even though the natural frequency is higher than the operating frequency because the excitation is of a low magnitude.

MISCELLANEOUS INDUSTRIAL EQUIPMENT

GEARS, BEARINGS, AND CAMS. Vibration resulting from the operation of bearings and gears usually has its source in friction between sliding members, in impacts between loose-fitting parts, and in turbulent action of lubricating fluids in bearings and gear boxes. Where gears are the major source of noise, the predominant frequency is the tooth frequency, that is, the product of the rotational frequency of the gear and the number of teeth on its periphery. Higher harmonics of the tooth frequency often are present, but subharmonics are seldom encountered. The vibration caused by bearings and fluid turbulence tends to lack discrete frequencies.

The method used to isolate the vibration originating in gears and bearings is of a noncritical nature. Adequate isolation generally can be attained with an isolator having a natural frequency as high as 25 cps. The forces are small and cause no appreciable overall motion of the mounted machine. Care should be taken to avoid resonance with any predominant frequency, such as the rotational or tooth frequency.

The operation of cams may involve either continuous sliding or rolling contact or momentary contact in the nature of an impact. In the former, the vibration resulting from cam action is very similar to that resulting from the operation of gears and bearings. Impact forces are different, and the required treatment depends upon the magnitude of the forces involved. In a loom, for example, the force exerted by the cam may be sufficiently great to displace the entire machine. The isolator then should be capable of

* In some large two-cylinder machines, the vertical inertia forces may form a couple about a transverse horizontal axis. Then the resulting motion of the machine may have a fore-and-aft component at the isolator locations. The application of isolators should make provision to accommodate this motion.

experiencing a deflection at least equal to this displacement. In other instances, the impacts associated with cam action are relatively small. For this latter condition, isolators designed in accordance with the principles of isolating gear and bearing vibration usually will give satisfactory results.

TRANSFORMERS. The principal source of vibration in a transformer is the fluctuation in the magnetic force resulting from the alternating current. Therefore, the vibration frequency is constant, generally two times the frequency of voltage variation. Since all the magnetic forces in a transformer are opposed by members which are structurally an integral part of the transformer, the forces transmitted externally of the transformer are small. This is in contrast to a motor, for example, which delivers a torque and exerts an equal reaction upon the motor supports.

The technique for isolating the structure-borne vibration from a transformer follows conventional principles of vibration isolation. The excitation frequency usually is high, and a relatively stiff isolator may be used. Experience with transformers, particularly in residential areas, shows that the major vibration problem resides in airborne rather than structure-borne vibration.

MAGNETICALLY ACTUATED DEVICES. This class of equipment includes relays, contactors, circuit breakers, etc., and commonly embodies an armature which is caused to move as a result of a magnetic force originating in a solenoid. If the solenoid is energized by alternating current, an alternating force similar to that created by a transformer exists and the same principles of isolation apply. There may be an additional problem resulting from the movement of the armature, particularly from the impact involved in stopping the armature. In most instances, the duration of the impact is quite short and a relatively stiff isolator can provide the required isolation at the high excitation frequencies. Magnetically operated devices sometimes disturb adjacent components of a delicate nature, and isolators are used to good advantage for supporting the devices.

FANS. It is common practice to distribute air throughout a heating and air-conditioning system by a "squirrel cage" fan driven by an electric motor. The rotor of the fan is a drum having a relatively large number of longitudinally extending blades arranged around the periphery. Air enters the fan by flowing axially into the drum, and is expelled radially between the blades into the duct system. There are two predominant sources of vibration: (1) rotor unbalance and (2) aerodynamic forces resulting from passage of the blades past the duct opening. The frequency of the former is the rotational speed of the rotor (rotor frequency); the frequency of the latter (blade frequency) is the product of rotor speed and number of blades on the rotor. It is conventional practice in the application of isolators to use the blade frequency as a basis for design. However, a large central air-conditioning system may employ a large fan operating at a relatively high speed and working against a pressure of 6 to 8 in. of water. Then the forces resulting from the rotor unbalance may become objectionable unless the isolators are designed on the basis of the rotor frequency.

Where the isolators are applied on the basis of the blade frequency, the application usually is noncritical because the forcing frequency is relatively high. For example, a fan may have 50 blades spaced around the periphery of the rotor. Then if the rotor operates at the typical speed of 900 rpm (15 cps), the forcing frequency is 750 cps. An isolator applied to have a natural frequency of 10 to 15 cps affords adequate isolation of vibration for most purposes. However, typical operating speeds for large fans are 1,000 to 1,600 rpm, and a natural frequency of the isolator as low as 4 cps may be required to attain adequate isolation of vibration at the rotor frequency. Such fans should be supported on concrete floor or roof panels at least 6 in. thick. For supports of less rigidity, it is advisable to mount the fan upon a floating concrete block which is in turn supported by isolators (see the section on *Isolated Foundations*). The mass of the block should be at least equal to the mass of the fan-and-motor assembly.

Smaller fans are constructed of relatively light sheet metal, and the motors are furnished separately. A conventional installation practice is to mount the fan and motor on an integral base fabricated from structural steel and equipped with vibration isolators. The rigidity of the base maintains the necessary alignment between motor and fan; it also adds such rigidity to the light sheet-metal housing of the fan as is required to attain effective vibration isolation. A typical fan and motor mounted upon an integral base are shown in Fig. 32.3. In another type of base, the fan is mounted upon a pair of rails and the motor upon another pair of rails; each set of rails is equipped with vibration isolators. Both types of bases are available commercially of the proper dimensions and equipped with isolators for use with designated types and sizes of fans and motors.

PUMPS. This class of equipment includes both centrifugal and reciprocating pumps. The former usually has a multivane rotor that operates at a relatively high speed. Vibration may be expected (1) at the rotational frequency, as a result of mass unbalance of the rotor, (2) at the vane frequency, as the moving vanes pass close to the fixed vanes, and (3) in a random manner, resulting from forces created by turbulent flow of liquid within the pump. Centrifugal pumps are quite adaptable to the application of vibration isolators because the operating speed is sufficiently high to justify relatively stiff isolators, and the oscillating forces are so small that the vibration amplitude of the pump body is not appreciable.

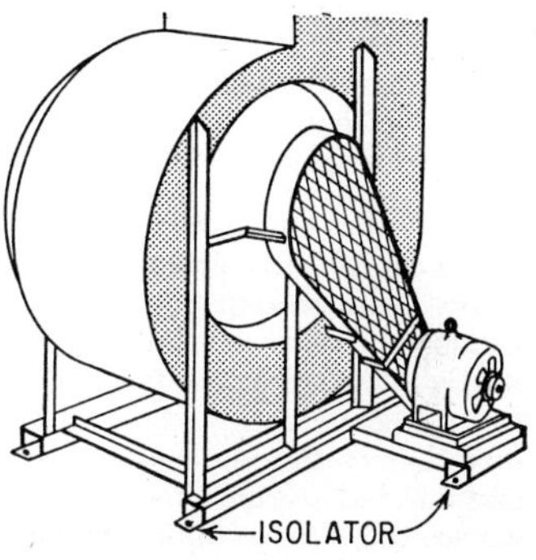

Reciprocating pumps are used for causing flow of liquid, for compressing air and gas, and for creating a vacuum. They are characteristically low-speed machines, and the number of cylinders often is small. The preceding discussion in this section on reciprocating machines is applicable generally. Unbalanced rotating and reciprocating parts, and torque impulses associated with the work being done on the fluid, are predominant sources of vibration.

FIG. 32.3. Squirrel cage fan with electric motor mounted upon integral base fabricated from structural steel and supported by vibration isolators. (*The Vibration Eliminator Co.*)

A large displacement of a reciprocating pump may occur when it is mounted upon vibration isolators. This may require that the pump be mounted upon a heavy base (e.g., a concrete block) to reduce its motion. Where the motion of the pump is large, flexible hose should be used instead of rigid pipe in all connections to the pump (see the section on *Flexible Conduits*). In other instances, it is expedient to forego the use of isolators or to use relatively stiff isolators adapted to isolate only the vibration resulting from gears and bearings, motor impulses, and hydrodynamic forces.

PRINTING PRESSES. Printing presses may be of the rotary or reciprocating type. In the former, the principal rotating parts are quite massive, but the unbalance usually is not great and the rotational frequency is relatively low. The principal need for isolation in a rotary press is related to the vibration originating in the action of gears, bearings, and cams. The frequencies associated with such vibration tend to be high, and adequate vibration isolation usually can be achieved with isolators having a natural frequency of 15 to 20 cps.

A large reciprocating printing press presents a more difficult problem. Characteristic features of such a press are:

1. Relatively large forces at low frequencies, as a result of the reciprocating motion of massive members. These forces may become sufficiently great to cause objectionable swaying of the building when the press is located on an upper floor.

2. Miscellaneous vibration and impacts resulting from the operation of gears, cams, bearings, and similar mechanisms. The magnitudes are usually small, but the number of occurrences is large and the total effect may be considerable.

3. A relatively nonrigid framework for supporting the mechanism. Many machines are fabricated from light and loosely joined structural members, and require a rigid mounting base for proper operation.

A printing press presents an unusual problem in that appreciable over-all movement of the press often cannot be tolerated because it interferes with the register of the paper moving through the press. This leaves the alternatives of either (1) using relatively stiff isolators which prevent appreciable movement of the press but transmit large forces to the building structure or (2) adding mass to the press in the form of a concrete foundation to limit the vibration amplitude and reduce the vibratory force transmitted to the building. The latter method is feasible only if the operating frequency of the press is approximately 5 cps or higher. For a lower operating frequency, the isolator would have a natural frequency so low that special consideration would be required to maintain the stability of the system. Some large presses operate at frequencies as low as 1 cps. Then there is little possibility of attaining isolation of the inertia forces resulting from the reciprocating motion of principal members of the press.

The most successful method of mounting a large printing press is to secure the press rigidly to a concrete foundation which is, in turn, supported by isolators. Two principal advantages of this type of installation are (1) the press acquires the rigid foundation which is required to maintain the alignment of structural members and (2) mass is provided whose inertia tends to reduce the vibratory motion of the press. Various concrete foundation designs are considered under *Isolated Foundations*.

LOOMS. One of the most troublesome machines with regard to vibration and shock is the cloth-weaving loom, illustrated schematically in Fig. 32.4. The *lay*, a relatively heavy member, is driven with a horizontally reciprocating motion by a pair of cranks and connecting rods. A shuttle is caused, by a mechanism described later, to travel alternately in opposite directions across the lay from one end to the other. The longitudinally extending warp thread is fed continuously in many parallel strands extending in the direction of the lay movement. Combinations of warp thread are alternately raised and lowered at each cycle of the lay by a harness not shown in the diagram. The shuttle carries the cross or filling thread between the raised and lowered strands of warp thread. After each passage of the shuttle, the lay presses or lays the last woven filling thread against the previously woven threads.

The shuttle, before it begins to move, rests in contact with a picker attached to the upper end of a picker stick. A picking roll mounted upon an arm carried by the bottom shaft strikes a picking shoe and causes the shaft that carries this shoe to rotate about its longitudinal axis. The resultant sudden angular movement of the picking sweep arm is transmitted to the picker stick through the sweep stick and lug strap. The sudden move-

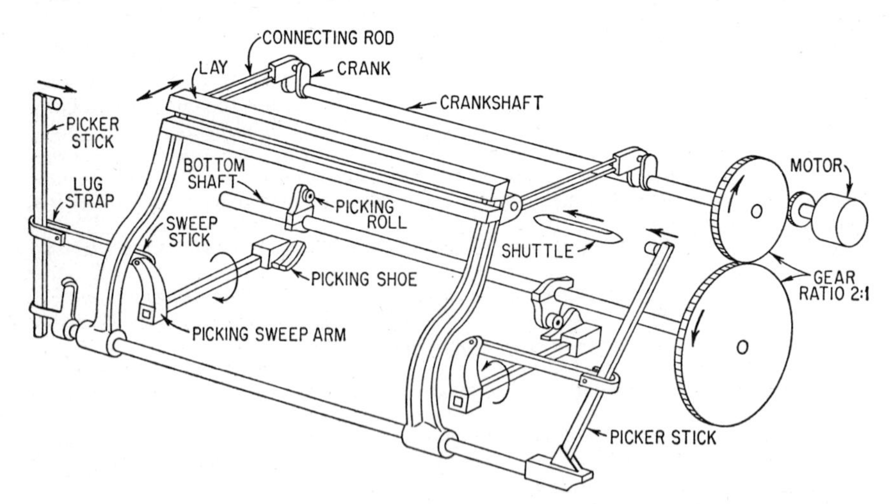

Fig. 32.4. Schematic diagram of a cloth-weaving loom indicating the principal moving parts responsible for the generation of shock and vibration.

ment of the picker stick propels the shuttle toward the opposite end of the lay. Upon arriving, it is first brought to rest and then propelled across the lay in the opposite direction to continue the cycle of operation.

The two principal sources of vibration and shock resulting from loom operation are as follows:

1. The inertia force created by the reciprocating motion of the lay is substantial. This is almost a pure harmonic force acting in a horizontal direction, and the reaction upon the frame of the loom is at the crankshaft. It is not feasible to isolate the vibration resulting from the lay motion; the frequency usually is approximately 3 to 4 cps in modern looms, and the stiffness of the isolators would be too low for effective operation of the loom. Furthermore, the mass of the lay is an appreciable proportion of the total mass of the loom; excessive displacement of the loom upon the vibration isolators would thus result. The forces generated by many hundreds of looms in a single building have caused excessive damage to mill buildings on occasion. Such damage has been counteracted by constructing steel reinforcements against the exterior walls of the buildings and by the installation of auxiliary mass dampers within the building [3] (see Chap. 6).

2. The force that propels the shuttle is in the nature of an impact. The complexity of the mechanism employed for this purpose makes the exact nature and direction of this impact uncertain. Although the picking action occurs with the same periodicity as the lay movement, the actual force endures for only a small fraction of this period, i.e., the excitation frequency is high. There appear to be some advantages to mounting looms upon relatively stiff isolators (approximately 30 to 40 cps), particularly if the looms are supported by a concrete floor. In addition to affording a degree of resilience in the support for the loom, the isolators facilitate the installation of looms by cementing procedures is common use to attach isolators to both the loom and the floor.

ISOLATED FOUNDATIONS

In special applications, it is not feasible to mount the machine directly upon the vibration isolators. Many such applications effectively employ a relatively heavy and rigid block, usually made of concrete, which is supported by suitable isolators and upon which the equipment to be isolated is placed. The use of such a massive block is desirable for a number of different reasons:

1. Some types of equipment do not operate properly unless supported by a rigid structure. This applies to certain types of machine tools whose inherent rigidity is not great and which maintain the prescribed accuracy only when rigidly supported. Another requirement for a rigid supporting structure involves such machinery as printing presses which often consist of articulated components wherein proper alignment of working parts can be maintained only by employing a rigid support.

2. If the machine to be mounted generates relatively large forces during its operation, the over-all movement of the machine tends to become excessive unless its effective mass is substantially increased. Such an increase in effective mass can be achieved by mounting the machine upon a concrete block. Diesel engines, forging hammers, vibration testing machines, etc., are effectively mounted in this manner.

3. In the application of vibration isolators where the excitation frequency is low, the most feasible method of achieving a low natural frequency with acceptable stability is to locate the isolators in the same horizontal plane as the center-of-gravity of the mounted machine. This is not generally possible with a machine intended to be supported only at its base. The machine can be mounted rigidly to a concrete block, and the assembly of machine and block supported by isolators located in the same horizontal plane as the center-of-gravity of the assembly.

4. Many types of equipment present relatively nonmassive mounting structures for the attachment of isolators. The low mechanical impedance of such structures makes it impossible to attain a high degree of vibration isolation without interposing a rigid structure having a relatively high mechanical impedance. A concrete block used to support the equipment provides a very large impedance at the interface with the isolator.

Concrete blocks adapted to meet the requirements outlined above may be installed either above the floor level or within a pit provided for this purpose. Isolators employed to support the blocks may be made of cork, rubber, steel springs, or other suitable material. The required size of the block depends upon the function of the equipment. The purpose of the block may be to provide rigidity, and the size is determined accordingly. However, the size of a block used to mount a forging machine is determined by the momentum available for transfer to the block and by the permissible motion of the machine. The desired natural frequency for the isolation system usually is established by operating characteristics of the mounted equipment, and the required properties of the supporting springs can be determined from the properties of resilient materials set forth in other chapters of this Handbook. Several typical installations of concrete blocks are illustrated and described in the following paragraphs.

One method of attaining a rigid and massive supporting structure is to cast a shallow concrete block which may be supported above the level of the floor surface, as illustrated in Fig. 32.5. Commercially available isolators can be used effectively with such a block, and particular advantages result from the use of isolators having a height adjustment feature to facilitate leveling of the block. Even though the block is relatively

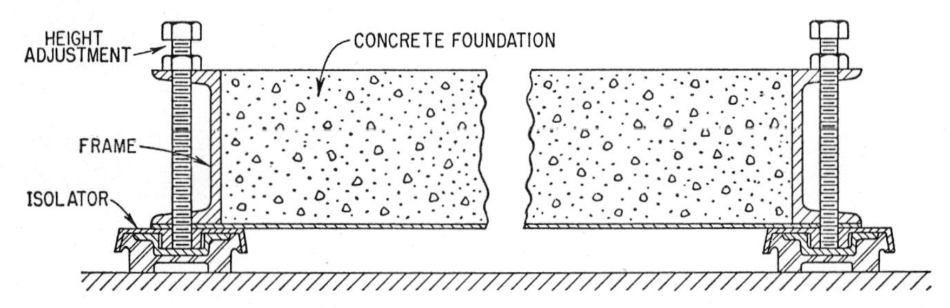

Fig. 32.5. Independent foundation supported upon vibration isolators and adapted to be placed above the floor level. (*Barry Wright Corp.*)

large, it can be moved from place to place with modern materials-handling equipment. A number of techniques have been developed to facilitate construction or to attain desired dynamic characteristics. For example, the block shown in Fig. 32.6 may be poured without a form, using the floor to close the lower side of the frame and covering the floor with waterproof paper before pouring the concrete. Upon hardening of the concrete, the block is lifted from the floor by turning the lifting screws.

In the arrangement shown in Fig. 32.7, the coil-spring isolators are placed upon pedestals located adjacent the edges of the block. A vertical rod extends downward from the strongback over the several springs and is attached at its lower end to a structural member embedded in the block. This pendulum-type suspension ensures stability of the system and makes possible a relatively low natural frequency in the horizontal mode of vibration.

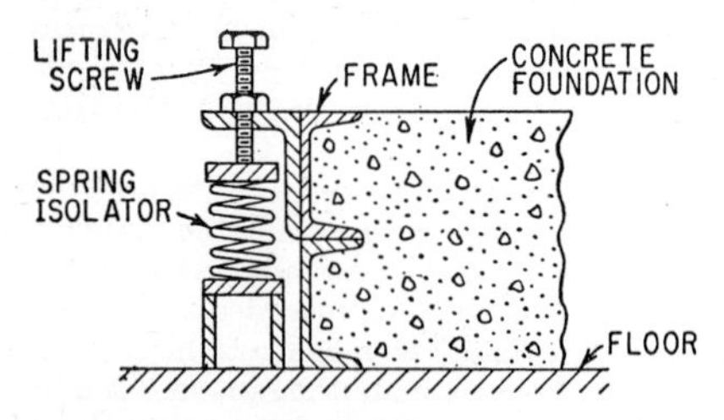

Fig. 32.6. Foundation similar to that shown in Fig. 32.5. The foundation may be constructed by covering the floor with waterproof paper, setting the frame upon the paper, and pouring concrete to fill the frame. After the concrete sets, the foundation is raised off the floor and made operative by turning the lifting screws. (*The Vibration Eliminator Co.*)

In some circumstances it is undesirable to have a block above the level of the floor. Then the block may be placed in a pit, as shown in Fig. 32.8, and often may be of substantially greater size. There are several supporting methods of constructing such blocks. One of these employs cork pads as the resilient means, as illustrated in Fig. 32.8. A pit of the required size is first provided and lined with concrete. The cork pads then are placed within the pit, the pads are covered with a layer of waterproof paper, and the concrete is poured into the form thus provided. The desired natural frequency can be attained by using cork pads of the necessary area and using inactive cork fill (e.g., free cork particles) in the spaces therebetween to fill out the form without adding appreciable stiffness.

Cork as used for isolating vibration is the bark of the cork tree; it may be applied in the form of (1) pads cut from the virgin cork or (2) pads made by compressing cork particles under high pressure and subsequently baking them with superheated steam. A trade name for this latter material is Vibracork.* Because of the limited dimensions of the cork tree, pads of virgin cork can be of only moderate size; however, pads of larger size can be assembled by gluing together several small pads. A representative value for the modulus of elasticity of cork in compression is 400 lb/in.^{2}; i.e., a 1-in. cube of cork deflects approximately 0.1 in. under a load of 40 lb. The natural frequency of Vibracork as a function of unit load is shown in Fig. 32.9[4] for pads 2 and 4 in. thick, and for several densities.

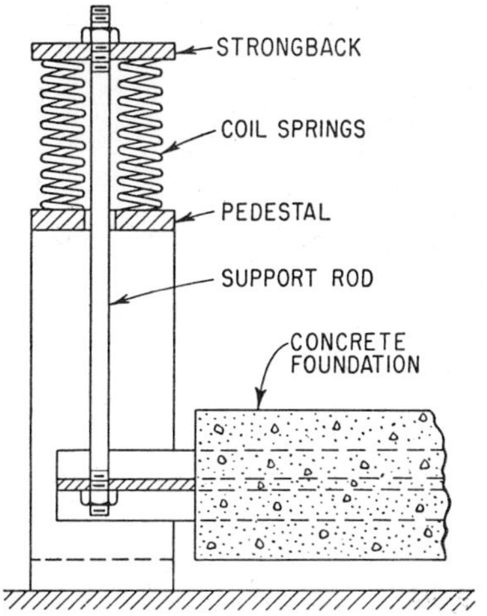

FIG. 32.7. Foundation similar to that shown in Fig. 32.5 but supported by coil springs located on pedestals arranged around the periphery of the foundation. The support rods afford pendulum action and make possible relatively low natural frequencies in horizontal modes. (*The Korfund Co., Inc.*)

Where the natural frequency of the foundation on the isolators is low and the static deflection is correspondingly large, steel springs must be used. In general, this results in a more expensive type of construction because forms are required for pouring the concrete block. A typical concrete-block installation employing steel springs is illustrated in Fig. 32.10. A structural-steel frame is embedded within the concrete block and protrudes from the sides thereof for attachment to isolators located within the pit. If the block is relatively small, it is possible to employ commercially available isolators. For more massive blocks, it is customary to design coil springs (see Chap. 34) specifically to the application and often to provide mechanical dampers in parallel with the springs.

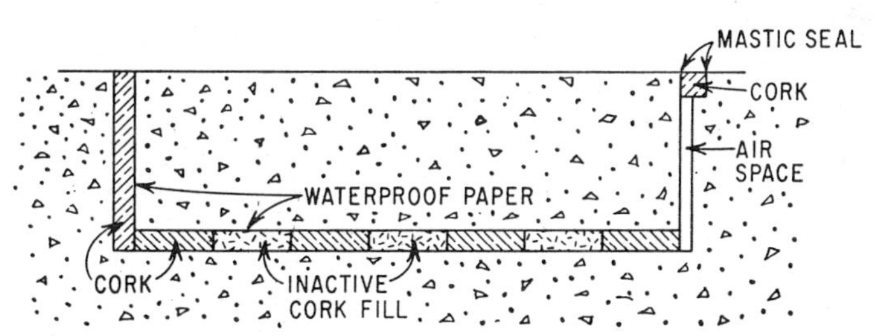

FIG. 32.8. Independent foundation flush with general floor level. Stiffness of cork support is determined by type of cork and area of cork pads; space between pads is filled with inactive cork fill, i.e., with cork particles of negligible stiffness. The foundation may be poured directly on the waterproof paper placed over the cork. Alternate arrangements are shown on opposite ends. At left, the cork serves as a form for the concrete; where less horizontal stiffness is required, removable forms may be used to provide construction shown at right. (*The Vibration Eliminator Co.*)

When the isolators are located substantially below the center-of-gravity of the block, some of the natural frequencies become relatively low. For example, see Fig. 30.18.

* Manufactured by Armstrong Cork Co., Lancaster, Pa.

A trend toward instability is thus introduced, and the effect becomes more important if the machine generates large horizontal forces during normal operation. This limitation can be minimized by installing the isolators in positions closer to the upper surface of the floating block, supported upon abutments extending inward from the walls of the pit. A more refined version of this concept is the T-shaped block illustrated in Fig. 32.11.[5] With such a design it is possible to place the isolators in the same horizontal plane as the

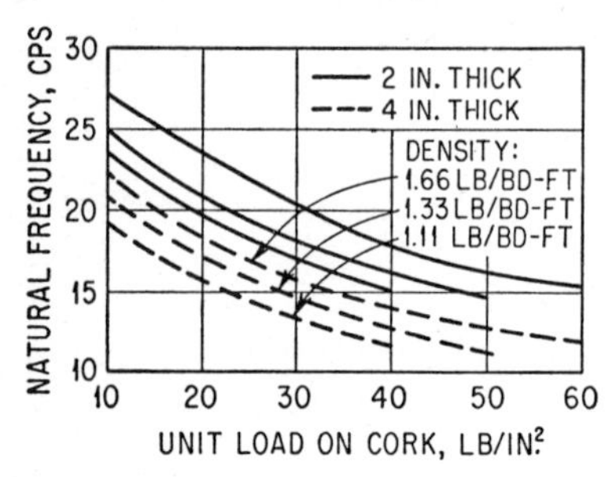

FIG. 32.9. Natural frequency of Vibracork as a function of the unit load (static) for pads of 2 and 4 in. thickness. Data are included for three densities of Vibracork expressed in units of pounds per board foot. A board foot is a quantity of material one foot square and one inch thick. (*Armstrong Cork Co.*[4])

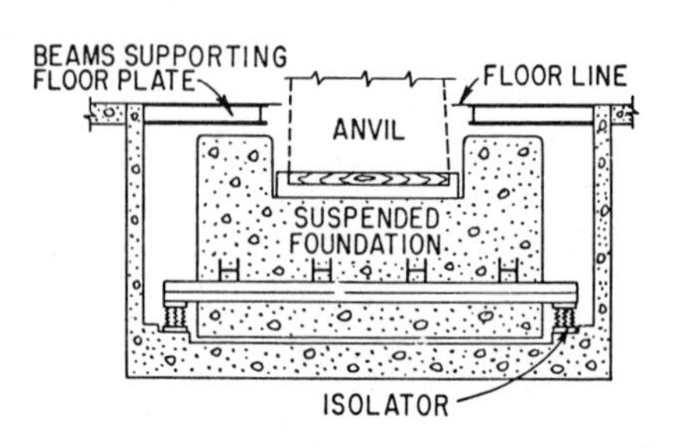

FIG. 32.10. Independent foundation adapted to be located in a pit; the upper surface of the foundation is flush with or below the floor level. The foundation is supported by isolators having coil springs as the principal resilient load-carrying means. (*The Korfund Co., Inc.*)

center-of-gravity and approximately to equalize the natural frequencies in all six modes of vibration. This requires coil springs having equal horizontal and vertical stiffnesses (see Chap. 34). If the distances between the springs when viewed in elevation from either of two sides are made equal to two times the respective radii of gyration, the vertical and rotational natural frequencies are equalized (see Fig. 30.18). The natural frequency in rotation about the vertical axis cannot be controlled independently, but usually is approximately equal to the natural frequencies in other modes.

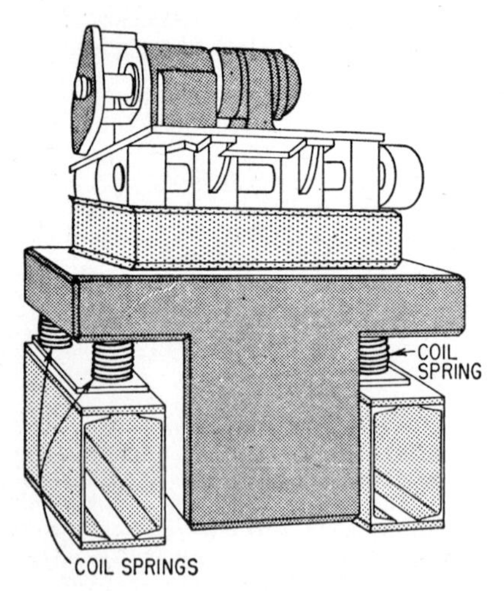

FIG. 32.11. Independent foundation supported by coil springs and arranged to equalize substantially the natural frequencies in all six natural modes of vibration.[5]

When the natural frequency of a concrete block on coil springs is relatively low, the possibility of instability is always present because the only horizontal constraint is that provided by the horizontal stiffness of the load-carrying springs. The stability of a spring in this sense is discussed in Chap. 34. The relatively great installation cost of a concrete block and the still greater cost of making modifications to an existing installation dictates a conservative design approach. If there are doubts concerning stability, additional stabilizing springs arranged to act horizontally may be added.

FLEXIBLE CONDUITS

The advantages attained through the use of vibration isolators may be lost if the mounted machine or equipment is attached to rigid piping, electrical conduit, or shafting. Conduits for supplying electric power or for conducting fluids should have flexible portions to prevent transmission of vibration. Rubber hose often is preferable but, if the temperature is too great or if chemically active fluids must be transmitted, metal hose or tubing must be used. The properties of flexible metal hose and tubing are discussed in the following paragraphs.

In the installation of flexible conduits, the following design principles are of importance:

1. The conduit should be attached to the mounted machine near the point of minimum vibratory movement This effects minimum vibration input and ensures minimum transmission of vibration through the conduit. A length of rigid conduit carried by the ma-

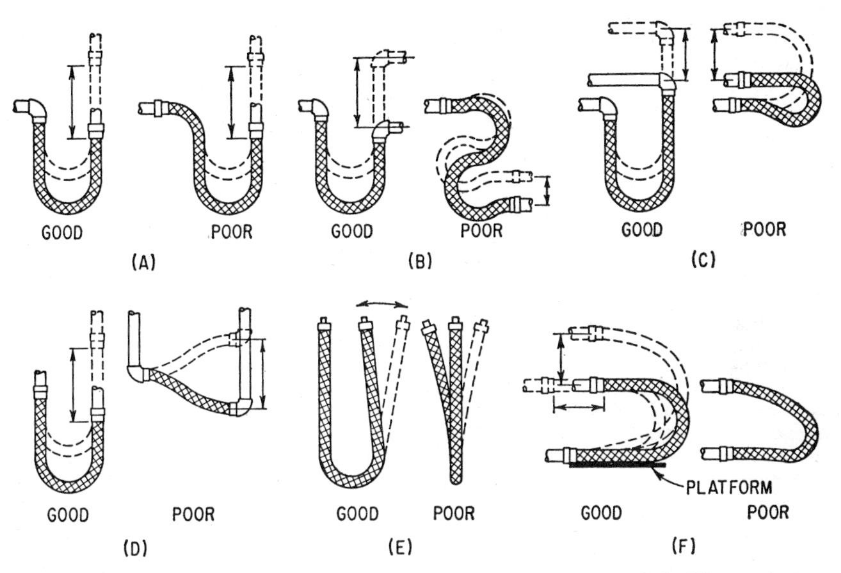

Fig. 32.12. Diagrams of preferred and nonpreferred arrangements of flexible conductors. Movement between the ends is distributed over the length of the conductor to minimize wear. (*P. H. Geiger.*[6])

chine may be desirable to enable the flexible conduit to be attached at the point of minimum movement.

2. The other end of the conduit should be attached to the most massive structure available. This tends to reduce the influence of transmitted vibration, and makes it possible to employ a more rigid conduit while achieving the desired isolation of vibration.

3. The conduit may be installed with relatively large loops where one end experiences large displacement relative to the other end or where a very low stiffness is required of the conduit. In general, the vibration amplitude is small and the conduit usually may be installed straight. The manufacturers of flexible conduit do not provide data on stiffness which make it possible to determine whether a loop is required for any particular application.

4. If a loop is used, the conduit should be arranged so that the vibratory movement is substantially distributed throughout its length. If the movement occurs only adjacent the ends of the conduit, excessive wear resulting from stress concentration at the end connections may shorten the life of the conduit. Figure 32.12 shows a number of preferred arrangements for ensuring distribution of movement throughout the length of conduit.[6]

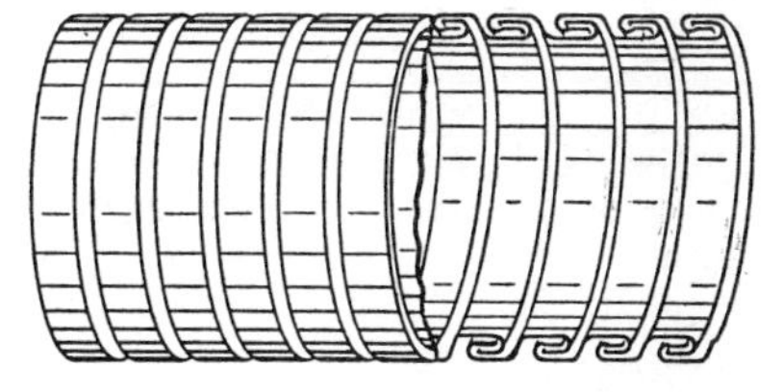

Fig. 32.13. Cutaway view of spiral-wound or interlocked type of flexible metallic hose. (*The American Brass Co.*)

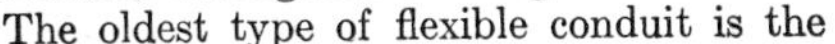

The oldest type of flexible conduit is the

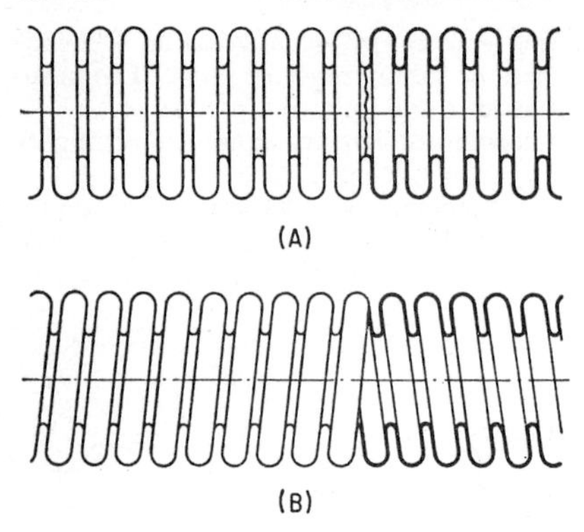

FIG. 32.14. Cutaway views of flexible seamless tubing of (*A*) annular and (*B*) spiral types. (*The American Brass Co.*)

strip wound hose illustrated in Fig. 32.13. This hose is made from a continuous coil of strip metal and embodies an interlocked construction with a packing wound continuously into the groove in the interlocked joint. Its flexibility is derived from the sliding action at the interlocked joint; its pressure-tight qualities are derived from the packing, which usually is asbestos. This type of hose is suitable for containing air, oil, water, and steam; in general it is unsatisfactory for handling gases at high pressure or highly volatile substances where the slightest leakage may create a dangerous condition. It is available in diameters from ½ to 6 in. and may be constructed from different alloys depending on service requirements.

A more modern type of flexible conduit is the corrugated tubing illustrated in Fig. 32.14. It is available with the corrugations having an annular configuration as illustrated in Fig. 32.14*A*, or a helical configuration as illustrated in Fig. 32.14*B*. It is made from seamless, welded, or soldered tubing in diameters from ⅛ to 4 in., depending upon the materials and styles offered by various manufacturers. Welded tubing in diameters as large as 20 in. also is available in corrugated form. Corrugated conduit commonly is made from bronze, steel, and stainless steel but is also available in a variety of alloys, including Monel, Super Nickel, Everdur, and Inconel. Corrugated seamless or welded tubing is not susceptible to leakage; therefore it is satisfactory for containing volatile substances and gases under pressure.

Flexible conduit as obtained from the manufacturer is normally equipped with integral fittings at the opposite ends. Lengths of tubing are available in standard sizes or to custom dimensions determined by the application. Corrugated tubing for pressure lines normally is enclosed within a flexible braid to prevent elongation and ultimate bursting of the tubing, the braid being attached to the end fittings. It is acceptable for use without a braid only for such applications as transmitting air or gas under low pressure.

Data on the stiffness of flexible conduit are not available from the manufacturers. A considerable background of experience has been accumulated, however, and good recommendations may be obtained regarding specific applications.*

VIBRATION THROUGH CRITICAL SPEED

It is inherent in vibration isolation that the natural frequency of the isolator system be lower than the frequency of the excitation. Where the machine supported by the isolator operates intermittently, a condition of resonance must exist momentarily during starting and stopping of the machine. The relatively great amplitude commonly associated with resonance does not occur instantaneously but rather requires a finite time to build up. If the forcing frequency is varied continuously as the machine starts or stops, the resonance condition may exist for such a short period of time that only a moderate amplitude builds up. This problem has been investigated for a linear system in which the forcing frequency varies uniformly through resonance,[7] and the results are summarized by the curves in Fig. 32.15. The ordinate in Fig. 32.15 is the maximum transmissibility; the abscissa is a ratio in which the numerator is the rate of change of forcing frequency in cycles per second per second while the denominator is the square of the natural frequency of the isolator expressed in cycles per second. From the curves for various values of the fraction of critical damping ζ, it is evident that the rate of change of forcing fre-

* Principal manufacturers of flexible hose and tubing include American Brass Company of Waterbury, Conn.; Flexonics Corporation of Maywood, Ill.; Pennsylvania Metallic Tubing Company of Philadelphia, Pa.; and Titeflex Corporation of Newark, N.J.

quency is of little importance with highly damped isolators but of considerable importance if the isolators are lightly damped.

For example, a machine operating at a frequency of 30 cps is mounted upon isolators whose natural frequency is 10 cps and whose fraction of critical damping is $\zeta = 0.025$. In starting the machine, the operating frequency increases from zero to 30 cps in 5 sec with a uniform rate of change of frequency. This rate of change is 30/5, or 6 cps/sec. For a natural frequency of 10 cps, the abscissa parameter is 6/100, or 0.06. Entering Fig. 32.15 at 0.06 on the horizontal scale, a value of 9.5 for maximum transmissibility

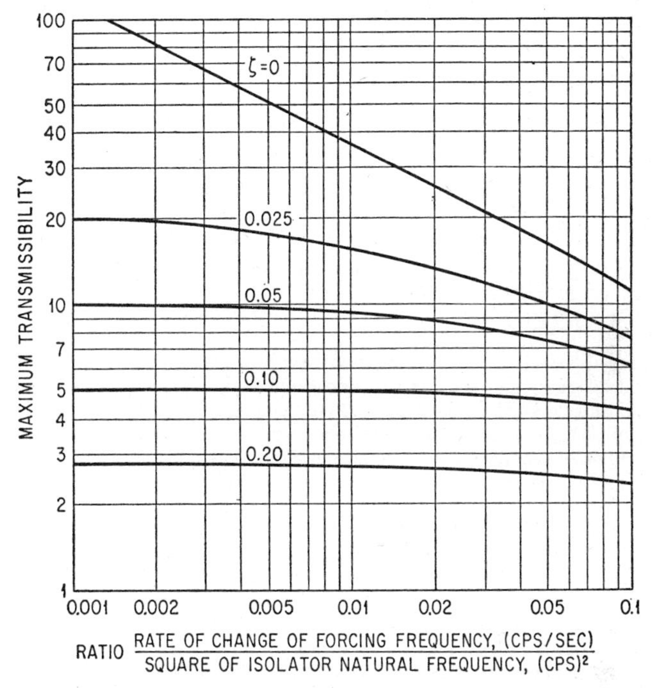

FIG. 32.15. Maximum transmissibility for a viscous-damped, single degree-of-freedom system when the forcing frequency changes at a constant rate in the region of the natural frequency of the system. (*After F. M. Lewis.*[7])

is obtained from the curve for $\zeta = 0.025$. This maximum transmissibility compares with a corresponding value of 20 which occurs with equal damping for continuous operation at resonance.

APPLICATIONS IN VEHICLES

This section discusses the application of isolators in particular vehicles. In general, the isolators are used to protect equipment from the shock and/or vibration that is characteristic of the environmental conditions in vehicles that travel in the air, on land, or on water. Many such requirements for isolators are related to military vehicles, or are influenced by similar requirements of a military nature. For example, many of the vibration isolators used in nonmilitary aircraft conform to specifications whose origin was in military equipment.

In the design and application of vibration isolators for military use, cognizance must be taken of the actual operating requirements as well as of the specifications which are applicable. Specifications are used extensively to specify the important characteristics of much military equipment. This is necessary because (1) the vehicle which ultimately will carry the equipment in question has not yet been built and its environmental characteristics can only be estimated, and (2) equipment frequently is designed for general use in many vehicles, the specification thus attempting to define a general environment which will apply to a number of vehicles. The specified tests therefore do not necessarily represent a reproduction of environmental conditions, but rather a simulation of environmental conditions taken as a group. The specification of tests to simulate an environment is discussed in Chap. 24.

Isolators serve an important function by reducing the importance of secondary or incidental aspects of laboratory tests. This is particularly significant where shock tests are involved. The specification defines the severity of the shock and its general characteristics; in general, the details of the shock motion are not specified and/or are not related to the actual environmental conditions being simulated. For example, the structure used to mount equipment on the shock testing machine generally has different dynamic characteristics than the structure used in a corresponding manner in service. To the extent that these structures are different, the test does not simulate the service condition. To equipment mounted upon isolators, such differences may be unimportant because the isolators "filter out" the vibration contributed by the mounting structures. By minimizing discrepancies between test and service conditions, isolators increase the extent to which the test simulates the service condition; i.e., the use of isolators increases the validity of laboratory tests.

AIRBORNE VEHICLES

This section discusses (1) the application of vibration and shock isolators to protect equipment installed in manned aircraft and guided missiles and (2) particular requirements of isolators for such applications. These classes of airborne vehicles encompass a wide range of sizes, operating characteristics, and tactical missions. Vehicles in the several classes have certain common characteristics that are relevant to the application of isolators:

1. The excitation involves vibration extending to relatively high frequencies, for any type of power plant. The vibration induced by rocket-type power plants and aerodynamic forces has pronounced random characteristics, whereas that originating in reciprocating engines and propellers is more likely to have pronounced periodic components. Significantly lower vibration frequencies characterize the vibration induced by reciprocating engines and propellers.
2. The shock occurring in airborne vehicles usually is not severe. In manned aircraft, the most severe sources of shock are landing and gunfire by the aircraft's own armament. In guided missiles, shock may result from action of the guidance system, separation of booster sections in flight, and handling prior to launch.
3. Guided missiles and certain types of manned aircraft characteristically operate in any attitude. Equipment which utilizes gravitational force to maintain the proper orientation of parts is subject to malfunction.
4. The airframe structure and airborne equipment tend to be as light as possible because unnecessary weight limits the payload and/or the flight range. A natural consequence of efforts to reduce weight is a reduction in stiffness of structural members, including those to which isolators are attached.

TEMPERATURE. Isolators frequently are required to operate at both high and low ambient temperatures. The minimum temperature varies somewhat from application to application; a representative temperature is −65°F (−54°C) although some applications require operation at lower temperatures. In many applications the temperature requirements are not as stringent, particularly for installation in low-performance aircraft or for isolators located within an air-conditioned space.

The minimum temperature at which an isolator is usable is the lowest temperature at which it maintains substantially the same stiffness as at room temperature. It is necessary to ensure that the natural frequency under low-temperature conditions remains at the design value. Isolators using springs made from natural or synthetic rubber are likely to have limited usefulness in applications involving very low ambient temperatures. Certain methods of evaluating the properties of rubber at low temperatures use a criterion based upon brittleness. A material may pass this test satisfactorily even though exhibiting a large increase in stiffness; thus, it is not suitable for use in isolators required to operate at low temperatures. Other tests evaluate the change in modulus (see Chap. 35).

For typical applications, a representative maximum ambient temperature is 175°F (79°C). Isolators sometimes are required to operate at much higher temperatures—particularly if located near the propulsion equipment or where they are subject to the influences of aerodynamic heating. The "thermal inertia" of an isolator may maintain the temperature at a relatively low ambient level even though a higher temperature exists for a short interval of time.

The representative specified temperature limits of −65°F to 175°F coincide approximately with the useful temperature range of natural and some synthetic rubbers. If interpreted literally and stringently, the specification may exclude such materials and require the substitution of other materials whose general suitability is somewhat lower. Thus, inferior equipment may result from undue emphasis on temperature limits. Silicone rubber especially compounded for use in isolators has many of the desirable properties of rubber and a substantially wider range of useful temperatures. Figure 35.3 gives data on the variation of stiffness (or modulus) with ambient temperature for several types of natural and synthetic rubber. Except for temperatures greater than several hundred degrees Fahrenheit, there are no limitations imposed by temperature on the use of metal springs in the form of coils, strips, or knitted wire.

INSTALLATION OF EQUIPMENT. In general, it is not feasible to attempt servicing or repair of equipment while installed in an aircraft or missile because (1) space is not available and (2) the vehicle is unavailable for use. Thus, equipment is designed for ready installation and removal using a minimum number of thumbscrews or other quick-disconnect fasteners. (See Chap. 43 for descriptions of such fasteners.) The isolators may be permanently attached to the equipment and removable from the airframe; however, a more common type of construction embodies isolators permanently installed in the vehicle. An intermediate structure (mounting tray or mounting rack) permanently attached to the isolators includes the quick-disconnect means for attachment of the equipment.

The mounting tray is a common source of difficulty. To some extent, it does not perform a function whose importance is proportionate to the added weight. The equipment frequently is attached to the tray at locations somewhat remote from the isolators; thus, the tray must have adequate strength and rigidity to transmit the necessary forces and to preserve the dynamic characteristics of the isolators. In some instances, several equipments are attached to a single tray that maintains the necessary interrelations of the equipments while supplying electric power, cooling air, etc. Trays adapted to mount several equipments offer a number of advantages: (1) The assembly of several equipments mounted together can be made to fit into an irregular-sized compartment more conveniently than several individual equipments, (2) the large assembly tends to be more stable on its isolators than individually isolated equipments, and (3) less total space is required for deflection of isolators because there are fewer isolators.

ACTIVE ISOLATORS. An active isolator is an isolator provided with a source of power and adapted to respond in a manner tailored to specific requirements. The design of active isolators is discussed in Chap. 33 with particular reference to applications in guided missiles and helicopters.

ENVIRONMENTAL CONDITIONS. Chapters 22 and 23 discuss the analysis of data defining environmental conditions, and the presentation of such data in a form suitable for further engineering use. Theoretical aspects of vibration isolation are discussed in Chap. 30; the theory of shock isolation is discussed in Chap. 31. Calculations, made on the basis of these chapters, indicate quantitatively the performance expected from isolators.

In general, the natural frequency of an isolator should be as low as possible without introducing undesirable characteristics. A low transmissibility, i.e., high efficiency of isolation, results from a natural frequency that is small relative to the forcing frequency. However, the maximum deflection of the isolator as a result of shock increases as the natural frequency decreases. It is common practice to equip isolators with relatively stiff snubbers or bumpers to limit such deflection. Repeated engagement with snubbers has the same effect as repeated applications of shock. This should be prevented by selecting the natural frequency and the clearance to the snubber so that snubber engagement seldom occurs during normal operation.

The objective of avoiding snubber engagement while maintaining the natural frequencies low can be best attained by selecting isolator locations to uncouple the natural modes of vibration (see Chaps. 3 and 30). Then the natural frequencies in the several modes each may be made approximately equal to an optimum value selected in accordance with isolation requirements and available clearances to the snubbers. Practical difficulties in achieving this objective may arise because many equipment designs do not contemplate the attachment of isolators in the locations necessary to uncouple natural modes of vibration. Then the desired result can be attained only by redesign of the equipment or by use of a mounting tray or brackets that place the isolators in optimum locations.

The value of natural frequency used in the design or selection of isolators depends upon the nature of environmental conditions, the type of equipment and the space available. In general, natural frequencies should be between 10 and 20 cps. The higher value of 20 cps is preferable for applications where space is limited because it tends to minimize deflection of the isolator as a result of shock. For effective use of such an isolator, both the equipment structure and the airframe structure to which the isolator is attached must be substantially more rigid than the isolator; otherwise, the deflection takes place in the structure and the isolator is not effective. There are many installations which do not meet this requirement. Then, isolators of lower natural frequency are required and greater deflection results under conditions of shock. Unfortunately, it is characteristic of many installations in aircraft and guided missiles that stiffnesses of structures dictate the use of isolators of low natural frequency but space limitations make it inevitable that snubber engagement occurs as a result of shock.

ROAD AND RAIL VEHICLES

The vibration that exists in a road vehicle may result from excitation by the power source or by irregularities in the surface of the road. The engine, usually an internal-combustion engine, operates at a relatively high frequency and usually is not an important source of vibration within the vehicle because the engine itself is supported by isolators. In exceptional cases, such as large trucks and army vehicles, vibration from the engine may be significant.

In the conventional road vehicle, road irregularities may excite each of three different elastic systems separately. The first system consists of the body of the vehicle and the main springs, a second system is formed by the unsprung weight (wheels, axles and drive mechanism) supported between the main springs and the elastic tires, and the third system is the local body structure which can experience structural vibration at the natural frequencies of the structures. The natural frequencies of these systems are primary factors in determining design requirements for isolators:

1. The natural frequency of the vehicle body on the main springs usually is between 1 and 4 cps, the higher value generally applying to military vehicles. It is not feasible to use isolators with a lower natural frequency than the natural frequency of the main-spring system. Therefore, it is unlikely that isolators can provide significant protection to equipment with respect to the over-all motion of the vehicle body.

2. The natural frequency of the unsprung weight (sometimes referred to as "wheel hop" frequency) usually is between 10 and 20 cps, the higher value generally applying to military vehicles. In general, it is not feasible to design isolators to have natural frequencies substantially lower than the natural frequency of the unsprung weight unless large clearances are available to permit isolator deflection without bottoming; thus, the isolators do not alleviate vibration from this source. A principal problem in the application of isolators is to avoid resonance with this source of vibration.

3. The natural frequencies of the structural members of the vehicle body usually are higher than the frequencies associated with the springing system. A principal function of isolators is to alleviate the effects of the resonance conditions that would develop when the natural frequency of an important member of the equipment coincides with a natural frequency of a structural member supporting the equipment.

Characteristics of isolators for vehicular applications can be determined more effectively if the principal natural frequencies of the vehicle are known. The natural frequency of the unsprung weight is of particular importance because (1) it borders on the frequency range typical of optimum isolator design, and (2) the vibration induced in the vehicle body by vibration of the unsprung weight may be of significant severity. If the natural frequency of the unsprung weight is known, the isolators should be applied so that the natural frequencies of the equipment in all modes of vibration are different than the frequency associated with the unsprung weight. It is necessary to consider all natural modes of vibration, both coupled and uncoupled, as set forth in Chap. 30. If the characteristics of the vehicle are not known, the isolators should be applied on the premise that the natural frequency of the unsprung weight may be as high as 20 cps; then the natural frequencies of the equipment-and-isolator assembly should have a minimum value of 25 cps. This objective is reasonably attainable only if the natural modes of vibration are substantially uncoupled; otherwise the maximum natural frequency becomes unreasonably high to attain a minimum natural frequency of 25 cps.

NAVAL SHIPS

Three principal types of shock and vibration on ships are significant in the design and application of isolators:

1. The predominant vibration of the ship is that induced by the propeller; the predominant frequency is the blade frequency, i.e., shaft speed times number of blades. In general, the blade frequency varies as the speed of the ship changes, from a minimum of near zero to a maximum generally considered to be approximately 20 cps. In exceptional instances, the maximum frequency may be substantially greater than 20 cps.

2. The predominant shock arises from enemy action, primarily underwater explosions. Secondary sources of shock reside in the ship's own armament. Severe shock on a ship is characterized by (*a*) suddenly acquired velocities of principal structural members of the ship with consequent displacements of large magnitude and (*b*) transient vibration of the structural members in response to the initial sudden shock motion. The general severity of the shock and the frequencies of the ensuing transient vibration are represented realistically by the corresponding characteristics of shock testing machines used for testing shipborne equipment (see Chap. 26).

3. The machinery operating on a ship creates an ambient vibration level which is not particularly objectionable on the ship but which radiates acoustic power into the water. In submarines and other ships that seek to avoid detection by underwater listening devices, the acoustic power radiated into the water constitutes a serious hazard.

The vibration induced by the propeller is of significant magnitude and relatively low frequency. Some types of shipborne equipment are capable of withstanding such vibration without alleviation; then the isolators may be designed with the intent only to isolate shock or they may be omitted entirely. Types of equipment that are unable to withstand the vibration are mounted upon isolators of relatively low natural frequency; such isolators must be well damped because they may become resonant with the shipboard vibration and must have relatively large clearances to prevent bottoming as a consequence of shock.

In the isolation of shock, it is important that the isolators satisfy the following characteristics: (*a*) an adequate stiffness and permissible deflection to respond to the maximum shock motion in a manner that reduces the severity of the shock as it is transmitted to the isolated equipment, (*b*) stiffness adapted to isolate the vibration created by the structural members of the ship in response to the shock, and (*c*) means to prevent excessive vibration of the equipment as a result of the propeller-induced vibration. The latter means may be either high damping in the isolators (so that the vibration severity is not excessive even though operating continuously at resonance) or, alternatively, a natural frequency greater than the maximum frequency of the propeller-induced vibration. The latter means is mechanically simpler insofar as isolator design is concerned but the relatively high stiffness then approaches the stiffness of many principal structural members of the equipment. Under these circumstances an isolator ceases to perform its intended function because the structure rather than the isolator deflects.

An isolator with adequate damping may have a lower natural frequency (in the range of frequencies of propeller-induced vibration); consequently the stiffness is lower and does not impose such a severe requirement on structural members of the equipment. For a given shock motion, characterized by a sudden movement of the ship's structure, the maximum deflection of an isolator varies inversely as its natural frequency. Thus, an isolator with a low natural frequency (and consequently high damping) must be relatively large in size to permit deflection to occur without bottoming. However, such an isolator may have the characteristics necessary to isolate propeller-induced vibration as well as to isolate shock.

An isolator used to reduce the level of acoustic power transmitted to the water should have a relatively low natural frequency. The efficiency of isolation increases as the natural frequency decreases; hence, an objective is to make the natural frequency as low as stability considerations permit. A natural frequency of 5 cps constitutes a reasonable design objective; from a vibration isolation requirement, an isolator with a substantially lower natural frequency would be desirable but may not be sufficiently stable for waterborne equipment (see Fig. 32.34 for a typical isolator). The limitation on isolation arises because the hull structure of a typical ship has many natural frequencies and corresponding modes of vibration. At these natural frequencies, it may experience vibration of objectionably large magnitude when subjected only to the relatively small forces transmitted through an isolator with a low stiffness. Chapter 30 discusses the use of isolators with a nonrigid structure and shows analytically that isolators may be ineffective at certain critical frequencies.

SPRING MATERIALS

This section discusses the application of spring materials to the design of isolators, supplementing the more fundamental discussions of metal springs in Chap. 34 and rubber springs in Chap. 35. Emphasis is given to the selection of spring materials and to forms of springs, considering required static deflection, stiffness in lateral directions, load-carrying capacity, environmental conditions, damping, weight and space limitations, linearity, and relation of dynamic stiffness to static stiffness.

GENERAL PROPERTIES

Many different materials are used as the resilient elements of isolators. These may be grouped somewhat arbitrarily as (1) natural and synthetic rubber (elastomers), (2) metal springs, and (3) a group of miscellaneous materials including cork, felt, sponge rubber, and various compounded materials. Each of these materials has its peculiar advantages and disadvantages. The materials listed in the third group generally are available only in slab form and are used principally to mount machinery. Isolators embodying steel springs and molded rubber parts, on the other hand, are used generally as component parts of machines as well as for the mounting of machinery.

Steel springs are commonly used where the static deflection is required to be great, where temperature or other environmental conditions make rubber unsuitable, and under some circumstances where a low-cost isolator is required. Molded rubber parts find wide application because they may be conveniently molded to many desired shapes and stiffnesses, embody more internal hysteresis than metal springs, often require a minimum of space and weight, and can be bonded to metallic inserts adapted for convenient attachment to the isolated structures.

Of the elastomers, natural rubber probably embodies the most favorable combination of mechanical properties, such as minimum drift, maximum tensile strength, and maximum elongation at failure; also, it tends to be the least expensive. Its usefulness is restricted somewhat by limited resistance to deterioration under the influence of hydrocarbons, ozone, and high ambient temperatures. Neoprene and buna N exhibit excellent resistance to hydrocarbons and ozone—buna N being particularly well suited to applications involving relatively high ambient temperatures. Buna S is a good general-purpose synthetic rubber for use in vibration isolators. Silicone rubber is a synthetic elastomer which is suitable for applications involving either high or low ambient temperatures at which no other elastomer can be used. It shows remarkable stability in important properties over a very wide temperature range, both above and below room temperature.

Felt, cork, sponge rubber, and compound materials usually are cut from large slabs and used flat, thereby lacking the ready adaptability of rubber parts molded to shape and frequently adhered to metal inserts for easy application. Cork frequently is used to support relatively large concrete foundations.* Felt sometimes is used under machinery; its application is convenient because it can be cemented in place and is generally unaffected by oil and other foreign substances. Glass fibers impregnated with a suitable binder, and knitted wire compressed to a specified shape and density may be used to form the resilient elements of vibration isolators. The latter is relatively immune to environmental influences, and has seen extensive use in military aircraft. Numerous other materials have been compounded in slab form and are used for vibration isolation, particularly in conjunction with machinery.

In the design of vibration isolators and in the selection of a resilient material, many practical factors must be evaluated. The principal factors discussed in this section are:

1. The static deflection of the isolator under the dead-weight load of the supported equipment
2. The isolator stiffness in the lateral directions relative to the stiffness in the direction of the dead-weight load
3. The load to be carried by the isolators
4. Extremes of temperature and other environmental conditions to which the isolators are subjected
5. Damping attainable from hysteresis of the load-carrying spring
6. Weight and space limitations applicable to the isolators
7. The effect of nonlinear elasticity upon the performance of the isolators
8. The relation between the static and dynamic stiffnesses of the isolator

The above factors are considered in detail in the following paragraphs.

STATIC DEFLECTION

The required static deflection under the dead-weight load determines to a considerable extent the type of elastic load-carrying element to be used in the isolators. Organic materials, such as rubber and cork, are capable of experiencing very large unit strains applied momentarily, but exhibit a pronounced tendency to drift or creep if large strains remain for an appreciable period of time. Metal springs, on the other hand, generally experience permanent deformation if the stress exceeds the yield stress, even momentarily, but show little drift or creep when the maximum stress is maintained below the yield stress.

Rubber is suited for use in isolators in which the deflection under the dead-weight load (static deflection) is moderately small but wherein the deflection may become significantly great momentarily. For example, an isolator used for the isolation of shock sometimes must sustain a large deflection for a short time. The unit strain in rubber (i.e.,

* Some relevant mechanical properties of cork are given in the section entitled *Isolated Foundations*.

the deflection divided by the thickness) under static conditions should not exceed 0.10 to 0.15 if the strain is compression, and should not exceed 0.25 to 0.50 if the strain is shear. The permissible dynamic strain is much larger. For example, the strain in compression may be as great as 0.50 to 0.75 without damage to the rubber although the stiffness of the isolator may become significantly greater at the extreme of strain (see Chap. 35). A dynamic strain of 2.0 to 3.0 in shear is permissible if the rubber is bonded adequately to supporting members.

It is advantageous to use metal springs when the deflection under the dead-weight load is large. In particular, a helical spring can be designed to sustain a relatively large deflection without exceeding the elastic limit of the material. (See Chap. 34 for a detailed discussion of metal springs.) Care should be exercised in the application of metal springs to ensure that the stress during dynamic deflection does not exceed the elastic limit.

STIFFNESS IN LATERAL DIRECTIONS

Effective vibration isolation requires the control of the natural frequencies, in the several degrees-of-freedom, of a rigid body on resilient supports. (See Chap. 30 for a detailed discussion of the basic principles.) Thus, it is necessary that the stiffnesses of the isolators in the directions of the several coordinate axes be controlled.

RUBBER SPRINGS. Rubber is a substantially incompressible material, i.e., it has a Poisson's ratio of approximately 0.5.* The stiffness of a rubber spring when strained in compression thus depends, to a considerable extent, upon the extent of the surface available for lateral expansion. The stiffness of the spring in shear, on the other hand, is substantially independent of the shape of the rubber member. As a very approximate indication of order of magnitude, it may be assumed for estimating purposes that the minimum likely compression stiffness of a given rubber spring is five times as great as the shear stiffness. The maximum compression stiffness may be several times as great as the minimum if lateral expansion of the rubber is constrained.

Rubber-pad Isolator. A vibration isolator formed of a pad of rubber has relatively low stiffness in the two shear directions, and a relatively high stiffness in the tension-compression direction. In general, the high stiffness leads to a high natural frequency in at least one mode with a consequent decrease in isolation effectiveness. In some applications, if the vibratory force or motion in a particular direction is negligible, it is possible to orient such a pad relative to the mounted equipment so that the large stiffness in compression is acceptable. Where this is not possible, means must be introduced to attain desired stiffnesses in the respective directions. Examples of rubber isolators which attain this objective are shown in Figs. 32.35 and 32.39. Another means of achieving the objective is to use a lever as discussed in the next section.

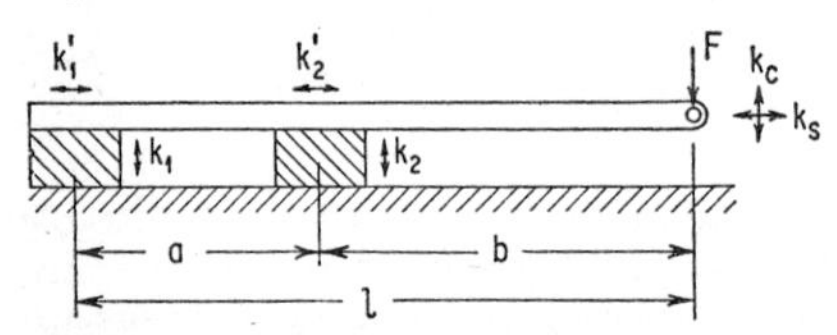

FIG. 32.16. Schematic diagram of isolator assembled from two rubber pads and a hinged bar. The force F acting on the isolator is applied normal to the length dimension of the bar through a hinged connection. The stiffnesses k_c, k_s are effective stiffnesses in the directions indicated, at point of application of force F.

Rubber Pads and Lever in Combination. Rubber pads can be used in conjunction with a lever, as illustrated in Fig. 32.16, to attain equal stiffnesses in orthogonal directions. The pads of stiffnesses k_1 and k_2 in compression are spaced apart a distance a and spanned by a massless bar of over-all length l. The corresponding shear stiffnesses of the rubber pads in directions perpendicular to the compression stiffnesses are k_1' and k_2', respectively. The force F is applied to a pivot at the extremity of the bar so that its direction is always parallel to the principal compression axis of the pads. The effective stiffness at the end of the bar is defined by the compression stiffness k_c and the shear stiffness k_s.

For the condition of static equilibrium, the following equations are written by tak-

* For a more exact value, see Chap. 35.

ing a summation of forces and a summation of moments about pad k_1:

$$\begin{aligned} F - k_1\delta_1 - k_2\delta_2 &= 0 \\ k_2\delta_2 a - Fl &= 0 \end{aligned} \tag{32.1}$$

where δ_1, δ_2 are the deflections of the respective pads. Solving Eqs. (32.1) simultaneously for the displacements δ_1 and δ_2,

$$\delta_1 = \frac{F}{k_1}\left(1 - \frac{l}{a}\right) \qquad \delta_2 = \frac{F}{k_2}\left(\frac{l}{a}\right) \tag{32.2}$$

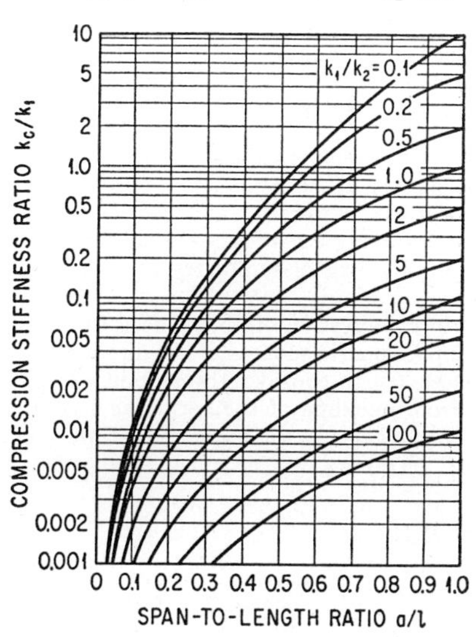

FIG. 32.17. Compression stiffness k_c of the isolator illustrated in Fig. 32.16.

Since the bar must translate and rotate as a rigid body, the deflection δ_F at point of application of force F is

$$\delta_F = \delta_1 + (\delta_2 - \delta_1)\left(\frac{l}{a}\right) = \left(\frac{l}{a}\right)\delta_2 + \left(1 - \frac{l}{a}\right)\delta_1 \tag{32.3}$$

where δ_F is the deflection at the pivot point due to the applied force F. Substituting Eq. (32.2) in Eq. (32.3), the effective compression stiffness k_c at point of application of force F is

$$k_c = \frac{F}{\delta_F} = \frac{k_1 k_2}{(l/a)^2 k_1 + (1 - l/a)^2 k_2} \tag{32.4}$$

Equation (32.4) may be written to define k_c with respect to k_1:

$$\frac{k_c}{k_1} = \frac{1}{(1 - l/a)^2 + (l/a)^2 (k_1/k_2)} \tag{32.5}$$

Equation (32.5) is shown graphically in Fig. 32.17. For small values of the span-to-length ratio a/l, the stiffness ratio is, approximately,

$$\frac{k_c}{k_1} \simeq \frac{(a/l)^2}{1 + k_1/k_2} \tag{32.6}$$

The shear stiffness of the system shown in Fig. 32.16 is

$$k_s = k_1' + k_2' \tag{32.7}$$

When the compression stiffness k_c is equal to the shear stiffness k_s, the expressions for these stiffnesses given by Eqs. (32.4) and (32.7) are equated to give the following relation:

$$\left(1 - \frac{l}{a}\right)^2 + \left(\frac{l}{a}\right)^2 \left(\frac{k_1}{k_2}\right) = \frac{1}{(k_1'/k_1) + (k_2/k_1)(k_2'/k_2)} \tag{32.8}$$

If the ratio of shear-to-compression stiffness k'/k is the same for both pads, Eq. (32.8) reduces to

$$\left(1 - \frac{l}{a}\right)^2 + \left(\frac{l}{a}\right)^2 \left(\frac{k_1}{k_2}\right) = \frac{k/k'}{1 + k_2/k_1} \tag{32.9}$$

The stiffness k_s also applies in a direction normal to the paper if the hinge has rigidity with respect to a vertical axis. Thus, using the relation of Eq. (32.8) or (32.9), an isolator can have equal stiffnesses in the directions of the three coordinate axes.

HELICAL SPRINGS. Commonly available data on the characteristics of helical springs apply to a load along the axis of symmetry, and usually neglect the stiffness in a lateral direction; this is significant for the design of vibration isolators. Data are included in Chap. 34 for determining the stiffness in the lateral direction when the deadweight load is applied along the axis of symmetry in such a manner as to compress the spring.

Helical springs loaded in compression often are used to isolate vibration of relatively low frequency. If the application is such as to require a relatively low stiffness in the lateral direction, it is necessary to consider the stability of the assembly. Instability is not likely to develop from buckling of the spring as a column but rather by lateral collapse of the entire assembly because the horizontal stiffness of the springs is not great enough to ensure stability. The stability of helical springs is considered in Chap. 34.

LOAD-CARRYING CAPACITY

Resilient materials strained in compression are most useful when the load is relatively great and the static deflection is relatively small. Unbonded resilient materials fall into this category. Such applications are difficult to design for a small load unless the required static deflection is relatively small. Otherwise, the small area and relatively great thickness tend to cause a condition of instability. To a considerable extent, this limitation may be overcome by using sponge rubber, a material of lower modulus. In general, where the load is relatively small, it is preferable to employ rubber springs that carry the load in shear, although this frequently introduces other problems related to the relatively great stiffness in one lateral direction.

ENVIRONMENTAL CONDITIONS

It is common for vibration isolators to be subjected to a wide range of severe environmental conditions. In military applications, a wide range of ambient temperatures is encountered in addition to ozone, hydraulic fluids, rocket fuel, etc. The stringency of environmental influences is not so great in industrial applications but includes lubricating oil, ozone, etc. Little difficulty is encountered in meeting most of the environmental requirements with metal springs. Organic materials are usually more susceptible to environmental influences. Their superior mechanical properties, such as lighter weight, smaller size, greater damping, and ability to store large quantities of energy under shock, constitute a large incentive to develop organic materials (e.g., natural and synthetic rubber) which are capable of meeting the environmental requirements of military applications.

In military applications, it is common to require that isolators remain operative at temperatures as low as −65°F. The maximum temperature varies widely, depending upon conditions. In many applications the maximum temperature is approximately 150°F but there are requirements as high as 400°F. Most natural and synthetic rubber compounds are usable at temperatures as high as 150°F. They sometimes have marginal acceptability at the minimum temperatures specified but often can be used with little difficulty under practical conditions.

Silicone rubbers offer substantially superior service at the low temperatures and usually can be used continuously at temperatures as high as approximately 300°F. There is no generally stated upper limit, depending upon the properties of the particular compound, the degree of deterioration which is permissible as a result of continued exposure at high temperatures, and the duration of the exposure. A temperature substantially greater than 300°F is permissible for several hours. The superior ability of silicone rubbers to withstand high and low temperatures is offset somewhat by inferior strength, tear resistance and abrasion resistance. Details of their characteristics are set forth in Chap. 35.

DAMPING

Damping is inherent in all resilient materials to some extent. For some purposes, the hysteresis in the load-carrying element affords adequate damping whereas, for other purposes, the addition of a separate damper is necessary. One of the principal uses for vibration isolators is in military equipment wherein the excitation covers a wide range of frequencies and may have random properties. Where the nature of the excitation is difficult to predict, it is desirable that the damping in the isolator be relatively great. It is also desirable that isolators embody substantial damping where it is known that they will be required to operate at resonance, even momentarily, as in the starting of a machine whose operating frequency is greater than the natural frequency of the machine on its isolators.

Many specifications used by the military services require, by imposing explicit or implicit limits on the maximum response of the system when vibrated at its resonant frequencies, that isolators have substantial damping. As a consequence of this requirement, isolators for military applications often include discrete dampers arranged in parallel with the load-carrying springs. Descriptions of such dampers and data for use in their design are included in a later section.

To determine whether auxiliary dampers are required, damping requirements should be compared with the damping available from internal hysteresis of the main load-carrying element. Internal hysteresis-damping of a particular spring is a function of the strain and the vibration frequency; it varies widely from one rubber compound to another. Typical values of damping are given in Figs. 32.18 and 32.19, expressed as the

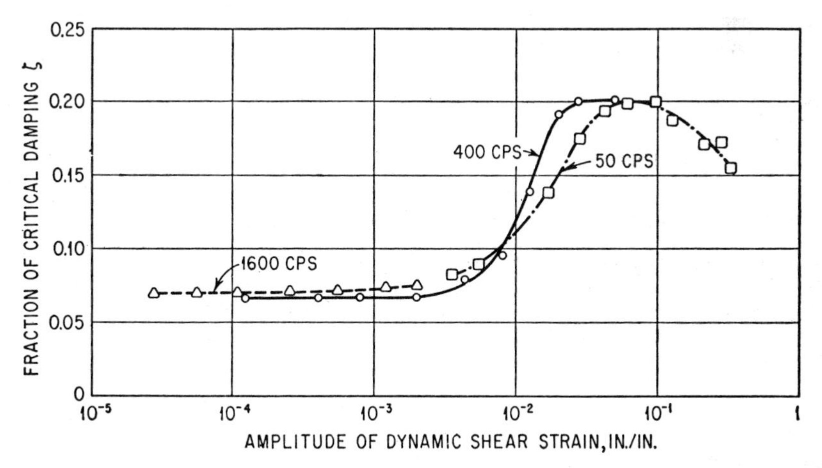

FIG. 32.18. Damping exhibited during vibration of a typical elastomeric material strained in shear, as a function of frequency and amplitude of dynamic shear strain. (*After G. W. Painter.*[8])

fraction of critical damping.[8] In general, the damping increases as the frequency increases. Damping in an isolator is of greatest significance at the resonant frequency. The data of Figs. 32.18 and 32.19 can be used to predict transmissibility at resonance by estimating the frequency and the strain amplitude; then the fraction of critical damping is obtained from the curves and used with Eq. (30.1) to calculate transmissibility at resonance.

When the excitation occurs at relatively high frequencies, the mass of the isolator spring becomes important because standing waves tend to develop in such springs. This results in a relatively high transmissibility at the standing-wave frequencies, as pointed out in Chap. 30. The influence of standing waves tends to increase as the mass

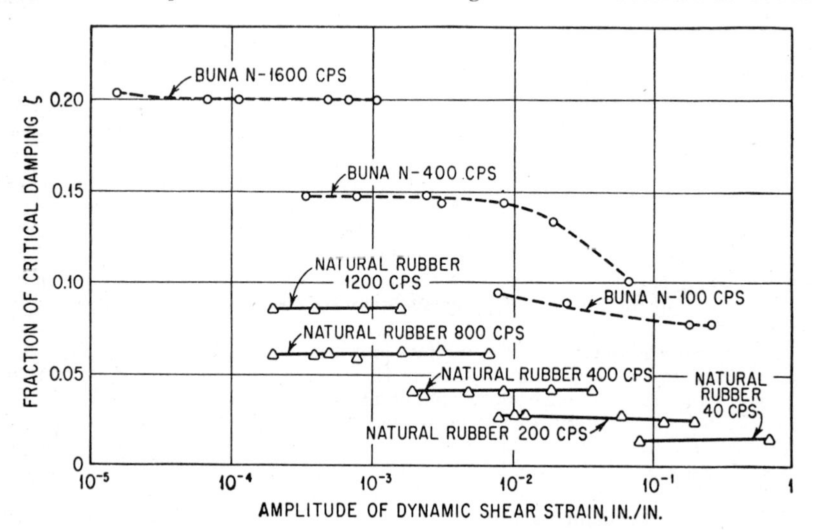

FIG. 32.19. Damping exhibited during vibration of buna N synthetic rubber and natural rubber strained in shear, as a function of frequency and amplitude of dynamic shear strain. (*After G. W. Painter.*[8])

of the spring increases and as its damping decreases. Rubber is less likely to cause trouble from standing waves than metal when measured by either of these criteria. If a metal spring is used in applications requiring the isolation of vibration at high frequencies, it is common to employ a rubber pad in series with the steel spring and, in some instances, to damp the vibration of the individual coils by the addition of suitable damping material.

WEIGHT AND SPACE LIMITATIONS

The amount of load-carrying resilient material required for a vibration isolator is determined by the quantity of energy to be stored. In the usual vibration-isolation condition, the vibration amplitude tends to be small relative to the static deflection and the amount of material may be calculated by equating the energy storage in the material of the isolators to the work done on the isolator system in deflecting the isolator. The work done on the isolator expressed in terms of the equipment weight W and static deflection δ_{st} is

$$\text{Work} = \frac{W\delta_{st}}{2} \tag{32.10}$$

The strain energy stored in a volume V of the resilient material of the isolator may be defined in terms of stress and strain or, alternatively, in terms of modulus and strain:

$$\text{Stored energy} = \int_V \frac{\sigma\epsilon}{2}\,dV = \int_V \frac{E\epsilon^2}{2}\,dV \tag{32.11}$$

If the strain is uniform over the volume of the resilient material, Eq. (32.11) may be integrated directly:

$$\text{Stored energy} = \frac{VE\epsilon^2}{2} \quad \text{in.-lb} \tag{32.12}$$

The volume of resilient material required for a given application is obtained by equating work and stored energy: *

$$\frac{V}{W} = \frac{\delta_{st}}{E\epsilon^2} \quad \text{in.}^3/\text{lb} \tag{32.13}$$

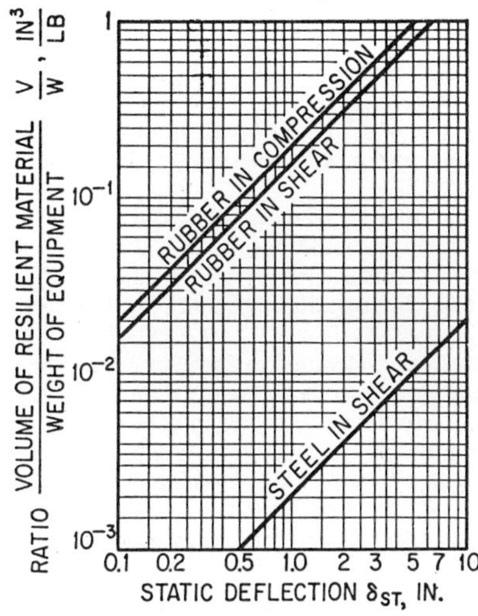

FIG. 32.20. Volume of resilient material required in an isolator with reference to the weight of the supported equipment, as a function of static deflection of the isolator. The curves are based upon compression and shear moduli of rubber, $E = 500$ lb/in.², $G = 100$ lb/in.², and upon allowable compression and shear strains (static) of $\epsilon = 0.1$ and $\gamma = 0.25$. Corresponding assumed values for steel are $G = 11.5 \times 10^6$ and $\gamma = 6.5 \times 10^{-3}$.

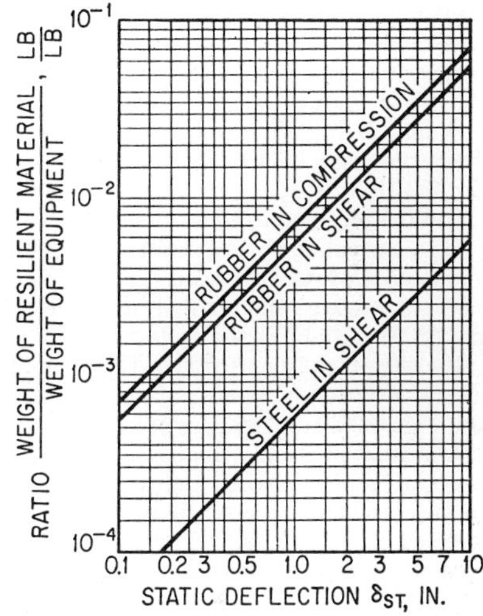

FIG. 32.21. Weight of resilient material required in an isolator with reference to the weight of the supported equipment, as a function of static deflection of the isolator. The curves are based upon compression and shear moduli of rubber, $E = 500$ lb/in.², $G = 100$ lb/in.², and upon allowable compression and shear strains (static) of $\epsilon = 0.1$ and $\gamma = 0.25$. Corresponding values for steel are $G = 11.5 \times 10^6$ and $\gamma = 6.5 \times 10^{-3}$. Weight density of rubber and steel are 0.035 and 0.285 lb/in.³, respectively.

The required volume of resilient material depends upon the values used for the modulus E and the strain ϵ (or G and γ for material strained in shear). For steel in torsion, $G = 11.5 \times 10^6$; a reasonable working stress for a typical spring material is 75,000 lb/in.² (strain $\gamma = 6.5 \times 10^{-3}$). Using these values for G and γ, the shear equivalent of Eq. (32.13) is shown by the "steel in shear" curve of Fig. 32.20. Corresponding results are

* For material strained in shear, the corresponding expression is

$$\frac{V}{W} = \frac{\delta_{st}}{G\gamma^2}$$

included in Fig. 32.21 in which the amount of spring material is expressed in terms of its weight.

Application of the results expressed by Eq. (32.13) to rubber springs is less straightforward because the value of E (and G) varies over a wide range, depending upon the compound, and the maximum permissible value of ϵ (and γ) is not as definite as with steel. Whereas the consequence of excessive strain in metal usually is permanent deformation, there is no immediate degradation of rubber as a result of a moderately large strain. If rubber is maintained under a constant strain of excessive magnitude for a long period of time, the load tends to relax; correspondingly, if excessive stress is maintained for a long period of time, the strain tends to increase continuously.

The modulus of elasticity of rubber in shear G varies between 50 lb/in.^{2} and 150 lb/in.^{2} for the range of rubber hardness commonly used in isolators. Corresponding values of the compression modulus depend upon the shape of the rubber member. Nominally, E may be taken as $5 \times G$; Chap. 35 includes a more detailed discussion of the properties of rubber. A very conservative strain for static loading is $\epsilon = 0.10$ and $\gamma = 0.25$. For applications in which some relaxation of load or increase in deflection with time (drift) is permissible, the strain may be as great as $\epsilon = 0.15$ and $\gamma = 0.50$. The relation of Eq. (32.13) is shown graphically in Figs. 32.20 and 32.21, using the reference values $E = 500$ lb/in.^{2}, $G = 100$ lb/in.^{2}, $\epsilon = 0.10$, and $\gamma = 0.25$. The values obtained from the curves may be adjusted to reflect other values of modulus or strain, as indicated by Eq. (32.13).

Rubber strained in shear must be bonded (i.e., adhered) to metal plates to transfer the load to the rubber. The stress at the bonded interface should not exceed 50 lb/in.^{2} For some types of rubber springs, notably those made from relatively stiff rubber, the permissible stress at the bonded interface rather than the relations set forth by Figs. 32.20 and 32.21 may limit the deflection.

If the resilient element is in the form of an elastomer strained in shear or compression, the distribution of strain throughout the volume tends to be uniform; then Eq. (32.13) and Figs. 32.20 and 32.21 can be used directly. It seldom is possible to attain a uniform distribution of strain throughout the volume of a metal spring; thus a greater volume of material is needed than indicated by Eq. (32.13). If the resilient element is a helical spring, the stress is essentially torsional and the strain distribution across the diameter is shown in Fig. 32.22*A*. For unit length normal to the paper, the energy stored per unit volume of spring material is found by substituting $\epsilon = \epsilon_0(R/R_0)$ in Eq. (32.11) and integrating:

$$\text{Stored energy} = \frac{1}{\pi R_0^2} \int_0^{2\pi} \int_0^{R_0} \frac{E\epsilon_0^2}{2} \left(\frac{R}{R_0}\right)^2 R\, dR\, d\theta = \frac{E\epsilon_0^2}{4} \qquad \text{in.-lb/in.}^3 \qquad (32.14)$$

By comparison of Eq. (32.14) with Eq. (32.12) when $V = 1$, a coil spring utilizes material to only 50 per cent efficiency.

A similar calculation can be made for a spring with the resilient material in bending, as shown in Fig. 32.22*B*. For a unit length of bar having a width normal to the paper equal to x_0, the energy storage per unit volume can be determined by substituting $\epsilon = \epsilon_0(2y/y_0)$ in Eq. (32.11) and integrating:

$$\text{Stored energy} = \frac{1}{x_0 y_0} \int_0^{x_0} \int_{-\frac{y_0}{2}}^{\frac{y_0}{2}} \frac{E\epsilon_0^2}{2} \left(\frac{2y}{y_0}\right) dy\, dx = \frac{E\epsilon_0^2}{6} \qquad \text{in.-lb/in.}^3 \qquad (32.15)$$

By comparison of Eq. (32.15) with Eq. (32.12) when $V = 1$, a spring in flexure utilizes its material to only 33 per cent efficiency. The attainment of an efficiency as high as this is contingent upon utilizing the material with the same efficiency along the length of the beam. For most common types of beam, this requires a variable cross section.[9]

The efficiency of a metal spring is increased if the strain can be distributed more uniformly over the volume of material. For a member in torsion, this can be done by using a tubular member, as shown in Fig. 32.22*C*. As the wall thickness approaches zero, the efficiency approaches that indicated by Eq. (32.12). Similarly, a flexural spring can attain a higher efficiency if constructed as a sandwich, as shown in Fig. 32.22*D*. The core material should be as light as possible, and must have a high modulus in shear lengthwise of the beam if the tension and compression stiffness of the outer plates is to be realized.

The curves of Figs. 32.20 and 32.21 apply only to the load-carrying element and do not include the weight of additional structure required to employ the springs. In general, rubber springs require substantially less additional structure than metal springs, and use

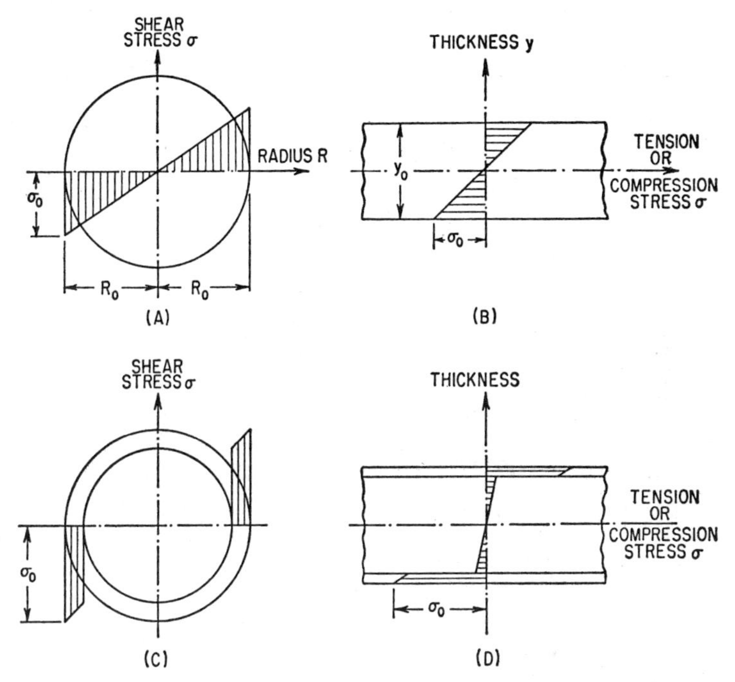

FIG. 32.22. Distribution of stress in cross sections of typical structural members: (*A*) solid circular bar in torsion; (*B*) bar with rectangular section in bending; (*C*) circular tube in torsion; (*D*) I beam in bending.

the spring material to a greater efficiency. If the additional weight which must be associated with a metal spring is large, the over-all installation may be heavier than one using a rubber spring even though the weight of the spring material is less.

NONLINEARITY

A nonlinear isolator often may be "linearized" for analysis of isolation if the vibration amplitude is small; nonlinear effects are important in vibration isolation if the amplitude is large or if there is a large sustained force. These conditions are discussed in Chap. 30. Shock isolation with nonlinear isolators is discussed in Chap. 31. Vibration of nonlinear systems in general is discussed in Chap. 4.

DYNAMIC STIFFNESS

When the main load-carrying spring is rubber or other organic material, the natural frequency calculated using the stiffness determined from a static force-deflection test of the spring almost invariably gives a value lower than that experienced during vibration. In other words, the dynamic modulus appears greater than the static modulus. The ratio of moduli seems to be approximately independent of the velocity of strain, and has

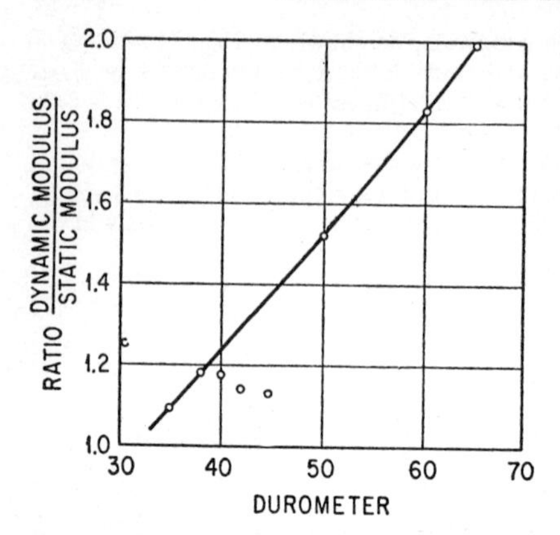

FIG. 32.23. Dynamic modulus of rubber with reference to the static modulus, as a function of durometer.[10]

a numerical value generally between one and two. This ratio increases substantially as the durometer increases, as indicated by the data presented in Fig. 32.23 from tests of a number of representative isolators selected at random.[10] The static and dynamic moduli for metal springs are approximately equal.

DAMPERS

Where adequate damping cannot be attained from internal hysteresis of the elastic load-carrying element, a discrete damper may be placed in parallel therewith. The degree of damping attainable in this manner is practically unlimited, depending only upon the space, cost, and mechanical complexity which can be tolerated. In general, discrete dampers intended for general applications are not commercially available although many commercially available vibration isolators embody damping means separate from the principal load-carrying spring.

FLUID FILM DAMPER (VISCOUS DAMPER)

The design of a viscous damper based upon shearing a film of fluid is straightforward conceptually, although the practical design problems may become difficult. While dampers of this type find rather wide acceptance in vibration-control techniques, the requirements usually can be met only by a design that is somewhat too elaborate for incorporating into a conventional vibration isolator intended for wide applications. There remain many uses for viscous dampers, however, and data upon which to base their design are included in this section.

In its simplest form, a viscous damper may be considered to consist of two spaced apart parallel plates with a film of viscous fluid interposed therebetween. One or both of the plates has a velocity v_1, v_2 in its respective plane. A damper of this type is illustrated in Fig. 32.24*A*. The area of each plate is S, and the thickness of viscous film is t. The coefficient of dynamic or "absolute" viscosity μ is the resistance offered by the viscous film of unit film thickness, unit plate area and unit differential velocity $v_1 - v_2$.

Dynamic viscosity is of fundamental significance in determining the characteristics of fluid film dampers of the type illustrated in Fig. 32.24*A*. When damping is achieved from the flow of fluid through a conduit, the Reynolds number is the criterion for laminar flow. The dynamic viscosity appears in the Reynolds number in combination with the mass density of the fluid, thus leading to the concept of kinematic viscosity $\nu = \mu/\rho$, where μ is dynamic viscosity and ρ is mass density.[11]

For purposes of damper design, dynamic viscosity is most conveniently expressed in units of lb-sec/in.^{2}, and kinematic viscosity in units of in.^{2}/sec. The mass density then

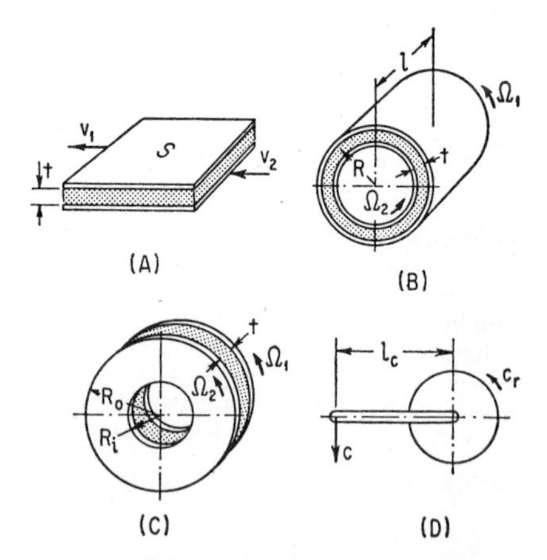

FIG. 32.24. Arrangements for obtaining damping from the shearing of a viscous fluid between solid members having relative velocity.

is in units of lb-sec^2/in.^{4} In the metric system, the unit of dynamic viscosity, one dyne-second per square centimeter, is customarily called a *poise*; the unit of kinematic viscosity, one square centimeter per second, is called a *stoke*. Commercially available fluids often are rated in units of centipoises or centistokes. For ready reference, conversion factors are as follows:

$$\text{One centipoise} = 1.45 \times 10^{-7} \text{ lb-sec/in.}^2$$

$$\text{One centistoke} = 6.44 \times 10^{-2} \text{ in.}^2\text{/sec}$$

Typical values of dynamic viscosity as a function of temperature are shown in Fig. 32.25 (also see Fig. 16.8) for guidance in the design of viscous dampers. Corresponding values of mass density for the fluids are given in Table 32.1, expressed in engineering units.

For the parallel plate damper illustrated in Fig. 32.24*A*, the resistance or drag between plates is directly proportional to the differential velocity $v_1 - v_2$. The damping coefficient is the force per unit of velocity as follows:

$$c = \frac{S\mu}{t} \quad \text{lb-sec/in.} \qquad (32.16)$$

where S is the area of each plate in square inches, t is the thickness of viscous film in inches, μ is the coefficient of dynamic viscosity in units of lb-sec/in.^{2}, and c is the damping coefficient in units of lb-sec/in. The value of c can be used directly in the differential equation of motion for a system embodying such a damper.

In a practical sense, it frequently is more convenient to attain viscous damping from the rotation of one cylinder within another. Then the film of viscous fluid is essentially the film of Fig. 32.24*A* curved into cylindrical form and interposed between the two cylinders, as illustrated in Fig. 32.24*B*. The damping is a resisting torque and is directly proportional to the differential rotational velocity of the concentric cylinders defined by $\Omega_1 - \Omega_2$ in units of radians per second. The damping coefficient is

$$c_r = \frac{2\pi R^3 l\mu}{t} \quad \frac{\text{lb-in.}}{\text{rad/sec}} \qquad (32.17)$$

where R is the radius to the center of the film in inches, l is the axial length of the film in inches, and t is the film thickness in inches. The damping coefficient c_r involves a torsional concept, and is the resistance offered by the film in units of pound-inches per unit of differential velocity expressed in radians per second.

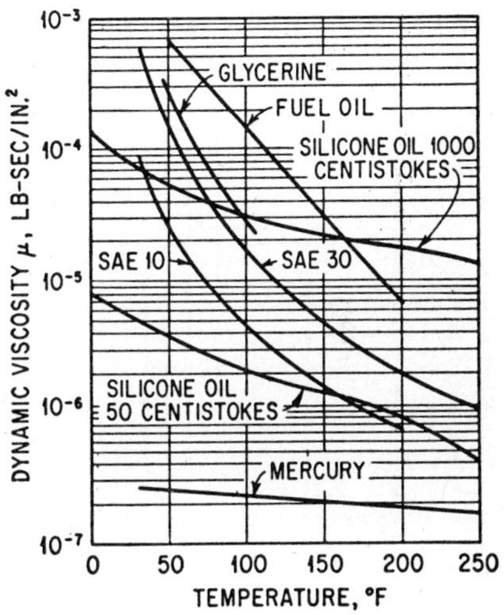

FIG. 32.25. Dynamic viscosity as a function of temperature for several fluids having possible use in viscous dampers.

Table 32.1. Mass Densities of Typical Fluids

Fluid	Mass density, lb-sec^2/in.
Glycerin	1.07×10^{-4}
Mercury	1.15×10^{-3}
Lubricating oil	7.22×10^{-5}
Fuel oil	8.10×10^{-5}
Silicone oil	8.23×10^{-5}

In another form of viscous damping involving rotational motion, the viscous film is interposed between two parallel discs rotating on the same axis, as illustrated in Fig. 32.24*C*. The portion of the film adjacent the periphery of the discs has a greater damping effect than that adjacent the axis because the differential velocity is greater and because the moment of the resultant force also is greater. The torque resulting from the damping force can be determined by integrating the force over the area of the discs. The coefficient of viscous damping, again involving a torsional concept with units equal to those of Eq. (32.17), is given by the following equation:

$$c_r = \frac{\pi\mu}{2t}(R_0^4 - R_i^4) \qquad \frac{\text{lb-in.}}{\text{rad/sec}} \tag{32.18}$$

where R_i and R_0, respectively, are the inner and outer radii of the discs in inches and t is the thickness of the viscous film in inches.

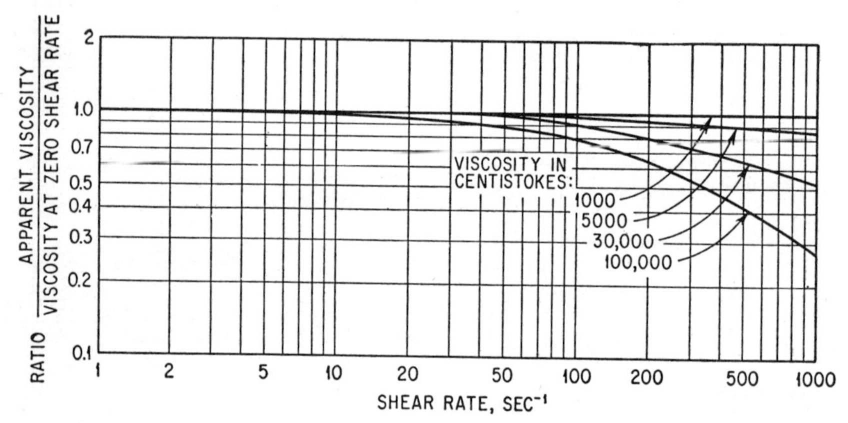

FIG. 32.26. Apparent viscosity of silicone oils as a function of the shear rate across the viscous film. (*W. J. Dugan.*[12])

Although practical considerations in the design and construction of viscous dampers frequently suggest constructions as shown schematically in Figs. 32.24*B* and 32.24*C*, it may be desirable to damp linear or translational vibration by adding a crank and connecting rod to the rotational damper as illustrated in Fig. 32.24*D*. If the length of the crank is l_c, the torsional damping coefficients given by Eqs. (32.17) and (32.18) may be transformed to the translational damping coefficient c by applying the following relation:

$$c = \frac{c_r}{l_c^2} \qquad \frac{\text{lb}}{\text{in./sec}} \tag{32.19}$$

In the design of a viscous damper, the calculations may indicate a damping force that is greater than can be justified by the shear *strength* of the viscous film. The probability of this limitation increases as the viscosity of the fluid increases. This effect is defined in Fig. 32.26 [12] for silicone fluids, indicating that the effective viscosity of a fluid tends to decrease as the shear rate increases. The relations indicated by the curves of Fig. 32.26 do not necessarily apply to all fluids, but are included here as a guide to the limitations that must be considered in the design of viscous dampers.

FLOW OF VISCOUS, INCOMPRESSIBLE FLUID THROUGH CONDUIT

When a viscous, incompressible fluid is forced through a pipe or conduit, the velocity distribution across the cross section varies from zero at the wall to a maximum at the center. This differential velocity involves shearing of the fluid, and the resistance force ideally is proportional to the velocity of the fluid. This ideal resistance characteristic can be attained only if the following requirements are met:

1. The velocity of flow must be low enough to ensure laminar rather than turbulent flow of the fluid.
2. The velocity of flow must be low enough to reduce the kinetic or orifice effect to a relatively small value.
3. The vibration frequency must not be too great; otherwise the inertia of the fluid tends to equalize the velocity across the conduit and shearing occurs at only the outer boundary of the conduit.

These effects are discussed separately in the following paragraphs.

The resistance offered by a viscous fluid in flowing through a conduit is defined in terms of the pressure in the fluid:[13]

$$p = \frac{8\pi\mu l v}{S} \qquad \frac{\text{lb/in.}^2}{\text{in./sec}} \tag{32.20}$$

where μ is the dynamic viscosity in units of lb-sec/in.^{2}, l is the length of the conduit in inches, v is the mean velocity of fluid flow in units of in./sec and S is the cross-section area of the conduit in square inches. Equation (32.20) should be used only where the Reynolds number is lower than approximately 2,000; for greater values, the flow is turbulent and does not develop the viscous resistance attainable from laminar flow. The limit on the Reynolds number is expressed as follows for a conduit of circular cross section:

$$\frac{vD\rho}{\mu} < 2{,}000 \tag{32.21}$$

where D is the diameter of the conduit in inches, v is the velocity of flow in units of in./sec, and ρ is the mass density of the liquid in units of lb-sec^2/in.^{4}

Damping attainable from flow of a compressible fluid through a conduit is discussed in Chap. 33.

The kinetic resistance in opposing the flow of liquid through a conduit corresponds to the pressure drop through an orifice. The resistance is proportional to the square of the velocity, in contrast to the resistance in viscous flow which is proportional to the first power of velocity. This method of damping is used in many applications; its principal limitation is the relatively large resistance developed at large values of velocity or frequency. The resistance is defined in terms of pressure in the fluid; it is independent of viscosity of the fluid, being dependent only upon its density and velocity as follows:

$$p = \frac{v^2\rho}{2} \qquad \text{lb/in.}^2 \tag{32.22}$$

where v is the velocity of fluid in the conduit in units of in./sec and ρ is the mass density of the fluid in units of lb-sec^2/in^4.

The expression for the damping coefficient in viscous flow, Eq. (32.20), applies only where the frequency is relatively low. The limitation on frequency is defined for a circular tube as follows:

$$\sqrt{\frac{R^2\rho\omega}{2\mu}} < 2 \tag{32.23}$$

where R is the radius of the tube in inches, ω is the vibration frequency in units of rad/sec and μ/ρ is the kinematic viscosity of the fluid in units of in.^{2}/sec. If the parameter given by Eq. (32.23) exceeds a value of approximately two, the coefficient given by Eq. (32.20) is modified by a frequency-dependent factor.[14] The corrected coefficient is defined with

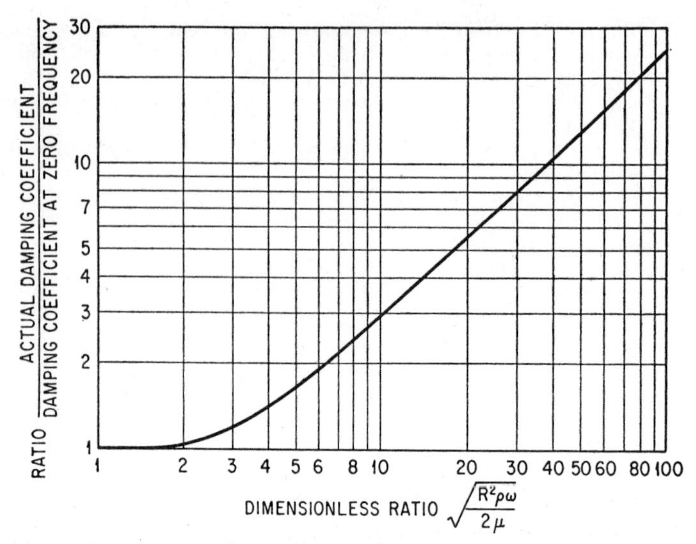

Fig. 32.27. Mass effect in a damper whose damping force is derived from the viscous drag of a fluid flowing through a conduit. (*C. K. Stedman.*[14])

reference to the coefficient for zero frequency given by Eq. (32.20), as a function of the parameter given by Eq. (32.23). This relation is shown graphically in Fig. 32.27.

Although the damping force that is attainable directly from the resistance offered by a viscous fluid in passing through a conduit is quite moderate, a very large mechanical advantage can be attained. This employs a piston and cylinder of relatively large area arranged so that motion of the piston forces fluid through the conduit, as shown schematically in Fig. 32.28. The velocity of fluid through the conduit is the piston velocity multiplied by the ratio of areas, and the force acting on the piston is the pressure of the fluid in the conduit multiplied by the piston area. Taking as a reference the force and velocity on the piston rod, the corresponding coefficient of viscous damping is obtained by extending Eq. (32.20) as follows:

$$c_p = \frac{8\pi\mu l}{S_c}\left(\frac{S_p}{S_c}\right)^2 \tag{32.24}$$

where S_p is the area of the piston and S_c is the area of the conduit. Care must be exercised in applying a damper as shown in Fig. 32.28 to be certain that the limits established by Eqs. (32.21) and (32.23) are not exceeded, the velocity v in Eq. (32.21) being the actual velocity of fluid in the conduit rather than the velocity of the piston.

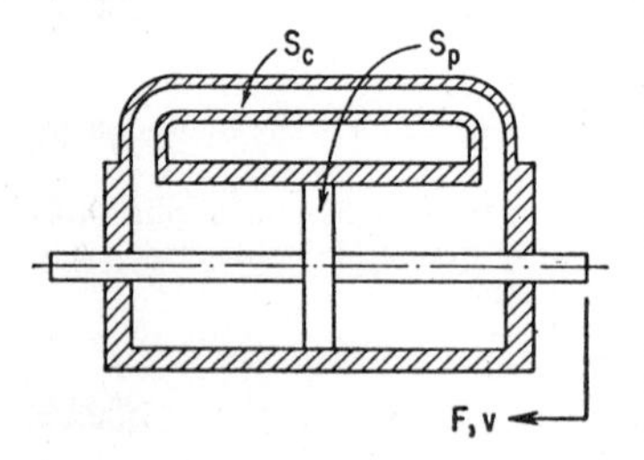

Fig. 32.28. Schematic diagram of a damper involving flow of fluid through a conduit, used in conjunction with a piston and cylinder to achieve a mechanical advantage for the damping force.

EDDY-CURRENT DAMPING

A damping force directly proportional to velocity can be attained by moving a short-circuited coil through a magnetic field. Both the direction of current flow in the conductor and the direction of motion of the conductor are perpendicular to the direction of the lines of flux of the magnetic field. A damper designed in accordance with this principle is illustrated in Fig. 32.29. It employs a permanent magnet cylindrical in form and polarized as indicated. Upper and lower flat plates and a cylindrical core constitute pole pieces which create a magnetic field extending radially through an annular air gap. The conductor forms a loop in this air gap, and is arranged to move vertically. The damping force acting on the conductor opposes its motion.

The damping force is directly proportional to the velocity of the conductor in the field, and is defined quantitatively by the coefficient of viscous damping c:[16]

$$c = \frac{5.6B^2lS}{\rho} \times 10^{-6} \qquad \frac{\text{lb}}{\text{in./sec}} \tag{32.25}$$

where B is the flux density in gauss, l is the length of the conductor in inches, S is the cross-sectional area of the conductor in square inches, and ρ is the resistivity of the conductor in ohms-circular mil per foot.* Equation (32.25) applies only if the entire conductor is in the magnetic field; the damping coefficient is a constant only if the vibration amplitude is small enough to permit the conductor to remain within a constant strength field continuously throughout the vibration cycle. The resistivity ρ of copper is approximately 10.37 ohms-circular mil per foot; a conservative design value for the flux density attainable from a permanent magnet is 1,000 gauss and a maximum value is approximately 7,000 gauss. In general, the damping attainable from eddy currents using a permanent magnet is useful only if both the mass and the natural frequency of the system are relatively low.

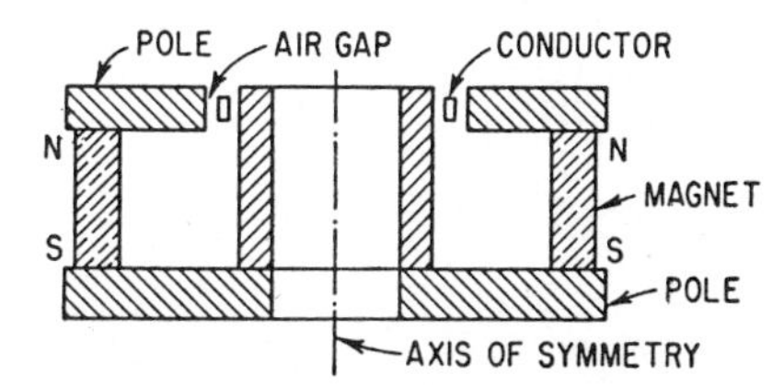

FIG. 32.29. Schematic diagram of a damper involving the induction of eddy currents in a short-circuited conductor.

COULOMB-FRICTION DAMPING

As pointed out in Chap. 30, the magnitude of the Coulomb-friction force is a constant independent of velocity or displacement, but having a direction that tends to oppose relative motion across the damper. A Coulomb-friction force generally results from the relative motion of two solid members held together under pressure, the motion to be damped being substantially parallel to the engaged surfaces. The magnitude of the damping force F_f is directly proportional to the coefficient of friction μ, the unit pressure P_n between the surfaces, and the area S of contact:

$$F_f = \mu S P_n \qquad \text{lb} \tag{32.26}$$

The coefficient of friction is a function of the materials at the interface where relative motion occurs. Typical values are given in Table 32.2. In general, the static coefficient of friction tends to be greater than the dynamic coefficient; i.e., the resistance offered by the damper decreases somewhat after a relative velocity has been attained. In steady-state vibration, the relative velocity becomes equal to zero twice during each cycle. The

Table 32.2. Representative Values for Coefficient of Friction

Materials	Coefficient of friction
Brass–steel	0.15
Steel–steel	0.15
Leather–steel	0.35
Nylon–metal	0.30
Teflon–metal	0.05

* A circular mil is the area of a circle whose diameter is 0.001 in. One circular mil = 7.85×10^{-7} in.^{2}

effective coefficient of friction thus falls between the extremes. The values of the coefficient given in Table 32.2 reflect this condition to some extent, and may be used for design purposes.

One of the problems encountered in the design of friction dampers is the mechanical complexity involved in attaining a structure capable of providing a damping force in three directions. Two flat plates held in contact provide a damping force along the two axes in the plane of the plates, but not perpendicular to the plates. Dampers operative along three axes are illustrated in Fig. 32.30*A* and *B*.* The former involves a central column encircled by a segmented damper held against the column by a contractive spring. The damper is located between cover plates; a flat spring is interposed between

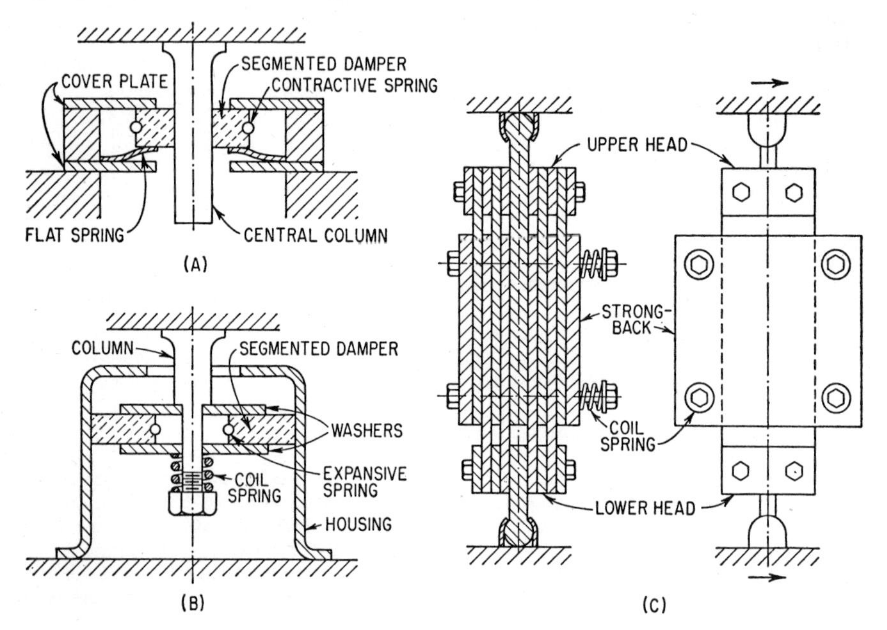

FIG. 32.30. Typical arrangements of mechanisms for producing Coulomb or dry friction damping. (*A: Lord Mfg. Co.*; *B and C: Barry Wright Corp.*)

the damper and one of the cover plates to hold the damper against the other cover plate. Damping of vertical motion is thus attained by friction between column and damper, whereas horizontal motion is damped by friction between damper and cover plate.

The arrangement shown in Fig. 32.30*B* is somewhat similar to that shown in Fig. 32.30*A*. The column carries two washers arranged on opposite sides of the segmented damper, the coil spring maintaining pressure at the engaging surfaces to damp horizontal motion. The expansive spring forces the damper segments against the housing, the relative motion at this interface providing the damping force that opposes vertical motion.

If the arrangement of isolators couples translational and rotational motion in certain modes, it may be possible to use a unidirectional damper placed so that it experiences relative motion in any natural mode of vibration of the mounted equipment. A damper adapted to such an application is shown in Fig. 32.30*C*. It is attached by universal joints at its opposite ends, and arranged to provide damping only along the vertical axis. It consists of an upper head and a lower head attached by the upper and lower universal joints between the isolated structures. Each head carries a number of spaced plates that alternate to provide a plurality of engaging surfaces. The assembly of alternating plates is held under pressure by the strongbacks and coil springs. In Eq. (32.26), the normal

* Certain features of these and similar dampers may be covered by United States patents.

pressure P_n is the total force exerted by the coil springs divided by the area of that portion of the strongback that bears upon a plate. The area S is the total area of interfaces in contact, a value that increases as the number of plates increases. It may be found convenient to use a relatively large number of plates to attain a large damping force.

In the design of dampers of the types shown in Fig. 32.30, it is desirable that the springs have a relatively low stiffness and a relatively large deflection to achieve the required normal pressure P_n. If the springs have these characteristics, the properties of the damper remain substantially unaffected by wear of the engaging surfaces. A damper with a relatively stiff spring, on the other hand, is likely to experience an appreciable change in its characteristics as wear occurs.

COMMERCIALLY AVAILABLE ISOLATORS

A catalog of commercially available vibration isolators must recognize the distinction between those intended for military use and those intended for civilian use, as well as the distinction between those of standard design and those of special design. In isolators intended for military use, specifications are important in defining both dimensions and performance. The many unusual equipments and unusual environmental conditions encourage the use of special isolators for military application. This is less true in nonmilitary applications where the greatest incentive to depart from the use of standard isolators is to achieve cost reduction by simplifying the over-all design, a result generally attainable only by use of special isolators in large quantity. To a considerable extent, the characteristics of isolators are determined by requirements of the application. This makes it convenient to separate commercially available isolators into several groups in accordance with their ultimate end use.

AIRCRAFT AND GUIDED MISSILES

Standard isolators used in aircraft, and to some extent in guided missiles, are closely controlled by applicable military specifications. These specifications define not only the size and mounting pattern of the respective isolators but also the load range for each size, the natural frequency when supporting the rated load, and the degree of damping required. The outline drawing of standard isolators is set forth in Fig. 32.31; dimensions and load ranges applicable to each size are given in Table 32.3. Names of principal manufacturers are included. The over-all load range for each size as given in the table is broken into increments as determined by each manufacturer, but these increments also are controlled to some extent by the weights of standard equipments.

Isolators of the types illustrated in Fig. 32.31 characteristically have natural frequencies of approximately 8 to 10 cps in vertical translation at rated load. These isolators are intended primarily for attachment to the bottom face of the mounted equipment, and generally are adapted to operate only in compression under the dead-weight load. The stiffness in directions of the horizontal axes is generally somewhat less than that in the direction of the vertical axis, to minimize the highest natural frequency in a coupled mode. This characteristic, together with the required damping capacity, is defined by the envelope of maximum permissible transmissibility included here as Fig. 32.32.[17] This envelope applies to transmissibility in all modes of vibration. In a general sense, the products of one manufacturer are equivalent to those of another in that they meet the same specification. The different manufacturers emphasize different materials and characteristic features, as explained in the commercial literature applying to the respective isolators and available from the manufacturers.

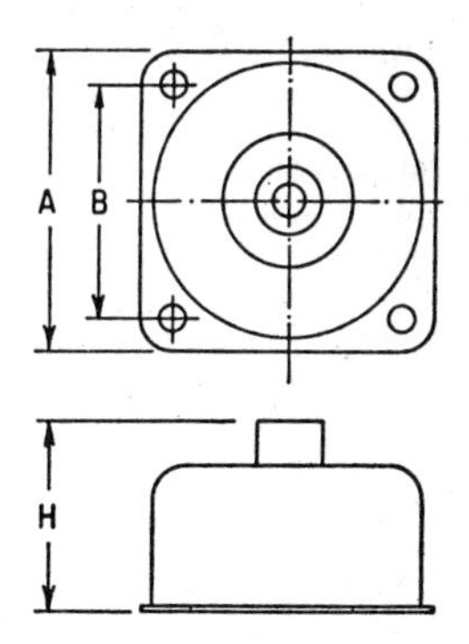

FIG. 32.31. Outline drawing of typical aircraft-type vibration isolator. Dimensions and load ranges are given in Table 32.3.

It may be desirable to use higher natural frequencies for iso-

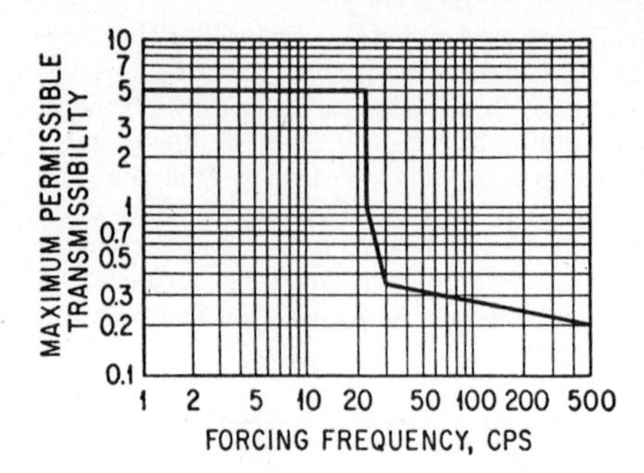

FIG. 32.32. Maximum permissible transmissibility for aircraft-type vibration isolators, in accordance with Military Specification MIL-C-172C.[17]

lators in aircraft and guided missiles than those suggested by Fig. 32.32. Several manufacturers offer isolators having approximately the dimensions shown in Fig. 32.31 and Table 32.3 but exhibiting natural frequencies of 20 to 30 cps at rated load.* The objective is to attain improved stability by decreasing the static deflection, this concept having considerable importance in fighter aircraft and guided missiles. In some instances, the loads indicated in Table 32.3 can be increased somewhat for a given isolator size when the natural frequency is increased.

Although the aircraft-type isolators illustrated in Fig. 32.31 find wide usage, there are many applications in aircraft and guided missiles in which the use of specially designed isolators is desirable. These special isolators cover a wide range of embodiments which cannot be effectively catalogued.

Table 32.3. Dimensions and Load Ranges for Isolators Shown in Fig. 32.31

Size	0	1	2	3
Width A, in.	1 11/32	1 3/4	2 3/8	3
Hole centers B, in.	1	1 3/8	1 15/16	2 1/2
Height H, in.:				
Low style	...	1 3/8	1 3/8	1 3/8
High style	...	1 9/16	1 9/16	1 9/16
Load range, lb (approximate)	0–5	1–10	2–43	20–86

SHIPBOARD APPLICATIONS

A type of isolator which has seen extensive use for the protection of shipborne equipment against shock is illustrated in Fig. 32.33. Dimensions and rated loads are given in Table 32.4. This isolator employs rubber loaded in compression for any direction of loading, thereby attaining a force-deflection curve that is devoid of excessive discontinuities. The mounted equipment is positively retained by interlocked metal members, and the stiffness along the several axes are approximately equal. Past practice in the application of isolators (e.g., the type shown in Fig. 32.33) for shipboard service has established the natural frequency somewhat greater than the forcing frequency, and stringent requirements for damping do not exist. Other considerations in the selection of isolators for shipboard use are discussed in the section entitled *Applications in Naval Ships.*

The effective isolation of noise in naval shipboard applications requires that the natural frequency of the isolator be as low as possible consistent with space limitations and stability of the mounted equipment. For such applications, the Navy Bureau of Ships has initiated the development of the rubber isolators shown in Fig. 32.34. The dimensions of the isolators are given in Table 32.5. When loaded with the rated loads indicated, the

* The axial and radial stiffnesses of these isolators are approximately equal, and the gravity load may be applied in any direction.

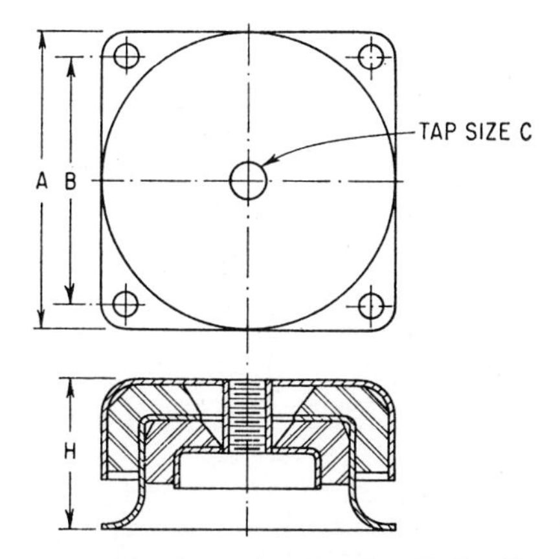

Fig. 32.33. Isolators having rubber in compression for all directions of loading. The stiffnesses along the several coordinate axes are approximately equal, and the minimum attainable natural frequency is relatively high. Dimensions and loads are given in Table 32.4.

Table 32.4. Dimensions and Load Ranges for Isolators Shown in Fig. 32.33

Type*	C-1000T	C-2000T	C-4000T	C-3000T
Width A, in.	2⅜	3	5¼	6⅞
Hole centers B, in.	1[15]⁄16	2½	4¼	5⅝
Tap size C	¼–20	⅜–16	⅝–11	⅝–11
Height H, in.	1⅛	1½	2½	3 9⁄16
Load* for natural frequency of 25 cps, lb	20–100	40–150	80–300	100–450

Manufacturer: Barry Wright Corp., Watertown, Mass.

* Particular load is indicated by digits substituted for zeros in type number; nature of center hole is indicated by a numerical suffix.

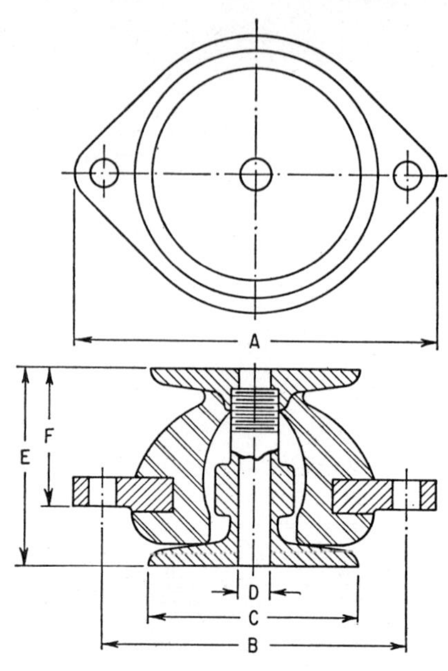

FIG. 32.34. Isolators designed to have equal stiffnesses in all directions and applied with natural frequencies as low as 6 cps. Dimensions and loads are given in Table 32.5.

Table 32.5. Dimensions and Load Ranges for Isolators Shown in Fig. 32.34

Type number	7E450	6E900	6E2000
Load range, lb	100–450	400–900	700–2,000
Dimension, in.			
A	6½	9¾	11⅜
B	5⅜	8¼	9⅝
C	3½	5½	6½
D	¾	1	1¼
E	3⅛	4¼	4 17/32
F	2⅛	3	3 3/16

Note: For approved manufacturers, see Qualified Products List of the Bureau of Ships.

isolators exhibit natural frequencies as low as 6 cps along any of the coordinate axes. Manufacture of these isolators must be in conformance with strict Navy requirements, and the isolators are available only from sources approved by the Bureau of Ships.

NONMILITARY

The influence of military specifications is seen in the dimensions of a group of isolators employing rubber springs and used predominantly for nonmilitary service. In many instances, these isolators initially had military applications and their dimensions reflect the requirements of such applications. A group of such isolators is shown in Fig. 32.35.

It has become common practice to interpose isolators between the floor and mounting feet of industrial machinery. Three distinct classes of isolators are used for such applications:

1. When the application is such as to require a static deflection approaching 1 in., it is conventional practice to employ an isolator embodying coil springs and auxiliary damping means. Commercially available isolators of this type are illustrated in Fig. 32.36.

2. When the application requirements call for a static deflection not exceeding approximately 0.1 in., it is common practice to employ isolators with springs made of organic material such as natural or synthetic rubber, cork, or compositions of other materials. It usually is not necessary that such isolators embody auxiliary dampers, because the spring exhibits adequate hysteresis damping. Several types of isolators in this class embody height adjustment features. Representative isolators are illustrated in Fig. 32.37.

3. The third class of isolators consists of pads of resilient material cut to proper size from larger pieces. It is not uncommon for applications of such pads to be on a cut-and-try basis, using the experience of the pad manufacturer.

Frequently it is desirable to suspend steam pipes, fans, ducts, acoustic ceilings, and similar equipment from overhead supports. Isolators adapted for this type of application are illustrated in Fig. 32.38.

In addition to the several classes of isolators intended specifically for designated applications, as described above, commercially available isolators include a relatively large number of miscellaneous types. Some of these, initially designed for specific purposes, find general application. Others were initially designed for general-purpose applications. A number of such isolators are illustrated in Fig. 32.39.

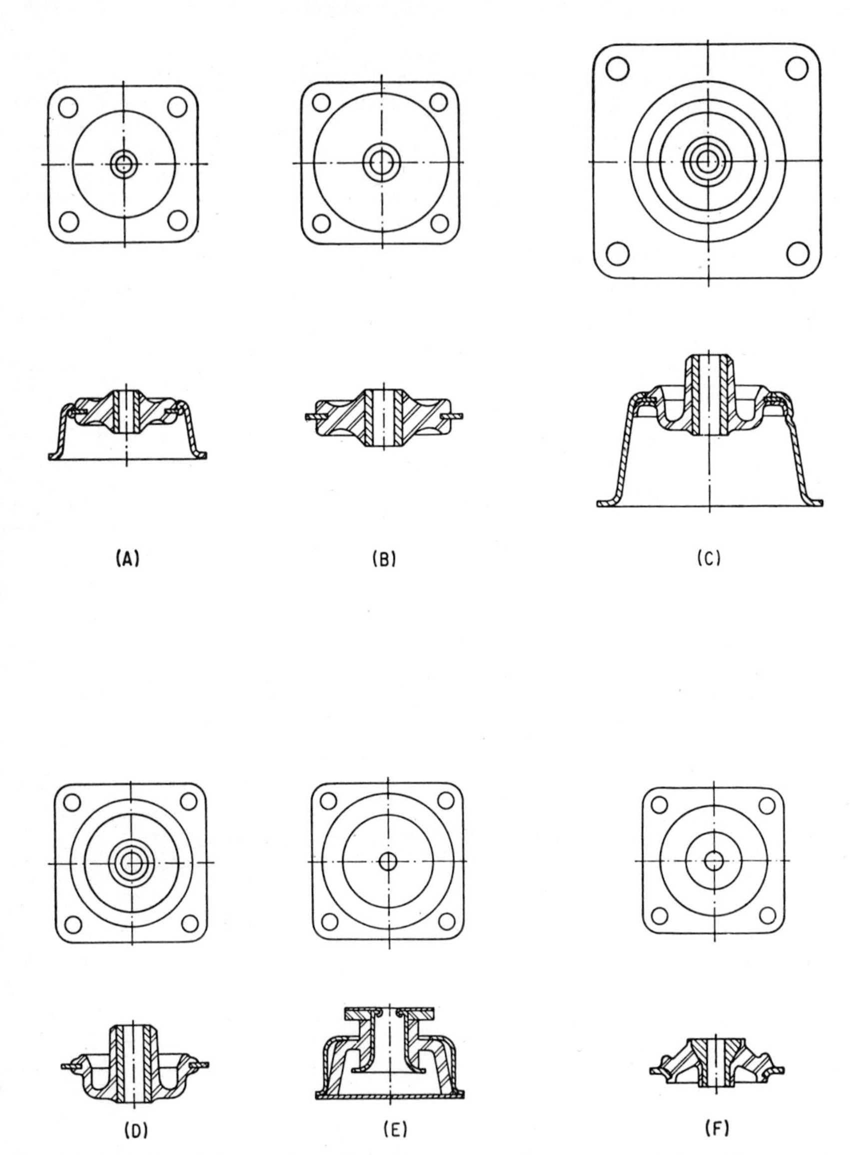

FIG. 32.35. Rubber isolators whose dimensions conform in some respects to military requirements but which see extensive nonmilitary use.

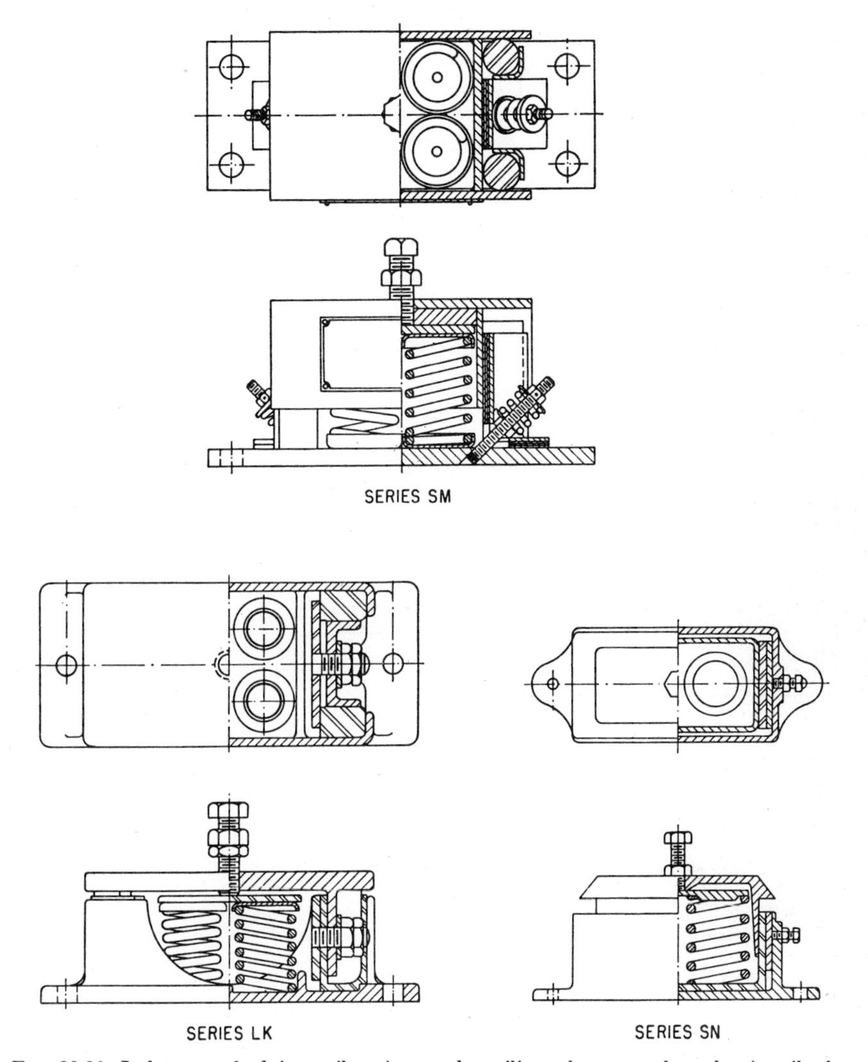

Fig. 32.36. Isolators embodying coil springs and auxiliary dampers adapted primarily for supporting industrial machinery.

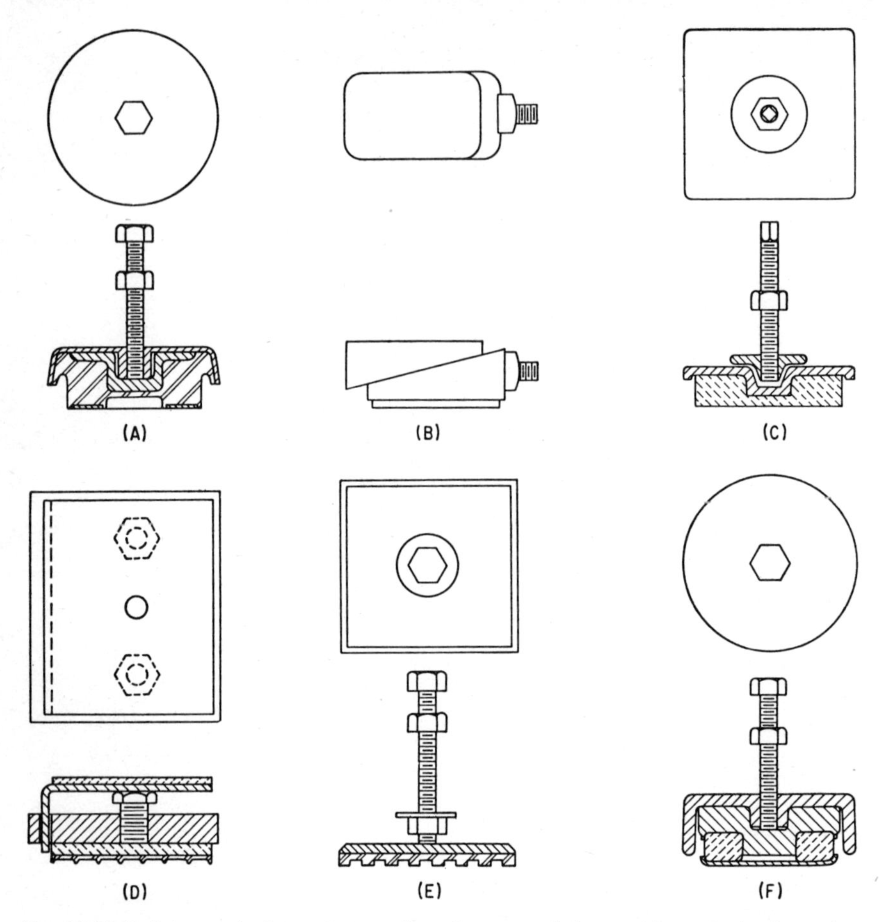

FIG. 32.37. Isolators embodying elastomeric and compounded materials as the load-carrying means, adapted primarily for supporting industrial machinery.

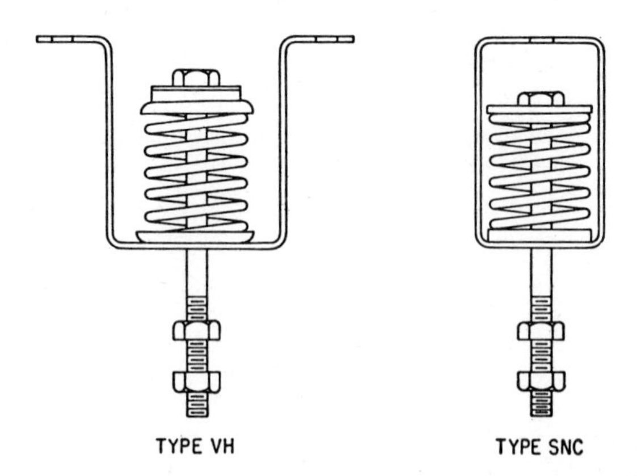

FIG. 32.38. Isolators for suspension of equipment from overhead support.

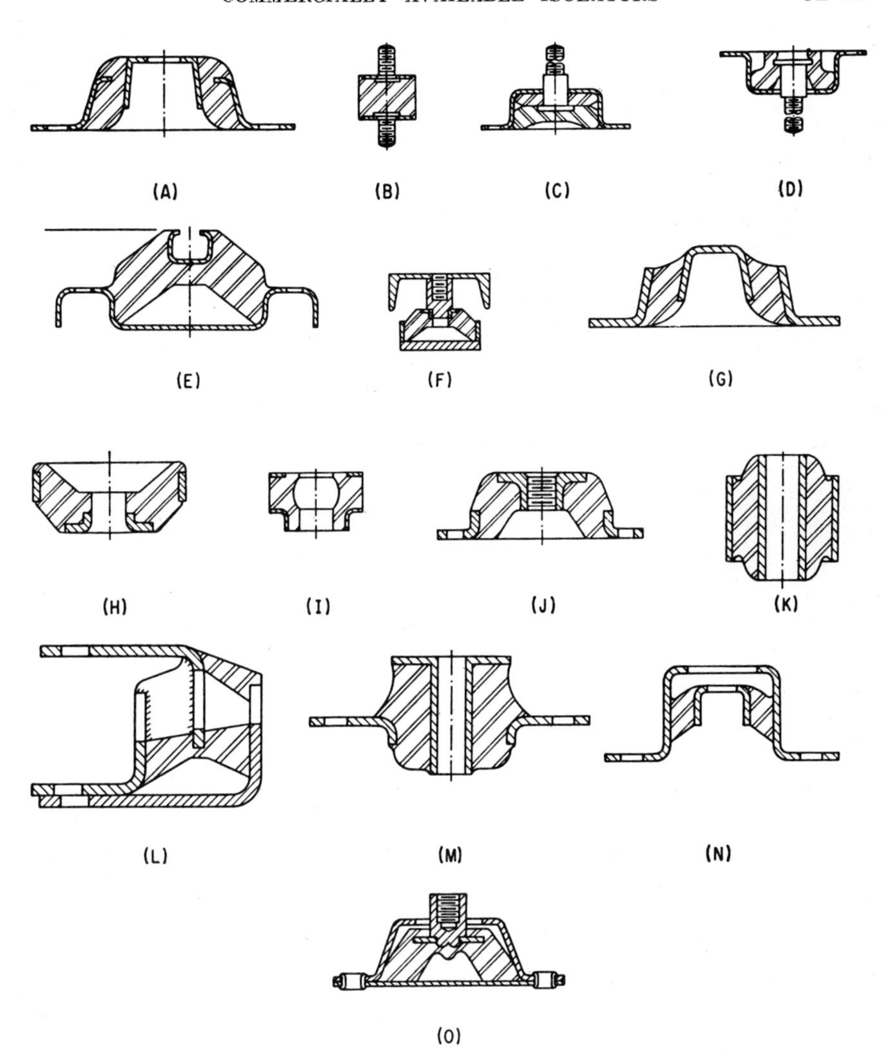

FIG. 32.39. Rubber isolators adapted for general nonmilitary use.

REFERENCES

1. Crede, C. E.: "Vibration and Shock Isolation," p. 290, John Wiley & Sons, Inc., New York, 1951.
2. Ref. 1, p. 291.
3. Crede, C. E.: *Trans. ASME*, **69**:937 (1947).
4. Ref. 1, p. 257.
5. Crede, C. E., and J. P. Walsh: *J. Appl. Mechanics*, **14**:A-7 (1947).
6. Geiger, P. H.: "Noise Reduction Manual," *Univ. Mich. Eng. Research Inst.*, 1953.
7. Lewis, F. M.: *Trans. ASME*, **54**:253 (1932).
8. Painter, G. W.: "Dynamic Properties of BTR Elastomer," *Paper* 83B, SAE National Aero. Meeting, Sept. 29 to Oct. 4, 1958.
9. Macduff, J. N., and R. P. Felgar: *ASME Trans.*, **79**:1459 (1957).
10. Ref. 1, p. 221.
11. Rouse, H.: "Elementary Mechanics of Fluids," p. 359, John Wiley & Sons, Inc., New York, 1946.
12. Dugan, W. J.: *Prod. Eng.*, July, 1954, p. 137.
13. Ref. 11, p. 162.
14. Stedman, C. K.: "Alternating Flow of Fluids in Tubes," *Statham Laboratories Notes*, No. 30, January, 1956.
15. Peterson, J. B.: "Damping Characteristics of Dashpots," *NACA Tech. Note* 830, 1941.
16. Hausmann, E., and E. P. Slack: "Physics," p. 426, D. Van Nostrand Company, Inc., Princeton, N.J., 1939.
17. Military Specification MIL-C-172C: Cases; Bases, Mounting; and Mounts, Vibration (For Use with Electronic Equipment in Aircraft), Dec. 8, 1958.

33

AIR SUSPENSION AND SERVO-CONTROLLED ISOLATION SYSTEMS

Richard D. Cavanaugh
Barry Wright Corporation

INTRODUCTION

This chapter discusses (1) vibration isolation using pneumatic springs as vibration isolators and (2) the application of servo control systems to vibration isolation. Servo control offers unique advantages in eliminating undesirable pseudostatic deflection of large magnitude while preserving the low natural frequency necessary for effective vibration isolation; it is particularly adapted to use with pneumatic springs. Additional advantages result from the general use of pneumatic springs, particularly where the load is large and the static deflection is great.

Vibration isolation is achieved conventionally by supporting a body on a linear spring or vibration isolator, as discussed in detail in Chap. 30. As a general rule, the natural frequency of the body-and-isolator system must be substantially lower than the excitation frequency. The deflection of the isolator under the deadweight load of the supported body is inversely proportional to the square of the natural frequency. Thus, the static deflection is large when the natural frequency is low. The combined requirement of large static deflection and large load-carrying capacity requires the storage of great quantities of potential energy in the isolator. The material used to construct mechanical springs generally has limited energy-storage capacity per unit of material. Therefore, springs which must carry large loads and provide large static deflections tend to become excessively bulky.

A spring which employs a compressible gas as its resilient element does not require a large static deflection; this is because the gas can be compressed to the pressure required to carry the load while maintaining the low stiffness necessary for vibration isolation. The energy-storage capacity of air is far greater per unit weight than that of mechanical spring materials, such as steel and rubber. The advantage of air is somewhat less than a comparison of energy-storage capacity per pound of material would indicate because the gas must be contained. However, if the load and static deflection are great, there usually is a large weight reduction resulting from the use of pneumatic springs.

If the isolator must operate over a relatively large load range, as in the springing system of an automobile, the requirement for a low natural frequency presents design problems which cannot be overcome by use of either the mechanical spring or the air spring alone. Because of the low stiffness of the spring, a small change in the weight of the supported body results in a large change in the static deflection of the spring. In many instances, this cannot be tolerated. For example, an automobile with a springing system having a natural frequency of 0.75 cps requires that the springs have a static deflection of 17.5 in. A 30 per cent change in the weight supported by the automobile suspension system results in a 5.25-in. change in its normal height. This change in position is too large to allow proper operation of the springs over the required load range, i.e., approximately equal clearance travel of the body in the up-and-down directions. A similar problem arises in rocket-propelled missiles where a substantially constant acceleration is sustained for an appreciably long time. This deflects an isolator in the same manner as would an increase in weight of the supported equipment. The clearance around such equipment must be large enough to provide for unobstructed displacement of the equipment under the influence of large steady acceleration experienced by the missile. Space on such missiles is extremely limited.

The large deflections experienced by conventional spring elements can be reduced by the use of a servomechanism which compensates for load changes, thereby maintaining the supported body at a given position with respect to its vehicle. Figure 33.1 illustrates such a device schematically. As force F is applied slowly, the spring k stretches and the supported body m moves downward. The relative motion between body m and the base causes the sensing device to respond, calling for power to be supplied to the servo actuator which moves the frame upward until the body m has returned to its original position with respect to the base. This position of equilibrium exists until another change in the force F occurs. The element which produces the change in position of the frame is a servomechanism arranged to provide integral feedback control of the relative displacement δ.

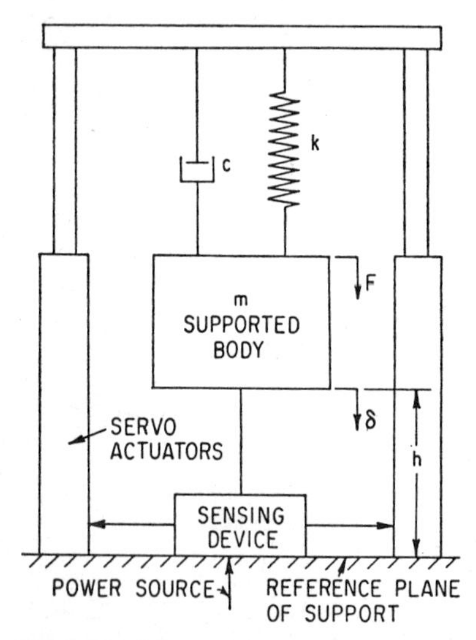

FIG. 33.1. Schematic diagram of a servo-controlled isolation system which maintains body m a fixed distance h from the reference plane of the support—irrespective of the steady force F applied to the body.

The isolation system illustrated in Fig. 33.1 always seeks as its equilibrium position a location at a distance h above the reference plane of the support, independent of the origin and magnitude of a steady force applied to the supported body. There is no change in the position of the body in response to a slowly applied load, whether the isolator natural frequency is 1 cps or 10 cps. Certain disadvantages attendant on the use of isolators of low natural frequency are thus removed. When the load is applied suddenly, however, the servomechanism may be unable to respond fast enough to compensate for the tendency of the supported body to change its position relative to the support; then the isolator can experience a significant deflection.

PNEUMATIC SPRING

The pneumatic spring provides a simple means for resiliently supporting large loads. Consequently, it finds application as the load-carrying element in a vibration isolation system. Referring to the pneumatic spring illustrated in Fig. 33.2A, the load F supported by the spring is the product of pressure P_i and area S, $F = P_iS$. For a given area, the pressure may be adjusted to carry any load within the strength limitations of the cylinder walls.

STIFFNESS OF PNEUMATIC SPRING

The stiffness is the derivative of the force F with respect to the displacement:

$$k = \frac{dF}{d\delta} = S\frac{dP}{d\delta}$$

The stiffness of the pneumatic spring of Fig. 33.2*A* is derived readily from the gas laws governing the pressure and volume relationship. Assuming adiabatic compression, the equation defining the pressure-volume relationship is

$$PV^n = P_iV_i^n \tag{33.1}$$

where P_i is the gas pressure at a reference displacement, V_i is the corresponding volume of the contained gas and n is the ratio of the specific heats of the gas ($n = 1.4$ for air). Written in terms of the force applied to the air spring, Eq. (33.1) becomes

$$\frac{F}{S}V^n = P_iV_i^n \tag{33.2}$$

The relationship between the volume V and the deflection δ of the spring of Fig. 33.2*A* is

$$V = V_i - S\delta$$

Substituting this expression into Eq. (33.2) and taking the derivative $dF/d\delta$, the stiffness k as a function of displacement δ at any position is

$$k = \frac{dF}{d\delta} = \frac{nP_iS^2}{V_i}\left(\frac{1}{1 - (S/V_i)\delta}\right)^{n+1}$$

If the change in volume is small relative to the initial volume V_i, $S\delta \ll V_i$, then the expression for stiffness reduces to

$$k = \frac{nP_iS^2}{V_i} \tag{33.3}$$

A constant area S is assumed in the above derivation. For cases where the effective load-supporting area of the pneumatic spring varies with the deflection δ; i.e., $S = f(\delta)$, the pressure P and area S both must be treated as variables. The resulting expression for stiffness is

$$k = \frac{dF}{d\delta} = S\frac{dP}{d\delta} + P\frac{dS}{d\delta}$$

Variation of the area S with deflection δ provides means of minimizing the nonlinearity inherent in the load-deflection characteristics of a constant-area pneumatic spring. Controlled variation of the load-area relationship also provides means for obtaining any desired nonlinearity other than that normally available in the constant-area pneumatic spring.

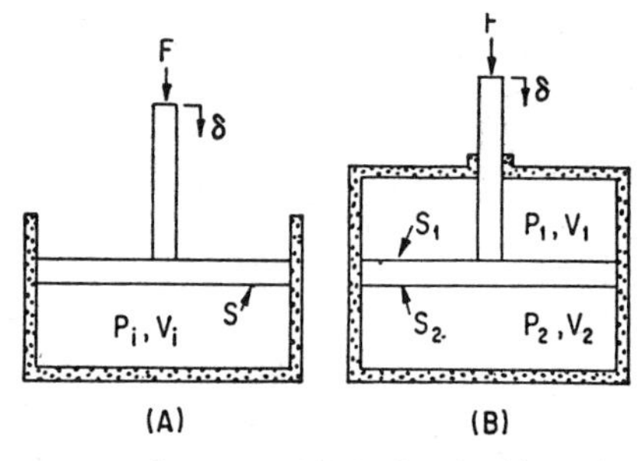

FIG. 33.2. Illustration of a single-acting (*A*) and a double-acting (*B*) pneumatic spring consisting of a piston and cylinder.

Variation in the load-supporting area of a pneumatic spring usually is obtained by means of a rolling diaphragm. Figure 33.3 illustrates a pneumatic spring typical of those used by the automotive industry in passenger car suspensions. The area of this particular spring is varied in such a manner that the load-deflection curve remains essentially linear over several inches of stroke.

DOUBLE-ACTING SPRING

The stiffness expression given by Eq. (33.3) is for a single-acting air spring as illustrated in Fig. 33.2*A*. The stiffness of the double-acting pneumatic spring illustrated in Fig. 33.2*B* is given by the expression

$$k = \frac{nP_1S_1^2}{V_1} + \frac{nP_2S_2}{V_2}$$

If the force on the pneumatic spring is zero when the piston is in its central position, $P_1 = P_2 = P_i$ and $V_1 = V_2 = V_i$. Then the stiffness is

$$k = \frac{2nP_iS^2}{V_i}$$

This equation assumes $S_1 = S_2 = S_i$; this is approximately correct when the piston rod area is small compared to the piston area.

The advantage of the double-acting spring of Fig. 33.2*B* over the single-acting spring of Fig. 33.2*A* is its ability to provide a restoring force in either the positive or negative direction. The single-acting spring is used in applications where the force acting on the spring is unidirectional. Two examples of such applications are automobile suspensions and vibration isolators for industrial machinery. An example of a double-acting pneumatic spring application is found in vibration-isolation systems for equipment used on rocket-propelled missiles which experience large sustained accelerations. The forces

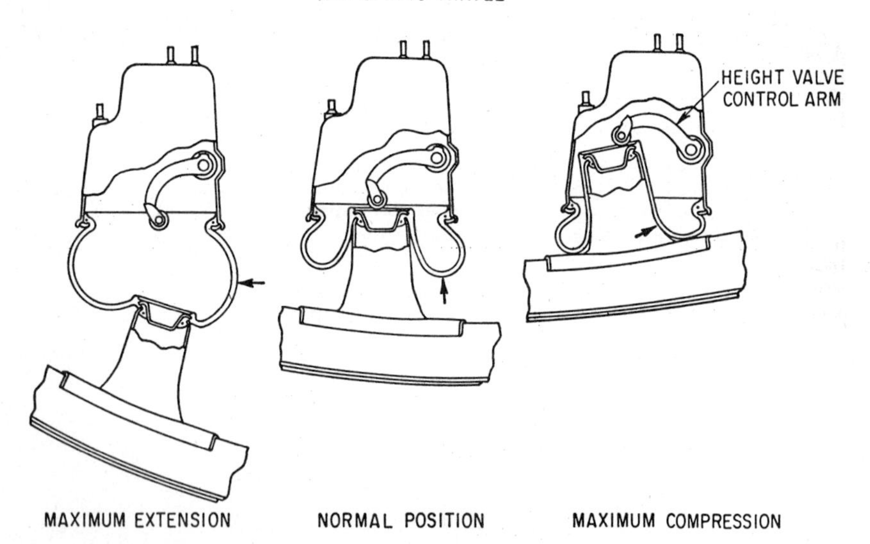

Fig. 33.3. Cross section of a pneumatic spring, typical of those used in automotive suspension systems. The heavy arrows indicate the rolling diaphragm which changes its maximum diameter with stroke, thereby providing a means of controlling the shape of the load-deflection curve. (*Courtesy of the Chevrolet Division of General Motors Corporation.*)

applied to the isolation systems in the latter applications may appear from either direction. It may be advantageous to achieve a double-acting pneumatic spring by employing two separate single-acting springs working in opposition to one another.

DAMPED PNEUMATIC SPRING

The above discussion of a closed cylinder pneumatic spring is limited to the stiffness of the spring. It is possible to design a pneumatic spring that is self-damped, as illustrated in Fig. 33.4. A self-damped pneumatic spring is more difficult to analyze than a closed cylinder spring because of the flow of gas which takes place between the cylinder and the surge tanks. This type of compressible flow can be analyzed by considering the rate of change of energy which occurs in the individual cylinder and tank volumes. Considering the cylinder volume, the rate at which energy leaves the cylinder is found by multiplying the total energy content per unit weight of gas by the weight rate of flow. The total energy content consists of the internal energy of the gas at a particular temperature and the energy gained or lost by the gas as a result of its compression or expansion. This is expressed as rate of change of enthalpy:

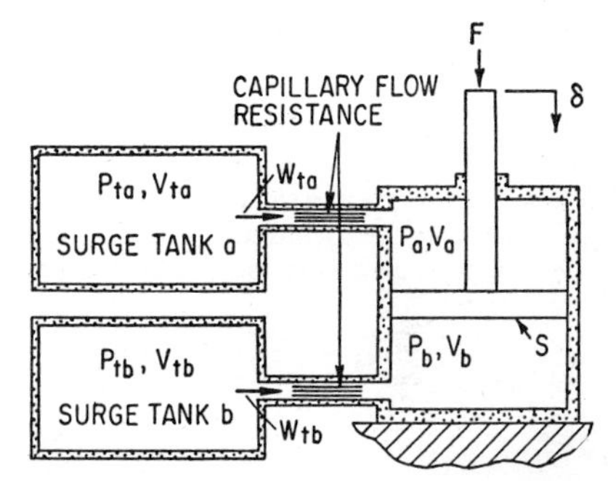

FIG. 33.4. A self-damped double-acting pneumatic spring in which the damping is achieved by placing a capillary flow resistance between the cylinder of the pneumatic spring and the connecting surge tanks.

$$\frac{dH}{dt} = \frac{dU}{dt} + \frac{dE}{dt} = Wc_pT_0$$

where dH/dt is the rate of change of enthalpy with time, dU/dt is the rate of change of internal energy, dE/dt is the rate at which work is performed on the gas, W is the flow of gas in unit weight per unit time, c_p is the specific heat of the gas at constant pressure, and T_0 is the absolute temperature of the gas. The rate of change of internal energy of the gas in the cylinder is the flow rate of the gas multiplied by its internal energy per pound:

$$\frac{dU}{dt} = Wc_vT_0 = \frac{gc_v}{R}\frac{d}{dt}(PV)$$

where dU/dt is the rate of change of internal energy of the gas, c_v is the specific heat of the gas at constant volume, R is the gas constant, g is the acceleration of gravity, P is the gas pressure, and V is the gas volume. The rate of energy change in the gas as a result of work done on or by the gas is

$$\frac{dE}{dt} = gP\frac{dV}{dt}$$

The expression for the rate of change of enthalpy thus may be written for the two ends of the cylinder:

$$W_{ta}c_pT_0 = \frac{gc_v}{R}\frac{d}{dt}(P_aV_a) + gP_a\frac{dV_a}{dt}$$

$$W_{tb}c_pT_0 = \frac{gc_v}{R}\frac{d}{dt}(P_bV_b) + gP_b\frac{dV_b}{dt}$$

where W_{ta} and W_{tb} are the weight rates of the gas flowing into the respective sides of the cylinder, P_a and P_b are the pressures on the respective sides of the cylinder, and V_a and V_b are the volumes on the respective sides of the cylinder.

Solving for W_{ta} and W_{tb}, and substituting $\frac{1}{c_p} + \frac{1}{nR} = \frac{1}{R}$ results in the following set of equations:

$$W_{ta} = \frac{g}{RT_0}\left[\frac{V_a}{n}\frac{dP_a}{dt} + P_a\frac{dV_a}{dt}\right]$$
$$W_{tb} = \frac{g}{RT_0}\left[\frac{V_b}{n}\frac{dP_b}{dt} + P_b\frac{dV_b}{dt}\right] \tag{33.4}$$

A corresponding set of equations may be written for the tank volumes. Inasmuch as the volume of the tank remains fixed, $\frac{dV_{ta}}{dt} = \frac{dV_{tb}}{dt} = 0$. Since the flow from the cylinders must equal the flow to the respective tanks,

$$W_{ta} = \frac{gV_{ta}}{nRT_0}\frac{dP_{ta}}{dt} \qquad W_{tb} = \frac{gV_{tb}}{nRT_0}\frac{dP_{tb}}{dt} \tag{33.5}$$

where P_{ta} and P_{tb} are the pressures in the respective tanks. The flow through the restriction between the cylinder and tank, assuming capillary type of flow; i.e., one in which the flow passage is very long when compared to the diameter of the passage, is defined by

$$W_{ta} = \frac{gC_r}{2RT_0}(P_a^2 - P_{ta}^2) \qquad W_{tb} = \frac{gC_r}{2RT_0}(P_b^2 - P_{tb}^2) \tag{33.6}$$

where

$$C_r = \frac{wt^3}{12\mu l} \tag{33.7}$$

and w, t, and l are width, thickness, and length, respectively, of the capillary passage, and μ is the dynamic viscosity of the gas in lb-sec/in.^{2}.

A capillary resistance is assumed since the relationship between the flow W_{ta}, W_{tb} through the resistance and the pressure drop $(P_a - P_{ta})$, $(P_b - P_{tb})$ across the resistance are linearized readily. Equation (33.6) may be written

$$W_{ta} = \frac{gC_r}{2RT_0}(P_a - P_{ta})(P_a + P_{ta})$$

$$W_{tb} = \frac{gC_r}{2RT_0}(P_b - P_{tb})(P_b + P_{tb})$$

For small motions $P_a + P_{ta} = 2P_i$ and $P_b + P_{tb} = 2P_i$, where P_i is the average pressure about which variations occur. Then the expression for the flow through the capillary resistance may be simplified:

$$W_{ta} = \frac{P_igC_r}{RT_0}(P_a - P_{ta}) \qquad W_{tb} = \frac{P_igC_r}{RT_0}(P_b - P_{tb}) \tag{33.8}$$

The cylinder volumes V_a and V_b are defined as functions of δ by

$$V_a = V_c + S\delta \qquad V_b = V_c - S\delta \tag{33.9}$$

where V_c is the volume of one side of the cylinder when the piston is in its neutral position.

The rates of change of volumes V_a and V_b are

$$\frac{dV_a}{dt} = S\frac{d\delta}{dt} \qquad \frac{dV_b}{dt} = -S\frac{d\delta}{dt} \tag{33.10}$$

Noting that positive values of δ result in a positive value for W_{ta} and a negative value of W_{tb}, the net rate of flow into the cylinder is found by substituting Eqs. (33.9) and

(33.10) into Eq. (33.4) and solving for $(W_{ta} - W_{tb})$:

$$(W_{ta} - W_{tb}) = \frac{g}{RT_0}\left[\frac{V_c}{n}\frac{d}{dt}(P_a - P_b) + \frac{S\delta}{n}\frac{d}{dt}(P_a + P_b) + (P_a + P_b)S\frac{d\delta}{dt}\right]$$

For small motion of the piston, P_a increases at nearly the same rate that P_b decreases; i.e., $(P_a + P_b) \simeq 2P_i$ and $d/dt\,(P_a + P_b) \simeq 0$. Using the differential operator notation $d/dt = D$, and introducing these approximations,

$$(W_{ta} - W_{tb}) = \frac{g}{RT_0}\left[\frac{V_c}{n}D(P_a - P_b) + 2P_iSD(\delta)\right] \tag{33.11}$$

Assuming both surge tanks to have the same volume, $V_{ta} = V_{tb} = V_t$, solving Eqs. (33.8) for P_{ta} and P_{tb}, substituting in Eq. (33.5), and solving for $(W_{ta} - W_{tb})$,

$$(W_{ta} - W_{tb}) = \frac{gV_t}{nRT_0}\left[\frac{D(P_a - P_b)}{1 + (V_t/nC_rP_i)D}\right] \tag{33.12}$$

Equating Eqs. (33.11) and (33.12), and solving for $(P_a - P_b)$ in terms of δ,

$$(P_a - P_b) = -\frac{2nP_iS\delta}{V_c\left[1 + \dfrac{V_t/V_c}{1 + (V_t/nC_rP_i)D}\right]}$$

The negative sign indicates that a positive deflection δ results in the pressure P_b being larger than P_a.

The external force on the piston is

$$F = (P_b - P_a)S = \frac{2nP_iS^2\delta}{V_c\left[1 + \dfrac{V_t/V_c}{1 + (V_t/nC_rP_i)D}\right]} \tag{33.13}$$

Dividing Eq. (33.13) by δ, the expression for the stiffness is

$$k = \frac{F}{\delta} = \frac{2nP_iS^2}{V_c}\left[\frac{1}{1 + \dfrac{V_t/V_c}{1 + (V_t/nC_rP_i)D}}\right] \tag{33.14}$$

SYSTEM WITH DAMPED SPRING. Using the expression for stiffness given by Eq. (33.14) and writing the equation of motion of the mass m of Fig. 33.5,

$$mD^2x = \frac{2nP_iS^2(u - x)}{V_c\left[1 + \dfrac{V_t/V_c}{1 + (V_t/nC_rP_i)D}\right]} \tag{33.15}$$

where x is the absolute displacement of the mass, u is the absolute displacement of the base, and δ is the relative displacement between mass and base. The following substitutions are made in Eq. (33.15):

$$\frac{V_t}{nC_rP_i} = 2\frac{c}{c_c}\frac{1}{\omega_n}; \qquad \frac{2nP_iS^2}{V_t} = k; \qquad \omega_n^2 = \frac{k}{m}; \qquad \frac{V_t}{V_c} = N \tag{33.16}$$

Solving for x/u, the absolute transmissibility,

$$\frac{x}{u} = \frac{2\dfrac{c}{c_c}\dfrac{D}{\omega_n} + 1}{2\dfrac{c}{c_c}\dfrac{1}{N}\dfrac{D^3}{\omega_n^3} + \left(\dfrac{N+1}{N}\right)\dfrac{D^2}{\omega_n^2} + 2\dfrac{c}{c_c}\dfrac{D}{\omega_n} + 1}$$

Substituting $j\omega = D$ and taking the square root of the sum of the squares of the real and imaginary parts:

$$\left|\frac{x_0}{u_0}\right| = T = \sqrt{\frac{1 + \left(2\frac{c}{c_c}\frac{\omega}{\omega_n}\right)^2}{\left(1 - \frac{N+1}{N}\frac{\omega^2}{\omega_n^2}\right)^2 + \left(2\frac{c}{c_c}\frac{\omega}{\omega_n}\right)^2\left(1 - \frac{1}{N}\frac{\omega^2}{\omega_n^2}\right)^2}} \tag{33.17}$$

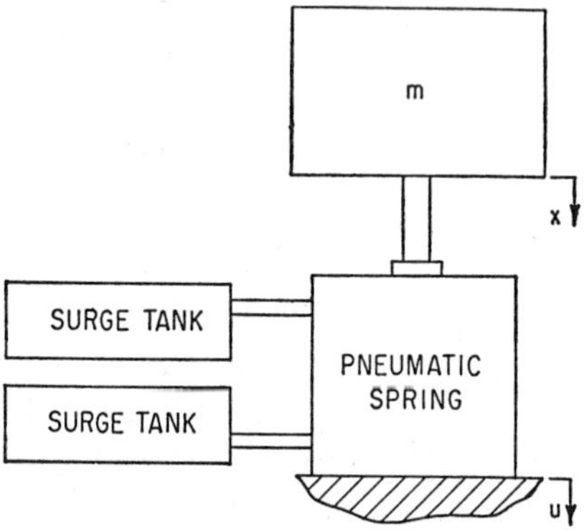

Fig. 33.5. Schematic diagram of a self-damped pneumatic-spring type isolation system in which the spring element is similar to that of Fig. 33.4. The system is excited by the motion $u = u_0 \sin \omega t$. The response motion is $x = x_0 \sin (\omega t + \theta)$.

EFFECT OF FREQUENCY. As the piston of the device illustrated in Fig. 33.4 is moved so that δ is positive, the gas in the B side of the cylinder is compressed and flows through the connecting pipe to the region of lower pressure in the surge tank. If the piston is left in its new position, the pressure in the tank eventually becomes equal to the pressure in the cylinder and the flow of air stops. If the piston now is moved back to its original position, the expansion of the air in the cylinder results in a lower pressure, and the air flows in the reverse direction until the pressures in the tank and cylinder again become equal. Similar conditions apply to the A side of the piston. Thus, the pressure developed when the piston is held displaced from its neutral position is a function of the combined volumes of the cylinder and tank. This then is also true when the piston moves with a low-frequency vibration. Then the expression for stiffness in Eq. (33.14) becomes

$$k_0 = \frac{2nP_iS^2}{V_t + V_c} \quad \text{lb/in.} \tag{33.18}$$

As the vibration frequency increases, the quantity of air which can move from one volume to the other decreases because there is less time for flow to take place. Eventually, for relatively high frequencies, essentially no flow can take place and the surge tank becomes effectively cut off from the cylinder volume. At high frequencies, the stiffness of the system is

$$k_\infty = \frac{2nP_iS^2}{V_c} \quad \text{lb/in.} \tag{33.19}$$

At low frequencies, the compression process tends to become isothermal rather than adiabatic. The parameter n which has a value of 1.4 for adiabatic conditions, has a value of 1.0 for isothermal conditions. For frequencies below approximately 3.0 cps, isothermal conditions exist unless special pains have been taken to insulate the system thermally. For thermally insulated systems, the transition to an isothermal process occurs at frequencies below 3.0 cps.

INFLUENCE OF FLOW RESTRICTION. Another effect which can be deduced from the physical considerations of the system is the influence of the flow restriction between the cylinder and tank. If the passage between the cylinder and tank is made large so that flow occurs without any restriction, $C_r = \infty$ and $c/c_c = 0$; i.e., large rates of flow W_{ta}, W_{tb} occur for very small pressure differentials $(P_a - P_{ta})$, $(P_b - P_{tb})$. Then the system should behave like a mass on an undamped spring—the spring having a stiffness as given by Eq. (33.18). The natural frequency of such a system is designated ω_0 and is expressed by

$$\omega_0 = \sqrt{\frac{2nP_iS^2}{(V_t + V_c)m}} = \omega_n \sqrt{\frac{N}{N+1}} \tag{33.20}$$

where N is the ratio V_t/V_c. The reference frequency ω_n is taken for convenience; physically it is the undamped natural frequency the system would have using an air spring with a volume V_t.

If the restriction is such that no flow can occur between the cylinder and tank volumes, $C_r = 0$ and $c/c_c = \infty$; again the system should behave like a mass on an undamped spring—the spring having a stiffness k_∞ as given by Eq. (33.19). The natural frequency of such a system is designated ω_∞ and is expressed by

$$\omega_\infty = \sqrt{\frac{2nP_iS^2}{V_cm}} = \omega_n \sqrt{N} \tag{33.21}$$

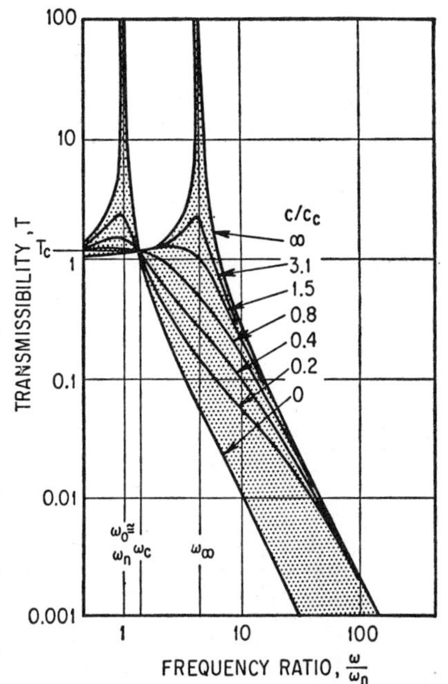

FIG. 33.6. Transmissibility curves of a self-damped pneumatic-spring type isolation system similar to that shown in Fig. 33.5 and having a volume ratio $N = V_t/V_c = 21$. These curves illustrate the effect of the value of the fraction of critical damping c/c_c on the system's response. The shaded area defines the envelope of the curves for all values of c/c_c and $N = 21$.

Having established that infinite response of the system occurs for the two extreme conditions of the flow restriction constant C_r, it follows that there must be an intermediate value of C_r which produces a minimum peak value for the transmissibility. A finite value of flow and a finite value of C_r will result in the dissipation of energy and consequently will reduce the response amplitude from infinity to some finite value.

Equation (33.17) is plotted in Fig. 33.6 for one value of N and several values of damping. If the damping is zero, $c/c_c = 0$ and Eq. (33.17) reduces to

$$T_0 = \frac{1}{1 - \dfrac{\omega^2}{{\omega_0}^2}} \tag{33.22}$$

Equation (33.22) defines the forced-vibration response of an undamped single degree-of-freedom system having a natural frequency ω_0.

If the damping is infinitely great, $c/c_c = \infty$ and Eq. (33.17) reduces to

$$T_\infty = \frac{1}{1 - \dfrac{\omega^2}{{\omega_\infty}^2}} \tag{33.23}$$

Equation (33.23) defines the forced-vibration response of an undamped single degree-of-freedom system having a natural frequency ω_∞. Equations (33.22) and (33.23) define the envelope of the responses for all values of damping, as depicted by the shaded area of Fig. 33.6. Since all the curves must remain in the shaded area, it is evident that they all must have the same value of T at the frequency ω_c as shown in Fig. 33.6. This frequency is called the *common transmissibility frequency;* it is found by equating Eqs. (33.22) and (33.23):

$$\omega_c = \omega_n \sqrt{\frac{2N}{2 + N}} \tag{33.24}$$

The value of T at this frequency, designated as T_c, is

$$T_c = \frac{2}{N} + 1 \tag{33.25}$$

If it is desired to have the lowest possible amplification at resonance for any given value

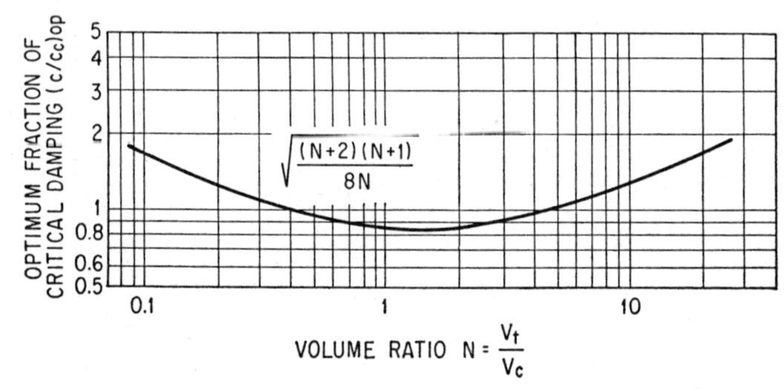

FIG. 33.7. Curve illustrating relationship of optimum fraction of critical damping $(c/c_c)_{op}$ to the volume ratio N for a self-damped pneumatic-spring isolation system as illustrated in Fig. 33.5.

of N, it is necessary to make the resonant frequency coincident with ω_c. The resonant frequency is found by differentiating Eq. (33.17) with respect to frequency and setting the derivative equal to zero. Substitution of the expression for ω_c from Eq. (33.24) in the expression for resonant frequency results in an equation which may be solved for the optimum value of damping as a function of N:

$$\left(\frac{c}{c_c}\right)_{op} = \sqrt{\frac{(N + 1)(N + 2)}{8N}} \tag{33.26}$$

where N is the ratio V_t/V_c of tank volume to the cylinder volume. This relation is shown graphically in Fig. 33.7.

Comparison of the system with pneumatic-spring and the conventional single degree-of-freedom isolation system (illustrated in Fig. 33.9 and discussed in Chap. 30) demonstrates a unique property of the pneumatic-spring system. As the excitation frequency ω approaches infinity, a conventional system has a transmissibility asymptote given by

$$T_i = \frac{2(c/c_c)}{\omega/\omega_n} \tag{33.27}$$

The pneumatic-spring system has a corresponding asymptote given by

$$T_p = \frac{1}{\dfrac{\omega^2}{N\omega_n^2}} = \frac{1}{\dfrac{\omega^2}{\omega_\infty^2}} \tag{33.28}$$

Equations (33.27) and (33.28) indicate the relative effectiveness of the two systems in isolating high-frequency vibration.

At high frequencies, the transmissibility of the pneumatic isolation system decreases at a rate proportional to the square of the forcing frequency, whereas the transmissibility of the conventional system decreases at a rate proportional to only the first power of the forcing frequency. Therefore, at high frequencies, the isolation efficiency of the pneumatic system is much greater than that of the conventional system.

SERVO-CONTROLLED ISOLATION SYSTEMS

Servo-controlled isolation systems are employed when the natural frequency of the isolation system must be relatively low—with the added requirement that the supported body be maintained at a relatively constant distance from the base to which it is attached. As described in the introduction, the position of the body, i.e., the deflection of the isolation system, can be controlled by means of a servomechanism. The servomechanism is made up of two basic components: (1) a sensing element which generates a signal proportional to the integral of the displacement of the body and (2) a motor, or actuator, which creates a force proportional to the signal from the sensing element. The force output of the motor, or actuator, then supplements the spring and damper force to help support the body. The combination of sensing element and actuator provides what is known as "integral control of displacement" in servomechanism terminology. It is not uncommon to place both the sensing element and the actuator in one physical assembly. The element, or elements, added to a conventional isolation system must have an over-all characteristic such that the output force is proportional to the integral of the displacement of the actuator. This corresponds to a spring which provides an output force proportional to the deflection of the spring or a damper which provides a force proportional to the first derivative of deflection of the damper. The individual elements which make up a servo-controlled isolation system, together with the equations which describe the force outputs of the element, are illustrated in Fig. 33.8. The schematic diagrams of the three elements as they are used in an active isolation system also are shown.

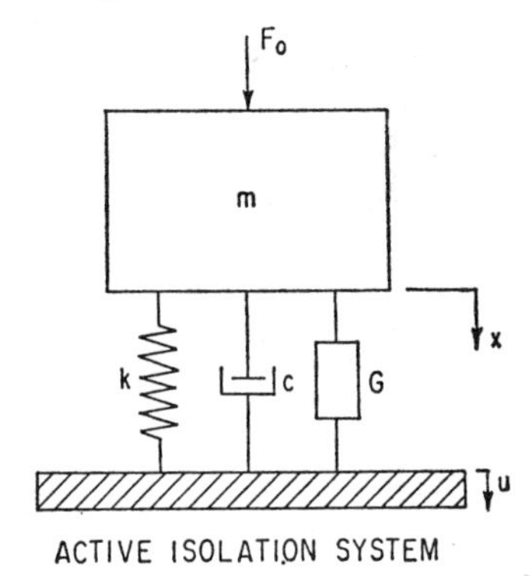

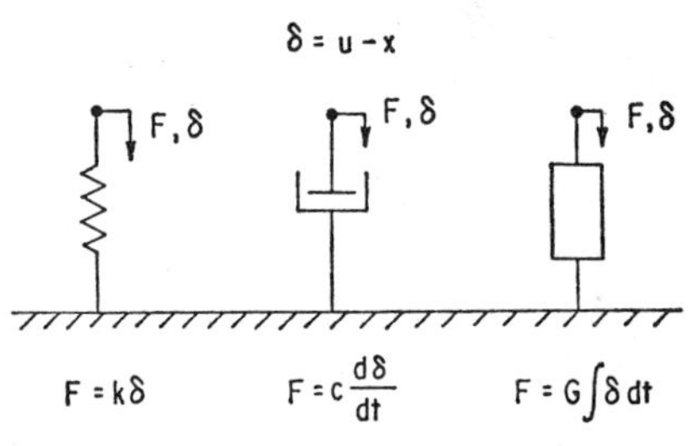

FIG. 33.8. Schematic diagram of an active isolation system. The constant G designates the gain of the active element. All the elements connecting the isolated mass to the base, together with their defining equations, appear in the lower portion of the figure. The system is excited by the motion $u = u_0 \sin \omega t$ and responds with a motion $x = x_0 \sin (\omega t + \theta)$. The relative motion δ between u and x is $\delta = u - x$. Excitation may also be caused by the force F_0.

Because integral feedback of displacement requires that energy be fed into the isolation system, it is possible to make the isolation system dynamically unstable by improper proportioning of the system constants. Therefore, one of the most important factors in achieving a satisfactory isolation system is the determination of the margin of dynamic stability of the system.

The elements which provide integral control of the displacement may take many forms. For example, the sensing element which determines the position of the isolated body may be a differential transformer which produces an electrical signal proportional to its displacement. The electrical signal then is amplified, integrated, and used to operate an electric motor. The differential transformer-integrator-motor system produces a force proportional to the integral of the signal from the transformer. Alternatively, the signal from the differential transformer may be used to operate a servo valve directly. the flow through the valve being proportional to the signal from the transformer. The servo valve then is used to control a hydraulic or pneumatic cylinder to produce the desired force. In this arrangement, the flow through the valve is proportional to the integral of the displacement and no electrical integration of the transformer signal is needed. It also is possible to operate the valve through a direct mechanical or hydraulic coupling whereby the motion of the mounted body with respect to its base is used directly to provide the desired valve motion. The possible combination of sensing elements and control devices is almost limitless. The choice of a suitable combination usually is dictated by the type of power available and the type of application, e.g., aircraft, missile, automotive, or industrial.

Active isolation systems usually are described by cubic or higher-order differential equations which, because of their complexity, make it difficult to visualize the effect of changes in the system constants on the performance of the system. In the case of a conventional isolation system, it is possible to determine many of the performance characteristics from the constants appearing in the differential equation. For example, the transmissibility T of the conventional system at resonance is approximately

$$T_r \simeq \frac{\sqrt{km}}{c} = \frac{1}{2(c/c_c)} \qquad \left[\frac{c}{c_c} < 0.2\right] \tag{33.29}$$

Similarly, the resonant frequency ω_r is approximately equal to the undamped natural frequency:

$$\omega_r \simeq \sqrt{\frac{k}{m}} \qquad \left[\frac{c}{c_c} < 0.2\right] \tag{33.30}$$

At high frequencies ($\omega \gg \omega_n$) the transmissibility of the conventional system approaches the asymptotic value

$$T_i = \frac{2(c/c_c)}{\omega/\omega_n} \qquad [\omega \gg \omega_n] \tag{33.31}$$

The transmissibility curve of a conventional system may be estimated from Eqs. (33.29) to (33.31) without plotting the transmissibility equation point by point. Somewhat similar relationships can be obtained for an active system if its equation of motion is not higher than third order. The most convenient way to obtain rules of thumb for design of an active system is to compare the characteristic properties of a conventional isolation system with those of the same isolation system having an element which provides integral feedback of displacement added in parallel with a spring and damper.

The element in an active isolation system that provides integral control of the displacement maintains the supported body at a constant distance from the base to which it is attached. When a step function of force is applied to the supported body, the response of the system gives a measure of the effectiveness of the element in performing the desired function. A comparison of the transient response of the active system, i.e., one having integral feedback of displacement, with that of the conventional passive system described in Chap. 30 illustrates the advantage obtained from the active system.

TRANSIENT RESPONSE

Referring to Fig. 33.9, the equation of motion for the mass of the passive isolation system is

$$m\frac{d^2x}{dt^2} + c\frac{dx}{dt} + kx = F(t) \tag{33.32}$$

where u is set equal to zero and $F(t)$ is a step function of force having a magnitude $F = F_0$ when $t > 0$ and $F = 0$ when $t < 0$. Writing the Laplace transform of Eq. (33.32),

$$\mathfrak{L}[x(t)] = X(s) = \frac{F_0}{ms}\frac{1}{[s^2 + (c/m)s + k/m]} \tag{33.33}$$

where $X(s)$ designates the Laplace transform of x, a function of time.* Letting $c/m = 2(c/c_c)$ and $k/m = \omega_n^2$, Eq. (33.33) may be written

$$X(s) = \frac{F_0}{ms}\frac{1}{[s^2 + 2(c/c_c)\omega_n s + \omega_n^2]} \tag{33.34}$$

Equation (33.34) is analogous to the transform equation

$$X(s) = \frac{F_0}{m}\frac{1}{s[(s+\alpha)^2 + \omega_d^2]}$$

where $\alpha = \omega_n c/c_c$ and $\omega_d = \omega_n\sqrt{1-(c/c_c)^2}$ is the damped natural frequency of the system. The inverse transform of Eq. (33.34) gives the solution:

$$x = \frac{F_0}{m\omega_n^2} + \frac{F_0}{m\omega_n^2}\cdot e^{-(c/c_c)\omega_n t}\sin(\omega_d t - \theta_1) \tag{33.35}$$

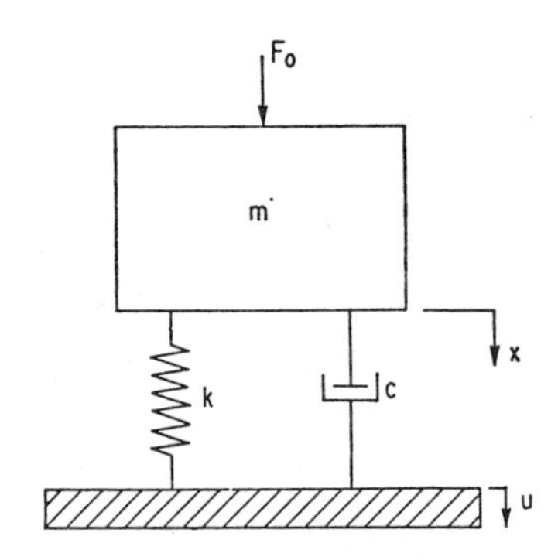

FIG. 33.9. Schematic diagram of a passive or conventional-type isolation system. The system is excited by the motion $u = u_0 \sin \omega t$ and responds with a motion $x = x_0 \sin(\omega t + \theta)$. The relative motion between u and x is $\delta = u - x$. Excitation may also be caused by the force F_0.

* Laplace transforms are discussed in detail in Chap. 8.

where $\theta_1 = \tan^{-1} \dfrac{\sqrt{1 - (c/c_c)^2}}{-c/c_c}$

$\omega_n = \sqrt{k/m}$ = undamped natural frequency, rad/sec

$\omega_d = \omega_n\sqrt{1 - (c/c_c)^2}$ = damped natural frequency, rad/sec

c/c_c = fraction of critical damping

$c_c = 2m\omega_n$ = critical damping

From Eq. (33.35), the mass takes a new position of static equilibrium at a distance $F_0/m\omega_n^2$ from the original position as $t \to \infty$. The first term $F_0/m\omega_n^2$ may be eliminated from Eq. (33.35) by adding an integral displacement control element to the isolation system to complement the spring force. This added element would produce a force proportional to the integral of the displacement x with respect to time. Referring to Fig. 33.8, the equation of motion for the system with the added element is

$$m\frac{d^2x}{dt^2} + c\frac{dx}{dt} + kx + G\int x\,dt = F(t)$$

This equation may be solved most easily for a step function excitation by the use of Laplace transform methods. If $F(t)$ is a force step function of magnitude F_0, the transform equation is

$$\mathfrak{L}[x(t)] = X(s) = \frac{F_0}{m} \cdot \frac{1}{s^3 + 2(c/c_c)\omega_n s^2 + \omega_n^2 s + (G/m\omega_n^3)\omega_n^3} \tag{33.36}$$

where the constants c/c_c, ω_n, m, and F_0 have the same meaning as in Eq. (33.34), and G is the gain of the integral feedback element in lb/in.-sec. Since the transform equation is a cubic, it is practical to solve only when numerical values are given. The equation can be written as

$$X(s) = \frac{F_0}{m}\,\frac{1}{(s + q\omega_p)(s^2 + 2p\omega_p + \omega_p^2)} \tag{33.37}$$

where q and p are dimensionless constants and ω_p is a characteristic angular frequency. Taking the inverse transform of Eq. (33.37),

$$x = \frac{F_0}{m\omega_n^2} \cdot \frac{\omega_n^2 e^{-q\omega_p t}}{\omega_p^2(q^2 - 2pq + 1)}$$

$$+ \frac{F_0}{m\omega_n^2} \cdot \frac{\omega_n^2 e^{-p\omega_p t}}{\omega_p^2[(1 - p^2)(q^2 - 2pq + 1)]^{1/2}} \sin\left(\omega_p\sqrt{1 - p^2}\,t - \theta_2\right) \tag{33.38}$$

$$\theta_2 = \tan^{-1}\frac{\sqrt{1 - p^2}}{q - p}$$

In considering the effect of the various parameters, it is convenient to rewrite Eq. (33.38) in the form:

$$x = Ae^{-q\omega_p t} + Be^{-p\omega_p t}\sin(\omega_d t - \theta_2) \tag{33.39}$$

The effect of the parameters p and q on the transient response can be visualized more readily if the terms of Eq. (33.39) are examined for their respective contributions to the total response of the system. The last term of Eq. (33.39) is simply a damped sinusoidal quantity having an initial amplitude B and frequency ω_d; it decays at a rate determined by the exponential $e^{-p\omega_p t}$. The first term of Eq. (33.39) is a nonoscillatory exponential having an initial value A which decays at a rate determined by the exponential $e^{-q\omega_p t}$. The first term of the equation forms a mean value about which the second (oscillatory) term varies. Therefore the product $p\omega_p$ determines how rapidly oscillations caused by a disturbing force decay, while $q\omega_p$ determines how rapidly the system is able to restore the mass m to its set position $x = 0$. Since both of the terms of Eq. (33.39) are multiplied by negative exponentials, the value of x approaches zero as t approaches infinity. Therefore, the position of the mass tends to remain a fixed distance from the base to which it is attached—regardless of the external forces acting upon the system.

Having established that the integral feedback control element performs the desired function in the active system (i.e., maintaining the isolated body in a given spatial relationship with respect to the base to which it is attached), the physical constants c/c_c, ω_n, and $G/m\omega_n^3$ now must be related to the constants p, q, and ω_p appearing in Eq. (33.37). This can be done by equating the denominators of Eqs. (33.36) and (33.37):

$$s^3 + 2\frac{c}{c_c}\omega_n s^2 + \omega_n^2 s + \frac{G}{m\omega_n^3}\omega_n^3 = (s + q\omega_p)(s^2 + 2p\omega_p + \omega_p^2) \tag{33.40}$$

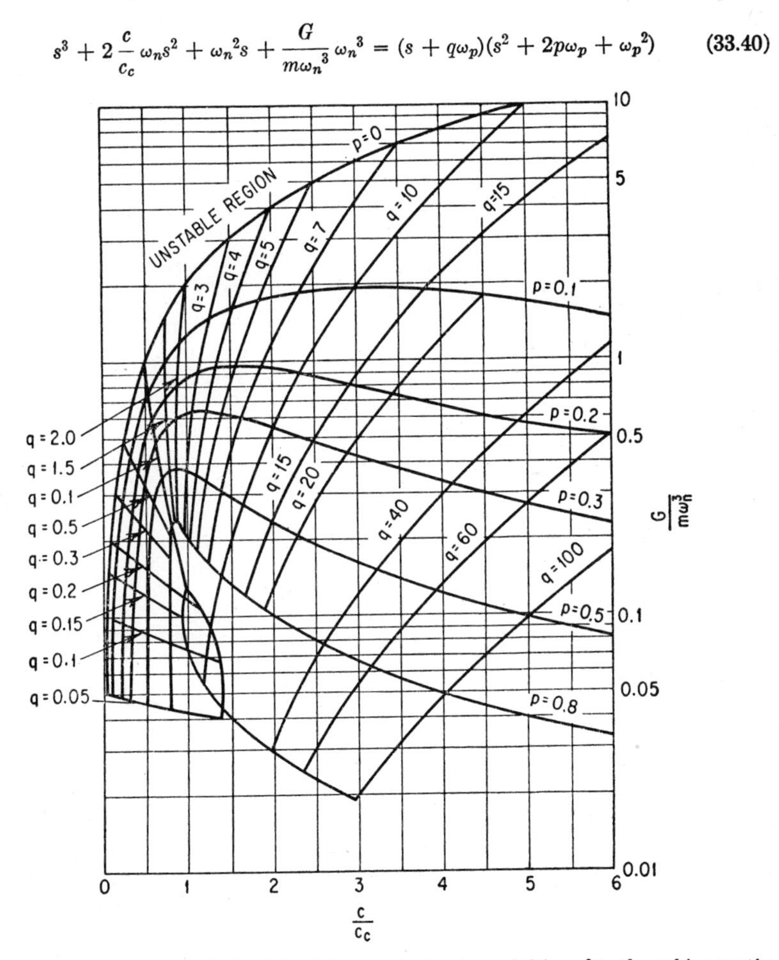

FIG. 33.10. Graph showing relationship of the constants c/c_c and $G/m\omega_n^3$ to the cubic equation root parameters p and q for the system illustrated in Fig. 33.8.

Equating like terms of s and solving for the various constants,

$$\frac{c}{c_c} = \frac{2p + q}{2(1 + 2pq)^{1/2}}; \qquad \frac{G}{m\omega_n^3} = \frac{q}{(1 + 2pq)^{3/2}}; \qquad \omega_n = \omega_p(1 + 2pq)^{1/2} \tag{33.41}$$

Comparing the form of the second-order root on the right-hand side of Eq. (33.40) with that of a conventional passive isolation system [see the denominator of Eq. (33.34)],

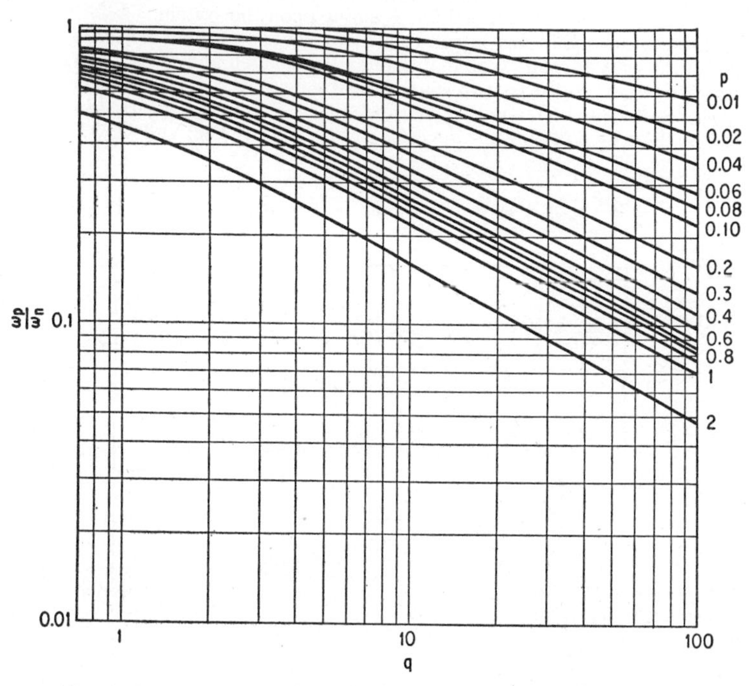

FIG. 33.11. Graph showing relationship of the constant ω_n to the cubic equation root parameters ω_p, p, and q for the system illustrated in Fig. 33.8.

it is evident that p in the second-order root of the active system equation is analogous to c/c_c in the passive system and that ω_p in the active system is analogous to ω_n in the passive system. The quantity q in the active system has no counterpart in the passive system since it is related to the gain of integral feedback element. The product $q\omega_p$ determines how rapidly the active system returns the mass to its undisturbed position. The larger the value of $q\omega_p$, the more rapid the return. Figure 33.10 shows the relationship of $G/m\omega_n^3$ and c/c_c to p and q. Figure 33.11 shows the relationship of ω_n and ω_p to p and q.

Figure 33.12 is a plot of Eqs. (33.35) and (33.38) for typical active and passive systems. Various combinations of fractions of critical damping (c/c_c) and dimensionless gain ($G/m\omega_n^3$) are used to illustrate their effect on the transient response. In general, the fraction of critical damping controls the amplitude of the oscillations. The time required for the system to reach its equilibrium position is an inverse function of the dimensionless gain. The optimum combination of damping and gain depends upon the combined transient and steady-state vibration environment which the system will experience.

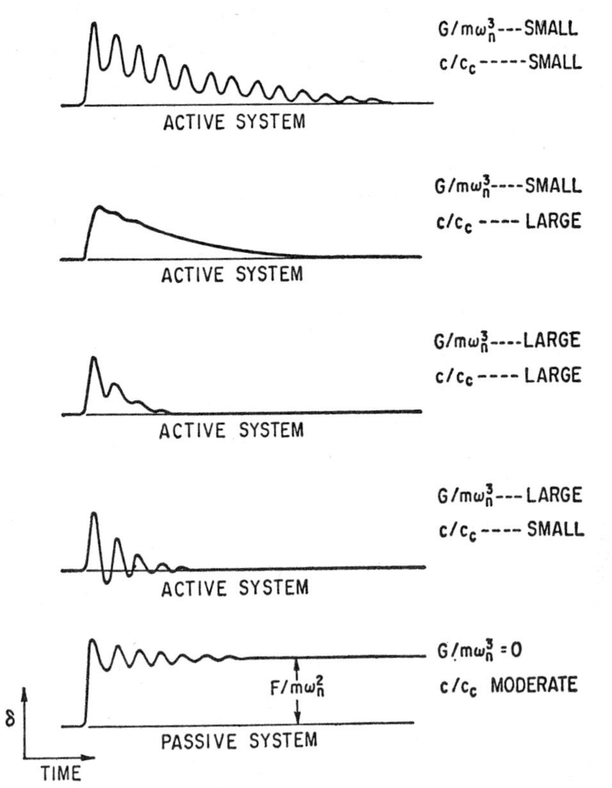

FIG. 33.12. Displacement time-histories for the active isolation system illustrated in Fig. 33.8 when subjected to a step force of magnitude F_0. The various magnitudes of gain and damping illustrate how each affects the transient response of the system.

STEADY-STATE RESPONSE

A comparison of the steady-state response of the active and passive systems illustrates some of the disadvantages associated with a servo-controlled active system. In Fig. 33.8, assume that $F(t) = 0$ and that the excitation is now caused by the motion u of the base. Then the equation of motion of the mounted body is

$$m\frac{d^2x}{dt^2} + c\frac{dx}{dt} + kx + G\int x\,dt = c\frac{du}{dt} + ku + G\int u\,dt \tag{33.42}$$

The solution of this equation may be expressed in terms of transmissibility

$$T = \left|\frac{x_0}{u_0}\right| = \sqrt{\frac{\left[\dfrac{G}{m\omega_n{}^3} - 2\left(\dfrac{c}{c_c}\right)\dfrac{\omega^2}{\omega_n{}^2}\right]^2 + \dfrac{\omega^2}{\omega_n{}^2}}{\dfrac{\omega^2}{\omega_n{}^2}\left[1 - \dfrac{\omega^2}{\omega_n{}^2}\right]^2 + \left[\dfrac{G}{m\omega_n{}^3} - 2\left(\dfrac{c}{c_c}\right)\dfrac{\omega^2}{\omega_n{}^2}\right]^2}} \tag{33.43}$$

Figure 33.13 is a plot of Eq. (33.43). The corresponding expression for the transmissibility of a conventional passive isolation system as illustrated in Fig. 33.9 is

$$T = \sqrt{\frac{1 + \left[2\left(\frac{c}{c_c}\right)\frac{\omega}{\omega_n}\right]^2}{\left(1 - \frac{\omega^2}{\omega_n^2}\right)^2 + \left[2\left(\frac{c}{c_c}\right)\frac{\omega}{\omega_n}\right]^2}} \tag{33.44}$$

Equations (33.43) and (33.44) both have the same characteristics at high excitation frequencies: As $\omega/\omega_n \to \infty$, $T \to \dfrac{2(c/c_c)}{\omega/\omega_n}$.

Therefore, the addition of the servo control element has no important effect on the capability of the system to isolate vibration of high frequency. However, the effect at resonance is significant, as shown in Fig. 33.13. As the gain of the servo control element is increased, the transmissibility of the system in the region of resonance increases. If the dimensionless gain of the servo control element for this particular system becomes larger than twice the damping ratio $[G/m\omega_n^3 > 2(c/c_c)]$, the system becomes unstable. Under these conditions, if the mass m receives the slightest disturbance, it continues to oscillate indefinitely. A more comprehensive discussion on stability is given in the section on *Stability of Active Systems.*

The approximate transmissibility at resonance for the active system having constants $G/m\omega_n^3 < 0.15$ and $c/c_c < 0.2$ is given by an expression similar to the one which applies to the passive system:

$$T \simeq \frac{1}{2p} \simeq \frac{1}{2(c/c_c)} \qquad \left[\frac{c}{c_c} < 0.2;\ \frac{G}{m\omega_n^3} < 0.15\right] \tag{33.45}$$

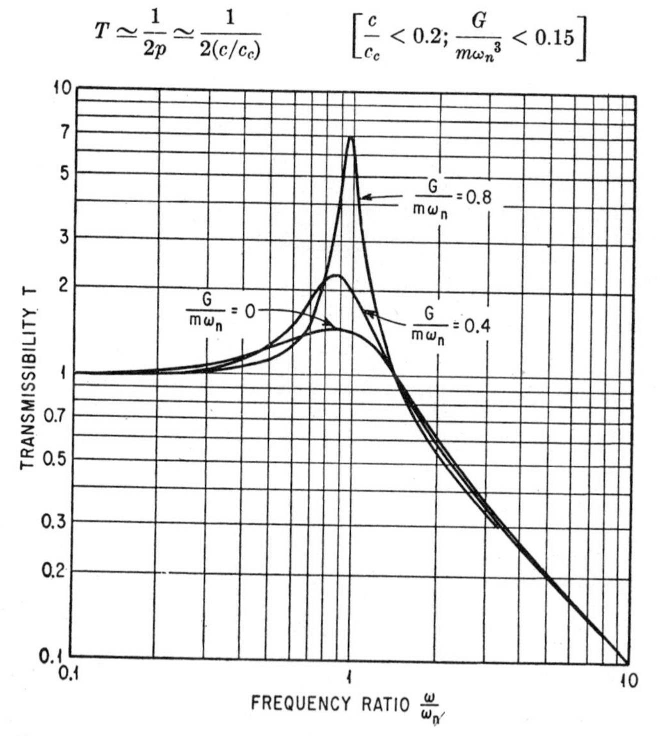

FIG. 33.13. Transmissibility T vs. frequency ratio ω/ω_n for the isolation system illustrated in Fig. 33.8. The fraction of critical damping for these curves is $c/c_c = 0.5$. The gain $G/m\omega_n^3$ is varied from 0 to 0.8 to illustrate the effect of gain on the transmissibility curves at resonance.

where p is determined from Fig. 33.10 for the given values of $G/m\omega_n^3$ and c/c_c. For these ranges, $G/m\omega_n^3 \simeq q$ and $c/c_c \simeq p$; the approximate resonant frequency is given by the same equation as that for the passive system:

$$\omega_p \simeq \sqrt{\frac{k}{m}} \qquad \left[\frac{c}{c_c} < 0.2; \frac{G}{m\omega_n^3} < 0.15\right] \tag{33.46}$$

The high-frequency asymptote of the active system is identical to that of the passive system:

$$T = \frac{2(c/c_c)}{\omega/\omega_n} \tag{33.47}$$

For values of $G/m\omega_n^3 > 0.15$ and of $c/c_c > 0.2$, each solution to the transmissibility equation must be considered separately because generalized results become complicated.

Values frequently used in practice for p and q are $p = 0.25$ and $q = 0.15$; then from Fig. 33.10, $c/c_c = 0.32$ and $G/m\omega_n^3 = 0.13$. These values result in a system which responds relatively rapidly to disturbances and yet does not tend to be dynamically unstable.

PNEUMATIC ISOLATION SYSTEM WITH SERVO CONTROL

This section includes an analysis of integral feedback control discussed in the preceding section as applied to isolators described in the section on *Pneumatic Springs*. The desired control consists of varying the pressure in the air spring so that the supported body has a given equilibrium position—regardless of the steady forces which act upon it. This type of control is obtained by a valve which is actuated by the position of the supported body; as forces are applied to the body, the resultant motion of the body opens a valve which allows air to escape from the cylinder on one side of the piston and to flow into the cylinder on the other side of the piston. The valve may be coupled to the body mechanically or may be actuated by an electrical signal which it receives from a transducer attached to the body. In either case, the characteristics of the valve, i.e., the rate of the gas flow in pounds per second per unit displacement of the body, must be known. Since the analysis is linearized, the valve characteristics used in the analysis also are assumed to be linear. The flow rate of air through the valve is proportional to the motion of the valve control element. Figure 33.14 illustrates the characteristics and shows a sectional drawing of a typical control valve; the application of the valve to a typical system is shown in Fig. 33.15.

For the flow rate to be a linear function of the valve opening when a compressible fluid is used, the upstream pressure must be equal to, or greater than, the pressure which causes choked flow conditions in the valve. Then the flow characteristics of the valve may be expressed by $W = G_v x_v$, where W is the weight rate of flow through the valve, x_v is the motion of the valve spool or control element with respect to the valve body, and G_v is the valve constant in pounds per second of gas flow per unit stroke of the control element of the valve.

The passive elements of the system shown in Fig. 33.15 are identical to those of Fig. 33.4. The flow equations [Eqs. (33.4)] must be modified, however, to accommodate the flow contributed by the valve. The net rate of flow into the cylinder is

$$\begin{aligned}(W_a - W_{ta}) &= \frac{g}{RT_0}\left[\frac{V_a}{n}D(P_a) + P_aD(V_a)\right] \\ (W_b - W_{tb}) &= \frac{g}{RT_0}\left[\frac{V_b}{n}D(P_b) - P_bD(V_b)\right]\end{aligned} \tag{33.48}$$

Flow W_a, W_b through the valve is related to the relative displacement $\delta = (u - x)$ by $W_b = +G\delta$, $W_b = -G\delta$, where the constant G defines the weight rate of flow through

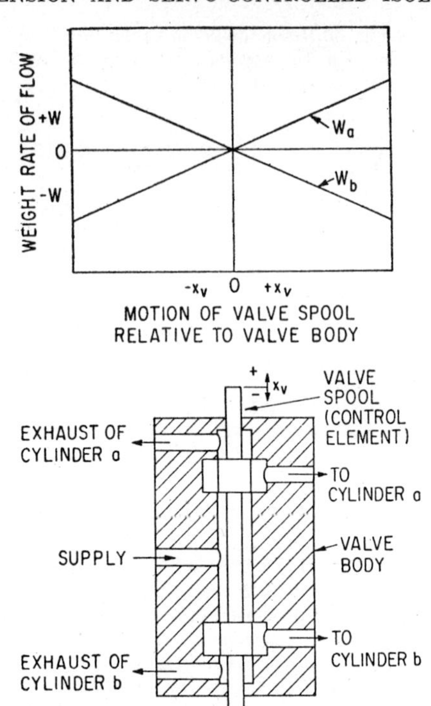

FIG. 33.14. Schematic diagram of a typical four-way peneumatic spool valve together with its weight rate of flow vs. spool-displacement characteristics.

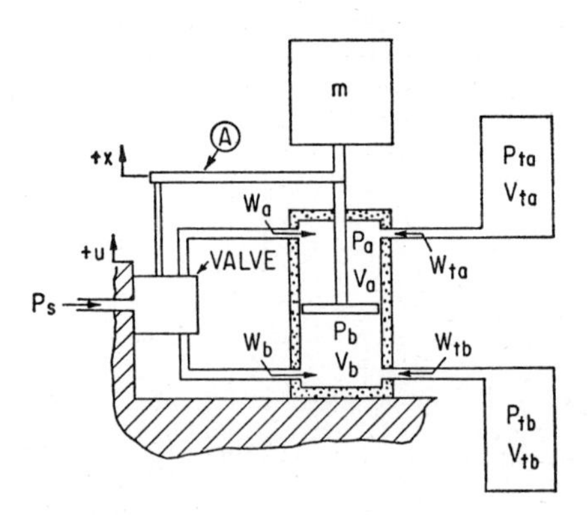

FIG. 33.15. Schematic diagram of an active isolation system employing a self-damped air spring and a pneumatic control valve. Motion of the arm A moves the control element of the valve to provide integral feedback of the displacement of mass m. The system is excited by the motion $u = u_0 \sin \omega t$ and responds with a motion $x = x_0 \sin (\omega t + \theta)$. The relative motion is $\delta = u - x$.

the valve with respect to the motion of the supported body:

$$G = G_v \frac{x_v}{\delta}$$

Proceeding in a manner similar to that used in Eq. (33.4), Eq. (33.48) leads to the result

$$\frac{x}{u} = \frac{2\frac{c}{c_c}\frac{D^2}{\omega_n^2} + \left[1 + 2\frac{c}{c_c}\cdot\frac{G}{Nm\omega_n^3}\right]\frac{D}{\omega_n} + \frac{G}{Nm\omega_n^3}}{2\frac{c}{c_c}\frac{D^4}{N\omega_n^4} + \left(\frac{N+1}{N}\right)\frac{D^3}{\omega_n^3} + 2\frac{c}{c_c}\frac{D^2}{\omega_n^2} + \left[1 + 2\frac{c}{c_c}\frac{G}{Nm\omega_n^3}\right]\frac{D}{\omega_n} + \frac{G}{Nm\omega_n^3}}$$

Noting that $D = j\omega$, the preceding equation may be written in the more familiar transmissibility form by taking the square root of the sum of the squares of the real and imaginary parts:

$$T = \sqrt{\frac{\left[\frac{G}{Nm\omega_n^3} - 2\left(\frac{c}{c_c}\right)\frac{\omega^2}{\omega_n^2}\right]^2 + \left[1 + 2\left(\frac{c}{c_c}\right)\frac{G}{Nm\omega_n^3}\right]^2\frac{\omega^2}{\omega_n^2}}{\left[\frac{G}{Nm\omega_n^3} - 2\left(\frac{c}{c_c}\right)\left(1 - \frac{\omega^2}{N\omega_n^2}\right)\frac{\omega^2}{\omega_n^2}\right]^2 + \left\{\left[1 + 2\left(\frac{c}{c_c}\right)\frac{G}{Nm\omega_n^3}\right]\frac{\omega}{\omega_n} - \frac{N+1}{N}\frac{\omega^3}{\omega_n^3}\right\}^2}} \tag{33.49}$$

As the frequency of excitation approaches infinity, Eq. (33.49) approaches the asymptotic value

$$T = \frac{1}{\omega^2/N\omega_n^2} \qquad [\omega \to \infty]$$

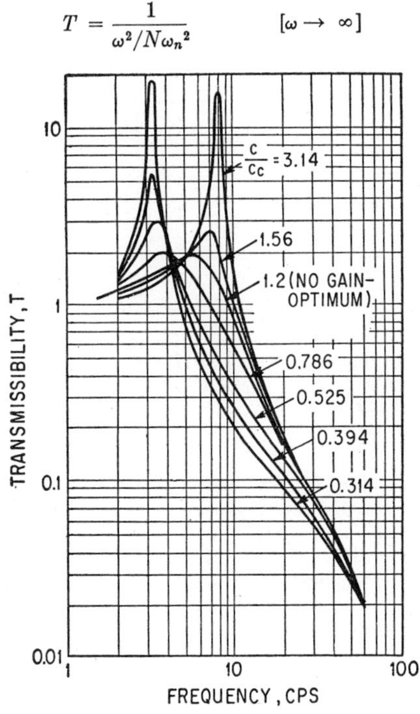

FIG. 33.16. Transmissibility curves of the active system illustrated in Fig. 33.15 for various values of fraction of critical damping. The gain $G/Nm\omega_n^3$ and the volume ratio $N = V_t/V_c$ are held constant.

This is the same asymptote as that of the passive system, as shown by Eq. (33.28) or Fig. 33.13. Therefore, the addition of integral feedback has no important effect on the isolation characteristics of the system at high frequency. This is a typical result of adding integral feedback. It tends to raise the value of the transmissibility function T only in the range near resonance.

Figure 33.16 shows graphically the transmissibility given by Eq. (33.49), and illustrates the effect of changes in damping. The curves do not have a common transmissibility point as do those for the passive system and the optimum fraction of critical damping for the active system is somewhat different from that of the passive system. For low values of valve gain, however, the values of optimum damping for the passive and active systems are approximately equal. For the particular values of N, G, m, and ω_n used to plot Fig. 33.16, the fraction of critical damping may be varied from $c/c_c = 0.78$ to

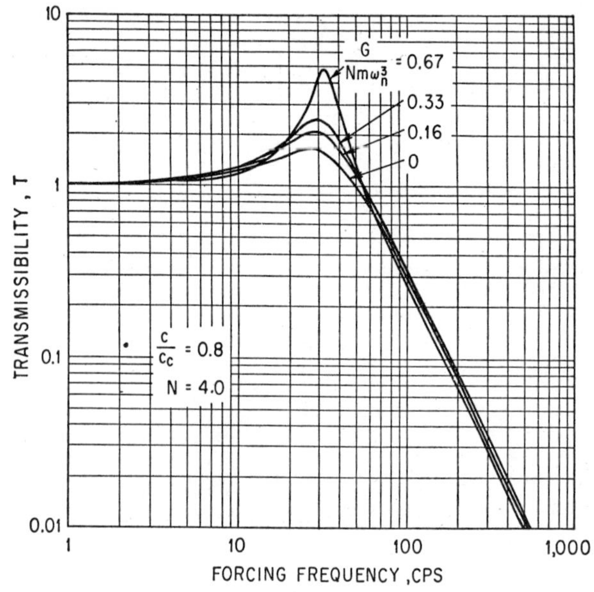

FIG. 33.17. Transmissibility curve of the active isolation system illustrated in Fig. 33.15; this illustrates the effect of increasing the gain $G/Nm\omega_n{}^3$ on the response of the system at resonance. The volume ratio $N = V_t/V_c$ is held at 4.0, and the fraction of critical damping c/c_c is 0.8 for this group of curves.

$c/c_c = 1.2$ with very little effect on the transmissibility of the isolation system at resonance. This is a desirable result since it allows broad tolerances on the design and manufacture of the components for a system with little chance of producing undesirable system performance. For N larger than approximately 3.0 and valve gains which are low enough to have the system well within the stable region, variations in the fraction of critical damping from the optimum value may be as large as 20 per cent without resulting in excessive transmissibility at resonance or loss of high-frequency isolation efficiency.

Figure 33.17 is a plot of Eq. (33.49) and illustrates the effect of the gain of the integral feedback loop in the pneumatic system. This is accomplished by increasing the value of G, the weight rate of gas flow per unit displacement of the suspended body. The variation of gain has little effect on the isolation properties of the suspension system in the region above resonance. At resonance, however, increased gain results in a higher transmissibility. This result could have been anticipated since the force resulting from integral feedback is 180° out-of-phase with the damping force; therefore, the over-all damping force is decreased. For low values of feedback gain, there is no significant change in the transmissibility curve of the isolation system. It is possible to introduce enough integral feedback gain to make the system unstable, as discussed in the section *Stability of Active Systems.*

STABILITY OF ACTIVE SYSTEMS

The addition of integral feedback control to a servo system tends to make the system less stable, i.e., more subject to self-excited oscillatory motion. It is possible to add enough integral feedback so that the system exhibits sustained oscillation with no input forces. This is an undesirable situation. Consider the system shown in Fig. 33.8; the equation of motion is

$$m\frac{d^2x}{dt^2} + c\frac{dx}{dt} + kx + G\int x\,dt = c\frac{du}{dt} + ku + G\int u\,dt \tag{33.50}$$

If u is defined by the expression $u = u_0 \sin \omega t$, Eq. (33.50) may be written

$$-m\omega^2 x_0 + jc\omega x_0 + kx_0 - j\frac{G}{\omega}x_0 = jc\omega u_0 + ku_0 - \frac{jG}{\omega}u_0$$

where x_0 is the amplitude of motion of the mass m and u_0 is the amplitude of the base. Solving for the ratio x_0/u_0 yields

$$\left|\frac{x_0}{u_0}\right| = \sqrt{\frac{1 + \left[2\left(\frac{c}{c_c}\right)\frac{\omega}{\omega_n} - \left(\frac{G}{m\omega_n{}^3}\right)\frac{1}{\omega/\omega_n}\right]^2}{\left(1 - \frac{\omega^2}{\omega_n{}^2}\right)^2 + \left[2\left(\frac{c}{c_c}\right)\frac{\omega}{\omega_n} - \left(\frac{G}{m\omega_n{}^3}\right)\frac{1}{\omega/\omega_n}\right]^2}} \tag{33.51}$$

When $\omega = \omega_n$, the first term of the denominator goes to zero. If the second term goes to zero at the same frequency, x_0 becomes infinite. By equating the second term to zero and letting $\omega = \omega_n$, the conditions which produce instability, i.e., an infinite amplitude of x_0, can be determined. Therefore, the condition of stability is

$$2\frac{c}{c_c} - \frac{G}{m\omega_n{}^3} \geq 0 \qquad \text{or} \qquad \frac{G}{m\omega_n{}^3} \leq 2\frac{c}{c_c} \tag{33.52}$$

The addition of integral feedback results in a force that is 180° out-of-phase with the damping force. Increasing the integral feedback force decreases the effective damping force. When G exceeds the value given in Eq. (33.52), the net result is a force which may be considered a negative damping force. This tends to excite oscillatory motion rather than to inhibit it.

Figure 33.18*A* illustrates the functional block diagram of a servo system. In such systems, r is referred to as the input function, d is the output function, and e is the error function or difference between r and d. The analogous quantities for the system illustrated in Fig. 33.18*A* are shown in Fig. 33.18*B* and are respectively, u, x, and δ where $\delta = (u - x)$. The relative displacement δ is defined as a function of ω for the system illustrated in Fig. 33.8 by

$$\left|\frac{\delta_0}{u_0}\right| = \sqrt{\frac{(\omega^2/\omega_n{}^2)^2}{(1 - \omega^2/\omega_n{}^2)^2 + \left[2(c/c_c)(\omega/\omega_n) - (G/m\omega_n{}^3)\left(\frac{1}{\omega/\omega_n}\right)\right]^2}} \tag{33.53}$$

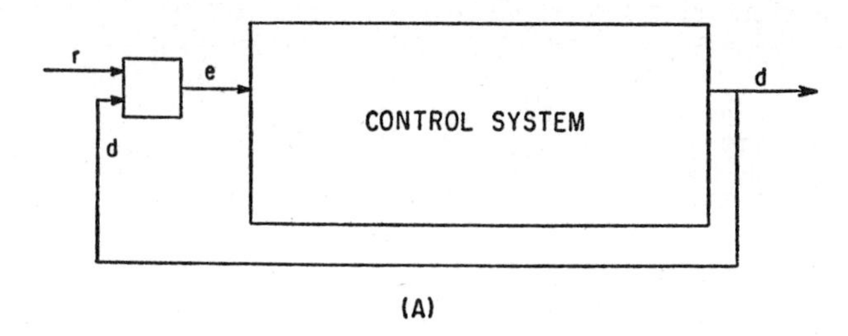

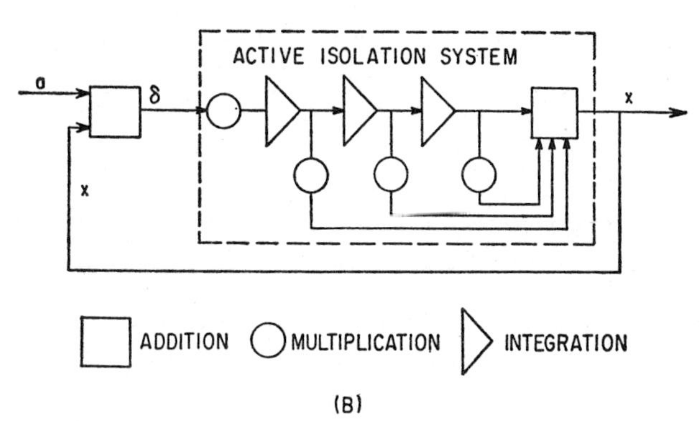

FIG. 33.18. Functional block diagram of a typical servo control system (A), illustrating its similarity to the active isolation system functional block diagram (B).

Equation (33.53) is the solution of Eq. (33.42) for the relative motion transmissibility $(u - x)/u = \delta/u$. Dividing Eq. (33.51) by Eq. (33.53), the magnitude of the ratio of the output function x to the error function δ is

$$\left|\frac{x_0}{\delta_0}\right| = \sqrt{\frac{1 + \left[2(c/c_c)(\omega/\omega_n) - (G/m\omega_n{}^3)\left(\frac{1}{\omega/\omega_n}\right)\right]^2}{(\omega/\omega_n)^4}}$$

$$\varphi = \tan^{-1}\frac{G/m\omega_n{}^3 - 2(c/c_c)(\omega/\omega_n)^2}{\omega/\omega_n}$$

The Bode criteria for stability states, in part, that the magnitude $\left|\frac{x_0}{\delta_0}\right|$ must be greater than 1.0 when the phase angle φ between x_0 and δ_0 is 180°. For the system under consideration, these conditions are met when $G/m\omega_n{}^3 < 2(c/c_c)$ [Eq. (33.52)].

The pneumatic isolation system with integral feedback of displacement presents a more complex stability problem since the equations of motion are of fourth order. Here the Bode methods are useful. The expression for x/δ of this system is

$$\frac{x}{\delta} = \frac{2\dfrac{c}{c_c}\dfrac{D^2}{\omega_n{}^2} + \left[1 + 2\dfrac{c}{c_c}\dfrac{G}{Nm\omega_n{}^3}\right]\dfrac{D}{\omega_n} + \dfrac{G}{Nm\omega_n{}^3}}{2\dfrac{c}{c_c}\dfrac{1}{N}\dfrac{D^4}{\omega_n{}^4} + \dfrac{N+1}{N}\dfrac{D^3}{\omega_n{}^3}} \qquad (33.54)$$

The magnitude of this expression is obtained by substituting $j\omega = D$ and taking the square root of the sum of the squares of the real and imaginary parts:

$$\left|\frac{x_0}{\delta_0}\right| = \sqrt{\frac{\left[\frac{G}{Nm\omega_n^3} - 2\frac{c}{c_c}\frac{\omega^2}{\omega_n^2}\right]^2 + \left[1 + 2\frac{c}{c_c}\frac{G}{Nm\omega_n^3}\right]^2\left(\frac{\omega}{\omega_n}\right)^2}{\left(\frac{\omega}{\omega_n}\right)^6\left\{\left[2\frac{c}{c_c}\frac{1}{N}\frac{\omega}{\omega_n}\right]^2 + \left(\frac{N+1}{N}\right)^2\right\}}} \tag{33.55}$$

The phase relationship between x and δ is

$$\varphi = \tan^{-1}\left[\frac{\left[1 + 2\frac{c}{c_c}\frac{G}{Nm\omega_n^3}\right]\frac{\omega}{\omega_n}}{\frac{G}{Nm\omega_n^3} - 2\frac{c}{c_c}\left(\frac{\omega}{\omega_n}\right)^2}\right] - \tan^{-1}\left[2\frac{c}{c_c}\left(\frac{1}{N+1}\right)\frac{\omega}{\omega_n}\right] - \frac{3}{2}\pi \tag{33.56}$$

Equation (33.54) may be plotted directly by the methods described in Appendix 33.1. Equations (33.55) and (33.56) are in a form more familiar to most engineers in the vibration field.

In the determination of stability, there are two critical frequencies in the magnitude-phase relationship which determine the margin of stability of any system. These frequencies are: (1) the gain crossover frequency, the frequency at which the magnitude ratio $\left|\frac{x_0}{\delta_0}\right| = 1.0$, and (2) phase crossover frequency, the frequency at which the phase angle $\varphi = -180°$. The value of the magnitude of $\left|\frac{x_0}{\delta_0}\right|$ at the phase crossover frequency is referred to as the *gain margin*. The phase difference, $\varphi - 180°$, at the gain crossover

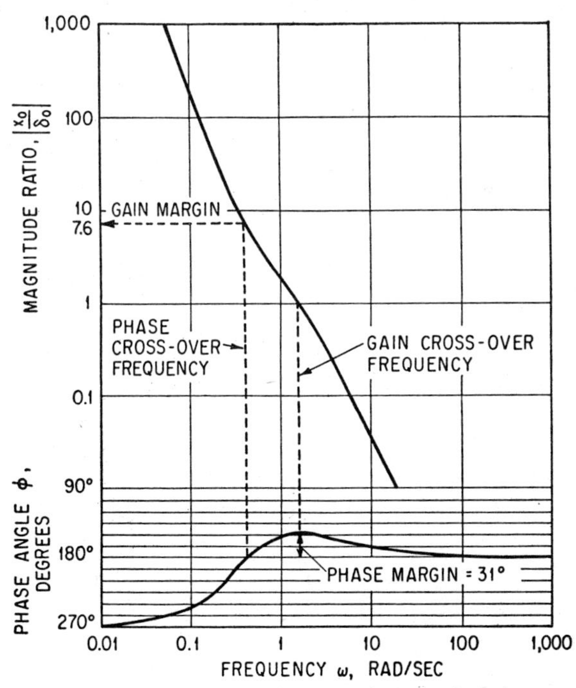

FIG. 33.19. Magnitude-phase plot of a typical pneumatic active isolation system, illustrating how the phase and gain margins are used in determining the dynamic stability of the system.

frequency is referred to as *phase margin*. The stability of the system increases as the gain and phase margins increase. For good stability the gain margin should be maintained at 2.5 or more and the phase margin at 30° or more.

For example, a magnitude-phase plot of a pneumatic system is shown in Fig. 33.19. The values of the various system constants are $m = 1.0$, $c/c_c = 1.0$, $\omega_n = 1.0$, $G = 1.0$, and $N = 4.0$. This system has adequate stability since the phase margin is 31° and the gain margin is 7.6.

If the gain G were increased by an appreciable amount, the phase and gain margins would both decrease and a less stable system would result. The phase margin is only slightly greater than the minimum dictated by good design practice. Therefore, additional gain would not be advisable.

REFERENCES

1. Crede, C. E.: "Vibration and Shock Isolation," pp. 32, 176, John Wiley & Sons, Inc., New York, 1951.
2. Polhemus, V. D., and L. J. Kehoe, Jr.: pt. 1; Cowin, F. H., and S. L. Milliken: pt. 2; *Trans. SAE*, **66**:346 (1958).
3. Dillman, O. D., and R. R. Love: *Trans. SAE*, **66**:295 (1958).
4. O'Shea, C. F.: *Trans. SAE*, **66**:475 (1958).
5. Hansen, K. H., J. F. Bertsch, and R. E. Denzer: *Trans. SAE*, **66**:483 (1958).
6. Perkins, R. W.: *Trans. SAE*, **66**:491 (1958).
7. Shearer, J. L.: *Trans. ASME*, **78**:233 (1956).
8. Shearer, J. L.: *Trans. ASME*, **79**:465 (1957).
9. Faires, V. M.: "Applied Thermodynamics," p. 16, The Macmillan Company, New York, 1948.
10. Ruzicka, J. E., and R. D. Cavanaugh: *Machine Design*, **30**:114 (1958).
11. Brown, G. S., and D. P. Campbell: "Principles of Servo-Mechanisms," pp. 59, 236, John Wiley & Sons, Inc., New York, 1948.
12. Gardner, M. F., and J. L. Barnes: "Transients in Linear Systems," vol. 1, p. 19, John Wiley & Sons, Inc., New York, 1942.
13. Oldenburger, R.: *Trans. ASME*, **76**:1155 (1954).
14. MacMillan, R. H.: "Frequency Response," p. 21, The Macmillan Company, New York, 1956.
15. Lauer, H., R. Lesnick, and L. Matson: "Servomechanism Fundamentals," 2d ed., p. 172, McGraw-Hill Book Company, Inc., New York, 1960.

34

MECHANICAL SPRINGS

Austin H. Church
New York University

SPRING TYPES

The more important types of mechanical springs used for shock and vibration isolation are: (1) Helical compression or tension springs made of bar stock or wire coiled into a helical form, the load being applied along the helix axis; in a compression spring, the helix is compressed; in a tension spring, it is extended. (2) Flat springs usually made of flat strips in a variety of forms. (3) Leaf springs made essentially of flat bars of varying lengths clamped together so as to obtain greater efficiency and resilience (automotive leaf springs); such springs may be full elliptic, semi-elliptic, or cantilever. (4) Belleville springs composed essentially of coned discs that may be stacked up to give a variety of load-deflection characteristics. This chapter describes the characteristics of such springs. A detailed discussion of the design of these springs is given in Ref. 1.

SPRING MATERIALS

PHYSICAL PROPERTIES AND COMPOSITIONS

For all ferrous materials used in metal springs (except stainless steel), the values of the modulus of elasticity E and modulus of rigidity G may be taken as 30×10^6 and 11.5×10^6 lb/in.^{2}, respectively, at room temperature. For type 302 stainless steel, the corresponding values are 28×10^6 and 10×10^6 lb/in.^{2}

Physical properties, chemical compositions, and SAE numbers for steel bars commonly used for springs are given in Table 34.1.[3,4] Physical properties include tensile and torsional ultimate strengths and elastic limits, modulus of elasticity, and shear modulus. Principal uses and special properties also are mentioned. Similar data are given in Table 34.2 for flat spring steels and in Table 34.3 for nonferrous spring materials. Additional data on material properties are given in Refs. 1 to 5, 24, and 25.

HELICAL TENSION OR COMPRESSION SPRINGS

Because of inherent advantages of low cost, compactness, and efficient use of material, helical tension or compression springs are widely used for shock and vibration control as well as for other purposes in machines.

Table 34.1. Properties and Compositions of Spring-steel Bars* (SAE† Committee[3] and Associated Spring Corp.[4])

Name and composition	Chemical composition (major elements), per cent	Tensile properties		Torsional properties		Principal uses and special properties
		Ultimate strength, 10^3 lb/in.^{2}	Elastic limit, 10^3 lb/in.^{2}	Ultimate strength, 10^3 lb/in.^{2}	Elastic limit, 10^3 lb/in.^{2}	
Carbon steel, SAE 1085	C 0.80–0.93 Mn 0.70–1.00	175 to 230	130 to 175	115 to 150	80 to 105	Used for hot coiled and flat springs; has good hardenability
Carbon steel, SAE 1095	C 0.90–1.03 Mn 0.30–0.50	170 to 220	125 to 170	110 to 145	75 to 100	Used for heavy hot coiled and flat springs; does not fully harden in oil quench when diameter is greater than 0.375 in.
Alloy steel, SAE 4068	C 0.63–0.70 Mn 0.75–1.00 Si 0.20–0.35 Mo 0.20–0.30	200 to 270	175 to 240	145 to 200	105 to 145	Used for hot coiled springs, also for hot formed leaf springs
Chrome-vanadium alloy steel, SAE 6150	C 0.48–0.53 Cr 0.80–1.10 Mn 0.70–0.90 V 0.15 min Si 0.20–0.35	200 to 250	180 to 230	140 to 175	100 to 130	Resists heat better than carbon steels
Chrome-silicon alloy steel, SAE 9254	C 0.50–0.60 Mn 0.50–0.80 Si 1.20–1.60 Cr 0.50–0.80	250 to 325	220 to 300	160 to 200	130 to 160	Resists heat well to 450° F; used at high stresses
Silicon-manganese alloy steel, SAE 9260	C 0.55–0.65 Mn 0.70–1.00 Si 1.80–2.20	200 to 250	180 to 230	140 to 175	100 to 130	

* Sizes available 0.375 to 2.25 in. in diameter. Modulus of elasticity $E = 30 \times 10^6$ lb/in.^{2}; shear modulus $G = 11.5 \times 10^6$ lb/in.^{2} (For hot-wound springs, use slightly lower values of E and G.) For larger sizes, SAE 8660 and 9262 may be used.

† Society of Automotive Engineers.

Table 34.2. Properties and Composition of Flat Cold-rolled Spring Steel* (SAE† Committee[3])

Name of material	Chemical composition (major elements), per cent	Tensile properties		Principal uses and special properties
		Ultimate strength, 10^3 lb/in.^{2}	Elastic limit, 10^3 lb/in.^{2}	
Flat spring steel, SAE 1050	C 0.48–0.55 Mn 0.60–0.90	160 to 270	120 to 210	Used for thin flat springs where severe forming is required
Flat spring steel, SAE 1060	C 0.54–0.66 Mn 0.60–0.90	160 to 270	120 to 210	Used for flat springs and stampings where higher carbon steel is not needed
Flat spring steel, SAE 1074	C 0.68–0.80 Mn 0.50–0.80	160 to 280	125 to 230	Very popular steel for miscellaneous flat springs Gives excellent life
Clock spring steel, SAE 1095	C 0.89–1.04 Mn 0.30–0.50	180 to 320	140 to 260	Supplied in heat-treated state Used for clock and motor springs and flat springs less than 0.06 in. thick

* Modulus of elasticity $E = 30 \times 10^6$ lb/in.^{2}
† Society of Automotive Engineers.

Table 34.3. Properties and Compositions of Nonferrous Spring Materials* (SAE Committee[3])

Name of material	Available wire sizes, in.	Typical chemical composition (major elem.), per cent	Tensile properties			Torsional properties			Principal uses and special properties
			Ultimate strength, 10^3 lb/in.^{2}	Elastic limit, 10^3 lb/in.^{2}	Young's modulus E, 10^6 lb/in.^{2}	Ultimate strength, 10^3 lb/in.^{2}	Elastic limit, 10^3 lb/in.^{2}	Shear modulus G, 10^6 lb/in.^{2}	
Spring brass	0.028 to 0.500	Cu 67.0 Zn 33.0	100 to 130	40 to 60	15	60 to 90	30 to 50	5.5	For low cost and low stresses For electrical conductivity
Nickel silver ASTM B206-54 (wire) ASTM B122-55T (strip)	0.020 to 0.250	Cu 55.0 Zn 27.0 Ni 18.0	130 to 150	80 to 100	16	80 to 100	60 to 70	5.5	Superior to brass Also used for its silver color
Phosphor bronze ASTM B159-54 (wire) ASTM B103-55 (strip)	0.028 to 0.500	Cu 90.0 min Sn 8.0 P 0.3 max	100 to 150	60 to 110	15	70 to 110	50 to 85	6.5	For electrical conductivity For higher stresses than brass
Silicon bronze	0.028 to 0.500	Cu 97.5 Si 2.5	100 to 150	60 to 110	15	70 to 110	50 to 85	6.5	Similar to phosphor bronze; less costly
Beryllium copper ASTM B197-52 (wire) ASTM B196-52 (rod) ASTM B194-55 (strip)	0.010 to 0.250	Cu 98.0 Be 2.0	160 to 200	110 to 150	19	100 to 130	70 to 100	7.0	High mechanical properties Excellent electrical conductor; low hysteresis Corrosion resistance like copper Properties based on age-hardened conditions
Monel	0.028 to 0.500	Ni 65.0 Cu 35.0	130 to 170	90 to 125	26	80 to 110	60 to 80	9.5	Fair electrical conductor; slightly magnetic, moderate stresses to 400°F
"K" Monel	0.028 to 0.500	Ni 66.0 Cu 31.0 Al 3.0	160 to 200	110 to 150	26	100 to 125	70 to 90	9.5	High stresses to 450°F Nonmagnetic to −150°F Properties based on age-hardened conditions
Permanickel	0.028 to 0.500	Ni 97.0 min Mn 0.4 Ti 0.4	180 to 230	130 to 170	30	110 to 150	80 to 110	11.0	High stresses to 550°F Properties based on age-hardened condition
Inconel	0.028 to 0.500	Ni 76.0 Cr 16.0 Fe 8.0	160 to 200	110 to 150	31	100 to 125	70 to 90	11.0	High stresses to 700°F Nonmagnetic to −40°F
Inconel X	0.028 to 0.500	Ni 70.0 min Cr 16.0 Fe 8.0 Ti 2.5	160 to 200	110 to 150	31	100 to 125	70 to 90	11.0	Can be used up to 900°F Properties based on age-hardened condition Nonmagnetic to −280°F
Ni-Span C †		Ni 42.0 Ti 2.3 Cr 5.2 Fe balance	200	110	27.5			9.4	Zero temperature coefficient of modulus −50 to 150°F

* All materials may be obtained either cold-rolled or cold-drawn and all are corrosion resistant.
† Data on this alloy from Associated Spring Co.[4]

BASIC FORMULAS—ROUND WIRE SPRINGS

STRESS. The uncorrected shearing stress in round wire helical springs loaded axially by a force F (Fig. 34.1) is

$$\tau = \frac{8FD}{\pi d^3} \tag{34.1}$$

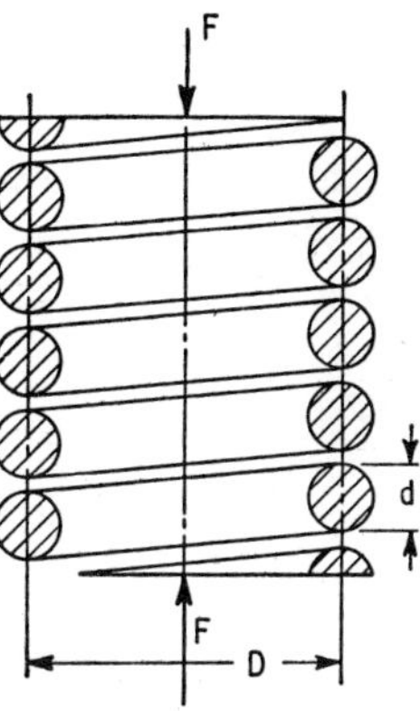

Fig. 34.1. Helical spring.

where d is the wire diameter, and D is the mean coil diameter. The maximum shearing stress in the round wire is

$$\tau_{\max} = \kappa \frac{8FD}{\pi d^3} \tag{34.2}$$

where κ is stress concentration factor known as "Wahl's factor,"[1] which is a function of the spring index, $C = D/d$.

$$\kappa = \frac{4C - 1}{4C - 4} + \frac{0.615}{C} \tag{34.3}$$

Since the value of κ ranges from about 1.1 when $C = 14$ to about 2.0 when $C = 2$, the value of the maximum shear stress $\tau_{\max}$ in the round wire is significantly affected by this factor.

DEFLECTION. The deflection of a round wire helical spring is calculated by considering the spring to be a straight bar under a torsion moment $FD/2$. The length l of this bar is approximately equal to πDn where n is the number of *active* coils in the spring. Calculating the angular deflection due to twisting of the bar and multiplying by the coil radius results in the following expression for deflection δ of the spring,

$$\delta = \frac{8FD^3n}{Gd^4} \tag{34.4}$$

where G is the modulus of rigidity in pounds per square inch.

The spring rate or stiffness k_y is

$$k_y = \frac{F}{\delta} = \frac{Gd^4}{8D^3n} \tag{34.5}$$

Fig. 34.2. Types of ends for compression springs. (*A*) Plain ends (coiled right-hand); (*B*) squared or closed ends, not ground; (*C*) plain ends, ground (coiled left-hand); (*D*) squared or closed ends, ground. (*After ASM Committee.*[2])

Although Eqs. (34.4) and (34.5) are derived on the basis of simplifying assumptions, they give sufficiently accurate results in most practical cases, provided the proper values of G and n are used.

Figure 34.2 shows common types of ends for compression springs.[2] If the ends are squared and ground, the number of active coils n used in Eq. (34.5) to calculate deflection may be taken equal to the total turns (tip to tip of bar) minus $1\frac{3}{4}$ turns.[1] For extension springs with full end loops, n may be taken as 1 plus the number of coils between points where the loops begin.

DYNAMICAL EFFECTS

A spring which is subject to rapid variation in load, such as a valve spring, may experience additional dynamical effects which often increase the stress seriously. Particularly severe effects occur when a harmonic in the motion of the spring end (or in the valve lift curve in the case of valve springs) has a frequency which coincides with one of the natural frequencies of the spring (corresponding to one of the modes of vibration). Harmonics of very high order may have to be considered in certain cases. Usually, the lowest natural frequency is of the most importance. For a spring compressed between parallel plates, the first mode of vibration (corresponding to the lowest natural frequency) consists of a vibratory motion of the middle part of the spring with the ends remaining stationary. The second mode of vibration (corresponding to a higher natural frequency) has a node in the middle of the spring, while maximum motion of the coils occurs at points one-fourth and three-fourths of the length distant from the end.

NATURAL FREQUENCIES. The lowest natural frequency for a helical spring between two parallel plates is

$$f_n = \frac{2d}{\pi D^2 n}\sqrt{\frac{Gg}{32\gamma}} \quad \text{cps} \tag{34.6}$$

where d = wire diameter, in.
D = coil diameter, in.
G = modulus of rigidity, lb/in.^{2}
g = acceleration of gravity = 386 in./sec.^{2}
γ = weight density of spring material, lb/in.^{3}
n = number of active coils

For steel springs, $G = 11.5 \times 10^6$ lb/in.^{2} and $\gamma = 0.285$ lb/in.^{3} The natural frequency for a steel spring between parallel plates is

$$f_n = \frac{14{,}100d}{D^2 n} \quad \text{cps} \tag{34.7}$$

Natural frequencies in the higher modes of vibration are 2, 3, 4, etc., times the frequency given by Eq. (34.7).

For a spring with one end free and the other fixed, the lowest natural frequency is equal to that of a similar spring twice as long but with both ends fixed. For such a spring, Eqs. (34.6) and (34.7) apply if n is taken as twice the actual number of active turns in the spring.

Example. A steel spring with $d = 0.3$ in., $D = 2$ in., and $n = 6$ is fixed at both ends. Using Eq. (34.7), the lowest natural frequency is

$$f_n = \frac{14{,}100 \times 0.3}{4 \times 6} = 176 \text{ cps}$$

In the second mode of vibration, the natural frequency of the spring is twice as great or 352 cps If such a spring were used as a valve spring, at an engine speed such as to give a frequency of one of the harmonic components of the valve-lift curve equal to 176 or 352 cps, excessive vibration would be expected.[1]

The following methods are useful in reducing stresses in valve springs due to dynamical effects: use of a high natural frequency to avoid resonance with lower harmonics; contour of driving cam shaped to reduce amplitudes of the harmonics that are of importance in the operating speed range; reducing pitch of coils near ends of the spring to change natural frequency by closing these coils during part of cycle of oscillation; and use of friction dampers pressing against center coils.

HELICAL EXTENSION SPRINGS

SPRING ENDS. Tension springs may be wound with a variety of ends, as indicated in Fig. 34.3. Some of these designs may result in considerably higher stress in the ends than are calculated from Eq. (34.3). For example, the design of Fig. 34.3*B*, consisting of a half loop turned up sharply, may have rather sharp curvature and severe stress concentration. This would result in a considerable decrease in strength, particularly under fatigue loading. Lower stress concentration effects are present in the full loop of Fig. 34.3*A*. If the hook is at the side (Fig. 34.3*C*), the moment arm of the load is much larger than that which would exist if a purely axial load were applied, thus resulting in a

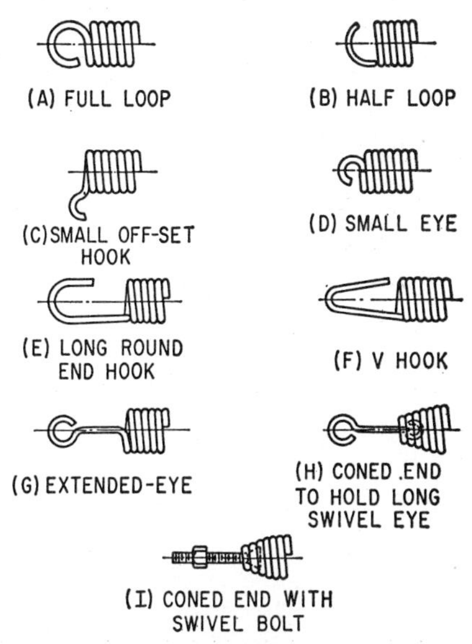

FIG. 34.3. End designs for extension springs.

large increase in stress. The coned-end designs shown in Fig. 34.3*H* and *I* have relatively low stress in the end turns. Threaded plugs which may be screwed to the ends of plain close-wound helical springs are used frequently.

HELICAL SPRINGS OF SQUARE OR RECTANGULAR BAR SECTION

Springs of square or rectangular bar section sometimes are used where it is desired to provide maximum energy storage within a limited space. In this manner, more material may be provided within a given outside diameter and length, particularly if the spring is coiled flatwise; hence, greater energy storage is possible. This advantage is partially offset by the fact that it is difficult to obtain material in square or rectangular bars whose quality is equal to that in round bars.

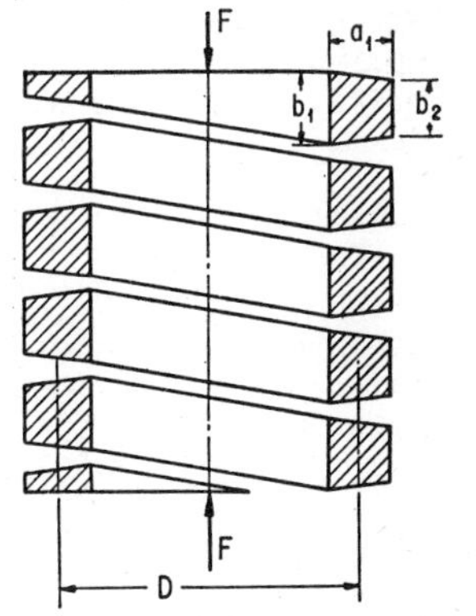

FIG. 34.4. Helical spring of square wire axially loaded. The wire becomes somewhat trapezoidal due to the coiling operation.

The uncorrected stress τ and deflection δ for a helical spring of square bar or wire section loaded axially by a force F (Fig. 34.4) are given by

$$\tau = \frac{2.4FD}{a^3} \quad \text{lb/in.}^2 \qquad (34.8)$$

$$\delta = \frac{5.58FD^3n}{Ga^4} \quad \text{in.} \qquad (34.9)$$

where D = mean coil diameter, in.
a = side of square section, in.
G = modulus of rigidity, lb/in.²
n = number of active coils

Theoretically, Eq. (34.9) is 2 to 4 per cent in error for indexes D/a between three and four; however, it is sufficiently accurate for most practical cases.

For square wire springs, the *corrected* stress is found by multiplying the stress of Eq. (34.8) by a factor κ':[22]

$$\kappa' = 1 + \frac{1.2}{C} + \frac{0.56}{C^2} + \frac{0.5}{C^3} \tag{34.10}$$

where $C = D/a$ is the spring index. The factor κ' is slightly less than the factor κ for round wire, Eq. (34.3).

Reference 23 discusses the design of helical springs of rectangular bar or wire.

FLAT SPRINGS

SIMPLE CANTILEVER SPRING

Springs made from flat strip or bar stock are included in this category, even though they may be formed into more or less complicated shapes. They often may be considered as simple cantilevers loaded by force F (Fig. 34.5). The deflection δ at the free end and the maximum bending stress σ are

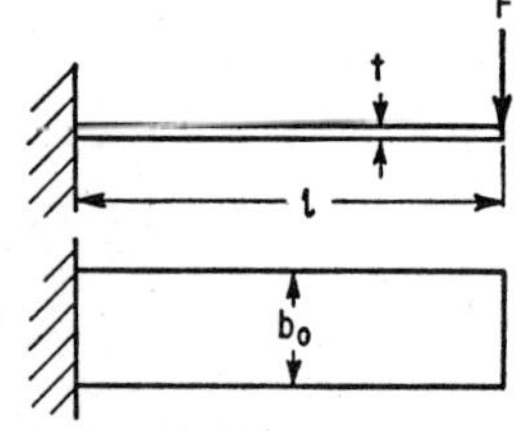

FIG. 34.5. Simple cantilever spring.

$$\delta = \frac{4Fl^3}{Eb_0t^3} \quad \text{in.} \tag{34.11}$$

$$\sigma = \frac{6Fl}{b_0t^2} \quad \text{lb/in.}^2 \tag{34.12}$$

where l = length, b_0 = width, and t = thickness as shown in Fig. 34.5 in inches, and E is Young's modulus in pounds per square inch.

Where the strip is wide compared to the thickness, as in most practical cases, slightly lower deflections than those calculated from Eq. (34.11) result. This difference may reach values up to 10 per cent, and occurs because lateral expansion or contraction of elements near the surface of the strip is prevented.

TRAPEZOIDAL PROFILE CANTILEVER SPRING

Somewhat more efficient use of material is obtained by the use of the cantilever spring of trapezoidal profile, as shown in Fig. 34.6. The maximum bending stress is given by Eq. (34.12), where b_0 is taken as the width at the built-in end. The deflection δ at the

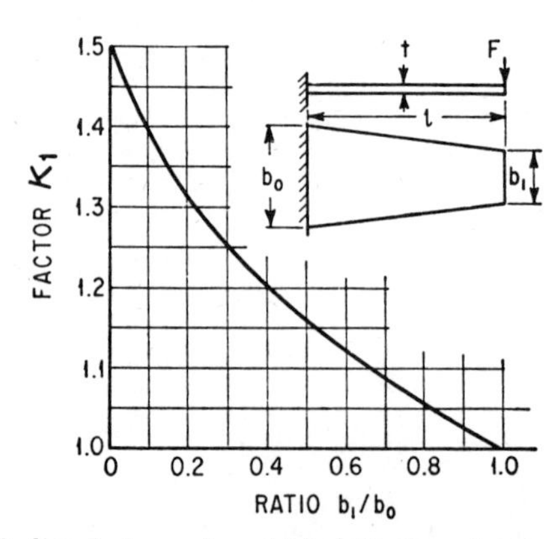

FIG. 34.6. Curve for finding factor κ_1 by which deflection of rectangular profile cantilever [Eq. (34.11)] should be multiplied to obtain deflection of trapezoidal profile spring.

free end is obtained by multiplying the deflection of the simple cantilever [Eq. (34.11)] by a factor κ_1 which depends on b_1/b_0 and is given by the curve of Fig. 34.6 where b_1 is the width at the loaded end.

LEAF SPRINGS

Leaf springs are somewhat less efficient in terms of energy storage capacity per pound of metal as compared with helical springs. However, leaf springs may be applied to function as structural members.[24] A typical semielliptic leaf spring is shown in Fig. 34.7; typical ends and center clamps are shown in Figs. 34.8 and 34.9. Rebound and alignment clips are illustrated in Fig. 34.10.

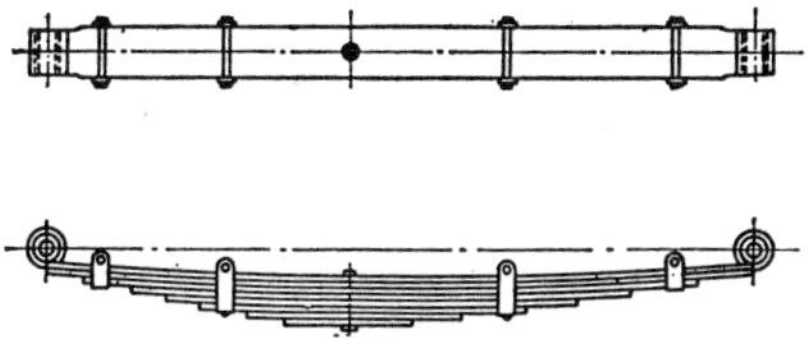

FIG. 34.7. Semielliptic leaf spring.

(A) UP-TURNED EYE
(D) BERLIN EYE
(B) PLAIN END MOUNTING
(E) REINFORCED EYE
(C) DOWN-TURNED
(F) SOLID EYE
(G) END BLOCK FOR SEMI-ELLIPTIC R. R. TYPE

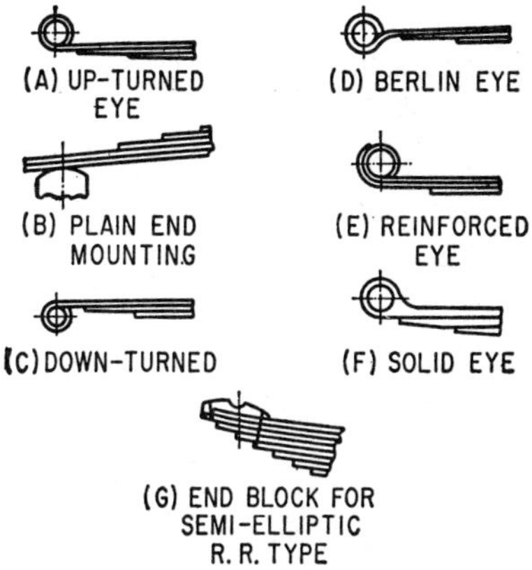

FIG. 34.8. Leaf-spring ends.

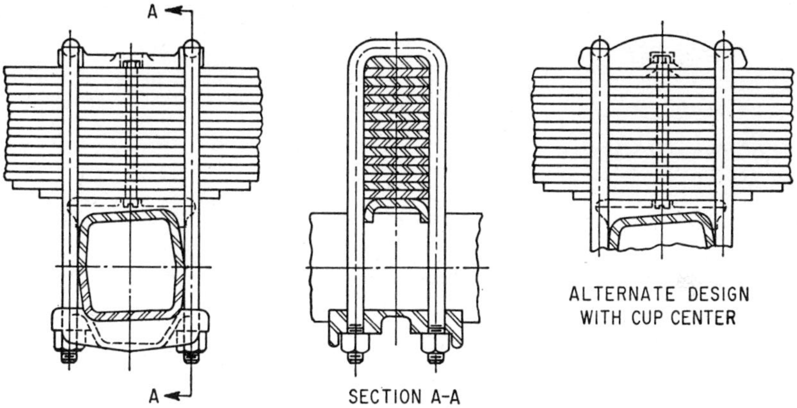

A
A
SECTION A-A
ALTERNATE DESIGN WITH CUP CENTER

FIG. 34.9. Center clamps for leaf springs.

(A) BOLT TYPE CLIP
(B) CLINCH TYPE CLIP

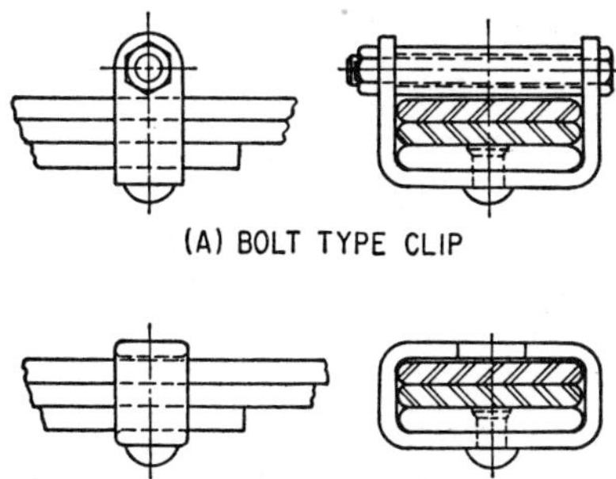

FIG. 34.10. Rebound and alignment clips for leaf springs.

DESIGN FORMULAS

The following formulas for leaf springs are based on the assumption of a beam of uniform strength.[24] This assumption requires that the total thickness of the spring decrease linearly from the center to each end; this is approximated by the arrangement of stepped leaves.

SYMMETRIC SEMIELLIPTIC LEAF SPRING. (Fig. 34.11*A*). The spring rate or stiffness is

$$k = \frac{8Enbt^3}{3l^3} \quad \text{lb/in.} \tag{34.13}$$

where b = width of leaves, in.
t = thickness of leaves, in.
l = maximum length, in.
n = number of leaves
E = Young's modulus, lb/in.²

The stress resulting from load F, assumed uniform along the spring, is

$$\sigma = \frac{3Fl}{2nbt^2} \quad \text{lb/in.}^2 \tag{34.14}$$

In actual springs where a center clamp is used, the length l in Eqs. (34.13) and (34.14) is taken as the total length minus the length of the center clamp.

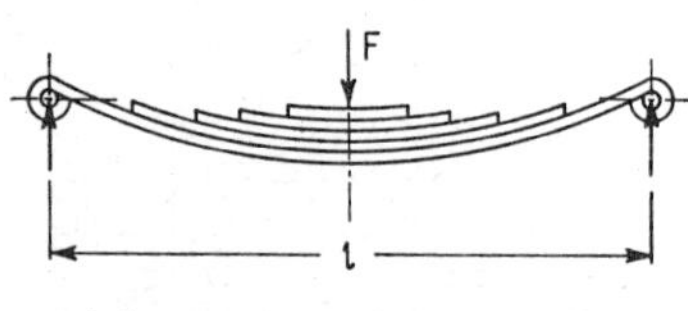

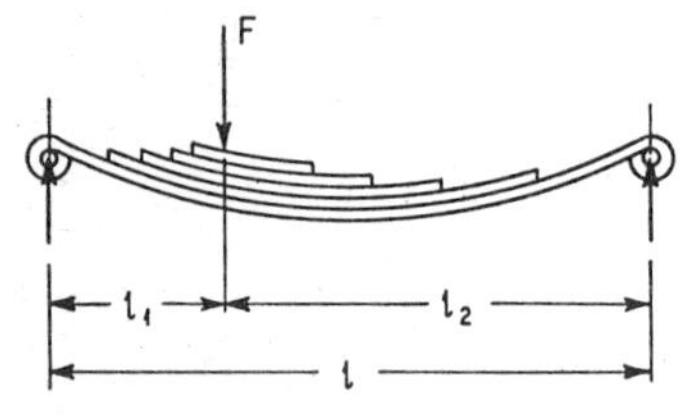

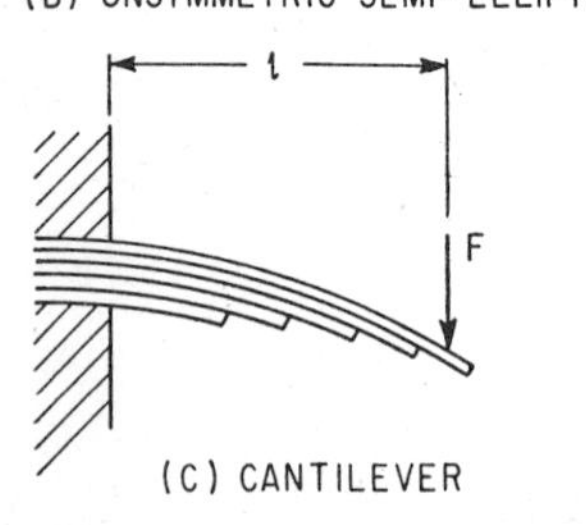

FIG. 34.11. Leaf springs showing types of symmetry and loading.

UNSYMMETRIC SEMIELLIPTIC LEAF SPRING. (Fig. 34.11*B*)

$$k = \frac{Ebnt^3l}{6l_1^2l_2^2} \quad \text{lb/in.} \tag{34.15}$$

$$\sigma = \frac{6Fl_1l_2}{lnbt^2} \quad \text{lb/in.}^2 \tag{34.16}$$

where l_1 and l_2 are defined in Fig. 34.11*B*.

CANTILEVER LEAF SPRING. (Fig. 34.11*C*)

$$k = \frac{Ebnt^3}{6l^3} \quad \text{lb/in.} \tag{34.17}$$

$$\sigma = \frac{6Fl}{nbt^2} \quad \text{lb/in.}^2 \tag{34.18}$$

where l is defined in Fig. 34.11*C*.

Equations (34.13) to (34.18), although based on the assumption of a beam of uniform strength (triangular profile), may be used for preliminary design calculations when the uniform strength condition is not fulfilled completely. Effects which may cause deviations from the ideal conditions assumed are: use of leaves of different thicknesses; use of more than one main leaf; interleaf friction; and use of spring shackles which result in angular loading at the ends. These effects may result in rather large deviations from the values calculated from the formulas.[24]

Materials commonly used for leaf springs include SAE 1085, 1095, 4068, 4161, 5160, 6150, and 9260. Physical properties and compositions of some of these materials are given in Table 34.1. Heat treatments giving Brinell hardness of 415 to 460 usually yield satisfactory results for spring steels. Shot-peening and cold-setting operations are desirable for satisfactory life in service.

CONED DISC (BELLEVILLE) SPRINGS

The coned or dished discs have diametral cross sections and loading, as indicated in Fig. 34.12. The shape of the force-deflection characteristic depends primarily on the ratio of the initial cone (or disc) height h to the thickness t. Some characteristics are indicated in Fig. 34.13 for various values of h/t, where the spring is supported so that it may deflect beyond the flattened position. For ratios of h/t approximately equal to 0.5, the curve approximates a straight line up to a deflection equal to half the thickness; for h/t equal to 1.5, the load is constant within a few per cent over a considerable range of deflection. Springs with ratios h/t approximating 1.5 are known as *constant-load* or *zero rate* springs. By stacking Belleville springs in parallel (Fig. 34.14*A*), the load capacity is increased while the series arrangement (Fig. 34.14*B*) gives increased deflection for a given load. The latter should not be used for ratios h/t greater than 1.3 since instability and an irregular force-deflection characteristic may result. Guides are advisable to prevent buckling or lateral deflection.

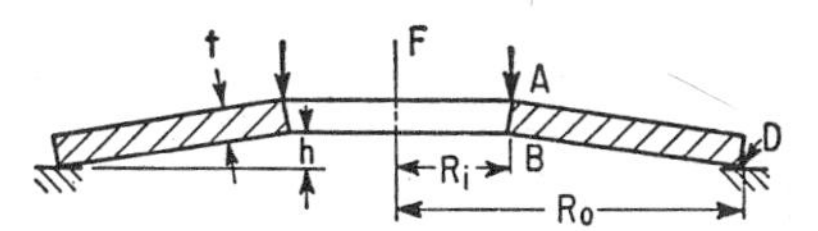

FIG. 34.12. Belleville spring.

Advantages of Belleville springs include: small space requirement in direction of load application; ability to carry lateral loads; characteristics variable by adding or removing discs. Disadvantages include nonuniformity of stress distribution, particularly for large ratios of outside to inside diameter.

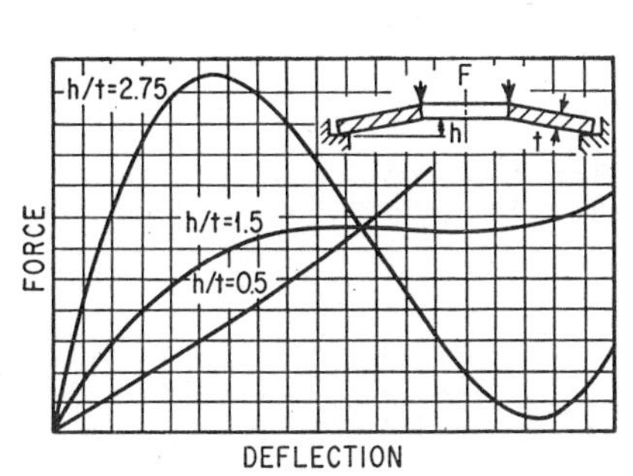

FIG. 34.13. Shapes of force-deflection diagrams for Belleville spring obtained by varying the ratio h/t (Fig. 34.12).

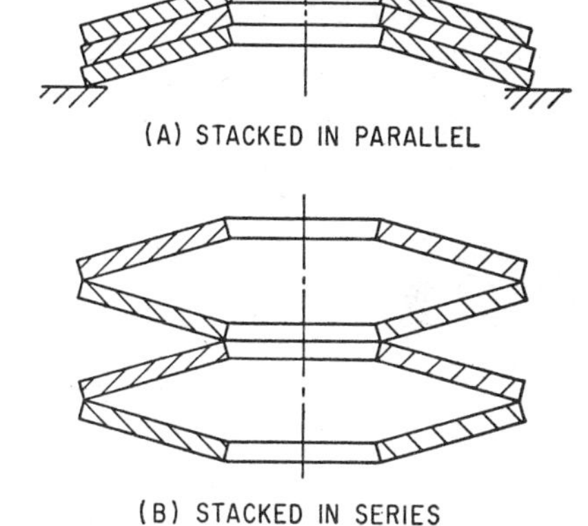

FIG. 34.14. Methods of stacking Belleville springs.

DESIGN FORMULAS BASED ON ELASTIC THEORY

The following analysis of stress and deflection in Belleville springs is based on elastic theory and on the assumption that radial cross sections of the spring do not distort during deflection.[25, 26] The results for deflection are in approximate agreement with available test data; however, agreement closer than ±5 per cent should not be expected because of friction and other effects not taken into account by the theory.

Load F at deflection δ is

$$F = \frac{1.1\delta EC}{R_0^2}\left[(h - \delta)\left(h - \frac{\delta}{2}\right)t + t^3\right] \quad \text{lb} \tag{34.19}$$

where R_o = outside radius, in.
h = height, in.
t = thickness, in.
δ = deflection, in.
E = Young's modulus, lb/in.^{2}
C = a factor depending on the ratio of outside radius R_0 to inside radius R_i and shown quantitatively in Fig. 34.15

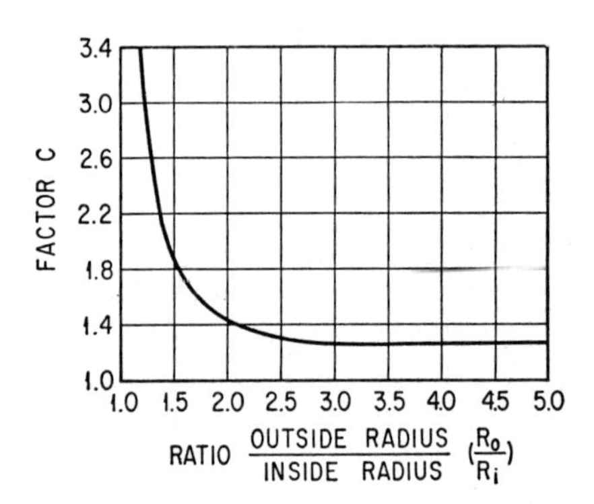

FIG. 34.15. Quantitative values of factor C in Eq. (34.19) et seq.

Equation (34.19) may be written

$$F = \frac{CC_1Et^4}{R_o^2} \quad \text{lb} \tag{34.20}$$

where the factor C_1, a function of both h/t and δ/t, is defined numerically in Fig. 34.16. Shapes of the curves for C_1 represent the shapes of the load-deflection diagrams for the various ratios h/t.

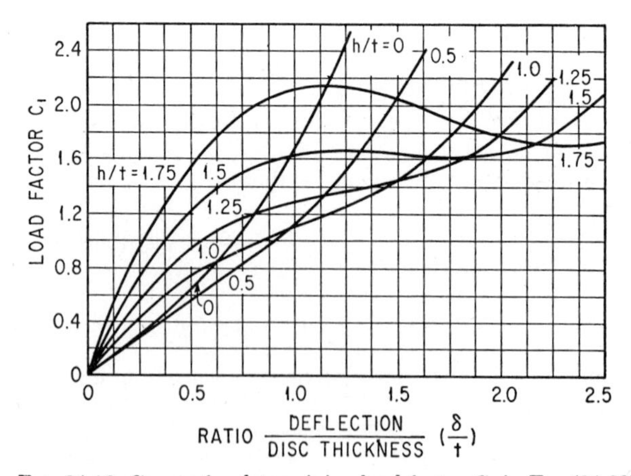

FIG. 34.16. Curves for determining load factor C_1 in Eq. (34.20).

REFERENCES

1. Wahl, A. M.: "Mechanical Springs," 2d ed., McGraw-Hill Book Company, New York, 1963.
2. "Metals Handbook," 8th ed., vol. 1, pp. 160–174, American Society for Metals, Cleveland, Ohio, 1961.
3. "Design and Application of Helical and Spiral Springs—SAE J795a," Society of Automotive Engineers, Warrendale, Pa., 1973.
4. "Design Handbook," Associated Spring Corp., Bristol, Conn., 1970.
5. Zimmerli, F. P.: *Metal Prog.*, **61**:(5, 6) 97 (May, 1952), **62**:(1) 84 (July, 1952).
6. Wahl, A. M.: *Trans. ASME*, APM 51-17, p. 191, 1929.
7. *The Mainspring*, Associated Spring Corp., December, 1959—January, 1960.
8. Zimmerli, F. P.: *Proc. ASTM*, **30**(2):356 (1930).
9. Johnson, W. R.: Paper presented at the Metals Congress, American Society for Metals, Cleveland, 1958.
10. Eakin, C. T.: *Met. Memo.* 1166, Westinghouse Materials Engineering Department, East Pittsburgh, Pa., 1948.
11. "Technical Bulletin T-35," International Nickel Co., New York, 1955.
12. "The Mainspring," Associated Spring Corp., November, 1957.
13. Meyers, O. G.: *Machine Design*, **23**:135 (1951).
14. Eakin, C. T.: *Met. Memo.* 1057, Westinghouse Materials Engineering Department, East Pittsburgh, Pa., 1946.
15. "Spring Design Data," Hunter Spring Co., Lansdale, Pa.
16. Edgerton, C. T.: *Trans. ASME*, **59**:609 (1937).
17. Keysor, H. C.: *Trans. ASME*, **62**:319 (1940).
18. Haringx, J. A.: *Philips Research Repts.*, **3**:401 (1948); **4**:49 (1949).
19. Crede, C. E.: "Vibration and Shock Isolation," John Wiley & Sons, Inc., New York, 1951.
20. Burdick, W. E., F. S. Chaplin, and W. L. Sheppard: *Trans. ASME*, **61**:623 (1939).
21. Ross, H. F.: *Trans. ASME*, **69**:727 (1947).
22. Göhner, O.: *Z. VDI*, **76**:269 (1932).
23. Liesecke, G.: *Z. VDI*, **77**:425, 892 (1933).
24. "Manual on Design and Application of Leaf Springs—SAE J788a," Society of Automotive Engineers, Warrendale, Pa., 1970.
25. "Manual on Design and Manufacture of Coned Disk Springs or Belleville Springs—SAE J798," Society of Automotive Engineers, Warrendale, Pa., 1971.
26. Almen, J. O., and A. Laszlo: *Trans. ASME*, **58**:303 (1936).

35

RUBBER SPRINGS

William A. Frye
Inland Manufacturing Division
General Motors Corporation

INTRODUCTION

Rubber may be defined as an elastomeric material which can be or already is modified to a state exhibiting small plastic flow, high elongation, and high speed of retraction. This definition includes natural rubber, which is made from the milky juice of the *Hevea brasiliensis* tree, and a large family of synthetic rubbers made of polymers and copolymers of various organic compounds. The adjectives *elastomeric* and *rubberlike* are terms often used to describe rubber or similar materials. The term *polymer* is used to designate rubber in the "raw" or uncompounded state. The designation and composition of some of the common polymers are shown in Table 35.1. The term *vulcanizate* indicates a vulcanized rubber compound.

Table 35.1. Designation and Composition of Common Polymers

Polymer designation*	Common name	Chemical composition
NR	Natural rubber	Natural polyisoprene
IR	Isoprene	Synthetic polyisoprene
SBR	SBR	Copolymer of butadiene and styrene
BR	Polybutadiene	Polybutadiene
IIR	Butyl	Copolymer of isobutylene and isoprene
CR	Neoprene	Polychloroprene
EPDM	EPDM	Terpolymer of ethylene, propylene, and a diene
PO	Propylene Oxide	Copolymer of allyl glycidyl ether and propylene oxide

* American Society for Testing Materials Method D1418-72.

This chapter describes the properties of rubber which are of importance in vibration- and shock-isolation applications, how these properties are influenced by environmental conditions, and presents basic information for the design of rubber vibration and shock isolators, i.e., rubber springs.

Rubber has a low modulus of elasticity, and is capable of sustaining deformation of as much as 1,000 per cent. After such deformation, it quickly and forcibly retracts to essentially its original dimensions. It is resilient, and yet exhibits internal damping. Rubber can be processed into a variety of shapes and can be adhered to metal inserts or mounting plates. It can be compounded to have widely varying properties. The load-deflection curve of a rubber isolator can be altered by changing its shape. Rubber will not corrode and normally requires no lubrication. Because of these properties, rubber is uniquely suited for application to isolation problems.

In spite of continued advances, rubber technology remains more of an art than a science, for a number of reasons: (1) Rubber itself is difficult to analyze and describe accurately because it is made up of very large and complicated organic molecules. (2) Commercial rubber compounds are made up of a great number of ingredients, many of which are also organic compounds. (3) Variations in processing methods cause variations in the properties of a rubber part. (4) Rubber manufacturing companies tend to keep compounding and processing information confidential.

The essential steps in producing a molded rubber part are (1) mastication of the polymer and compounding ingredients to produce an uncured compound, (2) cooling of this compound to prevent vulcanization while it is transported and temporarily stored at the molding location, (3) vulcanization or cure in the mold under heat and pressure, and (4) cooling and trimming.

PROPERTIES OF RUBBER

The physical properties of a vulcanized rubber specimen are dependent on the stiffness of the rubber and the size and shape of the test specimen. They are also dependent on the specimen temperature and rate of straining the specimen to an extent far greater than most materials. Therefore, it is convenient to consider the static properties of rubber and then the effect of environment and strain rate on these properties.

STATIC PHYSICAL PROPERTIES OF RUBBER

A typical tensile stress-strain curve for rubber is shown in Fig. 35.1*A*; similar curves for rubber in compression and in shear are given in Fig. 35.1*B* and *C*, respectively. Note

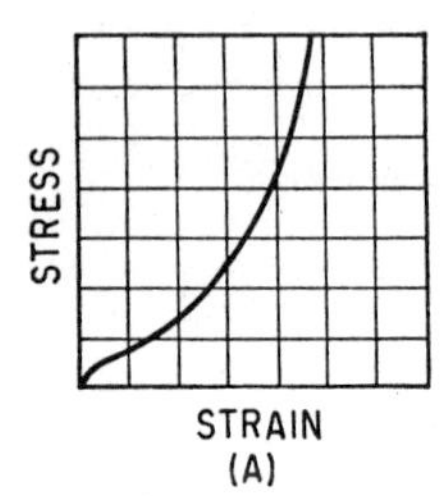

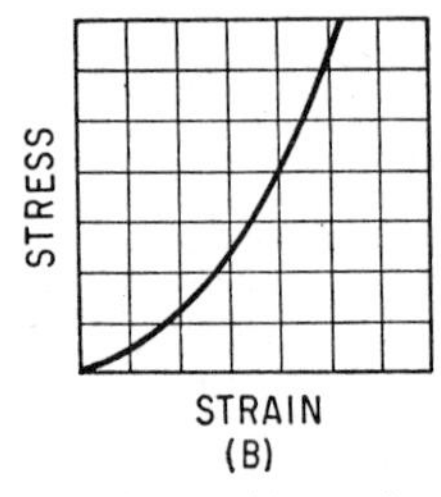

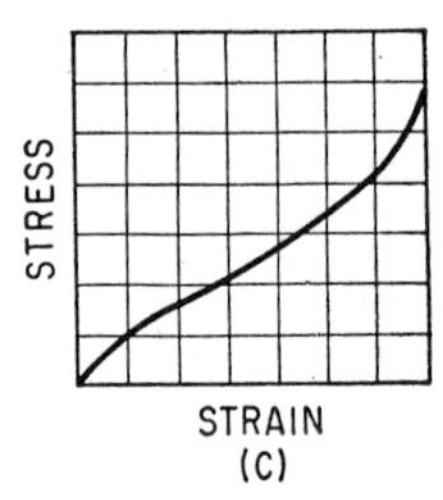

FIG. 35.1. Stress-strain curves for rubber in (*A*) tension, (*B*) compression, and (*C*) shear.

that in these curves Hooke's law of elasticity is not followed. This fact makes it difficult to specify the modulus of elasticity of the material, the property which is probably the most important for isolator design purposes. In addition, the compression stress-strain curve is greatly influenced by the *shape factor* of the specimen used, i.e., whether the specimen is relatively tall with a small cross section, or relatively short with a large cross section.* For these reasons, it is common to specify the durometer hardness of rubber compounds. The *durometer* is a pocket-type instrument which, when applied by hand, forces a small indentor into the rubber.

* Rubber is essentially an incompressible substance that deflects by changing shape rather than changing volume. Poisson's ratio is approximately 1/2. Values of Poisson's ratio to five significant figures are given in Ref. 25 for five compounds of natural and synthetic rubber. Average values for each compound are: natural rubber, 0.49935; SBR, 0.49894; butyl, 0.49691; Neoprene, 0.49950; and Nitrile, 0.49712.

The resistance to indentation is indicated by a pointer and a graduated scale.[1] The higher the reading, the harder the rubber. Rubber hardness is not important of itself; it is merely an approximate and convenient measurement which is related to modulus of elasticity and which is independent of a specimen shape factor. Because of the visco-elastic nature of rubber, a durometer reading reaches a maximum as soon as the instrument is completely applied to the specimen and then falls several points during the next 5 to 15 sec. The reading is also sensitive to indentor side loads caused by the manner in which the instrument is applied. These factors make it difficult to obtain good agreement between readings made in various laboratories.

OTHER BASIC PROPERTIES. Basic rubber properties which are often specified, in addition to *hardness*, are *tensile strength*, *ultimate elongation*, and *compression set.* Tensile strength and ultimate elongation are determined by pulling to destruction small dumbbell-shaped specimens which have been cut from laboratory cure sheets.[1] During this test, the tensile stress at a specified elongation may be determined. This value is called *modulus* and is specified as (for example) "modulus at 200 per cent" to avoid confusion with the general term "modulus of elasticity." Tensile test values for samples taken from finished articles may be somewhat different because of variance in processing and cures and because of the directional properties (anisotropy) of the material. Rubber seldom is stressed to a value approaching its ultimate tensile strength or elongation when used in an isolator. (Bond strength is usually the limiting factor.) Hence these properties are important only as they relate to fatigue strength, tear resistance, and similar properties. Rubber compounds commonly used for vibration isolators have tensile strengths ranging from 500 to 3,500 lb/in.^{2} (35 to 250 kg/cm^2). The ultimate elongation of such vulcanizates is from 150 to 600 per cent.

Compression set is determined by compressing a 1.129-in. (28.4-mm) diameter specimen which is 0.50 in. (12.7 mm) thick to a preset deflection and exposing it at an elevated temperature.[1] After exposure, the specimen is allowed to recover for one-half hour and measured. *Per cent compression set* is defined as the decrease in thickness divided by the original deflection. Typical rubber compounds used for vibration isolators have compression set values of from 10 to 50 per cent. The exposure may consist of 22 hr at 158°F (70°C), or 70 hr at 212°F (100°C), depending on the intended use of the isolator.

A *stress relaxation test* is used which utilizes the same size specimen as the above test. In this test, the force exerted by the specimen on the clamp is measured initially and after an exposure of 46 hr at 194°F (90°C).[1]

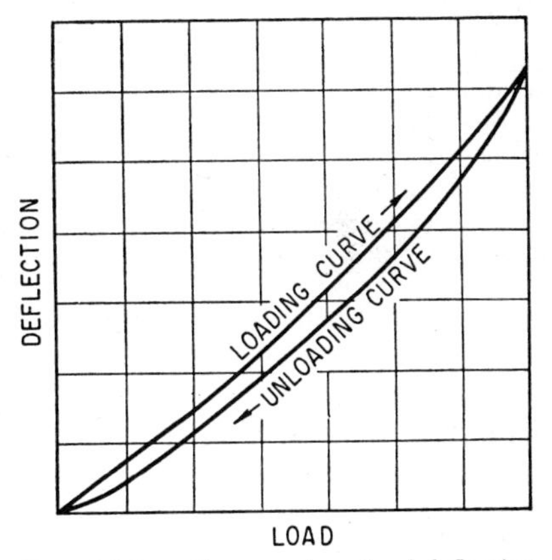

FIG. 35.2. Compression load-deflection curve for a rubber isolator. The data for this curve were obtained on the third loading cycle.

A "static" load-deflection curve for a rubber vibration isolator is shown in Fig. 35.2. The area between the loading and unloading curves represents hysteresis or damping. Both the stiffness and the hysteresis of the isolator depend on the specimen temperature and the rate of straining. It is also necessary to make three or more loading cycles in order to obtain a stable load-deflection curve. The flexing causes a slight softening of the material which is semipermanent in nature. This change is evidently caused by internal rearrangements of the carbon black structure which reverts to its original state after standing for several hours. It is common to specify that load-deflection readings be made at 73°F (23°C) and 0.125 in./min (0.318 cm/min) strain rate, flexing the specimen to the test load, or above, three times at those same conditions.

ENVIRONMENTAL FACTORS

AGING. The chemical reaction of vulcanization often continues at a very slow rate after a rubber part has been completed and placed in service. Therefore the hardness (stiffness) of rubber increases with time. At room temperature, the increase is quite rapid for the first few hours, and after a few days it drops to a very low value.

If the hardness of a typical vibration isolator were measured 3 days after vulcanization and again after a 10-year room-temperature exposure, a hardness increase of 5 durometer units might be found.

Indoor or shelf aging causes a small reduction in the tensile properties of rubber because the oxygen of the air attacks the molecular structure of the rubber. This effect is accelerated and magnified for test purposes by exposing tensile test specimens in the shape of dumbbells in a circulating air oven.[1] Exposures of 70 hr at 158°F (70°C), or at 212°F (100°C), are used—depending on the intended service. Measurements are made on aged and unaged samples, and the per cent reduction in tensile strength and in ultimate elongation as a result of the exposure are computed. Hardness increase is also measured in this test. Typical isolator rubber compounds may have a tensile strength reduction of as much as 25 per cent, ultimate elongation reduction of as high as 40 per cent, and a hardness increase of as much as 20 durometer units. Compounds of natural rubber and SBR normally used in vibration isolators are not suitable for continued use at service temperatures above 175°F (79°C). Chloroprene rubber compounds should not be used above 220°F (104°C). Propylene oxide and EPDM should not be used above 250°F (121°C). Continued exposure to high temperatures furthers the cure, promotes oxidation, and causes plasticizers and other ingredients to volatilize. Natural rubber is subject to *reversion*, in which the rubber becomes soft and tacky. Some of the other polymers undergo additional molecular "cross-linking" which causes a pronounced increase in hardness.

When rubber is exposed to direct weathering, it is exposed to oxygen, ultraviolet light, and to the leaching action of rain. The factors combine to promote oxidation. Sunlight entering the atmosphere generates ozone, which attacks the surface of the stressed rubber.[2] Ozone is especially deleterious to natural rubber, SBR (styrene-butadiene rubber), BR (butadiene rubber), and nitrile rubber (NBR) because it attacks the rubber molecule at the double bonds (the chemically unsaturated area) and produces surface cracks. These cracks can grow and cause complete failure of a rubber part if the part is continuously stressed. Other polymers have more inherent ozone resistance because they have fewer unsaturated areas in their molecular chains. The ozone resistance of the unsaturated polymers can be markedly improved by protective compounding ingredients. Ozone cracking will not occur if the rubber is unstressed. Although some reaction takes place between unstressed rubber and ozone, subsequent stretching will not reveal any cracks. There is a critical elongation at which ozone cracking is most severe. This elongation varies somewhat with the composition of the compound and the polymer used.[3] Threshold strains exist below which ozone damage is very small.[4] These strains are 7 to 9 per cent for natural rubber, SBR, and nitrile rubber compounds, 18 per cent for chloroprene rubber (CR), and 26 per cent for butyl rubber (IIR). Most polymers are not subject to ozone attack at temperatures below 0°F (−18°C).[5] Ozone is one constituent of "smog" and can occur in concentrations which are quite deleterious to rubber.

Ozone resistance usually is determined by exposing stressed specimens in a test chamber at 100°F (37.8°C) in an atmosphere of 50 parts of ozone per hundred million parts of air by volume.[1] Extruded specimens and actual parts such as weatherstrips are stressed by looping them around a 2-in.-diameter mandrel. Thinner samples are stressed by stretching them on a frame by an amount equal to 20 per cent of their lengths. The specimens are mounted, allowed to rest for 24 hr, and then exposed. They are examined frequently (perhaps every 2 hr) until the first cracking is noted and then are examined each day. The test is usually stopped after 3 days' exposure. At each examination, the specimens are rated for number and size of cracks. Vibration isolators normally are not subject to direct weathering, but sometimes operate near electrical equipment which may generate ozone by corona discharge. The correlation between ozone tests and actual weather exposure tests is poor. The actual exposure test normally is used for final evaluation and for acceptance tests, and the ozone test is used for development purposes.

IMMERSION IN FLUIDS. The properties of rubber are altered by exposure to fluids; the fluids present in vibration-isolation applications are usually petroleum oils or solvents. When immersed in these petroleum oils, natural rubber, SBR, isoprene, butadiene, butyl rubber, and silicone rubber soften, swell, and suffer a reduction in tensile strength and ultimate elongation. In a light petroleum oil, natural rubber might swell as much as 200 per cent and retain less than 20 per cent of its original tensile strength and ultimate elongation. Natural, SBR, and butyl rubbers usually are not tested for oil or solvent resistance. Two reference fuels and two reference oils have been established for immersion tests of other polymers. These tests are made by exposing tensile dumbbells and volume-swell specimens in the test fluid for 70 hr at 212°F (100°C) or 302°F (150°C), depending on the intended service. These samples, and unexposed duplicate samples, are then tested and per cent reduction in tensile strength and ultimate elongation are computed. Per cent volume swell and hardness change in durometer units are also found.[1] A typical chloroprene rubber compound, when exposed in a lubricating oil for 70 hr at 212°F (100°C), might suffer a 40 per cent reduction in tensile strength and a 30 per cent reduction in ultimate elongation. The volume might increase 15 per cent and the hardness might decrease 10 durometer units. Rubber compounds used for vibration isolators normally are not exposed to oil or solvents. It is usually more economical to locate the isolator in an area not exposed to these fluids, or to shield it, than to use an oil-resisting polymer.

LOW-TEMPERATURE PROPERTIES. When exposed to low temperatures, rubber becomes harder, stiffer, and less resilient. These changes are caused by changes in the mobility of the rubber molecule: (1) Simple temperature effects bring about an increase in stiffness and a decrease in resilience. They are caused by decreased thermal activity of the rubber molecule and are complete as soon as the specimen reaches thermal equilibrium. (2) Second-order transition of the rubber molecule also causes an increase in stiffness and a decrease in resilience; however, the change occurs in a short time at a definite temperature, which depends on both the polymer and the compounding ingredients. This transition results from a change in the type of motion of certain segments of the rubber molecule, and is accompanied by a change in the coefficient of thermal expansion and in the specific heat of the rubber. Normal rubber compounds are rendered unserviceable by simple temperature effects before the second-order transition effects become evident. (3) Crystallization causes hardening at temperatures above those necessary for second-order transition. It occurs over a long period of time, but is accelerated by stretching the specimen. It is accompanied by a decrease in volume and evolution of heat. Crystallization occurs most rapidly at an "optimum crystallization temperature" which is specific to the base polymer. The approximate optimum crystallization temperature for natural rubber is −13°F (−25°C); for chloroprene rubber it is 14°F (−10°C); and for SBR it is −49°F (−45°C). Butyl rubber does not crystallize unless stretched, and nitrile rubbers do not crystallize at all. Natural rubber, SBR, and silicone rubber have good low-temperature flexibility.

Figure 35.3 shows the effect of low temperature on the torsional modulus of elasticity of some tread-type rubber compounds. The modulus of each compound at various temperatures is shown relative to its own modulus at 77°F (25°C). Low-temperature stiffening is reduced by plasticizer addition to an extent which depends greatly on the amount and type of plasticizers used in the particular rubber compound. High plasticizer content generally has a detrimental effect on rubber-to-metal adhesion.

As a result of hysteresis, a rubber compound produces internal heat as it is dynamically strained, and its stiffness decreases. As the temperature of a rubber compound is lowered, hysteresis increases rapidly until the second-order transition temperature is reached. There is a simultaneous increase in stiffness. In a system which applies a constant dynamic force to a rubber isolator, there will be a "barrier" temperature below which the stiffness effect overcomes the hysteresis effect and the isolator will *not* warm up

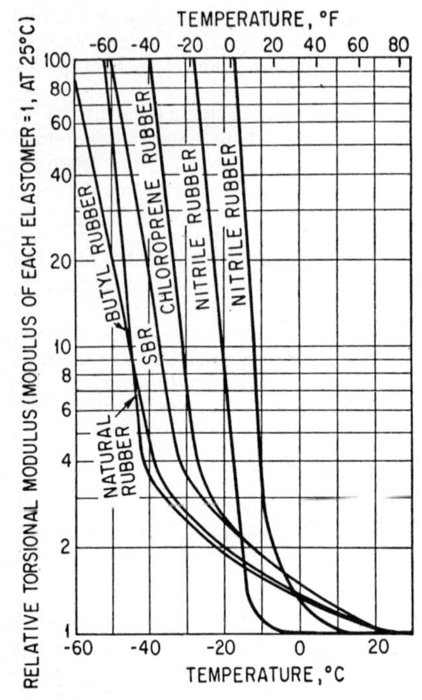

Fig. 35.3. Increase in torsional modulus of elasticity of various elastomers as a function of temperature. (*After Gehman.*[22])

as a result of hysteresis.[6] This barrier temperature depends greatly on the composition of the rubber compound, the shape of the isolator, the frequency and amplitude of vibration, and the heat losses to the surroundings. Barrier temperatures for typical compounds of several polymers are shown in Table 35.2.

Standard low-temperature tests for rubber include various methods of determining Young's modulus, hardness increase, and brittle point.[1] The method used to obtain the data in Fig. 35.3 utilizes a specimen 1.5 in. (38.2 mm) long, 0.125 in. (3.2 mm) wide, and 0.085 in. (2.2 mm) thick. The specimen is placed in an instrument containing one of three standard torsion spring wires. The twist of the specimen is determined at 73.4°F (23°C) and after 5 min at each test temperature. The temperatures used vary with the rubber being tested and may be as low as −94°F (−70°C) for natural rubber. The temperatures at which the relative torsional modulus of elasticity is 2, 5, 10, and 100 are determined from a plot of the data. In the brittleness method, samples 0.070 in. (1.8 mm) thick and 0.10 in. (2.5 mm) wide are immersed in a low-temperature bath and then impacted by a striker traveling at a velocity of 6.5 ft/sec (198 cm/sec). The test is repeated with new specimens at various temperatures until the temperature at which 50 per cent of the samples fail is determined.

Table 35.2. Rubber Mounting Barrier Temperature[6]

Polymer	Temperature	
	°F	°C
NR	−42	−41
SBR	−24	−31
CR	−18	−28
IIR	+2	−17
NBR	+16	−9

DRIFT. Drift is the characteristic of rubber by which it continues to deform after the initial deformation caused by an applied stress. *Relative drift* is defined as

$$d = \frac{\delta_t - \delta}{\delta} \tag{35.1}$$

where d is per cent relative drift at time t, δ is deflection at 1 min, and δ_t is deflection at time t. Drift is also called *creep* and *strain relaxation.* It is related to compression set and stress relaxation—all are manifestations of the flow properties of rubber. The theory of these effects has been developed by many investigators.[7–9] It is thought that the short-time effects are connected with the stability of the secondary molecular double bonds of the rubber compound, but that the long-time effects are associated with oxidation which causes scission of the molecular chain and chain cross-linking. Drift characteristics of several natural rubber compounds loaded in shear are shown in Fig. 35.4.

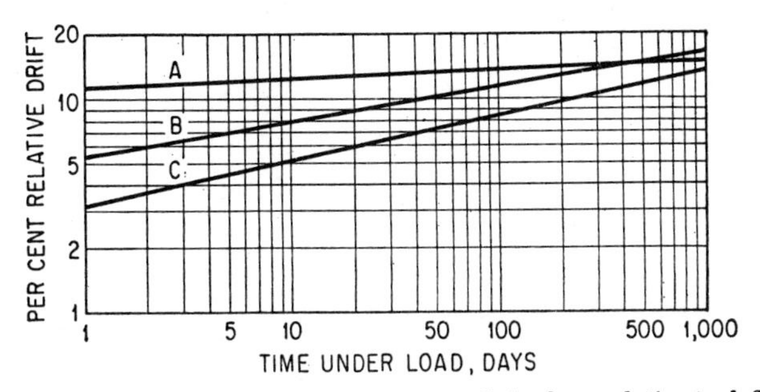

FIG. 35.4. Static drift of three natural rubber compounds in shear relative to deflection at one minute. The test specimens were subjected to a load of 60 lb/in.^{2} (4.2 kg/cm^2). (*After Morron.*[23])

Drift is quite dependent on the composition of the rubber compound and on the state of cure. Since the long-term effects are chemical in nature, temperature plays an important role. This effect is shown in Fig. 35.5.

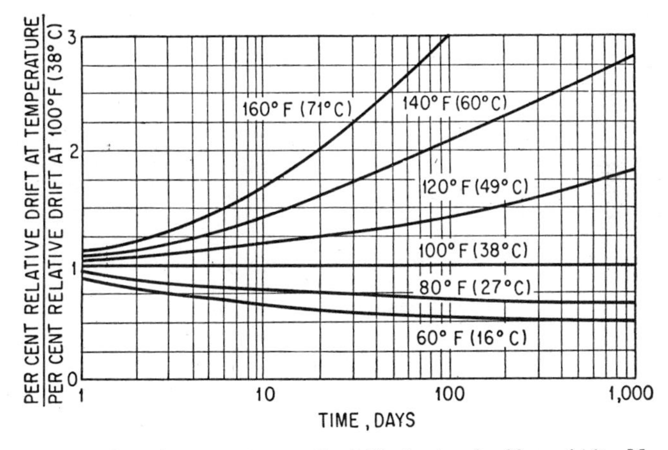

FIG. 35.5. The effect of temperature on the drift of natural rubber. (*After Morron.*[23])

Drift in compression is less than drift in shear, which in turn is less than drift in tension. Recommended maximum stress values normally are used to avoid excessive drift in an isolator.

Because of the long time required for actual drift tests, the compression set test is used to evaluate compounds.

POLYMER PROPERTIES. A qualitative comparison of the properties of various polymers is shown in Table 35.3. The composition of a rubber compound has a pro-

Table 35.3. Relative Properties

Poymer designa-tion	Common name	Shore A hardness range	Max. tensile strength		Com-pression set	Tear resist-ance	Resilience		Heat resistance	Outdoor aging resistance
			lb/in.²	kg/cm²			Room temp.	High temp.		
NR.......	Natural	30-100	4,000	280	Good	Good	High	High	Fair	Fair
SBR......	SBR	40-100	3,000	210	Good	Fair	Fairly high	Fairly high	Fair	Fair
CR.......	Neoprene	40-95	3,000	210	Poor (GN) Good (W)	Good	Fairly high	Fairly high	Good	Excellent
IIR.......	Butyl	40-75	2,000	140	Fair	Good	Low	Fairly high	Good	Good
EPDM...	EPDM	45-100	2,000	140	Fair	Fair	Fairly high	Fairly high	Excellent	Excellent
NBR.....	Nitrile	20-100	2,500	176	Good	Fair	Medium	Medium	Good to excellent	Poor
PO.......	Propylene Oxide	45-80	2,000	140	Fair	Fair	Fairly high	Fairly high	Good to excellent	Excellent
--	Thiokol	20-80	1,300	91	Poor	Good	Medium	Fairly high	Fair	Excellent
Si........	Silicone	20-90	1,000	70	Excellent	Poor	Fairly high	Medium	Excellent	Excellent
CSM.....	Hypalon	45-95	2,800	197	Fair	Fair to good	Fairly high	High	Excellent	Excellent
ACM.....	Polyacry-late	40-90	1,800	127	Good	Fair to poor	Low	High	Excellent	Excellent
FPM.....	Fluoro-rubber	60-90	3,000	210	Excellent	Fair to poor	Medium	Medium	Excellent	Excellent

* The relative ease of obtaining good adhesion to metal without employing costly metal treatments and elements.

nounced effect on its properties, and the comparisons shown are for each polymer as "normally" compounded.

The selection of the polymer to be used for any purpose is always a compromise. If high resilience, abrasion resistance, and good low-temperature flexibility are required, then natural rubber or isoprene is the best choice. If extreme temperature ranges are encountered, silicone rubber may be selected. If excellent weather resistance is of primary importance, EPDM or Hypalon might be selected. Butyl exhibits low permeability and low resilience at normal and low temperatures.

The cost of raw polymer does not represent the cost of a rubber compound in a finished part, because some rubber compounds require greater processing time and effort. Furthermore, the longer cures which are necessary for some polymers increase cost, and special ingredients necessary to modify the polymer properties to suit the application may be expensive.

DYNAMIC PROPERTIES OF RUBBER

MODULUS AND DAMPING. Rubber is not perfectly elastic—it exhibits internal damping and its stiffness tends to increase as the frequency of loading is increased. The action of rubber can be represented by an idealized mathematical model to which the measured performance can be compared. The idealized behavior most nearly approximating that of rubberlike materials is known as *linear viscoelastic behavior*. The mechanical model used for mathematical derivation is shown in Fig. 35.6. (Also see Chap. 29 for a discussion of electrical models.) If this model is subjected to a sinusoidal force of Various Polymers

Low temperature flexibility	Specific gravity	Abrasion resistance	Adhesion strength*	Flex life	Solvent resistance		Air permeability	Moisture resistance	Flame resistance
					Hydrocarbons	Oxygenated			
Excellent	0.93	Excellent	Excellent	Excellent	Poor	Good	Fairly high	Fair to good	Poor
Good	0.94	Excellent	Excellent	Good	Poor	Good	Fairly high	Good to excellent	Poor
Fair	1.23	Good	Good	Good	Good	Poor	Moderate	Fair to good	Excellent
Fair	0.92	Fair to good	Fair	Fair to good	Poor	Good	Very low	Excellent	Poor
Excellent	0.90	Fair to good	Fair	Fair	Poor	Good	Low	Excellent	Poor
Fair	1.00	Good	Good	Fair	Excellent	Poor	Low	Excellent	Poor
Excellent	1.01	Fair to good	Fair	Fair	Good	Poor	Moderate	Excellent	Poor
Good	1.34	Fair	Poor	....	Excellent	Good	Low	Fair	Poor
Excellent	0.95	Poor	Poor	Good	Fair	Good	Fairly high	Excellent	Fair
Fair	1.10	Excellent	Poor	Good	Good	Fair	Moderate	Excellent	Good
Poor	1.09	Good	Poor	Good	Excellent	Poor	Moderate	Fair	Poor
Fair	1.85	Good	Fair	Good	Excellent	Fair	Moderate	Excellent	Good

$F = F_0 \sin \omega t$, the response is given by[10]

$$x = \frac{F_0}{k_1 + k_2}\left[\sin \omega t + \frac{k_2 \sin \omega t}{k_1(1 + \omega^2\psi^2)} - \frac{k_2\omega\psi \cos \omega t}{k_1(1 + \omega^2\psi^2)}\right] \tag{35.2}$$

where k_1 is the stiffness of spring 1, k_2 is the stiffness of spring 2, c is the viscous damping coefficient, and

$$\psi = c\left(\frac{1}{k_1} + \frac{1}{k_2}\right) \tag{35.3}$$

The terms in brackets in Eq. (35.2) represent components of the model response. The first term represents the ordinary elastic response which is in-phase with the force and is independent of frequency; the second term is also in-phase with the force, but is frequency-dependent and represents the elastic component of viscoelastic response; the third term is frequency-dependent and is 90° out-of-phase with the force; it is responsible for the energy losses.

A number of methods have been devised for the measurement of the dynamic properties of vulcanizates and of isolators.[25] Such methods usually assume the simpler mechanical model shown in Fig. 35.7. The results are usually expressed as the "complex modulus" (i.e., the modulus of the entire model) and the damping "coefficient." They may also be expressed as the "storage mod-

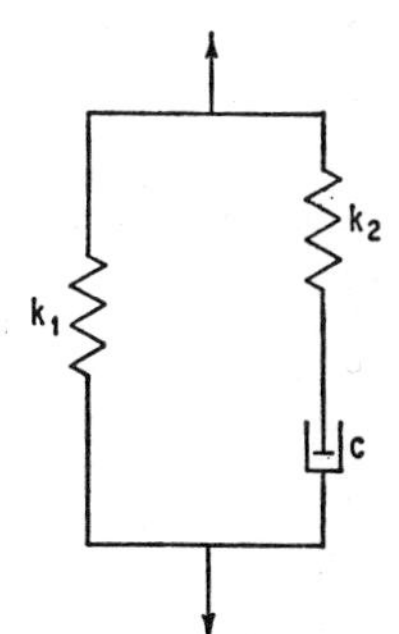

FIG. 35.6. Mechanical model for viscoelastic behavior of rubberlike materials with spring constants k_1 and k_2 and damping constant c.

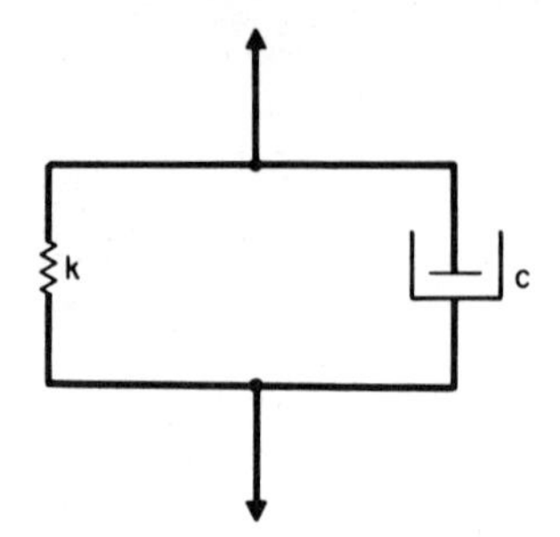

FIG. 35.7. Simplified mechanical model used for dynamic tests of isolators.

ulus" [the first two terms in Eq. (35.2)] and the "loss modulus" [the third term in (Eq. 35.2)]. Damping is also expressed as the loss tangent:

$$\tan \phi = \frac{\text{loss modulus}}{\text{storage modulus}} \tag{35.4}$$

EFFECT OF TEMPERATURE AND FREQUENCY

The dynamic properties of an elastomer are influenced by a number of factors. The general effects of temperature and frequency are shown in Fig. 35.8. (The storage modulus and loss tangent are shown on considerably different scales for clarity; the loss tangent is only a fraction of the storage modulus.) The exact shapes of these curves depend on the polymer and the specific composition used.

Figure 35.9 shows the response of typical isolator compounds made with several common polymers, as measured with the resonant beam apparatus.[25] These results, with the exception of IIR, represent a portion of the curve in Fig. 35.8*A* to the right of T_g.

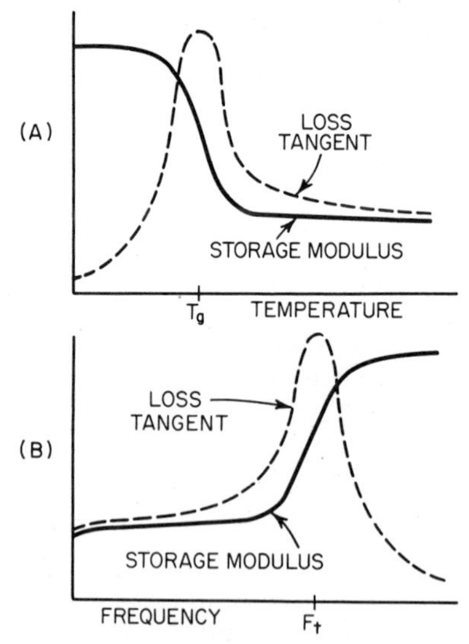

FIG. 35.8. Generalized response of vulcanizates to (*A*) temperature change at low frequency and (*B*) frequency change at normal temperature. T_g is the glass transition temperature (−99°F or −72.8°C for natural rubber). F_t is the transition frequency (10^6 to 10^8 cps).

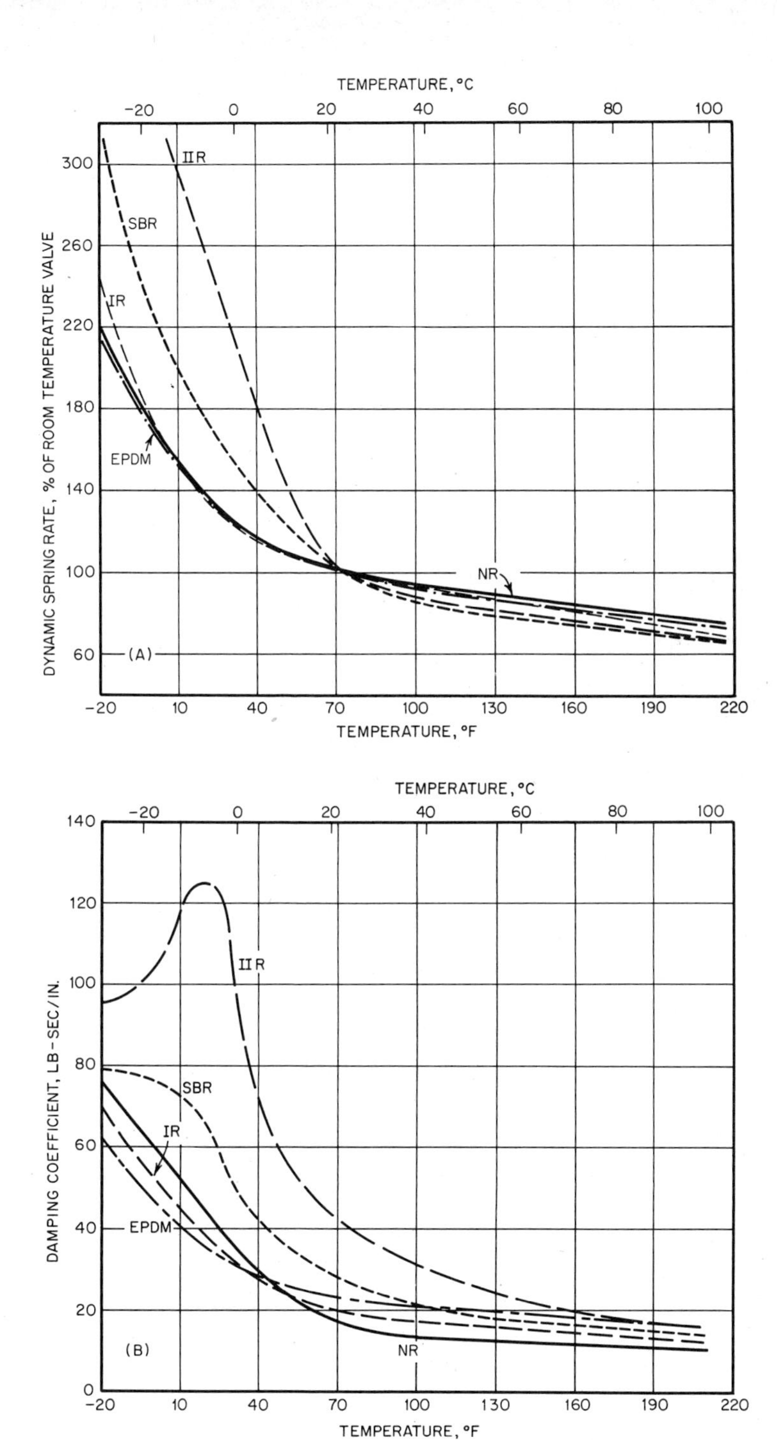

FIG. 35.9. The effect of temperature on (*A*) the dynamic spring rate and (*B*) the damping coefficient of typical isolator compounds using several polymers.

EFFECT OF AMPLITUDE

Most compounds for vibration isolators contain carbon black to give the compound the necessary strength. The dynamic properties of vulcanizates loaded with carbon black are quite dependent on the strain applied as those properties are measured. This effect is shown in Fig. 35.10. The high modulus at low strains is related to carbon black

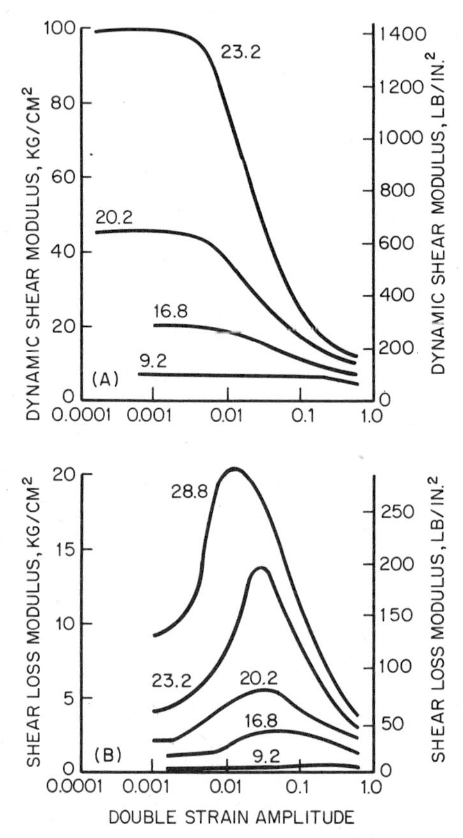

FIG. 35.10. The effect of amplitude of deformation on (*A*) the dynamic modulus and (*B*) the loss modulus of IIR vulcanizates. The numbers on the curves represent the percentage of carbon black by volume. (*After Payne.*[24])

structure and to the Van der Waals forces between the carbon black particles. For example, a specimen containing 23.2 per cent black has a modulus at 0.001 strain of 100 kg/cm². If that specimen is strained to 0.8 double amplitude, it has a modulus of 12 kg/cm². If it is then immediately tested at 0.001 strain, the modulus is still 12 kg/cm². If permitted to rest for over 24 hours, the carbon black structure will reform, and the modulus at 0.001 strain returns to nearly 100 kg/cm². These specific data are for carbon black in butyl rubber, but similar effects have been found with other reinforcing agents or fillers in other polymers.[24]

CRITICAL DAMPING

Representative values of the fraction of critical damping are given in Table 35.4. The range given for each polymer is necessary because of differences in properties brought about by hardness and by different compounding ingredients, and does not include the effect of temperature.

Table 35.4. Representative Values of Fraction of Critical Damping at Room Temperature

Polymer	*Fraction of critical damping c/c_c*
SBR	0.05–0.15
Natural rubber	0.01–0.08
Chloroprene rubber	0.03–0.08
Butyl rubber	0.05–0.50

FATIGUE. A rubber specimen subjected to a tensile stress approaching its tensile strength will continue to elongate with time and eventually will rupture. This process is known as *static fatigue;* it is the end result of the creep process. Dynamic fatigue occurs in a specimen subjected to an alternating stress centered about zero. In most vibration

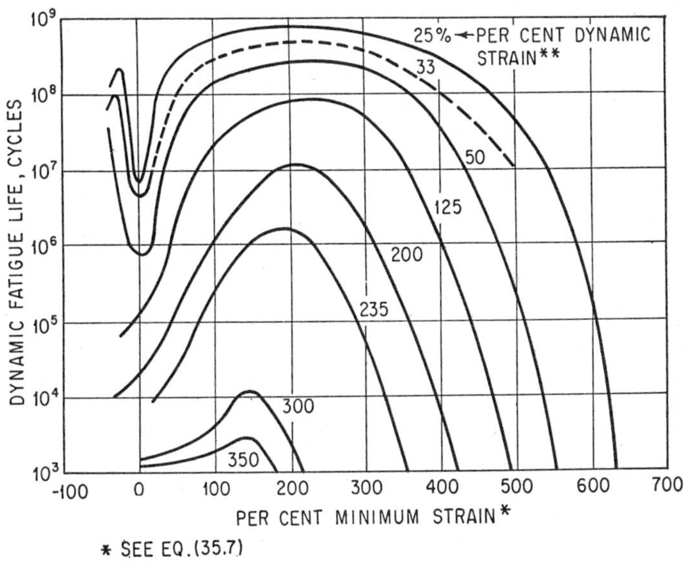

FIG. 35.11. The effect of strain on fatigue life of rubber specimens tested in tension and compression. (*After Caldwell, Merrill, Sloman, and Yost.*[12])

isolators, the fatigue process is some combination of static and dynamic fatigue. The results of an investigation of the dynamic fatigue characteristics of 50 durometer isolators strained in tension and compression are shown in Fig. 35.11. Dynamic fatigue life is plotted as a function of the per cent minimum strain, for fixed values of dynamic strain. The latter two parameters are computed in the following way. The per cent minimum strain is

$$\frac{l_{min} - l_0}{l_0}(100) \qquad \text{per cent} \tag{35.5}$$

and the per cent dynamic strain is

$$\frac{l_{max} - l_{min}}{l_0}(100) \qquad \text{per cent} \tag{35.6}$$

where l_0 is the unstrained length of the specimen, l_{min} is the minimum strained length, and l_{max} is the maximum strained length. Figure 35.11 shows that (1) for small strains there is a pronounced minimum in the fatigue life when the sample is returned to zero at the end of the stroke; (2) the point of maximum fatigue life shifts toward lower minimum strain as the dynamic strain is increased; and (3) fatigue life decreases as dynamic strain is increased. This latter effect may be caused partially by the greater heat generated. Specimens tested in shear also have a minimum life when returned to zero strain at one end of the stroke.[12] A summary of the fatigue life of shear specimens as a function of dynamic strain is given in Table 35.5.

Table 35.5. Fatigue Life in Cycles of Shear Specimens as a Function of Dynamic Strain for Various Lateral Strains[12]

Dynamic strain, per cent	Lateral strain		
	Zero	12.5 per cent compression	25 per cent tension
−25 to +25	7×10^6	20×10^6	12×10^6
0 to 50	1×10^6	2×10^6	2×10^6
75 to 125	15×10^6	2×10^6	40×10^6

Stress concentrations reduce the fatigue life of rubber isolators. Sudden changes in rubber section, sharp-edged inserts, protruding boltheads, and weld flash should be avoided. In many cases, fatigue failures will develop inside the rubber at some distance from the sharp edge of a metal insert.

DYNAMIC AGING. Dynamic stresses accelerate most aging effects, because dynamic strain exposes new areas of the molecular structure to attack. The cracks which form in a rubber flex test specimen are caused by a combination of dynamic fatigue and oxidation or ozone attack. One mechanism of protection against ozone is the inclusion of a wax in the rubber compound. The solubility of the wax is low and it *blooms* (i.e., migrates to the surface) and forms a protective coating. Such a compound has poor ozone resistance under dynamic conditions because the wax film is broken. A rubber compound containing wax usually is not used for bonded vibration isolators, since wax has a detrimental effect on rubber-to-metal adhesion. Flexible coatings are sometimes used when rubber isolators are exposed to abnormal ozone concentrations. Another method of surface protection is surface chlorination, in which chlorine is added at the molecular double bonds. This treatment also greatly reduces the coefficient of friction of the rubber on smooth, dry hard surfaces.

CLASSIFICATIONS AND SPECIFICATIONS

The ASTM Standards on Rubber Products[1] includes many test methods, specifications, recommended practices, and definitions covering the entire field of rubber products. The following are of particular interest in shock and vibration control.

ELASTOMERIC MATERIALS FOR AUTOMOTIVE APPLICATIONS

This classification system is identified as ASTM Designation D-2000 or SAE Standard J200. Although the word "automotive" is included in the title, this standard is widely used in many fields to specify the particular type of compound desired for a specific application. It enables the user to select a *type* of elastomer based on heat resistance and a *class* of elastomer based on oil resistance. Grade numbers allow specification of basic properties such as hardness and tensile strength, and suffix letters are added to signify special requirements such as compression set and adhesion strength. A tabular format is provided in order to guide the user to compounds which are commercially feasible and to prevent specification of combinations of properties which are impossible to obtain.

There are similar systems for specifying latex foam rubber (ASTM D-1055) and sponge rubber (ASTM D-1056).

FORCED VIBRATION TESTING OF VULCANIZATES

A recommended practice for forced vibration testing is given in ASTM D-2231. General background information specific to the design of isolators for automobiles is given in the proceedings of a symposium on vibration and noise in motor vehicles.[26]

DESIGN OF RUBBER ISOLATORS

METHODS OF LOADING RUBBER

COMPRESSION LOADING. Rubber is practically incompressible, having a bulk modulus of approximately 345,000 lb/in.^{2} (24,300 kg/cm^2) (Poisson's ratio approximately 1/2). Therefore when rubber is loaded in compression, the sides bulge out and the stress-strain curve increases in slope as the deflection is increased, as shown in Fig. 35.1*B*. The curves shown are for normal condition of no slippage at the loaded rubber face. If the loaded surface of the rubber is lubricated thoroughly, the stress-strain curve for small strains is fairly linear and the *compression modulus* is

$$E = \frac{F}{S_1}\left(\frac{h}{h - h_1}\right) \tag{35.7}$$

where E is the elastic modulus in compression in pounds per square inch, F is the compressive force in pounds; h is the unstrained height in inches; h_1 is the height in inches under a load of F in pounds, and S_1 is the area at that load in square inches. If S is the original area, then $Sh = S_1h_1$ (for an incompressible material), and the expression may be written as

$$E = \frac{F}{S}\left(\frac{h_1}{h - h_1}\right) \tag{35.8}$$

This relation is valid only for a specimen that is well lubricated and which is allowed to expand freely in the unloaded areas. Although rubber mountings usually are not used in such a manner, this scheme establishes a relation between hardness and modulus in compression which is independent of shape factor. This relationship is valuable as a design aid.

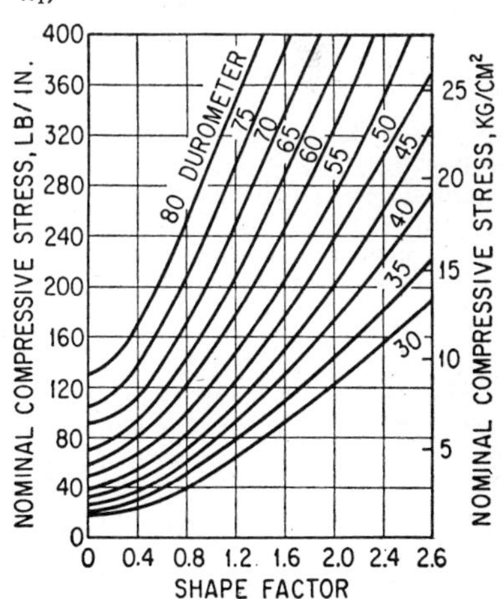

Fig. 35.12. Loads for 10 per cent deflection for rubber specimens having various hardness values and shape factors. (*After Kimmich.*[14])

The shape of the load-deflection curve of a dry rubber specimen depends on the relative dimensions of the specimen and on the hardness of the rubber compound. This is illustrated by the data of Fig. 35.12. The shape factor (along the abscissa) is defined as

$$\frac{\text{Loaded area}}{\text{Free area}}$$

For a rectangular specimen loaded vertically, the shape factor is the area of the loaded cross section divided by the area of the four sides.

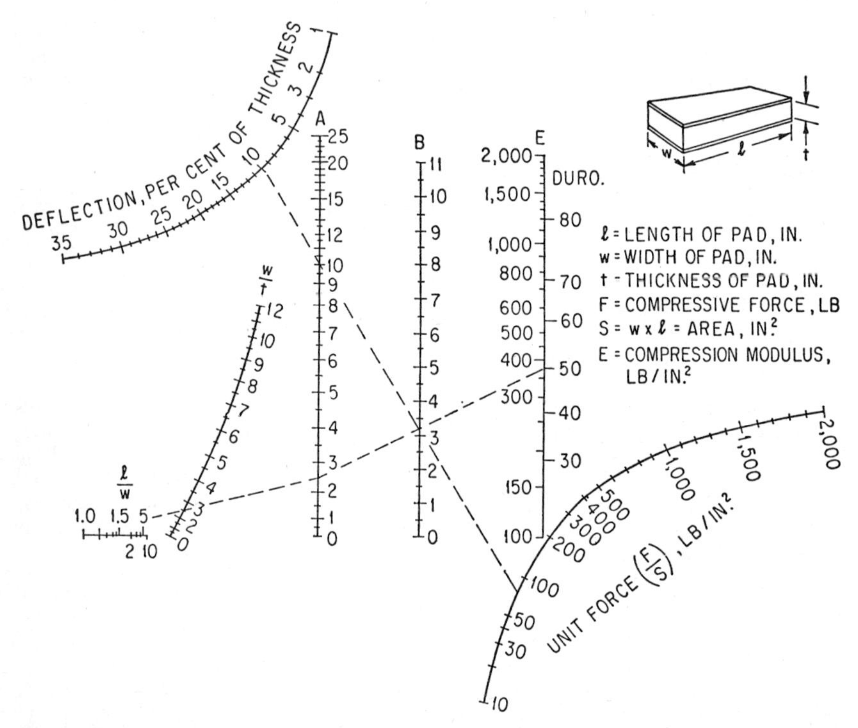

FIG. 35.13. Nomograph for determining relation of force and deflection for bonded rectangular rubber pads loaded in compression. A straight line is drawn intersecting the l/w and w/t scales at appropriate values. This determines a value of A, from which a second straight line is drawn to the E scale. To determine the deflection resulting from any unit force, a straight line is drawn from the force scale to the deflection scale, intersecting the B scale at the same point as the line from A to E. (*Courtesy of United States Rubber Company.*[18])

The nomograph shown in Fig. 35.13 applies to the design of bonded isolators loaded in compression. It covers the range of sizes and shapes normally used for mountings, and makes use of the compression modulus given by Eq. (35.8). The moduli shown in this figure were determined on the loading cycle with the samples at 70°F (21°C) after several loading and unloading cycles. Calculations made using this nomograph agree well with measured data.[14, 15, 16]

Rectangular Isolators. In Fig. 35.13, the A scale represents the ratio w/t for a square-shaped isolator. The E scale relates hardness and modulus; the B scale is merely a transfer device; and other quantities are defined in the legend. The nomograph is essentially an equation with six unknowns, l, w, t, F, E, and per cent compression. Given any five unknowns, the other can be found.

Example 35.1. Assume that a rubber mounting is 2 in. by 2 in. by 0.8 in. thick. Also assume that it is made of 50-durometer stock and that it is to have a compressive deflection of 10 per cent. What load will the mounting support?

$$\frac{l}{w} = 1 \quad \text{and} \quad \frac{w}{t} = 2.5$$

Connecting l/w and w/t yields a value on the A scale of 2.5. Connecting this value with the 50-durometer (375 lb/in.^{2}) value on the E scale gives a B-scale value of approximately 3.2. Connecting this value with the 10 per cent value yields a stress value of 80 lb/in.^{2} Since the mounting area S equals 4 in.^{2}, F = 320 lb.

Example 35.2. A machine weighing 12,000 lb is to be supported with four symmetrically located compression loaded isolators. Since the application involves infrequent vibration, a loading of 200 lb/in.^{2} and a deflection of 20 per cent are permissible (large static deflections contribute to high compression set; 20 per cent strain is the maximum recommended value). It is desired to use 60-durometer stock. What are the dimensions of square-shaped isolators to meet these requirements?

Connecting 200 lb/in.^{2} and 20 per cent yields a B-scale value of approximately 3.6. Connecting this value with 60 durometer (545 lb/in.^{2}) gives an A-scale value of approximately 2.

$$S = \frac{3{,}000}{200} = 15 \text{ in.}^2$$

Since $l = w$, $w^2 = 15$, and $w = 3.87$ in. From the nomograph for $l/w = 1$, $w/t = 2$, and $t = w/2 = 1.94$ in.

If a rectangular instead of square sandwich is desired, values of l/w and w/t can be obtained by cut-and-try alignments on those scales. This change would not affect the performance of the isolator, because the static deflection and pound per square inch loading would remain constant.

Cylindrical Isolators. For approximate calculations, using Fig. 35.13, a cylindrical isolator may be considered as a square isolator of the same thickness whose sides are equal to the cylinder diameter. Empirical determinations show that the A-scale value is actually 0.93 times the value for a square isolator because of the reduction of rubber area. When an approximation is made, the actual unit loading should be used for the F/S scale.

Triangular Isolators. A triangular isolator may be considered to be approximately equivalent to the largest cylindrical isolator of the same thickness which can be inscribed within it. For an equilateral triangle, the diameter of such a cylinder is

$$D = 0.577 \times \text{length of one side}$$

For any triangle, the diameter of the inscribed circle is

$$D = \frac{2}{\Sigma}\,[(\Sigma - x)(\Sigma - y)(\Sigma - z)]^{1/2}$$

where x, y, and z are the sides and

$$\Sigma = \tfrac{1}{2}(x + y + z)$$

Design Stress. The nomograph in Fig. 35.13 covers a wide range of possible designs. Designs which fall within its scope are not necessarily satisfactory for creep and flex life. Often there are unknown factors in a particular application such as actual operating temperature, magnitude of occasional overloads, amount of exposure to solvents, actual long-term creep rate of the compound used, etc. It is common practice to use a rule-of-thumb maximum design stress to cover these unknowns. A value of 90 lb/in.^{2} (6.3 kg/cm^2) is reasonable for an average compression isolator. This value can be modified to allow for operating conditions and expected life; for example, a value of 60 lb/in.^{2} (4.2 kg/cm^2) is used for some automotive engine mounts, and values as high as 400 lb/in.^{2} (28.1 kg/cm^2) are used.

The drift or the compression set of an isolator can cause misalignment of connecting shafts, pipes, etc. In addition, the reduction in thickness causes an increase in isolator stiffness which is not desired. Compression-loaded rubber isolators are used widely because the drift in compression is less than that in tension or shear and because the rubber-to-metal bond is not subjected to tensile loads.

TENSION LOADING. Rubber vibration isolators seldom are designed to operate primarily in tension because of the lower allowable bond stresses, increased importance

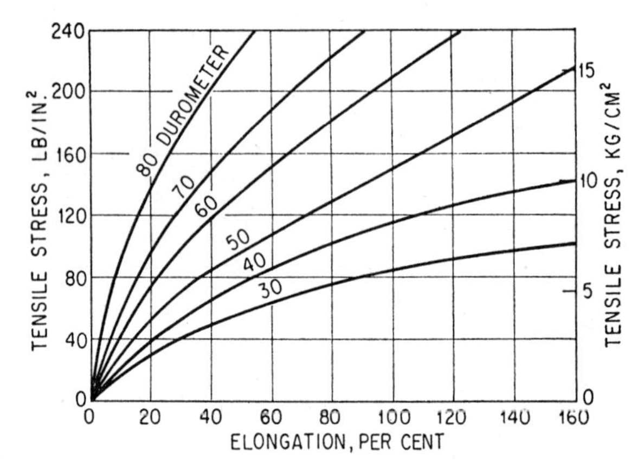

Fig. 35.14. Stress-strain curves for rubber of various hardnesses in tension. (*After U.S. Rubber Co.*[18])

of tear resistance, and lower creep limit. The relationships between tensile stress, tensile strain, and Shore "A" hardness for a typical compound are shown in Fig. 35.14. The curves cover only small strains. From these curves an approximate tension modulus can be computed for design purposes.

SHEAR LOADING. Rubber is often used in shear because large deflections in one direction can be obtained with good stability in the other directions and because the characteristics of a shear mounting can be predicted with greater ease. A typical shear

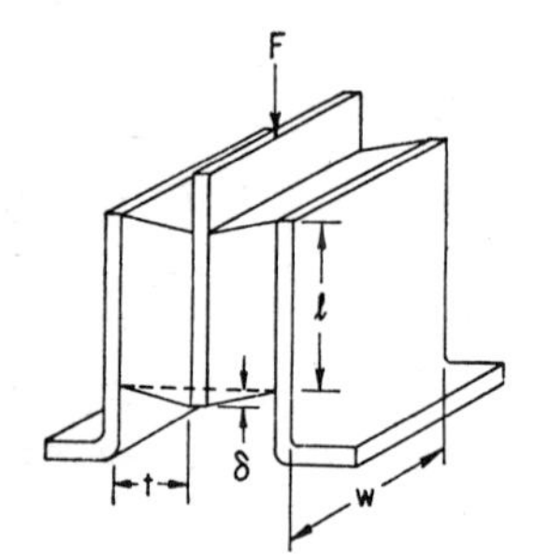

Fig. 35.15. A typical shear isolator, showing applied force F, length l, width w, and thickness t of rubber and deflection δ of isolator.

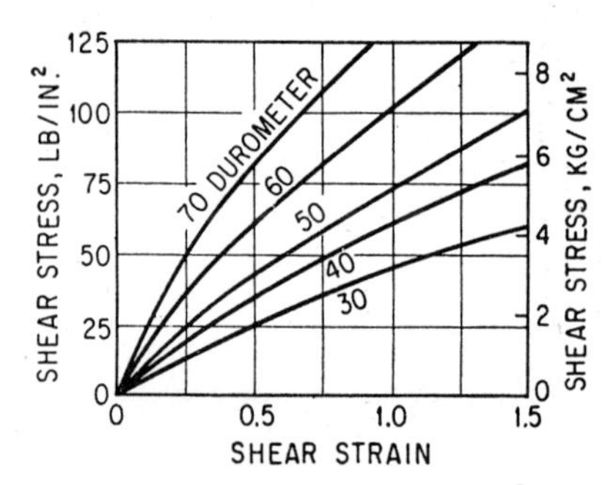

Fig. 35.16. Shear stress-strain curves for rubber of various hardnesses. (*After Crede.*[11])

mounting is shown in Fig. 35.15. Shear stress-strain curves for typical rubber compounds are shown in Fig. 35.16. The curves represent values taken on the loading cycle at 70°F (21°C) after several loading cycles. These curves are approximately linear for small strains. Because of this linearity and the fact that the stress-strain curves are not dependent on the shape factor, durometer hardness can be related to the shear modulus computed from the slope at the origin. This relation is shown in Fig. 35.17. With this relation the load-deflection characteristics of the shear mountings can be formulated as:

$$\frac{F}{S} = G\frac{\delta}{t}$$

where F = applied force, lb; S = area of rubber in shear ($2 \times l \times w$), in.^{2}; G = modulus of elasticity in shear, lb/in.^{2}; δ = deflection, in.; and t = rubber thickness, in.

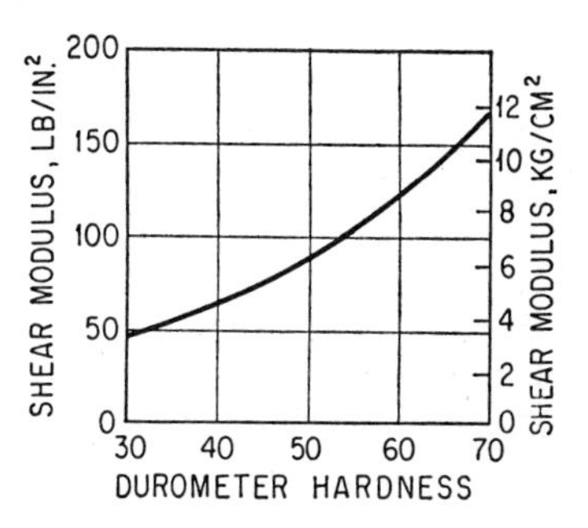

FIG. 35.17. Shear modulus of rubber for small deflections as a function of durometer hardness. (*After Crede.*[11])

Example 35.3. A double-shear sandwich, as in Fig. 35.15, supports a load of 500 lb. Each l and w are 3 in. and t is 0.75 in. The rubber is 40 durometer. Find the static deflection of the isolator.

The area is $S = 2 \times 3 \times 3 = 18$ in.^{2} Since two sides support the load, the shear stress is

$$\frac{F}{S} = \frac{500}{18} = 27.8 \text{ lb/in.}^2$$

The resulting shear strain may be found from Fig. 35.16 as 0.4. Thus the deflection is

$$\delta = 0.4 \times 0.75 = 0.3 \text{ in.}$$

Example 35.4. It is desired to support a 400-lb instrument with four single shear isolators, each of which is to have a static deflection of 20 per cent. If 50-durometer rubber is used, what size isolators are required?

The force per isolator F = 100 lb, and G is found from Fig. 35.17 as 88 lb/in.^{2} The strain $\delta/t = 0.2$ is given.

Thus

$$\frac{100}{S} = 88(0.2)$$

$$S = 5.68 \text{ in.}^2$$

Assume $l = w$; then $w = (5.68)^{1/2} = 2.38$ in. To determine the thickness of the sandwich, it is necessary to choose a static deflection commensurate with good stability, say ¼ in.

$$\frac{0.25}{t} = 0.2 \qquad t = 1.25 \text{ in.}$$

The bond stress is $100/5.68 = 17.6$ lb/in.^{2}, which is satisfactory.

Design Stress. As in the case of compression loading, it is usually necessary to use a recommended design stress figure. A design bond shear stress of 30 lb/in.^{2} (2.1 kg/cm^2) will be adequate for most situations. Some automotive engine mounts are designed at 20 lb/in.^{2} (1.4 kg/cm^2) and values as high as 100 lb/in.^{2} (7.0 kg/cm^2) are used.

If bond stress is *not* the limiting factor (i.e., if bond stress is quite low), a maximum shear strain of 50 per cent is recommended as a design figure.[11]

REFERENCES

1. American Society for Testing Materials, Standards on Rubber Products.
2. Crabtree, J., and A. R. Kemp: *Ind. Eng. Chem.*, **38**:278 (1946).
3. Newton, R. G. J.: *Rubber Research*, **14**:27 (1945).
4. Edwards, D. C., and E. B. Storey: *Trans. Inst. Rubber Ind.*, **31**:45 (1955).
5. Cuthbertson, G. R., and D. D. Dunnom: *Ind. Eng. Chem.*, **44**:834 (1952).
6. Morron, J. D., R. C. Knapp, E. F. Linhorst, and P. Viohl: *India Rubber World*, **110**:521 (1944).
7. Tobolsky, A. V., I. B. Prettyman, and J. H. Dillon: *J. Appl. Phys.*, **15**:380 (1944).
8. Peterson, L. E., R. L. Anthony, and E. Guth: *Ind. Eng. Chem.*, **34**:1349 (1942).

9. Beatty, J. R., and A. E. Juve: *India Rubber World*, **121**:537 (1950).
10. Gehman, S. D.: *Rubber Chem. and Technol.*, **30**:1202 (1957).
11. Crede, C. E.: "Vibration and Shock Isolation," John Wiley & Sons, Inc., New York, 1951.
12. Cadwell, S. M., R. A. Merrill, C. M. Sloman, and F. L. Yost: *Ind. Eng. Chem.*, **12**:19 (1940).
13. SAE (Society of Automotive Engineers) Handbook, 1959.
14. Kimmich, E. G.: *Rubber Chem. and Technol.*, **14**:407 (1941).
15. Smith, J. F. D.: *J. Appl. Mechanics*, A-13, March, 1938.
16. Chambers, G. G.: "The Development of a Rubber Engineering Design Handbook," General Motors Institute, 1957.
17. Smith, J. F. D.: *J. Appl. Mechanics*, **6**:159 (1939).
18. "Some Physical Properties of Rubber," United States Rubber Company, 1941.
19. Winspear, G. C. (ed.): "Vanderbilt Rubber Handbook," R. T. Vanderbilt Co., 1958.
20. Glossary of Terms Relating to Rubber and Rubber-like Materials, *ASTM Publ.* 184, 1956.
21. Whitby, G. S. (ed.): "Synthetic Rubber," John Wiley & Sons, Inc., New York, 1954.
22. Gehman, S. D., D. E. Woodford, and C. S. Wilkinson: *Ind. Eng. Chem.*, **39**:1108 (1947).
23. Morron, J. D.: *ASME Paper* 46-SA-18, presented June, 1946.
24. Payne, A. R., and R. E. Whittaker: *Rubber Chem. and Technol.*, **44**:440 (1971).
25. Hillberry, B. M.: "The Measurement of the Dynamic Properties of Elastomers and Elastomeric Mounts," Society of Automotive Engineers, January, 1973.
26. *Symp. Vibration and Noise in Motor Vehicles*, Automotive Division, Institute of Mechanical Engineers, London, 1972.

36

MATERIAL DAMPING AND SLIP DAMPING

L. E. Goodman
University of Minnesota

INTRODUCTION

The term *damping* as used in this chapter refers to the energy-dissipation properties of a material or system under cyclic stress but excludes energy-transfer devices such as dynamic absorbers. With this understanding of the meaning of the word, energy must be dissipated within the vibrating system. In most cases a conversion of mechanical energy to heat occurs. For convenience, damping is classified here as (1) material damping and (2) system damping. Material properties and the principles underlying measurement and prediction of damping magnitude are discussed in this chapter. For application to specific engineering problems, see Chap. 37.

MATERIAL DAMPING

Without a source of external energy, no real mechanical system maintains an undiminished amplitude of vibration. *Material damping* is a name for the complex physical effects that convert kinetic and strain energy in a vibrating mechanical system consisting of a volume of macrocontinuous (solid) matter into heat. Studies of material damping are employed in solid-state physics as guides to the internal structure of solids. The damping capacity of materials is also a significant property in the design of structures and mechanical devices; for example, in problems involving mechanical resonance and fatigue, shaft whirl, instrument hysteresis, and heating under cyclic stress. Three types of material that have been studied in detail are:

1. Viscoelastic materials. The idealized linear behavior generally assumed for this class of materials is amenable to the laws of superposition and other conventional rheological treatments including model analog analysis. In most cases linear (Newtonian) viscosity is considered to be the principal form of energy dissipation. Many polymeric materials as well as some other types of materials may be treated under this heading.

2. Structural metals and nonmetals. The linear dissipation functions generally assumed for the analysis of viscoelastic materials are not, as a rule, appropriate for structural materials. Significant nonlinearity characterizes structural materials, particularly at high levels of stress. A further complication arises from the fact that the stress and temperature histories may affect the material damping properties markedly; therefore, the concept of a stable material assumed in viscoelastic treatments may not be realistic for structural materials.

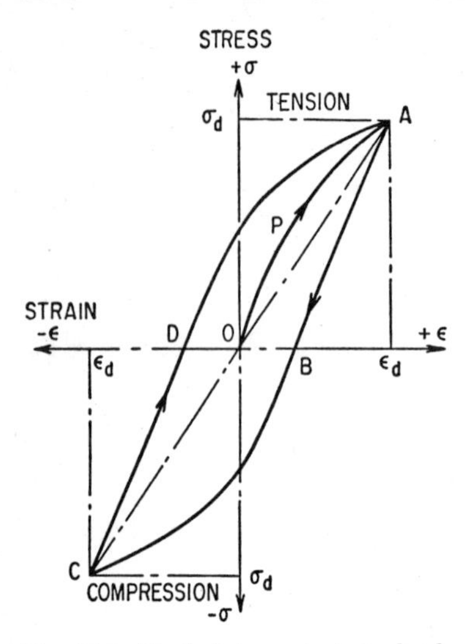

FIG. 36.1. Typical stress-strain (or load-deflection) hysteresis loop for a material under cyclic stress.

3. Surface coatings. The application of coatings to flat and curved surfaces to enhance energy dissipation by increasing the losses associated with fluid flow is a common device in acoustic noise control. These coatings also take advantage of material and interface damping through their bond with a structural material. They are treated in detail in Chap. 37.

Material damping of macrocontinuous media may be associated with such mechanisms as plastic slip or flow, magnetomechanical effects, dislocation movements, and inhomogeneous strain in fibrous materials. Under cyclic stress or strain these mechanisms lead to the formation of a stress-strain hysteresis loop of the type shown in Fig. 36.1. Since a variety of inelastic and anelastic mechanisms can be operative during cyclic stress, the unloading branch AB of the stress-strain curve falls below the initial loading branch OPA. Curves OPA and AB coincide only for a perfectly elastic material; such a material is never encountered in actual practice, even at very low stresses. The damping energy dissipated per unit volume during one stress cycle (between stress limits $\pm\sigma_d$ or strain limits $\pm\epsilon_d$) is equal to the area within the hysteresis loop $ABCDA$.

SLIP DAMPING

In contrast to material damping which occurs within a volume of solid material, slip damping arises from boundary shear effects at mating surfaces, or joints between distinguishable parts. Energy dissipation during cyclic shear strain at an interface may occur as a result of dry sliding (Coulomb friction), lubricated sliding (viscous forces), or cyclic strain in a separating adhesive (damping in viscoelastic layer between mating surfaces).

SIGNIFICANCE OF MECHANICAL DAMPING AS AN ENGINEERING PROPERTY

Large damping in a structural material may be either desirable or undesirable, depending on the engineering application at hand. For example, damping is a desirable property to the designer concerned with limiting the peak stresses and extending the fatigue life of structural elements and machine parts subjected to near-resonant cyclic forces or to suddenly applied forces. It is a desirable property if noise reduction is of importance. On the other hand, damping is undesirable if internal heating is to be avoided. It also can be a source of dynamic instability of rotating shafts and of error in sensitive instruments.

Resonant vibrations of large amplitude are encountered in a variety of modern devices, frequently causing rough and noisy operation and, in extreme cases, leading to seriously high repeated stresses. Various types of damping may be employed to minimize these resonant vibration amplitudes. Although special damping devices of the types described in Chap. 6 may be used to transfer energy from the system, there are many situations in which auxiliary dampers are not practical. Then accurate estimation of material and slip damping becomes important.

When an engineering structure is subjected to a harmonic exciting force $F_g \sin \omega t$ an induced force $F_d \sin(\omega t - \varphi)$ appears at the support. The ratio of the amplitudes, F_d/F_g, is a function of the exciting frequency ω. It is known as the vibration amplification factor. At resonance, when $\varphi = 90°$, this ratio becomes the resonance amplification factor[2] A_r:

$$A_r = \frac{F_d}{F_g} \tag{36.1}$$

This condition is pictured schematically in Fig. 36.2 for low, intermediate, and high damping (curves 3, 2, 1, respectively).

The magnitude of the resonance amplification factor varies over a wide range in engineering practice.[3] In laboratory tests, values as large as 1,000 have been observed. In actual engineering parts under high stress, a range of 500 to 10 is reasonably inclusive. These limits are exemplified by an airplane propeller, cyclically stressed in the fatigue range, which displayed a resonance amplification factor of 490, and a double leaf spring with optimum interface slip damping[5] which was observed to have a resonance amplification factor of 10. Because of the wide range of possible values of A_r, each case must be considered individually.

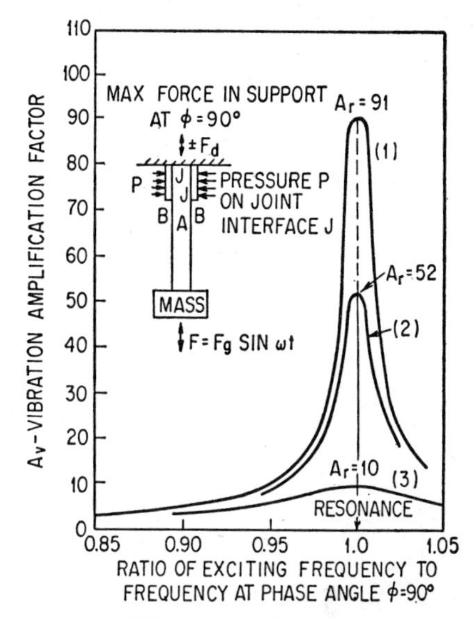

FIG. 36.2. Effect of material and slip damping on vibration amplification. Curve (1) illustrates case of small material and slip damping; (2) one damping is large while other is small; (3) both material and slip damping are large.

METHODS FOR MEASURING DAMPING PROPERTIES

STRESS-STRAIN (OR LOAD-DEFLECTION) HYSTERESIS LOOP

The hysteresis loop illustrated in Fig. 36.1 provides a direct and easily interpreted measure of damping energy. To determine damping at low stress levels requires instruments of extreme sensitivity. For example, the width (*DB* in Fig. 36.1) of the loop of chrome steel at an alternating direct-stress level of 103 MPa* is less than 2×10^{-6}. High-sensitivity and high-speed transducers and recording devices are required to attain sufficient accuracy for measurement of such strains. For metals in general extremely long gage lengths are required to measure damping in direct stress by the hysteresis loop method if the peak stress is less than about 60 per cent of the fatigue limit. Under torsional stress, however, greater sensitivity is possible and the hysteresis loop method is applicable to low stress work.

PROCEDURES EMPLOYING A VIBRATING SPECIMEN

The following methods of measuring damping utilize a vibrating system in which the deflected member, usually acting as a spring, serves as the specimen under test. For example, one end of the specimen may be fixed and the other end attached to a mass which is caused to vibrate; alternatively, a freely supported beam or a tuning fork may be used as the specimen vibrating system.[4] In any arrangement the damping is computed from the observed vibratory characteristics of the system.

In one class of these procedures the rate of decay of free damped vibration is measured. Typical vibration decay curves are shown in Fig. 36.3. The measure of damping

* 1 MPa = 10^6 N/m² = 146.5 lb/in² (103 MPa = 15,000 lb/in²).

usually used, the *logarithmic decrement*, is the natural logarithm of the ratio of any two successive amplitudes:

$$\Delta = \ln \frac{x_{n+1}}{x_n} \simeq \frac{\Delta x}{x_n} \tag{36.2}$$

The relation between logarithmic decrement and other units used to measure damping is given in Eq. (36.16). Vibration decay tests can be performed under a variety of stress and temperature conditions, and may utilize many different procedures for releasing the specimen and recording the vibration decay. It is essential to minimize loss of energy either to the specimen supports or in acoustic radiation.

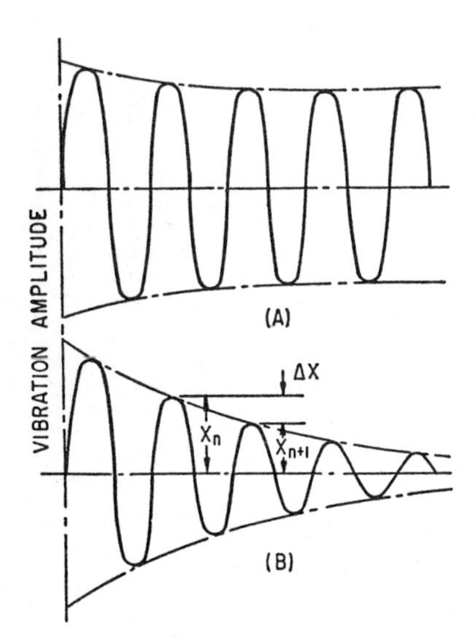

Fig. 36.3. Typical vibration decay curves: (*A*) low decay rate, small damping, and (*B*) high decay rate, large damping.

A second class of vibrating specimen procedures makes use of the fact, illustrated in Fig. 36.2, that higher damping is associated with a broader peak in the frequency response or resonance curve. If the exciting force is held constant and the exciting frequency varied, measurement of the steady-state amplitude of motion (or stress) yields a curve similar to those shown in Fig. 36.2. The damping is then determined by measuring the width of the curve at an amplification factor equal to $0.707A_r$. If a horizontal line drawn at this ordinate intercepts the resonance curve at frequency ratios f_1/f_n and f_2/f_n,

$$\Delta = \pi \left(\frac{f_2}{f_n} - \frac{f_1}{f_n} \right) \tag{36.3}$$

The quantity $(f_2 - f_1)$ is the *bandwidth at the half-power point*. This procedure has the advantage of requiring only steady-state tests. As in the case of the free-decay procedure, only the relative amplitude of the response need be measured. However, the procedure does impose a particular stress history. If the system behavior should be markedly nonlinear, the shape of the resonance curve will not be that assumed in the derivation of Eq. (36.3).[5]

If a system is operated exactly at resonance, the resonance amplification factor A_r is the ratio of the (induced) force F_d to the exciting force F_g [see Eq. (36.1)]. In direct application of this equation, F_g is usually made controllable and F_d computed from strain or displacement measurements. The principle has been applied to the measurement of damping in a large structure[6] and in simple test specimens. It can take account of high stress magnitude and of stress history as controlled variables. The natural frequency of vibration of a specimen can be altered so that damping as a function of frequency may be studied, but it is usually difficult to make such measurements over a wide frequency range. This technique requires accurately calibrated apparatus since measurements are absolute and not relative.

LATERAL DEFLECTION (OR LATERAL FORCE) OF ROTATING CANTILEVER METHOD

The principle of the lateral deflection method is illustrated in Fig. 36.4. If test specimen S is loaded by arm-weight combination A—W, the target T deflects vertically downward from position 1 to position 2. If the arm-specimen combination is rotated by

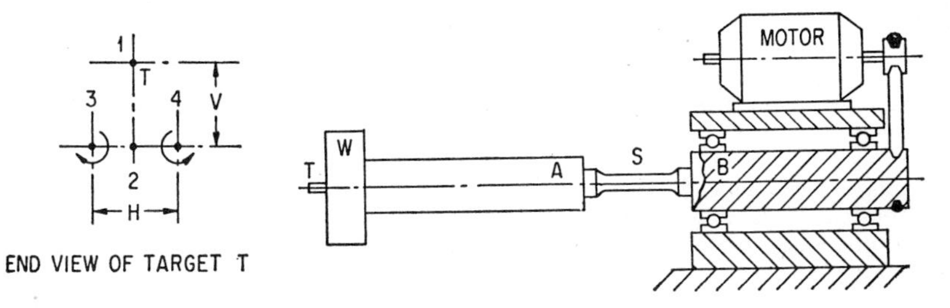

FIG. 36.4. Principle of rotating cantilever beam method for measuring damping.

spindle B, as in a rotating cantilever-beam fatigue test, target T moves from position 2 to position 3 for clockwise rotation. If the direction of rotation is counterclockwise, the target moves from position 3 horizontally to position 4. The horizontal traversal H is a direct measure of the total damping absorbed by the rotating system.[7]

A modification of the lateral deflection method is the *lateral force method.* The end of the rotating beam is confined and the lateral confining force is measured instead of the lateral deflection H. This modification is particularly useful for measurements of low modulus materials, such as plastic and viscoelastic materials.[8]

The advantages of the rotating cantilever beam method[9] are (1) the test variables, stress magnitude, stress history and frequency, may be easily and independently controlled so that this method is satisfactory for intermediate and high stress levels, and (2) it yields not only data on damping but also fatigue and elasticity properties.

The disadvantages of this method are (1) the tests are rather time-consuming, (2) accuracy is often questionable at low stress levels (below about 20 per cent of the fatigue limit) due to the small value of the horizontal traversal H, and (3) the method can be used under rotating-bending conditions only.

HIGH-FREQUENCY PULSE TECHNIQUES

A sequence of elastic pulses generated by a transducer such as a quartz crystal cemented to the front face of a specimen is reflected at the rear face and received again at the transducer. The frequencies are in the megacycle range. The velocity of such waves provides a measure of the elastic constants of the specimen; their decay rates provide a measure of the material damping.[10] This technique has been widely employed in the study of the viscoelastic properties of polymers and the elastic properties of crystals. So far as measurement of damping is concerned, it is open to the objection that the attenuation may be due to scattering by imperfections rather than to internal friction.

FUNDAMENTAL RELATIONSHIPS

Two general types of units[11] are used to specify the damping properties of structural materials: (1) the energy dissipated per cycle in a structural element or test specimen and (2) the ratio of this energy to a reference strain energy or elastic energy. Absolute damping energy units are:

D_0 = *total damping energy* dissipated by entire specimen or structural element per cycle of vibration, N.m/cycle

D_a = *average damping energy,* determined by dividing total damping energy D_0 by volume V_0 of specimen or structural element which is dissipating energy, $N.m/m^3$/cycle

D = *specific damping energy,* work dissipated per unit volume and per cycle at a point in the specimen, $N.m/m^3$/cycle

Of these absolute damping energy units, the total energy D_0 usually is of greatest interest to the engineer. The average damping energy D_a depends upon the shape of the specimen or structural element and upon the nature of the stress distribution in it, even though the specimens are made of the same material and have been subjected to the same stress distribution at the same temperature and frequency. Thus, quoted values of the average damping energy in the technical literature should be viewed with some reserve.

The specific damping energy D is the most fundamental of the three absolute units of damping since it depends only on the material in question and not on the shape, stress distribution, or volume of the vibrating element. However, most of the methods discussed previously for measuring damping properties yield data on total damping energy D_0 rather than on specific damping energy D. Therefore, the development of the relationships between these quantities assumes importance.

RELATIONSHIP BETWEEN D_0, D_a, AND D

If the specific damping energy is integrated throughout the stressed volume,

$$D_0 = \int_0^{V_0} D\, dV \tag{36.4}$$

This is a triple integral; $dV = dx\, dy\, dz$ and D is regarded as a function of the space coordinates x, y, z. If there is only one nonzero stress component the specific damping energy D may be considered a function of the stress level σ. Then

$$D_9 = \int_0^{\sigma_d} D \frac{dV}{d\sigma} d\sigma \tag{36.5}$$

In this integration V is the volume of material whose stress level is less than σ. The integration is a single integral and σ_d is the peak stress. The integrands may be put in dimensionless form by introducing D_d, the specific damping energy associated with the peak stress level reached anywhere in the specimen during the vibration (i.e., the value of D corresponding to $\sigma = \sigma_d$). Then

$$D_0 = D_d V_0 \alpha \tag{36.6}$$

where

$$\alpha = \int_0^1 \left(\frac{D}{D_d}\right) \frac{d(V/V_0)}{d(\sigma/\sigma_d)}\, d\left(\frac{\sigma}{\sigma_d}\right) \tag{36.7}$$

The average damping energy is

$$D_a = \frac{D_0}{V_0} = D_d \alpha \tag{36.8}$$

The relationship between the damping energies D_0, D_a, and D depends upon the dimensionless damping energy integral α. The integrand of α may be separated into two parts: (1) a damping function D/D_d which is a property of the material and (2) a volume-stress function $d(V/V_0)/d(\sigma/\sigma_d)$ which depends on the shape of the part and the stress distribution.

RELATIONSHIP BETWEEN SPECIFIC DAMPING ENERGY AND STRESS LEVEL

Before the damping function D/D_d can be determined, the specific damping energy D must be related to the stress level σ. Data of this type for typical engineering materials are given in Figs. 36.10 and 36.11. These results illustrate the fact that the damping-stress relationship for all materials cannot be expressed by one simple function. For a large number of structural materials in the low-intermediate stress region (up to 70 per cent of σ_e the fatigue strength at 2×10^7 cycles), the following relationship is reasonably satisfactory:

$$D = J\left(\frac{\sigma}{\sigma_e}\right)^n \tag{36.9}$$

Values of the constants J and n are given in Table 36.5 and Fig. 36.10. In general, $n = 2.0$ to 3.0 in the low-intermediate stress region but may be much larger at high stress levels. Where Eq. (36.9) is not applicable, as in the high stress regions of Figs. 36.10 and 36.11 or in the case of the 403 steel alloy of Fig. 36.9, analytical expressions are impractical and a graphical approach is more suitable for computation of α.

VOLUME-STRESS FUNCTION

The volume-stress function (V/V_0) may be visualized by referring to the dimensionless volume-stress curves shown in Fig. 36.5. The variety of specimen types included in

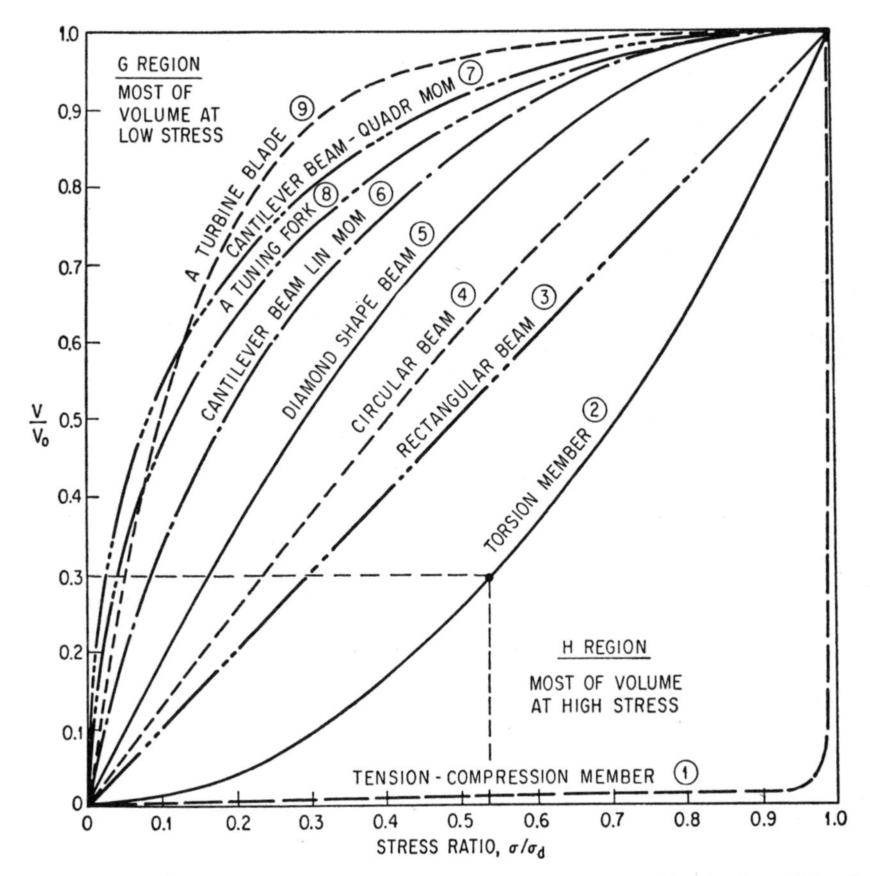

Fig. 36.5. Volume-stress functions for various types of parts. (See Table 36.1 for additional details on parts.)

this figure [tension-compression member (1) to turbine blade (9)] is representative of those encountered in practice. These curves give the fraction of the total volume which is stressed below a certain fraction of the peak stress. In a torsion member, for example, 30 per cent of the material is at a stress lower than 53 per cent of the peak stress. The volume-stress curves for a part having a reasonably uniform stress, i.e., having most of its volume stressed near the maximum stress, are in the region of this diagram labeled H. By contrast, curves for parts having a large stress gradient (such as a notched beam in

Table 36.1. Expressions and Values for α and β/α for Various Stress Distribution and Damping Functions

	Type of specimen and loading as designated in Fig. 36.1	Volume-stress function V/V_0	Dimensionless damping energy integral α for various damping functions: General case $D = f(\sigma)$	For special case $D = J(\sigma/\sigma_e)^n$: For any value of n	$n = 2.4$	$n = 8$	Dimensionless strain energy integral β	β/α if $n = 8$
1	Tension-compression member		1	1	1	1	1.0	1
2	Cylindrical torsion member or rotating beam	$\left(\frac{\sigma}{\sigma_d}\right)^2$	$\left[1 + \frac{\sigma_d}{2D_0}\frac{dD_0}{d\sigma_d}\right]^{-1}$	$\frac{2}{n+2}$	0.45	0.20	0.5	2.5
3	Rectangular beam under uniform bending	$\frac{\sigma}{\sigma_d}$	$\left[1 + \frac{\sigma_d}{D_0}\frac{dD_0}{d\sigma_d}\right]^{-1}$	$\frac{1}{n+1}$	0.29	0.11	0.33	3.0
4	Cylindrical beam under uniform bending	$\frac{2}{\pi}\left[\frac{\sigma}{\sigma_d}\sqrt{1-\left(\frac{\sigma}{\sigma_d}\right)^2}+\sin^{-1}\left(\frac{\sigma}{\sigma_d}\right)\right]$		$\frac{1}{\sqrt{\pi}}\frac{2}{n+2}\frac{\Gamma\left(\frac{n+1}{2}\right)}{\Gamma\left(\frac{n+2}{2}\right)}$	0.21	0.055	0.24	4.5
5	Diamond beam under uniform bending	$2\frac{\sigma}{\sigma_d} - \left(\frac{\sigma}{\sigma_d}\right)^2$	$\left[1 + \frac{2\sigma_d}{D_0}\frac{dD_0}{d\sigma_d} + \frac{\sigma_d^2}{2D_0}\frac{d^2D_0}{d\sigma_d^2}\right]^{-1}$	$\frac{2}{n^2+3n+2}$	0.13	0.022	0.17	7.7
6	Rectangular beam having bending moment shown: $M_x = \frac{x}{L}M_0$	$\frac{\sigma}{\sigma_d}\left[1 - \log_e\frac{\sigma}{\sigma_d}\right]$	$\left[1 + \frac{3\sigma_d}{D_0}\frac{dD_0}{d\sigma_d} + \frac{\sigma_d^2}{D_0}\frac{d^2D_0}{d\sigma_d^2}\right]^{-1}$	$\frac{1}{(n+1)^2}$	0.088	0.012	0.11	9.1
7	Rectangular beam having bending moment shown: $M_x = \left(\frac{x}{L}\right)^2 M_0$	$2\sqrt{\frac{\sigma}{\sigma_d}} - \frac{\sigma}{\sigma_d}$	$\left[1 + \frac{5\sigma_d}{D_0}\frac{dD_0}{d\sigma_d} + \frac{2\sigma_d^2}{D_0}\frac{d^2D_0}{d\sigma_d^2}\right]^{-1}$	$\frac{1}{2n^2+3n+1}$	0.051	0.0065	0.067	10.0
8	Tuning fork in bending	$K\frac{\sigma}{\sigma_d}\left[1 - \log_e\frac{\sigma}{\sigma_d}\right]$	$K\left[1 + \frac{3\sigma_d}{D_0}\frac{dD_0}{d\sigma_d} + \frac{\sigma_d^2}{D_0}\frac{d^2D_0}{d\sigma_d^2}\right]^{-1}$	$\frac{K}{(n+1)^2}$ For $K = 0.8$ →	0.091	0.0099	0.089	9.0

Note: $\beta/\alpha = 1$ for all cases if $n = 2$.

which very little volume is at the maximum stress and practically all of the volume is at a very low stress) are in the G region.

In order to illustrate representative values of α for several cases of engineering interest, the results of selected analytical and graphical computations[3] are summarized in Table 36.1 and in Fig. 36.6. In Fig. 36.6 the effect of the damping exponent n on the value

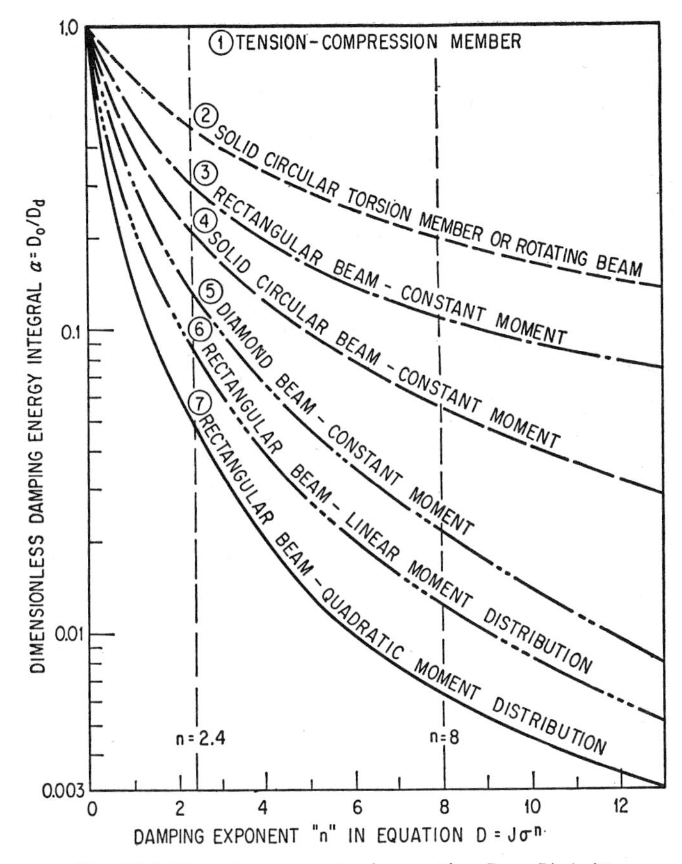

FIG. 36.6. Damping exponent n in equation $D = J(\sigma/\sigma_e)^n$.

of α for different types of representative specimens is illustrated. Note the wide range of α encountered for $n = 2.4$ (representative of many materials at low and intermediate stress) and for $n = 8$ (representative of materials at high stress, as shown in the next section).

RATIO OF DAMPING ENERGY TO STRAIN ENERGY

Owing to the complexity of the sources of material damping, the use of relative damping energy units does not produce all the advantages which might otherwise be associated with a nondimensional quantity. One motivation for the use of such units, however, is their direct relation to several conventional damping tests. The *logarithmic decrement* Δ is defined by Eq. (36.2). Other energy ratio units are tabulated and defined below. In this chapter, the energy ratio unit termed *loss factor* is used as the reference unit.

In defining the various energy ratio units, it is important to distinguish between loss factor η_s of a specimen or part (having variable stress distribution) and the loss factor η for a material (having uniform stress distribution). By definition the loss factor of a specimen (identified by subscript s) is

$$\eta_s = \frac{D_0}{2\pi W_0} \tag{36.10}$$

where the total damping D_0 in the specimen is given by Eq. (36.6). The total strain energy in the part is of the form

$$W_0 = \int_0^{V_0} \frac{1}{2}\left(\frac{\sigma^2}{E}\right) dV = \frac{1}{2}\left(\frac{\sigma_d^2}{E}\right) V_0\beta \tag{36.11}$$

where E denotes a modulus of elasticity and β is a dimensionless integral whose value depends upon the volume-stress function and the stress distribution:

$$\beta = \int_0^1 \left(\frac{\sigma}{\sigma_d}\right)^2 \frac{d(V/V_0)}{d(\sigma/\sigma_d)}\, d\left(\frac{\sigma}{\sigma_d}\right) \tag{36.12}$$

On substituting Eq. (36.6) and Eq. (36.11) in Eq. (36.10), it follows that

$$\eta_s = \frac{E}{\pi}\frac{D_d}{\sigma_d^2}\frac{\alpha}{\beta} \tag{36.13}$$

If the specimen has a uniform stress distribution, $\alpha = \beta = 1$ and the specimen loss factor η_s becomes the material loss factor η; in general, however,

$$\eta = \frac{ED_d}{\pi\sigma_d^2} = \eta_s\frac{\beta}{\alpha} \tag{36.14}$$

Other energy ratio (or relative energy) damping units in common use are defined below:

For specimens with variable stress distribution:

$$\eta_s = (\tan\phi)_s = \frac{\Delta_s}{\pi} = \frac{\psi_s}{\pi} = \left(\frac{\delta\omega}{\omega_n}\right)_s = \frac{1}{(A_r)_s} = \frac{1}{Q_s} = \frac{ED_d}{\pi\sigma_d^2}\left(\frac{\alpha}{\beta}\right) \tag{36.15}$$

For materials or specimens with uniform stress distribution:

$$\eta = \tan\phi = \frac{\Delta}{\pi} = \frac{\psi}{\pi} = \frac{\delta\omega}{\omega_n} = \frac{1}{A_r} = \frac{1}{Q} = Q^{-1} = \frac{ED_d}{\pi\sigma_d^2} \tag{36.16}$$

where η = loss factor of material = dissipation factor (high loss factor signifies high damping)

$\tan\phi$ = loss angle, where ϕ is phase angle by which strain lags stress in sinusoidal loading

$\psi = \pi\eta$ = specific damping capacity

$\delta\omega/\omega_n$ = (bandwidth at half-power point)/(natural frequency) [see Eq. (36.3)]

A_r = resonance amplification factor [see Eq. (36.1)]

$Q = 1/\eta$ = measure of the sharpness of a resonance peak and amplification produced by resonance

The material properties are related to the specimen properties as follows:

$$\psi = \psi_s\frac{\beta}{\alpha} \qquad \Delta = \Delta_s\frac{\beta}{\alpha} \qquad A_r = (A_r)_s\frac{\alpha}{\beta} \tag{36.17}$$

Thus, the various energy ratio units, as conventionally expressed for specimens, depend not only on the basic material properties D and E but also on β/α. The ratio β/α depends on the form of the damping-stress function and the stress distribution in the specimen. As in the case of average damping energy, D_a, the loss factor or the logarithmic decrement for specimens made from exactly the same material and exposed to the same stress range, frequency, temperature, and other test variables may vary significantly if the shape and stress distribution of the specimen are varied. Since data expressed as logarithmic decrement and similar energy ratio units reported in technical literature have been obtained on a variety of specimen types and stress distributions, any comparison of such data must be considered carefully. The ratio β/α may vary for specimens of exactly the same shape if made from materials having different damping-stress functions.[3] For different specimens made of exactly the same materials, the variation in β/α also may be large, as shown in Fig. 36.7. For example, for a material and

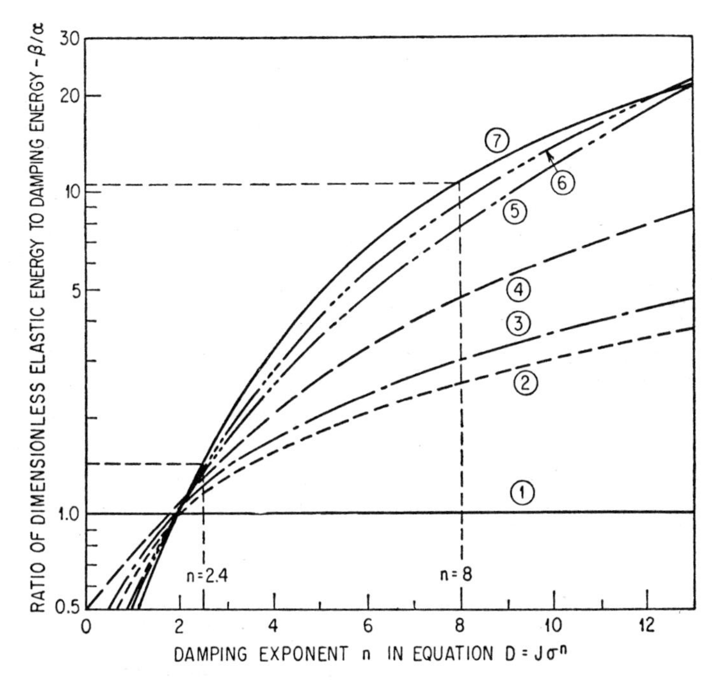

FIG. 36.7. Effect of damping exponent n on ratio β/α for $D = J\sigma^n$. Curves are (1) tension-compression member; (2) solid circular torsion member or rotating beam; (3) rectangular beam–constant moment; (4) solid circular beam–constant moment; (5) diamond beam–constant moment; (6) rectangular beam–linear moment distribution; and (7) rectangular beam–quadratic moment distribution.

stress region for which damping exponent $n = 2.4$ (characteristic of metals at low and intermediate stress), the value of β/α shown in Table 36.1 varies from 1 for a tension-compression member to 1.6 for a rectangular beam with quadratic moment distribution. If $n = 8$ (characteristic of materials at high stress), the variation is from 1 to 10, and larger for beams with higher stress gradient.

It is possible, for a variety of types of beams, to separate the ratio β/α into two factors:[12] (1) a cross-sectional shape factor K_c which quantitatively expresses the effect of stress distribution on a cross section, and (2) a longitudinal stress distribution factor K_s which expresses the effect of stress distribution along the length of the beam. Then

$$\frac{\beta}{\alpha} = K_s K_c \tag{36.18}$$

If material damping can be expressed as an exponential function of stress, as in Eq. (36.9), some significant generalizations can be made regarding the pronounced effect of the damping exponent n on each of these factors. Some of the results are shown in Fig. 36.8 for beams of constant cross-section. These curves indicate that high values of K_s and K_c are associated with a high damping exponent n, other factors being equal; K_c is high when very little material is near peak stress. For example, compare the diamond cross-section shape with the I beam, or compare the uniform stress beam with the cantilever.

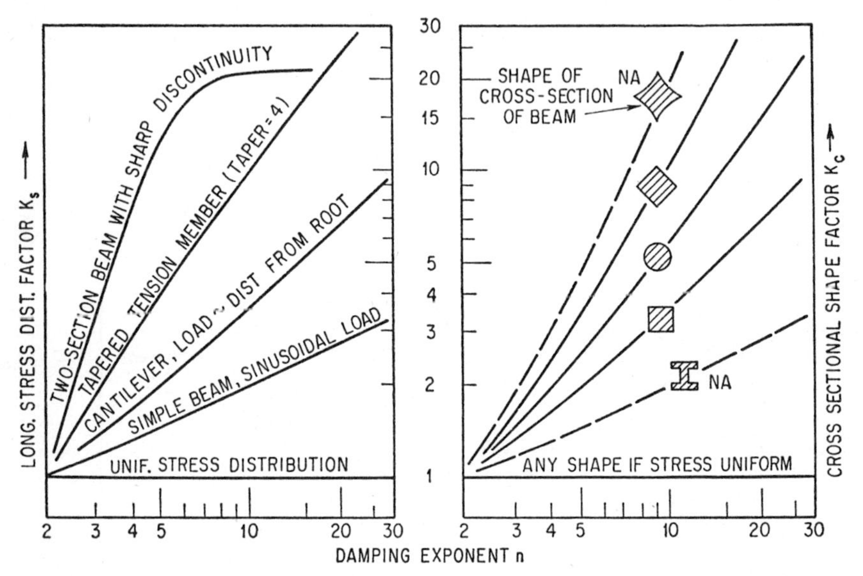

FIG. 36.8. Effect of damping exponent n on longitudinal stress distribution factor and cross-sectional shape factor of selected examples.

In much of the literature of damping, the existence of factors α and β (or K_s and K_c) is not recognized; the unstated assumption is that $\alpha = \beta = 1$. As discussed above, this assumption is true only for specimens under homogeneous stress.

Relative damping units such as logarithmic decrement depend on the ratio of two energies, the damping energy and the strain energy. Since strain energy increases with the square of the stress for reasonably linear materials, the logarithmic decrement remains constant with stress level and is independent of specimen shape and stress distribution only for materials whose damping energy also increases as the square of the stress [$n = 2$ in Eq. (36.9)]. For most materials at working stresses, n varies between 2 and 3 (see Fig. 36.10) but for some (Fig. 36.9) it is highly variable. In the high stress region, n lies in the range 8.0–20.0 (Fig. 36.10). In view of the broad range of materials and stresses encountered in design, the case $n = 2$ must be regarded as exceptional. Thus, logarithmic decrement is a variable rather than a "material constant." Its magnitude generally decreases significantly with stress amplitude. When referring to specimens such as beams in which all stresses between zero and some maximum stress occur simultaneously, the logarithmic decrement is an ambiguous average value associated with some mean stress. Published data require careful analysis before suitable comparisons can be made.

VISCOELASTIC MATERIALS

Some materials respond to load in a way that shows a pronounced influence of the rate of loading. Generally the strain is larger if the stress varies slowly than it is if the stress reaches its peak value swiftly. Among materials that exhibit this viscoelastic behavior are high polymers and metals at elevated temperatures as well as many glasses, rubbers, and plastics. As might be expected, these materials usually also exhibit creep, an increasing deformation under constant applied load.

When a sinusoidal exciting force is applied to a viscoelastic solid, the strain is observed to lag behind the stress. The phase angle between them, denoted φ, is the *loss angle*. The stress may be separated into two components, one in phase with the strain and one leading it by a quarter cycle. The magnitudes of these components depend upon the material and upon the exciting frequency, ω. For a specimen subject to homogeneous shear ($\alpha = \beta = 1$),

$$\gamma = \gamma_0 \sin \omega t \tag{36.19}$$

$$\sigma = \gamma_0[G'(\omega) \sin \omega t + G''(\omega) \cos \omega t] \tag{36.20}$$

This is a linear viscoelastic stress-strain law. The theory of linear viscoelasticity is the most thoroughly developed of viscoelastic theories. In Eq. (36.20) $G'(\omega)$ is known as the "storage modulus in shear" and $G''(\omega)$ is the "loss modulus in shear" (the symbols G_1 and G_2 are also widely used in the literature). The stiffness of the material depends on G' and the damping capacity on G''. In terms of these quantities the loss angle $\varphi = \tan^{-1} (G''/G')$. The *complex*, or *resultant*, modulus in shear is $G^* = G' + iG''$. In questions of stress analysis, complex moduli have the advantage that the form of Hooke's law is the same as in the elastic case except that the elastic constants are replaced by the corresponding complex moduli. Then a correspondence principle often makes it possible to adapt an existing elastic solution to the viscoelastic case. For details of viscoelastic stress analysis see Refs. 42 and 43.

The moduli of linear viscoelasticity are readily related to the specific damping energy D introduced previously. For a specimen in homogeneous shear of peak magnitude γ_0, the energy dissipated per cycle and per unit volume is

$$D = \int_0^{2\pi/\omega} \sigma \left(\frac{d\gamma}{dt}\right) dt \tag{36.21}$$

In view of Eqs. (36.19) and (36.20) this becomes

$$\begin{aligned} D &= \int_0^{2\pi/\omega} \gamma_0^2\omega(G' \sin \omega t + G'' \cos \omega t) \cos \omega t \, dt \\ &= \pi\gamma_0^2 G''(\omega) \end{aligned} \tag{36.22}$$

It is apparent from Eq. (36.22) that linear viscoelastic materials take the coefficient $\omega = 2$ in Eq. (36.9). These materials differ from metals, however, by having damping capacities that are strongly frequency- and temperature-sensitive.[16]

DAMPING PROPERTIES OF MATERIALS

The specific damping energy D dissipated in a material exposed to cyclic stress is affected by many factors. Some of the more important are:

1. Condition of the material
   a. In virgin state:
      Chemical composition; constitution (or structure) due to thermal and mechanical treatment; inhomogeneity effects
   b. During and after exposure to pretreatment, test, or service condition:
      Effect of stress and temperature histories on aging, precipitation, and other metallurgical solid-state transformations
2. State of internal stress
   a. Initially, due to surface-finishing operations (shot peening, rolling, case hardening)

   b. Changes caused by stress and temperature histories during test or service
3. Stress imposed by test or service conditions
   a. Type of stress (tension, compression, bending, shear, torsion)
   b. State of stress (uniaxial, biaxial, or triaxial)
   c. Stress-magnitude parameters, including mean stress and alternating components; loading spectrum if stress amplitude is not constant
   d. Characteristics of stress variations including frequency and waveform
   e. Environmental conditions:
      Temperature (magnitude and variation) and the surrounding medium and its (corrosive, erosive, and chemical) effects

Factors tabulated above, such as stress magnitude, history, and frequency, may be significant at one stress level or test condition and unimportant at another. The deformation mechanism that is operative governs sensitivity to the various factors tabulated.

Many types of inelastic mechanisms and hysteretic phenomena have been identified,[10, 14, 15] as shown in Table 36.2. The various damping phenomena and mechanisms may be classified under two main headings: *dynamic hysteresis* and *static hysteresis*.

Materials which display dynamic hysteresis (sometimes identified as viscoelastic, rheological, and rate-dependent hysteresis) have stress-strain laws which are describable by a differential equation containing stress, strain, and time derivatives of stress or strain. This differential equation need not be linear, though, to avoid mathematical complexity, much of existing theory is based on the linear viscoelastic law described in the previous section. One important type of dynamic hysteresis, a special case identified as *anelasticity* or *internal friction*, produces no permanent set after a long time. This means that if the load is suddenly removed at point B in Fig. 36.1, after cycle OAB, strain OB will gradually reduce to zero as the specimen recovers (or creeps negatively) from point B to point O.

A distinguishing characteristic of anelasticity and the more general case of viscoelastic damping is its dependence on time-derivative terms. The hysteresis loops tend to be elliptical in shape rather than pointed as in Fig. 36.1. Furthermore, the loop area is definitely related to the dynamic or cyclic nature of the loading and the area of the loop is dependent on frequency. In fact, the stress-strain curve for an ideally viscoelastic material becomes a single-valued curve (no hysteretic loop) if the cyclic stress is applied slowly enough to allow the material to be in complete equilibrium at all times (oscillation period very much longer than relaxation times). No hysteretic damping is produced by these mechanisms if the material is subjected to essentially static loading. Stated differently, the static hysteresis is zero.

Static hysteresis, by contrast, involves stress-strain laws which are insensitive to time, strain, or stress rate. The equilibrium value of strain is attained almost instantly for each value of stress and prior stress history (direction of loading, amplitudes, etc.), independent of loading rate. Hysteresis loops are pointed, as shown in Fig. 36.1, and if the stress is reduced to zero (point B) after cycle OAB, then OB remains as a permanent set or residual deformation. The two principal mechanisms which lead to static hysteresis are magnetostriction and plastic strain.

Table 36.2 also shows the simplest representative mechanical models for each of the behaviors classified. In these models k is a spring having linear elasticity (linear and single-valued stress-strain curve), C is a linear dashpot which produces a resisting force proportional to velocity, and D is a Coulomb friction unit which produces a constant force whenever slip occurs within the unit, the direction of the force being opposite to the direction of relative motion. More sophisticated models have been found to predict reliably the behavior of some materials, particularly polymeric materials.

Any one of the mechanisms to be discussed may dominate, depending on the stress level. For convenience, *low stress* is defined here as a (tension-compression) stress less than 1 per cent of the fatigue limit; *intermediate* stress levels are those between 1 per cent and 50 per cent of the fatigue limit of the material; and *high* stress levels are those exceeding 50 per cent of the fatigue limit.

Table 36.2. Classification of Types of Hysteretic Damping of Materials

	Types of material damping		
Name used here	Dynamic hysteresis	Static hysteresis	
Other names	Viscoelastic, rheological, and rate-dependent hysteresis	Plastic, plastic flow, plastic strain, and rate-independent hysteresis	
Nature of stress-strain laws	Essentially linear. Differential equation involving stress, strain, and their time derivatives	Essentially nonlinear, but excludes time derivatives of stress or strain	
Special cases and description	*Anelasticity*. Special because no permanent set after sufficient time. Called "internal friction"		
Simplest representative mechanical model	VOIGT UNIT: C, k, ±F (ANELASTICITY); MAXWELL UNIT: k, C, ±F	D, k, ±F	k, D, ±F
Frequency dependence	Critically at relaxation peaks	No, unless other mechanisms present	
Primary mechanisms	Solute atoms, grain boundaries. Micro- and macro-thermal and eddy currents. Molecular curling and uncurling in polymers.	Magnetoelasticity	Plastic strain
Value of n in $D = JS^n$	2	3—up to coercive force	2–3 up to σ_L 2 to >30 above σ_L
Variation of η with stress	No change, since $n-2 = 0$	Proportional to σ since $n-2 = 1$	Small increase up to σ_L Large increase above σ_L
Typical values for η	Anelasticity: <0.001 to 0.01 Viscoelasticity: <0.1 to >1.5	0.01 to 0.08	0.001 to 0.05 up to σ_L 0.001 to >0.1 above σ_L
Stress range of engineering importance	Anelasticity—low stress Viscoelasticity—all stresses	Low and medium. Sometimes high	Medium and high stress
Effect of fatigue cycles	No effect	No effect	No effect up to σ_L Large changes above σ_L
Effect of temperature	Critical effects near relaxation peaks	Damping disappears at Curie Temp.	Mixed. Depends on type of comparison
Effect of static preload		Large reduction for small coercive force	Either little effect or increase

DYNAMIC HYSTERESIS OF VISCOELASTIC MATERIALS

The linearity limits of a variety of plastics and rubbers are summarized in Table 36.3. While the stress limits are of the same order of magnitude for plastics and rubbers, the strain limits are much smaller for the former class of materials. Within these limits the dynamic storage and loss moduli of linear viscoelasticity may be used.

Table 36.3. Linearity Limits for a Variety of Plastics and Rubber

Material	Stress limit in creep MPa	Strain limit in relaxation
Polymethylmethacrylate	10	
Polystyrene	5	
Plasticized polyvinyl chloride	1	0.1–1.0%
Polythene	12	
Phenolic resins	10	
Polyisobutylene		50%
Natural rubber	1–10	100%
GR-S		100%

Note: 1 MPa = 10^6 N/m² = 146.5 lb/in.²

One distinguishing characteristic of the dynamic behavior of viscoelastic materials is a strong dependence on temperature and frequency.[16] At high frequencies (or low temperature) the storage modulus is large, the loss modulus is small, and the behavior resembles that of a stiff ideal material. This is known as the "glassy" region in which the "molecular curling and uncurling" cannot occur rapidly enough to follow the stress. Thus the material behaves essentially "elastically." At low frequencies (or high temperature) the storage modulus and the loss modulus are both small. This is the "rubbery" region in which the molecular curling and uncurling follow the stress in phase, resulting in an equilibrium condition not conducive to energy dissipation. At intermediate frequencies and temperatures there is a "transition" region in which the loss modulus is largest. In this region the molecular curling and uncurling[16] is out of phase with the cyclic stress and the resulting lag in the cyclic strain provides a mechanism for dissipating damping energy. The loss factor also shows a peak in this region although at a somewhat lower frequency than the peak in G''. Since the damping energy is proportional to G'', the specific damping curve also has its maximum in the transition region. Most engineering problems involving vibration are associated with the transition and glassy regions. In Table 36.4, values of G' and G'' are given for a variety of rubbers and plastics. In many cases the references from which these values are quoted contain additional useful information.

Metals at low stress exhibit certain properties that constitute dynamic hysteresis effects. Peaks are observed in curves of loss factors vs. frequency of excitation. For example, under conditions that maximize the internal friction associated with grain boundary effects polycrystalline aluminum will display a loss factor peak as high as $\eta = 0.09$. But for most metals, the peak values are less than 0.01. Although the rheological properties of metals at low stress can be described in terms of anelastic properties (rheology without permanent set), a more general approach which includes provisions for permanent set is required to specify the rheological properties of metals at high stress. This approach is best described in terms of static hysteresis.

Table 36.4. Typical Moduli of Viscoelastic Materials

(Two values are given: the upper value is G′; the bottom value is G″. Moduli units are megapascals, MPa)

Material	Ref.	Frequency, cps				Temperature, °C
		10	100	1,000	4,000	
Polyisobutylene	33		0.512	1.31	2.36	−60–100
			0.410	1.76	4.50	
M 169A Butyl gum	35		0.480	1.40	2.70	21–65
			0.502	1.32	2.88	
Du Pont fluoro rubber,	33		2.00	4.54	7.93	0–100
(Viton A)			1.60	8.41	27.0	
Silicon rubber gum	35		0.05	0.08		21–65
			0.02	0.04		
Natural rubber	36		0.33	0.50		25
			0.02	0.02		
3M tape No. 466	33		0.81	2.52	15.3	25
(adhesive)			0.95	4.59	13.0	
3M tape No. 435	33		0.28	0.55	0.87	−40–60
(sound damping tape)			0.16	0.37	0.63	
Natural rubber	37	3.91	4.91			−30–75
(tread stock)		0.68	0.97			
Thiokol M-5	37	7.86	8.34			−30–75
		3.91	10.27			
Natural gum	37	0.73				−30–75
(tread stock)		0.07				
Filled silicone rubber	35		2.00	2.50	3.41	21–65
			0.26	0.44	0.58	
Polyvinyl chloride	35		1.26	3.20	6.60	21–65
acetate			1.44	2.32	5.78	
X7 Polymerized tung oil	35			17.0	39.0	21–65
with polyoxane liquid				9.45	20.8	
Du Pont X7775 pyralin	38	4.50	12.0	45.0		−45–100
		2.51	9.45	28.3		
Polyvinyl butyral	38	30.0	200.0	600.0		−45–100
		3.1	12.5	37.6		
Polyvinyl chloride with	39		0.35	0.65		
dimethyl thianthrene			0.21	0.97		

Note: 1 MPa = 10^6 N/m² = 10^{-3} kN/mm² = 146.5 lb/in.²

STATIC HYSTERESIS

The metals used in engineering practice exhibit little internal damping at low stress levels. At intermediate and high stress levels, however, magnetostriction and plastic strain can introduce appreciable damping. The former effect is considered first.

Ferromagnetic metals have significantly higher damping at intermediate stress levels than do nonferromagnetic metals. This is because of the rotation of the magnetic domain vectors produced by the alternating stress field.[16, 17] If the specimen is magnetized to saturation, most of the damping disappears, indicating that it was due primarily to magnetoelastic hysteresis. Figure 36.9 shows the loss factor of three metals, each heat-treated for maximum damping. The damping of 403 steel (ferromagnetic material with 12% Cr and 5% Ni) is much higher than that of 310 steel (nonferromagnetic with 25% Cr and 20% Ni). Most structural metals at low and intermediate stress exhibit loss factors in the general range of 310 steel until the hysteresis produced by plastic strain becomes significant. The alloy Nivco 10[17] (approximately 72% Co and

23% Ni), developed to take maximum advantage of magnetoelastic hysteresis, displays significantly larger damping than other metals.

The damping energy dissipated by magnetoelastic hysteresis increases as the third power of stress up to a stress level governed by the magnetomechanical coercive force; thus, the loss factor should increase linearly with stress. Nivco 10 follows this relationship for the entire range of stress shown in Fig. 36.9. Beyond an alternating stress governed by the magnetomechanical coercive force, i.e., beyond approximately 34.5

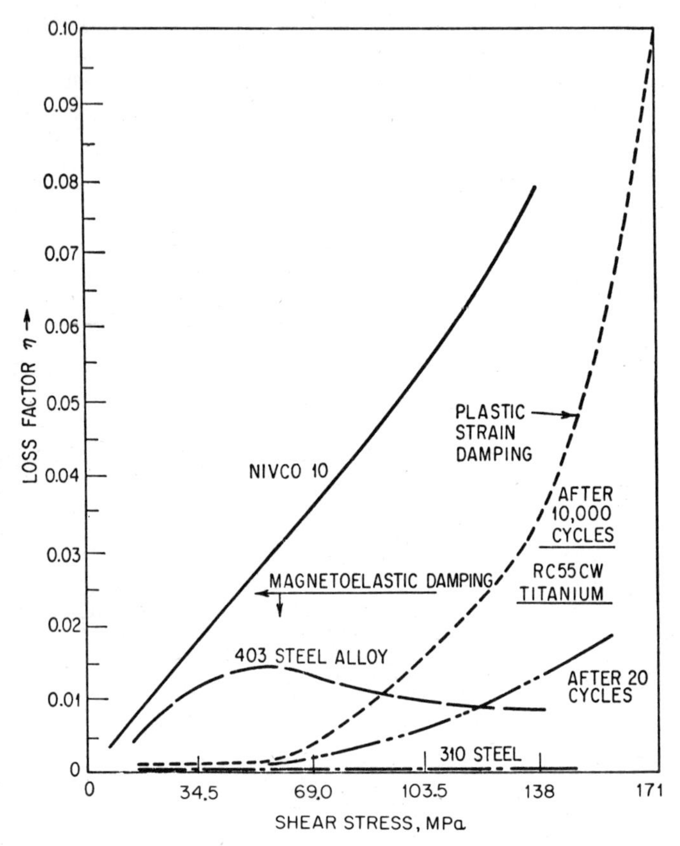

FIG. 36.9. Comparison of internal friction and damping values for different inelastic mechanisms.

MPa (5,000 lb/in²) for the 403 steel, the damping energy dissipated becomes constant. Since the elastic energy W_0 continues to increase as the square of the alternating stress, the value of loss factor (ratio of the two energies) decreases with the inverse square of stress. The curve for 403 steel in Fig. 36.9 at stresses between 62 MPa (9,000 lb/in²) and 103 MPa (15,000 lb/in²) demonstrates this behavior.

Magnetoelastic damping is independent of excitation frequency, at least in the frequency range that is of engineering interest. Magnetoelastic damping decreases only slightly with increasing temperature until the Curie temperature is reached, when it decreases rapidly to zero. Static stress superposed on alternating stress reduces magnetoelastic damping.[17, 18]

It is not entirely clear at this time what mechanisms are encompassed by the terms plastic strain, localized plastic deformation, crystal plasticity, and plastic flow in a range

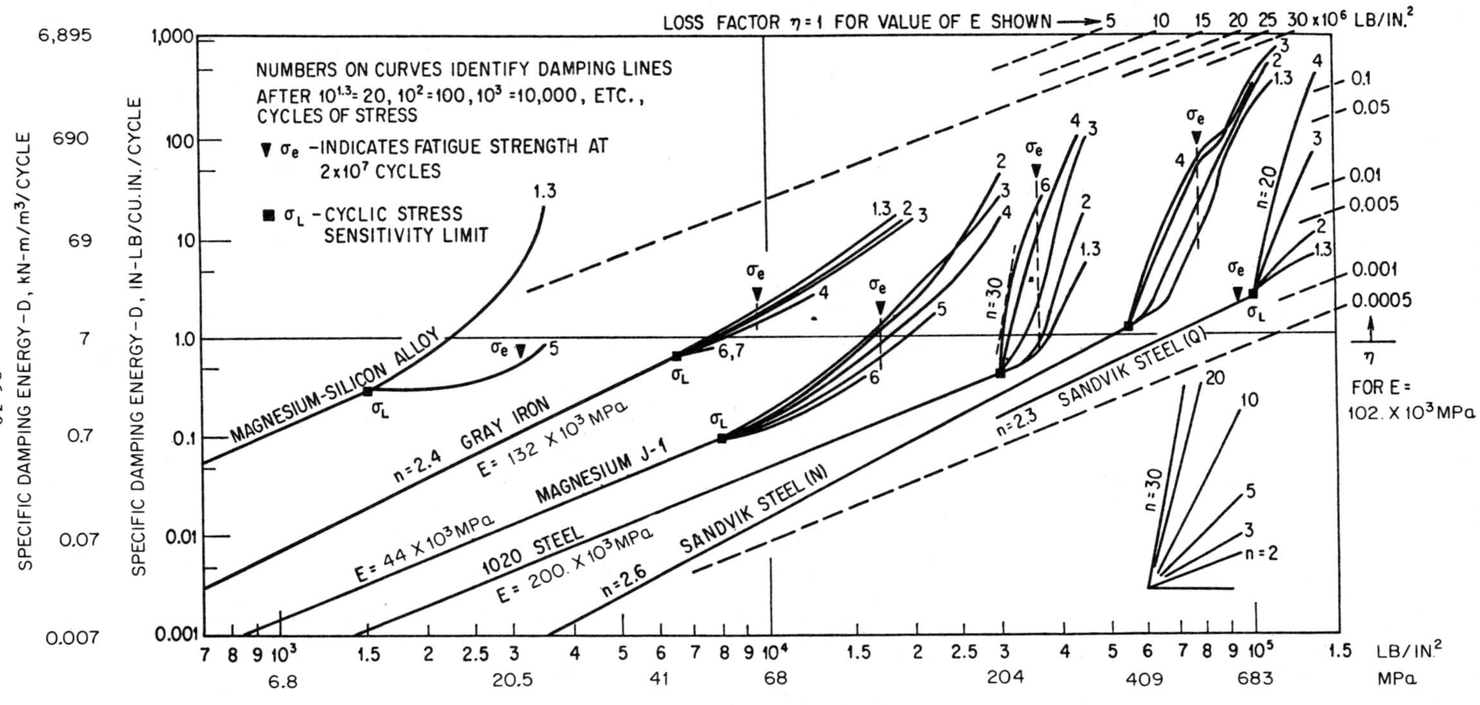

FIG. 36.10. Specific damping energy of various materials as a function of amplitude of reversed stress and number of fatigue cycles. Number of cycles is 10 to power indicated on curve. For example, a curved marked 3 is for 10^3 or 1,000 cycles. *Note:* 6.895 kN.m/m³ = 1 in.lb/in.³ and 1 MPa = 10^3 N/m² = 10^{-3} kN/mm² = 146.5 lb/in.²

of stress within the apparent elastic limit. On the microscopic scale, the inhomogeneity of stress distribution within crystals and the stress concentration at crystal boundary intersections produce local stress high enough to cause local plastic strain, even though the average (macroscopic) stress may be very low. The number and volume of local sites so affected probably increase rapidly with stress amplitude, particularly at stresses approaching the fatigue limit of a material. On the submicroscopic scale, the role of dislocations, their kind, number, dispersion, and lattice anchorage in the deformation process still remains to be determined. The processes involved in these various inelastic behaviors may be included under the general term "plastic strain."

At small and intermediate stress, the damping caused by plastic strain is small, probably of the same order as some of the internal friction peaks discussed previously and much smaller than magnetoelastic damping in many materials. In this stress region, damping generally is not affected by stress or strain history. However, as the stress is increased, the plastic strain mechanism becomes increasingly important and at stresses approaching the fatigue limit it begins to dominate as a damping mechanism. This is shown by the curves for titanium in Fig. 36.9.[18] In the region of high stress, microstructural changes and metallurgical instability appear to be initiated and promoted by cyclic stress. This occurs even though the stress amplitude may lie below the apparent elastic limit (that observed by conventional methods) and the fatigue limit of the material. This means that damping in the high stress region is a function not only of stress amplitude but also of stress history.

In Fig. 36.9, for example, the lower of the two curves for titanium indicates the damping of the virgin specimen and the upper curve gives the damping after 10,000 stress cycles.

The general position as regards stress history is given in Fig. 36.10. Below a certain peak stress, σ_L, known as the "cyclic stress sensitivity limit," the curve of damping vs. stress is a straight line on a log-log plot and displays no stress-history effect. The limit stress σ_L usually falls somewhat below the fatigue strength of the material. Above σ_L, stress-history effects appear; the curve labeled 1.3 indicates the damping energy after $10^{1.3} = 20$ cycles and the curve labeled 6 after 10^6 or 1 million cycles. To facilitate comparisons between the reference damping units, loss factor η and D under uniform stress ($\alpha/\beta = 1$), the loss factor also is plotted in Fig. 36.10. Since the relationship between D and η depends on the value of Young's modulus of elasticity E, a family of lines for the range of $E = 34 \times 10^3$ to 205.0×10^3 MPa (5×10^6 to 30×10^6 lb/in^2) is shown for $\eta = 1$. The lines for the other values of η correspond to a value of $E = 102.0 \times 10^3$ MPa (15×10^6 lb/in^2).

COMPARISON OF VARIOUS MATERIAL DAMPING MECHANISMS AND REPRESENTATIVE DATA FOR ENGINEERING MATERIALS

The general qualitative characteristics of the various types of damping are summarized in Table 36.2 by comparing the effects of different testing variables. The data tabulated indicate that, in general, anelastic mechanisms do not contribute significantly to total damping at intermediate and high stresses; in these regions magnetoelastic and plastic strain mechanisms probably are the most important from an engineering viewpoint.

Damping vs. stress ratio data have been determined for a variety of common structural materials at various temperatures.[2] Some of these data are listed in Table 36.5 (all tests at 0.33 cps). For a large variety of structural materials (not particularly selected for large magnetoelastic or plastic strain damping), the data are found to lie within a fairly well-established band shown in Fig. 36.11. The approximate geometric-mean curve is shown. Up to the fatigue limit, that is up to $\sigma_d = \sigma_e$, the specific damping energy D is given with sufficient accuracy by the expression

$$D = J\left(\frac{\sigma}{\sigma_e}\right)^{2.4} \qquad (36.23)$$

where $J = 6.8 \times 10^{-3}$

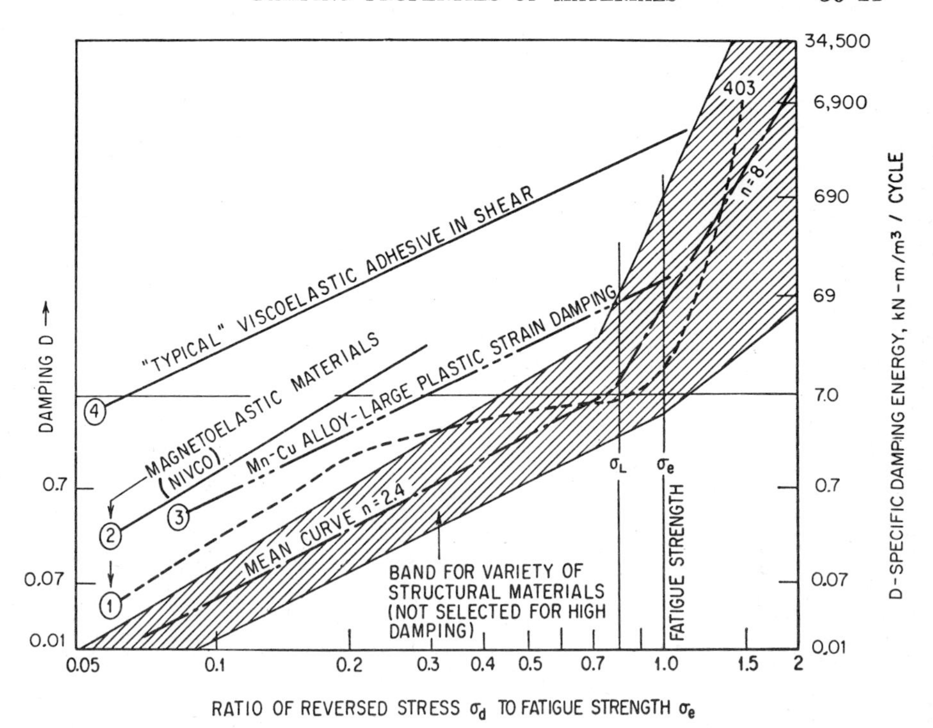

FIG. 36.11. Range of damping properties for a variety of structural materials. The shaded band defines the damping for most structural materials. 1 kN.m/m³ = 0.146 in.lb/in.³

if D is expressed in SI units of MN.m/m³/cycle and

$$J = 1.0$$

if D is expressed in units of in.lb/in.³/cycle.

The approximate bandwidth about this geometric mean curve for the various structural materials included in the band is as follows: from $\frac{1}{3}$ to 3 times the mean value at a stress ratio of 0.2 or less; from $\frac{1}{5}$ to 5 times at a ratio of 0.6; from $\frac{1}{10}$ to 10 times at a ratio of 1.0.

Also shown in Fig. 36.11 for comparison purposes are data for four materials having especially high damping. Materials 1 and 2 are the magnetoelastic alloys Nivco 10 and 403. Nivco 10 retains its high damping up to the stresses shown (data not available at higher stresses). However, the 403 alloy reaches its magnetoelastic peak at a stress ratio of approximately 0.2 and increases less rapidly beyond this point; when plastic strain damping becomes dominant (at stress ratio of approximately 0.8), damping increases very rapidly. By contrast, material 3, a manganese-copper alloy with large plastic strain damping, retains its high damping up to and beyond its fatigue strength.[20] Material 4 is a "typical" viscoelastic adhesive ($G'' = 0.95$ MPa $= 138$ lb/in²), assuming that the permissible cyclic shear strain is unity (experiments show that a shear strain of unity does not cause deterioration in this adhesive even after millions of cycles).[21] The magnetoelastic material has a damping thirty times as large as the average structural material in the stress range shown in Fig. 36.11, and the viscoelastic damping is over ten times as large as the magnetoelastic damping.

The range of D observed for common structural materials stressed at their fatigue limit is 0.003 to 0.7 MN.m/m³/cycle with a mean value of 0.05 (0.5 to 100 in. lb/in.³/cycle

Table 36.5. Static, Hysteretic, Elastic, and

Material*	Static properties			Fatigue behavior		
	Modulus of elasticity E, MPa 10^{-4}	Yield stress (0.2% offset), MPa	Tensile strength, MPa	Fatigue strength, σ_e, MPa	Cyclic stress sensitivity limit σ_L, MPa	Stress ratio σ_L/σ_e
N-155 (superalloy)	20.	410.	810.	360.	220.	0.62
Lapelloy (superalloy)	22.	764.	880.	490.	490.	1.00
Lapelloy (480°C)	17.5			270.	310.	1.14
RC 130B (titanium)	11.5	950.	1,040.	590.	650.	1.10
RC 130B (320°C)	9.9			430.	340.	0.81
Sandvik (O & T) steel	19.9	1,210.	1,400.	630.	680.	1.09
SAE 1020 steel	20.1	320.	490.	240.	200.	0.85
Gray iron	13.2		140.	65.	44.	0.69
24S-T4 aluminum	7.2	330.	500.	180.	160.	0.88
J-1 magnesium	4.4	230.	310.	120.	55.	0.47
Manganese-copper alloy		410.	610.	130.	120.	0.95

Note: 1 MPa = 10^6 N/m² = 146 lb/in.²
(Includes test temperature if above room temperature.)

with a mean value of 7). For materials stressed at a rate of 60 cps under uniform stress distribution (tension-compression), 16.4 cm³ (1 in³) of a typical material will safely absorb and dissipate 48 watts (0.064 hp). Some high damping materials can absorb almost 746 watts (1 hp) in the safe-stress range, assuming no significant frequency or stress-history effects.

SLIP DAMPING

INTRODUCTION

In some cases the hysteretic damping in a structural material is sufficient to keep resonant vibration stresses within reasonable limits. However, in many engineering designs, material damping is too small and structural damping must be considered. A structural damping mechanism which offers excellent potential for large energy dissipation is that associated with the interface shear at a structural joint.

The initial studies[22, 23, 24] on interface shear damping considered the case of Coulomb or dry friction. Under optimum pressure and geometry conditions, very large energy dissipation is possible at a joint interface. However, the application of the general concepts of optimum Coulomb interface damping to engineering structures introduces two new problems. First, if the configuration is optimum for maximum Coulomb damping, the resulting slip can lead to serious corrosion due to chafing; this may be worse than the high resonance amplification associated with small damping. Second, for many types of design configurations, the interface pressure or other design parameters must be carefully optimized initially and then accurately maintained during service; otherwise, a small shift from optimum conditions may lead to a pronounced reduction in total damping of the configuration.[22] Since it usually is difficult to maintain optimum pressure, particularly under fretting conditions, other types of interface treatment have been developed. One approach is to lubricate the interface surfaces. However, the maintenance of a lubricated surface often is difficult, particularly under the large normal pres-

Fatigue Properties of a Variety of Metals

Damping properties, kN.m/m³/cycle							
$D = J\left(\frac{\sigma}{\sigma_e}\right)^n$, $\sigma \leq \sigma_L$		D	D	D at $\sigma/\sigma_e = 1$		D at $\sigma/\sigma_e = 1.2$	
J	n, dimensionless	$\frac{\sigma}{\sigma_L} = 1$	$\frac{\sigma}{\sigma_e} = 0.6$	After $10^{1.3}$ cycles	After 10^6 cycles	After $10^{1.3}$ cycles	Maximum number of cycles
8.8	2.5	2.7	2.7	310.	170.	1,230.	1,500.
30.*	2.4*	10.9	4.0	11.	11.	55.	170.
24.	2.2	34.	8.2	26.	26.	41.	48.
14.	2.0	14.	4.4	12.	12.	18.	24.
17.	1.9	12.	6.1	13.	34.	30.	170.
16.	2.3	19.	5.5	16.	16.	31.	200.
4.3	2.0	3.1	1.6	4.5	140.	34.	680.
12.	2.4	4.5	3.4	14.	8.2	22.	16.
3.9	2.0	3.0	1.4	6.8	4.1	6.	15.
3.1	2.0	0.7	0.9	7.5	3.4	24.	7.
96.	2.8	82.	22.	89.	89.	170.	140.

Note: 1 kN.m/m³/cycle = 0.146 in.lb/in³/cycle.
* Up to σ = 96 MPa (14,000 lb/in²); at σ = 204 MPa (30,000 lb/in²) n = 1.5.

sure and shear sometimes necessary for high damping. Therefore, a more satisfactory form of interface treatment is an adhesive separator placed between mating surfaces at an interface. The function of the separating adhesive layer is to distort in shear and thus to dissipate energy with no significant Coulomb friction or sliding and therefore no fretting corrosion. The design of such layers is discussed in Chap. 37.

DAMPING BY SLIDING

The nature of interface shear damping can be explained by considering the behavior of two machine parts or structural elements which have been clamped together. The clamping force, whether it is the result of externally applied loads, of accelerations present in high-speed rotating machinery, or of a press fit, produces an interface common to the two parts. If an additional exciting force F_g is now gradually imposed, the two parts at first react as a single elastic body. There is shear on the interface but not enough to produce relative slip at any point. As F_g increases in magnitude, the resulting shearing traction at some places on the interface exceeds the limiting value permitted by the friction characteristics of the two mating surfaces. In these regions microscopic slip of adjacent points on opposite sides of the interface occurs. As a result, mechanical energy is converted into heat. If the mechanical energy is energy of free or forced vibration, damping occurs. The slipped region is local and does not, in general, extend over the entire interface. If it does extend over the entire interface, gross slip is said to occur. This usually is prevented by the geometry of the system.

The force-displacement relationship for systems with interface shear damping is shown in Fig. 36.12. Since there are many displacements which can be measured, the displacement which corresponds to the exciting force which acts on the system is taken as a basis. Then the product of displacement and exciting force, integrated over a complete cycle, is the work done by the exciting force and absorbed by the structural element. As shown in Fig. 36.13, there is an initial linear phase OP during which behavior is entirely elastic. This is followed, in general, by a nonlinear transition phase PB during which slip progresses across the contact area. The phase PB is nonlinear, not because of any plastic behavior, but simply because the specimen is changing in stiffness as slip progresses. After the nonlinear phase PB, there may be a second linear phase BC during which slip is present over the entire interface. The existence of such a phase requires some geometric constraint which prevents gross motion even after slip has progressed over the entire contact area. If no such constraint is provided, F_g cannot be allowed to exceed the value corresponding to point B. If it should exceed this gross value, slip would occur.

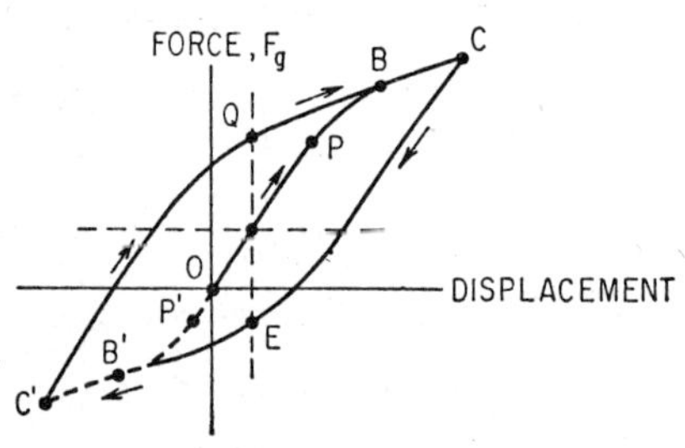

FIG. 36.12. Force-displacement hysteresis loop under Coulomb friction.

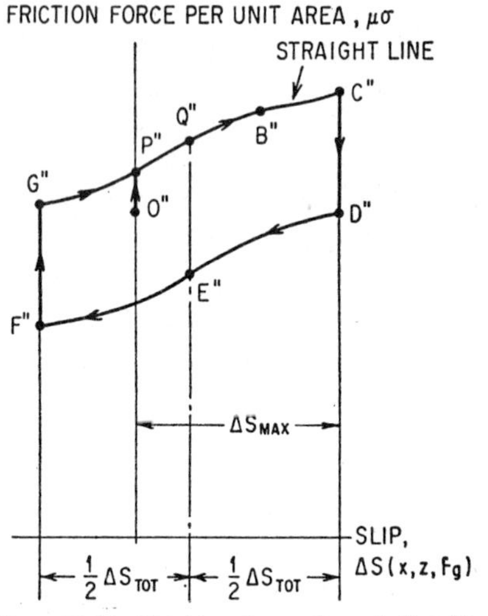

FIG. 36.13. Friction force-slip relationship under Coulomb friction.

If the clamping force itself does not produce any shear on the interface and if the exciting force does not affect the clamping pressure, the force-displacement curve is symmetrical about the origin O. These conditions are at least approximately fulfilled in many cases. If they are not fulfilled, the exciting force in one direction initiates slip at a different magnitude of load than the exciting force in the opposite direction. This is the case pictured in Fig. 36.12. With negative exciting force, slip is initiated at P' which corresponds to a force of considerably smaller magnitude than point P. However, the force-displacement curve is always symmetrical about the mid-point of PP' (intersection of dotted lines in Fig. 36.12).

The force-displacement curve has been followed from point O to point C. If now a reduction in the exciting force occurs, the curve proceeds from C in a direction parallel to its initial elastic phase. Eventually, as unloading proceeds, slip is initiated again. Its sense is now opposite to that which was produced by positive force. The curve continues to point B', where slip is complete, and then along a linear stretch to C', where the exciting force has its largest negative value. As the force reverses, the curve becomes again linear and parallel to OP. Slip eventually occurs again and covers the interface at B. The hysteresis loop is closed at C.

The energy dissipated in local slip can be found by computing the area enclosed by the force-displacement hysteresis loop. It usually is simpler, however, to determine the energy loss at a typical location on the interface by analysis, and then to integrate over the area of the interface. In this mode of procedure, interest centers on the frictional force per unit area $\mu\sigma$ and the relative displacement Δs of initially adjacent points on opposite sides of the interface.

The so-called "slip-curve" illustrating the relationship between $\mu\sigma$ and Δs is shown in Fig. 36.13. Before the exciting force is applied, conditions are represented by point O'' which corresponds to point O in Fig. 36.12. The initial elastic phase during which there is no slip is represented by $O''P''$ (note that the normal pressure σ may change during this phase). The phase during which slip occurs only over part of the interface is represented by the curved line $P''B''$; it corresponds to PB in Fig. 36.12. After slip has progressed over the entire interface, the normal force *vs.* relative-displacement relation is linear. This phase is represented by the curve $B''C''$ in Fig. 36.13 and by BC in Fig. 36.12. When the exciting force has reached its maximum value, a second nonslip phase $C''D''$ ensues. This is followed by slip along the curve $D''E''F''$ until the exciting force reaches its maximum negative value. As the exciting force completes its period, there is a nonslip phase $F''G''$ followed by slip along $G''C''$. The lengths $C''D''$ and $F''G''$ are equal and the curves $G''C''$ and $D''F''$ are congruent (F'' corresponds to C'' and D'' corresponds to G'').

If the point in question is at an element of area $dx\,dz$ of the xz interface, the energy dissipated in slip is proportional to the area enclosed by the slip curve. Because of the congruence of the curved portions of the diagram and the parallelism of the linear portions, this area can be expressed in terms of the total slip and the pressures at two instants during the loading cycle. Integrating over the entire interface,

$$D_0 = -\mu \iint [\sigma(E'') + \sigma(Q'')]\,\Delta s_{\text{tot}}\,dx\,dy \tag{36.24}$$

In this expression, the parameters $\sigma(E'')$ and $\sigma(Q'')$ and the total slip Δs_{tot} are functions of x and z. They are the normal stresses at points E'' and Q'' of Fig. 36.13, located midway between the vertical lines $G''F''$ and $C''D''$. Since the pressures σ are always compressive (negative) and the total slip is always taken as a positive quantity, the negative sign is required to ensure a positive energy dissipation. Equation (36.24) is of little engineering value in itself because the stresses are functions of F_g as well as of x and z. In many of the problems which are of design interest, however, the shear on the interface is produced primarily by the exciting force and not by the initial clamping pressure. Conversely, the clamping pressure is not greatly affected by the addition of the time-varying exciting force. Under these circumstances, the slip curve of Fig. 36.13, like the force-displacement curve of Fig. 36.12, is symmetric about the point O''. Points P'' and Q'' then coincide, and the mean ordinate of the slip curve is that corresponding to point O''. Then Eq. (36.24) reduces to

$$D_0 = -4\mu \iint \sigma(O'')\,\Delta s_{\max}\,dx\,dz \tag{36.25}$$

where $\sigma(O'')$ is the clamping stress corresponding to zero exciting force. It may be determined by any of the well-known methods of stress analysis. In most cases $\sigma(O'')$ can be determined without any reference to the existence of an interface. The term $\Delta s_{\max}$ represents the arc length of the maximum relative displacement, the so-called "scratch path." It is a function of the maximum value of F_g as well as of position on the interface. It may be inferred from Eq. (36.25) that energy dissipation due to interface shear is small both at very low clamping pressures and at very high ones. In the former case, $\sigma(O'') = 0$; in the latter case, $\Delta s_{\max} = 0$. It follows that for any distribution of clamping pressure there is an optimum intensity of clamping force at which the energy dissipation due to interface shear is a maximum. The maintenance of this optimum pressure is essential to the utilization of this form of damping. From the shape of the force-displacement curve *OPBC* shown in Fig. 36.12, it is evident that systems in which interface shear damping plays a significant role behave like softening springs. This means that instability and jump phenomena may occur at frequencies below the nominal resonant frequency.[5]

In the case of plane stress, the thickness of the material is t and Eq. (36.25) becomes

$$D_0 = -4\mu t \int \sigma(O'')\,\Delta s_{\max}\,dx \tag{36.26}$$

The slip can be related to stress through Hooke's law:

$$\Delta s = E^{-1} \int (\Delta\,\sigma_x)\,dx \tag{36.27}$$

This indicates that any discontinuity in displacement is associated with a discontinuity in the component of stress parallel to the interface. These displacement discontinuities due to slip are members of a class of generalized dislocations whose existence has been demonstrated theoretically.[25] If Eq. (36.27) is substituted in Eq. (36.26), the energy dissipation can be expressed in terms of stress alone:

$$D_0 = -4\mu E^{-1}t \int_0^l \sigma(O'') \left[\int_0^x (\Delta\, \sigma_x)_{\max}\, dx' \right] dx \tag{36.28}$$

The computation of energy dissipation per cycle D_0 is the first step in the prediction of the dynamic amplification factor to be expected in service. For interface shear damping, an elementary theory permits the dynamic amplification factor to be estimated even though the system behavior is nonlinear. The technique employs an averaging method. Denoting the displacement corresponding to the exciting force by the symbol v,

$$v = v_d \cos \omega t \qquad \text{and} \qquad F_g = F_m \cos(\omega t + \varphi) \tag{36.29}$$

where v_d is the peak dynamic displacement, F_m is the peak exciting force, and φ is the loss angle. One relationship between these quantities is obtained by making the average value of the virtual work vanish during each half-cycle of the steady-state forced vibration:

$$\int_0^{\pi/\omega} [mv + kv - F_g] \cos \omega t\, dt = 0 \tag{36.30}$$

In this integration, the stiffness k changes as slip progresses across the interface. If the hysteresis loop of Fig. 36.12 is replaced by a parallelogram, only two phases, elastic and fully slipped, need be considered. Denoting the stiffness (i.e., the ratio of exciting force to displacement) in the unslipped condition by the symbol k_e and the reduced stiffness in the fully slipped condition by the symbol k_s, the phase angle φ and the dynamic amplification factor A may be related by Eq. (36.30) to the duration of the elastic phase t':

$$\left(1 - \frac{k_s}{k_e}\right)(\omega t' + \sin \omega t') = \pi \left(\frac{m\omega^2 k_s}{k_e} + \frac{\cos \varphi}{A}\right) \tag{36.31}$$

where A is the conventional dynamic amplification factor, i.e., $A = v_d k_e / F_m$. The duration of the elastic phase is given by the first of Eqs. (36.29) with $v = v_d - 2v_s$, where v_s is the displacement at which slip first occurs. Then eliminating t' from Eq. (36.31):

$$\frac{\cos \varphi}{A} = \frac{1}{\pi}\left(1 - \frac{k_s}{k_e}\right)\left[2\frac{v_s k_e}{AF_m}\sqrt{1 + \frac{v_s k_e}{AF_m}} + \cos^{-1}\left(1 - 2\frac{v_s k_e}{AF_m}\right)\right] - \frac{m\omega^2 k_s}{k_e} \tag{36.32}$$

Equation (36.32) gives the relation between phase lag φ and amplification factor A. A second relationship between these quantities is found from the consideration that the energy dissipated during each half cycle of forced motion must be $D_0/2$:

$$\int_0^{\pi/\omega} F_g \frac{dv}{dt}\, dt = \tfrac{1}{2} D_0 \qquad \text{or} \qquad \sin \varphi = \frac{D_0 k_e}{\pi F_m{}^2 A} \tag{36.33}$$

Equations (36.32) and (36.33) serve to determine the dynamic amplification factor A, after D_0 has been estimated. Conversely, they serve to estimate the amount of energy which must be dissipated per cycle to produce a given reduction in the amplification factor by interface shear. A detailed analysis of response to a parallelogram hysteresis loop has been made.[40] Hysteresis loops other than parallelograms also have been studied.[26] At resonance, $\varphi = 90°$ and

$$A = A_r = \frac{D_0 k_e}{\pi F_m{}^2} \tag{36.34}$$

In general, the energy dissipation does not increase as rapidly as the square of the peak exciting force; consequently, the resonance amplification factor decreases as the exciting force increases. As a result, structures in which interface shear predominates tend to be self-limiting in their response to external excitation.

The foregoing discussion is based on the premise that changes in the exciting force do not materially affect the size of the contact area. There is an important class of problems for which this assumption is not valid, namely, those in which even the smallest exciting force produces some slip. An example of this type of joint is the press-fit bushing on a cylindrical shaft. If the ends of the shaft are subjected to a cyclic torque, part of this torque is transmitted to the bushing. Each part of the compound torque tube carries a moment proportional to its stiffness. Transmission of torque from the shaft to the bushing is effected by slip over the interface. The length of the slipped region grows in proportion to the applied torque. There is no initial elastic region such as OP or $O''P''$ in Figs. 36.12 and 36.13. If the peak value of the exciting torque is not too large, the fully slipped region BC or $B'C'$ in Fig. 36.12 never occurs. In these cases, Eqs. (36.31) to (36.34) are not applicable because there are no assignable constant values of k_s and k_e. A variety of simple cases of this type which occur in design practice have been analyzed.[27] They include the cylindrical shaft and bushing in tension and torsion, and the flexure of a beam with cover plate.

Another important case in which the smallest exciting force may produce slip arises in the contact of rounded solids. If these are pressed together by normal forces along the line joining their centers, a small contact region is formed. Subsequent application of a cyclic tangential force produces slip over a portion of the contact region even if the peak tangential force is not great enough to effect gross slip or sliding. This situation has been analyzed by Mindlin and by Deresiewicz.[45] Their conclusions have been verified experimentally by others.

REFERENCES

1. Phillips, E. M., and R. W. Weymouth: *SAE Paper* 438, January, 1955.
2. Lazan, B. J.: "Fatigue," chap. II, American Society for Metals, 1954.
3. Podnieks, E., and B. J. Lazan: *Wright Air Development Center Tech. Rept.* 55-284, August, 1955. Cochardt, A. W.: *J. Appl. Mechanics*, **21**:257 (1954).
4. Lazan, B. J.: *Trans. ASME*, **65**:87 (1943).
5. Goodman, L. E., and J. H. Klumpp: *Wright Air Development Center Tech. Rept.* 56-291, September, 1956.
6. Von Heydekampf, G. S.: *Proc. ASTM*, **31** (pt. II): 157 (1931).
7. Lazan, B. J.: *Trans. ASM*, **12**:499 (1950).
8. Maxwell, B.: *ASTM Bull.* 215, July, 1956, p. 76.
9. Lazan, B. J.: *Gen. Elec. Corp. Lab. Rept.* R 55GL129, January, 1955.
10. Nowick, A. S.: "Progress in Metal Physics," vol. 4, chap. I, p. 29, Interscience Publishers, Inc., New York, 1953.
11. Podnieks, E. R., and B. J. Lazan: *Wright Air Development Center Tech. Rept.* 56-44, 1956.
12. Lazan, B. J.: *J. Appl. Mechanics*, **20**:201 (1953).
13. Staverman, A. J., and F. Schwarzl: "Linear Deformation Behaviour of High Polymers," chap. I of "Theorie und molekulare Deutung technologischer Eigenschaften von hochpolymeren Werkstoffen," p. 71, Springer-Verlag, Berlin, 1956.
14. Zener, C.: "Elasticity and Anelasticity," University of Chicago Press, Chicago, 1948.
15. Wert, C.: "The Metallurgical Use of Anelasticity" in "Modern Research Techniques in Physical Metallurgy," American Society for Metals, Cleveland, Ohio, 1953.
16. Alfrey, T., Jr.: "Mechanical Behavior of High Polymers," Interscience Publishers, Inc., New York, 1948.
17. Cochardt, A.: *Scientific Paper* 8-0161-P7, Westinghouse Research Laboratories, West Pittsburgh, Pa., Nov. 6, 1956.
18. Person, N., and B. J. Lazan: *Proc. ASTM*, **56**:1399 (1956).
19. Demer, L. J., and B. J. Lazan: *Proc. ASTM*, **53**:839 (1953).
20. Torvik, P.: Appendix 72fg, *Status Rept.* 58-4 by B. J. Lazan, University of Minnesota, Wright Air Development Center, Dayton, Ohio, Contract AF-33(616)-2803, Dec. 31, 1958.
21. Whittier, J. S., and B. J. Lazan: Appendix B, *Prog. Rept.* 57-6, Wright Air Development Center, Dayton, Ohio, Contract AF-33(616)-2803, December, 1957.
22. Goodman, L. E., and J. H. Klumpp: *J. Appl. Mechanics*, **23**:421 (1956).
23. Lazan, B. J., and L. E. Goodman: "Shock and Vibration Instrumentation," p. 55, ASME, New York, 1956.
24. Pian, T. H. H., and F. C. Hallowell: *Proc. First U.S. Natl. Congr. Appl. Mechanics*, June, 1951, p. 97.

25. Bogdanoff, J. L.: *J. Appl. Phys.*, **21**:1258 (1950).
26. Rang, E.: *Wright Air Development Center Tech. Rept.* 59-121, February, 1959.
27. Goodman, L. E.: "A Review of Progress in Analysis of Interfacial Slip Damping," in "Structural Damping," American Society of Mechanical Engineers, New York, December, 1959.
28. Mentel, T. J.: *Wright Air Development Center Tech. Rept.* 58-547, December, 1958 (ASTIA Doc. 206,667).
29. Kerwin, E. M., Jr.: *J. Acoust. Soc. Amer.*, **31**:952 (1959).
30. Whittier, J. S.: *Wright Air Development Center Tech. Rept.* 58-568, May, 1959 (ASTIA Doc. 214,381).
31. Lazan, B. J.: *Proc. SESA*, **15**(1):1 (1957).
32. Kaelble, D. H.: private communication to the author.
33. Dalquist, C. A.: private communication to the author.
34. Fitzgerald, E., L. Grandine, and J. Ferry: *J. Appl. Phys.*, **24**:650 (1953).
35. Hopkins, I. L.: *Trans. ASME*, **73**:195 (1951).
36. Zapas, L. J., S. L. Shufler, and T. W. Dewitt: *J. Polymer Sci.*, **18**:245 (1955).
37. Phillipoff, W.: *J. Appl. Phys.*, **24**:685 (1953).
38. Rorden, H., and A. Grieco: *J. Appl. Phys.*, **22**:842 (1951).
39. Fitzgerald, E., and J. D. Ferry: *J. Colloid. Sci.*, **8**:1 (1953).
40. Caughey, T. K.: *J. Appl. Mechanics*, **27**:640 (1960).
41. Lazan, B. J.: "Damping of Materials and Members in Structural Mechanics," Pergamon Press, New York, 1968.
42. Flugge, W.: "Viscoelasticity," Blaisdell Publishing Company, a division of Ginn and Company, Waltham, Mass., 1967.
43. Lee, E. H.: Viscoelasticity, in W. Flugge (ed.), "Handbook of Engineering Mechanics," McGraw-Hill Book Company, New York, 1962.
44. Mindlin, R. D., W. F. Stubner, and H. L. Cooper: *Proc. Soc. Exp. Stress Anal.*, **5**(2), 69–87, 1948.
45. Deresiewicz, H.: Bodies in Contact with Applications to Granular Media, in G. Herrmann (ed.) "R. D. Mindlin and Applied Mechanics," Pergamon Press, New York, 1974.

37

VIBRATION CONTROL BY APPLIED DAMPING TREATMENTS

Robert Plunkett
University of Minnesota

PURPOSE OF SYSTEM DAMPING

Progress in mechanical design has been paced by the development of high-strength materials and efficient methods for fastening parts together. These have made possible designs which combine high energy densities, high speeds, and high stresses with very low mechanical system losses. Although this type of design is clearly useful, it does cause problems with dynamic stresses. It is fairly easy to design for steady dynamic stresses caused by centrifugal forces or impact but allows little room for amplification due to resonance.

All materials exhibit mechanical hysteresis under oscillatory strain, but most strong materials used for construction have a remarkably low loss factor for stresses less than half the yield. The actual mechanisms which lead to material damping are discussed later in this chapter, but if we eliminate such specific actions as thermoelastic and magnetoelastic damping, high strength steel and aluminum alloys normally have damping factors from 10^{-5} to 10^{-4}. If at the same time we use a design which eliminates joint damping by welding and then subject it to high dynamic loads with a broad frequency spectrum, we must deliberately add damping if we are to avoid excessive vibration. Fortunately it is possible to get surprisingly high system damping by the careful use of constrained and unconstrained damping treatments with an acceptable weight penalty and at low cost. The characteristics of most useful damping materials change markedly with frequency and temperature so that designs must be carefully made for each application. Chapter 36 discusses the mathematical analysis used for such design; this chapter considers the practical aspects. The various sections cover

1. How systems respond to different kinds of stimuli and effects of damping distribution
2. Physical mechanisms of material damping and effects of temperature and frequency
3. Mechanisms of system damping
4. Design configurations for system damping

Methods of controlling shock and vibration by the use of practical means of adding damping to a system are described. There are many additional sources of damping which are not discussed because they are not readily used in design or, like acoustic damping, are of limited applicability. Further information is contained in Ref. 23, which is a review of 272 published articles on damping methods, measurement, and analysis, in Ref. 22, which is a review of available information on material damping; in Ref. 12, which is a compilation of material damping properties; and in Ref. 1, which is a survey of the state of the art as of 1970.

EFFECTIVENESS OF SYSTEM DAMPING

EQUIVALENT DAMPING COEFFICIENTS

Classical vibration analysis was concerned with the oscillatory response of linear multiple degree-of-freedom or simple continuous systems excited by known periodic forces or motions. Viscous damping is defined for these systems by multiplying the velocity terms by constant coefficients. This type of equation may also be used with simple transient or random excitation. Structural damping, also called hysteretic or complex modulus damping, seems to give better agreement with the measured responses of large structures for typical low-frequency ranges. Either viscous or structural damping gives reasonable qualitative results for simple nonlinear systems.

The classical methods have been remarkably successful in spite of the fact that the damping or dissipative force is not really proportional to the relative velocity in any material, part, or system. The complex modulus or hysteretic description is reasonably valid for most metals, but pure material damping accounts for only a small fraction of the energy dissipation in most practical systems.

It has been shown[2] that for a single degree-of-freedom system with relatively small damping the motion is controlled by the mass and stiffness, so that one can calculate an equivalent linear damping constant; this technique has now been extended to multiple degree-of-freedom and continuous systems with the damping coefficient assumed to be a function of frequency in order to match the measured values. Even though this is a widely used and convenient technique, it is valid only if the damping forces are everywhere small in comparison with the spring and "inertia" forces.

Equivalent linear damping is found by assuming sinusoidal motion, calculating the energy dissipated by the actual nonlinear damping force acting through one cycle of the assumed motion, and finding the linear damping coefficient which would give the same energy dissipation. For a force which is a function of displacement only,

$$D_s = \int F\,dx = \int Fv\,dt$$

If x is a sinusoidal function of time,

$$x = a \cos \omega t$$

$$v = -\omega a \sin \omega t$$

Now if

$$F = -cv$$

$$D_s = \pi c \omega a^2$$

For example, if we have velocity-squared damping,

$$F = -c_1 v|v|$$

The velocity is multiplied by its absolute value to give the proper sign to the force.

$$\begin{aligned} D_s &= \int Fv\,dt \\ &= 4\int_0^{\pi/2} c_1\omega^2 a^3 \sin^3 (\omega t)\, d(\omega t) \\ &= \frac{8}{3} c_1 \omega^2 a^3 \end{aligned}$$

Equating

$$c_{\text{equiv}} = \frac{8}{3\pi}\omega a c_1 = \frac{8}{3\pi} c_1 v_{\max}$$

EFFECT OF DAMPING ON RESPONSE

If a single degree-of-freedom system is excited by a sinusoidal force or displacement at the frequency of maximum response, theory predicts the response to be inversely proportional to the damping coefficient for small damping. The basic definition of damping coefficient comes from the classical equation for the linear oscillator:

$$m\ddot{x} + c\dot{x} + kx = f(t) \tag{37.1}$$

This is usually put in nondimensional form as

$$\ddot{x} + 2\zeta\omega_n\dot{x} + \omega_n{}^2 = \frac{f(t)}{m} \tag{37.2}$$

where

$$\zeta = \frac{c}{c_c} \tag{37.3}$$

$$c_c = 2\sqrt{km} \tag{37.4}$$

$$\omega_n = \sqrt{\frac{k}{m}} \tag{37.5}$$

For steady-state sinusoidal motion, the maximum kinetic energy equals the maximum potential energy:

$$U = \tfrac{1}{2}mv_{\max}^2 = \tfrac{1}{2}m\omega_n{}^2x_{\max}^2 = \tfrac{1}{2}kx_{\max}^2 \tag{37.6}$$

The energy loss per cycle is

$$D_s = \pi c\omega_n x_{\max}^2 \tag{37.7}$$

Following the nomenclature of electronic circuit analysis, the quality factor is defined as

$$Q = \frac{D_s}{2\pi U} = \frac{1}{2\zeta} \tag{37.8}$$

and this is also the amplification factor at resonance for small damping ($Q > 2$).

If we use linear, structural damping with a complex spring constant, Eq. (37.1) becomes

$$m\ddot{x} + k'(1 + i\eta)x = f(t) \tag{37.9}$$

where the complex spring constant is defined as

$$k^* = k' + k'' = k'(1 + i\eta) \tag{37.10}$$

It can be shown that for sinusoidal motion

$$\eta = 2\zeta \tag{37.11}$$

If the right-hand side of Eq. (37.1) is set equal to zero, the amplitude of vibration decays exponentially:

$$x = x_0e^{-\zeta\omega_nt}\sin\left(\omega_n\sqrt{1 - \zeta^2}\,t\right) \tag{37.12}$$

The relative amplitude decrement per cycle is

$$\frac{x_{n+1}}{x_n} = e^{-\Delta} \approx (1 - \Delta) \tag{37.13}$$

where $\Delta = 2\pi\zeta$.

These relationships are shown in Table 37.1 along with some similar relationships for material volume damping. The transformations shown in this table are strictly valid only for small damping and single degree-of-freedom systems but they are useful for any system where the modal damping concept is valid (see Chap. 36).

If the right-hand side of Eq. (37.1) is a stationary, wide-band, random process with constant spectrum, the spectrum level of the response is also proportional to Q; but the rms level (a better measure of damage or noise) is proportional to $\sqrt{Q}$ (Chap. 11). If

Table 37.1. Classification of Recommended Damping Units, Their Interre Testing

(After

(1) Classification of type of damping units	(2) Preferred names	(3) Other names used	(4) Symbols and Units: Material	(4) Symbols and Units: Members	(5) Definition
Properties of materials and specimens					
A. Absolute energy units: Stored or dissipated per cycle of stress. Material properties: U and D give unit values. Specimen properties: U_s and D_s give total values for specimen	Unit elastic strain energy	Elastic energy Strain energy	U, in. lb/in.³ cycle		Area under σ_{mid}-ϵ curve from 0 to ϵ_{max}
	Total elastic strain energy			$U_s = UV_s\beta$, in. lb/cycle	Area under P_{mid}-X curve from 0 to P_{max}
	Unit damping energy	Specific damping Specific hysteresis	D, in. lb/in.³ cycle		Area within σ-ϵ hysteresis loop
	Total damping energy			$D_s = DV_s\alpha$, in. lb/cycle	Area within P-X hysteresis loop
B. Complex modulus notation: Appropriate for linear materials. Material properties: $E^* = E' + iE''$ Same for G^*, K^*, M^*. Properties of specimen, member, or total part, $k_s^* = k_s' + k_s''$	Complex Modulus		E^*, G^*, K^*, M^*, lb/in²	k_s^*, lb/in	$E^* = \|E^*\|e^{i\phi} = E'(1 + i\eta)$
	Absolute Modulus		$\|E^*\|$, $\|G^*\|$, $\|R^*\|$, $\|M^*\|$	$\|k_s^*\|$	$\|E^*\| = [E'^2 + E''^2]^{1/2}$
	Storage Modulus	Elastic modulus Real modulus	E', G', K', M', lb/in²	k_s', lb/in	$E' = \|E^*\| \cos\phi$
	Loss Modulus	Dissipation Modulus	E'', G'', K'', M'', lb/in²	k_s'', lb/in	$E'' = \|E^*\| \sin\phi$
C. Relative energy units: Dimensionless ratios of damping energy D and strain energy U	Loss Coefficient	Loss factor Damping factor	η	$\eta_s = \eta\dfrac{\alpha}{\beta}$	$\eta = \dfrac{D}{2\pi U}$; $\eta_s = \dfrac{D_s}{2\pi U_s}$
	Quality Factor	Storage coefficient Q Factor	Q	$Q_s = Q\dfrac{\beta}{\alpha}$	$Q = \dfrac{2\pi U}{D}$; $Q_s = \dfrac{2\pi U_s}{D_s}$
Damping properties of dynamic systems					
D. Temporal decay of free vibrations in members: $X_a(t) = X_a(0)e^{-\upsilon_t t}$; $\upsilon_t = \left(\frac{1}{t}\right)\ln\frac{X_a(0)}{X_a(t)}$; $\frac{(db)}{t} = 20\log\frac{X_a(0)}{X_a(t)} = 20\log e^{\upsilon_t t} = 8.68\upsilon_t t$			Measured quantities or properties of systems		How measured
	Logarithmic decrement	Decrement	Δ_t		$X_a(n+1) - X_a(n)e^{-\Delta_t}$; $\Delta_t = \ln\dfrac{X_a(n)}{X_a(n+1)}$
	Decay constant	Temporal decay constant	υ_t, (in/in)/sec.		$\upsilon_t = \dfrac{1}{t}\ln\dfrac{X_a(0)}{X_a(t)}$
		Decibel decay rate with time	Υ_t (db)/sec		$\Upsilon_t = 8.68\upsilon_t$
E. Spatial attenuation of waves in slender rods, beams, etc., having distributed parameters: $X_a(y) = X_a(0)e^{-\upsilon_y y}$; $\upsilon_y = \left(\frac{1}{y}\right)\ln\frac{X_a(0)}{X_a(y)}$; $\frac{(db)}{y} = 20\log\frac{X_a(0)}{X_a(y)} = 20\log e^{\upsilon_y y} = 8.68\upsilon_y y$	Logarithmic attenuation	Attenuation	Δ_y(or Δ_z)		$X_a(\lambda) - X_a(0)e^{-\Delta_x}$; $\Delta_y = \ln\dfrac{X_a(0)}{X_a(\lambda)}$
	Attenuation constant	Spatial attenuation constant	υ_y, (in/in) per in		$\upsilon_y = \dfrac{1}{y}\ln\dfrac{X_a(0)}{X_a(y)}$
		Decibel attenuation rate in space	Υ_y (db)/in		$\Upsilon_y = 8.68\upsilon_y$

F. Near-resonance response under sinusoidal loading:
Relations for single degree-of-freedom system (member-mass system) with linear properties.
Vibration phase angle ϕ: If exciting frequency much smaller than resonance ($\omega \ll \omega_n$, $\beta \approx 0$): $\phi = \arctan \eta_s$.
Resonance amplification factor A_r and bluntness: $1/A_r = B_s = B_p = (1/\sqrt{3}\,\omega_n)\,(\Delta\omega)_{0.5} = (1/\omega_n)\,(\Delta\omega)_{0.707}$

Notes:
1. For linear materials $\beta/\omega = 1$. Thus $\eta_s = \eta$, $Q_s = Q$, etc.
2. Some relations given above are accurate only if damping is small (say $\eta_s < 0.2$).
3. For longitudinal shear, torsion, and compressive nondispersive waves, $b = 1$. For simple bending waves, $b = 2$.

lations, and Their Relations to the Measured Quantities of Damping Systems

Lazan[6])

Relations of Properties of Linear Specimens to System Properties.

$$\eta_s = \frac{D_s}{2\pi U_s} = \frac{1}{Q_s} = \frac{1}{A_r} = \tan\delta = \frac{\Delta_t}{\pi} = \frac{\psi_s}{2\pi} = 2\zeta = \frac{k_s''}{k_s'}$$

Relations among unit properties of uniform materials

$E' = \frac{2U}{\epsilon^2}$		
$E'' = \frac{D}{\pi\epsilon^2}$		
$\eta = \frac{D}{2\pi U}$	$\eta = \frac{E''}{E'}$	
$Q = \frac{2\pi U}{D}$	$Q = \frac{E'}{E''}$	$Q = \frac{1}{\eta}$

Relations between system properties and properties of specimens — Relations among system properties for linear materials

$\frac{D_s}{U_s}$	$\frac{k_s''}{k_s'}$	η_s	Q_s		Δ_t	υ_t	Υ_t
$\Delta_t = \frac{D_s}{2U_s}$	$\Delta_t = \pi\frac{k_s''}{k_s'}$	$\Delta_t = \pi\eta_s$	$\Delta_t = \frac{\pi}{Q_s}$	Δ_t			
$\upsilon_t = \frac{\omega D_s}{4\pi U_s}$	$\upsilon_t = \frac{\omega}{2}\frac{k_s''}{k_s'}$	$\upsilon_t = \frac{\omega}{2}\eta_s$	$\upsilon_t = \frac{\omega}{2Q_s}$	υ_t	$\upsilon_t = \frac{\omega}{2\pi}\Delta_t$		
$\Upsilon_t = .69\omega\frac{D_s}{U_s}$	$\Upsilon_t = 4.35\omega\frac{k_s''}{k_s'}$	$\Upsilon_t = 4.35\omega\eta_s$	$\Upsilon_t = \frac{4.35\omega}{Q_s}$	Υ_t	$\Upsilon_t = 1.38\omega\Delta_t$	$\Upsilon_t = 8.68\upsilon_t$	
$\Delta_y = \frac{1}{2b}\frac{D_s}{U_s}$	$\Delta_y = \frac{\pi}{b}\frac{k_s''}{k_s'}$	$\Delta_y = \frac{\pi}{b}\eta_s$	$\Delta_y = \frac{\pi}{b}\frac{1}{Q_s}$	Δ_y	$\Delta_y = \frac{1}{b}\Delta_t$	$\Delta_y = \frac{2\pi}{b\omega}\upsilon_t$	$\Delta_y = \frac{0.73}{b\omega}\Upsilon_t$
$\upsilon_y = \frac{1}{2b\lambda}\frac{D_s}{U_s}$	$\upsilon_y = \frac{\pi}{b\lambda}\frac{k_s''}{k_s'}$	$\upsilon_y = \frac{\pi}{b\lambda}\eta_s$	$\upsilon_y = \frac{\pi}{b\lambda}\frac{1}{Q_s}$	υ_y	$\upsilon_y = \frac{1}{b\lambda}\Delta_t$	$\upsilon_y = \frac{2\pi}{b\lambda\omega}\upsilon_t$	$\upsilon_y = \frac{0.73}{b\lambda\omega}\Upsilon_t$
$\Upsilon_y = \frac{4.35}{b\lambda}\frac{D_s}{U_s}$	$\Upsilon_y = \frac{27.3}{b\lambda}\frac{k_s''}{k_s'}$	$\Upsilon_y = \frac{27.3}{b\lambda}\eta_s$	$\Upsilon_y = \frac{27.3}{b\lambda}\frac{1}{Q_s}$	Υ_y	$\Upsilon_y = \frac{8.68}{b\lambda}\Delta_t$	$\Upsilon_y = \frac{54.6}{b\lambda\omega}\upsilon_t$	$\Upsilon_y = \frac{2\pi}{b\lambda\omega}\Upsilon_t$

Equations expressed in terms of symbols indicated (see note *c*)

the system is given an abrupt velocity change (step-velocity shock), the maximum response is changed by less than a factor of 2 as the damping is increased from $\eta = 0$ to $\eta = 0.5$[3]. In spite of this, damping can have a very profound influence on the mean amplitude due to repeated impacts (such as hammering) or transient excitation of an oscillatory nature (such as earthquakes) (Fig. 37.1).

Damping controls the stability of linear self-excited systems (Chap. 5). The amplitude grows without limit for damping less than critical; self-excited vibrations are not possible for damping greater than critical. All real systems have a maximum amplitude when self-excited because they are at least slightly nonlinear, so that the usual experience with self-excited vibrations is that the amplitude is almost unaffected by increases in damping until the damping approaches the critical value and then the vibration decreases from almost full amplitude to an unmeasurably small level with a very small increase in damping factor.

The influence of damping factor on the response of even a simple quasi-linear single degree-of-freedom system thus depends on the type of excitation. It is even more complicated for multiple degree-of-freedom systems with some nonlinearity.

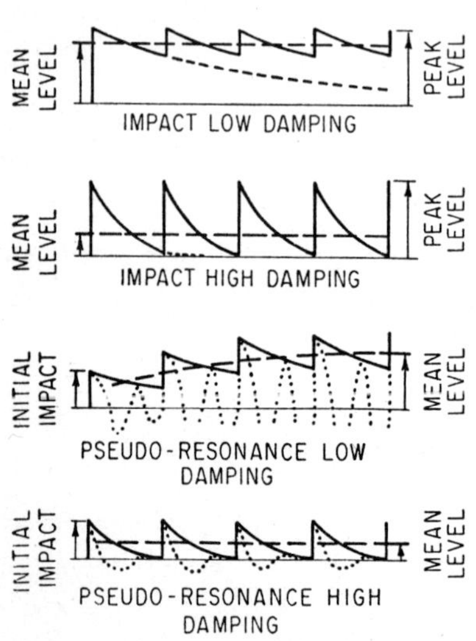

FIG. 37.1. Schematics illustrating the influence of damping in reducing the resonant noise of periodic impacts. In the bottom pair of traces, there is assumed to be an integral relation between natural period of vibration and periodicity of impact.

INSTRUMENTATION RESPONSE

The mechanical parts of most analog-type instruments such as pressure gages, mechanical vibrometers, speedometers, etc., are carefully designed to have linear response and only one important frequency. This is also true for electromechanical (galvanometers, accelerometers) and electroacoustic (microphones, loudspeakers) systems.

These systems must have either high damping or narrow-band excitation to avoid resonant amplitude distortion. Unless the damping is fairly close to $\zeta = 0.7$, there will also be phase distortion unless the maximum excitation frequency is less than one-third the resonant frequency.

DISTRIBUTION OF DAMPING

The concept of modal equivalent damping is valid if the response shape of the actual system does not differ appreciably from that which is computed for the undamped linear system. If, however, local damping forces are comparable to the effective spring forces at some particular frequency, the response shape can change and an increase in damping coefficient may cause an increase in dynamic response. Since a sufficiently small amount of damping must always cause a decrease in response and since the response is a continuous function of damping factor for a linear system, there is a specific value of local damping which causes a minimum dynamic response.[4] If the local damping is increased above this critical value, the effective system damping decreases and the system response increases but in a somewhat different vibration mode; this phenomenon is known as overdamping. It has been shown rigorously (see Ref. 40, Chap. 36) that damping distributed proportionately to stiffness or mass leads to modal damping with no interaction; in this case, overdamping cannot occur. Some limited studies[5] have shown that uniformly distributed damping is almost as efficient as optimally distributed damping.

Therefore in most cases surface damping treatments should be applied as uniformly as possible over the whole structure. The one exception to this philosophy is where the objective is only to damp lightly a large structure with a very high Q, such as a space vehicle in a high vacuum, in the most efficient manner with the least weight penalty. In this case the most efficient and effective damping mechanism is one or more damped dynamic absorbers,[3] but the design must be carefully done. It is extremely important to get the correct frequency, mass, location, and damping for the absorbers used.

NONLINEAR EFFECTS

Most nonlinear effects in damping are beneficial since the effective damping usually increases with increased stress or increased amplitude.[6] The only practical case where this does not happen is for Coulomb friction damping. It has been shown[7] that in the classical case of a single degree-of-freedom system with constant-force frictional damping, the effective damping decreases with increasing vibration amplitude up to the point where the damping force no longer controls the amplitude at all because the effective excitation force exceeds the friction force. To make the classical problem worse, friction force is actually dependent on motion history, temperature, and relative surface velocity; unfortunately it usually decreases to some minimum value under the influence of any of these parameters. Design based on energy dissipation due to deliberate slipping and dry friction should be very conservative and carefully tested under extreme conditions if it is to be successful. Cross slip can reduce the effective damping still further;[8] skidding of automobiles is a typical example of the interactive problems of cross slip. An automobile which is overbraked or overaccelerated first slips slightly in the traction direction and then slides very easily laterally; one which starts to slide laterally loses braking ability in the traction direction.

The effective stiffness of most mechanical systems is reduced if they are deformed beyond the elastic limit; this can have the effect of making the actual response to certain kinds of transient excitation greater than would normally be anticipated. Such behavior is fortunately rather rare if the amplitude predicted by linear analysis using the actual excitation is only slightly greater than that necessary to cause yielding. Since most designs are supposed to keep the system safely elastic, overloads are usually due to unexpectedly large excitation; unless the designer is incurably optimistic, nonlinear amplitude amplification effects are very unlikely.

Methods for finding the equivalent linear damping corresponding to inherently nonlinear damping were discussed in earlier sections of this chapter. Since the effective damping increases with amplitude for all but dry friction damping, the only requirement for conservative design is to be certain that the amplitude used for computing the equivalent damping is less than the maximum amplitude calculated by using it.

MATERIAL PROPERTIES

METALS

The damping parameter of interest for lightly damped vibratory systems is the loss factor, η, or other measures of the same property, such as fraction of critical damping, ζ, quality factor, Q, or log decrement, Δ (Table 37.1). Structural metals and alloys usually have very low damping ($\eta < 10^{-3}$) except for specifically developed high damping materials, such as certain manganese-copper,[9,10] iron-aluminum, and magnesium alloys.[11] The primary mechanism for damping in metals at stresses below the cyclic stress sensitivity limit is the energy loss associated with dislocation motion back and forth between grain boundaries. This effect is so remarkably small that η values for steel and duralumin greater than 10^{-4} at stresses less than one-quarter of yield are suspect as being influenced by the test procedure. Reference 12 gives a compilation of material damping data available in the open literature. In addition to higher damping for the special alloys noted above and castings of iron, aluminum, and magnesium, aluminum alloys in bending may show values for η approaching 0.01 due to thermoelastic heat transfer.[13]

Since the energy loss due to cyclic straining in most metals is very much smaller than that due to relative motion of the parts, the exact value of η for metals is seldom of any practical interest in design. Reference 6 shows how to find the contribution to system damping of a nonuniformly stressed material using a stress volume function. If a small portion such as a fillet is stressed so much that the dissipation per unit volume is very high, the volume involved is usually so small that the total energy dissipated is still negligible in comparison with the vibratory energy. As a result, it is extremely rare for energy loss or loss factor of structural metals to be of practical importance in controlling the effective damping of a mechanical system.

ELASTOMERS

If the system damping due to relative motion is insufficient to control the dynamic response, it is normally increased by adding a constrained or an unconstrained elastomeric damping coating. In Chapter 36, it is shown that the energy loss per unit volume of a material is equal to

$$D = \pi \gamma_0^2 G'' \tag{37.14}$$

$$D = \pi \gamma_0^2 \eta G' \tag{37.15}$$

η by itself is not a good measure of the damping effectiveness of such a coating material. If γ can be increased by suitable configurations or constraints (Ref. 14 and the last section of this chapter), the effective damping is greatly increased. Since all elastomers have moduli much lower than that in metals, the strain levels are usually controlled by the rest of the structure and not by the modulus of the elastomer. Therefore, the energy loss in the elastomer is normally proportional to the loss modulus G'', Eq. (37.14), or the product of the loss factor η and the storage modulus G', Eq. (37.15). For this reason, a material with a large loss factor and a small modulus is less effective than one with a smaller loss factor if the product leads to a larger loss modulus.

Both the storage modulus and the loss modulus of polymeric materials depend very markedly on temperature and vibration frequency. A master curve can be drawn for most materials showing this variation as a function of a single parameter involving both temperature and frequency.[15] The storage modulus and loss factor are typically shown as functions of the logarithm of the reduced frequency, as in Fig. 37.2. In this case the abscissa is defined as

$$x = \log \frac{f}{f_0} + \frac{Q}{RT} \tag{37.16}$$

where f = frequency
f_0 = reference frequency
Q = activation energy
R = gas constant
T = absolute temperature

The Arrhenius exponential relationship must hold over the temperature and frequency range of interest for this simple form to be valid. Where this is not strictly true, an apparent activation energy as a function of temperature may be used or one may simply write

$$x = \log \frac{f}{f_0} + \alpha(t) \tag{37.17}$$

where α is an experimentally determined function of T.

In any case, as seen from Fig. 37.2, most elastomers pass through three stages as the temperature is increased. At low temperature they are very stiff and lossless—the glassy region. The loss factor reaches a maximum at some intermediate critical temperature in the transition region. The storage modulus decreases monotonically with increasing temperature so that the loss modulus, the product of the two, has a maximum at a temperature somewhat lower than that for the peak loss factor. An increase in frequency of vibration of 10:1 will typically shift all curves to higher temperatures by 10 to 30°C depending upon the activation energy. If a given polymeric material has its

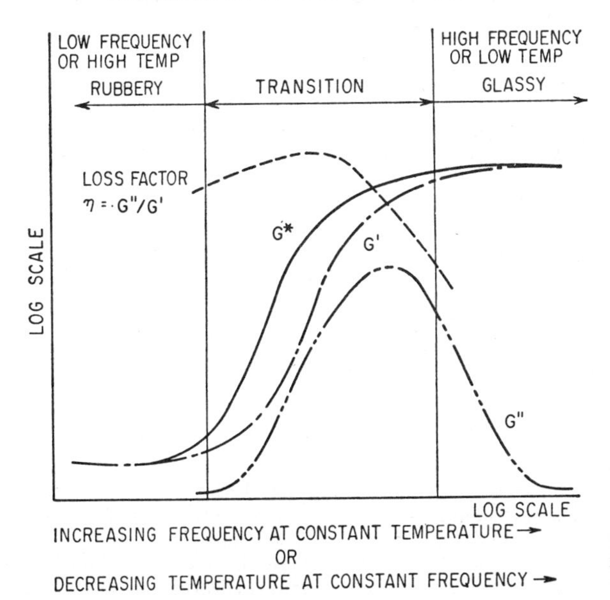

FIG. 37.2. Dependence of storage modulus, loss modulus, and loss factor η on temperature and frequency.

peak loss modulus at about room temperature (20°C) at 100 cps, it might peak at about 40°C at 1,000 cps. This could easily mean a decrease in damping effectiveness of 2:1 at room temperature for the higher frequency.

MECHANISMS OF SYSTEM DAMPING

TYPES OF SYSTEM DAMPING

As we have seen previously, any process by which vibratory energy at a given frequency can be removed from a system or changed to energy at another frequency is an effective damping process for the vibrational mode of the system corresponding to that frequency. Nonlinear damping mechanisms usually transform vibratory energy from one frequency to another, usually higher, where it is more easily dissipated. Linear mechanisms transmit the energy, at the same frequency, out of the system to some other place where they either radiate indefinitely or dissipate. Dissipative mechanisms transform the energy directly into heat. For design purposes it is simpler to classify damping mechanisms as material, radiation and relative motion damping rather than as linear or nonlinear behavior.

MATERIAL DAMPING

It was shown in the previous section that hysteretic dissipation of energy in metals is seldom an important factor in vibration control. The few exceptions are those in which strength, weight, or cost are not the controlling factors so that highly dissipative metals which are heavy or costly, such as lead, special manganese bronze, or cast metals (other than cast steel), may be used as structural elements. Lead occupies a special position since it has a density greater than most metals and creeps at room temperature. It may be used in compression when constrained by sandwiching it with stronger metals like steel or aluminum. It is also used for acoustical curtains where strength is not a primary factor.

Not all polymeric or other organic materials have high energy dissipation. Those which do usually have a rather low shear loss modulus (1 to 20 MPa or 150 to 3,000 psi) and a high loss factor (η = 0.1 to 2). The loss factor for simple stress (tension or bending) is only slightly less than that for shear[3], and Young's modulus is almost three times the shear modulus. Those organic materials which have a high modulus (over 50 MPa or 70,000 psi) are in the glassy region (Fig. 37.2) and have a very low loss factor so that they are no more effective in vibration control than are metals. The low modulus polymers and comonomers will survive vibratory shear strain at very high amplitudes; steady-state strains of 0.1 present no problem; and some soft materials will last indefinitely at shear strains greater than 1. In the next section we examine methods for inducing such large strains. These elastomeric materials are very temperature-sensitive and must be carefully compounded for the temperature and frequency range of interest. Fortunately, most practical applications for vibration control lie in the limited frequency range of 1 to 1,000 cps. If we choose a material for which the loss modulus peaks at the design temperature and 30 cps, it will normally drop off by about 2:1 at ±25°C or 10:1 in frequency or some combination of the two. Elastomeric materials can be designed to peak anywhere in the temperature range of −25 to 100°C with some experimental materials extending this range on both ends. They are usable to higher and lower temperatures at reduced effectiveness and some of them will survive short exposures to temperatures as high as 300°C. Since the properties are so sensitive to frequency, temperature, and aging, the designer should work from manufacturer's specification sheets and use the figures above only as a guide. *Sound and Vibration* magazine publishes an annual buyer's guide for materials (July issue) which should be consulted for current listings of manufacturers. Similar listings may also be found in more general annual buyer's guides such as *Sweet's*, *Aviation Week*, or *Science*.

Materials for vibration control have not been so fully investigated at higher temperatures. The inherent damping of most metals increases markedly as the temperatures rise into the creep range. It also appears that inorganic glasses such as porcelain enamels have high loss moduli (G'' of about 3 GPa or 400,000 psi) over a temperature range of 50 to 100°C at temperatures just below the melting point. Again, these materials may be designed to peak at any temperature from 400 to 1000°C. This application may develop rapidly over the next few years, so the designer should consult with the research departments of commercial manufacturers of porcelain enamel frits.

RADIATION DAMPING

One seldom designs for vibration control by radiating energy out of the system of interest. Nevertheless, the designer should be aware of these effects both because they contribute a large share of the damping in normal design and because their presence must be considered in deciding about the effectiveness of added damping. It can be shown that a traveling bending wave in an infinite lossless plate will not radiate energy below some critical (coincidence) frequency (1000 cps in air for 1 cm of steel). Above that frequency the energy radiated will increase approximately linearly with frequency. For thinner metal plates, the critical frequency increases inversely with the thickness, so that the direct radiation mechanism is of little importance for very large, flat, thin plates. This phenomenon is substantially affected by free edges and corners; in addition, the energy radiated is supplemented by the energy dissipated in the viscous flow of the fluid around the edges so that the damping factor of a plate with free edges vibrating in bending can easily reach values of 0.001 to 0.01 in either air or water. The radiation damping from cylinders and ovoids in rigid body and elastic vibration has been the subject of a large number of mainly mathematical technical papers which should be consulted for further information.

Another source of radiation damping is radiation between parts of a solid structure or from one structure to another. A cantilever beam, such as a turbine bucket on a wheel or a tall building on the earth, radiates elastic energy into its foundation as it vibrates. If the excitation is transient or if the foundation is very much larger than the vibrating structure and slightly dissipative, this energy is not reflected back and therefore damps the original vibration. This phenomenon also can account for damping factors of several tenths of a percent. Unless added damping can increase the energy loss by a measurable amount more than this, it may appear that no damping is actually being added. As one homely example of this phenomenon, if a steel ball bearing is dropped on a large mass of steel or rock (such as a foundation) from a height of several feet, the coefficient of restitution is about 0.96 even though the impact is purely elastic. The difference from 1.00 is accounted for by the energy radiated away from the point of impact.

RELATIVE MOTION DAMPING

By far the most important source of damping stems from relative motion between different parts of a large structure. We have already commented on the energy dissipation to be had from friction between parts sliding on each other. Even partial slip can dissipate enough energy to give a loss factor of several tenths of a percent. Partial slip occurs when two elastic members (all real materials are deformable) are subject to a vibratory shear force. Since the normal pressure on any but the most carefully ground and polished parts must decrease to zero at the edge of the area of contact, there will be some slip near the edges. The amount of energy dissipated depends in a very complicated and nonlinear fashion on both the normal and tangential forces; but for any given vibratory tangential force, there is one value of normal force which gives maximum energy dissipation or damping. A larger normal force will inhibit motion and a smaller one reduces the friction force. For the same reason, for any given normal force, there is one vibration amplitude for which the damping factor is a maximum. Any change in surface properties by fretting or corrosion usually tends to decrease the energy dissipation.

Impact is another source of energy dissipation. Once more, this mechanism decreases with increasing age due to wear and polishing of the parts. Impact dampers have been used in some applications to increase damping. Cantilever rods have been inserted along the length of turbine buckets to detune them and increase the damping by impact. Loose metal spheres have been built into aircraft control surfaces for damping purposes. Loose particles have been used in boring bars in machine tools for the same reason. Dry sand is sometimes used in hollow walls to make them deader. In all cases, great care must be taken to keep the sand or other particles from packing and losing their damping effect. There are as yet few rules that can be used in the design of impact dampers except to urge life tests for durability.

One of the most unusual sources of system damping is that due to pumping of air, water, or other fluids through highly constrained passages because of the change in spacing of the parts.[18] For example, if a rib is riveted or spot-welded to a plate, it will not make contact at all points. If the combination is then bent, the clearance between the two will change, pumping the surrounding fluid, air, or water through the narrow passage between them. The viscosity in the fluid will cause damping. If this viscosity changes, the damping will change. Once more there is a critical value; if the viscosity is greater than this, as in very cold oil, the motion will be prevented and there will be no dissipation. If the viscosity is decreased sufficiently, there will be no dissipation, regardless of the motion. In normal aerospace structures, this mechanism may account for damping factors of 0.001[18]; but this source of damping disappears as the structure moves into a hard vacuum. Likewise, if a design is changed from riveted or spot-welded to seam-welded or integrally machined, this source of damping is eliminated. Another similar source of damping comes from the sloshing of liquid in an elastic or moving container, especially with properly designed baffles. This source of damping disappears if the container is either completely full or completely empty.

All these mechanisms are very important for most structures. It is, however, very difficult to make a reasonable quantitative analysis of them and most of them are subject to change during use. Designers should understand their qualitative behavior so that they may use designs which take maximum advantage of naturally occurring mechanisms. It is also important to learn by experience the approximate magnitude of the energy dissipation due to them so that one is not trapped into using excessive effort to increase damping only 10 to 50 per cent by some of the more orthodox treatments mentioned later.

CONFIGURATIONAL DAMPING

Consider the use of high damping elastomeric materials for effectively damping a system. Since the loss modulus of the damping material is usually small in comparison with the storage modulus of the base material, we would like to devise some kinematic method for multiplying the shear strain in the elastomer. This procedure is permissible because elastomers will take high strains without failure.

The way in which we damp a structure and the effectiveness of the damping treatment depend very markedly on the details of the structural configuration. Some typical structures are (A) unstiffened beams and plates; (B) the skin rib type of construction used in aircraft skins, spacecraft launch vehicles, metal building, truck and trailer bodies, etc.; (C) integrally stiffened structures milled from solid metal plates; and (D) honeycomb plates and shells. These four classes are shown schematically in Fig. 37.3. Vibration response spectra typical of these structures are also shown. The frequencies corresponding to the various modes may either be very dense (B1) or widely separated (D or B2). The modal damping caused by a given treatment will depend very much on the frequency density and the structural stiffness.

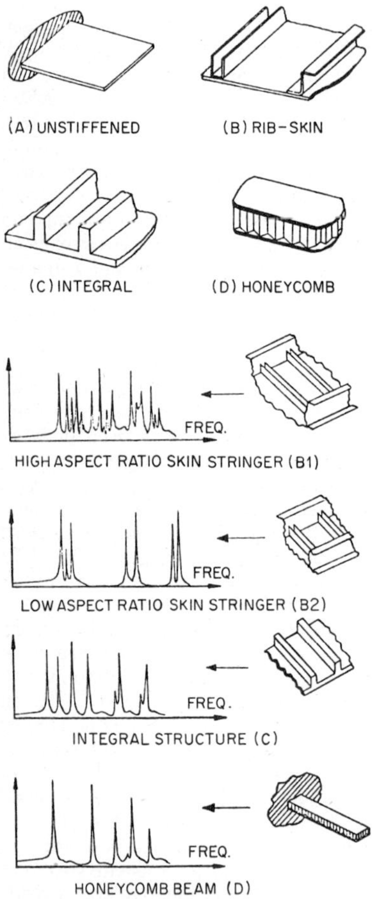

FIG. 37.3. Typical response spectra (sketches only). (*After D. I. G. Jones.*[14])

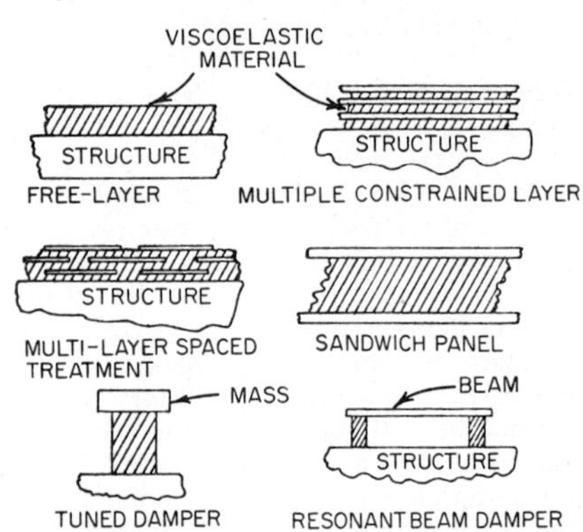

FIG. 37.4. Typical damping treatments. (*After D. I. G. Jones.*[14])

The choice of configuration depends on the amount of damping desired, cost of treatment, and allowable additional weight. The simplest, cheapest, and least-effective treatment is to spread a coating of a high loss elastomer on one side of a plate to be damped; this is called unconstrained or free damping layer (Fig. 37.4). The shear strain in the elastomer may be increased by the use of one or more constraining layers of a thin stiff material, usually metal. As the structure is strained, the constraining layer is not deformed, at least not as much. The relative movement between the structure and the constraining layer puts a large shear strain in the elastomer. This shear strain causes a shear stress, the shear stress causes axial stress in the constraining layer and for lengths much greater than some critical length, the increased length causes no shear in the elastomer. It may be shown that this characteristic length is:[21]

$$L_0^2 = t_1 t_2 \frac{E}{G} \tag{37.18}$$

where t_1 = thickness of constraining layer
t_2 = thickness of elastomer
E = Young's modulus of constraining layer
G = shear modulus of elastomer

If the structure is a flat plate in single-frequency bending vibration, the plate bends in a sinusoidal form with a wavelength λ equal to twice the distance between nodes. If this wavelength is short in comparison with the characteristic length, the elastomer is not very much constrained; if it is long, the constraining layer is stretched. The ratio of λ to L_0 is defined as the shear parameter g.

The most effective use of constraining layers is to build them up like bricks with each piece about three times the characteristic length,[21] which gives rise to a multilayer-spaced treatment.

Another approach is to use an elastomer for the filler of a sandwich panel. In this way we can introduce the lossy material at no increase in weight. If the filler is to be effective, the filler must be strained in shear. This means that the panel will be more flexible. It can be shown that optimum damping for a material loss constant $\eta = 1$ will be had with the configuration which cuts the stiffness in half. The strain in the elastomer can also be increased by using tuned dampers[3] or resonant beams (Fig. 37.5).

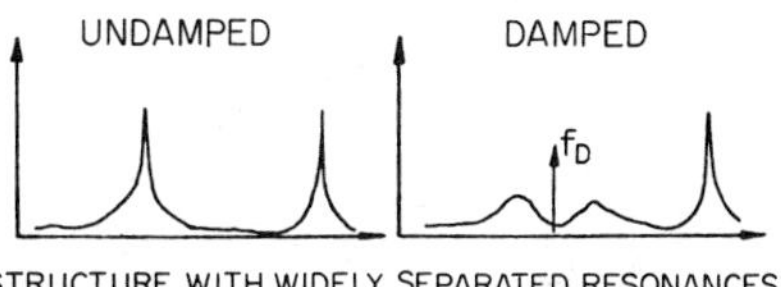

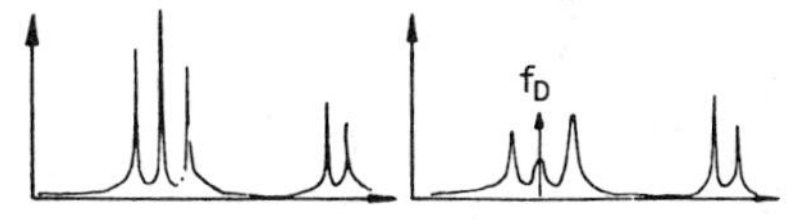

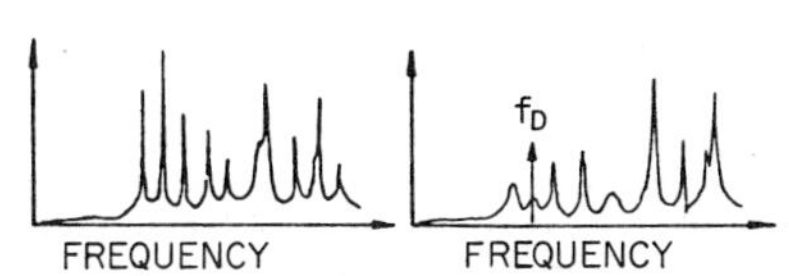

FIG. 37.5. Response spectra for structures with tuned dampers (sketches only). (*After D. I. G. Jones.*[14])

UNCONSTRAINED LAYERS

As discussed in earlier sections, molecular relaxation processes occur over certain temperature and frequency ranges during the vibration of so-called "linear viscoelastic materials." Much study has been devoted to the determination of the inherent dynamic stiffness moduli and damping parameters of these materials as a function of temperature and frequency of vibration,[15] so that it is of considerable interest to develop means of computation whereby the damping characteristics of treatments can be predicted from the material properties. Even though the dynamic moduli and loss factors of such materials are very sensitive functions of both temperature and frequency, good treatment-design predictions can be made employing universal functions that have been deduced for this class of materials.

THE HOMOGENEOUS ADHESIVE LAYER. The simplest analytical case of the computation of treatment performance from the material properties of viscoelastic high polymers is that of the homogeneous adhesive-layer treatment, namely, the uniform mastic application. Assuming that all losses occur due to extensional strain of the damping layer during flexure of the panel under treatment, the loss factor of the treated panel may be expressed in terms of the loss factor of the damping layer, the ratio of the Young's modulus of the damping layer to that of the panel under treatment, and the ratio of the thickness of the damping layer to that of the panel under treatment.[19] The graphical representation of the result is shown in Fig. 37.6, using the notation of a more general theory that treats the homogeneous single layer as a particular case. The relative damping factor of the treatment is plotted against the ratio between the treatment thickness and panel thickness. The validity of this result is limited to treatment thicknesses that are relatively small compared to the panel thickness and to damping materials for which the product of the loss factor and the stiffness of the damping layer is less than one-tenth the stiffness of the undamped panel. This means that the portions of the graph to the right of $h_2 = 1.0$ are of academic interest only and should not be used for design purposes. The result is derived for one-dimensional flexure of a uniform strip or straight-crested flexural waves in plates.

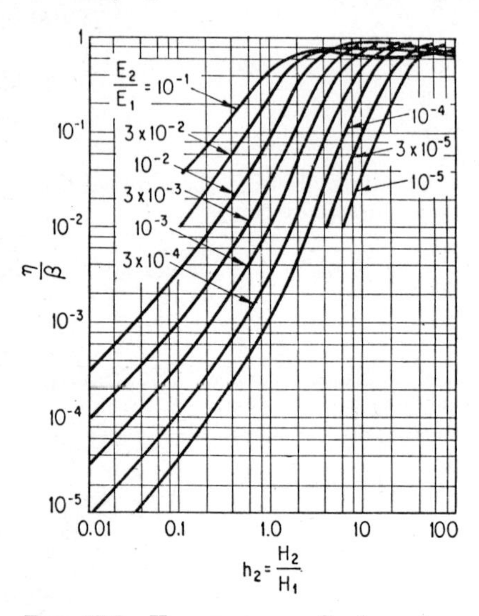

FIG. 37.6. Homogeneous adhesive viscoelastic layer theory: The relative damping factor η/β is the ratio between the loss factor of the treated panel η and the loss factor of the damping material β. The relative damping factor is plotted against the treatment thickness ratio h_2, the ratio between the treatment thickness H_2 and the panel thickness H_1. The parameter is the relative stiffness ratio between Young's modulus of the damping material E_2 and that of the panel material E_1. The subscript 1 refers to the panel, and the subscript 2 refers to the damping layer. (*After H. Oberst.*[19])

Observations from Theory. The following observations from the theory summarize its practical content. First, the loss factor of the panel treated with a mastic layer depends only on the thickness ratio between the treatment layer and the panel, rather than on the panel thickness itself. This establishes a scaling rule for comparisons of mastic layers with other classes of treatment in the laboratory. Thus the mastic thickness required on the test panel for valid ranking with other classes of material is established, once the treatment-weight limitation and panel thickness of the anticipated application have been specified. The form of the equation and slope of the curves in Fig. 37.6 show that the loss factor of the treated panel depends approximately on the square of the thickness ratio in the range of practical treatment thicknesses. Hence, the effect of the thickness of the laboratory test fixtures on the ranking of tested materials can be computed for comparison with other mechanisms of damping which may suffer less panel-thickness dependence.

Theory indicates that the optimum materials for damping must have both a high loss factor and high stiffness, because the damping factor of the treated panel depends on the loss modulus of the damping layer. The square-law dependence on thickness ratio favors application of all treatment on one side of the panel under treatment rather than dividing the weight of material between two layers on opposite sides.

Damping has been measured[14] as a function of temperature for roughly equal weights of a commercial free-layer damping material on the four structure types shown in Fig. 37.3. The thickness of the damping material was about 0.040 in. on structures A, B1, B2, and C and about 0.055 in. on D. It was attached in each case by means of a commercially available double-backed tape designed for this type of use. It will be seen (Fig. 37.7) that peak damping, whatever its specific value may be, always occurs near the same temperature, which lies near the center of the transition region for the damping material. Similar results will occur for other structures and damping materials. The exact temperature at which peak damping occurs depend on the geometry of the structure in a manner which can sometimes be calculated; but in its broad aspects, the main controlling factor is the transition temperature of the damping material itself. This fact is most important if one is designing a damping treatment for use under conditions where the temperature can vary significantly, as in aerospace applications and in the outdoor

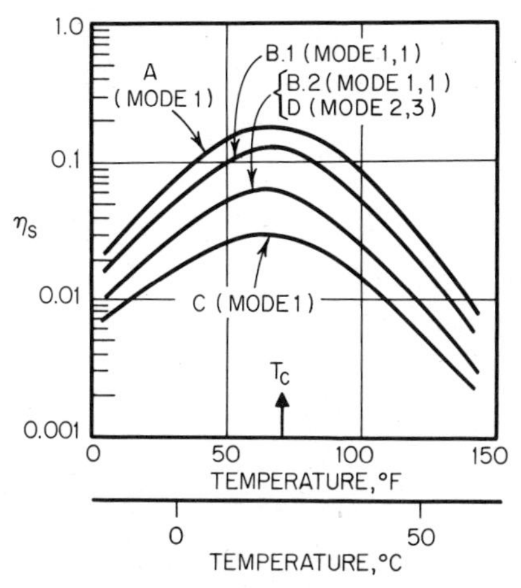

FIG. 37.7. Graphs of modal loss factor η_s vs. temperature for several structures with free layer damping. (*After D. I. G. Jones.*[14])

environment just about anywhere on earth. Internal environments in buildings may be less variable with modern heating and air-conditioning conditions, but even here some variability should be assumed.

CONSTRAINED DAMPING LAYERS

VIBRATION-DAMPING TAPES

The analysis of constrained damping layers in the form of homogeneous-adhesive tapes proceeds from certain dimensional restrictions and assumptions which are consistent with the construction of damping tapes as well as a broad class of larger-scale treatments. Among these restrictions are the following:[20] the composite bar (or plate with straight-crested flexural waves) must be thin compared with the bending wavelength so that only the damping layer is subject to shear distortion; the vibration amplitude must be small and the loss factor of the treated panel must be small; the loss modulus of the damping material must be small; the bending stiffness of the constraining layer must be small compared with that of the panel; and the Young's modulus of the damping layer must be restricted to a range between being small compared with that of the facing layers and being sufficiently large to make thickness changes in the damping layer negligible. The results of the analysis express the relative loss factor of the treated panel (the ratio between the loss factor of the treated panel η_s and the loss factor η_G of the complex shear modulus of the damping material) as a function of Young's moduli, thicknesses of the materials, and a dimensionless shear parameter g. The parameter g is the ratio between the bending wavelength and the shear length in the damping material. The shear length is the distance along the damping layer over which a local shear disturbance is reduced by a factor of $1/e$.

Engineering computations of the performance of damping tapes from shear-damping properties and dimensions of the adhesive are facilitated by families of reference curves computed for selected relative foil thicknesses. The extent of agreement between

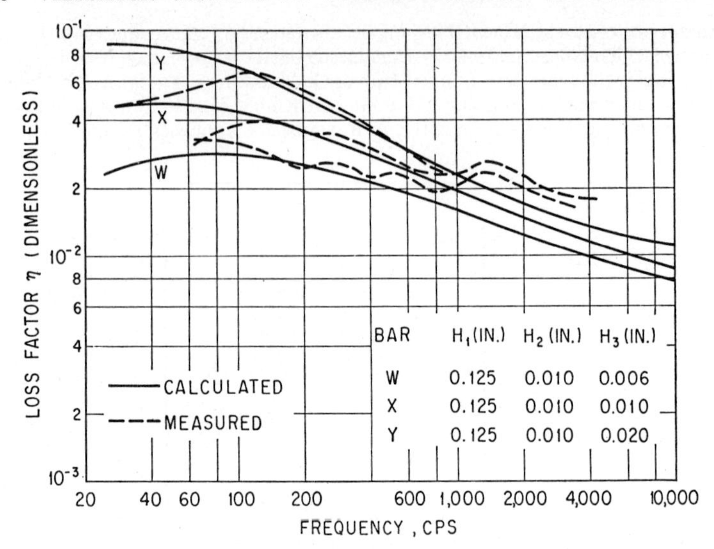

FIG. 37.8. Comparison of the measured and theoretical performance of constrained viscoelastic layers, showing the loss factor of the damped test bar at room temperature as a function of frequency for three different constraining layer thicknesses. H_1, H_2, and H_3 are the thicknesses of the bar, damping layer, and constraining layer, respectively. These results show the frequency dependence of the damping effectiveness of damping tapes. (*After E. M. Kerwin.*[20])

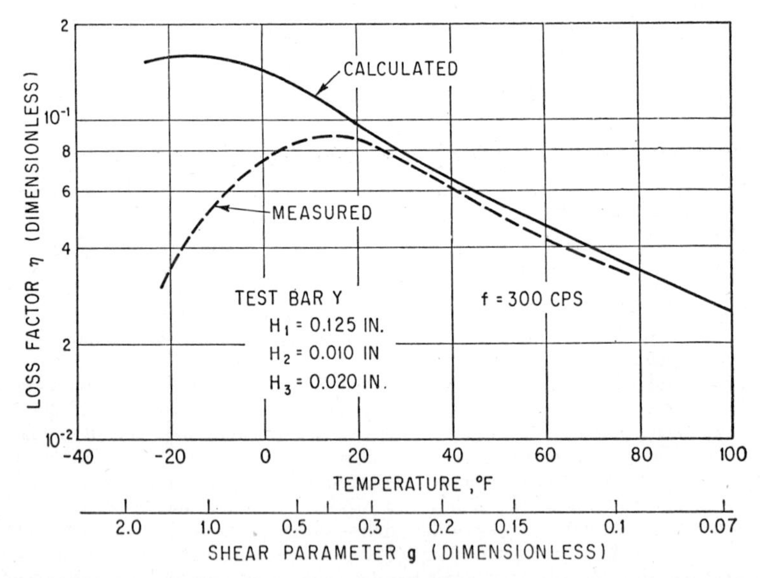

FIG. 37.9. Comparison of the measured and theoretical performance of constrained viscoelastic layers, showing the loss factor of a damped test bar at 300 cps as a function of temperature or shear parameter g (see text). H_1, H_2, and H_3 are the thicknesses of the bar, damping layer, and constraining layer, respectively. These results show the temperature dependence of the damping effectiveness of a damping tape. (*After E. M. Kerwin.*[20])

measured and calculated performance is shown in Fig. 37.8 for the case of three damping tapes applied to the same ⅛-in. bar with the same adhesive thickness of 0.010 in. but with different foil-facing thicknesses: 0.020, 0.010, and 0.006 in., respectively. The

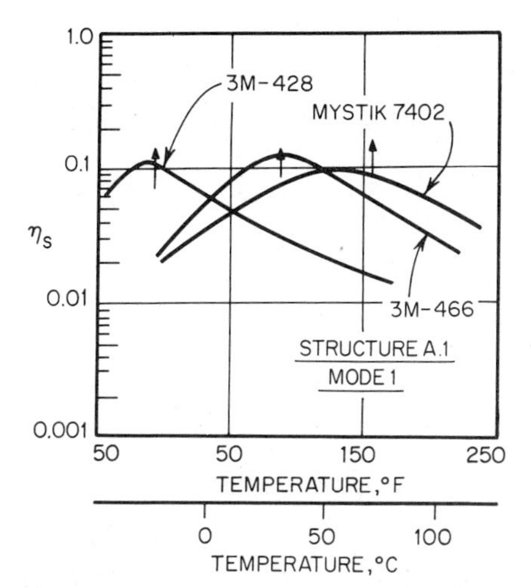

FIG. 37.10. Graphs of modal loss factor η_s vs. temperature for several multiple constrained-layer treatments. (*After D. I. G. Jones.*[14])

agreement is good over a wide range of frequency. The measured and calculated performance of one of the same tapes (0.020-in. foil backing) as a function of temperature at 300 cps is shown in Fig. 37.9. The poor agreement at low temperatures is explainable in terms of neglecting certain higher-order terms in order to faciliate computation.[20] The ability of the theory to predict shifts in the damping performance as a function of

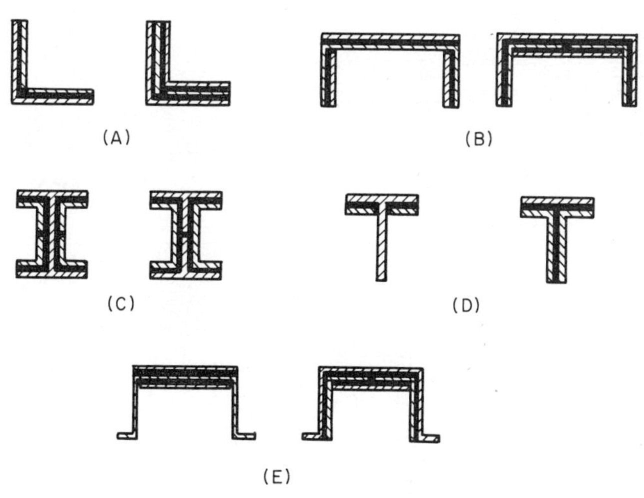

FIG. 37.11*A*. Cross sections of viscoelastic shear-damped composite structural beams of multilaminate construction: (*A*) angle, (*B*) channel, (*C*) I section, (*D*) T section, and (*E*) hat-section designs. (*After Ruzicka.*[25])

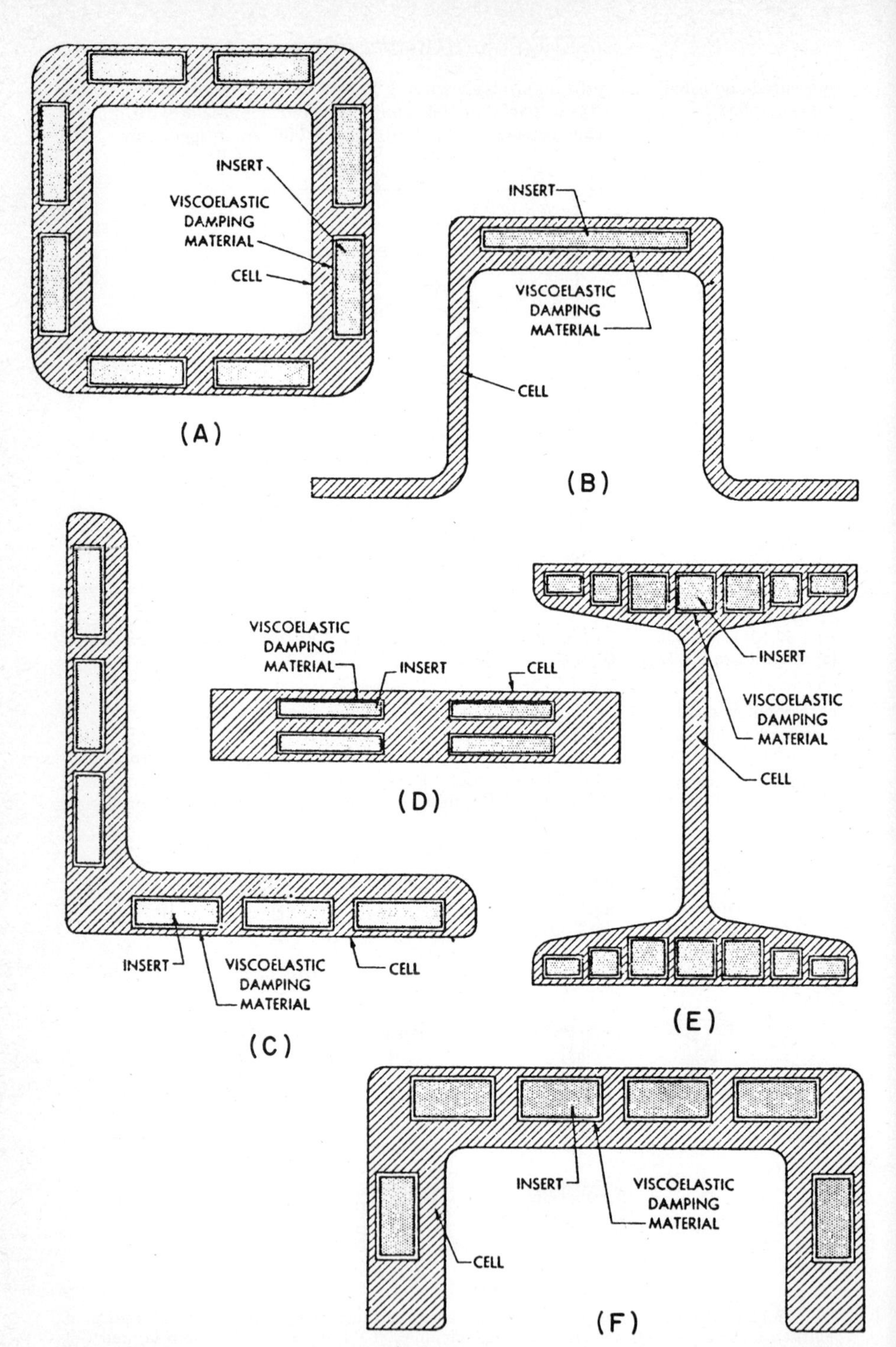

FIG. 37.11*B*. Cross sections of viscoelastic shear-damped composite structural beams of cell-insert construction: (*A*) square-tube, (*B*) hat-section, (*C*) angle, (*D*) flat-bar, (*E*) I-section, and (*F*) channel designs. (*After Ruzicka.*[25])

frequency when the same constrained layer is applied to bars of different thicknesses is also good. The scaling law involves two separate shifts of the damping-vs.-frequency characteristic: one in magnitude of damping, and the other in location along the frequency axis.

It is clearly possible to get system loss factors approaching 0.1 by optimal use of constrained elastomers. The importance of using the elastomer with the correct critical temperature is shown in Fig. 37.10. This shows the damping in a clamped-clamped beam with various five-layered treatments fully covering the surface and with adhesive thickness of 50 micron (μ) and aluminum constraining layer thickness of about 50μ. The peak loss modulus is not as significant as the transition temperature, since changes in loss modulus may be compensated for by changing the elastomer thickness.

GENERAL REFERENCES

As has been mentioned earlier, new damping materials are constantly being developed. Since their properties depend strongly on frequency and temperature, it is important to obtain the manufacturer's specifications. Manufacturers' names and addresses can be found in various annual buyers guides, including those issued by *Sweet's*, *Aviation Week*, and *Sound and Vibration* magazines. The journals of the Acoustical Society of America, The American Society of Mechanical Engineers, the Society of Environmental Engineers, the Institute of Noise Control Engineering and the other journals cited in the list of references are also fruitful sources of current information.

The *Shock and Vibration Digest*, a publication of SVIC, Naval Research Laboratory, Washington, D.C., is a unique source of reviews of the current literature as well as review articles in the field. Two review articles covering the technical literature were published in 1966; one covered the properties of materials, members, and composites[22] and the other covered the methods of measurement and analysis of system damping.[23]

References 24 and 25 are extensive compilations of the damping to be expected for structural composites with viscoelastic shear-damping mechanisms. The results are presented in dimensionless graphical form with supporting derivations and equations. Graphs are given for a wide range of viscoelastic loss factors and are easily interpreted for size and frequency. A representative but by no means exhaustive set of cross sections is shown in Fig. 37.11.

REFERENCES

1. Crandall, S. H.: *J. Sound Vibration*, **11**:3–18 (1970).
2. Jacobsen, L. S.: *Trans ASME*, APM-52-15 (1930).
3. Snowden, J. C.: "Vibration and Shock in Damped Mechanical Systems," John Wiley & Sons, Inc., New York, 1968.
4. Plunkett, R.: *J. Appl. Mech.*, **30**(9):70–75 (1963).
5. Plunkett, R.: *Shock and Vibration Bull.*, **42**(4, NRL):57–64 (1972).
6. Lazan, B. J.: "Damping of Materials and Members in Structural Mechanics," Pergamon Press, New York, 1968.
7. Den Hartog, J. P.: "Mechanical Vibrations," chap. 8, McGraw-Hill Book Company, New York, 1956.
8. Goodman, T. P.: *J. Eng. Ind.*, **85**:17 (1963).
9. Adams, R. D.: *J. Sound Vibration*, **23**:199–216 (1972).
10. Birchon, D.: *Eng.*, **222**:207–209 (1966).
11. Weismann, G. F., and W. Babington: *J. Environ. Sci.*, **9**(5):19–27 (1966).
12. Lee, L. T.: A Graphical Compilation of Damping Properties of Materials, AFML-TR-66-169, AFSC, Wright-Patterson, 1966.
13. Zener, C.: *Phys. Rev.*, **52**:230–235 (1937).
14. Jones, D. I. G.: *Sound and Vibration*, **6**(7):25 (1972).
15. Ferry, J. D.: "Viscoelastic Properties of Polymers," John Wiley & Sons, Inc., New York, 1970.
16. Sridharen, P., and R. Plunkett: *Trans ASME*, **96**(B 3):969–975 (1974).
17. Beranek, L. L. (ed.): "Noise and Vibration Control," chap. 11, McGraw-Hill Book Company, New York, 1971.

18. Ungar, E. E.: *J. Sound Vibration*, **26**:141–154 (1973).
19. Oberst, H.: *Acustica, 2, Akust. Beih.*, **4**:181 (1952).
20. Kerwin, E. M.: *J. Acoust. Soc. Am.*, **31**:952 (1959).
21. Plunkett, R., and C. T. Lee: *J. Acoust. Soc. Am.*, **48**(1):150–161 (1970).
22. Lazan, B. J.: Damping Properties of Materials, Members and Composites, in "Applied Mechanics Surveys," Spartan Books, Washington, D.C., 1966.
23. Plunkett, R.: Vibration Damping, in "Applied Mechanics Surveys," Spartan Books, Washington, D.C., 1966.
24. Derby, T. F., and J. E. Ruzicka: *NASA Rept.* CR-1269, 1969.
25. Ruzicka, J. E., et al.: *NASA Rept.* CR-742, 1967.

38

TORSIONAL VIBRATION IN RECIPROCATING MACHINES

Frank M. Lewis
Massachusetts Institute of Technology

INTRODUCTION

The crankshaft of a reciprocating engine and all the moving parts driven by it comprise a torsional elastic system. Such a system has several modes of free torsional oscillation. Each mode is characterized by a natural frequency and a pattern of relative amplitudes of parts of the system. The harmonic components of the driving torque excite vibration of the system in its fundamental modes. If the frequency of any harmonic component of the torque from a single cylinder is equal to or near the frequency of any fundamental mode of vibration, a condition of resonance exists and the engine is said to be running at a *critical speed.* Operation at such critical speeds can be very dangerous, resulting in fracture of the shafting; operation at other critical speeds may result in rapid wear of bearings, gear, and other parts, and may result in undesirable vibration of the engine and associated machinery.

The number of complete oscillations of the elastic system per unit revolution of the crankshaft is called the *order of a critical speed.* The orders of critical speeds that correspond to the harmonic components of the torque from the engine as a whole are called *major orders.* Critical speeds also can be excited which correspond to the harmonic components of the torque curve of a single cylinder. Since the fundamental period of the torque from a single cylinder in a four-cycle engine is 720°, the criticals in a four-cycle engine can be of ½, 1, 1½, 2, 2½, etc., order. In a two-cycle engine, only the criticals of 1, 2, 3, etc., order can exist. All criticals except those of the major orders are called *minor criticals,* a term which does not imply that they are necessarily unimportant.

Example 38.1. A six-cylinder four-cycle engine with 120° crank spacings has three equally spaced firing impulses per revolution. The major orders therefore are 3, 6, 9, 12, etc.

The critical speeds occur at

$$\frac{f_n}{q} \quad \text{rps} \tag{38.1}$$

where f_n is the natural frequency of one of the modes in cps and q is the order number of the critical. While many critical speeds exist in the operating range of an engine, only a few are likely to be of any importance.

A dynamic analysis of an engine consists of:

1. Calculation of the natural frequencies of the modes likely to be of importance. Generally, the calculation may be limited to the lowest mode or the lowest two modes, but in some complicated arrangements more modes may be needed.
2. Estimates of the vibration amplitude at the critical speeds which are in or near the operating range.
3. A study of remedial measures if any are needed.

CALCULATION OF THE NATURAL FREQUENCIES AND RELATIVE AMPLITUDE CURVES

The torsional elastic system of an engine and its associated machinery is a rather complicated arrangement of mass and elastic distribution. To make the system amenable to mathematical treatment, it is necessary to represent it by a simpler system that is substantially equivalent dynamically. The equivalent system generally used consists of concentrated rotors connected by massless, torsionally elastic springs. Rotors are placed at each crank center and at the center planes of actual flywheels, rotors, propellers, etc. In a variation of this procedure, the engine part of the system is replaced by a uniform distribution of mass and elasticity.[1,2,3]

The torsional calculation is made not for the engine alone but for the complete system, including all driven machinery. On the engine it usually is possible to consider such parts as camshafts, pumps, blowers, etc., to be detached from the engine if they are driven elastically, or as additional rigid masses at the point of attachment to the crankshaft if the drive is relatively rigid. If there is doubt, these parts should be included in the torsional calculation as elastically connected masses.

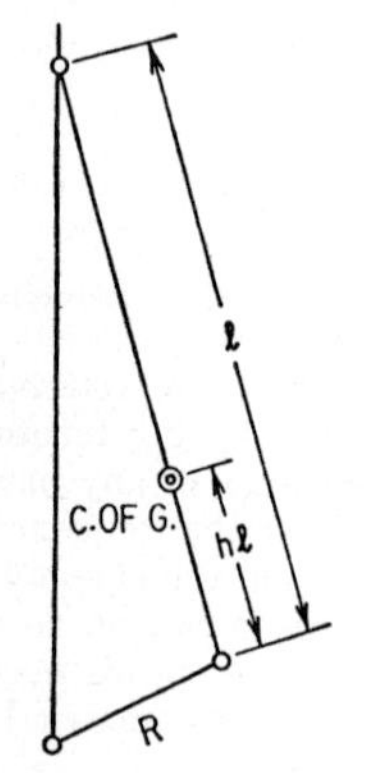

FIG. 38.1. Schematic diagram of crank and connecting rod.

CALCULATION OF POLAR MOMENTS OF INERTIA

The piston and connecting rod introduce a variable-mass problem whose solution is complex. The exact solution[4] shows that the effect of the piston and connecting rod can be closely approximated by representing them by a concentrated rotor of polar inertia J defined by

$$J = \left[\frac{W_p}{2} + W_c\left(1 - \frac{h}{2}\right)\right] R^2 \qquad \text{in.}^2\text{-lb} \qquad (38.2)$$

where W_p = weight of piston, piston pin, and cooling fluid, lb
W_c = weight of connecting rod, lb
h = fraction of rod length from the lower center to the center-of-gravity (see Fig. 38.1)
R = crank radius, in.

The polar inertias of the crank webs (see Fig. 38.2), the crankpin, and the journal sections are added to that given by Eq. (38.2). These polar inertias should be calculated with the best obtainable accuracy. The following procedure is recommended for calculation:

Let the crank web be intercepted by a series of concentric cylinders of radius R about the crank center, as shown in Fig. 38.2. The area of the intercept of each cylinder with the crank web is designated S, and a curve of R^2S is plotted as a function of R. Then the polar inertia of the crank web is defined by

$$J = \left(\gamma \int_{R_b}^{R_m} R^2 S\, dR\right) + J_b \qquad \text{in.}^2\text{-lb}$$

where γ = weight density of the crank web, lb/in.^{3}
R_m = maximum radius of the crank web, in.
R_b = radius of base cylinder (see Fig. 38.2), in.
J_b = polar inertia of portion of crank web within the base cylinder, in.^{2}-lb

The integral in the above expression for J is the area of the R^2S curve between the values of radii R_b, R_m.

For the crank web shown in Fig. 38.2, the area S may be defined as $S = bR\theta/57.3$, where θ is measured in degrees. Then the polar inertia may be expressed as

$$J = \left(\frac{\gamma}{57.3} \int_{R_b}^{R_m} bR^3\theta\, dR\right) + J_b \qquad \text{in.}^2\text{-lb}$$

The same procedure is used for calculating the polar inertia of propellers and other irregular parts. In a marine propeller of ogival sections, i.e., flat driving face, circular arc back, and elliptical developed outline (do not use for other shapes), the polar inertia (excluding hub) is given by

$$J = 0.0046\, nD^3bt \qquad \text{in.}^2\text{-lb}$$

where n = number of blades
D = diameter of propeller, in.
b = maximum blade width, in.
t = maximum blade thickness at one-half radius (axis to tip), in.

For propellers, pumps, hydraulic couplings, etc., an addition must be made for the virtual inertia of the entrained fluid. For marine propellers this is ordinarily assumed at 25 per cent of the propeller inertia. Virtual inertias for pumps and the like are not known accurately but can be assumed as if half the casing were filled with rotating fluid.

CALCULATION OF SHAFT STIFFNESS

The stiffness of the crankshaft is the most uncertain element in the torsional-vibration calculation. Shaft stiffness can be measured experimentally either by twisting a shaft with a known torque or from the observed values of the critical speeds in a running engine. Alternatively, it can be calculated by semiempirical formulas which have been proposed. Twenty such formulas are given in Ref. 23. The three given below are recommended. Referring to Figs. 38.2 and 38.3, l_e is the length of a solid shaft of diameter D_s equal in torsional stiffness to the section of crankshaft between crank centers. Then, from Ref. 5,

Wilson's Formula:

$$\frac{l_e}{D_s^4} = \frac{b + 0.4D_s}{D_s^4 - d_s^4} + \frac{a + 0.4D_c}{D_c^4 - d_c^4} + \frac{r - 0.2(D_s + D_c)}{hW^3} \qquad (38.3)$$

From Ref. 6,

Ziamanenko's Formula:

$$\frac{l_e}{D_s^4} = \frac{b + 0.6hD_s/b}{D_s^4 - d_s^4} + \frac{0.8a + 0.2(W/r)D_s}{D_c^4 - d_c^4} + \frac{r^{3/2}}{hW^3D_c^{1/2}} \qquad (38.4)$$

From Ref. 7,

Constant's Formula:

$$\frac{l_e}{D_s^4} = \frac{1}{\alpha_1\alpha_2\alpha_3\alpha_4}\left(\frac{b}{D_s^4 - d_s^4} + \frac{a}{D_c^4 - d_c^4} + \frac{0.94}{hW^3}\right) \qquad (38.5)$$

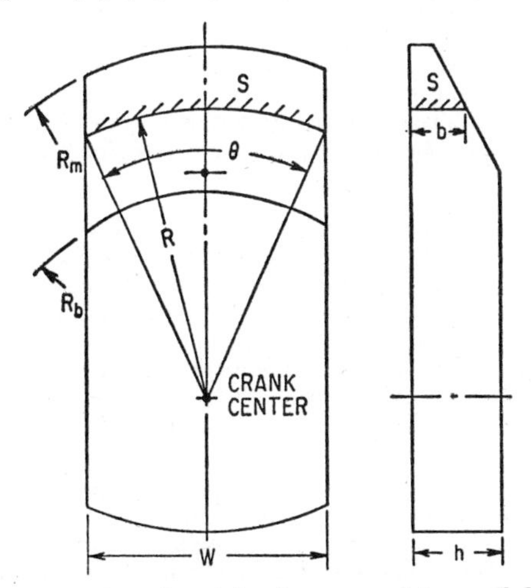

FIG. 38.2. View of crank web in plane normal to crankshaft axis.

where α_1, α_2, α_3, α_4 are modifying factors, determined as follows:

$$\alpha_1 = 1 - \frac{0.0825}{\sqrt{\dfrac{W_s - d_s}{2W_s} + \dfrac{W_c - d_c}{2W_c}} - 0.32} \tag{38.6}$$

If the shaft is solid, assume $\alpha_1 = 0.9$ instead of applying Eq. (38.6). The factor α_2 is a web-thickness modification determined as follows: If $4h/l$ is greater than ⅔, then $\alpha_2 = 1.666 - 4h/l$. If $4h/l < ⅔$, assume $\alpha_2 = 1$. The factor α_3 is a modification for web chamfering determined as follows: If the webs are chamfered, estimate α_3 by comparison with the cuts on Fig. 38.3:

Cut AB and $A'B'$, $\alpha_3 = 1.000$; cut CD alone, $\alpha_3 = 0.965$; cut CD and $C'D'$, $\alpha_3 = 0.930$; cut EF alone, $\alpha_3 = 0.950$; cut EF and $E'F'$, $\alpha_3 = 0.900$; if ends are square, $\alpha_3 = 1.010$.

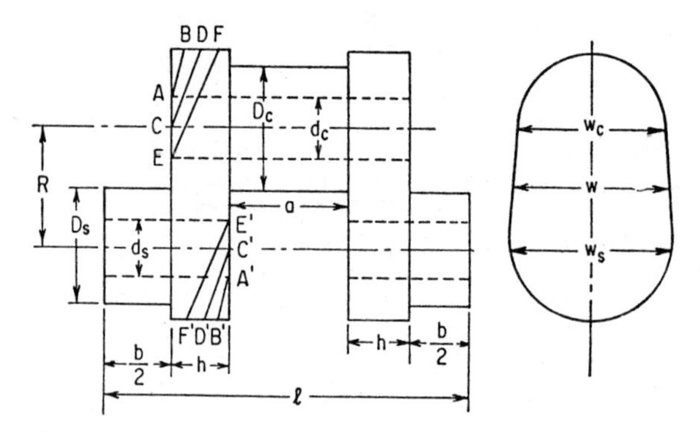

FIG. 38.3. Schematic diagram of one crank of a crankshaft.

The factor α_4 is a modification for bearing support given by

$$\alpha_4 = \frac{Al^3w}{D_c^4 - d_c^4} + B \tag{38.7}$$

For marine engine and large stationary engine shafts: $A = 0.0029$, $B = 0.91$
For auto and aircraft engine shafts: $A = 0.0100$, $B = 0.84$
If α_4 as given by Eq. (38.7) is less than 1.0, assume a value of 1.0.

The Constant formula, Eq. (38.5), is recommended for shafts with large bores and heavy chamfers.

THE STIFFNESS OF OTHER SHAFT ELEMENTS. The shafting of an engine system may contain elements such as changes of section, collars, shrunk and keyed armatures, etc., which require the exercise of judgment in the assessment of stiffness. For a change of section having a fillet radius equal to 10 per cent of the smaller diameter, the stiffness can be estimated by assuming that the smaller shaft is lengthened and the larger shaft is shortened by a length λ obtained from the curve of Fig. 38.4.[8] This also may be applied to flanges where D is the bolt diameter. The stiffening effect of thrust collars can be ignored.

The stiffness of shrunk and keyed parts is difficult to estimate as the stiffening effect depends to a large extent on the tightness of the shrunk fit and keying. The most reliable values of stiffness are obtained by neglecting the stiffening effect of an armature and assuming that the armature acts as a concentrated mass at the center of the shrunk or keyed fit. Some armature spiders and flywheels have considerable flexibility in their arms; the treatment of these is discussed under *Branched Systems*. The torsional stiffness of a tapered shaft of large diameter D_2 and small diameter D_1 is given by

$$k_r{}^* = \frac{3\pi G(D_2 - D_1)}{32l[(1/D_1{}^3) - (1/D_2{}^3)]} \tag{38.8}$$

Numerous types of torsionally elastic couplings can be constructed; Ref. 20 discusses some of these with their torsional properties.

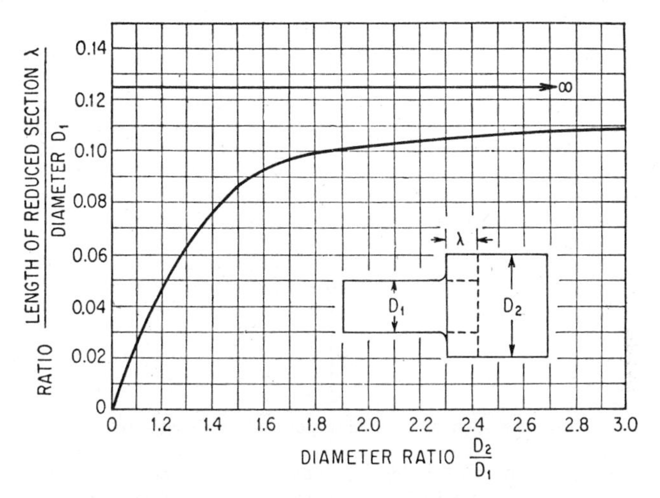

FIG. 38.4. Curve showing decrease of stiffness resulting from a change in shaft diameter. The stiffness of the shaft combination is the same as if shaft D_1 is lengthened by λ and D_2 is shortened by λ. (*F. Porter.*[8])

NATURAL FREQUENCY CALCULATIONS

If a system can be reduced to two rotors at opposite ends of a massless shaft, as shown in Fig. 38.5, the natural frequency is given by

$$f_n = \frac{1}{2\pi}\sqrt{\frac{(I_1 + I_2)k^*}{I_1 I_2}} \quad \text{cps} \tag{38.9}$$

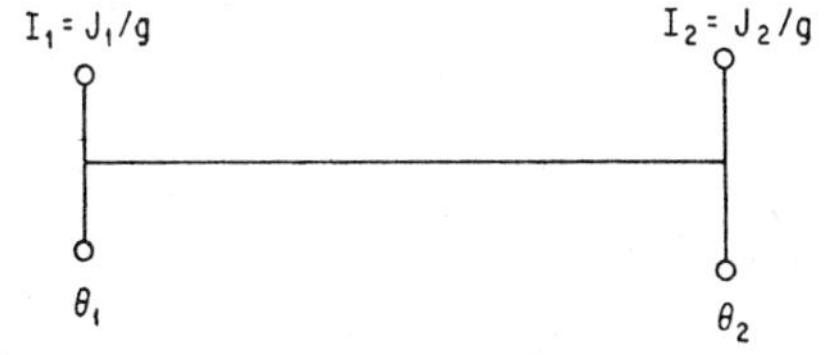

$I_1 = J_1/g$ $\qquad$ $I_2 = J_2/g$

θ_1 $\qquad$ θ_2

FIG. 38.5. Schematic diagram of shaft with two rotors.

The ratio of the vibration amplitudes is given by:

$$\frac{\theta_2}{\theta_1} = -\frac{I_1}{I_2}$$

* The symbol k is used in this chapter to denote torsional stiffness; subscripts are used to designate the particular shaft.

For a three-rotor system as shown in Fig. 38.6, the natural frequency is

$$f_n = \frac{1}{2\pi} \sqrt{A \pm \sqrt{A^2 - B}} \qquad \text{cps} \tag{38.10}$$

where
$$A = \frac{k_{12}(I_1 + I_2)}{2I_1I_2} + \frac{k_{23}(I_2 + I_3)}{2I_2I_3}$$

$$B = \frac{(I_1 + I_2 + I_3)k_{12}k_{23}}{I_1I_2I_3}$$

In Eqs. (38.9) and (38.10), the k's are torsional stiffness constants expressed in in.-lb/rad. The notation $k_{1,2}$ indicates that the constant applies to the shaft between rotors 1 and 2. The polar inertia J has units of lb-in.^{2} (often written $W\rho^2$ where W is the weight and ρ the radius of gyration), and $I = J/g$ is a polar moment of inertia having units of lb-in.-sec^2 where $g = 386$ in./sec.^{2} Any system of units, English or metric, can be used, provided that g is expressed in units consistent with the adopted units.

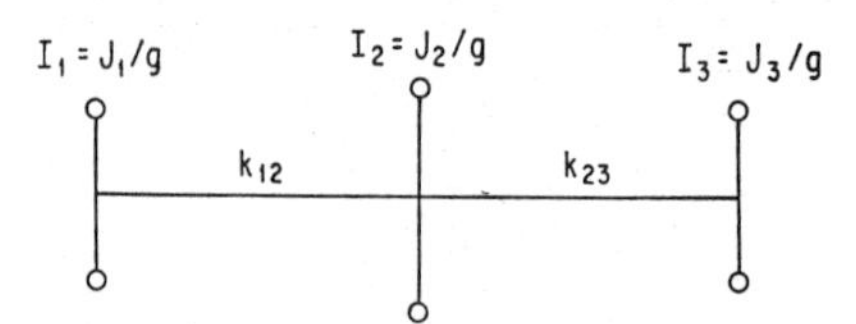

FIG. 38.6. Schematic diagram of shaft with three rotors.

The above formulas and all the developments for multimass torsional systems which follow apply also to systems with longitudinal motion, if the polar moments of inertia $I = J/g$ are replaced by the masses $m = W/g$ and the torsional stiffnesses are replaced by longitudinal stiffnesses.

If there are more than three masses the method of sequence calculation, commonly called the *Holzer table,*[9,10] can be used to calculate the natural frequencies. Other alternative methods can be employed but the sequence calculation has the advantages of extreme simplicity, ready extension to forced vibration, and ease of programming if a digital computer is available.* In this procedure, a frequency is assumed; then starting at one end of the system, a balance of torques and displacements is obtained, step by step. The final external torque required to achieve balance is called the *residual torque.* If this is zero, the assumed frequency is one of the natural frequencies of the system. The mode associated with this natural frequency is identified by the number of changes of sign of the amplitudes at the various rotors. A plot of residual torque vs. frequency yields all the natural frequencies of the system.

In many internal-combustion engines, the polar inertias of the rotors representing the cranks and pistons are equal, as are the stiffness constants between these rotors. Advantage can be taken of this fact to reduce the labor of computation considerably. Let I be one of the equal polar moments of inertia and k one of the equal spring constants. Divide all the polar moments of inertia by I, and all the spring constants by k. The new system is dynamically similar to the original one. It is called the *unity system.* Then if ω_n is the natural angular frequency of the actual system and ω_s that of the unity system,

$$\omega_n^2 = \omega_s^2 \frac{k}{I} \tag{38.11}$$

* This method was first proposed in 1912 by Gümbel[9] in graphical form; later by Holzer[10] in a tabular form. Thirty or more alternative methods have since been proposed, but the sequence calculation has the advantage of extreme simplicity, ready extension to forced vibration, and ease of programming if a digital computer is available. As given here, the original Holzer tabulation has been rearranged in a more consistent form.

This "unity form" of the calculation is a laborsaving technique which is useful only when there are a number of equal masses and/or stiffnesses in the system; whether calculations are made using the unity form or the original constants is a matter of choice.

Before starting the sequence calculation it is necessary to know approximately the frequencies of the desired modes. Usually not more than the two lowest are required. For an internal-combustion engine the following procedure is generally satisfactory. Add all the polar inertias within the engine part of the system and assume that 40 per cent of this is placed at the crank farthest from the flywheel or generator (No. 1 crank). Find the stiffness with reference to the flywheel, or the generator if no flywheel is fitted, to the No. 1 crank. If there is no flywheel, the system reduces to a two-rotor system, and Eq. (38.9) gives approximately the lowest natural frequency of the system. If there is a flywheel, the system reduces to one of three rotors; then Eq. (38.10) gives the approximate values of the two lowest natural frequencies of the system. For complicated systems involving numerous driven rotors no general rules can be given.

Example 38.2. Figure 38.7 (upper scale) shows the actual mass and elasticity distribution of an eight-cylinder, four-cycle engine. Four of the cranks are fitted with counterweights so that the crank inertias are not the same. Dividing the moments of inertia by 35.8 and the stiffness constants by 175.7×10^6, the constants for the unity system are obtained, as shown in Fig. 38.7 (lower scale). The sum of the inertia constants through rotor 9 is 7.152. The stiffness

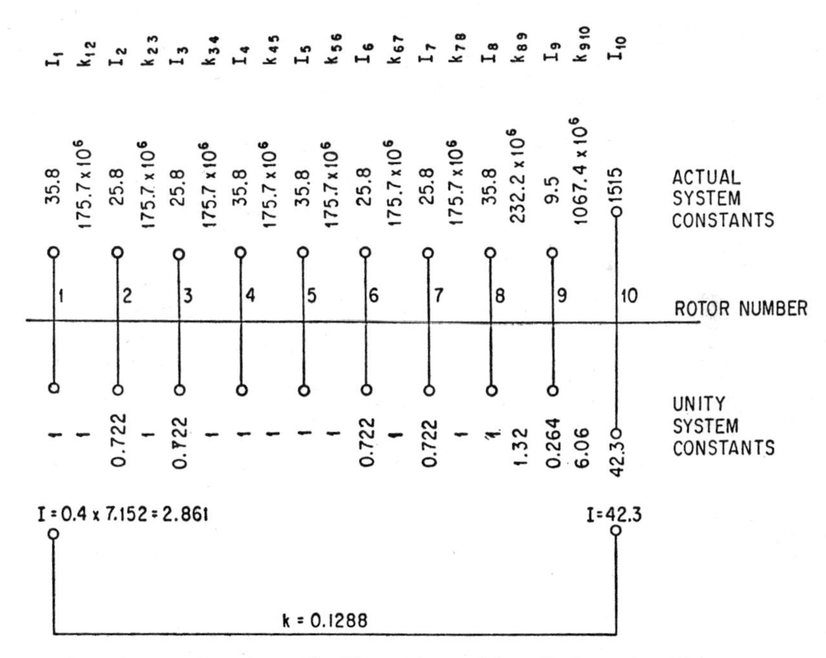

Fig. 38.7. Moments of inertia I and stiffness k for an eight-cylinder engine driving a generator. The constants for the actual system are shown in the upper scale and for the unity system in the lower scale.

of the shaft to rotor 10 is

$$\frac{1}{7 + \dfrac{1}{1.32} + \dfrac{1}{6.06}} = 0.126$$

The two-rotor system is therefore represented as in Fig. 38.5 and an approximate value of ω_s^2 is calculated from Eq. (38.9) where $I_1 = 0.40 \times 7.152 = 2.861$ and $I_2 = 42.3$:

$$\omega_s^2 = \frac{0.126 \times 45.16}{2.861 \times 42.3} = 0.047 \qquad (\text{rad/sec})^2$$

Table 38.1. Sequence Calculations for First Mode of System Shown in Fig. 38.7 ($\omega_s^2 = 0.047$)

Rotor number	I	$I\omega_s^2$	Deflection β	Torque M, in.-lb	k
1	1.000	0.0470	1.0000	0.0470	
			0.0470	0.0470	1.00
2	0.722	0.0339	0.9530	0.0323	
			0.0793	0.0793	1.00
3	0.722	0.0339	0.8737	0.0296	
			0.1089	0.1089	1.00
4	1.000	0.0470	0.7648	0.0360	
			0.1449	0.1449	1.00
5	1.000	0.0470	0.6199	0.0291	
			0.1740	0.1740	1.00
6	0.722	0.0339	0.4459	0.0151	
			0.1891	0.1891	1.00
7	0.722	0.0339	0.2568	0.0076	
			0.1967	0.1967	1.00
8	1.000	0.0470	0.0601	0.0028	
			0.1511	0.1995	1.32
9	0.265	0.0125	−0.0910	−0.0011	
			0.0327	0.1984	6.06
10	42.500	1.9975	−0.1237	−0.2471	
			Residual torque M_r =	−0.0487	

The sequence calculation is started with $\omega_s^2 = 0.047$. The calculation is arranged as shown in Table 38.1. The rotor numbers and their inertia constants I are placed on alternate lines. The β values opposite the rotor numbers define the first mode relative amplitude curve and the torques between rotor numbers are the shaft torques between the corresponding rotors. The intermediate lines are designated lines 1–2, lines 2–3, etc. The stiffness constants are placed on the intermediate lines in the last column. For the trial $\omega_s^2 = 0.047$, the $I\omega_s^2$ column is computed next. An amplitude of 1.0 is taken for rotor 1. Then the inertia reaction of rotor 1 is $0.047 \times 1.0 = 0.047$; it is placed in the M column on line 1. The torque in the shaft section between rotors 1 and 2 is equal to this inertia reaction and is placed in the M column in line 1–2. This torque divided by the stiffness, 1.00 in shaft section 1–2, is the relative deflection of rotors 1–2 and is placed in the β column of line 1–2. Subtracting this relative deflection 0.0470 from the deflection of rotor 1 gives the deflection 0.9530 of rotor 2. This deflection multiplied by 0.0339 ($I\omega_s^2$ for rotor 2) gives the inertia reaction 0.0323 of rotor 2. The inertia reaction of rotor 2 added to the torque in section 1–2 gives the torque in section 2–3. Thus, the table

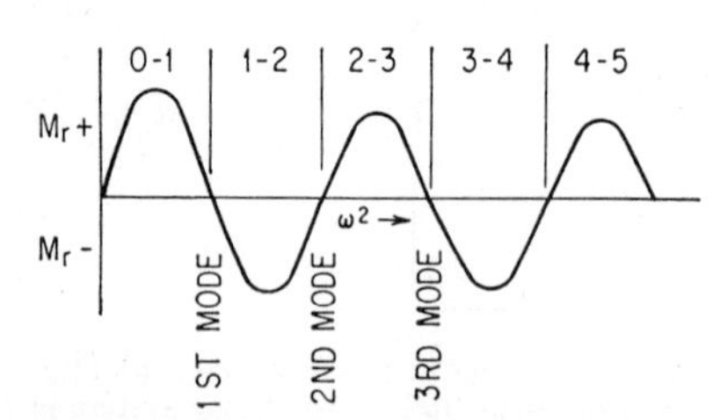

FIG. 38.8. Typical curve of residual torque M_r as a function of assumed frequency ω. The values at the top of the figure bracket are the numbers of change of sign of β that are set forth opposite the rotor numbers in the sequence calculation.

is carried through step by step. Note that in lines 8–9 and 9–10, k is no longer unity. The final figure in the torque column is called the *residual torque*. All additions and subtractions are made in the algebraic sense, i.e., on line 9 the deflection $-0.0910 = 0.0601 - 0.1511$.

The residual torque has the following physical significance: If a simple harmonic torque of amplitude M_r and angular frequency ω_s is applied to rotor 10 of the system, then all the rotors will oscillate with the amplitudes as given in the β column and the torques will be as given in the M column. If such a sequence calculation is made over a sufficient range of values of ω_s to cover all the natural frequencies, M_r will follow the general shape of the curve shown in Fig. 38.8. The number of crossings of zero amplitude, excluding the origin, is equal to the number of rotors minus one. The location of the trial values of ω_s^2 on this diagram can be determined from the sign of M_r and the number of changes of sign of the relative amplitudes β opposite the rotor numbers in the sequence table. In the above example, there is one change of sign of β and M_r is negative. Therefore the assumed value of ω_s^2 lies between the first and second modes.

To find the first mode, assume a new trial value of $\omega_s^2 = 0.045$. This leads to $M_r = +0.0253$. Interpolating for a zero M_r between $\omega_s^2 = 0.047$ and 0.045 yields 0.0457. Table 38.2 is constructed for this value and represents the first mode conditions with a frequency error of less than 1/1,000, a much higher accuracy than is justified by the data on which it is based.

The second mode frequency corresponding to the condition in which there are two nodes in the engine shaft is approximated by

$$\omega_s^2 = \frac{22}{n^2}$$

where n is the number of cylinders. In the example of Fig. 38.7:

$$\omega_s^2 = 22/64 = 0.345$$

By trial sequence calculation, this is found to be somewhat low, and a value which will make M_r nearly zero is $\omega_s^2 = 0.360$, as indicated in Table 38.3. The natural frequencies of the actual system are given by Eq. (38.11):

$$\omega_n^2 = \frac{175.7 \times 10^6 \omega_s^2}{35.8} = 4{,}910{,}000\omega_s^2$$

Table 38.2. Sequence Calculations for First Mode of System Shown in Fig. 38.7 ($\omega_s^2 = 0.0457$)

Rotor number	I	$I\omega_s^2$	Deflection β	Torque M, in.-lb	k
1	1.000	0.0457	1.0000	0.0457	
			0.0457	0.0457	1.00
2	0.722	0.0330	0.9543	0.0315	
			0.0772	0.0772	1.00
3	0.722	0.0330	0.8771	0.0290	
			0.1062	0.1062	1.00
4	1.000	0.0457	0.7709	0.0351	
			0.1413	0.1413	1.00
5	1.000	0.0457	0.6296	0.0288	
			0.1701	0.1701	1.00
6	0.722	0.0330	0.4595	0.0152	
			0.1853	0.1853	1.00
7	0.722	0.0330	0.2742	0.0090	
			0.1943	0.1943	1.00
8	1.000	0.0457	0.0799	0.0036	
			0.1499	0.1979	1.32
9	0.265	0.0121	−0.0700	−0.0008	
			0.0325	0.1971	6.06
10	42.500	1.9422	−0.1025	−0.1990	
			Residual torque M_r =	−0.0019	

Table 38.3. Sequence Calculations for Second Mode of System Shown in Fig. 38.7 ($\omega_s^2 = 0.360$)

Rotor number	I	$I\omega_s^2$	Deflection β	Torque M, in.-lb	k
1	1	0.036	1.0000	0.36	
			0.3600	0.36	1.00
2	0.722	0.260	0.6400	0.166	
			0.5260	0.526	1.00
3	0.722	0.260	0.1140	0.0296	
			0.5556	0.5556	1.00
4	1	0.36	−0.4416	−0.1580	
			0.3976	0.3976	1.00
5	1	0.36	−0.8392	−0.3020	
			0.0956	0.0956	1.00
6	0.722	0.260	−0.9348	−0.2430	
			−0.1474	−0.1474	1.00
7	0.722	0.260	−0.7874	−0.2045	
			−0.3519	−0.3519	1.00
8	1	0.36	−0.4355	−0.1570	
			−0.3850	−0.5089	1.32
9	0.265	0.095	−0.0505	−0.0048	
			−0.0848	−0.5137	6.06
10	42.3	15.23	+0.0343	0.5230	
			Residual torque	$M_r = +0.0093$	

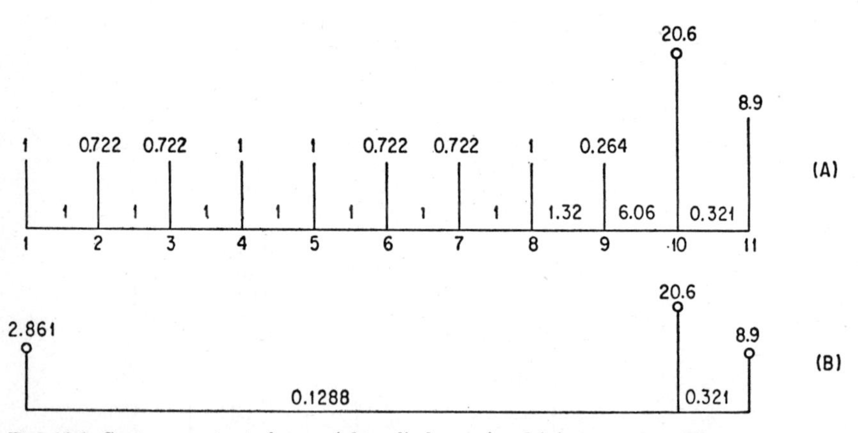

FIG. 38.9. System constants for an eight-cylinder engine driving a pump. The constants on the unity system for the actual engine are shown at A, and for the equivalent three-mass system at B.

Then the natural frequencies in the first and second modes are:

$$\text{1st mode: } \omega_s^2 = 0.0457;\ \omega^2 = 224{,}000;\ f = 75.3 \text{ cps or } 4{,}318 \text{ cpm}$$

$$\text{2nd mode: } \omega_s^2 = 0.3600;\ \omega^2 = 1{,}765{,}000;\ f = 212 \text{ cps or } 12{,}720 \text{ cpm}$$

The critical speeds then are given by Eq. (38.1) and expressed as follows:

Order	4	4½	5	5½	6	6½	7	7½	8	12
1-noded (rpm)....	1,125	1,000	900	820	750	692	643	600	562	375
2-noded (rpm)....	...	...	...	...	...	...	...	...	1,670	1,115

The maximum operating speed of the engine is 950 rpm. All the critical speeds in the above table may not be of importance.

The mass elasticity system of the above engine driving a pump is shown in Fig. 38.9*A*. Treating this as a three-mass system as shown in Fig. 38.9*B*, and then applying Eq. (38.10), the frequency constants of the unity system are approximately

$$\omega_s^2 = 0.0419 \text{ (1st mode)};\quad \omega_s^2 = 0.0609 \text{ (2nd mode)}$$

By trial sequence calculations, more exact values are found to be

$$\omega_s^2 = 0.0383 \text{ (1st mode)};\quad \omega_s^2 = 0.0633 \text{ (2nd mode)}$$

The calculations are shown in Tables 38.4 and 38.5. This case differs markedly from the first example calculated in that both the first and second modes are now of importance. The nodal point of the first mode is now in the pump shaft instead of the engine shaft, and the nodal points of the second mode are in the engine shaft and the pump shaft. The third mode nearly corresponds to the second mode of the previous case. A similar distribution of modes holds for marine-propeller-drive installations.

Table 38.4. Sequence Calculations for First Mode of System Shown in Fig. 38.9 ($\omega_s^2 = 0.0383$)

Rotor number	I	$I\omega_s^2$	Deflection β	Torque M, in.-lb	k
1	1	0.0383	1.0000	0.0383	
			0.0383	0.0383	1.00
2	0.722	0.0276	0.9617	0.0265	
			0.0648	0.0648	1.00
3	0.722	0.0276	0.8969	0.0247	
			0.0895	0.0895	1.00
4	1	0.0383	0.8074	0.0309	
			0.1204	0.1204	1.00
5	1	0.0383	0.6870	0.0263	
			0.1467	0.1467	1.00
6	0.722	0.0276	0.5403	0.0149	
			0.1616	0.1616	1.00
7	0.722	0.0276	0.3787	0.0104	
			0.1720	0.1720	1.00
8	1	0.0383	0.2067	0.0079	
			0.1360	0.1799	1.32
9	0.265	0.0102	0.0707	0.0007	
			0.0298	0.1806	6.06
10	20.6	0.788	0.0409	0.0322	
			0.6650	0.2128	0.321
11	8.9	0.341	−0.6241	−0.2130	
			Residual torque $M_r =$	−0.0002	

Table 38.5. Sequence Calculations for Second Mode of System Shown in Fig. 38.9 ($\omega_s^2 = 0.0633$)

Rotor number	I	$I\omega_s^2$	Deflection β	Torque M, in.-lb	k
1	1	0.0633	1.0000	0.0633	
			0.0633	0.0633	1.00
2	0.722	0.0457	0.9367	0.0428	
			0.1061	0.1061	1.00
3	0.722	0.0457	0.8306	0.0380	
			0.1441	0.1441	1.00
4	1	0.0633	0.6865	0.0435	
			0.1876	0.1876	1.00
5	1	0.0633	0.4989	0.0316	
			0.2192	0.2192	1.00
6	0.722	0.0457	0.2797	0.0128	
			0.2320	0.2320	1.00
7	0.722	0.0457	0.0477	0.0022	
			0.2342	0.2342	1.00
8	1	0.0633	−0.1865	−0.0118	
			0.1083	0.2224	1.32
9	0.265	0.0168	−0.3548	−0.0060	
			0.0336	0.2164	6.06
10	20.6	1.305	−0.3884	−0.5070	
			−0.9070	−0.2906	0.321
11	8.9	0.564	0.5180	0.2920	
			Residual torque M_r = 0.0014		

RESONANT AMPLITUDES

AMPLITUDE FORMULAS

The vibration amplitude at resonance is determined by the magnitude, points of application, and phase relations of the exciting torques produced by gas pressure and inertia, and the magnitudes and points of application of the damping torques.

Damping effects come from a variety of sources, such as pumping action in the engine bearings, hysteresis in the shafting and between fitted parts, energy absorbed in the engine frame and foundation, etc. In a few cases, notably with a marine propeller, the damping of the propeller predominates; this permits a rational calculation of the resonant amplitude. This is also possible when the engine is fitted with a damper. In the majority of applications, the uncertainties in the damping make a semiempirical approach necessary.

Both rational and empirical formulas for the resonant amplitudes of engines without dampers are based on the energy balance at resonance. It is assumed that the system vibrates in a normal mode and that the displacement is in a 90° phase relationship to the exciting and damping torques. Then the energy input by the exciting torques is equal to the energy output by the damping torques. Unless the damping is extremely large, this gives a very close approximation to the amplitude at resonance.

Figure 38.10 shows a curve of relative amplitude in the first mode of vibration. Assume that a cylinder acts at A. Let the actual amplitude at A be θ_a, and the amplitude relative to that of the No. 1 cylinder be β. The β values are taken from the column opposite each rotor number in the sequence calculation for the natural frequency. At a point such as B, where damping may be applied, let the actual amplitude be θ_d and the amplitude relative to the No. 1 cylinder be β_d.

The energy input to the system from the cylinder acting at A is

$$\pi M_e \theta_a \qquad \text{in.-lb/cycle}$$

and the energy output to the damper is

$$\pi c \omega \theta_d^2 \qquad \text{in.-lb/cycle}$$

where c * is the damping constant action of the damper at B. Equating input to output:

$$M_e \theta_a = c \omega \theta_d^2 \tag{38.12a}$$

Let θ' be the amplitude at the No. 1 cylinder produced by the cylinder acting at A. Then $\theta_a/\theta' = \beta$ and $\theta_d/\theta' = \beta_d$. Substituting in Eq. (38.12a) gives

$$\theta' = \frac{M_e \beta}{c \omega \beta_d^2} \tag{38.12b}$$

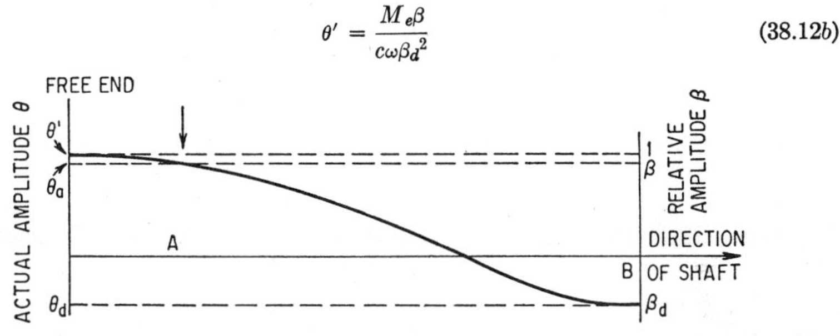

FIG. 38.10. Diagram of actual amplitude θ and relative amplitude β as a function of position along shaft. Excitation is at A, and B is the position where damping is applied. The No. 1 cylinder is at the free end of the crankshaft.

If all the cylinders act, and if damping is applied at a variety of points, the total amplitude at the No. 1 cylinder is

$$\theta = \Sigma \theta' = \frac{M_e \Sigma \beta}{\omega \Sigma c \beta_d^2} \tag{38.13}$$

where $\Sigma\beta$ is taken over the cylinders and $\Sigma c\beta_d^2$ is taken over the points at which damping is applied. This formula can be applied directly when the magnitude and points of application of the damping torques are known. For the great majority of applications, where the damping is unknown, a number of empirical formulas have been proposed with coefficients based on engine tests. These formulas may give an amplitude varying 30 per cent or more from test results if applied to a variety of engines. Better agreement should not be expected, for even identical engines may have amplitudes differing as much as 2 to 1, depending on length of service, bearing fits, mounting, variation in the harmonic excitation because of different combustion rates, and other unknown factors.[11]

Good results have been obtained using the Lewis formula [2]

$$M_m = \mathfrak{R} M_e \Sigma \beta \tag{38.14}$$

The maximum torque at resonance in any part of the system is M_m; the exciting torque per cylinder is M_e. The vector sum over the cylinders of the relative amplitudes as taken from the sequence calculation for a natural frequency (Table 38.2, for example) is $\Sigma\beta$. It is determined as follows.

For a four-cycle engine construct a phase diagram of the firing sequence in which 360° corresponds to a complete cycle of a single cylinder, or two revolutions. The phase relationship for a critical of order number q is obtained by multiplying the angles in this diagram by $2q$, with the No. 1 crank held fixed. The β values assigned to each direction then are obtained from the values corresponding to each cylinder in the β column of the sequence calculation. Then $\Sigma\beta$ is the vector sum. The summation extends only to those rotors on which exciting torques act.

* The symbol c is used in this chapter to denote a torsional damping coefficient.

In a two-cycle engine the β phase relations are determined by multiplying the crank diagram by q, holding the No. 1 cylinder fixed.

Table 38.6 shows the $\Sigma\beta$ phase diagrams and $\Sigma\beta$ values for the one-noded mode of the example of Table 38.2, with a firing sequence 1, 6, 2, 5, 8, 3, 7, 4. The firing sequence is drawn first; then the angles of this diagram are multiplied by 2, 3, 4, etc., in succeeding diagrams. After multiplication by eight for the fourth order, the diagrams repeat. Diagrams which are equidistant in order number from the 2, 6, 10, etc., orders are mirror images of each other and have the same $\Sigma\beta$. The numerical values of $\Sigma\beta$ in Table 38.6 have been obtained by calculation, summing the vertical and horizontal components.

The empirical factor $\mathfrak{R}$ is determined by the measurement of amplitudes in running engines. Representative values are:

Bore	*Stroke*	$\mathfrak{R}$
20 in. ×	24 in. or larger	50–60
8 in. ×	10 in.	40–50
4 in. ×	6 in. or smaller	35

The exciting torque per cylinder, M_e in Eq. (38.13), is composed of the sum of the torques produced by gas pressure, inertia force, gravity force, and friction force. The

Table 38.6. Phase Diagrams and Deflections for the Mode Calculated in Table 38.2

PHASE DIAGRAMS	$\Sigma\beta$	ORDERS
FIRING SEQUENCE 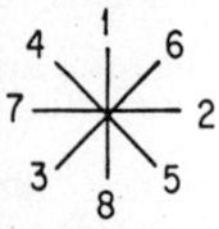 1 4 6 7 2 3 5 8	0.778	1/2, 4½, 8½, – – MIRROR IMAGE FOR 3½, 5½, 7½, – –
×2 OR 6 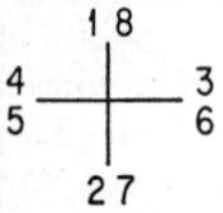 1 8 4 3 5 6 2 7	0.169	1, 5, 9, – MIRROR IMAGE FOR 3, 7, 11
×3 OR 5 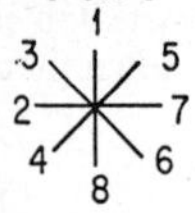 1 3 5 2 7 4 6 8	1.549	1½, 5½, 9½, + – MIRROR IMAGE FOR 2½, 6½, 10½, – –
×4 1,2,7,8  3,4,5,6	0.4287	2, 6, 10, – –
×8 1,2,3,4,5,6,7,8 	5.0455	4, 8, 12 MAJOR ORDERS

cn	β
1	1.0000
2	0.9543
3	0.8771
4	0.7709
5	0.6296
6	0.4595
7	0.2742
8	0.0799
	5.0455

gravity and friction torques are of negligible importance; and the inertia torque is of importance only for first-, second-, and third-order harmonic components.

INERTIA TORQUE. A harmonic analysis of the inertia torque of a cylinder is closely approximated by[2, 8]

$$M = \frac{W}{g}\,\Omega^2 r\left(\frac{\lambda}{4}\sin\theta - \frac{1}{2}\sin 2\theta - \frac{3}{4}\lambda\sin 3\theta - \frac{\lambda^2}{4}\sin 4\theta\cdots\right) \tag{38.15}$$

where $W = W_p + hW_c$, $\lambda = r/l$ } see Fig. 38.1 and Eq. (38.2)

and Ω = angular speed in radians/sec

It is usual to drop all terms above the third order.

GAS-PRESSURE TORQUE. A harmonic analysis of the turning effort curve yields the gas-pressure components of the exciting torque. The turning effort curve is obtained from the indicator card of the engine by the graphical construction shown in Fig. 38.11.

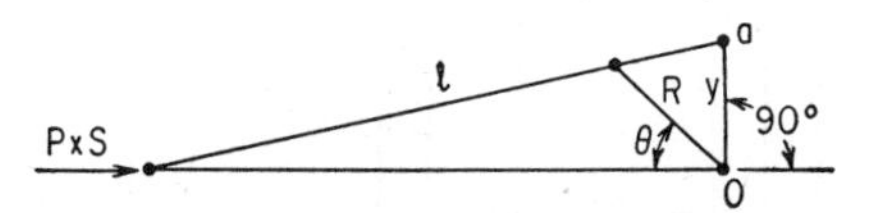

FIG. 38.11. Schematic diagram of crank and connecting rod used in plotting torque curve.

For a given crank angle θ, let the gas pressure on the piston be P. Erect a perpendicular to the line of action of the piston from the crank center, intersecting the line of the connecting rod. Let the intercept Oa on this perpendicular be y. Then the torque M for angle θ is given by

$$M = PSy \tag{38.16}$$

where S = piston area.

HARMONIC ANALYSIS. If an instrumental harmonic analyzer, such as the Coradi, is not available, the harmonic analysis can be made by any one of several rules based on selected ordinates. There are two types:

1. The Runge rules use the same set of ordinates for all harmonics. A forty-eight ordinate rule of this type is described in Refs. 1 and 12. It requires the scaling of forty-eight ordinates and numerous multiplications and additions, and is not recommended unless a digital computer is available.

2. The Fischer-Hinnen and Lewis rules use a different set of ordinates for each harmonic. They do not require the scaling of ordinates and numerical work is reduced to a single multiplication. A single harmonic component, which in many cases is all that is needed, can be obtained in a few minutes independently of all the others.

The Fischer-Hinnen rule is stated as follows: To determine the qth-order coefficients, divide the period into $4q$ intervals and erect ordinates at each division. Beginning at the left end designate them as y_0, y_1, y_2, etc., to y_{4q}. Then

$$a_q = \frac{1}{2q}\left\{(y_1 + y_5 + y_9 + \cdots + y_{4q-3}) - (y_3 + y_7 + \cdots + y_{4q-1})\right\} \qquad \text{(sine terms)}$$

$$b_q = \frac{1}{2q}\left\{(y_0 + y_4 + \cdots + y_{4q-4}) - (y_2 + y_6 + \cdots + y_{4q-2})\right\} \qquad \text{(cosine terms)}$$

This addition is accomplished most easily as follows: Adopt a length as standard for the period τ; 12 in. is convenient. Any curves to be analyzed should be plotted to this standard period. A series of charts is made on tracing cloth or transparent plastic, one for each order number. The ordinates are ruled on these charts, the plus ordinates in full lines and the minus ordinates in dashed lines. Such a chart, as ruled for the fourth-order coefficients, with the ordinates marked X, is shown in Fig. 38.12. It is laid over the curve to be analyzed so that the end points marked "sine" coincide with the ends of the curve period, and so that the base line of the chart is just below the lowest point of the curve, the exact position being immaterial. The positive ordinates are summed by marking them off in succession on one edge of a strip of paper; the negative ordinates are marked on the other edge. The harmonic coefficient is the difference of the sum of the positive and negative ordinates divided by the total number of ordinates $2q$. To obtain the cosine coefficients of the same order number, the chart is shifted until the points marked "cosine" coincide with the end points of the period; then the ordinates are summed as before.

In the analysis of the turning effort curve of a two-cycle engine, it is convenient to treat the compression portion by reversing it end to end and inverting. Sum the ordinates from this curve and add to the sum from the expansion portion. Thus, only the left half of the analysis chart is used. The divisor is $2q$.

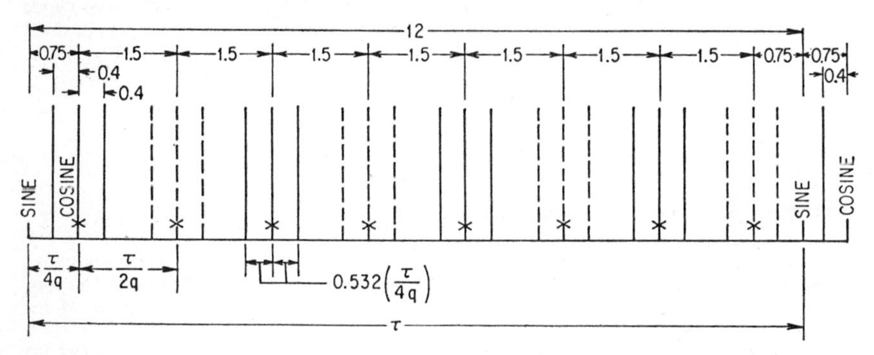

FIG. 38.12. Harmonic analysis chart, fourth-order spacing. For the Fischer-Hinnen rule use only the ordinates marked x; then a_n or $b_n = (1/2q)\Sigma$ ordinates, where full lines indicate positive values and the dashed lines negative values. For the three ordinate Lewis rule use all ordinates, full lines positive, dashed lines negative. Then a_q or $b_q = (2/3\pi q) \times \Sigma$ ordinates.

For a four-cycle engine the fundamental period is two revolutions, or 720°, but from 180 to 540° the gas-pressure torque is negligible. Charts of the type shown in Fig. 38.12 having a 12″ period are used for integral order coefficients; and an additional set, constructed according to the same rule, is required for the fractional order coefficients. The compression portion of the turning effort curve from 540 to 360° is inverted and reversed end to end and laid in the 0 to 180° interval so that the 720° point coincides with the 0° point; then the summation from this portion is added to the summation from the expansion portion. The divisor for the four-cycle engine is now $4q$ instead of $2q$ as for the two-cycle engine.

For the low order numbers the above rule makes the coefficients a function of a small number of ordinates and the accuracy may be low. The Lewis rules provide greater accuracy.[13] The three ordinate rule is shown in Fig. 38.12. Lay in additional ordinates $0.532/4q$ to the right and left of the ordinates marked x and sum as before. Then

$$a_q \text{ or } b_q = \frac{2}{3\pi q} \times \Sigma \text{ ordinates}$$

Rules for 5 and 7 ordinates per half of the interval $\tau/2q$ are also available [13] and have still higher accuracy. The rules are based on the assumption that within this interval the portion of the curve to be analyzed is a parabola of second, fourth, or sixth order for the 3, 5, and 7 ordinates rules, respectively. In most cases, this is a very close approximation, but no rule based on the use of a limited number of ordinates can be more than an approximation.

THE COEFFICIENTS OF STANDARD INDICATOR CARDS. In some cases it may be possible to use prepared analyses obtained from standard sets of indicator cards. One such set of diesel indicator cards and the harmonic analysis of the corresponding

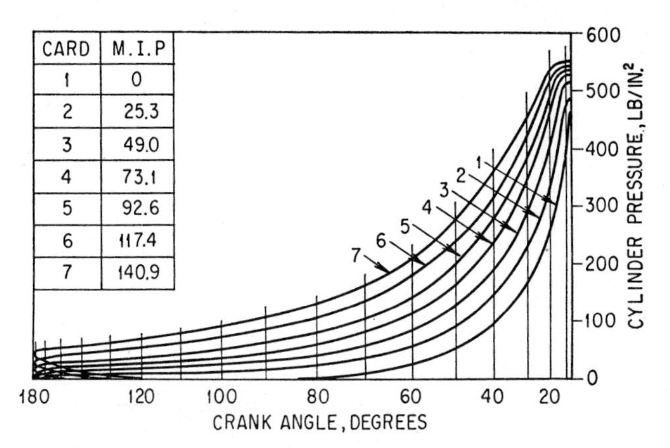

FIG. 38.13. Standard diesel indicator diagrams for engine with $l/r = 4.25$. The mean indicated pressure is shown in the inset table. (*F. M. Lewis.*[2])

turning effort curves are shown in Figs. 38.13 to 38.17. Except for the first, second, and third orders, the resultant harmonic coefficients are given by

$$h = \sqrt{a^2 + b^2}$$

Then the exciting torque per cylinder is

$$M_e = Srh \tag{38.17}$$

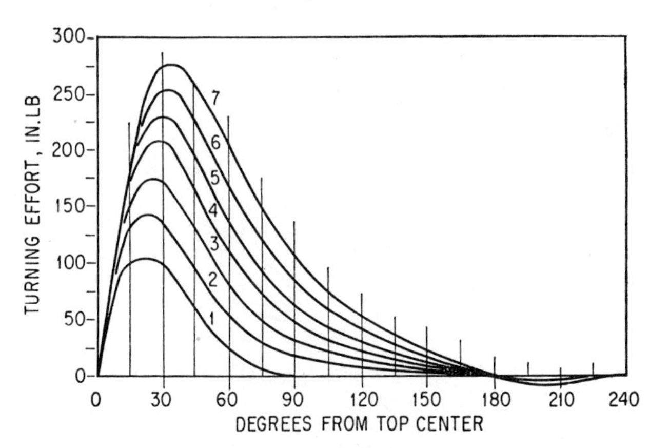

FIG. 38.14. Turning effort curves derived from indicator cards of Fig. 38.13 for a single-cylinder engine of 1-in.² piston area and 1-in. crank radius. (*F. M. Lewis.*[2])

The coefficients are for four-cycle engines; for two-cycle engines multiply by 2; there are no fractional orders.

For the first, second, and third orders, the sine inertia component of the turning effort curve is computed by Eq. (38.15) and added algebraically to the sine gas-pressure component. Then the resultant M_e is the square root of the sums of the squares of the sine and cosine components. Analyses for other standard diesel indicator cards are given in Ref. 14.

The indicator cards of modern high-speed diesel engines with supercharging may differ considerably from the standard cards of Fig. 38.13, and a full analysis may be necessary. Approximate values of the coefficients for such engines can be obtained by multiplying

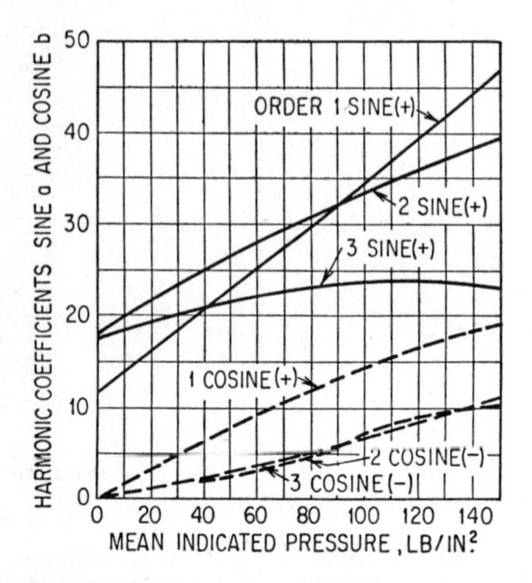

Fig. 38.15. Harmonic coefficients of sine and cosine terms for first, second, and third orders for indicator diagrams shown in Fig. 38.13. (*F. M. Lewis.*[2])

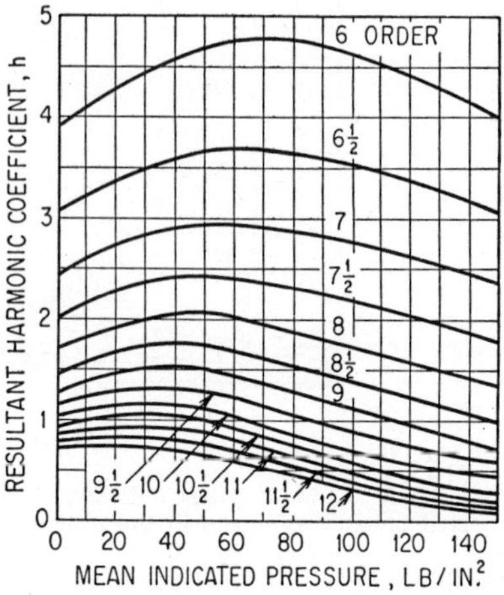

Fig. 38.16. Resultant harmonic coefficients for sixth and higher orders for indicator diagrams shown in Fig. 38.13. (*F. M. Lewis.*[2])

the coefficients of Figs. 38.15 to 38.17 by the ratio of the maximum combustion pressure to that shown by the cards of Fig. 38.13, with the additional multiplication by 2 for a two-cycle engine. The higher-order coefficients are extremely sensitive to the details of the fuel-injection and combustion process; unless good experimental indicator cards are available for harmonic analysis, it may not be possible to obtain accuracy in estimating exciting torques.

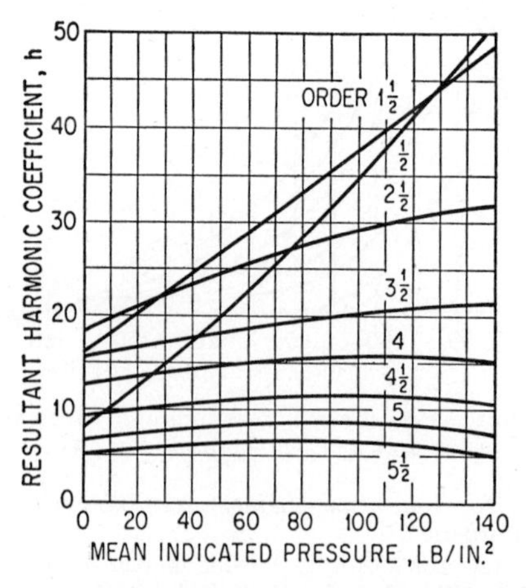

Fig. 38.17. Resultant harmonic coefficients for lower orders of indicator diagrams shown in Fig. 38.13. (*F. M. Lewis.*[2])

For a four-cycle gasoline engine the harmonic coefficients are more nearly proportional to the mean indicated pressure and are approximated by the values in Table 38.7.

Table 38.7. Harmonic Components of Gas-pressure Torque for a Four-cycle Gasoline Engine

Order	½	1	1½	2	2½	3	3½		
Resultant coefficient for h	34		31		17		10		
Sine coefficient for a		31		21		11.5			
Cosine coefficient for b		10		−4		4.5			
Order	4	4½	5	5½	6	6½	7	7½	
h	7.5	5.5	4	3.2	2.9	2.45	2	1.68	
Order	8	8½	9	9½	10	10½	11	11½	12
h	1.5	1.32	1.15	1.10	0.88	0.78	0.71	0.58	1.52

$$h,\ a,\ b = \frac{\text{mean indicated pressure}}{100} \times \text{coefficients above.}$$

$$M_e = Sr \times (h \text{ or } a \text{ or } b)$$

Table 38.8. Amplitude Estimates for First Mode of Diesel Engine Calculated in Table 38.2

Order	Rpm	$\Sigma\beta$	h	M_e, in.-lb	M_m, in.-lb	Stress	θ_1 rad
4	1,125	5.055	29.5	11,300	2,860,000	49,800	0.08257
4½	1,000	0.777	23.5	9,000	349,000	6,070	0.0105
5	900	0.164	18.5	7,080	58,000	1,005	
5½	820	1.54	15.0	5,730	441,000	7,700	0.0127
6	750	0.430	12.0	4,580	98,500	1,715	
6½	692	1.54	8.8	3,370	259,000	4,500	0.0075
7	643	0.164	6.5	2,490	20,400	355	
7½	600	0.777	5.5	2,110	81,900	656	
8	562	5.055	3.9	1,490	377,000	6,550	0.0108
12	375	5.055	1.4	537	136,000	2,380	0.0039
4	950	. . .	29.5	11,300	230,000	4,010	

Example 38.3. Table 38.8 gives the amplitude estimate for the first mode of the diesel engine, the frequency of which was calculated in Table 38.2. The engine is four-cycle, eight-cylinder with 9-in. bore by 12-in. stroke. The values of speed in rpm are as previously calculated, and $\Sigma\beta$ is taken from Table 38.6. The harmonic coefficients h are obtained from a harmonic analysis of an indicator card. Then the exciting torques are

$$M_e = \frac{\pi}{4} \times 9^2 \times 6 \times h = 383h \qquad \text{in.-lb}$$

For the maximum torques, M_m, a value of $\mathfrak{R} = 50$ was assumed in the equation

$$M_m = \mathfrak{R} M_e \Sigma\beta \qquad \text{in.-lb}$$

The crankpin diameter is 6½ in. so that the nominal stress is

$$\tau = \frac{16M_m}{\pi \times (6.5)^3} = 0.0174M_m \qquad \text{lb/in.}^2$$

From Table 38.2 (unity system) the torque in shaft section 8–9 is 0.1979 for unity amplitude at the No. 1 cylinder. Therefore the vibration amplitude at the No. 1 cylinder in the actual installation is

$$\theta = \frac{M_m}{0.1979 \times 175.7 \times 10^6} = \frac{M_m}{34.7 \times 10^6} \quad \text{radians}$$

TWO-ROTOR SYSTEMS

The lowest mode of vibration of some systems, particularly marine installations, can be approximated with a two-rotor system having an excitation M_e applied at one end and the damping at the other.

Referring to Fig. 38.18 the torque equation for rotor I_1 is

$$I_1\omega^2\theta_1 - k(\theta_1 - \theta_2) + M_e = 0 \tag{38.18}$$

The corresponding equation for rotor I_2 is

$$I_2\omega^2\theta_2 + k(\theta_1 - \theta_2) - jc\omega\theta_2 = 0 \tag{38.19}$$

The resonant frequency is given by

$$\omega^2 = \frac{k(I_1 + I_2)}{I_1 I_2}$$

The shaft torque is $M_{12} = k(\theta_1 - \theta_2)$. Solving the above equations, the amplitude of M_{12} at resonance is

$$|M_{12}| = k|\theta_1 - \theta_2| = M_e\frac{I_2}{I_1}\sqrt{1 + \frac{kI_2(I_1 + I_2)}{I_1 c^2}} \tag{38.20}$$

Since with usual damping the second term under the radical is large compared with unity, Eq. (38.20) reduces to

$$|M_{12}| \simeq \frac{M_e}{c}\frac{I_2}{I_1}\sqrt{(I_1 + I_2)\frac{I_2 k}{I_1}} \tag{38.21}$$

The torsional damping constant c of a marine propeller is a matter of some uncertainty. It is customary to use the "steady-state" value as derived in Ref. 3. This is approximated by

$$c = \frac{4M_{\text{mean}}}{\Omega} \quad \text{in.-lb/rad/sec}$$

where Ω = angular speed of shaft in radians per second. Considerations of oscillating airfoil theory indicate that this is too high and that a better value would be

$$c = \frac{2.3M_{\text{mean}}}{\Omega} \quad \text{in.-lb/rad/sec} \tag{38.22}$$

Equation (38.21) is applicable only where $I_1/I_2 > 1$ approximately. If used outside this range with other types of damping neglected, fictitiously large amplitudes will be obtained. Equation (38.21) gives the resonant amplitude, but the peak may not occur exactly at resonance. The complete amplitude curve is computed by the methods discussed in the section on *Nonresonant Amplitude*.

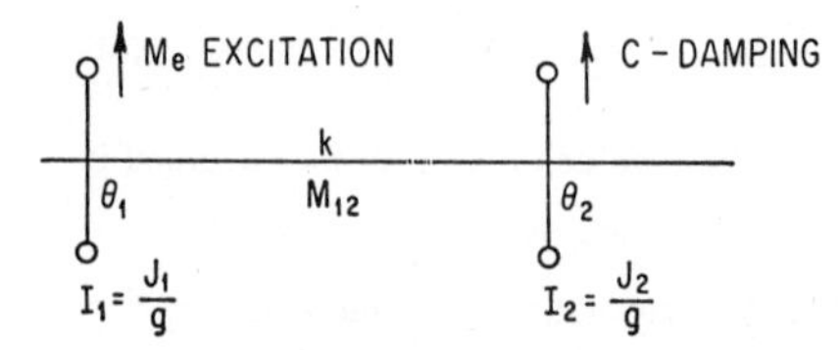

FIG. 38.18. Schematic diagram of shaft with two rotors showing positions of excitation and damping.

PERMISSIBLE AMPLITUDES

Failure caused by torsional vibration invariably initiates in fatigue cracks starting at points of stress concentration, such as the ends of keyway slots, the fillets at changes of shaft size, and, in particular, the oil holes in the crankshaft. Failure also may start at corrosion pits, such as may exist in marine shafting. At the shaft oil holes the cracks start on lines at 45° to the shaft axis and grow in a spiral pattern until failure occurs. Theoretically the stress at the edges of the oil holes is four times the mean shear stress in the shaft, and failure may be expected if this concentrated stress exceeds the fatigue limit of the material. The problem of estimating the stress required to cause failure is further complicated by the presence of the steady stress from the mean driving torque and the variable bending stresses.

In practice the severity of a critical speed is judged by the maximum nominal torsional stress:

$$\tau = \frac{16M_m}{\pi d^3}$$

where M_m is the torque amplitude from torsional vibration and d is the crankpin diameter.

U.S. MILITARY STANDARD

A military standard [22] issued by the U.S. Navy Department states that the limit of acceptable nominal torsional stress within the operating range is:

$$\tau = \frac{\text{ultimate tensile strength}}{25} \quad \text{for steel}$$

$$\tau = \frac{\text{torsional fatigue limit}}{6} \quad \text{for cast iron}$$

If the full-scale shaft has been given a fatigue test, then:

$$\tau = \frac{\text{torsional fatigue limit}}{2} \quad \text{for either material}$$

Such tests are rarely, if ever, possible.

For critical speeds below the operating range which are passed through in starting and stopping, the nominal torsional stress shall not exceed 1¾ times the above values.

Crankshaft steels which have ultimate tensile strengths between 75,000 and 115,000 lb/in.^{2} usually have * torsional stress limits of 3,000 to 4,600 lb/in.^{2}

For gear drives the vibratory torque across the gears, at any operating speed, shall not be greater than 75 per cent of the driving torque at the same speed, or 25 per cent of full-load torque, whichever is smaller.

GEARED SYSTEMS

The frequencies of systems containing gears may be calculated by two different procedures. Assume a system in which the engine speed is n times the speed of the driven equipment.

Procedure 1. The sequence calculation is started at the engine end and carried to the gear face. At this point a new tabulation is started in which the initial angular amplitude

* By the 1/25th rule.

is $1/n$ times the angular amplitude at the engine side of the gear face and in which the initial torque is n times that at the engine side of the gear face. The sequence then is carried through the remainder of the system.

Procedure 2. Multiply all the inertia and elastic constants on the driven side of the system by $1/n^2$, and calculate the system as if no gears exist. In any calculations involving damping constants on the driven side, these constants likewise are multiplied by $1/n^2$. Torques and deflections thus obtained on the driven side of this substitute system, when multiplied by n and $1/n$, respectively, give those in the actual geared system. Alternatively, the driven side may be used as the reference, multiplying the engine constants by n^2. Procedure 2 usually is preferred to Procedure 1.

BRANCHED SYSTEMS

Where two or more engines are geared to a common load, hydraulic or electrical couplings may be placed between the engines and the gears. These serve as disconnecting clutches; they also insulate the gears from any engine-produced vibration. This insulation is so perfect that the engine end of the system can be calculated as if terminating at the coupling gap. The damping effect of such couplings upon the vibration in the engine end of the system normally is quite small and should be disregarded in amplitude calculations.

The majority of applications without hydraulic or electrical couplings involve two identical engines. For such systems the modes of vibration are of two types:

1. The *opposite-phase* modes in which the engines vibrate against each other with a node at the gear. These are calculated for a single branch in the usual manner, terminating the calculation at the gear. The condition for a natural frequency is that $\beta = 0$ at the gear.

2. The *like-phase* modes in which the two engines vibrate in the same direction against the driven machinery. To calculate these frequencies, the inertia and stiffness constants of the engine side of one branch are doubled; then the calculation is made as if there were only a single engine. The condition for a natural frequency is zero residual torque at the end.

If the two identical engines rotating in the same direction are so phased that the same cranks are vertical simultaneously, all orders of the opposite-phase modes will be eliminated. The two engines can be so phased as to eliminate certain of the like-phase modes. For example, if the No. 1 cranks in the two branches are placed at an angle of 45° with respect to each other, the fourth, twelfth, twentieth, etc., orders, but no others, will be eliminated. If the engines are connected with clutches, these phasing possibilities cannot be utilized.

In the general case of nonidentical branches the calculation is made as follows: Reduce the system to a 1:1 gear ratio as described under *Geared Systems, Procedure* 2. Call the branches a and b. Make the sequence calculation for the a branch with initial amplitude $\beta = 1$, and for the b branch with the initial amplitude the algebraic unknown x. At the junction equate the amplitudes and find x. With this numerical value of the amplitude x substituted, the torques in the two branches and the torque of the gear are added; then the sequence calculation is continued through the last mass.

SINGLE-MEMBER BRANCH

The branch may consist of a single member elastically connected to the system. Examples of such a branch are a flywheel with appreciable flexibility in its spokes or an armature with flexibility in the spider. Let I be the moment of inertia of the flywheel rim and k the elastic constant of the connection. Then the flexibly mounted flywheel is equivalent to a rigid flywheel of moment of inertia

$$I' = \frac{I}{1 - I\omega^2/k} \tag{38.23}$$

This equivalent moment of inertia is negative if $\omega^2_n > k/I$, i.e., if the frequency of the system is higher than the natural frequency of the flywheel with the flywheel hub fixed.

AMPLITUDES AT NONRESONANT FREQUENCIES

A calculation of the nonresonant or "forced" vibration amplitude is required in some cases to define the range of the more severe critical speeds, particularly with geared drives; it also is required in the design of dampers. The calculation is readily made by an extension of the Holzer table sequence method illustrated in Table 38.1. In the sequence the initial amplitude is treated as an algebraic unknown θ. At each cylinder where an exciting torque acts, this torque is added. Assume first that there are no damping torques. Then the residual torque after the last rotor is of the form $a\theta + b$, where a and b are numerical constants resulting from the calculation. Since the residual torque is zero, $\theta = -b/a$.

Table 38.9. Relation between Parameters in Actual and Unity Systems

	Actual	Unity
Inertia constant......	$I = J/g$	1
Spring constant.......	k	1
Damping constant....	c	$c_s = \dfrac{c}{\sqrt{Ik}}$
Torques.............	M_e	1
Deflections...........	$\dfrac{M_e\theta}{k}$	θ

The amplitude and torque at any point of the system are found by substituting this numerical value of θ at the appropriate point in the tabulation. At frequencies well removed from resonance, damping has but little effect and can be neglected. Damping can be added to the system by treating it as an exciting torque equal to the imaginary quantity $-jc\omega\theta$, where c is the damping constant and θ is the amplitude at the point of application. Relative damping between two inertias can be treated as a spring of a stiffness constant equal to the imaginary quantity $+jc\omega$.

For the major critical speeds the exciting torques are all in phase and are real numbers. For the minor critical speeds the exciting torques are out-of-phase; they must be entered as complex numbers of amplitude and phase as determined from the phase diagram (discussed under *Resonant Amplitude Calculation*) for the critical speed of the order under consideration. With damping and/or out-of-phase exciting torques introduced, a and b in the equation $a\theta + b = 0$ are complex numbers, and θ must be entered as a complex number in the tabulation in order to determine the angle and torque at any point. The angles and torques are then of the form $r + js$, where r and s are numerical constants and the amplitudes are equal to $\sqrt{r^2 + s^2}$.

While the calculation can be made with the moments of inertia, stiffness constants, and exciting torques of the actual system, the unity form of the calculation used in frequency calculations [see Eq. (38.11)] saves considerable labor. In the forced-vibration calculation the amplitude of the exciting torque also is assumed to be unity. It is advisable, particularly when the calculation is made for damper design, to start the sequence at the flywheel end of the system. The introduction of exciting and damping terms thus is deferred as long as possible.

The relationship between the actual and the unity system is given in Table 38.9. Here $I = J/g$ is the moment of inertia by which all the moments of inertia are divided and k is the common stiffness constant by which all the stiffness constants are divided.

A damping constant c in the actual system becomes

$$c_s = \frac{c}{\sqrt{Ik}}$$

in the unity system. Exciting torques are assumed to have unity amplitude. After the torque in any part of the unity system is found, the torque in the actual system is determined by multiplying by M_e, and deflections in the actual system are obtained by multiplying those in the unity system by M_e/k.

Example 38.4. Determine the amplitude of the fourth-order vibration at the maximum speed of 950 rpm for the eight-cylinder, four-cycle engine driving a generator shown in Fig. 38.7.

The frequency in the unity system corresponding to fourth-order vibration at 950 rpm in the actual system is given by

$$\omega_s^2 = \left(\frac{2\pi \times 950 \times 4}{60}\right)^2 \times \frac{35.8}{175.7 \times 10^6} = 0.0321$$

$$\omega_s = 0.179$$

The calculation is carried out in Table 38.10. Assuming no damping, the residual torque is

$$M_r = 0.3946\theta + 5.852 = 0$$

from which $\theta = -14.85$.

The maximum torque occurs in section 8–9; it is given by

$$M_{8,9} = 1.3726\theta = -20.4$$

In section 6–7 the torque is given by

$$M_{6,7} = 1.326\theta + 1.977 = -17.73$$

Table 38.10. Sequence Calculations for Forced Vibration of Engine Generator Shown in Fig. 38.7 at Operating Speed of 950 rpm ($\omega_s^2 = 0.0321$)

Rotor number	I	$I\omega_s^2$	β		M, in.-lb		k
			Inertia $\theta \times$	Exciting	Inertia $\theta \times$	Exciting	
10	42.5	1.366	1.0000	. . .	1.366		
			0.226	. . .	1.366	. . .	6.06
9	0.265	0.0085	0.774	. . .	0.0066		
			1.042	. . .	1.3726	. . .	1.32
8	1	0.0321	−0.268	. . .	−0.0086		
			1.3640	1.000	1.3640	1.000	1.00
7	0.722	0.0231	−1.632	−1.000	−0.0377	−0.023	
			1.3263	1.977	1.3263	1.977	1.00
6	0.722	0.0231	−2.9583	−2.977	−0.0682	−0.069	
			1.2581	2.908	1.2581	2.908	1.00
5	1	0.0321	−4.2164	−5.885	−0.1350	−0.189	
			1.1231	3.719	1.1231	3.719	1.00
4	1	0.0321	−5.3395	−9.604	−0.1715	−0.308	
			0.9516	4.411	0.9516	4.411	1.00
3	0.722	0.0231	−6.2911	−14.015	−0.1452	−0.324	
			0.8064	5.087	0.8064	5.087	1.00
2	0.722	0.0231	−7.0975	−19.102	−0.1638	−0.441	
			0.6426	5.646	0.6426	5.646	1.00
1	1	0.0321	−7.7401	−24.748	−0.2480	−0.794	
					0.3946	5.852	

Fourth-order torque M_e in the actual engine is 11,300 in.-lb (Table 38.8); in the actual engine the fourth-order torque in section 8–9 is

$$M = 20.4 \times 11{,}300 = 231{,}000 \text{ in.-lb}$$

At the No. 1 cylinder the amplitude in the unity system is

$$-7.7401\theta - 24.748 = 90.3 \text{ radians}$$

In the actual system the corresponding amplitude is

$$\frac{90.3 \times 11{,}300}{175.7 \times 10^6} = 0.0058 \text{ radian}$$

The problem also can be solved by assuming that a concentrated damping of constant $c = 1{,}470$ in.-lb/rad acts at the No. 1 crank. In the unity system the corresponding damping constant is

$$c_s = \frac{1{,}470}{\sqrt{35.8 \times 175.7 \times 10^6}} = 0.0185$$

At the No. 1 crank the torque $-jc_s\omega_s\theta$ is added; this torque is

$$M = -j \times 0.0185 \times 0.175 \times (-7.7401\theta - 24.748) = j(0.0256\theta + 0.0819)$$

The resulting residual torque is given by

$$M_r = 0.3946\theta + 5.852 + j(0.0256\theta + 0.0819) = 0$$

Solving for θ:

$$\theta = \frac{-5.852 - 0.0819j}{0.3946 + 0.0256j}$$

The amplitude of the deflection is

$$|\theta| = \sqrt{\frac{(5.852)^2 + (0.0819)^2}{(0.3946)^2 + (0.0256)^2}} = 14.85$$

which is the same as if damping were not present; thus, at this speed the effect of the external damping is completely negligible.

MEASURES OF CONTROL

The various methods which are available for avoiding a critical speed or reducing the amplitude of vibration at the critical speed may be classified as:

1. Shifting the values of critical speeds by changes in mass and elasticity
2. Vector cancellation methods
3. Change in mass distribution to utilize the inherent damping in the system
4. Addition of dampers of various types

SHIFTING OF CRITICAL SPEEDS

If the stiffness of all the shafting in a system is increased in the ratio a, then all the frequencies will increase in the ratio $\sqrt{a}$, provided that there is no corresponding increase in the inertia. It is rarely possible to increase the crankshaft diameters on modern engines; in order to reduce bearing pressures, bearing diameters usually are made as large as practical. If bearing diameters are increased, the increase in the critical speed will be much smaller than indicated by the $\sqrt{a}$ ratio because a considerable increase in the inertia will accompany the increase in diameter. Changes in the stiffness of a system made near a nodal point will have maximum effect. Changes in inertia near a loop will have maximum effect while those near a node will have little effect.

By the use of elastic couplings it may be possible to place certain critical speeds below the operating speed where they are passed through only in starting and stopping; this leaves a clear range above the critical speed. This procedure must be used with caution because some critical speeds, for example the fourth order in an eight-cylinder, four-cycle engine, are so violent that it may be dangerous to pass through them. If the acceleration through the critical speed is sufficiently high, some reduction in amplitude may be attained,[16] but with a practical rate the reduction may not be large. The rate of deceleration when stopping is equally important. In some cases mechanical clutches disconnect the driven machinery from the engine until the engine has attained a speed above dangerous critical speeds. Elastic couplings may take many forms including helical springs arranged tangentially, flat leaf springs arranged longitudinally or radially, various arrangements using rubber, or small-diameter shaft sections of high tensile steel.[20]

VECTOR CANCELLATION METHODS

CHOICE OF CRANK ARRANGEMENT AND FIRING ORDER. The amplitude at certain minor critical speeds sometimes can be reduced by a suitable choice of crank arrangement and firing order (i.e., firing sequence). These fix the value of the vector sum $\Sigma\beta$ in Eq. (38.14), $M_m = \mathfrak{R} M_e \Sigma\beta$. But considerations of balance, bearing pressures, and internal bending moments restrict this freedom of choice. Also, an arrangement which decreases the amplitude at one order of critical speed invariably increases the amplitude at others. In four-cycle engines with an even number of cylinders, the amplitude at the half-order critical speeds is fixed by the firing order because this determines the $\Sigma\beta$ value. Tables 38.11 to 38.18 list the balance and torsional-vibration characteristics for the crank arrangements and firing orders, for six to ten cylinders, having the most desira-

Table 38.11. Balance and Torsional-vibration Characteristics for Six-cylinder, Four-cycle Engine Having 120° Crank Spacing

	$\Sigma\beta$ of orders					
Cranks at 1, 6 0° 2, 5 120° 3, 4 240°	½, 2½, 3½, 5½, 6½, 8½,	1½ 4½ 7½	1, 2 4, 5 7, 8	3 6 9	C_1	C_2
Firing sequence	. . .	. . .	. . .	12		
1, 5, 3, 6, 2, 4	0.577	1.5	0	3.5	0	0
1, 2, 4, 6, 5, 3	1.2	0.166	0	3.5	0	0

Values of 0 in the $\Sigma\beta$ column indicate small but not necessarily 0 values for actual β distributions.

Table 38.12. Balance and Torsional-vibration Characteristics for Six-cylinder, Four-cycle Engine Having 60° Crank Spacing

Firing sequence	$\Sigma\beta$ of orders				C_1	C_2
	1, 5, 7, 11	2, 4, 8, 10	3, 9	6, 12		
1, 5, 3, 4, 2, 6	0	0.577	1.5	3.5	0	$2\sqrt{3}$
1, 5, 3, 6, 2, 4	0.577	0	1.5	3.5	$2\sqrt{3}$	0

ble properties. All first- and second-order forces are balanced for these crank arrangements. The unbalanced couple coefficients C_1 and C_2 are defined as follows:

First-order unbalanced couple:

$$M_1 = \frac{C_1(W_p + hW_c)ra\Omega^2}{g} \quad \text{in.-lb} \tag{38.24}$$

where W_p = weight of piston, piston pin, and cooling fluid in pounds and W_c = weight of the connecting rod in pounds [see Eq. (38.2)].

Second-order unbalanced couple:

$$M_2 = \frac{C_2(W_p + hW_c)r^2a\Omega^2}{gl} \quad \text{in.-lb} \tag{38.25}$$

Ω = angular speed of shaft, rad/sec

a = cylinder spacing (assumed equal), in.

l = connecting-rod length, in.

Other terms are as defined with respect to Eq. (38.2).

The values of $\Sigma\beta$ are calculated by assuming $\beta = 1$ for the cylinder most remote from the flywheel, assuming $\beta = 1/n$ for the cylinder adjacent to the flywheel (where n is the number of cylinders), and assuming a linear variation of β therebetween. In any actual installation $\Sigma\beta$ must be calculated by taking β from the relative amplitude curve; however, if the $\Sigma\beta$ as determined above is small, it also will be small for the actual β distribution. These arrangements assume equal crank angles and firing intervals. The reverse arrangements (mirror images) have the same properties.

Table 38.13. Balance and Torsional-vibration Characteristics for Eight-cylinder, Four-cycle Engine Having 90° Crank Spacing

	$\Sigma\beta$· OF ORDERS					C_1	C_2
FIRING ORDER	1/2, 7½, 8½, 3½, 4½	1½, 5½, 6½	2, 6, 10	13, 58, 7	4,8		
1,6,2,5,8,3,7,4	0.745	1.44	0	0	4.5	0	0
1,6,2,4,8,3,7,5	0.686	1.48	0	0	4.5	0	0
1,3,2,5,8,6,7,4	1.48	0.686	0	0	4.5	0	0
6,3,5,7,8,6,4,2	1.74	0.176	0	0	4.5	0	0
1,7,4,3,8,2,5,6	0.176	1.74	0	0	4.5	0	0

CRANK ARRANGEMENT #1,2,3 #4,5

18, 45, 36, 27   18, 27, 36, 45

Table 38.14. Balance and Torsional-vibration Characteristics for Eight-cylinder, Two-cycle Engine Having 45° Crank Spacing

Firing order	$\Sigma\beta$ of orders					C_1	C_2
	1, 7, 9	2, 6, 10	3, 5, 11	4, 12	8, 16		
1, 8, 2, 6, 4, 5, 3, 7	0.056	0	0.79	2.0	4.5	0.448	0
1, 7, 4, 3, 8, 2, 5, 6	0.175	0	1.61	0	4.5	1.405	0
1, 6, 5, 2, 7, 4, 3, 8	0.112	0	1.58	0.5	4.5	0.897	0

Table 38.15. Balance and Torsional-vibration Characteristics for Seven-cylinder, Two- or Four-cycle Engine Having 51.4° Crank Spacing

No.	Crank sequence	$\Sigma\beta$ of orders				C_1	C_2
		1, 6, 7 2½ 4½ 9½	2, 5, 9 1½ 5½ 8½	3, 4, 10 ½ 6½ 7½	7 3½		
1	1, 6, 3, 4, 5, 2, 7	0.04	0.14	1.4	4	0.27	1.01
2	1, 5, 2, 4, 6, 3, 7	0.56	0.04	1.2	4	3.90	0.25
3	1, 4, 7, 2, 3, 5, 6	0.01	1.30	0.6	4	0.08	9.10
4	1, 5, 4, 3, 6, 2, 7	0.18	0.21	1.4	4	1.30	1.50
5	1, 6, 4, 3, 5, 2, 7	0.04	0.33	1.4	4	0.30	2.30

Table 38.16. Balance and Torsional-vibration Characteristics for Nine-cylinder, Two- or Four-cycle Engine Having 40° Crank Spacing

No.	Crank sequence	$\Sigma\beta$ of orders					C_1	C_2
		Order of critical { 1, 8, 10 3½ 5½	2, 7, 11 2½ 6½	3, 6, 12 1½ 7½	4, 5, 13 ½ 8½ 9½	9 4½ 13½		
1	1, 9, 2, 7, 4, 5, 6, 3, 8	0.022	0.055	0.019	1.8	5.0	0.2	0.5
2	1, 9, 2, 5, 7, 6, 3, 4, 8	0.10	0.19	0.94	1.8	5.0	0.8	1.5
3	1, 6, 8, 2, 4, 7, 5, 3, 9	0.10	0.48	1.77	0.91	5.0	0.8	3.8
4	1, 6, 7, 2, 5, 8, 3, 4, 9	0.11	0.14	1.95	0.6	5.0	0.9	1.1
5	1, 9, 2, 7, 5, 3, 6, 4, 8	0.14	0.35	0.51	1.7	5.0	1.1	2.8
6	1, 8, 7, 3, 2, 9, 4, 5, 6						0.028	8.5

Table 38.17. Balance and Torsional-vibration Characteristics for Ten-cylinder, Four-cycle Engine Having 72° Crank Spacing

For all orders given, $C_1 = C_2 = 0$
$\Sigma\beta = 0$ for all whole numbered orders except 5, 10, etc.

No.	Crank sequence	$\Sigma\beta$ of orders			
	Order of critical	½, 4½, 5½, 9½, 10½	1½, 3½, 6½, 8½	2½, 7½, 12½	5, 10, 15
1	1, 7, 9, 3, 6, 10, 4, 2, 8, 5	0.56	1.95	0.1	5.5
2	1, 3, 4, 5, 9, 10, 8, 7, 6, 2	1.95	0.56	0.1	5.5
3	1, 8, 5, 7, 2, 10, 3, 6, 4, 9	0.09	1.00	2.5	5.5
4	1, 7, 3, 9, 5, 10, 4, 8, 2, 6	1.00	0.09	2.5	5.5

Table 38.18. Balance and Torsional-vibration Characteristics for Ten-cylinder, Two-cycle Engine Having 36° Crank Spacing

No.	Crank sequence	$\Sigma\beta$ of order						C_1	C_2
		1, 9, 11	2, 8, 12	3, 7, 13	4, 6	5	10		
1	1, 9, 3, 7, 5, 6, 4, 8, 2, 10	0.00	0.10	0.00	1.11	2.8	5.5	0.00	0.90
2	1, 6, 10, 2, 4, 9, 5, 3, 7, 8	0.03	0.16	1.97	0.48	0.1	5.5	0.28	1.56
3	1, 9, 5, 6, 2, 10, 4, 3, 8, 7	0.03	0.03	0.53	0.80	1.5	5.5	0.33	0.28
4	1, 9, 6, 3, 5, 10, 2, 4, 8, 7	0.04	0.19	1.86	0.12	1.1	5.5	0.45	1.90
5	1, 7, 6, 4, 8, 2, 9, 3, 5, 10	0.05	0.22	0.65	1.90	0.3	5.5	0.51	2.18
6	1, 5, 9, 4, 3, 8, 7, 2, 6, 10	0.06	0.04	1.95	0.50	0.3	5.5	0.57	0.45
7	1, 6, 9, 2, 7, 4, 8, 3, 5, 10	0.06	0.08	1.42	1.56	0.5	5.5	0.60	0.80
8	1, 9, 4, 5, 8, 3, 6, 7, 2, 10	0.06	0.06	0.16	1.80	1.3	5.5	0.62	0.62
9	1, 10, 3, 5, 8, 4, 7, 2, 9, 6	0.07	0.26	0.98	1.76	0.1	5.5	0.65	2.56
10	1, 9, 4, 6, 3, 10, 2, 7, 5, 8	0.09	0.00	1.00	0.00	2.5	5.5	0.90	0.00
11	1, 6, 10, 2, 5, 9, 3, 4, 7, 8	0.10	0.20	1.87	0.72	0.3	5.5	0.98	2.00
12	1, 9, 4, 5, 8, 3, 7, 6, 2, 10	0.10	0.08	0.10	1.87	1.1	5.5	1.00	0.77

V-TYPE ENGINES. In V-type engines, it may be possible to choose an angle of the V which will cancel certain criticals. Letting ϕ be the V angle between cylinder banks, and q the order number of the critical, the general formula is

$$q\phi = 180°, 540°, 1080°, \text{etc.} \tag{38.26}$$

For example, in an eight-cylinder engine the eighth order is canceled at angles of 22½°, 67½°, 112½°, etc.

In four-cycle engines, ϕ is to be taken as the actual bank angle if the second-bank cylinders fire directly after the first and $360° + \phi$ if the second-bank cylinders omit a revolution before firing. In the latter case the cancellation formula is

$$(\phi + 360°)q = n \times 180° \tag{38.27}$$

where $n = 1, 3, 5$, etc. For example, to cancel a 4½ order critical the bank angle should be

$$\phi = \frac{180°}{4.5} = 40° \text{ for direct firing}$$

or

$$\phi = \frac{11 \times 180°}{4.5} - 360° = 80° \text{ for the } 360° \text{ delay}$$

CANCELLATION BY SHIFT OF THE NODE. If an engine can be arranged with approximately equal flywheel (or other rotors) at each end so that the node of a particular mode is at the center of the engine, $\Sigma\beta$ will cancel for the major orders of that mode. This procedure must be used with caution because the double flywheel arrangement may reduce the natural frequency in such a manner that low-order minor criticals of large amplitudes take the place of the canceled major criticals.

REDUCTION BY USE OF PROPELLER DAMPING IN MARINE INSTALLATIONS. From Eq. (38.21) it is evident that the torque amplitude in the shaft can be reduced below any desired level by making the flywheel moment of inertia I_1 of sufficient magnitude. The ratio of the propeller amplitude to the engine amplitude increases as the flywheel becomes larger; thus the effectiveness of the propeller as a damper is increased.

DAMPERS

Many arrangements of dampers can be employed (see Chap. 6). In each type there is a loose flywheel or inertia member which is coupled to the shaft by:

1. Coulomb friction (Lanchester damper)
2. Viscous fluid friction
3. Coulomb or viscous friction plus springs
4. Centrifugal force, equivalent to a spring having a constant proportional to the square of the speed (pendulum damper) (see Chap. 6)

Each of these types acts by generating torques in opposition to the exciting torques.

The Lanchester damper illustrated in Fig. 6.35 has been entirely superseded by designs in which fluid friction is utilized. In the Houdaille damper, Fig. 38.19, a flywheel is mounted in an oiltight case with small clearances; the case is filled with silicone fluid. The damping constant is

$$c = 2\pi\mu \left[\frac{r_2^3 b}{h_2} + \frac{1}{2} \frac{(r_2^4 - r_1^4)}{h_1} \right] \quad \text{in.-lb-sec} \tag{38.28}$$

where μ is the viscosity of the fluid and r_1, r_2, b, h_1 and h_2 are dimensions indicated in Fig. 38.19.

The paddle-type damper illustrated in Fig. 38.20 utilizes the engine lubricating oil supplied through the crankshaft. It has the damping constant:

$$c = \frac{3\mu d^2 (r_2^2 - r_1^2)^2 n}{h^3 \left[\frac{d}{b_1} + \frac{d}{b_3} + \frac{4(r_2 - r_1)}{b_1 + b_2} \right]} \quad \text{in.-lb-sec} \tag{38.29}$$

where n is the number of paddles, μ is the viscosity of the fluid and b_1, b_2, r_1, r_2 and d are dimensions indicated in Fig. 38.20. Other types of dampers are described in Ref. 5.

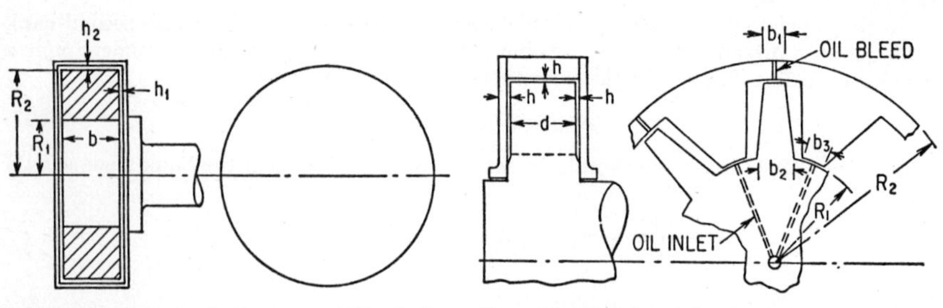

FIG. 38.19. Schematic diagram of Houdaille-type damper.

FIG. 38.20. Schematic diagram of paddle-type damper.

The effectiveness of these dampers may be increased somewhat by connecting the flywheel to the engine by a spring of proper stiffness, in addition to the fluid friction. In one form, Fig. 38.21, the connection is by rubber bonded between the flywheel and the shaft member. The rubber acts both as the spring and by hysteresis as the energy absorbing member. See Chaps. 32, 35 and 36 for discussions of damping in rubber. Dampers without and with springs are defined here as *untuned* and *tuned viscous dampers*, respectively.

RUBBER

INERTIA MEMBER

FIG. 38.21. Schematic diagram of bonded rubber damper.

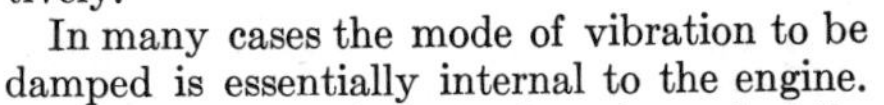

In many cases the mode of vibration to be damped is essentially internal to the engine. Then the damper is located at the end of the engine remote from the flywheel. If the mode to be damped is essentially one between driven masses, other locations may be desirable or necessary.

DESIGN OF THE UNTUNED VISCOUS DAMPER, EXACT PROCEDURE

The first step in the design procedure is to make a tentative assumption of the polar moment of inertia of the floating inertia member. If the damper is attached to the forward end of the crankshaft with the primary purpose of damping vibration in the engine, the size should be from 5 to 25 per cent, depending on the severity of the critical to be damped, of the total inertia in the engine part of the system, excluding the flywheel.

Usually it is advantageous to minimize the torque in a particular shaft section. This may be done as follows: For a series of frequencies plot the resonance curve of this torque, first without the floating damper mass and then with the damper mass locked to the damper hub. Plot the curves with all ordinates positive. The nature of such a plot is shown in Fig. 38.22. The point of intersection is called the *fixed point*. The plot is shown as if there were only one resonant frequency. Usually only one is of interest, and the curves are plotted in its vicinity. If the plot were extended, there would be a series of fixed points.

If a damping constant is assigned to the damper and the new resonance curve plotted, it will be similar to curve 3 in Fig. 38.22 and will pass through the fixed point. If there is no other damping in the system except that in the damper, all of the resonance curves will pass through the fixed points, independent of the value assigned to the damping constant.[17] Therefore, the amplitude at the fixed point is the lowest that can be obtained for the assumed damper size. If this amplitude is too large, it will be necessary to increase the damper size; if the amplitude is unnecessarily small, the damper size can be decreased. When a satisfactory size of damper has been selected, it is necessary to find the damping constant which will put the resonance curve through the fixed point with a zero slope. Assume a value of ω^2 slightly lower than its value at the fixed point, and compute the amplitude at that value of ω^2 with the damping constant c entered as an algebraic unknown. Equating this amplitude to that at the fixed point, the unknown damping constant c can be calculated. Repeat the calculation with a value of ω^2 higher than the fixed point value by the same increment. The mean of the two values of c thus obtained will be as close to the optimum value as construction of the damper will permit. In constructing these resonance curves, it is not necessary to con-

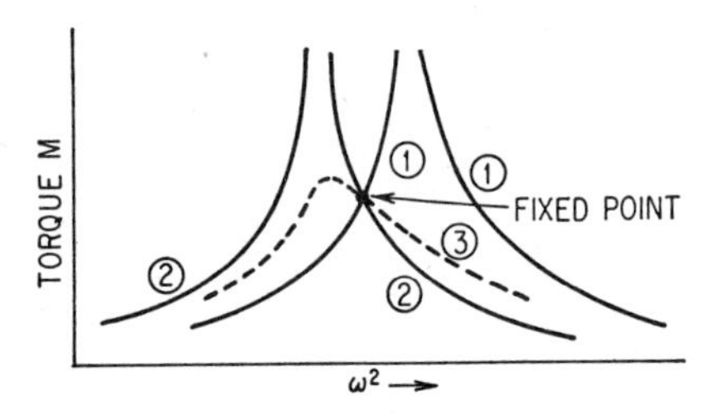

FIG. 38.22. Resonance curves for various conditions of auxiliary mass dampers: (1) damper free, $c = 0$; (2) damper locked, $c = \infty$; (3) auxiliary mass coupled to shaft by damping.

Table 38.19. Sequence Calculations for System of Fig. 38.7 with Damper at End of Crankshaft (ω_s^2 = 0.037)

Rotor number	I	$I\omega_s^2$	β Inertia $\theta \times$	β Exciting	M, in.-lb Inertia $\theta \times$	M, in.-lb Exciting	k
10	42.5	1.570	1.0000	...	1.570		
			0.259	...	1.570		6.06
9	0.265	0.0113	0.741	...	0.0083		
			1.191	...	1.5783		1.32
8	1	0.037	−0.450	...	−0.0166		
			1.5617	1.000	1.5617	1.000	1.00
7	0.722	0.0267	−2.0117	−1.000	−0.0538	−0.027	
			1.5079	1.973	1.5079	1.973	1.00
6	0.722	0.0267	−3.5196	−2.973	−0.0935	−0.080	
			1.4144	2.893	1.4144	2.893	1.00
5	1	0.037	−4.9340	−5.866	−0.1830	−0.217	
			1.2314	3.676	1.2314	3.676	1.00
4	1	0.037	−6.1654	−9.542	−0.2280	−0.352	
			1.0034	4.324	1.0034	4.324	1.00
3	0.722	0.0267	−7.1688	−13.866	−0.1910	−0.371	
			0.8124	4.953	0.8124	4.953	1.00
2	0.722	0.0267	−7.9812	−18.819	−0.2130	−0.513	
			0.5994	5.450	0.5994	5.450	1.00
1	1	0.037	−8.5806	−24.269	−0.3170	−0.898	
			0.1785	3.510	0.2824	5.552	11.58
a	0.42	0.0155	−8.7591	−27.779	−0.1350	−0.431	
					0.1474	5.121	
					0.2824	5.552	
$a + d$	1.13	0.0418	−8.7591	−27.779	−0.3670	−1.161	
					−0.0846	4.391	
			$-0.767j/c_s$	$-26.7j/c_s$	0.1474	5.121	$0.192c_sj$
d	0.71	0.0263	$\begin{Bmatrix}-8.7591 \\ +0.767j/c_s\end{Bmatrix}$	$\begin{Bmatrix}-27.779 \\ +26.7j/c_s\end{Bmatrix}$	$\begin{Bmatrix}-0.231 \\ +0.020j/c_s\end{Bmatrix}$	$\begin{Bmatrix}-0.730 \\ +0.703j/c_s\end{Bmatrix}$	
	$\omega_s = 0.192$				$\begin{Bmatrix}-0.084 \\ +0.020j/c_s\end{Bmatrix}$	$\begin{Bmatrix}4.39 \\ +0.703j/c_s\end{Bmatrix}$	

$$\theta = \frac{-4.39 - 0.703j/c_s}{-0.084 + 0.020j/c_s} \qquad |\theta|^2 = \frac{19.3 + 0.495/c_s^2}{0.0069 + 0.0004/c_s^2}$$

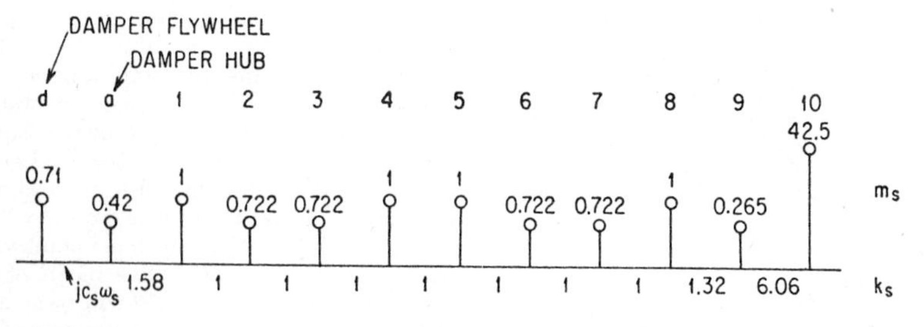

FIG. 38.23. Constants for system of Fig. 38.7 when damper is added at end of crankshaft.

Table 38.20. Summary of $\frac{1}{M_{9,10}}$ from Table 38.19 for Other Frequencies

ω_s^2	$\frac{1}{M_{9,10}}$	
	Free	Locked
0.035	−0.0261	0.000487
0.037	−0.01832	0.01225
0.038	−0.0137	0.0187

struct complete curves over a wide range of frequencies but only a short interval in the vicinity of the fixed point. Considerable labor is saved by making the calculation in the unity system with conversion to the actual system as a final step.

Figure 38.23 shows the constants in the unity system for the example of Fig. 38.7, when a damper is added at the end of the crankshaft. A floating rotor having 10 per cent of the total engine inertia is assumed; a is the fixed hub attached to the shaft. The "fixed point" is first located by calculating $M_{9,10}$ for $\omega_s^2 = 0.035$, 0.037, and 0.038. Table 38.19 gives the calculation for $\omega_s^2 = 0.037$.

In making the plot of the resonance curves, it is convenient to plot $1/M_{9,10}$ to avoid the resonant peak near $\omega_s^2 = 0.035$. Then $1/M_{9,10}$ is obtained from Table 38.19 as follows: Taking the case of the free damper, $\omega_s^2 = 0.037$. For zero residual torque: $0.1474\theta + 5.121 = 0$, or $\theta = -5.121/0.1474$.

The torque in the shaft between rotors 9 and 10 is $M_{9,10} = 1.57\theta$. The reciprocal is

$$\frac{1}{M_{9,10}} = \frac{0.1474}{5.121 \times 1.57} = 0.01832$$

Adding the damper d to the hub a, the corresponding value is

$$\frac{1}{M_{9,10}} = \frac{-0.0846}{4.391 \times 1.57} = -0.01225$$

Table 38.20 gives $1/M_{9,10}$ for the other frequencies. The reciprocal $1/M_{9,10}$ of the torque is plotted in Fig. 38.24 as a function of ω_s^2; all ordinates are plotted with positive sign. The intersection of the curves for free and locked damper is at $\omega^2 = 0.0375$; $1/M_{9,10} = 0.016$ in the unity system. In terms of the actual engine, this indicates that with an optimum damper the vibratory torque will be $1/0.016 = 62.5$ times the exciting torque of one cylinder. For the eighth-order critical, the resonant amplitude calculation gives $M_e = 1{,}490$; then, in the engine

$$M_{9,10} = 93{,}200 \text{ in.-lb}$$

The calculation for optimum damping constant c_s is continued in Table 38.19. The damping constant is treated as an algebraic unknown so that the damper "spring" constant is $jc_s\omega_s$. Now let $M_{9,10}$ at $\omega_s^2 = 0.037$ equal 62.5, its value at the fixed point. Then $|\theta| = 62.5/1.570 = 39.8$.

The damper "spring" constant $0.192c_sj$ is entered in the k column of Table 38.19 opposite the damper d. Then the relative amplitudes β in the "inertia" and "exciting" columns are the quotients M/k:

$$\frac{0.1474}{0.192c_sj} = -\frac{0.767j}{c_s}$$

$$\frac{5.121}{0.192c_sj} = -\frac{26.7j}{c_s}$$

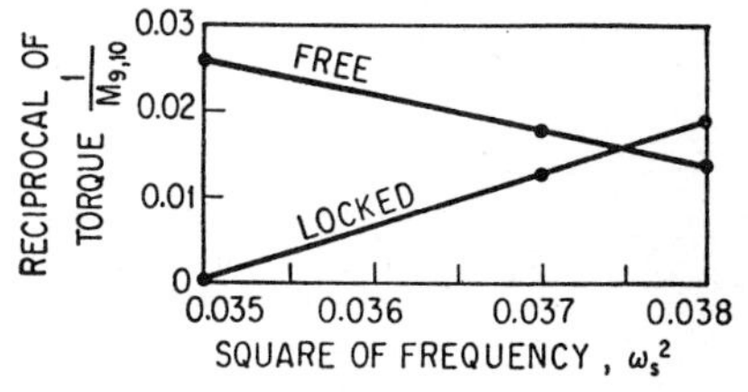

Fig. 38.24. Variation of torque in system of Fig. 38.23 when damper is either free or locked. The dots indicate computed points.

These values are entered in the β column and subtracted from the β values at rotor a, giving the β values at the damper rotor d. The amplitudes at this rotor multiplied by 0.0263 give the inertia torque reaction of damper rotor d. Adding this to the torque between a and d, the sum is equated to zero, and from this equation θ is solved for. Then the amplitude $|\theta|^2$ is obtained in the lower line of Table 38.19 by squaring and adding the real and imaginary terms of numerator and denominator. Substituting $\theta = 39.8$,

$$19.3 + \frac{0.495}{c_s^2} = (39.8^2)\left(0.0069 + \frac{0.0004}{c_s^2}\right)$$

Solving, $c_s^2 = 0.0172$, or $c_s = 0.131$. A similar calculation for $\omega_s^2 = 0.038$ leads to $c_s = 0.122$. A mean of these values is $c_s = 0.126$.

To check this value it is substituted in the $|\theta|^2$ equations of Table 38.19; then $M_{9,10}$ is obtained as follows:

$$M_{9,10} = 61.9 \text{ for } \omega_s^2 = 0.037$$

$$M_{9,10} = 61.3 \text{ for } \omega_s^2 = 0.038$$

Thus, the mean value of c_s places the fixed point very close to the maximum of the damped resonance curve.

It is of interest to find the fourth-order forced amplitude at 950 rpm with the damper in place. Table 38.21 shows the final items of Table 38.19 for the forced-vibration calculation with its continuation through the damper. The torque in section 9–10 is now 30.9 times the fourth-order exciting torque of one cylinder as against 20.3 without the damper (per Table 38.10); thus, the addition of the damper has increased the fourth-order stress by 50 per cent. This is because the addition of the damper inertia has lowered the frequency of the fourth-order critical, but the damper is of insufficient size to give any effective control over the fourth-order amplitude. It is extremely difficult to arrange a friction damper which will be effective for the lowest-order major criticals of four-cycle engines. For six- and eight-cylinder engines, the torque which must be developed in the damper is of the order of magnitude of the engine torque and requires very large dampers. In some cases a solution can be found with pendulum dampers (see Chap. 6).

Table 38.21. Final Items of Table 38.19 Continued through the Damper ($\omega_s^2 = 0.0321$)

Rotor	I	$I\omega_s^2$	β		M		k
			Inertia $\theta\times$	Exciting	Inertia $\theta\times$	Exciting	
1	1	0.0321	−7.7401	−24.748	−0.2480	−0.794	
			0.2490	3.710	0.3946	5.852	1.58
a	0.42	0.0135	−7.9891	−28.458	−0.1010	−0.384	
			$-13j$	$-242j$	0.2936	5.468	$0.0226j$
d	0.71	0.0228	−7.9891	−28.458	−0.162	−0.650	
			$+13j$	$+242j$	$0.296j$	$5.5j$	
				$RM \rightarrow$	0.1316	4.818	
					$0.296j$	$5.5j$	

$$|\theta| = \sqrt{\frac{(4.818)^2 + (5.5)^2}{(0.1316)^2 + (0.296)^2}} = 22.6$$

$$|M_{9,10}| \text{ (unity system)} = 22.6 \times 1.366 = 30.9$$

$$M_{9,10} \text{ (actual system)} = 30.9 \times 11{,}300 = 349{,}000 \text{ in.-lb}$$

In the actual engine the damper constant is

$$c = 0.131\sqrt{35.8 \times 175.7 \times 10^6} = 10{,}400 \text{ in.-lb/rad/sec}$$

The polar inertia of the floating member is

$$J = 35.8 \times 0.71 \times 386 = 9{,}800 \text{ in.}^2\text{-lb}$$

A damper of the Houdaille type having the dimensions $r_2 = 8.25$ in., $r_1 = 4.125$ in., and $b = 5$ in. will have the polar inertia 9,800 in.^{2}-lb. Viscosities in the range of 20,000 to 60,000 centipoises are used.

To convert to inch-pound-second units, centipoises are multiplied by 1.45×10^{-7}. Assuming 40,000 centipoises fluid, $\mu = 0.0058$ reyn, and equal clearances $h_1 = h$, the expression for clearances obtained from Eq. (38.29) is

$$h = \frac{2\pi}{c}\mu[r_2^3 b + \tfrac{1}{2}(r_2^4 - r_1^4)] \qquad (38.30)$$

which gives the damper clearance $h = 0.015$ in.

For a damper of the paddle type of Fig. 38.19, a diameter of 16.25 in. is taken together with the following dimensions: $r_2 = 7.16$ in., $r_1 = 3.58$ in., $d = 4.58$ in., $b_1 = b_2 = b_3 = 0.9$ in., and $n = 8$ pockets. The viscosity of SAE No. 30 oil at an operating temperature of 150°F is approximately 27 centipoises. Substituting in Eq. (38.30) and solving for h

$$h = 0.025 \text{ in. clearance}$$

To find the torque acting on the damper, refer to Table 38.20. Substituting $c_s = 0.126$ in the θ equation,

$$\theta = \frac{-4.39 - 5.58j}{-0.084 + 0.1586j}$$

But $M_{ad} = (0.1474\theta + 5.121)$ in.-lb reduces with the substitution of θ to

$$M_{ad} = \frac{1.079 - 0.013j}{0.084 - 0.1586j}$$

The absolute value is $|M_{ad}| = 6.01$, and in the actual system $|M_{ad}| = 6.01 \times 1{,}490 = 8{,}950$ in.-lb. Note that this is of the same order of magnitude as $M_e\Sigma\beta$. It is not necessarily the maximum torque in the damper, but is close to the maximum. To find the maximum it is necessary to make a similar calculation for several adjacent frequencies.

The oil pressure in the pockets produced by this torque is

$$\frac{8{,}950}{8 \times 5.37 \times 3.58 \times 4.58} = 12.8 \text{ lb/in.}^2$$

Oil must be supplied at a pressure in excess of this.

To find the oil pressure for the fourth-order critical at 950 rpm, M_{ad} is obtained from Table 38.21:

$$M_{ad} = 0.2936\theta + 5.468$$

where

$$\theta = \frac{-4.818 - 5.5j}{0.1316 + 0.296j}$$

Substituting this expression for θ in the expression for M_{ad},

$$M_{ad} = \frac{0.71}{0.1316 + 0.296j}$$

The absolute value is $|M_{ad}| = 2.18$. In the actual engine, the torque is $M_{\text{damper}} = 2.18 \times 11{,}300 = 24{,}600$ in.-lb, giving a pressure of 35.3 lb/in.^{2}

The above calculations are for a major critical with the excitations of the various cylinders all in-phase. In theory, for minor criticals where the excitations are out-of-phase, fixed points do not exist, and this procedure does not apply. An approximation to optimum conditions for this case is obtained by assuming that the single exciting torque $M_e\Sigma\beta$ acts at the No. 1 cylinder. This approximation can be used for major criticals as well.

TWO-MASS APPROXIMATION

If the system is replaced by a two-mass system in the manner utilized to make a first estimate (see Fig. 38.5 and the section *Natural Frequency Calculations*) of the one-noded mode, the results are further approximated by the following formulas:

For such a two-mass plus damper system the amplitude at the fixed point is given by [17]

$$M_{12} = M_e\Sigma\beta\left(\frac{2I_2 + I_d}{I_d}\right) \tag{38.31}$$

where $M_e = Srh$ is the exciting torque per cylinder [see Eq. (38.17)]. The optimum damping is

$$c = \left[\frac{kI_2I_d^2(2I_1 + 2I_2 + I_d)}{I_1(I_2 + I_d)(2I_2 + I_d)}\right]^{1/2} \quad \text{in.-lb/radian/sec} \tag{38.32}$$

where I_1 = polar moment of inertia for the flywheel or generator
I_2 = 40 per cent of the engine polar moment of inertia taken up to the flywheel
I_d = polar moment of inertia of the damper floating element
k = stiffness from the No.1 crank to the flywheel

Application to the engine discussed with reference to Table 38.21 gives

$$M_{9,10} = 74{,}300 \text{ in.-lb} \qquad \text{and} \qquad c = 9{,}700 \text{ in.-lb/radian/sec}$$

This compares with $M_{9,10} = 93{,}200$ and $c = 10{,}400$ as previously calculated.

TUNED VISCOUS DAMPERS

The procedure for the design of a tuned viscous damper is as follows:

1. Assume a polar inertia and a spring constant for the damper. As a first assumption, adjust the spring constant so that if f is the frequency of the mode to be suppressed and f_n is the natural frequency of the damper, assuming the hub as a fixed point,

$$\frac{f_n}{f} = 0.8$$

2. Plot the resonance curves of M for a particular section, first for the damper locked, then with zero damping but the damper spring in place. All ordinates are plotted positive. The curves have the general form of those shown in Fig. 38.25. They intersect in two fixed points through which all resonance curves pass, irrespective of the damping constant in the damper. If the fixed point a is higher than b, assume a lower constant for the damper spring and recalculate the M curve. If a is lower than b, do the reverse. Thus adjust the damper spring constant until a and b are of equal height. If this amplitude M is higher than desired, it is necessary to repeat the calculation with a larger damper.

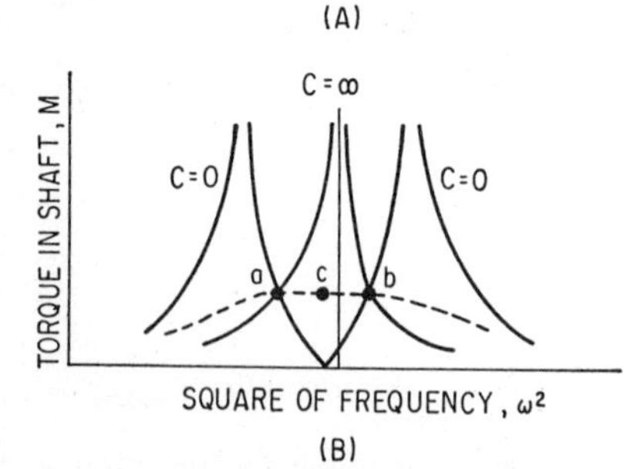

FIG. 38.25. Curves of torque vs. square of frequency for auxiliary mass damper.

With the spring and damper mass adjusted, a direct calculation (similar to that for the untuned damper) can be made to determine the damping constant c_r which will give the resonance curve the same ordinate at an intermediate frequency indicated by point c as at a and b. Figure 38.25*B* shows the resonance curve of an ideally adjusted damper.

3. For a range of frequencies, using the inertia, spring, and damping constants as determined above, compute the amplitude of the damper mass relative to its hub by a forced-vibration calculation. In this calculation the damper spring constant becomes the complex number $(k + jc\omega)$. The load for which the damper springs must be designed is k times the relative amplitude of the damper mass to its hub. The torque on the damper is approximately $M_e\Sigma\beta$.

PENDULUM DAMPERS

The principle of a pendulum damper is shown in Fig. 38.26*A*. (Also see Chap. 6.) The hole-pin construction usually used, which is equivalent to that of Fig. 38.26*A*, is shown in Fig. 38.26*B*. It is undesirable to have any friction in the damper. The damper produces an effect equivalent to a fixed flywheel, and the inertia of this flywheel is different for each order of vibration.

The design formulas for the pendulum damper are as follows:[3,18] If the length L is made equal to

$$L = \frac{R}{1 + q_o^2} \tag{38.33}$$

the damper is said to be tuned to order q_o. For excitation of q_o cycles per revolution, it will act as an infinite flywheel, keeping the shaft at its point of attachment to uniform rotation in so far as q_o order vibrations are concerned. But other orders of vibration may exist in the shaft.

If the shaft at the point of attachment of the damper is vibrating with order q and amplitude θ, the maximum link angle ψ (see Fig. 38.26) is

$$\psi = \frac{\theta q^2(1 + q_o^2)}{q_o^2 - q^2} \qquad \text{radians} \tag{38.34}$$

The torque exerted by a single element of the damper is

$$M = \left[\frac{WR^2}{1 - q^2/q_o^2} + J\right]\frac{q^2\Omega^2\theta}{g} \qquad \text{in.-lb} \tag{38.35}$$

where W is the weight of an element and J is the polar inertia of an element about its own center-of-gravity. The J term is equivalent to an addition to the damper hub. Dropping this term, the damper is equivalent to a flywheel of polar inertia

$$J_d = \frac{WR^2}{1 - q^2/q_o^2} \qquad \text{in.}^2\text{-lb} \tag{38.36}$$

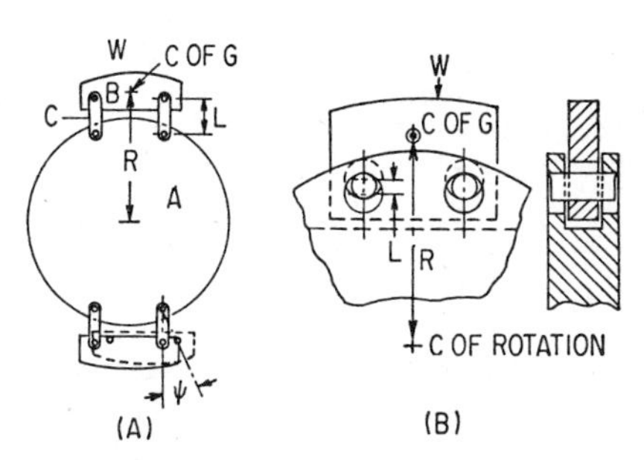

FIG. 38.26. Pendulum-type damper. The arrangement is shown in principle at *A* and the Chilton construction schematically at *B*.

For $q < q_o$ this is a positive flywheel, for $q = q_o$ an infinite flywheel, and for $q > q_o$ a negative flywheel. Omitting the J term and eliminating θ between Eqs. (38.32) and (38.33),

$$\psi = \frac{M(1 + q_o^2)g}{q_o^2 WR^2\Omega^2} \quad \text{radians} \tag{38.37}$$

APPLICATION TO A RADIAL AIRCRAFT ENGINE. Assume a nine-cylinder engine with the damper fitted to suppress the four and one-half order vibration. The damper would be tuned as an infinite flywheel with distance between hole centers

$$L = \frac{R}{1 + (4.5)^2}$$

In Eq. (38.35), M becomes the four and one-half order exciting torque. Then ψ is chosen as the permissible angle of pendulum swing, approximately 20 to 30°. The necessary polar inertia WR^2 of the damper is given by Eq. (38.35).

IN-LINE DIESEL ENGINE. As applied to a diesel engine, the above procedure is much more difficult. The exciting torques in diesel engines are nearly independent of speed. Hence from Eq. (38.37) it is evident that ψ will be inversely proportional to Ω^2. Thus for a variable-speed engine the damper size is fixed by the low-speed end of the range; if ψ is to be kept in the 20 to 30° limit, the size may be excessive. This difficulty usually can be overcome by tuning the damper as a negative flywheel, thus acting to raise the undesired critical above the operating range while keeping ψ to a reasonable limit at low speed. The procedure is as follows:

Assuming a damper size and a q/q_o ratio, a forced-vibration calculation is made starting at the flywheel end, for the maximum speed of the engine. In this calculation the damper is treated as a fixed flywheel of polar inertia $n\{[WR^2(1 - q^2/q_o^2)^{-1}] + J\}$ plus the inertia of the fixed carrier which supports the moving weights, where n is the number of weights. This calculation will yield θ, the amplitude at the damper hub, and the maximum torque in the engine shaft. Then ψ is given by Eq. (38.24). If either the shaft torque or the damper amplitude ψ is too large, it is necessary to increase the damper size and possibly adjust the q/q_o ratio as well. A similar check for ψ is made at the low-speed end of the range with further adjustment of WR^2 and q/q_o if necessary.

With a pendulum damper fitted, the equivalent inertia is different for each order of vibration so that each order has a different frequency. A damper tuned as a negative flywheel for one order becomes a positive flywheel for lower orders; thus, it reduces the frequencies of those orders, with possibly unfortunate results.

In in-line engines the application of a pendulum damper may be further complicated by the necessity of suppressing several orders of vibration, thus requiring several sets of damper weights. Alternatively, both a pendulum- and viscous-type damper may be fitted to an engine.

In general, the pendulum-type dampers are more expensive than the viscous types. Wear in the pins and their bushings changes the properties of the damper, thus requiring replacement of these parts at intervals.

The effect upon the fourth-order amplitude at 950 rpm of fitting a pendulum damper adjusted to act as a negative flywheel, for the generator-engine previously used, is calculated as follows: A damper is assumed which has four damper weights, each of 23.6 lb with their centers-of-gravity at a radius of 8.625 in. They will be tuned to $q_o = 3.7$ order. The length of link is calculated from Eq. (38.31):

$$l = \frac{8.625}{1 + (3.7)^2} = 0.59 \text{ in.}$$

For fourth-order vibration the four weights are equivalent to a flywheel of polar inertia calculated from Eq. (38.36):

$$J_d = \frac{4 \times 23.6 \times (8.625)^2}{1 - (4/3.7)^2} = -41{,}250 \text{ in.}^2\text{-lb}$$

The hub inertia is assumed to be 3,000 in.^{2}-lb; then the net inertia is −38,250 in.^{2}-lb. In the unity system the moment of inertia is

$$I_d = \frac{-38{,}250}{35.8 \times 386} = -2.77$$

Table 38.22 gives the last lines of Table 38.10 with the calculation continued through the damper. The fourth-order torque in the engine shaft becomes 107,300 in.-lb, as compared with 230,000 in.-lb before the damper was fitted.

The angle at the damper in the unity system is

$$\theta_d = -7.9891\theta - 28.458 = 32.042 \text{ rad}$$

In the actual system this corresponds to

$$\theta_d = \frac{32.042 \times 11{,}300}{175.7 \times 10} = 0.00206 \text{ rad}$$

The link angle is calculated from Eq. (38.34):

$$\psi = \frac{0.00206 \times 16(1 + 3.7^2)}{3.7^2 - 4^2} = 0.2 \text{ rad}$$

which is an acceptable amplitude.

A similar calculation is made at the low-speed end of the operating range. If the angle ψ is too large at the low-speed end it is necessary to adjust WR^2 or q_0, or both.

Table 38.22. Continuation of Table 38.10 to Include Pendulum-type Damper ($\omega_s^2 = 0.0321$)

Rotor	I	$I\omega_s^2$, $\omega_s^2 = 0.0321$	β		M		k
			Inertia $\theta \times$	Exciting	Inertia $\theta \times$	Exciting	
1	1.00	0.0321	−7.7401	−24.748	−0.2480	−7.94	
			0.2490	3.710	0.3946	5.852	1.58
d	−2.77	−0.0890	−7.9891	−28.458	+0.7110	2.540	
				$RM \rightarrow$	1.1056	8.392	

$$\theta = -8.392/1.1056 = -7.58$$

$$M_{9,10} \text{ (unity system)} = 1.366\theta = 10.38$$

$$M_{9,10} \text{ (actual system)} = 10.38 \times 11{,}300 = 107{,}300 \text{ in.-lb}$$

REFERENCES

1. Den Hartog, J.: "Mechanical Vibration," 4th ed., McGraw-Hill Book Company, Inc., New York, 1956.
2. Lewis, F. M.: *Trans. Soc. of Naval Arch. Marine Engrs.*, **33**:109 (1925).
3. Seward, H. L.: "Marine Engineering," vol. 2, chap. 2, The Society of Naval Architects and Marine Engineers, New York, 1944.
4. Biezeno and Grammel: "Engineering Dynamics" (translation), Blackie & Son, Ltd., Glasgow, 1946.
5. Wilson, W. K.: "Practical Solutions of Torsional Vibration Problems," John Wiley & Sons, Inc., New York, 1942.
6. Ziamenenko: *Engineers Digest*, **7** (11).
7. Constant, H.: *Brit. Aero. Res. Comm. R and M*, No. 1201, 1928.
8. Porter, F.: *Trans. ASME*, **50**: 8 (1928).

9. Gümbel: *Inst. Naval Architects* (*England*), **54**:219 (1912).
10. Holzer: "Die Berechnung der Drehschwingungen," Springer-Verlag, Berlin, 1921.
11. Draminsky, P.: *Proc. Inst. Mech. Engrs.* (*London*), **159**:416 (1948).
12. Klock, N.: *Trans. ASME*, **62**:A-148 (1940).
13. Lewis, F. M.: *Trans. ASME*, **57**:A-137 (1935).
14. Porter, F.: "Evaluation of Effects of Torsional Vibration," Society of Automotive Engineers, War Engineering Board; also *Trans. ASME*, **65**:A-33 (1943).
15. Dorey, S. F.: *Proc. Inst. Mech. Engrs.* (*London*), **159**:399 (1948).
16. Lewis, F. M.: *Trans. ASME*, **54**:253 (1932).
17. Lewis, F. M.: *Trans. ASME*, **78**:APM 377 (1955).
18. Zdanowich and T. S. Wilson: *Proc. Inst. Mech. Engs.* (*London*), **1943**:182 (1940).
19. Porter, F.: *Trans. ASME*, **75**:241 (1953).
20. Editorial Report on Flexible Couplings, *Engineers Digest*, **18**:271 (1957).
21. Roelig, H.: *ZVDI*, *Engineers Digest*, **87**:347 (1943); **1**:497 (1943).
22. U.S. Navy Department: "Military Standard Mechanical Vibrations of Mechanical Equipment," MIL-STD-167 (SHIPS).
23. Nesturides, E. J.: "A Handbook of Torsional Vibration," Cambridge University Press, 1958.

39

BALANCING OF ROTATING MACHINERY

Douglas Muster
University of Houston

Douglas G. Stadelbauer
Schenck Trebel Corporation

INTRODUCTION

The demanding requirements placed on modern rotating machines and equipment—for example, electric motors and generators, turbines, compressors, and blowers—have introduced a trend toward higher speeds and more stringent acceptable vibration levels. At lower speeds, the design of most rotors presents few problems which cannot be solved by relatively simple means, even for installations in vibration-sensitive environments. At higher speeds, which are sometimes in the range of tens of thousands of revolutions per minute, the design of rotors can be an engineering challenge which requires sophisticated solutions of interrelated problems in mechanical design, balancing procedures, bearing design, and the stability of the complete assembly. This has made balancing a first-order engineering problem from conceptual design through the final assembly and operation of modern machines.

This chapter describes some important aspects of balancing, such as the basic principles of the process by which an optimum state of balance is achieved in a rotor, balancing methods and machines, and definitions of balancing terms. The discussion is limited to those principles, methods, and procedures with which an engineer should be familiar in order to understand what is meant by "balancing."

In addition to unbalance, there are many possible sources of vibration in rotating machinery; some of them are related to or aggravated by unbalance so that, under appropriate conditions, they may be of paramount importance. However, this discussion is limited to the means by which the effect of once-per-revolution components of vibration (i.e., the effects due to mass unbalance) can be minimized.

DEFINITIONS[1]

Amount of Unbalance. The quantitative measure of unbalance in a rotor (referred to a plane) without referring to its angular position; obtained by taking the product of the unbalance mass and the distance of its center of gravity from the shaft axis.

Angle of Unbalance. Given a polar coordinate system fixed in a plane perpendicular to the shaft axis and rotating with the rotor, the polar angle at which an unbalance mass is located with reference to the given coordinate system.

Balance Quality. For rigid rotors, the product of the specific unbalance and the maximum service angular speed of the rotor.

Balancing. A procedure by which the mass distribution of a rotor is checked and, if necessary, adjusted in order to ensure that the vibration of the journals and/or forces on the bearings at a frequency corresponding to operational speed are within specified limits.

Balancing Machine. A machine that provides a measure of the unbalance in a rotor which can be used for adjusting the mass distribution of that rotor mounted on it so that once-per-revolution vibratory motion of the journals or force on the bearings can be reduced if necessary.

Bearing Support. The part, or series of parts, that transmits the load from the bearing to the main body of the structure.

Center-of-Gravity (Mass Center). The point in a body through which passes the resultant of the weights of its component particles for all orientations of the body with respect to a uniform gravitational field.

Compensating (Null Force) Balancing Machine. A balancing machine with a builtin calibrated force system which counteracts the unbalanced forces in the rotor.

Correction Plane Interference (Cross Effect). The change of balancing-machine indication at one correction plane of a given rotor, which is observed for a certain change of unbalance in the other correction plane.

Correction Plane Interference Ratios. The interference ratios (I_{AB}, I_{BA}) of two correction planes A and B of a given rotor are defined by the following relationships:

$$I_{AB} = \frac{U_{AB}}{U_{BB}}$$

where U_{AB} and U_{BB} are the unbalances referring to planes A and B, respectively, caused by the addition of a specified amount of unbalance in plane B; and

$$I_{BA} = \frac{U_{BA}}{U_{AA}}$$

where U_{BA} and U_{AA} are the unbalances referring to planes B and A, respectively, caused by the addition of a specified amount of unbalance in plane A.

Critical Speed. A characteristic speed such that the predominant response occurs at a resonance of the system. (In the case of a rotating system, the speed that corresponds to a resonance frequency of the system; for example, speed in revolutions per unit time equals the resonance frequency in cycles per unit time. Where there are several rotating systems, there will be several corresponding sets of critical speeds, one for each mode of the overall system.)

Couple Unbalance. That condition of unbalance for which the central principal axis intersects the shaft axis at the center of gravity.

Dynamic (Two-Plane) Balancing Machine. A centrifugal balancing machine that furnishes information for performing two-plane balancing.

Dynamic Unbalance. The condition in which the central principal axis is not coincident with the shaft axis.

Field Balancing Equipment. An assembly of measuring instruments for providing information for performing balancing operations on assembled machinery which is not mounted in a balancing machine.

Flexible Rotor. A rotor not satisfying the definition of a rigid rotor.

Flexural Critical Speed. A speed of a rotor at which there is maximum bending of the rotor and where flexure of the rotor is more significant than deflection of the bearings.

Flexural Principal Mode. For undamped rotor–bearing systems, that mode shape which the rotor takes up at one of the (rotor) flexural critical speeds.

High-speed Balancing (Relating to Flexible Rotors). A procedure of balancing at speeds where the rotor to be balanced cannot be considered rigid.

Initial Unbalance. Unbalance of any kind that exists in the rotor before balancing.

Journal Axis. The straight line joining the centroids of cross-sectional contours of the journal.

Low-speed Balancing (Relating to Flexible Rotors). A procedure of balancing at a speed where the rotor to be balanced can be considered rigid.

Minimum Achievable Residual Unbalance. The smallest value of residual unbalance that a balancing machine is capable of achieving.

Modal Balancing. A procedure for balancing flexible rotors in which balance corrections are made to reduce the amplitude of vibration in the separate significant principal flexural modes to within specified limits.

Multiplane Balancing. As applied to the balancing of flexible rotors, any balancing procedure that requires unbalance correction in more than two correction planes.

Perfectly Balanced Rotor. A rotor the mass distribution of which is such that it transmits no vibratory force or motion to its bearing as a result of centrifugal forces.

Permanent Calibration. That feature of a hard bearing balancing machine which permits calibration and setting the machine for any rotor within the capacity and speed range of the machine.

Plane Separation. Of a balancing machine, the operation of reducing the correction-plane interference ratio for a particular rotor.

Principal Inertia Axis. For each set of Cartesian coordinates at a given point, the values of the six moments of inertia of a body $I_{x_ix_j}$ $(i,j = 1, 2, 3)$ are in general unequal; for one such coordinate system the moments $I_{x_ix_j}$ $(i \neq j)$ vanish. The values of $I_{x_ix_j}$ $(i = j)$ for this particular coordinate system are called the principal moments of inertia and the corresponding coordinate directions are called the principal axes of inertia.

$$I_{x_ix_j} = \int_m x_ix_j \, dm \qquad \text{if } i \neq j$$

$$I_{x_ix_j} = \int_m (r^2 - x_i^2) \, dm \qquad \text{if } i = j$$

where $r^2 = x_1^2 + x_2^2 + x_1^2$ and x_ix_j are Cartesian coordinates.

Residual Unbalance. Unbalance of any kind that remains after balancing.

Resonance Balancing Machine. A balancing machine having an operating speed at the natural frequency of the suspension-and-rotor system.

Rigid Rotor. A rotor is considered rigid when it can be corrected in any two (arbitrarily selected) planes and, after the correction, its unbalance does not significantly exceed the balancing tolerances (relative to the shaft axis) at any speed up to maximum operating speed and when running under conditions which approximate closely those of the final supporting system.

Rotor. A body, capable of rotation, generally with journals which are supported by bearings.

Single-plane (Static) Balancing Machine. A gravitational or centrifugal balancing machine that provides information for accomplishing single-plane balancing.

Static Unbalance. That condition of unbalance for which the central principal axis is displaced only parallel to the shaft axis.

Unbalance. That condition which exists in a rotor when vibratory force or motion is imparted to its bearings as a result of centrifugal forces.

BASIC PRINCIPLES OF BALANCING

Descriptions of the behavior of rigid or flexible rotors are given as introductory material in standard vibration texts, in the references listed at the end of the chapter, and in the few books devoted to balancing. A similar description is included here for the purpose of examining the principles which govern the behavior of rotors as their speed of rotation is varied.

PERFECT BALANCE

Consider a rigid body which is rotating at a uniform speed about one of its three principal inertial axes. Suppose that the forces which cause the rotation and support the body are neglected; then it will rotate about this axis without wobbling, i.e., the principal axis (which is fixed in the body) coincides with a line fixed in space (Fig. 39.1). Now construct circular, concentric journals around the axis at the points where the axis protrudes from the body, i.e., on the stub shafts whose axes coincide with the principal axis. Since the axis does not wobble, the newly constructed journals also will not wobble. Next, place the journals in bearings which are circular and concentric to the principal axis (Fig. 39.2). It is assumed that there is no dynamic action of the elasticity

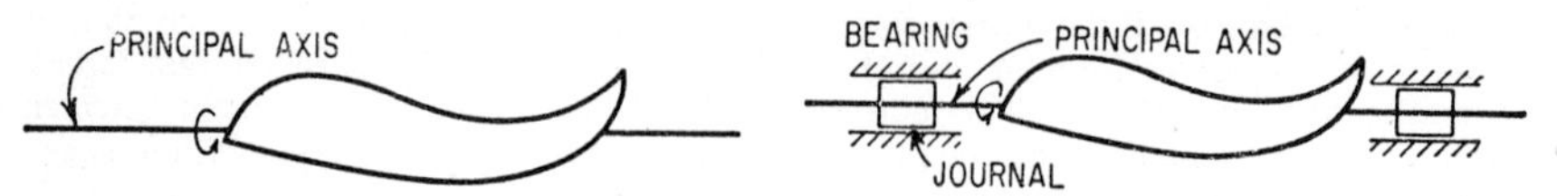

FIG. 39.1. Rigid body rotating about principal axis.

FIG. 39.2. Balanced rigid rotor.

of the rotor and the lubricant in the bearings. A rigid rotor constructed and supported in this manner will not wobble; the bearings will exert no forces other than those necessary to support the weight of the rotor. In this assembly, the radial distances between the center-of-gravity of the rotor and the journal and bearing axes are zero. This rotor is said to be *perfectly balanced.*

RIGID-ROTOR BALANCING—STATIC UNBALANCE

Rigid-rotor balancing is important because it comprises a substantial and probably the greater portion of balancing work done in industry. By far the greatest number of rotors manufactured and installed in equipment can be classified as "rigid" by definition. All balancing machines are designed to perform rigid-rotor balancing.*

Consider the case in which the journals are concentric to a line fixed in the body other than the principal axis, as illustrated in Fig. 39.3. In practice, with even the closest manufacturing tolerances, the journals are never concentric with the principal axis of the rotor. If concentric rigid bearings are placed around the journals, thus forcing the rotor to turn about the journal axis, a variable force is sensed at each bearing.

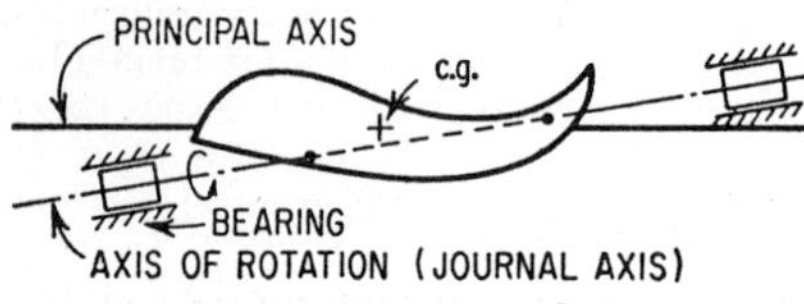

FIG. 39.3. Unbalanced rigid rotor.

The center-of-gravity is located on the principal axis, and is not on the axis of rotation (journal and bearing axes). From this it follows that there is a net radial force acting on the rotor which is due to centrifugal acceleration. The magnitude of this force is given by

$$F = m\epsilon\omega^2 \tag{39.1}$$

where m is the mass of the rotor, ϵ the eccentricity or radial distance of the center-of-gravity from the axis of rotation, and ω is the rotational speed in radians per second. Since the rotor is assumed to be rigid and thus not capable of distortion, this force is balanced by two reaction forces. There is one force at each bearing. Their algebraic sum is equal in magnitude and opposite in sense. The relative magnitudes of the two forces depend, in part, upon the axial position of each bearing with respect to the center-of-gravity of the rotor. In simplified form, this illustrates the "balancing problem." One must choose a practical method of constructing a perfectly balanced rotor from this unbalanced rotor.

The center-of-gravity may be moved to the journal (bearing) axis (or as close to this axis as is practical) in one of two ways. The journals may be modified so that the jour-

* Field balancing equipment is specifically excluded from this category since it is designed for use with both rigid and flexible rotors.

nal axis and an axis through the center-of-gravity are moved to essential coincidence. From theoretical considerations, this is a valid method of minimizing unbalance caused by the displacement of the center-of-gravity from the journal axis, but for practical reasons it is difficult to accomplish. Instead, it is easier to achieve a radial shift of the center-of-gravity by adding mass to or subtracting it from the mass of the rotor; this change in mass takes place in the longitudinal plane which includes the journal axis and the center-of-gravity. From Eq. (39.1), it follows that there can be no net radial force acting on the rotor at any speed of rotation if

$$m'r = m\epsilon \tag{39.2}$$

where m' is the mass added to or subtracted from that of the rotor and r is the radial distance to m'. There may be a *couple*, but there is no net *force*. Correspondingly, there can be no *net bearing reaction*. Any residual reactions sensed at the bearings would be due solely to the couple acting on the rotor.

If this rotor-bearing assembly were supported on a scale having a sufficiently rapid response to sense the change in force at the speed of rotation of the rotor, no fluctuations in the magnitude of the force would be observed. The scale would register only the dead weight of the rotor-bearing assembly.

This process of *effecting essential coincidence between the center-of-gravity of the rotor and the axis journal is called "single-plane (static) balancing."* The latter name for the process is more descriptive of the end result than the procedure that is followed.

If a rotor which is supported on two bearings has been balanced statically, the rotor will not rotate under the influence of gravity alone. It can be rotated to any position and, if left there, will remain in that position. However, if the rotor has not been balanced statically, then from any position in which the rotor is initially placed it will tend to turn to that position in which the center-of-gravity is lowest.

As indicated below, single-plane balancing can be accomplished most simply (but not necessarily with great accuracy) by supporting the rotor on flat, horizontal ways and allowing the center-of-gravity to seek its lowest position. It also can be accomplished in a centrifugal balancing machine by sensing and correcting for the unbalance force characterized by Eq. (39.1).

RIGID-ROTOR BALANCING—DYNAMIC UNBALANCE

When a rotor is balanced statically, the journal axis and principal inertial axis may not coincide; single-plane balancing ensures that the axes have only one common point, namely, the center-of-gravity. Thus, perfect balance is not achieved. To obtain perfect balance, the principal axis must be rotated about the center-of-gravity in the longitudinal plane characterized by the journal axis and the principal axis. This rotation can be accomplished by modifying the journals (but, as before, this is impractical) or by adding masses to or subtracting them from the mass of the rotor in the longitudinal plane characterized by the journal axis and the principal inertial axis. Although adding or subtracting a single mass may cause rotation of the principal axis relative to the journal axis, it also disturbs the static balance already achieved. From this it can be deduced that a couple must be applied to the rotor in the longitudinal plane. This is usually accomplished by adding or subtracting two masses of equal magnitude, one on each side of the principal axis (so as not to disturb the static balance) and one in each of two radial planes (so as to produce the necessary rotatory effect). Theoretically, it is not important which two radial planes are selected since the same rotatory effect can be achieved with appropriate masses, irrespective of the axial location of the two planes. Practically, the choice of suitable planes can be vital. Usually, it is best to select planes which are separated axially by as great a distance as possible in order to minimize the magnitude of the masses required.

The above process of *bringing the principal inertial axis of the rotor into essential coincidence with the journal axis is called "two-plane (dynamic) balancing."* If a rotor is balanced in two planes, then, by definition, it is balanced statically; however, the converse is not true.

FLEXIBLE-ROTOR BALANCING[2–4]

If the bearing supports are rigid, then the forces exerted on the bearings are due entirely to centrifugal forces caused by the residual unbalance. Dynamic action of the elasticity of the rotor and the lubricant in the bearings has been ignored.

The portion of the over-all problem in which the dynamic action and interaction of rotor elasticity, bearing elasticity, and damping are considered is called flexible rotor or modal balancing.

CRITICAL SPEED. Consider a long, slender rotor, as shown in Fig. 39.4. It represents the idealized form of a typical flexible rotor, such as a paper machinery roll or

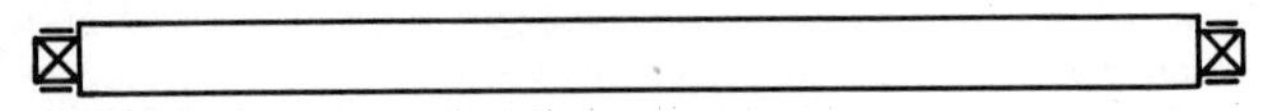

FIG. 39.4. Idealized flexible rotor.

turbogenerator rotor. Assuming further that all unbalances occurring along the rotor caused by machining tolerances, inhomogeneities of material, etc. are compensated by correction weights placed in the end faces of the rotor, and that the balancing is done at a low speed as if the rotor were a rigid body.

Assume there is no damping in the rotor or its bearing supports. Consider a thin slice of this rotor perpendicular to the journal axis (see Fig. 39.5*A*). This axis intersects the slice at its equilibrium center E when the rotor is not rotating, provided that deflection due to gravity forces is ignored. The center-of-gravity of the slice is displaced by δ from E due to an unbalance in the slice (caused by machining tolerances, inhomogeneity, etc., mentioned above) which was compensated by correction weights in the rotor's end planes. If the rotor starts to rotate about the journal axis with an angular speed ω, then the slice starts to rotate in its own plane at the same speed about an axis through E. Centrifugal force $me\omega^2$ is thus experienced by the slice. This force occurs in a direction perpendicular to the journal axis and may be accompanied by similarly caused forces at other cross sections along the rotor; such forces are likely to vary in magnitude and direction. They cause the rotor to bend, which in turn causes additional centrifugal forces and further bending of the rotor.

At every speed ω, equilibrium conditions require for one slice that the centrifugal and restoring forces be related by

$$m(\delta + x)\omega^2 = kx \tag{39.3}$$

where x is the deflection of the shaft (the radial distance between the equilibrium center and the journal axis) and k is the shaft stiffness (Fig. 39.5*B*). In Fig. 39.5, the centrifugal and restoring forces are plotted for various speeds ($\omega_1 < \omega_2 < \omega_3 < \omega_4 < \omega_5$). The point of intersection of the lines representing the two forces denotes the equilibrium condition for the rotor at the given speeds. For this ideal example, as the speed increases, the point which denotes equilibrium will move outward until, at say ω_3, a speed is reached at which there is no resulting force and the lines are parallel. Since equilibrium is not possible at this speed, it is called the critical speed. *The critical speed* ω_n *of a rotating system corresponds to a resonant frequency of the system.*

At speeds greater than ω_3 ($\equiv \omega_n$), the lines representing the centrifugal and restoring forces again intersect. As ω increases, the slope of the line $m\omega^2(x + \delta)$ increases correspondingly until, for speeds which are large, the deflection x approaches the value of δ, i.e., the rotor tends to rotate about its center-of-gravity.

UNBALANCE DISTRIBUTION. Apart from any special and obvious design features, the axial distribution of unbalance in the slices previously examined along any rotor is likely to be random. The distribution may be significantly influenced by the presence of large local unbalances arising from shrink-fitted discs, couplings, etc. The rotor may also have a substantial amount of initial bend which may produce effects similar to those due to unbalance. The method of construction can influence signifi-

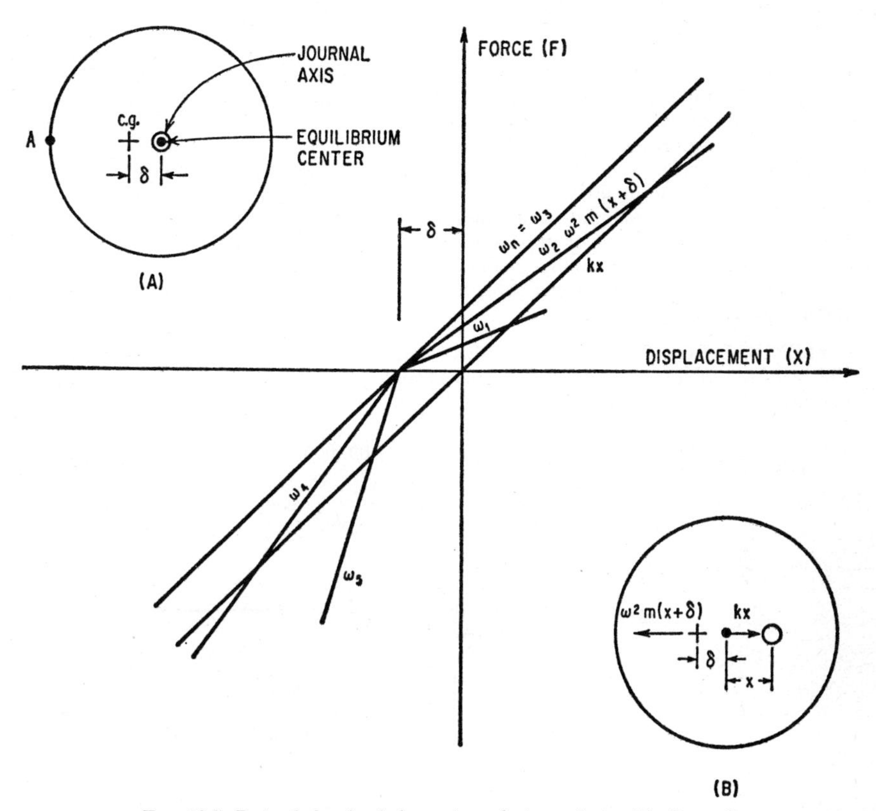

FIG. 39.5. Rotor behavior below, at, and above first critical speed.

cantly the magnitude and distribution of unbalance along a rotor. Rotors may be machined from a single forging or they may be constructed by fitting several components together. For example, jet-engine rotors are constructed by joining many shell and disc components, whereas alternator rotors are usually manufactured from a single piece of material, although they still have additional components fitted.

The unbalance distributions along two nominally identical rotors may be similar but rarely identical.

Contrary to the case of a rigid rotor, distribution of unbalance is significant in a flexible rotor because it determines the degree to which any bending or flexural mode of vibration is excited. The resulting modal shapes are reduced to acceptable levels by flexible-rotor balancing, also called "modal balancing."*

RESPONSE OF A FLEXIBLE ROTOR TO UNBALANCE. In common with all vibrating systems, rotor vibration is the sum of its modal components. For an *undamped* flexible rotor which rotates in flexible bearings, the flexural modes coincide with principal modes and are plane curves rotating about the axis of the bearing. For a *damped* flexible rotor, the flexural modes may be space (three-dimensional) curves rotating about the axis of the bearings. The damping forces also limit the flexural amplitudes at each critical speed. In many cases, however, the damped modes can be treated approximately as principal modes and hence regarded as rotating plane curves.

* All modal balancing is accomplished by multiplane corrections; however, multiplane balancing need not be modal balancing, since multiplane balancing refers only to unbalance correction in more than two planes.

The unbalance distribution along a rotor may be expressed in terms of modal components. The vibration in each mode is caused by the corresponding modal component of unbalance. Moreover, the response of the rotor in the vicinity of a critical speed is usually predominantly in the associated mode. The rotor modal response is a maximum at any rotor critical speed corresponding to that mode. Thus, when a rotor rotates at a speed near a critical speed, it is disposed to adopt a deflection shape corresponding to the mode associated with this critical speed. The degree to which large amplitudes of rotor deflection occur in these circumstances is determined by the modal component of unbalance and the amount of damping present in the rotor system.

If the modal component of unbalance is reduced by a number of discrete correction masses, then the corresponding modal component of vibration is similarly reduced. The reduction of the modal components of unbalance in this way forms the basis of the modal balancing technique.

FLEXIBLE-ROTOR MODE SHAPES. The stiffnesses of a rotor, its bearings, and the bearing supports affect critical speeds and therefore mode shapes in a complex manner. For example, Fig. 39.6 shows the effect of varying bearing and support stiff-

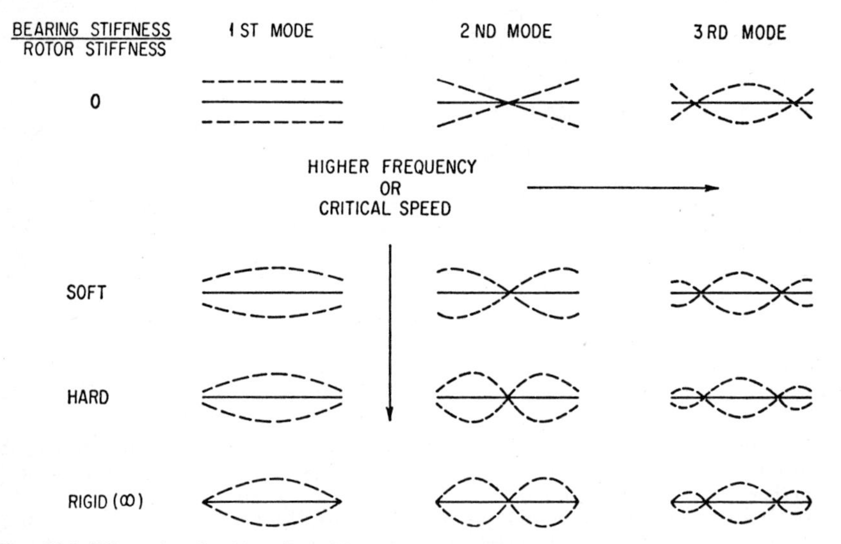

FIG. 39.6. Effect of ratio of bearing stiffness to rotor stiffness on mode shape at critical speeds.

ness relative to that of the rotor. The term "soft" or "hard" bearing is a relative one, since for different rotors the same bearing may appear to be either soft or hard. The schematic diagrams of the figure illustrate that the first critical speed of a rotor supported in a balancing machine having soft-spring-bearing supports occurs at a lower frequency and in an apparently different shape than that of the same rotor supported in a hard-bearing balancing machine where the bearing support stiffness approximates service conditions.

To evaluate whether a given rotor may require a flexible-rotor balancing procedure, the following rotor characteristics must be considered:

1. Rotor configuration and service speed.
2. Rotor design and manufacturing procedures. Rotors which are known to be flexible or unstable may still be capable of being balanced as rigid rotors.

ROTOR ELASTICITY TEST. This test is designed to determine if a rotor can be considered *rigid* for balancing purposes or if it must be treated as *flexible*. The test is carried out at service speed either under service conditions or in a high-speed, hard-bearing balancing machine whose support-bearing stiffness closely approximates that of the final supporting system. The rotor should first be balanced. A weight is then added in each end plane of the rotor near its journals; the two weights must be in the same angular position. During a subsequent test run, the vibration is measured at both bearings. Next, the rotor is stopped and the test weights are moved to the center of the rotor, or to a position where they are expected to cause the largest rotor distortion; in another run the vibration is again measured at the bearings. If the total of the first readings is designated x, and the total of the second readings y, then the ratio between $(y - x)/x$ should not exceed 0.2. Experience has shown that if this ratio is below 0.2, the rotor can be corrected satisfactorily at low speed by applying correction weights in two or three selected planes. Should the ratio exceed 0.2, the rotor usually must be checked at or near its service speed and corrected by a modal balancing technique.

HIGH-SPEED BALANCING MACHINES. Any technique of modal balancing requires a balancing machine having a variable balancing speed with a maximum speed at least equal to the maximum service speed of the flexible rotor. Such a machine must also have a drive-system power rating which takes into consideration not only acceleration of the rotor inertia but also windage losses and the energy required for a rotor to pass through a critical speed. For some rotors, windage is the major loss; such rotors may have to be run in vacuum chambers to reduce the fanlike action of the rotor and to prevent the rotor from becoming excessively hot. For high-speed balancing installations, appropriate controls and safety measures must be employed to protect the operator, the equipment, and the surrounding work areas.

MODAL (FLEXIBLE-ROTOR) BALANCING TECHNIQUES. Modal balancing consists essentially of a series of individual balancing operations performed at successively greater rotor speeds:

At a low speed, where the rotor is considered rigid.

At a speed whereby significant rotor deformation occurs in the mode of the first flexural critical speed. (This deformation may occur at speeds well below the critical speed.)

At a speed whereby significant rotor deformation occurs in the mode of the second flexural critical speed. (This applies only to rotors with a maximum service speed affected significantly by the mode shape of the second flexural critical speed.)

At a speed whereby significant rotor deformation occurs in the mode of the third critical speed, etc.

At the maximum service speed of the rotor.

The balancing of flexible rotors requires experience in determining the size of correction weights when the rotor runs in a flexible mode. The process is considerably more complex than standard low-speed balancing techniques used with rigid rotors. Primarily this is due to a shift of mass within the rotor, caused by shaft and/or body elasticity, asymmetric stiffness, thermal dissymmetry, incorrect centering of rotor mass and shifting of windings and associated components, magnetic forces, and fit tolerances and couplings.

Before starting the modal balancing procedure, the rotor temperature should be stabilized in the lower- or middle-speed range, until unbalance readings are repeatable. This preliminary warmup may take from a few minutes to several hours depending on the type of rotor, its dimensions, its mass, and its pretest condition.

Once the rotor is temperature-stabilized, the balancing process can begin. Several weight corrections in both end planes and along the rotor surface are required. Two axial planes, at right angles, are fixed in the rotor. Their line of intersection is the journal axis. Preferably, unbalance corrections should be made in these planes. This procedure permits all unbalance vectors (in a given correction plane) to be resolved into 90° components.

Modal balancing is performed in several discrete modes, each mode being associated with the speed range in which the rotor is deformed to the mode shape corresponding to a particular flexural critical speed. Figure 39.7 shows a rotor deformed in five of

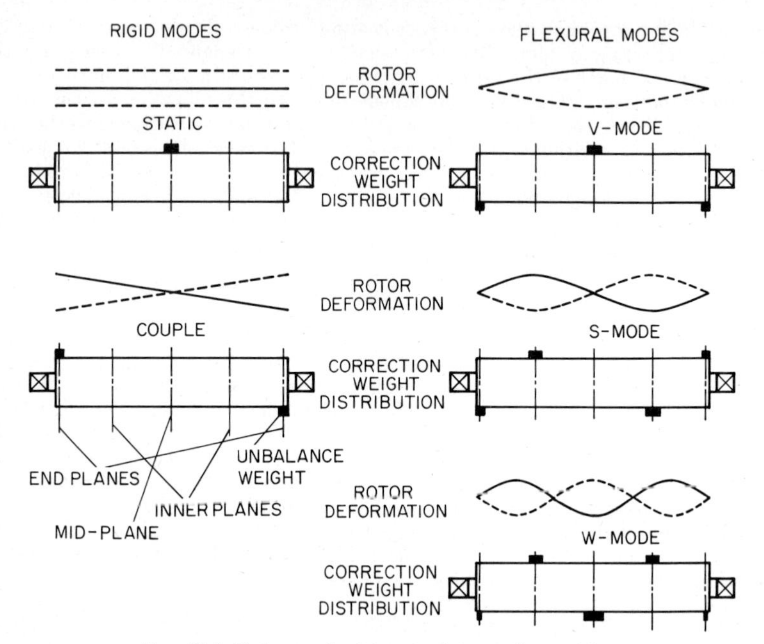

Fig. 39.7. Rotor mode shapes and correction weights.

the mode shapes of Fig. 39.6; the location of the weights which provide the proper correction for these mode shapes is indicated.

First, the rotor is rotated at a speed less than one-half the rotor's first flexural critical speed and balanced using a rigid-rotor balancing technique. Balancing corrections are performed at the end planes in the X and Y axes to reduce the original amount of unbalance to three or four times the final balancing tolerance.

Correction for the First Flexural Mode (V Mode). The balancing speed is increased until rotor deformation occurs, accompanied by a rapid increase in unbalance indication at the same angular position for both end planes. Unbalance corrections for this mode are made in at least three planes. Due to the bending of the rotor, the unbalance indication is not directly proportional to the correction to be applied. A new relationship between unbalance indication and corresponding correction weight must be established by test with trial weights. Such weights are first added in the correction plane nearest the middle of the rotor, in perpendicular components using the previously established axial planes fixed in the rotor. For large turbo-generator rotors the resultant of these components should be in the range of 30 to 60 oz-in./ton of rotor weight. Two additional corrections are added in the end planes diametrically opposite to the center weights, again in component form, each equal to one-half the magnitude of the corresponding center weights. This process may have to be repeated a number of times, each run reducing the magnitude of the weight applications until the residual unbalance is approximately 1 to 3 oz-in./ton of turbo-generator rotor weight. Then the speed is increased slowly to the maximum service speed; at the same time, the unbalance indicator is monitored. If an excessive unbalance indication is observed as the rotor passes through its first critical speed, further unbalance corrections are required in the V mode until the maximum service speed can be reached without an excessive unbalance indication. If a second flexural critical speed is observed before the maximum service speed is reached, the additional balancing operation in the S mode must be performed, as indicated below.

Correction for the Second Flexural Mode (S Mode). The rotor speed is increased until significant rotor deformation due to the second flexural mode is observed. This is indicated by a rapid increase in unbalance indication measured in the end planes at angular positions opposite to each other. Unbalance corrections for this S mode are made in at least four planes, as indicated in Fig. 39.7. The weights placed in the end correction planes, again in components, must be diametrically opposed; on the idealized symmetrical rotor, each end-plane weight must be equal to one-half the correction weight placed in one of the inner planes. Of primary concern is that the S-mode weight set will not have any influence on the previously corrected mode shapes. The correction weight in each inner plane must be diametrically opposed to its nearest end-plane correction weight. The procedure to determine the relationship between unbalance indication and required correction weight is similar to that used in the V-mode procedure, described above. The S-mode balancing process must be repeated until an acceptable residual unbalance is achieved.

Corrections for the Third Flexural Mode (W Mode). The rotor speed is increased further until significant rotor deformation due to the third flexural mode is observed. Corrections are made in the rotor with a five-weight set (shown in Fig. 39.7) and in a manner similar to that used in correcting for the first and second flexural modes.

If the service-speed range requires it, higher modes (those associated with the nth critical speed, for example) may have to be corrected as well. For each of these higher modes, a set of $(n + 2)$ correction weights is required.

Final Balancing at Service Speed. Final balancing takes place with the rotor at its service speed. Correction should be made only in the end planes. The final unbalance tolerance for large turbo-generators, for example, will normally be in the order of 1 oz-in./ton of rotor weight. If the rotor cannot be brought into proper balancing tolerances, the S-mode, V-mode, and W-mode corrections may require slight adjustment.

To achieve repeatability of the correction effects, the duration of each run must remain constant and the same balancing speed for each mode must be accurately maintained. Depending on the size of the rotor, the number of modes that must be corrected, and the ease with which weights can be applied, the entire process may take anywhere from 3 to 30 hours.

The relative position of the unbalance correction planes shown in Fig. 39.7 applies to symmetrical rotors only. Rotors with axial asymmetry generally require unsymmetrically spaced correction planes. Computer programs are avilable which facilitate the selection of the most appropriate correction planes and the computation of correction weights.[5,6] Other modal balancing techniques rely mostly on experience data available from previously manufactured rotors of the same type, or correct only for flexural modes if no low-speed balancing equipment is available.[7]

SOURCES OF UNBALANCE

Sources of unbalance in rotating machinery may be classified as resulting from

1. Dissymmetry
   (Core shifts in casting, rough surfaces on forgings, unsymmetrical configuration)
2. Nonhomogeneous material
   (Blowholes in rotor bars of electric motors, inclusions in rolled or forged materials, slag inclusions or variations in crystalline structure caused by variations in the density of the material)
3. Distortion at service speed
   (Blower blades in builtup design)
4. Eccentricity
   (Journals not concentric or circular, matching holes in builtup rotor not circular)
5. Misalignment of bearings
6. Shifting of parts due to plastic deformation of rotor part
   (Windings in rotor)
7. Hydraulic or aerodynamic unbalance
   (Cavitation or turbulence)
8. Thermal gradients
   (Steam-turbine rotors)

Often, balancing problems can be minimized by careful design in which unbalance is carefully controlled. When a part is to be balanced, large amounts of unbalance require large corrections. If such corrections are made by removal of material, additional cost is involved and part strength may be affected. If corrections are made by addition of material, cost is again a factor and space requirements for the added material may be a problem.

Manufacturing processes are the major source of unbalance. Unmachined portions of castings or forgings which cannot be made concentric and symmetrical with respect to the journal axis introduce substantial unbalance. Manufacturing tolerances and processes which permit any eccentricity or lack of squareness with respect to the journal axis are sources of unbalance. The tolerances, necessary for economical assembly of several elements of a rotor, permit radial displacement of parts of the assembly and thereby introduce unbalance.

Limitations imposed by design often introduce unbalance effects which cannot be corrected adequately by refinement in design. For example, electrical design limitations impose a requirement that one coil be at a greater radius than the others in a certain type of universal motor armature. It is impractical to design a compensating unbalance into the armature.

Fabricated parts, such as fans, often distort nonsymmetrically under service conditions. Design and economic considerations prevent the adaptation of methods which might eliminate this distortion and thereby reduce the resulting unbalance.

Ideally, rotating parts always should be designed for inherent balance, whether a balancing operation is to be performed or not. Where low service speeds are involved and the effects of a reasonable amount of unbalance can be tolerated, this practice may eliminate the need for balancing. In parts which require unbalanced masses for functional reasons, these masses often can be counterbalanced by designing for symmetry about the journal axis.

MOTIONS OF UNBALANCED ROTORS

In Fig. 39.8 a rotor is shown spinning freely in space. This corresponds to spinning above resonance in soft bearings. In Fig. 39.8*A* only static unbalance is present and the center line of the shaft sweeps out a cylindrical surface. Figure 39.8*B* illustrates the motion when only couple unbalance is present. In this case, the center line of the

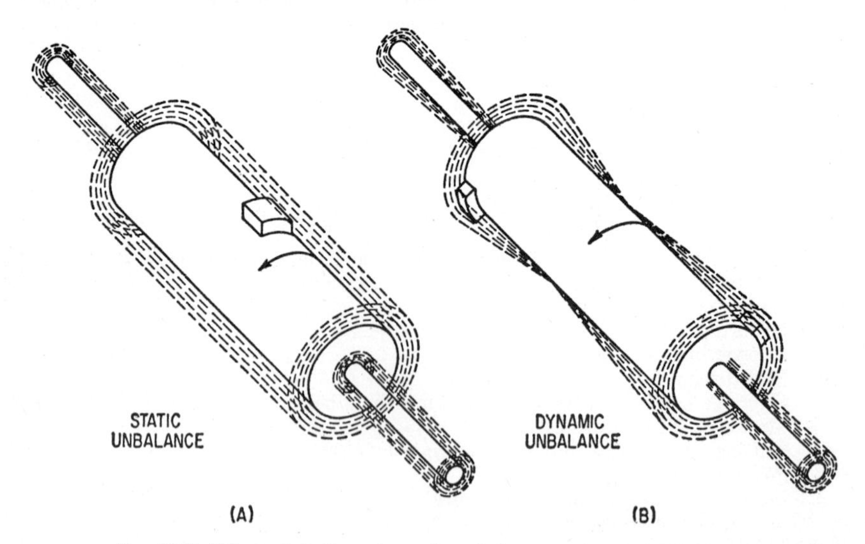

FIG. 39.8. Effect of static and couple unbalance on free rotor motion.

rotor shaft sweeps out two cones which have their apexes at the center-of-gravity of the rotor. The effect of combining these two types of unbalance when they occur in the same axial plane is to move the apex of the cones away from the center-of-gravity. In many cases, there will be no apex and the shaft will move in a more complex combination of the motions shown in Fig. 39.8.

OPERATING PRINCIPLES OF BALANCING MACHINES[8,9]

This section describes the basic operating principles and general features of the various types of balancing machines which are available commercially. With this type of information, it is possible to determine the basic type of machine required for a given application to stated specifications.

Every balancing machine must determine by some technique both the magnitude of a correction weight and its angular position in each of one, two, or more selected balancing planes. For single-plane balancing this can be done statically, but for two- or multiplane balancing it can be done only while the rotor is spinning. Finally, all machines must be able to resolve the unbalance readings, usually taken at the bearings, into equivalent corrections in each of the balancing planes.

On the basis of their method of operation, balancing machines and equipment can be grouped in two general categories:

1. Gravity balancing machines
2. Centrifugal balancing machines and field balancing equipment.

In the first category, advantage is taken of the fact that a body free to rotate always seeks that position in which its center-of-gravity is lowest. Gravity balancing machines, also called *nonrotating balancing machines*, include horizontal ways, knife-edges or roller arrangements, and vertical pendulum types. All are capable of only detecting and/or indicating static unbalance.

In the second category, the amplitude and phase of motions or reaction forces caused by once-per-revolution centrifugal forces resulting from unbalance are sensed, measured, and indicated by appropriate means. Field balancing equipment provides sensing and measuring instrumentation only; the necessary measurements for balancing a rotor are taken while the rotor runs in its own bearings and under its own power. However, on a centrifugal balancing machine, the rotor is supported by the machine and rotated around a horizontal or vertical axis by the machine's drive motor. Balancing-machine instrumentation differs in specific features from field balancing equipment which simplify the balancing process. A centrifugal balancing machine (also called a *rotating balancing machine*) is usually capable of measuring static unbalance (by means of a *single-plane rotating balancing machine*) or static *and* dynamic unbalance (by means of a *two-plane rotating balancing machine*). Only a two-plane rotating balancing machine can detect couple unbalance or dynamic unbalance.

GRAVITY BALANCING MACHINES

First, consider the simplest type of balancing—usually called "static" balancing, since the rotor is not spinning. In Fig. 39.9*A*, a disc-type rotor on a shaft is shown resting on knife-edges. The mass added to the disc at its rim represents a known unbalance. In this illustration, in Fig. 39.8, and in the illustrations which follow, the rotor is assumed to be balanced without this added unbalance weight. In order for this balancing procedure to work effectively, the knife-edges must be level, parallel, hard, and straight.

In operation, the heavier side of the disc will seek the lowest level—thus indicating the angular position of the unbalance. Then, the magnitude of the unbalance usually is determined by an empirical process, adding mass in the form of wax or putty to the light side of the disc until it is in balance, i.e., until the disc does not stop at the same angular position.

In Fig. 39.9*B* a set of balanced rollers or wheels is used in place of the knife-edges. These have the advantage of permitting the rotor to turn without, at the same time, moving laterally.

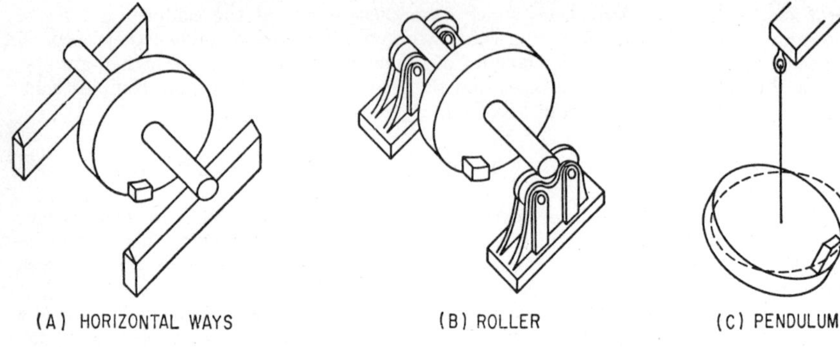

FIG. 39.9. Static (single-plane) balancing devices.

In Fig. 39.9*C*, a setup for another type of static, or "nonrotating," balancing procedure is shown. Here the disc to be balanced is supported by a flexible cable, fastened to a point on the disc which coincides with the center of the shaft and is slightly above the normal plane containing the center-of-gravity. As shown in Fig. 39.9*C*, the heavy side will tend to seek a lower level than the light side, thereby indicating the angular position of the unbalance. The disc can be balanced by adding weight to the diametrically opposed side of the disc until it hangs level. In this case, the center-of-gravity is moved until it is directly under the flexible support cable.

In modified versions of this setup, the cable is replaced by a hardened ball-and-socket arrangement (used on many automobile wheel balancers) or by a spherical air bearing (used on some industrial and aerospace balancers).

Static balancing is satisfactory for rotors having relatively low service speeds and axial lengths which are small in comparison with the rotor diameter. A preliminary static balance correction may be required on rotors having a combined unbalance so large that it is impossible in a dynamic, soft-bearing balancing machine to bring the rotor up to its proper balancing speed without damaging the machine. If the rotor is first balanced statically by one of the methods just outlined, it is usually possible to decrease the combined unbalance to the point where the rotor may be brought up to balancing speed and the residual unbalance measured. Such preliminary static correction is not required on hard-bearing balancing machines.

CENTRIFUGAL BALANCING MACHINES

The following procedures may be used to balance the rotor shown in Fig. 39.8*B*. First, select the planes in which the correction weights are to be added; these planes should be as far apart as possible and the weights should be added as far out from the shaft as feasible to minimize the size of the weights. Next, by a balancing technique, determine the size of the required correction weight and its angular position for each correction plane. To implement these procedures two types of machines, *soft-bearing* and *hard-bearing* balancing machines, which are described below, are employed.

SOFT-BEARING BALANCING MACHINES. Soft-bearing balancing machines permit the idealized free rotor motion illustrated in Fig. 39.8*B*; but on most machines the motion is restricted to a horizontal plane (as shown in Fig. 39.10). Furthermore, the bearings (and the directly attached components) vibrate in unison with the rotor, thus adding to its mass. The restriction on the vertical motion does not affect the amplitude of vibration in the horizontal plane, but the added mass of the bearings does. The greater the combined rotor-and-bearing mass, the smaller will be the displacement of the bearings, and the smaller will be the output of the devices which sense the unbalance.

Consider the following example. Assume a balanced disc (see Fig. 39.11) having a

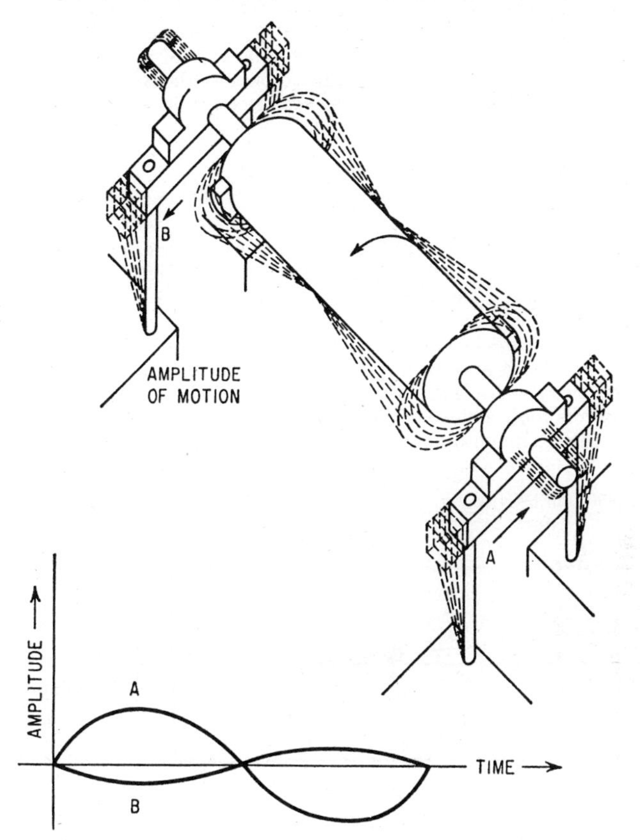

Fig. 39.10. Motion of unbalanced rotor and bearings in flexible-bearing, centrifugal balancing machine.

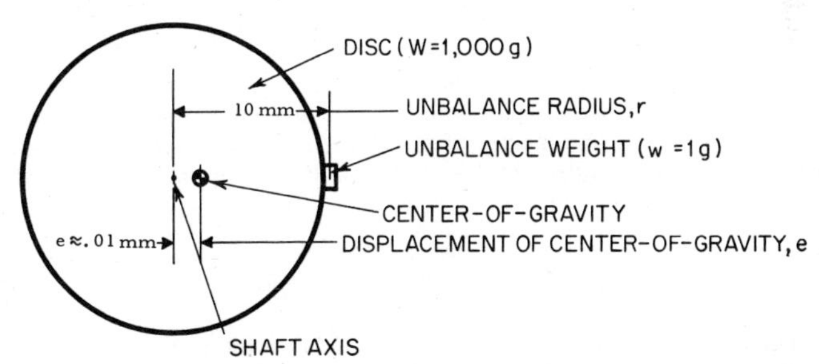

Fig. 39.11. Displacement of center-of-gravity because of unbalance.

weight W of 1,000 gm, rotating freely in space. An unbalance weight w of 1 gm is then added to the disc at a radius of 10 mm. The unbalance causes the center-of-gravity of the disc to be displaced from the journal axis by

$$e = \frac{wr}{W + w} = 0.00999 \text{ mm}$$

Since the addition of the weight of the unbalance to the rotor causes only an insignificant difference, the approximation $e \approx wr/W$ is generally used. Then $e \approx 001$ mm.

If the same disc with the same unbalance is rotated on a single-plane balancing machine having a bearing and bearing housing weight W' of 1,000 gm, the displacement of center-of-gravity will be significantly reduced because the bearing and housing weight is added to the weight W of the disc. The center-of-gravity of the combined vibrating components will now be displaced by

$$e' = \frac{wr}{W + W'} \approx 0.005 \text{ mm}$$

The conversion of unbalance into displacement of center-of-gravity as shown in the example above also holds true for rotors of greater axial length which normally require correction in two planes. However, such rotors are prone to have unbalance other than static unbalance, causing an inclination of the principal inertia axis from the journal axis. In turn, this results in a displacement of the principal inertia axis from the journal axis in the bearing planes of the rotor, causing the balancing machine bearings to vibrate.

To find the bearing displacement or bearing vibration amplitude resulting from a given unbalance is more involved than finding the center-of-gravity displacement, because other factors come into play, as is illustrated by Fig. 39.12.

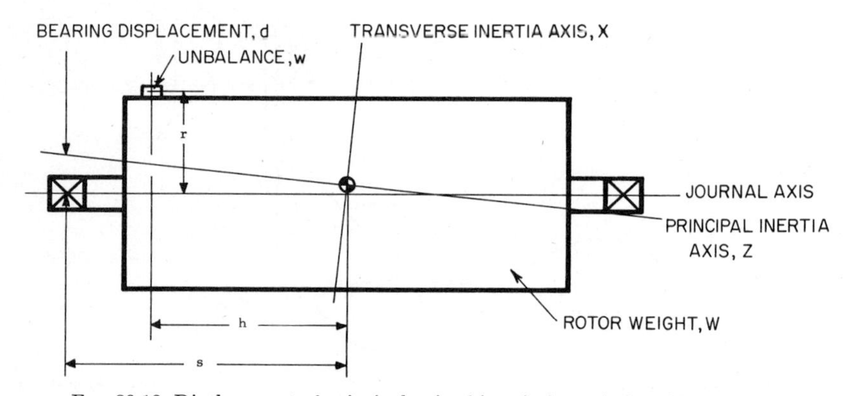

FIG. 39.12. Displacement of principal axis of inertia from shaft axis at bearing.

The weight and inertia of the balancing machine bearings and directly attached vibratory components are usually not known. In any case, they are usually small in relation to the weight and the inertia of the rotor and can generally be ignored. On this basis, the following formula may be used to find the approximate bearing displacement d:

$$d \approx -\frac{wr}{W} + \frac{wrhs}{g(I_x - I_z)}$$

where d = displacement at bearing of principal inertia axis from journal axis
r = distance from journal axis to unbalance weight
h = distance from center-of-gravity to unbalance plane
s = distance from center-of-gravity to bearing plane
g = gravitational constant
I_x = moment of inertia around transverse axis X
I_z = moment of inertia around principal axis Z

From the above can be seen that the relationship between bearing motion and unbalance in a soft-bearing balancing machine is complex. Therefore, a direct indication of unbalance can be obtained only after calibrating the indicating elements to a given rotor by use of test weights which produce a known amount of unbalance.

HARD-BEARING BALANCING MACHINES. Hard-bearing balancing machines are essentially of the same construction as soft-bearing balancing machines except that their bearing supports are significantly stiffer in the horizontal direction. This results in a horizontal critical speed for the machine which is several orders of magnitude greater than that for a comparable soft-bearing balancing machine. The hard-bearing balancing machine is designed to operate at speeds well below its horizontal critical speed. In this speed range, the output from the sensing elements attached to the balancing-machine bearing supports is directly proportional to the centrifugal force resulting from unbalance in the rotor. The output is not influenced by bearing mass, rotor weight, or inertia, so that a permanent relation between unbalance and sensing element output can be established. Unlike soft-bearing balancing machines, the use of test weights is not required to calibrate the machine for a given rotor.

MEASUREMENT OF AMOUNT AND ANGLE OF UNBALANCE. Both soft- and hard-bearing balancing machines use various types of sensing elements *at the rotor-bearing supports* to convert mechanical vibration into an electrical signal. On commercially available balancing machines, these sensing elements employed are usually velocity-type pickups, although certain hard-bearing balancing machines use magnetostrictive or piezoelectric pickups.

Three basic methods are used to obtain a reference signal by which the phase angle of the amount-of-unbalance indication signal may be correlated with the rotor. On end-drive machines (where the rotor is driven via a universal joint driver or similarly flexible coupling shaft) a phase reference generator, directly coupled to the balancing machine drive spindle, is used. On belt-drive machines (where the rotor is driven by a belt over the rotor periphery) or on air-drive or self-drive machines, a stroboscopic lamp flashing once per rotor revolution or a photoelectric cell is employed to obtain the phase reference. Both the stroboscopic light and photoelectric cell require a reference mark on the rotor to obtain the angular position of unbalance in the rotor. The output from the phase-reference sensor and the pickups at the rotor bearing supports are processed in various ways by different manufacturers. Generally, the processed signals result in an indication representing the amount of unbalance and its angular position. In Fig. 39.13 block diagrams are shown for typical balancing instrumentation. In Fig. 39.13*A* an indicating system is shown which uses switching between correction planes (i.e., single-channel instrumentation). This is generally employed on balancing machines with stroboscopic-angle indication and belt drive. In Fig. 39.13*B* an indicating system is shown with two-channel instrumentation. Combined indication of amount of unbalance and its angular position is provided on a vectormeter having an illuminated target projected on a screen. Displacement of the target from the central zero point provides a direct visual representation of the displacement of the principal inertia axis from the journal axis. Concentric circles on the screen indicate the amount of unbalance, and radial lines indicate its angular position.

INDICATED AND ACTUAL ANGLE OF UNBALANCE. An *unbalanced rotor* is a rotor in which the principal inertia axis does not coincide with the journal axis. When rotated in its bearings, an unbalanced rotor will cause periodic vibration of, and will exert a periodic force on, the rotor bearings and their supporting structure. If the structure is rigid, the force is larger than if the structure is flexible. In practice, supporting structures are neither rigid nor flexible but somewhere in between. The rotor-bearing support offers some restraint, forming a spring-mass system with damping having a single resonance frequency. When the rotor speed is below this frequency, the principal inertia axis of the rotor moves outward radially. This condition is illustrated in Fig. 39.14. If a soft pencil is held against the rotor, the so-called high spot is marked at the same angular position as that of the unbalance. When the rotor speed is increased, there is a small time lag between the instant at which the unbalance passes the pencil and the instant at which the rotor moves out enough to contact it. This is due to the damping in the system. The angle between these two points is called the "angle of lag." (See Fig. 39.14*B*.) As the rotor speed is increased further, resonance of the rotor and its supporting structure will occur; at this speed the angle of lag is

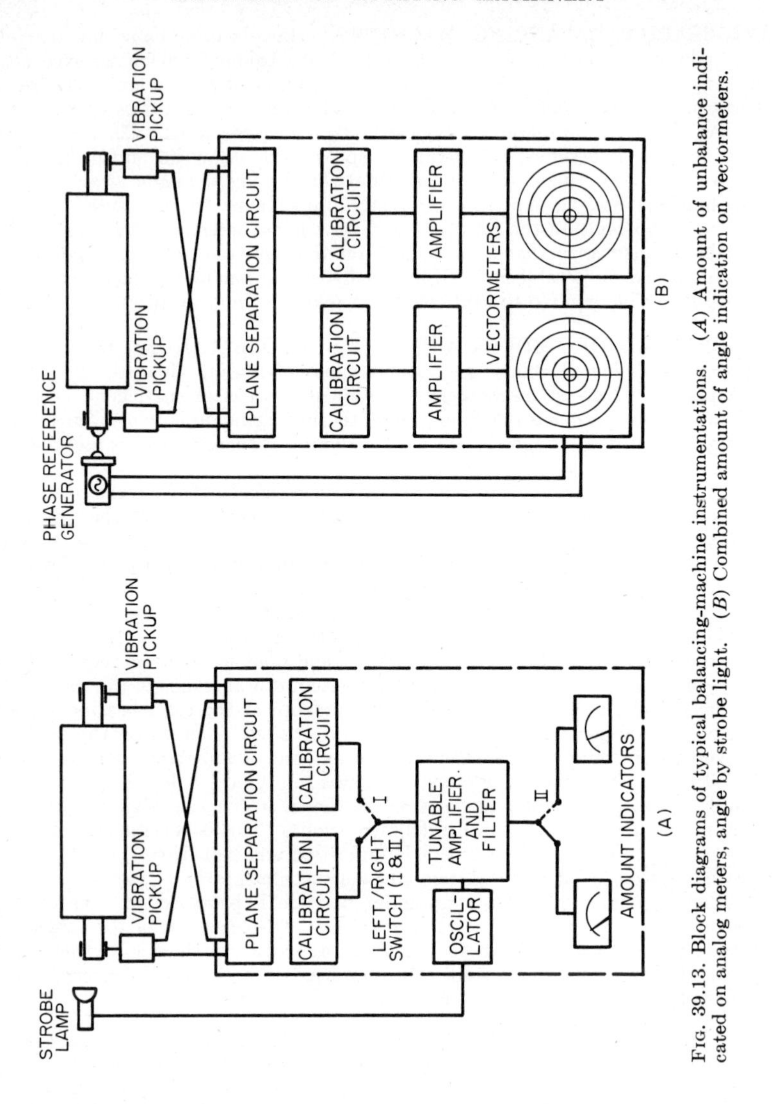

FIG. 39.13. Block diagrams of typical balancing-machine instrumentations. (*A*) Amount of unbalance indicated on analog meters, angle by strobe light. (*B*) Combined amount of angle indication on vectormeters.

90°. As the rotor passes through resonance there are large vibration amplitudes and the angle of lag changes rapidly. As the speed is increased to more than twice the critical speed, the angle of lag approaches 180°. At speeds greater than approximately twice the critical speed, the rotor tends to rotate about its principal inertia axis; the angle of lag (for all practical purposes) is 180°.

The changes in the relative position of pencil mark and unbalance as shown in Fig. 39.14 of a statically unbalanced rotor occurs in the same manner on a rotor with dynamic unbalance. However, the center-of-gravity shown in the illustrations then represents the position of the principal inertia axis in the plane at which the pencil is applied to the rotor shaft. Thus, the indicated angle of lag and displacement amplitude refer only to that particular plane and generally differ from any other plane in the rotor.

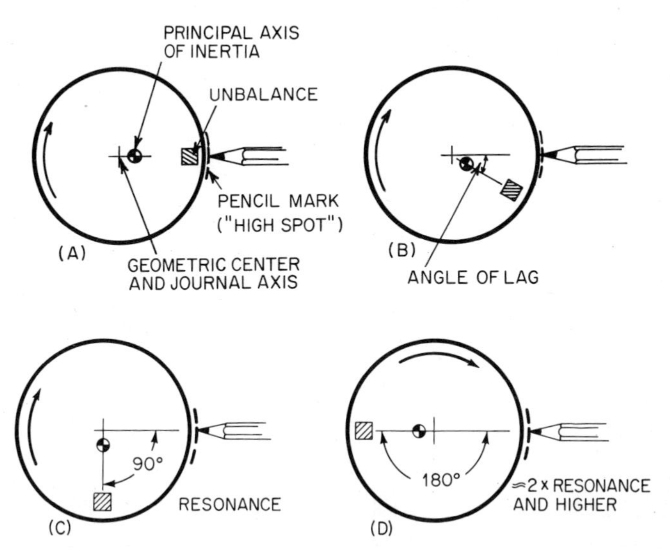

FIG. 39.14. A soft pencil is held against an unbalanced rotor. (*A*) A high spot is marked. (*B*) The angle of lag. Angle of lag between unbalance and high spot increases from 0° (*A*) to 180° (*D*) as rotor speed increases.

Angle of lag is shown as a function of rotational speed in Fig. 39.15: (*A*) for soft-bearing balancing machines whose balancing-speed ranges start at approximately twice the critical speed; and (*B*) for hard-bearing balancing machines. The effects of damping also are illustrated. Here the resonance frequency of the combined rotor-bearing support system is usually more than three times greater than the maximum balancing speed.

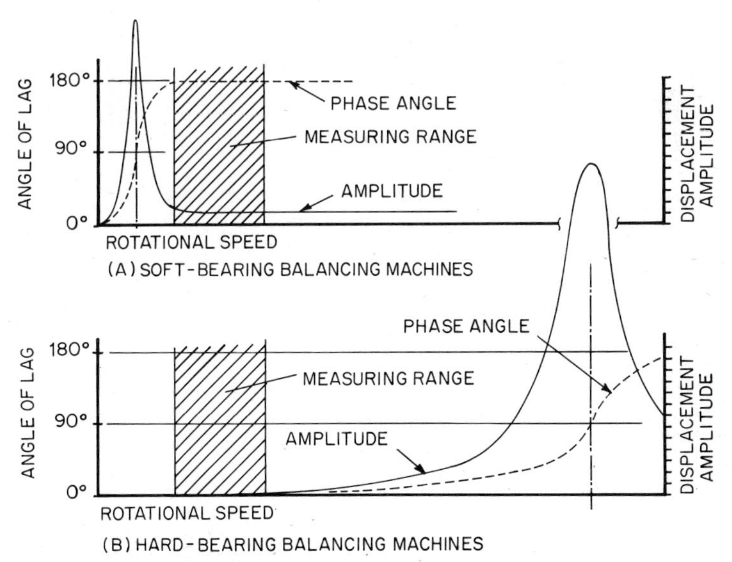

FIG. 39.15. Phase angle (angle of lag) and displacement amplitude vs. rotational speed in soft-bearing and hard-bearing balancing machines.

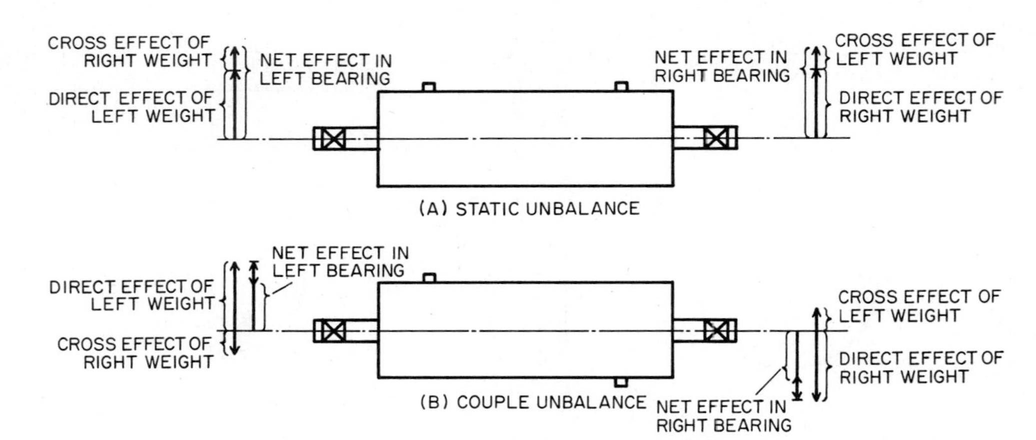

Fig. 39.16. Influence of cross effects in rotors with static and couple unbalance.

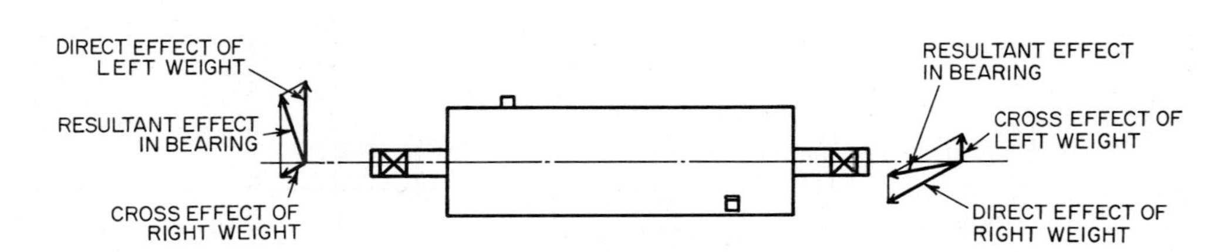

Fig. 39.17. Influence of cross effects in rotors with dynamic unbalance. All vectors seen from right side of rotor.

PLANE SEPARATION. Consider the rotor in Fig. 39.10 and assume that only the unbalance weight on the left is attached to the rotor. This weight causes not only the left bearing to vibrate but to a lesser degree the right. This influence is called "cross effect." If a second weight is attached in the right plane of the rotor as shown in Fig. 39.10, then the direct effect of the weight in the right plane combines with the cross effect of the weight in the left plane, resulting in a composite vibration of the right bearing. If the two unbalance weights are at the same angular position, the cross effect of one weight has the same angular position as the direct effect in the other rotor end plane; thus, their direct and cross effects are additive (Fig. 39.16*A*). If the two unbalance weights are 180° out of phase, their direct and cross effects are subtractive (Fig. 39.16*B*). In a hard-bearing balancing machine the additive or subtractive effect depends entirely on ratios between the axial positions of the correction planes and bearings. On a soft-bearing machine, this is not true, because the masses and inertias of the rotor and its bearings must be taken into account.

If the two unbalance weights on the rotor (Fig. 39.10) have an angular relationship other than 0 or 180°, then the cross effect in the right bearing has a different phase angle than the direct effect from the right weight. Addition or subtraction of these effects is vectorial. The net bearing vibration is equal to the resultant of the two vectors, as shown in Fig. 39.17. The phase angle indicated by the bearing vibration does not coincide with the angular position of either weight. This is the most common type of unbalance (dynamic unbalance of unknown amount and angular position). This interaction of direct and cross effects could cause the balancing process to be a trial-and-error procedure. To avoid this, balancing machines incorporate a feature called "plane separation" which eliminates the influence of cross effect.

Cross effect may be eliminated by supporting the rotor in a cradle which rests on a knife-edge and spring arrangement, as shown in Fig. 39.18. Either the bearing-support members of the cradle or the pivot point are movable so that one unbalance correction plane always can be brought into the plane of the knife-edge. Any unbalance in this plane is prevented from causing the cradle to vibrate. Unbalance in one end plane of the rotor is measured and corrected. The rotor is turned end for end, so that the knife-edge is in the plane of the first correction. Any vibration of the cradle is now due solely to unbalance present in the plane that was first over the knife-edge. Corrections are applied to this plane until the cradle ceases to vibrate. The rotor is now in balance. If it is again turned end for end, there will be no vibration. Mechanical plane separation cradles restrict the rotor length, diameter, and location of correction planes; thus, electronic devices are used to accomplish the function of plane separation.

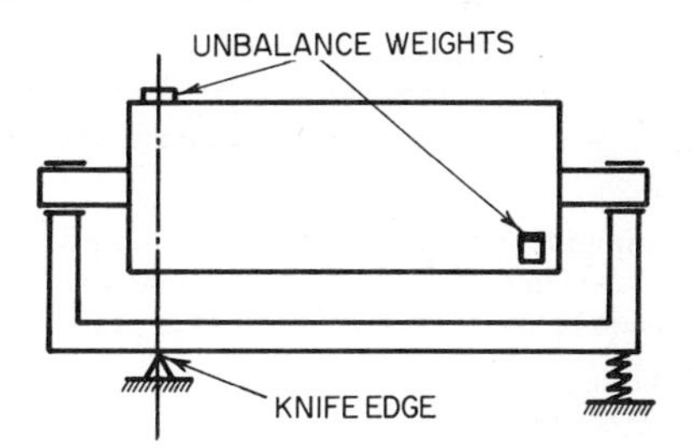

FIG. 39.18. Plane separation by mechanical means.

CLASSIFICATION OF CENTRIFUGAL BALANCING MACHINES

Centrifugal balancing machines may be categorized by the type of unbalance the machine is capable of indicating (static or dynamic), the attitude of the journal axis of the workpiece (vertical or horizontal), and the type of rotor-bearing-support system employed (soft- or hard-bearing). The four classes included in Table 39.1 are described below.

Class I: Trial-and-error Balancing Machines. Machines in this class are of the soft-bearing type. They do not indicate unbalance directly in weight units (such as ounces or grams in the actual correction planes) but indicate only displacement and/or velocity of vibration at the bearings. The instrumentation does not indicate the

Table 39.1. Classification of Balancing Machines

Principle employed	Unbalance indicated	Attitude of shaft axis	Type of machine	Available classes
Gravity (nonrotating)	Static (single-plane)	Vertical	Pendulum	Not classified
		Horizontal	Knife-edges	
			Roller sets	
Centrifugal (rotating)	Static (single-plane)	Vertical	Soft-bearing	
			Hard-bearing	
		Horizontal	Not commercially available	
	Dynamic* (two-plane); also suitable for static (single-plane)	Vertical	Soft-bearing	II, III
			Hard-bearing	III, IV
		Horizontal	Soft-bearing	I, II, III
			Hard-bearing	IV

* When suitably equipped, these machines may also be used for modal balancing of flexible rotors.

amount of weight which must be added or removed in each of the correction planes. Balancing with this type of machine involves a lengthy trial-and-error procedure for each rotor, even if it is one of a presumably identical series. The unbalance indication cannot be calibrated for specified correction planes because these machines do not have the feature of plane separation. Field balancing equipment usually falls into this class.

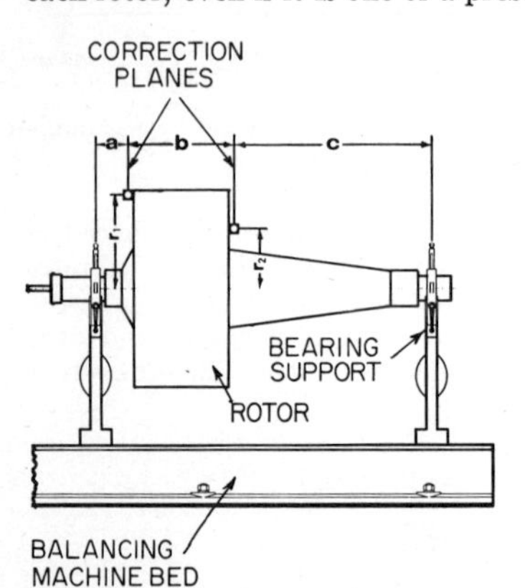

FIG. 39.19. A permanently calibrated balancing machine, showing five rotor dimensions used in computing unbalance. (See Class IV.)

Class II: Calibratable Balancing Machines Requiring a Balanced Prototype. Machines in this class are of the soft-bearing type using instrumentation which permits plane separation and calibration for a given rotor type, if a balanced master or prototype rotor is available. However, the same trial-and-error procedure as for class I machines is required for the first of a series of presumably identical rotors.

Class III: Calibratable Balancing Machines Not Requiring a Balanced Prototype. Machines in this class are of the soft-bearing type using instrumentation which includes an integral electronic unbalance compensator. Any (unbalanced) rotor may be used in place of a balanced master rotor. In turn, plane separation and calibration can be achieved for rotors without trial and error. This class includes soft-bearing machines with electrically driven shakers fitted to the vibratory part of their rotor supports.

Class IV: Permanently Calibrated Balancing Machines. Machines in this class are of the hard-bearing type. They are permanently calibrated by the manufacturer for all rotors falling within the weight and speed range of a given machine size. Unlike the machines in other classes, these machines indicate unbalance in the first run without individual rotor calibration. This is accomplished by the incorporation of an analog computer into the instrumentation associated with the machine. The following five rotor dimensions (see Fig. 39.19) are fed into the computer: distance from left correction plane to left support; distance between correction planes; distance from right correction plane to right support; and r_1 and r_2 radii of the correction weights in the left and right planes, respectively. The instrumentation then indicates the magnitude and angular position of the required correction weight for each of the two selected planes.

The null-force balancing machine is in this class. Although no longer manufactured, it is still widely used. It balances at the same speed as the natural frequency or resonance of its suspension system (including the rotor).

BALANCING-MACHINE EVALUATION[10]

To evaluate the suitability of a balancing machine for a given application, it is first necessary to establish a precise description of the required machine capacity and performance. Such description often becomes the basis for a balancing-machine purchase specification. It should contain details on the range of workpiece weight, the diameter, length, journal diameter, and service speed, and whether the rotors are rigid or flexible, their application, available line voltage, etc. Such information enables the machine

Table 39.2. Standards for Testing Balancing Machines

Application	Title	Issuer	Document no.
General industrial balancing machines	Balancing Machines–Description and Evaluation	International Standards Organization (ISO)	DIS 2953
Jet engine rotor balancing machines (for two-plane correction)	Balancing Equipment for Jet Engine Components, Compressor and Turbine, Rotating Type, for Measuring Unbalance in One or More Than One Transverse Planes	Society of Automotive Engineers, Inc. (SAE)	ARP 587
Jet engine rotor balancing machines (for single-plane correction)	Balancing Equipment for Jet Engine Components, Compressor and Turbine, Rotating Type, for Measuring Unbalance in One Transverse Plane	Society of Automotive Engineers, Inc. (SAE)	ARP 588
Gyroscope rotor balancing machines	Balancing Machine–Gyroscope Rotor	Defense General Supply Center, Richmond, Va.	FSN 6635–450–2208 NT
Field balancing equipment	Field Balancing Equipment–Description and Evaluation	International Standards Organization (ISO)	ISO 2371

vendor to propose a suitable machine. Next the vendor's proposal must be evaluated not only on compliance with the purchase specification but also on the operation of his machine and its features. In describing the machine, the vendor should conform with the applicable standards. Once the machine is purchased and ready for shipment, compliance with the purchase specification and vendor proposal should be verified. Depending on circumstances, such verification is usually repeated after installation of the machine at the buyer's facility.

Precise testing procedures vary for different fields of application. Table 39.2 lists a number of standards for testing balancing machines used in the United States and Canada.

UNBALANCE CORRECTION METHODS[11]

Corrections for rotor unbalance are made either by the addition of weight to the rotor or by the removal of material (and in some cases, by relocating the journal axis). The selected correction method should ensure that there is sufficient capacity to allow correction of the maximum unbalance which may occur. The ideal correction method permits reduction of the maximum initial unbalance to less than balancing tolerance in a single correction. However, this is often difficult to achieve. The more common methods, described below, e.g., drilling, usually permit a reduction of 10:1 in unbalance if carried out carefully. The addition of weight may achieve a reduction as great as 20:1 or higher, provided the weight and its position are closely controlled. If the method selected for reduction of maximum initial unbalance cannot be expected to bring the rotor within the permissible residual unbalance in a single correction, a preliminary correction is made. Then a second correction method is selected to reduce the remaining unbalance to less than its permissible value.

UNBALANCE CORRECTION BY THE ADDITION OF WEIGHT TO THE ROTOR

1. *The addition of wire solder.* It is difficult to apply the solder so that its center-of-gravity is at the desired correction location. Variations in diameter of the solder wire introduce errors in correction.
2. *The addition of bolted or riveted washers.* This method is used only where moderate balance quality is required.
3. *The addition of cast iron, lead, or lead weights.* Such weights, in incremental sizes, are used for unbalance correction.
4. *The addition of welded weights.* Resistance welding provides a means of attachment in which the total weight and center-of-gravity are changed only slightly.

UNBALANCE CORRECTION BY THE REMOVAL OF WEIGHT

1. *Drilling.* Material is removed from the rotor by a drill which penetrates the rotor to a measured depth, thereby removing the intended weight of material with a high degree of accuracy. A depth gage or limit switch can be provided on the drill spindle to ensure that the hole is drilled to the desired depth. This is probably the most effective method of unbalance correction.
2. *Milling, shaping, or fly cutting.* This method permits accurate removal of weight when the rotor surfaces, from which the depth of cut is measured, are machined surfaces and when means are provided for accurate measurement of cut with respect to those surfaces; used where relatively large corrections are required.
3. *Grinding.* In general, grinding is used as a trial-and-error method of correction. It is difficult to evaluate the actual weight of the material which is removed. This method is usually used only where the rotor design does not permit a more economical type of correction.

MASS CENTERING

A procedure known as "mass centering" is used to reduce unbalance effects in rotors. A rotor is mounted in a balanced cage or cradle which, in turn, is rotated in a balancing machine. The rotor is adjusted radially with respect to the cage, until the unbalance indication is zero; this provides a means for bringing the principal inertia axis of the rotor into essential coincidence with the journal axis of the balanced cage. Center drills (or other suitable tools guided along the axis of the cage) provide a means of establishing an axis in the rotor about which it is in balance. The beneficial effects of mass centering are reduced by any subsequent machining operations.

BALANCING OF ROTATING PARTS

MAINTENANCE AND PRODUCTION BALANCING MACHINES

Balancing machines of this type fall into three general categories: (1) universal balancing machines, (2) semiautomatic balancing machines, and (3) fully automatic balancing machines with automatic transfer of work. Each of these has been made in both the nonrotating and rotating types. The rotating type of balancer is available for rotors in which corrections for balance are required in either one or two planes.

Universal balancing machines are adaptable for balancing a considerable variety of sizes and types of rotors. These machines commonly have a capacity for balancing rotors whose weight varies as much as 100 to 1 from maximum to minimum. The elements of these machines are adapted easily to new sizes and types of rotors. The amount and location of unbalance are observed on suitable instrumentation by the machine operator as the machine performs its measuring functions. This category of machine is suitable for maintenance or job-shop balancing as well as for many small and medium lot-size production applications.

Semiautomatic balancing machines are of many types. They vary from an almost universal machine to an almost fully automatic machine. Machines in this category may perform automatically any one or all of the following functions in sequence or simultaneously: (1) retain the amount of unbalance indication for further reference, (2) retain the angular location of unbalance indication for further reference, (3) measure amount and position of unbalance, (4) couple the balancing-machine driver to the rotor, (5) initiate and stop rotation, (6) set the depth of a correction tool from the indication of amount of unbalance, (7) index the rotor to a desired position from indication of unbalance location, (8) apply correction of the proper magnitude at the indicated location, (9) inspect the residual unbalance after correction, and (10) uncouple the balancing-machine driver. Thus, the most complete semiautomatic balancing machine performs the complete balancing process and leaves only loading, unloading, and cycle initiation to the operator. Other semiautomatic balancing machines provide only means for retention of measurements to reduce operator fatigue and error. The equipment which is economically feasible on a semiautomatic balancing machine may be determined only from a study of the rotor to be balanced and the production requirements.

Fully automatic balancing machines with automatic transfer of the rotor are also available. These machines may be either single- or multiple-station machines. In either case, the parts to be balanced are brought to the balancing machine by conveyor, and balanced parts are taken away from the balancing machine by conveyor. All the steps of the balancing process and the required handling of the rotor are performed without an operator. These machines also may include means for inspecting the residual unbalance as well as monitoring means to ensure that the balance inspection operation is performed satisfactorily.

In single-station automatic balancing machines all functions of the balancing process (unbalance measurement, location, and correction) as well as inspection of the complete process are performed in a single checking at a single station. In a multiple-station machine, the individual steps of the balancing process may be done at individual stations. Automatic transfer is provided between stations at which the amount and location of unbalance are determined; then the correction for unbalance is applied; finally, the rotor is inspected for residual unbalance. Such machines generally have shorter cycle times than single-station machines.

FIELD BALANCING EQUIPMENT[12]

Many types of vibration indicators and measuring devices are available for field balancing operations. Although these devices are sometimes called "portable balancing machines," they never provide direct means for measuring the amount and location of the correction required to eliminate the vibration produced by the rotor at its supporting bearings. It is intended that these devices be used in the field to reduce or eliminate vibration produced by the rotating elements of a machine under service conditions. Basically, such a device consists of a combination of a transducer and meter which provides an indication proportional to the vibration magnitude. The vibration magnitude indicated may be displacement, velocity, or acceleration, depending on the type of transducer and readout system used. The transducer can be held by an operator, the hands of the operator providing a seismic mounting. A probe held against the vibrating machine is presumed to cause the transducer output to be proportional to the vibration of the machine. At frequencies below approximately 15 cps, it is almost impossible to hold the transducer sufficiently still to give stable readings. Frequently, the results obtained depend upon the technique of the operator; this can be shown by obtaining measurements of vibration magnitude on a machine with the transducer held with varying degrees of firmness.

Transducers having internal seismic mountings can be clamped firmly to the machine for measurement of vibration magnitude. This type of transducer should not be used where the frequency of the vibration being measured is less than three times the natural frequency of the transducer. The principles of vibration measurement are discussed more thoroughly in Chap. 12.

A transducer responds to all vibration to which it is subjected, within the useful frequency range of the transducer and associated instruments. The vibration detected on a machine may come through the floor from adjacent machines, may be caused by reciprocating forces or torques inherent in normal operation of the machine, or may be due to unbalances in different shafts or rotors in the machine. Location of the transducer on the axis of angular vibration of the machine can eliminate the effect of a reciprocating torque; however, the simple vibration indicator cannot discriminate between the other vibrations unless the magnitude at one frequency is considerably greater than the magnitude at other frequencies. For balancing it may be useful to have the magnitude indicated in displacement units.

Velocity and acceleration are functions of frequency as well as amplitude; therefore, they may be less convenient to use for reduction of unbalance effects where more than one frequency is present. Velocity- and acceleration-type transducers may be used if suitable integrating devices are introduced between the transducer and the meter. A suitable filter following the output of an electromechanical transducer may be introduced to attenuate frequencies other than the wanted frequency.

The approximate location of the unbalance may be determined by measuring the phase of the vibration. Phase of vibration may be measured by a stroboscopic lamp flashed each time the output of an electrical transducer changes polarity in a given direction. Phase also may be determined by use of a phase meter or by use of a wattmeter.

BALANCING OF ASSEMBLED MACHINES

The balancing of rotors assembled of two or more individually balanced parts and the balancing of rotors in complete machines are done frequently to obtain maximum reduction in vibration due to unbalance. In many cases the complete machine is run under service conditions during the balancing procedure.

The requirement for assembly balance often is made necessary by conditions dictated by machining operations and assembly procedures. For example, a balanced flywheel mounted on a balanced crankshaft may not produce a balanced assembly. When pistons and connecting rods are added to the above assembly, more unbalance is introduced. Such resultant unbalance effects can sometimes only be reduced by balancing the engine in assembly. The probable variation of unbalance in an assembly of balanced components is best determined by statistical methods.

Assemblies such as gyros, superchargers, and jet engines often run on antifriction bearings. The inner races of these bearings may not have concentric inside and outside surfaces. The eccentricity of the bearing races makes assembly balancing on the actual bearings desirable. In many cases such balancing is done with the stator supporting the antifriction bearings. This ensures that balance is achieved with the bearing race exactly in the position of final assembly.

PRACTICAL CONSIDERATIONS IN TOOLING A BALANCING MACHINE

SUPPORT OF THE ROTOR

The first consideration in tooling a balancing machine is the means for supporting the rotor. Various means are available, such as twin rollers, plain bearings, rolling element bearings (including slave bearings), gas bearings, nylon V blocks, etc. The most frequently used and easiest to adapt are twin rollers. A rotor should generally be supported at its journals to assure that balancing is carried out around the same axis on which it rotates in service.

Rotors which are normally supported at more than two journals may be balanced satisfactorily on only two journals provided that

1. All journal surfaces are concentric with respect to the axis determined by the two journals used for support in the balancing machine.
2. The rotor is rigid at the balancing speed when supported on only two bearings.
3. The rotor has equal stiffness in all radial planes when supported on only two journals.

If the other journal surfaces are not concentric with respect to the axis determined by the two supporting journals, the shaft should be straightened. If the rotor is not a rigid body or if it has unequal stiffness in different radial planes, the rotor should be supported in a (nonrotating) cradle at all journals during the balancing operation. This cradle should supply the stiffness usually supplied to the rotor by the machine in which it is used. The cradle should have minimum weight when used with a soft-bearing machine to permit maximum balancing sensitivity.

Rotors with stringent requirements for minimum residual unbalance and which run in antifriction bearings, should be balanced in the antifriction bearings which will ultimately support the rotor. Such rotors should be balanced either in (1) special machines where the antifriction bearings are aligned and the outer races held in half-shoe-type bearing supports, rigidly connected by tie bars, or (2) in standard machines having supports equipped with V-roller carriages.

Frequently, practical considerations make it necessary to remove antifriction bearings after balancing, to permit final assembly. If this cannot be avoided, the bearings should be match-marked to the rotor and returned to the location used while balancing. Antifriction bearings with considerable radial play or bearings with a quality less than ABEC (Annular Bearing Engineers Committee) Standard grade 3 tend to cause erratic indications of the balancing machine. In some cases the outer race can be clamped tightly enough to remove excessive radial play. Only indifferent balancing can be done when rotors are supported on bearings of a grade lower than ABEC 3.

When maintenance requires antifriction bearings to be changed occasionally on a rotor, it is best to balance the rotor on the journal on which the inner race of the antifriction bearings fits. The unbalance introduced by axis shift due to eccentricity of the inner race of the bearing then can be minimized by use of high-quality bearings to ensure minimum eccentricity.

BALANCING SPEED

The second consideration in tooling a balancing machine for a specific rotor is the balancing speed. For rigid rotors the balancing speed should be the lowest speed at which the balancing machine has the required sensitivity. Low speeds reduce the time for acceleration and deceleration of the rotor. If the rotor distorts nonsymmetrically at service speed, the balancing speed should be the same as the service speed. Rotors in which aerodynamic unbalance is present may require balancing under service conditions. Some machines show the effect of unbalance produced by varying electrical fields caused by changes in air gap and the like. Such disturbance can be reduced only by balancing (at service speed) if the disturbing frequency is identical to the service speed.

DRIVE FOR ROTOR

A final consideration in tooling a balancing machine for a specific rotor is the means for driving the rotor. For balancing rotors which do not have journals, the balancing machine may incorporate in its spindle the necessary journals; alternatively, an arbor may be used to provide the journal surfaces. An adapter must be provided to adapt the shaftless rotor to the balancing-machine spindle or arbor. This adapter should provide the following:

1. Rotor locating surfaces which are concentric and square with the spindle or arbor axis.
2. Locating surfaces which hold the rotor in the manner in which it is held in final assembly.
3. Locating surfaces which adjust the fit tolerance of the rotor to suit final assembly conditions.
4. A driver to ensure that a fixed angular relation is maintained between the rotor and the adapter.
5. Means for correcting unbalance in the adapter itself.

If the rotor to be tooled has its own journals, it may be driven through: (1) a universal joint or flexible coupling drive from one end of the rotor, (2) a pin-type drive from one end of the rotor, (3) a belt over the periphery of the rotor, or over a pulley attached to the rotor, or (4) air jets or other power means by which the rotor is normally driven in the final machine assembly.

The choice of universal joint or flexible coupling drive attached to one end of the rotor can affect the residual unbalance substantially. Careful attention must be given to the surfaces on the rotor to which the coupling is attached to ensure that the rotor journal axis and coupling are concentric (for example within 0.001 in. total indicator reading) when all fit tolerances and eccentricities have been considered. The weight of that part of the balancing machine drive which is supported by the rotor during the balancing operation, expressed in ounces (and, in this example multiplied by one-half of the total indicator reading or 0.0005 in.) must be less than the permissible residual unbalance in ounce-inches. Adjustable means must be provided in the coupling drive of the balancing machine to apply corrections for balance to the coupling. The adjustments may have to be effective in each of the correction planes of the rotor in an amount equal to at least twice the permissible residual unbalance. For convenience, the coupling should be movable to the rotor for attachment and away from the rotor for uncoupling.

The choice of a pin type of drive can seriously affect the residual unbalance. This drive generally takes the form of one or two driving pins in the balancing-machine face plate, and pins projecting from the adapter for the rotor toward the balancing-machine face plate. Slots or flexible members are used to transmit torque from the balancing-machine spindle through the pins. Misalignment of the rotor axis with respect to the spindle axis of the balancing machine introduces forces which have substantial effect on the indication of unbalance. The effect of misalignment on unbalance indication increases as the driving torque increases. Variations in the weight of successive rotors of the same type and wear of the balancing-machine members which support the rotor make it difficult to maintain alignment of the balancing-machine spindle axis and the rotor axis. For proper balancing, failure of the two axes to intersect by 0.005 in. at the point of the drive and failure of the two axes to be parallel within 0.005 in./ft should not introduce an error in amount of unbalance indication exceeding one-quarter of the permissible residual unbalance. The driver which attaches to the rotor must locate from surfaces of the rotor which are concentric with the journal axis because an accumulation of fit tolerances and eccentricities introduces an error in the result.

A belt drive can transmit only limited torque to the rotor. Driving belts must be extremely flexible and of uniform thickness. Driving pulleys attached to the rotor should be used only when it is impossible to transmit sufficient driving torque by running the belt over the rotor. Pulleys must be as light as possible, must be dynamically balanced, and should be mounted on surfaces of the rotor which are square and concentric with the journal axis. The belt drive should not cause disturbances in the unbalance indication exceeding one-quarter of the permissible residual unbalance. Rotors driven by belt should not drive components of the balancing machine by means of any mechanical connection.

The use of electrical means or air for driving rotors may influence the unbalance

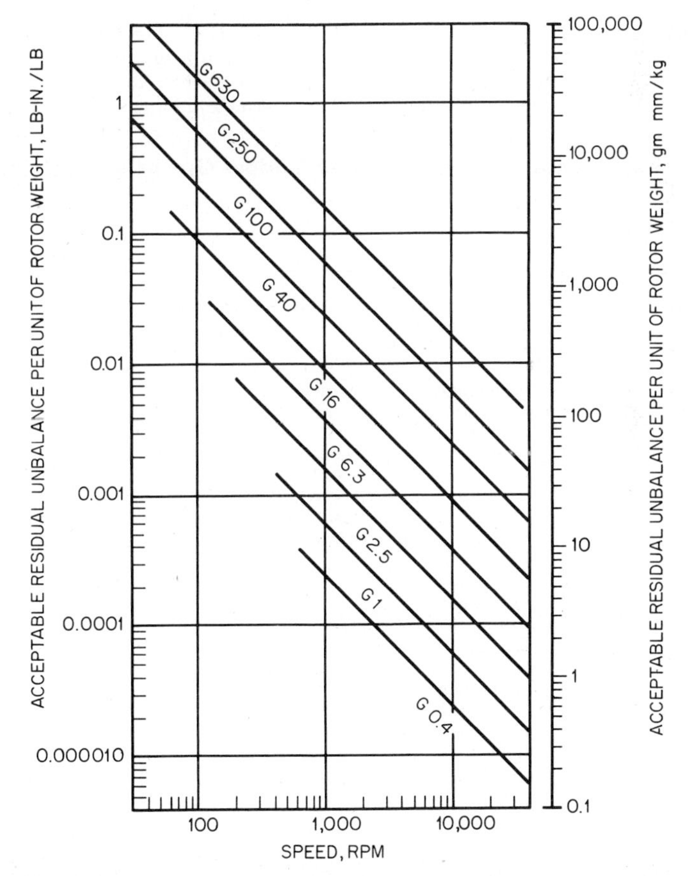

FIG. 39.20. Residual unbalance corresponding to various balancing quality grades, *G. Note:* 1 gm. mm/kg is equivalent to a displacement of the center-of-gravity of 0.001 mm = 39 μin.

readout. To avoid or minimize such influence, great care should be taken to bring in the power supply through very flexible leads, or have the airstream strike the rotor, at right angles to the direction of the vibration resulting from unbalance.

If the balancing machine incorporates filters tuned to a specific frequency only, it is essential that means be available to control the rotor speed to suit the filter setting.

BALANCE CRITERIA

Achieving better balance in rotors is primarily an art; therefore, it is often difficult for an engineer normally conversant with balancing methods and techniques to decide which particular balancing method to employ, the rotational speed for balancing to be used, and at what particular point in a production line the balancing procedure should be inserted. The appropriate choice of a balance criterion is likely to be an even greater problem.

A suitable criterion of the quality of balancing required would appear to be the running smoothness of the complete assembly; however, many other factors than unbalance contribute to uneven running of machines (for example, bearing dissymmetries, misalignment, aerodynamic and hydrodynamic effects, etc.). In addition, there is no simple relation between rotor unbalance and vibration amplitude measured on the bearing housing. Many factors, such as proximity of resonant frequencies, fit, and machining errors, may influence over-all vibration level considerably. Therefore, a measurement of the vibration amplitude will not indicate directly the magnitude of unbalance or whether an improved state of unbalance will cause the machine to run smoother. For certain classes of machines, particularly electric motors and large turbines and genera-

Table 39.3. Balance Quality Grades for Various Groups of Rigid Rotors[13]

Balance quality grade	Type of rotor
G4,000	Crankshaft drives of rigidly mounted slow marine diesel engines with uneven number of cylinders.
G1,600	Crankshaft drives of rigidly-mounted large two-cycle engines.
G630	Crankshaft drives of rigidly mounted large four-cycle engines; crankshaft drives of elastically mounted marine diesel engines.
G250	Crankshaft drives of rigidly mounted fast four-cylinder diesel engines.
G100	Crankshaft drives of fast diesel engines with six or more cylinders; complete engines (gasoline or diesel) for cars and trucks.
G40	Car wheels, wheel rims, wheel sets, drive shafts; crankshaft drives of elastically mounted fast four-cycle engines (gasoline or diesel) with six or more cylinders; crankshaft drives for engines of cars and trucks.
G16	Parts of agricultural machinery; individual components of engines (gasoline or diesel) for cars and trucks.
G6.3	Parts or process plant machines; marine main-turbine gears; centrifuge drums; fans; assembled aircraft gas-turbine rotors; fly wheels; pump impellers; machine-tool and general machinery parts; electrical armatures.
G2.5	Gas and steam turbines; rigid turbo-generator rotors; rotors; turbo-compressors; machine-tool drives; small electrical armatures; turbine-driven pumps.
G1	Tape recorder and phonograph drives; grinding-machine drives.
G0.4	Spindles, disks, and armatures of precision grinders; gyroscopes.

Note: In general, for rigid rotors with two correction planes, one-half the recommended residual unbalance is to be taken for each plane; these values apply usually for any two arbitrarily chosen planes, but the state of unbalance may be improved upon at the bearings; for disc-shaped rotors, the full recommended value holds for one plane.

tors, voluminous data have been collected which can be used as a guide for the establishment of criteria for such installations.

Table 39.3 and Fig. 39.20 show a classification system for various types of representative rotors, based on a document—ISO Standard 1940–1973, "Balance Quality of Rotating Rigid Bodies"—of the International Standards Organization. Balance quality grades are grouped according to numbers with the prefix G; the numbers associated with each group indicate the permissible acceptable residual unbalance per unit of rotor weight, expressed in English and SI units. The residual unbalance is equivalent to a displacement of the center-of-gravity. The recommended balance quality grades are based on experience with various types, sizes, and service speeds; they apply only to rotors which are rigid throughout their entire range of service speeds.

REFERENCES

1. "Balancing—Vocabulary," ISO 1925, 1st ed., 1974-11-01.
2. "Balancing Criteria for Flexible Rotors," ISO/TC 108/SC 1.
3. Federn, K. G.: Analysis of Balancing Procedure, lecture at General Electric Company Balancing Seminar, Schenectady, N.Y., May 3, 1956.
4. Federn, K. G.: *Werkstatt u. Betrieb*, **86**:243 (1953).
5. Giers, A.: "Rechnergestütztes Auswuchten elastischer Rotoren," Sonderdruck, June, 1974.
6. Rieger, N. F.: Balancing High-Speed Rotors to Reduce Vibration Levels, ASME Design Conference, Chicago, Ill., May 8, 1972.
7. Moore, L. S., and E. G. Dodd: *G.E.C. J.*, **31**(2) (1964).
8. Schneider, H.: "Auswuchttechnik," *VDI Taschenbücher* T29, Düsseldorf, Germany.
9. McQueary, D. E.: "Understanding Balancing Machines," American Machinist Special Rep. 656, June 11, 1973.
10. "Balancing Machines—Description and Evaluation," ISO 2953.
11. Senger, W. I.: *Machine Design*, **16**:101 (1944); **16**:131 (1944); **17**:127 (1945).
12. "Field Balancing Equipment—Description and Evaluation," ISO 2371, 1st ed., 1974.
13. "Balancing Quality of Rotating Rigid Bodies," ISO 1940, 1st ed., 1973-05-01, and ASA STD 2-1975 (ANSI S2.19-1975).

40

MACHINE-TOOL VIBRATION

J. G. Bollinger
University of Wisconsin, Madison

INTRODUCTION

The machining of metals and other materials is invariably accompanied by relative vibration between workpiece and tool. These vibrations are due to one or more of the following causes: (1) inhomogeneities in the workpiece material; (2) disturbances in the workpiece or tool drives, including the feed drive; (3) interrupted cutting; (4) vibration generated in other machines and transmitted through the foundation; (5) vibration generated by the cutting process itself (machine-tool chatter).

The tolerable level of relative vibration between tool and workpiece, i.e., the maximum amplitude and to some extent the frequency, depends on whether the machining process is of the roughing or finishing kind. In the roughing process the tolerable level is determined mainly by detrimental effects of the vibration on tool life (see *The Effect of Vibration on Tool Life*) and by the noise which is frequency generated. In finishing operations, surface finish and machining accuracy determine the tolerable level.

VIBRATION DUE TO INHOMOGENEITIES IN THE WORKPIECE

Hard spots in the material being worked impart small shocks to the tool and workpiece, as a result of which free vibrations are set up. If these transients are rapidly damped out, their effect is usually not serious; they simply form part of the general "background noise" encountered in making vibration measurements on machine tools. Cases in which transient disturbances do not decay but build up to vibrations of large amplitudes (as a result of dynamic instability) are of great practical importance, and are discussed later.

When machining under conditions resulting in discontinuous chip removal, the segmentation of chip elements results in a fluctuation of the cutting thrust. If the frequency of these fluctuations coincides with one of the natural frequencies of the structure, forced vibration of appreciable amplitude may be excited.[1] However, it is not clear whether the segmentation of the chip is a primary effect or whether it is produced by other vibration without which continuous chip flow would be encountered.

The breaking away of a built-up edge from the tool face also imparts impulses to the cutting tool which result in vibration. However, marks left by the built-up edge on the machined surface are far more pronounced than those caused by the ensuing vibration; it is probably for this reason that the built-up edge has not been studied from the vibration point of view. The built-up edge frequently accompanies certain types of vibration (chatter), and instances have been known when it disappeared as soon as the vibration was eliminated.

DISTURBANCES IN THE WORKPIECE AND TOOL DRIVE

Vibrations arise in drives that impart rotational motions (spindle and rotary table drives) and in drives that impart rectilinear motion (feed drives). Forced vibrations result from rotating unbalanced masses, gear and belt drives, bearing irregularities, and periodic but nonsinusoidal forces in drive motors. Free vibrations result from shocks caused by reciprocating unbalanced masses, a type of self-induced vibration frequently encountered with slide drives (stick-slip), and from spurious transient motions in the equipment.

VIBRATION CAUSED BY ROTATING UNBALANCED MEMBERS

When the speed of rotation of some unbalanced member falls near one of the natural frequencies of the machine-tool structure, forced vibration of large amplitude is induced which may affect both surface finish and tool life. This vibration can be eliminated by careful balancing, the procedure being basically similar to that described in Chap. 39.

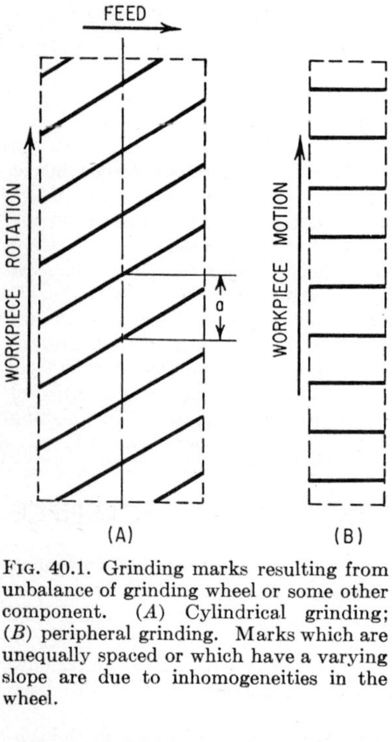

FEED

WORKPIECE ROTATION

a

WORKPIECE MOTION

(A) (B)

Fig. 40.1. Grinding marks resulting from unbalance of grinding wheel or some other component. (*A*) Cylindrical grinding; (*B*) peripheral grinding. Marks which are unequally spaced or which have a varying slope are due to inhomogeneities in the wheel.

When designing a new machine, a great deal of trouble can be forestalled by placing rotating components in a position in which the detrimental effect of their unbalance is likely to be relatively small. Motors should not be placed on the top of slender columns, and the plane of their unbalance should preferably be parallel to the plane of cutting. In some cases, vibration resulting from rotating unbalanced members can be eliminated by applying the usual vibration-isolation techniques (Chaps. 30 and 32); for example, the coolant pump may be supported by vibration isolators.[2]

Among all machining processes, grinding is most sensitive to vibration because of the high surface finish resulting from the operation. In cylindrical grinding, marks resulting from unbalance of the grinding wheel or of some other component are readily recognizable. They appear in the form of equally spaced, continuous spirals with a constant slope, as shown in Fig. 40.1*A*. From these marks, the machine component responsible for their existence is found by considering that its speed in rpm must be equal to $\pi DN/a$ where D is the workpiece diameter in inches, a is the pitch of the marks in inches, and N is the workpiece speed in rpm. An analogous procedure also can be applied to peripheral surface grinding. Marks produced in one pass of the wheel are shown in Fig. 40.1*B*. The speed of the responsible component in rpm is equal to the number of marks (produced in one pass) which fall into a distance equal to that traveled by the workpiece (or wheel) in 1 min.

MARKS CAUSED BY INHOMOGENEITIES IN THE GRINDING WHEEL. Although grinding marks usually indicate the presence of a vibration, this vibration may not necessarily be the primary cause of the marks. Hard spots on the cutting surface of the wheel result in similar, though generally less pronounced, marks. Grinding wheels usually are not of equal hardness throughout.[3] A hard region on the wheel circumference becomes rapidly glazed in use and establishes itself as a high spot on the wheel (since it retains the grains for a longer period than the softer parts). However, these high spots eventually break down or shift to other parts of the wheel; in cylindrical grinding, this manifests itself as a sudden change of the slope of the spiral marks. Marks which appear to be due to an unbalanced member rotating at two or three times the speed of the wheel and which are nonuniformly spaced are always due to two or three hard spots.

Owing to the sensitivity of high precision grinding to inhomogeneities in the grinding wheel, a vibration test has been suggested with which the presence of cracks and hard spots can be detected.[4] This test is based on the fact that in any elastic system of cylindrical symmetry (for instance, a disc or a cup wheel) the presence of imperfections (inhomogeneities or dimensional deviations from perfect cylindrical symmetry) results in a splitting up of the natural frequencies of the various bending modes of vibration. Each bending mode of vibration (specified by the number of nodal circles and nodal diameters) can be excited at two distinct natural frequencies. The difference between these natural frequencies is, to some extent, a measure of the imperfections present. In a perfect system the natural frequencies coincide.[5]

THE EFFECT OF VIBRATION ON THE WHEEL PROPERTIES. If vibration exists between wheel and workpiece, normal forces are produced which react on the wheel and tend to alter the wheel shape and/or its cutting properties.[6,7] In soft wheels the dominating influence of vibration appears to be inhomogeneous wheel wear, and in hard wheels inhomogeneous loading (i.e., packing of metal chips on and in crevasses between the grits). These effects result in an increased fluctuation of the normal force which produces further changes of the wheel properties. The over-all effect is that a vibration once initiated tends to grow.[8] When successive cuts or passes overlap, the inhomogeneous wear and loading of the wheel may cause a regenerative chatter effect which makes the cutting process dynamically unstable (see *Dynamic Stability*).

It is desirable to reduce vibration in so far as possible in the initial design. This necessitates careful balancing, well-adjusted bearings (see *Bearings*), a suitable belt (see *Belt Drives*), and a stiff machine construction (see *The Design of Vibration-free Machine Tools*).

GEAR DRIVES

All gear faults, eccentricities, pitch errors, profile errors, etc., produce nonuniform rotation which in some cases adversely affects surface finish, geometry, and possibly tool life. It is for this reason that in precision machines, where a high degree of surface finish is required, the workpiece or tool spindle usually is driven by belts. In this application, belts act like filters which suppress the high-frequency torsional oscillations.

Gear vibration[9,10] is an important subject, particularly in precision machinery. Gear cutting and grinding, where precise kinematic relations must be maintained and surface finish is important, is an area in which gear-system-induced vibrations are detrimental. The dynamic behavior of gear trains has been extensively investigated by computer simulation techniques[11] to gain insight into how gear accuracy, stiffness variation, and tooth geometry relate to resulting vibration. Dynamic tooth load resulting from vibration in gear trains is also an important consideration in gear wear and noise generation.[12]

Tests with a lathe having a headstock gear with a deliberately large eccentricity (0.1 mm) showed that when the center distance was adjusted to maintain the gear in mesh without backlash, large amplitudes of vibration were set up when the frequency of tooth engagement coincided nearly with the natural frequency of the work.[13] The amplitude of the forced vibration resulting from gear eccentricity can be decreased by increasing backlash between the gears.

BELT DRIVES

Belt drives may give rise to both torsional and rectilinear forced vibrations. Any variation of the effective belt radius, i.e., the radius of the neutral axis of the belt around the pulley axis, produces a variation of the belt tension and the belt velocity. This causes a variation of the bearing load and of the rotational velocity of the pulley.

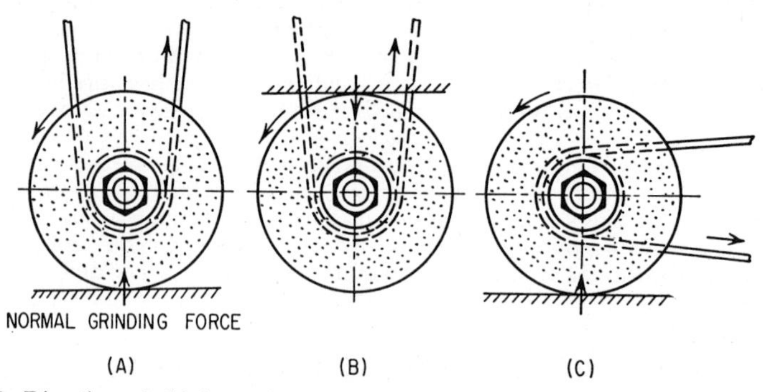

FIG. 40.2. Direction of driving belt and its influence on performance. (*A*) Vibration is minimized when belt tension and normal grinding force point in the same direction. (*B*) Large amplitudes may arise when the normal grinding force is substantially equal to the belt tension. (*C*) Vibration due to centrifugal force is likely to be caused by an unbalance of wheel. (*S. Doi.*[8])

The effective pulley radius can vary as a result of (1) form faults of the pulley itself (eccentric rim or groove) or (2) form and/or structural faults in the belt (variation of belt profile or inhomogeneity of belt material).[14] Tests have shown that from the point of view of vibration, flat belts are preferable to V belts, partly because of the smaller disturbances generated by them and partly because the disturbing force is almost independent of the belt tension.[14]

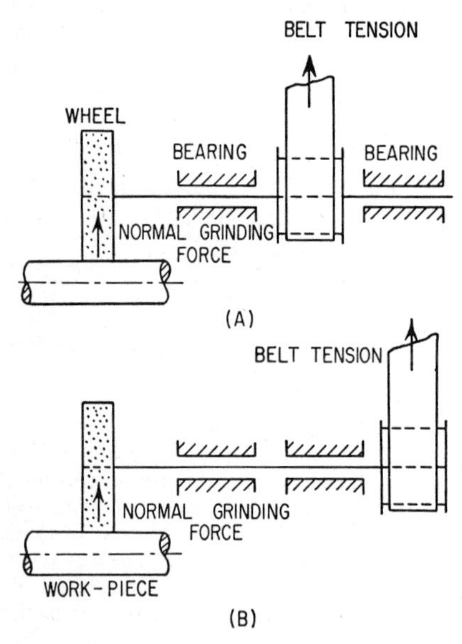

FIG. 40.3. Effect of relative position of grinding wheel, bearings, and driving pulley on grinding performance. (*A*) Driving pulley should be placed between bearings, as shown in (*A*). Arrangement shown in (*B*) is constructionally simpler but is more liable to cause trouble. (*S. Doi.*[8])

Grinding is particularly sensitive to disturbances caused by belts. Seamless silk belts or a direct motor drive to the main spindle is recommended for high-precision machines.[8] Attention also should be paid to the direction of the belt tension with respect to the normal grinding force. Vibration is minimized when the belt tension and the normal grinding force point in the same direction, as shown in Fig. 40.2*A*. The clearance between bearing and spindle is thus eliminated. With the arrangement shown in Fig. 40.2*B*, large amplitudes of vibration may arise when the normal grinding force is substantially equal to the belt tension and/or the peripheral surface of the wheel is nonuniform. Tests indicate that with the arrangement shown in Fig. 40.2*C*, vibration due to the centrifugal force is likely to be caused by an unbalance of the wheel.[8]

The spindle pulley should preferably be placed between the spindle bearings (Fig. 40.3*A*) and not at the end of the spindle (Fig. 40.3*B*); the latter usually is done to achieve constructional simplicity. With the arrangement of Fig. 40.3*B*, the wheel displacement attributable to the normal grinding force opposes that due to the belt tension; thus, conditions are basically similar to those of Fig. 40.2*B*.

BEARINGS

Dimensional inaccuracies of the components of ball or roller bearings and/or surface irregularities on the running surfaces (or the bearing housing) may give rise to vibration trouble in machines where high-quality surface finish is demanded. From the frequency of the vibration produced, it is sometimes possible to identify the component of the bearing responsible.[15, 16] For conventional bearings frequently used in machine tools, the outer race is stationary and the inner race rotates with n_i rpm; the cage velocity is of the order of $n_c \simeq 0.4n_i$, and that of the balls or rollers about $n_b \simeq 2.4n_i$. In some cases, a disturbing frequency of the order of $n_z = zn_c$ also can be detected where z is the number of rolling elements. This is the frequency with which successive rolling elements pass through the "loaded zone" of the bearing, which is determined by the direction of the load. These disturbing frequencies usually are not detectable with bearings having two rows of rolling elements, each unit of which lies halfway between units of the neighboring row.

From the point of view of vibration control, best results are achieved with bearings containing two rows, each having a large number of rolling elements and having a very small radial clearance. Where a large spindle stiffness is required, as in a lathe, the front spindle bearing may consist of two of such bearings, combined with a thrust bearing to take up the axial load acting on the spindle. Using such a bearing construction, a spindle supported only at its ends may have a greater stiffness and a higher natural frequency than a spindle supported in three places with single roller bearings.[17, 18]

The stiffness and damping of roller bearings increase with a decrease of the radial play. This is particularly pronounced when the radial play becomes negative, i.e., when the bearings are preloaded. Vibration is then reduced, and the surface finish and accuracy of the work improve. However, it is not clear whether preloading is practicable, because it is accompanied by an increase in heat generation and the likely decrease of bearing life.[19]

It is often claimed that sliding bearings have a greater damping capacity than roller bearings and are therefore superior with respect to vibration.[8] However, careful measurements show that the damping capacity of sliding bearings which have attained their steady working temperature is of the same order as that of roller bearings. As the temperature of the lubricating oil increases, the damping decreases. Consequently, vibration of undesirably large magnitude may arise only after the machine has been running for more than 30 min. The damping of sliding bearings is greater than that of roller bearings only if a highly viscous lubricating oil is used. Even under these conditions, the over-all damping effect is unlikely to be appreciable because the amplitude of vibration in the bearing is small.[16]

STICK-SLIP IN LINEAR DRIVES

The uniformity of feed motions is often disturbed by a phenomenon known as *stick-slip*.[20, 21] When motion of a tool support is initiated by engagement of a transmission, elements of the transmission suffer elastic deformation until the forces transmitted exceed the static frictional resistance of the tool support. Subsequently, the support commences to move, and the friction drops to its dynamic value. As a result of the drop of the friction force, the support receives a high acceleration and overshoots because of its inertia. At the end of the "jump," the transmission is wound up in the opposite sense; before any further motion can take place, this deformation must be unwound; this occurs during a period of standstill of the support. Subsequently, the phenomenon repeats itself. The physical sequence described falls into the category generally known as "relaxation oscillations"[23, 26] (Chap. 5).

The occurrence of stick-slip depends on the interaction of the following factors:[22] (1) the mass of the sliding body, (2) the drive stiffness, (3) the damping present in the drive,

(4) the sliding speed, (5) the surface roughness of the sliding surfaces, and (6) the lubricant used.[22] As a rule, it is encountered only at very low sliding speeds or when moving a mass of appreciable size a small distance from rest. Slide drives designed for stick-slip-free motion have small moving masses and a high drive stiffness. Excellent results also may be achieved by using cast iron and a suitable plastic material as mating surfaces.[24, 30] Pressurized slides are used with considerable success, particularly on servo-controlled machine tools. By keeping the oil film between the mating surfaces under a certain pressure, the possibility of mixed dry and viscous friction is eliminated and as a result stick-slip cannot arise. For many applications (measuring instruments, precision boring machines, etc.), slides in which sliding friction is replaced by rolling friction are used.[27–29]

VIBRATION DUE TO INTERRUPTED CUTTING

When machining interrupted surfaces, impulses of appreciable magnitude may be imparted to the tool which may lead to undesirable vibration. With single-pointed cutting tools, this vibration usually is not severe (although the tool may suffer) and is quickly damped out. In cylindrical grinding, when machining a workpiece which contains a slot or a keyway, visible marks frequently are observed near the "leaving edge" of the slot or keyway. These are due to a "bouncing" of the grinding wheel on the machined surface. They are eliminated or minimized by inserting a wooden plug or key into the recess.

Serious forced vibration owing to interrupted cutting is encountered when using cutting tools with multiple cutting edges, i.e., in the milling process.

VIBRATION TRANSMITTED FROM OTHER MACHINES

Vibration generated in machines (presses, machine tools, internal-combustion engines, compressors, etc.) is transmitted through the foundation to other machines and may set up forced vibration in these. In the workshop, the vibration of the floor contains a wide frequency spectrum;[31] it is almost inevitable that one of these frequencies should fall near a natural frequency of a particular machine tool. The amplitudes of this background vibration usually are small, and become troublesome only in the case of grinding machines and precision boring machines.

The isolation of vibration transmitted through the floor may be achieved by the customary methods of vibration isolation (Chaps. 30 and 32), i.e., the machine which generates the vibration or is to be protected from vibration is placed upon vibration isolators. In general, the first course is to be preferred, particularly when exciting forces result from presses or hammers.[32]

When applying vibration isolators to machine tools, some care must be exercised. The foundation constitutes the "end condition" of the machine-tool structure. Any alteration of the end condition affects the natural frequencies of the structure, the corresponding damping, equivalent stiffness, and the dynamic deflection curves (modes of vibration).[33–35]

In general, vibration isolators lower the natural frequencies, damping, and stiffness. The structure may thus become more susceptible to internal exciting forces, and its chatter behavior also may be affected in an undesirable way.[35] However, it is fortunate that only the lower modes of vibration are affected to an appreciable extent (primarily the rocking modes) and these do not play an important part in many machine tools (milling machines, grinding machines, etc.). In general, machine-tool structures which are very stiff by themselves (i.e., without being bolted down) can be placed on vibration isolators safely (milling machines, grinding machines, some lathes).[36, 37]

Isolators for medium loads usually contain rubber springs.[32, 38] Those suitable for very great loads (presses, hammers) are of all-metal construction, usually incorporating hydraulic or dry-friction dampers.[32, 39, 40] Sheets of cork, rubber, or plastic also have been used with good results as isolators.

The foundation of a machine tool has an important influence on the dynamic behavior of a machine structure due to excitations originating within the machine and those originating externally from the machine. This is particularly significant below 100 cps, as has been reported in studies directed at establishing important foundation parameters.[41]

MACHINE-TOOL CHATTER

INTRODUCTION

The cutting of metals is frequently accompanied by a violent vibration of workpiece and cutting tool known as machine-tool chatter. *Chatter* is a self-induced vibration which is caused not by external forces but which is induced and maintained by forces generated by the cutting process itself. It is highly detrimental to tool life[42–45] and surface finish,[44, 87] and is usually accompanied by considerable noise. Moreover, chatter also affects adversely the rate of production since, in many cases, its elimination can be achieved only by reducing the rate of metal removal.

Machine-tool chatter is characteristically erratic. Forced and free vibration effects in machine tools are usually detected in the development stage or during final inspection, and can be reduced or eliminated by known procedures. The inclination of a certain machine to chatter often remains unobserved in the plant of the machine-tool manufacturer; when finally encountered, its elimination from a particular machining process may be highly time-consuming and laborious.

A distinction can be drawn[46] between regenerative and nonregenerative chatter. The former occurs when there is an overlap in the process of performing successive cuts such that part of a previous cut surface is removed by a succeeding pass of the cutter. Under regenerative cutting, a displacement of the tool can result in a vibration of the tool relative to the workpiece resulting in a variation of the chip thickness. This in turn results in a variation in the cutting force during following revolutions. The regenerative chatter theory explains a wide variation of practical chatter situation in such operations as normal turning and milling.

Nonregenerative chatter is found in such operations as shaping, slothing, and screw-thread cutting. In this type of cutting, chatter has been explained through the principle of mode coupling.[8, 47–50] If a machine-tool vibration consists of two degrees-of-freedom, with the axes of major flexibilities normal and with a common mass, the dynamic motion can take an elliptical path. If the major axis of motion (axis with the greater compliance) lies within the angle formed by the total cutting force and the normal to the workpiece surface, energy can be transferred into the machine-tool structure, thus producing an effective negative damping. The width of cut for the threshold of stable operation is directly dependent upon the difference between the two principal stiffness values, and chatter tends to occur when the two principal stiffnesses are close in magnitude.

DYNAMIC STABILITY

Machine-tool chatter is essentially a problem of dynamic stability. A machine tool under vibration-free cutting conditions may be regarded as a dynamical system in steady-state motion. Systems of this kind may become dynamically unstable and break into oscillation around the steady motion. Instability is caused by an alteration of the cutting conditions produced by a disturbance of the cutting process (e.g., a hard spot in the material). As a result, a time-dependent thrust element dP is superimposed on the steady cutting thrust P. If this thrust element is such as to increase the original disturbance, oscillations will build up and the system is said to be unstable.

This chain of events is most easily investigated theoretically by considering that the incremental thrust element dP is a function not only of the original disturbance but also of the velocity of this disturbance. Forces which are dependent on the velocity of a displacement are damping forces; they are additive to or subtractive from the damping present in the system (e.g., structural damping, assumed to be viscous). When the damping due to dP is positive, the total damping (structural damping plus damping due to altered cutting conditions) also is positive and the system is stable. Any disturbance will then be damped out rapidly. However, the damping due to dP may be negative, in which case it will decrease the structural damping, which is always positive. If the negative damping due to dP predominates, the total damping is negative. Positive damping forces are energy absorbing. Negative damping forces feed energy into the system; when the total damping is negative, this energy is used for the maintenance of oscillations (chatter).

From the practical point of view, the fully developed chatter vibration (self-induced vibration) is of little interest. Production engineers are almost entirely concerned with conditions leading to chatter (dynamic instability). The build-up of chatter is very difficult to observe, and experimental work has to be carried out mainly under conditions which are only indirectly relevant to the problem being investigated. Experimental results obtained from fully developed chatter vibration may, in some instances, be not really relevant to the problem of dynamic stability.

The influence of the machine-tool structure on the dynamic stability of the cutting process is of great importance. This becomes clear by considering that with a structure (including tool and workpiece) of infinite stiffness, the cutting process could not be disturbed in the first place because hard spots, for example, would not be able to produce the deflections necessary to cause such a disturbance. Furthermore, it is clear that were the structural damping infinite, the total damping could not become negative and the cutting process would always be stable. This discussion indicates that an increase of structural stiffness and/or damping always has beneficial effects from the point of view of chatter.

In practically feasible machines, the interrelation between structural stiffness, damping, and dynamic stability is of considerable complexity. This is because machine-tool structures are systems with distributed mass, elasticity, and damping; their vibration is described by partial differential equations of intractable complexity. For this reason, the dynamic characteristics of the machine-tool structure are introduced into the analysis of dynamic stability by making certain simplifying assumptions. From this point of view, the existing theories of the dynamic stability of the cutting process can be divided into two groups. In the first group it is assumed that dynamic instability can arise only in one of the many modes of vibration of the structure. This assumption is legitimate when the modes of the structure are well separated and/or the geometry of the cutting tool does not permit the simultaneous excitation of several modes.

THE EFFECT OF VIBRATION ON TOOL LIFE

Inasmuch as the cutting speed and the chip cross section vary during vibration, it is to be expected that vibration affects tool life. The magnitude of this effect is unexpectedly large, even when impact loading of the tool is excluded. This has been shown by theoretical and experimental investigations of single-pointed cutting tools.[51, 43, 44, 52, 55]

Impact loading of the cutting edge may be very severe, and the elimination of vibration may multiply tool life 80 to 200 times.[45]

The life of face-mill blades may suffer considerably owing to torsional vibration executed by the cutter. A tool-life reduction of 30 per cent is not uncommon.[54] The torsional vibration need not necessarily be caused by dynamic instability of the cutting process but may be forced vibration, because of faulty gears or interrupted chip removal.

THE ELIMINATION OF VIBRATION

The vibration behavior of a machine tool can be improved by an increase of the equivalent static stiffness and the damping for the modes of vibration which result in a relative displacement between tool and workpiece. Furthermore, improvement can be achieved by appropriate choice of cutting conditions, tool design, and workpiece design. Stiffness and damping are important both for ordinary (free and forced) and self-induced (chatter) vibrations. Cutting conditions and tool design influence mainly self-induced vibration. In addition, the application of vibration dampers and absorbers has proven to be an effective technique for the solution of machine-vibration problems. These devices should be viewed as a functional part of a machine and not as an add-on to solve specific problems, although their past use has often taken that perspective.

THE ELIMINATION OF VIBRATION IN EXISTING MACHINES

Problems associated with free or forced vibration may be eliminated by getting rid of the source of the disturbance, isolating the source, absorbing the vibrational energy, or modifying the structure.

Cutting Conditions. In the elimination of chatter, cutting conditions are first altered. If this fails to achieve the desired result, an increase of stiffness between tool and workpiece and/or damping of disturbances impinging on the tool or improved tool design is resorted to.

Regenerative chatter usually is confined to definite speed ranges which are separated by chatter-free speeds. Other types of chatter are also usually confined to a certain speed range.[68] Thus, in some cases of regenerative chatter a small increase or decrease of speed may stabilize the cutting process. This is likely to be effective only when stiffness and/or damping are relatively large; otherwise the unstable speed ranges overlap. In most cases, very high cutting speeds are stable[48] and the consequent loss of tool life from speed may be compensated by the increase from elimination of vibration.

An increase of the feed rate is also beneficial in some types of machining (drilling, face milling, etc.). For the same cross-sectional area, narrow chips (high feed rate) are less likely to lead to chatter than wide chips (low feed rate) since the chip-thickness variation effect results in a relatively smaller variation of the cross-sectional area in the former (smaller dynamic cutting force).

Continuous variation of cutting conditions achieved through a programmed variation of spindle speed is a technique which has been investigated and under laboratory conditions appears to have some promise as a technique for chatter suppression.[87, 88]

Stiffness. Although the configuration of the structure is given in existing machines, the static and the equivalent static stiffness values may nevertheless vary within wide limits. High stiffness values are ensured by the use of steadies, by placing tool and workpiece in a position where the relative dynamic displacement between them is small (i.e., by placing them near the main column, etc.), by using rigid tools and clamps, by using jigs which rigidly clamp (and if necessary support) the workpiece, by clamping securely all parts of the machine which do not move with respect to each other, etc.[55]

In some cases chatter is eliminated by loosening the locks of slides. The conclusion often drawn from this is that flexible machines may be less inclined to chatter than stiff machines; i.e., rigidity sometimes is undesirable. When the rigidity of some machine element is reduced, greater damping may appear at the cutter; the increase of damping often outweighs the reduction in rigidity. Consequently, although a loss of rigidity is always undesirable, it may be tolerated when accompanied by a large increase of damping.

In some cases old machines are less likely to chatter than new machines of identical design. This may result from wear and tear of the slides which increases the damping and effects an improvement in performance. However, it would be wrong to conclude that lack of proper attention and maintenance is desirable. Proper attention to slides, bearings (minimum play), belts, etc., is necessary to satisfactory performance. It would be wrong also to conclude that a highly polluted workshop atmosphere is desirable because some new machines, even when not used but merely exposed to workshop dirt for a sufficiently long time, appear to improve in their chatter behavior. The explanation is that dirty slides increase the damping.

Damping. Damping should be increased without impairing the static stiffness and machining accuracy of the machine. This is achieved by the use of dampers. These are basically similar to those employed in other fields of vibration control (Chaps. 32 and 43). Dampers are effective only when placed in a position where vibration amplitudes are large.

The auxiliary mass absorber (Chap. 6) has been employed with considerable success on milling machines, radial drilling machines,[56] gear hobbing machines,[57] and grinding machines. An interesting design variant of this type of absorber is shown in Fig. 40.4. In this design a plastic ring element is used which combines both the elastic and the damping element of the absorber. The auxiliary mass may be attached to the top of a

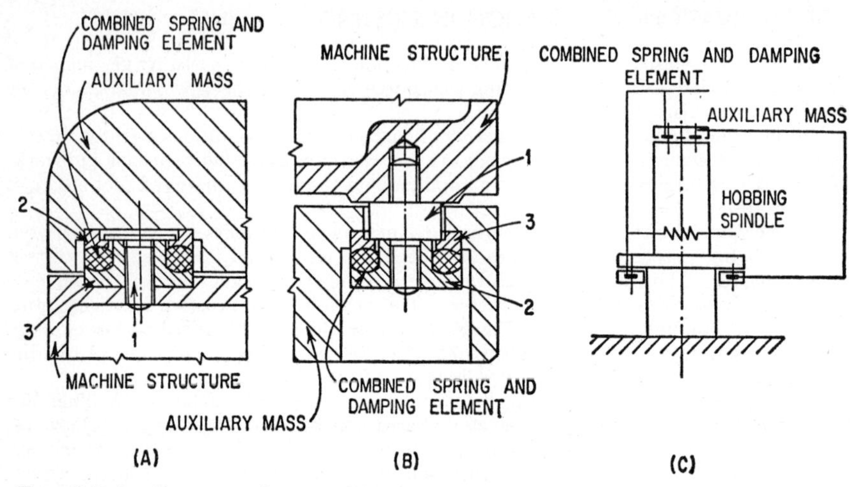

FIG. 40.4. Auxiliary mass damper with combined elastic and damping element. The combined element lies between two retainer rings, of which one (3) is attached with bolt 1 to the machine structure. The other ring (2) takes the weight of the auxiliary mass. (*A*) Arrangement when auxiliary mass is being supported. (*B*) Arrangement when auxiliary mass is being suspended. (*C*) Application of both types of arrangements to a hobbing machine. (*After F. Eisele and H. W. Lysen.*[57])

column (Fig. 40.4*C*), as shown in Fig. 40.4*A*. Alternatively, the auxiliary mass may be suspended on the underside of a table (Fig. 40.4*C*), using the design shown in Fig. 40.4*B*. In either case, several plastic ring elements may support one large auxiliary mass, as shown in Fig. 40.4*C*.[57]

A variation of the *Lanchester damper* (Chaps. 6 and 38) is frequently used in boring bars to good advantage.[58] This consists of an inertia weight fitted into a hole bored in the end of a quill, as shown in Fig. 40.5. To ensure effective operation, a relatively small radial clearance of about 0.001 to 0.006 in. must be provided, depending upon the diameter of the inertia weight. An axial clearance of about 0.006 to 0.010 in. is sufficient. A smooth surface finish of both plug and hole is desirable. The clearance values given refer to dry operation, using air as the damping medium. Oil also can be used as a damping medium but it does not necessarily result in an improved performance. When applying oil, clearance gaps larger than those stated above have to be ensured, depending on the viscosity of the oil. Further means for the suppression of boring-bar vibration are discussed under *Boring Bars with Improved Resistance to Vibration*.

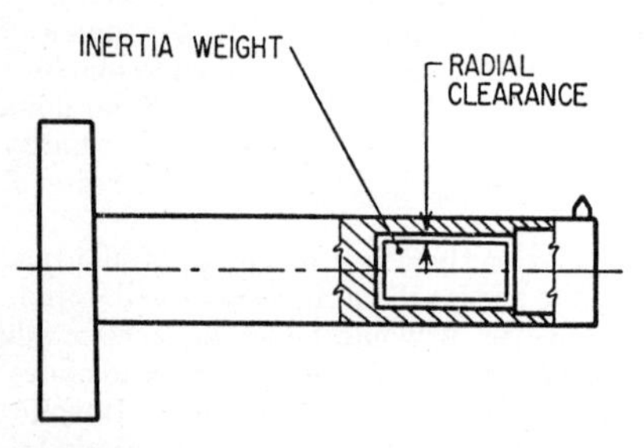

FIG. 40.5. Lanchester damper for the suppression of boring-bar vibration. (*After R. S. Hahn.*[58])

The impact damper shown in Fig. 40.6 also provides effective vibration control.[59] It consists of a bolt 1, a sleeve 2, a spring 3, and a cover 4. It is set on the lathe tool post or on the tool, at a point where vibration of large amplitude arises normally. In operation, the vibration of the tool results in an oscillatory motion of the inert mass 4 which, impinging on the bolt 1, absorbs energy. The important physical parameters of an impact damper are material, mass ratio, and stroke ratio (a function of the gap through which the free mass travels). Design charts have been prepared which facilitate the application of impact dampers to the solution of vibration problems found typically in machine tools.[60]

Dynamic absorbers can be designed to be self-optimizing. Thus, an absorber can be capable of self-adjustment of the spring rate to minimize vibration amplitude under changing excitation conditions.[61] The self-optimizing feature is achieved by placing vibration transducers on both the absorber mass and the main system. A control circuit measures the phase angle between the motions and activates a spring-modifying mechanism to maintain 90° phase difference between the two measured motions. It has been demonstrated that the 90° phase relationship guarantees minimum motion of the main vibrating mass.[61]

For the machining of long slender shafts, a hydraulically operated *support-vibration damper*[48, 63] can be used. This device consists of three cams which support the workpiece. The cams are hydraulically interconnected so that a floating support is ensured. Vibration energy is absorbed by viscous friction.

Tool Design. Sharp tools are more likely to chatter than slightly blunted tools. In the workshop, the cutting edge is often deliberately dulled by a slight honing. Consequently, a *vibro-damping bevel*[45] on the leading face of a lathe tool has been suggested. This bevel has a leading edge of −80° and a width of about 0.080 in. Tests show that the negative bevel does not in all cases eliminate vibration and that the life of the bevel is short.[64] Appreciably worn cutting edges cause violent chatter.[48]

Since narrow chips are less likely to lead to instability, a decrease of the approach angle of the cutting tool results in improved chatter behavior. With lathe tools, an increase of the rake angle results in improvement, but the influence of changes in the relief angle is relatively small.[48]

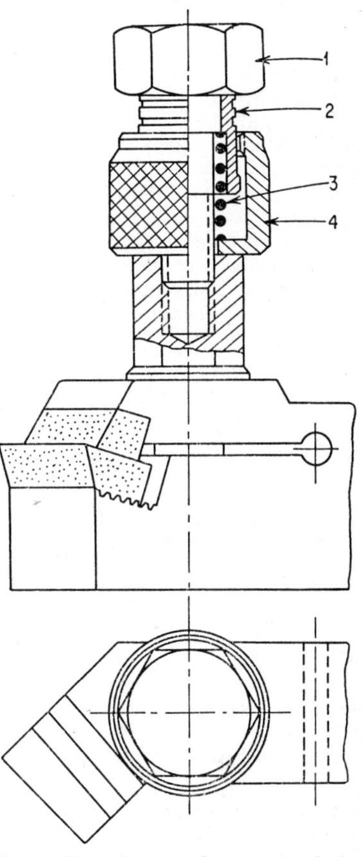

Fig. 40.6. Impact damper for lathe tools, consisting of bolt 1, sleeve 2, spring 3, and a cover 4. The oscillation of the inert mass 4, impinging on the bolt 1, absorbs vibratory energy. (*After D. I. Ryzhkov.*[59])

Theoretically, the tool shape should be such that any sudden increase of the feed rate meets the maximum possible resistance in the workpiece material. The penetration-rate coefficient Λ [see Eq. (40.1)] then is positive and large, and has an appreciable stabilizing influence. Such a stabilizing effect is produced by the chisel edge of a drill;[64, 65] it is not clear whether chisel-edge corrections, which decrease the static thrust, are also beneficial from the vibration point of view. Similarly, the wide face edge of certain face-mill blades prevents a "hacking-in" of these blades into the machined surface and appears to have a stabilizing influence.[46]

THE DESIGN OF VIBRATION-FREE MACHINE TOOLS. The following discussion is concerned mainly with problems of structural design, rather than with vibration sources.

Stiffness. In machine-tool practice, three different kinds of stiffness are considered.

1. *Static stiffness* k_s, defined as the ratio of the static force P_o, applied between tool and workpiece, to the resulting static deflection A_s between the points of application of force.

2. *Dynamic stiffness* k_d, defined as the ratio of the amplitude of a harmonic force P, applied between tool and workpiece, to the resulting resonance amplitude A_{max} between the points of application of force.

3. *Equivalent static stiffness* k, related to the dynamic stiffness by the expression $k = k_d Q$ where Q is the dynamic amplification factor.

A high static stiffness is necessary for machining accuracy. A high dynamic stiffness (i.e., large equivalent static stiffness and large damping) is necessary for satisfactory dynamic performance.

The stiffness of a structure is determined primarily by the stiffness of the most flexible component in the path of the force. Flexible members or "spots" show up in the static and dynamic deflection curves as sudden changes of their slope. In many cases the most flexible members are joints, i.e., bolted connections between relatively rigid elements such as column and bed, column and table, etc. Some points to be considered in the design of connections are illustrated in Fig. 40.7. To avoid bending of the flange in Fig. 40.7*A*, the

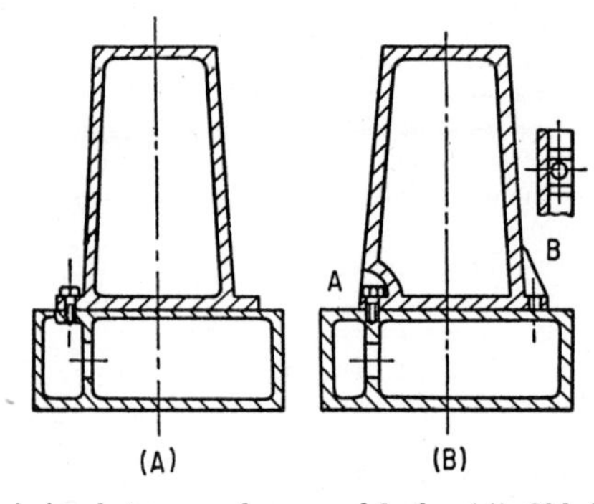

FIG. 40.7. Load transmission between column and bed. (*A*) Old design, relatively flexible owing to deformation of flange. (*B*) New design, bolt placed in a pocket (*A*) or flange stiffened with ribs on both sides of bolt (*B*). (*Afrer H. Opitz.*[65])

bolts should be placed in pockets or between ribs, as shown in Fig. 40.7*B*. Increasing the flange thickness does not necessarily increase the stiffness of the connection since this requires longer bolts which are more flexible. There is an optimum flange thickness (bolt length), the value of which depends on the elastic deformation in the vicinity of the connection. Deformation of the bed is minimized by placing ribs under connecting bolts.[65, 66, 68]

The efficiency of bolted connections, and other static and dynamic structural problems, is conveniently investigated by model analysis techniques.[65, 67, 69, 73, 74] Figure 40.8 shows the results of successive stages of a model experiment in which the effect of the design of bolt connections on the bending rigidity (X and Y directions) and the torsional rigidity of a column were investigated. The relative rigidities are shown by the length of bars. In the design of Fig. 40.8*A*, the connection consists of 12 bolts (⅝ in. in diameter) arranged in pairs along both sides of the column. In the design of Fig. 40.8*B*, the number of bolts is reduced to 10 and arranged as shown. With the addition of ribs, shown in succeeding figures, the bending stiffness in the direction X was raised by 40 per cent, in the direction Y by 45 per cent, and the torsional stiffness by 53 per cent, compared to the original design.[65, 67]

Openings in columns should be as small as possible. Figure 40.9 shows the loss of static flexural stiffness k_{sx}, k_{sy}, and torsional stiffness $k_{s\theta}$, and the decrease of the flexural natural frequencies f_x, f_y, resulting from the introduction of a hole in a box-type col-

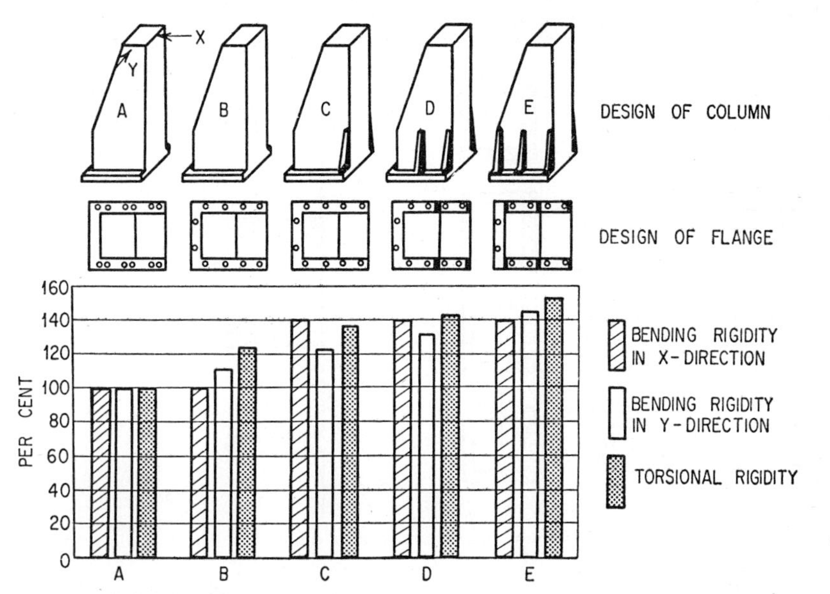

FIG. 40.8. Successive stages in the improvement of a flange connection. (*H. Opitz.*[65])

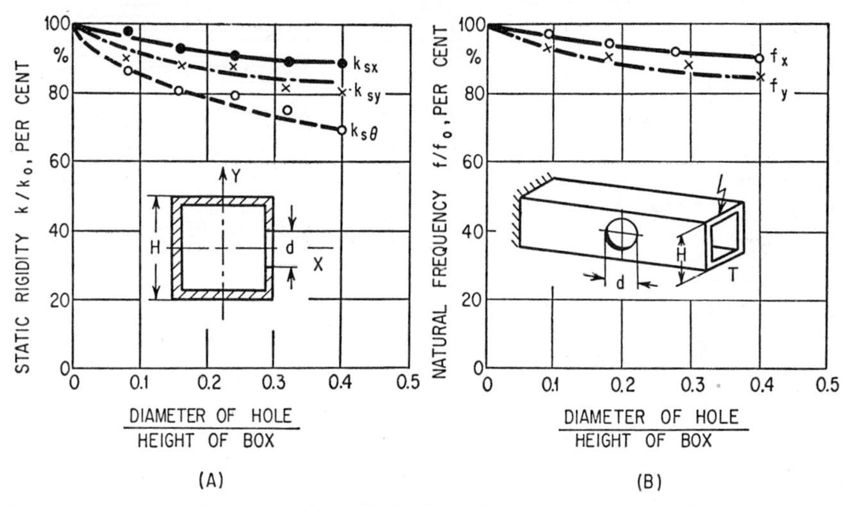

FIG. 40.9. Influence of a hole in the wall of a box column on the static stiffness and natural frequency. (*A*) Static stiffness; (*B*) natural frequency. (*H. Opitz.*[65])

umn.[65, 67] Smaller holes show relatively smaller decreases of stiffness and natural frequency than larger ones. The torsional rigidity $k_{s\theta}$ of a box-type column is particularly sensitive to openings, as shown in Fig. 40.10. Lids or doors used for covering these openings do not restore the stiffness. The influence of covers depends on their thickness, mode of attachment, and design, as shown in Fig. 40.11. However, covers may increase damping and thereby partly compensate for the detrimental effect of loss of stiffness.[65, 67]

Welded structural components not only are stiffer than cast-iron components but may also have a larger damping capacity.[70, 71] This is because welds are never perfect;

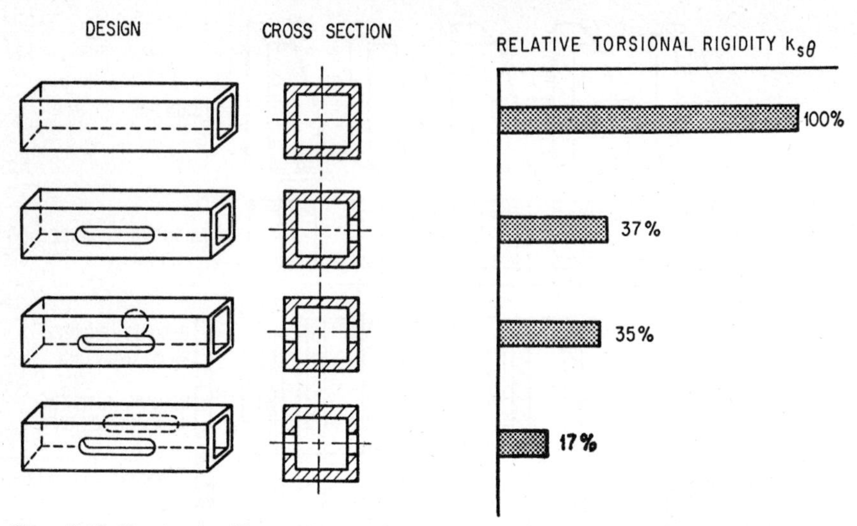

FIG. 40.10. Torsional stiffness of box columns with different holes in walls. (*H. Opitz.*[65])

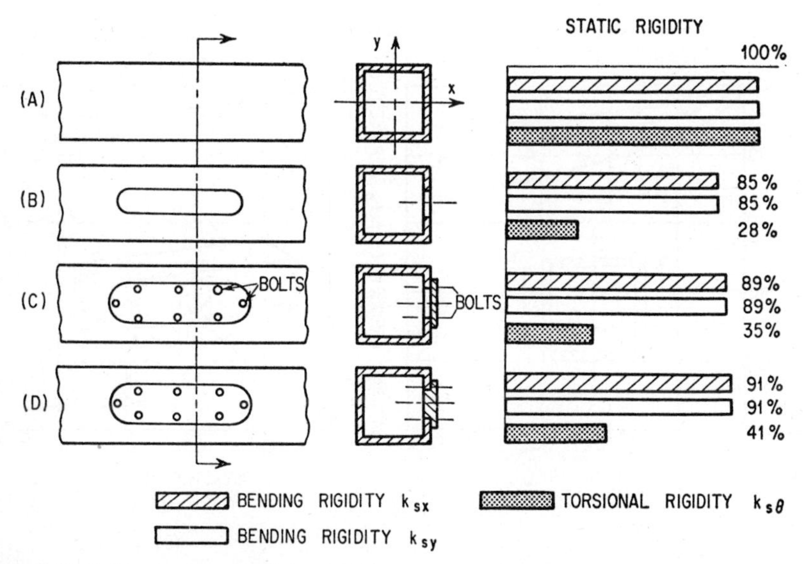

FIG. 40.11. Influence of cover plate and lid on static stiffness of box column. (*A*) Column without holes, (*B*) one hole uncovered, (*C*) hole covered with cover plate, and (*D*) hole covered with substantial lid, firmly attached. (*After H. Opitz.*[65])

consequently, rubbing takes place between joined members.[71] A considerable increase in damping can be achieved by using interrupted welds. This is of particular importance with welded ribs, which may be necessary not so much to increase rigidity but to prevent "drumming" (membrane vibration) of large unsupported areas.[65, 72, 73, 75]

The static and dynamic behavior of lathes is influenced significantly by the design of the spindle and its bearings. The static deflection of the spindle consists of two parts, X_1 and X_2, as shown in Fig. 40.12. The deflection X_1 corresponds to the deflection of a flexible beam on rigid supports, and X_2 corresponds to the deflection of a rigid beam on flexible supports which represent the flexibility of the bearings. The deflection of the spindle amounts to 50 to 70 per cent of the total deflection, and the bearings 30 to 50 per cent of the total depending on the relation of spindle cross section to bearing stiffness and span. The result is the same for roller or plain bearings of the same size. The stiffness of needle roller bearings is about twice that of ordinary roller bearings.[65, 76, 113]

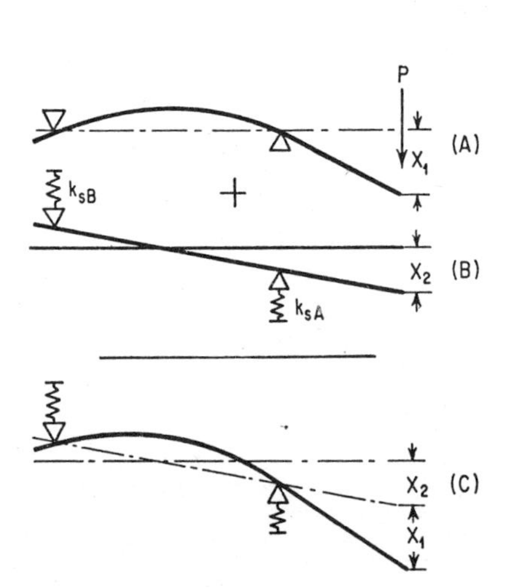

FIG. 40.12. Deflection of machine-tool spindle and bearings. A machine-tool spindle can be regarded as a beam on flexible supports. The total deflection under the force P consists of the sum of (A) the deflection X_1 of a flexible beam on rigid supports and (B) the deflection X_2 of a rigid beam on flexible supports. (*H. Opitz.*[65])

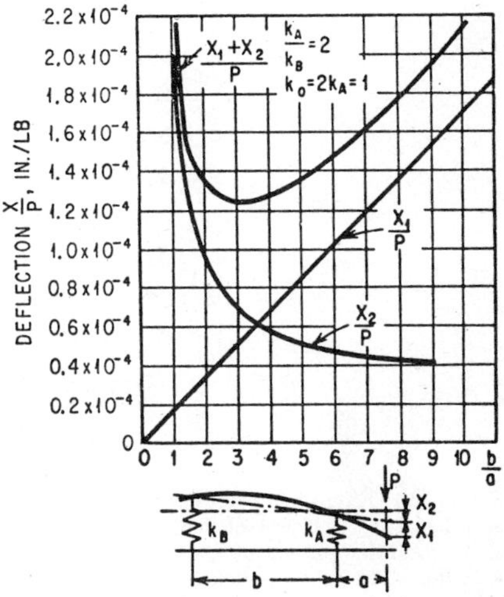

FIG. 40.13. Deflection of a beam on elastic supports as a function of the bearing distance. Bearing stiffness k_A and k_B, spindle stiffness k_o. (*After H. Opitz.*[65])

The distance between the bearings has considerable influence on the effective stiffness of the spindle, as shown in Fig. 40.13. The ordinate of the figure corresponds to the deflection in inches per pound and the abscissa represents the ratio of bearing distance b to cantilever length a. The straight line refers to the deflection of the spindle, and the hyperbola refers to the deflection of the bearings. The total deflection is obtained by the addition of the two curves; the minimum of the curve of total deflection corresponds to the optimum bearing distance. For a short cantilever length a, the optimum value of b/a lies between 3 and 5; for a long cantilever length a, b/a is optimum at 2.

The optimum bearing span for minimum deflection may be computed by the use of nomograms which have been prepared for a variable cross-section spindle, with overhanging end and supported by two bearings.[77]

It is often important to consider the dynamic behavior of a spindle before establishing an optimum bearing span. Maximizing the stiffness of a spindle at one point does not establish any dynamic properties. Care must be taken to investigate both bending and rocking modes of the spindle before accepting a final optimum span. For example, a large overhand on the rear of a spindle could produce an undesirable low-frequency rocking mode of the spindle even if the "optimum span" as defined previously were satisfied.[18]

The influence of the ratio of bore diameter to outside diameter on the stiffness of a hollow spindle is shown in Fig. 40.14. A 25 per cent decrease of stiffness occurs only at a diameter ratio of $d/D = 0.7$, where D is the outside diameter and d the bore diameter. This is important for the dynamic behavior of the spindle. A solid spindle has nearly the same stiffness, but a substantially greater mass. Consequently, the natural frequency of the solid spindle is considerably lower, which is undesirable.[65, 76, 18]

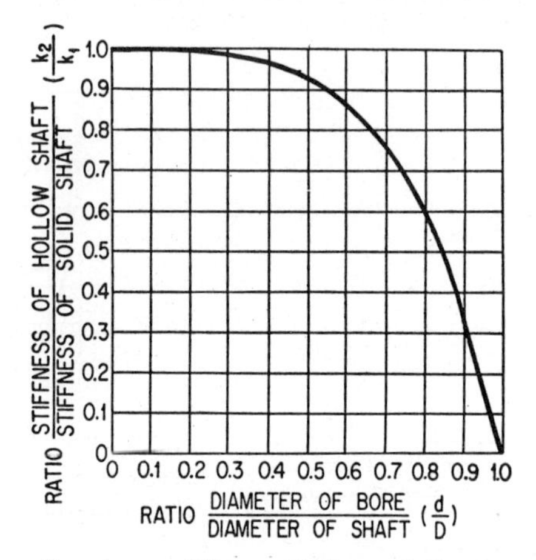

FIG. 40.14. Effect of bore diameter on stiffness of hollow spindle where k_1 = stiffness of solid spindle, k_2 = stiffness of hollow spindle, D = outer spindle diameter, d = bore diameter, J_2 = second moment of area of hollow spindle, and J_1 = second moment of area of solid spindle. The curve is defined by $k_2/k_1 = J_2/J_1 = 1 - (d/D).^4$ (*H. Opitz.*[65])

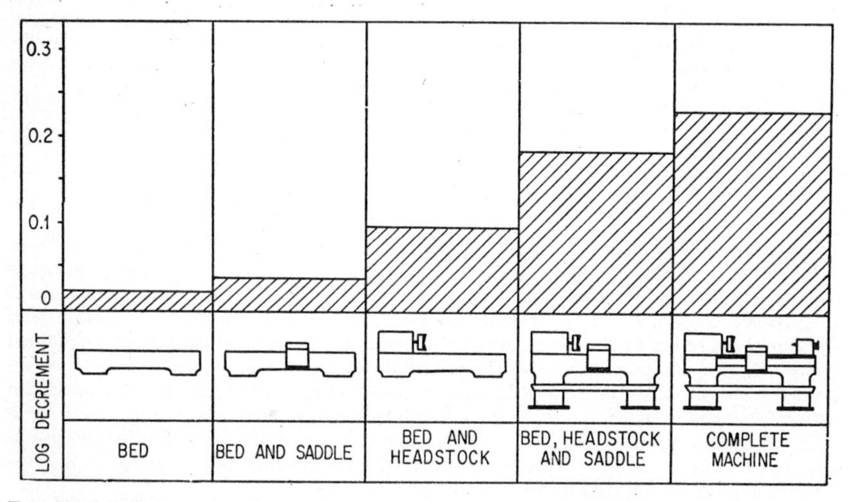

FIG. 40.15. Influence of various components on total damping of lathes. The major part of the damping is generated at the mating surfaces of the various components. (*K. Loewenfeld.*[78])

Damping. The over-all damping capacity of a structure is determined only to a small extent by the damping capacity of its individual components. The major part of the damping results from the interaction of joined components at slides or bolted joints. The interaction of the structure with the foundation also may produce considerable damping.[35] A qualitative picture of the influence of the various components of a lathe on the total damping is given in Fig. 40.15. The damping of the various modes of vibration differs appreciably; the values of the logarithmic decrement shown in the figure correspond to an average value for all the modes which play a significant part.[78]

The damping of various types of machine tool differs greatly, as indicated in Fig. 40.16. However, relatively little variation is found when comparing the damping of machines of the same type but different make.[78]

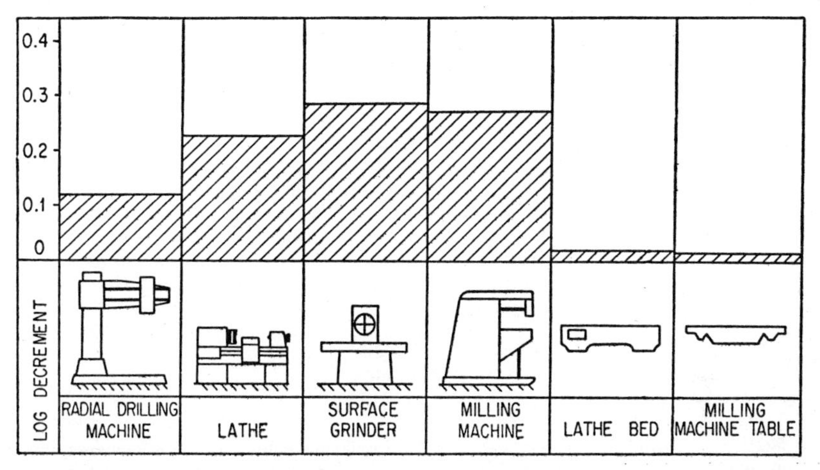

FIG. 40.16. Damping of machines and castings. Complete machines contain far greater damping than individual components. (*K. Loewenfeld.*[78])

Since damping in machine tools is due primarily to rubbing between mating surfaces, attempts have been made to increase this effect by attaching strips of sheet metal to the machine. Successive layers rub against each other when deformed and introduce damping. Sheets of plastic material may be placed between the strips. The design appears to be promising, particularly with welded structures where it suppresses "drumming" (membrane vibration) of unsupported walls.[79]

COMPUTER-AIDED DESIGN OF MACHINE TOOLS

As illustrated in the previous section, there is considerable data in the literature concerning the static and dynamic performance of machine-tool structures and elements as developed both analytically and experimentally. Experimental investigations have drawn upon both analytical computation and experimental measurement using actual machine elements and models of structures of machine elements. The desire to predict the static and dynamic behavior of a given specific machine-element design has led to the development of numerous computer techniques which employ both analog and digital computer systems.

USE OF ANALOG COMPUTER

The general-purpose analog computer is naturally suited for studying dynamic phenomena of machine elements. Fundamentally, any element or system which may be modeled mathematically by a differential equation or set of differential equations may be studied by electronic analog computation techniques. The problem of analyzing the dynamics of beams and structures may be handled in two ways. First, equations for the normal mode and lateral vibration may be set up, and a system of ordinary differential equations may be established whereby the independent variable is distance along the beam. The other approach is to establish a lump parameter model and replace the partial derivatives with finite differences. The latter technique offers several advantages for studying machine structures, in that variable cross sections, elastic supports, and the responses to forced excitations are readily handled.[80, 81]

An investigation of the static and dynamic behavior of the lathe spindle serves as an example of the application of the finite-difference techniques.[82] Once a model is functioning on the computer, it is possible to investigate a number of design concepts. For example, it is possible to optimize the bearing locations for maximum stiffness, explore the advantages of more than two supporting bearings, impose simulated cutting forces and investigate the chatter stability, and investigate vibration damping ideas.

USE OF DIGITAL COMPUTER

The utilization of the digital computer in analyzing the static and dynamic behavior of machine elements has become widespread in recent years. Lumped-parameter modeling for digital computation has been facilitated by the development of software programs for the stiffness matrix technique.[82] More recent utilization of finite-element methods has enhanced the computing capability and has extended the application of computers to investigating complex structural shapes and, in general, handling those problems which are not readily studied by beam methods.[83, 86]

REFERENCES

1. Landberg, P.: *Microtecnic*, **10**(5):219 (1956).
2. Saljé, E.: "Zweites Forschungs- und Konstruktionskolloquium Werkzeugmaschinen," p. 105, Vogel-Verlag, Coburg, 1955.
3. Späth, W.: *Werkstattstechnik und Maschinenbau*, **42**(2):59 (1952).
4. Rowe, R. G.: *Steel*, **126**(26):74, 84 (1950).
5. Tobias, S. A.: *Engineering*, **172**(4470):409 (1951).
6. Landberg, P.: *Microtecnic*, **11**(1):18 (1957).
7. Hahn, R. S.: *Trans. ASME*, Paper No. 58-A-97, 1958.
8. Doi, S.: *Trans. ASME*, **80**(1):133 (1958).
9. Strauch, H.: *Z. VDI*, **95**:159 (1953).
10. Harris, S. L.: *Proc. IME*, **172**(2):87 (1958).
11. Bollinger, J. G.: *Ind.-Anzeiger*, **46**:961 (June, 1963).
12. Bollinger, J. G., and M. Bosh: *Ind.-Anzeiger*, **86**(19):319 (March, 1964).
13. Doi, S.: *Mem. Fac. Eng., Nagoya Univ.*, **5**(2):179 (1953).
14. Lysen, H. W., and R. Schwaighofer: "Zweites Forschungs- und Konstruktionskolloquium Werkzeugmaschinen," pp. 121, 133, Vogel-Verlag, Coburg, 1955.
15. Schenk, O.: "Zweites Forschungs- und Konstruktionskolloquium Werkzeugmaschinen," p. 137, Vogel-Verlag, Coburg, 1955.
16. Lohmann, G.: "Zweites Forschungs- und Konstruktionskolloquium Werkzeugmaschinen," p. 139, Vogel-Verlag, Coburg, 1955.
17. Pittroff, H.: *Konstruktion*, **11**(3):106 (1959).
18. Bollinger, J. G., and G. Geiger: *Intern. J. Mach. Des. Res.*, **3**:193 (1964).
19. Piekenbrink, R.: "Drittes Forschungs- und Konstruktionskolloquium Werkzeugmaschinen und Betriebswissenschaft," vol. 2, p. D155, Vogel-Verlag, Coburg, 1958.
20. Bowden, F. P., and L. Leben: *Proc. Roy. Soc. (London)*, **169**:371 (1939).
21. Blok, H.: *SAE J.*, February, 1940.
22. Merchant, M. E.: "Characteristics of Typical Polar and Non-polar Lubricant Additives under Stick-slip Conditions," Meeting of the Society of Lubrication Engineers, Chicago, April, 1946.
23. Elyasberg, M. E.: *Stanki i Instrument*, **11**:1 (1951) and **12**:6 (1951); *Engineers' Digest*, **13**(9):298 (1952) and **13**(10):338 (1952).
24. Opitz, H., and E. Saljé: *Ind.-Anzeiger*, (63) (1952).
25. Beuerlein, P.: "Fortschrittliche Fertigung und moderne Werkzeugmaschinen," p. 117, Verlag W. Girardet, Essen, 1954.
26. Mikes, M.: *Czechoslovak Heavy Industry*, **1**(1):6 (1957).
27. Anon.: *Machinery (London)*, **89**(2292):912 (1956).
28. Galonska, D. A.: *Machinery (New York)*, **62**(8):159 (1956).
29. Havemeyer, H. R.: *Machine Design*, **29**(17):120 (1957).
30. Weiter, E. J., and A. O. Schmidt: *ASME* Paper No. 57-SA-100, 1957; *Engineers' Digest*, **18**(8):335 (1957).

31. Sadowy, M.: "Drittes Forschungs- und Konstruktionskolloquium Werkzeugmaschinen und Betriebswissenschaft," p. 127, Vogel-Verlag, Coburg, 1957.
32. Harris, C. M. (ed.): "Handbook of Noise Control," p. 13, McGraw-Hill Book Company, Inc., New York, 1957.
33. Sadowy, M.: "Drittes Forschungs- und Konstruktionskolloquium Werkzeugmaschinen und Betriebswissenschaft," p. 141, Vogel-Verlag, Coburg, 1957.
34. Tobias, S. A., and W. Fishwick: *Engineering*, **176**(4584):707 (1953).
35. Fishwick, W., and S. A. Tobias: *Engineering*, **185**(4808):568 (1958).
36. Eisele, F., and M. Sadowy: "Drittes Forschungs- und Konstruktionskolloquium Werkzeugmaschinen und Betriebswissenschaft," p. 167, Vogel-Verlag, Coburg, 1957.
37. Eisele, F., M. Sadowy, and E. Sun: "Drittes Forschungs- und Konstruktionskolloquium Werkzeugmaschinen und Betriebswissenschaft," pp. 163, 177, Vogel-Verlag, Coburg, 1957.
38. Anon: *Automobile Engr.*, **46**(4):150 (1956).
39. Nietsch, H. E.: *Tool Engr.*, **39**(1):105 (1957).
40. Hartz, H.: "Drittes Forschungs- und Konstruktionskolloquium Werkzeugmaschinen und Betriebswissenschaft," p. 153, Vogel-Verlag, Coburg, 1957.
41. Honshi, T.: *Ann. CIRP*, **22**:1 (1973).
42. Weilenmann, R.: *Werkstatt u. Betrieb.*, **87**(8):401 (1954).
43. Weilenmann, R.: *Werkstatt u. Betrieb.*, **89**(9):529 (1956).
44. Opitz, H., and E. Saljé: *Ind.-Anzeiger*, (45):690 (1954).
45. Klushin, M. I., and D. I. Ryzhkow: *Metalworking Production*, **100**:1153 (1956).
46. Tobias, S. A.: *Proc. IME*, **173**(18):474 (1959).
47. Saljé, E.: *Ind.-Anzeiger*, **36**:501 (1955).
48. Sokolowski, A. P.: "Präzision in der Metalbearbeitung," p. 345, VEB Verlag Technik, Berlin, 1955.
49. Hölken, W.: *Ind.-Anzeiger*, (1):1 (1958).
50. Lysen, H.: "Zweites Forschungs- und Konstruktionskolloquium Werkzeugmaschinen," p. 175, Vogel-Verlag, Coburg, 1955.
51. Tobias, S. A., and W. Fishwick: *Engineer*, **205**(5324):199 (1958) and **238**(5325):238 (1958).
52. Arnold, R. N.: *Proc. IME*, **154**:261 (1946).
53. Tourret, R.: "Performance of Metal-cutting Tools," p. 101, Butterworth & Co. (Publishers) Ltd., London, 1958.
54. Opitz, H.: *Microtecnic*, **12**(2):72 (1958).
55. Hinjink, J. A. W., and Vander Wolf, A. C. H.: *Ann. CIRP*, 22, 1, 123 (1971).
56. Galloway, D. F.: *Proc. IME*, **170**(6):207 (1956).
57. Eisele, F., and H. W. Lysen: "Zweites Forschungs- und Konstruktionskolloquium Werkzeugmaschinen," p. 89, Vogel-Verlag, Coburg, 1955.
58. Hahn, R. S.: *Trans. ASME*, **73**(4):331 (1951).
59. Ryzhkov, D. I.: *Stanki i Instrument*, (3):23 (1953); *Engineers' Digest*, **14**(7):246 (1953).
60. Pinotti, P. C., and M. M. Sadek: *Proc. 11th Intern. MTDR Conf.*, September, 1970.
61. Bonesho, J. A., and J. G. Bollinger: *Proc. 7th Intern. MTDR Conf.*, pp. 229–241, Pergamon Press, New York, 1968. (See also *Machine Design*, **40**, 123, Feb. 29, 1968.)
62. Sadowy, M.: "Zweites Forschungs- und Konstruktionskolloquium Werkzeugmaschinen," p. 167, Vogel-Verlag, Coburg, 1955.
63. Podporkin, V. G.: *Stanki i Instrument*, (7):12 (1953).
64. Sharin, Yu. S., and V. B. Serebrovsky: *Stanki i Instrument*, (7):21 (1955).
65. Opitz, H.: "Conference on Technology of Engineering Manufacture," Paper 7, The Institution of Mechanical Engineers, London, 1958.
66. Piekenbrink, R.: "Untersuchungen an Elementen im Kraftfluss der Werkzeugmaschinen, Forschungsbericht," Verlag W. Girardet, Essen, 1957.
67. Bielefeld, J.: "Untersuchungen an Elementen im Kraftfluss der Werkzeugmaschinen, 7. Forschungsbericht," Verlag W. Girardet, Essen, 1957; *Ind.-Anzeiger*, (63):953 (1957).
68. Hölken, W.: *Ind.-Anzeiger*, (62):898 (1955).
69. Saljé, E.: *Ind.-Anzeiger*, (62):894 (1955).
70. Heiss, A.: *VDI-Forschungsheft 429*, **16**(429):7 (1949/1950).
71. Kienzle, O., and H. Kettner: *Werkstattstechnik u. Werksleiter*, **33**(9):229 (1939).
72. Eisele, F., and H. Drumm: "Drittes Forschungs- und Konstruktionskolloquium Werkzeugmaschinen und Betriebswissenschaft," vol. 2, p. D125, Vogel-Verlag, Coburg, 1957.
73. Loewenfeld, K.: "Zweites Forschungs- und Konstruktionskolloquium Werkzeugmaschinen," pp. 59, 65, 73, 83, Vogel-Verlag, Coburg, 1955.
74. Bollinger, J. G., and B. C. Cuppan: ASME Paper 65-MD-13, (1965).
75. Loewenfeld, K.: "Drittes Forschungs- und Konstruktionskolloquium Werkzeugmaschinen und Betriebswissenshaft" (Vorberichtsheft), pp. 49, 63, Vogel-Verlag, Coburg, 1957.
76. Honrath, K.: "Untersuchungen an Elementen im Kraftfluss der Werkzeugmaschinen, 7. Forschungsbericht," Verlag W. Girardet, Essen, 1957.

77. Terman, T., and J. G. Bollinger: *Machine Design*, **37**, 159, May 27, 1965.
78. Loewenfeld, K.: "Zweites Forschungs- und Konstruktionskolloquium Werkzeugmaschinen," p. 117, Vogel-Verlag, Coburg, 1955.
79. Loewenfeld, K.: "Drittes Forschungs- und Konstruktionskolloquium Werkzeugmaschinen und Betriebswissenshaft" (Vorberichtsheft), p. 105, Vogel-Verlag, Coburg, 1957.
80. Cuppan, B. C., and J. G. Bollinger: *Proc. 7th Intern. MTDR Conf.*, 191, 1967.
81. Bollinger, J. G.: *Proc. CIRP Intern. Conf. Mfg. Tech.*, 77, 1967.
82. Cuppan, B. C., and J. G. Bollinger: *Ann. CIRP*, **17**:243 (1969).
83. Opitz, H., and R. Noppen: *Ann. CIRP*, **22**(2):227 (1973).
84. Hinduja, S., and A. Cowley: *Proc. 12th Intern. MTDR Conf.*, 455, 1971.
85. Yoshimura, M., and T. Hoshi: *12th Intern. MTDR Conf.*, 439, 1971.
86. Hinduja, S., and A. Cowley: *Ann. CIRP*, **21**(1):113 (1972).
87. Takemura, T., T. Kitamura, and T. Hoshi: *Ann. CIRP*, **22**(1):121 (1974).
88. Inamura, T., and Sata, T.: *Ann. CIRP*, **22**(1):119 (1974).
89. Eisele, P. T., and R. F. Griffin: *Trans. ASME*, **75**(10):1211 (1953).

41

PACKAGING DESIGN

Masaji T. Hatae
Rockwell International, Inc.

INTRODUCTION

Packaging is the technology of preparing an item for shipment in such a manner as to minimize the damage resulting from environmental hazards encountered during shipment. Such environmental hazards include moisture, temperature, dust, shock, vibration, etc. This chapter considers only the influence of shock and vibration.

The shock and vibration experienced by goods during shipment result from the vibration of a cargo-carrying vehicle, the shock resulting from the impacts of railroad cars, the shock caused by the handling (e.g., dropping) of packages, etc. Such goods are protected from damage by isolation. The theory and practice of vibration and shock isolation are discussed in Chaps. 30 to 33. This chapter discusses the application of the principles of isolation to the design of packages. In general, this consists of placing the packaged item within a container, and interposing resilient means between the item and the container to provide the necessary isolation. Such resilient means is known as *package cushioning*. The required degree of protection should be provided with minimum packaging and shipping costs through optimum combination of labor, materials, and package volume and weight. The means of protection may range from a simple corrugated carton to a more complex system such as a part "floated" in distributed cushioning material in a wooden box or by a spring suspension in a reusable metal drum.

In concept, the package cushioning is designed to protect an item of known strength from the known shock and vibration existing in the particular environment. Practically, package design cannot be pursued so rationally because the strength of equipment and the characteristics of the environment often are not known with the necessary accuracy. As a consequence, the strength of the equipment often must be inferred from exploratory tests or estimated on the basis of experience. Similarly, environmental conditions are simulated by simplified tests (e.g., drop tests) that represent environmental conditions of maximum expected severity. Then the design of the package cushioning proceeds in a straightforward manner using the methods and data set forth in this chapter.

FACTORS IN DESIGN OF PACKAGE CUSHIONING

This section discusses factors that are significant in the design of package cushioning.

FRAGILITY OF EQUIPMENT

The capability of an item to withstand shock and vibration is defined in terms of its "fragility." The term "fragility" is used as a quantitative index of the strength of the equipment when subjected to shock and vibration. The quantitative index is expressed as a maximum permissible acceleration $\ddot{x}_F$. In addition to the value of $\ddot{x}_F$, it is necessary to know the natural frequencies and damping characteristics of the equipment. With the foregoing information the package designer could proceed to select cushioning with the desired characteristics to protect the equipment during shipment. Unfortunately, it is difficult to determine this information for many types of equipment. As a consequence of this difficulty, several alternate procedures may be used to obtain information of use for package design.

ANALYTICAL METHODS. It is difficult to determine analytically (i.e., by theoretical analysis and calculation) the fragility of a complex device or structure. Sometimes it is possible to simplify or idealize the structure to a form that is more amenable to analysis, and the strengths of principal component structures may be calculated. This analytical basis for design is discussed in detail in Chap. 42.

LABORATORY TEST METHOD. Shock and vibration tests may be carried to failure in order to determine the fragility level of products. This method is particularly suited to products manufactured in large quantity where unit cost is small and tests-to-failure may be performed. Various types of laboratory shock test apparatus, such as drop test machines, hydraulic shock testing machines and pendulum impact machines, are available for testing a wide range of sizes and weights of equipment. These testing machines, which are described in Chap. 26, provide control of acceleration level, time duration of pulse, and shape of pulse. Laboratory vibration testing machines, described in Chap. 25, are available in various ranges of frequency and force output.

Shock tests-to-failure can provide data concerning the maximum allowable accelerations for particular input pulse shapes. These tests-to-failure define fragility levels for shock and vibration in terms of the characteristics of the testing machine. In general, such information is useful in the design of package cushioning if the testing machines are adjusted to have the same characteristics as the motion transmitted by the cushioning to the equipment or if the cushioning is made to have the characteristics of the testing machine.

Most products, particularly military equipment, are designed and tested according to requirements stipulated in contractual specifications; in general, such requirements relate to ultimate use rather than to packaging. The levels of shock and vibration to which the equipment is designed and tested most often represent the *minimum* acceptable resistance to damage; they do not represent the maximum level that the item can withstand without failure, nor are they generally related to the requirements for successful packaging. Strict adherence to these minimum values in package design often results in high costs of packaging and shipping.

ENGINEERING ESTIMATE. Field personnel experienced in the handling and operation of specific items (products) often can provide information on the ability of these items to withstand drops from various heights when protected by certain cushioning materials. These heights of drop can be converted into peak acceleration levels if the dynamic properties of the cushioning materials are known. Then the fragility is defined quantitatively for certain conditions. For example, assume that it has been determined that an item, blocked and braced in a corrugated carton, is able to withstand the shock that results from dropping the carton from a height of 12 in. Information on height of drop may be transformed to the parameters of rise time and peak acceleration by assuming a sinusoidal acceleration pulse. The applicable relations are given in Fig. 41.1.

Estimated rise times to maximum acceleration for various drop conditions are given in Table 41.1. Further refinements in the values may be made by actual instrumented drop tests using various types of containers and materials.

The fragility index of equipment as estimated from experience in dropping similar equipment applies only to the peak acceleration. The natural frequencies and damping

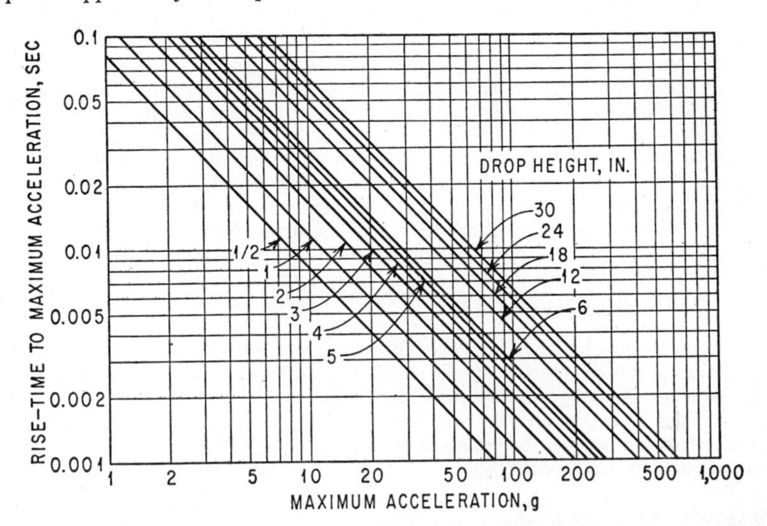

FIG. 41.1. Relation of drop height and rise time to maximum acceleration for a packaged item which experiences a half-sinusoidal pulse. The relation is defined mathematically as $\ddot{x}_1^2 = 0.0128ht_r^{-2}$, where $\ddot{x}_1$ = maximum acceleration, g; h = height of drop, in.; and t_r = rise time, sec.

present in the equipment are not indicated. In general, the natural frequencies of important components of the equipment are substantially greater than the natural frequency of the package cushioning; under these conditions, damage tends to be directly proportional to maximum acceleration and it is possible to use the maximum acceleration as an index of fragility (see Chap. 42).

Table 41.1. Approximate Sinusoidal Rise Time for Various Drop Conditions

Condition	*Rise time, sec*
(1) Unpackaged; metal container...	0.002
(2) Wooden box..................	0.004
(3) Carton......................	0.006
(4) Approx. 1 in. latex hair........	0.008
(5) Approx. 3 in. latex hair........	0.015

ENVIRONMENTAL CONDITIONS

In general, the shock received during handling (for example, from being dropped) is greater than that experienced in being transported in a vehicle; such shocks usually determine the severity of a shock test.

The standard shock test for packages is the drop test in which the package is dropped from a predetermined height onto a rigid floor. The height of drop is determined by the type of handling the package may receive during shipment. For example, packages from 0 to 50 lb may be considered within the "one man throwing limit"; i.e., packages of such weight may be thrown easily onto piles or in other ways severely mishandled due to their light weight. Packages weighing between 50 and 100 lb may be considered within a "one man carrying limit"; such packages are somewhat heavy to be thrown but can be carried and dropped from a height as great as shoulder height. A "two man dropping limit" may apply to a weight range between 100 to 300 lb; the corresponding drop height for this mode of handling may be waist height. A further range may be from 300 to 1,000 lb; packages in this range would be handled with light cranes or lift trucks and may be subjected to shock from excessive lifting or lowering velocities. Finally, very heavy packages weighing over 1,000 lb would be handled by heavier equipment with correspondingly more skill; any drops would be from very small heights. Similarly, the size of the package classifies the type of handling into one man, two men, light equipment, or heavy equipment with corresponding drop heights. Thus, the drop heights for package shock tests are derived from the type of handling to which the package is most likely to be subjected during a shipping cycle; the type of handling is dependent on the size and weight of the package.

In addition to drop heights that vary with package size and weight, another factor in shock testing is the orientation of the package at impact. For example, small, light-weight packages are likely to be subjected to free fall drops onto sides, edges, and corners of the package. Larger, heavier packages handled by light or heavy equipment are likely to encounter drops of the type where one end rests on the floor and the other end is dropped (bottom rotational drop).

PROPERTIES OF CUSHIONING MATERIALS

The static and dynamic properties of cushioning materials that are commonly used in package design are included below in the section *Properties of Cushioning Materials.*

SHOCK ISOLATION

In the analytical treatment of shock isolation in packages, the idealized system illustrated in Fig. 41.2*A* is considered. It consists of a heavy rigid outer container, a packaged item assumed to be a rigid body, and a shock-isolation medium having characteristics which are considered below. The container is dropped from a characteristic height h upon a rigid floor, and the protection afforded to the packaged item by the isolation medium is indicated by the maximum acceleration experienced by the item. For

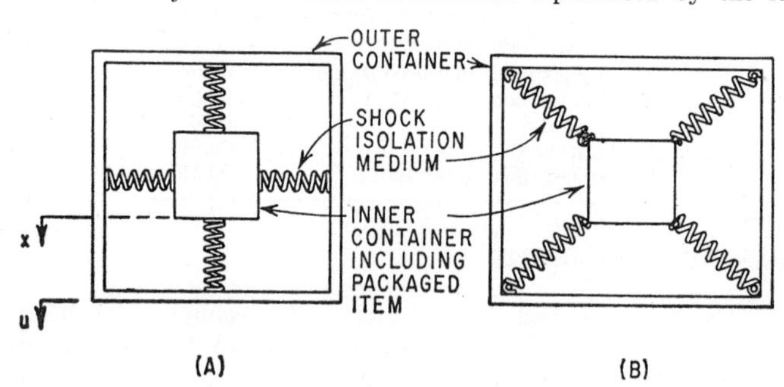

Fig. 41.2. Schematic arrangements of isolation media in shipping containers: A — axes of isolation medium perpendicular to faces of container; B — axes of isolation medium inclined to faces of container.

purposes of analysis, it is convenient to assume a medium having a linear stiffness k; then results are modified to include the effects of the nonlinearity in actual media.

The motion of the container as it falls freely from the height h is

$$u = \tfrac{1}{2} g t^2 \tag{41.1}$$

The equation of motion for the packaged item is

$$m\ddot{x} = mg + k\delta \tag{41.2}$$

where $\delta = u - x$ is the deflection of the isolation medium, m is the mass of the packaged item, and g is the acceleration of gravity. Substituting $u - x = \delta$ in Eq. (41.2) and

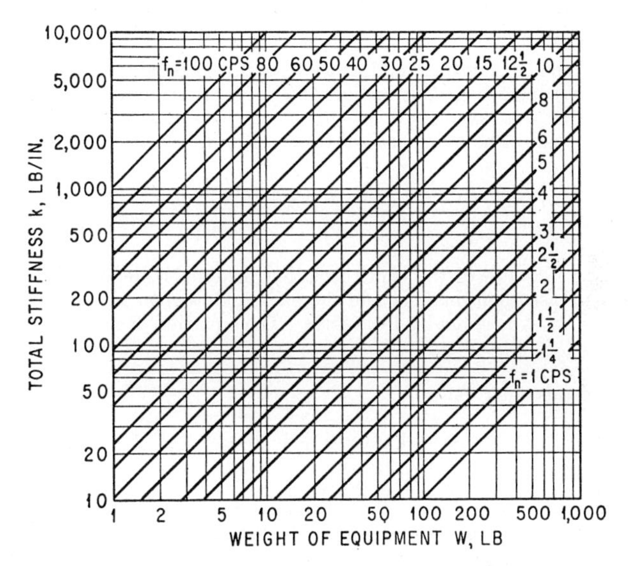

FIG. 41.3. Natural frequency of a package as a function of the weight of the packaged item and the stiffness of the isolation medium. These curves are defined by Eq. (41.5).

using the expression for u given by Eq. (41.1), the equation of motion of the packaged item is

$$\ddot{x} + \omega_n^2 x = g + \frac{\omega_n^2 g t^2}{2} \tag{41.3}$$

The solution of this equation is

$$x = A \sin \omega_n t + B \cos \omega_n t + \frac{g t^2}{2} \tag{41.4}$$

where $\omega_n = \sqrt{k/m}$ is the natural angular frequency in radians per second of the packaged item supported by the isolation medium. The natural frequency is expressed in units of cycles per second, as a function of the weight W of the packaged item:

$$f_n = \frac{1}{2\pi}\sqrt{\frac{kg}{W}} \tag{41.5}$$

This relation is shown graphically in Fig. 41.3.

The coefficients in Eq. (41.4) are evaluated from the conditions at time $t = 0$. At this instant, the packaged item is in equilibrium between the gravitational force and the upward force applied by the isolation medium; i.e., there is no downward acceleration until the container has traveled a sufficient distance to reduce the magnitude of the force applied by the isolation medium. Then, at time $t = 0$, $\ddot{x} = 0$ and Eq. (41.4) becomes

$$x = \frac{g}{\omega_n^2} \cos \omega_n t + \frac{gt^2}{2} \tag{41.6}$$

Thus, the packaged item falls at the same rate as the container, plus a superimposed oscillatory motion initiated at the moment of dropping. The velocity is found by differentiating Eq. (41.6):

$$\dot{x} = -\frac{g}{\omega_n} \sin \omega_n t + gt \tag{41.7}$$

The velocity of the packaged item after the container has fallen from height h is determined by substituting $t = \sqrt{2h/g}$ in Eq. (41.7). The maximum possible velocity of the packaged item as the downward velocity of the container is arrested occurs when $\sin \omega_n \sqrt{2h/g} = -1$. The magnitude of the maximum velocity is

$$\dot{x}_{\max} = \sqrt{2gh} + \frac{g}{\omega_n} \tag{41.8}$$

The resulting response of the shock-isolation medium is determined by considering the container to rest upon the floor and the packaged item to have an initial downward velocity $\dot{\delta}_{\max} = \dot{x}_{\max}$ as given by Eq. (41.8). The equation of motion for the packaged item is

$$m\ddot{\delta} = -k\delta + mg \tag{41.9}$$

The solution of this equation is

$$\delta = A \sin \omega_n t' + B \cos \omega_n t' + \frac{g}{\omega_n^2} \tag{41.10}$$

where $t' = 0$ at the moment that the downward velocity of the container is arrested. Initial conditions are $\delta = 0$, $\dot{\delta} = \dot{x}_{\max}$ from Eq. (41.8). Substituting these initial conditions, Eq. (41.10) becomes

$$\delta = \left(\frac{\sqrt{2gh}}{\omega_n} + \frac{g}{\omega_n^2}\right) \sin \omega_n t' + \frac{g}{\omega_n^2}(1 - \cos \omega_n t') \tag{41.11}$$

In general, the term g/ω_n^2 is relatively small in packaging applications. Then the term $g/\omega_n^2 \cos \omega_n t'$ may be neglected and the maximum deflection $\delta_{\max}$, at time $t' = \pi/(2\omega_n)$, is

$$\delta_{\max} = \left(\frac{\sqrt{2gh}}{\omega_n} + \frac{g}{\omega_n^2}\right) + \frac{g}{\omega_n^2} = \left(\frac{\sqrt{2gh}}{\omega_n} + \frac{2g}{\omega_n^2}\right) \quad \text{in.} \tag{41.12}$$

The maximum acceleration experienced by the packaged unit is found by differentiating Eq. (41.11) * twice with respect to time, neglecting the term $g(\cos \omega_n t')/\omega_n^2$ and taking the maximum value at time $t' = \pi/(2\omega_n)$:

$$\ddot{\delta}_{\max} = \omega_n \sqrt{2gh} + g \qquad \text{in./sec}^2 \tag{41.13}$$

The maximum acceleration $\ddot{\delta}_{\max}$ may be expressed as a dimensionless multiple $(\ddot{\delta}_g)_{\max}$ of

* In this portion of the analysis, the container is assumed to rest upon the floor; thus, $\ddot{\delta} = \ddot{x}$ is the acceleration of the packaged item.

the gravitational acceleration by dividing Eq. (41.13) by g:

$$(\ddot{\delta}_g)_{\max} = \frac{\omega_n \sqrt{2gh}}{g} + 1 \tag{41.14}$$

Solving Eq. (41.14) for the natural frequency ω_n and dividing the result by 2π to obtain units of cycles per second,

$$f_n = \frac{1}{2\pi}\sqrt{\frac{g[(\ddot{\delta}_g)_{\max} - 1]^2}{2h}} \quad \text{cps} \tag{41.15}$$

The relation of Eq. (41.15) is shown graphically by the solid lines in Fig. 41.4.

The relation between maximum deflection $\delta_{\max}$ and maximum acceleration $(\ddot{\delta}_g)_{\max}$ is obtained by combining Eqs. (41.12) and (41.14):

$$(\ddot{\delta}_g)_{\max} = \frac{2h + 3\dfrac{\sqrt{2gh}}{\omega_n} + \dfrac{2g}{\omega_n{}^2}}{\delta_{\max}} \tag{41.16}$$

For most packaging applications, the first term in the numerator of Eq. (41.16) is much larger than the other terms; thus, it is a reasonable approximation to neglect the latter and write Eq. (41.16) in the following form:

$$(\ddot{\delta}_g)_{\max} \simeq \frac{2h}{\delta_{\max}} \tag{41.17}$$

This relation is shown by the dotted lines in Fig. 41.4. For a successful package design, it is necessary that $\ddot{x}_g$ be less than $\ddot{x}_F$.

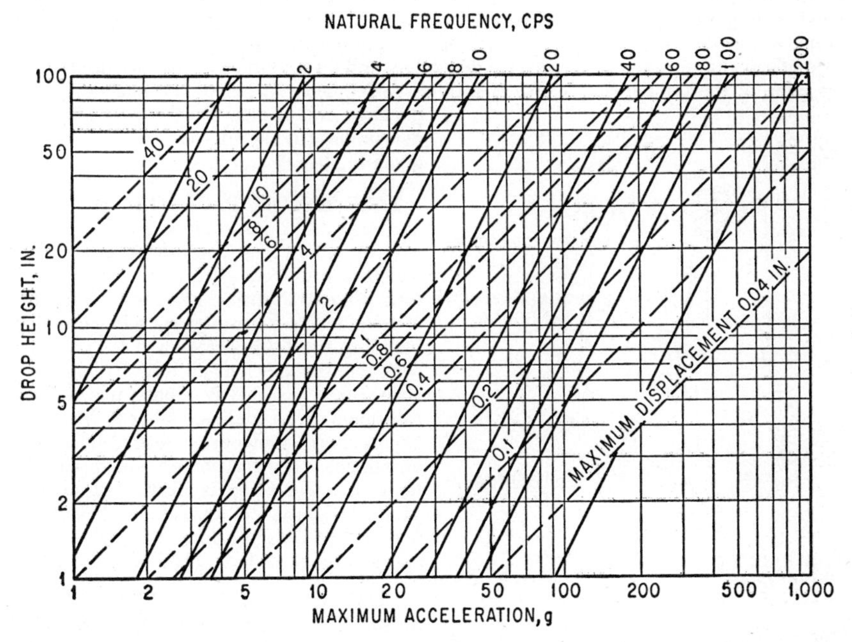

FIG. 41.4. Natural frequency and maximum displacement of the packaged item in terms of its maximum acceleration and the drop height of the package. These curves are based on Eqs. (41.15) and (41.17).

Example 41.1. Consider an item weighing 25 lb that is to be packaged in such a manner that it does not receive an acceleration greater than 30g ($\ddot{x}_F = 30g$) when the package is dropped from a height of 36 in. Determine the spring characteristics and maximum displacement required to attain the desired maximum acceleration, and indicate the natural frequency of the packaged item on the isolation system.

Entering Fig. 41.4 at a value of 30g on the abscissa and 36 in. on the ordinate, the following values are obtained by interpolation between pairs of solid and dotted diagonal lines:

From the solid lines:

$$\text{Natural frequency} = 10.7 \text{ cps}$$

From the dotted lines:

$$\text{Maximum spring deflection} = 2.4 \text{ in.}$$

Then, entering Fig. 41.3 at a value of 25 lb on the abscissa scale, and reading on the ordinate scale at the diagonal line for 10.7 cps:

$$\text{Total spring stiffness} = 300 \text{ lb/in.}$$

INFLUENCE OF DAMPING ON SHOCK ISOLATION

The presence of damping in the isolation medium affects the response to shock. This subject is considered in detail in Chap. 31.

APPLICATIONS OF SHOCK-ISOLATION THEORY

TENSION-SPRING SUSPENSION. When the springs are arranged parallel to the direction of motion, as shown in Fig. 41.2*A*, the stiffness indicated by Fig. 41.3 is used directly in the design of the springs (see Chap. 34 for design data on metal springs, and Chap. 35 for data on rubber springs). The total stiffness is divided by the number of springs to obtain the stiffness per spring. However, if the springs extend between the interior corners of the container and the exterior corners of the packaged item, as indicated in Fig. 41.2*B*, several additional factors must be considered in determining the spring characteristics: (1) the axes of the springs are not parallel with the deflection, (2) the angle of the spring axes varies with deflection, and (3) if tension springs are used, only a portion of the total spring length is active. A detailed analysis of the tension spring package and a step-by-step procedure to determine the spring characteristics are given in Refs. 1 and 2.

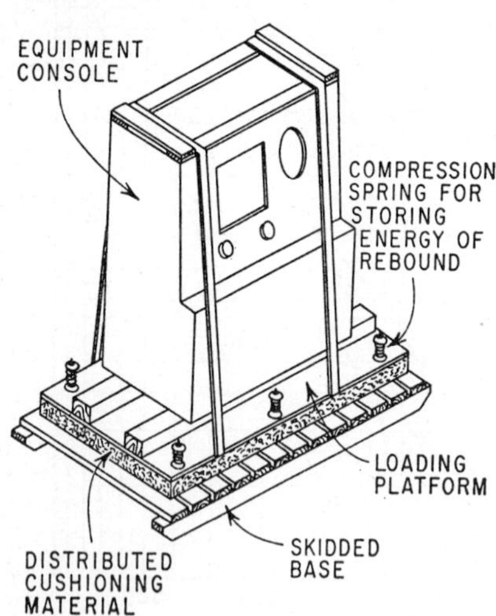

FIG. 41.5. Typical base-mounted package utilizing loading platform and skids. The sides and top of the container are mounted upon the base.

BASE-MOUNT PACKAGE. For large and heavy equipment, it is common practice to mount the equipment on a base which is provided with skids to allow for fork lift entry during handling. The use of a skidded base on a large container dictates handling and transportation of the package with the base down at all times. Then isolation need be provided only for bottom rotational drops, and side and end impacts, as discussed previously under shock environment. An isolation system applicable to such a condition is illustrated in Fig. 41.5. An analysis of a base-mount package with linear isolators for both end impact and rotational drop is included in Chap. 3.

CUSHION DESIGN USING FLAT SHEETS. A common method of packaging embodies the use of an isolation medium in sheet form of standard thicknesses; the sheet is cut into pads that are inserted between container and the packaged item on the several sides thereof. The theory of shock isolation with linear springs is applicable only for relatively small deflection of the isolation medium; for larger deflections, the stiffness becomes nonlinear and a different type of analysis must be used. In general, the stiffness cannot be defined analytically; * rather, the characteristics of the isolation medium are defined by stress-strain curves and by curves of maximum transmitted acceleration as a function of static loading stress for various heights of free fall of the container. These curves are obtained by physical tests of the material and are used to design the isolation medium, using the procedure described in the following paragraphs.

For a given isolation medium, the significant parameters in the design of a package are thickness of the medium, height of free fall of the package, unit loading on the medium, and maximum acceleration experienced by the packaged item. Unit loading is defined in terms of static conditions; it is expressed as a "static stress" representing the weight of the packaged item divided by the area of the medium in engagement with one side of the item. Isolation media are commercially available in a sequence of standard thicknesses; for example, thicknesses of 2, 4, and 6 in. are commonly available. Consequently, the design data are given for corresponding thicknesses. Similarly, the height of free fall is defined in discrete increments corresponding to accepted specification practices; the isolation at intermediate heights can be determined by interpolation. The shock isolation properties are defined by several families of curves. For example, Fig. 41.15 shows maximum acceleration of the packaged item as a function of static stress for several thicknesses of "Latex Hair, Firm" when the height of free fall is 12 in. (Corresponding data on the same material for other heights of free fall are given in Figs. 41.16 and 41.17.)

In the design of a package, data of the type given in Figs. 41.15 to 41.32 are applied by selecting the figure corresponding to the applicable height of free fall, entering the figure on the ordinate scale at a value of maximum acceleration corresponding to the fragility of the packaged item, and determining the required static loading from the curves representing various thicknesses of isolation medium. This determines the requirements of the design for purposes of shock isolation; other requirements are set forth below under *Vibration Isolation* and *Properties of Cushioning Materials.* The complete procedure for designing the isolation medium is discussed under *Design Technique—Nonlinear Cushioning.*

Comparison of Flat and Corner Drops. The selection of cushion dimensions is determined from the dynamic properties of cushioning materials. Such properties are determined by dropping an equipment mock-up flat onto cushion specimens. However, package tests require drops on corners and edges. The use of flat-drop data in cushion design has been justified by a series of tests summarized in Fig. 41.6.[3] These tests involved dropping a package consisting of a wooden block cushioned on all six sides and enclosed in an exterior container. Three accelerometers were mounted on the block with their directions of sensitivity along the three mutually perpendicular axes of symmetry of the block. The package was suspended above a rigid surface and its orientation was carefully measured. It was allowed to fall freely and the three accelerometer measurements were recorded. The first series of tests began with a flat drop. Subsequent drops from the same height were made with varying angles of orientation until a corner drop was reached. The angle of orientation was then further varied until the tests were concluded with a drop on an edge of the package. This procedure was repeated using 2- and 3-in. thick cushioning. The second series of tests began from a flat drop, continued to an edge, and concluded with a flat drop on the side 90° from the original impacting side. The acceleration measurements are plotted in Fig. 41.6 with a faired curve for each thickness. The accelerations are the vector resultants of the three accelerometer recordings.

VIBRATION ISOLATION. In general, the primary consideration in the design of a cushion system is the protection of the packaged item against shock. However, if the

* An analytical treatment of shock isolation wherein the characteristics of nonlinear isolators are defined analytically is included in Chap. 31.

item (or a component of the item) to be packaged has a resonant frequency falling within the range of vibration frequencies encountered during transportation, the vibration-isolation characteristics of the cushion system must be considered.

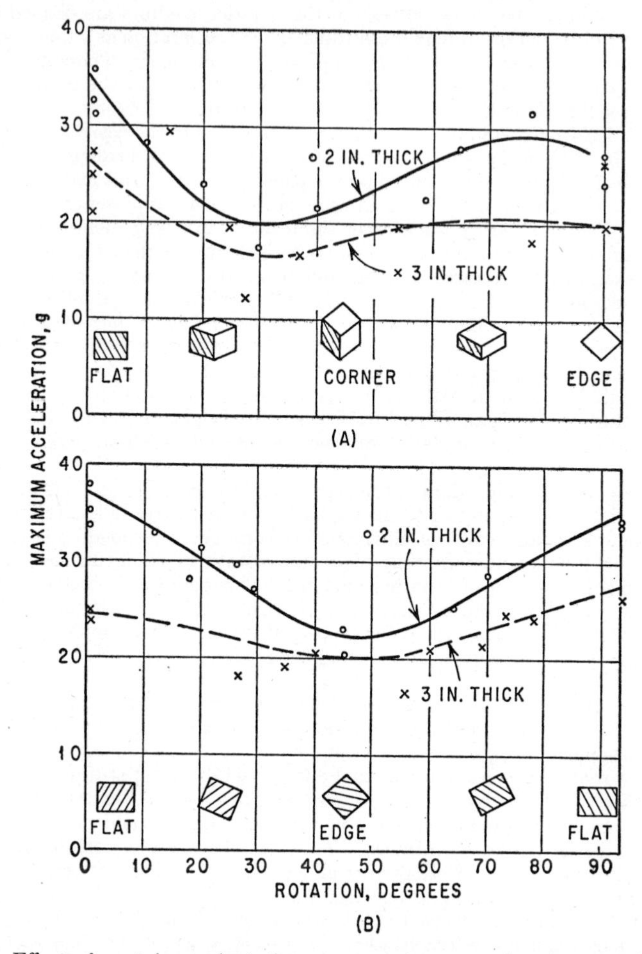

FIG. 41.6. Effect of container orientation upon maximum acceleration experienced by a packaged item during drop tests: *A* – sequence of orientations is flat to corner to edge; ***B*** – **sequence of orientations is flat to edge to flat.** (***L. W. Gammel and J. L. Gretz.***[3])

The theory of vibration isolation is discussed in detail in Chap. 30; equations for transmissibility with several types of damping are given in Table 30.2 and transmissibility with viscous damping is shown graphically in Fig. 2.17. It is highly desirable that the isolator have significant damping, so that the vibration amplitude does not become excessive if environmental conditions include vibration having a frequency equal to the natural frequency of the isolator. Furthermore, it is important that the natural frequency of the isolator be substantially less than the natural frequencies of vulnerable components of the equipment so that the excitation of the components is small.

Example 41.2. Consider the system shown schematically in Fig. 41.7 where m_1 represents the total mass of the packaged item, m_2 represents the mass of an element or component of the equipment (m_2 is assumed small relative to m_1), and $\ddot{u}$ is the acceleration of the vibration associated with the environmental condition. The environment is characterized by vibration having an acceleration amplitude of $1.3g$ at 6 cps (the natural frequency of the isolator) and an acceleration amplitude of $3g$ at 200 cps (the natural frequency of the element). The fragility $\ddot{x}_F$ of the critical element is $30g$ and its fraction of critical damping ζ_2 is 0.001. An undamped tension spring package with natural frequency f_1 has been designed. The fraction of critical damping ζ_1 for an undamped tension spring package has been experimentally determined to be 0.004.[4] Determine the response of the critical element at isolator resonance and element resonance.

Response of Element at Isolator Resonance. The transmissibility of the isolator at resonance is defined by $T_R = 1/(2\zeta)$ [see Eq. (30.1)]:

$$T_R = \frac{1}{2 \times 0.004} = 125$$

Then, the acceleration of m_1 at 6 cps is

$$\ddot{x}_1 = T_R \times \ddot{u} = 125 \times 1.3g = 163g$$

The acceleration $\ddot{x}_2$ of the element is approximately the same as that of the equipment; this is characteristic of elements of high natural frequency (relative to the natural frequency of the isolator).

Response of Element at Element Resonance. The transmissibility of the isolator at element resonance is $T = 0.0008$ [see Eq. (30.1)]. Then the acceleration of m_1 is

$$\ddot{x}_1 = T \times \ddot{u} = 0.0008 \times 3g = 0.0024g$$

The transmissibility of the element at resonance is $T_R = \frac{1}{2\zeta_2}$ [see Eq. (30.1)] $= 500$. Thus, the acceleration of the element is $\ddot{x}_2 = T_R \times \ddot{x}_1 = 500 \times 0.0024g = 1.2g$.

The results calculated above are tabulated in Table 41.2 and plotted in Fig. 41.8. Similar results are included for fractions of critical damping $\zeta_1 = 0.03$ and 0.10 characteristic of a damped spring suspension system. The addition of damping to the tension spring reduces the acceleration experienced by the component as a result of vibration at the resonant frequency of the spring but increases the acceleration resulting from vibration at the resonant frequency of the element.

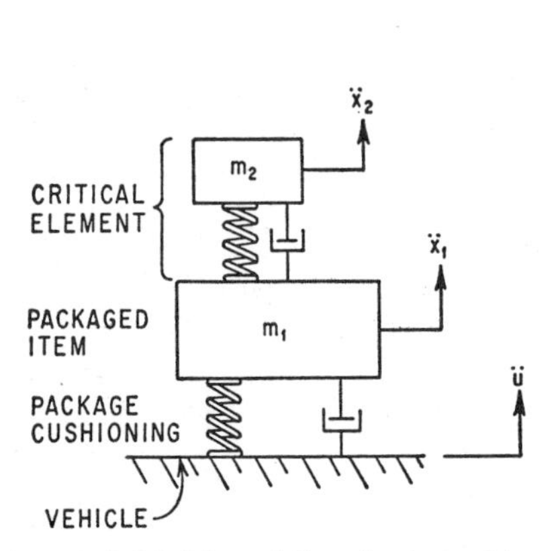

FIG. 41.7. Model consisting of a damped two degree-of-freedom system to represent a packaged item and a critical element thereof

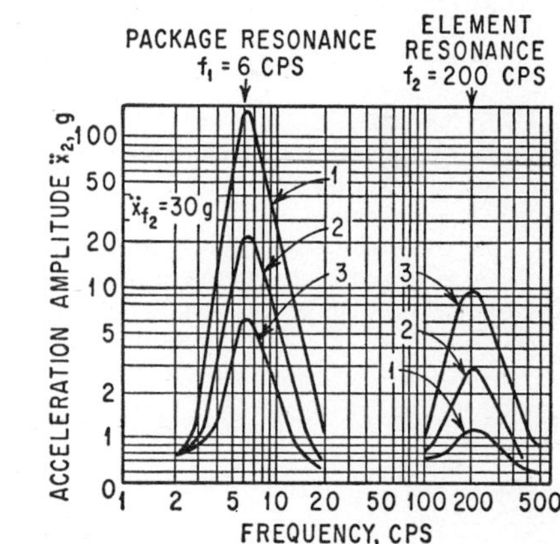

FIG. 41.8. Maximum acceleration amplitude $\ddot{x}_2$ of critical element m_2 in Fig. 41.7 for the conditions indicated in Example 41.2. The curves refer to different values of the fraction of critical damping ζ_1 for the package cushioning as follows: curve 1: $\zeta_1 = 0.004$; curve 2: $\zeta_1 = 0.03$; curve 3: $\zeta_1 = 0.10$.

Table 41.2. Element Acceleration for Tension Spring Suspension Package

(See Example 41.2)

Type of package	Fraction of critical damping	Response at package resonance				Response at element resonance			
		T_p	$\ddot{x}_1$, g	T_e	$\ddot{x}_2$, g	T_p	$\ddot{x}_1$, g	T_e	$\ddot{x}_2$, g
Undamped tension spring....	0.004	125	163	1	163	0.0008	0.0024	500	1.2
Damped tension spring....	0.03	16.7	21.7	1	21.7	0.002	0.006	500	3.0
Damped tension spring....	0.10	5.0	6.5	1	6.5	0.007	0.021	500	10.5

Damping of Tension Spring Package. One simple and economical means of providing damping is to stuff wool felt into the centers of the spring coils throughout the extended spring length. Fractions of critical damping in the range from 0.03 to 0.10 can be obtained by this method, indicating a range of maximum transmissibility from 16.7 to 5, respectively.

ANALYTICAL DETERMINATION OF NATURAL FREQUENCY. The relationships between transmissibility and frequency of vibration and between the natural frequency of the cushion system and the static stress on the cushion determine the degree of vibration isolation afforded by a cushion. Because of the many variables and unknowns involved in analyses of the materials commonly used as package cushioning, transmissibility data must be obtained empirically. For example, transmissibility curves for latex hair and polyester urethane foam are given in Figs. 41.37 and 41.38.

The natural frequency of a nonlinear isolator is

$$f_n = \frac{1}{2\pi}\sqrt{\frac{dF/d\delta}{W/g}} \qquad (41.18)$$

where f_n = natural frequency, cps
F = force, lb
δ = deflection, in.
W = weight supported by isolator, lb
g = gravitational acceleration = 386 in./sec^2

The force F is equal to the weight W and $dF/d\delta$ is the slope of force-displacement curve at the force $F = W$. These may be expressed in terms of stress and strain as follows:

$$\sigma = \frac{F}{S} = \text{stress, lb/in.}^2$$

$$\epsilon = \frac{\delta}{t} = \text{strain, in./in.}$$

where S = cushion area, in.^{2}
t = cushion thickness, in.,

Then Eq. (41.18) becomes

$$f_n = \frac{1}{2\pi}\sqrt{\frac{d\sigma/d\epsilon}{Wt/gS}} = 3.13\sqrt{\frac{d\sigma/d\epsilon}{Wt/S}} \qquad (41.19)$$

Applying the stress-strain relation $d\sigma/d\epsilon$ for a material, Eq. (41.19) gives the relationship between natural frequency f_n and static stress W/S for various cushion thicknesses t. Data giving this relationship for several package cushioning materials are presented in Figs. 41.39 through 41.41.

Equation (41.19) applies only for small vibration amplitudes because $d\sigma/d\epsilon$ does not remain constant throughout a vibration cycle of large amplitude; thus discrepancies between the actual and calculated natural frequencies may occur for large vibration amplitudes. Most materials also exhibit some difference between static and dynamic stiffness. Testing demonstrates fair correlation between actual and calculated values of natural frequency for small vibration displacement amplitudes; however, discrepancies may occur if large displacements are involved. Thus, the curves of static stress vs. strain should be used only to obtain the first approximation to the natural frequency of the isolator.

PROPERTIES OF CUSHIONING MATERIALS

This section considers cushioning materials such as latex hair, various plastic foams etc., which are supplied either in sheet form or in molded shapes; as package cushioning they are interposed between the container and the packaged item. Although such materials may exhibit linear force-deflection characteristics for small deflections, efficient package design involves large deflections and consequent nonlinearity of the cushioning materials.

MATERIAL CLASSIFICATION

All materials utilized to provide cushioning (i.e., shock and vibration isolation) in packaging may be classified as either of two general types—*elastic* (flexible or resilient) and *nonelastic* (crushable). Materials which, when deflected during impact, sustain no more than an arbitrarily selected amount of permanent deformation are placed in the first category; materials that sustain a greater amount are placed in the second category. Certain factors influence the choice of this arbitrarily selected level. One of the most important of these is that a sufficient thickness of material should be left after deformation to provide the required isolation on successive impacts. A permanent deformation of not more than 10 per cent after compression to a strain of 65 per cent is used as a criterion in the package design field.

Materials described in this section may be used to satisfy package design requirements other than those for cushioning, for example, to position an item within the package.

MATERIAL PROPERTIES

The fundamental property to be considered in application of a material in shock and vibration isolation is the manner in which it stores, absorbs, and dissipates energy. This characteristic is basic to the classification of materials set forth in the preceding paragraph.

Package cushioning may be designed analytically if an equation can be written for the force-deflection characteristics under dynamic conditions[1] (see Chap. 31). A quasi-analytical concept minimizes the ratio of transmitted force to stored energy.[5] The most directly useful method makes use of experimentally determined properties of the cushioning materials in a straightforward design procedure, as described in subsequent sections of this chapter.

PROPERTIES FOR ISOLATION. The properties of materials that are most useful in packaging design for shock isolation have been standardized as follows:

1. Compression stress vs. strain characteristics
2. Maximum acceleration vs. static stress characteristics, employing drop height and material thickness as parameters
3. Creep characteristics

Requirements covering these properties for elastic-type cushioning materials are set forth in a military specification on elastic-type cushioning materials.[6] Methods of testing for determination of these properties are detailed in Refs. 6 to 9.

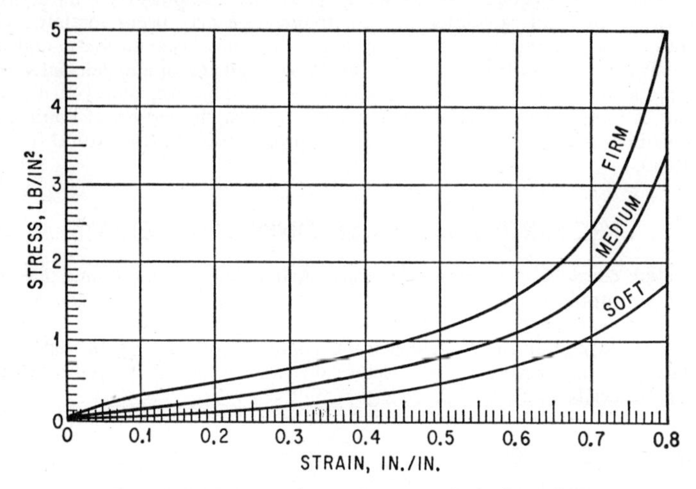

Fig. 41.9. Compression stress vs. strain for latex hair.

Data of the types listed above are presented in this chapter for a number of specific materials commonly used for package cushioning. Static stress vs. strain characteristics are given in Figs. 41.9 to 41.14; these data are used to determine the degree of static

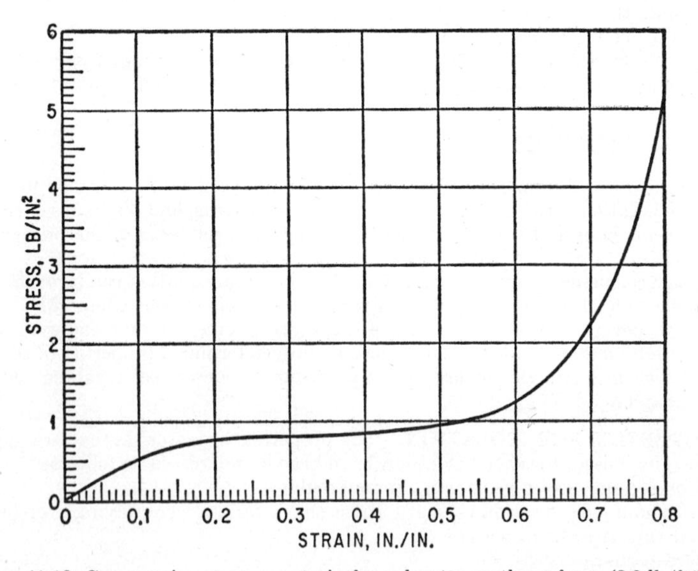

Fig. 41.10. Compression stress vs. strain for polyester urethane foam (2.2 lb/ft³).

compression of the cushion due to the weight of the item to be packaged. Maximum acceleration vs. static stress data are given in Figs. 41.15 to 41.32; the dynamic shock loading on the packaged item for various package drop heights can be determined from these data. Creep characteristics are given in Figs. 41.33 to 41.35; these data are used to determine the thickness loss of the cushion during storage. Because of wide differences in manufacturing processes and techniques associated with production of package cushioning materials, and the large number of manufacturers of each type of material, a great variation exists in the characteristics of commercially available materials. For this reason, the data presented here represent only an average for a particular material and condition.*

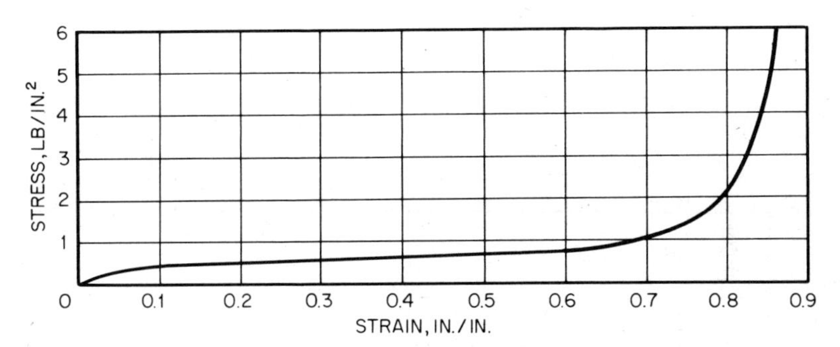

FIG. 41.11. Compression stress vs. strain for polyether urethane foam (2.0 lb/ft³).

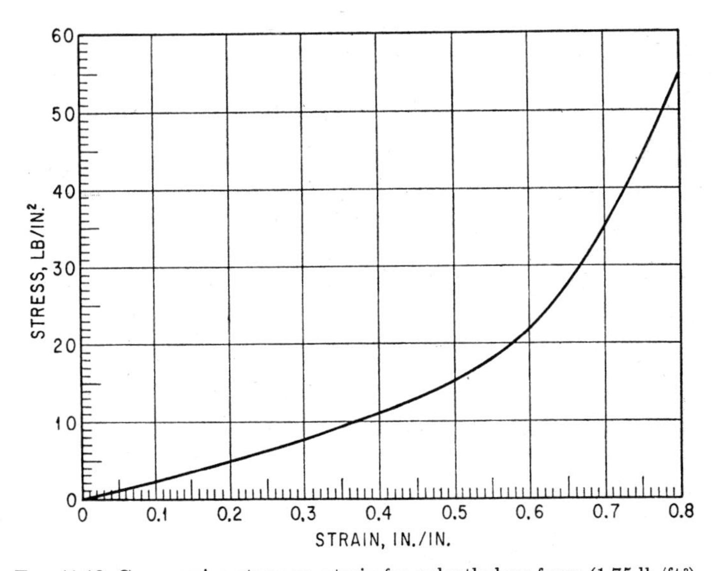

FIG. 41.12. Compression stress vs. strain for polyethylene foam (1.75 lb/ft³).

* The data used to plot Figs. 41.9 to 41.41 inclusive were made available through the courtesy of Rockwell International, Inc. and Ref. 10.

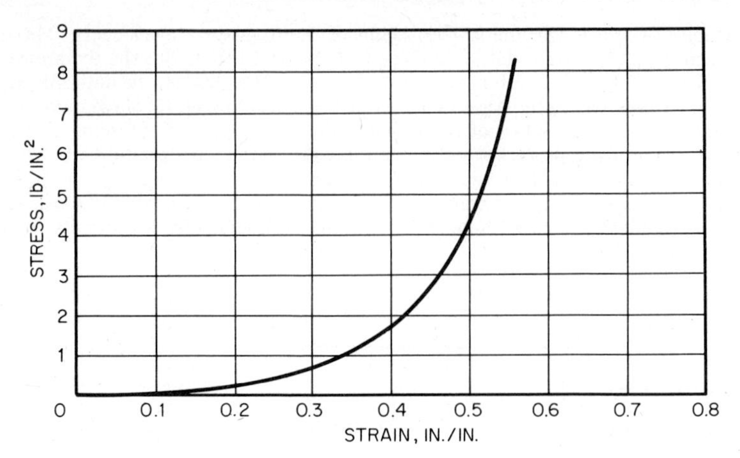

Fig. 41.13. Compression stress vs. strain for cellulose wadding.

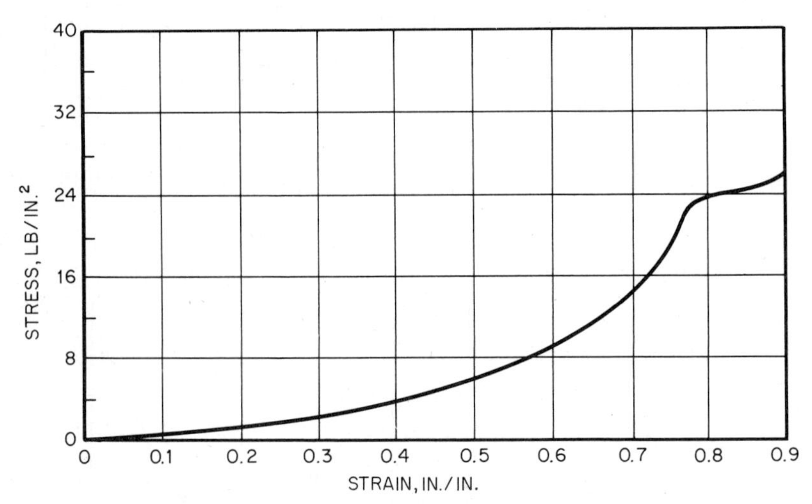

Fig. 41.14. Compression stress vs. strain for "AirCap" SD 240.

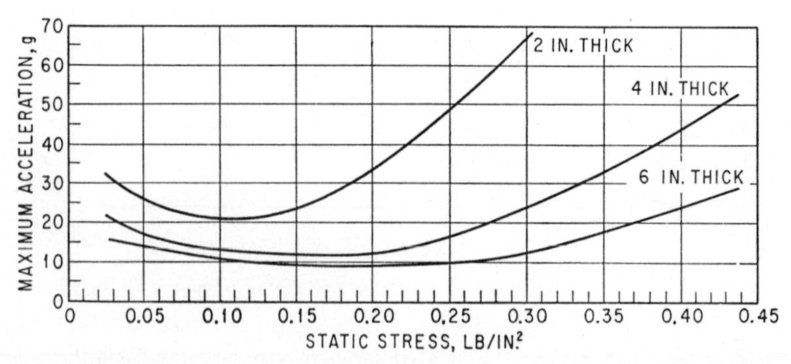

Fig. 41.15. Maximum acceleration vs. static stress for latex hair, firm—12-in. drop.

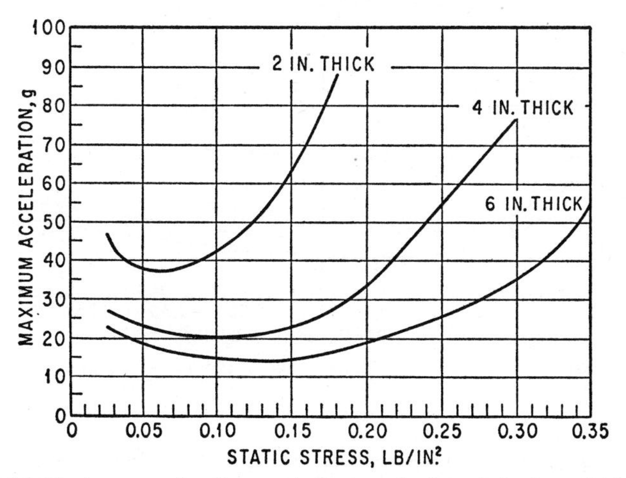

FIG. 41.16. Maximum acceleration vs. static stress for latex hair, firm—24-in. drop.

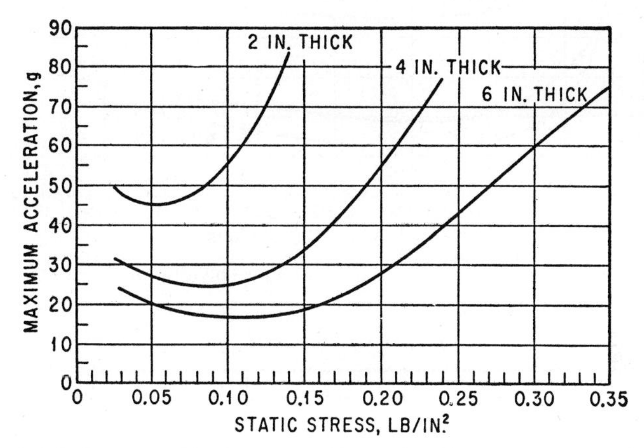

FIG. 41.17. Maximum acceleration vs. static stress for latex hair, firm—30-in. drop.

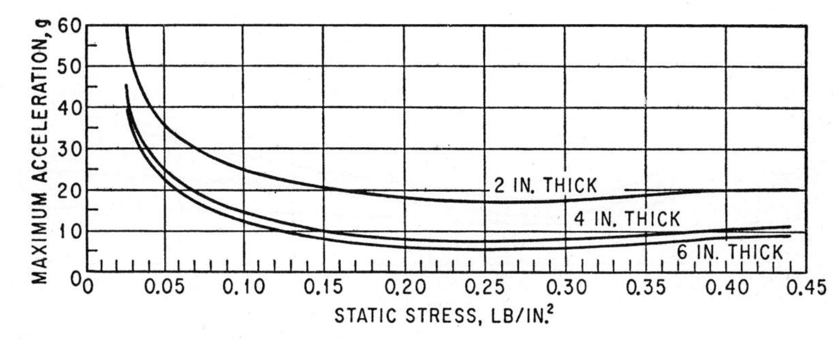

FIG. 41.18. Maximum acceleration vs. static stress for polyester urethane (2.2 lb/ft^3)—12-in. drop.

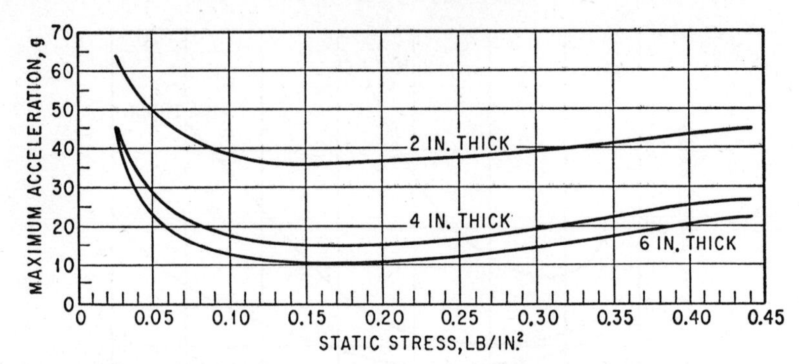

FIG. 41.19. Maximum acceleration vs. static stress for polyester urethane (2.2 lb/ft³)—24-in. drop.

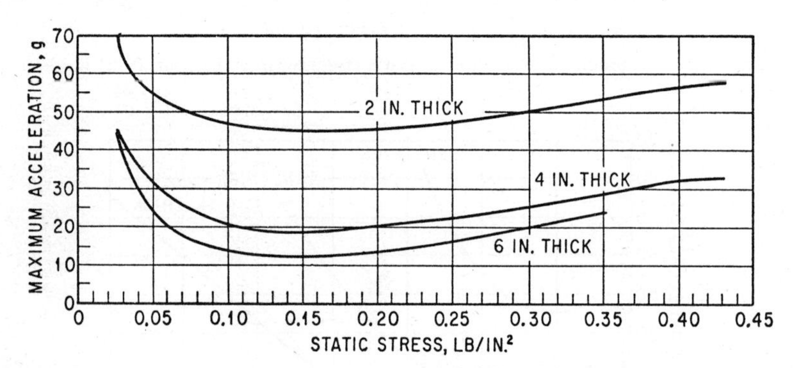

FIG. 41.20. Maximum acceleration vs. static stress for polyester urethane (2.2 lb/ft³)—30-in. drop.

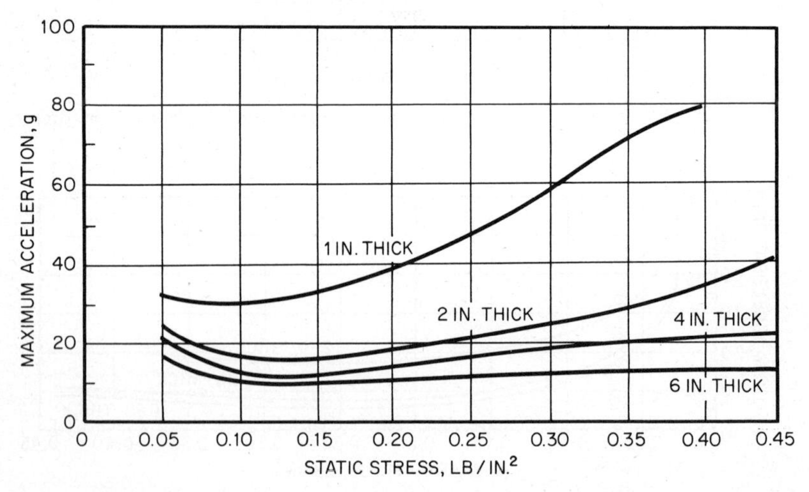

FIG. 41.21. Maximum acceleration vs. static stress for polyether urethane (2.0 lb/ft³)—12-in. drop.

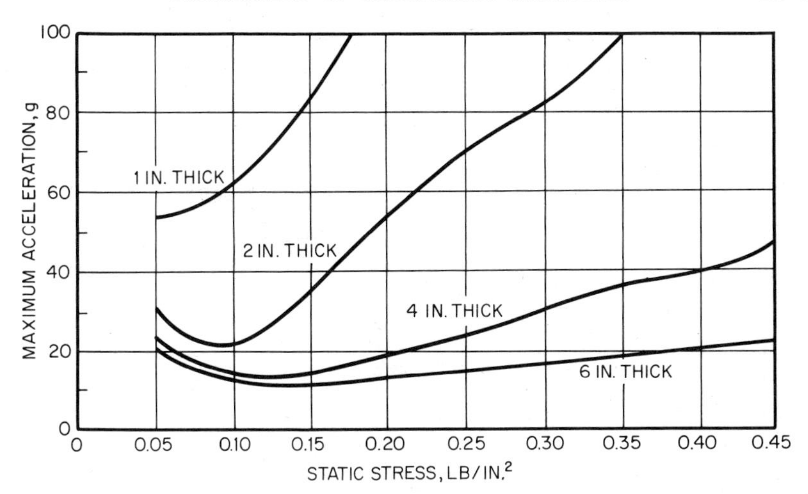

Fig. 41.22. Maximum acceleration vs. static stress for polyether urethane (2.0 lb/ft^3)—24-in. drop.

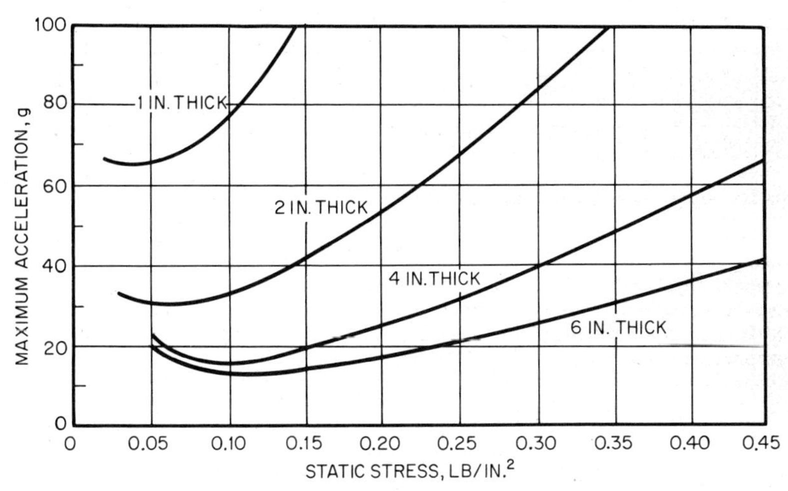

Fig. 41.23. Maximum acceleration vs. static stress for polyether urethane (2.0 lb/ft^3)—30-in. drop.

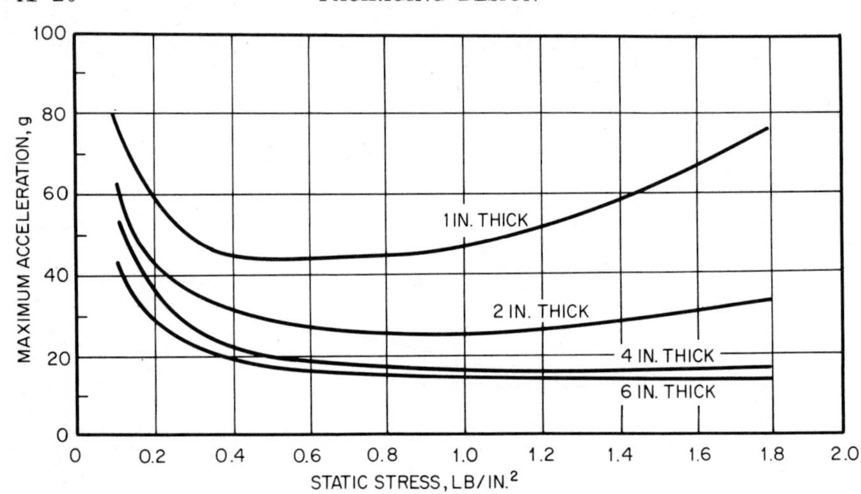

FIG. 41.24. Maximum acceleration vs. static stress for polyethylene foam (2.0 lb/ft³), 12-in. drop.

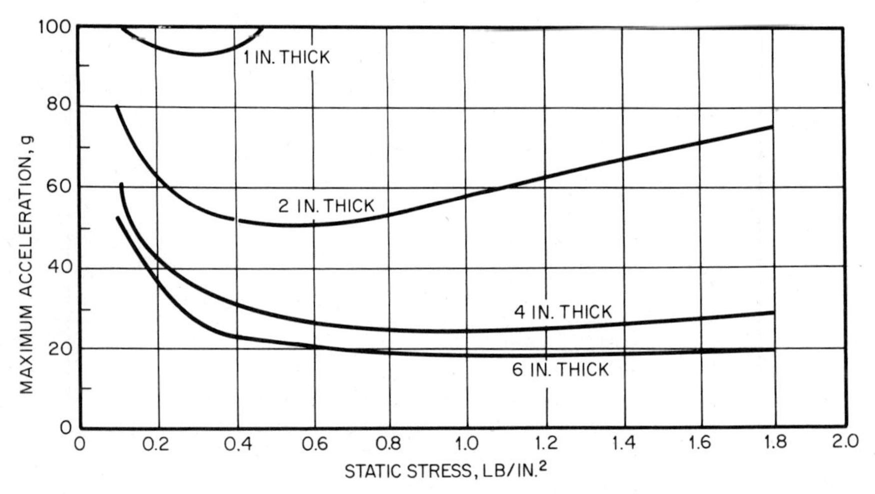

FIG. 41.25. Maximum acceleration vs. static stress for polyethylene foam (2.0 lb/ft³), 24-in. drop.

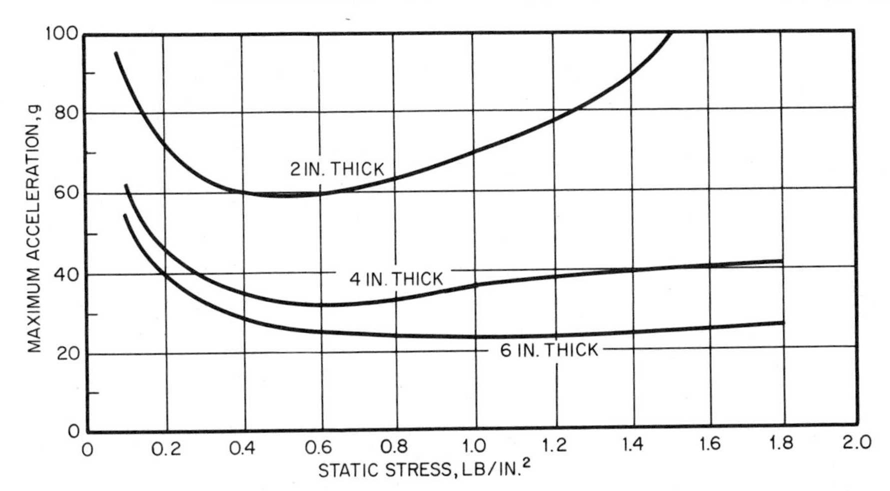

FIG. 41.26. Maximum acceleration vs. static stress for polyethylene foam (2.0 lb/ft^3), 30-in. drop.

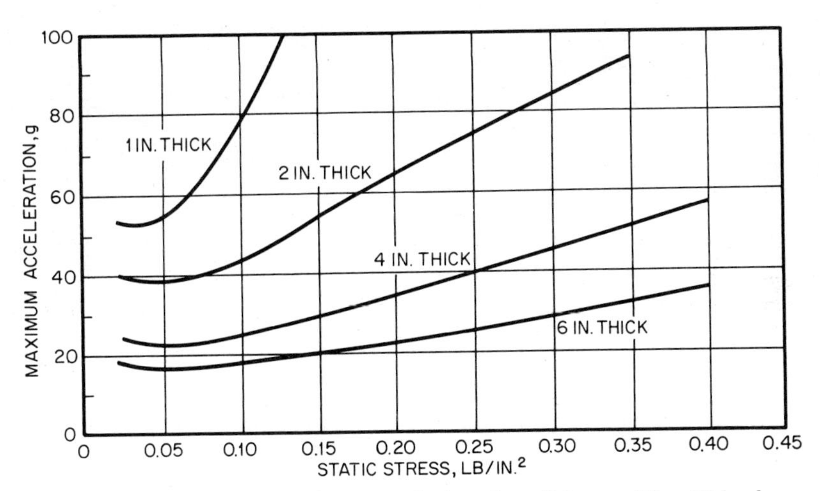

FIG. 41.27. Maximum acceleration vs. static stress for cellulose wadding, 12-in. drop.

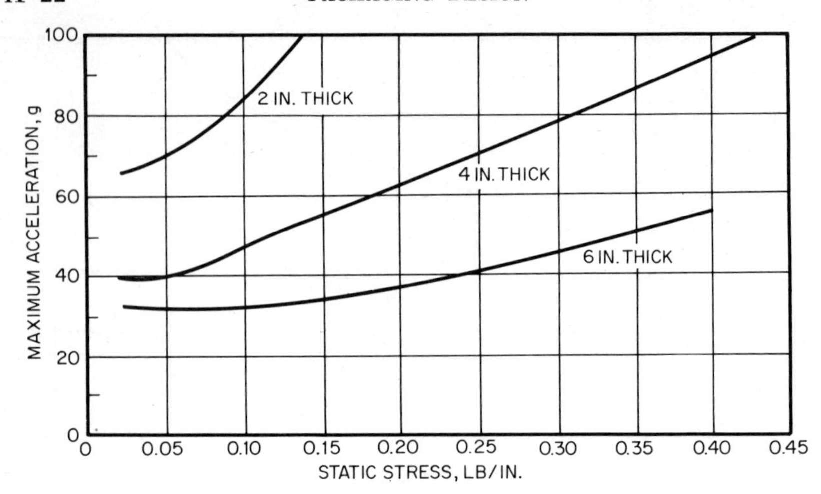

FIG. 41.28. Maximum acceleration vs. static stress for cellulose wadding, 24-in. drop.

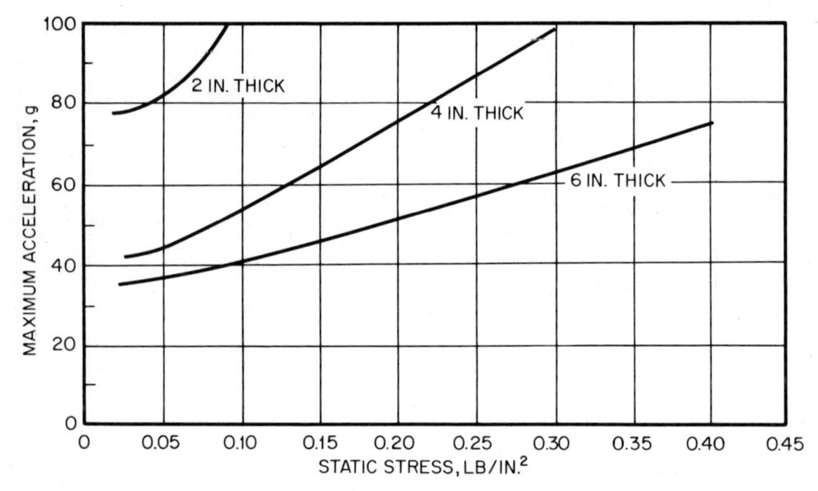

FIG. 41.29. Maximum acceleration vs. static stress for cellulose wadding, 30-in. drop.

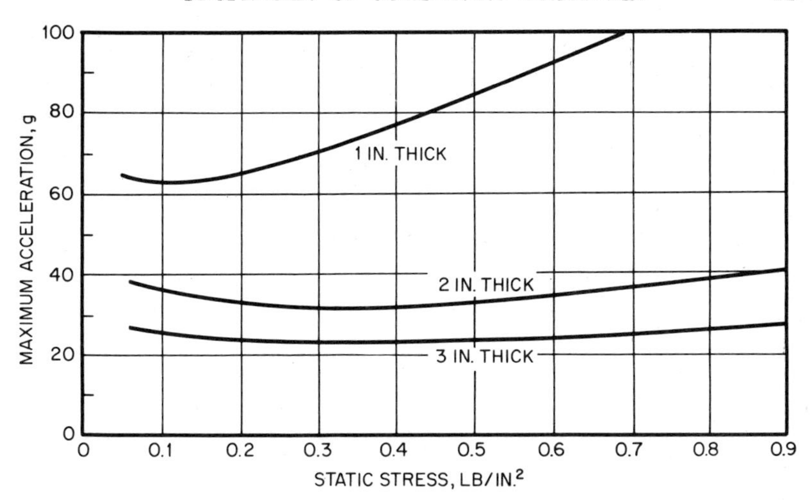

Fig. 41.30. Maximum acceleration vs. static stress for "AirCap" SD 240, 12-in. drop.

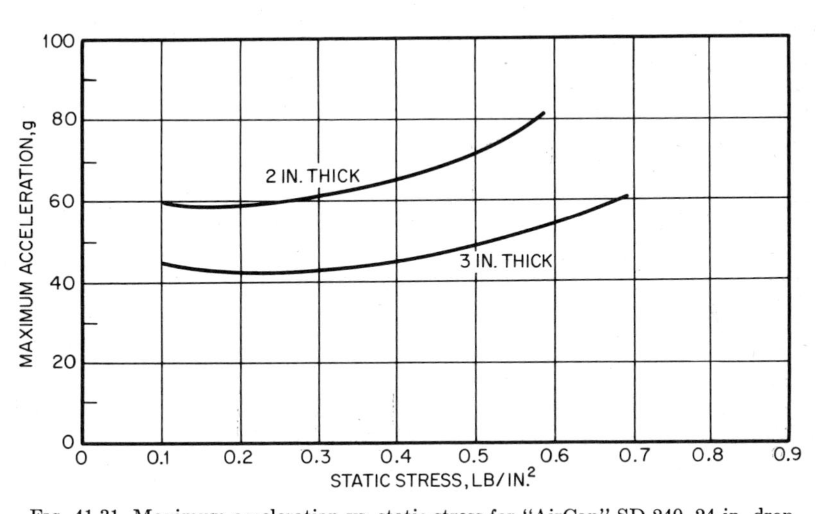

Fig. 41.31. Maximum acceleration vs. static stress for "AirCap" SD 240, 24-in. drop.

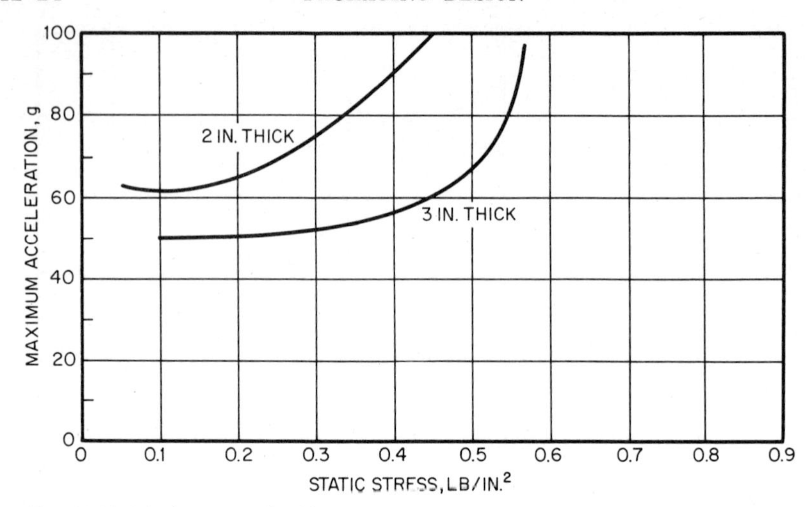

Fig. 41.32. Maximum acceleration vs. static stress for "AirCap" SD 240, 30-in. drop.

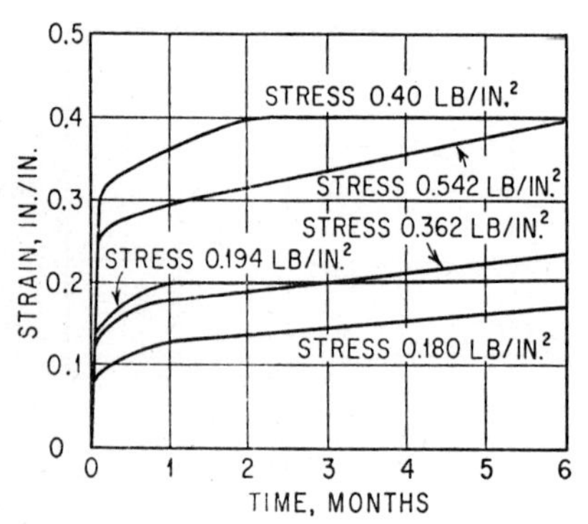

Fig. 41.33. Creep characteristics for latex hair, firm.

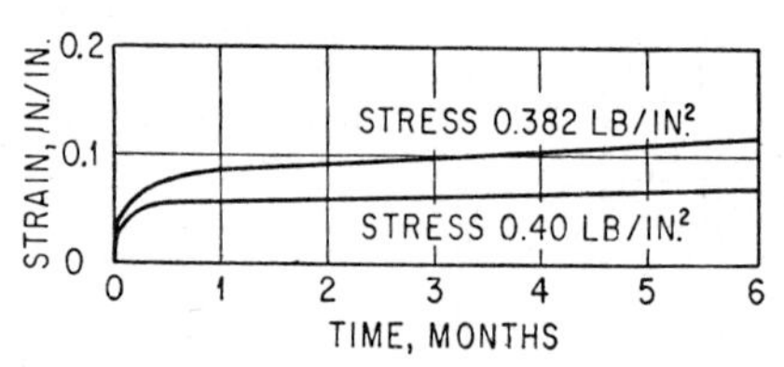

Fig. 41.34. Creep characteristics for polyester urethane (2.2 lb/ft³).

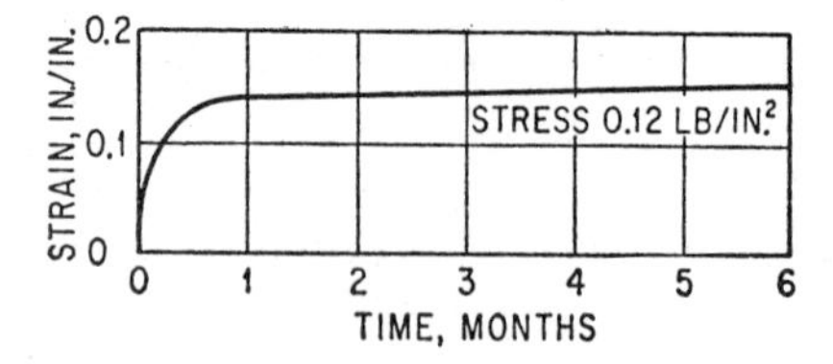

Fig. 41.35. Creep characteristics for polyether urethane (2.2 lb/ft³).

DESCRIPTION OF MATERIALS. This section includes a brief physical description of the various materials covered in this chapter, including their uses and characteristics.

Latex Hair. Curled animal hair (mostly from hogs but with some cattle tail and horse mane hair), bonded with natural latex (in molded shapes) or neoprene (in sheet form);

this material is most efficiently used to cushion items at static stresses up to approximately 0.3 lb/in.² and is available in various densities (normally referred to as soft, medium, and firm).

Urethane Foam. An open-cell plastic foam available in various densities, depending on the manufacturer; two forms of this material are available—one derived from a polyester and one from a polyether. They have somewhat different static stress-strain characteristics. Most commercially available materials are useful in approximately the same static stress range as latex hair, but some are available for use at a static stress as great as approximately 1.0 lb/in.²

Polyethylene Foam. A blown plastic foam, it is manufactured in various densities (approximately 2.0 to 9.0 lb/ft³). This is a stiff material but is very efficient in cushioning applications at static stresses from approximately 1.0 to 2.0 lb/in.²; in other applications the material can be cut very thin and used as a protective wrap.

Cellulose Wadding. A wood-fiber product most efficiently used as a protective wrap, but also used in some applications as cushioning where a static stress of less than 0.1 lb/in.² is required. For cushioning purposes, this material should not be used on extremely delicate items.

AirCap. A laminate of two layers of barrier-coated polyethylene film, one layer of which is embossed with rows of cells forming an air encapsulated cushioning material.

TEMPERATURE EFFECTS ON ISOLATION. The effects of temperature, both high and low, on the isolation characteristics of materials must be considered in the selection of the proper material for use as package cushioning. Most materials in common usage for package cushioning exhibit a marked increase in stiffness as the temperature decreases from 70°F (21.1°C), and a corresponding decrease in stiffness as the temperature is increased from 70°F. The variation in stiffness with temperature is considerably more pronounced in some materials than in others. For example, fibrous glass cushioning exhibits very little change in stiffness with temperature whereas the stiffness of polyester urethane foam increases rapidly as the temperature decreases. The change in the stiffness of latex-bound hair with variations in temperature may be considered approximately average for most cushioning materials. Because of wide variations in procedures and equipment used in testing to determine the effects of temperature on the isolation characteristics of cushioning materials, some published data of this type are not reliable. For this reason, only data on the effects of temperature on the stress vs. strain relationship for latex hair are presented here (see Fig. 41.36) as an indication of material performance.

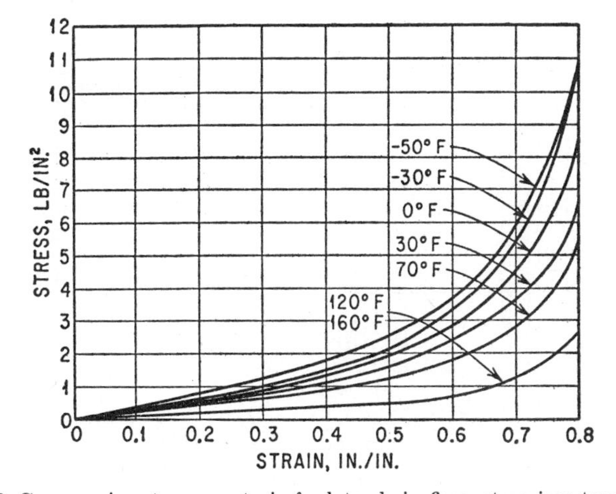

FIG. 41.36. Compression stress vs. strain for latex hair, firm, at various temperatures.

DESIGN TECHNIQUE—NONLINEAR CUSHIONING

The following procedure can be used as a guide in the development of design techniques utilizing distributed cushioning material for shock and vibration isolation:

1. Define the shock environment in terms of drop height; define the vibration environment in terms of frequency and amplitude.
2. Establish the fragility $\ddot{x}_F$ of the packaged item (see Chap. 42) and the characteristics (natural frequency and damping) of the critical components of the item.
3. Provide protection for the packaged item by means such as blocking and bracing of protruding parts, distribution of load to avoid stress concentrations, and preservation methods. (Interior containers adapted to house the packaged item may serve as protection when the item is rigidly secured within, thus approximating a single degree-of-freedom system.)
4. From the weight and size of the packaged item (or interior container plus contents), determine the static stress for each face; then from the maximum acceleration vs. static stress curves (Figs. 41.15 to 41.32) select the optimum material and the required thickness.
5. Check the static stress vs. strain curve (Figs. 41.9 to 41.14) for the applicable material to determine the static deflection.
6. Check the creep properties (Figs. 41.33 to 41.35) and, when applicable, make allowance for any loss in thickness during storage. The required initial thickness is

$$t_{\text{req}} = \frac{t_s}{1 - C} \tag{41.20}$$

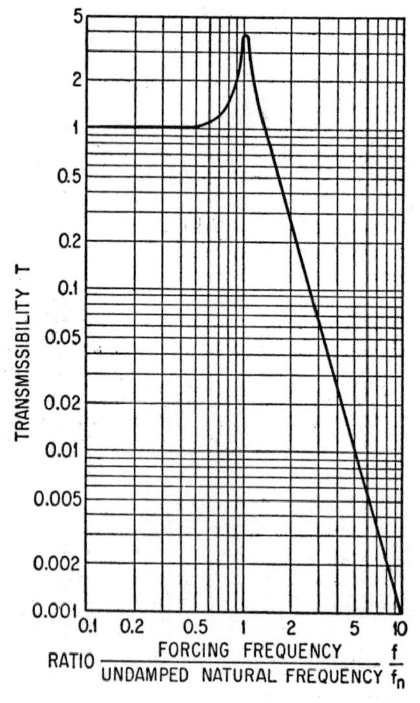

FIG. 41.37. Transmissibility curve for latex hair, firm.

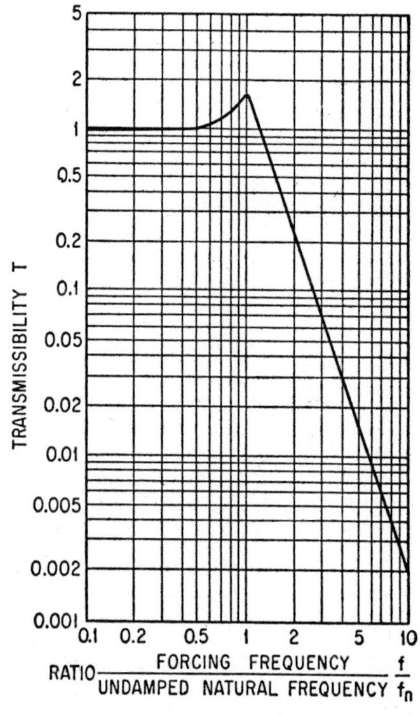

FIG. 41.38. Transmissibility curve for polyester urethane (2.2 lb/ft³).

where t_{req} = initial thickness required to compensate for creep, in.
t_s = thickness required for shock isolation, in.
C = drift, the ratio of thickness loss from creep to original thickness

7. To avoid localized buckling of the isolation medium, stability must be considered so that the cushion returns to its original position to provide protection for more than a single drop. A rule-of-thumb stability relation requires that

$$\frac{t}{\sqrt{S}} \leq \frac{4}{3} \qquad (41.21)$$

where t = thickness of cushion pad, in.
S = area of cushion pad, in.^{2}

8. For vibration isolation, determine the natural frequency of the isolation system (Figs. 41.39 to 41.41) and, from the transmissibility curves of Figs. 41.37 and 41.38, calculate the vibration of the packaged item at the resonant frequencies of the isolation system and the critical component.

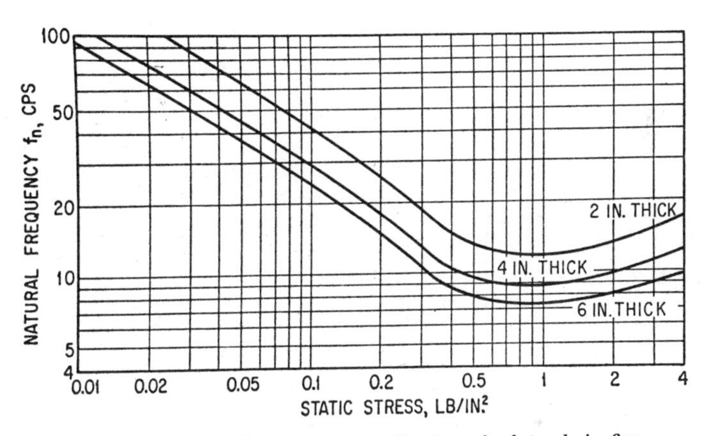

FIG. 41.39. Natural frequency vs. static stress for latex hair, firm.

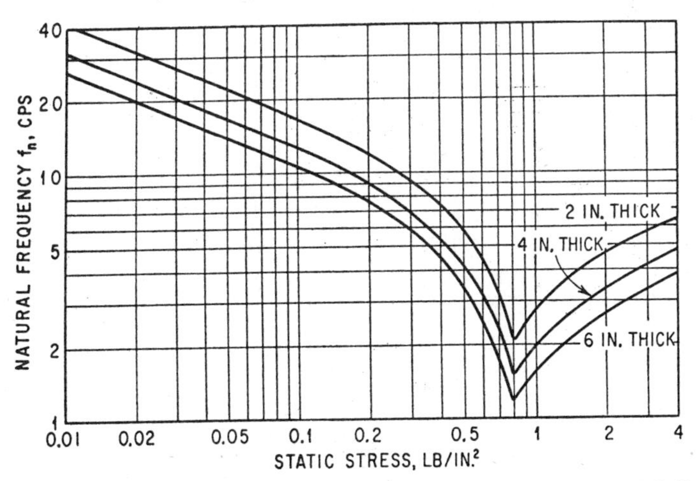

FIG. 41.40. Natural frequency vs. static stress for polyester urethane (2.2 lb/ft^3).

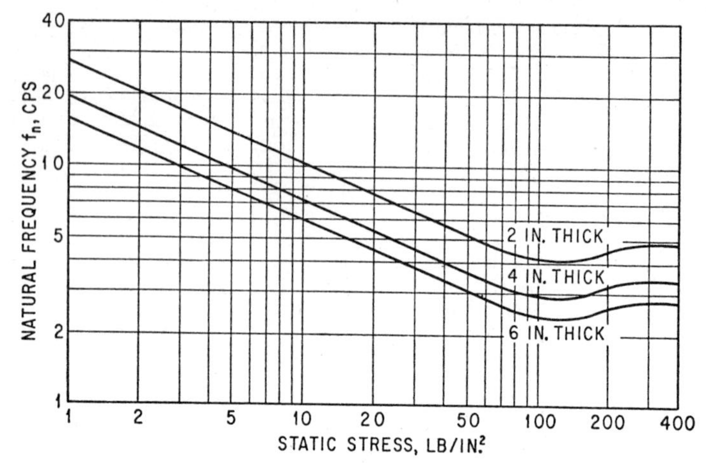

FIG. 41.41. Natural frequency vs. static stress for polyethylene foam (1.75 lb/ft³).

9. Select a method of cushioning to provide the exact amount of material required by the analysis. Wrapping a part with the material or use of other methods which provide a larger effective cushion area than required may be detrimental because the static stress is decreased and the maximum acceleration may be increased.

10. Select the outer container dimensions to provide the proper thickness of material and sufficient clearances to obtain the required deflection.

Example 41.3. A packaged item is capable of withstanding a maximum acceleration of $30g$ without damage. A critical component in this item has a natural frequency of 250 cps and an equivalent fraction of critical damping of 0.001. A criterion for rough handling has been established as a drop from a height of 30 in. The anticipated transportation environment has the following vibration characteristics: Displacement amplitude of 0.5 in. from 2 to 5 cps, and acceleration amplitude of $3g$ from 5 to 300 cps. The storage period is 6 months. An interior container with proper blocking and bracing, and necessary preservation requirements, has the following dimensions: Length, 24 in.; width, 18 in.; and height, 12 in. The gross weight of the interior container plus contents is 40 lb.

1. From the dimensions of the interior container and its gross weight, the available cushioning area and static stress for each container face are determined:

Container face	Available area, in.²	Static stress, lb/in.²
Top and bottom........	$24 \times 18 = 432$	0.0925
Sides..................	$24 \times 12 = 288$	0.139
Ends...................	$18 \times 12 = 216$	0.185

2. For this application, latex hair (firm) is to be used. From Fig. 41.17, for a 30-in. drop and $30g$ maximum acceleration, the thickness required is 3½ in. for top and bottom, 4 in. for the sides, and 5½ in. for the ends. Adjustment of the static stress on all faces to 0.0925 lb/in.² (total area on each face = 432 in.²) yields optimum design, resulting in a thickness of 3½ in. on each container face. The area on each face can be made equal by overlapping the cushioning materials in the manner illustrated in Fig. 41.42.

3. A static stress of 0.0925 lb/in.² produces a strain of approximately 0.025 (from Fig. 41.9), or a static deflection of 0.0875 in. This relatively small deflection does not significantly change the nominal dimensions of the package.

4. From Fig. 41.33, a static stress of 0.18 lb/in.² for a duration of 6 months results in an increase in strain of 0.167 in./in., i.e., in a thickness loss of approximately 16.7 per cent as

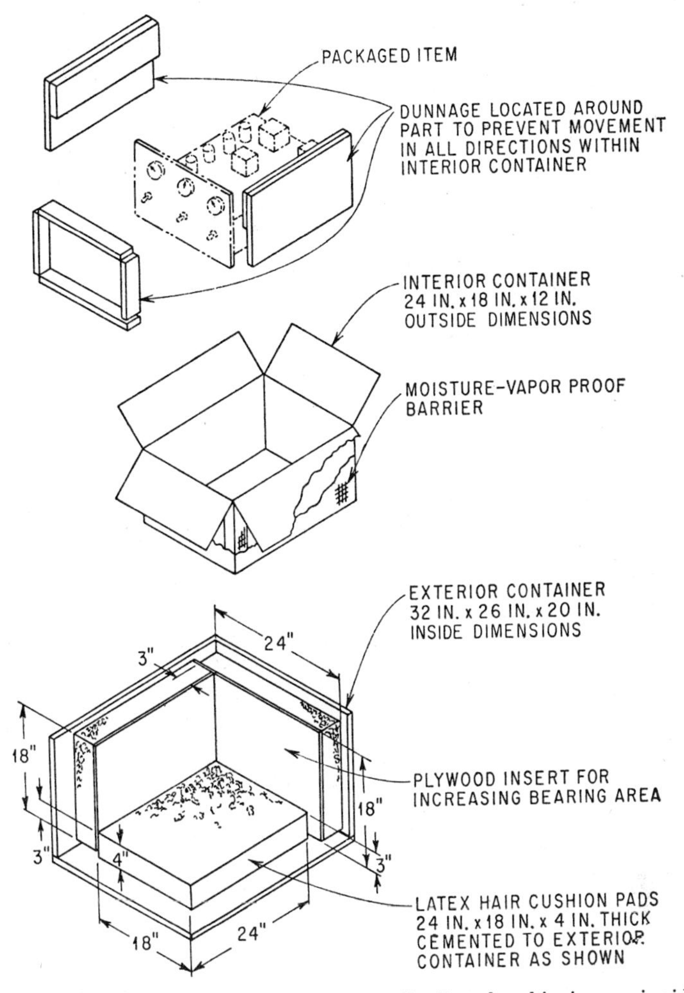

FIG. 41.42. Package cushioning design using distributed cushioning on six sides.

a result of creep. Assuming a linear relationship between stress and drift, the expected drift after 6 months under load at a static stress of 0.0925 lb/in.^{2} is approximately 8 per cent. Then, from Eq. (41.20), the initial thickness required to compensate for drift is

$$t_{\text{req}} = \frac{3.5}{1 - 0.08} = 3.8 \text{ in.}$$

Rounding this result off to the nearest nominal value, cushioning material 4 in. thick should be used.

5. Checking the stability by substituting the thickness and area in Eq. (41.21),

$$\frac{t}{\sqrt{S}} = \frac{4}{\sqrt{432}} = 0.193$$

This is less than 4⁄3, and the design is not susceptible to localized buckling.

6. From Fig. 41.39, the natural frequency of the isolation system for a static stress of 0.0925 lb/in.^{2} and 4-in. thickness is 30 cps.

7. At "package resonance," the impressed acceleration is $3g$ and (from Fig. 41.37) the interior container acceleration is $4 \times 3g = 12g$. The acceleration of the element also is $12g$. At resonance of the element, the acceleration of the interior container is $0.002 \times 3g = 0.006g$, and the acceleration of the element is $500 \times 0.006g = 3g$.

REFERENCES

1. Mindlin, R. D.: *Bell System Tech. J.*, **24**:353 (1945).
2. Franklin, P. E.: "Packaging for Shock and Vibration Protection with Tension Spring Suspensions," *Missile Division Rept.* AL-2608, Contract AF33(600)-28469, North American Aviation, Inc., 1957.
3. Gammell, L. W., and J. L. Gretz: "Report on Effect of Drop Test Orientation on Impact Accelerations," Physical Test Laboratory, Texfoam Division, B. F. Goodrich Sponge Products Division of B. F. Goodrich Company, Shelton, Conn., 1955.
4. Hatae, M. T.: "Shock and Vibration Tests of a Tension Spring Package," *Rept.* NA-54-1003, North American Aviation, Inc., 1954.
5. Janssen, R. R.: "A Method for the Proper Selection of a Package Cushion Material and Its Dimensions," *Rept.* NA-51-1004, North American Aviation, Inc., 1952.
6. "Military Specification—Cushioning Material, Elastic Type General," *USAF Specification* MIL-C-26861, 1959.
7. ASTM Method D 1372–64 Package Cushioning Materials, American Society for Testing Materials, 1964.
8. ASTM Method D 1596-64 Shock Absorbing Characteristics of Packaging Cushioning Materials, American Society for Testing Materials, 1964.
9. ASTM Method D 2221-68 Creep Properties of Package Cushioning Materials, American Society for Testing Materials, 1968.
10. "Military Standardization Handbook-Package Cushioning Design," *USAF Handbook* MIL-HDBK-304A, Sept. 25, 1974.

42

THEORY OF EQUIPMENT DESIGN

Edward G. Fischer
Westinghouse Electric Corporation

INTRODUCTION

The design of equipment to withstand dynamic loads is one of the more difficult and less developed aspects of shock and vibration control. The difficulty is inherent, in part, in the physical conditions because the nature and magnitude of such loads depend upon the dynamic characteristics of the equipment; thus, the loads cannot be determined until the equipment has been designed. Therefore, the designer must design for loads that are necessarily unknown. Such a problem is approached logically by first preparing an initial design based upon the best judgment of the designer. A subsequent analysis would show weaknesses and other undesirable features of the design, and point to desirable modifications. Thus, the design would be developed by cut-and-try methods.

Design by cut-and-try methods using only analytical procedures is not profitable in general because several important considerations are difficult to evaluate. For example, even though the dynamic loads could be determined by analysis, it is difficult to determine the resulting stresses if the equipment has an ordinary degree of complexity. Furthermore, it is difficult to evaluate the properties of materials under dynamic strain. Accordingly, it is preferred design practice to calculate strength requirements for only major and important structural members of the equipment. An initial prototype equipment is then constructed for test in which actual operating conditions preferably are duplicated. For military equipment, this is generally impractical and laboratory tests are used to simulate actual operating conditions. Laboratory testing and the simulation of actual conditions are discussed in Chap. 24. Neither designing nor testing to withstand dynamic loads is an exact science; hence, discrepancies frequently occur and must be resolved by experience and intuition. This chapter offers such guidance as the state of the art permits.

It is important to consider the design of the initial prototype from both the analytical and practical points of view. Much can be done at that stage of the design. Optimum stiffnesses can be determined for the interrelated structural members of the equipment with the objective of minimizing the stresses in such members. The forces experienced by components of the equipment also are minimized. The required strengths of these members then can be calculated. The succeeding sections of this chapter set forth procedures for determining strengths and stiffnesses of principal structural members, and outline applicable information on the selection and use of engineering materials in equipment that is required to withstand shock and vibration. A number of techniques, generally qualitative in nature, have been developed as a result of much practical experience in design. These are summarized in Chap. 43.

An important part of the design process is redesign to correct the deficiencies revealed by tests of the prototype. Such tests reveal not only the weaknesses of the equipment but also useful information, for example, on the natural frequencies of structural elements or the degree of damping present. With this information, new analyses can be carried out or original analyses modified using actual rather than assumed information.

Although methods for the design of equipment to withstand dynamic loads are not well developed in general, it is possible to apply known methods in particular cases. Some of these cases are discussed in other chapters. The design of internal-combustion engines and other types of rotating and reciprocating machinery is discussed in Chap. 38. Dynamic absorbers and auxiliary mass dampers are discussed in Chap. 6, and the use of damping materials is discussed in Chap. 37. An analysis of self-excited vibration is included in Chap. 5, and the isolation of vibration and shock is covered in Chaps. 30 to 33.

Although the connotation of the words "shock and vibration" is somewhat broader, this chapter is limited to the design of equipment that must withstand the motion of the support to which the equipment is attached. The motion may be of a continuing oscillatory nature, known broadly as vibration, or of a transient nature involving acceleration and/or displacements of large magnitude.

DYNAMIC LOAD

When an equipment is subjected to shock or vibration, the structural members of the equipment deflect or deform. The deflection is said to be caused by the *dynamic load;* i.e., the force required to cause structural members of the equipment to deflect in response to the motion imposed by the support. For practical purposes, the dynamic load may be considered as the static or dead-weight load multiplied by a *dynamic load factor*. The members must be designed to withstand, without failure, the stress imposed by the dynamic load. In general, the stress in the member subjected to a dynamic load is equal to that caused by a static load of the same magnitude. It does not necessarily follow, however, that a design stress which applies to static conditions also applies to dynamic conditions. For example, a structure designed to withstand only a static load may be satisfactory if the stress does not exceed a certain percentage of the ultimate stress. On the other hand, a structure may be unsatisfactory under a dynamic load inducing the same stress level because the structure may fail under repeated occurrences of the stress. In other words, if the dynamic load is repeated many times, a lower value of acceptable stress must be used to recognize the effect of fatigue of the material.

The magnitudes of the dynamic loads in an equipment subjected to shock and vibration usually are large relative to the static loads; thus, it is common practice in design to neglect static loads unless they are known to be significant relative to dynamic loads.

The physically observable effect of the application of a dynamic load is the deflection of a structural member. To require that the member have an acceptably low stress as a consequence of such deflection would introduce an uncommon concept in design because structures, in general, are designed to withstand designated *loads*. By considering the dynamic load to be equivalent to a static load that would cause the same deflection, the designer acquires a familiar working tool in that he can apply the principles of static loading. The process of designing for shock and vibration consists of (1) determining the magnitudes and directions of the dynamic loads and (2) designing the structure to withstand a hypothetical static load equivalent to the dynamic load, keeping in mind certain qualifying conditions related to the stress level and the distribution of stress.

METHODS OF ANALYSIS

USE OF MODELS

The first step in carrying out an analysis involving dynamic loading is to represent the structure by an *equivalent model* of rigid masses connected by massless springs and dampers. Theoretically, it is then possible to determine from any known excitation the motion of each mass and the consequent deflection of each spring. The force acting on each spring is known directly from its deflection and stiffness; thus, the springs can be designed with the strength required to withstand such forces. Therefore, it may be said that the equipment has adequate structural strength to withstand the excitation.

The type of model used for the analysis depends upon the accuracy required in the results and the effort that can be applied to the analysis. For example, consider the

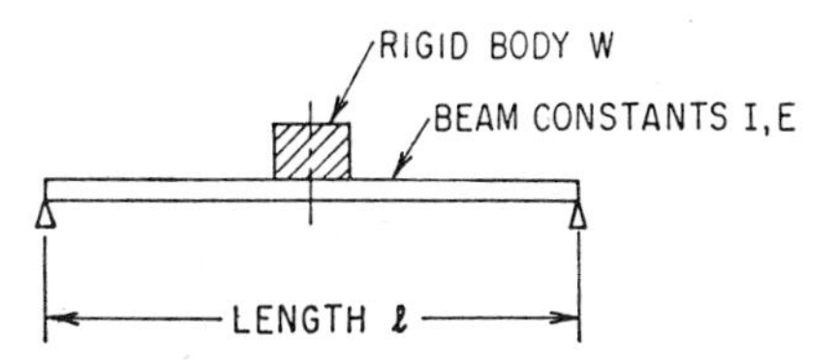

FIG. 42.1. Rigid body supported by an elastic beam with hinged ends.

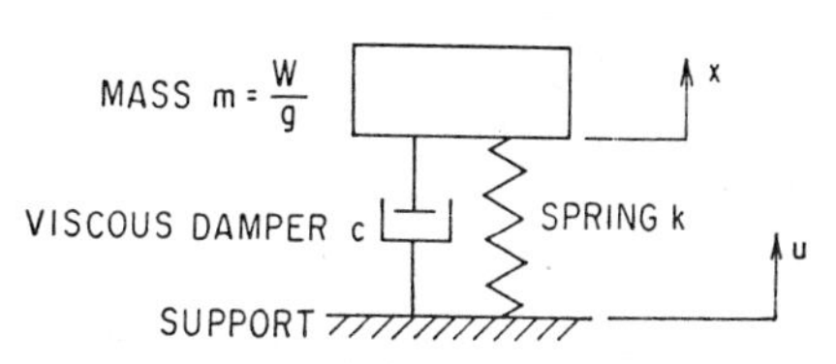

FIG. 42.2. Single degree-of-freedom system model of beam in Fig. 42.1, neglecting mass of beam.

problem of designing the simple beam supporting a rigid body of weight W as shown in Fig. 42.1. If the mass of the beam is small relative to that of the rigid body, a model that is adequate for most purposes is illustrated in Fig. 42.2 where the stiffness k and the damper c represent corresponding properties of the beam. However, if the mass of the beam is of the same order of magnitude as the mass of the supported body, parts of the beam tend to vibrate with respect to each other. This introduces dynamic loads that cannot be neglected if accurate results are desired. A model that includes the mass of the beam is shown in Fig. 42.3 where the sum of the incremental masses m' equals the total mass of the spring.

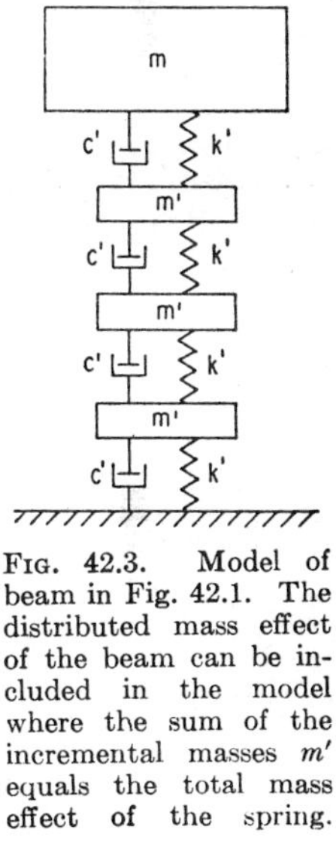

FIG. 42.3. Model of beam in Fig. 42.1. The distributed mass effect of the beam can be included in the model where the sum of the incremental masses m' equals the total mass effect of the spring.

A model may be constructed with any degree of complexity required. For example, the model of an automobile in Fig. 42.4 could be used to determine the effect of shock absorbers on riding comfort or to calculate the maximum force applied to the seat springs. In this instance, the discrete masses of the model correspond to similarly discrete bodies of the actual automobile. By contrast, a model suitable for investigating the strength and stiffness of the guided missile of Fig. 42.5*A* under bending loads would appear as in Fig. 42.5*B*. The total mass of the missile, distributed more or less uniformly along its length, is represented by the sum of the several rigid masses of the model; the stiffness of the missile in bending is represented by the stiffnesses in bending of the members connecting the masses of the model. As an objective in setting up the model, the natural frequencies of model and actual missile should agree.

In general, complex models are useful only for investigating the capability of a particular equipment to withstand a particular excitation. However, the development of high-speed digital computer facilities has made it practical to investigate the dynamic response of a large number of possible combinations of masses, springs, and dampers in a complex model and eventually draw general conclusions. It must be cautioned that in order for the model to be authentic, the computer programming should first demonstrate that it can reproduce the results of simple static deflection and snap-back natural frequency tests of the original physical system. There is almost no limit to the degree of sophistication (i.e., yielding elements, random excitation, etc.) that can be achieved with computer-aided analyses, provided that the results make sense in terms of engineering experience.

A model may be used to study equipment design problems arising from any type of dynamic excitation. The general designation "shock and vibration" used to describe an environment is separable for purposes of analysis into (1) periodic vibration, (2) transient vibration or shock, and (3) random vibration. The response of mechanical systems (a "system" as discussed generally in other chapters is equivalent to a "model"

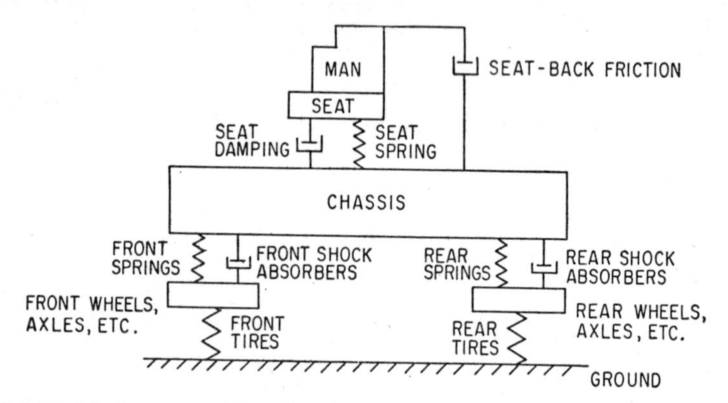

FIG. 42.4. Model of an automobile. The discrete masses of the model correspond to similarly discrete bodies of an actual automobile.

in this chapter) to these various classes of environmental conditions is discussed in several other chapters as follows:

Periodic vibration: Chaps. 2, 6, and 22.
Transient vibration or shock: Chaps. 8 and 23.
Random vibration: Chaps. 11 and 22.

The design of the model and the interpretation of the response are discussed in succeeding sections of this chapter.

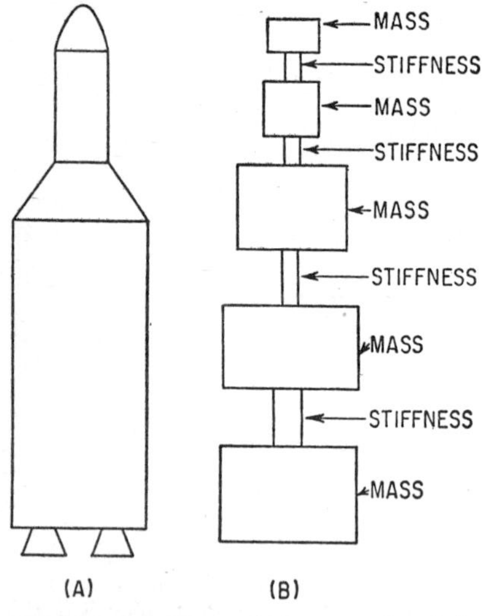

FIG. 42.5. (*A*) Schematic diagram of a missile with the total mass distributed more or less uniformly along its length. (*B*) Model of the missile wherein the total mass is subdivided into a series of masses connected by members having the appropriate stiffnesses in bending.

TYPES OF MODELS. The usefulness of a model for initial design purposes tends to increase as its simplicity increases. A model that attempts to simulate an equipment in detail cannot be constructed until after an equipment is designed; thus, it becomes the means to check a design rather than to implement the design initially. On the other hand, a less complex model can simulate primary or important structures in a more general sense and can be evaluated in more general terms. For example, the model shown in Fig. 42.2 is defined completely by its natural frequency and fraction of critical damping; its response to any given excitation can be expressed concisely in terms of these parameters. By contrast, the model shown in Fig. 42.4 involves five masses and five springs; the relatively large number of possible permutations demonstrates the impracticality of expressing the results concisely or in a form suitable for ready reference. Techniques for the application of models in design are described below under *Structural Design Considerations* for various types of excitation.

Single Degree-of-freedom Model. The model shown in Fig. 42.2 is a spring-mass-damper system where the rigid mass $m = W/g$ is assumed to move along a straight line. The mass can be located at any time t by the single coordinate x; therefore the system is referred to as having a single degree-of-freedom. When the mass is displaced from its equilibrium position, it has only one mode of vibration; this occurs in free vibration at the single natural frequency of the system. The massless spring has a linear force-deflection relationship defined by the spring constant k. Similarly, the massless damper exhibits a linear force-velocity relationship denoted by the damping coefficient c. The damping in a typical mechanical system usually is so small that it has a negligible effect on the value of the natural frequency. Hence, the free vibration of the mass in the single degree-of-freedom model can be considered to be a simple harmonic motion. The undamped natural frequency from Eq. (2.8) is

$$\omega_n = \sqrt{\frac{kg}{W}} = \sqrt{\frac{g}{\delta_{st}}} \quad \text{rad/sec} \tag{42.1}$$

where δ_{st} is the static deflection of the spring caused by the weight W.

The analogy between the structure of Fig. 42.1 and the model of Fig. 42.2 is established as follows: The rigid body of weight W is supported at the mid-position of the elastic beam of length l and moment of inertia (of the cross-sectional area) I; the modulus of elasticity of the beam material is E. The single degree-of-freedom model of Fig. 42.2 is applicable only when the weight of the beam is small compared to that of the rigid body W. The static deflection directly under the body W is[1] $\delta_{st} = Wl^3/48EI$. The spring constant of the elastic beam is $k = W/\delta_{st}$ which, when substituted into the first expression of Eq. (42.1), gives the natural frequency for lateral vibration of the beam carrying the weight W:

$$\omega_n = \sqrt{\frac{48EIg}{Wl^3}} \quad \text{rad/sec} \tag{42.2}$$

The model of Fig. 42.2 represents the beam of Fig. 42.1 when their natural frequencies are equal.

The model shown in Fig. 42.2 can be used to represent many different types of structures, generally beams or plates. The natural frequencies of such structures can be calculated from the analyses in Chap. 7 or the tables of formulas in Chap. 1. If the structure carries a load that may be considered a rigid body whose mass is large relative to the mass of the structure, the single degree-of-freedom model of Fig. 42.2 is likely to be quite satisfactory.

If the load carried by the structure is relatively small compared with the mass of the structure, the model of Fig. 42.2 cannot be used in evaluating the actual equipment when the higher modes of vibration are of interest. The multiple degree-of-freedom model based upon the normal mode concept (see *Multiple Degree-of-Freedom Model* below) then becomes applicable. However, if only the fundamental mode of vibration is of interest, a portion of the weight of the beam may be lumped with the load and treated as a single degree-of-freedom system (see Table 7.2).

Two Degree-of-freedom Model. When the equipment includes two structures joined together in such a way that the vibration of one is dependent upon the vibration of the other, the single degree-of-freedom model of Fig. 42.2 may not be appropriate. The two degree-of-freedom model of Fig. 42.6*A* then may be used. This simulates, for example, the structure of Fig. 42.7 where a load of mass m_2 supported by a beam of stiffness k_2 is mounted upon a load of mass m_1 that in turn is supported by a beam of stiffness k_1. One type of analysis applies if the mass m_2 is small relative to the mass m_1 (uncoupled two degree-of-freedom model), and another type of analysis applies if the two masses are of the same order of size (coupled two degree-of-freedom model). These two types of analysis are discussed separately under *Design for Vibration Conditions* and *Design for Shock Conditions*.

In the model of Fig. 42.6*A*, two spring-mass-damper systems are connected in a series system where the rigid masses m_2 and m_1 are both assumed to move along the same vertical line. The individual masses can be located at any time t by the two coordinates x_2 and x_1; thus, the system is referred to as having two degrees-of-freedom. The massless springs have linear force-deflection relationships denoted by the spring constants k_2 and k_1. Similarly, the massless dampers exhibit linear force-velocity relationships denoted by the damping coefficients c_2 and c_1. The system consisting of k_1, m_1, c_1 is designated

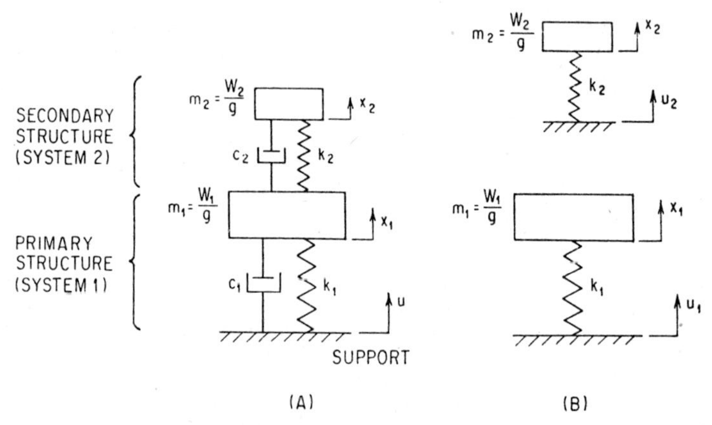

Fig. 42.6. (*A*) Two degree-of-freedom model consisting of a primary structure and a secondary structure, and (*B*) the uncoupled primary and secondary structures.

the *primary structure* because, generally, it represents a principal structure of the equipment or a major component thereof. Similarly, the system consisting of k_2, m_2, c_2 is designated the *secondary structure* because it represents a structure or component to which the excitation is transmitted by the primary structure. In some of the analyses that follow, the equations are simplified by omitting the terms for damping forces.

When the two masses m_2 and m_1 are arbitrarily displaced from their equilibrium positions and released, the system shown in Fig. 42.6*A* exhibits a complicated motion composed of two superimposed free vibrations at two different (*coupled*) natural frequencies. In the mode of vibration at the lower of these frequencies, the masses m_2

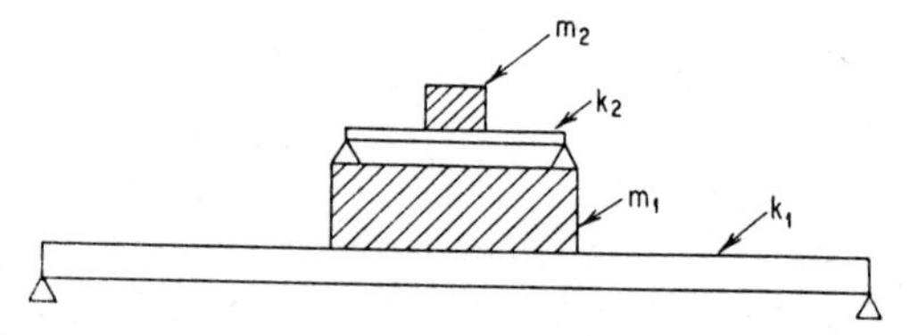

Fig. 42.7. Assembly of beams to which a two degree-of-freedom model is applicable.

and m_1 vibrate in phase; in the mode at the higher frequency of vibration, the two masses vibrate in opposition to each other. In general, the two natural frequencies can be determined as follows:[2] The differential equation of motion is derived for each mass in the system. Then, by means of the operational calculus, these equations are converted into two simultaneous algebraic equations. The determinant of the coefficients of x_2 and x_1 gives the characteristic equation which is solved for the unknown natural frequencies of the system. Sometimes it is more fruitful to discuss the analytical results in terms of the *uncoupled* natural frequencies defined according to Eq. (42.1) for the simple spring-mass models in Fig. 42.6*B*: $\omega_2 = \sqrt{k_2/m_2}$ and $\omega_1 = \sqrt{k_1/m_1}$. In the following analyses the uncoupled natural frequencies are used in place of the more complicated expressions for the actual natural frequencies.

Multiple Degree-of-freedom Model. Models with more than two degrees-of-freedom also may be used as a basis for design to withstand shock and vibration. Such models are used to represent equipment that may be idealized by several rigid masses and massless springs or, alternatively, by a continuous member whose vibration as an elastic body is significant. Response of such a model to excitation occurs in its normal modes; vibration in each of such modes occurs independently of vibration in the other normal modes and can be evaluated as the vibration of a single degree-of-freedom system. The fundamental concept of normal modes is discussed in Chap. 2. The vibration of an elastic body in terms of its normal modes is considered in Chap. 7.

DESIGN FOR VIBRATION CONDITIONS

The most important requirement in the design of equipment to withstand vibration is to avoid the large amplitude of vibration that accompanies a condition of resonance. Experience indicates that vibration troubles, whether leading to failure or improper operation of equipment or to personal discomfort, usually result from a condition of resonance. The excitation may be either that generated by normal operation of the machine or that coming from the environment in which the machine operates. The discussion in this chapter is limited to the design of equipment to withstand environmental conditions; a qualitative discussion of techniques to minimize the vibration generated by the machine is included in Chap. 43.

Techniques for minimizing the relatively large amplitude of vibration that occurs at the condition of resonance depend upon the nature of the excitation:

1. If the excitation is a continuing periodic vibration at a nonchanging frequency or frequencies, it is preferable that all natural frequencies of the equipment occur at other than the excitation frequencies.

2. If the excitation is (*a*) periodic vibration whose frequency varies between limits or (*b*) random vibration consisting of all frequencies within a defined band, it may not be feasible to design the equipment so that its natural frequencies do not coincide with excitation frequencies. Under these conditions, damping must be added to the vibrating structure.

SINGLE DEGREE-OF-FREEDOM MODEL. When the model of Fig. 42.2 is acted upon by a periodic excitation whose frequency coincides with the natural frequency of the model, the amplitude of vibration of the model is indicated quantitatively by Fig. 2.13. The relative motion or spring deflection is shown in Fig. 42.8. It usually is preferable to eliminate resonance by stiffening the spring k of the model; this increases the natural frequency and moves the operating condition to the left of the resonance peak in Fig. 42.8 where the spring deflection is small. In some instances, an increase in stiffness is not possible; a decrease in stiffness accomplishes the objective of avoiding resonance but may have the result not only of removing desirable rigidity from the equipment but also of making the spring deflection equal to or greater than the amplitude of vibration. The natural frequency of the model in Fig. 42.2 varies directly as the square root of the spring stiffness [see Eq. (42.1)]; consequently, a relatively large change in stiffness is required to effect an incremental change in natural frequency.

If the natural frequency of the structure cannot be modified or if the excitation covers a band of the frequency spectrum, it may be impossible to avoid a resonant condition. Under these circumstances, the desired low amplitude can be attained only by ensuring that the structures of the equipment are damped sufficiently. In general, metals commonly used in structures do not have sufficient hysteresis damping to maintain vibration amplitudes at an acceptable level if a resonant condition exists. Frequently, damping may be attained by use of bolted or riveted joints, or by use of damping materials in or on the structure. In extreme cases, discrete dampers must be added to achieve adequate control of vibration. These various techniques are discussed in Chap. 43 under *Damping of Structures*.

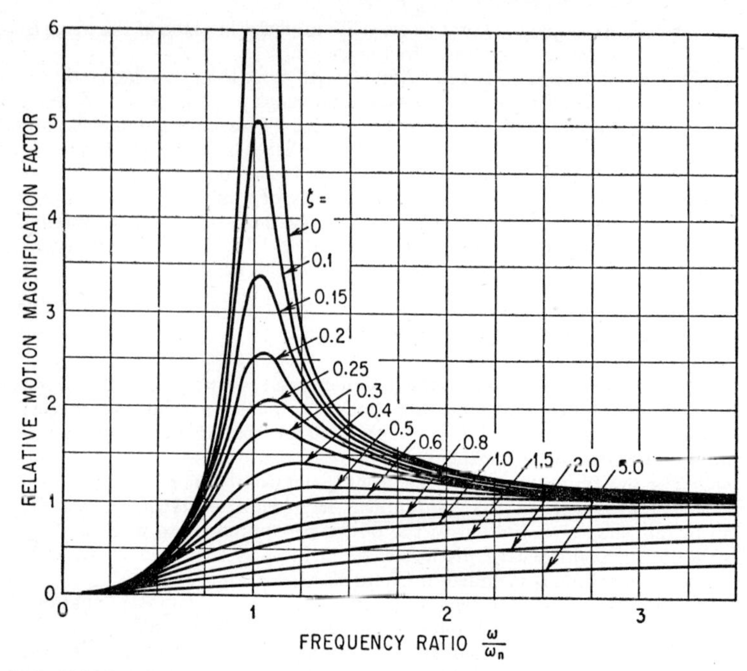

FIG. 42.8. Relative motion magnification factor for a single degree-of-freedom system with viscous damping. The forcing frequency is ω, the undamped natural frequency ω_n and the fraction of critical damping ζ.

UNCOUPLED TWO DEGREE-OF-FREEDOM MODEL. When the mass m_2 is small, it is unable to influence the motion of mass m_1. Then the vibratory motion of m_1 in system 1 of the model shown in Fig. 42.6*A* becomes the excitation for system 2, and the above discussion of the single degree-of-freedom model becomes applicable to evaluating the response of system 2. A condition of resonance of system 2 with its excitation should be avoided; i.e., the natural frequency of system 2 should be different from that of system 1. Preferably, the natural frequency of system 2 should be greater than the natural frequency of system 1; otherwise, the deflection of the spring k_2 may equal or exceed the vibratory displacement of the mass m_1. A system 2 with a natural frequency twice as great as that of system 1 experiences a vibration amplitude 1.35 times as great, as indicated in Fig. 2.13. This is a tolerable degree of amplification; it usually is achievable practically, and constitutes a good design objective. It has wide application to equipment that involves elastic systems supported on elastic systems, and should be adopted as a primary objective in the design of equipment to withstand vibration.

If the natural frequency criteria of the above paragraph can be met, damping requirements for system 2 may not be stringent. The excitation for system 2 is the vibration of system 1; the magnitude of such excitation is determined to a considerable extent by the damping in system 1. Thus, a basic principle of design where system 2 is of negligible mass is (*a*) to design system 2 to have a natural frequency at least twice as great as that of system 1 and (*b*) to introduce damping in system 1 to limit its amplitude of vibration if this is made necessary by the nature of the excitation. Methods for adding damping to structures are discussed in Chap. 43 under *Damping* in *Design of Structures.*

COUPLED TWO DEGREE-OF-FREEDOM MODEL. The analysis of an equipment represented by a two degree-of-freedom model as shown in Fig. 42.6*A* must consider coupling between the branches of the system when the two masses m_2, m_1 are of the same order of size. In other words, it cannot be considered that system 1 responds as a single degree-of-freedom system and imposes this response as the excitation for system 2. The spring k_2 and damper c_2 apply significant forces to system 1; this is knows as a *coupling effect* and causes the entire assembly to respond as a single system. The analysis of such a system is set forth in Chap. 6; it can be used to predict the response of the model to a known excitation. The effect of changes in the physical properties of the model can be determined and related to the actual equipment.

DESIGN FOR SHOCK CONDITIONS

The design of equipment to withstand shock involves the formulation of a model to represent the equipment and the determination of the response of the model to the shock. Theoretically, the design of equipment to withstand shock can proceed if the shock is defined as the time-history of the motion (displacement, velocity, or acceleration) of the support for the equipment. As pointed out below (see *Single Degree-of-Freedom Model*), the shock spectrum is an adequate definition of the shock motion for purposes of equipment design. Known methods can be applied to determine the maximum response of the model, and design requirements for the actual equipment can be inferred from this response. Considerations in selecting a model are discussed above under *Types of Models.*

SINGLE DEGREE-OF-FREEDOM MODEL. In the design of equipment to withstand shock, it is common practice to assume that failure occurs because the maximum allowable stress or acceleration is exceeded in an element of the equipment. Stress of this magnitude is assumed to lead to failure should it occur once.* The method of analysis involves a determination of the maximum stress in the model corresponding to the maximum deflection of spring k or maximum acceleration of mass m in Fig. 42.2.

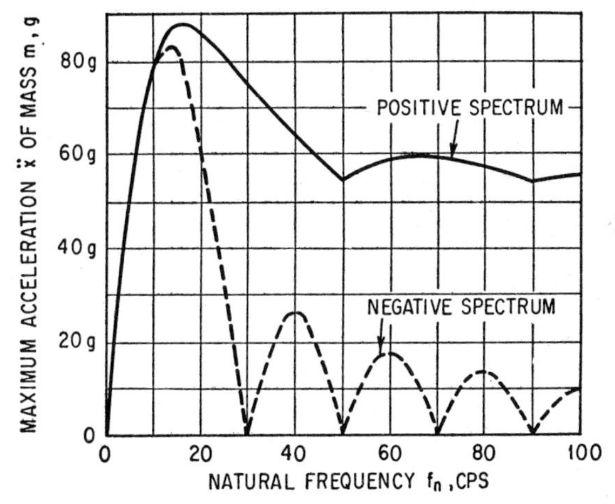

FIG. 42.9. Typical shock spectra for an undamped system. The solid line represents acceleration in the same direction as the principal excitation; the broken line represents acceleration in the opposite direction. (*After J. E. Ruzicka.*[3])

The Shock Spectrum. For a given shock motion, the maximum acceleration $\ddot{x}$ of mass m is a function of the natural frequency f_n of the model and its fraction of critical damping ζ. This maximum acceleration $\ddot{x}$ can be plotted as spectra of the natural frequency f_n with ζ as the parameter of the family of spectra. Typical spectra[3] are shown in Fig. 42.9. Each shock motion has characteristic shock spectra; this is discussed in Chap.

* In laboratory testing, consideration sometimes is given to the cumulative effect of repeated occurrences of stress. This concept also has application to the design of equipment; it is discussed in detail in Chap. 24.

23 where it is pointed out that the spectra constitute a convenient definition of a shock motion. An important aspect of such a definition is that it provides information directly useful for design purposes; i.e., a known value for the maximum acceleration of the mass of the single degree-of-freedom model makes possible the design of the spring. Methods for determining the shock spectrum and examples of spectra for typical excitations are given in Chaps. 8 and 23.

In using the spectra of Fig. 42.9 for design purposes, the chart is entered on the abscissa scale at the natural frequency of the model. The corresponding maximum acceleration of mass m, usually expressed as a multiple of the gravitational acceleration, is read from

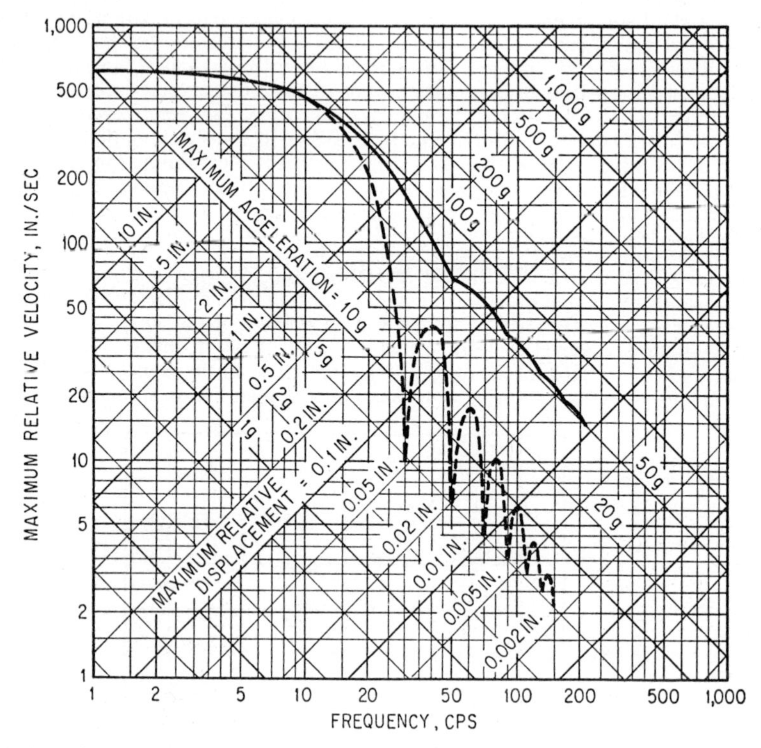

Fig. 42.10. Spectra of Fig. 42.9 redrawn to indicate maximum relative displacement and maximum relative velocity as well as maximum acceleration.

the ordinate scale. Inasmuch as the acceleration of the model under static conditions may be considered as $1g$, the ordinate indicates the factor by which the stress under static conditions must be multiplied to give the stress under conditions of shock. This maximum acceleration is approximately equal to the *equivalent static acceleration*,[4] a concept* that permits the application of design methods for static loads to the design of equipment required to withstand dynamic loading. The spectrum shown by solid lines in Fig. 42.9 represents acceleration in the same direction as the pulse; the spectrum shown by dotted lines represents accelerations in the opposite direction, usually caused by free vibration of the model after the excitation has disappeared.

In general, the spectrum has lower maximum values of acceleration as damping is added to the model. Therefore, the use of the spectrum for the undamped model is conservative in design work; it is not excessively conservative for spectra of the type

* See Chap. 23 for the distinction between maximum acceleration and equivalent static acceleration. For practical design purposes, the two may be considered equal.

shown in Fig. 42.9 where the excitation is a single pulse of acceleration. If the excitation is oscillatory, however, there is a tendency for a model whose natural frequency is equal to the frequency of oscillation to respond in a resonant manner; then the discrepancy between the spectra for damped and undamped models may be relatively great. If the spectra for the undamped model tend toward values that are great relative to spectra for damped models, a representative value of damping should be chosen as the basis of design. (See *Approximate Methods*, below, for examples of spectra for damped systems and spectra plotted to dimensionless coordinates.) Otherwise, the undamped model is preferable because the spectra are more readily computed.

The spectra of Fig. 42.9 may be redrawn, as shown in Fig. 42.10, to indicate maximum relative displacement and maximum relative velocity as well as maximum acceleration. Relative displacement is the deflection of spring k in Fig. 42.2. This information is of importance, for example, in determining whether the deflection of a shaft would be great enough to permit gears to become momentarily disengaged and then reengaged in a different position.

Approximate Methods. Ideally, the concept of designing to withstand shock contemplates the availability of a time-history of the shock motion, and the computation of spectra as shown in Fig. 42.10. Time-histories and/or facilities for computation sometimes are not available. It becomes instructive, then, to examine typical spectra and to compare them with the corresponding time-histories to determine trends that are useful for design purposes. It may be convenient to employ dimensionless or normalized parameters to evaluate important characteristics of the response; this leads directly to useful design methods.

Acceleration Pulse. An idealized version of a typical shock motion is the half-sinusoidal acceleration pulse defined by

$$\begin{aligned} \ddot{u} &= \ddot{u}_o \sin \omega_p t \qquad \left[0 \le t \le \frac{\pi}{\omega_p}\right] \\ \ddot{u} &= 0 \qquad \left[\frac{\pi}{\omega_p} \le t\right] \end{aligned} \tag{42.3}$$

where $\ddot{u}$ is the acceleration of the support for the model at any time t, $\ddot{u}_o$ is the peak acceleration, and ω_p is the angular frequency associated with the acceleration pulse. The time duration of the pulse is $\tau = \pi/\omega_p$. With the system initially at rest, the time-histories of acceleration, as given by Eq. (42.3), and velocity $\dot{u}$ and displacement u, as obtained from integration of Eq. (42.3), are shown in Fig. 42.11.

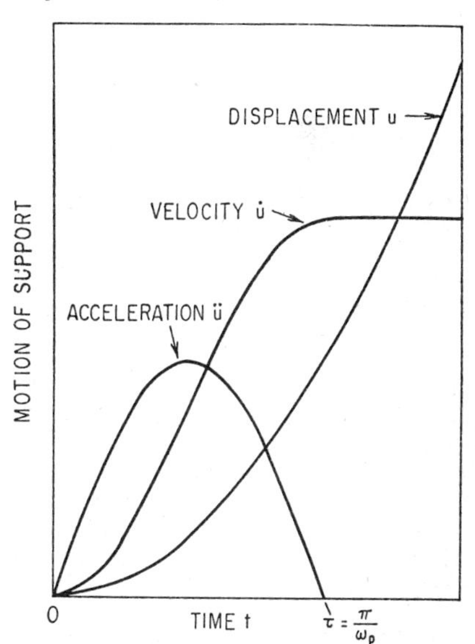

FIG. 42.11. Velocity and displacement as a function of time, corresponding to acceleration pulse defined by Eq. (42.3).

The transient response of the model of Fig. 42.2 to the pulse defined by Eq. (42.3) is illustrated in Fig. 42.12[5] where the acceleration $\ddot{x}$ of the mass m is plotted as a function of time from initiation of pulse $\ddot{u}$. Acceleration $\ddot{x}$ is plotted as a multiple of the peak pulse acceleration $\ddot{u}_o$, and time t is plotted as a multiple of the pulse duration τ. The half-sinusoidal acceleration pulse defined by Eq. (42.3) is the shaded area bounded by a dotted line.

The time-history of the response acceleration $\ddot{x}$ is plotted in Fig. 42.12 for three different values of the ratio of pulse duration to natural period of model: $\tau = 1\frac{0}{3}(2\pi/\omega_n)$; $(2\pi/\omega_n)$; and $\frac{1}{4}(2\pi/\omega_n)$, where ω_n is the natural frequency of the model [Eq. (42.1)]. The acceleration $\ddot{x}$ of the mass m can be equal to, greater than, or less than the peak acceleration of the support. In general, the acceleration of the support is not an indication of the forces induced in the model.

When the angular natural frequency of the model is appreciably larger than the frequency ω_p associated with the pulse, the response of the model approximates but does not exactly duplicate the acceleration pulse of the support. For example, when $\tau = {}^{10}\!/_{3}(2\pi/\omega_n)$ the response has superimposed small-amplitude vibration at the natural frequency ω_n of the model. The subsequent free vibration of the system represents the maximum *negative* response to the original positive (or upward) pulse. When $\tau = (2\pi/\omega_n)$ the maximum acceleration of the mass m is about 1.7 times the acceleration of

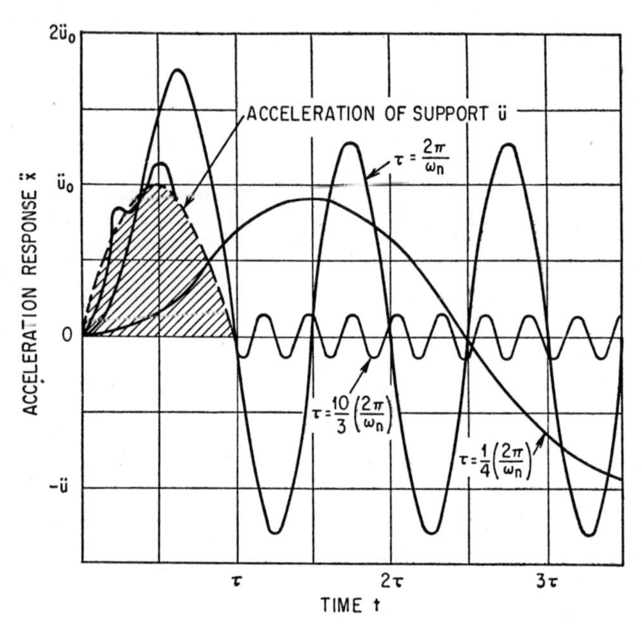

FIG. 42.12. Response of an undamped, single degree-of-freedom system to the half-sinusoidal pulse of acceleration shown shaded. Systems with different natural periods are included. (*J. M. Frankland.*[5])

the support. When $\tau = {}^{1}\!/_{4}(2\pi/\omega_n)$, the peak acceleration of the model occurs after the acceleration of the original pulse becomes zero and is about 0.9 times the peak acceleration of the pulse.

The shock response spectra for the half-sinusoidal acceleration pulse of Eq. (42.3) are shown in Fig. 42.13 for several values of the fraction of critical damping ζ [see Eq. (2.12)]. These spectra summarize the maximum values of the response for the time-histories of Fig. 42.12, for a wide range of frequency relations. The ordinate is the ratio of the peak acceleration $\ddot{x}_{\max}$ of the response to the peak acceleration $\ddot{u}_{\max}$ of the pulse where $\ddot{u}_{\max}$ is used in place of $\ddot{u}_0$ to suggest that this discussion is applicable to other than the sinusoidal pulse defined by Eq. (42.3). The numerical value of the ratio $\ddot{x}_{\max}/\ddot{u}_{\max}$ can be interpreted as the shock transmissibility since it represents both spring and damper forces. For very high values of damping, this system acts as a rigid connection so that the acceleration ratio quickly becomes unity below a frequency ratio of 1. For the curve with zero damping, the acceleration ratio can be interpreted as the shock amplification [dynamic load factor $= \omega_n^2(x - u)_{\max}/\ddot{u}_{\max}$], since it then represents only the spring force, or relative motion. When system damping is included, Fig. 8.43 shows the shock amplification factors which are somewhat different from shock transmissibility values. For very high values of damping (not shown in Fig. 8.43), the shock amplification factors can decrease below unity as the relative motion $x - u$ approaches zero.[3]

When the frequency ratio ω_n/ω_p is large, the acceleration ratio approaches unity; i.e., the maximum acceleration experienced by the model is approximately equal to that of the pulse. In other words, when a structure is stiff and the support motion is applied slowly, the maximum acceleration of the support can be used to determine the dynamic load imposed upon structures of the equipment. This relation is independent of the

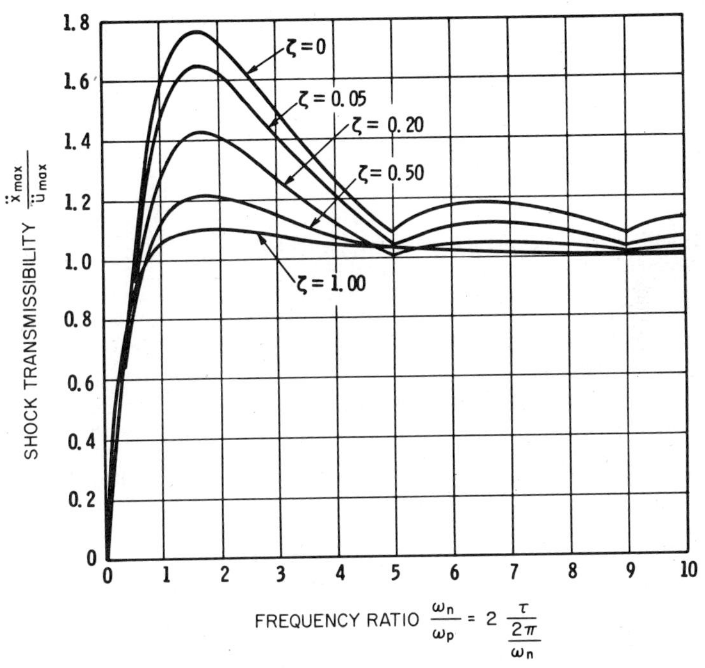

FIG. 42.13. Maximum dynamic response factor (shock transmissibility) for the half-sinusoidal pulse of acceleration. The fraction of critical damping in the system is indicated by ζ. (*After J. E. Ruzicka.*[3])

pulse shape; however, a pulse with superimposed high-frequency vibration cannot be interpreted in this manner because the superimposed vibration may induce a resonant response not contemplated by the preceding analysis. Furthermore, irregular pulses are sometimes made to appear smooth by limited high-frequency response of the instrumentation.

Starting Velocity. In Fig. 42.13 the equation of the tangent to the response spectrum for zero damping, at the origin, is

$$\frac{\ddot{x}_{\max}}{\ddot{u}_{\max}} = 2\left(\frac{\omega_n}{\omega_p}\right) \tag{42.4}$$

This straight line represents a good approximation to the spectrum for $\zeta = 0$ when $\omega_n/\omega_p \leq 0.50$. The velocity change $\dot{u}_o$ of the support is the integral of Eq. (42.3) with respect to time:

$$\dot{u}_o = \frac{2\ddot{u}_o}{\omega_p} \quad \text{in./sec} \tag{42.5}$$

Combining Eqs. (42.4) and (42.5) and letting $\ddot{u}_{\max} = \ddot{u}_o$,

$$\ddot{x}_{\max} = \omega_n \times \dot{u}_o \qquad \text{in./sec}^2 \qquad \left[\frac{\omega_n}{\omega_p} \leq 0.50\right] \tag{42.6}$$

where ω_n is in units of radians per second and $\dot{u}_o$ is in units of inches per second. Thus, the maximum acceleration $\ddot{x}_{\max}$ of the mass m in Fig. 42.2 is proportional to the magnitude of the velocity change $\dot{u}_o$ of the support and to the natural frequency ω_n of the model, provided the acceleration pulse duration τ is short compared to the natural period $2\pi/\omega_n$ of the model. This relation is convenient when the duration of the shock motion is small compared with the natural periods of primary structures; it finds particular application in simplified aspects of design using the single degree-of-freedom model, as well as in shock isolation (Chap. 31) and packaging (Chap. 41).

The maximum deflection $\delta_{\max}$ of a single degree-of-freedom model of natural frequency ω_n, when ω_n is small relative to $2\pi/\tau$, is

$$\delta_{\max} = \frac{\dot{u}_o}{\omega_n} \qquad \text{in.} \qquad \left[\frac{\omega_n}{\omega_p} \leq 0.50\right] \tag{42.7}$$

where $\dot{u}_o$ and ω_n have the units defined for Eq. (42.6).

In applying the relation of Eqs. (42.6) and (42.7), note that $\ddot{x}_{\max}$ and $\delta_{\max}$ are independent of the shape of the pulse and are determined solely by the area under the acceleration-time curve. If the excitation is irregular or vibratory, these relations are applicable if $\dot{u}_o$ is taken as the integral of the acceleration-time curve where $\omega_p = \pi/\tau$ and τ is the duration of the excitation.

UNCOUPLED TWO DEGREE-OF-FREEDOM MODEL. When the secondary structure of an equipment is small relative to the primary structure, the response of the model to a shock motion can be evaluated by first finding the response of the primary structure and using this response as the excitation for the secondary structure. For example, Fig. 42.12 gives several time-histories of the acceleration response of an undamped, single degree-of-freedom model to a half-period sinusoidal pulse of acceleration. The continuing periodic vibration of the primary structure appears to the secondary structure as steady-state vibration; thus, criteria established for design to withstand vibration constitute a good guide for design to withstand shock when the equipment can be represented by an uncoupled two degree-of-freedom model.

In accordance with the principle of design to withstand vibration, the secondary structure should have a natural frequency at least twice as great as that of the primary structure. The acceleration amplitude of the secondary structure then is 1.35 times as great as that of the primary structure (see *Design for Vibration Conditions—Uncoupled Two Degree-of-freedom Model*). The maximum acceleration of the primary structure is found by methods indicated under *Design for Shock Conditions—Single Degree-of-freedom Model.* The primary structure must be designed to withstand this acceleration; the secondary structure must be designed to withstand an acceleration 1.35 times as great.

If the natural frequencies of the primary and secondary structures are equal, the secondary structure may experience excessive vibration unless either the primary or secondary structure embodies significant damping. For example, Fig. 42.14 indicates the maximum acceleration of the secondary structure with reference to the maximum acceleration of the primary structure when the shock motion is a sudden velocity change $u = \dot{u}_0 t$ of the support for the primary structure.[6] When there is significant damping in the primary and/or secondary structure, the maximum acceleration of the secondary structure is relatively small.

COUPLED TWO DEGREE-OF-FREEDOM MODEL. When the mass of the secondary structure is of the same order of magnitude as that of the primary structure, it is necessary to consider the response of the composite model of primary and secondary structures. The response of this model to any excitation can be determined by known

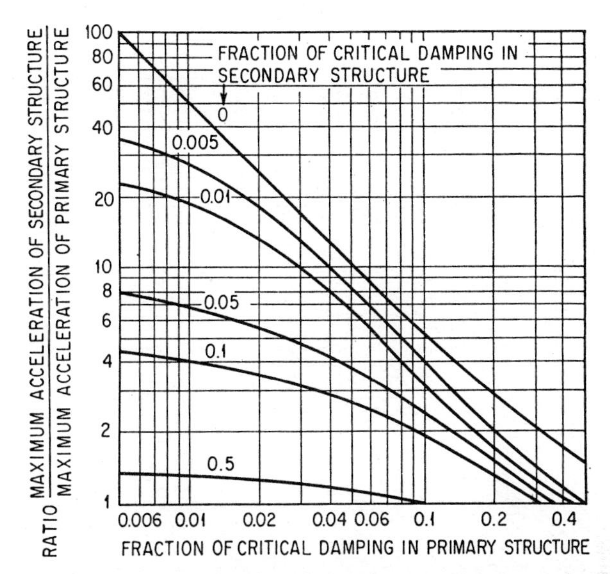

FIG. 42.14. Ratio of maximum acceleration of secondary structure to maximum acceleration of primary structure (see Fig. 42.6*A*) when support has motion $u = \dot{u}_o t$ and uncoupled natural frequencies of the structures are equal $\left(\text{i.e., } \omega_1 = \sqrt{\frac{k_1}{m_1}} = \omega_2 = \sqrt{\frac{k_2}{m_2}}\right)$. The mass m_2 of the secondary structure is negligible relative to the mass m_1 of the primary structure. (*C. E. Crede.*[6])

methods. For initial design work, it is convenient to generalize the excitation as a sudden motion change of the support, $u = \dot{u}_o t$. This is a good approximation when the velocity change takes place during a time interval that is small relative to the smallest natural period of the composite model.

The equations of motion for the system shown in Fig. 42.6*A*, when the excitation is a starting velocity $\dot{u}_o$ applied at the support, give results that can be expressed in terms of an equivalent starting velocity applied individually to each of the uncoupled spring-mass models in Fig. 42.6*B*; i.e., the maximum response of each model is equivalent to its maximum response as a part of the two degree-of-freedom model. The equivalent starting velocities are designated by $\dot{u}_2$ for the secondary structure and $\dot{u}_1$ for the primary structure. They are expressed in terms of the starting velocity $\dot{u}_o$ as:[7,8]

$$\frac{\dot{u}_2}{\dot{u}_o} = \frac{1}{\sqrt{\left(1 - \frac{\omega_2}{\omega_1}\right)^2 + \frac{m_2}{m_1}\left(\frac{\omega_2}{\omega_1}\right)^2}} \tag{42.8}$$

$$\frac{\dot{u}_1}{\dot{u}_o} = \frac{1 + \left(1 + \frac{m_2}{m_1}\right)\frac{\omega_2}{\omega_1}}{\sqrt{\left(1 + \frac{\omega_2}{\omega_1}\right)^2 + \frac{m_2}{m_1}\left(\frac{\omega_2}{\omega_1}\right)^2}} \tag{42.9}$$

where ω_1 and ω_2 are defined under *Two Degree-of-freedom Model.*

Figure 42.15 shows a contour chart of $\dot{u}_2/\dot{u}_o$ plotted against the stiffness ratio k_2/k_1, the mass ratio m_2/m_1, and the frequency ratio ω_2/ω_1. Figure 42.16 shows a contour chart of a factor D plotted against the same stiffness, mass, and frequency ratios. The

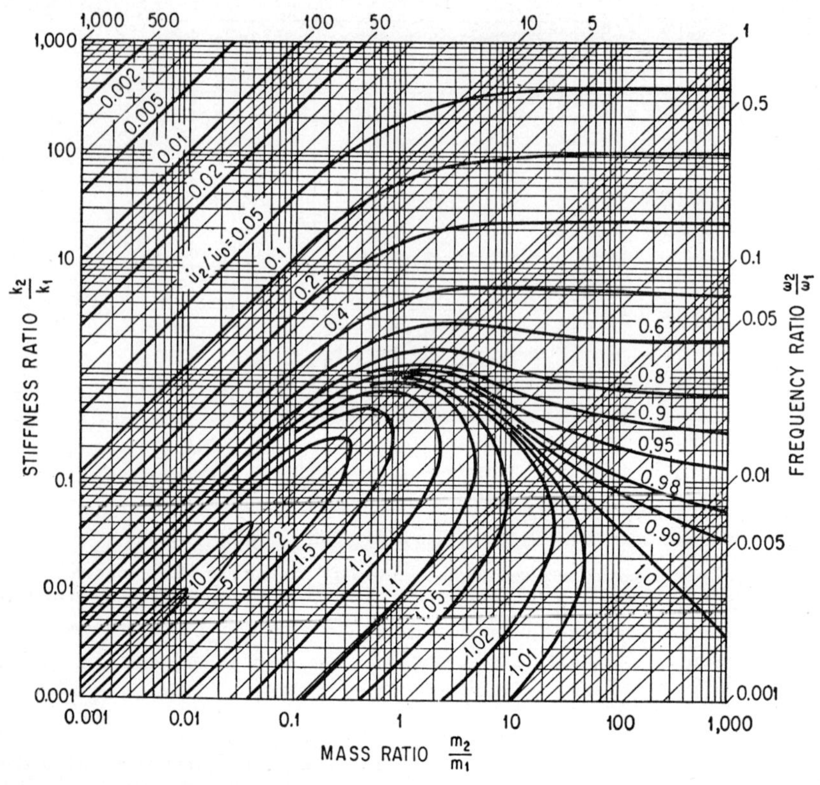

FIG. 42.15. Equivalent velocity change $\dot{u}_2$ for secondary structure relative to $\dot{u}_0$ where displacement of support for primary structure is $u = \dot{u}_0 t$. The natural frequencies are $\omega_1 = \sqrt{\frac{k_1}{m_1}}$, $\omega_2 = \sqrt{\frac{k_2}{m_2}}$ where m_1, m_2, k_1, k_2 are defined in Fig. 42.6A. (*R. B. McCalley, Jr.*[8])

factor D is related to the velocities $\dot{u}_1$ and $\dot{u}_o$ by

$$\frac{\dot{u}_1}{\dot{u}_o} = D\sqrt{1 + \frac{m_2}{m_1}} \tag{42.10}$$

The maximum relative motion $(x_1 - x_2)_{\max}$; i.e., the maximum deflection $(\delta_2)_{\max}$ of the spring k_2, is

$$(\delta_2)_{\max} = \frac{\dot{u}_2}{\omega_2} = \left(\frac{\dot{u}_o}{\omega_2}\right)\left(\frac{\dot{u}_2}{\dot{u}_o}\right) \quad \text{in.} \tag{42.11}$$

where $\dot{u}_2/\dot{u}_o$ is obtained from Fig. 42.15. The contour chart of Fig. 42.15 is particularly useful where the characteristics of the secondary structure are fixed and a primary structure must be selected to reduce the shock experienced by the secondary structure. The equivalent static acceleration of the secondary structure is

$$(\ddot{x}_2)_{\max} = \frac{k_2(\delta_2)_{\max}}{m_2} \quad \text{in./sec}^2 \tag{42.12}$$

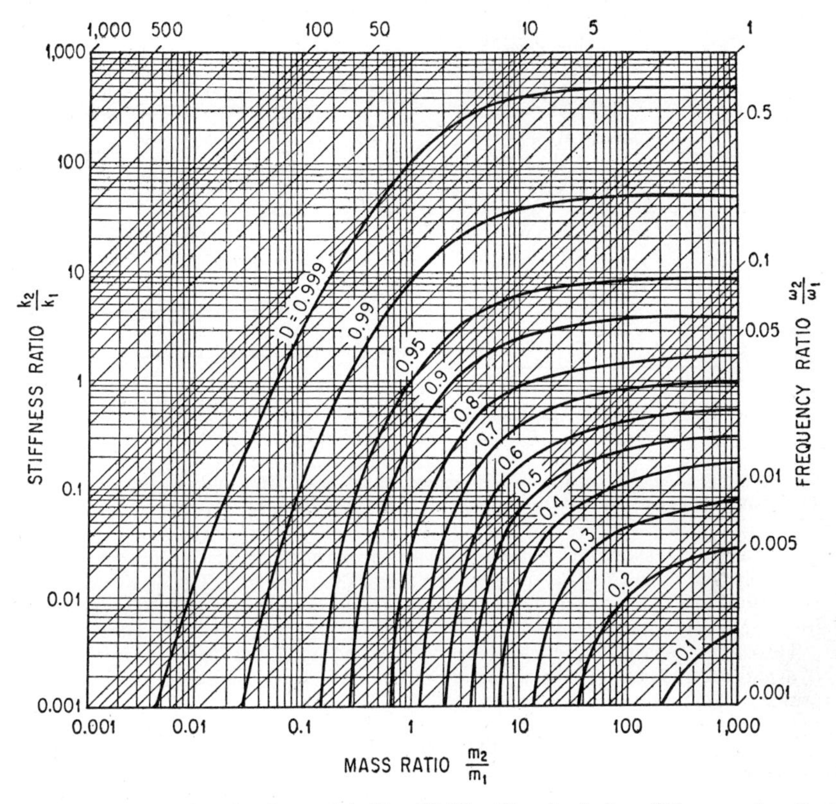

FIG. 42.16. Contour chart for factor D in Eq. (42.10). The physical conditions are described with reference to Fig. 42.15. (*R. B. McCalley, Jr.*[8])

The maximum relative motion $(u - x_1)_{max}$; i.e., the maximum deflection $(\delta_1)_{max}$ of the spring k_1, is

$$(\delta_1)_{max} = \frac{\dot{u}_1}{\omega_1} = \left(\frac{\dot{u}_o}{\omega_1}\right) D \sqrt{1 + \frac{m_2}{m_1}} \quad \text{in.} \tag{42.13}$$

where D is obtained from Fig. 42.16. The equivalent static acceleration of the primary structure is

$$(\ddot{x}_1)_{max} = \frac{k_1(\delta_1)_{max}}{m_1} \quad \text{in./sec}^2 \tag{42.14}$$

The response data given by the contour plots of Figs. 42.15 and 42.16, referring to the coupled two degree-of-freedom model of Fig. 42.6*A*, are shown in Fig. 42.17 as spectra of the frequency ratio ω_2/ω_1. Maximum deflection of the springs k_2, k_1 and the corresponding equivalent static acceleration for the respective structures are directly proportional, respectively, to the equivalent starting velocities $\dot{u}_2$, $\dot{u}_1$. These relations are defined by Eqs. (42.11) and (42.12); values for $\dot{u}_2$, $\dot{u}_1$ are given by the ordinates of Fig. 42.17.

Figure 42.17 illustrates the importance of maintaining appropriate relations between the natural frequencies ω_2 and ω_1:

1. If $\omega_2 = \omega_1$, the equivalent starting velocity of the secondary structure becomes great if its mass m_2 is small relative to the mass m_1 of the primary structure. This approaches the uncoupled model discussed in an earlier section.

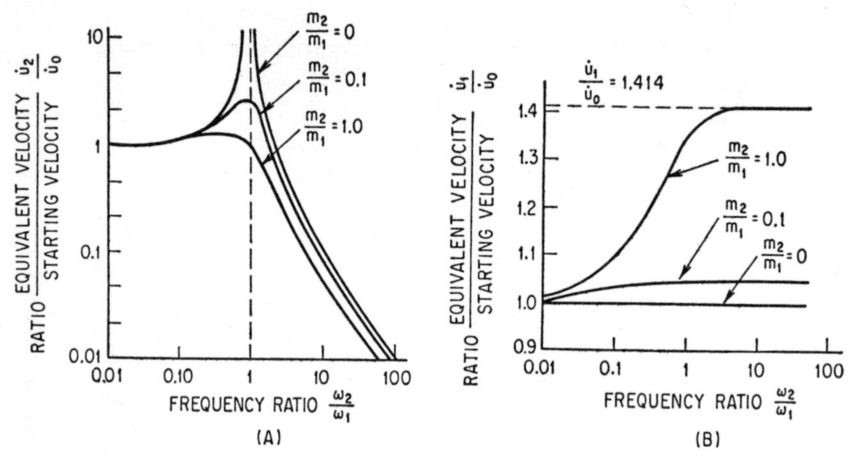

FIG. 42.17. Equivalent starting velocity ratios for (*A*) secondary and (*B*) primary structures. The motion of the support is $u = \dot{u}_0 t$. (*C. M. Friedrich.*[7])

2. If $\omega_2 > 2\omega_1$, the equivalent starting velocity of the secondary structure tends toward relatively low values for any ratio of the masses m_2 and m_1. The equivalent starting velocity of the primary structure tends toward larger though moderate values as the frequency ratio ω_2/ω_1 increases.

Thus, a design criterion for coupled as well as uncoupled two degree-of-freedom models is a natural frequency of the secondary structure at least twice as great as that of the primary structure.

EXAMPLES OF TYPICAL REQUIREMENTS. Shock testing machines play an important role in the design of equipment to withstand shock. Their importance is derived, in part, from the difficulty of applying the foregoing design principles to all details of a complex equipment because the equipment cannot be represented by a simple model. A shock test confirms calculations that were made and takes the place of calculations that could not be made. Furthermore, even though analytical techniques can be applied, the conditions of shock often are not defined well enough to serve as a basis for design. An equipment then is designed to withstand the shock created by a shock testing machine. In general, the shock created by the testing machine does not duplicate that experienced in service; however, it does represent an explicit requirement for design purposes and presumably takes into account the variances from one shock motion to another in actual service.

Shock testing machines are discussed, and several common machines are illustrated, in Chap. 26. The specification of laboratory tests to simulate actual service conditions is considered in Chap. 24. Most shock testing machines are adapted to several alternate modes of operation, at least to the extent of modifying the severity of the shock. Compilations of data are available to define the characteristics of the shock motion of commonly used shock testing machines. As examples of design requirements, typical shock spectra are shown in Fig. 42.18 for several machines. These spectra are useful directly with single degree-of-freedom models as a design procedure for equipment required to withstand the shock produced by the respective machines.

EFFECT OF EQUIPMENT WEIGHT. The weight of the equipment sometimes is a factor of importance when selecting strength requirements for the equipment. Heavy equipment often receives shock of less severity than light equipment because (1) sources of shock may involve only finite energy and have less effect upon the heavy equipment or (2) the structures that transmit the forces have finite strength and thereby limit the amount of energy that reaches the equipment. Many types of shock testing machines

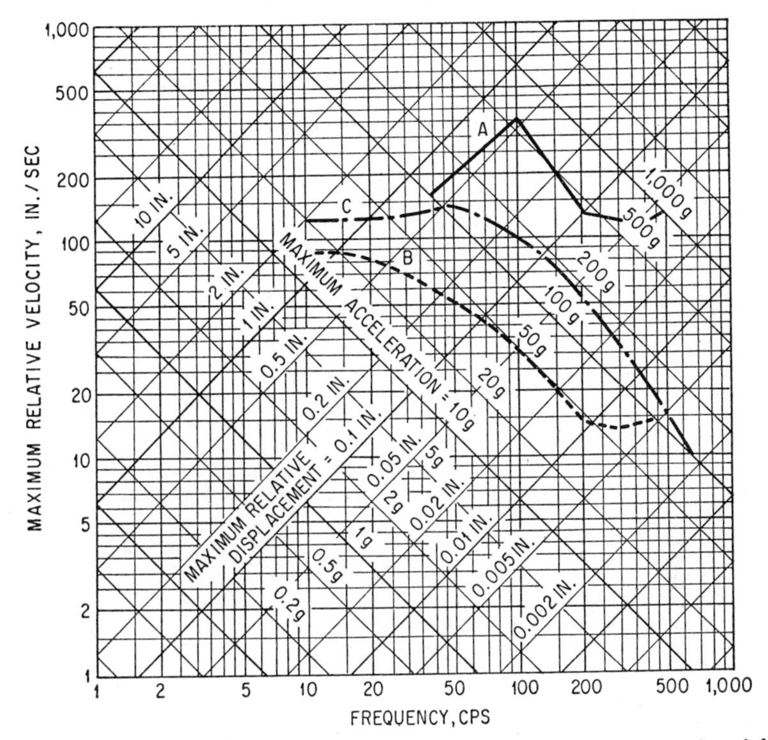

Fig. 42.18. Typical shock spectra for several shock testing machines as examples of design requirements: (*A*) MIL-S-901B, (*B*) MIL-S-4456 using sand, and (*C*) MIL-S-4456 using lead. The designation "MIL" denotes a military specification number.

recognize these limitations and deliver a shock motion whose severity tends to decrease with an increase in the weight of the equipment under test. For example, machines with swinging hammers tend to deliver a shock whose severity decreases as the equipment weight increases, even though the height of the hammer drop may be increased somewhat as the equipment weight increases.

Analytical methods for design of shock-resistant equipment employ one or two degree-of-freedom models that are tacitly assumed to be of negligible mass; i.e., the equipment offers no impedance to the shock motion. The effect of equipment weight in reducing the severity of shock is a factor of considerable significance; sometimes it is recognized empirically where design criteria are set forth. For example, the Bureau of Ships of the U.S. Navy specifies the strength requirements for bolts used to mount equipment on naval ships. The specification requires that the mounting bolts for light equipment be capable of withstanding a greater equivalent static acceleration than the bolts for mounting heavy equipment, as illustrated in Fig. 42.19. Strictly speaking, Fig. 42.19 applies only to mounting bolts and associated

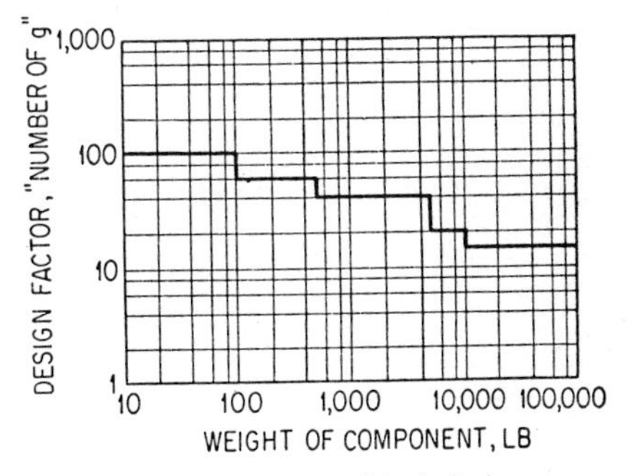

Fig. 42.19. Typical empirical design curve indicating required strength of mounting bolts as a function of equipment weight. This applies to some classes of naval service.

mounting brackets; generally, the trend may be recognized and applied to the over-all design where there is reason to believe that the shock severity would be decreased by the weight of the equipment.

STRUCTURAL DESIGN CONSIDERATIONS

ENERGY STORAGE CAPACITY

The deflection of structural members that characterizes the response of equipment to shock and vibration involves the temporary storage of strain energy in such members. In a given member, strain energy increases as stress increases. The severity of shock that an equipment may withstand without damage is reached when the stress in the most highly stressed member reaches the maximum allowable value. Any material in the structure that does not at the same time reach its maximum allowable stress may be considered as contributing to a loss in efficiency of design. This material not only tends to increase the cost and weight of the equipment but, more important, may increase the weight carried by other structures with a consequent increase in strength requirements.

Efficiency is determined by the type of loading, the constraints on the structure, and the shape of the cross section. For example, the potential energy stored in a uniform bar in pure tension is the product of the average force $F/2$ and the deflection δ:

$$V = \frac{F\delta}{2} = \frac{(\sigma S)(l\epsilon)}{2} \quad \text{in.-lb} \tag{42.15}$$

where σ = tensile stress, lb/in.²
S = cross-sectional area of the bar, in.²
l = length of the bar, in.
ϵ = strain, in./in.

The substitutions made in Eq. (42.15) are $F = \sigma S$ lb and $\delta = l\epsilon$ in. By making the additional substitutions $V = Sl$ in.³ and $\epsilon = \sigma_m/E$ in./in., Eq. (42.15) can be expressed as

$$\frac{F\delta}{2} = \eta \frac{\sigma_m{}^2 V}{2E} \quad \text{in.-lb} \tag{42.16}$$

where V = total volume of material being stressed, in.³
E = Young's modulus, lb/in.²
σ_m = maximum stress (uniform for bar in tension), lb/in.²
η = efficiency (0 to 1)

The efficiency η is the fraction of the total strain energy, based on a uniform stress σ_m in all elements, that is stored by the structure. Solving Eq. (42.16) for η,

$$\eta = \frac{F\delta E}{\sigma_m{}^2 V} \tag{42.17}$$

When each element in the structure has the same stress, the material is being used at maximum efficiency and $\eta = 1$.

For a beam in bending, the efficiency usually is a small fraction. For a cantilever beam with a concentrated load F at the free end, the deflection δ at the free end[1] is $Fl^3/3EI$ and the stress σ_m in the outermost fiber at the built-in end of the beam is Flc/I. The moment of inertia (of the cross-sectional area S) is $I = S\rho^2$ in.⁴, where ρ is the radius of gyration and c is the distance of the outer fiber from the neutral axis. Substituting in Eq. (42.17),

$$\eta = \frac{\rho^2}{3c^2} \tag{42.18}$$

Values of ρ and c are known for various cross sections.[9] For a cantilever beam with end load, the corresponding values of efficiency are $\eta = 1/12$ for a circular section, $1/9$ for a rectangular section, $1/6$ for a thin-walled tube, and about $1/3$ for an I beam. These increasing values reflect the fact that each successive shape of cross section further concentrates the material in the outermost fibers where the bending stress reaches a maximum.

The efficiency η depends not only upon the distribution of stress over the cross section but also upon the distribution along the length of the beam. The relatively low efficiency of only $\frac{1}{3}$ for the cantilever I beam occurs because the bending moment in a cantilever beam varies from a maximum at the built-in end to zero at the free end. By contrast, when a pure moment instead of a concentrated load is applied at the free end of the cantilever beam, the bending moment is uniform along the length of the beam and the efficiency increases by a factor of 3 to become $\eta = \rho^2/c^2$. This is approximately unity for the I beam.

The foregoing examples are included in Table 42.1 along with the efficiencies of other common structural members for various cross sections.[7] Column 1 gives the general

Table 42.1. Efficiencies of Common Structural Members as Defined by Eq. (42.17)

(*C. M. Friedrich.*[7])

Type of structure	Type of cross section		
	(1) Uniform	(2) Rectangular	(3) Circular
Bar in tension	1	1	1
Cantilever beam with concentrated load at end	$\rho^2/3c^2$	$\frac{1}{9}$	$\frac{1}{12}$
Simply supported beam with concentrated load at any point	$\rho^2/3c^2$	$\frac{1}{9}$	$\frac{1}{12}$
Built-in beam with concentrated load at center	$\rho^2/3c^2$	$\frac{1}{9}$	$\frac{1}{12}$
Simply supported beam with moment at center	$\rho^2/3c^2$	$\frac{1}{9}$	$\frac{1}{12}$
Built-in beam with moment at center	$\rho^2/2c^2$	$\frac{1}{6}$	$\frac{1}{8}$
Cantilever beam with moment at end	ρ^2/c^2	$\frac{1}{3}$	$\frac{1}{4}$
Simply supported beam with uniform load	$8\rho^2/15c^2$	$\frac{8}{45}$	$\frac{2}{15}$
Beam with one built-in end, one simply-supported end, load at center	$7\rho^2/27c^2$	$\frac{7}{81}$	$\frac{7}{108}$

ρ = radius of gyration of the beam cross section.
c = distance of outermost fibers from the neutral axis of the beam.

expression for efficiency applicable to any uniform cross section, for the particular type of loading and constraints on the structure. Columns 2 and 3 give the efficiencies for rectangular and circular sections, respectively, and are obtained by substituting appropriate values for ρ and c in the expressions of column 1.

ELASTIC CHARACTERISTICS OF BEAMS AND PLATES

Analytical methods of design to withstand shock and vibration make extensive use of models to determine the response of an equipment or structure to the shock and vibration. Generally, the model is an elastic system of concentrated masses, massless springs, and massless dampers whose natural frequencies can be related to the natural frequencies of the actual structure. The forces and deflections associated with the response of the model then are transformed to stresses and strains in the actual structure. The section of this chapter on *Models* shows the relation between a model and a simple beam with center load. This section extends the treatment to other types of beams, and includes analyses directed toward the ready calculation of bending moments and reaction forces.

UNIFORM BEAM WITH DISTRIBUTED LOAD. The spring stiffness for lateral deflection of a uniform beam can be generalized as[10]

$$k_b = n_b\left(\frac{EI}{l^3}\right) = n_b\left(\frac{ES\rho^2}{l^3}\right) \qquad \text{lb/in.} \tag{42.19}$$

where E = Young's modulus, lb/in.²
I = moment of inertia of cross section, in.⁴
S = cross-sectional area of beam, in.²
ρ = radius of gyration of cross section, in.
l = length of beam, in.

The numerical coefficient n_b depends upon the type of loading and the end constraints on the beam. For the beam shown in Fig. 42.1, the coefficient is $n_b = 48$.

The natural frequency of a uniform beam is[10]

$$\omega_n = \sqrt{\frac{EI}{\nu l^4}} = N_b\left(\frac{\rho}{l^2}\right)\sqrt{\frac{E}{\mu}} \qquad \text{rad/sec} \tag{42.20}$$

where ν = mass per unit length of beam, lb-sec²/in.²
μ = mass density of the material, lb-sec²/in.⁴

Hence, $\nu = \mu S$. The numerical coefficient N_b depends upon the end constraints of the beam. For the beam shown in Fig. 42.1 (without the rigid body W), $N_b = \pi^2$ for the lowest natural frequency. Values of N_b for higher natural frequencies, and for other constraints and types of loading, are given in Tables 7.3 and 7.5.

By considering a beam to be analogous to a single degree-of-freedom system having the stiffness k_b given by Eq. (42.19) and an effective mass m_e, Eq. (42.1) may be written as follows for the natural frequency of the beam:

$$\omega_n = \sqrt{\frac{k_b}{m_e}} \qquad \text{rad/sec} \tag{42.21}$$

Solving for m_e and substituting k_b and ω_n from Eqs. (42.19) and (42.20), respectively,

$$m_e = \left(\frac{n_b}{N_b{}^2}\right) m_b \qquad \text{lb-sec}^2\text{/in.} \tag{42.22}$$

where $m_b = \mu S l$ is the total mass of the beam.

UNIFORM BEAM WITH CONCENTRATED LOAD. The effects upon the natural frequency of the mass of both the rigid body W and the beam shown in Fig. 42.1 may be included by writing Eq. (42.21) as follows:

$$\omega_n{}' = \sqrt{\frac{k_b}{m_l + m_e}} \qquad \text{rad/sec} \tag{42.23}$$

where m_l is the mass of the load (rigid body W), m_e is given by Eq. (42.22), and k_b is given by Eq. (42.19). By using Eqs. (42.19) and (42.22), the natural frequency $\omega_n{}'$ can be expressed in terms of the dimensions and properties of the actual beam:

$$\omega_n{}' = \frac{\dfrac{\rho}{l^2}\sqrt{\dfrac{E}{\mu}}}{\xi_\delta} \qquad \text{rad/sec} \tag{42.24}$$

where

$$\xi_\delta = \sqrt{\frac{1}{N_b{}^2} + \frac{1}{n_b}\left(\frac{m_l}{m_b}\right)}$$

For the beam and load shown in Fig. 42.1, $\xi_\delta = \sqrt{0.0103 + 0.0208\, m_l/m_b}$.

When the duration τ of a shock motion is small relative to the natural period $2\pi/\omega_n'$ of the beam, the shock motion can be assumed to consist of the starting velocity $\dot{u}_0$ applied at the beam supports in a direction transverse to the beam. Then, Eq. (42.7) leads to the following expression for the maximum beam deflection δ_m at the concentrated load m_l:

$$\delta_m = \frac{\dot{u}_o}{\omega_n'} = \xi_\delta \dot{u}_o \left(\frac{l^2}{\rho}\right) \sqrt{\frac{\mu}{E}} \quad \text{in.} \tag{42.25}$$

The maximum bending moment M_m and the maximum force F_m at the end supports are determined from the maximum deflection:[1]

$$M_m = \xi_M \dot{u}_o S \rho \sqrt{E\mu} \quad \text{lb-in.} \tag{42.26}$$

$$F_m = \xi_F \dot{u}_o \left(\frac{S\rho}{l}\right) \sqrt{E\mu} \quad \text{lb} \tag{42.27}$$

Table 42.2 gives values of the dimensionless factors ξ_δ, ξ_M, and ξ_F to be used in Eqs. (42.25) to (42.27) for calculating the maximum deflection δ_m, the maximum bending

Table 42.2. Dimensionless Factors * in Eqs. (42.25) to (42.27) for Calculating the Response of Beams to a Velocity Shock $u = \dot{u}_o t$

(C. M. Friedrich.[7])

Type of structure	Maximum deflection ξ_δ Eq. (42.25)	Bending moment ξ_M Eq. (42.26)	End reaction ξ_F Eq. (42.27)
Simply supported beam with mass at center (deflection and moment at center, reaction at simple support)	$\sqrt{0.0103 + 0.0208\, m_l/m_b}$	$12\xi_\delta$	$24\xi_\delta$
Built-in beam with mass at center (deflection at center, moment and reaction at built-in end)	$\sqrt{0.00199 + 0.00521\, m_l/m_b}$	$24\xi_\delta$	$96\xi_\delta$
Cantilever beam with mass at free end (deflection at free end, moment and reaction at built-in end)	$\sqrt{0.0806 + 0.333\, m_l/m_b}$	$3\xi_\delta$	$3\xi_\delta$
Beam with one built-in end, one simply supported end, with mass at center (deflection near center, moment at built-in end), reaction at: Built-in end	$\sqrt{0.00422 + 0.00932\, m_l/m_b}$	$20.1\xi_\delta$	$73.6\xi_\delta$
Simple support	$\sqrt{0.00422 + 0.00932\, m_l/m_b}$	$20.1\xi_\delta$	$33.5\xi_\delta$

* The dimensionless factors depend upon the type of loading and the end supports for the beam as well as the following:

m_l = mass of load lb-sec²/in.

m_b = total mass of beam lb-sec²/in.

moment M_m, and the maximum end support force F_m, respectively, for several types of loading and end support.[7] The natural frequency ω_n' also can be found from Eq. (42.25):

$$\omega_n' = \frac{\dot{u}_o}{\delta_m} = \frac{\rho}{\xi_\delta l^2} \sqrt{\frac{E}{\mu}} \quad \text{rad/sec} \tag{42.28}$$

CIRCULAR PLATE WITH CONCENTRATED CENTER LOAD. The elastic behavior of plates is analogous to that of beams. For example, the stiffness of a uniform circular plate with a built-in circumferential edge and supporting a load at the center is[10]

$$k_p = n_p \left(\frac{Et^3}{R^2}\right) \quad \text{lb/in.} \tag{42.29}$$

where t is the thickness and R the radius of the plate, in inches. This expression is similar to Eq. (42.19). The constant n_p has a value of 4.60 when the plate is made of steel having a Poisson's ratio $\gamma = 0.3$. The expression for the natural frequency of the plate in its fundamental mode is[10]

$$\omega_n = N_p \left(\frac{t}{R^2}\right) \sqrt{\frac{E}{\mu}} \quad \text{rad/sec} \tag{42.30}$$

This expression is similar to Eq. (42.20). The constant N_p has a value of 3.09 when the plate is made of steel having a Poisson's ratio $\gamma = 0.3$. Finally, the expression for the natural frequency ω_n', which accounts for the equivalent mass m_e of the vibrating plate as well as the concentrated load m_l, is

$$\omega_n' = \frac{\dfrac{t}{R^2}\sqrt{\dfrac{E}{\mu}}}{\xi_\delta'} \quad \text{rad/sec} \tag{42.31}$$

where

$$\xi_\delta' = \sqrt{\frac{1}{N_p^2} + \left(\frac{\pi}{n_p}\right)\frac{m_l}{\mu t(\pi R^2)}}$$

Substituting the numerical values of N_p and n_p for a circular plate built in at the edge and carrying a concentrated load m_l at the center, $\xi_\delta' = \sqrt{0.1047 + 0.683\, m_l/(\mu t \pi R^2)}$. For plates with other edge conditions and other types of loading, procedures have been worked out for computing the natural frequencies.[11]

MECHANICAL PROPERTIES OF ENGINEERING MATERIALS

The mechanical properties of a material are determined by standardized measurements of the reactions of the material to specified physical conditions. Under the dynamic conditions of shock and vibration, materials are likely to react differently than under conventional static testing. Differences in behavior are ascribable to the rapid application and short duration of the loading, and to repeated loading.

The important static properties are ultimate tensile strength, yield strength, modulus of elasticity, elongation at failure, and reduction of area. Hence, a standard tensile test can be used for comparison of materials subject to static loading. This is acceptable if each material remains within its elastic limit and dynamic effects do not become important. However, under dynamic loading the yield strength and ultimate strength depend upon the strain rate which in turn depends upon the geometry of the structure and the type of loading. As a consequence, dynamic properties are not standardized easily. They usually are expressed as ultimate strength and yield strength variation with strain rate, impact strength variation with temperature, fatigue behavior, and energy dissipation by internal damping. A limited amount of information in these categories is available.[12]

Undoubtedly there are additional dynamic properties of importance that remain to be formulated and evaluated. In particular, more work must be done in determining material strength to resist rapid repeated loading at a stress slightly less than the yield strength. For the case of transient loading, there is doubt concerning the amount that the yield strength of a material increases under dynamic loading. Usually, it is advisable in the design of equipment to withstand shock and vibration to assume no increase in yield strength for any material.

STATIC PROPERTIES

A standard tensile test supplies a stress-strain curve which can be used to define many of the mechanical properties of a material. Hooke's law applies at stresses below the proportional limit. The extension e is directly proportional to the load F and to the length of the test bar l, and inversely proportional to the cross-sectional area S. A proportionality constant, the initial slope of the stress-strain curve, is called the modulus of elasticity or Young's modulus E:

$$e = \frac{Fl}{SE} \quad \text{in.} \tag{42.32}$$

For example, with a low- or medium-carbon steel, the extension e of a test bar with gage length $l = 2$ in. is about 0.002 in. at the proportional limit. With a Young's modulus for steel of $E = 30 \times 10^6$ lb/in.^{2}, Eq. (42.32) gives a tensile stress (force per unit area) of $F/S = 30{,}000$ lb/in.^{2}

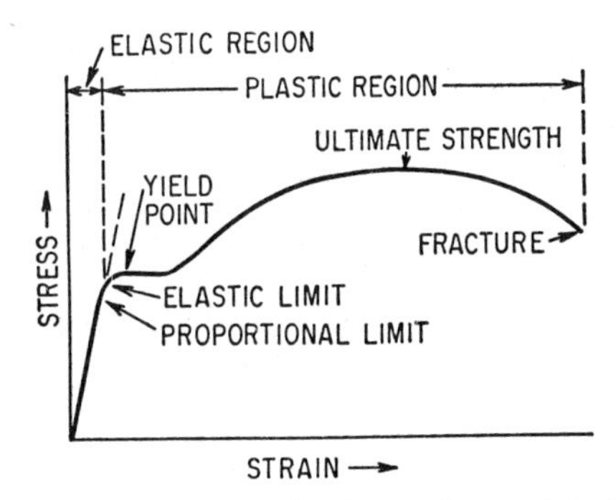

Fig. 42.20. Conventional stress-strain diagram for a metal with a yield point.

Most metals can be stressed slightly higher than the proportional limit without showing permanent deformation upon removal of the load. The maximum stress that can be attained without permanent deformation is called the elastic limit. Mild steel and a few rather uncommon metals exhibit a distinct yield point. This is defined as the stress at which permanent deformation begins suddenly and continues with no increase in stress. However, most materials continue to yield gradually after passing the elastic limit; there is no characteristic dip at strains somewhat greater than the yield point, as in the stress-strain curve of Fig. 42.20.

Various definitions of the yield condition have been proposed,[13] but they are not pertinent to the subsequent discussion of dynamic properties. In the United States of America the term *yield strength* has been adopted as standard.[14] The yield strength of a material is that stress which produces, on unloading, a permanent strain of 0.002 in./in. (0.2 per cent).

When the loading on a ductile test bar is increased beyond the yield strength, large permanent deformations occur and lead to actual fracture of the material. At the same time there is an appreciable reduction in cross-sectional area of the test bar at the section where fracture occurs. The shape of the stress-strain curve shown in Fig. 42.20 is the result of calculating an arbitrary nominal stress based on the original cross-sectional area. The ultimate strength is the nominal stress found by dividing the maximum load achieved in a tensile test by the original area of the test bar cross section.

If the tensile force on a test bar of ductile material (e.g., low-carbon steel) is continuously increased, the area of the bar at the cross section of eventual failure begins to

Table 42.3. Mechanical Properties of Typical Cast Irons

(A. Vallance and V. Doughtie.[16]*)*

Material	Ultimate strength		Endurance limit in reversed bending, lb/in.^{2}, σ_e	Brinell hardness number	Modulus of elasticity		Elongation in 2 in., %
	Tension, lb/in.^{2}, σ_u †	Compression, lb/in.^{2}, σ_u			Tension and compression, lb/in.^{2}, E	Shear lb/in.^{2}, G	
Gray, ordinary	18,000	80,000	9,000	100–150	10–12,000,000	4,000,000	0–1
Gray, good *	24,000 16,000	100,000	12,000	100–150	12,000,000	4,800,000	0–1
Gray, high grade	30,000	120,000	15,000	100–150	14,000,000	5,600,000	0–1
Malleable, S.A.E. 32510	50,000	120,000	25,000	100–145	23,000,000	9,200,000	10
Nickel alloys:							
Ni-0.75, C-3.40, Si-1.75, Mn-0.55 *	32,000 24,000	120,000	16,000	200 175	15,000,000	6,000,000	1–2
Ni-2.00, C-3.00, Si-1.10, Mn-0.80 *	40,000 31,000	155,000	20,000	220 200	20,000,000	8,000,000	1–2
Nickel-chromium alloys:							
Ni-0.75, Cr-0.30, C-3.40, Si-1.90, Mn-0.65	32,000	125,000	16,000	200	15,000,000	6,000,000	1–2
Ni-2.75, Cr-0.80, C-3.00, Si-1.25, Mn-0.60	45,000	160,000	22,000	300	20,000,000	8,000,000	1–2

* Upper figures refer to arbitration test bars. Lower figures refer to the center of 4-in. round specimens.
† *Flexure:* For cast irons in bending, the modulus of rupture may be taken as 1.75 σ_u (tension) for circular sections, 1.50 σ_u for rectangular sections and 1.25 σ_u for I and T sections.

Table 42.4. Mechanical Properties of Typical Carbon Steels

(A. Vallance and V. Doughtie.[16]*)*

Material	Ultimate strength		Yield strength		Endurance limit in reversed bending lb/in.^{2}, σ_e	Brinell hardness number	Modulus of elasticity		Elongation in 2 in., %
	Tension, lb/in.^{2}, σ_u	Shear, lb/in.^{2}, σ_u	Tension and compression, lb/in.^{2}, σ_y	Shear, lb/in.^{2}, σ_y			Tension and compression, lb/in.^{2}, E	Shear, lb/in.^{2}, G	
Wrought iron	48,000	50,000	27,000	30,000	25,000	100	28,000,000	11,200,000	30–40
Cast steel:									
Soft	60,000	42,000	27,000	16,000	26,000	110	30,000,000	12,000,000	22
Medium	70,000	49,000	31,500	19,000	30,000	120	30,000,000	12,000,000	18
Hard	80,000	56,000	36,000	21,000	34,000	130	30,000,000	12,000,000	15
SAE 1025:									
Annealed	67,000	41,000	34,000	20,000	29,000	120	30,000,000	12,000,000	26
Water-quenched *	78,000 90,000	55,000 63,000	41,000 58,000	24,000 34,000	43,000 50,000	159 183	30,000,000	12,000,000	35 27
SAE 1045:									
Annealed	85,000	60,000	45,000	26,000	42,000	140	30,000,000	12,000,000	20
Water-quenched *	95,000 120,000	67,000 84,000	60,000 90,000	35,000 52,000	53,000 67,000	197 248	30,000,000	12,000,000	28 15
Oil-quenched *	96,000 115,000	67,000 80,000	62,000 80,000	35,000 45,000	53,000 65,000	192 235	30,000,000	12,000,000	22 16
SAE 1095:									
Annealed	110,000	75,000	55,000	33,000	52,000	200	30,000,000	12,000,000	20
Oil-quenched *	130,000 188,000	85,000 120,000	66,000 130,000	39,000 75,000	68,000 100,000	300 380	30,000,000	11,500,000	16 10

* Upper figures: steel quenched and drawn to 1300°F. Lower figures: steel quenched and drawn to 800°F. Values for intermediate drawing temperatures may be approximated by direct interpolation.

decrease (neck down) as the maximum force is approached; then further extension takes place with decreasing force. This ability of the material to flow without immediate rupture is called ductility; it is defined as the per cent reduction of cross-sectional area measured at the section of fracture. Another measure of ductility is the per cent elongation of a 2-in. gage length; this depends upon the shape and size of the specimen (cross section and gage length).

Table 42.5. Mechanical Properties of Copper-Zinc Alloys (Brass)

(*A. Vallance and V. Doughtie.*[16])

Type of material	Ultimate strength	Yield strength	Endurance limit, lb/in.^{2}, σ_e	Brinell hard-ness number	Modulus of elasticity	Elon-gation in 2 in., %
	Tension, lb/in.^{2}, σ_u	Tension, lb/in.^{2}, σ_y			Tension and compression lb/in.^{2}, E	
Commercial bronze (90 Cu, 10 Zn):						
Rolled, hard	65,000	63,000	18,000	107	15,000,000	18
Rolled, soft	35,000	11,000	12,000	52	15,000,000	56
Forged, cold	40,000–65,000	25,000–61,000	12,000–16,000	62–102	15,000,000	55–20
Red brass (85 Cu, 15 Zn):						
Rolled, hard	75,000	72,000	20,000	126	15,000,000	18
Rolled, soft	37,000	14,000	14,000	54	15,000,000	55
Forged, cold	42,000–62,000	22,000–54,000	14,000–18,000	63–120	15,000,000	47–20
Low brass (80 Cu, 20 Zn):						
Rolled, hard	75,000	59,000	22,000	130	15,000,000	18
Rolled, soft	44,000	12,000	15,000	56	15,000,000	65
Forged, cold	47,000–80,000	20,000–65,000		63–133	15,000,000	30–15
Spring brass (75 Cu, 25 Zn):						
Hard	84,000	64,000	21,000	107 *	14,000,000	5
Soft	45,000	17,000	17,000	57 *	18,000,000	58
Cartridge brass (70 Cu, 30 Zn):						
Rolled, hard	100,000	75,000	22,000	154	15,000,000	14
Rolled, soft	48,000	30,000	17,000	70	15,000,000	55
Deep-drawing brass (68 Cu, 32 Zn):						
Strip, hard	85,000	79,000	21,000	106 *	15,000,000	3
Strip, soft	45,000	11,000	17,000	13 *	15,000,000	55
Muntz metal (60 Cu, 40 Zn):						
Rolled, hard	80,000	66,000	25,000	151	15,000,000	20
Rolled, soft	52,000	22,000	21,000	82	15,000,000	48
Tobin bronze (60 Cu, 39.25 Zn, 0.75 Sn):						
Hard	63,000	35,000	21,000	165	15,000,000	35
Soft	56,000	22,000		90	15,000,000	45
Manganese bronze (58 Cu, 40 Zn):						
Hard	75,000	45,000	20,000	110	15,000,000	20
Soft	60,000	30,000	16,000	90	15,000,000	30

* Rockwell hardness F.

Other static properties of materials find application in the design of equipment to withstand shock and vibration. The modulus of rigidity G is the ratio of shear stress to shear strain; it may be determined from the torsional stiffness of a thin-walled tube of the material. The value of G for steel is 12×10^6 lb/in.^{2} The modulus of volume expansion (bulk modulus) κ is the inverse of the fractional change in volume which accompanies the application of unit hydrostatic pressure. The value of κ for steel is 25×10^6 lb/in.^{2} The inverse of the modulus of volume expansion is the compressibility. Poisson's ratio γ is the ratio of the lateral unit strain to the normal unit strain in the elastic range of the material. This ratio evaluates the deformation of a material that occurs perpendicular to the direction of application of load. The value of γ for steel is 0.3. From geometrical considerations it is known that the three moduli of elasticity E, G, and κ, and Poisson's ratio γ, are related to each other.[15] More complete data on materials and their properties as used in machine and equipment design are compiled in available references.[16, 17] Values of the static properties of typical engineering materials are given in Tables 42.3 to 42.5.

DYNAMIC PROPERTIES

When a load is applied to a structure suddenly, the dynamic properties of the material may be significant. Dynamic properties can be considered as modifications of static

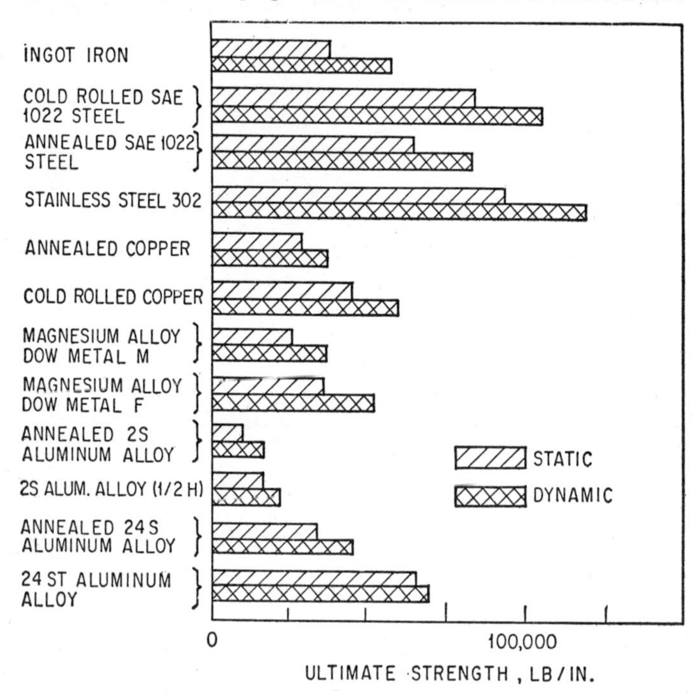

Fig. 42.21. Static and dynamic values of the ultimate strength of several metals where the dynamic strengths were obtained at impact velocities of 200 to 250 ft/sec. (*D. S. Clark and D. S. Wood.*[18])

properties. The most significant modifications are increases in ultimate strength and yield strength; other effects of some importance are an apparent lack of ductility at very high strain rates and delay in initiation of yielding under sudden application of load.

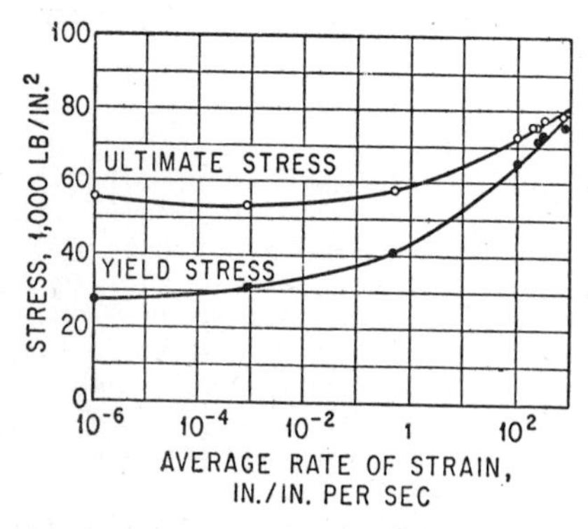

Fig. 42.22. Effect of strain rate on mechanical properties of mild steel. (*M. J. Manjoine.*[20])

STRAIN-RATE EFFECTS. The use of high-speed tensile testing machines shows that an increase in the rate of deformation raises both the yield strength and the ultimate strength of certain metals, as well as the entire stress level of the stress-strain curve. Figure 42.21 shows the static and dynamic values of the ultimate strengths of several metals where the dynamic strengths are determined at impact velocities of 200 to 250 ft/sec.[18] Impact tests of projectiles with striking velocities up to 2,750 ft/sec show as much as a 300 per cent increase in compressive strength for some types of steel.[19]

The influence of strain rate on the tensile properties of mild steel at room temperature is shown in Fig. 42.22. The marked difference between the yield stress and ultimate stress at low rates of strain disappears at high rates of strain.[20] Figure 42.22 also shows that the ultimate stress remains practically unchanged for strain rates below 1 in./in./sec; in this limited range the stress-strain curve of most engineering metals is not raised appreciably.[21] Mild steel is an exception in which the yield stress is influenced markedly by strain rate in the range from 0 to 1 in./in./sec.

TOUGHNESS AND DUCTILITY. It is useful to evaluate the total energy needed to fracture a test bar under tension; this energy is a measure of the toughness of the material. The area under the typical stress-strain diagram shown in Fig. 42.20 gives an approximate measure of the fracture energy per unit volume of material. However, the true fracture energy depends upon the true stress and true strain characteristics which take into account the nonuniform strain resulting from the reduction of area upon necking down of the test bar. Calculated values of the fracture energy for various metals are given in Table 42.6. Tough materials (e.g., wrought iron and low- or medium-carbon steel) exhibit high unit elongation and are considered to be ductile. By contrast, cast iron exhibits practically no elongation and is considered to be brittle. If only the elastic

Table 42.6. Fracture Energy or Toughness of Different Materials

(J. M. Lessells.[13])

Material	Condition	Yield strength, lb/in.^{2}, σ_y	Tensile strength lb/in.^{2}, σ_u	Unit elongation, in./in., ξ	Toughness or fracture energy, in.-lb/in.^{3}	Modulus of resilience, in.-lb/in.^{3}
Wrought iron.......	As received	24,000	47,000	0.50	17,700	7
Steel (0.13% C)....	As received	26,000	54,000	0.44	17,600	11
Steel (0.25% C)....	As received	44,000	76,000	0.36	21,600	24
Steel (0.53% C)....	Oil-quenched and drawn	86,000	134,000	0.11	12,000	100
Steel (1.2% C).....	Oil-quenched and drawn	130,000	180,000	0.09	10,800	280
Steel (spring).......	Oil-quenched and drawn	140,000	220,000	0.03	4,400	320
Cast iron..........	As received	...	20,000	0.005	70	1
Nickel cast iron.....	As received	20,000	50,000	0.10	3,500	9
Rolled bronze......	As received	40,000	65,000	0.20	10,500	60
Duralumin.........	Forged and heat-treated	30,000	52,000	0.25	10,200	17

strain energy up to the proportional limit is included, the resulting stored energy per unit volume is called the *modulus of resilience*. Values for this property also are given in Table 42.6.

CRITICAL STRAIN VELOCITY. When a large load is applied to a structure very suddenly, failure of the structure may occur with a relatively small over-all elongation. This has been interpreted as a brittle fracture, and it has been said that a material loses its ductility at high strain rates. However, an examination of the failure shows normal ductility (necking down) in a region close to the application of load. Large stresses are developed in this region by the inertia of the material remote from the application of the load, and failure occurs before the plastic stress waves are transmitted away from the point of load application. This effect is important only where loads are applied very suddenly, as in a direct hit by a projectile on armor plate. In general, equipment is mounted upon structures that are protected from direct hits; the resilience of such structures prevents a sufficiently rapid application of load for the above effect to be of significance in the design of equipment.

DELAY IN INITIATION OF YIELD. Yielding of a structure made of ductile material does not occur immediately upon sudden application of load. Rather, yielding occurs an instant of time later. This delay in initiation of yield is a function of the material, the stress level, and the temperature. Thus, a material may be stressed substantially above its yield strength for a short period of time without yielding. For mild steel at room temperature, the delay time is of the order of 0.001 sec. For repeated applications of load, the material has a memory; i.e., the durations of load are additive to determine the time of yielding. An equipment subjected to shock or vibration experiences an oscillatory stress pattern wherein the higher stresses occur repeatedly. The durations of these stresses add quickly to a time greater than the delay time for common materials; thus, the effect is of little significance in the design of equipment to withstand shock and vibration.

FATIGUE BEHAVIOR. The ability of a material to withstand repeated applications of stress is termed its endurance strength or fatigue strength. Endurance tests commonly are made by subjecting a test specimen to a stress pattern in which the stress varies sinusoidally with time. The test specimen usually* is a cantilever or simply supported beam with a dead-weight load to induce the desired stress; the cyclic stress pattern is attained by rotating the beam about its longitudinal axis. Rotation of the beam is continued until rupture occurs; this gives a point on a curve of stress vs. number of cycles to failure. The test is repeated using identical specimens with different stress, and the results are plotted as logarithm of stress vs. logarithm of number of cycles to failure. Typical curves are shown in Fig. 42.23.[22]

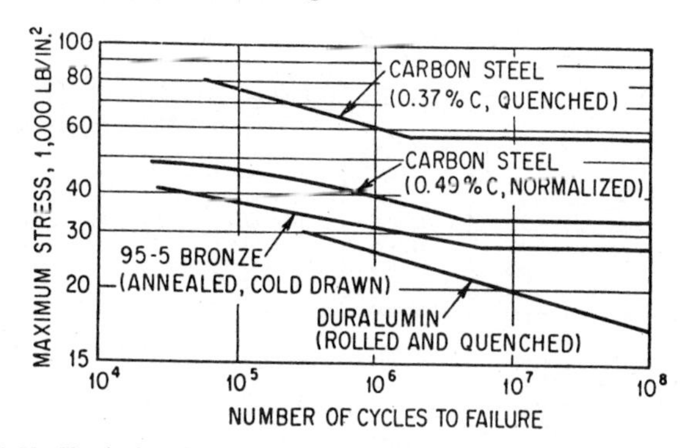

FIG. 42.23. Typical endurance curves. (*H. F. Moore and J. B. Kommers.*[22])

The sharp break and horizontal continuation of the curves in Fig. 42.23 indicate the stress level at which the material can endure an infinite number of stress reversals without failure. Sometimes, the curve gradually turns horizontal at its left-hand end. In other words, the stress required to cause failure after only a few cycles may be independent of the number of cycles.[23] Representative tensile and fatigue properties of steel are shown in Table 42.7. Similarly, Table 42.8 shows the tensile and fatigue properties of some nonferrous metals and alloys.

NOTCHED-BAR IMPACT TESTS. The Charpy and Izod notched-bar impact tests measure the energy absorbed by a standard specimen when it is fractured by a pendulum.[13] They are of value in showing quickly the effects of temperature, heat-treatment, and composition on the ductility of different metals. Typical values of impact energy are given in Table 42.6.

Repeated-impact notched-bar tests were at one time considered important but this practice has been discontinued. The notch sensitivity of the material influences the results which then become difficult to interpret as quantitative values of resistance to impact loads.[24] Figure 42.24, showing the results of such tests, is of general interest because the curves start with the impact strength for a single blow and follow the characteristic shape of conventional vibration endurance curves.[25]

* Numerous special types of fatigue tests are made, including alternating tension and compression, alternating torsion, alternating tension or torsion superimposed upon a non-varying component of tension or torsion, alternating combined tension and torsion, and many others.

Table 42.7. Tensile and Fatigue Properties of Steels

(*J. M. Lessells.*[13])

Material	State	Yield strength, lb/in.^{2}, σ_y	Tensile strength, lb/in.^{2}, σ_u	Elongation, %	Reduction of area, %	Endurance limit, lb/in.^{2}, σ_e	Ratio σ_e/σ_u
0.02% C	As received	19,000	42,400	48.3	76.2	26,000	0.61
Wrought iron	As received	29,600	47,000	35.0	29.0	23,000	0.49
0.24% C	As received	38,000	60,500	39.0	64.0	25,600	0.425
0.24% C	Water-quenched and drawn	45,600	67,000	38.0	71.0	30,200	0.45
0.37% C	Normalized	34,900	71,900	29.4	53.5	33,000	0.46
0.37% C	Water-quenched and drawn	63,100	94,200	25.0	63.0	45,000	0.476
0.52% C	Normalized	47,600	98,000	24.4	41.7	42,000	0.43
0.52% C	Water-quenched and drawn	84,300	111,400	21.9	56.6	55,000	0.48
0.93% C	Normalized	33,400	84,100	24.8	37.2	30,500	0.36
0.93% C	Oil-quenched and drawn	67,600	115,000	23.0	39.6	56,000	0.487
1.2% C	Normalized	60,700	116,900	7.9	11.6	50,000	0.43
1.2% C	Oil-quenched and drawn	130,000	180,000	9.0	15.2	92,000	0.51
0.31% C, 3.35% Ni	Normalized	53,500	104,000	23.0	45.0	49,500	0.47
0.31% C, 3.35% Ni	Oil-quenched and drawn	130,000	154,000	17.0	49.0	63,500	0.41
0.24% C, 3.3% Ni, 0.87% Cr	Oil-quenched and drawn	128,000	138,000	18.2	61.8	68,000	0.49

Table 42.8. Tensile and Fatigue Properties of Nonferrous Metals

(*J. M. Lessells.*[13])

Material	State	Tensile strength, lb/in.^{2}, σ_u	Endurance limit or fatigue strength, lb/in.^{2}, σ_e	N_1,* millions of cycles	N_2,† millions of cycles	Ratio σ_e/σ_u
Aluminum	...	22,600	10,500	100	6	0.46
Duralumin	Rolled	51,000	14,000	400	>400	0.27
Duralumin	Annealed	25,200	10,000	200	>200	0.40
Duralumin	Tempered	51,300	12,000	400	4½	0.24
Magnesium	Extruded	32,500	8,000	200	2	0.25
Magnesium alloy (4% Al)	...	35,200	12,000	600	½	0.34
Magnesium alloy (4% Al, 0.25% Mn)	...	39,000	15,000	100	1	0.38
Magnesium alloy (6.5% Al)	...	41,200	13,000	600	½	0.31
Magnesium alloy (6.5% Al, 0.25% Mn)	...	44,500	15,000	100	½	0.34
Magnesium alloy (10% Cu)	...	39,000	12,000	600	½	0.31
Electron metal	...	36,600	17,000	200	30	0.47
Copper	Annealed	32,400	10,000	500	20	0.31
Copper	Cold-drawn	56,200	10,000	500	>500	0.18
Brass (60–40)	Annealed	54,200	22,000	500	>500	0.44
Brass (60–40)	Cold-drawn	97,000	26,000	500	50	0.27
Naval brass	...	68,400	22,000	300	10	0.32
Aluminum bronze (10% Al)	As cast	59,200	23,000	60	3	0.39
Aluminum bronze (10% Al)	Heat-treated	77,800	27,000	40	1	0.35
Bronze (5% Sn)	Annealed	45,600	23,000	1000	10	0.50
Bronze (5% Sn)	Cold-drawn	85,000	27,000	500	50	0.32
Manganese bronze	As cast	70,000	17,000	150	20	0.24
Nickel	Annealed	70,000	28,000	100	50	0.40
Monel metal	Hot-rolled	90,000	32,000	450	>450	0.36

* N_1 = cycles on which σ_e is based.
† N_2 = cycles at which σ-N curve becomes and remains horizontal.

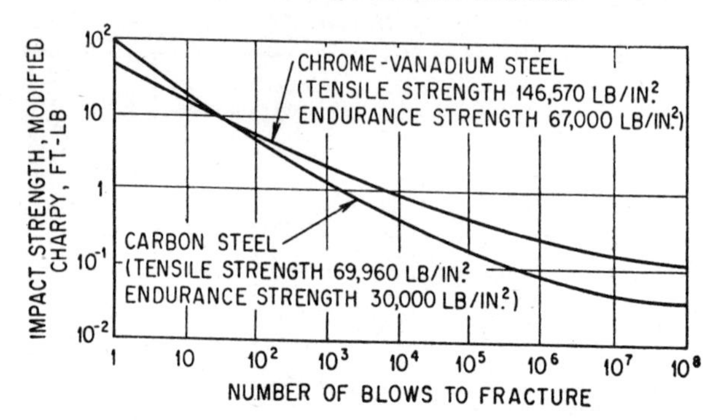

FIG. 42.24. Impact endurance curves. (*D. J. McAdam, Jr.*[25])

THE SIGNIFICANCE OF MATERIAL PROPERTIES IN DESIGN

The method of designing equipment to withstand shock and vibration follows conventional methods of design in that it involves (1) determination of the loads, (2) calculation of the resultant stresses, and (3) selection of a suitable material. The loads and the nominal stresses are determined most conveniently by constructing a hypothetical model of the equipment as described in preceding sections of this chapter. This section discusses some of the important considerations in adapting the results of the model study to the selection of appropriate materials. Significant properties of materials are set forth in the above section under *Mechanical Properties of Engineering Materials.*

If the most severe environmental condition encountered by an equipment is a shock experienced only once or a few times, the most important consideration is to design the equipment to withstand a sudden application of a large load with its resultant stresses. Equipment that must endure sustained vibration of large magnitude must withstand repeated application of load of possibly smaller magnitude than that experienced during shock. The relative importance of these types of loading must be determined for each application; however, they are discussed in the following paragraphs under *Strength Requirements* and *Fatigue Behavior.*[23]

STRENGTH REQUIREMENTS

The most important consideration involving ultimate strength is the attainment of a proper balance between stress and ductility. A model, in so far as it can represent an equipment effectively, indicates the maximum strain if the equipment characteristics are linear. Where nonductile materials are used or where permanent deformation of a structure cannot be tolerated, the equipment must be designed so that the strain does not exceed the strain at the elastic limit. Usually, permanent deformation of some structural members of an equipment is permissible. Permanent deformation of other members may not be permissible because it would cause misalignment of parts whose proper operation depends upon accurate alignment. A concept that has application to the design of yielding members is discussed in the following paragraphs.

Experience indicates that the yield strength of a part may be exceeded at the maximum strain experienced during shock, provided the material yields under the shock loading. This concept may be applied only if it is possible to specify a maximum plastic deflection of the structure which will not lead to malfunctioning of the equipment. It is difficult to determine the degree of deformation that occurs at stresses greater than the yield strength because the structure becomes nonlinear and the linear model theory is only approximately applicable. However, semi-empirical procedures have been developed to specify permissible stresses.

An empirical design procedure that has application in the design of members subjected to bending contemplates that some predetermined percentage of the cross-sectional area of the member has a calculated stress greater than the yield strength of the material.[26] It is necessary that the material have a high ductility. In a typical application of this concept, 25 per cent of the cross-sectional area of the beam has a stress greater than the yield strength. For example, applying the concept to a beam in bending, the stress at maximum deflection is calculated; if the beam is linear without regard to the magnitude of the stress, the distribution of stress over a rectangular cross section is as shown in Fig. 42.25*A*. The beam is designed so that the stress is above the yield strength over 25 per cent of the cross-sectional area. Then the actual stress distribution, assuming a stress-strain curve of the type shown in Fig. 42.20, is shown in Fig. 42.25*B*. The permanent deformation of the beam is less than the maximum deflection because 75 per cent of the material, adjacent to the neutral axis, remains elastic and tends to restore the beam to its original shape.

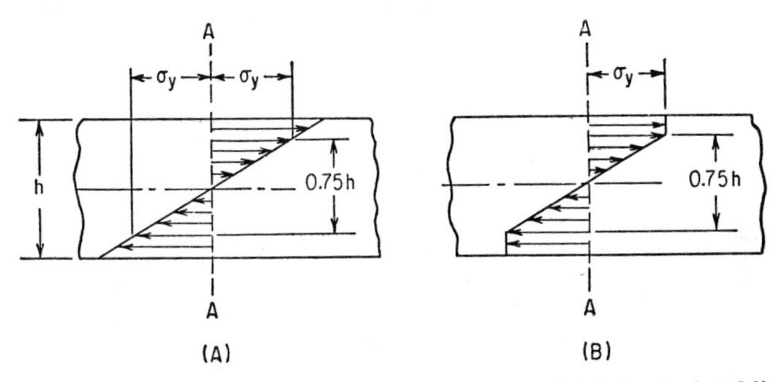

FIG. 42.25. Stress distribution in a beam subject to bending with (*A*) calculated linear stress distribution so that the stress is above the yield strength over 25 per cent of the cross-sectional area, and (*B*) the actual stress distribution if the material has a stress-strain curve similar to that shown in Fig. 42.20.

Another empirical design procedure establishes a maximum allowable stress under shock loading that depends upon the function served by a particular member of the equipment. When a material has sufficient ductility to exhibit an elongation at failure of at least 10 per cent in 2 in., the allowable stress σ_a based on permissible yielding can be expressed:

$$\sigma_a = \sigma_y + C(\sigma_u - \sigma_y) \qquad \text{lb/in.}^2 \tag{42.33}$$

where σ_y is the yield strength for a 0.2 per cent permanent extension, σ_u is the ultimate tensile stress, and the constant C depends upon the function served by a particular member. For bolts which are inaccessible for retightening, $C = 0$. Where dimensional stability is important, $C = 0.2$. If some yielding can be tolerated, as in piping, $C = 0.5$. For calculating the bearing stress between flat or mating surfaces or where permanent deformation can be tolerated, $C = 1.0$.

Although the yield strength and ultimate strength of mild steel show an increase as the rate of strain increases (Fig. 42.22), this effect is of very limited significance in the design of equipment to withstand shock and vibration. In general, a strain rate great enough to cause a significant increase in strength occurs only closely adjacent to a source of shock, as at the point of impact of a projectile with armor plate. Equipment seldom is subjected to shock of this character. In a typical installation, the structure interposed between the equipment and the source of shock is unable to transmit large forces suddenly enough to cause high strain rates at the equipment. Furthermore, the response of a structure to a shock is oscillatory; maximum strain rate occurs at zero strain, and vice versa. The data of Fig. 42.22 represent conditions where maximum stress and maximum rate of strain occur simultaneously; thus, they do not apply directly to the design of shock resistant equipment. The use of statically determined yield strength and ultimate strength for design purposes is a conservative but not overly conservative practice.

FATIGUE BEHAVIOR

Failure of equipment by repeated occurrence of stress usually is the result of a condition of resonance with consequent excessive stress. Several approaches for controlling the stress are suggested in Chap. 43. The effectiveness of such control determines the extent to which the fatigue properties of the material become significant. Most of the data defining the fatigue behavior of material are in the form of Fig. 42.23. In some applications, the stress on the ordinate of Fig. 42.23 may be applied directly to the calculated stress in a member of the equipment. Often, considerations that cannot be deduced directly from Fig. 42.23 become important: (1) the alternating stress may be superimposed upon a steady stress, (2) the stress may be greater than that calculated because of stress concentration, and (3) the stress amplitude may not be constant at each cycle of stress.

When a steady stress is combined with a variable stress, the resultant strength is determined from the Goodman or Gerber relations shown in Fig. 42.26[27] and from the applicable fatigue behavior of the type shown in Fig. 42.23. The abscissa in Fig. 42.26

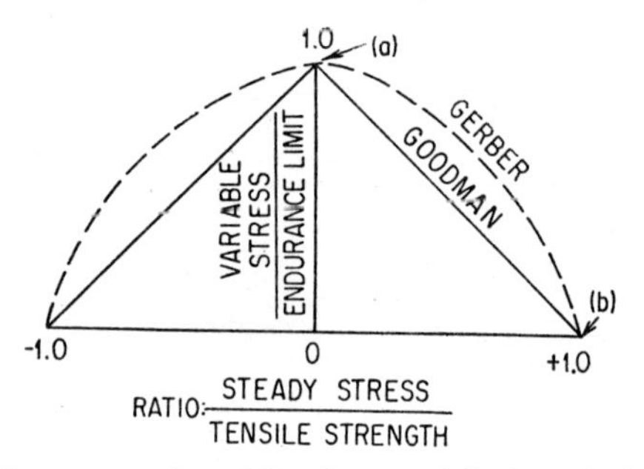

FIG. 42.26. Graphical representation of Goodman and Gerber relations. (*H. J. Gough.*[27])

is the ratio of the steady stress existing in the part to the tensile strength of the material. The ordinate is the ratio of the allowable variable stress in the part to the endurance limit found from Fig. 42.23. The limiting cases shown are as follows: (1) For a steady stress equal to zero, the allowable variable stress equals the endurance limit; and (2) for a variable stress equal to zero, the allowable steady stress is the tensile or compressive strength. The straight line indicates that for a steady stress equal to half the tensile strength, the allowable variable stress is only half the endurance limit. The Gerber relation is in good agreement with test data. However, the Goodman relation gives a smaller allowable stress and may be preferred for more conservative design.

The fatigue strength of a machine member is greatly affected by its geometric shape. Figure 42.27 shows typical values of stress concentration factor at the change in diameter of a circular rod subjected to a bending moment M_B.[28] The stress concentration factor can be interpreted as the ratio of actual maximum stress in the fillet to the nominal stress calculated on the basis of the diameter d. The stress concentration increases with the sharpness of the radius r, and depends upon the ratio of diameters D/d.

To provide the data required for the design of equipment subjected to loads of different magnitudes, endurance testing must be done with *different* stress levels applied to a single specimen before failure is reached. Such data are not available to any appreciable degree, and it becomes necessary to adapt data of the type shown in Fig. 42.23 to circumstances of varying stress amplitude. Several different cumulative damage criteria have been suggested. One hypothesis[29] assumes that the percentage of fatigue damage is a linear function of the number of cycles. For example, referring to Fig. 42.28,[30] if a specimen is subjected to a stress σ_1 for n_1 cycles, σ_2 for n_2 cycles, etc., failure tends to

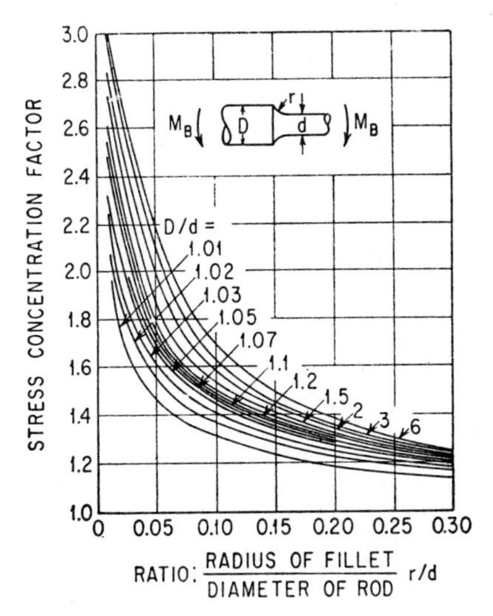

FIG. 42.27. Values of stress concentration factor K for fillets at change in diameter of a round rod in bending. (*R. E. Peterson.*[28])

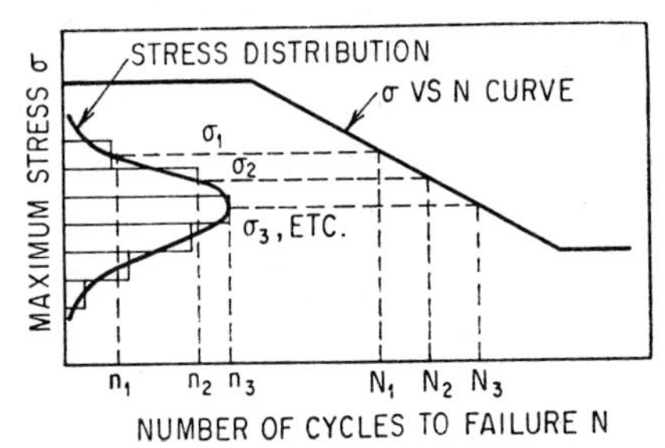

FIG. 42.28. Typical stress distribution and stress-cycle curve illustrating the application of Miner's hypothesis for cumulative fatigue damage. (*N. M. Newmark.*[30])

occur when

$$\frac{n_1}{N_1} + \frac{n_2}{N_2} + \frac{n_3}{N_3} + \cdots = 1 \tag{42.34}$$

where N_1, N_2, . . . are cycles-to-failure at stresses σ_1, σ_2, . . ., respectively. Limited experimental results indicate that the factor 1.0 in Eq. (42.34) varies between 0.3 and 3.0, depending on the order in which the stress cycles are applied and possibly on other influences. A somewhat more complicated hypothesis[31] assumes a nonlinear accumulation of damage.

REFERENCES

1. Roark, R. J.: "Formulas for Stress and Strain," 3d ed., p. 100, McGraw-Hill Book Company, Inc., New York, 1954.
2. Den Hartog, J. P.: "Mechanical Vibration," 3d ed., p. 105, McGraw-Hill Book Company, Inc., New York, 1947.
3. Ruzicka, J. E.: *Sound and Vibration*, **4**(9):10 September, 1970.
4. Walsh, J. P., and R. E. Blake: *Proc. SESA*, **6**(2):152 (1948).

5. Frankland, J. M.: *Proc. SESA*, **6**(2):11 (1948).
6. Crede, C. E., and M. C. Junger: "A Guide for Design of Shock Resistant Naval Equipment," U.S. Navy Department, Bureau of Ships, *NAVSHIPS* 230-660-3C, p. 17, 1949.
7. Friedrich, C. M.: "Shock Design Notes," unpublished report, Westinghouse Electric Corporation, Atomic Power Division, Pittsburgh, Pa., June, 1956.
8. McCalley, R. B., Jr.: *Supplement to Shock and Vibration Bull.* 23 (unclassified), Office of Secretary of Defense, June, 1956.
9. Ref. 1, p. 70.
10. Ref. 2, p. 457.
11. Stokey, W. F., and C. F. Zorowski: *J. Appl. Mechanics*, **E26**(2):210 (1959).
12. Rinehart, J. S., and J. Pearson: "Behavior of Metals Under Impulsive Loads," The Haddon Craftsmen, Inc., Scranton, Pa., 1954.
13. Lessells, J. M.: "Strength and Resistance of Metals," 1st ed., p. 7, John Wiley & Sons, Inc., New York, 1954.
14. ASTM Book of Standards, part 3, p. 117, 1958.
15. Sokolnikoff, I. S.: "Mathematical Theory of Elasticity," 2d ed., p. 65, McGraw-Hill Book Company, Inc., New York, 1956.
16. Vallance, A., and V. Doughtie: "Design of Machine Members," Chap. II, 3d ed., McGraw-Hill Book Company, Inc., New York, 1951.
17. Hoyt, S. L.: "Metal Data," 2d ed., Reinhold Publishing Corporation, New York, 1952.
18. Clark, D. S., and D. S. Wood: *Trans. ASM*, **42**:45 (1950).
19. Whiffin, A. C.: *Proc. Royal Society* (*London*), **194A**:300 (1948).
20. Manjoine, M. J.: *J. Appl. Mechanics*, **66**:A215 (1944).
21. MacGregor, C. W., and J. C. Fisher: *J. Appl. Mechanics*, **67**:A217 (1945).
22. Moore, H. F., and J. B. Kommers: "The Fatigue of Metals," Chap. VI, McGraw-Hill Book Company, Inc., New York, 1927.
23. Pope, J. A.: "Metal Fatigue," Chap. II, Chapman & Hall, Ltd., London, 1959.
24. Gillett, H. W.: *Steel*, Nov. 22, 1943, and Feb. 14, 1944.
25. McAdam, D. J., Jr.: *Proc. ASTM*, **23**:56-105 (1923).
26. Graziana, D. J.: *Prod. Eng.*, **44**:44, Sept. 29, 1958.
27. Gough, H. J.: "The Fatigue of Metals," D. Van Nostrand Company, Inc., Princeton, N.J., p. 69, 1924.
28. Peterson, R. E.: "Stress Concentration Design Factors," p. 75, John Wiley & Sons, Inc., New York, 1953.
29. Miner, M. A.: *J. Appl. Mechanics*, **67**:A159 (1945).
30. Murray, W. M.: "Fatigue and Fracture of Metals," p. 197, John Wiley & Sons, Inc., New York, 1952. (See Paper 10 by N. M. Newmark.)
31. Corten, H. T., and T. J. Dolan: *Intern. Conf. Fatigue of Metals, IME and ASME*, September, 1956.

43

PRACTICE OF EQUIPMENT DESIGN

Edward G. Fischer
Westinghouse Electric Corporation

Harold M. Forkois
U.S. Naval Research Laboratory

INTRODUCTION

Equipment of a type required to withstand shock and vibration generally consists of a housing or chassis to provide structural strength and an array of functional components supported by the chassis. A suitable design is characterized by (1) properly selected or designed components, (2) chassis mounting to minimize damage from shock and vibration, and (3) a chassis capable not only of withstanding shock and vibration but also of providing a degree of protection to the components. The design of chassis and other structures is discussed in Chap. 42 from a quantitative viewpoint; this chapter discusses (1) chassis design in a general or qualitative sense, (2) the selection and mounting of components and (3) design to minimize vibration. Many of the examples refer explicitly to electronic equipment but the general principles have wider application.

DESIGN OF STRUCTURES

One of the more troublesome problems in the design of equipment is attainment of the proper balance between flexibility and rigidity of structures. As shown analytically in Chap. 42, the maximum acceleration experienced by a component is determined primarily, for a particular shock motion, by the natural frequency of the structure supporting the component. The acceleration decreases as the natural frequency decreases. Thus, components that are susceptible to damage from shock may be given a degree of protection by mounting them on relatively flexible structures; however, if the equipment is required to withstand vibration as well as shock, it is possible that the flexibility introduced to attenuate the shock may lead to failure as a result of vibration.

Failure or damage from vibration usually is the result of a resonant condition; i.e., a structure or component has a natural frequency that coincides with a frequency of the forcing vibration. In some applications, e.g., on naval ships, the maximum frequency of the forcing vibration is relatively low and it becomes feasible in most instances to design structures whose natural frequencies are greater than the highest forcing frequency. On the other hand, the vibration in aircraft and missiles is characterized by relatively high frequencies; as a consequence, it is not feasible to design structures whose natural frequencies are higher than the forcing frequencies. A condition of resonance often must be tolerated. In extreme conditions, the effect of the resonant condition can be alleviated by the provision of damping or energy dissipation (see the section on *Damping*). In general, this is not necessary for structural members if the natural frequencies are chosen wisely.

Because of the many different types of components and equipments in common use and the wide range of environmental conditions encountered, it is not possible to establish specific design objectives for natural frequencies of structures. In general, a structure should be designed to be stiff rather than flexible because (1) most components of modern design are more shock-resistant than commonly anticipated and do not require the protection afforded by structures of low natural frequency, and (2) failure of structures by repeated reversal of stress during vibration is a common occurrence of damage. The stress in a structure tends to decrease when the natural frequency increases, provided the severity of the excitation and the fraction of critical damping for the structure remain unchanged. Furthermore, some environmental conditions embody a lower severity at the higher vibration frequencies; thus, a resonance at a high frequency may become less damaging. However, if a component is known to be vulnerable to damage by shock or if this is demonstrated during a shock test, the desired protection often may be attained by introducing a structure of lower flexibility. Such a structure may be the chassis of the equipment, a bracket to support a component, or an isolator (see Chaps. 30 to 33).

In typical electronic equipment, the electronic components are mounted upon a chassis to form a unit that has a discrete function. The unit is made small enough to permit ready maintenance and servicing. In some instances, the function of the unit may be complete; then the unit is considered a complete equipment and is provided with independent mounting means. In other instances, several units are required to fulfill a function; such units are mounted in a single cabinet and interconnected electrically and/or mechanically to perform a function. Particular design problems associated with the housing for the equipment and involving ability to withstand shock and vibration are (1) design of the unit chassis, (2) design of the cabinet, and (3) mounting of the unit chassis in the cabinet.[1]

CHASSIS DESIGN

In general, unit chassis are either (1) formed from sheet metal or (2) cast from a light metal (usually aluminum or magnesium) and machined to form intimate supports for the components. The former is more common and less costly to manufacture, but tends to

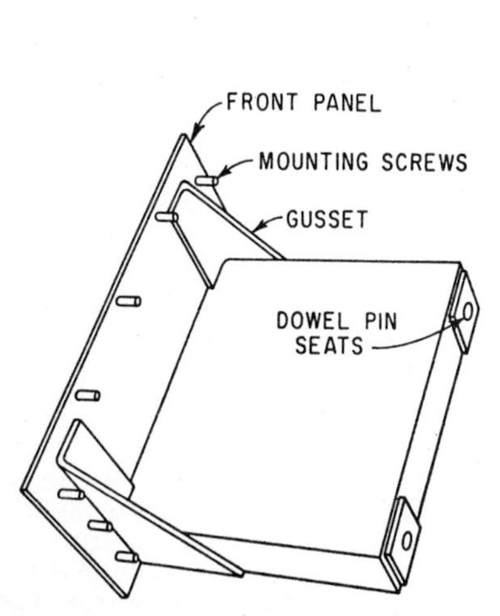

FIG. 43.1. Good chassis construction, showing front-panel mounting screws and hole-plate reinforcement for dowel pins in the rear. (*H. M. Forkois and K. E. Woodward.*[1])

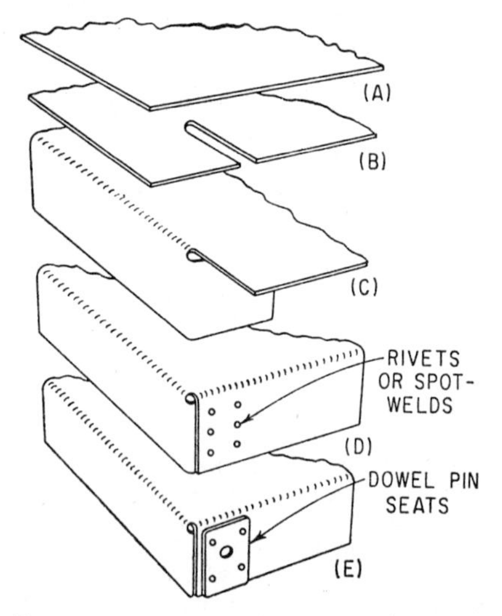

FIG. 43.2. A method of forming chassis which produces stiff corners. (*H. M. Forkois and K. E. Woodward.*[1])

lack adequate rigidity. The casting has the desirable property of relatively great rigidity, but refinements in design and manufacture are necessary to keep the weight within reasonable limits. Because of its rigidity, the cast chassis is not only less likely to fail as a result of vibration but it also minimizes damage from broken leads and collision of components that occur when the chassis is unduly flexible. The flexibility of structure required for protection from shock can be provided in the support for the chassis.

The most common type of sheet-metal chassis is a shallow rectangular "box" formed from a single piece of sheet metal; the front panel is added and rigidly attached to the "box" by gussets, as shown in Fig. 43.1. The chassis usually is oriented with the flanges or sides of the "box" extending downward. The sequence of illustrations in Fig. 43.2 shows one method of forming chassis corners. All bending radii should be greater than the thickness of the sheet to reduce the effects of stress concentration. The overlapped

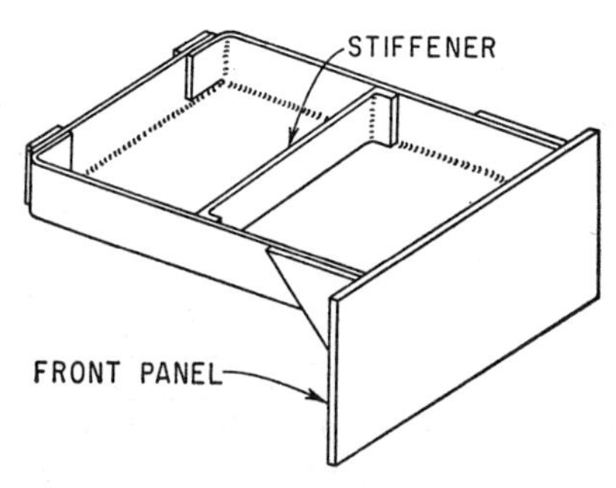

FIG. 43.3. View of underside of chassis showing stiffener added to support heavy component on upper side of chassis. (*H. M. Forkois and K. E. Woodward.*[1])

portions of the sheet can be joined by rivets or spot-welds (see *Methods of Construction*) to provide a stiff support for dowel pins or other attaching means. It is common practice to mount large components, such as transformers and electron tubes, on the upper side of the chassis; smaller components, such as resistors and capacitors together with substantially all wiring, are mounted on the lower side of the chassis.

Special care should be exercised to ensure maximum rigidity of the sheet-metal chassis. The overlapped flanges at the corners are areas of inherent rigidity and should be used to support the chassis. For example, the gussets for reinforcing the front panel and the plates to seat the dowel pins are attached at the corners as illustrated in Fig. 43.1. Where possible, heavy components such as transformers should be mounted adjacent the corners to attain a maximum ratio of chassis stiffness to component weight. If a heavy component cannot be mounted adjacent a corner, it should be mounted adjacent an edge; if it must be mounted near the center of the chassis, a stiffener should be provided on the lower side of the chassis, as shown in Fig. 43.3. This type of construction is necessary to obtain the required stiffness but is undesirable in that it limits the location of components and interferes with the wiring.

Where an equipment is comprised of a single chassis, it is common practice to provide an enclosure to protect the equipment during handling and during exposure to atmospheric contamination; such an enclosure is not intended as a support for the equipment and is not required to withstand the forces introduced by shock and vibration conditions. For example, in one type of application commonly used with vibration isolators in aircraft equipment, a supporting rack has forward-facing tapered pins adjacent the rear edge. These pins extend through clearance holes in the enclosure and engage dowel pin seats in the rear of the chassis, as illustrated in Fig. 43.1. Similarly, clamps at the forward edge of the rack engage the front panel of the chassis directly, and loads are not carried by the enclosure. Frequently, two or more chassis, each with its individual enclosure, are mounted independently upon one rack.

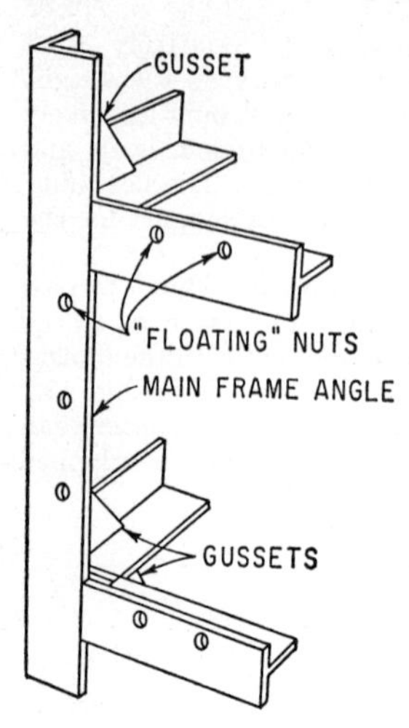

FIG. 43.4. Frame construction for cabinet showing clearance holes for drawer or chassis front-panel thumbscrews. The frame members are welded, riveted, or bolted together. (*H. M. Forkois and K. E. Woodward.*[1])

CABINET DESIGN

When several chassis are included in a single cabinet, it is common to design the cabinet to be both an enclosure and a structure capable of withstanding the forces introduced by shock and vibration conditions. The chassis usually must be readily removable from the cabinet for adjustment and maintenance. Suitable methods of mounting the chassis in the cabinet are discussed under *Chassis Mounting Methods*. If the objective of a rigid chassis is achieved, criteria to be applied in design of the cabinet are (1) optimum rigidity in view of shock and vibration conditions and (2) adequate structural strength. Suggestions for cabinet design are included in the following paragraphs.

A typical cabinet has horizontal dimensions corresponding to the width and depth of a chassis, and a height equal to the total heights of the chassis. The chassis are arranged one above the other, and are mounted independently to the cabinet. Except for very light equipment, the frame of the cabinet is assembled from structural steel or aluminum members welded, riveted, or bolted together, as illustrated in Fig. 43.4. Gussets are provided at the joints of structural members where they would not interfere with installation and removal of the chassis; they should be substantially the same thickness as the frame members. Generally, a cabinet constructed in this manner is notably lacking in rigidity, and all opportunities for stiffening should be employed. Usually, it is possible to cover the sides, back, top, and bottom with relatively light-gage sheet metal, securely attached to the structural members. Preferably, such sheets should be welded to the structural members to take full advantage of available rigidity, and they may be provided with welded or formed stiffeners, as illustrated in Figs. 43.5 and 43.6. It is difficult to achieve the desired rigidity in the front face of the cabinet because the requirement for removal of chassis leaves this face almost unstiffened. The several smaller front

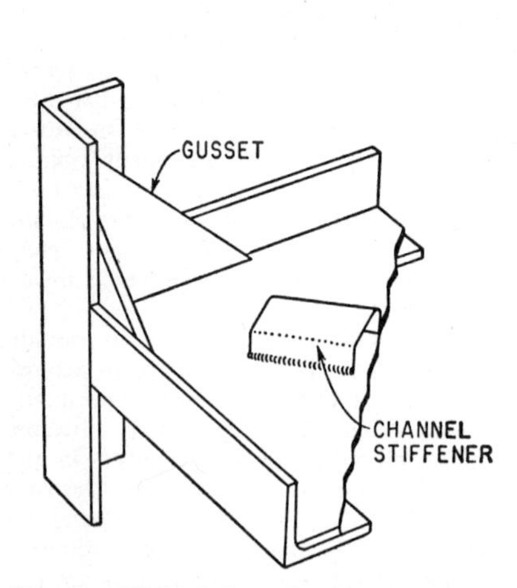

FIG. 43.5. Welded channel stiffener to increase cabinet rigidity. (*H. M. Forkois and K. E. Woodward.*[1])

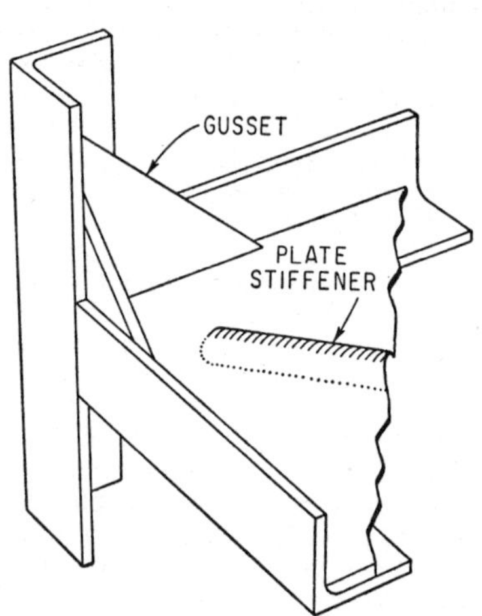

FIG. 43.6. Coined stiffener to increase cabinet rigidity. (*H. M. Forkois and K. E. Woodward.*[1])

panels of the individual chassis do not contribute as much stiffness as a single large sheet, and the screws and clamps commonly used to secure the chassis are not as effective as welding. This lack of rigidity is particularly noticeable if the equipment is supported at the bottom and braced to a bulkhead by a support adjacent only the upper rear edge.

A cabinet with several chassis commonly weighs several hundred pounds; the re-

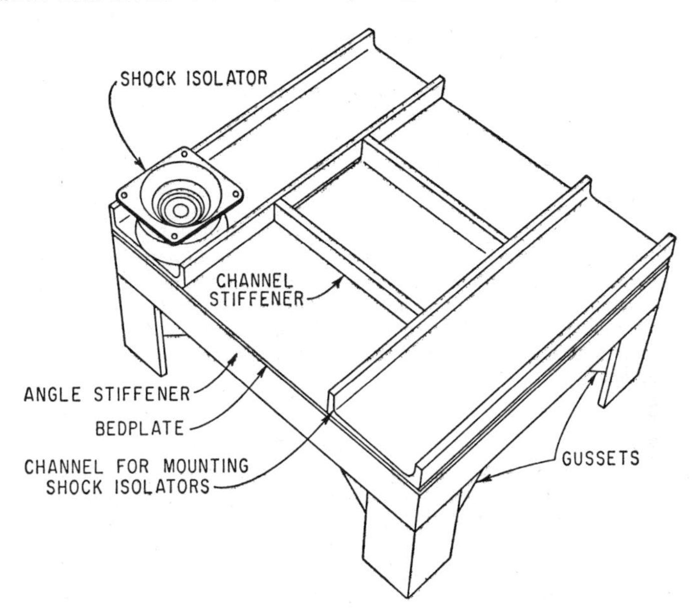

Fig. 43.7. Bottom view of cabinet construction showing channel stiffeners and shock isolator. (*H. M. Forkois and K. E. Woodward.*[1])

sultant loads at points of support may become very great under conditions of shock. Particular care in design is required to ensure that excessive deformation does not occur. For expedience of installation, it is preferable that the cabinet be mounted with a minimum of bolts. Usually it is necessary to provide additional structural members to distribute the forces from the bolts over much of the base area of the cabinet. Figure

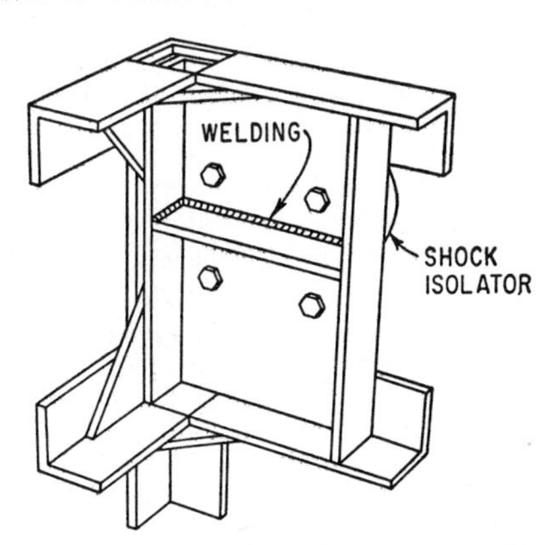

Fig. 43.8. Structural stiffeners for bulkhead attachment near top of cabinet. (*H. M. Forkois and K. E. Woodward.*[1])

43.7 is a bottom view of a cabinet showing an arrangement of channel stiffeners for distributing the forces. Such stiffeners are of particular convenience when used with shock isolators because the isolators nest in the channels, but they can also be used with direct mounting. Figure 43.8 shows a type of stiffener that may be used near the top of a cabinet for attachment to a bulkhead.

METHODS OF CONSTRUCTION

In selecting a method of construction, it is important to consider the suitability of the structure for use under conditions of shock and vibration. The greatest cause of damage in a structure subjected to vibration is the large vibration amplitude that tends to occur during vibration at resonance of lightly damped structures. The interface friction between bolted members or riveted members introduces damping to a significant degree (see Chap. 36). Therefore, bolted and riveted structures may be better adapted than welded structures to withstand vibration, particularly if damping is important. However, a welded structure may be designed with less stress concentration and with better ability to maintain alignment of component parts; thus, it may be the most suitable method of construction even though larger amplitudes of vibration result at resonance. A qualitative indication of damping in various structures is given in the section on *Damping;* important considerations in various types of construction are discussed in the following sections.[2,3]

WELDED JOINTS

Welded joints must be well designed, and effective quality control must be imposed upon the welding conditions. The most common defect is excessive stress concentration which leads to low fatigue strength and, consequently, to inferior capability of withstanding shock and vibration. Stress concentration can be minimized in design by reducing the number of welded lengths in intermittent welding. For example, individual welds in a series of intermittent welds should be at least 1½ in. long with at least 4 in. between welds. Internal crevices can be eliminated only by careful quality control to ensure full-depth welds with good fusion at the bottom of the welds. Welds of adequate quality can be made by either the electric arc or gas flame process. Subsequent heat-treatment to relieve residual stress tends to increase the fatigue strength.

SPOT-WELDED JOINTS. Spot welding is quick, easy, and economical but should be used only with caution when the welded structure may be subjected to shock and vibration. Basic strength members supporting relatively heavy components should not rely upon spot welding. However, spot welds can be used successfully to fasten a metal skin or covering to the structural framework. Even though improvements in spot welding techniques have increased the strength and fatigue properties, spot welds tend to be inherently weak because a high stress concentration exists in the junction between the two bonded materials when a tension stress exists at the weld. Spot-welded joints are satisfactory only if frequent tests are conducted to show that proper welding conditions exist. Quality can deteriorate rapidly with a change from proved welding methods, and such deterioration is difficult to detect by observation. However, accepted quality-control methods are available and should be followed stringently for all spot welding.

RIVETED JOINTS. Riveting is an acceptable method of joining structural members when riveted joints are properly designed and constructed. Rivets should be driven hot to avoid excessive residual stress concentration at the formed head and to ensure that the riveted members are tightly in contact. Cold-driven rivets are not suitable for use in structures subjected to shock and vibration, particularly rivets that are set by a single stroke of a press as contrasted to a peening operation. Cold-driven rivets have a relatively high probability of failure in tension because of residual stress concentration, and tend to spread between the riveted members with consequent lack of tightness in the joint. Joints in which slip develops exhibit a relatively low fatigue strength.

BOLTED JOINTS. Except for the welded joints of principal structures, the bolted joint is the most common type of joint. A bolted joint is readily detachable for changes in construction, and may be effected or modified with only a drill press and wrenches as equipment. However, bolts tend to loosen and require means to maintain tightness (see section on *Locking Devices*). Furthermore, bolts are not effective in maintaining alignment of bolted connections because slippage may occur at the joint; this can be prevented by using dowel pins in conjunction with bolts or by precision fitting the bolts; i.e., fitting the bolts tightly in the holes of the bolted members. Other characteristics of bolted joints are discussed in the following paragraphs.

BOLTS AND NUTS

Bolts and nuts are made by various methods from different materials. Grades of steel bolts as standardized by the SAE and the ASTM are shown in Table 43.1. The ordinary type (18-8) stainless steel bolts have a low yield point (about 25,000 lb/in.^{2}); thus, they tend to stretch and loosen under shock although they have a high ultimate strength. Where corrosion resistance as well as the ability to withstand shock is a consideration, bolts made from heat-treatable stainless-steel alloy having a high yield point should be

Table 43.1. Grades of Bolts (SAE and ASTM)

(*ASME Handbook.*[6])

SAE grade	ASTM designation	Tensile strength,* lb/in.^{2}	Proof load or yield strength,* lb/in.^{2}	Material
0	...	...	...	Low carbon
1	A307	55,000	...	Low carbon
2	...	69,000	55,000	Low carbon, stress-relieved
3	...	110,000	85,000	Medium carbon, cold-worked
5	...	120,000	85,000	Medium carbon, quenched and tempered
	A325	125,000	90,000	Medium carbon, quenched and tempered
6	...	140,000	110,000	Special medium carbon, quenched and tempered
7	...	130,000	105,000	Alloy steel, quenched and tempered
8	...	150,000	120,000	Alloy steel, quenched and tempered

* Properties for sizes ½ in. diameter and smaller. For complete data and for properties of larger bolts, see SAE Handbook and ASTM specifications.

Table 43.2. Comparison of Tensile Fatigue Strengths of Bolts with Rolled, Cut, and Milled Threads

(*H. Dinner and W. Felix.*[4])

Threading carried out by	Fatigue strength, lb/in.^{2} (pulsating tension)	
	Unaged specimen	Specimen aged at 482°F
Rolling	28,450	28,450
Cutting	17,070	
Milling	17,070	12,800

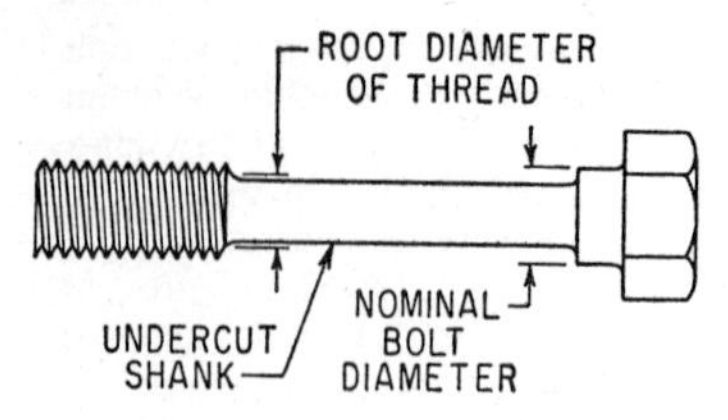

Fig. 43.9. Shank of bolt undercut slightly smaller than root diameter of thread to obtain uniform elongation under shock load.

used. The fatigue strength of bolts under vibratory loading can be improved by initial cold working of the surfaces. Cold working may result from rolling of thread roots, shot peening of shanks, or rolling of the fillet between head and shank.[4, 5] Benefits are derived from both the work hardening and the initial compression left in the surfaces. Table 43.2 gives data on the endurance limit of bolts in pulsating tension.[6]

The strength of a bolt when subjected to shock depends upon its capacity to absorb energy by elastic deflection. For optimum efficiency, all cross sections of the bolt should be designed to reach the yield stress simultaneously, where the stress is calculated by including stress concentrations. Figure 43.9 shows a bolt with the shank undercut slightly smaller than the root of the thread; therefore, the shank deflects equally throughout its length and stores energy without overstressing the material in the threads. This effect can be accentuated by increasing the length of the shank. The effect of a reduction in the shank diameter of bolts made from three classes of steel is shown in Table 43.3.[7] The composition and physical properties of the steels are given in the same table.

Stress raisers occur at abrupt changes in shape, such as shoulders, fillets, notches, grooves, and holes. They contribute to a lower endurance limit, and should be eliminated wherever possible. For example, Fig. 43.10*A* shows a stud with a smooth notch machined adjacent the last remaining thread. Maximum bending stress then occurs at the notch where the stress concentration is small, rather than at the root of a thread. The stud is closely fitted in the hole for a short length to reduce further the bending stress. Figure 43.10*B* shows a nut that is undercut so that the load is better distributed over several threads in the more flexible region adjacent the undercut. Plastic deformation of the first highly-loaded thread achieves somewhat the same result; this can be accomplished by a reverse taper or shorter pitch of the bolt threads relative to those of the nut. Subsequently, at lower bolt tensions, the load will be distributed more uniformly in the nut threads.[8, 9]

Bolts usually are available in both coarse and fine thread series. Coarse thread bolts have deeper threads and need not be manufactured to as close a tolerance as fine thread bolts where care must be exercised to ensure a full thread depth. Fine thread bolts have a smaller helix angle; thus, they have less tendency to loosen when vibration is present.

It is good practice in assembling equipment intended to withstand shock and vibration to tighten a bolt until some yielding occurs in the bottom threads, thus ensuring maximum tightness. This reduces the stress change in a bolt when the loading is tension-directed and increases the resistance to fatigue.[10] Bolts may be preloaded by being

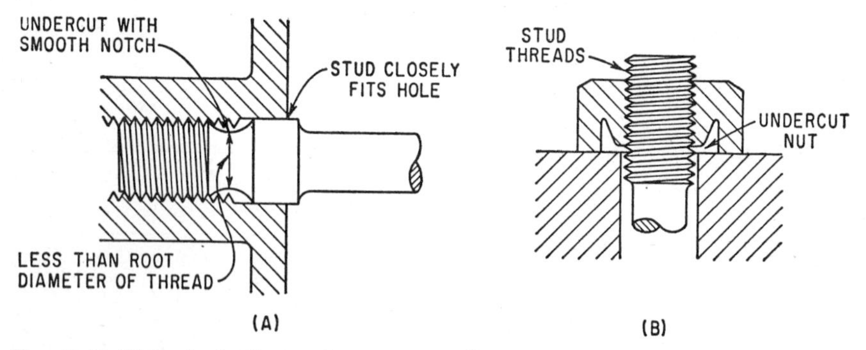

Fig. 43.10. Methods of relieving stress concentration at threads: (*A*) Smooth notch used to relieve bending stress at thread. (*B*) Undercut nut to attain more uniform stress distribution in bolt. (*R. A. MacGregor.*[8])

Table 43.3. Effect of Bolt Diameter and Material on Impact Fatigue Strength

(*W. Staedel.*[7])

Bolt design	Bolt body			Repeated impact fatigue limit					
	Diam., in.	Area, in.^{2}	Area, % root diam. area	Steel *A*		Steel *B*		Steel *C*	
				in.-lb	%	in.-lb	%	in.-lb	%
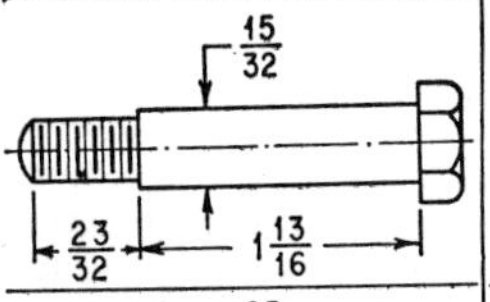 15/32; 23/32; 1 13/16	15/32	0.175	230	1.73	100	1.73	100		
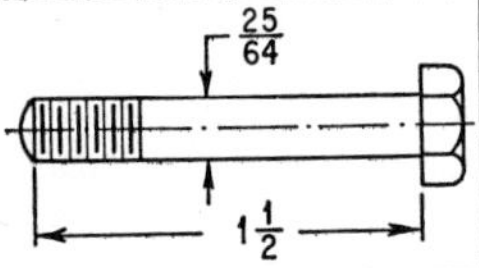 25/64; 1 1/2	25/64	0.122	160	2.60	150	2.26	130	~2.2	100
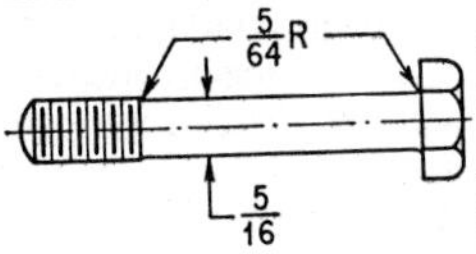 5/64 R; 5/16	5/16	0.076	100	4.34	250	2.78	160	~3.5	160
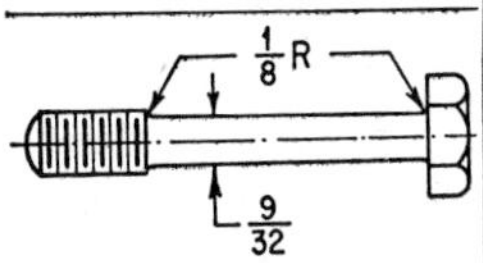 1/8 R; 9/32	9/32	0.060	78	5.21	300	2.95	170	5.21	240
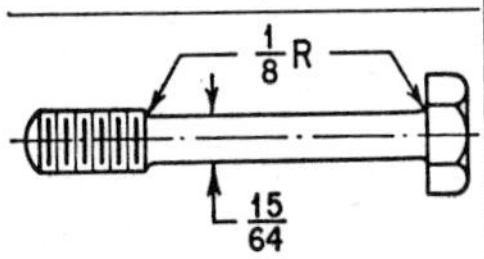 1/8 R; 15/64	15/64	0.044	57	6.51	375	3.65	210		

	Composition, %					Ultimate, lb/in.^{2}	Elong. %	Red. in area, %	Bolt * ultimate, lb
	C	Mn	P	S	Si				
Steel *A*.........	0.15	0.65	0.095	0.186	0.04	78,200	12.8	56.0	9,390
Steel *B*.........	0.25	0.75	0.019	0.024	0.25	96,400	12.8	54.0	11,800
Steel *C*.........	0.33	0.70	0.039	0.027	0.18	85,300–99,600	16–10	...	11,800

* Static tensile strength of actual bolt (for the second of the designs illustrated above) having 5/32-in. length of free threads under nut.

initially heated, then lightly tightened in place, and finally allowed to cool. It is more common to apply a known torque to the bolt head or nut with a special wrench, and estimate the preloading from the following equation:

$$M = 0.2DF \qquad \text{lb-in.} \tag{43.1}$$

where M = torque applied to bolt, lb-in.
D = diameter of bolt, in.
F = preload tension in bolt, lb

Relaxation of bolt tension usually occurs to some extent in service, and a periodic check on preloading is advisable to ensure tight joints.

LOCKING DEVICES. Locking devices[11] include cotter pins, friction nuts, and lock washers. The castellated nut and cotter pin, as well as the locking wire inserted through holes in bolts and nuts, are devices to prevent positively the rotation of the nut relative to the bolt. Such devices must be removed each time the nut is tightened or loosened, and permit adjustment of the nut only at the discrete increments of the locking holes. In another type of locking device, additional thread friction is introduced by the design of the nut; it is a less positive means to retard loosening. Methods of introducing the friction are (1) the bolt is forced through an unthreaded, nonmetallic insert forming part of the nut; (2) the hole in the nut is elliptical and appreciable torque must be applied to turn it on the round bolt; and (3) the inner face of the nut is concave so that when tightened against a shoulder it distorts and binds the threads. In general, means to introduce additional friction are satisfactory when newly installed but tend to deteriorate with repeated use.

The split-ring lock washer is an initially flat washer that has been split and then sprung out of its plane slightly. It is flattened when the nut is tightened, and the resultant friction between washer and nut tends to prevent loosening of the nut. Its resilience enables it to maintain the friction even though the bolt deforms somewhat as a result of shock. Split-ring lock washers may become brittle and fracture; they then fall out of place and a loose joint results. Another type of lock washer embodies a sawtooth edge with an axial set to the teeth; it is adapted to wedge between the nut and the shoulder against which the nut is seated. The teeth are relatively rigid and cannot compensate for the additional space resulting from elongation of the bolt or indentation of the bolted members. This type of lock washer is not suitable for use in equipment subjected to severe shock.

CHASSIS MOUNTING METHODS

To give accessibility to the component parts of an equipment, it is common practice to mount the chassis in the cabinet by hinges or rollers. This imposes the general requirements: (1) the chassis should be movable readily to positions of accessibility, (2) the operators should be protected while the chassis are being so adjusted, and (3) when the chassis are closed, they should be attached securely to the cabinet and should add as much stiffness as possible thereto. Particular considerations that are important with different types of chassis are set forth in the following paragraphs.

DOOR-MOUNTED CONSTRUCTION. Equipment sometimes is designed with many of the parts mounted on a door or on hinged panels to permit better accessibility for maintenance. Two undesirable features of this type of design are (1) the design moves the center-of-gravity of the equipment forward in the cabinet, and (2) it is difficult to stiffen large door panels adequately when they are heavily weighted. Hinges and their associated structures must be designed to support the door adequately in the open position. Dowel pins for the opening side of the door should be provided to relieve the door fasteners from loads when closed. To ensure personnel safety, the opening and closing of heavy doors should be controlled by pivoted lead screws with suitable hand cranks for operation.

RETRACTABLE CHASSIS. It is preferable to mount components on a retractable chassis or directly on the cabinet structure, rather than on a door or hinged panel. There are principal requirements for a satisfactory retractable chassis: (1) the chassis must be constrained from bouncing and (2) the equipment loads must be transferred directly to the frame of the cabinet. (For large and heavy chassis, the additional use of front-panel dowel pins may be desirable. These pins slide into corresponding receiving-hole plates, or receptacles, attached to the chassis.) The front of the chassis is bolted to the frame by screw-type fasteners located around the periphery of the front panel. To eliminate alignment difficulties, the front-panel fasteners screw into "floating" nuts attached to the frame. It is important to place these fasteners at regular intervals around the entire periphery of the front panel to develop maximum stiffness, particularly in the front-to-back direction.

Chassis Slides. Accessibility to the chassis for maintenance and troubleshooting is improved by the use of ball-bearing slides with built-in tilting features. Slides are avail-

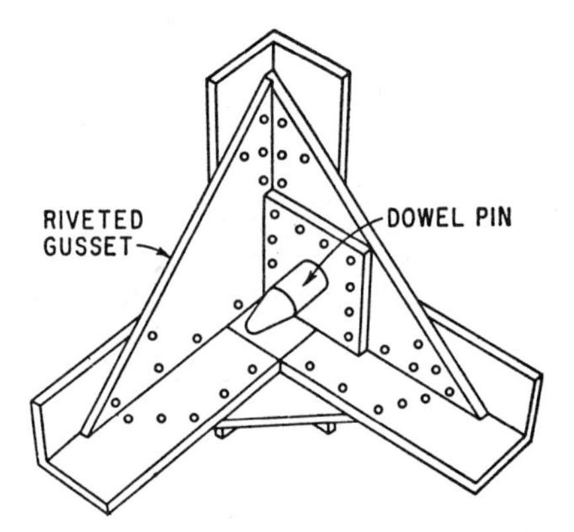

FIG. 43.11. Dowel pin arrangement for riveted or bolted gusset. (*H. M. Forkois and K. E. Woodward.*[1])

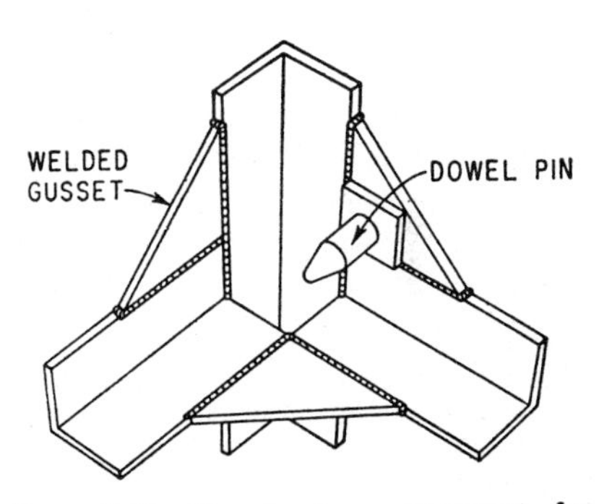

FIG. 43.12. Dowel pin arrangement for welded gusset. (*H. M. Forkois and K. E. Woodward.*[1])

able to permit tilting the chassis either up or down, and also to permit locking at a number of fixed angular positions. Since ball-bearing slides are not designed to carry large loads, it is important that the weight of the chassis in the locked-up position be supported entirely by the dowel pins and front-panel fasteners.

Dowel Pins. Dowel pins should be made of steel to avoid excessive abrasion. In general, they vary in diameter from ¼ to ½ in., and should be no longer than 1 in. to minimize bending stresses. The pin receptacle should also be made of steel with a close fit to the pin. It usually is better practice to locate the pin on the frame of the cabinet and the pin receptacle on the chassis because the frame is better able to support the pin as a cantilever. Figure 43.11 illustrates a pin attached to a riveted or bolted gusset, and Fig. 43.12 shows a pin installed on a welded gusset. Maintaining the necessary alignment of close-fitting dowels (maximum clearance on diametral fit of 0.005 in.) is difficult, particularly where interchangeability of chassis is a requirement. Figure 43.13 shows a dowel pin clamped to the frame and held in place by serrated flange and washer. A maximum adjustment of 3/16 in. in any direction is obtainable because of the oversize hole in the frame.

Quick-release Fasteners. Only a fraction of a turn is necessary to lock or release

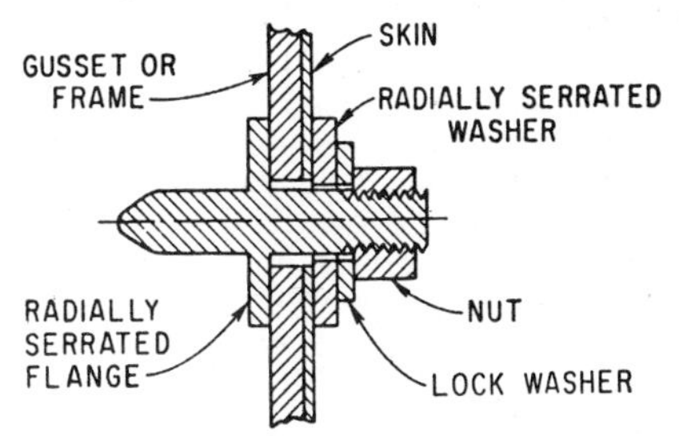

FIG. 43.13. Adjustable dowel pin. A maximum adjustment of 3/16 in. in any direction is obtainable as a result of the oversized hole in the frame. (*H. M. Forkois and K. E. Woodward.*[1])

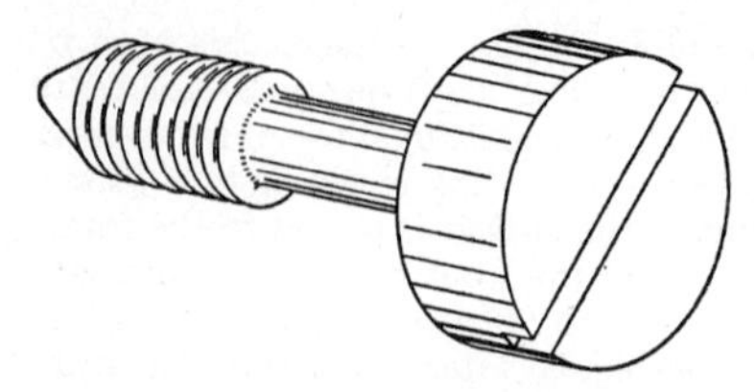

FIG. 43.14. Reduced-shank front-panel thumbscrew which is threaded into a threaded hole in the chassis until the threaded portion passes through the hole; the reduced shank portion then rests in the threaded hole in a captive position which permits turning freely to engage a threaded hole in the frame. (*H. M. Forkois and K. E. Woodward.*[1])

a device in which a spring element is used to maintain tension in the "locked-up" position. Shock and vibration forces of low magnitude tend to overcome the spring force and cause looseness in the assembly. Under shock, the spring or the cam rider (the member which engages the spring and through which the loads are transmitted from the captive mass to the support) may deform or fracture. Hence, quick-release fasteners should not be used to carry large loads but rather may be used to mount such items as inspection plates or covers.

Front-panel Thumbscrews. Figure 43.14 illustrates a captive fastener used to secure the chassis to the frame of the cabinet. It is knurled and slotted to facilitate loosening and tightening. The thumbscrew is threaded into a threaded hole in the chassis until the threaded portion passes through the hole; the reduced shank portion then rests in the threaded hole in a captive position which permits turning freely to engage a threaded hole in the frame. Reduced shank diameters vary according to design requirements but should not be less than $\frac{3}{16}$ in. for lightweight or $\frac{5}{16}$ in. for mediumweight equipment. Corresponding thread sizes are $\frac{1}{4}$ in.-20 and $\frac{3}{8}$ in.-16.

DAMPING

A direct approach to the reduction of vibration resulting from resonance is to supplement the inherent hysteresis damping of materials by other means to dissipate energy. A structure fabricated from a number of component parts exhibits a marked degree of energy dissipation compared with a similar structure formed in a single piece. The method of fabrication influences the degree of energy dissipation. Bolted and riveted joints are most effective in dissipating energy because they permit a limited slippage at the interfaces between members in contact while maintaining pressure at the interface. Welded joints also exhibit considerable damping, less than bolted or riveted joints but substantially more than solid materials. This apparently is the result of some slippage between members in contact not restrained fully by the weld.

ENERGY DISSIPATION. The damping or energy dissipation resulting from slippage of parts in a fabricated structure is termed *slip damping.* The analytical basis for this type of damping is established in Chap. 36, and applications of the theory are discussed. The considerations involved in attaining optimum conditions also are discussed. Many structures encountered in the design of equipment are irregular and complex, and the theory given in Chap. 36 may be applied only with considerable difficulty. Various types of structures have been found to exhibit damping properties that are known within wide and approximate limits. Knowledge of such properties in a quantitative sense would assist in the design of equipment having desirable energy dissipation.

Frictional effects in composite structures are not well adapted to analysis except under ideal conditions. As guidance for the designer, it is convenient to define the damping in terms of the *equivalent viscous damping.*[12] The inverse of the fraction of critical damping ζ is proportional to the amplification at resonance [13] and makes it possible to estimate the severity of a resonance. Table 43.4 gives typical values of the fraction of critical damping ζ found in various types of structures. The damping varies widely from structure to structure of a given type; for example, in a given structure, it varies significantly with a change in tightness of bolts. Thus, Table 43.4 should be interpreted only as an indication of the relative magnitudes of damping in different types of structures, and of the variation in damping that may be expected from a change in design.

Damping can be added to members in flexure by utilizing the shear stress at the neutral plane to deform a layer of viscous material and attain a relatively high degree of energy dissipation. This sometimes is known as *sandwich construction.* The analytical

Table 43.4. Typical Values of Damping for Different Methods of Construction

Method of construction	*Effective damping,** ζ
One-piece unit.	0.01
Welded assembly.	0.02
Riveted assembly.	0.04
Bolted assembly.	0.05

* ζ = fraction of critical damping [see Eq. (2.12); also see Chap. 42].

basis for an understanding of this concept is included in Chap. 36. The concept as applied to the design of structural members is illustrated in Fig. 43.15 where a cantilever beam of rectangular cross section (inset *A*) is compared with a sandwich-type beam (inset *B*) having the same length and natural frequency.[14] The viscoelastic layer provides a partial constraint in shear between the separated portions of the beam; thus, the natural frequency of the sandwich-type beam is lower than that of a solid beam having the same over-all dimensions, but greater than that of one-half of the sandwich-type beam. The effective value of the fraction of critical damping $\zeta = c/c_c$ for the beam depends upon the viscosity of the viscoelastic layer and decreases as the load carried by the beam increases.

The sandwich-type construction has inherent limitations. For a given over-all dimension, a sandwich-type structure is less rigid than a structure using solid members.

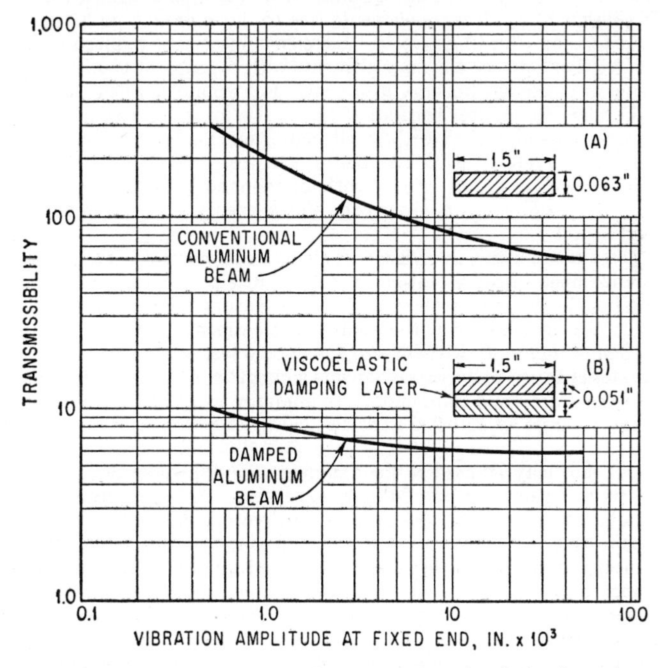

FIG. 43.15. Comparison of structural damping in a conventional aluminum beam with that in a damped aluminum beam having a viscoelastic damping layer. The beam is a cantilever 11 in. long with cross section (*A*) for conventional beam and (*B*) for damped beam. Natural frequency is 18 cps. Transmissibility is ratio of amplitude at free end to amplitude at fixed end, during vibration at resonance. (*J. E. Ruzicka.*[14])

From a fabrication viewpoint, a sandwich-type structure is not well adapted to welding; appropriate bolted connections preferably should include means to relieve the viscoelastic layer of bolt forces. Finally, the viscosity of the viscoelastic layer tends to vary with temperature. Nevertheless, a large increase in damping is attainable by use of sandwich-type construction and may justify use of the concept for certain applications.

Members that deflect in flexure also may be damped by the application of vibration damping material that dissipates energy; this concept is discussed in detail in Chap. 37. Its effectiveness is a function of the energy dissipated by the undercoating relative to the elastic energy of the structure being damped. For this reason, it finds principal application to relatively light structures of large area, such as hoods and trunk covers of automobiles or enclosures for household appliances.

When vibration of an equipment or structure is characterized by a large amplitude at a definable point, a possible method of reducing the vibration level involves the attachment of a damper. Such a device dissipates energy as it extends and compresses. A damper can be attached between a vibrating structure and a relatively fixed structure, or between two structures vibrating in opposite directions. The design of dampers is discussed in Chap. 32.

AUXILIARY MASS DAMPERS. A tuned damper or dynamic vibration absorber may be attached directly to the vibrating structure; it consists of a simple mass-spring system that may be damped or undamped. Reduction of vibration is achieved because the mass vibrates out of phase with the vibration to be eliminated; sometimes this is called *sympathetic vibration*. The theory of the tuned damper and a number of applications are discussed in Chap. 6; applications to internal-combustion engines are discussed in Chap. 38.

An auxiliary mass damper without a spring or damping is designated an *acceleration damper*. It consists of small mass elements free to move or slide in a sealed container that is attached to the vibrating system. Incipient vibration is reduced by the multiple collisions and friction between the mass elements with consequent transfer of momentum and conversion of mechanical energy into heat. For example, when electrical contacts are closed, they tend to chatter because of the elastic impact; to minimize the chatter, a small container partially filled with small mass elements is attached to the movable contact. The abrupt motion of the contactor arm tosses the powder about inside the container. The result is a dull impact without rebound, because the relative motion and friction of the mass elements absorb the elastic energy involved in the stress waves set up by the impact.[15]

Analogous results are attainable with a single massive member contained within a box attached to a member whose vibration is to be reduced. Analyses of this damper[16,17] indicate that its effectiveness is the result of momentum transfer induced by multiple impacts within the box; however, the analytical treatment is not sufficiently complete to permit design of a damper without supplementary test work.

An ideal material to use as the mass elements is tungsten powder. It has rough individual grains to create friction, is hard enough to minimize wear, and is sufficiently massive. A half-and-half mixture of 150–200 and 200–250 mesh powders is commercially available.*

Sand or steel shot may be used as the damping material where the quantity required would make tungsten powder prohibitively expensive. Some experimental results are available for the damping attainable from a flat, rectangular tray partially filled with steel shot of No 15 grit.[18] The tray was attached to a vibrating table and subjected to a vibratory motion in the lengthwise direction of the tray at a frequency of 13 cps. (The latter frequency happened to be of particular interest because it corresponded to the natural frequency of a radar antenna which was being set into vibration by gusts of wind.) The vibration table was put into motion by means of a rotating unbalance weight supported in bearings attached to the underside of the table. Various tray lengths and different weights of shot were tried; the width of the tray was fixed at 3 in.

* Westinghouse Electric Corporation, Lamp Division, Bloomfield, N.J., mixture specification 35-8-562.

Figure 43.16*A* shows that with a given rotating unbalance weight causing a harmonic forcing function $F = 1.7$ lb, the displacement amplitude could be reduced by adding more weight of damping material to the tray. At each of several different tray lengths an optimum weight of damping material was found beyond which the resulting displacement amplitude could no longer be reduced by the addition of damping material. The optimum weight of damping material was found to be larger as the length of the tray was increased. The results plotted in Fig. 43.16*A* indicate that energy was absorbed primarily because of sliding friction of the mass particles moving as a unit relative to the bottom and sides of the tray.

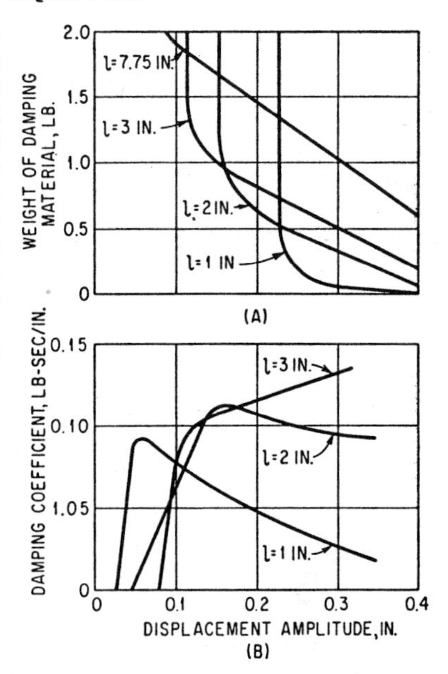

FIG. 43.16. Performance curves for steel shot damper attached to a table driven by a rotating unbalance at a frequency of 13 cps: (*A*) Variation of vibration amplitude as a function of weight of damping material for several lengths l of tray; driving force amplitude is 1.7 lb. (*B*) Damping coefficient as a function of vibration amplitude, calculated from driving force and vibration amplitude. Weight of damping material is 0.91 lb; the tray length $l = 3$ in. is divided into three 1-in. long compartments. (*G. O. Sankey.*[18])

Figure 43.16*B* shows the results when the tests were repeated for the same weight of damping material $W = 0.91$ lb but located in trays of various lengths. The magnitude of the rotating unbalance weight was adjusted successively to establish various displacement amplitudes while maintaining the frequency at 13 cps; the equivalent viscous damping coefficient was calculated at each amplitude. For the shorter tray lengths, the damping coefficient increased rapidly as the displacement amplitude increased; however, it reached a peak value and then decreased gradually as the displacement amplitude was further increased. The exception was the 3-in. long tray, which was partitioned into three 1-in. long compartments, and gave the most effective damping action. The rising characteristic for the 3-in. tray shown in Fig. 43.16*B* indicates that the damping coefficient continues to increase with displacement amplitude and indicates a safe, stable design for the damper. By using partitions to reduce the lengthwise freedom for relative motion, the sliding friction is supplemented by the more effective turbulence of the damping material. The increased energy absorption indicated by the larger damping coefficient was caused principally by the multiple, inelastic collisions. However, depending upon circumstances, the net damping is the result of the combined sliding friction and inelastic collisions of the mass elements.

MALFUNCTION OF EQUIPMENT

Failure of equipment as a result of shock and vibration may consist of (1) damage so severe that the ability of the equipment to perform its intended function is impaired permanently or (2) temporary disruption of normal operation in a manner permitting restoration of service by subsequent adjustment of the equipment or termination of the disturbance. For example, a common type of disruption involves excessive vibration of the elements within an electron tube; damage frequently does not occur but rather the electron tube generates spurious signals and is unable to perform its intended function. This difficulty is corrected by applying the principle set forth as a guide to the design of structures to withstand shock and vibration; i.e., a change in stiffness or damping of the electron tube elements is effected. However, another type of disruption may occur in a relay or circuit breaker where shock or vibration causes unintended and improper operation. Normal operation can be restored readily if the equipment is accessible to personnel for adjustment. Meanwhile, serious consequences may have developed from the disruption. The following paragraphs of this section discuss design principles that can be applied to minimize improper operation.

BALANCING OF MOVING MEMBERS

Where possible, pivoted or hinged members should be balanced statically to prevent maloperation. For example, Fig. 43.17 shows a circular rheostat with a single pivot arm

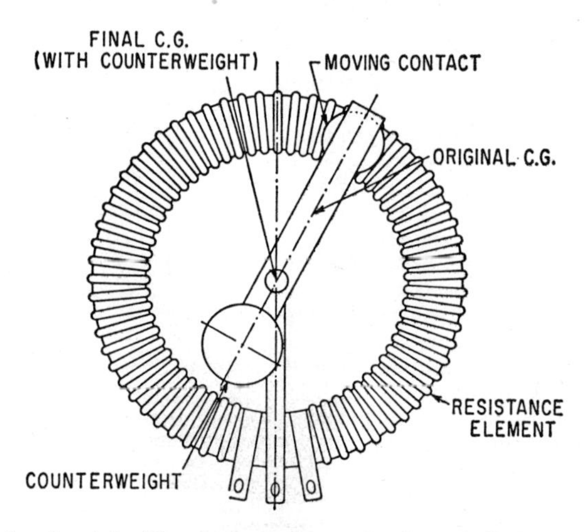

FIG. 43.17. Circular rheostat with a single pivot arm that is statically balanced. (*C. E. Crede and M. C. Junger.*[2])

that is statically balanced by the addition of a counterweight. Figure 43.18 shows a push-button station that is inherently unbalanced and vulnerable to maloperation as a result of accelerations along the line of motion of the push button. The pivoted arm and counterweight are added to prevent maloperation. The counterweight plus the assembly of push button with contacts are statically balanced about the fulcrum of the pivoted arm. Note that the statically balanced mechanisms shown in Figs. 43.17 and 43.18

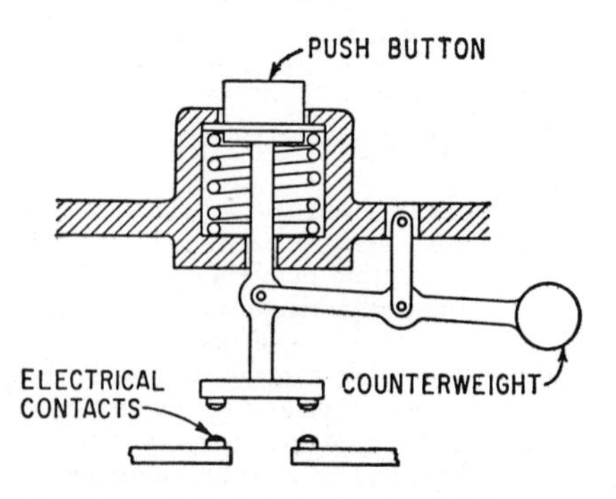

FIG. 43.18. Statically balanced push button. The center-of-gravity of entire linkage is at fulcrum of counterweight arm. (*W. P. Welch.*[3])

would tend to malfunction if subjected to a shock motion having a strong rotational component. In general, severe shock tends to involve translational motion primarily, and statically balanced mechanisms give adequate service.

In mechanisms involving more than one pivot, it is possible to make them unresponsive to rotational acceleration about any axis by arranging similar mechanisms to act in opposition or "back to back." Typical designs having this feature are described in Ref. 2.

MANUAL AND AUTOMATIC LATCHES

Locking devices, such as manual or automatic latches, represent the most direct means of preventing maloperation. For example, clamps can be added to hold a knife switch

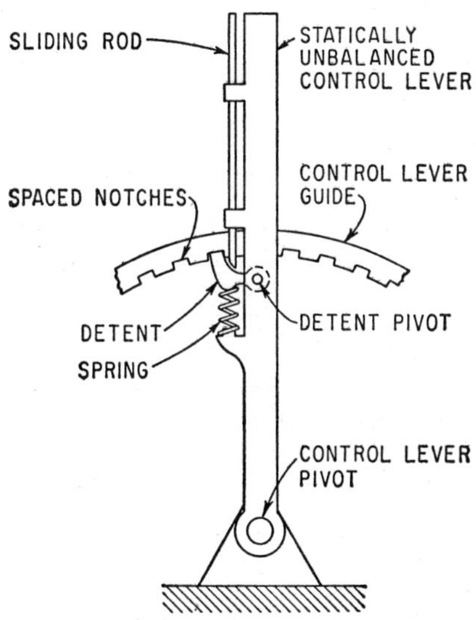

Fig. 43.19. Manual latch with control lever and detent that is not statically balanced. (*C. E. Crede and M. C. Junger.*[2])

blade in either the open or the closed position. Similarly, mechanisms can be used to block the tripping linkages of circuit breakers and contactors. Such devices are undesirable expedients, because normal operation of the knife switch is impeded and the protective feature of the circuit breaker is nullified. However, in some types of mechanisms, a manually operated latch can be used without impeding normal operation seriously. For example, Fig. 43.19 shows a control lever that is not statically balanced. Means are provided to latch it to a selected notch in a lever guide. A detent is pivoted to the lever and urged against the guide by a spring. A sliding rod carried by the lever is depressed by the operator to disengage the detent and allow the lever to be moved to another notch.

Latches may be actuated automatically to prevent maloperation without interfering with normal operation. Figure 43.20 shows a contactor locking device actuated by the electromagnet of the contactor. The contacts are closed by energizing the solenoid and causing the armature to pivot against the force of the tension spring. The magnetic flux from the solenoid rotates the latch against the force of the bias spring, thereby aligning the latch with the hole in the armature and allowing the armature to close in the normal manner. When the solenoid is deenergized, the armature is blocked open by the latch. This latch prevents accidental closing of the contacts as a result of acceleration toward the left; it does not prevent accidental opening due to acceleration toward the right. However, accidental opening is less likely because the magnetic force tends to hold the contacts closed.

Figure 43.21 shows a contactor with a set of armature latches which are shock-actuated to block the contacts in either the open or closed position. The contacts are shown in Fig. 43.21*A* in the open position and are subsequently closed by vertical upward motion of the armature. A pair of armature latches are mounted on pivots to the left of the

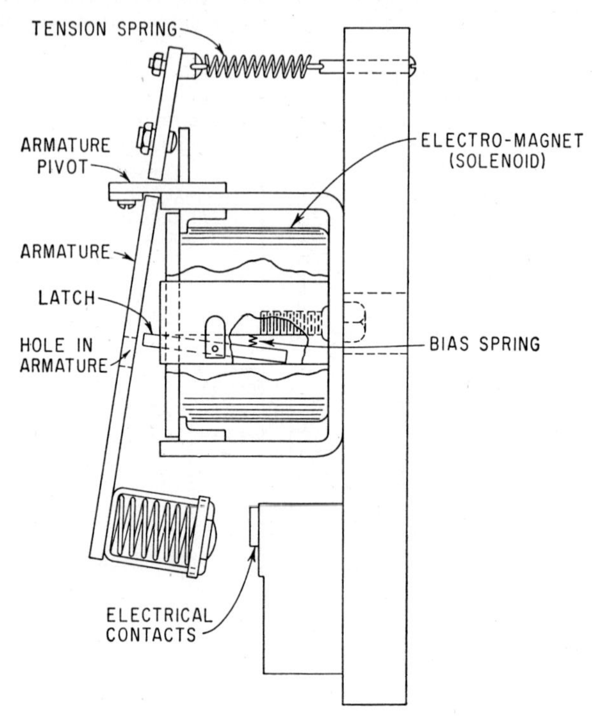

FIG. 43.20. Magnetically actuated latch to prevent contact closure under shock when solenoid is deenergized. (*W. P. Welch.*[3])

crossbar. The pivot for the "out" latch shown in Fig. 43.21*B* is located to the right of the center-of-gravity; thus, a downward shock causes the latch to rotate in a clockwise direction against the force of the tension spring and engage the crossbar. This prevents upward motion of the armature and accidental closing of the contacts. Similarly, the "in" latch would rotate in a counterclockwise direction if the contactor is closed as shown in Fig. 43.21*C* and receives an upward shock. Thus, the "in" latch prevents accidental opening of the contactor. This type of latch is successful only if the latches respond more rapidly than the armature to all shock motions that are likely to occur. Since it is impossible to ensure this, the shock-actuated latch involves a degree of unreliability.

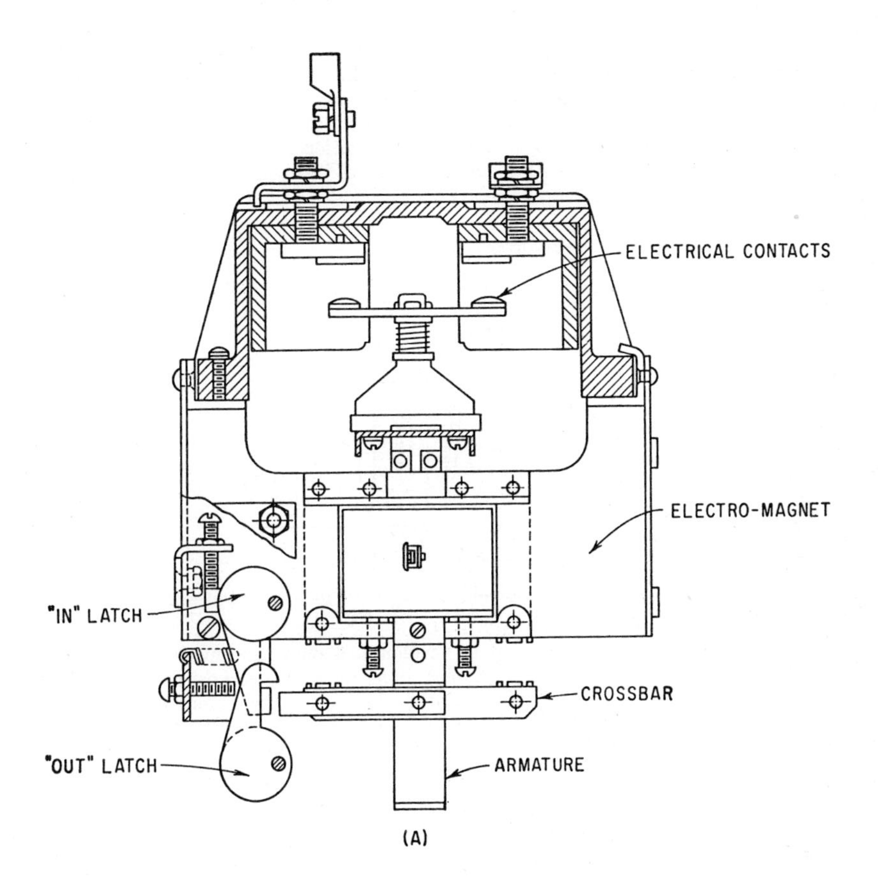

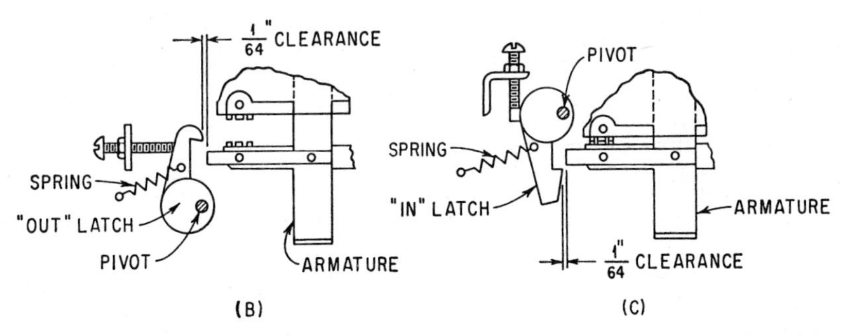

FIG. 43.21. (*A*) Contactor with shock-actuated antishock devices; (*B*) armature in position of "opened" contacts; (*C*) armature in position of "closed" contacts. (*W. P. Welch.*[3])

PRELOAD AND OVERCENTER

Electrical contacts can be prevented from chattering as a result of shock and vibration by interposing a spring that forces one contact against another in closed position. In the contact design shown in Fig. 43.22, the position of the contactor arm relative to the base is fixed when in closed position. This fixed position can maintain the compression spring in a deflected position with a consequent force being applied to the contacts.

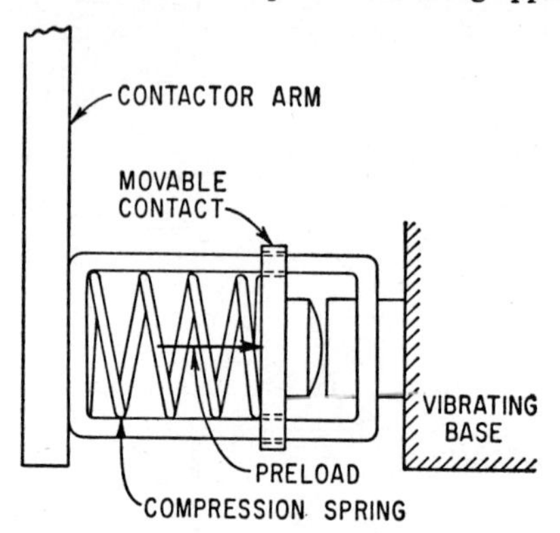

FIG. 43.22. Spring preloading of movable contact to prevent separation (opening) due to shock and vibration.

A contactor may be made resistant to accidental opening and closing by the use of preloading springs to hold a crank mechanism in an overcentered position, as illustrated in Fig. 43.23. In either extreme position of the mechanism, the crank is rotated beyond its dead center and the crank pin engages a stop. The mechanism is held in its open position shown by Fig. 43.23*A* by means of springs *A*. It is held in its closed position shown by Fig. 43.23*B* by means of springs *B*. In the event of either an upward or

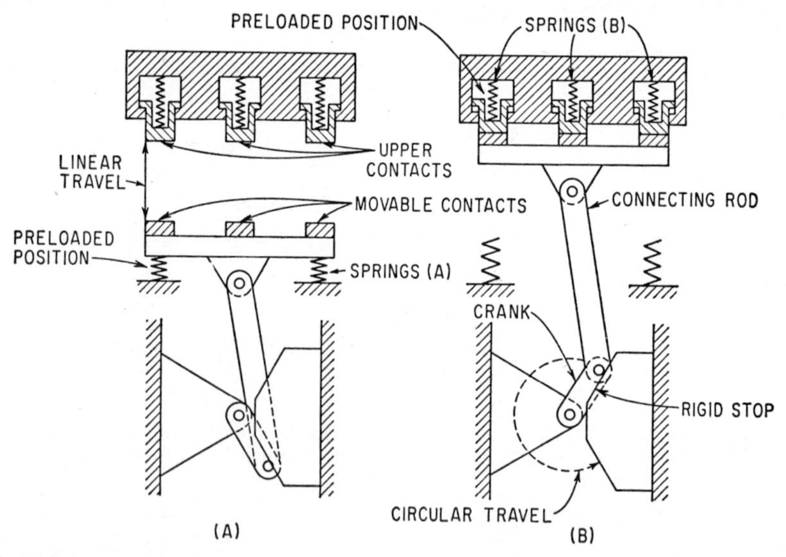

FIG. 43.23. Overcentered three-pole contactor in (*A*) open and (*B*) closed positions. (*C. E. Crede and M. C. Junger.*[2])

downward shock, the mechanism would respond only if the acceleration associated with the shock motion has the magnitude and direction necessary to move the crank over dead center against the opposing force of the preloading springs.

IRREVERSIBLE MECHANISMS

Shock resistance often can be obtained by employing an irreversible component or combination of elements. An irreversible component is one which responds only to a force or torque applied at the input side, remaining unresponsive to a force or torque applied at the output side. A common irreversible device is a screw and nut, in which the nut experiences translational motion as a result of a rotation of the screw. However, the screw cannot be made to rotate by applying an axial force to the nut unless the helix angle of the screw thread is made substantially larger than normally used. Similarly,

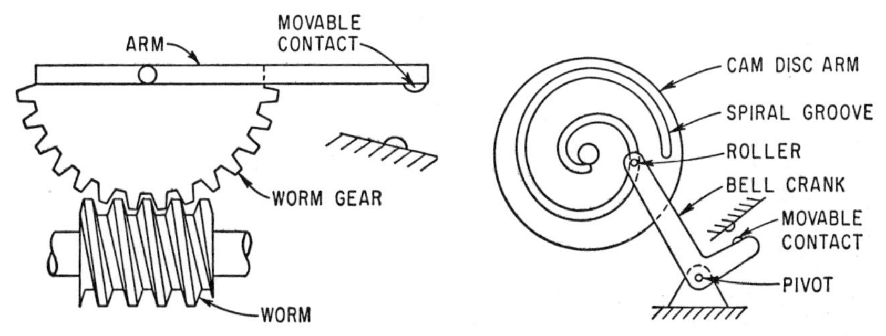

FIG. 43.24. Contactor actuated by worm and gear. (*C. E. Crede and M. C. Junger.*[2])

FIG. 43.25. Contactor actuated by spiral can. (*C. E. Crede and M. C. Junger.*[2])

in a worm-and-gear combination, the gear can be rotated by rotating the worm, but usually not the reverse. The condition for reversibility is a function of the helix angle and coefficient of friction at the sliding surfaces.

The application of a worm and gear to a contactor is illustrated in Fig. 43.24. The arm which carries the movable contact is secured to the same shaft as the worm gear. Movement of the contact arm can be attained only by rotation of the worm gear shaft. Note that it is not necessary for the contact arm to be statically balanced to attain resistance to shock.

A contactor which employs a spiral cam is illustrated in Fig. 43.25. The movable contact is carried on one arm of a bell crank. The opposite arm of the bell crank carries a roller which rides in a spiral groove in the disc. The cam is driven with a rotary motion by a motor which controls the action of the contactor. It is not necessary that the bell crank be statically balanced.

DESIGN OF COMPONENTS

A record of damage sustained by equipment during laboratory tests is significant in emphasizing the considerations that are important in attaining resistance to damage from shock and vibration. Compilations of damage experiences are available.[1] These show that failure of principal structures and mounting brackets is the most common form of damage. During shock, structures may not have sufficient strength to withstand the forces that are applied; during vibration, resonant conditions may develop and the relatively undamped structures may fail from fatigue. Failure of electrical leads from fatigue is common, often because the leads are used improperly to support resistors and capacitors. Failure of electron tubes is common, but only because they are used in large quantities in electronic equipment; failure in terms of percentage of tubes used is not large. On the other hand, a relatively large percentage of incandescent lamps and cathode-ray tubes experience failure. In some instances, failure may be ascribed to the inherent properties of the components that failed; in other instances, failure is the result of improper installation. This section discusses design and selection of components to withstand shock and vibration, and the following section discusses the mounting of such components.

BRITTLE MATERIALS

Components which embody brittle material having important functional roles, such as glass, ceramics, and jewels, present particular problems in the protection of such material from fracture under shock. General design principles for the protection of brittle materials are difficult to formulate; however, several examples are given to show the circumstances of typical damage occurrence and corrective measures that have proved effective.

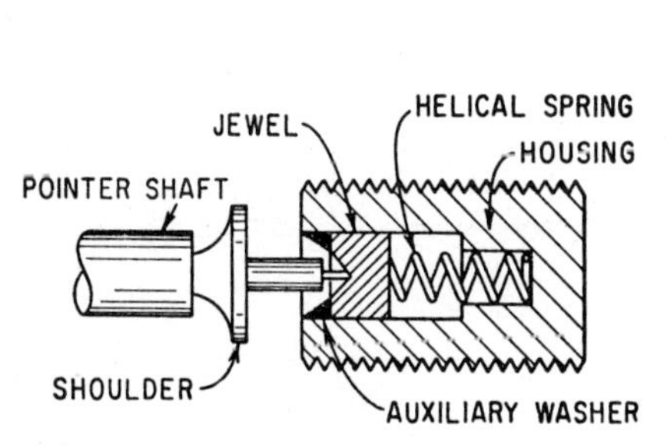

Fig. 43.26. Section through a flexibly mounted instrument bearing. (*R. D. Hickok.*[19])

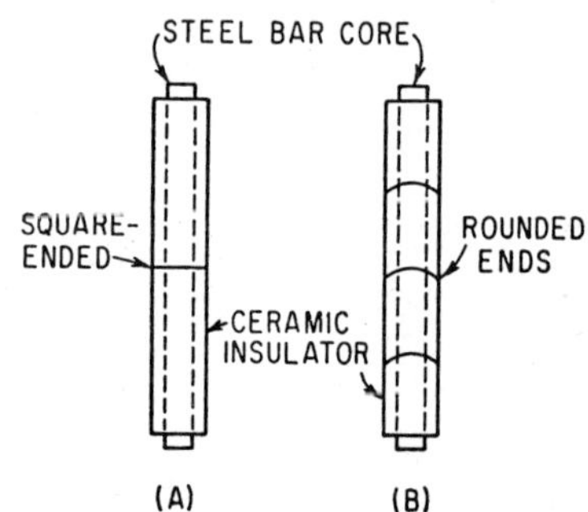

Fig. 43.27. Edgewound resistors showing long, square-ended ceramic sections (*A*) and short round-ended sections (*B*). The latter has improved resistance to shock and vibration.

JEWEL BEARINGS. Damage to brittle materials is likely to occur because an applied force is excessive and/or because it is applied too suddenly. For example, the pivot bearings used for the pointers of instruments usually are made of hard, brittle material that tends to fracture if a large force is applied to it suddenly. Figure 43.26 shows a spring mounting for the bearing to reduce the severity and suddenness of load application, thereby minimizing the possibility of blunting the pivot or splitting the jewel. The shoulder on the shaft is provided to limit the axial travel so that the bearing cannot bottom forcibly in its retainer.[19]

GLASS AND CERAMICS. A brittle member may be damaged because it is constrained to experience a large deflection as a consequence of its attachment to a large

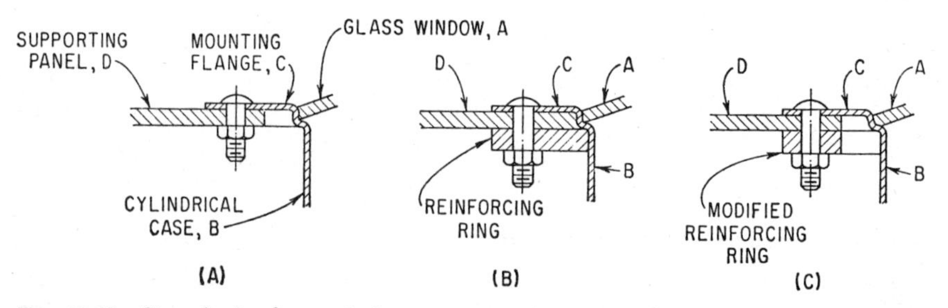

Fig. 43.28. Steps in development of correct strength and flexibility of the flange mounting for a glass window in a small instrument. (*C. E. Crede and M. C. Junger.*[2])

flexible member inherently capable of experiencing a greater deflection. For example, Fig. 43.27 shows an edge-wound resistor having a steel core capable of deflecting laterally when subjected to shock. The original long, square-ended ceramic sections shown in Fig. 43.27*A* tend to chip and split because they are incapable of participating in the lateral deflection of the core. Replacement of the original ceramic sections by shorter sections having rounded ends to permit knee action bending, as shown in Fig. 43.27*B*, increases the capability of the resistor to withstand shock.

Figure 43.28 shows the evolution in the design of a glass instrument window capable of withstanding severe shock. The glass window is sealed to a mounting flange, and must experience an equivalent distortion. In the design of Fig. 43.28*A*, the mounting flange was too flexible and its subsequent distortion under shock was greater than the glass could withstand. When the reinforcing ring shown in Fig. 43.28*B* was introduced to limit the distortion of the mounting flange, the glass failed because the forces transmitted by the rigid mounting arrangement to the glass were too great. A modified reinforcing ring, shown in Fig. 43.28*C*, reduced the magnitude of both the transmitted force and the mounting flange distortion, thereby minimizing damage to the glass window.

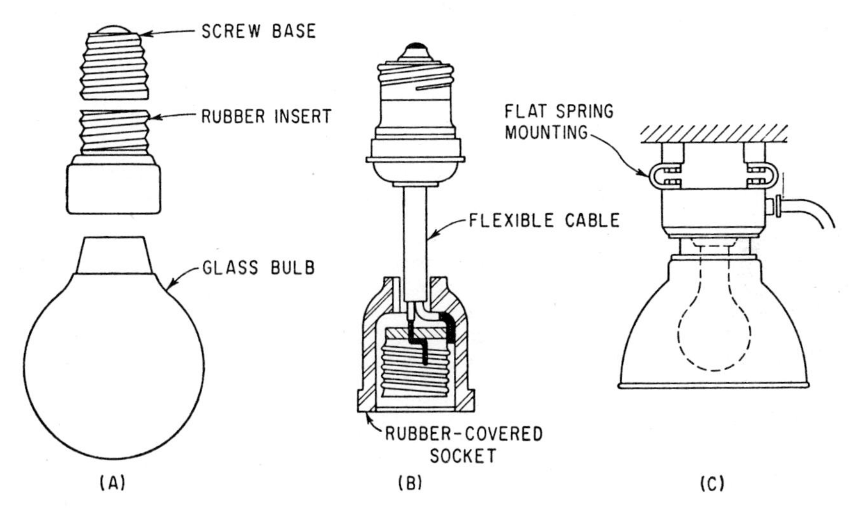

Fig. 43.29. Examples of shockproof designs and installations of incandescent lamps: (*A*) rubber insert between screw base and glass bulb; (*B*) rubber-covered socket and flexible cable; (*C*) flexible U-shaped flat spring mounting. (*C. E. Crede and M. C. Junger*;[2] *W. P. Welch.*[3])

Figure 43.29 illustrates several methods of protecting incandescent lamps. The lamp may embody a rubber insert which separates the screw base from the bulb and filament, as shown in Fig. 43.29*A*. Flexible electrical conductors extend through the rubber insert and connect the filament to the screw base. Alternatively, the socket may be flexibly mounted to permit the use of a conventional lamp. For example, a rubber-covered socket may be suspended by a short length of flexible cable from an overhead structure, as shown in Fig. 43.29*B*. Excessive swinging may be prevented by using a guard ring attached to the rubber socket. Where excessive relative motion between the lamp and fixture must be avoided, the entire lighting fixture may be mounted on flexible U-shaped flat springs, as shown in Fig. 43.29*C*.

MECHANICAL MEMBERS OF INHERENT FLEXIBILITY

Equipment that includes long shafts is particularly susceptible to malfunction or other impairment of operation as a result of lateral deflection of a shaft. A shaft with a long distance between supporting bearings is inherently capable of relatively large lateral

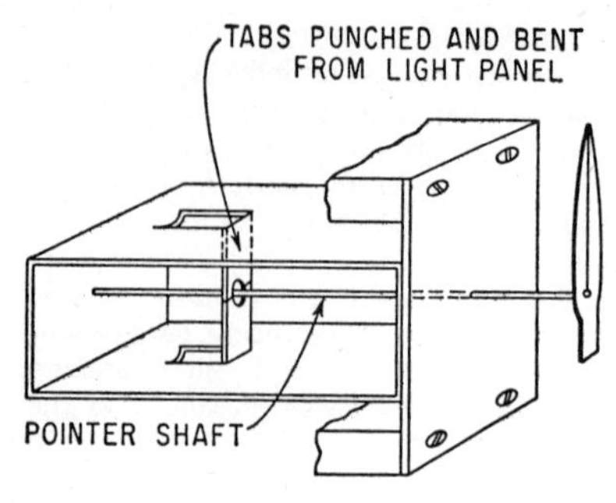

FIG. 43.30. Means to limit shaft deflection. (*C. E. Crede and M. C. Junger.*[2])

deflection without damage to the shaft itself; however, other damage may result from the deflection. For example, the shaft that carries the pointer of an electrical indicating instrument may deflect sufficiently to unseat the ends of the shaft from the jewel bearings. This type of damage may be minimized by the method illustrated in Fig. 43.30; it consists of tabs normally spaced from the shaft but arranged to prevent excessive deflection in event of severe shock. This concept may be applied to shafts having gears where a large lateral deflection of a shaft may cause temporary disengagement of gears, with the possibility of re-engagement in an incorrect relation.

In general, it is poor practice to rely upon friction to retain or position a component. However, the need for instant removal and replacement may justify the use of friction-type retainers if sufficient care has been used in design to ensure satisfactory operation. For example, fuse clips of the type shown in Fig. 43.31 must permit instant removal of the fuse and must provide high contact force to maintain a low electrical resistance at the connection. The clamping action must be tight enough to prevent a fuse from being dislodged by shock, yet loose enough to permit easy replacement of fuses. These results are achieved by refinement of design. The base of the clip is curved concave downward so that additional clamping action is obtained when the clip is screwed to a flat panel. The fuse cannot move lengthwise out of the clips because of the strips or lugs formed at the outer edges of the clips. Clips may be reinforced by additional spring retainers.

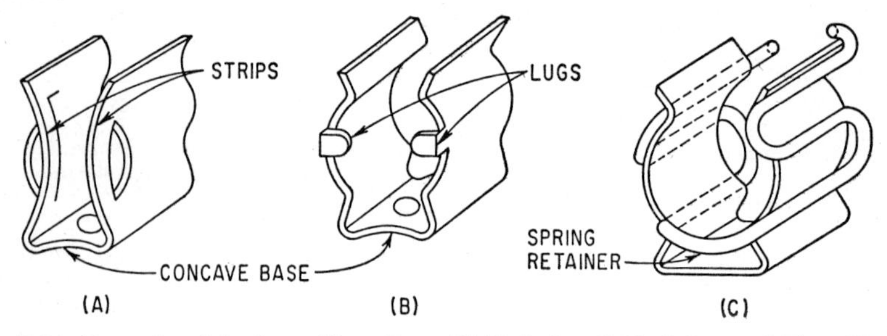

FIG. 43.31. Examples of shockproof fuse clips. (*C. E. Crede and M. C. Junger*; [2] *W. P. Welch.*[3])

RELAYS

A relay is a continuing source of potential difficulty when subjected to shock and vibration. Fundamentally, it includes a moving armature which introduces liability of malfunction as a result of gross movement of the armature. Furthermore, the electrical contacts are susceptible to chatter as a result of resonances of local structures and other dynamic effects. Relays designed to withstand shock and vibration give quite satisfactory service if the shock magnitude is relatively low and if the vibration frequencies are low. Higher levels of shock and higher frequencies of vibration tend to cause malfunction, but the limits of effective operation are difficult to define quantitatively. An indication of expected reliability under conditions of vibration is given in Fig. 43.32; the points on this chart are acceleration amplitudes at which malfunction occurred in a group of relays selected at random and subjected to sinusoidal vibration.[20]

Performance of relays when subjected to shock and vibration usually is better in the energized condition because the magnetic field of the coil tends to damp or restrain the motion of the armature. For example, in the case of the relay shown in Fig. 43.33, the duration of the disruption from shock, when the relay was energized, was in one case only 30 per cent of the duration of the corresponding disruption in the deenergized condition. This result was found by noting the time required for a shock-excited disruption of normal operation to disappear. Performance also is affected greatly by the direction of the shock. A shock motion parallel to a line through the axis or pivot points of the armature usually has the least disturbing effect. When the direction of the shock motion is known, it may be possible to gain shock resistance by the physical orientation of the relay.

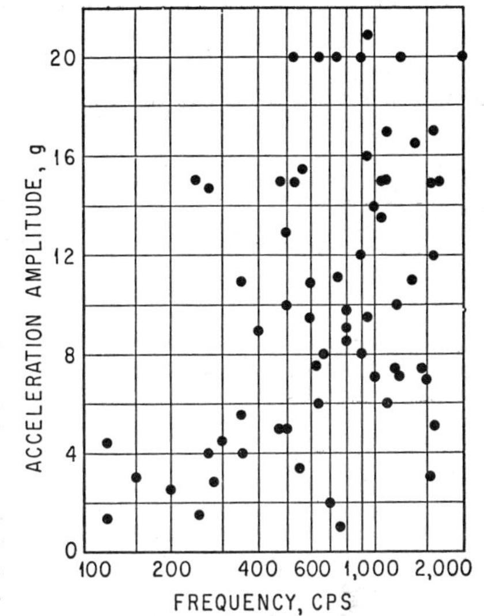

FIG. 43.32. Record of performance under sinusoidal vibration of a group of relays selected at random. The frequency and acceleration amplitude are indicated on the coordinate axes; each dot indicates malfunction of a relay. (*R. E. Barbiere and W. Hall.*[20])

Several types of relays are available, and exhibit varying degrees of reliability when subjected to shock and vibration. The clapper-type relay shown in Fig. 43.34 is inherently unbalanced; it is particularly susceptible to malfunction by movement of the armature when the excitation is along axis 1. To a somewhat less degree, it is vulnerable along axis 3. There is a tendency for chatter to develop at the contacts during vibration at relatively low frequencies.

Application of the concept of a balanced mechanism to the problem of relay design has resulted in the relays shown in Figs. 43.33 and 43.35, for example. In general, this type of relay resists improper operation even when subjected to severe shock; i.e., the balance of the armature prevents its motion. (These relays are not balanced with respect to rotational motion; see previous section on *Balancing of Moving Members.*) However, the problem of chatter at the contacts has not been solved with equal effectiveness. Tests indicate chatter at the contacts under conditions of severe shock and of vibration at high frequencies.

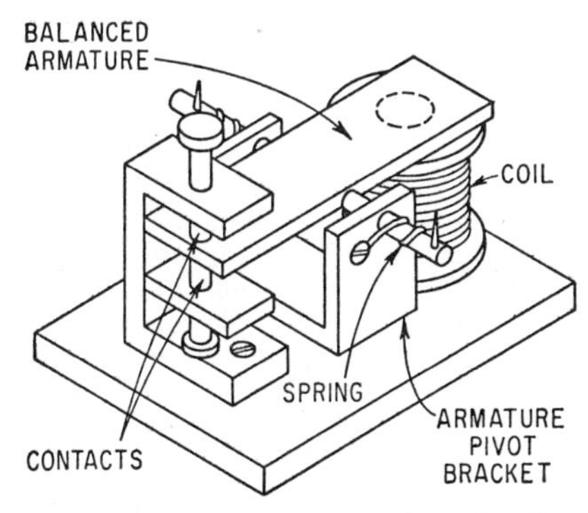

FIG. 43.33. Sensitive balanced relay. (*H. M. Forkois and K. E. Woodward.*[1])

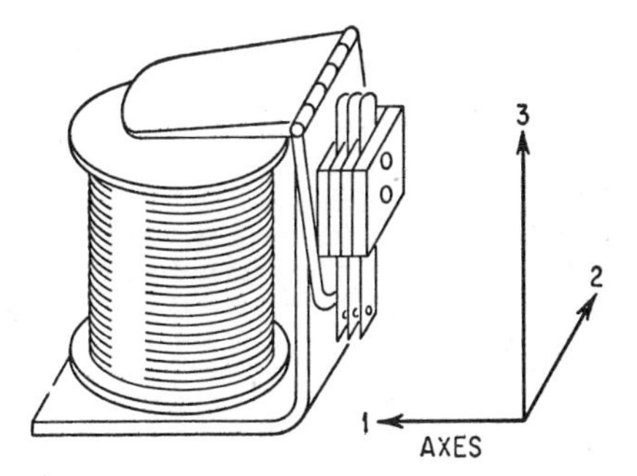

FIG. 43.34. Clapper-type relay. It is inherently unbalanced and is susceptible to shock along axes 1 and 3. (*R. E. Barbiere and W. Hall.*[20])

The plunger-type relay illustrated in Fig. 43.36 is inherently resistant to maloperation by movement of the armature when the excitation is along axes 1 and 2. Along axis 3, the magnetic forces acting on the armature are sufficient to withstand the imposed forces for moderate levels of shock and vibration. However, the relay is inherently unbalanced and susceptible to malfunction if the shock severity becomes sufficiently great. The contact support members in the plunger-type relay tend to be relatively short; as a consequence, tendency to chatter as a result of vibration is less than in some other types of relays.

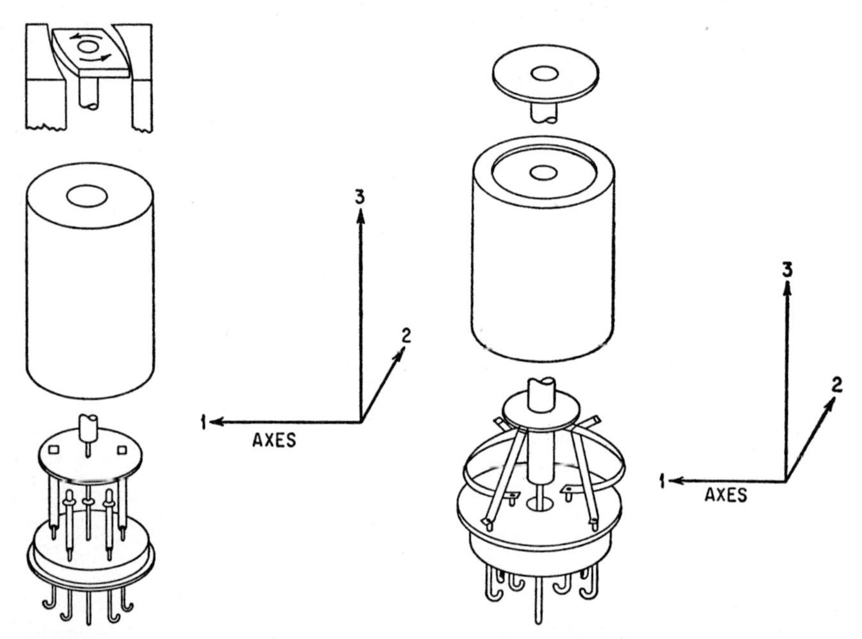

FIG. 43.35. Rotary-type relay. It is balanced to resist shock and vibration. (*R. E. Barbiere and W. Hall.*[20])

FIG. 43.36. Plunger-type relay. It is inherently unbalanced but resists mild shock and has little tendency toward contact chatter. (*R. E. Barbiere and W. Hall.*[20])

ELECTRON TUBES

Under conditions of shock, vibration, or acoustic noise, an electron tube may either fail completely or malfunction by acting as a transducer to create noise or by reacting to shock in such a way as to paralyze a circuit momentarily. Complete failure may embody breakage of the glass envelope, breakage of the filament, or distortion of the elements. Tube failures may be reduced by (1) improvement of the conventional tube to endure mechanical vibration, (2) development of an entirely new design which would be inherently more rugged, and (3) improved application of the available tubes.[20]

The conventional electron tube consists of several long, slender elements held close to one another by spacing discs, enclosed in a tubular envelope. The relative positions of the elements are critical; changes in spacing change the electrical characteristics of the tube. The internal structure of a tube has relatively little damping and is characteristically vulnerable to damage from vibration if a condition of resonance develops. Typical tube failures which result from shock and vibration are: broken filament, broken mica spacers, enlargement of holes in mica spacers, mounting pin breakages, loose metal particles, and cracked envelope or base seal. Envelope breakage due to shock is relatively uncommon, and usually results from stress concentration caused by inadequate envelope support. In general, low-frequency vibration has little effect on the average size electron tube since the natural frequencies of internal members are relatively high. Natural frequencies of internal members of typical tubes are approximately 400 to 600 cps, but large tubes may have elements with natural frequencies as low as 180 cps.

Ruggedized electron tubes provide more rugged supports for the elements and better fitting parts; materials and processing are more closely controlled in their manufacture. In general, they have improved resistance to shock and vibration.

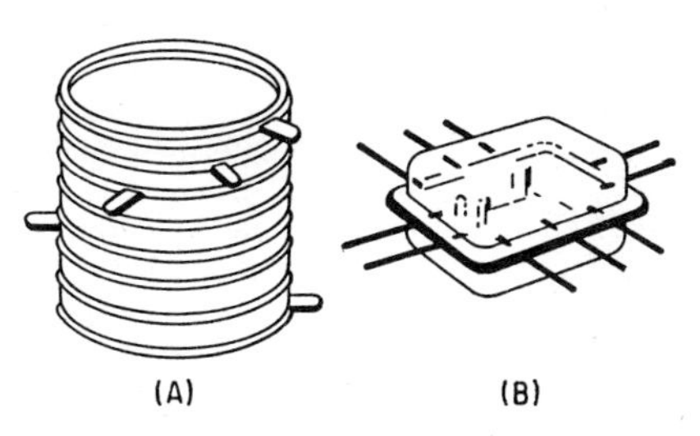

FIG. 43.37. Electron tube designs involving stacked ceramic plates. (*R. E. Barbiere and W. Hall.*[20])

More advanced concepts of tube design involving stacked ceramic plates, such as those shown in Fig. 43.37*A*, are the result primarily of an interest in automatic tube assembly. They are more resistant to shock and vibration than conventional tubes because of the compact design and the rigidity of the materials used. The small, box-shaped tube shown in Fig. 43.37*B* has all of its leads in one plane and is convenient for mounting in a circuit. Its size and shape indicate that it has desirable resistance to shock and vibration.[20]

SOLID-STATE DEVICES

The high reliability and compactness of solid-state devices make them particularly suitable for severe shock and vibration environments. In both military and industrial applications they have replaced relays, contactors, magnetic amplifiers, thyratrons, vacuum tubes, variable autotransformers, etc.[21]

Unlike electron tubes, they have no fragile filaments or glass envelopes and are rugged in construction and free from microphonics. The heart of the device is the disc of alternate layers of *N*- and *P*-type silicon, which may be brazed as shown in Fig. 43.38

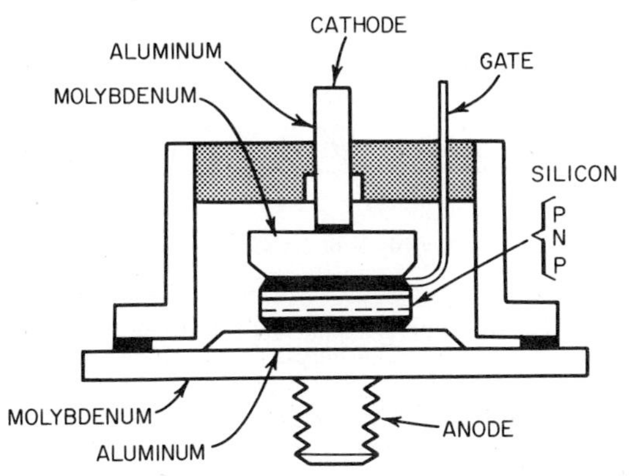

FIG. 43.38. Cross-sectional diagram of internal construction of a typical silicon-controlled rectifier (SCR).[21]

between heavy plates of molybdenum or tungsten. The latter materials have the same coefficient of expansion as the silicon and protect the fragile junctions against thermal and mechanical stresses.

MOUNTING OF COMPONENTS

CATHODE-RAY TUBES

Proper support and mounting of cathode-ray tubes to minimize damage from shock and vibration are important. Damage to glass is the most common type of damage. Breaks usually are confined to the neck of the tube where flaring toward the face begins. Internal elements may be deformed during shock with consequent permanent misalignment of the electron beam, although the tube continues to operate electrically. Where only friction positions the tube in its cradle, rotation of the tube as a result of shock may occur and cause the scope presentation to become canted. The most common type of damage as a result of vibration is fatigue of internal elements.

A cathode-ray tube must be supported by rubber of adequate firmness. Rubber not only absorbs part of the shock, but also accommodates some twisting of the supporting structure without damage to the tube, thereby protecting the brittle glass envelope from concentrated loads. Ruggedness appears to increase as envelope size decreases.

A recommended method for mounting a 5-in. cathode-ray tube is illustrated in Fig. 43.39; the method also is adaptable to larger tubes. A sheet-metal housing is formed

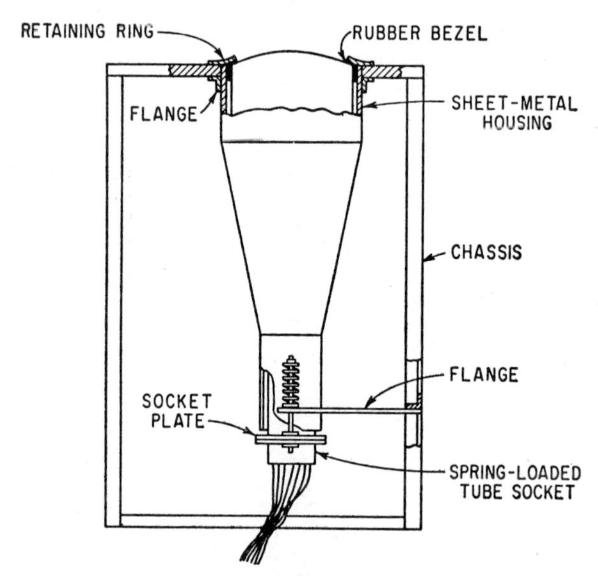

FIG. 43.39. Recommended mounting for 5-in. cathode-ray tube. (*H. M. Forkois and K. E. Woodward.*[1])

around the envelope of the tube and is supported at both ends. In the front, a circular flange is spot-welded to the housing and bolted to the front panel of the chassis; in the rear, another flange is fastened to the chassis structure. The housing serves two purposes: (1) it protects against stray magnetic and electrostatic fields and (2) it provides means of mechanical support. A bezel of firm rubber (not sponge rubber) is placed between the tube and the shield to prevent concentrated loading on the glass envelope and to reduce the intensity of the shock. It is important that the supporting chassis be stiff; lack of stiffness may cause the neck of the tube to be broken during shock. The tube is inserted neck first into the housing and is held in place by the rubber bezel. A suitably shaped metal retaining ring is bolted over the bezel. The socket for the tube rests against a socket plate mounted to the flange by a spring arranged to force the tube face against the rubber bezel. This accommodates variations in tube length.

ELECTRON TUBES

Tube clamps always are necessary to retain electron tubes in their sockets. The tube may be clamped either by its base or by its envelope. Base-gripping clamps usually are more rigid than other types and tend to transmit full shock forces directly to the tube. Thus, the glass envelope of the tube becomes a cantilever and the cement which secures the envelope to the base may fail. Clamps which grip the envelope are preferable to the base-gripping type because they tend to provide support along the length of the tube.

Figure 43.40 illustrates several types of satisfactory clamps that are used frequently

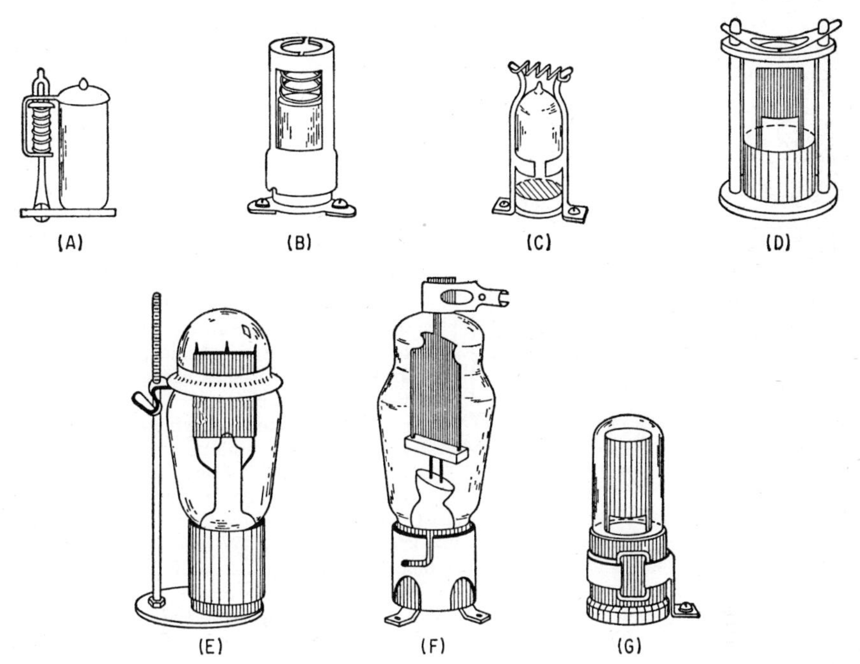

Fig. 43.40. Several popular types of electron-tube clamps. (*H. M. Forkois and K. E. Woodward.*[1])

in electronic equipment. Clamps illustrated in the top row are satisfactory for use with miniature tubes. The clamps shown in Figs. 43.40*D* and 43.40*G* are satisfactory for larger tubes. The clamp shown in Fig. 43.40*E* is satisfactory for medium-size tubes; styles with two posts are more satisfactory than those with one post as illustrated. The plate cap shown in Fig. 43.40*F* may cause failure of the tube unless it is light in weight. The ceramic cap shown in Fig. 43.41*A* may either damage the tube or be forcibly disconnected during severe shock because of its mass. The metal cap shown in Fig. 43.41*B* is preferable with respect to shock but lacks the insulating properties of ceramic. The plastic cap shown in Fig. 43.41*C* may be an optimum compromise between weight and insulating properties.

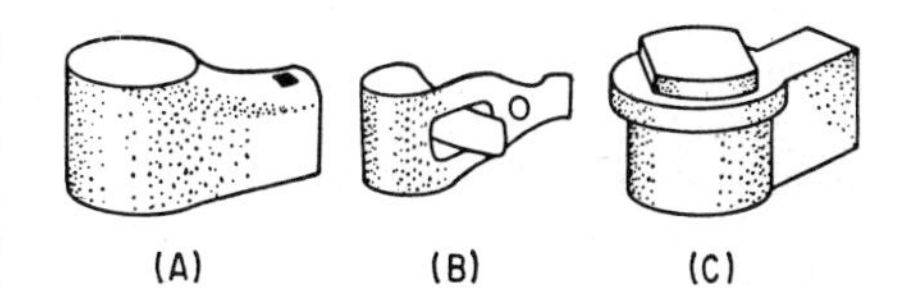

Fig. 43.41. Electron plate caps: (*A*) ceramic, (*B*) metallic, and (*C*) plastic. (*H. M. Forkois and K. E. Woodward.*[1])

TRANSISTORS AND THYRISTORS

When a second junction is added to a diode to provide power or voltage amplification, the solid-state device becomes known as a transistor.[21] Failure of transistors may occur because of improper mounting techniques.

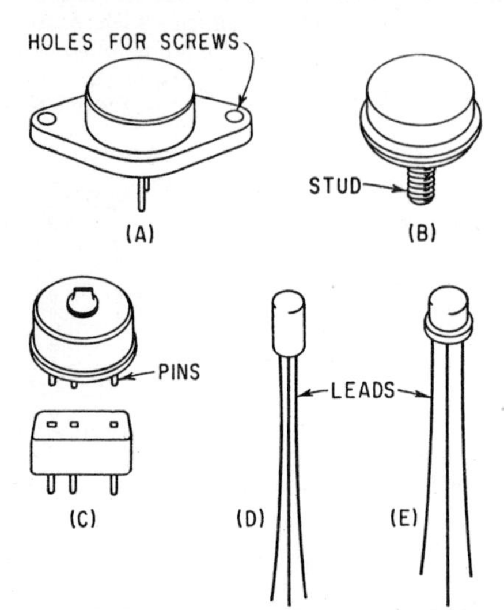

Fig. 43.42. Five types of transistors. They are attached to the chassis in various ways. (*R. E. Barbiere and W. Hall.*[20])

Figure 43.42 shows five types of transistors which are mounted in various ways. The transistor shown in Fig. 43.42*A* is held to the chassis with screws. The type shown in Fig. 43.42*B* has a threaded stud which screws into the chassis. These two types have a high degree of resistance to shock and vibration.

The transistors shown in the bottom row are not designed to be attached as securely as those shown above, and supplementary measures must be employed to ensure their resistance to shock and vibration. The transistor shown in Fig. 43.42*C* mounts in a socket in the same manner as an electron tube. To prevent ejection of the transistor from its socket, a strip may be extended across a row of plug-in transistors and be bolted to the chassis. Rubber grommets are used at contact of the strip with the transistor.

The transistors shown in Figs. 43.42*D* and 43.42*E* have no means of support except their leads. The long type shown in Fig. 43.42*D* usually is mounted on printed-wiring boards with the long side flat against the board, and is held to the board by spring clips or wire straps. The flat, button type shown in Fig. 43.42*E* may be used without retainers if mounted on a printed-wiring board with the leads holding the base tight against the board.[20]

When semiconductor materials are used in a series arrangment, the device becomes known as a thyristor (SCR, triac) and has control characteristics similar to those of a thyratron tube. Such devices with increased power ratings must be securely fastened by means of a base plate and screws indicated by Fig. 43.42*A*, or a copper stud shown in Fig. 43.42*B*, to a large area chassis. The latter acts as a heat sink, sometimes air- or water-cooled, and prevents possible thermal fatigue at the silicon interfaces.

TRANSFORMERS AND CHOKES

Difficulties involving transformers in a shock and vibration environment occur because of poor or inadequate provisions for mounting. Transformers are relatively heavy, but frequently are held by frail brackets or legs, or undersized bolts. Figures 43.43 to 43.46 illustrate some of the more common small transformer designs used in electronic equipments. The coil and core of the transformer shown in Fig. 43.43 are supported by cantilever brackets; this design often is not satisfactory because the resonant frequency is too low. A better design mounts the core and coil on symmetrical supports consisting of *Z*-section brackets, as shown in Fig. 43.44. Usually, no mechanical support is provided at the top of the core. A more adequate mounting could be obtained by supporting the top of the core, using the housing as a load-bearing structure. The transformer shown in Fig. 43.45 failed because the weight of the entire transformer assembly was taken at the base.

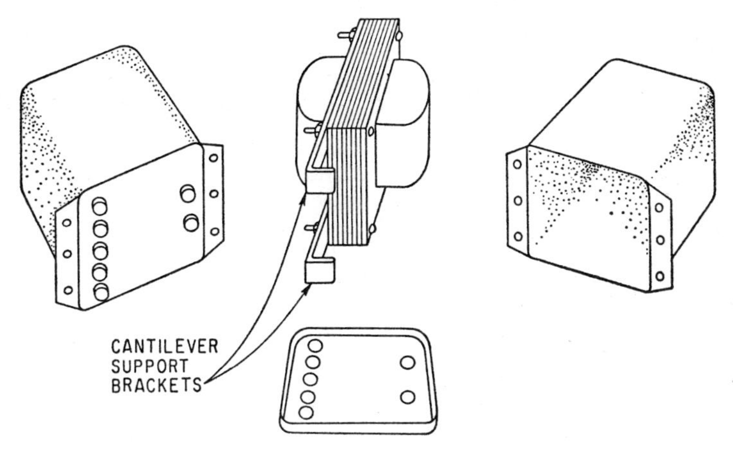

FIG. 43.43. Cantilever support for transformer coil and core which results in low resonant frequencies. (*H. M. Forkois and K. E. Woodward.*[1])

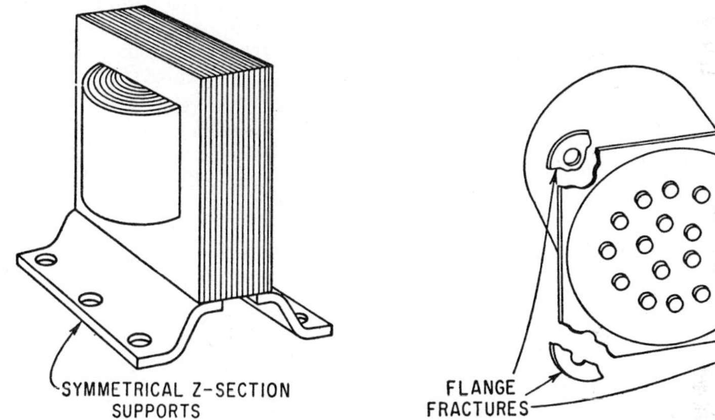

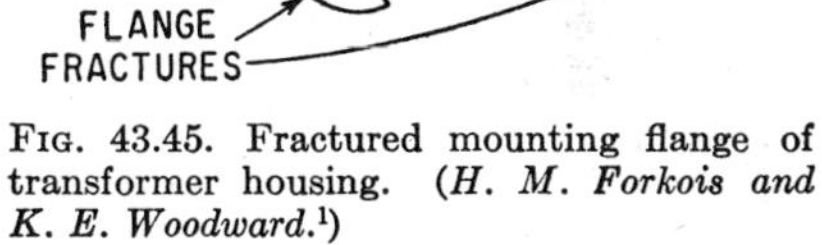

FIG. 43.44. Transformer mounting showing core secured to bottom Z brackets. (*H. M. Forkois and K. E. Woodward.*[1])

FIG. 43.45. Fractured mounting flange of transformer housing. (*H. M. Forkois and K. E. Woodward.*[1])

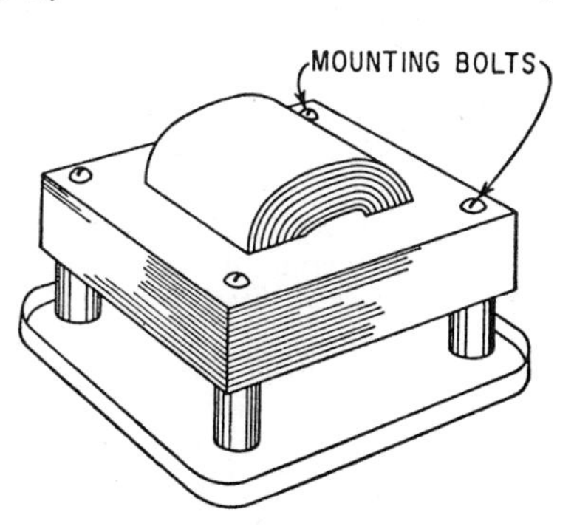

FIG. 43.46. Desirable transformer core mounting for shock and vibration environment showing mounting bolts passing through the core. (*H. M. Forkois and K. E. Woodward.*[1])

A preferred core mounting is shown in Fig. 43.46 where the mounting bolts for the transformer pass through the core itself. The housing can be supported by the same chassis mounting bolts; the housing then is not subjected to large loads and can be made from thin sheet metal because it serves only as a reservoir for the cooling or potting compound. It is difficult to devise adequate mountings for transformers having cores made of a strain sensitive material such as Hipersil because the cores cannot be drilled without adversely affecting the magnetic properties of the core material.

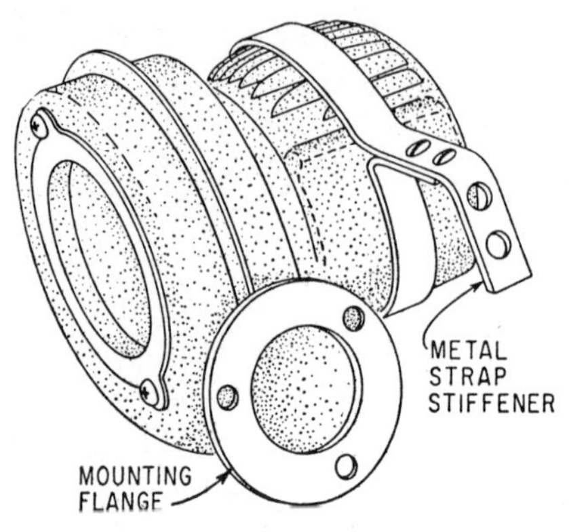

Fig. 43.47. Inadequate mounting flange arrangement for a centrifugal-type blower, improved by adding metal strap around motor body. (*H. M. Forkois and K. E. Woodward.*[1])

AIR BLOWERS

An air blower usually is sufficiently rugged to withstand shock and vibration of relatively great severity. However, care must be exercised in designing the support to ensure that it has adequate strength. Figure 43.47 illustrates a poor mounting of a centrifugal-type blower where the weight of the entire assembly is supported as a cantilever by the mounting flange surrounding the outer end of the delivery duct. Additional support should be provided for the motor, either by adding the strap shown in Fig. 43.47 or by using tapped holes in the motor housing. Vibration isolators should not be used to mount blowers unless the noise and vibration caused by the blower are known to be of great enough magnitude to cause difficulty.

CAPACITORS

Commercially available capacitors usually are sufficiently rugged to withstand shock and vibration, but failure may occur because means used to mount the capacitors to the chassis of the equipment may be inadequate. Some common methods of mounting various types of capacitors are illustrated in Fig. 43.48:

A. The capacitor is enclosed within a container which is provided with two attachable side plates, each of which includes a hooked end to engage the container and an opposite end fitted with anchor studs adapted to be secured to the chassis. Principal advantages of this design are (*a*) very little likelihood of damage to the container and (*b*) a design for the attachment which avoids stress concentration.

B. The cylindrical capacitor has a support rod extending along its axis of symmetry, and brackets that mount the capacitor by its support rod to the chassis. Care should be taken in mounting such a capacitor to ensure that the chassis is sufficiently rigid to avoid straining the capacitor because of chassis flexure.

C. The mounting lugs are designed so that the capacitor itself is not damaged by deformation of the mounting lugs. A capacitor with mounting lugs at the bottom is susceptible to damage unless properly designed. The failure illustrated in Fig. 43.49 shows the effects of excessive stress concentration at the mounting lugs. The capacitor shown in Fig. 43.50*B* exhibits desirable reinforcing of the mounting lugs; however, the design suggests damage to the container upon deformation of the lugs.

D. Small capacitors having their leads protruding from opposite ends often are mounted by soldering the leads to binding posts or terminals of adjacent components. This practice leads to frequent failure of leads unless precautions are taken. A preferred practice is to clamp the capacitor to the chassis with a strap at least partially encircling the capacitor, thereby relieving the leads of any load imposed by the weight of the capacitor.

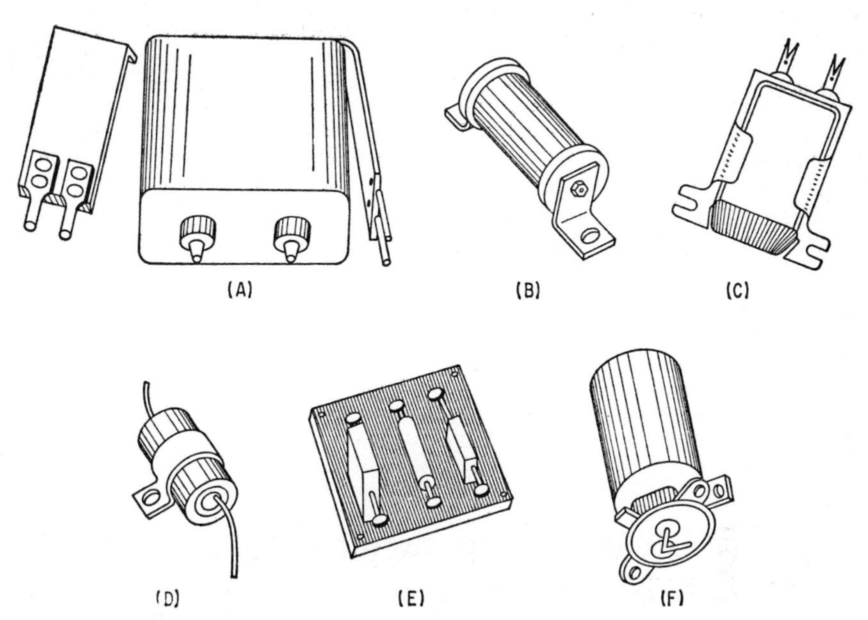

FIG. 43.48. Types of capacitors and mountings usually satisfactory for shock and vibration environment. (*H. M. Forkois and K. E. Woodward.*[1])

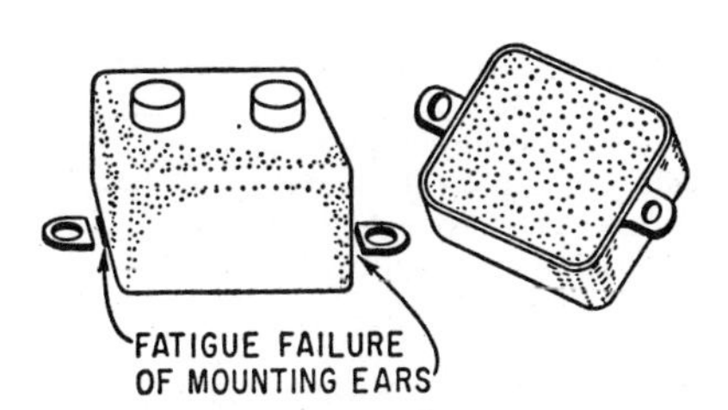

FIG. 43.49. Typical damage to a canned capacitor. (*H. M. Forkois and K. E. Woodward.*[1])

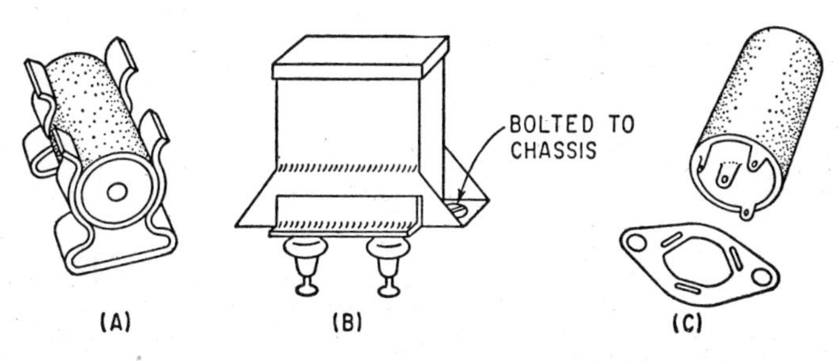

FIG. 43.50. Types of capacitor mountings which are less desirable for shock and vibration environment, especially in the larger sizes. (*H. M. Forkois and K. E. Woodward.*[1])

E. Small capacitors may be mounted by their leads successfully if (*a*) the leads are short and rigid enough to provide a high natural frequency and (*b*) the terminals to which the leads are attached are mounted to a rigid structure so that the leads are not stressed by flexure of the mounting structure.

F. Capacitors that are housed in cylindrical containers may be supported adequately by encircling the container with a strap which in turn is mounted to the chassis.

The capacitor mountings shown in Fig. 43.50*A* and *C* are somewhat marginal. Capacitors may be mounted in fuse clips as shown in Fig. 43.50*A* if the fuse clips are designed to retain the capacitors under conditions of severe shock. Examples of adequate fuse clips are shown in Fig. 43.31. The mounting shown in Fig. 43.50*C* exhibits a weakness where the tabs of the cylindrical container are soft-soldered to the mounting plate.

RESISTORS AND POTENTIOMETERS

Resistors perform exceptionally well under shock and vibration when proper mounting practices are followed. Resistor mountings are very similar to those for capacitors. Resistors of ⅓- to 2-watt rating may be supported by their leads, provided the leads are attached to a structure sufficiently rigid to prevent straining the leads by flexure. The length of lead at each end of the resistor should not exceed ⅜ in. Resistors larger than the 2-watt size should be body-mounted to avoid excessive loads on the leads.

WIRING AND CIRCUIT BOARDS

WIRES, CABLES, AND CONNECTORS

Wires, cables, and connectors are not in themselves susceptible to damage from shock and vibration. They can cause considerable difficulty, however, if improper methods are used in their application. Wires and cables have a high degree of flexibility as compared with other electronic component parts. This flexibility results from (1) the type of construction, (2) the kind of material, and (3) the high ratio of length to cross-sectional area. Because of this flexibility, the natural frequency of a span of wire or cable is low and may easily fall within the frequency range of existing vibration. The vibration of a wire or cable produces time-varying stresses which can lead to fatigue failure. The areas which usually are first to fail are the ends of the vibrating span, i.e., at the wire or cable terminations. Failures also may occur where the insulation rubs against an adjacent part, or where one conductor in a multiconductor cable rubs against another within the cable body. A cable passing through a hole in a partition should be protected by a grommet lining the hole.

Where several wires extend in the same direction, their stiffness may be increased by harnessing the wires with lacing cord to form an integral bundle. Harnessing also provides damping because of friction between the wires. The harnessed length should be supported adequately so that its weight will not load the wire termination points unduly. The relatively high degree of damping of stranded wire and cable is beneficial when they are subjected to flexing as a result of vibration[20] (see the discussion of *Cables* in Chap. 19).

MATERIALS AND CONSTRUCTION. Although all wire and cable are flexible when compared to other components, the degree of flexibility depends in part upon the material composition. The flexibility may vary widely with temperature, since many organic insulating materials stiffen considerably with decreased temperature and soften with increased temperature. Fluorinated hydrocarbons and silicone elastomers maintain flexibility at low temperatures better than most others.

LEAD LENGTH. The bending of a wire or cable may be reduced at its termination or support by shortening the length of lead between supports. This applies only in cases where the supports *do not* move relative to one another. The shortening of the span of wire or cable accomplishes two beneficial results: (1) it increases the natural frequency of the span and (2) it reduces the weight and inertial loads at each support. Where supports or terminals *do* move relative to one another, lead failures can be reduced considerably by allowing some slack in the lead. Lengthening a lead slightly (by about 20 per cent of its stretched length), as shown in Fig. 43.51, reduces the failure hazard considerably. The additional length, although lowering the natural frequency, prevents repeated straining of the lead. The use of stranded, insulated wire provides damping to limit the amplitude of vibration during resonant conditions.

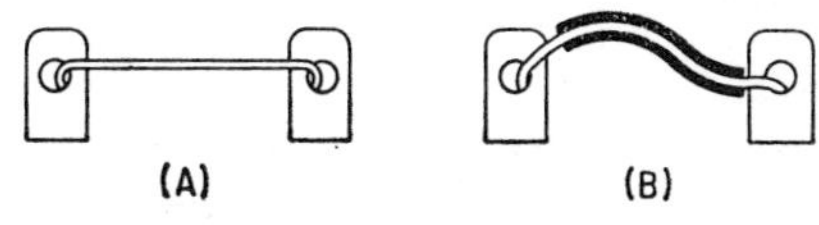

FIG. 43.51. Examples of jumper leads: (*A*) Tightly stretched solid wire liable to failure if terminals experience relative motion. (*B*) Slightly bowed stranded wire with insulation; the extra length prevents excessive stress in the wire, and the damping provided by friction and insulation prevent large amplitudes of vibration. (*R. E. Barbiere and W. Hall.*[20])

METHODS OF ATTACHMENT. Wire or cable failures due to shock and vibration usually take place where the lead is interrupted, either at a termination point or at a support along the lead. Failures occur primarily because these locations are subjected to the greatest bending stress. When wire is terminated by soldering, the area most susceptible to shock and vibration lies between the end of the insulation and the rigidly "frozen" section at the terminal post shown in Fig. 43.52. This portion of the wire is subjected to severe flexing and undergoes fatigue very quickly. Figure 43.52 indicates how solder flows up the strands of the conductor by capillary action and concentrates all bending action into a very small area of wire. Clamping the wire to a structure can restrict the motion and reduce damage from flexing at the termination, as shown in Fig. 43.53.

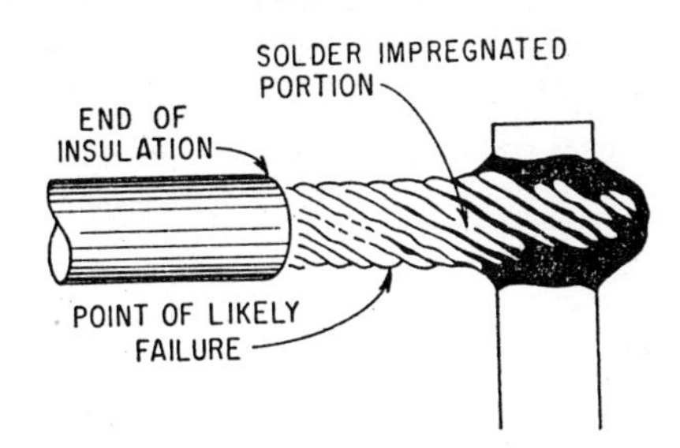

FIG. 43.52. View of cable end illustrating flow of solder along strands to form a rigid mass of cable. Failure is most likely to occur at edge of solder-impregnated portion. (*R. E. Barbiere and W. Hall.*[20])

When wire is terminated by an end support lug, the lug should grip both the insulation and the conductor. When cable is terminated by a connector, the connector should be attached to the cable so that it grips the cable jacket to restrict flexing of individual wires within the cable body. Attaching a sleeve as shown in Fig. 43.54 prevents severe cable bending at a termination point. Bending of individual wires can be restricted by potting the cable end permanently into the connector. The connections are made in a normal manner; then the connector body is filled with a self-curing plastic material to encapsulate the end of the cable completely. The disadvantage to potting a cable is that potted components are difficult to repair.

PRINTED WIRING BOARDS. Printed wiring boards usually require a stiffening structure to prevent low resonant frequencies. The stiffening may take the form of a metal chassis, conformal coating, or complete potting in plastic. Small wiring boards may be rigid enough without additional stiffening. To apply conformal coating, the side of the board having the component parts is dipped into a clear plastic material that hardens to form a rigid assembly. Resistors and capacitors are mounted to the board by leads, and the transformers and other heavy parts are screwed or riveted directly to the board. Printed circuit boards can also be made rigid by complete potting in plastic.

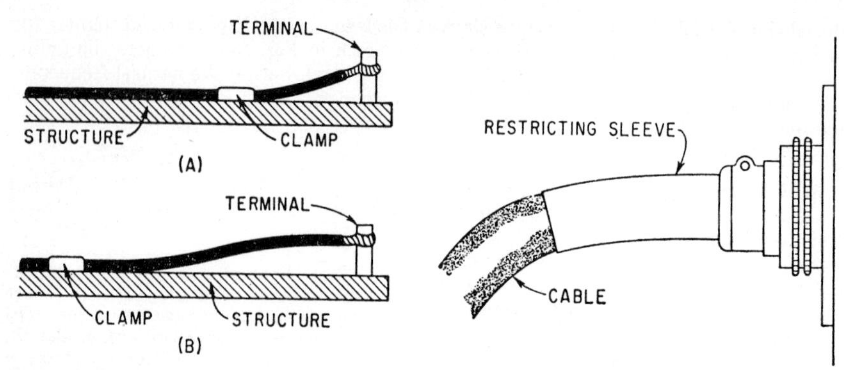

FIG. 43.53. Examples of preferred (*A*) and inferior (*B*) methods of clamping cable ends. The relatively great length of unsupported cable is liable to experience excessive vibration. (*R. E. Barbiere and W. Hall.*[20])

FIG. 43.54. Restricting sleeve to prevent severe cable bending at a termination point. (*R. E. Barbiere and W. Hall.*[20])

DESIGN TO MINIMIZE VIBRATION

This section discusses means to reduce vibration which may occur during normal operating conditions of certain types of equipment. It is possible to reduce vibration by (1) eliminating the forcing function by counterbalancing (see Chap. 39), (2) isolating the vibration with isolators (see Chaps. 30 to 33), (3) increasing the damping of structural members (see Chap. 37), or (4) adding tuned dampers (see Chap. 6). Examples of the application of these techniques are given below. Further examples are given in Chap. 40 relative to the design of machine tools.

BEARINGS

Vibration of machinery may result from sleeve or journal bearings, and from antifriction bearings using balls or rollers. The conventional oil-lubricated, babbitted bearings often are sufficiently quiet and vibration-free. An occasional instability in a lightly loaded bearing, sometimes referred to as "oil-whip," usually can be cured by relocating oil grooves, reducing bearing clearances, or using an oil of different viscosity. Sleeve bearings made from Teflon, nylon, or similar materials may be used when loads are light and extra quiet operation is required.

Antifriction bearings must have small machining tolerances and superior surface finishing to be quiet. Inaccuracies in machining may lead to vibration. For example, the small induction motor shown in Fig. 43.55*A* was equipped with precision ball bearings and vibrated at a number of relatively low harmonic speeds as the de-energized motor coasted from its normal operating speed to a standstill. The vibration was a maximum in the axial direction, the vibration frequency was approximately the same value regardless of the motor speed, and an audible hum accompanied the motion. The axial mode of the system consisted of the motor frame vibrating in opposition to the rotor; the ball bearings at either end of the rotor provided nonlinear spring effect, as shown schematically in Fig. 43.55*B*. Usually, the outer race of one antifriction bearing was tight in the motor end bracket; the other bearing had a sliding fit but was "seated" and carried a small thrust load. The vibration was caused by minor discrepancies in the outer race of the press-fit ball bearing. It was not practical to require more precise machining operations. Instead, a relatively loose sliding fit was specified on both ball bearings so that the motor frame and rotor were no longer "spring-coupled" and the axial resonance could not materialize. A spring of low stiffness, such as a wavy washer or a rubber O ring, may be added to achieve a more definite assembly of the axial system.[22]

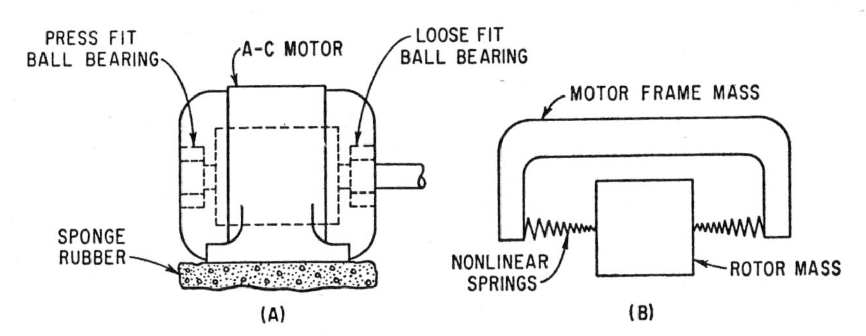

Fig. 43.55. A-C motor with (*A*) tight and loose-fit antifriction bearings, and (*B*) equivalent axial vibration system. (*E. G. Fischer.*[22])

GEARS

Gears are a common source of vibration. They can be fabricated from nylon, rawhide, or plastic-impregnated layers of textile material when loads are light and quiet operation is desired. A compliant material sometimes may be used to separate the tooth-carrying rim from the web and hub. Usually, improper design and inaccurate manufacture are major causes of gear noise.[23]

Friction is an important source of noise, and can be reduced by adequate lubrication. For example, a dangerous situation involving cracked turbine discs was discovered in a geared-turbine drive for a low-speed induced-draft fan on a steam boiler. Investigation revealed high-frequency torsional vibration of the turbine shaft and corresponding harmonic tooth spacing patterns in the high-speed pinion. An abrasive type of fly ash had leaked into the oil lubrication system for the gear case. In the presence of this "grinding compound," the natural mode of torsional vibration of the turbine shaft and discs provided a feed-back mechanism for accentuating the original hobbing errors in the pinion. A filter should have been used to ensure a supply of clean oil for the gears.

VIBRATION ISOLATION

Periodic forces sometimes are unavoidable in machinery, and it may be necessary to isolate the resultant vibration from the foundation or from other parts of the equipment. In electrical machines the predominant vibration and noise are caused by periodic magnetic forces in the air gap between the rotor and stator. The excitation forces can be minimized by enlarging the air gap and by skewing the rotor and stator slots. The natural frequencies for the pertinent rotor and stator modes of vibration should be outside the range of known excitation frequencies.[24] Resilient supports may be used to isolate any remaining structure-borne vibration from the foundation.

Figure 43.56 shows a flexible mounting system for a large two-pole generator, where the location of the flat isolation springs is based on a theoretical analysis of the stator core vibrating as a thick elastic ring. The pull of the magnetic field causes elastic deformation of the stator core laminations, and the resulting elliptic shape shown in Fig. 43.56*B* rotates with the salient magnetic poles. In general, both horizontal and vertical vibration will be produced at points on the (nonrotating) stator core. Three flat isolation springs shown in Fig. 43.56*A* are attached to the inner frame in order to support the core and to position it, securely, with a uniform air gap around the rotor. Some radial clearance is provided so that the inner frame can vibrate without touching the outer frame. The principal function of the flat spring supports is to prevent the core vibration from being transmitted to the outer frame and foundation.

One end of each flat spring is attached to the inner frame at a radial position corresponding to approximately $1\frac{1}{3}$ times the mean core radius; this is the point at which the tangential component of the vibratory motion in the stator core is a minimum, as

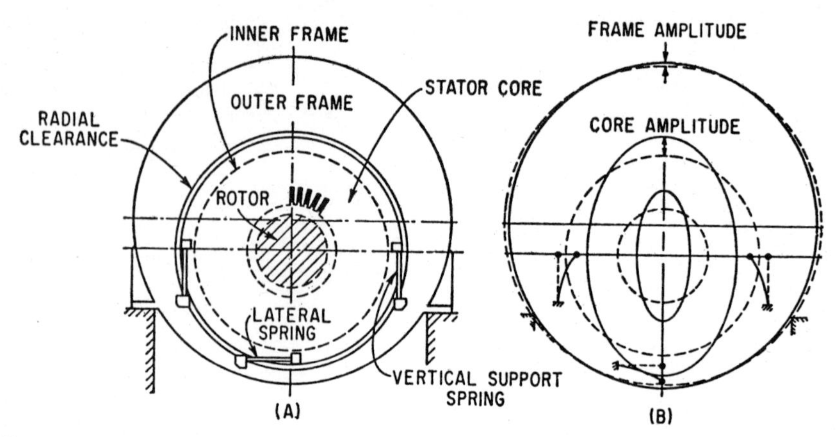

Fig. 43.56. Flexible spring mounting system for a two-pole generator: (A) generator core mounting, and (B) frame and core vibration patterns. (*A. C. Hagg, C. H. Janthey, and P. R. Heller.*[25])

shown by analysis. This position permits optimum design for vibration isolation, while also supporting the weight of the inner frame in a manner that resists rotation relative to the outer frame. The other end of each flat spring is attached close to the frame feet, where analysis indicates that the nodal points would appear for the principal elliptical mode of vibration in the outer frame.[25]

The effectiveness of a vibration isolator in the isolation of high-frequency vibration can be improved by loading the spring with a small mass to achieve the effect of two isolators in series.[26] For example, Fig. 43.57 shows a helical spring suspension for a motor-driven compressor of a household refrigerator; small weights are bonded to the suspension springs to improve the effectiveness of isolation and to the flexible discharge pipe to change the resonant frequency.

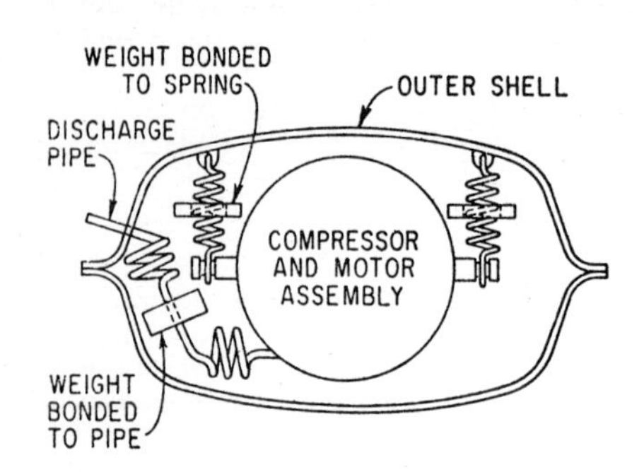

Fig. 43.57. Helical spring suspension showing small weights bonded to the springs to improve vibration isolation.

SPECIAL DAMPING DEVICES

The natural frequencies of structural members of machinery should differ appreciably from the normal speed of operation to avoid resonance. However, it may be necessary to accelerate a machine rotor through a relatively low critical frequency of its support while large unbalance forces are acting.[27] A special damping device can be installed to reduce the amplitude of vibration at resonance, and to reduce the power requirements for the drive motor. For example, Fig. 43.58 is a schematic diagram of a revolving-drum type of household washing machine where large unbalance forces are caused by the uneven distribution of the clothes being washed. The enclosure for the drum, together with the geared motor drive, is flexibly mounted to isolate the vibration from the laundry floor.

It is not practical to accelerate through the critical speed of a flexible suspension at such a high rate that vibration build-up is avoided. Hence, damping is required so that the vibration build-up at resonance is not excessive. The damping devices used in the washing machine consist of pads of brake-shoe material attached to metal straps in such a manner as to experience the relative motion between the vibrating drum enclosure and the stationary floor cabinet. Sometimes, when the machine ceases to operate, the dry rubbing friction in the damper pads causes an unpleasant "squealing" noise; this is an example of self-induced vibration.[28] Adding oil to the rubbing surfaces of the damper pads may decrease the coefficient of friction and destroy the effectiveness of the dampers. Thus, care should be exercised in selecting materials for friction damping to ensure quiet, effective operation.

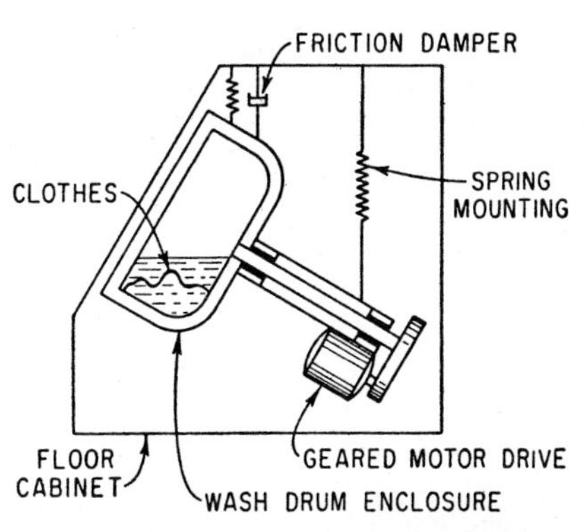

FIG. 43.58. Spring mounting for a revolving-drum type of household washing machine.

Where torsional vibration occurs in a rotating system, a tuned damper may be applied to the system to reduce the vibration. The theory of tuned dampers is given in Chap. 6, and the application to internal combustion engines is discussed in Chap. 38. Figure 43.59 is a schematic diagram of a damper that is capable of being tuned to the torsional vibration frequency of the system. The damper housing is keyed directly to the motor shaft, and includes a mass having an inner sleeve-type bearing. The mass oscillates torsionally between two sets of helical springs located at two positions around its outer periphery. Upon torsional vibration of the mass, oil is pumped through two orifices located between the two sets of springs. By adjusting the viscosity of the oil and the size of the orifice, the amount of damping can be optimized. The damper assembly is self-contained and permanently sealed.

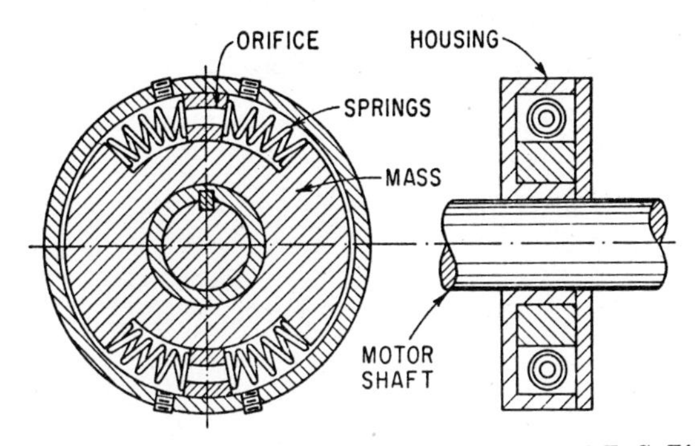

FIG. 43.59. Torsional vibration damper. (*A. M. Wahl and E. G. Fischer.*[29])

REFERENCES

1. Forkois, H. M., and K. E. Woodward: "Design of Shock- and Vibration-resistant Electronic Equipment for Shipboard Use," U.S. Navy Department, Bureau of Ships, *NAVSHIPS* 900, 185A, 1957. Also, *U.S. Naval Research Lab. Rpt.* 4789, 1956.
2. Crede, C. E., and M. C. Junger: "A Guide for the Design of Shock Resistant Naval Equipment," U.S. Navy Department, Bureau of Ships, *NAVSHIPS* 250-660-30, 1949.
3. Welch, W. P.: "Mechanical Shock on Naval Vessels," U.S. Navy Department, Bureau of Ships, *NAVSHIPS* 250-660-26, 1946.
4. Dinner, H., and W. Felix: *Engineers Digest*, **3**(2):85 (1946).
5. Almen, J. O.: *Prod. Eng.*, **22**(4):153 (1951).
6. ASME Handbook: "Metals Engineering—Design" McGraw-Hill Book Company, Inc., New York, 1953.
7. Staedel, W.: *Mitt. Materialprüfungsanstalt Berlin-Lichterfelde West*, **4** (1933).

8. MacGregor, R. A., W. S. Burn, and F. Bacon: *Trans. North East Coast Inst. Engrs. & Shipbuilders*, **51** (1935).
9. Battelle Memorial Institute: "Prevention of Failure of Metals under Repeated Stress," Appendix 10, John Wiley & Sons, Inc., New York, 1941.
10. Almen, J. O.: *Trans. SAE*, **52**(4):151 (1944). Also see *SAE Special Publication* 23.
11. Engineering Experiment Station, U.S. Navy Department, Annapolis, Md., Lock-Nut Test *Repts.* 7224, 7226, 9321, B-3984 and B-6180.
12. Jacobson, L. S.: *Trans. ASME*, APM-52-15, **52** (1930).
13. Den Hartog, J. P.: "Mechanical Vibrations," 4th ed, p. 87, McGraw-Hill Book Company, Inc., New York, 1956.
14. Ruzicka, J. E.: Paper No. 100Y, presented at *SAE National Aeronautic Meeting*, Oct. 5–9, 1959.
15. Johnson, W. V.: *Westinghouse Engineer*, May, 1942.
16. Grubin, C.: *Trans. ASME*, **78**:373 (1956).
17. Leiber, P., and D. P. Jensen: *Trans. ASME*, **67**:523 (1945).
18. Sankey, G. O.: "Some Experiments on a Particle or Shot Damper," unpublished memorandum, Westinghouse Research Laboratories, Pittsburgh, Pa., 1954.
19. Hickok, R. D., et al.: U.S. Patent 2,427,529, Sept. 16, 1947, for "Meter Pivot."
20. Barbiere, R. E., and W. Hall: "Electronic Designer's Shock and Vibration Guide for Airborne Applications," *WADC Tech. Rept.* 58-363, ASTIA No. AD 204095, December, 1958.
21. "Solid-State Power Circuits," RCA Designer's Handbook, Technical Series SP-52, RCA Solid State Division, Somerville, N.J., September, 1971.
22. Fischer, E. G.: *Conference Paper CP*58-200, *AIEE Winter Meeting*, New York, February 2–7, 1958.
23. Moeller, K. G. F.: "Gear Noise," chap. 23, in "Handbook of Noise Control," C. M. Harris (ed.), McGraw-Hill Book Company, Inc., New York, 1957.
24. Fehr, R. O., and D. F. Muster: "Electric Motor and Generator Noise," chap. 30, in "Handbook of Noise Control," C. M. Harris (ed.), McGraw-Hill Book Company, Inc., New York, 1957.
25. Hagg, A. C., C. H. Janthey, and P. R. Heller: *Proc. Midwest Power Conf., Chicago*, vol. 13, p. 145, 1951.
26. Himelblau, H., Jr.: *Prod. Eng.*, November, 1952.
27. Baker, J. G.: *J. Appl. Mechanics*, **61**:A-145 (1939).
28. Baker, J. G.: *J. Appl. Mechanics*, **55**:5 (1933).
29. Wahl, A. M., and E. G. Fischer: *J. Appl. Mechanics*, **64**:A-175 (1942).

44

EFFECTS OF SHOCK AND VIBRATION ON MAN

Henning E. von Gierke
Aerospace Medical Research Laboratory, USAF Dayton, O.

David E. Goldman
The Medical College of Pennsylvania and Hospital, Philadelphia, PA

INTRODUCTION*

This chapter considers the following problems: (1) the determination of the structure and properties of the human body regarded as a mechanical as well as a biological system, (2) the effects of shock and vibration forces on this system, (3) the protection required by the system under various exposure conditions, the means by which this protection is to be achieved, and (4) tolerance criteria for shock and vibration exposure. Man, as a mechanical system, is extremely complex and his mechanical properties readily undergo change. There is very little reliable information on the magnitude of the forces required to produce mechanical damage to the human body. To avoid damage to humans while obtaining such data, it is necessary to use experimental animals for most studies on mechanical injury. However, the data so obtained must be subjected to careful scrutiny to determine the degree of their applicability to humans, who differ from animals not only in size but in anatomical and physiological structure as well. Occasionally it is possible to obtain useful information from situations involving accidental injuries to man, but while the damage often can be assessed, the forces producing the damage cannot, and only rarely are useful data obtained in this way. It is also very difficult to obtain reliable data on the effects of mechanical forces on the performance of various tasks and on subjective responses to these forces largely because of the wide variation in the human being in both physical and behavioral respects. Measurement of some of the mechanical properties of man is, however, often practicable since only small forces are needed for such work. The difficulty here is in the variability and lability of the system and in the inaccessibility of certain structures.

One section of this chapter is devoted to a discussion of methods and instrumentation used for mechanical shock and vibration studies on man and animals. Subsequent sections deal with the mechanical structure of the body, the effects of shock and vibration forces on man, the methods and procedures for protection against these forces, and safety criteria.

DEFINITIONS AND CHARACTERIZATION OF FORCES

CHARACTERIZATION OF FORCES. Forces may be transmitted to the body through a gas, liquid, or solid. They may be diffuse or concentrated over a small area.

* The contributions of J. T. Shaffer, Aerospace Medical Research Laboratory, USAF, are gratefully acknowledged.

They may vary from tangential to normal, and may be applied in more than one direction. The shape of a solid body impinging on the surface of the human is as important as the position or shape of the human body itself. All these factors must be taken into account in comparing injuries produced by vehicle crashes, explosions, blows, vibration, etc. Laboratory studies often permit fairly accurate control of force application, but field situations are apt to be extremely complex. Therefore it is often very difficult to predict what will happen in the field on the basis of laboratory studies. It is equally difficult to interpret field observations without the benefit of laboratory studies.

SHOCK. The term "shock" is used differently in biology and medicine than in engineering; therefore one must be careful not to confuse the various meanings given to the term. In this chapter the term *shock* is used in its engineering sense as defined in Chap. 1 of this Handbook. A *shock wave* is a discontinuous pressure change propagated through a medium at velocity greater than that of sound in the medium. In general, forces reaching peak values in less than a few tenths of a second and of not more than a few seconds' duration may be considered as shock forces in relation to the human system.

The term *impact* (i.e., a *blow*) refers to a force applied when the human body comes into sudden contact with a solid body and when the momentum transfer is considerable, as in rapid deceleration in a vehicle crash or when a rapidly moving solid body strikes a human body.

VIBRATION. Biological systems may be influenced by vibration at all frequencies if the amplitude is sufficiently great. This chapter is concerned primarily with the frequency range from about 1 cps to several hundred cycles per second. (Studies at higher frequencies are very useful for the analysis of tissue characteristics.)

METHODS AND INSTRUMENTATION

Most quantitative investigations of the effects of shock and vibration on humans are conducted in the laboratory in controlled, simulated environments. Meaningful results can be obtained from such tests only if measurement methods and instrumentation are adapted to the particular properties of the biological system under investigation to ensure noninterference of the measurement with the system's behavior. This behavior may be physical, physiological, and psychological although these parameters should be studied separately if possible. The complexity of a living organism makes such separation, even assuming independent parameters, only an approximation at best. In many cases if extreme care is not exercised in planning and conducting the experiment, uncontrolled interaction between these parameters can lead to completely erroneous results. For example, the dynamic elasticity of tissue of a certain area of the body may depend on the simultaneous vibration excitation of other parts of the body; or the elasticity may change with the duration of the measurement since the subject's physiological response varies; or the elasticity may be influenced by the subject's psychological reaction to the test or to the measurement equipment.

Control of, and compensation for, the nonuniformity of living systems is absolutely essential because of the variation in size, shape, sensitivity, and responsiveness of people, and because these factors for a single subject vary with time, experience, and motivation. The use of adequate statistical experimental design is necessary and almost always requires a large number of observations and carefully arranged controls.

PHYSICAL MEASUREMENTS

In determining the effects of shock and vibration on man, the mechanical force environment to which the human body is exposed must be clearly defined. Force and vibration amplitudes should be specified for the area of contact with the body. Vibration measurements of the body's response should be made whenever possible by noncontact methods. X-ray methods can be used successfully to measure the displacement of internal organs. Optical, cinematographic, and stroboscopic observation can give the displacement amplitudes of larger parts of the body. Small vibrations sometimes can be measured without contact by capacitive probes located at small distance from the (grounded) body surface (see *Variable Capacitance Pickups*, Chap. 14). If vibration pickups in contact with the body are used, they must be small and light enough so as not to introduce a distorting mechanical load. This usually places a weight limitation on the pickup of a few grams or less, depending on the frequency range of interest and the effective mass to which the pickup is attached. Figure 44.1 illustrates the effect of mass and size on the response of accelerometers attached to the skin overlying soft tissue.[1] The lack of rigidity of the human body as a supporting structure makes measurements

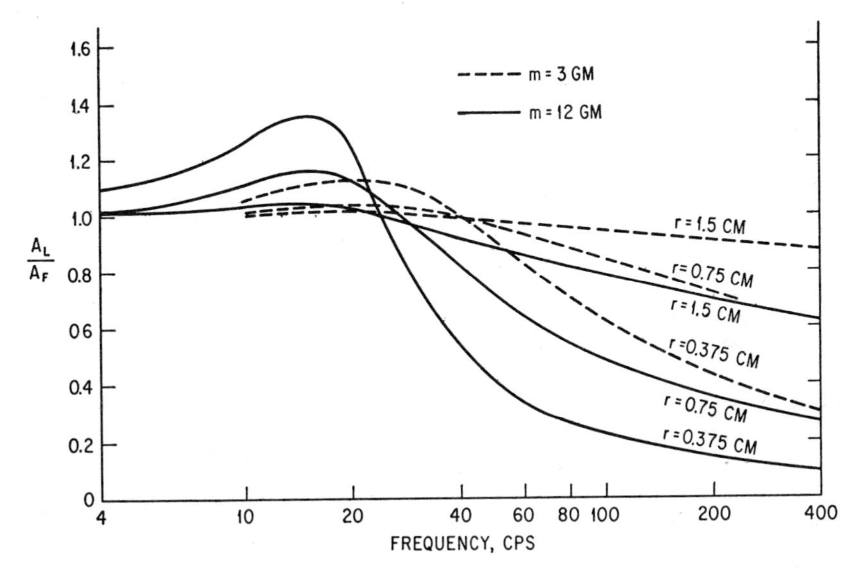

FIG. 44.1. Amplitude distortion due to accelerometers of different mass m and size which are attached to body surface over soft tissue of human subject exposed to vibration. The graph gives the ratio A_L/A_F of the response of the loaded to the unloaded surface for accelerometers having three different radii r. (*Values calculated from mechanical surface impedance data of Franke and von Gierke.*[1])

of acceleration usually preferable to those of velocity or displacement. The mechanical impedance of a sitting, standing, or supine subject is extremely useful for calculating the vibratory energy transmitted to the body by the vibrating structure.[2] The mechanical impedance of small areas of the body surface can be measured in many different ways (see Chap. 10), for example by vibrating pistons,[3] resonating rods, and acoustical impedance tubes.[4]

If the entire body is exposed to a pressure or blast wave in air or water, exact definition of the pressure environment is essential. The pressure distribution should be measured if possible.[5] If the environment deviates from free-field conditions it should be carefully specified because of its effect on peak pressure and pressure *vs.* time-history.[6]

BIOLOGICAL MEASUREMENTS

Instrumentation for the measurement of physiological properties such as blood pressure, respiration rate or depth, heart potentials, brain potentials, or galvanic skin response [7] must be carefully checked for freedom from artifacts when the subject, the instrument, or both are exposed to vibration, intense sound, or acceleration. For example, a bioelectronic harness,[8] such as is used in aviation medicine, is very useful for measurements of various physiological functions, but the proper functioning of instruments employed in such measurements must be verified for each environment.

If psychological experiments during exposure to the mechanical stimuli are planned, adherence to established procedures for such subjective tests is an absolute necessity to obtain valid results. The maintaining of a neutral situation with uniform motivation, subject instruction, and adequate statistical design of the experiment are some of the most important considerations. Care must be taken that the subject be not biased or influenced by environmental factors not purposely included in the test (e.g., the noise of a vibration table can act as a disturbing factor in a study of vibration effects).

SIMULATION OF MECHANICAL ENVIRONMENT

The desire to study the physical, physiological, and psychological responses of biological specimens in the laboratory, under well-controlled conditions, has led to the use of standard and specialized shock and vibration testing machines for experiments on man and animals. A summary of some machines employed in such tests is given in Fig. 44.2. An accurate simulation of the environmental conditions to which man is exposed frequently is not feasible for technical or economic reasons or may even be undesirable because of a need for more systematic investigation under somewhat simplified con-

TYPE OF MACHINE	APPLICATION OF FORCE	FORCE–TIME FUNCTION	FREQUENCY RANGE	MAXIMUM AMPLITUDE	REF NO.
SHAKE TABLE			MECHANICAL 0–50 CPS	UP TO 15g	10, 2
			ELECTRODYNAMIC 15–1,000 CPS		
VERTICAL ACCELERATOR			0–10 CPS	±10 FT, 3.7g PEAK	11
SHOCK MACHINE			DOWN TO T = 0.16 SEC, $\tau = 2.10^{-3}$ SEC	PEAK AMPLITUDE 10^{-2} TO 10^{-1} CM	12
HORIZONTAL OR VERTICAL DECELERATOR OR ACCELERATOR (SLEDS ON TRACKS, CAR, DROP-TOWER)			RATE OF ACCELERATION UP TO 1,400 g/SEC	40g PEAK	9, 13, 14, 15
BLAST TUBE FIELD EXPLOSION					11, 18, 5, 17
SIREN (AIRBORNE SOUND)			25–100,000 CPS	160–170 DB RE 0.0002 μBAR	
RESPIRATOR			0–15 CPS		19
VIBRATOR (SMALL PISTON)			0–10 MCPS		20, 3, 21
SHAKER ON CENTRIFUGE	ALTERNATING g FORCE IN ADDITION TO STATIC g FIELD		0–3 CPS		22
HEAD IMPACT MACHINE FOR DUMMY HEADS			DEPENDING ON STRUCTURE STRUCK	IMPACT VELOCITY 140 FT/SEC	23

FIG. 44.2. Summary of characteristics of shock and vibration machines used for human and animal experiments. Force-time functions are indicated schematically. Frequency range and maximum amplitudes refer to values used, not to capabilities of machines. References are to papers describing use of the machines for biological purposes.

ditions. Thus most investigations are limited to the study of a single degree-of-freedom at a time in which the human test specimen is vibrated only in one direction. Many fundamental studies are performed with sinusoidal forces. Usually mechanical and electrodynamic shake tables are employed for this purpose. Requirements for all shock and vibration machines used include: adequate safety precautions, safe and accurate control of the exposure, and sufficient load capacity for subject, seat, and instrumentation. Since the law of linear superposition is valid only in the linear physical domain, sinusoidal forces alone are not adequate for the study of nonlinear physical responses or physiological and psychological reactions to complex force functions. Therefore, some of the machines listed are designed uniquely for exposure of humans. One vertical accelerator, for example, employs a friction-drive mechanism to permit the simulation

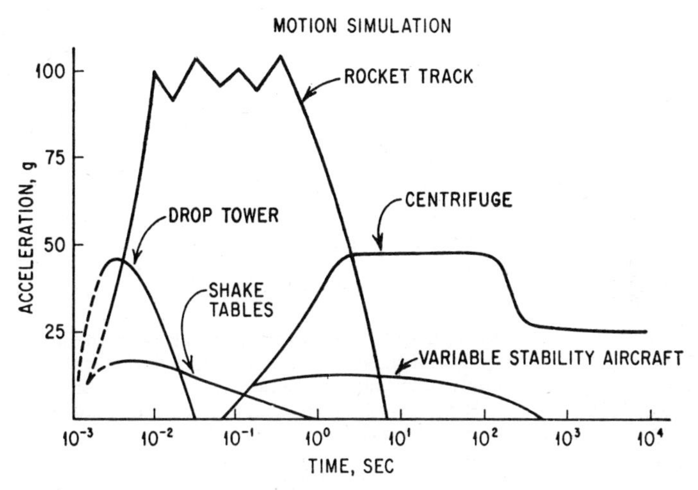

FIG. 44.3. Ranges of time and acceleration obtainable with certain high-acceleration devices for tests on man. (*After Clark and Gray.*[24])

of large amplitude sinusoidal and random vibrations such as those encountered in buffeting during low-altitude high-speed flight or anticipated during the launch and reentry phases of spacecraft.[11] This device can be programmed with acceleration recordings obtained under actual flight conditions. Other machines for the study of human tolerance to ejection from high-speed aircraft (ejection seat) have upward or downward acceleration tracks with sliding seats projected by explosive charges. Horizontal tracks with rocket-propelled sleds, which can be stopped by special braking mechanisms, have been used to study the effects of linear decelerations similar to those occurring in automobile or aircraft crashes.[9] Studies of combinations of static acceleration and vibration have been carried out by mounting vibrators on centrifuges. Blast tubes, sirens, and body respirators are used to study the body's response to pressure distributions surrounding it. At low frequencies, the respirator is valuable in studying the response of the lung-thorax system. Small vibrating pistons, which are available for a wide frequency range, have been used in investigating the mechanical impedance of small surfaces, the transmission of vibration and the physiological reaction to localized excitation. The range of some high-acceleration devices is shown in Fig. 44.3.

SIMULATION OF HUMAN SUBJECTS

The establishment of limits of human tolerance to mechanical forces, and the explanation of injuries produced when these limits are exceeded, frequently requires experimentation at various degrees of potential hazard. To avoid unnecessary risks to humans, animals are used first for detailed physiological studies. As a result of these studies, levels may be determined which are, with reasonable probability, safe for human subjects. However, such comparative experiments have obvious limitations. The different structure, size, and weight of most animals shift their response curves to mechanical forces into other frequency ranges and to other levels than those observed on humans. These differences must be considered in addition to the general and partially known physiological differences between species. For example, the natural frequency of the thorax-abdomen system of a human subject is between 3 and 4 cps; for a mouse the same resonance occurs between 18 and 25 cps. Therefore maximum effect and maximum damage occur at different vibration frequencies and different shock-time patterns in a mouse than in a human. However, studies on small animals are well worth making if care is taken in the interpretation of the data and if scaling laws are established. Dogs, pigs, and primates are used extensively in such tests.

Many kinematic processes, physical loadings, and gross destructive anatomical effects can be studied on dummies which approximate a human being in size, form, mobility, total weight, and weight distribution in body segments.[25] Several such dummies are commercially available. In contrast to those used only for load purposes, dummies simulating basic static and dynamic properties of the human body are called "anthropometric" or "anthropomorphic" dummies. They are used extensively in aviation and automotive crash research. In other studies dummies are used in place of human subjects to evaluate protective seats and harnesses. In such dummies, an attempt is made to match the "resiliency" of human flesh by some kind of padding. However, they are crude simulations at best and their dynamic mechanical properties are, if at all, only reasonably matched in a very narrow low-frequency range. This and the passiveness of such dummies constitute important mechanical differences between them and living subjects.

Efforts have been made to simulate the mechanical properties of the human head in order to study the physical phenomena occurring in the brain during crash conditions.[23] Although these head forms only approximate the human head, they are very useful in the evaluation of the protective features of crash, safety, and antibuffet helmets. Plastic head forms, conforming to standard head measurements, are designed to fracture in the same energy range as that established for the human head. A cranial vault is provided to house instrumentation and a simulated brain mass with comparable weight and consistency (a mixture of glycerin, ethylene glycol, etc.). The static properties of the skin and scalp tissue are simulated with polyvinyl foam.

The static and dynamic breaking strength of bones, ligaments, and muscles and the forces producing fractures in rapid decelerations have been studied frequently on cadaver material.[26, 27, 28] Extreme caution must be exercised in applying elastic and strength properties obtained in this way to a situation involving the living organism. The differences observed between properties of wet, dry, and embalmed materials are considerable;[29] changes in these properties also result in changes in the force distribution of a composite structure. Thus a multitude of physiological, anatomical, and physical factors must be considered for each specific situation, in which the use of animals, dummies, or cadavers as substitutes for live human subjects is planned.

PHYSICAL CHARACTERISTICS OF THE BODY

ANATOMY

Structurally, the body consists of a hard, bony skeleton whose pieces are held together by tough, fibrous ligaments and which is embedded in a highly organized mass of connective tissue and muscle. The soft visceral organs are contained within the rib cage and the abdominal cavity.

An outline of the skeleton is shown in Fig. 44.4. The slightly bowed vertebral column is the central structural element. It consists of a number of individual vertebrae which have roughly cylindrical load-bearing elements separated by fibrocartilaginous pads. Near the lower end several vertebrae are fused together to form the sacrum which is a tightly fitting part of the pelvic girdle. The skull rests at the top of the column and is held in place by muscle and connective tissue as well as by ligaments. At the bottom of each side of the pelvic girdle is a roughly hemispherical hollow into which the head of the femur fits. Below the femur are the tibia and fibula, which in turn rest on the ankle and foot bones.

The intervertebral discs are dense fibrocartilaginous pads. The hip, knee, and ankle joints have cartilaginous layers on their articular surfaces as do also the joints of the upper

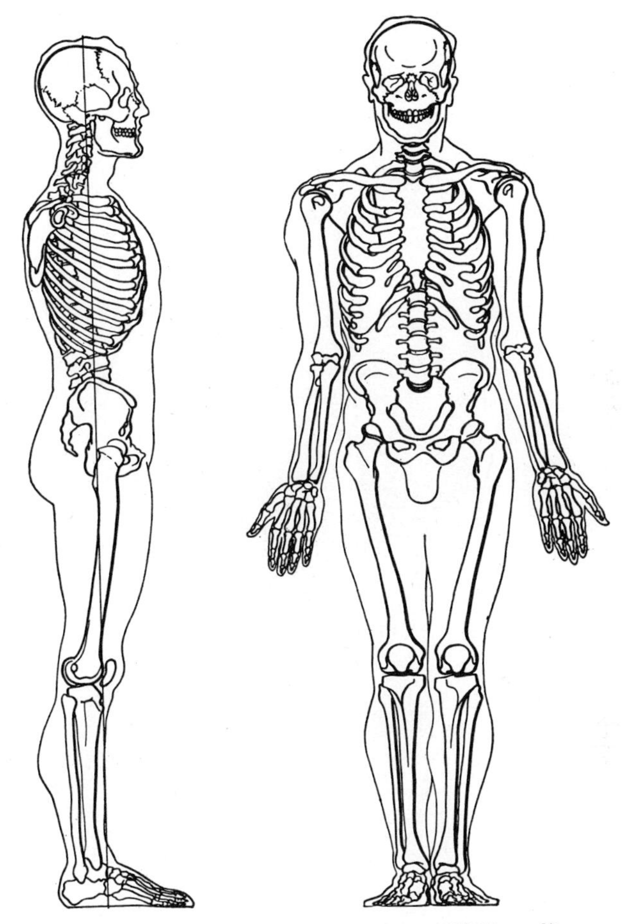

FIG. 44.4. Diagram of human skeleton. (*Goldman.*[34])

limbs. The foot has a tough connective tissue pad at the heel and an arrangement of bones which acts to distribute applied loads. All the joints are held together by ligaments which are flexible but relatively inextensible. These ligaments form a more or less crisscross lacing which permits movement of the joints in a suitable direction without stretching the ligaments themselves to an appreciable extent. The sacroiliac joint is held tightly and almost immovably. The rib cage and shoulder girdle are also dependent on muscle and connective tissue for their support.

In the "ideal" standing position, a plumb line through the center-of-gravity of the body passes through the lower lumbar and upper sacral vertebrae, slightly behind the hip joint sockets and a bit in front of the knee and ankle joints. Upward, the line passes in front of the thoracic curve of the vertebral column and through the support at the base of the skull. Vertical thrusts may be taken up by the compression of the joint pads and by bending of the vertebral column. There is often a slight forward turning moment at the pelvis, especially in older adults. Small deviations from postural symmetry may result in a markedly asymmetrical force distribution under the action of vertical thrusts.

The body musculature, supported from the skeleton by tendons and held together by a network of connective tissue fibers, forms a secondary supporting structure for the skeleton and joints. Fat and skin also contain connective tissue.

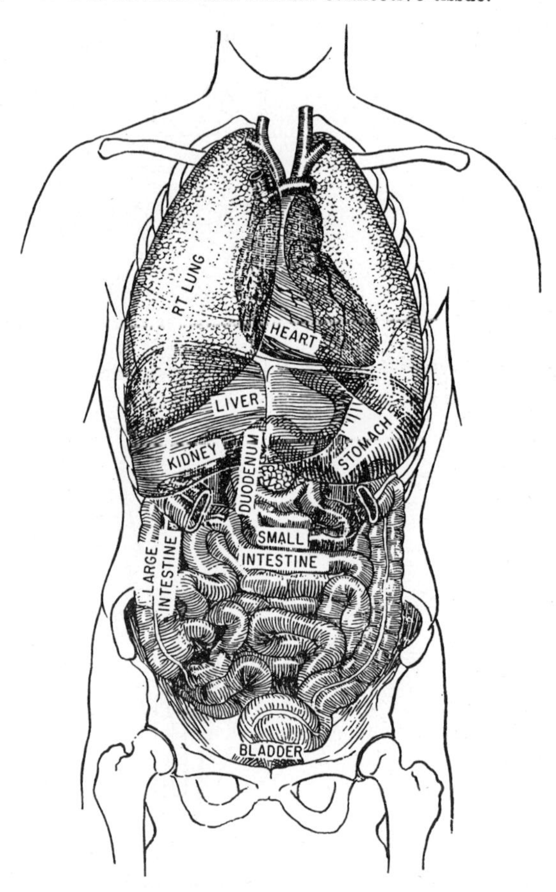

FIG. 44.5. Diagram of position of human viscera. (*Goldman.*[34])

In compression, soft tissues resemble water in their mechanical properties, but in shear they approximate stiff, nonlinear gels with internal losses.

The visceral organs (Fig. 44.5), contained within the thoracic cage and abdominal cavity, are soft tissue elements, separately encapsulated to slide freely over each other; they are supported individually by membranes and ligaments and collectively by the bone, muscle, and connective tissue surroundings. Their weights range from a fraction of an ounce to several pounds and most of the supporting membranes are quite flexible. The kidneys are embedded in fatty tissue and held by a connective tissue sheath in a depression in the posterior wall of the abdominal cavity. The stomach is supported by the esophagus; its displacement is restricted by the dome-shaped diaphragm, which is a large sheet of muscle separating the chest cavity from the abdominal cavity. The lungs, filled with tiny air sacs, are held in place by a combination of supports including a pressure differential. The heart is held in place by ligaments stretched longitudinally through the chest cavity and by large blood vessels. Considerable support also is provided by the diaphragm.

The brain and spinal cord have special protection. The brain is surrounded by liquid contained mostly in the subarachnoid space inside the skull. The spinal cord runs longitudinally through holes in the vertebrae lined with heavy ligaments and a membrane which forms a liquid-filled tube.

Fluid in the body consists of (1) 5 to 6 liters of blood in the heart, arteries, veins, and capillaries; (2) 100 to 150 cm^3 of cerebrospinal fluid surrounding the brain and spinal cord and also within the ventricular cavities of the brain; (3) the interstitial fluid found everywhere in the body as a bathing fluid for cells and tissues but nowhere in large reservoirs; and (4) liquid contained from time to time in the stomach, intestines, and bladder. Gas occurs in the sinuses of the skull, the oronasal cavity, the trachea, the lungs, and often in the stomach and intestines. The latter two organs often contain solid matter as well.

The cells of which the body is made have a wide variety of size and shape. Diameters range from about 0.001 in. to 0.01 in. Shapes may be spherical, disclike, columnar, flat or highly irregular. Many cells have filamentous processes projecting from them. The internal structure of cells is also very complex, containing salts, protein, carbohydrates, and many other substances. Approximately 60 to 80 per cent of the cell is water. Nuclei and other inclusion bodies are found. The rest of the cell is a viscous solution or gel with evidence of considerable submicroscopic structure.

Soft tissues consist of cells, held together by connective tissue and by intercellular links. Blood is a liquid containing nearly 50 per cent by volume of disc-like red cells, together with a few white cells. Soft tissues exhibit a wide variety of structures. Striped (voluntary) muscle consists of parallel bundles of long, thin cells which can be either relaxed or contracted. Control of contraction is provided by nerve fibers which act on small groups of muscle fibers. Thus a whole muscle can have graded degrees of contraction, even though the individual fibers have only two possible states. In this way the elastic stiffness of a muscle can vary widely. Smooth (involuntary) muscle occurs mostly in the walls of hollow organs such as the stomach, intestines, blood vessels, and other specialized organs. The heart consists of a specialized type of muscle fiber.

Nerve tissue is partly cellular (grey matter) or fiber (white matter). The latter contains considerable fatty material in the fiber sheaths.

The structure of bone is very complex. There is an outer layer of hard compact material underneath which is a layer of looser, spongelike bone so arranged as to produce a maximum of strength for commonly encountered stresses. The marrow of some bones contains blood-forming tissue.

The density of most soft tissue is between 1.0 and 1.2 with fatty tissue being somewhat lighter and bone somewhat heavier. Lung tissue is still lighter because of its air content.

PHYSICAL CONSTANTS AND MECHANICAL TRANSMISSION CHARACTERISTICS

USE OF THE PHYSICAL DATA. This section summarizes the passive mechanical responses of the human body and tissues exposed to vibration and impact. The data can be used to calculate quantitatively the transmission and dissipation of vibratory energy in human body tissue, to estimate vibration amplitudes and pressures at different locations of the body, and to predict the effectiveness of various protective measures. Table 44.1 lists some dynamic mechanical characteristics of the body and indicates some of the fields where these data find application. In cases where detailed quantitative investigations are lacking, the information may serve as a guide for the explanation of observed phenomena or for the prediction of results to be expected. Most physical characteristics of the human body presented in this section (except for the strength data) have been derived from the analysis of experimental data in which it is assumed that the body is a linear, passive mechanical system. This is an idealization which holds only for very small amplitudes. Therefore these data may not apply in analyses of mechanical injury to tissue. There is actually considerable nonlinearity of response well below amplitudes required for the production of damage. This is indicated, for example,

Table 44.1. Application of Mechanical Studies of Body

Dynamic mechanical quantity investigated	Field of application
Skull resonances and viscosity of brain tissue	Head injuries; bone-conduction hearing
Impedance of skull and mastoid	Matching and calibration of bone-conduction transducers; ear protection
Ultrasound transmission through skull and brain tissue	Brain tumor diagnosis; changes in central nervous system exposed to focused ultrasound
Sound transmission through skull and tissue	Bone-conduction hearing
Mechanical properties of outer, middle, and inner ear	Theory of hearing; correction of hearing deficiencies
Resonances of mouth, nasal, and pharyngeal cavities	Theory of speech generation; correction of speech deficiencies; oxygen mask design
Resonance of lower jaw	Bone-conduction hearing
Response of mouth-thorax system	Blast-wave injury; respirators
Propagation of pulse cardiac pressure	Circulatory physiology; hemodynamics
Heart sounds	Physiology of heart; diagnosis
Suspension of heart	Ballistocardiography; injury from severe vibrations and crash
Response of the thorax-abdominal mass system	Severe vibration and crash injury; crash protection
Impedance of subject sitting, standing, or lying on vibration platform	Isolation and protection against vibration and short-time accelerations; ballistocardiography
Impedance of body surface, surface wave velocity, sound velocity in tissue, absorption coefficient of body surface	Theory of energy transmission and attenuation in tissue; determination of tissue elasticity, viscosity and compressibility; determination of acoustic and vibratory energy entering the body; vibration isolation; design of vibration pickups; transfer of vibratory energy to inner organs and sensory receptors
Absorption of ultrasound in tissue	Theory of energy transmission on cellular basis; determination of doses for ultrasonic therapy

Table 44.2. Physical Properties of Human Tissue

	Tissue, soft	Bone, compact	
		Fresh	Embalmed, dry
Density, gm/cm^3	1–1.2	1.93–1.98	1.87
Young's modulus, dyne/cm^2	7.5×10^4	2.26×10^{11}	1.84×10^{11}
Volume compressibility,* dyne/cm^2	2.6×10^{10}	...	1.3×10^{11}
Shear elasticity,* dyne/cm^2	2.5×10^4	...	7.1×10^{10}
Shear viscosity,* dyne-sec/cm^2	1.5×10^2	...	....
Sound velocity, cm/sec	$1.5–1.6 \times 10^5$	3.36×10^5	....
Acoustic impedance, dyne-sec/cm^3	1.7×10^5	6×10^5	6×10^5
Tensile strength, dyne/cm^2	...	9.75×10^8	1.05×10^9
Shearing strength, dyne/cm^2, parallel	...	4.9×10^8	....
Shearing strength, dyne/cm^2, perpendicular	...	1.16×10^9	5.55×10^8

* Lamé elastic moduli.

by the data given in Fig. 44.6, which shows how the mechanical stiffness and resistance of soft tissue vary with static deflection. Bone behaves more or less like a normal solid; however, soft elastic tissues such as muscle, tendon, and connective tissue resemble elastomers with respect to their Young's modulus and S-shaped stress-strain relation. These properties have been studied in connection with the quasi-static pressure-volume relations of hollow organs such as arteries, the heart, and the urinary bladder,[30] assuming linear properties in studying dynamic responses. Then soft tissue can be described phenomenologically as a viscoelastic medium; plastic deformation need be considered only if injury occurs. Physical properties of human body tissue are summarized in Table 44.2.

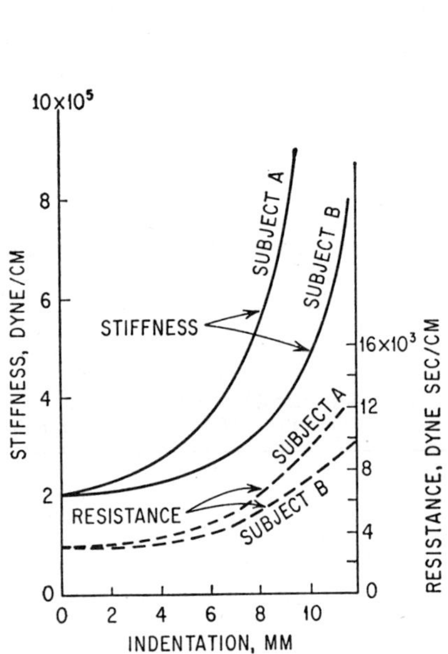

FIG. 44.6. Mechanical stiffness and resistance of soft tissue, per square centimeter, as a function of indentation (i.e., static deflection). The nonlinearity shows the effect of loading of body surface on surface impedance of soft tissues for two experimental human subjects *A* and *B*. (*After Franke.*[21])

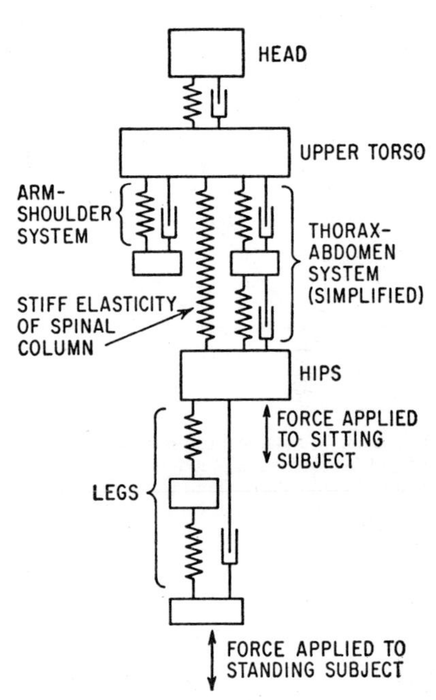

FIG. 44.7. Simplified mechanical system representing the human body—standing on a vertically vibrating platform at low frequencies. (*After Coermann et al.*[33])

The combination of soft tissue and bone in the structure of the body together with the body's geometric dimensions results in a system which exhibits different types of response to vibratory energy depending on the frequency range: At low frequencies (below approximately 100 cps), the body can be described for most purposes as a lumped parameter system; resonances occur due to the interaction of tissue masses with purely elastic structures. At higher frequencies, through the audio-frequency range and up to about 100,000 cps, the body behaves more as a complex distributed parameter system—the type of wave propagation (shear waves, surface waves, or compressional waves) being strongly influenced by boundaries and geometrical configurations.

THE LOW-FREQUENCY RANGE. Simple mechanical systems, such as the one shown in Fig. 44.7 for a standing man, are usually sufficient to describe and understand the important features of the response of the human body to low-frequency vibra-

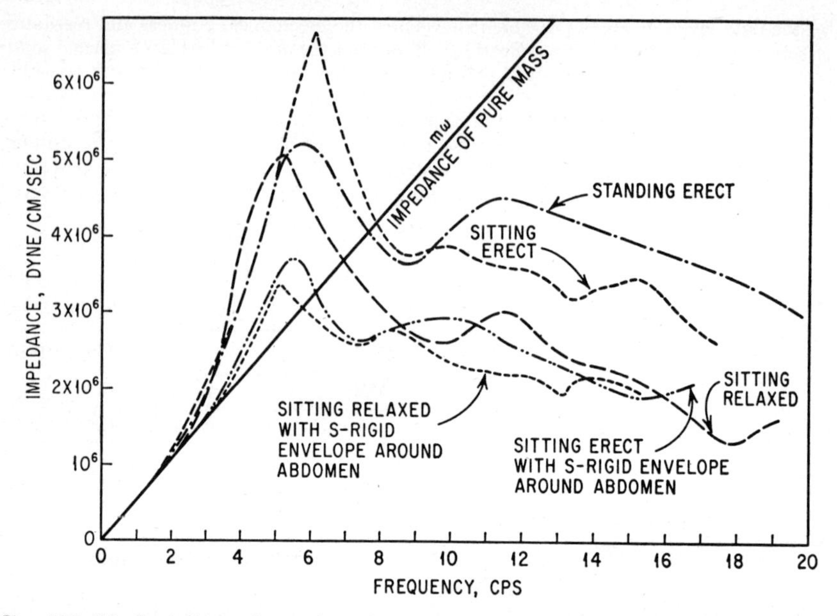

Fig. 44.8. Mechanical impedance of standing and sitting human subject vibrating in the direction of his longitudinal axis as a function of frequency. The effects of body posture and of a semirigid envelope around the abdomen are shown. The impedance of a mass m also is given. (*After Coermann* [32] *and Coermann et al.*[33])

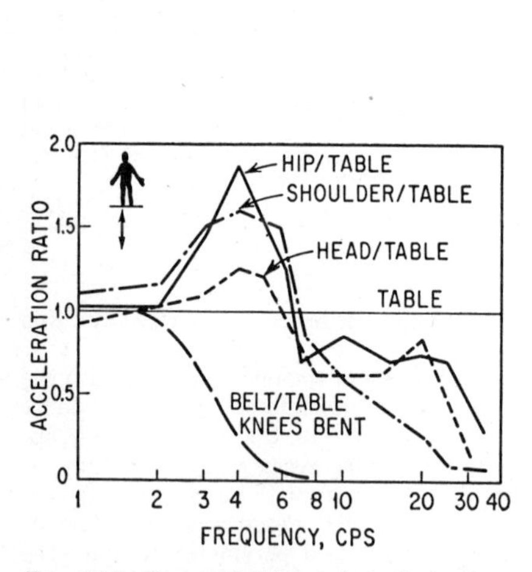

Fig. 44.9. Transmissibility of vertical vibration from table to various parts of the body of a standing human subject as a function of frequency. (*After Dieckmann;* [7] *data for transmission to belt, after Radke.*[35])

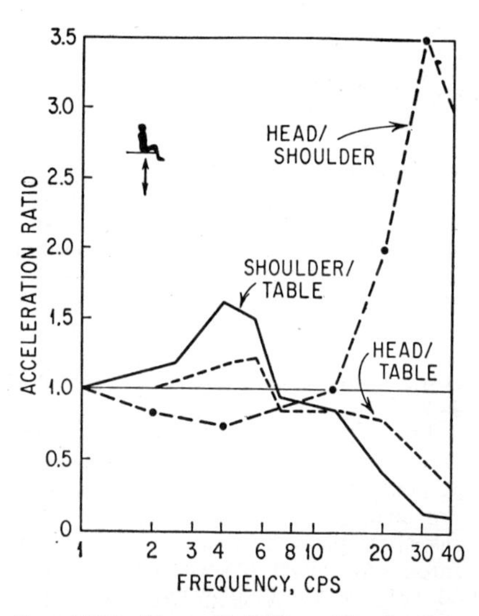

Fig. 44.10. Transmissibility of longitudinal vertical vibration from table to various parts of body of seated human subject as a function of frequency. (*After Dieckmann.*[7])

tions.[34, 36, 37] Nevertheless it is difficult to assign numerical values to the elements of the circuit, since they depend critically on the kind of excitation, the body type of the subject, and his position and muscle tone.

Subject Exposed to Vibrations in the Longitudinal Direction. The mechanical impedance of a man standing or sitting on a vertically vibrating platform is shown in Fig. 44.8. Below approximately 2 cps the body acts as a unit mass. For the sitting man, the first resonance is between 4 and 6 cps; for the standing man, resonance peaks occur at about 5 and 12 cps.[7, 32] The numerical value of the impedance together with its phase angle provides data for the calculation of the total energy transmitted to the subject.

The resonances at 4 to 6 cps and 10 to 14 cps are suggestive of mass-spring combinations of (1) the entire torso with the lower spine and pelvis, and (2) the upper torso with forward flexion movements of the upper vertebral column. The assumption that flexion of the upper vertebral column occurs is supported by observations of the transient response of the body to vertical impact loads and associated compression fractures. The greatest loads occur in the region of the eleventh thoracic to the second lumbar vertebra, which therefore can be assumed as the hinge area for flexion of the upper torso. Since the center-of-gravity of the upper torso is considerably forward of the spine, flexion movement will occur even if the force is applied parallel to the axis of the spine (also see Fig. 44.23). Changing the direction of the force so that it is applied at an angle with respect to the spine (for example, by tilting the torso forward) influences this effect considerably. Similarly the center-of-gravity of the head can be considerably in front of the neck joint which permits forward-backward motion. This situation results in forward-backward rotation of the head instead of pure vertical motion. Examples of relative amplitudes for different parts of the body when it is subject to vibration are shown in Fig. 44.9 for a standing subject, and in Fig. 44.10 for a sitting subject.[7] The curves show an amplification of motion in the region of resonance and a decrease at higher frequencies. The impedances and the transmissibility factors are changed considerably by individual differences in the body, the posture of the body, the type of support by a seat or back rest for a sitting subject, or by the state of the knee or ankle joints of a standing subject. The resonant frequencies remain relatively constant, whereas the transmissibility varies (for the condition of Fig. 44.10, values of transmissibility as high as 4 have been observed at 4 cps [35]). Above approximately 10 cps, vibration displacement amplitudes of the body are smaller than the amplitudes of the exciting table and they decrease continuously with increasing frequency. The attenuation of the vibrations transmitted from the table to the head is illustrated in Fig. 44.11. At 100 cps this attenuation is about 40 db. The attenuation along the body at 50 cps is shown in Fig. 44.12.

Between 20 and 30 cps the head exhibits a mechanical resonance, as indicated in Figs. 44.9 and 44.10. When subject to vibration in this range, the head displacement amplitude can exceed the shoulder amplitude by a factor of 3. This resonance is of importance in connection with the deterioration of visual acuity under the influence of vibration. Another frequency range of disturbances between 60 and 90 cps suggests an eyeball resonance.[36]

The mechanical impedance of the human body, lying on its back on a rigid surface and vibrating in the direction of its longitudinal axis, has been determined in connection with ballistocardiograph studies.[37] For tangential vibration, the total mass of the body behaves as a simple mass-spring system with the elasticity and resistance of the skin. For the average subject

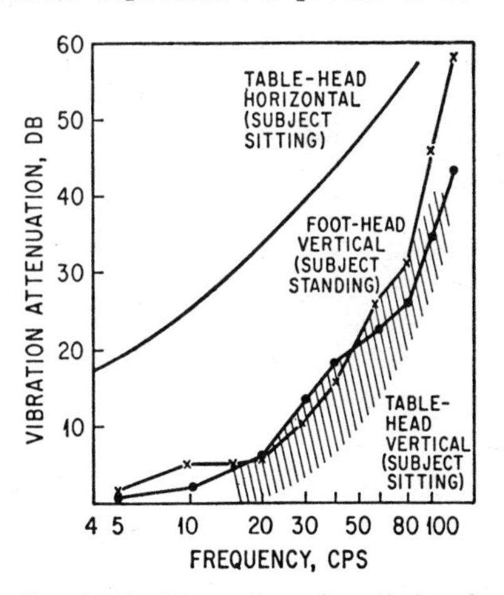

FIG. 44.11. Attenuation of vertical and horizontal vibration for standing and sitting human subjects. (*Continuous lines after von Békésy.*[2] *Shaded area is range of values for 10 subjects after Coermann.*[36])

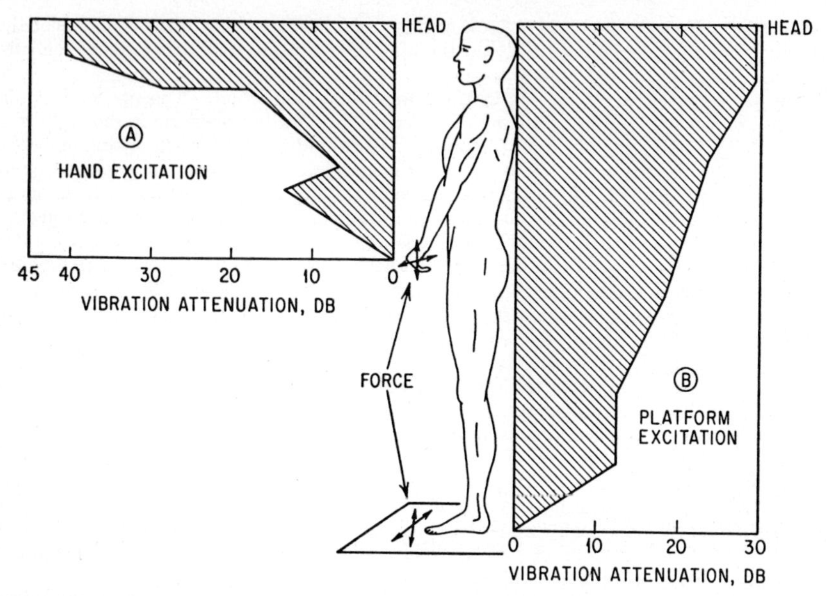

FIG. 44.12. Attenuation of vibration at 50 cps along human body. The attenuation is expressed in decibels below values at the point of excitation. For excitation of (*A*) hand, and (*B*) platform on which subject stands. (*After von Békésy.*[2])

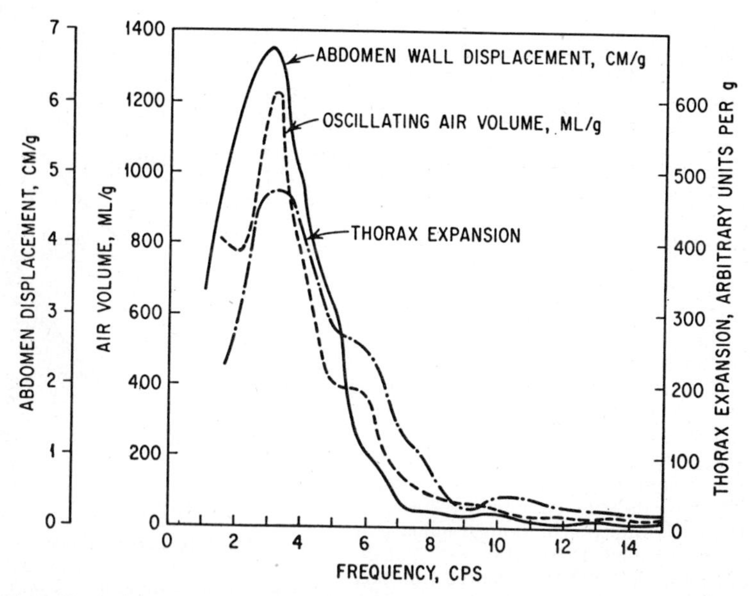

FIG. 44.13. Typical response curves of the thorax-abdomen system of a human subject in the supine position exposed to longitudinal vibrations. The displacement of the abdominal wall (2 in. below umbilicus), the air volume oscillating through the mouth, and the variations in thorax circumference are shown per *g* longitudinal acceleration. (*From Coermann et al.*[33])

the resonant frequency is between 3 and 3.5 cps, and the Q of the system is about 3. If the subject's motion is restricted by clamping the body at the feet and at the shoulders between plates connected with the table, the resonant frequency is shifted to approximately 9 cps and the Q is about 2.5.

One of the most important sub-systems of the body, which is excited in the standing and sitting position as well as in the lying position, is the thorax-abdomen system.[33,38] The abdominal viscera have a high mobility due to the very low stiffness of the diaphragm and the air volume of the lungs and the chest wall behind it. Under the influence of both longitudinal and transverse vibration of the torso, the abdominal mass vibrates in and out of the thoracic cage. Vibrations take place in other than the (longitudinal) direction of excitation; during the phase of the cycle when the abdominal contents swing toward the hips, the abdominal wall is stretched outward and the abdomen appears larger in volume; at the same time, the downward deflection of the diaphragm causes a decrease of the chest circumference. At the other end of the cycle the abdominal wall is pressed inward, the diaphragm upward and the chest wall is expanded. This periodic displacement of the abdominal viscera has a sharp resonance between 3 and 3.5 cps (Fig. 44.13). The oscillations of the abdominal mass are coupled with the air oscillations of the mouth-chest system.[19, 33] Measurements of the impedance of the latter system at the mouth (by applying oscillating air pressure to the mouth) show that the abdominal wall and the anterior chest wall respond to this pressure. The magnitude of the impedance is minimum and the phase angle is zero between 7 and 8 cps. The abdominal wall has a maximum response between 5 and 8 cps, the anterior chest wall between 7 and 11 cps. Vibration of the abdominal system resulting from exposure of a sitting or standing subject is detected clearly as modulation of the air flow velocity through the mouth (Fig. 44.13). (Therefore at large amplitudes of vibration, speech can be modulated at the exposure frequency.) A lumped parameter model of the thorax-abdomen-airway system is used successfully to explain and predict these detailed physiological responses (Fig. 44.14). The same model can also be used, when appropriately excited, to describe the effects of blast, infrasound, and chest impact and to derive curves of equal injury potential, i.e., tolerance curves. Appropriate scaling laws have been applied to it to make it useful for the interpretation and extrapolation of experimental animal data (von Gierke, 1971).

Subject Exposed to Vibrations in the Transverse Direction. The physical response of the body to transverse vibration—i.e., horizontal in the normal upright position—is quite different from that for vertical vibration. Instead of thrust forces acting primarily along the line of action of the force of gravity on the human body, they act at right angles to this line. Therefore the distribution of the body masses along this line is of the utmost importance. There is a greater difference in response between sitting and standing positions for transverse vibration than for vertical vibration where the supporting structure of the skeleton and the spine are designed for vertical loading.

Data of transmission of vibration along the body are given in Fig. 44.45.[39] For a standing subject, the displacement amplitudes of vibration of the hip, shoulder, and head are about 20 to 30 per cent of the table amplitude at 1 cps and decrease with increasing frequency. Relative maxima of shoulder and head amplitudes occur at 2 and 3 cps, respectively. The sitting subject exhibits amplification of the hip (1.5 cps) and head (2 cps) amplitudes. All critical resonant frequencies are between 1 and 3 cps. The transverse vibration patterns of the body can be described as standing waves, i.e., as a rough approximation one can compare the body with a rod in which transverse flexural waves are excited. Therefore there are nodal points on the body which become closer to the feet as the frequency of excitation increases, since the phase shift between all body parts and the table increases continuously with increasing frequency. At the first resonant frequency (1.5 cps), the head of the standing subject has a 180° phase shift with respect to the table; between 2 and 3 cps this phase shift is 360°.[39]

There are longitudinal head motions excited by the transverse vibration in addition to the transverse head motions shown in Fig. 44.15 and discussed above. The head performs a nodding motion due to the anatomy of the upper vertebrae and the location of the head's center-of-gravity. Above 5 cps, the head motion for the sitting and stand-

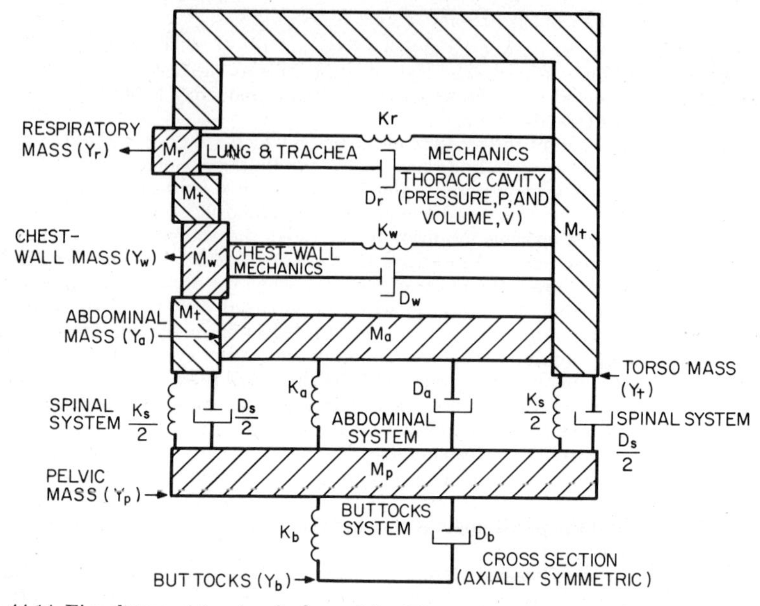

FIG. 44.14. Five-degree-of-freedom body model. The model is used to calculate body deformation (thorax compression, pressure in the lungs, airflow into and out of the lungs, diaphragm and abdominal mass movement) as a function of external longitudinal forces (vibration or impact) and pressure loads (blast, infrasonic acoustic loads). It has also been used to calculate thorax dynamics under impacts to the chest wall, M_w.

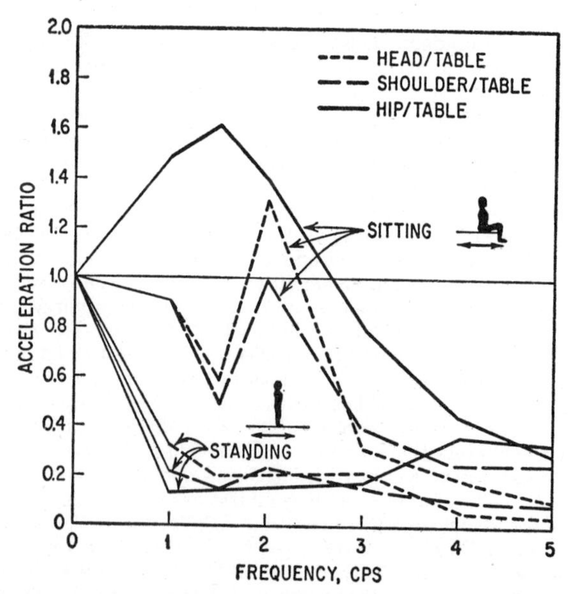

FIG. 44.15. Transmission of transverse, horizontal vibration from table to various parts of sitting and standing human subject. (*After Dieckmann.*[39])

ing subjects is predominantly vertical (about 10 to 30 per cent of the horizontal table motion).

Vibrations Transmitted from the Hand. In connection with studies on the use of vibrating hand tools, measurements have been made of vibration transmission from the hand to the arm and body.[20, 40, 41] The mechanical impedance measured on a hand grip for a specific condition representative of hand tool use is presented in Fig. 44.16. The impedance has one maximum in the range below 5 cps, probably determined by the natural frequencies for transverse excitation of the human body between 1 and 3 cps. A second strong maximum appears between 30 and 40 cps; the effective mass of the hand [approximately 2.2 lb (1 kg)] is here in resonance with the elasticity of the soft parts on the inside of the hand. This elasticity is estimated to be 2.10^{-8} cm/dyne. With a practical hand tool which operates between 40 and 50 cps, the vibration amplitude decreases from the palm to the back of the hand by 35 to 65 per cent. Further losses occur between

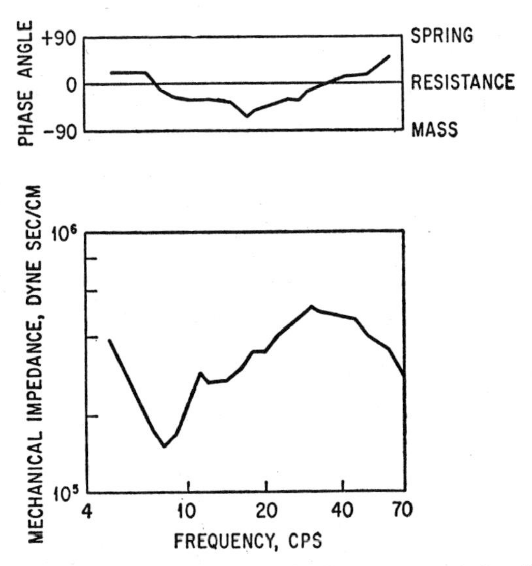

Fig. 44.16. Mechanical impedance and phase angle of arm measured at a vibrating hand grip. Elbow flexion 20 to 25°, static pressure on grip 22 lb. Measurements on one subject. (*After Dieckmann.*[40])

the hand and the elbow and the elbow and the shoulder. Figure 44.12, which shows this decrease of vibration amplitude from the hand to the head, indicates that the strongest attenuation occurs in the shoulder joint.

MIDDLE-FREQUENCY RANGE (WAVE PROPAGATION). Above about 100 cps, simple lumped parameter models become more and more unsatisfactory for describing vibration of tissue. At higher frequencies it is necessary to consider the tissue as a continuous medium for vibration propagation.

Skull Vibrations. The vibration pattern of the skull is approximately the same as the pattern of a spherical elastic shell.[42, 43] The nodal lines observed suggest that the fundamental resonant frequency is between 300 and 400 cps and that resonances for the higher modes are around 600 and 900 cps. The observed frequency ratio between the modes for the skull is approximately 1.7 while the theoretical ratio for a sphere is 1.5. From the observed resonances, the calculated value of the elasticity of skull bone (a value of Young's modulus = 1.4×10^{10} dynes/cm^2) agrees reasonably well with static test results on dry skull preparations, but is somewhat lower than the static test data obtained on the femur (Table 44.2). Mechanical impedances of small areas on the skull over the mastoid area[44] have been measured for practical problems (Table 44.1). The impedance of the skin lining in the auditory canal has been investigated and used in connection with studies on ear protectors.[45]

Vibration of the lower jaw with respect to the skull can be explained by a simple mass-spring system, which has a resonance, relative to the skull, between 100 and 200 cps.[46]

Impedance of Soft Human Tissue. Mechanical impedance measurements of small areas (1 to 17 cm^2) over soft human body tissue have been made with vibrating pistons between 10 and 20,000 cps. At low frequencies this impedance is a large elastic reactance. With increasing frequency the reactance decreases, becomes zero at a resonant frequency, and becomes a mass reactance with a further increase in frequency (Fig. 44.17).[3, 4, 47] These data cannot be explained by a simple lumped parameter model, but

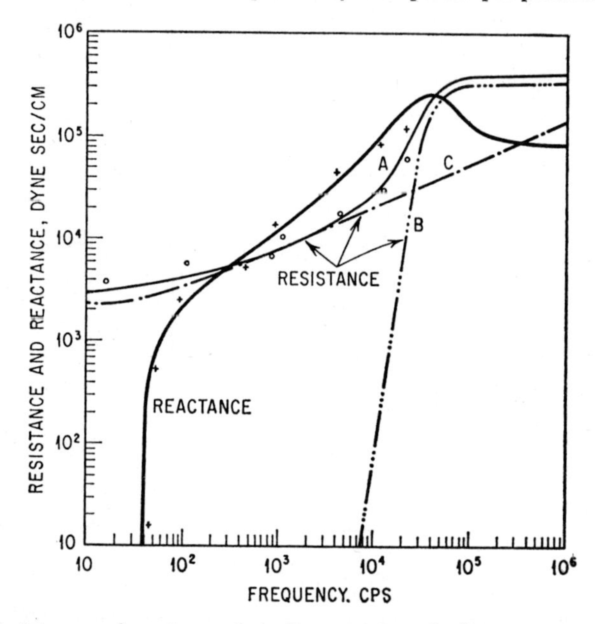

Fig. 44.17. Resistance and reactance of circular area, 2 cm in diameter, of soft tissue body surface as a function of frequency. Crosses and circles indicate measured values for reactance and resistance. Smooth curves calculated for 2-cm-diameter sphere vibrating in (*A*) viscoelastic medium with properties similar to soft tissue (parameters as in Table 44.2), (*B*) frictionless compressible fluid, and (*C*) incompressible viscous fluid. (*From von Gierke et al.*[47])

require a distributed parameter system including a viscoelastic medium—such as the tissue constitutes for this frequency range.[47, 48] The high viscosity of the medium makes possible the use of simplified theoretical assumptions, such as a homogeneous isotropic infinite medium and a vibrating sphere instead of a circular piston. The results of such a theory agree well with the measured characteristics. As a consequence it is possible to assign absolute values to the shear viscosity and the shear elasticity of soft tissue (Table 44.2). The theory together with the measurements show that, over the audiofrequency range, most of the vibratory energy is propagated through the tissue in the form of transverse shear waves—not in the form of longitudinal compression waves. The velocity of the shear waves is about 20 m/sec at 200 cps and increases approximately with the square root of the frequency. This may be compared with the constant sound velocity of about 1,500 m/sec for compressional waves. Some energy is propagated along the body surface in the form of surface waves which have been observed optically. Their velocity is of the same order as the velocity of shear waves.[47]

ULTRASONIC VIBRATION. Above several hundred thousand cycles per second, in the ultrasound range, most of the vibratory energy is propagated through tissue in the form of compressional waves; for these conditions, geometrical acoustics offer a good approximation for the description of their path. Since the tissue dimensions under consideration are almost always large compared to the wavelength (about 1.5 mm at 1 mcps) the mechanical impedance of the tissue is equal to the characteristic impedance, i.e., sound velocity times density. This value for soft tissue differs only slightly from the characteristic impedance of water.[49] The most important factor in this frequency range is the tissue viscosity, which brings about an increasing energy absorption with increasing frequency.[48] At very high frequencies this

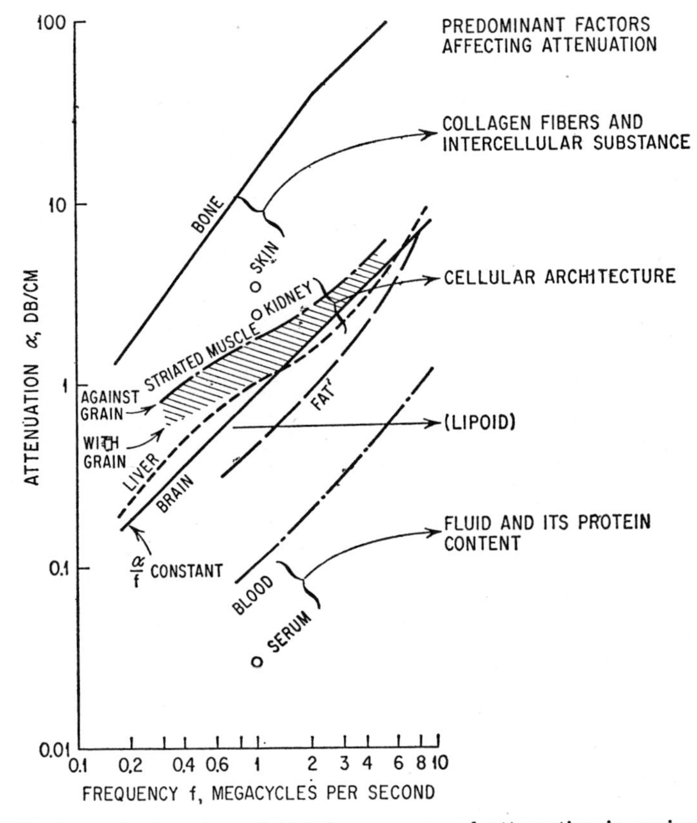

FIG. 44.18. Approximate values of high-frequency sound attenuation in various tissues. (*After Goldman and Hueter* [49] *and Dussik et al.*[50])

viscosity also generates shear waves at the boundaries of the medium, at the boundary of the acoustic beams, and in the areas of wave transition to media with somewhat different constants (e.g., boundary muscle to fat tissue, or soft tissue to bone). These shear waves are attenuated so rapidly that they are of no importance for energy transport but are noticeable as increased local absorption, i.e., heating.

From 500 kcps to 10 mcps the attenuation coefficient describing the decrease of the sound intensity in a plane ultrasound wave is only in fair agreement with the value one would calculate from the tissue viscosity measured in the audio-frequency range (Table 44.2). The tissue deviates in this frequency range from the behavior of a medium with constant viscosity. In Fig. 44.18, attenuation coefficients measured in different types of tissue are summarized.[49, 50] On this graph a slope α/f^2 = constant (where f is the frequency) is indicative of classical viscous absorption with constant shear viscosity. A small slope, or a change in slope, indicates a change in viscosity with frequency (relaxation phenomenon). The graph gives only a few examples and typical functions from a large body of attenuation data available. The absorption of most soft tissues is in the range from 0.5 to 2 db/cm/mcps. In order of increasing absorption are: brain tissue, liver tissue, striated muscle, smooth muscle, kidney, skin, and tendon. Bone has the highest value of absorption—approximately 10 db/cm. The ultrasonic absorption coefficient depends very much on the structural features of the tissue and might well aid in obtaining a clearer quantitative picture of the mechanical structure of cells. Interesting in this respect are the acoustic anisotropies, i.e., cases where attenuation depends on the direction of propagation: fiber anisotropy has been found in the collinear fibers of striated muscle (Fig. 44.18) and layer anisotropy can occur in structures consisting of parallel layers of different tissue types such as in the abdominal wall.

MECHANICAL DATA FROM SHOCK FORCES

Very little numerical data on mechanical characteristics of the body are available from studies on shock or impact forces. Some evidence of resonances has been noted[51] but much of this is better obtained from vibration studies. The application of mechanical data which has been obtained from studies on vibration to exposure to shock and impact is discussed in the section on mechanical damage. Mechanical responses to shock and impact are, in general, extremely difficult to analyze numerically for basic body characteristics.

EFFECTS OF SHOCK AND VIBRATION

The motions and mechanical stresses resulting from the application of mechanical forces to the human body have several possible effects: (1) the motion may interfere directly with physical activity; (2) there may be mechanical damage or destruction; (3) there may be secondary effects (including subjective phenomena) operating through biological receptors and transfer mechanisms, which produce changes in the organism. Thermal and chemical effects are usually unimportant.

MECHANICAL INTERFERENCE. There are many ways in which forces can be applied to the body. The effects of these forces are quite varied. Certain types of displacement, velocity, or acceleration (if of sufficient magnitude) can be very disturbing to sensory and neuromuscular activities such as reading instruments, or making fine adjustments of controls or of the position of the body. Speech communication may be rendered difficult. Very little is known about how much of what kind of motion interferes with particular activities. When available, such information has meaning only in terms of tolerances permitted for the activities in question. For example, the disturbance of visual acuity arising from body vibration is not only frequency-dependent[36] (Fig. 44.19) but is roughly proportional to the amplitude of the vibration. This disturbance may be controlled either by changing the frequency, reducing the amplitude, or by decreasing the acuity required for the given task. Mechanical motion resulting from occasional shocks usually will not interfere directly with most tasks unless timed critically with respect to some operation or unless repeated at very short intervals. Such interference with task performance as does arise from shocks is more apt to be the result of biological responses or of actual damage.

MECHANICAL DAMAGE. Of the many kinds and degrees of mechanical damage to the human body arising from the application of mechanical forces, certain kinds have been singled out for attention and study because they are particularly common, dangerous, or disturbing in some special way. Among these, short of actual destruction, are bone fracture, lung damage, injury to the inner wall of the intestine, brain injury, cardiac damage, ear damage, tearing or crushing of soft tissues, and certain special types of chronic injury such as tendon or joint strains and the "white finger" syndrome of vibrating tool operators.

BIOLOGICAL RESPONSES. Mechanical stresses and motions may stimulate various receptor organs in the skin and elsewhere or may excite parts of the nervous system directly. The result may be reflex activity or modification of it. The stimuli

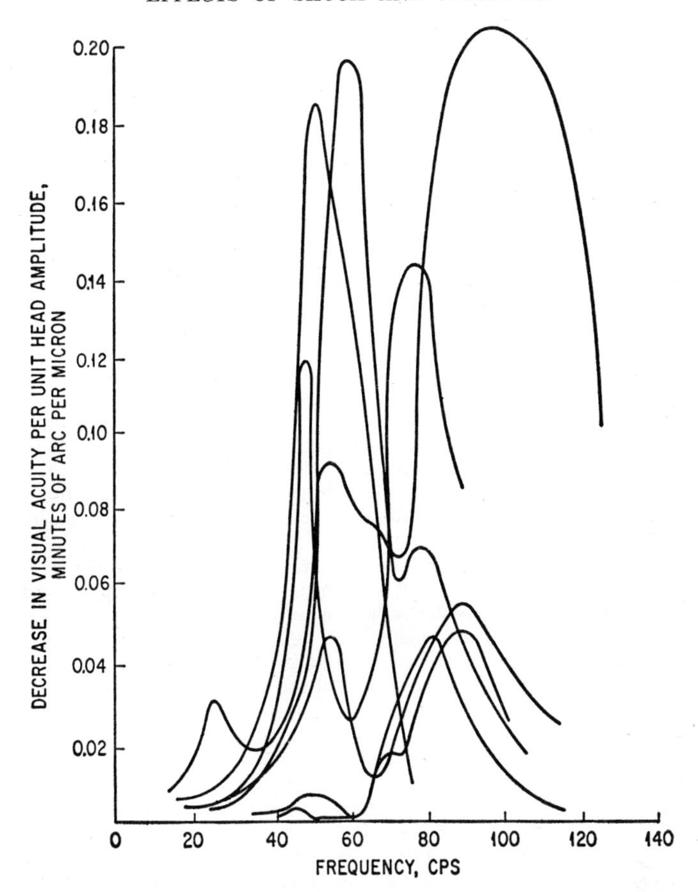

Fig. 44.19. Decrease of visual acuity of observers standing on vibrating platform. Each curve represents one subject. (*After Coermann.*[36])

initiate nervous system and hormonal activity which has a marked modifying action on many metabolic processes relating to food assimilation, muscular activity, reproductive activity, etc.[52] These changes are difficult to measure and correlate; they seem to differ in different species, at least in degree. Nevertheless, considerable indirect evidence exists for the reality of these response patterns and it usually is agreed that the general conclusions drawn from animal experimentation apply to man. Exposure to mechanical forces (when of great enough extent and duration) results, in part, in phenomena such as fatigue, changes in work capacity, ability to maintain attentiveness, etc. In response to acute stimulation, excitation of brain centers may produce emotional reactions such as fear or unpleasantness and lead to automatic or deliberate compensatory or protective behavior.

EFFECTS OF MECHANICAL VIBRATION

MECHANICAL DAMAGE. Damage is produced when the accelerative forces are high enough. However, experimental data can be obtained only from experiments on animals. Such data must be considered very carefully in applying the results to humans. Mice,[53] rats,[54] and cats[55, 56] have been killed by exposure to vibration. There is a definite frequency dependence of the lethal accelerations which coincides with the resonant displacement of the visceral organs, but which has been only partly established. Mice are killed at accelerations of 10 to 20*g* within a few minutes in the range 15 to 25 cps; above and below this frequency range, the survival time is longer. Rats and cats may be killed within 5 to 30 min at accelerations above about 10*g*. Post-mortem examination of these animals usually shows lung damage, often heart damage, and occasionally brain injury. The injuries to heart and lungs probably result from the beating of these organs against each other and against the rib cage. The brain injury, which is a superficial hemorrhage, may be due to relative motion of the brain within the skull, to mechanical action involving the blood vessels or sinuses directly, or to secondary mechanical effects. Tearing of intraabdominal membranes rarely occurs. Exposure for several minutes to peak acceleration of about 5*g* occasionally produces heart damage, as indicated by changes in the electrocardiogram.[56] An increase in body temperature is found on exposure to vibration. Since this occurs also in vibration experiments on dead animals it is probably mechanical in origin. Calculations of heat absorption based on body impedance data suggest that appreciable heat can be generated at large amplitudes. Exposure of monkeys to acceleration of 5*g* at a frequency of 10 and 20 cps for several hours seems to produce some damage to the vestibular system, but these findings require confirmation.[57] Observations on man[58] indicate that above about 3*g*, sharp pain in the chest may occur. Traces of blood occasionally have been found in the feces after exposure to acceleration of 6*g* at a frequency of 20 to 25 cps for about 15 min. This suggests mechanical damage to the intestine or rectum.

It is clear that several of the phenomena found in animals also must be possible in humans. Mechanical damage to heart and lungs, injury to the brain, tearing of membranes in the abdominal and chest cavities, as well as the above-mentioned intestinal injury, are all possible. Also it is likely that there is heating of the body when it is shaken. Acceleration-frequency curves for these effects have not been established. Because of the relatively greater visceral masses of the human, the minima of such curves, which would correspond to resonance ranges, must occur at relatively lower frequencies than in small animals. Subjective symptoms such as the occurrence of chest pain after exposure to acceleration of 3*g* (Table 44.3) may or may not be significant, although from the point of view of safety they must be taken seriously.

Chronic injuries may be produced by vibration exposure of long duration at levels which produce no apparent acute effects. In practice, such effects are usually found after exposure to repeated blows or to random jolts rather than to sinusoidal motion. When such blows are applied to the human body at relatively short intervals, the relation of the interval to tissue response times becomes very important. Exposure to such forces frequently occurs in connection with the riding of vehicles. Buffeting in aircraft or in high-speed small craft on the water, and shaking in heavy vehicles on rough surfaces, gives rise to irregular jolting motion. Acute injuries from exposure to these situations are rare, but complaints of discomfort and chronic minor injury are common. Truck and tractor drivers often have sacroiliac strain. Minor kidney injuries are occasionally suspected and, rarely, traces of blood may appear in the urine. The length of exposure and the details of the ways in which the body is supported play an important role.

Chronic injuries also are produced by localized vibration. A classical example of this is the pain and numbing of the fingers on exposure to cold which affects many people after several months of using such equipment as pneumatic hammers and drills or hand-held grinders or polishers. The heavier, slow-moving devices appear to produce more severe jolting. There is an extensive clinical literature on this subject[59] but little known of the mechanism of the injury or of the actual forces responsible. Many high-frequency components may be present.[41] The repeated insults to the tissues seem gradually to affect the capillaries and their nerve supply. Injuries resembling this have been produced in the feet of rats exposed to vibration at a frequency of 60 cps at an acceleration amplitude of 8 to 9*g* for 10 to 12 hr per day up to about 1,000 hr.[60]

PHYSIOLOGICAL RESPONSES. Rats exposed for 10 to 40 min a day for several months to vibration at a frequency of 12.5 cps at an acceleration amplitude of about 15*g* appear to show minor behavioral abnormalities.[61] The adrenal glands of rats show a rapid fall in ascorbic acid content on exposure to acceleration levels of a few tenths of a g unit at 5 to 10 cps,[62] and changes in the reproductive cycle and growth have been observed in rabbits exposed to vibration for several days.[64] Changes in respiration, heart activity and peripheral circulation have been observed in both man and animals as immediate and possibly transient responses to moderate vibration. Certain postural reflexes appear to be inhibited by vibratory motion.[64, 65]

SUBJECTIVE RESPONSES TO VIBRATION

The subjective responses of humans include perception, feelings of discomfort, apprehension, and pain. Since acceleration, frequency, mode of application, duration, and the situation of the subject are all involved, it is exceedingly difficult to find simple definitive ways of characterizing results. Early workers in this field[66–71] concerned themselves with whole-body exposure under conditions which were believed to be of practical interest. The results which they obtained were only very approximate since there appears to have been little control of subjective factors.

Generally there are three simple criteria for subjective responses to shock and vibration: the thresholds of perception, of unpleasantness, and of tolerance. The two latter are difficult to identify and reproduce, although agreement to within a factor of about 3 has been obtained. A compilation of these results based on exposures of about 5 to 20 min is given in Fig. 44.20.[72] For longer exposures, data are limited. Some information has been obtained on comfort and tolerance levels for aircraft pilots.[73] Very long exposure to vibration much above the level of perception seems to be irritating and fatiguing.

For short exposures, less than 5 min, a study has been carried out in the frequency range 1 to 15 cps.[74] Subjects were exposed to a specified acceleration amplitude until they could no longer tolerate it. They were then asked for their reactions and what their specific reason was for asking to be released. Table 44.3 shows a distribution of the major reasons for each of the frequencies used. Evidently no single criterion of tolerance was found although some reactions were more common than others. The estimated limits of tolerance for short exposures according to these criteria are given in Fig. 44.21.

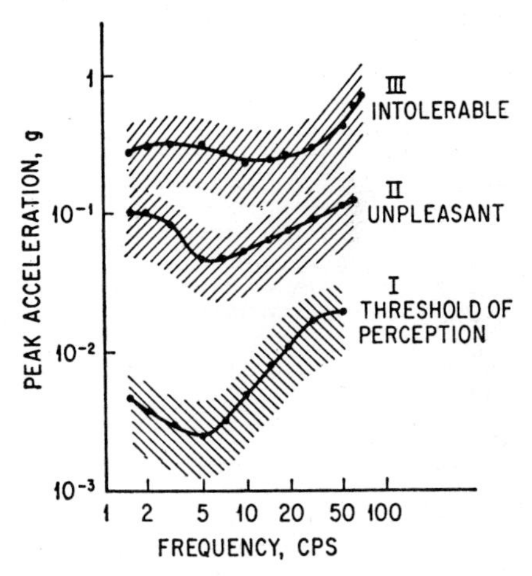

FIG. 44.20. Average peak accelerations at various frequencies at which subjects perceive vibration (I); find it unpleasant (II); or refuse to tolerate it further (III). Exposures of 5 to 20 min. Shaded areas are about one standard deviation on either side of mean. Data averaged from seven sources. (*Goldman.*[72])

EFFECTS OF MECHANICAL SHOCK

Mechanical shock includes several types of force application which have similar, though not identical, effects. Explosions, explosive compression or decompression, and impacts and blows from rapid changes in body velocity or from moving objects produce shock forces of importance. Major damage, short of complete tissue destruction, is usually to lungs, intestines, heart, or brain. Differences in injury patterns arise from differences in rates of loading, peak force, duration, localization of forces, etc.

BLAST AND SHOCK WAVES.[5, 17, 75, 110] The mechanical effects associated with rapid changes in environmental pressure are primarily localized to the vicinity of air-

Table 44.3. Criteria of Tolerance for Short Exposure to Vertical Vibration for Various Frequencies

(From Ziegenruecker and Magid;[74] *see Fig.* 44.21)

(Each cross indicates a decided comment from one of the ten human subjects used in this study as to his experiencing the symptom listed.)

Symptom → Freq. ↓	Abdominal pain	Chest pain	Testicular pain	Head symptoms	Dyspnea	Anxiety	General discomfort
1 cps					xxxxx xxx		xxx
2 cps					xxxxx xxx		xxxx
3 cps	xx	xx			xxxxx	x	xxxxx
4 cps	xx	xx		xx	xxx	xx	xxxxx
5 cps		xxxx				x	xxxxx x
6 cps	xxx	xxxx		x			xxxx
7 cps	xx	xxxxx	x	x			x
8 cps	x	xxxx		x		xx	xxx
9 cps	xx	xxxx			x		xxxxx
10 cps	x	x	xxx	xx		x	
15 cps							xxxxx xxx

filled cavities in the body, i.e., the lungs and the air-containing gastrointestinal tract. Here, heavy masses of blood or tissue border on light masses of air. The local impedance mismatch can lead to a relative tissue displacement, which is destructive, by several different mechanisms. Starting with very slow differential pressure changes, of approximately 1 sec duration or longer, dynamic mechanical effects are unimportant; the static pressure is responsible for destructive mechanical stress or physiological response. Such pressure-time functions occur with the explosive decompression of pressurized aircraft cabins at high altitude and occur with the slow response of very well-sealed shelters to blast waves. If the pressure rise times or fall times are shortened (roughly to the order of tenths of seconds), the dynamic response of the different resonating systems of the body becomes important, in particular the thorax-abdomen system of Fig. 44.14. The dynamic load factor of the specific blast disturbance under consideration determines the response. Available data for single pulse, "instantaneously rising" pressures suggest the existence of a minimum peak pressure which corresponds to natural frequencies for dogs of between 10 and 25 cps;[17] for humans this frequency is lower. Sensitivity curves for blast exposure, i.e., curves of equal injury potential (maximum tolerable level) for various species are shown in Fig. 44.22. The theoretical curves are obtained by means of the thorax model of Fig. 44.14 after application of appropriate scaling laws to account for the different species sizes (von Gierke, 1971). For pressures with total durations of milliseconds or less and much shorter rise times (duration of wave, short compared to the natural period of the responding tissue), the effect and destruction seem

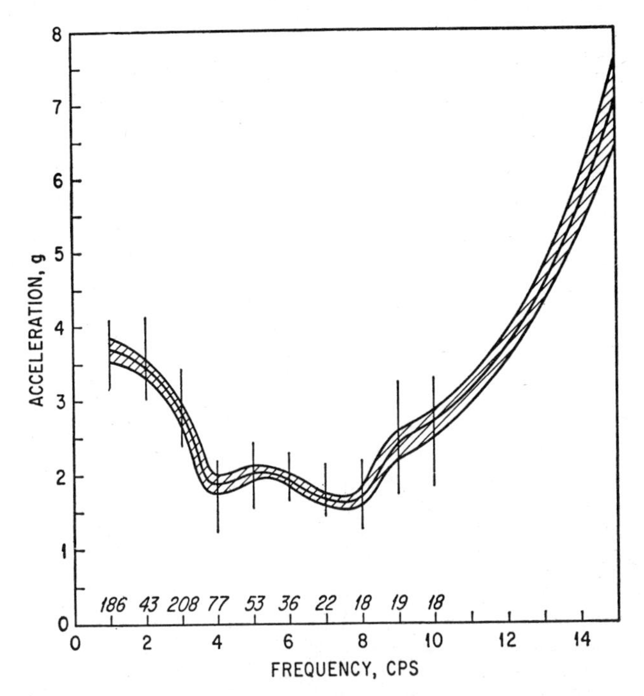

FIG. 44.21. Peak acceleration at various frequencies at which subjects refuse to tolerate further a short exposure (less than 5 min) to vertical vibration. The figures above the abscissa indicate the exposure time in seconds at the corresponding frequency. The shaded area has a width of one standard deviation on either side of the mean (10 subjects). (*Ziegenruecker and Magid.*[74])

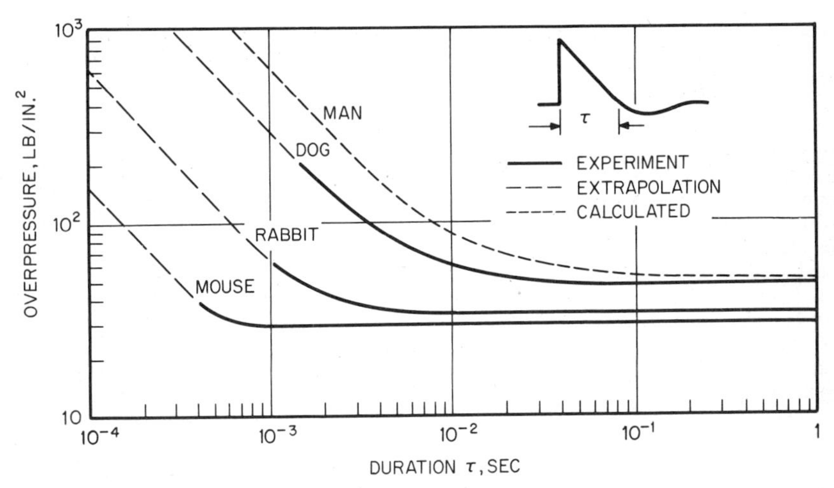

FIG. 44.22. Maximum tolerable blast overpressures for mouse, rabbit, dog, and man. The curve for man is calculated by means of a model of the type given in Fig. 44.14. Using the same model, dimensionally scaled to the animal sizes, results in curves matching closely the experimental data.

to depend primarily on the momentum of the shock wave. The mass m of an oscillatory system located in a wall or body surface, which is struck by a shock wave, is set into motion according to the relation $\int P_r\,dt = mv_0$, where P_r = reflected pressure at body surface, t = time, and v_0 = initial velocity. Experimental fatality curves on animals[5] generally show this dependence on momentum for short pressure phenomena (close to center of detonation) and the transition to a dependence on peak pressure for phenomena of long duration (far away from center). Fatal blast waves in air and water, for example, differ widely in peak pressure and duration (in air, 10 atm in excess of atmospheric pressure with a duration of 2.8 milliseconds, in water 135 atm in excess with a duration of 0.17 milliseconds) but their momenta are similar. In this most important range of short duration blasts, the mechanical effects are localized because of the short duration, i.e., the high-frequency content of the wave. The upper respiratory tract and bronchial tree, as well as the thorax and abdomen system, are too large and have resonant frequencies too low to be excited; there is no general compression or overexpansion of the thorax, which leads to pulmonary injury as in explosive decompression. The blast waves go directly through the thoracic wall, producing an impact or grazing blow. Inside the tissue, blast injury has three possible causes: (1) spalling effects, i.e., injuries caused by the tensile stresses arising from the reflection of the shock wave at the boundary between media with different propagation velocity [for example, subpleural pulmonary hemorrhages along the ribs]; (2) inertia effects which lead to different accelerations of adjacent tissues with various densities, when the shock wave passes simultaneously through these media; and (3) implosion of gas bubbles enclosed in a liquid. These phenomena are similar to the observations made with high-velocity missiles passing through water near air-containing tissues.[76, 77] The shock waves may produce not only pulmonary injuries but also hard, sharply circumscribed blows to the heart.

Of the injuries produced by exposure to high-explosive blast, lung hemorrhage is one of the most common. It may not of itself be fatal, since enough functional lung tissue may easily remain to permit marginal gas exchange. However, the rupture of the capillaries in the lung produces bleeding into the alveoli and tissue spaces, which can seriously hamper respiratory activity or produce various respiratory and cardiac reflexes. The heart rate is often very slow after a blast injury. Leakage of fluid through moderately injured, but not ruptured, capillaries may occur. There is also the possibility that air may enter the circulation to form bubbles or emboli and by reaching critical regions may impair fatally the heart or brain circulation, or produce secondary damage to other organs.[5] Fat emboli also may be formed and these, too, are capable of blocking vessels supplying vital parts. When gas pockets are present in the intestines, the shock may produce hemorrhage and in extreme cases rupture the intestinal wall itself.

The effects of underwater shock waves on man and animals are in general of the same kind as those produced by air blast. Differences which appear are those of magnitude and often depend on the mode of exposure of the body. A person in the water may, for example, be submerged from the waist down only; in this case, damage is practically confined to the lower half of the body so that intestinal, rather than lung, damage will occur.[78] Direct mechanical injury to the heart muscle and conducting mechanism is possible. Cerebral concussion resulting directly from exposure to shock waves is unusual. Neurological symptoms following exposure to blast, however, may include general depression of nervous activity sometimes to the point of abolition of certain reflexes. Psychological changes such as memory disturbances and abnormal emotional states are found sometimes. In extreme cases, there may be paralysis or muscular dysfunction. Unconsciousness and subsequent amnesia for events immediately preceding the injury result more commonly from blows to the head than from air blast. Recovery from minor concussion apparently may be complete, but repeated concussion may produce lasting damage.[79]

The ear is the part of the human body most sensitive to blast injury. Rupture of the tympanic membrane[17, 80] and injury to the conduction apparatus can occur singly or together with injury to the hair cells in the inner ear.[81] The two first-mentioned injuries may protect the inner ear through energy dissipation. The degree of injury depends on the frequency content of the blast pressure function. The fact that the ear's greatest mechanical sensitivity occurs at frequencies between 1,500 and 3,000 cps explains its vulnerability to short-duration blast waves. Peak pressures of only a few pounds per square inch can rupture the ear drum and still smaller pressures can damage the conducting mechanism and the inner ear. There are wide variations in individual susceptibility to these injuries.

IMPACTS, BLOWS, RAPID DECELERATION. This type of force is experienced in falls, in automobile or aircraft crashes, in parachute openings, in seat ejections for escape from high-speed military aircraft, and in many other situations. Interest in the body's responses to these forces centers on mechanical stress limits. The injuries which occur most often are bruises, tissue crushing, bone fracture, rupture of soft tissues and organs, and concussion. A bruise is a superficial area of slight tissue damage with rupture of the small blood vessels and accumulation of blood and fluid in and around the injured region. It is essentially a crushing injury produced by compression of the tissues, usually between the impinging solid and the underlying bone. It is extremely common and is readily repaired by the body itself. When the tissue is completely destroyed by crushing, the damage usually is irreparable. Bone fractures, like bruises, require that the forces be sustained long enough to produce appreciable displacements and properly concentrated stresses.

When soft tissues are displaced considerable distances by appropriate forces, so-called internal injuries, i.e., rupture of membranes or organ capsules, may take place. Such injuries are, in practice, more often produced by forces of relatively long duration and usually are dangerous.

Experiments have been carried out in which animals were embedded in plaster casts and then dropped about 20 ft onto a metal plate which had various degrees of cushioning. In this case the decelerative impact is well distributed. The degree of injury increases with the acceleration. Lung hemorrhage and laceration of liver and spleen capsules are most common. Damage to the diaphragm, the brain, and the bone structure occurs when the deceleration exceeds about $500g$ (velocity change 35 ft/sec).[82]

The obvious correlation between the response of the body system to continuous vibration and to spike and step-force functions may be used to guide and interpret experiments. The tissue areas stressed to maximum relative displacement at the various frequencies during steady-state excitation are naturally preferred target areas for injury under impact load if the force-time functions of the impacts have appreciable energy in these frequency bands, i.e., if the impact duration is of the same order of magnitude as the body's natural periods. If the impact exposure times are shorter, stress tolerance limits increase; if exposure times decrease to hundredths or thousandths of a second, the response becomes more and more limited and localized to the point of application of the force (blow). Elastic compression or injury will depend on the load distribution over the application area, i.e., the pressure, to which tissues are subjected. If tissue destruction or bone fracture occurs close to the area of application of the force these will absorb additional energy and protect deeper-seated tissues by reducing the peak force and spreading it over a longer period of time. An example is the fracture of foot and ankle of men standing on the deck of warships when an explosion occurs beneath. The support may be thrown upward with great momentum; if the velocity reaches 5 to 10 ft/sec (which corresponds, under these conditions, to an acceleration of several hundred g) fractures occur.[83] However, the energy absorption by the fracture protects structures of the body which are higher up.

If the force functions contain extremely high frequencies, the compression effects spread from the area of force application throughout the body as compression waves. If these are of sufficient amplitude they may cause considerable tissue disruption. Such compression waves are observed from the impact of high-velocity missiles.

If the exposure to the accelerating forces lasts long enough so that (as in most applications of interest) the whole body is displaced, exact measurement of the force applied to the body and of the direction and contact areas of application becomes of extreme importance. In studies of seat ejection, for example, a knowledge of seat acceleration alone is not sufficient for estimating responses. One must know the forces in those structures or restraining harnesses through which acceleration forces are transmitted. The location of the center-of-gravity of the various body parts such as arms, head, and upper torso must be known over the time of force application so that the resulting body motion and deformation can be analyzed and controlled for protection purposes. In addition to the primary displacements of body parts and organs, there are secondary forces from decelerations if, due to the large amplitudes, the motions of parts of the body are stopped suddenly by hitting other body parts. Examples occur in linear deceleration where, depending on the restraint, the head may be thrown forward until it hits the chest or, if only a lap belt is used, the upper torso may jackknife and the chest may hit the knees. There is always the additional possibility that the body may strike nearby objects, thus initiating a new impact deceleration history.

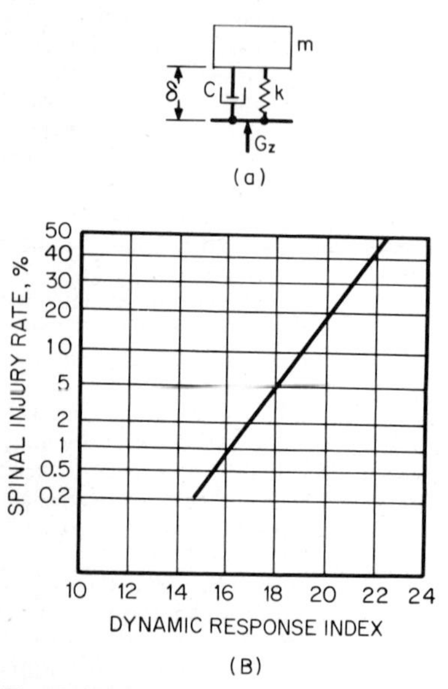

FIG. 44.23. Prediction of spinal compression fracture under longitudinal acceleration loads, G_z. (*A*) Simple model for the study of body ballistics and injury prediction. Deflection of the spine δ is used to represent stress in the vertebral column. (*B*) Probability of spinal compression injury as a function of dynamic response index (DRI) [DRI $= \omega_n{}^2\,\delta_{\max}/g$, where $\omega_n = (K/M)^{1/2}$, $\delta_{\max}$ = maximum deflection, and $g = 9.81$ m/sec²].

Longitudinal Acceleration. The study of positive longitudinal (headward) acceleration of short duration is connected closely with the development of upward ejection seats for escape from aircraft. Since the necessary ejection velocity of approximately 60 ft/sec and the available distance for the catapult guide rails of about 3 ft are determined by the aircraft without much leeway, the minimum acceleration required (step function) is approximately 18.6*g*. Since the high jolt of the instantaneous acceleration increase is undesirable because of the high dynamic load factor in this direction for the frequency range of body resonances (compare with Fig. 44.10), slower build-up of the acceleration with higher final acceleration is preferable to prevent injury. Investigations show that the body's ballistic response can be predicted by means of analog computations making use of the frequency-response characteristics of the body.[84] The simplest analog used for the study of headward accelerations is the single degree-of-freedom mechanical resonator composed of the lumped-parameter elements of a spring, mass, and damper. A diagram of this model is shown in Fig. 44.23*A*. The model is used to simulate the maximum stress developed within the vertebral column (the first failure mode in this direction) for any given impact environment. Use of the model provides a basis for a probabilistic approach to injury prediction and permits the evaluation of simple and complex acceleration waveforms in the same terms. The natural frequency, $\omega_n = \sqrt{K/M}$, for the model is 52.9 rad/sec and the damping ratio $c/2\sqrt{KM}$ is 0.224. The dynamic response index (DRI) is defined as $\omega_n{}^2\delta_{\max}/g$. It has been determined that a DRI value of 18 corresponds to a probability of spinal injury of 55 per cent—a value frequently used for design purposes. This relation between DRI and probability of spinal injury (Fig. 44.23*B*) is based on extensive strength data on the vertebral column and on the evaluation of military injury statistics. (For references, see Symposium on Biodynamic Models and their Applications, 1970, and von Gierke, 1971.) Although more refined models of the human spine are being used[85] (Stech and Payne, 1969, von Gierke, 1971), the simple approach shown in Fig. 44.23 is the one most frequently applied for practical design problems.

Actual control or prevention of injury in this direction is critically dependent on optimal body positioning and restraint to minimize unwanted and forceful flexion of the spinal column.[86] The fracture tolerance limits are influenced by age, physical condition, clothing, weight, and many other factors which detract from the optimum. If the tolerance limits are exceeded, fractures of the lumbar and thoracic vertebrae occur first. While in and of itself this injury may not be classified as severe, small changes in orientation may be enough to involve the spinal cord, an injury which is extremely severe and may be life-threatening. Neck injuries from headward accelerations appear to occur at considerably higher levels.

In general, for vertical crash loads the same considerations apply as discussed for seat ejections, although no control over the build-up time of the acceleration is possible and more sudden onsets must be expected.

For negative (tailward) acceleration (downward ejection) no firm point for application of the accelerating force is accessible as for positive acceleration. If the force is applied as usual through harness and belt at shoulder and groin,[87] the mobility of the shoulder girdle together with the elasticity of the belts results in a lower resonant frequency than the one observed in upward ejection. To avoid overshooting with standard harnesses, the acceleration rise time must be at least 0.15 sec. This type of impact can excite the thorax-abdomen system (Fig. 44.14). The diaphragm is pushed upward by the abdominal viscera; as a result, air rushes out of the lungs (if the glottis is open) or high pressures develop in the air passages.[88] Tolerance limits for negative acceleration probably are set by the compression load on the thoracic vertebrae, which are exposed to the load of the portion of the body below the chest. This load on the vertebrae is higher than that for the positive acceleration case due to the greater weight; therefore a tolerance limit for acceleration has been set at $13g$. Shoulder accelerations of $13g$ have been tolerated by human subjects without injury, when the load was divided between hips and shoulders.

Transverse Accelerations. The forward-facing and backward-facing seated positions are most frequently exposed to high transverse components of crash loads. Human tolerance to these forces has been studied extensively by volunteer tests on linear decelerators,[9] in automobile crashes,[15] and by the analysis of the records of accidental falls.[89] The results indicate the importance of distributing the decelerative forces or impact over as wide an area as possible. The tolerable levels of acceleration amplitude of well over $50g$ ($100g$ and over for falling flat on the back with minor injuries, 35 to $40g$ for 0.05 sec voluntary tolerance seated with restraining harness) are probably limited by injury to the brain. An indication that the latter might be sensitive to, and based on specific dynamic responses, is the fact that the tolerance limit depends strongly on the rise time of the acceleration. With rise times around 0.1 sec (rate of change of acceleration $500g$/sec), no overshooting of head and chest accelerations is observed, whereas faster rise times of around 0.03 sec (1,000 to $1{,}400g$/sec) result in overshooting of chest accelerations of 30 per cent (acceleration front to back) and even up to 70 per cent (acceleration back to front). All these results depend critically on the harness for fixation and the back support used[9] (see *Protection Methods and Procedures*, below). These dynamic load factors indicate a natural frequency of the body system between 10 and 20 cps. Impact of the heart against the chest wall is another possible injury discussed and noted in some animal experiments.[85, 90]

The head and neck supporting structures seem to be relatively tough.[9, 15] Injury seems to occur only upon backward flexion and extension of the neck ("whiplash") when the body is accelerated from back to front without head support, as is common in rear-end automobile collisions.

Head Impact. Injuries to the head, beyond superficial bruises and lacerations, usually consist of concussion or fracture of the skull. The symptoms produced by head impact range from pain and dizziness through disorientation and depression of function to unconsciousness and loss of memory for events immediately preceding the injury.[79] Head injuries usually occur from heavy blows by solid objects to the head, rather than by accelerative forces applied to the body. Since approximately 75 per cent of airplane crash fatalities result from head injuries, and the latter are without doubt of similar importance in many other types of accident (athletics, etc.), the mechanisms leading to head injury have been the object of a large number of investigations.[23, 86, 91]

Neck Response. The limitations for forward and lateral bending of the neck are, for practical purposes, the anterior chest wall and the shoulders, respectively. Since the head is almost entirely held by the neck muscles, the absence of their supporting action gives any blow to the head or neck a "flying start." As a consequence, fractures or dislocations are more apt to result. Dislocations involving the first and second vertebral joints usually are less severe if the odontoid process is fractured (less damage to the spinal cord). If this is not the case, the spinal cord may be severed or crushed. The latter is the essential mechanism in the hanging of criminals. The energy release required for this is around 1,000 ft-lb (150 kg-m). As animal experiments indicate, damage done to the cervical cord at the first vertebra also may play a role in the generation of one type of concussion which had been attributed primarily to brain damage.[77]

Head Response. The elastic shell of the skull is filled with nerve tissue, blood, and cerebrospinal fluid, which have about the same density. The compressibility of the brain substance is very small (like water), and its shear modulus is very low. The viscosity of the brain tissue is around 20 dyne-sec/cm^2. The reaction of the head to a blow is a function of the velocity, duration, and area of impact, and the transfer of momentum. Near the point of application of the blow there will be an indentation of the skull. This results in shear strains in the brain in a superficial region close to the dent. Compression waves emanate from this area, which have normally small amplitudes since the brain is nearly incompressible. In addition to the forces on the brain resulting from skull deformation there are acceleration forces, which also would act on a completely undeformable skull. The centrifugal forces and linear accelerations producing compressional strains are negligible compared to the shear strains produced by the rotational accelerations. The distribution of the shear strains over the brain has been studied on models[92] and the motion of the brain surface has been observed in animals with sections of skull replaced by Lucite.[93] The maximum strains are concentrated at regions where the skull has a good grip on the brain owing to inwardly projecting ridges, especially at the wing of the sphenoid bone of the skull. Shear strains also must be present throughout the brain and in the brain stem. Many investigators consider these shear strains, resulting from rotational accelerations due to a blow to the unsupported head, as the principal event leading to concussion. Blows to the supported, fixed head are supposed to produce concussion by compression of the skull and elevation of cerebrospinal fluid pressure. Despite the general acceptance of rotational acceleration as one of the main causes of concussion, experimental data on this quantity are almost completely missing and concussion thresholds are discussed in terms of "available energy" (which is usually not the energy transferred to the head) and impact velocity.

In general it can be assumed that a high-velocity projectile (for example, a bullet of 10 gm with a speed of 1,000 ft/sec) with its high kinetic energy and low momentum produces plainly visible injury to scalp, skull, and brain along its path. The high-frequency content of the impact is apt to produce compression waves which in the case of very high energies may conceivably lead to cavitation with resulting disruption of tissue.[94] Skull fracture is not a prerequisite for these compression waves. However, if the head hits a wall or another object whose mass is large compared to the head's mass, the local, visible damage is small and the damage due to rotational acceleration may be large. Blows to certain points, especially on the midline, produce no rotation. Blows to the chin upward and sideward produce rotation relatively easily ("knock-out" in boxing). Therefore, it is almost impossible to define a concussion velocity or energy. Velocities listed in the literature for concussion from impact of large masses range from 15 to 50 ft/sec. At impact velocities of about 30 ft/sec, approximately 200 in.-lb of energy is absorbed in 0.002 sec, resulting in an acceleration of the head of 47g. Impact energies for compression concussion are probably approximately in the same range.

Skull Response and Skull Fracture.[95] Impact studies on cadaver skulls with strain gages show that a hammer blow to the bone itself lasts 2.5 to 5 $\times$ 10^{-4} sec. After the initial impact the bone oscillates for 2 to 4 $\times$ 10^{-3} sec with a frequency of approximately 700 cps, which agrees with the fundamental frequency found with continuous period excitation.[43] Scalp, skin, and subcutaneous tissue reduce the energy applied to the bone. If the response of the skull to a blow exceeds the elastic deformation limit, skull fracture occurs. Impact by a high-velocity, blunt-shaped object results in localized circumscribed fracture and depression. Low-velocity blunt blows, insufficient to cause depression, occur frequently in falls and crashes. Stresscoat studies of such situations reveal a general deformation pattern;[95] an inward bent area of the skull surrounds the contact area and is followed at a considerable distance by an outbending of the skull. Sometimes fairly symmetrical undulating patterns are observed, occasionally with an outbending of the contrecoup type, i.e., an outbending in the skull area opposite the point of contact. Since the bone, a brittle material, fails in tension, most linear skull fractures originate in the outer surface of the outward bent area surrounding the indentation. From this area the crack may spread toward the center of impact, which rebounds immediately after the blow and becomes an area of high tensile stress. Propagation in the direction opposite to the center of the impact takes place also. The size of the fracture line depends on the energy expended. Given enough energy two, three, or more cracks appear all radiating from the center of the blow. The skull has both weak and strong areas, each impact area showing well-defined regions for the occurrence of the fracture lines.

The total energy required for skull fracture varies from 400 to 900 in.-lb, with an average often assumed to be 600 in.-lb. This energy is equivalent to the condition that the head hits a hard, flat surface after a free fall from a 5-ft height. Skull fractures

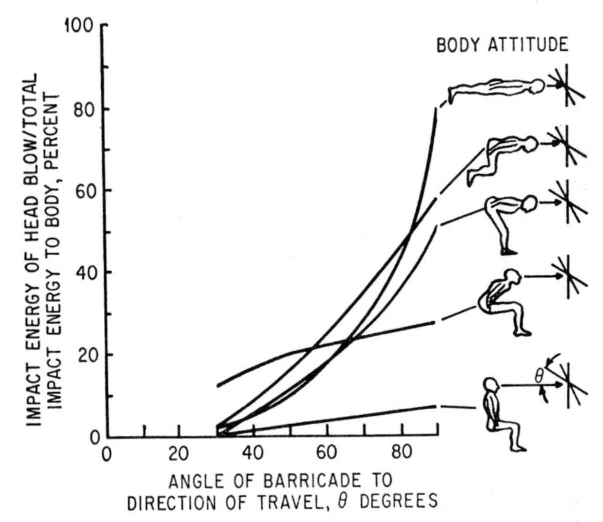

Fig. 44.24. Head impact energy as affected by body attitude and barricade angle. (*Data from dummy drop tests after Dye.*[96])

occurring when a batter is accidentally hit by a ball (5 oz) of high velocity (100 ft/sec) indicate that about the same energy (580 in.-lb) is required. Additional energy 10 to 20 per cent beyond the single linear fracture demolishes the skull completely. Dry skull preparations required only approximately 25 in.-lb for fracture. The reason for the large energy difference required in the two conditions is attributed mainly to the attenuating properties of the scalp.

For practical situations in automobile and aircraft crashes the form, elasticity, and plasticity of the object injuring the head is of extreme importance and determines its "head injury potential." For example, impact with a 90° sharp corner requires only a tenth of the energy for skull fracture (60 in.-lb) that impact with a hard, flat surface requires.[96] Head impact energies for various attitudes of the human body hitting a contact area at various angles are presented in Fig. 44.24. The conditions shown are representative of crash conditions involving unrestrained humans.

EFFECTS OF SHOCK AND VIBRATION ON TASK PERFORMANCE

Little is known about the effects of shock on task performance at accelerations which do not produce actual damage. Whole-body vibration affects task performance primarily through mechanical interference; performance capability can therefore be strongly influenced through vibration-insensitive human engineering of controls and displays. When mechanical interference is negligible, measurements of performance on very simple tasks have given equivocal or negative results.[36,97] The extensive literature in this field has recently been reviewed by Shoenberger (1972) and Hornick (1973). Experiments relating to task performance and fatigue are extraordinarily difficult to carry out properly, and the results have limited practical application, i.e., are not easily applied from one situation to another, even if they appear similar. General guidance with respect to the effects of vibration environments on task performance and the effect of fatigue is given in Fig. 44.31. Recent very promising approaches in this area attack the problem by measuring the changes in the human-operator control parameters caused by the environment using the methodologies of human manual-control theory.

PROTECTION METHODS AND PROCEDURES

Protection of man against mechanical forces is accomplished in two ways: (1) isolation to reduce transmission of the forces to the man and (2) increase of man's mechanical resistance to the forces. Isolation against shock and vibration is achieved if the natural frequency of the system to be isolated is lower than the exciting frequency at least by a factor of 2. Both linear and nonlinear resistive elements are used for damping the transmission system; irreversible resistive elements or energy-absorbing devices can be used once to change the time and amplitude pattern of impulsive forces. Human tolerance to mechanical forces is strongly influenced by selecting the proper body position with respect to the direction of forces to be expected. Man's resistance to mechanical forces also can be increased by proper distribution of the forces so that relative displacement of parts of the body is avoided as much as possible. This may be achieved by supporting the body over as wide an area as possible, preferably loading bony regions and thus making use of the rigidity available in the skeleton. Reinforcement of the skeleton is an important feature of seats designed to protect against crash loads. The flexibility of the body is reduced by fixation to the rigid seat structure. The mobility of various parts of the body, e.g., the abdominal mass, can be reduced by properly designed belts and suits. The factor of training and indoctrination is essential for the best use of protective equipment, for aligning the body in the least dangerous positions during intense vibration or crash exposure, and possibly for improving operator performance during vibration exposure. The latter type of training may be helpful in anticipating and preventing man-machine resonance effects and in reducing anxiety which might otherwise occur.

PROTECTION AGAINST VIBRATIONS

The transmission of vibration from a vehicle or platform to a man is reduced by mounting him on a spring or similar isolation device, such as an elastic cushion.[7,35] The degree of vibration isolation theoretically possible is limited, in the important resonance frequency range of the sitting man, by the fact that large static deflections of the man with the seat or into the seat cushion are undesirable. Large relative movements between operator and vehicle control interfere in many situations with man's performance. Therefore a compromise must be made. Cushions are used primarily for static comfort, but they are also effective in decreasing transmission of vibration above man's resonance range. They are ineffective in the resonance range and may even amplify the vibration in the subresonance range. In order to achieve effective isolation over the 2 to 5 cps range, the natural frequency of the man-cushion system should be reduced to 1 cps, i.e., the natural frequency should be small compared with the forcing frequency (see Chap. 30). This would require a cushion deflection of 10 in. (Fig. 44.25).

Examples of the mechanical impedance changes achieved for a human subject in the sitting position with various cushions are given in Fig. 44.26.[31] If a seat cushion without a back cushion is used (as is common in some tractor or vehicle arrangements), a condition known as "back scrub" (a backache) may result. Efforts of the operator to

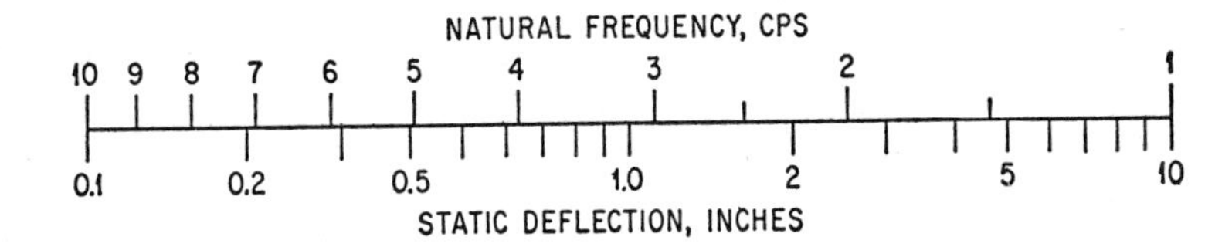

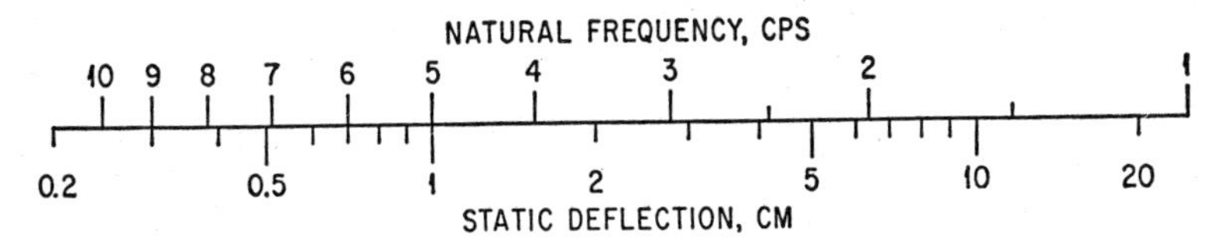

Fig. 44.25. Static deflection of a mass on a spring (such as a man on a seat cushion) as a function of the natural frequency of this linear system. To obtain a natural frequency for the man-seat system which is lower than the resonant frequency of the man alone (5 cps) requires a large static deflection.

wedge himself between the controls and the back of the seat often tend to accentuate this uncomfortable condition. In ordinary passenger seats, the seat cushion does not alter the resonant frequency of the man-seat system significantly, so that no isolation is achieved in the frequency range below 5 cps. Sometimes some amplification is unavoidable. Nevertheless, the damping properties of the seat cushion are very important in attenuating the frequencies above resonance, which may be more important in automobile or aircraft transportation (see Fig. 44.33), for example. For severe low-frequency vibration and shock conditions, such as occur in tractors and other field equipment, suspension of the whole seat is superior to the simple seat cushion. Hydraulic shock absorbers, rubber torsion bars, coil springs, and leaf springs all have been successfully used for suspension seats. The differences between these springs appear to be small. However, a seat that is guided so that it can move only in a linear direction seems to be more comfortable than one in which the seat simply pivots around a center of rotation (Fig. 44.27).[7] The latter situation produces an uncomfortable and fatiguing pitching motion. Suspension seats can be built which are capable of preloading for the operator's weight so as to maintain the static position of the seat and the natural frequency of the system at the desired value. Suspension seats for use on tractors and on similar vehicles are available which reduce the resonant frequency of the man-seat system from approximately 4 to 1 cps[35] (measured in terms of the ratio of the acceleration at the belt level of the man to the seat acceleration). The transmission ratio, which is between 2 and 2.5 at the resonant frequency for the man alone, is only 1.6 at 1 cps for the man on the suspension seat. It has not been possible to lower the resonant fre-

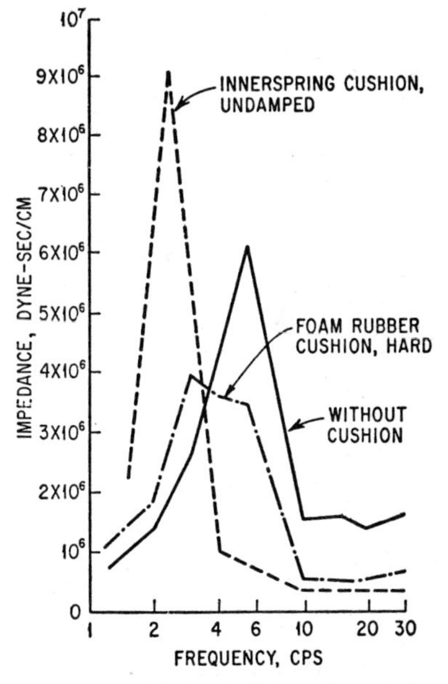

Fig. 44.26. Mechanical impedance of a sitting human subject with various cushions and without cushion (vertical, i.e., longitudinal vibrations). (*Dieckmann.*[31])

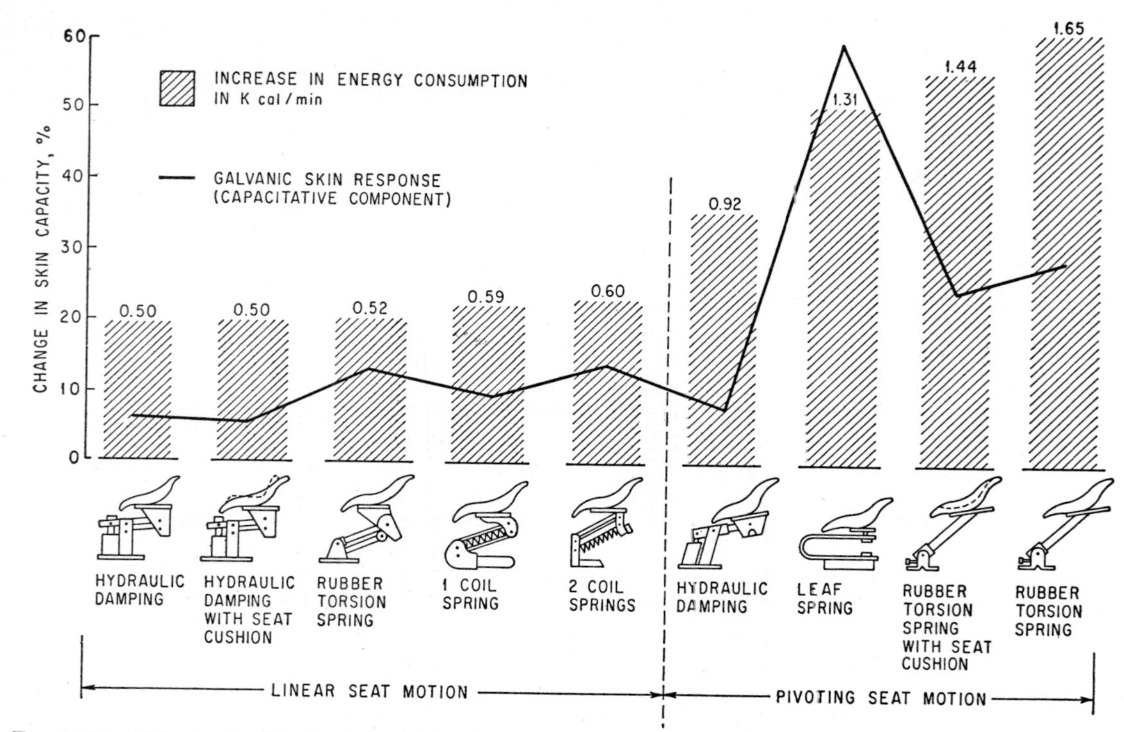

Fig. 44.27. Difference in vibration induced stress on a subject as a function of the seat design. Galvanic skin response and the increase in caloric energy consumption due to vibrations are plotted for the different seats indicated. The data represent averages over 10 tests of 15 min duration on five subjects with vibrations characteristic of tractor operation. The data prove the general superiority of seat designs which restrict seat vibrations to linear, compared to pivoting, motions. Subjective evaluations indicate a similar rank order as the physiological quantities in this graph. (*After Dieckmann.*[7])

quency below 2 to 3 cps, with a transmission ratio of 2.5 to 3, only with different types of foam and spring cushions. Therefore the use of elastic cushions alone results in an amplification of the vibration, as indicated by the impedance measurements of Fig. 44.26. These findings are confirmed not only by laboratory tests but also by field tests on tractors. (The superiority of man's legs to most seating devices, with respect to the transmission of vertical vibrations, is shown in Fig. 44.9.) Differences in positioning of the sitting subject also change the transmission as demonstrated by impedance measurements (Fig. 44.8).

For severe vibrations, close to or exceeding normal tolerance limits, such as those which may occur in military operations, special seats and restraints can be employed to provide maximum body support in all critical directions for the subject in the most advantageous position. In general, under these conditions, seat and restraint requirements are the same for vibration and rapidly applied accelerations. (A discussion of body restraints is given below.) Laboratory experiments[53] show that large protection can be achieved by the use of rigid or semirigid body enclosures. Immersion of the operator in a rigid, water-filled container with proper breathing provisions has been used in laboratory experiments to protect subjects against high, sustained static g loads.[98] This principle can be used to provide protection against high alternating loads.

The isolation of hand and arm against vibration from hand tools depends critically on the type, weight, and relative position of the tool. The model for the hand-arm system presented in Fig. 44.16 probably is adequate to estimate the effectiveness of isolation measures.

Since tolerance to continuous vibration depends critically on the exposure time, the control of the work hours in vibration environments or with vibrating tools is one of the most important protective measures. It can prevent cumulative permanent damage and reduce the possibility of accidents favored by vibration-aggravated fatigue.

PROTECTION AGAINST RAPIDLY APPLIED ACCELERATIONS (CRASH)

The study of automobile and aircraft crashes and of experiments with dummies and live subjects shows that complete body support and restraint of the extremities provide maximum protection against accelerating forces and give the best chance for survival.[9, 51] If the subject is restrained in the seat, he makes full use of the force moderation provided by the collapse of the vehicle structure, and he is protected against shifts in position which would injure him by bringing him in contact with interior surfaces of the cabin structure.[99, 100, 101] The decelerative load must be distributed over as wide a body area as possible to avoid force concentration with resulting bending moments and shearing effects. The load should be transmitted as directly as possible to the skeleton, preferably directly to the pelvic structure—not via the vertebral column.

Theoretically, a rigid envelope around the body will protect it to the maximum possible extent by preventing deformation. A body restrained to a rigid seat approximates such a condition; proper restraints against longitudinal acceleration shift part of the load of the shoulder girdle and arms from the spinal column to the back rest. Arm rests can remove the load of the arms from the shoulders.[51] Semirigid and elastic abdominal supports provide some protection against large abdominal displacements. The effectiveness of this principle has been shown by animal experiments[38] and by impedance measurements on human subjects (Fig. 44.8). Animals immersed in water, which distributes the load applied to the rigid container evenly over the body surface, or in rigid casts are able to survive acceleration loads many times their normal tolerance.

Many attempts have been made to incorporate energy-absorptive devices, either in a harness or in a seat with the intent to change the acceleration-time pattern by limiting peak accelerations. Parts of the seat or harness are designed with characteristics which become nonlinear at some given acceleration level. The benefits derived from such devices are usually small since little space for body or seat motion is available in airplanes or automobiles; furthermore, contact with interior cabin surfaces during the period of extension of the device is apt to result in more serious injury. For example (Fig. 44.28), consider an aircraft which is stopped in a crash from 100 mph in 5 ft; it is

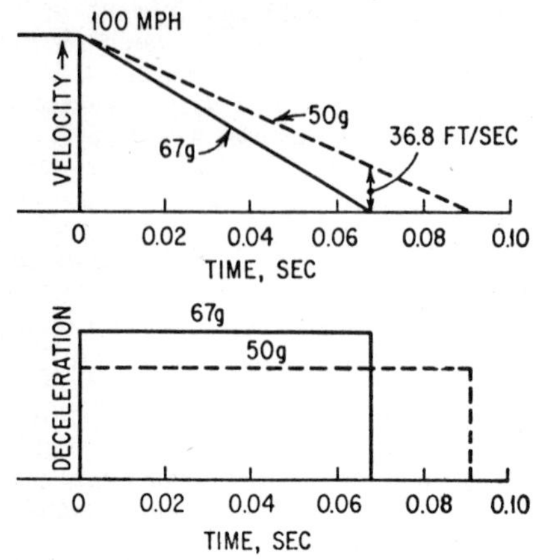

FIG. 44.28. Crash deceleration and velocity of an aircraft as a function of time. The aircraft comes to a complete stop from 100 mph within 5 ft. An energy-absorbing device limiting the maximum acceleration on the passenger to $50g$ is assumed. It would require a displacement of 19 in.

subjected to a constant deceleration of $67g$. An energy-absorptive device designed to elongate at $17g$ would require a displacement of 19 in. In traveling through this distance, the body or seat would be decelerated relative to the aircraft by $14.4g$ and would have a maximum velocity of 36.8 ft/sec relative to the aircraft structure. A head striking a solid surface with this velocity has many times the minimum energy required to fracture a skull. The available space for seat or passenger travel using the principle of energy absorption therefore must be considered carefully in the design. In the development of "catch-up" mechanisms for window washers and workers in similar situations, energy-absorptive elements in the form of undrawn nylon ropes may be applied. For seat belts and other crash restraint harnesses, extensible fabrics have been found to be extremely hazardous since their load characteristics cannot be sufficiently controlled.[102] Seats for jet airliners have been designed which have energy-absorptive mechanisms in the form of extendable rear-legs housing.[103] The maximum travel of the seats is 6 in.; their motion is designed to start between 9 and $12g$ horizontal load, depending on the floor strength. During motion, the legs pivot at the floor level—a feature considered to be beneficial if the floor wrinkles in the crash. Theoretically, such a seat can be exposed to a deceleration of $30g$ for 0.037 sec or $20g$ for 0.067 sec without transmitting a deceleration of more than $9g$ to the seat. However, the increase in exposure time must be considered as well as the reduction in peak acceleration. For very short exposure times where the body's tolerance probably is limited by the transferred momentum and not the peak acceleration, the benefits derived from reducing peak loads would disappear.

The high tolerance limits of the well-supported human body to decelerative forces (Figs. 44.34 to 44.41) suggest that in aircraft and other vehicles, seats, floors, and the whole inner structure surrounding crew and passengers should be designed to resist crash decelerations as near to $40g$ as weight or space limitations permit.[103] The structural members surrounding this inner compartment should be arranged so that their crushing reduces forces on the inner structure. Protruding and easily loosened objects should be avoided. To allow the best chance for survival, seats should also be stressed for dynamic loadings between 20 and $40g$. Civil Air Regulations require a minimum static strength of seats of $9g$.[104] A method for computing seat tolerance for typical survivable airplane crash decelerations is available for seats of conventional design.[105] It has been established that the passenger who is riding in a seat facing backward has a better chance to survive an abrupt crash deceleration since the impact forces are then more uniformly distributed over the body. Neck injury must be prevented by proper head support. Objections to riding backward on a railway or in a bus are minimized for air transportation because of the absence of disturbing motion of objects in the immediate field of view. Another consideration concerning the direction of passenger seats in aircraft stems from the fact that for a rearward-facing seat the center of passenger support is about 1 ft above the point where the seat belt would be attached for a forward-facing passenger. Consequently, the rearward-facing seat is subjected to a higher bending moment; in other words, for seats of the same weight the forward-facing seat will sustain higher crash forces without collapse. For the same seat weight, the rearward-facing seat will have approximately only half the design strength of the forward-facing seat and about one-third its natural frequency. A criterion for the selection of one type of seat over the other on the basis of allowable weight, seat tolerance to probable crash loads, and passenger injury has been proposed, but must await experiments for quantitative confirmation.[106]

Safety lap or seat belts are used to fix the occupants of aircraft or automobiles to their seats and to prevent their being hurled about within, or being ejected from, the car or aircraft. Their effectiveness has been proved by many laboratory tests and in actual crash accidents.[86] The belt load on the lower abdomen causes no severe intra-abdominal injury or injury to the lower spinal region—at least in most survivable crashes. A forward-facing passenger held by a seat belt flails about when suddenly decelerated; his hands, feet, and upper torso swing forward until his chest hits his knees or until the body is stopped in this motion by hitting other objects (back of seat in front, cabin wall, instrument panel, steering wheel, control stick, etc.). Since 15 to 18*g* longitudinal

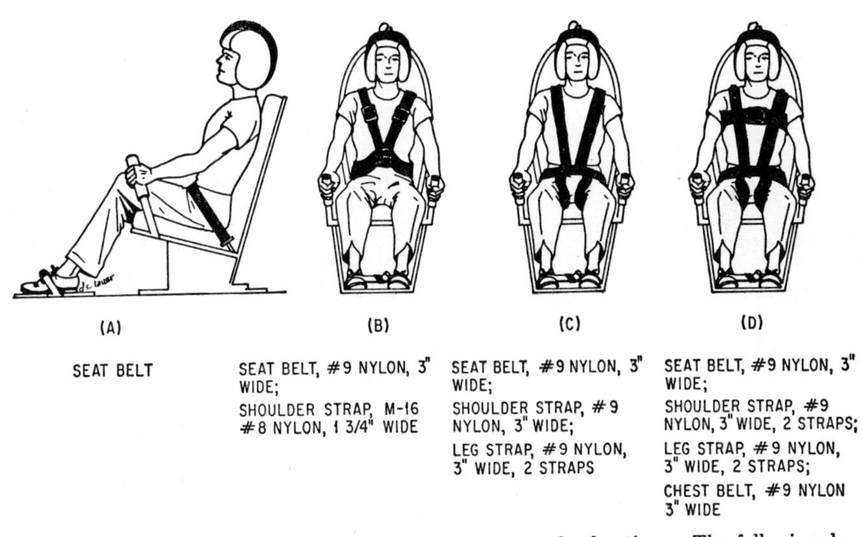

FIG. 44.29. Protective harnesses for rapid accelerations or decelerations. The following devices were evaluated in sled deceleration tests:
A. Seat belt for automobiles and commercial aviation.
B. Standard military lap and shoulder strap.
C. Like *B* but with thigh straps added to prevent headward rotation of the lap strap.
D. Like *C* but with chest strap added. (*Stapp*.[9])

deceleration can result in three times higher acceleration of the chest hitting the knees, this load appears to be about the limit a human can tolerate with a seat belt alone. Approximately the same limit is obtained when the head-neck structure is considered.

Increased safety in automobile as well as airplane crashes can be obtained by distributing the impact load over larger areas of the body and fixing the body more rigidly to the seat. Shoulder straps, thigh straps, chest straps, and hand holds are additional body supports used in experiments. They are illustrated in Fig. 44.29. Table 44.4 shows the desirability of these additional restraints to increase possible survivability to acceleration loads of various direction.[51] In airplane crashes, vertical and horizontal loads must be anticipated. In automobile crashes, horizontal loads are most likely. The effectiveness of adequately engineered shoulder or chest straps in automobile crashes is illustrated in Fig. 44.30. Lap straps always should be as tight as comfort will permit to exclude available slack. During forward movement, about 60 per cent of the body mass is restrained by the belt, and therefore represents the belt load. In order to reduce slack and elongation under load, it is recommended that an 8,000-lb loop strength, 3-in. wide nylon belt replace the 3,000- to 4,000-lb, 3-in. lap belt currently in wide use.

Table 44.4. Human-body Restraint and Possible Increased Impact Survivability
(*After Eiband*[51])

Direction of acceleration imposed on seated occupants	Conventional restraint	Possible survivability increases available by additional body supports *
Spineward:		
Crew	Lap strap Shoulder straps	**Forward facing:** (*a*) Thigh straps (Assume crew members will be performing emergency duties with hands and feet at impact.)
Passengers	Lap strap	**Forward facing:** (*a*) Shoulder straps (*b*) Thigh straps (*c*) Nonfailing arm rests (*d*) Suitable hand holds (*e*) Emergency toe straps in floor
Sternumward:		
Passengers only	Lap strap	**Aft facing:** (*a*) Nondeflecting seat back and (*b*) Integral, full-height head rest (*c*) Chest strap (axillary level) (*d*) Lateral head motion restricted by padded "winged back" (*e*) Leg and foot barriers (*f*) Arm rests and hand holds (prevent arm displacement beyond seat back)
Headward:		
Crew	Lap strap Shoulder straps	**Forward facing:** (*a*) Thigh straps (*b*) Chest strap (axillary level) (*c*) Full, integral head rest (Assume crew members will be performing emergency duties; extremity restraint useless.)
Passengers	Lap strap	**Forward facing:** (*a*) Shoulder straps (*b*) Thigh straps (*c*) Chest strap (axillary level) (*d*) Full, integral head rest (*e*) Nonfailing contoured arm rests (*f*) Suitable hand holds **Aft facing:** (*a*) Chest strap (axillary level) (*b*) Full, integral head rest (*c*) Nonfailing arm rests (*d*) Suitable hand holds
Tailward:		
Crew	Lap strap Shoulder straps	**Forward facing:** (*a*) Lap-belt tie-down strap (Assume crew members will be performing emergency duties; extremity restraint useless.)
Passengers	Lap strap	**Forward facing:** (*a*) Shoulder straps (*b*) Lap-belt tie-down strap (*c*) Hand holds (*d*) Emergency toe straps **Aft facing:** (*a*) Chest strap (axillary level) (*b*) Hand holds (*c*) Emergency toe straps
Berthed occupants	Lap strap	**Feet forward:** Full-support webbing net **Athwart ships:** Full-support webbing net

* Exposure to maximum tolerance limits (Figs. 44.34 to 44.41) requires straps exceeding conventional strap strength and width.

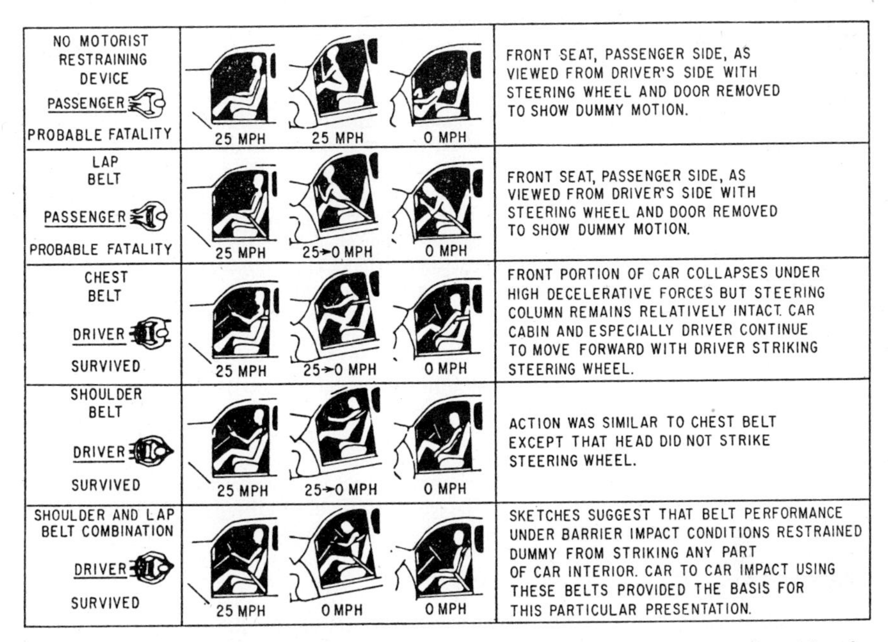
NO MOTORIST RESTRAINING DEVICE
PASSENGER
PROBABLE FATALITY
25 MPH
25 MPH
0 MPH
FRONT SEAT, PASSENGER SIDE, AS VIEWED FROM DRIVER'S SIDE WITH STEERING WHEEL AND DOOR REMOVED TO SHOW DUMMY MOTION.
LAP BELT
PASSENGER
PROBABLE FATALITY
25 MPH
25→0 MPH
0 MPH
FRONT SEAT, PASSENGER SIDE, AS VIEWED FROM DRIVER'S SIDE WITH STEERING WHEEL AND DOOR REMOVED TO SHOW DUMMY MOTION.
CHEST BELT
DRIVER
SURVIVED
25 MPH
25→0 MPH
0 MPH
FRONT PORTION OF CAR COLLAPSES UNDER HIGH DECELERATIVE FORCES BUT STEERING COLUMN REMAINS RELATIVELY INTACT. CAR CABIN AND ESPECIALLY DRIVER CONTINUE TO MOVE FORWARD WITH DRIVER STRIKING STEERING WHEEL.
SHOULDER BELT
DRIVER
SURVIVED
25 MPH
25→0 MPH
0 MPH
ACTION WAS SIMILAR TO CHEST BELT EXCEPT THAT HEAD DID NOT STRIKE STEERING WHEEL.
SHOULDER AND LAP BELT COMBINATION
DRIVER
SURVIVED
25 MPH
0 MPH
0 MPH
SKETCHES SUGGEST THAT BELT PERFORMANCE UNDER BARRIER IMPACT CONDITIONS RESTRAINED DUMMY FROM STRIKING ANY PART OF CAR INTERIOR. CAR TO CAR IMPACT USING THESE BELTS PROVIDED THE BASIS FOR THIS PARTICULAR PRESENTATION.

FIG. 44.30. Effect of varying safety-belt arrangements on driver and passenger for a 25-mph automobile collision with a fixed barrier. The sketches and evaluations are based on actual collision tests. (*UCLA-ITTE, Severy and Matthewson.*[99])

Double-thickness No. 9 undrawn nylon straps of 3 in. width are most satisfactory for all harnesses[50] with respect to strength, elongation, and supported surface area. If the upper torso is fixed to the back of the seat by any type of harness (shoulder harness, chest belt, etc.), the load on the seat is approximately the same for forward- and aft-facing seats. The difference between these seats with respect to crash tolerance as discussed above no longer exists. These body restraints for passenger and crew must be applied without creating excessive discomfort.[100, 107]

The dynamic properties of seat cushions are extremely important if an acceleration force is applied through the cushion to the body. In this case the steady-state response curve of the total man-seat system (Fig. 44.26) provides a clue to the possible dynamic load factors under impact. Overshooting should be avoided, at least for the most probable impact rise times. This problem has been studied in detail in connection with seat cushions used on upward ejection seats.[87] The ideal cushion is approached when its compression under static load spreads the load uniformly and comfortably over a wide area of the body and if almost full compression is reached under the normal weight. The impact acceleration then acts uniformly and almost directly on the body without intervening elastic elements. A slow-responding foam plastic, for example, with a thickness of 2 to 2.5 in. is very satisfactory in fulfilling these requirements.

PROTECTION AGAINST HEAD IMPACT

The impact-reducing properties of protective helmets are based on two principles: the distribution of the load over a large area of the skull and the interposition of energy-absorbing systems. The first principle is applied by using a hard shell, which is suspended by padding or support webbing at some distance from the head (5/8 to 3/4 in.). High local impact forces are distributed by proper supports over the whole side of the skull to which the blow is applied. Thus, skull injury from relatively small objects and projectiles can be avoided. However, tests usually show that contact padding alone over the skull results in most instances in greater load concentration, whereas helmets with web suspension distribute pressures uniformly.[23] Since helmets with contact padding usually permit less slippage of the helmet, a combination of web or strap suspension with contact padding is desirable. The shell itself must be as stiff as is compatible with weight considerations; when the shell is struck by a blow, its deflection must not be large enough to permit it to come in contact with the head. For light industrial safety hats, molded bakelite reinforced with steel wire, laminated bakelite, or high-strength aluminum alloy are used; for football helmets, combination rubber and plastic compounds molded in a single shell or a vulcanized fiber shell have been used; for antibuffet helmets molded fiberglass flock, molded cotton fabric, or cloth laminates, all impregnated with plastic are some of the most commonly used materials.

Padding materials can incorporate energy-absorptive features. Whereas foam rubber and felt are too elastic to absorb a blow, foam plastics like polystyrene or Ensolite result in lower transmitted accelerations.

Most helmets constitute compromises among several objectives such as pressurization, communication, temperature conditioning, minimum bulk and weight, visibility, protection against falling objects, etc.; usually, impact protection is but one of many design considerations. The protective effect of helmets against concussion and skull fracture has been proved in animal experiments[108] and is apparent from accident statistics. However, it is difficult to specify the exact physical conditions for a helmet which can provide optimum impact protection.

PROTECTION AGAINST BLAST WAVES

Individual protection against air blast waves is extremely difficult since only very thick protective covers can reduce the transmission of the blast energy significantly. Furthermore, not only the thorax but the whole trunk would require protection. In animal experiments, sponge-rubber wrappings and jackets of other elastic material have resulted in some reduction of blast injuries.[5] Enclosure of the animal in a metal cylinder with the head exposed to the blast wave has provided the best protection—short of complete enclosure of the animal. Therefore it is generally assumed that shelters are the only practical means of protecting humans against blast. They may be of either the open or closed type; both change the pressure environment. Changes in pressure rise time introduced by the door or other restricted openings are physiologically most important.[17]

HUMAN TOLERANCE CRITERIA FOR SHOCK AND VIBRATION EXPOSURE

Only approximate limits for man's safety and performance under field conditions can be given, since the exact physical mode of action of the environment varies with man's unpredictable position and motion, and since biological variations with respect to physical, physiological, and psychological reactions make such limits statistical in nature. Individual cases may deviate considerably from the average. Therefore biological criteria must be used with caution; they are summarized here as rough guides for engineering purposes. Tolerance criteria, as well as examples of field environment, indicate only orders of magnitude—not rigid limits. For critical safety problems, a detailed study of the literature or expert consultation is indispensable.

VIBRATION EXPOSURE

Many methods have been developed to assess man's reaction to vibration in a quantitative manner, but most of these are based on a limited number, specific types, or a specific interpretation of experiments and contradict each other to a certain degree.

Whole-Body Exposure. In the absence of any clearly defined vibration-induced occupational disease and recalling the wide variability of the effects of vibration on human performance, it is difficult to select firm rating procedures considering all factors influencing such rating (magnitude, frequency, time function, intermittency, direction of vibration, and position of the body and its coupling to the vibrating surface). In spite of these acknowledged difficulties, a general guide for the evaluation of human exposure to whole-body vibration has been developed and agreed upon as International Standard ISO 2631, 1974. This guide gives positive quantitative guidance with respect to safety and acceptability of whole-body vibration for the 1 to 80 cps frequency band and is recommended for the evaluation of vibration environments in transportation vehicles (air, land, and water) and in industry (factories, agriculture). This guide gives limits for the following items. (1) The preservation of health or safety (called "exposure limits"). These limits should not be exceeded without special justification and awareness of a health risk. (2) The preservation of working efficiency (called "fatigue, or decreased proficiency boundary"). It is obvious that this boundary cannot apply to all possible tasks; it applies to such typical tasks as flying an airplane or driving a tractor. (3) The preservation of comfort ("reduced comfort boundary"). All three boundaries are a function of the average daily exposure time: all boundaries are high for short exposures such as a minute per day and are lowest for 24 hr, i.e., continuous exposure. The "fatigue-decreased proficiency boundary" for longitudinal (head-to-foot or Z-axis) vibration is presented in Fig. 44.31A with daily exposure time as parameter. In Fig. 44.31B the same boundaries are shown as a function of exposure time. For transverse vibration (side-to-side or Y-axis and chest-to-back or X-axis) similar curves are shown in Fig. 44.31C and D. The difference in the frequency-dependence between Fig. 44.31A and C is a result of the changed transmissibility for the different axis (see Figs. 44.9 and 44.15). The boundaries are defined in terms of rms value for sinusoidal single frequency exposure or rms value in the third-octave bands for distributed vibra-

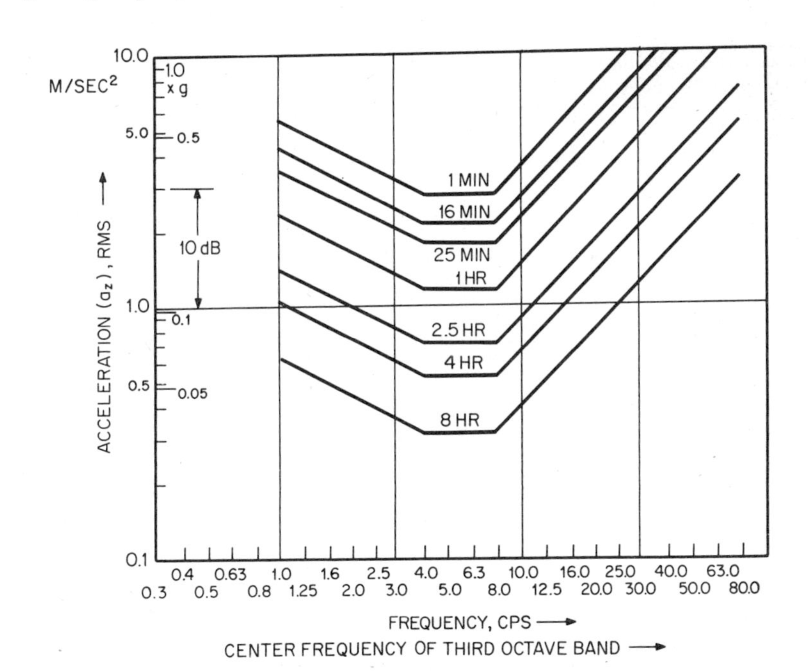

FIG. 44.31. Vibration exposure limits as a function of frequency and exposure time. (*Figures A to D are based on ISO Standard* 2631—1974. *Figures E and F are based on unpublished working papers of Subcommittee* 4 "*Effects of Shock and Vibration on Man*" *of ISO Technical Committee* 108 "*Mechanical Shock and Vibration.*") (*A*) Longitudinal whole-body acceleration limits: fatigue—decreased proficiency boundary.

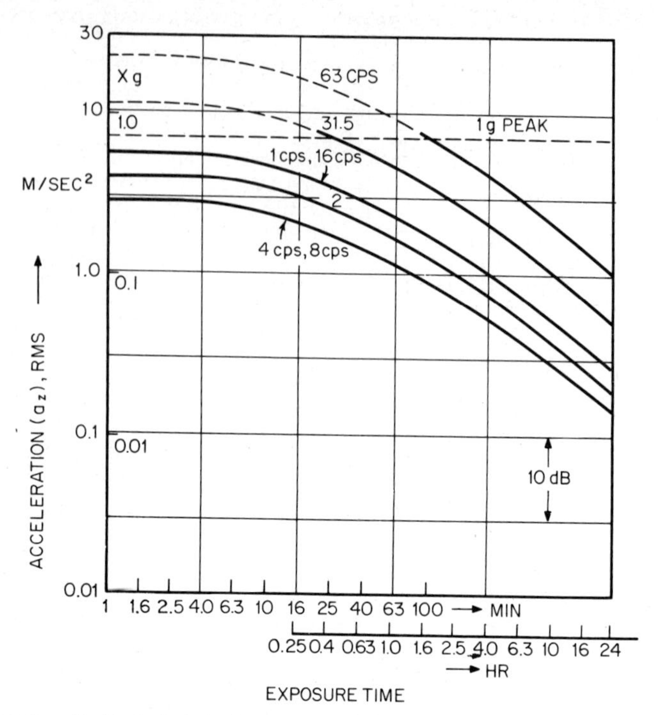

FIG. 44.31*B*. Longitudinal whole-body acceleration limits as a function of daily exposure time: fatigue-decreased proficiency boundary.

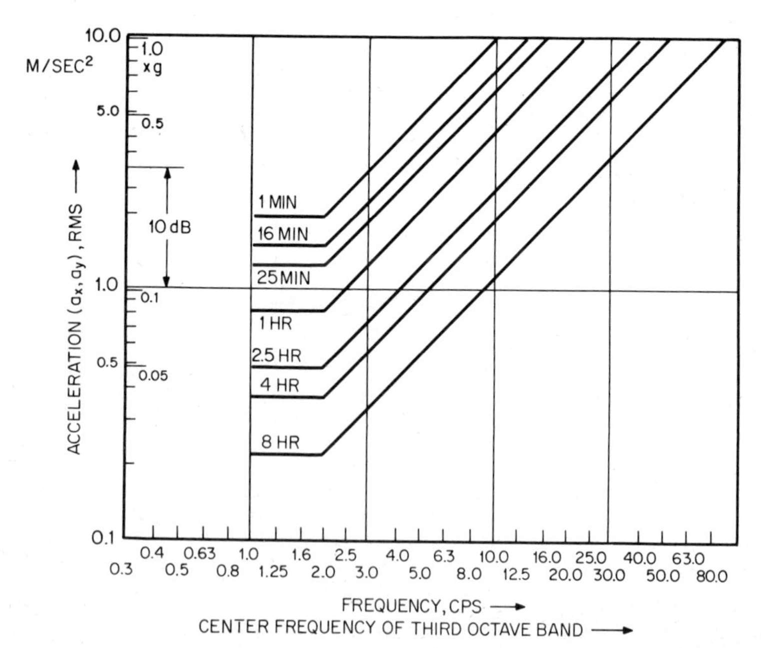

FIG. 44.31*C*. Transverse whole body acceleration limits: fatigue-decreased proficiency boundary.

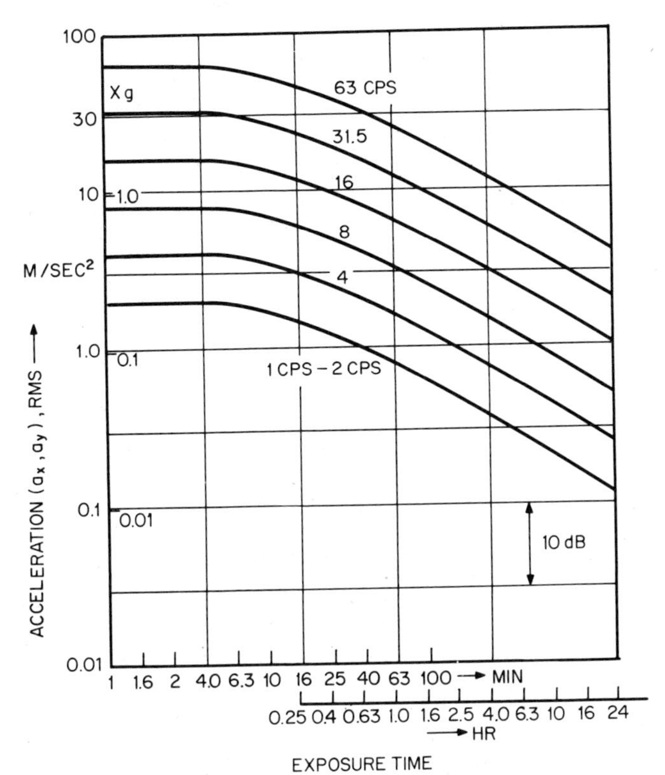

FIG. 44.31*D*. Transverse whole-body acceleration limits as a function of daily exposure time: fatigue-decreased proficiency boundary.

tion. For broad-band vibration, the rms value in each octave band should not exceed the limits shown; for vibration in more than one direction simultaneously, corresponding limits apply to each vectorial component in the three axes. For evaluating time-varying exposure patterns over a day's period, the standard proposes a method for arriving at an "equivalent exposure with constant vibration magnitude."

The "exposure limits" for health and safety reasons are obtained by raising the boundaries of Fig. 44.31 *A* to *D* by a factor of 2 (6 dB higher). To obtain the "reduced comfort boundaries," the values of the boundaries of Fig. 44.31 must be divided by a factor of 3.15 (10 dB lower). In transportation vehicles the reduced comfort boundary is related to such functions as reading, writing, and eating.

The standard also gives useful guidance for measuring and reporting the vibration environment when its evaluation with respect to its effects on man is of primary concern. Adherence to this standard will not only provide uniform evaluation of vibration environments but will also assist in collecting and evaluating more meaningful data and exposure histories than were previously available.

The ISO standard is only very tentative for vibrations having high crest factors (i.e., greater than 3). Very few studies exist for such vibrations.

The results of one study on single-shock acceleration pulses (such as floor vibrations near drop forges or similar equipment) are presented in Fig. 44.32.[12]

Extension of the curves in Fig. 44.31*A* to *D* to frequencies below 1 cps is extremely difficult because it is not possible to use the same categories of human-exposure criteria:

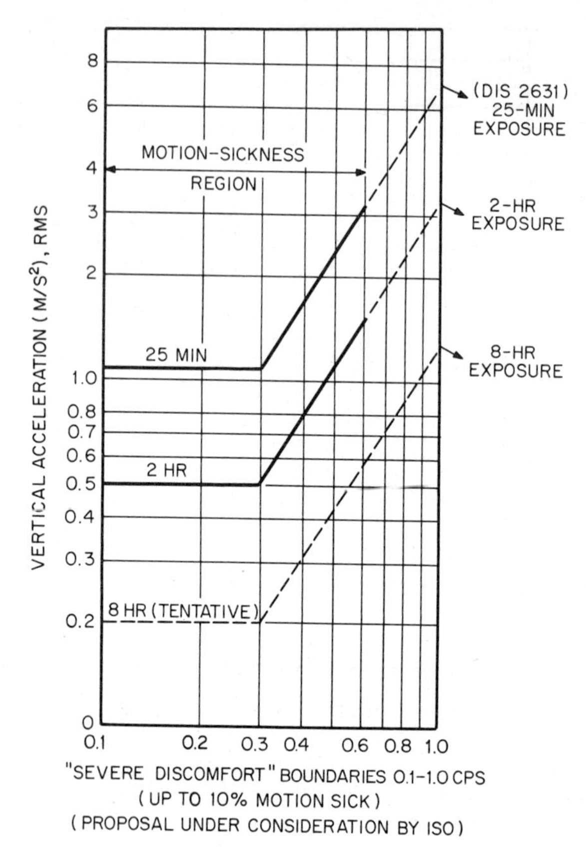

FIG. 44.31*E*. Severe discomfort boundaries for the frequency range of 0.1 to 1.0 cps (up to 10 per cent motion sickness).

motion sickness enters the picture, severly affecting comfort, work efficiency, and, under certain conditions, health and safety. Motion sickness also depends on other environmental factors in addition to the physical motion. It is very variable in its occurrence for the same subject and for different subjects, and it is, in general, very complex and hard to predict. The best tentative generalized guidance which can be provided at the present time is given in Fig. 44.31*E*.

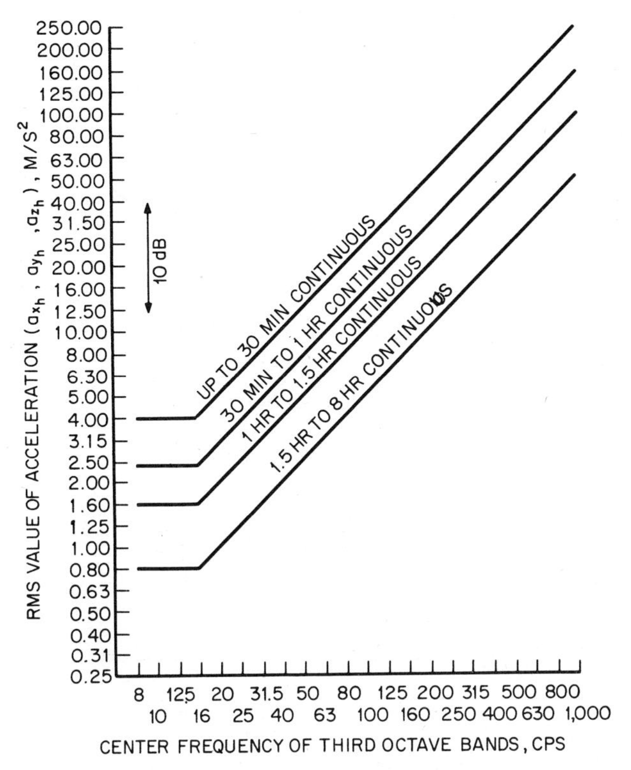

FIG. 44.31*F*. Vibration exposure boundaries for hand transmitted vibration.

Hand-Arm Vibrations. No generally accepted exposure limits for protecting against hand-transmitted vibration from tools such as pneumatic and electric hand tools used in the manufacturing, mining, or construction industries or chain saws used in forestry are available. Provisional exposure boundaries for routine daily exposures as a function of daily exposure time are provided in Fig. 44.31*F*. Adherence to the curves, which are based on the best data available, should reduce the risk of occupational vibration-induced disease to acceptable or negligible levels. The curves do not necessarily represent limits of safe exposure for all individuals and working conditions.

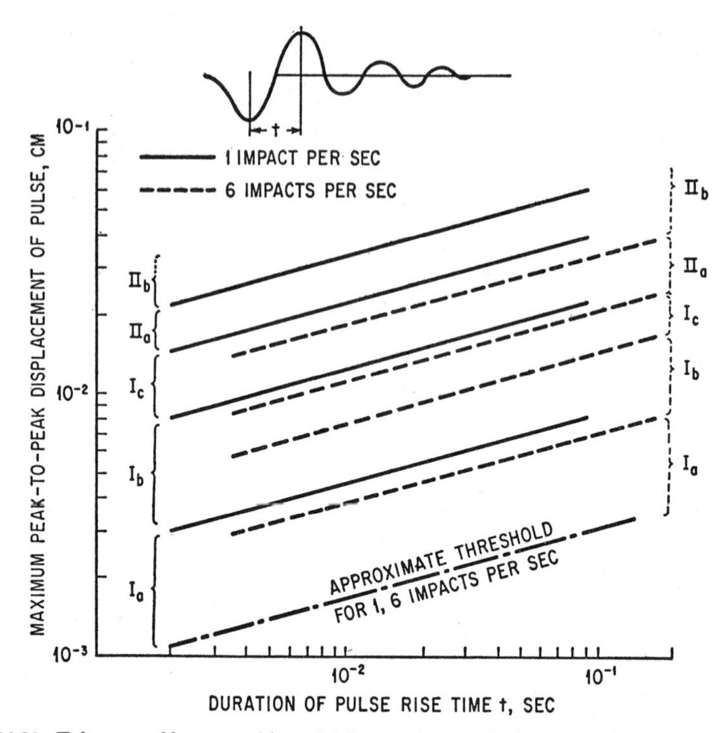

FIG. 44.32. Tolerance of human subjects in the standing or supine position to repetitive vertical impact pulses representative of impacts from pile drivers, heavy tools, heavy traffic, etc. Subjective reaction is plotted as a function of the maximum displacement of the initial pulse and its rise time. The numbers indicate the following reactions for the areas between the lines: I_a, threshold of perception; I_b, of easy perception; I_c, of strong perception, annoying; II_a, very unpleasant, potential danger for long exposures; II_b, extremely unpleasant, definitely dangerous. The decay process of the impact pulses was found to be of little practical significance. (*After Reiher and Meister.*[12])

(Biological effects of hand-transmitted vibration are very much influenced by the direction of the vibration transmitted to the hand, the method of working, proficiency, and the climatic conditions.) The acceleration values of Fig. 44.31*F* are not to be exceeded by accelerations measured at the handtool interface for any direction of a rectangular coordinate system.

Typical Vibration Environments. The range of vibration levels found in aircraft, trucks, and tractors is indicated in Fig. 44.33. Individual points represent flight vibration data obtained in various types of military aircraft.[73] The solid circles, squares, and triangles indicate vibrations at seat levels which were reported as excessive and undesirable in actual service; the open marks indicate conditions accepted as tolerable. The linearized dividing line between tolerable and excessive vibrations is the tolerance criterion, used as a long-time vibration tolerance criterion in military aviation before

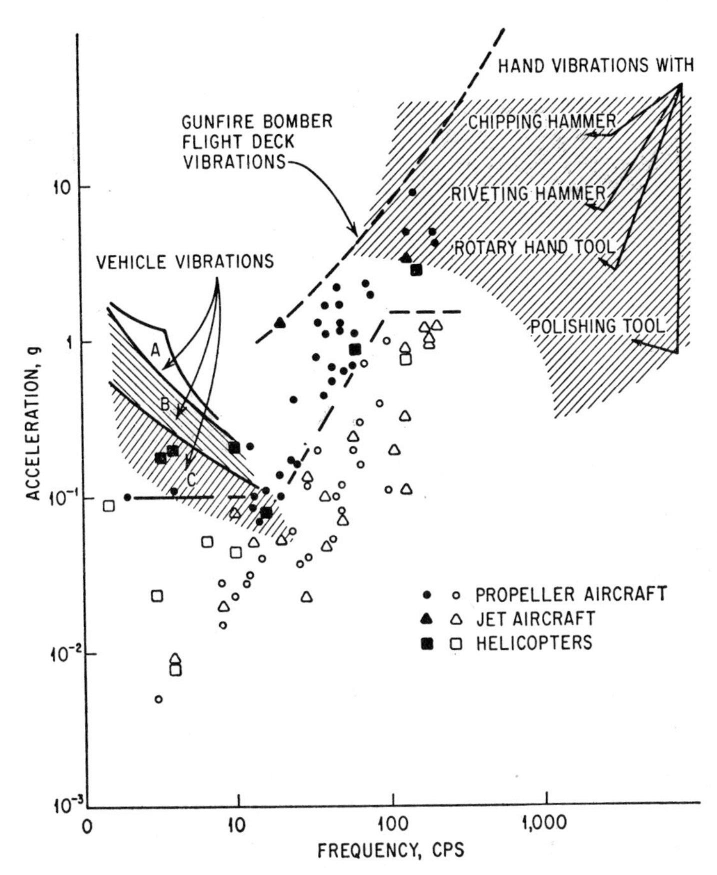

FIG. 44.33. Approximate vibration environments for automobiles, trucks, tractors, and aircraft. Peak values of acceleration are plotted as a function of frequency. Vibration levels observed on hands while operating various hand tools are also indicated. These levels are strongly dependent on tool design and type of operation. Most of the hand vibrations indicated are in the acceleration range at which chronic hand injuries may result. (*After Radke,*[35] *Dieckmann,*[7] *Getline,*[73] *and Agate and Druett,*[41] *and others.*)

the ISO standard (Fig. 44.31) was available. The areas indicated for truck and tractor vibration[7, 35] are the range for the respective vibration maxima and do not represent spectra. Most vehicles have very pronounced natural frequencies excited according to ground conditions. Very generally, rubber-tired farm tractors, as well as trucks, have the maximum of their vertical accelerations in the 2 to 6 cps range, whereas large rubber-tired earth-moving equipment is in the 1.5 to 3.5 cps range; crawler-type tractors may have higher frequencies—of the order of 5 cps.

DECELERATION EXPOSURE, CRASH, AND IMPACT

The approximate maximum tolerance limits to rapid decelerations applied to a sitting human subject are summarized in Figs. 44.34 to 44.41.[51] The data are compared and summarized on the basis of trapezoidal pulses on the seat in all four acceleration directions with respect to a sagittal plane through the body axis. The limits as shown are

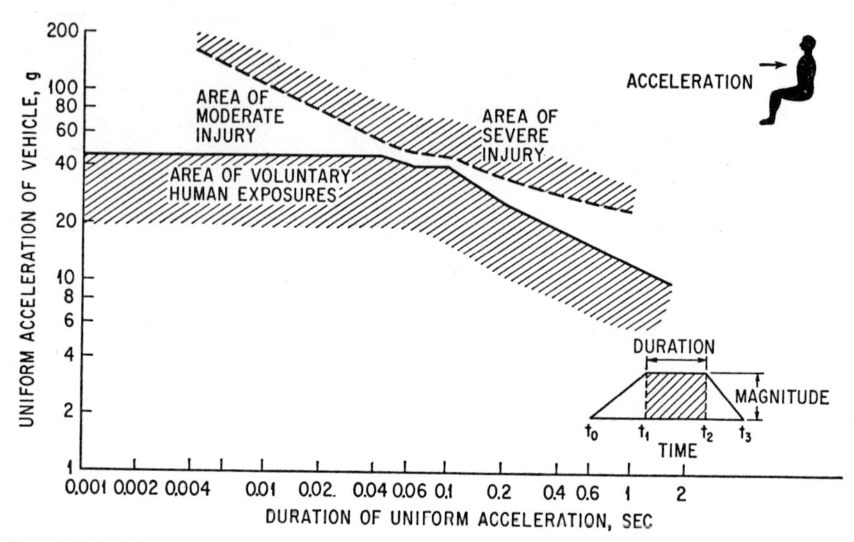

Fig. 44.34. Tolerance to spineward acceleration as a function of magnitude and duration of impulse. (*Eiband.*[51])

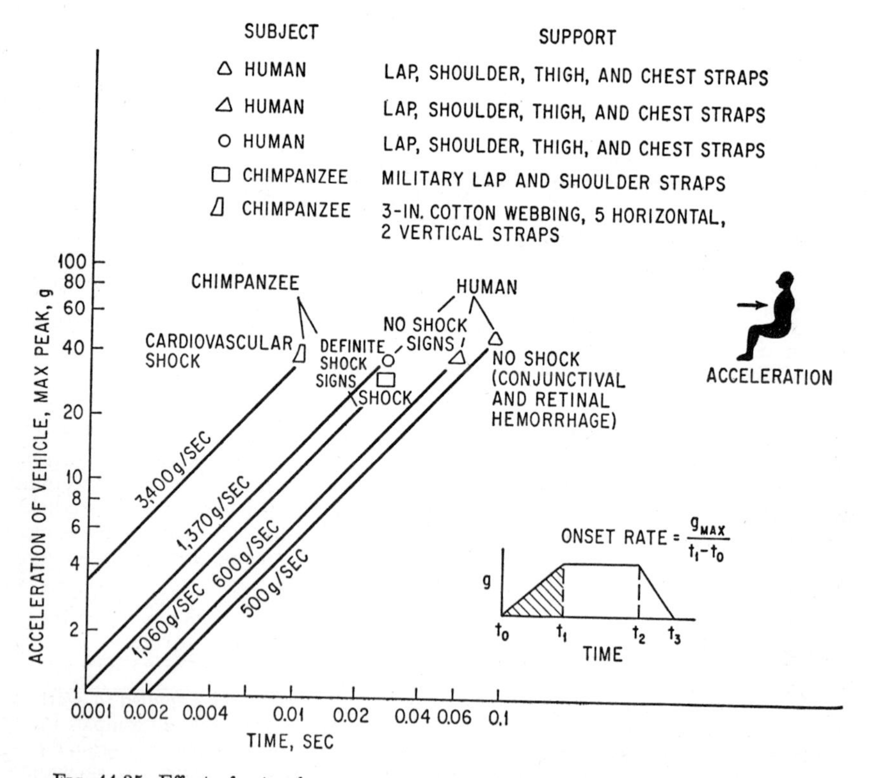

Fig. 44.35. Effect of rate of onset on spineward acceleration tolerance. (*Eiband.*[51])

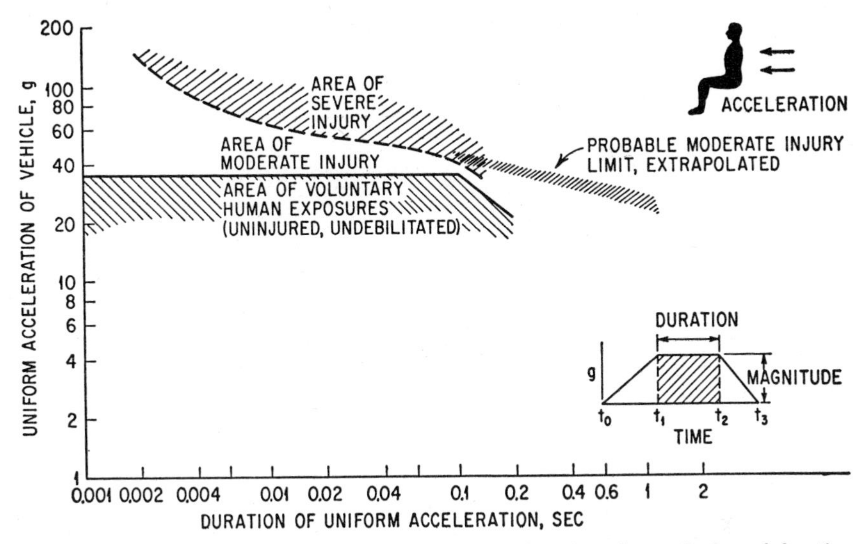

FIG. 44.36. Tolerance to sternumward acceleration as a function of magnitude and duration of impulse. (*Eiband.*[51])

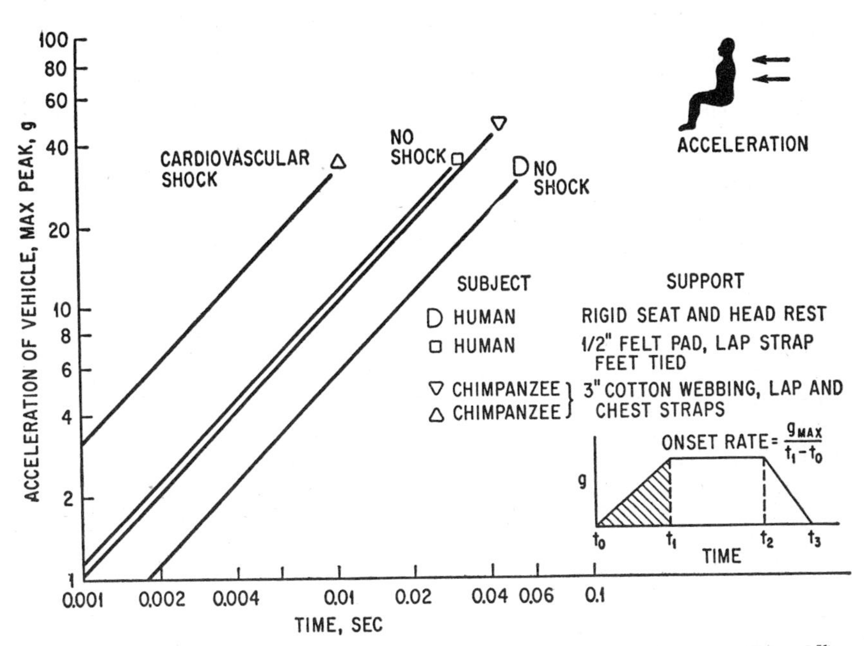

FIG. 44.37. Effect of rate of onset on sternumward acceleration tolerance. (*Eiband.*[51])

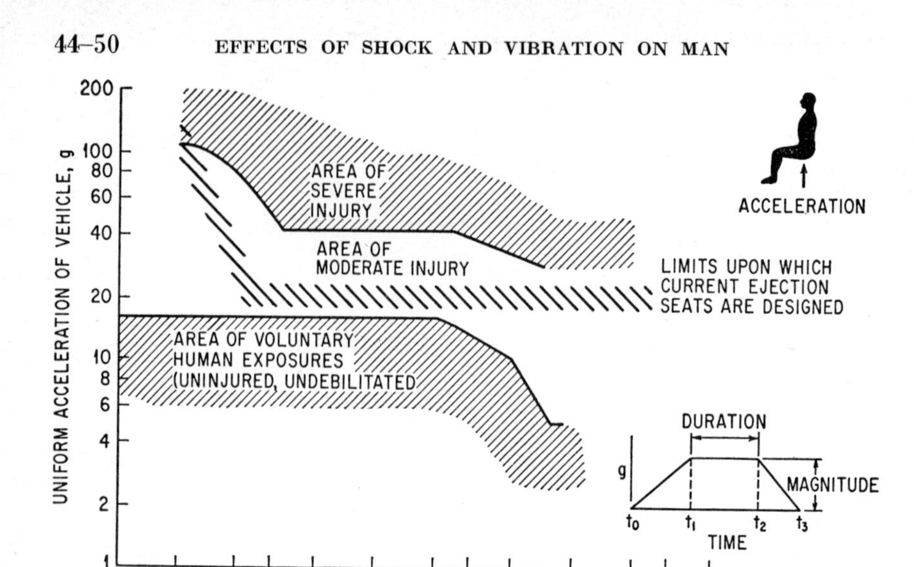

Fig. 44.38. Tolerance to headward acceleration as a function of magnitude and duration of impulse. (*Eiband.*[51])

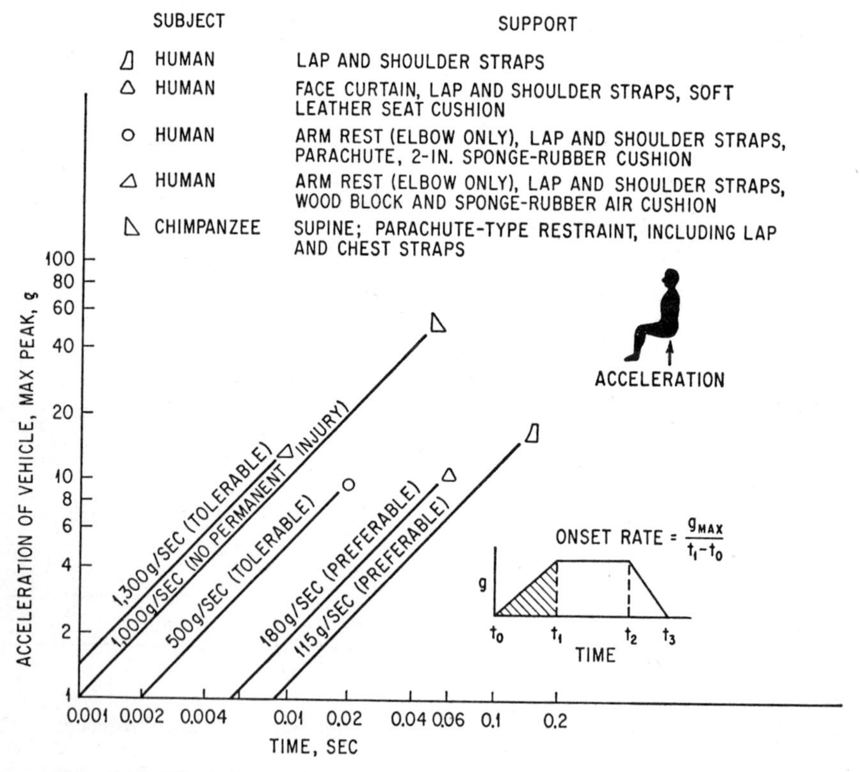

Fig. 44.39. Effect of rate of onset on headward acceleration tolerance. (*Eiband.*[51])

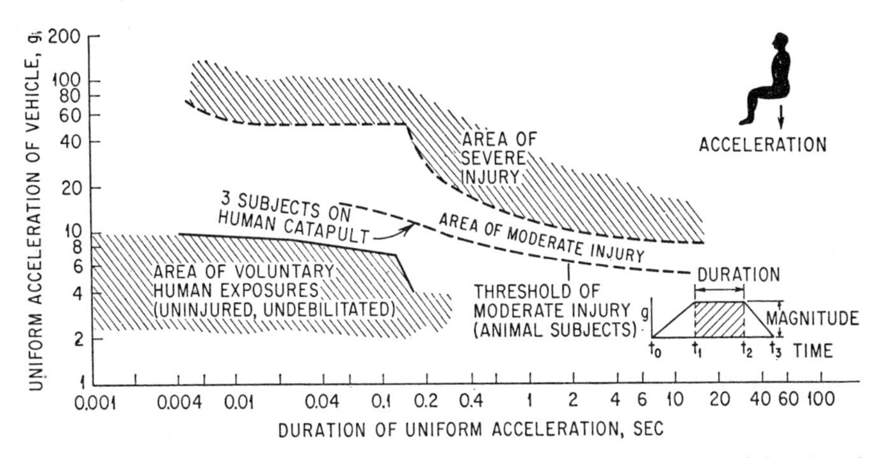

FIG. 44.40. Tolerance to tailward acceleration as a function of magnitude and duration of impulse. (*Eiband.*[51])

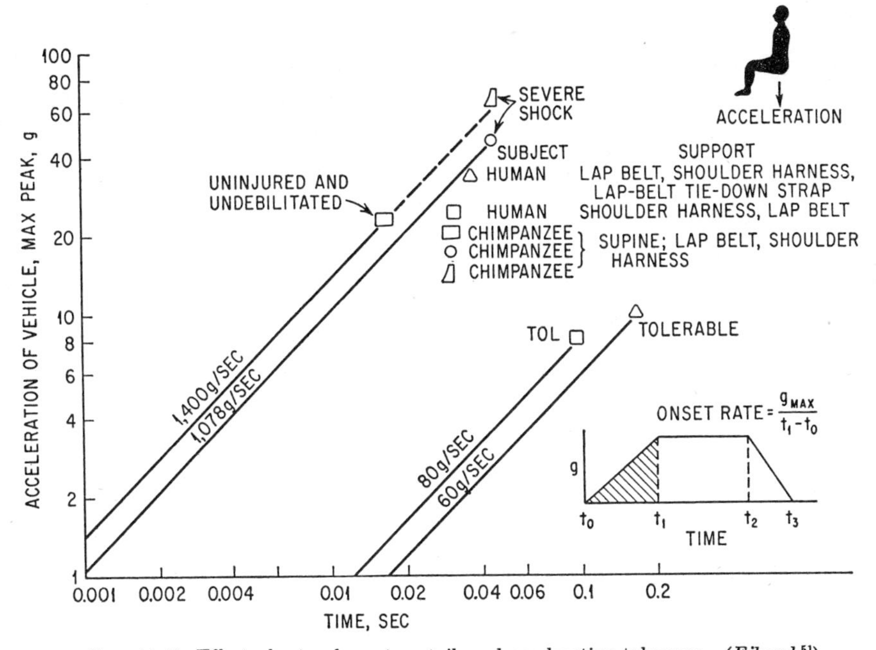

FIG. 44.41. Effect of rate of onset on tailward acceleration tolerance. (*Eiband.*[51])

based on experiments providing maximum body support (see Table 44.4), i.e., lap belt, shoulder harness, thigh and chest strap, and arm rests for the headward accelerations. The quantitative influence of the initial rate of change of acceleration is not too clearly established and not enough data are available for exact mathematical analysis of the influence of the total acceleration-time function. Although the separation of this function into duration of (uniform) acceleration and onset rate is not completely satisfying, it constitutes the most complete analysis of the experimental evidence available. Onset rates endured by various subjects therefore are summarized in separate graphs for the

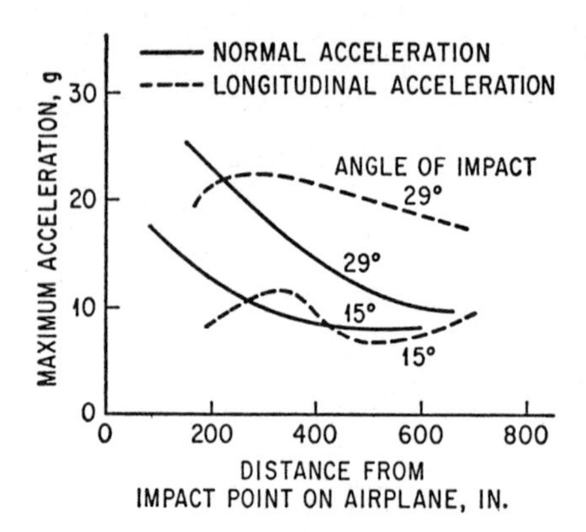

FIG. 44.42. Longitudinal and normal crash loads on a pressurized transport aircraft hitting the ground at 35 mph under an impact angle of 15° and 29°. Acceleration levels in the aircraft are shown as a function of the distance from the point of contact (nose). (*After Preston and Pesman.*[109])

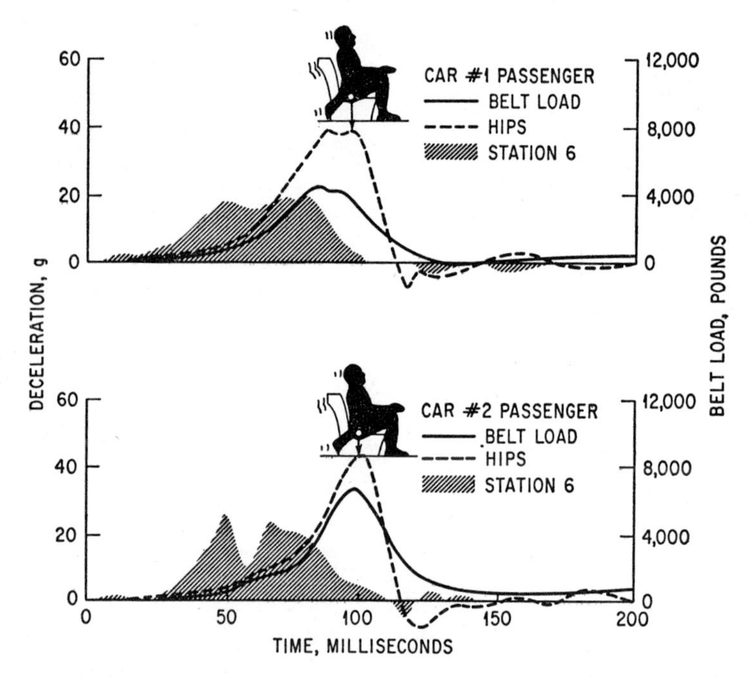

FIG. 44.43. Example of automobile head-on collision deceleration patterns as a function of time. The deceleration for the underbody under the seat (station 6) and the passenger's hip is plotted together with the load function of the seat belt. The data are for two cars engaged in experimental head-on collisions. (Impact speed 31 ft/sec. Kinetic energy of cars before impact approximately 45,000 ft-lb. Cars collapsed under impact approximately 1.7 ft.) (*Severy et al.*[100])

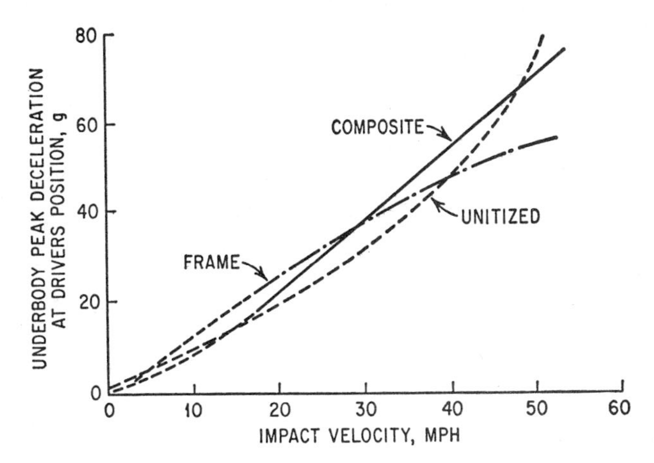

FIG. 44.44. Crash deceleration of the passenger compartment in head-on collisions of automobiles as a function of driving speed. The negligible difference between frame and unitized underbody construction also is shown. (*Severy et al.*[100])

different directions (Figs. 44.35, 44.37, 44.39, and 44.41). In applying these curves, caution must be exercised since they are based on well-designed body supports, minimum slack of the harnesses, heavy seat construction, young, healthy volunteer subjects, and subjects expecting the impact exposure. These curves constitute maxima in many respects, although further improvement in the protection methods certainly does not appear impossible. For example, the maximum limit for exposure to transverse front-to-back acceleration, as experienced in head-on automobile collisions, is indicated in Fig. 44.34 as between 40 and 50g for durations of less than 0.1 sec. For subjects without maximum upper torso restraint having only a lap belt or other types of abdominal restraints, this limit is estimated to be between 10 and 20g.

Approximate ranges for aircraft crash loads can be obtained from Fig. 44.42.[109] Horizontal crash loads, i.e., in the direction of the aircraft's longitudinal axis, increase with the crash angle to a maximum at 90°, whereas vertical loads reach their maximum approximately at 35°. The graph shows only one typical example; aircraft type, ground conditions, and point of initial crash contact have a strong influence in each individual case. For automobile head-on collisions, Fig. 44.43 shows typical deceleration patterns for the car structure under the seat and the passenger's hips; seat-belt loads are also indicated in the graph.[100] The two graphs in this figure are for two cars colliding with each other head-on. Figure 44.44 summarizes the results of many automobile crash

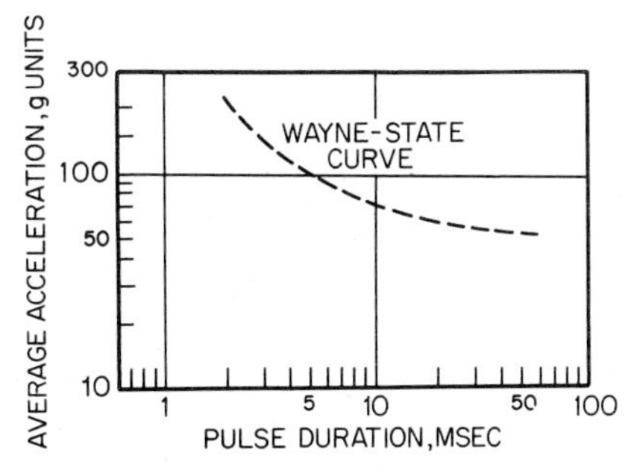

FIG. 44.45. Head-impact acceleration levels for the onset of cerebral concussion as a function of pulse duration. (*L. M. Patrick et al., Proc. Seventh Stapp Car Crash Conf., Ch. C. Thomas. Springfield, Ill., 1965.*)

experiments. The peak deceleration of the car body under the driver's compartment is plotted against the impact velocity. The difference in impact load between the frame and unitized underbody construction was negligible.

Examples of other types of short duration accelerations are given in Table 44.5.

With respect to head injury, the relation between trauma and mechanical insult is very complex and hard to compress into curves readily applicable for operational and/or design use. One curve frequently used for these purposes—and the one around which the performance specifications for most protective helmets (for example, ANSI Standard Z90.1—1971) are written—is presented in Fig. 44.45. It relates the onset of concussion to the magnitude and duration of the impact pulse to the head ("Wayne-State curve").

Table 44.5. Approximate Duration and Magnitude of Some Short-duration Acceleration Loads

Type of operation	Acceleration, *g*	Duration, sec
Elevators:		
Average in "fast service"	0.1–0.2	1–5
Comfort limit	0.3	
Emergency deceleration	2.5	
Public transit:		
Normal acceleration and deceleration	0.1–0.2	5
Emergency stop braking from 70 mph	0.4	2.5
Automobiles:		
Comfortable stop	0.25	5–8
Very undesirable	0.45	3–5
Maximum obtainable	0.7	3
Crash (potentially survivable)	20–100	<0.1
Aircraft:		
Ordinary take-off	0.5	>10
Catapult take-off	2.5–6	1.5
Crash landing (potentially survivable)	20–100	
Seat ejection	10–15	0.25
Man:		
Parachute opening, 40,000 ft	33	0.2–0.5
6,000 ft	8.5	0.5
Parachute landing	3–4	0.1–0.2
Fall into fireman's net	20	0.1
Approximate survival limit with well-distributed forces (fall into deep snow bank)	200	0.015–0.03
Head:		
Adult head falling from 6 ft onto hard surface	250	0.007
Voluntarily tolerated impact with protective headgear	18–23	0.02

REFERENCES

1. Franke, E. K., and H. E. von Gierke: Unpublished.
2. von Békésy, G.: *Akust. Z.*, **4**:360 (1939).
3. Franke, E. K.: *J. Appl. Physiol.*, **3**:582 (1951).
4. von Gierke, H. E.: *USAF Tech. Rept.* 6010, 1950.
5. German Aviation Medicine, World War II, vol. 2, Department of the Air Force, Government Printing Office, Washington, D.C., 1950.
6. Harris, C. M. (ed.): "Handbook of Noise Control," chaps. 16, 17, McGraw-Hill Book Company, Inc., New York, 1957.
7. Dieckmann, D.: *Intern. Z. angew. Physiol. einschl. Arbeitsphysiol.*, **16**:519 (1957).
8. Adams, O. S.: *USAF, WADC Tech. Rept.* 57-510, ASTIA Document AD 130983, 1958.
9. Stapp, J. P.: *USAF Tech. Rept.* 5915, pt. I, 1949, pt. II, 1951.
10. Forkois, H. M., and R. W. Conrad: *USNRL, Rept.* 4151, June, 1953.
11. von Gierke, H. E.: *Noise Control*, **5**:8 (1959).
12. Reiher, H., and F. J. Meister: *Forsch. Gebiete Ingenieurw.*, **3**:177 (1932).

13. Watts, D. T., et al.: *J. Av. Med.*, **18**:554 (1947).
14. Severy, D. M., and J. H. Matthewson: *Highway Research Board Bull.*, **91**:39 (1954).
15. Severy, D. M., and J. H. Matthewson: *Clin. Orthopaedics*, **8**:275 (1956).
16. Walker, R. Y., et al.: *USAMRL*, Ft. Knox, *Rept.* 321, November, 1957.
17. White, C. S., et al.: *AEC Civil Effects Test Group Rept.* WT-1179, Project 33.1, October, 1957.
18. Lee, R. N.: *U.S. Navy Mine Defense Lab.*, *Med. Research Rept. 2*, September, 1957.
19. Dubois, A. B., et al.: *J. Appl. Physiol.*, **8**:587 (1956).
20. von Békésy, G.: *Akust. Z.*, **4**:316 (1939).
21. Franke, E. K.: *USAF Tech. Rept.* 6469, April, 1951.
22. Hardy, J. D., and C. C. Clark: *Aero/Space Engineering*, **18**:48, June, 1959.
23. *Cornell Aeronaut. Lab. Rept.* 06-675-0-5, ONR project NM 118-389, April, 1951.
24. Clark, C. C., and F. Gray: *Proc. AGARD Aeromed. Panel*, Athens, 1959.
25. Hertzberg, H. T. E.: *USAF, WADC Tech. Rept.*, 56-30, May, 1958.
26. Evans, G. F., and M. Lebow: *J. Appl. Physiol.*, **3**:563 (1957).
27. Carothers, C. O., F. C. Smith, P. Calabresi: *USNMRI and NBS Rept.* NM 001 056.02.13, October, 1949.
28. Fasola, A. F., R. C. Baker, and F. A. Hitchcock: *USAF, WADC Tech. Rept.* 54-218, August, 1955.
29. Calabresi, P., and F. C. Smith: *USNMRI and NBS Memo Rept.* 51-2, February, 1951.
30. Stacy, R. W., et al.: "Essentials of Biological and Medical Physics," McGraw-Hill Book Company, Inc., New York, 1955.
31. Dieckmann, D.: *Intern. Z. angew. Physiol. einschl. Arbeitsphysiol.*, **17**:67 (1958).
32. Coermann, R. R.: *Human Factors*, **4**:227 (1962).
33. Coermann, R. R., et al.: *Aerospace Med.*, **31**:443 (1960).
34. Goldman, D. E.: In "Handbook of Noise Control," C. M. Harris (ed.), chap. 11, McGraw-Hill Book Company, Inc., New York, 1957.
35. Radke, A. O.: *Proc. ASME*, December, 1957.
36. Coermann, R. R.: *Jahrb. deut. Luftfahrtforschung*, **3**:111 (1938).
37. Von Wittern, W. W.: *Am. Heart J.*, **46**:705 (1953).
38. Roman, J., R. R. Coermann, and G. Ziegenruecker: *J. Av. Med.*, **29**:248 (1958).
39. Dieckmann, D.: *Intern. Z. angew. Physiol. einschl. Arbeitsphysiol.*, **17**:83 (1958).
40. Dieckmann, D.: *Intern. Z. angew. Physiol. einschl. Arbeitsphysiol.*, **17**:125 (1958).
41. Agate, J. N., and H. A. Druett: *Brit. J. Ind. Med.*, **4**:141 (1947).
42. von Békésy, G.: *J. Acoust. Soc. Amer.*, **20**:749 (1948).
43. Franke, E. K.: *USAF, WADC Tech. Rept.* 54-24, 1954.
44. Corliss, E. L., and W. Koidan: *J. Acoust. Soc. Amer.*, **27**:1164 (1955).
45. von Gierke, H. E.: *Noise Control*, **2**:37 (1956).
46. Franke, E. K., et al.: *J. Acoust. Soc. Amer.*, **27**:484 (1952).
47. von Gierke, H. E., et al.: *J. Appl. Physiol.*, **4**:886 (1952).
48. Oestreicher, H. L.: *J. Acoust. Soc. Amer.*, **23**:707 (1951).
49. Goldman, D. E., and T. F. Hueter: *J. Acoust. Soc. Amer.*, **28**:35 (1956).
50. Dussik, K. T., et al.: *Am. J. Phys. Med.*, **37**:107 (1958).
51. Eiband, M.: *NASA Memo* 5-10-59E.
52. Selye, H.: "Stress," Acta Inc., Montreal, 1950.
53. Roman, J.: *USAF, WADC Tech. Rept.* 58-107, April, 1958.
54. Schaefer, V. H., et al.: *USAMRL*, Ft. Knox, Ky., *Rept.* 390, June 20, 1959.
55. Fowler, R. C.: *Shock and Vibration Bull.* 22, *Suppl.*, Dept. of Defense, Washington, D.C., 1955.
56. Pape, R., et al.: *J. Appl. Physiol.*, **18**:1193 (1963).
57. Riopelle, A. J., M. Hines, and M. Lawrence: *USAMRL Rept.* 358, October, 1958.
58. White, D. C.: *USNADC*, Johnsville, Pa., *Rept.* NM 001 111 304.
59. Dart, E. E.: *Occup. Med.*, **1**:515 (1946).
60. Guillemin, V., and P. Wechsberg: *J. Av. Med.*, **24**:258 (1953).
61. Schaefer, V. H., et al.: *USAMRL*, Ft. Knox, Ky., *Rept.* 389, June 12, 1959.
62. Vollmer, E. P., and D. E. Goldman: Unpublished data.
63. Sueda, M.: *Mitt. med. Akad. Kioto*, **21**:1066 (1937).
64. Loeckle, W. E.: *Arbeitsphysiol.*, **13**:79 (1944).
65. Goldman, D. E.: *Am. J. Physiol.*, **155**:78 (1948).
66. Constant, H.: *J. Roy. Aeronaut. Soc.*, **36**:205 (1932).
67. Zand, S. J.: *SAE J.*, **31**:445 (1932).
68. Best, S. G.: *SAE J.*, **53**:648 (1945).
69. Reiher, H., and F. J. Meister: *Forsch. Gebiete Ingenieurw.*, **2**:381 (1931).
70. Jacklin, H. M., and G. J. Liddell: *Purdue Univ. Eng. Bull.*, *Research Series* 44, May, 1933.

71. von Békésy, G.: *Akust. Z.*, **4**:360 (1939).
72. Goldman, D. E.: *USNMRI Rept.* 1, NM 004 001, March, 1948.
73. Getline, G. L.: *Shock and Vibration Bull.* 22, *Suppl.*, Dept. of Defense, Washington, D.C., 1955.
74. Ziegenruecker, G., and E. B. Magid: *USAF, WADC Tech. Rept.* 59-18, 1959; also Magid, E. B., et al.: *Aerospace Med.*, **31**:915 (1960).
75. Clemedson, C. J.: *Physiol. Rev.*, **26**:336 (1956).
76. Harvey, E. N.: *Proc. Am. Phil. Soc.*, **92**:294 (1948).
77. Harvey, E. N., et al.: *Surgery*, **21**:218 (1947).
78. Greaves, F. C., et al.: *U.S. Naval Med. Bull.*, **41**:339 (1943).
79. Denny-Brown, D.: *Physiol. Rev.*, **25**:296 (1945).
80. Hoff, E. C., and L. J. Greenbaum: A Bibliographical Sourcebook of Compressed Air, Diving and Submarine Medicine, Dept. of the Navy, ONR and BuMed, vol. II, 1954.
81. Ruedi, L., and W. Furrer: "Das akustische Trauma," Karger, Basel, 1947.
82. Rushmer, R. F., E. L. Green, and H. D. Kingsley: *J. Av. Med.*, **17**:511 (1946).
83. Barr, J. S., R. H. Draeger, and W. W. Sager: *Mil. Surg.*, **91**:1 (1946).
84. Latham, F.: *Proc. Roy. Soc. (London)*, **B146**:121 (1957).
85. Hess, J. L., and C. F. Lombard: *J. Av. Med.*, **29**:66 (1958).
86. Pesman, G. J., and A. M. Eiband: *NACA Tech. Note* 3775, November, 1956.
87. Hecht, K. F., E. G. Sperry, F. J. Beaupre: *USAF, WADC Tech. Rept.* 53-443, November, 1953.
88. Shaw, R. S.: *J. Av. Med.*, **19**:39 (1948).
89. De Haven, H.: *War Med.*, **2**:586 (1942).
90. Joffe, M., and F. A. Hitchcock: *Fed. Proc.*, **8**:1 (1949).
91. Gross, G. A.: *USN BuAer Rept.* N1-R1, Bibliography on Head and Brain Injuries, January, 1955.
92. Holbourn, A. H. S.: *Lancet*, **2**:438 (1943).
93. Sheldon, C. H., R. H. Pudenz, and J. S. Restarski: *USNMRI Rept.* 2, Project X-182, January, 1946.
94. Ward, J. W., L. H. Montgomery, and S. L. Clark: *Science*, **107**:349 (1948).
95. Gurdjian, E. S., J. E. Webster, and H. R. Lissner: in "Medical Physics," Otto Glasser (ed.), Year Book Publishers, Inc., New York, 1950.
96. Dye, E. R.: *Clinical Orthopaedics*, **8**:305 (1956).
97. Mozell, M. M.: *USNADC Rept.* MA 5802, 1958.
98. Bondurant, S., et al.: *USAF, WADC Rept.* 58-156, 1958.
99. Severy, D. M., and J. H. Matthewson: *Proc. SAE*, January, 1956. *Trans. SAE*, **65**:70 (1957).
100. Severy, D. M., J. H. Matthewson, and A. W. Seigel: *Proc. SAE*, National Passenger Car, Body and Materials Meeting, Detroit, March, 1958.
101. De Haven, H.: *Aeronaut. Eng. Rev.*, **5**:11 (1946).
102. Bierman, H. R.: *USNMRI, Rept.* 11, Project X-630, April, 1949.
103. Hawthorne, R.: *Space/Aeronautics*, October, 1958.
104. Civil Air Regulations ISO-C39.
105. Pinkel, I. I., and E. G. Rosenbert: *NACA Rept.* 1332.
106. Pinkel, I. I.: *Proc. ASME Aviation Conf.*, Los Angeles, March, 1959.
107. Matthewson, J. H., and D. M. Severy: *Highway Research Board Bull.* 73.
108. Denny-Brown, D., and W. R. Russell: *Brain*, **64**:93 (1941).
109. Preston, G. M., and G. J. Pesman: *Proc. Inst. Aeronaut. Sci.*, January, 1958.
110. Lee, R. H., F. I. Whitten, and F. W. Brown: *U.S. Naval Mine Defense Lab., Med. Research Rept.* 4, NM 640 123, June, 1959.
111. Fibikar, R. T.: *Prod. Eng.*, **27**:177, November (1956).

ADDITIONAL REFERENCES

Coermann, R. R., et al.: The Passive Dynamic Mechanical Properties of the Human Thorax-Abdomen System and of the Whole Body System, *Aerospace Med.*, **31**:443 (1960).

Nixon, C. W.: Influence of Selected Vibrations on Speech: I. Range of 10 cps to 50 cps, *J. Aud. Res.*, **2**:247 (1962).

Coermann, R. R.: The Mechanical Impedance of the Human Body in Sitting and Standing Positions at Low Frequencies, *Human Factors*, **4**:227 (1962).

"Impact Acceleration Stress", *U.S. Natl. Acad. Sci. Natl. Res. Coun. Publ.* 997, 1962.

Clarke, N. P.: Acceleration, Vibration and Impact, *Bioastronautics—Fund. Prac. Prob.*, **17**:79 (1964).

von Gierke, H. E.: Transient Acceleration, Vibration and Noise Problems in Space Flight, in K. Schaefer (ed.) "Bioastronautics," The MacMillan Company, New York, 1964.

von Gierke, H. E.: Biodynamic Response of the Human Body, *Appl. Mech. Rev.* **17**:951 (1964).

Bowen, I. G., et al.: Biophysical Mechanisms and Scaling Procedures in Assessing Responses of the Thorax Energized by Air-Blast Over-pressures or by Nonpenetrating Missiles, *Ann. N.Y. Acad. Sci.*, **125**(1):122 (1968).
Harris, C. S., and R. W. Shoenberger: Effects of Frequency of Vibration on Human Performance, *J. Eng. Psychol.*, **5**(1):1 (1966).
"Ride and Vibration Data Manual," Society of Automotive Engineers, Information Report SAE J6a, 1965.
Liu, Y. K., and J. D. Murray: A Theoretical Study of the Effect of Impulses on the Human Torso, *Proc. ASME Symp. Biomechanics*, 1966.
Caveness, W. F., and A. E. Walker (eds.): "Head Injury," J. B. Lippincott Company, Philadelphia, 1966.
Brinkley, J. W.: Development of Aerospace Escape Systems, *Air Univ. Rev.*, 34 (July/August, 1968).
Martinez, J. L., and D. J. Garcia: A Model for Whiplash, *J. Biomechanics*, **1**:23 (1968).
Fletcher, E. R., and I. G. Bowen: Blast-Induced Translational Effects, *Ann. N.Y. Acad. Sci.*, **125**(1):378 (1968).
von Gierke, H. E.: Response of the Body to Mechanical Forces—an Overview, *Ann. N.Y. Acad. Sci.*, **125**(1):172 (1968).
Terry, C. T., and V. L. Roberts: A Viscoelastic Model of the Human Spine Subjected to $+G_z$ Acceleration, *J. Biomechanics*, **1**:161 (1968).
Stech, E. L., and P. R. Payne: "Dynamic Models of the Human Body," *Aerospace Med. Res. Lab.*, AMRL-TR-66-157, 1969.
Payne, P. R.: Injury Potential of Ejection Seat Cushions, *J. Aircraft* **6**:273 (1969).
Payne, P. R., and E. G. U. Band: "A Four-Degree-of-Freedom Lumped Parameter Model of the Seated Human Body," *Aerospace Med. Res. Lab.*, AMRL-TR-70-35, January, 1971.
von Gierke, H. E.: "Physiological and Performance Effects on the Aircrew During Low-Altitude, High-Speed Flight Missions," *Aerospace Med. Res. Lab.*, AMRL-TR-70-67 November, 1971.
Aerospace Med. Res. Lab., Symp. *Biodynamic Models and Their Applications*, AMRL-TR-71-29, December, 1971.
von Gierke, H. E.: "Biodynamic Models and Their Applications," *J. Acous. Soc. Amer.* **50**:1397 (1971).
Bartz, John A.: "A Three-Dimensional Computer Simulation of a Motor Vehicle Crash Victim," Calspan Corp. Rep. VJ-2978-V-1, July, 1971.
Hammond, R. A.: "Digital Simulation of an Inflatable Safety Restraint," Society Automotive Engineers preprint 710019, 1971.
Orne, D., and Y. K. Liu: A Mathematical Model of Spinal Response to Impact, *J. Biomechanics*, **4**:49 (1971).
Ommaya, A. K., and A. B. Hirsch: Tolerance for Cerebral Concussion from Head Impact and Whiplash in Primates, *J. Biomechanics*, **4**:13 (1971).
"Specifications for Protective Headgear for Vehicular Users," American National Standards Institute, ANSI S90.1, 1971.
Shoenberger, R. W.: Human Response to Whole-Body Vibration, *Perceptual and Motor Skills*, **34**:127 (Monograph Suppl. 1-V34) (1972).
Synder, R. G.: Impact, in "Bioastronautics Data Book," 2d ed., NASA SP-3006, 1973.
Hornick, R. J.: Vibration, in "Bioastronautics Data Book," 2d ed., NASA-3006, 1973.
Kenedi, R. M. (ed.): "Perspectives in Biomedical Engineering," MacMillan & Co., Ltd, London, 1973.
King, W. F. and A. J. Merts (eds.): "Human Impact Response," Plenum Press, New York, 1973.
"Guide for the Evaluation of Human Exposure to While-Body Vibration," International Standards Organization: ISO/IS 2631, 1974.
Proceedings of the First to Eighteenth (1956 to 1974) Stapp Car Crash Conferences, published each year by Society of Automotive Engineers, New York.
AGARD Conference Proceedings on "Vibration and Combined Stress in Advanced Systems," National Technical Information Service, Springfield, Va., 1975.
Brinkley, J. W., and H. E. von Gierke: Impact Acceleration, in "Foundations of Space Biology and Medicine," vol. II, pt. 3, Chap. 3 U.S. Academy of Sciences, Washington D.C., 1975.
von Gierke, H. E., C. W. Nixon and J. C. Guignard: Noise and Vibration, in "Foundations of Space Biology and Medicine," chap. 9, U.S. Academy of Sciences, Washington, D.C., 1975.
Taylor, W. (ed.): "The Vibration Syndrome," Academic Press, London, 1974.
von Gierke, H. E. (ed.): "Vibration and Combined Stress in Advanced Systems," *Proc. AGARD Conf.* Oslo, Norway, 1974, Nat. Tech. Info. Service, Springfield Va.
Saczalski, K., et al. (ed.): "Aircraft Crash Worthiness," University of Virginia Press, 1975.
U.S. Dept. of Transportation, 1975 Ride Quality Symposium, DOT-TSC-OSC-75-40, Nat. Tech. Info. Service, Springfield, Va.

INDEX